Advance Elements of Laser Circuits and Systems

Ofer Aluf

Advance Elements of Laser Circuits and Systems

Nonlinear Applications in Engineering

Springer

Ofer Aluf
Netanya, Israel

ISBN 978-3-030-64102-3 ISBN 978-3-030-64103-0 (eBook)
https://doi.org/10.1007/978-3-030-64103-0

This Springer imprint is published by the registered company Springer Nature Switzerland AG
The registered company address is: Gewerbestrasse 11, 6330 Cham, Switzerland

Preface

This book on laser circuits and systems: Nonlinearity applications in engineering covers and deals with two separate engineering and scientific areas and what between. It gives advance analysis methods for laser circuits and systems which represent many applications in engineering and science. Lasers circuits and systems come in many topological structures and represent many specific implementations which stand the target scientific and engineering features. The main types of lasers are categories by the lasing medium and are classified as either laser diode (semiconductor laser), Ti: Sapphire laser, ion-channel and log wavelength laser, solid-state laser, Nd: YAG, mid-infrared Q-switched microchip laser, gas laser, dual-wavelength laser, helium–neon laser, carbon dioxide laser, metal vapor laser, rare gas ion laser, excimer laser, chemical laser, and gas dynamic carbon dioxide laser. Additionally, there are dye laser, free-electron laser (FEL), X-ray laser, quantum cascade laser, lead salt laser, antimonide laser, and femtosecond laser. Ultrafast laser micromachining using picosecond or femtosecond laser can be used for the most challenging applications. Picosecond laser micromachining utilized a very short pulse laser (typically 8–12 picosecond duration), to prevent the formation of a heat-affected zone. Femtosecond laser micromachining system is used in ultrafast machining. Sometimes 20-fold reduction in pulse duration relative to a picosecond system gives advantages in terms of material interaction. By using picosecond or femtosecond laser, we can get a very high peak power density. These types of laser can be used to machine otherwise transparent materials such as silica or sapphire. Nonlinear absorption enables the light to interact with transparent materials, causing very strong, extremely localized heating and ablation. The strength of the laser ablation process results in a very low debris process. High-power (Peta watt, 10^{15} w) laser involves real-time controls and femtosecond precision timing systems. We can synchronize operation of lasers to specific beamline clock and get a broadband laser pulse stretchers with pulse compress. Laser diodes consist of a p–n diode with an active region where electrons and holes recombine resulting in light emission. In addition, a laser diode contains an optical cavity where stimulated emission takes place. The laser cavity consists of a waveguide terminated on each end by a mirror. There are some longitudinal modes to laser diode. Rate equations for each longitudinal mode (λ), with photon density S_λ and carrier density N_λ, are couple into specific mode. In the basic electrical equivalent circuit of a laser diode, the effects of spontaneous emission and self-pulsations are included. The self-pulsations are represented by a negative resistance in the model. Application of this model suggests purely electronic methods of suppressing relaxation oscillations in laser diodes. When dealing with MRI technology, needs are addressed by fiber optics MRI system. The laser diode circuitry is the component responsible for the conversion of MR signals from electrical to the optical state. The vertical-cavity surface-emitting laser (VCSEL) employs current-blocking layers to funnel the current to the active region located between the mirrors of the laser. Ti: Sapphire lasers are tunable lasers which emit red and near-infrared light in the range from 650 to 1100 nm. These lasers are mainly used in scientific research because of their tenability and their ability to generate ultrashort pulses. The ion-channel laser (ICL) is an ultracompact version of the free-electron laser (FEL), with the undulator replaced by an ion channel. Analysis of the resulting scaled equations shows a realistic prospect of building a compact ICL source for fundamental wavelength down to UV and harmonics potentially extending to X-rays. A laser is capable of working in the terahertz range (long-wavelength light from the far infrared to one millimeter). It enables to examine better than can be done using chemical analysis. Solid-state laser is a laser that uses a gain medium that is a solid, rather than a liquid such as in dye lasers or a gas as in gas lasers. Semiconductor-based lasers are also in the solid state but are generally considered as a separate class from solid-state lasers. Quantum composers laser features three series of diode pumped, solid state, Nd: YAG lasers. Gas laser is a laser in which an electric current is discharged through a gas to produce coherent light. Lasers emitting simultaneously at multiple wavelengths and can fit for wide applications in many fields such as environmental monitoring, laser radar, spectral analysis, THz research, etc. It is a good solution if you need a special measure compare to traditional laser systems like one wavelength laser. The laser circuits and systems analyze as linear and nonlinear dynamical systems and their dynamics under parameters variations. This book is aimed at newcomers to linear and nonlinear dynamics and chaos laser system and circuits. The presentation stresses analytical and numerical methods, concrete examples, and geometric intuition. The laser systems and circuits analysis is developed systematically, starting with first-order differential equations and their bifurcation, followed by phase plane analysis, limit cycles and their bifurcations, chaos, iterated maps, period doubling, renormalization and strange attractors. Additionally, the book dealing with delayed laser circuits which characterized by overall variables delayed with time. Each variable has specific delay parameter and can be inspect for dynamics. More realistic laser circuit's models should include some of the past states of laser circuits systems, that is, ideally, a real laser circuits should be modeled by differential equations with time delays. The use of delay differential equations (DDEs) in the modeling of laser type's circuit's dynamics is currently very active, largely due to progress achieved in the understanding of the dynamics of several classes of delayed differential equations and laser circuits and systems. This book is designed for advanced undergraduate or graduate students in electronics, physics, RF, and electronic engineering, mathematics who will interest in laser circuits dynamics and innovative analysis methods. It is also addressed to electrical and optic engineers, physics experts and researchers in physics, electronics, engineering, and mathematics who use dynamical systems as modeling tools in their studies. Therefore, only a moderate mathematical and electronic semiconductor background in geometry, linear algebra, analysis, and differential equations is required. Each book chapter includes various laser systems and circuits drawing and their equivalent analyses circuits. Laser circuits fixed points and stability analysis done by using much estimation. Various bifurcations laser systems and circuits are discussed.

In this book, we try to provide the reader with explicit procedures for application of laser systems and circuits mathematical representations to particular research problems. Special attention is given to numerical implementation of the developed techniques.

Let us briefly characterize the content of each chapter.

Chapter 1. **Dynamical and Nonlinearity of Laser Diode Circuits.** In this chapter, laser diode coupled delay rate equations, laser diode intrinsic and package electrical equivalents circuits, MRI system, and electron photon exchanges into VCSEL are discussed. The laser diode is a p–n diode with an active region where electrons and holes recombine resulting in light emission. The elements of the intrinsic laser equivalent circuit are derived from the coupled rate equations which describe the interplay between the injected carrier and photon densities in the active region of the laser diode. The parameter τ is the latent period which is related to parasitic effects of the stimulated emission process. We draw the laser diode intrinsic and package electrical equivalent circuits and analyze the stability under parameters variation. MRI system laser diode is one element of fiber optic MRI system. One of these technologies is phased array coils with high number of elements. These are multichannel coils with each channel. A miniature fiber optic transmission (FOT) system is used and it contains a laser diode connected to a photodetector with fiber optic cabling. We inspect the diode laser circuitry under variation of delay parameters which are related to circuit's internal circuitry microstrip lines. Vertical-cavity surface-emitting lasers (VCSELs) are a semiconductor lasers that have many advantages. Due to VCSEL active region parasitic effects, there is a time delay τ of the carrier–photon interaction time. The VCSEL rate equations include this time delay parameter, and under different values, there is a stability switching concerning the operation of VCSEL laser element.

Chapter 2. **Ti: Sapphire Laser Systems with Delay Parameters in Time Stability Analysis.** In this chapter, sapphire laser systems are discussed. Ti: Sapphire lase (Ti: Al_2O_3 laser, titanium-sapphire laser, or Ti: Sapphires) is a tunable laser which emit red and near-infrared light in the range from 650 nm to 1100 nm. In many scientific experiments, the Ti: Sapphire laser is used. It is very tenability and can generate ultrashort pulses. We use linearly polarized femtosecond laser pulses from a Ti: Sapphire laser to analyze the multiphoton ionization of D_2 (D_2 is a molecule of deuterium). There is a predictive model for the ionization rate of a diatomic molecule in the field of a short pulse with wavelength in the near-IR region. There are two equivalent models which give a good overlap with experimental results on the ionization of rate-gas atoms in the tunneling regime, interacting with femtosecond laser pulses. Shaping ultrashort pulses with a resolution approach 10 fs is done by a liquid-crystal spatial light modulator within reflective optics pulse-shaping apparatus. The principle is to use a spatial light modulator as a phase modulator. A spatial light modulator (SLM) is an object that imposes some form of spatially varying modulation on a beam of light. There is phase dispersion at higher orders, and especially cubic- and quartic-phase dispersions, stability analysis is done. Additionally, Ti: Sapphire laser can operate with mode of multiple pulse. It is mode locked by a semiconductor saturable absorber mirrors (SESAMs). The system (multipulse operation of Ti: Sapphire laser mode locked by an ion-implanted semiconductor saturable absorber mirror) differential equations are inspected for stability.

Chapter 3. **Ion Channel and Long-Wavelength Lasers, Nonlinearity Applications in Engineering**. In this chapter, ion-channel and long-wavelength laser's analysis is discussed. Ion-channel system includes relativistic electron beam propagating through plasma in the ion-focused regime. Growth is enhanced by optical guiding in the ion channel, which acts as a dielectric waveguide. It is a transport of relativistic electron beams (REBs) in plasmas and mechanism of ion focusing. The basic mechanism is laser-driven acceleration in ion channel. A long laser beam propagating through un-dense plasma produces a positively charged ion channel by expelling plasma electrons in the transverse direction. The perturbed equations of motion, average over betatron period, are inspected for stability and stability switching. The ion-channel laser perturbed system equations of motion are inspected for behavior analysis. A semiconductor laser (Laser diode) is a device that causes laser oscillation by the flowing an electric current to semiconductor. Practically, the mechanism of light emission is the same as a lighting-emitting diode (LED). The optical gain is produced in a semiconductor material. The nonlinear dynamics of the lasing nodes are described mathematically by multimode rate equations of the photon number S_p of the lasing modes and the injected electron number N and analysis is done.

Chapter 4. **Solid State Laser Nonlinearity Applications in Engineering**. In this chapter, solid-state lasers are analyzed for best performances. A solid-state laser is a laser that uses a gain medium that is a solid. It is very different from liquid laser such as in dye lasers or a gas lasers. Lasers are commonly designed by the type of lasing material employed: Solid-state lasers have lasing material distributed in a solid matrix such as the ruby or neodymium "YAG" lasers. Solid-state lasers use optical pumping and such pump sources are relatively cheap and can provide very high powers. High-power diode lasers array can be utilized to pump solid-state laser materials efficiently. Laser materials with high intrinsic quantum efficiencies and an improved scalable cavity design make high output powers. The powerful and compact laser sources have fundamental research application such as X-ray, plasma, and higher-harmonic generation. The simplest laser model of a single-mode laser with homogeneously laser medium and saturate absorber is simulated and analyzed for stability behavior. In many applications, we use laser diode-pumped microchip solid-state laser (DPSSLs). One application is a self-aligned optical feedback vibrometry technique. The laser diode-pumped microchip solid-state laser is useful for compact, durable, coherent light sources. In laser, we have the phenomena of light injection-induced. It is injection locking and return light-induced instabilities. Important is the basic properties of laser dynamics as well as their practical importance. One group of lasers is Nd stoichiometric lasers. Stoichiometric laser materials are pure chemical compounds capable of emitting coherent light in the undiluted state as opposed to conventional laser materials where the active ions or molecules are dispersed in a host. We characterize the light injection model for rotating-wave approximation fields and inspect stability. A combination of gain switching with passive Q-switching of a miniature diode-pumped solid-state laser is done. A composite pumping pulse, consisting of a long, low-intensity pulse and a following short, high-intensity pulse, is used to reduce the timing jitter. The simple model of a passively Q-switched laser is presented and inspected for stability.

Chapter 5. **Nd: YAG, Mid-Infrared and Q-Switched Microchip Lasers Stability Analysis**. In this chapter, Nd: YAG, mid-infrared, and Q-switched lasers are analyzed for best performances and stability is discussed. A diode-pumped Nd: YAG laser is Q-switched by a GaAs saturable absorber and intra-cavity frequency doubled by a KTP crystal. At pump power, the device produces high quality pulses at specific pulse repetition rate. We characterize the dynamics of the pulse formation by rate equations and an energy-level model which accounts for the various energy-transfer process in GaAs. A diode-dumped Nd: YAG laser is Q-switched by a GaAs saturable absorber and intra-cavity frequency doubled by a KTP crystal. The dynamics of a pulse formation can be described by rate equations and an energy-level model which accounts for the various energy-transfer processes in GaAs and stability is discussed. Semiconductor saturable absorber mirror (SESAM) is a key component of ultrafast passive mode-locked laser sources. The key SESAM parameters are saturation fluence, modulation depth, and non-saturable losses are measurable with a high accuracy. We control these parameters to obtain stable pulse generation for a given laser. The model function for the nonlinear reflectivity is based on a simple two-level travelling wave system and the behavior is analyzed. Ion-doped crystalline laser mainly operates in the mid-IR spectral range between 2 and 5 μm. Types are rare-earth and transition-metal-based ionic crystals, and color-center lasers. They are compact all-solid-state room-temperature tunable sources, belonging to class of vibrionic lasers. The four-level scheme which characterized by balance equation for the population of the upper laser level n and dynamic is inspected.

Chapter 6. **Gas Laser Systems Stability Analysis Under Parameters Variation**. In this chapter, gas laser systems are discussed and there stability behavior is analyzed. Lasing from molecular nitrogen is used in many scientific and industrial applications. The discharge pumped nitrogen laser, operating in a broad range of gas pressures, from several mill bars to the atmospheric pressure, and repetition rates from several hertz to several kilo hertz. It is robust source of high-power near-UV radiation. Achieving nitrogen lasing via remote excitation would pave the way to many potential applications. It is remotely initiated lasing from molecular gases by femtosecond filaments. The nitrogen laser emits radiation at 337 nm, and its active medium is gaseous nitrogen. The gas is confined within a pressure vessel, usually at a total pressure between 101 and 2700 Pa. The nitrogen laser generates pulses of duration between 300 ps (at atmospheric pressure) and 10 ns (at 2700 Pa). The repetition rate changes from 1 to 100 Hz. The dynamical behavior of nitrogen laser is discussed. Metal vapor lasers are devices in which the lasing medium is a vapor of metal atoms or ions, sometimes mixed with another gas. Metal vapor lasers use a variety of metal types to generate a variety of laser lines for applications throughout the electromagnetic spectrum. A self-consistent model copper vapor laser (CVL) is very helpful for characterization of the copper laser and its implementation. It uses vapors of copper as the lasing medium in a three-level laser. It can produce green laser light (510.6 nm) and yellow laser light (578.2 nm). The model for a high repetition rate copper vapor laser is described. Equations for the discharge pulse, laser pulse, and inter-pulse afterglow are discussed and analyzed for stability.

Chapter 7. **Dual-Wavelength Laser Systems Stability Analysis Under Parameters Variation (I)**. In this chapter, filters systems in many circuits are inspected for dynamical behavior and stability analysis. Dual-wavelength lasers are not common in scientific facilities but some applications need to use them. Dual-wave lasers demonstrate simultaneous dual-wavelength lasing and outcome of operating in multiple wavelength is possible. A four-energy-level system of dual-wavelength Ti: Sapphire laser is characterized by rate equation. Two sub-resonators structure gives dual-wavelength laser. The Ti: Sapphire laser system simultaneously generates sequence of femtosecond pulses at two independent wavelength regions. It shows a high degree of synchronization and reduced relative timing jitter. The analytic model is analyzed for stability. A vertical external-cavity surface-emitting laser (VECSEL) is a semiconductor laser based on a surface-emitting semiconductor gain chip and a laser resonator. VECSELs generate high optical power in diffraction limited beams and have wavelength versatility. VECSEL unit is constructed from a semiconductor gain chip and an external laser resonator, and additionally, there are arrangements for pumping and cooling. Er-doped fiber lasers (EDFLs) are a erbium-doped fiber amplifier (EDFA) which operate in the particular regime where coherent oscillation of ASE occurs due to feedbacks mean. EDFLs are used as sources for coherent light signal generation, while EDFAs are used as wave-wave amplifiers for coherent light signal generation. EDFLs can be pumped with compact, efficient laser diodes. A compact high-power yellow pulsed laser uses an intra-cavity sum-frequency mixing in a diode-end-pumped Q-switched Nd: YVO_4 dual-wavelength laser. A three-mirror configuration forming two separate laser cavities is used to optimize the gain match for simultaneous dual-wavelength emission in Q-switched operation. A diode-pumped Nd-doped laser is a compact all-solid-state sources, in the blue, green, and red spectral regions by use of intra-cavity frequency doubling. These sources cover the region from 550 to 650 nm. The dynamic of yellow pulsed laser is discussed for stability.

Chapter 8. **Dual-Wavelength Laser Systems Stability Analysis Under Parameters Variation (II)**. In this, we discuss a dual-wavelength laser system and stability analysis under parameters variation. The dynamic responses of the photon and carrier densities in an asymmetric dual quantum well laser exhibit wavelength switching. The laser consists of two different quantum wells isolated by a high and/or thick barrier layer that results in an inhomogeneous carrier injection into the two wells. The self-pulsing of the Tm^{3+}-doped silica fiber laser is operated near 2 μm. There are various self-pulsing regimes for a range of pumping rates where the fiber is end-pumped with a high-power Nd: YAG laser which operated at 1.319 μm in a linear bidirectional cavity. There are a rich variety of nonlinear phenomena, ranging from self-pulsing to self-Q-switching and to a modulated quasi-cw wave. Quantum cascade (QC) lasers are semiconductor devices in which radiation is generated by electronic transitions within an artificial crystal, heterostructure crystal. This heterostructure consists of alternating layers of two semiconductor materials, with the layer thickness determining the electronic states inside the crystal, and thereby, the frequency of the emitted light as well as the electrical transport. The operation of QC lasers in the range of frequencies 1–10 THz is demonstrated by employing super lattice active material and developing a waveguide concept based on interface modes called surface Plasmon.

Chapter 9. **Laser Circuits and Systems Bifurcation Behaviors—Investigation, Comparison, and Conclusions**. In this chapter, we summary the main topics regarding laser circuits and systems, inspect behavior, dynamics, stability, comparison, and conclusion. Laser systems and circuits are an integrated part of every industrial and scientific system. There are many laser source types which characterized by different parameters. Plasma physics is also an area which is integrated with laser systems and circuits. Laser systems and circuits can be represented by set of nonlinear differential equations with delays (DDEs) or without delays. The dynamic of laser systems and circuits is analyzed by nonlinear dynamic, fixed points, bifurcations; BOA, stability, and stability switching are inspected. Plasma diagnostic is also a very important in every laser system which has characterized by overall parameters and different diagnostic methods. Some laser systems and circuits models may involve some delay-dependent parameters. The presence of laser systems and circuits parameters often greatly complicates the task of an analytical study of such models. The laser systems and circuits stability of a given steady state is simply determined by the graphs of some functions of τ which can be expressed explicitly. Most application problems, we look at one such function and locate its zeros. This function has number of zeros, providing thresholds for stability switches. As time delay increases, stability changes from stable to unstable to stable, implying that a large delay can be stabilizing.

Netanya, Israel Ofer Aluf

Introduction

The laser stands for light amplifications by stimulated emission of radiation. We get laser by some excited atoms which emit photons then in turn stimulated other atoms to emit photons. To make a powerful laser, we trap atoms between two mirrors. This bounces the photons back and forth, increasing the stimulation of other atoms. The light from a laser is made up of one color (one wavelength). All the waves in light from a laser travel in the same direction, making a concentrated beam. The lasers have implementation like DVD players, medicine, eye surgery, space exploration, to drill holes in diamonds, communication by fiber optics, Internet and TV, etc. The electrons in atoms are special glasses, crystals, or gases absorb energy from an electrical current or another laser and become "excited." The excited electrons move from a lower energy orbit to a higher-energy orbit around the atom's nucleus. When they return to their normal or "ground" state, the electrons emit photons (particles of light). All photons are at the same wavelength and are "coherent," meaning the crests and troughs of the light waves are all in lockstep. In contrast, ordinary visible light comprises multiple wavelengths and is not coherent. The particular wavelength of light is determined by the amount of energy released when the excited electron drops to a lower orbit and the laser light is directional. Whereas a laser generates a very tight beam, a flash light product light that is diffuse. Because laser light is coherent, it stays focused for vast distances, even to the moon and back. Modern lasers can produce pulses that are billions of times more powerful. Some lasers, such as ruby lasers, emit short pulses of light. Others, like helium–neon gas lasers or liquid dye lasers, emit light that is continuous. The ruby laser emits pulses of light lasting only billionths of a second. Laser light does not need to be visible and can be invisible infrared light and then pass through special optics that convert them to visible green light and then to visible, high-energy, ultraviolet light for optimum interaction with the target. There are much kind of lasers we can analyze and simulate the dynamics and stability, laser diodes, Ti: Sapphire laser systems, ion-channel and long-wavelength lasers, solid-state laser, Nd: YAG, mid-infrared, and Q-switched microchip lasers, gas lasers, and dual-wavelength laser systems. In laser diode circuits, we can characterize the laser diode intrinsic and package electrical equivalent circuit and analyze the stability under parameters variations. Each laser system or circuit can be representing by set of differential equations which depend on laser system's variable parameters. The laser system differential equations are related to the rate equations most of the times, or other process that happened in a particular laser type. The investigation of laser systems or circuits differential equation bifurcation theory, includes the study of possible changes in the structure of the orbits of a differential equation depending on variable parameters. The book illustrates certain observations and analyzes local bifurcations of an appropriate arbitrary scalar differential equation. Since the implicit function theorem is the main ingredient used in these generalizations, then it includes a precise statement of the theorem. Additional analyze the bifurcations of a laser system's differential equation on the circle. Bifurcation behavior of specific differential equations can be encapsulated in certain pictures called bifurcation diagrams. This is done for optimization of lasers systems parameters and to get the best performance.

Dynamics (chaos, fractals) change with laser systems that evolution in time. There are two types of dynamical systems: differential equations and iterated map (difference equations). Differential equation is described the evolution of a systems in continues time. Iterated map is arising in problems where the time is discrete. Differential equations can be divided to two main groups, ordinary differential equations and partial differential equations. Differential equation system can be represented as the below set.

$$\frac{d\xi_1}{dt} = f_1(\xi_1, \ldots, \xi_n); \ \frac{d\xi_2}{dt} = f_2(\xi_1, \ldots, \xi_n); \ \ldots; \ \frac{d\xi_n}{dt} = f_n(\xi_1, \ldots, \xi_n); \ \dot{\xi}_i = \frac{d\xi_i}{dt}$$

Some of laser systems or circuits can be represented as equation in dimensions "one." Basic notions of laser system and circuit stability and bifurcations of vector fields are easily explained for scalar autonomous equations dimension one—because their flows are determined from the equilibrium points. Numerical solutions of each equations lead to scalar maps and show some of the "anomalies" albeit profound and exciting that may arise when numerical approximation is poor period doubling bifurcation, chaos, etc. Laser system rate equations can be turned to the dynamics and bifurcations of periodic solutions of non-autonomous equations with periodic coefficients dimension one and one half, where scalar maps reappear naturally as Poincare maps. Laser system schematic's investigation the dynamics of planar autonomous equations—dimension two—where, in addition to equilibria, new dynamical behavior, such as periodic and homoclinic orbits, appears. Laser systems and circuit schematics stability of an equilibrium point is inspected. The subtle topological aspects of linear systems as well as the standard theory of Liapunov functions are discussed. Center manifolds and the method of Liapunov Schmidt are done to make a reduction to a scalar autonomous equation. Periodic orbit—Poincare—Andronov—Hopf bifurcation—and its analysis can be reduced to that of a non-autonomous periodic equation. Some of laser system's rate differential equations or other laser phenomenon differential equations are delayed in time. The laser system rate differential equations variables delay in time are due to parasitic phenomenon inside the laser. The laser system rate delay differential equations in population dynamics incorporate a time delays and are the result of the existence of some stage structure.

The laser system through stage survival rate is often a function of time delays, and the models may involve some delay-dependent parameters. The laser system analysis combines graphical information with analytical work to effectively analyze the local stability of some models involving delay-dependent parameters. The stability of a given steady state is simply determined by the graphs of some functions of τ which can be expressed explicitly and thus can be easily depicted by MATLAB software. We look at function and locate its zeros. This function often has only two zeros, providing thresholds for stability switches. The common scenario in laser system rate equation is that as time delay increases, stability changes from stable to unstable to stable, implying that a large delay can be stabilizing. We incorporate of a time delay which is the result of the existence of some stage structure. There are practical guidelines that combine graphical information with analytical to effectively analyze the local stability of laser models involving delay-dependent parameters. Generally, we analyze the occurrence of any possible stability switching resulting from the increase of value of the time delay τ for the general laser system characteristic equation $D(\lambda, \tau) = 0$.

$$D(\lambda, \tau) = P_n(\lambda, \tau) + Q_m(\lambda, \tau) \cdot e^{-\lambda \cdot \tau}; \; P_n(\lambda, \tau) = \sum_{k=0}^{n} p_k(\tau) \cdot \lambda^k; \; Q_m(\lambda, \tau) = \sum_{k=0}^{m} q_k(\tau) \cdot \lambda^k$$

$n, m \in \mathbb{N}_0$; $n > m$ and $p_k(\cdot), q_k(\cdot) : \mathbb{R}_{+0} \to \mathbb{R}$ are continuous and differentiable function of τ.

Contents

Chapter 1
Dynamical and Nonlinearity of Laser Diode Circuits

Laser diode consists of a p-n diode with an active region where electrons and holes recombine resulting in light emission. The laser diode contains an optical cavity where stimulated emission takes place. The laser cavity consists of a waveguide terminated on each end by a mirror. Other names to Laser Diode (LD) are injection laser diode or ILD. The laser diode is the type of laser produced with a wide range of uses that include fiber optic communications, barcode readers, laser pointers, etc.

The laser diode complete equivalent circuit is separated into two parts. The first part represents the intrinsic electrical equivalent circuit of the laser chip. The second part is the electrical equivalent circuit of the package including the major parasitic elements. The elements of the intrinsic laser equivalent circuit are derived from the coupled rate equations which describe the interplay between the injected carrier and photon densities in the active region of the laser diode. Coupled rate equations parasitic effects of the stimulated emission process is talent period τ. The investigation of our laser diode coupled delay rate equations, differential equations is based on bifurcation theory, a study of possible changes in the structure of the orbits of a delay differential equations (DDEs) depending on τ parameter. Additionally, we check the laser diode intrinsic and package electrical equivalent circuit's stability switching for parameters variation. An optical system which capable of transmitting MRI signals consist laser diode circuitry, responsible for the conversion of MR signal from the electrical to the optical state. We investigate the dynamic and stability of MRI system laser diode circuit under parameters variation. Laser diode transmitter RF section circuit contains 5 O microstrip, DC bias circuit, RFC, and additional discrete components, input signal is RF. The stability of the circuit is analyzed under parameters variation. The vertical-cavity surface-emitting laser (VCSEL) is a type of semiconductor laser diode with laser beam emission perpendicular from the top surface. VCSEL applications include fiber optic communications, precision sensing, etc. There are two main VCSEL's variables, Carrier density in the active region N(t) and Photon density in the active region S(t). Due to VCSEL active region parasitic effects, there is a time delay (τ) of the carrier-photon interaction time. The VCSEL rate equations with delay parameter (τ) of the carrier-photon interaction time are

O. Aluf, *Advance Elements of Laser Circuits and Systems*,
https://doi.org/10.1007/978-3-030-64103-0_1

analysed for stability switching under variation of (τ) parameter. We review the VCSEL structure, characteristics and inspect deeply the main variables and parameters. First is to write the VCSEL rate equations with delay parameter (τ) of the carrier-photon interaction time. We combine graphical information with analytical work to effectively study the local stability of the VCSEL model involving delay dependent parameters [1–7].

1.1 Laser Diode Coupled Delay Rate Equations Stability Analysis

The laser diode consists of a p-n diode with an active region where electrons and holes recombine resulting in light emission. A laser diode contains an optical cavity where stimulated emission takes place. The laser cavity consists of a waveguide terminated on each end by a mirror. Photons, which are emitted into the waveguide, can travel back and forth in this waveguide provided they are reflected at the mirrors. The distance between the two mirrors is the cavity length, labeled L. The light in the waveguide is amplified by stimulated emission. Stimulated emission is a process where a photon triggers the radiative recombination of an electron and hole creating an additional photon with the same energy and phase as the incident photon. This "cloning" of photons results in a coherent beam. The stimulated emission process yields an increase in photons as they travel along the waveguide. The laser diode lasing condition is as follow, when combined with the waveguide losses, stimulated emission yields a net gain per unit length g. The number of photons can therefore be maintained if the roundtrip amplification in a cavity of length, L, including the partial reflection at the mirrors with reflectivity R_1 and R_2 equals unity ($R_1 \cdot R_2 \cdot \exp(2 \cdot g \cdot L) = 1$). In case roundtrip amplification $R_1 \cdot R_2 \cdot \exp(2 \cdot g \cdot L) < 1$ is less than one the number of photons steadily decreases. If the roundtrip amplification ($R_1 \cdot R_2 \cdot \exp(2 \cdot g \cdot L) > 1$), the number of photons increases as the photons travel back and forth in the cavity and no steady state value would be obtained. The gain required for lasing equals: $g = \frac{1}{2 \cdot L} \cdot \ln\left(\frac{1}{R_1 \cdot R_2}\right)$. The process is as follow: initially, the gain is negative if no current is applied to the laser diode as absorption dominates in the waveguide. As the laser current is increased, the absorption first decreases and gain increases. The current for which the gain satisfies the lasing condition is the threshold current of the laser I_{th}. When below the threshold current very little light is emitted by the laser. If we applied current larger than the threshold current, the output power, P_{out} increases linearity with the applied current. The output power equals: $P_{out} = \eta \cdot \frac{h \cdot \nu}{q} \cdot (I - I_{th})$ where $h \cdot \nu$ is the energy per photon. The factor η indicates that only a fraction of the generated photons contribute to the output power of the laser as photons are partially lost through the other mirror and through the waveguide. The laser diode consists of a cavity, which is the region between two mirrors with reflectivity R_1 and R_2, and a gain medium (quantum wall). The optical mode originates in spontaneous emission, which is confined to the cavity by the waveguide. The optical mode is amplified by the gain medium and partially reflected by the mirrors. The modal gain depends on the gain of the medium, multiplied with the overlap between the gain medium and the optical mode (confinement factor Γ). The modal gain is $g(N) \cdot \Gamma$. Lasing occurs when the round trip optical gain equals the losses. Laser with modal gain $g(N) \cdot \Gamma$ and waveguide lossα, this condition implies: $R_1 \cdot R_2 \cdot \exp[2 \cdot (g(N) - \alpha) \cdot L] = 1$ where L is the length of the cavity. The distributed loss of the mirrors is $\frac{1}{L} \cdot \ln\left(\frac{1}{\sqrt{R_1 \cdot R_2}}\right)$.

We need to analyze the dynamical of a double heterojunction laser diode circuits using scattering parameters [1]. The laser diode is highly coherent light output that can be modulated to carry coded information at a high density and speed along a fiber optic cable. The modulation response characteristics of laser diode is studied to find the optimum operation conditions. The modulation response of the laser diode is determined by solving the rate equations numerically. The disadvantage that it does not take into account the influence of the package parasitic and device circuit interactions on the modulation response. The solution is to use a circuit analysis based on the complete electrical circuit model of the conventional double heterostructure lasers or quantum well lasers. The complete equivalent circuit of a laser diode can be separated into two parts. The first part is the intrinsic electrical equivalent circuit of the laser chip and the second part is the electrical equivalent circuit of the package including the major parasitic elements. The elements of the intrinsic laser equivalent circuit are derived from the coupled rate equations which describe the interplay between the injected carrier and photon densities in the active region of the laser diode. We define the parasitic effects of the stimulated emission process as a latent period τ. The laser diode coupled single-mode delayed differential rate equations are given by the following DDEs (Delay Differential Equations), $N_e = N_e(t)$; $N_{ph}=N_{ph}(t)$; $N_i = N_i(t)$:

$$\frac{dN_e(t)}{dt} = \frac{1}{q \cdot a \cdot d} - A \cdot (N_e(t-\tau) - N_{om}) \cdot N_{ph}(t-\tau) - \frac{N_e(t)}{\tau_s}$$

$$\frac{dN_{ph}(t)}{dt} = A \cdot (N_e(t-\tau) - N_{om}) \cdot N_{ph}(t-\tau) - \frac{N_{ph}(t)}{\tau_{ph}} + \beta \cdot \frac{N_e(t)}{\tau_s}$$

where variable N_{ph} is the photon density. Variable N_e is the electron density, q is the electronic charge, d is the thickness of the active region, a is the area of the diode contact stripe, I is the injected current, N_{om} is the minimum electron density to obtain a positive gain, A is a constant related to the stimulated emission process, τ_s is the spontaneous emission lifetime, τ_{ph} is the photon lifetime and β is the fraction of the spontaneous emission that is coupled to the lasing mode. The model differential equations assume that the inversion is homogenous and the gain is linear in the difference between N_e and N_{om}. The carrier density N_e is a function of the junction voltage, where N_i is the intrinsic carrier density and V is the junction voltage. The carrier density equation: $N_e = N_i \cdot \exp\left(\frac{q \cdot V}{2 \cdot k \cdot T}\right)$; $N_e(t-\tau) = N_i(t-\tau) \cdot \exp\left(\frac{q \cdot V}{2 \cdot k \cdot T}\right)$

$$N_e(t) = N_i(t) \cdot \exp\left(\frac{q \cdot V}{2 \cdot k \cdot T}\right); \quad \frac{dN_e(t)}{dt} = \frac{dN_i(t)}{dt} \cdot \exp\left(\frac{q \cdot V}{2 \cdot k \cdot T}\right)$$

$$\frac{dN_i(t)}{dt} = \frac{1}{q \cdot a \cdot d} \cdot \exp\left(-\frac{q \cdot V}{2 \cdot k \cdot T}\right) - A \cdot \exp\left(-\frac{q \cdot V}{2 \cdot k \cdot T}\right) \cdot \left(N_i(t-\tau) \cdot \exp\left(\frac{q \cdot V}{2 \cdot k \cdot T}\right) - N_{om}\right) \cdot N_{ph}(t-\tau) - \frac{N_i(t)}{\tau_s}$$

$$\frac{dN_{ph}(t)}{dt} = A \cdot \left(N_i(t-\tau) \cdot \exp\left(\frac{q \cdot V}{2 \cdot k \cdot T}\right) - N_{om}\right) \cdot N_{ph}(t-\tau) - \frac{N_{ph}(t)}{\tau_{ph}} + \frac{\beta}{\tau_s} \cdot N_i(t) \cdot \exp\left(\frac{q \cdot V}{2 \cdot k \cdot T}\right)$$

We get two possible sets of differential equations:
No.1: $\frac{dN_e(t)}{dt} = \frac{1}{q \cdot a \cdot d} - A \cdot (N_e(t-\tau) - N_{om}) \cdot N_{ph}(t-\tau) - \frac{N_e(t)}{\tau_s}$

$$\frac{dN_{ph}(t)}{dt} = A \cdot (N_e(t-\tau) - N_{om}) \cdot N_{ph}(t-\tau) - \frac{N_{ph}(t)}{\tau_{ph}} + \beta \cdot \frac{N_e(t)}{\tau_s}$$

To find the equilibrium points (fixed points) of the Laser diode coupled delay rate equation: $\lim_{t\to\infty} N_e(t-\tau) = \lim_{t\to\infty} N_e(t)$; $\lim_{t\to\infty} N_{ph}(t-\tau) = \lim_{t\to\infty} N_{ph}(t)$; $t \gg \tau$; $t - \tau \approx t \, \forall t \to \infty$.

$$\frac{dN_e(t)}{dt} = 0; \quad \frac{dN_{ph}(t)}{dt} = 0; \quad \frac{1}{q \cdot a \cdot d} - A \cdot \left(N_e^* - N_{om}\right) \cdot N_{ph}^* - \frac{N_e^*}{\tau_s} = 0$$

$$A \cdot \left(N_e^* - N_{om}\right) \cdot N_{ph}^* - \frac{N_{ph}^*}{\tau_{ph}} + \beta \cdot \frac{N_e^*}{\tau_s} = 0 \Rightarrow A \cdot \left(N_e^* - N_{om}\right) \cdot N_{ph}^* = \frac{N_{ph}^*}{\tau_{ph}} - \beta \cdot \frac{N_e^*}{\tau_s}$$

$$\frac{1}{q \cdot a \cdot d} - A \cdot \left(N_e^* - N_{om}\right) \cdot N_{ph}^* - \frac{N_e^*}{\tau_s} = 0 \Rightarrow A \cdot \left(N_e^* - N_{om}\right) \cdot N_{ph}^* = \frac{1}{q \cdot a \cdot d} - \frac{N_e^*}{\tau_s}$$

We can write the equation: $\frac{N_{ph}^*}{\tau_{ph}} - \beta \cdot \frac{N_e^*}{\tau_s} = \frac{1}{q \cdot a \cdot d} - \frac{N_e^*}{\tau_s}$

$$\frac{N_e^*}{\tau_s}(\beta - 1) = \frac{N_{ph}^*}{\tau_{ph}} - \frac{1}{q \cdot a \cdot d} \Rightarrow N_e^* = \frac{\tau_s}{(\beta - 1)} \cdot \left[\frac{N_{ph}^*}{\tau_{ph}} - \frac{1}{q \cdot a \cdot d}\right]$$

$$\frac{1}{q \cdot a \cdot d} - A \cdot \left(\frac{\tau_s}{(\beta - 1)} \cdot \left[\frac{N_{ph}^*}{\tau_{ph}} - \frac{1}{q \cdot a \cdot d}\right] - N_{om}\right) \cdot N_{ph}^* - \frac{1}{\tau_s} \cdot$$

$$\frac{\tau_s}{(\beta-1)}\cdot\left[\frac{N_{ph}^*}{\tau_{ph}}-\frac{1}{q\cdot a\cdot d}\right]=0$$

$$\frac{1}{q\cdot a\cdot d}-\frac{A\cdot\tau_s\cdot\left(N_{ph}^*\right)^2}{(\beta-1)\cdot\tau_{ph}}+A\cdot\left[\frac{\tau_s}{(\beta-1)}\cdot\frac{1}{q\cdot a\cdot d}+N_{om}\right]\cdot N_{ph}^*$$
$$-\frac{1}{(\beta-1)}\cdot\frac{N_{ph}^*}{\tau_{ph}}+\frac{1}{(\beta-1)}\cdot\frac{1}{q\cdot a\cdot d}=0$$

$$-\frac{A\cdot\tau_s\cdot\left(N_{ph}^*\right)^2}{(\beta-1)\cdot\tau_{ph}}+\left\{A\cdot\left[\frac{\tau_s}{(\beta-1)}\cdot\frac{1}{q\cdot a\cdot d}+N_{om}\right]-\frac{1}{(\beta-1)}\cdot\frac{1}{\tau_{ph}}\right\}\cdot N_{ph}^*$$
$$+\frac{1}{q\cdot a\cdot d}+\frac{1}{(\beta-1)}\cdot\frac{1}{q\cdot a\cdot d}=0$$

$$\frac{A\cdot\tau_s\cdot\left(N_{ph}^*\right)^2}{(1-\beta)\cdot\tau_{ph}}+\left\{A\cdot\left[\frac{\tau_s}{(\beta-1)}\cdot\frac{1}{q\cdot a\cdot d}+N_{om}\right]\frac{1}{(\beta-1)}\cdot\frac{1}{\tau_{ph}}\right\}\cdot N_{ph}^*$$
$$+\frac{1}{q\cdot a\cdot d}\cdot\left[1-\frac{1}{(1-\beta)}\right]=0$$

$$\frac{A\cdot\tau_s\cdot\left(N_{ph}^*\right)^2}{(1-\beta)\cdot\tau_{ph}}+\left\{A\cdot\left[-\frac{\tau_s}{(1-\beta)}\cdot\frac{1}{q\cdot a\cdot d}+N_{om}\right]+\frac{1}{(1-\beta)}\cdot\frac{1}{\tau_{ph}}\right\}\cdot N_{ph}^*$$
$$+\frac{1}{q\cdot a\cdot d}\cdot\left[1-\frac{1}{(1-\beta)}\right]=0$$

$$\frac{A\cdot\tau_s\cdot\left(N_{ph}^*\right)^2}{(1-\beta)\cdot\tau_{ph}}+\left\{\frac{1}{(1-\beta)}\cdot\frac{1}{\tau_{ph}}-A\cdot\left[\frac{\tau_s}{(1-\beta)}\cdot\frac{1}{q\cdot a\cdot d}-N_{om}\right]\right\}\cdot N_{ph}^*$$
$$+\frac{1}{q\cdot a\cdot d}\cdot\left[1-\frac{1}{(1-\beta)}\right]=0$$

$$\left(N_{ph}^*\right)^2\cdot\Xi_1+N_{ph}^*\cdot\Xi_2+\Xi_3=0;\;N_{ph}^*=\frac{-\Xi_2\pm\sqrt{\Xi_2^2-4\cdot\Xi_1\cdot\Xi_3}}{2\cdot\Xi_1};\;N_{ph}^{*\,(1)}$$
$$=\frac{-\Xi_2+\sqrt{\Xi_2^2-4\cdot\Xi_1\cdot\Xi_3}}{2\cdot\Xi_1}$$

$$N_{ph}^{*\,(2)}=\frac{-\Xi_2-\sqrt{\Xi_2^2-4\cdot\Xi_1\cdot\Xi_3}}{2\cdot\Xi_1};\;(N_e^*)^{(1)}$$
$$=\frac{\tau_s}{(\beta-1)}\cdot\left[\frac{1}{\tau_{ph}}\cdot\left(\frac{-\Xi_2+\sqrt{\Xi_2^2-4\cdot\Xi_1\cdot\Xi_3}}{2\cdot\Xi_1}\right)-\frac{1}{q\cdot a\cdot d}\right]$$

$$(N_e^*)^{(2)} = \frac{\tau_s}{(\beta - 1)} \cdot \left[\frac{1}{\tau_{ph}} \cdot \left(\frac{-\Xi_2 - \sqrt{\Xi_2^2 - 4 \cdot \Xi_1 \cdot \Xi_3}}{2 \cdot \Xi_1} \right) - \frac{1}{q \cdot a \cdot d} \right]$$

The first fixed point:

$$E^{(1)}\left(N_{ph}^{*}{}^{(1)}, (N_e^*)^{(1)}\right) = \left(\frac{-\Xi_2 + \sqrt{\Xi_2^2 - 4 \cdot \Xi_1 \cdot \Xi_3}}{2 \cdot \Xi_1}, \frac{\tau_s}{(\beta - 1)} \cdot \left[\frac{1}{\tau_{ph}} \cdot \left(\frac{-\Xi_2 + \sqrt{\Xi_2^2 - 4 \cdot \Xi_1 \cdot \Xi_3}}{2 \cdot \Xi_1} \right) - \frac{1}{q \cdot a \cdot d} \right] \right)$$

The second fixed point:

$$E^{(2)}\left(N_{ph}^{*}{}^{(2)}, (N_e^*)^{(2)}\right) = \left(\frac{-\Xi_2 - \sqrt{\Xi_2^2 - 4 \cdot \Xi_1 \cdot \Xi_3}}{2 \cdot \Xi_1}, \frac{\tau_s}{(\beta - 1)} \cdot \left[\frac{1}{\tau_{ph}} \cdot \left(\frac{-\Xi_2 - \sqrt{\Xi_2^2 - 4 \cdot \Xi_1 \cdot \Xi_3}}{2 \cdot \Xi_1} \right) - \frac{1}{q \cdot a \cdot d} \right] \right)$$

Stability analysis: The standard local stability analysis about any one of the equilibrium point of Laser diode delayed rate equations consists in adding to its coordinates $[N_{ph}\ N_e]$ arbitrarily small increments of exponential form $[n_{ph}\ n_e]\cdot e^{\lambda \cdot t}$, and retaining the first order terms in n_{ph}, n_e. The system of two homogeneous equations leads to polynomial characteristics equation in the eigenvalue λ. The polynomial characteristic equations accept by set the below photon density and electron density equations. The photon density and electron density fixed values with arbitrarily small increments of exponential form $[n_{ph}\ n_e]\cdot e^{\lambda \cdot t}$ are: i = 0 (first fixed point), i = 1 (second fixed point), i = 2 (third fixed point), etc.

$$N_{ph}(t) = N_{ph}^{(i)} + n_{ph} \cdot e^{\lambda \cdot t};\ N_e(t) = N_e^{(i)} + n_e \cdot e^{\lambda \cdot t};\ N_{ph}(t - \tau) = N_{ph}^{(i)} + n_{ph} \cdot e^{\lambda \cdot (t - \tau)}$$

$$N_e(t - \tau) = N_e^{(i)} + n_e \cdot e^{\lambda \cdot (t - \tau)};\ \frac{dN_{ph}(t)}{dt} = n_{ph} \cdot \lambda \cdot e^{\lambda \cdot t};\ \frac{dN_e(t)}{dt} = n_e \cdot \lambda \cdot e^{\lambda \cdot t}$$

$$\frac{dN_{ph}(t - \tau)}{dt} = n_{ph} \cdot \lambda \cdot e^{\lambda \cdot t} \cdot e^{-\lambda \cdot \tau};\ \frac{dN_e(t - \tau)}{dt} = n_e \cdot \lambda \cdot e^{\lambda \cdot t} \cdot e^{-\lambda \cdot \tau}$$

We choose the above expressions for our $N_{ph}(t)$, $N_e(t)$ as small displacement $[n_{ph}\ n_e]$ from the system fixed points at time t = 0: $N_{ph}(t = 0) = N_{ph}^{(i)} + n_{ph}$; $N_e(t = 0) = N_e^{(i)} + n_e$ for $t > 0,\ \lambda < 0$ the selected fixed point is stable otherwise $t > 0,\ \lambda > 0$ is unstable. Our system tends to the selected fixed point exponentially for $t > 0,\ \lambda < 0$ otherwise go away from the selected fixed point exponentially. The eigenvalue λ parameter establishes if the fixed point is stable or

Table 1.1 N_{ph}, N_e expressions for eigenvalue λ and time parameter

	$\lambda < 0$	$\lambda > 0$
$t = 0$	$N_{ph}(t=0) = N_{ph}^{(i)} + n_{ph}$ $N_e(t=0) = N_e^{(i)} + n_e$	$N_{ph}(t=0) = N_{ph}^{(i)} + n_{ph}$ $N_e(t=0) = N_e^{(i)} + n_e$
$t > 0$	$N_{ph}(t) = N_{ph}^{(i)} + n_{ph} \cdot e^{-\lvert\lambda\rvert \cdot t}$ $N_e(t) = N_e^{(i)} + n_e \cdot e^{-\lvert\lambda\rvert \cdot t}$	$N_{ph}(t) = N_{ph}^{(i)} + n_{ph} \cdot e^{\lvert\lambda\rvert \cdot t}$ $N_e(t) = N_e^{(i)} + n_e \cdot e^{\lvert\lambda\rvert \cdot t}$
$t \to \infty$	$N_{ph}(t \to \infty) = N_{ph}^{(i)}$ $N_e(t \to \infty) = N_e^{(i)}$	$N_{ph}(t \to \infty) \simeq n_{ph} \cdot e^{\lvert\lambda\rvert \cdot t}$ $N_e(t \to \infty) \simeq n_e \cdot e^{\lvert\lambda\rvert \cdot t}$

unstable, additionally his absolute value |λ| establishes the speed of flow toward or away from the selected fixed point [8, 9] (Table 1.1).

The speeds of flow toward or away from the selected fixed point for Laser diode delayed rate equations:

$$\frac{dN_e(t)}{dt} = \frac{1}{q \cdot a \cdot d} - A \cdot (N_e(t-\tau) - N_{om}) \cdot N_{ph}(t-\tau) - \frac{N_e(t)}{\tau_s}$$

$$n_e \cdot \lambda \cdot e^{\lambda \cdot t} = \frac{1}{q \cdot a \cdot d} - A \cdot \left(N_e^{(i)} + n_e \cdot e^{\lambda \cdot t} \cdot e^{-\lambda \cdot \tau} - N_{om}\right) \cdot \left[N_{ph}^{(i)} + n_{ph} \cdot e^{\lambda \cdot t} \cdot e^{-\lambda \cdot \tau}\right] - \frac{1}{\tau_s} \cdot \left[N_e^{(i)} + n_e \cdot e^{\lambda \cdot t}\right]$$

$$n_e \cdot \lambda \cdot e^{\lambda \cdot t} = \frac{1}{q \cdot a \cdot d} - A \cdot \left(\left[N_e^{(i)} - N_{om}\right] + n_e \cdot e^{\lambda \cdot t} \cdot e^{-\lambda \cdot \tau}\right) \cdot \left(N_{ph}^{(i)} + n_{ph} \cdot e^{\lambda \cdot t} \cdot e^{-\lambda \cdot \tau}\right) - \frac{1}{\tau_s} \cdot N_e^{(i)} - \frac{1}{\tau_s} \cdot n_e \cdot e^{\lambda \cdot t}$$

$$\begin{aligned} n_e \cdot \lambda \cdot e^{\lambda \cdot t} = &\frac{1}{q \cdot a \cdot d} \\ &- A \cdot \left[N_e^{(i)} - N_{om}\right] \cdot N_{ph}^{(i)} - A \cdot \left[N_e^{(i)} - N_{om}\right] \cdot n_{ph} \cdot e^{\lambda \cdot t} \cdot e^{-\lambda \cdot \tau} \\ &- A \cdot N_{ph}^{(i)} \cdot n_e \cdot e^{\lambda \cdot t} \cdot e^{-\lambda \cdot \tau} - A \cdot n_e \cdot n_{ph} \cdot e^{\lambda \cdot t} \cdot e^{-\lambda \cdot \tau} \cdot e^{\lambda \cdot t} \cdot e^{-\lambda \cdot \tau} \\ &- \frac{1}{\tau_s} \cdot N_e^{(i)} - \frac{1}{\tau_s} \cdot n_e \cdot e^{\lambda \cdot t} \end{aligned}$$

$$\begin{aligned} n_e \cdot \lambda \cdot e^{\lambda \cdot t} = &\frac{1}{q \cdot a \cdot d} - A \cdot \left[N_e^{(i)} - N_{om}\right] \cdot N_{ph}^{(i)} - \frac{1}{\tau_s} \cdot N_e^{(i)} \\ &- A \cdot \left[N_e^{(i)} - N_{om}\right] \cdot n_{ph} \cdot e^{\lambda \cdot t} \cdot e^{-\lambda \cdot \tau} \\ &- A \cdot N_{ph}^{(i)} \cdot n_e \cdot e^{\lambda \cdot t} \cdot e^{-\lambda \cdot \tau} - A \cdot n_e \cdot n_{ph} \cdot e^{2 \cdot \lambda \cdot t} \cdot e^{-2 \cdot \lambda \cdot \tau} \\ &- \frac{1}{\tau_s} \cdot n_e \cdot e^{\lambda \cdot t} \end{aligned}$$

$n_e \cdot n_{ph}$ multiplication is very small $n_e \cdot n_{ph} \to \varepsilon$

$$\begin{aligned} n_e \cdot \lambda \cdot e^{\lambda \cdot t} = {} & \frac{1}{q \cdot a \cdot d} - A \cdot \left[N_e^{(i)} - N_{om}\right] \cdot N_{ph}^{(i)} - \frac{1}{\tau_s} \cdot N_e^{(i)} \\ & - A \cdot \left[N_e^{(i)} - N_{om}\right] \cdot n_{ph} \cdot e^{\lambda \cdot t} \cdot e^{-\lambda \cdot \tau} \\ & - A \cdot N_{ph}^{(i)} \cdot n_e \cdot e^{\lambda \cdot t} \cdot e^{-\lambda \cdot \tau} - \frac{1}{\tau_s} \cdot n_e \cdot e^{\lambda \cdot t} \end{aligned}$$

At fixed point: $\frac{1}{q \cdot a \cdot d} - A \cdot \left[N_e^{(i)} - N_{om}\right] \cdot N_{ph}^{(i)} - \frac{1}{\tau_s} \cdot N_e^{(i)} = 0$

$$\begin{aligned} n_e \cdot \lambda \cdot e^{\lambda \cdot t} = {} & -A \cdot \left[N_e^{(i)} - N_{om}\right] \cdot n_{ph} \cdot e^{\lambda \cdot t} \cdot e^{-\lambda \cdot \tau} \\ & - A \cdot N_{ph}^{(i)} \cdot n_e \cdot e^{\lambda \cdot t} \cdot e^{-\lambda \cdot \tau} - \frac{1}{\tau_s} \cdot n_e \cdot e^{\lambda \cdot t} \end{aligned}$$

Dividing the two sides of the above equation by $e^{\lambda \cdot t}$ term

$$-A \cdot \left[N_e^{(i)} - N_{om}\right] \cdot n_{ph} \cdot e^{-\lambda \cdot \tau} + n_e \left\{-A \cdot N_{ph}^{(i)} \cdot e^{-\lambda \cdot \tau} - \frac{1}{\tau_s} - \lambda\right\} = 0$$

$$\frac{dN_{ph}(t)}{dt} = A \cdot (N_e(t - \tau) - N_{om}) \cdot N_{ph}(t - \tau) - \frac{N_{ph}(t)}{\tau_{ph}} + \beta \cdot \frac{N_e(t)}{\tau_s}$$

$$\begin{aligned} n_{ph} \cdot \lambda \cdot e^{\lambda \cdot t} = {} & A \cdot \left(N_e^{(i)} + n_e \cdot e^{\lambda \cdot (t - \tau)} - N_{om}\right) \cdot \left(N_{ph}^{(i)} + n_{ph} \cdot e^{\lambda \cdot (t - \tau)}\right) \\ & - \frac{1}{\tau_{ph}} \cdot \left(N_{ph}^{(i)} + n_{ph} \cdot e^{\lambda \cdot t}\right) + \beta \cdot \frac{1}{\tau_s} \cdot \left(N_e^{(i)} + n_e \cdot e^{\lambda \cdot t}\right) \end{aligned}$$

$$\begin{aligned} n_{ph} \cdot \lambda \cdot e^{\lambda \cdot t} = {} & A \cdot \left(\left[N_e^{(i)} - N_{om}\right] + n_e \cdot e^{\lambda \cdot (t - \tau)}\right) \cdot \left(N_{ph}^{(i)} + n_{ph} \cdot e^{\lambda \cdot (t - \tau)}\right) \\ & - \frac{1}{\tau_{ph}} \cdot N_{ph}^{(i)} - \frac{1}{\tau_{ph}} \cdot n_{ph} \cdot e^{\lambda \cdot t} + \beta \cdot \frac{1}{\tau_s} \cdot N_e^{(i)} + \beta \cdot \frac{1}{\tau_s} \cdot n_e \cdot e^{\lambda \cdot t} \end{aligned}$$

$$\begin{aligned} n_{ph} \cdot \lambda \cdot e^{\lambda \cdot t} = {} & A \cdot \left[N_e^{(i)} - N_{om}\right] \cdot N_{ph}^{(i)} \\ & + A \cdot \left[N_e^{(i)} - N_{om}\right] \cdot n_{ph} \cdot e^{\lambda \cdot (t - \tau)} + A \cdot N_{ph}^{(i)} \cdot n_e \cdot e^{\lambda \cdot (t - \tau)} \\ & + A \cdot n_e \cdot n_{ph} \cdot e^{2 \cdot \lambda \cdot (t - \tau)} - \frac{1}{\tau_{ph}} \cdot N_{ph}^{(i)} - \frac{1}{\tau_{ph}} \cdot n_{ph} \cdot e^{\lambda \cdot t} \\ & + \beta \cdot \frac{1}{\tau_s} \cdot N_e^{(i)} + \beta \cdot \frac{1}{\tau_s} \cdot n_e \cdot e^{\lambda \cdot t} \end{aligned}$$

$n_e \cdot n_{ph}$ Multiplication is very small $n_e \cdot n_{ph} \to \varepsilon$

$$\begin{aligned} n_{ph} \cdot \lambda \cdot e^{\lambda \cdot t} = {} & A \cdot \left[N_e^{(i)} - N_{om}\right] \cdot N_{ph}^{(i)} \\ & + A \cdot \left[N_e^{(i)} - N_{om}\right] \cdot n_{ph} \cdot e^{\lambda \cdot (t - \tau)} + A \cdot N_{ph}^{(i)} \cdot n_e \cdot e^{\lambda \cdot (t - \tau)} \end{aligned}$$

$$-\frac{1}{\tau_{ph}} \cdot N_{ph}^{(i)} - \frac{1}{\tau_{ph}} \cdot n_{ph} \cdot e^{\lambda \cdot t} + \beta \cdot \frac{1}{\tau_s} \cdot N_e^{(i)} + \beta \cdot \frac{1}{\tau_s} \cdot n_e \cdot e^{\lambda \cdot t}$$

$$\begin{aligned} n_{ph} \cdot \lambda \cdot e^{\lambda \cdot t} &= A \cdot \left[N_e^{(i)} - N_{om}\right] \cdot N_{ph}^{(i)} \\ &- \frac{1}{\tau_{ph}} \cdot N_{ph}^{(i)} + \beta \cdot \frac{1}{\tau_s} \cdot N_e^{(i)} \\ &+ A \cdot \left[N_e^{(i)} - N_{om}\right] \cdot n_{ph} \cdot e^{\lambda \cdot (t-\tau)} + A \cdot N_{ph}^{(i)} \cdot n_e \cdot e^{\lambda \cdot (t-\tau)} \\ &- \frac{1}{\tau_{ph}} \cdot n_{ph} \cdot e^{\lambda \cdot t} + \beta \cdot \frac{1}{\tau_s} \cdot n_e \cdot e^{\lambda \cdot t} \end{aligned}$$

At fixed point: $A \cdot \left[N_e^{(i)} - N_{om}\right] \cdot N_{ph}^{(i)} - \frac{1}{\tau_{ph}} \cdot N_{ph}^{(i)} + \beta \cdot \frac{1}{\tau_s} \cdot N_e^{(i)} = 0$

$$\begin{aligned} n_{ph} \cdot \lambda \cdot e^{\lambda \cdot t} &= A \cdot \left[N_e^{(i)} - N_{om}\right] \cdot n_{ph} \cdot e^{\lambda \cdot t} \cdot e^{-\lambda \cdot \tau} \\ &+ A \cdot N_{ph}^{(i)} \cdot n_e \cdot e^{\lambda \cdot t} \cdot e^{-\lambda \cdot \tau} - \frac{1}{\tau_{ph}} \cdot n_{ph} \cdot e^{\lambda \cdot t} + \beta \cdot \frac{1}{\tau_s} \cdot n_e \cdot e^{\lambda \cdot t} \end{aligned}$$

Dividing the two sides of the above equation by $e^{\lambda \cdot t}$ term

$$n_{ph} \cdot \lambda = A \cdot \left[N_e^{(i)} - N_{om}\right] \cdot n_{ph} \cdot e^{-\lambda \cdot \tau} + A \cdot N_{ph}^{(i)} \cdot n_e \cdot e^{-\lambda \cdot \tau} - \frac{1}{\tau_{ph}} \cdot n_{ph} + \beta \cdot \frac{1}{\tau_s} \cdot n_e$$

$$n_{ph} \cdot \left(A \cdot \left[N_e^{(i)} - N_{om}\right] \cdot e^{-\lambda \cdot \tau} - \frac{1}{\tau_{ph}} - \lambda\right) + n_e \cdot \left(A \cdot N_{ph}^{(i)} \cdot e^{-\lambda \cdot \tau} + \beta \cdot \frac{1}{\tau_s}\right) = 0$$

We can summary our system arbitrarily small increments equation:

$$\begin{aligned} n_{ph} \cdot &\left(A \cdot \left[N_e^{(i)} - N_{om}\right] \cdot e^{-\lambda \cdot \tau} - \frac{1}{\tau_{ph}} - \lambda\right) \\ &+ n_e \cdot \left(A \cdot N_{ph}^{(i)} \cdot e^{-\lambda \cdot \tau} + \beta \cdot \frac{1}{\tau_s}\right) = 0 \end{aligned}$$

$$-A \cdot \left[N_e^{(i)} - N_{om}\right] \cdot e^{-\lambda \cdot \tau} \cdot n_{ph} + n_e \left\{-A \cdot N_{ph}^{(i)} \cdot e^{-\lambda \cdot \tau} - \frac{1}{\tau_s} - \lambda\right\} = 0$$

We get the Laser diode delayed rate equations eigenvalue matrix:

$$\begin{pmatrix} A \cdot \left[N_e^{(i)} - N_{om}\right] \cdot e^{-\lambda \cdot \tau} - \frac{1}{\tau_{ph}} - \lambda & A \cdot N_{ph}^{(i)} \cdot e^{-\lambda \cdot \tau} + \beta \cdot \frac{1}{\tau_s} \\ -A \cdot \left[N_e^{(i)} - N_{om}\right] \cdot e^{-\lambda \cdot \tau} & -A \cdot N_{ph}^{(i)} \cdot e^{-\lambda \cdot \tau} - \frac{1}{\tau_s} - \lambda \end{pmatrix} \cdot \begin{pmatrix} n_{ph} \\ n_e \end{pmatrix} = \begin{pmatrix} 0 \\ 0 \end{pmatrix}$$

$$(A - \lambda \cdot I) = \begin{pmatrix} A \cdot \left[N_e^{(i)} - N_{om}\right] \cdot e^{-\lambda \cdot \tau} - \frac{1}{\tau_{ph}} - \lambda & A \cdot N_{ph}^{(i)} \cdot e^{-\lambda \cdot \tau} + \beta \cdot \frac{1}{\tau_s} \\ -A \cdot \left[N_e^{(i)} - N_{om}\right] \cdot e^{-\lambda \cdot \tau} & -A \cdot N_{ph}^{(i)} \cdot e^{-\lambda \cdot \tau} - \frac{1}{\tau_s} - \lambda \end{pmatrix}$$

$$\det(A - \lambda \cdot I) = 0 \Rightarrow \det\begin{pmatrix} A \cdot \left[N_e^{(i)} - N_{om}\right] \cdot e^{-\lambda \cdot \tau} - \frac{1}{\tau_{ph}} - \lambda & A \cdot N_{ph}^{(i)} \cdot e^{-\lambda \cdot \tau} + \beta \cdot \frac{1}{\tau_s} \\ -A \cdot \left[N_e^{(i)} - N_{om}\right] \cdot e^{-\lambda \cdot \tau} & -A \cdot N_{ph}^{(i)} \cdot e^{-\lambda \cdot \tau} - \frac{1}{\tau_s} - \lambda \end{pmatrix} = 0$$

$$\left(A \cdot \left[N_e^{(i)} - N_{om}\right] \cdot e^{-\lambda \cdot \tau} - \frac{1}{\tau_{ph}} - \lambda\right) \cdot \left(-A \cdot N_{ph}^{(i)} \cdot e^{-\lambda \cdot \tau} - \frac{1}{\tau_s} - \lambda\right) + A \cdot \left[N_e^{(i)} - N_{om}\right] \cdot e^{-\lambda \cdot \tau} \cdot \left(A \cdot N_{ph}^{(i)} \cdot e^{-\lambda \cdot \tau} + \beta \cdot \frac{1}{\tau_s}\right) = 0$$

$$\begin{aligned} &-A^2 \cdot N_{ph}^{(i)} \cdot \left[N_e^{(i)} - N_{om}\right] \cdot e^{-2 \cdot \lambda \cdot \tau} - \frac{1}{\tau_s} \cdot A \cdot \left[N_e^{(i)} - N_{om}\right] \cdot e^{-\lambda \cdot \tau} \\ &-\lambda \cdot A \cdot \left[N_e^{(i)} - N_{om}\right] \cdot e^{-\lambda \cdot \tau} \\ &+\frac{1}{\tau_{ph}} \cdot A \cdot N_{ph}^{(i)} \cdot e^{-\lambda \cdot \tau} + \frac{1}{\tau_{ph} \cdot \tau_s} + \frac{1}{\tau_{ph}} \cdot \lambda + \lambda \cdot A \cdot N_{ph}^{(i)} \cdot e^{-\lambda \cdot \tau} \\ &+\lambda \cdot \frac{1}{\tau_s} + \lambda^2 \\ &+A \cdot \left[N_e^{(i)} - N_{om}\right] \cdot A \cdot N_{ph}^{(i)} \cdot e^{-2 \cdot \lambda \cdot \tau} \\ &+A \cdot \left[N_e^{(i)} - N_{om}\right] \cdot \beta \cdot \frac{1}{\tau_s} \cdot e^{-\lambda \cdot \tau} = 0 \end{aligned}$$

$$\begin{aligned} &-A^2 \cdot N_{ph}^{(i)} \cdot \left[N_e^{(i)} - N_{om}\right] \cdot e^{-2 \cdot \lambda \cdot \tau} + A^2 \cdot \left[N_e^{(i)} - N_{om}\right] \cdot N_{ph}^{(i)} \cdot e^{-2 \cdot \lambda \cdot \tau} \\ &+\left(A \cdot \left[N_e^{(i)} - N_{om}\right] \cdot \frac{1}{\tau_s} \cdot (\beta - 1) + \frac{1}{\tau_{ph}} \cdot A \cdot N_{ph}^{(i)} + \lambda \cdot A \cdot \left[N_{ph}^{(i)} - N_e^{(i)} + N_{om}\right]\right) \cdot e^{-\lambda \cdot \tau} \\ &+\frac{1}{\tau_{ph} \cdot \tau_s} + \left(\frac{1}{\tau_{ph}} + \frac{1}{\tau_s}\right) \cdot \lambda + \lambda^2 = 0 \end{aligned}$$

$$-A^2 \cdot N_{ph}^{(i)} \cdot \left[N_e^{(i)} - N_{om}\right] \cdot e^{-2 \cdot \lambda \cdot \tau} + A^2 \cdot \left[N_e^{(i)} - N_{om}\right] \cdot N_{ph}^{(i)} \cdot e^{-2 \cdot \lambda \cdot \tau} = 0$$

$$\frac{1}{\tau_{ph} \cdot \tau_s} + \left(\frac{1}{\tau_{ph}} + \frac{1}{\tau_s}\right) \cdot \lambda + \lambda^2 + \left\{A \cdot \left([N_e^{(i)} - N_{om}] \cdot \frac{1}{\tau_s} \cdot (\beta - 1) + \frac{1}{\tau_{ph}} \cdot N_{ph}^{(i)}\right) + \lambda \cdot A \cdot [N_{ph}^{(i)} - N_e^{(i)} + N_{om}]\right\} \cdot e^{-\lambda \cdot \tau} = 0$$

$$D(\lambda, \tau) = P_n(\lambda, \tau) + Q_m(\lambda, \tau) \cdot e^{-\lambda \cdot \tau}$$

$$n = 2;\; m = 1;\; n < m;\; P_n(\lambda, \tau) = \frac{1}{\tau_{ph} \cdot \tau_s} + \left(\frac{1}{\tau_{ph}} + \frac{1}{\tau_s}\right) \cdot \lambda + \lambda^2$$

$$Q_m(\lambda, \tau) = A \cdot \left(\left[N_e^{(i)} - N_{om}\right] \cdot \frac{1}{\tau_s} \cdot (\beta - 1) + \frac{1}{\tau_{ph}} \cdot N_{ph}^{(i)}\right) + \lambda \cdot A \cdot \left[N_{ph}^{(i)} - N_e^{(i)} + N_{om}\right]$$

The expression for $P_n(\lambda, \tau)$: $P_n(\lambda, \tau) = \sum_{k=0}^{n=2} \lambda^k \cdot p_k(\tau)$

$$p_0(\tau) = \frac{1}{\tau_{ph} \cdot \tau_s};\; p_1(\tau) = \frac{1}{r_{ph}} + \frac{1}{\tau_s};\; p_2(\tau) = 1$$

The expression for $Q_m(\lambda, \tau)$: $Q_m(\lambda, \tau) = \sum_{k=0}^{m=1} \lambda^k \cdot q_k(\tau)$

$$q_0(\tau) = A \cdot \left(\left[N_c^{(i)} - N_{om}\right] \cdot \frac{1}{\tau_s} \cdot (\beta - 1) + \frac{1}{\tau_{ph}} \cdot N_{ph}^{(i)}\right)$$

$$q_1(\tau) = A \cdot \left[N_{ph}^{(i)} - N_e^{(i)} + N_{om}\right]$$

The coefficients p_i, q_j are not depend on τ parameter; i = 0,1, n ; j = 0, m

The homogeneous system for N_{ph}, N_e leads to a characteristic equation for the eigenvalue λ having the form $P(\lambda, \tau) + Q(\lambda, \tau) \cdot e^{-\lambda t} = 0$; $P(\lambda) = \sum_{j=0}^{2} a_j \cdot \lambda^j$

$Q(\lambda) = \sum_{j=0}^{1} c_j \cdot \lambda^j$. The coefficients $\{a_j(q_i, q_1, \tau), c_j(q_i, q_k, \tau)\} \in \Gamma$ depend on q_i, q_k and delay τ. The parameters q_i, q_k are any Laser diode delayed rate equation's parameters, other parameters kept as a constant.

$$a_0 = \frac{1}{\tau_{ph} \cdot \tau_s}; \quad a_1 = \frac{1}{\tau_{ph}} + \frac{1}{\tau_s}; \quad a_2 = 1$$

$$c_0 = A \cdot \left(\left[N_c^{(i)} - N_{om}\right] \cdot \frac{1}{\tau_s} \cdot (\beta - 1) + \frac{1}{\tau_{ph}} \cdot N_{ph}^{(i)}\right)$$

$$c_1 = A \cdot \left[N_{ph}^{(i)} - N_e^{(n)} + N_{om}\right]$$

Unless strictly necessary, the designation of the varied arguments (q_i, q_k) will subsequently be omitted from P, Q, a_j, and c_j. The coefficients a_j, c_j are continuous and differentiable functions of their arguments, and direct substitution shows that $a_0 + c_0 \neq 0$ for $q_1, q_k \in \mathbb{R}_+$; that is, $\lambda = 0$ is not of $P(\lambda) + Q(\lambda) \cdot e^{-\lambda \cdot \tau} = 0$ Furthermore P(λ), Q(λ) are analytic functions of λ, for which the following requirements of the analysis [2, 4] can also be verified in the present case:

(a) If $\lambda = i \cdot \omega$; $\omega \in \Gamma$ then $P(i \cdot \omega) + Q(i \cdot \omega) \neq 0$.
(b) If $|Q(\lambda)/P(\lambda)|$ is bounded for $|\lambda| \to \infty$; $\text{Re}\lambda \geq 0$. No roots bifurcation from ∞.
(c) $F(\omega) = |P(i \cdot \omega)|^2 - |Q(i \cdot \omega)|^2$ has a finite number of zeros. Indeed, this is a polynomial in ω.
(d) Each positive root $\omega(q_i, q_k)$ of $F(\omega) = 0$ is continuous and differentiable with respect to q_i, q_k.

We assume that $P_n(\lambda, \tau)$ and $Q_m(\lambda, \tau)$ cannot have common imaginary roots. That is for any real number ω: $P_n(\lambda = i \cdot \omega, \tau) + Q_m(\lambda = i \cdot \omega, \tau) \neq 0$.

$$P_n(\lambda = i \cdot \omega, \tau) = \frac{1}{\tau_{ph} \cdot \tau_s} - \omega^2 + \left(\frac{1}{\tau_{ph}} + \frac{1}{\tau_s}\right) \cdot i \cdot \omega$$

$$Q_m(\lambda = i \cdot \omega, \tau) = A \cdot \left(\left[N_e^{(i)} - N_{om}\right] \cdot \frac{1}{\tau_s} \cdot (\beta - 1) + \frac{1}{\tau_{ph}} \cdot N_{ph}^{(i)}\right)$$

$$+ i \cdot \omega \cdot A \cdot \left[N_{ph}^{(i)} - N_e^{(i)} + N_{om}\right]$$

$$P_n(\lambda = i \cdot \omega, \tau) + Q_m(\lambda = i \cdot \omega, \tau)$$
$$= \frac{1}{\tau_{ph} \cdot \tau_s} + A \cdot \left(\left[N_e^{(i)} - N_{om}\right] \cdot \frac{1}{\tau_s} \cdot (\beta - 1) + \frac{1}{\tau_{ph}} \cdot N_{ph}^{(i)}\right) - \omega^2$$
$$+ \left\{\left(\frac{1}{\tau_{ph}} + \frac{1}{\tau_s}\right) + A \cdot \left(N_{ph}^{(i)} - N_e^{(i)} + N_{om}\right)\right\} \cdot i \cdot \omega \neq 0$$

$$|P(i \cdot \omega, \tau)|^2 = \left[\frac{1}{\tau_{ph} \cdot \tau_s} - \omega^2\right]^2 + \left(\frac{1}{\tau_{ph}} + \frac{1}{\tau_s}\right)^2 \cdot$$
$$\omega^2 = \frac{1}{(\tau_{ph} \cdot \tau_s)^2} + \left[\left(\frac{1}{\tau_{ph}} + \frac{1}{\tau_s}\right)^2 - 2 \cdot \frac{1}{\tau_{ph} \cdot \tau_s}\right] \cdot \omega^2 + \omega^4$$

$$|Q(i \cdot \omega, \tau)|^2 = A^2 \cdot \left(\left[N_e^{(i)} - N_{om}\right] \cdot \frac{1}{\tau_s} \cdot (\beta - 1) + \frac{1}{\tau_{ph}} \cdot N_{ph}^{(i)}\right)^2$$
$$+ \omega^2 \cdot A^2 \cdot \left[N_{ph}^{(i)} - N_e^{(i)} + N_{om}\right]^2$$

$$F(\omega, \tau) = |P(i \cdot \omega)|^2 - |Q(i \cdot \omega)|^2 = \frac{1}{(\tau_{ph} \cdot \tau_s)^2}$$
$$- A^2 \cdot \left(\left[N_e^{(i)} - N_{om}\right] \cdot \frac{1}{\tau_s} \cdot (\beta - 1) + \frac{1}{\tau_{ph}} \cdot N_{ph}^{(i)}\right)^2$$
$$+ \left\{\left[\left(\frac{1}{\tau_{ph}} + \frac{1}{\tau_s}\right)^2 - 2 \cdot \frac{1}{\tau_{ph} \cdot \tau_s}\right] - A^2 \cdot \left[N_{ph}^{(i)} - N_e^{(i)} + N_{om}\right]^2\right\} \cdot \omega^2 + \omega^4$$

We define the following parameters for simplicity: $\prod_0, \prod_2, \prod_4$

$$\prod\nolimits_0 = \frac{1}{(\tau_{ph} \cdot \tau_s)^2} - A^2 \cdot \left(\left[N_e^{(i)} - N_{om}\right] \cdot \frac{1}{\tau_s} \cdot (\beta - 1) + \frac{1}{\tau_{ph}} \cdot N_{ph}^{(i)}\right)^2$$

$$\prod\nolimits_2 = \left[\left(\frac{1}{\tau_{ph}} + \frac{1}{\tau_s}\right)^2 - 2 \cdot \frac{1}{\tau_{ph} \cdot \tau_s}\right] - A^2 \cdot \left[N_{ph}^{(i)} - N_e^{(i)} + N_{om}\right]^2; \ \prod\nolimits_4 = 1$$

Hence $F(\omega, \tau) = 0$ implies $\sum_{k=0}^{2} \prod_{2 \cdot k} \cdot \omega^{2 \cdot k} = 0$ and its roots are given by solving the polynomial.

$$P_R(i \cdot \omega, \tau) = \frac{1}{\tau_{ph} \cdot \tau_s} - \omega^2; \ P_I(i \cdot \omega, \tau) = \left(\frac{1}{\tau_{ph}} + \frac{1}{\tau_s}\right) \cdot \omega$$

$$Q_R(i\cdot\omega,\tau) = A\cdot\left(\left[N_e^{(i)} - N_{om}\right]\cdot\frac{1}{\tau_s}\cdot(\beta-1) + \frac{1}{\tau_{ph}}\cdot N_{ph}^{(i)}\right)$$

$$Q_I(i\cdot\omega,\tau) = \omega\cdot A\cdot\left[N_{ph}^{(i)} - N_e^{(i)} + N_{om}\right]$$

$$\sin\theta(\tau) = \frac{-P_R(i\cdot\omega,\tau)\cdot Q_I(i\cdot\omega,\tau) + P_I(i\cdot\omega,\tau)\cdot Q_R(i\cdot\omega,\tau)}{|Q(i\cdot\omega,\tau)|^2}$$

$$\cos\theta(\tau) = -\frac{P_R(i\cdot\omega,\tau)\cdot Q_R(i\cdot\omega,\tau) + P_I(i\cdot\omega,\tau)\cdot Q_I(i\cdot\omega,\tau)}{|Q(i\cdot\omega,\tau)|^2}$$

We use different parameters terminology from our last characteristics parameters definition:

$$k\to j;\; p_k(\tau)\to a_j;\; q_k(\tau)\to c_j;\; n=2$$
$$m=1;\; n>;\; m;\; P_n(\lambda,\tau)\to P(\lambda);\; Q_m(\lambda,\tau)\to Q(\lambda)$$

$$P(\lambda) = \sum_{j=0}^{2} a_j\cdot\lambda^j;\; Q(\lambda) = \sum_{j=0}^{1} c_j\cdot\lambda^j;\; P(\lambda)$$
$$= a_0 + a_1\cdot\lambda + a_2\cdot\lambda^2;\; Q(\lambda) = c_0 + c_1\cdot\lambda$$

$n, m \in \mathbb{N}_0$; $n > m$ and $a_j, c_j : \mathbb{R}_{0+} \to \mathbb{R}$ are continuous and differentiable function of τ such that $a_0 + c_0 \neq 0$. In the following "-" denotes complex and conjugate. $P(\lambda)$, $Q(\lambda)$ are analytic functions of λ and differentiable in τ. The coefficients $a_j(N_{om}, \tau_s, \tau_{ph}, \beta, \ldots) \in \mathbb{R}$ and $c_j(N_{om}, \tau_s, \tau_{ph}, \beta, \ldots) \in \mathbb{R}$ depend on Laser diode delayed rate equations systems $N_{om}, \tau_s, \tau_{ph}, \beta, \ldots$ values. Unless strictly necessary, the designation of the varied arguments: $(N_{om}, \tau_s, \tau_{ph}, \beta, \tau, \ldots)$ will subsequently be omitted from P, Q, a_j, c_j. The coefficients a_j, c_j are continuous and differentiable functions of their arguments and direct substitution show that $a_0 + c_0 \neq 0$ [2, 3].

$$a_0 = \frac{1}{\tau_{ph}\cdot\tau_s};\; c_0 = A\cdot\left(\left[N_e^{(i)} - N_{om}\right]\cdot\frac{1}{\tau_s}\cdot(\beta-1) + \frac{1}{\tau_{ph}}\cdot N_{ph}^{(i)}\right)$$

$$\frac{1}{\tau_{ph}\cdot\tau_s} + A\cdot\left(\left[N_e^{(i)} - N_{om}\right]\cdot\frac{1}{\tau_s}\cdot(\beta-1) + \frac{1}{\tau_{ph}}\cdot N_{ph}^{(i)}\right) \neq 0$$

$\forall N_{om}, \tau_s, \tau_{ph}, \beta, \tau, \ldots \in \mathbb{R}_+$ i.e. $\lambda = 0$ is not a root of the characteristic equation. Furthermore $P(\lambda)$, $Q(\lambda)$ are analytic functions of λ for which the following requirements of the analysis (see [2], Sect. 1.3.4) can also be verified in the present case.

(a) If $\lambda = i\cdot\omega$; $\omega \in \mathbb{R}$ then $P(i\cdot\omega) + Q(i\cdot\omega) \neq 0$, i.e. P and Q have no common imaginary roots. This condition was verified numerically in the entire $N_{om}, \tau_s, \tau_{ph}, \beta, \tau, \ldots$ domain of interest.

(b) $|\frac{P(\lambda)}{Q(\lambda)}|$ is bounded for $|\lambda| \to \infty$; $\mathrm{Re}(\lambda) \geq 0$. No roots bifurcation from ∞. Indeed, in the limit $|\frac{Q(\lambda)}{P(\lambda)}| = |\frac{c_0+c_1\cdot\lambda}{a_0+a_1\cdot\lambda+a_2\cdot\lambda^2}|$.
(c) The following expression exists: $F(\omega,\tau) = |P(i\cdot\omega)|^2 - |Q(i\cdot\omega)|^2 = \sum_{k=0}^{2}\prod_{2\cdot k}\cdot\omega^{2\cdot k}$ has at most a finite number of zeros. Indeed, this is a polynomial in ω (degree in ω^4).
(d) Each positive root $\omega(N_{om},\tau_s,\tau_{ph},\beta,\tau,\ldots)$ of $F(\omega) = 0$ is continuous and differentiable with respect to $N_{om},\tau_s,\tau_{ph},\beta,\tau,\ldots$ The condition can only be assessed numerically.

In addition, since the coefficients in P and Q are real, we have $\overline{P(-i\cdot\omega)} = P(i\cdot\omega)$ $\overline{Q(-i\cdot\omega)} = Q(i\cdot\omega)$ thus, $\omega > 0$ maybe on eigenvalue of characteristic equations. The analysis consists in identifying the roots of the characteristic equation situated on the imaginary axis of the complex λ - plane, whereby increasing the parameters: $N_{om},\tau_s,\tau_{ph},\beta,\tau,\ldots$ Re(λ) may, at the crossing, change its sign from (−) to (+), i.e. from a stable focus $E^{(1)}\left(N_{ph}^{*\,(1)},(N_e^*)^{(1)}\right)$ or $E^{(2)}\left(N_{ph}^{*\,(2)},(N_e^*)^{(2)}\right)$ to an unstable one or vice versa. This feature may be further assessed by examining the sign of the partial derivatives with respect to $N_{om},\tau_s,\tau_{ph},\beta,\tau,\ldots$ and system parameters.

$$\Lambda^{-1}(N_{om}) = \left(\frac{\partial\mathrm{Re}\lambda}{\partial N_{om}}\right)_{\lambda=i\omega};\ a,d,q,A,\tau_s,\tau_{ph},\beta,\tau = const$$

$$\Lambda^{-1}(a) = \left(\frac{\partial\mathrm{Re}\lambda}{\partial a}\right)_{\lambda=i\omega};\ N_{om},d,q,A,\tau_s,\tau_{ph},\beta,\tau = const$$

$$\Lambda^{-1}(d) = \left(\frac{\partial\mathrm{Re}\lambda}{\partial d}\right)_{\lambda=i\omega};\ a,N_{om},q,A,\tau_s,\tau_{ph},\beta,\tau = const$$

$$\Lambda^{-1}(q) = \left(\frac{\partial\mathrm{Re}\lambda}{\partial q}\right)_{\lambda=i\omega};\ a,d,N_{om},A,\tau_s,\tau_{ph},\beta,\tau = const$$

$$\Lambda^{-1}(A) = \left(\frac{\partial\mathrm{Re}\lambda}{\partial A}\right)_{\lambda=i\omega};\ a,d,q,N_{om},\tau_s,\tau_{ph},\beta,\tau = const$$

$$\Lambda^{-1}(\tau_s) = \left(\frac{\partial\mathrm{Re}\lambda}{\partial \tau_s}\right)_{\lambda=i\omega};\ a,d,q,A,N_{om},\tau_{ph},\beta,\tau = const$$

$$\Lambda^{-1}(\tau_{ph}) = \left(\frac{\partial\mathrm{Re}\lambda}{\partial \tau_{ph}}\right)_{\lambda=i\omega};\ a,d,q,A,\tau_s,N_{om},\beta,\tau = const$$

$$\Lambda^{-1}(\beta) = \left(\frac{\partial\mathrm{Re}\lambda}{\partial \beta}\right)_{\lambda=i\omega};\ a,d,q,A,\tau_s,\tau_{ph},N_{om},\tau = const$$

$$\Lambda^{-1}(\tau) = \left(\frac{\partial\mathrm{Re}\lambda}{\partial \tau}\right)_{\lambda=i\omega};\ a,d,q,A,\tau_s,\tau_{ph},\beta,N_{om} = const$$

$$P(\lambda) = P_R(\lambda) + i \cdot P_I(\lambda); \;\; Q(\lambda) = Q_R(\lambda) + i \cdot Q_I(\lambda)$$

When writing and inserting $\lambda = i \cdot \omega$ into Laser diode delayed rate equations characteristic equation ω must satisfy the following equations.

$$\sin(\omega \cdot \tau) = g(\omega) = \frac{-P_R(i \cdot \omega) \cdot Q_I(i \cdot \omega) + P_I(i \cdot \omega) \cdot Q_R(i \cdot \omega)}{|Q(i \cdot \omega)|^2}$$

$$\cos(\omega \cdot \tau) = h(\omega) = -\frac{P_R(i \cdot \omega) \cdot Q_R(i \cdot \omega) + P_I(i \cdot \omega) \cdot Q_I(i \cdot \omega)}{|Q(i \cdot \omega)|^2}$$

where $|Q(i \cdot \omega)|^2 \neq 0$ in view of requirement (a) above, and $(g, h) \in \mathbb{R}$. Furthermore, it follows above $\sin(\omega \cdot \tau)$ and $\cos(\omega \cdot \tau)$ equations that, by squaring and adding the sides, ω must be a positive root of $F(\omega) = |P(i \cdot \omega)|^2 - |Q(i \cdot \omega)|^2 = 0$. Note:$F(\omega)$ is dependent on τ. Now it is important to notice that if $\tau \notin I$, assume that $I \subseteq \mathbb{R}_{0+}$ is the set where $\omega(\tau)$ is a positive root of $F(\omega)$ and for $\tau \notin I$,$\omega(\tau)$ is not defined. Then for all τ in I, $\omega(\tau)$ is satisfied that $F(\omega) = 0$. Then there are no positive $\omega(\tau)$ solutions for $F(\omega, \tau) = 0$, and we can not have stability switches. For $\tau \in I$ where $\omega(\tau)$ is a positive solution of $F(\omega, \tau) = 0$, we can define the angle $\theta(\tau) \in [0, 2 \cdot \pi]$ as the solution of as the solution of $\sin\theta(\tau) = \ldots; \;\; \cos\theta(\tau) = \ldots$.

$$\sin\theta(\tau) = \frac{-P_R(i \cdot \omega) \cdot Q_I(i \cdot \omega) + P_I(i \cdot \omega) \cdot Q_R(i \cdot \omega)}{|Q(i \cdot \omega)|^2}$$

$$\cos\theta(\tau) = -\frac{P_R(i \cdot \omega) \cdot Q_R(i \cdot \omega) + P_I(i \cdot \omega) \cdot Q_I(i \cdot \omega)}{|Q(i \cdot \omega)|^2}$$

And the relation between the argument $\theta(\tau)$ and $\tau \cdot \omega(\tau)$ for $\tau \in I$ must be described $\omega(\tau) \cdot \tau = \theta(\tau) + 2 \cdot n \cdot \pi \; \forall n \in \mathbb{N}_0$. Hence we can define the maps $\tau_n : I \to \mathbb{R}_{0+}$ given by $\tau_n(\tau) = \frac{\theta(\tau)+2 \cdot n \cdot \pi}{\omega(\tau)}; \;\; n \in \mathbb{N}_0; \;\; \tau \in I$. Let us introduce the functions: $I \to \mathbb{R} \; S_n(\tau) = \tau - \tau_n(\tau); \;\; \tau \in I; \;\; n \in \mathbb{N}_0$ that is continuous and differentiable in τ. In the following, the subscripts $\lambda, \omega, a, d, q, A, \tau_s, \tau_{ph}, \beta, \ldots$ indicate the corresponding partial derivatives. Lut us first concentrate on $\Lambda(x)$, remember in $\lambda(a, d, q, A, \tau_s, \tau_{ph}, \beta, \ldots)$ and $\omega(a, d, q, A, \tau_s, \tau_{ph}, \beta, \ldots)$ and keeping all parameters except one (x) and τ. The derivation closely follows that in the reference [BK]. Differentiating Laser diode delayed rate equations characteristic equation $P(\lambda) + Q(\lambda) \cdot e^{-\lambda \cdot \tau} = 0$ with respect to specific parameter (x), and inverting the derivative, for convenience, one calculates: $x = a, d, q, A, \tau_s, \tau_{ph}, \beta, \ldots$

$$\left(\frac{\partial \lambda}{\partial x}\right)^{-1} = \frac{-P_\lambda(\lambda, x) \cdot Q(\lambda, x) + Q_\lambda(\lambda, x) \cdot P(\lambda, x) - \tau \cdot P(\lambda, x) \cdot Q(\lambda, x)}{P_x(\lambda, x) \cdot Q(\lambda, x) - Q_x(\lambda, x) \cdot P(\lambda, x)}$$

where $P_\lambda = \frac{\partial P}{\partial \lambda}, \ldots$. etc., substituting $\lambda = i \cdot \omega$ and bearing $\overline{P(-i \cdot \omega)} = P(i \cdot \omega); \;\; \overline{Q(-i \cdot \omega)} = Q(i \cdot \omega)$

$i \cdot P_{\lambda}(i \cdot \omega) = P_{\omega}(i \cdot \omega)$; $i \cdot Q_{\lambda}(i \cdot \omega) = Q_{\omega}(i \cdot \omega)$ and that on the surface $|P(i \cdot \omega)|^2 = |Q(i \cdot \omega)|^2$, one obtains:

$$\left(\frac{\partial \lambda}{\partial x}\right)^{-1} |_{\lambda = i \cdot \omega}$$
$$= \frac{i \cdot P_{\omega}(i \cdot \omega, x) \cdot \overline{P(i \cdot \omega, x)} + i \cdot Q_{\lambda}(i \cdot \omega, x) \cdot \overline{Q(\lambda, x)} - \tau \cdot |P(i \cdot \omega, x)|^2}{P_x(i \cdot \omega, x) \cdot \overline{P(i \cdot \omega, x)} - Q_x(i \cdot \omega, x) \cdot \overline{Q(i \cdot \omega, x)}}$$

Upon separating into real and imaginary parts, with $P = P_R + i \cdot P_I$; $Q = Q_R + i \cdot Q_I$

$$P_{\omega} = P_{R\omega} + i \cdot P_{I\omega}; \ Q_{\omega} = Q_{R\omega} + i \cdot Q_{I\omega}$$
$$P_x = P_{Rx} + i \cdot P_{Ix}; \ Q_x = Q_{Rx} + i \cdot Q_{Ix}; \ P^2 = P_R^2 + P_I^2$$

When (x) can be any Laser diode delayed rate equations parameters $a, d, q, A, \tau_s, \tau_{ph}, \beta, \ldots$ and time delay τ etc. where for convenience, we have dropped the arguments $(i \cdot \omega, x)$ and where $F_{\omega} = 2 \cdot [(P_{R\omega} \cdot P_R + P_{I\omega} \cdot P_I) - (Q_{R\omega} \cdot Q_R + Q_{I\omega} \cdot Q_I)]$

$F_x = 2 \cdot [(P_{Rx} \cdot P_R + P_{Ix} \cdot P_I) - (Q_{Rx} \cdot Q_R + Q_{Ix} \cdot Q_I)]$; $\omega_x = \frac{-F_x}{F_{\omega}}$. We define U and V:

$$U = (P_R \cdot P_{I\omega} - P_I \cdot P_{R\omega}) - (Q_R \cdot Q_{I\omega} - Q_I \cdot Q_{R\omega})$$
$$V = (P_R \cdot P_{Ix} - P_I \cdot P_{Rx}) - (Q_R \cdot Q_{Ix} - Q_I \cdot Q_{Rx})$$

We choose our specific parameter as time delay $x = \tau$

$$P_R = \frac{1}{\tau_{ph} \cdot \tau_s} - \omega^2; \ P_I = \left(\frac{1}{\tau_{ph}} + \frac{1}{\tau_s}\right) \cdot \omega$$
$$Q_R = A \cdot \left(\left[N_e^{(i)} - N_{om}\right] \cdot \frac{1}{\tau_s} \cdot (\beta - 1) + \frac{1}{\tau_{ph}} \cdot N_{ph}^{(i)}\right)$$

$$Q_I = \omega \cdot A \cdot \left[N_{ph}^{(i)} - N_e^{(i)} + N_{om}\right]; \ P_{R\omega} = -2 \cdot \omega$$

$$P_{I\omega} = \left(\frac{1}{\tau_{ph}} + \frac{1}{\tau_s}\right); \ Q_{R\omega} = 0; \ Q_{I\omega} = A \cdot \left[N_{ph}^{(i)} - N_e^{(i)} + N_{om}\right]$$

$$P_{R\tau} = 0; \ P_{I\tau} = 0; \ Q_{R\tau} = 0; \ Q_{I\tau} = 0; \ P_{R\omega} \cdot$$
$$P_R = -2 \cdot \omega \cdot \left(\frac{1}{\tau_{ph} \cdot \tau_s} - \omega^2\right); \ Q_{R\omega} \cdot Q_R = 0$$

$$F_{\tau} = 2 \cdot \left[(P_{R\tau} \cdot P_R + P_{I\tau} \cdot P_I) - (Q_{R\tau} \cdot Q_R + Q_{Iz} \cdot Q_I)\right] = 0$$

$$P_R \cdot P_{I\omega} = \left(\frac{1}{\tau_{ph} \cdot \tau_s} - \omega^2\right) \cdot \left(\frac{1}{\tau_{ph}} + \frac{1}{\tau_s}\right); \; P_I \cdot$$

$$P_{R\omega} = -2 \cdot \omega^2 \cdot \left(\frac{1}{\tau_{ph}} + \frac{1}{\tau_s}\right)$$

$$Q_R \cdot Q_{I\omega} = A^2 \cdot \left(\left[N_e^{(i)} - N_{om}\right] \cdot \frac{1}{\tau_s} \cdot (\beta - 1) + \frac{1}{\tau_{ph}} \cdot N_{ph}^{(i)}\right) \cdot \left[N_{ph}^{(i)} - N_e^{(i)} + N_{om}\right]$$

$$V = (P_R \cdot P_{I\tau} - P_I \cdot P_{R\tau}) - (Q_R \cdot Q_{I\tau} - Q_I \cdot Q_{R\tau}) = 0$$

$F(\omega, \tau) = 0$. Differentiating with respect to τ and we get

$$F_\omega \cdot \frac{\partial \omega}{\partial \tau} + F_\tau = 0; \; \tau \in I \Rightarrow \frac{\partial \omega}{\partial \tau}$$

$$= -\frac{F_\tau}{F_\omega}; \; \Lambda^{-1}(\tau) = \left(\frac{\partial \mathrm{Re}\lambda}{\partial \tau}\right)_{\lambda - i\omega}; \; \frac{\partial \omega}{\partial \tau} = \omega_\tau = -\frac{F_z}{F_\omega}$$

$$\Lambda^{-1}(\tau) = \mathrm{Re}\left\{\frac{-2 \cdot (U + \tau \cdot |P|^2) + i \cdot F_\omega}{F_\tau + i \cdot 2 \cdot (V + \omega \cdot |P|^2)}\right\}$$

$$sign\{\Lambda^{-1}(\tau)\} = sign\left\{\left(\frac{\partial \mathrm{Re}\lambda}{\partial \tau}\right)_{\lambda = i\omega}\right\}$$

$$sign\{\Lambda^{-1}(\tau)\} = sign\{F_\omega\} \cdot sign\left\{\frac{V + \frac{\partial \omega}{\partial \tau} \cdot U}{|P|^2} + \omega + \frac{\partial \omega}{\partial \tau} \cdot \tau\right\}$$

We shall presently examine the possibility of stability transitions (bifurcations) Laser diode delayed rate equations system, about the equilibrium point $E^{(*)}\left(N_{ph}^*, \left(N_e^*\right)\right)$ as a result of a variation of delay parameter τ. The analysis consists in identifying the roots of our system characteristics equation situated on the imaginary axis of the complex λ-plane where by increasing the delay parameter τ, Re λ may at the crossing, changes its sign from − to +, i.e. from a stable focus $E^{(*)}$ to an unstable one, or vice versa. This feature may be further assessed by examining the sign of the partial derivatives with respect to τ [4].

$$\Lambda^{-1}(\tau) = \left(\frac{\partial \mathrm{Re}\lambda}{\partial \tau}\right)_{\lambda = i \cdot \omega}; \; a, d, q, A, \tau_s, \tau_{ph}, \beta, N_{om} = const$$

$$U = (P_R \cdot P_{I\omega} - P_I \cdot P_{R\omega}) - (Q_R \cdot Q_{I\omega} - Q_I \cdot Q_{R\omega})$$

$$U = \left\{ \left(\frac{1}{\tau_{ph} \cdot \tau_s} - \omega^2 \right) \cdot \left(\frac{1}{\tau_{ph}} + \frac{1}{\tau_s} \right) + \left(\frac{1}{\tau_{ph}} + \frac{1}{\tau_s} \right) \cdot 2 \cdot \omega^2 \right\}$$
$$- \left\{ A \cdot \left(\left[N_e^{(i)} - N_{om} \right] \cdot \frac{1}{\tau_s} \cdot (\beta - 1) + \frac{1}{\tau_{ph}} \cdot N_{ph}^{(i)} \right) \cdot A \cdot \left[N_{ph}^{(i)} - N_e^{(i)} + N_{om} \right] \right\}$$

We get the expression for $F(\omega, \tau)$ Laser diode delayed rate equations parameter values. We find those ω, τ values which fulfill $F(\omega, \tau) = 0$. We ignore negative, complex, and imaginary values of ω for specific τ values. Parameter τ has possible values [0.001…10] second ($\tau \in [0.001 \ldots 10]$). It can be express by 3D function $F(\omega, \tau) = 0$. We plot the stability switch diagram based on different delay values of our Laser diode delayed rate equations system.

$$\Lambda^{-1}(\tau) = \left(\frac{\partial \mathrm{Re}\lambda}{\partial \tau} \right)_{\lambda = i \cdot \omega} = \mathrm{Re} \left\{ \frac{-2 \cdot \left[U + \tau \cdot |P|^2 \right] + i \cdot F_\omega}{F_\tau + 2 \cdot i \cdot \left[V + \omega \cdot |P|^2 \right]} \right\}; \ \Lambda^{-1}(\tau)$$
$$= \left(\frac{\partial \mathrm{Re}\lambda}{\partial \tau} \right)_{\lambda = i \cdot \omega} = \frac{2 \cdot \left\{ F_\omega \cdot \left(V + \omega \cdot P^2 \right) - F_\tau \cdot \left(U + \tau \cdot P^2 \right) \right\}}{F_\tau^2 + 4 \cdot \left(V + \omega \cdot P^2 \right)^2}$$

The stability switch occurs only on those delay values (τ) which fit the equation: $\tau = \frac{\theta_+(\tau)}{\omega_+(\tau)}$ and $\theta_+(\tau)$ is the solution of $\sin\theta(\tau) = \ldots; \ \cos(\tau) = \ldots$ when $\omega = \omega_+(\tau)$ if only ω_+ is feasible. Additionally, when all Laser diode delayed rate equations parameters are known the stability switch due to various time delay values τ is described in the following expression:

$$sign\left\{ \Lambda^{-1}(\tau) \right\} = sign\{ F_\omega(\omega(\tau), \tau) \} \cdot$$
$$sign\left\{ \tau \cdot \omega_\tau(\omega(\tau)) + \omega(\tau) + \frac{U \cdot (\omega(\tau)) \cdot \omega_\tau(\omega(\tau)) + V \cdot (\omega(\tau))}{|P(\omega(\tau))|^2} \right\}$$

Remark We know $F(\omega, \tau) = 0$ implies its roots $\omega_i(\tau)$ and finding those delays values τ which ω_i is feasible. There are τ values which give complex ω_i or imaginary number, then unable to analyse stability. F function is independent on τ parameter $F(\omega, \tau) = 0$.

The results: We find those ω, τ values which fulfill $F(\omega, \tau) = 0$. We ignore negative, complex and imaginary values of ω.

Next is to find those ω, τ values which fulfill $\sin\theta(\tau) = \ldots; \ \cos(\tau) = \ldots$ then

$$\sin(\omega \cdot \tau) = \frac{-P_R \cdot Q_I + P_I \cdot Q_R}{|Q|^2}; \ \cos(\omega \cdot \tau)$$
$$= -\frac{(P_R \cdot Q_R + P_I \cdot Q_I}{|Q|^2})|Q|^2 = Q_R^2 + Q_I^2; \ |P|^2 = P_R^2 + P_I^2$$

We can plot the stability switch diagram based on $g(\tau)$ function behavior.

$$g(\tau) = \Lambda^{-1}(\tau) = \left(\frac{\partial \mathrm{Re}\lambda}{\partial \tau}\right)_{\lambda = i\cdot\omega}; \; g(\tau)$$

$$= \Lambda^{-1}(\tau) = \frac{2\cdot\left[F_\omega\cdot\left(V+\omega\cdot P^2\right) - F_\tau\cdot\left(U+\tau\cdot P^2\right)\right]}{F_\tau^2 + 4\cdot\left(V+\omega\cdot P^2\right)^2}$$

$$sign[g(\tau)] = sign[\Lambda^{-1}(\tau)] = sign\left[\left(\frac{\partial \mathrm{Re}\lambda}{\partial \tau}\right)_{\lambda=i\cdot\omega}\right]$$

$$sign[g(\tau)] = sign\left\{\frac{2\cdot[F_\omega\cdot(V+\omega\cdot P^2) - F_\tau\cdot(U+\tau\cdot P^2)]}{F_\tau^2 + 4\cdot(V+\omega\cdot P^2)^2}\right\}$$

$$F_\tau^2 + 4\cdot\left(V+\omega\cdot P^2\right)^2 > 0; \; sign\left[\Lambda^{-1}(\tau)\right] = sign\left\{F_\omega\cdot\left(V+\omega\cdot P^2\right) - F_\tau\cdot\left(U+\tau\cdot P^2\right)\right\}$$

$$sign\left[\Lambda^{-1}(\tau)\right] = sign\left\{F_\omega\cdot\left[\left(V+\omega\cdot P^2\right) - \frac{F_\tau}{F_\omega}\cdot\left(U+\tau\cdot P^2\right)\right]\right\}; \; \omega_\tau = -\frac{F_\tau}{F_\omega}; \; \omega_\tau = \left(\frac{\partial\omega}{\partial\tau}\right)^{-1} = -\frac{\partial F/\partial\omega}{\partial F/\partial\tau}$$

$$sign\left[\Lambda^{-1}(\tau)\right] = sign\left\{F_\omega\cdot\left[V+\omega\cdot P^2 + \omega_\tau\cdot\left(U+\tau\cdot P^2\right)\right]\right\}$$

$$sign\left[\Lambda^{-1}(\tau)\right] = sign\left\{F_\omega\cdot\left[V+\omega_\tau\cdot U + \omega\cdot P^2 + \omega_\tau\cdot\tau\cdot P^2\right]\right\}$$

$$sign\left[\Lambda^{-1}(\tau)\right] = sign\left\{F_\omega\cdot P^2\cdot\left[\frac{(V+\omega_\tau\cdot U)}{P^2} + \omega + \omega_\tau\cdot\tau\right]\right\}; \; sign\left(P^2\right) > 0$$

$$sign\left[\Lambda^{-1}(\tau)\right] = sign\left\{[F_\omega]\cdot\left[\frac{(V+\omega_\tau\cdot U)}{P^2} + \omega + \omega_\tau\cdot\tau\right]\right\}$$

$$sign\left[\Lambda^{-1}(\tau)\right] = sign[F_\omega]\cdot sign\left[\frac{(V+\omega_\tau\cdot U)}{P^2} + \omega + \omega_\tau\cdot\tau\right]$$

$$sign\left[\Lambda^{-1}(\tau)\right] = sign[F_\omega]\cdot sign\left[\frac{V+\omega_\tau\cdot U}{P^2} + \omega + \omega_\tau\cdot\tau\right]$$

$$V = 0 \Rightarrow sign\left[\Lambda^{-1}(\tau)\right] = sign[F_\omega]\cdot sign\left[\frac{\omega_\tau\cdot U}{P^2} + \omega + \omega_\tau\cdot\tau\right]$$

$$F_\tau = 0; \; F_\omega \neq 0 \Rightarrow \omega_\tau = -\frac{F_\tau}{F_\omega} = 0 \Rightarrow sign\left[\Lambda^{-1}(\tau)\right] = sign[F_\omega]\cdot sign[\omega]$$

$$F_\omega = 2\cdot[(P_{R\omega}\cdot P_R + P_{I\omega}\cdot P_I) - (Q_{R\omega}\cdot Q_R + Q_{I\omega}\cdot Q_I)]$$

Table 1.2 Sign of $\Lambda^{-1}(\tau)$ Vs sign of $sign[F_\omega]$ and $sign[\omega]$

$sign[F_\omega]$	$sign[\omega]$	$sign[\Lambda^{-1}(\tau)]$
$\pm$	$\pm$	+
$\pm$	$\mp$	−

Table 1.3 System cases behavior for $sign[\Lambda^{-1}(\tau)]$sign

Case	Crossing behavior
$sign[\Lambda^{-1}(\tau)] > 0$	Crossing proceeds from (−) to (+) respectively (stable to unstable)
$sign[\Lambda^{-1}(\tau)] < 0$	Crossing proceeds from (+) to (−) respectively (unstable to stable)

$$P_{R\omega} \cdot P_R = -2 \cdot \omega \cdot \left(\frac{1}{\tau_{ph} \cdot \tau_s} - \omega^2\right);\; P_{I\omega} \cdot P_I = \left(\frac{1}{\tau_{ph}} + \frac{1}{\tau_s}\right)^2 \cdot \omega;\; Q_{R\omega} \cdot Q_R = 0$$

$$Q_{I\omega} \cdot Q_I = A^2 \cdot \left[N_{ph}^{(i)} - N_e^{(i)} + N_{om}\right]^2 \cdot \omega$$

$$F_\omega = 2 \cdot [P_{R\omega} \cdot P_R + P_{I\omega} \cdot P_I - Q_{I\omega} \cdot Q_I]$$
$$= -2 \cdot \omega \cdot \left\{2 \cdot \left(\frac{1}{\tau_{ph} \cdot \tau_s} - \omega^2\right) + \left(\frac{1}{\tau_{ph}} + \frac{1}{\tau_s}\right)^2 + A^2 \cdot \left[N_{ph}^{(i)} - N_e^{(i)} + N_{om}\right]^2\right\}$$

$$U = \left\{\left(\frac{1}{\tau_{ph} \cdot \tau_s} - \omega^2\right) \cdot \left(\frac{1}{\tau_{ph}} + \frac{1}{\tau_s}\right) + \left(\frac{1}{\tau_{ph}} + \frac{1}{\tau_s}\right) \cdot 2 \cdot \omega^2\right\}$$
$$- \left\{A \cdot \left(\left[N_e^{(i)} - N_{om}\right] \cdot \frac{1}{\tau_s} \cdot (\beta - 1) + \frac{1}{\tau_{ph}} \cdot N_{ph}^{(i)}\right) \cdot A \cdot \left[N_{ph}^{(i)} - N_e^{(i)} + N_{om}\right]\right\}$$

We check the sign of $\Lambda^{-1}(\tau)$ according the following rule (Table 1.2).
We can summary our system cases behavior in the following table (Table 1.3).
No. 2:

$$\frac{dN_i(t)}{dt} = \frac{1}{q \cdot a \cdot d} \cdot \exp\left(-\frac{q \cdot V}{2 \cdot k \cdot T}\right) - A \cdot \exp\left(-\frac{q \cdot V}{2 \cdot k \cdot T}\right) \cdot$$
$$\left(N_i(t-\tau) \cdot \exp\left(\frac{q \cdot V}{2 \cdot k \cdot T}\right) - N_{om}\right) \cdot N_{ph}(t-\tau) - \frac{N_i(t)}{\tau_s}$$

$$\frac{dN_{ph}(t)}{dt} = A \cdot \left(N_i(t-\tau) \cdot \exp\left(\frac{q \cdot V}{2 \cdot k \cdot T}\right) - N_{om}\right) \cdot$$
$$N_{ph}(t-\tau) - \frac{N_{ph}(t)}{\tau_{ph}} + \frac{\beta}{\tau_s} \cdot N_i(t) \cdot \exp\left(\frac{q \cdot V}{2 \cdot k \cdot T}\right)$$

Remark It is reader exercise to analyze the second possible set of differential equations. To find the equilibrium points (fixed points), stability analysis, Laser diode delayed rate equations eigenvalue matrix, and stability switching inspection.

1.2 Laser Diode Intrinsic and Package Electrical Equivalent Circuits Stability Analysis for Parameters Variation

The complete equivalent circuit of a laser diode can be separated into two parts. The first part represents the intrinsic electrical equivalent circuit of the laser chip itself. The second part is the electrical equivalent circuit of the package including the major parasitic elements. The elements of the intrinsic laser equivalent circuit are derived from the coupled rate equations which describe the interplay between the injected carrier and photon densities in the active region of the laser diode. We describe the laser diode coupled single-mode rate equation parameters. N_{ph} is the photon density, N_e is the electron density, q is the electronic charge, d is the thickness of the active region, a is the area of the diode contact stripe, I is the injected current, N_{om} is the minimum electron density required to obtain a positive gain, A is a constant related to the stimulated emission process, τ_S is the spontaneous emission lifetime, τ_{ph} is the photon lifetime and β is the fraction of the spontaneous emission that is coupled to the lasing mode. We describe the carrier density as a function of the junction voltage ($N_e = N_i \cdot e^{\frac{q \cdot V}{2 \cdot k \cdot T}}$; N_i is the intrinsic carrier density and V is the junction voltage). We define R_d as the differential resistance of the laser diode and n_{ph}^0, n_e^0 and n_{om} are the normalized steady-state values of N_{ph}^0, N_e^0 and N_{om}, respectively [1].

$$R_d = \frac{2 \cdot k \cdot T}{q} \cdot \frac{1}{I_d}; \; I_d = \frac{N_e^0 \cdot a \cdot q \cdot d}{\tau_s}; \; R_d = \frac{2 \cdot k \cdot T}{q^2} \cdot \frac{\tau_s}{N_e^0 \cdot a \cdot d}$$

I_d is the normalized current.

$$n_{ph}^0 = A \cdot \tau_S \cdot N_{ph}^0; \; n_e^0 = A \cdot \tau_{ph} \cdot N_e^0; \; n_{om} = A \cdot \tau_{ph} \cdot N_{om}$$

The basic intrinsic electrical equivalent circuit of the laser diode is describe below (Fig. 1.1).

V_{sup} is the power source that drive the laser diode with its parasitic resistance R_p. R_{se} is the series resistance with the inductor (L_i).

$$R_i = \frac{R_d}{n_{ph}^0 + 1}; \; R_i = \frac{2 \cdot k \cdot T}{q^2} \cdot \frac{\tau_s}{N_e^0 \cdot a \cdot d} \cdot \frac{1}{\left(A \cdot \tau_S \cdot N_{ph}^0 + 1\right)}$$

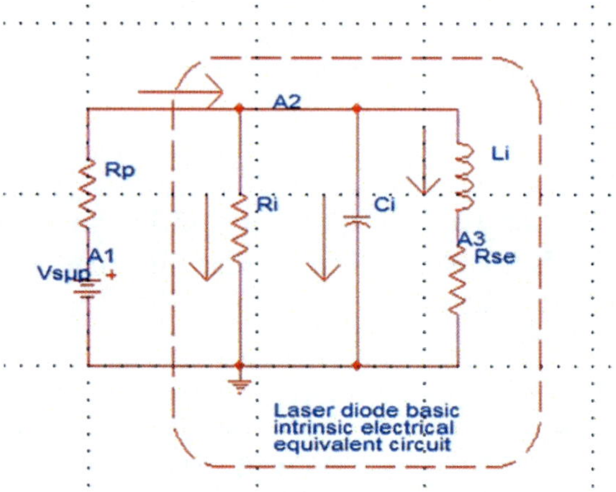

Fig. 1.1 Basic intrinsic equivalent circuit of laser diode

$$L_i = \frac{R_d \cdot \tau_{ph}}{\left[\left(n_{ph}^0 + \beta\right) \cdot \left(n_e^0 - n_{om}\right)\right]} \approx \frac{R_d \cdot \tau_{ph}}{n_{ph}^0}; \; L_i \approx \frac{2 \cdot k \cdot T \cdot \tau_{ph}}{A \cdot N_{ph}^0 \cdot q^2 \cdot N_e^0 \cdot a \cdot d}$$

$$C_i = \frac{\tau_s}{R_d}; \; C_i = \frac{N_e^0 \cdot a \cdot q^2 \cdot d}{2 \cdot k \cdot T}$$

$$R_{se} = \beta \cdot R_d \cdot \frac{n_e^0}{n_{ph}^0 \cdot \left(n_{ph}^0 + \beta\right) \cdot \left(n_e^0 - n_{om}\right)} \approx \beta \cdot R_d \cdot \frac{n_e^0}{\left(n_{ph}^0\right)^2}$$

$$R_{se} \approx \beta \cdot \frac{2 \cdot k \cdot T}{q^2} \cdot \frac{\tau_{ph}}{a \cdot d \cdot A \cdot \tau_s \cdot \left(N_{ph}^0\right)^2}$$

The resistance R_i including the differential resistance (R_d) of the laser diode models and is damping due to the spontaneous and stimulated recombination terms in the rate equations. The resistance R_{se} models damping due to spontaneous emission coupled into the lasing mode. Resistances R_i and R_{se} are responsible for the damping of the electro-optical resonance. The capacitance C_i represents the active layer diffusion capacitance of the laser diode. The inductance L_i arises from the small signal analysis of the rate equations and represents the resonance phenomenon of the laser diode with the capacitance C_i. The resonant frequency (f_r) of the circuit is

$$f_r = \frac{1}{2 \cdot \pi} \cdot \frac{1}{\sqrt{L_i \cdot C_i}}; \; f_r = \frac{1}{2 \cdot \pi} \cdot \sqrt{\frac{A \cdot N_{ph}^0}{\tau_{ph}}}; \; L_i \cdot C_i = \frac{\tau_{ph}}{A \cdot N_{ph}^0}$$

We can increase the value of the resonant frequency by decreasing the values of L_i and C_i since the values of L_i and C_i are depend on the laser diode parameters, the resonant frequency can be increased by decreasing the photon lifetime (τ_{ph}), increasing the photon density (n_{ph}^0) (or increasing the N_{ph}^0), or by increasing the gain coefficient A, that is, increasing electron density n_e^0. There is a close relationship between the circuit components of the equivalent circuit and the laser diode parameters. We derive circuit differential equations.

$$I_{R_p} = I_{R_i} + I_{C_i} + I_{L_i};\ I_{L_i} = I_{R_{se}};\ V_{A_2} = V_{R_i} = V_{C_i} = V_{L_i} + V_{R_{se}};\ V_{A_3} = V_{R_{se}}$$

$$V_{A_2} = I_{R_i} \cdot R_i;\ I_{C_i} = C_i \cdot \frac{dV_{A_2}}{dt};\ V_{L_i} = L_i \cdot \frac{dI_{L_i}}{dt}$$

$$V_{R_{se}} = I_{R_{se}} \cdot R_{se} = I_{L_i} \cdot R_{se};\ I_{C_i} \cdot \frac{1}{C_i} = \frac{dV_{A_2}}{dt}$$

$$V_{A_2} = \frac{1}{C_i} \cdot \int I_{C_i} \cdot dt;\ V_{A_2} = L_i \cdot \frac{dI_{L_i}}{dt}$$

$$+ I_{L_i} \cdot R_{se};\ \frac{1}{C_i} \cdot \int I_{C_i} \cdot dt = L_i \cdot \frac{dI_{L_i}}{dt} + I_{L_i} \cdot R_{se}$$

$$\frac{d}{dt}\left\{\frac{1}{C_i} \cdot \int I_{C_i} \cdot dt = L_i \cdot \frac{dI_{L_i}}{dt} + I_{L_i} \cdot R_{se}\right\} \Rightarrow \frac{1}{C_i} \cdot I_{C_i} = L_i \cdot \frac{d^2 I_{L_i}}{dt^2} + \frac{dI_{L_i}}{dt} \cdot R_{se}$$

$$I_{R_p} = \frac{V_{\text{sup}} - V_{A_2}}{R_p};\ I_{R_p} = I_{R_i} + I_{C_i} + I_{L_i} \Rightarrow \frac{V_{\text{sup}} - V_{A_2}}{R_p}$$

$$= \frac{V_{A_2}}{R_i} + C_i \cdot \frac{dV_{A_2}}{dt} + I_{L_i};\ I_{L_i} = \frac{1}{L_i} \cdot \int V_{L_i} \cdot dt$$

$$V_{L_i} = V_{A_2} - V_{R_{se}} = V_{A_2} - I_{L_i} \cdot R_{se};\ I_{L_i}$$

$$= \frac{1}{L_i} \cdot \int \left(V_{A_2} - I_{L_i} \cdot R_{se}\right) \cdot dt;\ I_{L_i} = \frac{1}{L_i} \cdot \int V_{A_2} \cdot dt - \frac{R_{se}}{L_i} \cdot \int I_{L_i} \cdot dt$$

$$\frac{V_{\text{sup}} - V_{A_2}}{R_p} = \frac{V_{A_2}}{R_i} + C_i \cdot \frac{dV_{A_2}}{dt} + \frac{1}{L_i} \cdot \int V_{A_2} \cdot dt - \frac{R_{se}}{L_i} \cdot \int I_{L_i} \cdot dt$$

$$\frac{d}{dt}\left\{\frac{V_{\text{sup}} - V_{A_2}}{R_p} = \frac{V_{A_2}}{R_i} + C_i \cdot \frac{dV_{A_2}}{dt} + \frac{1}{L_i} \cdot \int V_{A_2} \cdot dt - \frac{R_{se}}{L_i} \cdot \int I_{L_i} \cdot dt\right\}$$

$$\frac{d}{dt} V_{\text{sup}} \to \varepsilon (= 0)$$

$$-\frac{1}{R_p} \cdot \frac{dV_{A_2}}{dt} = \frac{1}{R_i} \cdot \frac{dV_{A_2}}{dt} + C_i \cdot \frac{d^2 V_{A_2}}{dt^2} + \frac{1}{L_i} \cdot V_{A_2} - \frac{R_{se}}{L_i} \cdot I_{L_i}$$

$$\frac{1}{C_i} \cdot I_{C_i} = L_i \cdot \frac{d^2 I_{L_i}}{dt^2} + \frac{dI_{L_i}}{dt} \cdot R_{se} \Rightarrow \frac{dV_{A_2}}{dt} = L_i \cdot \frac{d^2 I_{L_i}}{dt^2} + \frac{dI_{L_i}}{dt} \cdot R_{se}$$

We define new variables for simplicity: $Y_1 = \frac{dV_{A_2}}{dt};\ \frac{dI_{L_i}}{dt} = Y_2$

$$Y_1 = L_i \cdot \frac{dY_2}{dt} + Y_2 \cdot R_{se}; \quad -\frac{1}{R_p} \cdot Y_1 = \frac{1}{R_i} \cdot Y_1 + C_i \cdot \frac{dY_1}{dt} + \frac{1}{L_i} \cdot V_{A_2} - \frac{R_{se}}{L_i} \cdot I_{L_i}$$

We can summary our system differential equation:

$$\frac{dY_1}{dt} = -Y_1 \cdot \left(\frac{1}{R_p} + \frac{1}{R_i}\right) \cdot \frac{1}{C_i} - \frac{1}{L_i \cdot C_i} \cdot V_{A_2} + \frac{R_{se}}{L_i \cdot C_i} \cdot I_{L_i}; \quad \frac{dY_2}{dt} = \frac{1}{L_i} Y_1 - Y_2 \cdot \frac{R_{se}}{L_i}$$

$$\frac{dV_{A_2}}{dt} = Y_1; \quad \frac{dI_{L_i}}{dt} = Y_2; \quad V_{A_2} = V_{A_2}(t); \quad I_{L_i} = I_{L_i}(t); \quad Y_1 = Y_1(t); \quad Y_2 = Y_2(t)$$

We define four functions: $f_i = f_i\left(V_{A_2}, I_{L_i}, Y_1, Y_2\right) \forall i = 1, 2, 3, 4$

$$\frac{dY_1}{dt} = f_1\left(V_{A_2}, I_{L_i}, Y_1, Y_2\right); \quad \frac{dY_2}{dt} = f_2\left(V_{A_2}, I_{L_i}, Y_1, Y_2\right)$$
$$\frac{dV_{A_2}}{dt} = f_3\left(V_{A_2}, I_{L_i}, Y_1, Y_2\right); \quad \frac{dI_{L_i}}{dt} = f_4\left(V_{A_2}, I_{L_i}, Y_1, Y_2\right)$$

$$f_1\left(V_{A_2}, I_{L_i}, Y_1, Y_2\right) = -Y_1 \cdot \left(\frac{1}{R_p} + \frac{1}{R_i}\right) \cdot \frac{1}{C_i} - \frac{1}{L_i \cdot C_i} \cdot V_{A_2} + \frac{R_{se}}{L_i \cdot C_i} \cdot I_{L_i}$$

$$f_2\left(V_{A_2}, I_{L_i}, Y_1, Y_2\right) = \frac{1}{L_i} Y_1 - Y_2 \cdot \frac{R_{se}}{L_i}; \quad f_3\left(V_{A_2}, I_{L_i}, Y_1, Y_2\right)$$
$$= Y_1; \quad f_4\left(V_{A_2}, I_{L_i}, Y_1, Y_2\right) = Y_2$$

To find system fixed points (equilibrium points), we set

$$\frac{dY_1}{dt} = 0; \quad \frac{dY_2}{dt} = 0; \quad \frac{dV_{A_2}}{dt} = 0; \quad \frac{dI_{L_i}}{dt} = 0; \quad \frac{dV_{A_2}}{dt} = 0 \Rightarrow Y_1^* = 0; \quad \frac{dI_{L_i}}{dt} = 0 \Rightarrow Y_2^* = 0$$

$$\frac{dY_1}{dt} = 0 \Rightarrow -Y_1^* \cdot \left(\frac{1}{R_p} + \frac{1}{R_i}\right) \cdot \frac{1}{C_i} - \frac{1}{L_i \cdot C_i} \cdot V_{A_2}^*$$
$$+ \frac{R_{se}}{L_i \cdot C_i} \cdot I_{L_i}^* = 0 \longrightarrow Y_1^* = 0 R_{se} \cdot I_{L_i}^* = V_{A_2}^*$$

System fixed point: $E^*\left(V_{A_2}^*, I_{L_i}^*, Y_1^*, Y_2^*\right) = \left(R_{se} \cdot I_{L_i}^*, I_{L_i}^*, 0, 0\right)$

Stability analysis: The standard local stability analysis about one of the equilibrium points of laser diode basic intrinsic electrical equivalent circuit consists in adding to coordinate $[V_{A_2}\ I_{L_i}\ Y_1\ Y_2]$ arbitrarily small increment of exponential form $v_{A_2} \cdot e^{\lambda \cdot t}$, $i_{L_i} \cdot e^{\lambda \cdot t}$, $y_1 \cdot e^{\lambda \cdot t}$, $y_2 \cdot e^{\lambda \cdot t}$ respectively and retaining the first order terms in $[v_{A_2}, i_{L_i}, y_1, y_2]$. The system of one homogeneous equation leads to a polynomial characteristic equation in the eigenvalue λ. The polynomial characteristic equation accept by set the below laser diode intrinsic electrical equivalent circuit respect to time into one system equation. Our system fixed points/values with arbitrarily small increment of the exponential form $v_{A_2} \cdot e^{\lambda \cdot t}$, $i_{L_i} \cdot e^{\lambda \cdot t}$, $y_1 \cdot e^{\lambda \cdot t}$, $y_2 \cdot e^{\lambda \cdot t}$ are i = 0 (first fixed point), i = 1 (second fixed point), and i = 2 (third fixed point) [8, 9].

$$V_{A_2}(t) = V_{A_2}^{(i)} + v_{A_2} \cdot e^{\lambda \cdot t}; \ I_{L_i}(t) = I_{L_i}^{(i)} + i_{L_i} \cdot e^{\lambda \cdot t}; \ Y_1(t) = Y_1^{(i)} + y_1 \cdot e^{\lambda \cdot t}; \ Y_2(t) = Y_2^{(i)} + y_2 \cdot e^{\lambda \cdot t}$$

$$\frac{dV_{A_2}(t)}{dt} = v_{A_2} \cdot \lambda \cdot e^{\lambda \cdot t}; \ \frac{dI_{L_i}(t)}{dt} = i_{L_i} \cdot \lambda \cdot e^{\lambda \cdot t}; \ \frac{dY_1(t)}{dt} = y_1 \cdot \lambda \cdot e^{\lambda \cdot t}; \ \frac{dY_2(t)}{dt} = y_2 \cdot \lambda \cdot e^{\lambda \cdot t}$$

We choose the above expressions for our $V_{A_2}(t)$, $I_{L_i}(t)$, $Y_1(t)$, and $Y_2(t)$ as small displacement $[v_{A_2}, i_{L_i}, y_1, y_2]$ from our laser diode intrinsic equation circuit fixed points at time t = 0. For $\lambda < 0, t > 0$ the selected fixed point is stable otherwise $\lambda > 0, t > 0$ is unstable. Our laser diode intrinsic equation circuit tends to the selected fixed point exponentially for $\lambda < 0, t > 0$, otherwise go away from the selected fixed point exponentially. Eigenvalue λ is the parameter which establish if the fixed point is stable or unstable, additionally his absolute value $|\lambda|$ establish the speed of the flow toward or away from the selected fixed point. The laser diode intrinsic electrical equivalent circuit variables for negative eigenvalue ($\lambda < 0$) are as follow (Table 1.4).

The laser diode intrinsic electrical equivalent circuit variables for positive eigenvalue ($\lambda > 0$) are as follow (Table 1.5).

$\frac{dV_{A_2}}{dt} = Y_1; \ v_{A_2} \cdot \lambda \cdot e^{\lambda \cdot t} = Y_1^{(i)} + y_1 \cdot e^{\lambda \cdot t}$. At fixed point $Y_1^{(i)} = 0; \ -v_{A_2} \cdot \lambda + y_1 = 0$

$\frac{dI_{L_i}}{dt} = Y_2; \ i_{L_i} \cdot \lambda \cdot e^{\lambda \cdot t} = Y_2^{(i)} + y_2 \cdot e^{\lambda \cdot t}$. At fixed point $Y_2^{(i)} = 0; \ -i_{L_i} \cdot \lambda + y_2 = 0$

$$\frac{dY_1}{dt} = -Y_1 \cdot \left(\frac{1}{R_p} + \frac{1}{R_i}\right) \cdot \frac{1}{C_i} - \frac{1}{L_i \cdot C_i} \cdot V_{A_2} + \frac{R_{se}}{L_i \cdot C_i} \cdot I_{L_i}$$

$$y_1 \cdot \lambda \cdot e^{\lambda \cdot t} = -\left(Y_1^{(i)} + y_1 \cdot e^{\lambda \cdot t}\right) \cdot \left(\frac{1}{R_p} + \frac{1}{R_i}\right) \cdot \frac{1}{C_i} - \frac{1}{L_i \cdot C_i} \cdot \left(V_{A_2}^{(i)} + v_{A_2} \cdot e^{\lambda \cdot t}\right)$$

Table 1.4 System variables for $\lambda < 0$ and time parameter

Time (t)	$\lambda < 0$
t = 0	$V_{A_2}(t=0) = V_{A_2}^{(i)} + v_{A_2}; \ I_{L_i}(t=0) = I_{L_i}^{(i)} + i_{L_i}$
	$Y_1(t=0) = Y_1^{(i)} + y_1; \ Y_2(t=0) = Y_2^{(i)} + y_2$
t > 0	$V_{A_2}(t) = V_{A_2}^{(i)} + v_{A_2} \cdot e^{-\lvert\lambda\rvert \cdot t}; \ I_{L_i}(t) = I_{L_i}^{(i)} + i_{L_i} \cdot e^{-\lvert\lambda\rvert \cdot t}$
	$Y_1(t) = Y_1^{(i)} + y_1 \cdot e^{-\lvert\lambda\rvert \cdot t}; \ Y_2(t) = Y_2^{(i)} + y_2 \cdot e^{-\lvert\lambda\rvert \cdot t}$
t > 0; t → ∞	$V_{A_2}(t \to \infty) = V_{A_2}^{(i)}; \ I_{L_i}(t \to \infty) = I_{L_i}^{(i)}$
	$Y_1(t \to \infty) = Y_1^{(i)}; \ Y_2(t \to \infty) = Y_2^{(i)}$

Table 1.5 System variables for $\lambda > 0$ and time parameter

Time (t)	$\lambda > 0$
t = 0	$V_{A_2}(t=0) = V_{A_2}^{(i)} + v_{A_2};\ I_{L_i}(t=0) = I_{L_i}^{(i)} + i_{L_i}$
	$Y_1(t=0) = Y_1^{(i)} + y_1;\ Y_2(t=0) = Y_2^{(i)} + y_2$
t > 0	$V_{A_2}(t) = V_{A_2}^{(i)} + v_{A_2} \cdot e^{\vert\lambda\vert \cdot t};\ I_{L_i}(t) = I_{L_i}^{(i)} + i_{L_i} \cdot e^{\vert\lambda\vert \cdot t}$
	$Y_1(t) = Y_1^{(i)} + y_1 \cdot e^{\vert\lambda\vert \cdot t};\ Y_2(t) = Y_2^{(i)} + y_2 \cdot e^{\vert\lambda\vert \cdot t}$
t > 0; t → ∞	$V_{A_2}(t \to \infty) = v_{A_2} \cdot e^{\vert\lambda\vert \cdot t};\ I_{L_i}(t \to \infty) = i_{L_i} \cdot e^{\vert\lambda\vert \cdot t}$
	$Y_1(t \to \infty) = y_1 \cdot e^{\vert\lambda\vert \cdot t};\ Y_2(t \to \infty) = y_2 \cdot e^{\vert\lambda\vert \cdot t}$

$$+ \frac{R_{se}}{L_i \cdot C_i} \cdot \left(I_{L_i}^{(i)} + i_{L_i} \cdot e^{\lambda \cdot t}\right)$$

$$y_1 \cdot \lambda \cdot e^{\lambda \cdot t} = -Y_1^{(i)} \cdot \left(\frac{1}{R_p} + \frac{1}{R_i}\right) \cdot \frac{1}{C_i} - \frac{1}{L_i \cdot C_i} \cdot V_{A_2}^{(i)} + \frac{R_{se}}{L_i \cdot C_i} \cdot I_{L_i}^{(i)} - y_1 \cdot \left(\frac{1}{R_p} + \frac{1}{R_i}\right) \cdot \frac{1}{C_i} \cdot e^{\lambda \cdot t} - v_{A_2} \cdot \frac{1}{L_i \cdot C_i} \cdot e^{\lambda \cdot t} + i_{L_i} \cdot \frac{R_{se}}{L_i \cdot C_i} \cdot e^{\lambda \cdot t}$$

At fixed point $-Y_1^{(i)} \cdot \left(\frac{1}{R_p} + \frac{1}{R_i}\right) \cdot \frac{1}{C_i} - \frac{1}{L_i \cdot C_i} \cdot V_{A_2}^{(i)} + \frac{R_{se}}{L_i \cdot C_i} \cdot I_{L_i}^{(i)} = 0$

$$-y_1 \cdot \left(\frac{1}{R_p} + \frac{1}{R_i}\right) \cdot \frac{1}{C_i} - y_1 \cdot \lambda - v_{A_2} \cdot \frac{1}{L_i \cdot C_i} + i_{L_i} \cdot \frac{R_{se}}{L_i \cdot C_i} = 0$$

$$\frac{dY_2}{dt} = \frac{1}{L_i} Y_1 - Y_2 \cdot \frac{R_{se}}{L_i};\ y_2 \cdot \lambda \cdot e^{\lambda \cdot t} = \frac{1}{L_i} \cdot \left(Y_1^{(i)} + y_1 \cdot e^{\lambda \cdot t}\right) - \left(Y_2^{(i)} + y_2 \cdot e^{\lambda \cdot t}\right) \cdot \frac{R_{se}}{L_i}$$

$$y_2 \cdot \lambda \cdot e^{\lambda \cdot t} = \frac{1}{L_i} \cdot Y_1^{(i)} - Y_2^{(i)} \cdot \frac{R_{se}}{L_i} + y_1 \cdot \frac{1}{L_i} \cdot e^{\lambda \cdot t} - y_2 \cdot \frac{R_{se}}{L_i} \cdot e^{\lambda \cdot t}$$

At fixed point $\frac{1}{L_i} \cdot Y_1^{(i)} - Y_2^{(i)} \cdot \frac{R_{se}}{L_i} = 0;\ y_1 \cdot \frac{1}{L_i} - y_2 \cdot \frac{R_{se}}{L_i} - y_2 \cdot \lambda = 0.$
We can summary our laser diode intrinsic electrical equivalent circuit arbitrarily small increment equations:

$$-v_{A_2} \cdot \lambda + y_1 = 0;\ -i_{L_i} \cdot \lambda + y_2 = 0;\ -y_1 \cdot \left(\frac{1}{R_p} + \frac{1}{R_i}\right) \cdot \frac{1}{C_i} - y_1 \cdot \lambda - v_{A_2} \cdot \frac{1}{L_i \cdot C_i} + i_{L_i} \cdot \frac{R_{se}}{L_i \cdot C_i} = 0$$

$$y_1 \cdot \frac{1}{L_i} - y_2 \cdot \frac{R_{se}}{L_i} - y_2 \cdot \lambda = 0$$

The arbitrarily small increments Jacobian of our laser diode intrinsic electrical equivalent circuit is as follow:

$$\begin{pmatrix} \Xi_{11} & \ldots & \Xi_{14} \\ \vdots & \ddots & \vdots \\ \Xi_{41} & \cdots & \Xi_{44} \end{pmatrix} \cdot \begin{pmatrix} v_{A_2} \\ i_{L_i} \\ y_1 \\ y_2 \end{pmatrix} = 0; \ A - \lambda \cdot I = \begin{pmatrix} \Xi_{11} & \ldots & \Xi_{14} \\ \vdots & \ddots & \vdots \\ \Xi_{41} & \cdots & \Xi_{44} \end{pmatrix}$$

$$\Xi_{11} = -\lambda; \ \Xi_{12} = 0; \ \Xi_{13} = 1; \ \Xi_{14} = 0; \ \Xi_{21} = 0; \ \Xi_{22} = -\lambda; \ \Xi_{23} = 0; \ \Xi_{24} = 1$$

$$\Xi_{31} = -\frac{1}{L_i \cdot C_i}; \ \Xi_{32} = \frac{R_{se}}{L_i \cdot C_i}; \ \Xi_{33} = -\left(\frac{1}{R_p} + \frac{1}{R_i}\right) \cdot \frac{1}{C_i} - \lambda; \ \Xi_{34} = 0; \ \Xi_{41} = 0$$

$$\Xi_{42} = 0; \ \Xi_{43} = \frac{1}{L_i}; \ \Xi_{44} = -\frac{R_{se}}{L_i} - \lambda$$

$$\det(A - \lambda \cdot I) = 0; \ \det \begin{pmatrix} \Xi_{11} & \ldots & \Xi_{14} \\ \vdots & \ddots & \vdots \\ \Xi_{41} & \cdots & \Xi_{44} \end{pmatrix} = 0$$

$$\det(A - \lambda \cdot I) = -\lambda \cdot \det \begin{pmatrix} -\lambda & 0 & 1 \\ \frac{R_{se}}{L_i \cdot C_i} & -\left(\frac{1}{R_p} + \frac{1}{R_i}\right) \cdot \frac{1}{C_i} - \lambda & 0 \\ 0 & \frac{1}{L_i} & -\frac{R_{se}}{L_i} - \lambda \end{pmatrix}$$

$$+ \det \begin{pmatrix} 0 & -\lambda & 1 \\ -\frac{1}{L_i \cdot C_i} & \frac{R_{se}}{L_i \cdot C_i} & 0 \\ 0 & 0 & -\frac{R_{se}}{L_i} - \lambda \end{pmatrix}$$

$$\det(A - \lambda \cdot I) = -\lambda \cdot \left\{ -\lambda \cdot \left[\left(-\left(\frac{1}{R_p} + \frac{1}{R_i}\right) \cdot \frac{1}{C_i} - \lambda \right) \cdot \left(-\frac{R_{se}}{L_i} - \lambda \right) \right] + \frac{R_{se}}{L_i^2 \cdot C_i} \right\}$$
$$+ \lambda \cdot \left\{ \left(-\frac{1}{L_i \cdot C_i} \right) \cdot \left(-\frac{R_{se}}{L_i} - \lambda \right) \right\}$$

$$\det(A - \lambda \cdot I) = -\lambda \cdot \left\{ -\lambda \cdot \left[\left(\left(\frac{1}{R_p} + \frac{1}{R_i}\right) \cdot \frac{1}{C_i} + \lambda \right) \cdot \left(\frac{R_{se}}{L_i} + \lambda \right) \right] + \frac{R_{se}}{L_i^2 \cdot C_i} \right\}$$
$$+ \lambda \cdot \left\{ \left(\frac{1}{L_i \cdot C_i} \right) \cdot \left(\frac{R_{se}}{L_i} + \lambda \right) \right\}$$

$$\det(A - \lambda \cdot I) = -\lambda \cdot \left\{ -\lambda \cdot \left[\left(\frac{1}{R_p} + \frac{1}{R_i}\right) \cdot \frac{1}{C_i} \cdot \frac{R_{se}}{L_i} + \left[\left(\frac{1}{R_p} + \frac{1}{R_i}\right) \cdot \frac{1}{C_i} + \frac{R_{se}}{L_i} \right] \right.\right.$$
$$\left.\left. \cdot \lambda + \lambda^2 \right] + \frac{R_{se}}{L_i^2 \cdot C_i} \right\} + \lambda \cdot \left(\frac{1}{L_i \cdot C_i} \right) \cdot \frac{R_{se}}{L_i} + \left(\frac{1}{L_i \cdot C_i} \right) \cdot \lambda^2$$

$$\det(A - \lambda \cdot I) = \lambda^2 \cdot \left(\frac{1}{R_p} + \frac{1}{R_i}\right) \cdot \frac{1}{C_i} \cdot \frac{R_{se}}{L_i} + \left[\left(\frac{1}{R_p} + \frac{1}{R_i}\right) \cdot \frac{1}{C_i} + \frac{R_{se}}{L_i} \right] \cdot$$

$$\lambda^3 + \lambda^4 - \lambda \cdot \frac{R_{se}}{L_i^2 \cdot C_i} + \lambda \cdot \left(\frac{1}{L_i \cdot C_i}\right) \cdot \frac{R_{se}}{L_i} + \left(\frac{1}{L_i \cdot C_i}\right) \cdot \lambda^2 - \lambda \cdot \frac{R_{se}}{L_i^2 \cdot C_i} + \lambda \cdot \left(\frac{1}{L_i \cdot C_i}\right) \cdot \frac{R_{se}}{L_i} = 0$$

$$\det(A - \lambda \cdot I) = \lambda^4 + \lambda^3 \cdot \left[\left(\frac{1}{R_p} + \frac{1}{R_i}\right) \cdot \frac{1}{C_i} + \frac{R_{se}}{L_i}\right] + \lambda^2 \cdot \left[\left(\frac{1}{R_p} + \frac{1}{R_i}\right) \cdot \frac{1}{C_i} \cdot \frac{R_{se}}{L_i} + \frac{1}{L_i \cdot C_i}\right]$$

$$\det(A - \lambda \cdot I) = 0 \Rightarrow \lambda^4 + \lambda^3 \cdot \left[\left(\frac{1}{R_p} + \frac{1}{R_i}\right) \cdot \frac{1}{C_i} + \frac{R_{se}}{L_i}\right] + \lambda^2 \cdot \left[\left(\frac{1}{R_p} + \frac{1}{R_i}\right) \cdot \frac{1}{C_i} \cdot \frac{R_{se}}{L_i} + \frac{1}{L_i \cdot C_i}\right] = 0$$

$$\lambda^2 \cdot \left\{\lambda^2 + \lambda \cdot \left[\left(\frac{1}{R_p} + \frac{1}{R_i}\right) \cdot \frac{1}{C_i} + \frac{R_{se}}{L_i}\right] + \left[\left(\frac{1}{R_p} + \frac{1}{R_i}\right) \cdot \frac{1}{C_i} \cdot \frac{R_{se}}{L_i} + \frac{1}{L_i \cdot C_i}\right]\right\} = 0$$

$$\sum_{k=0}^{4} \lambda^k \cdot \prod_k = 0; \; \prod_4 = 1; \; \prod_3 = \left(\frac{1}{R_p} + \frac{1}{R_i}\right) \cdot \frac{1}{C_i} + \frac{R_{se}}{L_i}; \; \prod_2 = \left(\frac{1}{R_p} + \frac{1}{R_i}\right) \cdot \frac{1}{C_i} \cdot \frac{R_{se}}{L_i} + \frac{1}{L_i \cdot C_i}; \; \prod_1 = 0; \; \prod_0 = 0$$

We have three possible cases:

Case A: $\lambda^2 = 0; \; \lambda^2 + \lambda \cdot \left[\left(\frac{1}{R_p} + \frac{1}{R_i}\right) \cdot \frac{1}{C_i} + \frac{R_{se}}{L_i}\right] + \left[\left(\frac{1}{R_p} + \frac{1}{R_i}\right) \cdot \frac{1}{C_i} \cdot \frac{R_{se}}{L_i} + \frac{1}{L_i \cdot C_i}\right] \neq 0.$

Case B: $\lambda^2 \neq 0; \; \lambda^2 + \lambda \cdot \left[\left(\frac{1}{R_p} + \frac{1}{R_i}\right) \cdot \frac{1}{C_i} + \frac{R_{se}}{L_i}\right] + \left[\left(\frac{1}{R_p} + \frac{1}{R_i}\right) \cdot \frac{1}{C_i} \cdot \frac{R_{se}}{L_i} + \frac{1}{L_i \cdot C_i}\right] = 0.$

Case C: $\lambda^2 = 0; \; \lambda^2 + \lambda \cdot \left[\left(\frac{1}{R_p} + \frac{1}{R_i}\right) \cdot \frac{1}{C_i} + \frac{R_{se}}{L_i}\right] + \left[\left(\frac{1}{R_p} + \frac{1}{R_i}\right) \cdot \frac{1}{C_i} \cdot \frac{R_{se}}{L_i} + \frac{1}{L_i \cdot C_i}\right] = 0.$

We analyze cases B, C since the eigenvale's quadratic equation is equal to zero. Totally we get four (4) eigenvalues for our laser diode basic intrinsic electrical equivalent circuit's characteristic equation.

Case B: $\lambda_1 \neq 0; \; \lambda_2 \neq 0; \; \lambda^2 + \lambda \cdot \left[\left(\frac{1}{R_p} + \frac{1}{R_i}\right) \cdot \frac{1}{C_i} + \frac{R_{se}}{L_i}\right] + \left[\left(\frac{1}{R_p} + \frac{1}{R_i}\right) \cdot \frac{1}{C_i} \cdot \frac{R_{se}}{L_i} + \frac{1}{L_i \cdot C_i}\right] = 0.$

Case C: $\lambda_1 = 0$; $\lambda_2 \neq 0$ or $\lambda_1 \neq 0$; $\lambda_2 = 0$ or $\lambda_1 = 0$; $\lambda_2 = 0$ and $\lambda^2 + \lambda \cdot \left[\left(\frac{1}{R_p} + \frac{1}{R_i}\right) \cdot \frac{1}{C_i} + \frac{R_{se}}{L_i}\right] + \left[\left(\frac{1}{R_p} + \frac{1}{R_i}\right) \cdot \frac{1}{C_i} \cdot \frac{R_{se}}{L_i} + \frac{1}{L_i \cdot C_i}\right] = 0$.

Eigenvalues stability discussion: Our laser diode intrinsic equation circuit system involving N variables (N > 2, N = 4), the characteristic equation is of degree N = 4 ($\sum_{k=0}^{4} \lambda^k \cdot \prod_k = 0$) and must often be solved numerically. Expect in some particular cases, such an equation has (N = 4) distinct roots that can be real or complex. These values are the eigenvalues of the 4 × 4 Jacobian matrix A. The general rule is that the Steady State (SS) is stable if there is no eigenvalue with positive real part. It is sufficient that one eigenvalue is positive for the steady state to be unstable. Our 4-variables (V_{A_2} I_{L_i} Y_1 Y_2) laser diode intrinsic equation circuit system has four eigenvalues. The type of behavior can be characterized as a function of the position of these eigenvalues in the Re/Im plane. Five non-degenerated cases can be distinguished: (1) the four eigenvalues are real and negative (stable steady state), (2) the four eigenvalues are real, three of them are negative (unstable steady state), (3) and (4) two eigenvalues are complex conjugates with a negative real part and the other eigenvalues are real and negative (stable steady state), two cases can be distinguished depending on the relative value of the real part of the complex eigenvalues and of the real one, (5) two eigenvalues are complex conjugates with a negative real part and at least one eigenvalue is positive (unstable steady state). For each of the five non-degenerated cases, if at least one of the eigenvalues is equal to zero then system's fixed points are on a line [3, 4].

The complete intrinsic electrical equivalent circuit of the laser diode includes terms which are added to the rate equations for explaining the different phenomena appearing in the light output of the laser diodes and the different geometry lasers. The complete intrinsic electrical equivalent circuit includes another capacitor parallel to C_i for explaining the space-charge effects in the active region of the laser diode. The name of the capacitance is as the space-charge capacitance C_{sc}, of the active region. Additionally there is a non-uniform electron density which arrives at an equivalent circuit containing a series resistor R_{se1}, with the inductor. The equation of C_{sc} is given by $C_{sc} = C_{sc(0)} \cdot \left(1 - \frac{V}{V_D}\right)^{-\frac{1}{2}}$ where $C_{sc(0)}$ is the zero-bias space-charge capacitance. V_D is the heterojunction built-in potential (typical value 1.65 v). The resistance R_{se1} is given by $R_{se1} = \frac{(\Gamma \cdot A)^2 \cdot \tau_s \cdot (N_e^0 - N_{om})}{2 \cdot \left[1 + \left(\frac{2 \cdot \pi \cdot L_{eff}}{w}\right)^2\right]} \cdot N_{ph}^0 \cdot L_i$ where Γ is the optical confinement factor normal to the junction plane, L_{eff} is the effective carrier diffusion length and w is the width of the active layer. Resistor R_{se1} models the lateral carrier diffusion and the optical confinement in the active region. The complete intrinsic electrical equivalent circuit of the laser diode (Fig. 1.2) which takes into account all the major factors affecting the dynamic response of the laser diode is shown below [1].

$$V_{A_2} = V_{C_{sc}} = V_{C_i} = V_{R_i}; \quad V_{A_4} = V_{R_{se}}; \quad V_{A_3} - V_{A_4} = V_{L_i}; \quad V_{A_2} - V_{A_3} = V_{R_{se1}}; \quad \frac{V_{\text{sup}} - V_{A_2}}{R_p} = I_{R_p}$$

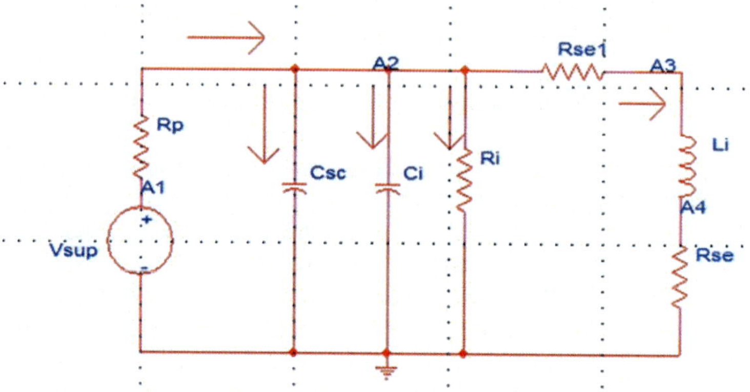

Fig. 1.2 Complete intrinsic electrical equivalent circuit of the laser diode

$$I_{C_{sc}} = C_{sc} \cdot \frac{dV_{A_2}}{dt}; \; I_{C_i} = C_i \cdot \frac{dV_{A_2}}{dt}$$

$$I_{R_i} = \frac{V_{A_2}}{R_i}; \; I_{R_{se1}} = I_{L_i} = I_{R_{se}}; \; I_{R_{se1}} = \frac{V_{A_2} - V_{A_3}}{R_{se1}}$$

$$V_{L_i} = V_{A_3} - V_{A_4} = L_i \cdot \frac{dI_{L_i}}{dt}; \; I_{R_{se}} = \frac{V_{A_4}}{R_{se}}; \; I_{R_p} = I_{C_{sc}} + I_{C_i} + I_{R_i} + I_{R_{se1}}$$

$$\frac{V_{sup} - V_{A_2}}{R_p} = C_{sc} \cdot \frac{dV_{A_2}}{dt} + C_i \cdot \frac{dV_{A_2}}{dt} + \frac{V_{A_2}}{R_i} + \frac{V_{A_2} - V_{A_3}}{R_{se1}}; \; V_{A_2} = V_{R_{se1}} + V_{L_i} + V_{R_{se}}$$

$$V_{A_2} = I_{L_i} \cdot R_{se1} + L_i \cdot \frac{dI_{L_i}}{dt} + I_{L_i} \cdot R_{se}; \; \frac{V_{A_2} - V_{A_3}}{R_{se1}} = \frac{V_{A_4}}{R_{se}} = \frac{1}{L_i} \cdot \int (V_{A_3} - V_{A_4}) \cdot dt$$

$$\frac{V_{sup} - V_{A_2}}{R_p} = (C_{sc} + C_i) \cdot \frac{dV_{A_2}}{dt} + \left(\frac{1}{R_i} + \frac{1}{R_{se1}}\right) \cdot V_{A_2} - \frac{V_{A_3}}{R_{se1}}$$

$$\frac{V_{A_2} - V_{A_3}}{R_{se1}} = \frac{1}{L_i} \cdot \int (V_{A_3} - V_{A_4}) \cdot dt \Rightarrow \left(\frac{dV_{A_2}}{dt} - \frac{dV_{A_3}}{dt}\right) \cdot \frac{1}{R_{se1}} = \frac{1}{L_i} \cdot (V_{A_3} - V_{A_4})$$

$$\frac{V_{A_4}}{R_{se}} = \frac{1}{L_i} \cdot \int (V_{A_3} - V_{A_4}) \cdot dt \Rightarrow \frac{1}{R_{se}} \cdot \frac{dV_{A_4}}{dt} = \frac{1}{L_i} \cdot (V_{A_3} - V_{A_4})$$

$$\frac{1}{R_{se}} \cdot \frac{dV_{A_4}}{dt} = \frac{1}{L_i} \cdot V_{A_3} - \frac{1}{L_i} \cdot V_{A_4}; \; \frac{1}{R_{se1}} \cdot \frac{dV_{A_2}}{dt} - \frac{1}{R_{se1}} \cdot \frac{dV_{A_3}}{dt} = \frac{1}{L_i} \cdot V_{A_3} - \frac{1}{L_i} \cdot V_{A_4}$$

$$\frac{1}{R_{se1}} \cdot \frac{dV_{A_3}}{dt} = \frac{1}{R_{se1}} \cdot \frac{dV_{A_2}}{dt} - \frac{1}{L_i} \cdot V_{A_3} + \frac{1}{L_i} \cdot V_{A_4} \Rightarrow \frac{dV_{A_3}}{dt} = \frac{dV_{A_2}}{dt} - \frac{R_{se1}}{L_i} \cdot V_{A_3} + \frac{R_{se1}}{L_i} \cdot V_{A_4}$$

$$\frac{V_{\sup}}{R_p} - \frac{V_{A_2}}{R_p} = (C_{sc} + C_i) \cdot \frac{dV_{A_2}}{dt} + \left(\frac{1}{R_i} + \frac{1}{R_{se1}}\right) \cdot V_{A_2} - \frac{V_{A_3}}{R_{se1}}$$

$$\frac{d}{dt}\left\{\frac{V_{\sup}}{R_p} - \frac{V_{A_2}}{R_p} = (C_{sc} + C_i) \cdot \frac{dV_{A_2}}{dt} + \left(\frac{1}{R_i} + \frac{1}{R_{se1}}\right) \cdot V_{A_2} - \frac{V_{A_3}}{R_{se1}}\right\}; \quad \frac{dV_{\sup}}{dt} \to \varepsilon$$

$$\frac{V_{\sup}}{R_p} = (C_{sc} + C_i) \cdot \frac{dV_{A_2}}{dt} + \left(\frac{1}{R_i} + \frac{1}{R_{se1}} + \frac{1}{R_p}\right) \cdot V_{A_2} - \frac{V_{A_3}}{R_{se1}}$$

$$\frac{d}{dt}\{\ldots\} \Rightarrow (C_{sc} + C_i) \cdot \frac{d^2 V_{A_2}}{dt^2} + \left(\frac{1}{R_i} + \frac{1}{R_{se1}} + \frac{1}{R_p}\right) \cdot \frac{dV_{A_2}}{dt} - \frac{1}{R_{se1}} \frac{dV_{A_3}}{dt} = 0$$

$$(C_{sc} + C_i) \cdot \frac{d^2 V_{A_2}}{dt^2} + \left(\frac{1}{R_i} + \frac{1}{R_{se1}} + \frac{1}{R_p}\right) \cdot \frac{dV_{A_2}}{dt} - \frac{1}{R_{se1}} \cdot \left(\frac{dV_{A_2}}{dt} - \frac{R_{se1}}{L_i} \cdot V_{A_3} + \frac{R_{se1}}{L_i} \cdot V_{A_4}\right) = 0$$

$$(C_{sc} + C_i) \cdot \frac{d^2 V_{A_2}}{dt^2} + \left(\frac{1}{R_i} + \frac{1}{R_{se1}} + \frac{1}{R_p}\right) \cdot \frac{dV_{A_2}}{dt} - \frac{1}{R_{se1}} \cdot \frac{dV_{A_2}}{dt} + \frac{1}{L_i} \cdot V_{A_3} - \frac{1}{L_i} \cdot V_{A_4} = 0$$

$$(C_{sc} + C_i) \cdot \frac{d^2 V_{A_2}}{dt^2} + \left(\frac{1}{R_i} + \frac{1}{R_p}\right) \cdot \frac{dV_{A_2}}{dt} + \frac{1}{L_i} \cdot V_{A_3} - \frac{1}{L_i} \cdot V_{A_4} = 0$$

We define new variable: $Y = \frac{dV_{A_2}}{dt}; \ \frac{dY}{dt} = \frac{d^2 V_{A_2}}{dt^2}$

$$\frac{dV_{A_3}}{dt} = Y - \frac{R_{se1}}{L_i} \cdot V_{A_3} + \frac{R_{se1}}{L_i} \cdot V_{A_4}; \quad \frac{dV_{A_4}}{dt} = \frac{R_{se}}{L_i} \cdot V_{A_3} - \frac{R_{se}}{L_i} \cdot V_{A_4}$$

$$(C_{sc} + C_i) \cdot \frac{dY}{dt} + \left(\frac{1}{R_i} + \frac{1}{R_p}\right) \cdot Y + \frac{1}{L_i} \cdot V_{A_3} - \frac{1}{L_i} \cdot V_{A_4} = 0 \Rightarrow (C_{sc} + C_i) \cdot \frac{dY}{dt} = -\left(\frac{1}{R_i} + \frac{1}{R_p}\right) \cdot Y - \frac{1}{L_i} \cdot V_{A_3} + \frac{1}{L_i} \cdot V_{A_4}$$

$$\frac{dY}{dt} = -\frac{1}{(C_{sc} + C_i)} \cdot \left(\frac{1}{R_i} + \frac{1}{R_p}\right) \cdot Y - \frac{1}{L_i \cdot (C_{sc} + C_i)} \cdot V_{A_3} + \frac{1}{L_i \cdot (C_{sc} + C_i)} \cdot V_{A_4}$$

We can summary our system differential equations: $Y = Y(t); \ V_{A_3} = V_{A_3}(t); \ V_{A_4} = V_{A_4}(t)$

$$\frac{dY}{dt} = -\frac{1}{(C_{sc} + C_i)} \cdot \left(\frac{1}{R_i} + \frac{1}{R_p}\right) \cdot Y$$

$$-\frac{1}{L_i\cdot(C_{sc}+C_i)}\cdot V_{A_3}+\frac{1}{L_i\cdot(C_{sc}+C_i)}\cdot V_{A_4}$$

$$\frac{dV_{A_3}}{dt}=Y-\frac{R_{se1}}{L_i}\cdot V_{A_3}+\frac{R_{se1}}{L_i}\cdot V_{A_4};\ \frac{dV_{A_4}}{dt}$$
$$=\frac{R_{se}}{L_i}\cdot V_{A_3}-\frac{R_{se}}{L_i}\cdot V_{A_4}$$

We define three functions: $g_i=g_i\left(Y,V_{A_3},V_{A_4}\right)\forall i=1,2,3$

$$\frac{dY}{dt}=g_1\left(Y,V_{A_3},V_{A_4}\right);\ \frac{dV_{A_3}}{dt}$$
$$=g_2\left(Y,V_{A_3},V_{A_4}\right);\ \frac{dV_{A_4}}{dt}=g_3\left(Y,V_{A_3},V_{A_4}\right)$$

$$g_1\left(Y,V_{A_3},V_{A_4}\right)=-\frac{1}{(C_{sc}+C_i)}\cdot\left(\frac{1}{R_i}+\frac{1}{R_p}\right)\cdot$$
$$Y-\frac{1}{L_i\cdot(C_{sc}+C_i)}\cdot V_{A_3}+\frac{1}{L_i\cdot(C_{sc}+C_i)}\cdot V_{A_4}$$

$$g_2\left(Y,V_{A_3},V_{A_4}\right)=Y-\frac{R_{se1}}{L_i}\cdot V_{A_3}+\frac{R_{se1}}{L_i}\cdot V_{A_4};\ g_3\left(Y,V_{A_3},V_{A_4}\right)$$
$$=\frac{R_{se}}{L_i}\cdot V_{A_3}-\frac{R_{se}}{L_i}\cdot V_{A_4}$$

At fixed points: $\frac{dY}{dt}=0;\ \frac{dV_{A_3}}{dt}=0;\ \frac{dV_{A_4}}{dt}=0$

$$\frac{dV_{A_4}}{dt}=0\Rightarrow V^*_{A_3}=V^*_{A_4};\ \frac{dV_{A_3}}{dt}=0\Rightarrow Y^*-\frac{R_{se1}}{L_i}\cdot V^*_{A_3}$$
$$+\frac{R_{se1}}{L_i}\cdot V^*_{A_3}=0\Rightarrow Y^*=0$$

$$E^*\left(Y^*,V^*_{A_3},V^*_{A_4}\right)=\left(0,V^*_{A_3},V^*_{A_3}\right)=\left(0,V^*_{A_4},V^*_{A_4}\right)$$

Stability analysis: The standard local stability analysis about one of the equilibrium points of laser diode complete intrinsic electrical equivalent circuit consists in adding to coordinate $[Y\ V_{A_3}\ V_{A_4}]$ arbitrarily small increment of exponential form $y\cdot e^{\lambda\cdot t}$, $v_{A_3}\cdot e^{\lambda\cdot t}$, $v_{A_4}\cdot e^{\lambda\cdot t}$ Respectively and retaining the first order terms in $[y,v_{A_3},v_{A_4}]$. The system of one homogeneous equation leads to a polynomial characteristic equation in the eigenvalue λ. The polynomial characteristic equation accept by set the below laser diode complete intrinsic electrical equivalent circuit respect to time into one system equation. Our system fixed points/values with arbitrarily small increment of the exponential form $y\cdot e^{\lambda\cdot t}$, $v_{A_3}\cdot e^{\lambda\cdot t}$, $v_{A_4}\cdot e^{\lambda\cdot t}$ are i = 0 (first fixed point), i = 1 (second fixed point), and i = 2 (third fixed point).

$$Y(t) = Y^{(i)} + y \cdot e^{\lambda \cdot t};\ V_{A_3}(t) = V_{A_3}^{(i)} + v_{A_3} \cdot e^{\lambda \cdot t};\ V_{A_4}(t) = V_{A_4}^{(i)} + v_{A_4} \cdot e^{\lambda \cdot t};\ \frac{dY(t)}{dt} = y \cdot \lambda \cdot e^{\lambda \cdot t}$$

$$\frac{dV_{A_3}(t)}{dt} = v_{A_3} \cdot \lambda \cdot e^{\lambda \cdot t};\ \frac{dV_{A_4}(t)}{dt} = v_{A_4} \cdot \lambda \cdot e^{\lambda \cdot t}$$

We choose the above expressions for our $Y(t)$, $V_{A_3}(t)$, and $V_{A_4}(t)$ as small displacement $[y, v_{A_3}, v_{A_4}]$ from our laser diode complete intrinsic equation circuit fixed points at time t = 0. For $\lambda < 0, t > 0$ the selected fixed point is stable otherwise $\lambda > 0, t > 0$ is unstable. Our laser diode complete intrinsic equation circuit tends to the selected fixed point exponentially for $\lambda < 0, t > 0$, otherwise go away from the selected fixed point exponentially. Eigenvalue λ is the parameter which establish if the fixed point is stable or unstable, additionally his absolute value $|\lambda|$ establish the speed of the flow toward or away from the selected fixed point. The laser diode complete intrinsic electrical equivalent circuit variables for negative eigenvalue ($\lambda < 0$) are as follow (Table 1.6).

The laser diode complete intrinsic electrical equivalent circuit variables for positive eigenvalue ($\lambda > 0$) are as follow (Table 1.7).

$$\frac{dY}{dt} = -\frac{1}{(C_{sc} + C_i)} \cdot \left(\frac{1}{R_i} + \frac{1}{R_p}\right) \cdot Y - \frac{1}{L_i \cdot (C_{sc} + C_i)} \cdot V_{A3} + \frac{1}{L_i \cdot (C_{sc} + C_i)} \cdot V_{A4}$$

Table 1.6 System variables for $\lambda < 0$ and time parameter

Time (t)	$\lambda < 0$
t = 0	$Y(t=0) = Y^{(i)} + y;\ V_{A3}(t=0) = V_{A_3}^{(i)} + v_{A3}$ $V_{A4}(t=0) = V_{A_4}^{(i)} + v_{A4}$
t > 0	$Y(t) = Y^{(i)} + y \cdot e^{-\lvert\lambda\rvert \cdot t};\ V_{A3}(t) = V_{A_3}^{(i)} + v_{A3} \cdot e^{-\lvert\lambda\rvert \cdot t}$ $V_{A4}(t) = V_{A_4}^{(i)} + v_{A4} \cdot e^{-\lvert\lambda\rvert \cdot t}$
t > 0; t → ∞	$Y(t \to \infty) = Y^{(i)};\ V_{A3}(t \to \infty) = V_{A_3}^{(i)};\ V_{A4}(t \to \infty) = V_{A4}^{(i)}$

Table 1.7 System variables for $\lambda > 0$ and time parameter

Time (t)	$\lambda > 0$
t = 0	$Y(t=0) = Y^{(i)} + y;\ V_{A3}(t=0) = V_{A_3}^{(i)} + v_{A3}$ $V_{A4}(t=0) = V_{A_4}^{(i)} + v_{A4}$
t > 0	$Y(t) = Y^{(i)} + y \cdot e^{\lvert\lambda\rvert \cdot t};\ V_{A3}(t) = V_{A_3}^{(i)} + v_{A3} \cdot e^{\lvert\lambda\rvert \cdot t}$ $V_{A4}(t) = V_{A_4}^{(i)} + v_{A4} \cdot e^{\lvert\lambda\rvert \cdot t}$
t > 0; t → ∞	$Y(t \to \infty) = y \cdot e^{\lvert\lambda\rvert \cdot t};\ V_{A3}(t \to \infty) = v_{A_3} \cdot e^{\lvert\lambda\rvert \cdot t};\ V_{A4}(t \to \infty) = v_{A_4} \cdot e^{\lvert\lambda\rvert \cdot t}$

$$y \cdot \lambda \cdot e^{\lambda \cdot t} = -\frac{1}{(C_{sc} + C_i)} \cdot \left(\frac{1}{R_i} + \frac{1}{R_p}\right) \cdot \left(Y^{(i)} + y \cdot e^{\lambda \cdot t}\right) - \frac{1}{L_i \cdot (C_{sc} + C_i)} \cdot \left(V_{A_3}^{(i)} + v_{A_3} \cdot e^{\lambda \cdot t}\right) + \frac{1}{L_i \cdot (C_{sc} + C_i)} \cdot \left(V_{A_4}^{(i)} + v_{A_4} \cdot e^{\lambda \cdot t}\right)$$

$$y \cdot \lambda \cdot e^{\lambda \cdot t} = -\frac{1}{(C_{sc} + C_i)} \cdot \left(\frac{1}{R_i} + \frac{1}{R_p}\right) \cdot Y^{(i)} - \frac{1}{L_i \cdot (C_{sc} + C_i)} \cdot V_{A_3}^{(i)} + \frac{1}{L_i \cdot (C_{sc} + C_i)} \cdot V_{A_4}^{(i)} - \frac{1}{(C_{sc} + C_i)} \cdot \left(\frac{1}{R_i} + \frac{1}{R_p}\right) \cdot y \cdot e^{\lambda \cdot t} - \frac{1}{L_i \cdot (C_{sc} + C_i)} \cdot v_{A_3} \cdot e^{\lambda \cdot t} + \frac{1}{L_i \cdot (C_{sc} + C_i)} \cdot v_{A_4} \cdot e^{\lambda \cdot t}$$

At fixed point: $-\frac{1}{(C_{sc} + C_i)} \cdot \left(\frac{1}{R_i} + \frac{1}{R_p}\right) \cdot Y^{(i)} - \frac{1}{L_i \cdot (C_{sc} + C_i)} \cdot V_{A_3}^{(i)} + \frac{1}{L_i \cdot (C_{sc} + C_i)} \cdot V_{A_4}^{(i)} = 0$

$$-\frac{1}{(C_{sc} + C_i)} \cdot \left(\frac{1}{R_i} + \frac{1}{R_p}\right) \cdot y - y \cdot \lambda - \frac{1}{L_i \cdot (C_{sc} + C_i)} \cdot v_{A_3} + \frac{1}{L_i \cdot (C_{sc} + C_i)} \cdot v_{A_4} = 0$$

$$\frac{dV_{A_3}}{dt} = Y - \frac{R_{se1}}{L_i} \cdot V_{A_3} + \frac{R_{se1}}{L_i} \cdot V_{A_4}$$

$$v_{A_3} \cdot \lambda \cdot e^{\lambda \cdot t} = Y^{(i)} + y \cdot e^{\lambda \cdot t} - \frac{R_{se1}}{L_i} \cdot \left(V_{A_3}^{(i)} + v_{A_3} \cdot e^{\lambda \cdot t}\right) + \frac{R_{se1}}{L_i} \cdot \left(V_{A_4}^{(i)} + v_{A_4} \cdot e^{\lambda \cdot t}\right)$$

$$v_{A_3} \cdot \lambda \cdot e^{\lambda \cdot t} = Y^{(i)} - \frac{R_{se1}}{L_i} \cdot V_{A_3}^{(i)} + \frac{R_{se1}}{L_i} \cdot V_{A_4}^{(i)} + y \cdot e^{\lambda \cdot t} - \frac{R_{se1}}{L_i} \cdot v_{A_3} \cdot e^{\lambda \cdot t} + \frac{R_{se1}}{L_i} \cdot v_{A_4} \cdot e^{\lambda \cdot t}$$

At fixed point: $Y^{(i)} - \frac{R_{se1}}{L_i} \cdot V_{A_3}^{(i)} + \frac{R_{se1}}{L_i} \cdot V_{A_4}^{(i)} = 0$

$$y - \frac{R_{se1}}{L_i} \cdot v_{A_3} - v_{A_3} \cdot \lambda + \frac{R_{se1}}{L_i} \cdot v_{A_4} = 0$$

$$\frac{dV_{A_4}}{dt} = \frac{R_{se}}{L_i} \cdot V_{A_3} - \frac{R_{se}}{L_i} \cdot V_{A_4}; \; v_{A_4} \cdot \lambda \cdot e^{\lambda \cdot t} = \frac{R_{se}}{L_i} \cdot \left(V_{A_3}^{(i)} + v_{A_3} \cdot e^{\lambda \cdot t}\right) - \frac{R_{se}}{L_i} \cdot \left(V_{A_4}^{(i)} + v_{A_4} \cdot e^{\lambda \cdot t}\right)$$

$$v_{A_4} \cdot \lambda \cdot e^{\lambda \cdot t} = \frac{R_{se}}{L_i} \cdot V_{A_3}^{(i)} - \frac{R_{se}}{L_i} \cdot V_{A_4}^{(i)} + \frac{R_{se}}{L_i} \cdot v_{A_3} \cdot e^{\lambda \cdot t} - \frac{R_{se}}{L_i} \cdot v_{A_4} \cdot e^{\lambda \cdot t}$$

At fixed point: $\frac{R_{se}}{L_i} \cdot V_{A_3}^{(i)} - \frac{R_{se}}{L_i} \cdot V_{A_4}^{(i)} = 0$

$$\frac{R_{se}}{L_i} \cdot v_{A_3} - \frac{R_{se}}{L_i} \cdot v_{A_4} - v_{A_4} \cdot \lambda = 0$$

We can summary our laser diode complete intrinsic electrical equivalent circuit arbitrarily small increment equations:

$$-\frac{1}{(C_{sc} + C_i)} \cdot \left(\frac{1}{R_i} + \frac{1}{R_p}\right) \cdot y - y \cdot \lambda - \frac{1}{L_i \cdot (C_{sc} + C_i)} \cdot v_{A_3} + \frac{1}{L_i \cdot (C_{sc} + C_i)} \cdot v_{A_4} = 0$$

$$y - \frac{R_{sel}}{L_i} \cdot v_{A_3} - v_{A_3} \cdot \lambda + \frac{R_{sel}}{L_i} \cdot v_{A_4} = 0; \quad \frac{R_{se}}{L_i} \cdot v_{A_3} - \frac{R_{se}}{L_i} \cdot v_{A_4} - v_{A_4} \cdot \lambda = 0$$

The arbitrarily small increments Jacobian of our laser diode complete intrinsic electrical equivalent circuit is as follow:

$$\begin{pmatrix} -\frac{1}{(C_{sc}+C_i)} \cdot \left(\frac{1}{R_i} + \frac{1}{R_p}\right) - \lambda & -\frac{1}{L_i \cdot (C_{sc}+C_i)} & \frac{1}{L_i \cdot (C_{sc}+C_i)} \\ 1 & -\frac{R_{sel}}{L_i} - \lambda & \frac{R_{sel}}{L_i} \\ 0 & \frac{R_{se}}{L_i} & -\frac{R_{se}}{L_i} - \lambda \end{pmatrix} \cdot \begin{pmatrix} y \\ v_{A_3} \\ v_{A_4} \end{pmatrix} = 0$$

$$A - \lambda \cdot I = \begin{pmatrix} -\frac{1}{(C_{sc}+C_i)} \cdot \left(\frac{1}{R_i} + \frac{1}{R_p}\right) - \lambda & -\frac{1}{L_i \cdot (C_{sc}+C_i)} & \frac{1}{L_i \cdot (C_{sc}+C_i)} \\ 1 & -\frac{R_{sel}}{L_i} - \lambda & \frac{R_{sel}}{L_i} \\ 0 & \frac{R_{se}}{L_i} & -\frac{R_{se}}{L_i} - \lambda \end{pmatrix}$$

$$\det(A - \lambda \cdot I) = 0$$

$$\det(A - \lambda \cdot I) = \left[-\frac{1}{(C_{sc} + C_i)} \cdot \left(\frac{1}{R_i} + \frac{1}{R_p}\right) - \lambda\right] \cdot \det\begin{pmatrix} -\frac{R_{sel}}{L_i} - \lambda & \frac{R_{sel}}{L_i} \\ \frac{R_{se}}{L_i} & -\frac{R_{se}}{L_i} - \lambda \end{pmatrix}$$
$$+ \frac{1}{L_i \cdot (C_{sc} + C_i)} \cdot \det\begin{pmatrix} 1 & \frac{R_{sel}}{L_i} \\ 0 & -\frac{R_{se}}{L_i} - \lambda \end{pmatrix}$$
$$+ \frac{1}{L_i \cdot (C_{sc} + C_i)} \cdot \det\begin{pmatrix} 1 & -\frac{R_{sel}}{L_i} - \lambda \\ 0 & \frac{R_{se}}{L_i} \end{pmatrix}$$

$$\det(A - \lambda \cdot I) = \left[-\frac{1}{(C_{sc} + C_i)} \cdot \left(\frac{1}{R_i} + \frac{1}{R_p}\right) - \lambda\right] \cdot$$

$$\left\{\left(-\frac{R_{se1}}{L_i}-\lambda\right)\cdot\left(-\frac{R_{se}}{L_i}-\lambda\right)-\frac{R_{se}}{L_i}\cdot\frac{R_{se1}}{L_i}\right\}$$
$$+\frac{1}{L_i\cdot(C_{sc}+C_i)}\cdot\left(-\frac{R_{se}}{L_i}-\lambda\right)+\frac{1}{L_i\cdot(C_{sc}+C_i)}\cdot\frac{R_{se}}{L_i}$$

$$\det(A-\lambda\cdot I)=-\left[\frac{1}{(C_{sc}+C_i)}\cdot\left(\frac{1}{R_i}+\frac{1}{R_p}\right)+\lambda\right]\cdot$$
$$\left\{\left(\frac{R_{se1}}{L_i}+\lambda\right)\cdot\left(\frac{R_{se}}{L_i}+\lambda\right)-\frac{R_{se}\cdot R_{se1}}{L_i^2}\right\}$$
$$-\frac{1}{L_i\cdot(C_{sc}+C_i)}\cdot\left(\frac{R_{se}}{L_i}+\lambda\right)+\frac{1}{L_i\cdot(C_{sc}+C_i)}\cdot\frac{R_{se}}{L_i}$$

$$\det(A-\lambda\cdot I)=-\left[\frac{1}{(C_{sc}+C_i)}\cdot\left(\frac{1}{R_i}+\frac{1}{R_p}\right)+\lambda\right]\cdot$$
$$\left[\lambda^2+\lambda\cdot\left(\frac{R_{se1}}{L_i}+\frac{R_{se}}{L_i}\right)\right]-\frac{1}{L_i\cdot(C_{sc}+C_i)}\cdot\lambda$$

$$\det(A-\lambda\cdot I)=-\lambda^3-\lambda^2\cdot\left[\frac{1}{(C_{sc}+C_i)}\cdot\left(\frac{1}{R_i}+\frac{1}{R_p}\right)+\left(\frac{R_{se1}}{L_i}+\frac{R_{se}}{L_i}\right)\right]$$
$$-\lambda\cdot\left[\left(\frac{R_{se1}}{L_i}+\frac{R_{se}}{L_i}\right)\cdot\left(\frac{1}{R_i}+\frac{1}{R_p}\right)\cdot\frac{1}{(C_{sc}+C_i)}+\frac{1}{L_i\cdot(C_{sc}+C_i)}\right]$$

$$\det(A-\lambda\cdot I)=-\lambda\cdot\left\{\lambda^2+\lambda\cdot\left[\frac{1}{(C_{sc}+C_i)}\cdot(\frac{1}{R_i}+\frac{1}{R_p})+(\frac{R_{se1}}{L_i}+\frac{R_{se}}{L_i})\right]\right.$$
$$\left.+\left[\left(\frac{R_{se1}}{L_i}+\frac{R_{se}}{L_i}\right)\cdot\left(\frac{1}{R_i}+\frac{1}{R_p}\right)\cdot\frac{1}{(C_{sc}+C_i)}+\frac{1}{L_i\cdot(C_{sc}+C_i)}\right]\right\}$$

$$\sum_{k=0}^{3}\lambda^k\cdot\prod\nolimits_k=0;\ \prod\nolimits_3=-1;\ \prod\nolimits_2=-\left[\frac{1}{(c_{sc}+c_i)}\cdot\left(\frac{1}{r_i}+\frac{1}{r_p}\right)+\left(\frac{r_{se1}}{l_i}+\frac{r_{se}}{l_i}\right)\right]$$
$$\prod\nolimits_1=-\left[\left(\frac{r_{se1}}{l_i}+\frac{r_{se}}{l_i}\right)\cdot\left(\frac{1}{r_i}+\frac{1}{r_p}\right)\cdot\frac{1}{(c_{sc}+c_i)}+\frac{1}{l_i\cdot(c_{sc}+c_i)}\right];\ \prod\nolimits_0=0$$

We have three possible cases:

Case A: $\lambda_1=0$

$$\lambda^2+\lambda\cdot\left[\frac{1}{(C_{sc}+C_i)}\cdot\left(\frac{1}{R_i}+\frac{1}{R_p}\right)+\left(\frac{R_{se1}}{L_i}+\frac{R_{se}}{L_i}\right)\right]$$
$$+\left[\left(\frac{R_{se1}}{L_i}+\frac{R_{se}}{L_i}\right)\cdot(\frac{1}{R_i}+\frac{1}{R_p})\cdot\frac{1}{(C_{sc}+C_i)}+\frac{1}{L_i\cdot(C_{sc}+C_i)}\right]\neq 0$$

Case B: $\lambda_1\neq 0$

$$\lambda^2+\lambda\cdot\left[\frac{1}{(C_{sc}+C_i)}\cdot\left(\frac{1}{R_i}+\frac{1}{R_p}\right)+\left(\frac{R_{se1}}{L_i}+\frac{R_{se}}{L_i}\right)\right]$$

$$+\left[\left(\frac{R_{sel}}{L_i}+\frac{R_{se}}{L_i}\right)\cdot\left(\frac{1}{R_i}+\frac{1}{R_p}\right)\cdot\frac{1}{(C_{sc}+C_i)}+\frac{1}{L_i\cdot(C_{sc}+C_i)}\right]=0$$

Case C: $\lambda_1 = 0$

$$\lambda^2+\lambda\cdot\left[\frac{1}{(C_{sc}+C_i)}\cdot\left(\frac{1}{R_i}+\frac{1}{R_p}\right)+\left(\frac{R_{sel}}{L_i}+\frac{R_{se}}{L_i}\right)\right]$$
$$+\left[\left(\frac{R_{sel}}{L_i}+\frac{R_{se}}{L_i}\right)\cdot\left(\frac{1}{R_i}+\frac{1}{R_p}\right)\cdot\frac{1}{(C_{sc}+C_i)}+\frac{1}{L_i\cdot(C_{sc}+C_i)}\right]=0$$

We analyze case B, C since the eigenvale's quadratic equation is equal to zero. Totally we get three (3) eigenvalues for our complete intrinsic electrical equivalent circuit of the laser diode characteristic equation.

Eigenvalues stability discussion: Our complete intrinsic electrical equivalent circuit of the laser diode system involving N variables (N > 2, N = 3), the characteristic equation is of degree N = 3 ($\sum_{k=0}^{3}\lambda^k\cdot\prod_k = 0$) and must often be solved numerically. Expect in some particular cases, such an equation has (N = 3) distinct roots that can be real or complex. These values are the eigenvalues of the 3 × 3 Jacobian matrix A. The general rule is that the Steady State (SS) is stable if there is no eigenvalue with positive real part. It is sufficient that one eigenvalue is positive for the steady state to be unstable. Our 3-variables ($Y\ V_{A_3}\ V_{A_4}$) complete intrinsic electrical equivalent circuit of the laser diode system has three eigenvalues. The type of behavior can be characterized as a function of the position of these eigenvalues in the Re/Im plane. Five non-degenerated cases can be distinguished: (1) the three eigenvalues are real and negative (stable steady state), (2) the three eigenvalues are real, two of them are negative (unstable steady state), (3) and (4) one eigenvalue is complex conjugates with a negative real part and the other eigenvalues are real and negative (stable steady state), two cases can be distinguished depending on the relative value of the real part of the complex eigenvalues and of the real one, (5) one eigenvalue is complex conjugates with a negative real part and at least one eigenvalue is positive (unstable steady state). For each of the five non-degenerated cases, if at least one of the eigenvalues is zero then system's fixed points are on a line [10, 11].

1.3 MRI System Laser Diode Circuitry Dynamic and Stability Analysis

Fiber optics MRI system addresses all the needs of MRI technology. One of these technologies is phased array coils with high number of elements. The phase array is a type of MRI coil that allows parallel imaging. These are multi-channel coils with each channel having its own receiver when channel system increases the number of receiver channels over traditional MRI coils. Fast scan time is the most obvious benefit of phase array. The fact is by using multiple receiver coils in parallel scan time in Fourier imaging can be considerably reduced. In order to minimize the total size of system interconnections and improve the system safety we use a miniature fiber optic transmission (FOT) system. The TOF system consists of a receiver coil with active detuning, a low-noise preamplifier, and a laser diode connected to a photodetector with fiber optic cabling. The laser diode circuitry is the component responsible for the conversion of MR signals from electrical to the optical state. The circuitry is placed after the preamplifiers in the RF signal path of the fiber-optic MRI probe. The circuit is preceded by preamplifiers from the fact of purely noise performance. Diode laser circuitry is integrated electrical-optical-electrical (E-O-E) conversion process and is associated with a loss in the transmitted power. Adding preamplifiers provide that the gain is sufficient enough to move the signal away from the noise floor even after attenuation. The laser diode circuitry consists of an L-match section, a Bias-T, and the laser diode itself (Fig. 1.3). The L-match is responsible for the matching of the low impedance laser diode to the high impedance output of the preamplifier. The Bias-T combines the RF signal with the appropriate DC current injection which required biasing the laser. The laser diode holds the micro-lens to couple the generated optical energy to the fiber. Due to laser diode circuit microstrip parasitic effects there are instabilities on the circuit operation. The circuit microstrip parasitic effects are represented by delay lines elements which delay the current in time that flow through them [6]. We consider ideal delay lines then the voltages that fall are neglected ($V_{\tau_i} \to \varepsilon \,\forall\, i = 1, 2$).

The voltage on the laser diode is V and the current is I ($V = f(I)$). We consider ideal delay lines $V_{\tau_i} \to \varepsilon \,\forall\, i = 1, 2$ and the delay is in the currents. We write circuit equations:

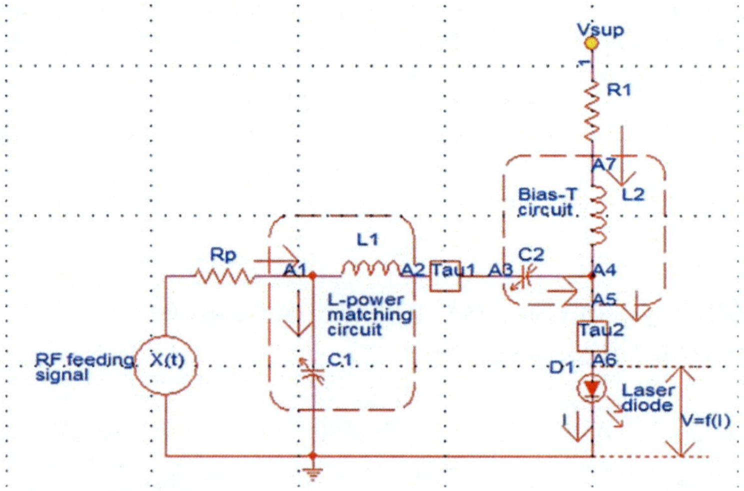

Fig. 1.3 MRI system laser diode circuit with delay lines

$$V_{A_5} = V_{A_4};\; V_{A_2} = V_{A_3};\; V_{A_5} = V_{A_6};\; I_{R_p} = \frac{X(t) - V_{A_1}}{R_p}$$

$$I_{C_1} = C_1 \cdot \frac{dV_{A_1}}{dt};\; V_{A_1} - V_{A_2} = V_{L_1} = L_1 \cdot \frac{dI_{L_1}}{dt}$$

$$I_{C_2}(t) = I_{L_1}(t - \tau_1);\; I_{C_2} = C_2 \cdot \frac{d\left(V_{A_3} - V_{A_4}\right)}{dt}$$

$$V_{L_2} = V_{A_7} - V_{A_4} = L_2 \cdot \frac{dI_{L_2}}{dt};\; I_{R_1} = I_{L_2};\; I_{L_1} + I_{C_1} = I_{R_p}$$

$$I_{R_1} = \frac{V_{\text{sup}} - V_{A_7}}{R_1};\; V_{A_6} = V = f(I);\; I(t) = I_{L_2}(t - \tau_2) + I_{C_2}(t - \tau_2)$$

$$I(t) = I_{L_2}(t - \tau_2) + I_{L_1}\left(t - \sum_{k=1}^{2} \tau_k\right)$$

$$I_{R_p} = \frac{X(t) - V_{A_1}}{R_p} \Rightarrow I_{R_p} \cdot R_p = X(t) - V_{A_1};\; V_{A_1} = X(t) - I_{R_p} \cdot R_p$$

$$I_{C_1} = C_1 \cdot \frac{dV_{A_1}}{dt} = C_1 \cdot \frac{d}{dt}\left[X(t) - I_{R_p} \cdot R_p\right]$$

$$I_{R_p} - I_{L_1} = C_1 \cdot \frac{dX(t)}{dt} - C_1 \cdot R_p \cdot \frac{dI_{R_p}}{dt};\; V_{A_1} - V_{A_2} = V_{L_1} = L_1 \cdot \frac{dI_{L_1}}{dt} \Rightarrow V_{A_2} = V_{A_1} - L_1 \cdot \frac{dI_{L_1}}{dt}$$

$$V_{A_2} = V_{A_3} = X(t) - I_{R_p} \cdot R_p - L_1 \cdot \frac{dI_{L_1}}{dt}$$

$$I_{C_2} = C_2 \cdot \frac{d\left(V_{A_3} - V_{A_4}\right)}{dt} \Rightarrow V_{A_3} - V_{A_4} = \frac{1}{C_2} \cdot \int I_{C_2} \cdot dt$$

$$V_{A_4} = V_{A_3} - \frac{1}{C_2} \cdot \int I_{C_2} \cdot dt;\; V_{A_4} = X(t) - I_{R_p} \cdot R_p - L_1 \cdot \frac{dI_{L_1}}{dt} - \frac{1}{C_2} \cdot \int I_{C_2} \cdot dt$$

$$V_{A_7} - V_{A_4} = L_2 \cdot \frac{dI_{L_2}}{dt} \Rightarrow V_{A_7} = V_{A_4} + L_2 \cdot \frac{dI_{L_2}}{dt}$$

$$V_{A_7} = X(t) - I_{R_p} \cdot R_p - L_1 \cdot \frac{dI_{L_1}}{dt} - \frac{1}{C_2} \cdot \int I_{C_2} \cdot dt + L_2 \cdot \frac{dI_{L_2}}{dt}$$

$$I_{R_1} = \frac{V_{\text{sup}} - V_{A_7}}{R_1} = \frac{V_{\text{sup}}}{R_1} - \frac{1}{R_1} \cdot V_{A_7}$$

$$I_{R_1} = \frac{V_{\text{sup}}}{R_1} - \frac{1}{R_1} \cdot \left[X(t) - I_{R_p} \cdot R_p - L_1 \cdot \frac{dI_{L_1}}{dt} - \frac{1}{C_2} \cdot \int I_{C_2} \cdot dt + L_2 \cdot \frac{dI_{L_2}}{dt}\right]$$

$$\frac{d}{dt}\left\{I_{R_1} = \frac{V_{\text{sup}}}{R_1} - \frac{1}{R_1} \cdot \left[X(t) - I_{R_p} \cdot R_p - L_1 \cdot \frac{dI_{L_1}}{dt} - \frac{1}{C_2} \cdot \int I_{C_2} \cdot dt + L_2 \cdot \frac{dI_{L_2}}{dt}\right]\right\}$$

$$\frac{dI_{R_1}}{dt} = \frac{dV_{\text{sup}}}{dt} \cdot \frac{1}{R_1} - \frac{1}{R_1} \cdot \left[\frac{dX(t)}{dt} - \frac{dI_{R_p}}{dt} \cdot R_p - L_1 \cdot \frac{d^2 I_{L_1}}{dt^2} - \frac{1}{C_2} \cdot I_{C_2} + L_2 \cdot \frac{d^2 I_{L_2}}{dt^2}\right]$$

$$\frac{dV_{\text{sup}}}{dt} \to \varepsilon; \quad \frac{dI_{R_1}}{dt} = -\frac{1}{R_1} \cdot \left[\frac{dX(t)}{dt} - \frac{dI_{R_p}}{dt} \cdot R_p - L_1 \cdot \frac{d^2 I_{L_1}}{dt^2} - \frac{1}{C_2} \cdot I_{C_2} + L_2 \cdot \frac{d^2 I_{L_2}}{dt^2}\right]$$

$$I_{C_2}(t) = I_{L_1}(t - \tau_1) \& I_{R_1} = I_{L_2}; \quad \frac{dI_{L_2}}{dt} = -\frac{1}{R_1} \cdot \left[\frac{dX(t)}{dt} - \frac{dI_{R_p}}{dt} \cdot R_p - L_1 \cdot \frac{d^2 I_{L_1}}{dt^2} - \frac{1}{C_2} \cdot I_{L_1}(t - \tau_1) + L_2 \cdot \frac{d^2 I_{L_2}}{dt^2}\right]$$

$$V_{A_5} = V_{A_6} = V = f(I); \quad V_{A_5} = V_{A_4} = V = f(I); \quad I(t) = I_{L_2}(t - \tau_2) + I_{C_2}(t - \tau_2)$$

$$V_{A_4} = V \Rightarrow V = f\left(I_{L_2}(t - \tau_2) + I_{C_2}(t - \tau_2)\right)$$

$$V = f\left(I_{L_2}(t - \tau_2) + I_{L_1}\left(t - \sum_{k=1}^{2} \tau_k\right)\right)$$

$$V_{A_4} = X(t) - I_{R_p} \cdot R_p - L_1 \cdot \frac{dI_{L_1}}{dt} - \frac{1}{C_2} \cdot \int I_{C_2} \cdot dt; \quad V_{A_4} = f(I)$$

$$X(t) - I_{R_p} \cdot R_p - L_1 \cdot \frac{dI_{L_1}}{dt} - \frac{1}{C_2} \cdot \int I_{C_2} \cdot dt = f\left(I_{L_2}(t - \tau_2) + I_{L_1}\left(t - \sum_{k=1}^{2} \tau_k\right)\right)$$

$$X(t) - I_{R_p} \cdot R_p - L_1 \cdot \frac{dI_{L_1}}{dt} - \frac{1}{C_2} \cdot \int I_{L_1}(t - \tau_1) \cdot dt = f\left(I_{L_2}(t - \tau_2) + I_{L_1}\left(t - \sum_{k=1}^{2} \tau_k\right)\right)$$

We can summary our system differential equations:

$$I_{R_p} - I_{L_1} = C_1 \cdot \frac{dX(t)}{dt} - C_1 \cdot R_p \cdot \frac{dI_{R_p}}{dt}$$

$$\frac{dI_{L_2}}{dt} = -\frac{1}{R_1} \cdot \left[\frac{dX(t)}{dt} - \frac{dI_{R_p}}{dt} \cdot R_p - L_1 \cdot \frac{d^2 I_{L_1}}{dt^2} - \frac{1}{C_2} \cdot I_{L_1}(t - \tau_1) + L_2 \cdot \frac{d^2 I_{L_2}}{dt^2}\right]$$

$$X(t) - I_{R_p} \cdot R_p - L_1 \cdot \frac{dI_{L_1}}{dt} - \frac{1}{C_2} \cdot \int I_{L_1}(t - \tau_1) \cdot dt = f\left(I_{L_2}(t - \tau_2) + I_{L_1}\left(t - \sum_{k=1}^{2} \tau_k\right)\right)$$

By differentiation of the above equation we get the following differential equation:

$$\frac{dX(t)}{dt} - \frac{dI_{R_p}}{dt} \cdot R_p - L_1 \cdot \frac{d^2 I_{L_1}}{dt^2} - \frac{1}{C_2} \cdot I_{L_1}(t - \tau_1) = \frac{d}{dt} f\left(I_{L_2}(t - \tau_2) + I_{L_1}\left(t - \sum_{k=1}^{2} \tau_k\right)\right)$$

The above system differential equations are sophisticated to solve analytically and it is recommended to solve numerically by using MATLAB software or other software.

We use for our analysis the intrinsic electrical equivalent circuit of the laser diode (Fig. 1.4). It is reader exercise to use the complete intrinsic electrical equivalent circuit of the laser diode which takes into accounts all the major factors affecting the laser diode's dynamic response. There is a close relationship between the intrinsic electrical equivalent circuit of the laser diode components and the laser diode parameters.

$$I_{R_p} = \frac{X(t) - V_{A_1}}{R_p};\; I_{C_1} = C_1 \cdot \frac{dV_{C_1}}{dt} = C_1 \cdot \frac{dV_{A_1}}{dt}$$

$$V_{C_1} = V_{A_1};\; V_{A_1} - V_{A_2} = L_1 \cdot \frac{dI_{L_1}}{dt};\; V_{A_2} = V_{A_3}$$

$$I_{C_2}(t) = I_{L_1}(t - \tau_1);\; V_{C_2} = V_{A_3} - V_{A_4};\; V_{L_1} = V_{A_1} - V_{A_2};\; I_{C_2} = C_2 \cdot \frac{d}{dt}(V_{A_3} - V_{A_4})$$

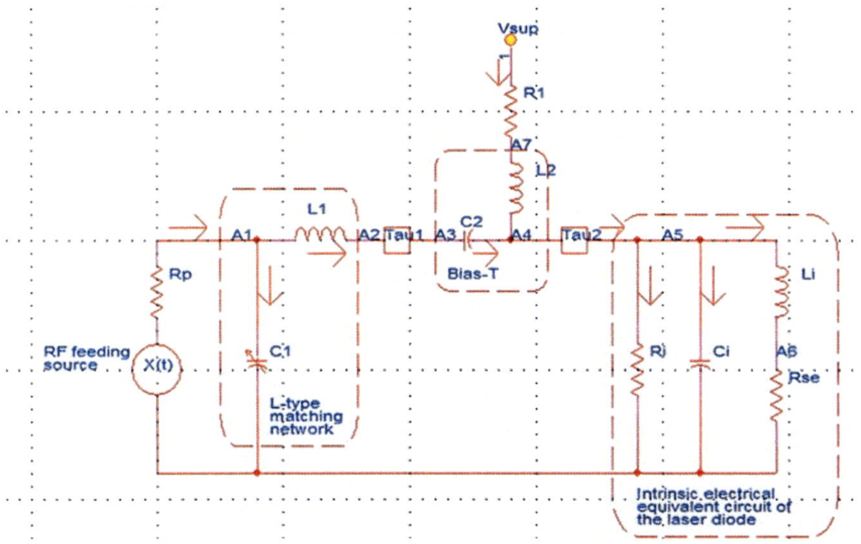

Fig. 1.4 MRI system laser diode, intrinsic electrical equivalent circuit of the laser diode

$$I_{R_1} = \frac{V_{\text{sup}} - V_{A_7}}{R_1};\ V_{L_2} = V_{A_7} - V_{A_4} = L_2 \cdot \frac{dI_{L_2}}{dt};\ I_{R_1} = I_{L_2};\ V_{A_4} = V_{A_5};\ I_{L_i} = I_{R_{se}}$$

$$I_{L_2}(t - \tau_2) + I_{C_2}(t - \tau_2) = I_{R_i} + I_{C_i} + I_{L_i};\ I_{R_i} = \frac{V_{A_5}}{R_i}$$

$$I_{C_i} = C_i \cdot \frac{dV_{A_5}}{dt};\ V_{R_i} = V_{A_5};\ V_{C_i} = V_{A_5}$$

$$V_{A_5} - V_{A_6} = L_i \cdot \frac{dI_{L_i}}{dt};\ I_{R_{se}} = \frac{V_{A_6}}{R_{se}};\ I_{R_p} = I_{L_1} + I_{C_1};\ I_{L_2}(t - \tau_2) + I_{L_1}\left(t - \sum_{k=1}^{2} \tau_k\right) = I_{R_i} + I_{C_i} + I_{L_i}$$

$$I_{R_p} = \frac{X(t) - V_{A_1}}{R_p} \Rightarrow I_{R_p} \cdot R_p = X(t) - V_{A_1};\ V_{A_1} = X(t) - I_{R_p} \cdot R_p$$

$$I_{C_1} = C_1 \cdot \frac{dV_{A_1}}{dt} = C_1 \cdot \frac{d}{dt}\left[X(t) - I_{R_p} \cdot R_p\right]$$

$$I_{C_1} = C_1 \cdot \frac{dX(t)}{dt} - R_p \cdot C_1 \cdot \frac{dI_{R_p}}{dt};\ V_{A_1} - V_{A_2} = L_1 \cdot \frac{dI_{L_1}}{dt} \Rightarrow V_{A_2} = V_{A_1} - L_1 \cdot \frac{dI_{L_1}}{dt};\ V_{A_2} = V_{A_3}$$

$$V_{A_3} = V_{A_2} = X(t) - I_{R_p} \cdot R_p - L_1 \cdot \frac{dI_{L_1}}{dt};\ I_{C_2} = C_2 \cdot \frac{d}{dt}\left(V_{A_3} - V_{A_4}\right) \Rightarrow V_{A_3} - V_{A_4} = \frac{1}{C_2} \cdot \int I_{C_2} \cdot dt$$

$$V_{A_4} = V_{A_3} - \frac{1}{C_2} \cdot \int I_{C_2} \cdot dt;\ V_{A_4} = X(t) - I_{R_p} \cdot R_p - L_1 \cdot \frac{dI_{L_1}}{dt} - \frac{1}{C_2} \cdot \int I_{C_2} \cdot dt$$

$$I_{R_1} = \frac{V_{\text{sup}} - V_{A_7}}{R_1} \Rightarrow I_{R_1} \cdot R_1 = V_{\text{sup}} - V_{A_7}$$

$$V_{A_7} = V_{\text{sup}} - I_{R_1} \cdot R_1;\ V_{A_7} - V_{A_4} = L_2 \cdot \frac{dI_{L_2}}{dt}$$

$$V_{\text{sup}} - I_{R_1} \cdot R_1 - \left[X(t) - I_{R_p} \cdot R_p - L_1 \cdot \frac{dI_{L_1}}{dt} - \frac{1}{C_2} \cdot \int I_{C_2} \cdot dt\right] = L_2 \cdot \frac{dI_{L_2}}{dt}$$

$$\frac{d}{dt}\left\{V_{\text{sup}} - I_{R_1} \cdot R_1 - \left[X(t) - I_{R_p} \cdot R_p - L_1 \cdot \frac{dI_{L_1}}{dt} - \frac{1}{C_2} \cdot \int I_{C_2} \cdot dt\right] = L_2 \cdot \frac{dI_{L_2}}{dt}\right\}$$

$$\frac{dV_{\text{sup}}}{dt} - \frac{dI_{R_1}}{dt} \cdot R_1 - \frac{dX(t)}{dt} + \frac{dI_{R_p}}{dt} \cdot R_p + L_1 \cdot \frac{d^2 I_{L_1}}{dt^2} + \frac{1}{C_2} \cdot I_{C_2} = L_2 \cdot \frac{d^2 I_{L_2}}{dt^2}; \ \frac{dV_{\text{sup}}}{dt} \to \varepsilon$$

$$-\frac{dI_{R_1}}{dt} \cdot R_1 - \frac{dX(t)}{dt} + \frac{dI_{R_p}}{dt} \cdot R_p + L_1 \cdot \frac{d^2 I_{L_1}}{dt^2} + \frac{1}{C_2} \cdot I_{C_2} = L_2 \cdot \frac{d^2 I_{L_2}}{dt^2}; \ I_{C_2}(t) = I_{L_1}(t - \tau_1)$$

$$-\frac{dI_{R_1}}{dt} \cdot R_1 - \frac{dX(t)}{dt} + \frac{dI_{R_p}}{dt} \cdot R_p + L_1 \cdot \frac{d^2 I_{L_1}}{dt^2} + \frac{1}{C_2} \cdot I_{L_1}(t - \tau_1) = L_2 \cdot \frac{d^2 I_{L_2}}{dt^2}$$

$$I_{R_i} \cdot R_i = V_{A_5}; \ V_{A_5} = V_{A_4} = X(t) - I_{R_p} \cdot R_p - L_1 \cdot \frac{dI_{L_1}}{dt} - \frac{1}{C_2} \cdot \int I_{C_2} \cdot dt; \ I_{C_i} = C_i \cdot \frac{dV_{A_5}}{dt}$$

$$I_{C_i} = C_i \cdot \frac{d}{dt}\left[X(t) - I_{R_p} \cdot R_p - L_1 \cdot \frac{dI_{L_1}}{dt} - \frac{1}{C_2} \cdot \int I_{C_2} \cdot dt\right]$$

$$I_{C_i} = C_i \cdot \left[\frac{dX(t)}{dt} - \frac{dI_{R_p}}{dt} \cdot R_p - L_1 \cdot \frac{d^2 I_{L_1}}{dt^2} - \frac{1}{C_2} \cdot I_{C_2}\right]$$

$$I_{C_i} = C_i \cdot \left[\frac{dX(t)}{dt} - \frac{dI_{R_p}}{dt} \cdot R_p - L_1 \cdot \frac{d^2 I_{L_1}}{dt^2} - \frac{1}{C_2} \cdot I_{L_1}(t - \tau_1)\right]$$

$$V_{A_6} = I_{R_{se}} \cdot R_{se}; \ V_{A_5} - V_{A_6} = L_i \cdot \frac{dI_{L_i}}{dt}$$

$$X(t) - I_{R_p} \cdot R_p - L_1 \cdot \frac{dI_{L_1}}{dt} - \frac{1}{C_2} \cdot \int I_{C_2} \cdot dt - I_{R_{se}} \cdot R_{se} = L_i \cdot \frac{dI_{L_i}}{dt}$$

$$\frac{d}{dt}\left\{X(t) - I_{R_p} \cdot R_p - L_1 \cdot \frac{dI_{L_1}}{dt} - \frac{1}{C_2} \cdot \int I_{C_2} \cdot dt - I_{R_{se}} \cdot R_{se} = L_i \cdot \frac{dI_{L_i}}{dt}\right\}$$

$$\frac{dX(t)}{dt} - \frac{dI_{R_p}}{dt} \cdot R_p - L_1 \cdot \frac{d^2 I_{L_1}}{dt^2} - \frac{1}{C_2} \cdot I_{C_2} - \frac{dI_{R_{se}}}{dt} \cdot R_{se} = L_i \cdot \frac{d^2 I_{L_i}}{dt^2}; \ I_{C_2}(t) = I_{L_1}(t - \tau_1)$$

$$\frac{dX(t)}{dt} - \frac{dI_{R_p}}{dt} \cdot R_p - L_1 \cdot \frac{d^2 I_{L_1}}{dt^2} - \frac{1}{C_2} \cdot I_{L_1}(t - \tau_1)$$

$$-\frac{dI_{R_{se}}}{dt}\cdot R_{se}=L_i\cdot\frac{d^2I_{L_i}}{dt^2};\ I_{R_{se}}=I_{L_i}$$

$$\frac{dX(t)}{dt}-\frac{dI_{R_p}}{dt}\cdot R_p-L_1\cdot\frac{d^2I_{L_1}}{dt^2}-\frac{1}{C_2}\cdot I_{L_1}(t-\tau_1)-\frac{dI_{L_i}}{dt}\cdot R_{se}=L_i\cdot\frac{d^2I_{L_i}}{dt^2}$$

$$I_{R_i}\cdot R_i=V_{A_5};\ I_{R_i}\cdot R_i=X(t)-I_{R_p}\cdot R_p-L_1\cdot\frac{dI_{L_1}}{dt}-\frac{1}{C_2}\cdot\int I_{C_2}\cdot dt$$

$$\frac{d}{dt}\left\{I_{R_i}\cdot R_i=X(t)-I_{R_p}\cdot R_p-L_1\cdot\frac{dI_{L_1}}{dt}-\frac{1}{C_2}\cdot\int I_{C_2}\cdot dt\right\}$$

$$\frac{dI_{R_i}}{dt}\cdot R_i=\frac{dX(t)}{dt}-\frac{dI_{R_p}}{dt}\cdot R_p-L_1\cdot\frac{d^2I_{L_1}}{dt^2}-\frac{1}{C_2}\cdot I_{C_2}$$

$$\frac{dI_{R_i}}{dt}\cdot R_i=\frac{dX(t)}{dt}-\frac{dI_{R_p}}{dt}\cdot R_p-L_1\cdot\frac{d^2I_{L_1}}{dt^2}-\frac{1}{C_2}\cdot I_{L_1}(t-\tau_1)$$

$$I_{R_i}\cdot R_i=V_{A_5};\ V_{A_5}=\frac{1}{C_i}\cdot\int I_{C_i}\cdot dt;\ I_{R_i}\cdot R_i=\frac{1}{C_i}\cdot\int I_{C_i}\cdot dt;\ \frac{d}{dt}\left\{I_{R_i}\cdot R_i=\frac{1}{C_i}\cdot\int I_{C_i}\cdot dt\right\}$$

$$\frac{dI_{R_i}}{dt}\cdot R_i=\frac{1}{C_i}\cdot I_{C_i}\Rightarrow\frac{dI_{R_i}}{dt}=\frac{1}{C_i\cdot R_i}\cdot I_{C_i};\ I_{R_{se}}=I_{L_i}$$

We can summary our MRI system laser diode circuitry differential equations:

1. $-\frac{dI_{R_1}}{dt}\cdot R_1-\frac{dX(t)}{dt}+\frac{dI_{R_p}}{dt}\cdot R_p+L_1\cdot\frac{d^2I_{L_1}}{dt^2}+\frac{1}{C_2}\cdot I_{L_1}(t-\tau_1)=L_2\cdot\frac{d^2I_{L_2}}{dt^2}$

2. $I_{C_i}=C_i\cdot\left[\frac{dX(t)}{dt}-\frac{dI_{R_p}}{dt}\cdot R_p-L_1\cdot\frac{d^2I_{L_1}}{dt^2}-\frac{1}{C_2}\cdot I_{L_1}(t-\tau_1)\right]$

3. $\frac{dX(t)}{dt}-\frac{dI_{R_p}}{dt}\cdot R_p-L_1\cdot\frac{d^2I_{L_1}}{dt^2}-\frac{1}{C_2}\cdot I_{L_1}(t-\tau_1)-\frac{dI_{L_i}}{dt}\cdot R_{se}=L_i\cdot\frac{d^2I_{L_i}}{dt^2}$

4. $\frac{dI_{R_i}}{dt}\cdot R_i=\frac{dX(t)}{dt}-\frac{dI_{R_p}}{dt}\cdot R_p-L_1\cdot\frac{d^2I_{L_1}}{dt^2}-\frac{1}{C_2}\cdot I_{L_1}(t-\tau_1);\ \frac{dI_{R_i}}{dt}=\frac{1}{C_i\cdot R_i}\cdot I_{C_i}$

Additionally KCL at nodes A_1 and A_5: $I_{R_p}=I_{L_1}+I_{C_1};\ I_{L_2}(t-\tau_2)+I_{L_1}\left(t-\sum_{k=1}^{2}\tau_k\right)=I_{R_i}+I_{C_i}+I_{L_i}$

For simplicity we define the following function in time:

$$\xi(t,\tau_1) = -\frac{dX(t)}{dt} + \frac{dI_{R_p}}{dt} \cdot R_p + L_1 \cdot \frac{d^2 I_{L_1}}{dt^2} + \frac{1}{C_2} \cdot I_{L_1}(t-\tau_1)$$

We can summary again our system differential equations:

$$-\frac{dI_{R_1}}{dt} \cdot R_1 + \xi(t,\tau_1) = L_2 \cdot \frac{d^2 I_{L_2}}{dt^2};\ I_{C_i}$$
$$= -C_i \cdot \xi(t,\tau_1);\ -\xi(t,\tau_1) - \frac{dI_{L_i}}{dt} \cdot R_{se} = L_i \cdot \frac{d^2 I_{L_i}}{dt^2}$$

$$\frac{dI_{R_i}}{dt} \cdot R_i = -\xi(t,\tau_1);\ \frac{dI_{R_i}}{dt} = \frac{1}{C_i \cdot R_i} \cdot I_{C_i}$$

$$I_{R_p} = I_{L_1} + I_{C_1};\ I_{L_2}(t-\tau_2) + I_{L_1}\left(t - \sum_{k=1}^{2}\tau_k\right) = I_{R_i} + I_{C_i} + I_{L_i}$$

We can write differential equation (4) as follow:

$$L_1 \cdot \frac{d^2 I_{L_1}}{dt^2} = -\frac{dI_{R_i}}{dt} \cdot R_i + \frac{dX(t)}{dt} - \frac{dI_{R_p}}{dt} \cdot R_p - \frac{1}{C_2} \cdot I_{L_1}(t-\tau_1)$$

4. $\rightarrow 1.;\ -\frac{dI_{R_1}}{dt} \cdot R_1 - \frac{dI_{R_i}}{dt} \cdot R_i = L_2 \cdot \frac{d^2 I_{L_2}}{dt^2}$

$$I_{L_2}(t-\tau_2) + I_{L_1}\left(t - \sum_{k=1}^{2}\tau_k\right) = I_{R_i} + I_{C_i} + I_{L_i} \Rightarrow I_{C_i}$$
$$= I_{L_2}(t-\tau_2) + I_{L_1}\left(t - \sum_{k=1}^{2}\tau_k\right) - I_{R_i} - I_{L_i}$$

2. $I_{C_i} = C_i \cdot \left[\frac{dX(t)}{dt} - \frac{dI_{R_p}}{dt} \cdot R_p - L_1 \cdot \frac{d^2 I_{L_1}}{dt^2} - \frac{1}{C_2} \cdot I_{L_1}(t-\tau_1)\right]$

$$I_{L_2}(t-\tau_2) + I_{L_1}\left(t - \sum_{k=1}^{2}\tau_k\right) - I_{R_i} - I_{L_i}$$
$$= C_i \cdot \left[\frac{dX(t)}{dt} - \frac{dI_{R_p}}{dt} \cdot R_p - L_1 \cdot \frac{d^2 I_{L_1}}{dt^2} - \frac{1}{C_2} \cdot I_{L_1}(t-\tau_1)\right]$$

3. $\rightarrow \frac{dX(t)}{dt} - \frac{dI_{R_p}}{dt} \cdot R_p - L_1 \cdot \frac{d^2 I_{L_1}}{dt^2} - \frac{1}{C_2} \cdot I_{L_1}(t-\tau_1) - \frac{dI_{L_i}}{dt} \cdot R_{se} = L_i \cdot \frac{d^2 I_{L_i}}{dt^2}$

$$\frac{dX(t)}{dt} - \frac{dI_{R_p}}{dt} \cdot R_p - \left[-\frac{dI_{R_i}}{dt} \cdot R_i + \frac{dX(t)}{dt} - \frac{dI_{R_p}}{dt} \cdot R_p - \frac{1}{C_2} \cdot I_{L_1}(t - \tau_1) \right]$$
$$- \frac{1}{C_2} \cdot I_{L_1}(t - \tau_1) - \frac{dI_{L_i}}{dt} \cdot R_{se} = L_i \cdot \frac{d^2 I_{L_i}}{dt^2}$$

$$\frac{dI_{R_i}}{dt} \cdot R_i - \frac{dI_{L_i}}{dt} \cdot R_{se} = L_i \cdot \frac{d^2 I_{L_i}}{dt^2}$$

$$\frac{dI_{R_i}}{dt} = \frac{1}{C_i \cdot R_i} \cdot I_{C_i} \Rightarrow \frac{dI_{R_i}}{dt}$$
$$= \frac{1}{C_i \cdot R_i} \cdot \left[I_{L_2}(t - \tau_2) + I_{L_1}\left(t - \sum_{k=1}^{2} \tau_k\right) - I_{R_i} - I_{L_i} \right]$$

We can summary our system differential equations:

$$L_2 \cdot \frac{d^2 I_{L_2}}{dt^2} = -\frac{dI_{R_1}}{dt} \cdot R_1 - \frac{dI_{R_i}}{dt} \cdot R_i$$

$$\frac{dX(t)}{dt} - \frac{dI_{R_p}}{dt} \cdot R_p - L_1 \cdot \frac{d^2 I_{L_1}}{dt^2} - \frac{1}{C_2} \cdot I_{L_1}(t - \tau_1)$$
$$= \frac{1}{C_i} \cdot I_{L_2}(t - \tau_2) + \frac{1}{C_i} \cdot I_{L_1}\left(t - \sum_{k=1}^{2} \tau_k\right) - \frac{1}{C_i} \cdot I_{R_i} - \frac{1}{C_i} \cdot I_{L_i}$$

$$L_i \cdot \frac{d^2 I_{L_i}}{dt^2} = \frac{dI_{R_i}}{dt} \cdot R_i - \frac{dI_{L_i}}{dt} \cdot R_{se} \Rightarrow L_i \cdot \frac{d^2 I_{L_i}}{dt^2}$$
$$= \frac{1}{C_i} \cdot \left[I_{L_2}(t - \tau_2) + I_{L_1}\left(t - \sum_{k=1}^{2} \tau_k\right) - I_{R_i} - I_{L_i} \right] - \frac{dI_{L_i}}{dt} \cdot R_{se}$$

$$\frac{dI_{R_i}}{dt} = \frac{1}{C_i \cdot R_i} \cdot \left[I_{L_2}(t - \tau_2) + I_{L_1}\left(t - \sum_{k=1}^{2} \tau_k\right) - I_{R_i} - I_{L_i} \right]$$

&&

$$I_{R_1} = I_{L_2} \Rightarrow L_2 \cdot \frac{d^2 I_{L_2}}{dt^2} = -\frac{dI_{L_2}}{dt} \cdot R_1 - \frac{dI_{R_i}}{dt} \cdot R_i$$

$$L_2 \cdot \frac{d^2 I_{L_2}}{dt^2} = -\frac{dI_{L_2}}{dt} \cdot R_1 - \frac{1}{C_i} \cdot \left[I_{L_2}(t - \tau_2) + I_{L_1}\left(t - \sum_{k=1}^{2} \tau_k\right) - I_{R_i} - I_{L_i} \right]$$

$$\frac{dX(t)}{dt} - \frac{dI_{R_p}}{dt} \cdot R_p - L_1 \cdot \frac{d^2 I_{L_1}}{dt^2} - \frac{1}{C_2} \cdot I_{L_1}(t - \tau_1)$$

$$= \frac{1}{C_i} \cdot I_{L_2}(t - \tau_2) + \frac{1}{C_i} \cdot I_{L_1}\left(t - \sum_{k=1}^{2} \tau_k\right) - \frac{1}{C_i} \cdot I_{R_i} - \frac{1}{C_i} \cdot I_{L_i}$$

$$L_i \cdot \frac{d^2 I_{L_i}}{dt^2} = \frac{1}{C_i} \cdot \left[I_{L_2}(t - \tau_2) + I_{L_1}\left(t - \sum_{k=1}^{2} \tau_k\right) - I_{R_i} - I_{L_i}\right] - \frac{dI_{L_i}}{dt} \cdot R_{se}$$

$$\frac{dI_{R_i}}{dt} = \frac{1}{C_i \cdot R_i} \cdot \left[I_{L_2}(t - \tau_2) + I_{L_1}\left(t - \sum_{k=1}^{2} \tau_k\right) - I_{R_i} - I_{L_i}\right]$$

We can summary our system differential equation in this stage:

$$L_2 \cdot \frac{d^2 I_{L_2}}{dt^2} = -\frac{dI_{L_2}}{dt} \cdot R_1 - \frac{1}{C_i} \cdot \left[I_{L_2}(t - \tau_2) + I_{L_1}\left(t - \sum_{k=1}^{2} \tau_k\right) - I_{R_i} - I_{L_i}\right]$$

$$\frac{dX(t)}{dt} - \frac{dI_{R_p}}{dt} \cdot R_p - L_1 \cdot \frac{d^2 I_{L_1}}{dt^2} - \frac{1}{C_2} \cdot I_{L_1}(t - \tau_1)$$

$$= \frac{1}{C_i} \cdot I_{L_2}(t - \tau_2) + \frac{1}{C_i} \cdot I_{L_1}\left(t - \sum_{k=1}^{2} \tau_k\right) - \frac{1}{C_i} \cdot I_{R_i} - \frac{1}{C_i} \cdot I_{L_i}$$

$$L_i \cdot \frac{d^2 I_{L_i}}{dt^2} = \frac{1}{C_i} \cdot \left[I_{L_2}(t - \tau_2) + I_{L_1}\left(t - \sum_{k=1}^{2} \tau_k\right) - I_{R_i} - I_{L_i}\right] - \frac{dI_{L_i}}{dt} \cdot R_{se}$$

$$\frac{dI_{R_i}}{dt} = \frac{1}{C_i \cdot R_i} \cdot \left[I_{L_2}(t - \tau_2) + I_{L_1}\left(t - \sum_{k=1}^{2} \tau_k\right) - I_{R_i} - I_{L_i}\right]$$

&&&

$$I_{C_1} = C_1 \cdot \frac{dX(t)}{dt} - R_p \cdot C_1 \cdot \frac{dI_{R_p}}{dt} \Rightarrow \frac{dI_{R_p}}{dt}$$

$$= \frac{1}{R_p} \cdot \frac{dX(t)}{dt} - \frac{1}{R_p \cdot C_1} \cdot I_{C_1}; \; I_{C_1} = I_{R_p} - I_{L_1}$$

$$\frac{dI_{R_p}}{dt} = \frac{1}{R_p} \cdot \frac{dX(t)}{dt} - \frac{1}{R_p \cdot C_1} \cdot \left(I_{R_p} - I_{L_1}\right)$$

$$\frac{dX(t)}{dt} - \frac{dI_{R_p}}{dt} \cdot R_p - L_1 \cdot \frac{d^2 I_{L_1}}{dt^2} - \frac{1}{C_2} \cdot I_{L_1}(t - \tau_1)$$

$$= \frac{1}{C_i} \cdot I_{L_2}(t - \tau_2) + \frac{1}{C_i} \cdot I_{L_1}\left(t - \sum_{k=1}^{2} \tau_k\right) - \frac{1}{C_i} \cdot I_{R_i} - \frac{1}{C_i} \cdot I_{L_i}$$

$$\frac{dX(t)}{dt} - \left[\frac{dX(t)}{dt} - \frac{1}{C_1} \cdot \left(I_{R_p} - I_{L_1}\right)\right] - L_1 \cdot \frac{d^2 I_{L_1}}{dt^2} - \frac{1}{C_2} \cdot I_{L_1}(t - \tau_1) = \frac{1}{C_i} \cdot I_{L_2}(t - \tau_2)$$

$$+\frac{1}{C_i}\cdot I_{L_1}\left(t-\sum_{k=1}^{2}\tau_k\right)-\frac{1}{C_i}\cdot I_{R_i}-\frac{1}{C_i}\cdot I_{L_i}$$

$$\frac{1}{C_1}\cdot I_{R_p}-\frac{1}{C_1}\cdot I_{L_1}-L_1\cdot\frac{d^2I_{L_1}}{dt^2}-\frac{1}{C_2}\cdot I_{L_1}(t-\tau_1)$$
$$=\frac{1}{C_i}\cdot I_{L_2}(t-\tau_2)+\frac{1}{C_i}\cdot I_{L_1}\left(t-\sum_{k=1}^{2}\tau_k\right)-\frac{1}{C_i}\cdot I_{R_i}-\frac{1}{C_i}\cdot I_{L_i}$$

We can summary our system differential equation at this stage:

1. $L_2\cdot\frac{d^2I_{L_2}}{dt^2}=-\frac{dI_{L_2}}{dt}\cdot R_1-\frac{1}{C_i}\cdot\left[I_{L_2}(t-\tau_2)+I_{L_1}\left(t-\sum_{k=1}^{2}\tau_k\right)-I_{R_i}-I_{L_i}\right]$

2. $\frac{1}{C_1}\cdot I_{R_p}-\frac{1}{C_1}\cdot I_{L_1}-L_1\cdot\frac{d^2I_{L_1}}{dt^2}-\frac{1}{C_2}\cdot I_{L_1}(t-\tau_1)$
$=\frac{1}{C_i}\cdot I_{L_2}(t-\tau_2)+\frac{1}{C_i}\cdot I_{L_1}\left(t-\sum_{k=1}^{2}\tau_k\right)-\frac{1}{C_i}\cdot I_{R_i}-\frac{1}{C_i}\cdot I_{L_i}$

3. $L_i\cdot\frac{d^2I_{L_i}}{dt^2}=\frac{1}{C_i}\cdot\left[I_{L_2}(t-\tau_2)+I_{L_1}\left(t-\sum_{k=1}^{2}\tau_k\right)-I_{R_i}-I_{L_i}\right]-\frac{dI_{L_i}}{dt}\cdot R_{se}$

4. $\frac{dI_{R_i}}{dt}=\frac{1}{C_i\cdot R_i}\cdot\left[I_{L_2}(t-\tau_2)+I_{L_1}\left(t-\sum_{k=1}^{2}\tau_k\right)-I_{R_i}-I_{L_i}\right]$

5. $\frac{dI_{R_p}}{dt}=\frac{1}{R_p}\cdot\frac{dX(t)}{dt}-\frac{1}{R_p\cdot C_1}\cdot\left(I_{R_p}-I_{L_1}\right)$

Our above system variables are $I_{L_2}, I_{R_i}, I_{L_i}, I_{R_p}, I_{L_1}$

We define the following new variables: $\frac{d^2I_{L_1}}{dt^2}=\frac{dY_1}{dt}$; $\frac{dI_{L_1}}{dt}=Y_1$; $\frac{d^2I_{L_2}}{dt^2}=\frac{dY_2}{dt}$; $\frac{dI_{L_2}}{dt}=Y_2$

$$\frac{d^2I_{L_i}}{dt^2}=\frac{dY_3}{dt};\ \frac{dI_{L_i}}{dt}=Y_3$$

We can write our system differential equation with the new variables:

$$L_2\cdot\frac{dY_2}{dt}=-Y_2\cdot R_1-\frac{1}{C_i}\cdot\left[I_{L_2}(t-\tau_2)+I_{L_1}\left(t-\sum_{k=1}^{2}\tau_k\right)-I_{R_i}-I_{L_i}\right]$$

$$\frac{1}{C_1}\cdot I_{R_p}-\frac{1}{C_1}\cdot I_{L_1}-L_1\cdot\frac{dY_1}{dt}-\frac{1}{C_2}\cdot I_{L_1}(t-\tau_1)$$

$$= \frac{1}{C_i} \cdot I_{L_2}(t - \tau_2) + \frac{1}{C_i} \cdot I_{L_1}\left(t - \sum_{k=1}^{2} \tau_k\right) - \frac{1}{C_i} \cdot I_{R_i} - \frac{1}{C_i} \cdot I_{L_i}$$

$$L_i \cdot \frac{dY_3}{dt} = \frac{1}{C_i} \cdot \left[I_{L_2}(t - \tau_2) + I_{L_1}\left(t - \sum_{k=1}^{2} \tau_k\right) - I_{R_i} - I_{L_i}\right]$$

$$- Y_3 \cdot R_{se}; \quad \frac{dI_{L_1}}{dt} = Y_1; \quad \frac{dI_{L_2}}{dt} = Y_2$$

$$\frac{dI_{R_i}}{dt} = \frac{1}{C_i \cdot R_i} \cdot \left[I_{L_2}(t - \tau_2) + I_{L_1}\left(t - \sum_{k=1}^{2} \tau_k\right) - I_{R_i} - I_{L_i}\right]; \quad \frac{dI_{L_i}}{dt} = Y_3$$

$$\frac{dI_{R_p}}{dt} = \frac{1}{R_p} \cdot \frac{dX(t)}{dt} - \frac{1}{R_p \cdot C_1} \cdot (I_{R_p} - I_{L_1})$$

Finally, we summary our system differential equations:

$$\frac{dY_2}{dt} = -Y_2 \cdot \frac{R_1}{L_2} - \frac{1}{C_i \cdot L_2} \cdot \left[I_{L_2}(t - \tau_2) + I_{L_1}\left(t - \sum_{k=1}^{2} \tau_k\right) - I_{R_i} - I_{L_i}\right]$$

$$\frac{dY_1}{dt} = \frac{1}{C_1 \cdot L_1} \cdot I_{R_p} - \frac{1}{C_1 \cdot L_1} \cdot I_{L_1} - \frac{1}{C_2 \cdot L_1} \cdot I_{L_1}(t - \tau_1)$$

$$- \frac{1}{C_i \cdot L_1} \cdot I_{L_1}\left(t - \sum_{k=1}^{2} \tau_k\right) - \frac{1}{C_i \cdot L_1} \cdot I_{L_2}(t - \tau_2)$$

$$+ \frac{1}{C_i \cdot L_1} \cdot I_{R_i} + \frac{1}{C_i \cdot L_1} \cdot I_{L_i}$$

$$\frac{dY_3}{dt} = \frac{1}{L_i \cdot C_i} \cdot \left[I_{L_2}(t - \tau_2) + I_{L_1}\left(t - \sum_{k=1}^{2} \tau_k\right) - I_{R_i} - I_{L_i}\right]$$

$$- Y_3 \cdot \frac{R_{se}}{L_i}; \quad \frac{dI_{L_1}}{dt} = Y_1; \quad \frac{dI_{L_2}}{dt} = Y_2$$

$$\frac{dI_{R_i}}{dt} = \frac{1}{C_i \cdot R_i} \cdot \left[I_{L_2}(t - \tau_2) + I_{L_1}\left(t - \sum_{k=1}^{2} \tau_k\right) - I_{R_i} - I_{L_i}\right]; \quad \frac{dI_{L_i}}{dt} = Y_3$$

$$\frac{dI_{R_p}}{dt} = \frac{1}{R_p} \cdot \frac{dX(t)}{dt} - \frac{1}{R_p \cdot C_1} \cdot (I_{R_p} - I_{L_1})$$

Our above system variables are $I_{L_2}, I_{R_i}, I_{L_i}, I_{R_p}, I_{L_1}, Y_1, Y_2, Y_3$

To find the equilibrium point (fixed points) of our MRI system Laser diode circuitry by using intrinsic electrical equivalent circuit of laser diode.

$$\lim_{t\to\infty} I_{L_1}(t-\tau_1) = I_{L_1}(t);\ \lim_{t\to\infty} I_{L_2}(t-\tau_2)$$
$$= I_{L_2}(t);\ \lim_{t\to\infty} I_{L_1}\left(t-\sum_{k=1}^{2}\tau_k\right) = I_{L_1}(t);\ t \gg \tau_1;\ t \gg \tau_2$$

And setting $\frac{dI_{L_1}}{dt}=0;\ \frac{dI_{L_2}}{dt}=0;\ \frac{dI_{L_i}}{dt}=0;\ \frac{dI_{R_i}}{dt}=0;\ \frac{dI_{R_p}}{dt}=0;\ \frac{dY_i}{dt}=0\,\forall i=1,2,3$

$$t \gg \sum_{k=1}^{2}\tau_k;\ \frac{dI_{L_1}}{dt}=0 \Rightarrow Y_1^*=0;\ \frac{dI_{L_2}}{dt}=0 \Rightarrow Y_2^*=0;\ \frac{dI_{L_i}}{dt}=0 \Rightarrow Y_3^*=0$$

$$-\frac{1}{C_i\cdot L_2}\cdot\left[I_{L_2}^*+I_{L_1}^*-I_{R_i}^*-I_{L_i}^*\right]=0 \Rightarrow I_{L_2}^*+I_{L_1}^*-I_{R_i}^*-I_{L_i}^*=0$$

$$\frac{1}{C_1}\cdot I_{R_p}^*-\left[\frac{1}{C_1}+\frac{1}{C_2}+\frac{1}{C_i}\right]\cdot I_{L_1}^*-\frac{1}{C_i}\cdot I_{L_2}^*+\frac{1}{C_i}\cdot I_{R_i}^*+\frac{1}{C_i}\cdot I_{L_i}^*=0$$

$$I_{L_2}^*+I_{L_1}^*-I_{R_i}^*-I_{L_i}^*=0;\ \frac{dX(t)}{dt}-\frac{1}{C_1}\cdot\left(I_{R_p}^*-I_{L_1}^*\right)=0$$

We consider our RF feeding source as $X(t)=\Gamma+\xi(t);\ \Gamma \gg \xi(t)$

$$\frac{dX(t)}{dt}=\left(\frac{d\Gamma}{dt}\to\varepsilon\right)+\frac{d\xi(t)}{dt};\ \frac{d\xi(t)}{dt}\to\varepsilon;\ \frac{dX(t)}{dt}\to\varepsilon$$

$$\left(\frac{dX(t)}{dt}\to\varepsilon\right)-\frac{1}{C_1}\cdot\left(I_{R_p}^*-I_{L_1}^*\right)=0 \Rightarrow I_{R_p}^*$$
$$=I_{L_1}^*;\ -\left[\frac{1}{C_2}+\frac{1}{C_i}\right]\cdot I_{L_1}^*-\frac{1}{C_i}\cdot I_{L_2}^*+\frac{1}{C_i}\cdot I_{R_i}^*+\frac{1}{C_i}\cdot I_{L_i}^*=0$$

$$I_{L_2}^*+I_{L_1}^*-I_{R_i}^*-I_{L_i}^*=0 \Rightarrow I_{L_i}^*=I_{L_2}^*+I_{L_1}^*-I_{R_i}^*$$
$$-\frac{1}{C_2}\cdot I_{L_1}^*+\frac{1}{C_i}\cdot I_{R_i}^*-\frac{1}{C_i}\cdot I_{R_i}^*=0$$

$$-\frac{1}{C_2}\cdot I_{L_1}^*=0 \Rightarrow -I_{L_1}^*=0 \Rightarrow I_{L_1}^*=0;\ I_{R_p}^*=0;\ I_{L_2}^*$$
$$-I_{R_i}^*-I_{L_i}^*=0 \Rightarrow I_{L_2}^*=I_{R_i}^*+I_{L_i}^*$$

We can define our fixed points (equilibrium points):

$$E^*\left(I_{L_1}^*, I_{L_2}^*, I_{R_i}^*, I_{R_p}^*, I_{L_i}^*, Y_1^*, Y_2^*, Y_3^*\right)=\left(0, I_{R_i}^*+I_{L_i}^*, I_{R_i}^*, 0, I_{L_i}^*, 0, 0, 0\right)$$

First we analyze our system stability for no delays: $\tau_k\to\varepsilon\,\forall k=1,2;\ \tau_1+\tau_2\to\varepsilon$

We get the following set of system differential equations:

$$\frac{dY_2}{dt} = -Y_2 \cdot \frac{R_1}{L_2} - \frac{1}{C_i \cdot L_2} \cdot [I_{L_2} + I_{L_1} - I_{R_i} - I_{L_i}]$$

$$\frac{dY_1}{dt} = \frac{1}{C_1 \cdot L_1} \cdot I_{R_p} - \frac{1}{L_1} \cdot \left[\frac{1}{C_1} + \frac{1}{C_2} + \frac{1}{C_i}\right] \cdot I_{L_1} - \frac{1}{C_i \cdot L_1} \cdot I_{L_2}$$
$$+ \frac{1}{C_i \cdot L_1} \cdot I_{R_i} + \frac{1}{C_i \cdot L_1} \cdot I_{L_i}$$

$$\frac{dY_3}{dt} = \frac{1}{L_i \cdot C_i} \cdot [I_{L_2} + I_{L_1} - I_{R_i} - I_{L_i}]$$
$$- Y_3 \cdot \frac{R_{se}}{L_i}; \ \frac{dI_{L_1}}{dt} = Y_1; \ \frac{dI_{L_2}}{dt} = Y_2$$

$$\frac{dI_{R_i}}{dt} = \frac{1}{C_i \cdot R_i} \cdot [I_{L_2} + I_{L_1} - I_{R_i} - I_{L_i}]; \ \frac{dI_{L_i}}{dt}$$
$$= Y_3; \ \frac{dI_{R_p}}{dt} = -\frac{1}{R_p \cdot C_1} \cdot (I_{R_p} - I_{L_1}); \ \frac{dX(t)}{dt} \to \varepsilon$$

Stability analysis: The standard local stability analysis about any one of the equilibrium points of MRI system Laser diode circuitry consists in adding to its coordinates $[I_{L_1} I_{L_2}\ I_{R_i}\ I_{R_p} I_{L_i}\ Y_1\ Y_2\ Y_3]$ arbitrarily small increments of exponential form $[i_{L_1}\, i_{L_2}\, i_{R_i}\, i_{R_p}\, i_{L_i}\ y_1\ y_2\ y_3] \cdot e^{\lambda \cdot t}$, and retaining the first order terms in $i_{L_1}\, i_{L_2}\, i_{R_i}\, i_{R_p}\, i_{L_i}\ y_1\ y_2\ y_3$. The system of eight homogeneous equations leads to a polynomial characteristics equation in the eigenvalue λ. The polynomial characteristics equation accept by set the below equations [8]. The MRI system Laser diode circuitry fixed values with arbitrarily small increments of exponential form $[i_{L_1}\, i_{L_2}\, i_{R_i}\, i_{R_p}\, i_{L_i}\ y_1\ y_2\ y_3] \cdot e^{\lambda \cdot t}$ are $i = 0$ (first fixed point) $i = 1$ (second fixed point), $i = 2$ (third fixed point), etc.,

$$I_{L_1}(t) = I_{L_1}^{(i)} + i_{L_1} \cdot e^{\lambda \cdot t}; \ I_{L_2}(t) = I_{L_2}^{(i)} + i_{L_2} \cdot e^{\lambda \cdot t}; \ I_{R_i}(t)$$
$$= I_{R_i}^{(i)} + i_{R_i} \cdot e^{\lambda \cdot t}; \ I_{R_p}(t) = I_{R_p}^{(i)} + i_{R_p} \cdot e^{\lambda \cdot t}$$

$$Y_1(t) = Y_1^{(i)} + y_1 \cdot e^{\lambda \cdot t}; \ Y_2(t) = Y_2^{(i)} + y_2 \cdot e^{\lambda \cdot t}; \ Y_3(t)$$
$$= Y_3^{(i)} + y_3 \cdot e^{\lambda \cdot t}; \ \frac{dI_{L_1}(t)}{dt} = i_{L_1} \cdot \lambda \cdot e^{\lambda \cdot t}$$

$$\frac{dI_{L_2}(t)}{dt} = i_{L_2} \cdot \lambda \cdot e^{\lambda \cdot t}; \ \frac{dI_{R_i}(t)}{dt} = i_{R_i} \cdot \lambda \cdot e^{\lambda \cdot t}; \ \frac{dI_{R_p}(t)}{dt}$$
$$= i_{R_p} \cdot \lambda \cdot e^{\lambda \cdot t}; \ \frac{dY_1(t)}{dt} = y_1 \cdot \lambda \cdot e^{\lambda \cdot t}$$

$$\frac{dY_2(t)}{dt} = y_2 \cdot \lambda \cdot e^{\lambda \cdot t}; \quad \frac{dY_3(t)}{dt} = y_3 \cdot \lambda \cdot e^{\lambda \cdot t}; \; I_{L_i}(t) = I_{L_i}^{(i)} + i_{L_i} \cdot e^{\lambda \cdot t}; \quad \frac{dI_{L_i}(t)}{dt} = i_{L_i} \cdot \lambda \cdot e^{\lambda \cdot t}$$

We choose the above expressions for our $I_{L_1}(t)$, $I_{L_2}(t)$, $I_{R_i}(t)$, $I_{R_p}(t)$, $I_{L_i}(t)$, $Y_1(t)$, $Y_2(t)$, $Y_3(t)$ as small displacement $\left[i_{L_1}\, i_{L_2}\, i_{R_i}\, i_{R_p}\, i_{L_i}\, y_1\, y_2\, y_3\right]$ from the system fixed points at time t = 0

$$I_{L_1}(t=0) = I_{L_1}^{(i)} + i_{L_1}; \; I_{L_2}(t=0) = I_{L_2}^{(i)} + i_{L_2}; \; I_{R_i}(t=0) = I_{R_i}^{(i)} + i_{R_i}; \; I_{R_p}(t=0) = I_{R_p}^{(i)} + i_{R_p}$$

$$Y_1(t=0) = Y_1^{(i)} + y_1; \; Y_2(t=0) = Y_2^{(i)} + y_2; \; Y_3(t=0) = Y_3^{(i)} + y_3$$

For t > 0, $\lambda < 0$ the selected fixed point is stable otherwise t > 0, $\lambda > 0$ is unstable. Our system tends to the selected fixed point exponentially for t > 0, $\lambda < 0$ otherwise go away from the selected fixed point exponentially. The eigenvalue λ is the parameter which if the selected fixed point is stable or unstable, additionally his absolute value $|\lambda|$ establish the speed of flow toward or away from the selected fixed point.

$$\frac{dY_2}{dt} = -Y_2 \cdot \frac{R_1}{L_2} - \frac{1}{C_i \cdot L_2} \cdot \left[I_{L_2} + I_{L_1} - I_{R_i} - I_{L_i}\right]$$

$$y_2 \cdot \lambda \cdot e^{\lambda \cdot t} = -\left[Y_2^{(i)} + y_2 \cdot e^{\lambda \cdot t}\right] \cdot \frac{R_1}{L_2} - \frac{1}{C_i \cdot L_2} \cdot \left[I_{L_2}^{(i)} + i_{L_2} \cdot e^{\lambda \cdot t} + I_{L_1}^{(i)} + i_{L_1} \cdot e^{\lambda \cdot t} - I_{R_i}^{(i)} - i_{R_i} \cdot e^{\lambda \cdot t} - I_{L_i}^{(i)} - i_{L_i} \cdot e^{\lambda \cdot t}\right]$$

$$y_2 \cdot \lambda \cdot e^{\lambda \cdot t} = -Y_2^{(i)} \cdot \frac{R_1}{L_2} - y_2 \cdot \frac{R_1}{L_2} \cdot e^{\lambda \cdot t} - \left[\frac{1}{C_i \cdot L_2} \cdot I_{L_2}^{(i)} + \frac{1}{C_i \cdot L_2} \cdot I_{L_1}^{(i)} - \frac{1}{C_i \cdot L_2} \cdot I_{R_i}^{(i)} - \frac{1}{C_i \cdot L_2} \cdot I_{L_i}^{(i)} + \frac{1}{C_i \cdot L_2} \cdot i_{L_2} \cdot e^{\lambda \cdot t} + \frac{1}{C_i \cdot L_2} \cdot i_{L_1} \cdot e^{\lambda \cdot t} - \frac{1}{C_i \cdot L_2} \cdot i_{R_i} \cdot e^{\lambda \cdot t} - \frac{1}{C_i \cdot L_2} \cdot i_{L_i} \cdot e^{\lambda \cdot t}\right]$$

$$y_2 \cdot \lambda \cdot e^{\lambda \cdot t} = -Y_2^{(i)} \cdot \frac{R_1}{L_2} - \left[\frac{1}{C_i \cdot L_2} \cdot I_{L_2}^{(i)} + \frac{1}{C_i \cdot L_2} \cdot I_{L_1}^{(i)} - \frac{1}{C_i \cdot L_2} \cdot I_{R_i}^{(i)} - \frac{1}{C_i \cdot L_2} \cdot I_{L_i}^{(i)}\right] - y_2 \cdot \frac{R_1}{L_2} \cdot e^{\lambda \cdot t} - \left[\frac{1}{C_i \cdot L_2} \cdot i_{L_2} \cdot e^{\lambda \cdot t} + \frac{1}{C_i \cdot L_2} \cdot i_{L_1} \cdot e^{\lambda \cdot t} - \frac{1}{C_i \cdot L_2} \cdot i_{R_i} \cdot e^{\lambda \cdot t} - \frac{1}{C_i \cdot L_2} \cdot i_{L_i} \cdot e^{\lambda \cdot t}\right]$$

At fixed point $-Y_2^{(i)} \cdot \frac{R_1}{L_2} - \left[\frac{1}{C_i \cdot L_2} \cdot I_{L_2}^{(i)} + \frac{1}{C_i \cdot L_2} \cdot I_{L_1}^{(i)} - \frac{1}{C_i \cdot L_2} \cdot I_{R_i}^{(i)} - \frac{1}{C_i \cdot L_2} \cdot I_{L_i}^{(i)}\right] = 0$

$$-y_2 \cdot \frac{R_1}{L_2} - y_2 \cdot \lambda - \frac{1}{C_i \cdot L_2} \cdot i_{L_2} - \frac{1}{C_i \cdot L_2} \cdot i_{L_1} + \frac{1}{C_i \cdot L_2} \cdot i_{R_i} + \frac{1}{C_i \cdot L_2} \cdot i_{L_i} = 0$$

$$\frac{dY_1}{dt} = \frac{1}{C_1 \cdot L_1} \cdot I_{R_p} - \frac{1}{L_1} \cdot \left[\frac{1}{C_1} + \frac{1}{C_2} + \frac{1}{C_i}\right] \cdot I_{L_1} - \frac{1}{C_i \cdot L_1} \cdot I_{L_2} + \frac{1}{C_i \cdot L_1} \cdot I_{R_i} + \frac{1}{C_i \cdot L_1} \cdot I_{L_i}$$

$$y_1 \cdot \lambda \cdot e^{\lambda \cdot t} = \frac{1}{C_1 \cdot L_1} \cdot \left[I_{R_p}^{(i)} + i_{R_p} \cdot e^{\lambda \cdot t}\right] - \frac{1}{L_1} \cdot \left[\frac{1}{C_1} + \frac{1}{C_2} + \frac{1}{C_i}\right] \cdot \left[I_{L_1}^{(i)} + i_{L_1} \cdot e^{\lambda \cdot t}\right] - \frac{1}{C_i \cdot L_1} \cdot \left[I_{L_2}^{(i)} + i_{L_2} \cdot e^{\lambda \cdot t}\right] + \frac{1}{C_i \cdot L_1} \cdot \left[I_{R_i}^{(i)} + i_{R_i} \cdot e^{\lambda \cdot t}\right] + \frac{1}{C_i \cdot L_1} \cdot \left[I_{L_i}^{(i)} + i_{L_i} \cdot e^{\lambda \cdot t}\right]$$

$$y_1 \cdot \lambda \cdot e^{\lambda \cdot t} = \frac{1}{C_1 \cdot L_1} \cdot I_{R_p}^{(i)} - \frac{1}{L_1} \cdot \left[\frac{1}{C_1} + \frac{1}{C_2} + \frac{1}{C_i}\right] \cdot I_{L_1}^{(i)} - \frac{1}{C_i \cdot L_1} \cdot I_{L_2}^{(i)} + \frac{1}{C_i \cdot L_1} \cdot I_{R_i}^{(i)} + \frac{1}{C_i \cdot L_1} \cdot I_{L_i}^{(i)} + i_{R_p} \cdot \frac{1}{C_1 \cdot L_1} \cdot e^{\lambda \cdot t} - i_{L_1} \cdot \frac{1}{L_1} \cdot \left[\frac{1}{C_1} + \frac{1}{C_2} + \frac{1}{C_i}\right] \cdot e^{\lambda \cdot t} - i_{L_2} \cdot \frac{1}{C_i \cdot L_1} \cdot e^{\lambda \cdot t} + i_{R_i} \cdot \frac{1}{C_i \cdot L_1} \cdot e^{\lambda \cdot t} + i_{L_i} \cdot \frac{1}{C_i \cdot L_1} \cdot e^{\lambda \cdot t}$$

At fixed point $\frac{1}{C_1 \cdot L_1} \cdot I_{R_p}^{(i)} - \frac{1}{L_1} \cdot \left[\frac{1}{C_1} + \frac{1}{C_2} + \frac{1}{C_i}\right] \cdot I_{L_1}^{(i)} - \frac{1}{C_i \cdot L_1} \cdot I_{L_2}^{(i)} + \frac{1}{C_i \cdot L_1} \cdot I_{R_i}^{(i)} + \frac{1}{C_i \cdot L_1} \cdot I_{L_i}^{(i)} = 0$

$$-y_1 \cdot \lambda + i_{R_p} \cdot \frac{1}{C_1 \cdot L_1} - i_{L_1} \cdot \frac{1}{L_1} \cdot \left[\frac{1}{C_1} + \frac{1}{C_2} + \frac{1}{C_i}\right] - i_{L_2} \cdot \frac{1}{C_i \cdot L_1} + i_{R_i} \cdot \frac{1}{C_i \cdot L_1} + i_{L_i} \cdot \frac{1}{C_i \cdot L_1} = 0$$

$$\frac{dY_3}{dt} = \frac{1}{L_i \cdot C_i} \cdot \left[I_{L_2} + I_{L_1} - I_{R_i} - I_{L_i}\right] - Y_3 \cdot \frac{R_{se}}{L_i}$$

$$y_3 \cdot \lambda \cdot e^{\lambda \cdot t} = \frac{1}{L_i \cdot C_i} \cdot \left(I_{L_2}^{(i)} + i_{L_2} \cdot e^{\lambda \cdot t} + I_{L_1}^{(i)} + i_{L_1} \cdot e^{\lambda \cdot t} - \left[I_{R_i}^{(i)} + i_{R_i} \cdot e^{\lambda \cdot t}\right] - \left[I_{L_i}^{(i)} + i_{L_i} \cdot e^{\lambda \cdot t}\right]\right) - \left[Y_3^{(i)} + y_3 \cdot e^{\lambda \cdot t}\right] \cdot \frac{R_{se}}{L_i}$$

$$y_3 \cdot \lambda \cdot e^{\lambda \cdot t} = \frac{1}{L_i \cdot C_i} \cdot \left(I_{L_2}^{(i)} + I_{L_1}^{(i)} - I_{R_i}^{(i)} - I_{L_i}^{(i)} + i_{L_2} \cdot e^{\lambda \cdot t} \right.$$
$$\left. + i_{L_1} \cdot e^{\lambda \cdot t} - i_{R_i} \cdot e^{\lambda \cdot t} - i_{L_i} \cdot e^{\lambda \cdot t} \right) - Y_3^{(i)} \cdot \frac{R_{se}}{L_i} - y_3 \cdot \frac{R_{se}}{L_i} \cdot e^{\lambda \cdot t}$$

$$y_3 \cdot \lambda \cdot e^{\lambda \cdot t} = \frac{1}{L_i \cdot C_i} \cdot (I_{L_2}^{(i)} + I_{L_1}^{(i)} - I_{R_i}^{(i)} - I_{L_i}^{(i)}) - Y_3^{(i)} \cdot \frac{R_{se}}{L_i}$$
$$+ \frac{1}{L_i \cdot C_i} \cdot (i_{L_2} \cdot e^{\lambda \cdot t} + i_{L_1} \cdot e^{\lambda \cdot t} - i_{R_i} \cdot e^{\lambda \cdot t}$$
$$- i_{L_i} \cdot e^{\lambda \cdot t}) - y_3 \cdot \frac{R_{se}}{L_i} \cdot e^{\lambda \cdot t}$$

At fixed point $\frac{1}{L_i \cdot C_i} \cdot \left(I_{L_2}^{(i)} + I_{L_1}^{(i)} - I_{R_i}^{(i)} - I_{L_i}^{(i)} \right) - Y_3^{(i)} \cdot \frac{R_{se}}{L_i} = 0$

$$\frac{1}{L_i \cdot C_i} \cdot \left(i_{L_2} + i_{L_1} - i_{R_i} - i_{L_i} \right) - y_3 \cdot \frac{R_{se}}{L_i} - y_3 \cdot \lambda = 0$$

$$\frac{dI_{L_1}}{dt} = Y_1; \; i_{L_1} \cdot \lambda \cdot e^{\lambda \cdot t} = Y_1^{(i)} + y_1 \cdot e^{\lambda \cdot t}; \; Y_1^{(i)} = 0; \; -i_{L_1} \cdot \lambda + y_1 = 0$$

$$\frac{dI_{L_2}}{dt} = Y_2; \; i_{L_2} \cdot \lambda \cdot e^{\lambda \cdot t} = Y_2^{(i)} + y_2 \cdot e^{\lambda \cdot t}; \; Y_2^{(i)} = 0; \; -i_{L_2} \cdot \lambda + y_2 = 0$$

$$\frac{dI_{R_i}}{dt} = \frac{1}{C_i \cdot R_i} \cdot \left[I_{L_2} + I_{L_1} - I_{R_i} - I_{L_i} \right] \Rightarrow \frac{dI_{R_i}}{dt} = \frac{1}{C_i \cdot R_i} \cdot \left[\sum_{k=1}^{2} I_{L_k} - I_{R_i} - I_{L_i} \right]$$

$$i_{R_i} \cdot \lambda \cdot e^{\lambda \cdot t} = \frac{1}{C_i \cdot R_i} \cdot$$
$$\left[I_{L_1}^{(i)} + i_{L_1} \cdot e^{\lambda \cdot t} + I_{L_2}^{(i)} + i_{L_2} \cdot e^{\lambda \cdot t} - \left(I_{R_i}^{(i)} + i_{R_i} \cdot e^{\lambda \cdot t} \right) - \left(I_{L_i}^{(i)} + i_{L_i} \cdot e^{\lambda \cdot t} \right) \right]$$

$$i_{R_i} \cdot \lambda \cdot e^{\lambda \cdot t} = \frac{1}{C_i \cdot R_i} \cdot \left[I_{L_1}^{(i)} + I_{L_2}^{(i)} - I_{R_i}^{(i)} - I_{L_i}^{(i)} \right]$$
$$+ \frac{1}{C_i \cdot R_i} \cdot \left[i_{L_1} \cdot e^{\lambda \cdot t} + i_{L_2} \cdot e^{\lambda \cdot t} - i_{R_i} \cdot e^{\lambda \cdot t} - i_{L_i} \cdot e^{\lambda \cdot t} \right]$$

At fixed point
$$\frac{1}{C_i \cdot R_i} \cdot \left[I_{L_1}^{(i)} + I_{L_2}^{(i)} - I_{R_i}^{(i)} - I_{L_i}^{(i)} \right] = 0; \; -i_{R_i} \cdot \lambda$$
$$+ \frac{1}{C_i \cdot R_i} \cdot \left[i_{L_1} + i_{L_2} - i_{R_i} - i_{L_i} \right] = 0$$

$$\frac{dI_{L_i}}{dt} = Y_3; \; i_{L_i} \cdot \lambda \cdot e^{\lambda \cdot t} = Y_3^{(i)} + y_3 \cdot e^{\lambda \cdot t}; \; Y_3^{(i)} = 0; \; -i_{L_i} \cdot \lambda + y_3 = 0$$

$$\frac{dI_{R_p}}{dt} = -\frac{1}{R_p \cdot C_1} \cdot (I_{R_p} - I_{L_1}); \; i_{R_p} \cdot \lambda \cdot e^{\lambda \cdot t}$$
$$= -\frac{1}{R_p \cdot C_1} \cdot \left(I_{R_p}^{(i)} + i_{R_p} \cdot e^{\lambda \cdot t} - \left[I_{L_1}^{(i)} + i_{L_1} \cdot e^{\lambda \cdot t}\right]\right)$$

$$i_{R_p} \cdot \lambda \cdot e^{\lambda \cdot t} = -\frac{1}{R_p \cdot C_1} \cdot \left(I_{R_p}^{(i)} - I_{L_1}^{(i)}\right) - \frac{1}{R_p \cdot C_1} \cdot \left(i_{R_p} \cdot e^{\lambda \cdot t} - i_{L_1} \cdot e^{\lambda \cdot t}\right)$$

At fixed point $-\frac{1}{R_p \cdot C_1} \cdot \left(I_{R_p}^{(i)} - I_{L_1}^{(i)}\right) = 0; \; -i_{R_p} \cdot \lambda - \frac{1}{R_p \cdot C_1} \cdot \left(i_{R_p} - i_{L_1}\right) = 0$
We can summary our system arbitrarily small increments equations:

$$- y_1 \cdot \lambda + i_{R_p} \cdot \frac{1}{C_1 \cdot L_1} - i_{L_1} \cdot \frac{1}{L_1} \cdot \left[\frac{1}{C_1} + \frac{1}{C_2} + \frac{1}{C_i}\right]$$
$$- i_{L_2} \cdot \frac{1}{C_i \cdot L_1} + i_{R_i} \cdot \frac{1}{C_i \cdot L_1} + i_{L_i} \cdot \frac{1}{C_i \cdot L_1} = 0$$

$$- y_2 \cdot \frac{R_1}{L_2} - y_2 \cdot \lambda - \frac{1}{C_i \cdot L_2} \cdot i_{L_2} - \frac{1}{C_i \cdot L_2} \cdot i_{L_1}$$
$$+ \frac{1}{C_i \cdot L_2} \cdot i_{R_i} + \frac{1}{C_i \cdot L_2} \cdot i_{L_i} = 0$$

$$\frac{1}{L_i \cdot C_i} \cdot i_{L_2} + \frac{1}{L_i \cdot C_i} \cdot i_{L_1} - \frac{1}{L_i \cdot C_i} \cdot i_{R_i} - \frac{1}{L_i \cdot C_i} \cdot i_{L_i} - y_3 \cdot \frac{R_{se}}{L_i}$$
$$- y_3 \cdot \lambda = 0; \; -i_{L_1} \cdot \lambda + y_1 = 0; \; -i_{L_2} \cdot \lambda + y_2 = 0$$

$$- i_{R_i} \cdot \lambda + \frac{1}{C_i \cdot R_i} \cdot i_{L_1} + \frac{1}{C_i \cdot R_i} \cdot i_{L_2}$$
$$- \frac{1}{C_i \cdot R_i} \cdot i_{R_i} - \frac{1}{C_i \cdot R_i} \cdot i_{L_i} = 0$$

$$-i_{L_i} \cdot \lambda + y_3 = 0; \; -i_{R_p} \cdot \lambda - \frac{1}{R_p \cdot C_1} \cdot i_{R_p} + \frac{1}{R_p \cdot C_1} \cdot i_{L_1} = 0$$

The small increments Jacobian of our MRI system Laser diode circuitry system is as below:

$$\begin{pmatrix} \Xi_{11} & \ldots & \Xi_{18} \\ \vdots & \ddots & \vdots \\ \Xi_{81} & \cdots & \Xi_{88} \end{pmatrix} \cdot \begin{pmatrix} y_1 \\ y_2 \\ y_3 \\ i_{L_1} \\ i_{L_2} \\ i_{R_i} \\ i_{L_i} \\ i_{R_p} \end{pmatrix} = 0; \;\; \Xi_{11} = -\lambda; \;\; \Xi_{12} = \Xi_{13} = 0; \;\; \Xi_{14}$$

$$= -\frac{1}{L_1} \cdot \left[\frac{1}{C_1} + \frac{1}{C_2} + \frac{1}{C_i}\right]; \;\; \Xi_{15} = -\frac{1}{C_i \cdot L_1}$$

$$\Xi_{16} = \frac{1}{C_i \cdot L_1}; \;\; \Xi_{17} = \frac{1}{C_i \cdot L_1}; \;\; \Xi_{18} = \frac{1}{C_1 \cdot L_1}; \;\; \Xi_{21} = 0; \;\; \Xi_{22}$$
$$= -\lambda - \frac{R_1}{L_2}; \;\; \Xi_{23} = 0; \;\; \Xi_{24} = -\frac{1}{C_i \cdot L_2}; \;\; \Xi_{25} = -\frac{1}{C_i \cdot L_2}$$

$$\Xi_{26} = \frac{1}{C_i \cdot L_2}; \;\; \Xi_{27} = \frac{1}{C_i \cdot L_2}; \;\; \Xi_{28} = 0; \;\; \Xi_{31} = \Xi_{32} = 0; \;\; \Xi_{33}$$
$$= -\frac{R_{se}}{L_i} - \lambda; \;\; \Xi_{34} = \Xi_{35} = \frac{1}{C_i \cdot L_i}$$

$$\Xi_{36} = \Xi_{37} = -\frac{1}{C_i \cdot L_i}; \;\; \Xi_{38} = 0; \;\; \Xi_{41} = 1; \;\; \Xi_{42} = \Xi_{43} = 0; \;\; \Xi_{44}$$
$$= -\lambda; \;\; \Xi_{45} = \ldots = \Xi_{48} = 0$$

$$\Xi_{51} = 0; \;\; \Xi_{52} = 1; \;\; \Xi_{53} = \Xi_{54} = 0; \;\; \Xi_{55} = -\lambda; \;\; \Xi_{56} = \Xi_{57}$$
$$= \Xi_{58} = 0; \;\; \Xi_{61} = \Xi_{62} = \Xi_{63} = 0$$

$$\Xi_{64} = \Xi_{65} = \frac{1}{C_i \cdot R_i}; \;\; \Xi_{66} = -\lambda - \frac{1}{C_i \cdot R_i}; \;\; \Xi_{67}$$
$$= -\frac{1}{C_i \cdot R_i}; \;\; \Xi_{68} = 0; \;\; \Xi_{71} = \Xi_{72} = 0; \;\; \Xi_{73} = 1$$

$$\Xi_{74} = \Xi_{75} = \Xi_{76} = 0; \;\; \Xi_{77} = -\lambda; \;\; \Xi_{78} = 0; \;\; \Xi_{81}$$
$$= \Xi_{82} = \Xi_{83} = 0; \;\; \Xi_{84} = \frac{1}{R_p \cdot C_1}; \;\; \Xi_{85} = \Xi_{86} = \Xi_{87} = 0$$

$$\Xi_{88} = -\lambda - \frac{1}{R_p \cdot C_1}$$

$$A - \lambda \cdot I = \begin{pmatrix} \Xi_{11} & \ldots & \Xi_{18} \\ \vdots & \ddots & \vdots \\ \Xi_{81} & \cdots & \Xi_{88} \end{pmatrix}; \ \det(A - \lambda \cdot I)$$

$$= 0 \Rightarrow \sum_{k=0}^{8} \lambda^k \cdot f_k(L_1, L_2, C_i, R_i, R_p, C_1, R_{se}, L_i, \ldots) = 0$$

We get zeros of the polynomial $\sum_{k=0}^{8} \lambda^k \cdot f_k(L_1, L_2, C_i, R_i, R_p, C_1, R_{se}, L_i, \ldots) = 0$ with real coefficients $f_1, f_2, \ldots$ which are functions of our system parameters$L_1, L_2, C_i, \ldots$. The characteristic equation is a polynomial of degree $N = 8;\ \sum_{k=0}^{8} \lambda^k \cdot f_k(L_1, L_2, \ldots) = 0$.

Eigenvalue stability discussion: Our MRI system Laser diode circuitry system involving N variables (N > 2, N = 8), the characteristic equation is of degree N = 8 and must often be solved numerically. Expect in some particular cases, such an equation has (N = 8) distinct roots that can be real or complex. These values are the eigenvalues of the 8x8 Jacobian matrix (A). The general rule is that the Steady State (SS) is stable if there is no eigenvalue with positive real part. It is sufficient that one eigenvalue is positive for the steady state to be unstable. Our 8-variables I_{L_1} I_{L_2} I_{R_i} $I_{R_p} I_{L_i}$ Y_1 Y_2 Y_3 system has eight eigenvalues. The type of behavior can be characterized as a function of the position of these eigenvalues in the Re/Im plane. Five non-degenerated cases can be distinguished: (1) the eight eigenvalues are real and negative (stable steady state), (2) the eight eigenvalues are real, seven of them are negative (unstable steady state), (3) and (4) at least two eigenvalues are complex conjugates with a negative real part and the other are negative (stable steady state), two cases can be distinguished depending on the relative value of the real part of the complex eigenvalues and of the real one, (5) at least two eigenvalues are complex conjugates with a negative real part and the other, one of them real is positive (unstable steady state) [12].

Second we analyze our system stability for $\tau_1 = \tau,\ \tau_2 = 0;\ \tau_1 = 0,\ \tau_2 = \tau;\ \tau_1 = \tau_2 = \tau$. We choose the first case $\tau_1 = \tau,\ \tau_2 = 0$. We summary out system differential equations ($\tau_1 = \tau,\ \tau_2 = 0;\ \sum_{k=1}^{2} \tau_k = \tau$):

$$\frac{dY_2}{dt} = -Y_2 \cdot \frac{R_1}{L_2} - \frac{1}{C_i \cdot L_2} \cdot [I_{L_2}(t) + I_{L_1}(t - \tau) - I_{R_i} - I_{L_i}]$$

$$\frac{dY_1}{dt} = \frac{1}{C_1 \cdot L_1} \cdot I_{R_p} - \frac{1}{C_1 \cdot L_1} \cdot I_{L_1} - \frac{1}{L_1} \cdot \left[\frac{1}{C_2} + \frac{1}{C_i}\right] \cdot I_{L_1}(t - \tau)$$
$$- \frac{1}{C_i \cdot L_1} \cdot I_{L_2}(t) + \frac{1}{C_i \cdot L_1} \cdot I_{R_i} + \frac{1}{C_i \cdot L_1} \cdot I_{L_i}$$

$$\frac{dY_3}{dt} = \frac{1}{L_i \cdot C_i} \cdot [I_{L_2}(t) + I_{L_1}(t - \tau) - I_{R_i} - I_{L_i}]$$
$$- Y_3 \cdot \frac{R_{se}}{L_i};\ \frac{dI_{L_1}}{dt} = Y_1;\ \frac{dI_{L_2}}{dt} = Y_2$$

$$\frac{dI_{R_i}}{dt} = \frac{1}{C_i \cdot R_i} \cdot \left[I_{L_2}(t) + I_{L_1}(t-\tau) - I_{R_i} - I_{L_i}\right]; \quad \frac{dI_{L_i}}{dt} = Y_3$$

$$\frac{dI_{R_p}}{dt} = \frac{1}{R_p} \cdot \frac{dX(t)}{dt} - \frac{1}{R_p \cdot C_1} \cdot \left(I_{R_p} - I_{L_1}\right)$$

$$I_{L_1} = I_{L_1}(t); \; I_{L_2} = I_{L_2}(t); \; I_{R_i} = I_{R_i}(t); \; I_{L_i} = I_{L_i}(t); \; I_{R_p} = I_{R_p}(t); \; Y_k = Y_k(t) \, \forall \, k = 1, 2, 3$$

We already define the MRI system Laser diode circuitry fixed values with arbitrarily small increments of exponential form $\left[i_{L_1}\, i_{L_2}\, i_{R_i}\, i_{R_p}\, i_{L_i}\, y_1\, y_2\, y_3\right] \cdot e^{\lambda \cdot t}$ (i = 0 (first fixed point), i = 1 (second fixed point), i = 2 (third fixed point), etc.) and complete it.

$$I_{L_1}(t-\tau) = I_{L_1}^{(i)} + i_{L_1} \cdot e^{\lambda \cdot (t-\tau)} \Rightarrow I_{L_1}(t-\tau) = I_{L_1}^{(i)} + i_{L_1} \cdot e^{\lambda \cdot t} \cdot e^{-\lambda \cdot \tau}; \quad \frac{dI_{L_1}(t-\tau)}{dt} = i_{L_1} \cdot \lambda \cdot e^{\lambda \cdot t} \cdot e^{-\lambda \cdot \tau}$$

$$\frac{dY_1}{dt} = \frac{1}{C_1 \cdot L_1} \cdot I_{R_p} - \frac{1}{C_1 \cdot L_1} \cdot I_{L_1} - \frac{1}{L_1} \cdot \left[\frac{1}{C_2} + \frac{1}{C_i}\right] \cdot I_{L_1}(t-\tau) - \frac{1}{C_i \cdot L_1} \cdot I_{L_2}(t) + \frac{1}{C_i \cdot L_1} \cdot I_{R_i} + \frac{1}{C_i \cdot L_1} \cdot I_{L_i}$$

$$y_1 \cdot \lambda \cdot e^{\lambda \cdot t} = \frac{1}{C_1 \cdot L_1} \cdot \left[I_{R_p}^{(i)} + i_{R_p} \cdot e^{\lambda \cdot t}\right] - \frac{1}{C_1 \cdot L_1} \cdot \left[I_{L_1}^{(i)} + i_{L_1} \cdot e^{\lambda \cdot t}\right] - \frac{1}{L_1} \cdot \left[\frac{1}{C_2} + \frac{1}{C_i}\right] \cdot \left[I_{L_1}^{(i)} + i_{L_1} \cdot e^{\lambda \cdot (t-\tau)}\right] - \frac{1}{C_i \cdot L_1} \cdot \left[I_{L_2}^{(i)} + i_{L_2} \cdot e^{\lambda \cdot t}\right] + \frac{1}{C_i \cdot L_1} \cdot \left[I_{R_i}^{(i)} + i_{R_i} \cdot e^{\lambda \cdot t}\right] + \frac{1}{C_i \cdot L_1} \cdot \left[I_{L_i}^{(i)} + i_{L_i} \cdot e^{\lambda \cdot t}\right]$$

$$y_1 \cdot \lambda \cdot e^{\lambda \cdot t} = \frac{1}{C_1 \cdot L_1} \cdot I_{R_p}^{(i)} - \frac{1}{C_1 \cdot L_1} \cdot I_{L_1}^{(i)} - \frac{1}{L_1} \cdot \left[\frac{1}{C_2} + \frac{1}{C_i}\right] \cdot I_{L_1}^{(i)} - \frac{1}{C_i \cdot L_1} \cdot I_{L_2}^{(i)} + \frac{1}{C_i \cdot L_1} \cdot I_{R_i}^{(i)} + \frac{1}{C_i \cdot L_1} \cdot I_{L_i}^{(i)} + \frac{1}{C_1 \cdot L_1} \cdot i_{R_p} \cdot e^{\lambda \cdot t} - \frac{1}{C_1 \cdot L_1} \cdot i_{L_1} \cdot e^{\lambda \cdot t} - \frac{1}{L_1} \cdot \left[\frac{1}{C_2} + \frac{1}{C_i}\right] \cdot i_{L_1} \cdot e^{\lambda \cdot (t-\tau)} - \frac{1}{C_i \cdot L_1} \cdot i_{L_2} \cdot e^{\lambda \cdot t} + \frac{1}{C_i \cdot L_1} \cdot i_{R_i} \cdot e^{\lambda \cdot t} + \frac{1}{C_i \cdot L_1} \cdot i_{L_i} \cdot e^{\lambda \cdot t}$$

At fixed point
$$\frac{1}{C_1 \cdot L_1} \cdot I_{R_p}^{(i)} - \frac{1}{C_1 \cdot L_1} \cdot I_{L_1}^{(i)} - \frac{1}{L_1} \cdot \left[\frac{1}{C_2} + \frac{1}{C_i}\right] \cdot I_{L_1}^{(i)} - \frac{1}{C_i \cdot L_1} \cdot I_{L_2}^{(i)} + \frac{1}{C_i \cdot L_1} \cdot I_{R_i}^{(i)} + \frac{1}{C_i \cdot L_1} \cdot I_{L_i}^{(i)} = 0$$

$$- y_1 \cdot \lambda - \frac{1}{L_1} \cdot \left(\frac{1}{C_1} + \left[\frac{1}{C_2} + \frac{1}{C_i}\right] \cdot e^{-\lambda \cdot \tau}\right) \cdot i_{L_1} - \frac{1}{C_i \cdot L_1} \cdot i_{L_2} + \frac{1}{C_1 \cdot L_1} \cdot i_{R_p} + \frac{1}{C_i \cdot L_1} \cdot i_{R_i} + \frac{1}{C_i \cdot L_1} \cdot i_{L_i} = 0$$

$$\frac{dY_2}{dt} = -Y_2 \cdot \frac{R_1}{L_2} - \frac{1}{C_i \cdot L_2} \cdot [I_{L_2}(t) + I_{L_1}(t - \tau) - I_{R_i} - I_{L_i}]$$

$$y_2 \cdot \lambda \cdot e^{\lambda \cdot t} = -\left[Y_2^{(i)} + y_2 \cdot e^{\lambda \cdot t}\right] \cdot \frac{R_1}{L_2} - \frac{1}{C_i \cdot L_2} \cdot \left[I_{L_2}^{(i)} + i_{L_2} \cdot e^{\lambda \cdot t} + I_{L_1}^{(i)} + i_{L_1} \cdot e^{\lambda \cdot (t-\tau)} - I_{R_i}^{(i)} - i_{R_i} \cdot e^{\lambda \cdot t} - I_{L_i}^{(i)} - i_{L_i} \cdot e^{\lambda \cdot t}\right]$$

$$y_2 \cdot \lambda \cdot e^{\lambda \cdot t} = -Y_2^{(i)} \cdot \frac{R_1}{L_2} - \frac{1}{C_i \cdot L_2} \cdot \left[I_{L_2}^{(i)} + I_{L_1}^{(i)} - I_{R_i}^{(i)} - I_{L_i}^{(i)}\right] - y_2 \cdot \frac{R_1}{L_2} \cdot e^{\lambda \cdot t} - \frac{1}{C_i \cdot L_2} \cdot \left[i_{L_2} \cdot e^{\lambda \cdot t} + i_{L_1} \cdot e^{\lambda \cdot (t-\tau)} - i_{R_i} \cdot e^{\lambda \cdot t} - i_{L_i} \cdot e^{\lambda \cdot t}\right]$$

At fixed point $-Y_2^{(i)} \cdot \frac{R_1}{L_2} - \frac{1}{C_i \cdot L_2} \cdot \left[I_{L_2}^{(i)} + I_{L_1}^{(i)} - I_{R_i}^{(i)} - I_{L_i}^{(i)}\right] = 0$

$$- y_2 \cdot \frac{R_1}{L_2} - y_2 \cdot \lambda - \frac{1}{C_i \cdot L_2} \cdot i_{L_2} - i_{L_1} \cdot \frac{1}{C_i \cdot L_2} \cdot e^{-\lambda \cdot \tau} + \frac{1}{C_i \cdot L_2} \cdot i_{R_i} + \frac{1}{C_i \cdot L_2} \cdot i_{L_i} = 0$$

$$\frac{dY_3}{dt} = \frac{1}{L_i \cdot C_i} \cdot [I_{L_2}(t) + I_{L_1}(t - \tau) - I_{R_i} - I_{L_i}] - Y_3 \cdot \frac{R_{se}}{L_i}$$

$$y_3 \cdot \lambda \cdot e^{\lambda \cdot t} = \frac{1}{L_i \cdot C_i} \cdot \left[I_{L_2}^{(i)} + i_{L_2} \cdot e^{\lambda \cdot t} + I_{L_1}^{(i)} + i_{L_1} \cdot e^{\lambda \cdot (t-\tau)} - I_{R_i}^{(i)} - i_{R_i} \cdot e^{\lambda \cdot t} - I_{L_i}^{(i)} - i_{L_i} \cdot e^{\lambda \cdot t}\right] - Y_3^{(i)} \cdot \frac{R_{se}}{L_i} - y_3 \cdot \frac{R_{se}}{L_i} \cdot e^{\lambda \cdot t}$$

$$y_3 \cdot \lambda \cdot e^{\lambda \cdot t} = \frac{1}{L_i \cdot C_i} \cdot \left[I_{L_2}^{(i)} + I_{L_1}^{(i)} - I_{R_i}^{(i)} - I_{L_i}^{(i)}\right]$$

$$- Y_3^{(i)} \cdot \frac{R_{se}}{L_i} + \frac{1}{L_i \cdot C_i} \cdot \left[i_{L_2} \cdot e^{\lambda \cdot t} + i_{L_1} \cdot e^{\lambda \cdot (t-\tau)} - i_{R_i} \cdot e^{\lambda \cdot t} - i_{L_i} \cdot e^{\lambda \cdot t} \right]$$
$$- y_3 \cdot \frac{R_{se}}{L_i} \cdot e^{\lambda \cdot t}$$

At fixed point $\frac{1}{L_i \cdot C_i} \cdot \left[I_{L_2}^{(i)} + I_{L_1}^{(i)} - I_{R_i}^{(i)} - I_{L_i}^{(i)} \right] - Y_3^{(i)} \cdot \frac{R_{se}}{L_i} = 0$

$$- y_3 \cdot \lambda - y_3 \cdot \frac{R_{se}}{L_i} + \frac{1}{L_i \cdot C_i} \cdot i_{L_2} + i_{L_1} \cdot \frac{1}{L_i \cdot C_i} \cdot e^{-\lambda \cdot \tau}$$
$$- \frac{1}{L_i \cdot C_i} \cdot i_{R_i} - \frac{1}{L_i \cdot C_i} \cdot i_{L_i} = 0$$

$$\frac{dI_{L_1}}{dt} = Y_1; \; i_{L_1} \cdot \lambda \cdot e^{\lambda \cdot t} = Y_1^{(i)} + y_1 \cdot e^{\lambda \cdot t}; \; Y_1^{(i)}$$
$$= 0; \; -i_{L_1} \cdot \lambda + y_1 = 0$$

$$\frac{dI_{L_2}}{dt} = Y_2; \; i_{L_2} \cdot \lambda \cdot e^{\lambda \cdot t} = Y_2^{(i)} + y_2 \cdot e^{\lambda \cdot t}; \; Y_2^{(i)}$$
$$= 0; \; -i_{L_2} \cdot \lambda + y_2 = 0$$

$$\frac{dI_{R_i}}{dt} = \frac{1}{C_i \cdot R_i} \cdot \left[I_{L_2}(t) + I_{L_1}(t - \tau) - I_{R_i} - I_{L_i} \right]$$

$$i_{R_i} \cdot \lambda \cdot e^{\lambda \cdot t} = \frac{1}{C_i \cdot R_i} \cdot$$
$$\left[I_{L_2}^{(i)} + i_{L_2} \cdot e^{\lambda \cdot t} + I_{L_1}^{(i)} + i_{L_1} \cdot e^{\lambda \cdot t} \cdot e^{-\lambda \cdot \tau} - I_{R_i}^{(i)} \right.$$
$$\left. - i_{R_i} \cdot e^{\lambda \cdot t} - I_{L_i}^{(i)} - i_{L_i} \cdot e^{\lambda \cdot t} \right]$$

$$i_{R_i} \cdot \lambda \cdot e^{\lambda \cdot t} = \frac{1}{C_i \cdot R_i} \cdot \left[I_{L_2}^{(i)} + I_{L_1}^{(i)} - I_{R_i}^{(i)} - I_{L_i}^{(i)} \right]$$
$$+ \frac{1}{C_i \cdot R_i} \cdot \left[i_{L_2} \cdot e^{\lambda \cdot t} + i_{L_1} \cdot e^{\lambda \cdot t} \cdot e^{-\lambda \cdot \tau} - i_{R_i} \cdot e^{\lambda \cdot t} - i_{L_i} \cdot e^{\lambda \cdot t} \right]$$

At fixed point $\frac{1}{C_i \cdot R_i} \cdot \left[I_{L_2}^{(i)} + I_{L_1}^{(i)} - I_{R_i}^{(i)} - I_{L_i}^{(i)} \right] = 0$

$$- \frac{1}{C_i \cdot R_i} \cdot i_{R_i} - i_{R_i} \cdot \lambda + \frac{1}{C_i \cdot R_i} \cdot i_{L_2}$$
$$+ i_{L_1} \cdot \frac{1}{C_i \cdot R_i} \cdot e^{-\lambda \cdot \tau} - \frac{1}{C_i \cdot R_i} \cdot i_{L_i} = 0$$

$$\frac{dI_{L_i}}{dt} = Y_3; \; i_{L_i} \cdot \lambda \cdot e^{\lambda \cdot t} = Y_3^{(i)} + y_3 \cdot e^{\lambda \cdot t}; \; Y_3^{(i)}$$

$$= 0; \ -i_{L_i} \cdot \lambda + y_3 = 0$$

$$\frac{dI_{R_p}}{dt} = -\frac{1}{R_p \cdot C_1} \cdot (I_{R_p} - I_{L_1}); \ i_{R_p} \cdot \lambda \cdot e^{\lambda \cdot t}$$
$$= -\frac{1}{R_p \cdot C_1} \cdot \left(I_{R_p}^{(i)} + i_{R_p} \cdot e^{\lambda \cdot t} - \left[I_{L_1}^{(i)} + i_{L_1} \cdot e^{\lambda \cdot t} \right] \right)$$

$$i_{R_p} \cdot \lambda \cdot e^{\lambda \cdot t} = -\frac{1}{R_p \cdot C_1} \cdot \left(I_{R_p}^{(i)} - I_{L_1}^{(i)} \right) - \frac{1}{R_p \cdot C_1} \cdot \left(i_{R_p} \cdot e^{\lambda \cdot t} - i_{L_1} \cdot e^{\lambda \cdot t} \right)$$

At fixed point $-\frac{1}{R_p \cdot C_1} \cdot \left(I_{R_p}^{(i)} - I_{L_1}^{(i)} \right) = 0; \ -i_{R_p} \cdot \lambda - \frac{1}{R_p \cdot C_1} \cdot \left(i_{R_p} - i_{L_1} \right) = 0$

We can summary our system arbitrarily small increments equations with delay $\left(\tau_1 = \tau, \ \tau_2 = 0; \ \sum_{k=1}^{2} \tau_k = \tau \right)$:

$$- y_1 \cdot \lambda - \frac{1}{L_1} \cdot \left(\frac{1}{C_1} + \left[\frac{1}{C_2} + \frac{1}{C_i} \right] \cdot e^{-\lambda \cdot \tau} \right) \cdot i_{L_1}$$
$$- \frac{1}{C_i \cdot L_1} \cdot i_{L_2} + \frac{1}{C_1 \cdot L_1} \cdot i_{R_p} + \frac{1}{C_i \cdot L_1} \cdot i_{R_i} + \frac{1}{C_i \cdot L_1} \cdot i_{L_i} = 0$$

$$- y_2 \cdot \frac{R_1}{L_2} - y_2 \cdot \lambda - \frac{1}{C_i \cdot L_2} \cdot i_{L_2}$$
$$- i_{L_1} \cdot \frac{1}{C_i \cdot L_2} \cdot e^{-\lambda \cdot \tau} + \frac{1}{C_i \cdot L_2} \cdot i_{R_i} + \frac{1}{C_i \cdot L_2} \cdot i_{L_i} = 0$$

$$- y_3 \cdot \lambda - y_3 \cdot \frac{R_{se}}{L_i} + \frac{1}{L_i \cdot C_i} \cdot i_{L_2} + i_{L_1} \cdot \frac{1}{L_i \cdot C_i} \cdot e^{-\lambda \cdot \tau}$$
$$- \frac{1}{L_i \cdot C_i} \cdot i_{R_i} - \frac{1}{L_i \cdot C_i} \cdot i_{L_i} = 0; \ -i_{L_1} \cdot \lambda + y_1 = 0; \ -i_{L_2} \cdot \lambda + y_2 = 0$$

$$- \frac{1}{C_i \cdot R_i} \cdot i_{R_i} - i_{R_i} \cdot \lambda + \frac{1}{C_i \cdot R_i} \cdot i_{L_2} + i_{L_1} \cdot \frac{1}{C_i \cdot R_i} \cdot e^{-\lambda \cdot \tau}$$
$$- \frac{1}{C_i \cdot R_i} \cdot i_{L_i} = 0; \ -i_{L_i} \cdot \lambda + y_3 = 0$$

$$-i_{R_p} \cdot \lambda - \frac{1}{R_p \cdot C_1} \cdot i_{R_p} + \frac{1}{R_p \cdot C_1} \cdot i_{L_1} = 0$$

The small increments Jacobian of our MRI system Laser diode circuitry system with delay is as below $\left(\tau_1 = \tau, \ \tau_2 = 0; \ \sum_{k=1}^{2} \tau_k = \tau \right)$:

$$\begin{pmatrix} \Upsilon_{11} & \ldots & \Upsilon_{18} \\ \vdots & \ddots & \vdots \\ \Upsilon_{81} & \cdots & \Upsilon_{88} \end{pmatrix} \cdot \begin{pmatrix} y_1 \\ y_2 \\ y_3 \\ i_{L_1} \\ i_{L_2} \\ i_{R_i} \\ i_{L_i} \\ i_{R_p} \end{pmatrix} = 0; \ \Upsilon_{11} = -\lambda; \ \Upsilon_{12} = \Upsilon_{13} = 0; \ \Upsilon_{14}$$

$$= -\frac{1}{L_1} \cdot \left(\frac{1}{C_1} + \left[\frac{1}{C_2} + \frac{1}{C_i} \right] \cdot e^{-\lambda \cdot \tau} \right); \ \Upsilon_{15} = -\frac{1}{C_i \cdot L_1}$$

$$\Upsilon_{16} = \frac{1}{C_i \cdot L_1}; \ \Upsilon_{17} = \frac{1}{C_i \cdot L_1}; \ \Upsilon_{18} = \frac{1}{C_1 \cdot L_1}; \ \Upsilon_{21} = 0; \ \Upsilon_{22} = -\frac{R_1}{L_2} - \lambda; \ \Upsilon_{23} = 0; \ \Upsilon_{24} = -\frac{1}{C_i \cdot L_2} \cdot e^{-\lambda \cdot \tau}$$

$$\Upsilon_{25} = -\frac{1}{C_i \cdot L_2}; \ \Upsilon_{26} = \frac{1}{C_i \cdot L_2}; \ \Upsilon_{27} = \frac{1}{C_i \cdot L_2}; \ \Upsilon_{28} = 0; \ \Upsilon_{31} = \Upsilon_{32} = 0; \ \Upsilon_{33} = -\lambda - \frac{R_{se}}{L_i}$$

$$\Upsilon_{34} = \frac{1}{L_i \cdot C_i} \cdot e^{-\lambda \cdot \tau}; \ \Upsilon_{35} = \frac{1}{L_i \cdot C_i}; \ \Upsilon_{36} = -\frac{1}{L_i \cdot C_i}; \ \Upsilon_{37} = -\frac{1}{L_i \cdot C_i}; \ \Upsilon_{38} = 0; \ \Upsilon_{41} = 1; \ \Upsilon_{42} = \Upsilon_{43} = 0$$

$$\Upsilon_{44} = -\lambda; \ \Upsilon_{45} = \Upsilon_{46} = \Upsilon_{47} = \Upsilon_{48} = 0; \ \Upsilon_{51} = 0; \ \Upsilon_{52} = 1; \ \Upsilon_{53} = \Upsilon_{54} = 0; \ \Upsilon_{55} = -\lambda; \ \Upsilon_{56} = \Upsilon_{57} = \Upsilon_{58} = 0$$

$$\Upsilon_{61} = \Upsilon_{62} = \Upsilon_{63} = 0; \ \Upsilon_{64} = \frac{1}{C_i \cdot R_i} \cdot e^{-\lambda \cdot \tau}; \ \Upsilon_{65} = \frac{1}{C_i \cdot R_i}; \ \Upsilon_{66} = -\frac{1}{C_i \cdot R_i} - \lambda; \ \Upsilon_{67} = -\frac{1}{C_i \cdot R_i}$$

$$\Upsilon_{68} = 0; \ \Upsilon_{71} = \Upsilon_{72} = 0; \ \Upsilon_{73} = 1; \ \Upsilon_{74} = \Upsilon_{75} = \Upsilon_{76} = 0; \ \Upsilon_{77} = -\lambda; \ \Upsilon_{78} = 0$$

$$\Upsilon_{81} = \Upsilon_{82} = \Upsilon_{83} = 0; \ \Upsilon_{84} = \frac{1}{R_p \cdot C_1}; \ \Upsilon_{85} = \Upsilon_{86} = \Upsilon_{87} = 0; \ \Upsilon_{88} = -\lambda - \frac{1}{R_p \cdot C_1}$$

$$A-\lambda\cdot I=\begin{pmatrix}\Upsilon_{11} & \ldots & \Upsilon_{18}\\ \vdots & \ddots & \vdots\\ \Upsilon_{81} & \cdots & \Upsilon_{88}\end{pmatrix};\ \det(A-\lambda\cdot I)=0$$

$$\sigma_1=-\frac{1}{L_1};\ \sigma_2=\frac{1}{C_1};\ \sigma_3=\frac{1}{C_2}+\frac{1}{C_i}$$

$$\sigma_4=-\frac{1}{C_i\cdot L_1};\ \sigma_5=\frac{1}{C_i\cdot L_1};\ \sigma_6=\frac{1}{C_1\cdot L_1};\ \sigma_7=-\frac{R_1}{L_2}$$
$$\sigma_8=-\frac{1}{C_i\cdot L_2};\ \sigma_9=\frac{1}{C_i\cdot L_2}$$

$$\sigma_{10}=\frac{1}{L_i\cdot C_i};\ \sigma_{11}=-\frac{1}{L_i\cdot C_i};\ \sigma_{12}=\frac{1}{C_i\cdot R_i}$$
$$\sigma_{13}=-\frac{1}{C_i\cdot R_i};\ \sigma_{14}=\frac{1}{R_p\cdot C_1};\ \sigma_{15}=\frac{R_{se}}{L_i}$$

We need to get the characteristic equation for the case $\tau_1=\tau,\ \tau_2=0;\ \sum_{k=1}^{2}\tau_k=\tau$ and discuss stability. We study the occurrence of any possible stability switching, resulting from the increase of the value of the time delay τ for the general characteristic equation $D(\lambda,\tau)$. The expression for the characteristic equation is in the form $D(\lambda,\tau)=P_n(\lambda,\tau)+Q_m(\lambda,\tau)\cdot e^{-\lambda\cdot\tau};\ n,m\in\mathbb{N}_0;\ n>m$.

$$\Upsilon_{11}=-\lambda;\ \Upsilon_{12}=\Upsilon_{13}=0;\ \Upsilon_{14}=\sigma_1\cdot\left(\sigma_2+\sigma_3\cdot e^{-\lambda\cdot\tau}\right);\ \Upsilon_{15}=\sigma_4$$

$$\Upsilon_{16}=\sigma_5;\ \Upsilon_{17}=\sigma_5;\ \Upsilon_{18}=\sigma_6;\ \Upsilon_{21}=0;\ \Upsilon_{22}=\sigma_7-\lambda;\ \Upsilon_{23}=0;\ \Upsilon_{24}=\sigma_8\cdot e^{-\lambda\cdot\tau}$$

$$\Upsilon_{25}=\sigma_8;\ \Upsilon_{26}=\sigma_9;\ \Upsilon_{27}=\sigma_9;\ \Upsilon_{28}=0;\ \Upsilon_{31}=\Upsilon_{32}=0;\ \Upsilon_{33}=-\lambda-\sigma_{15}$$

$$\Upsilon_{34}=\sigma_{10}\cdot e^{-\lambda\cdot\tau};\ \Upsilon_{35}=\sigma_{10};\ \Upsilon_{36}=\sigma_{11};\ \Upsilon_{37}=\sigma_{11};\ \Upsilon_{38}=0;\ \Upsilon_{41}=1;\ \Upsilon_{42}=\Upsilon_{43}=0$$

$$\Upsilon_{44}=-\lambda;\ \Upsilon_{45}=\Upsilon_{46}=\Upsilon_{47}=\Upsilon_{48}=0;\ \Upsilon_{51}=0;\ \Upsilon_{52}=1;\ \Upsilon_{53}=\Upsilon_{54}$$
$$=0;\ \Upsilon_{55}=-\lambda;\ \Upsilon_{56}=\Upsilon_{57}=\Upsilon_{58}=0$$

$$\Upsilon_{61}=\Upsilon_{62}=\Upsilon_{63}=0;\ \Upsilon_{64}=\sigma_{12}\cdot e^{-\lambda\cdot\tau};\ \Upsilon_{65}=\sigma_{12};\ \Upsilon_{66}=\sigma_{13}-\lambda;\ \Upsilon_{67}=\sigma_{13}$$

$$\Upsilon_{68}=0;\ \Upsilon_{71}=\Upsilon_{72}=0;\ \Upsilon_{73}=1;\ \Upsilon_{74}=\Upsilon_{75}=\Upsilon_{76}=0;\ \Upsilon_{77}=-\lambda;\ \Upsilon_{78}=0$$

$$\Upsilon_{81}=\Upsilon_{82}=\Upsilon_{83}=0;\ \Upsilon_{84}=\sigma_{14};\ \Upsilon_{85}=\Upsilon_{86}=\Upsilon_{87}=0;\ \Upsilon_{88}=-\lambda-\sigma_{14}$$

Table 1.8 Matrix $A - \lambda \cdot I$ (8x8), matrix elements by rows and columns

$-\lambda$	0	0	$\sigma_1 \cdot (\sigma_2 + \sigma_3 \cdot e^{-\lambda \cdot \tau})$	σ_4	σ_5	σ_5	σ_6
0	$\sigma_7 - \lambda$	0	$\sigma_8 \cdot e^{-\lambda \cdot \tau}$	σ_8	σ_9	σ_9	0
0	0	$-\lambda - \sigma_{15}$	$\sigma_{10} \cdot e^{-\lambda \cdot \tau}$	σ_{10}	σ_{11}	σ_{11}	0
1	0	0	$-\lambda$	0	0	0	0
0	1	0	0	$-\lambda$	0	0	0
0	0	0	$\sigma_{12} \cdot e^{-\lambda \cdot \tau}$	σ_{12}	$\sigma_{13} - \lambda$	σ_{13}	0
0	0	1	0	0	0	$-\lambda$	0
0	0	0	σ_{14}	0	0	0	$-\lambda - \sigma_{14}$

$$\det(A - \lambda \cdot I) = (-\lambda) \cdot \det(A_1) - \sigma_1 \cdot \left(\sigma_2 + \sigma_3 \cdot e^{-\lambda \cdot \tau}\right) \cdot \det(A_2) + \sigma_4 \cdot \det(A_3)$$
$$- \sigma_5 \cdot \det(A_4) + \sigma_5 \cdot \det(A_5) - \sigma_6 \cdot \det(A_6)$$

Matrix $A - \lambda \cdot I$ is 8x8 dimensions. Matrix $A_1, A_2, \ldots, A_6$ are 7×7 dimension.

Matrix $A - \lambda \cdot I$ (8x8) determinant: matrix elements by rows and columns (Table 1.8)

$$\det(A - \lambda \cdot I) = -\lambda \cdot \det(A_1) - \sigma_1 \cdot \left(\sigma_2 + \sigma_3 \cdot e^{-\lambda \cdot \tau}\right) \cdot \det(A_2) + \sigma_4 \cdot \det(A_3)$$
$$- \sigma_5 \cdot \det(A_4) + \sigma_5 \cdot \det(A_5) - \sigma_6 \cdot \det(A_6)$$

Matrix A_1 (7×7) determinant: matrix elements by rows and columns (Table 1.9)

$$\det(A_1) = (\sigma_7 - \lambda) \cdot \det(A_{11}) + \sigma_8 \cdot e^{-\lambda \cdot \tau} \cdot \det(A_{12}) - \sigma_8 \cdot \det(A_{13})$$
$$+ \sigma_9 \cdot \det(A_{14}) - \sigma_9 \cdot \det(A_{15})$$

Matrix A_{11} (6×6) determinant: matrix elements by rows and columns (Table 1.10).

$$\det(A_{11}) = (-\lambda - \sigma_{15}) \cdot \det(A_{111}) - \sigma_{10} \cdot e^{-\lambda \cdot \tau} \cdot \det(A_{112}) + \sigma_{10} \cdot \det(A_{113})$$
$$- \sigma_{11} \cdot \det(A_{114}) + \sigma_{11} \cdot \det(A_{115})$$

Table 1.9 Matrix A_1 (7×7) elements by rows and columns

$\sigma_7 - \lambda$	0	$\sigma_8 \cdot e^{-\lambda \cdot \tau}$	σ_8	σ_9	σ_9	0
0	$-\lambda - \sigma_{15}$	$\sigma_{10} \cdot e^{-\lambda \cdot \tau}$	σ_{10}	σ_{11}	σ_{11}	0
0	0	$-\lambda$	0	0	0	0
1	0	0	$-\lambda$	0	0	0
0	0	$\sigma_{12} \cdot e^{-\lambda \cdot \tau}$	σ_{12}	$\sigma_{13} - \lambda$	σ_{13}	0
0	1	0	0	0	$-\lambda$	0
0	0	σ_{14}	0	0	0	$-\lambda - \sigma_{14}$

Table 1.10 Matrix A_{11} (6 × 6) elements by rows and columns

$-\lambda-\sigma_{15}$	$\sigma_{10}\cdot e^{-\lambda\cdot\tau}$	σ_{10}	σ_{11}	σ_{11}	0
0	$-\lambda$	0	0	0	0
0	0	$-\lambda$	0	0	0
0	$\sigma_{12}\cdot e^{-\lambda\cdot\tau}$	σ_{12}	$\sigma_{13}-\lambda$	σ_{13}	0
1	0	0	0	$-\lambda$	0
0	σ_{14}	0	0	0	$-\lambda-\sigma_{14}$

Table 1.11 Matrix A_{111} (5 × 5) elements by rows and columns

$-\lambda$	0	0	0	0
0	$-\lambda$	0	0	0
$\sigma_{12}\cdot e^{-\lambda\cdot\tau}$	σ_{12}	$\sigma_{13}-\lambda$	σ_{13}	0
0	0	0	$-\lambda$	0
σ_{14}	0	0	0	$-\lambda-\sigma_{14}$

Matrix A_{111} (5 × 5) determinant: matrix elements by rows and columns (Table 1.11).

$$\det(A_{111}) = \lambda^4\cdot(\sigma_{13}-\lambda)\cdot(\lambda+\sigma_{14})$$

Matrix A_{112} (5 × 5) determinant: matrix elements by rows and columns (Table 1.12)

$$\det(A_{112}) = 0$$

Matrix A_{113} (5 × 5) determinant: matrix elements by rows and columns (Table 1.13).

$$\det(A_{113}) = 0$$

Matrix A_{114} (5 × 5) determinant: matrix elements by rows and columns (Table 1.14).

Table 1.12 Matrix A_{112} (5 × 5) elements by rows and columns

0	0	0	0	0
0	$-\lambda$	0	0	0
0	σ_{12}	$\sigma_{13}-\lambda$	σ_{13}	0
1	0	0	$-\lambda$	0
0	0	0	0	$-\lambda-\sigma_{14}$

Table 1.13 Matrix A_{113} (5 × 5) elements by rows and columns

0	$-\lambda$	0	0	0
0	0	0	0	0
0	$\sigma_{12} \cdot e^{-\lambda \cdot \tau}$	$\sigma_{13} - \lambda$	σ_{13}	0
1	0	0	$-\lambda$	0
0	σ_{14}	0	0	$-\lambda - \sigma_{14}$

Table 1.14 Matrix A_{114} (5 × 5) elements by rows and columns

0	$-\lambda$	0	0	0
0	0	$-\lambda$	0	0
0	$\sigma_{12} \cdot e^{-\lambda \cdot \tau}$	σ_{12}	σ_{13}	0
1	0	0	$-\lambda$	0
0	σ_{14}	0	0	$-\lambda - \sigma_{14}$

$$\det(A_{114}) = \lambda^2 \cdot \sigma_{13} \cdot (\lambda + \sigma_{14})$$

Matrix A_{115} (5 × 5) determinant: matrix elements by rows and columns (Table 1.15).

$$\det(A_{115}) = \lambda^2 \cdot (\sigma_{13} - \lambda) \cdot (\lambda + \sigma_{14})$$

Matrix A_{12} (6 × 6) determinant: matrix elements by rows and columns (Table 1.16).

Table 1.15 Matrix A_{115} (5 × 5) elements by rows and columns

0	$-\lambda$	0	0	0
0	0	$-\lambda$	0	0
0	$\sigma_{12} \cdot e^{-\lambda \cdot \tau}$	σ_{12}	$\sigma_{13} - \lambda$	0
1	0	0	0	0
0	σ_{14}	0	0	$-\lambda - \sigma_{14}$

Table 1.16 Matrix A_{12} (6 × 6) elements by rows and columns

0	$-\lambda - \sigma_{15}$	σ_{10}	σ_{11}	σ_{11}	0
0	0	0	0	0	0
1	0	$-\lambda$	0	0	0
0	0	σ_{12}	$\sigma_{13} - \lambda$	σ_{13}	0
0	1	0	0	$-\lambda$	0
0	0	0	0	0	$-\lambda - \sigma_{14}$

Table 1.17 Matrix A_{13} (6 × 6) elements by rows and columns

0	$-\lambda-\sigma_{15}$	$\sigma_{10}\cdot e^{-\lambda\cdot\tau}$	σ_{11}	σ_{11}	0
0	0	$-\lambda$	0	0	0
1	0	0	0	0	0
0	0	$\sigma_{12}\cdot e^{-\lambda\cdot\tau}$	$\sigma_{13}-\lambda$	σ_{13}	0
0	1	0	0	$-\lambda$	0
0	0	σ_{14}	0	0	$-\lambda-\sigma_{14}$

$$\det(A_{12})=0$$

Matrix A_{13} (6 × 6) determinant: matrix elements by rows and columns (Table 1.17).

$$\det(A_{13})=-(-\lambda-\sigma_{15})\cdot\det(A_{131})+\sigma_{10}\cdot e^{-\lambda\cdot\tau}\cdot\det(A_{132})$$
$$-\sigma_{11}\cdot\det(A_{133})+\sigma_{11}\cdot\det(A_{134})$$

Matrix A_{131} (5 × 5) determinant: matrix elements by rows and columns (Table 1.18).

$$\det(A_{131})=\lambda^2\cdot(\sigma_{13}-\lambda)\cdot(\lambda+\sigma_{14})$$

Matrix A_{132} (5 × 5) determinant: matrix elements by rows and columns (Table 1.19).

$$\det(A_{132})=0$$

Table 1.18 Matrix A_{131} (5 × 5) elements by rows and columns

0	$-\lambda$	0	0	0
1	0	0	0	0
0	$\sigma_{12}\cdot e^{-\lambda\cdot\tau}$	$\sigma_{13}-\lambda$	σ_{13}	0
0	0	0	$-\lambda$	0
0	σ_{14}	0	0	$-\lambda-\sigma_{14}$

Table 1.19 Matrix A_{132} (5 × 5) elements by rows and columns

0	0	0	0	0
1	0	0	0	0
0	0	$\sigma_{13}-\lambda$	σ_{13}	0
0	1	0	$-\lambda$	0
0	0	0	0	$-\lambda-\sigma_{14}$

Table 1.20 Matrix A_{133} (5 × 5) elements by rows and columns

0	0	$-\lambda$	0	0
1	0	0	0	0
0	0	$\sigma_{12} \cdot e^{-\lambda \cdot \tau}$	σ_{13}	0
0	1	0	$-\lambda$	0
0	0	σ_{14}	0	$-\lambda - \sigma_{14}$

Matrix A_{133} (5 × 5) determinant: matrix elements by rows and columns (Table 1.20).

$$\det(A_{133}) = -\lambda \cdot \sigma_{13} \cdot (\lambda + \sigma_{14})$$

Matrix A_{134} (5 × 5) determinant: matrix elements by rows and columns (Table 1.21).

$$\det(A_{134}) = -\lambda \cdot (\sigma_{13} - \lambda) \cdot (\lambda + \sigma_{14})$$

Matrix A_{14} (6 × 6) determinant: matrix elements by rows and columns (Table 1.22).

$$\det(A_{14}) = -(-\lambda - \sigma_{15}) \cdot \det(A_{141}) + \sigma_{10} \cdot e^{-\lambda \cdot \tau} \cdot \det(A_{142}) - \sigma_{10} \cdot \det(A_{143}) + \sigma_{11} \cdot \det(A_{144})$$

Table 1.21 Matrix A_{134} (5 × 5) elements by rows and columns

0	0	$-\lambda$	0	0
1	0	0	0	0
0	0	$\sigma_{12} \cdot e^{-\lambda \cdot \tau}$	$\sigma_{13} - \lambda$	0
0	1	0	0	0
0	0	σ_{14}	0	$-\lambda - \sigma_{14}$

Table 1.22 Matrix A_{14} (6 × 6) elements by rows and columns

0	$-\lambda - \sigma_{15}$	$\sigma_{10} \cdot e^{-\lambda \cdot \tau}$	σ_{10}	σ_{11}	0
0	0	$-\lambda$	0	0	0
1	0	0	$-\lambda$	0	0
0	0	$\sigma_{12} \cdot e^{-\lambda \cdot \tau}$	σ_{12}	σ_{13}	0
0	1	0	0	$-\lambda$	0
0	0	σ_{14}	0	0	$-\lambda - \sigma_{14}$

Table 1.23 Matrix A_{141} (5 × 5) elements by rows and columns

0	$-\lambda$	0	0	0
1	0	$-\lambda$	0	0
0	$\sigma_{12} \cdot e^{-\lambda \cdot \tau}$	σ_{12}	σ_{13}	0
0	0	0	$-\lambda$	0
0	σ_{14}	0	0	$-\lambda - \sigma_{14}$

Matrix A_{141} (5 × 5) determinant: matrix elements by rows and columns (Table 1.23).

$$\det(A_{141}) = \lambda^2 \cdot \sigma_{12} \cdot (\lambda + \sigma_{14})$$

Matrix A_{142} (5 × 5) determinant: matrix elements by rows and columns (Table 1.24).

$$\det(A_{142}) = 0$$

Matrix A_{143} (5 × 5) determinant: matrix elements by rows and columns (Table 1.25).

$$\det(A_{143}) = -\lambda \cdot \sigma_{13} \cdot (\lambda + \sigma_{14})$$

Matrix A_{144} (5 × 5) determinant: matrix elements by rows and columns (Table 1.26).

$$\det(A_{144}) = -\lambda \cdot \sigma_{12} \cdot (\lambda + \sigma_{14})$$

Table 1.24 Matrix A_{142} (5 × 5) elements by rows and columns

0	0	0	0	0
1	0	$-\lambda$	0	0
0	0	σ_{12}	σ_{13}	0
0	1	0	$-\lambda$	0
0	0	0	0	$-\lambda - \sigma_{14}$

Table 1.25 Matrix A_{143} (5 × 5) elements by rows and columns

0	0	$-\lambda$	0	0
1	0	0	0	0
0	0	$\sigma_{12} \cdot e^{-\lambda \cdot \tau}$	σ_{13}	0
0	1	0	$-\lambda$	0
0	0	σ_{14}	0	$-\lambda - \sigma_{14}$

Table 1.26 Matrix A_{144} (5 × 5) elements by rows and columns

0	0	$-\lambda$	0	0
1	0	0	$-\lambda$	0
0	0	$\sigma_{12} \cdot e^{-\lambda \cdot \tau}$	σ_{12}	0
0	1	0	0	0
0	0	σ_{14}	0	$-\lambda - \sigma_{14}$

Table 1.27 Matrix A_{15} (6 × 6) elements by rows and columns

0	$-\lambda - \sigma_{15}$	$\sigma_{10} \cdot e^{-\lambda \cdot \tau}$	σ_{10}	σ_{11}	0
0	0	$-\lambda$	0	0	0
1	0	0	$-\lambda$	0	0
0	0	$\sigma_{12} \cdot e^{-\lambda \cdot \tau}$	σ_{12}	$\sigma_{13} - \lambda$	0
0	1	0	0	0	0
0	0	σ_{14}	0	0	$-\lambda - \sigma_{14}$

Matrix A_{15} (6 × 6) determinant: matrix elements by rows and columns (Table 1.27)

$$\det(A_{15}) = -(-\lambda - \sigma_{15}) \cdot \det(A_{151}) + \sigma_{10} \cdot e^{-\lambda \cdot \tau} \cdot \det(A_{152}) - \sigma_{10} \cdot \det(A_{153}) + \sigma_{11} \cdot \det(A_{154})$$

Matrix A_{151} (5 × 5) determinant: matrix elements by rows and columns (Table 1.28)

$$\det(A_{151}) = 0$$

Matrix A_{152} (5 × 5) determinant: matrix elements by rows and columns (Table 1.29).

$$\det(A_{152}) = 0$$

Matrix A_{153} (5 × 5) determinant: matrix elements by rows and columns (Table 1.30).

Table 1.28 Matrix A_{151} (5 × 5) elements by rows and columns

0	$-\lambda$	0	0	0
1	0	$-\lambda$	0	0
0	$\sigma_{12} \cdot e^{-\lambda \cdot \tau}$	σ_{12}	$\sigma_{13} - \lambda$	0
0	0	0	0	0
0	σ_{14}	0	0	$-\lambda - \sigma_{14}$

Table 1.29 Matrix A_{152} (5 × 5) elements by rows and columns

0	0	0	0	0
1	0	$-\lambda$	0	0
0	0	σ_{12}	$\sigma_{13} - \lambda$	0
0	1	0	0	0
0	0	0	0	$-\lambda - \sigma_{14}$

Table 1.30 Matrix A_{153} (5 × 5) elements by rows and columns

0	0	$-\lambda$	0	0
1	0	0	0	0
0	0	$\sigma_{12} \cdot e^{-\lambda \cdot \tau}$	$\sigma_{13} - \lambda$	0
0	1	0	0	0
0	0	σ_{14}	0	$-\lambda - \sigma_{14}$

$$\det(A_{153}) = -\lambda \cdot (\sigma_{13} - \lambda) \cdot (\lambda + \sigma_{14})$$

Matrix A_{154} (5 × 5) determinant: matrix elements by rows and columns (Table 1.31).

$$\det(A_{154}) = -\lambda \cdot \sigma_{12} \cdot (\lambda + \sigma_{14})$$

Matrix A_2 (7 × 7) determinant: matrix elements by rows and columns (Table 1.32).

Table 1.31 Matrix A_{154} (5 × 5) elements by rows and columns

0	0	$-\lambda$	0	0
1	0	0	$-\lambda$	0
0	0	$\sigma_{12} \cdot e^{-\lambda \cdot \tau}$	σ_{12}	0
0	1	0	0	0
0	0	σ_{14}	0	$-\lambda - \sigma_{14}$

Table 1.32 Matrix A_2 (7 × 7) elements by rows and columns

0	$\sigma_7 - \lambda$	0	σ_8	σ_9	σ_9	0
0	0	$-\lambda - \sigma_{15}$	σ_{10}	σ_{11}	σ_{11}	0
1	0	0	0	0	0	0
0	1	0	$-\lambda$	0	0	0
0	0	0	σ_{12}	$\sigma_{13} - \lambda$	σ_{13}	0
0	0	1	0	0	$-\lambda$	0
0	0	0	0	0	0	$-\lambda - \sigma_{14}$

Table 1.33 Matrix A_{21} (6 × 6) elements by rows and columns

0	$-\lambda-\sigma_{15}$	σ_{10}	σ_{11}	σ_{11}	0
1	0	0	0	0	0
0	0	$-\lambda$	0	0	0
0	0	σ_{12}	$\sigma_{13}-\lambda$	σ_{13}	0
0	1	0	0	$-\lambda$	0
0	0	0	0	0	$-\lambda-\sigma_{14}$

$$\det(A_2) = -(\sigma_7-\lambda)\cdot\det(A_{21}) - \sigma_8\cdot\det(A_{22}) + \sigma_9\cdot\det(A_{23}) - \sigma_9\cdot\det(A_{24})$$

Matrix A_{21} (6 × 6) determinant: matrix elements by rows and columns (Table 1.33).

$$\det(A_{21}) = (\lambda+\sigma_{15})\cdot\det(A_{211}) + \sigma_{10}\cdot\det(A_{212}) - \sigma_{11}\cdot\det(A_{213}) + \sigma_{11}\cdot\det(A_{214})$$

Matrix A_{211} (5 × 5) determinant: matrix elements by rows and columns (Table 1.34).

$$\det(A_{211}) = -\lambda^2\cdot(\sigma_{13}-\lambda)\cdot(\lambda+\sigma_{14})$$

Matrix A_{212} (5 × 5) determinant: matrix elements by rows and columns (Table 1.35).

$$\det(A_{212}) = 0$$

Table 1.34 Matrix A_{211} (5 × 5) elements by rows and columns

1		0	0	0	0
0		$-\lambda$	0	0	0
0		σ_{12}	$\sigma_{13}-\lambda$	σ_{13}	0
0		0	0	$-\lambda$	0
0		0	0	0	$-\lambda-\sigma_{14}$

Table 1.35 Matrix A_{212} (5 × 5) elements by rows and columns

1	0	0	0	0
0	0	0	0	0
0	0	$\sigma_{13}-\lambda$	σ_{13}	0
0	1	0	$-\lambda$	0
0	0	0	0	$-\lambda-\sigma_{14}$

Table 1.36 Matrix A_{213} (5 × 5) elements by rows and columns

1	0	0	0	0
0	0	$-\lambda$	0	0
0	0	σ_{12}	σ_{13}	0
0	1	0	$-\lambda$	0
0	0	0	0	$-\lambda-\sigma_{14}$

Matrix A_{213} (5 × 5) determinant: matrix elements by rows and columns (Table 1.36).

$$\det(A_{213}) = \lambda \cdot \sigma_{13} \cdot (\lambda + \sigma_{14})$$

Matrix A_{214} (5 × 5) determinant: matrix elements by rows and columns (Table 1.37).

$$\det(A_{214}) = \lambda \cdot (\sigma_{13} - \lambda) \cdot (\lambda + \sigma_{14})$$

Matrix A_{22} (6 × 6) determinant: matrix elements by rows and columns (Table 1.38).

$$\det(A_{22}) = (-\lambda - \sigma_{15}) \cdot \det(A_{221}) - \sigma_{11} \cdot \det(A_{222}) + \sigma_{11} \cdot \det(A_{223})$$

Matrix A_{221} (5 × 5) determinant: matrix elements by rows and columns (Table 1.39).

Table 1.37 Matrix A_{214} (5 × 5) elements by rows and columns

1	0	0	0	0
0	0	$-\lambda$	0	0
0	0	σ_{12}	$\sigma_{13}-\lambda$	0
0	1	0	0	0
0	0	0	0	$-\lambda-\sigma_{14}$

Table 1.38 Matrix A_{22} (6 × 6) elements by rows and columns

0	0	$-\lambda-\sigma_{15}$	σ_{11}	σ_{11}	0
1	0	0	0	0	0
0	1	0	0	0	0
0	0	0	$\sigma_{13}-\lambda$	σ_{13}	0
0	0	1	0	$-\lambda$	0
0	0	0	0	0	$-\lambda-\sigma_{14}$

Table 1.39 Matrix A_{221} (5 × 5) elements by rows and columns

1	0	0	0	0
0	1	0	0	0
0	0	$\sigma_{13} - \lambda$	σ_{13}	0
0	0	0	$-\lambda$	0
0	0	0	0	$-\lambda - \sigma_{14}$

Table 1.40 Matrix A_{222} (5 × 5) elements by rows and columns

1	0	0	0	0
0	1	0	0	0
0	0	0	σ_{13}	0
0	0	1	$-\lambda$	0
0	0	0	0	$-\lambda - \sigma_{14}$

$$\det(A_{221}) = \lambda \cdot (\sigma_{13} - \lambda) \cdot (\lambda + \sigma_{14})$$

Matrix A_{222} (5 × 5) determinant: matrix elements by rows and columns (Table 1.40).

$$\det(A_{222}) = \sigma_{13} \cdot (\lambda + \sigma_{14})$$

Matrix A_{223} (5 × 5) determinant: matrix elements by rows and columns (Table 1.41).

$$\det(A_{223}) = (\sigma_{13} - \lambda) \cdot (\lambda + \sigma_{14})$$

Matrix A_{23} (6 × 6) determinant: matrix elements by rows and columns (Table 1.42).

$$\det(A_{23}) = (-\lambda - \sigma_{15}) \cdot \det(A_{231}) - \sigma_{10} \cdot \det(A_{232}) + \sigma_{11} \cdot \det(A_{233})$$

Matrix A_{231} (5 × 5) determinant: matrix elements by rows and columns (Table 1.43).

Table 1.41 Matrix A_{223} (5 × 5) elements by rows and columns

1	0	0	0	0
0	1	0	0	0
0	0	0	$\sigma_{13} - \lambda$	0
0	0	1	0	0
0	0	0	0	$-\lambda - \sigma_{14}$

Table 1.42 Matrix A_{23} (6 × 6) elements by rows and columns

0	0	$-\lambda - \sigma_{15}$	σ_{10}	σ_{11}	0
1	0	0	0	0	0
0	1	0	$-\lambda$	0	0
0	0	0	σ_{12}	σ_{13}	0
0	0	1	0	$-\lambda$	0
0	0	0	0	0	$-\lambda - \sigma_{14}$

Table 1.43 Matrix A_{231} (5 × 5) elements by rows and columns

1	0	0	0	0
0	1	$-\lambda$	0	0
0	0	σ_{12}	σ_{13}	0
0	0	0	$-\lambda$	0
0	0	0	0	$-\lambda - \sigma_{14}$

$$\det(A_{231}) = \lambda \cdot \sigma_{12} \cdot (\lambda + \sigma_{14})$$

Matrix A_{232} (5 × 5) determinant: matrix elements by rows and columns (Table 1.44).

$$\det(A_{232}) = \sigma_{13} \cdot (\lambda + \sigma_{14})$$

Matrix A_{233} (5 × 5) determinant: matrix elements by rows and columns (Table 1.45).

Table 1.44 Matrix A_{232} (5 × 5) elements by rows and columns

1	0	0	0	0
0	1	0	0	0
0	0	0	σ_{13}	0
0	0	1	$-\lambda$	0
0	0	0	0	$-\lambda - \sigma_{14}$

Table 1.45 Matrix A_{233} (5 × 5) elements by rows and columns

1	0	0	0	0
0	1	0	$-\lambda$	0
0	0	0	σ_{12}	0
0	0	1	0	0
0	0	0	0	$-\lambda - \sigma_{14}$

Table 1.46 Matrix A_{24} (6 × 6) elements by rows and columns

0	0	$-\lambda - \sigma_{15}$	σ_{10}	σ_{11}	0
1	0	0	0	0	0
0	1	0	$-\lambda$	0	0
0	0	0	σ_{12}	$\sigma_{13} - \lambda$	0
0	0	1	0	0	0
0	0	0	0	0	$-\lambda - \sigma_{14}$

$$\det(A_{233}) = \sigma_{12} \cdot (\lambda + \sigma_{14})$$

Matrix A_{24} (6 × 6) determinant: matrix elements by rows and columns (Table 1.46).

$$\det(A_{24}) = (-\lambda - \sigma_{15}) \cdot \det(A_{241}) - \sigma_{10} \cdot \det(A_{242}) + \sigma_{11} \cdot \det(A_{243})$$

Matrix A_{241} (5 × 5) determinant: matrix elements by rows and columns (Table 1.47).

$$\det(A_{241}) = -\lambda \cdot (\sigma_{13} - \lambda) \cdot (\lambda + \sigma_{14})$$

Matrix A_{242} (5 × 5) determinant: matrix elements by rows and columns (Table 1.48).

$$\det(A_{242}) = (\sigma_{13} - \lambda) \cdot (\lambda + \sigma_{14})$$

Table 1.47 Matrix A_{241} (5 × 5) elements by rows and columns

1	0	0	0	0
0	1	$-\lambda$	0	0
0	0	σ_{12}	$\sigma_{13} - \lambda$	0
0	0	0	0	0
0	0	0	0	$-\lambda - \sigma_{14}$

Table 1.48 Matrix A_{242} (5 × 5) elements by rows and columns

1	0	0	0	0
0	1	0	0	0
0	0	0	$\sigma_{13} - \lambda$	0
0	0	1	0	0
0	0	0	0	$-\lambda - \sigma_{14}$

Table 1.49 Matrix A_{243} (5 × 5) elements by rows and columns

1	0	0	0	0
0	1	0	$-\lambda$	0
0	0	0	σ_{12}	0
0	0	1	0	0
0	0	0	0	$-\lambda - \sigma_{14}$

Matrix A_{243} (5 × 5) determinant: matrix elements by rows and columns (Table 1.49).

$$\det(A_{243}) = \sigma_{12} \cdot (\lambda + \sigma_{14})$$

Matrix A_3 (7 × 7) determinant: matrix elements by rows and columns (Table 1.50).

$$\det(A_3) = -(\sigma_7 - \lambda) \cdot \det(A_{31}) - \sigma_8 \cdot e^{-\lambda \cdot \tau} \cdot \det(A_{32}) + \sigma_9 \cdot \det(A_{33}) - \sigma_9 \cdot \det(A_{34})$$

Matrix A_{31} (6 × 6) determinant: matrix elements by rows and columns (Table 1.51).

$$\det(A_{31}) = 0$$

Table 1.50 Matrix A_3 (7 × 7) elements by rows and columns

0	$\sigma_7 - \lambda$	0	$\sigma_8 \cdot e^{-\lambda \cdot \tau}$	σ_9	σ_9	0
0	0	$-\lambda - \sigma_{15}$	$\sigma_{10} \cdot e^{-\lambda \cdot \tau}$	σ_{11}	σ_{11}	0
1	0	0	$-\lambda$	0	0	0
0	1	0	0	0	0	0
0	0	0	$\sigma_{12} \cdot e^{-\lambda \cdot \tau}$	$\sigma_{13} - \lambda$	σ_{13}	0
0	0	1	0	0	$-\lambda$	0
0	0	0	σ_{14}	0	0	$-\lambda - \sigma_{14}$

Table 1.51 Matrix A_{31} (6 × 6) elements by rows and columns

0	$-\lambda - \sigma_{15}$	$\sigma_{10} \cdot e^{-\lambda \cdot \tau}$	σ_{11}	σ_{11}	0
1	0	$-\lambda$	0	0	0
0	0	0	0	0	0
0	0	$\sigma_{12} \cdot e^{-\lambda \cdot \tau}$	$\sigma_{13} - \lambda$	σ_{13}	0
0	1	0	0	$-\lambda$	0
0	0	σ_{14}	0	0	$-\lambda - \sigma_{14}$

Table 1.52 Matrix A_{32} (6 × 6) elements by rows and columns

0	0	$-\lambda - \sigma_{15}$	σ_{11}	σ_{11}	0
1	0	0	0	0	0
0	1	0	0	0	0
0	0	0	$\sigma_{13} - \lambda$	σ_{13}	0
0	0	1	0	$-\lambda$	0
0	0	0	0	0	$-\lambda - \sigma_{14}$

Matrix A_{32} (6 × 6) determinant: matrix elements by rows and columns (Table 1.52).

$$\det(A_{32}) = (-\lambda - \sigma_{15}) \cdot \det(A_{321}) - \sigma_{11} \cdot \det(A_{322}) + \sigma_{11} \cdot \det(A_{323})$$

Matrix A_{321} (5 × 5) determinant: matrix elements by rows and columns (Table 1.53).

$$\det(A_{321}) = \lambda \cdot (\sigma_{13} - \lambda) \cdot (\lambda + \sigma_{14})$$

Matrix A_{322} (5 × 5) determinant: matrix elements by rows and columns (Table 1.54).

$$\det(A_{322}) = \sigma_{13} \cdot (\lambda + \sigma_{14})$$

Matrix A_{323} (5 × 5) determinant: matrix elements by rows and columns (Table 1.55).

Table 1.53 Matrix A_{321} (5 × 5) elements by rows and columns

1	0	0	0	0
0	1	0	0	0
0	0	$\sigma_{13} - \lambda$	σ_{13}	0
0	0	0	$-\lambda$	0
0	0	0	0	$-\lambda - \sigma_{14}$

Table 1.54 Matrix A_{322} (5 × 5) elements by rows and columns

1	0	0	0	0
0	1	0	0	0
0	0	0	σ_{13}	0
0	0	1	$-\lambda$	0
0	0	0	0	$-\lambda - \sigma_{14}$

Table 1.55 Matrix A_{323} (5 × 5) elements by rows and columns

1	0	0	0	0
0	1	0	0	0
0	0	0	$\sigma_{13} - \lambda$	0
0	0	1	0	0
0	0	0	0	$-\lambda - \sigma_{14}$

Table 1.56 Matrix A_{33} (6 × 6) elements by rows and columns

0	0	$-\lambda - \sigma_{15}$	$\sigma_{10} \cdot e^{-\lambda \cdot \tau}$	σ_{11}	0
1	0	0	$-\lambda$	0	0
0	1	0	0	0	0
0	0	0	$\sigma_{12} \cdot e^{-\lambda \cdot \tau}$	σ_{13}	0
0	0	1	0	$-\lambda$	0
0	0	0	σ_{14}	0	$-\lambda - \sigma_{14}$

$$\det(A_{323}) = (\sigma_{13} - \lambda) \cdot (\lambda + \sigma_{14})$$

Matrix A_{33} (6 × 6) determinant: matrix elements by rows and columns (Table 1.56).

$$\det(A_{33}) = (-\lambda - \sigma_{15}) \cdot \det(A_{331}) - \sigma_{10} \cdot e^{-\lambda \cdot \tau} \cdot \det(A_{332}) + \sigma_{11} \cdot \det(A_{333})$$

Matrix A_{331} (5 × 5) determinant: matrix elements by rows and columns (Table 1.57).

$$\det(A_{331}) = \sigma_{12} \cdot e^{-\lambda \cdot \tau} \cdot \lambda \cdot (\lambda + \sigma_{14})$$

Matrix A_{332} (5 × 5) determinant: matrix elements by rows and columns (Table 1.58).

$$\det(A_{332}) = \sigma_{13} \cdot (\lambda + \sigma_{14})$$

Table 1.57 Matrix A_{331} (5 × 5) elements by rows and columns

1	0	$-\lambda$	0	0
0	1	0	0	0
0	0	$\sigma_{12} \cdot e^{-\lambda \cdot \tau}$	σ_{13}	0
0	0	0	$-\lambda$	0
0	0	σ_{14}	0	$-\lambda - \sigma_{14}$

Table 1.58 Matrix A_{332} (5 × 5) elements by rows and columns

1	0	0	0	0
0	1	0	0	0
0	0	0	σ_{13}	0
0	0	1	$-\lambda$	0
0	0	0	0	$-\lambda - \sigma_{14}$

Table 1.59 Matrix A_{333} (5 × 5) elements by rows and columns

1	0	0	$-\lambda$	0
0	1	0	0	0
0	0	0	$\sigma_{12} \cdot e^{-\lambda \cdot \tau}$	0
0	0	1	0	0
0	0	0	σ_{14}	$-\lambda - \sigma_{14}$

Matrix A_{333} (5 × 5) determinant: matrix elements by rows and columns (Table 1.59).

$$\det(A_{333}) = \sigma_{12} \cdot e^{-\lambda \cdot \tau} \cdot (\lambda + \sigma_{14})$$

Matrix A_{34} (6 × 6) determinant: matrix elements by rows and columns (Table 1.60).

$$\det(A_{34}) = (-\lambda - \sigma_{15}) \cdot \det(A_{341}) - \sigma_{10} \cdot e^{-\lambda \cdot \tau} \cdot \det(A_{342}) + \sigma_{11} \cdot \det(A_{343})$$

Matrix A_{341} (5 × 5) determinant: matrix elements by rows and columns (Table 1.61).

$$\det(A_{341}) = 0$$

Matrix A_{342} (5 × 5) determinant: matrix elements by rows and columns (Table 1.62).

Table 1.60 Matrix A_{34} (6 × 6) elements by rows and columns

0	0	$-\lambda - \sigma_{15}$	$\sigma_{10} \cdot e^{-\lambda \cdot \tau}$	σ_{11}	0
1	0	0	$-\lambda$	0	0
0	1	0	0	0	0
0	0	0	$\sigma_{12} \cdot e^{-\lambda \cdot \tau}$	$\sigma_{13} - \lambda$	0
0	0	1	0	0	0
0	0	0	σ_{14}	0	$-\lambda - \sigma_{14}$

Table 1.61 Matrix A_{341} (5 × 5) elements by rows and columns

1	0	$-\lambda$	0	0
0	1	0	0	0
0	0	$\sigma_{12} \cdot e^{-\lambda \cdot \tau}$	$\sigma_{13} - \lambda$	0
0	0	0	0	0
0	0	σ_{14}	0	$-\lambda - \sigma_{14}$

Table 1.62 Matrix A_{342} (5 × 5) elements by rows and columns

1	0	0	0	0
0	1	0	0	0
0	0	0	$\sigma_{13} - \lambda$	0
0	0	1	0	0
0	0	0	0	$-\lambda - \sigma_{14}$

$$\det(A_{342}) = (\sigma_{13} - \lambda) \cdot (\lambda + \sigma_{14})$$

Matrix A_{343} (5 × 5) determinant: matrix elements by rows and columns (Table 1.63).

$$\det(A_{343}) = \sigma_{12} \cdot e^{-\lambda \cdot \tau} \cdot (\lambda + \sigma_{14})$$

Matrix A_4 (7 × 7) determinant: matrix elements by rows and columns (Table 1.64).

Table 1.63 Matrix A_{343} (5 × 5) elements by rows and columns

1	0	0	$-\lambda$	0
0	1	0	0	0
0	0	0	$\sigma_{12} \cdot e^{-\lambda \cdot \tau}$	0
0	0	1	0	0
0	0	0	σ_{14}	$-\lambda - \sigma_{14}$

Table 1.64 Matrix A_4 (7 × 7) elements by rows and columns

0	$\sigma_7 - \lambda$	0	$\sigma_8 \cdot e^{-\lambda \cdot \tau}$	σ_8	σ_9	0
0	0	$-\lambda - \sigma_{15}$	$\sigma_{10} \cdot e^{-\lambda \cdot \tau}$	σ_{10}	σ_{11}	0
1	0	0	$-\lambda$	0	0	0
0	1	0	0	$-\lambda$	0	0
0	0	0	$\sigma_{12} \cdot e^{-\lambda \cdot \tau}$	σ_{12}	σ_{13}	0
0	0	1	0	0	$-\lambda$	0
0	0	0	σ_{14}	0	0	$-\lambda - \sigma_{14}$

Table 1.65 Matrix A_{41} (6 × 6) elements by rows and columns

0	$-\lambda - \sigma_{15}$	$\sigma_{10} \cdot e^{-\lambda \cdot \tau}$	σ_{10}	σ_{11}	0
1	0	$-\lambda$	0	0	0
0	0	0	$-\lambda$	0	0
0	0	$\sigma_{12} \cdot e^{-\lambda \cdot \tau}$	σ_{12}	σ_{13}	0
0	1	0	0	$-\lambda$	0
0	0	σ_{14}	0	0	$-\lambda - \sigma_{14}$

$$\det(A_4) = -(\sigma_7 - \lambda) \cdot \det(A_{41}) - \sigma_8 \cdot e^{-\lambda \cdot \tau} \cdot \det(A_{42}) + \sigma_8 \cdot \det(A_{43}) - \sigma_9 \cdot \det(A_{44})$$

Matrix A_{41} (6 × 6) determinant: matrix elements by rows and columns (Table 1.65).

$$\det(A_{41}) = -(-\lambda - \sigma_{15}) \cdot \det(A_{411})$$
$$+ \sigma_{10} \cdot e^{-\lambda \cdot \tau} \cdot \det(A_{412}) - \sigma_{10} \cdot \det(A_{413}) + \sigma_{11} \cdot \det(A_{414})$$

Matrix A_{411} (5 × 5) determinant: matrix elements by rows and columns (Table 1.66).

$$\det(A_{411}) = \lambda^2 \cdot \sigma_{12} \cdot e^{-\lambda \cdot \tau} \cdot (\lambda + \sigma_{14})$$

Matrix A_{412} (5 × 5) determinant: matrix elements by rows and columns (Table 1.67).

$$\det(A_{412}) = \lambda \cdot \sigma_{13} \cdot (\lambda + \sigma_{14})$$

Table 1.66 Matrix A_{411} (5 × 5) elements by rows and columns

1	$-\lambda$	0	0	0
0	0	$-\lambda$	0	0
0	$\sigma_{12} \cdot e^{-\lambda \cdot \tau}$	σ_{12}	σ_{13}	0
0	0	0	$-\lambda$	0
0	σ_{14}	0	0	$-\lambda - \sigma_{14}$

Table 1.67 Matrix A_{412} (5 × 5) elements by rows and columns

1	0	0	0	0
0	0	$-\lambda$	0	0
0	0	σ_{12}	σ_{13}	0
0	1	0	$-\lambda$	0
0	0	0	0	$-\lambda - \sigma_{14}$

Table 1.68 Matrix A_{413} (5 × 5) elements by rows and columns

1	0	$-\lambda$	0	0
0	0	0	0	0
0	0	$\sigma_{12} \cdot e^{-\lambda \cdot \tau}$	σ_{13}	0
0	1	0	$-\lambda$	0
0	0	σ_{14}	0	$-\lambda - \sigma_{14}$

Matrix A_{413} (5 × 5) determinant: matrix elements by rows and columns (Table 1.68).

$$\det(A_{413}) = 0$$

Matrix A_{414} (5 × 5) determinant: matrix elements by rows and columns (Table 1.69).

$$\det(A_{414}) = -\lambda \cdot \sigma_{12} \cdot e^{-\lambda \cdot \tau} \cdot (\lambda + \sigma_{14})$$

Matrix A_{42} (6 × 6) determinant: matrix elements by rows and columns (Table 1.70).

$$\det(A_{42}) = (-\lambda - \sigma_{15}) \cdot \det(A_{421}) - \sigma_{10} \cdot \det(A_{422}) + \sigma_{11} \cdot \det(A_{423})$$

Matrix A_{421} (5 × 5) determinant: matrix elements by rows and columns (Table 1.71).

Table 1.69 Elements by rows and columns

1	0	$-\lambda$	0	0
0	0	0	$-\lambda$	0
0	0	$\sigma_{12} \cdot e^{-\lambda \cdot \tau}$	σ_{12}	0
0	1	0	0	0
0	0	σ_{14}	0	$-\lambda - \sigma_{14}$

Table 1.70 Matrix A_{42} (6 × 6) elements by rows and columns

0	0	$-\lambda - \sigma_{15}$	σ_{10}	σ_{11}	0
1	0	0	0	0	0
0	1	0	$-\lambda$	0	0
0	0	0	σ_{12}	σ_{13}	0
0	0	1	0	$-\lambda$	0
0	0	0	0	0	$-\lambda - \sigma_{14}$

Table 1.71 Matrix A_{421} (5 × 5) elements by rows and columns

1	0	0	0	0
0	1	$-\lambda$	0	0
0	0	σ_{12}	σ_{13}	0
0	0	0	$-\lambda$	0
0	0	0	0	$-\lambda - \sigma_{14}$

Table 1.72 Matrix A_{422} (5 × 5) elements by rows and columns

1	0	0	0	0
0	1	0	0	0
0	0	0	σ_{13}	0
0	0	1	$-\lambda$	0
0	0	0	0	$-\lambda - \sigma_{14}$

$$\det(A_{421}) = \lambda \cdot \sigma_{12} \cdot (\lambda + \sigma_{14})$$

Matrix A_{422} (5 × 5) determinant: matrix elements by rows and columns (Table 1.72).

$$\det(A_{422}) = \sigma_{13} \cdot (\lambda + \sigma_{14})$$

Matrix A_{423} (5 × 5) determinant: matrix elements by rows and columns (Table 1.73).

$$\det(A_{423}) = \sigma_{12} \cdot (\lambda + \sigma_{14})$$

Matrix A_{43} (6 × 6) determinant: matrix elements by rows and columns (Table 1.74).

$$\det(A_{43}) = (-\lambda - \sigma_{15}) \cdot \det(A_{431}) - \sigma_{10} \cdot e^{-\lambda \cdot \tau} \cdot \det(A_{432}) + \sigma_{11} \cdot \det(A_{433})$$

Matrix A_{431} (5 × 5) determinant: matrix elements by rows and columns (Table 1.75).

Table 1.73 Matrix A_{423} (5 × 5) elements by rows and columns

1	0	0	0	0
0	1	0	$-\lambda$	0
0	0	0	σ_{12}	0
0	0	1	0	0
0	0	0	0	$-\lambda - \sigma_{14}$

Table 1.74 Matrix A_{43} (6 × 6) elements by rows and columns

0	0	$-\lambda - \sigma_{15}$	$\sigma_{10} \cdot e^{-\lambda \cdot \tau}$	σ_{11}	0
1	0	0	$-\lambda$	0	0
0	1	0	0	0	0
0	0	0	$\sigma_{12} \cdot e^{-\lambda \cdot \tau}$	σ_{13}	0
0	0	1	0	$-\lambda$	0
0	0	0	σ_{14}	0	$-\lambda - \sigma_{14}$

Table 1.75 Matrix A_{431} (5 × 5) elements by rows and columns

1	0	$-\lambda$	0	0
0	1	0	0	0
0	0	$\sigma_{12} \cdot e^{-\lambda \cdot \tau}$	σ_{13}	0
0	0	0	$-\lambda$	0
0	0	σ_{14}	0	$-\lambda - \sigma_{14}$

$$\det(A_{431}) = \sigma_{12} \cdot e^{-\lambda \cdot \tau} \cdot \lambda \cdot (\lambda + \sigma_{14})$$

Matrix A_{432} (5 × 5) determinant: matrix elements by rows and columns (Table 1.76).

$$\det(A_{432}) = \sigma_{13} \cdot (\lambda + \sigma_{14})$$

Matrix A_{433} (5 × 5) determinant: matrix elements by rows and columns (Table 1.77).

Table 1.76 Matrix A_{432} (5 × 5) elements by rows and columns

1	0	0	0	0
0	1	0	0	0
0	0	0	σ_{13}	0
0	0	1	$-\lambda$	0
0	0	0	0	$-\lambda - \sigma_{14}$

Table 1.77 Matrix A_{433} (5 × 5) elements by rows and columns

1	0	0	$-\lambda$	0
0	1	0	0	0
0	0	0	$\sigma_{12} \cdot e^{-\lambda \cdot \tau}$	0
0	0	1	0	0
0	0	0	σ_{14}	$-\lambda - \sigma_{14}$

Table 1.78 Matrix A_{44} (6 × 6) elements by rows and columns

0	0	$-\lambda - \sigma_{15}$	$\sigma_{10} \cdot e^{-\lambda \cdot \tau}$	σ_{10}	0
1	0	0	$-\lambda$	0	0
0	1	0	0	$-\lambda$	0
0	0	0	$\sigma_{12} \cdot e^{-\lambda \cdot \tau}$	σ_{12}	0
0	0	1	0	0	0
0	0	0	σ_{14}	0	$-\lambda - \sigma_{14}$

$$\det(A_{433}) = \sigma_{12} \cdot e^{-\lambda \cdot \tau} \cdot (\lambda + \sigma_{14})$$

Matrix A_{44} (6 × 6) determinant: matrix elements by rows and columns (Table 1.78).

$$\det(A_{44}) = (-\lambda - \sigma_{15}) \cdot \det(A_{441}) - \sigma_{10} \cdot e^{-\lambda \cdot \tau} \cdot \det(A_{442}) + \sigma_{10} \cdot \det(A_{443})$$

Matrix A_{441} (5 × 5) determinant: matrix elements by rows and columns (Table 1.79).

$$\det(A_{441}) = 0$$

Matrix A_{442} (5 × 5) determinant: matrix elements by rows and columns (Table 1.80).

$$\det(A_{442}) = \sigma_{12} \cdot (\lambda + \sigma_{14})$$

Table 1.79 Matrix A_{441} (5 × 5) elements by rows and columns

1	0	$-\lambda$	0	0
0	1	0	$-\lambda$	0
0	0	$\sigma_{12} \cdot e^{-\lambda \cdot \tau}$	σ_{12}	0
0	0	0	0	0
0	0	σ_{14}	0	$-\lambda - \sigma_{14}$

Table 1.80 Matrix A_{442} (5 × 5) elements by rows and columns

1	0	0	0	0
0	1	0	$-\lambda$	0
0	0	0	σ_{12}	0
0	0	1	0	0
0	0	0	0	$-\lambda - \sigma_{14}$

Table 1.81 Matrix A_{443} (5 × 5) elements by rows and columns

1	0	0	$-\lambda$	0
0	1	0	0	0
0	0	0	$\sigma_{12} \cdot e^{-\lambda \cdot \tau}$	0
0	0	1	0	0
0	0	0	σ_{14}	$-\lambda - \sigma_{14}$

Matrix A_{443} (5 × 5) determinant: matrix elements by rows and columns (Table 1.81).

$$\det(A_{443}) = \sigma_{12} \cdot e^{-\lambda \cdot \tau} \cdot (\lambda + \sigma_{14})$$

Matrix A_5 (7 × 7) determinant: matrix elements by rows and columns (Table 1.82).

$$\det(A_5) = -(\sigma_7 - \lambda) \cdot \det(A_{51}) - \sigma_8 \cdot e^{-\lambda \cdot \tau} \cdot \det(A_{52}) + \sigma_8 \cdot \det(A_{53}) - \sigma_9 \cdot \det(A_{54})$$

Matrix A_{51} (6 × 6) determinant: matrix elements by rows and columns (Table 1.83).

$$\det(A_{51}) = -(-\lambda - \sigma_{15}) \cdot \det(A_{511}) + \sigma_{10} \cdot e^{-\lambda \cdot \tau} \cdot \det(A_{512}) - \sigma_{10} \cdot \det(A_{513}) + \sigma_{11} \cdot \det(A_{514})$$

Table 1.82 Matrix A_5 (7 × 7) elements by rows and columns

0	$\sigma_7 - \lambda$	0	$\sigma_8 \cdot e^{-\lambda \cdot \tau}$	σ_8	σ_9	0
0	0	$-\lambda - \sigma_{15}$	$\sigma_{10} \cdot e^{-\lambda \cdot \tau}$	σ_{10}	σ_{11}	0
1	0	0	$-\lambda$	0	0	0
0	1	0	0	$-\lambda$	0	0
0	0	0	$\sigma_{12} \cdot e^{-\lambda \cdot \tau}$	σ_{12}	$\sigma_{13} - \lambda$	0
0	0	1	0	0	0	0
0	0	0	σ_{14}	0	0	$-\lambda - \sigma_{14}$

Table 1.83 Matrix A_{51} (6 × 6) elements by rows and columns

0	$-\lambda - \sigma_{15}$	$\sigma_{10} \cdot e^{-\lambda \cdot \tau}$	σ_{10}	σ_{11}	0
1	0	$-\lambda$	0	0	0
0	0	0	$-\lambda$	0	0
0	0	$\sigma_{12} \cdot e^{-\lambda \cdot \tau}$	σ_{12}	$\sigma_{13} - \lambda$	0
0	1	0	0	0	0
0	0	σ_{14}	0	0	$-\lambda - \sigma_{14}$

Table 1.84 Matrix A_{511} (5 × 5) elements by rows and columns

1	$-\lambda$	0	0	0
0	0	$-\lambda$	0	0
0	$\sigma_{12} \cdot e^{-\lambda \cdot \tau}$	σ_{12}	$\sigma_{13} - \lambda$	0
0	0	0	0	0
0	σ_{14}	0	0	$-\lambda - \sigma_{14}$

Matrix A_{511} (5 × 5) determinant: matrix elements by rows and columns (Table 1.84).

$$\det(A_{511}) = 0$$

Matrix A_{512} (5 × 5) determinant: matrix elements by rows and columns (Table 1.85).

$$\det(A_{512}) = \lambda \cdot (\sigma_{13} - \lambda) \cdot (\lambda + \sigma_{14})$$

Matrix A_{513} (5 × 5) determinant: matrix elements by rows and columns (Table 1.86).

$$\det(A_{513}) = 0$$

Matrix A_{514} (5 × 5) determinant: matrix elements by rows and columns (Table 1.87).

$$\det(A_{514}) = -\lambda \cdot \sigma_{12} \cdot e^{-\lambda \cdot \tau} \cdot (\lambda + \sigma_{14})$$

Table 1.85 Matrix A_{512} (5 × 5) elements by rows and columns

1	0	0	0	0
0	0	$-\lambda$	0	0
0	0	σ_{12}	$\sigma_{13} - \lambda$	0
0	1	0	0	0
0	0	0	0	$-\lambda - \sigma_{14}$

Table 1.86 Matrix A_{513} (5 × 5) elements by rows and columns

1	0	$-\lambda$	0	0
0	0	0	0	0
0	0	$\sigma_{12} \cdot e^{-\lambda \cdot \tau}$	$\sigma_{13} - \lambda$	0
0	1	0	0	0
0	0	σ_{14}	0	$-\lambda - \sigma_{14}$

Table 1.87 Matrix A_{514} (5 × 5) elements by rows and columns

1	0	$-\lambda$	0	0
0	0	0	$-\lambda$	0
0	0	$\sigma_{12} \cdot e^{-\lambda \cdot \tau}$	σ_{12}	0
0	1	0	0	0
0	0	σ_{14}	0	$-\lambda - \sigma_{14}$

Table 1.88 Matrix A_{52} (6 × 6) elements by rows and columns

0	0	$-\lambda - \sigma_{15}$	σ_{10}	σ_{11}	0
1	0	0	0	0	0
0	1	0	$-\lambda$	0	0
0	0	0	σ_{12}	$\sigma_{13} - \lambda$	0
0	0	1	0	0	0
0	0	0	0	0	$-\lambda - \sigma_{14}$

Matrix A_{52} (6 × 6) determinant: matrix elements by rows and columns (Table 1.88).

$$\det(A_{52}) = (-\lambda - \sigma_{15}) \cdot \det(A_{521}) - \sigma_{10} \cdot \det(A_{522}) + \sigma_{11} \cdot \det(A_{523})$$

Matrix A_{521} (5 × 5) determinant: matrix elements by rows and columns (Table 1.89).

$$\det(A_{521}) = 0$$

Matrix A_{522} (5 × 5) determinant: matrix elements by rows and columns (Table 1.90).

$$\det(A_{522}) = (\sigma_{13} - \lambda) \cdot (\lambda + \sigma_{14})$$

Matrix A_{523} (5 × 5) determinant: matrix elements by rows and columns (Table 1.91).

Table 1.89 Matrix A_{521} (5 × 5) elements by rows and columns

1	0	0	0	0
0	1	$-\lambda$	0	0
0	0	σ_{12}	$\sigma_{13} - \lambda$	0
0	0	0	0	0
0	0	0	0	$-\lambda - \sigma_{14}$

Table 1.90 Matrix A_{522} (5 × 5) elements by rows and columns

1	0	0	0	0
0	1	0	0	0
0	0	0	$\sigma_{13} - \lambda$	0
0	0	1	0	0
0	0	0	0	$-\lambda - \sigma_{14}$

Table 1.91 Matrix A_{523} (5 × 5) elements by rows and columns

1	0	0	0	0
0	1	0	$-\lambda$	0
0	0	0	σ_{12}	0
0	0	1	0	0
0	0	0	0	$-\lambda - \sigma_{14}$

$$\det(A_{523}) = \sigma_{12} \cdot (\lambda + \sigma_{14})$$

Matrix A_{53} (6 × 6) determinant: matrix elements by rows and columns (Table 1.92).

$$\det(A_{53}) = (-\lambda - \sigma_{15}) \cdot \det(A_{531}) - \sigma_{10} \cdot e^{-\lambda \cdot \tau} \cdot \det(A_{532}) + \sigma_{11} \cdot \det(A_{533})$$

Matrix A_{531} (5 × 5) determinant: matrix elements by rows and columns (Table 1.93).

Table 1.92 Matrix A_{53} (6 × 6) elements by rows and columns

0	0	$-\lambda - \sigma_{15}$	$\sigma_{10} \cdot e^{-\lambda \cdot \tau}$	σ_{11}	0
1	0	0	$-\lambda$	0	0
0	1	0	0	0	0
0	0	0	$\sigma_{12} \cdot e^{-\lambda \cdot \tau}$	$\sigma_{13} - \lambda$	0
0	0	1	0	0	0
0	0	0	σ_{14}	0	$-\lambda - \sigma_{14}$

Table 1.93 Matrix A_{531} (5 × 5) elements by rows and columns

1	0	$-\lambda$	0	0
0	1	0	0	0
0	0	$\sigma_{12} \cdot e^{-\lambda \cdot \tau}$	$\sigma_{13} - \lambda$	0
0	0	0	0	0
0	0	σ_{14}	0	$-\lambda - \sigma_{14}$

Table 1.94 Matrix A_{532} (5 × 5) elements by rows and columns

1	0	0	0	0
0	1	0	0	0
0	0	0	$\sigma_{13} - \lambda$	0
0	0	1	0	0
0	0	0	0	$-\lambda - \sigma_{14}$

$$\det(A_{531}) = 0$$

Matrix A_{532} (5 × 5) determinant: matrix elements by rows and columns (Table 1.94).

$$\det(A_{532}) = (\sigma_{13} - \lambda) \cdot (\lambda + \sigma_{14})$$

Matrix A_{533} (5 × 5) determinant: matrix elements by rows and columns (Table 1.95).

$$\det(A_{533}) = \sigma_{12} \cdot e^{-\lambda \cdot \tau} \cdot (\lambda + \sigma_{14})$$

Matrix A_{54} (6 × 6) determinant: matrix elements by rows and columns (Table 1.96).

$$\det(A_{54}) = -(\lambda + \sigma_{15}) \cdot \det(A_{541}) - \sigma_{10} \cdot e^{-\lambda \cdot \tau} \cdot \det(A_{542}) + \sigma_{10} \cdot \det(A_{543})$$

Table 1.95 Matrix A_{533} (5 × 5) elements by rows and columns

1	0	0	$-\lambda$	0
0	1	0	0	0
0	0	0	$\sigma_{12} \cdot e^{-\lambda \cdot \tau}$	0
0	0	1	0	0
0	0	0	σ_{14}	$-\lambda - \sigma_{14}$

Table 1.96 Matrix A_{54} (6 × 6) elements by rows and columns

0	0	$-\lambda - \sigma_{15}$	$\sigma_{10} \cdot e^{-\lambda \cdot \tau}$	σ_{10}	0
1	0	0	$-\lambda$	0	0
0	1	0	0	$-\lambda$	0
0	0	0	$\sigma_{12} \cdot e^{-\lambda \cdot \tau}$	σ_{12}	0
0	0	1	0	0	0
0	0	0	σ_{14}	0	$-\lambda - \sigma_{14}$

Table 1.97 Matrix A_{541} (5 × 5) elements by rows and columns

1	0	$-\lambda$	0	0
0	1	0	$-\lambda$	0
0	0	$\sigma_{12} \cdot e^{-\lambda \cdot \tau}$	σ_{12}	0
0	0	0	0	0
0	0	σ_{14}	0	$-\lambda - \sigma_{14}$

Matrix A_{541} (5 × 5) determinant: matrix elements by rows and columns (Table 1.97).

$$\det(A_{541}) = 0$$

Matrix A_{542} (5 × 5) determinant: matrix elements by rows and columns (Table 1.98).

$$\det(A_{542}) = \sigma_{12} \cdot (\lambda + \sigma_{14})$$

Matrix A_{543} (5 × 5) determinant: matrix elements by rows and columns (Table 1.99).

$$\det(A_{543}) = \sigma_{12} \cdot e^{-\lambda \cdot \tau} \cdot (\lambda + \sigma_{14})$$

Matrix A_6 (7 × 7) determinant: matrix elements by rows and columns (Table 1.100).

$$\det(A_6) = -(\sigma_7 - \lambda) \cdot \det(A_{61}) - \sigma_8 \cdot e^{-\lambda \cdot \tau} \cdot \det(A_{62})$$

Table 1.98 Matrix A_{542} (5 × 5) elements by rows and columns

1	0	0	0	0
0	1	0	$-\lambda$	0
0	0	0	σ_{12}	0
0	0	1	0	0
0	0	0	0	$-\lambda - \sigma_{14}$

Table 1.99 Matrix A_{543} (5 × 5) elements by rows and columns

1	0	0	$-\lambda$	0
0	1	0	0	0
0	0	0	$\sigma_{12} \cdot e^{-\lambda \cdot \tau}$	0
0	0	1	0	0
0	0	0	σ_{14}	$-\lambda - \sigma_{14}$

Table 1.100 Matrix A_6 (7 × 7) elements by rows and columns

0	$\sigma_7 - \lambda$	0	$\sigma_8 \cdot e^{-\lambda \cdot \tau}$	σ_8	σ_9	σ_9
0	0	$-\lambda - \sigma_{15}$	$\sigma_{10} \cdot e^{-\lambda \cdot \tau}$	σ_{10}	σ_{11}	σ_{11}
1	0	0	$-\lambda$	0	0	0
0	1	0	0	$-\lambda$	0	0
0	0	0	$\sigma_{12} \cdot e^{-\lambda \cdot \tau}$	σ_{12}	$\sigma_{13} - \lambda$	σ_{13}
0	0	1	0	0	0	$-\lambda$
0	0	0	σ_{14}	0	0	0

$$+ \sigma_8 \cdot \det(A_{63}) - \sigma_9 \cdot \det(A_{64}) + \sigma_9 \cdot \det(A_{65})$$

Matrix A_{61} (6 × 6) determinant: matrix elements by rows and columns (Table 1.101).

$$\det(A_{61}) = -(-\lambda - \sigma_{15}) \cdot \det(A_{611}) + \sigma_{10} \cdot e^{-\lambda \cdot \tau} \cdot \det(A_{612}) - \sigma_{10} \cdot \det(A_{613}) + \sigma_{11} \cdot \det(A_{614}) - \sigma_{11} \cdot \det(A_{615})$$

Matrix A_{611} (5 × 5) determinant: matrix elements by rows and columns (Table 1.102).

$$\det(A_{611}) = -\lambda^2 \cdot \sigma_{14} \cdot (\sigma_{13} - \lambda)$$

Matrix A_{612} (5 × 5) determinant: matrix elements by rows and columns (Table 1.103).

Table 1.101 Matrix A_{61} (6 × 6) elements by rows and columns

0	$-\lambda - \sigma_{15}$	$\sigma_{10} \cdot e^{-\lambda \cdot \tau}$	σ_{10}	σ_{11}	σ_{11}
1	0	$-\lambda$	0	0	0
0	0	0	$-\lambda$	0	0
0	0	$\sigma_{12} \cdot e^{-\lambda \cdot \tau}$	σ_{12}	$\sigma_{13} - \lambda$	σ_{13}
0	1	0	0	0	$-\lambda$
0	0	σ_{14}	0	0	0

Table 1.102 Matrix A_{611} (5 × 5) elements by rows and columns

1	$-\lambda$	0	0	0
0	0	$-\lambda$	0	0
0	$\sigma_{12} \cdot e^{-\lambda \cdot \tau}$	σ_{12}	$\sigma_{13} - \lambda$	σ_{13}
0	0	0	0	$-\lambda$
0	σ_{14}	0	0	0

Table 1.103 Matrix A_{612} (5 × 5) elements by rows and columns

1	0	0	0	0
0	0	$-\lambda$	0	0
0	0	σ_{12}	$\sigma_{13} - \lambda$	σ_{13}
0	1	0	0	$-\lambda$
0	0	0	0	0

Table 1.104 Matrix A_{613} (5 × 5) elements by rows and columns

1	0	$-\lambda$	0	0
0	0	0	0	0
0	0	$\sigma_{12} \cdot e^{-\lambda \cdot \tau}$	$\sigma_{13} - \lambda$	σ_{13}
0	1	0	0	$-\lambda$
0	0	σ_{14}	0	0

$$\det(A_{612}) = 0$$

Matrix A_{613} (5 × 5) determinant: matrix elements by rows and columns (Table 1.104).

$$\det(A_{613}) = 0$$

Matrix A_{614} (5 × 5) determinant: matrix elements by rows and columns (Table 1.105).

$$\det(A_{614}) = -\lambda \cdot \sigma_{14} \cdot \sigma_{13}$$

Matrix A_{615} (5 × 5) determinant: matrix elements by rows and columns (Table 1.106).

$$\det(A_{615}) = -\lambda \cdot \sigma_{14} \cdot (\sigma_{13} - \lambda)$$

Matrix A_{62} (6 × 6) determinant: matrix elements by rows and columns (Table 1.107).

Table 1.105 Matrix A_{614} (5 × 5) elements by rows and columns

1	0	$-\lambda$	0	0
0	0	0	$-\lambda$	0
0	0	$\sigma_{12} \cdot e^{-\lambda \cdot \tau}$	σ_{12}	σ_{13}
0	1	0	0	$-\lambda$
0	0	σ_{14}	0	0

Table 1.106 Matrix A_{615} (5 × 5) elements by rows and columns

1	0	$-\lambda$	0	0
0	0	0	$-\lambda$	0
0	0	$\sigma_{12} \cdot e^{-\lambda \cdot \tau}$	σ_{12}	$\sigma_{13} - \lambda$
0	1	0	0	0
0	0	σ_{14}	0	0

Table 1.107 Matrix A_{62} (6 × 6) elements by rows and columns

0	0	$-\lambda - \sigma_{15}$	σ_{10}	σ_{11}	σ_{11}
1	0	0	0	0	0
0	1	0	$-\lambda$	0	0
0	0	0	σ_{12}	$\sigma_{13} - \lambda$	σ_{13}
0	0	1	0	0	$-\lambda$
0	0	0	0	0	0

$$\det(A_{62}) = 0$$

Matrix A_{63} (6 × 6) determinant: matrix elements by rows and columns (Table 1.108).

$$\det(A_{63}) = (-\lambda - \sigma_{15}) \cdot \det(A_{631}) - \sigma_{10} \cdot e^{-\lambda \cdot \tau} \cdot \det(A_{632}) + \sigma_{11} \cdot \det(A_{633}) - \sigma_{11} \cdot \det(A_{634})$$

Matrix A_{631} (5 × 5) determinant: matrix elements by rows and columns (Table 1.109).

$$\det(A_{631}) = -\lambda \cdot \sigma_{14} \cdot (\sigma_{13} - \lambda)$$

Matrix A_{632} (5 × 5) determinant: matrix elements by rows and columns (Table 1.110)

Table 1.108 Matrix A_{63} (6 × 6) elements by rows and columns

0	0	$-\lambda - \sigma_{15}$	$\sigma_{10} \cdot e^{-\lambda \cdot \tau}$	σ_{11}	σ_{11}
1	0	0	$-\lambda$	0	0
0	1	0	0	0	0
0	0	0	$\sigma_{12} \cdot e^{-\lambda \cdot \tau}$	$\sigma_{13} - \lambda$	σ_{13}
0	0	1	0	0	$-\lambda$
0	0	0	σ_{14}	0	0

Table 1.109 Matrix A_{631} (5 × 5) elements by rows and columns

1	0	$-\lambda$	0	0
0	1	0	0	0
0	0	$\sigma_{12} \cdot e^{-\lambda \cdot \tau}$	$\sigma_{13} - \lambda$	σ_{13}
0	0	0	0	$-\lambda$
0	0	σ_{14}	0	0

Table 1.110 Matrix A_{632} (5 × 5) elements by rows and columns

1	0	0	0	0
0	1	0	0	0
0	0	0	$\sigma_{13} - \lambda$	σ_{13}
0	0	1	0	$-\lambda$
0	0	0	0	0

$$\det(A_{632}) = 0$$

Matrix A_{633} (5 × 5) determinant: matrix elements by rows and columns (Table 1.111).

$$\det(A_{633}) = \sigma_{13} \cdot \sigma_{14}$$

Matrix A_{634} (5 × 5) determinant: matrix elements by rows and columns (Table 1.112)

Table 1.111 Matrix A_{633} (5 × 5) elements by rows and columns

1	0	0	$-\lambda$	0
0	1	0	0	0
0	0	0	$\sigma_{12} \cdot e^{-\lambda \cdot \tau}$	σ_{13}
0	0	1	0	$-\lambda$
0	0	0	σ_{14}	0

Table 1.112 Matrix A_{634} (5 × 5) elements by rows and columns

1	0	0	$-\lambda$	0
0	1	0	0	0
0	0	0	$\sigma_{12} \cdot e^{-\lambda \cdot \tau}$	$\sigma_{13} - \lambda$
0	0	1	0	0
0	0	0	σ_{14}	0

Table 1.113 Matrix A_{64} (6 × 6) elements by rows and columns

0	0	$-\lambda-\sigma_{15}$	$\sigma_{10}\cdot e^{-\lambda\cdot\tau}$	σ_{10}	σ_{11}
1	0	0	$-\lambda$	0	0
0	1	0	0	$-\lambda$	0
0	0	0	$\sigma_{12}\cdot e^{-\lambda\cdot\tau}$	σ_{12}	σ_{13}
0	0	1	0	0	$-\lambda$
0	0	0	σ_{14}	0	0

$$\det(A_{634}) = \sigma_{14}\cdot(\sigma_{13}-\lambda)$$

Matrix A_{64} (6 × 6) determinant: matrix elements by rows and columns (Table 1.113).

$$\det(A_{64}) = (-\lambda-\sigma_{15})\cdot\det(A_{641}) - \sigma_{10}\cdot e^{-\lambda\cdot\tau}\cdot\det(A_{642}) + \sigma_{10}\cdot\det(A_{643}) - \sigma_{11}\cdot\det(A_{644})$$

Matrix A_{641} (5 × 5) determinant: matrix elements by rows and columns (Table 1.114)

$$\det(A_{641}) = -\lambda\cdot\sigma_{14}\cdot\sigma_{12}$$

Matrix A_{642} (5 × 5) determinant: matrix elements by rows and columns (Table 1.115).

$$\det(A_{642}) = 0$$

Table 1.114 Matrix A_{641} (5 × 5) elements by rows and columns

1	0	$-\lambda$	0	0
0	1	0	$-\lambda$	0
0	0	$\sigma_{12}\cdot e^{-\lambda\cdot\tau}$	σ_{12}	σ_{13}
0	0	0	0	$-\lambda$
0	0	σ_{14}	0	0

Table 1.115 Matrix A_{642} (5 × 5) elements by rows and columns

1	0	0	0	0
0	1	0	$-\lambda$	0
0	0	0	σ_{12}	σ_{13}
0	0	1	0	$-\lambda$
0	0	0	0	0

Table 1.116 Matrix A_{643} (5 × 5) elements by rows and columns

1	0	0	$-\lambda$	0
0	1	0	0	0
0	0	0	$\sigma_{12} \cdot e^{-\lambda \cdot \tau}$	σ_{13}
0	0	1	0	$-\lambda$
0	0	0	σ_{14}	0

Matrix A_{643} (5 × 5) determinant: matrix elements by rows and columns (Table 1.116).

$$\det(A_{643}) = \sigma_{13} \cdot \sigma_{14}$$

Matrix A_{644} (5 × 5) determinant: matrix elements by rows and columns (Table 1.117).

$$\det(A_{644}) = \sigma_{12} \cdot \sigma_{14}$$

Matrix A_{65} (6 × 6) determinant: matrix elements by rows and columns (Table 1.118)

$$\det(A_{65}) = (-\lambda - \sigma_{15}) \cdot \det(A_{651}) - \sigma_{10} \cdot e^{-\lambda \cdot \tau} \cdot \det(A_{652}) + \sigma_{10} \cdot \det(A_{653}) - \sigma_{11} \cdot \det(A_{654})$$

Table 1.117 Matrix A_{644} (5 × 5) elements by rows and columns

1	0	0	$-\lambda$	0
0	1	0	0	$-\lambda$
0	0	0	$\sigma_{12} \cdot e^{-\lambda \cdot \tau}$	σ_{12}
0	0	1	0	0
0	0	0	σ_{14}	0

Table 1.118 Matrix A_{65} (6 × 6) elements by rows and columns

0	0	$-\lambda - \sigma_{15}$	$\sigma_{10} \cdot e^{-\lambda \cdot \tau}$	σ_{10}	σ_{11}
1	0	0	$-\lambda$	0	0
0	1	0	0	$-\lambda$	0
0	0	0	$\sigma_{12} \cdot e^{-\lambda \cdot \tau}$	σ_{12}	$\sigma_{13} - \lambda$
0	0	1	0	0	0
0	0	0	σ_{14}	0	0

Table 1.119 Matrix A_{651} (5 × 5) elements by rows and columns

1	0	$-\lambda$	0	0
0	1	0	$-\lambda$	0
0	0	$\sigma_{12} \cdot e^{-\lambda \cdot \tau}$	σ_{12}	$\sigma_{13} - \lambda$
0	0	0	0	0
0	0	σ_{14}	0	0

Table 1.120 Matrix A_{652} (5 × 5) elements by rows and columns

1	0	0	0	0
0	1	0	$-\lambda$	0
0	0	0	σ_{12}	$\sigma_{13} - \lambda$
0	0	1	0	0
0	0	0	0	0

Matrix A_{651} (5 × 5) determinant: matrix elements by rows and columns (Table 1.119).

$$\det(A_{651}) = 0$$

Matrix A_{652} (5 × 5) determinant: matrix elements by rows and columns (Table 1.120).

$$\det(A_{652}) = 0$$

Matrix A_{653} (5 × 5) determinant: matrix elements by rows and columns (Table 1.121).

$$\det(A_{653}) = \sigma_{44} \cdot (\sigma_{13} - \lambda)$$

Matrix A_{654} (5 × 5) determinant: matrix elements by rows and columns (Table 1.122).

$$\det(A_{654}) = \sigma_{12} \cdot \sigma_{14}$$

Table 1.121 Matrix A_{653} (5 × 5) elements by rows and columns

1	0	0	$-\lambda$	0
0	1	0	0	0
0	0	0	$\sigma_{12} \cdot e^{-\lambda \cdot \tau}$	$\sigma_{13} - \lambda$
0	0	1	0	0
0	0	0	σ_{14}	0

Table 1.122 Matrix A_{654} (5 × 5) elements by rows and columns

1	0	0	$-\lambda$	0
0	1	0	0	$-\lambda$
0	0	0	$\sigma_{12} \cdot e^{-\lambda \cdot \tau}$	σ_{12}
0	0	1	0	0
0	0	0	σ_{14}	0

The determinants and sub determinants structure is as follow (Table 1.123).

$$\det(A_{11}) = \lambda^5 - \lambda^4 \cdot (\sigma_{13} - \sigma_{15} + \sigma_{11} - \sigma_{14}) - \lambda^3 \cdot [\sigma_{15} \cdot \sigma_{13} + (\sigma_{13} - \sigma_{15} + \sigma_{11}) \cdot \sigma_{14}] - \lambda^2 \cdot \sigma_{14} \cdot \sigma_{15} \cdot \sigma_{13}$$

$$\Gamma_0 = 1; \ \Gamma_1 = -(\sigma_{13} - \sigma_{15} + \sigma_{11} - \sigma_{14}); \ \Gamma_2 = -[\sigma_{15} \cdot \sigma_{13} + (\sigma_{13} - \sigma_{15} + \sigma_{11}) \cdot \sigma_{14}]; \ \Gamma_3 = -\sigma_{14} \cdot \sigma_{15} \cdot \sigma_{13}$$

$$\det(A_{11}) = \sum_{k=2}^{5} \lambda^k \cdot \Gamma_{(5-k)} = \lambda^5 \cdot \Gamma_0 + \lambda^4 \cdot \Gamma_1 + \lambda^3 \cdot \Gamma_2 + \lambda^2 \cdot \Gamma_3$$

$$\det(A_{13}) = -\lambda^5 + \lambda^4 \cdot (\sigma_{13} - \sigma_{15} - \sigma_{14}) + \lambda^3 \cdot [\sigma_{15} \cdot \sigma_{13} + \sigma_{11} + (\sigma_{13} - \sigma_{15}) \cdot \sigma_{14}] + \lambda^2 \cdot (\sigma_{15} \cdot \sigma_{13} + \sigma_{11}) \cdot \sigma_{14}$$

$$\Gamma_4 = -1; \ \Gamma_5 = \sigma_{13} - \sigma_{15} - \sigma_{14}; \ \Gamma_6 = \sigma_{15} \cdot \sigma_{13} + \sigma_{11} + (\sigma_{13} - \sigma_{15}) \cdot \sigma_{14}; \ \Gamma_7 = (\sigma_{15} \cdot \sigma_{13} + \sigma_{11}) \cdot \sigma_{14}$$

$$\det(A_{13}) = \sum_{k=2}^{5} \lambda^k \cdot \Gamma_{(9-k)} = \lambda^2 \cdot \Gamma_7 + \lambda^3 \cdot \Gamma_6 + \lambda^4 \cdot \Gamma_5 + \lambda^5 \cdot \Gamma_4$$

$$\det(A_{14}) = \lambda^4 \cdot \sigma_{12} + \lambda^3 \cdot (\sigma_{12} \cdot \sigma_{15} + \sigma_{12} \cdot \sigma_{14}) + \lambda^2 \cdot (\sigma_{10} \cdot \sigma_{13} - \sigma_{11} \cdot \sigma_{12} + \sigma_{12} \cdot \sigma_{15} \cdot \sigma_{14}) + \lambda \cdot (\sigma_{10} \cdot \sigma_{13} - \sigma_{11} \cdot \sigma_{12}) \cdot \sigma_{14}$$

$$\Gamma_8 = \sigma_{12}; \ \Gamma_9 = \sigma_{12} \cdot \sigma_{15} + \sigma_{12} \cdot \sigma_{14}; \ \Gamma_{10} = \sigma_{10} \cdot \sigma_{13} - \sigma_{11} \cdot \sigma_{12} + \sigma_{12} \cdot \sigma_{15} \cdot \sigma_{14}$$
$$\Gamma_{11} = (\sigma_{10} \cdot \sigma_{13} - \sigma_{11} \cdot \sigma_{12}) \cdot \sigma_{14}$$

$$\det(A_{14}) = \sum_{k=1}^{4} \lambda^k \cdot \Gamma_{(12-k)} = \lambda^4 \cdot \Gamma_8 + \lambda^3 \cdot \Gamma_9 + \lambda^2 \cdot \Gamma_{10} + \lambda \cdot \Gamma_{11}$$

Table 1.123 Determinants and sub determinants structure

$\det(A-\lambda\cdot I)=-\lambda\cdot\det(A_1)-\sigma_1\cdot(\sigma_2+\sigma_3\cdot e^{-\lambda\cdot\tau})\cdot\det(A_2)+\sigma_4\cdot\det(A_3)$ $-\sigma_5\cdot\det(A_4)+\sigma_5\cdot\det(A_5)-\sigma_6\cdot\det(A_6)$				
$\det(A_1)=(\sigma_7-\lambda)\cdot\det(A_{11})+\sigma_8\cdot e^{-\lambda\cdot\tau}\cdot\det(A_{12})-\sigma_8\cdot\det(A_{13})+\sigma_9\cdot\det(A_{14})-\sigma_9\cdot\det(A_{15})$				
$\det(A_{11})$ $=(-\lambda-\sigma_{15})\cdot\det(A_{111})$ $-\sigma_{10}\cdot e^{-\lambda\cdot\tau}\cdot\det(A_{112})$ $+\sigma_{10}\cdot\det(A_{113})$ $-\sigma_{11}\cdot\det(A_{114})$ $+\sigma_{11}\cdot\det(A_{115})$	$\det(A_{12})=0$	$\det(A_{13})=-(-\lambda$ $-\sigma_{15})\cdot\det(A_{131})$ $+\sigma_{10}\cdot e^{-\lambda\cdot\tau}\cdot\det(A_{132})$ $-\sigma_{11}\cdot\det(A_{133})$ $+\sigma_{11}\cdot\det(A_{134})$	$\det(A_{14})=-(-\lambda-\sigma_{15})$ $\cdot\det(A_{141})$ $+\sigma_{10}\cdot e^{-\lambda\cdot\tau}\cdot\det(A_{142})$ $-\sigma_{10}\cdot\det(A_{143})$ $+\sigma_{11}\cdot\det(A_{144})$	$\det(A_{15})=-(-\lambda$ $-\sigma_{15})\cdot\det(A_{151})$ $+\sigma_{10}\cdot e^{-\lambda\cdot\tau}\cdot\det(A_{152})$ $-\sigma_{10}\cdot\det(A_{153})$ $+\sigma_{11}\cdot\det(A_{154})$
$\det(A_{111})=\lambda^4\cdot(\sigma_{13}$ $-\lambda)\cdot(\lambda+\sigma_{14})$ $\det(A_{112})=0$ $\det(A_{113})=0$ $\det(A_{114})=\lambda^2$ $\cdot\sigma_{13}\cdot(\lambda+\sigma_{14})$ $\det(A_{115})=\lambda^2\cdot(\sigma_{13}$ $-\lambda)\cdot(\lambda+\sigma_{14})$		$\det(A_{131})=\lambda^2\cdot(\sigma_{13}$ $-\lambda)\cdot(\lambda+\sigma_{14})$ $\det(A_{132})=$ 0 $\det(A_{133})=-\lambda$ $\cdot\sigma_{13}\cdot(\lambda+\sigma_{14})$ $\det(A_{134})=-\lambda\cdot(\sigma_{13}$ $-\lambda)\cdot(\lambda+\sigma_{14})$	$\det(A_{141})=\lambda^2$ $\cdot\sigma_{12}\cdot(\lambda+\sigma_{14})$ $\det(A_{142})=0$ $\det(A_{143})=-\lambda$ $\cdot\sigma_{13}\cdot(\lambda+\sigma_{14})$ $\det(A_{144})=-\lambda$ $\cdot\sigma_{12}\cdot(\lambda+\sigma_{14})$	$\det(A_{151})=0$ $\det(A_{152})=0$ $\det(A_{153})=-\lambda\cdot(\sigma_{13}$ $-\lambda)\cdot(\lambda+\sigma_{14})$ $\det(A_{154})=-\lambda$ $\cdot\sigma_{12}\cdot(\lambda+\sigma_{14})$

$$\det(A_{15}) = -\lambda^3 \cdot \sigma_{10} + \lambda^2 \cdot [\sigma_{10} \cdot (\sigma_{13} - \sigma_{14}) - \sigma_{12} \cdot \sigma_{11}] + \lambda \cdot (\sigma_{10} \cdot \sigma_{13} \cdot \sigma_{14} - \sigma_{11} \cdot \sigma_{12} \cdot \sigma_{14})$$

$$\Gamma_{12} = -\sigma_{10};\ \Gamma_{13} = \sigma_{10} \cdot (\sigma_{13} - \sigma_{14}) - \sigma_{12} \cdot \sigma_{11};\ \Gamma_{14} = \sigma_{10} \cdot \sigma_{13} \cdot \sigma_{14} - \sigma_{11} \cdot \sigma_{12} \cdot \sigma_{14}$$

$$\det(A_{15}) = \sum_{k=1}^{3} \lambda^k \cdot \Gamma_{(15-k)} = \lambda \cdot \Gamma_{14} + \lambda^2 \cdot \Gamma_{13} + \lambda^3 \cdot \Gamma_{12}$$

$$\det(A_1) = (\sigma_7 - \lambda) \cdot \det(A_{11}) + \sigma_8 \cdot e^{-\lambda \cdot \tau} \cdot \det(A_{12}) - \sigma_8 \cdot \det(A_{13}) + \sigma_9 \cdot \det(A_{14}) - \sigma_9 \cdot \det(A_{15})$$

$$\det(A_1)|_{\det(A_{12})=0} = (\sigma_7 - \lambda) \cdot \det(A_{11}) - \sigma_8 \cdot \det(A_{13}) + \sigma_9 \cdot \det(A_{14}) - \sigma_9 \cdot \det(A_{15})$$

$$\begin{aligned}\det(A_1)|_{\det(A_{12})=0} = {} & (\sigma_7 - \lambda) \cdot \left(\lambda^5 \cdot \Gamma_0 + \lambda^4 \cdot \Gamma_1 + \lambda^3 \cdot \Gamma_2 + \lambda^2 \cdot \Gamma_3\right) \\ & - \sigma_8 \cdot \left(\lambda^2 \cdot \Gamma_7 + \lambda^3 \cdot \Gamma_6 + \lambda^4 \cdot \Gamma_5 + \lambda^5 \cdot \Gamma_4\right) \\ & + \sigma_9 \cdot \left(\lambda^4 \cdot \Gamma_8 + \lambda^3 \cdot \Gamma_9 + \lambda^2 \cdot \Gamma_{10} + \lambda \cdot \Gamma_{11}\right) \\ & - \sigma_9 \cdot \left(\lambda \cdot \Gamma_{14} + \lambda^2 \cdot \Gamma_{13} + \lambda^3 \cdot \Gamma_{12}\right)\end{aligned}$$

We define for simplicity the following sub determinants products:

$$\det(A_1) = .\det(A_1)|_{\det(A_{12})=0} = \det(A_1)^{(1)} - \det(A_1)^{(2)} + \det(A_1)^{(3)} - \det(A_1)^{(4)}$$

$$\det(A_1)^{(1)} = -\lambda^6 \cdot \Gamma_0 + \lambda^5 \cdot (\sigma_7 \cdot \Gamma_0 - \Gamma_1) + \lambda^4 \cdot (\sigma_7 \cdot \Gamma_1 - \Gamma_2) + \lambda^3 \cdot (\sigma_7 \cdot \Gamma_2 - \Gamma_3) + \lambda^2 \cdot \sigma_7 \cdot \Gamma_3$$

$$\det(A_1)^{(2)} = \lambda^2 \cdot \sigma_8 \cdot \Gamma_7 + \lambda^3 \cdot \sigma_8 \cdot \Gamma_6 + \lambda^4 \cdot \sigma_8 \cdot \Gamma_5 + \lambda^5 \cdot \sigma_8 \cdot \Gamma_4$$

$$\det(A_1)^{(3)} = \lambda^4 \cdot \sigma_9 \cdot \Gamma_8 + \lambda^3 \cdot \sigma_9 \cdot \Gamma_9 + \lambda^2 \cdot \sigma_9 \cdot \Gamma_{10} + \lambda \cdot \sigma_9 \cdot \Gamma_{11}$$

$$\det(A_1)^{(4)} = \lambda \cdot \sigma_9 \cdot \Gamma_{14} + \lambda^2 \cdot \sigma_9 \cdot \Gamma_{13} + \lambda^3 \cdot \sigma_9 \cdot \Gamma_{12}$$

$$\begin{aligned}\det(A_1) = {} & \det(A_1)|_{\det(A_2)=0} = -\lambda^6 \cdot \Gamma_0 + \lambda^5 \cdot (\sigma_7 \cdot \Gamma_0 - \Gamma_1 - \sigma_8 \cdot \Gamma_4) \\ & + \lambda^4 \cdot (\sigma_7 \cdot \Gamma_1 - \Gamma_2 + \sigma_9 \cdot \Gamma_8 - \sigma_8 \cdot \Gamma_5) \\ & + \lambda^3 \cdot (\sigma_7 \cdot \Gamma_2 - \Gamma_3 - \sigma_8 \cdot \Gamma_6 + \sigma_9 \cdot \Gamma_9 - \sigma_9 \cdot \Gamma_{12})\end{aligned}$$

$$+\lambda^2 \cdot (\sigma_7 \cdot \Gamma_3 - \sigma_8 \cdot \Gamma_7 + \sigma_9 \cdot \Gamma_{10} - \sigma_9 \cdot \Gamma_{13}) + \lambda \cdot (\sigma_9 \cdot \Gamma_{11} - \sigma_9 \cdot \Gamma_{14})$$

$$\Omega_6 = -\Gamma_0;\ \Omega_5 = \sigma_7 \cdot \Gamma_0 - \Gamma_1 - \sigma_8 \cdot \Gamma_4;\ \Omega_4 = \sigma_7 \cdot \Gamma_1 - \Gamma_2 + \sigma_9 \cdot \Gamma_8 - \sigma_8 \cdot \Gamma_5$$

$$\Omega_3 = \sigma_7 \cdot \Gamma_2 - \Gamma_3 - \sigma_8 \cdot \Gamma_6 + \sigma_9 \cdot \Gamma_9 - \sigma_9 \cdot \Gamma_{12};\ \Omega_2 = \sigma_7 \cdot \Gamma_3 - \sigma_8 \cdot \Gamma_7 + \sigma_9 \cdot \Gamma_{10} - \sigma_9 \cdot \Gamma_{13}$$

$$\Omega_1 = \sigma_9 \cdot \Gamma_{11} - \sigma_9 \cdot \Gamma_{14}$$

$$\det(A_1) = \det(A_1)|_{\det(A_2)=0} = \sum_{k=1}^{6} \lambda^k \cdot \Omega_t = \lambda^6 \cdot \Omega_6 + \lambda^5 \cdot \Omega_5 + \lambda^4 \cdot \Omega_4 + \lambda^3 \cdot \Omega_3 + \lambda^2 \cdot \Omega_2 + \lambda \cdot \Omega_1$$

Table 1.124.

$$\det(A_{21}) = \lambda^5 + \lambda^4 \cdot (\sigma_{14} - \sigma_{13} + \sigma_{15}) - \lambda^3 \cdot (\sigma_{15} \cdot \sigma_{13} + \sigma_{11} + \sigma_{14} \cdot (\sigma_{13} - \sigma_{15})) - \lambda^2 \cdot \sigma_{14} \cdot (\sigma_{15} \cdot \sigma_{13} + \sigma_{11})$$

$$\det(A_{21}) = \sum_{k=2}^{5} \lambda^k \cdot T_k = \lambda^2 \cdot T_2 + \lambda^3 \cdot T_3 + \lambda^4 \cdot T_4 + \lambda^5 \cdot T_5;\ T_2 = -\sigma_{14} \cdot (\sigma_{15} \cdot \sigma_{13} + \sigma_{11})$$

$$T_3 = -(\sigma_{15} \cdot \sigma_{13} + \sigma_{11} + \sigma_{14} \cdot (\sigma_{13} - \sigma_{15}))$$
$$T_4 = \sigma_{14} - \sigma_{13} + \sigma_{15};\ T_5 = 1$$

$$\det(A_{22}) = \lambda^4 + \lambda^3 \cdot (\sigma_{15} - \sigma_{13} + \sigma_{14}) + \lambda^2 \cdot (\sigma_{14} \cdot (\sigma_{15} - \sigma_{13}) - \sigma_{15} \cdot \sigma_{13} - \sigma_{11}) - \lambda \cdot \sigma_{14} \cdot (\sigma_{15} \cdot \sigma_{13} + \sigma_{11})$$

$$\det(A_{22}) = \sum_{k=1}^{4} \lambda^k \cdot \mathrm{H}_k = \lambda^4 \cdot \mathrm{H}_4 + \lambda^3 \cdot \mathrm{H}_3 + \lambda^2 \cdot \mathrm{H}_2 + \lambda \cdot \mathrm{H}_1$$

$$\mathrm{H}_4 = 1;\ \mathrm{H}_3 = \sigma_{15} - \sigma_{13} + \sigma_{14};\ \mathrm{H}_2 = \sigma_{14} \cdot (\sigma_{15} - \sigma_{13}) - \sigma_{15} \cdot \sigma_{13} - \sigma_{11};\ \mathrm{H}_1 = -\sigma_{14} \cdot (\sigma_{15} \cdot \sigma_{13} + \sigma_{11})$$

$$\det(A_{23}) = -\lambda^3 \cdot \sigma_{12} - \lambda^2 \cdot \sigma_{12} \cdot (\sigma_{15} + \sigma_{14}) + \lambda \cdot (\sigma_{11} \cdot \sigma_{12} - \sigma_{10} \cdot \sigma_{13} - \sigma_{12} \cdot \sigma_{15} \cdot \sigma_{14}) + (\sigma_{11} \cdot \sigma_{12} - \sigma_{10} \cdot \sigma_{13}) \cdot \sigma_{14}$$

Table 1.124 Determinants and sub determinants structure

$\det(A-\lambda\cdot I)=-\lambda\cdot\det(A_1)-\sigma_1\cdot(\sigma_2+\sigma_3\cdot e^{-\lambda\cdot\tau})\cdot\det(A_2)+\sigma_4\cdot\det(A_3)$ $-\sigma_5\cdot\det(A_4)+\sigma_5\cdot\det(A_5)-\sigma_6\cdot\det(A_6)$			
$\det(A_2)=-(\sigma_7-\lambda)\cdot\det(A_{21})-\sigma_8\cdot\det(A_{22})+\sigma_9\cdot\det(A_{23})-\sigma_9\cdot\det(A_{24})$			
$\det(A_{21})=(\lambda$ $+\sigma_{15})\cdot\det(A_{211})$ $+\sigma_{10}\cdot\det(A_{212})$ $-\sigma_{11}\cdot\det(A_{213})$ $+\sigma_{11}\cdot\det(A_{214})$	$\det(A_{22})=(-\lambda$ $-\sigma_{15})\cdot\det(A_{221})$ $-\sigma_{11}\cdot\det(A_{222})$ $+\sigma_{11}\cdot\det(A_{223})$	$\det(A_{23})=(-\lambda$ $-\sigma_{15})\cdot\det(A_{231})$ $-\sigma_{10}\cdot\det(A_{232})$ $+\sigma_{11}\cdot\det(A_{233})$	$\det(A_{24})=(-\lambda$ $-\sigma_{15})\cdot\det(A_{241})$ $-\sigma_{10}\cdot\det(A_{242})$ $+\sigma_{11}\cdot\det(A_{243})$
$\det(A_{211})=-\lambda^2\cdot(\sigma_{13}$ $-\lambda)\cdot(\lambda+\sigma_{14})$ $\det(A_{212})=0$ $\det(A_{213})=\lambda\cdot\sigma_{13}$ $\cdot(\lambda+\sigma_{14})$ $\det(A_{214})=\lambda\cdot(\sigma_{13}-\lambda)\cdot(\lambda+\sigma_{14})$	$\det(A_{221})=\lambda\cdot(\sigma_{13}$ $-\lambda)\cdot(\lambda+\sigma_{14})$ $\det(A_{222})=\sigma_{13}\cdot(\lambda$ $+\sigma_{14})$ $\det(A_{223})=(\sigma_{13}-\lambda)\cdot(\lambda+\sigma_{14})$	$\det(A_{231})=\lambda\cdot\sigma_{12}$ $\cdot(\lambda+\sigma_{14})$ $\det(A_{232})=\sigma_{13}\cdot(\lambda+\sigma_{14})$ $\det(A_{233})=\sigma_{12}\cdot(\lambda+\sigma_{14})$	$\det(A_{241})=-\lambda\cdot(\sigma_{13}$ $-\lambda)\cdot(\lambda+\sigma_{14})$ $\det(A_{242})=(\sigma_{13}$ $-\lambda)\cdot(\lambda+\sigma_{14})$ $\det(A_{243})=\sigma_{12}\cdot(\lambda+\sigma_{14})$

$$\det(A_{23}) = \sum_{k=0}^{3} \lambda^k \cdot \mathrm{B}_k = \lambda^3 \cdot \mathrm{B}_3 + \lambda^2 \cdot \mathrm{B}_2 + \lambda \cdot \mathrm{B}_1 + \mathrm{B}_0; \; \mathrm{B}_3 = -\sigma_{12}$$

$$\mathrm{B}_2 = -\sigma_{12} \cdot (\sigma_{15} + \sigma_{14})$$

$$\mathrm{B}_1 = \sigma_{11} \cdot \sigma_{12} - \sigma_{10} \cdot \sigma_{13} - \sigma_{12} \cdot \sigma_{15} \cdot \sigma_{14}; \; \mathrm{B}_0 = (\sigma_{11} \cdot \sigma_{12} - \sigma_{10} \cdot \sigma_{13}) \cdot \sigma_{14}$$

$$\begin{aligned}\det(A_{24}) = &-\lambda^4 + \lambda^3 \cdot (\sigma_{13} - \sigma_{15} - \sigma_{14}) + \lambda^2 \cdot (\sigma_{15} \cdot \sigma_{13} + \sigma_{10} + \sigma_{14} \cdot (\sigma_{13} - \sigma_{15})) \\ &+ \lambda \cdot (\sigma_{11} \cdot \sigma_{12} - \sigma_{10} \cdot \sigma_{13} + \sigma_{14} \cdot (\sigma_{15} \cdot \sigma_{13} + \sigma_{10})) \\ &+ \sigma_{14} \cdot (\sigma_{11} \cdot \sigma_{12} - \sigma_{10} \cdot \sigma_{13})\end{aligned}$$

$$\det(A_{24}) = \sum_{k=0}^{4} \lambda^k \cdot \mathrm{A}_k = \lambda^4 \cdot \mathrm{A}_4 + \lambda^3 \cdot \mathrm{A}_3 + \lambda^2 \cdot \mathrm{A}_2 + \lambda \cdot \mathrm{A}_1 + \mathrm{A}_0; \; \mathrm{A}_4 = -1$$

$$\mathrm{A}_3 = \sigma_{13} - \sigma_{15} - \sigma_{14}; \; \mathrm{A}_2 = \sigma_{15} \cdot \sigma_{13} + \sigma_{10} + \sigma_{44} \cdot (\sigma_{13} - \sigma_{15})$$

$$\mathrm{A}_1 = \sigma_{11} \cdot \sigma_{12} - \sigma_{10} \cdot \sigma_{13} + \sigma_{14} \cdot (\sigma_{15} \cdot \sigma_{13} + \sigma_{10}); \; \mathrm{A}_0 = \sigma_{14} \cdot (\sigma_{11} \cdot \sigma_{12} - \sigma_{10} \cdot \sigma_{13})$$

$$\det(A_2) = -(\sigma_7 - \lambda) \cdot \det(A_{21}) - \sigma_8 \cdot \det(A_2) + \sigma_9 \cdot \det(A_{23}) - \sigma_9 \cdot \det(A_{24})$$

$$\begin{aligned}\det(A_2) = &-(\sigma_7 - \lambda) \cdot \left(\lambda^2 \cdot T_2 + \lambda^3 \cdot T_3 + \lambda^4 \cdot T_4 + \lambda^5 \cdot T_5\right) \\ &- \sigma_8 \cdot \left(\lambda^4 \cdot \mathrm{H}_4 + \lambda^3 \cdot \mathrm{H}_3 + \lambda^2 \cdot \mathrm{H}_2 + \lambda \cdot \mathrm{H}_1\right) \\ &+ \sigma_9 \cdot \left(\lambda^3 \cdot \mathrm{B}_3 + \lambda^2 \cdot \mathrm{B}_2 + \lambda \cdot \mathrm{B}_1 + \mathrm{B}_0\right) \\ &- \sigma_9 \cdot \left(\lambda^4 \cdot \mathrm{A}_4 + \lambda^3 \cdot \mathrm{A}_3 + \lambda^2 \cdot \mathrm{A}_2 + \lambda \cdot \mathrm{A}_1 + \mathrm{A}_0\right)\end{aligned}$$

$$\begin{aligned}\det(A_2) = \sum_{k=0}^{6} \lambda^k \cdot \Xi_k = &\lambda^6 \cdot \Xi_6 + \lambda^5 \cdot \Xi_5 + \lambda^4 \cdot \Xi_4 \\ &+ \lambda^3 \cdot \Xi_3 + \lambda^2 \cdot \Xi_2 + \lambda \cdot \Xi_1 + \Xi_0\end{aligned}$$

$$\Xi_6 = \mathrm{T}_5; \; \Xi_5 = \mathrm{T}_4 - \mathrm{T}_5 \cdot \sigma_7; \; \Xi_4 = \mathrm{T}_3 - \mathrm{T}_4 \cdot \sigma_7 - \sigma_8 \cdot \mathrm{H}_4 - \mathrm{A}_4 \cdot \sigma_9$$

$$\begin{aligned}\Xi_3 = \mathrm{T}_2 - \mathrm{T}_3 \cdot \sigma_7 - \sigma_8 \cdot \mathrm{H}_3 - \mathrm{A}_3 \cdot \sigma_9 + \mathrm{B}_3 \cdot \sigma_9; \; \Xi_2 = -\mathrm{T}_2 \cdot \sigma_7 - \sigma_8 \cdot \mathrm{H}_2 \\ + \mathrm{B}_2 \cdot \sigma_9 - \mathrm{A}_2 \cdot \sigma_9\end{aligned}$$

$$\Xi_1 = -\mathrm{H}_1 \cdot \sigma_8 + \mathrm{B}_1 \cdot \sigma_9 - \mathrm{A}_1 \cdot \sigma_9; \; \Xi_0 = \sigma_9 \cdot (\mathrm{B}_0 - \mathrm{A}_0)$$

(Table 1.125)

Table 1.125 Determinants and sub determinants structure

$\det(A - \lambda \cdot I) = -\lambda \cdot \det(A_1) - \sigma_1 \cdot (\sigma_2 + \sigma_3 \cdot e^{-\lambda \cdot \tau}) \cdot \det(A_2) + \sigma_4 \cdot \det(A_3) - \sigma_5 \cdot \det(A_4) + \sigma_5 \cdot \det(A_5) - \sigma_6 \cdot \det(A_6)$			
$\det(A_3) = -(\sigma_7 - \lambda) \cdot \det(A_{31}) - \sigma_8 \cdot e^{-\lambda \cdot \tau} \cdot \det(A_{32}) + \sigma_9 \cdot \det(A_{33}) - \sigma_9 \cdot \det(A_{34})$			
$\det(A_{31}) = 0$	$\det(A_{32}) = (-\lambda - \sigma_{15}) \cdot \det(A_{321}) - \sigma_{11} \cdot \det(A_{322}) + \sigma_{11} \cdot \det(A_{323})$	$\det(A_{33}) = (-\lambda - \sigma_{15}) \cdot \det(A_{331}) - \sigma_{10} \cdot e^{-\lambda \cdot \tau} \cdot \det(A_{332}) + \sigma_{11} \cdot \det(A_{333})$	$\det(A_{34}) = (-\lambda - \sigma_{15}) \cdot \det(A_{341}) - \sigma_{10} \cdot e^{-\lambda \cdot \tau} \cdot \det(A_{342}) + \sigma_{11} \cdot \det(A_{343})$
	$\det(A_{321}) = \lambda \cdot (\sigma_{13} - \lambda) \cdot (\lambda + \sigma_{14})$ $\det(A_{322}) = \sigma_{13} \cdot (\lambda + \sigma_{14})$ $\det(A_{323}) = (\sigma_{13} - \lambda) \cdot (\lambda + \sigma_{14})$	$\det(A_{331}) = \sigma_{12} \cdot e^{-\lambda \cdot \tau} \cdot \lambda \cdot (\lambda + \sigma_{14})$ $\det(A_{332}) = \sigma_{13} \cdot (\lambda + \sigma_{14})$ $\det(A_{333}) = \sigma_{12} \cdot e^{-\lambda \cdot \tau} \cdot (\lambda + \sigma_{14})$	$\det(A_{341}) = 0$ $\det(A_{342}) = (\sigma_{13} - \lambda) \cdot (\lambda + \sigma_{14})$ $\det(A_{343}) = \sigma_{12} \cdot e^{-\lambda \cdot \tau} \cdot (\lambda + \sigma_{14})$

$$\det(A_{31}) = 0$$

$$\det(A_{32}) = (-\lambda - \sigma_{15}) \cdot \det(A_{321}) - \sigma_{11} \cdot \det(A_{322}) + \sigma_{11} \cdot \det(A_{323})$$

$$\det(A_{32}) = (-\lambda - \sigma_{15}) \cdot \lambda \cdot (\sigma_{13} - \lambda) \cdot (\lambda + \sigma_{14}) - \sigma_{11} \cdot \sigma_{13} \cdot (\lambda + \sigma_{14}) + \sigma_{11} \cdot (\sigma_{13} - \lambda) \cdot (\lambda + \sigma_{14})$$

$$\det(A_{32}) = \lambda^4 + \lambda^3 \cdot (\sigma_{14} - \sigma_{13} + \sigma_{15}) - \lambda^2 \cdot (\sigma_{15} \cdot \sigma_{13} + \sigma_{11} + \sigma_{14} \cdot (\sigma_{13} - \sigma_{15})) - \lambda \cdot \sigma_{14} \cdot (\sigma_{15} \cdot \sigma_{13} + \sigma_{11})$$

$$\det(A_{32}) = \sum_{k=1}^{4} \lambda^k \cdot P_k = \lambda \cdot P_1 + \lambda^2 \cdot P_2 + \lambda^3 \cdot P_3 + \lambda^4 \cdot P_4; \; P_1 = -\sigma_{14} \cdot (\sigma_{15} \cdot \sigma_{13} + \sigma_{11})$$

$$\mathrm{P}_2 = -(\sigma_{15} \cdot \sigma_{13} + \sigma_{11} + \sigma_{14} \cdot (\sigma_{13} - \sigma_{15})); \; \mathrm{P}_3 = \sigma_{14} - \sigma_{13} + \sigma_{15}; \; \mathrm{P}_4 = 1$$

$$\det(A_{33}) = (-\lambda - \sigma_{15}) \cdot \det(A_{331}) - \sigma_{10} \cdot e^{-\lambda \cdot \tau} \cdot \det(A_{332}) + \sigma_{11} \cdot \det(A_{333})$$

$$\det(A_{33}) = (-\lambda - \sigma_{15}) \cdot \sigma_{12} \cdot e^{-\lambda \cdot \tau} \cdot \lambda \cdot (\lambda + \sigma_{14}) - \sigma_{10} \cdot e^{-\lambda \cdot \tau} \cdot \sigma_{13} \cdot (\lambda + \sigma_{14}) + \sigma_{11} \cdot \sigma_{12} \cdot e^{-\lambda \cdot \tau} \cdot (\lambda + \sigma_{14})$$

$$\det(A_{33}) = e^{-\lambda \cdot \tau} \cdot \left[-\lambda^3 \cdot \sigma_{12} - \lambda^2 \cdot \sigma_{12} \cdot (\sigma_{15} + \sigma_{14}) + \lambda \cdot (\sigma_{11} \cdot \sigma_{12} - \sigma_{10} \cdot \sigma_{13} - \sigma_{15} \cdot \sigma_{12} \cdot \sigma_{14}) + (\sigma_{11} \cdot \sigma_{12} - \sigma_{10} \cdot \sigma_{13}) \cdot \sigma_{14} \right]$$

$$\det(A_{33}) = e^{-\lambda \cdot \tau} \cdot \sum_{k=0}^{3} \lambda^k \cdot Y_k = e^{-\lambda \cdot \tau} \cdot \left(Y_0 + \lambda \cdot Y_1 + \lambda^2 \cdot Y_2 + \lambda^3 \cdot Y_3 \right)$$

$$Y_0 = (\sigma_{11} \cdot \sigma_{12} - \sigma_{10} \cdot \sigma_{13}) \cdot \sigma_{14}; \; Y_1 = \sigma_{11} \cdot \sigma_2 - \sigma_{10} \cdot \sigma_3 - \sigma_{15} \cdot \sigma_{12} \cdot \sigma_{14}$$
$$Y_2 = -\sigma_{12} \cdot (\sigma_5 + \sigma_4)$$

$$Y_3 = -\sigma_{12}$$

$$\det(A_{34}) = (-\lambda - \sigma_{15}) \cdot \det(A_{341}) - \sigma_{10} \cdot e^{-\lambda \cdot \tau} \cdot \det(A_{34}) + \sigma_{11} \cdot \det(A_{343})$$

$$\det(A_4) = -\sigma_{10} \cdot e^{-\lambda \cdot \tau} \cdot (\sigma_{13} - \lambda) \cdot (\lambda + \sigma_{14}) + \sigma_{11} \cdot \sigma_1 \cdot e^{-\lambda \cdot \tau} \cdot (\lambda + \sigma_{14})$$

$$\det(A_{34}) = e^{-\lambda\cdot\tau}\cdot$$
$$\left[\lambda^2\cdot\sigma_{10}+\lambda\cdot(\sigma_{11}\cdot\sigma_{12}-\sigma_{10}\cdot\sigma_{13}+\sigma_{14}\cdot\sigma_{10})+\sigma_{14}\cdot(\sigma_{11}\cdot\sigma_{12}-\sigma_{10}\cdot\sigma_{13})\right]$$

$$\det(A_{34}) = e^{-\lambda\cdot\tau}\cdot\sum_{k=0}^{2}\lambda^k\cdot \mathrm{I}_k = e^{-\lambda\cdot\tau}\cdot\left(\mathrm{I}_0+\mathrm{I}_1\cdot\lambda+\mathrm{I}_2\cdot\lambda^2\right)$$
$$\mathrm{I}_0 = \sigma_{14}\cdot(\sigma_{11}\cdot\sigma_{12}-\sigma_{10}\cdot\sigma_{13})$$

$$\mathrm{I}_1 = \sigma_{11}\cdot\sigma_{12}-\sigma_{10}\cdot\sigma_{13}+\sigma_{14}\cdot\sigma_{10};\;\; \mathrm{I}_2=\sigma_{10}$$

$$\det(A_3) = -(\sigma_7-\lambda)\cdot\det(A_{31})-\sigma_8\cdot e^{-\lambda\cdot\tau}\cdot\det(A_{32})$$
$$+\sigma_9\cdot\det(A_{33})-\sigma_9\cdot\det(A_{34})$$

$$\det(A_3)|_{\det(A_{31})=0} = -\sigma_8\cdot e^{-\lambda\cdot\tau}\cdot\left(\sum_{k=1}^{4}\lambda^k\cdot \mathrm{P}_k\right)$$
$$+\sigma_9\cdot e^{-\lambda\cdot\tau}\cdot\left(\sum_{k=0}^{3}\lambda^k\cdot Y_k\right)-\sigma_9\cdot e^{-\lambda\cdot\tau}\cdot\left(\sum_{k=0}^{2}\lambda^k\cdot \mathrm{I}_k\right)$$

$$\det(A_3) = \det(A_3)|_{\mathrm{der}\,(A_{31})=0}$$
$$= e^{-\lambda\cdot\tau}\cdot\left\{-\sigma_8\cdot\left(\sum_{k=1}^{4}\lambda^k\cdot P_k\right)+\sigma_9\cdot\left(\sum_{k=0}^{3}\lambda^k\cdot Y_k\right)-\sigma_9\cdot\left(\sum_{k=0}^{2}\lambda^k\cdot \mathrm{I}_k\right)\right\}$$

$$\det(A_3) = e^{-\lambda\cdot\tau}\cdot\{-\lambda^4\cdot \mathrm{P}_4\cdot\sigma_8+\lambda^3\cdot(Y_3\cdot\sigma_9-\mathrm{P}_3\cdot\sigma_8)$$
$$+\lambda^2\cdot(Y_2\cdot\sigma_9-\mathrm{P}_2\cdot\sigma_8-\sigma_9\cdot \mathrm{I}_2)$$
$$+\lambda\cdot(Y_1\cdot\sigma_9-\mathrm{P}_1\cdot\sigma_8-\sigma_9\cdot \mathrm{I}_1)+\sigma_9\cdot(Y_0-\mathrm{I}_0)$$

$$\det(A_3) = e^{-\lambda\cdot\tau}\cdot\left(\sum_{k=0}^{4}\lambda^k\cdot\Upsilon_k\right);\;\; \Upsilon_0=\sigma_9\cdot(Y_0-\mathrm{I}_0);\;\; \Upsilon_1$$
$$= Y_1\cdot\sigma_9-\mathrm{P}_1\cdot\sigma_8-\sigma_9\cdot \mathrm{I}_1$$

$$\Upsilon_2 = Y_2\cdot\sigma_9-\mathrm{P}_2\cdot\sigma_8-\sigma_9\cdot \mathrm{I}_2;\;\; \Upsilon_3=Y_3\cdot\sigma_9-\mathrm{P}_3\cdot\sigma_8;\;\; \Upsilon_4=-\mathrm{P}_4\cdot\sigma_8$$

Table 1.126.

$$\det(A_{41}) = e^{-\lambda\cdot\tau}\cdot[\lambda^4\cdot\sigma_{12}+\lambda^3\cdot\sigma_{12}(\cdot\sigma_{15}+\sigma_{14})+\lambda^2\cdot$$
$$(\sigma_{10}\cdot\sigma_{13}-\sigma_{11}\cdot\sigma_{12}+\sigma_{12}\cdot\sigma_{15}\cdot\sigma_{14})+\lambda\cdot(\sigma_{10}\cdot\sigma_{13}-\sigma_{11}-\sigma_{12})\cdot\sigma_{14}]$$

Table 1.126 Determinants and sub determinants structure

$\det(A - \lambda \cdot I) = -\lambda \cdot \det(A_1) - \sigma_1 \cdot (\sigma_2 + \sigma_3 \cdot e^{-\lambda \cdot \tau}) \cdot \det(A_2) + \sigma_4 \cdot \det(A_3) - \sigma_5 \cdot \det(A_4) + \sigma_5 \cdot \det(A_5) - \sigma_6 \cdot \det(A_6)$			
$\det(A_4) = -(\sigma_7 - \lambda) \cdot \det(A_{41}) - \sigma_8 \cdot e^{-\lambda \cdot \tau} \cdot \det(A_{42}) + \sigma_8 \cdot \det(A_{43}) - \sigma_9 \cdot \det(A_{44})$			
$\det(A_{41}) = -(-\lambda - \sigma_{15}) \cdot \det(A_{411}) + \sigma_{10} \cdot e^{-\lambda \cdot \tau} \cdot \det(A_{412}) - \sigma_{10} \cdot \det(A_{413}) + \sigma_{11} \cdot \det(A_{414})$	$\det(A_{42}) = (-\lambda - \sigma_{15}) \cdot \det(A_{421}) - \sigma_{10} \cdot \det(A_{422}) + \sigma_{11} \cdot \det(A_{423})$	$\det(A_{43}) = (-\lambda - \sigma_{15}) \cdot \det(A_{431}) - \sigma_{10} \cdot e^{-\lambda \cdot \tau} \cdot \det(A_{432}) + \sigma_{11} \cdot \det(A_{433})$	$\det(A_{44}) = (-\lambda - \sigma_{15}) \cdot \det(A_{441}) - \sigma_{10} \cdot e^{-\lambda \cdot \tau} \cdot \det(A_{442}) + \sigma_{10} \cdot \det(A_{443})$
$\det(A_{411}) = \lambda^2 \cdot \sigma_{12} \cdot e^{-\lambda \cdot \tau} \cdot (\lambda + \sigma_{14})$ $\det(A_{412}) = \lambda \cdot \sigma_{13} \cdot (\lambda + \sigma_{14})$ $\det(A_{413}) = 0$ $\det(A_{414}) = -\lambda \cdot \sigma_{12} \cdot e^{-\lambda \cdot \tau} \cdot (\lambda + \sigma_{14})$	$\det(A_{421}) = \lambda \cdot \sigma_{12} \cdot (\lambda + \sigma_{14})$ $\det(A_{422}) = \sigma_{13} \cdot (\lambda + \sigma_{14})$ $\det(A_{423}) = \sigma_{12} \cdot (\lambda + \sigma_{14})$	$\det(A_{431}) = \sigma_{12} \cdot e^{-\lambda \cdot \tau} \cdot \lambda \cdot (\lambda + \sigma_{14})$ $\det(A_{432}) = \sigma_{13} \cdot (\lambda + \sigma_{14})$ $\det(A_{433}) = \sigma_{12} \cdot e^{-\lambda \cdot \tau} \cdot (\lambda + \sigma_{14})$	$\det(A_{441}) = 0$ $\det(A_{442}) = \sigma_{12} \cdot (\lambda + \sigma_{14})$ $\det(A_{443}) = \sigma_{12} \cdot e^{-\lambda \cdot \tau} \cdot (\lambda + \sigma_{14})$

$$\det(A_{41}) = e^{-\lambda\cdot\tau}\cdot\left(\sum_{k=1}^{4}\lambda^k\cdot E_k\right) = e^{-\lambda\cdot\tau}\cdot\left(\lambda^4\cdot E_4+\lambda^3\cdot E_3+\lambda^2\cdot E_2+\lambda\cdot E_1\right)$$

$$E_1 = (\sigma_{10}\cdot\sigma_{13}-\sigma_{11}\cdot\sigma_{12})\cdot\sigma_{14};\;\; E_2 = \sigma_{10}\cdot\sigma_{13}-\sigma_{11}\cdot\sigma_{12}+\sigma_{12}\cdot\sigma_{15}\cdot\sigma_{14}$$

$$E_3 = \sigma_{12}(\cdot\sigma_{15}+\sigma_{14});\;\; E_4 = \sigma_{12}$$

$$\begin{aligned}\det(A_{42}) &= -\lambda^3\cdot\sigma_{12}-\lambda^2\cdot\sigma_{12}-(\sigma_{15}+\sigma_{14})\\ &\quad+\lambda\cdot(\sigma_{11}\cdot\sigma_{12}-\sigma_{10}\cdot\sigma_{13}-\sigma_{15}\cdot\sigma_{12}\cdot\sigma_{14})\\ &\quad+(\sigma_{11}\cdot\sigma_{12}-\sigma_{10}\cdot\sigma_{13})\cdot\sigma_{14}\end{aligned}$$

$$\begin{aligned}\det(A_{42}) &= \sum_{i=0}^{3}\lambda^k\cdot T_k = \lambda^3\cdot T_3+\lambda^2\cdot T_2+\lambda\cdot T_1+T_0;\;\; T_3\\ &= -\sigma_{12};\;\; T_2 = -\sigma_{12}\cdot(\sigma_{15}+\sigma_{14})\end{aligned}$$

$$T_1 = \sigma_{11}\cdot\sigma_{12}-\sigma_{10}\cdot\sigma_{13}-\sigma_{15}\cdot\sigma_{12}\cdot\sigma_{14};\;\; T_0 = (\sigma_{11}\cdot\sigma_{12}-\sigma_{10}\cdot\sigma_{13})\cdot\sigma_{14}$$

$$\begin{aligned}&\det(A_{43}) = e^{-\lambda z}\cdot\left[-\lambda^3\cdot\sigma_{12}-\lambda^2\cdot\sigma_{12}\cdot(\sigma_{15}+\sigma_{14})\right.\\ &\quad\left.+\lambda\cdot(\sigma_{11}\cdot\sigma_{12}-\sigma_{10}\cdot\sigma_{13}-\sigma_{12}\cdot\sigma_{13}\cdot\sigma_{14})+\sigma_{14}\cdot(\sigma_{11}\cdot\sigma_{12}-\sigma_{10}\cdot\sigma_{13})\right]\end{aligned}$$

$$\begin{aligned}\det(A_{43}) &= e^{-\lambda\cdot\tau}\cdot\left(\sum_{k=0}^{3}\lambda^k\cdot L_k\right)\\ &= e^{-\lambda\cdot z}\cdot\left(\lambda^3\cdot L_3+\lambda^2\cdot L_2+\lambda\cdot L_1+L_0\right)\end{aligned}$$

$$L_3 = -\sigma_{12};\;\; L_2 = -\sigma_{12}\cdot(\sigma_{15}+\sigma_{14});\;\; L_1 = \sigma_{11}\cdot\sigma_{12}-\sigma_{10}\cdot\sigma_{13}-\sigma_{12}\cdot\sigma_{15}-\sigma_{14}$$

$$L_0 = \sigma_{14}\cdot(\sigma_{11}\cdot\sigma_{12}-\sigma_{10}\cdot\sigma_{13})$$

$$\det(A_{44}) = 0$$

$$\begin{aligned}\det(A_4) &= -(\sigma_7-\lambda)\cdot\det(A_{41})-\sigma_8\cdot e^{-\lambda\cdot\tau}\cdot\det(A_{42})\\ &\quad+\sigma_8\cdot\det(A_{43})-\sigma_9\cdot\det(A_{44})\end{aligned}$$

$$\det(A_4) = -(\sigma_7-\lambda)\cdot e^{-\lambda\cdot\tau}\cdot\left(\sum_{i=1}^{4}\lambda^k\cdot E_k\right)$$

$$- \sigma_8 \cdot e^{-\lambda \cdot \tau} \cdot \left(\sum_{k=0}^{3} \lambda^k \cdot T_k \right) + \sigma_8 \cdot e^{-\lambda \cdot \tau} \cdot \left(\sum_{k=0}^{3} \lambda^k \cdot L_k \right)$$

$$\det(A_4) = e^{-\lambda \cdot \tau} \cdot \left\{ (-\sigma_7 + \lambda) \cdot \left(\sum_{k=1}^{4} \lambda^k \cdot E_k \right) + \sigma_8 \cdot \left[\left(\sum_{k=0}^{3} \lambda^k \cdot L_k \right) - \left(\sum_{k=0}^{3} \lambda^k \cdot T_k \right) \right] \right\}$$

$$\begin{aligned} \det(A_4) = e^{-\lambda \cdot \tau} \cdot \{ & \lambda^5 \cdot E_4 + \lambda^4 \cdot [E_3 + \sigma_8 \cdot (L_3 - T_3) - E_4 \cdot \sigma_7] \\ & + \lambda^3 \cdot [E_2 + \sigma_8 \cdot (L_2 - T_2) - \sigma_7 \cdot (E_3 + \sigma_8 \cdot (L_3 - T_3))] \\ & + \lambda^2 \cdot [E_1 + \sigma_8 \cdot (L_1 - T_1) - \sigma_7 \cdot (E_2 + \sigma_8 \cdot (L_2 - T_2))] \\ & + \lambda \cdot [\sigma_8 \cdot (L_0 - T_0) - \sigma_7 \cdot (E_1 + \sigma_8 \cdot (L_1 - T_1))] - \sigma_7 \cdot \sigma_8 \cdot (L_0 - T_0) \} \end{aligned}$$

$$\det(A_4) = e^{-\lambda \cdot \tau} \cdot \left(\sum_{k=0}^{5} \lambda^k \cdot \chi_k \right); \;\; \chi_5 = E_4; \;\; \chi_4 = E_3 + \sigma_8 \cdot (L_3 - T_3) - E_4 \cdot \sigma_7$$

$$\chi_3 = E_2 + \sigma_8 \cdot (L_2 - T_2) - \sigma_7 \cdot (E_3 + \sigma_8 - (L_3 - T_3)); \;\; \chi_2 = E_1 + \sigma_8 - (L_1 - T_1) - \sigma_7 \cdot (E_2 + \sigma_8 \cdot (L_2 - T_2))$$

$$\chi_1 = \sigma_8 \cdot (L_0 - T_0) - \sigma_7 \cdot (E_1 + \sigma_8 - (L_1 - T_1)); \;\; \chi_0 = -\sigma_7 \cdot \sigma_8 \cdot (L_0 - T_0)$$

Table 1.127.

$$\begin{aligned} \det(A_{51}) = e^{-\lambda \cdot \tau} \cdot \{ & -\sigma_{10} \cdot \lambda^3 + \lambda^2 \cdot (\sigma_{10} \cdot \sigma_{13} - \sigma_{12} \cdot \sigma_{11} - \sigma_{10} \cdot \sigma_{14}) \\ & + \lambda \cdot \sigma_{14} \cdot (\sigma_{10} \cdot \sigma_{13} - \sigma_{12} \cdot \sigma_{11}) \} \end{aligned}$$

$$\begin{aligned} \det(A_{51}) &= e^{-\lambda \cdot \tau} \cdot \left(\sum_{k=1}^{3} \lambda^k \cdot G_k \right) \\ &= e^{-\lambda \cdot \tau} \cdot \left(\lambda \cdot G_1 + \lambda^2 \cdot G_2 + \lambda^3 \cdot G_3 \right); \;\; G_3 = -\sigma_{10} \end{aligned}$$

$$G_2 = \sigma_{10} \cdot \sigma_{13} - \sigma_{12} \cdot \sigma_{11} - \sigma_{10} \cdot \sigma_{14}; \;\; G_1 = \sigma_{14} \cdot (\sigma_{10} \cdot \sigma_{13} - \sigma_{12} \cdot \sigma_{11})$$

$$\begin{aligned} \det(A_{52}) = & \sigma_{10} \cdot \lambda^2 + \lambda \cdot (\sigma_{11} \cdot \sigma_{12} - \sigma_{10} \cdot \sigma_{13} + \sigma_{14} \cdot \sigma_{10}) \\ & + \sigma_{14} \cdot (\sigma_{11} \cdot \sigma_{12} - \sigma_{10} \cdot \sigma_{13}) \end{aligned}$$

$$\det(A_{52}) = \sum_{i=0}^{2} \lambda^k \cdot I_k = \lambda^2 \cdot I_2 + \lambda \cdot I_1 + I_0; \;\; I_0 = \sigma_{14} \cdot (\sigma_{11} \cdot \sigma_{12} - \sigma_{10} \cdot \sigma_{13})$$

$$I_1 = \sigma_{11} \cdot \sigma_{12} - \sigma_{10} \cdot \sigma_{13} + \sigma_{14} \cdot \sigma_{10}; \;\; I_2 = \sigma_{10}$$

Table 1.127 Determinants and sub determinants structure

$\det(A - \lambda \cdot I) = -\lambda \cdot \det(A_1) - \sigma_1 \cdot \left(\sigma_2 + \sigma_3 \cdot e^{-\lambda \cdot \tau}\right) \cdot \det(A_2) + \sigma_4 \cdot \det(A_3)$ $- \sigma_5 \cdot \det(A_4) + \sigma_5 \cdot \det(A_5) - \sigma_6 \cdot \det(A_6)$			
$\det(A_5) = -(\sigma_7 - \lambda) \cdot \det(A_{51}) - \sigma_8 \cdot e^{-\lambda \cdot \tau} \cdot \det(A_{52}) + \sigma_8 \cdot \det(A_{53}) - \sigma_9 \cdot \det(A_{54})$			
$\det(A_{51}) = -(-\lambda$ $-\sigma_{15}) \cdot \det(A_{511})$ $+ \sigma_{10} \cdot e^{-\lambda \cdot \tau} \cdot \det(A_{512})$ $- \sigma_{10} \cdot \det(A_{513})$ $+\sigma_{11} \cdot \det(A_{514})$	$\det(A_{52}) = (-\lambda$ $-\sigma_{15}) \cdot \det(A_{521})$ $- \sigma_{10} \cdot \det(A_{522})$ $+ \sigma_{11} \cdot \det(A_{523})$	$\det(A_{53}) = (-\lambda$ $-\sigma_{15}) \cdot \det(A_{531})$ $- \sigma_{10} \cdot e^{-\lambda \cdot \tau} \cdot \det(A_{532})$ $+ \sigma_{11} \cdot \det(A_{33})$	$\det(A_{54}) = -(\lambda$ $+\sigma_{15}) \cdot \det(A_{541})$ $- \sigma_{10} \cdot e^{-\lambda \cdot \tau} \cdot \det(A_{542})$ $+ \sigma_{10} \cdot \det(A_{543})$
$\det(A_{511}) = 0$ $\det(A_{512}) = \lambda \cdot (\sigma_{13}$ $- \lambda) \cdot (\lambda + \sigma_{14})$ $\det(A_{513}) = 0$ $\det(A_{514}) = -\lambda \cdot \sigma_{12}$ $\cdot e^{-\lambda \cdot \tau} \cdot (\lambda + \sigma_{14})$	$\det(A_{521}) = 0$ $\det(A_{522}) = (\sigma_{13}$ $- \lambda) \cdot (\lambda + \sigma_{14})$ $\det(A_{523}) = \sigma_{12}$ $(\lambda + \sigma_{14})$	$\det(A_{531}) = 0$ $\det(A_{332}) = (\sigma_{13}$ $- \lambda) \cdot (\lambda + \sigma_{14})$ $\det(A_{533}) = \sigma_{12}$ $\cdot e^{-\lambda \cdot \tau} \cdot (\lambda + \sigma_{14})$	$\det(A_{541}) = 0$ $\det(A_{542}) = \sigma_{12} \cdot (\lambda$ $+\sigma_{14})$ $\det(A_{543}) = \sigma_{12} \cdot e^{-\lambda \cdot \tau}$ $(\lambda + \sigma_{14})$

$$\det(A_{53}) = e^{-\lambda\cdot\tau}\cdot\{\sigma_{10}\cdot\lambda^2+\lambda\cdot(\sigma_{11}\cdot\sigma_{12}-\sigma_{10}\cdot\sigma_{13}+\sigma_{10}\cdot\sigma_{14}) +\sigma_{14}\cdot(\sigma_{11}\cdot\sigma_{12}-\sigma_{10}\cdot\sigma_{13})\}$$

$$\det(A_{33}) = e^{-\lambda\cdot\tau}\cdot\left(\sum_{k=0}^{2}\lambda^k\cdot J_k\right) = e^{-\lambda\cdot\tau}\cdot\left(\lambda^2\cdot J_2+\lambda\cdot J_1+J_0\right)$$

$$J_0=\sigma_{14}\cdot(\sigma_{11}\cdot\sigma_{12}-\sigma_{10}\cdot\sigma_{13})$$

$$J_1=\sigma_{11}\cdot\sigma_{12}-\sigma_{10}\cdot\sigma_{13}+\sigma_{10}\cdot\sigma_{14};\;\; J_2=\sigma_{10};\;\; \det(A_{54})=0$$

$$\det(A_5)=-(\sigma_7-\lambda)\cdot\det(A_{51})-\sigma_8\cdot e^{-\lambda\cdot\tau}\cdot\det(A_{52})+\sigma_8\cdot\det(A_{53})-\sigma_9\cdot\det(A_{54})$$

$$\det(A_5)=-(\sigma_7-\lambda)\cdot e^{-\lambda\cdot\tau}\cdot\left(\lambda\cdot G_1+\lambda^2\cdot G_2+\lambda^3\cdot G_3\right) - \sigma_8\cdot e^{-\lambda\cdot\tau}\cdot\left(\lambda^2\cdot I_2+\lambda\cdot I_1+I_0\right) + \sigma_8\cdot e^{-\lambda\cdot\tau}\cdot\left(\lambda^2\cdot J_2+\lambda\cdot J_1+J_0\right)$$

Since $J_2 = I_2;\; J_1 = I_1;\; J_0 = I_0$ then $\det(A_5) = -(\sigma_7-\lambda)\cdot e^{-\lambda\cdot\tau}\cdot\left(\lambda\cdot G_1+\lambda^2\cdot G_2+\lambda^3\cdot G_3\right)$

$$\det(A_5)=e^{-\lambda\cdot\tau}\cdot\left[\lambda^4\cdot G_3+\lambda^3\cdot(G_2-\sigma_7\cdot G_3)+\lambda^2\cdot(G_1-G_2\cdot\sigma_7)-\lambda\cdot G_1\cdot\sigma_7\right]$$

$$\det(A_5)=e^{-\lambda\cdot\tau}\cdot\left(\sum_{k=1}^{4}\lambda^k\cdot t_k\right)$$

$$t_4=G_3;\;\; t_3=G_2-\sigma_7\cdot G_3;\;\; t_2=G_1-G_2\cdot\sigma_7;\;\; t_1=-G_1\cdot\sigma_7$$

Table 1.128.

$$\det(A_{61})=-(-\lambda-\sigma_{15})\cdot\det(A_{611})+\sigma_{10}\cdot e^{-\lambda\cdot\tau}\cdot\det(A_{612})-\sigma_{10}\cdot\det(A_{613}) +\sigma_{11}\cdot\det(A_{614})-\sigma_{11}\cdot\det(A_{615})$$

$$\det(A_{61})=\lambda^4\cdot\sigma_{14}+\lambda^3\cdot\sigma_{14}\cdot(\sigma_{15}-\sigma_{13})-\lambda^2\cdot\sigma_{14}\cdot(\sigma_{15}\cdot\sigma_{13}+\sigma_{11})$$

$$\det(A_{61})=\sum_{k=2}^{4}\lambda^k\cdot V_k=\lambda^2\cdot V_2+\lambda^3\cdot V_3+\lambda^4\cdot V_4;\;\; V_2=-\sigma_{14}\cdot(\sigma_{15}\cdot\sigma_{13}+\sigma_{11})$$

$$V_3=\sigma_{14}\cdot(\sigma_{15}-\sigma_{13});\;\; V_4=\sigma_{14}$$

$$\det(A_{63})=(-\lambda-\sigma_{15})\cdot\det(A_{631})-\sigma_{10}\cdot e^{-\lambda\cdot\tau}\cdot\det(A_{632}) +\sigma_{11}\cdot\det(A_{633})-\sigma_{11}\cdot\det(A_{634})$$

Table 1.128 Determinants and sub determinants structure

$\det(A-\lambda\cdot I)=-\lambda\cdot\det(A_1)-\sigma_1\cdot(\sigma_2+\sigma_3\cdot e^{-\lambda\cdot\tau})\cdot\det(A_2)+\sigma_4\cdot\det(A_3)$ $-\sigma_5\cdot\det(A_4)+\sigma_5\cdot\det(A_5)-\sigma_6\cdot\det(A_6);\;\det(A_{62})=0$			
$\det(A_6)=-(\sigma_7-\lambda)\cdot\det(A_{61})-\sigma_8\cdot e^{-\lambda\cdot\tau}\cdot\det(A_{62})+\sigma_8\cdot\det(A_{63})-\sigma_9\cdot\det(A_{64})+\sigma_9\cdot\det(A_{65})$			
$\det(A_{61})=-(-\lambda$ $-\sigma_{15})\cdot\det(A_{611})$ $+\sigma_{10}\cdot e^{-\lambda\cdot\tau}\cdot\det(A_{612})$ $-\sigma_{10}\cdot\det(A_{613})$ $+\sigma_{11}\cdot\det(A_{614})$ $-\sigma_{11}\cdot\det(A_{615})$	$\det(A_{63})=(-\lambda$ $-\sigma_{15})\cdot\det(A_{631})$ $-\sigma_{10}\cdot e^{-\lambda\cdot\tau}\cdot\det(A_{632})$ $+\sigma_{11}\cdot\det(A_{633})$ $-\sigma_{11}\cdot\det(A_{634})$	$\det(A_{64})=(-\lambda$ $-\sigma_{15})\cdot\det(A_{641})$ $-\sigma_{10}\cdot e^{-\lambda\cdot\tau}\cdot\det(A_{642})$ $+\sigma_{10}\cdot\det(A_{643})$ $-\sigma_{11}\cdot\det(A_{644})$	$\det(A_{65})=(-\lambda$ $-\sigma_{15})\cdot\det(A_{651})$ $-\sigma_{10}\cdot e^{-\lambda\cdot\tau}\cdot\det(A_{652})$ $+\sigma_{10}\cdot\det(A_{653})$ $-\sigma_{11}\cdot\det(A_{654})$
$\det(A_{611})=-\lambda^2$ $\cdot\sigma_{14}\cdot(\sigma_{13}-\lambda)$ $\det(A_{612})=0$ $\det(A_{613})=0$ $\det(A_{614})=-\lambda\cdot\sigma_{14}\cdot\sigma_{13}$ $\det(A_{615})=-\lambda\cdot\sigma_{14}\cdot(\sigma_{13}-\lambda)$	$\det(A_{631})=-\lambda\cdot\sigma_{14}\cdot(\sigma_{13}-\lambda)$ $\det(A_{632})=0$ $\det(A_{633})=\sigma_{13}\cdot\sigma_{14}$ $\det(A_{634})=\sigma_{14}\cdot(\sigma_{13}-\lambda)$	$\det(A_{641})=-\lambda\cdot\sigma_{14}\cdot\sigma_{12}$ $\det(A_{642})=0$ $\det(A_{643})=\sigma_{13}\cdot\sigma_{14}$ $\det(A_{644})=\sigma_{12}\cdot\sigma_{14}$	$\det(A_{651})=0$ $\det(A_{652})=0$ $\det(A_{653})=\sigma_{14}\cdot(\sigma_{13}-\lambda)$ $\det(A_{654})=\sigma_{12}\cdot\sigma_{14}$

$$\det(A_{63}) = -\lambda^3 \cdot \sigma_{14} + \lambda^2 \cdot \sigma_{14} \cdot (\sigma_{13} - \sigma_{15}) + \lambda \cdot \sigma_{14} \cdot (\sigma_{15} \cdot \sigma_{13} + \sigma_{11})$$

$$\det(A_{63}) = \sum_{k=1}^{3} \lambda^k \cdot W_k = \lambda \cdot W_1 + \lambda^2 \cdot W_2 + \lambda^3 \cdot W_3; \ W_1 = \sigma_{14} \cdot (\sigma_{15} \cdot \sigma_{13} + \sigma_{11})$$

$$W_2 = \sigma_{14} \cdot (\sigma_{13} - \sigma_{15}); \ W_3 = -\sigma_{14}$$

$$\det(A_{64}) = (-\lambda - \sigma_{15}) \cdot \det(A_{641}) - \sigma_{10} \cdot e^{-\lambda \cdot \tau} \cdot \det(A_{642}) + \sigma_{10} \cdot \det(A_{643}) - \sigma_{11} \cdot \det(A_{644})$$

$$\det(A_{64}) = \lambda^2 \cdot \sigma_{14} \cdot \sigma_{12} + \lambda \cdot \sigma_{13} \cdot \sigma_{14} \cdot \sigma_{12} + \sigma_{14} \cdot (\sigma_{10} \cdot \sigma_{13} - \sigma_{11} \cdot \sigma_{12})$$

$$\det(A_{64}) = \sum_{k=0}^{2} \lambda^k \cdot X_i = X_0 + \lambda \cdot X_1 + \lambda^2 \cdot X_2; \ X_0 = \sigma_{14} \cdot (\sigma_{10} \cdot \sigma_{13} - \sigma_{11} \cdot \sigma_{12})$$

$$X_1 = \sigma_{15} \cdot \sigma_{14} \cdot \sigma_{12}; \ X_2 = \sigma_{14} \cdot \sigma_{12}$$

$$\det(A_{65}) = (-\lambda - \sigma_{15}) \cdot \det(A_{651}) - \sigma_{10} \cdot e^{-\lambda \cdot \tau} \cdot \det(A_{652}) + \sigma_{10} \cdot \det(A_{653}) - \sigma_{11} \cdot \det(A_{654})$$

$$\det(A_{65}) = -\lambda \cdot \sigma_{10} \cdot \sigma_{14} + \sigma_{14} \cdot (\sigma_{10} \cdot \sigma_{13} - \sigma_{11} \cdot \sigma_{12}); \ \det(A_{65}) = \sum_{k=0}^{1} \lambda^k \cdot Z_k = Z_0 + \lambda \cdot Z_1$$

$$Z_0 = \sigma_{14} \cdot (\sigma_{10} \cdot \sigma_{13} - \sigma_{11} \cdot \sigma_{12}); \ Z_1 = -\sigma_{10} \cdot \sigma_{14}$$

$$\det(A_6) = -(\sigma_7 - \lambda) \cdot \det(A_{61}) + \sigma_8 \cdot \det(A_{63}) - \sigma_9 \cdot \det(A_{64}) + \sigma_9 \cdot \det(A_{65})$$

$$\det(A_6) = \lambda^5 \cdot V_4 + \lambda^4 \cdot (V_3 - \sigma_7 \cdot V_4) + \lambda^3 \cdot (V_2 - \sigma_7 \cdot V_3 + \sigma_8 \cdot W_3) + \lambda^2 \cdot (-\sigma_7 \cdot V_2 + \sigma_8 \cdot W_2 - \sigma_9 \cdot X_2) + \lambda \cdot (W_1 \cdot \sigma_8 - X_1 \cdot \sigma_9 + \sigma_9 \cdot Z_1) + (\sigma_9 \cdot Z_0 - X_0 \cdot \sigma_9)$$

$$\det(A_6) = \sum_{k=0}^{5} \lambda^k \cdot \mathrm{O}_k = \mathrm{O}_0 + \lambda \cdot \mathrm{O}_1 + \lambda^2 \cdot \mathrm{O}_2 + \lambda^3 \cdot \mathrm{O}_3 + \lambda^4 \cdot \mathrm{O}_4 + \lambda^5 \cdot \mathrm{O}_5$$

$$\mathrm{O}_o = \sigma_9 \cdot Z_0 - X_0 \cdot \sigma_9; \ \mathrm{O}_1 = W_1 \cdot \sigma_8 - X_1 \cdot \sigma_9 + \sigma_9 \cdot Z_1$$

$$\text{O}_2 = -\sigma_7 \cdot V_2 + \sigma_8 - W_2 - \sigma_9 \cdot X_2$$

$$\text{O}_3 = V_2 - \sigma_7 \cdot V_3 + \sigma_8 \cdot W_3; \ \text{O}_4 = V_3 - \sigma_7 \cdot V_4; \ \text{O}_5 = V_4$$

We can summary our intermediate results:

$$\det(A_1) = \det(A_1)|_{\det(A_1)=0} = \sum_{k=1}^{6} \lambda^k \cdot \Omega_k$$
$$= \lambda^6 \cdot \Omega_6 + \lambda^5 \cdot \Omega_5 + \lambda^4 \cdot \Omega_4 + \lambda^3 \cdot \Omega_3 + \lambda^2 \cdot \Omega_2 + \lambda \cdot \Omega_1$$

$$\det(A_2) = \sum_{k=0}^{6} \lambda^k \cdot \Xi_k = \lambda^6 \cdot \Xi_6 + \lambda^5 \cdot \Xi_5 + \lambda^4 \cdot \Xi_4 + \lambda^3 \cdot \Xi_3$$
$$+ \lambda^2 \cdot \Xi_2 + \lambda \cdot \Xi_1 + \Xi_0$$

$$\det(A_3) = e^{-\lambda \cdot \tau} \cdot \left(\sum_{k=0}^{4} \lambda^k \cdot \Upsilon_k \right) = e^{-\lambda \cdot \tau} \cdot$$
$$\left(\Upsilon_0 + \lambda \cdot Y_1 + \lambda^2 \cdot \Upsilon_2 + \lambda^3 \cdot \Upsilon_3 + \lambda^4 \cdot \Upsilon_4 \right)$$

$$\det(A_4) = e^{-\lambda \cdot \tau} \cdot \left(\sum_{k=0}^{5} \lambda^k \cdot \chi_k \right)$$
$$= e^{-\lambda \cdot \tau} \cdot \left(\chi_0 + \lambda \cdot \chi_1 + \lambda^2 \cdot \chi_3 + \lambda^3 \cdot \chi_3 + \lambda^4 \cdot \chi_4 + \lambda^5 \cdot \chi_5 \right)$$

$$\det(A_5) = e^{-\lambda \cdot \tau} \cdot \left(\sum_{k=1}^{4} \lambda^k \cdot t_k \right) = e^{-\lambda \cdot \tau} \cdot$$
$$\left[\lambda^4 \cdot G_3 + \lambda^3 \cdot (G_2 - \sigma_7 \cdot G_3) + \lambda^2 \cdot (G_1 - G_2 \cdot \sigma_7) - \lambda \cdot G_1 \cdot \sigma_7 \right]$$

$$\det(A_6) = \sum_{k=0}^{5} \lambda^k \cdot \text{O}_k = \text{O}_0 + \lambda \cdot \text{O}_1 + \lambda^2 \cdot \text{O}_2 + \lambda^3 \cdot \text{O}_3 + \lambda^4 \cdot \text{O}_4 + \lambda^5 \cdot \text{O}_5$$

Finally we get the expression for $\det(A - \lambda \cdot I)$.

$$\det(A - \lambda \cdot I) = (-\lambda) \cdot \det(A_1) - \sigma_1 \cdot \left(\sigma_2 + \sigma_3 \cdot e^{-\lambda \cdot \tau} \right) \cdot \det(A_2) + \sigma_4 \cdot \det(A_3)$$
$$- \sigma_5 \cdot \det(A_4) + \sigma_5 \cdot \det(A_5) - \sigma_6 \cdot \det(A_6)$$

$$\det(A - \lambda \cdot I) = (-\lambda) \cdot \left(\sum_{k=1}^{6} \lambda^k - \Omega_k \right) - \sigma_1 \cdot \left(\sigma_2 + \sigma_3 \cdot e^{-\lambda \cdot \tau} \right) \cdot \left(\sum_{k=0}^{6} \lambda^i \cdot \Xi_k \right)$$

$$+\sigma_4 \cdot e^{-\lambda\cdot\tau} \cdot \left(\sum_{k=0}^{4} \lambda^k \cdot \Upsilon_k\right)$$
$$-\sigma_5 \cdot e^{-\lambda\cdot\tau} \cdot \left(\sum_{k=0}^{5} \lambda^k \cdot \chi_k\right)$$
$$+\sigma_5 \cdot e^{-\lambda\cdot\tau} \cdot \left(\sum_{k=1}^{4} \lambda^k - t_k\right) - \sigma_6 \cdot \left(\sum_{k=0}^{5} \lambda^k \cdot \mathrm{O}_k\right)$$

$$\det(A - \lambda \cdot I) = (-\lambda) \cdot \left(\sum_{k=1}^{6} \lambda^k \cdot \Omega_k\right) - \sigma_1 \cdot \sigma_2 \cdot \left(\sum_{k=0}^{6} \lambda^k \cdot \Xi_k\right)$$
$$-\sigma_6 \cdot \left(\sum_{k=0}^{5} \lambda^k \cdot \mathrm{O}_k\right)$$
$$+\left\{\sigma_4 \cdot \left(\sum_{k=0}^{4} \lambda^k \cdot \Upsilon_k\right) - \sigma_1 \cdot \sigma_3 \cdot \left(\sum_{k=0}^{6} \lambda^k \cdot \Xi_k\right)\right.$$
$$\left. -\sigma_5 \cdot \left(\sum_{k=0}^{5} \lambda^k \cdot \chi_i\right) + \sigma_5 \cdot \left(\sum_{k=1}^{4} \lambda^k \cdot t_k\right)\right\} \cdot e^{-\lambda\cdot\tau}$$

We get our MRI system laser diode equivalent circuit characteristic equation:

$$D(\lambda, \tau) = P_n(\lambda, \tau) + Q_m(\lambda, \tau) \cdot e^{-\lambda\cdot\tau}; \; P_n(\lambda, \tau) = (-\lambda) \cdot \left(\sum_{k=1}^{6} \lambda^k \cdot \Omega_k\right)$$
$$-\sigma_1 \cdot \sigma_2 \cdot \left(\sum_{k=0}^{6} \lambda^k \cdot \Xi_k\right) - \sigma_6 \cdot \left(\sum_{k=0}^{5} \lambda^k \cdot \mathrm{O}_k\right)$$
$$Q_m(\lambda, \tau) = \sigma_4 \cdot \left(\sum_{k=0}^{4} \lambda^k \cdot \Upsilon_k\right) - \sigma_1 \cdot \sigma_3 \cdot \left(\sum_{k=0}^{6} \lambda^k \cdot \Xi_k\right)$$
$$-\sigma_5 \cdot \left(\sum_{k=0}^{5} \lambda^k - \chi_i\right) + \sigma_5 \cdot \left(\sum_{k=1}^{4} \lambda^k \cdot t_k\right)$$

$$P_n(\lambda, \tau) = -\lambda^7 \cdot \Omega_6 - \lambda^6 \cdot (\Omega_3 + \sigma_1 \cdot \sigma_2 \cdot \Xi_6)$$
$$-\lambda^5 \cdot (\Omega_4 + \Xi_3 \cdot \sigma_1 \cdot \sigma_2 + \mathrm{O}_5 \cdot \sigma_6)$$
$$-\lambda^4 \cdot (\Omega_3 + \sigma_1 \cdot \sigma_2 \cdot \Xi_4 + \sigma_6 \cdot \mathrm{O}_4)$$
$$-\lambda^3 \cdot (\Omega_2 + \Xi_3 \cdot \sigma_1 \cdot \sigma_2 + \sigma_6 \cdot \mathrm{O}_3)$$
$$-\lambda^2 \cdot (\Omega_1 + \sigma_1 \cdot \sigma_2 \cdot \Xi_2 + \mathrm{O}_2 \cdot \sigma_6)$$
$$-\lambda \cdot (\Xi_1 \cdot \sigma_1 \cdot \sigma_2 + \mathrm{O}_1 \cdot \sigma_6) - (\sigma_1 \cdot \sigma_2 \cdot \Xi_0 + \mathrm{O}_0 \cdot \sigma_6)$$

$$\begin{aligned}Q_m(\lambda,\tau) = &-\sigma_1\cdot\sigma_3\cdot\Xi_6 - \lambda^6 - \lambda^5 - (\Xi_5\cdot\sigma_1\cdot\sigma_2 + \chi_5\cdot\sigma_5)\\ &+\lambda^4\cdot(\Upsilon_4\cdot\sigma_4 - \sigma_1\cdot\sigma_3\cdot\Xi_4 - \chi_4\cdot\sigma_5 + \sigma_5\cdot t_4)\\ &+\lambda^3\cdot(\Upsilon_3\cdot\sigma_4 - \sigma_1\cdot\sigma_3\cdot\Xi_3 - \chi_3\cdot\sigma_5 + \sigma_5\cdot t_3)\\ &+\lambda^2\cdot(\Upsilon_2\cdot\sigma_4 - \sigma_1\cdot\sigma_3\cdot\Xi_2 - \chi_2\cdot\sigma_3 + t_2\cdot\sigma_5)\\ &+\lambda\cdot(\Upsilon_1\cdot\sigma_4 - \sigma_1\cdot\sigma_3\cdot\Xi_1 - \chi_1\cdot\sigma_5 + \sigma_5\cdot t_1)\\ &+(\sigma_4\cdot\Upsilon_0 - \sigma_1\cdot\sigma_3 - \sigma_5\cdot\chi_0)\end{aligned}$$

We get the characteristics equation for stability analysis. We study the occurrence of any possible stability switching, resulting from the increase of the value of the time delay τ or any other circuit parameter for the general characteristic equation $D(\lambda,\tau)$.

$$D(\lambda,\tau) = P_n(\lambda,\tau) + Q_m(\lambda,\tau)\cdot e^{-\lambda\cdot\tau};\ n,m\in\mathbb{N}_0;\ n>m, n=7, m=6$$

$$\begin{aligned}Q_m(\lambda,\tau) &= \sum_{k=0}^{m} q_k(\tau)\cdot\lambda^k = q_0(\tau) + q_1(\tau)\cdot\lambda + \ldots + q_6(\tau)\cdot\lambda^6;\ q_0(\tau)\\ &= \sigma_4\cdot\Upsilon_0 - \sigma_1\cdot\sigma_3 - \sigma_5\cdot\chi_0\end{aligned}$$

$$\begin{aligned}q_1(\tau) &= \Upsilon_1\cdot\sigma_4 - \sigma_1\cdot\sigma_3\cdot\Xi_1 - \chi_1\cdot\sigma_5 + \sigma_5\cdot t_1;\ q_2(\tau)\\ &= \Upsilon_2\cdot\sigma_4 - \sigma_1\cdot\sigma_3\cdot\Xi_2 - \chi_2\cdot\sigma_5 + t_2\cdot\sigma_5\end{aligned}$$

$$\begin{aligned}q_3(\tau) &= \Upsilon_3\cdot\sigma_4 - \sigma_1\cdot\sigma_3\cdot\Xi_3 - \chi_3\cdot\sigma_5 + \sigma_5\cdot t_3;\ q_4(\tau)\\ &= \Upsilon_4\cdot\sigma_4 - \sigma_1\cdot\sigma_3\cdot\Xi_4 - \chi_4\cdot\sigma_5 + \sigma_5\cdot t_4\end{aligned}$$

$$q_5(\tau) = -(\Xi_5\cdot\sigma_1\cdot\sigma_2 + \chi_5\cdot\sigma_5);\ q_6(\tau) = -\sigma_1\cdot\sigma_3\cdot\Xi_6$$

$$\begin{aligned}P_n(\lambda,\tau) &= \sum_{k=0}^{n} p_k(\tau)\cdot\lambda^k = p_0(\tau) + p_1(\tau)\cdot\lambda + \ldots + p_7(\tau)\cdot\lambda^7;\ p_0(\tau)\\ &= -(\sigma_1\cdot\sigma_2\cdot\Xi_0 + \mathrm{O}_0\cdot\sigma_6)\end{aligned}$$

$$p_1(\tau) = -(\Xi_1\cdot\sigma_1\cdot\sigma_2 + \mathrm{O}_1\cdot\sigma_6);\ p_2(\tau) = -(\Omega_1 + \sigma_1\cdot\sigma_2\cdot\Xi_2 + \mathrm{O}_2\cdot\sigma_6)$$

$$p_3(\tau) = -(\Omega_2 + \Xi_3\cdot\sigma_1\cdot\sigma_2 + \sigma_6\cdot\mathrm{O}_3);\ p_4(\tau) = -(\Omega_3 + \sigma_1\cdot\sigma_2\cdot\Xi_4 + \sigma_6\cdot\mathrm{O}_4)$$

$$p_5(\tau) = -(\Omega_4 + \Xi_5\cdot\sigma_1\cdot\sigma_2 + \mathrm{O}_5\cdot\sigma_6);\ p_6(\tau) = -(\Omega_5 + \sigma_1\cdot\sigma_2\cdot\Xi_6);\ p_7(\tau) = -\Omega_6$$

The homogeneous system for $Y_1, Y_2, Y_3, I_{L_1}, I_{L_2}, I_{R_i}, I_{L_i}, I_{R_p}$ leads to characteristics equation for the eigenvalue λ having the form $P(\lambda,\tau) + Q(\lambda,\tau) \cdot e^{-\lambda\cdot\tau} = 0$; $Q(\lambda,\tau) = \sum_{j=0}^{6} c_j \cdot \lambda^j$

$P(\lambda,\tau) = \sum_{j=0}^{7} a_j \cdot \lambda^j$. The coefficients $\{a_j(q_i, q_k, \tau), c_j(q_i, q_k, \tau)\} \in \mathbb{R}$ depend on q_i, q_k and delay τ. q_i, q_k are any MRI system laser diode equivalent circuit parameters, other parameters kept as constant.

$$a_0 = -(\sigma_1 \cdot \sigma_2 \cdot \Xi_0 + \mathrm{O}_0 \cdot \sigma_6);\ a_1 = -(\Xi_1 \cdot \sigma_1 \cdot \sigma_2 + \mathrm{O}_1 \cdot \sigma_6);\ a_2 = -(\Omega_1 + \sigma_1 \cdot \sigma_2 \cdot \Xi_2 + \mathrm{O}_2 \cdot \sigma_6)$$

$$a_2 = -(\Omega_1 + \sigma_1 \cdot \sigma_2 \cdot \Xi_2 + \mathrm{O}_2 \cdot \sigma_6);\ a_3 = -(\Omega_2 + \Xi_3 \cdot \sigma_1 \cdot \sigma_2 + \sigma_6 \cdot \mathrm{O}_3)$$

$$a_4 = -(\Omega_3 + \sigma_1 \cdot \sigma_2 \cdot \Xi_4 + \sigma_6 \cdot \mathrm{O}_4);\ a_5 = -(\Omega_4 + \Xi_5 \cdot \sigma_1 \cdot \sigma_2 + \mathrm{O}_5 \cdot \sigma_6)$$

$$a_6 = -(\Omega_5 + \sigma_1 \cdot \sigma_2 \cdot \Xi_6);\ a_7 = -\Omega_6;\ c_0 = \sigma_4 \cdot \Upsilon_0 - \sigma_1 \cdot \sigma_3 - \sigma_5 \cdot \gamma_0$$

$$c_1 = \Upsilon_1 \cdot \sigma_4 - \sigma_1 \cdot \sigma_3 \cdot \Xi_1 - \chi_1 \cdot \sigma_5 + \sigma_5 \cdot t_1;\ c_2 = \Upsilon_2 \cdot \sigma_4 - \sigma_1 \cdot \sigma_3 \cdot \Xi_2 - \chi_2 \cdot \sigma_5 + t_2 \cdot \sigma_5$$

$$c_3 = \Upsilon_3 \cdot \sigma_4 - \sigma_1 \cdot \sigma_3 \cdot \Xi_3 - \gamma_3 \cdot \sigma_5 + \sigma_5 \cdot t_3;\ c_4 = \Upsilon_4 \cdot \sigma_4 - \sigma_1 \cdot \sigma_3 \cdot \Xi_4 - \chi_4 \cdot \sigma_5 + \sigma_5 \cdot t_4$$

$$c_5 = -(\Xi_5 \cdot \sigma_1 \cdot \sigma_2 + \chi_5 \cdot \sigma_5);\ c_6 = -\sigma_1 \cdot \sigma_3 \cdot \Xi_6$$

Unless strictly necessary, the designation of varied arguments (q_i, q_k) will subsequently be omitted from P, Q, a_j, and c_j. The coefficients a_j, c_j are continuous, and differentiable functions of their arguments, and direct substitution shows that $a_0 + c_0 \neq 0; -(\sigma_1 \cdot \sigma_2 \cdot \Xi_0 + \mathrm{O}_0 \cdot \sigma_6) + \sigma_4 \cdot \Upsilon_0 - \sigma_1 \cdot \sigma_3 - \sigma_5 \cdot \chi_0 \neq 0$ for $q_i, q_k \in \mathbb{R}_+$; that is $\lambda = 0$ is not of $P(\lambda,\tau) + Q(\lambda,\tau) \cdot e^{-\lambda\cdot\tau} = 0$; $P(\lambda) = P(\lambda,\tau)$; $Q(\lambda) = Q(\lambda,\tau)$ [2, 3]. Furthermore, $P(\lambda)$, $Q(\lambda)$ are analytic functions of λ, for which the following requirements of the analysis [2, 4] can also be verified in the present case:

1. If $\lambda = i \cdot \omega, \omega \in \mathbb{R}$ then $P(i \cdot \omega) + Q(i \cdot \omega) \neq 0$.
2. If $|Q(\lambda)/P(\lambda)|$ is bounded for $|\lambda| \to \infty$; $\operatorname{Re}\lambda \geq 0$. No roots bifurcation from ∞.
3. $F(\omega) = |P(i \cdot \omega)|^2 - |Q(i \cdot \omega)|^2$ has a finite number of zeros. Indeed, this is polynomial in ω.
4. Each positive root $\omega(q_i, q_k)$ of $F(\omega) = 0$ is continuous and differentiable with respect to q_i, q_k.

We assume that $P_n(\lambda, \tau)$ and $Q_m(\lambda, \tau)$ cannot have common imaginary roots. That is for any real number ω. $P_n(\lambda = i \cdot \omega, \tau) + Q_m(\lambda = i \cdot \omega, \tau) \neq 0$

$$P_n(\lambda = i \cdot \omega, \tau) = i \cdot \{\omega^7 \cdot \Omega_8 - \omega^5 \cdot (\Omega_4 + \Xi_5 \cdot \sigma_1 \cdot \sigma_2 + O_5 \cdot \sigma_6) + \omega^3 \cdot (\Omega_2 + \Xi_3 \cdot \sigma_1 \cdot \sigma_2 + \sigma_6 \cdot O_3) - \omega \cdot (\Xi_1 \cdot \sigma_1 \cdot \sigma_2 + O_1 \cdot \sigma_6)\} + \omega^6 \cdot (\Omega_5 + \sigma_1 \cdot \sigma_2 \cdot \Xi_6) - \omega^4 \cdot (\Omega_3 + \sigma_1 \cdot \sigma_2 \cdot \Xi_4 + \sigma_6 \cdot O_4) + \omega^2 \cdot (\Omega_1 + \sigma_1 \cdot \sigma_2 \cdot \Xi_2 + O_2 \cdot \sigma_6) - (\sigma_1 \cdot \sigma_2 \cdot \Xi_0 + O_0 \cdot \sigma_6)$$

$$Q_m(\lambda = i \cdot \omega, \tau) = \omega^6 \cdot \sigma_1 \cdot \sigma_3 \cdot \Xi_6 + \omega^4 \cdot (\Upsilon_4 \cdot \sigma_4 - \sigma_1 \cdot \sigma_3 \cdot \Xi_4 - \chi_4 \cdot \sigma_5 + \sigma_5 \cdot t_4) - \omega^2 \cdot (\Upsilon_2 \cdot \sigma_4 - \sigma_1 \cdot \sigma_3 \cdot \Xi_2 - \chi_2 \cdot \sigma_5 + t_2 \cdot \sigma_5) + (\sigma_4 \cdot \Upsilon_0 - \sigma_1 \cdot \sigma_3 - \sigma_5 \cdot \chi_0) + i \cdot \{-\omega^5 \cdot (\Xi_5 \cdot \sigma_1 \cdot \sigma_2 + \chi_5 \cdot \sigma_5) - \omega^3 \cdot (\Upsilon_3 \cdot \sigma_4 - \sigma_1 \cdot \sigma_3 \cdot \Xi_3 - \chi_3 \cdot \sigma_5 + \sigma_5 \cdot t_3) + \omega \cdot (\Upsilon_1 \cdot \sigma_4 - \sigma_1 \cdot \sigma_3 \cdot \Xi_1 - \chi_1 \cdot \sigma_5 + \sigma_5 \cdot t_1)\}$$

$$\lambda^k = i^k \cdot \omega^k;\ i^k = \begin{cases} i & \forall k = 4 \cdot n - 3 \\ -1 & \forall k = 4 \cdot n - 2 \\ -i & \forall k = 4 \cdot n - 1 \\ 1 & \forall k = 4 \cdot n \end{cases};\ n \in \mathbb{R}_0;\ n = 1, 2, 3, \ldots$$

$$Q_m(\lambda = i \cdot \omega, \tau) = Q_m(i \cdot \omega, \tau) = Q(i \cdot \omega, \tau)$$

$$P_n(\lambda = i \cdot \omega, \tau) + Q_m(\lambda = i \cdot \omega, \tau) = \omega^6 \cdot [\Omega_5 + \sigma_1 \cdot \sigma_2 \cdot \Xi_6 + \sigma_1 \cdot \sigma_3 \cdot \Xi_6] + \omega^4 \cdot [\Upsilon_4 \cdot \sigma_4 - \sigma_1 \cdot \sigma_3 \cdot \Xi_4 - \chi_4 \cdot \sigma_5 + \sigma_5 \cdot t_4 - \Omega_3 - \sigma_1 \cdot \sigma_2 \cdot \Xi_4 - \sigma_6 \cdot O_4] + \omega^2 \cdot [\Omega_1 + \sigma_1 \cdot \sigma_2 \cdot \Xi_2 + O_2 \cdot \sigma_6 - (\Upsilon_2 \cdot \sigma_4 - \sigma_1 \cdot \sigma_3 \cdot \Xi_2 - \chi_2 \cdot \sigma_5 + t_2 \cdot \sigma_5)] + (\sigma_4 \cdot \Upsilon_0 - \sigma_1 \cdot \sigma_3 - \sigma_3 \cdot \chi_0) - \sigma_1 \cdot \sigma_2 \cdot \Xi_0 - O_0 \cdot \sigma_6 + i \cdot \{\omega^7 \cdot \Omega_6 - \omega^5 \cdot [(\Omega_4 + \Xi_5 \cdot \sigma_1 \cdot \sigma_2 + O_5 \cdot \sigma_6) + (\Xi_5 \cdot \sigma_1 \cdot \sigma_2 + \chi_5 \cdot \sigma_5)] + \omega^3 \cdot [(\Omega_2 + \Xi_3 \cdot \sigma_1 \cdot \sigma_2 + \sigma_6 \cdot O_3) - (\Upsilon_3 \cdot \sigma_4 - \sigma_1 \cdot \sigma_3 \cdot \Xi_3 - \chi_3 \cdot \sigma_5 + \sigma_5 \cdot t_3)] + \omega \cdot [(\Upsilon_1 \cdot \sigma_4 - \sigma_1 \cdot \sigma_3 \cdot \Xi_1 - \chi_1 \cdot \sigma_5 + \sigma_5 \cdot t_1) - (\Xi_1 \cdot \sigma_1 \cdot \sigma_2 + O_1 \cdot \sigma_6)]\} \neq 0$$

$$P_n(\lambda = i \cdot \omega, \tau) = P_n(i \cdot \omega, \tau) = P(i \cdot \omega, \tau)$$

$$|P(i \cdot \omega, \tau)|^2 = \{\omega^6 \cdot (\Omega_5 + \sigma_1 \cdot \sigma_2 \cdot \Xi_6) - \omega^4 \cdot (\Omega_3 + \sigma_1 \cdot \sigma_2 \cdot \Xi_4 + \sigma_6 \cdot O_4) + \omega^2 \cdot (\Omega_1 + \sigma_1 \cdot \sigma_2 \cdot \Xi_2 + O_2 \cdot \sigma_6) - (\sigma_1 \cdot \sigma_2 \cdot \Xi_0 + O_0 \cdot \sigma_6)\}^2 + \{\omega^7 \cdot \Omega_6 - \omega^5 \cdot (\Omega_4 + \Xi_5 \cdot \sigma_1 \cdot \sigma_2 + O_5 \cdot \sigma_6) + \omega^3 \cdot (\Omega_2 + \Xi_3 \cdot \sigma_1 \cdot \sigma_2 + \sigma_6 \cdot O_3) - \omega \cdot (\Xi_1 \cdot \sigma_1 \cdot \sigma_2 + O_1 \cdot \sigma_6)\}^2$$

$$|Q(i \cdot \omega, \tau)|^2 = \{\omega^6 \cdot \sigma_1 \cdot \sigma_3 \cdot \Xi_6 + \omega^4 \cdot (\Upsilon_4 \cdot \sigma_4 - \sigma_1 \cdot \sigma_3 \cdot \Xi_4 - \chi_4 \cdot \sigma_5 + \sigma_5 \cdot t_4) - \omega^2 \cdot (\Upsilon_2 \cdot \sigma_4 - \sigma_1 \cdot \sigma_3 \cdot \Xi_2 - \chi_2 \cdot \sigma_5 + t_2 \cdot \sigma_5) + (\sigma_4 \cdot \Upsilon_0 - \sigma_1 \cdot \sigma_3 - \sigma_5 \cdot \chi_0)\}^2$$

$$+\{-\omega^5\cdot(\Xi_5\cdot\sigma_1\cdot\sigma_2+\chi_5\cdot\sigma_5)-\omega^3\cdot(\Upsilon_3\cdot\sigma_4-\sigma_1\cdot\sigma_3\cdot\Xi_3-\chi_3\cdot\sigma_5+\sigma_5\cdot t_3)$$
$$+\omega\cdot(\Upsilon_1\cdot\sigma_4-\sigma_1\cdot\sigma_3\cdot\Xi_1-\chi_1\cdot\sigma_5+\sigma_5\cdot t_1)\}^2$$

$$F(\omega,\tau)=P(i\cdot\omega,\tau)|^2-|Q(i\cdot\omega,\tau)|^2=\sum_{k=0}^{7}\Pi_{2.k}\cdot\omega^{2\cdot k}$$

Hence $F(\omega,\tau)=0$ implies $\sum_{k=0}^{7}\Pi_{2.k}\cdot\omega^{2.k}=0$ and its roots are given by solving the above polynomial.

Remark It is the reader exercise to find the $\Pi_{2\cdot k};\ k=0,1,\ldots,7$ expressions.

$$P_R(i\cdot\omega,\tau)=\omega^6\cdot(\Omega_5+\sigma_1\cdot\sigma_2\cdot\Xi_6)-\omega^4\cdot(\Omega_3+\sigma_1\cdot\sigma_2\cdot\Xi_4+\sigma_6\cdot O_4)$$
$$+\omega^2\cdot(\Omega_1+\sigma_1\cdot\sigma_2\cdot\Xi_2+O_2\cdot\sigma_6)-(\sigma_1\cdot\sigma_2\cdot\Xi_0+O_0\cdot\sigma_6)$$

$$P_I(i\cdot\omega,\tau)=\omega^7\cdot\Omega_6-\omega^5\cdot(\Omega_4+\Xi_5\cdot\sigma_1\cdot\sigma_2+O_5\cdot\sigma_6)$$
$$+\omega^3\cdot(\Omega_2+\Xi_3\cdot\sigma_1\cdot\sigma_2+\sigma_6\cdot O_3)-\omega\cdot(\Xi_1\cdot\sigma_1\cdot\sigma_2+O_1\cdot\sigma_6)$$

$$Q_R(i\cdot\omega,\tau)=\omega^6\cdot\sigma_1\cdot\sigma_3\cdot\Xi_6+\omega^4\cdot(\Upsilon_4\cdot\sigma_4-\sigma_1\cdot\sigma_3\cdot\Xi_4-\chi_4\cdot\sigma_5+\sigma_5\cdot t_4)$$
$$-\omega^2\cdot(\Upsilon_2\cdot\sigma_4-\sigma_1\cdot\sigma_3\cdot\Xi_2-\chi_2\cdot\sigma_5+t_2\cdot\sigma_5)$$
$$+(\sigma_4\cdot\Upsilon_0-\sigma_1\cdot\sigma_3-\sigma_5\cdot\chi_0)$$

$$Q_I(i\cdot\omega,\tau)=-\omega^5\cdot(\Xi_5\cdot\sigma_1\cdot\sigma_2+\chi_5\cdot\sigma_5)$$
$$-\omega^3\cdot(\Upsilon_3\cdot\sigma_4-\sigma_1\cdot\sigma_3\cdot\Xi_3-\chi_3\cdot\sigma_5+\sigma_5\cdot t_3)$$
$$+\omega\cdot(\Upsilon_1\cdot\sigma_4-\sigma_1\cdot\sigma_3\cdot\Xi_1-\chi_1\cdot\sigma_5+\sigma_5\cdot t_1)$$

$$\sin\theta(\tau)=\frac{-P_R(i\cdot\omega,\tau)\cdot Q_I(i\cdot\omega,\tau)+P_I(i\cdot\omega,\tau)\cdot Q_R(i\cdot\omega,\tau)}{|Q(i\cdot\omega,\tau)|^2}$$

$$\cos\theta(\tau)=-\frac{P_R(i\cdot\omega,\tau)\cdot Q_R(i\cdot\omega,\tau)+P_I(i\cdot\omega,\tau)\cdot Q_I(i\cdot\omega,\tau)}{|Q(i\cdot\omega,\tau)|^2}$$

We use different parameters terminology from our last characteristics parameters definition: $k\to j;\ p_k(\tau)\to a_j;\ q_k(\tau)\to c_j;\ n=7;\ m=6;\ n>m$.

$$P_n(\lambda,\tau)\to P(\lambda);\ Q_m(\lambda,\tau)\to Q(\lambda);\ P(\lambda)=\sum_{j=0}^{7}a_j\cdot\lambda^j;\ Q(\lambda)=\sum_{j=0}^{6}c_j\cdot\lambda^j$$

$$P(\lambda) = -\lambda^7 \cdot \Omega_6 - \lambda^6 \cdot (\Omega_5 + \sigma_1 \cdot \sigma_2 \cdot \Xi_6) - \lambda^5 \cdot (\Omega_4 + \Xi_5 \cdot \sigma_1 \cdot \sigma_2 + O_5 \cdot \sigma_6)$$
$$- \lambda^4 \cdot (\Omega_3 + \sigma_1 \cdot \sigma_2 \cdot \Xi_4 + \sigma_6 \cdot O_4) - \lambda^3 \cdot (\Omega_2 + \Xi_3 \cdot \sigma_1 \cdot \sigma_2 + \sigma_6 \cdot O_3)$$
$$- \lambda^2 \cdot (\Omega_1 + \sigma_1 \cdot \sigma_2 \cdot \Xi_2 + O_2 \cdot \sigma_6) - \lambda \cdot (\Xi_1 \cdot \sigma_1 \cdot \sigma_2 + O_1 \cdot \sigma_6)$$
$$- (\sigma_1 \cdot \sigma_2 \cdot \Xi_0 + O_0 \cdot \sigma_6)$$

$$Q(\lambda) = -\sigma_1 \cdot \sigma_3 \cdot \Xi_6 \cdot \lambda^6 - \lambda^5 \cdot (\Xi_5 \cdot \sigma_1 \cdot \sigma_2 + \chi_5 \cdot \sigma_5)$$
$$+ \lambda^4 \cdot (\Upsilon_4 \cdot \sigma_4 - \sigma_1 \cdot \sigma_3 \cdot \Xi_4 - \chi_4 \cdot \sigma_5 + \sigma_5 \cdot t_4)$$
$$+ \lambda^3 \cdot (\Upsilon_3 \cdot \sigma_4 - \sigma_1 \cdot \sigma_3 \cdot \Xi_3 - \chi_3 \cdot \sigma_5 + \sigma_5 \cdot t_3)$$
$$+ \lambda^2 \cdot (\Upsilon_2 \cdot \sigma_4 - \sigma_1 \cdot \sigma_3 \cdot \Xi_2 - \chi_2 \cdot \sigma_5 + t_2 \cdot \sigma_5)$$
$$+ \lambda \cdot (\Upsilon_1 \cdot \sigma_4 - \sigma_1 \cdot \sigma_3 \cdot \Xi_1 - \chi_1 \cdot \sigma_5 + \sigma_5 \cdot t_1)$$
$$+ (\sigma_4 \cdot \Upsilon_0 - \sigma_1 \cdot \sigma_3 - \sigma_5 \cdot \chi_0)$$

$n, m \in \mathbb{N}_0$; $n > m$ and $a_j, c_j : \mathbb{R}_{+0} \to \mathbb{R}$ are continuous and differentiable function of τ such that $a_0 + c_0 \neq 0$. In the following "–" denotes complex and conjugate. P(λ), Q(λ) are analytic function in λ and differentiable in τ. The coefficients $a_j(Y_1, Y_2, Y_3, I_{L_1}, \tau, \ldots) \in \mathbb{R}$ and $c_j(Y_1, Y_2, Y_3, I_{L_1}, \tau, \ldots) \in \mathbb{R}$ depend on MRI system laser diode equivalent circuit's $Y_1, Y_2, Y_3, I_{L_1}, \tau, \ldots$ values. Unless strictly necessary, the designation of the varied arguments: will subsequently be omitted P, Q, a_j, c_j. The coefficients a_j, c_j are continuous and differentiable functions of their arguments, and direct substitution shows that $a_0 + c_0 \neq 0$.

$$- (\sigma_1 \cdot \sigma_2 \cdot \Xi_0 + O_0 \cdot \sigma_6) + \sigma_4 \cdot \Upsilon_0 - \sigma_1 \cdot \sigma_3$$
$$- \sigma_5 \cdot \chi_0 \neq 0 \; \forall \; Y_1, Y_2, Y_3, I_{L_1}, \tau, \ldots \in \mathbb{R}_+$$

i.e. $\lambda = 0$ is not a root of the characteristic equation. Furthermore $P(\lambda)$, $Q(\lambda)$ are analytic functions of λ for which the following requirements of the analysis (see Kuang [2], Sect. 3.4) can also be verified in the present case.

(a) If $\lambda = i \cdot \omega$; $\omega \in \mathbb{R}$ then $P(i \cdot \omega) + Q(i \cdot \omega) \neq 0$, i.e. P and Q have no common imaginary roots. This condition was verified numerically in the entire $(Y_1, Y_2, Y_3, I_{L_1}, \tau, \ldots)$ domain of interest.

(b) $\left|\frac{P(\lambda)}{Q(\lambda)}\right|$ is bounded for $|\lambda| \to \infty$; $\operatorname{Re}\lambda \geq 0$. No roots bifurcation from ∞. Indeed, in the limit: $\left|\frac{Q(\lambda)}{P(\lambda)}\right| = \left|\frac{q_0(\tau)+q_1(\tau)\cdot\lambda+\ldots+q_6(\tau)\cdot\lambda^6}{p_0(\tau)+p_1(\tau)\cdot\lambda+\ldots+p_7(\tau)\cdot\lambda^7}\right|$.

(c) The following expressions exist: $F(\omega) = P(i \cdot \omega)|^2 - |Q(i \cdot \omega)|^2$

$$F(\omega, \tau) = |P(i \cdot \omega, \tau)|^2 - |Q(i \cdot \omega, \tau)|^2 = \sum_{i=0}^{7} \prod\nolimits_{2 \cdot k} \cdot \omega^{2 \cdot k}$$

Has at most a finite number of zeros. Indeed, this is a polynomial in ω (Degree in ω^{14}).

(d) Each positive root $\omega(Y_1, Y_2, Y_3, I_{L_1}, \tau, \ldots)$ of $F(\omega) = 0$ is continuous and differentiable with respect to $Y_1, Y_2, Y_3, I_{L_1}, \tau, \ldots$. The condition can only be assessed numerically.

In addition, since the coefficients in P and Q are real, we have $\overline{P(-i \cdot \omega)} = P(i \cdot \omega)$ and $\overline{Q(-i \cdot \omega)} = Q(i \cdot \omega)$ thus, $\omega > 0$ maybe on eigenvalue of characteristic equation. The analysis consists in identifying the roots of the characteristic equation situated on the imaginary axis of the complex λ plan, whereby increasing the parameters:$Y_1, Y_2, Y_3, I_{L_1}, \tau, \ldots$, Reλ may at the crossing change its sign from (−) to (+), i.e. from a stable focus $E^*\left(I_{L_1}^*, I_{L_2}^*, I_{R_i}^*, I_{R_p}^*, I_{L_i}^*, Y_1^*, Y_2^*, Y_3^*\right) = \left(0, I_{R_i}^* + I_{L_i}^*, I_{R_i}^*, 0, I_{L_i}^*, 0, 0, 0\right)$ to an unstable one, or vice versa. This feature may be further assessed by examining the sign of the partial derivatives with respect to $Y_1, Y_2, Y_3, I_{L_1}, \tau, \ldots$ and system parameters.

$$\Lambda^{-1}(\tau) = \left(\frac{\partial \mathrm{Re}\lambda}{\partial \tau}\right)_{\lambda = i \cdot \omega} ; \quad R_p, C_1, L_1, C_2, R_1, L_2, R_i, C_i, L_i, R_{se} = \mathrm{const}$$

$$\Lambda^{-1}(\tau) = \left(\frac{\partial \mathrm{Re}\lambda}{\partial R_p}\right)_{\lambda = i \cdot \omega} ; \quad \tau, C_1, L_1, C_2, R_1, L_2, R_i, C_i, L_i, R_{se} = \mathrm{const}$$

$$\Lambda^{-1}(\tau) = \left(\frac{\partial \mathrm{Re}\lambda}{\partial C_1}\right)_{\lambda = i \cdot \omega} ; \quad R_p, \tau, L_1, C_2, R_1, L_2, R_i, C_i, L_i, R_{se} = \mathrm{const}$$

$$\Lambda^{-1}(\tau) = \left(\frac{\partial \mathrm{Re}\lambda}{\partial L_1}\right)_{\lambda = i \cdot \omega} ; \quad R_p, C_1, \tau, C_2, R_1, L_2, R_i, C_i, L_i, R_{se} = \mathrm{const}$$

$$\Lambda^{-1}(\tau) = \left(\frac{\partial \mathrm{Re}\lambda}{\partial C_2}\right)_{\lambda = i \cdot \omega} ; \quad R_p, C_1, L_1, \tau, R_1, L_2, R_i, C_i, L_i, R_{se} = \mathrm{const}$$

$$\Lambda^{-1}(\tau) = \left(\frac{\partial \mathrm{Re}\lambda}{\partial R_1}\right)_{\lambda = i \cdot \omega} ; \quad R_p, C_1, L_1, C_2, \tau, L_2, R_i, C_i, L_i, R_{se} = \mathrm{const}$$

$$\Lambda^{-1}(\tau) = \left(\frac{\partial \mathrm{Re}\lambda}{\partial L_2}\right)_{\lambda = i \cdot \omega} ; \quad R_p, C_1, L_1, C_2, R_1, \tau, R_i, C_i, L_i, R_{se} = \mathrm{const}$$

$$\Lambda^{-1}(\tau) = \left(\frac{\partial \mathrm{Re}\lambda}{\partial R_i}\right)_{\lambda = i \cdot \omega} ; \quad R_p, C_1, L_1, C_2, R_1, L_2, \tau, C_i, L_i, R_{se} = \mathrm{const}$$

$$\Lambda^{-1}(\tau) = \left(\frac{\partial \mathrm{Re}\lambda}{\partial C_i}\right)_{\lambda = i \cdot \omega} ; \quad R_p, C_1, L_1, C_2, R_1, L_2, R_i, \tau, L_i, R_{se} = \mathrm{const}$$

$$\Lambda^{-1}(\tau) = \left(\frac{\partial \mathrm{Re}\lambda}{\partial L_i}\right)_{\lambda = i \cdot \omega} ; \quad R_p, C_1, L_1, C_2, R_1, L_2, R_i, C_i, \tau, R_{se} = \mathrm{const}$$

$$\Lambda^{-1}(\tau) = \left(\frac{\partial \mathrm{Re}\lambda}{\partial R_{se}}\right)_{\lambda=i\cdot\omega}; \; R_p, C_1, L_1, C_2, R_1, L_2, R_i, C_i, L_i, \tau = \text{const}$$

$$P(\lambda) = P_R(\lambda) + i \cdot P_I(\lambda); \; Q(\lambda) = Q_R(\lambda) + i \cdot Q_I(\lambda)$$

When writing and inserting $\lambda = i \cdot \omega$ into MRI system laser diode circuit's characteristic equation, ω must satisfy the following equations:

$$\sin(\omega \cdot \tau) = g(\omega) = \frac{-P_R(i\cdot\omega)\cdot Q_I(i\cdot\omega) + P_I(i\cdot\omega)\cdot Q_R(i\cdot\omega)}{|Q(i\cdot\omega)|^2}$$

$$\cos(\omega \cdot \tau) = h(\omega) = -\frac{P_R(i\cdot\omega)\cdot Q_R(i\cdot\omega) + P_I(i\cdot\omega)\cdot Q_I(i\cdot\omega)}{|Q(i\cdot\omega)|^2}$$

where $|Q(i \cdot \omega, \tau)|^2 \neq 0$ in view of requirement (a) above, and $(g, h) \in \mathbb{R}$. Furthermore, it follows above $\sin(\omega \cdot \tau)$ and $\cos(\omega \cdot \tau)$ equations that, by squaring and adding the sides, ω must be a positive root for $F(\omega) = P(i \cdot \omega)|^2 - |Q(i \cdot \omega)|^2$. Note: $F(\omega)$ is dependent onτ. Now it is important to notice that if $\tau \notin I$ (assume that $I \subseteq \mathbb{R}_{+0}$ is the set where $\omega(\tau)$ is a positive root of $F(\omega)$ and for $\tau \notin I$, $\omega(\tau)$ is not defined. Then for all τ in I, $\omega(\tau)$ is satisfied that $F(\omega, \tau) = 0$. Then there are no positive $\omega(\tau)$ solutions for $F(\omega, \tau) = 0$, and we cannot have stability switches. For $\tau \in I$ where $\omega(\tau)$ is a positive solution of $F(\omega, \tau) = 0$, we can define the angle $\theta(\tau) \in [0, 2 \cdot \pi]$ as the solution of $\sin\theta(\tau) = \ldots$; $\cos\theta(\tau) = \ldots$ and the relation between the arguments $\theta(\tau)$ and $\tau \cdot \omega(\tau)$ for $\tau \in I$ must be describing below [10].

$$\sin\theta(\tau) = \frac{-P_R(i\cdot\omega)\cdot Q_I(i\cdot\omega) + P_I(i\cdot\omega)\cdot Q_R(i\cdot\omega)}{|Q(i\cdot\omega)|^2}$$

$$\cos\theta(\tau) = -\frac{P_R(i\cdot\omega)\cdot Q_R(i\cdot\omega) + P_I(i\cdot\omega)\cdot Q_I(i\cdot\omega)}{|Q(i\cdot\omega)|^2}$$

$$\omega(\tau)\cdot\tau = \theta(\tau) + 2\cdot n\cdot\pi \;\forall\, n \in \mathbb{N}_0$$

Hence we can define the maps $\tau_n : I \to \mathbb{R}_{+0}$ given by $\tau_n(\tau) = \frac{\theta(\tau)+2\cdot n\cdot\pi}{\omega(\tau)}$; $n \in \mathbb{N}_0, \tau \in I$.

Let us introduce the functions: $I \to \mathbb{R}$; $S_n(\tau) = \tau - \tau_n(\tau), \tau \in I, n \in \mathbb{N}_0$ that is continuous and differentiable in τ. In the following, the subscripts $\lambda, \omega, R_p, C_1, L_1, \ldots$ indicate the corresponding partial derivatives. Let us first concentrate on Λ(x), remember in $\lambda(\omega, R_p, C_1, L_1, \ldots)$ and $\omega(R_p, C_1, L_1, C_2, \ldots)$, keeping all parameters except one (x) andτ. The derivation closely follows that in reference [BK]. Differentiating MRI system Laser diode circuit characteristic equation $P(\lambda) + Q(\lambda) \cdot e^{-\lambda\cdot\tau} = 0$ with respect to specific parameter (x), and inverting the derivative, for convenience, one calculates: $x = R_p, C_1, L_1, C_2, \ldots$

$$\left(\frac{\partial\lambda}{\partial x}\right)^{-1} = \frac{-P_\lambda(\lambda,x)\cdot Q(\lambda,x) + Q_\lambda(\lambda,x)\cdot P(\lambda,x) - \tau\cdot P(\lambda,x)\cdot Q(\lambda,x)}{P_x(\lambda,x)\cdot Q(\lambda,x) - Q_x(\lambda,x)\cdot P(\lambda,x)}$$

where $P_\lambda = \frac{\partial P}{\partial\lambda}$; $Q_\lambda = \frac{\partial Q}{\partial\lambda}$; $P_x = \frac{\partial P}{\partial x}$; $Q_x = \frac{\partial Q}{\partial x}$, substituting $\lambda = i\cdot\omega$ and bearing $\overline{P(-i\cdot\omega)} = P(i\cdot\omega)$; $\overline{Q(-i\cdot\omega)} = Q(i\cdot\omega)$; $i\cdot P_\lambda(i\cdot\omega) = P_\omega(i\cdot\omega)$; $i\cdot Q_\lambda(i\cdot\omega) = Q_\omega(i\cdot\omega)$

And that on the surface $|P(i\cdot\omega)|^2 = |Q(i\cdot\omega)|^2$, one obtains:

$$\left.\left(\frac{\partial\lambda}{\partial x}\right)^{-1}\right|_{\lambda=i\cdot\omega} = \frac{i\cdot P_\omega(i\cdot\omega,x)\cdot\overline{P(i\cdot\omega,x)} + i\cdot Q_\lambda(i\cdot\omega,x)\cdot\overline{Q(\lambda,x)} - \tau\cdot|P(i\cdot\omega,x)|^2}{P_x(i\cdot\omega,x)\cdot\overline{P(i\cdot\omega,x)} - Q_x(i\cdot\omega,x)\cdot Q(i\cdot\omega,x)}$$

Upon separating into real and imaginary parts, with $P = P_R + i\cdot P_I$; $Q = Q_R + i\cdot Q_I$

$$P_\omega = P_{R\omega} + i\cdot P_{I\omega};\ Q_\omega = Q_{R\omega} + i\cdot Q_{I\omega}$$
$$P_x = P_{Rx} + i\cdot P_{Ix};\ Q_x = Q_{Rx} + i\cdot Q_{Ix};\ P^2 = P_R^2 + P_I^2$$

When (x) can be any MRI system laser diode circuit's parameter $R_p, C_1, L_1, C_2, \ldots$ and time delayτ, etc., Where for convenience, we have dropped the arguments(i,ω,x), and where $F_\omega = 2\cdot[(P_{R\omega}\cdot P_R + P_{I\omega}\cdot P_I) - (Q_{R\omega}\cdot Q_R + Q_{I\omega}\cdot Q_I)]$; $\omega_x = -\frac{F_x}{F_\omega}$
$F_x = 2\cdot[(P_{Rx}\cdot P_R + P_{Ix}\cdot P_I) - (Q_{Rx}\cdot Q_R + Q_{Ix}\cdot Q_I)]$. We define U and V functions:

$$U = (P_R\cdot P_{I\omega} - P_I\cdot P_{R\omega}) - (Q_R\cdot Q_{I\omega} - Q_I\cdot Q_{R\omega})$$
$$V = (P_R\cdot P_{Ix} - P_I\cdot P_{Rx}) - (Q_R\cdot Q_{Ix} - Q_I\cdot Q_{Rx})$$

We choose our specific parameter as time delay x = τ.

$$P_R = \omega^6\cdot(\Omega_5 + \sigma_1\cdot\sigma_2\cdot\Xi_6) - \omega^4\cdot(\Omega_3 + \sigma_1\cdot\sigma_2\cdot\Xi_4 + \sigma_6\cdot O_4)$$
$$+ \omega^2\cdot(\Omega_1 + \sigma_1\cdot\sigma_2\cdot\Xi_2 + O_2\cdot\sigma_6) - (\sigma_1\cdot\sigma_2\cdot\Xi_0 + O_0\cdot\sigma_6)$$

$$P_I = \omega^7\cdot\Omega_6 - \omega^5\cdot(\Omega_4 + \Xi_5\cdot\sigma_1\cdot\sigma_2 + O_5\cdot\sigma_6)$$
$$+ \omega^3\cdot(\Omega_2 + \Xi_3\cdot\sigma_1\cdot\sigma_2 + \sigma_6\cdot O_3)$$
$$- \omega\cdot(\Xi_1\cdot\sigma_1\cdot\sigma_2 + O_1\cdot\sigma_6)$$

$$Q_R = \omega^6\cdot\sigma_1\cdot\sigma_3\cdot\Xi_6 + \omega^4\cdot(\Upsilon_4\cdot\sigma_4 - \sigma_1\cdot\sigma_3\cdot\Xi_4 - \chi_4\cdot\sigma_5 + \sigma_5\cdot l_4)$$
$$- \omega^2\cdot(\Upsilon_2\cdot\sigma_4 - \sigma_1\cdot\sigma_3\cdot\Xi_2 - \chi_2\cdot\sigma_5 + c_2\cdot\sigma_5) + (\sigma_4\cdot\Upsilon_0 - \sigma_1\cdot\sigma_3 - \sigma_5\cdot\chi_0)$$

$$Q_I = -\omega^5 \cdot (\Xi_5 \cdot \sigma_1 \cdot \sigma_2 + \chi_5 \cdot \sigma_5) - \omega^3 \cdot (\Upsilon_3 \cdot \sigma_4 - \sigma_1 \cdot \sigma_3 \cdot \Xi_3 - \chi_3 \cdot \sigma_5 + \sigma_5 \cdot t_3)$$
$$+ \omega \cdot (\Upsilon_1 \cdot \sigma_4 - \sigma_1 \cdot \sigma_3 \cdot \Xi_1 - \chi_1 \cdot \sigma_5 + \sigma_5 \cdot t_1)$$

$$P_{R\omega} = 6 \cdot \omega^5 \cdot (\Omega_5 + \sigma_1 \cdot \sigma_2 \cdot \Xi_6) - 4 \cdot \omega^3 \cdot (\Omega_3 + \sigma_1 \cdot \sigma_2 \cdot \Xi_4 + \sigma_6 \cdot \mathrm{O}_4)$$
$$+ 2 \cdot \omega \cdot (\Omega_1 + \sigma_1 \cdot \sigma_2 \cdot \Xi_2 + \mathrm{O}_2 \cdot \sigma_6)$$

$$P_{I\omega} = 7 \cdot \omega^6 \cdot \Omega_6 - 5 \cdot \omega^4 \cdot (\Omega_4 + \Xi_5 \cdot \sigma_1 \cdot \sigma_2 + \mathrm{O}_5 \cdot \sigma_6)$$
$$+ 3 \cdot \omega^2 \cdot (\Omega_2 + \Xi_3 \cdot \sigma_1 \cdot \sigma_2 + \sigma_6 \cdot \mathrm{O}_3)$$
$$- (\Xi_1 \cdot \sigma_1 \cdot \sigma_2 + \mathrm{O}_1 \cdot \sigma_6)$$

$$Q_{R\omega} = 6 \cdot \omega^5 \cdot \sigma_1 \cdot \sigma_3 \cdot \Xi_6 + 4 \cdot \omega^3 \cdot (\Upsilon_4 \cdot \sigma_4 - \sigma_1 \cdot \sigma_3 \cdot \Xi_4 - \chi_4 \cdot \sigma_5 + \sigma_5 \cdot t_4)$$
$$- 2 \cdot \omega \cdot (\Upsilon_2 \cdot \sigma_4 - \sigma_1 \cdot \sigma_3 \cdot \Xi_2 - \chi_2 \cdot \sigma_5 + t_2 \cdot \sigma_5)$$

$$Q_{I\omega} = -5 \cdot \omega^4 \cdot (\Xi_5 \cdot \sigma_1 \cdot \sigma_2 + \chi_5 \cdot \sigma_5)$$
$$- 3 \cdot \omega^2 \cdot (\Upsilon_3 \cdot \sigma_4 - \sigma_1 \cdot \sigma_3 \cdot \Xi_3 - \chi_3 \cdot \sigma_5 + \sigma_5 \cdot l_3)$$
$$+ (\Upsilon_1 \cdot \sigma_4 - \sigma_1 \cdot \sigma_3 \cdot \Xi_1 - \chi_1 \cdot \sigma_5 + \sigma_5 \cdot c_1)$$

$$P_{R\tau} = 0; \quad P_{I\tau} = 0; \quad \mathrm{Q}_{R\tau} = 0; \quad Q_{I\tau} = 0; \quad \omega_\tau = -\frac{F_\tau}{F_\omega}$$

$$P_{R\omega} \cdot P_R = \{6 \cdot \omega^5 \cdot (\Omega_5 + \sigma_1 \cdot \sigma_2 \cdot \Xi_6) - 4 \cdot \omega^3 \cdot (\Omega_3 + \sigma_1 \cdot \sigma_2 \cdot \Xi_4 + \sigma_6 \cdot \mathrm{O}_4)$$
$$+ 2 \cdot \omega \cdot (\Omega_1 + \sigma_1 \cdot \sigma_2 \cdot \Xi_2 + \mathrm{O}_2 \cdot \sigma_6)\}$$
$$\cdot \left\{\omega^6 \cdot (\Omega_5 + \sigma_1 \cdot \sigma_2 \cdot \Xi_6) - \omega^4 \cdot (\Omega_3 + \sigma_1 \cdot \sigma_2 \cdot \Xi_4 + \sigma_6 \cdot \mathrm{O}_4)\right.$$
$$+ \omega^2 \cdot (\Omega_1 + \sigma_1 \cdot \sigma_2 \cdot \Xi_2 + \mathrm{O}_2 \cdot \sigma_6)$$
$$\left. - (\sigma_1 \cdot \sigma_2 \cdot \Xi_0 + \mathrm{O}_0 \cdot \sigma_6)\right\}$$

$$Q_{R\omega} \cdot Q_R = \left\{6 \cdot \omega^5 \cdot \sigma_1 \cdot \sigma_3 \cdot \Xi_6 + 4 \cdot \omega^3 \cdot (\Upsilon_4 \cdot \sigma_4 - \sigma_1 \cdot \sigma_3 \cdot \Xi_4 - \chi_4 \cdot \sigma_5 + \sigma_5 \cdot t_4)\right.$$
$$\left. - 2 \cdot \omega \cdot (\Upsilon_2 \cdot \sigma_4 - \sigma_1 \cdot \sigma_3 \cdot \Xi_2 - \chi_2 \cdot \sigma_5 + t_2 \cdot \sigma_5)\right\} \cdot$$
$$\left\{\omega^6 \cdot \sigma_1 \cdot \sigma_3 \cdot \Xi_6 + \omega^4 \cdot (\Upsilon_4 \cdot \sigma_4\right.$$
$$-\sigma_1 \cdot \sigma_3 \cdot \Xi_4 - \chi_4 \cdot \sigma_5 + \sigma_5 \cdot t_4) - \omega^2 \cdot$$
$$(\Upsilon_2 \cdot \sigma_4 - \sigma_1 \cdot \sigma_3 \cdot \Xi_2 - \chi_2 \cdot \sigma_5 + t_2 \cdot \sigma_5)$$
$$\left. +(\sigma_4 \cdot \Upsilon_0 - \sigma_1 \cdot \sigma_3 - \sigma_5 \cdot \chi_0)\right\}$$

$$Q_R \cdot Q_{I\omega} = \left\{\omega^6 \cdot \sigma_1 \cdot \sigma_3 \cdot \Xi_6 + \omega^4 \cdot (\Upsilon_4 \cdot \sigma_4 - \sigma_1 \cdot \sigma_3 \cdot \Xi_4 - \chi_4 \cdot \sigma_5 + \sigma_5 \cdot t_4)\right.$$
$$\left. -\omega^2 \cdot (\Upsilon_2 \cdot \sigma_4 - \sigma_1 \cdot \sigma_3 \cdot \Xi_2 - \chi_2 \cdot \sigma_5 + t_2 \cdot \sigma_5) + (\sigma_4 \cdot \Upsilon_0 - \sigma_1 \cdot \sigma_3 - \sigma_5 \cdot \chi_0)\right\}$$
$$\cdot \left\{-5 \cdot \omega^4 \cdot (\Xi_5 \cdot \sigma_1 \cdot \sigma_2 + \chi_5 \cdot \sigma_5) - 3 \cdot \omega^2 \cdot (\Upsilon_3 \cdot \sigma_4 - \sigma_1 \cdot \sigma_3 \cdot \Xi_3 - \chi_3 \cdot \sigma_5 + \sigma_5 \cdot t_3)\right.$$

$$+(\Upsilon_1 \cdot \sigma_4 - \sigma_1 \cdot \sigma_3 \cdot \Xi_1 - \chi_1 \cdot \sigma_5 + \sigma_5 \cdot t_1)\}$$

$$Q_I \cdot Q_{R\omega} = \left\{-\omega^5 \cdot (\Xi_5 \cdot \sigma_1 \cdot \sigma_2 + \chi_5 \cdot \sigma_5) - \omega^3 \cdot (\Upsilon_3 \cdot \sigma_4 - \sigma_1 \cdot \sigma_3 \cdot \Xi_3 - \chi_3 \cdot \sigma_5 + \sigma_5 \cdot t_3)\right.$$
$$+\omega \cdot (\Upsilon_1 \cdot \sigma_4 - \sigma_1 \cdot \sigma_3 \cdot \Xi_1 - \chi_1 \cdot \sigma_5 + \sigma_5 \cdot t_1)\} \cdot \left\{6 \cdot \omega^5 \cdot \sigma_1 \cdot \sigma_3 \cdot \Xi_6 + 4 \cdot \omega^3 \cdot (\Upsilon_4 \cdot \sigma_4\right.$$
$$-\sigma_1 \cdot \sigma_3 \cdot \Xi_4 - \chi_4 \cdot \sigma_5 + \sigma_5 \cdot l_4) - 2 \cdot \omega \cdot (\Upsilon_2 \cdot \sigma_4 - \sigma_1 \cdot \sigma_3 \cdot \Xi_2 - \chi_2 \cdot \sigma_5 + t_2 \cdot \sigma_5)\}$$

$$V = (P_R \cdot P_{I\tau} - P_I \cdot P_{R\tau}) - (Q_R \cdot Q_{I\tau} - Q_I \cdot Q_{R\tau}) = 0; \;\; F(\omega, \tau) = 0$$

Differentiating with respect to τ and we get $F_\omega \cdot \frac{\partial \omega}{\partial \tau} + F_\tau = 0; \;\; \tau \in I \Rightarrow \frac{\partial \omega}{\partial \tau} = -\frac{F_\tau}{F_\omega}$

$$\Lambda^{-1}(\tau) = \left(\frac{\partial \mathrm{Re}\lambda}{\partial \tau}\right)_{\lambda = i \cdot \omega}; \;\; \frac{\partial \omega}{\partial \tau} = \omega_\tau = -\frac{F_\tau}{F_\omega}; \;\; \Lambda^{-1}(\tau)$$
$$= \mathrm{Re}\left\{\frac{-2 \cdot [U + \tau \cdot |P|^2] + i \cdot F_\omega}{F_\tau + i \cdot 2 \cdot [V + \omega \cdot |P|^2]}\right\}$$

$$sign\{\Lambda^{-1}(\tau)\} = sign\left\{\left(\frac{\partial \mathrm{Re}\lambda}{\partial \tau}\right)_{\lambda = i \cdot \omega}\right\}; \;\; sign\{\Lambda^{-1}(\tau)\}$$
$$= sign\{F_\omega\} \cdot sign\left\{\frac{V + \frac{\partial \omega}{\partial \tau} \cdot U}{|P|^2} + \omega + \frac{\partial \omega}{\partial \tau} \cdot \tau\right\}$$

We shall presently examine the possibility of stability transitions (bifurcations) MRI system laser diode circuit, about the equilibrium point$E^*\left(I_L^*, I_{L_2}^*, I_{R_i}^*, I_{R_p}^*, I_{L_i}^*, Y_1^*, Y_2^*, Y_3^*\right)$.

$E^*\left(I_{L_1}^*, I_{L_2}^*, I_{R_i}^*, I_{R_p}^*, I_{L_i}^*, Y_1^*, Y_2^*, Y_3^*\right) = \left(0, I_{R_i}^* + I_{L_i}^*, I_{R_i}^*, 0, I_{L_i}^*, 0, 0, 0\right)$ as a result of a variation of delay parameter τ. The analysis consists in identifying the roots of our system characteristic equation situated on the imaginary axis of the complex λ plane, Where by increasing the delay parameter τ, Reλ may at the crossing, changes its sign from "−" to " + ", i.e. from a stable focus E^* to an unstable one, or vice versa. This feature may be further assessed by examining the sign of the partial derivatives with respect toτ.

$$\Lambda^{-1}(\tau) = \left(\frac{\partial \mathrm{Re}\lambda}{\partial C_1}\right)_{\lambda = i \cdot \omega}; \;\; R_p, \tau, L_1, C_2, R_1, L_2, R_i, C_i, L_i, R_{se} = \text{const}$$

$$U = (P_R \cdot P_{I\omega} - P_I \cdot P_{R\omega}) - (Q_R \cdot Q_{I\omega} - Q_I \cdot Q_{R\omega})$$

After getting the expression for $F(\omega, \tau)$ MRI system laser diode circuit's parameter values, we find those ω, τ values which fulfill $F(\omega, \tau) = 0$. We ignore negative, complex, and imaginary values of ω for specific τ values $\tau \in [0.001 \ldots 10]$, it can be express by 3D function $F(\omega, \tau) = 0$. The next step is to plot the stability switch diagram based on different delay values of our system.

$$\Lambda^{-1}(\tau) = \left(\frac{\partial \mathrm{Re}\lambda}{\partial \tau}\right)_{\lambda=i\cdot\omega} = \mathrm{Re}\left\{\frac{-2\cdot[U+\tau\cdot|P|^2]+i\cdot F_\omega}{F_\tau+i\cdot 2\cdot[V+\omega\cdot|P|^2]}\right\};\ \Lambda^{-1}(\tau)$$
$$= \frac{2\cdot\{F_\omega\cdot(V+\omega\cdot P^2)-F_\tau\cdot(U+\tau\cdot P^2)}{F_\tau^2+4\cdot(V+\omega\cdot P^2)^2}$$

The stability switch occurs only on those delay values (τ) which fit the equation: $\tau = \frac{\theta_+(\tau)}{\omega_+(\tau)}$ and $\theta_+(\tau)$ is the solution of $\sin\theta(\tau) = \ldots$; $\cos\theta(\tau) = \ldots$ when $\omega = \omega_+(\tau)$ if only ω_+ is feasible. Additionally, when all MRI system laser diode circuit parameters are known and the stability switch due to various time delay valuesτ is describe in the following expression:

$$sign\{\Lambda^{-1}(\tau)\} = sign\{F_\omega(\omega(\tau),\tau)\}\cdot$$
$$sign\left\{\tau\cdot\omega_\tau(\omega(\tau))+\omega(\tau)+\frac{U(\omega(\tau))\cdot\omega_\tau(\omega(\tau))+V(\omega(\tau))}{|P(\omega(\tau))|^2}\right\}$$

Remark We know $F(\omega,\tau) = 0$ implies its roots $\omega_i(\tau)$ and finding those delays values τ which ω_i is feasible. There are τ values which give complex ω_i or imaginary number, then unable to analyze stability. F function is independent on τ parameter $F(\omega,\tau) = 0$.

The results: we find those ω, τ values which fulfill $F(\omega,\tau) = 0$. We ignore negative, complex, and imaginary values ofω. $F(\omega) = 0 \Rightarrow \omega_1, \omega_2, \ldots$ and next is to find those ω, τ values which fulfill $\sin\theta(\tau) = \ldots$; $\cos\theta(\tau) = \ldots$. Finally, we plot the stability which diagram.

$$g(\tau) = \Lambda^{-1}(\tau) = \left(\frac{\partial \mathrm{Re}\lambda}{\partial \tau}\right)_{\lambda=i\cdot\omega};\ g(\tau)$$
$$= \frac{2\cdot\{F_\omega\cdot(V+\omega\cdot P^2)-F_\tau\cdot(U+\tau\cdot P^2)\}}{F_\tau^2+4\cdot(V+\omega\cdot P^2)^2}$$

$$sign[g(\tau)] = sign[\Lambda^{-1}(\tau)] = sign\left[\left(\frac{\partial \mathrm{Re}\lambda}{\partial \tau}\right)_{\lambda=i\cdot\omega}\right]$$
$$= sign\left[\frac{2\cdot\{F_\omega\cdot(V+\omega\cdot P^2)-F_\tau\cdot(U+\tau\cdot P^2)\}}{F_\tau^2+4\cdot(V+\omega\cdot P^2)^2}\right]$$

$$F_\tau^2+4\cdot(V+\omega\cdot P^2)^2 > 0$$
$$sign[\Lambda^{-1}(\tau)] = sign\{F_\omega\cdot(V+\omega\cdot P^2)-F_\tau\cdot(U+\tau\cdot P^2)\}$$

$$sign[\Lambda^{-1}(\tau)] = sign\left\{[F_\omega]\cdot\left[(V+\omega\cdot P^2)-\frac{F_\tau}{F_\omega}\cdot(U+\tau\cdot P^2)\right]\right\};\ \omega_\tau = -\frac{F_\tau}{F_\omega};\ \omega_\tau$$
$$= \left(\frac{\partial\omega}{\partial\tau}\right)^{-1} = -\frac{\partial F/\partial\omega}{\partial F/\partial\tau}$$

Table 1.129 Sign of $\Lambda^{-1}(\tau)$ for stability switching

$sign[F_{\omega}]$	$sign\left[\frac{V+\omega_{\tau}\cdot U}{P^2}+\omega+\omega_{\tau}\cdot\tau\right]$	$sign[\Lambda^{-1}(\tau)]$
$\pm$	$\pm$	+
$\pm$	$\mp$	−

$$sign[\Lambda^{-1}(\tau)] = sign\{[F_{\omega}]\cdot[V+\omega_{\tau}\cdot U+\omega\cdot P^2+\omega_{\tau}\cdot\tau\cdot P^2]\}$$

$$sign[\Lambda^{-1}(\tau)] = sign\left\{[F_{\omega}]\cdot\left[\frac{1}{P^2}\right]\cdot\left[\frac{V+\omega_{\tau}\cdot U}{P^2}+\omega+\omega_{\tau}\cdot\tau\right]\right\}$$

$$sign\left[\frac{1}{P^2}\right] > 0 \Rightarrow sign[\Lambda^{-1}(\tau)] = sign\left\{[F_{\omega}]\cdot\left[\frac{V+\omega_{\tau}\cdot U}{P^2}+\omega+\omega_{\tau}\cdot\tau\right]\right\}$$

$$sign[\Lambda^{-1}(\tau)] = sign[F_{\omega}]\cdot sign\left[\frac{V+\omega_{\tau}\cdot U}{P^2}+\omega+\omega_{\tau}\cdot\tau\right]$$

$$F_{\omega} = 2\cdot[(P_{R\omega}\cdot P_R+P_{I\omega}\cdot P_I)-(Q_{R\omega}\cdot Q_R+Q_{I\omega}\cdot Q_I)]$$

We check the sign of $\Lambda^{-1}(\tau)$ according the following rule (Table 1.129).

If $sign[\Lambda^{-1}(\tau)] > 0$ then the crossing proceeds from (−) to (+) respectively (stable to unstable). If $\text{sign}[\Lambda^{-1}(\tau)] < 0$ then the crossing proceeds from (+) to (−) respectively (unstable to stable). The stability switching can occur for specific ω values.

1.4 Dynamic of Electron-Photon Exchanges into VCSEL, Stability Optimization Under Delayed Carrier-Photon Interaction in Time

In this sub chapter, we discuss the dynamic of electron-photon exchanges into VCSEL, stability optimization under delayed carrier-photon interaction in time. Vertical-cavity surface-emitting lasers (VCSELs) employ current-blocking layers to funnel the current to the active region located between the mirrors of the laser. The implementation depth is limited so that the lateral resistance can become substantial for devices with large-area contacts. First we describe the VCSEL DC characteristics. Above the diode threshold potential (typically 1.5–1.6 v), the VCSEL current has an almost linear dependence on the applied forward-bias voltage, specified by a resistance R_d. Lasing occurs when the injected current exceeds the VCSEL's threshold current I_{th}, and the optical output power is linearly proportional to the input current by the slope efficiency factor η. VCSELs typically have thresholds currents below 1 mA and slope efficiencies of 0.3–0.5 mW/mA. At higher current densities (above

10 mA), the slope efficiency decreases due to self-heating effects in the VCSEL. VCSELs can be designed to lase light of a variety of Datacom wavelengths. GaAs-based VCSELs operating at 850 nm designed for optical interconnects and short distances links are the most mature and common VCSEL technology. Longer wavelength GaAs and InP-based VCSELs lasing at 1.3 and 1.55 μm have developed for use in long-distance links due to lower loss and fiber dispersion at those wavelengths. The VCSEL structure is formed by two horizontal electrically conductive distributed Bragg reflector (DBR) surrounding an optical gain region. The highly reflective DBR mirrors provide the necessary optical feedback for lasing to occur. The gain region of most high speed VCSELs is constructed of several thin semiconductor layers that form multiple quanta well (MQW). The VCSEL's optical characteristics and the emitted wavelength are thus determined by the gain profile of the quantum well material and the geometry of the DBR. VCSEL optical response is regulated by two coupled differential equations governing the carrier (N) and photon (S) density in the active region. When the laser is turned on, a current I_j flows into the laser's gain region volume V_{qw}, increasing the carrier density. Photon generation begin as spontaneous emission (β), until the carrier density exceeds the threshold level, stimulated emission occurs, determined by the gain coefficient G. The time difference between the current being applied and the light emission is the laser's turn on delay. N(t) and S(t) are the densities in the VCSEL active region of the carrier and photon respectively. Due to VCSEL active region parasitic effects, there is a time delay (τ) of the carrier-photon interaction time. We introduce the VCSEL carrier-photon interaction active region volume (Fig. 1.5) [7, 13].

Terminology:

N—Carrier density in the active region.
S—Photon density in the active region.
N_0—Carrier density in the active region at $t = 0$.
η—Injection efficiency.
τ_n—Carrier recombination lifetime.
τ_p—Photon lifetime.
G—Gain coefficient.
N_{tr}—Carrier's transparency number.
β—Spontaneous emission coupling coefficient.
ε—Gain compression factor.
I—VCSEL's injection current.
I_{off}—VCSEL's offset current.

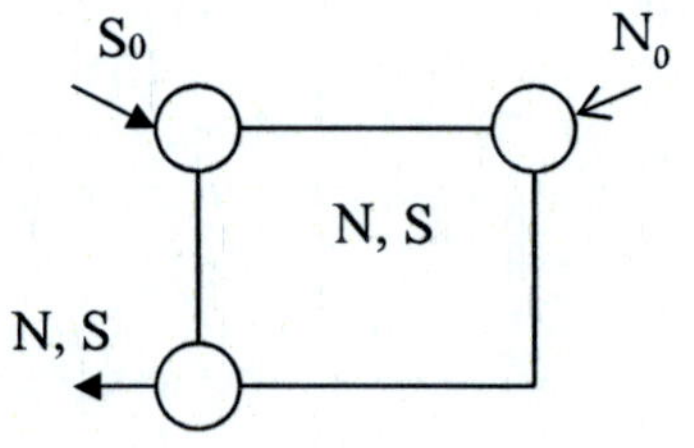

Fig. 1.5 VCSEL carrier-photon interaction active region volume

τ—Time delay due to VCSEL active region parasitic effects.

The VCSEL rate equation with delay parameter (τ) of the carrier-photon interaction time:

$$\frac{dN(t)}{dt} = \frac{\eta \cdot \left(I - I_{off}\right)}{q} - \frac{N(t)}{\tau_n} - \frac{G \cdot (N(t-\tau) - N_{tr}) \cdot S(t-\tau)}{1 + \varepsilon \cdot S(t-\tau)}; \ \frac{dS(t)}{dt}$$
$$= -\frac{S(t)}{\tau_p} + \frac{\beta \cdot N(t)}{\tau_n} + \frac{G \cdot (N(t-\tau) - N_{tr}) \cdot S(t-\tau)}{1 + \varepsilon \cdot S(t-\tau)}$$

$$N = N(t); \ S = S(t)$$

Optical output power $P_0 = k \cdot S$; k-scaling factor accounting for the output coupling efficiency of the VCSEL.

$$T = T_0 + (I \cdot V - P_0) \cdot R_{th} - \tau_{th} \cdot \frac{dT}{dt}$$

R_{th}—VCSEL's thermal impedance (which related the change in device temperature to the power dissipated as heat).

τ_{th}—Thermal time constant (which is necessary to account for the nonzero response time of the device temperature, observed to be on the order of 1 μs).

T_0—Ambient temperature.

V—Laser voltage.

Under DC condition $\frac{dT}{dt}$ term disappears. Expression $(I \cdot V - P_0)$ models the power dissipated in the VCSEL, where we assume that any power not carried in the optical output is dissipated as heat in the device. We describe the key VCSEL parameters as functions of temperature, in particular the laser gain and the thermal physics of the device. We partition the thermal threshold current into a constant value of threshold current I_{tho} plus an empirical offset current $I_{off}(T)$.

$$P_o = \eta \cdot \left(I - I_{tho} - I_{off}(T)\right)$$

All static thermal effects are accounted for via the offset current model. This offset current using a polynomial function of temperature.

$$I_{off}(T) = \sum_{k=0}^{\infty} a_k \cdot T^k = a_0 + a_1 \cdot T + a_2 \cdot T^2 + a_3 \cdot T^3 + a_4 \cdot T^4 + \ldots$$

It fit of the threshold current $I_{th}(T)$ to the expression $I_{tho} + I_{off}(T)$. The equilibrium points (fixed points) of our system $\frac{dN}{dt} = 0$; $\frac{dS}{dt} = 0$; $\lim_{t \to \infty} S(t - \tau) = S$; $\lim_{t \to \infty} N(t - \tau) = N$. We define as VCSEL fixed points: $N^{(j)}, S^{(j)}$; $j = 0, 1, 2, \ldots$ index of VCSEL fixed point is (j).

$$\frac{\eta \cdot (I - I_{off})}{q} - \frac{N^{(j)}}{\tau_n} - \frac{G \cdot (N^{(j)} - N_{tr}) \cdot S^{(j)}}{1 + \varepsilon \cdot S^{(j)}} = 0; \quad -\frac{S^{(j)}}{\tau_p} + \frac{\beta \cdot N^{(j)}}{\tau_n} + \frac{G \cdot (N^{(j)} - N_n) \cdot S^{(j)}}{1 + \varepsilon \cdot S^{(j)}} = 0$$

$$N^{(j)} = \frac{\tau_n}{\tau_p \cdot (\beta - 1)} \cdot S^{(j)} - \frac{\tau_n \cdot \eta \cdot (I - I_{off})}{(\beta - 1) \cdot q}; \quad A = \frac{\tau_n}{\tau_p \cdot (\beta - 1)}$$

$$B = \frac{\tau_n \cdot \eta \cdot (I - I_{off})}{(\beta - 1) \cdot q}$$

$$A = A(\tau_n, \tau_p, \beta); \quad B = B(\tau_n, \eta, \beta, I, I_{off}); \quad N^{(j)} = A \cdot S^{(j)} - B$$

$$[S^{(j)}]^2 \cdot \left[\frac{A \cdot \varepsilon}{\tau_n} + G \cdot A\right] + S^{(j)} \cdot \left[\frac{A}{\tau_n} - G \cdot B - G \cdot N_{tr} - \left(\frac{\eta \cdot (I - I_{off})}{q} + \frac{B}{\tau_n}\right) \cdot \varepsilon\right] - \left[\frac{\eta \cdot (I - I_{off})}{q} + \frac{B}{\tau_n}\right] = 0$$

$$\frac{A \cdot \varepsilon}{\tau_n} = \frac{\varepsilon}{\tau_p \cdot (\beta - 1)}; \quad \frac{A}{\tau_n} = \frac{1}{\tau_p \cdot (\beta - 1)}; \quad \frac{B}{\tau_n} = \frac{\eta \cdot (I - I_{off})}{(\beta - 1) \cdot q}$$

$$[S^{(j)}]^2 \cdot \left[\frac{\varepsilon + G \cdot \tau_n}{\tau_p \cdot (\beta - 1)}\right] + S^{(j)} \cdot \left[\frac{1}{\tau_p \cdot (\beta - 1)} - G \cdot \frac{\tau_n \cdot \eta \cdot (I - I_{off})}{(\beta - 1) \cdot q} - G \cdot N_{tr} - \frac{\eta \cdot (I - I_{off})}{q} \cdot \frac{\beta}{(\beta - 1)} \cdot \varepsilon\right] - \frac{\eta \cdot (I - I_{off})}{q} \cdot \frac{\beta}{(\beta - 1)} = 0$$

$$\Xi_1 = \frac{\varepsilon + G \cdot \tau_n}{\tau_p \cdot (\beta - 1)}; \quad \Xi_3 = -\frac{\eta \cdot (I - I_{off})}{q} \cdot \frac{\beta}{(\beta - 1)}; \quad \Xi_2 = \frac{1}{\tau_p \cdot (\beta - 1)} - G \cdot \frac{\tau_n \cdot \eta \cdot (I - I_{off})}{(\beta - 1) \cdot q} G \cdot N_{tr} - \frac{\eta \cdot (I - I_{off})}{q} \cdot \frac{\beta}{(\beta - 1)} \cdot \varepsilon$$

$$[S^{(j)}]^2 \cdot \Xi_1 + S^{(j)} \cdot \Xi_2 + \Xi_3 = 0; \quad N^{(j)} \geq 0; \quad S^{(j)} \geq 0$$

$$S^{(j)} = \frac{-\Xi_2 \pm \sqrt{\Xi_2^2 - 4 \cdot \Xi_1 \cdot \Xi_3}}{2 \cdot \Xi_1}; \quad S^{(j)} = \frac{-\Xi_2}{2 \cdot \Xi_1} \pm \frac{\sqrt{\Xi_2^2 - 4 \cdot \Xi_1 \cdot \Xi_3}}{2 \cdot \Xi_1}; \quad N^{(j)} = -\left(\frac{\Xi_2 \cdot A}{2 \cdot \Xi_1} + B\right) \pm \frac{A \cdot \sqrt{\Xi_2^2 - 4 \cdot \Xi_1 \cdot \Xi_3}}{2 \cdot \Xi_1}$$

We can check our Electron-photon exchanges into VCSEL system for $\tau \neq 0$; $\tau > 0$.The standard local stability analysis about any one of the equilibrium points of the Electron-photon exchanges into VCSEL system consists in adding to coordinate

[S, N] arbitrarily small increments of exponential form $[s, n] \cdot e^{\lambda t}$ and retaining the first order terms in S, N. The system of two homogeneous equations leads to a polynomial characteristic equation in the eigenvalues. The polynomial characteristic equations accept by set the below carrier density in the active region (N) and photon density in the active region (S) and derivatives with respect to time into VCSEL rate equation with delay parameter (τ) of the carrier-photon interaction time. Electron-photon exchanges into VCSEL system fixed values with arbitrarily small increaments of exponential form $[s, n] \cdot e^{\lambda t}$ are j = 0 (first fixed point), j = 1 (second fixed point), j = 2 (third fixed point), etc., $\left(S(t) = S^{(j)} + s \cdot e^{\lambda \cdot t};\ N(t) = N^{(j)} + n \cdot e^{\lambda \cdot t}\right)$. We choose these expressions for ourselves S(t), N(t) as a small displacement [s, n] from Electron-photon exchanges into VCSEL system fixed points in time t = 0.

$$S(t = 0) = S^{(j)} + s;\ N(t = 0) = N^{(j)} + n$$

For $t > 0;\ \lambda < 0$ the selected fixed point is stable otherwise $t > 0;\ \lambda > 0$ is unstable. Our system tends to the selected fixed point exponentially for $t > 0;\ \lambda < 0$ otherwise go away from the selected fixed point (equilibrium point) exponentially. Eigenvalue λ parameter is established if the fixed point is stable or unstable; additionally, his absolute value |λ| establishes the speed of flow toward or away from the selected fixed point. The speed of flow toward or away from the selected fixed point for VCSEL rate equations with delay parameter (τ) system with respect to time are as follow: $\frac{dS(t)}{dt} = \lambda \cdot s \cdot e^{\lambda \cdot t};\ \frac{dN(t)}{dt} = \lambda \cdot n \cdot e^{\lambda \cdot t}$.

Additionally

$$S(t - \tau) = S^{(j)} + s \cdot e^{\lambda(t-\tau)} = S^{(j)} + s \cdot e^{\lambda \cdot t} \cdot e^{-\lambda \cdot \tau};\ N(t - \tau) = N^{(j)} + n \cdot e^{\lambda \cdot (t-\tau)} = N^{(j)} + n \cdot e^{\lambda \cdot t} \cdot e^{-\lambda \cdot \tau}$$

$$\frac{dS(t - \tau)}{dt} = \lambda \cdot s \cdot e^{i \cdot t} \cdot e^{-\lambda \cdot \tau};\ \frac{dN(t - \tau)}{dt} = \lambda \cdot n \cdot e^{\lambda \cdot t} \cdot e^{-\lambda \cdot \tau}.$$

$$\frac{dN(t)}{dt} = \frac{\eta \cdot \left(I - I_{off}\right)}{q} - \frac{N(t)}{\tau_n} - \frac{G \cdot (N(t - \tau) - N_{tr}) \cdot S(t - \tau)}{1 + \varepsilon \cdot S(t - \tau)}$$

$$\lambda \cdot n \cdot e^{\lambda \cdot t} = \frac{\eta \cdot \left(I - I_{off}\right)}{q} - \frac{\left(N^{(j)} + n \cdot e^{\lambda \cdot t}\right)}{\tau_n} - \frac{G \cdot \left(N^{(j)} + n \cdot e^{\lambda \cdot t} \cdot e^{-\lambda \cdot \tau} - N_{tr}\right) \cdot \left(S^{(j)} + s \cdot e^{\lambda \cdot t} \cdot e^{-\lambda \cdot \tau}\right)}{1 + \varepsilon \cdot \left(S^{(j)} + s \cdot e^{\lambda \cdot t} \cdot e^{-\lambda \cdot \tau}\right)}$$

$$\lambda \cdot n \cdot e^{\lambda \cdot t} = \frac{\eta \cdot \left(I - I_{off}\right)}{q} - \frac{N^{(j)}}{\tau_n} - \frac{n \cdot e^{\lambda \cdot t}}{\tau_n} - \frac{G \cdot \left(N^{(j)} + n \cdot e^{\lambda \cdot t} \cdot e^{-\lambda \cdot \tau} - N_{tr}\right) \cdot \left(S^{(j)} + s \cdot e^{\lambda \cdot t} \cdot e^{-\lambda \cdot \tau}\right)}{1 + \varepsilon \cdot \left(S^{(j)} + s \cdot e^{\lambda \cdot t} \cdot e^{-\lambda \cdot \tau}\right)}$$

$$\begin{aligned}&\left(N^{(j)}+n\cdot e^{\lambda\cdot t}\cdot e^{-\lambda\cdot\tau}-N_{tr}\right)\cdot\left(S^{(j)}+s\cdot e^{\lambda\cdot t}\cdot e^{-\lambda\cdot\tau}\right)\\&\quad=N^{(j)}\cdot S^{(j)}+N^{(j)}\cdot s\cdot e^{\lambda\cdot t}\cdot e^{-\lambda\cdot\tau}\\&\quad+n\cdot e^{\lambda\cdot t}\cdot e^{-\lambda\cdot\tau}\cdot S^{(j)}+n\cdot s\cdot e^{2\cdot\lambda\cdot t}\cdot e^{-2\cdot\lambda\cdot\tau}\\&\quad-N_{tr}\cdot S^{(j)}-N_{tr}\cdot s\cdot e^{\lambda\cdot t}\cdot e^{-\lambda\cdot\tau}\end{aligned}$$

$$\lambda\cdot n\cdot e^{\lambda\cdot t}=\frac{\eta\cdot\left(I-I_{off}\right)}{q}-\frac{N^{(j)}}{\tau_n}-\frac{n\cdot e^{\lambda\cdot t}}{\tau_n}-\frac{\begin{array}{l}G\cdot\left\{N^{(j)}\cdot S^{(j)}+N^{(j)}\cdot s\cdot e^{\lambda\cdot t}\cdot e^{-\lambda\cdot\tau}+n\cdot e^{\lambda\cdot t}\cdot e^{-\lambda\cdot\tau}\cdot S^{(j)}\right.\\\left.+n\cdot s\cdot e^{2\cdot\lambda\cdot t}\cdot e^{-2\cdot\lambda\cdot\tau}-N_{tr}\cdot S^{(j)}-N_{tr}\cdot s\cdot e^{\lambda\cdot t}\cdot e^{-i\cdot\tau}\right\}\end{array}}{1+\varepsilon\cdot\left(S^{(j)}+s\cdot e^{\lambda\cdot t}\cdot e^{-\lambda\cdot\tau}\right)}$$

$$n\cdot s\to 0\Rightarrow n\cdot s\cdot e^{2\cdot\lambda\cdot t}\cdot e^{-2\cdot\lambda\cdot\tau}\to 0$$

$$\lambda\cdot n\cdot e^{\lambda\cdot t}=\frac{\eta\cdot\left(I-I_{off}\right)}{q}-\frac{N^{(j)}}{\tau_n}-\frac{n\cdot e^{\lambda\cdot t}}{\tau_n}-\frac{\begin{array}{l}G\cdot\left\{N^{(j)}\cdot S^{(j)}+N^{(j)}\cdot s\cdot e^{\lambda\cdot t}\cdot e^{-\lambda\cdot\tau}+n\cdot e^{\lambda\cdot t}\cdot e^{-\lambda\cdot\tau}\cdot S^{(j)}\right.\\\left.-N_{tr}\cdot S^{(j)}-N_{tr}\cdot s\cdot e^{\lambda\cdot t}\cdot e^{-\lambda\cdot\tau}\right\}\end{array}}{1+\varepsilon\cdot\left(S^{(j)}+s\cdot e^{\lambda\cdot t}\cdot e^{-\lambda\cdot\tau}\right)}$$

$$\begin{aligned}\lambda\cdot n\cdot e^{\lambda\cdot t}&=\frac{\eta\cdot\left(I-I_{off}\right)}{q}-\frac{N^{(j)}}{\tau_n}-\frac{n\cdot e^{\lambda\cdot t}}{\tau_n}\\&\quad-\frac{\begin{array}{l}G\cdot\left\{N^{(j)}\cdot S^{(j)}+\left(N^{(j)}\cdot s+n\cdot S^{(j)}\right)\cdot e^{\lambda\cdot t}\cdot e^{-\lambda\cdot\tau}\right.\\\left.-N_{tr}\cdot S^{(j)}-N_{tr}\cdot s\cdot e^{\lambda\cdot t}\cdot e^{-\lambda\cdot\tau}\right\}\end{array}}{\left(1+\varepsilon\cdot S^{(j)}\right)+\varepsilon\cdot s\cdot e^{\lambda\cdot t}\cdot e^{-\lambda\cdot\tau}}\\&\quad\cdot\left\{\frac{\left(1+\varepsilon\cdot S^{(j)}\right)-\varepsilon\cdot s\cdot e^{\lambda\cdot t}\cdot e^{-\lambda\cdot\tau}}{\left(1+\varepsilon\cdot S^{(j)}\right)-\varepsilon\cdot s\cdot e^{\lambda\cdot t}\cdot e^{-\lambda\cdot\tau}}\right\}\end{aligned}$$

$$\lambda\cdot n\cdot e^{\lambda\cdot t}=\frac{\eta\cdot\left(I-I_{off}\right)}{q}-\frac{N^{(j)}}{\tau_n}-\frac{n\cdot e^{\lambda\cdot t}}{\tau_n}-\frac{\begin{array}{l}G\cdot\left\{N^{(j)}\cdot S^{(j)}+\left(N^{(j)}\cdot s+n\cdot S^{(j)}\right)\cdot e^{\lambda\cdot t}\cdot e^{-\lambda\cdot\tau}-N_{tr}\cdot S^{(j)}\right.\\\left.-N_{tr}\cdot s\cdot e^{i\cdot t}\cdot e^{-\lambda\cdot\tau}\right\}\cdot\left\{\left(1+\varepsilon\cdot S^{(j)}\right)-\varepsilon\cdot s\cdot e^{\lambda\cdot t}\cdot e^{-\lambda\cdot\tau}\right\}\end{array}}{\left(1+\varepsilon\cdot S^{(j)}\right)^2-\varepsilon^2\cdot s^2\cdot e^{2\cdot\lambda\cdot t}\cdot e^{-2\cdot\lambda\cdot\tau}}$$

$$s^2\to 0$$

$$\lambda\cdot n\cdot e^{\lambda\cdot t}=\frac{\eta\cdot\left(I-I_{off}\right)}{q}-\frac{N^{(j)}}{\tau_n}-\frac{n\cdot e^{\lambda\cdot t}}{\tau_n}$$

$$-\frac{\begin{array}{l}G \cdot \{N^{(j)} \cdot S^{(j)} + (N^{(j)} \cdot s + n \cdot S^{(j)}) \cdot e^{\lambda \cdot t} \cdot e^{-\lambda \cdot \tau} - N_{tr} \cdot S^{(j)} \\ -N_{tr} \cdot s \cdot e^{\lambda \cdot t} \cdot e^{-\lambda \cdot \tau}\} \cdot \{(1 + \varepsilon \cdot S^{(j)}) - \varepsilon \cdot s \cdot e^{\lambda \cdot t} \cdot e^{-\lambda \cdot \tau}\}\end{array}}{(1 + \varepsilon \cdot S^{(j)})^2}$$

$$\begin{aligned}&\{N^{(j)} \cdot S^{(j)} + (N^{(j)} \cdot s + n \cdot S^{(j)}) \cdot e^{\lambda \cdot t} \cdot e^{-\lambda \cdot \tau} - N_{tr} \cdot S^{(j)} - N_{tr} \cdot s \cdot e^{\lambda \cdot t} \cdot e^{-\lambda \cdot \tau}\} \cdot \\ &\quad \{(1 + \varepsilon \cdot S^{(j)}) - \varepsilon \cdot s \cdot e^{\lambda \cdot t} \cdot e^{-\lambda \cdot \tau} \\ &\quad = N^{(j)} \cdot S^{(j)} \cdot (1 + \varepsilon \cdot S^{(j)}) \\ &\quad + (N^{(j)} \cdot s + n \cdot S^{(j)}) \cdot (1 + \varepsilon \cdot S^{(j)}) \cdot e^{\lambda \cdot t} \cdot e^{-\lambda \cdot \tau} - N_{tr} \cdot S^{(j)} \cdot (1 + \varepsilon \cdot S^{(j)}) \\ &\quad - N_{tr} \cdot (1 + \varepsilon \cdot S^{(j)}) \cdot s \cdot e^{\lambda \cdot t} \cdot e^{-\lambda \cdot \tau} - N^{(j)} \cdot S^{(j)} \cdot \varepsilon \cdot s \cdot e^{\lambda \cdot t} \cdot e^{-\lambda \cdot \tau} \\ &\quad - (N^{(j)} \cdot s^2 + n \cdot s \cdot S^{(j)}) \cdot e^{\lambda \cdot t} \cdot e^{-\lambda \cdot \tau} \cdot \varepsilon \cdot e^{\lambda \cdot t} \cdot e^{-\lambda \cdot \tau} \\ &\quad + N_{tr} \cdot S^{(j)} \cdot \varepsilon \cdot s \cdot e^{\lambda \cdot t} \cdot e^{-\lambda \cdot \tau} + N_{tr} \cdot s^2 \cdot e^{2 \cdot \lambda \cdot t} \cdot e^{-2 \cdot \lambda \cdot \tau} \cdot \varepsilon\end{aligned}$$

$$s^2 \to 0; \; n \cdot s \to 0$$

$$\begin{aligned}&\{N^{(j)} \cdot S^{(j)} + (N^{(j)} \cdot s + n \cdot S^{(j)}) \cdot e^{\lambda \cdot t} \cdot e^{-\lambda \cdot \tau} - N_p \cdot S^{(j)} - N_{tr} \cdot s \cdot e^{\lambda \cdot t} \cdot e^{-\lambda \cdot \tau}\} \cdot \\ &\{(1 + \varepsilon \cdot S^{(j)}) - \varepsilon \cdot s \cdot e^{\lambda \cdot t} \cdot e^{-\lambda \cdot \tau}\} \\ &= N^{(j)} \cdot S^{(j)} \cdot (1 + \varepsilon \cdot S^{(j)}) + (N^{(j)} \cdot s + n \cdot S^{(j)}) \cdot (1 + \varepsilon \cdot S^{(j)}) \cdot e^{\lambda \cdot t} \cdot e^{-\lambda \cdot \tau} \\ &- N_{tr} \cdot S^{(j)} \cdot (1 + \varepsilon \cdot S^{(j)}) \\ &\quad - N_n \cdot (1 + \varepsilon \cdot S^{(j)}) \cdot s \cdot e^{\lambda \cdot t} \cdot e^{-\lambda \cdot \tau} - N^{(j)} \cdot S^{(j)} \cdot \varepsilon \cdot s \cdot e^{\lambda \cdot t} \cdot e^{-\lambda \cdot \tau} \\ &+ N_p \cdot S^{(j)} \cdot \varepsilon \cdot s \cdot e^{\lambda \cdot t} \cdot e^{-\lambda \cdot \tau}\end{aligned}$$

$$\lambda \cdot n \cdot e^{\lambda \cdot t} = \frac{\eta \cdot (I - I_{off})}{q} - \frac{N^{(j)}}{\tau_n} - \frac{n \cdot e^{\lambda \cdot t}}{\tau_n} - \frac{\begin{array}{l}G \cdot \{N^{(j)} \cdot S^{(j)} \cdot (1 + \varepsilon \cdot S^{(j)}) + (N^{(j)} \cdot s + n \cdot S^{(j)}) \\ \cdot (1 + \varepsilon \cdot S^{(j)}) \cdot e^{\lambda \cdot t} \cdot e^{-\lambda \cdot \tau} - N_{tr} \cdot S^{(j)} \cdot (1 + \varepsilon \cdot S^{(j)}) \\ -N_{tr} \cdot (1 + \varepsilon \cdot S^{(j)}) \cdot s \cdot e^{\lambda \cdot t} \cdot e^{-\lambda \cdot \tau} - N^{(j)} \cdot S^{(j)} \cdot \varepsilon \cdot s \cdot e^{\lambda \cdot t} \cdot e^{-\lambda \cdot \tau} \\ +N_{tr} \cdot S^{(j)} \cdot \varepsilon \cdot s \cdot e^{\lambda \cdot t} \cdot e^{-\lambda \cdot \tau}\}\end{array}}{(1 + \varepsilon \cdot S^{(j)})^2}$$

$$\lambda \cdot n \cdot e^{\lambda \cdot t} = \frac{\eta \cdot (I - I_{off})}{q} - \frac{N^{(j)}}{\tau_n} - \frac{n \cdot e^{\lambda \cdot t}}{\tau_n} - \frac{\begin{array}{l}G \cdot \{[N^{(j)} \cdot S^{(j)} + (N^{(j)} \cdot s + n \cdot S^{(j)}) \cdot e^{\lambda \cdot t} \cdot e^{-\lambda \cdot \tau} - N_{tr} \cdot S^{(j)} \\ -N_{tr} \cdot s \cdot e^{\lambda t} \cdot e^{-\lambda \cdot \tau}] \cdot (1 + \varepsilon \cdot S^{(j)}) - N^{(j)} \cdot S^{(j)} \cdot \varepsilon \cdot s \cdot e^{\lambda \cdot t} \cdot e^{-\lambda \cdot \tau} \\ +N_{tr} \cdot S^{(j)} \cdot \varepsilon \cdot s \cdot e^{\lambda \cdot t} \cdot e^{-\lambda \cdot \tau}\}\end{array}}{(1 + \varepsilon \cdot S^{(j)})^2}$$

$$\lambda \cdot n \cdot e^{\lambda \cdot t} = \frac{\eta \cdot (I - I_{off})}{q} - \frac{N^{(j)}}{\tau_n} - \frac{n \cdot e^{\lambda \cdot t}}{\tau_n}$$
$$- \frac{G \cdot \left[N^{(j)} \cdot S^{(j)} + \left(N^{(j)} \cdot s + n \cdot S^{(j)}\right) \cdot e^{\lambda \cdot t} \cdot e^{-\lambda \cdot \tau} - N_p \cdot S^{(j)} - N_{tr} \cdot s \cdot e^{\lambda \cdot t} \cdot e^{-\lambda \cdot \tau}\right] \cdot \left(1 + \varepsilon \cdot S^{(j)}\right)}{\left(1 + \varepsilon \cdot S^{(j)}\right)^2}$$
$$- \frac{G \cdot \left\{-N^{(j)} \cdot S^{(j)} \cdot \varepsilon \cdot s \cdot e^{\lambda \cdot t} \cdot e^{-\lambda \cdot \tau} + N_o \cdot S^{(j)} \cdot \varepsilon \cdot s \cdot e^{\lambda \cdot t} \cdot e^{-\lambda \cdot \tau}\right\}}{\left(1 + \varepsilon \cdot S^{(j)}\right)^2}$$

$$\lambda \cdot n \cdot e^{\lambda \cdot t} = \frac{\eta \cdot (I - I_{off})}{q} - \frac{N^{(j)}}{\tau_n} - \frac{n \cdot e^{i \cdot t}}{\tau_n}$$
$$- \frac{G \cdot \left[\left(N^{(j)} - N_{tr}\right) \cdot S^{(j)} + \left(N^{(j)} \cdot s + n \cdot S^{(j)}\right) \cdot e^{\lambda \cdot t} \cdot e^{-\lambda \cdot \tau} - N_n \cdot s \cdot e^{\lambda \cdot t} \cdot e^{-\lambda \cdot \tau}\right]}{\left(1 + \varepsilon \cdot S^{(j)}\right)}$$
$$- \frac{G \cdot \left\{-N^{(j)} \cdot S^{(j)} \cdot \varepsilon \cdot s \cdot e^{\lambda \cdot t} \cdot e^{-\lambda \cdot \tau} + N_{tr} \cdot S^{(j)} \cdot \varepsilon \cdot s \cdot e^{\lambda \cdot t} \cdot e^{-\lambda \cdot \tau}\right\}}{\left(1 + \varepsilon \cdot S^{(j)}\right)^2}$$

$$\lambda \cdot n \cdot e^{\lambda \cdot t} = \frac{\eta \cdot (I - I_{off})}{q} - \frac{N^{(j)}}{\tau_n} - \frac{G \cdot \left(N^{(j)} - N_{tr}\right) \cdot S^{(j)}}{\left(1 + \varepsilon \cdot S^{(j)}\right)} - \frac{n \cdot e^{\lambda \cdot t}}{\tau_n}$$
$$- \frac{G \cdot \left[\left(N^{(j)} \cdot s + n \cdot S^{(j)}\right) - N_{tr} \cdot s\right]}{\left(1 + \varepsilon \cdot S^{(j)}\right)} \cdot e^{\lambda \cdot t} \cdot e^{-\lambda \cdot \tau}$$
$$- \frac{G \cdot \left\{N_{tr} - N^{(j)}\right\}}{\left(1 + \varepsilon \cdot S^{(j)}\right)^2} \cdot S^{(j)} \cdot \varepsilon \cdot s \cdot e^{\lambda \cdot t} \cdot e^{-\lambda \cdot \tau}$$

At fixed points: $\lim_{t \to \infty} N(t - \tau) = N(t)$; $\lim_{t \to \infty} S(t - \tau) = S(t)$; t $\gg \tau \Rightarrow t - \tau \approx t$

$$\frac{\eta \cdot (I - I_{off})}{q} - \frac{N^{(j)}}{\tau_n} - \frac{G \cdot \left(N^{(j)} - N_r\right) \cdot S^{(j)}}{\left(1 + \varepsilon \cdot S^{(j)}\right)} = 0$$

$$\lambda \cdot n = -\frac{n}{\tau_n} - \frac{G \cdot \left[\left(N^{(j)} \cdot s + n \cdot S^{(j)}\right) - N_{tr} \cdot s\right]}{\left(1 + \varepsilon \cdot S^{(j)}\right)} \cdot e^{-\lambda \cdot \tau}$$
$$- \frac{G \cdot \left\{N_{tr} - N^{(j)}\right\}}{\left(1 + \varepsilon \cdot S^{(j)}\right)^2} \cdot S^{(j)} \cdot \varepsilon \cdot s \cdot e^{-\lambda \cdot \tau}$$

$$-\lambda \cdot n - \frac{n}{\tau_n} - \frac{G \cdot \left[\left(N^{(j)} \cdot s + n \cdot S^{(j)}\right) - N_{tr} \cdot s\right]}{\left(1 + \varepsilon \cdot S^{(j)}\right)} \cdot e^{-\lambda \cdot \tau}$$
$$- \frac{G \cdot \left\{N_{tr} - N^{(j)}\right\}}{\left(1 + \varepsilon \cdot S^{(j)}\right)^2} \cdot S^{(j)} \cdot \varepsilon \cdot s \cdot e^{-\lambda \cdot \tau} = 0$$

$$\left[-\lambda - \left(\frac{1}{\tau_n} + \frac{G \cdot S^{(j)}}{(1 + \varepsilon \cdot S^{(j)})} \cdot e^{-\lambda \cdot \tau}\right)\right] \cdot n - \frac{G \cdot (N^{(j)} - N_{tr})}{(1 + \varepsilon \cdot S^{(j)})} \cdot$$
$$\left[1 - \frac{S^{(j)} \cdot \varepsilon}{(1 + \varepsilon \cdot S^{(j)})}\right] \cdot s \cdot e^{-\lambda \cdot \tau} = 0$$

$$\frac{dS(t)}{dt} = -\frac{S(t)}{\tau_p} + \frac{\beta \cdot N(t)}{\tau_n} + \frac{G \cdot (N(t - \tau) - N_{tr}) \cdot S(t - \tau)}{1 + \varepsilon \cdot S(t - \tau)}$$

$$\lambda \cdot s \cdot e^{\lambda \cdot t} = -\frac{S^{(j)} + s \cdot e^{\lambda \cdot t}}{\tau_p} + \frac{\beta \cdot (N^{(j)} + n \cdot e^{\lambda t})}{\tau_n}$$
$$+ \frac{G \cdot (N^{(j)} + n \cdot e^{\lambda \cdot t} \cdot e^{-\lambda \cdot \tau} - N_{tr}) \cdot (S^{(j)} + s \cdot e^{\lambda \cdot t} \cdot e^{-\lambda \cdot \tau})}{1 + \varepsilon \cdot (S^{(j)} + s \cdot e^{\lambda \cdot t} \cdot e^{-\lambda \cdot \tau})}$$

$$\lambda \cdot s \cdot e^{\lambda \cdot t} = -\frac{S^{(j)}}{\tau_p} + \frac{\beta \cdot N^{(j)}}{\tau_n} - \frac{s \cdot e^{\lambda \cdot t}}{\tau_p} + \frac{\beta \cdot n \cdot e^{\lambda \cdot t}}{\tau_n}$$
$$+ \frac{G \cdot (N^{(j)} + n \cdot e^{\lambda \cdot t} \cdot e^{-\lambda \cdot \tau} - N_{tr}) \cdot (S^{(j)} + s \cdot e^{\lambda \cdot t} \cdot e^{-\lambda \cdot \tau})}{1 + \varepsilon \cdot (S^{(j)} + s \cdot e^{\lambda \cdot t} \cdot e^{-\lambda \cdot \tau})}$$

We already got the expression for $\frac{G \cdot (N^{(j)} + n \cdot e^{\lambda \cdot t} \cdot e^{-\lambda \cdot \tau} - N_n) \cdot (S^{(j)} + s \cdot e^{\lambda \cdot t} \cdot e^{-\lambda \cdot \tau})}{1 + \varepsilon \cdot (S^{(j)} + s \cdot e^{\lambda \cdot t} \cdot e^{-\lambda \cdot \tau})}$

$$\frac{G \cdot (N^{(j)} + n \cdot e^{\lambda \cdot t} \cdot e^{-\lambda \cdot \tau} - N_{tr}) \cdot (S^{(j)} + s \cdot e^{\lambda \cdot t} \cdot e^{-\lambda \cdot \tau})}{1 + \varepsilon \cdot (S^{(j)} + s \cdot e^{\lambda \cdot t} \cdot e^{-\lambda \cdot \tau})}$$
$$= \frac{\begin{array}{l} G \cdot \{N^{(j)} \cdot S^{(j)} + (N^{(j)} \cdot s + n \cdot S^{(j)}) \cdot e^{\lambda \cdot t} \cdot e^{-\lambda \cdot \tau} - N_{tr} \cdot S^{(j)} \\ -N_{tr} \cdot s \cdot e^{\lambda \cdot t} \cdot e^{-\lambda \cdot \tau}\} \cdot \{(1 + \varepsilon \cdot S^{(j)}) - \varepsilon \cdot s \cdot e^{\lambda \cdot t} \cdot e^{-\lambda \cdot \tau}\} \end{array}}{(1 + \varepsilon \cdot S^{(j)})^2}$$

$$s^2 \to 0; \quad n \cdot s \to 0$$

$$\frac{G \cdot (N^{(j)} + n \cdot e^{i \cdot t} \cdot e^{-\lambda \cdot \tau} - N_{tr}) \cdot (S^{(j)} + s \cdot e^{\lambda \cdot t} \cdot e^{-\lambda \cdot \tau})}{1 + \varepsilon \cdot (S^{(j)} + s \cdot e^{\lambda \cdot t} \cdot e^{-\lambda \cdot \tau})}$$
$$= \frac{\begin{array}{l} G \cdot \{N^{(j)} \cdot S^{(j)} \cdot (1 + \varepsilon \cdot S^{(j)}) + (N^{(j)} \cdot s + n \cdot S^{(j)}) \\ (1 + \varepsilon \cdot S^{(j)}) \cdot e^{\lambda \cdot t} \cdot e^{-\lambda \cdot \tau} - N_{tr} \cdot S^{(j)} \cdot (1 + \varepsilon \cdot S^{(j)}) \\ -N_{tr} \cdot (1 + \varepsilon \cdot S^{(j)}) \cdot s \cdot e^{i \cdot t} \cdot e^{-\lambda \cdot \tau} - N^{(j)} \cdot S^{(j)} \cdot \varepsilon \cdot s \cdot e^{\lambda \cdot t} \cdot e^{-\lambda \cdot t} \\ +N_o \cdot S^{(j)} \cdot \varepsilon \cdot s \cdot e^{\lambda \cdot t} \cdot e^{-\lambda \cdot \tau}\} \end{array}}{(1 + \varepsilon \cdot S^{(j)})^2}$$

$$\frac{G \cdot (N^{(j)} + n \cdot e^{\lambda \cdot t} \cdot e^{-\lambda \cdot \tau} - N_r) \cdot (S^{(j)} + s \cdot e^{\lambda \cdot t} \cdot e^{-\lambda \cdot \tau})}{1 + \varepsilon \cdot (S^{(j)} + s \cdot e^{\lambda \cdot t} \cdot e^{-\lambda \cdot \tau})}$$

$$= \frac{G \cdot \left[\left(N^{(j)} - N_{tr}\right) \cdot S^{(j)} + \left(N^{(j)} \cdot s + n \cdot S^{(j)}\right) \cdot e^{\lambda \cdot t} \cdot e^{-\lambda \cdot t} - N_{tr} \cdot s \cdot e^{\lambda \cdot t} \cdot e^{-\lambda \cdot \tau} \right]}{\left(1 + \varepsilon \cdot S^{(j)}\right)} + \frac{G \cdot \left\{ -N^{(j)} \cdot S^{(j)} \cdot \varepsilon \cdot s \cdot e^{\lambda \cdot t} \cdot e^{-\lambda \cdot \tau} + N_{tr} \cdot S^{(j)} \cdot \varepsilon \cdot s \cdot e^{\lambda \cdot t} \cdot e^{-\lambda \cdot \tau} \right\}}{\left(1 + \varepsilon \cdot S^{(j)}\right)^2}$$

$$\frac{G \cdot \left(N^{(j)} + n \cdot e^{\lambda t} \cdot e^{-\lambda \cdot \tau} - N_n\right) \cdot \left(S^{(j)} + s \cdot e^{\lambda \cdot t} \cdot e^{-\lambda \cdot \tau}\right)}{1 + \varepsilon \cdot \left(S^{(j)} + s \cdot e^{\lambda \cdot t} \cdot e^{-\lambda \cdot \tau}\right)} = \frac{G \cdot \left(N^{(j)} - N_{tr}\right) \cdot S^{(j)}}{\left(1 + \varepsilon \cdot S^{(j)}\right)} + \frac{G \cdot \left[\left(N^{(j)} \cdot s + n \cdot S^{(j)}\right) - N_{tr} \cdot s \right]}{\left(1 + \varepsilon \cdot S^{(j)}\right)} \cdot e^{\lambda \cdot t} \cdot e^{-\lambda \cdot \tau} + \frac{G \cdot \left\{ N_{tr} - N^{(j)} \right\}}{\left(1 + \varepsilon \cdot S^{(j)}\right)^2} \cdot S^{(j)} \cdot \varepsilon \cdot s \cdot e^{\lambda \cdot t} \cdot e^{-\lambda \cdot \tau}$$

$$\lambda \cdot s \cdot e^{\lambda \cdot t} = -\frac{S^{(j)}}{\tau_p} + \frac{\beta \cdot N^{(j)}}{\tau_n} - \frac{s \cdot e^{\lambda \cdot t}}{\tau_p} + \frac{\beta \cdot n \cdot e^{\lambda \cdot t}}{\tau_n} + \frac{G \cdot \left(N^{(j)} - N_{tr}\right) \cdot S^{(j)}}{\left(1 + \varepsilon \cdot S^{(j)}\right)} + \frac{G \cdot \left[\left(N^{(j)} \cdot s + n \cdot S^{(j)}\right) - N_{tr} \cdot s \right]}{\left(1 + \varepsilon \cdot S^{(j)}\right)} \cdot e^{\lambda \cdot t} \cdot e^{-\lambda \cdot \tau} + \frac{G \cdot \left\{ N_{tr} - N^{(j)} \right\}}{\left(1 + \varepsilon \cdot S^{(j)}\right)^2} \cdot S^{(j)} \cdot \varepsilon \cdot s \cdot e^{\lambda \cdot t} \cdot e^{-\lambda \cdot \tau}$$

$$\lambda \cdot s \cdot e^{\lambda \cdot t} = -\frac{S^{(j)}}{\tau_p} + \frac{\beta \cdot N^{(j)}}{\tau_n} + \frac{G \cdot \left(N^{(j)} - N_{tr}\right) \cdot S^{(j)}}{\left(1 + \varepsilon \cdot S^{(j)}\right)} + \frac{\beta \cdot n \cdot e^{\lambda \cdot t}}{\tau_n} - \frac{s \cdot e^{\lambda \cdot t}}{\tau_p} + \frac{G \cdot \left[\left(N^{(j)} \cdot s + n \cdot S^{(j)}\right) - N_{tr} \cdot s \right]}{\left(1 + \varepsilon \cdot S^{(j)}\right)} \cdot e^{\lambda \cdot t} \cdot e^{-\lambda \cdot \tau} + \frac{G \cdot \left\{ N_{tr} - N^{(j)} \right\}}{\left(1 + \varepsilon \cdot S^{(j)}\right)^2} \cdot S^{(j)} \cdot \varepsilon \cdot s \cdot e^{\lambda \cdot t} \cdot e^{-\lambda \cdot \tau}$$

At fixed points: $\lim_{t \to \infty} N(t - \tau) = N(t)$; $\lim_{t \to \infty} S(t - \tau) = S(t)$; t $\gg \tau \Rightarrow t - \tau \approx t$

$$-\frac{S^{(j)}}{\tau_p} + \frac{\beta \cdot N^{(j)}}{\tau_n} + \frac{G \cdot \left(N^{(j)} - N_{tr}\right) \cdot S^{(j)}}{\left(1 + \varepsilon \cdot S^{(j)}\right)} = 0$$

$$\lambda \cdot s \cdot e^{\lambda \cdot t} = \frac{\beta \cdot n \cdot e^{\lambda \cdot t}}{\tau_n} - \frac{s \cdot e^{\lambda \cdot t}}{\tau_p} + \frac{G \cdot \left[\left(N^{(j)} \cdot s + n \cdot S^{(j)}\right) - N_{tr} \cdot s \right]}{\left(1 + \varepsilon \cdot S^{(j)}\right)} \cdot e^{\lambda \cdot t} \cdot e^{-\lambda \cdot \tau} + \frac{G \cdot \left\{ N_{tr} - N^{(j)} \right\}}{\left(1 + \varepsilon \cdot S^{(j)}\right)^2} \cdot S^{(j)} \cdot \varepsilon \cdot s \cdot e^{\lambda \cdot t} \cdot e^{-\lambda \cdot \tau}$$

$$\lambda \cdot s = \frac{\beta \cdot n}{\tau_n} - \frac{s}{\tau_p} + \frac{G \cdot \left[\left(N^{(j)} \cdot s + n \cdot S^{(j)}\right) - N_{tr} \cdot s\right]}{\left(1 + \varepsilon \cdot S^{(j)}\right)} \cdot e^{-\lambda \cdot \tau} + \frac{G \cdot \left\{N_{tr} - N^{(j)}\right\}}{\left(1 + \varepsilon \cdot S^{(j)}\right)^2} \cdot S^{(j)} \cdot \varepsilon \cdot s \cdot e^{-\lambda \cdot \tau}$$

$$\left[\frac{\beta}{\tau_n} + \frac{G \cdot S^{(j)}}{\left(1 + \varepsilon \cdot S^{(j)}\right)} \cdot e^{-\lambda \cdot \tau}\right] \cdot n + \left[-\lambda - \frac{1}{\tau_p} + \left(1 - \frac{S^{(j)} \cdot \varepsilon}{\left(1 + \varepsilon \cdot S^{(j)}\right)}\right) \cdot \frac{G \cdot \left(N^{(j)} - N_{tr}\right)}{\left(1 + \varepsilon \cdot S^{(j)}\right)} \cdot e^{-\lambda \cdot \tau}\right] \cdot s = 0$$

We can summarize our last results:

$$\left[-\lambda - \left(\frac{1}{\tau_n} + \frac{G \cdot S^{(j)}}{\left(1 + \varepsilon \cdot S^{(j)}\right)} \cdot e^{-\lambda \cdot \tau}\right)\right] \cdot n - \frac{G \cdot \left(N^{(j)} - N_{tr}\right)}{\left(1 + \varepsilon \cdot S^{(j)}\right)} \cdot \left[1 - \frac{S^{(j)} \cdot \varepsilon}{\left(1 + \varepsilon \cdot S^{(j)}\right)}\right] \cdot s \cdot e^{-\lambda \cdot \tau} = 0$$

$$\left[\frac{\beta}{\tau_n} + \frac{G \cdot S^{(j)}}{\left(1 + \varepsilon \cdot S^{(j)}\right)} \cdot e^{-\lambda \cdot \tau}\right] \cdot n + \left[-\lambda - \frac{1}{\tau_p} + \left(1 - \frac{S^{(j)} \cdot \varepsilon}{\left(1 + \varepsilon \cdot S^{(j)}\right)}\right) \cdot \frac{G \cdot \left(N^{(j)} - N_{tr}\right)}{\left(1 + \varepsilon \cdot S^{(j)}\right)} \cdot e^{-\lambda \cdot \tau}\right] \cdot s = 0$$

The small increments Jacobian of our Electron-photon exchanges into VCSEL system is as follow:

$$\begin{pmatrix} -\lambda - \left(\frac{1}{\tau_n} + \frac{G \cdot S^{(j)}}{(1+\varepsilon \cdot S^{(j)})} \cdot e^{-\lambda \cdot \tau}\right) & -\frac{G \cdot \left(N^{(j)} - N_{tr}\right)}{(1+\varepsilon \cdot S^{(j)})} \cdot \left[1 - \frac{S^{(j)} \cdot \varepsilon}{(1+\varepsilon \cdot S^{(j)})}\right] \cdot e^{-\lambda \cdot t} \\ \frac{\beta}{\tau_n} + \frac{G \cdot S^{(j)}}{(1+\varepsilon \cdot S^{(j)})} \cdot e^{-\lambda \cdot \tau} & -\lambda - \frac{1}{\tau_p} + \left(1 - \frac{S^{(j)} \cdot \varepsilon}{(1+\varepsilon \cdot S^{(j)})}\right) \cdot \frac{G \cdot \left(N^{(j)} - N_{tr}\right)}{(1+\varepsilon \cdot S^{(j)})} \cdot e^{-\lambda \cdot \tau} \end{pmatrix} \cdot \begin{pmatrix} n \\ s \end{pmatrix} = 0$$

$$(A - \lambda \cdot I) = \begin{pmatrix} -\lambda - \left(\frac{1}{\tau_n} + \frac{G \cdot S^{(j)}}{(1+\varepsilon \cdot S^{(j)})} \cdot e^{-\lambda \cdot \tau}\right) & -\frac{G \cdot (N^{(j)} - N_{tr})}{(1+\varepsilon \cdot S^{(j)})} \cdot \left[1 - \frac{S^{(j)} \cdot \varepsilon}{(1+\varepsilon \cdot S^{(j)})}\right] \cdot e^{-\lambda \cdot t} \\ \frac{\beta}{\tau_n} + \frac{G \cdot S^{(j)}}{(1+\varepsilon \cdot S^{(j)})} \cdot e^{-\lambda \cdot \tau} & -\lambda - \frac{1}{\tau_p} + \left(1 - \frac{S^{(j)} \cdot \varepsilon}{(1+\varepsilon \cdot S^{(j)})}\right) \cdot \frac{G \cdot (N^{(j)} - N_{tr})}{(1+\varepsilon \cdot S^{(j)})} \cdot e^{-\lambda \cdot \tau} \end{pmatrix}$$

$$(A - \lambda \cdot I) = 0$$

$$\Rightarrow \det \begin{pmatrix} -\lambda - \left(\frac{1}{\tau_n} + \frac{G \cdot S^{(j)}}{(1+\varepsilon \cdot S^{(j)})} \cdot e^{-\lambda \cdot \tau}\right) & -\frac{G \cdot (N^{(j)} - N_{tr})}{(1+\varepsilon \cdot S^{(j)})} \cdot [1 - \frac{S^{(j)} \cdot \varepsilon}{(1+\varepsilon \cdot S^{(j)})}] \cdot e^{-\lambda \cdot \tau} \\ \frac{\beta}{\tau_n} + \frac{G \cdot S^{(j)}}{(1+\varepsilon \cdot S^{(j)})} \cdot e^{-\lambda \cdot \tau} & -\lambda - \frac{1}{\tau_p} + (1 - \frac{S^{(j)} \cdot \varepsilon}{(1+\varepsilon \cdot S^{(j)})}) \cdot \frac{G \cdot (N^{(j)} - N_{tr})}{(1+\varepsilon \cdot S^{(j)})} \cdot e^{-\lambda \cdot \tau} \end{pmatrix} = 0$$

For simplicity we define the function: $g_1(S^{(j)}, N^{(j)}) = \left(1 - \frac{S^{(j)} \cdot \varepsilon}{(1+\varepsilon \cdot S^{(j)})}\right) \cdot \frac{G \cdot (N^{(j)} - N_{tr})}{(1+\varepsilon \cdot S^{(j)})}$

$$g_2(S^{(j)}, N^{(j)}) = \frac{G \cdot S^{(j)}}{(1 + \varepsilon \cdot S^{(j)})}$$

$$\det(A - \lambda \cdot I) = 0$$

$$\Rightarrow \det\begin{pmatrix} -\lambda - \left(\frac{1}{\tau_n} + g_2(S^{(j)}, N^{(j)}) \cdot e^{-\lambda \cdot \tau}\right) & -g_1(S^{(j)}, N^{(j)}) \cdot e^{-\lambda \cdot \tau} \\ \frac{\beta}{\tau_n} + g_2(S^{(j)}, N^{(j)}) \cdot e^{-\lambda \cdot \tau} & -\lambda - \frac{1}{\tau_p} + g_1(S^{(j)}, N^{(j)}) \cdot e^{-\lambda \cdot \tau} \end{pmatrix} = 0$$

$$\left\{-\lambda - \left(\frac{1}{\tau_n} + g_2(S^{(j)}, N^{(j)}) \cdot e^{-\lambda \cdot \tau}\right)\right\} \cdot \left\{-\lambda - \frac{1}{\tau_p} + g_1(S^{(j)}, N^{(j)}) \cdot e^{-\lambda \cdot \tau}\right\}$$
$$- \left\{\frac{\beta}{\tau_n} + g_2(S^{(j)}, N^{(j)}) \cdot e^{-\lambda \cdot \tau}\right\} \cdot \left\{-g_1(S^{(j)}, N^{(j)}) \cdot e^{-\lambda \cdot \tau}\right\} = 0$$

$$\lambda^2 + \lambda \cdot \frac{1}{\tau_p} - \lambda \cdot g_1(S^{(j)}, N^{(j)}) \cdot e^{-\lambda \cdot \tau} + \frac{1}{\tau_n} \cdot \lambda$$
$$+ \frac{1}{\tau_n \cdot \tau_p} - \frac{1}{\tau_n} \cdot g_1(S^{(j)}, N^{(j)}) \cdot e^{-\lambda \cdot \tau}$$
$$+ g_2(S^{(j)}, N^{(j)}) \cdot e^{-\lambda \cdot \tau} \cdot \lambda + g_2(S^{(j)}, N^{(j)}) \cdot e^{-\lambda \cdot \tau} \cdot$$
$$\frac{1}{\tau_p} - g_2(S^{(j)}, N^{(j)}) \cdot g_1(S^{(j)}, N^{(j)}) \cdot e^{-2 \cdot \lambda \cdot \tau}$$
$$+ \frac{\beta}{\tau_n} \cdot g_1(S^{(j)}, N^{(j)}) \cdot e^{-\lambda \cdot \tau} + g_2(S^{(j)}, N^{(j)}) \cdot$$
$$g_1(S^{(j)}, N^{(j)}) \cdot e^{-2 \cdot \lambda \cdot \tau} = 0$$

$$\lambda^2 + \lambda \cdot \left(\frac{1}{\tau_p} + \frac{1}{\tau_n}\right) + \frac{1}{\tau_n \cdot \tau_p} + [\{g_2(S^{(j)}, N^{(j)}) - g_1(S^{(j)}, N^{(j)})\} \cdot \lambda$$
$$+ g_2(S^{(j)}, N^{(j)}) \cdot \frac{1}{\tau_p} + \frac{\beta}{\tau_n} \cdot g_1(S^{(j)}, N^{(j)}) - \frac{1}{\tau_n} \cdot g_1(S^{(j)}, N^{(j)})\Big] \cdot e^{-\lambda \cdot \tau} = 0$$

The characteristic equation $D(\lambda, \tau)$ and we study the occurrence of any possible stability switching, resulting from the increase of the value of the time delay τ for general characteristic equation $D(\lambda, \tau)$ [3, 4].

$$D(\lambda, \tau) = P_n(\lambda, \tau) + Q_m(\lambda, \tau) \cdot e^{-\tau \cdot \lambda}; \; n, m \in \mathbb{N}_0; \; n > m; \; n = 2; \; m = 1$$

$$P_n(\lambda, \tau) = \lambda^2 + \lambda \cdot \left(\frac{1}{\tau_p} + \frac{1}{\tau_n}\right) + \frac{1}{\tau_n \cdot \tau_p}$$

$$Q_m(\lambda, \tau) = \{g_2(S^{(j)}, N^{(j)}) - g_1(S^{(j)}, N^{(j)})\} \cdot \lambda$$

$$+ g_2(S^{(j)}, N^{(j)}) \cdot \frac{1}{\tau_p} + \frac{\beta}{\tau_n} \cdot g_1(S^{(j)}, N^{(j)}) - \frac{1}{\tau_n} \cdot g_1(S^{(j)}, N^{(j)})$$

$$P_n(\lambda, \tau) = \sum_{k=0}^{n=2} P_k(\tau) \cdot \lambda^k = P_0(\tau) + P_1(\tau) \cdot \lambda + P_2(\tau) \cdot \lambda^2; \; P_0(\tau) = \frac{1}{\tau_n \cdot \tau_p}; \; P_1(\tau) = \frac{1}{\tau_p} + \frac{1}{\tau_n}$$

$$P_2(\tau) = 1; \; Q_m(\lambda, \tau) = \sum_{k=0}^{m=1} q_k(\tau) \cdot \lambda^k = q_0(\tau) + q_1(\tau) \cdot \lambda$$

$$q_0(\tau) = g_2(S^{(j)}, N^{(j)}) \cdot \frac{1}{\tau_p} + \frac{\beta}{\tau_n} \cdot g_1(S^{(j)}, N^{(j)}) - \frac{1}{\tau_n} \cdot g_1(S^{(j)}, N^{(j)}); \; q_1(\tau) = g_2(S^{(j)}, N^{(j)}) - g_1(S^{(j)}, N^{(j)})$$

The homogeneous system for S, N leads to a characteristic equation for the eigenvalue λ having the form

$$P(\lambda, \tau) + Q(\lambda, \tau) \cdot e^{-\tau \cdot \lambda} = 0; \; P(\lambda) = \sum_{j=0}^{2} a_j \cdot \lambda^j; \; Q(\lambda) = \sum_{j=0}^{1} c_j \cdot \lambda^j$$

The coefficients $\{a_j(q_i, q_k, \tau), c_j(q_i, q_k, \tau)\} \in \mathbb{R}$ depend on q_i, q_k and delay τ. Parameters q_i, q_k are any electron-photon exchanges into VCSEL system's global parameters, other parameters kept as a constant.

$$a_0 = \frac{1}{\tau_n \cdot \tau_p}; \; a_1 = \frac{1}{\tau_p} + \frac{1}{\tau_n}; \; a_2 = 1$$

$$c_0 = g_2(S^{(j)}, N^{(j)}) \cdot \frac{1}{\tau_p} + \frac{\beta}{\tau_n} \cdot g_1(S^{(j)}, N^{(j)}) - \frac{1}{\tau_n} \cdot g_1(S^{(j)}, N^{(j)}); \; c_1 = g_2(S^{(j)}, N^{(j)}) - g_1(S^{(j)}, N^{(j)})$$

Unless strictly necessary, the designation of the varied arguments (q_i, q_k) will subsequently be omitted from P, Q, a_j, and c_j. The coefficients a_j, c_j are continuous, and differentiable functions of their arguments, and direct substitution shows that $a_0 + c_0 \neq 0$ for $q_i, q_k \in \mathbb{R}_+$; that is, $\lambda = 0$ is not of $P(\lambda) + Q(\lambda) \cdot e^{-\tau \cdot \lambda} = 0$.

$$a_0 + c_0 = \frac{1}{\tau_n \cdot \tau_p} + g_2(S^{(j)}, N^{(j)}) \cdot \frac{1}{\tau_p} + \frac{\beta}{\tau_n} \cdot g_1(S^{(j)}, N^{(j)}) - \frac{1}{\tau_n} \cdot g_1(S^{(j)}, N^{(j)})$$

Furthermore, $P(\lambda)$, $Q(\lambda)$ are analytic functions of λ, for which the following requirements of the analysis can be verified in the present case.

1. If $\lambda = i \cdot \omega;\ \omega \in \mathbb{R}$ then $P(i \cdot \omega) + Q(i \cdot \omega) \neq 0$.
2. If $\left|\frac{Q(\lambda)}{P(\lambda)}\right|$ is bounded for $|\lambda| \to \infty;\ \mathrm{Re}\lambda \geq 0$. No roots bifurcation from ∞.
3. $F(\omega) = |P(i \cdot \omega)|^2 - |Q(i \cdot \omega)|^2$ has a finite number of zeros. Indeed, this is a polynomial in ω.
4. Each positive root $\omega(q_i, q_k)$ of $F(\omega) = 0$ is continuous and differentiable with respect to q_i, q_k.

We assume that $P_n(\lambda, \tau)$ and $Q_m(\lambda, \tau)$ cannot have common imaginary roots. That is for any real numberω :$P_n(\lambda = i \cdot \omega, \tau) + Q_m(\lambda = i \cdot \omega, \tau) \neq 0$.

$$P_n(\lambda = i \cdot \omega, \tau) = -\omega^2 + \frac{1}{\tau_n \cdot \tau_p} + i \cdot \omega \cdot \left(\frac{1}{\tau_p} + \frac{1}{\tau_n}\right)$$

$$Q_m(\lambda = i \cdot \omega, \tau) = i \cdot \omega \cdot \left\{g_2(S^{(j)}, N^{(j)}) - g_1(S^{(j)}, N^{(j)})\right\} + g_2(S^{(j)}, N^{(j)}) \cdot \frac{1}{\tau_p} + \frac{\beta}{\tau_n} \cdot g_1(S^{(j)}, N^{(j)}) - \frac{1}{\tau_n} \cdot g_1(S^{(j)}, N^{(j)})$$

$$\begin{aligned} &P_n(\lambda = i \cdot \omega, \tau) + Q_m(\lambda = i \cdot \omega, \tau) \\ &= -\omega^2 + \frac{1}{\tau_n \cdot \tau_p} + g_2(S^{(j)}, N^{(j)}) \cdot \frac{1}{\tau_p} + [\beta - 1] \cdot \frac{1}{\tau_n} \cdot g_1(S^{(j)}, N^{(j)}) \\ &\quad + i \cdot \omega \cdot \left[\left(\frac{1}{\tau_p} + \frac{1}{\tau_n}\right) + g_2(S^{(j)}, N^{(j)}) - g_1(S^{(j)}, N^{(j)})\right] \end{aligned}$$

$$|P(i \cdot \omega, \tau)|^2 = \left[-\omega^2 + \frac{1}{\tau_n \cdot \tau_p}\right]^2 + \omega^2 \cdot \left(\frac{1}{\tau_p} + \frac{1}{\tau_n}\right)^2 = \omega^4 - 2 \cdot \omega^2 \cdot \frac{1}{\tau_n \cdot \tau_p} + \frac{1}{\tau_n^2 \cdot \tau_p^2} + \omega^2 \cdot \left(\frac{1}{\tau_p} + \frac{1}{\tau_n}\right)^2$$

$$|P(i \cdot \omega, \tau)|^2 = \omega^4 + \omega^2 \cdot \left[\left(\frac{1}{\tau_p} + \frac{1}{\tau_n}\right)^2 - 2 \cdot \frac{1}{\tau_n \cdot \tau_p}\right] + \frac{1}{\tau_n^2 \cdot \tau_p^2}$$

$$|P(i \cdot \omega, \tau)|^2 = \omega^4 + \omega^2 \cdot \left(\frac{1}{\tau_p^2} + \frac{1}{\tau_n^2}\right) + \frac{1}{\tau_n^2 \cdot \tau_p^2}$$

$$|Q(i \cdot \omega, \tau)|^2 = \omega^2 \cdot \left[g_2(S^{(j)}, N^{(j)}) - g_1(S^{(j)}, N^{(j)})\right]^2 + \left[g_2(S^{(j)}, N^{(j)}) \cdot \frac{1}{\tau_p} + (\beta - 1) \cdot \frac{1}{\tau_n} \cdot g_1(S^{(j)}, N^{(j)})\right]^2$$

$$F(\omega, \tau) = P(i \cdot \omega, \tau)|^2 - |Q(i \cdot \omega, \tau)|^2$$

$$= \omega^4 + \omega^2 \cdot \left\{ \left(\frac{1}{\tau_p^2} + \frac{1}{\tau_n^2} \right) - [g_2(S^{(j)}, N^{(j)}) - g_1(S^{(j)}, N^{(j)})]^2 \right\}$$

$$+ \frac{1}{\tau_n^2 \cdot \tau_p^2} - \left[g_2(S^{(j)}, N^{(j)}) \cdot \frac{1}{\tau_p} + (\beta - 1) \cdot \frac{1}{\tau_n} \cdot g_1(S^{(j)}, N^{(j)}) \right]^2$$

We define the following parameters for simplicity: Π_0, Π_2, Π_4.

$$\Pi_0 = \frac{1}{\tau_n^2 \cdot \tau_p^2} - \left[g_2(S^{(j)}, N^{(j)}) \cdot \frac{1}{\tau_p} + (\beta - 1) \cdot \frac{1}{\tau_n} \cdot g_1(S^{(j)}, N^{(j)}) \right]^2$$

$$\Pi_2 = \left(\frac{1}{\tau_p^2} + \frac{1}{\tau_n^2} \right) - [g_2(S^{(j)}, N^{(j)}) - g_1(S^{(j)}, N^{(j)})]^2; \; \Pi_4 = 1$$

Hence $F(\omega, \tau) = 0$ implies $\sum_{k=0}^{2} \Pi_{2,k} \cdot \omega^{2 \cdot k} = 0$ and its roots are given by solving the above polynomial.

$$P_R(i \cdot \omega, \tau) = -\omega^2 + \frac{1}{\tau_n \cdot \tau_p}; \; P_I(i \cdot \omega, \tau) = \omega \cdot \left(\frac{1}{\tau_p} + \frac{1}{\tau_n} \right)$$

$$Q_R(i \cdot \omega, \tau) = g_2(S^{(j)}, N^{(j)}) \cdot \frac{1}{\tau_p} + (\beta - 1) \cdot \frac{1}{\tau_n} \cdot g_1(S^{(j)}, N^{(j)})$$

$$Q_I(i \cdot \omega, \tau) = \omega \cdot [g_2(S^{(j)}, N^{(j)}) - g_1(S^{(j)}, N^{(j)})]$$

$$\sin \theta(\tau) = \frac{-P_R(i \cdot \omega, \tau) \cdot Q_I(i \cdot \omega, \tau) + P_I(i \cdot \omega, \tau) \cdot Q_R(i \cdot \omega, \tau)}{|Q(i \cdot \omega, \tau)|^2}$$

$$\cos \theta(\tau) = -\frac{P_R(i \cdot \omega, \tau) \cdot Q_R(i \cdot \omega, \tau) + P_I(i \cdot \omega, \tau) \cdot Q_I(i \cdot \omega, \tau)}{|Q(i \cdot \omega, \tau)|^2}$$

We use different parameters terminology from our last characteristics parameter definition:

$$k \to j; \; p_k(\tau) \to a_j; \; q_k(\tau) \to c_j; \; n = 2; \; m = 1; \; n > m; \; P_n(\lambda, \tau) \to P(\lambda); \; Q_m(\lambda, \tau) \to Q(\lambda)$$

$$P(\lambda) = \sum_{j=0}^{2} a_j \cdot \lambda^j = a_o + a_1 \cdot \lambda + a_2 \cdot \lambda^2; \; Q(\lambda) = \sum_{j=0}^{1} c_j \cdot \lambda^j = c_0 + c_1 \cdot \lambda$$

$n, m \in \mathbb{N}_0$; $n > m$; $a_j, c_j : \mathbb{R}_{0+} \to \mathbb{R}$ are continuous and differentiable function of τ such that $a_0 + c_0 \neq 0$. In the following "-" denotes complex and conjugate. $P(\lambda)$, $Q(\lambda)$ are analytic functions in λ and differentiable in τ. The coefficients $a_j(\eta, \tau_n, \tau_p, G, \beta, \varepsilon, \ldots) \in \mathbb{R}$ and $c_j(\eta, \tau_n, \tau_p, G, \beta, \varepsilon, \ldots) \in \mathbb{R}$ depend on electron-photon exchanges into VCSEL system's $\eta, \tau_n, \tau_p, G, \beta, \varepsilon, \ldots$ values. Unless strictly necessary, the designation of the varied arguments: $\eta, \tau_n, \tau_p, G, \beta, \varepsilon, \ldots$ will subsequently be omitted from P, Q, a_j, c_j. The coefficients a_j, c_j are continuous, and differentiable functions of their argument, and direct substitution show that $a_0 + c_0 \neq 0$ [2, 3].

$$a_0 = \frac{1}{\tau_n \cdot \tau_p};\; c_0 = g_2\left(S^{(j)}, N^{(j)}\right) \cdot \frac{1}{\tau_p} + \frac{\beta}{\tau_n} \cdot g_1\left(S^{(j)}, N^{(j)}\right) - \frac{1}{\tau_n} \cdot g_1\left(S^{(j)}, N^{(j)}\right)$$

$$\frac{1}{\tau_n \cdot \tau_p} + g_2\left(S^{(j)}, N^{(j)}\right) \cdot \frac{1}{\tau_p} + \frac{\beta}{\tau_n} \cdot g_1\left(S^{(j)}, N^{(j)}\right)$$
$$- \frac{1}{\tau_n} \cdot g_1\left(S^{(j)}, N^{(j)}\right) \neq 0 \;\forall\; \eta, \tau_n, \tau_p, G, \beta, \varepsilon, \ldots$$

i.e. $\lambda = 0$ is not a root of the characteristic equation. Furthermore $P(\lambda)$, $Q(\lambda)$ are analytic functions of λ for which the following requirements of the analysis scan also be verified in the present case.

(a) If $\lambda = i \cdot \omega;\; \omega \in \mathbb{R}$ then $P(i \cdot \omega) + Q(i \cdot \omega) \neq 0$, i.e. P and Q have no common imaginary roots. This condition was verified numerically in the entire $\eta, \tau_n, \tau_p, G, \beta, \varepsilon, \ldots$ domain of interest.
(b) $\frac{P(\lambda)}{Q(\lambda)}$ is bounded for $|\lambda| \to \infty$; $\mathrm{Re}\lambda \geq 0$. No roots bifurcation from ∞. Indeed, in the limit:

$$\left\{g_2\left(S^{(j)}, N^{(j)}\right) - g_1\left(S^{(j)}, N^{(j)}\right)\right\} \cdot \lambda + g_2\left(S^{(j)}, N^{(j)}\right) \cdot \frac{1}{\tau_p}$$
$$\text{in the limit}: \left|\frac{Q(\lambda)}{P(\lambda)}\right| = \left|\frac{+\frac{\beta}{\tau_n} \cdot g_1\left(S^{(j)}, N^{(j)}\right) - \frac{1}{\tau_n} \cdot g_1\left(S^{(j)}, N^{(j)}\right)}{\lambda^2 + \lambda \cdot \left(\frac{1}{\tau_p} + \frac{1}{\tau_n}\right) + \frac{1}{\tau_n \cdot \tau_p}}\right|$$

(c) The following expressions exist: $F(\omega, \tau) = P(i \cdot \omega, \tau)|^2 - |Q(i \cdot \omega, \tau)|^2$ $F(\omega, \tau) = P(i \cdot \omega, \tau)|^2 - |Q(i \cdot \omega, \tau)|^2 = \sum_{k=0}^{2} \Pi_{2 \cdot k} \cdot \omega^{2 \cdot k}$ Has at most a finite number of zeros. Indeed, this is a polynomial in ω (degree in ω^4).
(d) Each positive root $\omega(\eta, \tau_n, \tau_p, G, \beta, \varepsilon, \tau, \ldots)$ of $F(\omega) = 0$ is continuous and differentiable with respect to $\eta, \tau_n, \tau_p, G, \beta, \varepsilon, \tau, \ldots$. The condition can only be assessed numerically.

In addition, since the coefficients in P and Q are real, we have $\overline{P(-i \cdot \omega)} = P(i \cdot \omega)$, and $\overline{Q(-i \cdot \omega)} = Q(i \cdot \omega)$ thus, $\omega > 0$ may be on eigenvalue of characteristic equations. The analysis consists in identifying the roots of the characteristic equation situated on the imaginary axis of the complex λ - plane, whereby increasing the parameters: $\eta, \tau_n, \tau_p, G, \beta, \varepsilon, \tau, \ldots$ Re λ may, at the crossing change its sign from (−) to (+), i.e. from a stable focus:

$$E^{(j)}(N^{(j)}, S^{(j)})$$
$$= \left(-\left(\frac{\Xi_2 \cdot A}{2 \cdot \Xi_1} + B \right) \pm \frac{A \cdot \sqrt{\Xi_2^2 - 4 \cdot \Xi_1 \cdot \Xi_3}}{2 \cdot \Xi_1}, \frac{-\Xi_2}{2 \cdot \Xi_1} \pm \frac{\sqrt{\Xi_2^2 - 4 \cdot \Xi_1 \cdot \Xi_3}}{2 \cdot \Xi_1} \right)$$

to an unstable one, or vice versa. This feature may be further assessed by examining the sign of the partial derivatives with respect to $\eta, \tau_n, \tau_p, G, \beta, \varepsilon, \tau, \ldots$ and system parameters.

$$\Lambda^{-1}(\tau) = \left(\frac{\partial \mathrm{Re}\lambda}{\partial \tau} \right)_{\lambda = i \cdot \omega} ; \ \eta, \tau_n, \tau_p, G, \beta, \varepsilon, \ldots$$

$$\Lambda^{-1}(\eta) = \left(\frac{\partial \mathrm{Re}\lambda}{\partial \eta} \right)_{\lambda = i \cdot \omega} ; \ \tau_n, \tau_p, G, \beta, \varepsilon, \tau, \ldots$$

$$\Lambda^{-1}(\tau_n) = \left(\frac{\partial \mathrm{Re}\lambda}{\partial \tau_n} \right)_{\lambda = i \cdot \omega} ; \ \eta, \tau_p, G, \beta, \varepsilon, \tau, \ldots$$

$$\Lambda^{-1}(\tau_p) = \left(\frac{\partial \mathrm{Re}\lambda}{\partial \tau_p} \right)_{\lambda = i \cdot \omega} ; \ \eta, \tau_n, G, \beta, \varepsilon, \tau, \ldots$$

$$\Lambda^{-1}(G) = \left(\frac{\partial \mathrm{Re}\lambda}{\partial G} \right)_{\lambda = i \cdot \omega} ; \ \eta, \tau_n, \tau_p, \beta, \varepsilon, \tau, \ldots$$

$$\Lambda^{-1}(\beta) = \left(\frac{\partial \mathrm{Re}\lambda}{\partial \beta} \right)_{\lambda = i \cdot \omega} ; \ \eta, \tau_n, \tau_p, G, \varepsilon, \tau, \ldots$$

$$\Lambda^{-1}(\varepsilon) = \left(\frac{\partial \mathrm{Re}\lambda}{\partial \varepsilon} \right)_{\lambda = i \cdot \omega} ; \ \eta, \tau_n, \tau_p, G, \beta, \tau, \ldots$$

$$P(\lambda) = P_R(\lambda) + i \cdot P_I(\lambda); \ \ Q(\lambda) = Q_R(\lambda) + i \cdot Q_I(\lambda)$$

When writing and inserting $\lambda = i \cdot \omega; \ \omega \in \mathbb{R}$ into electron-photon exchanges into VCSEL system's characteristic equation, ω must satisfy the following equations:

$$\sin(\omega \cdot \tau) = g(\omega) = \frac{-P_R(i \cdot \omega) \cdot Q_I(i \cdot \omega) + P_I(i \cdot \omega) \cdot Q_R(i \cdot \omega)}{|Q(i \cdot \omega)|^2}$$

$$\cos(\omega \cdot \tau) = h(\omega) = -\frac{P_R(i \cdot \omega) \cdot Q_R(i \cdot \omega) + P_I(i \cdot \omega) \cdot Q_I(i \cdot \omega)}{|Q(i \cdot \omega)|^2}$$

where $|Q(i \cdot \omega)|^2 \neq 0$ in the view of requirement (a) above, and $(g, h) \in \mathbb{R}$. Furthermore, it follows above $\sin(\omega \cdot \tau)$ and $\cos(\omega \cdot \tau)$ equations that by squaring and adding the sides, ω must be a positive root of $F(\omega, \tau) = P(i \cdot \omega, \tau)|^2 - |Q(i \cdot \omega, \tau)|^2 = 0$. Note: In case that $F(\omega)$ is dependent on τ. It is important to notice that if $\tau \notin I$

(assume that $I \subseteq \mathbb{R}_{+0}$ is the set where $\omega(\tau)$ is a positive root of $F(\omega)$ and for, $\tau \notin I$, $\omega(\tau)$ is not defined). Then for all $\tau \in I, \omega(\tau)$ is satisfied that $F(\omega, \tau) = 0$. Then there are no positive $\omega(\tau)$ solutions for $F(\omega, \tau) = 0$, and we cannot have stability switches [5]. For $\tau \in I$ where $\omega(\tau)$ is a positive solution of $F(\omega, \tau) = 0$, we can define the angel $\theta(\tau) \in [0, 2 \cdot \pi]$ as the solution of $\sin\theta(\tau) = \ldots;\ \cos\theta(\tau) = \ldots$

$$\sin\theta(\tau) = \frac{-P_R(i \cdot \omega) \cdot Q_I(i \cdot \omega) + P_I(i \cdot \omega) \cdot Q_R(i \cdot \omega)}{|Q(i \cdot \omega)|^2}$$

$$\cos\theta(\tau) = -\frac{P_R(i \cdot \omega) \cdot Q_R(i \cdot \omega) + P_I(i \cdot \omega) \cdot Q_I(i \cdot \omega)}{|Q(i \cdot \omega)|^2}$$

And the relation between the argument $\theta(\tau)$ and $\tau \cdot \omega(\tau)$ for $\tau \in I$ must be $\omega(\tau) \cdot \tau = \theta(\tau) + 2 \cdot n \cdot \pi\ \forall\, n \in \mathbb{R}_0$. Hence we can define the maps $\tau_n : I \to \mathbb{R}_{+0}$ given by $\tau_n(\tau) = \frac{\theta(\tau)+2\cdot n\cdot\pi}{\omega(\tau)};\ n \in \mathbb{R}_0;\ \tau \in I$. Let us introduce the functions: $I \to \mathbb{R}$

$S_n(\tau) = \tau - \tau_n(\tau);\ \tau \in I;\ n \in \mathbb{R}_0$ that is continuous and differentiable in τ. In the following, the subscripts $\lambda, \omega, \eta, \tau_n, \tau_p, G, \beta, \varepsilon, \ldots$ indicate the corresponding partial derivatives. Let us first concentrate on $\Lambda(x)$, remember in $\lambda(\eta, \tau_n, \tau_p, G, \beta, \varepsilon, \ldots)$ and $\omega(\eta, \tau_n, \tau_p, G, \beta, \varepsilon, \ldots)$, and keeping all parameters except one (x) and τ. Differentiating electron-photon exchanges into VCSEL system's characteristic equation $P(\lambda) + Q(\lambda) \cdot e^{-\lambda\cdot\tau} = 0$ with respect to specific parameter (x), and inverting the derivative, for convenience, one calculates: $x = \eta, \tau_n, \tau_p, G, \beta, \varepsilon, \ldots$

$$\left(\frac{\partial\lambda}{\partial x}\right)^{-1} = \frac{-P_\lambda(\lambda, x) \cdot Q(\lambda, x) + Q_\lambda(\lambda, x) \cdot P(\lambda, x) - \tau \cdot P(\lambda, x) \cdot Q(\lambda, x)}{P_x(\lambda, x) \cdot Q(\lambda, x) - Q_x(\lambda, x) \cdot P(\lambda, x)}$$

Where $P_\lambda(\lambda, x) = \frac{\partial P(\lambda,x)}{\partial\lambda};\ Q_\lambda(\lambda, x) = \frac{\partial Q(\lambda,x)}{\partial\lambda};\ P_x(\lambda, x) = \frac{\partial P(\lambda,x)}{\partial x};\ Q_x(\lambda, x) = \frac{\partial Q(\lambda,x)}{\partial x}$

Substituting $\lambda = i \cdot \omega$ and bearing $\overline{P(-i \cdot \omega)} = P(i \cdot \omega);\ \overline{Q(-i \cdot \omega)} = Q(i \cdot \omega);$ $i \cdot P_\lambda(i \cdot \omega) = P_\omega(i \cdot \omega)$

$i \cdot Q_\lambda(i \cdot \omega) = Q_\omega(i \cdot \omega)$ and that on the surface $|P(i \cdot \omega)|^2 = |Q(i \cdot \omega)|^2$, one obtains:

$$\left.\left(\frac{\partial\lambda}{\partial x}\right)^{-1}\right|_{\lambda=i\cdot\omega} = \left(\frac{i \cdot P_\omega(i \cdot \omega, x) \cdot \overline{P(i \cdot \omega, x)} + i \cdot Q_\lambda(i \cdot \omega, x) \cdot \overline{Q(\lambda, x)} - \tau \cdot |P(i \cdot \omega, x)|^2}{P_x(i \cdot \omega, x) \cdot \overline{P(i \cdot \omega, x)} - Q_x(i \cdot \omega, x) \cdot \overline{Q(i \cdot \omega, x)}}\right)$$

Upon separating into real and imaginary parts, with $P = P_R + i \cdot P_I;\ Q = Q_R + i \cdot Q_I$

$$P_\omega = P_{R\omega} + i \cdot P_{I\omega};\ Q_\omega = Q_{R\omega} + i \cdot Q_{I\omega}$$

$$P_x = P_{Rx} + i \cdot P_{Ix};\ Q_x = Q_{Rx} + i \cdot Q_{Ix};\ P^2 = P_R^2 + P_I^2$$

When (x) can be any electron-photon exchanges into VCSEL system's parameter $\eta, \tau_n, \tau_p, G, \beta, \varepsilon, \ldots$ and time delayτ, etc. Where for convenient, we have dropped the arguments $(i \cdot \omega, x)$ and where $F_\omega = 2 \cdot [(P_{R\omega} \cdot P_R + P_{I\omega} \cdot P_I) - (Q_{R\omega} \cdot Q_R + Q_{I\omega} \cdot Q_I)]$ $F_x = 2 \cdot [(P_{Rx} \cdot P_R + P_{Ix} \cdot P_I) - (Q_{Rx} \cdot Q_R + Q_{Ix} \cdot Q_I)]$; $\omega_x = -\frac{F_x}{F_\omega}$. We define U and V:

$$U = (P_R \cdot P_{I\omega} - P_I \cdot P_{R\omega}) - (Q_R \cdot Q_{I\omega} - Q_I \cdot Q_{R\omega})$$
$$V = (P_R \cdot P_{Ix} - P_I \cdot P_{Rx}) - (Q_R \cdot Q_{Ix} - Q_I \cdot Q_{Rx})$$

We choose our specific parameter as time delay $x = \tau$.

$$P_R = -\omega^2 + \frac{1}{\tau_n \cdot \tau_p};\ P_I = \omega \cdot \left(\frac{1}{\tau_p} + \frac{1}{\tau_n}\right)$$
$$Q_R = g_2\left(S^{(j)}, N^{(j)}\right) \cdot \frac{1}{\tau_p} + (\beta - 1) \cdot \frac{1}{\tau_n} \cdot g_1\left(S^{(j)}, N^{(j)}\right)$$
$$Q_I = \omega \cdot \left[g_2\left(S^{(j)}, N^{(j)}\right) - g_1\left(S^{(j)}, N^{(j)}\right)\right];\ P_{R\omega} = -2 \cdot \omega;\ P_{I\omega} = \frac{1}{\tau_p} + \frac{1}{\tau_n};\ Q_{R\omega} = 0$$

$$Q_{I\omega} = g_2\left(S^{(j)}, N^{(j)}\right) - g_1\left(S^{(j)}, N^{(j)}\right);\ P_{R\tau} = 0$$
$$P_{I\tau} = 0;\ Q_{R\tau} = 0;\ Q_{I\tau} = 0;\ \omega_\tau = -\frac{F_\tau}{F_\omega}$$

$$P_{R\omega} \cdot P_R = -2 \cdot \omega \cdot \left(-\omega^2 + \frac{1}{\tau_n \cdot \tau_p}\right);\ \mathrm{Q}_{R\omega} \cdot Q_R = 0;\ F_\tau$$
$$= 2 \cdot [(P_{R\tau} \cdot P_R + P_{I\tau} \cdot P_I) - (Q_{R\tau} \cdot Q_R + Q_{I\tau} \cdot Q_I)] = 0$$

$$P_{I\omega} \cdot P_I = \omega \cdot \left(\frac{1}{\tau_p} + \frac{1}{\tau_n}\right)^2;\ Q_{I\omega} \cdot$$
$$Q_I = \omega \cdot \left[g_2\left(S^{(j)}, N^{(j)}\right) - g_1\left(S^{(j)}, N^{(j)}\right)\right]^2$$

$$P_R \cdot P_{I\omega} = \left[-\omega^2 + \frac{1}{\tau_n \cdot \tau_p}\right] \cdot \left[g_2\left(S^{(j)}, N^{(j)}\right) - g_1\left(S^{(j)}, N^{(j)}\right)\right];\ P_I \cdot P_{R\omega}$$
$$= -2 \cdot \omega^2 \cdot \left(\frac{1}{\tau_p} + \frac{1}{\tau_n}\right)$$

$$Q_R \cdot Q_{I\omega} = \left[g_2\left(S^{(j)}, N^{(j)}\right) \cdot \frac{1}{\tau_p} + (\beta - 1) \cdot \frac{1}{\tau_n} \cdot g_1\left(S^{(j)}, N^{(j)}\right)\right] \cdot$$
$$\left[g_2\left(S^{(j)}, N^{(j)}\right) - g_1\left(S^{(j)}, N^{(j)}\right)\right]$$

$Q_I \cdot Q_{R\omega} = 0;\ V = 0$. $F(\omega, \tau) = 0$, differentiating with respect to τ and we get

$$F_\omega \cdot \frac{\partial \omega}{\partial \tau} + F_\tau = 0;\ \tau \in I \Rightarrow \frac{\partial \omega}{\partial \tau} = -\frac{F_\tau}{F_\omega};\ \Lambda^{-1}(\tau)$$
$$= \left(\frac{\partial \mathrm{Re}\lambda}{\partial \tau}\right)_{\lambda = i\cdot\omega};\ \frac{\partial \omega}{\partial \tau} = \omega_\tau = -\frac{F_\tau}{F_\omega}$$
$$\Lambda^{-1}(\tau) = \mathrm{Re}\left\{\frac{-2\cdot\left[U + \tau\cdot|P|^2 + i\cdot F_\omega\right.}{F_\tau + i\cdot 2\cdot\left[V + \omega\cdot|P|^2\right]}\right\};\ \mathrm{sign}\left\{\Lambda^{-1}(\tau)\right\} = \mathrm{sign}\left\{\left(\frac{\partial \mathrm{Re}\lambda}{\partial \tau}\right)_{\lambda = i\cdot\omega}\right\}$$

$$\mathrm{sign}\left\{\Lambda^{-1}(\tau)\right\} = \mathrm{sign}\{F_\omega\}\cdot\mathrm{sign}\left\{\frac{V + \frac{\partial\omega}{\partial\tau}\cdot U}{|P|^2} + \omega + \frac{\partial \omega}{\partial \tau}\cdot\tau\right\}$$

We shall presently examine the possibility of stability transitions (bifurcations) electron-photon exchanges into VCSEL system, about the equilibrium points

$$E^{(j)}\left(N^{(j)}, S^{(j)}\right)$$
$$= \left(-\left(\frac{\Xi_2\cdot A}{2\cdot\Xi_1} + B\right) \pm \frac{A\cdot\sqrt{\Xi_2^2 - 4\cdot\Xi_1\cdot\Xi_3}}{2\cdot\Xi_1}, \frac{-\Xi_2}{2\cdot\Xi_1} \pm \frac{\sqrt{\Xi_2^2 - 4\cdot\Xi_1\cdot\Xi_3}}{2\cdot\Xi_1}\right)$$

$$j = 0, 1, 2, \ldots$$

as a result of a variation of delay parameterτ. The analysis consists in identifying the roots of our system characteristic equation situated on the imaginary axis of the complex λ—plane, where by increasing the delay parameterτ, Re λ may at the crossing, changes its sign from – to +, i.e. from a stable focus $E(*)$ to an unstable one, or vice versa. This feature may be further assessed by examining the sign of the partial derivatives with respect to τ.

$$\Lambda^{-1}(\tau) = \left(\frac{\partial \mathrm{Re}\lambda}{\partial \tau}\right)_{\lambda = i\cdot\omega};\ \eta, \tau_n, \tau_p, G, \beta, \varepsilon, \ldots$$

$$U = (P_R\cdot P_{I\omega} - P_I\cdot P_{R\omega}) - (Q_R\cdot Q_{I\omega} - Q_I\cdot Q_{R\omega})$$

$$U = \left(\left[-\omega^2 + \frac{1}{\tau_n\cdot\tau_p}\right]\cdot\left[g_2\left(S^{(j)}, N^{(j)}\right) - g_1\left(S^{(j)}, N^{(j)}\right)\right] + 2\cdot\omega^2\cdot\left(\frac{1}{\tau_p} + \frac{1}{\tau_n}\right)\right)$$
$$-\left[g_2\left(S^{(j)}, N^{(j)}\right)\cdot\frac{1}{\tau_p} + (\beta - 1)\cdot\frac{1}{\tau_n}\cdot g_1\left(S^{(j)}, N^{(j)}\right)\right]\cdot$$
$$\left[g_2\left(S^{(j)}, N^{(j)}\right) - g_1\left(S^{(j)}, N^{(j)}\right)\right]$$

Then we get the expression for $F(\omega, \tau)$ electron-photon exchanges into VCSEL system's parameter values. We find those ω, τ values which fullfill $F(\omega, \tau) = 0$.

We ignore negative, complex, and imaginary values of ω for specific τ values. In our case $F(\omega, \tau)$ is independent on τ values. We can express by 3D or 2D function $F(\omega, \tau) = 0$. We plot the stability switch diagram based on different delay values of our electron-photon exchanges into VCSEL system.

$$\Lambda^{-1}(\tau) = \left(\frac{\partial \mathrm{Re}\lambda}{\partial \tau}\right)_{\lambda = i\cdot\omega} = \mathrm{Re}\left\{\frac{-2\cdot\left[U + \tau\cdot|P|^2\right] + i\cdot F_\omega}{F_\tau + 2\cdot i\cdot\left[V + \omega\cdot|P|^2\right]}\right\}$$

$$\Lambda^{-1}(\tau) = \left(\frac{\partial \mathrm{Re}\lambda}{\partial \tau}\right)_{\lambda = i\cdot\omega} = \frac{2\cdot\left\{F_\omega\cdot\left(V + \omega\cdot P^2\right) - F_\tau\cdot\left(U + \tau\cdot P^2\right)\right\}}{F_\tau^2 + 4\cdot\left(V + \omega\cdot P^2\right)^2}$$

The stability switch occurs only on those delay values (τ) which fit the equation: $\tau = \frac{\theta_+(\tau)}{\omega_+(\tau)}$ and $\theta_+(\tau)$ is the solution of $\sin\theta(\tau) = \ldots;\ \cos\theta(\tau) = \ldots$ when $\omega = \omega_+(\tau)$ if only ω_+ is feasible. Additionally, when all electron-photon exchanges into VCSEL system's parameters are known and the stability switch due to various time delay values τ is described in the following expression:

$$\mathrm{sign}\left\{\Lambda^{-1}(\tau)\right\} = \mathrm{sign}\{F_\omega(\omega(\tau), \tau)\}\cdot$$
$$\mathrm{sign}\left\{\tau\cdot\omega_\tau(\omega(\tau)) + \omega(\tau) + \frac{U(\omega(\tau))\cdot\omega_\tau(\omega(\tau)) + V(\omega(\tau))}{|P(\omega(\tau))|^2}\right.$$

Remark We know $F(\omega, \tau) = 0$ implies its roots $\omega_i(\tau)$ and finding those delay values τ which ω_i is feasible. There are τ values which give complex ω_i or imaginary number, then unable to analyse stability. F function is independent on τ parameter values $F(\omega, \tau) = 0$.

The results: We find those ω, τ values which fullfil $F(\omega, \tau) = 0$. We ignore negative, complex and imaginary values of ω.
Next is to find those ω, τ values which fullfil $\sin\theta(\tau) = \ldots;\ \cos\theta(\tau) = \ldots$

$$\sin(\omega\cdot\tau) = \frac{-P_R\cdot Q_I + P_I\cdot Q_R}{|Q|^2};\ \cos(\omega\cdot\tau) = -\frac{(P_R\cdot Q_R + P_I\cdot Q_I)}{|Q|^2};\ |Q|^2 = Q_R^2 + Q_I^2$$

Finally, we plot the stability switch diagram$g(\tau) = \Lambda^{-1}(\tau) = \left(\frac{\partial \mathrm{Re}\lambda}{\partial \tau}\right)_{\lambda = i\cdot\omega}$

$$g(\tau) = \Lambda^{-1}(\tau) = \frac{2\cdot\left\{F_\omega\cdot\left(V + \omega\cdot P^2\right) - F_\tau\cdot\left(U + \tau\cdot P^2\right)\right\}}{F_\tau^2 + 4\cdot\left(V + \omega\cdot P^2\right)^2}$$

$$\mathrm{sign}[g(\tau)] = \mathrm{sign}\left[\Lambda^{-1}(\tau)\right] = \mathrm{sign}\left[\frac{2\cdot\left\{F_\omega\cdot\left(V + \omega\cdot P^2\right) - F_\tau\cdot\left(U + \tau\cdot P^2\right)\right\}}{F_\tau^2 + 4\cdot\left(V + \omega\cdot P^2\right)^2}\right]$$

$$F_{\tau}^{2}+4\cdot\left(V+\omega\cdot P^{2}\right)^{2}>0;\ \operatorname{sign}\left[\Lambda^{-1}(\tau)\right]$$
$$=\operatorname{sign}\left[F_{\omega}\cdot\left(V+\omega\cdot P^{2}\right)-F_{\tau}\cdot\left(U+\tau\cdot P^{2}\right)\right]$$

$$\operatorname{sign}\left[\Lambda^{-1}(\tau)\right]=\operatorname{sign}\left\{[F_{\omega}]\cdot\left[\left(V+\omega\cdot P^{2}\right)-\frac{F_{\tau}}{F_{\omega}}\cdot\left(U+\tau\cdot P^{2}\right)\right]\right\}$$

$$\omega_{\tau}=-\frac{F_{\tau}}{F_{\omega}};\ \omega_{\tau}=\left(\frac{\partial\omega}{\partial\tau}\right)^{-1}=-\frac{\partial F/\partial\omega}{\partial F/\partial\tau}$$

$$\operatorname{sign}\left[\Lambda^{-1}(\tau)\right]=\operatorname{sign}\left\{[F_{\omega}]\cdot\left[V+\omega_{\tau}\cdot U+\omega\cdot P^{2}+\omega_{\tau}\cdot\tau\cdot P^{2}\right]\right\}$$

$$\operatorname{sign}\left[\Lambda^{-1}(\tau)\right]=\operatorname{sign}\left\{[F_{\omega}]\cdot\left[P^{2}\right]\cdot\left[\frac{V+\omega_{\tau}\cdot U}{P^{2}}+(\omega+\omega_{\tau}\cdot\tau)\right]\right\}$$

$$\operatorname{sign}\left[P^{2}\right]>0\Rightarrow\operatorname{sign}\left[\Lambda^{-1}(\tau)\right]=\operatorname{sign}\left\{[F_{\omega}]\cdot\left[\frac{V+\omega_{\tau}\cdot U}{P^{2}}+(\omega+\omega_{\tau}\cdot\tau)\right]\right\}$$

$$\operatorname{sign}\left[\Lambda^{-1}(\tau)\right]=\operatorname{sign}[F_{\omega}]\cdot\operatorname{sign}\left[\frac{V+\omega_{\tau}\cdot U}{P^{2}}+\omega+\omega_{\tau}\cdot\tau\right]$$

$$F_{\omega}=2\cdot[(P_{R\omega}\cdot P_{R}+P_{I\omega}\cdot P_{I})-(Q_{R\omega}\cdot Q_{R}+Q_{I\omega}\cdot Q_{I})]$$

$$F_{\omega}=2\cdot\left[\left(-2\cdot\omega\cdot\left(-\omega^{2}+\frac{1}{\tau_{n}\cdot\tau_{p}}\right)+\omega\cdot\left(\frac{1}{\tau_{p}}+\frac{1}{\tau_{n}}\right)^{2}\right)\right.$$
$$\left.-\left(\omega\cdot\left[g_{2}\left(S^{(j)},N^{(j)}\right)-g_{1}\left(S^{(j)},N^{(j)}\right)\right]^{2}\right)\right]$$

$$F_{\omega}=2\cdot\left[2\cdot\omega\cdot\left(\omega^{2}-\frac{1}{\tau_{n}\cdot\tau_{p}}\right)+\omega\cdot\left(\frac{1}{\tau_{p}}+\frac{1}{\tau_{n}}\right)^{2}\right.$$
$$\left.-\omega\cdot\left[g_{2}\left(S^{(j)},N^{(j)}\right)-g_{1}\left(S^{(j)},N^{(j)}\right)\right]^{2}\right]$$

We check the sign of $\Lambda^{-1}(\tau)$, if $\operatorname{sign}\left[\Lambda^{-1}(\tau)\right]>0$ then the crossing proceeds from (−) to (+) respectively (stable to unstable). If $\operatorname{sign}\left[\Lambda^{-1}(\tau)\right]<0$ then the crossing proceeds from (+) to (−) respectively (unstable to stable).

1.5 Questions

1. We have Laser diode coupled delay rate equations which characterize our system. variable N_{ph} is the photon density. Variable N_e is the electron density, q is the electronic charge, d is the thickness of the active region, a is the area of

the diode contact stripe, I is the injected current, N_{om} is the minimum electron density to obtain a positive gain, A is a constant related to the stimulated emission process, τ_s is the spontaneous emission lifetime, τ_{ph} is the photon lifetime and β is the fraction of the spontaneous emission that is coupled to the lasing mode. The carrier density N_e is a function of the junction voltage, where N_i is the intrinsic carrier density and V is the junction voltage.

$$\frac{dN_i(t)}{dt} = \frac{1}{q \cdot a \cdot d} \cdot \exp\left(-\frac{q \cdot V}{2 \cdot k \cdot T}\right) - A \cdot \exp\left(-\frac{q \cdot V}{2 \cdot k \cdot T}\right) \cdot \left(N_i(t-\tau) \cdot \exp\left(\frac{q \cdot V}{2 \cdot k \cdot T}\right) - N_{om}\right) \cdot N_{ph}(t-\tau) - \frac{N_i(t)}{\tau_s}$$

$$\frac{dN_{ph}(t)}{dt} = A \cdot \left(N_i(t-\tau) \cdot \exp\left(\frac{q \cdot V}{2 \cdot k \cdot T}\right) - N_{om}\right) \cdot N_{ph}(t-\tau) - \frac{N_{ph}(t)}{\tau_{ph}} + \frac{\beta}{\tau_s} \cdot N_i(t) \cdot \exp\left(\frac{q \cdot V}{2 \cdot k \cdot T}\right)$$

(second possible sets of differential equations)
The model differential equations assume that the inversion is homogenous and the gain is linear in the difference between N_e and N_{om}. The carrier density N_e is a function of the junction voltage, where N_i is the intrinsic carrier density and V is the junction voltage. The carrier density equation: $N_e = N_i \cdot \exp\left(\frac{q \cdot V}{2 \cdot k \cdot T}\right)$; $N_e(t-\tau) = N_i(t-\tau) \cdot \exp\left(\frac{q \cdot V}{2 \cdot k \cdot T}\right)$. We define the parasitic effects of the stimulated emission process as a latent period τ.

1.1 Find system equilibrium points (fixed points) and explain the meaning.
1.2 Discuss stability and stability switching of our system (τ variation).
1.3 We define our τ_s (spontaneous emission lifetime) and τ_{ph} (photon lifetime) parameters as a function of our latent period parameter τ; $\tau_s = \tau^2 - \Gamma_1 \cdot \tau$, $\tau_{ph} = \sqrt{\tau^2 - \Gamma_2 \cdot \tau}$; $\Gamma_1, \Gamma_2 \in \mathbb{R}$. Find system fixed points and discuss stability/stability switching for different values of τ parameter ($\tau > 0$).
1.4 What are the restrictions on Γ_1, Γ_2; $\Gamma_1, \Gamma_2 \in \mathbb{R}$ parameters which are related to τ_s, τ_{ph} expressions (see 1.3)? ($\tau_s > 0, \tau_{ph} > 0$).
1.5 Discuss our system stability and stability switching for the following cases: (1) $\Gamma_1 > 0$; $\Gamma_2 = 0$ (2) $\Gamma_1 = 0$; $\Gamma_2 > 0$ (3) $\Gamma_1 = 0$; $\Gamma_2 = 0, \Gamma_1, \Gamma_2 \in \mathbb{R}$.

2. The dynamic behavior of a laser is determined by the interaction of the intracavity light field with the gain medium. Essentially, the intracavity laser power can grow or decay exponentially according to the difference between gain and resonator losses, whereas the rate of change in the gain is determined by stimulated and spontaneous emission (and possibly by other effects such as quenching and energy transfer). With certain approximations, including a not too high laser

gain, the dynamics of the intracavity optical power P and the gain coefficient g in a continuous-wave laser can be described with the following coupled delay differential equations. Due to the delay of the optical power P which amplified by gain coefficient g, the multiplication element is $P(t-\tau)\cdot g(t-\tau)$.

$$\frac{dP}{dt}=\frac{g(t-\tau)\cdot P(t-\tau)}{T_R}-\frac{I\cdot P(t)}{T_R};\ \frac{dg}{dt}=-\frac{(g-g_{ss})}{\tau_g}-\frac{g(t-\tau)\cdot P(t-\tau)}{E_{sat}}$$

where T_R is the cavity round-trip time, I the cavity loss, g_{ss} the small-signal gain (for a given pump intensity), τ_g the gain relaxation time (often close to the upper-state lifetime), and E_{sat} the saturation energy of the gain medium.

2.1. Find system equilibrium points (fixed points) and explain the Meaning.
2.2 Discuss stability and stability switching of our system for different values of τ parameter.
2.3 We define τ_g (gain relaxation time) as a function of τ and Ω; $\Omega\in\mathbb{R}$ parameters, $\tau_g=\sqrt{\tau-\Omega\cdot\sqrt{\tau}}$. Find system fixed points and discuss stability/stability switching for different values of τ parameter ($\tau>0$).
2.4 What are the restrictions on Ω; $\Omega\in\mathbb{R}$ parameter which is related to τ_g expression (See 2.3)? ($\tau_g>0$).
2.5 We define τ_g (gain relaxation time) as a function of $\tau$and Ω; $\Omega\in\mathbb{R}$ parameters, $\tau_g=\sqrt{\tau-\sqrt{\Omega\cdot\sqrt{\tau}}}$. Find system fixed points and discuss stability/stability switching for different values of τ parameter ($\tau>0$).

3. We have system of back to back connection laser diodes (D1, D2) which are attached to fiber optic. The supply signal (square wave alternates between V_{sup1} and V_{sup2}, $V_{sup1}>0$; $V_{sup2}<0$; $|V_{sup1}|>|V_{sup2}|$) is connected through resistor R_p to the set of two laser diodes.

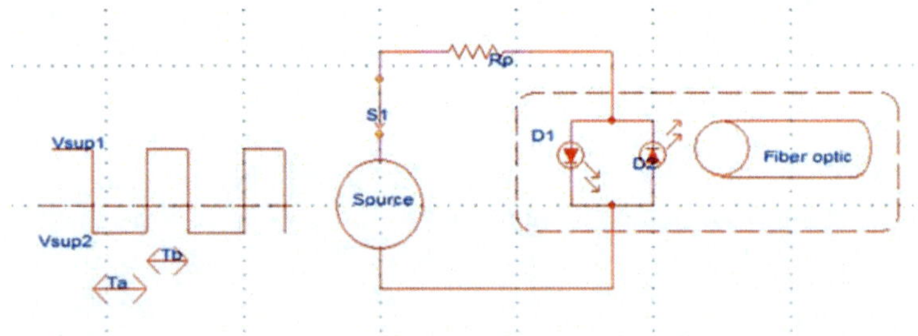

The durations of the positive voltage pulse and the negative voltage pulse are T_b and T_a respectively ($T_a\neq T_b$; $T_a>0$; $T_b>0$). At t=0 switch S1 changes his position from OFF to ON state. Investigate the operation of the system when you use the basic intrinsic equivalent circuit of the laser diode. Some of the laser diodes (D1 and D2) parameters are not the same (see below list):

$$R_{i-D1} = \frac{2 \cdot k \cdot T}{q^2} \cdot \frac{\tau_{s-D1}}{N_e^0 \cdot a \cdot d} \cdot \frac{1}{(A \cdot \tau_{S-D1} \cdot N_{ph}^0 + 1)}$$

$$R_{i-D2} = \frac{2 \cdot k \cdot T}{q^2} \cdot \frac{\tau_{s-D2}}{N_e^0 \cdot a \cdot d} \cdot \frac{1}{(A \cdot \tau_{S-D2} \cdot N_{ph}^0 + 1)}$$

$$L_{i-D1} \approx \frac{2 \cdot k \cdot T \cdot \tau_{ph-D1}}{A \cdot N_{ph}^0 \cdot q^2 \cdot N_e^0 \cdot a \cdot d}; \; L_{i-D2} \approx \frac{2 \cdot k \cdot T \cdot \tau_{ph-D2}}{A \cdot N_{ph}^0 \cdot q^2 \cdot N_e^0 \cdot a \cdot d}$$

$$C_{i-D1} = C_{i-D2} = C_i = \frac{N_e^0 \cdot a \cdot q^2 \cdot d}{2 \cdot k \cdot T}$$

$$R_{se-D1} = \beta \cdot R_d \cdot \frac{n_e^0}{n_{ph}^0 \cdot (n_{ph}^0 + \beta) \cdot (n_e^0 - n_{om-D1})}$$

$$R_{se-D1} = \beta \cdot R_d \cdot \frac{n_e^0}{n_{ph}^0 \cdot (n_{ph}^0 + \beta) \cdot (n_e^0 - n_{om-D2})}$$

3.1 Represent our system by set of differential equations (separate between the operation of D1 and D2 diodes).
3.2 Find system fixed points (equilibrium points). Discuss how your results are affected by the diodes different basic intrinsic equivalent circuit's parameters?
3.3 Discuss our system stability and bifurcation for different diode's basic intrinsic equivalent circuit's parameters. Is there any stability switching for different diode's parameter values?
3.4 We change the direction of D1, how it influences our system dynamic? Find fixed points and discuss stability.
3.5 We change the direction of D2, how it influences our system dynamic? Find fixed points and discuss stability.

4. A semiconductor laser is biased with current I above threshold. The steady state values of carrier and photon densities are n and n_p, respectively. Since $I > I_{th}$, $n \approx n_{th}$. Suppose at time t = 0, the carrier density is suddenly increased from n to $n + \Delta n$. The new value of the carrier density is not the steady state value and we need to improve it, if at all, the steady state value is recovered. The carrier density $n + \Delta n$ is greater than n_{th}, and, consequently, the gain $\widetilde{g}$ is greater than the threshold gain $\widetilde{g_{th}}$. In steady state, can never be greater than $\widetilde{g_{th}}$, but this restriction does not hold in non-steady state situations. Another way is to perturbe the photon density at time t = 0 to $n_p + \Delta n_p$. If the carrier or photon densities in a laser are distributed from their steady state values than it is important to know if these quantities return to their steady state values. If they do the laser is stable. If they don't, the laser is unstable. Inspecting the recovery dynamics associated with such carrier Density or photon density perturbations help for understanding the laser dynamic. The coupled differential equations for the perturbations in the carrier and photons densities constitute a second order linear system. Due to laser operation parasitic effects there is

a delay in $\Delta n_p(t)$ and $\Delta n(t)$ variables in time ($\Delta n_p(t) \to \Delta n_p(t-\tau_1)$ and $\Delta n(t) \to \Delta n(t-\tau_2)$; $\tau_1 > 0$; $\tau_2 > 0$; $\tau_1 \neq \tau_2$). The coupled equations give the following identical second order delay differential equations for the perturbations (relaxation oscillations).

$$\frac{d^2\Delta n_p(t)}{dt^2} + \gamma \cdot \frac{d\Delta n_p(t)}{dt} + \omega_R^2 \cdot \Delta n_p(t-\tau_1) = 0; \ \omega_R = \sqrt{\frac{1}{\tau_{st}\cdot\tau_p}}; \ \gamma = \frac{1}{\tau_r} + \frac{1}{\tau_{st}}$$

$$\frac{d^2\Delta n(t)}{dt^2} + \gamma \cdot \frac{d\Delta n(t)}{dt} + \omega_R^2 \cdot \Delta n(t-\tau_2) = 0$$

τ_{st} differential stimulated emission time.
τ_r differential recombination time.
τ_p cavity photon lifetime.
n_{sp} spontaneous emission factor.
$\widetilde{g}$ gain which is related to the net stimulate rate.
$\widetilde{g_{th}}$ threshold gain.
N_p total photon number.
n_p average photon density.
Δn_p photon density perturbed additive term in time.
Δn carrier density perturbed additive term in time.

The carrier density at which the gain $\widetilde{g}$ equals the threshold gain $\widetilde{g_{th}}$ ($\widetilde{g} = \widetilde{g_{th}}$) is called the threshold carrier density.

4.1 Find system fixed points (equilibrium points).
4.2 Discuss system stability and stability switching for the following cases: (1) $\tau_1 = \tau; \ \tau_2 = 0$ (2) $\tau_1 = 0; \ \tau_2 = \tau$ (3) $\tau_1 = \tau; \ \tau_2 = \tau$.
4.3 What happened if the differential stimulated emission time (τ_{st}) and differential recombination time (τ_r) are very big ($\tau_{st} \to \infty; \ \tau_r \to \infty$)? How the system's dynamic changes? Discuss stability and stability switching in that case.
4.4 What happened if the cavity photon lifetime τ_p is very big ($\tau_p \to \infty$)? How the system's dynamic changes? Discuss stability and stability switching in that case.
4.5 The second derivative in time of $\Delta n_p(t)$ variable is neglected ($\frac{d^2\Delta n_p(t)}{dt^2} \to \varepsilon$). How the system dynamic is changed? Find fixed points and discuss stability and stability switching for the following cases: (1) $\tau_1 = \tau; \ \tau_2 = 0$ (2) $\tau_1 = 0; \ \tau_2 = \tau$ (3) $\tau_1 = \tau; \ \tau_2 = \tau$.

5. We have laser diode which an electric current input is converted to an output of photons. The time-dependent relation between the input electric current and output photon is described by a pair of equations describing the time evolution of photon and carrier densities inside the laser medium. This pair of equations are the laser rate equations, we analyze the "kinetics" by the coupled rate equations.

The differential equations are local photon and injected carrier conservation equations. We define some parameters:
z: spatial dimension along the length of the laser (distance along the active medium with z=0 at the center of the laser).
R: laser reflectors of (power) reflectivities placed at $z = \pm\frac{L}{2}$.
X^+, X^-: forward (X^+) and backward (X^-) propagating photon densities (which are proportional to the light intensities).
N: local carrier densities.
N_{tr}: electron density where the semiconductor medium becomes transparent.
C:group velocity of the waveguide mode.
A: gain constant in $s^{-1}/$(unit carrier density), differential optical gain.
β: fraction of spontaneous emission entering the lasing mode.
τ_s: spontaneous recombination lifetime of the carrier.
J: uniform pump current density.
τ: delay parameter (related to$N^*(t-\tau)\cdot P^*(t-\tau)$).
e: electronic charge.
d:thickness of the active region in which the carriers are confined.
P: total local photon density and the boundary conditions have been used ($P = X^+ + X^-$).
"*": denotes the spatial average $\int_{-L/2}^{+L/2}\frac{dz}{dL}$.
f_1, f_2: factor functions ($f_1 = \frac{(N\cdot P)^*}{N^*\cdot P^*}$, $f_2 = \frac{P\cdot(\frac{L}{2})}{P^*\cdot(1+R)}$). If the following conditions are satisfied $f_1 = 1$; $f_2 = -\frac{1}{2}\cdot\frac{\ln R}{(1-R)}$, we can recognize the following rate equations: $N^* = N^*(t)$; $P^* = P^*(t)$

$$\frac{dP^*}{dt} = A\cdot f_1\cdot N^*(t-\tau)\cdot P^*(t-\tau) - 2\cdot c\cdot(1-R)\cdot f_2\cdot\frac{P^*}{L} + 2\cdot\beta\cdot\frac{N^*}{\tau_s}$$

$$\frac{dN^*}{dt} = \frac{J}{e\cdot d} - \frac{N^*}{\tau_s} - A\cdot f_1\cdot N^*(t-\tau)\cdot P^*(t-\tau)$$

The first of these conditions ($f_1 = 1$) requires, for the quantities N and P, that the spatial average of the product equals the product of the spatial averages. This condition is not satisfied in general, but it will be true if the electron density N is uniform, as in the case when R approaches unity. The second condition ($f_2 = -\frac{1}{2}\cdot\frac{\ln R}{(1-R)}$) requires the photon loss rate to be inversely proportional to the conventional photon lifetime. It will also be satisfied if R is very close to unity. Due to laser's parasitic effects there is a delay in time of τ which is related to the multiplication element ($N^*(t)\cdot P^*(t) \to N^*(t-\tau)\cdot P^*(t-\tau)$).

5.1 Find system fixed points (equilibrium points).
5.2 Discuss stability and stability switching for different values of τ parameter.
5.3 Discuss system stability and stability switching if the group velocity of the waveguide mode c is a function of parameter τ and fundamental group velocity of the waveguide mode c_0; $c_0 \in \mathbb{R}_+$ $c = c_0\cdot\tau^2 + \sqrt{\tau}$.

5.4 Discuss system stability and stability switching if the fraction of spontaneous emission entering the lasing modeβ is a function of parameter τ, $\beta = \sqrt{\Gamma_0 \cdot \sqrt{\tau} + \tau^3}$; $\Gamma_0 \in \mathbb{R}$. What are the restrictions on$\Gamma_0$parameter if we know that$\beta > 0$.

5.5 If we can neglect our uniform pump current density (J), $J \to \varepsilon$; How the system dynamic change? Find fixed points and discuss stability and stability switching for different values of τparameter.

6. We have a system of two laser diodes (D1, D2) which are attached to fiber optic. There are two power sources in our system; one is fix DC source (V_{sup}) and the other is a square wave which is alternated between $V_{\text{sup}\,1}$and$V_{\text{sup}\,2}$($V_{\text{sup}} > V_{\text{sup}\,1}$; $V_{\text{sup}\,1} > 0$; $V_{\text{sup}\,2} < 0$; $|V_{\text{sup}}| > |V_{\text{sup}\,2}|$, $|V_{\text{sup}\,1}| \neq |V_{\text{sup}\,2}|$). At t = 0 switch S1 changes his position from OFF to ON state. Investigate the operation of the system when you use the complete intrinsic electrical equivalent circuit of the laser diode. Some of the laser diodes (D1 and D2) parameters are not the same (see below list):

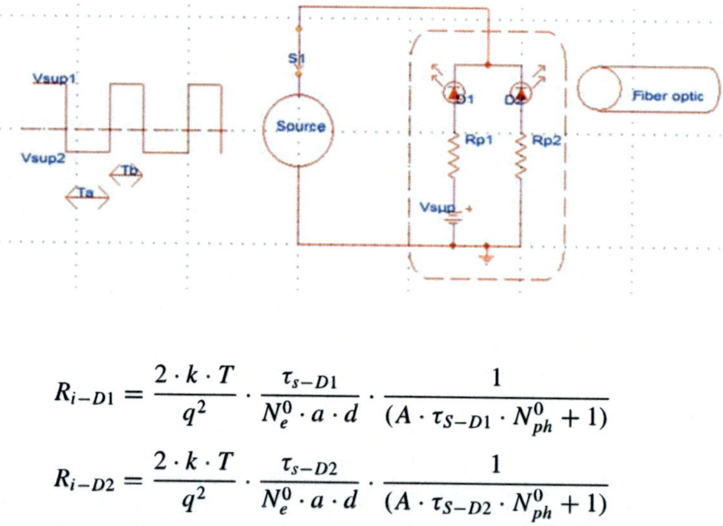

$$R_{i-D1} = \frac{2 \cdot k \cdot T}{q^2} \cdot \frac{\tau_{s-D1}}{N_e^0 \cdot a \cdot d} \cdot \frac{1}{(A \cdot \tau_{S-D1} \cdot N_{ph}^0 + 1)}$$

$$R_{i-D2} = \frac{2 \cdot k \cdot T}{q^2} \cdot \frac{\tau_{s-D2}}{N_e^0 \cdot a \cdot d} \cdot \frac{1}{(A \cdot \tau_{S-D2} \cdot N_{ph}^0 + 1)}$$

$$L_{i-D1} \approx \frac{2 \cdot k \cdot T \cdot \tau_{ph-D1}}{A \cdot N_{ph}^0 \cdot q^2 \cdot N_e^0 \cdot a \cdot d}; \quad L_{i-D2} \approx \frac{2 \cdot k \cdot T \cdot \tau_{ph-D2}}{A \cdot N_{ph}^0 \cdot q^2 \cdot N_e^0 \cdot a \cdot d}$$

$$C_{i-D1} = C_{i-D2} = C_i = \frac{N_e^0 \cdot a \cdot q^2 \cdot d}{2 \cdot k \cdot T}$$

$$R_{se-D1} = \beta \cdot R_d \cdot \frac{n_e^0}{n_{ph}^0 \cdot (n_{ph}^0 + \beta) \cdot (n_e^0 - n_{om-D1})}$$

$$R_{se-D1} = \beta \cdot R_d \cdot \frac{n_e^0}{n_{ph}^0 \cdot (n_{ph}^0 + \beta) \cdot (n_e^0 - n_{om-D2})}$$

C_{sc} and R_{se1}elements have the same value for D1 and D2.

6.1 Represent our system by set of differential equations (saperate between the operation of D1 and D2 diodes).
6.2 Find system fixed points (equilibrium points). Discuss how your results are affected by the diodes different complete intrinsic electrical equivalent circuit's parameters?
6.3 Discuss our system stability and bifurcation for different diode's complete intrinsic electrical equivalent circuit's parameters. Is there any stability switching for different diode's parameter values?
6.4 We change the direction of D1, how it influences our system dynamic? Find fixed points and discuss stability.
6.5 We change the direction of D2, how it influences our system dynamic? Find fixed points and discuss stability.

7. We have MRI system laser diode circuit with delay lines. The laser diode circuitry consists of T—matching network (C_a, C_b, L_1), a Bias—T (C_1, L_2) match network, and laser diode itself. The T—matching network is responsible for the matching of the low impedance laser diode to the high impedance output of the pre amplifier. The Bias—T combines the RF signal with the appropriate DC current injection which required biasing the laser. The laser diode holds the micro-lens to couple the generated optical energy to the fiber. Due to laser diode circuit microstrip parasitic effects there are instabilities on the circuit operation. The circuit microstrip parasitic effects are represented by delay lines elements which delay the current in time that flow through them. We consider ideal delay lines then the voltages that fall are neglected ($V_{\tau_i} \to \varepsilon \, \forall i = 1, 2, 3$).

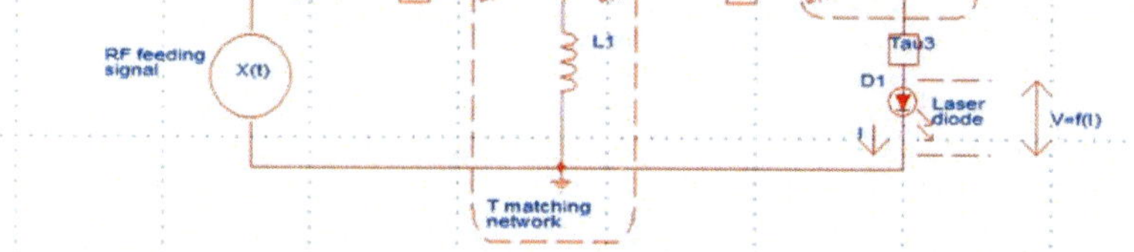

7.1 Find system fixed points (equilibrium points).
7.2 Find system delay differential equations (DDEs) and Jacobian.

7.3 Discuss stability and stability switching for different value of τ parameter, cases: (1) $\tau_1 = \tau;\ \tau_2 = 0;\ \tau_3 = 0$ (2) $\tau_1 = 0;\ \tau_2 = \tau;\ \tau_3 = 0$ (3) $\tau_1 = \tau;\ \tau_2 = \tau;\ \tau_3 = 0$ (4) $\tau_1 = 0;\ \tau_2 = \tau;\ \tau_3 = \tau$ (5) $\tau_1 = \tau;\ \tau_2 = 0;\ \tau_3 = \tau$ (6) $\tau_1 = \tau;\ \tau_2 = \tau;\ \tau_3 = \tau$.

7.4 We replace the elements in T matching network to L_a, L_b, C_2, find fixed points and discuss stability and stability switching for the cases in (7.3).

7.5 We replace system T—matching network with PI—matching network (C_a, C_b, L_1). Find system fixed points and discuss stability and stability switching for different value of τ parameter, cases: (1) $\tau_1 = \tau;\ \tau_2 = 0;\ \tau_3 = 0$ (2) $\tau_1 = 0;\ \tau_2 = \tau;\ \tau_3 = 0$ (3) $\tau_1 = \tau;\ \tau_2 = \tau;\ \tau_3 = 0$ (4) $\tau_1 = 0;\ \tau_2 = \tau;\ \tau_3 = \tau$ (5) $\tau_1 = \tau;\ \tau_2 = 0;\ \tau_3 = \tau$ (6) $\tau_1 = \tau;\ \tau_2 = \tau;\ \tau_3 = \tau$.

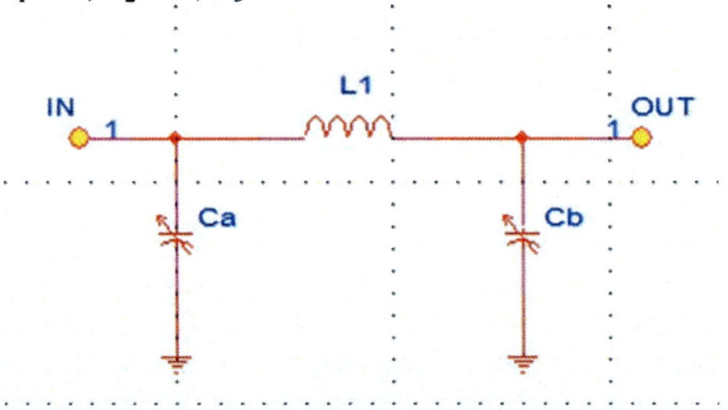

8. The MRI signals, after being coupled to the optical beam in the fiber, are then transmitted to the receiver. The main responsibility of the receiver is to convert optical signals back to electricity, and it uses a special photo detector (photo diode). An analysis of the laser beam that falls on the Photo Diode (PD) reveals that the envelope of the beam consists of two components, the one generated by MRI signals and the one generated by the DC biasing current of the laser diode. The current that the photo detector generates is the sum of these components converted. As the demodulation of the optical carrier is performed internally by the photo detector, the only circuitry needed is a Bias—T to separate these two components. The schematics of the receiver is as follow.

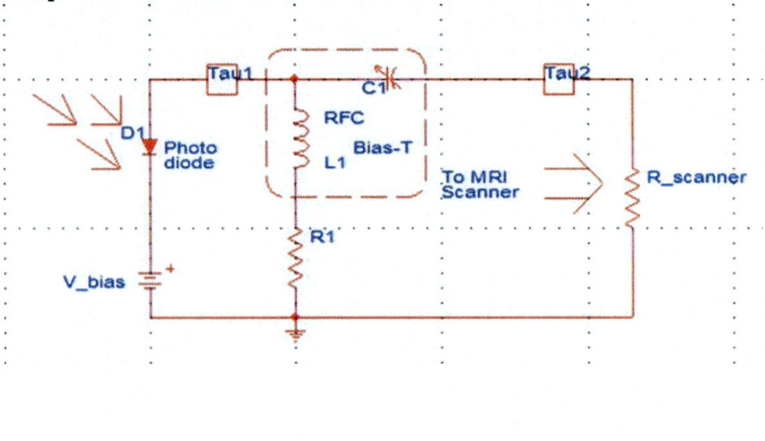

Circuit parasitic effects are represented by two delay lines Tau1, Tau2, which $V_{\tau_1}, V_{\tau_2} \to \varepsilon$ (delay on the current which flow through τ_1, τ_2). The Bias—T circuit (C_1, L_1), separates DC component of the signal from radio frequency component. The biasing circuit for photo diode is V_bias, R_1. The photo diode used is high speed diode which is a 100 um diameter InGaAs photodetector in a plastic housing with an SC receptacle. MRI scanner is represented by R_scanner. The photo diode equivalent circuit is shown below.

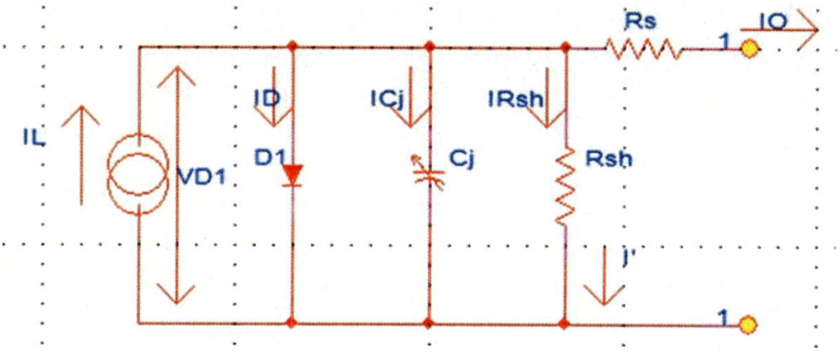

IL—current generated by the incident light and its proportional to the amount of light, I_D—diode current, C_j—junction capacitance, R_{sh}—shunt resistance, R_s—series resistance, $I' = I_{RSH}$—shunt resistance current, V_{D1}—voltage across the diode, I_{out}—output current, V_o—output voltage. Using the above equivalent circuit by neglecting the current through junction capaciatance, the output current I_0 is given as follow:

$$I_{out} = I_L - I_0 \cdot (e^{\frac{V_{D1}}{V_t}} - 1) - C_j \cdot \frac{dV_{D1}}{dt} - \frac{V_{D1}}{R_{sh}}; \; V_t = \frac{k \cdot T}{q},$$

I_0 photodiode reverse saturation current,
q electron charge,
k Boltzmann's constant,
T absolute temperature of the photodiode

$$C_j \cdot \frac{dV_{D1}}{dt} \to \varepsilon \Rightarrow I_{out} = I_L - I_0 \cdot (e^{\frac{V_{D1}}{V_t}} - 1) - \frac{V_{D1}}{R_{sh}}.$$

8.1 Find system Delay Differential Equations (DDEs) and fixed points.
8.2 Discuss stability and stability switching for different values of τ parameter (Cases: (1) $\tau_1 = \tau; \; \tau_2 = 0$ (2) $\tau_1 = 0; \; \tau_2 = \tau$ (3) $\tau_1 = \tau; \; \tau_2 = \tau$).
8.3 We change the value of resistor R_1, $R_1 \to R_1'; \; R_1' = \sqrt{\Gamma \cdot R_1}; \; \Gamma \in \mathbb{R}_+$. Discuss stability and stability switching for different values of Γ parameter (τ is constant).

8.4 How system dynamic is changed if we disconnect photo diode equivalent circuit's R_{sh} resistor? Find fixed points and discuss stability and stability switching for the following cases: (1) $\tau_1 = \tau;\ \tau_2 = 0$ (2) $\tau_1 = 0;\ \tau_2 = \tau$ (3) $\tau_1 = \tau;\ \tau_2 = \tau$.

8.5 We change the polarity of the photo diode (D_1), How the dynamic of the circuit is change? Is it possible to get the right outcome? Explain deeply your answer.

9. We have a system of Vertical-Cavity Surface-emitting laser (VCSELs) which employ current-blocking layers to funnel the current to the active region located between the mirrors of the laser. The VCSEL carrier-photon interaction active region volume is describe in the next figure.

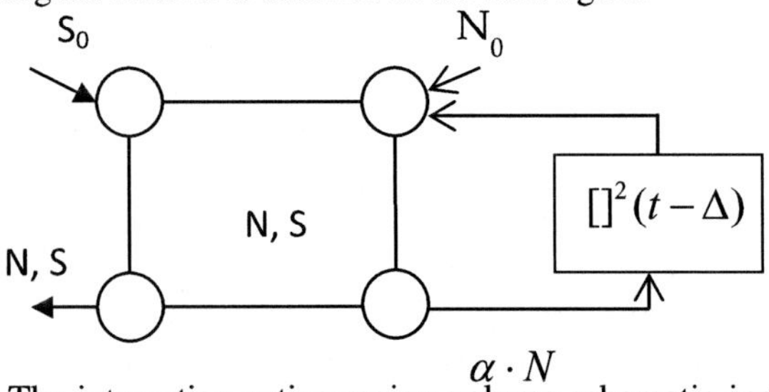

The interaction active region volume schematic includes additionally mechanism which takes some carriers (carriers (α factor, $0 < \alpha < 1$) in the active region volume) are amplified by squaring and reinject into the active region volume. The time takes that mechanism to happened is Δsecond. The VCSEL rate equations with delay parameter (τ) of the carrier-photon interaction time and delay parameter Δ of the additional mechanism:

$$\frac{dN(t)}{dt} = \frac{\eta \cdot (I - I_{off})}{q} - \frac{N(t)}{\tau_n} - \frac{G \cdot (N(t-\tau) - N_{tr}) \cdot S(t-\tau)}{1 + \varepsilon \cdot S(t-\tau)} - \alpha \cdot N(t) + \alpha^2 \cdot N^2(t - \Delta)$$

$$\frac{dS(t)}{dt} = -\frac{S(t)}{\tau_p} + \frac{\beta \cdot N(t)}{\tau_n} + \frac{G \cdot (N(t-\tau) - N_{tr}) \cdot S(t-\tau)}{1 + \varepsilon \cdot S(t-\tau)}$$

- N Carrier density in the active region.
- S Photon density in the active region.
- S_0 Photon density in the active region at t = 0.
- N_0 Carrier density in the active region at t = 0.
- η Injection efficiency.
- τ_n Carrier recombination lifetime.
- τ_p Photon lifetime.
- G Gain coefficient.
- N_{tr} Carrier's transparency number.
- β Spontaneous emission coupling coefficient.

ε Gain compression factor.
I VCSEL's injection current.
I_{off} VCSEL's offset current.
τ Time delay due to VCSEL active region parasitic effects.
Δ Time delay which related to the additional mechanism.

9.1 Find system fixed points and draw functions of each fixed point for different values of αparameter ($0 < \alpha < 1$).
9.2 Discuss stability and stability switching for different values of Δ parameter (other parameters are constant).
9.3 Discuss stability and stability switching for different values of τ parameter (other parameters are constant).
9.4 We change the additional mechanism so some carriers (carriers (αfactor, $0 < \alpha < 1$) in the active region volume) are amplified by tripling ($[]^3(t - \Delta)$) and reinject into the active region volume. Find fixed points and discuss stability and stability switching for different values of $\tau$and Δparameters.
9.5 We change the additional mechanism so some carriers (carriers (αfactor, $0 < \alpha < 1$) in the active region volume) are attenuated by square root ($\sqrt{[]}(t - \Delta)$) and reinject into the active region volume. Find fixed points and discuss stability and stability switching for different values of $\tau$and Δparameters.

10. Vertical cavity surface-emitting laser (VCSEL's) are the $\sqrt[4]{}\sqrt[4]{}$ transmitters in short distance optical communication system such as LAN's, where 1-2Gb/s signal must be carried 0–3 km over multimode fiber. Laser diode rate-equation parameters are extracted for simulation of on-off keyed digital communication links. The extraction procedure uses measurements of the current voltage light characteristic, the AC small-signal response above threshold, and the turn-on delay due to an isolated pulse. Many characteristics of laser diode behavior can be described by a set of rate equations, which may be written as

$$\frac{dN}{dt} = \frac{\eta_i \cdot I}{q \cdot N_\omega \cdot V_{act}} - R(N) - \Gamma \cdot v_{gr} \cdot G(N, S) \cdot S$$

$$\frac{dS}{dt} = N_\omega \cdot R_\beta(N) + N_\omega \cdot \Gamma \cdot v_{gr} \cdot G(N, S) \cdot S - \frac{S}{\tau_p}$$

where N is the carrier density, I is the injection current, and S is the normalized photon density $\frac{S_{tot}}{V_{act}}$, with S_{tot} being the photon number. This definition provides symmetry in the rate-equation stimulated emission terms. η_i is the current injection efficiency, q is the electron charge, N_ω the number of QW's (Quantum Well's), V_{act} is the volume of a single QW, $R(N)$ is the carrier recombination rate, Γ is the carrier confinement factor, v_{gr} is the velocity of the light in the cavity, $G(N, S)$ is the gain function, including gain compression, $R_\beta(N)$ is the coupled spontaneous emission, and τ_p is the photon lifetime. The output power is thus $\frac{\eta_c \cdot h \cdot c \cdot \Gamma \cdot S \cdot V_{act}}{\lambda \cdot \tau_p}$, where η_c is the output coupling efficiency, and $\frac{h \cdot c}{\lambda}$ is the photon energy. The recombination rate $R(N)$ may be written $A \cdot N + B \cdot N^2 + C \cdot N^3$. When B and C are set to zero, linearized form $\frac{N}{\tau_n}$ often used when simple rate equations are desired. The coupled spontaneous emission $R_\beta(N)$ may be written as $\beta_B \cdot B \cdot N^2$ or $\beta_A \cdot A \cdot N$ when the linearized recombination rate is used. The gain function may be written as $G(N, S) = \frac{G(N)}{1+\varepsilon \cdot \Gamma \cdot S}$ where $\varepsilon$is the gain suppression factor. The gain confinement factor Γappears in the gain suppression term because the photon density $\frac{S_{tot}}{V_{\text{mod}\,e}}$is equal to $\Gamma \cdot S$ where $V_{\text{mod}\,e}$is the volume of the optical mode. The gain carrier dependence be written as $G(N) = G_0 \cdot \ln\{\frac{R(N)}{R(N_{th})}\}$. The gain function may be linearized about the transparency carrier density, yielding $G(N) = m \cdot G_0 \cdot (\frac{N}{N_0} - 1)$, where m is a linearization constant. There are three model variations which define $R(N)$ and $G(N, S)$, as shown in the below table.

	Model 1	Model 2	Model 3
$R(N)$	$A \cdot N + B \cdot N^2 + C \cdot N^3$	$A \cdot N + B \cdot N^2 + C \cdot N^3$	$A \cdot N$
$R_\beta(N)$	$\beta_B \cdot B \cdot N^2$	$\beta_B \cdot B \cdot N^2$	$\beta_A \cdot A \cdot N$
$G(N)$	$G_0 \cdot \ln\left\{\frac{R(N)}{R(N_{th})}\right\}$	$m \cdot G_0 \cdot \left(\frac{N}{N_0} - 1\right)$	$G_0 \cdot \ln\left\{\frac{R(N)}{R(N_{th})}\right\}$

10.1 Find system fixed points and jacobian for all models (1, 2, 3).
10.2 Discuss stability for each system differential equations models (1, 2, 3). Classify fixed points for different parameter's values.
10.3 The carrier confinement factor $\Gamma = 0$, How the system dynamic is change? Find fixed points and discuss stability.
10.4 Due to parasitic effects there is a delay τ in the multiplication Elements $G(N(t), S(t)) \cdot S(t) \to G(N(t-\tau), S(t-\tau)) \cdot S(t-\tau)$, discuss stability and stability switching for different values of τ parameter.
10.5 The number of QW's (Quantum Well's) $N_\omega = 1$, how the system dynamic is changed? Discuss stability for that case.

References

1. M.S. Ozyazici, The complete electrical equivalent circuit of a double heterojunction laser diode using scattering parameters. J. Optoelectron Adv Mater **6**(4), 1243–1253 (2004)
2. Y. Kuang, *Delay Differential Equations with Applications in Population Dynamics* (Academic Press, Boston, 1993)
3. E. Beretta, Y. Kuang, Geometric stability switch criteria in delay differential systems with delay dependent parameters. SIAM J. Math. Anal. **33**, 1144–1165 (2002)
4. J. Kuang, Y. Cong. *Stability of Numerical methods for Delay Differential Equations.* Elsevier Science (2007)
5. B. Balachandran, T. Kalmár-Nagy D.E. Gilsinn, Delay *Differential Equations: Recent Advances and New Directions (Hardcover).* 1st edn. (Springer; 2009)

6. O. Gokalp Memis, Y. Eryaman, O. Aytur, E. Atalar, Miniaturized fiber-optic transmission system for MRI signals. Magn. Reson. Med. **59**,165–173 (2008)
7. P.V. Mena, J.J. Morikuni, S.-M. Kang, A.V. Harton, K.W. Wyatt, A simple rate-equation-based thermal VCSEL model. J. Lightwave Technol. **17**(5) (1999)
8. S.H. Strogatz, *Nonlinear Dynamics and Chaos.* Westview Press
9. J.C. Sprott, Simplifications of the lorenz attractor. Nonlinear Dynam. Psychol. Life Sci. **13**(3), 271–278
10. G. Stepan, T. Insperger, R. Szalai, Delay, Parametric excitation, and the nonlinear dynamics of cutting process. Int. J. Bifurcat. Chaos, **15**(9), 2783-2798 (2005)
11. R. Rand, A. Barcilon, T. Morrison, Parametric resonance of hopf bifurcation. Nonlinear Dynam. **39**, 411–421 (2005)
12. H.D.I. Abarbanel, M. I. Rabinovich, M. M. Sushchik, *Introduction to Nonlinear Dynamics for Physicists* (World Scientific, 1993)
13. X. Shi, C. Qi, G. Wang, J. Hu, Rate-equation-based VCSEL model and simulation. Comput-Aided Des Comput Graph (2009)

Chapter 2
Ti: Sapphire Laser Systems with Delay Parameters in Time Stability Analysis

Ti: Sapphire lase (Ti: Al_2O_3 laser, titanium-sapphire laser, or Ti: Sapphires) is a tunable laser which emit red and near infrared light in the range from 650 to 1100 nm. In many scientific experiments the Ti: Sapphire laser is used. It is very tenability and can generate ultrashort pulses. The lasing medium is Titanium-sapphire and the crystal of sapphire (Al_2O_3) that is doped with titanium ions. In many systems the Ti: Sapphire laser is pumped with another laser with a wavelength of 514–532 nm. Another laser can be argon-ion laser (514.5 nm) and frequency doubled Nd:YAG, Nd:YLF and Nd:YVO lasers (527–532 nm). The most efficient operation of Ti: Sapphire laser is at wavelengths near 800 nm. Additionally, the other widely use of Titanium-doped sapphire is transition-metal-doped gain medium for tunable lasers and femto second solid-state lasers. Sapphire has an excellent thermal conductivity, and alleviating thermal effects for high laser powers and intensities. The T_i^{3+} ion has a very large gain bandwidth which allowing the generation of a very short pulses and wide wavelength tenability. The maximum gain and laser efficiency are obtained around 800 nm. The T_i^{3+} doping concentration has to be kept fairly low because otherwise no good crystal is possible. The upper-state lifetime of Ti: Sapphire is short (3.2 us), and the saturation power is very high. The Ti: Sapphire pump intensity needs to be high and we get a pump source with high beam quality. The Ti: Sapphire has relatively high laser cross sections, which reduces the tendency of Ti: Sapphire laser for Q-switching instabilities. Ultrashort pulses from Ti: Sapphire laser can be generated with passive mode locking, in the form of kerr lens mode locking (KLM). A pulse duration around 100 fsec is easily achieved and is typical for commercial devices. Typical output powers of mode-locked Ti: Sapphire lasers are of the order 0.3–1 w, whereas continuous-wave versions sometimes generate several watts. A typical pulse repetition rate is 80 MHz, and devices with multi-gigahertz repetition rates are available and use as frequency comb sources. Ti: Sapphire is also used for multi-pass amplifiers and regenerative amplifiers. Nonlinear frequency conversion is used to extend further the range of emission wavelengths of Ti: Sapphire laser system. Linearly polarized femtosecond laser pulses from Ti: sapphire laser is related to multiphoton ionization of D_2 molecule. The rate of ionization of the molecule

O. Aluf, *Advance Elements of Laser Circuits and Systems*,
https://doi.org/10.1007/978-3-030-64103-0_2

is much lower than that of the atom. The interaction of strong laser fields with diatomic molecules characterized by the following processes: non-sequential ionization of molecules, high-order harmonic generation and fragmentation of diatomic molecules. They are closely related to the tunneling ionization (TI) of the molecules to a singly charged ion. There are two aspects which are related to finding a predictive model for the ionization rate of a diatomic molecule in the field of a short pulse with wavelength in the near-IR region. The first one is related to the fact that there was no experimentally proven model capable of providing a quantitative fit with the data on rare-gas atoms. The second is related to the difficulty in modeling the TI rate of molecules (orientation of the neutral molecule in the laser field). The second order differential equation for the rotation of the molecule with reduced mass, polarizability and inter-nuclear distance in the field of laser with frequency (ω) and field strength (F) has a stability switching under parameter variation and it is inspected. Ultra-broad bandwidth pulses from a Ti: sapphire laser is characterized by effects of higher-order phase dispersion on the fidelity of the shaped pulse. The phase dispersion at higher orders cubic phase and quartic phase dispersions. The phase dispersion differential equations are inspected for stability under parameters variation. Semiconductor saturable-absorber mirror mode-locked Ti: sapphire soliton laser is operated in the multiple-pulse regime. The stability of the pair of solitons is bounded by two limits. A stability criterion from the evolution of pulse energies as well as the gain is inspected in a set of differential equations. We analyze energies and gain delayed in time stability analysis due to parasitic effects [1–4].

2.1 Linearly Polarized Femtosecond Laser Pulses from Ti: Sapphire Laser Stability Analysis Under Parameters Variation

We use linearly polarized femtosecond laser pulses from a Ti: Sapphire laser to analyze the multiphoton ionization of D_2 (D_2 is a molecule of Deuterium). Deuterium (Hydrogen-2, Symbol D or ^2H) is one of two stable isotopes of hydrogen. The nucleus of deuterium, called a deuteron, contains one proton and one neutron, where the far more common protium has no neutron in the nucleus. We can compare between two mechanisms, one is the ion signal versus intensity curve of the singly charged molecular ion, second is the ion signal versus the intensity curve of the A_r atom with similar ionization potential. The ionization energy or ionization potential is the energy necessary to remove an electron from the neutral atom. It is a minimum for the alkali metals which have a single electron outside a closed shell. It generally increases across a row on the periodic maximum for the noble gases which have closed shells. Comparing between two mechanisms gives the outcome that the rate of ionization of the molecule is much lower than that of atom. There are two-centre nature of the potential which produces the barrier for tunneling, and the orientation of the molecular axis with respect to the laser polarization. A molecule can have

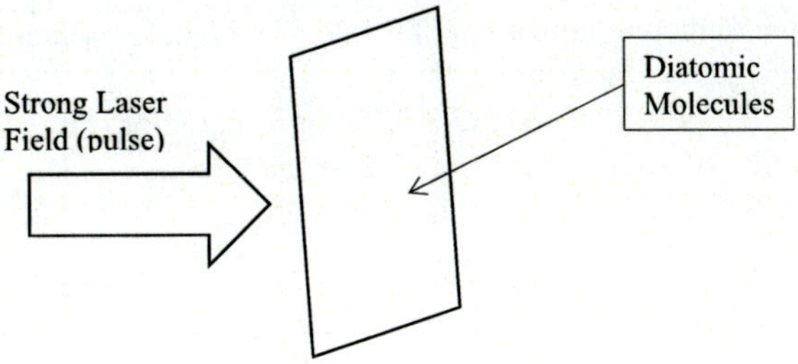

Fig. 2.1 Structure of strong laser field pulse and diatomic molecules experiment

more than one symmetry axis; the one with the highest n is called the principal axis, and by convention is aligned with the z-axis in a Cartesian coordinate system. Plane of symmetry: a plane of reflection through which an identical copy of the original molecule is generated. The effective barrier is equivalent to that produced by an atom with an effective charge of less than unity [1].

{Rate of ionization of the molecule} $\ll$ Rate of ionization of an atom}.

We investigate the interaction of strong laser fields with diatomic molecules (Fig. 2.1). There are some mechanism's processes, first the non-sequential ionization of molecules, second the high-order harmonic generation, and third the fragmentation of diatomic molecules. Diatomic molecules are molecules composed of only two atoms, of the same or different chemical elements (The prefix di- is of Greek origin, meaning "two"). If a diatomic molecule consists of two atoms of the same element, such as hydrogen (H_2) or oxygen (O_2), then it is said to be homo nuclear. All processes are related to the Tunneling Ionization (TI) of the molecule to a singly charged ion. Tunnel ionization is a process in which electrons in an atom (or a molecule) pass through the potential barrier and escape from the atom (or molecule). In an intense electric field, the potential barrier of an atom (molecule) is distorted drastically. Therefore, the length of the barrier that electrons have to pass decreases and electrons can escape from the atom (molecule) easily. We can consider that the ionization rate of the diatomic molecules aligned in the direction of the field. It should be enhanced compared to an atom with a similar ionization potential. There is an impact on fragmentation of molecular ions which involve the ionization of aligned species. The total rate of tunneling ionization (TI) is affected by the fact that at the first stage of ionization the neutral molecules are randomly oriented with respect to the polarization of the laser. The rate of multiphoton ionization of an N_2 molecule is suppressed compared to that of an A_r atom with the same ionization potential. This suppression is more pronounced in the case of ionization of D_2 compared to A_r. The random orientation of the molecules gives the result of suppression. There is an analytic expression for the single-electron wave function of the D_2 molecule and we predict the rate of tunneling ionization (TI) of this molecule. There is a predictive model for the ionization rate of a diatomic molecule in the field of a short pulse with wavelength in the near-IR region. There are two equivalent models which give a good overlap with experimental results on the ionization of rate-gas atoms in the tunneling regime, interacting with femtosecond laser pulses. There is a difficult in modeling the tunneling ionization (TI) rate of molecules which is related to the orientation of the neutral molecule in the laser field.

There is a second order differential equation, gives the rotation of a molecule with reduced mass (μ), polarization (α) and inter nuclear distance (R) in the field of a laser with frequency (ω) and field strength (F).

$$\frac{d^2\theta}{d\tau^2} + \left(\frac{\omega_0}{\omega}\right)^2 \cdot \cos^2(\tau) \cdot \sin(2 \cdot \theta) = 0; \tau = \omega \cdot t; \omega_0^2 = \frac{\alpha \cdot F^2}{\mu \cdot R^2}; d\tau^2 = d(\omega \cdot t)^2 = \omega^2 \cdot dt^2$$

θ Angle between the polarization of the laser and inter nuclear axis.
ω_0 Establish the rotation dynamic of the molecule.
$\frac{\omega_0}{\omega}$ Quantity determines the rotation dynamics of the molecules.
θ_0 Molecule initially angle with the direction of laser polarization.
μ reduced mass of the molecule.
α Polarization factor (polarizability).
R Inter nuclear distance in the field of laser.
ω Laser frequency (we consider it as a constant in our system).
ω_0 Calculated parameter $\left(\omega_0^2 = \frac{\alpha \cdot F^2}{\mu \cdot R^2}; \omega_0 = \frac{F}{R} \cdot \sqrt{\frac{\alpha}{\mu}}\right)$.

($\omega_0 = 7.9 \times 10^{-4}$au (Atomic units (au or a.u.) form a system of natural units which is especially convenient for atomic physics calculations) for D_2 molecule at a peak intensity of 10^{14} W/cm^2).

$\frac{2 \cdot \pi \cdot \omega}{\omega_0}$ The division between the period of oscillation for the molecule around the polarization axis and the laser period $\left(T_{\text{molecule}} = \frac{2 \cdot \pi \cdot \omega}{\omega_0} \cdot T_{\text{Laser}}\right)$.

T_{molecule} Period of oscillation for the molecule around the polarization axis.
T_{Laser} Laser periodLaser period.

We can write our system second order differential equation:

$$\frac{1}{\omega^2} \cdot \frac{d^2\theta}{dt^2} + \left(\frac{\omega_0}{\omega}\right)^2 \cdot \cos^2(\omega \cdot t) \cdot \sin(2 \cdot \theta) = 0; y = \frac{d\theta}{dt}$$

$$\frac{1}{\omega^2} \cdot \frac{dy}{dt} + \left(\frac{\omega_0}{\omega}\right)^2 \cdot \cos^2(\omega \cdot t) \cdot \sin(2 \cdot \theta) = 0; \frac{d\theta}{dt} = y$$

We can write our system differential equations:

$$\frac{dy}{dt} = -\omega^2 \cdot \left(\frac{\omega_0}{\omega}\right)^2 \cdot \cos^2(\omega \cdot t) \cdot \sin(2 \cdot \theta); \frac{d\theta}{dt} = y$$

$$\frac{dy}{dt} = -\omega_0^2 \cdot \cos^2(\omega \cdot t) \cdot \sin(2 \cdot \theta); \frac{d\theta}{dt} = y$$

The rotation dynamics of the molecule is determined by the quantity $\frac{\omega_0}{\omega}$. The molecule initially making an angel θ_0 with the direction of laser polarization will oscillate around the polarization axis with a period of $\frac{2\cdot\pi\cdot\omega}{\omega_0}$ times the laser period. The molecule in the Ti: Sapphire laser field should be considered as randomly oriented. TI is defined as the tunneling ionization of the molecules to a charged ion. The interaction of strong laser fields with diatomic molecules involves the following processes:

1. Non sequential ionization of molecules.
2. High-order harmonic generation.
3. Fragmentation of diatomic molecules.

Predictive mode for ionization of diatomic molecules, short laser pulse with wavelength in the infra-red (IR) region is constructed from

1. No experimental model which fit with data on rare-gas atom.
2. Orientation of the neutral molecule in the laser field.

Typical experiment is to use Ti: Sapphire femtosecond laser at 800 nm with pulse length of 200 fsec which hit A_r and D_2 and cause to ionization. These two particles have similar ionization potential (D_2 15.46 eV and A_r 15.76 eV). We investigate the dynamic and stability of our system for two cases: (1) Regular differential equations (2) Delay Differential Equations (DDEs)—parasitic effects which are related to laser diatomic molecules interaction.

Case (1): Our femtosecond laser pulses from a Ti: Sapphire interact with diatomic molecule—differential equations for the rotation of a molecule (with reduced mass (μ), polarization (α) and inter nuclear distance (R) in the field of a laser with frequency (ω) and field strength (F)).

$$\frac{dy}{dt} = -\omega_0^2 \cdot \cos^2(\omega \cdot t) \cdot \sin(2 \cdot \theta); \ \frac{d\theta}{dt} = y; \ \omega_0^2 = \frac{\alpha \cdot F^2}{\mu \cdot R^2}$$

$$\frac{dy}{dt} = -\frac{\alpha \cdot F^2}{\mu \cdot R^2} \cdot \cos^2(\omega \cdot t) \cdot \sin(2 \cdot \theta); \ \frac{d\theta}{dt} = y$$

At fixed points $\frac{d\theta}{dt} = 0; \frac{dy}{dt} = 0$ then $y^* = 0; -\frac{\alpha \cdot F^2}{\mu \cdot R^2} \cdot \cos^2(\omega \cdot t) \cdot \sin(2 \cdot \theta^*) = 0$.

$$\frac{\alpha \cdot F^2}{\mu \cdot R^2} \neq 0; \cos^2(\omega \cdot t) \neq 0 \Rightarrow \sin(2 \cdot \theta^*) = 0; 2 \cdot \theta^* = k \cdot \pi \Rightarrow \theta^* = k \cdot \frac{\pi}{2} \ \forall\, k = \ldots, -2, -1, 0, +1, +2, \ldots$$

System fixed points: $E^*(\theta^*, y^*) = \left(k \cdot \frac{\pi}{2}, 0\right) \ \forall\, k = \ldots, -2, -1, 0, +1, +2, \ldots$

Stability analysis: The standard local stability analysis about any one of the equilibrium points (fixed points) of diatomic molecule—differential equations for the rotation of a molecule system consists in adding to coordinates $[\theta\ y]$ arbitrary small increments of exponential form $[\theta\ y] \cdot e^{\lambda \cdot t}$, and retaining the first order terms in $\theta\ y$. The system of two homogeneous equations leads to a polynomial characteristics equation in the eigenvalue λ. The polynomial characteristics equations accept by set of below Angle between the polarization of the laser and inter nuclear axis and Angle between the polarization of the laser and inter nuclear axis derivative respect to the time into diatomic molecule—system differential equations for the rotation of a molecule. Diatomic molecule – differential equations for the rotation of a molecule system fixed values with arbitrarily small increments of exponential form $[\theta\ y] \cdot e^{\lambda \cdot t}$ are: i = 0 (first fixed point), i = 1 (second fixed point), i = 2 (third fixed point), etc.

$$\theta(t) = \theta^{(i)} + \theta \cdot e^{\lambda \cdot t};\ y(t) = y^{(i)} + y \cdot e^{\lambda \cdot t}$$

We choose the above expressions for our $\theta(t)$, $y(t)$ as small displacement $[\theta\ y]$ from the system fixed points at time t = 0 ($\theta(t = 0) = \theta^{(i)} + \theta$; $y(t = 0) = y^{(i)} + y$). For t > 0; $\lambda < 0$ the selected fixed point is stable otherwise t > 0; $\lambda > 0$ is unstable. Our system tends to the selected fixed point exponentially for t > 0; $\lambda < 0$ otherwise go away from the selected fixed point exponentially. Eigenvalue λ is the parameter which establish if the fixed point is stable or unstable, additionally his absolute value $|\lambda|$ establish the speed of flow toward or away from the selected fixed point. The next table describes diatomic molecule—differential equations for the rotation of a molecule system variables for different eigenvalue λ and t values (Table 2.1). The speeds of flow toward or away from the selected fixed point diatomic molecule—differential equations for the rotation of a molecule system angle (θ) between the polarization of the laser and inter nuclear axis and angle (θ) derivative respect to time are as follow:

$$\frac{d\theta(t)}{dt} = \lambda \cdot \theta \cdot e^{\lambda \cdot t};\ \frac{dy(t)}{dt} = \lambda \cdot y \cdot e^{\lambda \cdot t}$$

$$\frac{dy}{dt} = -\frac{\alpha \cdot F^2}{\mu \cdot R^2} \cdot \cos^2(\omega \cdot t) \cdot \sin(2 \cdot \theta) \Rightarrow \lambda \cdot y \cdot e^{\lambda \cdot t}$$

Table 2.1 Molecular system's angle (θ) and angle derivative for positive ($\lambda > 0$) and negative ($\lambda < 0$) eigenvalues

	$\lambda < 0$	$\lambda > 0$
t = 0	$\theta(t = 0) = \theta^{(i)} + \theta$ $y(t = 0) = y^{(i)} + y$	$\theta(t = 0) = \theta^{(i)} + \theta$ $y(t = 0) = y^{(i)} + y$
t > 0	$\theta(t) = \theta^{(i)} + \theta \cdot e^{-\lvert\lambda\rvert \cdot t}$ $y(t) = y^{(i)} + y \cdot e^{-\lvert\lambda\rvert \cdot t}$	$\theta(t) = \theta^{(i)} + \theta \cdot e^{\lvert\lambda\rvert \cdot t}$ $y(t) = y^{(i)} + y \cdot e^{\lvert\lambda\rvert \cdot t}$
$t \to \infty$	$\theta(t \to \infty) = \theta^{(i)}$ $y(t \to \infty) = y^{(i)}$	$\theta(t \to \infty) \approx \theta \cdot e^{\lambda \cdot t}$ $y(t \to \infty) \approx y \cdot e^{\lambda \cdot t}$

$$= -\frac{\alpha \cdot F^2}{\mu \cdot R^2} \cdot \cos^2(\omega \cdot t) \cdot \sin(2 \cdot [\theta^{(i)} + \theta \cdot e^{\lambda \cdot t}])$$

$$\sin(2 \cdot [\theta^{(i)} + \theta \cdot e^{\lambda \cdot t}]) = 2 \cdot \sin(\theta^{(i)} + \theta \cdot e^{\lambda \cdot t}) \cdot \cos(\theta^{(i)} + \theta \cdot e^{\lambda \cdot t})$$

$$\sin(\theta^{(i)} + \theta \cdot e^{\lambda \cdot t}) = \sin(\theta^{(i)}) \cdot \cos(\theta \cdot e^{\lambda \cdot t}) + \cos(\theta^{(i)}) \cdot \sin(\theta \cdot e^{\lambda \cdot t})$$

$$\cos(\theta^{(i)} + \theta \cdot e^{\lambda \cdot t}) = \cos(\theta^{(i)}) \cdot \cos(\theta \cdot e^{\lambda \cdot t}) - \sin(\theta^{(i)}) \cdot \sin(\theta \cdot e^{\lambda \cdot t})$$

$$\sin(2 \cdot [\theta^{(i)} + \theta \cdot e^{\lambda \cdot t}]) = 2 \cdot \sin(\theta^{(i)} + \theta \cdot e^{\lambda \cdot t}) \cdot \cos(\theta^{(i)} + \theta \cdot e^{\lambda \cdot t})$$

$$= 2 \cdot \{\sin(\theta^{(i)}) \cdot \cos(\theta \cdot e^{\lambda \cdot t}) + \cos(\theta^{(i)}) \cdot \sin(\theta \cdot e^{\lambda \cdot t})\} \cdot \{\cos(\theta^{(i)}) \cdot \cos(\theta \cdot e^{\lambda \cdot t})$$
$$- \sin(\theta^{(i)}) \cdot \sin(\theta \cdot e^{\lambda \cdot t})\}$$

$$= 2 \cdot \sin(\theta^{(i)}) \cdot \cos(\theta \cdot e^{\lambda \cdot t}) \cdot \cos(\theta^{(i)}) \cdot \cos(\theta \cdot e^{\lambda \cdot t})$$
$$- 2 \cdot \sin(\theta^{(i)}) \cdot \cos(\theta \cdot e^{\lambda \cdot t}) \cdot \sin(\theta^{(i)}) \cdot \sin(\theta \cdot e^{\lambda \cdot t})$$
$$+ 2 \cdot \cos(\theta^{(i)}) \cdot \sin(\theta \cdot e^{\lambda \cdot t}) \cdot \cos(\theta^{(i)}) \cdot \cos(\theta \cdot e^{\lambda \cdot t})$$
$$- 2 \cdot \cos(\theta^{(i)}) \cdot \sin(\theta \cdot e^{\lambda \cdot t}) \cdot \sin(\theta^{(i)}) \cdot \sin(\theta \cdot e^{\lambda \cdot t})$$

$$\sin(2 \cdot [\theta^{(i)} + \theta \cdot e^{\lambda \cdot t}]) = \sin(2 \cdot \theta^{(i)}) \cdot \cos^2(\theta \cdot e^{\lambda \cdot t}) - \sin(2 \cdot \theta \cdot e^{\lambda \cdot t}) \cdot \sin^2(\theta^{(i)})$$
$$+ \sin(2 \cdot \theta \cdot e^{\lambda \cdot t}) \cdot \cos^2(\theta^{(i)}) - \sin(2 \cdot \theta^{(i)}) \cdot \sin^2(\theta \cdot e^{\lambda \cdot t})$$

$$\sin(2 \cdot [\theta^{(i)} + \theta \cdot e^{\lambda \cdot t}]) = \sin(2 \cdot \theta^{(i)}) \cdot [\cos^2(\theta \cdot e^{\lambda \cdot t}) - \sin^2(\theta \cdot e^{\lambda \cdot t})]$$
$$+ \sin(2 \cdot \theta \cdot e^{\lambda \cdot t}) \cdot [\cos^2(\theta^{(i)}) - \sin^2(\theta^{(i)})]$$

$$\cos^2(\theta \cdot e^{\lambda \cdot t}) - \sin^2(\theta \cdot e^{\lambda \cdot t}) = \cos(2 \cdot \theta \cdot e^{\lambda \cdot t}); \cos^2(\theta^{(i)}) - \sin^2(\theta^{(i)}) = \cos(2 \cdot \theta^{(i)})$$

$$\sin(2 \cdot [\theta^{(i)} + \theta \cdot e^{\lambda \cdot t}]) = \sin(2 \cdot \theta^{(i)}) \cdot \cos(2 \cdot \theta \cdot e^{\lambda \cdot t}) + \sin(2 \cdot \theta \cdot e^{\lambda \cdot t}) \cdot \cos(2 \cdot \theta^{(i)})$$

$$\theta^* = \theta^{(i)} = k \cdot \frac{\pi}{2} \, \forall k = \ldots, -2, -1, 0, +1, +2, \ldots; 2 \cdot \theta^{(i)}$$
$$= k \cdot \pi \, \forall k = \ldots, -2, -1, 0, +1, +2, \ldots$$

$$\sin(2 \cdot \theta^{(i)}) = \sin(k \cdot \pi) = 0; \cos(2 \cdot \theta^{(i)}) = \cos(k \cdot \pi) = (-1)^k$$

$$\sin(2 \cdot [\theta^{(i)} + \theta \cdot e^{\lambda \cdot t}]) = \sin(2 \cdot \theta \cdot e^{\lambda \cdot t}) \cdot (-1)^k \, \forall k = \ldots, -2, -1, 0, +1, +2, \ldots$$

$$\xi(\theta, \lambda) = 2 \cdot \theta \cdot e^{\lambda \cdot t}; \sin(2 \cdot \theta \cdot e^{\lambda \cdot t}) = \sin(\xi(\theta, \lambda))$$
$$= \sum_{n=0}^{\infty} \frac{(-1)^n}{(2 \cdot n + 1)!} \cdot [\xi(\theta, \lambda)]^{2 \cdot n + 1}$$

$$\sin(\xi(\theta, \lambda)) = \sum_{n=0}^{\infty} \frac{(-1)^n}{(2 \cdot n + 1)!} \cdot [\xi(\theta, \lambda)]^{2 \cdot n + 1} = \xi(\theta, \lambda)$$
$$- \frac{[\xi(\theta, \lambda)]^3}{3!} + \frac{[\xi(\theta, \lambda)]^5}{5!} - \frac{[\xi(\theta, \lambda)]^7}{7!} + \cdots$$

$$\sin(\xi(\theta, \lambda)) = \sum_{n=0}^{\infty} \frac{(-1)^n}{(2 \cdot n + 1)!} \cdot [\xi(\theta, \lambda)]^{2 \cdot n + 1} = 2 \cdot \theta \cdot e^{\lambda \cdot t}$$
$$- \frac{[2 \cdot \theta \cdot e^{\lambda \cdot t}]^3}{3!} + \frac{[2 \cdot \theta \cdot e^{\lambda \cdot t}]^5}{5!} - \frac{[2 \cdot \theta \cdot e^{\lambda \cdot t}]^7}{7!} + \cdots$$

$$\sin(\xi(\theta,\lambda)) = \sum_{n=0}^{\infty} \frac{(-1)^n}{(2\cdot n+1)!}\cdot[\xi(\theta,\lambda)]^{2\cdot n+1} = 2\cdot\theta\cdot e^{\lambda\cdot t}$$
$$-\frac{2^3\cdot\theta^3\cdot e^{3\cdot\lambda\cdot t}}{3!} + \frac{2^5\cdot\theta^5\cdot e^{5\cdot\lambda\cdot t}}{5!} - \frac{2^7\cdot\theta^7\cdot e^{7\cdot\lambda\cdot t}}{7!} + \cdots$$

We consider $\theta^m \to \varepsilon\ \forall\ m \geq 2$ then $\frac{2^3\cdot\theta^3\cdot e^{3\cdot\lambda\cdot t}}{3!} + \frac{2^5\cdot\theta^5\cdot e^{5\cdot\lambda\cdot t}}{5!} - \frac{2^7\cdot\theta^7\cdot e^{7\cdot\lambda\cdot t}}{7!} + \cdots \to \varepsilon$

$$\sin(\xi(\theta,\lambda)) = \sum_{n=0}^{\infty} \frac{(-1)^n}{(2\cdot n+1)!}\cdot[\xi(\theta,\lambda)]^{2\cdot n+1} \approx 2\cdot\theta\cdot e^{\lambda\cdot t}$$
$$\sin(2\cdot[\theta^{(i)}+\theta\cdot e^{\lambda\cdot t}]) = 2\cdot\theta\cdot e^{\lambda\cdot t}\cdot(-1)^k\ \forall k = \cdots, -2, -1, 0, +1, +2, \ldots$$
$$\lambda\cdot y\cdot e^{\lambda\cdot t} = -\frac{\alpha\cdot F^2}{\mu\cdot R^2}\cdot\cos^2(\omega\cdot t)\cdot\sin(2\cdot[\theta^{(i)}+\theta\cdot e^{\lambda\cdot t}])$$
$$= -\frac{\alpha\cdot F^2}{\mu\cdot R^2}\cdot\cos^2(\omega\cdot t)\cdot 2\cdot\theta\cdot e^{\lambda\cdot t}\cdot(-1)^k$$
$$-\lambda\cdot y - \frac{\alpha\cdot F^2}{\mu\cdot R^2}\cdot\cos^2(\omega\cdot t)\cdot 2\cdot\theta\cdot(-1)^k = 0$$
$$\frac{d\theta}{dt} = y \Rightarrow \lambda\cdot\theta\cdot e^{\lambda\cdot t} = y^{(i)} + y\cdot e^{\lambda\cdot t};\ y^* = y^{(i)} = 0;\ -\lambda\cdot\theta + y = 0$$

We can summary our diatomic molecule—differential equations for the rotation of a molecule system's arbitrarily small increments equations:

$$-\lambda\cdot y - \frac{\alpha\cdot F^2}{\mu\cdot R^2}\cdot\cos^2(\omega\cdot t)\cdot 2\cdot\theta\cdot(-1)^k = 0;\ -\lambda\cdot\theta + y = 0\,\forall k$$
$$= \ldots, -2, -1, 0, +1, +2, \ldots$$
$$\begin{pmatrix} -\lambda & -\frac{\alpha\cdot F^2}{\mu\cdot R^2}\cdot\cos^2(\omega\cdot t)\cdot 2\cdot(-1)^k \\ 1 & -\lambda \end{pmatrix}\cdot\begin{pmatrix} y \\ \theta \end{pmatrix} = 0$$
$$A - \lambda\cdot I = \begin{pmatrix} -\lambda & -\frac{\alpha\cdot F^2}{\mu\cdot R^2}\cdot\cos^2(\omega\cdot t)\cdot 2\cdot(-1)^k \\ 1 & -\lambda \end{pmatrix};\ \det(A-\lambda\cdot I) = 0$$
$$\lambda^2 + \frac{\alpha\cdot F^2}{\mu\cdot R^2}\cdot\cos^2(\omega\cdot t)\cdot 2\cdot(-1)^k = 0\,\forall k = \ldots, -2, -1, 0, +1, +2, \ldots$$
$$\lambda^2 = -\frac{\alpha\cdot F^2}{\mu\cdot R^2}\cdot\cos^2(\omega\cdot t)\cdot 2\cdot(-1)^k \Rightarrow \lambda_{1,2} = \pm\sqrt{-2\cdot\frac{\alpha\cdot F^2}{\mu\cdot R^2}\cdot\cos^2(\omega\cdot t)\cdot(-1)^k}$$

$\mu > 0$ (Reduced mass of the molecule), $\alpha > 0$ (Polarization factor (polarizability)).

$$2\cdot\frac{\alpha\cdot F^2}{\mu\cdot R^2}\cdot\cos^2(\omega\cdot t) > 0;\ \lambda_{1,2} = \pm\frac{F}{R}\cdot\sqrt{2\cdot\frac{\alpha}{\mu}}\cdot\sqrt{-\cos^2(\omega\cdot t)\cdot(-1)^k}$$
$$\frac{F}{R}\cdot\sqrt{2\cdot\frac{\alpha}{\mu}} > 0\,\lambda_{1,2} = \pm\frac{F}{R}\cdot\sqrt{2\cdot\frac{\alpha}{\mu}}\cdot\sqrt{\cos^2(\omega\cdot t)}\cdot\sqrt{-(-1)^k}$$

We define $u = \cos(\omega \cdot t)$; $u^2 = \cos^2(\omega \cdot t)$. $\sqrt{u^2} = |u|$, if we cannot be sure that u is positive.

$$\sqrt{[\cos(\omega \cdot t)]^2} = |\cos(\omega \cdot t)| = \begin{cases} \cos(\omega \cdot t)\ \forall\ 0 \le \omega \cdot t \le \frac{\pi}{2} \\ -\cos(\omega \cdot t)\ \forall\ \frac{\pi}{2} \le \omega \cdot t \le \pi \end{cases}$$

$$\lambda_{1,2} = \pm\frac{F}{R} \cdot \sqrt{2 \cdot \frac{\alpha}{\mu}} \cdot |\cos(\omega \cdot t)| \cdot \sqrt{-(-1)^k};\ \lambda_{1,2} = \pm\frac{F}{R} \cdot \sqrt{2 \cdot \frac{\alpha}{\mu}} \cdot |\cos(\omega \cdot t)| \cdot j \cdot \sqrt{(-1)^k}$$

We consider in our analysis $k \ge 0$ (k is natural number $k \in \mathbb{N}$).

$$\sqrt{(-1)^k} = \begin{cases} 1\ \forall\ k = 0, 2, 4, \ldots \\ j\ \forall\ k = 1, 3, 5, \ldots \end{cases}$$

$$\lambda_{1,2} = \begin{cases} \pm\frac{F}{R} \cdot \sqrt{2 \cdot \frac{\alpha}{\mu}} \cdot |\cos(\omega \cdot t)| \cdot j\ \forall\ k = 0, 2, 4, \ldots \\ \pm\frac{F}{R} \cdot \sqrt{2 \cdot \frac{\alpha}{\mu}} \cdot |\cos(\omega \cdot t)| \cdot j^2\ \forall\ k = 1, 3, 5, \ldots \end{cases}$$

$$\lambda_{1,2} = \begin{cases} \pm j \cdot \frac{F}{R} \cdot \sqrt{2 \cdot \frac{\alpha}{\mu}} \cdot |\cos(\omega \cdot t)|\ \forall\ k = 0, 2, 4, \ldots \\ \mp\frac{F}{R} \cdot \sqrt{2 \cdot \frac{\alpha}{\mu}} \cdot |\cos(\omega \cdot t)|\ \forall\ k = 1, 3, 5, \ldots \end{cases}$$

If the eigenvalues are pure imaginary $\lambda_{1,2} = \pm j \cdot \frac{F}{R} \cdot \sqrt{2 \cdot \frac{\alpha}{\mu}} \cdot |\cos(\omega \cdot t)|\ \forall\ k = 0, 2, 4, \ldots$ then all the solutions are periodic with period $T = \frac{2 \cdot \pi}{\frac{F}{R} \cdot \sqrt{2 \cdot \frac{\alpha}{\mu}} \cdot |\cos(\omega \cdot t)|}$. The oscillations have fixed amplitudes and fixed point is a center. If $\lambda_1 \cdot \lambda_2 < 0$, in our case: $\lambda_{1,2} = \mp\frac{F}{R} \cdot \sqrt{2 \cdot \frac{\alpha}{\mu}} \cdot |\cos(\omega \cdot t)|\ \forall\ k = 1, 3, 5, \ldots$ then eigenvalues are real and have opposite signs; hence the fixed point is a saddle point. The next figure illustrates the system fixed points (equilibrium points) phase portrait and classification (Fig. 2.2) [5, 6].

Case (2): Our femtosecond laser pulses from a Ti: Sapphire interact with diatomic molecule—delay differential equations for the rotation of a molecule (with reduced mass (μ), polarization (α) and inter nuclear distance (R) in the field of a laser with frequency (ω) and field strength (F)). Due to parasitic effects there is a delay on angle between the polarization of the laser and inter nuclear axis ($\theta(t) \to \theta(t - \tau_2)$) and y parameter ($y(t) \to y(t - \tau_1)$), there is no effect of delay on θ, y variables derivative in time.

$$\frac{dy(t)}{dt} = -\omega_0^2 \cdot \cos^2(\omega \cdot t) \cdot \sin(2 \cdot \theta(t - \tau_2));\ \frac{d\theta(t)}{dt} = y(t - \tau_1);\ \omega_0^2 = \frac{\alpha \cdot F^2}{\mu \cdot R^2}$$

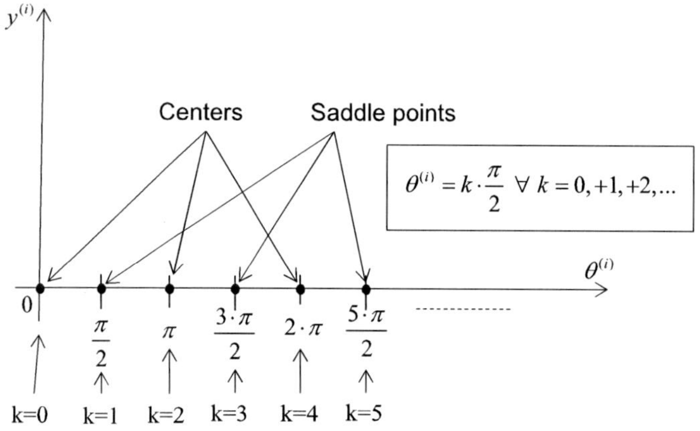

Fig. 2.2 System fixed points (equilibrium points) phase portrait and classification

$$\frac{dy(t)}{dt} = -\frac{\alpha \cdot F^2}{\mu \cdot R^2} \cdot \cos^2(\omega \cdot t) \cdot \sin(2 \cdot \theta(t - \tau_2)); \; \frac{d\theta(t)}{dt} = y(t - \tau_1)$$

At fixed points $\frac{d\theta}{dt} = 0$; $\frac{dy}{dt} = 0$; $\lim_{t \to \infty} y(t - \tau_1) = y(t)$; $\lim_{t \to \infty} \theta(t - \tau_2) = \theta(t)$; $t \gg \tau_1, \tau_2$ then $y^* = 0$; $-\frac{\alpha \cdot F^2}{\mu \cdot R^2} \cdot \cos^2(\omega \cdot t) \cdot \sin(2 \cdot \theta^*) = 0$

$$\frac{\alpha \cdot F^2}{\mu \cdot R^2} \neq 0; \cos^2(\omega \cdot t) \neq 0 \Rightarrow \sin(2 \cdot \theta^*) = 0; 2 \cdot \theta^* = k \cdot \pi \Rightarrow \theta^* = k \cdot \frac{\pi}{2} \forall k = \ldots, -2, -1, 0, +1, +2, \ldots$$

System fixed points: $E^*(\theta^*, y^*) = \left(k \cdot \frac{\pi}{2}, 0\right) \; \forall \; k = \ldots, -2, -1, 0, +1, +2, \ldots$

Diatomic molecule—differential equations for the rotation of a molecule system fixed values with arbitrarily small increments of exponential form $[\theta \;\; y] \cdot e^{\lambda \cdot t}$ are: i = 0 (first fixed point), i = 1 (second fixed point), i = 2 (third fixed point), etc.

$$\theta(t - \tau_2) = \theta^{(i)} + \theta \cdot e^{\lambda \cdot (t - \tau_2)}; \; y(t - \tau_1) = y^{(i)} + y \cdot e^{\lambda \cdot (t - \tau_1)}; \; \frac{d\theta(t)}{dt} = \lambda \cdot \theta \cdot e^{\lambda \cdot t}; \; \frac{dy(t)}{dt} = \lambda \cdot y \cdot e^{\lambda \cdot t}$$

We get two delayed differential equations respect to coordinates $[\theta \;\; y]$ arbitrarily small increments of exponential $[\theta \;\; y] \cdot e^{\lambda \cdot t}$.

$$\lambda \cdot y \cdot e^{\lambda \cdot t} = -\frac{\alpha \cdot F^2}{\mu \cdot R^2} \cdot \cos^2(\omega \cdot t) \cdot \sin(2 \cdot [\theta^{(i)} + \theta \cdot e^{\lambda \cdot (t - \tau_2)}]); \; \lambda \cdot \theta \cdot e^{\lambda \cdot t} = y^{(i)} + y \cdot e^{\lambda \cdot (t - \tau_1)}$$

$$\lambda \cdot y \cdot e^{\lambda \cdot t} = -\frac{\alpha \cdot F^2}{\mu \cdot R^2} \cdot \cos^2(\omega \cdot t) \cdot \sin(2 \cdot [\theta^{(i)} + \theta \cdot e^{\lambda \cdot t} \cdot e^{-\lambda \cdot \tau_2}]); \lambda \cdot \theta \cdot e^{\lambda \cdot t}$$
$$= y^{(i)} + y \cdot e^{\lambda \cdot t} \cdot e^{-\lambda \cdot \tau_1}$$

We have three possible cases: (1) $\tau_1 = \tau_A; \tau_2 = 0$ (2) $\tau_1 = 0; \tau_2 = \tau_A$ (3) $\tau_1 = \tau_A; \tau_2 = \tau_A$.

We analyze the stability of our system for the second case: $\tau_1 = 0; \tau_2 = \tau_A$.

$$\lambda \cdot y \cdot e^{\lambda \cdot t} = -\frac{\alpha \cdot F^2}{\mu \cdot R^2} \cdot \cos^2(\omega \cdot t) \cdot \sin(2 \cdot [\theta^{(i)} + \theta \cdot e^{\lambda \cdot t} \cdot e^{-\lambda \cdot \tau_A}]); \lambda \cdot \theta \cdot e^{\lambda \cdot t}$$
$$= y^{(i)} + y \cdot e^{\lambda \cdot t}$$
$$\sin(2 \cdot [\theta^{(i)} + \theta \cdot e^{\lambda \cdot t}]) = 2 \cdot \sin(\theta^{(i)} + \theta \cdot e^{\lambda \cdot t} \cdot e^{-\lambda \cdot \tau_A}) \cdot \cos(\theta^{(i)} + \theta \cdot e^{\lambda \cdot t} \cdot e^{-\lambda \cdot \tau_A})$$
$$\sin(\theta^{(i)} + \theta \cdot e^{\lambda \cdot t}) = \sin(\theta^{(i)}) \cdot \cos(\theta \cdot e^{\lambda \cdot t} \cdot e^{-\lambda \cdot \tau_A}) + \cos(\theta^{(i)}) \cdot \sin(\theta \cdot e^{\lambda \cdot t} \cdot e^{-\lambda \cdot \tau_A})$$
$$\cos(\theta^{(i)} + \theta \cdot e^{\lambda \cdot t}) = \cos(\theta^{(i)}) \cdot \cos(\theta \cdot e^{\lambda \cdot t} \cdot e^{-\lambda \cdot \tau_A}) - \sin(\theta^{(i)}) \cdot \sin(\theta \cdot e^{\lambda \cdot t} \cdot e^{-\lambda \cdot \tau_A})$$

$$\sin(2 \cdot [\theta^{(i)} + \theta \cdot e^{\lambda \cdot t}]) = 2 \cdot \sin(\theta^{(i)} + \theta \cdot e^{\lambda \cdot t} \cdot e^{-\lambda \cdot \tau_A}) \cdot \cos(\theta^{(i)} + \theta \cdot e^{\lambda \cdot t} \cdot e^{-\lambda \cdot \tau_A})$$
$$= 2 \cdot \{\sin(\theta^{(i)}) \cdot \cos(\theta \cdot e^{\lambda \cdot t} \cdot e^{-\lambda \cdot \tau_A}) + \cos(\theta^{(i)}) \cdot$$
$$\sin(\theta \cdot e^{\lambda \cdot t} \cdot e^{-\lambda \cdot \tau_A})\} \cdot \{\cos(\theta^{(i)}) \cdot \cos(\theta \cdot e^{\lambda \cdot t} \cdot e^{-\lambda \cdot \tau_A})$$
$$- \sin(\theta^{(i)}) \cdot \sin(\theta \cdot e^{\lambda \cdot t} \cdot e^{-\lambda \cdot \tau_A})\} = 2 \cdot \sin(\theta^{(i)}) \cdot$$
$$\cos(\theta \cdot e^{\lambda \cdot t} \cdot e^{-\lambda \cdot \tau_A}) \cdot \cos(\theta^{(i)}) \cdot \cos(\theta \cdot e^{\lambda \cdot t} \cdot e^{-\lambda \cdot \tau_A})$$
$$- 2 \cdot \sin(\theta^{(i)}) \cdot \cos(\theta \cdot e^{\lambda \cdot t} \cdot e^{-\lambda \cdot \tau_A}) \cdot \sin(\theta^{(i)}) \cdot \sin(\theta \cdot e^{\lambda \cdot t} \cdot e^{-\lambda \cdot \tau_A})$$
$$+ 2 \cdot \cos(\theta^{(i)}) \cdot \sin(\theta \cdot e^{\lambda \cdot t} \cdot e^{-\lambda \cdot \tau_A}) \cdot \cos(\theta^{(i)}) \cdot \cos(\theta \cdot e^{\lambda \cdot t} \cdot e^{-\lambda \cdot \tau_A})$$
$$- 2 \cdot \cos(\theta^{(i)}) \cdot \sin(\theta \cdot e^{\lambda \cdot t} \cdot e^{-\lambda \cdot \tau_A}) \cdot \sin(\theta^{(i)}) \cdot \sin(\theta \cdot e^{\lambda \cdot t} \cdot e^{-\lambda \cdot \tau_A})$$

$$\sin(2 \cdot [\theta^{(i)} + \theta \cdot e^{\lambda \cdot t}]) = \sin(2 \cdot \theta^{(i)}) \cdot \cos^2(\theta \cdot e^{\lambda \cdot t} \cdot e^{-\lambda \cdot \tau_A})$$
$$- \sin(2 \cdot \theta \cdot e^{\lambda \cdot t} \cdot e^{-\lambda \cdot \tau_A}) \cdot \sin^2(\theta^{(i)})$$
$$+ \sin(2 \cdot \theta \cdot e^{\lambda \cdot t} \cdot e^{-\lambda \cdot \tau_A}) \cdot \cos^2(\theta^{(i)}) - \sin(2 \cdot \theta^{(i)}) \cdot \sin^2(\theta \cdot e^{\lambda \cdot t} \cdot e^{-\lambda \cdot \tau_A})$$
$$\sin(2 \cdot [\theta^{(i)} + \theta \cdot e^{\lambda \cdot t}]) = \sin(2 \cdot \theta^{(i)}) \cdot [\cos^2(\theta \cdot e^{\lambda \cdot t} \cdot e^{-\lambda \cdot \tau_A}) - \sin^2(\theta \cdot e^{\lambda \cdot t} \cdot e^{-\lambda \cdot \tau_A})]$$
$$+ \sin(2 \cdot \theta \cdot e^{\lambda \cdot t} \cdot e^{-\lambda \cdot \tau_A}) \cdot [\cos^2(\theta^{(i)}) - \sin^2(\theta^{(i)})]$$
$$\cos^2(\theta \cdot e^{\lambda \cdot t} \cdot e^{-\lambda \cdot \tau_A}) - \sin^2(\theta \cdot e^{\lambda \cdot t} \cdot e^{-\lambda \cdot \tau_A}) = \cos(2 \cdot \theta \cdot e^{\lambda \cdot t} \cdot e^{-\lambda \cdot \tau_A}); \cos^2(\theta^{(i)})$$
$$- \sin^2(\theta^{(i)}) = \cos(2 \cdot \theta^{(i)})$$
$$\sin(2 \cdot [\theta^{(i)} + \theta \cdot e^{\lambda \cdot t} \cdot e^{-\lambda \cdot \tau_A}]) = \sin(2 \cdot \theta^{(i)}) \cdot \cos(2 \cdot \theta \cdot e^{\lambda \cdot t} \cdot e^{-\lambda \cdot \tau_A})$$
$$+ \sin(2 \cdot \theta \cdot e^{\lambda \cdot t} \cdot e^{-\lambda \cdot \tau_A}) \cdot \cos(2 \cdot \theta^{(i)})$$
$$\theta^* = \theta^{(i)} = k \cdot \frac{\pi}{2} \; \forall k = \ldots, -2, -1, 0, +1, +2, \ldots; 2 \cdot \theta^{(i)}$$
$$= k \cdot \pi \; \forall k = \ldots, -2, -1, 0, +1, +2, \ldots$$
$$\sin(2 \cdot \theta^{(i)}) = \sin(k \cdot \pi) = 0; \cos(2 \cdot \theta^{(i)}) = \cos(k \cdot \pi) = (-1)^k$$
$$\sin(2 \cdot [\theta^{(i)} + \theta \cdot e^{\lambda \cdot t} \cdot e^{-\lambda \cdot \tau_A}]) = \sin(2 \cdot \theta \cdot e^{\lambda \cdot t} \cdot e^{-\lambda \cdot \tau_A}) \cdot (-1)^k \; \forall k$$
$$= \ldots, -2, -1, 0, +1, +2, \ldots$$

$$\xi(\theta,\lambda,\tau_A) = 2\cdot\theta\cdot e^{\lambda\cdot t}\cdot e^{-\lambda\cdot\tau_A};\ \sin(2\cdot\theta\cdot e^{\lambda\cdot t}e^{-\lambda\cdot\tau_A}) = \sin(\xi(\theta,\lambda,\tau_A))$$
$$= \sum_{n=0}^{\infty}\frac{(-1)^n}{(2\cdot n+1)!}\cdot[\xi(\theta,\lambda,\tau_A)]^{2\cdot n+1}$$

$$\sin(\xi(\theta,\lambda,\tau_A)) = \sum_{n=0}^{\infty}\frac{(-1)^n}{(2\cdot n+1)!}\cdot[\xi(\theta,\lambda,\tau_A)]^{2\cdot n+1}$$
$$= \xi(\theta,\lambda,\tau_A) - \frac{[\xi(\theta,\lambda,\tau_A)]^3}{3!}$$
$$+ \frac{[\xi(\theta,\lambda,\tau_A)]^5}{5!} - \frac{[\xi(\theta,\lambda,\tau_A)]^7}{7!} + \cdots$$

$$\sin(\xi(\theta,\lambda,\tau_A)) = \sum_{n=0}^{\infty}\frac{(-1)^n}{(2\cdot n+1)!}\cdot[\xi(\theta,\lambda,\tau_A)]^{2\cdot n+1}$$
$$= 2\cdot\theta\cdot e^{\lambda\cdot t}\cdot e^{-\lambda\cdot\tau_A} - \frac{[2\cdot\theta\cdot e^{\lambda\cdot t}\cdot e^{-\lambda\cdot\tau_A}]^3}{3!}$$
$$+ \frac{[2\cdot\theta\cdot e^{\lambda\cdot t}\cdot e^{-\lambda\cdot\tau_A}]^5}{5!} - \frac{[2\cdot\theta\cdot e^{\lambda\cdot t}\cdot e^{-\lambda\cdot\tau_A}]^7}{7!} + \cdots$$

$$\sin(\xi(\theta,\lambda,\tau_A)) = \sum_{n=0}^{\infty}\frac{(-1)^n}{(2\cdot n+1)!}\cdot[\xi(\theta,\lambda,\tau_A)]^{2\cdot n+1}$$
$$= 2\cdot\theta\cdot e^{\lambda\cdot t}\cdot e^{-\lambda\cdot\tau_A} - \frac{2^3\cdot\theta^3\cdot e^{3\cdot\lambda\cdot t}\cdot e^{-3\cdot\lambda\cdot\tau_A}}{3!}$$
$$+ \frac{2^5\cdot\theta^5\cdot e^{5\cdot\lambda\cdot t}\cdot e^{-5\cdot\lambda\cdot\tau_A}}{5!} - \frac{2^7\cdot\theta^7\cdot e^{7\cdot\lambda\cdot t}\cdot e^{-7\cdot\lambda\cdot\tau_A}}{7!} + \cdots$$

We consider $\theta^m \to \varepsilon\ \forall\ m \geq 2$ then

$$\frac{2^3\cdot\theta^3\cdot e^{3\cdot\lambda\cdot t}\cdot e^{-3\cdot\lambda\cdot\tau_A}}{3!} + \frac{2^5\cdot\theta^5\cdot e^{5\cdot\lambda\cdot t}\cdot e^{-5\cdot\lambda\cdot\tau_A}}{5!} - \frac{2^7\cdot\theta^7\cdot e^{7\cdot\lambda\cdot t}\cdot e^{-7\cdot\lambda\cdot\tau_A}}{7!} + \cdots \to \varepsilon$$

$$\sin(\xi(\theta,\lambda,\tau_A)) = \sum_{n=0}^{\infty}\frac{(-1)^n}{(2\cdot n+1)!}\cdot[\xi(\theta,\lambda,\tau_A)]^{2\cdot n+1} \approx 2\cdot\theta\cdot e^{\lambda\cdot t}\cdot e^{-\lambda\cdot\tau_A}$$

$$\sin(2\cdot[\theta^{(i)} + \theta\cdot e^{\lambda\cdot t}\cdot e^{-\lambda\cdot\tau_A}]) = 2\cdot\theta\cdot e^{\lambda\cdot t}\cdot e^{-\lambda\cdot\tau_A}\cdot(-1)^k\ \forall k$$
$$= \ldots, -2, -1, 0, +1, +2, \ldots$$

$$\lambda\cdot y\cdot e^{\lambda\cdot t} = -\frac{\alpha\cdot F^2}{\mu\cdot R^2}\cdot\cos^2(\omega\cdot t)\cdot\sin(2\cdot[\theta^{(i)} + \theta\cdot e^{\lambda\cdot t}\cdot e^{-\lambda\cdot\tau_A}])$$
$$= -\frac{\alpha\cdot F^2}{\mu\cdot R^2}\cdot\cos^2(\omega\cdot t)\cdot 2\cdot\theta\cdot e^{\lambda\cdot t}\cdot e^{-\lambda\cdot\tau_A}\cdot(-1)^k$$
$$-\lambda\cdot y - \frac{\alpha\cdot F^2}{\mu\cdot R^2}\cdot\cos^2(\omega\cdot t)\cdot 2\cdot\theta\cdot e^{-\lambda\cdot\tau_A}\cdot(-1)^k = 0$$

$$\frac{d\theta}{dt} = y \Rightarrow \lambda\cdot\theta\cdot e^{\lambda\cdot t} = y^{(i)} + y\cdot e^{\lambda\cdot t};\ y^* = y^{(i)} = 0;\ -\lambda\cdot\theta + y = 0$$

$$\begin{pmatrix} -\lambda - \frac{\alpha \cdot F^2}{\mu \cdot R^2} \cdot \cos^2(\omega \cdot t) \cdot 2 \cdot e^{-\lambda \cdot \tau_A} \cdot (-1)^k \\ 1 & -\lambda \end{pmatrix} \cdot \begin{pmatrix} y \\ \theta \end{pmatrix} = 0$$

$$A - \lambda \cdot I = \begin{pmatrix} -\lambda - \frac{\alpha \cdot F^2}{\mu \cdot R^2} \cdot \cos^2(\omega \cdot t) \cdot 2 \cdot e^{-\lambda \cdot \tau_A} \cdot (-1)^k \\ 1 & -\lambda \end{pmatrix}; \det(A - \lambda \cdot I) = 0$$

$$\lambda^2 + \frac{\alpha \cdot F^2}{\mu \cdot R^2} \cdot \cos^2(\omega \cdot t) \cdot 2 \cdot e^{-\lambda \cdot \tau_A} \cdot (-1)^k = 0 \; \forall \; k$$
$$= \ldots, -2, -1, 0, +1, +2, \ldots$$

$$D(\lambda, \tau_A) = \lambda^2 + \frac{\alpha \cdot F^2}{\mu \cdot R^2} \cdot \cos^2(\omega \cdot t) \cdot 2 \cdot e^{-\lambda \cdot \tau_A} \cdot (-1)^k = 0 \; \forall \; k$$
$$= \ldots, -2, -1, 0, +1, +2, \ldots$$

We study the occurrence of any possible stability switching resulting from the increase the value of time delay τ_A for the general characteristic equation $D(\lambda, \tau_A)$. If we choose τ_A parameter then, $D(\lambda, \tau_A) = P_n(\lambda, \tau_A) + Q_m(\lambda, \tau_A) \cdot e^{-\lambda \cdot \tau_A}$.

The expression for $P_n(\lambda, \tau_A)$ is

$$P_n(\lambda, \tau_A) = \sum_{k=0}^{n} p_k(\tau_A) \cdot \lambda^k = p_0(\tau_A) + p_1(\tau_A) \cdot \lambda + p_2(\tau_A) \cdot \lambda^2 + \cdots$$

The expression for $Q_m(\lambda, \tau_A)$ is

$$Q_m(\lambda, \tau_A) = \sum_{k=0}^{m} q_k(\tau_A) \cdot \lambda^k = q_0(\tau_A) + q_1(\tau_A) \cdot \lambda + q_2(\tau_A) \cdot \lambda^2 + \cdots$$

The general characteristic equation $D(\lambda, \tau_A)$ is as follow:

$$D(\lambda, \tau_A) = \lambda^2 + \frac{\alpha \cdot F^2}{\mu \cdot R^2} \cdot \cos^2(\omega \cdot t) \cdot 2 \cdot (-1)^k \cdot e^{-\lambda \cdot \tau_A}$$
$$= 0 \, \forall \, k = \ldots, -2, -1, 0, +1, +2, \ldots$$
$$D(\lambda, \tau_A) = P_n(\lambda, \tau_A) + Q_m(\lambda, \tau_A) \cdot e^{-\lambda \cdot \tau_A}$$
$$n = 2; m = 0; n > m$$

The expression for $P_n(\lambda, \tau_A)$: $P_{n=2}(\lambda, \tau_A) = \sum_{k=0}^{n=2} p_k(\tau_A) \cdot \lambda^k = p_0(\tau_A) + p_1(\tau_A) \cdot \lambda + p_2(\tau_A) \cdot \lambda^2$.

$$p_0(\tau_A) = 0; \; p_1(\tau_A) = 0; \; p_2(\tau_A) = 1$$

The expression for $Q_m(\lambda, \tau_A)$: $Q_{m=0}(\lambda, \tau_A) = \sum_{k=0}^{m=0} q_k(\tau_A) \cdot \lambda^k = q_0(\tau_A)$

$$q_0(\tau_A) = \frac{\alpha \cdot F^2}{\mu \cdot R^2} \cdot \cos^2(\omega \cdot t) \cdot 2 \cdot (-1)^k \; \forall \, k = \ldots, -2, -1, 0, +1, +2, \ldots$$

The homogeneous system for θ y leads to a characteristic equation for the eigenvalue λ having the form $P(\lambda) + Q(\lambda) \cdot e^{-\lambda \cdot \tau_A}$; $P(\lambda) = \sum_{j=0}^{2} a_j \cdot \lambda^j$; $Q(\lambda) = \sum_{j=0}^{0} c_j \cdot \lambda^j$ and the coefficients $\{a_j(q_i, q_k), c_j(q_i, q_k)\} \in \mathbb{R}$ depend on q_i, q_k but not on τ_A, q_i, q_k are any molecule—differential equations for the rotation of a molecule system's parameters, other parameters keep as a constant.

$$a_0 = 0;\, a_1 = 0;\, a_2 = 1;\, c_0 = \frac{\alpha \cdot F^2}{\mu \cdot R^2} \cdot \cos^2(\omega \cdot t) \cdot 2 \cdot (-1)^k \, \forall k = \ldots, -2, -1, 0, +1, +2, \ldots$$

Unless strictly necessary, the designation of the variation arguments (q_i, q_k) will subsequently be omitted from P, Q, a_j,c_j. The coefficients a_j and c_j are continuous, and differential functions of their arguments, and direct substitution shows that

$$a_0 + c_0 = \frac{\alpha \cdot F^2}{\mu \cdot R^2} \cdot \cos^2(\omega \cdot t) \cdot 2 \cdot (-1)^k \; \forall \, k = \ldots, -2, -1, 0, +1, +2, \ldots$$

$$a_0 + c_0 \neq 0 \; \forall \, q_i, q_k \in \mathbb{R}_+;\, \omega \cdot t \neq \pi \cdot \left(\frac{1}{2} + l\right);\, l = 0, 1, 2, \ldots;\, l \in \mathbb{N}_0$$

$\lambda = 0$ is not a of $P(\lambda) + Q(\lambda) \cdot e^{-\lambda \cdot \tau_A} = 0$. Furthermore $P(\lambda)$, $Q(\lambda)$ are analytic functions of λ for which the following requirements of the analysis (BK] can also be verified in the present case:

1. If $\lambda = i \cdot \omega$; $\omega \in \mathbb{R}$, then $P(i \cdot \omega) + Q(i \cdot \omega) \neq 0$.
2. $\left|\frac{Q(\lambda)}{P(\lambda)}\right|$ is bounded for $|\lambda| \to \infty$, $\operatorname{Re}\lambda \geq 0$. No roots bifurcation from ∞.
3. $F(\omega) = |P(i \cdot \omega)|^2 - |Q(i \cdot \omega)|^2$ has a finite number of zeros. Indeed, this is a polynomial in ω.
4. Each positive root $\omega(q_i, q_k)$ of $F(\omega) = 0$ is continuous and differentiable respect to q_i, q_k.

Remark Please do not confuse ω which is related to our stability analysis and ω which is Laser frequency (we consider it as a constant in our system)

We assume that $P_n(\lambda, \tau_A) = P_n(\lambda)$ and $Q_m(\lambda, \tau_A) = Q_m(\lambda)$ cannot have common imaginary roots. That is for any real number ω: $P_n(\lambda = i \cdot \omega, \tau_A) + Q_m(\lambda = i \cdot \omega, \tau_A) \neq 0$.

$$P_n(\lambda = i \cdot \omega, \tau_A) = -\omega^2;\, Q_m(\lambda = i \cdot \omega, \tau_A) = \frac{\alpha \cdot F^2}{\mu \cdot R^2} \cdot \cos^2(\omega \cdot t) \cdot 2 \cdot (-1)^k \; \forall \, k = \ldots, -2, -1, 0, +1, +2, \ldots$$

$$P_n(\lambda = i \cdot \omega, \tau_A) + Q_m(\lambda = i \cdot \omega, \tau_A) = -\omega^2 + \frac{\alpha \cdot F^2}{\mu \cdot R^2} \cdot \cos^2(\omega \cdot t) \cdot$$
$$2 \cdot (-1)^k \neq 0 \; \forall \, k = \ldots, -2, -1, 0, +1, +2, \ldots$$
$$|P_n(\lambda = i \cdot \omega, \tau_A)|^2 = \omega^4; \, |Q_m(\lambda = i \cdot \omega, \tau_A)|^2$$
$$= \frac{\alpha^2 \cdot F^4}{\mu^2 \cdot R^4} \cdot \cos^4(\omega \cdot t) \cdot 4 \cdot (-1)^{2 \cdot k}$$
$$F(\omega) = |P(i \cdot \omega)|^2 - |Q(i \cdot \omega)|^2$$
$$= \omega^4 - \frac{\alpha^2 \cdot F^4}{\mu^2 \cdot R^4} \cdot \cos^4(\omega \cdot t) \cdot 4 \cdot (-1)^{2 \cdot k}$$

Since $(-1)^{2 \cdot k} = 1; \forall k = \ldots, -2, -1, 0, +1, +2, \ldots$ then

$$F(\omega) = |P(i \cdot \omega)|^2 - |Q(i \cdot \omega)|^2 = \omega^4 - \frac{\alpha^2 \cdot F^4}{\mu^2 \cdot R^4} \cdot \cos^4(\omega \cdot t) \cdot 4$$

Remark Please note that $F(\omega)$ is a function and F is one of system's parameter.

$$F(\omega) = |P(i \cdot \omega)|^2 - |Q(i \cdot \omega)|^2 = \sum_{k=0}^{2} \Xi_{2 \cdot k} \cdot \omega^{2 \cdot k}$$
$$= \Xi_4 \cdot \omega^4 + \Xi_2 \cdot \omega^2 + \Xi_0$$
$$\Xi_4 = 1; \; \Xi_2 = 0; \; \Xi_0 = -\frac{\alpha^2 \cdot F^4}{\mu^2 \cdot R^4} \cdot \cos^4(\omega \cdot t) \cdot 4 \cdot (-1)^{2 \cdot k}; \; \forall \, k$$
$$= \ldots, -2, -1, 0, +1, +2, \ldots$$

Since $(-1)^{2 \cdot k} = 1; \; \forall \; k = \ldots, -2, -1, 0, +1, +2, \ldots$ then $\Xi_0 = -\frac{\alpha^2 \cdot F^4}{\mu^2 \cdot R^4} \cdot \cos^4(\omega \cdot t) \cdot 4$.

Hence $F(\omega) = 0$ implies $\sum_{k=0}^{2} \Xi_{2 \cdot k} \cdot \omega^{2 \cdot k} = 0$ and its roots are given by solving the above polynomial. Furthermore $P_R(i \cdot \omega, \tau_A) = -\omega^2$; $P_I(i \cdot \omega, \tau_A) = 0$; $Q_I(i \cdot \omega, \tau_A) = 0$.

$$Q_R(i \cdot \omega, \tau_A) = \frac{\alpha \cdot F^2}{\mu \cdot R^2} \cdot \cos^2(\omega \cdot t) \cdot 2 \cdot (-1)^k \; \forall \, k = \ldots, -2, -1, 0, +1, +2, \ldots$$

Hence $\sin\theta(\tau_A) = \frac{-P_R(i \cdot \omega, \tau_A) \cdot Q_I(i \cdot \omega, \tau_A) + P_I(i \cdot \omega, \tau_A) \cdot Q_R(i \cdot \omega, \tau_A)}{|Q(i \cdot \omega, \tau_A)|^2}$

And $\cos\theta(\tau_A) = -\frac{P_R(i \cdot \omega, \tau_A) \cdot Q_R(i \cdot \omega, \tau_A) + P_I(i \cdot \omega, \tau_A) \cdot Q_I(i \cdot \omega, \tau_A)}{|Q(i \cdot \omega, \tau_A)|^2}$

$$\sin\theta(\tau_A) = 0; \cos\theta(\tau_A) = -\frac{-\omega^2 \cdot \frac{\alpha \cdot F^2}{\mu \cdot R^2} \cdot \cos^2(\omega \cdot t) \cdot 2 \cdot (-1)^k}{\frac{\alpha^2 \cdot F^4}{\mu^2 \cdot R^4} \cdot \cos^4(\omega \cdot t) \cdot 4 \cdot (-1)^{2 \cdot k}}$$

Since $(-1)^{2\cdot k}= 1;\ \forall k = \ldots, -2, -1, 0, +1, +2, \ldots$ then

$$\cos\theta(\tau_A) = -\frac{-\omega^2 \cdot \frac{\alpha\cdot F^2}{\mu\cdot R^2} \cdot \cos^2(\omega\cdot t)\cdot 2\cdot(-1)^k}{\frac{\alpha^2\cdot F^4}{\mu^2\cdot R^4}\cdot\cos^4(\omega\cdot t)\cdot 4}$$

$$\cos\theta(\tau_A) = \frac{\mu\cdot R^2}{2\cdot\alpha\cdot F^2}\cdot\frac{\omega^2}{\cos^2(\omega\cdot t)}\cdot(-1)^k$$

We can summary our results for $\sin\theta(\tau_A), \cos\theta(\tau_A)$:

$$\sin\theta(\tau_A) = 0; \cos\theta(\tau_A) = \frac{\mu\cdot R^2}{2\cdot\alpha\cdot F^2}\cdot\frac{\omega^2}{\cos^2(\omega\cdot t)}\cdot(-1)^k\,\forall k$$
$$= \ldots, -2, -1, 0, +1, +2, \ldots$$

Which jointly with $F(\omega) = 0 \Rightarrow \sum_{k=0}^{2} \Xi_{2\cdot k}\cdot\omega^{2\cdot k} = 0$ that are continuous and differentiable in τ_A based on Lemma 1.1. Hence we use Theorem 1.2 and this prove the Theorem 1.3. We use different parameters terminology from our last characteristic parameters definition $k \to j;\ p_k(\tau_A) \to a_j;\ q_k(\tau_A) \to c_j;\ n = 2;\ m = 0;\ n > m$ additionally

$$P_n(\lambda,\tau_A) \to P(\lambda,\tau_A);\ Q_m(\lambda,\tau_A) \to Q(\lambda,\tau_A) \text{ then } P(\lambda,\tau_A)$$
$$= \sum_{j=0}^{2} a_j\cdot\lambda^j;\ Q(\lambda,\tau_A) = \sum_{j=0}^{2} c_j\cdot\lambda^j.$$
$$P(\lambda) = \lambda^2;\ Q(\lambda) = \frac{\alpha\cdot F^2}{\mu\cdot R^2}\cdot\cos^2(\omega\cdot t)\cdot 2\cdot(-1)^k$$
$$= 0\,\forall k = \ldots, -2, -1, 0, +1, +2, \ldots$$

$n, m \in \mathbb{N}_0;\ n > m$ and $a_j, c_j : \mathbb{R}_{+0} \to \mathbb{R}$ are continuous and differentiable function of τ_A such that $a_0 + c_0 \neq 0$. In the following "—" denotes complex and conjugate. $P(\lambda), Q(\lambda)$ are analytic functions in λ and differentiable in τ_A. The coefficients: $\{a_j(\alpha, F, \mu, R, \ldots)\ \&\ c_j(\alpha, F, \mu, R, \ldots)\} \in \mathbb{R}$; depend on differential equations for the rotation of a molecule system's parameters $\alpha, F, \mu, R, \ldots$. Unless strictly necessary, the designation of the variation arguments. These parameters $\alpha, F, \mu, R, \ldots$ will subsequently be omitted from P, Q, a_j, c_j. The coefficients a_j, c_j are continuous (for c_0 coefficient you need to choose k as an odd or even numbers to keep continuity), and differentiable functions of their arguments, and direct substitution shows that $a_0 + c_0 \neq 0\ \forall\ \alpha, F, \mu, R, \ldots \in \mathbb{R}_+$ [7, 8, 9].

$$a_0 = 0;\ a_1 = 0;\ a_2 = 1;\ c_0 = \frac{\alpha\cdot F^2}{\mu\cdot R^2}\cdot\cos^2(\omega\cdot t)\cdot 2\cdot(-1)^k\ \forall k$$
$$= \ldots, -2, -1, 0, +1, +2, \ldots$$

$\frac{\alpha \cdot F^2}{\mu \cdot R^2} \cdot \cos^2(\omega \cdot t) \cdot 2 \cdot (-1)^k \neq 0 \ \forall \ k = \ldots, -2, -1, 0, +1, +2, \ldots; \alpha, F, \mu, R, \ldots \in \mathbb{R}_+$ For $\omega \cdot t \neq \pi \cdot \left(\frac{1}{2} + l\right); l = 0, 1, 2, \ldots; l \in \mathbb{N}_0$ i.e. $\lambda = 0$ is not a root of characteristic equation. Furthermore $P(\lambda)$, $Q(\lambda)$ are analytic functions of λ for which the following requirements of the analysis (see [7], Sect. 3.4) can also verified in the present case.

1. If $\lambda = i \cdot \omega; \omega \in \mathbb{R}$, then $P(i \cdot \omega) + Q(i \cdot \omega) \neq 0$. i.e. P and Q have no common imaginary roots. This condition can be verified numerically in the entire (differential equations for the rotation of a molecule system's parameters $\alpha, F, \mu, R, \ldots$) domain of interest.
2. $\left|\frac{Q(\lambda)}{P(\lambda)}\right|$ is bounded for $|\lambda| \to \infty$, $\mathrm{Re}\,\lambda \geq 0$. No roots bifurcation from ∞.

$$\left|\frac{Q(\lambda)}{P(\lambda)}\right| = \left|\frac{\frac{\alpha \cdot F^2}{\mu \cdot R^2} \cdot \cos^2(\omega \cdot t) \cdot 2 \cdot (-1)^k}{\lambda^2}\right| \forall \, k = \ldots, -2, -1, 0, +1, +2, \ldots$$

3. $F(\omega) = |P(i \cdot \omega)|^2 - |Q(i \cdot \omega)|^2$ has a finite number of zeros. Indeed, this is a polynomial in ω.

$$F(\omega) = |P(i \cdot \omega)|^2 - |Q(i \cdot \omega)|^2 = \omega^4 - \frac{\alpha^2 \cdot F^4}{\mu^2 \cdot R^4} \cdot \cos^4(\omega \cdot t) \cdot 4$$

4. Each positive root $\omega(\alpha, F, \mu, R, \ldots \in \mathbb{R}_+)$ of $F(\omega) = 0$ is continuous and differentiable respect to $\alpha, F, \mu, R, \ldots \in \mathbb{R}_+$. This condition can be assessed numerically.

In addition, since the coefficients in P and Q are real, we have, $\overline{P(-i \cdot \omega)} = P(i \cdot \omega)$ and $\overline{Q(-i \cdot \omega)} = Q(i \cdot \omega)$ thus $\lambda = i \cdot \omega; \omega > 0$ may be on eigenvalue of characteristic equation. The analysis consists in identifying the roots of characteristic equation situated on the imaginary axis of the complex λ-plane, where by increasing the parameters $\alpha, F, \mu, R, \tau_A \ldots$, $\mathrm{Re}\,\lambda$ may, at the crossing, change its sign from (−) to (+), i.e. from a stable focus $E^*(\theta^*, y^*) = \left(k \cdot \frac{\pi}{2}, 0\right) \forall \, k = \ldots, -2, -1, 0, +1, +2, \ldots$ to an unstable one, or vice versa. This feature may be further assessed by examining the sign of the partial derivatives with respect to $\alpha, F, \mu, R, \tau_A \ldots$ parameters, $\Lambda^{-1}(F) = \left(\frac{\partial \mathrm{Re}\lambda}{\partial F}\right)_{\lambda = i \cdot \omega}$ when other parameters are constant $(\alpha, \mu, R, \tau_A \ldots)$.

$$\Lambda^{-1}(R) = \left(\frac{\partial \mathrm{Re}\lambda}{\partial R}\right)_{\lambda = i \cdot \omega}; \alpha, \mu, F, \tau_A \ldots = const$$

$$\Lambda^{-1}(\alpha) = \left(\frac{\partial \mathrm{Re}\lambda}{\partial \alpha}\right)_{\lambda = i \cdot \omega}; R, \mu, F, \tau_A \ldots = const$$

$$\Lambda^{-1}(\mu) = \left(\frac{\partial \mathrm{Re}\lambda}{\partial \mu}\right)_{\lambda=i\cdot\omega}; R, \alpha, F, \tau_A \ldots = const$$

$$\Lambda^{-1}(\tau_A) = \left(\frac{\partial \mathrm{Re}\lambda}{\partial \tau_A}\right)_{\lambda=i\cdot\omega}; R, \alpha, F, \mu \ldots = const$$

When writing $P(\lambda) = P_R(\lambda) + i \cdot P_I(\lambda)$ and $Q(\lambda) = Q_R(\lambda) + i \cdot Q_I(\lambda)$, and inserting $\lambda = i \cdot \omega$; $\omega > 0$ into differential equations for the rotation of a molecule system's characteristic equation ω must satisfy the following:

$$\sin(\omega \cdot \tau_A) = g(\omega) = \frac{-P_R(i\cdot\omega)\cdot Q_I(i\cdot\omega) + P_I(i\cdot\omega)\cdot Q_R(i\cdot\omega)}{|Q(i\cdot\omega)|^2}$$

$$\cos(\omega \cdot \tau_A) = h(\omega) = -\frac{P_R(i\cdot\omega)\cdot Q_R(i\cdot\omega) + P_I(i\cdot\omega)\cdot Q_I(i\cdot\omega)}{|Q(i\cdot\omega)|^2}$$

where $|Q(i \cdot \omega)|^2 \neq 0$ in the view of requirement (a) above, $(g, h) \in \mathbb{R}$. Furthermore, it follows above equations $\sin(\omega \cdot \tau_A)$ and $\cos(\omega \cdot \tau_A)$ that, by squaring and adding sides, ω must be positive root of $F(\omega) = |P(i\cdot\omega)|^2 - |Q(i\cdot\omega)|^2 = 0$. Note $F(\omega)$ is independent of τ_A. Now it is important to notice that if $\tau_A \notin I$ (assume that $I \subseteq \mathbb{R}_{+0}$ is the set where $\omega(\tau_A)$ is a positive root of $F(\omega)$ and for $\tau_A \notin I$; $\omega(\tau_A)$ is not define. Then for all τ_A in I, $\omega(\tau_A)$ is satisfies that $F(\omega, \tau_A) = 0$). Then there are no positive $\omega(\tau_A)$ solutions for $F(\omega, \tau_A) = 0$, and we cannot have stability switches. For any $\tau_A \in I$, where $\omega(\tau_A)$ is a positive solution of $F(\omega, \tau_A) = 0$, we can define the angle $\theta(\tau_A) \in [0, 2\cdot\pi]$ as the solution of $\sin\theta(\tau_A) = \ldots$; $\cos\theta(\tau_A) = \ldots$ and the relation between the argument $\theta(\tau_A)$ and $\omega(\tau_A)\cdot\tau_A$ for $\tau_A \in I$ must be $\omega(\tau_A)\cdot\tau_A = \theta(\tau_A) + n\cdot 2\cdot\pi\ \forall n \in \mathbb{N}_0$. Hence we can define the maps $\tau_{An}: I \to \mathbb{R}_{+0}$ given by $\tau_{An}(\tau_A) = \frac{\theta(\tau_A)+n\cdot 2\cdot\pi}{\omega(\tau_A)}$; $n \in \mathbb{N}_0$; $\tau_A \in I$. Let us introduce the functions $I \to \mathbb{R}$; $S_n(\tau) = \tau_A - \tau_{An}(\tau_A)$; $\tau_A \in I$; $n \in \mathbb{N}_0$ that are continuous and differentiable in τ. In the following, the subscripts $\lambda, \omega, \alpha, \mu, R, \tau_A \ldots$ indicate the corresponding partial derivatives. Let us first concentrate on $\Lambda(x)$, remember in $\lambda(\alpha, \mu, R, \tau_A \ldots \in \mathbb{R}_+)$; $\omega(\alpha, \mu, R, \tau_A \ldots \in \mathbb{R}_+)$ and keeping all parameters except one (x) and τ_A. The deviation closely follows that in reference [BK]. Differentiating differential equations for the rotation of a molecule system's characteristic equation $P(\lambda) + Q(\lambda)\cdot e^{-\lambda\cdot\tau_A} = 0$ with respect to specific parameter (x), and inverting the derivative, for convenience, one calculates:

Remark $x = \alpha, \mu, R, F, \ldots \in \mathbb{R}_+$

$$\left(\frac{\partial\lambda}{\partial x}\right)^{-1} = \frac{-P_\lambda(\lambda, x)\cdot Q(\lambda, x) + Q_\lambda(\lambda, x)\cdot P(\lambda, x) - \tau_A\cdot P(\lambda, x)\cdot Q(\lambda, x)}{P_x(\lambda, x)\cdot Q(\lambda, x) - Q_x(\lambda, x)\cdot P(\lambda, x)}$$

where $P_\lambda = \frac{\partial P}{\partial \lambda}$; $Q_\lambda = \frac{\partial Q}{\partial \lambda}, \ldots$ etc. Substituting $\lambda = i\cdot\omega$ and bearing $\overline{P(-i\cdot\omega)} = P(i\cdot\omega)$; $\overline{Q(-i\cdot\omega)} = Q(i\cdot\omega)$ then $i\cdot P_\lambda(i\cdot\omega) = P_\omega(i\cdot\omega)$; $i\cdot Q_\lambda(i\cdot\omega) = Q_\omega(i\cdot\omega)$ that on the surface $|P(i\cdot\omega)|^2 = |Q(i\cdot\omega)|^2$, one obtains

$$\left(\frac{\partial\lambda}{\partial x}\right)^{-1}|_{\lambda=i\cdot\omega}$$
$$=\frac{i\cdot P_{\omega}(i\cdot\omega,x)\cdot\overline{P(i\cdot\omega,x)}+i\cdot Q_{\lambda}(i\cdot\omega,x)\cdot\overline{Q(\lambda,x)}-\tau_A\cdot|P(i\cdot\omega,x)|^2}{P_x(i\cdot\omega,x)\cdot\overline{P(i\cdot\omega,x)}-Q_x(i\cdot\omega,x)\cdot\overline{Q(i\cdot\omega,x)}}$$

Upon separating into real and imaginary parts, with $P = P_R + i\cdot P_I$; $Q = Q_R + i\cdot Q_I$.

$$P_{\omega}=P_{R\omega}+i\cdot P_{I\omega};\ Q_{\omega}=Q_{R\omega}+i\cdot Q_{I\omega};\ P_x=P_{Rx}+i\cdot P_{Ix};\ Q_x$$
$$=Q_{Rx}+i\cdot Q_{Ix};\ P^2=P_R^2+P_I^2$$

When (x) can be any system parameters $\alpha, \mu, R, F\ldots$ and delay parameter τ_A, etc. Where for convenience, we have dropped the arguments $(i\cdot\omega, x)$, and where $F_{\omega}=2\cdot[(P_{R\omega}\cdot P_R+P_{I\omega}\cdot P_I)-(Q_{R\omega}\cdot Q_R+Q_{I\omega}\cdot Q_I)]$

$$F_x=2\cdot[(P_{Rx}\cdot P_R+P_{Ix}\cdot P_I)-(Q_{Rx}\cdot Q_R+Q_{Ix}\cdot Q_I)];\ \omega_x=-\frac{F_x}{F_{\omega}}$$

We define U and V:

$$U=(P_R\cdot P_{I\omega}-P_I\cdot P_{R\omega})-(Q_R\cdot Q_{I\omega}-Q_I\cdot Q_{R\omega})$$

$$V=(P_R\cdot P_{Ix}-P_I\cdot P_{Rx})-(Q_R\cdot Q_{Ix}-Q_I\cdot Q_{Rx})$$

We choose our specific parameter as time delay τ_A.
Furthermore $P_R(i\cdot\omega,\tau_A)=-\omega^2$; $P_I(i\cdot\omega,\tau_A)=0$; $Q_I(i\cdot\omega,\tau_A)=0$

$$Q_R(i\cdot\omega,\tau_A)=\frac{\alpha\cdot F^2}{\mu\cdot R^2}\cdot\cos^2(\omega\cdot t)\cdot 2\cdot(-1)^k\,\forall k=\ldots,-2,-1,0,+1,+2,\ldots$$

Remark Please do not confuse ω which is related to our stability analysis and ω which is Laser frequency (we consider it as a constant in our system).

$$P_{R\omega}(i\cdot\omega,\tau_A)=-2\cdot\omega;\ P_{I\omega}(i\cdot\omega,\tau_A)=0;\ Q_{R\omega}(i\cdot\omega,\tau_A)$$
$$=0;\ Q_{I\omega}(i\cdot\omega,\tau_A)=0$$
$$P_{R\tau_A}(i\cdot\omega,\tau_A)=0;\ P_{I\tau_A}(i\cdot\omega,\tau_A)=0;\ Q_{R\tau_A}(i\cdot\omega,\tau_A)$$
$$=0;\ Q_{I\tau_A}(i\cdot\omega,\tau_A)=0$$
$$V=(P_R\cdot P_{I\tau_A}-P_I\cdot P_{R\tau_A})-(Q_R\cdot Q_{I\tau_A}-Q_I\cdot Q_{R\tau_A})=0$$
$$P_R\cdot P_{I\omega}=0;\ P_I\cdot P_{R\omega}=0;\ Q_R\cdot Q_{I\omega}=0;\ Q_I\cdot Q_{R\omega}=0\Rightarrow U$$
$$=0;\ F_{\omega}=4\cdot\omega^3;\ F_{\omega}\neq 0$$

$$F_{\tau_A} = 2 \cdot [(P_{R\tau_A} \cdot P_R + P_{I\tau_A} \cdot P_I) - (Q_{R\tau_A} \cdot Q_R + Q_{I\tau_A} \cdot Q_I)] = 0; \omega_{\tau_A} = -\frac{F_{\tau_A}}{F_\omega} = 0$$

$$F_\omega \cdot \frac{\partial \omega}{\partial \tau_A} + F_{\tau_A} = 0; \tau_A \in I; \frac{\partial \omega}{\partial \tau_A} = -\frac{F_{\tau_A}}{F_\omega}$$

$$\Lambda^{-1}(\tau_A) = \left(\frac{\partial \mathrm{Re}\lambda}{\partial \tau_A}\right)_{\lambda = i\cdot\omega}; \Lambda^{-1}(\tau_A) = \mathrm{Re}\left\{\frac{-2 \cdot [U + \tau_A \cdot |P|^2] + i \cdot F_\omega}{F_{\tau_A} + i \cdot 2 \cdot [V + \omega \cdot |P|^2}\right\}; \frac{\partial \omega}{\partial \tau_A} = \omega_{\tau_A} = -\frac{F_{\tau_A}}{F_\omega}$$

$$sign\{\Lambda^{-1}(\tau_A)\} = sign\left\{\left(\frac{\partial \mathrm{Re}\lambda}{\partial \tau_A}\right)_{\lambda = i\cdot\omega}\right\}$$

$$sign\{\Lambda^{-1}(\tau_A)\} = sign\{F_\omega\} \cdot sign\left\{\tau_A \cdot \frac{\partial \omega}{\partial \tau_A} + \omega + \frac{U \cdot \frac{\partial \omega}{\partial \tau_A} + V}{|P|^2}\right\}$$

$$sign\{\Lambda^{-1}(\tau_A)\} = sign\{4 \cdot \omega^3\} \cdot sign\{\omega\}$$

We shall presently examine the possibility of stability transitions (bifurcations) in a differential equations for the rotation of a molecule system, about the equilibrium points $E^*(\theta^*, y^*) = \left(k \cdot \frac{\pi}{2}, 0\right)\ \forall\ k = \ldots, -2, -1, 0, +1, +2, \ldots$ as a result of a variation of delay parameter τ_A. The analysis consists in identifying the roots of our system characteristic equation situated on the imaginary axis of the complex λ-plane where by increasing the delay parameter τ_A, $\mathrm{Re}\,\lambda$ may at the crossing, change its sign from $-$ to $+$, i.e. from a stable focus $E^*(\theta^*, y^*)$ to unstable one, or vice versa. This feature may be further assessed by examining the sign of the partial derivatives with respect to τ_A, $sign\{\Lambda^{-1}(\tau_A)\} = sign\left\{\left(\frac{\partial \mathrm{Re}\lambda}{\partial \tau_A}\right)_{\lambda = i\cdot\omega}\right\}$ when other parameters are constant where $\omega \in \mathbb{R}_+$. We need to plot the stability switch diagram based on different delay values of our system. Since it is very complex function we recommended to solve it numerically rather than analytic.

$$\Lambda^{-1}(\tau_A) = \left(\frac{\partial \mathrm{Re}\lambda}{\partial \tau_A}\right)_{\lambda = i\cdot\omega} = \mathrm{Re}\left\{\frac{-2 \cdot [U + \tau_A \cdot |P|^2] + i \cdot F_\omega}{F_{\tau_A} + i \cdot 2 \cdot [V + \omega \cdot |P|^2}\right\}$$

$$\Lambda^{-1}(\tau_A) = \left(\frac{\partial \mathrm{Re}\lambda}{\partial \tau_A}\right)_{\lambda = i\cdot\omega} = \frac{2 \cdot \{F_\omega \cdot (V + \omega \cdot P^2) - F_{\tau_A} \cdot (U + \tau_A \cdot P^2)\}}{F_{\tau_A}^2 + 4 \cdot (V + \omega \cdot P^2)^2}$$

The stability switch occurs only on those delay values τ_A which fit the equation: $\tau_A = \frac{\theta_+(\tau_A)}{\omega_+(\tau_A)}$ and $\theta_+(\tau_A)$ is the solution of $\sin\theta(\tau_A) = \ldots; \cos\theta(\tau_A) = \ldots$ when $\omega = \omega_+(\tau_A)$ if only ω_+ is feasible [10, 11]. Additionally when all differential equations for the rotation of a molecule system parameters are known and the stability switch due to various time delay values τ_A is describe in the following expression:

$$sign\{\Lambda^{-1}(\tau_A)\} = sign\{F_\omega(\omega(\tau_A), \tau_A)\} \cdot sign\{\tau_A \cdot \omega_{\tau_A}(\omega(\tau_A)) + \omega(\tau_A)$$

$$+\frac{U(\omega(\tau_A))\cdot\omega_{\tau_A}(\omega(\tau_A))+V(\omega(\tau_A))}{|P(\omega(\tau_A))|^2}\Bigg\}$$

Remark We know $F(\omega,\tau_A)=0$ implies it roots $\omega_i(\tau_A)$ and finding those delays values τ_A which ω_i is feasible. There are τ_A values which ω_i are complex or imaginary numbers, then unable to analyze stability.

Lemma 1.1 *Assume that $\omega(\tau)$ is a positive real root $F(\omega,\tau)=0$ defined for $\tau\in I$, which is continuous and differentiable. Assume further that if $\lambda=i\cdot\omega$; $\omega\in\mathbb{R}$, then $P_n(i\cdot\omega,\tau)+Q_n(i\cdot\omega,\tau)\neq 0$; $\tau\in\mathbb{R}$ hold true. The functions $S_n(\tau)$; $n\in\mathbb{N}_0$, are continuous and differentiable on I.*

Theorem 1.2 *Assume that $\omega(\tau)$ is a positive real root $F(\omega,\tau)=0$ defined for $\tau\in I$; $I\subseteq\mathbb{R}_{+0}$, and some $\tau^*\in I$; $S_n(\tau^*)=0$ for some $n\in\mathbb{N}_0$ then a pair of simple conjugate pure imaginary roots $\lambda_+(\tau^*)=i\cdot\omega(\tau^*)$; $\lambda_-(\tau^*)=-i\cdot\omega(\tau^*)$ of $D(\lambda,\tau)=0$ exist at $\tau=\tau^*$ which crosses the imaginary axis from left to right if $\delta(\tau^*)>0$ and cross the imaginary axis from right to left if $\delta(\tau^*)<0$ where*

$$\delta(\tau^*)=sign\left\{\frac{d\,\mathrm{Re}\,\lambda}{d\tau}|_{\lambda=i\cdot\omega}\right\}=sign\{F_\omega(\omega(\tau^*),\tau^*)\}\cdot sign\left\{\frac{dS_n(\tau)}{d\tau}|_{\tau=\tau^*}\right\}$$

Theorem 1.3 *The characteristic equation has a pair of simple and conjugate pure imaginary roots $\lambda=\pm\omega(\tau^*)$, $\omega(\tau^*)$ real at $\tau^*\in I$ if $S_n(\tau^*)=\tau^*-\tau_n(\tau^*)=0$ for some $n\in\mathbb{N}_0$.*

If $\omega(\tau^*)=\omega_+(\tau^*)$, this pair of simple conjugate pure imaginary roots crosses the imaginary axis from left to right if $\delta_+(\tau^*)>0$ and crosses the imaginary axis from right to left if $\delta_+(\tau^*)<0$ where $\delta_+(\tau^*)=sign\{\frac{d\,\mathrm{Re}\,\lambda}{d\tau}|_{\lambda=i\cdot\omega_+(\tau^*)}\}$.

$$\delta_+(\tau^*)=sign\left\{\frac{d\,\mathrm{Re}\,\lambda}{d\tau}|_{\lambda=i\cdot\omega_+(\tau^*)}\right\}=sign\left\{\frac{dS_n(\tau)}{d\tau}|_{\tau=\tau^*}\right\}$$

If $\omega(\tau^*)=\omega_-(\tau^*)$, this pair of simple conjugate pure imaginary roots cross the imaginary axis from left to right, if $\delta_-(\tau^*)>0$ and crosses the imaginary axis from right to left. If $\delta_-(\tau^*)<0$ where $\delta_-(\tau^*)=sign\{\frac{d\,\mathrm{Re}\,\lambda}{d\tau}|_{\lambda=i\cdot\omega_-(\tau^*)}\}=-sign\{\frac{dS_n(\tau)}{d\tau}|_{\tau=\tau^*}\}$.

If $\omega_+(\tau^*)=\omega_-(\tau^*)=\omega(\tau^*)$ then $\Delta(\tau^*)=0$ and $sign\{\frac{d\,\mathrm{Re}\,\lambda}{d\tau}|_{\lambda=i\cdot\omega(\tau^*)}\}=0$, the same is true when $S_n'(\tau^*)=0$ the following result can be useful in identifying values of τ where stability switches happened.

Remark Lemma 1.1 and Theorems 1.2 and 1.3, In our analysis we discuss delay parameter τ_A.

2.2 Ultraboard-Bandwidth Pulse from a Ti: Sapphire Laser Stability Analysis Under Parameters Variation

We can shape ultrashort pulses with a resolution approach 10 fsec by a liquid-crystal spatial light modulator within reflective optics pulse-shaping apparats. The principle is to use a spatial light modulator as a phase modulator. A Spatial Light Modulator (SLM) is an object that imposes some form of spatially varying modulation on a beam of light. Usually, and SLM modulates the intensity of the light beam. It is possible to produce devices that modulate the phase of the beam or both the intensity and the phase simultaneously. SLMs are used extensively in holographic data storage setups to encode information into a laser beam. Liquid crystal SLMs can help to solve problems related to laser micro particle manipulation. In this case spiral beam parameters can be changed dynamically. The outcome is a variety of complex ultrafast waveforms, including odd pulses, high repetition rate pulse trains, and asymmetric pulse trains. There is a limitation of shaping ultra-broad-bandwidth pulses. The synthesis of general waveforms may require compensating for nonlinear spatial dispersion of frequency in the masking plane. The implementation of cubic and quartic-phase modulations of the pulse helps to compensate for large amounts of high-order phase dispersion. One of the powerful methods for synthesizing complex femtosecond optical waveforms is pulse shaping within a dispersion-free grating-lens apparatus. The optical pulse is first transformed from the time domain into the frequency domain. We modify the spectrum of the pulse by a frequency dependent complex linear filter. When a filter is a linear filter (but not necessarily time-invariant), and its input is a complex signal, then, by linearity, If the filter is real, then filtering of complex signals can be carried out by simply performing real filtering on the real and imaginary parts separately (thereby avoiding complex arithmetic) (Fig. 2.3). Finally, the pulse is transformed back into the time domain. We get output pulse which as a temporal field profile that is essentially the Fourier transform of the frequency filter imposed on the electric-field spectrum of the pulse [2, 12].

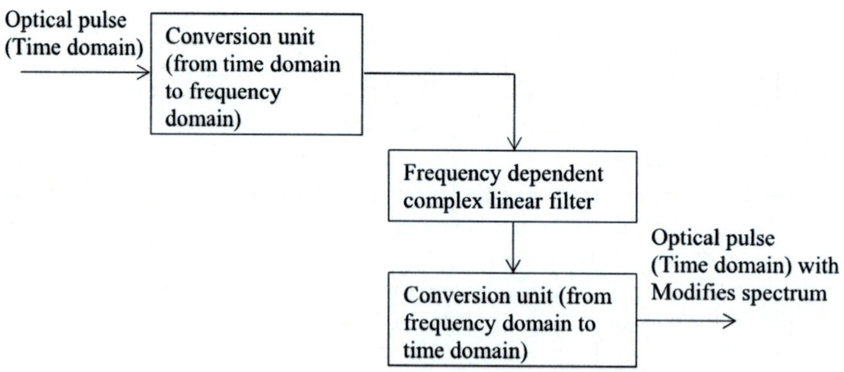

Fig. 2.3 Spectrum modify procedure of the pulse by a frequency dependent complex linear filter

There are many applications which use shape pulses, observation of the fundamental dark soliton in optical fiber and mode-selective excitation of coherent photons. By femtosecond pulse shaping- we dynamic control of molecular and chemical system. The result is system which driven into user-specific final quantum state by use of tailored optical fields to interact coherently with and to manipulate the dynamic evolution of the system. Other areas which can be implemented are solid state electronics and opto electronics. Mainly the pulses can be shaped by dynamically reconfigurable filters (masks), which is a good method. The method includes the use of a single liquid-crystal modulator within a pulse shaper. It can alter either pulse or amplitude independently, the use of two liquid-crystal modulators within a modified pulse shaper, which provide independent control of both phase and amplitude. We use acousto-optic modulator within a pulse shaping apparatus, which simultaneously modulate phase and amplitude. Spectral holographic techniques are complex optical processing and waveform synthesis by convolution and time-to-space conversion. Shaped pulses are influenced by quality of the input pulses which were generated by the use of spatial soliton pulse compression techniques. The mode locked Ti: Sapphire lasers extend these methods to pulses that approach 10 fsec in duration and, to explore the limits imposed by short-duration, ultra-broad bandwidth optical pulses. It is possible to produce pulses less than 10 fsec in duration directly from a laser oscillator. It is a challenge to extend spectral-domain pulse-shaping techniques to ultra-broad bandwidth pulses and investigate the limitations inherent in tailoring ultra-short wave form. The schematic of the ultra-broad bandwidth pulse-shaping apparatus is described below (Fig. 2.4).

The diffraction grating is an optical component with a periodic structure that splits and diffracts light into several beams (red and blue in our case) travelling in different directions. The directions of these beams depend on the spacing of the grating and the wavelength of the light and the grating acts as the dispersive element. The spherical mirror is a mirror which has the shape of a piece cut out of a spherical surface. There are two types of spherical mirrors: concave and convex. The concave mirror is a spherical mirror who's reflecting surface faces towards the center of the sphere and convex mirror is a spherical mirror with reflecting surface curved outwards. In our system we use a concave mirror as the first reflector of the light comes from the first grating and direct to Liquid Crystal SLM and as the second reflector from the light comes from Liquid Crystal SLM to the second grating. Concave mirrors have a curved surface with a center of curvature equidistant from every point on the mirror's surface. An object beyond the center of curvature forms a real and inverted image between the focal point and the center of curvature.

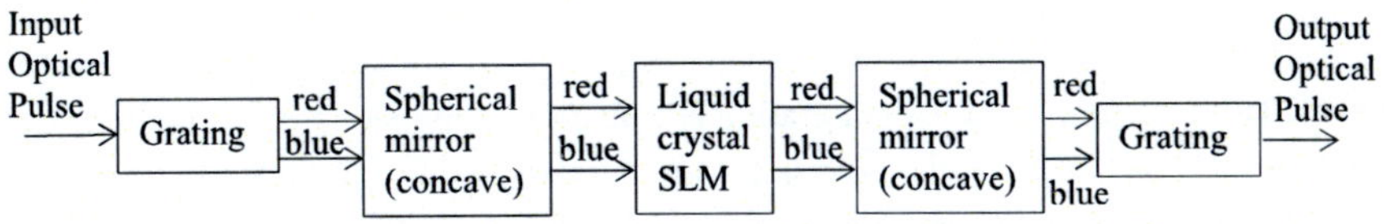

Fig. 2.4 Ultra-broad bandwidth pulse-shaping apparatus schematic

Our pulse-shaping system is performed with one-dimensional liquid-crystal SLM. The system consists of a pair of gratings placed at the focal planes of a unit magnification confocal pair of concave focal-length gold spherical mirrors. Mid way through the apparatus the optical frequencies are spatially separated, with a linear spatial dispersion given by $\frac{dx}{d\lambda} \approx \frac{f}{d \cdot \cos\theta_d}$ ($\frac{dx}{d\lambda}$—linear spatial dispersion, f—focal length, d—grating period, θ_d—diffraction angle, x—position of each frequency component in the masking plane). The compensation of cubic and quartic phase dispersions is very important in our analysis. It is a compensation of high-order phase dispersion in pulse compression and pulse-amplification. The propagation of ultra-board bandwidth pulses through dispersive media results in severe pulse broadening as frequency dependent phase shifts accumulate. For optical pulses with spectra not too broad $\left(\frac{\Delta\omega}{\omega_0} \ll 1\right)$, the phase of an optical pulse propagating through dispersive media can be expressed by Taylor series:

$$\Phi(\omega) = \Phi(\omega_0) + \Phi^{(1)}(\omega) \cdot (\omega - \omega_0) + \frac{1}{2!} \cdot \Phi^{(2)}(\omega) \cdot (\omega - \omega_0)^2 + \frac{1}{3!} \cdot \Phi^{(3)}(\omega) \cdot (\omega - \omega_0)^3 + \frac{1}{4!} \cdot \Phi^{(4)}(\omega) \cdot (\omega - \omega_0)^4 + \cdots$$

where ω_0 is the center frequency of the pulse and $\Phi^{(n)}(\omega) = \left(\frac{d^n\Phi}{d\omega^n}\right)_{\omega=\omega_0}$ is the nth-order derivative of the phase evaluated at ω_0. The group delay and the linear dispersion of an optical pulse are given by the linear term $\Phi^{(1)}(\omega) = \frac{d\Phi}{d\omega}|_{\omega=\omega_0}$ and the quadratic term $\Phi^{(2)}(\omega) = \frac{d^2\Phi}{d\omega^2}|_{\omega=\omega_0}$ respectively. In optics the group delay is the rate of change of the total phase shift with respect to angular frequency. We can get from our system variety of temporal waveforms by using ultra-board bandwidth pulses coupled with programmable pulse shaping techniques. There are limitations which ultra-board bandwidths impose on pulse shaping. Large bandwidths affect the quality of shaped pulses:

1. Higher order phase dispersion of the pulse during propagation through the shaper, which can lead to pulse broadening and loss of temporal resolution.
2. Higher order dispersion of the frequency components in the Fourier plane of the shaper, which can lead to deviations from the ideal pulse shape.
3. Amplitude and phase filtering of the pulse from other optical elements (gratings, mirrors) in the shaper, which lead to pulse broadening.

There are effects of higher-order phase dispersion on the fidelity of the shaped pulse. Any residual phase dispersion can be automatically compensated for the adjustment of the phases of individual modulator pixels. Any material (SLM) contributions to positive second order group velocity dispersion (GVD) within the shaper can be canceled if the position of the final grating is adjusted. Dispersion is the phenomenon in which the phase velocity of a wave depends on its frequency. Group velocity dispersion (GVD) is a characteristic of a dispersive medium, used most often to determine how the medium will affect the duration of an optical pulse traveling through it. There is phase dispersion at higher orders, specially cubic and quartic phase dispersions.

Both the material and the grating contribution can be analyze and we can estimate the material contributions of quadratic cubic and quartic phase distortions by the fiber propagation parameter $\beta(\omega) = \frac{\Phi(\omega)}{L}$, where L is the propagation distance then $\beta^{(n)}(\omega) = \frac{\Phi^{(n)}(\omega)}{L} = \left(\frac{d^n\beta}{d\omega^n}\right)_{\omega=\omega_0}$. The derivatives $\beta^{(n)}(\omega)$ are given in terms of the index of refraction.

$$\beta^{(2)}(\omega) = \frac{\lambda^3}{2\cdot\pi\cdot c^2}\cdot\frac{d^2n}{d\lambda^2}; \beta^{(3)}(\omega) = -\frac{\lambda^4}{4\pi^2c^3}\cdot\left(3\cdot\frac{d^2n}{d\lambda^2}+\lambda\cdot\frac{d^3n}{d\lambda^3}\right)$$

$$\beta^{(4)}(\omega) = \frac{\lambda^5}{\pi^3\cdot c^4}\cdot\left(\frac{3}{4}\cdot\frac{d^2n}{d\lambda^2}+\lambda\cdot\frac{d^3n}{d\lambda^3}+\frac{\lambda^2}{4}\cdot\frac{d^4n}{d\lambda^4}\right)$$

The refraction index (n) is dependent on the wavelength (λ), $n = n(\lambda)$. The refraction index (n) is equal to the speed of light divided by the wave's speed v. The wave speed v is related to both the frequency f and the wavelength λ ($v = \lambda\cdot f$). The wavelength in a medium is $\lambda = \frac{v}{f}$ and the wavelength in vacuum is $\lambda_0 = \frac{c}{f}$. The frequency (f) does not change as light moves from one medium to another.

$$n(\lambda) = \frac{\lambda_0}{\lambda}; n = \frac{c/f}{v/f} = \frac{c}{v}; n(\lambda) = \frac{c/f}{\lambda}$$

$$\beta^{(2)}(\omega) = \frac{\Phi^{(2)}(\omega)}{L} = \left(\frac{d^2\beta}{d\omega^2}\right)_{\omega=\omega_0}; \beta^{(3)}(\omega)$$

$$= \frac{\Phi^{(3)}(\omega)}{L} = \left(\frac{d^3\beta}{d\omega^3}\right)_{\omega=\omega_0}; \beta^{(4)}(\omega)$$

$$= \frac{\Phi^{(4)}(\omega)}{L} = \left(\frac{d^4\beta}{d\omega^4}\right)_{\omega=\omega_0}$$

We take the second, third, and fourth derivative of propagation parameter ($\beta(\omega)$) in ω as a constant at $\omega = \omega_0$. We define new variables: Y_1, Y_2, Y_3, Y_4.

$$Y_1 = \frac{dn}{d\lambda}, Y_2 = \frac{dY_1}{d\lambda} = \frac{d^2n}{d\lambda^2}, Y_3 = \frac{dY_2}{d\lambda} = \frac{d^3n}{d\lambda^3}, Y_4 = \frac{dY_3}{d\lambda} = \frac{d^4n}{d\lambda^4}$$

$$\beta^{(2)}(\omega) = \frac{\lambda^3}{2\cdot\pi\cdot c^2}\cdot\frac{dY_1}{d\lambda}; \beta^{(3)}(\omega) = -\frac{\lambda^4}{4\pi^2c^3}\cdot\left(3\cdot Y_2+\lambda\cdot\frac{dY_2}{d\lambda}\right)$$

$$\beta^{(4)}(\omega) = \frac{\lambda^5}{\pi^3\cdot c^4}\cdot\left(\frac{3}{4}\cdot Y_2+\lambda\cdot Y_3+\frac{\lambda^2}{4}\cdot\frac{dY_3}{d\lambda}\right)$$

$$\beta^{(2)}(\omega) = \frac{\lambda^3}{2\cdot\pi\cdot c^2}\cdot\frac{dY_1}{d\lambda} \Rightarrow \frac{dY_1}{d\lambda} = \frac{2\cdot\pi\cdot c^2}{\lambda^3}\cdot\beta^{(2)}(\omega)$$

$$\beta^{(3)}(\omega) = -\frac{\lambda^4}{4\pi^2c^3}\cdot\left(3\cdot Y_2+\lambda\cdot\frac{dY_2}{d\lambda}\right) \Rightarrow 3\cdot Y_2+\lambda\cdot\frac{dY_2}{d\lambda}$$

$$= -\frac{4\pi^2c^3}{\lambda^4}\cdot\beta^{(3)}(\omega)$$

$$\frac{dY_2}{d\lambda} = -\frac{4\pi^2 c^3}{\lambda^5} \cdot \beta^{(3)}(\omega) - \frac{3}{\lambda} \cdot Y_2$$

$$\beta^{(4)}(\omega) = \frac{\lambda^5}{\pi^3 \cdot c^4} \cdot \left(\frac{3}{4} \cdot Y_2 + \lambda \cdot Y_3 + \frac{\lambda^2}{4} \cdot \frac{dY_3}{d\lambda}\right) \Rightarrow \frac{3}{4} \cdot Y_2$$

$$+ \lambda \cdot Y_3 + \frac{\lambda^2}{4} \cdot \frac{dY_3}{d\lambda} = \frac{\pi^3 \cdot c^4}{\lambda^5} \cdot \beta^{(4)}(\omega)$$

$$\frac{3}{4} \cdot Y_2 + \lambda \cdot Y_3 + \frac{\lambda^2}{4} \cdot \frac{dY_3}{d\lambda} = \frac{\pi^3 \cdot c^4}{\lambda^5} \cdot \beta^{(4)}(\omega) \Rightarrow \frac{dY_3}{d\lambda}$$

$$= \frac{4}{\lambda^2} \cdot \frac{\pi^3 \cdot c^4}{\lambda^5} \cdot \beta^{(4)}(\omega) - \frac{3}{\lambda^2} \cdot Y_2 - \frac{4}{\lambda} \cdot Y_3$$

We can summary our system set of differential equations with the new variables:

$$\frac{dY_1}{d\lambda} = \frac{2 \cdot \pi \cdot c^2}{\lambda^3} \cdot \beta^{(2)}(\omega); \quad \frac{dY_2}{d\lambda} = -\frac{4 \cdot \pi^2 \cdot c^3}{\lambda^5} \cdot \beta^{(3)}(\omega) - \frac{3}{\lambda} \cdot Y_2$$

$$\frac{dY_3}{d\lambda} = \frac{4 \cdot \pi^3 \cdot c^4}{\lambda^7} \cdot \beta^{(4)}(\omega) - \frac{3}{\lambda^2} \cdot Y_2 - \frac{4}{\lambda} \cdot Y_3$$

$$\frac{dY_1}{d\lambda} = \frac{2 \cdot \pi \cdot c^2}{\lambda^3} \cdot \beta^{(2)}(\omega) \Rightarrow Y_2 = \frac{2 \cdot \pi \cdot c^2}{\lambda^3} \cdot \beta^{(2)}(\omega); \lambda$$

$$= \frac{\sqrt[3]{2 \cdot \pi \cdot c^2 \cdot \beta^{(2)}(\omega)}}{\sqrt[3]{Y_2}}$$

$$\frac{1}{\lambda} = \frac{\sqrt[3]{Y_2}}{\sqrt[3]{2 \cdot \pi \cdot c^2 \cdot \beta^{(2)}(\omega)}}; \quad \frac{1}{\lambda^5} = \frac{(\sqrt[3]{Y_2})^5}{(\sqrt[3]{2 \cdot \pi \cdot c^2 \cdot \beta^{(2)}(\omega)})^5}$$

$$\frac{1}{\lambda^2} = \frac{(\sqrt[3]{Y_2})^2}{(\sqrt[3]{2 \cdot \pi \cdot c^2 \cdot \beta^{(2)}(\omega)})^2}$$

$$\frac{1}{\lambda^7} = \frac{(\sqrt[3]{Y_2})^7}{(\sqrt[3]{2 \cdot \pi \cdot c^2 \cdot \beta^{(2)}(\omega)})^7}$$

We get the following two system differential equations:

$$\frac{dY_2}{d\lambda} = -\frac{4 \cdot \pi^2 \cdot c^3 \cdot (\sqrt[3]{Y_2})^5}{(\sqrt[3]{2 \cdot \pi \cdot c^2 \cdot \beta^{(2)}(\omega)})^5} \cdot \beta^{(3)}(\omega)$$

$$- \frac{3 \cdot \sqrt[3]{Y_2}}{\sqrt[3]{2 \cdot \pi \cdot c^2 \cdot \beta^{(2)}(\omega)}} \cdot Y_2$$

$$\frac{dY_3}{d\lambda} = \frac{4 \cdot \pi^3 \cdot c^4 \cdot (\sqrt[3]{Y_2})^7}{(\sqrt[3]{2 \cdot \pi \cdot c^2 \cdot \beta^{(2)}(\omega)})^7} \cdot \beta^{(4)}(\omega)$$

$$- \frac{3 \cdot (\sqrt[3]{Y_2})^2}{(\sqrt[3]{2 \cdot \pi \cdot c^2 \cdot \beta^{(2)}(\omega)})^2} \cdot Y_2 - \frac{4 \cdot \sqrt[3]{Y_2}}{\sqrt[3]{2 \cdot \pi \cdot c^2 \cdot \beta^{(2)}(\omega)}} \cdot Y_3$$

We can write our system's two differential equations:

$$\frac{dY_2}{d\lambda} = -\frac{4\cdot\pi^2\cdot c^3\cdot\beta^{(3)}(\omega)}{(\sqrt[3]{2\cdot\pi\cdot c^2\cdot\beta^{(2)}(\omega)})^5}\cdot(Y_2)^{\frac{5}{3}} - \frac{3}{\sqrt[3]{2\cdot\pi\cdot c^2\cdot\beta^{(2)}(\omega)}}\cdot(Y_2)^{\frac{4}{3}}$$

$$\frac{dY_3}{d\lambda} = \frac{4\cdot\pi^3\cdot c^4\cdot\beta^{(4)}(\omega)}{(\sqrt[3]{2\cdot\pi\cdot c^2\cdot\beta^{(2)}(\omega)})^7}\cdot(Y_2)^{\frac{7}{3}} - \frac{3}{(\sqrt[3]{2\cdot\pi\cdot c^2\cdot\beta^{(2)}(\omega)})^2}\cdot(Y_2)^{\frac{5}{3}} - \frac{4}{\sqrt[3]{2\cdot\pi\cdot c^2\cdot\beta^{(2)}(\omega)}}\cdot(Y_2)^{\frac{1}{3}}\cdot Y_3$$

At fixed points: $\frac{dY_2}{d\lambda} = 0$; $\frac{dY_3}{d\lambda} = 0$.

$$\frac{dY_2}{d\lambda} = 0 \Rightarrow -\frac{4\cdot\pi^2\cdot c^3\cdot\beta^{(3)}(\omega)}{(\sqrt[3]{2\cdot\pi\cdot c^2\cdot\beta^{(2)}(\omega)})^5}\cdot(Y_2^*)^{\frac{5}{3}} - \frac{3}{\sqrt[3]{2\cdot\pi\cdot c^2\cdot\beta^{(2)}(\omega)}}\cdot(Y_2^*)^{\frac{4}{3}} = 0$$

$$-(Y_2^*)^{\frac{4}{3}}\cdot\left\{\frac{4\cdot\pi^2\cdot c^3\cdot\beta^{(3)}(\omega)}{(\sqrt[3]{2\cdot\pi\cdot c^2\cdot\beta^{(2)}(\omega)})^5}\cdot(Y_2^*)^{\frac{1}{3}} + \frac{3}{\sqrt[3]{2\cdot\pi\cdot c^2\cdot\beta^{(2)}(\omega)}}\right\} = 0$$

Option 1: $Y_2^* = 0$
Option 2: $\frac{4\cdot\pi^2\cdot c^3\cdot\beta^{(3)}(\omega)}{(\sqrt[3]{2\cdot\pi\cdot c^2\cdot\beta^{(2)}(\omega)})^5}\cdot(Y_2^*)^{\frac{1}{3}} + \frac{3}{\sqrt[3]{2\cdot\pi\cdot c^2\cdot\beta^{(2)}(\omega)}} = 0.$

$$(Y_2^*)^{\frac{1}{3}} = -\frac{3}{\sqrt[3]{2\cdot\pi\cdot c^2\cdot\beta^{(2)}(\omega)}}\cdot\frac{\left(\sqrt[3]{2\cdot\pi\cdot c^2\cdot\beta^{(2)}(\omega)}\right)^5}{4\cdot\pi^2\cdot c^3\cdot\beta^{(3)}(\omega)}$$

$$(Y_2^*)^{\frac{1}{3}} = -\frac{3\cdot\left(\sqrt[3]{2\cdot\pi\cdot c^2\cdot\beta^{(2)}(\omega)}\right)^4}{4\cdot\pi^2\cdot c^3\cdot\beta^{(3)}(\omega)}$$

$$(Y_2^*)^{\frac{1}{3}} = -\frac{3\cdot(2\cdot\pi\cdot c^2\cdot\beta^{(2)}(\omega))^{\frac{4}{3}}}{4\cdot\pi^2\cdot c^3\cdot\beta^{(3)}(\omega)}$$

$$(Y_2^*)^{\frac{1}{3}} = -\frac{3\cdot 2^{\frac{4}{3}}\cdot\pi^{\frac{4}{3}}\cdot c^{\frac{8}{3}}\cdot[\beta^{(2)}(\omega)]^{\frac{4}{3}}}{4\cdot\pi^2\cdot c^3\cdot\beta^{(3)}(\omega)}$$

$$(Y_2^*)^{\frac{1}{3}} = -\frac{3\cdot 2^{\frac{4}{3}}}{4}\cdot\pi^{(\frac{4}{3}-2)}\cdot c^{(\frac{8}{3}-3)}\cdot\frac{[\beta^{(2)}(\omega)]^{\frac{4}{3}}}{\beta^{(3)}(\omega)}$$

$$(Y_2^*)^{\frac{1}{3}} = -\frac{3\cdot 2^{\frac{4}{3}}}{4}\cdot\pi^{(\frac{4}{3}-2)}\cdot c^{(\frac{8}{3}-3)}\cdot\frac{[\beta^{(2)}(\omega)]^{\frac{4}{3}}}{\beta^{(3)}(\omega)}$$

$$(Y_2^*)^{\frac{1}{3}} = -\frac{3\cdot 2^{\frac{4}{3}}}{4}\cdot\pi^{-\frac{2}{3}}\cdot c^{-\frac{1}{3}}\cdot\frac{[\beta^{(2)}(\omega)]^{\frac{4}{3}}}{\beta^{(3)}(\omega)}$$

$$Y_2^* = \left\{-\frac{3\cdot 2^{\frac{4}{3}}}{4}\cdot\pi^{-\frac{2}{3}}\cdot c^{-\frac{1}{3}}\cdot\frac{[\beta^{(2)}(\omega)]^{\frac{4}{3}}}{\beta^{(3)}(\omega)}\right\}^3;\ Y_2^* = -\frac{3^3\cdot 2^4}{4^3}\cdot\pi^{-2}\cdot c^{-1}\cdot\frac{[\beta^{(2)}(\omega)]^4}{[\beta^{(3)}(\omega)]^3}$$

$$\frac{dY_3}{d\lambda} = 0 \Rightarrow \frac{4\cdot\pi^3\cdot c^4\cdot\beta^{(4)}(\omega)}{\left(\sqrt[3]{2\cdot\pi\cdot c^2\cdot\beta^{(2)}(\omega)}\right)^7}\cdot(Y_2^*)^{\frac{7}{3}} - \frac{3}{\left(\sqrt[3]{2\cdot\pi\cdot c^2\cdot\beta^{(2)}(\omega)}\right)^2}\cdot(Y_2^*)^{\frac{5}{3}}$$

$$-\frac{4}{\sqrt[3]{2\cdot\pi\cdot c^2\cdot\beta^{(2)}(\omega)}}\cdot(Y_2^*)^{\frac{1}{3}}\cdot Y_3^* = 0$$

$$\frac{4}{\sqrt[3]{2\cdot\pi\cdot c^2\cdot\beta^{(2)}(\omega)}}\cdot(Y_2^*)^{\frac{1}{3}}\cdot Y_3^* = \frac{4\cdot\pi^3\cdot c^4\cdot\beta^{(4)}(\omega)}{(\sqrt[3]{2\cdot\pi\cdot c^2\cdot\beta^{(2)}(\omega)})^7}\cdot(Y_2^*)^{\frac{7}{3}}$$

$$-\frac{3}{(\sqrt[3]{2\cdot\pi\cdot c^2\cdot\beta^{(2)}(\omega)})^2}\cdot(Y_2^*)^{\frac{5}{3}}$$

$$Y_3^* = \frac{\frac{4\cdot\pi^3\cdot c^4\cdot\beta^{(4)}(\omega)}{\left(\sqrt[3]{2\cdot\pi\cdot c^2\cdot\beta^{(2)}(\omega)}\right)^7}\cdot(Y_2^*)^{\frac{7}{3}} - \frac{3}{\left(\sqrt[3]{2\cdot\pi\cdot c^2\cdot\beta^{(2)}(\omega)}\right)^2}\cdot(Y_2^*)^{\frac{5}{3}}}{\frac{4}{\sqrt[3]{2\cdot\pi\cdot c^2\cdot\beta^{(2)}(\omega)}}\cdot(Y_2^*)^{\frac{1}{3}}}$$

$$Y_3^* = \frac{\left\{\frac{4\cdot\pi^3\cdot c^4\cdot\beta^{(4)}(\omega)}{\left(\sqrt[3]{2\cdot\pi\cdot c^2\cdot\beta^{(2)}(\omega)}\right)^7}\cdot(Y_2^*)^{\frac{6}{3}} - \frac{3}{\left(\sqrt[3]{2\cdot\pi\cdot c^2\cdot\beta^{(2)}(\omega)}\right)^2}\cdot(Y_2^*)^{\frac{4}{3}}\right\}\cdot(Y_2^*)^{\frac{1}{3}}}{\frac{4}{\sqrt[3]{2\cdot\pi\cdot c^2\cdot\beta^{(2)}(\omega)}}\cdot(Y_2^*)^{\frac{1}{3}}}$$

$$Y_3^* = \frac{\left\{\frac{4\cdot\pi^3\cdot c^4\cdot\beta^{(4)}(\omega)}{\left(\sqrt[3]{2\cdot\pi\cdot c^2\cdot\beta^{(2)}(\omega)}\right)^7}\cdot(Y_2^*)^{\frac{6}{3}} - \frac{3}{\left(\sqrt[3]{2\cdot\pi\cdot c^2\cdot\beta^{(2)}(\omega)}\right)^2}\cdot(Y_2^*)^{\frac{4}{3}}\right\}}{\frac{4}{\sqrt[3]{2\cdot\pi\cdot c^2\cdot\beta^{(2)}(\omega)}}}; \frac{4}{\sqrt[3]{2\cdot\pi\cdot c^2\cdot\beta^{(2)}(\omega)}} \neq 0$$

Option 1: $Y_2^* = 0 \Rightarrow Y_3^* = 0$
Option 2: $Y_2^* = -\frac{3^3\cdot 2^4}{4^3}\cdot\pi^{-2}\cdot c^{-1}\cdot\frac{[\beta^{(2)}(\omega)]^4}{[\beta^{(3)}(\omega)]^3}$; $Y_2^* = -\frac{3^3}{2^2}\cdot\pi^{-2}\cdot c^{-1}\cdot\frac{[\beta^{(2)}(\omega)]^4}{[\beta^{(3)}(\omega)]^3}$

$$Y_3^* = \frac{1}{4}\cdot\left\{\frac{4\cdot\pi^3\cdot c^4\cdot\beta^{(4)}(\omega)}{\left(\sqrt[3]{2\cdot\pi\cdot c^2\cdot\beta^{(2)}(\omega)}\sqrt[3]{2\cdot\pi\cdot c^2\cdot\beta^{(2)}(\omega)}\right)^7}\cdot(Y_2^*)^{\frac{6}{3}}\right.$$

$$\left.-\frac{3}{\left(\sqrt[3]{2\cdot\pi\cdot c^2\cdot\beta^{(2)}(\omega)}\right)^2}\cdot(Y_2^*)^{\frac{4}{3}}\right\}\cdot\sqrt[3]{2\cdot\pi\cdot c^2\cdot\beta^{(2)}(\omega)}$$

$$\left(\sqrt[3]{2\cdot\pi\cdot c^2\cdot\beta^{(2)}(\omega)}\right)^6 = 2^2\cdot\pi^2\cdot c^4\cdot[\beta^{(2)}(\omega)]^2; \sqrt[3]{2\cdot\pi\cdot c^2\cdot\beta^{(2)}(\omega)}$$

$$= 2^{\frac{1}{3}}\cdot\pi^{\frac{1}{3}}\cdot c^{\frac{2}{3}}\cdot[\beta^{(2)}(\omega)]^{\frac{1}{3}}$$

$$Y_3^* = \frac{1}{4}\cdot\left\{\frac{4\cdot\pi^3\cdot c^4\cdot\beta^{(4)}(\omega)}{2^2\cdot\pi^2\cdot c^4\cdot[\beta^{(2)}(\omega)]^2}\cdot(Y_2^*)^{\frac{6}{3}} - \frac{3}{2^{\frac{1}{3}}\cdot\pi^{\frac{1}{3}}\cdot c^{\frac{2}{3}}\cdot[\beta^{(2)}(\omega)]^{\frac{1}{3}}}\cdot(Y_2^*)^{\frac{4}{3}}\right\}$$

$$Y_3^* = \frac{1}{4}\cdot\left\{\frac{\pi\cdot\beta^{(4)}(\omega)}{[\beta^{(2)}(\omega)]^2}\cdot(Y_2^*)^2 - \frac{3}{2^{\frac{1}{3}}\cdot\pi^{\frac{1}{3}}\cdot c^{\frac{2}{3}}\cdot[\beta^{(2)}(\omega)]^{\frac{1}{3}}}\cdot(Y_2^*)^{\frac{4}{3}}\right\}$$

$$(Y_2^*)^2 = \left(-\frac{3^3}{2^2}\cdot\pi^{-2}\cdot c^{-1}\cdot\frac{[\beta^{(2)}(\omega)]^4}{[\beta^{(3)}(\omega)]^3}\right)^2 = \frac{3^6}{2^6}\cdot\pi^{-4}\cdot c^{-2}\cdot\frac{[\beta^{(2)}(\omega)]^8}{[\beta^{(3)}(\omega)]^6}$$

$$(Y_2^*)^{\frac{4}{3}} = \left(-\frac{3^3}{2^2}\cdot\pi^{-2}\cdot c^{-1}\cdot\frac{[\beta^{(2)}(\omega)]^4}{[\beta^{(3)}(\omega)]^3}\right)^{\frac{4}{3}}$$

$$(Y_2^*)^{\frac{4}{3}} = \frac{3^4}{2^{\frac{8}{3}}}\cdot\pi^{-\frac{8}{3}}\cdot c^{-\frac{4}{3}}\cdot\frac{[\beta^{(2)}(\omega)]^{\frac{16}{3}}}{[\beta^{(3)}(\omega)]^4}$$

$$Y_3^* = \frac{1}{4}\cdot\left\{\frac{\pi\cdot\beta^{(4)}(\omega)}{[\beta^{(2)}(\omega)]^2}\cdot\frac{3^6}{2^6}\cdot\pi^{-4}\cdot c^{-2}\cdot\frac{[\beta^{(2)}(\omega)]^8}{[\beta^{(3)}(\omega)]^6}\right.$$
$$\left.-\frac{3}{2^{\frac{1}{3}}\cdot\pi^{\frac{1}{3}}\cdot c^{\frac{2}{3}}\cdot[\beta^{(2)}(\omega)]^{\frac{1}{3}}}\cdot\frac{3^4}{2^{\frac{8}{3}}}\cdot\pi^{-\frac{8}{3}}\cdot c^{-\frac{4}{3}}\cdot\frac{[\beta^{(2)}(\omega)]^{\frac{16}{3}}}{[\beta^{(3)}(\omega)]^4}\right\}$$

$$Y_3^* = \frac{1}{4}\cdot\frac{3^5}{2^3}\cdot\frac{1}{\pi^3\cdot c^2}\cdot\left\{\frac{3}{2^3}\cdot\frac{[\beta^{(2)}(\omega)]^6}{[\beta^{(3)}(\omega)]^6}\cdot\beta^{(4)}(\omega)-\frac{[\beta^{(2)}(\omega)]^5}{[\beta^{(3)}(\omega)]^4}\right\}$$

We can summary our system fixed points in Table 2.2.

$$\beta^{(k)}(\omega) = \frac{\Phi^{(k)}(\omega)}{L} = \left(\frac{d^k\beta}{d\omega^k}\right)_{\omega=\omega_0};\ k = 2, 3, 4$$

Stability analysis: Our ultra-board bandwidth pulse from a Ti: Sapphire laser system, derivatives of propagation parameter $\beta^{(n)}(\omega)$ in terms of the index of refraction $n = n(\lambda)$ derivatives $\left(\frac{d^k n}{d\lambda^k}; k = 2, 3, 4\right)$ are presented by the following differential equations $\left(Y_2 = \frac{dY_1}{d\lambda} = \frac{d^2n}{d\lambda^2}, Y_3 = \frac{dY_2}{d\lambda} = \frac{d^3n}{d\lambda^3}\right)$:

$$\frac{dY_2}{d\lambda} = -\frac{4\cdot\pi^2\cdot c^3\cdot\beta^{(3)}(\omega)}{(\sqrt[3]{2\cdot\pi\cdot c^2\cdot\beta^{(2)}(\omega)})^5}\cdot(Y_2)^{\frac{5}{3}} - \frac{3}{\sqrt[3]{2\cdot\pi\cdot c^2\cdot\beta^{(2)}(\omega)}}\cdot(Y_2)^{\frac{4}{3}}$$

$$\frac{dY_3}{d\lambda} = \frac{4\cdot\pi^3\cdot c^4\cdot\beta^{(4)}(\omega)}{(\sqrt[3]{2\cdot\pi\cdot c^2\cdot\beta^{(2)}(\omega)})^7}\cdot(Y_2)^{\frac{7}{3}}$$

Table 2.2 System (derivatives $\beta^{(n)}(\omega)$ which given in terms of the index of refraction) fixed points

Fixed point	Y_2^*	Y_3^*
First fixed point	0	0
Second fixed point	$-\frac{3^3}{2^2}\cdot\pi^{-2}\cdot c^{-1}\cdot\frac{[\beta^{(2)}(\omega)]^4}{[\beta^{(3)}(\omega)]^3}$	$\frac{1}{4}\cdot\frac{3^5}{2^3}\cdot\frac{1}{\pi^3\cdot c^2}\cdot\left\{\frac{3}{2^3}\cdot\frac{[\beta^{(2)}(\omega)]^6}{[\beta^{(3)}(\omega)]^6}\cdot\beta^{(4)}(\omega)-\frac{[\beta^{(2)}(\omega)]^5}{[\beta^{(3)}(\omega)]^4}\right\}$

$$-\frac{3}{(\sqrt[3]{2\cdot\pi\cdot c^2\cdot\beta^{(2)}(\omega)})^2}\cdot(Y_2)^{\frac{5}{3}}-\frac{4}{\sqrt[3]{2\cdot\pi\cdot c^2\cdot\beta^{(2)}(\omega)}}\cdot(Y_2)^{\frac{1}{3}}\cdot Y_3$$

We use linearization technique and we can approximate the phase portrait near a fixed point by that of a corresponding linear system. We consider our system $\frac{dY_2}{d\lambda}=f_1(Y_2,Y_3)$; $\frac{dY_3}{d\lambda}=f_2(Y_2,Y_3)$.

$$f_1(Y_2,Y_3)=-\frac{4\cdot\pi^2\cdot c^3\cdot\beta^{(3)}(\omega)}{(\sqrt[3]{2\cdot\pi\cdot c^2\cdot\beta^{(2)}(\omega)})^5}\cdot(Y_2)^{\frac{5}{3}}$$
$$-\frac{3}{\sqrt[3]{2\cdot\pi\cdot c^2\cdot\beta^{(2)}(\omega)}}\cdot(Y_2)^{\frac{4}{3}}$$
$$f_2(Y_2,Y_3)=\frac{4\cdot\pi^3\cdot c^4\cdot\beta^{(4)}(\omega)}{\left(\sqrt[3]{2\cdot\pi\cdot c^2\cdot\beta^{(2)}(\omega)}\right)^7}\cdot(Y_2)^{\frac{7}{3}}$$
$$-\frac{3}{(\sqrt[3]{2\cdot\pi\cdot c^2\cdot\beta^{(2)}(\omega)})^2}\cdot(Y_2)^{\frac{5}{3}}-\frac{4}{\sqrt[3]{2\cdot\pi\cdot c^2\cdot\beta^{(2)}(\omega)}}\cdot(Y_2)^{\frac{1}{3}}\cdot Y_3$$

And we suppose that (Y_2^*,Y_3^*) is a fixed point, $f_1(Y_2^*,Y_3^*)=0$; $f_2(Y_2^*,Y_3^*)=0$. Let $u=Y_2-Y_2^*$; $v=Y_3-Y_3^*$ denote the components of a small disturbance from the fixed point [5, 6, 13]. We need to inspect whether the disturbance grows or decays, we need to derive differential equations for u and v. $\frac{du}{d\lambda}=\frac{dY_2}{d\lambda}$ since Y_2^* is a constant then

$$Y_2=u+Y_2^*;\ Y_3=v+Y_3^*;\ \frac{du}{d\lambda}=\frac{dY_2}{d\lambda}=f_1(u+Y_2^*,v+Y_3^*)$$
$$\frac{du}{d\lambda}=f_1(u+Y_2^*,v+Y_3^*)=f_1(Y_2^*,Y_3^*)+u\cdot\frac{\partial f_1}{dY_2}+v\cdot\frac{\partial f_1}{dY_3}+O(u^2,v^2,u\cdot v)$$
$$\frac{du}{d\lambda}=f_1(u+Y_2^*,v+Y_3^*)=u\cdot\frac{\partial f_1}{dY_2}+v\cdot\frac{\partial f_1}{dY_3}+O(u^2,v^2,u\cdot v),\ \text{since } f_1(Y_2^*,Y_3^*)=0$$

The partial derivatives are to be evaluated at the fixed point (Y_2^*,Y_3^*); thus they are numbers, not functions. The $O(u^2,v^2,u\cdot v)$ denotes quadratic terms in u and v. Since u and v are small, these quadratic terms are extremely small. Similarly we find

$$Y_2=u+Y_2^*;\ Y_3=v+Y_3^*;\ \frac{dv}{d\lambda}=\frac{dY_3}{d\lambda}=f_2(u+Y_2^*,v+Y_3^*)$$
$$\frac{dv}{d\lambda}=f_2(u+Y_2^*,v+Y_3^*)=f_2(Y_2^*,Y_3^*)$$
$$+u\cdot\frac{\partial f_2}{dY_2}+v\cdot\frac{\partial f_2}{dY_3}+O(u^2,v^2,u\cdot v)$$

$$\frac{dv}{d\lambda}=f_2(u+Y_2^*,v+Y_3^*)=u\cdot\frac{\partial f_2}{dY_2}+v\cdot\frac{\partial f_2}{dY_3}+O(u^2,v^2,u\cdot v),\ \text{since } f_2(Y_2^*,Y_3^*)=0$$

$$\begin{pmatrix} \frac{du}{d\lambda} \\ \frac{dv}{d\lambda} \end{pmatrix} = \begin{pmatrix} \frac{\partial f_1}{\partial Y_2} & \frac{\partial f_1}{\partial Y_3} \\ \frac{\partial f_2}{\partial Y_2} & \frac{\partial f_2}{\partial Y_3} \end{pmatrix} \cdot \begin{pmatrix} u \\ v \end{pmatrix} + \text{quadratic terms}$$

The Jacobian matrix at fixed point (Y_2^*, Y_3^*): $A = \begin{pmatrix} \frac{\partial f_1}{\partial Y_2} & \frac{\partial f_1}{\partial Y_3} \\ \frac{\partial f_2}{\partial Y_2} & \frac{\partial f_2}{\partial Y_3} \end{pmatrix}_{(Y_2^*, Y_3^*)}$.

The quadratic terms are tiny and we can neglect them. We get the linearized system.

$$\begin{pmatrix} \frac{du}{d\lambda} \\ \frac{dv}{d\lambda} \end{pmatrix} = \begin{pmatrix} \frac{\partial f_1}{\partial Y_2} & \frac{\partial f_1}{\partial Y_3} \\ \frac{\partial f_2}{\partial Y_2} & \frac{\partial f_2}{\partial Y_3} \end{pmatrix} \cdot \begin{pmatrix} u \\ v \end{pmatrix}$$

$$\frac{\partial f_1(Y_2, Y_3)}{\partial Y_2} = -\frac{4 \cdot \pi^2 \cdot c^3 \cdot \beta^{(3)}(\omega)}{\left(\sqrt[3]{2 \cdot \pi \cdot c^2 \cdot \beta^{(2)}(\omega)}\right)^5} \cdot \frac{5}{3} \cdot (Y_2)^{\frac{2}{3}}$$

$$- \frac{3}{\sqrt[3]{2 \cdot \pi \cdot c^2 \cdot \beta^{(2)}(\omega)}} \cdot \frac{4}{3} \cdot (Y_2)^{\frac{1}{3}}; \; \frac{\partial f_1(Y_2, Y_3)}{\partial Y_3} = 0$$

$$\frac{\partial f_2(Y_2, Y_3)}{\partial Y_2} = \frac{4 \cdot \pi^3 \cdot c^4 \cdot \beta^{(4)}(\omega)}{\left(\sqrt[3]{2 \cdot \pi \cdot c^2 \cdot \beta^{(2)}(\omega)}\right)^7} \cdot \frac{7}{3} (Y_2)^{\frac{4}{3}}$$

$$- \frac{3}{\left(\sqrt[3]{2 \cdot \pi \cdot c^2 \cdot \beta^{(2)}(\omega)}\right)^2} \cdot \frac{5}{3} \cdot (Y_2)^{\frac{2}{3}}$$

$$- \frac{4}{\sqrt[3]{2 \cdot \pi \cdot c^2 \cdot \beta^{(2)}(\omega)}} \cdot \frac{1}{3} \cdot (Y_2)^{-\frac{2}{3}} \cdot Y_3$$

$$\frac{\partial f_2(Y_2, Y_3)}{\partial Y_3} = -\frac{4}{\sqrt[3]{2 \cdot \pi \cdot c^2 \cdot \beta^{(2)}(\omega)}} \cdot (Y_2)^{\frac{1}{3}}$$

Case A: First fixed point $E^{(0)}(Y_2^{(0)}, Y_3^{(0)}) = (0, 0)$.

$$A = \begin{pmatrix} \frac{\partial f_1}{\partial Y_2} & \frac{\partial f_1}{\partial Y_3} \\ \frac{\partial f_2}{\partial Y_2} & \frac{\partial f_2}{\partial Y_3} \end{pmatrix}_{(Y_2^*, Y_3^*)=(0,0)} = \begin{pmatrix} 0 & 0 \\ 0 & 0 \end{pmatrix}$$

Then we have a plane of fixed points.

Case B: Second fixed point

$$E^{(1)}(Y_2^{(1)}, Y_3^{(1)}) = \left(-\frac{3^3}{2^2}\cdot\pi^{-2}\cdot c^{-1}\cdot\frac{[\beta^{(2)}(\omega)]^4}{[\beta^{(3)}(\omega)]^3}, \frac{1}{4}\cdot\frac{3^5}{2^3}\cdot\frac{1}{\pi^3\cdot c^2}\cdot \left\{\frac{3}{2^3}\cdot\frac{[\beta^{(2)}(\omega)]^6}{[\beta^{(3)}(\omega)]^6}\cdot\beta^{(4)}(\omega) - \frac{[\beta^{(2)}(\omega)]^5}{[\beta^{(3)}(\omega)]^4}\right\}\right)$$

$$A = \begin{pmatrix} \frac{\partial f_1}{\partial Y_2} & \frac{\partial f_1}{\partial Y_3} \\ \frac{\partial f_2}{\partial Y_2} & \frac{\partial f_2}{\partial Y_3} \end{pmatrix}_{(Y_2^{(1)}, Y_3^{(1)})}$$

$$= \begin{pmatrix} \frac{\partial f_1}{\partial Y_2} & \frac{\partial f_1}{\partial Y_3} \\ \frac{\partial f_2}{\partial Y_2} & \frac{\partial f_2}{\partial Y_3} \end{pmatrix}_{(Y_2^{(1)}, Y_3^{(1)}) = \left(-\frac{3^3}{2^2}\cdot\pi^{-2}\cdot c^{-1}\cdot\frac{[\beta^{(2)}(\omega)]^4}{[\beta^{(3)}(\omega)]^3}, \frac{1}{4}\cdot\frac{3^5}{2^3}\cdot\frac{1}{\pi^3\cdot c^2}\cdot\left\{\frac{3}{2^3}\cdot\frac{[\beta^{(2)}(\omega)]^6}{[\beta^{(3)}(\omega)]^6}\cdot\beta^{(4)}(\omega) - \frac{[\beta^{(2)}(\omega)]^5}{[\beta^{(3)}(\omega)]^4}\right\}\right)}$$

$$\frac{\partial f_1(Y_2, Y_3)}{\partial Y_2}|_{(Y_2^{(1)}, Y_3^{(1)})} = -\frac{4\cdot\pi^2\cdot c^3\cdot\beta^{(3)}(\omega)}{\left(\sqrt[3]{2\cdot\pi\cdot c^2\cdot\beta^{(2)}(\omega)}\right)^5}\cdot \frac{5}{3}\cdot\left(-\frac{3^3}{2^2}\cdot\pi^{-2}\cdot c^{-1}\cdot\frac{[\beta^{(2)}(\omega)]^4}{[\beta^{(3)}(\omega)]^3}\right)^{\frac{2}{3}} - \frac{3}{\sqrt[3]{2\cdot\pi\cdot c^2\cdot\beta^{(2)}(\omega)}}\cdot\frac{4}{3}\cdot\left(-\frac{3^3}{2^2}\cdot\pi^{-2}\cdot c^{-1}\cdot\frac{[\beta^{(2)}(\omega)]^4}{[\beta^{(3)}(\omega)]^3}\right)^{\frac{1}{3}}$$

$$\frac{\partial f_1(Y_2, Y_3)}{\partial Y_3}|_{(Y_2^{(1)}, Y_3^{(1)})} = 0$$

$$\frac{\partial f_2(Y_2, Y_3)}{\partial Y_2}|_{(Y_2^{(1)}, Y_3^{(1)})} = \frac{4\cdot\pi^3\cdot c^4\cdot\beta^{(4)}(\omega)}{\left(\sqrt[3]{2\cdot\pi\cdot c^2\cdot\beta^{(2)}(\omega)}\right)^7}\cdot\frac{7}{3}\left(-\frac{3^3}{2^2}\cdot\pi^{-2}\cdot c^{-1}\cdot\frac{[\beta^{(2)}(\omega)]^4}{[\beta^{(3)}(\omega)]^3}\right)^{\frac{4}{3}} - \frac{3}{(\sqrt[3]{2\cdot\pi\cdot c^2\cdot\beta^{(2)}(\omega)})^2}\cdot\frac{5}{3}\cdot\left(-\frac{3^3}{2^2}\cdot\pi^{-2}\cdot c^{-1}\cdot\frac{[\beta^{(2)}(\omega)]^4}{[\beta^{(3)}(\omega)]^3}\right)^{\frac{2}{3}} - \frac{4}{\sqrt[3]{2\cdot\pi\cdot c^2\cdot\beta^{(2)}(\omega)}}\cdot\frac{1}{3}\cdot\left(-\frac{3^3}{2^2}\cdot\pi^{-2}\cdot c^{-1}\cdot\frac{[\beta^{(2)}(\omega)]^4}{[\beta^{(3)}(\omega)]^3}\right)^{-\frac{2}{3}}\cdot \left(\frac{1}{4}\cdot\frac{3^5}{2^3}\cdot\frac{1}{\pi^3\cdot c^2}\cdot\left\{\frac{3}{2^3}\cdot\frac{[\beta^{(2)}(\omega)]^6}{[\beta^{(3)}(\omega)]^6}\cdot\beta^{(4)}(\omega) - \frac{[\beta^{(2)}(\omega)]^5}{[\beta^{(3)}(\omega)]^4}\right\}\right)$$

$$\frac{\partial f_2(Y_2, Y_3)}{\partial Y_3}_{(Y_2^{(1)}, Y_3^{(1)})} = -\frac{4}{\sqrt[3]{2\cdot\pi\cdot c^2\cdot\beta^{(2)}(\omega)}}\cdot\left(-\frac{3^3}{2^2}\cdot\pi^{-2}\cdot c^{-1}\cdot\frac{[\beta^{(2)}(\omega)]^4}{[\beta^{(3)}(\omega)]^3}\right)^{\frac{1}{3}}$$

We define the following parameters:

$$\Xi_1 = \frac{\partial f_1(Y_2, Y_3)}{\partial Y_2}|_{(Y_2^{(1)}, Y_3^{(1)})};\ \Xi_2 = \frac{\partial f_1(Y_2, Y_3)}{\partial Y_3}|_{(Y_2^{(1)}, Y_3^{(1)})}$$

$$\Xi_3 = \frac{\partial f_2(Y_2, Y_3)}{\partial Y_2}|_{(Y_2^{(1)}, Y_3^{(1)})}$$

$$\Xi_4 = \frac{\partial f_2(Y_2, Y_3)}{\partial Y_3}_{(Y_2^{(1)}, Y_3^{(1)})}$$

$$A = \begin{pmatrix} \frac{\partial f_1}{\partial Y_2} & \frac{\partial f_1}{\partial Y_3} \\ \frac{\partial f_2}{\partial Y_2} & \frac{\partial f_2}{\partial Y_3} \end{pmatrix}_{(Y_2^{(1)}, Y_3^{(1)})} = \begin{pmatrix} \Xi_1 & \Xi_2 \\ \Xi_3 & \Xi_4 \end{pmatrix} = \begin{pmatrix} \Xi_1 & 0 \\ \Xi_3 & \Xi_4 \end{pmatrix}$$

$$A - \lambda \cdot I = \begin{pmatrix} \frac{\partial f_1}{\partial Y_2} & \frac{\partial f_1}{\partial Y_3} \\ \frac{\partial f_2}{\partial Y_2} & \frac{\partial f_2}{\partial Y_3} \end{pmatrix}_{(Y_2^{(1)}, Y_3^{(1)})} - \lambda \cdot \begin{pmatrix} 1 & 0 \\ 0 & 1 \end{pmatrix} = \begin{pmatrix} \Xi_1 - \lambda & \Xi_2 \\ \Xi_3 & \Xi_4 - \lambda \end{pmatrix}$$
$$= \begin{pmatrix} \Xi_1 - \lambda & 0 \\ \Xi_3 & \Xi_4 - \lambda \end{pmatrix}$$

$$\det(A - \lambda \cdot I) = (\Xi_1 - \lambda) \cdot (\Xi_4 - \lambda) = \Xi_1 \cdot \Xi_4 - \Xi_1 \cdot \lambda - \Xi_4 \cdot \lambda + \lambda^2$$
$$= \lambda^2 - \lambda \cdot (\Xi_1 + \Xi_4) + \Xi_1 \cdot \Xi_4$$

$$\det(A - \lambda \cdot I) = 0 \Rightarrow \lambda^2 - \lambda \cdot (\Xi_1 + \Xi_4) + \Xi_1 \cdot \Xi_4 = 0; \tau = \text{trace}(A)$$
$$= \Xi_1 + \Xi_4; \Delta = \det(A) = \Xi_1 \cdot \Xi_4$$

$$\lambda_1 = \frac{\tau + \sqrt{\tau^2 - 4 \cdot \Delta}}{2} = \frac{\Xi_1 + \Xi_4 + \sqrt{(\Xi_1 + \Xi_4)^2 - 4 \cdot \Xi_1 \cdot \Xi_4}}{2}$$

$$\lambda_2 = \frac{\tau - \sqrt{\tau^2 - 4 \cdot \Delta}}{2} = \frac{\Xi_1 + \Xi_4 - \sqrt{(\Xi_1 + \Xi_4)^2 - 4 \cdot \Xi_1 \cdot \Xi_4}}{2}$$

$$\tau = \text{trace}(A) = -\frac{4 \cdot \pi^2 \cdot c^3 \cdot \beta^{(3)}(\omega)}{\left(\sqrt[3]{2 \cdot \pi \cdot c^2 \cdot \beta^{(2)}(\omega)}\right)^5} \cdot \frac{5}{3} \cdot \left(-\frac{3^3}{2^2} \cdot \pi^{-2} \cdot c^{-1} \cdot \frac{[\beta^{(2)}(\omega)]^4}{[\beta^{(3)}(\omega)]^3}\right)^{\frac{2}{3}}$$
$$- \frac{3}{\sqrt[3]{2 \cdot \pi \cdot c^2 \cdot \beta^{(2)}(\omega)}} \cdot \frac{4}{3} \cdot \left(-\frac{3^3}{2^2} \cdot \pi^{-2} \cdot c^{-1} \cdot \frac{[\beta^{(2)}(\omega)]^4}{[\beta^{(3)}(\omega)]^3}\right)^{\frac{1}{3}}$$
$$- \frac{4}{\sqrt[3]{2 \cdot \pi \cdot c^2 \cdot \beta^{(2)}(\omega)}} \cdot \left(-\frac{3^3}{2^2} \cdot \pi^{-2} \cdot c^{-1} \cdot \frac{[\beta^{(2)}(\omega)]^4}{[\beta^{(3)}(\omega)]^3}\right)^{\frac{1}{3}}$$

$$\Delta = \det(A) = \left\{-\frac{4 \cdot \pi^2 \cdot c^3 \cdot \beta^{(3)}(\omega)}{(\sqrt[3]{2 \cdot \pi \cdot c^2 \cdot \beta^{(2)}(\omega)})^5} \cdot \frac{5}{3} \cdot \left(-\frac{3^3}{2^2} \cdot \pi^{-2} \cdot c^{-1} \cdot \frac{[\beta^{(2)}(\omega)]^4}{[\beta^{(3)}(\omega)]^3}\right)^{\frac{2}{3}}\right.$$
$$\left.- \frac{3}{\sqrt[3]{2 \cdot \pi \cdot c^2 \cdot \beta^{(2)}(\omega)}} \cdot \frac{4}{3} \cdot \left(-\frac{3^3}{2^2} \cdot \pi^{-2} \cdot c^{-1} \cdot \frac{[\beta^{(2)}(\omega)]^4}{[\beta^{(3)}(\omega)]^3}\right)^{\frac{1}{3}}\right\}$$
$$\cdot \left\{-\frac{4}{\sqrt[3]{2 \cdot \pi \cdot c^2 \cdot \beta^{(2)}(\omega)}} \cdot \left(-\frac{3^3}{2^2} \cdot \pi^{-2} \cdot c^{-1} \cdot \frac{[\beta^{(2)}(\omega)]^4}{[\beta^{(3)}(\omega)]^3}\right)^{\frac{1}{3}}\right\}$$

Table 2.3 System (derivatives $\beta^{(n)}(\omega)$ which given in terms of the index of refraction) classification of the second fixed point

Condition	Outcome
$\Delta < 0$	Eigenvalues are real and have opposite sign; hence the fixed point is a saddle point
$\Delta > 0$	Eigenvalues are either real with same sign (nodes—satisfy $\tau^2 - 4 \cdot \Delta > 0$), or complex conjugate (spiral and centers). Spirals satisfy $\tau^2 - 4 \cdot \Delta < 0$
$\Delta = 0$	At least one of the eigenvalues is zero. There is either a whole line of fixed points or a plane of fixed points if $A = 0$
$\tau^2 - 4 \cdot \Delta = 0$	Star nodes and degenerate nodes live on this parabola; it is the borderline between nodes and spirals
$\tau < 0$	Both eigenvalues have negative real parts and the fixed point is stable (related to nodes and spirals)
$\tau > 0$	Unstable spiral and nodes
$\tau = 0$	Neutrally stable centers live on the borderline, eigenvalues are purely imaginary

Classification of the second fixed point (Table 2.3).

2.3 Multipulse Operation of a Ti: Sapphire Laser Mode Semiconductor Saturable-Absorbed Mirror Pulse Energies and Gain Delayed in Time Stability Analysis

Ti: Sapphire laser can be operates with mode of multiple-pulse. It is mode locked by a semiconductor saturable absorber mirrors (SESAM's). First, saturable absorption is a property of materials where the absorption of the light decreases with increasing light intensity. Saturable absorbers are useful in laser cavities and the key parameters for a saturable absorber are its wavelength range where it absorbs and how fast it recovers which it dynamic response, saturation intensity which is the intensity level or pulse energy it saturates. We can describe it by the absorption rate (A) equation: $A = \frac{\alpha}{1+\frac{I}{I_0}}$, A—absorption rate, I—light intensity,α—linear absorption, I_0—saturation intensity. A semiconductor saturable absorber mirror (SESAM) is a mirror structure with an incorporated saturable absorber in semiconductor technology. It is used for the generation of ultrashort pulses by passive mode locking of various types of lasers. The SESAM contains a semiconductor Bragg mirror and near the surface a single quantum well absorber layer. The Bragg mirror material has larger hand gap energy and no absorption occurs in that region. It is a saturable Bragg reflector (SBRs). We can calculate the penetration of the optical field into a SESAM by matrix method which is applied to other types of dielectric mirrors. The design of the structure influences the bandwidth and the chromatic dispersion. The absorber layer is placed in an anti-node of the electric field and the results are maximum saturable absorption and the smallest possible saturation fluency. Another option is to use multiple absorber layers which give us high modulation depth. The multiple absorber layers are placed in separate anti-nodes or near one anti-node. Figure 2.5 describes the structure of typical SESAM at specific wavelength. On a GaAs substrate, a GaAs/AlGaAs Bragg mirror is grown within the top layers, there is an InGaAs quantum well.

The most important characteristics of a SESAM are modulation depth, saturation fluence, recovery time and non saturable losses. The modulation depth is the maximum nonlinear change in reflectivity and it is depends on the thickness and material of the absorber, optical wavelength, and the degree of optical field penetration into the absorber structure. The saturation fluence is the fluence of an incident short pulse which is required for causing significant absorption saturation. The recovery time is the exponential time constant of absorption recovery after a saturating pulse. In every SESAM operation there are some non saturable losses, which are unwanted. They lead to device heating while not contributing to the pulse shaping. The non saturable losses tend to be higher for SESAMs with a larger modulation depth and faster recovery [3, 14]. Basically, soliton laser is a mode-locked laser using pulse compression and solitons in a single mode fiber to force the laser itself to produce pulses of a well-defined shape and width. Pulse compression and solitons in single mode fibers result from the interaction of nonlinearity with "negative" group velocity dispersion. Operation of soliton lasers can be in multiple pulse modes. A higher order soliton is a soliton pulse the energy of which is higher than that of a fundamental soliton by a factor which is the square of an integer number. The temporal shape of such a pulse is not constant, but rather varies periodically during propagation. The soliton period is the period of this evolution. We can breakup of single pulses into multiple pulses by Kerr lens mode-locked Ti:Sapphire laser or by semiconductor saturable absorber mirrors (SESAM's). We get pulses with larger spacing compare to single-pulse width. Ti: Sapphire laser includes titanium-doped sapphire crystal which characterized by the broad gain profile and excellent physical properties. The typical mode-locking mechanism for most Ti: Sapphire lasers is soft-aperture Kerr-lens mode-locking (KLM). The Kerr effect of the gain crystal results in intensity

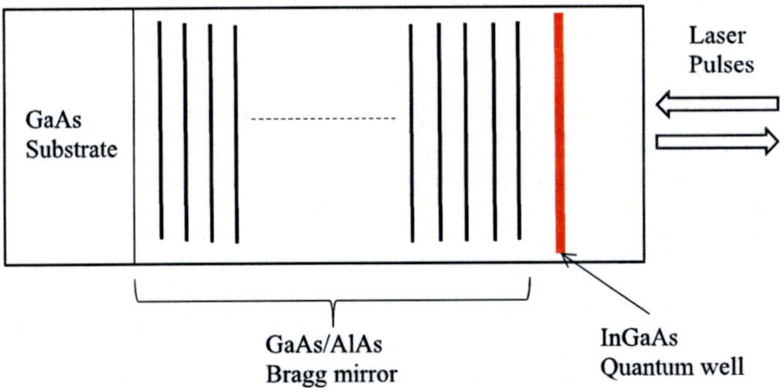

Fig. 2.5 Structure of semiconductor saturable absorber mirror

dependent gain distribution and beam overlapping of the laser and pump beams. The short pulse experiences a higher gain than the multimode continuous-wave (CW) signals, resulting in stable mode-locking. The Ti:Sapphire crystal works as not only a gain crystal to supply energy but also a Kerr medium to start mode-locking. There is a limited controllability of the cavity features, such as the pump threshold to start mode-locking which depends on the nonlinearities inside the cavity, the stable regions to maintain mode-locking whose gap depends on the asymmetry of the cavity. The laser cavity experienced multiple pulses, and production of high-repetition rate soliton pulse sources for optical fiber communication systems. Erbium-doped fiber lasers are characterized by harmonic mode locking. C_r^{4+} Yag Bulk laser incorporated the same characteristic of harmonic mode locking. Harmonic mode locking is a technique for achieving higher pulse repetition rates and is mainly applied to actively mode locked lasers. The principle is that multiple pulses circulate in the laser resonator with equal spacing. We achieve it by driving the modulator of an actively mode-locked laser with a harmonic of the resonator's round-trip frequency. There is no proving that the pulses have equal pulse energies. The pulses are not always mutually phase coherent. The nonlinear polarization rotation or a semiconductor Bragg reflector is used as the passive mode-locking mechanisms. The passive mode locking is a technique of mode locking, based on a saturable absorber inside the laser resonator. The passive mode locking can be achieved by incorporating a saturable absorber with suitable properties into the laser resonator. The incorporation of the absorber may lead to passive Q switching, to Q—switching mode locking, or some noisy mode of operation, if the absorber properties are not appropriate. The steady state is considered, where a short pulse is already circulating in the laser resonator, it is assumed that there is a single circulating pulse, and a fast absorber. We analyze the dynamic and stability of the laser by using nonlinear dynamic and possible numerical simulations. It is interested to analyze the stability of a soliton in the laser when subject to perturbations from a filter and an absorber. We can define some system parameters:

T	Time parameter.
T_R	Round trip time.
g	Saturable gain.
q	Saturable absorption.
T_g	Saturable gain recovery time.
T_a	Saturable absorption recovery time (absorber recovery time).
$P_g \cdot T_g$; E_α	Saturation energies.
g_0; q_0	Unsaturated values of gain and absorption.
l_0	Linear loss (output coupling).
$\beta = \frac{1}{\Omega^2}$	Effect of the birefringent filter.
Ω	HWHM of the filter.
κ	Self-phase-modulation parameter (calculated from the nonlinear refractive index Nonlinear refractive index n_2 of sapphire.
A_{eff}	Effective mode area inside the crystal.
l_L	Total length of the Ti:Sapphire crystal per round trip $\left(\kappa = \frac{2\times\pi\cdot n_2\cdot l_L}{\lambda_0\cdot A_{eff}}\right)$.

ψ	Electric field envelope that describes the pulses on two time scales.
Q	Total intra cavity energy $\left(Q = \int \lvert\psi\rvert^2 \cdot dt\right)$.
Q_a, Q_b	Pulse energies.
r	Ratio of round-trip time to soliton period as a measure of the strength of solitonic pulse formation $\left(r = \frac{Q^2 \cdot \kappa^2}{2\times\pi\cdot\lvert\beta_2\rvert} = \frac{Q\cdot\kappa}{\pi\cdot t_0}\right)$.
l_s	Total energy loss of soliton ($l_s = \delta + \alpha$).
δ	Filter loss (energy loss due to filter).
α	Absorber loss (energy loss due to absorber).
l_c	Loss experienced by the continuum.

The state which the growth of dispersive continuum radiation is effectively suppressed happened when $l_s < l_c$; $l_s = \delta + \alpha$. We can consider that apart from filter and absorber perturbations, the discreteness of all effects like self-phase modulation, dispersion, gain, etc., causes shedding of continuum by the soliton. The discreteness is not covered by the current model but will become important for cases when $r \approx 1$ or larger ($r > 1$). We can write the laser in-equilibrium condition of energy balance as $g_0 - l_0 - \delta - \alpha = 0$. The laser adjust it saturated gain (g) according the values of δ, α. The frame defined by the two lines separated by q_0 therefore moves with respect to the zero-gain line. If the line $g - l_0 - q_0$ crosses the zero line, because of an increase in δ, continuum growth is possible. It is true also for soliton mode-locking schemes that use ultrafast saturable absorbers such as the Kerr lens nonlinear polarization rotation. The onset of continuum growth of (δ (filter loss) for instance) is the critical condition that destabilizes the soliton pulse in the laser cavity. We define (ε) parameter which characterize the absorber, $\varepsilon = 0$ for the ideally fast absorber, $\varepsilon \neq 0$ for slow absorber (growth occurs slightly earlier, indicated by $\varepsilon \neq 0$, but physical process is the same). The stability of the pair of solitons is bounded by two limits. The first limit, toward smaller $\lvert\beta_2\rvert$, is that of renewed shedding and growth of the continuum if $(\alpha_2 + \delta_2 + \varepsilon)_{Q2} > q_0$, and similarly to the single-soliton case. The process drives the transition from double to triple pulse. The second limit, toward larger $\lvert\beta_2\rvert$, is not absolute since the pulse pair always appears to experience lower loss than a single pulse with the same total energy, because of the dominant influence of the filter loss. We derive a stability criterion from the evolution of pulse energies Q_a and Q_b respectively as well as the gain (g)—soliton perturbation theory.

Mainly we can present the system generalized complex Ginzburg-Landau equation (GCGLE) which is the master equation to describe the dynamical effects of laser by means of numerical simulations [15, 16].

$$i \cdot T_R \cdot \psi_T - \frac{\beta_2}{2} \cdot \frac{\partial^2 \psi}{\partial t^2} + \kappa \cdot \lvert\psi\rvert^2 \cdot \psi = i \cdot (g - q - l_0) \cdot \psi + i \cdot \beta \cdot \frac{\partial^2 \psi}{\partial t^2}$$

And the solution we get by neglecting the perturbations on the above equation's right hand side, it is known as the Non Linear Schrodinger Equation (NLSE), which has the first-order soliton solution.

$$\psi = \sqrt{\frac{\kappa \cdot Q^2}{4 \cdot |\beta_2|}} \cdot \operatorname{sech}\left(\frac{\kappa \cdot Q}{2 \times |\beta_2|} \cdot t\right) \cdot \exp\left(i \cdot \frac{Q^2 \cdot \kappa^2}{8 \cdot |\beta_2| \cdot T_R} \cdot T\right)$$

We can consider our system differential equations when the perturbations on the right hand side are not neglected.

$$T_R \cdot \frac{dQ_a}{dT} = 2 \cdot [g \cdot (Q_a + Q_b) - l_0 - \delta_a - \alpha_a] \cdot Q_a$$
$$T_R \cdot \frac{dQ_b}{dT} = 2 \cdot [g \cdot (Q_a + Q_b) - l_0 - \delta_b - \alpha_b] \cdot Q_b$$
$$\frac{dg}{dT} = -\frac{(g - g_0)}{T_g} - \frac{(Q_a + Q_b) \cdot g}{P_g \cdot T_g \cdot T_R}$$

δ_a, δ_b Pulse filter losses (a, b).
α_a, α_b Pulse absorber losses (a, b).

The filter loss and absorber loss are given by the following expression: $l_s = \delta + \alpha$.

$$l_s = \frac{\beta}{3 \cdot t_0^2} + \frac{1}{2 \cdot t_0} \cdot \int_{-\infty}^{+\infty} \operatorname{sech}^2\left(\frac{t}{t_0}\right) \cdot q(t) \cdot dt; \; t_0 = \frac{2 \cdot |\beta_2|}{\kappa \cdot Q_{a,b}} = \frac{\tau_{FWHM}}{1.763}$$

We analyze the dynamic and stability of our system by using nonlinear dynamic. Our system variables are pulse energies: Q_a, Q_b.

Our system (Multi pulse operation of Ti: Sapphire laser mode locked by an ion-implanted semiconductor saturable-absorber mirror) differential equations when the perturbations on the right hand side are not neglected ($Q_a = Q_a(T); Q_b = Q_b(T); g = g(T)$).

$$\frac{dQ_a}{dT} = 2 \cdot \frac{1}{T_R} \cdot [g \cdot (Q_a + Q_b) - l_0 - \delta_a - \alpha_a] \cdot Q_a$$

$$\frac{dQ_b}{dT} = 2 \cdot \frac{1}{T_R} \cdot [g \cdot (Q_a + Q_b) - l_0 - \delta_b - \alpha_b] \cdot Q_b; \; \frac{dg}{dT} = -\frac{(g - g_0)}{T_g} - \frac{(Q_a + Q_b) \cdot g}{P_g \cdot T_g \cdot T_R}$$

At fixed points: $\frac{dQ_a}{dT} = 0$; $\frac{dQ_b}{dT} = 0$; $\frac{dg}{dT} = 0$.

$$2 \cdot \frac{1}{T_R} \cdot [g^* \cdot (Q_a^* + Q_b^*) - l_0 - \delta_a - \alpha_a] \cdot Q_a^* = 0$$

$$2 \cdot \frac{1}{T_R} \cdot [g^* \cdot (Q_a^* + Q_b^*) - l_0 - \delta_b - \alpha_b] \cdot Q_b^* = 0$$

$$-\frac{(g^* g_0)}{T_g}-\frac{(Q_a^*+Q_b^*)\cdot g^*}{P_g\cdot T_g\cdot T_R}=0$$

$$\frac{dQ_a}{dT}=0\Rightarrow 2\cdot\frac{1}{T_R}\cdot[g^*\cdot(Q_a^*+Q_b^*)-l_0-\delta_a-\alpha_a]\cdot Q_a^*=0$$

Case 1: $Q_a^*=0$; Case 2: $g^*\cdot(Q_a^*+Q_b^*)-l_0-\delta_a-\alpha_a=0\Rightarrow Q_a^*=\frac{1}{g^*}\cdot(l_0+\delta_a+\alpha_a)-Q_b^*$

Case 1: $Q_a^*=0$; $\frac{dQ_b}{dT}=0\Rightarrow 2\cdot\frac{1}{T_R}\cdot[g^*\cdot Q_b^*-l_0-\delta_b-\alpha_b]\cdot Q_b^*=0$

Case 1.1: $Q_a^*=0$; $Q_b^*=0$.; Case 1.2: $Q_a^*=0$; $g^*\cdot Q_b^*-l_0-\delta_b-\alpha_b=0$ $\Rightarrow Q_b^*=\frac{1}{g^*}(l_0+\delta_b+\alpha_b)$

Case 1.1: $Q_a^*=0$; $Q_b^*=0$; $\frac{dg}{dT}=0\Rightarrow-\frac{(g^*-g_0)}{T_g}=0\Rightarrow g^*=g_0$

First fixed point: $E^{(0)}(Q_a^{(0)},Q_b^{(0)},g^{(0)})=(0,0,g_0)$.

Case 1.2: $Q_a^*=0$; $Q_b^*=\frac{1}{g^*}\cdot(l_0+\delta_b+\alpha_b)$

$$\frac{dg}{dT}=0\Rightarrow-\frac{(g^*-g_0)}{T_g}-\frac{\frac{1}{g^*}\cdot(l_0+\delta_b+\alpha_b)\cdot g^*}{P_g\cdot T_g\cdot T_R}=0$$

$$-(g^*-g_0)-\frac{(l_0+\delta_b+\alpha_b)}{P_g\cdot T_R}=0$$

$$-(g^*-g_0)-\frac{(l_0+\delta_b+\alpha_b)}{P_g\cdot T_R}=0\Rightarrow g^*-g_0$$

$$=-\frac{(l_0+\delta_b+\alpha_b)}{P_g\cdot T_R};\ g^*=g_0-\frac{(l_0+\delta_b+\alpha_b)}{P_g\cdot T_R}$$

$$Q_b^*=\frac{1}{g^*}\cdot(l_0+\delta_b+\alpha_b)=\frac{1}{\left[g_0-\frac{(l_0+\delta_b+\alpha_b)}{P_g\cdot T_R}\right]}\cdot(l_0+\delta_b+\alpha_b)$$

Second fixed point:

$$E^{(1)}(Q_a^{(1)},Q_b^{(1)},g^{(1)})=\left(0,\frac{1}{\left[g_0-\frac{(l_0+\delta_b+\alpha_b)}{P_g\cdot T_R}\right]}\cdot(l_0+\delta_b+\alpha_b),g_0-\frac{(l_0+\delta_b+\alpha_b)}{P_g\cdot T_R}\right)$$

Case 2: $Q_a^*=\frac{1}{g^*}\cdot(l_0+\delta_a+\alpha_a)-Q_b^*$

$$\frac{dQ_b}{dT}=0\Rightarrow 2\cdot\frac{1}{T_R}\cdot[g^*\cdot(Q_a^*+Q_b^*)-l_0-\delta_b-\alpha_b]\cdot Q_b^*=0$$

$$\frac{dQ_b}{dT}=0\Rightarrow 2\cdot\frac{1}{T_R}\cdot[g^*\cdot(\frac{1}{g^*}\cdot(l_0+\delta_a+\alpha_a)-Q_b^*+Q_b^*)-l_0-\delta_b-\alpha_b]\cdot Q_b^*=0$$

$$2\cdot\frac{1}{T_R}\cdot[\delta_a+\alpha_a-\delta_b-\alpha_b]\cdot Q_b^*=0;\ 2\cdot\frac{1}{T_R}\cdot[\delta_a+\alpha_a-\delta_b-\alpha_b]\neq 0\Rightarrow Q_b^*=0$$

$$\frac{dg}{dT} = 0 \Rightarrow -\frac{(g^* - g_0)}{T_g} - \frac{(Q_a^* + Q_b^*) \cdot g^*}{P_g \cdot T_g \cdot T_R} = 0$$

$$-\frac{(g^* - g_0)}{T_g} - \frac{\left(\frac{1}{g^*} \cdot (l_0 + \delta_a + \alpha_a) - Q_b^* + Q_b^*\right) \cdot g^*}{P_g \cdot T_g \cdot T_R}$$

$$= 0 \Rightarrow -(g^* - g_0) - \frac{(l_0 + \delta_a + \alpha_a)}{P_g \cdot T_R} = 0$$

$$g^* - g_0 = -\frac{(l_0 + \delta_a + \alpha_a)}{P_g \cdot T_R} \Rightarrow g^* = g_0 - \frac{(l_0 + \delta_a + \alpha_a)}{P_g \cdot T_R}$$

$$Q_a^* = \frac{1}{g^*} \cdot (l_0 + \delta_a + \alpha_a) - Q_b^* = \frac{1}{\left[g_0 - \frac{(l_0+\delta_a+\alpha_a)}{P_g \cdot T_R}\right]} \cdot (l_0 + \delta_a + \alpha_a)$$

Third fixed point:

$$E^{(2)}(Q_a^{(2)}, Q_b^{(2)}, g^{(2)}) = \left(\frac{1}{\left[g_0 - \frac{(l_0+\delta_a+\alpha_a)}{P_g \cdot T_R}\right]} \cdot (l_0 + \delta_a + \alpha_a), 0, g_0 - \frac{(l_0 + \delta_a + \alpha_a)}{P_g \cdot T_R}\right)$$

We can summary our system fixed points:

$$E^{(0)}(Q_a^{(0)}, Q_b^{(0)}, g^{(0)}) = (0, 0, g_0)$$

$$E^{(1)}(Q_a^{(1)}, Q_b^{(1)}, g^{(1)}) = \left(0, \frac{1}{\left[g_0 - \frac{(l_0+\delta_b+\alpha_b)}{P_g \cdot T_R}\right]} \cdot (l_0 + \delta_b + \alpha_b), g_0 - \frac{(l_0 + \delta_b + \alpha_b)}{P_g \cdot T_R}\right)$$

$$E^{(2)}(Q_a^{(2)}, Q_b^{(2)}, g^{(2)}) = \left(\frac{1}{\left[g_0 - \frac{(l_0+\delta_a+\alpha_a)}{P_g \cdot T_R}\right]} \cdot (l_0 + \delta_a + \alpha_a), 0, g_0 - \frac{(l_0 + \delta_a + \alpha_a)}{P_g \cdot T_R}\right)$$

The system is not linear and we need to implement linearization technique and then to approximate the phase portrait near a fixed points by that of a corresponding linear system.

$$\frac{dQ_a}{dT} = 2 \cdot \frac{1}{T_R} \cdot [g \cdot (Q_a + Q_b) - l_0 - \delta_a - \alpha_a] \cdot Q_a$$

$$\frac{dQ_b}{dT} = 2 \cdot \frac{1}{T_R} \cdot [g \cdot (Q_a + Q_b) - l_0 - \delta_b - \alpha_b] \cdot Q_b; \ \frac{dg}{dT}$$

$$= -\frac{(g - g_0)}{T_g} - \frac{(Q_a + Q_b) \cdot g}{P_g \cdot T_g \cdot T_R}$$

We define three functions: $f_1(Q_a, Q_b, g) = 2 \cdot \frac{1}{T_R} \cdot [g \cdot (Q_a + Q_b) - l_0 - \delta_a - \alpha_a] \cdot Q_a$

$$f_2(Q_a, Q_b, g) = 2 \cdot \frac{1}{T_R} \cdot [g \cdot (Q_a + Q_b) - l_0 - \delta_b - \alpha_b] \cdot Q_b$$

$$f_3(Q_a, Q_b, g) = -\frac{(g - g_0)}{T_g} - \frac{(Q_a + Q_b) \cdot g}{P_g \cdot T_g \cdot T_R}$$

$$\frac{dQ_a}{dT} = f_1(Q_a, Q_b, g); \frac{dQ_b}{dT} = f_2(Q_a, Q_b, g); \frac{dg}{dT} = f_3(Q_a, Q_b, g)$$

And suppose that $(Q_a^{(*)}, Q_b^{(*)}, g^{(*)})$ is specific system fixed point.

$$f_1(Q_a^{(*)}, Q_b^{(*)}, g^{(*)}) = 0; f_2(Q_a^{(*)}, Q_b^{(*)}, g^{(*)}) = 0; f_3(Q_a^{(*)}, Q_b^{(*)}, g^{(*)}) = 0$$

Let $u = Q_a - Q_a^*; v = Q_b - Q_b^*; w = g - g^*$.

Denote the components of a small disturbance from the fixed point. To see whether the disturbance grows or decays, we need to derive differential equations for u, v, w.

$$\frac{du}{dT} = \frac{dQ_a}{dT}; \frac{dv}{dT} = \frac{dQ_b}{dT}; \frac{dw}{dT} = \frac{dg}{dT}; Q_a^*, Q_b^*, g^* - \text{const}$$

And by substitution:

$$\frac{du}{dT} = \frac{dQ_a}{dT} = f_1(Q_a^* + u, Q_b^* + v, g^* + w)$$
$$\frac{dv}{dT} = \frac{dQ_b}{dT} = f_2(Q_a^* + u, Q_b^* + v, g^* + w)$$

$$\frac{dw}{dT} = \frac{dg}{dT} = f_3(Q_a^* + u, Q_b^* + v, g^* + w)$$

And by Taylor series expansion:

$$\frac{du}{dT} = \frac{dQ_a}{dT} = f_1(Q_a^*, Q_b^*, g^*) + u \cdot \frac{\partial f_1}{\partial Q_a} + v \cdot \frac{\partial f_1}{\partial Q_b} + w \cdot \frac{\partial f_1}{\partial g} + O(u^2, v^2, w^2, u \cdot v \cdot w)$$

$$\frac{dv}{dT} = \frac{dQ_b}{dT} = f_2(Q_a^*, Q_b^*, g^*) + u \cdot \frac{\partial f_2}{\partial Q_a} + v \cdot \frac{\partial f_2}{\partial Q_b} + w \cdot \frac{\partial f_2}{\partial g} + O(u^2, v^2, w^2, u \cdot v \cdot w)$$

$$\frac{dw}{dT} = \frac{dg}{dT} = f_3(Q_a^*, Q_b^*, g^*) + u \cdot \frac{\partial f_3}{\partial Q_a} + v \cdot \frac{\partial f_3}{\partial Q_b} + w \cdot \frac{\partial f_3}{\partial g} + O(u^2, v^2, w^2, u \cdot v \cdot w)$$

Since $f_1(Q_a^*, Q_b^*, g^*) = 0; f_2(Q_a^*, Q_b^*, g^*). = 0; f_3(Q_a^*, Q_b^*, g^*) = 0.$

$$\frac{du}{dT} = \frac{dQ_a}{dT} = u \cdot \frac{\partial f_1}{\partial Q_a} + v \cdot \frac{\partial f_1}{\partial Q_b} + w \cdot \frac{\partial f_1}{\partial g} + O(u^2, v^2, w^2, u \cdot v \cdot w)$$

$$\frac{dv}{dT} = \frac{dQ_b}{dT} = u \cdot \frac{\partial f_2}{\partial Q_a} + v \cdot \frac{\partial f_2}{\partial Q_b} + w \cdot \frac{\partial f_2}{\partial g} + O(u^2, v^2, w^2, u \cdot v \cdot w)$$

$$\frac{dw}{dT} = \frac{dg}{dT} = u \cdot \frac{\partial f_3}{\partial Q_a} + v \cdot \frac{\partial f_3}{\partial Q_b} + w \cdot \frac{\partial f_3}{\partial g} + O(u^2, v^2, w^2, u \cdot v \cdot w)$$

To simplify the notation, we have written $\frac{\partial f_1}{\partial Q_a}, \frac{\partial f_1}{\partial Q_b}, \frac{\partial f_1}{\partial g}, \frac{\partial f_2}{\partial Q_a} \cdot \frac{\partial f_2}{\partial Q_b}, \frac{\partial f_2}{\partial g}$. And $\frac{\partial f_3}{\partial Q_a}, \frac{\partial f_3}{\partial Q_b}, \frac{\partial f_3}{\partial g}$. These partial derivatives are to be evaluated at the fixed point (Q_a^*, Q_b^*, g^*); thus they are numbers, no functions. The shorthand $O(u^2, v^2, w^2, u \cdot v \cdot w)$ denotes quadratic terms in u, v, w. Since u, v, w are small, these quadratic terms are extremely small [5, 6].

$$\frac{du}{dT} = u \cdot \frac{\partial f_1}{\partial Q_a} + v \cdot \frac{\partial f_1}{\partial Q_b} + w \cdot \frac{\partial f_1}{\partial g} + O(u^2, v^2, w^2, u \cdot v \cdot w)$$

$$\frac{dv}{dT} = u \cdot \frac{\partial f_2}{\partial Q_a} + v \cdot \frac{\partial f_2}{\partial Q_b} + w \cdot \frac{\partial f_2}{\partial g} + O(u^2, v^2, w^2, u \cdot v \cdot w)$$

$$\frac{dw}{dT} = u \cdot \frac{\partial f_3}{\partial Q_a} + v \cdot \frac{\partial f_3}{\partial Q_b} + w \cdot \frac{\partial f_3}{\partial g} + O(u^2, v^2, w^2, u \cdot v \cdot w)$$

$$\begin{pmatrix} \frac{du}{dT} \\ \frac{dv}{dT} \\ \frac{dw}{dT} \end{pmatrix} = \begin{pmatrix} \frac{\partial f_1}{\partial Q_a} & \frac{\partial f_1}{\partial Q_b} & \frac{\partial f_1}{\partial g} \\ \frac{\partial f_2}{\partial Q_a} & \frac{\partial f_2}{\partial Q_b} & \frac{\partial f_2}{\partial g} \\ \frac{\partial f_3}{\partial Q_a} & \frac{\partial f_3}{\partial Q_b} & \frac{\partial f_3}{\partial g} \end{pmatrix} \cdot \begin{pmatrix} u \\ v \\ w \end{pmatrix} + \text{quadratic terms}$$

$$A = \begin{pmatrix} \frac{\partial f_1}{\partial Q_a} & \frac{\partial f_1}{\partial Q_b} & \frac{\partial f_1}{\partial g} \\ \frac{\partial f_2}{\partial Q_a} & \frac{\partial f_2}{\partial Q_b} & \frac{\partial f_2}{\partial g} \\ \frac{\partial f_3}{\partial Q_a} & \frac{\partial f_3}{\partial Q_b} & \frac{\partial f_3}{\partial g} \end{pmatrix}_{(Q_a^*, Q_b^*, g^*)}$$

It is called the Jacobian matrix A at the fixed points (Q_a^*, Q_b^*, g^*). Since the quadratic terms are tiny, it is tempting to neglect them altogether. We get our linearized system whose dynamics can be analyzed by the methods of classification of linear systems.

$$\begin{pmatrix} \frac{du}{dT} \\ \frac{dv}{dT} \\ \frac{dw}{dT} \end{pmatrix} = \begin{pmatrix} \frac{\partial f_1}{\partial Q_a} & \frac{\partial f_1}{\partial Q_b} & \frac{\partial f_1}{\partial g} \\ \frac{\partial f_2}{\partial Q_a} & \frac{\partial f_2}{\partial Q_b} & \frac{\partial f_2}{\partial g} \\ \frac{\partial f_3}{\partial Q_a} & \frac{\partial f_3}{\partial Q_b} & \frac{\partial f_3}{\partial g} \end{pmatrix} \cdot \begin{pmatrix} u \\ v \\ w \end{pmatrix}$$

$$f_1(Q_a, Q_b, g) = 2 \cdot \frac{1}{T_R} \cdot [g \cdot (Q_a^2 + Q_b \cdot Q_a) - (l_0 + \delta_a + \alpha_a) \cdot Q_a]$$

$$\frac{\partial f_1}{\partial Q_a} = 2 \cdot \frac{1}{T_R} \cdot [g \cdot (2 \cdot Q_a + Q_b) - (l_0 + \delta_a + \alpha_a)]$$
$$\frac{\partial f_1}{\partial Q_b} = 2 \cdot \frac{1}{T_R} \cdot g \cdot Q_a; \frac{\partial f_1}{\partial g} = 2 \cdot \frac{1}{T_R} \cdot (Q_a^2 + Q_b \cdot Q_a)$$

$$f_2(Q_a, Q_b, g) = 2 \cdot \frac{1}{T_R} \cdot [g \cdot (Q_a \cdot Q_b + Q_b^2) - (l_0 + \delta_b + \alpha_b) \cdot Q_b]$$

$$\frac{\partial f_2}{\partial Q_a} = 2 \cdot \frac{1}{T_R} \cdot g \cdot Q_b; \frac{\partial f_2}{\partial Q_b} = 2 \cdot \frac{1}{T_R} \cdot [g \cdot (Q_a + 2 \cdot Q_b)$$
$$- (l_0 + \delta_b + \alpha_b)]; \frac{\partial f_2}{\partial g} = 2 \cdot \frac{1}{T_R} \cdot (Q_a \cdot Q_b + Q_b^2)$$

$$f_3(Q_a, Q_b, g) = -\frac{(g - g_0)}{T_g} - \frac{(Q_a + Q_b) \cdot g}{P_g \cdot T_g \cdot T_R}$$
$$\frac{\partial f_3}{\partial Q_a} = -\frac{g}{P_g \cdot T_g \cdot T_R}; \frac{\partial f_3}{\partial Q_b} = -\frac{g}{P_g \cdot T_g \cdot T_R}$$
$$\frac{\partial f_3}{\partial g} = -\frac{1}{T_g} - \frac{(Q_a + Q_b)}{P_g \cdot T_g \cdot T_R} = -\frac{1}{T_g} \cdot \left[1 + \frac{(Q_a + Q_b)}{P_g \cdot T_R}\right]$$

We can summary our Jacobian matrix A elements expressions:

$$\frac{\partial f_1}{\partial Q_a} = 2 \cdot \frac{1}{T_R} \cdot [g \cdot (2 \cdot Q_a + Q_b) - (l_0 + \delta_a + \alpha_a)]$$
$$\frac{\partial f_1}{\partial Q_b} = 2 \cdot \frac{1}{T_R} \cdot g \cdot Q_a; \frac{\partial f_1}{\partial g} = 2 \cdot \frac{1}{T_R} \cdot (Q_a^2 + Q_b \cdot Q_a)$$
$$\frac{\partial f_2}{\partial Q_a} = 2 \cdot \frac{1}{T_R} \cdot g \cdot Q_b; \frac{\partial f_2}{\partial Q_b} = 2 \cdot \frac{1}{T_R} \cdot [g \cdot (Q_a + 2 \cdot Q_b) - (l_0 + \delta_b + \alpha_b)]$$
$$\frac{\partial f_2}{\partial g} = 2 \cdot \frac{1}{T_R} \cdot (Q_a \cdot Q_b + Q_b^2)$$
$$\frac{\partial f_3}{\partial Q_a} = -\frac{g}{P_g \cdot T_g \cdot T_R}; \frac{\partial f_3}{\partial Q_b} = -\frac{g}{P_g \cdot T_g \cdot T_R}$$
$$\frac{\partial f_3}{\partial g} = -\frac{1}{T_g} \cdot \left[1 + \frac{(Q_a + Q_b)}{P_g \cdot T_R}\right]$$

Jacobian matrix A at the fixed points (Q_a^*, Q_b^*, g^*):
First fixed point: $E^{(0)}(Q_a^{(0)}, Q_b^{(0)}, g^{(0)}) = (0, 0, g_0)$.

$$\frac{\partial f_1}{\partial Q_a}|_{E^{(0)}(Q_a^{(0)}, Q_b^{(0)}, g^{(0)}) = (0,0,g_0)} = -2 \cdot \frac{1}{T_R} \cdot (l_0 + \delta_a + \alpha_a)$$

$$\frac{\partial f_1}{\partial Q_b}|_{E^{(0)}(Q_a^{(0)},Q_b^{(0)},g^{(0)})=(0,0,g_0)} = 0$$
$$\frac{\partial f_1}{\partial g}|_{E^{(0)}(Q_a^{(0)},Q_b^{(0)},g^{(0)})=(0,0,g_0)} = 0$$
$$\frac{\partial f_2}{\partial Q_a}|_{E^{(0)}(Q_a^{(0)},Q_b^{(0)},g^{(0)})=(0,0,g_0)} = 0$$

$$\frac{\partial f_2}{\partial Q_b}|_{E^{(0)}(Q_a^{(0)},Q_b^{(0)},g^{(0)})=(0,0,g_0)} = -2\cdot\frac{1}{T_R}\cdot(l_0+\delta_b+\alpha_b)$$
$$\frac{\partial f_2}{\partial g}_{E^{(0)}(Q_a^{(0)},Q_b^{(0)},g^{(0)})=(0,0,g_0)} = 0$$
$$\frac{\partial f_3}{\partial Q_a}|_{E^{(0)}(Q_a^{(0)},Q_b^{(0)},g^{(0)})=(0,0,g_0)} = -\frac{g_0}{P_g\cdot T_g\cdot T_R}$$
$$\frac{\partial f_3}{\partial Q_b}|_{E^{(0)}(Q_a^{(0)},Q_b^{(0)},g^{(0)})=(0,0,g_0)} = -\frac{g_0}{P_g\cdot T_g\cdot T_R}$$

$$\frac{\partial f_3}{\partial g}|_{E^{(0)}(Q_a^{(0)},Q_b^{(0)},g^{(0)})=(0,0,g_0)} = -\frac{1}{T_g}$$

$$A = \begin{pmatrix} \frac{\partial f_1}{\partial Q_a} & \frac{\partial f_1}{\partial Q_b} & \frac{\partial f_1}{\partial g} \\ \frac{\partial f_2}{\partial Q_a} & \frac{\partial f_2}{\partial Q_b} & \frac{\partial f_2}{\partial g} \\ \frac{\partial f_3}{\partial Q_a} & \frac{\partial f_3}{\partial Q_b} & \frac{\partial f_3}{\partial g} \end{pmatrix}_{E^{(0)}(Q_a^{(0)},Q_b^{(0)},g^{(0)})=(0,0,g_0)}$$
$$= \begin{pmatrix} -2\cdot\frac{1}{T_R}\cdot(l_0+\delta_a+\alpha_a) & 0 & 0 \\ 0 & -2\cdot\frac{1}{T_R}\cdot(l_0+\delta_b+\alpha_b) & 0 \\ -\frac{g_0}{P_g\cdot T_g\cdot T_R} & -\frac{g_0}{P_g\cdot T_g\cdot T_R} & -\frac{1}{T_g} \end{pmatrix}$$

Remark As long as the fixed point for the linearized system in not one of the borderline cases, we can neglect the quadratic terms. Then the linearized system gives a qualitatively correct picture of the phase portrait near specific fixed point $(Q_a^{(*)}, Q_b^{(*)}, g^{(*)})$ [5, 6].

$$A-\lambda\cdot I = \begin{pmatrix} -2\cdot\frac{1}{T_R}\cdot(l_0+\delta_a+\alpha_a)-\lambda & 0 & 0 \\ 0 & -2\cdot\frac{1}{T_R}\cdot(l_0+\delta_b+\alpha_b)-\lambda & 0 \\ -\frac{g_0}{P_g\cdot T_g\cdot T_R} & -\frac{g_0}{P_g\cdot T_g\cdot T_R} & -\frac{1}{T_g}-\lambda \end{pmatrix}$$

$$\det(A-\lambda\cdot I) = -[2\cdot\frac{1}{T_R}\cdot(l_0+\delta_a+\alpha_a)+\lambda]\cdot$$
$$\det\begin{pmatrix} -\left[2\cdot\frac{1}{T_R}\cdot(l_0+\delta_b+\alpha_b)+\lambda\right] & 0 \\ -\frac{g_0}{P_g\cdot T_g\cdot T_R} & -\left[\frac{1}{T_g}+\lambda\right] \end{pmatrix}$$

$$\det(A-\lambda\cdot I) = -\left[2\cdot\frac{1}{T_R}\cdot(l_0+\delta_a+\alpha_a)+\lambda\right]\cdot$$

$$\left[2 \cdot \frac{1}{T_R} \cdot (l_0 + \delta_b + \alpha_b) + \lambda\right] \cdot \left[\frac{1}{T_g} + \lambda\right]$$

$$\det(A - \lambda \cdot I) = 0 \Rightarrow -\left[2 \cdot \frac{1}{T_R} \cdot (l_0 + \delta_a + \alpha_a) + \lambda\right] \cdot$$
$$\left[2 \cdot \frac{1}{T_R} \cdot (l_0 + \delta_b + \alpha_b) + \lambda\right] \cdot \left[\frac{1}{T_g} + \lambda\right] = 0$$

We get three eigenvalues: $\lambda_1 = -2 \cdot \frac{1}{T_R} \cdot (l_0 + \delta_a + \alpha_a)$; $\lambda_2 = -2 \cdot \frac{1}{T_R} \cdot (l_0 + \delta_b + \alpha_b)$; $\lambda_3 = -\frac{1}{T_g}$.

Our system's typical modeling parameters are as follow:

$T_R = 12.5$ns (Round trip time), $l_0 = 0.08$ (Total linear loss per round trip), $T_g = 2.5$us (Upper state lifetime of Ti: Sapphire), $4.5 \times 10^{-3} < \delta_a + \alpha_a < 5.2 \times 10^{-3}$ and $4.5 \times 10^{-3} < \delta_b + \alpha_b < 5.2 \times 10^{-3}$ for soliton operation time 0.6–2 ps.

Three eigenvalues are real and negative numbers and the first fixed point is stable.

Second fixed point:

$$E^{(1)}(Q_a^{(1)}, Q_b^{(1)}, g^{(1)}) = \left(0, \left[\frac{(l_0 + \delta_b + \alpha_b)}{\left[g_0 - \frac{(l_0+\delta_b+\alpha_b)}{P_g \cdot T_R}\right]}\right], \left[g_0 - \frac{(l_0 + \delta_b + \alpha_b)}{P_g \cdot T_R}\right]\right)$$

$$\frac{\partial f_1}{\partial Q_a}|_{E^{(1)}(Q_a^{(1)}, Q_b^{(1)}, g^{(1)})} = 2 \cdot \frac{1}{T_R} \cdot \left[\left[g_0 - \frac{(l_0 + \delta_b + \alpha_b)}{P_g \cdot T_R}\right] \cdot \left[\frac{(l_0 + \delta_b + \alpha_b)}{\left[g_0 - \frac{(l_0+\delta_b+\alpha_b)}{P_g \cdot T_R}\right]}\right]\right.$$
$$\left. -(l_0 + \delta_a + \alpha_a)\right]$$

$$\frac{\partial f_1}{\partial Q_b}|_{E^{(1)}(Q_a^{(1)}, Q_b^{(1)}, g^{(1)})} = 0; \frac{\partial f_1}{\partial g}|_{E^{(1)}(Q_a^{(1)}, Q_b^{(1)}, g^{(1)})} = 0$$

$$\frac{\partial f_2}{\partial Q_a}|_{E^{(1)}(Q_a^{(1)}, Q_b^{(1)}, g^{(1)})} = 2 \cdot \frac{1}{T_R} \cdot \left[g_0 - \frac{(l_0 + \delta_b + \alpha_b)}{P_g \cdot T_R}\right] \cdot \left[\frac{(l_0 + \delta_b + \alpha_b)}{\left[g_0 - \frac{(l_0+\delta_b+\alpha_b)}{P_g \cdot T_R}\right]}\right]$$

$$\frac{\partial f_2}{\partial Q_b}|_{E^{(1)}(Q_a^{(1)}, Q_b^{(1)}, g^{(1)})} = 2 \cdot \frac{1}{T_R} \cdot \left[2 \cdot \left[g_0 - \frac{(l_0 + \delta_b + \alpha_b)}{P_g \cdot T_R}\right] \cdot \left[\frac{(l_0 + \delta_b + \alpha_b)}{\left[g_0 - \frac{(l_0+\delta_b+\alpha_b)}{P_g \cdot T_R}\right]}\right]\right.$$
$$\left. -(l_0 + \delta_b + \alpha_b)\right]$$

$$\frac{\partial f_2}{\partial g}|_{E^{(1)}(Q_a^{(1)}, Q_b^{(1)}, g^{(1)})} = 2 \cdot \frac{1}{T_R} \cdot \left[\frac{(l_0 + \delta_b + \alpha_b)}{\left[g_0 - \frac{(l_0+\delta_b+\alpha_b)}{P_g \cdot T_R}\right]}\right]^2$$

$$\frac{\partial f_3}{\partial Q_a}|_{E^{(1)}(Q_a^{(1)},Q_b^{(1)},g^{(1)})} = -\left\{\frac{\left[g_0 - \frac{(l_0+\delta_b+\alpha_b)}{P_g\cdot T_R}\right]}{P_g\cdot T_g\cdot T_R}\right\}$$

$$\frac{\partial f_3}{\partial Q_b}|_{E^{(1)}(Q_a^{(1)},Q_b^{(1)},g^{(1)})} = -\frac{\left[g_0 - \frac{(l_0+\delta_b+\alpha_b)}{P_g\cdot T_R}\right]}{P_g\cdot T_g\cdot T_R}$$

$$\frac{\partial f_3}{\partial g}|_{E^{(1)}(Q_a^{(1)},Q_b^{(1)},g^{(1)})} = -\frac{1}{T_g}\cdot\left[1+\frac{\left(\frac{(l_0+\delta_b+\alpha_b)}{\left[g_0-\frac{(l_0+\delta_b+\alpha_b)}{P_g\cdot T_R}\right]}\right)}{P_g\cdot T_R}\right]$$

We use the identification to simplify our above expressions.

$$g_0 - \frac{(l_0+\delta_b+\alpha_b)}{P_g\cdot T_R} = \frac{g_0\cdot P_g\cdot T_R-(l_0+\delta_b+\alpha_b)}{P_g\cdot T_R}$$

$$\frac{1}{g_0 - \frac{(l_0+\delta_b+\alpha_b)}{P_g\cdot T_R}} = \frac{P_g\cdot T_R}{g_0\cdot P_g\cdot T_R-(l_0+\delta_b+\alpha_b)}$$

We get the following Jacobian matrix A's elements at the second fixed point:

$$\frac{\partial f_1}{\partial Q_a}|_{E^{(1)}(Q_a^{(1)},Q_b^{(1)},g^{(1)})} = 2\cdot\frac{1}{T_R}\cdot(\delta_b+\alpha_b-\delta_a-\alpha_a)$$

$$\frac{\partial f_1}{\partial Q_b}|_{E^{(1)}(Q_a^{(1)},Q_b^{(1)},g^{(1)})} = 0;\ \frac{\partial f_1}{\partial g}|_{E^{(1)}(Q_a^{(1)},Q_b^{(1)},g^{(1)})} = 0$$

$$\frac{\partial f_2}{\partial Q_a}|_{E^{(1)}(Q_a^{(1)},Q_b^{(1)},g^{(1)})} = 2\cdot\frac{1}{T_R}\cdot(l_0+\delta_b+\alpha_b)$$

$$\frac{\partial f_2}{\partial Q_b}|_{E^{(1)}(Q_a^{(1)},Q_b^{(1)},g^{(1)})} = 2\cdot\frac{1}{T_R}\cdot(l_0+\delta_b+\alpha_b)$$

$$\frac{\partial f_2}{\partial g}|_{E^{(1)}(Q_a^{(1)},Q_b^{(1)},g^{(1)})} = 2\cdot\frac{1}{T_R}\cdot\left[\frac{(l_0+\delta_b+\alpha_b)\cdot P_g\cdot T_R}{g_0\cdot P_g\cdot T_R-(l_0+\delta_b+\alpha_b)}\right]^2$$

$$\frac{\partial f_3}{\partial Q_a}|_{E^{(1)}(Q_a^{(1)},Q_b^{(1)},g^{(1)})} = -\left\{\frac{g_0\cdot P_g\cdot T_R-(l_0+\delta_b+\alpha_b)}{T_g\cdot P_g^2\cdot T_R^2}\right\}$$

$$\frac{\partial f_3}{\partial Q_b}|_{E^{(1)}(Q_a^{(1)},Q_b^{(1)},g^{(1)})} = -\frac{g_0\cdot P_g\cdot T_R-(l_0+\delta_b+\alpha_b)}{T_g\cdot T_R^2\cdot P_g^2}$$

$$\frac{\partial f_3}{\partial g}|_{E^{(1)}(Q_a^{(1)},Q_b^{(1)},g^{(1)})} = -\frac{1}{T_g}\cdot\left[1+\frac{(l_0+\delta_b+\alpha_b)}{g_0\cdot P_g\cdot T_R-(l_0+\delta_b+\alpha_b)}\right]$$

$$A = \begin{pmatrix} \frac{\partial f_1}{\partial Q_a} & \frac{\partial f_1}{\partial Q_b} & \frac{\partial f_1}{\partial g} \\ \frac{\partial f_2}{\partial Q_a} & \frac{\partial f_2}{\partial Q_b} & \frac{\partial f_2}{\partial g} \\ \frac{\partial f_3}{\partial Q_a} & \frac{\partial f_3}{\partial Q_b} & \frac{\partial f_3}{\partial g} \end{pmatrix}_{E^{(1)}(Q_a^{(1)}, Q_b^{(1)}, g^{(1)})}$$

$$= \begin{pmatrix} 2 \cdot \frac{1}{T_R} \cdot (\delta_b + \alpha_b - \delta_a - \alpha_a) & 0 & 0 \\ 2 \cdot \frac{1}{T_R} \cdot (l_0 + \delta_b + \alpha_b) & 2 \cdot \frac{1}{T_R} \cdot (l_0 + \delta_b + \alpha_b) & 2 \cdot \frac{1}{T_R} \cdot \left[\frac{(l_0+\delta_b+\alpha_b) \cdot P_g \cdot T_R}{g_0 \cdot P_g \cdot T_R - (l_0+\delta_b+\alpha_b)}\right]^2 \\ -\frac{g_0 \cdot P_g \cdot T_R - (l_0+\delta_b+\alpha_b)}{T_g \cdot P_g^2 \cdot T_R^2} & -\frac{g_0 \cdot P_g \cdot T_R - (l_0+\delta_b+\alpha_b)}{T_g \cdot T_R^2 \cdot P_g^2} & -\frac{1}{T_g} \cdot \left[1 + \frac{(l_0+\delta_b+\alpha_b)}{g_0 \cdot P_g \cdot T_R - (l_0+\delta_b+\alpha_b)}\right] \end{pmatrix}$$

$$A - \lambda \cdot I = \begin{pmatrix} \frac{\partial f_1}{\partial Q_a} & \frac{\partial f_1}{\partial Q_b} & \frac{\partial f_1}{\partial g} \\ \frac{\partial f_2}{\partial Q_a} & \frac{\partial f_2}{\partial Q_b} & \frac{\partial f_2}{\partial g} \\ \frac{\partial f_3}{\partial Q_a} & \frac{\partial f_3}{\partial Q_b} & \frac{\partial f_3}{\partial g} \end{pmatrix}_{E^{(1)}(Q_a^{(1)}, Q_b^{(1)}, g^{(1)})} - \lambda \cdot \begin{pmatrix} 1 & 0 & 0 \\ 0 & 1 & 0 \\ 0 & 0 & 1 \end{pmatrix}$$

$$= \begin{pmatrix} 2 \cdot \frac{1}{T_R} \cdot (\delta_b + \alpha_b - \delta_a - \alpha_a) - \lambda & 0 & 0 \\ 2 \cdot \frac{1}{T_R} \cdot (l_0 + \delta_b + \alpha_b) & 2 \cdot \frac{1}{T_R} \cdot (l_0 + \delta_b + \alpha_b) - \lambda & 2 \cdot \frac{1}{T_R} \cdot \left[\frac{(l_0+\delta_b+\alpha_b) \cdot P_g \cdot T_R}{g_0 \cdot P_g \cdot T_R - (l_0+\delta_b+\alpha_b)}\right]^2 \\ -\frac{g_0 \cdot P_g \cdot T_R - (l_0+\delta_b+\alpha_b)}{T_g \cdot P_g^2 \cdot T_R^2} & -\frac{g_0 \cdot P_g \cdot T_R - (l_0+\delta_b+\alpha_b)}{T_g \cdot T_R^2 \cdot P_g^2} & -\frac{1}{T_g} \cdot \left[1 + \frac{(l_0+\delta_b+\alpha_b)}{g_0 \cdot P_g \cdot T_R - (l_0+\delta_b+\alpha_b)}\right] - \end{pmatrix}$$

$$\det(A - \lambda \cdot I) = 0$$

$$\left[2 \cdot \frac{1}{T_R} \cdot (\delta_b + \alpha_b - \delta_a - \alpha_a) - \lambda\right] \cdot \left\{\left[2 \cdot \frac{1}{T_R} \cdot (l_0 + \delta_b + \alpha_b) - \lambda\right] \cdot \left[-\frac{1}{T_g} \cdot \left[1 + \frac{(l_0 + \delta_b + \alpha_b)}{g_0 \cdot P_g \cdot T_R - (l_0 + \delta_b + \alpha_b)}\right] - \lambda\right] - \left[-\frac{g_0 \cdot P_g \cdot T_R - (l_0 + \delta_b + \alpha_b)}{T_g \cdot T_R^2 \cdot P_g^2}\right] \cdot \left[2 \cdot \frac{1}{T_R} \cdot \left[\frac{(l_0 + \delta_b + \alpha_b) \cdot P_g \cdot T_R}{g_0 \cdot P_g \cdot T_R - (l_0 + \delta_b + \alpha_b)}\right]^2\right]\right\} = 0$$

$$\left[2 \cdot \frac{1}{T_R} \cdot (\delta_b + \alpha_b - \delta_a - \alpha_a) - \lambda\right] \cdot \left\{\left[2 \cdot \frac{1}{T_R} \cdot (l_0 + \delta_b + \alpha_b) - \lambda\right] \cdot \left[-\frac{1}{T_g} \cdot \left[1 + \frac{(l_0 + \delta_b + \alpha_b)}{g_0 \cdot P_g \cdot T_R - (l_0 + \delta_b + \alpha_b)}\right] - \lambda\right] + 2 \cdot \frac{1}{T_R \cdot T_g} \cdot \left[\frac{(l_0 + \delta_b + \alpha_b)^2}{g_0 \cdot P_g \cdot T_R - (l_0 + \delta_b + \alpha_b)}\right]\right\} = 0$$

We get eigenvalue characteristic equation $\sum_{k=0}^{3} \lambda^k \cdot \Upsilon_k = 0$.

Remark it is reader task to find the exact expression for Υ_k; $k = 0, 1, 2, 3$. We get three eigenvalues $\lambda_1, \lambda_2, \lambda_3$.

Eigenvalues stability discussion: Our system (Multi pulse operation of Ti: Sapphire laser mode locked by an ion-implanted semiconductor saturable-absorber mirror), the characteristic equation is of degree N = 3 $\left(\sum_{k=0}^{3} \lambda^k \cdot \Upsilon_k = 0\right)$ and must often be solved numerically. Expect in some particular cases, such an equation has (N = 3) distinct roots that can be real or complex. These values are the eigenvalues of the 3 × 3 Jacobian matrix (A). The general rule is that the Steady State (SS) is stable if there is no eigenvalue with positive real part. It is sufficient that one eigenvalue is positive for the steady state to be unstable. Our 3-variables (Q_a, Q_b, g) system has three eigenvalues. The type of behavior can be characterized as a function of the position of these eigenvalues in the Re/Im plane. Five non-degenerated cases can be distinguished: (1) the three eigenvalues are real and negative (stable steady state), (2) the three eigenvalues are real, two of them are negative (unstable steady state), (3) and (4) one eigenvalue is complex with a negative real part and the other eigenvalues are real and negative (stable steady state), two cases can be distinguished depending on the relative value of the real part of the complex eigenvalues and of the real one, (5) one eigenvalue is complex with a negative real part and at least one eigenvalue is positive (unstable steady state) [5, 6].

Third fixed point:

$$E^{(2)}(Q_a^{(2)}, Q_b^{(2)}, g^{(2)}) = \left(\frac{(l_0 + \delta_a + \alpha_a)}{\left[g_0 - \frac{(l_0+\delta_a+\alpha_a)}{P_g \cdot T_R}\right]}, 0, g_0 - \frac{(l_0 + \delta_a + \alpha_a)}{P_g \cdot T_R} \right)$$

$$\frac{\partial f_1}{\partial Q_a}|_{E^{(2)}(Q_a^{(2)},Q_b^{(2)},g^{(2)})} = 2 \cdot \frac{1}{T_R} \cdot \left[2 \cdot \left[g_0 - \frac{(l_0 + \delta_a + \alpha_a)}{P_g \cdot T_R}\right] \cdot \left[\frac{(l_0 + \delta_a + \alpha_a)}{\left[g_0 - \frac{(l_0+\delta_a+\alpha_a)}{P_g \cdot T_R}\right]}\right] - (l_0 + \delta_a + \alpha_a)\right]$$

$$\frac{\partial f_1}{\partial Q_b}|_{E^{(2)}(Q_a^{(2)},Q_b^{(2)},g^{(2)})} = 2 \cdot \frac{1}{T_R} \cdot \left[g_0 - \frac{(l_0 + \delta_a + \alpha_a)}{P_g \cdot T_R}\right] \cdot \left(\frac{(l_0 + \delta_a + \alpha_a)}{\left[g_0 - \frac{(l_0+\delta_a+\alpha_a)}{P_g \cdot T_R}\right]}\right)$$

$$\frac{\partial f_1}{\partial g}|_{E^{(2)}(Q_a^{(2)},Q_b^{(2)},g^{(2)})} = 2 \cdot \frac{1}{T_R} \cdot \left[\frac{(l_0 + \delta_a + \alpha_a)}{\left[g_0 - \frac{(l_0+\delta_a+\alpha_a)}{P_g \cdot T_R}\right]}\right]^2; \frac{\partial f_2}{\partial Q_a}|_{E^{(1)}(Q_a^{(2)},Q_b^{(2)},g^{(2)})} = 0$$

$$\frac{\partial f_2}{\partial Q_b}|_{E^{(2)}(Q_a^{(2)},Q_b^{(2)},g^{(2)})} = 2 \cdot \frac{1}{T_R} \cdot \left\{\left[g_0 - \frac{(l_0 + \delta_a + \alpha_a)}{P_g \cdot T_R}\right] \cdot \left[\frac{(l_0 + \delta_a + \alpha_a)}{\left[g_0 - \frac{(l_0+\delta_a+\alpha_a)}{P_g \cdot T_R}\right]}\right] - (l_0 + \delta_b + \alpha_b)\right\}$$

$$\frac{\partial f_2}{\partial g}|_{E^{(2)}(Q_a^{(2)},Q_b^{(2)},g^{(2)})} = 0; \frac{\partial f_3}{\partial Q_a}|_{E^{(2)}(Q_a^{(2)},Q_b^{(2)},g^{(2)})} = -\frac{\left[g_0 - \frac{(l_0+\delta_a+\alpha_a)}{P_g \cdot T_R}\right]}{P_g \cdot T_g \cdot T_R}$$

$$\frac{\partial f_3}{\partial Q_b}|_{E^{(2)}(Q_a^{(2)},Q_b^{(2)},g^{(2)})} = -\frac{\left[g_0 - \frac{(l_0+\delta_a+\alpha_a)}{P_g \cdot T_R}\right]}{P_g \cdot T_g \cdot T_R}$$

$$\frac{\partial f_3}{\partial g}|_{E^{(2)}(Q_a^{(2)},Q_b^{(2)},g^{(2)})} = -\frac{1}{T_g} \cdot \left[1 + \frac{\left(\frac{(l_0+\delta_a+\alpha_a)}{\left[g_0 - \frac{(l_0+\delta_a+\alpha_a)}{P_g \cdot T_R}\right]}\right)}{P_g \cdot T_R}\right]$$

We use the identification to simplify our above expressions.

$$g_0 - \frac{(l_0 + \delta_a + \alpha_a)}{P_g \cdot T_R} = \frac{g_0 \cdot P_g \cdot T_R - (l_0 + \delta_a + \alpha_a)}{P_g \cdot T_R}$$

$$\frac{1}{g_0 - \frac{(l_0+\delta_a+\alpha_a)}{P_g \cdot T_R}} = \frac{P_g \cdot T_R}{g_0 \cdot P_g \cdot T_R - (l_0 + \delta_a + \alpha_a)}$$

We get the following Jacobian matrix A's elements at the third fixed point:

$$\frac{\partial f_1}{\partial Q_a}|_{E^{(2)}(Q_a^{(2)},Q_b^{(2)},g^{(2)})} = 2 \cdot \frac{1}{T_R} \cdot [l_0 + \delta_a + \alpha_a]$$

$$\frac{\partial f_1}{\partial Q_b}|_{E^{(2)}(Q_a^{(2)},Q_b^{(2)},g^{(2)})} = 2 \cdot \frac{1}{T_R} \cdot (l_0 + \delta_a + \alpha_a)$$

$$\frac{\partial f_1}{\partial g}|_{E^{(2)}(Q_a^{(2)},Q_b^{(2)},g^{(2)})} = 2 \cdot \frac{1}{T_R} \cdot \left[\frac{(l_0 + \delta_a + \alpha_a) \cdot P_g \cdot T_R}{g_0 \cdot P_g \cdot T_R - (l_0 + \delta_a + \alpha_a)}\right]^2$$

$$\frac{\partial f_2}{\partial Q_a}|_{E^{(2)}(Q_a^{(2)},Q_b^{(2)},g^{(2)})} = 0$$

$$\frac{\partial f_2}{\partial Q_b}|_{E^{(2)}(Q_a^{(2)},Q_b^{(2)},g^{(2)})} = 2 \cdot \frac{1}{T_R} \cdot (\delta_a + \alpha_a - \delta_b - \alpha_b)$$

$$\frac{\partial f_2}{\partial g}|_{E^{(2)}(Q_a^{(2)},Q_b^{(2)},g^{(2)})} = 0$$

$$\frac{\partial f_3}{\partial Q_a}|_{E^{(2)}(Q_a^{(2)},Q_b^{(2)},g^{(2)})} = -\frac{g_0 \cdot P_g \cdot T_R - (l_0 + \delta_a + \alpha_a)}{T_g \cdot P_g^2 \cdot T_R^2}$$

$$\frac{\partial f_3}{\partial Q_b}|_{E^{(2)}(Q_a^{(2)},Q_b^{(2)},g^{(2)})} = -\frac{g_0 \cdot P_g \cdot T_R - (l_0 + \delta_a + \alpha_a)}{T_g \cdot P_g^2 \cdot T_R^2}$$

$$\frac{\partial f_3}{\partial g}|_{E^{(2)}(Q_a^{(2)},Q_b^{(2)},g^{(2)})} = -\frac{1}{T_g} \cdot \left[1 + \frac{(l_0 + \delta_a + \alpha_a)}{g_0 \cdot P_g \cdot T_R - (l_0 + \delta_a + \alpha_a)}\right]$$

$$A = \begin{pmatrix} \frac{\partial f_1}{\partial Q_a} & \frac{\partial f_1}{\partial Q_b} & \frac{\partial f_1}{\partial g} \\ \frac{\partial f_2}{\partial Q_a} & \frac{\partial f_2}{\partial Q_b} & \frac{\partial f_2}{\partial g} \\ \frac{\partial f_3}{\partial Q_a} & \frac{\partial f_3}{\partial Q_b} & \frac{\partial f_3}{\partial g} \end{pmatrix}_{E^{(2)}(Q_a^{(2)}, Q_b^{(2)}, g^{(2)})}$$

$$= \begin{pmatrix} 2 \cdot \frac{1}{T_R} \cdot [l_0 + \delta_a + \alpha_a] & 2 \cdot \frac{1}{T_R} \cdot (l_0 + \delta_a + \alpha_a) & 2 \cdot \frac{1}{T_R} \cdot \left[\frac{(l_0+\delta_a+\alpha_a) \cdot P_g \cdot T_R}{g_0 \cdot P_g \cdot T_R - (l_0+\delta_a+\alpha_a)}\right]^2 \\ 0 & 2 \cdot \frac{1}{T_R} \cdot (\delta_a + \alpha_a - \delta_b - \alpha_b) & 0 \\ -\frac{g_0 \cdot P_g \cdot T_R - (l_0+\delta_a+\alpha_a)}{T_g \cdot P_g^2 \cdot T_R^2} & -\frac{g_0 \cdot P_g \cdot T_R - (l_0+\delta_a+\alpha_a)}{T_g \cdot P_g^2 \cdot T_R^2} & -\frac{1}{T_g} \cdot \left[1 + \frac{(l_0+\delta_a+\alpha_a)}{g_0 \cdot P_g \cdot T_R - (l_0+\delta_a+\alpha_a)}\right] \end{pmatrix}$$

$$A - \lambda \cdot I = \begin{pmatrix} \frac{\partial f_1}{\partial Q_a} & \frac{\partial f_1}{\partial Q_b} & \frac{\partial f_1}{\partial g} \\ \frac{\partial f_2}{\partial Q_a} & \frac{\partial f_2}{\partial Q_b} & \frac{\partial f_2}{\partial g} \\ \frac{\partial f_3}{\partial Q_a} & \frac{\partial f_3}{\partial Q_b} & \frac{\partial f_3}{\partial g} \end{pmatrix}_{E^{(2)}(Q_a^{(2)}, Q_b^{(2)}, g^{(2)})} - \lambda \cdot \begin{pmatrix} 1 & 0 & 0 \\ 0 & 1 & 0 \\ 0 & 0 & 1 \end{pmatrix}$$

$$= \begin{pmatrix} 2 \cdot \frac{1}{T_R} \cdot [l_0 + \delta_a + \alpha_a] - \lambda & 2 \cdot \frac{1}{T_R} \cdot (l_0 + \delta_a + \alpha_a) & 2 \cdot \frac{1}{T_R} \cdot \left[\frac{(l_0+\delta_a+\alpha_a) \cdot P_g \cdot T_R}{g_0 \cdot P_g \cdot T_R - (l_0+\delta_a+\alpha_a)}\right]^2 \\ 0 & 2 \cdot \frac{1}{T_R} \cdot (\delta_a + \alpha_a - \delta_b - \alpha_b) - \lambda & 0 \\ -\frac{g_0 \cdot P_g \cdot T_R - (l_0+\delta_a+\alpha_a)}{T_g \cdot P_g^2 \cdot T_R^2} & -\frac{g_0 \cdot P_g \cdot T_R - (l_0+\delta_a+\alpha_a)}{T_g \cdot P_g^2 \cdot T_R^2} & -\frac{1}{T_g} \cdot \left[1 + \frac{(l_0+\delta_a+\alpha_a)}{g_0 \cdot P_g \cdot T_R - (l_0+\delta_a+\alpha_a)}\right] - \end{pmatrix}$$

$$\det(A - \lambda \cdot I) = 0$$

$$\left(2 \cdot \frac{1}{T_R} \cdot [l_0 + \delta_a + \alpha_a] - \lambda\right) \cdot \left\{\left[2 \cdot \frac{1}{T_R} \cdot (\delta_a + \alpha_a - \delta_b - \alpha_b) - \lambda\right] \right.$$
$$\left. \cdot \left[-\frac{1}{T_g} \cdot \left[1 + \frac{(l_0 + \delta_a + \alpha_a)}{g_0 \cdot P_g \cdot T_R - (l_0 + \delta_a + \alpha_a)}\right] - \lambda\right]\right\}$$
$$+ 2 \cdot \frac{1}{T_R} \cdot \left[\frac{(l_0 + \delta_a + \alpha_a) \cdot P_g \cdot T_R}{g_0 \cdot P_g \cdot T_R - (l_0 + \delta_a + \alpha_a)}\right]^2 \cdot \left\{-\left[-\frac{g_0 \cdot P_g \cdot T_R - (l_0 + \delta_a + \alpha_a)}{T_g \cdot P_g^2 \cdot T_R^2}\right] \right.$$
$$\left. \cdot \left[2 \cdot \frac{1}{T_R} \cdot (\delta_a + \alpha_a - \delta_b - \alpha_b) - \lambda\right]\right\} = 0$$

$$\left(2 \cdot \frac{1}{T_R} \cdot [l_0 + \delta_a + \alpha_a] - \lambda\right) \cdot \left\{\left[2 \cdot \frac{1}{T_R} \cdot (\delta_a + \alpha_a - \delta_b - \alpha_b) - \lambda\right] \right.$$
$$\left. \cdot \left[-\frac{1}{T_g} \cdot \left[1 + \frac{(l_0 + \delta_a + \alpha_a)}{g_0 \cdot P_g \cdot T_R - (l_0 + \delta_a + \alpha_a)}\right] - \lambda\right]\right\}$$
$$+ 2 \cdot \frac{1}{T_R \cdot T_g} \cdot \frac{(l_0 + \delta_a + \alpha_a)^2}{[g_0 \cdot P_g \cdot T_R - (l_0 + \delta_a + \alpha_a)]}$$
$$\cdot \left[2 \cdot \frac{1}{T_R} \cdot (\delta_a + \alpha_a - \delta_b - \alpha_b) - \lambda\right] = 0$$

We get eigenvalue characteristic equation $\sum_{k=0}^{3} \lambda^k \cdot \Upsilon_k = 0$.

Remark it is reader task to find the exact expression for Υ_k; $k = 0, 1, 2, 3$. We get three eigenvalues $\lambda_1, \lambda_2, \lambda_3$.

Eigenvalues stability discussion: Our system (Multi pulse operation of Ti:Sapphire laser mode locked by an ion-implanted semiconductor saturable-absorber mirror), the characteristic equation is of degree N = 3 $\left(\sum_{k=0}^{3} \lambda^k \cdot \Upsilon_k = 0\right)$ and must often be solved numerically. Expect in some particular cases, such an equation has (N = 3) distinct roots that can be real or complex. These values are the eigenvalues of the 3 × 3 Jacobian matrix (A). The general rule is that the Steady State (SS) is stable if there is no eigenvalue with positive real part. It is sufficient that one eigenvalue is positive for the steady state to be unstable. Our 3-variables (Q_a, Q_b, g) system has three eigenvalues. The type of behavior can be characterized as a function of the position of these eigenvalues in the Re/Im plane. Five non-degenerated cases can be distinguished: (1) the three eigenvalues are real and negative (stable steady state), (2) the three eigenvalues are real, two of them are negative (unstable steady state), (3) and (4) one eigenvalue is complex with a negative real part and the other eigenvalues are real and negative (stable steady state), two cases can be distinguished depending on the relative value of the real part of the complex eigenvalues and of the real one, (5) one eigenvalue is complex with a negative real part and at least one eigenvalue is positive (unstable steady state) [5, 6].

2.4 Questions

1. We have a system of stronger laser field pulse and diatomic molecules. The system is characterized by a second order differential equation, gives the rotation of a molecule with reduced mass (μ), polarization (α) and inter nuclear distance (R) in the field of a laser with frequency (ω) and field strength (F). The field strength (F) is dependent on the angle between the polarization of the laser and inter nuclear axis (θ) according to the following function $F = \Gamma_0 + \Gamma_1 \cdot \sin^2(\theta)$ $\Gamma_0, \Gamma_1 \in \mathbb{R}_+$. Our system second order differential equation:

$$\frac{d^2\theta}{d\tau^2} + (\frac{\omega_0}{\omega})^2 \cdot \cos^2(\tau) \cdot \sin(2 \cdot \theta) = 0$$

$$\tau = \omega \cdot t; \omega_0^2 = \frac{\alpha \cdot F^2}{\mu \cdot R^2}; d\tau^2 = d(\omega \cdot t)^2 = \omega^2 \cdot dt^2$$

θ	Angle between the polarization of the laser and inter nuclear axis.
ω_0	Establish the rotation dynamic of the molecule.
$\frac{\omega_0}{\omega}$	Quantity determines the rotation dynamics of the molecules.
θ_0	Molecule initially angle with the direction of laser polarization.
μ	reduced mass of the molecule.
α	Polarization factor (polarizability).
R	Inter nuclear distance in the field of laser.

ω Laser frequency (we consider it as a constant in our system).
ω_0 Calculated parameter $\left(\omega_0^2 = \frac{\alpha \cdot F^2}{\mu \cdot R^2}; \omega_0 = \frac{F}{R} \cdot \sqrt{\frac{\alpha}{\mu}}\right)$

($\omega_0 = 7.9 \times 10^{-4}$au (Atomic units (au or a.u.) form a system of natural units which is especially convenient for atomic physics calculations) for D_2 molecule at a peak intensity of 10^{14}W/cm^2).

1.1 Find possible fixed points in our system.
1.2 Discuss stability of our system by using stability analysis about any one of the equilibrium points (fixed points) of diatomic molecule-differential equations for the rotation of a molecule system.
1.3 Plot our system fixed points on the phase portrait and classify them.
1.4 Discuss the case when there is a delay on angle between the polarization of the laser and inter nuclear axis $\theta(t) \to \theta(t - \Delta)$. Find dynamics, stability behavior, and stability switching of our system.

2. We have a system of stronger laser field pulse and diatomic molecules. The system is characterized by a second order differential equation, gives the rotation of a molecule with reduced mass (μ), polarization (α) and inter nuclear distance (R) in the field of a laser with frequency (ω) and field strength (F). There are interferences in our system which dependent on the angle between the polarization of the laser and inter nuclear axis (θ) according to the function $f(\theta) = |\sqrt[3]{\theta} + \cos(\theta)|$. Our system second order differential equation (with interferences function):

$$\frac{d^2\theta}{d\tau^2} + \left(\frac{\omega_0}{\omega}\right)^2 \cdot f(\theta) \cdot \cos^2(\tau) \cdot \sin(2 \cdot \theta) = 0$$

$$\tau = \omega \cdot t; \omega_0^2 = \frac{\alpha \cdot F^2}{\mu \cdot R^2}; d\tau^2 = d(\omega \cdot t)^2 = \omega^2 \cdot dt^2$$

θ Angle between the polarization of the laser and inter nuclear axis.
ω_0 Establish the rotation dynamic of the molecule.
$\frac{\omega_0}{\omega}$ Quantity determines the rotation dynamics of the molecules.
θ_0 Molecule initially angle with the direction of laser polarization.
μ reduced mass of the molecule.
α Polarization factor (polarizability).
R Inter nuclear distance in the field of laser.
ω Laser frequency (we consider it as a constant in our system).
ω_0 Calculated parameter $\left(\omega_0^2 = \frac{\alpha \cdot F^2}{\mu \cdot R^2}; \omega_0 = \frac{F}{R} \cdot \sqrt{\frac{\alpha}{\mu}}\right)$.

($\omega_0 = 7.9 \times 10^{-4}$au (Atomic units (au or a.u.) form a system of natural units which is especially convenient for atomic physics calculations) for D_2 molecule at a peak intensity of 10^{14} W/cm^2).

2.1 Find possible fixed points in our system.

2.2 Discuss stability of our system by using stability analysis about any one of the equilibrium points (fixed points) of diatomic molecule-differential equations for the rotation of a molecule system.
2.3 Plot our system fixed points on the phase portrait and classify them.
2.4 Discuss the case when there is a delay on angle between the polarization of the laser and inter nuclear axis$\theta(t) \rightarrow \theta(t-\Delta)$. Find dynamics, stability behavior, and stability switching of our system.

3. We have a system of stronger laser field pulse and diatomic molecules. The system is characterized by a second order differential equation, gives the rotation of a molecule with reduced mass (μ), polarization (α) and inter nuclear distance (R) in the field of a laser with frequency (ω) and field strength (F). The polarization parameter (α) is dependent on the angle between the polarization of the laser and inter nuclear axis (θ) according to the following function $\alpha = |\sqrt{\theta} + \sin(\theta)|$. Additionally, there are interferences in our system which dependent on the angle between the polarization of the laser and inter nuclear axis (θ) according to the function $f(\theta) = |\sqrt[5]{\theta} + \sin(\theta)|$. Our system second order differential equation (with interferences function):

$$\frac{d^2\theta}{d\tau^2} + \left(\frac{\omega_0}{\omega}\right)^2 \cdot f(\theta) \cdot \cos^2(\tau) \cdot \sin(2 \cdot \theta) = 0$$

$$\tau = \omega \cdot t;\ \omega_0^2 = \frac{\alpha \cdot F^2}{\mu \cdot R^2};\ d\tau^2 = d(\omega \cdot t)^2 = \omega^2 \cdot dt^2$$

θ Angle between the polarization of the laser and inter nuclear axis.
ω_0 Establish the rotation dynamic of the molecule.
$\frac{\omega_0}{\omega}$ Quantity determines the rotation dynamics of the molecules.
θ_0 Molecule initially angle with the direction of laser polarization.
μ reduced mass of the molecule.
α Polarization factor (polarizability).
R Inter nuclear distance in the field of laser.
ω Laser frequency (we consider it as a constant in our system).
ω_0 Calculated parameter $\left(\omega_0^2 = \frac{\alpha \cdot F^2}{\mu \cdot R^2};\ \omega_0 = \frac{F}{R} \cdot \sqrt{\frac{\alpha}{\mu}}\right)$.

($\omega_0 = 7.9 \times 10^{-4}$au (Atomic units (au or a.u.) form a system of natural units which is especially convenient for atomic physics calculations) for D_2 molecule at a peak intensity of 10^{14} W/cm^2).

3.1 Find possible fixed points in our system.
3.2 Discuss stability of our system by using stability analysis about any one of the equilibrium points (fixed points) of diatomic molecule-differential equations for the rotation of a molecule system.
3.3 Plot our system fixed points on the phase portrait and classify them.

3.4 Discuss the case when there is a delay on angle between the polarization of the laser and inter nuclear axis $\theta(t) \to \theta(t-\Delta)$. Find dynamics, stability behavior, and stability switching of our system.

4. We have a system of Ultra-broad bandwidth pulse-shaping apparatus which includes two gratings, two spherical mirrors, and crystal SLM (Spatial Light Modulator). Any material (SLM) contributions to positive second order group velocity dispersion (GVD) within the shaper can be canceled if the position of the final grating is adjusted. There is phase dispersion at higher orders, specially cubic and quartic phase dispersions. Fiber propagation parameter is $\beta(\omega) = \frac{\Phi(\omega)}{L}$, where L is the propagation distance then $\beta^{(n)}(\omega) = \frac{\Phi^{(n)}(\omega)}{L} = \left(\frac{d^n\beta}{d\omega^n}\right)_{\omega=\omega_0}$.
The derivatives $\beta^{(n)}(\omega)$are given in terms of the index of refraction.

$$\beta^{(2)}(\omega) = \frac{\lambda^3}{2\cdot\pi\cdot c^2}\cdot\frac{d^2n}{d\lambda^2};\ \beta^{(3)}(\omega) = -\frac{\lambda^4}{4\pi^2c^3}\cdot\left(3\cdot\frac{d^2n}{d\lambda^2} + \lambda\cdot\frac{d^3n}{d\lambda^3}\right)$$

$$\beta^{(4)}(\omega) = \frac{\lambda^5}{\pi^3\cdot c^4}\cdot\left(\frac{3}{4}\cdot\frac{d^2n}{d\lambda^2} + \lambda\cdot\frac{d^3n}{d\lambda^3} + \frac{\lambda^2}{4}\cdot\frac{d^4n}{d\lambda^4}\right)$$

We take the second, third, and fourth derivative of propagation parameter $(\beta(\omega))$ in ω as a constant at $\omega = \omega_0$. We define new variables: Y_1, Y_2, Y_3, Y_4.

$$Y_1 = \frac{dn}{d\lambda},\ Y_2 = \frac{dY_1}{d\lambda} = \frac{d^2n}{d\lambda^2},\ Y_3 = \frac{dY_2}{d\lambda} = \frac{d^3n}{d\lambda^3},\ Y_4 = \frac{dY_3}{d\lambda} = \frac{d^4n}{d\lambda^4}$$

Due to interferences in our system there are shifting in new variables Y_2, Y_3, $Y_2(\lambda) \to Y_2(\lambda - \lambda_A)$; $Y_3(\lambda) \to Y_3(\lambda - \lambda_B)$ and there is no effect of delay on the derivatives of these variables $\left(\frac{dY_2}{d\lambda}, \frac{dY_3}{d\lambda}\right)$.

4.1 Find our system final two delay differential equations (DDEs).
4.2 Find our system fixed points and how there are dependent on system parameters?
4.3 Discuss the stability of our system by the standard local stability analysis about any one of the equilibrium points (fixed points) for the following cases: (a) $\lambda_A = 0; \lambda_B = \lambda_\Delta$(b) $\lambda_A = \lambda_\Delta; \lambda_B = 0$.
4.4 Find the small increment Jacobian of our system.
4.5 Discuss stability switching for different values of λ_Δ parameter (cases (a), (b)).

5. We have a system of Ultra-broad bandwidth pulse-shaping apparatus which includes two gratings, two spherical mirrors, and crystal SLM (Spatial Light Modulator). Any material (SLM) contributions to positive second order group velocity dispersion (GVD) within the shaper can be canceled if the position of the final grating is adjusted. There is phase dispersion at higher orders, specially

cubic and quartic phase dispersions. Fiber propagation parameter is $\beta(\omega) = \frac{\Phi(\omega)}{L}$, where L is the propagation distance then $\beta^{(n)}(\omega) = \frac{\Phi^{(n)}(\omega)}{L} = \left(\frac{d^n\beta}{d\omega^n}\right)_{\omega=\omega_0}$. The derivatives $\beta^{(n)}(\omega)$ are given in terms of the index of refraction. Due to other source of probe laser in our system there is a polynomial representation of the third, fourth, and fifth power of wavelength variable in our derivatives of $\beta^{(n)}(\omega)$ in terms of index of refraction differential equations.

$$\beta^{(2)}(\omega) = \frac{\sum_{k=0}^{3}\Gamma_k\cdot\lambda^k}{2\cdot\pi\cdot c^2}\cdot\frac{d^2n}{d\lambda^2}$$

$$\beta^{(3)}(\omega) = -\frac{\sum_{k=0}^{4}\Gamma_k\cdot\lambda^k}{4\pi^2c^3}\cdot\left(3\cdot\frac{d^2n}{d\lambda^2}+\lambda\cdot\frac{d^3n}{d\lambda^3}\right)$$

$$\beta^{(4)}(\omega) = \frac{\sum_{k=0}^{5}\Gamma_k\cdot\lambda^k}{\pi^3\cdot c^4}\cdot\left(\frac{3}{4}\cdot\frac{d^2n}{d\lambda^2}+\lambda\cdot\frac{d^3n}{d\lambda^3}+\frac{\lambda^2}{4}\cdot\frac{d^4n}{d\lambda^4}\right)$$

$$\Gamma_0, \Gamma_1, \ldots, \Gamma_5 \in \mathbb{R}_+$$

We take the second, third, and fourth derivative of propagation parameter ($\beta(\omega)$) in ω as a constant at$\omega = \omega_0$. We define new variables: Y_1, Y_2, Y_3, Y_4.

$$Y_1 = \frac{dn}{d\lambda}, Y_2 = \frac{dY_1}{d\lambda} = \frac{d^2n}{d\lambda^2}, Y_3 = \frac{dY_2}{d\lambda} = \frac{d^3n}{d\lambda^3}, Y_4 = \frac{dY_3}{d\lambda} = \frac{d^4n}{d\lambda^4}$$

5.1 Find system fixed point, How they dependent on system parameters?
5.2 Use linearization technique and try to approximate the phase portrait near a fixed point by that of a corresponding linear system.
5.3 Find the Jacobian matrix at each system's fixed point.
5.4 Classify each fixed point according to eigenvalues and discuss stability.
5.5 How system's parameters $\Gamma_0, \Gamma_1, \ldots, \Gamma_5 \in \mathbb{R}_+$ establish our system stability? Discuss each parameter in details.

6. We have a system of Ultra-broad bandwidth pulse-shaping apparatus which includes two gratings, two spherical mirrors, and crystal SLM (Spatial Light Modulator). Any material (SLM) contributions to positive second order group velocity dispersion (GVD) within the shaper can be canceled if the position of the final grating is adjusted. There is phase dispersion at higher orders, specially cubic and quartic phase dispersions. Fiber propagation parameter is $\beta(\omega) = \frac{\Phi(\omega)}{L}$, where L is the propagation distance then $\beta^{(n)}(\omega) = \frac{\Phi^{(n)}(\omega)}{L} = (\frac{d^n\beta}{d\omega^n})_{\omega=\omega_0}$. The derivatives $\beta^{(n)}(\omega)$ are given in terms of the index of refraction. Due to other source of probe laser in our system there is a polynomial representation of the third, fourth, and fifth power of wavelength variable in our derivatives of

$\beta^{(n)}(\omega)$ in terms of index of refraction differential equations.

$$\beta^{(2)}(\omega) = \frac{\sum_{k=0}^{3} \Gamma_k \cdot \lambda^k}{2 \cdot \pi \cdot c^2} \cdot \frac{d^2 n}{d\lambda^2}; \beta^{(3)}(\omega)$$
$$= -\frac{\sum_{k=0,k\neq 2}^{4} \Gamma_k \cdot \lambda^k}{4\pi^2 c^3} \cdot \left(3 \cdot \frac{d^2 n}{d\lambda^2} + \lambda \cdot \frac{d^3 n}{d\lambda^3}\right)$$
$$\beta^{(4)}(\omega) = \frac{\sum_{k=0,k\neq 3}^{5} \Gamma_k \cdot \lambda^k}{\pi^3 \cdot c^4}$$
$$\cdot \left(\frac{3}{4} \cdot \frac{d^2 n}{d\lambda^2} + \lambda \cdot \frac{d^3 n}{d\lambda^3} + \frac{\lambda^2}{4} \cdot \frac{d^4 n}{d\lambda^4}\right); \Gamma_0, \Gamma_1, \ldots, \Gamma_5 \in \mathbb{R}_+$$

We take the second, third, and fourth derivative of propagation parameter ($\beta(\omega)$) in ω as a constant at$\omega = \omega_0$. We define new variables: Y_1, Y_2, Y_3, Y_4.

$$Y_1 = \frac{dn}{d\lambda}, Y_2 = \frac{dY_1}{d\lambda} = \frac{d^2 n}{d\lambda^2}, Y_3 = \frac{dY_2}{d\lambda} = \frac{d^3 n}{d\lambda^3}, Y_4 = \frac{dY_3}{d\lambda} = \frac{d^4 n}{d\lambda^4}$$

Additionally, due to interferences in our system there are shifting in new variables $Y_2, Y_3, Y_2(\lambda) \to Y_2(\lambda - \lambda_A); Y_3(\lambda) \to Y_3(\lambda - \lambda_B)$and there is no effect of delay on the derivatives of these variables $\left(\frac{dY_2}{d\lambda}, \frac{dY_3}{d\lambda}\right)$.

6.1 Find our system final two delay differential equations (DDEs).
6.2 Find our system fixed points and how there are dependent on system parameters?
6.3 Discuss the stability of our system by the standard local stability analysis about any one of the equilibrium points (fixed points) for the following cases: (a) $\lambda_A = 0; \lambda_B = \lambda_\Delta$ (b) $\lambda_A = \lambda_\Delta; \lambda_B = 0$.
6.4 Find the small increment Jacobian of our system.
6.5 Discuss stability switching for different values of λ_Δ parameter (cases (a), (b)).

7. We have a system of Ti: Sapphir laser mode semiconductor saturable absorber mirror (SESAM) that operate in a multipulse mode. We need to analyze the stability of a soliton in the laser when subject to perturbations from a filter and an absorber. The system generalized complex Ginzburg-Landau equation (GCGLE) which is the master equation to describe the dynamical effects of laser by means of numerical simulations: $i \cdot T_R \cdot \psi_T - \frac{\beta_2}{2} \cdot \frac{\partial^2 \psi}{\partial t^2} + \kappa \cdot |\psi|^2 \cdot \psi = i \cdot (g - q - l_0) \cdot \psi + i \cdot \beta \cdot \frac{\partial^2 \psi}{\partial t^2}$. We can consider our system delay differential equations (DDEs) when the perturbations on the right hand side are not neglected.

$$T_R \cdot \frac{dQ_a}{dT} = 2 \cdot [g(T) \cdot (Q_a(T-\tau_1) + Q_b(T-\tau_2)) - l_0 - \delta_a - \alpha_a] \cdot Q_a(T-\tau_1)$$

$$T_R \cdot \frac{dQ_b}{dT} = 2 \cdot [g(T) \cdot (Q_a(T-\tau_1) + Q_b(T-\tau_2)) - l_0 - \delta_b - \alpha_b] \cdot Q_b(T-\tau_2)$$

$$\frac{dg}{dT} = -\frac{(g(T)-g_0)}{T_g} - \frac{(Q_a(T-\tau_1)+Q_b(T-\tau_2)) \cdot g(T)}{P_g \cdot T_g \cdot T_R}$$

τ_1, τ_2—Time delay parameters which are related to pulse energies, we get it for our system when the semiconductor saturable absorber mirror (SESAM) is not ideal in his functionality, δ_a, δ_b—Pulse filter losses (a, b), α_a, α_b—Pulse absorber losses (a, b), Q_a, Q_b—Pulse energies, l_0—Linear loss (output coupling), T_R—Round trip time, T_g—Saturable gain recovery time, $P_g \cdot T_g$—Saturation energies, g, Saturable gain, g_0—Unsaturated values of gain, T—Time parameter.

7.1 Find system fixed points and how they change for different values of system parameter (plot graphs of the behavior).
7.2 Discuss the stability of our system by the standard local stability analysis about any one of the equilibrium points (fixed points) for the following cases: (a) $\tau_1 = 0; \tau_2 = \Delta$ (b) $\tau_1 = \Delta; \tau_2 = 0$.
7.3 Find the small increment Jacobian of our system and characteristic equation $D(\lambda, \Delta)$.
7.4 Discuss stability switching for different values of Δ parameter (cases (a), (b)).

8. We have a system of Ti: Sapphir laser mode semiconductor saturable absorber mirror (SESAM) that operate in a multipulse mode. We need to analyze the stability of a soliton in the laser when subject to perturbations from a filter and an absorber. The system generalized complex Ginzburg-Landau equation (GCGLE) which is the master equation to describe the dynamical effects of laser by means of numerical simulations: $i \cdot T_R \cdot \psi_T - \frac{\beta_2}{2} \cdot \frac{\partial^2 \psi}{\partial t^2} + \kappa \cdot |\psi|^2 \cdot \psi = i \cdot (g - q - l_0) \cdot \psi + i \cdot \beta \cdot \frac{\partial^2 \psi}{\partial t^2}$. We can consider our system delay differential equations (DDEs) when the perturbations on the right hand side are not neglected.

$$T_R \cdot \frac{dQ_a}{dT} = 2 \cdot [g(T-\tau) \cdot (Q_a(T) + Q_b(T)) - l_0 - \delta_a - \alpha_a] \cdot Q_a(T)$$

$$T_R \cdot \frac{dQ_b}{dT} = 2 \cdot [g(T-\tau) \cdot (Q_a(T) + Q_b(T)) - l_0 - \delta_b - \alpha_b] \cdot Q_b(T)$$

$$\frac{dg}{dT} = -\frac{(g(T-\tau)-g_0)}{T_g} - \frac{(Q_a(T)+Q_b(T)) \cdot g(T-\tau)}{P_g \cdot T_g \cdot T_R}$$

τ—Time delay parameters which is related to Saturable gain (g): we get it for our system when the semiconductor saturable absorber mirror (SESAM) is not exactly the same as our original design, δ_a, δ_b—Pulse filter losses (a, b), α_a, α_b—Pulse absorber losses (a, b), Q_a, Q_b—Pulse energies, l_0—Linear loss (output coupling), T_R—Round trip time, T_g—Saturable gain recovery time, $P_g \cdot T_g$—Saturation energies, g—Saturable gain. g_0—Unsaturated values of gain, T—Time parameter.

8.1 Find system fixed points and how they change for different values of system parameter (plot graphs of the behavior).
8.2 Discuss the stability of our system by the standard local stability analysis about any one of the equilibrium points (fixed points) for different value of time delay parameter (τ).
8.3 Find the small increment Jacobian of our system and characteristic equation $D(\lambda, \tau)$.
8.4 Discuss stability switching for different values ofτparameter.

9. Semiconductor ring lasers (SRLs) are miniature ring laser devices with potential applications in optoelectronics, photonics and all-optical circuits. Semiconductor ring lasers are literally ring-shaped optical waveguides with a lasing medium. They have the ability to trap light in a ring, and recirculate it continuously as long as they remain powered. SRL have considerable gain and a low threshold pump power, making it a highly efficient laser system. The light was confined only to the film and did not reflect off of the outer surface of the optical fiber. There are nonlinear behaviors in semiconductor ring lasers (SRLs). A multimode model has been developed for linear and nonlinear interactions between modes in an SRL lasing uni-directionally. Heterodyne detection has been used to make high resolution measurements of the lasing spectra of an SRL in which the individual resonances associated with the coupled eigenvalues can be observed. There are number of key parameters characterizing the coupling mechanisms in the device and the semiconductor gain medium. There is an analytical model for mode interactions in an SRL. The analytic model helps us to extract key parameters characterizing the linear and nonlinear coupling mechanisms in the semiconductor laser medium. The analysis of nonlinear effects in a near single-mode laser can be simplified by considering only interactions pumped by the lasing mode. The analyses ignore the effect of the cavity resonance on the mode gain and analyze the interaction between a strong pump field at ω_0 and a pair of low intensity fields at $\omega_1 = \omega_0 + \Omega$ and $\omega_1 = \omega_0 - \Omega$. The refractive index dispersion is usually specified in terms of the first and second order variation in refractive index n with wavelength or refractive index variation in terms of ω. We can then express the first and second dispersion of $\chi_b^{'}$ in terms of n ($\chi_b^{'} = \chi_b^{'}(n)$). χ_b^1 accounts for the linear gain and refractive index of the material. χ_b^1 varies with frequency due to gain and refractive index dispersion.

$$\chi_b^{1'} = n_b^2 = n_0^2 - 2 \cdot n_0 \cdot \frac{\Delta\omega}{\omega_0} \cdot \lambda \cdot \frac{dn}{d\lambda} + \frac{(\Delta\omega)^2}{\omega_0^2} \cdot \left[\left(\lambda \cdot \frac{dn}{d\lambda}\right)^2 + 2 \cdot n_0 \cdot \lambda^2 \cdot \frac{d^2 n}{d\lambda^2}\right]$$

$$\chi_b^{1'} = n_b^2 = \chi_0^{1'} + \Delta\omega \cdot \frac{d\chi^{1'}}{d\omega} + (\Delta\omega)^2 \cdot \frac{d^2\chi^{1'}}{d\omega^2}$$

The only variables in our system are $n = n(\lambda)$; $\chi^{1'} = \chi^{1'}(\omega)$. We consider the parameters of our system: $\chi_b^{1'}, n_b, n_0, \chi_0^{1'}, \Delta\omega, \omega_0$.

9.1 Find system fixed points and how they change for different values of system parameter (plot graphs of the behavior).
9.2 Find the small increment Jacobian of our system and characteristic equation.
9.3 Use linearization technique (if you need) and try to approximate the phase portrait near a fixed point by that of a corresponding linear system.
9.4 Classify each fixed point according to eigenvalues and discuss stability.
9.5 How system parameters influence our system stability.

10. There is an important directional mode switching in semiconductor ring lasers through optical injection co-propagating with the lasing mode. It based on the particular structure of a two dimensional asymptotic phase space. Semiconductor Ring Lasers (SRLs) have Bi-stability, possibility of operation in either of the two counter propagating directions. The switching of the direction of operation of a SRL consists of injecting an optical signal counter propagating to the lasing mode. Once the signal is removed, the SRL remains stable in this direction. There is a scheme based on injection from only one side of the SRL, which would remove the need to have an injection laser at both sides of the SRL in order to switch back and forth between the two directional sides. The analysis of directional switching within the existing models is complicated due to the large number of dynamical variables. The analysis is based on an asymptotic reduction to two dimensions of a general rate equation model for a SRL. The switching scheme relies on the special topology of the invariant manifolds of the system. The SRL is operated in a single longitudinal, single-transverse mode. The rate equations model for that device is as follow:

$$\frac{dE_{1,2}}{dt} = \kappa \cdot (1 + i \cdot \alpha) \cdot [N \cdot (1 - s \cdot |E_{1,2}|^2 - c \cdot |E_{2,1}|^2) - 1] \cdot E_{1,2} - k \cdot e^{i \cdot \phi_k} \cdot E_{2,1} + \frac{1}{\kappa \cdot \tau_{in}} \cdot F_{1,2}(t)$$

Hint: The above differential equation is actually two differential equations: first, is with the first index and second is with the second index (Related to $E_i, F_i; i = 1, 2$).

$$\frac{dE_1}{dt} = \kappa \cdot (1 + i \cdot \alpha) \cdot [N \cdot (1 - s \cdot |E_1|^2 - c \cdot |E_2|^2) - 1] \cdot E_1$$
$$- k \cdot e^{i \cdot \phi_k} \cdot E_2 + \frac{1}{\kappa \cdot \tau_{in}} \cdot F_1(t)$$
$$\frac{dE_2}{dt} = \kappa \cdot (1 + i \cdot \alpha) \cdot [N \cdot (1 - s \cdot |E_2|^2 - c \cdot |E_1|^2) - 1] \cdot E_2$$
$$- k \cdot e^{i \cdot \phi_k} \cdot E_1 + \frac{1}{\kappa \cdot \tau_{in}} \cdot F_2(t)$$
$$\frac{dN}{dt} = \gamma \cdot [\mu - N - N \cdot (1 - s \cdot |E_1|^2 - c \cdot |E_2|^2) \cdot |E_1|^2$$
$$- N \cdot (1 - s \cdot |E_2|^2 - c \cdot |E_1|^2) \cdot |E_2|^2]$$

E_1, E_2—Slowly varying envelopes of the counter-propagating fields, N—Carrier inversion density, κ—Field decay rate, γ—Carrier decay rate, α—Linewidth enhancement factor, μ—Renormalized injection current with $\mu \approx 0$ at transparency and $\mu \approx 1$ at lasing threshold. s, c: parameters which self and cross saturation effects are added phenomenologically and are molded by them, k—Amplitude that characterized the localized reflections result in a linear coupling between the two fields, ϕ_k—Phase shift that characterized the localized reflections result in a linear coupling between the two fields, τ_{in}—Flight time in the ring cavity, $|E_{inj}|^2$, Injected power, Δ—Detuning between both lasers.

Remark Reflection of the counter-propagating modes occurs where ring cavity and Coupling waveguide meet and can also occur at the end facets of the coupling waveguide. Two counter-propagating modes are considered to saturate both their own and each other's gain due to spectral hole burning effects. The term $F_{1,2}(t) = E_{inj} \cdot e^{i \cdot \Delta \cdot t}$ represents the optically injected field in one of the two modes.

The variables of our system are $E_1 = E_1(t)$; $E_2 = E_2(t)$; $N = N(t)$.

10.1 Find system fixed points and how they change for different values of system parameter (plot graphs of the behavior).
10.2 Find the small increment Jacobian of our system and characteristic equation.
10.3 Use linearization technique (if you need) and try to approximate the phase portrait near a fixed point by that of a corresponding linear system.
10.4 Classify each fixed point according to eigenvalues and discuss stability.
10.5 How system parameters influence our system stability.

Remark The rate equations model for that device include complex functions, so you need to analyze them differently than a regular real functions.

References

1. A. Talebpour, S. Larochelle, S.L. Chin, Suppressed tunneling ionization of the D_2 molecule in an intense Ti: sapphire laser pulse. J. Phys. B: At. Mol. Opt. Phys. **31**, L49–L58 (1998)
2. A. Efimov, C. Schaffer, D. H. Reitze, Programmable shaping of Ultrabroad-bandwidth pulses from a Ti: sapphire laser. J. Opt. Soc. Am. **12**(10) (1995)
3. M.J. Lederer, B. Luther-Davies, H.H. Tan, C. Jagadish, N.N. Akhmediev, J.M. Soto-Crespo, Multipulse operation of a Ti: sapphire laser mode locked by an ion-implanted semiconductor saturable absorber mirror. J. Opt. Soc. Am. B **16**(6) (1999)
4. F. Song, J.Q. Yao, D.W. Zhou, J.Y. Qiao, G.Y. Zhang, J.G. Tian, Rate-equation theory and experimental research on dual-wavelength operation of a Ti: sapphire laser. Appl. Phys. B **72**, 605–610 (2001)
5. S.H. Strogatz, *Nonlinear Dynamics and Chaos*. Westview Press
6. S. Wiggins, Introduction to applied nonlinear dynamical systems and chaos. Text Appl. Math. (Hardcover)
7. Y. Kuang, *Delay Differential Equations with Applications in Population Dynamics* (Academic Press, Boston, 1993)
8. E. Beretta, Y. Kuang, Geometric stability switch criteria in delay differential systems with delay dependent parameters. SIAM J. Math. Anal. **33**, 1144–1165 (2002)
9. J. Kuang, Y. Cong, Stability of numerical methods for delay differential equations. Elsev. Sci. (2007)
10. E. Beretta, Y. Kuang, Geometric stability switch criteria in delay differential systems with delay dependent parameters. SIAM J. Math. Anal. **33**(5), 1144–1165 (Published electronically February 14, 2002)
11. Y. Kuang, Delay differential equations with applications in population dynamics. Math. Sci. Eng. **191** (1993 by Academic Press, Inc)
12. J.A. Davis, D.E. McNamara, D.M. Cottrell, T. Sonehara, Two-dimensional polarization encoding with a phase-only liquid-crystal spatial light modulator. Appl. Opt. **39**(10) (2000)
13. J. Guckenheimer, Nonlinear oscillations, dynamical systems, and bifurcations of vector fields. Appl. Math. Sci. **42**
14. X. Han, H. Zeng, Kerr-lens mode-locked Ti: sapphire laser with an additional intra cavity nonlinear medium. Opt. Express, **16**(23) (2008)
15. W. Van Saarloos, P.C. Hohenberg, Pulses and fronts in the complex Ginzburg-Landau equation near a subcritical bifurcation. Phys. Rev. Lett. **64**(7) (1990)
16. N. N. Akhmediev, A. Ankiewicz, Multi soliton solutions of the complex Ginzburg-Landau equation. Phys. Rev. Lett. **79**(21) (1997)

Chapter 3
Ion Channel and Long-Wavelength Lasers, Nonlinearity Applications in Engineering

Ion-channel system includes relativistic electron beam propagating through plasma in the ion-focused regime. Growth is enhanced by optical guiding in the ion channel, which acts as a dielectric waveguide. It is a transport of relativistic electron beams (REBs) in plasmas and mechanism of ion-focusing. The propagation in plasmas of short pulse, low emittance relativistic electron beams is related to plasma lens, the continuous plasma focus, the plasma wake field accelerator, and the beat wave accelerator. Beam injection and extraction from a plasma module is important in order to produce high quality electron beams with a plasma accelerator. There is a need for proper matching conditions to focus the incoming high brightness beam down to few microns size and to capture a high divergent beam at the exit without loss of beam quality. Plasma-based lenses provide focusing gradients of the order of kT/m with radially symmetric focusing. It gives compact and affordable alternative to permanent magnets in the design of transport lines. There is a coherent radiation from intense, relativistic electron beams in un-magnetized plasmas. The ion channel "free electron" laser (ICL), consisting of an intense, relativistic beam of electrons injected into a plasma less dense than the beam. The ion channel FEL (Free Electron Laser) makes use of ion-focusing to transport the beam, and a resonance, akin to that of the planer wiggler FEL, to produce coherent radiation. The ion channel acts as an optical fiber with step index of refraction. For high current, dielectric guiding eliminates the usual constraint that the Rayleigh length must be longer than the gain length. Rayleigh length or Rayleigh range is the distance along the propagation direction of a beam from the waist to the place where the area of the cross section is doubled. The three-dimensional free electron laser (FEL) power gain length's increase due to the collective effects of an ultra-relativistic electron beam, geometric transverse wake-field, coherent synchrotron radiation, and micro bunching instability. The gain length is affected by an increase of the electron beam projected emittance. It is done even though the slice local emittance is preserved. The ion channel FEL consists of a tank of neutral gas, from centimeters to meters in length, through which a plasma column millimeters in width is produced by an ionizing laser pulse. The relativistic electron beam (REB) is injected within less than a recombination time, and with

O. Aluf, *Advance Elements of Laser Circuits and Systems*,
https://doi.org/10.1007/978-3-030-64103-0_3

proper matching, as in a continuous plasma focus. It is propagating in the axial (+z) direction. As the beam head propagates through the plasma, it continuously expels plasma electrons from the beam volume, leaving fixed the relatively immobile ions to provide focusing for the remainder of the beam. The transverse force on the beam due to self-fields is much less than the transverse electric field due to the ion charges. The expulsion of plasma electrons produces a cylindrical channel occupied by un-neutralized ions and this is the "ion channel". We consider the motion in the ion-channel of a single electron, subject to an electromagnetic wave linearly polarized in the y direction. A Betatron is a type of cyclic particle accelerator which is a transformer with a torus-shaped vacuum tube as its secondary coil. An alternating current in the primary coils accelerates electrons in the vacuum around a circular path. The betatron period is a definition which is related to our ion-channel laser system. We get the perturbed equations of motion when averaging over the betatron period. Longitudinal modes in Long-wavelength lasers have mode-competition phenomena which we need to measure. The model behavior of an InGaAsP-InP Fabry Perot laser emitting in a long wavelength region is analyzed and stability is inspected. Mode-competition phenomena induce quasi-periodic hopping among several longitudinal modes, which reveal multimode like output spectra as the time-averaged spectra in long-wavelength lasers. There are variations of photon number and their frequency spectra in addition to the longitudinal mode spectra. The mechanism of the predicted multi-mode like oscillation is analyzed and theoretical model, variation of the photon number S_p, injected electron number N by set of differential equations. There is influence of instantaneous mode competition on the dynamics of semiconductor lasers. The nonlinear dynamics of the lasing modes are described mathematically by multimode rate equations of the photon number S_p of the lasing modes and the injected electron number N [1], 2, 3].

3.1 Ion Channel Laser Perturbed Differential Equations of Motion Stability Analysis

The basic mechanism is laser-driven acceleration in ion-channel. A long laser beam propagating through un-dense plasma produces a positively charged ion channel by expelling plasma electrons in the transverse direction. There is a dynamics of an electron in a resulting two-dimensional channel under the action of the laser field and the transverse electric field of the channel. We can enhance the axial momentum via amplification of betatron oscillations. Oscillations can be parametrically amplified when the betatron frequency, which increases with the wave amplitude, becomes comparable to the frequency of its modulations. There is a non-inertial (accelerated/decelerated) relativistic axial motion induced by the wave regardless of the angle between the laser electric field and the field of the channel and caused to modulation. For specific density, there is a well pronounced wave amplitude threshold above which the maximum electron energy is considerably enhanced. The energy enhancement we get is accompanied by spectrum broadening. The regime of laser interaction with the target electrons is greatly influenced by the density of the target. The interaction takes place only at the surface of the target if the density is over-critical (solid density targets). The laser beam is able to go through the target if the density is sub-critical, as in the case of gas jets. The low-density targets have extended interaction length with the laser and a higher energy gain by the electrons. A sub-critical plasma layer occurs also with solid-density targets where a significant pre-pulse is present. A pre-pulse can create a transparent pre-plasma, extending many wavelengths from the target surface along the beam path. The main pulse interacts with low-density plasma before reaching the target. We consider laser interaction with low density sub-critical plasma in order to see if such interactions can generate hot electrons in addition to the ones produced at the critical surface. The electron acceleration mechanism in this regime depends on laser pulse duration. A laser-produced wake field, the wake field mechanism becomes less efficient for longer laser pulses. For long laser beam, its ponder motive pressure tends to expel plasma electrons from the beam in the transverse direction. The plasmas generate a counter-acting electric field via charge separation. The laser can create a positively charged channel with a quasi-static transverse electric field evolving on an ion time scale. There are dynamics of an electron irradiated by a linearly polarized electromagnetic wave in a two-dimensional steady-state ion channel. The relativistic electron beam propagating through plasma in the ion-focused regime exhibits an electromagnetic instability at a resonant frequency$\omega \sim 2 \cdot \gamma^2 \cdot \omega_p$ (γ—Lorentz factor, ω_p—beam plasma radial frequency). There is an optical guiding in the ion channel, which acts as a dielectric waveguide, with fiber parameters $V \sim 2 \cdot \sqrt{\frac{I}{I_A}}$ (I—Peak beam current, I_A—Alfven current). The Alfven current I_A is a useful scaling parameter for intense beam studies. The magnetic deflection of electrons in a charge-neutral beam defines an upper limit on the transportable current. When a high-current beam enters plasma, plasma particles move to cancel the beam generated electric field. There is a response of a plasma to a steady-state beam and characterized plasma Debye length. Plasmas are collections of ions and electrons governed by long-range electromagnetic interactions. Usually, the densities of ions and electrons, n_e, and n_i, are almost equal and the mixture is space-charge neutralized. On plasma transport of electron beams the beam density n_b is much smaller than the plasma density $n_e \ldots (n_b \ll n_e)$. The beam drives out a few of the plasma electrons to achieve complete space-charge neutralization, where; $n_i = n_e + n_b$. Relativistic electron beams generate strong magnetic focusing forces. The beams are self-contained with only partial space-charge neutralization. The transport of such beams in low-density plasmas ($n_b > n_e$) is of considerable interest. The beam expels all the low-energy plasma electrons. The kinetic energy of plasma particles affects how they respond to beams. The velocity dispersion of plasma ions and electrons as Maxwell distributions with temperatures T_i and T_e is presented. The plasma particles have average drift velocities: V_i and V_e. Unconfined plasmas are stable against velocity space instabilities if $|V_i, |V_e| \ll \sqrt{\frac{2 \cdot k \cdot T_e}{m_e}}$. The condition holds when moderate current beams propagate through dense plasmas. High-current pulsed electron beams can induce a large plasma electron drift velocity, leading to a two stream instability and rapid plasma heating. Plasmas used for electron beam transport can have low temperature ($k \cdot T_e < 100\,\text{eV}$), and large spatial extent. The methods for beam plasma generation include laser ionization, pulsed plasma guns, low-energy electron discharges or collisions of beam particles with a background gas. Beam-generated plasmas are often not fully ionized-neutral atoms are present in the beam volume. Atoms may influence the beam and plasma responses through collisions. There is a possible simple model to calculate the steady-state response of plasma to an injected beam. The beam enters unconfined plasma of infinite dimension with no included magnetic field. Low-energy electrons respond rapidly to the presence of the beam while the massive ions respond slowly. Assuming that beam pulse is long enough for plasma electrons to adjust to a modified equilibrium. The ion channel laser (ICL), consisting of an intense, relativistic beam of electrons injected into a plasma less dense than the beam. The ion channel laser (ICL) makes use of ion-focusing to transport the beam, and a resonance, akin to that of the planar Wiggler FEL, to produce coherent radiation. A uniform static axial magnetic field is induced in addition to the ion electro static field and Wiggler magnetic field, this is a possibility. The evolution of electron beam in the tapered planar Wiggler field is with self-electric and self-magnetic fields. Free-electron laser (FEL) operation often requires sufficiently large gain, which increase when the beam current is increased. In the high-current regime, the electron motion can be altered by the self-generated field effects. There is evolution of electron beam in the tapered planar Wiggler field with self-electric and self-magnetic fields. The ion channel laser (ICL) consists of a tank of neutral gas, through which a plasma column millimeters in width is produced by a laser pulse. A relativistic electron beams (REBs) is injected, propagating in the axial (+z) direction. It is done within less than a recombination time, and with proper matching. A relativistic electron beam, propagating through an under dense plasma ($n_b > n_e$), expels plasma electron from the beam volume and beyond to produce an "ion-channel", which then focusses the beam, and causes it to radiate. The structure of relativistic electron beam which injected and propagating in the axial (+z) direction is as follow (Fig. 3.1).

The beam head is propagated through the plasma and expels plasma electrons from the beam volume. It is leaving fixed the relatively immobile ions to provide focusing for the remainder of the beam. Assumption: The transverse force on the beam due to self-fields is much less than the transverse electric fields due to the ion-charge. The plasma densities possibilities are as follow (Table 3.1).

The expulsion of plasma electron produces a cylindrical channel of radius b, $b \approx a \cdot \sqrt{\frac{n_b}{n_p}}$ occupied by un-neutralized ions and this is the "ion-channel". We define τ_r as the current rise time. The large radial plasma oscillations are not excited as plasma electrons are ejected from the channel and this requires $\omega_p \cdot \tau_r \gg 1$; $\omega_p^2 = \frac{4 \cdot \pi \cdot n_p \cdot e^2}{m}$.

E—Electron charge, m—Electron mass, ω_p—Plasma frequency, ω_b —beam-plasma frequency, η—Overlap integral of the mode and beam radial profiles, τ_r—Current rise time, n_p—Plasma density prior to channel formation, a—Relativistic electron beams (REBs) cylindrical channel radius, b—A cylindrical channel radius

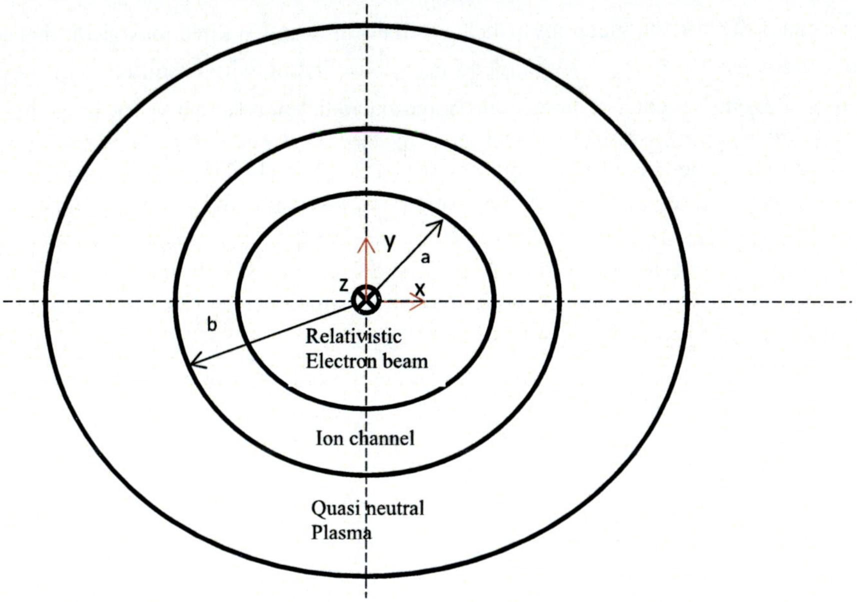

Figure 3.1 Structure of relativistic electron beam which injected and propagating in the axial (+z)

Table 3.1 System plasma densities possibilities, in term of n_p and n_b

Condition	Meaning (outcomes)
$n_p \gg \frac{n_b}{\gamma^2}$	Transverse force on the beam due to self-fields is much less than the transverse electric field due to the ion charge (n_p—Plasma density prior to channel formation, n_b—Peak beam density, γ—Lorentz factor)
$n_b > n_p$	The generalization to an over dense plasma

which expulsion of plasma electron happened, $_c$—Speed of light in vacuum, I—Peak beam current, I_A—Alfven current, τ—Beam length, τ_i—Time for the ions to collapse inward due to the radial electric field of the beam, ω_β—Betatron frequency which electrons oscillate in x and y, ω_1—Up shifted frequency which electrons are oscillating, ω—Frequency of radiation in the forward (+z) direction in the lab frame, A, φ - Eikonal amplitude and phase, respectively, A_y—Vector potential at y direction, p_x, p_y, p_z—Canonical momenta in x, y and z, respectively, V_z—Betatron averaged drift velocity in z, ε_n—rms normalized emittance, $\Delta\omega$—Detuning parameter, k_β—Betatron angular wave number, k_z—Wave number in z direction, ψ—Phase variable, a_β—Betatron parameter which is analogous to the wiggler parameter in an Free Electron Laser (FEL), q_x, q_y, q_z—System's variables in x, y, and z directions, θ_x, θ_y—System variables in x and y directions. The beam length (τ) is short compare to the time for ions to collapse inward due to the radial electric field of the beam (τ_i), $\tau < \tau_i$; $\tau_i \sim \frac{1}{4} \cdot (\frac{b}{c}) \cdot \sqrt{\frac{m_i}{m}} \cdot \sqrt{\frac{I_A}{I}} > \tau$. Additionally $I = \pi \cdot a^2 \cdot n_b \cdot e \cdot c$; $I_A = \frac{m \cdot c^3}{e}$, we can write for the electrons which oscillate in x and y directions at the betatron frequency$\omega_\beta \sim \omega_p \cdot \sqrt{\frac{m\cdot c}{2\cdot p_z}}$. Electrons are oscillating with upshifted frequency $\omega_1 \sim \gamma\cdot\omega_\beta$ in the center of momentum frame and radiate incoherently. The frequency of radiation in the forward (+z) direction, in the lab frame is $\omega \sim 2\cdot\gamma\cdot\omega_1 \sim 2\cdot\gamma^2\cdot\omega_\beta$. The motion in the "ion-channel" of a single electron is dependent on the electromagnetic wave linearity polarized in the y direction. We define the vector potentialA_y, $A_y = (\frac{mc^2}{e})\cdot A\cdot\sin(\xi)$; $\xi = k_z\cdot z - \omega\cdot t + \varphi$. We define A, φ as Eikonal amplitude and phase, respectively (the two parameters are related to Eikonal approximation method useful in wave scattering equation which occur in our system). The beam-plasma frequencyω_b, $\omega_b^2 = \frac{4\cdot\pi\cdot n_b\cdot e^2}{m}$ $\omega_b = 2\cdot e\cdot\sqrt{\frac{\pi\cdot n_b}{m}}$. Assumption: The an-harmonic in the transverse motion and second order terms in A are neglected, as well as derivatives of the slowly varying Eikonal quantities, and transverse gradients. We relate to the canonical momenta in x, y and z, respectively p_x, p_y, p_z, such that $p_x = m\cdot c\cdot q_x\cdot\sin(\theta_x)$; $p_y = m\cdot c\cdot q_y\cdot\sin(\theta_y)$; $p_z = m\cdot c\cdot q_z$. If the Eikonal amplitude$A$, $A = 0$, q_x, q_y are constants then $\frac{d\theta_x}{dt} = \omega_\beta$; $\frac{d\theta_y}{dt} = \omega_\beta$. The phase variable ψ, $\psi = \theta_y + \xi$; $\psi = \theta_y + k_z\cdot z - \omega\cdot t + \varphi$, average over the betatron period. The detuning parameter $\Delta\omega$, $\Delta\omega = k_z\cdot V_z - \omega + \omega_\beta$, where V_z is betatron-average drift velocity in z, $\frac{V_z}{c} = 1 - \frac{(2+q_x^2+q_y^2)}{4\cdot q_z^2}$. The a_β, $a_\beta^2 = \frac{q_x^2+q_y^2}{2}$; $a_\beta = \sqrt{\frac{q_x^2+q_y^2}{2}}$ is analogous to the wiggler parameter in an FEL. We can consider for the round beam (Fig. 3.1) a_β, $a_\beta = \frac{q_z\cdot k_\beta\cdot a}{\sqrt{2}} a_\beta = \frac{\sqrt{2^3}\cdot\varepsilon_n}{a}$ (initially the same for each particle). Additionally, $\varepsilon_n = \frac{q_z\cdot k_\beta\cdot a^2}{4}$; $k_\beta = \frac{\omega_\beta}{c}$. The perturbed equations of motion derived (average over betatron period) [1]:

$$\frac{d\psi}{dt} = k_z\cdot V_z - \omega + \frac{d\theta_y}{dt} + \frac{d\varphi}{dt} + \frac{1}{2}\cdot k_z\cdot c\cdot\frac{q_y}{q_z^2}\cdot A\cdot\cos(\psi)$$

$$\frac{d\theta_y}{dt} = \omega_\beta\cdot[1 - \frac{1}{2\cdot q_y}\cdot A\cdot\cos(\psi)]\ ;\ \frac{dq_z}{dt} = -\frac{1}{2}\cdot k_z\cdot c\cdot\frac{q_y}{q_z}\cdot A\cdot\sin(\psi)$$

$$\frac{dq_y}{dt} = -\frac{1}{2}\cdot(\omega_\beta + \frac{1}{4}\cdot k_z\cdot c\cdot\frac{q_y^2}{q_z^2})\cdot A\cdot\sin(\psi)$$

$$\frac{dq_x}{dt} = -\frac{1}{8}\cdot k_z\cdot c\cdot\frac{q_y\cdot q_x}{q_z^2}\cdot A\cdot\sin(\psi)$$

We consider that the derivative of Eikonal phase in time $\frac{d\varphi}{dt}$ is very small $\frac{d\varphi}{dt}\to 0$. We get the following perturbed equations of motion (average over betatron period):

$$\frac{d\psi}{dt} = k_z\cdot V_z - \omega + \omega_\beta\cdot[1 - \frac{1}{2\cdot q_y}\cdot A\cdot\cos(\psi)] + (\frac{d\varphi}{dt}\to 0)$$
$$+ \frac{1}{2}\cdot k_z\cdot c\cdot\frac{q_y}{q_z^2}\cdot A\cdot\cos(\psi)$$

$$\frac{dq_z}{dt} = -\frac{1}{2} \cdot k_z \cdot c \cdot \frac{q_y}{q_z} \cdot A \cdot \sin(\psi);$$

$$\frac{dq_y}{dt} = -\frac{1}{2} \cdot (\omega_\beta + \frac{1}{4} \cdot k_z \cdot c \cdot \frac{q_y^2}{q_z^2}) \cdot A \cdot \sin(\psi)$$

$$\frac{dq_x}{dt} = -\frac{1}{8} \cdot k_z \cdot c \cdot \frac{q_y \cdot q_x}{q_z^2} \cdot A \cdot \sin(\psi)$$

Our system variables are ψ, q_x, q_y, q_z. $\frac{V_z}{c} = 1 - \frac{(2+q_x^2+q_y^2)}{4 \cdot q_z^2} \Rightarrow V_z = c \cdot [1 - \frac{(2+q_x^2+q_y^2)}{4 \cdot q_z^2}]$.

We get the following perturbed equations of motion (average over betatron period):

$$\frac{d\psi}{dt} = k_z \cdot c \cdot [1 - \frac{(2 + q_x^2 + q_y^2)}{4 \cdot q_z^2}] - \omega + \omega_\beta \cdot [1 - \frac{1}{2 \cdot q_y} \cdot A \cdot \cos(\psi)] + \frac{1}{2} \cdot k_z \cdot c \cdot \frac{q_y}{q_z^2} \cdot A \cdot \cos(\psi)$$

$$\frac{dq_z}{dt} = -\frac{1}{2} \cdot k_z \cdot c \cdot \frac{q_y}{q_z} \cdot A \cdot \sin(\psi) \; ; \; \frac{dq_y}{dt} = -\frac{1}{2} \cdot (\omega_\beta + \frac{1}{4} \cdot k_z \cdot c \cdot \frac{q_y^2}{q_z^2}) \cdot A \cdot \sin(\psi)$$

$$\frac{dq_x}{dt} = -\frac{1}{8} \cdot k_z \cdot c \cdot \frac{q_y \cdot q_x}{q_z^2} \cdot A \cdot \sin(\psi)$$

At fixed points: $\frac{d\psi}{dt} = 0$; $\frac{dq_z}{dt} = 0$; $\frac{dq_y}{dt} = 0$; $\frac{dq_x}{dt} = 0$

$$\frac{dq_x}{dt} = 0 \Rightarrow -\frac{1}{8} \cdot k_z \cdot c \cdot \frac{q_y^* \cdot q_x^*}{(q_z^*)^2} \cdot A \cdot \sin(\psi^*) = 0$$

Case 1 $q_y^* = 0$

$$q_y^* = 0; \frac{dq_y}{dt} = 0 \Rightarrow -\frac{1}{2} \cdot \omega_\beta \cdot A \cdot \sin(\psi^*) = 0; \omega_\beta \cdot A \neq 0$$
$$\Rightarrow \sin(\psi^*) = 0$$
$$\Rightarrow \psi^* = n \cdot \pi \, \forall n = 0, 1, 2, \ldots; n \in \mathbb{N}_0$$

$$\frac{d\psi}{dt} = 0 \Rightarrow k_z \cdot c \cdot [1 - \frac{(2 + (q_x^*)^2)}{4 \cdot (q_z^*)^2}] - \omega + \omega_\beta \cdot [1 - \frac{1}{2 \cdot (q_y^* = 0)} \cdot A \cdot \cos(\psi^*)] = 0$$

$$\cos(\psi^*) = \pm\sqrt{1 - \sin^2(\psi^*)};\ \sin(\psi^*) = 0 \Rightarrow \cos(\psi^*) = \pm 1$$

$$\frac{d\psi}{dt} = 0 \Rightarrow k_z \cdot c \cdot [1 - \frac{(2 + (q_x^*)^2)}{4 \cdot (q_z^*)^2}] - \omega + \omega_\beta \cdot [1 - \frac{1}{2 \cdot (q_y^* = 0)} \cdot A \cdot (\pm 1)] = 0$$

$$\omega_\beta > 0;\ \omega_\beta \cdot [1 - \frac{1}{2 \cdot (q_y^* = 0)} \cdot A \cdot (\pm 1)] \to \mp\infty$$

$$k_z \cdot c \cdot [1 - \frac{(2 + (q_x^*)^2)}{4 \cdot (q_z^*)^2}] - \omega \to \pm\infty$$

$$[k_z \cdot c - \omega] - k_z \cdot c \cdot \frac{(2 + (q_x^*)^2)}{4 \cdot (q_z^*)^2} \to \pm\infty$$

$$\Rightarrow -k_z \cdot c \cdot \frac{(2 + (q_x^*)^2)}{4 \cdot (q_z^*)^2} \to \pm\infty - [k_z \cdot c - \omega] \overset{k_z \cdot c - \omega \ll 1}{\approx} \pm\infty$$

$$k_z \cdot c \cdot \frac{(2 + (q_x^*)^2)}{4 \cdot (q_z^*)^2} \to \mp\infty \Rightarrow q_z^* = 0$$

The first fixed point: $E^{(0)}(\psi^*, q_x^*, q_y^*, q_z^*) = (n \cdot \pi, q_x^*, 0, 0);\ n \in \mathbb{N}_0;\ q_x^* \in \mathbb{R}$.
Remark The q_x^* values we choose establish first fixed point coordinate's values.
Case 2 $q_x^* = 0$

$$q_x^* = 0;\ \frac{dq_z}{dt} = 0 \Rightarrow -\frac{1}{2} \cdot k_z \cdot c \cdot \frac{q_y^*}{q_z^*} \cdot A \cdot \sin(\psi^*) = 0$$

$$\frac{dq_y}{dt} = 0 \Rightarrow -\frac{1}{2} \cdot (\omega_\beta + \frac{1}{4} \cdot k_z \cdot c \cdot \frac{(q_y^*)^2}{(q_z^*)^2}) \cdot A \cdot \sin(\psi^*) = 0$$

$$\frac{d\psi}{dt} = 0 \Rightarrow k_z \cdot c \cdot [1 - \frac{(2 + (q_y^*)^2)}{4 \cdot (q_z^*)^2}] - \omega + \omega_\beta \cdot [1 - \frac{1}{2 \cdot q_y^*} \cdot A \cdot \cos(\psi^*)] + \frac{1}{2} \cdot k_z \cdot c \cdot \frac{q_y^*}{(q_z^*)^2} \cdot A \cdot \cos(\psi^*) = 0$$

Case 2.1 $q_x^* = 0;\ \frac{dq_z}{dt} = 0 \Rightarrow -\frac{1}{2} \cdot k_z \cdot c \cdot \frac{q_y^*}{q_z^*} \cdot A \cdot \sin(\psi^*) = 0 \Rightarrow \sin(\psi^*) = 0$

$$\sin(\psi^*) = 0 \Rightarrow \psi^* = n \cdot \pi \; \forall n = 0, 1, 2, \ldots; n \in \mathbb{N}_0$$

$$\cos(\psi^*) = \pm\sqrt{1 - \sin^2(\psi^*)}; \; \sin(\psi^*) = 0 \Rightarrow \cos(\psi^*) = \pm 1$$

$$k_z \cdot c \cdot [1 - \frac{(2 + (q_y^*)^2)}{4 \cdot (q_z^*)^2}] - \omega + \omega_\beta \cdot [1 - \frac{1}{2 \cdot q_y^*} \cdot A \cdot \cos(\psi^*)] + \frac{1}{2} \cdot k_z \cdot c \cdot \frac{q_y^*}{(q_z^*)^2} \cdot A \cdot \cos(\psi^*) = 0$$

$$k_z \cdot c \cdot [1 - \frac{(2 + (q_y^*)^2)}{4 \cdot (q_z^*)^2}] + [\omega_\beta - \omega] + \{\frac{1}{2} \cdot k_z \cdot c \cdot \frac{q_y^*}{(q_z^*)^2} - \frac{\omega_\beta}{2 \cdot q_y^*}\} \cdot A \cdot \cos(\psi^*) = 0$$

$$k_z \cdot c \cdot [1 - \frac{(2 + (q_y^*)^2)}{4 \cdot (q_z^*)^2}] + [\omega_\beta - \omega] + \{\frac{1}{2} \cdot k_z \cdot c \cdot \frac{q_y^*}{(q_z^*)^2} - \frac{\omega_\beta}{2 \cdot q_y^*}\} \cdot (\pm 1) = 0$$

Case 2.1.1

$$k_z \cdot c \cdot [1 - \frac{(2 + (q_y^*)^2)}{4 \cdot (q_z^*)^2}] + [\omega_\beta - \omega] + \{\frac{1}{2} \cdot k_z \cdot c \cdot \frac{q_y^*}{(q_z^*)^2} - \frac{\omega_\beta}{2 \cdot q_y^*}\} \cdot (+1) = 0$$

$$k_z \cdot c - \frac{k_z \cdot c \cdot (2 + (q_y^*)^2)}{4 \cdot (q_z^*)^2} + [\omega_\beta - \omega] + \frac{1}{2} \cdot k_z \cdot c \cdot \frac{q_y^*}{(q_z^*)^2} - \frac{\omega_\beta}{2 \cdot q_y^*} = 0$$

$$\frac{1}{2} \cdot k_z \cdot c \cdot \frac{q_y^*}{(q_z^*)^2} - \frac{k_z \cdot c \cdot (2 + (q_y^*)^2)}{4 \cdot (q_z^*)^2} - \frac{\omega_\beta}{2 \cdot q_y^*} = \omega - \omega_\beta - k_z \cdot c$$

$$\frac{2 \cdot k_z \cdot c \cdot (q_y^*)^2 - k_z \cdot c \cdot (2 + (q_y^*)^2) \cdot q_y^* - \omega_\beta \cdot 2 \cdot (q_z^*)^2}{4 \cdot (q_z^*)^2 \cdot q_y^*} = (\omega - \omega_\beta - k_z \cdot c)$$

$$2 \cdot k_z \cdot c \cdot (q_y^*)^2 - k_z \cdot c \cdot (2 + (q_y^*)^2) \cdot q_y^* - \omega_\beta \cdot 2 \cdot (q_z^*)^2 = 4 \cdot (q_z^*)^2 \cdot q_y^* \cdot (\omega - \omega_\beta - k_z \cdot c)$$

$$2 \cdot k_z \cdot c \cdot (q_y^*)^2 - k_z \cdot c \cdot (2 + (q_y^*)^2) \cdot q_y^* = \omega_\beta \cdot 2 \cdot (q_z^*)^2 + 4 \cdot (q_z^*)^2 \cdot q_y^* \cdot (\omega - \omega_\beta - k_z \cdot c)$$

$$2 \cdot k_z \cdot c \cdot (q_y^*)^2 - k_z \cdot c \cdot (2 + (q_y^*)^2) \cdot q_y^*$$
$$= 2 \cdot (q_z^*)^2 \cdot [\omega_\beta + 2 \cdot q_y^* \cdot (\omega - \omega_\beta - k_z \cdot c)]$$

$$2 \cdot k_z \cdot c \cdot (q_y^*)^2 - k_z \cdot c \cdot 2 \cdot q_y^* - k_z \cdot c \cdot (q_y^*)^3$$
$$= 2 \cdot (q_z^*)^2 \cdot [\omega_\beta + 2 \cdot q_y^* \cdot (\omega - \omega_\beta - k_z \cdot c)]$$

$$2 \cdot k_z \cdot c \cdot (q_y^*)^2 - k_z \cdot c \cdot 2 \cdot q_y^* - k_z \cdot c \cdot (q_y^*)^3$$
$$= 2 \cdot (q_z^*)^2 \cdot [\omega_\beta + 2 \cdot q_y^* \cdot (\omega - \omega_\beta - k_z \cdot c)]$$

$$2 \cdot (q_z^*)^2 \cdot [\omega_\beta + 2 \cdot q_y^* \cdot (\omega - \omega_\beta - k_z \cdot c)]$$
$$= k_z \cdot c \cdot q_y^* \cdot [2 \cdot q_y^* - 2 - (q_y^*)^2]$$

$$(q_z^*)^2 = \frac{k_z \cdot c \cdot q_y^* \cdot [2 \cdot q_y^* - 2 - (q_y^*)^2]}{2 \cdot [\omega_\beta + 2 \cdot q_y^* \cdot (\omega - \omega_\beta - k_z \cdot c)]}$$

$$q_z^* = \pm \sqrt{\frac{k_z \cdot c \cdot q_y^* \cdot [2 \cdot q_y^* - 2 - (q_y^*)^2]}{2 \cdot [\omega_\beta + 2 \cdot q_y^* \cdot (\omega - \omega_\beta - k_z \cdot c)]}}$$

The second fixed point:

$$E^{(1)}(\psi^*, q_x^*, q_y^*, q_z^*) = (n \cdot \pi, 0, q_y^*, \pm \sqrt{\frac{k_z \cdot c \cdot q_y^* \cdot [2 \cdot q_y^* - 2 - (q_y^*)^2]}{2 \cdot [\omega_\beta + 2 \cdot q_y^* \cdot (\omega - \omega_\beta - k_z \cdot c)]}})$$
$$n \in \mathbb{N}_0;\ q_y^* \in \mathbb{R}$$

Remark The q_y^* values we choose establish q_z^* fixed point values.

Case 2.1.2

$$k_z \cdot c \cdot \left[1 - \frac{(2 + (q_y^*)^2)}{4 \cdot (q_z^*)^2}\right] + [\omega_\beta - \omega] + \left\{\frac{1}{2} \cdot k_z \cdot c \cdot \frac{q_y^*}{(q_z^*)^2} - \frac{\omega_\beta}{2 \cdot q_y^*}\right\} \cdot (-1) = 0$$

$$k_z \cdot c \cdot \left[1 - \frac{(2 + (q_y^*)^2)}{4 \cdot (q_z^*)^2}\right] + [\omega_\beta - \omega] - \frac{1}{2} \cdot k_z \cdot c \cdot \frac{q_y^*}{(q_z^*)^2} + \frac{\omega_\beta}{2 \cdot q_y^*} = 0$$

$$k_z \cdot c - k_z \cdot c \cdot \frac{(2 + (q_y^*)^2)}{4 \cdot (q_z^*)^2} + [\omega_\beta - \omega] - \frac{1}{2} \cdot k_z \cdot c \cdot \frac{q_y^*}{(q_z^*)^2} + \frac{\omega_\beta}{2 \cdot q_y^*} = 0$$

$$[k_z \cdot c + \omega_\beta - \omega] - k_z \cdot c \cdot \frac{(2 + (q_y^*)^2)}{4 \cdot (q_z^*)^2} - \frac{1}{2} \cdot k_z \cdot c \cdot \frac{q_y^*}{(q_z^*)^2} + \frac{\omega_\beta}{2 \cdot q_y^*} = 0$$

$$[k_z \cdot c + \omega_\beta - \omega] + \frac{-q_y^* \cdot k_z \cdot c \cdot (2 + (q_y^*)^2) - 2 \cdot q_y^* \cdot k_z \cdot c \cdot q_y^* + 2 \cdot (q_z^*)^2 \cdot \omega_\beta}{4 \cdot (q_z^*)^2 \cdot q_y^*} = 0$$

$$- q_y^* \cdot k_z \cdot c \cdot (2 + (q_y^*)^2) - 2 \cdot q_y^* \cdot k_z \cdot c \cdot q_y^* + 2 \cdot (q_z^*)^2 \cdot \omega_\beta = [-k_z \cdot c - \omega_\beta + \omega] \cdot 4 \cdot (q_z^*)^2 \cdot q_y^*$$

$$- q_y^* \cdot k_z \cdot c \cdot (2 + (q_y^*)^2) - 2 \cdot q_y^* \cdot k_z \cdot c \cdot q_y^* = 2 \cdot (q_z^*)^2 \cdot \{[-k_z \cdot c - \omega_\beta + \omega] \cdot 2 \cdot q_y^* - \omega_\beta\}$$

$$k_z \cdot c \cdot q_y^* \cdot [-(2 + (q_y^*)^2) - 2 \cdot q_y^*] = 2 \cdot (q_z^*)^2 \cdot \{[-k_z \cdot c - \omega_\beta + \omega] \cdot 2 \cdot q_y^* - \omega_\beta\}$$

$$(q_z^*)^2 = \frac{k_z \cdot c \cdot q_y^* \cdot [-(2 + (q_y^*)^2) - 2 \cdot q_y^*]}{2 \cdot \{[-k_z \cdot c - \omega_\beta + \omega] \cdot 2 \cdot q_y^* - \omega_\beta\}}$$

$$q_z^* = \pm \sqrt{\frac{k_z \cdot c \cdot q_y^* \cdot [-(2 + (q_y^*)^2) - 2 \cdot q_y^*]}{2 \cdot \{[-k_z \cdot c - \omega_\beta + \omega] \cdot 2 \cdot q_y^* - \omega_\beta\}}}$$

The third fixed point:

$$E^{(2)}(\psi^*, q_x^*, q_y^*, q_z^*) = (n \cdot \pi, 0, q_y^*, \pm \sqrt{\frac{k_z \cdot c \cdot q_y^* \cdot [-(2 + (q_y^*)^2) - 2 \cdot q_y^*]}{2 \cdot \{[-k_z \cdot c - \omega_\beta + \omega] \cdot 2 \cdot q_y^* - \omega_\beta\}}})$$

$$n \in \mathbb{N}_0; q_y^* \in \mathbb{R}$$

Remark The q_y^* values we choose establish q_z^* fixed point values.

Case 2.2 $q_x^* = 0; \frac{dq_z}{dt} = 0 \Rightarrow -\frac{1}{2} \cdot k_z \cdot c \cdot \frac{q_y^*}{q_z^*} \cdot A \cdot \sin(\psi^*) = 0 \Rightarrow q_y^* = 0$

$$\frac{dq_y}{dt} = 0 \Rightarrow -\frac{1}{2} \cdot (\omega_\beta + \frac{1}{4} \cdot k_z \cdot c \cdot \frac{(q_y^*)^2}{(q_z^*)^2}) \cdot A \cdot \sin(\psi^*)|_{q_y^*=0} = -\frac{1}{2} \cdot \omega_\beta \cdot A \cdot \sin(\psi^*) = 0$$

$$-\frac{1}{2}\cdot\omega_{\beta}\cdot A\cdot\sin(\psi^{*})=0;\ \omega_{\beta}\cdot A\neq 0$$
$$\sin(\psi^{*})=0\Rightarrow\psi^{*}=n\cdot\pi\ \forall n=0,1,2,...;\ n\in\mathbb{N}_{0}$$

$$\cos(\psi^{*})=\pm\sqrt{1-\sin^{2}(\psi^{*})};\ \sin(\psi^{*})=0\Rightarrow\cos(\psi^{*})=\pm 1$$

$$\frac{d\psi}{dt}=0\Rightarrow k_{z}\cdot c\cdot\left[1-\frac{(2+(q_{y}^{*})^{2})}{4\cdot(q_{z}^{*})^{2}}\right]-\omega+\omega_{\beta}\cdot[1-\frac{1}{2\cdot q_{y}^{*}}\cdot A\cdot\cos(\psi^{*})]+\frac{1}{2}\cdot k_{z}\cdot c\cdot\frac{q_{y}^{*}}{(q_{z}^{*})^{2}}\cdot A\cdot\cos(\psi^{*})=0$$

$$q_{x}^{*}=0;\ q_{y}^{*}=0\Rightarrow k_{z}\cdot c\cdot[1-\frac{1}{2\cdot(q_{z}^{*})^{2}}]-\omega+\omega_{\beta}\cdot[1-\frac{1}{2\cdot(q_{y}^{*}=0)}\cdot A\cdot\cos(\psi^{*})]=0$$

$$k_{z}\cdot c-k_{z}\cdot c\cdot\frac{1}{2\cdot(q_{z}^{*})^{2}}-\omega+\omega_{\beta}-\frac{\omega_{\beta}}{2\cdot(q_{y}^{*}=0)}\cdot A\cdot\cos(\psi^{*})=0$$

$$[k_{z}\cdot c-\omega+\omega_{\beta}]-k_{z}\cdot c\cdot\frac{1}{2\cdot(q_{z}^{*})^{2}}-\frac{\omega_{\beta}}{2\cdot(q_{y}^{*}=0)}\cdot A\cdot\cos(\psi^{*})=0$$

$$k_{z}\cdot c\cdot\frac{1}{2\cdot(q_{z}^{*})^{2}}=[k_{z}\cdot c-\omega+\omega_{\beta}]-\frac{\omega_{\beta}}{2\cdot(q_{y}^{*}=0)}\cdot A\cdot(\pm 1)$$
$$\frac{\omega_{\beta}}{2\cdot(q_{y}^{*}=0)}\cdot A\cdot(\pm 1)\rightarrow\pm\infty$$

$$|\frac{\omega_{\beta}}{2\cdot(q_{y}^{*}=0)}\cdot A\cdot(\pm 1)|\gg k_{z}\cdot c-\omega+\omega_{\beta}$$
$$k_{z}\cdot c\cdot\frac{1}{2\cdot(q_{z}^{*})^{2}}=-\frac{\omega_{\beta}}{2\cdot(q_{y}^{*}=0)}\cdot A\cdot(\pm 1)$$
$$k_{z}\cdot c\cdot\frac{1}{2\cdot(q_{z}^{*})^{2}}\rightarrow\mp\infty\Rightarrow q_{z}^{*}\rightarrow 0$$

The fourth fixed point: $E^{(3)}(\psi^{*},q_{x}^{*},q_{y}^{*},q_{z}^{*})=(n\cdot\pi,0,0,0);\ n\in\mathbb{N}_{0}$.
Case 3 $\sin(\psi^{*})=0\Rightarrow\psi^{*}=n\cdot\pi\ \forall n=0,1,2,\ldots;\ n\in\mathbb{N}_{0}$

$$\cos(\psi^{*})=\pm\sqrt{1-\sin^{2}(\psi^{*})};\ \sin(\psi^{*})=0\Rightarrow\cos(\psi^{*})=\pm 1$$

$$\frac{d\psi}{dt} = 0 \Rightarrow k_z \cdot c \cdot [1 - \frac{(2 + (q_y^*)^2 + (q_x^*)^2)}{4 \cdot (q_z^*)^2}] - \omega + \omega_\beta \cdot [1 - \frac{1}{2 \cdot q_y^*} \cdot A \cdot \cos(\psi^*)] + \frac{1}{2} \cdot k_z \cdot c \cdot \frac{q_y^*}{(q_z^*)^2} \cdot A \cdot \cos(\psi^*) = 0$$

$$[k_z \cdot c - \omega + \omega_\beta] - \frac{k_z \cdot c \cdot (2 + (q_y^*)^2 + (q_x^*)^2)}{4 \cdot (q_z^*)^2} + \frac{1}{2} \cdot [k_z \cdot c \cdot \frac{q_y^*}{(q_z^*)^2} - \frac{\omega_\beta}{q_y^*}] \cdot A \cdot \cos(\psi^*) = 0$$

$$[k_z \cdot c - \omega + \omega_\beta] - \frac{k_z \cdot c \cdot (2 + (q_y^*)^2 + (q_x^*)^2)}{4 \cdot (q_z^*)^2} + \frac{1}{2} \cdot [k_z \cdot c \cdot \frac{q_y^*}{(q_z^*)^2} - \frac{\omega_\beta}{q_y^*}] \cdot A \cdot (\pm 1) = 0$$

$$-\frac{k_z \cdot c \cdot (2 + (q_y^*)^2 + (q_x^*)^2)}{4 \cdot (q_z^*)^2} + \frac{1}{2} \cdot [k_z \cdot c \cdot \frac{q_y^*}{(q_z^*)^2} - \frac{\omega_\beta}{q_y^*}] \cdot A \cdot (\pm 1) = [-k_z \cdot c + \omega - \omega_\beta]$$

Case 3.1 $$-\frac{k_z \cdot c \cdot (2 + (q_y^*)^2 + (q_x^*)^2)}{4 \cdot (q_z^*)^2} + \frac{1}{2} \cdot [k_z \cdot c \cdot \frac{q_y^*}{(q_z^*)^2} - \frac{\omega_\beta}{q_y^*}] \cdot A = [-k_z \cdot c + \omega - \omega_\beta]$$

$$-\frac{k_z \cdot c \cdot (2 + (q_y^*)^2 + (q_x^*)^2)}{4 \cdot (q_z^*)^2} + k_z \cdot c \cdot \frac{q_y^* \cdot A}{2 \cdot (q_z^*)^2} - \frac{\omega_\beta \cdot A}{2 \cdot q_y^*} = [-k_z \cdot c + \omega - \omega_\beta]$$

$$\frac{-k_z \cdot c \cdot (2 + (q_y^*)^2 + (q_x^*)^2) \cdot q_y^* + k_z \cdot c \cdot q_y^* \cdot A \cdot 2 \cdot q_y^* - \omega_\beta \cdot A \cdot 2 \cdot (q_z^*)^2}{4 \cdot (q_z^*)^2 \cdot q_y^*} = [-k_z \cdot c + \omega - \omega_\beta]$$

$$-k_z \cdot c \cdot (2 + (q_y^*)^2 + (q_x^*)^2) \cdot q_y^* + k_z \cdot c \cdot q_y^* \cdot A \cdot 2 \cdot q_y^* - \omega_\beta \cdot A \cdot 2 \cdot (q_z^*)^2 = 4 \cdot (q_z^*)^2 \cdot q_y^* \cdot [-k_z \cdot c + \omega - \omega_\beta]$$

$$-k_z \cdot c \cdot (2 + (q_y^*)^2 + (q_x^*)^2) \cdot q_y^* + k_z \cdot c \cdot q_y^* \cdot A \cdot 2 \cdot q_y^* = 4 \cdot (q_z^*)^2 \cdot q_y^* \cdot [-k_z \cdot c + \omega - \omega_\beta] + \omega_\beta \cdot A \cdot 2 \cdot (q_z^*)^2$$

$$2 \cdot (q_z^*)^2 \cdot \{2 \cdot q_y^* \cdot [-k_z \cdot c + \omega - \omega_\beta] + \omega_\beta \cdot A\}$$
$$= -k_z \cdot c \cdot (2 + (q_y^*)^2 + (q_x^*)^2) \cdot q_y^* + k_z \cdot c \cdot q_y^* \cdot A \cdot 2 \cdot q_y^*$$

$$(q_z^*)^2 = \frac{-k_z \cdot c \cdot (2 + (q_y^*)^2 + (q_x^*)^2) \cdot q_y^* + k_z \cdot c \cdot q_y^* \cdot A \cdot 2 \cdot q_y^*}{2 \cdot \{2 \cdot q_y^* \cdot [-k_z \cdot c + \omega - \omega_\beta] + \omega_\beta \cdot A\}}$$

$$q_z^* = \pm\sqrt{\frac{-k_z \cdot c \cdot (2 + (q_y^*)^2 + (q_x^*)^2) \cdot q_y^* + k_z \cdot c \cdot q_y^* \cdot A \cdot 2 \cdot q_y^*}{2 \cdot \{2 \cdot q_y^* \cdot [-k_z \cdot c + \omega - \omega_\beta] + \omega_\beta \cdot A\}}}$$

The fifth fixed point: $q_x^* \in \mathbb{R}; q_y^* \in \mathbb{R}$

$$E^{(4)}(\psi^*, q_x^*, q_y^*, q_z^*)$$
$$= (n \cdot \pi, q_x^*, q_y^*, \pm\sqrt{\frac{-k_z \cdot c \cdot (2 + (q_y^*)^2 + (q_x^*)^2) \cdot q_y^* + k_z \cdot c \cdot q_y^* \cdot A \cdot 2 \cdot q_y^*}{2 \cdot \{2 \cdot q_y^* \cdot [-k_z \cdot c + \omega - \omega_\beta] + \omega_\beta \cdot A\}}})$$
$n \in \mathbb{N}_0$

Remark The q_x^*, q_y^* values we choose establish q_z^* fixed point values.

Case 3.2 $$-\frac{k_z \cdot c \cdot (2 + (q_y^*)^2 + (q_x^*)^2)}{4 \cdot (q_z^*)^2} + \frac{1}{2} \cdot [k_z \cdot c \cdot \frac{q_y^*}{(q_z^*)^2} - \frac{\omega_\beta}{q_y^*}] \cdot A \cdot (-1)$$
$$= [-k_z \cdot c + \omega - \omega_\beta]$$

$$-\frac{k_z \cdot c \cdot (2 + (q_y^*)^2 + (q_x^*)^2)}{4 \cdot (q_z^*)^2} + \frac{1}{2} \cdot [-k_z \cdot c \cdot \frac{q_y^*}{(q_z^*)^2} + \frac{\omega_\beta}{q_y^*}] \cdot A$$
$$= [-k_z \cdot c + \omega - \omega_\beta]$$

$$-\frac{k_z \cdot c \cdot (2 + (q_y^*)^2 + (q_x^*)^2)}{4 \cdot (q_z^*)^2} - k_z \cdot c \cdot \frac{q_y^* \cdot A}{2 \cdot (q_z^*)^2} + \frac{\omega_\beta \cdot A}{2 \cdot q_y^*}$$
$$= [-k_z \cdot c + \omega - \omega_\beta]$$

$$\frac{-k_z \cdot c \cdot (2 + (q_y^*)^2 + (q_x^*)^2) \cdot q_y^* - k_z \cdot c \cdot q_y^* \cdot A \cdot 2 \cdot q_y^* + \omega_\beta \cdot A \cdot 2 \cdot (q_z^*)^2}{4 \cdot (q_z^*)^2 \cdot q_y^*}$$
$$= [-k_z \cdot c + \omega - \omega_\beta]$$

$$-k_z \cdot c \cdot (2 + (q_y^*)^2 + (q_x^*)^2) \cdot q_y^* - k_z \cdot c \cdot q_y^* \cdot A \cdot 2 \cdot q_y^* + \omega_\beta \cdot A \cdot 2 \cdot (q_z^*)^2$$
$$= 4 \cdot (q_z^*)^2 \cdot q_y^* \cdot [-k_z \cdot c + \omega - \omega_\beta]$$

$$-k_z \cdot c \cdot (2 + (q_y^*)^2 + (q_x^*)^2) \cdot q_y^* - k_z \cdot c \cdot q_y^* \cdot A \cdot 2 \cdot q_y^*$$

$$= 4 \cdot (q_z^*)^2 \cdot q_y^* \cdot [-k_z \cdot c + \omega - \omega_\beta] - \omega_\beta \cdot A \cdot 2 \cdot (q_z^*)^2$$

$$2 \cdot (q_z^*)^2 \cdot \{2 \cdot q_y^* \cdot [-k_z \cdot c + \omega - \omega_\beta] - \omega_\beta \cdot A\}$$
$$= -k_z \cdot c \cdot (2 + (q_y^*)^2 + (q_x^*)^2) \cdot q_y^* - k_z \cdot c \cdot q_y^* \cdot A \cdot 2 \cdot q_y^*$$

$$(q_z^*)^2 = \frac{-k_z \cdot c \cdot (2 + (q_y^*)^2 + (q_x^*)^2) \cdot q_y^* - k_z \cdot c \cdot q_y^* \cdot A \cdot 2 \cdot q_y^*}{2 \cdot \{2 \cdot q_y^* \cdot [-k_z \cdot c + \omega - \omega_\beta] - \omega_\beta \cdot A\}}$$

$$q_z^* = \pm\sqrt{\frac{-k_z \cdot c \cdot (2 + (q_y^*)^2 + (q_x^*)^2) \cdot q_y^* - k_z \cdot c \cdot q_y^* \cdot A \cdot 2 \cdot q_y^*}{2 \cdot \{2 \cdot q_y^* \cdot [-k_z \cdot c + \omega - \omega_\beta] - \omega_\beta \cdot A\}}}$$

The six fixed point: $q_x^* \in \mathbb{R}; q_y^* \in \mathbb{R}$

$$E^{(4)}(\psi^*, q_x^*, q_y^*, q_z^*)$$
$$= (n \cdot \pi, q_x^*, q_y^*, \pm\sqrt{\frac{-k_z \cdot c \cdot (2 + (q_y^*)^2 + (q_x^*)^2) \cdot q_y^* - k_z \cdot c \cdot q_y^* \cdot A \cdot 2 \cdot q_y^*}{2 \cdot \{2 \cdot q_y^* \cdot [-k_z \cdot c + \omega - \omega_\beta] - \omega_\beta \cdot A\}}})$$

$n \in \mathbb{N}_0$

Remark The q_x^*, q_y^* values we choose establish q_z^* fixed point values.

We can summary the process to find system possible fixed points by the following flowchart.

(*) $\frac{dq_x}{dt} = -\frac{1}{8} \cdot k_z \cdot c \cdot \frac{q_y \cdot q_x}{q_z^2} \cdot A \cdot \sin(\psi)$

(**) $\frac{dq_z}{dt} = -\frac{1}{2} \cdot k_z \cdot c \cdot \frac{q_y}{q_z} \cdot A \cdot \sin(\psi)$;

(***) $\frac{dq_y}{dt} = -\frac{1}{2} \cdot (\omega_\beta + \frac{1}{4} \cdot k_z \cdot c \cdot \frac{q_y^2}{q_z^2}) \cdot A \cdot \sin(\psi)$

(****)
$$\frac{d\psi}{dt} = k_z \cdot c \cdot [1 - \frac{(2 + q_x^2 + q_y^2)}{4 \cdot q_z^2}] - \omega + \omega_\beta \cdot [1 - \frac{1}{2 \cdot q_y} \cdot A \cdot \cos(\psi)] + \frac{1}{2} \cdot k_z \cdot c \cdot \frac{q_y}{q_z^2} \cdot A \cdot \cos(\psi)$$

At fixed points: $\frac{dq_x}{dt} = 0; \frac{dq_z}{dt} = 0; \frac{dq_y}{dt} = 0; \frac{d\psi}{dt} = 0.$

We can summary our system fixed points (Fig. 3.2a–e):

The first fixed point: $E^{(0)}(\psi^*, q_x^*, q_y^*, q_z^*) = (n \cdot \pi, q_x^*, 0, 0); n \in \mathbb{N}_0; q_x^* \in \mathbb{R}$.

The second fixed point:

$$E^{(1)}(\psi^*, q_x^*, q_y^*, q_z^*) = (n \cdot \pi, 0, q_y^*, \pm\sqrt{\frac{k_z \cdot c \cdot q_y^* \cdot [2 \cdot q_y^* - 2 - (q_y^*)^2]}{2 \cdot [\omega_\beta + 2 \cdot q_y^* \cdot (\omega - \omega_\beta - k_z \cdot c)]}})$$

$n \in \mathbb{N}_0; q_y^* \in \mathbb{R}$

The third fixed point:

$$E^{(2)}(\psi^*, q_x^*, q_y^*, q_z^*) = (n \cdot \pi, 0, q_y^*, \pm\sqrt{\frac{k_z \cdot c \cdot q_y^* \cdot [-(2 + (q_y^*)^2) - 2 \cdot q_y^*]}{2 \cdot \{[-k_z \cdot c - \omega_\beta + \omega] \cdot 2 \cdot q_y^* - \omega_\beta\}}})$$

$$n \in \mathbb{N}_0;\ q_y^* \in \mathbb{R}$$

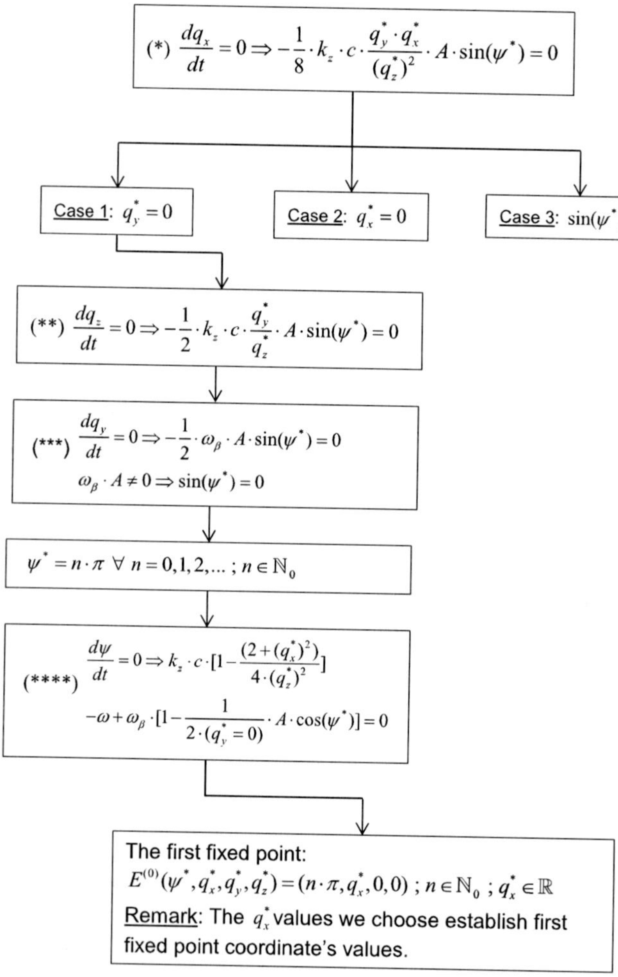

Figure 3.2 **a** Process finding system fixed points (flow chart), Case 1, **b** Process finding system fixed points (flow chart), Case 2, **c** Process finding system fixed points (flow chart), Case 2.1, **d** Process finding system fixed points (flow chart), Case 2.2, **e** Process finding system fixed points (flow chart), Case 3

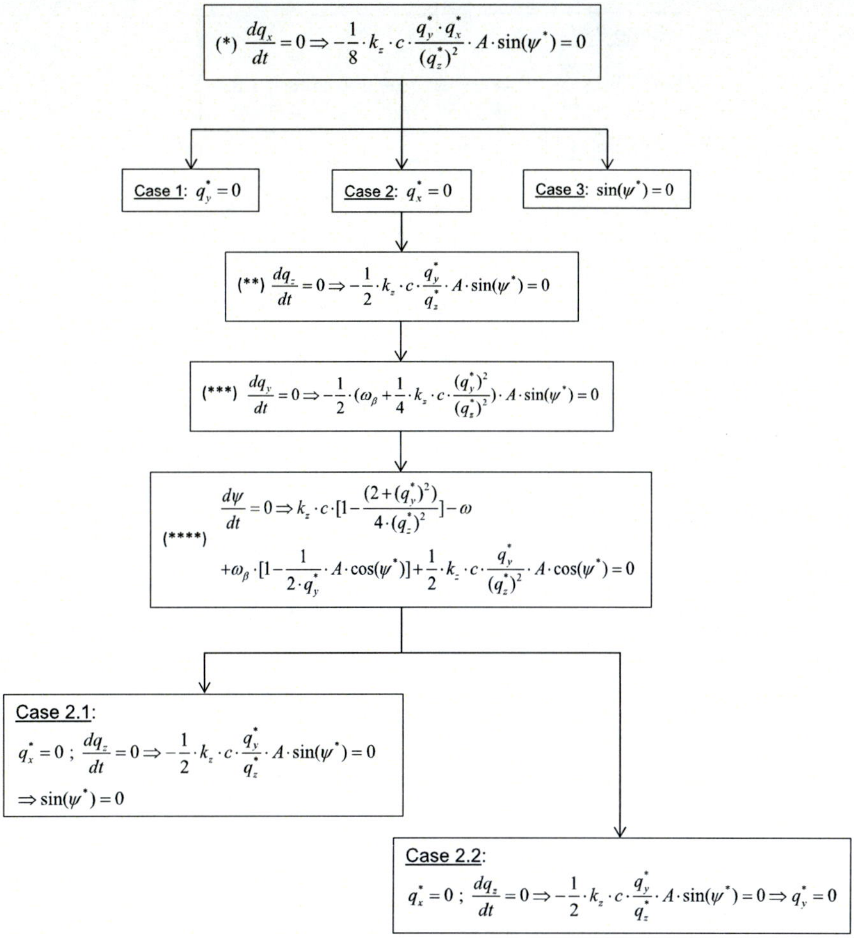

Figure 3.2 (continued)

Case 2.1:

$$q_x^* = 0 \; ; \; \frac{dq_z}{dt} = 0 \Rightarrow -\frac{1}{2} \cdot k_z \cdot c \cdot \frac{q_y^*}{q_z^*} \cdot A \cdot \sin(\psi^*) = 0$$

$$\Rightarrow \sin(\psi^*) = 0$$

↓

$$\sin(\psi^*) = 0 \Rightarrow \psi^* = n \cdot \pi \; \forall \; n = 0,1,2,\ldots \; ; \; n \in \mathbb{N}_0$$

$$\cos(\psi^*) = \pm\sqrt{1 - \sin^2(\psi^*)} \; ; \; \sin(\psi^*) = 0 \Rightarrow \cos(\psi^*) = \pm 1$$

↓

$$(****) \; k_z \cdot c \cdot [1 - \frac{(2 + (q_y^*)^2)}{4 \cdot (q_z^*)^2}] + [\omega_\beta - \omega] + \{\frac{1}{2} \cdot k_z \cdot c \cdot \frac{q_y^*}{(q_z^*)^2} - \frac{\omega_\beta}{2 \cdot q_y^*}\} \cdot (\pm 1) = 0$$

↓

Case 2.1.1:

$$k_z \cdot c \cdot [1 - \frac{(2 + (q_y^*)^2)}{4 \cdot (q_z^*)^2}] + [\omega_\beta - \omega] + [\frac{1}{2} \cdot k_z \cdot c \cdot \frac{q_y^*}{(q_z^*)^2} - \frac{\omega_\beta}{2 \cdot q_y^*}] = 0$$

↓

The second fixed point:

$$E^{(1)}(\psi^*, q_x^*, q_y^*, q_z^*) = (n \cdot \pi, 0, q_y^*, \pm\sqrt{\frac{k_z \cdot c \cdot q_y^* \cdot [2 \cdot q_y^* - 2 - (q_y^*)^2]}{2 \cdot [\omega_\beta + 2 \cdot q_y^* \cdot (\omega - \omega_\beta - k_z \cdot c)]}})$$

$$n \in \mathbb{N}_0 \; ; \; q_y^* \in \mathbb{R}$$

Case 2.1.2:

$$k_z \cdot c \cdot [1 - \frac{(2 + (q_y^*)^2)}{4 \cdot (q_z^*)^2}] + [\omega_\beta - \omega] - [\frac{1}{2} \cdot k_z \cdot c \cdot \frac{q_y^*}{(q_z^*)^2} - \frac{\omega_\beta}{2 \cdot q_y^*}] = 0$$

↓

The third fixed point:

$$E^{(2)}(\psi^*, q_x^*, q_y^*, q_z^*) = (n \cdot \pi, 0, q_y^*, \pm\sqrt{\frac{k_z \cdot c \cdot q_y^* \cdot [-(2 + (q_y^*)^2) - 2 \cdot q_y^*]}{2 \cdot \{[-k_z \cdot c - \omega_\beta + \omega] \cdot 2 \cdot q_y^* - \omega_\beta\}}})$$

$$n \in \mathbb{N}_0 \; ; \; q_y^* \in \mathbb{R}$$

Figure 3.2 (continued)

Case 2.2:

$$q_x^* = 0 \; ; \; \frac{dq_z}{dt} = 0 \Rightarrow -\frac{1}{2}\cdot k_z \cdot c \cdot \frac{q_y^*}{q_z^*}\cdot A \cdot \sin(\psi^*) = 0 \Rightarrow q_y^* = 0$$

↓

$$\frac{dq_y}{dt} = 0 \Rightarrow -\frac{1}{2}\cdot(\omega_\beta + \frac{1}{4}\cdot k_z \cdot c \cdot \frac{(q_y^*)^2}{(q_z^*)^2})\cdot A \cdot \sin(\psi^*)|_{q_y^*=0}$$

$$= -\frac{1}{2}\cdot \omega_\beta \cdot A \cdot \sin(\psi^*) = 0$$

↓

$$-\frac{1}{2}\cdot \omega_\beta \cdot A \cdot \sin(\psi^*) = 0 \; ; \; \omega_\beta \cdot A \neq 0$$

$$\sin(\psi^*) = 0 \Rightarrow \psi^* = n \cdot \pi \;\; \forall \; n = 0,1,2,\ldots \; ; \; n \in \mathbb{N}_0$$

$$\cos(\psi^*) = \pm\sqrt{1-\sin^2(\psi^*)} \; ; \; \sin(\psi^*) = 0 \Rightarrow \cos(\psi^*) = \pm 1$$

↓

$$k_z \cdot c \cdot \frac{1}{2\cdot(q_z^*)^2} \to \mp\infty \Rightarrow q_z^* \to 0$$

↓

The fourth fixed point:

$$E^{(3)}(\psi^*, q_x^*, q_y^*, q_z^*) = (n\cdot\pi, 0, 0, 0) \; ; \; n \in \mathbb{N}_0$$

Figure 3.2 (continued)

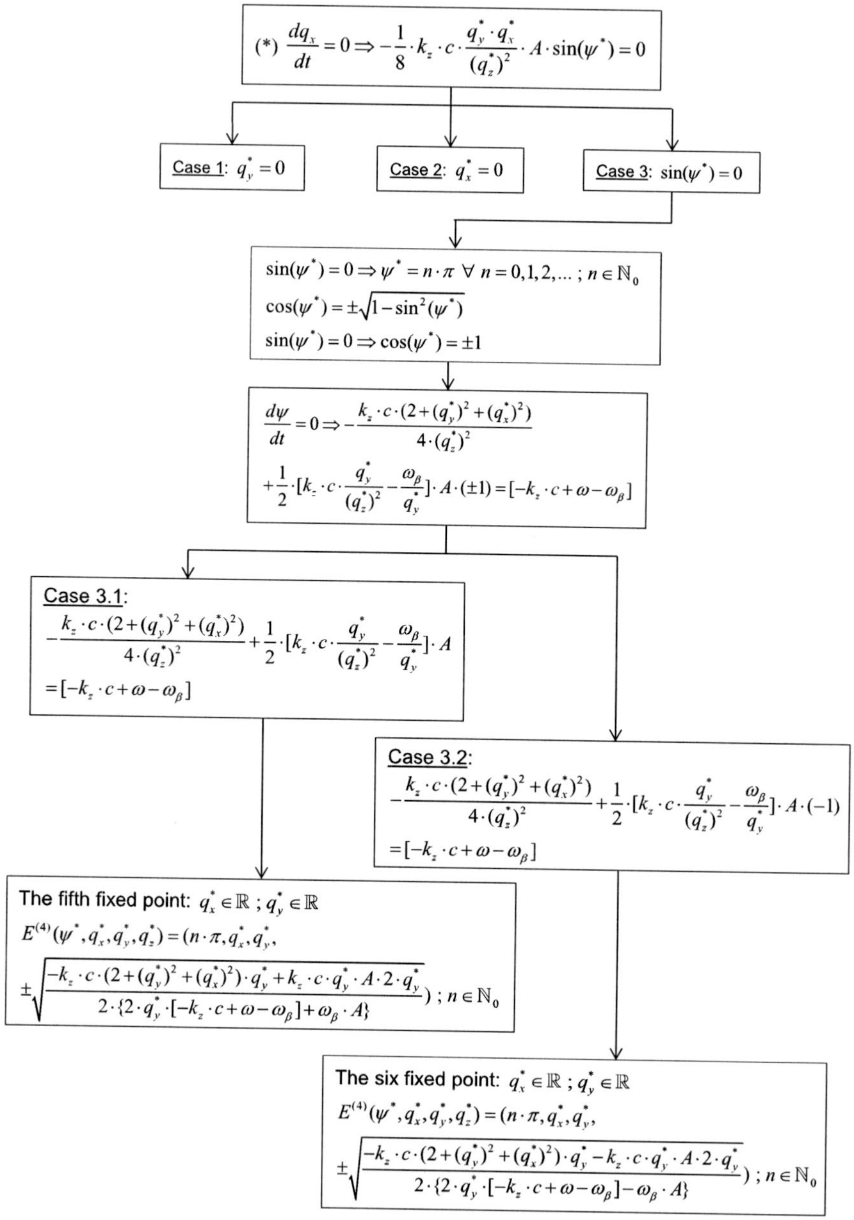

Figure 3.2 (continued)

The fourth fixed point: $E^{(3)}(\psi^*, q_x^*, q_y^*, q_z^*) = (n \cdot \pi, 0, 0, 0); n \in \mathbb{N}_0$.
The fifth fixed point: $q_x^* \in \mathbb{R};\ q_y^* \in \mathbb{R}$

$$E^{(4)}(\psi^*, q_x^*, q_y^*, q_z^*) = (n \cdot \pi, q_x^*, q_y^*, \pm\sqrt{\frac{-k_z \cdot c \cdot (2 + (q_y^*)^2 + (q_x^*)^2) \cdot q_y^* + k_z \cdot c \cdot q_y^* \cdot A \cdot 2 \cdot q_y^*}{2 \cdot \{2 \cdot q_y^* \cdot [-k_z \cdot c + \omega - \omega_\beta] + \omega_\beta \cdot A\}}})$$

$n \in \mathbb{N}_0$

The six fixed point: $q_x^* \in \mathbb{R}; q_y^* \in \mathbb{R}$

$$E^{(4)}(\psi^*, q_x^*, q_y^*, q_z^*) = (n \cdot \pi, q_x^*, q_y^*, \pm\sqrt{\frac{-k_z \cdot c \cdot (2 + (q_y^*)^2 + (q_x^*)^2) \cdot q_y^* - k_z \cdot c \cdot q_y^* \cdot A \cdot 2 \cdot q_y^*}{2 \cdot \{2 \cdot q_y^* \cdot [-k_z \cdot c + \omega - \omega_\beta] - \omega_\beta \cdot A\}}})$$

$n \in \mathbb{N}_0$

Stability analysis: The standard local stability analysis about any one of the equilibrium points of Ion channel laser perturbed system consists in adding to its coordinated $[\psi\ q_x\ q_y\ q_z]$ arbitrarily small increments of exponential terms $[\psi, q_x, q_y, q_z] \cdot e^{\lambda \cdot t}$, and retaining the first order terms in $\psi\ q_x\ q_y\ q_z$. The system of four homogeneous equations leads to a polynomial characteristic equation in the eigenvalueλ. The polynomial characteristic equation accepts by set the Ion channel laser perturbed system equations. The Ion channel laser perturbed system fixed values with arbitrarily small increments of exponential form $[\psi, q_x, q_y, q_z] \cdot e^{\lambda \cdot t}$ are; i = 0 (first fixed point), i = 1 (second fixed point), i = 2 (third fixed point), etc.,

$$\psi(t) = \psi^{(i)} + \psi \cdot e^{\lambda \cdot t}; q_x(t) = q_x^{(i)} + q_x \cdot e^{\lambda \cdot t}$$
$$q_y(t) = q_y^{(i)} + q_y \cdot e^{\lambda \cdot t}; q_z(t) = q_z^{(i)} + q_z \cdot e^{\lambda \cdot t}$$

$$\frac{d\psi(t)}{dt} = \psi \cdot \lambda \cdot e^{\lambda \cdot t}; \frac{dq_x(t)}{dt} = q_x \cdot \lambda \cdot e^{\lambda \cdot t}$$
$$\frac{dq_y(t)}{dt} = q_y \cdot \lambda \cdot e^{\lambda \cdot t}; \frac{dq_z(t)}{dt} = q_z \cdot \lambda \cdot e^{\lambda \cdot t}$$

&&&&

$$\frac{d\psi}{dt} = k_z \cdot c \cdot [1 - \frac{(2 + q_x^2 + q_y^2)}{4 \cdot q_z^2}] - \omega + \omega_\beta \cdot [1 - \frac{1}{2 \cdot q_y} \cdot A \cdot \cos(\psi)] + \frac{1}{2} \cdot k_z \cdot c \cdot \frac{q_y}{q_z^2} \cdot A \cdot \cos(\psi)$$

$$\psi \cdot \lambda \cdot e^{\lambda \cdot t} = k_z \cdot c \cdot \left[1 - \frac{(2 + [q_x^{(i)} + q_x \cdot e^{\lambda \cdot t}]^2 + [q_y^{(i)} + q_y \cdot e^{\lambda \cdot t}]^2)}{4 \cdot [q_z^{(i)} + q_z \cdot e^{\lambda \cdot t}]^2}\right]$$
$$- \omega + \omega_\beta \cdot \left\{1 - \frac{1}{2 \cdot [q_y^{(i)} + q_y \cdot e^{\lambda \cdot t}]} \cdot A \cdot \cos(\psi^{(i)} + \psi \cdot e^{\lambda \cdot t})\right\}$$
$$+ \frac{1}{2} \cdot k_z \cdot c \cdot \frac{[q_y^{(i)} + q_y \cdot e^{\lambda \cdot t}]}{[q_z^{(i)} + q_z \cdot e^{\lambda \cdot t}]^2} \cdot A \cdot \cos(\psi^{(i)} + \psi \cdot e^{\lambda \cdot t})$$

We take each part from the above expression and develop it mathematically differently.

$$\frac{(2 + [q_x^{(i)} + q_x \cdot e^{\lambda \cdot t}]^2 + [q_y^{(i)} + q_y \cdot e^{\lambda \cdot t}]^2)}{4 \cdot [q_z^{(i)} + q_z \cdot e^{\lambda \cdot t}]^2}$$

(1)
$$= \frac{(2 + (q_x^{(i)})^2 + 2 \cdot q_x^{(i)} \cdot q_x \cdot e^{\lambda \cdot t} + (q_x)^2 \cdot e^{2 \cdot \lambda \cdot t} + (q_y^{(i)})^2 + 2 \cdot q_y^{(i)} \cdot q_y \cdot e^{\lambda \cdot t} + (q_y)^2 \cdot e^{2 \cdot \lambda \cdot t})}{4 \cdot [(q_z^{(i)})^2 + 2 \cdot q_z^{(i)} \cdot q_z \cdot e^{\lambda \cdot t} + (q_z)^2 \cdot e^{2 \cdot \lambda \cdot t}]}$$

We consider $(q_x)^2$, $(q_y)^2$, and $(q_z)^2$ are very small and tend to zero, then:$(q_x)^2 \to \varepsilon$, $(q_y)^2 \to \varepsilon$, and $(q_z)^2 \to \varepsilon$.

$$\frac{(2 + [q_x^{(i)} + q_x \cdot e^{\lambda \cdot t}]^2 + [q_y^{(i)} + q_y \cdot e^{\lambda \cdot t}]^2)}{4 \cdot [q_z^{(i)} + q_z \cdot e^{\lambda \cdot t}]^2} = \frac{(2 + (q_x^{(i)})^2 + (q_y^{(i)})^2 + 2 \cdot q_x^{(i)} \cdot q_x \cdot e^{\lambda \cdot t} + 2 \cdot q_y^{(i)} \cdot q_y \cdot e^{\lambda \cdot t})}{4 \cdot [(q_z^{(i)})^2 + 2 \cdot q_z^{(i)} \cdot q_z \cdot e^{\lambda \cdot t}]}$$

$$\frac{(2 + (q_x^{(i)})^2 + (q_y^{(i)})^2 + 2 \cdot q_x^{(i)} \cdot q_x \cdot e^{\lambda \cdot t} + 2 \cdot q_y^{(i)} \cdot q_y \cdot e^{\lambda \cdot t})}{4 \cdot [(q_z^{(i)})^2 + 2 \cdot q_z^{(i)} \cdot q_z \cdot e^{\lambda \cdot t}]} \cdot \frac{[(q_z^{(i)})^2 - 2 \cdot q_z^{(i)} \cdot q_z \cdot e^{\lambda \cdot t}]}{[(q_z^{(i)})^2 - 2 \cdot q_z^{(i)} \cdot q_z \cdot e^{\lambda \cdot t}]}$$

$$= \frac{\begin{array}{l}[2 + (q_x^{(i)})^2 + (q_y^{(i)})^2] \cdot (q_z^{(i)})^2 - [2 + (q_x^{(i)})^2 + (q_y^{(i)})^2] \cdot 2 \cdot q_z^{(i)} \cdot q_z \cdot e^{\lambda \cdot t} \\ + [2 \cdot q_x^{(i)} \cdot q_x \cdot e^{\lambda \cdot t} + 2 \cdot q_y^{(i)} \cdot q_y \cdot e^{\lambda \cdot t}] \cdot (q_z^{(i)})^2 \\ - [2 \cdot q_x^{(i)} \cdot q_x \cdot e^{\lambda \cdot t} + 2 \cdot q_y^{(i)} \cdot q_y \cdot e^{\lambda \cdot t}] \cdot 2 \cdot q_z^{(i)} \cdot q_z \cdot e^{\lambda \cdot t}\end{array}}{4 \cdot [(q_z^{(i)})^4 - 4 \cdot (q_z^{(i)})^2 \cdot (q_z)^2 \cdot e^{2 \cdot \lambda \cdot t}]}$$

$$= \frac{\begin{array}{l}[2 + (q_x^{(i)})^2 + (q_y^{(i)})^2] \cdot (q_z^{(i)})^2 - [2 + (q_x^{(i)})^2 + (q_y^{(i)})^2] \cdot 2 \cdot q_z^{(i)} \cdot q_z \cdot e^{\lambda \cdot t} \\ + 2 \cdot [q_x^{(i)} \cdot q_x + q_y^{(i)} \cdot q_y] \cdot (q_z^{(i)})^2 \cdot e^{\lambda \cdot t} - 2 \cdot [q_x^{(i)} \cdot q_x + q_y^{(i)} \cdot q_y] \cdot 2 \cdot q_z^{(i)} \cdot q_z \cdot e^{2 \cdot \lambda \cdot t}\end{array}}{4 \cdot [(q_z^{(i)})^4 - 4 \cdot (q_z^{(i)})^2 \cdot (q_z)^2 \cdot e^{2 \cdot \lambda \cdot t}]}$$

$$= \frac{\begin{array}{l}[2 + (q_x^{(i)})^2 + (q_y^{(i)})^2] \cdot (q_z^{(i)})^2 - [2 + (q_x^{(i)})^2 + (q_y^{(i)})^2] \cdot 2 \cdot q_z^{(i)} \cdot q_z \cdot e^{\lambda \cdot t} \\ + 2 \cdot [q_x^{(i)} \cdot q_x + q_y^{(i)} \cdot q_y] \cdot (q_z^{(i)})^2 \cdot e^{\lambda \cdot t} - 2 \cdot [q_x^{(i)} \cdot q_x \cdot q_z + q_y^{(i)} \cdot q_y \cdot q_z] \cdot 2 \cdot q_z^{(i)} \cdot e^{2 \cdot \lambda \cdot t}\end{array}}{4 \cdot [(q_z^{(i)})^4 - 4 \cdot (q_z^{(i)})^2 \cdot (q_z)^2 \cdot e^{2 \cdot \lambda \cdot t}]}$$

We consider $q_x \cdot q_z \approx 0$; $q_y \cdot q_z \approx 0$; $(q_z)^2 \approx 0$ then

$$\frac{(2 + [q_x^{(i)} + q_x \cdot e^{\lambda \cdot t}]^2 + [q_y^{(i)} + q_y \cdot e^{\lambda \cdot t}]^2)}{4 \cdot [q_z^{(i)} + q_z \cdot e^{\lambda \cdot t}]^2} = \frac{\begin{array}{c}\{[2 + (q_x^{(i)})^2 + (q_y^{(i)})^2] + 2 \cdot [q_x^{(i)} \cdot q_x + q_y^{(i)} \cdot q_y] \cdot e^{\lambda \cdot t}\} \cdot (q_z^{(i)})^2 \\ -[2 + (q_x^{(i)})^2 + (q_y^{(i)})^2] \cdot 2 \cdot q_z^{(i)} \cdot q_z \cdot e^{\lambda \cdot t}\end{array}}{4 \cdot (q_z^{(i)})^4}$$

$$\frac{(2 + [q_x^{(i)} + q_x \cdot e^{\lambda \cdot t}]^2 + [q_y^{(i)} + q_y \cdot e^{\lambda \cdot t}]^2)}{4 \cdot [q_z^{(i)} + q_z \cdot e^{\lambda \cdot t}]^2} = \frac{[2 + (q_x^{(i)})^2 + (q_y^{(i)})^2]}{4 \cdot (q_z^{(i)})^2} + \frac{2 \cdot q_x^{(i)}}{4 \cdot (q_z^{(i)})^2} \cdot q_x \cdot e^{\lambda \cdot t} + \frac{2 \cdot q_y^{(i)}}{4 \cdot (q_z^{(i)})^2} \cdot q_y \cdot e^{\lambda \cdot t} - \frac{[2 + (q_x^{(i)})^2 + (q_y^{(i)})^2] \cdot 2}{4 \cdot (q_z^{(i)})^3} \cdot q_z \cdot e^{\lambda \cdot t}$$

(2)
$$\frac{1}{2 \cdot [q_y^{(i)} + q_y \cdot e^{\lambda \cdot t}]} = \frac{1}{2 \cdot [q_y^{(i)} + q_y \cdot e^{\lambda \cdot t}]} \cdot \frac{[q_y^{(i)} - q_y \cdot e^{\lambda \cdot t}]}{[q_y^{(i)} - q_y \cdot e^{\lambda \cdot t}]} = \frac{q_y^{(i)} - q_y \cdot e^{\lambda \cdot t}}{2 \cdot [(q_y^{(i)})^2 - (q_y)^2 \cdot e^{2 \cdot \lambda \cdot t}]} \overset{(q_y)^2 \approx 0}{=} \frac{q_y^{(i)} - q_y \cdot e^{\lambda \cdot t}}{2 \cdot (q_y^{(i)})^2} = \frac{1}{2 \cdot q_y^{(i)}} - \frac{1}{2 \cdot (q_y^{(i)})^2} \cdot q_y \cdot e^{\lambda \cdot t}$$

(3)
$$\frac{[q_y^{(i)} + q_y \cdot e^{\lambda \cdot t}]}{[q_z^{(i)} + q_z \cdot e^{\lambda \cdot t}]^2} = \frac{[q_y^{(i)} + q_y \cdot e^{\lambda \cdot t}]}{(q_z^{(i)})^2 + 2 \cdot q_z^{(i)} \cdot q_z \cdot e^{\lambda \cdot t} + (q_z)^2 \cdot e^{2 \cdot \lambda \cdot t}} \overset{(q_z)^2 \approx 0}{=} \frac{[q_y^{(i)} + q_y \cdot e^{\lambda \cdot t}]}{(q_z^{(i)})^2 + 2 \cdot q_z^{(i)} \cdot q_z \cdot e^{\lambda \cdot t}}$$

$$\frac{[q_y^{(i)} + q_y \cdot e^{\lambda \cdot t}]}{[q_z^{(i)} + q_z \cdot e^{\lambda \cdot t}]^2} \overset{(q_z)^2 \approx 0}{=} \frac{[q_y^{(i)} + q_y \cdot e^{\lambda \cdot t}]}{(q_z^{(i)})^2 + 2 \cdot q_z^{(i)} \cdot q_z \cdot e^{\lambda \cdot t}} \cdot \frac{[(q_z^{(i)})^2 - 2 \cdot q_z^{(i)} \cdot q_z \cdot e^{\lambda \cdot t}]}{[(q_z^{(i)})^2 - 2 \cdot q_z^{(i)} \cdot q_z \cdot e^{\lambda \cdot t}]} = \frac{\begin{array}{c}q_y^{(i)} \cdot (q_z^{(i)})^2 - 2 \cdot q_z^{(i)} \cdot q_y^{(i)} \cdot q_z \cdot e^{\lambda \cdot t} \\ +(q_z^{(i)})^2 \cdot q_y \cdot e^{\lambda \cdot t} - 2 \cdot q_z^{(i)} \cdot q_z \cdot q_y \cdot e^{2 \cdot \lambda \cdot t}\end{array}}{(q_z^{(i)})^4 - 4 \cdot (q_z^{(i)})^2 \cdot (q_z)^2 \cdot e^{2 \cdot \lambda \cdot t}}$$

$$\frac{[q_y^{(i)} + q_y \cdot e^{\lambda \cdot t}]}{[q_z^{(i)} + q_z \cdot e^{\lambda \cdot t}]^2} \overset{(q_z)^2 \approx 0}{=} \frac{q_y^{(i)} \cdot (q_z^{(i)})^2 - 2 \cdot q_z^{(i)} \cdot q_y^{(i)} \cdot q_z \cdot e^{\lambda \cdot t} + (q_z^{(i)})^2 \cdot q_y \cdot e^{\lambda \cdot t} - 2 \cdot q_z^{(i)} \cdot q_z \cdot q_y \cdot e^{2 \cdot \lambda \cdot t}}{(q_z^{(i)})^4 - 4 \cdot (q_z^{(i)})^2 \cdot (q_z)^2 \cdot e^{2 \cdot \lambda \cdot t}}$$

Table 3.2 System's first differential equation elements by adding variables their arbitrarily small increments of exponential term

	Expression	Final expression
1	$\frac{(2+[q_x^{(i)}+q_x\cdot e^{\lambda\cdot t}]^2+[q_y^{(i)}+q_y\cdot e^{\lambda\cdot t}]^2)}{4\cdot[q_z^{(i)}+q_z\cdot e^{\lambda\cdot t}]^2}$	$\frac{[2+(q_x^{(i)})^2+(q_y^{(i)})^2]}{4\cdot(q_z^{(i)})^2}+\frac{2\cdot q_x^{(i)}}{4\cdot(q_z^{(i)})^2}\cdot q_x\cdot e^{\lambda\cdot t}+\frac{2\cdot q_y^{(i)}}{4\cdot(q_z^{(i)})^2}\cdot q_y\cdot e^{\lambda\cdot t}-\frac{[2+(q_x^{(i)})^2+(q_y^{(i)})^2]\cdot 2}{4\cdot(q_z^{(i)})^3}\cdot q_z\cdot e^{\lambda\cdot t}$
2	$\frac{1}{2\cdot[q_y^{(i)}+q_y\cdot e^{\lambda\cdot t}]}$	$\frac{1}{2\cdot q_y^{(i)}}-\frac{1}{2\cdot(q_y^{(i)})^2}\cdot q_y\cdot e^{\lambda\cdot t}$
3	$\frac{[q_y^{(i)}+q_y\cdot e^{\lambda\cdot t}]}{[q_z^{(i)}+q_z\cdot e^{\lambda\cdot t}]^2}$	$\frac{q_y^{(i)}}{(q_z^{(i)})^2}-\frac{2\cdot q_y^{(i)}}{(q_z^{(i)})^3}\cdot q_z\cdot e^{\lambda\cdot t}+\frac{1}{(q_z^{(i)})^2}\cdot q_y\cdot e^{\lambda\cdot t}$

We assume that $q_z\cdot q_y\approx 0$; $(q_z)^2\approx 0$ then

$$\frac{[q_y^{(i)}+q_y\cdot e^{\lambda\cdot t}]}{[q_z^{(i)}+q_z\cdot e^{\lambda\cdot t}]^2}\overset{(q_z)^2\approx 0}{=}\frac{q_y^{(i)}\cdot(q_z^{(i)})^2-2\cdot q_z^{(i)}\cdot q_y^{(i)}\cdot q_z\cdot e^{\lambda\cdot t}+(q_z^{(i)})^2\cdot q_y\cdot e^{\lambda\cdot t}}{(q_z^{(i)})^4}$$
$$=\frac{q_y^{(i)}}{(q_z^{(i)})^2}-\frac{2\cdot q_y^{(i)}}{(q_z^{(i)})^3}\cdot q_z\cdot e^{\lambda\cdot t}+\frac{1}{(q_z^{(i)})^2}\cdot q_y\cdot e^{\lambda\cdot t}$$

We can summary all algebraic manipulation in the next table (Table 3.2).

$$\cos(\psi^{(i)}+\psi\cdot e^{\lambda\cdot t})=\cos(\psi^{(i)})\cdot\cos(\psi\cdot e^{\lambda\cdot t})-\sin(\psi^{(i)})\cdot\sin(\psi\cdot e^{\lambda\cdot t})$$

We use Taylor series to represent the functions $\cos(\psi\cdot e^{\lambda\cdot t})$ and $\sin(\psi\cdot e^{\lambda\cdot t})$ as an infinite sum of terms that are calculated from the values of the function's derivatives at a single point. The Taylor series are concentrated near to zero since $(\psi\to 0)\cdot e^{\lambda\cdot t}$ which is called a Maclaurin series. The functions are approximated by using a finite number of terms of their Taylor series. It gives quantitative estimates on the error introduced by the use of such an approximations.

$$\cos(\psi\cdot e^{\lambda\cdot t})=\sum_{n=0}^{\infty}\frac{(-1)^n}{(2\cdot n)!}\cdot\psi^{(2\cdot n)}\cdot e^{2\cdot n\cdot\lambda\cdot t}$$
$$=1-\frac{\psi^2\cdot e^{2\cdot\lambda\cdot t}}{2!}+\frac{\psi^4\cdot e^{4\cdot\lambda\cdot t}}{4!}$$
$$-\ldots\ \forall\psi\cdot e^{\lambda\cdot t}$$

Since $\psi^2\approx 0$; $\psi^4\approx 0$; .... then $\cos(\psi\cdot e^{\lambda\cdot t})\approx 1$

$$\sin(\psi\cdot e^{\lambda\cdot t})=\sum_{n=0}^{\infty}\frac{(-1)^n}{(2\cdot n+1)!}\cdot\psi^{(2\cdot n+1)}\cdot e^{(2\cdot n+1)\cdot\lambda\cdot t}$$
$$=\psi\cdot e^{\lambda\cdot t}-\frac{\psi^3\cdot e^{3\cdot\lambda\cdot t}}{3!}+\frac{\psi^5\cdot e^{5\cdot\lambda\cdot t}}{5!}$$
$$-\ldots\ \forall\psi\cdot e^{\lambda\cdot t}$$

Since $\psi^3 \approx 0; \psi^5 \approx 0; \ldots$ then $\sin(\psi \cdot e^{\lambda \cdot t}) \approx \psi \cdot e^{\lambda \cdot t}$

$$\cos(\psi^{(i)} + \psi \cdot e^{\lambda \cdot t}) \approx \cos(\psi^{(i)}) - \sin(\psi^{(i)}) \cdot \psi \cdot e^{\lambda \cdot t}$$

$$\sin(\psi^{(i)} + \psi \cdot e^{\lambda \cdot t}) = \sin(\psi^{(i)}) \cdot \cos(\psi \cdot e^{\lambda \cdot t}) + \cos(\psi^{(i)}) \cdot \sin(\psi \cdot e^{\lambda \cdot t})$$

$$\sin(\psi^{(i)} + \psi \cdot e^{\lambda \cdot t}) \approx \sin(\psi^{(i)}) + \cos(\psi^{(i)}) \cdot \psi \cdot e^{\lambda \cdot t}$$

$$\begin{aligned} \psi \cdot \lambda \cdot e^{\lambda \cdot t} = k_z \cdot c \cdot & \left[1 - \frac{(2 + [q_x^{(i)} + q_x \cdot e^{\lambda \cdot t}]^2 + [q_y^{(i)} + q_y \cdot e^{\lambda \cdot t}]^2)}{4 \cdot [q_z^{(i)} + q_z \cdot e^{\lambda \cdot t}]^2} \right] \\ & - \omega + \omega_\beta \cdot \left\{ 1 - \frac{1}{2 \cdot [q_y^{(i)} + q_y \cdot e^{\lambda \cdot t}]} \cdot A \cdot \cos(\psi^{(i)} + \psi \cdot e^{\lambda \cdot t}) \right\} \\ & + \frac{1}{2} \cdot k_z \cdot c \cdot \frac{[q_y^{(i)} + q_y \cdot e^{\lambda \cdot t}]}{[q_z^{(i)} + q_z \cdot e^{\lambda \cdot t}]^2} \cdot A \cdot \cos(\psi^{(i)} + \psi \cdot e^{\lambda \cdot t}) \end{aligned}$$

$$\begin{aligned} \psi \cdot \lambda \cdot e^{\lambda \cdot t} & = k_z \cdot c \cdot [1 - \{ \frac{[2 + (q_x^{(i)})^2 + (q_y^{(i)})^2]}{4 \cdot (q_z^{(i)})^2} \\ & + \frac{2 \cdot q_x^{(i)}}{4 \cdot (q_z^{(i)})^2} \cdot q_x \cdot e^{\lambda \cdot t} + \frac{2 \cdot q_y^{(i)}}{4 \cdot (q_z^{(i)})^2} \cdot q_y \cdot e^{\lambda \cdot t} \\ & - \frac{[2 + (q_x^{(i)})^2 + (q_y^{(i)})^2] \cdot 2}{4 \cdot (q_z^{(i)})^3} \cdot q_z \cdot e^{\lambda \cdot t} \}] - \omega \\ & + \omega_\beta \cdot \{ 1 - [\frac{1}{2 \cdot q_y^{(i)}} - \frac{1}{2 \cdot (q_y^{(i)})^2} \cdot q_y \cdot e^{\lambda \cdot t}] \\ & \cdot A \cdot [\cos(\psi^{(i)}) - \sin(\psi^{(i)}) \cdot \psi \cdot e^{\lambda \cdot t}] \} \\ & + \frac{1}{2} \cdot k_z \cdot c \cdot \{ \frac{q_y^{(i)}}{(q_z^{(i)})^2} - \frac{2 \cdot q_y^{(i)}}{(q_z^{(i)})^3} \cdot q_z \cdot e^{\lambda \cdot t} \\ & + \frac{1}{(q_z^{(i)})^2} \cdot q_y \cdot e^{\lambda \cdot t} \} \cdot A \cdot [\cos(\psi^{(i)}) - \sin(\psi^{(i)}) \cdot \psi \cdot e^{\lambda \cdot t}] \end{aligned}$$

$$\begin{aligned} & \psi \cdot \lambda \cdot e^{\lambda \cdot t} \\ & = k_z \cdot c \cdot [1 - \frac{[2 + (q_x^{(i)})^2 + (q_y^{(i)})^2]}{4 \cdot (q_z^{(i)})^2}] \\ & + k_z \cdot c \cdot [\frac{[2 + (q_x^{(i)})^2 + (q_y^{(i)})^2] \cdot 2}{4 \cdot (q_z^{(i)})^3} \cdot q_z \cdot e^{\lambda \cdot t} \\ & - \frac{2 \cdot q_x^{(i)}}{4 \cdot (q_z^{(i)})^2} \cdot q_x \cdot e^{\lambda \cdot t} - \frac{2 \cdot q_y^{(i)}}{4 \cdot (q_z^{(i)})^2} \cdot q_y \cdot e^{\lambda \cdot t}] \end{aligned}$$

$$- \omega + \omega_{\beta} \cdot \{1 - (\frac{1}{2 \cdot q_y^{(i)}} \cdot A \cdot [\cos(\psi^{(i)}) - \sin(\psi^{(i)}) \cdot \psi \cdot e^{\lambda \cdot t}]$$
$$- \frac{1}{2 \cdot (q_y^{(i)})^2} \cdot q_y \cdot e^{\lambda \cdot t} \cdot A \cdot [\cos(\psi^{(i)}) - \sin(\psi^{(i)}) \cdot \psi \cdot e^{\lambda \cdot t}])\}$$
$$+ \frac{1}{2} \cdot k_z \cdot c \cdot \{\frac{q_y^{(i)}}{(q_z^{(i)})^2}\} \cdot A \cdot [\cos(\psi^{(i)}) - \sin(\psi^{(i)}) \cdot \psi \cdot e^{\lambda \cdot t}]$$
$$+ \frac{1}{2} \cdot k_z \cdot c \cdot \{\frac{1}{(q_z^{(i)})^2} \cdot q_y \cdot e^{\lambda \cdot t} - \frac{2 \cdot q_y^{(i)}}{(q_z^{(i)})^3} \cdot q_z \cdot e^{\lambda \cdot t}\}$$
$$\cdot A \cdot [\cos(\psi^{(i)}) - \sin(\psi^{(i)}) \cdot \psi \cdot e^{\lambda \cdot t}]$$

$$\psi \cdot \lambda \cdot e^{\lambda \cdot t} = k_z \cdot c \cdot [1 - \frac{[2 + (q_x^{(i)})^2 + (q_y^{(i)})^2]}{4 \cdot (q_z^{(i)})^2}]$$
$$+ k_z \cdot c \cdot [\frac{[2 + (q_x^{(i)})^2 + (q_y^{(i)})^2] \cdot 2}{4 \cdot (q_z^{(i)})^3} \cdot q_z$$
$$\cdot e^{\lambda \cdot t} - \frac{2 \cdot q_x^{(i)}}{4 \cdot (q_z^{(i)})^2} \cdot q_x \cdot e^{\lambda \cdot t} - \frac{2 \cdot q_y^{(i)}}{4 \cdot (q_z^{(i)})^2} \cdot q_y \cdot e^{\lambda \cdot t}] - \omega$$
$$+ \omega_{\beta} \cdot \{1 - (\frac{1}{2 \cdot q_y^{(i)}} \cdot A \cdot \cos(\psi^{(i)}) - \frac{1}{2 \cdot q_y^{(i)}} \cdot A \cdot \sin(\psi^{(i)})$$
$$\cdot \psi \cdot e^{\lambda \cdot t} - \frac{1}{2 \cdot (q_y^{(i)})^2} \cdot q_y \cdot e^{\lambda \cdot t} \cdot A \cdot [\cos(\psi^{(i)}) - \sin(\psi^{(i)}) \cdot \psi \cdot e^{\lambda \cdot t}])\}$$
$$+ \frac{1}{2} \cdot k_z \cdot c \cdot \{\frac{q_y^{(i)}}{(q_z^{(i)})^2}\} \cdot A \cdot \cos(\psi^{(i)}) - \frac{1}{2} \cdot k_z$$
$$\cdot c \cdot \{\frac{q_y^{(i)}}{(q_z^{(i)})^2}\} \cdot A \cdot \sin(\psi^{(i)}) \cdot \psi \cdot e^{\lambda \cdot t}$$
$$+ \frac{1}{2} \cdot k_z \cdot c \cdot \{\frac{1}{(q_z^{(i)})^2} \cdot q_y \cdot e^{\lambda \cdot t} - \frac{2 \cdot q_y^{(i)}}{(q_z^{(i)})^3} \cdot q_z \cdot e^{\lambda \cdot t}\}$$
$$\cdot A \cdot [\cos(\psi^{(i)}) - \sin(\psi^{(i)}) \cdot \psi \cdot e^{\lambda \cdot t}]$$

$$\psi \cdot \lambda \cdot e^{\lambda \cdot t} = k_z \cdot c \cdot [1 - \frac{[2 + (q_x^{(i)})^2 + (q_y^{(i)})^2]}{4 \cdot (q_z^{(i)})^2}]$$
$$+ k_z \cdot c \cdot [\frac{[2 + (q_x^{(i)})^2 + (q_y^{(i)})^2] \cdot 2}{4 \cdot (q_z^{(i)})^3} \cdot q_z \cdot e^{\lambda \cdot t}$$
$$- \frac{2 \cdot q_x^{(i)}}{4 \cdot (q_z^{(i)})^2} \cdot q_x \cdot e^{\lambda \cdot t} - \frac{2 \cdot q_y^{(i)}}{4 \cdot (q_z^{(i)})^2} \cdot q_y \cdot e^{\lambda \cdot t}] - \omega$$

$$+\omega_\beta \cdot \{1 - \frac{1}{2 \cdot q_y^{(i)}} \cdot A \cdot \cos(\psi^{(i)}) + \frac{1}{2 \cdot q_y^{(i)}} \cdot A \cdot \sin(\psi^{(i)})$$

$$\cdot \psi \cdot e^{\lambda \cdot t} + \frac{1}{2 \cdot (q_y^{(i)})^2} \cdot q_y \cdot e^{\lambda \cdot t} \cdot A \cdot [\cos(\psi^{(i)}) - \sin(\psi^{(i)}) \cdot \psi \cdot e^{\lambda \cdot t}])\}$$

$$+ \frac{1}{2} \cdot k_z \cdot c \cdot \{\frac{q_y^{(i)}}{(q_z^{(i)})^2}\} \cdot A \cdot \cos(\psi^{(i)}) - \frac{1}{2}$$

$$\cdot k_z \cdot c \cdot \{\frac{q_y^{(i)}}{(q_z^{(i)})^2}\} \cdot A \cdot \sin(\psi^{(i)}) \cdot \psi \cdot e^{\lambda \cdot t}$$

$$+ \frac{1}{2} \cdot k_z \cdot c \cdot \{\frac{1}{(q_z^{(i)})^2} \cdot q_y \cdot e^{\lambda \cdot t} - \frac{2 \cdot q_y^{(i)}}{(q_z^{(i)})^3} \cdot q_z \cdot e^{\lambda \cdot t}\}$$

$$\cdot A \cdot [\cos(\psi^{(i)}) - \sin(\psi^{(i)}) \cdot \psi \cdot e^{\lambda \cdot t}]$$

$$\psi \cdot \lambda \cdot e^{\lambda \cdot t} = k_z \cdot c \cdot \left[1 - \frac{[2 + (q_x^{(i)})^2 + (q_y^{(i)})^2]}{4 \cdot (q_z^{(i)})^2}\right]$$

$$- \omega + \omega_\beta \cdot \{1 - \frac{1}{2 \cdot q_y^{(i)}} \cdot A \cdot \cos(\psi^{(i)})\}$$

$$+ \frac{1}{2} \cdot k_z \cdot c \cdot \left\{\frac{q_y^{(i)}}{(q_z^{(i)})^2}\right\} \cdot A \cdot \cos(\psi^{(i)})$$

$$+ k_z \cdot c \cdot \left[\frac{[2 + (q_x^{(i)})^2 + (q_y^{(i)})^2] \cdot 2}{4 \cdot (q_z^{(i)})^3} \cdot q_z \cdot e^{\lambda \cdot t}\right.$$

$$\left. - \frac{2 \cdot q_x^{(i)}}{4 \cdot (q_z^{(i)})^2} \cdot q_x \cdot e^{\lambda \cdot t} - \frac{2 \cdot q_y^{(i)}}{4 \cdot (q_z^{(i)})^2} \cdot q_y \cdot e^{\lambda \cdot t}\right]$$

$$+ \omega_\beta \cdot \left\{\frac{1}{2 \cdot q_y^{(i)}} \cdot A \cdot \sin(\psi^{(i)}) \cdot \psi \cdot e^{\lambda \cdot t} + \frac{1}{2 \cdot (q_y^{(i)})^2} \cdot q_y\right.$$

$$\cdot e^{\lambda \cdot t} \cdot A \cdot [\cos(\psi^{(i)}) - \sin(\psi^{(i)}) \cdot \psi \cdot e^{\lambda \cdot t}]\}$$

$$- \frac{1}{2} \cdot k_z \cdot c \cdot \left\{\frac{q_y^{(i)}}{(q_z^{(i)})^2}\right\} \cdot A \cdot \sin(\psi^{(i)}) \cdot \psi \cdot e^{\lambda \cdot t}$$

$$+ \frac{1}{2} \cdot k_z \cdot c \cdot \{\frac{1}{(q_z^{(i)})^2} \cdot q_y \cdot e^{\lambda \cdot t} - \frac{2 \cdot q_y^{(i)}}{(q_z^{(i)})^3} \cdot q_z \cdot e^{\lambda \cdot t}\}$$

$$\cdot A \cdot [\cos(\psi^{(i)}) - \sin(\psi^{(i)}) \cdot \psi \cdot e^{\lambda \cdot t}]$$

At fixed points: $\frac{d\psi}{dt} = 0$

$$k_z \cdot c \cdot \left[1 - \frac{[2 + (q_x^{(i)})^2 + (q_y^{(i)})^2]}{4 \cdot (q_z^{(i)})^2}\right] - \omega + \omega_\beta$$

$$\cdot \left\{1 - \frac{1}{2 \cdot q_y^{(i)}} \cdot A \cdot \cos(\psi^{(i)})\right\} + \frac{1}{2} \cdot k_z \cdot c \cdot \{\frac{q_y^{(i)}}{(q_z^{(i)})^2}\} \cdot A \cdot \cos(\psi^{(i)}) = 0$$

$$\psi \cdot \lambda \cdot e^{\lambda \cdot t} = k_z \cdot c \cdot \left[\frac{[2 + (q_x^{(i)})^2 + (q_y^{(i)})^2] \cdot 2}{4 \cdot (q_z^{(i)})^3} \cdot q_z \cdot e^{\lambda \cdot t}\right.$$

$$\left. - \frac{2 \cdot q_x^{(i)}}{4 \cdot (q_z^{(i)})^2} \cdot q_x \cdot e^{\lambda \cdot t} - \frac{2 \cdot q_y^{(i)}}{4 \cdot (q_z^{(i)})^2} \cdot q_y \cdot e^{\lambda \cdot t}\right]$$

$$+ \omega_\beta \cdot \left\{\frac{1}{2 \cdot q_y^{(i)}} \cdot A \cdot \sin(\psi^{(i)}) \cdot \psi \cdot e^{\lambda \cdot t} + \frac{1}{2 \cdot (q_y^{(i)})^2}\right.$$

$$\cdot q_y \cdot e^{\lambda \cdot t} \cdot A \cdot [\cos(\psi^{(i)}) - \sin(\psi^{(i)}) \cdot \psi \cdot e^{\lambda \cdot t}]\}$$

$$- \frac{1}{2} \cdot k_z \cdot c \cdot \left\{\frac{q_y^{(i)}}{(q_z^{(i)})^2}\right\} \cdot A \cdot \sin(\psi^{(i)}) \cdot \psi \cdot e^{\lambda \cdot t}$$

$$+ \frac{1}{2} \cdot k_z \cdot c \cdot \left\{\frac{1}{(q_z^{(i)})^2} \cdot q_y \cdot e^{\lambda \cdot t} - \frac{2 \cdot q_y^{(i)}}{(q_z^{(i)})^3} \cdot q_z \cdot e^{\lambda \cdot t}\right\}$$

$$\cdot A \cdot [\cos(\psi^{(i)}) - \sin(\psi^{(i)}) \cdot \psi \cdot e^{\lambda \cdot t}]$$

$$\psi \cdot \lambda \cdot e^{\lambda \cdot t} = k_z \cdot c \cdot [\frac{[2 + (q_x^{(i)})^2 + (q_y^{(i)})^2] \cdot 2}{4 \cdot (q_z^{(i)})^3} \cdot q_z \cdot e^{\lambda \cdot t}$$

$$- \frac{2 \cdot q_x^{(i)}}{4 \cdot (q_z^{(i)})^2} \cdot q_x \cdot e^{\lambda \cdot t} - \frac{2 \cdot q_y^{(i)}}{4 \cdot (q_z^{(i)})^2} \cdot q_y \cdot e^{\lambda \cdot t}]$$

$$+ \omega_\beta \cdot \{\frac{1}{2 \cdot q_y^{(i)}} \cdot A \cdot \sin(\psi^{(i)}) \cdot \psi \cdot e^{\lambda \cdot t} + \frac{1}{2 \cdot (q_y^{(i)})^2}$$

$$\cdot q_y \cdot e^{\lambda \cdot t} \cdot A \cdot \cos(\psi^{(i)}) - \frac{1}{2 \cdot (q_y^{(i)})^2} \cdot e^{\lambda \cdot t}$$

$$\cdot A \cdot \sin(\psi^{(i)}) \cdot q_y \cdot \psi \cdot e^{\lambda \cdot t}\}$$

$$- \frac{1}{2} \cdot k_z \cdot c \cdot \{\frac{q_y^{(i)}}{(q_z^{(i)})^2}\} \cdot A \cdot \sin(\psi^{(i)}) \cdot \psi \cdot e^{\lambda \cdot t}$$

$$+ \frac{1}{2} \cdot k_z \cdot c \cdot \{\frac{1}{(q_z^{(i)})^2} \cdot q_y \cdot e^{\lambda \cdot t} \cdot A \cdot \cos(\psi^{(i)})$$

$$- \frac{1}{(q_z^{(i)})^2} \cdot e^{\lambda \cdot t} \cdot A \cdot \sin(\psi^{(i)}) \cdot \psi \cdot q_y \cdot e^{\lambda \cdot t}$$

$$- \frac{2 \cdot q_y^{(i)}}{(q_z^{(i)})^3} \cdot q_z \cdot e^{\lambda \cdot t} \cdot A \cdot \cos(\psi^{(i)}) + \frac{2 \cdot q_y^{(i)}}{(q_z^{(i)})^3}$$

$$\cdot e^{\lambda \cdot t} \cdot A \cdot \sin(\psi^{(i)}) \cdot q_z \cdot \psi \cdot e^{\lambda \cdot t}\}$$

Assumption: $q_y q \cdot \psi \approx 0$; $q_z \cdot \psi \approx 0$

$$\psi \cdot \lambda \cdot e^{\lambda \cdot t} = k_z \cdot c \cdot [\frac{[2 + (q_x^{(i)})^2 + (q_y^{(i)})^2] \cdot 2}{4 \cdot (q_z^{(i)})^3} \cdot q_z \cdot e^{\lambda \cdot t}$$
$$- \frac{2 \cdot q_x^{(i)}}{4 \cdot (q_z^{(i)})^2} \cdot q_x \cdot e^{\lambda \cdot t} - \frac{2 \cdot q_y^{(i)}}{4 \cdot (q_z^{(i)})^2} \cdot q_y \cdot e^{\lambda \cdot t}]$$
$$+ \omega_\beta \cdot \{\frac{1}{2 \cdot q_y^{(i)}} \cdot A \cdot \sin(\psi^{(i)}) \cdot \psi \cdot e^{\lambda \cdot t}$$
$$+ \frac{1}{2 \cdot (q_y^{(i)})^2} \cdot q_y \cdot e^{\lambda \cdot t} \cdot A \cdot \cos(\psi^{(i)})\}$$
$$- \frac{1}{2} \cdot k_z \cdot c \cdot \{\frac{q_y^{(i)}}{(q_z^{(i)})^2}\} \cdot A \cdot \sin(\psi^{(i)}) \cdot \psi \cdot e^{\lambda \cdot t}$$
$$+ \frac{1}{2} \cdot k_z \cdot c \cdot \{\frac{1}{(q_z^{(i)})^2} \cdot q_y \cdot e^{\lambda \cdot t} \cdot A \cdot \cos(\psi^{(i)})$$
$$- \frac{2 \cdot q_y^{(i)}}{(q_z^{(i)})^3} \cdot q_z \cdot e^{\lambda \cdot t} \cdot A \cdot \cos(\psi^{(i)})\}$$

$$\psi \cdot \lambda \cdot e^{\lambda \cdot t} = k_z \cdot c \cdot [\frac{[2 + (q_x^{(i)})^2 + (q_y^{(i)})^2] \cdot 2}{4 \cdot (q_z^{(i)})^3} \cdot q_z$$
$$- \frac{2 \cdot q_x^{(i)}}{4 \cdot (q_z^{(i)})^2} \cdot q_x - \frac{2 \cdot q_y^{(i)}}{4 \cdot (q_z^{(i)})^2} \cdot q_y] \cdot e^{\lambda \cdot t}$$
$$+ \omega_\beta \cdot \{\frac{1}{2 \cdot q_y^{(i)}} \cdot A \cdot \sin(\psi^{(i)}) \cdot \psi + \frac{1}{2 \cdot (q_y^{(i)})^2}$$
$$\cdot q_y \cdot A \cdot \cos(\psi^{(i)})\} \cdot e^{\lambda \cdot t}$$
$$- \frac{1}{2} \cdot k_z \cdot c \cdot \{\frac{q_y^{(i)}}{(q_z^{(i)})^2}\} \cdot A \cdot \sin(\psi^{(i)}) \cdot \psi \cdot e^{\lambda \cdot t}$$
$$+ \frac{1}{2} \cdot k_z \cdot c \cdot \{\frac{1}{(q_z^{(i)})^2} \cdot q_y \cdot A \cdot \cos(\psi^{(i)})$$
$$- \frac{2 \cdot q_y^{(i)}}{(q_z^{(i)})^3} \cdot q_z \cdot A \cdot \cos(\psi^{(i)})\} \cdot e^{\lambda \cdot t}$$

Divide the two sides of the above equation by $e^{\lambda \cdot t}$ term.

$$\psi \cdot \lambda = k_z \cdot c \cdot \left[\frac{[2 + (q_x^{(i)})^2 + (q_y^{(i)})^2] \cdot 2}{4 \cdot (q_z^{(i)})^3} \cdot q_z \right.$$

$$- \frac{2 \cdot q_x^{(i)}}{4 \cdot (q_z^{(i)})^2} \cdot q_x - \frac{2 \cdot q_y^{(i)}}{4 \cdot (q_z^{(i)})^2} \cdot q_y]$$
$$+ \omega_\beta \cdot \{ \frac{1}{2 \cdot q_y^{(i)}} \cdot A \cdot \sin(\psi^{(i)}) \cdot \psi$$
$$+ \frac{1}{2 \cdot (q_y^{(i)})^2} \cdot q_y \cdot A \cdot \cos(\psi^{(i)})\}$$
$$- \frac{1}{2} \cdot k_z \cdot c \cdot \{ \frac{q_y^{(i)}}{(q_z^{(i)})^2} \} \cdot A \cdot \sin(\psi^{(i)}) \cdot \psi$$
$$+ \frac{1}{2} \cdot k_z \cdot c \cdot \{ \frac{1}{(q_z^{(i)})^2} \cdot q_y \cdot A \cdot \cos(\psi^{(i)})$$
$$- \frac{2 \cdot q_y^{(i)}}{(q_z^{(i)})^3} \cdot q_z \cdot A \cdot \cos(\psi^{(i)})\}$$

$$\psi \cdot \lambda = k_z \cdot c \cdot \frac{[2 + (q_x^{(i)})^2 + (q_y^{(i)})^2] \cdot 2}{4 \cdot (q_z^{(i)})^3} \cdot q_z$$
$$- k_z \cdot c \cdot \frac{2 \cdot q_x^{(i)}}{4 \cdot (q_z^{(i)})^2} \cdot q_x - k_z \cdot c \cdot \frac{2 \cdot q_y^{(i)}}{4 \cdot (q_z^{(i)})^2} \cdot q_y$$
$$+ \omega_\beta \cdot \frac{1}{2 \cdot q_y^{(i)}} \cdot A \cdot \sin(\psi^{(i)}) \cdot \psi$$
$$+ \omega_\beta \cdot \frac{1}{2 \cdot (q_y^{(i)})^2} \cdot q_y \cdot A \cdot \cos(\psi^{(i)})$$
$$- \frac{1}{2} \cdot k_z \cdot c \cdot \{ \frac{q_y^{(i)}}{(q_z^{(i)})^2} \} \cdot A \cdot \sin(\psi^{(i)}) \cdot \psi$$
$$+ \frac{1}{2} \cdot k_z \cdot c \cdot \frac{1}{(q_z^{(i)})^2} \cdot q_y \cdot A \cdot \cos(\psi^{(i)})$$
$$- \frac{1}{2} \cdot k_z \cdot c \cdot \frac{2 \cdot q_y^{(i)}}{(q_z^{(i)})^3} \cdot q_z \cdot A \cdot \cos(\psi^{(i)})$$

$$\psi \cdot \lambda = [\omega_\beta \cdot \frac{1}{2 \cdot q_y^{(i)}} \cdot A \cdot \sin(\psi^{(i)}) - \frac{1}{2} \cdot k_z \cdot c \cdot \{ \frac{q_y^{(i)}}{(q_z^{(i)})^2} \}$$
$$\cdot A \cdot \sin(\psi^{(i)})] \cdot \psi - k_z \cdot c \cdot \frac{2 \cdot q_x^{(i)}}{4 \cdot (q_z^{(i)})^2} \cdot q_x$$
$$+ [\omega_\beta \cdot \frac{1}{2 \cdot (q_y^{(i)})^2} \cdot A \cdot \cos(\psi^{(i)}) + \frac{1}{2} \cdot k_z \cdot c$$
$$\cdot \frac{1}{(q_z^{(i)})^2} \cdot A \cdot \cos(\psi^{(i)}) - k_z \cdot c \cdot \frac{2 \cdot q_y^{(i)}}{4 \cdot (q_z^{(i)})^2}] \cdot q_y$$

$$+[k_z \cdot c \cdot \frac{[2 + (q_x^{(i)})^2 + (q_y^{(i)})^2] \cdot 2}{4 \cdot (q_z^{(i)})^3}$$

$$-\frac{1}{2} \cdot k_z \cdot c \cdot \frac{2 \cdot q_y^{(i)}}{(q_z^{(i)})^3} \cdot A \cdot \cos(\psi^{(i)})] \cdot q_z$$

$$\Bigg[\omega_\beta \cdot \frac{1}{2 \cdot q_y^{(i)}} \cdot A \cdot \sin(\psi^{(i)}) - \frac{1}{2} \cdot k_z \cdot c \cdot \{\frac{q_y^{(i)}}{(q_z^{(i)})^2}\}$$

$$\cdot A \cdot \sin(\psi^{(i)})] \cdot \psi - \psi \cdot \lambda - k_z \cdot c \cdot \frac{2 \cdot q_x^{(i)}}{4 \cdot (q_z^{(i)})^2} \cdot q_x$$

$$+[\omega_\beta \cdot \frac{1}{2 \cdot (q_y^{(i)})^2} \cdot A \cdot \cos(\psi^{(i)}) + \frac{1}{2} \cdot k_z$$

$$\cdot c \cdot \frac{1}{(q_z^{(i)})^2} \cdot A \cdot \cos(\psi^{(i)}) - k_z \cdot c \cdot \frac{2 \cdot q_y^{(i)}}{4 \cdot (q_z^{(i)})^2}] \cdot q_y$$

$$+\Bigg[k_z \cdot c \cdot \frac{[2 + (q_x^{(i)})^2 + (q_y^{(i)})^2] \cdot 2}{4 \cdot (q_z^{(i)})^3}$$

$$-\frac{1}{2} \cdot k_z \cdot c \cdot \frac{2 \cdot q_y^{(i)}}{(q_z^{(i)})^3} \cdot A \cdot \cos(\psi^{(i)})] \cdot q_z = 0$$

We define the following global parameters for simplicity.

$$\Xi_{11} = \omega_\beta \cdot \frac{1}{2 \cdot q_y^{(i)}} \cdot A \cdot \sin(\psi^{(i)}) - \frac{1}{2} \cdot k_z \cdot c \cdot \left\{\frac{q_y^{(i)}}{(q_z^{(i)})^2}\right\}$$

$$\cdot A \cdot \sin(\psi^{(i)}); \; \Xi_{12} = -k_z \cdot c \cdot \frac{2 \cdot q_x^{(i)}}{4 \cdot (q_z^{(i)})^2}$$

$$\Xi_{13} = \omega_\beta \cdot \frac{1}{2 \cdot (q_y^{(i)})^2} \cdot A \cdot \cos(\psi^{(i)}) + \frac{1}{2} \cdot k_z$$

$$\cdot c \cdot \frac{1}{(q_z^{(i)})^2} \cdot A \cdot \cos(\psi^{(i)}) - k_z \cdot c \cdot \frac{2 \cdot q_y^{(i)}}{4 \cdot (q_z^{(i)})^2}$$

$$\Xi_{14} = k_z \cdot c \cdot \frac{[2 + (q_x^{(i)})^2 + (q_y^{(i)})^2] \cdot 2}{4 \cdot (q_z^{(i)})^3}$$

$$-\frac{1}{2} \cdot k_z \cdot c \cdot \frac{2 \cdot q_y^{(i)}}{(q_z^{(i)})^3} \cdot A \cdot \cos(\psi^{(i)})$$

$$\Xi_{11} \cdot \psi - \psi \cdot \lambda + \Xi_{12} \cdot q_x + \Xi_{13} \cdot q_y + \Xi_{14} \cdot q_z = 0$$

$$\frac{dq_x}{dt} = -\frac{1}{8} \cdot k_z \cdot c \cdot \frac{q_y \cdot q_x}{q_z^2} \cdot A \cdot \sin(\psi)$$

$$q_x \cdot \lambda \cdot e^{\lambda \cdot t} = -\frac{1}{8} \cdot k_z \cdot c \cdot \frac{(q_y^{(i)} + q_y \cdot e^{\lambda \cdot t}) \cdot (q_x^{(i)} + q_x \cdot e^{\lambda \cdot t})}{(q_z^{(i)} + q_z \cdot e^{\lambda \cdot t})^2} \cdot A \cdot \sin(\psi)$$

$$\sin(\psi) = \sin(\psi^{(i)} + \psi \cdot e^{\lambda \cdot t}) \approx \sin(\psi^{(i)}) + \cos(\psi^{(i)}) \cdot \psi \cdot e^{\lambda \cdot t}$$

$$q_x \cdot \lambda \cdot e^{\lambda \cdot t} = -\frac{1}{8} \cdot k_z \cdot c \cdot \frac{(q_y^{(i)} + q_y \cdot e^{\lambda \cdot t}) \cdot (q_x^{(i)} + q_x \cdot e^{\lambda \cdot t})}{(q_z^{(i)} + q_z \cdot e^{\lambda \cdot t})^2} \cdot A \cdot [\sin(\psi^{(i)}) + \cos(\psi^{(i)}) \cdot \psi \cdot e^{\lambda \cdot t}]$$

$$\frac{(q_y^{(i)} + q_y \cdot e^{\lambda \cdot t}) \cdot (q_x^{(i)} + q_x \cdot e^{\lambda \cdot t})}{(q_z^{(i)} + q_z \cdot e^{\lambda \cdot t})^2} = \frac{q_y^{(i)} \cdot q_x^{(i)} + q_y^{(i)} \cdot q_x \cdot e^{\lambda \cdot t} + q_x^{(i)} \cdot q_y \cdot e^{\lambda \cdot t} + q_y \cdot q_x \cdot e^{2 \cdot \lambda \cdot t}}{(q_z^{(i)})^2 + 2 \cdot q_z^{(i)} \cdot q_z \cdot e^{\lambda \cdot t} + (q_z)^2 \cdot e^{2 \cdot \lambda \cdot t}}$$

Assumption $q_y \cdot q_x \approx 0$; $(q_z)^2 \approx 0$ then

$$\frac{(q_y^{(i)} + q_y \cdot e^{\lambda \cdot t}) \cdot (q_x^{(i)} + q_x \cdot e^{\lambda \cdot t})}{(q_z^{(i)} + q_z \cdot e^{\lambda \cdot t})^2} = \frac{q_y^{(i)} \cdot q_x^{(i)} + q_y^{(i)} \cdot q_x \cdot e^{\lambda \cdot t} + q_x^{(i)} \cdot q_y \cdot e^{\lambda \cdot t}}{(q_z^{(i)})^2 + 2 \cdot q_z^{(i)} \cdot q_z \cdot e^{\lambda \cdot t}}$$

$$\frac{q_y^{(i)} \cdot q_x^{(i)} + q_y^{(i)} \cdot q_x \cdot e^{\lambda \cdot t} + q_x^{(i)} \cdot q_y \cdot e^{\lambda \cdot t}}{(q_z^{(i)})^2 + 2 \cdot q_z^{(i)} \cdot q_z \cdot e^{\lambda \cdot t}} \cdot \frac{[(q_z^{(i)})^2 - 2 \cdot q_z^{(i)} \cdot q_z \cdot e^{\lambda \cdot t}]}{[(q_z^{(i)})^2 - 2 \cdot q_z^{(i)} \cdot q_z \cdot e^{\lambda \cdot t}]}$$

$$= \frac{\begin{array}{l} q_y^{(i)} \cdot q_x^{(i)} \cdot (q_z^{(i)})^2 - q_y^{(i)} \cdot q_x^{(i)} \cdot 2 \cdot q_z^{(i)} \cdot q_z \cdot e^{\lambda \cdot t} \\ +(q_z^{(i)})^2 \cdot q_y^{(i)} \cdot q_x \cdot e^{\lambda \cdot t} - q_y^{(i)} \cdot 2 \cdot q_z^{(i)} \cdot q_z \cdot q_x \cdot e^{2 \cdot \lambda \cdot t} \\ +(q_z^{(i)})^2 \cdot q_x^{(i)} \cdot q_y \cdot e^{\lambda \cdot t} - q_x^{(i)} \cdot 2 \cdot q_z^{(i)} \cdot q_z \cdot q_y \cdot e^{2 \cdot \lambda \cdot t} \end{array}}{(q_z^{(i)})^4 - 4 \cdot (q_z^{(i)})^2 \cdot (q_z)^2 \cdot e^{2 \cdot \lambda \cdot t}}$$

Assumption $q_z \cdot q_x \approx 0$; $q_z \cdot q_y \approx 0$; $(q_z)^2$ then

$$\frac{(q_y^{(i)} + q_y \cdot e^{\lambda \cdot t}) \cdot (q_x^{(i)} + q_x \cdot e^{\lambda \cdot t})}{(q_z^{(i)} + q_z \cdot e^{\lambda \cdot t})^2} = \frac{\begin{array}{l} q_y^{(i)} \cdot q_x^{(i)} \cdot (q_z^{(i)})^2 - q_y^{(i)} \cdot q_x^{(i)} \cdot 2 \cdot q_z^{(i)} \cdot q_z \cdot e^{\lambda \cdot t} \\ +(q_z^{(i)})^2 \cdot q_y^{(i)} \cdot q_x \cdot e^{\lambda \cdot t} + (q_z^{(i)})^2 \cdot q_x^{(i)} \cdot q_y \cdot e^{\lambda \cdot t} \end{array}}{(q_z^{(i)})^4}$$

$$\frac{(q_y^{(i)} + q_y \cdot e^{\lambda \cdot t}) \cdot (q_x^{(i)} + q_x \cdot e^{\lambda \cdot t})}{(q_z^{(i)} + q_z \cdot e^{\lambda \cdot t})^2} = \frac{q_y^{(i)} \cdot q_x^{(i)}}{(q_z^{(i)})^2} - \frac{q_y^{(i)} \cdot q_x^{(i)} \cdot 2}{(q_z^{(i)})^3} \cdot q_z \cdot e^{\lambda \cdot t} + \frac{q_y^{(i)}}{(q_z^{(i)})^2} \cdot q_x \cdot e^{\lambda \cdot t} + \frac{q_x^{(i)}}{(q_z^{(i)})^2} \cdot q_y \cdot e^{\lambda \cdot t}$$

$$q_x \cdot \lambda \cdot e^{\lambda \cdot t} = -\frac{1}{8} \cdot k_z \cdot c \cdot [\frac{q_y^{(i)} \cdot q_x^{(i)}}{(q_z^{(i)})^2} - \frac{q_y^{(i)} \cdot q_x^{(i)} \cdot 2}{(q_z^{(i)})^3} \cdot q_z \cdot e^{\lambda \cdot t} + \frac{q_y^{(i)}}{(q_z^{(i)})^2} \cdot q_x \cdot e^{\lambda \cdot t} + \frac{q_x^{(i)}}{(q_z^{(i)})^2} \cdot q_y \cdot e^{\lambda \cdot t}] \cdot A \cdot [\sin(\psi^{(i)}) + \cos(\psi^{(i)}) \cdot \psi \cdot e^{\lambda \cdot t}]$$

$$q_x \cdot \lambda \cdot e^{\lambda \cdot t} = [-\frac{1}{8} \cdot k_z \cdot c \cdot A \cdot \frac{q_y^{(i)} \cdot q_x^{(i)}}{(q_z^{(i)})^2} + \frac{1}{8} \cdot k_z \cdot c \cdot A \cdot \frac{q_y^{(i)} \cdot q_x^{(i)} \cdot 2}{(q_z^{(i)})^3} \cdot q_z \cdot e^{\lambda \cdot t} - \frac{1}{8} \cdot k_z \cdot c \cdot A \cdot \frac{q_y^{(i)}}{(q_z^{(i)})^2} \cdot q_x \cdot e^{\lambda \cdot t} - \frac{1}{8} \cdot k_z \cdot c \cdot A \cdot \frac{q_x^{(i)}}{(q_z^{(i)})^2} \cdot q_y \cdot e^{\lambda \cdot t}] \cdot [\sin(\psi^{(i)}) + \cos(\psi^{(i)}) \cdot \psi \cdot e^{\lambda \cdot t}]$$

$$q_x \cdot \lambda \cdot e^{\lambda \cdot t} = [-\frac{1}{8} \cdot k_z \cdot c \cdot A \cdot \frac{q_y^{(i)} \cdot q_x^{(i)}}{(q_z^{(i)})^2} + \frac{1}{8} \cdot k_z \cdot c \cdot A \cdot \frac{q_y^{(i)} \cdot q_x^{(i)} \cdot 2}{(q_z^{(i)})^3} \cdot q_z \cdot e^{\lambda \cdot t} - \frac{1}{8} \cdot k_z \cdot c \cdot A \cdot \frac{q_y^{(i)}}{(q_z^{(i)})^2} \cdot q_x \cdot e^{\lambda \cdot t} - \frac{1}{8} \cdot k_z \cdot c \cdot A \cdot \frac{q_x^{(i)}}{(q_z^{(i)})^2} \cdot q_y \cdot e^{\lambda \cdot t}] \cdot \sin(\psi^{(i)}) + [-\frac{1}{8} \cdot k_z \cdot c \cdot A \cdot \frac{q_y^{(i)} \cdot q_x^{(i)}}{(q_z^{(i)})^2} + \frac{1}{8} \cdot k_z \cdot c \cdot A \cdot \frac{q_y^{(i)} \cdot q_x^{(i)} \cdot 2}{(q_z^{(i)})^3} \cdot q_z \cdot e^{\lambda \cdot t} - \frac{1}{8} \cdot k_z \cdot c \cdot A \cdot \frac{q_y^{(i)}}{(q_z^{(i)})^2} \cdot q_x \cdot e^{\lambda \cdot t} - \frac{1}{8} \cdot k_z \cdot c \cdot A \cdot \frac{q_x^{(i)}}{(q_z^{(i)})^2} \cdot q_y \cdot e^{\lambda \cdot t}] \cdot \cos(\psi^{(i)}) \cdot \psi \cdot e^{\lambda \cdot t}$$

$$q_x \cdot \lambda \cdot e^{\lambda \cdot t} = [-\frac{1}{8} \cdot k_z \cdot c \cdot A \cdot \frac{q_y^{(i)} \cdot q_x^{(i)}}{(q_z^{(i)})^2} + \frac{1}{8} \cdot k_z \cdot c \cdot A \cdot \frac{q_y^{(i)} \cdot q_x^{(i)} \cdot 2}{(q_z^{(i)})^3}$$

$$\cdot q_z \cdot e^{\lambda \cdot t} - \frac{1}{8} \cdot k_z \cdot c \cdot A \cdot \frac{q_y^{(i)}}{(q_z^{(i)})^2} \cdot q_x \cdot e^{\lambda \cdot t}$$
$$- \frac{1}{8} \cdot k_z \cdot c \cdot A \cdot \frac{q_x^{(i)}}{(q_z^{(i)})^2} \cdot q_y \cdot e^{\lambda \cdot t}] \cdot \sin(\psi^{(i)})$$
$$+ [-\frac{1}{8} \cdot k_z \cdot c \cdot A \cdot \frac{q_y^{(i)} \cdot q_x^{(i)}}{(q_z^{(i)})^2} \cdot \psi + \frac{1}{8} \cdot k_z \cdot c \cdot A$$
$$\cdot \frac{q_y^{(i)} \cdot q_x^{(i)} \cdot 2}{(q_z^{(i)})^3} \cdot q_z \cdot \psi \cdot e^{\lambda \cdot t} - \frac{1}{8} \cdot k_z \cdot c \cdot A \cdot \frac{q_y^{(i)}}{(q_z^{(i)})^2} \cdot q_x \cdot \psi \cdot e^{\lambda \cdot t}$$
$$- \frac{1}{8} \cdot k_z \cdot c \cdot A \cdot \frac{q_x^{(i)}}{(q_z^{(i)})^2} \cdot q_y \cdot \psi \cdot e^{\lambda \cdot t}] \cdot \cos(\psi^{(i)}) \cdot e^{\lambda \cdot t}$$

Assumption: $q_x \cdot \psi \approx 0; q_z \cdot \psi \approx 0; q_y \cdot \psi \approx 0$ then

$$q_x \cdot \lambda \cdot e^{\lambda \cdot t} = [-\frac{1}{8} \cdot k_z \cdot c \cdot A \cdot \frac{q_y^{(i)} \cdot q_x^{(i)}}{(q_z^{(i)})^2} + \frac{1}{8} \cdot k_z \cdot c \cdot A$$
$$\cdot \frac{q_y^{(i)} \cdot q_x^{(i)} \cdot 2}{(q_z^{(i)})^3} \cdot q_z \cdot e^{\lambda \cdot t} - \frac{1}{8} \cdot k_z \cdot c \cdot A \cdot \frac{q_y^{(i)}}{(q_z^{(i)})^2} \cdot q_x \cdot e^{\lambda \cdot t}$$
$$- \frac{1}{8} \cdot k_z \cdot c \cdot A \cdot \frac{q_x^{(i)}}{(q_z^{(i)})^2} \cdot q_y \cdot e^{\lambda \cdot t}] \cdot \sin(\psi^{(i)})$$
$$+ [-\frac{1}{8} \cdot k_z \cdot c \cdot A \cdot \frac{q_y^{(i)} \cdot q_x^{(i)}}{(q_z^{(i)})^2} \cdot \psi] \cdot \cos(\psi^{(i)}) \cdot e^{\lambda \cdot t}$$

$$q_x \cdot \lambda \cdot e^{\lambda \cdot t} = -\frac{1}{8} \cdot k_z \cdot c \cdot A \cdot \frac{q_y^{(i)} \cdot q_x^{(i)}}{(q_z^{(i)})^2} \cdot \sin(\psi^{(i)})$$
$$+ \frac{1}{8} \cdot k_z \cdot c \cdot A \cdot \frac{q_y^{(i)} \cdot q_x^{(i)} \cdot 2}{(q_z^{(i)})^3} \cdot q_z \cdot e^{\lambda \cdot t} \cdot \sin(\psi^{(i)})$$
$$- \frac{1}{8} \cdot k_z \cdot c \cdot A \cdot \frac{q_y^{(i)}}{(q_z^{(i)})^2} \cdot q_x \cdot e^{\lambda \cdot t} \cdot \sin(\psi^{(i)})$$
$$- \frac{1}{8} \cdot k_z \cdot c \cdot A \cdot \frac{q_x^{(i)}}{(q_z^{(i)})^2} \cdot q_y \cdot e^{\lambda \cdot t} \cdot \sin(\psi^{(i)})$$
$$+ [-\frac{1}{8} \cdot k_z \cdot c \cdot A \cdot \frac{q_y^{(i)} \cdot q_x^{(i)}}{(q_z^{(i)})^2} \cdot \psi] \cdot \cos(\psi^{(i)}) \cdot e^{\lambda \cdot t}$$

At fixed point $\frac{dq_x}{dt} = 0; -\frac{1}{8} \cdot k_z \cdot c \cdot A \cdot \frac{q_y^{(i)} \cdot q_x^{(i)}}{(q_z^{(i)})^2} \cdot \sin(\psi^{(i)}) = 0$ then

$$q_x \cdot \lambda \cdot e^{\lambda \cdot t} = \frac{1}{8} \cdot k_z \cdot c \cdot A \cdot \frac{q_y^{(i)} \cdot q_x^{(i)} \cdot 2}{(q_z^{(i)})^3} \cdot q_z \cdot e^{\lambda \cdot t} \cdot \sin(\psi^{(i)})$$

$$-\frac{1}{8}\cdot k_z\cdot c\cdot A\cdot\frac{q_y^{(i)}}{(q_z^{(i)})^2}\cdot q_x\cdot e^{\lambda\cdot t}\cdot\sin(\psi^{(i)})$$
$$-\frac{1}{8}\cdot k_z\cdot c\cdot A\cdot\frac{q_x^{(i)}}{(q_z^{(i)})^2}\cdot q_y\cdot e^{\lambda\cdot t}\cdot\sin(\psi^{(i)})$$
$$+[-\frac{1}{8}\cdot k_z\cdot c\cdot A\cdot\frac{q_y^{(i)}\cdot q_x^{(i)}}{(q_z^{(i)})^2}\cdot\psi]\cdot\cos(\psi^{(i)})\cdot e^{\lambda\cdot t}$$

We divide the two sides of the above expression by $e^{\lambda\cdot t}$ term.

$$q_x\cdot\lambda=\frac{1}{8}\cdot k_z\cdot c\cdot A\cdot\frac{q_y^{(i)}\cdot q_x^{(i)}\cdot 2}{(q_z^{(i)})^3}\cdot q_z\cdot\sin(\psi^{(i)})$$
$$-\frac{1}{8}\cdot k_z\cdot c\cdot A\cdot\frac{q_y^{(i)}}{(q_z^{(i)})^2}\cdot q_x\cdot\sin(\psi^{(i)})$$
$$-\frac{1}{8}\cdot k_z\cdot c\cdot A\cdot\frac{q_x^{(i)}}{(q_z^{(i)})^2}\cdot q_y\cdot\sin(\psi^{(i)})$$
$$-\frac{1}{8}\cdot k_z\cdot c\cdot A\cdot\frac{q_y^{(i)}\cdot q_x^{(i)}}{(q_z^{(i)})^2}\cdot\psi\cdot\cos(\psi^{(i)})$$

$$-\frac{1}{8}\cdot k_z\cdot c\cdot A\cdot\frac{q_y^{(i)}\cdot q_x^{(i)}}{(q_z^{(i)})^2}\cdot\cos(\psi^{(i)})\cdot\psi$$
$$-\frac{1}{8}\cdot k_z\cdot c\cdot A\cdot\frac{q_y^{(i)}}{(q_z^{(i)})^2}\cdot\sin(\psi^{(i)})\cdot q_x-q_x\cdot\lambda$$
$$-\frac{1}{8}\cdot k_z\cdot c\cdot A\cdot\frac{q_x^{(i)}}{(q_z^{(i)})^2}\cdot\sin(\psi^{(i)})\cdot q_y$$
$$+\frac{1}{8}\cdot k_z\cdot c\cdot A\cdot\frac{q_y^{(i)}\cdot q_x^{(i)}\cdot 2}{(q_z^{(i)})^3}\cdot\sin(\psi^{(i)})\cdot q_z=0$$

We define the following global parameters for simplicity

$$\Xi_{21}=-\frac{1}{8}\cdot k_z\cdot c\cdot A\cdot\frac{q_y^{(i)}\cdot q_x^{(i)}}{(q_z^{(i)})^2}\cdot\cos(\psi^{(i)})$$
$$\Xi_{22}=-\frac{1}{8}\cdot k_z\cdot c\cdot A\cdot\frac{q_y^{(i)}}{(q_z^{(i)})^2}\cdot\sin(\psi^{(i)})$$
$$\Xi_{23}=-\frac{1}{8}\cdot k_z\cdot c\cdot A\cdot\frac{q_x^{(i)}}{(q_z^{(i)})^2}\cdot\sin(\psi^{(i)})$$
$$\Xi_{24}=\frac{1}{8}\cdot k_z\cdot c\cdot A\cdot\frac{q_y^{(i)}\cdot q_x^{(i)}\cdot 2}{(q_z^{(i)})^3}\cdot\sin(\psi^{(i)})$$

$$\Xi_{21}\cdot\psi+\Xi_{22}\cdot q_x-q_x\cdot\lambda+\Xi_{23}\cdot q_y+\Xi_{24}\cdot q_z=0$$

$$\frac{dq_y}{dt}=-\frac{1}{2}\cdot(\omega_\beta+\frac{1}{4}\cdot k_z\cdot c\cdot\frac{q_y^2}{q_z^2})\cdot A\cdot\sin(\psi)$$

$$q_y\cdot\lambda\cdot e^{\lambda\cdot t}=-\frac{1}{2}\cdot(\omega_\beta+\frac{1}{4}\cdot k_z\cdot c\cdot\frac{(q_y^{(i)}+q_y\cdot e^{\lambda\cdot t})^2}{(q_z^{(i)}+q_z\cdot e^{\lambda\cdot t})^2})\cdot A\cdot\sin(\psi)$$

$$\sin(\psi)=\sin(\psi^{(i)}+\psi\cdot e^{\lambda\cdot t})\approx\sin(\psi^{(i)})+\cos(\psi^{(i)})\cdot\psi\cdot e^{\lambda\cdot t}$$

$$q_y\cdot\lambda\cdot e^{\lambda\cdot t}=-\frac{1}{2}\cdot(\omega_\beta+\frac{1}{4}\cdot k_z\cdot c\cdot\frac{(q_y^{(i)}+q_y\cdot e^{\lambda\cdot t})^2}{(q_z^{(i)}+q_z\cdot e^{\lambda\cdot t})^2})\cdot A\cdot[\sin(\psi^{(i)})+\cos(\psi^{(i)})\cdot\psi\cdot e^{\lambda\cdot t}]$$

$$\frac{(q_y^{(i)}+q_y\cdot e^{\lambda\cdot t})^2}{(q_z^{(i)}+q_z\cdot e^{\lambda\cdot t})^2}=\frac{(q_y^{(i)})^2+2\cdot q_y^{(i)}\cdot q_y\cdot e^{\lambda\cdot t}+(q_y)^2\cdot e^{2\cdot\lambda\cdot t}}{(q_z^{(i)})^2+2\cdot q_z^{(i)}\cdot q_z\cdot e^{\lambda\cdot t}+(q_z)^2\cdot e^{2\cdot\lambda\cdot t}}|_{(q_y)^2\approx 0;\ (q_z)^2\approx 0}\approx\frac{(q_y^{(i)})^2+2\cdot q_y^{(i)}\cdot q_y\cdot e^{\lambda\cdot t}}{(q_z^{(i)})^2+2\cdot q_z^{(i)}\cdot q_z\cdot e^{\lambda\cdot t}}$$

$$\frac{(q_y^{(i)}+q_y\cdot e^{\lambda\cdot t})^2}{(q_z^{(i)}+q_z\cdot e^{\lambda\cdot t})^2}\approx\frac{[(q_y^{(i)})^2+2\cdot q_y^{(i)}\cdot q_y\cdot e^{\lambda\cdot t}]}{[(q_z^{(i)})^2+2\cdot q_z^{(i)}\cdot q_z\cdot e^{\lambda\cdot t}]}\cdot\frac{[(q_z^{(i)})^2-2\cdot q_z^{(i)}\cdot q_z\cdot e^{\lambda\cdot t}]}{[(q_z^{(i)})^2-2\cdot q_z^{(i)}\cdot q_z\cdot e^{\lambda\cdot t}]}$$

$$=\frac{(q_y^{(i)})^2\cdot(q_z^{(i)})^2-(q_y^{(i)})^2\cdot 2\cdot q_z^{(i)}\cdot q_z\cdot e^{\lambda\cdot t}+2\cdot q_y^{(i)}\cdot q_y\cdot e^{\lambda\cdot t}\cdot(q_z^{(i)})^2-4\cdot q_y^{(i)}\cdot q_z^{(i)}\cdot q_z\cdot q_y\cdot e^{2\cdot\lambda\cdot t}}{(q_z^{(i)})^4-4\cdot(q_z^{(i)})^2\cdot(q_z)^2\cdot e^{2\cdot\lambda\cdot t}}$$

Assumption: $q_z\cdot q_y\approx 0;\ (q_z)^2\approx 0$ then

$$\frac{(q_y^{(i)}+q_y\cdot e^{\lambda\cdot t})^2}{(q_z^{(i)}+q_z\cdot e^{\lambda\cdot t})^2}\approx\frac{(q_y^{(i)})^2\cdot(q_z^{(i)})^2-(q_y^{(i)})^2\cdot 2\cdot q_z^{(i)}\cdot q_z\cdot e^{\lambda\cdot t}+2\cdot q_y^{(i)}\cdot q_y\cdot e^{\lambda\cdot t}\cdot(q_z^{(i)})^2}{(q_z^{(i)})^4}$$

$$\frac{(q_y^{(i)}+q_y\cdot e^{\lambda\cdot t})^2}{(q_z^{(i)}+q_z\cdot e^{\lambda\cdot t})^2}\approx\frac{(q_y^{(i)})^2}{(q_z^{(i)})^2}-\frac{(q_y^{(i)})^2\cdot 2}{(q_z^{(i)})^3}\cdot q_z\cdot e^{\lambda\cdot t}+\frac{2\cdot q_y^{(i)}}{(q_z^{(i)})^2}\cdot q_y\cdot e^{\lambda\cdot t}$$

$$q_y\cdot\lambda\cdot e^{\lambda\cdot t}=-\frac{1}{2}\cdot(\omega_\beta+\frac{1}{4}\cdot k_z\cdot c\cdot[\frac{(q_y^{(i)})^2}{(q_z^{(i)})^2}-\frac{(q_y^{(i)})^2\cdot 2}{(q_z^{(i)})^3}\cdot q_z\cdot e^{\lambda\cdot t}$$

$$+\frac{2\cdot q_y^{(i)}}{(q_z^{(i)})^2}\cdot q_y\cdot e^{\lambda\cdot t}])\cdot A\cdot[\sin(\psi^{(i)})+\cos(\psi^{(i)})\cdot\psi\cdot e^{\lambda\cdot t}]$$

$$q_y\cdot\lambda\cdot e^{\lambda\cdot t}=-\frac{1}{2}\cdot(\omega_\beta+\frac{1}{4}\cdot k_z\cdot c\cdot\frac{(q_y^{(i)})^2}{(q_z^{(i)})^2}-\frac{1}{4}\cdot k_z\cdot c\cdot\frac{(q_y^{(i)})^2\cdot 2}{(q_z^{(i)})^3}\cdot q_z\cdot e^{\lambda\cdot t}$$
$$+\frac{1}{4}\cdot k_z\cdot c\cdot\frac{2\cdot q_y^{(i)}}{(q_z^{(i)})^2}\cdot q_y\cdot e^{\lambda\cdot t})\cdot A\cdot[\sin(\psi^{(i)})+\cos(\psi^{(i)})\cdot\psi\cdot e^{\lambda\cdot t}]$$

$$q_y\cdot\lambda\cdot e^{\lambda\cdot t}=-\frac{1}{2}\cdot([\omega_\beta+\frac{1}{4}\cdot k_z\cdot c\cdot\frac{(q_y^{(i)})^2}{(q_z^{(i)})^2}]+[\frac{1}{4}\cdot k_z\cdot c\cdot\frac{2\cdot q_y^{(i)}}{(q_z^{(i)})^2}\cdot q_y\cdot e^{\lambda\cdot t}$$
$$-\frac{1}{4}\cdot k_z\cdot c\cdot\frac{(q_y^{(i)})^2\cdot 2}{(q_z^{(i)})^3}\cdot q_z\cdot e^{\lambda\cdot t}])\cdot A\cdot[\sin(\psi^{(i)})+\cos(\psi^{(i)})\cdot\psi\cdot e^{\lambda\cdot t}]$$

$$q_y\cdot\lambda\cdot e^{\lambda\cdot t}=-\frac{1}{2}\cdot\left[\omega_\beta+\frac{1}{4}\cdot k_z\cdot c\cdot\frac{(q_y^{(i)})^2}{(q_z^{(i)})^2}\right]\cdot A\cdot[\sin(\psi^{(i)})$$
$$+\cos(\psi^{(i)})\cdot\psi\cdot e^{\lambda\cdot t}]-\frac{1}{2}\cdot\left[\frac{1}{4}\cdot k_z\cdot c\cdot\frac{2\cdot q_y^{(i)}}{(q_z^{(i)})^2}\cdot q_y\cdot e^{\lambda\cdot t}\right.$$
$$\left.-\frac{1}{4}\cdot k_z\cdot c\cdot\frac{(q_y^{(i)})^2\cdot 2}{(q_z^{(i)})^3}\cdot q_z\cdot e^{\lambda\cdot t}\right]\cdot A\cdot[\sin(\psi^{(i)})$$
$$+\cos(\psi^{(i)})\cdot\psi\cdot e^{\lambda\cdot t}]$$

$$q_y\cdot\lambda\cdot e^{\lambda\cdot t}=-\frac{1}{2}\cdot[\omega_\beta+\frac{1}{4}\cdot k_z\cdot c\cdot\frac{(q_y^{(i)})^2}{(q_z^{(i)})^2}]\cdot A\cdot\sin(\psi^{(i)})$$
$$-\frac{1}{2}\cdot\left[\omega_\beta+\frac{1}{4}\cdot k_z\cdot c\cdot\frac{(q_y^{(i)})^2}{(q_z^{(i)})^2}\right]\cdot A\cdot\cos(\psi^{(i)})$$
$$\cdot\psi\cdot e^{\lambda\cdot t}-\frac{1}{2}\cdot\left[\frac{1}{4}\cdot k_z\cdot c\cdot\frac{2\cdot q_y^{(i)}}{(q_z^{(i)})^2}\cdot q_y\cdot e^{\lambda\cdot t}\right.$$
$$\left.-\frac{1}{4}\cdot k_z\cdot c\cdot\frac{(q_y^{(i)})^2\cdot 2}{(q_z^{(i)})^3}\cdot q_z\cdot e^{\lambda\cdot t}\right]\cdot A$$
$$\cdot[\sin(\psi^{(i)})+\cos(\psi^{(i)})\cdot\psi\cdot e^{\lambda\cdot t}]$$

At fixed points $\frac{dq_y}{dt}=0\Rightarrow-\frac{1}{2}\cdot\left[\omega_\beta+\frac{1}{4}\cdot k_z\cdot c\cdot\frac{(q_y^{(i)})^2}{(q_z^{(i)})^2}\right]\cdot A\cdot\sin(\psi^{(i)})=0$ then

$$q_y\cdot\lambda\cdot e^{\lambda\cdot t}=-\frac{1}{2}\cdot\left[\omega_\beta+\frac{1}{4}\cdot k_z\cdot c\cdot\frac{(q_y^{(i)})^2}{(q_z^{(i)})^2}\right]\cdot A\cdot\cos(\psi^{(i)})$$

$$\cdot \psi \cdot e^{\lambda \cdot t} - \frac{1}{2} \cdot \left[\frac{1}{4} \cdot k_z \cdot c \cdot \frac{2 \cdot q_y^{(i)}}{(q_z^{(i)})^2} \cdot q_y \cdot e^{\lambda \cdot t} - \frac{1}{4} \cdot k_z \cdot c \cdot \frac{(q_y^{(i)})^2 \cdot 2}{(q_z^{(i)})^3} \cdot q_z \cdot e^{\lambda \cdot t}\right] \cdot A \cdot [\sin(\psi^{(i)}) + \cos(\psi^{(i)}) \cdot \psi \cdot e^{\lambda \cdot t}]$$

$$q_y \cdot \lambda \cdot e^{\lambda \cdot t} = -\frac{1}{2} \cdot [\omega_\beta + \frac{1}{4} \cdot k_z \cdot c \cdot \frac{(q_y^{(i)})^2}{(q_z^{(i)})^2}] \cdot A \cdot \cos(\psi^{(i)}) \cdot \psi \cdot e^{\lambda \cdot t} - \frac{1}{2} \cdot [A \cdot \frac{1}{4} \cdot k_z \cdot c \cdot \frac{2 \cdot q_y^{(i)}}{(q_z^{(i)})^2} \cdot q_y \cdot e^{\lambda \cdot t} - A \cdot \frac{1}{4} \cdot k_z \cdot c \cdot \frac{(q_y^{(i)})^2 \cdot 2}{(q_z^{(i)})^3} \cdot q_z \cdot e^{\lambda \cdot t}] \cdot \sin(\psi^{(i)}) - \frac{1}{2} \cdot [A \cdot \frac{1}{4} \cdot k_z \cdot c \cdot \frac{2 \cdot q_y^{(i)}}{(q_z^{(i)})^2} \cdot q_y \cdot e^{\lambda \cdot t} - A \cdot \frac{1}{4} \cdot k_z \cdot c \cdot \frac{(q_y^{(i)})^2 \cdot 2}{(q_z^{(i)})^3} \cdot q_z \cdot e^{\lambda \cdot t}] \cdot \cos(\psi^{(i)}) \cdot \psi \cdot e^{\lambda \cdot t} q_y$$

$$\cdot \lambda \cdot e^{\lambda \cdot t} = -\frac{1}{2} \cdot [\omega_\beta + \frac{1}{4} \cdot k_z \cdot c \cdot \frac{(q_y^{(i)})^2}{(q_z^{(i)})^2}] \cdot A \cdot \cos(\psi^{(i)}) \cdot \psi \cdot e^{\lambda \cdot t} - \frac{1}{2} \cdot [A \cdot \frac{1}{4} \cdot k_z \cdot c \cdot \frac{2 \cdot q_y^{(i)}}{(q_z^{(i)})^2} \cdot q_y - A \cdot \frac{1}{4} \cdot k_z \cdot c \cdot \frac{(q_y^{(i)})^2 \cdot 2}{(q_z^{(i)})^3} \cdot q_z] \cdot \sin(\psi^{(i)}) \cdot e^{\lambda \cdot t} - \frac{1}{2} \cdot [A \cdot \frac{1}{4} \cdot k_z \cdot c \cdot \frac{2 \cdot q_y^{(i)}}{(q_z^{(i)})^2} \cdot q_y \cdot \psi \cdot e^{\lambda \cdot t} - A \cdot \frac{1}{4} \cdot k_z \cdot c \cdot \frac{(q_y^{(i)})^2 \cdot 2}{(q_z^{(i)})^3} \cdot q_z \cdot \psi \cdot e^{\lambda \cdot t}] \cdot \cos(\psi^{(i)}) \cdot e^{\lambda \cdot t}$$

Assumption $q_y \cdot \psi \approx 0; q_z \cdot \psi \approx 0$ then

$$q_y \cdot \lambda \cdot e^{\lambda \cdot t} = -\frac{1}{2} \cdot \left[\omega_\beta + \frac{1}{4} \cdot k_z \cdot c \cdot \frac{(q_y^{(i)})^2}{(q_z^{(i)})^2}\right] \cdot A \cdot \cos(\psi^{(i)}) \cdot \psi \cdot e^{\lambda \cdot t} - \frac{1}{2} \cdot \left[A \cdot \frac{1}{4} \cdot k_z \cdot c \cdot \frac{2 \cdot q_y^{(i)}}{(q_z^{(i)})^2} \cdot q_y\right.$$

$$-A \cdot \frac{1}{4} \cdot k_z \cdot c \cdot \frac{(q_y^{(i)})^2 \cdot 2}{(q_z^{(i)})^3} \cdot q_z \Bigg] \cdot \sin(\psi^{(i)}) \cdot e^{\lambda \cdot t}$$

We divide the two sides of the above equation by $e^{\lambda \cdot t}$ term.

$$q_y \cdot \lambda = -\frac{1}{2} \cdot [\omega_\beta + \frac{1}{4} \cdot k_z \cdot c \cdot \frac{(q_y^{(i)})^2}{(q_z^{(i)})^2}] \cdot A \cdot \cos(\psi^{(i)}) \cdot \psi$$
$$-\frac{1}{2} \cdot [A \cdot \frac{1}{4} \cdot k_z \cdot c \cdot \frac{2 \cdot q_y^{(i)}}{(q_z^{(i)})^2} \cdot q_y - A \cdot \frac{1}{4}$$
$$\cdot k_z \cdot c \cdot \frac{(q_y^{(i)})^2 \cdot 2}{(q_z^{(i)})^3} \cdot q_z] \cdot \sin(\psi^{(i)})$$

$$q_y \cdot \lambda = -\frac{1}{2} \cdot [\omega_\beta + \frac{1}{4} \cdot k_z \cdot c \cdot \frac{(q_y^{(i)})^2}{(q_z^{(i)})^2}] \cdot A \cdot \cos(\psi^{(i)})$$
$$\cdot \psi - \frac{1}{8} \cdot A \cdot k_z \cdot c \cdot \frac{2 \cdot q_y^{(i)}}{(q_z^{(i)})^2} \cdot \sin(\psi^{(i)}) \cdot q_y$$
$$+\frac{1}{8} \cdot A \cdot k_z \cdot c \cdot \frac{(q_y^{(i)})^2 \cdot 2}{(q_z^{(i)})^3} \cdot \sin(\psi^{(i)}) \cdot q_z$$

$$-\frac{1}{2} \cdot [\omega_\beta + \frac{1}{4} \cdot k_z \cdot c \cdot \frac{(q_y^{(i)})^2}{(q_z^{(i)})^2}] \cdot A \cdot \cos(\psi^{(i)})$$
$$\cdot \psi - \frac{1}{8} \cdot A \cdot k_z \cdot c \cdot \frac{2 \cdot q_y^{(i)}}{(q_z^{(i)})^2} \cdot \sin(\psi^{(i)}) \cdot q_y - q_y \cdot \lambda$$
$$+\frac{1}{8} \cdot A \cdot k_z \cdot c \cdot \frac{(q_y^{(i)})^2 \cdot 2}{(q_z^{(i)})^3} \cdot \sin(\psi^{(i)}) \cdot q_z = 0$$

We define the following global parameters for simplicity

$$\Xi_{31} = -\frac{1}{2} \cdot \left[\omega_\beta + \frac{1}{4} \cdot k_z \cdot c \cdot \frac{(q_y^{(i)})^2}{(q_z^{(i)})^2}\right] \cdot A \cdot \cos(\psi^{(i)})$$

$$\Xi_{32} = 0; \Xi_{33} = -\frac{1}{8} \cdot A \cdot k_z \cdot c \cdot \frac{2 \cdot q_y^{(i)}}{(q_z^{(i)})^2} \cdot \sin(\psi^{(i)}$$

$$\Xi_{34} = \frac{1}{8} \cdot A \cdot k_z \cdot c \cdot \frac{(q_y^{(i)})^2 \cdot 2}{(q_z^{(i)})^3} \cdot \sin(\psi^{(i)})$$

$$\Xi_{31} \cdot \psi + \Xi_{32} \cdot q_x + \Xi_{33} \cdot q_y - q_y \cdot \lambda + \Xi_{34} \cdot q_z = 0$$

$$\frac{dq_z}{dt} = -\frac{1}{2} \cdot k_z \cdot c \cdot \frac{q_y}{q_z} \cdot A \cdot \sin(\psi)$$

$$q_z \cdot \lambda \cdot e^{\lambda \cdot t} = -\frac{1}{2} \cdot k_z \cdot c \cdot \frac{q_y^{(i)} + q_y \cdot e^{\lambda \cdot t}}{q_z^{(i)} + q_z \cdot e^{\lambda \cdot t}} \cdot A \cdot \sin(\psi)$$

$$\sin(\psi) = \sin(\psi^{(i)} + \psi \cdot e^{\lambda \cdot t}) \approx \sin(\psi^{(i)}) + \cos(\psi^{(i)}) \cdot \psi \cdot e^{\lambda \cdot t}$$

$$q_z \cdot \lambda \cdot e^{\lambda \cdot t} = -\frac{1}{2} \cdot k_z \cdot c \cdot [\frac{q_y^{(i)} + q_y \cdot e^{\lambda \cdot t}}{q_z^{(i)} + q_z \cdot e^{\lambda \cdot t}}] \cdot A \cdot [\sin(\psi^{(i)}) + \cos(\psi^{(i)}) \cdot \psi \cdot e^{\lambda \cdot t}]$$

$$\frac{q_y^{(i)} + q_y \cdot e^{\lambda \cdot t}}{q_z^{(i)} + q_z \cdot e^{\lambda \cdot t}} = \frac{[q_y^{(i)} + q_y \cdot e^{\lambda \cdot t}]}{[q_z^{(i)} + q_z \cdot e^{\lambda \cdot t}]} \cdot \frac{[q_z^{(i)} - q_z \cdot e^{\lambda \cdot t}]}{[q_z^{(i)} - q_z \cdot e^{\lambda \cdot t}]}$$
$$= \frac{q_y^{(i)} \cdot q_z^{(i)} - q_y^{(i)} \cdot q_z \cdot e^{\lambda \cdot t} + q_z^{(i)} \cdot q_y \cdot e^{\lambda \cdot t} - q_y \cdot q_z \cdot e^{2 \cdot \lambda \cdot t}}{(q_z^{(i)})^2 - (q_z)^2 \cdot e^{2 \cdot \lambda \cdot t}}$$

Assumption $q_y \cdot q_z \approx 0$; $(q_z)^2 \approx 0$ then

$$\frac{q_y^{(i)} + q_y \cdot e^{\lambda \cdot t}}{q_z^{(i)} + q_z \cdot e^{\lambda \cdot t}} \approx \frac{q_y^{(i)} \cdot q_z^{(i)} - q_y^{(i)} \cdot q_z \cdot e^{\lambda \cdot t} + q_z^{(i)} \cdot q_y \cdot e^{\lambda \cdot t}}{(q_z^{(i)})^2}$$
$$= \frac{q_y^{(i)}}{q_z^{(i)}} - \frac{q_y^{(i)}}{(q_z^{(i)})^2} \cdot q_z \cdot e^{\lambda \cdot t} + \frac{1}{q_z^{(i)}} \cdot q_y \cdot e^{\lambda \cdot t}$$

$$q_z \cdot \lambda \cdot e^{\lambda \cdot t} = -\frac{1}{2} \cdot k_z \cdot c \cdot [\frac{q_y^{(i)}}{q_z^{(i)}} - \frac{q_y^{(i)}}{(q_z^{(i)})^2} \cdot q_z \cdot e^{\lambda \cdot t} + \frac{1}{q_z^{(i)}} \cdot q_y \cdot e^{\lambda \cdot t}] \cdot A \cdot [\sin(\psi^{(i)}) + \cos(\psi^{(i)}) \cdot \psi \cdot e^{\lambda \cdot t}]$$

$$q_z \cdot \lambda \cdot e^{\lambda \cdot t} = -\frac{1}{2} \cdot k_z \cdot c \cdot [\frac{q_y^{(i)}}{q_z^{(i)}} - \frac{q_y^{(i)}}{(q_z^{(i)})^2} \cdot q_z \cdot e^{\lambda \cdot t} + \frac{1}{q_z^{(i)}} \cdot q_y \cdot e^{\lambda \cdot t}] \cdot A \cdot \sin(\psi^{(i)})$$
$$- \frac{1}{2} \cdot k_z \cdot c \cdot [\frac{q_y^{(i)}}{q_z^{(i)}} - \frac{q_y^{(i)}}{(q_z^{(i)})^2} \cdot q_z \cdot e^{\lambda \cdot t} + \frac{1}{q_z^{(i)}} \cdot q_y \cdot e^{\lambda \cdot t}] \cdot A \cdot \cos(\psi^{(i)}) \cdot \psi \cdot e^{\lambda \cdot t}$$

$$q_z \cdot \lambda \cdot e^{\lambda \cdot t} = -\frac{1}{2} \cdot k_z \cdot c \cdot \frac{q_y^{(i)}}{q_z^{(i)}} \cdot A \cdot \sin(\psi^{(i)})$$

$$+\frac{1}{2}\cdot k_z\cdot c\cdot\frac{q_y^{(i)}}{(q_z^{(i)})^2}\cdot q_z\cdot e^{\lambda\cdot t}\cdot A\cdot\sin(\psi^{(i)})$$
$$-\frac{1}{2}\cdot k_z\cdot c\cdot\frac{1}{q_z^{(i)}}\cdot q_y\cdot e^{\lambda\cdot t}\cdot A\cdot\sin(\psi^{(i)})$$
$$-\frac{1}{2}\cdot k_z\cdot c\cdot[\frac{q_y^{(i)}}{q_z^{(i)}}\cdot\psi-\frac{q_y^{(i)}}{(q_z^{(i)})^2}\cdot q_z\cdot\psi\cdot e^{\lambda\cdot t}$$
$$+\frac{1}{q_z^{(i)}}\cdot q_y\cdot\psi\cdot e^{\lambda\cdot t}]\cdot A\cdot\cos(\psi^{(i)})\cdot e^{\lambda\cdot t}$$

Assumption $q_z\cdot\psi\approx 0$; $q_y\cdot\psi\approx 0$ then

$$q_z\cdot\lambda\cdot e^{\lambda\cdot t}=-\frac{1}{2}\cdot k_z\cdot c\cdot\frac{q_y^{(i)}}{q_z^{(i)}}\cdot A\cdot\sin(\psi^{(i)})$$
$$+\frac{1}{2}\cdot k_z\cdot c\cdot\frac{q_y^{(i)}}{(q_z^{(i)})^2}\cdot q_z\cdot e^{\lambda\cdot t}\cdot A\cdot\sin(\psi^{(i)})$$
$$-\frac{1}{2}\cdot k_z\cdot c\cdot\frac{1}{q_z^{(i)}}\cdot q_y\cdot e^{\lambda\cdot t}\cdot A\cdot\sin(\psi^{(i)})$$
$$-\frac{1}{2}\cdot k_z\cdot c\cdot\frac{q_y^{(i)}}{q_z^{(i)}}\cdot A\cdot\cos(\psi^{(i)})\cdot e^{\lambda\cdot t}\cdot\psi$$

At fixed point $\frac{dq_z}{dt}=0\Rightarrow-\frac{1}{2}\cdot k_z\cdot c\cdot\frac{q_y^{(i)}}{q_z^{(i)}}\cdot\sin(\psi^{(i)})=0$ then

$$q_z\cdot\lambda\cdot e^{\lambda\cdot t}=\frac{1}{2}\cdot k_z\cdot c\cdot\frac{q_y^{(i)}}{(q_z^{(i)})^2}\cdot q_z\cdot e^{\lambda\cdot t}\cdot A\cdot\sin(\psi^{(i)})$$
$$-\frac{1}{2}\cdot k_z\cdot c\cdot\frac{1}{q_z^{(i)}}\cdot q_y\cdot e^{\lambda\cdot t}\cdot A\cdot\sin(\psi^{(i)})$$
$$-\frac{1}{2}\cdot k_z\cdot c\cdot\frac{q_y^{(i)}}{q_z^{(i)}}\cdot A\cdot\cos(\psi^{(i)})\cdot e^{\lambda\cdot t}\cdot\psi$$

We divide the two sides of the above equation by $e^{\lambda\cdot t}$ term.

$$q_z\cdot\lambda=\frac{1}{2}\cdot k_z\cdot c\cdot\frac{q_y^{(i)}}{(q_z^{(i)})^2}\cdot A\cdot\sin(\psi^{(i)})\cdot q_z$$
$$-\frac{1}{2}\cdot k_z\cdot c\cdot\frac{1}{q_z^{(i)}}\cdot A\cdot\sin(\psi^{(i)})\cdot q_y$$
$$-\frac{1}{2}\cdot k_z\cdot c\cdot\frac{q_y^{(i)}}{q_z^{(i)}}\cdot A\cdot\cos(\psi^{(i)})\cdot\psi$$

$$-\frac{1}{2}\cdot k_z\cdot c\cdot\frac{q_y^{(i)}}{q_z^{(i)}}\cdot A\cdot\cos(\psi^{(i)})\cdot\psi$$
$$-\frac{1}{2}\cdot k_z\cdot c\cdot\frac{1}{q_z^{(i)}}\cdot A\cdot\sin(\psi^{(i)})\cdot q_y$$
$$+\frac{1}{2}\cdot k_z\cdot c\cdot\frac{q_y^{(i)}}{(q_z^{(i)})^2}\cdot A\cdot\sin(\psi^{(i)})\cdot q_z-q_z\cdot\lambda=0$$

We define the following global parameters for simplicity

$$\Xi_{41}=-\frac{1}{2}\cdot k_z\cdot c\cdot\frac{q_y^{(i)}}{q_z^{(i)}}\cdot A\cdot\cos(\psi^{(i)})$$
$$\Xi_{42}=0;\ \Xi_{43}=-\frac{1}{2\cdot q_z^{(i)}}\cdot k_z\cdot c\cdot A\cdot\sin(\psi^{(i)})$$
$$\Xi_{44}=\frac{1}{2}\cdot k_z\cdot c\cdot\frac{q_y^{(i)}}{(q_z^{(i)})^2}\cdot A\cdot\sin(\psi^{(i)})$$
$$\Xi_{41}\cdot\psi+\Xi_{42}\cdot q_x+\Xi_{43}\cdot q_y+\Xi_{44}\cdot q_z-\lambda\cdot q_z=0$$

We can summary our Ion channel laser perturbed system arbitrarily small increments equations:

$$\Xi_{11}\cdot\psi-\psi\cdot\lambda+\Xi_{12}\cdot q_x+\Xi_{13}\cdot q_y+\Xi_{14}\cdot q_z=0$$
$$\Xi_{21}\cdot\psi+\Xi_{22}\cdot q_x-q_x\cdot\lambda+\Xi_{23}\cdot q_y+\Xi_{24}\cdot q_z=0$$

$$\Xi_{31}\cdot\psi+\Xi_{32}\cdot q_x+\Xi_{33}\cdot q_y-q_y\cdot\lambda+\Xi_{34}\cdot q_z=0$$
$$\Xi_{41}\cdot\psi+\Xi_{42}\cdot q_x+\Xi_{43}\cdot q_y+\Xi_{44}\cdot q_z-\lambda\cdot q_z=0$$

$$I=\begin{pmatrix}1&\dots&0\\ \vdots&\ddots&\vdots\\ 0&\cdots&1\end{pmatrix};\ A=\begin{pmatrix}\Xi_{11}&\dots&\Xi_{14}\\ \vdots&\ddots&\vdots\\ \Xi_{41}&\cdots&\Xi_{44}\end{pmatrix}$$

$$A-\lambda\cdot I=\begin{pmatrix}\Xi_{11}&\dots&\Xi_{14}\\ \vdots&\ddots&\vdots\\ \Xi_{41}&\cdots&\Xi_{44}\end{pmatrix}-\begin{pmatrix}1&\dots&0\\ \vdots&\ddots&\vdots\\ 0&\cdots&1\end{pmatrix}\cdot\lambda$$

$$\{\begin{pmatrix}\Xi_{11}&\dots&\Xi_{14}\\ \vdots&\ddots&\vdots\\ \Xi_{41}&\cdots&\Xi_{44}\end{pmatrix}-\begin{pmatrix}1&\dots&0\\ \vdots&\ddots&\vdots\\ 0&\cdots&1\end{pmatrix}\cdot\lambda\}\cdot\begin{pmatrix}\psi\\ q_x\\ q_y\\ q_z\end{pmatrix}=0\ ;\ \det(A-\lambda\cdot I)=0$$

$$\det\{\begin{pmatrix} \Xi_{11} & \dots & \Xi_{14} \\ \vdots & \ddots & \vdots \\ \Xi_{41} & \cdots & \Xi_{44} \end{pmatrix} - \begin{pmatrix} 1 & \dots & 0 \\ \vdots & \ddots & \vdots \\ 0 & \cdots & 1 \end{pmatrix} \cdot \lambda\} = 0;\ \Xi_{32} = 0;\ \Xi_{42} = 0$$

$$\begin{pmatrix} \Xi_{11} & \dots & \Xi_{14} \\ \vdots & \ddots & \vdots \\ \Xi_{41} & \cdots & \Xi_{44} \end{pmatrix} - \begin{pmatrix} 1 & \dots & 0 \\ \vdots & \ddots & \vdots \\ 0 & \cdots & 1 \end{pmatrix} \cdot \lambda$$

$$= \begin{pmatrix} (\Xi_{11} - \lambda) & \Xi_{12} & \Xi_{13} & \Xi_{14} \\ \Xi_{21} & (\Xi_{22} - \lambda) & \Xi_{23} & \Xi_{24} \\ \Xi_{31} & 0 & (\Xi_{33} - \lambda) & \Xi_{34} \\ \Xi_{41} & 0 & \Xi_{43} & (\Xi_{44} - \lambda) \end{pmatrix}$$

$$\det(A - \lambda \cdot I) = \det \begin{pmatrix} (\Xi_{11} - \lambda) & \Xi_{12} & \Xi_{13} & \Xi_{14} \\ \Xi_{21} & (\Xi_{22} - \lambda) & \Xi_{23} & \Xi_{24} \\ \Xi_{31} & 0 & (\Xi_{33} - \lambda) & \Xi_{34} \\ \Xi_{41} & 0 & \Xi_{43} & (\Xi_{44} - \lambda) \end{pmatrix}$$
$$= 0$$

$$\det(A - \lambda \cdot I) = (\Xi_{11} - \lambda) \cdot \det \begin{pmatrix} (\Xi_{22} - \lambda) & \Xi_{23} & \Xi_{24} \\ 0 & (\Xi_{33} - \lambda) & \Xi_{34} \\ 0 & \Xi_{43} & (\Xi_{44} - \lambda) \end{pmatrix}$$
$$- \Xi_{12} \cdot \det \begin{pmatrix} \Xi_{21} & \Xi_{23} & \Xi_{24} \\ \Xi_{31} & (\Xi_{33} - \lambda) & \Xi_{34} \\ \Xi_{41} & \Xi_{43} & (\Xi_{44} - \lambda) \end{pmatrix}$$
$$+ \Xi_{13} \cdot \det \begin{pmatrix} \Xi_{21} & (\Xi_{22} - \lambda) & \Xi_{24} \\ \Xi_{31} & 0 & \Xi_{34} \\ \Xi_{41} & 0 & (\Xi_{44} - \lambda) \end{pmatrix}$$
$$- \Xi_{14} \cdot \det \begin{pmatrix} \Xi_{21} & (\Xi_{22} - \lambda) & \Xi_{23} \\ \Xi_{31} & 0 & (\Xi_{33} - \lambda) \\ \Xi_{41} & 0 & \Xi_{43} \end{pmatrix}$$

No.1

$$\det \begin{pmatrix} (\Xi_{22} - \lambda) & \Xi_{23} & \Xi_{24} \\ 0 & (\Xi_{33} - \lambda) & \Xi_{34} \\ 0 & \Xi_{43} & (\Xi_{44} - \lambda) \end{pmatrix}$$
$$= (\Xi_{22} - \lambda) \cdot \det \begin{pmatrix} (\Xi_{33} - \lambda) & \Xi_{34} \\ \Xi_{43} & (\Xi_{44} - \lambda) \end{pmatrix}$$
$$= (\Xi_{22} - \lambda) \cdot \{(\Xi_{33} - \lambda) \cdot (\Xi_{44} - \lambda) - \Xi_{43} \cdot \Xi_{34}\}$$

$$\det\begin{pmatrix} (\Xi_{22}-\lambda) & \Xi_{23} & \Xi_{24} \\ 0 & (\Xi_{33}-\lambda) & \Xi_{34} \\ 0 & \Xi_{43} & (\Xi_{44}-\lambda) \end{pmatrix}$$
$$= (\Xi_{22}-\lambda)\cdot\{(\Xi_{33}-\lambda)\cdot(\Xi_{44}-\lambda) - \Xi_{43}\cdot\Xi_{34}\}$$
$$= (\Xi_{22}-\lambda)\cdot\{\lambda^2 - (\Xi_{33}+\Xi_{44})\cdot\lambda + (\Xi_{33}\cdot\Xi_{44} - \Xi_{43}\cdot\Xi_{34})\}$$

$$\det\begin{pmatrix} (\Xi_{22}-\lambda) & \Xi_{23} & \Xi_{24} \\ 0 & (\Xi_{33}-\lambda) & \Xi_{34} \\ 0 & \Xi_{43} & (\Xi_{44}-\lambda) \end{pmatrix}$$
$$= (\Xi_{22}-\lambda)\cdot\{\lambda^2 - (\Xi_{33}+\Xi_{44})\cdot\lambda + (\Xi_{33}\cdot\Xi_{44} - \Xi_{43}\cdot\Xi_{34})\}$$
$$= \lambda^2\cdot\Xi_{22} - (\Xi_{33}+\Xi_{44})\cdot\Xi_{22}\cdot\lambda + (\Xi_{33}\cdot\Xi_{44} - \Xi_{43}\cdot\Xi_{34})\cdot\Xi_{22}$$
$$- \lambda^3 + (\Xi_{33}+\Xi_{44})\cdot\lambda^2 - (\Xi_{33}\cdot\Xi_{44} - \Xi_{43}\cdot\Xi_{34})\cdot\lambda$$

$$\det\begin{pmatrix} (\Xi_{22}-\lambda) & \Xi_{23} & \Xi_{24} \\ 0 & (\Xi_{33}-\lambda) & \Xi_{34} \\ 0 & \Xi_{43} & (\Xi_{44}-\lambda) \end{pmatrix}$$
$$= -\lambda^3 + [\Xi_{22} + (\Xi_{33}+\Xi_{44})]\cdot\lambda^2$$
$$- [(\Xi_{33}+\Xi_{44})\cdot\Xi_{22} + (\Xi_{33}\cdot\Xi_{44} - \Xi_{43}\cdot\Xi_{34})]\cdot\lambda$$
$$+ (\Xi_{33}\cdot\Xi_{44} - \Xi_{43}\cdot\Xi_{34})\cdot\Xi_{22}$$

We define for simplicity the following system global parameters:

$$\Upsilon_1 = \Xi_{22} + (\Xi_{33}+\Xi_{44});\ \Upsilon_2$$
$$= -[(\Xi_{33}+\Xi_{44})\cdot\Xi_{22} + (\Xi_{33}\cdot\Xi_{44} - \Xi_{43}\cdot\Xi_{34})]$$
$$\Upsilon_3 = (\Xi_{33}\cdot\Xi_{44} - \Xi_{43}\cdot\Xi_{34})\cdot\Xi_{22}$$

$$\det\begin{pmatrix} (\Xi_{22}-\lambda) & \Xi_{23} & \Xi_{24} \\ 0 & (\Xi_{33}-\lambda) & \Xi_{34} \\ 0 & \Xi_{43} & (\Xi_{44}-\lambda) \end{pmatrix} = -\lambda^3 + \Upsilon_1\cdot\lambda^2 + \Upsilon_2\cdot\lambda + \Upsilon_3$$

No.2

$$\det\begin{pmatrix} \Xi_{21} & \Xi_{23} & \Xi_{24} \\ \Xi_{31} & (\Xi_{33}-\lambda) & \Xi_{34} \\ \Xi_{41} & \Xi_{43} & (\Xi_{44}-\lambda) \end{pmatrix}$$
$$= \Xi_{21}\cdot\det\begin{pmatrix} (\Xi_{33}-\lambda) & \Xi_{34} \\ \Xi_{43} & (\Xi_{44}-\lambda) \end{pmatrix} - \Xi_{23}\cdot\det\begin{pmatrix} \Xi_{31} & \Xi_{34} \\ \Xi_{41} & (\Xi_{44}-\lambda) \end{pmatrix}$$

$$+ \Xi_{24} \cdot \det\begin{pmatrix} \Xi_{31} & (\Xi_{33} - \lambda) \\ \Xi_{41} & \Xi_{43} \end{pmatrix}$$

$$\det\begin{pmatrix} \Xi_{21} & \Xi_{23} & \Xi_{24} \\ \Xi_{31} & (\Xi_{33} - \lambda) & \Xi_{34} \\ \Xi_{41} & \Xi_{43} & (\Xi_{44} - \lambda) \end{pmatrix}$$
$$= \Xi_{21} \cdot [(\Xi_{33} - \lambda) \cdot (\Xi_{44} - \lambda) - \Xi_{43} \cdot \Xi_{34}]$$
$$- \Xi_{23} \cdot [\Xi_{31} \cdot (\Xi_{44} - \lambda) - \Xi_{41} \cdot \Xi_{34}]$$
$$+ \Xi_{24} \cdot [\Xi_{31} \cdot \Xi_{43} - \Xi_{41} \cdot (\Xi_{33} - \lambda)]$$

$$\det\begin{pmatrix} \Xi_{21} & \Xi_{23} & \Xi_{24} \\ \Xi_{31} & (\Xi_{33} - \lambda) & \Xi_{34} \\ \Xi_{41} & \Xi_{43} & (\Xi_{44} - \lambda) \end{pmatrix}$$
$$= \Xi_{21} \cdot [\Xi_{33} \cdot \Xi_{44} - \Xi_{33} \cdot \lambda - \Xi_{44} \cdot \lambda + \lambda^2 - \Xi_{43} \cdot \Xi_{34}]$$
$$- \Xi_{23} \cdot [\Xi_{31} \cdot \Xi_{44} - \Xi_{31} \cdot \lambda - \Xi_{41} \cdot \Xi_{34}]$$
$$+ \Xi_{24} \cdot [\Xi_{31} \cdot \Xi_{43} - \Xi_{41} \cdot \Xi_{33} + \Xi_{41} \cdot \lambda]$$

$$\det\begin{pmatrix} \Xi_{21} & \Xi_{23} & \Xi_{24} \\ \Xi_{31} & (\Xi_{33} - \lambda) & \Xi_{34} \\ \Xi_{41} & \Xi_{43} & (\Xi_{44} - \lambda) \end{pmatrix}$$
$$= \Xi_{21} \cdot \Xi_{33} \cdot \Xi_{44} - \Xi_{21} \cdot \Xi_{33} \cdot \lambda - \Xi_{21} \cdot \Xi_{44} \cdot \lambda$$
$$+ \Xi_{21} \cdot \lambda^2 - \Xi_{21} \cdot \Xi_{43} \cdot \Xi_{34}$$
$$- \Xi_{23} \cdot (\Xi_{31} \cdot \Xi_{44} - \Xi_{41} \cdot \Xi_{34}) + \Xi_{23} \cdot \Xi_{31} \cdot \lambda$$
$$+ \Xi_{24} \cdot (\Xi_{31} \cdot \Xi_{43} - \Xi_{41} \cdot \Xi_{33}) + \Xi_{24} \cdot \Xi_{41} \cdot \lambda$$

$$\det\begin{pmatrix} \Xi_{21} & \Xi_{23} & \Xi_{24} \\ \Xi_{31} & (\Xi_{33} - \lambda) & \Xi_{34} \\ \Xi_{41} & \Xi_{43} & (\Xi_{44} - \lambda) \end{pmatrix}$$
$$= \Xi_{21} \cdot \lambda^2 + [\Xi_{23} \cdot \Xi_{31} + \Xi_{24} \cdot \Xi_{41}$$
$$- \Xi_{21} \cdot \Xi_{33} - \Xi_{21} \cdot \Xi_{44}] \cdot \lambda$$
$$+ \Xi_{21} \cdot \Xi_{33} \cdot \Xi_{44} - \Xi_{21} \cdot \Xi_{43} \cdot \Xi_{34} - \Xi_{23}$$
$$\cdot (\Xi_{31} \cdot \Xi_{44} - \Xi_{41} \cdot \Xi_{34}) + \Xi_{24} \cdot (\Xi_{31} \cdot \Xi_{43} - \Xi_{41} \cdot \Xi_{33})$$

$$\Upsilon_4 = \Xi_{21};\ \Upsilon_5 = \Xi_{23} \cdot \Xi_{31} + \Xi_{24} \cdot \Xi_{41} - \Xi_{21} \cdot \Xi_{33} - \Xi_{21} \cdot \Xi_{44}$$
$$\Upsilon_6 = \Xi_{21} \cdot \Xi_{33} \cdot \Xi_{44} - \Xi_{21} \cdot \Xi_{43} \cdot \Xi_{34} - \Xi_{23} \cdot (\Xi_{31} \cdot \Xi_{44} - \Xi_{41} \cdot \Xi_{34})$$
$$+ \Xi_{24} \cdot (\Xi_{31} \cdot \Xi_{43} - \Xi_{41} \cdot \Xi_{33})$$

$$\det\begin{pmatrix} \Xi_{21} & \Xi_{23} & \Xi_{24} \\ \Xi_{31} & (\Xi_{33}-\lambda) & \Xi_{34} \\ \Xi_{41} & \Xi_{43} & (\Xi_{44}-\lambda) \end{pmatrix} = \Upsilon_4\cdot\lambda^2 + \Upsilon_5\cdot\lambda + \Upsilon_6$$

No.3

$$\det\begin{pmatrix} \Xi_{21} & (\Xi_{22}-\lambda) & \Xi_{24} \\ \Xi_{31} & 0 & \Xi_{34} \\ \Xi_{41} & 0 & (\Xi_{44}-\lambda) \end{pmatrix}$$
$$= \Xi_{21}\cdot\det\begin{pmatrix} 0 & \Xi_{34} \\ 0 & (\Xi_{44}-\lambda) \end{pmatrix} - (\Xi_{22}-\lambda)\cdot\det\begin{pmatrix} \Xi_{31} & \Xi_{34} \\ \Xi_{41} & (\Xi_{44}-\lambda) \end{pmatrix}$$
$$+ \Xi_{24}\cdot\det\begin{pmatrix} \Xi_{31} & 0 \\ \Xi_{41} & 0 \end{pmatrix}$$
$$= -(\Xi_{22}-\lambda)\cdot[\Xi_{31}\cdot(\Xi_{44}-\lambda) - \Xi_{41}\cdot\Xi_{34}]$$

$$\det\begin{pmatrix} \Xi_{21} & (\Xi_{22}-\lambda) & \Xi_{24} \\ \Xi_{31} & 0 & \Xi_{34} \\ \Xi_{41} & 0 & (\Xi_{44}-\lambda) \end{pmatrix}$$
$$= \Xi_{21}\cdot\det\begin{pmatrix} 0 & \Xi_{34} \\ 0 & (\Xi_{44}-\lambda) \end{pmatrix} - (\Xi_{22}-\lambda)\cdot\det\begin{pmatrix} \Xi_{31} & \Xi_{34} \\ \Xi_{41} & (\Xi_{44}-\lambda) \end{pmatrix}$$
$$+ \Xi_{24}\cdot\det\begin{pmatrix} \Xi_{31} & 0 \\ \Xi_{41} & 0 \end{pmatrix}$$
$$= (-\Xi_{22}+\lambda)\cdot[(\Xi_{31}\cdot\Xi_{44} - \Xi_{41}\cdot\Xi_{34}) - \Xi_{31}\cdot\lambda]$$

$$\det\begin{pmatrix} \Xi_{21} & (\Xi_{22}-\lambda) & \Xi_{24} \\ \Xi_{31} & 0 & \Xi_{34} \\ \Xi_{41} & 0 & (\Xi_{44}-\lambda) \end{pmatrix}$$
$$= \Xi_{21}\cdot\det\begin{pmatrix} 0 & \Xi_{34} \\ 0 & (\Xi_{44}-\lambda) \end{pmatrix} - (\Xi_{22}-\lambda)\cdot\det\begin{pmatrix} \Xi_{31} & \Xi_{34} \\ \Xi_{41} & (\Xi_{44}-\lambda) \end{pmatrix}$$
$$+ \Xi_{24}\cdot\det\begin{pmatrix} \Xi_{31} & 0 \\ \Xi_{41} & 0 \end{pmatrix} = -\Xi_{31}\cdot\lambda^2 + [\Xi_{22}\cdot\Xi_{31}$$
$$+ (\Xi_{31}\cdot\Xi_{44} - \Xi_{41}\cdot\Xi_{34})]\cdot\lambda - \Xi_{22}\cdot(\Xi_{31}\cdot\Xi_{44} - \Xi_{41}\cdot\Xi_{34})$$

$$\Upsilon_7 = -\Xi_{31};\ \Upsilon_8 = \Xi_{22}\cdot\Xi_{31} + (\Xi_{31}\cdot\Xi_{44} - \Xi_{41}\cdot\Xi_{34})$$
$$\Upsilon_9 = -\Xi_{22}\cdot(\Xi_{31}\cdot\Xi_{44} - \Xi_{41}\cdot\Xi_{34})$$

$$\det\begin{pmatrix} \Xi_{21} & (\Xi_{22}-\lambda) & \Xi_{24} \\ \Xi_{31} & 0 & \Xi_{34} \\ \Xi_{41} & 0 & (\Xi_{44}-\lambda) \end{pmatrix} = \Upsilon_7\cdot\lambda^2+\Upsilon_8\cdot\lambda+\Upsilon_9$$

No.4

$$\det\begin{pmatrix} \Xi_{21} & (\Xi_{22}-\lambda) & \Xi_{23} \\ \Xi_{31} & 0 & (\Xi_{33}-\lambda) \\ \Xi_{41} & 0 & \Xi_{43} \end{pmatrix}$$
$$= \Xi_{21}\cdot\det\begin{pmatrix} 0 & (\Xi_{33}-\lambda) \\ 0 & \Xi_{43} \end{pmatrix} - (\Xi_{22}-\lambda)\cdot\det\begin{pmatrix} \Xi_{31} & (\Xi_{33}-\lambda) \\ \Xi_{41} & \Xi_{43} \end{pmatrix}$$
$$+ \Xi_{23}\cdot\det\begin{pmatrix} \Xi_{31} & 0 \\ \Xi_{41} & 0 \end{pmatrix}$$

$$\det\begin{pmatrix} \Xi_{21} & (\Xi_{22}-\lambda) & \Xi_{23} \\ \Xi_{31} & 0 & (\Xi_{33}-\lambda) \\ \Xi_{41} & 0 & \Xi_{43} \end{pmatrix} = -(\Xi_{22}-\lambda)\cdot[\Xi_{31}\cdot\Xi_{43}-\Xi_{41}\cdot(\Xi_{33}-\lambda)]$$

$$\det\begin{pmatrix} \Xi_{21} & (\Xi_{22}-\lambda) & \Xi_{23} \\ \Xi_{31} & 0 & (\Xi_{33}-\lambda) \\ \Xi_{41} & 0 & \Xi_{43} \end{pmatrix}$$
$$= (-\Xi_{22}+\lambda)\cdot[(\Xi_{31}\cdot\Xi_{43}-\Xi_{41}\cdot\Xi_{33})+\Xi_{41}\cdot\lambda)]$$

$$\det\begin{pmatrix} \Xi_{21} & (\Xi_{22}-\lambda) & \Xi_{23} \\ \Xi_{31} & 0 & (\Xi_{33}-\lambda) \\ \Xi_{41} & 0 & \Xi_{43} \end{pmatrix}$$
$$= \Xi_{41}\cdot\lambda^2+[(\Xi_{31}\cdot\Xi_{43}-\Xi_{41}\cdot\Xi_{33})-\Xi_{22}\cdot\Xi_{41}]\cdot\lambda$$
$$- \Xi_{22}\cdot(\Xi_{31}\cdot\Xi_{43}-\Xi_{41}\cdot\Xi_{33})$$

$$\Upsilon_{10}=\Xi_{41};\ \Upsilon_{11}=(\Xi_{31}\cdot\Xi_{43}-\Xi_{41}\cdot\Xi_{33})-\Xi_{22}\cdot\Xi_{41}$$
$$\Upsilon_{12}=-\Xi_{22}\cdot(\Xi_{31}\cdot\Xi_{43}-\Xi_{41}\cdot\Xi_{33})$$

$$\det\begin{pmatrix} \Xi_{21} & (\Xi_{22}-\lambda) & \Xi_{23} \\ \Xi_{31} & 0 & (\Xi_{33}-\lambda) \\ \Xi_{41} & 0 & \Xi_{43} \end{pmatrix} = \lambda^2\cdot\Upsilon_{10}+\lambda\cdot\Upsilon_{11}+\Upsilon_{12}$$

We can summary our last results in the Table 3.3.3.

$$\det(A-\lambda\cdot I) = (\Xi_{11}-\lambda)\cdot(-\lambda^3+\Upsilon_1\cdot\lambda^2+\Upsilon_2\cdot\lambda+\Upsilon_3)$$
$$- \Xi_{12}\cdot(\Upsilon_4\cdot\lambda^2+\Upsilon_5\cdot\lambda+\Upsilon_6)$$

Table 3.3 System's Jacobian matrix determinant elements

No	Determinant	Equivalent expression
1	$\det\begin{pmatrix} (\Xi_{22}-\lambda) & \Xi_{23} & \Xi_{24} \\ 0 & (\Xi_{33}-\lambda) & \Xi_{34} \\ 0 & \Xi_{43} & (\Xi_{44}-\lambda) \end{pmatrix}$	$-\lambda^3 + \Upsilon_1 \cdot \lambda^2 + \Upsilon_2 \cdot \lambda + \Upsilon_3$
2	$\det\begin{pmatrix} \Xi_{21} & \Xi_{23} & \Xi_{24} \\ \Xi_{31} & (\Xi_{33}-\lambda) & \Xi_{34} \\ 0 & \Xi_{43} & (\Xi_{44}-\lambda) \end{pmatrix}$	$\Upsilon_4 \cdot \lambda^2 + \Upsilon_5 \cdot \lambda + \Upsilon_6$
3	$\det\begin{pmatrix} \Xi_{21} & (\Xi_{22}-\lambda) & \Xi_{24} \\ \Xi_{31} & 0 & \Xi_{34} \\ 0 & 0 & (\Xi_{44}-\lambda) \end{pmatrix}$	$\Upsilon_7 \cdot \lambda^2 + \Upsilon_8 \cdot \lambda + \Upsilon_9$
4	$\det\begin{pmatrix} \Xi_{21} & (\Xi_{22}-\lambda) & \Xi_{23} \\ \Xi_{31} & 0 & (\Xi_{33}-\lambda) \\ 0 & 0 & \Xi_{43} \end{pmatrix}$	$\lambda^2 \cdot \Upsilon_{10} + \lambda \cdot \Upsilon_{11} + \Upsilon_{12}$

$$+ \Xi_{13} \cdot (\Upsilon_7 \cdot \lambda^2 + \Upsilon_8 \cdot \lambda + \Upsilon_9)$$
$$- \Xi_{14} \cdot (\lambda^2 \cdot \Upsilon_{10} + \lambda \cdot \Upsilon_{11} + \Upsilon_{12})$$

$$\begin{aligned}\det(A - \lambda \cdot I) = & -\lambda^3 \cdot \Xi_{11} + \Upsilon_1 \cdot \Xi_{11} \cdot \lambda^2 + \Upsilon_2 \cdot \Xi_{11} \cdot \lambda \\ & + \Upsilon_3 \cdot \Xi_{11} + \lambda^4 - \Upsilon_1 \cdot \lambda^3 \\ & - \Upsilon_2 \cdot \lambda^2 - \Upsilon_3 \cdot \lambda - \Xi_{12} \cdot \Upsilon_4 \cdot \lambda^2 \\ & - \Xi_{12} \cdot \Upsilon_5 \cdot \lambda - \Xi_{12} \cdot \Upsilon_6 \\ & + \Xi_{13} \cdot \Upsilon_7 \cdot \lambda^2 + \Xi_{13} \cdot \Upsilon_8 \cdot \lambda + \Xi_{13} \cdot \Upsilon_9 \\ & - \lambda^2 \cdot \Xi_{14} \cdot \Upsilon_{10} - \lambda \cdot \Xi_{14} \cdot \Upsilon_{11} - \Xi_{14} \cdot \Upsilon_{12})\end{aligned}$$

$$\begin{aligned}\det(A - \lambda \cdot I) = \lambda^4 & - (\Xi_{11} + \Upsilon_1) \cdot \lambda^3 + (\Upsilon_1 \cdot \Xi_{11} - \Upsilon_2 \\ & - \Xi_{12} \cdot \Upsilon_4 + \Xi_{13} \cdot \Upsilon_7 - \Xi_{14} \cdot \Upsilon_{10}) \cdot \lambda^2 \\ & + (\Upsilon_2 \cdot \Xi_{11} - \Upsilon_3 - \Xi_{12} \cdot \Upsilon_5 + \Xi_{13} \cdot \Upsilon_8 - \Xi_{14} \cdot \Upsilon_{11}) \\ & \cdot \lambda + (\Upsilon_3 \cdot \Xi_{11} - \Xi_{12} \cdot \Upsilon_6 + \Xi_{13} \cdot \Upsilon_9 - \Xi_{14} \cdot \Upsilon_{12})\end{aligned}$$

We define new global parameters: $\Omega_4 = 1$; $\Omega_3 = -(\Xi_{11} + \Upsilon_1)$

$$\Omega_2 = \Upsilon_1 \cdot \Xi_{11} - \Upsilon_2 - \Xi_{12} \cdot \Upsilon_4 + \Xi_{13} \cdot \Upsilon_7 - \Xi_{14} \cdot \Upsilon_{10}$$
$$\Omega_1 = \Upsilon_2 \cdot \Xi_{11} - \Upsilon_3 - \Xi_{12} \cdot \Upsilon_5 + \Xi_{13} \cdot \Upsilon_8 - \Xi_{14} \cdot \Upsilon_{11}$$

$$\Omega_0 = \Upsilon_3 \cdot \Xi_{11} - \Xi_{12} \cdot \Upsilon_6 + \Xi_{13} \cdot \Upsilon_9 - \Xi_{14} \cdot \Upsilon_{12}$$

$$\det(A - \lambda \cdot I) = \sum_{k=0}^{4} \Omega_k \cdot \lambda^k; \, det(A - \lambda \cdot I) = 0 \Rightarrow \sum_{k=0}^{4} \Omega_k \cdot \lambda^k = 0$$

Eigenvalues stability discussion: Our Ion channel laser perturbed system involving N variables (N > 2, N = 4), the characteristic equation is of degree N = 4 ($\sum_{k=0}^{4} \Omega_k \cdot \lambda^k = 0$) and must often be solve numerically. Except in some particular cases, such an equation has (N = 4) distinct roots that can be real or complex. These values are the eigenvalues of the 4 × 4 Jacobian matrix (A). The general rule is that the Steady State (SS) is stable if there is no eigenvalue with positive real part. It is sufficient that one eigenvalue is positive for the steady state to be unstable. Our 4-variables ($\psi \; q_x \; q_y \; q_z$) Ion channel laser perturbed system has four eigenvalues. The type of behavior can be characterized as a function of the position of these eigenvalues in the Re/Im plane. Five non-degenerated cases can be distinguished: (1) the four eigenvalues are real and negative (stable steady state), (2) the four eigenvalues are real, three of them are negative (unstable steady state), (3) and (4) two eigenvalues are complex conjugates with a negative real part and the other eigenvalues are real and negative (stable steady state), two cases can be distinguished depending on the relative value of the real part of the complex eigenvalues and of the real one, (5) two eigenvalues are complex conjugates with a negative real part and at least one eigenvalue is positive (unstable steady state).

3.2 Ion Channel Laser Perturbed Differential Equations of Motion Stability Optimization Under Delayed Variables in Time

We get already (3.1) the Ion channel laser perturbed system equations of motion (average over betatron period):

$$\frac{d\psi}{dt} = k_z \cdot c \cdot [1 - \frac{(2 + q_x^2 + q_y^2)}{4 \cdot q_z^2}] - \omega + \omega_\beta$$
$$\cdot [1 - \frac{1}{2 \cdot q_y} \cdot A \cdot \cos(\psi)] + \frac{1}{2} \cdot k_z$$
$$\cdot c \cdot \frac{q_y}{q_z^2} \cdot A \cdot \cos(\psi)$$

$$\frac{dq_z}{dt} = -\frac{1}{2} \cdot k_z \cdot c \cdot \frac{q_y}{q_z} \cdot A \cdot \sin(\psi)$$

$$\frac{dq_y}{dt} = -\frac{1}{2} \cdot (\omega_\beta + \frac{1}{4} \cdot k_z \cdot c \cdot \frac{q_y^2}{q_z^2}) \cdot A \cdot \sin(\psi)$$

$$\frac{dq_x}{dt} = -\frac{1}{8} \cdot k_z \cdot c \cdot \frac{q_y \cdot q_x}{q_z^2} \cdot A \cdot \sin(\psi)$$

The mechanism is which a laser beam propagating through un-dense plasma produces a positively charged ion channels by expelling plasma electrons in the transverse direction. Basically the relativistic electron beam is injected and propagates in the axial (+z). The beam head is propagated through the plasma and expels plasma electrons from the beam volume. The result is focusing for the remainder of the beam. Self- focusing of intense laser pulses in plasmas can also be achieved by the expulsion of plasma electrons (cavitation) caused by the extreme ponderomotive force of a focused laser pulse. It is optical guiding of high intensity laser pulses in under dense plasmas and the creation of a plasma channel that guides subsequently injected laser pulses. The radial electrostatic force from charge separation expels the plasma ions from the electron cavitation regions. Due to their inertia, the ions continue to drift radially outward even after the passage of the laser pulse and a plasma channel is formed. The formation of a plasma channel can be most easily attributed to the effects of ponderomotive forces associated with the intense pump laser pulse as it propagates through the plasma. Initially, the pump laser pulses exert a ponderomotive force on the plasma electrons and expel them radially. There are four variables in our system ψ(phase variable in time) and q_x,q_y,q_z are system variables in x, y, and z directions respectively. We assume that electrons with $q_y < A$(A—Eikonal amplitude) contribute negligibly to the amplification, and that $a_\beta < 1$. Additionally,$\Delta\omega \ll \omega_\beta$, or $\omega - k_z \cdot V_z \sim \omega_\beta$. The fast wave, with $a_\beta \ll 1$, this corresponds to $\omega \sim 2 \cdot \gamma^2 \cdot \omega_\beta$. The mechanism which a laser beam propagating through un-dense plasma is not ideal and there is phase variable (ψ) delay in time, $\psi(t) \to \psi(t - \tau)$ but the derivative of the phase in time ($\frac{d\psi(t)}{dt}$) is not effected [1]. We can define q_x=$q_x(t)$; q_y=$q_y(t)$; $q_z = q_z(t)$

$$\frac{d\psi(t)}{dt} = k_z \cdot c \cdot [1 - \frac{(2 + q_x^2 + q_y^2)}{4 \cdot q_z^2}] - \omega + \omega_\beta$$
$$\cdot [1 - \frac{1}{2 \cdot q_y} \cdot A \cdot \cos(\psi(t - \tau))]$$
$$+ \frac{1}{2} \cdot k_z \cdot c \cdot \frac{q_y}{q_z^2} \cdot A \cdot \cos(\psi(t - \tau))$$

$$\frac{dq_z(t)}{dt} = -\frac{1}{2} \cdot k_z \cdot c \cdot \frac{q_y}{q_z} \cdot A \cdot \sin(\psi(t - \tau))$$

$$\frac{dq_y(t)}{dt} = -\frac{1}{2} \cdot (\omega_\beta + \frac{1}{4} \cdot k_z \cdot c \cdot \frac{q_y^2}{q_z^2}) \cdot A \cdot \sin(\psi(t - \tau))$$

$$\frac{dq_x(t)}{dt} = -\frac{1}{8}\cdot k_z\cdot c\cdot\frac{q_y\cdot q_x}{q_z^2}\cdot A\cdot\sin(\psi(t-\tau))$$

At fixed points $\lim_{t\to\infty}\psi(t-\tau) = \psi(t); t \gg \tau \Rightarrow t-\tau \approx t\forall\tau \in \mathbb{R}_+$ then our system fixed points are the same as we find (3.1). The Ion channel laser perturbed system fixed values with arbitrarily small increments of exponential form $[\psi, q_x, q_y, q_z]\cdot e^{\lambda\cdot t}$ are; i = 0 (first fixed point), i = 1 (second fixed point), i = 2 (third fixed point), etc.,

$$\psi(t) = \psi^{(i)} + \psi\cdot e^{\lambda\cdot t}$$
$$\psi(t-\tau) = \psi^{(i)} + \psi\cdot e^{\lambda\cdot(t-\tau)} = \psi^{(i)} + \psi\cdot e^{\lambda\cdot t}\cdot e^{-\lambda\cdot\tau}$$
$$q_x(t) = q_x^{(i)} + q_x\cdot e^{\lambda\cdot t}$$

$$q_y(t) = q_y^{(i)} + q_y\cdot e^{\lambda\cdot t}; q_z(t) = q_z^{(i)} + q_z\cdot e^{\lambda\cdot t}$$

$$\frac{d\psi(t)}{dt} = \psi\cdot\lambda\cdot e^{\lambda\cdot t}; \frac{dq_x(t)}{dt} = q_x\cdot\lambda\cdot e^{\lambda\cdot t}$$
$$\frac{dq_y(t)}{dt} = q_y\cdot\lambda\cdot e^{\lambda\cdot t}; \frac{dq_z(t)}{dt} = q_z\cdot\lambda\cdot e^{\lambda\cdot t}$$

&&&&

$$\frac{d\psi}{dt} = k_z\cdot c\cdot[1-\frac{(2+q_x^2+q_y^2)}{4\cdot q_z^2}]$$
$$-\omega+\omega_\beta\cdot[1-\frac{1}{2\cdot q_y}\cdot A\cdot\cos(\psi(t-\tau))]$$
$$+\frac{1}{2}\cdot k_z\cdot c\cdot\frac{q_y}{q_z^2}\cdot A\cdot\cos(\psi(t-\tau))$$

$$\psi\cdot\lambda\cdot e^{\lambda\cdot t} = k_z\cdot c\cdot[1-\frac{(2+[q_x^{(i)}+q_x\cdot e^{\lambda\cdot t}]^2+[q_y^{(i)}+q_y\cdot e^{\lambda\cdot t}]^2)}{4\cdot[q_z^{(i)}+q_z\cdot e^{\lambda\cdot t}]^2}]-\omega$$
$$+\omega_\beta\cdot\{1-\frac{1}{2\cdot[q_y^{(i)}+q_y\cdot e^{\lambda\cdot t}]}\cdot A\cdot\cos(\psi^{(i)}+\psi\cdot e^{\lambda\cdot t}\cdot e^{-\lambda\cdot\tau})\}$$
$$+\frac{1}{2}\cdot k_z\cdot c\cdot\frac{[q_y^{(i)}+q_y\cdot e^{\lambda\cdot t}]}{[q_z^{(i)}+q_z\cdot e^{\lambda\cdot t}]^2}\cdot A\cdot\cos(\psi^{(i)}+\psi\cdot e^{\lambda\cdot t}\cdot e^{-\lambda\cdot\tau})$$

We take each part from the above expression and develop it mathematically differently.

$$\frac{(2+[q_x^{(i)}+q_x\cdot e^{\lambda\cdot t}]^2+[q_y^{(i)}+q_y\cdot e^{\lambda\cdot t}]^2)}{4\cdot[q_z^{(i)}+q_z\cdot e^{\lambda\cdot t}]^2}$$

(1)
$$= \frac{\begin{array}{l}(2+(q_x^{(i)})^2+2\cdot q_x^{(i)}\cdot q_x\cdot e^{\lambda\cdot t}+(q_x)^2\cdot e^{2\cdot\lambda\cdot t}\\ +(q_y^{(i)})^2+2\cdot q_y^{(i)}\cdot q_y\cdot e^{\lambda\cdot t}+(q_y)^2\cdot e^{2\cdot\lambda\cdot t})\end{array}}{4\cdot[(q_z^{(i)})^2+2\cdot q_z^{(i)}\cdot q_z\cdot e^{\lambda\cdot t}+(q_z)^2\cdot e^{2\cdot\lambda\cdot t}]}$$

We consider $(q_x)^2$, $(q_y)^2$, and $(q_z)^2$ are very small and tend to zero, then:$(q_x)^2 \to \varepsilon$, $(q_y)^2 \to \varepsilon$, and $(q_z)^2 \to \varepsilon$.

$$\frac{(2+[q_x^{(i)}+q_x\cdot e^{\lambda\cdot t}]^2+[q_y^{(i)}+q_y\cdot e^{\lambda\cdot t}]^2)}{4\cdot[q_z^{(i)}+q_z\cdot e^{\lambda\cdot t}]^2} = \frac{(2+(q_x^{(i)})^2+(q_y^{(i)})^2+2\cdot q_x^{(i)}\cdot q_x\cdot e^{\lambda\cdot t}+2\cdot q_y^{(i)}\cdot q_y\cdot e^{\lambda\cdot t})}{4\cdot[(q_z^{(i)})^2+2\cdot q_z^{(i)}\cdot q_z\cdot e^{\lambda\cdot t}]}$$

$$\frac{(2+(q_x^{(i)})^2+(q_y^{(i)})^2+2\cdot q_x^{(i)}\cdot q_x\cdot e^{\lambda\cdot t}+2\cdot q_y^{(i)}\cdot q_y\cdot e^{\lambda\cdot t})}{4\cdot[(q_z^{(i)})^2+2\cdot q_z^{(i)}\cdot q_z\cdot e^{\lambda\cdot t}]} \cdot \frac{[(q_z^{(i)})^2-2\cdot q_z^{(i)}\cdot q_z\cdot e^{\lambda\cdot t}]}{[(q_z^{(i)})^2-2\cdot q_z^{(i)}\cdot q_z\cdot e^{\lambda\cdot t}]}$$

$$= \frac{\begin{array}{l}[2+(q_x^{(i)})^2+(q_y^{(i)})^2]\cdot(q_z^{(i)})^2-[2+(q_x^{(i)})^2+(q_y^{(i)})^2]\cdot 2\cdot q_z^{(i)}\cdot q_z\cdot e^{\lambda\cdot t}\\ +[2\cdot q_x^{(i)}\cdot q_x\cdot e^{\lambda\cdot t}+2\cdot q_y^{(i)}\cdot q_y\cdot e^{\lambda\cdot t}]\cdot(q_z^{(i)})^2\\ -[2\cdot q_x^{(i)}\cdot q_x\cdot e^{\lambda\cdot t}+2\cdot q_y^{(i)}\cdot q_y\cdot e^{\lambda\cdot t}]\cdot 2\cdot q_z^{(i)}\cdot q_z\cdot e^{\lambda\cdot t}\end{array}}{4\cdot[(q_z^{(i)})^4-4\cdot(q_z^{(i)})^2\cdot(q_z)^2\cdot e^{2\cdot\lambda\cdot t}]}$$

$$= \frac{\begin{array}{l}[2+(q_x^{(i)})^2+(q_y^{(i)})^2]\cdot(q_z^{(i)})^2-[2+(q_x^{(i)})^2+(q_y^{(i)})^2]\cdot 2\cdot q_z^{(i)}\cdot q_z\cdot e^{\lambda\cdot t}\\ +2\cdot[q_x^{(i)}\cdot q_x+q_y^{(i)}\cdot q_y]\cdot(q_z^{(i)})^2\cdot e^{\lambda\cdot t}-2\cdot[q_x^{(i)}\cdot q_x+q_y^{(i)}\cdot q_y]\cdot 2\cdot q_z^{(i)}\cdot q_z\cdot e^{2\cdot\lambda\cdot t}\end{array}}{4\cdot[(q_z^{(i)})^4-4\cdot(q_z^{(i)})^2\cdot(q_z)^2\cdot e^{2\cdot\lambda\cdot t}]}$$

$$= \frac{\begin{array}{l}[2+(q_x^{(i)})^2+(q_y^{(i)})^2]\cdot(q_z^{(i)})^2-[2+(q_x^{(i)})^2+(q_y^{(i)})^2]\cdot 2\cdot q_z^{(i)}\cdot q_z\cdot e^{\lambda\cdot t}\\ +2\cdot[q_x^{(i)}\cdot q_x+q_y^{(i)}\cdot q_y]\cdot(q_z^{(i)})^2\cdot e^{\lambda\cdot t}-2\cdot[q_x^{(i)}\cdot q_x\cdot q_z+q_y^{(i)}\cdot q_y\cdot q_z]\cdot 2\cdot q_z^{(i)}\cdot e^{2\cdot\lambda\cdot t}\end{array}}{4\cdot[(q_z^{(i)})^4-4\cdot(q_z^{(i)})^2\cdot(q_z)^2\cdot e^{2\cdot\lambda\cdot t}]}$$

We consider $q_x\cdot q_z\approx 0$; $q_y\cdot q_z\approx 0$; $(q_z)^2\approx 0$ then

$$\frac{(2+[q_x^{(i)}+q_x\cdot e^{\lambda\cdot t}]^2+[q_y^{(i)}+q_y\cdot e^{\lambda\cdot t}]^2)}{4\cdot[q_z^{(i)}+q_z\cdot e^{\lambda\cdot t}]^2} = \frac{\begin{array}{l}\{[2+(q_x^{(i)})^2+(q_y^{(i)})^2]+2\cdot[q_x^{(i)}\cdot q_x+q_y^{(i)}\cdot q_y]\cdot e^{\lambda\cdot t}\}\cdot(q_z^{(i)})^2\\ -[2+(q_x^{(i)})^2+(q_y^{(i)})^2]\cdot 2\cdot q_z^{(i)}\cdot q_z\cdot e^{\lambda\cdot t}\end{array}}{4\cdot(q_z^{(i)})^4}$$

$$\frac{(2+[q_x^{(i)}+q_x\cdot e^{\lambda\cdot t}]^2+[q_y^{(i)}+q_y\cdot e^{\lambda\cdot t}]^2)}{4\cdot[q_z^{(i)}+q_z\cdot e^{\lambda\cdot t}]^2}$$

$$= \frac{[2 + (q_x^{(i)})^2 + (q_y^{(i)})^2]}{4 \cdot (q_z^{(i)})^2} + \frac{2 \cdot q_x^{(i)}}{4 \cdot (q_z^{(i)})^2} \cdot q_x \cdot e^{\lambda \cdot t} + \frac{2 \cdot q_y^{(i)}}{4 \cdot (q_z^{(i)})^2} \cdot q_y \cdot e^{\lambda \cdot t} - \frac{[2 + (q_x^{(i)})^2 + (q_y^{(i)})^2] \cdot 2}{4 \cdot (q_z^{(i)})^3} \cdot q_z \cdot e^{\lambda \cdot t}$$

(2)

$$\frac{1}{2 \cdot [q_y^{(i)} + q_y \cdot e^{\lambda \cdot t}]} = \frac{1}{2 \cdot [q_y^{(i)} + q_y \cdot e^{\lambda \cdot t}]} \cdot \frac{[q_y^{(i)} - q_y \cdot e^{\lambda \cdot t}]}{[q_y^{(i)} - q_y \cdot e^{\lambda \cdot t}]} = \frac{q_y^{(i)} - q_y \cdot e^{\lambda \cdot t}}{2 \cdot [(q_y^{(i)})^2 - (q_y)^2 \cdot e^{2 \cdot \lambda \cdot t}]}$$

$$\overset{(q_y)^2 \approx 0}{=} \frac{q_y^{(i)} - q_y \cdot e^{\lambda \cdot t}}{2 \cdot (q_y^{(i)})^2} = \frac{1}{2 \cdot q_y^{(i)}} - \frac{1}{2 \cdot (q_y^{(i)})^2} \cdot q_y \cdot e^{\lambda \cdot t}$$

(3)

$$\frac{[q_y^{(i)} + q_y \cdot e^{\lambda \cdot t}]}{[q_z^{(i)} + q_z \cdot e^{\lambda \cdot t}]^2} = \frac{[q_y^{(i)} + q_y \cdot e^{\lambda \cdot t}]}{(q_z^{(i)})^2 + 2 \cdot q_z^{(i)} \cdot q_z \cdot e^{\lambda \cdot t} + (q_z)^2 \cdot e^{2 \cdot \lambda \cdot t}} \overset{(q_z)^2 \approx 0}{=} \frac{[q_y^{(i)} + q_y \cdot e^{\lambda \cdot t}]}{(q_z^{(i)})^2 + 2 \cdot q_z^{(i)} \cdot q_z \cdot e^{\lambda \cdot t}}$$

$$\frac{[q_y^{(i)} + q_y \cdot e^{\lambda \cdot t}]}{[q_z^{(i)} + q_z \cdot e^{\lambda \cdot t}]^2} \overset{(q_z)^2 \approx 0}{=} \frac{[q_y^{(i)} + q_y \cdot e^{\lambda \cdot t}]}{(q_z^{(i)})^2 + 2 \cdot q_z^{(i)} \cdot q_z \cdot e^{\lambda \cdot t}} \cdot \frac{[(q_z^{(i)})^2 - 2 \cdot q_z^{(i)} \cdot q_z \cdot e^{\lambda \cdot t}]}{[(q_z^{(i)})^2 - 2 \cdot q_z^{(i)} \cdot q_z \cdot e^{\lambda \cdot t}]}$$

$$= \frac{q_y^{(i)} \cdot (q_z^{(i)})^2 - 2 \cdot q_z^{(i)} \cdot q_y^{(i)} \cdot q_z \cdot e^{\lambda \cdot t} + (q_z^{(i)})^2 \cdot q_y \cdot e^{\lambda \cdot t} - 2 \cdot q_z^{(i)} \cdot q_z \cdot q_y \cdot e^{2 \cdot \lambda \cdot t}}{(q_z^{(i)})^4 - 4 \cdot (q_z^{(i)})^2 \cdot (q_z)^2 \cdot e^{2 \cdot \lambda \cdot t}}$$

$$\frac{[q_y^{(i)} + q_y \cdot e^{\lambda \cdot t}]}{[q_z^{(i)} + q_z \cdot e^{\lambda \cdot t}]^2} \overset{(q_z)^2 \approx 0}{=} \frac{q_y^{(i)} \cdot (q_z^{(i)})^2 - 2 \cdot q_z^{(i)} \cdot q_y^{(i)} \cdot q_z \cdot e^{\lambda \cdot t} + (q_z^{(i)})^2 \cdot q_y \cdot e^{\lambda \cdot t} - 2 \cdot q_z^{(i)} \cdot q_z \cdot q_y \cdot e^{2 \cdot \lambda \cdot t}}{(q_z^{(i)})^4 - 4 \cdot (q_z^{(i)})^2 \cdot (q_z)^2 \cdot e^{2 \cdot \lambda \cdot t}}$$

We assume that $q_z \cdot q_y \approx 0$; $(q_z)^2 \approx 0$ then

$$\frac{[q_y^{(i)} + q_y \cdot e^{\lambda \cdot t}]}{[q_z^{(i)} + q_z \cdot e^{\lambda \cdot t}]^2} \overset{(q_z)^2 \approx 0}{=} \frac{q_y^{(i)} \cdot (q_z^{(i)})^2 - 2 \cdot q_z^{(i)} \cdot q_y^{(i)} \cdot q_z \cdot e^{\lambda \cdot t} + (q_z^{(i)})^2 \cdot q_y \cdot e^{\lambda \cdot t}}{(q_z^{(i)})^4}$$

$$= \frac{q_y^{(i)}}{(q_z^{(i)})^2} - \frac{2 \cdot q_y^{(i)}}{(q_z^{(i)})^3} \cdot q_z \cdot e^{\lambda \cdot t} + \frac{1}{(q_z^{(i)})^2} \cdot q_y \cdot e^{\lambda \cdot t}$$

$$\cos(\psi^{(i)} + \psi \cdot e^{\lambda \cdot (t-\tau)}) = \cos(\psi^{(i)}) \cdot \cos(\psi \cdot e^{\lambda \cdot (t-\tau)}) - \sin(\psi^{(i)}) \cdot \sin(\psi \cdot e^{\lambda \cdot (t-\tau)})$$

$$\cos(\psi^{(i)} + \psi \cdot e^{\lambda \cdot t} \cdot e^{-\lambda \cdot \tau}) = \cos(\psi^{(i)}) \cdot \cos(\psi \cdot e^{\lambda \cdot t} \cdot e^{-\lambda \cdot \tau}) - \sin(\psi^{(i)}) \cdot \sin(\psi \cdot e^{\lambda \cdot t} \cdot e^{-\lambda \cdot \tau})$$

We use Taylor series to represent the functions $\cos(\psi \cdot e^{\lambda \cdot (t-\tau)})$ and $\sin(\psi \cdot e^{\lambda \cdot (t-\tau)})$ as an infinite sum of terms that are calculated from the values of the function's derivatives at a single point. The Taylor series are concentrated near to zero since $(\psi \to 0) \cdot e^{\lambda \cdot (t-\tau)}$ which is called a Maclaurin series. The functions are approximated by using a finite number of terms of their Taylor series. It gives quantitative estimates on the error introduced by the use of such an approximations.

$$\cos(\psi \cdot e^{\lambda \cdot (t-\tau)}) = \sum_{n=0}^{\infty} \frac{(-1)^n}{(2 \cdot n)!} \cdot \psi^{(2 \cdot n)} \cdot e^{2 \cdot n \cdot \lambda \cdot (t-\tau)} = 1 - \frac{\psi^2 \cdot e^{2 \cdot \lambda \cdot (t-\tau)}}{2!} + \frac{\psi^4 \cdot e^{4 \cdot \lambda \cdot (t-\tau)}}{4!} - \ldots \; \forall \psi \cdot e^{\lambda \cdot (t-\tau)}$$

Since $\psi^2 \approx 0$; $\psi^4 \approx 0$; .... then $\cos(\psi \cdot e^{\lambda \cdot (t-\tau)}) \approx 1$

$$\sin(\psi \cdot e^{\lambda \cdot (t-\tau)}) = \sum_{n=0}^{\infty} \frac{(-1)^n}{(2 \cdot n + 1)!} \cdot \psi^{(2 \cdot n+1)} \cdot e^{(2 \cdot n+1) \cdot \lambda \cdot (t-\tau)} = \psi \cdot e^{\lambda \cdot (t-\tau)} - \frac{\psi^3 \cdot e^{3 \cdot \lambda \cdot (t-\tau)}}{3!} + \frac{\psi^5 \cdot e^{5 \cdot \lambda \cdot (t-\tau)}}{5!} - \ldots . \; \forall \psi \cdot e^{\lambda \cdot (t-\tau)}$$

Since $\psi^3 \approx 0$; $\psi^5 \approx 0$; .... then $\sin(\psi \cdot e^{\lambda \cdot (t-\tau)}) \approx \psi \cdot e^{\lambda \cdot (t-\tau)}$

$$\cos(\psi^{(i)} + \psi \cdot e^{\lambda \cdot t}) \approx \cos(\psi^{(i)}) - \sin(\psi^{(i)}) \cdot \psi \cdot e^{\lambda \cdot (t-\tau)}$$

$$\sin(\psi^{(i)} + \psi \cdot e^{\lambda \cdot (t-\tau)}) = \sin(\psi^{(i)}) \cdot \cos(\psi \cdot e^{\lambda \cdot (t-\tau)}) + \cos(\psi^{(i)}) \cdot \sin(\psi \cdot e^{\lambda \cdot (t-\tau)})$$

$$\sin(\psi^{(i)} + \psi \cdot e^{\lambda \cdot (t-\tau)}) \approx \sin(\psi^{(i)}) + \cos(\psi^{(i)}) \cdot \psi \cdot e^{\lambda \cdot (t-\tau)}$$

$$\psi \cdot \lambda \cdot e^{\lambda \cdot t} = k_z \cdot c \cdot [1 - \frac{(2 + [q_x^{(i)} + q_x \cdot e^{\lambda \cdot t}]^2 + [q_y^{(i)} + q_y \cdot e^{\lambda \cdot t}]^2)}{4 \cdot [q_z^{(i)} + q_z \cdot e^{\lambda \cdot t}]^2}] - \omega + \omega_\beta \cdot \{1 - \frac{1}{2 \cdot [q_y^{(i)} + q_y \cdot e^{\lambda \cdot t}]} \cdot A \cdot \cos(\psi^{(i)} + \psi \cdot e^{\lambda \cdot (t-\tau)})\} + \frac{1}{2} \cdot k_z \cdot c \cdot \frac{[q_y^{(i)} + q_y \cdot e^{\lambda \cdot t}]}{[q_z^{(i)} + q_z \cdot e^{\lambda \cdot t}]^2} \cdot A \cdot \cos(\psi^{(i)} + \psi \cdot e^{\lambda \cdot (t-\tau)})$$

$$\psi \cdot \lambda \cdot e^{\lambda \cdot t} = k_z \cdot c \cdot [1 - \{\frac{[2 + (q_x^{(i)})^2 + (q_y^{(i)})^2]}{4 \cdot (q_z^{(i)})^2}$$
$$+ \frac{2 \cdot q_x^{(i)}}{4 \cdot (q_z^{(i)})^2} \cdot q_x \cdot e^{\lambda \cdot t} + \frac{2 \cdot q_y^{(i)}}{4 \cdot (q_z^{(i)})^2} \cdot q_y \cdot e^{\lambda \cdot t}$$
$$- \frac{[2 + (q_x^{(i)})^2 + (q_y^{(i)})^2] \cdot 2}{4 \cdot (q_z^{(i)})^3} \cdot q_z \cdot e^{\lambda \cdot t}\}] - \omega$$
$$+ \omega_\beta \cdot \{1 - [\frac{1}{2 \cdot q_y^{(i)}} - \frac{1}{2 \cdot (q_y^{(i)})^2} \cdot q_y \cdot e^{\lambda \cdot t}]$$
$$\cdot A \cdot [\cos(\psi^{(i)}) - \sin(\psi^{(i)}) \cdot \psi \cdot e^{\lambda \cdot (t-\tau)}]\}$$
$$+ \frac{1}{2} \cdot k_z \cdot c \cdot \{\frac{q_y^{(i)}}{(q_z^{(i)})^2} - \frac{2 \cdot q_y^{(i)}}{(q_z^{(i)})^3} \cdot q_z \cdot e^{\lambda \cdot t}$$
$$+ \frac{1}{(q_z^{(i)})^2} \cdot q_y \cdot e^{\lambda \cdot t}\} \cdot A \cdot [\cos(\psi^{(i)})$$
$$- \sin(\psi^{(i)}) \cdot \psi \cdot e^{\lambda \cdot (t-\tau)}]$$

$$\psi \cdot \lambda \cdot e^{\lambda \cdot t} = k_z \cdot c \cdot [1 - \frac{[2 + (q_x^{(i)})^2 + (q_y^{(i)})^2]}{4 \cdot (q_z^{(i)})^2}]$$
$$+ k_z \cdot c \cdot [\frac{[2 + (q_x^{(i)})^2 + (q_y^{(i)})^2] \cdot 2}{4 \cdot (q_z^{(i)})^3} \cdot q_z \cdot e^{\lambda \cdot t}$$
$$- \frac{2 \cdot q_x^{(i)}}{4 \cdot (q_z^{(i)})^2} \cdot q_x \cdot e^{\lambda \cdot t} - \frac{2 \cdot q_y^{(i)}}{4 \cdot (q_z^{(i)})^2} \cdot q_y \cdot e^{\lambda \cdot t}] - \omega$$
$$+ \omega_\beta \cdot \{1 - (\frac{1}{2 \cdot q_y^{(i)}} \cdot A \cdot [\cos(\psi^{(i)}) - \sin(\psi^{(i)})$$
$$\cdot \psi \cdot e^{\lambda \cdot (t-\tau)}] - \frac{1}{2 \cdot (q_y^{(i)})^2} \cdot q_y \cdot e^{\lambda \cdot t} \cdot A \cdot [\cos(\psi^{(i)})$$
$$- \sin(\psi^{(i)}) \cdot \psi \cdot e^{\lambda \cdot (t-\tau)}])\}$$
$$+ \frac{1}{2} \cdot k_z \cdot c \cdot \{\frac{q_y^{(i)}}{(q_z^{(i)})^2}\} \cdot A \cdot [\cos(\psi^{(i)}) - \sin(\psi^{(i)}) \cdot \psi \cdot e^{\lambda \cdot (t-\tau)}]$$
$$+ \frac{1}{2} \cdot k_z \cdot c \cdot \{\frac{1}{(q_z^{(i)})^2} \cdot q_y \cdot e^{\lambda \cdot t} - \frac{2 \cdot q_y^{(i)}}{(q_z^{(i)})^3} \cdot q_z \cdot e^{\lambda \cdot t}\}$$
$$\cdot A \cdot [\cos(\psi^{(i)}) - \sin(\psi^{(i)}) \cdot \psi \cdot e^{\lambda \cdot (t-\tau)}]$$

$$\psi \cdot \lambda \cdot e^{\lambda \cdot t} = k_z \cdot c \cdot [1 - \frac{[2 + (q_x^{(i)})^2 + (q_y^{(i)})^2]}{4 \cdot (q_z^{(i)})^2}]$$

$$
\begin{aligned}
&+ k_z \cdot c \cdot [\frac{[2 + (q_x^{(i)})^2 + (q_y^{(i)})^2] \cdot 2}{4 \cdot (q_z^{(i)})^3} \cdot q_z \cdot e^{\lambda \cdot t} \\
&- \frac{2 \cdot q_x^{(i)}}{4 \cdot (q_z^{(i)})^2} \cdot q_x \cdot e^{\lambda \cdot t} - \frac{2 \cdot q_y^{(i)}}{4 \cdot (q_z^{(i)})^2} \cdot q_y \cdot e^{\lambda \cdot t}] - \omega \\
&+ \omega_\beta \cdot \{1 - (\frac{1}{2 \cdot q_y^{(i)}} \cdot A \cdot \cos(\psi^{(i)}) - \frac{1}{2 \cdot q_y^{(i)}} \\
&\cdot A \cdot \sin(\psi^{(i)}) \cdot \psi \cdot e^{\lambda \cdot (t-\tau)} - \frac{1}{2 \cdot (q_y^{(i)})^2} \cdot q_y \cdot e^{\lambda \cdot t} \\
&\cdot A \cdot [\cos(\psi^{(i)}) - \sin(\psi^{(i)}) \cdot \psi \cdot e^{\lambda \cdot (t-\tau)}])\} \\
&+ \frac{1}{2} \cdot k_z \cdot c \cdot \{\frac{q_y^{(i)}}{(q_z^{(i)})^2}\} \cdot A \cdot \cos(\psi^{(i)}) - \frac{1}{2} \cdot k_z \\
&\cdot c \cdot \{\frac{q_y^{(i)}}{(q_z^{(i)})^2}\} \cdot A \cdot \sin(\psi^{(i)}) \cdot \psi \cdot e^{\lambda \cdot (t-\tau)} \\
&+ \frac{1}{2} \cdot k_z \cdot c \cdot \{\frac{1}{(q_z^{(i)})^2} \cdot q_y \cdot e^{\lambda \cdot t} - \frac{2 \cdot q_y^{(i)}}{(q_z^{(i)})^3} \cdot q_z \cdot e^{\lambda \cdot t}\} \\
&\cdot A \cdot [\cos(\psi^{(i)}) - \sin(\psi^{(i)}) \cdot \psi \cdot e^{\lambda \cdot (t-\tau)}]
\end{aligned}
$$

$$
\begin{aligned}
\psi \cdot \lambda \cdot e^{\lambda \cdot t} = {} & k_z \cdot c \cdot [1 - \frac{[2 + (q_x^{(i)})^2 + (q_y^{(i)})^2]}{4 \cdot (q_z^{(i)})^2}] \\
&+ k_z \cdot c \cdot [\frac{[2 + (q_x^{(i)})^2 + (q_y^{(i)})^2] \cdot 2}{4 \cdot (q_z^{(i)})^3} \cdot q_z \cdot e^{\lambda \cdot t} \\
&- \frac{2 \cdot q_x^{(i)}}{4 \cdot (q_z^{(i)})^2} \cdot q_x \cdot e^{\lambda \cdot t} - \frac{2 \cdot q_y^{(i)}}{4 \cdot (q_z^{(i)})^2} \cdot q_y \cdot e^{\lambda \cdot t}] - \omega \\
&+ \omega_\beta \cdot \{1 - \frac{1}{2 \cdot q_y^{(i)}} \cdot A \cdot \cos(\psi^{(i)}) + \frac{1}{2 \cdot q_y^{(i)}} \\
&\cdot A \cdot \sin(\psi^{(i)}) \cdot \psi \cdot e^{\lambda \cdot (t-\tau)} + \frac{1}{2 \cdot (q_y^{(i)})^2} \cdot q_y \cdot e^{\lambda \cdot t} \\
&\cdot A \cdot [\cos(\psi^{(i)}) - \sin(\psi^{(i)}) \cdot \psi \cdot e^{\lambda \cdot (t-\tau)}])\} \\
&+ \frac{1}{2} \cdot k_z \cdot c \cdot \{\frac{q_y^{(i)}}{(q_z^{(i)})^2}\} \cdot A \cdot \cos(\psi^{(i)}) - \frac{1}{2} \cdot k_z \\
&\cdot c \cdot \{\frac{q_y^{(i)}}{(q_z^{(i)})^2}\} \cdot A \cdot \sin(\psi^{(i)}) \cdot \psi \cdot e^{\lambda \cdot (t-\tau)} \\
&+ \frac{1}{2} \cdot k_z \cdot c \cdot \{\frac{1}{(q_z^{(i)})^2} \cdot q_y \cdot e^{\lambda \cdot t} - \frac{2 \cdot q_y^{(i)}}{(q_z^{(i)})^3} \cdot q_z \cdot e^{\lambda \cdot t}\} \\
&\cdot A \cdot [\cos(\psi^{(i)}) - \sin(\psi^{(i)}) \cdot \psi \cdot e^{\lambda \cdot (t-\tau)}]
\end{aligned}
$$

$$\psi \cdot \lambda \cdot e^{\lambda \cdot t} = k_z \cdot c \cdot [1 - \frac{[2 + (q_x^{(i)})^2 + (q_y^{(i)})^2]}{4 \cdot (q_z^{(i)})^2}]$$
$$- \omega + \omega_\beta \cdot \{1 - \frac{1}{2 \cdot q_y^{(i)}} \cdot A \cdot \cos(\psi^{(i)})\}$$
$$+ \frac{1}{2} \cdot k_z \cdot c \cdot \{\frac{q_y^{(i)}}{(q_z^{(i)})^2}\} \cdot A \cdot \cos(\psi^{(i)})$$
$$+ k_z \cdot c \cdot [\frac{[2 + (q_x^{(i)})^2 + (q_y^{(i)})^2] \cdot 2}{4 \cdot (q_z^{(i)})^3} \cdot q_z \cdot e^{\lambda \cdot t}$$
$$- \frac{2 \cdot q_x^{(i)}}{4 \cdot (q_z^{(i)})^2} \cdot q_x \cdot e^{\lambda \cdot t} - \frac{2 \cdot q_y^{(i)}}{4 \cdot (q_z^{(i)})^2} \cdot q_y \cdot e^{\lambda \cdot t}]$$
$$+ \omega_\beta \cdot \{\frac{1}{2 \cdot q_y^{(i)}} \cdot A \cdot \sin(\psi^{(i)}) \cdot \psi \cdot e^{\lambda \cdot (t - \tau)}$$
$$+ \frac{1}{2 \cdot (q_y^{(i)})^2} \cdot q_y \cdot e^{\lambda \cdot t} \cdot A \cdot [\cos(\psi^{(i)}) - \sin(\psi^{(i)}) \cdot \psi \cdot e^{\lambda \cdot (t - \tau)}]\}$$
$$- \frac{1}{2} \cdot k_z \cdot c \cdot \{\frac{q_y^{(i)}}{(q_z^{(i)})^2}\} \cdot A \cdot \sin(\psi^{(i)}) \cdot \psi \cdot e^{\lambda \cdot (t - \tau)}$$
$$+ \frac{1}{2} \cdot k_z \cdot c \cdot \{\frac{1}{(q_z^{(i)})^2} \cdot q_y \cdot e^{\lambda \cdot t} - \frac{2 \cdot q_y^{(i)}}{(q_z^{(i)})^3} \cdot q_z \cdot e^{\lambda \cdot t}\}$$
$$\cdot A \cdot [\cos(\psi^{(i)}) - \sin(\psi^{(i)}) \cdot \psi \cdot e^{\lambda \cdot (t - \tau)}]$$

At fixed points: $\frac{d\psi}{dt} = 0$

$$k_z \cdot c \cdot \left[1 - \frac{[2 + (q_x^{(i)})^2 + (q_y^{(i)})^2]}{4 \cdot (q_z^{(i)})^2}\right] - \omega + \omega_\beta$$
$$\left\{\cdot 1 - \frac{1}{2 \cdot q_y^{(i)}} \cdot A \cdot \cos(\psi^{(i)})\right\} + \frac{1}{2} \cdot k_z \cdot c \cdot \left\{\frac{q_y^{(i)}}{(q_z^{(i)})^2}\right\}$$
$$\cdot A \cdot \cos(\psi^{(i)}) = 0$$

$$\psi \cdot \lambda \cdot e^{\lambda \cdot t} = k_z \cdot c \cdot \left[\frac{[2 + (q_x^{(i)})^2 + (q_y^{(i)})^2] \cdot 2}{4 \cdot (q_z^{(i)})^3} \cdot q_z \cdot e^{\lambda \cdot t}\right.$$
$$\left. - \frac{2 \cdot q_x^{(i)}}{4 \cdot (q_z^{(i)})^2} \cdot q_x \cdot e^{\lambda \cdot t} - \frac{2 \cdot q_y^{(i)}}{4 \cdot (q_z^{(i)})^2} \cdot q_y \cdot e^{\lambda \cdot t}\right]$$
$$+ \omega_\beta \cdot \left\{\frac{1}{2 \cdot q_y^{(i)}} \cdot A \cdot \sin(\psi^{(i)}) \cdot \psi \cdot e^{\lambda \cdot (t - \tau)} + \frac{1}{2 \cdot (q_y^{(i)})^2} \cdot q_y\right.$$

$$\cdot e^{\lambda \cdot t} \cdot A \cdot [\cos(\psi^{(i)}) - \sin(\psi^{(i)}) \cdot \psi \cdot e^{\lambda \cdot (t-\tau)}]\}$$

$$- \frac{1}{2} \cdot k_z \cdot c \cdot \left\{ \frac{q_y^{(i)}}{(q_z^{(i)})^2} \right\} \cdot A \cdot \sin(\psi^{(i)}) \cdot \psi \cdot e^{\lambda \cdot (t-\tau)}$$

$$+ \frac{1}{2} \cdot k_z \cdot c \cdot \left\{ \frac{1}{(q_z^{(i)})^2} \cdot q_y \cdot e^{\lambda \cdot t} - \frac{2 \cdot q_y^{(i)}}{(q_z^{(i)})^3} \cdot q_z \cdot e^{\lambda \cdot t} \right\}$$

$$\cdot A \cdot [\cos(\psi^{(i)}) - \sin(\psi^{(i)}) \cdot \psi \cdot e^{\lambda \cdot (t-\tau)}]$$

$$\psi \cdot \lambda \cdot e^{\lambda \cdot t} = k_z \cdot c \cdot \left[\frac{[2 + (q_x^{(i)})^2 + (q_y^{(i)})^2] \cdot 2}{4 \cdot (q_z^{(i)})^3} \cdot q_z \right.$$

$$\left. \cdot e^{\lambda \cdot t} - \frac{2 \cdot q_x^{(i)}}{4 \cdot (q_z^{(i)})^2} \cdot q_x \cdot e^{\lambda \cdot t} - \frac{2 \cdot q_y^{(i)}}{4 \cdot (q_z^{(i)})^2} \cdot q_y \cdot e^{\lambda \cdot t} \right]$$

$$+ \omega_\beta \cdot \left\{ \frac{1}{2 \cdot q_y^{(i)}} \cdot A \cdot \sin(\psi^{(i)}) \cdot \psi \cdot e^{\lambda \cdot (t-\tau)} \right.$$

$$+ \frac{1}{2 \cdot (q_y^{(i)})^2} \cdot q_y \cdot e^{\lambda \cdot t} \cdot A \cdot \cos(\psi^{(i)})$$

$$\left. - \frac{1}{2 \cdot (q_y^{(i)})^2} \cdot e^{\lambda \cdot t} \cdot A \cdot \sin(\psi^{(i)}) \cdot q_y \cdot \psi \cdot e^{\lambda \cdot (t-\tau)} \right\}$$

$$- \frac{1}{2} \cdot k_z \cdot c \cdot \left\{ \frac{q_y^{(i)}}{(q_z^{(i)})^2} \right\} \cdot A \cdot \sin(\psi^{(i)}) \cdot \psi \cdot e^{\lambda \cdot (t-\tau)}$$

$$+ \frac{1}{2} \cdot k_z \cdot c \cdot \left\{ \frac{1}{(q_z^{(i)})^2} \cdot q_y \cdot e^{\lambda \cdot t} \cdot A \cdot \cos(\psi^{(i)}) \right.$$

$$- \frac{1}{(q_z^{(i)})^2} \cdot e^{\lambda \cdot t} \cdot A \cdot \sin(\psi^{(i)}) \cdot \psi \cdot q_y \cdot e^{\lambda \cdot (t-\tau)}$$

$$- \frac{2 \cdot q_y^{(i)}}{(q_z^{(i)})^3} \cdot q_z \cdot e^{\lambda \cdot t} \cdot A \cdot \cos(\psi^{(i)})$$

$$\left. + \frac{2 \cdot q_y^{(i)}}{(q_z^{(i)})^3} \cdot e^{\lambda \cdot t} \cdot A \cdot \sin(\psi^{(i)}) \cdot q_z \cdot \psi \cdot e^{\lambda \cdot (t-\tau)} \right\}$$

Assumption: $q_y \cdot \psi \approx 0; q_z \cdot \psi \approx 0$

$$\psi \cdot \lambda \cdot e^{\lambda \cdot t} = k_z \cdot c \cdot [\frac{[2 + (q_x^{(i)})^2 + (q_y^{(i)})^2] \cdot 2}{4 \cdot (q_z^{(i)})^3} \cdot q_z \cdot e^{\lambda \cdot t}$$

$$- \frac{2 \cdot q_x^{(i)}}{4 \cdot (q_z^{(i)})^2} \cdot q_x \cdot e^{\lambda \cdot t} - \frac{2 \cdot q_y^{(i)}}{4 \cdot (q_z^{(i)})^2} \cdot q_y \cdot e^{\lambda \cdot t}]$$

$$+\omega_{\beta}\cdot\{\frac{1}{2\cdot q_y^{(i)}}\cdot A\cdot\sin(\psi^{(i)})\cdot\psi\cdot e^{\lambda\cdot(t-\tau)}$$

$$+\frac{1}{2\cdot(q_y^{(i)})^2}\cdot q_y\cdot e^{\lambda\cdot t}\cdot A\cdot\cos(\psi^{(i)})\}$$

$$-\frac{1}{2}\cdot k_z\cdot c\cdot\{\frac{q_y^{(i)}}{(q_z^{(i)})^2}\}\cdot A\cdot\sin(\psi^{(i)})\cdot\psi\cdot e^{\lambda\cdot(t-\tau)}$$

$$+\frac{1}{2}\cdot k_z\cdot c\cdot\{\frac{1}{(q_z^{(i)})^2}\cdot q_y\cdot e^{\lambda\cdot t}\cdot A\cdot\cos(\psi^{(i)})$$

$$-\frac{2\cdot q_y^{(i)}}{(q_z^{(i)})^3}\cdot q_z\cdot e^{\lambda\cdot t}\cdot A\cdot\cos(\psi^{(i)})\}$$

$$\psi\cdot\lambda\cdot e^{\lambda\cdot t}=k_z\cdot c\cdot[\frac{[2+(q_x^{(i)})^2+(q_y^{(i)})^2]\cdot 2}{4\cdot(q_z^{(i)})^3}\cdot q_z$$

$$-\frac{2\cdot q_x^{(i)}}{4\cdot(q_z^{(i)})^2}\cdot q_x-\frac{2\cdot q_y^{(i)}}{4\cdot(q_z^{(i)})^2}\cdot q_y]\cdot e^{\lambda\cdot t}$$

$$+\omega_{\beta}\cdot\{\frac{1}{2\cdot q_y^{(i)}}\cdot A\cdot\sin(\psi^{(i)})\cdot\psi\cdot e^{-\lambda\cdot\tau}$$

$$+\frac{1}{2\cdot(q_y^{(i)})^2}\cdot q_y\cdot A\cdot\cos(\psi^{(i)})\}\cdot e^{\lambda\cdot t}$$

$$-\frac{1}{2}\cdot k_z\cdot c\cdot\{\frac{q_y^{(i)}}{(q_z^{(i)})^2}\}\cdot A\cdot\sin(\psi^{(i)})\cdot\psi\cdot e^{\lambda\cdot t}\cdot e^{-\lambda\cdot\tau}$$

$$+\frac{1}{2}\cdot k_z\cdot c\cdot\{\frac{1}{(q_z^{(i)})^2}\cdot q_y\cdot A\cdot\cos(\psi^{(i)})$$

$$-\frac{2\cdot q_y^{(i)}}{(q_z^{(i)})^3}\cdot q_z\cdot A\cdot\cos(\psi^{(i)})\}\cdot e^{\lambda\cdot t}$$

Divide the two sides of the above equation by $e^{\lambda\cdot t}$ term.

$$\psi\cdot\lambda=k_z\cdot c\cdot\left[\frac{[2+(q_x^{(i)})^2+(q_y^{(i)})^2]\cdot 2}{4\cdot(q_z^{(i)})^3}\cdot q_z\right.$$

$$\left.-\frac{2\cdot q_x^{(i)}}{4\cdot(q_z^{(i)})^2}\cdot q_x-\frac{2\cdot q_y^{(i)}}{4\cdot(q_z^{(i)})^2}\cdot q_y\right]$$

$$+\omega_{\beta}\cdot\left\{\frac{1}{2\cdot q_y^{(i)}}\cdot A\cdot\sin(\psi^{(i)})\cdot\psi\cdot e^{-\lambda\cdot\tau}\right.$$

$$\left.+\frac{1}{2\cdot(q_y^{(i)})^2}\cdot q_y\cdot A\cdot\cos(\psi^{(i)})\right\}$$

$$-\frac{1}{2}\cdot k_z\cdot c\cdot\left\{\frac{q_y^{(i)}}{(q_z^{(i)})^2}\right\}\cdot A\cdot\sin(\psi^{(i)})\cdot\psi\cdot e^{-\lambda\cdot\tau}$$
$$+\frac{1}{2}\cdot k_z\cdot c\cdot\left\{\frac{1}{(q_z^{(i)})^2}\cdot q_y\cdot A\cdot\cos(\psi^{(i)})\right.$$
$$\left.-\frac{2\cdot q_y^{(i)}}{(q_z^{(i)})^3}\cdot q_z\cdot A\cdot\cos(\psi^{(i)})\right\}$$

$$\psi\cdot\lambda=k_z\cdot c\cdot\frac{[2+(q_x^{(i)})^2+(q_y^{(i)})^2]\cdot 2}{4\cdot(q_z^{(i)})^3}\cdot q_z$$
$$-k_z\cdot c\cdot\frac{2\cdot q_x^{(i)}}{4\cdot(q_z^{(i)})^2}\cdot q_x-k_z\cdot c\cdot\frac{2\cdot q_y^{(i)}}{4\cdot(q_z^{(i)})^2}\cdot q_y$$
$$+\omega_\beta\cdot\frac{1}{2\cdot q_y^{(i)}}\cdot A\cdot\sin(\psi^{(i)})\cdot\psi\cdot e^{-\lambda\cdot\tau}$$
$$+\omega_\beta\cdot\frac{1}{2\cdot(q_y^{(i)})^2}\cdot q_y\cdot A\cdot\cos(\psi^{(i)})$$
$$-\frac{1}{2}\cdot k_z\cdot c\cdot\{\frac{q_y^{(i)}}{(q_z^{(i)})^2}\}\cdot A\cdot\sin(\psi^{(i)})\cdot\psi\cdot e^{-\lambda\cdot\tau}$$
$$+\frac{1}{2}\cdot k_z\cdot c\cdot\frac{1}{(q_z^{(i)})^2}\cdot q_y\cdot A\cdot\cos(\psi^{(i)})$$
$$-\frac{1}{2}\cdot k_z\cdot c\cdot\frac{2\cdot q_y^{(i)}}{(q_z^{(i)})^3}\cdot q_z\cdot A\cdot\cos(\psi^{(i)})$$

$$\psi\cdot\lambda=[\omega_\beta\cdot\frac{1}{2\cdot q_y^{(i)}}\cdot A\cdot\sin(\psi^{(i)})-\frac{1}{2}\cdot k_z\cdot c\cdot\{\frac{q_y^{(i)}}{(q_z^{(i)})^2}\}$$
$$\cdot A\cdot\sin(\psi^{(i)})]\cdot e^{-\lambda\cdot\tau}\cdot\psi-k_z\cdot c\cdot\frac{2\cdot q_x^{(i)}}{4\cdot(q_z^{(i)})^2}\cdot q_x$$
$$+[\omega_\beta\cdot\frac{1}{2\cdot(q_y^{(i)})^2}\cdot A\cdot\cos(\psi^{(i)})-k_z\cdot c\cdot\frac{2\cdot q_y^{(i)}}{4\cdot(q_z^{(i)})^2}$$
$$+\frac{1}{2}\cdot k_z\cdot c\cdot\frac{1}{(q_z^{(i)})^2}\cdot A\cdot\cos(\psi^{(i)})]\cdot q_y$$
$$+[k_z\cdot c\cdot\frac{[2+(q_x^{(i)})^2+(q_y^{(i)})^2]\cdot 2}{4\cdot(q_z^{(i)})^3}$$
$$-\frac{1}{2}\cdot k_z\cdot c\cdot\frac{2\cdot q_y^{(i)}}{(q_z^{(i)})^3}\cdot A\cdot\cos(\psi^{(i)})]\cdot q_z$$

$$[\omega_\beta\cdot\frac{1}{2\cdot q_y^{(i)}}\cdot A\cdot\sin(\psi^{(i)})-\frac{1}{2}\cdot k_z\cdot c\cdot\{\frac{q_y^{(i)}}{(q_z^{(i)})^2}\}\cdot A$$

$$\cdot \sin(\psi^{(i)})] \cdot e^{-\lambda\cdot\tau} \cdot \psi - \lambda \cdot \psi - k_z \cdot c \cdot \frac{2 \cdot q_x^{(i)}}{4 \cdot (q_z^{(i)})^2} \cdot q_x$$

$$+ [\omega_\beta \cdot \frac{1}{2 \cdot (q_y^{(i)})^2} \cdot A \cdot \cos(\psi^{(i)}) - k_z \cdot c \cdot \frac{2 \cdot q_y^{(i)}}{4 \cdot (q_z^{(i)})^2}$$

$$+ \frac{1}{2} \cdot k_z \cdot c \cdot \frac{1}{(q_z^{(i)})^2} \cdot A \cdot \cos(\psi^{(i)})] \cdot q_y$$

$$+ [k_z \cdot c \cdot \frac{[2 + (q_x^{(i)})^2 + (q_y^{(i)})^2] \cdot 2}{4 \cdot (q_z^{(i)})^3} - \frac{1}{2} \cdot k_z \cdot c$$

$$\cdot \frac{2 \cdot q_y^{(i)}}{(q_z^{(i)})^3} \cdot A \cdot \cos(\psi^{(i)})] \cdot q_z = 0$$

We define the following global parameters for simplicity.

$$\Xi_{11} = \omega_\beta \cdot \frac{1}{2 \cdot q_y^{(i)}} \cdot A \cdot \sin(\psi^{(i)})$$
$$- \frac{1}{2} \cdot k_z \cdot c \cdot \{\frac{q_y^{(i)}}{(q_z^{(i)})^2}\} \cdot A \cdot \sin(\psi^{(i)})$$

$$\Xi_{12} = -k_z \cdot c \cdot \frac{2 \cdot q_x^{(i)}}{4 \cdot [(q_z^{(i)})^2}$$

$$\Xi_{13} = \omega_\beta \cdot \frac{1}{2 \cdot (q_y^{(i)})^2} \cdot A \cdot \cos(\psi^{(i)})$$
$$+ \frac{1}{2} \cdot k_z \cdot c \cdot \frac{1}{(q_z^{(i)})^2} \cdot A \cdot \cos(\psi^{(i)})$$
$$- k_z \cdot c \cdot \frac{2 \cdot q_y^{(i)}}{4 \cdot [(q_z^{(i)})^2}$$

$$\Xi_{14} = k_z \cdot c \cdot \frac{[2 + (q_x^{(i)})^2 + (q_y^{(i)})^2] \cdot 2}{4 \cdot (q_z^{(i)})^3}$$
$$- \frac{1}{2} \cdot k_z \cdot c \cdot \frac{2 \cdot q_y^{(i)}}{(q_z^{(i)})^3} \cdot A \cdot \cos(\psi^{(i)})$$

$$\Xi_{11} \cdot e^{-\lambda\cdot\tau} \cdot \psi - \psi \cdot \lambda + \Xi_{12} \cdot q_x + \Xi_{13} \cdot q_y + \Xi_{14} \cdot q_z = 0$$

$$\frac{dq_x}{dt} = -\frac{1}{8} \cdot k_z \cdot c \cdot \frac{q_y \cdot q_x}{q_z^2} \cdot A \cdot \sin(\psi(t - \tau))$$

$$q_x \cdot \lambda \cdot e^{\lambda \cdot t} = -\frac{1}{8} \cdot k_z \cdot c \cdot \frac{(q_y^{(i)} + q_y \cdot e^{\lambda \cdot t}) \cdot (q_x^{(i)} + q_x \cdot e^{\lambda \cdot t})}{(q_z^{(i)} + q_z \cdot e^{\lambda \cdot t})^2} \cdot A \cdot \sin(\psi(t - \tau))$$

$$\sin(\psi(t - \tau)) = \sin(\psi^{(i)} + \psi \cdot e^{\lambda \cdot (t-\tau)}) \approx \sin(\psi^{(i)}) + \cos(\psi^{(i)}) \cdot \psi \cdot e^{\lambda \cdot (t-\tau)}$$

$$q_x \cdot \lambda \cdot e^{\lambda \cdot t} = -\frac{1}{8} \cdot k_z \cdot c \cdot \frac{(q_y^{(i)} + q_y \cdot e^{\lambda \cdot t}) \cdot (q_x^{(i)} + q_x \cdot e^{\lambda \cdot t})}{(q_z^{(i)} + q_z \cdot e^{\lambda \cdot t})^2} \cdot A \cdot [\sin(\psi^{(i)}) + \cos(\psi^{(i)}) \cdot \psi \cdot e^{\lambda \cdot (t-\tau)}]$$

$$\frac{(q_y^{(i)} + q_y \cdot e^{\lambda \cdot t}) \cdot (q_x^{(i)} + q_x \cdot e^{\lambda \cdot t})}{(q_z^{(i)} + q_z \cdot e^{\lambda \cdot t})^2} = \frac{q_y^{(i)} \cdot q_x^{(i)} + q_y^{(i)} \cdot q_x \cdot e^{\lambda \cdot t} + q_x^{(i)} \cdot q_y \cdot e^{\lambda \cdot t} + q_y \cdot q_x \cdot e^{2 \cdot \lambda \cdot t}}{(q_z^{(i)})^2 + 2 \cdot q_z^{(i)} \cdot q_z \cdot e^{\lambda \cdot t} + (q_z)^2 \cdot e^{2 \cdot \lambda \cdot t}}$$

Assumption $q_y \cdot q_x \approx 0$; $(q_z)^2 \approx 0$ then

$$\frac{(q_y^{(i)} + q_y \cdot e^{\lambda \cdot t}) \cdot (q_x^{(i)} + q_x \cdot e^{\lambda \cdot t})}{(q_z^{(i)} + q_z \cdot e^{\lambda \cdot t})^2} = \frac{q_y^{(i)} \cdot q_x^{(i)} + q_y^{(i)} \cdot q_x \cdot e^{\lambda \cdot t} + q_x^{(i)} \cdot q_y \cdot e^{\lambda \cdot t}}{(q_z^{(i)})^2 + 2 \cdot q_z^{(i)} \cdot q_z \cdot e^{\lambda \cdot t}}$$

$$\frac{q_y^{(i)} \cdot q_x^{(i)} + q_y^{(i)} \cdot q_x \cdot e^{\lambda \cdot t} + q_x^{(i)} \cdot q_y \cdot e^{\lambda \cdot t}}{(q_z^{(i)})^2 + 2 \cdot q_z^{(i)} \cdot q_z \cdot e^{\lambda \cdot t}} \cdot \frac{[(q_z^{(i)})^2 - 2 \cdot q_z^{(i)} \cdot q_z \cdot e^{\lambda \cdot t}]}{[(q_z^{(i)})^2 - 2 \cdot q_z^{(i)} \cdot q_z \cdot e^{\lambda \cdot t}]}$$

$$= \frac{\begin{array}{l} q_y^{(i)} \cdot q_x^{(i)} \cdot (q_z^{(i)})^2 - q_y^{(i)} \cdot q_x^{(i)} \cdot 2 \cdot q_z^{(i)} \cdot q_z \cdot e^{\lambda \cdot t} \\ +(q_z^{(i)})^2 \cdot q_y^{(i)} \cdot q_x \cdot e^{\lambda \cdot t} - q_y^{(i)} \cdot 2 \cdot q_z^{(i)} \cdot q_z \cdot q_x \cdot e^{2 \cdot \lambda \cdot t} \\ +(q_z^{(i)})^2 \cdot q_x^{(i)} \cdot q_y \cdot e^{\lambda \cdot t} - q_x^{(i)} \cdot 2 \cdot q_z^{(i)} \cdot q_z \cdot q_y \cdot e^{2 \cdot \lambda \cdot t} \end{array}}{(q_z^{(i)})^4 - 4 \cdot (q_z^{(i)})^2 \cdot (q_z)^2 \cdot e^{2 \cdot \lambda \cdot t}}$$

Assumption $q_z \cdot q_x \approx 0$; $q_z \cdot q_y \approx 0$; $(q_z)^2$ then

$$\frac{(q_y^{(i)} + q_y \cdot e^{\lambda \cdot t}) \cdot (q_x^{(i)} + q_x \cdot e^{\lambda \cdot t})}{(q_z^{(i)} + q_z \cdot e^{\lambda \cdot t})^2} = \frac{\begin{array}{l} q_y^{(i)} \cdot q_x^{(i)} \cdot (q_z^{(i)})^2 - q_y^{(i)} \cdot q_x^{(i)} \cdot 2 \cdot q_z^{(i)} \cdot q_z \cdot e^{\lambda \cdot t} \\ +(q_z^{(i)})^2 \cdot q_y^{(i)} \cdot q_x \cdot e^{\lambda \cdot t} + (q_z^{(i)})^2 \cdot q_x^{(i)} \cdot q_y \cdot e^{\lambda \cdot t} \end{array}}{(q_z^{(i)})^4}$$

$$\frac{(q_y^{(i)} + q_y \cdot e^{\lambda \cdot t}) \cdot (q_x^{(i)} + q_x \cdot e^{\lambda \cdot t})}{(q_z^{(i)} + q_z \cdot e^{\lambda \cdot t})^2} = \frac{q_y^{(i)} \cdot q_x^{(i)}}{(q_z^{(i)})^2} - \frac{q_y^{(i)} \cdot q_x^{(i)} \cdot 2}{(q_z^{(i)})^3} \cdot q_z \cdot e^{\lambda \cdot t}$$

$$+\frac{q_y^{(i)}}{(q_z^{(i)})^2}\cdot q_x\cdot e^{\lambda\cdot t}+\frac{q_x^{(i)}}{(q_z^{(i)})^2}\cdot q_y\cdot e^{\lambda\cdot t}$$

$$q_x\cdot\lambda\cdot e^{\lambda\cdot t}=-\frac{1}{8}\cdot k_z\cdot c\cdot[\frac{q_y^{(i)}\cdot q_x^{(i)}}{(q_z^{(i)})^2}$$
$$-\frac{q_y^{(i)}\cdot q_x^{(i)}\cdot 2}{(q_z^{(i)})^3}\cdot q_z\cdot e^{\lambda\cdot t}+\frac{q_y^{(i)}}{(q_z^{(i)})^2}\cdot q_x\cdot e^{\lambda\cdot t}$$
$$+\frac{q_x^{(i)}}{(q_z^{(i)})^2}\cdot q_y\cdot e^{\lambda\cdot t}]\cdot A\cdot[\sin(\psi^{(i)})$$
$$+\cos(\psi^{(i)})\cdot\psi\cdot e^{\lambda\cdot(t-\tau)}]$$

$$q_x\cdot\lambda\cdot e^{\lambda\cdot t}=[-\frac{1}{8}\cdot k_z\cdot c\cdot A\cdot\frac{q_y^{(i)}\cdot q_x^{(i)}}{(q_z^{(i)})^2}+\frac{1}{8}\cdot k_z\cdot c$$
$$\cdot A\cdot\frac{q_y^{(i)}\cdot q_x^{(i)}\cdot 2}{(q_z^{(i)})^3}\cdot q_z\cdot e^{\lambda\cdot t}-\frac{1}{8}\cdot k_z\cdot c\cdot A\cdot\frac{q_y^{(i)}}{(q_z^{(i)})^2}\cdot q_x\cdot e^{\lambda\cdot t}$$
$$-\frac{1}{8}\cdot k_z\cdot c\cdot A\cdot\frac{q_x^{(i)}}{(q_z^{(i)})^2}\cdot q_y\cdot e^{\lambda\cdot t}]\cdot[\sin(\psi^{(i)})$$
$$+\cos(\psi^{(i)})\cdot\psi\cdot e^{\lambda\cdot(t-\tau)}]$$

$$q_x\cdot\lambda\cdot e^{\lambda\cdot t}=[-\frac{1}{8}\cdot k_z\cdot c\cdot A\cdot\frac{q_y^{(i)}\cdot q_x^{(i)}}{(q_z^{(i)})^2}+\frac{1}{8}\cdot k_z\cdot c\cdot A$$
$$\cdot\frac{q_y^{(i)}\cdot q_x^{(i)}\cdot 2}{(q_z^{(i)})^3}\cdot q_z\cdot e^{\lambda\cdot t}-\frac{1}{8}\cdot k_z\cdot c\cdot A\cdot\frac{q_y^{(i)}}{(q_z^{(i)})^2}\cdot q_x\cdot e^{\lambda\cdot t}$$
$$-\frac{1}{8}\cdot k_z\cdot c\cdot A\cdot\frac{q_x^{(i)}}{(q_z^{(i)})^2}\cdot q_y\cdot e^{\lambda\cdot t}]\cdot\sin(\psi^{(i)})$$
$$+[-\frac{1}{8}\cdot k_z\cdot c\cdot A\cdot\frac{q_y^{(i)}\cdot q_x^{(i)}}{(q_z^{(i)})^2}+\frac{1}{8}\cdot k_z\cdot c\cdot A$$
$$\cdot\frac{q_y^{(i)}\cdot q_x^{(i)}\cdot 2}{(q_z^{(i)})^3}\cdot q_z\cdot e^{\lambda\cdot t}-\frac{1}{8}\cdot k_z\cdot c\cdot A\cdot\frac{q_y^{(i)}}{(q_z^{(i)})^2}\cdot q_x\cdot e^{\lambda\cdot t}$$
$$-\frac{1}{8}\cdot k_z\cdot c\cdot A\cdot\frac{q_x^{(i)}}{(q_z^{(i)})^2}\cdot q_y\cdot e^{\lambda\cdot t}]\cdot\cos(\psi^{(i)})\cdot\psi\cdot e^{\lambda\cdot(t-\tau)}$$

$$q_x\cdot\lambda\cdot e^{\lambda\cdot t}=\left[-\frac{1}{8}\cdot k_z\cdot c\cdot A\cdot\frac{q_y^{(i)}\cdot q_x^{(i)}}{(q_z^{(i)})^2}+\frac{1}{8}\cdot k_z\cdot c\cdot A\right.$$
$$\cdot\frac{q_y^{(i)}\cdot q_x^{(i)}\cdot 2}{(q_z^{(i)})^3}\cdot q_z\cdot e^{\lambda\cdot t}-\frac{1}{8}\cdot k_z\cdot c\cdot A\cdot\frac{q_y^{(i)}}{(q_z^{(i)})^2}\cdot q_x\cdot e^{\lambda\cdot t}$$

$$-\frac{1}{8}\cdot k_z\cdot c\cdot A\cdot\frac{q_x^{(i)}}{(q_z^{(i)})^2}\cdot q_y\cdot e^{\lambda\cdot t}\Bigg]\cdot\sin(\psi^{(i)})$$

$$+\Bigg[-\frac{1}{8}\cdot k_z\cdot c\cdot A\cdot\frac{q_y^{(i)}\cdot q_x^{(i)}}{(q_z^{(i)})^2}\cdot\psi+\frac{1}{8}\cdot k_z\cdot c\cdot A$$

$$\cdot\frac{q_y^{(i)}\cdot q_x^{(i)}\cdot 2}{(q_z^{(i)})^3}\cdot q_z\cdot\psi\cdot e^{\lambda\cdot t}-\frac{1}{8}\cdot k_z\cdot c\cdot A\cdot\frac{q_y^{(i)}}{(q_z^{(i)})^2}\cdot q_x\cdot\psi\cdot e^{\lambda\cdot t}$$

$$-\frac{1}{8}\cdot k_z\cdot c\cdot A\cdot\frac{q_x^{(i)}}{(q_z^{(i)})^2}\cdot q_y\cdot\psi\cdot e^{\lambda\cdot t}\Bigg]\cdot\cos(\psi^{(i)})\cdot e^{\lambda\cdot(t-\tau)}$$

Assumption: $q_x\cdot\psi\approx 0$; $q_z\cdot\psi\approx 0$; $q_y\cdot\psi\approx 0$ then

$$q_x\cdot\lambda\cdot e^{\lambda\cdot t}=\Bigg[-\frac{1}{8}\cdot k_z\cdot c\cdot A\cdot\frac{q_y^{(i)}\cdot q_x^{(i)}}{(q_z^{(i)})^2}+\frac{1}{8}\cdot k_z$$

$$\cdot c\cdot A\cdot\frac{q_y^{(i)}\cdot q_x^{(i)}\cdot 2}{(q_z^{(i)})^3}\cdot q_z\cdot e^{\lambda\cdot t}$$

$$-\frac{1}{8}\cdot k_z\cdot c\cdot A\cdot\frac{q_y^{(i)}}{(q_z^{(i)})^2}\cdot q_x\cdot e^{\lambda\cdot t}$$

$$-\frac{1}{8}\cdot k_z\cdot c\cdot A\cdot\frac{q_x^{(i)}}{(q_z^{(i)})^2}\cdot q_y\cdot e^{\lambda\cdot t}\Bigg]\cdot\sin(\psi^{(i)})$$

$$+\Bigg[-\frac{1}{8}\cdot k_z\cdot c\cdot A\cdot\frac{q_y^{(i)}\cdot q_x^{(i)}}{(q_z^{(i)})^2}\cdot\psi\Bigg]\cdot\cos(\psi^{(i)})\cdot e^{\lambda\cdot(t-\tau)}$$

$$q_x\cdot\lambda\cdot e^{\lambda\cdot t}=-\frac{1}{8}\cdot k_z\cdot c\cdot A\cdot\frac{q_y^{(i)}\cdot q_x^{(i)}}{(q_z^{(i)})^2}\cdot\sin(\psi^{(i)})$$

$$+\frac{1}{8}\cdot k_z\cdot c\cdot A\cdot\frac{q_y^{(i)}\cdot q_x^{(i)}\cdot 2}{(q_z^{(i)})^3}\cdot q_z\cdot e^{\lambda\cdot t}\cdot\sin(\psi^{(i)})$$

$$-\frac{1}{8}\cdot k_z\cdot c\cdot A\cdot\frac{q_y^{(i)}}{(q_z^{(i)})^2}\cdot q_x\cdot e^{\lambda\cdot t}\cdot\sin(\psi^{(i)})$$

$$-\frac{1}{8}\cdot k_z\cdot c\cdot A\cdot\frac{q_x^{(i)}}{(q_z^{(i)})^2}\cdot q_y\cdot e^{\lambda\cdot t}\cdot\sin(\psi^{(i)})$$

$$+[-\frac{1}{8}\cdot k_z\cdot c\cdot A\cdot\frac{q_y^{(i)}\cdot q_x^{(i)}}{(q_z^{(i)})^2}\cdot\psi]\cdot\cos(\psi^{(i)})\cdot e^{\lambda\cdot(t-\tau)}$$

At fixed point $\frac{dq_x}{dt}=0$; $-\frac{1}{8}\cdot k_z\cdot c\cdot A\cdot\frac{q_y^{(i)}\cdot q_x^{(i)}}{(q_z^{(i)})^2}\cdot\sin(\psi^{(i)})=0$ then

$$q_x \cdot \lambda \cdot e^{\lambda \cdot t} = \frac{1}{8} \cdot k_z \cdot c \cdot A \cdot \frac{q_y^{(i)} \cdot q_x^{(i)} \cdot 2}{(q_z^{(i)})^3} \cdot q_z \cdot e^{\lambda \cdot t} \cdot \sin(\psi^{(i)})$$
$$- \frac{1}{8} \cdot k_z \cdot c \cdot A \cdot \frac{q_y^{(i)}}{(q_z^{(i)})^2} \cdot q_x \cdot e^{\lambda \cdot t} \cdot \sin(\psi^{(i)})$$
$$- \frac{1}{8} \cdot k_z \cdot c \cdot A \cdot \frac{q_x^{(i)}}{(q_z^{(i)})^2} \cdot q_y \cdot e^{\lambda \cdot t} \cdot \sin(\psi^{(i)})$$
$$+ [- \frac{1}{8} \cdot k_z \cdot c \cdot A \cdot \frac{q_y^{(i)} \cdot q_x^{(i)}}{(q_z^{(i)})^2} \cdot \psi] \cdot \cos(\psi^{(i)}) \cdot e^{\lambda \cdot (t - \tau)}$$

We divide the two sides of the above expression by $e^{\lambda \cdot t}$ term.

$$q_x \cdot \lambda = \frac{1}{8} \cdot k_z \cdot c \cdot A \cdot \frac{q_y^{(i)} \cdot q_x^{(i)} \cdot 2}{(q_z^{(i)})^3} \cdot q_z \cdot \sin(\psi^{(i)})$$
$$- \frac{1}{8} \cdot k_z \cdot c \cdot A \cdot \frac{q_y^{(i)}}{(q_z^{(i)})^2} \cdot q_x \cdot \sin(\psi^{(i)})$$
$$- \frac{1}{8} \cdot k_z \cdot c \cdot A \cdot \frac{q_x^{(i)}}{(q_z^{(i)})^2} \cdot q_y \cdot \sin(\psi^{(i)})$$
$$- \frac{1}{8} \cdot k_z \cdot c \cdot A \cdot \frac{q_y^{(i)} \cdot q_x^{(i)}}{(q_z^{(i)})^2} \cdot \psi \cdot \cos(\psi^{(i)}) \cdot e^{-\lambda \cdot \tau}$$

$$- \frac{1}{8} \cdot k_z \cdot c \cdot A \cdot \frac{q_y^{(i)} \cdot q_x^{(i)}}{(q_z^{(i)})^2} \cdot \cos(\psi^{(i)}) \cdot e^{-\lambda \cdot \tau} \cdot \psi$$
$$- \frac{1}{8} \cdot k_z \cdot c \cdot A \cdot \frac{q_y^{(i)}}{(q_z^{(i)})^2} \cdot \sin(\psi^{(i)}) \cdot q_x - q_x \cdot \lambda$$
$$- \frac{1}{8} \cdot k_z \cdot c \cdot A \cdot \frac{q_x^{(i)}}{(q_z^{(i)})^2} \cdot \sin(\psi^{(i)}) \cdot q_y$$
$$+ \frac{1}{8} \cdot k_z \cdot c \cdot A \cdot \frac{q_y^{(i)} \cdot q_x^{(i)} \cdot 2}{(q_z^{(i)})^3} \cdot \sin(\psi^{(i)}) \cdot q_z = 0$$

We define the following global parameters for simplicity

$$\Xi_{21} = - \frac{1}{8} \cdot k_z \cdot c \cdot A \cdot \frac{q_y^{(i)} \cdot q_x^{(i)}}{(q_z^{(i)})^2} \cdot \cos(\psi^{(i)})$$

$$\Xi_{22} = - \frac{1}{8} \cdot k_z \cdot c \cdot A \cdot \frac{q_y^{(i)}}{(q_z^{(i)})^2} \cdot \sin(\psi^{(i)})$$

$$\Xi_{23} = - \frac{1}{8} \cdot k_z \cdot c \cdot A \cdot \frac{q_x^{(i)}}{(q_z^{(i)})^2} \cdot \sin(\psi^{(i)})$$

$$\Xi_{24} = \frac{1}{8} \cdot k_z \cdot c \cdot A \cdot \frac{q_y^{(i)} \cdot q_x^{(i)} \cdot 2}{(q_z^{(i)})^3} \cdot \sin(\psi^{(i)})$$

$$\Xi_{21} \cdot \psi \cdot e^{-\lambda \cdot \tau} + \Xi_{22} \cdot q_x - q_x \cdot \lambda + \Xi_{23} \cdot q_y + \Xi_{24} \cdot q_z = 0$$

$$\frac{dq_y}{dt} = -\frac{1}{2} \cdot (\omega_\beta + \frac{1}{4} \cdot k_z \cdot c \cdot \frac{q_y^2}{q_z^2}) \cdot A \cdot \sin(\psi(t - \tau))$$

$$q_y \cdot \lambda \cdot e^{\lambda \cdot t} = -\frac{1}{2} \cdot (\omega_\beta + \frac{1}{4} \cdot k_z \cdot c \cdot \frac{(q_y^{(i)} + q_y \cdot e^{\lambda \cdot t})^2}{(q_z^{(i)} + q_z \cdot e^{\lambda \cdot t})^2}) \cdot A \cdot \sin(\psi(t - \tau))$$

$$\sin(\psi(t - \tau)) = \sin(\psi^{(i)} + \psi \cdot e^{\lambda \cdot (t - \tau)}) \approx \sin(\psi^{(i)}) + \cos(\psi^{(i)}) \cdot \psi \cdot e^{\lambda \cdot (t - \tau)}$$

$$q_y \cdot \lambda \cdot e^{\lambda \cdot t} = -\frac{1}{2} \cdot (\omega_\beta + \frac{1}{4} \cdot k_z \cdot c \cdot \frac{(q_y^{(i)} + q_y \cdot e^{\lambda \cdot t})^2}{(q_z^{(i)} + q_z \cdot e^{\lambda \cdot t})^2}) \cdot A \cdot [\sin(\psi^{(i)}) + \cos(\psi^{(i)}) \cdot \psi \cdot e^{\lambda \cdot (t - \tau)}]$$

$$\frac{(q_y^{(i)} + q_y \cdot e^{\lambda \cdot t})^2}{(q_z^{(i)} + q_z \cdot e^{\lambda \cdot t})^2} = \frac{(q_y^{(i)})^2 + 2 \cdot q_y^{(i)} \cdot q_y \cdot e^{\lambda \cdot t} + (q_y)^2 \cdot e^{2 \cdot \lambda \cdot t}}{(q_z^{(i)})^2 + 2 \cdot q_z^{(i)} \cdot q_z \cdot e^{\lambda \cdot t} + (q_z)^2 \cdot e^{2 \cdot \lambda \cdot t}} \Big|_{(q_y)^2 \approx 0;\, (q_z)^2 \approx 0} \approx \frac{(q_y^{(i)})^2 + 2 \cdot q_y^{(i)} \cdot q_y \cdot e^{\lambda \cdot t}}{(q_z^{(i)})^2 + 2 \cdot q_z^{(i)} \cdot q_z \cdot e^{\lambda \cdot t}}$$

$$\frac{(q_y^{(i)} + q_y \cdot e^{\lambda \cdot t})^2}{(q_z^{(i)} + q_z \cdot e^{\lambda \cdot t})^2} \approx \frac{[(q_y^{(i)})^2 + 2 \cdot q_y^{(i)} \cdot q_y \cdot e^{\lambda \cdot t}]}{[(q_z^{(i)})^2 + 2 \cdot q_z^{(i)} \cdot q_z \cdot e^{\lambda \cdot t}]} \cdot \frac{[(q_z^{(i)})^2 - 2 \cdot q_z^{(i)} \cdot q_z \cdot e^{\lambda \cdot t}]}{[(q_z^{(i)})^2 - 2 \cdot q_z^{(i)} \cdot q_z \cdot e^{\lambda \cdot t}]}$$

$$= \frac{\begin{array}{c}(q_y^{(i)})^2 \cdot (q_z^{(i)})^2 - (q_y^{(i)})^2 \cdot 2 \cdot q_z^{(i)} \cdot q_z \cdot e^{\lambda \cdot t} \\ +2 \cdot q_y^{(i)} \cdot q_y \cdot e^{\lambda \cdot t} \cdot (q_z^{(i)})^2 - 4 \cdot q_y^{(i)} \cdot q_z^{(i)} \cdot q_z \cdot q_y \cdot e^{2 \cdot \lambda \cdot t}\end{array}}{(q_z^{(i)})^4 - 4 \cdot (q_z^{(i)})^2 \cdot (q_z)^2 \cdot e^{2 \cdot \lambda \cdot t}}$$

Assumption: $q_z \cdot q_y \approx 0$; $(q_z)^2 \approx 0$ then

$$\frac{(q_y^{(i)} + q_y \cdot e^{\lambda \cdot t})^2}{(q_z^{(i)} + q_z \cdot e^{\lambda \cdot t})^2} \approx \frac{(q_y^{(i)})^2 \cdot (q_z^{(i)})^2 - (q_y^{(i)})^2 \cdot 2 \cdot q_z^{(i)} \cdot q_z \cdot e^{\lambda \cdot t} + 2 \cdot q_y^{(i)} \cdot q_y \cdot e^{\lambda \cdot t} \cdot (q_z^{(i)})^2}{(q_z^{(i)})^4}$$

$$\frac{(q_y^{(i)} + q_y \cdot e^{\lambda \cdot t})^2}{(q_z^{(i)} + q_z \cdot e^{\lambda \cdot t})^2} \approx \frac{(q_y^{(i)})^2}{(q_z^{(i)})^2} - \frac{(q_y^{(i)})^2 \cdot 2}{(q_z^{(i)})^3} \cdot q_z \cdot e^{\lambda \cdot t} + \frac{2 \cdot q_y^{(i)}}{(q_z^{(i)})^2} \cdot q_y \cdot e^{\lambda \cdot t}$$

$$q_y \cdot \lambda \cdot e^{\lambda \cdot t} = -\frac{1}{2} \cdot (\omega_\beta + \frac{1}{4} \cdot k_z \cdot c \cdot [\frac{(q_y^{(i)})^2}{(q_z^{(i)})^2} - \frac{(q_y^{(i)})^2 \cdot 2}{(q_z^{(i)})^3} \cdot q_z \cdot e^{\lambda \cdot t}$$

$$+\frac{2\cdot q_y^{(i)}}{(q_z^{(i)})^2}\cdot q_y\cdot e^{\lambda\cdot t}]) \cdot A\cdot[\sin(\psi^{(i)})+\cos(\psi^{(i)})\cdot\psi\cdot e^{\lambda\cdot(t-\tau)}]$$

$$q_y\cdot\lambda\cdot e^{\lambda\cdot t} = -\frac{1}{2}\cdot(\omega_\beta+\frac{1}{4}\cdot k_z\cdot c\cdot\frac{(q_y^{(i)})^2}{(q_z^{(i)})^2}$$
$$-\frac{1}{4}\cdot k_z\cdot c\cdot\frac{(q_y^{(i)})^2\cdot 2}{(q_z^{(i)})^3}\cdot q_z\cdot e^{\lambda\cdot t}$$
$$+\frac{1}{4}\cdot k_z\cdot c\cdot\frac{2\cdot q_y^{(i)}}{(q_z^{(i)})^2}\cdot q_y\cdot e^{\lambda\cdot t})\cdot A\cdot[\sin(\psi^{(i)})$$
$$+\cos(\psi^{(i)})\cdot\psi\cdot e^{\lambda\cdot(t-\tau)}]$$

$$q_y\cdot\lambda\cdot e^{\lambda\cdot t} = -\frac{1}{2}\cdot([\omega_\beta+\frac{1}{4}\cdot k_z\cdot c\cdot\frac{(q_y^{(i)})^2}{(q_z^{(i)})^2}]+[\frac{1}{4}\cdot k_z\cdot c\cdot\frac{2\cdot q_y^{(i)}}{(q_z^{(i)})^2}\cdot q_y\cdot e^{\lambda\cdot t}$$
$$-\frac{1}{4}\cdot k_z\cdot c\cdot\frac{(q_y^{(i)})^2\cdot 2}{(q_z^{(i)})^3}\cdot q_z\cdot e^{\lambda\cdot t}])\cdot A\cdot[\sin(\psi^{(i)})$$
$$+\cos(\psi^{(i)})\cdot\psi\cdot e^{\lambda\cdot(t-\tau)}]$$

$$q_y\cdot\lambda\cdot e^{\lambda\cdot t} = -\frac{1}{2}\cdot[\omega_\beta+\frac{1}{4}\cdot k_z\cdot c\cdot\frac{(q_y^{(i)})^2}{(q_z^{(i)})^2}]\cdot A\cdot[\sin(\psi^{(i)})$$
$$+\cos(\psi^{(i)})\cdot\psi\cdot e^{\lambda\cdot(t-\tau)}]-\frac{1}{2}\cdot[\frac{1}{4}\cdot k_z\cdot c\cdot\frac{2\cdot q_y^{(i)}}{(q_z^{(i)})^2}\cdot q_y\cdot e^{\lambda\cdot t}$$
$$-\frac{1}{4}\cdot k_z\cdot c\cdot\frac{(q_y^{(i)})^2\cdot 2}{(q_z^{(i)})^3}\cdot q_z\cdot e^{\lambda\cdot t}]\cdot A\cdot[\sin(\psi^{(i)})$$
$$+\cos(\psi^{(i)})\cdot\psi\cdot e^{\lambda\cdot(t-\tau)}]$$

$$q_y\cdot\lambda\cdot e^{\lambda\cdot t} = -\frac{1}{2}\cdot[\omega_\beta+\frac{1}{4}\cdot k_z\cdot c\cdot\frac{(q_y^{(i)})^2}{(q_z^{(i)})^2}]\cdot A\cdot\sin(\psi^{(i)})$$
$$-\frac{1}{2}\cdot[\omega_\beta+\frac{1}{4}\cdot k_z\cdot c\cdot\frac{(q_y^{(i)})^2}{(q_z^{(i)})^2}]\cdot A\cdot\cos(\psi^{(i)})\cdot\psi$$
$$\cdot e^{\lambda\cdot(t-\tau)}-\frac{1}{2}\cdot[\frac{1}{4}\cdot k_z\cdot c\cdot\frac{2\cdot q_y^{(i)}}{(q_z^{(i)})^2}\cdot q_y\cdot e^{\lambda\cdot t}$$
$$-\frac{1}{4}\cdot k_z\cdot c\cdot\frac{(q_y^{(i)})^2\cdot 2}{(q_z^{(i)})^3}\cdot q_z\cdot e^{\lambda\cdot t}]\cdot A\cdot[\sin(\psi^{(i)})$$
$$+\cos(\psi^{(i)})\cdot\psi\cdot e^{\lambda\cdot(t-\tau)}]$$

At fixed points $\frac{dq_y}{dt}=0\Rightarrow -\frac{1}{2}\cdot[\omega_\beta+\frac{1}{4}\cdot k_z\cdot c\cdot\frac{(q_y^{(i)})^2}{(q_z^{(i)})^2}]\cdot A\cdot\sin(\psi^{(i)})=0$ then

$$q_y \cdot \lambda \cdot e^{\lambda \cdot t} = -\frac{1}{2} \cdot [\omega_\beta + \frac{1}{4} \cdot k_z \cdot c \cdot \frac{(q_y^{(i)})^2}{(q_z^{(i)})^2}] \cdot A$$
$$\cdot \cos(\psi^{(i)}) \cdot \psi \cdot e^{\lambda \cdot (t-\tau)} - \frac{1}{2} \cdot [\frac{1}{4} \cdot k_z \cdot c \cdot \frac{2 \cdot q_y^{(i)}}{(q_z^{(i)})^2} \cdot q_y \cdot e^{\lambda \cdot t}$$
$$- \frac{1}{4} \cdot k_z \cdot c \cdot \frac{(q_y^{(i)})^2 \cdot 2}{(q_z^{(i)})^3} \cdot q_z \cdot e^{\lambda \cdot t}] \cdot A \cdot [\sin(\psi^{(i)})$$
$$+ \cos(\psi^{(i)}) \cdot \psi \cdot e^{\lambda \cdot (t-\tau)}]$$

$$q_y \cdot \lambda \cdot e^{\lambda \cdot t} = -\frac{1}{2} \cdot [\omega_\beta + \frac{1}{4} \cdot k_z \cdot c \cdot \frac{(q_y^{(i)})^2}{(q_z^{(i)})^2}] \cdot A \cdot \cos(\psi^{(i)}) \cdot \psi \cdot e^{\lambda \cdot (t-\tau)}$$
$$- \frac{1}{2} \cdot [A \cdot \frac{1}{4} \cdot k_z \cdot c \cdot \frac{2 \cdot q_y^{(i)}}{(q_z^{(i)})^2} \cdot q_y \cdot e^{\lambda \cdot t}$$
$$- A \cdot \frac{1}{4} \cdot k_z \cdot c \cdot \frac{(q_y^{(i)})^2 \cdot 2}{(q_z^{(i)})^3} \cdot q_z \cdot e^{\lambda \cdot t}] \cdot \sin(\psi^{(i)})$$
$$- \frac{1}{2} \cdot [A \cdot \frac{1}{4} \cdot k_z \cdot c \cdot \frac{2 \cdot q_y^{(i)}}{(q_z^{(i)})^2} \cdot q_y \cdot e^{\lambda \cdot t}$$
$$- A \cdot \frac{1}{4} \cdot k_z \cdot c \cdot \frac{(q_y^{(i)})^2 \cdot 2}{(q_z^{(i)})^3} \cdot q_z \cdot e^{\lambda \cdot t}] \cdot \cos(\psi^{(i)}) \cdot \psi \cdot e^{\lambda \cdot (t-\tau)}$$

$$q_y \cdot \lambda \cdot e^{\lambda \cdot t} = -\frac{1}{2} \cdot [\omega_\beta + \frac{1}{4} \cdot k_z \cdot c \cdot \frac{(q_y^{(i)})^2}{(q_z^{(i)})^2}] \cdot A \cdot \cos(\psi^{(i)}) \cdot \psi \cdot e^{\lambda \cdot (t-\tau)}$$
$$- \frac{1}{2} \cdot [A \cdot \frac{1}{4} \cdot k_z \cdot c \cdot \frac{2 \cdot q_y^{(i)}}{(q_z^{(i)})^2} \cdot q_y$$
$$- A \cdot \frac{1}{4} \cdot k_z \cdot c \cdot \frac{(q_y^{(i)})^2 \cdot 2}{(q_z^{(i)})^3} \cdot q_z] \cdot \sin(\psi^{(i)}) \cdot e^{\lambda \cdot t}$$
$$- \frac{1}{2} \cdot [A \cdot \frac{1}{4} \cdot k_z \cdot c \cdot \frac{2 \cdot q_y^{(i)}}{(q_z^{(i)})^2} \cdot q_y \cdot \psi \cdot e^{\lambda \cdot t}$$
$$- A \cdot \frac{1}{4} \cdot k_z \cdot c \cdot \frac{(q_y^{(i)})^2 \cdot 2}{(q_z^{(i)})^3} \cdot q_z \cdot \psi \cdot e^{\lambda \cdot t}] \cdot \cos(\psi^{(i)}) \cdot e^{\lambda \cdot (t-\tau)}$$

Assumption $q_y \cdot \psi \approx 0; q_z \cdot \psi \approx 0$ then

$$q_y \cdot \lambda \cdot e^{\lambda \cdot t} = -\frac{1}{2} \cdot [\omega_\beta + \frac{1}{4} \cdot k_z \cdot c \cdot \frac{(q_y^{(i)})^2}{(q_z^{(i)})^2}]$$
$$\cdot A \cdot \cos(\psi^{(i)}) \cdot \psi \cdot e^{\lambda \cdot (t-\tau)}$$

$$-\frac{1}{2}\cdot[A\cdot\frac{1}{4}\cdot k_z\cdot c\cdot\frac{2\cdot q_y^{(i)}}{(q_z^{(i)})^2}\cdot q_y$$
$$-A\cdot\frac{1}{4}\cdot k_z\cdot c\cdot\frac{(q_y^{(i)})^2\cdot 2}{(q_z^{(i)})^3}\cdot q_z]\cdot\sin(\psi^{(i)})\cdot e^{\lambda\cdot t}$$

We divide the two sides of the above equation by $e^{\lambda\cdot t}$ term.

$$q_y\cdot\lambda=-\frac{1}{2}\cdot[\omega_\beta+\frac{1}{4}\cdot k_z\cdot c\cdot\frac{(q_y^{(i)})^2}{(q_z^{(i)})^2}]$$
$$\cdot A\cdot\cos(\psi^{(i)})\cdot\psi\cdot e^{-\lambda\cdot\tau}$$
$$-\frac{1}{2}\cdot[A\cdot\frac{1}{4}\cdot k_z\cdot c\cdot\frac{2\cdot q_y^{(i)}}{(q_z^{(i)})^2}\cdot q_y$$
$$-A\cdot\frac{1}{4}\cdot k_z\cdot c\cdot\frac{(q_y^{(i)})^2\cdot 2}{(q_z^{(i)})^3}\cdot q_z]\cdot\sin(\psi^{(i)})$$

$$q_y\cdot\lambda=-\frac{1}{2}\cdot[\omega_\beta+\frac{1}{4}\cdot k_z\cdot c\cdot\frac{(q_y^{(i)})^2}{(q_z^{(i)})^2}]\cdot A\cdot\cos(\psi^{(i)})\cdot\psi\cdot e^{-\lambda\cdot\tau}$$
$$-\frac{1}{8}\cdot A\cdot k_z\cdot c\cdot\frac{2\cdot q_y^{(i)}}{(q_z^{(i)})^2}\cdot\sin(\psi^{(i)})\cdot q_y$$
$$+\frac{1}{8}\cdot A\cdot k_z\cdot c\cdot\frac{(q_y^{(i)})^2\cdot 2}{(q_z^{(i)})^3}\cdot\sin(\psi^{(i)})\cdot q_z$$

$$-\frac{1}{2}\cdot[\omega_\beta+\frac{1}{4}\cdot k_z\cdot c\cdot\frac{(q_y^{(i)})^2}{(q_z^{(i)})^2}]\cdot A\cdot\cos(\psi^{(i)})\cdot\psi\cdot e^{-\lambda\cdot\tau}$$
$$-\frac{1}{8}\cdot A\cdot k_z\cdot c\cdot\frac{2\cdot q_y^{(i)}}{(q_z^{(i)})^2}\cdot\sin(\psi^{(i)})\cdot q_y-q_y\cdot\lambda$$
$$+\frac{1}{8}\cdot A\cdot k_z\cdot c\cdot\frac{(q_y^{(i)})^2\cdot 2}{(q_z^{(i)})^3}\cdot\sin(\psi^{(i)})\cdot q_z=0$$

We define the following global parameters for simplicity

$$\Xi_{31}=-\frac{1}{2}\cdot\left[\omega_\beta+\frac{1}{4}\cdot k_z\cdot c\cdot\frac{(q_y^{(i)})^2}{(q_z^{(i)})^2}\right]\cdot A\cdot\cos(\psi^{(i)})$$

$$\Xi_{32}=0;\ \Xi_{33}=-\frac{1}{8}\cdot A\cdot k_z\cdot c\cdot\frac{2\cdot q_y^{(i)}}{(q_z^{(i)})^2}\cdot\sin(\psi^{(i)})$$

$$\Xi_{34}=\frac{1}{8}\cdot A\cdot k_z\cdot c\cdot\frac{(q_y^{(i)})^2\cdot 2}{(q_z^{(i)})^3}\cdot\sin(\psi^{(i)})$$

$$\Xi_{31} \cdot \psi \cdot e^{-\lambda \cdot \tau} + \Xi_{32} \cdot q_x + \Xi_{33} \cdot q_y - q_y \cdot \lambda + \Xi_{34} \cdot q_z = 0$$

$$\frac{dq_z}{dt} = -\frac{1}{2} \cdot k_z \cdot c \cdot \frac{q_y}{q_z} \cdot A \cdot \sin(\psi(t - \tau))$$

$$q_z \cdot \lambda \cdot e^{\lambda \cdot t} = -\frac{1}{2} \cdot k_z \cdot c \cdot \frac{q_y^{(i)} + q_y \cdot e^{\lambda \cdot t}}{q_z^{(i)} + q_z \cdot e^{\lambda \cdot t}} \cdot A \cdot \sin(\psi(t - \tau))$$

$$\sin(\psi(t - \tau)) = \sin(\psi^{(i)} + \psi \cdot e^{\lambda \cdot (t - \tau)}) \approx \sin(\psi^{(i)}) + \cos(\psi^{(i)}) \cdot \psi \cdot e^{\lambda \cdot (t - \tau)}$$

$$q_z \cdot \lambda \cdot e^{\lambda \cdot t} = -\frac{1}{2} \cdot k_z \cdot c \cdot \left[\frac{q_y^{(i)} + q_y \cdot e^{\lambda \cdot t}}{q_z^{(i)} + q_z \cdot e^{\lambda \cdot t}}\right] \cdot A \cdot [\sin(\psi^{(i)}) + \cos(\psi^{(i)}) \cdot \psi \cdot e^{\lambda \cdot (t - \tau)}]$$

$$\begin{aligned} &\frac{q_y^{(i)} + q_y \cdot e^{\lambda \cdot t}}{q_z^{(i)} + q_z \cdot e^{\lambda \cdot t}} \\ &= \frac{[q_y^{(i)} + q_y \cdot e^{\lambda \cdot t}]}{[q_z^{(i)} + q_z \cdot e^{\lambda \cdot t}]} \cdot \frac{[q_z^{(i)} - q_z \cdot e^{\lambda \cdot t}]}{[q_z^{(i)} - q_z \cdot e^{\lambda \cdot t}]} \\ &= \frac{q_y^{(i)} \cdot q_z^{(i)} - q_y^{(i)} \cdot q_z \cdot e^{\lambda \cdot t} + q_z^{(i)} \cdot q_y \cdot e^{\lambda \cdot t} - q_y \cdot q_z \cdot e^{2 \cdot \lambda \cdot t}}{(q_z^{(i)})^2 - (q_z)^2 \cdot e^{2 \cdot \lambda \cdot t}} \end{aligned}$$

Assumption $q_y \cdot q_z \approx 0$; $(q_z)^2 \approx 0$ then

$$\begin{aligned} \frac{q_y^{(i)} + q_y \cdot e^{\lambda \cdot t}}{q_z^{(i)} + q_z \cdot e^{\lambda \cdot t}} &\approx \frac{q_y^{(i)} \cdot q_z^{(i)} - q_y^{(i)} \cdot q_z \cdot e^{\lambda \cdot t} + q_z^{(i)} \cdot q_y \cdot e^{\lambda \cdot t}}{(q_z^{(i)})^2} \\ &= \frac{q_y^{(i)}}{q_z^{(i)}} - \frac{q_y^{(i)}}{(q_z^{(i)})^2} \cdot q_z \cdot e^{\lambda \cdot t} + \frac{1}{q_z^{(i)}} \cdot q_y \cdot e^{\lambda \cdot t} \end{aligned}$$

$$\begin{aligned} q_z \cdot \lambda \cdot e^{\lambda \cdot t} = -\frac{1}{2} \cdot k_z \cdot c \cdot &\left[\frac{q_y^{(i)}}{q_z^{(i)}} - \frac{q_y^{(i)}}{(q_z^{(i)})^2} \cdot q_z \cdot e^{\lambda \cdot t} \right. \\ &\left. + \frac{1}{q_z^{(i)}} \cdot q_y \cdot e^{\lambda \cdot t}\right] \cdot A \cdot [\sin(\psi^{(i)}) \\ &+ \cos(\psi^{(i)}) \cdot \psi \cdot e^{\lambda \cdot (t - \tau)}] \end{aligned}$$

$$q_z \cdot \lambda \cdot e^{\lambda \cdot t} = -\frac{1}{2} \cdot k_z \cdot c \cdot \left[\frac{q_y^{(i)}}{q_z^{(i)}} - \frac{q_y^{(i)}}{(q_z^{(i)})^2} \cdot q_z \cdot e^{\lambda \cdot t}\right.$$

$$+\frac{1}{q_z^{(i)}} \cdot q_y \cdot e^{\lambda \cdot t}\Bigg] \cdot A \cdot \sin(\psi^{(i)})$$

$$-\frac{1}{2} \cdot k_z \cdot c \cdot \left[\frac{q_y^{(i)}}{q_z^{(i)}} - \frac{q_y^{(i)}}{(q_z^{(i)})^2} \cdot q_z \cdot e^{\lambda \cdot t}\right.$$

$$\left. + \frac{1}{q_z^{(i)}} \cdot q_y \cdot e^{\lambda \cdot t}\right] \cdot A \cdot \cos(\psi^{(i)}) \cdot \psi \cdot e^{\lambda \cdot (t-\tau)}$$

$$\begin{aligned} q_z \cdot \lambda \cdot e^{\lambda \cdot t} = & -\frac{1}{2} \cdot k_z \cdot c \cdot \frac{q_y^{(i)}}{q_z^{(i)}} \cdot A \cdot \sin(\psi^{(i)}) \\ & + \frac{1}{2} \cdot k_z \cdot c \cdot \frac{q_y^{(i)}}{(q_z^{(i)})^2} \cdot q_z \cdot e^{\lambda \cdot t} \cdot A \cdot \sin(\psi^{(i)}) \\ & - \frac{1}{2} \cdot k_z \cdot c \cdot \frac{1}{q_z^{(i)}} \cdot q_y \cdot e^{\lambda \cdot t} \cdot A \cdot \sin(\psi^{(i)}) \\ & - \frac{1}{2} \cdot k_z \cdot c \cdot [\frac{q_y^{(i)}}{q_z^{(i)}} \cdot \psi - \frac{q_y^{(i)}}{(q_z^{(i)})^2} \cdot q_z \cdot \psi \cdot e^{\lambda \cdot t} \\ & + \frac{1}{q_z^{(i)}} \cdot q_y \cdot \psi \cdot e^{\lambda \cdot t}] \cdot A \cdot \cos(\psi^{(i)}) \cdot e^{\lambda \cdot (t-\tau)} \end{aligned}$$

Assumption $q_z \cdot \psi \approx 0; q_y \cdot \psi \approx 0$ then

$$\begin{aligned} q_z \cdot \lambda \cdot e^{\lambda \cdot t} = & -\frac{1}{2} \cdot k_z \cdot c \cdot \frac{q_y^{(i)}}{q_z^{(i)}} \cdot A \cdot \sin(\psi^{(i)}) \\ & + \frac{1}{2} \cdot k_z \cdot c \cdot \frac{q_y^{(i)}}{(q_z^{(i)})^2} \cdot q_z \cdot e^{\lambda \cdot t} \cdot A \cdot \sin(\psi^{(i)}) \\ & - \frac{1}{2} \cdot k_z \cdot c \cdot \frac{1}{q_z^{(i)}} \cdot q_y \cdot e^{\lambda \cdot t} \cdot A \cdot \sin(\psi^{(i)}) \\ & - \frac{1}{2} \cdot k_z \cdot c \cdot [\frac{q_y^{(i)}}{q_z^{(i)}} \cdot \psi] \cdot A \cdot \cos(\psi^{(i)}) \cdot e^{\lambda t} \cdot e^{-\lambda \cdot \tau} \end{aligned}$$

At fixed point $\frac{dq_z}{dt} = 0 \Rightarrow -\frac{1}{2} \cdot k_z \cdot c \cdot \frac{q_y^{(i)}}{q_z^{(i)}} \cdot A \cdot \sin(\psi^{(i)}) = 0$ then

$$\begin{aligned} q_z \cdot \lambda \cdot e^{\lambda \cdot t} = & \frac{1}{2} \cdot k_z \cdot c \cdot \frac{q_y^{(i)}}{(q_z^{(i)})^2} \cdot q_z \cdot e^{\lambda \cdot t} \cdot A \cdot \sin(\psi^{(i)}) \\ & - \frac{1}{2} \cdot k_z \cdot c \cdot \frac{1}{q_z^{(i)}} \cdot q_y \cdot e^{\lambda \cdot t} \cdot A \cdot \sin(\psi^{(i)}) \\ & - \frac{1}{2} \cdot k_z \cdot c \cdot [\frac{q_y^{(i)}}{q_z^{(i)}} \cdot \psi] \cdot A \cdot \cos(\psi^{(i)}) \cdot e^{\lambda t} \cdot e^{-\lambda \cdot \tau} \end{aligned}$$

We divide the two sides of the above equation by $e^{\lambda \cdot t}$ term.

$$\begin{aligned} q_z \cdot \lambda = & \frac{1}{2} \cdot k_z \cdot c \cdot \frac{q_y^{(i)}}{(q_z^{(i)})^2} \cdot q_z \cdot A \cdot \sin(\psi^{(i)}) \\ & - \frac{1}{2} \cdot k_z \cdot c \cdot \frac{1}{q_z^{(i)}} \cdot q_y \cdot A \cdot \sin(\psi^{(i)}) \\ & - \frac{1}{2} \cdot k_z \cdot c \cdot [\frac{q_y^{(i)}}{q_z^{(i)}} \cdot \psi] \cdot A \cdot \cos(\psi^{(i)}) \cdot e^{-\lambda \cdot \tau} \end{aligned}$$

$$\begin{aligned} & - \frac{1}{2} \cdot k_z \cdot c \cdot \frac{q_y^{(i)}}{q_z^{(i)}} \cdot A \cdot \cos(\psi^{(i)}) \cdot e^{-\lambda \cdot \tau} \cdot \psi \\ & - \frac{1}{2} \cdot k_z \cdot c \cdot \frac{1}{q_z^{(i)}} \cdot A \cdot \sin(\psi^{(i)}) \cdot q_y \\ & + \frac{1}{2} \cdot k_z \cdot c \cdot \frac{q_y^{(i)}}{(q_z^{(i)})^2} \cdot A \cdot \sin(\psi^{(i)}) \cdot q_z - q_z \cdot \lambda = 0 \end{aligned}$$

We define the following global parameters for simplicity

$$\Xi_{41} = -\frac{1}{2} \cdot k_z \cdot c \cdot \frac{q_y^{(i)}}{q_z^{(i)}} \cdot A \cdot \cos(\psi^{(i)})$$

$$\Xi_{42} = 0; \ \Xi_{43} = -\frac{1}{2 \cdot q_z^{(i)}} \cdot k_z \cdot c \cdot A \cdot \sin(\psi^{(i)})$$

$$\Xi_{44} = \frac{1}{2} \cdot k_z \cdot c \cdot \frac{q_y^{(i)}}{(q_z^{(i)})^2} \cdot A \cdot \sin(\psi^{(i)})$$

$$\Xi_{41} \cdot e^{-\lambda \cdot \tau} \cdot \psi + \Xi_{42} \cdot q_x + \Xi_{43} \cdot q_y + \Xi_{44} \cdot q_z - \lambda \cdot q_z = 0$$

We can summary our Ion channel laser perturbed system arbitrarily small increments equations:

$$\Xi_{11} \cdot e^{-\lambda \cdot \tau} \cdot \psi - \psi \cdot \lambda + \Xi_{12} \cdot q_x + \Xi_{13} \cdot q_y + \Xi_{14} \cdot q_z = 0$$
$$\Xi_{21} \cdot \psi \cdot e^{-\lambda \cdot \tau} + \Xi_{22} \cdot q_x - q_x \cdot \lambda + \Xi_{23} \cdot q_y + \Xi_{24} \cdot q_z = 0$$

$$\Xi_{31} \cdot \psi \cdot e^{-\lambda \cdot \tau} + \Xi_{32} \cdot q_x + \Xi_{33} \cdot q_y - q_y \cdot \lambda + \Xi_{34} \cdot q_z = 0$$
$$\Xi_{41} \cdot e^{-\lambda \cdot \tau} \cdot \psi + \Xi_{42} \cdot q_x + \Xi_{43} \cdot q_y + \Xi_{44} \cdot q_z - \lambda \cdot q_z = 0$$

The small increments Jacobian of Ion channel laser perturbed system is as follow:

$$(A - \lambda \cdot I) = \begin{pmatrix} (\Xi_{11} \cdot e^{-\lambda \cdot \tau} - \lambda) & \Xi_{12} & \Xi_{13} & \Xi_{14} \\ \Xi_{21} \cdot e^{-\lambda \cdot \tau} & (\Xi_{22} - \lambda) & \Xi_{23} & \Xi_{24} \\ \Xi_{31} \cdot e^{-\lambda \cdot \tau} & 0 & (\Xi_{33} - \lambda) & \Xi_{34} \\ \Xi_{41} \cdot e^{-\lambda \cdot \tau} & 0 & \Xi_{43} & (\Xi_{44} - \lambda) \end{pmatrix}$$

$$\det(A - \lambda \cdot I) = 0\ ;\ D = D(\lambda, \tau)$$

$$\det(A - \lambda \cdot I) = (\Xi_{11} \cdot e^{-\lambda \cdot \tau} - \lambda) \cdot \det \begin{pmatrix} (\Xi_{22} - \lambda) & \Xi_{23} & \Xi_{24} \\ 0 & (\Xi_{33} - \lambda) & \Xi_{34} \\ 0 & \Xi_{43} & (\Xi_{44} - \lambda) \end{pmatrix}$$
$$- \Xi_{12} \cdot \det \begin{pmatrix} \Xi_{21} \cdot e^{-\lambda \cdot \tau} & \Xi_{23} & \Xi_{24} \\ \Xi_{31} \cdot e^{-\lambda \cdot \tau} & (\Xi_{33} - \lambda) & \Xi_{34} \\ \Xi_{41} \cdot e^{-\lambda \cdot \tau} & \Xi_{43} & (\Xi_{44} - \lambda) \end{pmatrix}$$
$$+ \Xi_{13} \cdot \det \begin{pmatrix} \Xi_{21} \cdot e^{-\lambda \cdot \tau} & (\Xi_{22} - \lambda) & \Xi_{24} \\ \Xi_{31} \cdot e^{-\lambda \cdot \tau} & 0 & \Xi_{34} \\ \Xi_{41} \cdot e^{-\lambda \cdot \tau} & 0 & (\Xi_{44} - \lambda) \end{pmatrix}$$
$$- \Xi_{14} \cdot \det \begin{pmatrix} \Xi_{21} \cdot e^{-\lambda \cdot \tau} & (\Xi_{22} - \lambda) & \Xi_{23} \\ \Xi_{31} \cdot e^{-\lambda \cdot \tau} & 0 & (\Xi_{33} - \lambda) \\ \Xi_{41} \cdot e^{-\lambda \cdot \tau} & 0 & \Xi_{43} \end{pmatrix}$$

No.1

$$\det \begin{pmatrix} (\Xi_{22} - \lambda) & \Xi_{23} & \Xi_{24} \\ 0 & (\Xi_{33} - \lambda) & \Xi_{34} \\ 0 & \Xi_{43} & (\Xi_{44} - \lambda) \end{pmatrix}$$
$$= (\Xi_{22} - \lambda) \cdot \det \begin{pmatrix} (\Xi_{33} - \lambda) & \Xi_{34} \\ \Xi_{43} & (\Xi_{44} - \lambda) \end{pmatrix}$$
$$= (\Xi_{22} - \lambda) \cdot \{(\Xi_{33} - \lambda) \cdot (\Xi_{44} - \lambda) - \Xi_{43} \cdot \Xi_{34}\}$$

$$\det \begin{pmatrix} (\Xi_{22} - \lambda) & \Xi_{23} & \Xi_{24} \\ 0 & (\Xi_{33} - \lambda) & \Xi_{34} \\ 0 & \Xi_{43} & (\Xi_{44} - \lambda) \end{pmatrix}$$
$$= (\Xi_{22} - \lambda) \cdot \{(\Xi_{33} - \lambda) \cdot (\Xi_{44} - \lambda) - \Xi_{43} \cdot \Xi_{34}\}$$
$$= (\Xi_{22} - \lambda) \cdot \{\lambda^2 - (\Xi_{33} + \Xi_{44}) \cdot \lambda + (\Xi_{33} \cdot \Xi_{44} - \Xi_{43} \cdot \Xi_{34})\}$$

$$\det \begin{pmatrix} (\Xi_{22} - \lambda) & \Xi_{23} & \Xi_{24} \\ 0 & (\Xi_{33} - \lambda) & \Xi_{34} \\ 0 & \Xi_{43} & (\Xi_{44} - \lambda) \end{pmatrix}$$
$$= (\Xi_{22} - \lambda) \cdot \{\lambda^2 - (\Xi_{33} + \Xi_{44}) \cdot \lambda + (\Xi_{33} \cdot \Xi_{44} - \Xi_{43} \cdot \Xi_{34})\}$$

$$= \lambda^2 \cdot \Xi_{22} - (\Xi_{33} + \Xi_{44}) \cdot \Xi_{22} \cdot \lambda + (\Xi_{33} \cdot \Xi_{44} - \Xi_{43} \cdot \Xi_{34}) \cdot \Xi_{22} - \lambda^3 + (\Xi_{33} + \Xi_{44}) \cdot \lambda^2 - (\Xi_{33} \cdot \Xi_{44} - \Xi_{43} \cdot \Xi_{34}) \cdot \lambda$$

$$\det\begin{pmatrix} (\Xi_{22} - \lambda) & \Xi_{23} & \Xi_{24} \\ 0 & (\Xi_{33} - \lambda) & \Xi_{34} \\ 0 & \Xi_{43} & (\Xi_{44} - \lambda) \end{pmatrix} = -\lambda^3 + [\Xi_{22} + (\Xi_{33} + \Xi_{44})] \cdot \lambda^2 - [(\Xi_{33} + \Xi_{44}) \cdot \Xi_{22} + (\Xi_{33} \cdot \Xi_{44} - \Xi_{43} \cdot \Xi_{34})] \cdot \lambda + (\Xi_{33} \cdot \Xi_{44} - \Xi_{43} \cdot \Xi_{34}) \cdot \Xi_{22}$$

We define for simplicity the following system global parameters:

$$\Upsilon_1 = \Xi_{22} + (\Xi_{33} + \Xi_{44})$$
$$\Upsilon_2 = -[(\Xi_{33} + \Xi_{44}) \cdot \Xi_{22} + (\Xi_{33} \cdot \Xi_{44} - \Xi_{43} \cdot \Xi_{34})]$$
$$\Upsilon_3 = (\Xi_{33} \cdot \Xi_{44} - \Xi_{43} \cdot \Xi_{34}) \cdot \Xi_{22}$$

$$\det\begin{pmatrix} (\Xi_{22} - \lambda) & \Xi_{23} & \Xi_{24} \\ 0 & (\Xi_{33} - \lambda) & \Xi_{34} \\ 0 & \Xi_{43} & (\Xi_{44} - \lambda) \end{pmatrix} = -\lambda^3 + \Upsilon_1 \cdot \lambda^2 + \Upsilon_2 \cdot \lambda + \Upsilon_3$$

No.2

$$\det\begin{pmatrix} \Xi_{21} \cdot e^{-\lambda \cdot \tau} & \Xi_{23} & \Xi_{24} \\ \Xi_{31} \cdot e^{-\lambda \cdot \tau} & (\Xi_{33} - \lambda) & \Xi_{34} \\ \Xi_{41} \cdot e^{-\lambda \cdot \tau} & \Xi_{43} & (\Xi_{44} - \lambda) \end{pmatrix} = \Xi_{21} \cdot e^{-\lambda \cdot \tau} \cdot \det\begin{pmatrix} (\Xi_{33} - \lambda) & \Xi_{34} \\ \Xi_{43} & (\Xi_{44} - \lambda) \end{pmatrix} - \Xi_{23} \cdot \det\begin{pmatrix} \Xi_{31} \cdot e^{-\lambda \cdot \tau} & \Xi_{34} \\ \Xi_{41} \cdot e^{-\lambda \cdot \tau} & (\Xi_{44} - \lambda) \end{pmatrix} + \Xi_{24} \cdot \det\begin{pmatrix} \Xi_{31} \cdot e^{-\lambda \cdot \tau} & (\Xi_{33} - \lambda) \\ \Xi_{41} \cdot e^{-\lambda \cdot \tau} & \Xi_{43} \end{pmatrix}$$

$$\det\begin{pmatrix} \Xi_{21} \cdot e^{-\lambda \cdot \tau} & \Xi_{23} & \Xi_{24} \\ \Xi_{31} \cdot e^{-\lambda \cdot \tau} & (\Xi_{33} - \lambda) & \Xi_{34} \\ \Xi_{41} \cdot e^{-\lambda \cdot \tau} & \Xi_{43} & (\Xi_{44} - \lambda) \end{pmatrix} = \Xi_{21} \cdot e^{-\lambda \cdot \tau} \cdot [(\Xi_{33} - \lambda) \cdot (\Xi_{44} - \lambda) - \Xi_{43} \cdot \Xi_{34}] - \Xi_{23} \cdot [\Xi_{31} \cdot e^{-\lambda \cdot \tau} \cdot (\Xi_{44} - \lambda) - \Xi_{41} \cdot e^{-\lambda \cdot \tau} \cdot \Xi_{34}] + \Xi_{24} \cdot [\Xi_{31} \cdot e^{-\lambda \cdot \tau} \cdot \Xi_{43} - \Xi_{41} \cdot e^{-\lambda \cdot \tau} \cdot (\Xi_{33} - \lambda)]$$

$$\det\begin{pmatrix} \Xi_{21}\cdot e^{-\lambda\cdot\tau} & \Xi_{23} & \Xi_{24} \\ \Xi_{31}\cdot e^{-\lambda\cdot\tau} & (\Xi_{33}-\lambda) & \Xi_{34} \\ \Xi_{41}\cdot e^{-\lambda\cdot\tau} & \Xi_{43} & (\Xi_{44}-\lambda) \end{pmatrix}$$
$$= \Xi_{21}\cdot e^{-\lambda\cdot\tau}\cdot[\Xi_{33}\cdot\Xi_{44}-\Xi_{43}\cdot\Xi_{34}-\lambda\cdot(\Xi_{33}+\Xi_{44})+\lambda^2]$$
$$-\Xi_{23}\cdot[(\Xi_{31}\cdot\Xi_{44}-\Xi_{41}\cdot\Xi_{34})-\Xi_{31}\cdot\lambda]\cdot e^{-\lambda\cdot\tau}$$
$$+\Xi_{24}\cdot[(\Xi_{31}\cdot\Xi_{43}-\Xi_{41}\cdot\Xi_{33})+\Xi_{41}\cdot\lambda]\cdot e^{-\lambda\cdot\tau}$$

$$\det\begin{pmatrix} \Xi_{21}\cdot e^{-\lambda\cdot\tau} & \Xi_{23} & \Xi_{24} \\ \Xi_{31}\cdot e^{-\lambda\cdot\tau} & (\Xi_{33}-\lambda) & \Xi_{34} \\ \Xi_{41}\cdot e^{-\lambda\cdot\tau} & \Xi_{43} & (\Xi_{44}-\lambda) \end{pmatrix}$$
$$= \{\Xi_{21}\cdot[\Xi_{33}\cdot\Xi_{44}-\Xi_{43}\cdot\Xi_{34}-\lambda\cdot(\Xi_{33}+\Xi_{44})+\lambda^2]$$
$$-\Xi_{23}\cdot[(\Xi_{31}\cdot\Xi_{44}-\Xi_{41}\cdot\Xi_{34})-\Xi_{31}\cdot\lambda]$$
$$+\Xi_{24}\cdot[(\Xi_{31}\cdot\Xi_{43}-\Xi_{41}\cdot\Xi_{33})+\Xi_{41}\cdot\lambda]\}\cdot e^{-\lambda\cdot\tau}$$

$$\det\begin{pmatrix} \Xi_{21}\cdot e^{-\lambda\cdot\tau} & \Xi_{23} & \Xi_{24} \\ \Xi_{31}\cdot e^{-\lambda\cdot\tau} & (\Xi_{33}-\lambda) & \Xi_{34} \\ \Xi_{41}\cdot e^{-\lambda\cdot\tau} & \Xi_{43} & (\Xi_{44}-\lambda) \end{pmatrix}$$
$$= \{\Xi_{21}\cdot(\Xi_{33}\cdot\Xi_{44}-\Xi_{43}\cdot\Xi_{34})-\lambda\cdot\Xi_{21}\cdot(\Xi_{33}+\Xi_{44})$$
$$+\Xi_{21}\cdot\lambda^2-\Xi_{23}\cdot(\Xi_{31}\cdot\Xi_{44}-\Xi_{41}\cdot\Xi_{34})+\Xi_{23}\cdot\Xi_{31}\cdot\lambda$$
$$+\Xi_{24}\cdot(\Xi_{31}\cdot\Xi_{43}-\Xi_{41}\cdot\Xi_{33})+\Xi_{24}\cdot\Xi_{41}\cdot\lambda\}\cdot e^{-\lambda\cdot\tau}$$

$$\det\begin{pmatrix} \Xi_{21}\cdot e^{-\lambda\cdot\tau} & \Xi_{23} & \Xi_{24} \\ \Xi_{31}\cdot e^{-\lambda\cdot\tau} & (\Xi_{33}-\lambda) & \Xi_{34} \\ \Xi_{41}\cdot e^{-\lambda\cdot\tau} & \Xi_{43} & (\Xi_{44}-\lambda) \end{pmatrix}$$
$$= \{\Xi_{21}\cdot\lambda^2+[\Xi_{23}\cdot\Xi_{31}+\Xi_{24}\cdot\Xi_{41}-\Xi_{21}\cdot(\Xi_{33}+\Xi_{44})]\cdot\lambda$$
$$+\Xi_{21}\cdot(\Xi_{33}\cdot\Xi_{44}-\Xi_{43}\cdot\Xi_{34})-\Xi_{23}\cdot(\Xi_{31}\cdot\Xi_{44}-\Xi_{41}\cdot\Xi_{34})$$
$$+\Xi_{24}\cdot(\Xi_{31}\cdot\Xi_{43}-\Xi_{41}\cdot\Xi_{33})\}\cdot e^{-\lambda\cdot\tau}$$

$$\Upsilon_4=\Xi_{21};\ \Upsilon_5=\Xi_{23}\cdot\Xi_{31}+\Xi_{24}\cdot\Xi_{41}-\Xi_{21}\cdot(\Xi_{33}+\Xi_{44})$$
$$\Upsilon_6=\Xi_{21}\cdot(\Xi_{33}\cdot\Xi_{44}-\Xi_{43}\cdot\Xi_{34})-\Xi_{23}\cdot(\Xi_{31}\cdot\Xi_{44}-\Xi_{41}\cdot\Xi_{34})$$
$$+\Xi_{24}\cdot(\Xi_{31}\cdot\Xi_{43}-\Xi_{41}\cdot\Xi_{33})$$

$$\det\begin{pmatrix} \Xi_{21}\cdot e^{-\lambda\cdot\tau} & \Xi_{23} & \Xi_{24} \\ \Xi_{31}\cdot e^{-\lambda\cdot\tau} & (\Xi_{33}-\lambda) & \Xi_{34} \\ \Xi_{41}\cdot e^{-\lambda\cdot\tau} & \Xi_{43} & (\Xi_{44}-\lambda) \end{pmatrix} = \{\Upsilon_4\cdot\lambda^2+\Upsilon_5\cdot\lambda+\Upsilon_6\}\cdot e^{-\lambda\cdot\tau}$$

No.3

$$\det\begin{pmatrix} \Xi_{21}\cdot e^{-\lambda\cdot\tau} & (\Xi_{22}-\lambda) & \Xi_{24} \\ \Xi_{31}\cdot e^{-\lambda\cdot\tau} & 0 & \Xi_{34} \\ \Xi_{41}\cdot e^{-\lambda\cdot\tau} & 0 & (\Xi_{44}-\lambda) \end{pmatrix}$$
$$= \Xi_{21}\cdot e^{-\lambda\cdot\tau}\cdot\det\begin{pmatrix} 0 & \Xi_{34} \\ 0 & (\Xi_{44}-\lambda) \end{pmatrix}$$
$$-(\Xi_{22}-\lambda)\cdot\det\begin{pmatrix} \Xi_{31}\cdot e^{-\lambda\cdot\tau} & \Xi_{34} \\ \Xi_{41}\cdot e^{-\lambda\cdot\tau} & (\Xi_{44}-\lambda) \end{pmatrix}$$
$$+\Xi_{24}\cdot\det\begin{pmatrix} \Xi_{31}\cdot e^{-\lambda\cdot\tau} & 0 \\ \Xi_{41}\cdot e^{-\lambda\cdot\tau} & 0 \end{pmatrix}$$

$$\det\begin{pmatrix} \Xi_{21}\cdot e^{-\lambda\cdot\tau} & (\Xi_{22}-\lambda) & \Xi_{24} \\ \Xi_{31}\cdot e^{-\lambda\cdot\tau} & 0 & \Xi_{34} \\ \Xi_{41}\cdot e^{-\lambda\cdot\tau} & 0 & (\Xi_{44}-\lambda) \end{pmatrix} = -(\Xi_{22}-\lambda)\cdot[\Xi_{31}\cdot e^{-\lambda\cdot\tau}\cdot(\Xi_{44}-\lambda)-\Xi_{41}\cdot e^{-\lambda\cdot\tau}\cdot\Xi_{34}]$$

$$\det\begin{pmatrix} \Xi_{21}\cdot e^{-\lambda\cdot\tau} & (\Xi_{22}-\lambda) & \Xi_{24} \\ \Xi_{31}\cdot e^{-\lambda\cdot\tau} & 0 & \Xi_{34} \\ \Xi_{41}\cdot e^{-\lambda\cdot\tau} & 0 & (\Xi_{44}-\lambda) \end{pmatrix} = -(\Xi_{22}-\lambda)\cdot[\Xi_{31}\cdot\Xi_{44}-\Xi_{41}\cdot\Xi_{34}-\Xi_{31}\cdot\lambda]\cdot e^{-\lambda\cdot\tau}$$

$$\det\begin{pmatrix} \Xi_{21}\cdot e^{-\lambda\cdot\tau} & (\Xi_{22}-\lambda) & \Xi_{24} \\ \Xi_{31}\cdot e^{-\lambda\cdot\tau} & 0 & \Xi_{34} \\ \Xi_{41}\cdot e^{-\lambda\cdot\tau} & 0 & (\Xi_{44}-\lambda) \end{pmatrix}$$
$$= \{-\Xi_{22}\cdot(\Xi_{31}\cdot\Xi_{44}-\Xi_{41}\cdot\Xi_{34})+\Xi_{22}\cdot\Xi_{31}\cdot\lambda + (\Xi_{31}\cdot\Xi_{44}-\Xi_{41}\cdot\Xi_{34})\cdot\lambda-\Xi_{31}\cdot\lambda^2\}\cdot e^{-\lambda\cdot\tau}$$

$$\det\begin{pmatrix} \Xi_{21}\cdot e^{-\lambda\cdot\tau} & (\Xi_{22}-\lambda) & \Xi_{24} \\ \Xi_{31}\cdot e^{-\lambda\cdot\tau} & 0 & \Xi_{34} \\ \Xi_{41}\cdot e^{-\lambda\cdot\tau} & 0 & (\Xi_{44}-\lambda) \end{pmatrix}$$
$$= \{-\Xi_{31}\cdot\lambda^2+[\Xi_{22}\cdot\Xi_{31}+(\Xi_{31}\cdot\Xi_{44}-\Xi_{41}\cdot\Xi_{34})]\cdot\lambda - \Xi_{22}\cdot(\Xi_{31}\cdot\Xi_{44}-\Xi_{41}\cdot\Xi_{34})\}\cdot e^{-\lambda\cdot\tau}$$

$$\Upsilon_7 = -\Xi_{31};\ \Upsilon_8 = \Xi_{22}\cdot\Xi_{31}+(\Xi_{31}\cdot\Xi_{44}-\Xi_{41}\cdot\Xi_{34})$$
$$\Upsilon_9 = -\Xi_{22}\cdot(\Xi_{31}\cdot\Xi_{44}-\Xi_{41}\cdot\Xi_{34})$$

$$\det\begin{pmatrix} \Xi_{21}\cdot e^{-\lambda\cdot\tau} & (\Xi_{22}-\lambda) & \Xi_{24} \\ \Xi_{31}\cdot e^{-\lambda\cdot\tau} & 0 & \Xi_{34} \\ 0 & 0 & (\Xi_{44}-\lambda) \end{pmatrix} = \{\Upsilon_7\cdot\lambda^2+\Upsilon_8\cdot\lambda+\Upsilon_9\}\cdot e^{-\lambda\cdot\tau}$$

No.4

$$\det\begin{pmatrix} \Xi_{21}\cdot e^{-\lambda\cdot\tau} & (\Xi_{22}-\lambda) & \Xi_{23} \\ \Xi_{31}\cdot e^{-\lambda\cdot\tau} & 0 & (\Xi_{33}-\lambda) \\ \Xi_{41}\cdot e^{-\lambda\cdot\tau} & 0 & \Xi_{43} \end{pmatrix}$$

$$= \Xi_{21}\cdot e^{-\lambda\cdot\tau}\cdot\det\begin{pmatrix} 0 & (\Xi_{33}-\lambda) \\ 0 & \Xi_{43} \end{pmatrix}$$

$$-(\Xi_{22}-\lambda)\cdot\det\begin{pmatrix} \Xi_{31}\cdot e^{-\lambda\cdot\tau} & (\Xi_{33}-\lambda) \\ \Xi_{41}\cdot e^{-\lambda\cdot\tau} & \Xi_{43} \end{pmatrix}$$

$$+\Xi_{23}\cdot\det\begin{pmatrix} \Xi_{31}\cdot e^{-\lambda\cdot\tau} & 0 \\ \Xi_{41}\cdot e^{-\lambda\cdot\tau} & 0 \end{pmatrix}$$

$$\det\begin{pmatrix} \Xi_{21}\cdot e^{-\lambda\cdot\tau} & (\Xi_{22}-\lambda) & \Xi_{23} \\ \Xi_{31}\cdot e^{-\lambda\cdot\tau} & 0 & (\Xi_{33}-\lambda) \\ \Xi_{41}\cdot e^{-\lambda\cdot\tau} & 0 & \Xi_{43} \end{pmatrix} = -(\Xi_{22}-\lambda)\cdot[\Xi_{31}\cdot\Xi_{43} - \Xi_{41}\cdot(\Xi_{33}-\lambda)]\cdot e^{-\lambda\cdot\tau}$$

$$\det\begin{pmatrix} \Xi_{21}\cdot e^{-\lambda\cdot\tau} & (\Xi_{22}-\lambda) & \Xi_{23} \\ \Xi_{31}\cdot e^{-\lambda\cdot\tau} & 0 & (\Xi_{33}-\lambda) \\ \Xi_{41}\cdot e^{-\lambda\cdot\tau} & 0 & \Xi_{43} \end{pmatrix} = -(\Xi_{22}-\lambda)\cdot[(\Xi_{31}\cdot\Xi_{43} - \Xi_{41}\cdot\Xi_{33}) + \Xi_{41}\cdot\lambda]\cdot e^{-\lambda\cdot\tau}$$

$$\det\begin{pmatrix} \Xi_{21}\cdot e^{-\lambda\cdot\tau} & (\Xi_{22}-\lambda) & \Xi_{23} \\ \Xi_{31}\cdot e^{-\lambda\cdot\tau} & 0 & (\Xi_{33}-\lambda) \\ \Xi_{41}\cdot e^{-\lambda\cdot\tau} & 0 & \Xi_{43} \end{pmatrix} = \{\Xi_{41}\cdot\lambda^2 + [(\Xi_{31}\cdot\Xi_{43} - \Xi_{41}\cdot\Xi_{33}) - \Xi_{22}\cdot\Xi_{41}]\cdot\lambda - \Xi_{22}\cdot(\Xi_{31}\cdot\Xi_{43} - \Xi_{41}\cdot\Xi_{33})\}\cdot e^{-\lambda\cdot\tau}$$

$$\Upsilon_{10} = \Xi_{41};\ \Upsilon_{11} = (\Xi_{31}\cdot\Xi_{43} - \Xi_{41}\cdot\Xi_{33}) - \Xi_{22}\cdot\Xi_{41}$$
$$\Upsilon_{12} = -\Xi_{22}\cdot(\Xi_{31}\cdot\Xi_{43} - \Xi_{41}\cdot\Xi_{33})$$

$$\det\begin{pmatrix} \Xi_{21}\cdot e^{-\lambda\cdot\tau} & (\Xi_{22}-\lambda) & \Xi_{23} \\ \Xi_{31}\cdot e^{-\lambda\cdot\tau} & 0 & (\Xi_{33}-\lambda) \\ \Xi_{41}\cdot e^{-\lambda\cdot\tau} & 0 & \Xi_{43} \end{pmatrix} = \{\Upsilon_{10}\cdot\lambda^2 + \Upsilon_{11}\cdot\lambda + \Upsilon_{12}\}\cdot e^{-\lambda\cdot\tau}$$

We can summary our last results in the following table (Table 3.4).

$$\det(A-\lambda\cdot I) = (\Xi_{11}\cdot e^{-\lambda\cdot\tau} - \lambda)\cdot(-\lambda^3 + \Upsilon_1\cdot\lambda^2 + \Upsilon_2\cdot\lambda + \Upsilon_3) - \Xi_{12}\cdot(\Upsilon_4\cdot\lambda^2 + \Upsilon_5\cdot\lambda + \Upsilon_6)\cdot e^{-\lambda\cdot\tau}$$

Table 3.4 System's Jacobian matrix determinant elements (with delay parameter τ)

No	Determinant	Equivalent expression
1	$\det\begin{pmatrix} (\Xi_{22}-\lambda) & \Xi_{23} & \Xi_{24} \\ 0 & (\Xi_{33}-\lambda) & \Xi_{34} \\ 0 & \Xi_{43} & (\Xi_{44}-\lambda) \end{pmatrix}$	$-\lambda^3 + \Upsilon_1\cdot\lambda^2 + \Upsilon_2\cdot\lambda + \Upsilon_3$
2	$\det\begin{pmatrix} \Xi_{21}\cdot e^{-\lambda\cdot\tau} & \Xi_{23} & \Xi_{24} \\ \Xi_{31}\cdot e^{-\lambda\cdot\tau} & (\Xi_{33}-\lambda) & \Xi_{34} \\ \Xi_{41}\cdot e^{-\lambda\cdot\tau} & \Xi_{43} & (\Xi_{44}-\lambda) \end{pmatrix}$	$(\Upsilon_4\cdot\lambda^2 + \Upsilon_5\cdot\lambda + \Upsilon_6)\cdot e^{-\lambda\cdot\tau}$
3	$\det\begin{pmatrix} \Xi_{21}\cdot e^{-\lambda\cdot\tau} & (\Xi_{22}-\lambda) & \Xi_{24} \\ \Xi_{31}\cdot e^{-\lambda\cdot\tau} & 0 & \Xi_{34} \\ \Xi_{41}\cdot e^{-\lambda\cdot\tau} & 0 & (\Xi_{44}-\lambda) \end{pmatrix}$	$(\Upsilon_7\cdot\lambda^2 + \Upsilon_8\cdot\lambda + \Upsilon_9)\cdot e^{-\lambda\cdot\tau}$
4	$\det\begin{pmatrix} \Xi_{21}\cdot e^{-\lambda\cdot\tau} & (\Xi_{22}-\lambda) & \Xi_{23} \\ \Xi_{31}\cdot e^{-\lambda\cdot\tau} & 0 & (\Xi_{33}-\lambda) \\ \Xi_{41}\cdot e^{-\lambda\cdot\tau} & 0 & \Xi_{43} \end{pmatrix}$	$(\Upsilon_{10}\cdot\lambda^2 + \Upsilon_{11}\cdot\lambda + \Upsilon_{12})\cdot e^{-\lambda\cdot\tau}$

$$
\begin{aligned}
&+ \Xi_{13}\cdot(\Upsilon_7\cdot\lambda^2 + \Upsilon_8\cdot\lambda + \Upsilon_9)\cdot e^{-\lambda\cdot\tau} \\
&- \Xi_{14}\cdot(\Upsilon_{10}\cdot\lambda^2 + \Upsilon_{11}\cdot\lambda + \Upsilon_{12})\cdot e^{-\lambda\cdot\tau}
\end{aligned}
$$

$$
\begin{aligned}
\det(A-\lambda\cdot I) = {} & \Xi_{11}\cdot(-\lambda^3 + \Upsilon_1\cdot\lambda^2 + \Upsilon_2\cdot\lambda + \Upsilon_3) \\
& \cdot e^{-\lambda\cdot\tau} - \lambda\cdot(-\lambda^3 + \Upsilon_1\cdot\lambda^2 + \Upsilon_2\cdot\lambda + \Upsilon_3) \\
& - (\Xi_{12}\cdot\Upsilon_4\cdot\lambda^2 + \Xi_{12}\cdot\Upsilon_5\cdot\lambda + \Xi_{12}\cdot\Upsilon_6)\cdot e^{-\lambda\cdot\tau} \\
& + (\Xi_{13}\cdot\Upsilon_7\cdot\lambda^2 + \Xi_{13}\cdot\Upsilon_8\cdot\lambda + \Xi_{13}\cdot\Upsilon_9)\cdot e^{-\lambda\cdot\tau} \\
& - (\Xi_{14}\cdot\Upsilon_{10}\cdot\lambda^2 + \Xi_{14}\cdot\Upsilon_{11}\cdot\lambda + \Xi_{14}\cdot\Upsilon_{12})\cdot e^{-\lambda\cdot\tau}
\end{aligned}
$$

$$
\begin{aligned}
\det(A-\lambda\cdot I) = {} & (\lambda^4 - \Upsilon_1\cdot\lambda^3 - \Upsilon_2\cdot\lambda^2 - \Upsilon_3\cdot\lambda) \\
& + (-\Xi_{11}\cdot\lambda^3 + \Xi_{11}\cdot\Upsilon_1\cdot\lambda^2 + \Xi_{11}\cdot\Upsilon_2\cdot\lambda + \Xi_{11}\cdot\Upsilon_3)\cdot e^{-\lambda\cdot\tau} \\
& - (\Xi_{12}\cdot\Upsilon_4\cdot\lambda^2 + \Xi_{12}\cdot\Upsilon_5\cdot\lambda + \Xi_{12}\cdot\Upsilon_6)\cdot e^{-\lambda\cdot\tau} \\
& + (\Xi_{13}\cdot\Upsilon_7\cdot\lambda^2 + \Xi_{13}\cdot\Upsilon_8\cdot\lambda + \Xi_{13}\cdot\Upsilon_9)\cdot e^{-\lambda\cdot\tau} \\
& - (\Xi_{14}\cdot\Upsilon_{10}\cdot\lambda^2 + \Xi_{14}\cdot\Upsilon_{11}\cdot\lambda + \Xi_{14}\cdot\Upsilon_{12})\cdot e^{-\lambda\cdot\tau}
\end{aligned}
$$

$$
\begin{aligned}
\det(A-\lambda\cdot I) = {} & (\lambda^4 - \Upsilon_1\cdot\lambda^3 - \Upsilon_2\cdot\lambda^2 - \Upsilon_3\cdot\lambda) \\
& + (-\Xi_{11}\cdot\lambda^3 + \Xi_{11}\cdot\Upsilon_1\cdot\lambda^2 + \Xi_{11}\cdot\Upsilon_2\cdot\lambda + \Xi_{11}\cdot\Upsilon_3 \\
& - \Xi_{12}\cdot\Upsilon_4\cdot\lambda^2 - \Xi_{12}\cdot\Upsilon_5\cdot\lambda - \Xi_{12}\cdot\Upsilon_6 \\
& + \Xi_{13}\cdot\Upsilon_7\cdot\lambda^2 + \Xi_{13}\cdot\Upsilon_8\cdot\lambda + \Xi_{13}\cdot\Upsilon_9
\end{aligned}
$$

$$- \Xi_{14} \cdot \Upsilon_{10} \cdot \lambda^2 - \Xi_{14} \cdot \Upsilon_{11} \cdot \lambda - \Xi_{14} \cdot \Upsilon_{12}) \cdot e^{-\lambda \cdot \tau}$$

$$\begin{aligned} \det(A - \lambda \cdot I) = {} & (\lambda^4 - \Upsilon_1 \cdot \lambda^3 - \Upsilon_2 \cdot \lambda^2 - \Upsilon_3 \cdot \lambda) \\ & + (-\Xi_{11} \cdot \lambda^3 + [\Xi_{11} \cdot \Upsilon_1 - \Xi_{12} \cdot \Upsilon_4 \\ & + \Xi_{13} \cdot \Upsilon_7 - \Xi_{14} \cdot \Upsilon_{10}] \cdot \lambda^2 + [\Xi_{11} \cdot \Upsilon_2 \\ & - \Xi_{12} \cdot \Upsilon_5 + \Xi_{13} \cdot \Upsilon_8 - \Xi_{14} \cdot \Upsilon_{11}] \cdot \lambda + \Xi_{11} \\ & \cdot \Upsilon_3 - \Xi_{12} \cdot \Upsilon_6 + \Xi_{13} \cdot \Upsilon_9 - \Xi_{14} \cdot \Upsilon_{12}) \cdot e^{-\lambda \cdot \tau} \end{aligned}$$

The characteristics equation for our system is as follow:

$$\begin{aligned} D(\lambda, \tau) = {} & (\lambda^4 - \Upsilon_1 \cdot \lambda^3 - \Upsilon_2 \cdot \lambda^2 - \Upsilon_3 \cdot \lambda) \\ & + (-\Xi_{11} \cdot \lambda^3 + [\Xi_{11} \cdot \Upsilon_1 - \Xi_{12} \cdot \Upsilon_4 \\ & + \Xi_{13} \cdot \Upsilon_7 - \Xi_{14} \cdot \Upsilon_{10}] \cdot \lambda^2 + [\Xi_{11} \cdot \Upsilon_2 \\ & - \Xi_{12} \cdot \Upsilon_5 + \Xi_{13} \cdot \Upsilon_8 - \Xi_{14} \cdot \Upsilon_{11}] \cdot \lambda + \Xi_{11} \\ & \cdot \Upsilon_3 - \Xi_{12} \cdot \Upsilon_6 + \Xi_{13} \cdot \Upsilon_9 - \Xi_{14} \cdot \Upsilon_{12}) \cdot e^{-\lambda \cdot \tau} \end{aligned}$$

We study the occurrence of any possible stability switching resulting from the increase the value of time delay τ for general characteristic equation$D(\lambda, \tau)$. We choose τ parameter then $D(\lambda, \tau) = P_n(\lambda, \tau) + Q_m(\lambda, \tau) \cdot e^{-\lambda \cdot \tau}$. The expression for $P_n(\lambda, \tau)$ is

$$P_n(\lambda, \tau) = \sum_{k=0}^{n} p_k(\tau) \cdot \lambda^k = p_0(\tau) + p_1(\tau) \cdot \lambda + p_2(\tau) \cdot \lambda^2 + \ldots$$

The expression for $Q_m(\lambda, \tau)$ is

$$Q_m(\lambda, \tau) = \sum_{k=0}^{m} q_k(\tau) \cdot \lambda^k = q_0(\tau) + q_1(\tau) \cdot \lambda + q_2(\tau) \cdot \lambda^2 + \ldots$$

$$D(\lambda, \tau) = P_n(\lambda, \tau) + Q_m(\lambda, \tau) \cdot e^{-\lambda \cdot \tau}; n = 4; m = 3; n > m$$

The expression for $P_n(\lambda, \tau)$:

$$\begin{aligned} P_{n=4}(\lambda, \tau) & = \sum_{k=0}^{n=4} p_k(\tau) \cdot \lambda^k \\ & = p_0(\tau) + p_1(\tau) \cdot \lambda + p_2(\tau) \cdot \lambda^2 \\ & \quad + p_3(\tau) \cdot \lambda^3 + p_4(\tau) \cdot \lambda^4 \end{aligned}$$

$$p_0(\tau) = 0; p_1(\tau) = -\Upsilon_3; p_2(\tau) = -\Upsilon_2; p_3(\tau) = -\Upsilon_1; p_4(\tau) = 1$$

The expression for $Q_m(\lambda, \tau)$:

$$Q_{m=3}(\lambda,\tau) = \sum_{k=0}^{m=3} q_k(\tau)\cdot\lambda^k = q_0(\tau) + q_1(\tau)\cdot\lambda + q_2(\tau)\cdot\lambda^2 + q_3(\tau)\cdot\lambda^3$$

$$q_0(\tau) = \Xi_{11}\cdot\Upsilon_3 - \Xi_{12}\cdot\Upsilon_6 + \Xi_{13}\cdot\Upsilon_9 - \Xi_{14}\cdot\Upsilon_{12}$$
$$q_1(\tau) = \Xi_{11}\cdot\Upsilon_2 - \Xi_{12}\cdot\Upsilon_5 + \Xi_{13}\cdot\Upsilon_8 - \Xi_{14}\cdot\Upsilon_{11}$$

$$q_2(\tau) = \Xi_{11}\cdot\Upsilon_1 - \Xi_{12}\cdot\Upsilon_4 + \Xi_{13}\cdot\Upsilon_7 - \Xi_{14}\cdot\Upsilon_{10};\ q_3(\tau) = -\Xi_{11}$$

The homogeneous system for $\psi\ q_x\ q_y\ q_z$ leads to a characteristic equation for the eigenvalue λ having the form $P(\lambda) + Q(\lambda)\cdot e^{-\lambda\cdot\tau}$; $P(\lambda) = \sum_{j=0}^{4} a_j\cdot\lambda^j$; $Q(\lambda) = \sum_{j=0}^{3} c_j\cdot\lambda^j$ and the coefficients $\{a_j(q_i, q_k), c_j(q_i, q_k)\} \in \mathbb{R}$ depend on q_i, q_k but not on τ,q_i, q_k are any Ion channel laser perturbed system's parameters, other parameters keep as a constant [4], 5, 6.

$$a_0(\tau) = 0;\ a_1(\tau) = -\Upsilon_3;\ a_2(\tau) = -\Upsilon_2;\ a_3(\tau) = -\Upsilon_1;\ a_4(\tau) = 1$$

$$c_0(\tau) = \Xi_{11}\cdot\Upsilon_3 - \Xi_{12}\cdot\Upsilon_6 + \Xi_{13}\cdot\Upsilon_9 - \Xi_{14}\cdot\Upsilon_{12}$$
$$c_1(\tau) = \Xi_{11}\cdot\Upsilon_2 - \Xi_{12}\cdot\Upsilon_5 + \Xi_{13}\cdot\Upsilon_8 - \Xi_{14}\cdot\Upsilon_{11}$$

$$c_2(\tau) = \Xi_{11}\cdot\Upsilon_1 - \Xi_{12}\cdot\Upsilon_4 + \Xi_{13}\cdot\Upsilon_7 - \Xi_{14}\cdot\Upsilon_{10};\ c_3(\tau) = -\Xi_{11}$$

Unless strictly necessary, the designation of the variation arguments (q_i, q_k) will subsequently be omitted from P, Q, a_j, and c_j. The coefficients, a_j, and c_j are continuous and differential functions of their arguments, and direct substitution shows that $a_0 + c_0 \neq 0 \forall q_i, q_k \in \mathbb{R}_+$. $\lambda = 0$ is not a of $P(\lambda) + Q(\lambda)\cdot e^{-\lambda\cdot\tau} = 0$. Furthermore,$P(\lambda)$, $Q(\lambda)$ are analytic functions of λ, for which the following requirements of the analysis [BK] can also be verified:

(a) If $\lambda = i\cdot\omega;\ \omega\in\mathbb{R}$ then $P(i\cdot\omega) + Q(i\cdot\omega) \neq 0$.
(b) $|\frac{Q(\lambda)}{P(\lambda)}|$ is bounded for $|\lambda| \to \infty$; $\mathrm{Re}\lambda \geq 0$. No roots bifurcation from ∞.
(c) $F(\omega) = |P(i\cdot\omega)|^2 - |Q(i\cdot\omega)|^2$ has a finite number of zeros. Indeed, this is a polynomial in ω.
(d) Each positive root $\omega(q_i, q_k)$ of $F(\omega) = 0$ is continuous and differentiable respect to q_i, q_k.

We assume that $P_n(\lambda, \tau) = P_n(\lambda)$ and $Q_m(\lambda, \tau) = Q_m(\lambda)$ cannot have common imaginary roots. That is for any real numberω: $P_n(\lambda = i\cdot\omega, \tau) + Q_m(\lambda = i\cdot\omega, \tau) \neq$

0.

$$P_n(\lambda = i\cdot\omega,\tau) = -\Upsilon_3\cdot i\cdot\omega + \Upsilon_2\cdot\omega^2 + i\cdot\Upsilon_1\cdot\omega^3 + \omega^4$$
$$= i\cdot(-\Upsilon_3\cdot\omega + \Upsilon_1\cdot\omega^3) + \Upsilon_2\cdot\omega^2 + \omega^4$$

$$Q_m(\lambda = i\cdot\omega,\tau) = \Xi_{11}\cdot\Upsilon_3 - \Xi_{12}\cdot\Upsilon_6 + \Xi_{13}\cdot\Upsilon_9 - \Xi_{14}\cdot\Upsilon_{12}$$
$$+ (\Xi_{11}\cdot\Upsilon_2 - \Xi_{12}\cdot\Upsilon_5 + \Xi_{13}\cdot\Upsilon_8 - \Xi_{14}\cdot\Upsilon_{11})\cdot i\cdot\omega$$
$$- (\Xi_{11}\cdot\Upsilon_1 - \Xi_{12}\cdot\Upsilon_4 + \Xi_{13}\cdot\Upsilon_7 - \Xi_{14}\cdot\Upsilon_{10})\cdot\omega^2$$
$$+ \Xi_{11}\cdot i\cdot\omega^3$$

$$Q_m(\lambda = i\cdot\omega,\tau) = \Xi_{11}\cdot\Upsilon_3 - \Xi_{12}\cdot\Upsilon_6 + \Xi_{13}\cdot\Upsilon_9 - \Xi_{14}\cdot\Upsilon_{12}$$
$$- (\Xi_{11}\cdot\Upsilon_1 - \Xi_{12}\cdot\Upsilon_4 + \Xi_{13}\cdot\Upsilon_7 - \Xi_{14}\cdot\Upsilon_{10})\cdot\omega^2$$
$$+ ((\Xi_{11}\cdot\Upsilon_2 - \Xi_{12}\cdot\Upsilon_5 + \Xi_{13}\cdot\Upsilon_8 - \Xi_{14}\cdot\Upsilon_{11})\cdot\omega$$
$$+ \Xi_{11}\cdot\omega^3)\cdot i$$

$$P_n(\lambda = i\cdot\omega,\tau) + Q_m(\lambda = i\cdot\omega,\tau) = i\cdot(-\Upsilon_3\cdot\omega + \Upsilon_1\cdot\omega^3) + \Upsilon_2\cdot\omega^2 + \omega^4$$
$$+ \Xi_{11}\cdot\Upsilon_3 - \Xi_{12}\cdot\Upsilon_6 + \Xi_{13}\cdot\Upsilon_9 - \Xi_{14}$$
$$\cdot\Upsilon_{12} - (\Xi_{11}\cdot\Upsilon_1 - \Xi_{12}\cdot\Upsilon_4 + \Xi_{13}\cdot\Upsilon_7$$
$$- \Xi_{14}\cdot\Upsilon_{10})\cdot\omega^2 + ((\Xi_{11}\cdot\Upsilon_2 - \Xi_{12}\cdot\Upsilon_5$$
$$+ \Xi_{13}\cdot\Upsilon_8 - \Xi_{14}\cdot\Upsilon_{11})\cdot\omega + \Xi_{11}\cdot\omega^3)\cdot i \neq 0$$

$$P_n(\lambda = i\cdot\omega,\tau) + Q_m(\lambda = i\cdot\omega,\tau)$$
$$= [(\Xi_{11}\cdot\Upsilon_2 - \Xi_{12}\cdot\Upsilon_5 + \Xi_{13}\cdot\Upsilon_8 - \Xi_{14}\cdot\Upsilon_{11} - \Upsilon_3)\cdot\omega$$
$$+ (\Xi_{11} + \Upsilon_1)\cdot\omega^3]\cdot i + \omega^4 + [\Upsilon_2 - (\Xi_{11}\cdot\Upsilon_1 - \Xi_{12}\cdot\Upsilon_4$$
$$+ \Xi_{13}\cdot\Upsilon_7 - \Xi_{14}\cdot\Upsilon_{10})]\cdot\omega^2 + \Xi_{11}\cdot\Upsilon_3 - \Xi_{12}\cdot\Upsilon_6$$
$$+ \Xi_{13}\cdot\Upsilon_9 - \Xi_{14}\cdot\Upsilon_{12} \neq 0$$

$$P_n(\lambda = i\cdot\omega,\tau) = i\cdot(-\Upsilon_3\cdot\omega + \Upsilon_1\cdot\omega^3) + \Upsilon_2\cdot\omega^2 + \omega^4$$

$$Q_m(\lambda = i\cdot\omega,\tau) = (\Xi_{11}\cdot\Upsilon_3 - \Xi_{12}\cdot\Upsilon_6 + \Xi_{13}\cdot\Upsilon_9 - \Xi_{14}\cdot\Upsilon_{12})$$
$$- (\Xi_{11}\cdot\Upsilon_1 - \Xi_{12}\cdot\Upsilon_4 + \Xi_{13}\cdot\Upsilon_7 - \Xi_{14}\cdot\Upsilon_{10})\cdot\omega^2$$
$$+ ((\Xi_{11}\cdot\Upsilon_2 - \Xi_{12}\cdot\Upsilon_5 + \Xi_{13}\cdot\Upsilon_8 - \Xi_{14}\cdot\Upsilon_{11})$$
$$\cdot\omega + \Xi_{11}\cdot\omega^3)\cdot i$$

We define for simplicity new parameters which are related to $Q_m(\lambda = i\cdot\omega,\tau) =$ …. expression.

$$\Omega_1 = \Xi_{11} \cdot \Upsilon_3 - \Xi_{12} \cdot \Upsilon_6 + \Xi_{13} \cdot \Upsilon_9 - \Xi_{14} \cdot \Upsilon_{12}$$
$$\Omega_2 = \Xi_{11} \cdot \Upsilon_1 - \Xi_{12} \cdot \Upsilon_4 + \Xi_{13} \cdot \Upsilon_7 - \Xi_{14} \cdot \Upsilon_{10}$$

$$\Omega_3 = \Xi_{11} \cdot \Upsilon_2 - \Xi_{12} \cdot \Upsilon_5 + \Xi_{13} \cdot \Upsilon_8 - \Xi_{14} \cdot \Upsilon_{11}$$
$$Q_m(\lambda = i \cdot \omega, \tau) = \Omega_1 - \Omega_2 \cdot \omega^2 + (\Omega_3 \cdot \omega + \Xi_{11} \cdot \omega^3) \cdot i$$

$$|P_n(\lambda = i \cdot \omega, \tau)|^2 = (\Upsilon_2 \cdot \omega^2 + \omega^4)^2 + (-\Upsilon_3 \cdot \omega + \Upsilon_1 \cdot \omega^3)^2$$

$$|P_n(\lambda = i \cdot \omega, \tau)|^2 = \omega^8 + (2 \cdot \Upsilon_2 + \Upsilon_1^2) \cdot \omega^6 + (\Upsilon_2^2 - 2 \cdot \Upsilon_3 \cdot \Upsilon_1) \cdot \omega^4 + \Upsilon_3^2 \cdot \omega^2$$

$$|Q_m(\lambda = i \cdot \omega, \tau)|^2 = (\Omega_1 - \Omega_2 \cdot \omega^2)^2 + (\Omega_3 \cdot \omega + \Xi_{11} \cdot \omega^3)^2$$

$$|Q_m(\lambda = i \cdot \omega, \tau)|^2 = \Xi_{11}^2 \cdot \omega^6 + (\Omega_2^2 + 2 \cdot \Omega_3 \cdot \Xi_{11}) \cdot \omega^4 + (\Omega_3^2 - 2 \cdot \Omega_1 \cdot \Omega_2) \cdot \omega^2 + \Omega_1^2$$

$$\begin{aligned} F(\omega) &= |P(i \cdot \omega, \tau)|^2 - |Q(i \cdot \omega, \tau)|^2 \\ &= \omega^8 + (2 \cdot \Upsilon_2 + \Upsilon_1^2) \cdot \omega^6 + (\Upsilon_2^2 - 2 \cdot \Upsilon_3 \cdot \Upsilon_1) \cdot \omega^4 + \Upsilon_3^2 \cdot \omega^2 \\ &\quad - [\Xi_{11}^2 \cdot \omega^6 + (\Omega_2^2 + 2 \cdot \Omega_3 \cdot \Xi_{11}) \cdot \omega^4 \\ &\quad + (\Omega_3^2 - 2 \cdot \Omega_1 \cdot \Omega_2) \cdot \omega^2 + \Omega_1^2] \end{aligned}$$

$$\begin{aligned} F(\omega) &= |P(i \cdot \omega, \tau)|^2 - |Q(i \cdot \omega, \tau)|^2 \\ &= \omega^8 + [2 \cdot \Upsilon_2 + \Upsilon_1^2 - \Xi_{11}^2] \cdot \omega^6 \\ &\quad + [\Upsilon_2^2 - 2 \cdot \Upsilon_3 \cdot \Upsilon_1 - \Omega_2^2 - 2 \cdot \Omega_3 \cdot \Xi_{11}] \cdot \omega^4 \\ &\quad + [\Upsilon_3^2 - \Omega_3^2 + 2 \cdot \Omega_1 \cdot \Omega_2] \cdot \omega^2 - \Omega_1^2 \end{aligned}$$

$$\Pi_8 = 1; \Pi_6 = 2 \cdot \Upsilon_2 + \Upsilon_1^2 - \Xi_{11}^2$$
$$\Pi_4 = \Upsilon_2^2 - 2 \cdot \Upsilon_3 \cdot \Upsilon_1 - \Omega_2^2 - 2 \cdot \Omega_3 \cdot \Xi_{11}$$

$$\Pi_2 = \Upsilon_3^2 - \Omega_3^2 + 2 \cdot \Omega_1 \cdot \Omega_2; \Pi_0 = -\Omega_1^2$$
$$F(\omega) = |P(i \cdot \omega, \tau)|^2 - |Q(i \cdot \omega, \tau)|^2 = \sum_{k=0}^{4} \Pi_k \cdot \omega^{2 \cdot k}$$

Hence $F(\omega) = 0$ implies $\sum_{k=0}^{4} \Pi_k \cdot \omega^{2 \cdot k} = 0$ and its roots are given by solving the above polynomial. Furthermore $P_R(i \cdot \omega, \tau) = \Upsilon_2 \cdot \omega^2 + \omega^4$; $P_I(i \cdot \omega, \tau) = -\Upsilon_3 \cdot \omega + \Upsilon_1 \cdot \omega^3$

$$Q_R(i \cdot \omega, \tau) = \Omega_1 - \Omega_2 \cdot \omega^2; \; Q_I(i \cdot \omega, \tau) = \Omega_3 \cdot \omega + \Xi_{11} \cdot \omega^3$$

Hence $\sin\theta(\tau) = \dfrac{-P_R(i \cdot \omega, \tau) \cdot Q_I(i \cdot \omega, \tau) + P_I(i \cdot \omega, \tau) \cdot Q_R(i \cdot \omega, \tau)}{|Q(i \cdot \omega, \tau)|^2}$.

And $\cos\theta(\tau) = -\dfrac{P_R(i \cdot \omega, \tau) \cdot Q_R(i \cdot \omega, \tau) + P_I(i \cdot \omega, \tau) \cdot Q_I(i \cdot \omega, \tau)}{|Q(i \cdot \omega, \tau)|^2}$

$$\sin\theta(\tau) = \frac{-[\Upsilon_2 \cdot \omega^2 + \omega^4] \cdot [\Omega_3 \cdot \omega + \Xi_{11} \cdot \omega^3] + [-\Upsilon_3 \cdot \omega + \Upsilon_1 \cdot \omega^3] \cdot [\Omega_1 - \Omega_2 \cdot \omega^2]}{\Xi_{11}^2 \cdot \omega^6 + (\Omega_2^2 + 2 \cdot \Omega_3 \cdot \Xi_{11}) \cdot \omega^4 + (\Omega_3^2 - 2 \cdot \Omega_1 \cdot \Omega_2) \cdot \omega^2 + \Omega_1^2}$$
$$\cos\theta(\tau) = -\frac{[\Upsilon_2 \cdot \omega^2 + \omega^4] \cdot [\Omega_1 - \Omega_2 \cdot \omega^2] + [-\Upsilon_3 \cdot \omega + \Upsilon_1 \cdot \omega^3] \cdot [\Omega_3 \cdot \omega + \Xi_{11} \cdot \omega^3]}{\Xi_{11}^2 \cdot \omega^6 + (\Omega_2^2 + 2 \cdot \Omega_3 \cdot \Xi_{11}) \cdot \omega^4 + (\Omega_3^2 - 2 \cdot \Omega_1 \cdot \Omega_2) \cdot \omega^2 + \Omega_1^2}$$

Which jointly with $F(\omega) = 0 \Rightarrow \sum_{k=0}^{4} \Pi_k \cdot \omega^{2 \cdot k} = 0$ that are continuous and differentiable in τ based on Lemma 1.1 (see Subchapter 2.1). Hence we use Theorem 1.2 (see Subchapter 2.1) and this proves the Theorem 1.3 (see Subchapter 2.1).

We use different parameters terminology from our last characteristics parameters definition $k \to j$; $p_k(\tau) \to a_j$; $q_k(\tau) \to c_j$; $n = 4$; $m = 3$; $n > m$ additionally $P_n(\lambda, \tau) \to P(\lambda, \tau)$.

And $Q_m(\lambda, \tau) \to Q(\lambda, \tau)$ then $P(\lambda, \tau) = \sum_{j=0}^{4} a_j \cdot \lambda^j$; $Q(\lambda, \tau) = \sum_{j=0}^{3} c_j \cdot \lambda^j$.

$$P(\lambda) = \lambda^4 - \Upsilon_1 \cdot \lambda^3 - \Upsilon_2 \cdot \lambda^2 - \Upsilon_3 \cdot \lambda$$

$$\begin{aligned} Q(\lambda) = & -\Xi_{11} \cdot \lambda^3 + [\Xi_{11} \cdot \Upsilon_1 - \Xi_{12} \cdot \Upsilon_4 + \Xi_{13} \cdot \Upsilon_7 - \Xi_{14} \cdot \Upsilon_{10}] \\ & \cdot \lambda^2 + [\Xi_{11} \cdot \Upsilon_2 - \Xi_{12} \cdot \Upsilon_5 + \Xi_{13} \cdot \Upsilon_8 - \Xi_{14} \cdot \Upsilon_{11}] \cdot \lambda \\ & + \Xi_{11} \cdot \Upsilon_3 - \Xi_{12} \cdot \Upsilon_6 + \Xi_{13} \cdot \Upsilon_9 - \Xi_{14} \cdot \Upsilon_{12} \end{aligned}$$

$n, m \in \mathbb{N}_0$; $n > m$ and $a_j, c_j : \mathbb{R}_{+0} \to \mathbb{R}$ are continuous and differentiable function of τ such that $a_0 + c_0 \neq 0$. In the following "$-$" denotes complex and conjugate. $P(\lambda)$, $Q(\lambda)$ are analytic functions in λ and differentiable in λ. The coefficients:

$$\{a_j(k_z, \omega, \omega_\beta, A, \ldots) \;\&\; c_j(k_z, \omega, \omega_\beta, A, \upsilon \ldots)\} \in \mathbb{R}$$

Depend on Ion channel laser perturbed system's parameters $k_z, \omega, \omega_\beta, A, \tau, \ldots$ values. Unless strictly necessary, the designation of the variation arguments. System parameters $k_z, \omega, \omega_\beta, A, \tau, \ldots$ will subsequently be omitted from P, Q, a_j, c_j. The coefficients a_j, c_j are continuous, and differentiable functions of their arguments.

Direct substitution shows that $a_0 + c_0 \neq 0$.

$$a_0 + c_0 = \Xi_{11} \cdot \Upsilon_3 - \Xi_{12} \cdot \Upsilon_6 + \Xi_{13} \cdot \Upsilon_9 - \Xi_{14} \cdot \Upsilon_{12} \neq 0$$

i.e. $\lambda = 0$ is not a root of characteristic equation. Furthermore $P(\lambda)$, $Q(\lambda)$ are analytic function of λ for which the following requirements of the analysis (see kuang, 1993, Sect. 3.4) can also be verified in the present case.

(a) If $\lambda = i \cdot \omega; \omega \in \mathbb{R}$ then $P(i \cdot \omega) + Q(i \cdot \omega) \neq 0$, i.e. P and Q have no common imaginary roots. This condition was verified numerically in the entire ($k_z, \omega, \omega_\beta, A, \ldots$ system parameters) domain of interest.

(b) $|\frac{Q(\lambda)}{P(\lambda)}|$ is bounded for $|\lambda| \to \infty$; $\mathrm{Re}\lambda \geq 0$. No roots bifurcation from ∞. Indeed, in the limit:

$$|\frac{Q(\lambda)}{P(\lambda)}| = |\frac{\begin{array}{c} -\Xi_{11} \cdot \lambda^3 + [\Xi_{11} \cdot \Upsilon_1 - \Xi_{12} \cdot \Upsilon_4 + \Xi_{13} \cdot \Upsilon_7 - \Xi_{14} \cdot \Upsilon_{10}] \cdot \lambda^2 \\ + [\Xi_{11} \cdot \Upsilon_2 - \Xi_{12} \cdot \Upsilon_5 + \Xi_{13} \cdot \Upsilon_8 - \Xi_{14} \cdot \Upsilon_{11}] \cdot \lambda \\ + \Xi_{11} \cdot \Upsilon_3 - \Xi_{12} \cdot \Upsilon_6 + \Xi_{13} \cdot \Upsilon_9 - \Xi_{14} \cdot \Upsilon_{12} \end{array}}{\lambda^4 - \Upsilon_1 \cdot \lambda^3 - \Upsilon_2 \cdot \lambda^2 - \Upsilon_3 \cdot \lambda}|$$

(c) $F(\omega) = |P(i \cdot \omega)|^2 - |Q(i \cdot \omega)|^2$

$$\begin{aligned} F(\omega) &= |P(i \cdot \omega, \tau)|^2 - |Q(i \cdot \omega, \tau)|^2 \\ &= \omega^8 + [2 \cdot \Upsilon_2 + \Upsilon_1^2 - \Xi_{11}^2] \cdot \omega^6 \\ &\quad + [\Upsilon_2^2 - 2 \cdot \Upsilon_3 \cdot \Upsilon_1 - \Omega_2^2 - 2 \cdot \Omega_3 \cdot \Xi_{11}] \cdot \omega^4 \\ &\quad + [\Upsilon_3^2 - \Omega_3^2 + 2 \cdot \Omega_1 \cdot \Omega_2] \cdot \omega^2 - \Omega_1^2 \end{aligned}$$

Has at most a finite number of zeros. Indeed, this is a polynomial in ω(degree in ω^8).

(d) Each positive rootω ($k_z, \omega, \omega_\beta, A, \ldots$ system parameters) of $F(\omega) = 0$ is continuous and differentiable with respect to $k_z, \omega, \omega_\beta, A, \ldots$ system parameters. This condition can only be assessed numerically.

In addition, since the coefficients in P and Q are real, we have $\overline{P(-i \cdot \omega)} = P(i \cdot \omega)$ and $\overline{Q(-i \cdot \omega)} = Q(i \cdot \omega)$ thus $\lambda = i \cdot \omega$; $\omega > 0$ may be on eigenvalue of characteristic equation. The analysis consists in identifying the roots of characteristic equation situated on the imaginary axis of the complex λ-plane, where by increasing the parameters $k_z, \omega, \omega_\beta, A, \ldots$.

Re λ may, at the crossing, change its sign from (−) to (+), i.e. from a stable focus $E^{(k)}(\psi^{(k)}, q_x^{(k)}, q_y^{(k)}, q_z^{(k)})$; $k = 0, 1, 2, 3, 4$ to an unstable one, or vice versa. This feature may be further assessed by examining the sign of the partial derivatives with respect to $k_z, \omega, \omega_\beta, A, \tau, \ldots$ parameters. $\Lambda^{-1}(k_z) = (\frac{\partial \mathrm{Re}\lambda}{\partial k_z})_{\lambda = i \cdot \omega}, \omega, \omega_\beta, A, \tau, \ldots$, Ion channel laser perturbed system parameters are constant [4, 5].

$$\Lambda^{-1}(\omega) = (\frac{\partial \mathrm{Re}\lambda}{\partial \omega})_{\lambda = i \cdot \omega}; k_z, \omega_\beta, A, \tau, \ldots = const$$

$$\Lambda^{-1}(\omega_\beta) = (\frac{\partial \mathrm{Re}\lambda}{\partial \omega_\beta})_{\lambda=i\cdot\omega}; k_z, \omega, A, \tau, \ldots = const$$

$$\Lambda^{-1}(A) = (\frac{\partial \mathrm{Re}\lambda}{\partial A})_{\lambda=i\cdot\omega}; k_z, \omega, \omega_\beta, \tau, \ldots = const$$

$$\Lambda^{-1}(\tau) = (\frac{\partial \mathrm{Re}\lambda}{\partial \tau})_{\lambda=i\cdot\omega}; k_z, \omega, \omega_\beta, A, \ldots = const$$

Remark We have two ω parameters in our analysis, one is the system variable (ω—Frequency of radiation in the forward (+z) direction in the lab frame) and the second is related to the general geometric criterion for stability (see the article, Edoardo Beretta and Yang Kuang, Geometric stability switch criteria in delay differential systems with delay dependent parameters, SIAM J. Math. Anal, 2001), $\omega \in \mathbb{R}_+$.

When writing $P(\lambda) = P_R(\lambda)+i\cdot P_I(\lambda)$; $Q(\lambda) = Q_R(\lambda)+i\cdot Q_I(\lambda)$, and inserting $\lambda = i\cdot\omega$ into Ion channel laser perturbed system's characteristic equation, ω must satisfy the following:

$$\sin(\omega\cdot\tau) = g(\omega) = \frac{-P_R(i\cdot\omega)\cdot Q_I(i\cdot\omega) + P_I(i\cdot\omega)\cdot Q_R(i\cdot\omega)}{|Q(i\cdot\omega)|^2}$$

$$\cos(\omega\cdot\tau) = h(\omega) = -\frac{P_R(i\cdot\omega)\cdot Q_R(i\cdot\omega) + P_I(i\cdot\omega)\cdot Q_I(i\cdot\omega)}{|Q(i\cdot\omega)|^2}$$

where $|Q(i\cdot\omega)|^2 \neq 0$ in view of requirement (a) above; $(g, h) \in \mathbb{R}$. Furthermore, it follows above equations $\sin(\omega\cdot\tau)$ and $\cos(\omega\cdot\tau)$ that, by squaring and adding the sides, ω must be a positive root of $F(\omega) = |P(i\cdot\omega)|^2 - |Q(i\cdot\omega)|^2 = 0$. Note $F(\omega)$ is dependent on τ. Now it is important to notice that if $\tau \notin I$ (assume that $I \subseteq \mathbb{R}_{+0}$ is the set where $\omega(\tau)$ is a positive root of $F(\omega)$ and for $\tau \notin I$; $\omega(\tau)$ is not define. Then for all τ in I, $\omega(\tau)$ is satisfies that $F(\omega,\tau) = 0$. Then there are no positive $\omega(\tau)$ solutions for $F(\omega,\tau) = 0$, and we cannot have stability switches. For any $\tau \in I$, where $\omega(\tau)$ is a positive solution of $F(\omega,\tau) = 0$ we can define the angle $\theta(\tau) \in [0, 2\cdot\pi]$ as the solution of

$$\sin\theta(\tau) = \frac{-P_R(i\cdot\omega)\cdot Q_I(i\cdot\omega) + P_I(i\cdot\omega)\cdot Q_R(i\cdot\omega)}{|Q(i\cdot\omega)|^2}$$

$$\cos\theta(\tau) = -\frac{P_R(i\cdot\omega)\cdot Q_R(i\cdot\omega) + P_I(i\cdot\omega)\cdot Q_I(i\cdot\omega)}{|Q(i\cdot\omega)|^2}$$

And the relation between the argument $\theta(\tau)$ and $\omega(\tau)\cdot\tau$ for $\tau \in I$ must be $\omega(\tau)\cdot\tau = \theta(\tau) + 2\cdot n\cdot\pi\ \forall n \in \mathbb{N}_0$. Hence we can define the maps $\tau_n : I \to \mathbb{R}_{+0}$ given by $\tau_n(\tau) = \frac{\theta(\tau)+2\cdot n\cdot\pi}{\omega(\tau)}$; $n \in \mathbb{N}_0$; $\tau \in I$. Let us introduce the functions $I \to \mathbb{R}$; $S_n(\tau) = \tau - \tau_n(\tau)$ $\tau \in I$; $n \in \mathbb{N}_0$ that are continuous and differentiable in τ. In the following, the subscripts $\lambda, \omega, k_z, \omega_\beta, A, \ldots$Ion channel laser perturbed system parameters indicate the corresponding partial derivatives. Let us first concentrate on $\Lambda(x)$, remember in $\lambda(k_z, \omega_\beta, A, \ldots; \in \mathbb{R}_+)$; $\omega(k_z, \omega_\beta, A, \ldots\ ; \in \mathbb{R}_+)$ and keeping all parameters except one (x) and τ. The derivation closely follows that in reference [BK]. Differentiating Ion channel laser perturbed system characteristic equation $P(\lambda) + Q(\lambda) \cdot e^{-\lambda \cdot \tau}$ with respect to specific parameter (x), and inverting the derivative, for convenience, one calculates:

Remark $x = k_z, \omega_\beta, A, \ldots; \in \mathbb{R}_+$

$$\left(\frac{\partial \lambda}{\partial x}\right)^{-1} = \frac{-P_\lambda(\lambda, x) \cdot Q(\lambda, x) + Q_\lambda(\lambda, x) \cdot P(\lambda, x) - \tau \cdot P(\lambda, x) \cdot Q(\lambda, x)}{P_x(\lambda, x) \cdot Q(\lambda, x) - Q_x(\lambda, x) \cdot P(\lambda, x)}$$

where, $P_\lambda = \frac{\partial P}{\partial \lambda}, \ldots$ etc. Substituting $\lambda = i \cdot \omega$, and bearing $\overline{P(-i \cdot \omega)} = P(i \cdot \omega)$; $\overline{Q(-i \cdot \omega)} = Q(i \cdot \omega)$ then $i \cdot P_\lambda(i \cdot \omega) = P_\omega(i \cdot \omega)$ and $i \cdot Q_\lambda(i \cdot \omega) = Q_\omega(i \cdot \omega)$ that on the surface $|P(i \cdot \omega)|^2 = |Q(i \cdot \omega)|^2$, one obtains.

$$\left(\frac{\partial \lambda}{\partial x}\right)^{-1}\Big|_{\lambda = i \cdot \omega} = \frac{i \cdot P_\omega(i \cdot \omega, x) \cdot \overline{P(i \cdot \omega, x)} + i \cdot Q_\lambda(i \cdot \omega, x) \cdot \overline{Q(\lambda, x)} - \tau \cdot |P(i \cdot \omega, x)|^2}{P_x(i \cdot \omega, x) \cdot \overline{P(i \cdot \omega, x)} - Q_x(i \cdot \omega, x) \cdot \overline{Q(i \cdot \omega, x)}}$$

Upon separating into real and imaginary parts, with $P = P_R + i \cdot P_I$; $Q = Q_R + i \cdot Q_I$

$$P_\omega = P_{R\omega} + i \cdot P_{I\omega};\ Q_\omega = Q_{R\omega} + i \cdot Q_{I\omega}$$
$$P_x = P_{Rx} + i \cdot P_{Ix};\ Q_x = Q_{Rx} + i \cdot Q_{Ix};\ P^2 = P_R^2 + P_I^2$$

When (x) can be any Ion channel laser perturbed system parameters $k_z, \omega_\beta, A, \ldots$ and time delay τ etc. Where for convenience, we have dropped the arguments $(i \cdot \omega, x)$, and where

$$F_\omega = 2 \cdot [(P_{R\omega} \cdot P_R + P_{I\omega} \cdot P_I) - (Q_{R\omega} \cdot Q_R + Q_{I\omega} \cdot Q_I)]$$
$$F_x = 2 \cdot [(P_{Rx} \cdot P_R + P_{Ix} \cdot P_I) - (Q_{Rx} \cdot Q_R + Q_{Ix} \cdot Q_I)]$$

And $\omega_x = -\frac{F_\lambda}{F_\omega}$.

$$U = (P_R \cdot P_{I\omega} - P_I \cdot P_{R\omega}) - (Q_R \cdot Q_{I\omega} - Q_I \cdot Q_{R\omega})$$
$$V = (P_R \cdot P_{Ix} - P_I \cdot P_{Rx}) - (Q_R \cdot Q_{Ix} - Q_I \cdot Q_{Rx})$$

We choose our specific parameter as time delay $x = \tau$.

$$P_R(i \cdot \omega, \tau) = \Upsilon_2 \cdot \omega^2 + \omega^4;\ P_I(i \cdot \omega, \tau) = -\Upsilon_3 \cdot \omega + \Upsilon_1 \cdot \omega^3$$

$$Q_R(i\cdot\omega,\tau)=\Omega_1-\Omega_2\cdot\omega^2;\ Q_I(i\cdot\omega,\tau)=\Omega_3\cdot\omega+\Xi_{11}\cdot\omega^3$$

$$P_{R\omega}(i\cdot\omega,\tau)=2\cdot\Upsilon_2\cdot\omega+4\cdot\omega^3;\ P_{I\omega}(i\cdot\omega,\tau)=-\Upsilon_3+3\cdot\Upsilon_1\cdot\omega^2$$

$$Q_{R\omega}(i\cdot\omega,\tau)=-2\cdot\Omega_2\cdot\omega;\ Q_{I\omega}(i\cdot\omega,\tau)=\Omega_3+3\cdot\Xi_{11}\cdot\omega^2$$

$$P_{R\tau}(i\cdot\omega,\tau)=0;\ P_{I\tau}(i\cdot\omega,\tau)=0$$
$$Q_{R\tau}(i\cdot\omega,\tau)=0;\ Q_{I\tau}(i\cdot\omega,\tau)=0$$

$$P_{R\omega}\cdot P_R=(2\cdot\Upsilon_2\cdot\omega+4\cdot\omega^3)\cdot(\Upsilon_2\cdot\omega^2+\omega^4)$$
$$P_{I\omega}\cdot P_I=(-\Upsilon_3+3\cdot\Upsilon_1\cdot\omega^2)\cdot(-\Upsilon_3\cdot\omega+\Upsilon_1\cdot\omega^3)$$

$$\omega_\tau=-\frac{F_\tau}{F_\omega};\ Q_{R\omega}\cdot Q_R=-2\cdot\Omega_2\cdot\omega\cdot(\Omega_1-\Omega_2\cdot\omega^2)$$
$$Q_{I\omega}\cdot Q_I=(\Omega_3+3\cdot\Xi_{11}\cdot\omega^2)\cdot(\Omega_3\cdot\omega+\Xi_{11}\cdot\omega^3)$$

$$F_\tau=2\cdot[(P_{R\tau}\cdot P_R+P_{I\tau}\cdot P_I)-(Q_{R\tau}\cdot Q_R+Q_{I\tau}\cdot Q_I)]=0$$

$$P_R\cdot P_{I\omega}=(\Upsilon_2\cdot\omega^2+\omega^4)\cdot(-\Upsilon_3+3\cdot\Upsilon_1\cdot\omega^2)$$
$$Q_I\cdot Q_{R\omega}=(\Omega_3\cdot\omega+\Xi_{11}\cdot\omega^3)\cdot(-2\cdot\Omega_2\cdot\omega)$$

$$V=(P_R\cdot P_{I\tau}-P_I\cdot P_{R\tau})-(Q_R\cdot Q_{I\tau}-Q_I\cdot Q_{R\tau})=0$$

Differentiating with respect to τ and we get $F_\omega\cdot\frac{\partial\omega}{\partial\tau}+F_\tau=0;\ \tau\in I\Rightarrow\frac{\partial\omega}{\partial\tau}=-\frac{F_\tau}{F_\omega}$

$$\Lambda^{-1}(\tau)=\left(\frac{\partial\mathrm{Re}\lambda}{\partial\tau}\right)_{\lambda=i\cdot\omega}$$
$$\Lambda^{-1}(\tau)=\mathrm{Re}\left\{\frac{-2\cdot[U+\tau\cdot|P|^2]+i\cdot F_\omega}{F_\tau+i\cdot 2\cdot[V+\omega\cdot|P|^2]}\right\}$$
$$\frac{\partial\omega}{\partial\tau}=\omega_\tau=-\frac{F_\tau}{F_\omega}$$

$$sign\{\Lambda^{-1}(\tau)\}=sign\left\{\left(\frac{\partial\mathrm{Re}\lambda}{\partial\tau}\right)_{\lambda=i\cdot\omega}\right\}$$
$$sign\{\Lambda^{-1}(\tau)\}=sign\{F_\omega\}\cdot sign\left\{\tau\cdot\frac{\partial\omega}{\partial\tau}+\omega+\frac{U\cdot\frac{\partial\omega}{\partial\tau}+V}{|P|^2}\right\}$$

We shall presently examine the possibility of stability transitions (bifurcations) in a Ion channel laser perturbed system, about the equilibrium point$E^{(k)}(\psi^{(k)}, q_x^{(k)}, q_y^{(k)}, q_z^{(k)}); k = 0, 1, 2, 3, 4$ as a result of a variation of delay parameterτ. The analysis consists in identifying the roots of our system characteristic equation situated on the imaginary axis of the complex λ-plane where by increasing the delay parameterτ, Re λ may at the crossing, change its sign from "- " to " + ", i.e. from a stable focus $E^{(*)}$ to an unstable one, or vice versa. This feature may be further assessed by examining the sign of the partial derivatives with respect to τ. $\Lambda^{-1}(\tau) = (\frac{\partial \mathrm{Re}\lambda}{\partial \tau})_{\lambda = i\cdot\omega}, k_z, \omega_\beta, A, \ldots$ system parameters are constant where $\omega \in \mathbb{R}_+$. We need to plot the stability switch diagram based on different delay values of our Ion channel laser perturbed system. Since it is a very complex function we recommend to solve it numerically rather than analytic.

$$\Lambda^{-1}(\tau) = \left(\frac{\partial \mathrm{Re}\lambda}{\partial \tau}\right)_{\lambda = i\cdot\omega} = \mathrm{Re}\left\{\frac{-2\cdot[U + \tau\cdot|P|^2] + i\cdot F_\omega}{F_\tau + i\cdot 2\cdot[V + \omega\cdot|P|^2]}\right\}$$

$$\Lambda^{-1}(\tau) = \left(\frac{\partial \mathrm{Re}\lambda}{\partial \tau}\right)_{\lambda = i\cdot\omega} = \frac{2\cdot\{F_\omega\cdot(V + \omega\cdot P^2) - F_\tau\cdot(U + \tau\cdot P^2)\}}{F_\tau^2 + 4\cdot(V + \omega\cdot P^2)^2}$$

The stability switch occurs only on those delay values (τ) which fit the equation: $\tau = \frac{\theta_+(\tau)}{\omega_+(\tau)}$ and $\theta_+(\tau)$ is the solution of $\sin\theta(\tau) = \ldots; \cos\theta(\tau) = \ldots$ when $\omega = \omega_+(\tau)$ if only ω_+ is feasible. Additionally when all Ion channel laser perturbed system parameters are known and the stability switch due to various time delay values τ is describe in the following expression:

$$sign\{\Lambda^{-1}(\tau)\} = sign\{F_\omega(\omega(\tau), \tau)\} \cdot sign\{\tau\cdot\omega_\tau(\omega(\tau)) + \omega(\tau) + \frac{U(\omega(\tau))\cdot\omega_\tau(\omega(\tau)) + V(\omega(\tau))}{|P(\omega(\tau))|^2}$$

Remark we know $F(\omega, \tau) = 0$ implies it roots $\omega_i(\tau)$ and finding those delays values τ which ω_i is feasible. There are τ values which ω_i are complex or imaginary numbers, then unable to analyze stability.

3.3 Mode-Competition Phenomena in Long-Wavelength Lasers Instability in Time

A semiconductor laser (LD) is a device that causes laser oscillation by the flowing an electric current to semiconductor. Practically, the mechanism of light emission is the same as a Lighting-emitting diode (LED). The optical gain is produced in a semiconductor material. By establishing the desired wavelength and other properties such as modulation speed we define the material. It is bulk semiconductor, quantum hetero structure. The pumping mechanism is done electrically or optically (disk laser). The laser light is produced when electrons and photons interact in a p-n junction. The concentrated light beam is produced by "pumping" the light emitted from atoms repeatedly, between two mirrors. The semiconductor laser produces coherent radiation (wave are all at the same frequency and phase) in the visible or IR spectrum when current passes through it. Semiconductor lasers are widely using in optical fiber communication systems and optical disk system, optical subscriber networks. The mode competition phenomena are the phenomenon that different laser resonator modes experience laser amplification in the same gain medium, leading to cross-saturation effects. The resonator modes are characterized by the light circulating in an optical resonator. We characterize the light in each mode by parameters (electric field amplitude, optical phase, apart from optical frequency and polarization). The overall number of parameters is small in the lasing process. The different modes of a laser resonator experience optical amplification in the same gain medium. Different modes experience amplification in the same gain medium which leads to cross saturation effects. A stimulated emission by one mode causes gain saturation for itself (self- saturation) and for other modes. The power distribution over several modes is unstable. We need to investigate the influence of instantaneous mode-competition phenomena on the dynamics of semiconductor lasers. The system analysis and stability is based on the multimode rate equations which are superposed by Langevin noise sources that account for the intrinsic fluctuations associated with the spontaneous emission. The Langevin noise sources derive the laser rate equations keeping their cross-correlations satisfied. They are time-varying profiles of the fluctuating photon and carrier numbers and the instantaneous shift of the oscillating frequency. The gain saturation effects cause competition phenomena among lasing modes are based on a self-consistent model. The mode-competition phenomena are caused by nonlinear gain suppression effects and are characterized by the corresponding saturation coefficients. The gain suppression effect among different lasing modes is stronger than the suppression effect for an identical mode when the laser supports only the fundamental transverse mode. The latter effect is called self-saturation, and the former one is called cross-saturation. The semiconductor lasers show several types of extra noise, such as hopping noise, and optical feedback noise. These types of noise are results of dynamic behavior of the mode-competition phenomena. There are cases on mode-competition phenomena, the intensity fluctuations were assumed to be not sufficiently large to change the operation state. The fluctuations of mode intensities can change the state of operation as pointed out in. The model is the multimode rate equations in which both gain saturation effects and Langevin noise sources are presented. The nonlinear dynamics of the lasing nodes are described mathematically by multimode rate equations of the photon number S_p of the lasing modes and the injected electron number N [7].

$$\frac{dS_p}{dt} = (G_p - G_{th}) \cdot S_p + \frac{a \cdot \xi \cdot N}{V \cdot \{[\frac{2 \cdot (\lambda_p - \lambda_0)}{\delta\lambda}]^2 + 1\}} + F_p(t)$$

$$\frac{dN}{dt} = -\sum_{p} A_p \cdot S_p - \frac{N}{\tau_s} + \frac{I}{e} + F_N(t)$$

$$C_p \cdot \frac{N}{\tau_s} = \frac{a \cdot \xi \cdot N}{V \cdot \{[\frac{2 \cdot (\lambda_p - \lambda_0)}{\delta\lambda}]^2 + 1\}} \Rightarrow C_p$$
$$= \frac{a \cdot \xi \cdot \tau_s}{V \cdot \{[\frac{2 \cdot (\lambda_p - \lambda_0)}{\delta\lambda}]^2 + 1\}}$$

Hint: The terms $F_p(t)$ and $F_N(t)$ are Langevin noise sources and are added to the rate equations to trigger fluctuations in the modal photon numbers and electron number.

$$G_p = A_p - B \cdot S_p - \sum_{q \neq p} (D_{p(q)} + H_{p(q)}) \cdot S_q$$

$$A_p = \frac{a \cdot \xi}{V} \cdot \{N - N_g - b \cdot V \cdot (\lambda_p - \lambda_0)^2\}$$

$$B = \frac{9}{2} \cdot \frac{\pi \cdot c}{\varepsilon_0 \cdot n_r^2 \cdot \hbar \cdot \lambda_0} \cdot (\frac{\xi \cdot \tau_{in}}{V})^2 \cdot a \cdot |R_{cv}|^2 \cdot (N - N_s)$$
$$B = B(N);\ A_p = A_p(N)$$

$$D_{p(q)} = \frac{4}{3} \cdot \frac{B}{(\frac{2 \cdot \pi \cdot c \cdot \tau_{in}}{\lambda_p^2})^2 \cdot (\lambda_p - \lambda_q)^2 + 1}$$

$$D_{p(q)} = \frac{4}{3} \cdot \frac{\frac{9}{2} \cdot \frac{\pi \cdot c}{\varepsilon_0 \cdot n_r^2 \cdot \hbar \cdot \lambda_0} \cdot (\frac{\xi \cdot \tau_{in}}{V})^2 \cdot a \cdot |R_{cv}|^2 \cdot (N - N_s)}{(\frac{2 \cdot \pi \cdot c \cdot \tau_{in}}{\lambda_p^2})^2 \cdot (\lambda_p - \lambda_q)^2 + 1}$$

$$H_{p(q)} = \frac{3}{4} \cdot (\frac{a \cdot \xi}{V})^2 \cdot (N - N_g) \cdot \left\{ \frac{\frac{1}{\tau_s} + \frac{3}{2} \cdot \frac{a \cdot \xi}{V} \cdot S + \alpha \cdot \frac{2 \cdot \pi \cdot c}{\lambda_p^2} \cdot (\lambda_p - \lambda_q)}{(\frac{1}{\tau_s} + \frac{3}{2} \cdot \frac{a \cdot \xi}{V} \cdot S)^2 + (\frac{2 \cdot \pi \cdot c}{\lambda_p^2})^2 \cdot (\lambda_p - \lambda_q)^2} \right\}$$

$$D_{p(q)} = D_{p(q)}(p, q, N);\ H_{p(q)} = H_{p(q)}(p, q, N)$$

Index $p = 0, \pm 1, \pm 2, \ldots, \pm M$; $M \in \mathbb{N}_0$ is the mode number, G_p is the optical gain of mode p. The mode p = 0 with wavelength λ_0 is assumed to be at the center of the spectral profile of the gain. The lasing modes on the long-wavelength side of the central mode of the gain profile $\lambda_p > \lambda_0$ are indicated with positive numbers, $0 < p \leq M$, while the modes on the shorter side, $\lambda_p < \lambda_0$, are indicated with negative numbers $-M \leq p < 0$; $\lambda_p = \lambda_0 + p \cdot \Delta\lambda$.

Where, $\Delta\lambda = \frac{\lambda_0^2}{n_r \cdot L}$; $\lambda_p = \lambda_0 + p \cdot \frac{\lambda_0^2}{n_r \cdot L}$. n_r—Refractive index of the active region, L—Length of the active region, $\Delta\lambda$—Mode wavelength spacing, $A_p \cdot S_p$—Photon number after amplification (A_p—linear gain coefficient) at mode p (Fig. 3.3).

G_{th} is the threshold gain level, and is determined by the loss coefficient κ of the laser and mirror loss, $G_{th} = \frac{c}{n_r} \cdot [\kappa + \frac{1}{2 \cdot L} \cdot \ln(\frac{1}{R_f \cdot R_b})]$ where R_f and R_b are power

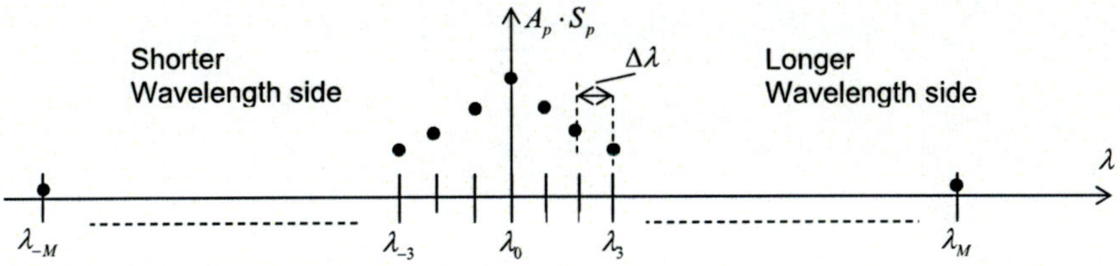

Fig. 3.3 Typical spectral distribution of photon number after amplification for different wavelengths at mode (p)

reflectivity at the front and back facets, respectively. τ_s—Lifetime of the electron for spontaneous emission, e—electron charge, I—Injection current, B—self saturation coefficient, I_{th}—Threshold current level, $D_{p(q)}$, $H_{p(q)}$—Coefficients for cross saturation effects of gain, a —Slope (or tangential) coefficient to give the local linear gain, b- Coefficient giving the wavelength dispersion of the linear gain, N_g—Transparent electron number, τ_{in}—Intra-band relaxation time of the electron wave, R_{cv}—Dipole moment, N_s—Electron number characterization the saturation coefficient (B), c—Speed of light in free space, $S = \sum_p S_p$—Total photon number summed for all existing modes, $\delta\lambda$ —Half width of the spontaneous emission profile, α—Linewidth enhancement factor, $F_p(t)$, $F_N(t)$—Generating term of the fluctuation caused by the spontaneous emission (Langevin noise sources), V—Volume of the active region, ξ—Field confinement factor into the active region, $D_{p(q)}$ —Symmetric cross saturation coefficient, n—Injected density of electrons ($n = \frac{N}{V}$), C_p—Spontaneous emission factor. Assumption for large model separation: $\frac{1}{\tau_s} + \frac{3}{2} \cdot \frac{a \cdot \xi}{V} \cdot S \ll \frac{2 \cdot \pi \cdot c}{\lambda_p^2} \cdot |\lambda_q - \lambda_p|$ and we get the reduce expression for $H_{p(q)}$, $H_{p(q)} \approx \frac{3 \cdot \lambda_p^2}{8 \cdot \pi \cdot c} \cdot (\frac{a \cdot \xi}{V})^2 \cdot \frac{\alpha \cdot (N - N_g)}{(\lambda_q - \lambda_p)}$. Function coefficient $D_{p(q)} \sim \frac{1}{(\lambda_p - \lambda_q)^2}$ gives the similar value if $\lambda_p > \lambda_q$ or $\lambda_p < \lambda_q$, then is called the symmetric cross saturation coefficient. On the other hand, the reduced function $H_{p(q)}$, $H_{p(q)} \sim \frac{1}{\lambda_q - \lambda_p}$ is inversely proportional to $\lambda_q - \lambda_p . H_{p(q)}$ is the asymmetric cross saturation coefficient. Wavelengths λ_p, λ_q are at two modes p and q respectively ($p \neq q$). Function $H_{p(q)}$, $H_{p(q)} = \begin{cases} > 0 \, \lambda_p < \lambda_q \\ < 0 \, \lambda_p > \lambda_q \end{cases}$.

We can summary the asymmetric cross saturation coefficient $H_{p(q)}$ sign for λ_p, λ_q conditions and the related meaning in the next table (Table 3.5).

Table 3.5 Asymmetric cross saturation coefficient $H_{p(q)}$ sign for λ_p, λ_q conditions and the related meaning

λ_p, λ_q condition	$H_{p(q)}$ value	Meaning
$\lambda_p < \lambda_q$	$H_{p(q)} > 0$	Suppress the lasing gain for the modes on the short wavelength side
$\lambda_p > \lambda_q$	$H_{p(q)} < 0$	Enhance the lasing gain for the modes on the longer wavelength side

$$\sum_{q \neq p} (D_{p(q)} + H_{p(q)}) \cdot S_q|_{\lambda_p < \lambda_q} > \sum_{q \neq p} (D_{p(q)} + H_{p(q)}) \cdot S_q|_{\lambda_p > \lambda_q}$$

$$\underset{\text{(Supress lasing gain)}}{G_p|_{\lambda_p < \lambda_q}} < \underset{\text{(Enhance lasing gain)}}{G_p|_{\lambda_p > \lambda_q}}$$

The periodic mode hopping ($0 \to 1 \to 2 \to 3 \to 0 \to \ldots$) is caused by the asymmetric cross saturation function $H_{p(q)}$. The following flowchart shows our laser periodic mode hopping (Fig. 3.4).

We need to analyze our system stability and stability switching for parameters variation. We take some assumptions before doing system analysis [2, 3, 7].

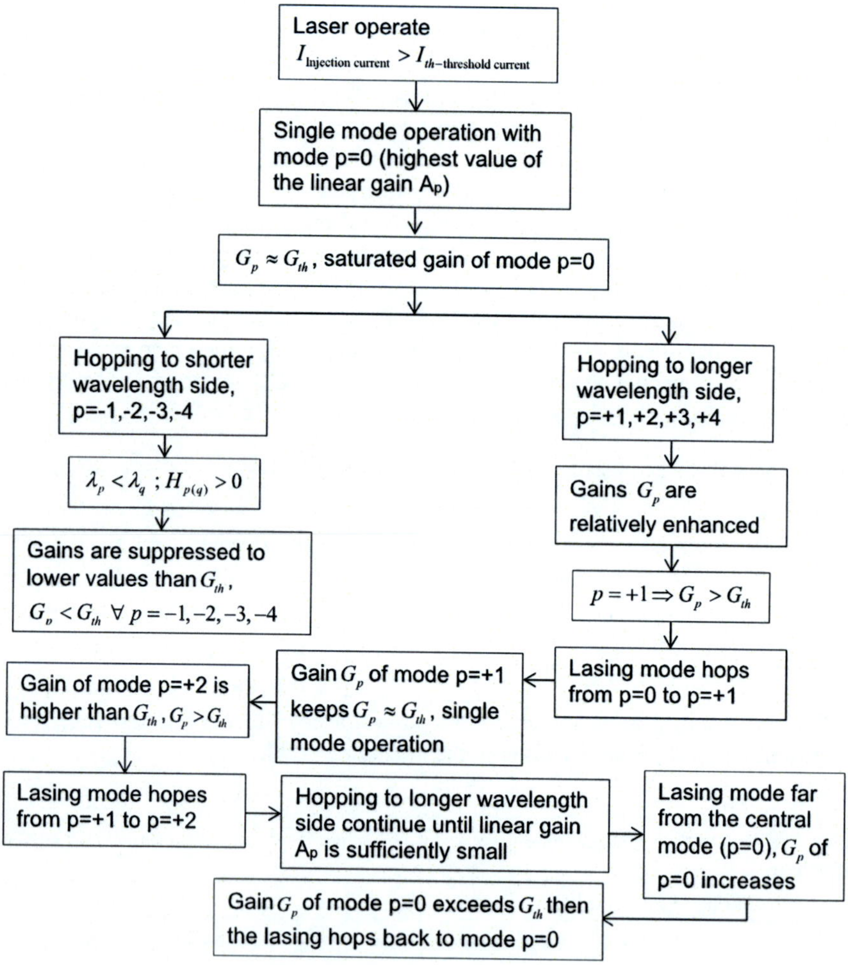

Fig. 3.4 Schematic for the rotating effect of the mode hopping among several longitudinal modes

Assumption A Terms $F_p(t)$ and $F_N(t)$ (Langevin noise sources) are very small compare to other elements on the rate equations.
$(G_p - G_{th}) \cdot S_p + \frac{a \cdot \xi \cdot N}{V \cdot \{[\frac{2 \cdot (\lambda_p - \lambda_0)}{\delta\lambda}]^2 + 1\}} \gg F_p(t); -\sum_p A_p \cdot S_p - \frac{N}{\tau_s} + \frac{I}{e} \gg F_N(t)$ Then we can write our reduced rate equations:

$$\frac{dS_p}{dt} = (G_p - G_{th}) \cdot S_p + \frac{a \cdot \xi \cdot N}{V \cdot \{[\frac{2 \cdot (\lambda_p - \lambda_0)}{\delta\lambda}]^2 + 1\}}$$

$$\frac{dN}{dt} = -\sum_p A_p \cdot S_p - \frac{N}{\tau_s} + \frac{I}{e}$$

$$D_{p(q)} = D_{p(q)}(p, q, N); H_{p(q)} = H_{p(q)}(p, q, N)$$
$$G_p = G_p(p, q, N); B = B(N); A_p = A_p(N)$$

$$G_p = A_p - B \cdot S_p - \sum_{q \neq p} (D_{p(q)} + H_{p(q)}) \cdot S_q$$

$$G_p = A_p(N) - B(N) \cdot S_p - \sum_{q \neq p} \{D_{p(q)}(p, q, N) + H_{p(q)}(p, q, N)\} \cdot S_q$$

Assumption B Large model separation: $\frac{1}{\tau_s} + \frac{3}{2} \cdot \frac{a \cdot \xi}{V} \cdot S \ll \frac{2 \cdot \pi \cdot c}{\lambda_p^2} \cdot |\lambda_q - \lambda_p|$ then.

$$H_{p(q)} \approx \frac{3 \cdot \lambda_p^2}{8 \cdot \pi \cdot c} \cdot (\frac{a \cdot \xi}{V})^2 \cdot \frac{\alpha \cdot (N - N_g)}{(\lambda_q - \lambda_p)}; H_{p(q)} = H_{p(q)}(p, q, N)$$

Assumption C Laser operates on a periodic mode hopping ($0 \to 1 \to 2 \to 3 \to 0 \to \ldots$) then $\sum_p A_p \cdot S_p = \sum_{p=0}^{3} A_p \cdot S_p; \Omega_0 = \sum_{p=0}^{3} A_p \cdot S_p; \Omega_0 \in \mathbb{R}_+$. The total photon number summed for existing modes $S = \sum_p S_p = \sum_{p=0}^{3} S_p; \Omega_1 = \sum_{p=0}^{3} S_p; \Omega_1 \in \mathbb{R}_+$.

Assumption D Laser modes are on the longer wavelength side and there is a periodically hopping ($0 \to 1 \to 2 \to 3 \to 0 \to \ldots$), increase the lasing gain $G_p(\lambda_p > \lambda_q, H_{p(q)} < 0)$. We define new variable $\Delta\lambda_{q-p} = \lambda_q - \lambda_p; q \neq p; \Delta\lambda_{q-p} < 0; \Delta\lambda_{q-p} \in \mathbb{R}$ and we define parameters $\Delta\lambda_{q-p=0} = \lambda_q - \lambda_0; \Delta\lambda_{q-p=1} = \lambda_q - \lambda_1; \Delta\lambda_{q-p=2} = \lambda_q - \lambda_2; \Delta\lambda_{q-p=3} = \lambda_q - \lambda_3$ $\Delta\lambda_{q-p=0} = \Delta\lambda_{q-p=0}(q); \Delta\lambda_{q-p=1} = \Delta\lambda_{q-p=1}(q); \Delta\lambda_{q-p=2} = \Delta\lambda_{q-p=2}(q); \Delta\lambda_{q-p=3} = \Delta\lambda_{q-p=3}(q), \Delta\lambda_{q-p=0}, \Delta\lambda_{q-p=1}, \Delta\lambda_{q-p=2}, \Delta\lambda_{q-p=3} < 0$(Index q is fixed).

$$H_{p=0} \approx \frac{3 \cdot \lambda_0^2}{8 \cdot \pi \cdot c} \cdot (\frac{a \cdot \xi}{V})^2 \cdot \frac{\alpha \cdot (N - N_g)}{(\lambda_q - \lambda_0)}$$

$$H_{p=1} \approx \frac{3 \cdot \lambda_1^2}{8 \cdot \pi \cdot c} \cdot (\frac{a \cdot \xi}{V})^2 \cdot \frac{\alpha \cdot (N - N_g)}{(\lambda_q - \lambda_1)}$$

$$H_{p=2} \approx \frac{3 \cdot \lambda_2^2}{8 \cdot \pi \cdot c} \cdot (\frac{a \cdot \xi}{V})^2 \cdot \frac{\alpha \cdot (N - N_g)}{(\lambda_q - \lambda_2)}$$

$$H_{p=3} \approx \frac{3 \cdot \lambda_3^2}{8 \cdot \pi \cdot c} \cdot (\frac{a \cdot \xi}{V})^2 \cdot \frac{\alpha \cdot (N - N_g)}{(\lambda_q - \lambda_3)}$$

$$H_{p=0} \approx \frac{3 \cdot \lambda_0^2}{8 \cdot \pi \cdot c} \cdot (\frac{a \cdot \xi}{V})^2 \cdot \frac{\alpha}{\Delta\lambda_{q-p=0}} \cdot (N - N_g)$$

$$H_{p=1} \approx \frac{3 \cdot \lambda_1^2}{8 \cdot \pi \cdot c} \cdot (\frac{a \cdot \xi}{V})^2 \cdot \frac{\alpha}{\Delta\lambda_{q-p=1}} \cdot (N - N_g)$$

$$H_{p=2} \approx \frac{3 \cdot \lambda_2^2}{8 \cdot \pi \cdot c} \cdot (\frac{a \cdot \xi}{V})^2 \cdot \frac{\alpha}{\Delta\lambda_{q-p=2}} \cdot (N - N_g)$$

$$H_{p=3} \approx \frac{3 \cdot \lambda_3^2}{8 \cdot \pi \cdot c} \cdot (\frac{a \cdot \xi}{V})^2 \cdot \frac{\alpha}{\Delta\lambda_{q-p=3}} \cdot (N - N_g)$$

We define new global parameters: $\Psi_0 = \frac{3 \cdot \lambda_0^2}{8 \cdot \pi \cdot c} \cdot (\frac{a \cdot \xi}{V})^2 \cdot \frac{\alpha}{\Delta\lambda_{q-p=0}}$

$$\Psi_1 = \frac{3 \cdot \lambda_1^2}{8 \cdot \pi \cdot c} \cdot (\frac{a \cdot \xi}{V})^2 \cdot \frac{\alpha}{\Delta\lambda_{q-p=1}}$$

$$\Psi_2 = \frac{3 \cdot \lambda_2^2}{8 \cdot \pi \cdot c} \cdot (\frac{a \cdot \xi}{V})^2 \cdot \frac{\alpha}{\Delta\lambda_{q-p=2}}$$

$$\Psi_3 = \frac{3 \cdot \lambda_3^2}{8 \cdot \pi \cdot c} \cdot (\frac{a \cdot \xi}{V})^2 \cdot \frac{\alpha}{\Delta\lambda_{q-p=3}}$$

$$H_{p=0} \approx \Psi_0 \cdot (N - N_g);\ H_{p=1} \approx \Psi_1 \cdot (N - N_g)$$
$$H_{p=2} \approx \Psi_2 \cdot (N - N_g);\ H_{p=3} \approx \Psi_3 \cdot (N - N_g)$$

$$D_{p(q)} = \frac{4}{3} \cdot \frac{\frac{9}{2} \cdot \frac{\pi \cdot c}{\varepsilon_0 \cdot n_r^2 \cdot \hbar \cdot \lambda_0} \cdot (\frac{\xi \cdot \tau_{in}}{V})^2 \cdot a \cdot |R_{cv}|^2 \cdot (N - N_s)}{(\frac{2 \cdot \pi \cdot c \cdot \tau_{in}}{\lambda_p^2})^2 \cdot (\lambda_p - \lambda_q)^2 + 1}$$

$$\Delta\lambda_{p-q} = \lambda_p - \lambda_q;\ \Delta\lambda_{p-q} > 0$$

$$\Delta\lambda_{p=0-q} = \lambda_0 - \lambda_q;\ \Delta\lambda_{p=1-q} = \lambda_1 - \lambda_q$$
$$\Delta\lambda_{p=2-q} = \lambda_2 - \lambda_q;\ \Delta\lambda_{p=3-q} = \lambda_3 - \lambda_q$$

$$D_{p(q)} = \frac{4}{3} \cdot \left[\frac{\frac{9}{2} \cdot \frac{\pi \cdot c}{\varepsilon_0 \cdot n_r^2 \cdot \hbar \cdot \lambda_0} \cdot (\frac{\xi \cdot \tau_{in}}{V})^2 \cdot a \cdot |R_{cv}|^2}{(\frac{2 \cdot \pi \cdot c \cdot \tau_{in}}{\lambda_p^2})^2 \cdot (\Delta\lambda_{p-q})^2 + 1} \right] \cdot (N - N_s)$$

$$D_{p=0} = \frac{4}{3} \cdot \left[\frac{\frac{9}{2} \cdot \frac{\pi \cdot c}{\varepsilon_0 \cdot n_r^2 \cdot \hbar \cdot \lambda_0} \cdot (\frac{\xi \cdot \tau_{in}}{V})^2 \cdot a \cdot |R_{cv}|^2}{(\frac{2 \cdot \pi \cdot c \cdot \tau_{in}}{\lambda_p^2})^2 \cdot (\Delta\lambda_{p=0-q})^2 + 1} \right] \cdot (N - N_s)$$

$$D_{p=1} = \frac{4}{3} \cdot \left[\frac{\frac{9}{2} \cdot \frac{\pi \cdot c}{\varepsilon_0 \cdot n_r^2 \cdot \hbar \cdot \lambda_0} \cdot (\frac{\xi \cdot \tau_{in}}{V})^2 \cdot a \cdot |R_{cv}|^2}{(\frac{2 \cdot \pi \cdot c \cdot \tau_{in}}{\lambda_p^2})^2 \cdot (\Delta\lambda_{p=1-q})^2 + 1} \right] \cdot (N - N_s)$$

$$D_{p=2} = \frac{4}{3} \cdot \left[\frac{\frac{9}{2} \cdot \frac{\pi \cdot c}{\varepsilon_0 \cdot n_r^2 \cdot \hbar \cdot \lambda_0} \cdot (\frac{\xi \cdot \tau_{in}}{V})^2 \cdot a \cdot |R_{cv}|^2}{(\frac{2 \cdot \pi \cdot c \cdot \tau_{in}}{\lambda_p^2})^2 \cdot (\Delta\lambda_{p=2-q})^2 + 1} \right] \cdot (N - N_s)$$

$$D_{p=3} = \frac{4}{3} \cdot \left[\frac{\frac{9}{2} \cdot \frac{\pi \cdot c}{\varepsilon_0 \cdot n_r^2 \cdot \hbar \cdot \lambda_0} \cdot (\frac{\xi \cdot \tau_{in}}{V})^2 \cdot a \cdot |R_{cv}|^2}{(\frac{2 \cdot \pi \cdot c \cdot \tau_{in}}{\lambda_p^2})^2 \cdot (\Delta\lambda_{p=3-q})^2 + 1} \right] \cdot (N - N_s)$$

We define new global parameters: $\Omega_0 = \frac{4}{3} \cdot [\frac{\frac{9}{2} \cdot \frac{\pi \cdot c}{\varepsilon_0 \cdot n_r^2 \cdot \hbar \cdot \lambda_0} \cdot (\frac{\xi \cdot \tau_{in}}{V})^2 \cdot a \cdot |R_{cv}|^2}{(\frac{2 \cdot \pi \cdot c \cdot \tau_{in}}{\lambda_p^2})^2 \cdot (\Delta\lambda_{p=0-q})^2 + 1}]$

$$\Omega_1 = \frac{4}{3} \cdot [\frac{\frac{9}{2} \cdot \frac{\pi \cdot c}{\varepsilon_0 \cdot n_r^2 \cdot \hbar \cdot \lambda_0} \cdot (\frac{\xi \cdot \tau_{in}}{V})^2 \cdot a \cdot |R_{cv}|^2}{(\frac{2 \cdot \pi \cdot c \cdot \tau_{in}}{\lambda_p^2})^2 \cdot (\Delta\lambda_{p=1-q})^2 + 1}]$$

$$\Omega_2 = \frac{4}{3} \cdot [\frac{\frac{9}{2} \cdot \frac{\pi \cdot c}{\varepsilon_0 \cdot n_r^2 \cdot \hbar \cdot \lambda_0} \cdot (\frac{\xi \cdot \tau_{in}}{V})^2 \cdot a \cdot |R_{cv}|^2}{(\frac{2 \cdot \pi \cdot c \cdot \tau_{in}}{\lambda_p^2})^2 \cdot (\Delta\lambda_{p=2-q})^2 + 1}]$$

$$\Omega_3 = \frac{4}{3} \cdot [\frac{\frac{9}{2} \cdot \frac{\pi \cdot c}{\varepsilon_0 \cdot n_r^2 \cdot \hbar \cdot \lambda_0} \cdot (\frac{\xi \cdot \tau_{in}}{V})^2 \cdot a \cdot |R_{cv}|^2}{(\frac{2 \cdot \pi \cdot c \cdot \tau_{in}}{\lambda_p^2})^2 \cdot (\Delta\lambda_{p=3-q})^2 + 1}]$$

$$D_{p=0} = \Omega_0 \cdot (N - N_s);\ D_{p=1} = \Omega_1 \cdot (N - N_s)$$

$$D_{p=2} = \Omega_2 \cdot (N - N_s);\ D_{p=3} = \Omega_3 \cdot (N - N_s)$$

$$\sum_{q \neq p} \{D_{p(q)}(p, q, N) + H_{p(q)}(p, q, N)\} \cdot S_q = \sum_{\substack{p=0 \\ p \neq q}}^{3} \{D_{p(q)}(p, q, N) + H_{p(q)}(p, q, N)\} \cdot S_q$$

$$\sum_{q \neq p} \{D_{p(q)}(p, q, N) + H_{p(q)}(p, q, N)\} \cdot S_q = \{\Omega_0 \cdot (N - N_s) + \Omega_1 \cdot (N - N_s) + \Omega_2 \cdot (N - N_s) + \Omega_3 \cdot (N - N_s) + \Psi_0 \cdot (N - N_g) + \Psi_1 \cdot (N - N_g)$$

$$+ \Psi_2 \cdot (N - N_g) + \Psi_3 \cdot (N - N_g)\} \cdot S_q$$

$$\sum_{q \neq p} \{D_{p(q)}(p, q, N) + H_{p(q)}(p, q, N)\} \cdot S_q$$
$$= \{(N - N_s) \cdot \sum_{k=0}^{3} \Omega_k + (N - N_g) \cdot \sum_{k=0}^{3} \Psi_k\} \cdot S_q$$

We can rewrite our system rate equations again based on assumptions A – D.

$$\frac{dS_p}{dt} = (G_p - G_{th}) \cdot S_p + \frac{a \cdot \xi \cdot N}{V \cdot \{[\frac{2 \cdot (\lambda_p - \lambda_0)}{\delta\lambda}]^2 + 1\}}$$
$$\frac{dN}{dt} = -\Omega_0 - \frac{N}{\tau_s} + \frac{I}{e}$$

$$G_p = A_p(N) - B(N) \cdot S_p - \sum_{q \neq p} \{D_{p(q)}(p, q, N) + H_{p(q)}(p, q, N)\} \cdot S_q$$

$$G_p = \frac{a \cdot \xi}{V} \cdot \{N - N_g - b \cdot V \cdot (\lambda_p - \lambda_0)^2\}$$
$$- \frac{9}{2} \cdot \frac{\pi \cdot c}{\varepsilon_0 \cdot n_r^2 \cdot \hbar \cdot \lambda_0} \cdot (\frac{\xi \cdot \tau_{in}}{V})^2 \cdot a \cdot |R_{cv}|^2 \cdot (N - N_s) \cdot S_p$$
$$- \{(N - N_s) \cdot \sum_{k=0}^{3} \Omega_k + (N - N_g) \cdot \sum_{k=0}^{3} \Psi_k\} \cdot S_q; \; G_p = G_p(N, S_p)$$

Our system rate equations variables are S_p and N. At fixed points (equilibrium points) $\frac{dS_p}{dt} = 0$; $\frac{dN}{dt} = 0$.

$$(G_p(N^*, S_p^*) - G_{th}) \cdot S_p^* + \frac{a \cdot \xi \cdot N^*}{V \cdot \{[\frac{2 \cdot (\lambda_p - \lambda_0)}{\delta\lambda}]^2 + 1\}} = 0;$$
$$- \Omega_0 - \frac{N^*}{\tau_s} + \frac{I}{e} = 0$$

$$-\Omega_0 - \frac{N^*}{\tau_s} + \frac{I}{e} = 0 \Rightarrow N^* = (\frac{I}{e} - \Omega_0) \cdot \tau_s$$

$$G_p(N^*, S_p^*) = \frac{a \cdot \xi}{V} \cdot \{N^* - N_g - b \cdot V \cdot (\lambda_p - \lambda_0)^2\}$$
$$- \frac{9}{2} \cdot \frac{\pi \cdot c}{\varepsilon_0 \cdot n_r^2 \cdot \hbar \cdot \lambda_0} \cdot (\frac{\xi \cdot \tau_{in}}{V})^2 \cdot a \cdot |R_{cv}|^2 \cdot (N^* - N_s) \cdot S_p^*$$
$$- \{(N^* - N_s) \cdot \sum_{k=0}^{3} \Omega_k + (N^* - N_g) \cdot \sum_{k=0}^{3} \Psi_k\} \cdot S_q$$

$$G_p(N^* = (\frac{I}{e} - \Omega_0) \cdot \tau_s, S_p^*)$$
$$= \frac{a \cdot \xi}{V} \cdot \{(\frac{I}{e} - \Omega_0) \cdot \tau_s - N_g - b \cdot V \cdot (\lambda_p - \lambda_0)^2\}$$
$$- \frac{9}{2} \cdot \frac{\pi \cdot c}{\varepsilon_0 \cdot n_r^2 \cdot \hbar \cdot \lambda_0} \cdot (\frac{\xi \cdot \tau_{in}}{V})^2 \cdot a \cdot |R_{cv}|^2 \cdot \{(\frac{I}{e} - \Omega_0) \cdot \tau_s - N_s\} \cdot S_p^*$$
$$- \{[(\frac{I}{e} - \Omega_0) \cdot \tau_s - N_s] \cdot \sum_{k=0}^{3} \Omega_k + (N^* - N_g) \cdot \sum_{k=0}^{3} \Psi_k\} \cdot S_q$$

We define for simplicity new global parameters: $\Xi_1 = \frac{a \cdot \xi}{V} \cdot \{(\frac{I}{e} - \Omega_0) \cdot \tau_s - N_g - b \cdot V \cdot (\lambda_p - \lambda_0)^2\}$

$$\Xi_2 = \frac{9}{2} \cdot \frac{\pi \cdot c}{\varepsilon_0 \cdot n_r^2 \cdot \hbar \cdot \lambda_0} \cdot (\frac{\xi \cdot \tau_{in}}{V})^2 \cdot a \cdot |R_{cv}|^2 \cdot \{(\frac{I}{e} - \Omega_0) \cdot \tau_s - N_s\}$$

$$\Xi_3 = \{[(\frac{I}{e} - \Omega_0) \cdot \tau_s - N_s] \cdot \sum_{k=0}^{3} \Omega_k + (N^* - N_g) \cdot \sum_{k=0}^{3} \Psi_k\} \cdot S_q$$

$$G_p(N^* = (\frac{I}{e} - \Omega_0) \cdot \tau_s, S_p^*) = \Xi_1 - \Xi_2 \cdot S_p^* - \Xi_3$$
$$G_p(N^* = (\frac{I}{e} - \Omega_0) \cdot \tau_s, S_p^*) = (\Xi_1 - \Xi_3) - \Xi_2 \cdot S_p^*$$

$$(G_p(N^*, S_p^*) - G_{th}) \cdot S_p^* + \frac{a \cdot \xi \cdot N^*}{V \cdot \{[\frac{2 \cdot (\lambda_p - \lambda_0)}{\delta\lambda}]^2 + 1\}} = 0$$

$$((\Xi_1 - \Xi_3) - \Xi_2 \cdot S_p^* - G_{th}) \cdot S_p^* + \frac{a \cdot \xi \cdot [(\frac{I}{e} - \Omega_0) \cdot \tau_s]}{V \cdot \{[\frac{2 \cdot (\lambda_p - \lambda_0)}{\delta\lambda}]^2 + 1\}} = 0$$

$$(\Xi_1 - \Xi_3 - G_{th}) \cdot S_p^* - \Xi_2 \cdot (S_p^*)^2 + \frac{a \cdot \xi \cdot [(\frac{I}{e} - \Omega_0) \cdot \tau_s]}{V \cdot \{[\frac{2 \cdot (\lambda_p - \lambda_0)}{\delta\lambda}]^2 + 1\}} = 0$$

$$\Xi_2 \cdot (S_p^*)^2 - (\Xi_1 - \Xi_3 - G_{th}) \cdot S_p^* - \frac{a \cdot \xi \cdot [(\frac{I}{e} - \Omega_0) \cdot \tau_s]}{V \cdot \{[\frac{2 \cdot (\lambda_p - \lambda_0)}{\delta\lambda}]^2 + 1\}} = 0$$

$$S_{p(1)}^* = \frac{(\Xi_1 - \Xi_3 - G_{th}) + \sqrt{(\Xi_1 - \Xi_3 - G_{th})^2 + 4 \cdot \Xi_2 \cdot \frac{a \cdot \xi \cdot [(\frac{I}{e} - \Omega_0) \cdot \tau_s]}{V \cdot \{[\frac{2 \cdot (\lambda_p - \lambda_0)}{\delta\lambda}]^2 + 1\}}}}{2 \cdot \Xi_2}$$

$$S_{p(2)}^* = \frac{(\Xi_1 - \Xi_3 - G_{th}) - \sqrt{(\Xi_1 - \Xi_3 - G_{th})^2 + 4 \cdot \Xi_2 \cdot \frac{a \cdot \xi \cdot [(\frac{I}{e} - \Omega_0) \cdot \tau_s]}{V \cdot \{[\frac{2 \cdot (\lambda_p - \lambda_0)}{\delta\lambda}]^2 + 1\}}}}{2 \cdot \Xi_2}$$

We have two fixed points in our system.

$$E^{(0)}(S_p^{(0)}, N^{(0)}) = \left(\frac{(\Xi_1 - \Xi_3 - G_{th}) + \sqrt{(\Xi_1 - \Xi_3 - G_{th})^2 + 4 \cdot \Xi_2 \cdot \frac{a \cdot \xi \cdot [(\frac{I}{e} - \Omega_0) \cdot \tau_s]}{V \cdot \{[\frac{2 \cdot (\lambda_p - \lambda_0)}{\delta\lambda}]^2 + 1\}}}}{2 \cdot \Xi_2}, (\frac{I}{e} - \Omega_0) \cdot \tau_s\right)$$

$$E^{(1)}(S_p^{(1)}, N^{(1)}) = \left(\frac{(\Xi_1 - \Xi_3 - G_{th}) - \sqrt{(\Xi_1 - \Xi_3 - G_{th})^2 + 4 \cdot \Xi_2 \cdot \frac{a \cdot \xi \cdot [(\frac{I}{e} - \Omega_0) \cdot \tau_s]}{V \cdot \{[\frac{2 \cdot (\lambda_p - \lambda_0)}{\delta\lambda}]^2 + 1\}}}}{2 \cdot \Xi_2}, (\frac{I}{e} - \Omega_0) \cdot \tau_s\right)$$

Stability analysis: The standard local stability analysis about any one of the equilibrium points (fixed points) of long-wavelength semiconductor laser rate equations consists in adding to its coordinated $[S_p\ N]$ arbitrarily small increments of exponential terms $[s_p\ n] \cdot e^{\lambda \cdot t}$, and retaining the first order terms in $s_p\ n$. The system of two homogeneous equations leads to a polynomial characteristic equation in the eigenvalue λ. The polynomial characteristic equation accepts by set the long-wavelength semiconductor laser rate equations. The long-wavelength semiconductor laser rate equations fixed values with arbitrarily small increments of exponential form $[s_p\ n] \cdot e^{\lambda \cdot t}$ are i = 0 (first fixed point), i = 1 (second fixed point), i = 2 (third fixed point), etc. [8–11].

Hint: please take care that we have one variable with two meanings, n—Injected density of electrons ($n = \frac{N}{V}$) and stability analysis arbitrarily small increment n.

$$S_p(t) = S_p^{(i)} + s_p \cdot e^{\lambda \cdot t};\ N(t) = N^{(i)} + n \cdot e^{\lambda \cdot t}$$

$$\frac{dS_p(t)}{dt} = s_p \cdot \lambda \cdot e^{\lambda \cdot t};\ \frac{dN(t)}{dt} = n \cdot \lambda \cdot e^{\lambda \cdot t}$$

#######

$$\frac{dS_p}{dt} = (G_p - G_{th}) \cdot S_p + \frac{a \cdot \xi \cdot N}{V \cdot \{[\frac{2 \cdot (\lambda_p - \lambda_0)}{\delta\lambda}]^2 + 1\}}$$

$$G_p = \frac{a \cdot \xi}{V} \cdot \{N - N_g - b \cdot V \cdot (\lambda_p - \lambda_0)^2\} - \frac{9}{2} \cdot \frac{\pi \cdot c}{\varepsilon_0 \cdot n_r^2 \cdot \hbar \cdot \lambda_0} \cdot (\frac{\xi \cdot \tau_{in}}{V})^2 \cdot a \cdot |R_{cv}|^2 \cdot (N - N_s) \cdot S_p$$

$$- \{(N - N_s) \cdot \sum_{k=0}^{3} \Omega_k + (N - N_g) \cdot \sum_{k=0}^{3} \Psi_k\} \cdot S_q; G_p = G_p(N, S_p)$$

$$G_p = \frac{a \cdot \xi}{V} \cdot \{N^{(i)} + n \cdot e^{\lambda \cdot t} - N_g - b \cdot V \cdot (\lambda_p - \lambda_0)^2\}$$
$$- \frac{9}{2} \cdot \frac{\pi \cdot c}{\varepsilon_0 \cdot n_r^2 \cdot \hbar \cdot \lambda_0} \cdot (\frac{\xi \cdot \tau_{in}}{V})^2 \cdot a \cdot |R_{cv}|^2$$
$$\cdot (N^{(i)} + n \cdot e^{\lambda \cdot t} - N_s) \cdot (S_p^{(i)} + s_p \cdot e^{\lambda \cdot t})$$
$$- \{(N^{(i)} + n \cdot e^{\lambda \cdot t} - N_s) \cdot \sum_{k=0}^{3} \Omega_k$$
$$+ (N^{(i)} + n \cdot e^{\lambda \cdot t} - N_g) \cdot \sum_{k=0}^{3} \Psi_k\} \cdot S_q$$

$$G_p = \frac{a \cdot \xi}{V} \cdot \{N^{(i)} - N_g - b \cdot V \cdot (\lambda_p - \lambda_0)^2\} + \frac{a \cdot \xi}{V} \cdot n \cdot e^{\lambda \cdot t}$$
$$- \frac{9}{2} \cdot \frac{\pi \cdot c}{\varepsilon_0 \cdot n_r^2 \cdot \hbar \cdot \lambda_0} \cdot (\frac{\xi \cdot \tau_{in}}{V})^2 \cdot a \cdot |R_{cv}|^2 \cdot ([N^{(i)} - N_s]$$
$$\cdot S_p^{(i)} + [N^{(i)} - N_s] \cdot s_p \cdot e^{\lambda \cdot t} + S_p^{(i)} \cdot n \cdot e^{\lambda \cdot t} + n \cdot s_p \cdot e^{2 \cdot \lambda \cdot t})$$
$$- \{[N^{(i)} - N_s] \cdot \sum_{k=0}^{3} \Omega_k + n \cdot e^{\lambda \cdot t} \cdot \sum_{k=0}^{3} \Omega_k + [N^{(i)} - N_g]$$
$$\cdot \sum_{k=0}^{3} \Psi_k + n \cdot e^{\lambda \cdot t} \cdot \sum_{k=0}^{3} \Psi_k\} \cdot S_q$$

$$G_p = \frac{a \cdot \xi}{V} \cdot \{N^{(i)} - N_g - b \cdot V \cdot (\lambda_p - \lambda_0)^2\}$$
$$- \frac{9}{2} \cdot \frac{\pi \cdot c}{\varepsilon_0 \cdot n_r^2 \cdot \hbar \cdot \lambda_0} \cdot (\frac{\xi \cdot \tau_{in}}{V})^2 \cdot a \cdot |R_{cv}|^2 \cdot ([N^{(i)} - N_s] \cdot S_p^{(i)})$$
$$- \{[N^{(i)} - N_s] \cdot \sum_{k=0}^{3} \Omega_k + [N^{(i)} - N_g] \cdot \sum_{k=0}^{3} \Psi_k\} \cdot S_q + \frac{a \cdot \xi}{V} \cdot n \cdot e^{\lambda \cdot t}$$
$$- \frac{9}{2} \cdot \frac{\pi \cdot c}{\varepsilon_0 \cdot n_r^2 \cdot \hbar \cdot \lambda_0} \cdot (\frac{\xi \cdot \tau_{in}}{V})^2 \cdot a \cdot |R_{cv}|^2 \cdot ([N^{(i)} - N_s] \cdot s_p$$
$$\cdot e^{\lambda \cdot t} + S_p^{(i)} \cdot n \cdot e^{\lambda \cdot t} + n \cdot s_p \cdot e^{2 \cdot \lambda \cdot t})$$
$$- \{\sum_{k=0}^{3} \Omega_k + \sum_{k=0}^{3} \Psi_k\} \cdot S_q \cdot n \cdot e^{\lambda \cdot t}; n \cdot s_p \approx 0$$

$$G_p = \frac{a \cdot \xi}{V} \cdot \{N^{(i)} - N_g - b \cdot V \cdot (\lambda_p - \lambda_0)^2\}$$

$$- \frac{9}{2} \cdot \frac{\pi \cdot c}{\varepsilon_0 \cdot n_r^2 \cdot \hbar \cdot \lambda_0} \cdot (\frac{\xi \cdot \tau_{in}}{V})^2 \cdot a \cdot |R_{cv}|^2 \cdot ([N^{(i)} - N_s] \cdot S_p^{(i)})$$

$$- \{[N^{(i)} - N_s] \cdot \sum_{k=0}^{3} \Omega_k + [N^{(i)} - N_g] \cdot \sum_{k=0}^{3} \Psi_k\} \cdot S_q + \frac{a \cdot \xi}{V} \cdot n \cdot e^{\lambda \cdot t}$$

$$- \frac{9}{2} \cdot \frac{\pi \cdot c}{\varepsilon_0 \cdot n_r^2 \cdot \hbar \cdot \lambda_0} \cdot (\frac{\xi \cdot \tau_{in}}{V})^2 \cdot a \cdot |R_{cv}|^2 \cdot ([N^{(i)} - N_s]$$

$$\cdot s_p \cdot e^{\lambda \cdot t} + S_p^{(i)} \cdot n \cdot e^{\lambda \cdot t}) - \{\sum_{k=0}^{3} \Omega_k + \sum_{k=0}^{3} \Psi_k\} \cdot S_q \cdot n \cdot e^{\lambda \cdot t}$$

$$G_p = \frac{a \cdot \xi}{V} \cdot \{N^{(i)} - N_g - b \cdot V \cdot (\lambda_p - \lambda_0)^2\}$$

$$- \frac{9}{2} \cdot \frac{\pi \cdot c}{\varepsilon_0 \cdot n_r^2 \cdot \hbar \cdot \lambda_0} \cdot (\frac{\xi \cdot \tau_{in}}{V})^2 \cdot a \cdot |R_{cv}|^2 \cdot ([N^{(i)} - N_s] \cdot S_p^{(i)})$$

$$- \{[N^{(i)} - N_s] \cdot \sum_{k=0}^{3} \Omega_k + [N^{(i)} - N_g] \cdot \sum_{k=0}^{3} \Psi_k\} \cdot S_q$$

$$+ [\frac{a \cdot \xi}{V} - \frac{9}{2} \cdot \frac{\pi \cdot c}{\varepsilon_0 \cdot n_r^2 \cdot \hbar \cdot \lambda_0} \cdot (\frac{\xi \cdot \tau_{in}}{V})^2 \cdot a \cdot |R_{cv}|^2 \cdot S_p^{(i)}$$

$$- \{\sum_{k=0}^{3} \Omega_k + \sum_{k=0}^{3} \Psi_k\} \cdot S_q] \cdot n \cdot e^{\lambda \cdot t}$$

$$- \frac{9}{2} \cdot \frac{\pi \cdot c}{\varepsilon_0 \cdot n_r^2 \cdot \hbar \cdot \lambda_0} \cdot (\frac{\xi \cdot \tau_{in}}{V})^2 \cdot a \cdot |R_{cv}|^2 \cdot [N^{(i)} - N_s] \cdot s_p \cdot e^{\lambda \cdot t}$$

We define two new global parameters for simplicity Φ_1, Φ_2.

$$\Phi_1 = [\frac{a \cdot \xi}{V} - \frac{9}{2} \cdot \frac{\pi \cdot c}{\varepsilon_0 \cdot n_r^2 \cdot \hbar \cdot \lambda_0} \cdot (\frac{\xi \cdot \tau_{in}}{V})^2 \cdot a \cdot |R_{cv}|^2 \cdot S_p^{(i)}$$

$$- \{\sum_{k=0}^{3} \Omega_k + \sum_{k=0}^{3} \Psi_k\} \cdot S_q]$$

$$\Phi_2 = \frac{9}{2} \cdot \frac{\pi \cdot c}{\varepsilon_0 \cdot n_r^2 \cdot \hbar \cdot \lambda_0} \cdot (\frac{\xi \cdot \tau_{in}}{V})^2 \cdot a \cdot |R_{cv}|^2 \cdot [N^{(i)} - N_s]$$

$$G_p = \frac{a \cdot \xi}{V} \cdot \{N^{(i)} - N_g - b \cdot V \cdot (\lambda_p - \lambda_0)^2\}$$

$$- \frac{9}{2} \cdot \frac{\pi \cdot c}{\varepsilon_0 \cdot n_r^2 \cdot \hbar \cdot \lambda_0} \cdot (\frac{\xi \cdot \tau_{in}}{V})^2 \cdot a \cdot |R_{cv}|^2 \cdot ([N^{(i)} - N_s] \cdot S_p^{(i)})$$

$$- \{[N^{(i)} - N_s] \cdot \sum_{k=0}^{3} \Omega_k + [N^{(i)} - N_g] \cdot \sum_{k=0}^{3} \Psi_k\} \cdot S_q$$

$$+ \Phi_1 \cdot n \cdot e^{\lambda \cdot t} - \Phi_2 \cdot s_p \cdot e^{\lambda \cdot t}$$

$$G_p(N^*, S_p^*) = \frac{a \cdot \xi}{V} \cdot \{N^{(i)} - N_g - b \cdot V \cdot (\lambda_p - \lambda_0)^2\}$$
$$- \frac{9}{2} \cdot \frac{\pi \cdot c}{\varepsilon_0 \cdot n_r^2 \cdot \hbar \cdot \lambda_0} \cdot (\frac{\xi \cdot \tau_{in}}{V})^2 \cdot a \cdot |R_{cv}|^2$$
$$\cdot ([N^{(i)} - N_s] \cdot S_p^{(i)})$$

$$G_p(S_p, N) = G_p(N^*, S_p^*) + \Phi_1 \cdot n \cdot e^{\lambda \cdot t} - \Phi_2 \cdot s_p \cdot e^{\lambda \cdot t}$$

$$\frac{dS_p}{dt} = (G_p - G_{th}) \cdot S_p + \frac{a \cdot \xi \cdot N}{V \cdot \{[\frac{2 \cdot (\lambda_p - \lambda_0)}{\delta\lambda}]^2 + 1\}}$$

$$s_p \cdot \lambda \cdot e^{\lambda \cdot t} = (G_p(S_p, N) - G_{th}) \cdot (S_p^{(i)} + s_p \cdot e^{\lambda \cdot t}) + \frac{a \cdot \xi \cdot (N^{(i)} + n \cdot e^{\lambda \cdot t})}{V \cdot \{[\frac{2 \cdot (\lambda_p - \lambda_0)}{\delta\lambda}]^2 + 1\}}$$

$$s_p \cdot \lambda \cdot e^{\lambda \cdot t} = (G_p(N^*, S_p^*) + \Phi_1 \cdot n \cdot e^{\lambda \cdot t} - \Phi_2 \cdot s_p \cdot e^{\lambda \cdot t} - G_{th})$$
$$\cdot (S_p^{(i)} + s_p \cdot e^{\lambda \cdot t}) + \frac{a \cdot \xi \cdot (N^{(i)} + n \cdot e^{\lambda \cdot t})}{V \cdot \{[\frac{2 \cdot (\lambda_p - \lambda_0)}{\delta\lambda}]^2 + 1\}}$$

$$s_p \cdot \lambda \cdot e^{\lambda \cdot t} = ([G_p(N^*, S_p^*) - G_{th}] + \Phi_1 \cdot n \cdot e^{\lambda \cdot t} - \Phi_2 \cdot s_p \cdot e^{\lambda \cdot t}) \cdot (S_p^{(i)} + s_p \cdot e^{\lambda \cdot t})$$
$$+ \frac{a \cdot \xi \cdot N^{(i)}}{V \cdot \{[\frac{2 \cdot (\lambda_p - \lambda_0)}{\delta\lambda}]^2 + 1\}} + \frac{a \cdot \xi \cdot n \cdot e^{\lambda \cdot t}}{V \cdot \{[\frac{2 \cdot (\lambda_p - \lambda_0)}{\delta\lambda}]^2 + 1\}}$$

$$s_p \cdot \lambda \cdot e^{\lambda \cdot t} = [G_p(N^*, S_p^*) - G_{th}] \cdot S_p^{(i)} + \Phi_1 \cdot S_p^{(i)} \cdot n \cdot e^{\lambda \cdot t} - \Phi_2 \cdot S_p^{(i)} \cdot s_p \cdot e^{\lambda \cdot t}$$
$$+ [G_p(N^*, S_p^*) - G_{th}] \cdot s_p \cdot e^{\lambda \cdot t} + \Phi_1 \cdot n \cdot s_p \cdot e^{2 \cdot \lambda \cdot t} - \Phi_2 \cdot [s_p]^2 \cdot e^{2 \cdot \lambda \cdot t}$$
$$+ \frac{a \cdot \xi \cdot N^{(i)}}{V \cdot \{[\frac{2 \cdot (\lambda_p - \lambda_0)}{\delta\lambda}]^2 + 1\}} + \frac{a \cdot \xi \cdot n \cdot e^{\lambda \cdot t}}{V \cdot \{[\frac{2 \cdot (\lambda_p - \lambda_0)}{\delta\lambda}]^2 + 1\}}$$

$$s_p \cdot \lambda \cdot e^{\lambda \cdot t} = [G_p(N^*, S_p^*) - G_{th}] \cdot S_p^{(i)} + \frac{a \cdot \xi \cdot N^{(i)}}{V \cdot \{[\frac{2 \cdot (\lambda_p - \lambda_0)}{\delta\lambda}]^2 + 1\}}$$
$$+ \Phi_1 \cdot S_p^{(i)} \cdot n \cdot e^{\lambda \cdot t} - \Phi_2 \cdot S_p^{(i)} \cdot s_p \cdot e^{\lambda \cdot t}$$
$$+ [G_p(N^*, S_p^*) - G_{th}] \cdot s_p \cdot e^{\lambda \cdot t} + \Phi_1 \cdot n \cdot s_p \cdot e^{2 \cdot \lambda \cdot t}$$
$$- \Phi_2 \cdot [s_p]^2 \cdot e^{2 \cdot \lambda \cdot t} + \frac{a \cdot \xi}{V \cdot \{[\frac{2 \cdot (\lambda_p - \lambda_0)}{\delta\lambda}]^2 + 1\}} \cdot n \cdot e^{\lambda \cdot t}$$

At fixed points $[G_p(N^*, S_p^*) - G_{th}] \cdot S_p^{(i)} + \frac{a \cdot \xi \cdot N^{(i)}}{V \cdot \{[\frac{2 \cdot (\lambda_p - \lambda_0)}{\delta\lambda}]^2 + 1\}} = 0$

$$s_p \cdot \lambda \cdot e^{\lambda \cdot t} = \Phi_1 \cdot S_p^{(i)} \cdot n \cdot e^{\lambda \cdot t} - \Phi_2 \cdot S_p^{(i)} \cdot s_p \cdot e^{\lambda \cdot t}$$
$$+ [G_p(N^*, S_p^*) - G_{th}] \cdot s_p \cdot e^{\lambda \cdot t} + \Phi_1 \cdot n \cdot s_p \cdot e^{2 \cdot \lambda \cdot t}$$

$$- \Phi_2 \cdot [s_p]^2 \cdot e^{2 \cdot \lambda \cdot t} + \frac{a \cdot \xi}{V \cdot \{[\frac{2 \cdot (\lambda_p - \lambda_0)}{\delta \lambda}]^2 + 1\}} \cdot n \cdot e^{\lambda \cdot t}$$

Assumption: $n \cdot s_p \approx 0; [s_p]^2 \approx 0$

$$\begin{aligned} s_p \cdot \lambda \cdot e^{\lambda \cdot t} &= \Phi_1 \cdot S_p^{(i)} \cdot n \cdot e^{\lambda \cdot t} - \Phi_2 \cdot S_p^{(i)} \cdot s_p \cdot e^{\lambda \cdot t} \\ &+ [G_p(N^*, S_p^*) - G_{th}] \cdot s_p \cdot e^{\lambda \cdot t} \\ &+ \frac{a \cdot \xi}{V \cdot \{[\frac{2 \cdot (\lambda_p - \lambda_0)}{\delta \lambda}]^2 + 1\}} \cdot n \cdot e^{\lambda \cdot t} \end{aligned}$$

By dividing the two sides of the above equation by $e^{\lambda \cdot t}$ term we get.

$$\begin{aligned} &- s_p \cdot \lambda + [G_p(N^*, S_p^*) - G_{th} - \Phi_2 \cdot S_p^{(i)}] \cdot s_p \\ &+ [\frac{a \cdot \xi}{V \cdot \{[\frac{2 \cdot (\lambda_p - \lambda_0)}{\delta \lambda}]^2 + 1\}} + \Phi_1 \cdot S_p^{(i)}] \cdot n = 0 \\ &\frac{dN}{dt} = -\Omega_0 - \frac{N}{\tau_s} + \frac{I}{e} \\ &n \cdot \lambda \cdot e^{\lambda \cdot t} = -\Omega_0 - \frac{N^{(i)} + n \cdot e^{\lambda \cdot t}}{\tau_s} + \frac{I}{e} \\ &n \cdot \lambda \cdot e^{\lambda \cdot t} = -\Omega_0 - \frac{N^{(i)}}{\tau_s} + \frac{I}{e} - \frac{n \cdot e^{\lambda \cdot t}}{\tau_s} \end{aligned}$$

At fixed points$-\Omega_0 - \frac{N^{(i)}}{\tau_s} + \frac{I}{e} = 0$ then $-n \cdot \lambda - \frac{n}{\tau_s} = 0$.

We summary our long-wavelength semiconductor laser rate small increments equations:

$$\begin{aligned} &- s_p \cdot \lambda + [G_p(N^*, S_p^*) - G_{th} - \Phi_2 \cdot S_p^{(i)}] \cdot s_p \\ &+ [\frac{a \cdot \xi}{V \cdot \{[\frac{2 \cdot (\lambda_p - \lambda_0)}{\delta \lambda}]^2 + 1\}} + \Phi_1 \cdot S_p^{(i)}] \cdot n = 0 \\ &- n \cdot \lambda - \frac{n}{\tau_s} = 0 \end{aligned}$$

$$\begin{pmatrix} -\lambda + [G_p(N^*, S_p^*) - G_{th} - \Phi_2 \cdot S_p^{(i)}] & [\frac{a \cdot \xi}{V \cdot \{[\frac{2 \cdot (\lambda_p - \lambda_0)}{\delta \lambda}]^2 + 1\}} + \Phi_1 \cdot S_p^{(i)}] \\ 0 & -\lambda - \frac{1}{\tau_s} \end{pmatrix} \cdot \begin{pmatrix} s_p \\ n \end{pmatrix} = 0$$

$$(A - \lambda \cdot I) = \begin{pmatrix} -\lambda + [G_p(N^*, S_p^*) - G_{th} - \Phi_2 \cdot S_p^{(i)}] & [\frac{a \cdot \xi}{V \cdot \{[\frac{2 \cdot (\lambda_p - \lambda_0)}{\delta \lambda}]^2 + 1\}} + \Phi_1 \cdot S_p^{(i)}] \\ 0 & -\lambda - \frac{1}{\tau_s} \end{pmatrix}$$

$$\det(A - \lambda \cdot I) = 0 \Rightarrow (-\lambda + [G_p(N^*, S_p^*) - G_{th} - \Phi_2 \cdot S_p^{(i)}]) \cdot (-\lambda - \frac{1}{\tau_s}) = 0$$

The characteristic equation of our system:

$$\lambda^2 + \lambda \cdot (\frac{1}{\tau_s} - [G_p(N^*, S_p^*) - G_{th} - \Phi_2 \cdot S_p^{(i)}])$$
$$- [G_p(N^*, S_p^*) - G_{th} - \Phi_2 \cdot S_p^{(i)}] \cdot \frac{1}{\tau_s} = 0$$

$$\lambda^2 - \lambda \cdot (-\frac{1}{\tau_s} + [G_p(N^*, S_p^*) - G_{th} - \Phi_2 \cdot S_p^{(i)}])$$
$$+ [G_{th} + \Phi_2 \cdot S_p^{(i)} - G_p(N^*, S_p^*)] \cdot \frac{1}{\tau_s} = 0$$

$$A = \begin{pmatrix} [G_p(N^*, S_p^*) - G_{th} - \Phi_2 \cdot S_p^{(i)}] & [\frac{a \cdot \xi}{V \cdot \{[\frac{2 \cdot (\lambda_p - \lambda_0)}{\delta\lambda}]^2 + 1\}} + \Phi_1 \cdot S_p^{(i)}] \\ 0 & -\frac{1}{\tau_s} \end{pmatrix}$$

We define Δ_1 as Trace (A) and Δ_2 as Det (A) elements for stability analysis.

$$\Delta_1 = -\frac{1}{\tau_s} + G_p(N^*, S_p^*) - G_{th} - \Phi_2 \cdot S_p^{(i)}$$

$$\Delta_2 = [G_{th} + \Phi_2 \cdot S_p^{(i)} - G_p(N^*, S_p^*)] \cdot \frac{1}{\tau_s}$$

$$\lambda_{1,2} = \frac{1}{2}(\Delta_1 \pm \sqrt{\Delta_1^2 - 4 \cdot \Delta_2})$$

$$\lambda_1 = \frac{-\frac{1}{\tau_s} + [G_p(N^*, S_p^*) + G_{th} + \Phi_2 \cdot S_p^{(i)}] + \sqrt{\begin{array}{c}(-\frac{1}{\tau_s} + G_p(N^*, S_p^*) - G_{th} - \Phi_2 \cdot S_p^{(i)})^2 \\ +4 \cdot [G_p(N^*, S_p^*) - G_{th} - \Phi_2 \cdot S_p^{(i)}] \cdot \frac{1}{\tau_s}\end{array}}}{2}$$

$$\lambda_2 = \frac{-\frac{1}{\tau_s} + [G_p(N^*, S_p^*) + G_{th} + \Phi_2 \cdot S_p^{(i)}] - \sqrt{\begin{array}{c}(-\frac{1}{\tau_s} + G_p(N^*, S_p^*) - G_{th} - \Phi_2 \cdot S_p^{(i)})^2 \\ +4 \cdot [G_p(N^*, S_p^*) - G_{th} - \Phi_2 \cdot S_p^{(i)}] \cdot \frac{1}{\tau_s}\end{array}}}{2}$$

Are the solutions of the quadratic equation in λ, eigenvalues dependent only on the trace and determinant of the matrix A. We need to classify system fixed points. The following formulas exist:

$$\lambda_{1,2} = \frac{-\frac{1}{\tau_s} + [G_p(N^*, S_p^*) + G_{th} + \Phi_2 \cdot S_p^{(i)}] \pm \sqrt{\begin{array}{l}(-\frac{1}{\tau_s} + G_p(N^*, S_p^*) - G_{th} - \Phi_2 \cdot S_p^{(i)})^2 \\ +4 \cdot [G_p(N^*, S_p^*) - G_{th} - \Phi_2 \cdot S_p^{(i)}] \cdot \frac{1}{\tau_s}\end{array}}}{2}$$

$\Delta_1 = \sum_{k=1}^{2} \lambda_k$; $\Delta_2 = \prod_{k=1}^{2} \lambda_k$. If $\Delta_2 = [G_{th} + \Phi_2 \cdot S_p^{(i)} - G_p(N^*, S_p^*)] \cdot \frac{1}{\tau_s} < 0$, the eigenvalues are real and have opposite signs; hence the fixed point is a saddle point.

If $\Delta_2 = [G_{th} + \Phi_2 \cdot S_p^{(i)} - G_p(N^*, S_p^*)] \cdot \frac{1}{\tau_s} > 0$, the eigenvalues are either real with the same sign (nodes), or complex conjugate (spiral and center). Nodes satisfy $\Delta_1^2 - 4 \cdot \Delta_2 > 0$, $(-\frac{1}{\tau_s} + G_p(N^*, S_p^*) - G_{th} - \Phi_2 \cdot S_p^{(i)})^2 + 4 \cdot [G_p(N^*, S_p^*) - G_{th} - \Phi_2 \cdot S_p^{(i)}] \cdot \frac{1}{\tau_s} > 0$ And spirals satisfy $\Delta_1^2 - 4 \cdot \Delta_2 < 0$; $(-\frac{1}{\tau_s} + G_p(N^*, S_p^*) - G_{th} - \Phi_2 \cdot S_p^{(i)})^2 + 4 \cdot [G_p(N^*, S_p^*) - G_{th} - \Phi_2 \cdot S_p^{(i)}] \cdot \frac{1}{\tau_s} < 0$. The parabola $\Delta_1^2 - 4 \cdot \Delta_2 = 0$ is the borderline between nodes and spirals; star nodes and degenerate nodes live on this parabola. The stability of the nodes and spirals is determined by Δ_1. When $\Delta_1 < 0$, both eigenvalues have negative real parts, the fixed point is stable. Unstable spirals and nodes have $\Delta_1 > 0$. Neutrally stable centers live on the borderline$\Delta_1 = 0$, where the eigenvalues are purely imaginary. If $\Delta_2 = 0$ then at least one of the eigenvalues is zero. Then the origin is not an isolated fixed point. There is either a whole line of fixed points, or a plane of fixed points, if $A = 0$. Saddle points, nodes, and spirals are the major types of fixed points; they occur in large open regions of the (Δ_1, Δ_2) plane. Centers, stars, degenerate nodes, and non-isolated fixed points are borderline cases that occur along curves in the (Δ_1, Δ_2) plane [10].

3.4 Questions

1. We have Ion-channel Laser system that is characterized by set of perturbed differential equations. Due to additional external laser source which radiate inside our main Ion-channel path, there is additional element ($\Gamma_\Delta \cdot tg(\psi)$; $\Gamma_\Delta \in \mathbb{R}_+$) in phase change in time ($\frac{d\psi}{dt} = ...$). All system parameters are as describe on subchapter (3.1).

$$\frac{d\psi}{dt} = k_z \cdot c \cdot \left[1 - \frac{(2 + q_x^2 + q_y^2)}{4 \cdot q_z^2}\right] - \omega + \omega_\beta \cdot \left[1 - \frac{1}{2 \cdot q_y} \cdot A \cdot \cos(\psi)\right] + \frac{1}{2} \cdot k_z \cdot c \cdot \frac{q_y}{q_z^2} \cdot A \cdot \cos(\psi) + \Gamma_\Delta \cdot tg(\psi)$$

$$\frac{dq_z}{dt} = -\frac{1}{2} \cdot k_z \cdot c \cdot \frac{q_y}{q_z} \cdot A \cdot \sin(\psi); \quad \frac{dq_y}{dt} = -\frac{1}{2} \cdot (\omega_\beta + \frac{1}{4} \cdot k_z \cdot c \cdot \frac{q_y^2}{q_z^2}) \cdot A \cdot \sin(\psi)$$

$$\frac{dq_x}{dt} = -\frac{1}{8} \cdot k_z \cdot c \cdot \frac{q_y \cdot q_x}{q_z^2} \cdot A \cdot \sin(\psi)$$

1.1 Find system fixed points; plot the possible graphs that show how they change for different values of system parameters.

1.2 Plot the appropriate flow charts that show how you find system fixed points.

1.3 Discuss stability and stability switching for different values of Γ_Δ; $\Gamma_\Delta \in \mathbb{R}_+$ parameter.

1.4 Parameter $\Gamma_\Delta = \sqrt{|\psi| \cdot \sqrt{\psi^3}}$, Find system fixed points and discuss stability.

1.5 We change the additional element to $\Gamma_\Delta \cdot tg(\sqrt{\psi \cdot \sqrt{\psi + 1}})$; $\Gamma_\Delta \in \mathbb{R}_+$, find System fixed points and discuss stability and stability switching for different values of Γ_Δ; $\Gamma_\Delta \in \mathbb{R}_+$.

2. We have Ion-channel Laser system that is characterized by set of perturbed differential equations. Due to additional interference source which radiate inside our main Ion-channel path, there is additional element ($\Gamma_\Delta \cdot \sec(\psi)$; $\Gamma_\Delta \in \mathbb{R}_+$) in q_z variable change in time ($\frac{dq_z}{dt} = ...$). All system parameters are as describe on subchapter (3.1).

$$\frac{d\psi}{dt} = k_z \cdot c \cdot [1 - \frac{(2 + q_x^2 + q_y^2)}{4 \cdot q_z^2}] - \omega + \omega_\beta \cdot [1 - \frac{1}{2 \cdot q_y} \cdot A \cdot \cos(\psi)] + \frac{1}{2} \cdot k_z \cdot c \cdot \frac{q_y}{q_z^2} \cdot A \cdot \cos(\psi)$$

$$\frac{dq_z}{dt} = -\frac{1}{2} \cdot k_z \cdot c \cdot \frac{q_y}{q_z} \cdot A \cdot \sin(\psi) + \Gamma_\Delta \cdot \sec(\psi)$$

$$\frac{dq_y}{dt} = -\frac{1}{2} \cdot (\omega_\beta + \frac{1}{4} \cdot k_z \cdot c \cdot \frac{q_y^2}{q_z^2}) \cdot A \cdot \sin(\psi)$$

$$\frac{dq_x}{dt} = -\frac{1}{8} \cdot k_z \cdot c \cdot \frac{q_y \cdot q_x}{q_z^2} \cdot A \cdot \sin(\psi)$$

2.1 Find system fixed points; plot the possible graphs that show how they change for different values of system parameters.

2.2 Plot the appropriate flow charts that show how you find system fixed points.

2.3 Discuss stability and stability switching for different values of Γ_Δ; $\Gamma_\Delta \in \mathbb{R}_+$ parameter.

2.4 Parameter $\Gamma_\Delta = \sqrt{|\psi| \cdot \sqrt{\psi^5}}$, Find system fixed points and discuss stability.

2.5 We change the additional element to $\Gamma_\Delta \cdot tg(\sqrt{\psi \cdot \sqrt{\psi^2 + 1}})$; $\Gamma_\Delta \in \mathbb{R}_+$, find System fixed points and discuss stability and stability switching for different values of Γ_Δ; $\Gamma_\Delta \in \mathbb{R}_+$.

3. We have Ion-channel Laser system that is characterized by set of perturbed differential equations. Due to additional interference source which radiate inside our main Ion-channel path, there is additional element ($\Gamma_\Delta \cdot \text{cosec}(\sqrt{\psi})$; $\Gamma_\Delta \in \mathbb{R}_+$) in q_x variable change in time ($\frac{dq_x}{dt} = \ldots\ldots$). All system parameters are as describe on subchapter (3.1).

$$\frac{d\psi}{dt} = k_z \cdot c \cdot [1 - \frac{(2 + q_x^2 + q_y^2)}{4 \cdot q_z^2}] - \omega + \omega_\beta \cdot \left[1 - \frac{1}{2 \cdot q_y} \cdot A \cdot \cos(\psi)\right] + \frac{1}{2} \cdot k_z \cdot c \cdot \frac{q_y}{q_z^2} \cdot A \cdot \cos(\psi)$$

$$\frac{dq_z}{dt} = -\frac{1}{2} \cdot k_z \cdot c \cdot \frac{q_y}{q_z} \cdot A \cdot \sin(\psi)$$

$$\frac{dq_y}{dt} = -\frac{1}{2} \cdot (\omega_\beta + \frac{1}{4} \cdot k_z \cdot c \cdot \frac{q_y^2}{q_z^2}) \cdot A \cdot \sin(\psi)$$

$$\frac{dq_x}{dt} = -\frac{1}{8} \cdot k_z \cdot c \cdot \frac{q_y \cdot q_x}{q_z^2} \cdot A \cdot \sin(\psi) + \Gamma_\Delta \cdot \text{cosec}(\sqrt{\psi})$$

3.1 Find system fixed points; plot the possible graphs that show how they change for different values of system parameters.

3.2 Plot the appropriate flow charts that show how you find system fixed points.

3.3 Discuss stability and stability switching for different values of Γ_Δ; $\Gamma_\Delta \in \mathbb{R}_+$ parameter.

3.4 Parameter $\Gamma_\Delta = \sqrt{|\psi| \cdot \psi}$, Find system fixed points and discuss stability.

3.5 We change the additional element to $\Gamma_\Delta \cdot tg(\sqrt{\psi \cdot \sqrt{\psi^3 + 1}})$; $\Gamma_\Delta \in \mathbb{R}_+$, find System fixed points and discuss stability and stability switching for different values of Γ_Δ; $\Gamma_\Delta \in \mathbb{R}_+$.

4. We have Ion-channel Laser system that is characterized by set of perturbed differential equations. The mechanism is which a laser beam propagating through un-dense plasma produces a positively charged ion channels by expelling plasma electrons in the transverse direction. Due to interference source which radiate inside our main Ion-channel path, there is additional element ($\Gamma_\Delta \cdot \sec(\sqrt{\psi})$; $\Gamma_\Delta \in \mathbb{R}_+$) in q_x variable change in time ($\frac{dq_x}{dt} = \ldots\ldots$). Additionally, the mechanism is not ideal and there is q_x variable delay in time, $q_x(t) \to q_x(t - \tau)$ but the derivative of the $q_x(t)$ in time ($\frac{dq_x(t)}{dt}$) is not effected. We can define $q_x = q_x(t)$; $q_y = q_y(t)$; $q_z = q_z(t)$.

$$\frac{d\psi}{dt} = k_z \cdot c \cdot \left[1 - \frac{(2 + q_x^2(t - \tau) + q_y^2)}{4 \cdot q_z^2}\right] - \omega + \omega_\beta \cdot [1 - \frac{1}{2 \cdot q_y} \cdot A \cdot \cos(\psi)]$$

$$+\frac{1}{2}\cdot k_z\cdot c\cdot\frac{q_y}{q_z^2}\cdot A\cdot\cos(\psi)$$

$$\frac{dq_z}{dt}=-\frac{1}{2}\cdot k_z\cdot c\cdot\frac{q_y}{q_z}\cdot A\cdot\sin(\psi)$$

$$\frac{dq_y}{dt}=-\frac{1}{2}\cdot(\omega_\beta+\frac{1}{4}\cdot k_z\cdot c\cdot\frac{q_y^2}{q_z^2})\cdot A\cdot\sin(\psi)$$

$$\frac{dq_x}{dt}=-\frac{1}{8}\cdot k_z\cdot c\cdot\frac{q_y\cdot q_x(t-\tau)}{q_z^2}\cdot A\cdot\sin(\psi)+\Gamma_\Delta\cdot\sec(\sqrt{\psi})$$

4.1 Find system fixed points; plot the possible graphs that show how they change for different values of system parameters.

4.2 Plot the appropriate flow charts that show how you find system fixed points.

4.3 Discuss stability and stability switching for different values of $\tau;\tau\in\mathbb{R}_+$ parameter.

4.4 Parameter $\Gamma_\Delta=\sqrt{|\psi|\cdot\psi}$, Find system fixed points and discuss stability for different values of $\tau;\tau\in\mathbb{R}_+$.

4.5 We change the additional element to $\Gamma_\Delta\cdot tg(\sqrt{\psi^2\cdot\sqrt{\psi^3+1}});\Gamma_\Delta\in\mathbb{R}_+$, find System fixed points and discuss stability and stability switching for different values of $\Gamma_\Delta;\Gamma_\Delta\in\mathbb{R}_+$ (parameter $\tau;\tau\in\mathbb{R}_+$ is fix).

5. We have Ion-channel Laser system that is characterized by set of perturbed differential equations. The mechanism is which a laser beam propagating through un-dense plasma produces a positively charged ion channels by expelling plasma electrons in the transverse direction. Due to interference source which radiate inside our main Ion-channel path, there is additional element ($\Gamma_\Delta\cdot\sqrt{\sec(\sqrt{\psi})};\Gamma_\Delta\in\mathbb{R}_+$) in q_z variable change in time ($\frac{dq_z}{dt}=\ldots..$). Additionally, the mechanism is not ideal and there is q_z variable delay in time, $q_z(t)\to q_z(t-\tau)$ but the derivative of the $q_z(t)$ in time ($\frac{dq_z(t)}{dt}$) is not effected. We can define $q_x=q_x(t);q_y=q_y(t);q_z=q_z(t)$.

$$\frac{d\psi}{dt}=k_z\cdot c\cdot[1-\frac{(2+q_x^2+q_y^2)}{4\cdot q_z^2(t-\tau)}]-\omega+\omega_\beta\cdot[1-\frac{1}{2\cdot q_y}\cdot A\cdot\cos(\psi)]$$

$$+\frac{1}{2}\cdot k_z\cdot c\cdot\frac{q_y}{q_z^2(t-\tau)}\cdot A\cdot\cos(\psi)$$

$$\frac{dq_z}{dt}=-\frac{1}{2}\cdot k_z\cdot c\cdot\frac{q_y}{q_z(t-\tau)}\cdot A\cdot\sin(\psi)+\Gamma_\Delta\cdot\sqrt{\sec(\sqrt{\psi})}$$

$$\frac{dq_y}{dt}=-\frac{1}{2}\cdot\left(\omega_\beta+\frac{1}{4}\cdot k_z\cdot c\cdot\frac{q_y^2}{q_z^2(t-\tau)}\right)\cdot A\cdot\sin(\psi)$$

$$\frac{dq_x}{dt} = -\frac{1}{8} \cdot k_z \cdot c \cdot \frac{q_y \cdot q_x}{q_z^2(t-\tau)} \cdot A \cdot \sin(\psi)$$

5.1 Find system fixed points; plot the possible graphs that show how they change for different values of system parameters.

5.2 Plot the appropriate flow charts that show how you find system fixed points.

5.3 Discuss stability and stability switching for different values of $\tau; \tau \in \mathbb{R}_+$ parameter.

5.4 Parameter $\Gamma_\Delta = \sqrt{|\psi| \cdot \sqrt[3]{\psi}}$, Find system fixed points and discuss stability for different values of $\tau; \tau \in \mathbb{R}_+$.

5.5 We change the additional element to $\Gamma_\Delta \cdot \cos(\sqrt{\psi \cdot \sqrt{\psi^3 + 1}}); \Gamma_\Delta \in \mathbb{R}_+$, find System fixed points and discuss stability and stability switching for different values of $\Gamma_\Delta; \Gamma_\Delta \in \mathbb{R}_+$ (parameter $\tau; \tau \in \mathbb{R}_+$ is fix).

6. We have Ion-channel Laser system that is characterized by set of perturbed differential equations. The mechanism is which a laser beam propagating through un-dense plasma produces a positively charged ion channels by expelling plasma electrons in the transverse direction. Due to interference source which radiate inside our main Ion-channel path, there is additional element ($\Gamma_\Delta \cdot \sqrt{\mathrm{cosec}(\sqrt{\psi})}; \Gamma_\Delta \in \mathbb{R}_+$) in q_y variable change in time ($\frac{dq_y}{dt} = \ldots..$). Additionally, the mechanism is not ideal and there is q_z variable delay in time, $q_y(t) \to q_y(t-\tau)$ but the derivative of the $q_y(t)$ in time ($\frac{dq_y(t)}{dt}$) is not effected. We can define $q_x = q_x(t); q_y = q_y(t); q_z = q_z(t)$.

$$\frac{d\psi}{dt} = k_z \cdot c \cdot \left[1 - \frac{(2 + q_x^2 + q_y^2(t-\tau))}{4 \cdot q_z^2}\right]$$

$$- \omega + \omega_\beta \cdot [1 - \frac{1}{2 \cdot q_y(t-\tau)} \cdot A \cdot \cos(\psi)]$$

$$+ \frac{1}{2} \cdot k_z \cdot c \cdot \frac{q_y(t-\tau)}{q_z^2} \cdot A \cdot \cos(\psi)$$

$$\frac{dq_z}{dt} = -\frac{1}{2} \cdot k_z \cdot c \cdot \frac{q_y(t-\tau)}{q_z} \cdot A \cdot \sin(\psi)$$

$$\frac{dq_x}{dt} = -\frac{1}{8} \cdot k_z \cdot c \cdot \frac{q_y(t-\tau) \cdot q_x}{q_z^2} \cdot A \cdot \sin(\psi)$$

$$\frac{dq_y}{dt} = -\frac{1}{2} \cdot \left(\omega_\beta + \frac{1}{4} \cdot k_z \cdot c \cdot \frac{q_y^2(t-\tau)}{q_z^2}\right) \cdot A \cdot \sin(\psi) + \Gamma_\Delta \cdot \sqrt{\mathrm{cosec}(\sqrt{\psi})}$$

6.1 Find system fixed points; plot the possible graphs that show how they change for different values of system parameters.

6.2 Plot the appropriate flow charts that show how you find system fixed points.

6.3 Discuss stability and stability switching for different values of $\tau;\tau \in \mathbb{R}_+$ parameter.

6.4 Parameter $\Gamma_\Delta = \sqrt{|\psi| \cdot \sqrt[5]{\psi}}$, Find system fixed points and discuss stability for different values of $\tau;\tau \in \mathbb{R}_+$.

6.5 We change the additional element to $\Gamma_\Delta \cdot \cos(\sqrt{\psi \cdot \sqrt{\psi^2 + 1}})$; $\Gamma_\Delta \in \mathbb{R}_+$, find System fixed points and discuss stability and stability switching for different values of $\Gamma_\Delta;\Gamma_\Delta \in \mathbb{R}_+$ (parameter $\tau;\tau \in \mathbb{R}_+$ is fix).

7. The simple case of two-mode competition describes the characterization of the mode-competition phenomena with the gain saturation effects. We describe the effect of the Langevin noise sources on the mode competition phenomena by considering the two modes p and q as follow.

$$\frac{dS_p}{dt} = [A_p - B \cdot S_p - \{D_{p(q)} + H_{p(q)}\} \cdot S_q - G_{th}] \cdot S_p + C_p \cdot \frac{N}{\tau_s} + F_p(t)$$

$$\frac{dS_q}{dt} = [A_q - B \cdot S_q - \{D_{q(p)} + H_{q(p)}\} \cdot S_p - G_{th}] \cdot S_q + C_q \cdot \frac{N}{\tau_s} + F_q(t)$$

$$\frac{dN}{dt} = -\sum_p A_p \cdot S_p - \sum_q A_q \cdot S_q - \frac{N}{\tau_s} + \frac{I}{e} + F_N(t)$$

where $\lambda_q > \lambda_p$ and there is a competition phenomenon between the two modes.

We consider $F_p(t) \to \varepsilon$; $F_q(t) \to \varepsilon$(neglected functions), $F_N(t) \to \varepsilon$ all above differential equation's parameters are as describe in (3.3). The assumptions are the same as (3.3) for $D_{p(q)}$, $H_{p(q)}$ and $D_{q(p)}$, $H_{q(p)}$ respectively.

7.1 Find system fixed points and Basin of Attraction (BOA) for each fixed point.

7.2 Discuss the case when the injection current I is far above the threshold current level I_{th}(a bi-stable state), and when the injection current I is near to I_{th} (a near-threshold stable multimode state).

7.3 Discuss the effect of the asymmetric gain saturation in our system.

7.4 Discuss stability and stability switching for different values of system.

parameters.

8. The simple case of two-mode competition describes the characterization of the mode-competition phenomena with the gain saturation effects. We describe the effect of the Langevin noise sources on the mode competition phenomena by considering the two modes p and q as follow. The two-mode competition mechanism is not ideal

and there is a delay in time on S_p, S_q variables, $S_p(t) \to S_p(t-\tau_1)$, $S_q(t) \to S_q(t-\tau_2)$ but the derivatives are not affected.

$$\frac{dS_p}{dt} = [A_p - B\cdot S_p(t-\tau_1) - \{D_{p(q)} + H_{p(q)}\}\cdot S_q(t-\tau_2) - G_{th}]\cdot S_p(t-\tau_1) + C_p\cdot\frac{N}{\tau_s} + F_p(t)$$
$$\frac{dS_q}{dt} = [A_q - B\cdot S_q(t-\tau_2) - \{D_{q(p)} + H_{q(p)}\}\cdot S_p(t-\tau_1) - G_{th}]\cdot S_q(t-\tau_2) + C_q\cdot\frac{N}{\tau_s} + F_q(t)$$

$$\frac{dN}{dt} = -\sum_p A_p\cdot S_p(t-\tau_1) - \sum_q A_q\cdot S_q(t-\tau_2) - \frac{N}{\tau_s} + \frac{I}{e} + F_N(t)$$

where $\lambda_q > \lambda_p$ and there is a competition phenomenon between the two modes.

We consider $F_p(t) \to \varepsilon$; $F_q(t) \to \varepsilon$(neglected functions), $F_N(t) \to \varepsilon$ all above differential equation's parameters are as describe in (3.3). The assumptions are the same as (3.3) for $D_{p(q)}$, $H_{p(q)}$ and $D_{q(p)}$, $H_{q(p)}$ respectively.

8.1 Find system fixed points and Basin of Attraction (BOA) for each fixed point.

8.2 Discuss the stability and stability switching for different values of τ; Case A $\tau_1 = \tau$; $\tau_2 = 0$.

8.3 Discuss the stability and stability switching for different values of τ; Case B $\tau_1 = 0$; $\tau_2 = \tau$.

8.4 Discuss the stability and stability switching for different values of τ; Case C $\tau_1 = \tau$; $\tau_2 = \tau$.

9. We have Ion-Channel Laser (ICL) system which relies on the injection of a relativistic electron beam in an Ion-Channel (IC) to create a coherent and highly amplified radiation source. The Ion-Channel (IC) can be produced in a plasma based wake field accelerator in the blowout or bubble regime. While propagating in plasma, a laser pulse or a particle beam pushes the electrons off-axis and lets an Ion-Channel in its wake. Analysis beam assumptions: No.1, Pierce parameter ρ is much smaller than one ($\rho \ll 1$), FEL-like amplification, No.2, In an ICL the wiggler parameter K depends on the electron properties; it is different value for each electron. The beam energy spread and beam Wiggler parameter spread are limited and satisfy: $\frac{\Delta\gamma}{\gamma} < \frac{2}{3}\cdot\rho$; $\frac{\Delta K}{K} < (\frac{2+K^2}{2\cdot K^2})\cdot\rho$; γ- Lorentz factor,K—Wiggler parameter,ρ—Pierce parameter, γ_η—Electron Lorentz factor after its interaction with the wave, $\bar{z}$—Longitudinal position of the electron averaged over one betatron oscillation. The interaction between the electron beam and the radiation can lead to the amplification of the radiation and the interaction with the wave leads to the following DDEs (Delay Differential Equations) of motion for the electron in the (ϕ, η) phase space. τ_1 and τ_2 time delay parameters are due to non- ideal interaction with the wave process.

$$\frac{d\phi}{dt} = \left(\frac{4+K^2}{8\cdot\gamma_0^2}\right)\cdot\eta(t-\tau_1); \quad \frac{d\eta}{dt} = \frac{A_1\cdot K\cdot[JJ]}{2\cdot\gamma_0^2}\cdot\cos[\phi(t-\tau_2) + \psi_1]$$

$$[JJ] = J_0 \cdot \left(\frac{K^2}{4 + 2 \cdot K^2}\right) - J_1 \cdot \left(\frac{K^2}{4 + 2 \cdot K^2}\right)$$

J_0 and J_1 are the Bessel functions. Assumption: J_0 and J_1 Bessel functions are constant for our system. The beam of electrons is bunched by the Electro-Magnetic (EM) wave at the phase $\phi = -\psi_1 + \frac{\pi}{2} + 2 \cdot m \cdot \pi$; m is integer which leads to a bunching at the position $r = r_0 \cdot \sin(k_1 \cdot \overline{z} - \omega_1 \cdot t + \psi_1)$. ϕ—Electron phase, η—Relative electron in the EM wave, A_1 and ψ_1 are respectively the wave amplitude and phase. The electron oscillates in the (x, z) plane. K—Wiggler parameter. You need to consider that $\eta = \frac{\gamma_\eta - \gamma_0}{\gamma_0} \Rightarrow \gamma_0 = \frac{\gamma_\eta}{\eta+1}$. Hint: In an Ion-Channel Laser (ICL), the K parameter depends on r_0 and γ_0 which can be different for each electron and the radiation wavelength spread can be induced by both the beam energy spread and K spread.

9.1 Find ICL system fixed points (ϕ^*, η^*) and plot them as a function of K parameter.

9.2 Discuss stability and stability switching of our system for different values of τ parameter, Cases: (1) $\tau_1 = \tau; \tau_2 = 0$ (2) $\tau_1 = 0; \tau_2 = \tau$ (3) $\tau_1 = \tau; \tau_2 = \tau$.

9.3 There is a special case that parameter K (Wiggler parameter) is dependent on ϕ variable according the following function, $K = \Gamma \cdot \sec(\phi); \Gamma \in \mathbb{R}_+$. Find system fixed points and plot the graphs (ϕ^*, η^*) for different value of Γ parameter.

Discuss Stability and Stability Switching (9.3), $\Lambda^{-1}(\tau); sign\Lambda^{-1}(\tau)$ for the Following Cases: (1) $\tau_1 = \tau; \tau_2 = 0$(2) $\tau_1 = 0; \tau_2 = \tau$(3) $\tau_1 = \tau; \tau_2 = \tau$.

10. In an ICL, K is a function of γ, $K = K(\gamma)$. K is a function of time. We define K_0 and the time-dependent longitudinal momentum p_z and the maximum radius r_m such that $K_0 = K(t = 0); p_z(t = 0) = p_0; r_m(t = 0) = r_0; \gamma_0 = \sqrt{1 + p_0^2}$. In presence of an EM wave, the energy and momentum change of an electron following betatron motion in the (x, z) plan is given by the following set of differential equations:

$$\frac{d\gamma}{dt} = \beta_r \cdot \alpha - \beta_r \cdot \frac{K \cdot (2 + K^2)}{4 \cdot \gamma^2} \cdot \cos(\theta_r)$$

$$\frac{dp_r}{dt} = (1 - \beta_z) \cdot \alpha - \frac{K \cdot (2 + K^2)}{4 \cdot \gamma^2} \cdot \cos(\theta_r)$$

$$\frac{dK}{dt} = (\frac{1 + K^2}{2 \cdot \gamma^2}) \cdot \alpha \cdot \sin(\theta_r); \frac{dp_z}{dt} = \beta_r \cdot \alpha$$

where $\alpha = A_1 \cdot \sin(k_1 \cdot z - \omega_1 \cdot t + \psi_1)$.$A_1$ and ψ_1 are respectively the wave amplitude and phase.

Hint: Betatron, a type of particle accelerator that uses the electric field induced by a varying magnetic field to accelerate electrons (beta particles) to high speeds in a circular orbit. ... Modern compact betatron designs are used to produce high-energy X-ray beams for a variety of applications.

β_r and β_z are the normalized transverse and longitudinal electron velocities ($\beta_r = \frac{v_r}{c}$; $\beta_z = \frac{v_z}{c}$).v_r and v_z are the transverse and longitudinal electron velocities in transverse and z direction respectively. $r = r_m \cdot \cos(\theta_r)$; $p_r = K \cdot \sin(\theta_r)$.

$K = r_m \cdot k_p \cdot \sqrt{\frac{\gamma}{2}}$; $K = \sqrt{p_r^2 + \frac{\gamma}{2} \cdot r^2 \cdot k_p^2}$. γ—Lorentz factor,K—Wiggler parameter,θ_r—Transverse angle of electron velocity vector.

10.1 Find system fixed points (equilibrium points) and plot fixed points (γ^*, p_r^*, K^*) graphs for different values of β_r, β_z parameters (normalized transverse and longitudinal electron velocities).

10.2 Discuss stability of our system and plot the relevant graphs.

10.3 Due to some interference we get delay in time (τ) of K parameter, $K(t) \to K(t - \tau)$. The derivative of K in time ($\frac{dK}{dt}$) is not effected by τ parameter. Discuss stability and stability switching for different values of τ parameter.

10.4 Unexpected event happened in our system. It is represented by additional element to $\frac{d\gamma}{dt} = ...$ differential equation ($\frac{d\gamma}{dt} = ... + \Omega \cdot \text{cosec}(\theta_r)$, $\Omega \in \mathbb{R}_+$. Find system fixed points and discuss stability and stability switching for parameters variation.

References

1. D.H. Whittum, A.M. Sessler, *The Ion-Channel Laser*, Dept. of Physics, Lawrence Berkeley Lab, Berkeley, California
2. M. Yamada, W. Ishimori, H. Sakaguchi, M. Ahmed, Time-dependent measurement of the Mode-competition phenomena among longitudinal modes in Long-wavelength lasers. IEEE J. Quantum Electronics **39**(12) (2003)
3. M. Ahmed, M. Yamada, Influence of Instantaneous mode competition on the dynamics of semiconductor lasers. IEEE J. Quantum Electronics **38**(6) (2002)
4. Y. Kuang, *Delay Differential Equations with Applications in Population Dynamics* (Academic Press, Boston, 1993).
5. E. Beretta, Y. Kuang, Geometric stability switch criteria in delay Differential systems with delay dependent parameters. SIAMJ. Math. Anal. **33**, 1144–1165 (2002)
6. B. Balachandran, T. Kalmár-Nagy, D.E. Gilsinn, *Delay Differential Equations: Recent Advances and New Directions (Hardcover)*. Springer, 1 edn (March 5, 2009)
7. M. Yamada, Theory of mode competition noise in semiconductor injection laser. IEEE J. Quantum Electronics **QE-22**(7) (1986)
8. Y.A. Kuznetsov, Elelments of applied bifurcation theory, in *Applied Mathematical Sciences*
9. J.K. Hale. Dynamics and bifurcations, in *Texts in Applied Mathematics*, vol. 3
10. S.H. Strogatz, *Nonlinear Dynamics and Chaos*. Westview Press
11. S. Wiggins, Introduction to applied nonlinear dynamical systems and chaos, in *Text in Applied Mathematics (Hardcover)*

Chapter 4
Solid State Laser Nonlinearity Applications in Engineering

There are many different types of lasers. The laser medium can be solid, gas, liquid or semiconductor. A solid-state laser is a laser that uses a gain medium that is a solid. It is very different from liquid laser such as in dye lasers or a gas lasers. Lasers are commonly designed by the type of lasing material employed: solid-state lasers have lasing material distributed in a solid matrix such as the ruby or neodymium "YAG" lasers. Solid-state lasers use optical pumping and such pump sources are relatively cheap and can provide very high powers. It has low power efficiency, moderate life-time, and strong thermal effects such as thermal lensing in the gain medium. Laser diodes are very often used for pumping solid-state lasers. Lasers use crystals like Nd:YAG crystal. The Neodymium doped Yttrium Aluminum Garnet-Nd:YAG laser crystal is the most popular lasing media for solid-state lasers. The gain medium in solid-state laser is a solid rather than a liquid laser. Semiconductor-based lasers are also in the solid-state, but are generally considered as a separate class from solid-state lasers. The solid-state lasers gain media can be crystals or glasses doped with rare earth or transition metal ions, or semiconductor lasers. Ion-doped solid-state lasers (Doped insulator lasers) can be made in the form of bulk lasers, fiber lasers or other types of waveguide lasers. Solid-state lasers may generate output powers between a few mill-watts and in high-power versions many kilowatts. Solid-state lasers are optically pumped with flash lamps or arc lamps. Such pump sources are good high power supply and lead to a low power efficiency, moderate lifetime, and strong thermal effects such as thermal lensing in the gain medium. Laser diodes are frequently used for pumping solid-state lasers. Diode-pumped solid-state lasers (DPSS lasers, all-solid-state lasers) have advantages as compact setup, long lifetime, and good beam quality. The laser transitions of rare-earth or transition-metal-doped crystal or glasses are normally weakly allowed transitions. The transitions are with very low oscillator strength, which leads to long radiative upper-state lifetime and consequently to good energy storage, with upper-state life-times of microseconds to milliseconds. The long upper-state lifetimes makes solid-state lasers very suitable for Q switching: the laser crystal can easily store an amount of energy which, when released in the form of a nanosecond pulse, leads to a peak power which is orders of

O. Aluf, *Advance Elements of Laser Circuits and Systems*,
https://doi.org/10.1007/978-3-030-64103-0_4

magnitude above the achievable average power. In mode-locked operation, solid-state lasers can generate ultrashort pulses with durations measured in picoseconds, with passive mode locking. They have a tendency for Q-switching instabilities, if there are not suppressed with suitable measures. Types of solid-state lasers: Small diode-pumped Nd:YAG (YAG lasers) or Nd:YVO_4 lasers (vanadate lasers) and Q-switched versions generate pulses with durations of a few nanoseconds, micro-joule pulse energies and peak powers of many kilowatts. Other types of solid-state laser are unidirectional ring lasers (single-frequency operation) or microchip lasers which allows for operation with very small linewidth in the lower kilohertz region. Larger lasers in side-pumped or end-pumped configurations, having the geometry of rod lasers, slab lasers, or thin-disk lasers, are suitable for output powers up to several kilo-watts. Particularly thin-disk lasers offer very high beam quality and a high power efficiency. An appropriately designed semiconductor saturate absorber device, the anti-resonant Fabry–Perot saturable absorber, can reliably start and sustain stable mode locking of solid state lasers such as Nd:YAG, Nd:YLF, Nd:Glass, Cr:LiSAF, and Ti:sapphire lasers. Solid-state lasers with long upper-state lifetimes, previous attempts to produce self-starting passive mode locking with saturate absorbers was accompanied by self Q-switching. There is a criteria that characterize the dynamic behavior of solid-state lasers in the important regimes of Q-switching, mode-locked Q-switching, and continuous-wave mode locking in the picosecond and femtosecond range for the pulse width. The rate equations of the laser are presented and the dynamics is analyze, semiconductor devices. By using laser-diode-pumped microchip solid-state laser we get nanometer vibration analysis of a target. It has been demonstrated by a self-aligned optical feedback vibrometry technique. The laser output waveform, which modulated through interference between a lasing field and an extremely week frequency-modulated (FM) feedback field is analyzed by the Hilbert transformation to yield the vibration waveform of the target. A simulation and analysis is done on the model differential equations, that is, laser with frequency-modulated optical feedback system. Light-injection-induced phenomena in laser, such as injection locking and return-light-induced instabilities, are profound interest because of the basic properties of laser dynamics as well as their practical importance. In Nd stoichiometric lasers the resonant excitation of periodic oscillations and chaotic oscillation in relaxation oscillation frequency regions, resulting from Dopper-shifted light injection from a moving light-scattering object. The phenomenon is the injection-induced modulation of lasers. The application of this phenomenon is to ultra-high sensitivity Dopper-shift measurements in wide velocity regions. A theoretical model that includes multimode effect is inspected and analyze for stability. The light-injection model differential equations for rotating-wave approximation fields are given by introducing injection field into spatially-hole-burned N-mode lasers.

A novel scheme that combines gain switching with passive Q switching of a miniature diode-pumped solid-state laser is analyze and simulate for stability inspection. A composite pumping pulse, consisting of a long, low-intensity pulse and a short, high-intensity pulse, is used to reduce the timing Jitter. There is a simple model of a Q-switched laser. The effective pumping rate of the upper laser level is $P_{eff} = \frac{\eta \cdot P}{h \cdot \nu_L \cdot V}$, where P, the electrical power input; η is the pumping efficiency, which is the product of laser-diode efficiency and optical coupling efficiency; ν_L is the laser frequency; V is the laser mode volume, and h is the Planck constant. The differential equations of the population density of the upper laser level, N_G, and the photon density in the cavity are presented and analyze by using nonlinear dynamics for stability and stability switching under system parameter's variation [1–4].

4.1 Solid State Laser Controlled by Semiconductor Devices Stability Analysis

High power diode lasers array can be utilized to pump solid-state laser materials efficiently. Laser materials with high intrinsic quantum efficiencies and an improved scalable cavity design make high output powers. The powerful and compact laser sources have fundamental research application such as X-ray, plasma, and higher-harmonic generation, etc. The target is to get high laser peak power and ultrashort optical pulses and control the laser dynamics by various Q-switching and continuous-wave (CW) mode locking techniques. It is controlling the laser dynamics with compact scalable passive semiconductor devices. The semiconductor device is a saturate absorber. A saturate absorber can be used to passively Q-switch or mode-lock laser. Third regime is laser which is Q-switch-mode-locked. Q-switching is a technique for obtaining energetic short pulses from laser by modulating the intra-cavity losses and thus the Q-factor of the laser resonator. It generates nanosecond pulses of high energy and peak power with solid-state bulk lasers. Q-switch pulse generation: (1) the resonator losses are kept at a high level and the energy fed into the gain medium by the pumping mechanism accumulates there. The stored energy can be a multiple of the saturation energy (2) the losses are reduced to a small value, and the power of the laser radiation builds up in the laser resonator (3) the temporary integrated intra-cavity power reaches the order of the saturation energy of the gain medium and the gain starts to be saturated. When the gain equals the remaining (low) resonator losses then the peak of the pulse is reached. The large intra-cavity power is presented and there is a depletion of the stored energy during the time where the power decays. The pulse duration achieved with Q-switching is typically in the nanosecond range and above the resonator round-trip time. In the most cases, Q-switched lasers generate regular pulse trains via repetitive Q-switching. Basically Q-switching (giant pulse formation, Q-spooling) is a technique by which a laser can be made to produce a pulsed output beam. The production of light pulses is with high peak power (higher than if it be produced by the same laser operating in continuous wave mode). Compare to mode locking, Q-switching leads to much lower pulse repetition rates, higher pulse energies and longer pulse duration. The solid-state lasers (Nd:YAG or Nd:YLF) can be passively CW mode locked using saturate absorbers without Q-switching or at least Q-switched mode locking. The parameters of saturate absorber are recovery time of the absorption and the saturation energy, pure passive Q-switching or pure CW mode locking, Q-switched mode locking. There is a full control over the saturate absorber device parameters and well-matched optimum values for a given solid-state laser material. Semiconductor absorbers have an intrinsic bit emporia pulse response, and the scales being given by intra band carrier-carrier scattering and thermal process. The absorber's performances are ideal for the mode locking of solid-state lasers in the femtosecond and picosecond regime. The basic laser models describe the dynamic of solid-state laser. Based on the feature of the laser dynamics, there are criteria that characterize the laser dynamics with respect to Q-switching, mode-locked Q-switching, and self-starting of mode locking by a saturate absorber. A saturate absorber is an optical component with a certain optical loss, which is reduced at high optical intensities. The reduction can occur in a medium with absorbing dopant ions, when a strong optical intensity leads to depletion of the ground state of these ions. Artificial saturate absorbers are where there is no real absorption, but an optical loss which decreases for increasing optical power. The main applications of saturate absorbers are passive mode locking and Q-switching of lasers. Saturate absorbers are also useful for purposes of nonlinear filtering outside laser resonators like cleaning up pulse shapes, and in optical signal processing. The saturate absorbers with relaxation times (picosecond range) possibly generate stable picosecond pulses if the gain bandwidth is broad enough. A picosecond absorber can generate femtosecond pulses of soliton like pulse shaping is additionally employed. Additionally there is a soliton mode locking by a slow saturate absorber. This mode locking technique does not impose any requirements on the cavity design or operation of the cavity close to its stability limit. The soliton mode locking is a stable phenomenon, which allows for a scalable resonator design. The semiconductor elements with the required dynamical behavior can be fabricated due to the fact of existing Molecular Beam Epitaxy (MBE) technology. Molecular beam epitaxy (MBE) is an evaporation process performed in an ultra-high vacuum for the deposition of compounds of extreme regularity of layer thickness and composition from well-controlled deposition rates. We can explain molecular beam epitaxy (MBE) as an atomic layer by atomic layer crystal growth technique, based on reaction of molecular or atomic beams with a heated crystalline substrate, performed in an ultra-high vacuum (UHV) environment. The term "molecular beam" describes a unidirectional kinematic flow of atoms or molecules with no collisions among them, as opposed to a viscous, fluid-like flow. The term "epitaxy" is akin or upon, and taxi-meaning arrangement or order. Epitaxy refers to ordered growth of one crystalline layer on another crystalline layer, with the same or related crystal arrangement. The semiconductors tend to have optical loss, low a saturation intensity, and low a damage threshold for typical solid-state lasers then some semiconductor material parameters are not well matched to the characteristics required for solid-state lasers. Anti-resonant Fabry–Perot saturate absorber (A-FPSA) integrates the semiconductor saturate absorber inside a Fabry–Perot cavity operated at anti resonance with a free spectral range larger than the gain bandwidth of the solid-state laser. This can be possible solution to unmatched parameters of the semiconductor material. We can investigate and inspect the dynamics of the absorber within a simple two-level rate equation model. A large variety of dynamical effects occurring in a laser is sufficient to model the semiconductor absorber as an equivalent two-level

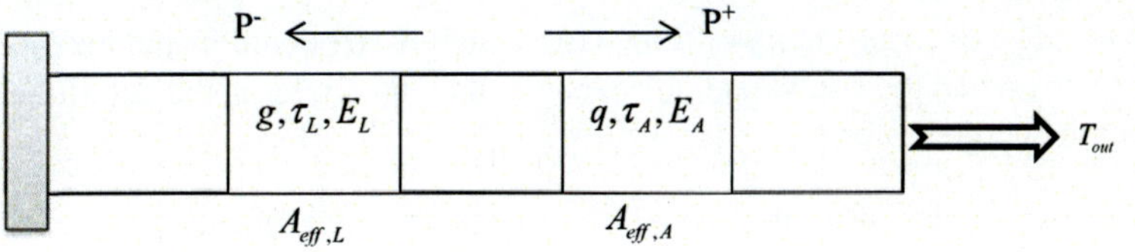

Fig. 4.1 Solid-state laser by semiconductor devices model corresponding to the rate Differential equation

system. We can then determine and modify the key parameters of the equivalent two-level model and get a stable system. There are a system parameter variation ranges for CW Q-switching and CW mode locking and extension to the case of Q-switched mode locking [1].

The simplest laser model of a single-mode laser with homogeneously laser medium and saturate absorber is as shown in the next figure (Fig. 4.1).

T_{out}—Small output coupling (laser power within one round trip does not change very much), P^{+}, P^{-}—Twice circulating power (related field density), neglecting standing-wave effects in the cavity [1].

Model assumptions: (1) the transverse relaxation times of the equivalent two-level models for the laser gain medium and for the saturate absorber are much faster than any other dynamics in the system. The outcome is that the model can be described by the rate equations to describe the laser dynamic (2) the changes in the laser intensity, gain, and saturate absorption on a time scale of the order of the round-trip time (T_R) in the cavity are small [1].

$$\frac{dP}{dt} = \frac{2}{T_R} \cdot (g - l - q) \cdot P$$
$$\frac{dg}{dt} = -\frac{(g - g_0)}{T_L \cdot T_R} - \frac{g \cdot P}{E_L}; \frac{dq}{dt} = -\frac{(q - q_0)}{T_A \cdot T_R} - \frac{q \cdot P}{E_A}$$
$$P = P(t); g = g(t); q = q(t)$$

T_R—Round-trip time in cavity, P—Laser power, g—Gain per round trip, l—Linear losses per round trip, q—Saturate losses per round trip, g_0—Small-signal gain per round trip, q_0—Un saturate but saturate losses per round trip, T_L—Upper state lifetime of the gain medium (normalized to the round trip time of the cavity); $T_L = \frac{\tau_L}{T_R}$, T_A—Absorber recovery time (normalized to the round trip time of the cavity); $T_A = \frac{\tau_A}{T_R}$, τ_L—Upper state lifetime of the gain medium, τ_A—Absorber recovery time, E_L—Saturation energy of the gain; $E_L = \frac{h \cdot \nu \cdot A_{eff,L}}{2 \cdot \sigma_L}$, E_A—Saturation energy of the absorber; $E_A = \frac{h \cdot \nu \cdot A_{eff,A}}{2 \cdot \sigma_A}$.

Remark (1): the factor of 2 in the definition of the saturation energies is due to averaging over the standing wave effects in a linear cavity.

Remark (2): For a ring cavity or if the laser is in pulsed operation and the media are much shorter than the equivalent pulse length, then the factor in the definition of the saturation energies is one.

The laser model of a single-mode laser with homogeneously broadband laser medium and saturate absorber is not ideal and there are interferences. We present these interferences by time delay parameters τ_1, τ_2, τ_3 for laser power, gain per round trip and saturate losses per round trip in time respectively. We can write our system Delay Differential Equations (DDEs): $P(t) \to P(t-\tau_1)$; $g(t) \to g(t-\tau_2)$; $q(t) \to q(t-\tau_3)$

$$\frac{dP}{dt} = \frac{2}{T_R} \cdot [g(t-\tau_2) - l - q(t-\tau_3)] \cdot P(t-\tau_1)$$
$$\frac{dg}{dt} = -\frac{(g(t-\tau_2) - g_0)}{T_L \cdot T_R} - \frac{g(t-\tau_2) \cdot P(t-\tau_1)}{E_L}$$
$$\frac{dq}{dt} = -\frac{(q(t-\tau_3) - q_0)}{T_A \cdot T_R} - \frac{q(t-\tau_3) \cdot P(t-\tau_1)}{E_A}$$

We consider that time delay parameters τ_1, τ_2, τ_3 do not affect the laser power, gain per round trip and saturate losses per round trip derivatives in time $\left(\frac{dP}{dt}, \frac{dg}{dt}, \frac{dq}{dt}\right)$.

We have possible seven cases for system stability analysis: (1) $\tau_1 = \tau; \tau_2 = 0; \tau_3 = 0$ (2) $\tau_1 = 0; \tau_2 = \tau; \tau_3 = 0$ (3) $\tau_1 = \tau; \tau_2 = \tau; \tau_3 = 0$ (4) $\tau_1 = 0; \tau_2 = 0; \tau_3 = \tau$ (5) (6) $\tau_1 = 0; \tau_2 = \tau; \tau_3 = \tau$ (7) $\tau_1 = \tau; \tau_2 = \tau; \tau_3 = \tau$.

We analyze the stability of our system for the first case $\tau_1 = \tau; \tau_2 = 0; \tau_3 = 0$. We get the following system DDEs:

$$\frac{dP}{dt} = \frac{2}{T_R} \cdot [g(t) - l - q(t)] \cdot P(t-\tau);$$
$$\frac{dg}{dt} = -\frac{(g(t) - g_0)}{T_L \cdot T_R} - \frac{g(t) \cdot P(t-\tau)}{E_L}$$
$$\frac{dq}{dt} = -\frac{(q(t) - q_0)}{T_A \cdot T_R} - \frac{q(t) \cdot P(t-\tau)}{E_A}$$

At fixed points (equilibrium points): $\frac{dP}{dt} = 0; \frac{dg}{dt} = 0; \frac{dq}{dt} = 0; \lim_{t\to\infty} P(t-\tau) = P(t)$

$$\frac{2}{T_R} \cdot [g^* - l - q^*] \cdot P^* = 0; \; -\frac{(g^* - g_0)}{T_L \cdot T_R} - \frac{g^* \cdot P^*}{E_L} = 0$$
$$-\frac{(q^* - q_0)}{T_A \cdot T_R} - \frac{q^* \cdot P^*}{E_A} = 0$$

Case (1): $\frac{2}{T_R} \cdot [g^* - l - q^*] \cdot P^* = 0$.

Case (1.1): $\frac{2}{T_R} \cdot [g^* - l - q^*] \neq 0$; $P^* = 0$ ten we get $-\frac{(g^* - g_0)}{T_L \cdot T_R} = 0; -\frac{(q^* - q_0)}{T_A \cdot T_R} = 0$

$$T_L \cdot T_R \neq 0;\, T_A \cdot T_R \neq 0;\, g^* - g_0 = 0 \Rightarrow g^* = g_0;\, q^* - q_0 = 0 \Rightarrow q^* = q_0$$

The first fixed point: $E^{(0)}(P^{(0)}, g^{(0)}, q^{(0)}) = (0, g_0, q_0)$.
Case (1.2): $\frac{2}{T_R} \cdot [g^* - l - q^*] = 0;\, P^* \neq 0;\, T_R \neq 0$ then $g^* = l + q^*$

$$(*) - \frac{(l + q^* - g_0)}{T_L \cdot T_R} - \frac{(l + q^*) \cdot P^*}{E_L} = 0 \; (**) - \frac{(q^* - q_0)}{T_A \cdot T_R} - \frac{q^* \cdot P^*}{E_A} = 0$$

$$(**) \rightarrow -\frac{(q^* - q_0)}{T_A \cdot T_R} - \frac{q^* \cdot P^*}{E_A} = 0 \Rightarrow -\left[\frac{(q^* - q_0) \cdot E_A + q^* \cdot P^* \cdot T_A \cdot T_R}{T_A \cdot T_R \cdot E_A}\right]$$
$$= 0;\, T_A \cdot T_R \cdot E_A \neq 0$$

$$(q^* - q_0) \cdot E_A + q^* \cdot P^* \cdot T_A \cdot T_R = 0 \Rightarrow q^* \cdot (E_A + P^* \cdot T_A \cdot T_R)$$
$$- q_0 \cdot E_A = 0;\, q^* = \frac{q_0 \cdot E_A}{E_A + P^* \cdot T_A \cdot T_R}$$

Submission to equation

$$(*) : -\frac{(l + q^* - g_0)}{T_L \cdot T_R} - \frac{(l + q^*) \cdot P^*}{E_L} = 0$$
$$-\frac{(l + \frac{q_0 \cdot E_A}{E_A + P^* \cdot T_A \cdot T_R} - g_0)}{T_L \cdot T_R} - \frac{(l + \frac{q_0 \cdot E_A}{E_A + P^* \cdot T_A \cdot T_R}) \cdot P^*}{E_L} = 0$$
$$-\frac{(l - g_0)}{T_L \cdot T_R} - \frac{q_0 \cdot E_A}{(E_A + P^* \cdot T_A \cdot T_R) \cdot T_L \cdot T_R}$$
$$-\frac{P^*}{E_L} \cdot \left(l + \frac{q_0 \cdot E_A}{E_A + P^* \cdot T_A \cdot T_R}\right) = 0 - \frac{q_0 \cdot E_A}{(E_A + P^* \cdot T_A \cdot T_R) \cdot T_L \cdot T_R}$$
$$-\frac{P^*}{E_L} \cdot \left(l + \frac{q_0 \cdot E_A}{E_A + P^* \cdot T_A \cdot T_R}\right) = \frac{(l - g_0)}{T_L \cdot T_R} - \frac{q_0 \cdot E_A}{(E_A + P^* \cdot T_A \cdot T_R) \cdot T_L \cdot T_R}$$
$$-\frac{P^*}{E_L} \cdot \left(\frac{l \cdot (E_A + P^* \cdot T_A \cdot T_R) + q_0 \cdot E_A}{E_A + P^* \cdot T_A \cdot T_R}\right) = \frac{(l - g_0)}{T_L \cdot T_R} \frac{q_0 \cdot E_A}{(E_A + P^* \cdot T_A \cdot T_R) \cdot T_L \cdot T_R}$$
$$+\frac{P^*}{E_L} \cdot \left(\frac{l \cdot (E_A + P^* \cdot T_A \cdot T_R) + q_0 \cdot E_A}{E_A + P^* \cdot T_A \cdot T_R}\right)$$
$$= \frac{(g_0 - l)}{T_L \cdot T_R} \frac{q_0 \cdot E_A \cdot E_L + P^* \cdot [l \cdot (E_A + P^* \cdot T_A \cdot T_R) + q_0 \cdot E_A] \cdot T_L \cdot T_R}{(E_A + P^* \cdot T_A \cdot T_R) \cdot T_L \cdot T_R \cdot E_L}$$
$$= \frac{(g_0 - l)}{T_L \cdot T_R} q_0 \cdot E_A \cdot E_L + P^* \cdot [l \cdot (E_A + P^* \cdot T_A \cdot T_R) + q_0 \cdot E_A] \cdot T_L \cdot T_R$$
$$= \frac{(g_0 - l)}{T_L \cdot T_R} \cdot (E_A + P^* \cdot T_A \cdot T_R) \cdot T_L \cdot T_R \cdot E_L$$
$$q_0 \cdot E_A \cdot E_L + P^* \cdot [l \cdot E_A + l \cdot P^* \cdot T_A \cdot T_R + q_0 \cdot E_A] \cdot T_L \cdot T_R$$
$$= (g_0 - l) \cdot (E_A + P^* \cdot T_A \cdot T_R) \cdot E_L$$
$$q_0 \cdot E_A \cdot E_L + P^* \cdot [(l + q_0) \cdot E_A + l \cdot P^* \cdot T_A \cdot T_R] \cdot T_L \cdot T_R$$
$$= (g_0 - l) \cdot (E_A \cdot E_L + P^* \cdot T_A \cdot T_R \cdot E_L)$$
$$q_0 \cdot E_A \cdot E_L + P^* \cdot [(l + q_0) \cdot E_A \cdot T_L \cdot T_R + l \cdot P^* \cdot T_A \cdot T_R \cdot T_L \cdot T_R]$$
$$= (g_0 - l) \cdot E_A \cdot E_L + P^* \cdot (g_0 - l) \cdot T_A \cdot T_R \cdot E_L$$
$$q_0 \cdot E_A \cdot E_L + P^* \cdot (l + q_0) \cdot E_A \cdot T_L \cdot T_R + (P^*)^2 \cdot l \cdot T_A \cdot T_R \cdot T_L \cdot T_R$$

$$= (g_0 - l) \cdot E_A \cdot E_L + P^* \cdot (g_0 - l) \cdot T_A \cdot T_R \cdot E_L$$

$$(P^*)^2 \cdot l \cdot T_A \cdot T_L \cdot T_R^2 + P^* \cdot [(l + q_0) \cdot E_A \cdot T_L \cdot T_R$$
$$- (g_0 - l) \cdot T_A \cdot T_R \cdot E_L] + q_0 \cdot E_A \cdot E_L - (g_0 - l) \cdot E_A \cdot E_L = 0$$

We define for simplicity the following global parameters:

$$\Xi_1 = l \cdot T_A \cdot T_L \cdot T_R^2;\ \Xi_2 = (l + q_0) \cdot E_A \cdot T_L \cdot T_R - (g_0 - l) \cdot T_A \cdot T_R \cdot E_L$$
$$\Xi_3 = q_0 \cdot E_A \cdot E_L - (g_0 - l) \cdot E_A \cdot E_L$$

$$(P^*)^2 \cdot \Xi_1 + P^* \cdot \Xi_2 + \Xi_3 = 0 \Rightarrow P^* = \frac{-\Xi_2 \pm \sqrt{\Xi_2^2 - 4 \cdot \Xi_1 \cdot \Xi_3}}{2 \cdot \Xi_1}$$

P^* Define two possible P variable fixed points ($P^{(1)}, P^{(2)}$).

$$P^{(1)} = \frac{-\Xi_2 + \sqrt{\Xi_2^2 - 4 \cdot \Xi_1 \cdot \Xi_3}}{2 \cdot \Xi_1};\ P^{(2)} = \frac{-\Xi_2 - \sqrt{\Xi_2^2 - 4 \cdot \Xi_1 \cdot \Xi_3}}{2 \cdot \Xi_1}$$

Then we get two possible q variable fixed points ($q^{(1)}, q^{(2)}$).

$$q^* = \frac{q_0 \cdot E_A}{E_A + P^* \cdot T_A \cdot T_R} \Rightarrow$$
$$q^{(1)} = \frac{q_0 \cdot E_A}{E_A + P^{(1)} \cdot T_A \cdot T_R};\ q^{(2)} = \frac{q_0 \cdot E_A}{E_A + P^{(2)} \cdot T_A \cdot T_R}$$
$$q^{(1)} = \frac{q_0 \cdot E_A}{E_A + [\frac{-\Xi_2 + \sqrt{\Xi_2^2 - 4 \cdot \Xi_1 \cdot \Xi_3}}{2 \cdot \Xi_1}] \cdot T_A \cdot T_R}$$
$$q^{(2)} = \frac{q_0 \cdot E_A}{E_A + [\frac{-\Xi_2 - \sqrt{\Xi_2^2 - 4 \cdot \Xi_1 \cdot \Xi_3}}{2 \cdot \Xi_1}] \cdot T_A \cdot T_R}$$

Then we get two possible g variable fixed points ($g^{(1)}, g^{(2)}$).

$$g^* = l + q^* \Rightarrow g^{(1)} = l + q^{(1)};\ g^{(2)} = l + q^{(2)};\ g^{(1)}$$
$$= l + \frac{q_0 \cdot E_A}{E_A + P^{(1)} \cdot T_A \cdot T_R};\ g^{(2)} = l + \frac{q_0 \cdot E_A}{E_A + P^{(2)} \cdot T_A \cdot T_R}$$
$$g^{(1)} = l + \frac{q_0 \cdot E_A}{E_A + \left[\frac{-\Xi_2 + \sqrt{\Xi_2^2 - 4 \cdot \Xi_1 \cdot \Xi_3}}{2 \cdot \Xi_1}\right] \cdot T_A \cdot T_R}$$
$$g^{(2)} = l + \frac{q_0 \cdot E_A}{E_A + \left[\frac{-\Xi_2 - \sqrt{\Xi_2^2 - 4 \cdot \Xi_1 \cdot \Xi_3}}{2 \cdot \Xi_1}\right] \cdot T_A \cdot T_R}$$

The second fixed point:

$$E^{(1)}(P^{(1)}, g^{(1)}, q^{(1)}) = \left(\frac{-\Xi_2 + \sqrt{\Xi_2^2 - 4 \cdot \Xi_1 \cdot \Xi_3}}{2 \cdot \Xi_1}, \right.$$

$$l + \frac{q_0 \cdot E_A}{E_A + \left[\frac{-\Xi_2 + \sqrt{\Xi_2^2 - 4 \cdot \Xi_1 \cdot \Xi_3}}{2 \cdot \Xi_1}\right] \cdot T_A \cdot T_R},$$

$$\left. \frac{q_0 \cdot E_A}{E_A + \left(\frac{-\Xi_2 + \sqrt{\Xi_2^2 - 4 \cdot \Xi_1 \cdot \Xi_3}}{2 \cdot \Xi_1}\right) \cdot T_A \cdot T_R} \right)$$

The third fixed point:

$$E^{(2)}(P^{(2)}, g^{(2)}, q^{(2)}) = \left(\frac{-\Xi_2 - \sqrt{\Xi_2^2 - 4 \cdot \Xi_1 \cdot \Xi_3}}{2 \cdot \Xi_1}, \right.$$

$$l + \frac{q_0 \cdot E_A}{E_A + \left[\frac{-\Xi_2 - \sqrt{\Xi_2^2 - 4 \cdot \Xi_1 \cdot \Xi_3}}{2 \cdot \Xi_1}\right] \cdot T_A \cdot T_R},$$

$$\left. \frac{q_0 \cdot E_A}{E_A + \left[\frac{-\Xi_2 - \sqrt{\Xi_2^2 - 4 \cdot \Xi_1 \cdot \Xi_3}}{2 \cdot \Xi_1}\right] \cdot T_A \cdot T_R} \right)$$

Stability analysis: The standard local stability analysis about any one of the equilibrium points of the Solid-state laser by semiconductor devices model corresponding to the rate differential equation system consists in adding to coordinate $[P, g, q]$ arbitrarily small increments of exponential form $[p, g, q] \cdot e^{\lambda \cdot t}$ and retaining the first order terms in P, g, q. The system of three homogeneous equations leads to a polynomial characteristic equation in the eigenvalues. The polynomial characteristic equations accept by set the below variables (P, g, q) and variables derivatives with respect to time into system rate equations [5–7]. System fixed values with arbitrarily small increments of exponential form $[p, g, q] \cdot e^{\lambda \cdot t}$ are: j = 0 (first fixed point), j = 1 (second fixed point), j = 2 (third fixed point), etc.

$$P(t) = P^{(j)} + p \cdot e^{\lambda \cdot t}; \; g(t) = g^{(j)} + g \cdot e^{\lambda \cdot t}; \; q(t) = q^{(j)} + q \cdot e^{\lambda \cdot t}; \; P(t - \tau) = P^{(j)} + p \cdot e^{\lambda \cdot (t - \tau)}$$

$$\frac{dP(t)}{dt} = \lambda \cdot p \cdot e^{\lambda \cdot t}; \; \frac{dg(t)}{dt} = \lambda \cdot g \cdot e^{\lambda \cdot t}; \; \frac{dq(t)}{dt} = \lambda \cdot q \cdot e^{\lambda \cdot t}$$

We choose these expressions for ourselves $P(t), g(t), q(t)$ as a small displacement $[p, g, q]$ from the system fixed points in time t = 0.

$$P(t=0) = P^{(j)} + p;\ g(t=0) = g^{(j)} + g;\ q(t=0) = q^{(j)} + q$$

$$\frac{dP}{dt} = \frac{2}{T_R} \cdot [g(t) - l - q(t)] \cdot P(t - \tau)$$

$$\lambda \cdot p \cdot e^{\lambda \cdot t} = \frac{2}{T_R} \cdot [g^{(j)} + g \cdot e^{\lambda \cdot t} - l - (q^{(j)} + q \cdot e^{\lambda \cdot t})] \cdot [P^{(j)} + p \cdot e^{\lambda \cdot (t-\tau)}]$$

$$\lambda \cdot p \cdot e^{\lambda \cdot t} = \frac{2}{T_R} \cdot [(g^{(j)} - l - q^{(j)}) + g \cdot e^{\lambda \cdot t} - q \cdot e^{\lambda \cdot t}] \cdot [P^{(j)} + p \cdot e^{\lambda \cdot (t-\tau)}]$$

$$\lambda \cdot p \cdot e^{\lambda \cdot t} = \frac{2}{T_R} \cdot [(g^{(j)} - l - q^{(j)}) \cdot P^{(j)}$$
$$+ (g^{(j)} - l - q^{(j)}) \cdot p \cdot e^{\lambda \cdot (t-\tau)} + P^{(j)} \cdot g \cdot e^{\lambda \cdot t}$$
$$+ g \cdot p \cdot e^{\lambda \cdot (t-\tau)} \cdot e^{\lambda \cdot t} - P^{(j)} \cdot q \cdot e^{\lambda \cdot t} - q \cdot p \cdot e^{\lambda \cdot (t-\tau)} \cdot e^{\lambda \cdot t}]$$

Assumption: $g \cdot p \approx 0;\ q \cdot p \approx 0$

$$\lambda \cdot p \cdot e^{\lambda \cdot t} = \frac{2}{T_R} \cdot [(g^{(j)} - l - q^{(j)}) \cdot P^{(j)} + (g^{(j)} - l - q^{(j)}) \cdot p \cdot e^{\lambda \cdot (t-\tau)}$$
$$+ P^{(j)} \cdot g \cdot e^{\lambda \cdot t} - P^{(j)} \cdot q \cdot e^{\lambda \cdot t}] \lambda \cdot p \cdot e^{\lambda \cdot t}$$
$$= \frac{2}{T_R} \cdot [(g^{(j)} - l - q^{(j)}) \cdot P^{(j)}]$$
$$+ \frac{2}{T_R} \cdot [(g^{(j)} - l - q^{(j)}) \cdot p \cdot e^{-\lambda \cdot \tau} + P^{(j)} \cdot g - P^{(j)} \cdot q] \cdot e^{\lambda \cdot t}$$

At fixed points: $\frac{2}{T_R} \cdot [(g^{(j)} - l - q^{(j)}) \cdot P^{(j)}] = 0.$

$$\lambda \cdot p \cdot e^{\lambda \cdot t} = \frac{2}{T_R} \cdot [(g^{(j)} - l - q^{(j)}) \cdot p \cdot e^{-\lambda \cdot \tau} + P^{(j)} \cdot g - P^{(j)} \cdot q] \cdot e^{\lambda \cdot t}$$

Divide the two sides of the above equation by $e^{\lambda \cdot t}$ term.

$$\lambda \cdot p = \frac{2}{T_R} \cdot [(g^{(j)} - l - q^{(j)}) \cdot p \cdot e^{-\lambda \cdot \tau} + P^{(j)} \cdot g - P^{(j)} \cdot q]$$

$$\left[\frac{2}{T_R} \cdot (g^{(j)} - l - q^{(j)}) \cdot e^{-\lambda \cdot \tau} - \lambda\right] \cdot p + \frac{2}{T_R} \cdot P^{(j)} \cdot g - \frac{2}{T_R} \cdot P^{(j)} \cdot q = 0$$

$$\frac{dg}{dt} = -\frac{(g(t) - g_0)}{T_L \cdot T_R} - \frac{g(t) \cdot P(t - \tau)}{E_L}$$

$$\lambda \cdot g \cdot e^{\lambda \cdot t} = -\frac{(g^{(j)} + g \cdot e^{\lambda \cdot t} - g_0)}{T_L \cdot T_R} - \frac{(g^{(j)} + g \cdot e^{\lambda \cdot t}) \cdot [P^{(j)} + p \cdot e^{\lambda \cdot (t-\tau)}]}{E_L}$$

$$\lambda \cdot g \cdot e^{\lambda \cdot t} = -\frac{(g^{(j)} - g_0)}{T_L \cdot T_R} - \frac{g \cdot e^{\lambda \cdot t}}{T_L \cdot T_R}$$

$$-\frac{[g^{(j)}\cdot P^{(j)}+g^{(j)}\cdot p\cdot e^{\lambda\cdot(t-\tau)}+P^{(j)}\cdot g\cdot e^{\lambda\cdot t}+g\cdot p\cdot e^{\lambda\cdot(t-\tau)}\cdot e^{\lambda\cdot t}]}{E_L}$$

$$\lambda\cdot g\cdot e^{\lambda\cdot t}=-\frac{(g^{(j)}-g_0)}{T_L\cdot T_R}-\frac{g^{(j)}\cdot P^{(j)}}{E_L}-\frac{g\cdot e^{\lambda\cdot t}}{T_L\cdot T_R}$$
$$-\frac{[g^{(j)}\cdot p\cdot e^{\lambda\cdot(t-\tau)}+P^{(j)}\cdot g\cdot e^{\lambda\cdot t}+g\cdot p\cdot e^{\lambda\cdot(t-\tau)}\cdot e^{\lambda\cdot t}]}{E_L}$$

Assumption: $g\cdot p\approx 0$

$$\lambda\cdot g\cdot e^{\lambda\cdot t}=-\frac{(g^{(j)}-g_0)}{T_L\cdot T_R}-\frac{g^{(j)}\cdot P^{(j)}}{E_L}-\frac{g}{T_L\cdot T_R}\cdot e^{\lambda\cdot t}$$
$$-\frac{[g^{(j)}\cdot p\cdot e^{-\lambda\cdot\tau}+P^{(j)}\cdot g]}{E_L}\cdot e^{\lambda\cdot t}$$

At fixed points: $-\frac{(g^{(j)}-g_0)}{T_L\cdot T_R}-\frac{g^{(j)}\cdot P^{(j)}}{E_L}=0$.

$$\lambda\cdot g\cdot e^{\lambda\cdot t}=-\frac{g}{T_L\cdot T_R}\cdot e^{\lambda\cdot t}-\frac{[g^{(j)}\cdot p\cdot e^{-\lambda\cdot\tau}+P^{(j)}\cdot g]}{E_L}\cdot e^{\lambda\cdot t}$$

Divide the two sides of the above equation by $e^{\lambda\cdot t}$ term.

$$\lambda\cdot g=-\frac{g}{T_L\cdot T_R}-\frac{g^{(j)}\cdot e^{-\lambda\cdot\tau}}{E_L}\cdot p-\frac{P^{(j)}}{E_L}\cdot g\Rightarrow-\frac{g^{(j)}\cdot e^{-\lambda\cdot\tau}}{E_L}$$
$$\cdot p+\left(-\frac{1}{T_L\cdot T_R}-\frac{P^{(j)}}{E_L}-\lambda\right)\cdot g=0\frac{dq}{dt}$$
$$=-\frac{(q(t)-q_0)}{T_A\cdot T_R}-\frac{q(t)\cdot P(t-\tau)}{E_A}\lambda\cdot q\cdot e^{\lambda\cdot t}$$
$$=-\frac{(q^{(j)}+q\cdot e^{\lambda\cdot t}-q_0)}{T_A\cdot T_R}-\frac{(q^{(j)}+q\cdot e^{\lambda\cdot t})\cdot[P^{(j)}+p\cdot e^{\lambda\cdot(t-\tau)}]}{E_A}$$
$$\lambda\cdot q\cdot e^{\lambda\cdot t}=-\frac{q^{(j)}-q_0}{T_A\cdot T_R}-\frac{q\cdot e^{\lambda\cdot t}}{T_A\cdot T_R}$$
$$-\frac{[q^{(j)}\cdot P^{(j)}+q^{(j)}\cdot p\cdot e^{\lambda\cdot(t-\tau)}+P^{(j)}\cdot q\cdot e^{\lambda\cdot t}+q\cdot p\cdot e^{\lambda\cdot(t-\tau)}\cdot e^{\lambda\cdot t}]}{E_A}$$

Assumption: $q\cdot p\approx 0$

$$\lambda\cdot q\cdot e^{\lambda\cdot t}=-\frac{q^{(j)}-q_0}{T_A\cdot T_R}-\frac{q\cdot e^{\lambda\cdot t}}{T_A\cdot T_R}$$
$$-\frac{[q^{(j)}\cdot P^{(j)}+q^{(j)}\cdot p\cdot e^{\lambda\cdot(t-\tau)}+P^{(j)}\cdot q\cdot e^{\lambda\cdot t}]}{E_A}$$

$$\lambda\cdot q\cdot e^{\lambda\cdot t}=-\frac{q^{(j)}-q_0}{T_A\cdot T_R}-\frac{q\cdot e^{\lambda\cdot t}}{T_A\cdot T_R}-\frac{q^{(j)}\cdot P^{(j)}}{E_A}$$

$$-\frac{[q^{(j)} \cdot p \cdot e^{\lambda \cdot (t-\tau)} + P^{(j)} \cdot q \cdot e^{\lambda \cdot t}]}{E_A} \lambda \cdot q \cdot e^{\lambda \cdot t} = -\frac{q^{(j)} - q_0}{T_A \cdot T_R}$$
$$-\frac{q^{(j)} \cdot P^{(j)}}{E_A} - \frac{q}{T_A \cdot T_R} \cdot e^{\lambda \cdot t} - \frac{[q^{(j)} \cdot p \cdot e^{-\lambda \cdot \tau} + P^{(j)} \cdot q]}{E_A} \cdot e^{\lambda \cdot t}$$

At fixed points: $-\frac{q^{(j)} - q_0}{T_A \cdot T_R} - \frac{q^{(j)} \cdot P^{(j)}}{E_A} = 0$.

$$\lambda \cdot q \cdot e^{\lambda \cdot t} = -\frac{q}{T_A \cdot T_R} \cdot e^{\lambda \cdot t} - \frac{[q^{(j)} \cdot p \cdot e^{-\lambda \cdot \tau} + P^{(j)} \cdot q]}{E_A} \cdot e^{\lambda \cdot t}$$

Divide the two sides of the above equation by $e^{\lambda \cdot t}$ term.

$$\lambda \cdot q = -\frac{q}{T_A \cdot T_R} - \frac{[q^{(j)} \cdot p \cdot e^{-\lambda \cdot \tau} + P^{(j)} \cdot q]}{E_A} \Rightarrow -\frac{q^{(j)} \cdot e^{-\lambda \cdot \tau}}{E_A} \cdot p$$
$$+ \left(-\frac{1}{T_A \cdot T_R} - \frac{P^{(j)}}{E_A} - \lambda\right) \cdot q = 0$$

We can summarize our last results:

$$\left[\frac{2}{T_R} \cdot (g^{(j)} - l - q^{(j)}) \cdot e^{-\lambda \cdot \tau} - \lambda\right] \cdot p + \frac{2}{T_R} \cdot P^{(j)} \cdot g - \frac{2}{T_R} \cdot P^{(j)} \cdot q = 0$$
$$-\frac{g^{(j)} \cdot e^{-\lambda \cdot \tau}}{E_L} \cdot p + \left(-\frac{1}{T_L \cdot T_R} - \frac{P^{(j)}}{E_L} - \lambda\right) \cdot g = 0$$
$$-\frac{q^{(j)} \cdot e^{-\lambda \cdot \tau}}{E_A} \cdot p + \left(-\frac{1}{T_A \cdot T_R} - \frac{P^{(j)}}{E_A} - \lambda\right) \cdot q = 0$$

The small increments Jacobian of our system is as follow:

$$\begin{pmatrix} \frac{2}{T_R}\cdot(g^{(j)}-l-q^{(j)})\cdot e^{-\lambda\cdot\tau}-\lambda & \frac{2}{T_R}\cdot P^{(j)} & -\frac{2}{T_R}\cdot P^{(j)} \\ -\frac{g^{(j)}}{E_L}\cdot e^{-\lambda\cdot\tau} & -\frac{1}{T_L\cdot T_R}-\frac{P^{(j)}}{E_L}-\lambda & 0 \\ -\frac{q^{(j)}}{E_A}\cdot e^{-\lambda\cdot\tau} & 0 & -\frac{1}{T_A\cdot T_R}-\frac{P^{(j)}}{E_A}-\lambda \end{pmatrix}\cdot\begin{pmatrix} p \\ g \\ q \end{pmatrix}=0$$

$$(A-\lambda\cdot I)=\begin{pmatrix} \frac{2}{T_R}\cdot(g^{(j)}-l-q^{(j)})\cdot e^{-\lambda\cdot\tau}-\lambda & \frac{2}{T_R}\cdot P^{(j)} & -\frac{2}{T_R}\cdot P^{(j)} \\ -\frac{g^{(j)}}{E_L}\cdot e^{-\lambda\cdot\tau} & -\frac{1}{T_L\cdot T_R}-\frac{P^{(j)}}{E_L}-\lambda & 0 \\ -\frac{q^{(j)}}{E_A}\cdot e^{-\lambda\cdot\tau} & 0 & -\frac{1}{T_A\cdot T_R}-\frac{P^{(j)}}{E_A}-\lambda \end{pmatrix}$$

$$\det(A-\lambda\cdot I)=0$$

$$\det\begin{pmatrix} \frac{2}{T_R}\cdot(g^{(j)}-l-q^{(j)})\cdot e^{-\lambda\cdot\tau}-\lambda & \frac{2}{T_R}\cdot P^{(j)} & -\frac{2}{T_R}\cdot P^{(j)} \\ -\frac{g^{(j)}}{E_L}\cdot e^{-\lambda\cdot\tau} & -\frac{1}{T_L\cdot T_R}-\frac{P^{(j)}}{E_L}-\lambda & 0 \\ -\frac{q^{(j)}}{E_A}\cdot e^{-\lambda\cdot\tau} & 0 & -\frac{1}{T_A\cdot T_R}-\frac{P^{(j)}}{E_A}-\lambda \end{pmatrix}=0$$

$$\det\begin{pmatrix} \frac{2}{T_R}\cdot(g^{(j)}-l-q^{(j)})\cdot e^{-\lambda\cdot\tau}-\lambda & \frac{2}{T_R}\cdot P^{(j)} & -\frac{2}{T_R}\cdot P^{(j)} \\ -\frac{g^{(j)}}{E_L}\cdot e^{-\lambda\cdot\tau} & -\frac{1}{T_L\cdot T_R}-\frac{P^{(j)}}{E_L}-\lambda & 0 \\ -\frac{q^{(j)}}{E_A}\cdot e^{-\lambda\cdot\tau} & 0 & -\frac{1}{T_A\cdot T_R}-\frac{P^{(j)}}{E_A}-\lambda \end{pmatrix}=$$

$$[\frac{2}{T_R}\cdot(g^{(j)}-l-q^{(j)})\cdot e^{-\lambda\cdot\tau}-\lambda]\cdot\det\begin{pmatrix} -\frac{1}{T_L\cdot T_R}-\frac{P^{(j)}}{E_L}-\lambda & 0 \\ 0 & -\frac{1}{T_A\cdot T_R}-\frac{P^{(j)}}{E_A}-\lambda \end{pmatrix}$$

$$-\frac{2}{T_R}\cdot P^{(j)}\cdot\det\begin{pmatrix} -\frac{g^{(j)}}{E_L}\cdot e^{-\lambda\cdot\tau} & 0 \\ -\frac{q^{(j)}}{E_A}\cdot e^{-\lambda\cdot\tau} & -\frac{1}{T_A\cdot T_R}-\frac{P^{(j)}}{E_A}-\lambda \end{pmatrix}-\frac{2}{T_R}\cdot P^{(j)}\cdot\det\begin{pmatrix} -\frac{g^{(j)}}{E_L}\cdot e^{-\lambda\cdot\tau} & -\frac{1}{T_L\cdot T_R}-\frac{P^{(j)}}{E_L}-\lambda \\ -\frac{q^{(j)}}{E_A}\cdot e^{-\lambda\cdot\tau} & 0 \end{pmatrix}$$

$$\det(A-\lambda\cdot I)=\left[\frac{2}{T_R}\cdot(g^{(j)}-l-q^{(j)})\cdot e^{-\lambda\cdot\tau}-\lambda\right]$$
$$\cdot\left[\frac{1}{T_L\cdot T_R}+\frac{P^{(j)}}{E_L}+\lambda\right]\cdot\left[\frac{1}{T_A\cdot T_R}+\frac{P^{(j)}}{E_A}+\lambda\right]$$
$$-\frac{2}{T_R}\cdot P^{(j)}\cdot\frac{g^{(j)}}{E_L}\cdot\left[\frac{1}{T_A\cdot T_R}+\frac{P^{(j)}}{E_A}+\lambda\right]\cdot e^{-\lambda\cdot\tau}$$
$$+\frac{2}{T_R}\cdot P^{(j)}\cdot\frac{q^{(j)}}{E_A}\cdot\left[\frac{1}{T_L\cdot T_R}+\frac{P^{(j)}}{E_L}+\lambda\right]\cdot e^{-\lambda\cdot\tau}$$

$$\det(A-\lambda\cdot I)=\left[\frac{2}{T_R}\cdot(g^{(j)}-l-q^{(j)})\cdot e^{-\lambda\cdot\tau}-\lambda\right]\cdot$$
$$\left[\left(\frac{1}{T_L\cdot T_R}+\frac{P^{(j)}}{E_L}\right)+\lambda\right]\cdot\left[\left(\frac{1}{T_A\cdot T_R}+\frac{P^{(j)}}{E_A}\right)+\lambda\right]$$
$$-\frac{2}{T_R}\cdot P^{(j)}\cdot\frac{g^{(j)}}{E_L}\cdot\left(\frac{1}{T_A\cdot T_R}+\frac{P^{(j)}}{E_A}\right)\cdot e^{-\lambda\cdot\tau}$$
$$-\frac{2}{T_R}\cdot P^{(j)}\cdot\frac{g^{(j)}}{E_L}\cdot\lambda\cdot e^{-\lambda\cdot\tau}$$
$$+\frac{2}{T_R}\cdot P^{(j)}\cdot\frac{q^{(j)}}{E_A}\cdot\left(\frac{1}{T_L\cdot T_R}+\frac{P^{(j)}}{E_L}\right)\cdot e^{-\lambda\cdot\tau}$$
$$+\frac{2}{T_R}\cdot P^{(j)}\cdot\frac{q^{(j)}}{E_A}\cdot\lambda\cdot e^{-\lambda\cdot\tau}$$

$$\det(A-\lambda\cdot I)=\left[\frac{2}{T_R}\cdot(g^{(j)}-l-q^{(j)})\cdot e^{-\lambda\cdot\tau}-\lambda\right]$$
$$\cdot\left[\left(\frac{1}{T_L\cdot T_R}+\frac{P^{(j)}}{E_L}\right)\cdot\left(\frac{1}{T_A\cdot T_R}+\frac{P^{(j)}}{E_A}\right)\right.$$
$$\left.+\left(\frac{1}{T_L\cdot T_R}+\frac{1}{T_A\cdot T_R}+\frac{P^{(j)}}{E_A}+\frac{P^{(j)}}{E_L}\right)\cdot\lambda+\lambda^2\right]$$
$$-\frac{2}{T_R}\cdot P^{(j)}\cdot\frac{g^{(j)}}{E_L}\cdot\left(\frac{1}{T_A\cdot T_R}+\frac{P^{(j)}}{E_A}\right)\cdot e^{-\lambda\cdot\tau}$$
$$-\frac{2}{T_R}\cdot P^{(j)}\cdot\frac{g^{(j)}}{E_L}\cdot\lambda\cdot e^{-\lambda\cdot\tau}$$
$$+\frac{2}{T_R}\cdot P^{(j)}\cdot\frac{q^{(j)}}{E_A}\cdot\left(\frac{1}{T_L\cdot T_R}+\frac{P^{(j)}}{E_L}\right)\cdot e^{-\lambda\cdot\tau}$$
$$+\frac{2}{T_R}\cdot P^{(j)}\cdot\frac{q^{(j)}}{E_A}\cdot\lambda\cdot e^{-\lambda\cdot\tau}$$

#####

$$\left[\frac{2}{T_R}\cdot(g^{(j)}-l-q^{(j)})\cdot e^{-\lambda\cdot\tau}-\lambda\right]$$
$$\cdot\left[\left(\frac{1}{T_L\cdot T_R}+\frac{P^{(j)}}{E_L}\right)\cdot\left(\frac{1}{T_A\cdot T_R}+\frac{P^{(j)}}{E_A}\right)\right.$$
$$\left.+\left(\frac{1}{T_L\cdot T_R}+\frac{1}{T_A\cdot T_R}+\frac{P^{(j)}}{E_A}+\frac{P^{(j)}}{E_L}\right)\cdot\lambda+\lambda^2\right]$$
$$=\left(\frac{1}{T_L\cdot T_R}+\frac{P^{(j)}}{E_L}\right)\cdot\left(\frac{1}{T_A\cdot T_R}+\frac{P^{(j)}}{E_A}\right)\cdot\frac{2}{T_R}\cdot(g^{(j)}-l-q^{(j)})\cdot e^{-\lambda\cdot\tau}$$
$$+\frac{2}{T_R}\cdot(g^{(j)}-l-q^{(j)})\cdot e^{-\lambda\cdot\tau}\cdot\left(\frac{1}{T_L\cdot T_R}+\frac{1}{T_A\cdot T_R}+\frac{P^{(j)}}{E_A}+\frac{P^{(j)}}{E_L}\right)\cdot\lambda$$
$$+\frac{2}{T_R}\cdot(g^{(j)}-l-q^{(j)})\cdot e^{-\lambda\cdot\tau}\cdot\lambda^2$$
$$-\lambda\cdot\left(\frac{1}{T_L\cdot T_R}+\frac{P^{(j)}}{E_L}\right)\cdot\left(\frac{1}{T_A\cdot T_R}+\frac{P^{(j)}}{E_A}\right)$$
$$-\left(\frac{1}{T_L\cdot T_R}+\frac{1}{T_A\cdot T_R}+\frac{P^{(j)}}{E_A}+\frac{P^{(j)}}{E_L}\right)\cdot\lambda^2-\lambda^3$$

$$\left[\frac{2}{T_R}\cdot(g^{(j)}-l-q^{(j)})\cdot e^{-\lambda\cdot\tau}-\lambda\right]$$
$$\cdot\left[\left(\frac{1}{T_L\cdot T_R}+\frac{P^{(j)}}{E_L}\right)\cdot\left(\frac{1}{T_A\cdot T_R}+\frac{P^{(j)}}{E_A}\right)\right.$$
$$\left.+\left(\frac{1}{T_L\cdot T_R}+\frac{1}{T_A\cdot T_R}+\frac{P^{(j)}}{E_A}+\frac{P^{(j)}}{E_L}\right))\lambda+\lambda^2\right]$$
$$=\left\{\left(\frac{1}{T_L\cdot T_R}+\frac{P^{(j)}}{E_L}\right)\cdot\left(\frac{1}{T_A\cdot T_R}+\frac{P^{(j)}}{E_A}\right)\cdot\frac{2}{T_R}\cdot(g^{(j)}-l-q^{(j)})\right.$$
$$+\frac{2}{T_R}\cdot(g^{(j)}-l-q^{(j)})\cdot\left(\frac{1}{T_L\cdot T_R}+\frac{1}{T_A\cdot T_R}+\frac{P^{(j)}}{E_A}+\frac{P^{(j)}}{E_L}\right)\cdot\lambda$$
$$\left.+\frac{2}{T_R}\cdot(g^{(j)}-l-q^{(j)})\cdot\lambda^2\right\}\cdot e^{-\lambda\cdot\tau}$$
$$-\lambda\cdot\left(\frac{1}{T_L\cdot T_R}+\frac{P^{(j)}}{E_L}\right)\cdot\left(\frac{1}{T_A\cdot T_R}+\frac{P^{(j)}}{E_A}\right)$$
$$-\left(\frac{1}{T_L\cdot T_R}+\frac{1}{T_A\cdot T_R}+\frac{P^{(j)}}{E_A}+\frac{P^{(j)}}{E_L}\right)\cdot\lambda^2-\lambda^3$$

#####

$$\det(A-\lambda\cdot I)=\left\{\left(\frac{1}{T_L\cdot T_R}+\frac{P^{(j)}}{E_L}\right)\cdot\left(\frac{1}{T_A\cdot T_R}+\frac{P^{(j)}}{E_A}\right)\cdot\frac{2}{T_R}\cdot(g^{(j)}-l-q^{(j)})\right.$$

$$+\frac{2}{T_R}\cdot(g^{(j)}-l-q^{(j)})\cdot\left(\frac{1}{T_L\cdot T_R}+\frac{1}{T_A\cdot T_R}+\frac{P^{(j)}}{E_A}+\frac{P^{(j)}}{E_L}\right)\cdot\lambda$$

$$+\frac{2}{T_R}\cdot(g^{(j)}-l-q^{(j)})\cdot\lambda^2\Bigg\}\cdot e^{-\lambda\cdot\tau}$$

$$-\lambda\cdot\left(\frac{1}{T_L\cdot T_R}+\frac{P^{(j)}}{E_L}\right)\cdot\left(\frac{1}{T_A\cdot T_R}+\frac{P^{(j)}}{E_A}\right)$$

$$-\left(\frac{1}{T_L\cdot T_R}+\frac{1}{T_A\cdot T_R}+\frac{P^{(j)}}{E_A}+\frac{P^{(j)}}{E_L}\right)\cdot\lambda^2-\lambda^3$$

$$-\frac{2}{T_R}\cdot P^{(j)}\cdot\frac{g^{(j)}}{E_L}\cdot\left(\frac{1}{T_A\cdot T_R}+\frac{P^{(j)}}{E_A}\right)\cdot e^{-\lambda\cdot\tau}$$

$$-\frac{2}{T_R}\cdot P^{(j)}\cdot\frac{g^{(j)}}{E_L}\cdot\lambda\cdot e^{-\lambda\cdot\tau}$$

$$+\frac{2}{T_R}\cdot P^{(j)}\cdot\frac{q^{(j)}}{E_A}\cdot\left(\frac{1}{T_L\cdot T_R}+\frac{P^{(j)}}{E_L}\right)\cdot e^{-\lambda\cdot\tau}$$

$$+\frac{2}{T_R}\cdot P^{(j)}\cdot\frac{q^{(j)}}{E_A}\cdot\lambda\cdot e^{-\lambda\cdot\tau}$$

$$\det(A-\lambda\cdot I)=\Bigg\{\left(\frac{1}{T_L\cdot T_R}+\frac{P^{(j)}}{E_L}\right)\cdot\left(\frac{1}{T_A\cdot T_R}+\frac{P^{(j)}}{E_A}\right)\cdot\frac{2}{T_R}\cdot(g^{(j)}-l-q^{(j)})$$

$$+\frac{2}{T_R}\cdot(g^{(j)}-l-q^{(j)})\cdot\left(\frac{1}{T_L\cdot T_R}+\frac{1}{T_A\cdot T_R}+\frac{P^{(j)}}{E_A}+\frac{P^{(j)}}{E_L}\right)\cdot\lambda$$

$$+\frac{2}{T_R}\cdot(g^{(j)}-l-q^{(j)})\cdot\lambda^2\Bigg\}\cdot e^{-\lambda\cdot\tau}$$

$$-\lambda\cdot\left(\frac{1}{T_L\cdot T_R}+\frac{P^{(j)}}{E_L}\right)\cdot\left(\frac{1}{T_A\cdot T_R}+\frac{P^{(j)}}{E_A}\right)$$

$$-\left(\frac{1}{T_L\cdot T_R}+\frac{1}{T_A\cdot T_R}+\frac{P^{(j)}}{E_A}+\frac{P^{(j)}}{E_L}\right)\cdot\lambda^2-\lambda^3$$

$$+\Bigg\{\frac{2}{T_R}\cdot P^{(j)}\cdot\frac{q^{(j)}}{E_A}\cdot\left(\frac{1}{T_L\cdot T_R}+\frac{P^{(j)}}{E_L}\right)$$

$$-\frac{2}{T_R}\cdot P^{(j)}\cdot\frac{g^{(j)}}{E_L}\cdot\left(\frac{1}{T_A\cdot T_R}+\frac{P^{(j)}}{E_A}\right)\Bigg\}\cdot e^{-\lambda\cdot\tau}$$

$$+\left\{\frac{2}{T_R}\cdot P^{(j)}\cdot\frac{q^{(j)}}{E_A}-\frac{2}{T_R}\cdot P^{(j)}\cdot\frac{g^{(j)}}{E_L}\right\}\cdot\lambda\cdot e^{-\lambda\cdot\tau}$$

$$\det(A-\lambda\cdot I)=\Bigg[\left(\frac{1}{T_L\cdot T_R}+\frac{P^{(j)}}{E_L}\right)\cdot\left(\frac{1}{T_A\cdot T_R}+\frac{P^{(j)}}{E_A}\right)\cdot\frac{2}{T_R}\cdot(g^{(j)}-l-q^{(j)})$$

$$+\frac{2}{T_R}\cdot P^{(j)}\cdot\frac{q^{(j)}}{E_A}\cdot\left(\frac{1}{T_L\cdot T_R}+\frac{P^{(j)}}{E_L}\right)$$

$$-\frac{2}{T_R}\cdot P^{(j)}\cdot\frac{g^{(j)}}{E_L}\cdot\left(\frac{1}{T_A\cdot T_R}+\frac{P^{(j)}}{E_A}\right)$$

$$+\left\{\frac{2}{T_R}\cdot(g^{(j)}-l-q^{(j)})\cdot\left(\frac{1}{T_L\cdot T_R}+\frac{1}{T_A\cdot T_R}+\frac{P^{(j)}}{E_A}+\frac{P^{(j)}}{E_L}\right)\right.$$
$$\left.+\frac{2}{T_R}\cdot P^{(j)}\cdot\frac{q^{(j)}}{E_A}-\frac{2}{T_R}\cdot P^{(j)}\cdot\frac{g^{(j)}}{E_L}\right\}\cdot\lambda$$
$$+\frac{2}{T_R}\cdot(g^{(j)}-l-q^{(j)})\cdot\lambda^2]\cdot e^{-\lambda\cdot\tau}$$
$$-\lambda\cdot\left(\frac{1}{T_L\cdot T_R}+\frac{P^{(j)}}{E_L}\right)\cdot\left(\frac{1}{T_A\cdot T_R}+\frac{P^{(j)}}{E_A}\right)$$
$$-\left(\frac{1}{T_L\cdot T_R}+\frac{1}{T_A\cdot T_R}+\frac{P^{(j)}}{E_A}+\frac{P^{(j)}}{E_L}\right)\cdot\lambda^2-\lambda^3$$

$$D(\lambda,\tau)=\left[\left(\frac{1}{T_L\cdot T_R}+\frac{P^{(j)}}{E_L}\right)\cdot\left(\frac{1}{T_A\cdot T_R}+\frac{P^{(j)}}{E_A}\right)\cdot\frac{2}{T_R}\cdot(g^{(j)}-l-q^{(j)})\right.$$
$$+\frac{2}{T_R}\cdot P^{(j)}\cdot\frac{q^{(j)}}{E_A}\cdot\left(\frac{1}{T_L\cdot T_R}+\frac{P^{(j)}}{E_L}\right)$$
$$-\frac{2}{T_R}\cdot P^{(j)}\cdot\frac{g^{(j)}}{E_L}\cdot\left(\frac{1}{T_A\cdot T_R}+\frac{P^{(j)}}{E_A}\right)$$
$$+\left\{\frac{2}{T_R}\cdot(g^{(j)}-l-q^{(j)})\cdot\left(\frac{1}{T_L\cdot T_R}+\frac{1}{T_A\cdot T_R}+\frac{P^{(j)}}{E_A}+\frac{P^{(j)}}{E_L}\right)\right.$$
$$\left.+\frac{2}{T_R}\cdot P^{(j)}\cdot\frac{q^{(j)}}{E_A}-\frac{2}{T_R}\cdot P^{(j)}\cdot\frac{g^{(j)}}{E_L}\right\}\cdot\lambda$$
$$\left.+\frac{2}{T_R}\cdot(g^{(j)}-l-q^{(j)})\cdot\lambda^2\right]\cdot e^{-\lambda\cdot\tau}$$
$$-\lambda\cdot\left(\frac{1}{T_L\cdot T_R}+\frac{P^{(j)}}{E_L}\right)\cdot\left(\frac{1}{T_A\cdot T_R}+\frac{P^{(j)}}{E_A}\right)$$
$$-\left(\frac{1}{T_L\cdot T_R}+\frac{1}{T_A\cdot T_R}+\frac{P^{(j)}}{E_A}+\frac{P^{(j)}}{E_L}\right)\cdot\lambda^2-\lambda^3$$

$$D(\lambda,\tau)=P_n(\lambda,\tau)+Q_m(\lambda,\tau)\cdot e^{-\lambda\cdot\tau};\ n,m\in\mathbb{N}_0;\ n>m$$

$$P_n(\lambda,\tau)=-\lambda\cdot\left(\frac{1}{T_L\cdot T_R}+\frac{P^{(j)}}{E_L}\right)\cdot\left(\frac{1}{T_A\cdot T_R}+\frac{P^{(j)}}{E_A}\right)$$
$$-\left(\frac{1}{T_L\cdot T_R}+\frac{1}{T_A\cdot T_R}+\frac{P^{(j)}}{E_A}+\frac{P^{(j)}}{E_L}\right)\cdot\lambda^2-\lambda^3$$

$$Q_m(\lambda,\tau)=\left(\frac{1}{T_L\cdot T_R}+\frac{P^{(j)}}{E_L}\right)\cdot\left(\frac{1}{T_A\cdot T_R}+\frac{P^{(j)}}{E_A}\right)\cdot\frac{2}{T_R}\cdot(g^{(j)}-l-q^{(j)})$$
$$+\frac{2}{T_R}\cdot P^{(j)}\cdot\frac{q^{(j)}}{E_A}\cdot\left(\frac{1}{T_L\cdot T_R}+\frac{P^{(j)}}{E_L}\right)$$

$$-\frac{2}{T_R}\cdot P^{(j)}\cdot\frac{g^{(j)}}{E_L}\cdot\left(\frac{1}{T_A\cdot T_R}+\frac{P^{(j)}}{E_A}\right)$$
$$+\left\{\frac{2}{T_R}\cdot(g^{(j)}-l-q^{(j)})\cdot\left(\frac{1}{T_L\cdot T_R}+\frac{1}{T_A\cdot T_R}+\frac{P^{(j)}}{E_A}+\frac{P^{(j)}}{E_L}\right)\right.$$
$$\left.+\frac{2}{T_R}\cdot P^{(j)}\cdot\frac{q^{(j)}}{E_A}-\frac{2}{T_R}\cdot P^{(j)}\cdot\frac{g^{(j)}}{E_L}\right\}\cdot\lambda$$
$$+\frac{2}{T_R}\cdot(g^{(j)}-l-q^{(j)})\cdot\lambda^2$$

$$D(\lambda,\tau)=P_n(\lambda,\tau)+Q_m(\lambda,\tau)\cdot e^{-\lambda\cdot\tau};\ n=3,m=2\in\mathbb{N}_0;\ n>m$$

$$P_n(\lambda,\tau)=\sum_{k=0}^{n=3}p_k(\tau)\cdot\lambda^k=p_0(\tau)+p_1(\tau)\cdot\lambda+p_2(\tau)\cdot\lambda^2+p_3(\tau)\cdot\lambda^3$$

$$p_0(\tau)=0;\ p_1(\tau)=-\left(\frac{1}{T_L\cdot T_R}+\frac{P^{(j)}}{E_L}\right)\cdot\left(\frac{1}{T_A\cdot T_R}+\frac{P^{(j)}}{E_A}\right);\ p_2(\tau)$$
$$=-\left(\frac{1}{T_L\cdot T_R}+\frac{1}{T_A\cdot T_R}+\frac{P^{(j)}}{E_A}+\frac{P^{(j)}}{E_L}\right)p_3(\tau)=-1$$

$$Q_m(\lambda,\tau)=\sum_{k=0}^{n=2}q_k(\tau)\cdot\lambda^k=q_0(\tau)+q_1(\tau)\cdot\lambda+q_2(\tau)\cdot\lambda^2$$

$$q_0(\tau)=\left(\frac{1}{T_L\cdot T_R}+\frac{P^{(j)}}{E_L}\right)\cdot\left(\frac{1}{T_A\cdot T_R}+\frac{P^{(j)}}{E_A}\right)\cdot\frac{2}{T_R}\cdot(g^{(j)}-l-q^{(j)})$$
$$+\frac{2}{T_R}\cdot P^{(j)}\cdot\frac{q^{(j)}}{E_A}\cdot\left(\frac{1}{T_L\cdot T_R}+\frac{P^{(j)}}{E_L}\right)$$
$$-\frac{2}{T_R}\cdot P^{(j)}\cdot\frac{g^{(j)}}{E_L}\cdot\left(\frac{1}{T_A\cdot T_R}+\frac{P^{(j)}}{E_A}\right)$$

$$q_1(\tau)=\frac{2}{T_R}\cdot(g^{(j)}-l-q^{(j)})\cdot\left(\frac{1}{T_L\cdot T_R}+\frac{1}{T_A\cdot T_R}+\frac{P^{(j)}}{E_A}+\frac{P^{(j)}}{E_L}\right)$$
$$+\frac{2}{T_R}\cdot P^{(j)}\cdot\frac{q^{(j)}}{E_A}-\frac{2}{T_R}\cdot P^{(j)}\cdot\frac{g^{(j)}}{E_L}$$
$$q_2(\tau)=\frac{2}{T_R}\cdot(g^{(j)}-l-q^{(j)})$$

The homogeneous system for P, g, q leads to a characteristic equation for the λ; having the form $P(\lambda,\tau)+Q(\lambda,\tau)\cdot e^{-\lambda\cdot\tau}=0$; $P(\lambda)=\sum_{j=0}^{3}a_j\cdot\lambda^j$; $Q(\lambda)=\sum_{j=0}^{2}c_j\cdot\lambda^j$. The coefficients $\{a_j(q_i,q_k,\tau),c_j(q_i,q_k,\tau)\}\in\mathbb{R}$ depend on q_i, q_k

and delayτ. q_i, q_k are any Solid-state laser by semiconductor devices model corresponding to the rate differential equation system parameters, other parameters kept as a constant [5–7].

Hint: We use "P" as the system Laser power variable and as a part of stability analysis criteria parameter ($P(\lambda, \tau)$... characteristic equation). Reader must differentiate between the two meanings. Additionally, we use "q" as the system saturate losses per round trip and as a part of stability analysis criteria parameter ($q_0, q_1, \ldots$ in characteristic equation).

$$a_0 = 0;\ a_1 = -\left(\frac{1}{T_L \cdot T_R} + \frac{P^{(j)}}{E_L}\right) \cdot \left(\frac{1}{T_A \cdot T_R} + \frac{P^{(j)}}{E_A}\right)$$

$$a_2 = -\left(\frac{1}{T_L \cdot T_R} + \frac{1}{T_A \cdot T_R} + \frac{P^{(j)}}{E_A} + \frac{P^{(j)}}{E_L}\right)$$

$$a_3 = -1$$

$$c_0 = \left(\frac{1}{T_L \cdot T_R} + \frac{P^{(j)}}{E_L}\right) \cdot \left(\frac{1}{T_A \cdot T_R} + \frac{P^{(j)}}{E_A}\right) \cdot \frac{2}{T_R} \cdot (g^{(j)} - l - q^{(j)})$$
$$+ \frac{2}{T_R} \cdot P^{(j)} \cdot \frac{q^{(j)}}{E_A} \cdot \left(\frac{1}{T_L \cdot T_R} + \frac{P^{(j)}}{E_L}\right)$$
$$- \frac{2}{T_R} \cdot P^{(j)} \cdot \frac{g^{(j)}}{E_L} \cdot \left(\frac{1}{T_A \cdot T_R} + \frac{P^{(j)}}{E_A}\right)$$

$$c_1 = \frac{2}{T_R} \cdot (g^{(j)} - l - q^{(j)}) \cdot \left(\frac{1}{T_L \cdot T_R} + \frac{1}{T_A \cdot T_R} + \frac{P^{(j)}}{E_A} + \frac{P^{(j)}}{E_L}\right)$$
$$+ \frac{2}{T_R} \cdot P^{(j)} \cdot \frac{q^{(j)}}{E_A} - \frac{2}{T_R} \cdot P^{(j)} \cdot \frac{g^{(j)}}{E_L}$$

$$c_2 = \frac{2}{T_R} \cdot (g^{(j)} - l - q^{(j)})$$

Unless strictly necessary, the designation of the varied arguments (q_i, q_k) will subsequently be omitted from P, Q, a_j, c_j. The coefficients a_j, c_j are continuous and differentiable functions of their arguments, and direct substitution shows that $a_0 + c_0 \neq 0$ for $q_i, q_k \in \mathbb{R}_+$; that is, $\lambda = 0$ is not of $P(\lambda) + Q(\lambda) \cdot e^{-\lambda \cdot \tau} = 0$. Furthermore $P(\lambda), Q(\lambda)$ are analytic functions of λ, for which the following requirements of the analysis [5] can also be verified in the present case:

(a) If $\lambda = i \cdot \omega;\ \omega \in \mathbb{R}$, then $P(i \cdot \omega) + Q(i \cdot \omega) \neq 0$.
(b) If $|\frac{Q(\lambda)}{P(\lambda)}|$ is bounded for $|\lambda| \to \infty$; $\mathrm{Re}\lambda \geq 0$. No roots bifurcation from ∞.
(c) $F(\omega) = |P(i \cdot \omega)|^2 - |Q(i \cdot \omega)|^2$ has a finite number of zeros. Indeed, this is a polynomial in ω.
(d) Each positive root $\omega(q_i, q_k)$ of $F(\omega) = 0$ is continuous and differentiable with respect to q_i, q_k.

We assume that $P_n(\lambda, \tau)$ and $Q_m(\lambda, \tau)$ cannot have common imaginary roots. That is for any real numberω: $P_n(\lambda = i \cdot \omega, \tau) + Q_m(\lambda = i \cdot \omega, \tau) \neq 0$.

$$P_n(\lambda = i \cdot \omega, \tau) = -i \cdot \omega \cdot \left(\frac{1}{T_L \cdot T_R} + \frac{P^{(j)}}{E_L}\right) \cdot \left(\frac{1}{T_A \cdot T_R} + \frac{P^{(j)}}{E_A}\right)$$
$$+ \left(\frac{1}{T_L \cdot T_R} + \frac{1}{T_A \cdot T_R} + \frac{P^{(j)}}{E_A} + \frac{P^{(j)}}{E_L}\right) \cdot \omega^2 + i \cdot \omega^3$$

$$P_n(\lambda = i \cdot \omega, \tau) = i \cdot \left[\omega^3 - \omega \cdot \left(\frac{1}{T_L \cdot T_R} + \frac{P^{(j)}}{E_L}\right) \cdot \left(\frac{1}{T_A \cdot T_R} + \frac{P^{(j)}}{E_A}\right)\right]$$
$$+ \left(\frac{1}{T_L \cdot T_R} + \frac{1}{T_A \cdot T_R} + \frac{P^{(j)}}{E_A} + \frac{P^{(j)}}{E_L}\right) \cdot \omega^2$$

$$Q_m(\lambda = i \cdot \omega, \tau) = \left(\frac{1}{T_L \cdot T_R} + \frac{P^{(j)}}{E_L}\right) \cdot \left(\frac{1}{T_A \cdot T_R} + \frac{P^{(j)}}{E_A}\right) \cdot \frac{2}{T_R} \cdot (g^{(j)} - l - q^{(j)})$$
$$+ \frac{2}{T_R} \cdot P^{(j)} \cdot \frac{q^{(j)}}{E_A} \cdot \left(\frac{1}{T_L \cdot T_R} + \frac{P^{(j)}}{E_L}\right)$$
$$- \frac{2}{T_R} \cdot P^{(j)} \cdot \frac{g^{(j)}}{E_L} \cdot \left(\frac{1}{T_A \cdot T_R} + \frac{P^{(j)}}{E_A}\right) - \frac{2}{T_R} \cdot (g^{(j)} - l - q^{(j)}) \cdot \omega^2$$
$$+ \left\{\frac{2}{T_R} \cdot (g^{(j)} - l - q^{(j)}) \cdot \left(\frac{1}{T_L \cdot T_R} + \frac{1}{T_A \cdot T_R} + \frac{P^{(j)}}{E_A} + \frac{P^{(j)}}{E_L}\right)\right.$$
$$\left. + \frac{2}{T_R} \cdot P^{(j)} \cdot \frac{q^{(j)}}{E_A} - \frac{2}{T_R} \cdot P^{(j)} \cdot \frac{g^{(j)}}{E_L}\right\} \cdot \omega \cdot i$$

$$P_n(\lambda = i \cdot \omega, \tau) + Q_m(\lambda = i \cdot \omega, \tau)$$
$$= i \cdot \left[\omega^3 - \omega \cdot (\frac{1}{T_L \cdot T_R} + \frac{P^{(j)}}{E_L}) \cdot (\frac{1}{T_A \cdot T_R} + \frac{P^{(j)}}{E_A})\right]$$
$$+ \left(\frac{1}{T_L \cdot T_R} + \frac{1}{T_A \cdot T_R} + \frac{P^{(j)}}{E_A} + \frac{P^{(j)}}{E_L}\right) \cdot \omega^2$$
$$+ \left(\frac{1}{T_L \cdot T_R} + \frac{P^{(j)}}{E_L}\right) \cdot \left(\frac{1}{T_A \cdot T_R} + \frac{P^{(j)}}{E_A}\right) \cdot \frac{2}{T_R} \cdot (g^{(j)} - l - q^{(j)})$$
$$+ \frac{2}{T_R} \cdot P^{(j)} \cdot \frac{q^{(j)}}{E_A} \cdot \left(\frac{1}{T_L \cdot T_R} + \frac{P^{(j)}}{E_L}\right)$$
$$- \frac{2}{T_R} \cdot P^{(j)} \cdot \frac{g^{(j)}}{E_L} \cdot \left(\frac{1}{T_A \cdot T_R} + \frac{P^{(j)}}{E_A}\right) - \frac{2}{T_R} \cdot (g^{(j)} - l - q^{(j)}) \cdot \omega^2$$
$$+ \left\{\frac{2}{T_R} \cdot (g^{(j)} - l - q^{(j)}) \cdot \left(\frac{1}{T_L \cdot T_R} + \frac{1}{T_A \cdot T_R} + \frac{P^{(j)}}{E_A} + \frac{P^{(j)}}{E_L}\right)\right.$$
$$\left. + \frac{2}{T_R} \cdot P^{(j)} \cdot \frac{q^{(j)}}{E_A} - \frac{2}{T_R} \cdot P^{(j)} \cdot \frac{g^{(j)}}{E_L}\right\} \cdot i \cdot \omega$$

$$
\begin{aligned}
&P_n(\lambda = i \cdot \omega, \tau) + Q_m(\lambda = i \cdot \omega, \tau) \\
&= i \cdot \left[\omega^3 - \omega \cdot \left(\frac{1}{T_L \cdot T_R} + \frac{P^{(j)}}{E_L}\right) \cdot \left(\frac{1}{T_A \cdot T_R} + \frac{P^{(j)}}{E_A}\right)\right. \\
&+ \left\{\frac{2}{T_R} \cdot (g^{(j)} - l - q^{(j)}) \cdot \left(\frac{1}{T_L \cdot T_R} + \frac{1}{T_A \cdot T_R} + \frac{P^{(j)}}{E_A} + \frac{P^{(j)}}{E_L}\right)\right\} \\
&\left.+ \frac{2}{T_R} \cdot P^{(j)} \cdot \frac{q^{(j)}}{E_A} - \frac{2}{T_R} \cdot P^{(j)} \cdot \frac{g^{(j)}}{E_L}\right\} \cdot \omega \\
&+ \left(\frac{1}{T_L \cdot T_R} + \frac{P^{(j)}}{E_L}\right) \cdot \left(\frac{1}{T_A \cdot T_R} + \frac{P^{(j)}}{E_A}\right) \cdot \frac{2}{T_R} \cdot (g^{(j)} - l - q^{(j)}) \\
&+ \frac{2}{T_R} \cdot P^{(j)} \cdot \frac{q^{(j)}}{E_A} \cdot \left(\frac{1}{T_L \cdot T_R} + \frac{P^{(j)}}{E_L}\right) \\
&- \frac{2}{T_R} \cdot P^{(j)} \cdot \frac{g^{(j)}}{E_L} \cdot \left(\frac{1}{T_A \cdot T_R} + \frac{P^{(j)}}{E_A}\right) \\
&+ \left[\frac{1}{T_L \cdot T_R} + \frac{1}{T_A \cdot T_R} + \frac{P^{(j)}}{E_A} + \frac{P^{(j)}}{E_L}\right. \\
&\left.- \frac{2}{T_R} \cdot (g^{(j)} - l - q^{(j)})\right] \cdot \omega^2
\end{aligned}
$$

$$
\begin{aligned}
|P(i \cdot \omega, \tau)|^2 &= \left[\omega^3 - \omega \cdot \left(\frac{1}{T_L \cdot T_R} + \frac{P^{(j)}}{E_L}\right) \cdot \left(\frac{1}{T_A \cdot T_R} + \frac{P^{(j)}}{E_A}\right)\right]^2 \\
&+ \left(\frac{1}{T_L \cdot T_R} + \frac{1}{T_A \cdot T_R} + \frac{P^{(j)}}{E_A} + \frac{P^{(j)}}{E_L}\right)^2 \cdot \omega^4
\end{aligned}
$$

$$
\begin{aligned}
|P(i \cdot \omega, \tau)|^2 = \omega^6 + \omega^4 \cdot &\left[\left(\frac{1}{T_L \cdot T_R} + \frac{1}{T_A \cdot T_R} + \frac{P^{(j)}}{E_A} + \frac{P^{(j)}}{E_L}\right)^2\right. \\
&\left.-2 \cdot \left(\frac{1}{T_L \cdot T_R} + \frac{P^{(j)}}{E_L}\right) \cdot \left(\frac{1}{T_A \cdot T_R} + \frac{P^{(j)}}{E_A}\right)\right] \\
&+ \omega^2 \cdot \left(\frac{1}{T_L \cdot T_R} + \frac{P^{(j)}}{E_L}\right)^2 \cdot \left(\frac{1}{T_A \cdot T_R} + \frac{P^{(j)}}{E_A}\right)^2
\end{aligned}
$$

$$
\begin{aligned}
|Q(i \cdot \omega, \tau)|^2 = &\left\{\left(\frac{1}{T_L \cdot T_R} + \frac{P^{(j)}}{E_L}\right) \cdot \left(\frac{1}{T_A \cdot T_R} + \frac{P^{(j)}}{E_A}\right) \cdot \frac{2}{T_R} \cdot (g^{(j)} - l - q^{(j)})\right. \\
&+ \frac{2}{T_R} \cdot P^{(j)} \cdot \frac{q^{(j)}}{E_A} \cdot \left(\frac{1}{T_L \cdot T_R} + \frac{P^{(j)}}{E_L}\right) \\
&\left.- \frac{2}{T_R} \cdot P^{(j)} \cdot \frac{g^{(j)}}{E_L} \cdot \left(\frac{1}{T_A \cdot T_R} + \frac{P^{(j)}}{E_A}\right) - \frac{2}{T_R} \cdot (g^{(j)} - l - q^{(j)}) \cdot \omega^2\right\}^2 \\
&+ \left\{\frac{2}{T_R} \cdot (g^{(j)} - l - q^{(j)}) \cdot \left(\frac{1}{T_L \cdot T_R} + \frac{1}{T_A \cdot T_R} + \frac{P^{(j)}}{E_A} + \frac{P^{(j)}}{E_L}\right)\right.
\end{aligned}
$$

$$+\frac{2}{T_R} \cdot P^{(j)} \cdot \frac{q^{(j)}}{E_A} - \frac{2}{T_R} \cdot P^{(j)} \cdot \frac{g^{(j)}}{E_L}\Bigg\} \cdot \omega\Bigg\}^2$$

$$|Q(i \cdot \omega, \tau)|^2 = \left[\left(\frac{1}{T_L \cdot T_R} + \frac{P^{(j)}}{E_L}\right) \cdot \left(\frac{1}{T_A \cdot T_R} + \frac{P^{(j)}}{E_A}\right)\right.$$
$$\cdot \frac{2}{T_R} \cdot (g^{(j)} - l - q^{(j)}) + \frac{2}{T_R} \cdot P^{(j)} \cdot \frac{q^{(j)}}{E_A} \cdot \left(\frac{1}{T_L \cdot T_R} + \frac{P^{(j)}}{E_L}\right)$$
$$\left. - \frac{2}{T_R} \cdot P^{(j)} \cdot \frac{g^{(j)}}{E_L} \cdot \left(\frac{1}{T_A \cdot T_R} + \frac{P^{(j)}}{E_A}\right)\right]^2$$
$$- 2 \cdot \left[\left(\frac{1}{T_L \cdot T_R} + \frac{P^{(j)}}{E_L}\right) \cdot \left(\frac{1}{T_A \cdot T_R} + \frac{P^{(j)}}{E_A}\right)\right.$$
$$\cdot \frac{2}{T_R} \cdot (g^{(j)} - l - q^{(j)})$$
$$+ \frac{2}{T_R} \cdot P^{(j)} \cdot \frac{q^{(j)}}{E_A} \cdot \left(\frac{1}{T_L \cdot T_R} + \frac{P^{(j)}}{E_L}\right)$$
$$\left. - \frac{2}{T_R} \cdot P^{(j)} \cdot \frac{g^{(j)}}{E_L} \cdot \left(\frac{1}{T_A \cdot T_R} + \frac{P^{(j)}}{E_A}\right)\right]$$
$$\cdot \frac{2}{T_R} \cdot (g^{(j)} - l - q^{(j)}) \cdot \omega^2$$
$$+ (\frac{2}{T_R})^2 \cdot (g^{(j)} - l - q^{(j)})^2 \cdot \omega^4$$
$$+ \left[\frac{2}{T_R} \cdot (g^{(j)} - l - q^{(j)}) \cdot \left(\frac{1}{T_L \cdot T_R} + \frac{1}{T_A \cdot T_R} + \frac{P^{(j)}}{E_A} + \frac{P^{(j)}}{E_L}\right)\right.$$
$$\left. + \frac{2}{T_R} \cdot P^{(j)} \cdot \frac{q^{(j)}}{E_A} - \frac{2}{T_R} \cdot P^{(j)} \cdot \frac{g^{(j)}}{E_L}\right]^2 \cdot \omega^2$$

$$|Q(i \cdot \omega, \tau)|^2 = \left[\left(\frac{1}{T_L \cdot T_R} + \frac{P^{(j)}}{E_L}\right) \cdot \left(\frac{1}{T_A \cdot T_R} + \frac{P^{(j)}}{E_A}\right) \cdot \frac{2}{T_R} \cdot (g^{(j)} - l - q^{(j)})\right.$$
$$+ \frac{2}{T_R} \cdot P^{(j)} \cdot \frac{q^{(j)}}{E_A} \cdot \left(\frac{1}{T_L \cdot T_R} + \frac{P^{(j)}}{E_L}\right)$$
$$\left. - \frac{2}{T_R} \cdot P^{(j)} \cdot \frac{g^{(j)}}{E_L} \cdot \left(\frac{1}{T_A \cdot T_R} + \frac{P^{(j)}}{E_A}\right)\right]^2$$
$$+ \left\{\left[\frac{2}{T_R} \cdot (g^{(j)} - l - q^{(j)}) \cdot \left(\frac{1}{T_L \cdot T_R} + \frac{1}{T_A \cdot T_R} + \frac{P^{(j)}}{E_A} + \frac{P^{(j)}}{E_L}\right)\right.\right.$$
$$\left. + \frac{2}{T_R} \cdot P^{(j)} \cdot \frac{q^{(j)}}{E_A} - \frac{2}{T_R} \cdot P^{(j)} \cdot \frac{g^{(j)}}{E_L}\right]^2$$
$$- 2 \cdot \left[\left(\frac{1}{T_L \cdot T_R} + \frac{P^{(j)}}{E_L}\right) \cdot \left(\frac{1}{T_A \cdot T_R} + \frac{P^{(j)}}{E_A}\right) \cdot \frac{2}{T_R} \cdot (g^{(j)} - l - q^{(j)})\right.$$

$$+\frac{2}{T_R}\cdot P^{(j)}\cdot\frac{q^{(j)}}{E_A}\cdot\left(\frac{1}{T_L\cdot T_R}+\frac{P^{(j)}}{E_L}\right)$$

$$-\frac{2}{T_R}\cdot P^{(j)}\cdot\frac{g^{(j)}}{E_L}\cdot\left(\frac{1}{T_A\cdot T_R}+\frac{P^{(j)}}{E_A}\right)\Bigg]\cdot\frac{2}{T_R}\cdot(g^{(j)}-l-q^{(j)})\Bigg\}\cdot\omega^2$$

$$+\left(\frac{2}{T_R}\right)^2\cdot(g^{(j)}-l-q^{(j)})^2\cdot\omega^4$$

$$F(\omega,\tau)=|P(i\cdot\omega,\tau)|^2-|Q(i\cdot\omega,\tau)|^2=\omega^6$$

$$+\omega^4\cdot\Bigg[\left(\frac{1}{T_L\cdot T_R}+\frac{1}{T_A\cdot T_R}+\frac{P^{(j)}}{E_A}+\frac{P^{(j)}}{E_L}\right)^2$$

$$-2\cdot\left(\frac{1}{T_L\cdot T_R}+\frac{P^{(j)}}{E_L}\right)\cdot\left(\frac{1}{T_A\cdot T_R}+\frac{P^{(j)}}{E_A}\right)\Bigg]$$

$$+\omega^2\cdot\left(\frac{1}{T_L\cdot T_R}+\frac{P^{(j)}}{E_L}\right)^2\cdot\left(\frac{1}{T_A\cdot T_R}+\frac{P^{(j)}}{E_A}\right)^2$$

$$\Bigg[-\left(\frac{1}{T_L\cdot T_R}+\frac{P^{(j)}}{E_L}\right)\cdot\left(\frac{1}{T_A\cdot T_R}+\frac{P^{(j)}}{E_A}\right)\cdot\frac{2}{T_R}\cdot(g^{(j)}-l-q^{(j)})$$

$$+\frac{2}{T_R}\cdot P^{(j)}\cdot\frac{q^{(j)}}{E_A}\cdot\left(\frac{1}{T_L\cdot T_R}+\frac{P^{(j)}}{E_L}\right)$$

$$-\frac{2}{T_R}\cdot P^{(j)}\cdot\frac{g^{(j)}}{E_L}\cdot\left(\frac{1}{T_A\cdot T_R}+\frac{P^{(j)}}{E_A}\right)\Bigg]^2$$

$$-\Bigg\{\Bigg[\frac{2}{T_R}\cdot(g^{(j)}-l-q^{(j)})\cdot\left(\frac{1}{T_L\cdot T_R}+\frac{1}{T_A\cdot T_R}+\frac{P^{(j)}}{E_A}+\frac{P^{(j)}}{E_L}\right)$$

$$+\frac{2}{T_R}\cdot P^{(j)}\cdot\frac{q^{(j)}}{E_A}-\frac{2}{T_R}\cdot P^{(j)}\cdot\frac{g^{(j)}}{E_L}\Bigg]^2$$

$$-2\cdot\Bigg[\left(\frac{1}{T_L\cdot T_R}+\frac{P^{(j)}}{E_L}\right)\cdot\left(\frac{1}{T_A\cdot T_R}+\frac{P^{(j)}}{E_A}\right)\cdot\frac{2}{T_R}\cdot(g^{(j)}-l-q^{(j)})$$

$$+\frac{2}{T_R}\cdot P^{(j)}\cdot\frac{q^{(j)}}{E_A}\cdot\left(\frac{1}{T_L\cdot T_R}+\frac{P^{(j)}}{E_L}\right)$$

$$-\frac{2}{T_R}\cdot P^{(j)}\cdot\frac{g^{(j)}}{E_L}\cdot\left(\frac{1}{T_A\cdot T_R}+\frac{P^{(j)}}{E_A}\right)\Bigg]\cdot\frac{2}{T_R}\cdot(g^{(j)}-l-q^{(j)})\Bigg\}\cdot\omega^2$$

$$-\left(\frac{2}{T_R}\right)^2\cdot(g^{(j)}-l-q^{(j)})^2\cdot\omega^4$$

$$F(\omega,\tau)=|P(i\cdot\omega,\tau)|^2-|Q(i\cdot\omega,\tau)|^2=\omega^6+\omega^4$$

$$\cdot\Bigg[\left(\frac{1}{T_L\cdot T_R}+\frac{1}{T_A\cdot T_R}+\frac{P^{(j)}}{E_A}+\frac{P^{(j)}}{E_L}\right)^2$$

$$-2\cdot\left(\frac{1}{T_L\cdot T_R}+\frac{P^{(j)}}{E_L}\right)\cdot\left(\frac{1}{T_A\cdot T_R}+\frac{P^{(j)}}{E_A}\right)$$
$$\left.-\left(\frac{2}{T_R}\right)^2\cdot(g^{(j)}-l-q^{(j)})^2\right]$$
$$+\omega^2\cdot\left[\left(\frac{1}{T_L\cdot T_R}+\frac{P^{(j)}}{E_L}\right)^2\cdot\left(\frac{1}{T_A\cdot T_R}+\frac{P^{(j)}}{E_A}\right)^2\right.$$
$$-\left\{\left[\frac{2}{T_R}\cdot(g^{(j)}-l-q^{(j)})\cdot\left(\frac{1}{T_L\cdot T_R}+\frac{1}{T_A\cdot T_R}+\frac{P^{(j)}}{E_A}+\frac{P^{(j)}}{E_L}\right)\right.\right.$$
$$\left.+\frac{2}{T_R}\cdot P^{(j)}\cdot\frac{q^{(j)}}{E_A}-\frac{2}{T_R}\cdot P^{(j)}\cdot\frac{g^{(j)}}{E_L}\right]^2$$
$$-2\cdot\left[\left(\frac{1}{T_L\cdot T_R}+\frac{P^{(j)}}{E_L}\right)\cdot\left(\frac{1}{T_A\cdot T_R}+\frac{P^{(j)}}{E_A}\right)\cdot\frac{2}{T_R}\cdot(g^{(j)}-l-q^{(j)})\right.$$
$$+\frac{2}{T_R}\cdot P^{(j)}\cdot\frac{q^{(j)}}{E_A}\cdot\left(\frac{1}{T_L\cdot T_R}+\frac{P^{(j)}}{E_L}\right)$$
$$\left.\left.\left.-\frac{2}{T_R}\cdot P^{(j)}\cdot\frac{g^{(j)}}{E_L}\cdot\left(\frac{1}{T_A\cdot T_R}+\frac{P^{(j)}}{E_A}\right)\right]\cdot\frac{2}{T_R}\cdot(g^{(j)}-l-q^{(j)})\right\}\right]$$
$$-\left[\left(\frac{1}{T_L\cdot T_R}+\frac{P^{(j)}}{E_L}\right)\cdot\left(\frac{1}{T_A\cdot T_R}+\frac{P^{(j)}}{E_A}\right)\cdot\frac{2}{T_R}\cdot(g^{(j)}-l-q^{(j)})\right.$$
$$+\frac{2}{T_R}\cdot P^{(j)}\cdot\frac{q^{(j)}}{E_A}\cdot\left(\frac{1}{T_L\cdot T_R}+\frac{P^{(j)}}{E_L}\right)$$
$$\left.-\frac{2}{T_R}\cdot P^{(j)}\cdot\frac{g^{(j)}}{E_L}\cdot\left(\frac{1}{T_A\cdot T_R}+\frac{P^{(j)}}{E_A}\right)\right]^2$$

We define the following parameters for simplicity: $\Pi_0, \Pi_2, \Pi_4, \Pi_6$.

$$\Pi_0=-\left[\left(\frac{1}{T_L\cdot T_R}+\frac{P^{(j)}}{E_L}\right)\cdot\left(\frac{1}{T_A\cdot T_R}+\frac{P^{(j)}}{E_A}\right)\cdot\frac{2}{T_R}\cdot(g^{(j)}-l-q^{(j)})\right.$$
$$+\frac{2}{T_R}\cdot P^{(j)}\cdot\frac{q^{(j)}}{E_A}\cdot\left(\frac{1}{T_L\cdot T_R}+\frac{P^{(j)}}{E_L}\right)$$
$$\left.-\frac{2}{T_R}\cdot P^{(j)}\cdot\frac{g^{(j)}}{E_L}\cdot\left(\frac{1}{T_A\cdot T_R}+\frac{P^{(j)}}{E_A}\right)\right]^2$$

$$\Pi_2=\left(\frac{1}{T_L\cdot T_R}+\frac{P^{(j)}}{E_L}\right)^2\cdot\left(\frac{1}{T_A\cdot T_R}+\frac{P^{(j)}}{E_A}\right)^2$$
$$-\left\{\left[\frac{2}{T_R}\cdot(g^{(j)}-l-q^{(j)})\cdot\left(\frac{1}{T_L\cdot T_R}+\frac{1}{T_A\cdot T_R}+\frac{P^{(j)}}{E_A}+\frac{P^{(j)}}{E_L}\right)\right.\right.$$

$$+\frac{2}{T_R}\cdot P^{(j)}\cdot\frac{q^{(j)}}{E_A}-\frac{2}{T_R}\cdot P^{(j)}\cdot\frac{g^{(j)}}{E_L}\Bigg]^2$$

$$-2\cdot\Bigg[\left(\frac{1}{T_L\cdot T_R}+\frac{P^{(j)}}{E_L}\right)\cdot\left(\frac{1}{T_A\cdot T_R}+\frac{P^{(j)}}{E_A}\right)\cdot\frac{2}{T_R}\cdot(g^{(j)}-l-q^{(j)})$$

$$+\frac{2}{T_R}\cdot P^{(j)}\cdot\frac{q^{(j)}}{E_A}\cdot\left(\frac{1}{T_L\cdot T_R}+\frac{P^{(j)}}{E_L}\right)$$

$$-\frac{2}{T_R}\cdot P^{(j)}\cdot\frac{g^{(j)}}{E_L}\cdot\left(\frac{1}{T_A\cdot T_R}+\frac{P^{(j)}}{E_A}\right)\Bigg]\cdot\frac{2}{T_R}\cdot(g^{(j)}-l-q^{(j)})\Bigg\}$$

$$\Pi_4=\left(\frac{1}{T_L\cdot T_R}+\frac{1}{T_A\cdot T_R}+\frac{P^{(j)}}{E_A}+\frac{P^{(j)}}{E_L}\right)^2-2\cdot\left(\frac{1}{T_L\cdot T_R}+\frac{P^{(j)}}{E_L}\right)$$

$$\cdot\left(\frac{1}{T_A\cdot T_R}+\frac{P^{(j)}}{E_A}\right)-\left(\frac{2}{T_R}\right)^2\cdot(g^{(j)}-l-q^{(j)})^2\Pi_6=1$$

Hence $F(\omega,\tau)$ implies $\sum_{k=0}^{3}\Pi_{2\cdot k}\cdot\omega^{2\cdot k}=0$ and its roots are given by solving the above polynomial.

$$P_R(i\cdot\omega,\tau)=\left(\frac{1}{T_L\cdot T_R}+\frac{1}{T_A\cdot T_R}+\frac{P^{(j)}}{E_A}+\frac{P^{(j)}}{E_L}\right)\cdot\omega^2$$

$$P_I(i\cdot\omega,\tau)=\omega^3-\omega\cdot\left(\frac{1}{T_L\cdot T_R}+\frac{P^{(j)}}{E_L}\right)\cdot\left(\frac{1}{T_A\cdot T_R}+\frac{P^{(j)}}{E_A}\right)$$

$$Q_R(i\cdot\omega,\tau)=\left(\frac{1}{T_L\cdot T_R}+\frac{P^{(j)}}{E_L}\right)\cdot\left(\frac{1}{T_A\cdot T_R}+\frac{P^{(j)}}{E_A}\right)\cdot\frac{2}{T_R}\cdot(g^{(j)}-l-q^{(j)})$$

$$+\frac{2}{T_R}\cdot P^{(j)}\cdot\frac{q^{(j)}}{E_A}\cdot\left(\frac{1}{T_L\cdot T_R}+\frac{P^{(j)}}{E_L}\right)$$

$$-\frac{2}{T_R}\cdot P^{(j)}\cdot\frac{g^{(j)}}{E_L}\cdot\left(\frac{1}{T_A\cdot T_R}+\frac{P^{(j)}}{E_A}\right)-\frac{2}{T_R}\cdot(g^{(j)}-l-q^{(j)})\cdot\omega^2$$

$$Q_I(i\cdot\omega,\tau)=\Bigg\{\frac{2}{T_R}\cdot(g^{(j)}-l-q^{(j)})\cdot\left(\frac{1}{T_L\cdot T_R}+\frac{1}{T_A\cdot T_R}+\frac{P^{(j)}}{E_A}+\frac{P^{(j)}}{E_L}\right)$$

$$+\frac{2}{T_R}\cdot P^{(j)}\cdot\frac{q^{(j)}}{E_A}-\frac{2}{T_R}\cdot P^{(j)}\cdot\frac{g^{(j)}}{E_L}\Bigg\}\cdot\omega$$

$$\sin\theta(\tau)=\frac{-P_R(i\cdot\omega,\tau)\cdot Q_I(i\cdot\omega,\tau)+P_I(i\cdot\omega,\tau)\cdot Q_R(i\cdot\omega,\tau)}{|Q(i\cdot\omega,\tau)|^2}$$

$$\cos\theta(\tau)=-\frac{P_R(i\cdot\omega,\tau)\cdot Q_R(i\cdot\omega,\tau)+P_I(i\cdot\omega,\tau)\cdot Q_i(i\cdot\omega,\tau)}{|Q(i\cdot\omega,\tau)|^2}$$

We use different parameters terminology from our last characteristics parameters definition:

$$k \to j;\ p_k(\tau) \to a_j;\ q_k(\tau) \to c_j;\ n = 3;\ m = 2;\ n > m;\ P_n(\lambda, \tau) \to P(\lambda);\ Q_m(\lambda, \tau) \to Q(\lambda)$$

$$P(\lambda) = \sum_{j=0}^{3} a_j \cdot \lambda^j;\ Q(\lambda) = \sum_{j=0}^{2} c_j \cdot \lambda^j;\ P(\lambda) = a_0 + a_1 \cdot \lambda + a_2 \cdot \lambda^2 + a_3 \cdot \lambda^3;\ Q(\lambda) = c_0 + c_1 \cdot \lambda + c_2 \cdot \lambda^2$$

$n, m \in \mathbb{N}_0$; $n > m$ and $a_j, c_j : \mathbb{R}_{+0} \to \mathbb{R}$ are continuous and differentiable function of τ such that $a_0 + c_0 \neq 0$. In the following "$-$" denotes complex and conjugate. $P(\lambda)$, $Q(\lambda)$ are analytic functions in λ and differentiable in τ. The coefficients $a_j(T_L, T_R, T_A, E_L, E_A, l, \tau, \ldots) \in \mathbb{R}$.

And $c_j(T_L, T_R, T_A, E_L, E_A, l, \tau, \ldots) \in \mathbb{R}$ depend on systems $T_L, T_R, T_A, E_L, E_A, l, \tau, \ldots$ values. Unless strictly, the designation of the varied arguments: $(T_L, T_R, T_A, E_L, E_A, l, \tau, \ldots)$ will subsequently be omitted from P, Q, a_j, c_j. The coefficients a_j, c_j are continuous and differentiable functions of their arguments, and direct substitution shows that $a_0 + c_0 \neq 0$.

$$a_0 = 0;\ c_0 = \left(\frac{1}{T_L \cdot T_R} + \frac{P^{(j)}}{E_L}\right) \cdot \left(\frac{1}{T_A \cdot T_R} + \frac{P^{(j)}}{E_A}\right) \cdot \frac{2}{T_R} \cdot (g^{(j)} - l - q^{(j)}) + \frac{2}{T_R} \cdot P^{(j)} \cdot \frac{q^{(j)}}{E_A} \cdot \left(\frac{1}{T_L \cdot T_R} + \frac{P^{(j)}}{E_L}\right) - \frac{2}{T_R} \cdot P^{(j)} \cdot \frac{g^{(j)}}{E_L} \cdot \left(\frac{1}{T_A \cdot T_R} + \frac{P^{(j)}}{E_A}\right)$$

$$\left(\frac{1}{T_L \cdot T_R} + \frac{P^{(j)}}{E_L}\right) \cdot \left(\frac{1}{T_A \cdot T_R} + \frac{P^{(j)}}{E_A}\right) \cdot \frac{2}{T_R} \cdot (g^{(j)} - l - q^{(j)}) + \frac{2}{T_R} \cdot P^{(j)} \cdot \frac{q^{(j)}}{E_A} \cdot \left(\frac{1}{T_L \cdot T_R} + \frac{P^{(j)}}{E_L}\right) - \frac{2}{T_R} \cdot P^{(j)} \cdot \frac{g^{(j)}}{E_L} \cdot \left(\frac{1}{T_A \cdot T_R} + \frac{P^{(j)}}{E_A}\right) \neq 0$$

$\forall\, T_L, T_R, T_A, E_L, E_A, l, \tau, \ldots \in \mathbb{R}_+$ i.e. $\lambda = 0$ is not a root of the characteristic equation. Furthermore $P(\lambda)$, $Q(\lambda)$ are analytic functions of λ for which the following requirements of the analysis (see Kuang 1993, Sect. 3.4) can be also be verified in the present case.

(a) If $\lambda = i \cdot \omega$; $\omega \in \mathbb{R}$ then $P(i \cdot \omega) + Q(i \cdot \omega) \neq 0$, i.e. P and Q have no common imaginary roots. This condition was verified numerically in the entire $(T_L, T_R, T_A, E_L, E_A, l, \tau, \ldots)$ domain of interest.
(b) $\left|\frac{P(\lambda)}{Q(\lambda)}\right|$ is bounded for $|\lambda| \to \infty$; $\mathrm{Re}\lambda \geq 0$. No roots bifurcation from ∞. Indeed, in the limit: $\left|\frac{Q(\lambda)}{P(\lambda)}\right| = \left|\frac{c_0 + c_1 \cdot \lambda + c_2 \cdot \lambda^2}{a_0 + a_1 \cdot \lambda + a_2 \cdot \lambda^2 + a_3 \cdot \lambda^3}\right|$.

(c) The following expressions exist: $F(\omega) = |P(i\cdot\omega)|^2 - |Q(i\cdot\omega)|^2$ $F(\omega,\tau) = |P(i\cdot\omega,\tau)|^2 - |Q(i\cdot\omega,\tau)|^2 = \sum_{k=0}^{3}\Pi_{2\cdot k}\cdot\omega^{2\cdot k} = 0$. Has at most a finite number of zeros. Indeed, this is a polynomial in ω (degree in ω^6).

(d) Each positive root $\omega(T_L, T_R, T_A, E_L, E_A, l, \tau, \ldots)$ of $F(\omega) = 0$ is continuous and differentiable with respect to $T_L, T_R, T_A, E_L, E_A, l, \tau, \ldots$ the condition can only be assessed numerically.

In addition, since the coefficients in P and Q are real, we have $\overline{P(-i\cdot\omega)} = P(i\cdot\omega)$, and $\overline{Q(-i\cdot\omega)} = Q(i\cdot\omega)$ thus, $\omega > 0$ maybe on eigenvalue of characteristic equations. The analysis consists in identifying the roots of the characteristic equation situated on the imaginary axis of the complex λ—plane, whereby increasing the parameters: $T_L, T_R, T_A, E_L, E_A, l, \tau, \ldots$ $\mathrm{Re}\,\lambda$ may, at the crossing, change its sign from (−) to (+), i.e. from a stable focus $E^{(*)}(P^{(*)}, g^{(*)}, q^{(*)})$ to an unstable one, or vice versa. This feature may be further assessed by examining the sign of the partial derivatives with respect to $T_L, T_R, T_A, E_L, E_A, l, \tau, \ldots$ and system parameters [5–7].

$$\Lambda^{-1}(T_L) = \left(\frac{\partial \mathrm{Re}\lambda}{\partial T_L}\right)_{\lambda=i\cdot\omega};\ T_R, T_A, E_L, E_A, l, \tau, \ldots = const$$

$$\Lambda^{-1}(T_A) = \left(\frac{\partial \mathrm{Re}\lambda}{\partial T_A}\right)_{\lambda=i\cdot\omega};\ T_L, T_R, E_L, E_A, l, \tau, \ldots = const$$

$$\Lambda^{-1}(E_L) = \left(\frac{\partial \mathrm{Re}\lambda}{\partial E_L}\right)_{\lambda=i\cdot\omega};\ T_L, T_R, T_A, E_A, l, \tau, \ldots = const$$

$$\Lambda^{-1}(E_A) = \left(\frac{\partial \mathrm{Re}\lambda}{\partial E_A}\right)_{\lambda=i\cdot\omega};\ T_L, T_R, T_A, E_L, l, \tau, \ldots = const$$

$$\Lambda^{-1}(l) = \left(\frac{\partial \mathrm{Re}\lambda}{\partial l}\right)_{\lambda=i\cdot\omega};\ T_L, T_R, T_A, E_L, E_A, \tau, \ldots = const$$

$$\Lambda^{-1}(\tau) = \left(\frac{\partial \mathrm{Re}\lambda}{\partial \tau}\right)_{\lambda=i\cdot\omega};\ T_L, T_R, T_A, E_L, E_A, l, \ldots = const$$

$$P(\lambda) = P_R(\lambda) + i\cdot P_I(\lambda);\ Q(\lambda) = Q_R(\lambda) + i\cdot Q_I(\lambda)$$

When writing and inserting $\lambda = i\cdot\omega$ into system's characteristic equation ω must satisfy the following equations.

$$\sin(\omega\cdot\tau) = g(\omega) = \frac{-P_R(i\cdot\omega)\cdot Q_I(i\cdot\omega) + P_I(i\cdot\omega)\cdot Q_R(i\cdot\omega)}{|Q(i\cdot\omega)|^2}$$

$$\cos(\omega\cdot\tau) = h(\omega) = -\frac{P_R(i\cdot\omega)\cdot Q_R(i\cdot\omega) + P_I(i\cdot\omega)\cdot Q_I(i\cdot\omega)}{|Q(i\cdot\omega)|^2}$$

where $|Q(i\cdot\omega)|^2 \neq 0$ in view of requirement (a) above, and $(g, h) \in \mathbb{R}$. Furthermore, it follows above $\sin(\omega\cdot\tau)$ and $\cos(\omega\cdot\tau)$ equation that, by squaring and adding the sides, ω must be a positive root of $F(\omega) = |P(i\cdot\omega)|^2 - |Q(i\cdot\omega)|^2 = 0$. Note: $F(\omega)$ is dependent on τ. Now it is important to notice that if $\tau \notin I$(assume that

$I \subseteq \mathbb{R}_{+0}$ is the set where $\omega(\tau)$ is a positive root of $F(\omega)$ and for, $\tau \notin I, \omega(\tau)$ is not defined. Then for all τ in I, $\omega(\tau)$ is satisfied that $F(\omega, \tau) = 0$. Then there are no positive $\omega(\tau)$solutions for $F(\omega, \tau) = 0$, and we cannot have stability switches. For $\tau \in I$ where $\omega(\tau)$ is a positive solution of $F(\omega, \tau) = 0$, we can define the angle $\theta(\tau) \in [0, 2 \cdot \pi]$ as the solution of $\sin(\theta) = \cdots; \cos(\theta) = \cdots$

$$\sin\theta(\tau) = \frac{-P_R(i\cdot\omega)\cdot Q_I(i\cdot\omega) + P_I(i\cdot\omega)\cdot Q_R(i\cdot\omega)}{|Q(i\cdot\omega)|^2}$$

$$\cos\theta(\tau) = -\frac{P_R(i\cdot\omega)\cdot Q_R(i\cdot\omega) + P_I(i\cdot\omega)\cdot Q_I(i\cdot\omega)}{|Q(i\cdot\omega)|^2}$$

And the relation between the argument $\theta(\tau)$ and $\tau \cdot \omega(\tau)$ for $\tau \in I$ must be as describe $\tau\cdot\omega(\tau) = \theta(\tau) + 2\cdot n\cdot\pi \ \forall \mathbb{N}_0$, hence we can define the maps $\tau_n : I \to \mathbb{R}_{+0}$ given by $\tau_n(\tau) = \frac{\theta(\tau)+2\cdot n\cdot\pi}{\omega(\tau)}$; $n \in \mathbb{N}_0$; $\tau \in I$. Let us introduce the functions:$I \to \mathbb{R}$; $S_n(\tau) = \tau - \tau_n(\tau)$.

$\tau \in I; n \in \mathbb{N}_0$ that is continuous and differentiable in τ. In the following, the subscribes $\lambda, T_L, T_R, T_A, E_L, E_A, l, \ldots$ Indicate the corresponding partial derivatives. Let us first concentrate on $\Lambda(x)$, remember in $\lambda(T_L, T_R, T_A, E_L, E_A, l, \ldots)$ and $\omega(\lambda, T_L, T_R, T_A, E_L, E_A, l, \ldots)$, and keeping all parameters except one (x) and τ. The derivation closely follows that in reference [BK]. Differentiating system characteristic equation $P(\lambda) + Q(\lambda) \cdot e^{-\lambda\cdot\tau} = 0$ with respect to specific parameter (x), and inverting the derivative, for convenience, one calculates: $x = T_L, T_R, T_A, E_L, E_A, l, \ldots$

$$\left(\frac{\partial\lambda}{\partial x}\right)^{-1} = \frac{-P_\lambda(\lambda,x)\cdot Q(\lambda,x) + Q_\lambda(\lambda,x)\cdot P(\lambda,x) - \tau\cdot P(\lambda,x)\cdot Q(\lambda,x)}{P_x(\lambda,x)\cdot Q(\lambda,x) - Q_x(\lambda,x)\cdot P(\lambda,x)}$$

where $P_\lambda(\lambda, x) = \frac{\partial P}{\partial\lambda}$; $Q_\lambda(\lambda, x) = \frac{\partial Q}{\partial\lambda}$; $P_x(\lambda, x) = \frac{\partial P}{\partial x}$; $Q_x(\lambda, x) = \frac{\partial Q}{\partial x}$, substituting $\lambda = i\cdot\omega$ and bearing $\overline{P(-i\cdot\omega)} = P(i\cdot\omega)$; $\overline{Q(-i\cdot\omega)} = Q(i\cdot\omega)$; $i\cdot P_\lambda(i\cdot\omega) = P_\omega(i\cdot\omega)$; $i\cdot Q_\lambda(i\cdot\omega) = Q_\omega(i\cdot\omega)$ and that on the surface $|P(i\cdot\omega)|^2 = |Q(i\cdot\omega)|^2$, one obtain:

$$\left(\frac{\partial\lambda}{\partial x}\right)^{-1}\Big|_{\lambda=i\cdot\omega}$$
$$= \frac{i\cdot P_\omega(i\cdot\omega,x)\cdot\overline{P(i\cdot\omega,x)} + i\cdot Q_\lambda(i\cdot\omega,x)\cdot\overline{Q(\lambda,x)} - \tau\cdot|P(i\cdot\omega,x)|^2}{P_x(i\cdot\omega,x)\cdot\overline{P(i\cdot\omega,x)} - Q_x(i\cdot\omega,x)\cdot\overline{Q(i\cdot\omega,x)}}$$

Upon separating into real and imaginary parts, with $P = P_R + i\cdot P_I$; $Q = Q_R + i\cdot Q_I$ and $P_\omega = P_{R\omega} + i\cdot P_{I\omega}$; $Q_\omega = Q_{R\omega} + i\cdot Q_{I\omega}$; $Q_x = Q_{Rx} + i\cdot Q_{Ix}$; $P^2 = P_R^2 + P_I^2$. When (x) can be any system parameter's $T_L, T_R, T_A, E_L, E_A, l, \ldots$ and time delayτ. Where for convenience, we have dropped the arguments $(i\cdot\omega, x)$, and where $F_\omega = 2\cdot[(P_{R\omega}\cdot P_R + P_{I\omega}\cdot P_I) - (Q_{R\omega}\cdot Q_R + Q_{I\omega}\cdot Q_I)]$, $F_x = 2\cdot[(P_{Rx}\cdot P_R + P_{Ix}\cdot P_I) - (Q_{Rx}\cdot Q_R + Q_{Ix}\cdot Q_I)]$; $\omega_x = -\frac{F_x}{F_\omega}$. We define U and V:

$$U = (P_R \cdot P_{I\omega} - P_I \cdot P_{R\omega}) - (Q_R \cdot Q_{I\omega} - Q_I \cdot Q_{R\omega})$$
$$V = (P_R \cdot P_{Ix} - P_I \cdot P_{Rx}) - (Q_R \cdot Q_{Ix} - Q_I \cdot Q_{Rx})$$

We choose our specific parameters as time delay $x = \tau$.

$$P_R = \left(\frac{1}{T_L \cdot T_R} + \frac{1}{T_A \cdot T_R} + \frac{P^{(j)}}{E_A} + \frac{P^{(j)}}{E_L}\right) \cdot \omega^2$$
$$P_I = \omega^3 - \omega \cdot \left(\frac{1}{T_L \cdot T_R} + \frac{P^{(j)}}{E_L}\right) \cdot \left(\frac{1}{T_A \cdot T_R} + \frac{P^{(j)}}{E_A}\right)$$

$$Q_R = \left(\frac{1}{T_L \cdot T_R} + \frac{P^{(j)}}{E_L}\right) \cdot \left(\frac{1}{T_A \cdot T_R} + \frac{P^{(j)}}{E_A}\right) \cdot \frac{2}{T_R} \cdot (g^{(j)} - l - q^{(j)}) + \frac{2}{T_R} \cdot P^{(j)} \cdot \frac{q^{(j)}}{E_A} \cdot \left(\frac{1}{T_L \cdot T_R} + \frac{P^{(j)}}{E_L}\right) - \frac{2}{T_R} \cdot P^{(j)} \cdot \frac{g^{(j)}}{E_L} \cdot \left(\frac{1}{T_A \cdot T_R} + \frac{P^{(j)}}{E_A}\right) - \frac{2}{T_R} \cdot (g^{(j)} - l - q^{(j)}) \cdot \omega^2$$

$$Q_I = \left[\frac{2}{T_R} \cdot (g^{(j)} - l - q^{(j)}) \cdot \left(\frac{1}{T_L \cdot T_R} + \frac{1}{T_A \cdot T_R} + \frac{P^{(j)}}{E_A} + \frac{P^{(j)}}{E_L}\right) + \frac{2}{T_R} \cdot P^{(j)} \cdot \frac{q^{(j)}}{E_A} - \frac{2}{T_R} \cdot P^{(j)} \cdot \frac{g^{(j)}}{E_L}\right] \cdot \omega$$

$$P_{R\omega} = 2 \cdot \left(\frac{1}{T_L \cdot T_R} + \frac{1}{T_A \cdot T_R} + \frac{P^{(j)}}{E_A} + \frac{P^{(j)}}{E_L}\right) \cdot \omega; \; P_{I\omega} = 2 \cdot \omega^2 - \left(\frac{1}{T_L \cdot T_R} + \frac{P^{(j)}}{E_L}\right) \cdot \left(\frac{1}{T_A \cdot T_R} + \frac{P^{(j)}}{E_A}\right)$$

$$Q_{R\omega} = -\frac{4}{T_R} \cdot (g^{(j)} - l - q^{(j)}) \cdot \omega; \; P_{R\tau} = 0; \; P_{I\tau} = 0$$
$$Q_{R\tau} = 0; \; Q_{I\tau} = 0; \; \omega_\tau = -\frac{F_\tau}{F_\omega}$$

$$Q_{I\omega} = \frac{2}{T_R} \cdot (g^{(j)} - l - q^{(j)}) \cdot \left(\frac{1}{T_L \cdot T_R} + \frac{1}{T_A \cdot T_R} + \frac{P^{(j)}}{E_A} + \frac{P^{(j)}}{E_L}\right) + \frac{2}{T_R} \cdot P^{(j)} \cdot \frac{q^{(j)}}{E_A} - \frac{2}{T_R} \cdot P^{(j)} \cdot \frac{g^{(j)}}{E_L}$$

$$P_{R\omega} \cdot P_R = 2 \cdot (\frac{1}{T_L \cdot T_R} + \frac{1}{T_A \cdot T_R} + \frac{P^{(j)}}{E_A} + \frac{P^{(j)}}{E_L})^2 \cdot \omega^3$$

$$Q_{R\omega} \cdot Q_R = -\left[\frac{4}{T_R} \cdot (g^{(j)} - l - q^{(j)}) \cdot \omega\right] \cdot \left[\left(\frac{1}{T_L \cdot T_R} + \frac{P^{(j)}}{E_L}\right)\right.$$

$$\cdot\left(\frac{1}{T_A\cdot T_R}+\frac{P^{(j)}}{E_A}\right)\cdot\frac{2}{T_R}\cdot(g^{(j)}-l-q^{(j)})$$
$$+\frac{2}{T_R}\cdot P^{(j)}\cdot\frac{q^{(j)}}{E_A}\cdot\left(\frac{1}{T_L\cdot T_R}+\frac{P^{(j)}}{E_L}\right)$$
$$\left.-\frac{2}{T_R}\cdot P^{(j)}\cdot\frac{g^{(j)}}{E_L}\cdot\left(\frac{1}{T_A\cdot T_R}+\frac{P^{(j)}}{E_A}\right)-\frac{2}{T_R}\cdot(g^{(j)}-l-q^{(j)})\cdot\omega^2\right]$$

$$Q_R\cdot Q_{I\omega}=\left[\left(\frac{1}{T_L\cdot T_R}+\frac{P^{(j)}}{E_L}\right)\cdot\left(\frac{1}{T_A\cdot T_R}+\frac{P^{(j)}}{E_A}\right)\cdot\frac{2}{T_R}\cdot(g^{(j)}-l-q^{(j)})\right.$$
$$+\frac{2}{T_R}\cdot P^{(j)}\cdot\frac{q^{(j)}}{E_A}\cdot\left(\frac{1}{T_L\cdot T_R}+\frac{P^{(j)}}{E_L}\right)$$
$$-\frac{2}{T_R}\cdot P^{(j)}\cdot\frac{g^{(j)}}{E_L}\cdot\left(\frac{1}{T_A\cdot T_R}+\frac{P^{(j)}}{E_A}\right)$$
$$\left.-\frac{2}{T_R}\cdot(g^{(j)}-l-q^{(j)})\cdot\omega^2\right]\cdot$$
$$\left[\frac{2}{T_R}\cdot(g^{(j)}-l-q^{(j)})\cdot\left(\frac{1}{T_L\cdot T_R}+\frac{1}{T_A\cdot T_R}+\frac{P^{(j)}}{E_A}+\frac{P^{(j)}}{E_L}\right)\right.$$
$$\left.+\frac{2}{T_R}\cdot P^{(j)}\cdot\frac{q^{(j)}}{E_A}-\frac{2}{T_R}\cdot P^{(j)}\cdot\frac{g^{(j)}}{E_L}\right]$$

$$Q_I\cdot Q_{R\omega}=-\left[\frac{2}{T_R}\cdot(g^{(j)}-l-q^{(j)})\right.$$
$$\cdot\left(\frac{1}{T_L\cdot T_R}+\frac{1}{T_A\cdot T_R}+\frac{P^{(j)}}{E_A}+\frac{P^{(j)}}{E_L}\right)+\frac{2}{T_R}\cdot P^{(j)}\cdot\frac{q^{(j)}}{E_A}$$
$$\left.-\frac{2}{T_R}\cdot P^{(j)}\cdot\frac{g^{(j)}}{E_L}\right]\cdot\frac{4}{T_R}\cdot(g^{(j)}-l-q^{(j)})\cdot\omega^2$$

$$V=(P_R\cdot P_{I\tau}-P_I\cdot P_{R\tau})-(Q_R\cdot Q_{I\tau}-Q_I\cdot Q_{R\tau})=0\Rightarrow F(\omega,\tau)=0$$

Differentiating with respect to τ and we get $F_\omega\cdot\frac{\partial\omega}{\partial\tau}+F_\tau=0;\ \tau\in I\Rightarrow\frac{\partial\omega}{\partial\tau}=-\frac{F_\tau}{F_\omega}$

$$\Lambda^{-1}(\tau)=\left(\frac{\partial\mathrm{Re}\lambda}{\partial\tau}\right)_{\lambda=i\cdot\omega};\quad\frac{\partial\omega}{\partial\tau}=\omega_\tau$$
$$=-\frac{F_\tau}{F_\omega};\ \Lambda^{-1}(\tau)=\mathrm{Re}\left\{\frac{-2\cdot[U+\tau\cdot|P|^2]+i\cdot F_\omega}{F_\tau+i\cdot 2\cdot[V+\omega\cdot|P|^2]}\right\}$$
$$sign\{\Lambda^{-1}(\tau)\}=sign\left\{\left(\frac{\partial\mathrm{Re}\lambda}{\partial\tau}\right)_{\lambda=i\cdot\omega}\right\}$$
$$sign\{\Lambda^{-1}(\tau)\}=sign\{F_\omega\}\cdot sign\left\{\frac{V+\frac{\partial\omega}{\partial\tau}\cdot U}{|P|^2}+\omega+\frac{\partial\omega}{\partial\tau}\cdot\tau\right\}$$

We shall presently examine the possibility of stability transitions (bifurcations) of Solid-state laser by semiconductor devices model corresponding to the rate differential equation system, about the equilibrium point $E^{(*)}(P^{(*)}, g^{(*)}, q^{(*)})$ as a result of a variation of delay parameterτ. The analysis consists in identifying the roots of our system characteristic equation situated on the imaginary axis of the complexλ—plane where by increasing the delay parameterτ, $\mathrm{Re}\lambda$ may at the crossing, changes its sign from "−" to " + ", i.e. from a stable focus $E^{(*)}$ to an unstable one, or vice versa. This feature may be further assessed by examining the sign of the partial derivatives with respect to τ.

$$\Lambda^{-1}(\tau) = \left(\frac{\partial \mathrm{Re}\lambda}{\partial \tau}\right)_{\lambda = i \cdot \omega}; T_L, T_R, T_A, E_L, E_A, l, \ldots = const$$

$$U = (P_R \cdot P_{I\omega} - P_I \cdot P_{R\omega}) - (Q_R \cdot Q_{I\omega} - Q_I \cdot Q_{R\omega})$$

$$P_R \cdot P_{I\omega} = \left[\frac{1}{T_L \cdot T_R} + \frac{1}{T_A \cdot T_R} + \frac{P^{(j)}}{E_A} + \frac{P^{(j)}}{E_L}\right] \cdot \left[2 \cdot \omega^2 - \left(\frac{1}{T_L \cdot T_R} + \frac{P^{(j)}}{E_L}\right) \cdot \left(\frac{1}{T_A \cdot T_R} + \frac{P^{(j)}}{E_A}\right)\right] \cdot \omega^2$$

$$P_I \cdot P_{R\omega} = 2 \cdot \omega^2 \cdot \left[\omega^2 - \left(\frac{1}{T_L \cdot T_R} + \frac{P^{(j)}}{E_L}\right) \cdot \left(\frac{1}{T_A \cdot T_R} + \frac{P^{(j)}}{E_A}\right)\right] \cdot \left[\frac{1}{T_L \cdot T_R} + \frac{1}{T_A \cdot T_R} + \frac{P^{(j)}}{E_A} + \frac{P^{(j)}}{E_L}\right]$$

$$Q_R \cdot Q_{I\omega} = \left\{\left(\frac{1}{T_L \cdot T_R} + \frac{P^{(j)}}{E_L}\right) \cdot \left(\frac{1}{T_A \cdot T_R} + \frac{P^{(j)}}{E_A}\right) \cdot \frac{2}{T_R} \cdot (g^{(j)} - l - q^{(j)}) + \frac{2}{T_R} \cdot P^{(j)} \cdot \frac{q^{(j)}}{E_A} \cdot \left(\frac{1}{T_L \cdot T_R} + \frac{P^{(j)}}{E_L}\right) - \frac{2}{T_R} \cdot P^{(j)} \cdot \frac{g^{(j)}}{E_L} \cdot \left(\frac{1}{T_A \cdot T_R} + \frac{P^{(j)}}{E_A}\right) - \frac{2}{T_R} \cdot (g^{(j)} - l - q^{(j)}) \cdot \omega^2\right\} \cdot \left\{\frac{2}{T_R} \cdot (g^{(j)} - l - q^{(j)}) \cdot \left(\frac{1}{T_L \cdot T_R} + \frac{1}{T_A \cdot T_R} + \frac{P^{(j)}}{E_A} + \frac{P^{(j)}}{E_L}\right) + \frac{2}{T_R} \cdot P^{(j)} \cdot \frac{q^{(j)}}{E_A} - \frac{2}{T_R} \cdot P^{(j)} \cdot \frac{g^{(j)}}{E_L}\right\}$$

$$Q_I \cdot Q_{R\omega} = -\left[\frac{2}{T_R} \cdot (g^{(j)} - l - q^{(j)}) \cdot \left(\frac{1}{T_L \cdot T_R} + \frac{1}{T_A \cdot T_R} + \frac{P^{(j)}}{E_A} + \frac{P^{(j)}}{E_L}\right) + \frac{2}{T_R} \cdot P^{(j)} \cdot \frac{q^{(j)}}{E_A}\right.$$

$$-\frac{2}{T_R}\cdot P^{(j)}\cdot\frac{g^{(j)}}{E_L}\bigg]\cdot\frac{4}{T_R}\cdot(g^{(j)}-l-q^{(j)})\cdot\omega^2$$

$$U=\left(\left[\frac{1}{T_L\cdot T_R}+\frac{1}{T_A\cdot T_R}+\frac{P^{(j)}}{E_A}+\frac{P^{(j)}}{E_L}\right]\cdot[2\cdot\omega^2\right.$$

$$-\left(\frac{1}{T_L\cdot T_R}+\frac{P^{(j)}}{E_L}\right)\cdot\left(\frac{1}{T_A\cdot T_R}+\frac{P^{(j)}}{E_A}\right)\bigg]\cdot\omega^2$$

$$-2\cdot\omega^2\cdot\left[\omega^2-\left(\frac{1}{T_L\cdot T_R}+\frac{P^{(j)}}{E_L}\right)\cdot\left(\frac{1}{T_A\cdot T_R}+\frac{P^{(j)}}{E_A}\right)\right.$$

$$\frac{1}{T_L\cdot T_R}+\frac{1}{T_A\cdot T_R}+\frac{P^{(j)}}{E_A}+\frac{P^{(j)}}{E_L}\bigg]\cdot$$

$$-\left(\left\{\left(\frac{1}{T_L\cdot T_R}+\frac{P^{(j)}}{E_L}\right)\cdot\left(\frac{1}{T_A\cdot T_R}+\frac{P^{(j)}}{E_A}\right)\right.\right.$$

$$\cdot\frac{2}{T_R}\cdot(g^{(j)}-l-q^{(j)})+\frac{2}{T_R}\cdot P^{(j)}\cdot\frac{q^{(j)}}{E_A}\cdot\left(\frac{1}{T_L\cdot T_R}+\frac{P^{(j)}}{E_L}\right)$$

$$-\frac{2}{T_R}\cdot P^{(j)}\cdot\frac{g^{(j)}}{E_L}\cdot\left(\frac{1}{T_A\cdot T_R}+\frac{P^{(j)}}{E_A}\right)$$

$$-\frac{2}{T_R}\cdot(g^{(j)}-l-q^{(j)})\cdot\omega^2\bigg\}\cdot\bigg\{\frac{2}{T_R}\cdot(g^{(j)}-l-q^{(j)})\cdot\left(\frac{1}{T_L\cdot T_R}+\frac{1}{T_A\cdot T_R}+\frac{P^{(j)}}{E_A}\right.$$

$$\left.+\frac{P^{(j)}}{E_L}\right)+\frac{2}{T_R}\cdot P^{(j)}\cdot\frac{q^{(j)}}{E_A}-\frac{2}{T_R}\cdot P^{(j)}\cdot\frac{g^{(j)}}{E_L}\bigg\}$$

$$+[\frac{2}{T_R}\cdot(g^{(j)}-l-q^{(j)})\cdot\left(\frac{1}{T_L\cdot T_R}+\frac{1}{T_A\cdot T_R}+\frac{P^{(j)}}{E_A}+\frac{P^{(j)}}{E_L}\right)$$

$$+\frac{2}{T_R}\cdot P^{(j)}\cdot\frac{q^{(j)}}{E_A}\frac{2}{T_R}\cdot P^{(j)}\cdot\frac{g^{(j)}}{E_L}\bigg]\cdot\frac{4}{T_R}\cdot(g^{(j)}-l-q^{(j)})\cdot\omega^2)$$

Then we get the expression for $F(\omega,\tau)$ system parameter values. We find those ω,τ values which fulfill $F(\omega,\tau)=0$. We ignore, complex, and imaginary values of ω for specific τ values, $\tau\in[0.001\ \ldots\ 10]$ second. We can plot the stability switch diagram based on different delay values of our system.

$$\Lambda^{-1}(\tau)=\left(\frac{\partial\mathrm{Re}\lambda}{\partial\tau}\right)_{\lambda=i\cdot\omega}=\mathrm{Re}\left\{\frac{-2\cdot[U+\tau\cdot|P|^2+i\cdot F_\omega}{F_\tau+2\cdot i\cdot[V+\omega\cdot|P|^2]}\right\}$$

$$\Lambda^{-1}(\tau)=\left(\frac{\partial\mathrm{Re}\lambda}{\partial\tau}\right)_{\lambda=i\cdot\omega}=\frac{2\cdot\{F_\omega\cdot(V+\omega\cdot P^2)-F_\tau\cdot(U+\tau\cdot P^2)\}}{F_\tau^2+4\cdot(V+\omega\cdot P^2)^2}$$

The stability switch occurs only on those delay values (τ) which fit the equation: $\tau=\frac{\theta_+(\tau)}{\omega_+(\tau)}$ and $\theta_+(\tau)$ is the solution of $\sin\theta(\tau)=\cdots;\cos\theta(\tau)=\cdots$ when $\omega=\omega_+(\tau)$ if only ω_+ is feasible. Additionally, when system parameters are known and the stability switch due to various time delay values τ is described in the following expression:

$$sign\{\Lambda^{-1}(\tau)\} = sign\{F_{\omega}(\omega(\tau),\tau)\}\cdot sign\{\tau\cdot\omega_{\tau}(\omega(\tau)) + \omega(\tau) + \frac{U(\omega(\tau))\cdot\omega_{\tau}(\omega(\tau)) + V(\omega(\tau))}{|P(\omega(\tau))|^2}\}$$

Remark We know $F(\omega,\tau) = 0$ implies its roots $\omega_i(\tau)$ and finding those values τ which ω_i is feasible. There are τ values which give complex ω_i or imaginary number, then unable to analyze stability. F function is independent on τ the parameter $F(\omega,\tau) = 0$.

The results: We find those ω,τ values which fulfill $F(\omega,\tau) = 0$. We ignore negative, complex, and imaginary values of ω.

Next is to find those ω,τ values which fulfill $\sin\theta(\tau) = \cdots$; $\cos\theta(\tau) = \cdots$ meaning $\sin(\omega\cdot\tau) = \frac{-P_R\cdot Q_I + P_I\cdot Q_R}{|Q|^2}$; $\cos(\omega\cdot\tau) = -\frac{(P_R\cdot Q_R + P_I\cdot Q_I)}{|Q|^2}$; $|Q|^2 = Q_R^2 + Q_I^2$.

Finally, we plot the stability switch diagram $g(\tau) = \Lambda^{-1}(\tau) = \left(\frac{\partial \mathrm{Re}\lambda}{\partial\tau}\right)_{\lambda=i\cdot\omega}$.

$$g(\tau) = \Lambda^{-1}(\tau) = \left(\frac{\partial \mathrm{Re}\lambda}{\partial\tau}\right)_{\lambda=i\cdot\omega} = \frac{2\cdot\{F_{\omega}\cdot(V+\omega\cdot P^2) - F_{\tau}\cdot(U+\tau\cdot P^2)\}}{F_{\tau}^2 + 4\cdot(V+\omega\cdot P^2)^2}$$

$$sign[g(\tau)] = sign[\Lambda^{-1}(\tau)] = sign\left[\left(\frac{\partial \mathrm{Re}\lambda}{\partial\tau}\right)_{\lambda=i\cdot\omega}\right]$$

$$= sign\left[\frac{2\cdot\{F_{\omega}\cdot(V+\omega\cdot P^2) - F_{\tau}\cdot(U+\tau\cdot P^2)\}}{F_{\tau}^2 + 4\cdot(V+\omega\cdot P^2)^2}\right]$$

$$F_{\tau}^2 + 4\cdot(V+\omega\cdot P^2)^2 >; sign[\Lambda^{-1}(\tau)]$$

$$= sign[F_{\omega}\cdot(V+\omega\cdot P^2) - F_{\tau}\cdot(U+\tau\cdot P^2)]$$

$$sign[\Lambda^{-1}(\tau)] = sign\left\{[F_{\omega}]\cdot\left[(V+\omega\cdot P^2) - \frac{F_{\tau}}{F_{\omega}}\cdot(U+\tau\cdot P^2)\right]\right\}; \omega_{\tau}$$

$$= -\frac{F_{\tau}}{F_{\omega}}; \omega_{\tau} = \left(\frac{\partial\omega}{\partial\tau}\right)^{-1} = -\frac{\partial F/\partial\omega}{\partial F/\partial\tau}$$

$$sign[\Lambda^{-1}(\tau)] = sign\{[F_{\omega}]\cdot[(V+\omega\cdot P^2) + \omega_{\tau}\cdot(U+\tau\cdot P^2)]\}$$

$$sign[\Lambda^{-1}(\tau)] = sign\{[F_{\omega}]\cdot[V+\omega_{\tau}\cdot U + \omega\cdot P^2 + \omega_{\tau}\cdot\tau\cdot P^2]\}$$

$$sign[\Lambda^{-1}(\tau)] = sign\left\{[F_{\omega}]\cdot\left[\frac{(V+\omega_{\tau}\cdot U)}{P^2}\cdot P^2 + (\omega+\omega_{\tau}\cdot\tau)\cdot P^2\right]\right\}$$

$$sign[\Lambda^{-1}(\tau)] = sign\left\{[F_{\omega}]\cdot[P^2]\cdot\left[\frac{V+\omega_{\tau}\cdot U}{P^2} + \omega + \omega_{\tau}\cdot\tau\right]\right\}$$

$$sign[P^2] > 0 \Rightarrow sign[\Lambda^{-1}(\tau)] = sign\left\{[F_{\omega}]\cdot\left[\frac{V+\omega_{\tau}\cdot U}{P^2} + \omega + \omega_{\tau}\cdot\tau\right]\right\}$$

$$sign[P^2] > 0 \Rightarrow sign[\Lambda^{-1}(\tau)] = sign[F_{\omega}] \cdot sign\left[\frac{V+\omega_{\tau}\cdot U}{P^2} + \omega + \omega_{\tau}\cdot\tau\right]$$

$$F_{\omega} = 2\cdot[(P_{R\omega}\cdot P_R + P_{I\omega}\cdot P_I) - (Q_{R\omega}\cdot Q_R + Q_{I\omega}\cdot Q_I)]$$

We check the sign of $\Lambda^{-1}(\tau)$ according the following rule (Table 4.1).

Table. 4.1 Solid-state laser by semiconductor devices model corresponding to the rate differential equation system sign of $sign[\Lambda^{-1}(\tau)]$

$sign[F_\omega]$	$sign[\frac{V+\omega_\tau \cdot U}{P^2}+\omega+\omega_\tau\cdot\tau]$	$sign[\Lambda^{-1}(\tau)]$
$\pm$	$\pm$	$+$
$\pm$	$\pm$	$-$

If $sign[\Lambda^{-1}(\tau)] > 0$; then the crossing proceeds from (−) to (+) respectively (stable to unstable). If $sign[\Lambda^{-1}(\tau)] < 0$; then the crossing proceeds from (+) to (−) respectively (unstable to stable) [7–9].

Appendix A

A. Appendix A1 (Lemma 1.1)

Assume that $\omega(\tau)$ is a positive and real root of $F(\omega,\tau)=0$ are defined for $\tau\in I$, which is continuous and differentiable. Assume further that if $\lambda=i\cdot\omega$; $\omega\in\mathbb{R}$, then $P_n(i\cdot\omega,\tau)+Q_n(i\cdot\omega,\tau)\neq 0$; $\tau\in\mathbb{R}$ hold true. Then the functions $S_n(\tau)$; $n\in\mathbb{N}_0$, are continuous and differentiable on I.

B. Appendix A_2 (Theorem 1.2)

Assume that $\omega(\tau)$ is a positive real root of $F(\omega,\tau)=0$ defined for $\tau\in I$, $I\subseteq\mathbb{R}_{+0}$ and at some $\tau^*\in I$, $S_n(\tau^*)=0$. For some $n\in\mathbb{N}_0$ then a pair of simple conjugate pure imaginary roots $\lambda_+(\tau^*)=i\cdot\omega(\tau^*)$ and $\lambda_-(\tau^*)=-i\cdot\omega(\tau^*)$ of $D(\lambda,\tau)=0$ exists at $\tau=\tau^*$ which crosses the imaginary axis from left to right if $\delta(\tau^*)>0$ and cross the imaginary axis from right to left if $\delta(\tau^*)<0$ where $\delta(\tau^*)=sign\{\frac{d\mathrm{Re}\lambda}{d\tau}|_{\lambda=i\cdot\omega(\tau^*)}\}=sign\{F_\omega(\omega(\tau^*),\tau^*)\}\cdot sign\left\{\frac{dS_n(\tau)}{d\tau}|_{\tau=\tau^*}\right\}$.
The theorem becomes $sign\{\frac{d\mathrm{Re}\lambda}{d\tau}|_{\lambda=i\cdot\omega\pm}\}=sign\{\pm\Delta^{1/2}\}\cdot sign\left\{\frac{dS_n(\tau)}{d\tau}|_{\tau=\tau^*}\right\}$.

C. Appendix A_3 (Theorem 1.3)

The characteristic equation has a pair of simple and conjugate pure imaginary roots$\lambda=\pm\omega(\tau^*)$, $\omega(\tau^*)$ real at $\tau^*\in I$ if $S_n(\tau^*)=\tau^*-\tau_n(\tau^*)=0$ for some $n\in\mathbb{N}_0$. If $\omega(\tau^*)=\omega_+(\tau^*)$, this pair of simple conjugate pure imaginary roots crosses the imaginary axis from left to right if $\delta_+(\tau^*)>0$ and crosses the imaginary axis from right to left if $\delta_+(\tau^*)<0$ where $\delta_+(\tau^*)=sign\{\frac{d\mathrm{Re}\lambda}{d\tau}|_{\lambda=i\cdot\omega_+(\tau^*)}\}=sign\left\{\frac{dS_n(\tau)}{d\tau}|_{\tau=\tau^*}\right\}$. If $\omega(\tau^*)=\omega_-(\tau^*)$, this pair of simple conjugate pure imaginary roots cross the imaginary axis from left to right if $\delta_-(\tau^*)>0$ and crosses the imaginary axis from right to left if $\delta_-(\tau^*)<0$ where $\delta_-(\tau^*)=sign\{\frac{d\mathrm{Re}\lambda}{d\tau}|_{\lambda=i\cdot\omega_-(\tau^*)}\}$ and $\delta_-(\tau^*)=-sign\left\{\frac{dS_n(\tau)}{d\tau}|_{\tau=\tau^*}\right\}$. If $\omega_+(\tau^*)=\omega_-(\tau^*)=\omega(\tau^*)$ then $\Delta(\tau^*)=0$ and $sign\{\frac{d\mathrm{Re}\lambda}{d\tau}|_{\lambda=i\cdot\omega(\tau^*)}\}=0$ the same is true when $S_n(\tau^*)=0$. The following result can be useful in identifying values of τ where stability switches happened.

D. Appendix A_4 (Theorem 1.4)

Assume that for all $\tau \in I$, $\omega(\tau)$ is defined as a function of $F(\omega, \tau) = 0$ then $\delta_{\pm}(\tau) = sign\{\pm\Delta^{1/2}(\tau)\} \cdot sign\{D_{\pm}(\tau)\}$.

4.2 Nanometer-Vibration Measurement with Microchip Solid-State Laser Instability Under Parameters Variation

In many applications we use laser-diode-pumped microchip solid-state laser (DPSSLs). One application is a self-aligned optical feedback vibrometry technique [2]. It can help to understand the nanometer vibration analysis. The laser-diode-pumped microchip solid-state laser is useful for compact, durable, coherent light sources. If we use external optical feedback light we get an ultrahigh sensitivity response of DPSSL. They have short photon lifetimes compared with the population lifetimes that are inherent in microchip lasers. The results are in such significant applications as self-aligned laser Doppler velocimetry, vibrometry, and imaging. The laser acts as a high-efficiency mixer oscillator and a shot-noise-limited quantum detector. In laser-diode-pumped microchip solid-state lasers the effective long-term interference between a lasing field and a coherent component of a scattered field that is fed back to the micro cavity laser resonator occurs because of the high degree of coherence of the lasing field. The experimental system configuration of self-mixing laser Doppler vibrometry and optical micro-phony include the following elements: Laser diode (LD), Anamorphic prism pairs (APPs), Microscope objective lens (OLs), Glass-plate beam splitter (BS), Photodiode receiver (PD), RF spectrum analyzer (SA), Variable attenuator (VA), Digital oscilloscope (DO), Frequency demodulator circuit (DC), and Personal computer (PC). It is c plate Nd direct compound $LiNdP_4O_{12}$ (LNP) laser with a thick plane-parallel Fabry–perot-cavity. An end surface is coated to be trans-missive at the laser-diode pump wavelength and highly reflective at specific lasing wavelength. The other surface is coated to be 1% transmissivity at the lasing wavelength. The collimated laser-diode pump light is passed through anamorphic prism pairs, which transformed the elliptical beam into a circular beam, and it is focused onto LNP crystal by a microscope objective lens of 20× magnification. An anamorphic prism pair is a pair of prisms which is usually used for reshaping the profile of a laser beam. In our case the elliptical beam from the laser diode is transformed into a beam with circular cross-section, by using a prism pair as a beam expander for only one direction. The principle of beam shaping with anamorphic prism pairs is not based on focusing effects (i.e. changes of wave front curvature), but rather an changed of the beam radius for reflection at plat prism interfaces. Such changes occur at the interfaces of any prism, because the angle of the beam against the surface-normal direction is different inside and outside according to Snell's law. In the case of a symmetric beam path, where the beam angles against input and output face of the prism are identical, the two changes in beam radius cancel each other. Therefore, one has to use an asymmetric configuration. A single prism is sufficient for changing the beam radius in one direction, but it also changes the beam direction. By using anamorphic prism pair, one can obtain an output beam with an unchanged direction, only a position offset. The two prisms are oriented such that they change the beam radius in the same direction. The overall magnification is then square of the refractive index, or the inverse of that. If the mentioned beam offset needs to be avoided, one may use a combination of four prisms. In an anamorphic prism the output beam is substantially narrower than the input beam. For example an anamorphic prism pair with refractive index 1.5, where Brewster's angle is used on one side of each prism, and normal incidence on the other one. Two parallel beams passing through the prisms. The distance changes, and likewise their beam radii in the direction of the plane are changed. The prism pair thus works as a beam expander if the input beam comes from the left side. The beam radius in the direction perpendicular to the drawing plane is not changed. In our system a part ($\approx$96%) of the output light is shifted by two Acoustic Optic Modulators (AOMs) and impinged upon a speaker that had an Al-coated surface with an average roughness of few hundred micro meters that is placed $\approx$1 m from the laser. An acoustic optic modulator (AOM) is a device which can be used for controlling the power, frequency or spatial direction of a laser beam with an electrical drive signal. It is based on the acousto-optic effect. The modification of the refractive index is by the oscillating mechanical pressure of a sound wave. The key element of an Acoustic optical modulator (AOM) is a transparent crystal through which the light propagates. A piezo electric traducer attached to the crystal is used to excite a sound wave with a frequency of the order of 100 MHz. The light can then experience Bragg diffraction at the traveling periodic refractive index grating generated by the sound wave; therefore, AOMs are Bragg cells. The optical frequency of the beam is increased or decreased by the frequency of the sound wave which is depending on the propagation direction of the acoustic wave relative to the beam and propagates in a slightly different direction. The frequency and direction of the scattered beam can be controlled via the frequency of the sound wave, whereas the acoustic power is the control for the optical powers. The acoustic wave may be absorbed at the other end of the crystal. It is possible to achieve a broad modulation bandwidth of many megahertz. In the experiment we use two AOMs to achieve appropriate frequency shifts. By changing the modulation frequencies of upshifted and downshifted AOMs, we shift the optical carrier frequency by specific frequency after a round trip for the feedback field. AOMs can also be used for cavity dumping of solid-state lasers, generating either nanosecond or ultrashort pulses. Active mode locking is often performed with an AOM for modulating the resonator losses at the round-trip frequency or a multiple thereof (Fig. 4.2). In our system another part ($\approx$4%) of the output light is detected by an InGaAs photo receiver. The additional units which used are frequency demodulation circuit, a digital oscilloscope connected to a Personal Computer (PC), and RF spectrum analyzer (Fig. 4.3). Microscope objective lens: The objective is the optical element that gathers light from the object being observed and focused the light rays to introduce a real image [2]. Objectives can be

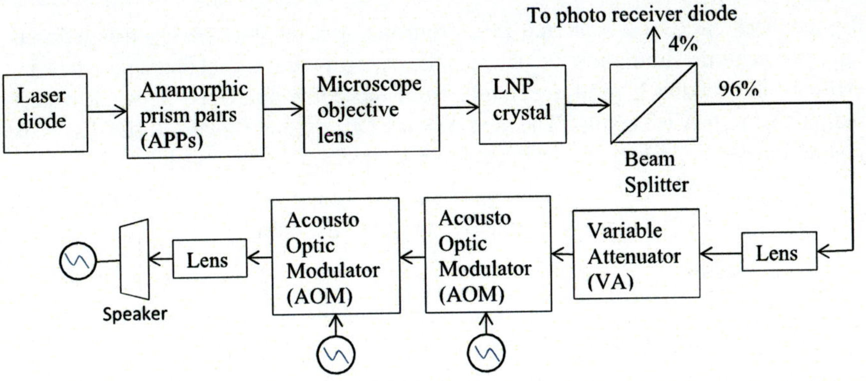

Fig. 4.2 Experimental configuration of self-mixing laser-Doppler vibrometry and optical micro phony (flow chart for 96% of the output light)

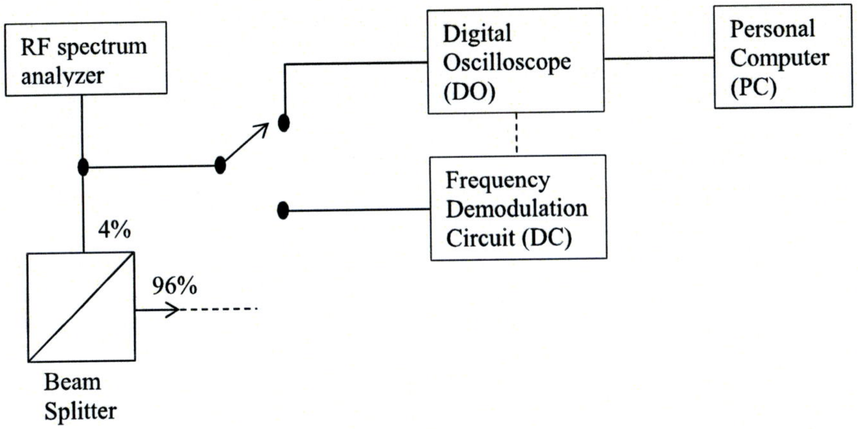

Fig. 4.3 Experimental configuration of self-mixing laser-Doppler vibrometry and optical micro phony (flow chart for 4% of the output light)

a single lens or mirror or combinations of several optical elements. f_m—Modulation frequency $\left(f_m = \frac{\omega_m}{2\cdot\pi}\right)$, β—Frequency modulation index, $A_{v,m}$—Maximum vibration amplitude $\left(A_{v,m} = \frac{\lambda\cdot\beta}{2\cdot\pi}\right)$, V_a—Applied voltage, $A_v(t)$—Temporal evolutions of nanometer vibration (vibration amplitude), $\Delta\Phi(t)$—Phase difference between the reference (carrier) signed and the modulated signal ($A_v(t) = \lambda \cdot \Delta\Phi(t)$), Φ_A—Analytic phase, related to analytic signal V_A, $\langle V_A\rangle$—Analytic signal time average, $V_A(t) = I(t) + i \cdot I_H(t)$, $V_A(t) - \langle V_A\rangle = R_A(t) \cdot e^{i\cdot\Phi_A(t)}$, $I(t)$—Time series of the scalar intensity, I_H—Hilbert transform.

There are modulated output waveforms and the corresponding vibration waveforms for different feedback ratios. The variable optical attenuator (VA) is used to change the feedback ratio. There is an analytic and numerical simulation to measure the feedback ratio of a scattered field from the speaker into the laser experimentally. By analytic simulation we produce the experimental signal characteristics and estimate the feedback ratio accurately from the correspondence between experimental and numerical results. The model Delay Differential Equations (DDEs) is the simulation, that is, laser with frequency-modulated optical feedback [2].

$E(t)$—Normalized field amplitude, $N(t)$—Normalized excess population inversion, ω—Relative pump power normalized by the threshold $\left(\omega = \frac{P}{P_{th}}\right)$, K—Population to photon lifetime ration $\left(K = \frac{\tau}{\tau_p}\right)$, $\phi(t)$—Phase of the lasing field, $\psi(t)$—Phase difference between the lasing and the feedback fields, $\Delta\Omega$—Normalized frequency shift,$\Delta\Omega = (\omega_i - \omega_0) \cdot \tau_p$.

(τ_p—photon lifetime), m—Amplitude feedback coefficient, Ω_m—Normalized modulation frequency ($\Omega_m = \omega_m \cdot \tau_p$), t—Time normalized by the photon lifetime, t_D—Delay time normalized by the photon lifetime, ε—Spontaneous emission rate, $\xi(t)$—Gaussian white noise with zero mean. The value $\langle \xi(t) \cdot \xi(t') \rangle = \delta(t - t')$ is δ correlated in time. $\left(\sqrt{2 \cdot \varepsilon \cdot [N(t) + 1]}\right) \cdot \xi(t)$ is the quantum spontaneous emission noise.

The model DDEs (lasers with frequency-modulated optical feedback):

$$\frac{dN(t)}{dt} = K \cdot \{\omega - 1 - N(t) - [1 + 2 \cdot N(t)] \cdot E^2(t)\}$$

$$N = N(t);\ E = E(t);\ \phi = \phi(t)$$

$$\frac{dE(t)}{dt} = N(t) \cdot E(t) + m \cdot E(t - t_D) \times \cos\psi(t) + \left(\sqrt{2 \cdot \varepsilon \cdot [N(t) + 1]}\right) \cdot \xi(t)$$

$$\frac{d\phi(t)}{dt} = \left[\frac{E(t - t_D)}{E(t)}\right] \cdot \sin\psi(t);\ \psi(t) = \Delta\Omega \cdot t + \beta \cdot \sin(\Omega_m \cdot t) - \phi(t) + \phi(t - t_D) - \left(\Omega_0 + \frac{\Delta\Omega}{2}\right) \cdot t_D$$

At fixed points $\frac{dN(t)}{dt} = 0$; $\frac{dE(t)}{dt} = 0$; $\frac{d\phi(t)}{dt} = 0$; $\lim_{t\to\infty} E(t - t_D) = E(t)$; $\lim_{t\to\infty} \phi(t - t_D) = \phi(t)$.

At fixed points: $t \gg t_D \Rightarrow t - t_D \approx t$; $\psi^*(t) = \psi(t, \phi^*)$

$$\psi^* = \Delta\Omega \cdot t + \beta \cdot \sin(\Omega_m \cdot t) - \left(\Omega_0 + \frac{\Delta\Omega}{2}\right) \cdot t_D;\ \psi^* = \Delta\Omega \cdot t + \beta \cdot \sin(\Omega_m \cdot t) - \Omega_0 \cdot t_D - \frac{\Delta\Omega}{2} \cdot t_D$$

$$\psi^* = \Delta\Omega \cdot \left(t - \frac{1}{2} \cdot t_D\right) + \beta \cdot \sin(\Omega_m \cdot t) - \Omega_0 \cdot t_D;\ t \gg \frac{1}{2} \cdot t_D \Rightarrow t - \frac{1}{2} \cdot t_D \approx t$$

$$\psi^* \approx \Delta\Omega \cdot t + \beta \cdot \sin(\Omega_m \cdot t) - \Omega_0 \cdot t_D; \frac{d\phi(t)}{dt}$$
$$= 0 \Rightarrow \sin \psi^* = 0; \psi^* = n \cdot \pi \, \forall n = 0, 1, 2, \ldots$$

$$n \in \mathbb{N}_0; n \cdot \pi = \Delta\Omega \cdot t + \beta \cdot \sin(\Omega_m \cdot t) - \Omega_0 \cdot t_D$$

White Gaussian Noise (WGN): Gaussian white noise with zero mean ($\xi(t)$). White Gaussian Noise (WGN) is a basic noise model used in information theory to mimic the effect of many random processes that occur in nature. It is additive because it is added to any noise that might be intrinsic to the information system. The noise is called "white" because it has the same intensity at every frequency. It called "white" when it contains all visible frequency. Gaussian White Noise (GWN) is a stationary and ergodic random process with zero mean that is defined by the fundamental property that two values of GWN are statistically independent no matter how close they are in time. The power spectrum of GWN is constant over all frequencies, hence the name "white noise", in analogy to the white light that contains all visible wavelength with the same power. The additional adjective "Gaussian" in GWN indicates that the amplitude distribution of the white noise signal is Gaussian-like the independent steps in Brownian motion. The most common approximation for GWN is the band-limited GWN signal that has the "Sinc function" as autocorrelation function. Non-Gaussian white-noise signals have symmetric amplitude probability density functions and may exhibits, practical advantages in certain applications over band-limited GWN. High-order autocorrelation functions of GWN signals have a specific structure that is suitable for nonlinear system identification. Specially, all the odd-order autocorrelation functions are uniformly zero and the even-order ones can be expressed in terms of sums of products of delta functions.

Assumption: The quantum spontaneous emission noise is very small and negligible $(\sqrt{2 \cdot \varepsilon \cdot [N(t) + 1]}) \cdot \xi(t) \ll 1; (\sqrt{2 \cdot \varepsilon \cdot [N(t) + 1]}) \cdot \xi(t) \to 0$.

Then $\frac{dN}{dt} = 0 \Rightarrow K \cdot \{\omega - 1 - N^* - [1 + 2 \cdot N^*] \cdot (E^*)^2\} = 0$

$$\frac{dE}{dt} = 0 \Rightarrow N^* \cdot E^* + m \cdot E^* \times \cos(\psi^*) = 0; N^* \cdot E^* + m \cdot E^* \times \cos(\psi^*) = 0$$
$$\sin \psi^* = 0 \Rightarrow \cos \psi^* = \pm 1; n = 0, 1, 2, \ldots; n \in \mathbb{N}_0; \psi^* = n \cdot \pi; \cos \psi^* = (-1)^n$$
$$\frac{dE}{dt} = 0 \Rightarrow N^* \cdot E^* + m \cdot E^* \times \cos(\psi^*) = 0; N^* \cdot E^*$$
$$+ m \cdot E^* \cdot (-1)^n = 0 \, \forall n = 0, 1, 2, \ldots$$
$$\frac{dN}{dt} = 0 \Rightarrow K \cdot \{\omega - 1 - N^* - [1 + 2 \cdot N^*] \cdot (E^*)^2\} = 0$$
$$K \neq 0 \Rightarrow \omega - 1 - N^* - [1 + 2 \cdot N^*] \cdot (E^*)^2 = 0$$

We get two fixed points equations: $N^* \cdot E^* + m \cdot E^* \cdot (-1)^n = 0 \, \forall n = 0, 1, 2, \ldots$

$$\omega - 1 - N^* - [1 + 2 \cdot N^*] \cdot (E^*)^2 = 0$$

&&&

$$\omega - 1 - (E^*)^2 = N^* \cdot [1 + 2 \cdot (E^*)^2];\; N^* = \frac{\omega - 1 - (E^*)^2}{1 + 2 \cdot (E^*)^2}$$

$$N^* \cdot E^* + m \cdot E^* \cdot (-1)^n$$
$$= 0 \Rightarrow \left[\frac{\omega - 1 - (E^*)^2}{1 + 2 \cdot (E^*)^2}\right] \cdot E^* + m \cdot E^* \cdot (-1)^n = 0$$
$$E^* \cdot \left[\frac{\omega - 1 - (E^*)^2}{1 + 2 \cdot (E^*)^2} + m \cdot (-1)^n\right] = 0$$

Case 1: $E^* = 0$ then $N^* = \omega - 1$. First fixed point: $(N^{(0)}, E^{(0)}, \phi^{(0)}) = (\omega - 1, 0, \phi^*);\ \phi^* \in \mathbb{R}$.

Case 2: $\frac{\omega-1-(E^*)^2}{1+2\cdot(E^*)^2} + m \cdot (-1)^n = 0$

$$\frac{\omega - 1 - (E^*)^2}{1 + 2 \cdot (E^*)^2} + m \cdot (-1)^n = 0 \Rightarrow \frac{\omega - 1 - (E^*)^2}{1 + 2 \cdot (E^*)^2}$$
$$= -m \cdot (-1)^n \Rightarrow \omega - 1 - (E^*)^2 = -m \cdot (-1)^n \cdot [1 + 2 \cdot (E^*)^2]$$
$$(E^*)^2 \cdot [m \cdot (-1)^n \cdot 2 - 1] = 1 - \omega - m \cdot (-1)^n;\ (E^*)^2$$
$$= \frac{1 - \omega - m \cdot (-1)^n}{m \cdot (-1)^n \cdot 2 - 1};\ E^* = \pm\sqrt{\frac{1 - \omega - m \cdot (-1)^n}{m \cdot (-1)^n \cdot 2 - 1}}$$

Case 2.1: $E^* = \sqrt{\frac{1-\omega-m\cdot(-1)^n}{m\cdot(-1)^n\cdot 2-1}}$

$$N^* = \frac{\omega - 1 - (E^*)^2}{1 + 2 \cdot (E^*)^2} \Rightarrow N^* = \frac{\omega - 1 - \left(\sqrt{\frac{1-\omega-m\cdot(-1)^n}{m\cdot(-1)^n\cdot 2-1}}\right)^2}{1 + 2 \cdot \left(\sqrt{\frac{1-\omega-m\cdot(-1)^n}{m\cdot(-1)^n\cdot 2-1}}\right)^2}$$

$$N^* = \frac{\omega - 1 - \left[\frac{1-\omega-m\cdot(-1)^n}{m\cdot(-1)^n\cdot 2-1}\right]}{1 + 2 \cdot \left[\frac{1-\omega-m\cdot(-1)^n}{m\cdot(-1)^n\cdot 2-1}\right]}$$
$$N^* = \frac{(\omega - 1) \cdot (m \cdot (-1)^n \cdot 2 - 1) - (1 - \omega - m \cdot (-1)^n)}{m \cdot (-1)^n \cdot 2 - 1 + 2 \cdot (1 - \omega - m \cdot (-1)^n)}$$
$$N^* = \frac{[\omega \cdot 2 - 1] \cdot m \cdot (-1)^n}{1 - 2 \cdot \omega};\ N^* = -m \cdot (-1)^n$$

Second fixed point: $(N^{(1)}, E^{(1)}, \phi^{(1)}) = \left(-m\cdot(-1)^n, \sqrt{\frac{1-\omega-m\cdot(-1)^n}{m\cdot(-1)^n\cdot 2-1}}, \phi^*\right);\ \phi^* \in \mathbb{R}$.

Case 2.2: $E^* = -\sqrt{\frac{1-\omega-m\cdot(-1)^n}{m\cdot(-1)^n\cdot 2-1}}$

$$N^* = \frac{\omega - 1 - (E^*)^2}{1 + 2 \cdot (E^*)^2} \Rightarrow N^* = \frac{\omega - 1 - \left(-\sqrt{\frac{1-\omega-m\cdot(-1)^n}{m\cdot(-1)^n\cdot 2-1}}\right)^2}{1 + 2 \cdot \left(-\sqrt{\frac{1-\omega-m\cdot(-1)^n}{m\cdot(-1)^n\cdot 2-1}}\right)^2}$$

$$N^* = \frac{\omega - 1 - \left[\frac{1-\omega-m\cdot(-1)^n}{m\cdot(-1)^n\cdot 2-1}\right]}{1 + 2 \cdot \left[\frac{1-\omega-m\cdot(-1)^n}{m\cdot(-1)^n\cdot 2-1}\right]};$$

$$N^* = \frac{[\omega \cdot 2 - 1] \cdot m \cdot (-1)^n}{1 - 2 \cdot \omega}; \; N^* = -m \cdot (-1)^n$$

Third fixed point: $(N^{(2)}, E^{(2)}, \phi^{(2)}) = \left(-m \cdot (-1)^n, -\sqrt{\frac{1-\omega-m\cdot(-1)^n}{m\cdot(-1)^n\cdot 2-1}}, \phi^*\right); \phi^* \in \mathbb{R}$.

We can summary our system fixed points:

First fixed point: $(N^{(0)}, E^{(0)}, \phi^{(0)}) = (\omega - 1, 0, \phi^*); \phi^* \in \mathbb{R}; \forall n = 0, 1, 2, \ldots; n \in \mathbb{N}_0$.

Second fixed point: $(N^{(1)}, E^{(1)}, \phi^{(1)}) = \left(-m \cdot (-1)^n, \sqrt{\frac{1-\omega-m\cdot(-1)^n}{m\cdot(-1)^n\cdot 2-1}}, \phi^*\right); \phi^* \in \mathbb{R}$.

Third fixed point: $(N^{(2)}, E^{(2)}, \phi^{(2)}) = \left(-m \cdot (-1)^n, -\sqrt{\frac{1-\omega-m\cdot(-1)^n}{m\cdot(-1)^n\cdot 2-1}}, \phi^*\right); \phi^* \in \mathbb{R}$.

Stability analysis: The standard local stability analysis about any one of the equilibrium points of lasers with frequency-modulated optical feedback system consists in adding to coordinates $[N\ E\ \phi]$ arbitrarily small increments of exponential form $[n\ \mathrm{E}\ \phi] \cdot e^{\lambda \cdot t}$, and retaining the first order terms in $N\ E\ \phi$. The system of three homogenous equations leads to a polynomial characteristics equation in the eigenvalues λ. The polynomial characteristics equations accept by set the below variables and variables derivative respect to time into lasers with frequency-modulated optical feedback system equations. System fixed values with arbitrarily small increments of exponential form $[n\ E\ \phi] \cdot e^{\lambda \cdot t}$ are i = 0 (first fixed point), i = 1 (second fixed point), i = 2 (third fixed point), etc. [5–7].

$$N(t) = N^{(i)} + n \cdot e^{\lambda \cdot t}; \; E(t) = E^{(i)} + E \cdot e^{\lambda \cdot t};$$

$$\phi(t) = \phi^{(i)} + \phi \cdot e^{\lambda \cdot t}; \; E(t - t_D) = E^{(i)} + E \cdot e^{\lambda \cdot (t - t_D)}$$

$$\phi(t - t_D) = \phi^{(i)} + \phi \cdot e^{\lambda \cdot (t - t_D)}; \; \frac{dN(t)}{dt} = \lambda \cdot n \cdot e^{\lambda \cdot t}$$

$$\frac{dE(t)}{dt} = \lambda \cdot E \cdot e^{\lambda \cdot t}; \; \frac{d\phi(t)}{dt} = \lambda \cdot \phi \cdot e^{\lambda \cdot t}$$

$$\frac{dN(t)}{dt} = K \cdot \{\omega - 1 - N(t) - [1 + 2 \cdot N(t)] \cdot E^2(t)\}$$

$$\lambda \cdot n \cdot e^{\lambda \cdot t} = K \cdot \{\omega - 1 - (N^{(i)} + n \cdot e^{\lambda \cdot t}) - [1 + 2 \cdot (N^{(i)} + n \cdot e^{\lambda \cdot t})] \cdot (E^{(i)} + E \cdot e^{\lambda \cdot t})^2\}$$

$$\lambda \cdot n \cdot e^{\lambda \cdot t} = K \cdot \{\omega - 1 - N^{(i)} - n \cdot e^{\lambda \cdot t} - [1 + (2 \cdot N^{(i)} + 2 \cdot n \cdot e^{\lambda \cdot t})] \cdot ([E^{(i)}]^2 + 2 \cdot E^{(i)} \cdot E \cdot e^{\lambda \cdot t} + E^2 \cdot e^{2 \cdot \lambda \cdot t})\}$$

Assumption: $E^2 \approx 0$

$$\lambda \cdot n \cdot e^{\lambda \cdot t} = K \cdot \{\omega - 1 - N^{(i)} - n \cdot e^{\lambda \cdot t}$$
$$3 - [(1 + 2 \cdot N^{(i)}) + 2 \cdot n \cdot e^{\lambda \cdot t}] \cdot ([E^{(i)}]^2 + 2 \cdot E^{(i)} \cdot E \cdot e^{\lambda \cdot t})\}$$
$$\lambda \cdot n \cdot e^{\lambda \cdot t} = K \cdot \{\omega - 1 - N^{(i)} - n \cdot e^{\lambda \cdot t}$$
$$- [(1 + 2 \cdot N^{(i)}) \cdot [E^{(i)}]^2 + (1 + 2 \cdot N^{(i)}) \cdot 2 \cdot E^{(i)} \cdot E \cdot e^{\lambda \cdot t}$$
$$+ [E^{(i)}]^2 \cdot 2 \cdot n \cdot e^{\lambda \cdot t} + 4 \cdot E^{(i)} \cdot n \cdot E \cdot e^{2 \cdot \lambda \cdot t}]\}$$

Assumption: $n \cdot E \approx 0$

$$\lambda \cdot n \cdot e^{\lambda \cdot t} = K \cdot \{\omega - 1 - N^{(i)} - (1 + 2 \cdot N^{(i)}) \cdot [E^{(i)}]^2\}$$
$$- K \cdot n \cdot e^{\lambda \cdot t} + K \cdot (1 + 2 \cdot N^{(i)}) \cdot 2 \cdot E^{(i)} \cdot E \cdot e^{\lambda \cdot t}$$
$$+ K \cdot [E^{(i)}]^2 \cdot 2 \cdot n \cdot e^{\lambda \cdot t}$$

At fixed points: $K \cdot \{\omega - 1 - N^{(i)} - (1 + 2 \cdot N^{(i)}) \cdot [E^{(i)}]^2\} = 0$.

$$\lambda \cdot n \cdot e^{\lambda \cdot t} = -K \cdot n \cdot e^{\lambda \cdot t} + K \cdot (1 + 2 \cdot N^{(i)}) \cdot 2 \cdot E^{(i)} \cdot E \cdot e^{\lambda \cdot t} + K \cdot [E^{(i)}]^2 \cdot 2 \cdot n \cdot e^{\lambda \cdot t}$$
$$- \lambda \cdot n \cdot e^{\lambda \cdot t} - K \cdot n \cdot e^{\lambda \cdot t} + K \cdot [E^{(i)}]^2 \cdot 2 \cdot n \cdot e^{\lambda \cdot t}$$
$$+ K \cdot (1 + 2 \cdot N^{(i)}) \cdot 2 \cdot E^{(i)} \cdot E \cdot e^{\lambda \cdot t} = 0$$

Divide the above equation by $e^{\lambda \cdot t}$ term.

$$(-\lambda - K + K \cdot [E^{(i)}]^2 \cdot 2) \cdot n + K \cdot (1 + 2 \cdot N^{(i)}) \cdot 2 \cdot E^{(i)} \cdot E = 0$$
$$(-\lambda + K \cdot \{2 \cdot [E^{(i)}]^2 - 1\}) \cdot n + 2 \cdot K \cdot (1 + 2 \cdot N^{(i)}) \cdot E^{(i)} \cdot E = 0$$
$$\frac{dE(t)}{dt} = N(t) \cdot E(t) + m \cdot E(t - t_D) \times \cos\psi(t) + \left(\sqrt{2 \cdot \varepsilon \cdot [N(t) + 1]}\right) \cdot \xi(t)$$

Assumption: The quantum spontaneous emission noise is very small and negligible $(\sqrt{2 \cdot \varepsilon \cdot [N(t) + 1]}) \cdot \xi(t) \ll 1$; $(\sqrt{2 \cdot \varepsilon \cdot [N(t) + 1]}) \cdot \xi(t) \to 0$.

$$\frac{dE(t)}{dt} = N(t) \cdot E(t) + m \cdot E(t - t_D) \times \cos\psi(t)$$
$$\lambda \cdot E \cdot e^{\lambda \cdot t} = (N^{(i)} + n \cdot e^{\lambda \cdot t}) \cdot (E^{(i)} + E \cdot e^{\lambda \cdot t})$$
$$+ m \cdot (E^{(i)} + E \cdot e^{\lambda \cdot (t - t_D)}) \times \cos\{\Delta\Omega \cdot t + \beta \cdot \sin(\Omega_m \cdot t)$$
$$\left. -\phi(t) + \phi(t - t_D) - (\Omega_0 + \frac{\Delta\Omega}{2}) \cdot t_D \right\}$$
$$\lambda \cdot E \cdot e^{\lambda \cdot t} = (N^{(i)} + n \cdot e^{\lambda \cdot t}) \cdot (E^{(i)} + E \cdot e^{\lambda \cdot t})$$
$$+ m \cdot (E^{(i)} + E \cdot e^{\lambda \cdot (t - t_D)}) \times \cos\{\Delta\Omega \cdot t + \beta \cdot \sin(\Omega_m \cdot t)$$

$$-(\phi^{(i)} + \phi \cdot e^{\lambda \cdot t}) + (\phi^{(i)} + \phi \cdot e^{\lambda \cdot (t-t_D)}) - (\Omega_0 + \frac{\Delta\Omega}{2}) \cdot t_D \Big\}$$

First we do algebraic manipulation for the cos{...} expression.

$$\cos\Big\{\Delta\Omega \cdot t + \beta \cdot \sin(\Omega_m \cdot t) - (\phi^{(i)} + \phi \cdot e^{\lambda \cdot t}) + (\phi^{(i)} + \phi \cdot e^{\lambda \cdot (t-t_D)}) - \left(\Omega_0 + \frac{\Delta\Omega}{2}\right) \cdot t_D\Big\}$$

$$\cos\Big\{\Delta\Omega \cdot t + \beta \cdot \sin(\Omega_m \cdot t) - \phi^{(i)} - \phi \cdot e^{\lambda \cdot t} + \phi^{(i)} + \phi \cdot e^{\lambda \cdot (t-t_D)} - \Omega_0 \cdot t_D - \frac{\Delta\Omega}{2} \cdot t_D\Big\}$$

$$\cos\Big\{\Delta\Omega \cdot t + \beta \cdot \sin(\Omega_m \cdot t) - \phi \cdot e^{\lambda \cdot t} + \phi \cdot e^{\lambda \cdot (t-t_D)} - \Omega_0 \cdot t_D - \frac{\Delta\Omega}{2} \cdot t_D\Big\}$$

$$\cos\Big\{[\Delta\Omega \cdot t + \beta \cdot \sin(\Omega_m \cdot t) - \Omega_0 \cdot t_D] + \phi \cdot e^{\lambda \cdot (t-t_D)} - \phi \cdot e^{\lambda \cdot t} - \frac{\Delta\Omega}{2} \cdot t_D\Big\}$$

$$\cos\Big\{[\Delta\Omega \cdot t + \beta \cdot \sin(\Omega_m \cdot t) - \Omega_0 \cdot t_D] + \left[\phi \cdot e^{\lambda \cdot t} \cdot (e^{-\lambda \cdot t_D} - 1) - \frac{\Delta\Omega}{2} \cdot t_D\right]\Big\}$$

We can write our second differential equation by adding to coordinates $[N\ E\ \phi]$ arbitrarily small increments of exponential form$[n\ \mathrm{E}\ \phi] \cdot e^{\lambda \cdot t}$

$$\begin{aligned}\lambda \cdot E \cdot e^{\lambda \cdot t} &= (N^{(i)} + n \cdot e^{\lambda \cdot t}) \cdot (E^{(i)} + E \cdot e^{\lambda \cdot t}) \\ &\quad + m \cdot (E^{(i)} + E \cdot e^{\lambda \cdot (t-t_D)}) \\ &\quad \times \cos\Big\{[\Delta\Omega \cdot t + \beta \cdot \sin(\Omega_m \cdot t) - \Omega_0 \cdot t_D] \\ &\quad + \left[\phi \cdot e^{\lambda \cdot t} \cdot (e^{-\lambda \cdot t_D} - 1) - \frac{\Delta\Omega}{2} \cdot t_D\right]\Big)\Big\}\end{aligned}$$

&&&

$$\begin{aligned}&\cos\Big\{\Big[[\Delta\Omega \cdot t + \beta \cdot \sin(\Omega_m \cdot t) - \Omega_0 \cdot t_D] + \left[\phi \cdot e^{\lambda \cdot t} \cdot (e^{-\lambda \cdot t_D} - 1) - \frac{\Delta\Omega}{2} \cdot t_D\right]\Big]\Big\} \\ &= \cos[\Delta\Omega \cdot t + \beta \cdot \sin(\Omega_m \cdot t) - \Omega_0 \cdot t_D] \cdot \cos\left[\phi \cdot e^{\lambda \cdot t} \cdot (e^{-\lambda \cdot t_D} - 1) - \frac{\Delta\Omega}{2} \cdot t_D\right] \\ &\quad - \sin[\Delta\Omega \cdot t + \beta \cdot \sin(\Omega_m \cdot t) - \Omega_0 \cdot t_D] \cdot \sin\left[\phi \cdot e^{\lambda \cdot t} \cdot (e^{-\lambda \cdot t_D} - 1) - \frac{\Delta\Omega}{2} \cdot t_D\right]\end{aligned}$$

$$\cos\left[\phi \cdot e^{\lambda \cdot t} \cdot (e^{-\lambda \cdot t_D} - 1) - \frac{\Delta\Omega}{2} \cdot t_D\right]$$

$$= \cos[\phi \cdot e^{\lambda \cdot t} \cdot (e^{-\lambda \cdot t_D} - 1)] \cdot \cos\left[\frac{\Delta\Omega}{2} \cdot t_D\right]$$
$$+ \sin[\phi \cdot e^{\lambda \cdot t} \cdot (e^{-\lambda \cdot t_D} - 1)] \cdot \sin\left[\frac{\Delta\Omega}{2} \cdot t_D\right]$$

$$\sin\left[\phi \cdot e^{\lambda \cdot t} \cdot (e^{-\lambda \cdot t_D} - 1) - \frac{\Delta\Omega}{2} \cdot t_D\right]$$
$$= \sin[\phi \cdot e^{\lambda \cdot t} \cdot (e^{-\lambda \cdot t_D} - 1)] \cdot \cos\left[\frac{\Delta\Omega}{2} \cdot t_D\right]$$
$$- \cos[\phi \cdot e^{\lambda \cdot t} \cdot (e^{-\lambda \cdot t_D} - 1)] \cdot \sin\left[\frac{\Delta\Omega}{2} \cdot t_D\right]$$

$$\cos\left\{[\Delta\Omega \cdot t + \beta \cdot \sin(\Omega_m \cdot t) - \Omega_0 \cdot t_D] + \left[\phi \cdot e^{\lambda \cdot t} \cdot (e^{-\lambda \cdot t_D} - 1) - \frac{\Delta\Omega}{2} \cdot t_D\right]\right\}$$
$$= \cos[\Delta\Omega \cdot t + \beta \cdot \sin(\Omega_m \cdot t) - \Omega_0 \cdot t_D] \cdot \{\cos[\phi \cdot e^{\lambda \cdot t} \cdot (e^{-\lambda \cdot t_D} - 1)]$$
$$\cdot \cos\left[\frac{\Delta\Omega}{2} \cdot t_D\right] + \sin[\phi \cdot e^{\lambda \cdot t} \cdot (e^{-\lambda \cdot t_D} - 1)] \cdot \sin\left[\frac{\Delta\Omega}{2} \cdot t_D\right]\bigg\}$$
$$- \sin[\Delta\Omega \cdot t + \beta \cdot \sin(\Omega_m \cdot t) - \Omega_0 \cdot t_D] \cdot \left\{\sin[\phi \cdot e^{\lambda \cdot t} \cdot (e^{-\lambda \cdot t_D} - 1)] \cdot \cos\left[\frac{\Delta\Omega}{2} \cdot t_D\right]\right.$$
$$\left. - \cos[\phi \cdot e^{\lambda \cdot t} \cdot (e^{-\lambda \cdot t_D} - 1)] \cdot \sin\left[\frac{\Delta\Omega}{2} \cdot t_D\right]\right\}$$

Assumption (A):

$$\xi_1(\phi, \lambda, t_D) = \phi \cdot e^{\lambda \cdot t} \cdot (e^{-\lambda \cdot t_D} - 1); \sin[\phi \cdot e^{\lambda \cdot t} \cdot (e^{-\lambda \cdot t_D} - 1)]$$
$$= \sin(\xi_1(\phi, \lambda, t_D)) = \sum_{n=0}^{\infty} \frac{(-1)^n}{(2 \cdot n + 1)!} \cdot [\xi_1(\phi, \lambda, t_D)]^{2 \cdot n + 1}$$

$$\sin(\xi_1(\phi, \lambda, t_D)) = \sum_{n=0}^{\infty} \frac{(-1)^n}{(2 \cdot n + 1)!} \cdot [\xi_1(\phi, \lambda, t_D)]^{2 \cdot n + 1}$$
$$= \xi_1(\phi, \lambda, t_D) - \frac{[\xi_1(\phi, \lambda, t_D)]^3}{3!}$$
$$+ \frac{[\xi_1(\phi, \lambda, t_D)]^5}{5!} - \frac{[\xi_1(\phi, \lambda, t_D)]^7}{7!} + \cdots$$

$$\sin(\xi_1(\phi, \lambda, t_D)) = \sum_{n=0}^{\infty} \frac{(-1)^n}{(2 \cdot n + 1)!} \cdot [\xi_1(\phi, \lambda, t_D)]^{2 \cdot n + 1}$$
$$= \phi \cdot e^{\lambda \cdot t} \cdot (e^{-\lambda \cdot t_D} - 1) - \frac{[\phi \cdot e^{\lambda \cdot t} \cdot (e^{-\lambda \cdot t_D} - 1)]^3}{3!}$$
$$+ \frac{[\phi \cdot e^{\lambda \cdot t} \cdot (e^{-\lambda \cdot t_D} - 1)]^5}{5!} - \frac{[\phi \cdot e^{\lambda \cdot t} \cdot (e^{-\lambda \cdot t_D} - 1)]^7}{7!} + \cdots$$

$$\sin(\xi_1(\phi,\lambda,t_D)) = \sum_{n=0}^{\infty}\frac{(-1)^n}{(2\cdot n+1)!}\cdot[\xi_1(\phi,\lambda,t_D)]^{2\cdot n+1}$$
$$= \phi\cdot e^{\lambda\cdot t}\cdot(e^{-\lambda\cdot t_D}-1) - \frac{\phi^3\cdot e^{3\cdot\lambda\cdot t}\cdot(e^{-\lambda\cdot t_D}-1)^3}{3!}$$
$$+\frac{\phi^5\cdot e^{5\cdot\lambda\cdot t}\cdot(e^{-\lambda\cdot t_D}-1)^5}{5!} - \frac{\phi^7\cdot e^{7\cdot\lambda\cdot t}\cdot(e^{-\lambda\cdot t_D}-1)^7}{7!}+\cdots$$

We consider $\phi^m \to \varepsilon\,\forall\, m \geq 2$ then

$$-\frac{\phi^3\cdot e^{3\cdot\lambda\cdot t}\cdot(e^{-\lambda\cdot t_D}-1)^3}{3!} + \frac{\phi^5\cdot e^{5\cdot\lambda\cdot t}\cdot(e^{-\lambda\cdot t_D}-1)^5}{5!}$$
$$-\frac{\phi^7\cdot e^{7\cdot\lambda\cdot t}\cdot(e^{-\lambda\cdot t_D}-1)^7}{7!}+\cdots\to\varepsilon$$

$$\sin(\xi_1(\phi,\lambda,t_D)) = \sum_{n=0}^{\infty}\frac{(-1)^n}{(2\cdot n+1)!}\cdot[\xi_1(\phi,\lambda,t_D)]^{2\cdot n+1} \approx \phi\cdot e^{\lambda\cdot t}\cdot(e^{-\lambda\cdot t_D}-1)$$

Assumption (B):

$$\xi_2(\Delta\Omega,t_D) = \frac{\Delta\Omega}{2}\cdot t_D;\ \sin\left[\frac{\Delta\Omega}{2}\cdot t_D\right]$$
$$= \sin[\xi_2(\Delta\Omega,t_D)] = \sum_{n=0}^{\infty}\frac{(-1)^n}{(2\cdot n+1)!}\cdot[\xi_2(\Delta\Omega,t_D)]^{2\cdot n+1}$$

$$\sin(\xi_2(\Delta\Omega,t_D)) = \sum_{n=0}^{\infty}\frac{(-1)^n}{(2\cdot n+1)!}\cdot[\xi_2(\Delta\Omega,t_D)]^{2\cdot n+1}$$
$$= \xi_2(\Delta\Omega,t_D) - \frac{[\xi_2(\Delta\Omega,t_D)]^3}{3!}$$
$$+\frac{[\xi_2(\Delta\Omega,t_D)]^5}{5!} - \frac{[\xi_2(\Delta\Omega,t_D)]^7}{7!}+\cdots$$

$$\sin(\xi_2(\Delta\Omega,t_D)) = \sum_{n=0}^{\infty}\frac{(-1)^n}{(2\cdot n+1)!}\cdot[\xi_2(\Delta\Omega,t_D)]^{2\cdot n+1}$$
$$= \frac{\Delta\Omega}{2}\cdot t_D - \frac{\left[\frac{\Delta\Omega}{2}\cdot t_D\right]^3}{3!} + \frac{\left[\frac{\Delta\Omega}{2}\cdot t_D\right]^5}{5!} - \frac{\left[\frac{\Delta\Omega}{2}\cdot t_D\right]^7}{7!}+\cdots$$
$$\sin(\xi_2(\Delta\Omega,t_D)) = \sum_{n=0}^{\infty}\frac{(-1)^n}{(2\cdot n+1)!}\cdot[\xi_2(\Delta\Omega,t_D)]^{2\cdot n+1}$$
$$= \frac{\Delta\Omega}{2}\cdot t_D - \frac{\frac{(\Delta\Omega)^3}{2^3}\cdot t_D^3}{3!} + \frac{\frac{(\Delta\Omega)^5}{2^5}\cdot t_D^5}{5!} - \frac{\frac{(\Delta\Omega)^7}{2^7}\cdot t_D^7}{7!}+\cdots$$

We consider $t_D^m \to \varepsilon \, \forall \, m \geq 2$ then $-\frac{\frac{(\Delta\Omega)^3}{2^3} \cdot t_D^3}{3!} + \frac{\frac{(\Delta\Omega)^5}{2^5} \cdot t_D^5}{5!} - \frac{\frac{(\Delta\Omega)^7}{2^7} \cdot t_D^7}{7!} + \ldots \to \varepsilon$

$$\sin(\xi_2(\Delta\Omega, t_D)) = \sum_{n=0}^{\infty} \frac{(-1)^n}{(2 \cdot n + 1)!} \cdot [\xi_2(\Delta\Omega, t_D)]^{2 \cdot n+1} \approx \frac{\Delta\Omega}{2} \cdot t_D$$

Assumption (C):

$$\xi_1(\phi, \lambda, t_D) = \phi \cdot e^{\lambda \cdot t} \cdot (e^{-\lambda \cdot t_D} - 1); \cos[\phi \cdot e^{\lambda \cdot t} \cdot (e^{-\lambda \cdot t_D} - 1)]$$
$$= \cos(\xi_1(\phi, \lambda, t_D)) = \sum_{n=0}^{\infty} \frac{(-1)^n}{(2 \cdot n)!} \cdot [\xi_1(\phi, \lambda, t_D)]^{2 \cdot n}$$
$$\cos(\xi_1(\phi, \lambda, t_D)) = \sum_{n=0}^{\infty} \frac{(-1)^n}{(2 \cdot n)!} \cdot [\xi_1(\phi, \lambda, t_D)]^{2 \cdot n}$$
$$= 1 - \frac{\phi^2 \cdot e^{2 \cdot \lambda \cdot t} \cdot (e^{-\lambda \cdot t_D} - 1)^2}{2!} + \frac{\phi^4 \cdot e^{4 \cdot \lambda \cdot t} \cdot (e^{-\lambda \cdot t_D} - 1)^4}{4!} - \ldots$$
$$\forall \phi \cdot e^{\lambda \cdot t} \cdot (e^{-\lambda \cdot t_D} - 1)$$

We consider $\phi^m \to \varepsilon \, \forall \, m \geq 2$ then $-\frac{\phi^2 \cdot e^{2 \cdot \lambda \cdot t} \cdot (e^{-\lambda \cdot t_D} - 1)^2}{2!} + \frac{\phi^4 \cdot e^{4 \cdot \lambda \cdot t} \cdot (e^{-\lambda \cdot t_D} - 1)^4}{4!} - \ldots \to \varepsilon$

$$\cos(\xi_1(\phi, \lambda, t_D)) = \sum_{n=0}^{\infty} \frac{(-1)^n}{(2 \cdot n)!} \cdot [\xi_1(\phi, \lambda, t_D)]^{2 \cdot n} \approx 1$$

Assumption (D):

$$\xi_2(\Delta\Omega, t_D) = \frac{\Delta\Omega}{2} \cdot t_D; \cos\left[\frac{\Delta\Omega}{2} \cdot t_D\right]$$
$$= \cos(\xi_2(\Delta\Omega, t_D)) = \sum_{n=0}^{\infty} \frac{(-1)^n}{(2 \cdot n)!} \cdot [\xi_2(\Delta\Omega, t_D)]^{2 \cdot n}$$
$$\cos(\xi_2(\Delta\Omega, t_D)) = \sum_{n=0}^{\infty} \frac{(-1)^n}{(2 \cdot n)!} \cdot \left[\frac{\Delta\Omega}{2} \cdot t_D\right]^{2 \cdot n}$$
$$= 1 - \frac{\frac{(\Delta\Omega)^2}{2^2} \cdot t_D^2}{2!} + \frac{\frac{(\Delta\Omega)^4}{2^4} \cdot t_D^4}{4!} - \ldots \; \forall \frac{\Delta\Omega}{2} \cdot t_D$$

We consider $t_D^m \to \varepsilon \, \forall \, m \geq 2$ then $-\frac{\frac{(\Delta\Omega)^2}{2^2} \cdot t_D^2}{2!} + \frac{\frac{(\Delta\Omega)^4}{2^4} \cdot t_D^4}{4!} - \ldots \to \varepsilon$

Table. 4.2 lasers with frequency-modulated optical feedback system assumptions A–D

Assumption A	$\sin(\xi_1(\phi,\lambda,t_D))=\sum_{n=0}^{\infty}\frac{(-1)^n}{(2\cdot n+1)!}\cdot[\xi_1(\phi,\lambda,t_D)]^{2\cdot n+1}\approx\phi\cdot e^{\lambda\cdot t}\cdot(e^{-\lambda\cdot t_D}-1)$
Assumption B	$\sin(\xi_2(\Delta\Omega,t_D))=\sum_{n=0}^{\infty}\frac{(-1)^n}{(2\cdot n+1)!}\cdot[\xi_2(\Delta\Omega,t_D)]^{2\cdot n+1}\approx\frac{\Delta\Omega}{2}\cdot t_D$
Assumption C	$\cos(\xi_1(\phi,\lambda,t_D))=\sum_{n=0}^{\infty}\frac{(-1)^n}{(2\cdot n)!}\cdot[\xi_1(\phi,\lambda,t_D)]^{2\cdot n}\approx 1$
Assumption D	$\cos(\xi_2(\Delta\Omega,t_D))=\sum_{n=0}^{\infty}\frac{(-1)^n}{(2\cdot n)!}\cdot\left[\frac{\Delta\Omega}{2}\cdot t_D\right]^{2\cdot n}\approx 1$

$$\cos(\xi_2(\Delta\Omega,t_D))=\sum_{n=0}^{\infty}\frac{(-1)^n}{(2\cdot n)!}\cdot\left[\frac{\Delta\Omega}{2}\cdot t_D\right]^{2\cdot n}\approx 1$$

We can summary assumptions A–D in the next table (Table 4.2):
We implement the above assumptions (A–D) in the below expression.

$$\begin{aligned}&\cos\{[\Delta\Omega\cdot t+\beta\cdot\sin(\Omega_m\cdot t)-\Omega_0\cdot t_D]\\&+\left[\phi\cdot e^{\lambda\cdot t}\cdot(e^{-\lambda\cdot t_D}-1)-\frac{\Delta\Omega}{2}\cdot t_D\right])\}\\&=\cos[\Delta\Omega\cdot t+\beta\cdot\sin(\Omega_m\cdot t)-\Omega_0\cdot t_D]\\&\cdot\{\cos[\phi\cdot e^{\lambda\cdot t}\cdot(e^{-\lambda\cdot t_D}-1)]\cdot\cos\left[\frac{\Delta\Omega}{2}\cdot t_D\right]\\&+\sin[\phi\cdot e^{\lambda\cdot t}\cdot(e^{-\lambda\cdot t_D}-1)]\cdot\sin\left[\frac{\Delta\Omega}{2}\cdot t_D\right]\}\\&-\sin[\Delta\Omega\cdot t+\beta\cdot\sin(\Omega_m\cdot t)-\Omega_0\cdot t_D]\\&\cdot\{\sin[\phi\cdot e^{\lambda\cdot t}\cdot(e^{-\lambda\cdot t_D}-1)]\cdot\cos\left[\frac{\Delta\Omega}{2}\cdot t_D\right]\\&-\cos[\phi\cdot e^{\lambda\cdot t}\cdot(e^{-\lambda\cdot t_D}-1)]\cdot\sin\left[\frac{\Delta\Omega}{2}\cdot t_D\right]\}\end{aligned}$$

$$\begin{aligned}&\cos\{[\Delta\Omega\cdot t+\beta\cdot\sin(\Omega_m\cdot t)-\Omega_0\cdot t_D]\\&+\left[\phi\cdot e^{\lambda\cdot t}\cdot(e^{-\lambda\cdot t_D}-1)-\frac{\Delta\Omega}{2}\cdot t_D\right])\}\\&=\cos[\Delta\Omega\cdot t+\beta\cdot\sin(\Omega_m\cdot t)-\Omega_0\cdot t_D]\\&\cdot\left[1+\phi\cdot e^{\lambda\cdot t}\cdot(e^{-\lambda\cdot t_D}-1)\cdot\frac{\Delta\Omega}{2}\cdot t_D\right]\\&-\sin[\Delta\Omega\cdot t+\beta\cdot\sin(\Omega_m\cdot t)-\Omega_0\cdot t_D]\\&\cdot\left[\phi\cdot e^{\lambda\cdot t}\cdot(e^{-\lambda\cdot t_D}-1)-\frac{\Delta\Omega}{2}\cdot t_D\right]\end{aligned}$$

We implement our last expression in the following formula

$$\lambda \cdot E \cdot e^{\lambda \cdot t} = (N^{(i)} + n \cdot e^{\lambda \cdot t}) \cdot (E^{(i)} + E \cdot e^{\lambda \cdot t})$$
$$+ m \cdot (E^{(i)} + E \cdot e^{\lambda \cdot (t - t_D)})$$
$$\times \cos\{[\Delta\Omega \cdot t + \beta \cdot \sin(\Omega_m \cdot t) - \Omega_0 \cdot t_D]$$
$$+ \left[\phi \cdot e^{\lambda \cdot t} \cdot (e^{-\lambda \cdot t_D} - 1) - \frac{\Delta\Omega}{2} \cdot t_D\right]\Big)\Big\}$$

$$\lambda \cdot E \cdot e^{\lambda \cdot t} = (N^{(i)} + n \cdot e^{\lambda \cdot t}) \cdot (E^{(i)} + E \cdot e^{\lambda \cdot t}) + m \cdot (E^{(i)} + E \cdot e^{\lambda \cdot (t - t_D)})$$
$$\times \{\cos[\Delta\Omega \cdot t + \beta \cdot \sin(\Omega_m \cdot t) - \Omega_0 \cdot t_D]$$
$$\cdot \left[1 + \phi \cdot e^{\lambda \cdot t} \cdot (e^{-\lambda \cdot t_D} - 1) \cdot \frac{\Delta\Omega}{2} \cdot t_D\right]$$
$$- \sin[\Delta\Omega \cdot t + \beta \cdot \sin(\Omega_m \cdot t) - \Omega_0 \cdot t_D]$$
$$\cdot \left[\phi \cdot e^{\lambda \cdot t} \cdot (e^{-\lambda \cdot t_D} - 1) - \frac{\Delta\Omega}{2} \cdot t_D\right]\Big\}$$

$$\lambda \cdot E \cdot e^{\lambda \cdot t} = N^{(i)} \cdot E^{(i)} + N^{(i)} \cdot E \cdot e^{\lambda \cdot t} + E^{(i)} \cdot n \cdot e^{\lambda \cdot t}$$
$$+ n \cdot E \cdot e^{2 \cdot \lambda \cdot t} + [m \cdot E^{(i)} + m \cdot E \cdot e^{\lambda \cdot (t - t_D)}]$$
$$\times \{\cos[\Delta\Omega \cdot t + \beta \cdot \sin(\Omega_m \cdot t) - \Omega_0 \cdot t_D]$$
$$\cdot \left[1 + \phi \cdot e^{\lambda \cdot t} \cdot (e^{-\lambda \cdot t_D} - 1) \cdot \frac{\Delta\Omega}{2} \cdot t_D\right]$$
$$- \sin[\Delta\Omega \cdot t + \beta \cdot \sin(\Omega_m \cdot t) - \Omega_0 \cdot t_D]$$
$$\cdot \left[\phi \cdot e^{\lambda \cdot t} \cdot (e^{-\lambda \cdot t_D} - 1) - \frac{\Delta\Omega}{2} \cdot t_D\right]\Big\}$$

Assumption: $n \cdot E \approx 0$

$$\lambda \cdot E \cdot e^{\lambda \cdot t} = N^{(i)} \cdot E^{(i)} + N^{(i)} \cdot E \cdot e^{\lambda \cdot t} + E^{(i)} \cdot n \cdot e^{\lambda \cdot t}$$
$$+ [m \cdot E^{(i)} + m \cdot E \cdot e^{\lambda \cdot (t - t_D)}]$$
$$\times \{\cos[\Delta\Omega \cdot t + \beta \cdot \sin(\Omega_m \cdot t) - \Omega_0 \cdot t_D]$$
$$\cdot \left[1 + \phi \cdot e^{\lambda \cdot t} \cdot (e^{-\lambda \cdot t_D} - 1) \cdot \frac{\Delta\Omega}{2} \cdot t_D\right]$$
$$- \sin[\Delta\Omega \cdot t + \beta \cdot \sin(\Omega_m \cdot t) - \Omega_0 \cdot t_D]$$
$$\cdot \left[\phi \cdot e^{\lambda \cdot t} \cdot (e^{-\lambda \cdot t_D} - 1) - \frac{\Delta\Omega}{2} \cdot t_D\right]\Big\}$$

$$\lambda \cdot E \cdot e^{\lambda \cdot t} = N^{(i)} \cdot E^{(i)} + N^{(i)} \cdot E \cdot e^{\lambda \cdot t} + E^{(i)} \cdot n \cdot e^{\lambda \cdot t} + [m \cdot E^{(i)} + m \cdot E \cdot e^{\lambda \cdot (t - t_D)}]$$
$$\times \{\cos[\Delta\Omega \cdot t + \beta \cdot \sin(\Omega_m \cdot t) - \Omega_0 \cdot t_D]$$
$$+ \phi \cdot e^{\lambda \cdot t} \cdot (e^{-\lambda \cdot t_D} - 1) \cdot \frac{\Delta\Omega}{2} \cdot t_D \cdot \cos[\Delta\Omega \cdot t + \beta \cdot \sin(\Omega_m \cdot t) - \Omega_0 \cdot t_D]$$
$$- \sin[\Delta\Omega \cdot t + \beta \cdot \sin(\Omega_m \cdot t) - \Omega_0 \cdot t_D] \cdot \left[\phi \cdot e^{\lambda \cdot t} \cdot (e^{-\lambda \cdot t_D} - 1) - \frac{\Delta\Omega}{2} \cdot t_D\right]\Big\}$$

$$\lambda \cdot E \cdot e^{\lambda \cdot t} = N^{(i)} \cdot E^{(i)} + N^{(i)} \cdot E \cdot e^{\lambda \cdot t} + E^{(i)} \cdot n \cdot e^{\lambda \cdot t} + m \cdot E^{(i)}$$
$$\times \{\cos[\Delta\Omega \cdot t + \beta \cdot \sin(\Omega_m \cdot t) - \Omega_0 \cdot t_D]$$
$$+ \phi \cdot e^{\lambda \cdot t} \cdot (e^{-\lambda \cdot t_D} - 1) \cdot \frac{\Delta\Omega}{2} \cdot t_D \cdot \cos[\Delta\Omega \cdot t + \beta \cdot \sin(\Omega_m \cdot t) - \Omega_0 \cdot t_D]$$
$$- \sin[\Delta\Omega \cdot t + \beta \cdot \sin(\Omega_m \cdot t) - \Omega_0 \cdot t_D] \cdot [\phi \cdot e^{\lambda \cdot t} \cdot (e^{-\lambda \cdot t_D} - 1)$$
$$\left. - \frac{\Delta\Omega}{2} \cdot t_D] \right\} + m \cdot E \cdot e^{\lambda \cdot (t - t_D)}$$
$$\times \{\cos[\Delta\Omega \cdot t + \beta \cdot \sin(\Omega_m \cdot t) - \Omega_0 \cdot t_D]$$
$$+ \phi \cdot e^{\lambda \cdot t} \cdot (e^{-\lambda \cdot t_D} - 1) \cdot \frac{\Delta\Omega}{2} \cdot t_D \cdot \cos[\Delta\Omega \cdot t + \beta \cdot \sin(\Omega_m \cdot t) - \Omega_0 \cdot t_D]$$
$$\left. - \sin[\Delta\Omega \cdot t + \beta \cdot \sin(\Omega_m \cdot t) - \Omega_0 \cdot t_D] \cdot \left[\phi \cdot e^{\lambda \cdot t} \cdot (e^{-\lambda \cdot t_D} - 1) - \frac{\Delta\Omega}{2} \cdot t_D\right]\right\}$$

$$\lambda \cdot E \cdot e^{\lambda \cdot t} = N^{(i)} \cdot E^{(i)} + N^{(i)} \cdot E \cdot e^{\lambda \cdot t} + E^{(i)} \cdot n \cdot e^{\lambda \cdot t}$$
$$+ m \cdot E^{(i)} \times \cos[\Delta\Omega \cdot t + \beta \cdot \sin(\Omega_m \cdot t) - \Omega_0 \cdot t_D]$$
$$+ m \cdot E^{(i)} \times \phi \cdot e^{\lambda \cdot t} \cdot (e^{-\lambda \cdot t_D} - 1) \cdot \frac{\Delta\Omega}{2} \cdot t_D \cdot \cos[\Delta\Omega \cdot t$$
$$+ \beta \cdot \sin(\Omega_m \cdot t) - \Omega_0 \cdot t_D]$$
$$- m \cdot E^{(i)} \times \sin[\Delta\Omega \cdot t + \beta \cdot \sin(\Omega_m \cdot t)$$
$$- \Omega_0 \cdot t_D] \cdot \left[\phi \cdot e^{\lambda \cdot t} \cdot (e^{-\lambda \cdot t_D} - 1) - \frac{\Delta\Omega}{2} \cdot t_D\right] + m \cdot E \cdot e^{\lambda \cdot (t - t_D)}$$
$$\times \{\cos[\Delta\Omega \cdot t + \beta \cdot \sin(\Omega_m \cdot t) - \Omega_0 \cdot t_D]$$
$$+ \phi \cdot e^{\lambda \cdot t} \cdot (e^{-\lambda \cdot t_D} - 1) \cdot \frac{\Delta\Omega}{2} \cdot t_D \cdot \cos[\Delta\Omega \cdot t$$
$$+ \beta \cdot \sin(\Omega_m \cdot t) - \Omega_0 \cdot t_D]$$
$$- \sin[\Delta\Omega \cdot t + \beta \cdot \sin(\Omega_m \cdot t)$$
$$\left. - \Omega_0 \cdot t_D] \cdot \left[\phi \cdot e^{\lambda \cdot t} \cdot (e^{-\lambda \cdot t_D} - 1) - \frac{\Delta\Omega}{2} \cdot t_D\right]\right\}$$

At fixed points: $N^{(i)} \cdot E^{(i)} + m \cdot E^{(i)} \times \cos[\Delta\Omega \cdot t + \beta \cdot \sin(\Omega_m \cdot t) - \Omega_0 \cdot t_D] = 0$.

Remark $\psi^* \approx \Delta\Omega \cdot t + \beta \cdot \sin(\Omega_m \cdot t) - \Omega_0 \cdot t_D$; $\frac{d\phi(t)}{dt} = 0 \Rightarrow \sin\psi^* = 0$; $\psi^* = n \cdot \pi \; \forall n = 0, 1, 2, \ldots$

$$\frac{dE}{dt} = 0 \Rightarrow N^* \cdot E^* + m \cdot E^* \times \cos(\psi^*)$$
$$= 0; n \in \mathbb{N}_0; n \cdot \pi = \Delta\Omega \cdot t + \beta \cdot \sin(\Omega_m \cdot t) - \Omega_0 \cdot t_D$$

$$\lambda \cdot E \cdot e^{\lambda \cdot t} = N^{(i)} \cdot E \cdot e^{\lambda \cdot t} + E^{(i)} \cdot n \cdot e^{\lambda \cdot t} + m \cdot E^{(i)} \times \phi \cdot e^{\lambda \cdot t} \cdot (e^{-\lambda \cdot t_D} - 1)$$
$$\cdot \frac{\Delta\Omega}{2} \cdot t_D \cdot \cos[\Delta\Omega \cdot t + \beta \cdot \sin(\Omega_m \cdot t) - \Omega_0 \cdot t_D]$$
$$- m \cdot E^{(i)} \times \sin[\Delta\Omega \cdot t + \beta \cdot \sin(\Omega_m \cdot t) - \Omega_0 \cdot t_D]$$

$$\cdot\left[\phi\cdot e^{\lambda\cdot t}\cdot(e^{-\lambda\cdot t_D}-1)-\frac{\Delta\Omega}{2}\cdot t_D\right]+m\cdot E\cdot e^{\lambda\cdot(t-t_D)}$$
$$\times\left\{\cos[\Delta\Omega\cdot t+\beta\cdot\sin(\Omega_m\cdot t)-\Omega_0\cdot t_D]+\phi\cdot e^{\lambda\cdot t}\cdot(e^{-\lambda\cdot t_D}-1)\right.$$
$$\cdot\frac{\Delta\Omega}{2}\cdot t_D\cdot\cos[\Delta\Omega\cdot t+\beta\cdot\sin(\Omega_m\cdot t)-\Omega_0\cdot t_D]$$
$$\left.-\sin[\Delta\Omega\cdot t+\beta\cdot\sin(\Omega_m\cdot t)-\Omega_0\cdot t_D]\cdot\left[\phi\cdot e^{\lambda\cdot t}\cdot(e^{-\lambda\cdot t_D}-1)-\frac{\Delta\Omega}{2}\cdot t_D\right]\right\}$$

$$n\cdot\pi=\Delta\Omega\cdot t+\beta\cdot\sin(\Omega_m\cdot t)-\Omega_0\cdot t_D\Rightarrow\sin[\Delta\Omega\cdot t+\beta\cdot\sin(\Omega_m\cdot t)-\Omega_0\cdot t_D]=\sin[n\cdot\pi]=0$$
$$n\cdot\pi=\Delta\Omega\cdot t+\beta\cdot\sin(\Omega_m\cdot t)-\Omega_0\cdot t_D\Rightarrow\cos[\Delta\Omega\cdot t+\beta\cdot\sin(\Omega_m\cdot t)-\Omega_0\cdot t_D]=\cos[n\cdot\pi]=(-1)^n=0\,\forall n=0,1,2,\ldots$$

Then we can write our second differential equation by adding to coordinates $[N\ E\ \phi]$ arbitrarily small increments of exponential form $[n\ \mathrm{E}\ \phi]\cdot e^{\lambda\cdot t}$ as follow.

$$\lambda\cdot E\cdot e^{\lambda\cdot t}=N^{(i)}\cdot E\cdot e^{\lambda\cdot t}+E^{(i)}\cdot n\cdot e^{\lambda\cdot t}+m\cdot E^{(i)}\times\phi\cdot e^{\lambda\cdot t}\cdot(e^{-\lambda\cdot t_D}-1)\cdot\frac{\Delta\Omega}{2}\cdot t_D\cdot(-1)^n+m\cdot E\cdot e^{\lambda\cdot(t-t_D)}\times\{(-1)^n+\phi\cdot e^{\lambda\cdot t}\cdot(e^{-\lambda\cdot t_D}-1)\cdot\frac{\Delta\Omega}{2}\cdot t_D\cdot(-1)^n\}\,\forall n=0,1,2,\ldots$$

$$\lambda\cdot E\cdot e^{\lambda\cdot t}=N^{(i)}\cdot E\cdot e^{\lambda\cdot t}+E^{(i)}\cdot n\cdot e^{\lambda\cdot t}+m\cdot E^{(i)}\times\phi\cdot e^{\lambda\cdot t}\cdot(e^{-\lambda\cdot t_D}-1)\cdot\frac{\Delta\Omega}{2}\cdot t_D\cdot(-1)^n+m\cdot E\cdot e^{\lambda\cdot(t-t_D)}\times\{1+\phi\cdot e^{\lambda\cdot t}\cdot(e^{-\lambda\cdot t_D}-1)\cdot\frac{\Delta\Omega}{2}\cdot t_D\}\cdot(-1)^n\,\forall n=0,1,2,\ldots$$

$$\lambda\cdot E\cdot e^{\lambda\cdot t}=N^{(i)}\cdot E\cdot e^{\lambda\cdot t}+E^{(i)}\cdot n\cdot e^{\lambda\cdot t}+m\cdot E^{(i)}\cdot\phi\cdot e^{\lambda\cdot t}\cdot(e^{-\lambda\cdot t_D}-1)\cdot\frac{\Delta\Omega}{2}\cdot t_D\cdot(-1)^n+\left\{m\cdot E\cdot e^{\lambda\cdot(t-t_D)}+m\cdot E\cdot\phi\cdot e^{\lambda\cdot(t-t_D)}\cdot e^{\lambda\cdot t}\cdot(e^{-\lambda\cdot t_D}-1)\cdot\frac{\Delta\Omega}{2}\cdot t_D\right\}\cdot(-1)^n\,\forall n=0,1,2,\ldots$$

Assumption: $E\cdot\phi\approx 0$

$$\lambda\cdot E\cdot e^{\lambda\cdot t}=N^{(i)}\cdot E\cdot e^{\lambda\cdot t}+E^{(i)}\cdot n\cdot e^{\lambda\cdot t}$$

$$+m\cdot E^{(i)}\cdot\phi\cdot e^{\lambda\cdot t}\cdot(e^{-\lambda\cdot t_D}-1)\cdot\frac{\Delta\Omega}{2}\cdot t_D\cdot(-1)^n$$
$$+m\cdot E\cdot e^{\lambda\cdot(t-t_D)}\cdot(-1)^n\,\forall n=0,1,2,\ldots$$

$$-\lambda\cdot E\cdot e^{\lambda\cdot t}+N^{(i)}\cdot E\cdot e^{\lambda\cdot t}+E^{(i)}\cdot n\cdot e^{\lambda\cdot t}$$
$$+m\cdot E^{(i)}\cdot\phi\cdot e^{\lambda\cdot t}\cdot(e^{-\lambda\cdot t_D}-1)\cdot\frac{\Delta\Omega}{2}\cdot t_D\cdot(-1)^n$$
$$+m\cdot E\cdot e^{\lambda\cdot t}\cdot e^{-\lambda\cdot t_D}\cdot(-1)^n=0\,\forall n=0,1,2,\ldots$$

$$E^{(i)}\cdot n\cdot e^{\lambda\cdot t}-\lambda\cdot E\cdot e^{\lambda\cdot t}+N^{(i)}\cdot E\cdot e^{\lambda\cdot t}+m\cdot E\cdot e^{\lambda\cdot t}\cdot e^{-\lambda\cdot t_D}\cdot(-1)^n$$
$$+m\cdot E^{(i)}\cdot\phi\cdot e^{\lambda\cdot t}\cdot(e^{-\lambda\cdot t_D}-1)\cdot\frac{\Delta\Omega}{2}\cdot t_D$$
$$\cdot(-1)^n=0\,\forall n=0,1,2,\ldots$$

Divide the two side of the above equation by $e^{\lambda\cdot t}$ term, we get the expression:

$$E^{(i)}\cdot n+[N^{(i)}+m\cdot e^{-\lambda\cdot t_D}\cdot(-1)^n-\lambda]\cdot E$$
$$+m\cdot E^{(i)}\cdot(e^{-\lambda\cdot t_D}-1)\cdot\frac{\Delta\Omega}{2}\cdot t_D\cdot(-1)^n\cdot\phi=0\,\forall n=0,1,2,\ldots$$

We can write our Third differential equation by adding to coordinates $[N\ E\ \phi]$ arbitrarily small increments of exponential form $[n\ \mathrm{E}\ \phi]\cdot e^{\lambda\cdot t}$

$$\frac{d\phi(t)}{dt}=\left[\frac{E(t-t_D)}{E(t)}\right]\cdot\sin\psi(t);\ \psi(t)=\Delta\Omega\cdot t$$
$$+\beta\cdot\sin(\Omega_m\cdot t)-\phi(t)+\phi(t-t_D)-\left(\Omega_0+\frac{\Delta\Omega}{2}\right)\cdot t_D$$

$$\lambda\cdot\phi\cdot e^{\lambda\cdot t}=\left[\frac{E^{(i)}+E\cdot e^{\lambda\cdot(t-t_D)}}{E^{(i)}+E\cdot e^{\lambda\cdot t}}\right]\cdot\sin[\Delta\Omega\cdot t+\beta\cdot\sin(\Omega_m\cdot t)$$
$$-\phi^{(i)}-\phi\cdot e^{\lambda\cdot t}+\phi^{(i)}+\phi\cdot e^{\lambda\cdot(t-t_D)}-\left(\Omega_0+\frac{\Delta\Omega}{2}\right)\cdot t_D]$$

$$\lambda\cdot\phi\cdot e^{\lambda\cdot t}=\left[\frac{E^{(i)}+E\cdot e^{\lambda\cdot(t-t_D)}}{E^{(i)}+E\cdot e^{\lambda\cdot t}}\right]\cdot\sin[\Delta\Omega\cdot t$$
$$+\beta\cdot\sin(\Omega_m\cdot t)-\phi\cdot e^{\lambda\cdot t}+\phi\cdot e^{\lambda\cdot(t-t_D)}-\left(\Omega_0+\frac{\Delta\Omega}{2}\right)\cdot t_D]$$

$$\lambda\cdot\phi\cdot e^{\lambda\cdot t}=\left[\frac{E^{(i)}+E\cdot e^{\lambda\cdot(t-t_D)}}{E^{(i)}+E\cdot e^{\lambda\cdot t}}\right]\cdot\left[\frac{E^{(i)}-E\cdot e^{\lambda\cdot t}}{E^{(i)}-E\cdot e^{\lambda\cdot t}}\right]$$
$$\cdot\sin\left[\Delta\Omega\cdot t+\beta\cdot\sin(\Omega_m\cdot t)-\phi\cdot e^{\lambda\cdot t}+\phi\cdot e^{\lambda\cdot(t-t_D)}-\left(\Omega_0+\frac{\Delta\Omega}{2}\right)\cdot t_D\right]$$

$$\lambda\cdot\phi\cdot e^{\lambda\cdot t}=\left[\frac{E^{(i)}+E\cdot e^{\lambda\cdot(t-t_D)}}{(E^{(i)})^2-E^2\cdot e^{2\cdot\lambda\cdot t}}\right]\cdot[E^{(i)}-E\cdot e^{\lambda\cdot t}]$$

$$\cdot \sin\left[\Delta\Omega \cdot t + \beta \cdot \sin(\Omega_m \cdot t) - \phi \cdot e^{\lambda \cdot t} + \phi \cdot e^{\lambda \cdot (t - t_D)} - \left(\Omega_0 + \frac{\Delta\Omega}{2}\right) \cdot t_D\right]$$

$$\lambda \cdot \phi \cdot e^{\lambda \cdot t} = \left[\frac{(E^{(i)})^2 - E^{(i)} \cdot E \cdot e^{\lambda \cdot t} + E^{(i)} \cdot E \cdot e^{\lambda \cdot (t - t_D)} - E^2 \cdot e^{\lambda \cdot (t - t_D)} \cdot e^{\lambda \cdot t}}{(E^{(i)})^2 - E^2 \cdot e^{2 \cdot \lambda \cdot t}}\right]$$
$$\cdot \sin[\Delta\Omega \cdot t + \beta \cdot \sin(\Omega_m \cdot t)$$
$$-\phi \cdot e^{\lambda \cdot t} + \phi \cdot e^{\lambda \cdot (t - t_D)} - \left(\Omega_0 + \frac{\Delta\Omega}{2}\right) \cdot t_D\Big]$$

Assumption: $E^2 \approx 0$

$$\lambda \cdot \phi \cdot e^{\lambda \cdot t} = \left[\frac{(E^{(i)})^2 - E^{(i)} \cdot E \cdot e^{\lambda \cdot t} + E^{(i)} \cdot E \cdot e^{\lambda \cdot (t - t_D)}}{(E^{(i)})^2}\right] \cdot \sin[\Delta\Omega \cdot t$$
$$+\beta \cdot \sin(\Omega_m \cdot t) - \phi \cdot e^{\lambda \cdot t} + \phi \cdot e^{\lambda \cdot (t - t_D)} - \left(\Omega_0 + \frac{\Delta\Omega}{2}\right) \cdot t_D\Big]$$

$$\lambda \cdot \phi \cdot e^{\lambda \cdot t} = \left[\frac{(E^{(i)})^2 - E^{(i)} \cdot E \cdot e^{\lambda \cdot t} + E^{(i)} \cdot E \cdot e^{\lambda \cdot (t - t_D)}}{(E^{(i)})^2}\right]$$
$$\cdot \sin\left[\Delta\Omega \cdot t + \beta \cdot \sin(\Omega_m \cdot t) + \phi \cdot e^{\lambda \cdot t} \cdot (e^{-\lambda \cdot t_D} - 1) - \Omega_0 \cdot t_D - \frac{\Delta\Omega}{2} \cdot t_D\right]$$

$$\lambda \cdot \phi \cdot e^{\lambda \cdot t} = \left(1 - \frac{E \cdot e^{\lambda \cdot t}}{E^{(i)}} + \frac{E \cdot e^{\lambda \cdot (t - t_D)}}{E^{(i)}}\right) \cdot \sin[\Delta\Omega \cdot t + \beta \cdot \sin(\Omega_m \cdot t)$$
$$+\phi \cdot e^{\lambda \cdot t} \cdot (e^{-\lambda \cdot t_D} - 1) - \Omega_0 \cdot t_D - \frac{\Delta\Omega}{2} \cdot t_D\Big]$$

$$\lambda \cdot \phi \cdot e^{\lambda \cdot t} = \left(1 - \frac{E \cdot e^{\lambda \cdot t}}{E^{(i)}} + \frac{E \cdot e^{\lambda \cdot (t - t_D)}}{E^{(i)}}\right) \cdot \sin[(\Delta\Omega \cdot t$$
$$+\beta \cdot \sin(\Omega_m \cdot t) - \Omega_0 \cdot t_D) + \left(\phi \cdot e^{\lambda \cdot t} \cdot (e^{-\lambda \cdot t_D} - 1) - \frac{\Delta\Omega}{2} \cdot t_D\right)\Big]$$

&&&

$$\sin[(\Delta\Omega \cdot t + \beta \cdot \sin(\Omega_m \cdot t) - \Omega_0 \cdot t_D)$$
$$+\left(\phi \cdot e^{\lambda \cdot t} \cdot (e^{-\lambda \cdot t_D} - 1) - \frac{\Delta\Omega}{2} \cdot t_D\right)\Big]$$
$$= \sin(\Delta\Omega \cdot t + \beta \cdot \sin(\Omega_m \cdot t) - \Omega_0 \cdot t_D)$$
$$\cdot \cos\left(\phi \cdot e^{\lambda \cdot t} \cdot (e^{-\lambda \cdot t_D} - 1) - \frac{\Delta\Omega}{2} \cdot t_D\right)$$
$$+ \cos(\Delta\Omega \cdot t + \beta \cdot \sin(\Omega_m \cdot t) - \Omega_0 \cdot t_D)$$
$$\cdot \sin\left(\phi \cdot e^{\lambda \cdot t} \cdot (e^{-\lambda \cdot t_D} - 1) - \frac{\Delta\Omega}{2} \cdot t_D\right)$$

&&&

Table. 4.3 lasers with frequency-modulated optical feedback system assumptions A–D

Assumption A	$\xi_1(\phi, \lambda, t_D) = \phi \cdot e^{\lambda \cdot t} \cdot (e^{-\lambda \cdot t_D} - 1)$
	$\sin(\xi_1(\phi, \lambda, t_D)) = \sum_{n=0}^{\infty} \frac{(-1)^n}{(2 \cdot n+1)!} \cdot [\xi_1(\phi, \lambda, t_D)]^{2 \cdot n+1} \approx \phi \cdot e^{\lambda \cdot t} \cdot (e^{-\lambda \cdot t_D} - 1)$
Assumption B	$\xi_2(\Delta\Omega, t_D) = \frac{\Delta\Omega}{2} \cdot t_D$
	$\sin(\xi_2(\Delta\Omega, t_D)) = \sum_{n=0}^{\infty} \frac{(-1)^n}{(2 \cdot n+1)!} \cdot [\xi_2(\Delta\Omega, t_D)]^{2 \cdot n+1} \approx \frac{\Delta\Omega}{2} \cdot t_D$
Assumption C	$\xi_1(\phi, \lambda, t_D) = \phi \cdot e^{\lambda \cdot t} \cdot (e^{-\lambda \cdot t_D} - 1)$
	$\cos(\xi_1(\phi, \lambda, t_D)) = \sum_{n=0}^{\infty} \frac{(-1)^n}{(2 \cdot n)!} \cdot [\xi_1(\phi, \lambda, t_D)]^{2 \cdot n} \approx 1$
Assumption D	$\xi_2(\Delta\Omega, t_D) = \frac{\Delta\Omega}{2} \cdot t_D$
	$\cos(\xi_2(\Delta\Omega, t_D)) = \sum_{n=0}^{\infty} \frac{(-1)^n}{(2 \cdot n)!} \cdot \left[\frac{\Delta\Omega}{2} \cdot t_D\right]^{2 \cdot n} \approx 1$

$$\cos\left[\phi \cdot e^{\lambda \cdot t} \cdot (e^{-\lambda \cdot t_D} - 1) - \frac{\Delta\Omega}{2} \cdot t_D\right]$$
$$= \cos[\phi \cdot e^{\lambda \cdot t} \cdot (e^{-\lambda \cdot t_D} - 1)] \cdot \cos\left[\frac{\Delta\Omega}{2} \cdot t_D\right]$$
$$+ \sin[\phi \cdot e^{\lambda \cdot t} \cdot (e^{-\lambda \cdot t_D} - 1)] \cdot \sin\left[\frac{\Delta\Omega}{2} \cdot t_D\right]$$

$$\sin\left[\phi \cdot e^{\lambda \cdot t} \cdot (e^{-\lambda \cdot t_D} - 1) - \frac{\Delta\Omega}{2} \cdot t_D\right]$$
$$= \sin[\phi \cdot e^{\lambda \cdot t} \cdot (e^{-\lambda \cdot t_D} - 1)] \cdot \cos\left[\frac{\Delta\Omega}{2} \cdot t_D\right]$$
$$- \cos[\phi \cdot e^{\lambda \cdot t} \cdot (e^{-\lambda \cdot t_D} - 1)] \cdot \sin\left[\frac{\Delta\Omega}{2} \cdot t_D\right]$$

We already got the following results by assumptions (Table 4.3).
We implement the above assumptions (A–D) in the below expression.

$$\cos\left[\phi \cdot e^{\lambda \cdot t} \cdot (e^{-\lambda \cdot t_D} - 1) - \frac{\Delta\Omega}{2} \cdot t_D\right] = 1 + \phi \cdot e^{\lambda \cdot t} \cdot (e^{-\lambda \cdot t_D} - 1) \cdot \frac{\Delta\Omega}{2} \cdot t_D$$
$$\sin\left[\phi \cdot e^{\lambda \cdot t} \cdot (e^{-\lambda \cdot t_D} - 1) - \frac{\Delta\Omega}{2} \cdot t_D\right] = \phi \cdot e^{\lambda \cdot t} \cdot (e^{-\lambda \cdot t_D} - 1) - \frac{\Delta\Omega}{2} \cdot t_D$$
$$\sin\left[(\Delta\Omega \cdot t + \beta \cdot \sin(\Omega_m \cdot t) - \Omega_0 \cdot t_D) + \left(\phi \cdot e^{\lambda \cdot t} \cdot (e^{-\lambda \cdot t_D} - 1) - \frac{\Delta\Omega}{2} \cdot t_D\right)\right]$$
$$= \sin(\Delta\Omega \cdot t + \beta \cdot \sin(\Omega_m \cdot t) - \Omega_0 \cdot t_D) \cdot \left[1 + \phi \cdot e^{\lambda \cdot t} \cdot (e^{-\lambda \cdot t_D} - 1) \cdot \frac{\Delta\Omega}{2} \cdot t_D\right]$$
$$+ \cos(\Delta\Omega \cdot t + \beta \cdot \sin(\Omega_m \cdot t) - \Omega_0 \cdot t_D) \cdot \left[\phi \cdot e^{\lambda \cdot t} \cdot (e^{-\lambda \cdot t_D} - 1) - \frac{\Delta\Omega}{2} \cdot t_D\right]$$

Implement it in the third differential equation arbitrarily small increments, we get.

$$\lambda \cdot \phi \cdot e^{\lambda \cdot t} = \left(1 - \frac{E \cdot e^{\lambda \cdot t}}{E^{(i)}} + \frac{E \cdot e^{\lambda \cdot (t - t_D)}}{E^{(i)}}\right)$$
$$\cdot \sin[(\Delta\Omega \cdot t + \beta \cdot \sin(\Omega_m \cdot t) - \Omega_0 \cdot t_D)$$
$$+ \left(\phi \cdot e^{\lambda \cdot t} \cdot (e^{-\lambda \cdot t_D} - 1) - \frac{\Delta\Omega}{2} \cdot t_D\right)\Bigg]\lambda \cdot \phi \cdot e^{\lambda \cdot t}$$
$$= \left(1 - \frac{E \cdot e^{\lambda \cdot t}}{E^{(i)}} + \frac{E \cdot e^{\lambda \cdot (t - t_D)}}{E^{(i)}}\right)$$
$$\cdot \{\sin(\Delta\Omega \cdot t + \beta \cdot \sin(\Omega_m \cdot t) - \Omega_0 \cdot t_D)$$
$$\cdot \left[1 + \phi \cdot e^{\lambda \cdot t} \cdot (e^{-\lambda \cdot t_D} - 1) \cdot \frac{\Delta\Omega}{2} \cdot t_D\right]$$
$$+ \cos(\Delta\Omega \cdot t + \beta \cdot \sin(\Omega_m \cdot t)$$
$$-\Omega_0 \cdot t_D) \cdot \left[\phi \cdot e^{\lambda \cdot t} \cdot (e^{-\lambda \cdot t_D} - 1) - \frac{\Delta\Omega}{2} \cdot t_D\right]\Bigg\}$$

$$\lambda \cdot \phi \cdot e^{\lambda \cdot t} = \left(1 - \frac{E \cdot e^{\lambda \cdot t}}{E^{(i)}} + \frac{E \cdot e^{\lambda \cdot (t - t_D)}}{E^{(i)}}\right) \cdot \left[1 + \phi \cdot e^{\lambda \cdot t} \cdot (e^{-\lambda \cdot t_D} - 1) \cdot \frac{\Delta\Omega}{2} \cdot t_D\right]$$
$$\cdot \sin(\Delta\Omega \cdot t + \beta \cdot \sin(\Omega_m \cdot t) - \Omega_0 \cdot t_D)$$
$$+ \left(1 - \frac{E \cdot e^{\lambda \cdot t}}{E^{(i)}} + \frac{E \cdot e^{\lambda \cdot (t - t_D)}}{E^{(i)}}\right) \cdot \left[\phi \cdot e^{\lambda \cdot t} \cdot (e^{-\lambda \cdot t_D} - 1) - \frac{\Delta\Omega}{2} \cdot t_D\right]$$
$$\cdot \cos(\Delta\Omega \cdot t + \beta \cdot \sin(\Omega_m \cdot t) - \Omega_0 \cdot t_D)$$

$$\lambda \cdot \phi \cdot e^{\lambda \cdot t} = \left(1 + \frac{E \cdot e^{\lambda \cdot t}}{E^{(i)}} \cdot [e^{-\lambda \cdot t_D} - 1]\right) \cdot \left[1 + \phi \cdot e^{\lambda \cdot t} \cdot (e^{-\lambda \cdot t_D} - 1) \cdot \frac{\Delta\Omega}{2} \cdot t_D\right]$$
$$\cdot \sin(\Delta\Omega \cdot t + \beta \cdot \sin(\Omega_m \cdot t) - \Omega_0 \cdot t_D)$$
$$+ \left(1 + \frac{E \cdot e^{\lambda \cdot t}}{E^{(i)}} \cdot [e^{-\lambda \cdot t_D} - 1]\right) \cdot \left[\phi \cdot e^{\lambda \cdot t} \cdot (e^{-\lambda \cdot t_D} - 1) - \frac{\Delta\Omega}{2} \cdot t_D\right]$$
$$\cdot \cos(\Delta\Omega \cdot t + \beta \cdot \sin(\Omega_m \cdot t) - \Omega_0 \cdot t_D)$$

$$\lambda \cdot \phi \cdot e^{\lambda \cdot t} = \Bigg\{1 + \phi \cdot e^{\lambda \cdot t} \cdot (e^{-\lambda \cdot t_D} - 1) \cdot \frac{\Delta\Omega}{2} \cdot t_D + \frac{E \cdot e^{\lambda \cdot t}}{E^{(i)}} \cdot [e^{-\lambda \cdot t_D} - 1]$$
$$+ \frac{E \cdot \phi \cdot e^{\lambda \cdot t}}{E^{(i)}} \cdot [e^{-\lambda \cdot t_D} - 1] \cdot e^{\lambda \cdot t} \cdot (e^{-\lambda \cdot t_D} - 1) \cdot \frac{\Delta\Omega}{2} \cdot t_D\Bigg\}$$
$$\cdot \sin(\Delta\Omega \cdot t + \beta \cdot \sin(\Omega_m \cdot t) - \Omega_0 \cdot t_D)$$
$$+ \Bigg\{\phi \cdot e^{\lambda \cdot t} \cdot (e^{-\lambda \cdot t_D} - 1) - \frac{\Delta\Omega}{2} \cdot t_D$$
$$+ \frac{E \cdot \phi \cdot e^{\lambda \cdot t}}{E^{(i)}} \cdot [e^{-\lambda \cdot t_D} - 1] \cdot e^{\lambda \cdot t} \cdot (e^{-\lambda \cdot t_D} - 1)$$
$$- \frac{E \cdot e^{\lambda \cdot t}}{E^{(i)}} \cdot [e^{-\lambda \cdot t_D} - 1] \cdot \frac{\Delta\Omega}{2} \cdot t_D\Bigg\}$$

$$\cdot \cos(\Delta\Omega \cdot t + \beta \cdot \sin(\Omega_m \cdot t) - \Omega_0 \cdot t_D)$$

Assumption: $E \cdot \phi \approx 0$

$$\lambda \cdot \phi \cdot e^{\lambda \cdot t} = \{1 + \phi \cdot e^{\lambda \cdot t} \cdot (e^{-\lambda \cdot t_D} - 1) \cdot \frac{\Delta\Omega}{2} \cdot t_D + \frac{E \cdot e^{\lambda \cdot t}}{E^{(i)}} \cdot [e^{-\lambda \cdot t_D} - 1]\} \cdot \sin(\Delta\Omega \cdot t + \beta \cdot \sin(\Omega_m \cdot t) - \Omega_0 \cdot t_D) + \left\{\phi \cdot e^{\lambda \cdot t} \cdot (e^{-\lambda \cdot t_D} - 1) - \frac{\Delta\Omega}{2} \cdot t_D - \frac{E \cdot e^{\lambda \cdot t}}{E^{(i)}} \cdot [e^{-\lambda \cdot t_D} - 1] \cdot \frac{\Delta\Omega}{2} \cdot t_D\right\} \cdot \cos(\Delta\Omega \cdot t + \beta \cdot \sin(\Omega_m \cdot t) - \Omega_0 \cdot t_D)$$

$$\lambda \cdot \phi \cdot e^{\lambda \cdot t} = \sin(\Delta\Omega \cdot t + \beta \cdot \sin(\Omega_m \cdot t) - \Omega_0 \cdot t_D) + \left\{\phi \cdot e^{\lambda \cdot t} \cdot (e^{-\lambda \cdot t_D} - 1) \cdot \frac{\Delta\Omega}{2} \cdot t_D + \frac{E \cdot e^{\lambda \cdot t}}{E^{(i)}} \cdot [e^{-\lambda \cdot t_D} - 1]\right\} \cdot \sin(\Delta\Omega \cdot t + \beta \cdot \sin(\Omega_m \cdot t) - \Omega_0 \cdot t_D) + \left\{\phi \cdot e^{\lambda \cdot t} \cdot (e^{-\lambda \cdot t_D} - 1) - \frac{\Delta\Omega}{2} \cdot t_D - \frac{E \cdot e^{\lambda \cdot t}}{E^{(i)}} \cdot [e^{-\lambda \cdot t_D} - 1] \cdot \frac{\Delta\Omega}{2} \cdot t_D\right\} \cdot \cos(\Delta\Omega \cdot t + \beta \cdot \sin(\Omega_m \cdot t) - \Omega_0 \cdot t_D)$$

At fixed point: $\frac{E^*}{E^*} \cdot \sin(\psi^*) = 0 \Rightarrow \sin(\psi^*) = 0$; $\psi^* \approx \Delta\Omega \cdot t + \beta \cdot \sin(\Omega_m \cdot t) - \Omega_0 \cdot t_D$ then

$$\lambda \cdot \phi \cdot e^{\lambda \cdot t} = \{\phi \cdot e^{\lambda \cdot t} \cdot (e^{-\lambda \cdot t_D} - 1) \cdot \frac{\Delta\Omega}{2} \cdot t_D + \frac{E \cdot e^{\lambda \cdot t}}{E^{(i)}} \cdot [e^{-\lambda \cdot t_D} - 1]\} \cdot \sin(\Delta\Omega \cdot t + \beta \cdot \sin(\Omega_m \cdot t) - \Omega_0 \cdot t_D) + \{\phi \cdot e^{\lambda \cdot t} \cdot (e^{-\lambda \cdot t_D} - 1) - \frac{\Delta\Omega}{2} \cdot t_D - \frac{E \cdot e^{\lambda \cdot t}}{E^{(i)}} \cdot [e^{-\lambda \cdot t_D} - 1] \cdot \frac{\Delta\Omega}{2} \cdot t_D\} \cdot \cos(\Delta\Omega \cdot t + \beta \cdot \sin(\Omega_m \cdot t) - \Omega_0 \cdot t_D)$$

$$\frac{d\phi(t)}{dt} = 0 \Rightarrow \sin\psi^* = 0;\ \psi^* = n \cdot \pi\ \forall n = 0, 1, 2, \ldots$$

$$n \in \mathbb{N}_0;\ n \cdot \pi = \Delta\Omega \cdot t + \beta \cdot \sin(\Omega_m \cdot t) - \Omega_0 \cdot t_D$$

$$\sin(n \cdot \pi) = \sin(\Delta\Omega \cdot t + \beta \cdot \sin(\Omega_m \cdot t) - \Omega_0 \cdot t_D) = 0$$

$$\sin\psi^* = 0 \Rightarrow \cos\psi^* = \pm 1;\ n = 0, 1, 2, \ldots;\ n \in \mathbb{N}_0;\ \psi^* = n \cdot \pi;\ \cos\psi^* = (-1)^n \cos(n \cdot \pi) = \cos(\Delta\Omega \cdot t + \beta \cdot \sin(\Omega_m \cdot t) - \Omega_0 \cdot t_D) = (-1)^n \lambda \cdot \phi \cdot e^{\lambda \cdot t} = \left\{\phi \cdot e^{\lambda \cdot t} \cdot (e^{-\lambda \cdot t_D} - 1) - \frac{\Delta\Omega}{2} \cdot t_D - \frac{E \cdot e^{\lambda \cdot t}}{E^{(i)}} \cdot [e^{-\lambda \cdot t_D} - 1] \cdot \frac{\Delta\Omega}{2} \cdot t_D\right\} \cdot (-1)^n$$

Assumption: $\phi \cdot e^{\lambda \cdot t} \cdot (e^{-\lambda \cdot t_D} - 1) - \frac{E \cdot e^{\lambda \cdot t}}{E^{(i)}} \cdot [e^{-\lambda \cdot t_D} - 1] \cdot \frac{\Delta\Omega}{2} \cdot t_D \gg \frac{\Delta\Omega}{2} \cdot t_D$ $(e^{-\lambda \cdot t_D} - 1) \cdot e^{\lambda \cdot t} \cdot \left\{\phi - \frac{E}{E^{(i)}} \cdot \frac{\Delta\Omega}{2} \cdot t_D\right\} \gg \frac{\Delta\Omega}{2} \cdot t_D; \phi \gg E \cdot t_D \cdot \frac{1}{E^{(i)}} \cdot \frac{\Delta\Omega}{2}$ then

$$\phi \cdot e^{\lambda \cdot t} \cdot (e^{-\lambda \cdot t_D} - 1) - \frac{\Delta\Omega}{2} \cdot t_D - \frac{E \cdot e^{\lambda \cdot t}}{E^{(i)}} \cdot [e^{-\lambda \cdot t_D} - 1] \cdot \frac{\Delta\Omega}{2} \cdot t_D$$
$$\approx \phi \cdot e^{\lambda \cdot t} \cdot (e^{-\lambda \cdot t_D} - 1) - \frac{E \cdot e^{\lambda \cdot t}}{E^{(i)}} \cdot [e^{-\lambda \cdot t_D} - 1] \cdot \frac{\Delta\Omega}{2} \cdot t_D$$
$$\lambda \cdot \phi \cdot e^{\lambda \cdot t} = \left\{\phi \cdot (e^{-\lambda \cdot t_D} - 1) - \frac{E}{E^{(i)}} \cdot [e^{-\lambda \cdot t_D} - 1] \cdot \frac{\Delta\Omega}{2} \cdot t_D\right\} \cdot e^{\lambda \cdot t} \cdot (-1)^n$$

Divide the above expression by $e^{\lambda \cdot t}$ term.

$$-\frac{1}{E^{(i)}} \cdot [e^{-\lambda \cdot t_D} - 1] \cdot \frac{\Delta\Omega}{2} \cdot t_D \cdot (-1)^n \cdot E + [(e^{-\lambda \cdot t_D} - 1) \cdot (-1)^n - \lambda] \cdot \phi = 0$$

We can summary our arbitrary small increments equations.

$$(-\lambda + K \cdot \{2 \cdot [E^{(i)}]^2 - 1\}) \cdot n + 2 \cdot K \cdot (1 + 2 \cdot N^{(i)}) \cdot E^{(i)} \cdot E = 0$$
$$E^{(i)} \cdot n + [N^{(i)} + m \cdot e^{-\lambda \cdot t_D} \cdot (-1)^n - \lambda] \cdot E + m \cdot E^{(i)} \cdot (e^{-\lambda \cdot t_D} - 1)$$
$$\cdot \frac{\Delta\Omega}{2} \cdot t_D \cdot (-1)^n \cdot \phi = 0 \; \forall n = 0, 1, 2, \ldots$$
$$-\frac{1}{E^{(i)}} \cdot [e^{-\lambda \cdot t_D} - 1] \cdot \frac{\Delta\Omega}{2} \cdot t_D \cdot (-1)^n \cdot E + [(e^{-\lambda \cdot t_D} - 1)$$
$$\cdot (-1)^n - \lambda] \cdot \phi = 0 \; \forall n = 0, 1, 2, \ldots$$

The small increments Jacobian of our lasers with frequency-modulated optical feedback system is as follow:

$$\begin{pmatrix} -\lambda + K \cdot \{2 \cdot [E^{(i)}]^2 - 1\} & 2 \cdot K \cdot (1 + 2 \cdot N^{(i)}) \cdot E^{(i)} & 0 \\ E^{(i)} & N^{(i)} + m \cdot e^{-\lambda \cdot t_D} \cdot (-1)^n - \lambda & m \cdot E^{(i)} \cdot (e^{-\lambda \cdot t_D} - 1) \cdot \frac{\Delta\Omega}{2} \cdot t_D \cdot (-1)^n \\ 0 & -\frac{1}{E^{(i)}} \cdot [e^{-\lambda \cdot t_D} - 1] \cdot \frac{\Delta\Omega}{2} \cdot t_D \cdot (-1)^n & (e^{-\lambda \cdot t_D} - 1) \cdot (-1)^n - \lambda \end{pmatrix} \cdot \begin{pmatrix} n \\ E \\ \phi \end{pmatrix} = 0$$

$$A - \lambda \cdot I = \begin{pmatrix} -\lambda + K \cdot \{2 \cdot [E^{(i)}]^2 - 1\} & 2 \cdot K \cdot (1 + 2 \cdot N^{(i)}) \cdot E^{(i)} & 0 \\ E^{(i)} & N^{(i)} + m \cdot e^{-\lambda \cdot t_D} \cdot (-1)^n - \lambda & m \cdot E^{(i)} \cdot (e^{-\lambda \cdot t_D} - 1) \cdot \frac{\Delta\Omega}{2} \cdot t_D \cdot (-1)^n \\ 0 & -\frac{1}{E^{(i)}} \cdot [e^{-\lambda \cdot t_D} - 1] \cdot \frac{\Delta\Omega}{2} \cdot t_D \cdot (-1)^n & (e^{-\lambda \cdot t_D} - 1) \cdot (-1)^n - \lambda \end{pmatrix}$$

We inspect the occurrence of any possible stability switching resulting from the increase of value of time delay t_D(delay time normalized by the photon lifetime) for lasers with frequency-modulated optical feedback system general characteristic equation $D(\lambda, t_D) = \det|A - \lambda \cdot I|$.

$$D(\lambda,t_D)=\det|A-\lambda\cdot I|=$$

$$(-\lambda+K\cdot\{2\cdot[E^{(i)}]^2-1\})\cdot\det\begin{pmatrix} N^{(i)}+m\cdot e^{-\lambda\cdot t_D}\cdot(-1)^n-\lambda & m\cdot E^{(i)}\cdot(e^{-\lambda\cdot t_D}-1)\cdot\frac{\Delta\Omega}{2}\cdot t_D\cdot(-1)^n \\ -\frac{1}{E^{(i)}}\cdot[e^{-\lambda\cdot t_D}-1]\cdot\frac{\Delta\Omega}{2}\cdot t_D\cdot(-1)^n & (e^{-\lambda\cdot t_D}-1)\cdot(-1)^n-\lambda \end{pmatrix}$$

$$-2\cdot K\cdot(1+2\cdot N^{(i)})\cdot E^{(i)}\cdot\det\begin{pmatrix} E^{(i)} & m\cdot E^{(i)}\cdot(e^{-\lambda\cdot t_D}-1)\cdot\frac{\Delta\Omega}{2}\cdot t_D\cdot(-1)^n \\ 0 & (e^{-\lambda\cdot t_D}-1)\cdot(-1)^n-\lambda \end{pmatrix}$$

$$\begin{aligned} D(\lambda,t_D) &= \det|A-\lambda\cdot I| \\ &= (-\lambda+K\cdot\{2\cdot[E^{(i)}]^2-1\}) \\ &\quad\cdot\{N^{(i)}+m\cdot e^{-\lambda\cdot t_D}\cdot(-1)^n-\lambda]\cdot[(e^{-\lambda\cdot t_D}-1)\cdot(-1)^n-\lambda] \\ &\quad-\left[-\frac{1}{E^{(i)}}\cdot[e^{-\lambda\cdot t_D}-1]\cdot\frac{\Delta\Omega}{2}\cdot t_D\cdot(-1)^n\right] \\ &\quad\cdot\left[m\cdot E^{(i)}\cdot(e^{-\lambda\cdot t_D}-1)\cdot\frac{\Delta\Omega}{2}\cdot t_D\cdot(-1)^n\right]\Bigg\} \\ &\quad-2\cdot K\cdot(1+2\cdot N^{(i)})\cdot E^{(i)}\cdot E^{(i)}\cdot[(e^{-\lambda\cdot t_D}-1)\cdot(-1)^n-\lambda] \end{aligned}$$

$$\begin{aligned} D(\lambda,t_D) &= \det|A-\lambda\cdot I| \\ &= (-\lambda+K\cdot\{2\cdot[E^{(i)}]^2-1\}) \\ &\quad\cdot\{[N^{(i)}+m\cdot e^{-\lambda\cdot t_D}\cdot(-1)^n-\lambda]\cdot[(e^{-\lambda\cdot t_D}-1)\cdot(-1)^n-\lambda] \\ &\quad+[e^{-\lambda\cdot t_D}-1]^2\cdot\left(\frac{\Delta\Omega}{2}\right)^2\cdot m\cdot t_D^2\cdot(-1)^n\cdot(-1)^n\} \\ &\quad-2\cdot K\cdot(1+2\cdot N^{(i)})\cdot(E^{(i)})^2\cdot[(e^{-\lambda\cdot t_D}-1)\cdot(-1)^n-\lambda] \end{aligned}$$

<u>Hint:</u> $(-1)^n\cdot(-1)^n=1\,\forall n\in\mathbb{N}_0$

$$\begin{aligned} D(\lambda,t_D) &= \det|A-\lambda\cdot I| \\ &= (-\lambda+K\cdot\{2\cdot[E^{(i)}]^2-1\}) \\ &\quad\cdot\Big\{[N^{(i)}+m\cdot e^{-\lambda\cdot t_D}\cdot(-1)^n-\lambda]\cdot[(e^{-\lambda\cdot t_D}-1)\cdot(-1)^n-\lambda] \\ &\quad+[e^{-\lambda\cdot t_D}-1]^2\cdot\left(\frac{\Delta\Omega}{2}\right)^2\cdot m\cdot t_D^2\Big\} \\ &\quad-2\cdot K\cdot(1+2\cdot N^{(i)})\cdot(E^{(i)})^2\cdot[(e^{-\lambda\cdot t_D}-1)\cdot(-1)^n-\lambda] \end{aligned}$$

$$\begin{aligned} D(\lambda,t_D) &= \det|A-\lambda\cdot I| \\ &= (-\lambda+K\cdot\{2\cdot[E^{(i)}]^2-1\}) \end{aligned}$$

$$\cdot \{[N^{(i)} + m \cdot e^{-\lambda \cdot t_D} \cdot (-1)^n - \lambda] \cdot [e^{-\lambda \cdot t_D} \cdot (-1)^n - (-1)^n - \lambda]$$
$$+ [e^{-2 \cdot \lambda \cdot t_D} - 2 \cdot e^{-\lambda \cdot t_D} + 1] \cdot \left(\frac{\Delta\Omega}{2}\right)^2 \cdot m \cdot t_D^2 \Big\}$$
$$- 2 \cdot K \cdot (1 + 2 \cdot N^{(i)}) \cdot (E^{(i)})^2 \cdot [(e^{-\lambda \cdot t_D} - 1) \cdot (-1)^n - \lambda]$$

$$D(\lambda, t_D) = \det |A - \lambda \cdot I|$$
$$= (-\lambda + K \cdot \{2 \cdot [E^{(i)}]^2 - 1\}) \cdot \{N^{(i)} \cdot e^{-\lambda \cdot t_D} \cdot (-1)^n$$
$$- N^{(i)} \cdot (-1)^n - N^{(i)} \cdot \lambda + e^{-\lambda \cdot t_D} \cdot m \cdot e^{-\lambda \cdot t_D} \cdot (-1)^n$$
$$\cdot (-1)^n - m \cdot e^{-\lambda \cdot t_D} \cdot (-1)^n \cdot (-1)^n$$
$$- \lambda \cdot m \cdot e^{-\lambda \cdot t_D} \cdot (-1)^n - \lambda \cdot e^{-\lambda \cdot t_D} \cdot (-1)^n$$
$$+ \lambda \cdot (-1)^n + \lambda^2 + [e^{-2 \cdot \lambda \cdot t_D} - 2 \cdot e^{-\lambda \cdot t_D} + 1] \cdot \left(\frac{\Delta\Omega}{2}\right)^2 \cdot m \cdot t_D^2 \Big\}$$
$$- 2 \cdot K \cdot (1 + 2 \cdot N^{(i)}) \cdot (E^{(i)})^2 \cdot [(e^{-\lambda \cdot t_D} - 1) \cdot (-1)^n - \lambda]$$

Hint: $(-1)^n \cdot (-1)^n = 1 \, \forall n \in \mathbb{N}_0$

$$D(\lambda, t_D) = \det |A - \lambda \cdot I|$$
$$= (-\lambda + K \cdot \{2 \cdot [E^{(i)}]^2 - 1\}) \cdot \{N^{(i)} \cdot e^{-\lambda \cdot t_D} \cdot (-1)^n$$
$$- N^{(i)} \cdot (-1)^n - N^{(i)} \cdot \lambda + m \cdot e^{-\lambda \cdot t_D} \cdot e^{-\lambda \cdot t_D}$$
$$- m \cdot e^{-\lambda \cdot t_D} - \lambda \cdot m \cdot e^{-\lambda \cdot t_D} \cdot (-1)^n$$
$$- \lambda \cdot e^{-\lambda \cdot t_D} \cdot (-1)^n + \lambda \cdot (-1)^n + \lambda^2$$
$$+ [e^{-2 \cdot \lambda \cdot t_D} - 2 \cdot e^{-\lambda \cdot t_D} + 1] \cdot \left(\frac{\Delta\Omega}{2}\right)^2 \cdot m \cdot t_D^2 \Big\}$$
$$- 2 \cdot K \cdot (1 + 2 \cdot N^{(i)}) \cdot (E^{(i)})^2 \cdot [e^{-\lambda \cdot t_D} \cdot (-1)^n - (-1)^n - \lambda]$$

$$D(\lambda, t_D) = \det |A - \lambda \cdot I|$$
$$= (-\lambda + K \cdot \{2 \cdot [E^{(i)}]^2 - 1\}) \cdot \{N^{(i)} \cdot e^{-\lambda \cdot t_D} \cdot (-1)^n - m \cdot e^{-\lambda \cdot t_D}$$
$$- \lambda \cdot m \cdot e^{-\lambda \cdot t_D} \cdot (-1)^n - \lambda \cdot e^{-\lambda \cdot t_D} \cdot (-1)^n$$
$$- N^{(i)} \cdot (-1)^n + \lambda \cdot (-1)^n - N^{(i)} \cdot \lambda + \lambda^2 + m \cdot e^{-2 \cdot \lambda \cdot t_D}$$
$$+ [e^{-2 \cdot \lambda \cdot t_D} - 2 \cdot e^{-\lambda \cdot t_D} + 1] \cdot \left(\frac{\Delta\Omega}{2}\right)^2 \cdot m \cdot t_D^2 \Big\}$$
$$- 2 \cdot K \cdot (1 + 2 \cdot N^{(i)}) \cdot (E^{(i)})^2 \cdot e^{-\lambda \cdot t_D} \cdot (-1)^n$$
$$+ 2 \cdot K \cdot (1 + 2 \cdot N^{(i)}) \cdot (E^{(i)})^2 \cdot (-1)^n$$
$$+ 2 \cdot K \cdot (1 + 2 \cdot N^{(i)}) \cdot (E^{(i)})^2 \cdot \lambda$$

$$\begin{aligned}
D(\lambda, t_D) &= \det|A - \lambda \cdot I| \\
&= (-\lambda + K \cdot \{2 \cdot [E^{(i)}]^2 - 1\}) \cdot \{N^{(i)} \cdot e^{-\lambda \cdot t_D} \cdot (-1)^n - m \cdot e^{-\lambda \cdot t_D} \\
&\quad - \lambda \cdot m \cdot e^{-\lambda \cdot t_D} \cdot (-1)^n - \lambda \cdot e^{-\lambda \cdot t_D} \cdot (-1)^n \\
&\quad - 2 \cdot e^{-\lambda \cdot t_D} \cdot \left(\frac{\Delta\Omega}{2}\right)^2 \cdot m \cdot t_D^2 \\
&\quad + \left(\frac{\Delta\Omega}{2}\right)^2 \cdot m \cdot t_D^2 - N^{(i)} \cdot (-1)^n + \lambda \cdot (-1)^n \\
&\quad - N^{(i)} \cdot \lambda + \lambda^2 + m \cdot e^{-2 \cdot \lambda \cdot t_D} + e^{-2 \cdot \lambda \cdot t_D} \cdot \left(\frac{\Delta\Omega}{2}\right)^2 \cdot m \cdot t_D^2 \Big\} \\
&\quad - 2 \cdot K \cdot (1 + 2 \cdot N^{(i)}) \cdot (E^{(i)})^2 \cdot e^{-\lambda \cdot t_D} \cdot (-1)^n \\
&\quad + 2 \cdot K \cdot (1 + 2 \cdot N^{(i)}) \cdot (E^{(i)})^2 \cdot (-1)^n \\
&\quad + 2 \cdot K \cdot (1 + 2 \cdot N^{(i)}) \cdot (E^{(i)})^2 \cdot \lambda
\end{aligned}$$

$$\begin{aligned}
D(\lambda, t_D) &= \det|A - \lambda \cdot I| \\
&= (-\lambda + K \cdot \{2 \cdot [E^{(i)}]^2 - 1\}) \cdot \{[N^{(i)} \cdot (-1)^n - m \\
&\quad - \lambda \cdot m \cdot (-1)^n - \lambda \cdot (-1)^n - 2 \cdot \left(\frac{\Delta\Omega}{2}\right)^2 \\
&\quad \cdot m \cdot t_D^2] \cdot e^{-\lambda \cdot t_D} + \left(\frac{\Delta\Omega}{2}\right)^2 \cdot m \cdot t_D^2 - N^{(i)} \cdot (-1)^n \\
&\quad + [(-1)^n - N^{(i)}] \cdot \lambda + \lambda^2 + m \cdot e^{-2 \cdot \lambda \cdot t_D} \cdot \left[1 + \left(\frac{\Delta\Omega}{2}\right)^2 \cdot t_D^2\right]\Big\} \\
&\quad - 2 \cdot K \cdot (1 + 2 \cdot N^{(i)}) \cdot (E^{(i)})^2 \cdot e^{-\lambda \cdot t_D} \cdot (-1)^n \\
&\quad + 2 \cdot K \cdot (1 + 2 \cdot N^{(i)}) \cdot (E^{(i)})^2 \cdot (-1)^n 4 \\
&\quad + 2 \cdot K \cdot (1 + 2 \cdot N^{(i)}) \cdot (E^{(i)})^2 \cdot \lambda
\end{aligned}$$

Assumption: $K = \frac{\tau}{\tau_p}$ is the population-to-photon lifetime ratio ($\tau_p \approx 600$ ps) then typical value for K is $K = 2 \times 10^5$. $\Delta\Omega = (\omega_i - \omega_0) \cdot \tau_p$ is the normalized frequency shift (τ_p is the photon lifetime). m is the amplitude feedback coefficient, $m \approx 10^{-5}$ then $\left(\frac{\Delta\Omega}{2}\right)^2 \cdot t_D^2 \ll 1$, $1 + \left(\frac{\Delta\Omega}{2}\right)^2 \cdot t_D^2\Big] \approx 1$; and $m \cdot e^{-2 \cdot \lambda \cdot t_D} \cdot \left[1 + \left(\frac{\Delta\Omega}{2}\right)^2 \cdot t_D^2\right]$ is very small compare to other elements in the above expression and we can ignore it and assume as a very small expression value. Then we get.

$$\begin{aligned}
D(\lambda, t_D) &= \det|A - \lambda \cdot I| \\
&= (-\lambda + K \cdot \{2 \cdot [E^{(i)}]^2 - 1\}) \cdot \{[N^{(i)} \cdot (-1)^n - m
\end{aligned}$$

$$-\lambda \cdot m \cdot (-1)^n - \lambda \cdot (-1)^n - 2 \cdot \left(\frac{\Delta\Omega}{2}\right)^2 \cdot m \cdot t_D^2\Bigg] \cdot e^{-\lambda \cdot t_D}$$
$$+ \left(\frac{\Delta\Omega}{2}\right)^2 \cdot m \cdot t_D^2 - N^{(i)} \cdot (-1)^n$$
$$+ [(-1)^n - N^{(i)}] \cdot \lambda + \lambda^2\} - 2 \cdot K$$
$$\cdot (1 + 2 \cdot N^{(i)}) \cdot (E^{(i)})^2 \cdot e^{-\lambda \cdot t_D} \cdot (-1)^n$$
$$+ 2 \cdot K \cdot (1 + 2 \cdot N^{(i)}) \cdot (E^{(i)})^2 \cdot (-1)^n$$
$$+ 2 \cdot K \cdot (1 + 2 \cdot N^{(i)}) \cdot (E^{(i)})^2 \cdot \lambda$$

$$D(\lambda, t_D) = \det|A - \lambda \cdot I| = -\lambda \cdot \Big\{\Big[N^{(i)} \cdot (-1)^n - m$$
$$-\lambda \cdot m \cdot (-1)^n - \lambda \cdot (-1)^n - 2 \cdot \left(\frac{\Delta\Omega}{2}\right)^2 \cdot m \cdot t_D^2\Bigg]$$
$$\cdot e^{-\lambda \cdot t_D} + \left(\frac{\Delta\Omega}{2}\right)^2 \cdot m \cdot t_D^2 - N^{(i)} \cdot (-1)^n$$
$$+[(-1)^n - N^{(i)}] \cdot \lambda + \lambda^2\Big\} + K \cdot (2 \cdot [E^{(i)}]^2 - 1) \cdot \{[N^{(i)} \cdot (-1)^n - m$$
$$- \lambda \cdot m \cdot (-1)^n - \lambda \cdot (-1)^n - 2 \cdot \left(\frac{\Delta\Omega}{2}\right)^2$$
$$\cdot m \cdot t_D^2] \cdot e^{-\lambda \cdot t_D} + \left(\frac{\Delta\Omega}{2}\right)^2 \cdot m \cdot t_D^2 - N^{(i)} \cdot (-1)^n$$
$$+ [(-1)^n - N^{(i)}] \cdot \lambda + \lambda^2\}$$
$$- 2 \cdot K \cdot (1 + 2 \cdot N^{(i)}) \cdot (E^{(i)})^2 \cdot e^{-\lambda \cdot t_D} \cdot (-1)^n$$
$$+ 2 \cdot K \cdot (1 + 2 \cdot N^{(i)}) \cdot (E^{(i)})^2 \cdot (-1)^n$$
$$+ 2 \cdot K \cdot (1 + 2 \cdot N^{(i)}) \cdot (E^{(i)})^2 \cdot \lambda$$

$$D(\lambda, t_D) = \det|A - \lambda \cdot I| = -\Big\{\lambda \cdot [N^{(i)} \cdot (-1)^n$$
$$- m - \lambda \cdot m \cdot (-1)^n - \lambda \cdot (-1)^n$$
$$-2 \cdot \left(\frac{\Delta\Omega}{2}\right)^2 \cdot m \cdot t_D^2\Bigg] \cdot e^{-\lambda \cdot t_D} + \lambda \cdot \left(\frac{\Delta\Omega}{2}\right)^2 \cdot m \cdot t_D^2$$
$$-\lambda \cdot N^{(i)} \cdot (-1)^n + [(-1)^n - N^{(i)}] \cdot \lambda^2 + \lambda^3\Big\}$$
$$+ \{K \cdot (2 \cdot [E^{(i)}]^2 - 1)$$
$$\cdot \Big[(N^{(i)} \cdot (-1)^n - m) - \lambda \cdot (-1)^n$$
$$\cdot (m + 1) - 2 \cdot \left(\frac{\Delta\Omega}{2}\right)^2 \cdot m \cdot t_D^2\Bigg] \cdot e^{-\lambda \cdot t_D}$$

$$
\begin{aligned}
&+ K \cdot (2 \cdot [E^{(i)}]^2 - 1) \cdot \left(\frac{\Delta\Omega}{2}\right)^2 \cdot m \cdot t_D^2 \\
&- K \cdot (2 \cdot [E^{(i)}]^2 - 1) \cdot N^{(i)} \cdot (-1)^n \\
&+ K \cdot (2 \cdot [E^{(i)}]^2 - 1) \cdot [(-1)^n - N^{(i)}] \\
&\cdot \lambda + K \cdot (2 \cdot [E^{(i)}]^2 - 1) \cdot \lambda^2\} \\
&- 2 \cdot K \cdot (1 + 2 \cdot N^{(i)}) \cdot (E^{(i)})^2 \cdot e^{-\lambda \cdot t_D} \cdot (-1)^n \\
&+ 2 \cdot K \cdot (1 + 2 \cdot N^{(i)}) \cdot (E^{(i)})^2 \cdot (-1)^n \\
&+ 2 \cdot K \cdot (1 + 2 \cdot N^{(i)}) \cdot (E^{(i)})^2 \cdot \lambda
\end{aligned}
$$

$$
\begin{aligned}
D(\lambda, t_D) = \det|A - \lambda \cdot I| = \Big[-\lambda \cdot \Big(N^{(i)} \\
&\cdot (-1)^n - m - 2 \cdot \left(\frac{\Delta\Omega}{2}\right)^2 \cdot m \cdot t_D^2\Big) \\
&+ \lambda^2 \cdot m \cdot (-1)^n + \lambda^2 \cdot (-1)^n\Big] \cdot e^{-\lambda \cdot t_D} \\
&- \lambda \cdot \left(\frac{\Delta\Omega}{2}\right)^2 \cdot m \cdot t_D^2 + \lambda \cdot N^{(i)} \cdot (-1)^n - [(-1)^n - N^{(i)}] \cdot \lambda^2 - \lambda^3 \\
&+ \Big[K \cdot (2 \cdot [E^{(i)}]^2 - 1) \cdot (N^{(i)} \cdot (-1)^n - m) \\
&- K \cdot (2 \cdot [E^{(i)}]^2 - 1) \cdot \lambda \cdot (-1)^n \cdot (m + 1) \\
&- K \cdot (2 \cdot [E^{(i)}]^2 - 1) \cdot 2 \cdot \left(\frac{\Delta\Omega}{2}\right)^2 \cdot m \cdot t_D^2\Big] \cdot e^{-\lambda \cdot t_D} \\
&+ K \cdot (2 \cdot [E^{(i)}]^2 - 1) \cdot \left(\frac{\Delta\Omega}{2}\right)^2 \cdot m \cdot t_D^2 \\
&- K \cdot (2 \cdot [E^{(i)}]^2 - 1) \cdot N^{(i)} \cdot (-1)^n \\
&+ K \cdot (2 \cdot [E^{(i)}]^2 - 1) \cdot [(-1)^n - N^{(i)}] \\
&\cdot \lambda + K \cdot (2 \cdot [E^{(i)}]^2 - 1) \cdot \lambda^2 \\
&- 2 \cdot K \cdot (1 + 2 \cdot N^{(i)}) \cdot (E^{(i)})^2 \cdot e^{-\lambda \cdot t_D} \cdot (-1)^n \\
&+ 2 \cdot K \cdot (1 + 2 \cdot N^{(i)}) \cdot (E^{(i)})^2 \cdot (-1)^n \\
&+ 2 \cdot K \cdot (1 + 2 \cdot N^{(i)}) \cdot (E^{(i)})^2 \cdot \lambda
\end{aligned}
$$

$$
\begin{aligned}
D(\lambda, t_D) = \det|A - \lambda \cdot I| = \Big[-\lambda \cdot (N^{(i)} \cdot (-1)^n \\
&- m - 2 \cdot \left(\frac{\Delta\Omega}{2}\right)^2 \cdot m \cdot t_D^2) + \lambda^2 \cdot m \cdot (-1)^n + \lambda^2 \cdot (-1)^n] \cdot e^{-\lambda \cdot t_D} \\
&+ [K \cdot (2 \cdot [E^{(i)}]^2 - 1) \cdot (N^{(i)} \cdot (-1)^n - m) \\
&- K \cdot (2 \cdot [E^{(i)}]^2 - 1) \cdot \lambda \cdot (-1)^n \cdot (m + 1)
\end{aligned}
$$

$$
\begin{aligned}
&- K \cdot (2 \cdot [E^{(i)}]^2 - 1) \cdot 2 \cdot \left(\frac{\Delta\Omega}{2}\right)^2 \cdot m \cdot t_D^2] \cdot e^{-\lambda \cdot t_D} \\
&- 2 \cdot K \cdot (1 + 2 \cdot N^{(i)}) \cdot (E^{(i)})^2 \cdot (-1)^n \cdot e^{-\lambda \cdot t_D} \\
&+ K \cdot (2 \cdot [E^{(i)}]^2 - 1) \cdot \left(\frac{\Delta\Omega}{2}\right)^2 \cdot m \cdot t_D^2 - K \cdot (2 \cdot [E^{(i)}]^2 - 1) \\
&\cdot N^{(i)} \cdot (-1)^n + 2 \cdot K \cdot (1 + 2 \cdot N^{(i)}) \cdot (E^{(i)})^2 \cdot (-1)^n \\
&+ \lambda \cdot N^{(i)} \cdot (-1)^n + K \cdot (2 \cdot [E^{(i)}]^2 - 1) \cdot [(-1)^n - N^{(i)}] \\
&\cdot \lambda + 2 \cdot K \cdot (1 + 2 \cdot N^{(i)}) \cdot (E^{(i)})^2 \cdot \lambda - \lambda \cdot \left(\frac{\Delta\Omega}{2}\right)^2 \cdot m \cdot t_D^2 \\
&+ K \cdot (2 \cdot [E^{(i)}]^2 - 1) \cdot \lambda^2 - [(-1)^n - N^{(i)}] \cdot \lambda^2 - \lambda^3
\end{aligned}
$$

$$
\begin{aligned}
D(\lambda, t_D) = \det|A - \lambda \cdot I| = {} & [-\lambda \cdot (N^{(i)} \cdot (-1)^n - m - 2 \\
& \cdot \left(\frac{\Delta\Omega}{2}\right)^2 \cdot m \cdot t_D^2 \Bigg) + \lambda^2 \cdot m \cdot (-1)^n + \lambda^2 \cdot (-1)^n] \cdot e^{-\lambda \cdot t_D} \\
& + [K \cdot (2 \cdot [E^{(i)}]^2 - 1) \cdot (N^{(i)} \cdot (-1)^n - m) \\
& - K \cdot (2 \cdot [E^{(i)}]^2 - 1) \cdot \lambda \cdot (-1)^n \cdot (m + 1) \\
& -K \cdot (2 \cdot [E^{(i)}]^2 - 1) \cdot 2 \cdot \left(\frac{\Delta\Omega}{2}\right)^2 \cdot m \cdot t_D^2 \Bigg] \cdot e^{-\lambda \cdot t_D} \\
& - 2 \cdot K \cdot (1 + 2 \cdot N^{(i)}) \cdot (E^{(i)})^2 \cdot (-1)^n \cdot e^{-\lambda \cdot t_D} \\
& +K \cdot (2 \cdot [E^{(i)}]^2 - 1) \cdot \left(\frac{\Delta\Omega}{2}\right)^2 \cdot m \cdot t_D^2 - K \cdot (2 \cdot [E^{(i)}]^2 - 1 \Bigg) \\
& \cdot N^{(i)} \cdot (-1)^n + 2 \cdot K \cdot (1 + 2 \cdot N^{(i)}) \cdot (E^{(i)})^2 \cdot (-1)^n \\
& + \{N^{(i)} \cdot (-1)^n + K \cdot (2 \cdot [E^{(i)}]^2 - 1) \cdot [(-1)^n - N^{(i)}] \\
& +2 \cdot K \cdot (1 + 2 \cdot N^{(i)}) \cdot (E^{(i)})^2 - \left(\frac{\Delta\Omega}{2}\right)^2 \cdot m \cdot t_D^2 \Bigg\} \cdot \lambda \\
& + \{K \cdot (2 \cdot [E^{(i)}]^2 - 1) - [(-1)^n - N^{(i)}]\} \cdot \lambda^2 - \lambda^3
\end{aligned}
$$

$$
\begin{aligned}
D(\lambda, t_D) = \det|A - \lambda \cdot I| = {} & \{K \cdot (2 \cdot [E^{(i)}]^2 - 1) \\
& \cdot (N^{(i)} \cdot (-1)^n - m) - K \cdot (2 \cdot [E^{(i)}]^2 - 1) \cdot 2 \cdot \left(\frac{\Delta\Omega}{2}\right)^2 \cdot m \cdot t_D^2 \\
& - 2 \cdot K \cdot (1 + 2 \cdot N^{(i)}) \cdot (E^{(i)})^2 \cdot (-1)^n \\
& - [(N^{(i)} \cdot (-1)^n - m - 2 \cdot \left(\frac{\Delta\Omega}{2}\right)^2 \cdot m \cdot t_D^2) \\
& + K \cdot (2 \cdot [E^{(i)}]^2 - 1) \cdot (-1)^n \cdot (m + 1)]
\end{aligned}
$$

$$\cdot \lambda + [m \cdot (-1)^n + (-1)^n] \cdot \lambda^2\} \cdot e^{-\lambda \cdot t_D}$$
$$+ K \cdot (2 \cdot [E^{(i)}]^2 - 1) \cdot \left(\frac{\Delta\Omega}{2}\right)^2 \cdot m \cdot t_D^2 - K \cdot (2 \cdot [E^{(i)}]^2 - 1)$$
$$\cdot N^{(i)} \cdot (-1)^n + 2 \cdot K \cdot (1 + 2 \cdot N^{(i)}) \cdot (E^{(i)})^2 \cdot (-1)^n$$
$$+ \{N^{(i)} \cdot (-1)^n + K \cdot (2 \cdot [E^{(i)}]^2 - 1) \cdot [(-1)^n - N^{(i)}]$$
$$+ 2 \cdot K \cdot (1 + 2 \cdot N^{(i)}) \cdot (E^{(i)})^2 - \left(\frac{\Delta\Omega}{2}\right)^2 \cdot m \cdot t_D^2\} \cdot \lambda$$
$$+ \{K \cdot (2 \cdot [E^{(i)}]^2 - 1) - [(-1)^n - N^{(i)}]\} \cdot \lambda^2 - \lambda^3$$

We define for simplicity new global parameters:

$$\Upsilon_1(n) = K \cdot (2 \cdot [E^{(i)}]^2 - 1) \cdot (N^{(i)} \cdot (-1)^n - m); \Upsilon_2 = K \cdot (2 \cdot [E^{(i)}]^2 - 1) \cdot 2 \cdot \left(\frac{\Delta\Omega}{2}\right)^2 \cdot m$$
$$\Upsilon_3(n) = 2 \cdot K \cdot (1 + 2 \cdot N^{(i)}) \cdot (E^{(i)})^2 \cdot (-1)^n; \Upsilon_4(n) = N^{(i)} \cdot (-1)^n - m; \Upsilon_5(n) = K \cdot (2 \cdot [E^{(i)}]^2 - 1) \cdot (-1)^n \cdot (m+1)$$
$$\Upsilon_6(n) = m \cdot (-1)^n + (-1)^n; \Upsilon_7 = K \cdot (2 \cdot [E^{(i)}]^2 - 1) \cdot \left(\frac{\Delta\Omega}{2}\right)^2 \cdot m; \Upsilon_8(n) = K \cdot (2 \cdot [E^{(i)}]^2 - 1) \cdot N^{(i)} \cdot (-1)^n$$
$$\Upsilon_9(n) = 2 \cdot K \cdot (1 + 2 \cdot N^{(i)}) \cdot (E^{(i)})^2 \cdot (-1)^n; \Upsilon_{11}(n) = K \cdot (2 \cdot [E^{(i)}]^2 - 1) - [(-1)^n - N^{(i)}]$$
$$\Upsilon_{10}(n) = N^{(i)} \cdot (-1)^n + K \cdot (2 \cdot [E^{(i)}]^2 - 1) \cdot [(-1)^n - N^{(i)}] + 2 \cdot K \cdot (1 + 2 \cdot N^{(i)}) \cdot (E^{(i)})^2$$
$$\Upsilon_1 = \Upsilon_1(n); \Upsilon_3 = \Upsilon_3(n); \Upsilon_4 = \Upsilon_4(n); \Upsilon_5 = \Upsilon_5(n); \Upsilon_6 = \Upsilon_6(n); \Upsilon_8 = \Upsilon_8(n); \Upsilon_9 = \Upsilon_9(n)$$
$$\Upsilon_{11} = \Upsilon_{11}(n); \Upsilon_{10} = \Upsilon_{10}(n)$$

$$D(\lambda, t_D) = \det|A - \lambda \cdot I| = \left\{\Upsilon_1 - \Upsilon_2 \cdot t_D^2 - \Upsilon_3 - \left[\Upsilon_4 - 2 \cdot \left(\frac{\Delta\Omega}{2}\right)^2 \cdot m \cdot t_D^2 + \Upsilon_5\right] \cdot \lambda + \Upsilon_6 \cdot \lambda^2\right\} \cdot e^{-\lambda \cdot t_D}$$
$$+ \Upsilon_7 \cdot t_D^2 - \Upsilon_8 + \Upsilon_9 + \left[\Upsilon_{10} - \left(\frac{\Delta\Omega}{2}\right)^2 \cdot m \cdot t_D^2\right] \cdot \lambda$$
$$+ \Upsilon_{11} \cdot \lambda^2 - \lambda^3$$

Table. 4.4 $\Upsilon_k(n); k = 1, 2, \ldots, 11$ expression for n parameter values ($n = 0, 1, 2, \ldots$)

$\Upsilon_k(n); k = 1, 2, \ldots, 11$	$n = 0, 2, 4, 6, 8, \ldots$(n is even number)	$n = 1, 3, 5, 7, 9, \ldots$(n is odd number)
$\Upsilon_1(n)$	$K \cdot (2 \cdot [E^{(i)}]^2 - 1) \cdot (N^{(i)} - m)$	$-K \cdot (2 \cdot [E^{(i)}]^2 - 1) \cdot (N^{(i)} + m)$
$\Upsilon_2(n)$	$K \cdot (2 \cdot [E^{(i)}]^2 - 1) \cdot 2 \cdot \left(\frac{\Delta\Omega}{2}\right)^2 \cdot m$	$K \cdot (2 \cdot [E^{(i)}]^2 - 1) \cdot 2 \cdot \left(\frac{\Delta\Omega}{2}\right)^2 \cdot m$
$\Upsilon_3(n)$	$2 \cdot K \cdot (1 + 2 \cdot N^{(i)}) \cdot (E^{(i)})^2$	$-2 \cdot K \cdot (1 + 2 \cdot N^{(i)}) \cdot (E^{(i)})^2$
$\Upsilon_4(n)$	$N^{(i)} - m$	$-(N^{(i)} + m)$
$\Upsilon_5(n)$	$K \cdot (2 \cdot [E^{(i)}]^2 - 1) \cdot (m + 1)$	$-K \cdot (2 \cdot [E^{(i)}]^2 - 1) \cdot (m + 1)$
$\Upsilon_6(n)$	$m + 1$	$-(m + 1)$
$\Upsilon_7(n)$	$K \cdot (2 \cdot [E^{(i)}]^2 - 1) \cdot \left(\frac{\Delta\Omega}{2}\right)^2 \cdot m$	$K \cdot (2 \cdot [E^{(i)}]^2 - 1) \cdot \left(\frac{\Delta\Omega}{2}\right)^2 \cdot m$
$\Upsilon_8(n)$	$K \cdot (2 \cdot [E^{(i)}]^2 - 1) \cdot N^{(i)}$	$-K \cdot (2 \cdot [E^{(i)}]^2 - 1) \cdot N^{(i)}$
$\Upsilon_9(n)$	$2 \cdot K \cdot (1 + 2 \cdot N^{(i)}) \cdot (E^{(i)})^2$	$-2 \cdot K \cdot (1 + 2 \cdot N^{(i)}) \cdot (E^{(i)})^2$
$\Upsilon_{10}(n)$	$N^{(i)} + K \cdot (2 \cdot [E^{(i)}]^2 - 1) \cdot [1 - N^{(i)}]$ $+2 \cdot K \cdot (1 + 2 \cdot N^{(i)}) \cdot (E^{(i)})^2$	$-N^{(i)} + K \cdot (2 \cdot [E^{(i)}]^2 - 1) \cdot [-1 - N^{(i)}]$ $+2 \cdot K \cdot (1 + 2 \cdot N^{(i)}) \cdot (E^{(i)})^2$
$\Upsilon_{11}(n)$	$K \cdot (2 \cdot [E^{(i)}]^2 - 1) - [1 - N^{(i)}]$	$K \cdot (2 \cdot [E^{(i)}]^2 - 1) + [1 + N^{(i)}]$

We can summary our $\Upsilon_k(n); k = 1, 2, \ldots, 11$ expression for n parameter values ($n = 0, 1, 2, \ldots$) in the next table (Table 4.4):

$$\begin{aligned} D(\lambda, t_D) = \det|A - \lambda \cdot I| &= [\Upsilon_7 \cdot t_D^2 - \Upsilon_8 + \Upsilon_9] \\ &+ [\Upsilon_{10} - \left(\frac{\Delta\Omega}{2}\right)^2 \cdot m \cdot t_D^2] \cdot \lambda + \Upsilon_{11} \cdot \lambda^2 - \lambda^3 \\ &+ \{[\Upsilon_1 - \Upsilon_2 \cdot t_D^2 - \Upsilon_3] \\ &- \left[\Upsilon_4 - 2 \cdot \left(\frac{\Delta\Omega}{2}\right)^2 \cdot m \cdot t_D^2 + \Upsilon_5\right] \cdot \lambda + \Upsilon_6 \cdot \lambda^2\Big\} \cdot e^{-\lambda \cdot t_D} \end{aligned}$$

$$\begin{aligned} D(\lambda, t_D) &= P_n(\lambda, t_D) + Q_m(\lambda, t_D) \\ &\cdot e^{-\lambda \cdot t_D}; n, m \in \mathbb{N}_0; n > m; n = 3; m = 2 \end{aligned}$$

$$\begin{aligned} P_{n=3}(\lambda, t_D) &= \Upsilon_7 \cdot t_D^2 - \Upsilon_8 + \Upsilon_9 + \left[\Upsilon_{10} - \left(\frac{\Delta\Omega}{2}\right)^2 \cdot m \cdot t_D^2\right] \\ &\cdot \lambda + \Upsilon_{11} \cdot \lambda^2 - \lambda^3 Q_{m=2}(\lambda, t_D) = \Upsilon_1 - \Upsilon_2 \cdot t_D^2 - \Upsilon_3 \\ &- \left[\Upsilon_4 - 2 \cdot \left(\frac{\Delta\Omega}{2}\right)^2 \cdot m \cdot t_D^2 + \Upsilon_5\right] \cdot \lambda + \Upsilon_6 \cdot \lambda^2 \end{aligned}$$

$$P_{n=3}(\lambda, t_D) = \sum_{k=0}^{3} p_k(t_D) \cdot \lambda^k = p_0(t_D) + p_1(t_D) \cdot \lambda + p_2(t_D) \cdot \lambda^2 + p_3(t_D) \cdot \lambda^3$$

$$p_0(t_D) = \Upsilon_7 \cdot t_D^2 - \Upsilon_8 + \Upsilon_9; \; p_1(t_D) = \Upsilon_{10} - \left(\frac{\Delta\Omega}{2}\right)^2 \cdot m \cdot t_D^2; \; p_2(t_D) = \Upsilon_{11}; \; p_3(t_D) = -1$$

$$Q_{m=2}(\lambda, t_D) = \sum_{k=0}^{2} q_k(t_D) \cdot \lambda^k = q_0(t_D) + q_1(t_D) \cdot \lambda + q_2(t_D) \cdot \lambda^2$$

$$q_0(t_D) = \Upsilon_1 - \Upsilon_2 \cdot t_D^2 - \Upsilon_3; \; q_1(t_D) = -\left[\Upsilon_4 - 2 \cdot \left(\frac{\Delta\Omega}{2}\right)^2 \cdot m \cdot t_D^2 + \Upsilon_5\right]; \; q_2(t_D) = \Upsilon_6$$

The homogeneous system for $N\,E\,\phi$ leads to a characteristic equation for the eigenvalue λ having the form $P(\lambda, t_D) + Q(\lambda, t_D) \cdot e^{-\lambda \cdot t_D}$; $P(\lambda, t_D) = \sum_{j=0}^{3} a_j \cdot \lambda^j$; $Q(\lambda, t_D) = \sum_{j=0}^{2} c_j \cdot \lambda^j$. The coefficients $\{a_j(q_i, q_k, t_D), c_j(q_i, q_k, t_D)\} \in \mathbb{R}$ depend on q_i, q_k and delay$t_D.q_i, q_k$ are any lasers with frequency-modulated optical feedback system's parameters, other parameters kept as a constant [5–7].

$$a_0(t_D) = \Upsilon_7 \cdot t_D^2 - \Upsilon_8 + \Upsilon_9; \; a_1(t_D) = \Upsilon_{10} - \left(\frac{\Delta\Omega}{2}\right)^2 \cdot m \cdot t_D^2; \; a_2(t_D) = \Upsilon_{11}; \; a_3(t_D) = -1$$

$$c_0(t_D) = \Upsilon_1 - \Upsilon_2 \cdot t_D^2 - \Upsilon_3; \; c_1(t_D) = -\left[\Upsilon_4 - 2 \cdot \left(\frac{\Delta\Omega}{2}\right)^2 \cdot m \cdot t_D^2 + \Upsilon_5\right]; \; c_2(t_D) = \Upsilon_6$$

Unless strictly necessary, the designation of the varied arguments (q_i, q_k) will subsequently be omitted from P, Q, a_j, c_j. The coefficients a_j, c_j are continuous, and differentiable functions of their arguments and direct substitution show that $a_0 + c_0 \neq 0$ for $q_i, q_k \in \mathbb{R}_+$; that is, $\lambda = 0$ is not of $P(\lambda, t_D) + Q(\lambda, t_D) \cdot e^{-\lambda \cdot t_D} = 0$. Furthermore,$P(\lambda, t_D)$, $Q(\lambda, t_D)$.

Are analytic functions of λ, for which the following requirements of the analysis [5] can also be verified in the present case:

(a) If $\lambda = i \cdot \omega$; $\omega \in \mathbb{R}$, then $P(i \cdot \omega) + Q(i \cdot \omega) \neq 0$.
(b) If $|\frac{Q(\lambda, t_D)}{P(\lambda, t_D)}|$ is bounded for $|\lambda| \to \infty$; $\mathrm{Re}\lambda \geq 0$. No roots bifurcation from ∞.
(c) $F(\omega) = |P(i \cdot \omega)|^2 - |Q(i \cdot \omega)|^2$ has a finite number of zeros. Indeed, this is a polynomial in ω.
(d) Each positive root $\omega(q_i, q_k)$ of $F(\omega) = 0$ is continuous and differentiable with respect to q_i, q_k.

We assume that $P_n(\lambda, t_D)$, $Q_m(\lambda, t_D)$ cannot have common imaginary roots. That is for any real number ω: $P_n(\lambda = i \cdot \omega, t_D) + Q_m(\lambda = i \cdot \omega, t_D) \neq 0$.

$$P_n(\lambda = i \cdot \omega, t_D) = \Upsilon_7 \cdot t_D^2 - \Upsilon_8 + \Upsilon_9$$

$$+\left[\Upsilon_{10}-\left(\frac{\Delta\Omega}{2}\right)^2\cdot m\cdot t_D^2\right]\cdot i\cdot\omega-\Upsilon_{11}\cdot\omega^2-i\cdot\omega^3$$

$$P_n(\lambda=i\cdot\omega,t_D)=\left(\Upsilon_7\cdot t_D^2-\Upsilon_8+\Upsilon_9\right)$$
$$-\Upsilon_{11}\cdot\omega^2+i\cdot\left(\left[\Upsilon_{10}-\left(\frac{\Delta\Omega}{2}\right)^2\cdot m\cdot t_D^2\right]\cdot\omega-\omega^3\right)$$

$$Q_m(\lambda=i\cdot\omega,t_D)=\left[\Upsilon_1-\Upsilon_2\cdot t_D^2-\Upsilon_3\right]-\Upsilon_6\cdot\omega^2$$
$$-\left[\Upsilon_4-2\cdot\left(\frac{\Delta\Omega}{2}\right)^2\cdot m\cdot t_D^2+\Upsilon_5\right]\cdot i\cdot\omega$$

$$P_n(\lambda=i\cdot\omega,t_D)+Q_m(\lambda=i\cdot\omega,t_D)$$
$$=\left(\Upsilon_7\cdot t_D^2-\Upsilon_8+\Upsilon_9+\Upsilon_1-\Upsilon_2\cdot t_D^2-\Upsilon_3\right)-(\Upsilon_{11}+\Upsilon_6)\cdot\omega^2$$
$$+i\cdot\left(\left[\Upsilon_{10}-\left(\frac{\Delta\Omega}{2}\right)^2\cdot m\cdot t_D^2\right.\right.$$
$$\left.\left.-\Upsilon_4+2\cdot\left(\frac{\Delta\Omega}{2}\right)^2\cdot m\cdot t_D^2-\Upsilon_5\right]\cdot\omega-\omega^3\right)$$

$$|P(i\cdot\omega,t_D)|^2=\left[\left(\Upsilon_7\cdot t_D^2-\Upsilon_8+\Upsilon_9\right)-\Upsilon_{11}\cdot\omega^2\right]^2$$
$$+\left[\left(\Upsilon_{10}-\left(\frac{\Delta\Omega}{2}\right)^2\cdot m\cdot t_D^2\right)\cdot\omega-\omega^3\right]^2$$

$$|P(i\cdot\omega,t_D)|^2=\left(\Upsilon_7\cdot t_D^2-\Upsilon_8+\Upsilon_9\right)^2$$
$$+\left[\left(\Upsilon_{10}-\left(\frac{\Delta\Omega}{2}\right)^2\cdot m\cdot t_D^2\right)^2\right.$$
$$\left.-2\cdot\left(\Upsilon_7\cdot t_D^2-\Upsilon_8+\Upsilon_9\right)\cdot\Upsilon_{11}\right]\cdot\omega^2$$
$$+\left[\Upsilon_{11}^2-2\cdot\left(\Upsilon_{10}-\left(\frac{\Delta\Omega}{2}\right)^2\cdot m\cdot t_D^2\right)\right]\cdot\omega^4+\omega^6$$

$$|Q(i\cdot\omega,t_D)|^2=\left(\left[\Upsilon_1-\Upsilon_2\cdot t_D^2-\Upsilon_3\right]-\Upsilon_6\cdot\omega^2\right)^2$$
$$+\left[\Upsilon_4-2\cdot\left(\frac{\Delta\Omega}{2}\right)^2\cdot m\cdot t_D^2+\Upsilon_5\right]^2\cdot\omega^2$$

$$|Q(i\cdot\omega,t_D)|^2=\left[\Upsilon_1-\Upsilon_2\cdot t_D^2-\Upsilon_3\right]^2$$
$$+\left(\left[\Upsilon_4-2\cdot\left(\frac{\Delta\Omega}{2}\right)^2\cdot m\cdot t_D^2+\Upsilon_5\right]^2\right.$$
$$\left.-2\cdot\left[\Upsilon_1-\Upsilon_2\cdot t_D^2-\Upsilon_3\right]\cdot\Upsilon_6\right)\cdot\omega^2+\Upsilon_6^2\cdot\omega^4$$

$$F(\omega,t_D)=|P(i\cdot\omega,t_D)|^2-|Q(i\cdot\omega,t_D)|^2$$

$$= (\Upsilon_7 \cdot t_D^2 - \Upsilon_8 + \Upsilon_9)^2 - [\Upsilon_1 - \Upsilon_2 \cdot t_D^2 - \Upsilon_3]^2$$
$$+ \left[\left(\Upsilon_{10} - \left(\frac{\Delta\Omega}{2} \right)^2 \cdot m \cdot t_D^2 \right)^2 - 2 \cdot (\Upsilon_7 \cdot t_D^2 - \Upsilon_8 + \Upsilon_9) \cdot \Upsilon_{11} \right] \cdot \omega^2$$
$$- \left(\left[\Upsilon_4 - 2 \cdot \left(\frac{\Delta\Omega}{2} \right)^2 \cdot m \cdot t_D^2 + \Upsilon_5 \right]^2 \right.$$
$$\left. - -2 \cdot [\Upsilon_1 - \Upsilon_2 \cdot t_D^2 - \Upsilon_3] \cdot \Upsilon_6 \right) \cdot \omega^2$$
$$+ \left[\Upsilon_{11}^2 - 2 \cdot \left(\Upsilon_{10} - \left(\frac{\Delta\Omega}{2} \right)^2 \cdot m \cdot t_D^2 \right) \right] \cdot \omega^4 - \Upsilon_6^2 \cdot \omega^4 + \omega^6$$

$$F(\omega, t_D) = |P(i \cdot \omega, t_D)|^2 - |Q(i \cdot \omega, t_D)|^2$$
$$= (\Upsilon_7 \cdot t_D^2 - \Upsilon_8 + \Upsilon_9)^2 - [\Upsilon_1 - \Upsilon_2 \cdot t_D^2 - \Upsilon_3]^2$$
$$+ [(\Upsilon_{10} - \left(\frac{\Delta\Omega}{2} \right)^2 \cdot m \cdot t_D^2)^2 - 2 \cdot (\Upsilon_7 \cdot t_D^2 - \Upsilon_8 + \Upsilon_9)$$
$$\cdot \Upsilon_{11} - \left(\left[\Upsilon_4 - 2 \cdot \left(\frac{\Delta\Omega}{2} \right)^2 \cdot m \cdot t_D^2 + \Upsilon_5 \right]^2 \right.$$
$$\left. -2 \cdot [\Upsilon_1 - \Upsilon_2 \cdot t_D^2 - \Upsilon_3] \cdot \Upsilon_6 \right)] \cdot \omega^2$$
$$+ \left(\left[\Upsilon_{11}^2 - 2 \cdot \left(\Upsilon_{10} - \left(\frac{\Delta\Omega}{2} \right)^2 \cdot m \cdot t_D^2 \right) \right] - \Upsilon_6^2 \right) \cdot \omega^4 + \omega^6$$

We define the following parameters for simplicity: $\Pi_0, \Pi_2, \Pi_4, \Pi_6$.

$$\Pi_0 = (\Upsilon_7 \cdot t_D^2 - \Upsilon_8 + \Upsilon_9)^2 - [\Upsilon_1 - \Upsilon_2 \cdot t_D^2 - \Upsilon_3]^2$$
$$\Pi_2 = \left(\Upsilon_{10} - \left(\frac{\Delta\Omega}{2} \right)^2 \cdot m \cdot t_D^2 \right)^2$$
$$- 2 \cdot (\Upsilon_7 \cdot t_D^2 - \Upsilon_8 + \Upsilon_9) \cdot \Upsilon_{11} - \left(\left[[\Upsilon_4 - 2 \cdot \left(\frac{\Delta\Omega}{2} \right)^2 \cdot m \cdot t_D^2 + \Upsilon_5 \right]^2 \right.$$
$$\left. -2 \cdot [\Upsilon_1 - \Upsilon_2 \cdot t_D^2 - \Upsilon_3] \cdot \Upsilon_6 \right)$$
$$\Pi_4 = \Upsilon_{11}^2 - 2 \cdot \left(\Upsilon_{10} - \left(\frac{\Delta\Omega}{2} \right)^2 \cdot m \cdot t_D^2 \right) - \Upsilon_6^2; \ \Pi_6 = 1$$

Hence $F(\omega, t_D) = 0$ implies $\sum_{k=0}^{3} \Pi_{2 \cdot k} \cdot \omega^{2 \cdot k} = 0$ and its roots are given by solving the above polynomial.

$$Q_R(i\cdot\omega,\tau) = \left(\frac{1}{T_L\cdot T_R}+\frac{P^{(j)}}{E_L}\right)\cdot\left(\frac{1}{T_A\cdot T_R}+\frac{P^{(j)}}{E_A}\right)\cdot\frac{2}{T_R}\cdot(g^{(j)}-l-q^{(j)})$$
$$+\frac{2}{T_R}\cdot P^{(j)}\cdot\frac{q^{(j)}}{E_A}\cdot\left(\frac{1}{T_L\cdot T_R}+\frac{P^{(j)}}{E_L}\right)$$
$$-\frac{2}{T_R}\cdot P^{(j)}\cdot\frac{g^{(j)}}{E_L}\cdot\left(\frac{1}{T_A\cdot T_R}+\frac{P^{(j)}}{E_A}\right)-\frac{2}{T_R}\cdot(g^{(j)}-l-q^{(j)})\cdot\omega^2$$
$$Q_I(i\cdot\omega,\tau) = \left\{\frac{2}{T_R}\cdot(g^{(j)}-l-q^{(j)})\cdot\left(\frac{1}{T_L\cdot T_R}+\frac{1}{T_A\cdot T_R}+\frac{P^{(j)}}{E_A}+\frac{P^{(j)}}{E_L}\right)\right.$$
$$\left.+\frac{2}{T_R}\cdot P^{(j)}\cdot\frac{q^{(j)}}{E_A}-\frac{2}{T_R}\cdot P^{(j)}\cdot\frac{g^{(j)}}{E_L}\right\}\cdot\omega$$
$$\sin\theta(\tau) = \frac{-P_R(i\cdot\omega,\tau)\cdot Q_I(i\cdot\omega,\tau)+P_I(i\cdot\omega,\tau)\cdot Q_R(i\cdot\omega,\tau)}{|Q(i\cdot\omega,\tau)|^2}$$
$$\cos\theta(\tau) = -\frac{P_R(i\cdot\omega,\tau)\cdot Q_R(i\cdot\omega,\tau)+P_I(i\cdot\omega,\tau)\cdot Q_i(i\cdot\omega,\tau)}{|Q(i\cdot\omega,\tau)|^2}$$

We use different parameters terminology from our last characteristics parameters definition: $k \to j;\ p_k(t_D) \to a_j;\ q_k(t_D) \to c_j;\ n = 3;\ m = 2;\ n > m;\ a_j = a_j(t_D);\ c_j = c_j(t_D)$

$$P_n(\lambda,t_D) \to P(\lambda,t_D);\ Q_m(\lambda,t_D) \to Q(\lambda,t_D);\ P(\lambda,t_D)$$
$$= \sum_{j=0}^{3} a_j\cdot\lambda^j;\ Q(\lambda,t_D) = \sum_{j=0}^{2} c_j\cdot\lambda^j$$

$$P(\lambda,t_D) = \sum_{j=0}^{3} a_j\cdot\lambda^j = a_0 + a_1\cdot\lambda + a_2\cdot\lambda^2$$
$$+ a_3\cdot\lambda^3;\ Q(\lambda,t_D) = \sum_{j=0}^{2} c_j\cdot\lambda^j = c_0 + c_1\cdot\lambda + c_2\cdot\lambda^2$$

$n, m \in \mathbb{N}_0;\ n > m$ and $a_j, c_j: \mathbb{R}_{+0} \to \mathbb{R}$ are continuous and differentiable function of t_D such that $a_0 + c_0 \neq 0$. In the following "–" denotes complex and conjugate. $P(\lambda,t_D)$, $Q(\lambda,t_D)$ are analytic functions in λ and differentiable in t_D. The coefficients $a_j(K,\omega,\varepsilon,\Delta\Omega,\beta,t_D,\ldots) \in \mathbb{R}$ and $c_j(K,\omega,\varepsilon,\Delta\Omega,\beta,t_D,\ldots) \in \mathbb{R}$ depend on lasers with frequency-modulated optical feedback system's $K, \omega, \varepsilon, \Delta\Omega, \beta, \Omega_m, \Omega_0, \Delta\Omega, t_D, \ldots$ values.

Remark we use ω as system relative pump power normalized by the threshold $\left(\omega = \frac{P}{P_{th}}\right)$ and also for Geometric stability switch criteria ω parameter ($\lambda = i\cdot\omega;\ \omega \in \mathbb{R}$). Reader must differentiate between the two utilizations. Additionally ε is used as a system spontaneous emission rate and as a small positive infinitesimal quantity in math.

Unless strictly necessary, the designation of the varied arguments: $(K, \omega, \varepsilon, \Delta\Omega, \beta, t_D, \ldots)$ will subsequently be omitted from P, Q, a_j, c_j are continuous and differentiable functions of their arguments, and direct substitution shows that $a_0 + c_0 \neq 0$.

$$a_0(t_D) = \Upsilon_7 \cdot t_D^2 - \Upsilon_8 + \Upsilon_9;\ c_0(t_D) = \Upsilon_1 - \Upsilon_2 \cdot t_D^2 - \Upsilon_3$$
$$a_0 + c_0 = -\Upsilon_8 + \Upsilon_9 + \Upsilon_1 - \Upsilon_3 + (\Upsilon_7 - \Upsilon_2) \cdot t_D^2 \neq 0;\ t_D^2$$
$$\neq \frac{\Upsilon_8 - \Upsilon_9 - \Upsilon_1 + \Upsilon_3}{\Upsilon_7 - \Upsilon_2};\ t_D \neq \sqrt{\frac{\Upsilon_8 - \Upsilon_9 - \Upsilon_1 + \Upsilon_3}{\Upsilon_7 - \Upsilon_2}}$$

$\forall K, \omega, \varepsilon, \Delta\Omega, \beta, \Omega_m, \Omega_0, \Delta\Omega, t_D, \ldots \in \mathbb{R}_+$ i.e. $\lambda = 0$ is not a root of the characteristic equation. Furthermore $P(\lambda, t_D)$, $Q(\lambda, t_D)$ are analytic functions of λ for which the following requirements of the analysis (see Kuang 1993, Sect. 3.4) can also be verified in the present case.

(a) If $\lambda = i \cdot \omega; \omega \in \mathbb{R}$ then $P(i \cdot \omega) + Q(i \cdot \omega) \neq 0$ i.e. P and Q have no common imaginary roots. This condition was verified numerically in the entire $(K, \omega, \varepsilon, \Delta\Omega, t_D, \ldots)$ domain of interest.

(b) $\left|\frac{P(\lambda, t_D)}{Q(\lambda, t_D)}\right|$ is bounded for $|\lambda| \to \infty$; $\mathrm{Re}\lambda \geq 0$. No roots bifurcation from ∞. Indeed, in the limit:

$$\left|\frac{P(\lambda, t_D)}{Q(\lambda, t_D)}\right| = \left|\frac{\Upsilon_7 \cdot t_D^2 - \Upsilon_8 + \Upsilon_9 + [\Upsilon_{10} - \left(\frac{\Delta\Omega}{2}\right)^2 \cdot m \cdot t_D^2] \cdot \lambda + \Upsilon_{11} \cdot \lambda^2 - \lambda^3}{\Upsilon_1 - \Upsilon_2 \cdot t_D^2 - \Upsilon_3 - [\Upsilon_4 - 2 \cdot \left(\frac{\Delta\Omega}{2}\right)^2 \cdot m \cdot t_D^2 + \Upsilon_5] \cdot \lambda + \Upsilon_6 \cdot \lambda^2}\right|$$

(c) The following expressions exist: $F(\omega, t_D) = |P(i \cdot \omega, t_D)|^2 - |Q(i \cdot \omega, t_D)|^2$

$$F(\omega, t_D) = |P(i \cdot \omega, t_D)|^2 - |Q(i \cdot \omega, t_D)|^2 = \sum_{k=0}^{3} \Pi_{2 \cdot k} \cdot \omega^{2 \cdot k}$$

Has at most a finite number of zeros. Indeed, this is a polynomial in ω (degree in ω^6).

(d) Each positive root $\omega(K, \omega, \varepsilon, \Delta\Omega, t_D, \ldots)$ of $F(\omega, t_D) = 0$ is continuous and differentiable with respect to $K, \omega, \varepsilon, \Delta\Omega, t_D, \ldots$ The condition can only be assessed numerically.

In addition, since the coefficients P and Q are real, we have $\overline{P(-i \cdot \omega)} = P(i \cdot \omega)$ and $\overline{Q(-i \cdot \omega)} = Q(i \cdot \omega)$ thus, $\omega > 0$ maybe on eigenvalue of characteristic equations. The analysis consists in identifying the roots of the characteristic equation situated on the imaginary axis of the complex λ—plane, whereby increasing the parameters: $K, \omega, \varepsilon, \Delta\Omega, \beta, \Omega_m, \Omega_0, \Delta\Omega, t_D, \ldots$ Re λ may, at the crossing, change its sign from (−) to (+), i.e. from stable focus($N^{(k)}$, $E^{(k)}$, $\phi^{(k)}$); $k = 0, 1, 2$ to an unstable one, or vice versa. This feature may be further assessed by examining the sign of the partial derivatives with respect to $K, \omega, \varepsilon, \Delta\Omega, \beta, \Omega_m, t_D, \ldots$ and system parameters [5–7].

$$\Lambda^{-1}(t_D) = (\frac{\partial \mathrm{Re}\lambda}{\partial t_D})_{\lambda = i\cdot\omega};\ K, \omega, \varepsilon, \Delta\Omega, \beta, \Omega_m, \Omega_0, \Delta\Omega, \ldots = const$$
$$\Lambda^{-1}(K) = (\frac{\partial \mathrm{Re}\lambda}{\partial K})_{\lambda = i\cdot\omega};\ t_D, \omega, \varepsilon, \Delta\Omega, \beta, \Omega_m, \Omega_0, \Delta\Omega, \ldots = const$$
$$\Lambda^{-1}(\omega) = (\frac{\partial \mathrm{Re}\lambda}{\partial \omega})_{\lambda = i\cdot\omega};\ K, t_D, \varepsilon, \Delta\Omega, \beta, \Omega_m, \Omega_0, \Delta\Omega, \ldots = const$$
$$\Lambda^{-1}(\varepsilon) = (\frac{\partial \mathrm{Re}\lambda}{\partial \varepsilon})_{\lambda = i\cdot\omega};\ K, \omega, t_D, \Delta\Omega, \beta, \Omega_m, \Omega_0, \Delta\Omega, \ldots = const$$
$$\Lambda^{-1}(\Delta\Omega) = (\frac{\partial \mathrm{Re}\lambda}{\partial \Delta\Omega})_{\lambda = i\cdot\omega};\ K, \omega, \varepsilon, t_D, \beta, \Omega_m, \Omega_0, \Delta\Omega, \ldots = const$$
$$\Lambda^{-1}(\beta) = (\frac{\partial \mathrm{Re}\lambda}{\partial \beta})_{\lambda = i\cdot\omega};\ K, \omega, \varepsilon, \Delta\Omega, t_D, \Omega_m, \Omega_0, \Delta\Omega, \ldots = const$$
$$\Lambda^{-1}(\Omega_m) = (\frac{\partial \mathrm{Re}\lambda}{\partial \Omega_m})_{\lambda = i\cdot\omega};\ K, \omega, \varepsilon, \Delta\Omega, \beta, t_D, \Omega_0, \Delta\Omega, \ldots = const$$
$$\Lambda^{-1}(\Omega_0) = (\frac{\partial \mathrm{Re}\lambda}{\partial \Omega_0})_{\lambda = i\cdot\omega};\ K, \omega, \varepsilon, \Delta\Omega, \beta, \Omega_m, t_D, \Delta\Omega, \ldots = const$$
$$\Lambda^{-1}(\Delta\Omega) = (\frac{\partial \mathrm{Re}\lambda}{\partial \Delta\Omega})_{\lambda = i\cdot\omega};\ K, \omega, \varepsilon, \Delta\Omega, \beta, \Omega_m, \Omega_0, t_D, \ldots = const$$
$$P(\lambda) = P_R(\lambda) + i \cdot P_I(\lambda);\ Q(\lambda) = Q_R(\lambda) + i \cdot Q_I(\lambda)$$

When writing and inserting $\lambda = i \cdot \omega$ into lasers with frequency-modulated optical feedback system's characteristic equation ω must satisfy the following equations.

$$\sin(\omega \cdot t_D) = g(\omega) = \frac{-P_R(i\cdot\omega, t_D)\cdot Q_I(i\cdot\omega, t_D) + P_I(i\cdot\omega, t_D)\cdot Q_R(i\cdot\omega, t_D)}{|Q(i\cdot\omega, t_D)|^2}$$
$$\cos(\omega \cdot t_D) = h(\omega) = -\frac{P_R(i\cdot\omega, t_D)\cdot Q_R(i\cdot\omega, t_D) + P_I(i\cdot\omega, t_D)\cdot Q_I(i\cdot\omega, t_D)}{|Q(i\cdot\omega, t_D)|^2}$$

where $|Q(i\cdot\omega, t_D)|^2 \neq 0$ in view of requirement (a) above, and $(g, h) \in \mathbb{R}$. Furthermore, it follows above $\sin(\omega\cdot t_D)$ and $\cos(\omega\cdot t_D)$ equations that, by squaring and adding the sides, ω must be a positive root of $F(\omega, t_D) = |P(i\cdot\omega, t_D)|^2 - |Q(i\cdot\omega, t_D)|^2 = 0$. Note: $F(\omega, t_D)$ is dependent on t_D. Now it is important to notice that if $t_D \notin I$(assume that $I \subseteq \mathbb{R}_{+0}$ is the set where $\omega(t_D)$ is a positive root of $F(\omega, t_D)$ and for, $t_D \notin I$, $\omega(t_D)$ is not defined. Then for all t_D in I, $\omega(t_D)$ is satisfied that $F(\omega, t_D) = 0$. Then there are no positive $\omega(t_D)$ solutions for $F(\omega, t_D) = 0$, and we cannot have stability switches. For $t_D \in I$ where $\omega(t_D)$ is a positive solution of $F(\omega, t_D) = 0$, we can define the angle $\theta(t_D) \in [0, 2\cdot\pi]$ as the solution of $\sin\theta(t_D) = \ldots$; $\cos\theta(t_D) = \ldots$.

$$\sin\theta(t_D) = \frac{-P_R(i\cdot\omega,t_D)\cdot Q_I(i\cdot\omega,t_D) + P_I(i\cdot\omega,t_D)\cdot Q_R(i\cdot\omega,t_D)}{|Q(i\cdot\omega,t_D)|^2}$$

$$\cos\theta(t_D) = -\frac{P_R(i\cdot\omega,t_D)\cdot Q_R(i\cdot\omega,t_D) + P_I(i\cdot\omega,t_D)\cdot Q_I(i\cdot\omega,t_D)}{|Q(i\cdot\omega,t_D)|^2}$$

And the relation between the argument $\theta(t_D)$ and $t_D\cdot\omega(t_D)$ for $t_D \in I$ must be as describe below.

$$\omega(t_D)\cdot t_D = \theta(t_D) + 2\cdot n\cdot\pi\ \forall n \in \mathbb{N}_0$$

Hence we can define the maps $t_{D(n)} : I \to \mathbb{R}_{+0}$ given by $t_{D(n)}(t_D) = \frac{\theta(t_D)+2\cdot n\cdot\pi}{\omega(t_D)}$; $n \in \mathbb{N}_0$; $t_D \in I$. Let us introduce the functions: $I \to \mathbb{R}$; $S_n(t_D) = t_D - t_{D(n)}(t_D)$; $t_D \in I$; $n \in \mathbb{N}_0$, that is continuous and differentiable in t_D. In the following, the subscripts $\lambda, K, \omega, \varepsilon, \Delta\Omega, \beta, \Omega_m, \ldots$ indicate the corresponding partial derivatives. Let us first concentrate on $\Lambda(x)$, remember in $\lambda(K,\omega,\varepsilon,\Delta\Omega,\beta,\Omega_m,\ldots)$ and $\omega(K,\omega,\varepsilon,\Delta\Omega,\beta,\Omega_m,\ldots)$, and keeping all parameters except one (x) and t_D. The derivation closely follows that in reference [BK], Differentiating system characteristic equation $P(\lambda,t_D) + Q(\lambda,t_D)\cdot e^{-\lambda\cdot t_D} = 0$ with respect to specific parameter (x), and inverting the derivative, for convenience, one calculate: $x = K, \omega, \varepsilon, \Delta\Omega, \beta, \Omega_m, \ldots$

$$\left(\frac{\partial\lambda}{\partial x}\right)^{-1} = \frac{-P_\lambda(\lambda,x)\cdot Q(\lambda,x) + Q_\lambda(\lambda,x)\cdot P(\lambda,x) - t_D\cdot P(\lambda,x)\cdot Q(\lambda,x)}{P_x(\lambda,x)\cdot Q(\lambda,x) - Q_x(\lambda,x)\cdot P(\lambda,x)}$$

where $P_\lambda = \frac{\partial P}{\partial\lambda}, \ldots$ etc., substituting $\lambda = i\cdot\omega$ and bearing $\overline{P(-i\cdot\omega)} = P(i\cdot\omega)$; $\overline{Q(-i\cdot\omega)} = Q(i\cdot\omega)$ and $i\cdot P_\lambda(i\cdot\omega) = P_\omega(i\cdot\omega)$; $i\cdot Q_\lambda(i\cdot\omega) = Q_\omega(i\cdot\omega)$ and that on the surface $|P(i\cdot\omega,t_D)|^2 = |Q(i\cdot\omega,t_D)|^2$, one obtains:

$$\left(\frac{\partial\lambda}{\partial x}\right)^{-1}|_{\lambda=i\cdot\omega} = \left(\frac{i\cdot P_\omega(i\cdot\omega,x)\cdot\overline{P(i\cdot\omega,x)} + i\cdot Q_\lambda(i\cdot\omega,x)\cdot\overline{Q(\lambda,x)} - t_D\cdot|P(i\cdot\omega,x)|^2}{P_x(i\cdot\omega,x)\cdot\overline{P(i\cdot\omega,x)} - Q_x(i\cdot\omega,x)\cdot\overline{Q(i\cdot\omega,x)}}\right)$$

Upon separating into real and imaginary parts, with $P = P_R + i\cdot P_I$; $Q = Q_R + i\cdot Q_I$ and $P_\omega = P_{R\omega} + i\cdot P_{I\omega}$; $Q_\omega = Q_{R\omega} + i\cdot Q_{I\omega}$; $P_x = P_{Rx} + i\cdot P_{Ix}$; $Q_x = Q_{Rx} + i\cdot Q_{Ix}$; $P^2 = P_R^2 + P_I^2$. When (x) can be and lasers with frequency-modulated optical feedback system's parameters $K, \omega, \varepsilon, \Delta\Omega, \beta, \Omega_m, \Omega_0, \Delta\Omega, \ldots$ and the time delayt_D etc. Where for convenience, we dropped the arguments $(i\cdot\omega, x)$, and where $F_\omega = 2\cdot[(P_{R\omega}\cdot P_R + P_{I\omega}\cdot P_I) - (Q_{R\omega}\cdot Q_R + Q_{I\omega}\cdot Q_I)]$ and $F_x = 2\cdot[(P_{Rx}\cdot P_R + P_{Ix}\cdot P_I) - (Q_{Rx}\cdot Q_R + Q_{Ix}\cdot Q_I)]$; $\omega_x = -\frac{F_x}{F_\omega}$. We define U and V:

$U = (P_R\cdot P_{I\omega} - P_I\cdot P_{R\omega}) - (Q_R\cdot Q_{I\omega} - Q_I\cdot Q_{R\omega})$; $V = (P_R\cdot P_{Ix} - P_I\cdot P_{Rx}) - (Q_R\cdot Q_{Ix} - Q_I\cdot Q_{Rx})$.

We choose our specific parameter as time delay $x = t_D$.

$$P_R(i \cdot \omega, t_D) = \left(\Upsilon_7 \cdot t_D^2 - \Upsilon_8 + \Upsilon_9\right) - \Upsilon_{11} \cdot \omega^2$$

$$P_I(i \cdot \omega, t_D) = \left[\Upsilon_{10} - \left(\frac{\Delta\Omega}{2}\right)^2 \cdot m \cdot t_D^2\right] \cdot \omega - \omega^3$$

$$Q_R(i \cdot \omega, t_D) = \left[\Upsilon_1 - \Upsilon_2 \cdot t_D^2 - \Upsilon_3\right] - \Upsilon_6 \cdot \omega^2$$

$$Q_I(i \cdot \omega, t_D) = -\left[\Upsilon_4 - 2 \cdot \left(\frac{\Delta\Omega}{2}\right)^2 \cdot m \cdot t_D^2 + \Upsilon_5\right]] \cdot \omega$$

$$P_R = P_R(i \cdot \omega, t_D);\ P_I = P_I(i \cdot \omega, t_D);\ Q_R$$
$$= Q_R(i \cdot \omega, t_D);\ Q_I = Q_I(i \cdot \omega, t_D)$$

$$P_{R\omega} = -2 \cdot \Upsilon_{11} \cdot \omega;\ P_{I\omega} = \left[\Upsilon_{10} - \left(\frac{\Delta\Omega}{2}\right)^2 \cdot m \cdot t_D^2\right]$$
$$-3 \cdot \omega^2;\ Q_{R\omega} = -2 \cdot \Upsilon_6 \cdot \omega;\ Q_{I\omega} = -\left[\Upsilon_4 - 2 \cdot \left(\frac{\Delta\Omega}{2}\right)^2 \cdot m \cdot t_D^2 + \Upsilon_5\right]$$

$$P_{Rt_D} = 2 \cdot \Upsilon_7 \cdot t_D;\ P_{It_D} = -2 \cdot \left(\frac{\Delta\Omega}{2}\right)^2 \cdot m \cdot t_D \cdot \omega;\ Q_{Rt_D}$$
$$= -2 \cdot \Upsilon_2 \cdot t_D;\ Q_{It_D} = 4 \cdot \left(\frac{\Delta\Omega}{2}\right)^2 \cdot m \cdot t_D \cdot \omega$$

$$P_{R\omega} \cdot P_R = -2 \cdot \Upsilon_{11} \cdot \omega \cdot \left[\left(\Upsilon_7 \cdot t_D^2 - \Upsilon_8 + \Upsilon_9\right) - \Upsilon_{11} \cdot \omega^2\right]$$
$$Q_{R\omega} \cdot Q_R = -2 \cdot \Upsilon_6 \cdot \omega \cdot \left(\left[\Upsilon_1 - \Upsilon_2 \cdot t_D^2 - \Upsilon_3\right] - \Upsilon_6 \cdot \omega^2\right])$$

$$\begin{aligned} F_{t_D} &= 2 \cdot [(P_{Rt_D} \cdot P_R + P_{It_D} \cdot P_I) - (Q_{Rt_D} \cdot Q_R + Q_{It_D} \cdot Q_I)] \\ &= 2 \cdot \left[\left(2 \cdot \Upsilon_7 \cdot t_D \cdot \left[\left(\Upsilon_7 \cdot t_D^2 - \Upsilon_8 + \Upsilon_9\right) - \Upsilon_{11} \cdot \omega^2\right]\right.\right. \\ &\quad -2 \cdot \left(\frac{\Delta\Omega}{2}\right)^2 \cdot m \cdot t_D \cdot \omega \cdot \left.\left\{\left[\Upsilon_{10} - \left(\frac{\Delta\Omega}{2}\right)^2 \cdot m \cdot t_D^2\right] \cdot \omega - \omega^3\right\}\right) \\ &\quad - (-2 \cdot \Upsilon_2 \cdot t_D \cdot \{[\Upsilon_1 - \Upsilon_2 \cdot t_D^2 - \Upsilon_3] - \Upsilon_6 \cdot \omega^2\} \\ &\quad \left.\left. -4 \cdot \left(\frac{\Delta\Omega}{2}\right)^2 \cdot m \cdot t_D \cdot \omega \cdot \left[\Upsilon_4 - 2 \cdot \left(\frac{\Delta\Omega}{2}\right)^2 \cdot m \cdot t_D^2 + \Upsilon_5\right] \cdot \omega\right)\right] \end{aligned}$$

$$\begin{aligned} Q_R \cdot Q_{I\omega} &= -\left(\left[\Upsilon_1 - \Upsilon_2 \cdot t_D^2 - \Upsilon_3\right]\right. \\ &\quad \left. -\Upsilon_6 \cdot \omega^2\right) \cdot \left[\Upsilon_4 - 2 \cdot \left(\frac{\Delta\Omega}{2}\right)^2 \cdot m \cdot t_D^2 + \Upsilon_5\right] \end{aligned}$$

$$Q_I \cdot Q_{R\omega} = \left[\Upsilon_4 - 2 \cdot \left(\frac{\Delta\Omega}{2}\right)^2 \cdot m \cdot t_D^2 + \Upsilon_5\right] \cdot 2 \cdot \Upsilon_6 \cdot \omega^2$$

$$V = (P_R \cdot P_{It_D} - P_I \cdot P_{Rt_D}) - (Q_R \cdot Q_{It_D} - Q_I \cdot Q_{Rt_D})$$

$$= \left(-[(\Upsilon_7 \cdot t_D^2 - \Upsilon_8 + \Upsilon_9) - \Upsilon_{11} \cdot \omega^2] \cdot 2 \cdot \left(\frac{\Delta\Omega}{2}\right)^2 \cdot m \cdot t_D \cdot \omega \right.$$

$$\left. - \left\{ \left[\Upsilon_{10} - \left(\frac{\Delta\Omega}{2}\right)^2 \cdot m \cdot t_D^2 \right] \cdot \omega - \omega^3 \right\} \cdot 2 \cdot \Upsilon_7 \cdot t_D \right)$$

$$- \left(\{[\Upsilon_1 - \Upsilon_2 \cdot t_D^2 - \Upsilon_3] - \Upsilon_6 \cdot \omega^2\} \cdot 4 \cdot \left(\frac{\Delta\Omega}{2}\right)^2 \cdot m \cdot t_D \cdot \omega \right.$$

$$\left. - \left[\Upsilon_4 - 2 \cdot \left(\frac{\Delta\Omega}{2}\right)^2 \cdot m \cdot t_D^2 + \Upsilon_5 \right] \cdot \omega \cdot 2 \cdot \Upsilon_2 \cdot t_D \right)$$

$F(\omega, t_D) = 0$. Differentiating with respect to t_D and we get $F_\omega \cdot \frac{\partial \omega}{\partial t_D} + F_{t_D} = 0$.

$$t_D \in I \Rightarrow \frac{\partial \omega}{\partial t_D} = -\frac{F_{t_D}}{F_\omega}; \ \Lambda^{-1}(t_D)$$

$$= \left(\frac{\partial \mathrm{Re}\lambda}{\partial t_D}\right)_{\lambda = i \cdot \omega}; \ \frac{\partial \omega}{\partial t_D} = \omega_{t_D} = -\frac{F_{t_D}}{F_\omega}$$

$$\Lambda^{-1}(t_D) = \mathrm{Re}\left\{ \frac{-2 \cdot (U + t_D \cdot |P|^2) + i \cdot F_\omega}{F_{t_D} + i \cdot 2 \cdot (V + \omega \cdot |P|^2)} \right\}$$

$$sign\{\Lambda^{-1}(t_D)\} = sign\left\{ \left(\frac{\partial \mathrm{Re}\lambda}{\partial t_D}\right)_{\lambda = i \cdot \omega} \right\}$$

$$sign\{\Lambda^{-1}(t_D)\} = sign\{F_\omega\} \cdot sign\left\{ \frac{V + \frac{\partial \omega}{\partial t_D} \cdot U}{|P|^2} + \omega + \frac{\partial \omega}{\partial t_D} \cdot t_D \right\}$$

We shall presently examine the possibility of stability transitions (bifurcations) lasers with frequency-modulated optical feedback system, about the equilibrium point $(N^{(k)}, E^{(k)}, \phi^{(k)})$; $k = 0, 1, 2$ as a result of a variation of delay parametert_D. The analysis consists in identifying the roots of our system characteristic equation situated on the imaginary axis of the complex λ-plane where by increasing the delay parametert_D, Reλ may at the crossing, changes its sign from "−" to "+", i.e. from a stable focus $(N^{(k)}, E^{(k)}, \phi^{(k)})$; $k = 0, 1, 2$ to an unstable one, or vice versa. This feature may be further assessed by examining the sign of the partial derivatives with respect to t_D.

Hint: t_D is delay time normalized by the photon lifetime (delay parameter).

$$\Lambda^{-1}(t_D) = \left(\frac{\partial \mathrm{Re}\lambda}{\partial t_D}\right)_{\lambda = i \cdot \omega}; \ \Lambda^{-1}(t_D) = \left(\frac{\partial \mathrm{Re}\lambda}{\partial t_D}\right)_{\lambda = i \cdot \omega};$$

$$K, \omega, \varepsilon, \Delta\Omega, \beta, \Omega_m, \Omega_0, \Delta\Omega, \ldots = const$$

$$U = (P_R \cdot P_{I\omega} - P_I \cdot P_{R\omega}) - (Q_R \cdot Q_{I\omega} - Q_I \cdot Q_{R\omega})$$

$$U = [(\Upsilon_7 \cdot t_D^2 - \Upsilon_8 + \Upsilon_9) - \Upsilon_{11} \cdot \omega^2]$$
$$\cdot \left\{ \left[\Upsilon_{10} - \left(\frac{\Delta\Omega}{2}\right)^2 \cdot m \cdot t_D^2 \right] - 3 \cdot \omega^2 \right\}$$
$$- \left\{ - \left[\Upsilon_{10} - \left(\frac{\Delta\Omega}{2}\right)^2 \cdot m \cdot t_D^2 \right] \cdot \omega - \omega^3 \right\} \cdot 2 \cdot \Upsilon_{11} \cdot \omega)$$
$$- (-\{[\Upsilon_1 - \Upsilon_2 \cdot t_D^2 - \Upsilon_3] - \Upsilon_6 \cdot \omega^2\}$$
$$\cdot \left[\Upsilon_4 - 2 \cdot \left(\frac{\Delta\Omega}{2}\right)^2 \cdot m \cdot t_D^2 + \Upsilon_5 \right]$$
$$- \left[\Upsilon_4 - 2 \cdot \left(\frac{\Delta\Omega}{2}\right)^2 \cdot m \cdot t_D^2 + \Upsilon_5 \right] \cdot \omega \cdot 2 \cdot \Upsilon_6 \cdot \omega \Big)$$

Then we get the expression for $F(\omega, t_D)$ lasers with frequency-modulated optical feedback system parameters values. We find those ω, t_D values which fulfil $F(\omega, t_D) = 0$. We ignore negative, complex, and imaginary values of ω for specific t_D values. We can be express by 3D function $F(\omega, t_D) = 0$. We plot the stability switch diagram based on different delay values of our system.

$$\Lambda^{-1}(t_D) = \left(\frac{\partial \mathrm{Re}\lambda}{\partial t_D}\right)_{\lambda = i\cdot\omega} = \mathrm{Re}\left\{\frac{-2 \cdot (U + t_D \cdot |P|^2) + i \cdot F_\omega}{F_{t_D} + i \cdot 2 \cdot (V + \omega \cdot |P|^2)}\right\}$$
$$\Lambda^{-1}(t_D) = \left(\frac{\partial \mathrm{Re}\lambda}{\partial t_D}\right)_{\lambda = i\cdot\omega} = \frac{2 \cdot \{F_\omega \cdot (V + \omega \cdot P^2) - F_{t_D} \cdot (U + t_D \cdot P^2)\}}{F_{t_D}^2 + 4 \cdot (V + \omega \cdot P^2)^2}$$

The stability switch occurs only on those delay values (t_D) which fit the equation: $t_D = \frac{\theta_+(t_D)}{\omega_+(t_D)}$ and $\theta_+(t_D)$ is the solution of $\sin\theta(t_D) = \cdots$; $\cos\theta(t_D) = \cdots$ when $\omega = \omega_+(t_D)$ if only ω_+ is feasible. Additionally, when all lasers with frequency-modulated optical feedback system parameters are known and the stability switch due to various time delay values t_D is described in the following expression:

$$sign\{\Lambda^{-1}(t_D)\} = sign\{F_\omega(\omega(t_D), t_D)\}$$
$$\cdot sign\left\{t_D \cdot \omega_{t_D}(\omega(t_D)) + \omega(t_D) + \frac{U(\omega(t_D)) \cdot \omega_{t_D}(\omega(t_D)) + V(\omega(t_D))}{|P(\omega(t_D))|^2}\right\}$$

Remark we know $F(\omega, t_D) = 0$ implies its roots $\omega_i(t_D)$ and finding those delay values t_D which ω_i is feasible. There are t_D values which give complex ω_i or imaginary number, then unable to analyze stability. $F(\omega, t_D)$ function is dependent on t_D the parameter $F(\omega, t_D) = 0$. The results: We find those ω, t_D values which fulfill $F(\omega, t_D) = 0$. We ignore negative, complex, and imaginary values of ω.

Next is to find those ω, t_D values which fulfill $\sin\theta(t_D) = \cdots$; $\cos\theta(t_D) = \cdots$

$$\sin(\omega \cdot t_D) = \frac{-P_R \cdot Q_I + P_I \cdot Q_R}{|Q|^2}; \cos(\omega \cdot t_D) = -\frac{(P_R \cdot Q_R + P_I \cdot Q_I)}{|Q|^2}; |Q|^2 = Q_R^2 + Q_I^2$$

Finally, we plot the stability switch diagram$g(t_D) = \Lambda^{-1}(t_D) = \left(\frac{\partial \mathrm{Re}\lambda}{\partial t_D}\right)_{\lambda = i \cdot \omega}$.

$$g(t_D) = \Lambda^{-1}(t_D) = \left(\frac{\partial \mathrm{Re}\lambda}{\partial t_D}\right)_{\lambda = i \cdot \omega} = \frac{2 \cdot \{F_\omega \cdot (V + \omega \cdot P^2) - F_{t_D} \cdot (U + t_D \cdot P^2)\}}{F_{t_D}^2 + 4 \cdot (V + \omega \cdot P^2)^2}$$

$$sign\{g(t_D)\} = sign\{\Lambda^{-1}(t_D)\} = sign\left\{\left(\frac{\partial \mathrm{Re}\lambda}{\partial t_D}\right)_{\lambda = i \cdot \omega}\right\} = sign\left\{\frac{2 \cdot \{F_\omega \cdot (V + \omega \cdot P^2) - F_{t_D} \cdot (U + t_D \cdot P^2)\}}{F_{t_D}^2 + 4 \cdot (V + \omega \cdot P^2)^2}\right\}$$

$$F_{t_D}^2 + 4 \cdot (V + \omega \cdot P^2)^2 > 0$$

$$sign\{\Lambda^{-1}(t_D)\} = sign\{F_\omega \cdot (V + \omega \cdot P^2) - F_{t_D} \cdot (U + t_D \cdot P^2)\}$$

$$sign\{\Lambda^{-1}(t_D)\} = sign\left\{[F_\omega] \cdot \left[(V + \omega \cdot P^2) - \frac{F_{t_D}}{F_\omega} \cdot (U + t_D \cdot P^2)\right]\right\}$$

$$\omega_{t_D} = -\frac{F_{t_D}}{F_\omega}; \omega_{t_D} = \left(\frac{\partial \omega}{\partial t_D}\right)^{-1} = -\frac{\partial F / \partial \omega}{\partial F / \partial t_D}$$

$$sign\{\Lambda^{-1}(t_D)\} = sign\{[F_\omega] \cdot [V + \omega_{t_D} \cdot U + \omega \cdot P^2 + \omega_{t_D} \cdot t_D \cdot P^2]\}$$

$$sign\{\Lambda^{-1}(t_D)\} = sign\left\{[F_\omega] \cdot [P^2] \cdot \left[\frac{V + \omega_{t_D} \cdot U}{P^2} + \omega_{t_D} + \omega_{t_D} \cdot t_D\right]\right\}$$

$$sign[P^2] > 0 \Rightarrow sign\{\Lambda^{-1}(t_D)\} = sign\left\{[F_\omega] \cdot \left[\frac{V + \omega_{t_D} \cdot U}{P^2} + \omega + \omega_{t_D} \cdot t_D\right]\right\}$$

$$sign\{\Lambda^{-1}(t_D)\} = sign[F_\omega] \cdot sign\left[\frac{V + \omega_{t_D} \cdot U}{P^2} + \omega + \omega_{t_D} \cdot t_D\right]$$

$$F_\omega = 2 \cdot [(P_{R\omega} \cdot P_R + P_{I\omega} \cdot P_I) - (Q_{R\omega} \cdot Q_R + Q_{I\omega} \cdot Q_I)]$$

$$F_\omega = 2 \cdot [(-2 \cdot \Upsilon_{11} \cdot \omega \cdot \{(\Upsilon_7 \cdot t_D^2 - \Upsilon_8 + \Upsilon_9) - \Upsilon_{11} \cdot \omega^2\} + \left\{\left[\Upsilon_{10} - \left(\frac{\Delta\Omega}{2}\right)^2 \cdot m \cdot t_D^2\right] - 3 \cdot \omega^2\right\} \cdot \left\{\left[\Upsilon_{10} - \left(\frac{\Delta\Omega}{2}\right)^2 \cdot m \cdot t_D^2\right] \cdot \omega - \omega^3\right\}) - (-2 \cdot \Upsilon_6 \cdot \omega \cdot \{[\Upsilon_1 - \Upsilon_2 \cdot t_D^2 - \Upsilon_3] - \Upsilon_6 \cdot \omega^2\}$$

Table. 4.5 lasers with frequency-modulated optical feedback system sign of $\Lambda^{-1}(t_D)$.

$sign[F_\omega]$	$sign[\frac{V+\omega_{t_D}\cdot U}{P^2}+\omega+\omega_{t_D}\cdot t_D]$	$sign\{\Lambda^{-1}(t_D)\}$
±	±	+
±	±	−

$$+\left[\Upsilon_4-2\cdot\left(\frac{\Delta\Omega}{2}\right)^2\cdot m\cdot t_D^2+\Upsilon_5\right]^2\cdot\omega\Bigg)\Bigg]$$

We check the sign of $\Lambda^{-1}(t_D)$ according the following rule (Table 4.5):

If $sign\{\Lambda^{-1}(t_D)\}>0$ then the crossing proceeds from (−) to (+) respectively (stable to unstable). If $sign\{\Lambda^{-1}(t_D)\}<0$ then the crossing proceeds from (+) to (−) respectively (unstable to stable).

4.3 Doppler-Shift with a Microchip Solid State Laser Stability Analysis Under Parameters Variation

In laser we have the phenomena of light-injection-induced. It is injection locking and return-light-induced instabilities. Important is the basic properties of laser dynamics as well as their practical importance. One group of lasers is Nd stoichiometric lasers. Stoichiometric laser materials are pure chemical compounds capable of emitting coherent light in the undiluted state as opposed to conventional laser materials where the active ions or molecules are dispersed in a host. One chemical compound is Neodymium compounds. The stoichiometric laser materials can store several kilo-joules of energy per cm^3 for times ranging from 100 ns to 10 ms. Continuous room temperature laser operation can be achieved with just a few hundred microwatts of pump power. Optical gains up to 10 dB per optical wavelength can be expected. Stoichiometric systems are related to spectroscopic properties. Threshold, pump power, and dynamical behavior (relaxation oscillations) can be calculated including the effects of anisotropy, saturation, inversion profile, reabsorption, pump profile, and mode profile effects that are important for materials with a high concentration of active ions. Structural aspects such as phase transitions, Ferro elasticity, ionic spacing, probability density of rare earth electrons, and energy transfer can be discussed using NdP_5O_{14}. Hybrid exchange can be a possible mechanism for energy transfer between rare earth ions. Their wave functions overlap appreciably in stoichiometric materials. There is a resonant excitation of periodic oscillations and chaotic oscillations in relaxation oscillation, resulting from Doppler-shifted light injection from a moving light-scattering object. It is injection-induced modulation of lasers phenomenon. One application of this phenomenon is ultra-high sensitivity Doppler-shift measurements in wide velocity regions. Light waves from a moving source experience the Doppler Effect to result in either a red shift or blue shift in the light's frequency. Light waves do not require a medium for travel. This is one effect on light which shows that its source is moving with respect to the observer; its color changes. The Doppler Effect can be observed to occur with all types of waves and light waves. The relativistic Doppler Effect is the change in frequency (and wavelength) of light, caused by the relative motion of the source and the observer, when taking into account effects described by the special theory of relativity. The Doppler Effect is caused when the source of a waveform light-send out waves at a regular rate or frequency, but there is a constant relative motion between the source and observer, causing the observed frequency to change. Doppler Effect Redshift and Blue shift describe how light shifts toward shorter or longer wavelengths as objects move closer or farther away from us. When an object moves away from us, the light is shifted to the red end of the spectrum, as its wavelengths get longer. The model includes multimode effects which are in correlation with the experimental results. Our experimental system includes the following elements: (1) An argon laser which served as a pump laser to NLP laser, (2) $LiNdP_4O_{12}$ (LNP) crystal which is coated with dielectric mirrors, (3) Infrared transmission filter (IR), (4) Beam splitter element (BS), (5) InGaAs photodiode (~120 MHz bandwidth), (6) Variable Attenuator (VA), (7) Lens to focus the main beam on a rotation circular paper sheet stuck to a metal plate which is attached to the rotating arm. Argon ion laser is powerful gas laser, which typically generate multiple watts of optical power in a green or blue output beam with high beam quality. The core component of an argon ion laser is an argon-filed tube, made of beryllium oxide ceramics, in which an intense electrical discharge between two hollow electrodes generates plasma with a high density of argon (Ar^+) ions. A solenoid around the tube can be used for generating a magnetic field, which increases the output power by better confining the plasma. Multi-watt argon ion lasers can be used for pumping titanium-sapphire lasers and dye lasers, or for laser light activities. They are rivaled by frequency-doubled diode-pumped solid-state lasers. The latter are far more power efficient and have longer lifetimes. Argon tubes have a limited lifetime of the order of a few thousand hours. An argon laser may thus be preferable if it is used only during a limited number of hours, whereas a diode-pumped solid-state laser is the better solution for reliable and efficient long-term operation. The argon laser pumps the LNP laser which emits light through infrared transmission filter. Then the light goes through Beam Splitter (BS) which splits the light to the InGaAs photo diode and to Variable Attenuator (VA), lens and finally hit paper sheet placed on rotating arm. The LNP laser radiation is linearly polarized along the pseudo-orthorhombic c axis. Pseudo-orthorhombic shape is having unequal axes at right angles (Fig. 4.4). Orthorhombic crystal system is orthorhombic lattices result from stretching a cubic lattice along two of its orthogonal pairs by two different factors, resulting in a rectangular prism with a rectangular base [3, 10. Two-longitudinal-mode oscillation is obtained above $w = \frac{P}{P_{th}} = 1.09$ (P_{th} is threshold pump power) and three-longitudinal-mode oscillation is obtained above $w = \frac{P}{P_{th}} = 1.7$. Part of the output light is used for monitoring output waveforms and power spectra with an InGaAs photo diode (120 MHz bandwidth). The impinged laser power on the paper sheet is changed by a Variable Attenuator (VA). Our system

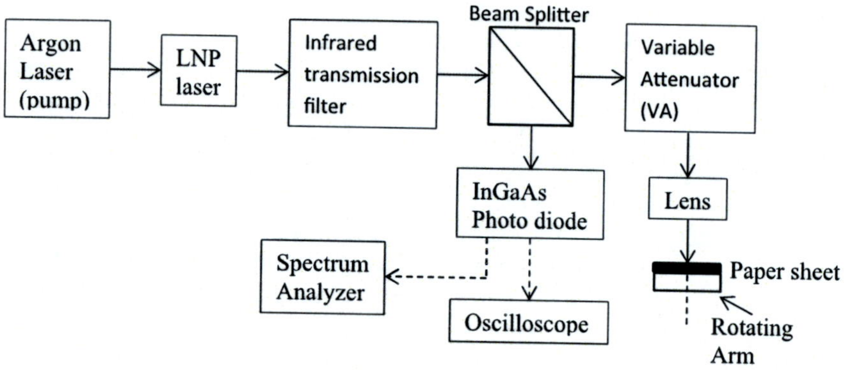

Fig. 4.4 Experimental structure of Doppler-shift with Microchip solid-state laser

observes the injection induce modulation, even the fact that the impinged light is attenuated [11].

The system phenomenon can be inspected in terms of injection-induced intensity modulation resulting from interference between oscillating field $\widetilde{E}_i$ and Doppler shifted field $\widetilde{E_{i,s}}$ injected into the laser cavity, where "i" denotes the model index. Laser cavity is a means of optical confinement intended to increase the gain length of radiation prior to emission from the device. The means of optical confinement used to increase gain path length vary depending upon the properties of the beam desired within the lasing medium. High light intensities occur within a laser cavity and dielectric mirrors coated for the lasing wavelength are used. The position and curvature of the optical cavity elements is alerted in order to optimize the laser performance as needed for a particular application. A similar phenomenon is expected outside the injection-locking region in single-mode lasers. The injection-locking properties of a semiconductor laser can be analyze and the injected carrier density dependent refractive index in the action region.

We can characterize the light-injection model for rotating-wave approximation fields by the following set of differential equations. It is done by introducing injection fields into spatially-hole-burned N-mode lasers.

$$\frac{dN_0}{dt} = P - \frac{N_0}{\tau} - \sum_{k=1}^{N} B_k \cdot S_k \cdot \left(N_0 - \frac{N_k}{2}\right); \frac{dN_i}{dt}$$

$$= -\frac{N_i}{\tau} + B_i \cdot N_0 \cdot S_i - N_i \cdot \sum_{k=1}^{N} B_k \cdot S_k$$

$$\frac{dE_i}{dt} = \left[B_i \cdot (N_0 - \frac{N_i}{2}) - \frac{1}{\tau_p}\right] \cdot \left(\frac{E_i}{2}\right) + \left(\frac{\kappa}{\tau_l}\right) \cdot E_{i,s} \cdot \cos[\psi_i(t) + \phi_{i,s}(t)]$$

$$\frac{d\psi_i}{dt} = \omega_i - \omega_{i,s} - \left(\frac{\kappa}{\tau_l}\right) \cdot \left(\frac{E_{i,s}}{E_i}\right) \cdot \sin[\psi_i(t) + \phi_{i,s}(t)]; i = 1, 2, \ldots, N$$

N_0- Space average of population inversion density, N_i—First spatial Fourier component of population inversion for the i – th mode, i—Modal index, P—Pump power, τ—Population lifetime, τ_p—Photon lifetime, κ—Effective amplitude transmission coefficient of the output mirror, τ_l—Round trip time in the cavity, ψ_i—Phase different between the two fields, $\phi_{i,s}$—Express phase fluctuations of scattered light, $\widetilde{E}_i$, E_i—Lasing field, $\widetilde{E_{i,s}}$—Scattered field, B_i—Gain, S_i—Photon density, $S_i = E_i^2$; $E_{i,s}^2(t) \propto E_i(t - t_d)$; t_d—Delay time, $\frac{E_{i,s}}{E_i} < 10^{-4}$(extremely small). In the lock-in region: $\frac{\Delta\omega}{2\cdot\pi} = \left(\frac{\kappa}{\tau_l}\right)\cdot\left(\frac{E_{i,s}}{E_i}\right) < 100\,\text{kHz}$. Where "$i$" denotes the model index (laser mode number). N—Number of relaxation oscillations which each mode may exhibit.

Hint: The phase interaction between longitudinal modes are omitted since the longitudinal mode spacing is extremely large compared with the Doppler-shift frequencies and the lasing time scales.

$w = \frac{P}{P_{th,1}}$; $P_{th,1}$ is the first lasing mode pump threshold, n_0, n_i—are the space average and the first Fourier components of population inversion normalized by the mode threshold value $\left(N_0 - \frac{N_1}{2}\right)_{th}$. s_i—Photon density normalized by the steady state S_1value at $w = 2$. γ_i—Gain ratio to the first lasing mode $\left(\left(\gamma_i = \frac{B_i}{B}\right)\right)$, $m = \frac{2\cdot E_{i,s}}{\kappa\cdot E_i}$; $(\omega_i - \omega_{i,s})\cdot\tau \equiv \Omega_D$ (time is scaled by τ), $K = \frac{\tau}{\tau_p}$. ω_D—Doppler-shift frequency. We define the relaxation oscillation frequency f_{RO} as $f_{RO} = \frac{\omega_{RO}}{2\cdot\pi} = \left(\frac{1}{2\cdot\pi}\right)\cdot\sqrt{\frac{(w-1)}{\tau\cdot\tau_p}}|_{\tau=120\,\mu\text{s},\tau_p=310\,\text{ps}} = 540\,\text{kHz}$. $k\cdot v$—Angular frequency of Doppler shift and increases with increasing angular moving velocity v and with decreasing the scattering angle θ_s, where θ_s is the angle between the laser axis and velocity vector. λ—Oscillation wavelength.

Assumption: The delay time t_d is extremely small compared with modulation time scales.

The simplest case of our system is when the number of relaxation oscillations which each mode exhibits is N = 2 and N = 3. The system same collective behavior is featuring antiphase dynamics for larger number of modes. We analyze the system for the simplest case N = 2 and get differential equations [3, 10].

$$\frac{dN_0}{dt} = P - \frac{N_0}{\tau} - \sum_{k=1}^{N=2} B_k\cdot S_k\cdot\left(N_0 - \frac{N_k}{2}\right);$$

$$\frac{dN_i}{dt} = -\frac{N_i}{\tau} + B_i\cdot N_0\cdot S_i - N_i\cdot\sum_{k=1}^{N=2} B_k\cdot S_k$$

$$\frac{dE_i}{dt} = \left[B_i\cdot\left(N_0 - \frac{N_i}{2}\right) - \frac{1}{\tau_p}\right]\cdot\left(\frac{E_i}{2}\right) + \left(\frac{\kappa}{\tau_l}\right)\cdot E_{i,s}\cdot\cos[\psi_i(t) + \phi_{i,s}(t)]$$

$$\frac{d\psi_i}{dt} = \omega_i - \omega_{i,s} - \left(\frac{\kappa}{\tau_l}\right)\cdot\left(\frac{E_{i,s}}{E_i}\right)\cdot\sin[\psi_i(t) + \phi_{i,s}(t)];\ i = 1, 2, \ldots, N$$

We write the above differential equations for $N = 2$.

$$\frac{dN_0}{dt} = P - \frac{N_0}{\tau} - B_1 \cdot S_1 \cdot \left(N_0 - \frac{N_1}{2}\right) - B_2 \cdot S_2 \cdot \left(N_0 - \frac{N_2}{2}\right)$$
$$\frac{dN_i}{dt} = -\frac{N_i}{\tau} + B_i \cdot N_0 \cdot S_i - N_i \cdot (B_1 \cdot S_1 + B_2 \cdot S_2)$$
$$\frac{dE_i}{dt} = \left[B_i \cdot \left(N_0 - \frac{N_i}{2}\right) - \frac{1}{\tau_p}\right] \cdot \left(\frac{E_i}{2}\right) + \left(\frac{\kappa}{\tau_l}\right) \cdot E_{i,s} \cdot \cos[\psi_i(t) + \phi_{i,s}(t)]$$
$$\frac{d\psi_i}{dt} = \omega_i - \omega_{i,s} - \left(\frac{\kappa}{\tau_l}\right) \cdot \left(\frac{E_{i,s}}{E_i}\right) \cdot \sin[\psi_i(t) + \phi_{i,s}(t)];\ i = 1, 2, \ldots, N$$

At fixed points: $\frac{dN_0}{dt} = 0$; $\frac{dN_i}{dt} = 0$; $\frac{dE_i}{dt} = 0$; $\frac{d\psi_i}{dt} = 0$

$$\frac{dN_0}{dt} = 0 \Rightarrow P - \frac{N_0^*}{\tau} - B_1 \cdot S_1 \cdot \left(N_0^* - \frac{N_1}{2}\right) - B_2 \cdot S_2 \cdot \left(N_0^* - \frac{N_2}{2}\right) = 0$$
$$- N_0^* \cdot \left(\frac{1}{\tau} + B_1 \cdot S_1 + B_2 \cdot S_2\right) + P + \sum_{k=1}^{2} B_k \cdot S_k \cdot \frac{N_k}{2} = 0;\ N_0^* = \frac{P + \sum_{k=1}^{2} B_k \cdot S_k \cdot \frac{N_k}{2}}{\left(\frac{1}{\tau} + \sum_{k=1}^{2} B_k \cdot S_k\right)}$$
$$\frac{dN_i}{dt} = 0 \Rightarrow -\frac{N_i^*}{\tau} + B_i \cdot N_0^* \cdot S_i - N_i^* \cdot \sum_{k=1}^{2} B_k \cdot S_k = 0;\ N_i^* \cdot \left(\frac{1}{\tau} + \sum_{k=1}^{2} B_k \cdot S_k\right) = B_i \cdot N_0^* \cdot S_i$$
$$N_i^* = \frac{B_i \cdot N_0^* \cdot S_i}{\left(\frac{1}{\tau} + \sum_{k=1}^{2} B_k \cdot S_k\right)};$$
$$N_i^* = \frac{B_i \cdot S_i \cdot \left[\frac{P + \sum_{k=1}^{2} B_k \cdot S_k \cdot \frac{N_k}{2}}{\left(\frac{1}{\tau} + \sum_{k=1}^{2} B_k \cdot S_k\right)}\right]}{\left(\frac{1}{\tau} + \sum_{k=1}^{2} B_k \cdot S_k\right)} = \frac{B_i \cdot S_i \cdot \left(P + \sum_{k=1}^{2} B_k \cdot S_k \cdot \frac{N_k}{2}\right)}{\left(\frac{1}{\tau} + \sum_{k=1}^{2} B_k \cdot S_k\right)^2}$$
$$\frac{dE_i}{dt} = 0 \Rightarrow \left[B_i \cdot \left(N_0^* - \frac{N_i^*}{2}\right) - \frac{1}{\tau_p}\right] \cdot \left(\frac{E_i^*}{2}\right)$$

$$+\left(\frac{\kappa}{\tau_l}\right)\cdot E_{i,s}\cdot\cos\left[\psi_i^*+\phi_{i,s}(t)\right]=0$$

$$E_i^*=-\frac{2\cdot\left(\frac{\kappa}{\tau_l}\right)\cdot E_{i,s}\cdot\cos\left[\psi_i^*+\phi_{i,s}(t)\right]}{\left[B_i\cdot\left(N_0^*-\frac{N_i^*}{2}\right)-\frac{1}{\tau_p}\right]}\Rightarrow E_i^*$$

$$=\frac{2\cdot\left(\frac{\kappa}{\tau_l}\right)\cdot E_{i,s}\cdot\cos\left[\psi_i^*+\phi_{i,s}(t)\right]}{\left[\frac{1}{\tau_p}-B_i\cdot\left(N_0^*-\frac{N_i^*}{2}\right)\right]}$$

$$E_i^*=\frac{2\cdot\left(\frac{\kappa}{\tau_l}\right)\cdot E_{i,s}\cdot\cos\left[\psi_i^*+\phi_{i,s}(t)\right]}{\frac{1}{\tau_p}-B_i\cdot\left(\frac{P+\sum_{k=1}^{2}B_k\cdot S_k\cdot\frac{N_k}{2}}{\left(\frac{1}{\tau}+\sum_{k=1}^{2}B_k\cdot S_k\right)}-\frac{1}{2}\cdot\left[\frac{B_i\cdot S_i\cdot\left(P+\sum_{k=1}^{2}B_k\cdot S_k\cdot\frac{N_k}{2}\right)}{\left(\frac{1}{\tau}+\sum_{k=1}^{2}B_k\cdot S_k\right)^2}\right]\right)}$$

$$E_i^*=\frac{2\cdot\left(\frac{\kappa}{\tau_l}\right)\cdot E_{i,s}\cdot\cos\left[\psi_i^*+\phi_{i,s}(t)\right]}{\frac{1}{\tau_p}-B_i\cdot\left(\frac{P+\sum_{k=1}^{2}B_k\cdot S_k\cdot\frac{N_k}{2}}{\left(\frac{1}{\tau}+\sum_{k=1}^{2}B_k\cdot S_k\right)}\right)\cdot\left(1-\frac{1}{2}\cdot\left[\frac{B_i\cdot S_i}{\left(\frac{1}{\tau}+\sum_{k=1}^{2}B_k\cdot S_k\right)}\right]\right)}$$

$$\frac{d\psi_i}{dt}=0\Rightarrow\omega_i-\omega_{i,s}-\left(\frac{\kappa}{\tau_l}\right)\cdot\left(\frac{E_{i,s}}{E_i^*}\right)\cdot\sin\left[\psi_i^*+\phi_{i,s}(t)\right]=0;\ i=1,2,\ldots,N$$

$$\left(\frac{\kappa}{\tau_l}\right)\cdot\left(\frac{E_{i,s}}{E_i^*}\right)\cdot\sin[\psi_i^*+\phi_{i,s}(t)]=\omega_i-\omega_{i,s}\Rightarrow\frac{E_{i,s}}{E_i^*}$$

$$=\frac{\omega_i-\omega_{i,s}}{\left(\frac{\kappa}{\tau_l}\right)\cdot\sin[\psi_i^*+\phi_{i,s}(t)]};\ E_i^*=\frac{E_{i,s}\cdot\left(\frac{\kappa}{\tau_l}\right)\cdot\sin[\psi_i^*+\phi_{i,s}(t)]}{\omega_i-\omega_{i,s}}$$

Then we get the following equation for fixed point's coordinate ψ_i^*:

$$\frac{2\cdot\left(\frac{\kappa}{\tau_l}\right)\cdot E_{i,s}\cdot\cos[\psi_i^*+\phi_{i,s}(t)]}{\frac{1}{\tau_p}-B_i\cdot\left(\frac{P+\sum_{k=1}^{2}B_k\cdot S_k\cdot\frac{N_k}{2}}{\left(\frac{1}{\tau}+\sum_{k=1}^{2}B_k\cdot S_k\right)}\right)\cdot\left(1-\frac{1}{2}\cdot\left[\frac{B_i\cdot S_i}{\left(\frac{1}{\tau}+\sum_{k=1}^{2}B_k\cdot S_k\right)}\right]\right)}$$

$$=\frac{E_{i,s}\cdot\left(\frac{\kappa}{\tau_l}\right)\cdot\sin[\psi_i^*+\phi_{i,s}(t)]}{\omega_i-\omega_{i,s}}$$

The above equation needs to solve numerically for finding ψ_i^* then submit the value in one of $E_i^*=\ldots$ equations.

Our system fixed points:

$$\left(N_0^*, N_i^*, E_i^*, \psi_i^*\right) = \left(\frac{P + \sum_{k=1}^{2} B_k \cdot S_k \cdot \frac{N_k}{2}}{\left(\frac{1}{\tau} + \sum_{k=1}^{2} B_k \cdot S_k\right)}, \frac{B_i \cdot S_i \cdot \left(P + \sum_{k=1}^{2} B_k \cdot S_k \cdot \frac{N_k}{2}\right)}{\left(\frac{1}{\tau} + \sum_{k=1}^{2} B_k \cdot S_k\right)^2}, \frac{2 \cdot \left(\frac{\kappa}{\tau_l}\right) \cdot E_{i,s} \cdot \cos[\psi_i^* + \phi_{i,s}(t)]}{\left[\frac{1}{\tau_p} - B_i \cdot \left(N_0^* - \frac{N_i^*}{2}\right)\right]}, \psi_i^* \right)$$

Stability analysis: The standard local stability analysis about any one of the equilibrium points of the light-injection model for rotating-wave approximation fields differential equation system consists in adding to coordinate $[N_0, N_i, E_i, \psi_i]$ arbitrarily small increments of exponential form $[n_0, n_i, E_i, \psi_i] \cdot e^{\lambda \cdot t}$ and retaining the first order terms in N_0, N_i, E_i, ψ_i. The system of four homogeneous equations leads to a polynomial characteristic equation in the eigenvalues. The polynomial characteristic equations accept by set the below variables (N_0, N_i, E_i, ψ_i) and variables derivatives with respect to time into system equations [12–14]. System fixed values with arbitrarily small increments of exponential form $[n_0, n_i, E_i, \psi_i] \cdot e^{\lambda \cdot t}$ are: j = 0 (first fixed point), j = 1 (second fixed point), j = 2 (third fixed point), etc. $N_0 = N_0(t); N_i = N_i(t); E_i = E_i(t); \psi_i = \psi_i(t)$

$$N_0(t) = N_0^{(j)} + n_0 \cdot e^{\lambda \cdot t}; N_i(t) = N_i^{(j)} + n_i \cdot e^{\lambda \cdot t}; E_i(t) = E_i^{(j)} + E_i \cdot e^{\lambda \cdot t}; \psi_i(t) = \psi_i^{(j)} + \psi_i \cdot e^{\lambda \cdot t}$$

$$\frac{dN_0(t)}{dt} = n_0 \cdot \lambda \cdot e^{\lambda \cdot t}; \frac{dN_i(t)}{dt} = n_i \cdot \lambda \cdot e^{\lambda \cdot t}; \frac{dE_i(t)}{dt} = E_i \cdot \lambda \cdot e^{\lambda \cdot t}; \frac{d\psi_i(t)}{dt} = \psi_i \cdot \lambda \cdot e^{\lambda \cdot t}$$

We choose these expressions for ourselves $N_0(t), N_i(t), E_i(t), \psi_i(t)$ as a small displacement $[n_0, n_i, E_i, \psi_i]$ from the system fixed points in time t = 0.

$$N_0(t=0) = N_0^{(j)} + n_0; N_i(t=0) = N_i^{(j)} + n_i; E_i(t=0) = E_i^{(j)} + E_i; \psi_i(t=0) = \psi_i^{(j)} + \psi_i$$

$$\frac{dN_0}{dt} = P - \frac{N_0}{\tau} - B_1 \cdot S_1 \cdot (N_0 - \frac{N_1}{2}) - B_2 \cdot S_2 \cdot (N_0 - \frac{N_2}{2}) \frac{dN_0}{dt} = P - \frac{N_0}{\tau} - N_0 \cdot \left(\sum_{k=1}^{2} B_k \cdot S_k\right) + \sum_{k=1}^{2} B_k \cdot S_k \cdot \frac{N_k}{2}; \frac{dN_0}{dt}$$

$$= P - N_0 \cdot \left[\frac{1}{\tau} + \left(\sum_{k=1}^{2} B_k \cdot S_k\right)\right] + \sum_{k=1}^{2} B_k \cdot S_k \cdot \frac{N_k}{2}$$

$$n_0 \cdot \lambda \cdot e^{\lambda \cdot t} = P - [N_0^{(j)} + n_0 \cdot e^{\lambda \cdot t}] \cdot \left[\frac{1}{\tau} + \left(\sum_{k=1}^{2} B_k \cdot S_k\right)\right] + \sum_{k=1}^{2} B_k \cdot S_k \cdot \frac{N_k}{2}$$

$$n_0 \cdot \lambda \cdot e^{\lambda \cdot t} = P - N_0^{(j)} \cdot \left[\frac{1}{\tau} + \left(\sum_{k=1}^{2} B_k \cdot S_k\right)\right]$$
$$- n_0 \cdot \left[\frac{1}{\tau} + \left(\sum_{k=1}^{2} B_k \cdot S_k\right)\right] \cdot e^{\lambda \cdot t} + \sum_{k=1}^{2} B_k \cdot S_k \cdot \frac{N_k}{2}$$

At fixed points:

$$P - N_0^{(j)} \cdot \left[\frac{1}{\tau} + \left(\sum_{k=1}^{2} B_k \cdot S_k\right)\right] + \sum_{k=1}^{2} B_k \cdot S_k \cdot \frac{N_k}{2}$$
$$= 0 - n_0 \cdot \lambda - n_0 \cdot \left[\frac{1}{\tau} + \left(\sum_{k=1}^{2} B_k \cdot S_k\right)\right]$$
$$= 0 \frac{dN_i}{dt} = -\frac{N_i}{\tau} + B_i \cdot N_0 \cdot S_i - N_i \cdot (B_1 \cdot S_1 + B_2 \cdot S_2)$$

$$n_i \cdot \lambda \cdot e^{\lambda \cdot t} = -\frac{\left(N_i^{(j)} + n_i \cdot e^{\lambda \cdot t}\right)}{\tau} + B_i \cdot \left(N_0^{(j)} + n_0 \cdot e^{\lambda \cdot t}\right)$$
$$\cdot S_i - \left(N_i^{(j)} + n_i \cdot e^{\lambda \cdot t}\right) \cdot (B_1 \cdot S_1 + B_2 \cdot S_2) n_i \cdot \lambda \cdot e^{\lambda \cdot t}$$
$$= -\frac{N_i^{(j)}}{\tau} + B_i \cdot N_0^{(j)} \cdot S_i - N_i^{(j)} \cdot (B_1 \cdot S_1 + B_2 \cdot S_2) - \frac{n_i \cdot e^{\lambda \cdot t}}{\tau}$$
$$- n_i \cdot e^{\lambda \cdot t} \cdot (B_1 \cdot S_1 + B_2 \cdot S_2) + B_i \cdot n_0 \cdot e^{\lambda \cdot t} \cdot S_i$$

At fixed points: $-\frac{N_i^{(j)}}{\tau} + B_i \cdot N_0^{(j)} \cdot S_i - N_i^{(j)} \cdot (B_1 \cdot S_1 + B_2 \cdot S_2) = 0$

$$B_i \cdot S_i \cdot n_0 - n_i \cdot \lambda - \frac{n_i}{\tau} - n_i \cdot (B_1 \cdot S_1 + B_2 \cdot S_2) = 0$$

$$\frac{dE_i}{dt} = \left[B_i \cdot \left(N_0 - \frac{N_i}{2}\right) - \frac{1}{\tau_p}\right] \cdot \left(\frac{E_i}{2}\right) + \left(\frac{\kappa}{\tau_l}\right)$$
$$\cdot E_{i,s} \cdot \cos[\psi_i(t) + \phi_{i,s}(t)]$$

$$E_i \cdot \lambda \cdot e^{\lambda \cdot t} = \left[B_i \cdot \left(N_0^{(j)} + n_0 \cdot e^{\lambda \cdot t} - \frac{[N_i^{(j)} + n_i \cdot e^{\lambda \cdot t}]}{2}\right) - \frac{1}{\tau_p}\right]$$

$$\cdot \left(\frac{E_i^{(j)} + E_i \cdot e^{\lambda \cdot t}}{2} \right) + \left(\frac{\kappa}{\tau_l} \right) \cdot E_{i,s}$$

$$\cdot \cos[\psi_i^{(j)} + \psi_i \cdot e^{\lambda \cdot t} + \phi_{i,s}(t)]$$

First we do algebraic manipulation for the cos{…} expression

$$\cos[\psi_i^{(j)} + \psi_i \cdot e^{\lambda \cdot t} + \phi_{i,s}(t)] = \cos[\psi_i^{(j)} + \phi_{i,s}(t) + \psi_i \cdot e^{\lambda \cdot t}]$$
$$\cos[\psi_i^{(j)} + \phi_{i,s}(t) + \psi_i \cdot e^{\lambda \cdot t}] = \cos[\psi_i^{(j)} + \phi_{i,s}(t)]$$
$$\cdot \cos[\psi_i \cdot e^{\lambda \cdot t}] - \sin[\psi_i^{(j)} + \phi_{i,s}(t)] \cdot \sin[\psi_i \cdot e^{\lambda \cdot t}]$$

Assumption
(A): $\xi_1(\psi_i, \lambda) = \psi_i \cdot e^{\lambda \cdot t}$; $\cos[\xi_1(\psi_i, \lambda)] = \cos[\psi_i \cdot e^{\lambda \cdot t}]$.

$$\cos(\xi_1(\psi_i, \lambda)) = \sum_{n=0}^{\infty} \frac{(-1)^n}{(2 \cdot n)!} \cdot [\xi_1(\psi_i, \lambda)]^{2 \cdot n}; \cos(\xi_1(\psi_i, \lambda))$$
$$= 1 - \frac{\psi_i^2 \cdot e^{2 \cdot \lambda \cdot t}}{2!} + \frac{\psi_i^4 \cdot e^{4 \cdot \lambda \cdot t}}{4!} - \cdots \ \forall \ \psi_i \cdot e^{\lambda \cdot t}$$

We consider $\psi_i^m \to \varepsilon \, \forall m \geq 2$ then $-\frac{\psi_i^2 \cdot e^{2 \cdot \lambda \cdot t}}{2!} + \frac{\psi_i^4 \cdot e^{4 \cdot \lambda \cdot t}}{4!} - \ldots \to \varepsilon$

$$\cos(\xi_1(\psi_i, \lambda)) = \sum_{n=0}^{\infty} \frac{(-1)^n}{(2 \cdot n)!} \cdot [\xi_1(\psi_i, \lambda)]^{2 \cdot n} \approx 1$$

Assumption
(B): $\xi_1(\psi_i, \lambda) = \psi_i \cdot e^{\lambda \cdot t}$; $\sin[\xi_1(\psi_i, \lambda)] = \sin[\psi_i \cdot e^{\lambda \cdot t}]$.

$$\sin(\xi_1(\psi_i, \lambda)) = \sum_{n=0}^{\infty} \frac{(-1)^n}{(2 \cdot n + 1)!} \cdot [\xi_1(\psi_i, \lambda)]^{2 \cdot n + 1}$$
$$= \xi_1(\psi_i, \lambda) - \frac{[\xi_1(\psi_i, \lambda)]^3}{3!} + \frac{[\xi_1(\psi_i, \lambda)]^5}{5!} - \frac{[\xi_1(\psi_i, \lambda)]^7}{7!} + \cdots$$
$$\sin(\xi_1(\psi_i, \lambda)) = \sum_{n=0}^{\infty} \frac{(-1)^n}{(2 \cdot n + 1)!} \cdot [\xi_1(\psi_i, \lambda)]^{2 \cdot n + 1}$$
$$= \psi_i \cdot e^{\lambda \cdot t} - \frac{[\psi_i \cdot e^{\lambda \cdot t}]^3}{3!} + \frac{[\psi_i \cdot e^{\lambda \cdot t}]^5}{5!} - \frac{[\psi_i \cdot e^{\lambda \cdot t}]^7}{7!} + \cdots$$
$$\sin(\xi_1(\psi_i, \lambda)) = \sum_{n=0}^{\infty} \frac{(-1)^n}{(2 \cdot n + 1)!} \cdot [\xi_1(\psi_i, \lambda)]^{2 \cdot n + 1}$$
$$= \psi_i \cdot e^{\lambda \cdot t} - \frac{\psi_i^3 \cdot e^{3 \cdot \lambda \cdot t}}{3!} + \frac{\psi_i^5 \cdot e^{5 \cdot \lambda \cdot t}}{5!} - \frac{\psi_i^7 \cdot e^{7 \cdot \lambda \cdot t}}{7!} + \cdots$$

We consider $\psi_i^m \to \varepsilon\, \forall m \geq 2$ then $-\frac{\psi_i^3 \cdot e^{3\cdot\lambda\cdot t}}{3!} + \frac{\psi_i^5 \cdot e^{5\cdot\lambda\cdot t}}{5!} - \frac{\psi_i^7 \cdot e^{7\cdot\lambda\cdot t}}{7!} + \cdots \to \varepsilon$

$$\sin(\xi_1(\psi_i, \lambda)) = \sum_{n=0}^{\infty} \frac{(-1)^n}{(2 \cdot n + 1)!} \cdot [\xi_1(\psi_i, \lambda)]^{2\cdot n+1} \approx \psi_i \cdot e^{\lambda\cdot t}$$

Implementing our last results in cos{...} expression

$$\cos\left[\psi_i^{(j)} + \phi_{i,s}(t) + \psi_i \cdot e^{\lambda\cdot t}\right] = \cos\left[\psi_i^{(j)} + \phi_{i,s}(t)\right] - \sin\left[\psi_i^{(j)} + \phi_{i,s}(t)\right] \cdot \psi_i \cdot e^{\lambda\cdot t}$$

Then we get

$$E_i \cdot \lambda \cdot e^{\lambda\cdot t} = \left[B_i \cdot \left(N_0^{(j)} + n_0 \cdot e^{\lambda\cdot t} - \frac{[N_i^{(j)} + n_i \cdot e^{\lambda\cdot t}]}{2}\right) - \frac{1}{\tau_p}\right] \cdot \left(\frac{E_i^{(j)} + E_i \cdot e^{\lambda\cdot t}}{2}\right) + \left(\frac{\kappa}{\tau_l}\right) \cdot E_{i,s} \cdot (\cos[\psi_i^{(j)} + \phi_{i,s}(t)] - \sin[\psi_i^{(j)} + \phi_{i,s}(t)] \cdot \psi_i \cdot e^{\lambda\cdot t})$$

$$E_i \cdot \lambda \cdot e^{\lambda\cdot t} = \left[B_i \cdot \left(N_0^{(j)} - \frac{N_i^{(j)}}{2} + n_0 \cdot e^{\lambda\cdot t} - \frac{n_i \cdot e^{\lambda\cdot t}}{2}\right) - \frac{1}{\tau_p}\right] \cdot \left(\frac{E_i^{(j)}}{2} + \frac{E_i \cdot e^{\lambda\cdot t}}{2}\right) + \left(\frac{\kappa}{\tau_l}\right) \cdot E_{i,s} \cdot \cos[\psi_i^{(j)} + \phi_{i,s}(t)] - \left(\frac{\kappa}{\tau_l}\right) \cdot E_{i,s} \cdot \sin[\psi_i^{(j)} + \phi_{i,s}(t)] \cdot \psi_i \cdot e^{\lambda\cdot t}$$

$$E_i \cdot \lambda \cdot e^{\lambda\cdot t} = \left[\left\{B_i \cdot \left(N_0^{(j)} - \frac{N_i^{(j)}}{2}\right) - \frac{1}{\tau_p}\right\} + B_i \cdot \left(n_0 - \frac{n_i}{2}\right) \cdot e^{\lambda\cdot t}\right] \cdot \left(\frac{E_i^{(j)}}{2} + \frac{E_i \cdot e^{\lambda\cdot t}}{2}\right) + \left(\frac{\kappa}{\tau_l}\right) \cdot E_{i,s} \cdot \cos[\psi_i^{(j)} + \phi_{i,s}(t)] - \left(\frac{\kappa}{\tau_l}\right) \cdot E_{i,s} \cdot \sin[\psi_i^{(j)} + \phi_{i,s}(t)] \cdot \psi_i \cdot e^{\lambda\cdot t}$$

$$E_i \cdot \lambda \cdot e^{\lambda\cdot t} = \left[\left\{B_i \cdot \left(N_0^{(j)} - \frac{N_i^{(j)}}{2}\right) - \frac{1}{\tau_p}\right\} + B_i \cdot \left(n_0 - \frac{n_i}{2}\right) \cdot e^{\lambda\cdot t}\right] \cdot \frac{E_i^{(j)}}{2}$$

$$+\left[\left\{B_i\cdot(N_0^{(j)}-\frac{N_i^{(j)}}{2})-\frac{1}{\tau_p}\right\}+B_i\cdot(n_0-\frac{n_i}{2})\cdot e^{\lambda\cdot t}\right]\cdot\frac{E_i\cdot e^{\lambda\cdot t}}{2}$$
$$+\left(\frac{\kappa}{\tau_l}\right)\cdot E_{i,s}\cdot\cos[\psi_i^{(j)}+\phi_{i,s}(t)]-\left(\frac{\kappa}{\tau_l}\right)\cdot E_{i,s}$$
$$\cdot\sin[\psi_i^{(j)}+\phi_{i,s}(t)]\cdot\psi_i\cdot e^{\lambda\cdot t}$$

$$E_i\cdot\lambda\cdot e^{\lambda\cdot t}=\left\{B_i\cdot\left(N_0^{(j)}-\frac{N_i^{(j)}}{2}\right)-\frac{1}{\tau_p}\right\}\cdot\frac{E_i^{(j)}}{2}+B_i\cdot\left(n_0-\frac{n_i}{2}\right)\cdot e^{\lambda\cdot t}\cdot\frac{E_i^{(j)}}{2}$$
$$+\left\{B_i\cdot\left(N_0^{(j)}-\frac{N_i^{(j)}}{2}\right)-\frac{1}{\tau_p}\right\}\cdot\frac{E_i\cdot e^{\lambda\cdot t}}{2}$$
$$+B_i\cdot\left(n_0-\frac{n_i}{2}\right)\cdot e^{\lambda\cdot t}\cdot\frac{E_i\cdot e^{\lambda\cdot t}}{2}+\left(\frac{\kappa}{\tau_l}\right)\cdot E_{i,s}\cdot\cos[\psi_i^{(j)}+\phi_{i,s}(t)]$$
$$-\left(\frac{\kappa}{\tau_l}\right)\cdot E_{i,s}\cdot\sin[\psi_i^{(j)}+\phi_{i,s}(t)]\cdot\psi_i\cdot e^{\lambda\cdot t}$$

$$E_i\cdot\lambda\cdot e^{\lambda\cdot t}=\left[B_i\cdot\left(N_0^{(j)}-\frac{N_i^{(j)}}{2}\right)-\frac{1}{\tau_p}\right]\cdot\frac{E_i^{(j)}}{2}+B_i\cdot\left(n_0-\frac{n_i}{2}\right)\cdot\frac{E_i^{(j)}}{2}\cdot e^{\lambda\cdot t}$$
$$+\left[B_i\cdot\left(N_0^{(j)}-\frac{N_i^{(j)}}{2}\right)-\frac{1}{\tau_p}\right]\cdot E_i\cdot\frac{e^{\lambda\cdot t}}{2}$$
$$+B_i\cdot\left(E_i\cdot n_0-\frac{E_i\cdot n_i}{2}\right)\cdot\frac{e^{2\cdot\lambda\cdot t}}{2}$$
$$+\left(\frac{\kappa}{\tau_l}\right)\cdot E_{i,s}\cdot\cos[\psi_i^{(j)}+\phi_{i,s}(t)]$$
$$-\left(\frac{\kappa}{\tau_l}\right)\cdot E_{i,s}\cdot\sin[\psi_i^{(j)}+\phi_{i,s}(t)]\cdot\psi_i\cdot e^{\lambda\cdot t}$$

Assumption: $E_i\cdot n_0\approx 0$; $E_i\cdot n_i\approx 0$

$$E_i\cdot\lambda\cdot e^{\lambda\cdot t}=\left[B_i\cdot\left(N_0^{(j)}-\frac{N_i^{(j)}}{2}\right)-\frac{1}{\tau_p}\right]\cdot\frac{E_i^{(j)}}{2}+\left(\frac{\kappa}{\tau_l}\right)\cdot E_{i,s}\cdot\cos\left[\psi_i^{(j)}+\phi_{i,s}(t)\right]$$
$$+B_i\cdot\left(n_0-\frac{n_i}{2}\right)\cdot\frac{E_i^{(j)}}{2}\cdot e^{\lambda\cdot t}+\left[B_i\cdot\left(N_0^{(j)}-\frac{N_i^{(j)}}{2}\right)-\frac{1}{\tau_p}\right]\cdot E_i\cdot\frac{e^{\lambda\cdot t}}{2}$$
$$-\left(\frac{\kappa}{\tau_l}\right)\cdot E_{i,s}\cdot\sin[\psi_i^{(j)}+\phi_{i,s}(t)]\cdot\psi_i\cdot e^{\lambda\cdot t}$$

At fixed points: $\left[B_i\cdot\left(N_0^{(j)}-\frac{N_i^{(j)}}{2}\right)-\frac{1}{\tau_p}\right]\cdot\frac{E_i^{(j)}}{2}+\left(\frac{\kappa}{\tau_l}\right)\cdot E_{i,s}\cdot\cos[\psi_i^{(j)}+\phi_{i,s}(t)]=0.$

$$E_i \cdot \lambda \cdot e^{\lambda \cdot t} = B_i \cdot \left(n_0 - \frac{n_i}{2}\right) \cdot \frac{E_i^{(j)}}{2} \cdot e^{\lambda \cdot t} + \left[B_i \cdot \left(N_0^{(j)} - \frac{N_i^{(j)}}{2}\right) - \frac{1}{\tau_p}\right] \cdot E_i \cdot \frac{e^{\lambda \cdot t}}{2}$$

$$- \left(\frac{\kappa}{\tau_l}\right) \cdot E_{i,s} \cdot \sin[\psi_i^{(j)} + \phi_{i,s}(t)] \cdot \psi_i \cdot e^{\lambda \cdot t}$$

$$E_i \cdot \lambda = B_i \cdot \left(n_0 - \frac{n_i}{2}\right) \cdot \frac{E_i^{(j)}}{2} + \left[B_i \cdot \left(N_0^{(j)} - \frac{N_i^{(j)}}{2}\right) - \frac{1}{\tau_p}\right]$$

$$\cdot E_i \cdot \frac{1}{2} - \left(\frac{\kappa}{\tau_l}\right) \cdot E_{i,s} \cdot \sin\left[\psi_i^{(j)} + \phi_{i,s}(t)\right] \cdot \psi_i$$

$$- E_i \cdot \lambda + B_i \cdot n_0 \cdot \frac{E_i^{(j)}}{2} - B_i \cdot \frac{n_i}{2} \cdot \frac{E_i^{(j)}}{2} + \left[B_i \cdot \left(N_0^{(j)} - \frac{N_i^{(j)}}{2}\right)\right.$$

$$\left. - \frac{1}{\tau_p}\right] \cdot E_i \cdot \frac{1}{2} - \left(\frac{\kappa}{\tau_l}\right) \cdot E_{i,s} \cdot \sin\left[\psi_i^{(j)} + \phi_{i,s}(t)\right] \cdot \psi_i = 0 \; B_i$$

$$\cdot \frac{E_i^{(j)}}{2} \cdot n_0 - B_i \cdot \frac{E_i^{(j)}}{4} \cdot n_i - E_i \cdot \lambda + \frac{1}{2} \cdot \left[B_i \cdot \left(N_0^{(j)} - \frac{N_i^{(j)}}{2}\right) - \frac{1}{\tau_p}\right]$$

$$\cdot E_i - \left(\frac{\kappa}{\tau_l}\right) \cdot E_{i,s} \cdot \sin\left[\psi_i^{(j)} + \phi_{i,s}(t)\right] \cdot \psi_i = 0$$

$$\frac{d\psi_i}{dt} = \omega_i - \omega_{i,s} - \left(\frac{\kappa}{\tau_l}\right) \cdot \left(\frac{E_{i,s}}{E_i}\right) \cdot \sin[\psi_i(t) + \phi_{i,s}(t)]\psi_i \cdot \lambda \cdot e^{\lambda \cdot t}$$

$$= \omega_i - \omega_{i,s} - \left(\frac{\kappa}{\tau_l}\right) \cdot \left(\frac{E_{i,s}}{E_i^{(j)} + E_i \cdot e^{\lambda \cdot t}}\right)$$

$$\cdot \sin\left[\psi_i^{(j)} + \psi_i \cdot e^{\lambda \cdot t} + \phi_{i,s}(t)\right] \psi_i \cdot \lambda \cdot e^{\lambda \cdot t} = \omega_i - \omega_{i,s} - \left(\frac{\kappa}{\tau_l}\right)$$

$$\cdot \left(\frac{E_{i,s}}{\left[E_i^{(j)} + E_i \cdot e^{\lambda \cdot t}\right]} \cdot \left[\frac{E_i^{(j)} - E_i \cdot e^{\lambda \cdot t}}{E_i^{(j)} - E_i \cdot e^{\lambda \cdot t}}\right]\right) \cdot \sin\left[\psi_i^{(j)} + \psi_i \cdot e^{\lambda \cdot t} + \phi_{i,s}(t)\right]$$

$$\psi_i \cdot \lambda \cdot e^{\lambda \cdot t} = \omega_i - \omega_{i,s} - \left(\frac{\kappa}{\tau_l}\right) \cdot \frac{E_{i,s} \cdot \left[E_i^{(j)} - E_i \cdot e^{\lambda \cdot t}\right]}{\left[\left(E_i^{(j)}\right)^2 - E_i^2 \cdot e^{2 \cdot \lambda \cdot t}\right]}$$

$$\cdot \sin\left[\psi_i^{(j)} + \psi_i \cdot e^{\lambda \cdot t} + \phi_{i,s}(t)\right]$$

Assumption: $E_i^2 \approx 0$

$$\psi_i \cdot \lambda \cdot e^{\lambda \cdot t} = \omega_i - \omega_{i,s} - \left(\frac{\kappa}{\tau_l}\right)$$

$$\cdot \frac{E_{i,s} \cdot [E_i^{(j)} - E_i \cdot e^{\lambda \cdot t}]}{(E_i^{(j)})^2} \cdot \sin[\psi_i^{(j)} + \psi_i \cdot e^{\lambda \cdot t} + \phi_{i,s}(t)]$$

$$\psi_i \cdot \lambda \cdot e^{\lambda \cdot t} = \omega_i - \omega_{i,s} - \left(\frac{\kappa}{\tau_l}\right) \cdot \left[\frac{E_{i,s}}{E_i^{(j)}} - \frac{E_{i,s}}{(E_i^{(j)})^2} \cdot E_i \cdot e^{\lambda \cdot t}\right]$$

$$\cdot \sin[\psi_i^{(j)} + \psi_i \cdot e^{\lambda \cdot t} + \phi_{i,s}(t)]$$

First we do algebraic manipulation for the sin{...} expression.

$$\sin[\psi_i^{(j)} + \psi_i \cdot e^{\lambda \cdot t} + \phi_{i,s}(t)] = \sin[\psi_i^{(j)} + \phi_{i,s}(t) + \psi_i \cdot e^{\lambda \cdot t}]$$

$$\sin[\psi_i^{(j)} + \phi_{i,s}(t) + \psi_i \cdot e^{\lambda \cdot t}] = \sin(\psi_i^{(j)} + \phi_{i,s}(t)) \cdot \cos(\psi_i \cdot e^{\lambda \cdot t})$$
$$+ \cos(\psi_i^{(j)} + \phi_{i,s}(t)) \cdot \sin(\psi_i \cdot e^{\lambda \cdot t})$$

Based on assumptions A and B:$\xi_1(\psi_i, \lambda) = \psi_i \cdot e^{\lambda \cdot t}$

$$\cos(\xi_1(\psi_i, \lambda)) = \cos(\psi_i \cdot e^{\lambda \cdot t}) = \sum_{n=0}^{\infty} \frac{(-1)^n}{(2 \cdot n)!} \cdot [\xi_1(\psi_i, \lambda)]^{2 \cdot n} \approx 1$$

$$\sin(\xi_1(\psi_i, \lambda)) = \sin(\psi_i \cdot e^{\lambda \cdot t}) = \sum_{n=0}^{\infty} \frac{(-1)^n}{(2 \cdot n + 1)!} \cdot [\xi_1(\psi_i, \lambda)]^{2 \cdot n+1} \approx \psi_i$$
$$\cdot e^{\lambda \cdot t} \sin[\psi_i^{(j)} + \phi_{i,s}(t) + \psi_i \cdot e^{\lambda \cdot t}] = \sin(\psi_i^{(j)} + \phi_{i,s}(t)) + \cos(\psi_i^{(j)} + \phi_{i,s}(t)) \cdot \psi_i$$
$$\cdot e^{\lambda \cdot t} \psi_i \cdot \lambda \cdot e^{\lambda \cdot t} = \omega_i - \omega_{i,s} - \left(\frac{\kappa}{\tau_l}\right) \cdot \left[\frac{E_{i,s}}{E_i^{(j)}} - \frac{E_{i,s}}{(E_i^{(j)})^2} \cdot E_i \cdot e^{\lambda \cdot t}\right]$$
$$\cdot [\sin(\psi_i^{(j)} + \phi_{i,s}(t)) + \cos(\psi_i^{(j)} + \phi_{i,s}(t)) \cdot \psi_i \cdot e^{\lambda \cdot t}]$$

$$\psi_i \cdot \lambda \cdot e^{\lambda \cdot t} = \omega_i - \omega_{i,s} - \left(\frac{\kappa}{\tau_l}\right)$$
$$\cdot \left[\frac{E_{i,s}}{E_i^{(j)}} \cdot \sin(\psi_i^{(j)} + \phi_{i,s}(t)) + \frac{E_{i,s}}{E_i^{(j)}} \cdot \cos(\psi_i^{(j)} + \phi_{i,s}(t)) \cdot \psi_i \cdot e^{\lambda \cdot t}\right.$$
$$- \frac{E_{i,s}}{(E_i^{(j)})^2} \cdot E_i \cdot e^{\lambda \cdot t} \cdot \sin(\psi_i^{(j)} + \phi_{i,s}(t))$$
$$\left. - \frac{E_{i,s}}{(E_i^{(j)})^2} \cdot E_i \cdot \psi_i \cdot e^{2 \cdot \lambda \cdot t} \cdot \cos(\psi_i^{(j)} + \phi_{i,s}(t))\right]$$

Assumption: $E_i \cdot \psi_i \approx 0$

$$\psi_i \cdot \lambda \cdot e^{\lambda \cdot t} = \omega_i - \omega_{i,s} - \left(\frac{\kappa}{\tau_l}\right) \cdot \left[\frac{E_{i,s}}{E_i^{(j)}} \cdot \sin(\psi_i^{(j)} + \phi_{i,s}(t))\right.$$
$$+ \frac{E_{i,s}}{E_i^{(j)}} \cdot \cos(\psi_i^{(j)} + \phi_{i,s}(t)) \cdot \psi_i \cdot e^{\lambda \cdot t}$$
$$\left. - \frac{E_{i,s}}{(E_i^{(j)})^2} \cdot E_i \cdot e^{\lambda \cdot t} \cdot \sin(\psi_i^{(j)} + \phi_{i,s}(t))\right]$$

$$\psi_i \cdot \lambda \cdot e^{\lambda \cdot t} = \omega_i - \omega_{i,s} - \left(\frac{\kappa}{\tau_l}\right) \cdot \frac{E_{i,s}}{E_i^{(j)}} \cdot \sin(\psi_i^{(j)} + \phi_{i,s}(t))$$
$$- \left(\frac{\kappa}{\tau_l}\right) \cdot \frac{E_{i,s}}{E_i^{(j)}} \cdot \cos(\psi_i^{(j)} + \phi_{i,s}(t)) \cdot \psi_i \cdot e^{\lambda \cdot t}$$

$$+\left(\frac{\kappa}{\tau_l}\right)\cdot\frac{E_{i,s}}{(E_i^{(j)})^2}\cdot E_i\cdot e^{\lambda\cdot t}\cdot\sin(\psi_i^{(j)}+\phi_{i,s}(t))$$

At fixed points: $\omega_i-\omega_{i,s}-(\frac{\kappa}{\tau_l})\cdot\frac{E_{i,s}}{E_i^{(j)}}\cdot\sin(\psi_i^{(j)}+\phi_{i,s}(t))=0.$

$$\left(\frac{\kappa}{\tau_l}\right)\cdot\frac{E_{i,s}}{(E_i^{(j)})^2}\cdot\sin(\psi_i^{(j)}+\phi_{i,s}(t))\cdot E_i-\psi_i\cdot\lambda$$
$$-\left(\frac{\kappa}{\tau_l}\right)\cdot\frac{E_{i,s}}{E_i^{(j)}}\cdot\cos(\psi_i^{(j)}+\phi_{i,s}(t))\cdot\psi_i=0$$

We can summarize our last arbitrarily small increments system equations:

$$-n_0\cdot\lambda-n_0\cdot\left[\frac{1}{\tau}+\left(\sum_{k=1}^{2}B_k\cdot S_k\right)\right]=0$$

$$B_i\cdot S_i\cdot n_0-n_i\cdot\lambda-\frac{n_i}{\tau}-n_i\cdot(B_1\cdot S_1+B_2\cdot S_2)=0$$

$$B_i\cdot\frac{E_i^{(j)}}{2}\cdot n_0-B_i\cdot\frac{E_i^{(j)}}{4}\cdot n_i-E_i\cdot\lambda+\frac{1}{2}$$
$$\cdot\left[B_i\cdot(N_0^{(j)}-\frac{N_i^{(j)}}{2})-\frac{1}{\tau_p}\right]\cdot E_i-\left(\frac{\kappa}{\tau_l}\right)$$
$$\cdot E_{i,s}\cdot\sin[\psi_i^{(j)}+\phi_{i,s}(t)]\cdot\psi_i=0$$

$$\left(\frac{\kappa}{\tau_l}\right)\cdot\frac{E_{i,s}}{(E_i^{(j)})^2}\cdot\sin(\psi_i^{(j)}+\phi_{i,s}(t))\cdot E_i-\psi_i\cdot\lambda$$
$$-(\frac{\kappa}{\tau_l})\cdot\frac{E_{i,s}}{E_i^{(j)}}\cdot\cos(\psi_i^{(j)}+\phi_{i,s}(t))\cdot\psi_i=0$$

The small increments Jacobian of our system is as follow:Jacobian

$$\begin{pmatrix} -\lambda-[\frac{1}{\tau}+(\sum_{k=1}^{2}B_k\cdot S_k)] & 0 & 0 & 0 \\ B_i\cdot S_i & -\lambda-\frac{1}{\tau}-(B_1\cdot S_1+B_2\cdot S_2) & 0 & 0 \\ B_i\cdot\frac{E_i^{(j)}}{2} & -B_i\cdot\frac{E_i^{(j)}}{4} & -\lambda+\frac{1}{2}\cdot[B_i\cdot(N_0^{(j)}-\frac{N_i^{(j)}}{2})-\frac{1}{\tau_p}] & -(\frac{\kappa}{\tau_l})\cdot E_{i,s}\cdot\sin[\psi_i^{(j)}+\phi_{i,s}(t)] \\ 0 & 0 & (\frac{\kappa}{\tau_l})\cdot\frac{E_{i,s}}{(E_i^{(j)})^2}\cdot\sin(\psi_i^{(j)}+\phi_{i,s}(t)) & -\lambda-(\frac{\kappa}{\tau_l})\cdot\frac{E_{i,s}}{E_i^{(j)}}\cdot\cos(\psi_i^{(j)}+\phi_{i,s}(t)) \end{pmatrix}\cdot\begin{pmatrix} n_0 \\ n_i \\ E_i \\ \psi_i \end{pmatrix}=\begin{pmatrix} 0 \\ 0 \\ 0 \\ 0 \end{pmatrix}$$

$$(A-\lambda\cdot I)=\begin{pmatrix} -\lambda-[\frac{1}{\tau}+(\sum_{k=1}^{2}B_k\cdot S_k)] & 0 & 0 & 0 \\ B_i\cdot S_i & -\lambda-\frac{1}{\tau}-(B_1\cdot S_1+B_2\cdot S_2) & 0 & 0 \\ B_i\cdot\frac{E_i^{(j)}}{2} & -B_i\cdot\frac{E_i^{(j)}}{4} & -\lambda+\frac{1}{2}\cdot[B_i\cdot(N_0^{(j)}-\frac{N_i^{(j)}}{2})-\frac{1}{\tau_p}] & -(\frac{\kappa}{\tau_l})\cdot E_{i,s}\cdot\sin[\psi_i^{(j)}+\phi_{i,s}(t)] \\ 0 & 0 & (\frac{\kappa}{\tau_l})\cdot\frac{E_{i,s}}{(E_i^{(j)})^2}\cdot\sin(\psi_i^{(j)}+\phi_{i,s}(t)) & -\lambda-(\frac{\kappa}{\tau_l})\cdot\frac{E_{i,s}}{E_i^{(j)}}\cdot\cos(\psi_i^{(j)}+\phi_{i,s}(t)) \end{pmatrix}$$

$$\det(A-\lambda\cdot I)=0$$

$$\det(A-\lambda\cdot I)=$$

$$\det\begin{pmatrix} -\lambda-[\frac{1}{\tau}+(\sum_{k=1}^{2}B_k\cdot S_k)] & 0 & 0 & 0 \\ B_i\cdot S_i & -\lambda-\frac{1}{\tau}-(B_1\cdot S_1+B_2\cdot S_2) & 0 & 0 \\ B_i\cdot\frac{E_i^{(j)}}{2} & -B_i\cdot\frac{E_i^{(j)}}{4} & -\lambda+\frac{1}{2}\cdot[B_i\cdot(N_0^{(j)}-\frac{N_i^{(j)}}{2})-\frac{1}{\tau_p}] & -(\frac{\kappa}{\tau_l})\cdot E_{i,s}\cdot\sin[\psi_i^{(j)}+\phi_{i,s}(t)] \\ 0 & 0 & (\frac{\kappa}{\tau_l})\cdot\frac{E_{i,s}}{(E_i^{(j)})^2}\cdot\sin(\psi_i^{(j)}+\phi_{i,s}(t)) & -\lambda-(\frac{\kappa}{\tau_l})\cdot\frac{E_{i,s}}{E_i^{(j)}}\cdot\cos(\psi_i^{(j)}+\phi_{i,s}(t)) \end{pmatrix}$$

$$\det(A-\lambda\cdot I)$$

$$=(-\lambda-[\frac{1}{\tau}+(\sum_{k=1}^{2}B_k\cdot S_k)])\cdot\det\begin{pmatrix} -\lambda-\frac{1}{\tau}-(B_1\cdot S_1+B_2\cdot S_2) & 0 & 0 \\ -B_i\cdot\frac{E_i^{(j)}}{4} & -\lambda+\frac{1}{2}\cdot[B_i\cdot(N_0^{(j)}-\frac{N_i^{(j)}}{2})-\frac{1}{\tau_p}] & -(\frac{\kappa}{\tau_l})\cdot E_{i,s}\cdot\sin[\psi_i^{(j)}+\phi_{i,s}(t)] \\ 0 & (\frac{\kappa}{\tau_l})\cdot\frac{E_{i,s}}{(E_i^{(j)})^2}\cdot\sin(\psi_i^{(j)}+\phi_{i,s}(t)) & -\lambda-(\frac{\kappa}{\tau_l})\cdot\frac{E_{i,s}}{E_i^{(j)}}\cdot\cos(\psi_i^{(j)}+\phi_{i,s}(t)) \end{pmatrix}$$

$$\begin{aligned} \det(A-\lambda\cdot I) = {} & \left(\lambda^2+2\cdot\lambda\cdot\left[\frac{1}{\tau}+\left(\sum_{k=1}^{2}B_k\cdot S_k\right)\right]+\left[\frac{1}{\tau}+\left(\sum_{k=1}^{2}B_k\cdot S_k\right)\right]^2\right) \\ & \cdot\left[\left\{-\lambda+\frac{1}{2}\cdot\left[B_i\cdot(N_0^{(j)}-\frac{N_i^{(j)}}{2})-\frac{1}{\tau_p}\right]\right\}\right. \\ & \cdot\left\{-\lambda-\left(\frac{\kappa}{\tau_l}\right)\cdot\frac{E_{i,s}}{E_i^{(j)}}\cdot\cos(\psi_i^{(j)}+\phi_{i,s}(t))\right\} \\ & -\left\{\left(\frac{\kappa}{\tau_l}\right)\cdot\frac{E_{i,s}}{(E_i^{(j)})^2}\cdot\sin(\psi_i^{(j)}+\phi_{i,s}(t))\right\} \\ & \left.\cdot\left\{-\left(\frac{\kappa}{\tau_l}\right)\cdot E_{i,s}\cdot\sin\left[\psi_i^{(j)}+\phi_{i,s}(t)\right]\right\}\right] \end{aligned}$$

Part I:

$$\left\{-\lambda + \frac{1}{2}\cdot\left[B_i\cdot\left(N_0^{(j)} - \frac{N_i^{(j)}}{2}\right) - \frac{1}{\tau_p}\right]\right\}\cdot\left\{-\lambda - \left(\frac{\kappa}{\tau_l}\right)\cdot\frac{E_{i,s}}{E_i^{(j)}}\cdot\cos\left(\psi_i^{(j)} + \phi_{i,s}(t)\right)\right\}$$

$$\left\{-\lambda + \frac{1}{2}\cdot\left[B_i\cdot\left(N_0^{(j)} - \frac{N_i^{(j)}}{2}\right) - \frac{1}{\tau_p}\right]\right\}\cdot\left\{-\lambda - \left(\frac{\kappa}{\tau_l}\right)\cdot\frac{E_{i,s}}{E_i^{(j)}}\right.$$
$$\left.\cdot\cos\left(\psi_i^{(j)} + \phi_{i,s}(t)\right)\right\} = \lambda^2 + \lambda\cdot\left(\frac{\kappa}{\tau_l}\right)\cdot\frac{E_{i,s}}{E_i^{(j)}}\cdot\cos\left(\psi_i^{(j)} + \phi_{i,s}(t)\right)$$
$$-\lambda\cdot\frac{1}{2}\cdot\left[B_i\cdot\left(N_0^{(j)} - \frac{N_i^{(j)}}{2}\right) - \frac{1}{\tau_p}\right] - \frac{1}{2}\cdot\left[B_i\cdot\left(N_0^{(j)} - \frac{N_i^{(j)}}{2}\right) - \frac{1}{\tau_p}\right]$$
$$\cdot\left(\frac{\kappa}{\tau_l}\right)\cdot\frac{E_{i,s}}{E_i^{(j)}}\cdot\cos\left(\psi_i^{(j)} + \phi_{i,s}(t)\right)$$

$$\left\{-\lambda + \frac{1}{2}\cdot\left[B_i\cdot\left(N_0^{(j)} - \frac{N_i^{(j)}}{2}\right) - \frac{1}{\tau_p}\right]\right\}\cdot\left\{-\lambda - \left(\frac{\kappa}{\tau_l}\right)\cdot\frac{E_{i,s}}{E_i^{(j)}}\cdot\cos\left(\psi_i^{(j)} + \phi_{i,s}(t)\right)\right\}$$
$$= \lambda^2 + \lambda\cdot\left\{\left(\frac{\kappa}{\tau_l}\right)\cdot\frac{E_{i,s}}{E_i^{(j)}}\cdot\cos\left(\psi_i^{(j)} + \phi_{i,s}(t)\right)\right.$$
$$\left. - \frac{1}{2}\cdot\left[B_i\cdot\left(N_0^{(j)} - \frac{N_i^{(j)}}{2}\right) - \frac{1}{\tau_p}\right]\right\}$$
$$- \frac{1}{2}\cdot\left[B_i\cdot\left(N_0^{(j)} - \frac{N_i^{(j)}}{2}\right) - \frac{1}{\tau_p}\right]\cdot\left(\frac{\kappa}{\tau_l}\right)$$
$$\cdot\frac{E_{i,s}}{E_i^{(j)}}\cdot\cos\left(\psi_i^{(j)} + \phi_{i,s}(t)\right)$$

Part II:

$$\left\{\left(\frac{\kappa}{\tau_l}\right)\cdot\frac{E_{i,s}}{\left(E_i^{(j)}\right)^2}\cdot\sin\left(\psi_i^{(j)} + \phi_{i,s}(t)\right)\right\}\cdot\left\{-\left(\frac{\kappa}{\tau_l}\right)\cdot E_{i,s}\cdot\sin\left[\psi_i^{(j)} + \phi_{i,s}(t)\right]\right\}$$

$$\left\{\left(\frac{\kappa}{\tau_l}\right)\cdot\frac{E_{i,s}}{\left(E_i^{(j)}\right)^2}\cdot\sin\left(\psi_i^{(j)} + \phi_{i,s}(t)\right)\right\}\cdot\left\{-\left(\frac{\kappa}{\tau_l}\right)\cdot E_{i,s}\cdot\sin\left[\psi_i^{(j)} + \phi_{i,s}(t)\right]\right\}$$
$$= -\left(\frac{\kappa}{\tau_l}\right)^2\cdot\left(\frac{E_{i,s}}{E_i^{(j)}}\right)^2\cdot\sin^2\left[\psi_i^{(j)} + \phi_{i,s}(t)\right]$$

Implementing Part I, II in system eigenvalue equation:

$$\det(A-\lambda\cdot I)=\left(\lambda^2+2\cdot\lambda\cdot\left[\frac{1}{\tau}+\left(\sum_{k=1}^{2}B_k\cdot S_k\right)\right]+\left[\frac{1}{\tau}+\left(\sum_{k=1}^{2}B_k\cdot S_k\right)\right]^2\right)$$
$$\cdot\left\{\lambda^2+\lambda\cdot\left\{\left(\frac{\kappa}{\tau_l}\right)\cdot\frac{E_{i,s}}{E_i^{(j)}}\cdot\cos\left(\psi_i^{(j)}+\phi_{i,s}(t)\right)\right.\right.$$
$$\left.-\frac{1}{2}\cdot\left[B_i\cdot\left(N_0^{(j)}-\frac{N_i^{(j)}}{2}\right)-\frac{1}{\tau_p}\right]\right\}-\frac{1}{2}\cdot\left[B_i\cdot\left(N_0^{(j)}-\frac{N_i^{(j)}}{2}\right)-\frac{1}{\tau_p}\right]$$
$$\cdot\left(\frac{\kappa}{\tau_l}\right)\cdot\frac{E_{i,s}}{E_i^{(j)}}\cdot\cos\left(\psi_i^{(j)}+\phi_{i,s}(t)\right)$$
$$\left.+\left(\frac{\kappa}{\tau_l}\right)^2\cdot\left(\frac{E_{i,s}}{E_i^{(j)}}\right)^2\cdot\sin^2\left[\psi_i^{(j)}+\phi_{i,s}(t)\right]\right\}$$

We define for simplicity the following global parameters:

$$\Xi_1=2\cdot\left[\frac{1}{\tau}+\left(\sum_{k=1}^{2}B_k\cdot S_k\right)\right];\ \Xi_2=\left[\frac{1}{\tau}+\left(\sum_{k=1}^{2}B_k\cdot S_k\right)\right]^2$$
$$\Xi_3=\left(\frac{\kappa}{\tau_l}\right)\cdot\frac{E_{i,s}}{E_i^{(j)}}\cdot\cos\left(\psi_i^{(j)}+\phi_{i,s}(t)\right)-\frac{1}{2}\cdot\left[B_i\cdot\left(N_0^{(j)}-\frac{N_i^{(j)}}{2}\right)-\frac{1}{\tau_p}\right]$$
$$\Xi_4=\frac{1}{2}\cdot\left[B_i\cdot\left(N_0^{(j)}-\frac{N_i^{(j)}}{2}\right)-\frac{1}{\tau_p}\right]\cdot\left(\frac{\kappa}{\tau_l}\right)\cdot\frac{E_{i,s}}{E_i^{(j)}}$$
$$\cdot\cos\left(\psi_i^{(j)}+\phi_{i,s}(t)\right)+\left(\frac{\kappa}{\tau_l}\right)^2\cdot\left(\frac{E_{i,s}}{E_i^{(j)}}\right)^2\cdot\sin^2\left[\psi_i^{(j)}+\phi_{i,s}(t)\right]$$

$$\det(A-\lambda\cdot I)=(\lambda^2+\lambda\cdot\Xi_1+\Xi_2)\cdot(\lambda^2+\lambda\cdot\Xi_3-\Xi_4)=\lambda^4+\lambda^3\cdot\Xi_3$$
$$-\lambda^2\cdot\Xi_4+\lambda^3\cdot\Xi_1+\lambda^2\cdot\Xi_1\cdot\Xi_3-\lambda\cdot\Xi_1\cdot\Xi_4$$
$$+\lambda^2\cdot\Xi_2+\lambda\cdot\Xi_3\cdot\Xi_2-\Xi_4\cdot\Xi_2$$

$$\det(A-\lambda\cdot I)=(\lambda^2+\lambda\cdot\Xi_1+\Xi_2)\cdot(\lambda^2+\lambda\cdot\Xi_3-\Xi_4)$$
$$=\lambda^4+\lambda^3\cdot(\Xi_3+\Xi_1)+\lambda^2(\Xi_1\cdot\Xi_3+\Xi_2-\Xi_4)$$
$$+\lambda\cdot(\Xi_3\cdot\Xi_2-\Xi_1\cdot\Xi_4)-\Xi_4\cdot\Xi_2$$

$$\det(A-\lambda\cdot I)=\sum_{k=0}^{4}\lambda^k\cdot\Upsilon_k;\ \Upsilon_4=1;\ \Upsilon_3$$
$$=\Xi_3+\Xi_1;\ \Upsilon_2=\Xi_1\cdot\Xi_3+\Xi_2-\Xi_4$$

$$\Upsilon_1 = \Xi_3 \cdot \Xi_2 - \Xi_1 \cdot \Xi_4;\ \Upsilon_0 = -\Xi_4 \cdot \Xi_2$$

$$\Upsilon_3 = \Xi_3 + \Xi_1 = \left(\frac{\kappa}{\tau_l}\right) \cdot \frac{E_{i,s}}{E_i^{(j)}} \cdot \cos\left(\psi_i^{(j)} + \phi_{i,s}(t)\right) - \frac{1}{2} \cdot \left[B_i \cdot \left(N_0^{(j)} - \frac{N_i^{(j)}}{2}\right) - \frac{1}{\tau_p}\right] + 2 \cdot \left[\frac{1}{\tau} + \left(\sum_{k=1}^{2} B_k \cdot S_k\right)\right]$$

$$\Upsilon_2 = \Xi_1 \cdot \Xi_3 + \Xi_2 - \Xi_4 = 2 \cdot \left[\frac{1}{\tau} + \left(\sum_{k=1}^{2} B_k \cdot S_k\right)\right] \cdot \left\{\left(\frac{\kappa}{\tau_l}\right) \cdot \frac{E_{i,s}}{E_i^{(j)}} \cdot \cos\left(\psi_i^{(j)} + \phi_{i,s}(t)\right) - \frac{1}{2} \cdot \left[B_i \cdot (N_0^{(j)} - \frac{N_i^{(j)}}{2}) - \frac{1}{\tau_p}\right]\right\} + \left[\frac{1}{\tau} + \left(\sum_{k=1}^{2} B_k \cdot S_k\right)\right]^2$$
$$\left\{-\frac{1}{2} \cdot \left[B_i \cdot \left(N_0^{(j)} - \frac{N_i^{(j)}}{2}\right) - \frac{1}{\tau_p}\right] \cdot \left(\frac{\kappa}{\tau_l}\right) \cdot \frac{E_{i,s}}{E_i^{(j)}} \cdot \cos\left(\psi_i^{(j)} + \phi_{i,s}(t)\right) + \left(\frac{\kappa}{\tau_l}\right)^2 \cdot \left(\frac{E_{i,s}}{E_i^{(j)}}\right)^2 \cdot \sin^2\left[\psi_i^{(j)} + \phi_{i,s}(t)\right]\right\}$$

$$\Upsilon_1 = \Xi_3 \cdot \Xi_2 - \Xi_1 \cdot \Xi_4 = \left\{\left(\frac{\kappa}{\tau_l}\right) \cdot \frac{E_{i,s}}{E_i^{(j)}} \cdot \cos\left(\psi_i^{(j)} + \phi_{i,s}(t)\right) - \frac{1}{2} \cdot \left[B_i \cdot \left(N_0^{(j)} - \frac{N_i^{(j)}}{2}\right) - \frac{1}{\tau_p}\right]\right\} \cdot \left[\frac{1}{\tau} + \left(\sum_{k=1}^{2} B_k \cdot S_k\right)\right]^2$$
$$- 2 \cdot \left[\frac{1}{\tau} + \left(\sum_{k=1}^{2} B_k \cdot S_k\right)\right] \cdot \left\{\frac{1}{2} \cdot \left[B_i \cdot \left(N_0^{(j)} - \frac{N_i^{(j)}}{2}\right) - \frac{1}{\tau_p}\right] \cdot \left(\frac{\kappa}{\tau_l}\right) \cdot \frac{E_{i,s}}{E_i^{(j)}} \cdot \cos\left(\psi_i^{(j)} + \phi_{i,s}(t)\right) + \left(\frac{\kappa}{\tau_l}\right)^2 \cdot \left(\frac{E_{i,s}}{E_i^{(j)}}\right)^2 \cdot \sin^2\left[\psi_i^{(j)} + \phi_{i,s}(t)\right]\right\}$$

$$\Upsilon_0 = -\Xi_4 \cdot \Xi_2 = -\left\{\frac{1}{2} \cdot \left[B_i \cdot \left(N_0^{(j)} - \frac{N_i^{(j)}}{2}\right) - \frac{1}{\tau_p}\right] \cdot \left(\frac{\kappa}{\tau_l}\right) \cdot \frac{E_{i,s}}{E_i^{(j)}} \cdot \cos\left(\psi_i^{(j)} + \phi_{i,s}(t)\right)\right.$$

$$+\left(\frac{\kappa}{\tau_l}\right)^2 \cdot \left(\frac{E_{i,s}}{E_i^{(j)}}\right)^2 \cdot \sin^2\left[\psi_i^{(j)} + \phi_{i,s}(t)\right]\Bigg\} \cdot \left[\frac{1}{\tau} + \left(\sum_{k=1}^{2} B_k \cdot S_k\right)\right]^2$$

Eigenvalues stability discussion: Our light-injection model for rotating-wave approximation fields system involving N variables (N > 2, N = 4, arbitrarily small increments), the characteristic equation is of degree N = 4 and must often be solved numerically. Except in some particular cases, such an equation has (N = 4) distinct roots that can be real or complex. These values are the eigenvalues of the (4 × 4) Jacobian matrix (A). The general rule is that the light-injection model for rotating-wave approximation fields system is stable if there is no eigenvalue with positive real part. It is sufficient that one eigenvalue is positive for the steady state to be unstable. Our 4 variables (n_0, n_i, E_i, ψ_i) system has four eigenvalues (four system's arbitrarily small increments). The type of behavior can be characterized as a function of the position of these eigenvalues in the Re/Im plane. Five non-degenerate cases can be distinguished: (1) the four eigenvalues are real and negative (stable steady state), (2) the four eigenvalues are real, at least one of them is positive (unstable steady state), (3) and (4) two eigenvalues are complex conjugates with a negative real part and other eigenvalues real are negative (stable steady state), two cases can be distinguished depending on the relative value of the real part of the complex eigenvalues and of the real one, (5) two eigenvalues are complex conjugates with a negative real part and at least one of the other eigenvalues real is positive (unstable steady state) [12, 13].

$$\det(A - \lambda \cdot I) = \sum_{k=0}^{4} \lambda^k \cdot \Upsilon_k; \det(A - \lambda \cdot I) = 0 \Rightarrow \sum_{k=0}^{4} \lambda^k \cdot \Upsilon_k = 0$$

Remark We use N as the Number of relaxation oscillations which each mode may exhibit, and also as system number of variables (eigenvalue stability discussion). Reader must consider it.

4.4 Q-Switched Diode-Pumped Solid-State Laser Stability Under Parameters Variation

Our system implements and combines gain switching with passive Q switching of a miniature diode-pumped solid-state laser. A composite pumping pulse, consisting of a long, low-intensity pulse and a following short, high-intensity pulse, and it is used to reduce the timing jitter. A greater-than-tenfold reduction in timing jitter is exists. Compact all-solid-state lasers play a crucial role in applications as laser radars and range finders for smart munitions, for which precision, light weight, low power consumption, and ruggedness are achieved simultaneously. Many applications require powerful nanosecond pulses whose firing times can be controlled as well.

These pulses can be obtained by Q switching. Passive Q switching with saturate absorbers has the advantage of ruggedness, simplicity, compactness, and good control of transverse mode structure. Q switching is a method for obtaining energetic pulses from lasers by modulating the intra cavity losses. Q switching is a technique for obtaining energetic short pulses from a laser by modulating the intra cavity losses and thus the Q factor of the laser resonator. The technique is mainly applied for the generation of nanosecond pulses of high energy and peak power with solid-state bulk lasers. A generation of a Q switched pulse follows the following steps: (1) initially, the resonator losses are kept at a high level. As lasing cannot occur at that time, the energy fed into the gain medium by the pumping mechanism accumulates there. The amount of stored energy is limited only by spontaneous emission, in other cases with strong enough gain by the onset of lasing or strong ASE, or by the pump energy available. The stored energy can be a multiple of the saturation energy, (2) the losses are suddenly reduced to a small value, and the power of the laser radiation builds up very quickly in the laser resonator. The process starts with noise from spontaneous emission, which is amplified to macroscopic power levels within hundreds or thousands of resonator roundtrips, (3) once, the temporally integrated intra cavity power has reached the order of the saturation energy of the gain medium, the gain starts to be saturated. The peak of the pulse is reached when the gain equals the remaining resonator losses. The large intra cavity power present at that time leads to further depletion of the stored energy during the time where the power decays. The pulse duration achieved with Q switching is typically in nanosecond range, and usually well above the resonator round-trip time. The energy of the generated pulse is typically higher than the saturation energy of the gain medium and can be in the mill joule range even for small lasers. The peak power can be orders of magnitude higher than the power which is achievable in continuous-wave operation.

Regularly, Q-switched lasers generate regular pulse trains via repetitive Q switching. The pulse repetition rate is typically in the range from 1 to 100 kHz, sometimes higher. Passively Q-switched microchip lasers have reached pulse durations from below 1nsec and repetition rates up to several MHz, whereas large laser systems can deliver pulses with many kilojoules of energy and durations in the nanosecond range. Q switched lasers are lasers which applied Q-switching technique [4].

Active Q switching: The losses are modulated with an active control element (active Q switch). The pulse is formed shortly after an electrical trigger signal arrives. There are mechanical Q-switches such as spinning mirrors, used as end mirrors of laser resonators. The achieved pulse energy and pulse duration depends on the energy stored in the gain medium (pump power and the pulse repetition rate). The pulse repetition rate of an actively Q-switched laser can be controlled via the modulator. Higher repetition rates typically lead to lower pulse energies, if the pump power is kept constant. At the same time, the pulses then become longer, as the initial laser gain become lower. Once the pulse period exceeds the upper-state lifetime, increasing losses via spontaneous emission limit the possible pulse energy. The duration of the generated pulses is at least of the order of the resonator round-trip time, and often substantially longer than that, if the laser gain and/or the resonator losses are low. For high pulse repetition rates, it can be difficult to obtain very short pulses. Cavity dumping method can solve the problem. Once most of the stored energy has been transferred into the circulating pulse, the energy is suddenly released with the cavity dumper, which is a fast optical switch. The optical energy in the resonator can be extracted within one resonator round-trip time, independent of the time required for pulse build-up.

Passive Q switching: For passive Q switching (self Q switching), the losses are automatically modulated with a saturate absorber. The pulse is formed as soon as the energy stored in the gain medium has reached a sufficiently high level. In many cases, the pulse energy and duration are then fixed, and changes of the pump power only influence the pulse repetition rate. However, the performance of a passively Q switched laser can be substantially degraded by timing jitter. The jitter is caused by fluctuations (temperature, pump power and wavelength, and intra cavity loss, within the system). Timing jitter in a Q-switched laser is inevitable because the first photon of the oscillation mode comes from spontaneous emission of the gain medium. This initial timing jitter imposes a fundamental lower limit on the timing jitter of all Q-switched lasers. A second limitation derives from the inevitable fluctuations of the pump. Temperature fluctuation, which affects pump power and wavelength and the absorption coefficient of the gain medium, usually contributes to the long-term drift of the repetition rate.

An actively Q-switched solid-state laser has substantially less jitter but requires a number of additional components, which increases the size, weight, complexity, and ultimately the cost of the laser. We discuss a novel scheme to reduce timing jitter that combines gain switching with passive Q switching of a miniature diode-pumped solid-state laser by use of a composite pumping pulse that consists of a long, low-intensity pulse and a following short, high intensity pulse. The simple model of a passively Q-switched laser is presented. The effective pumping rate of the upper laser level is $p_{eff} = \frac{\eta \cdot P}{h \cdot v_L \cdot V}$, where P is the electrical power input; η is the pumping efficiency, which is the product of laser-diode efficiency and optical coupling efficiency; v_L is the laser frequency; V is the laser mode volume, and h is the Planck constant. Then the delay differential equations (DDEs) of our system is for population density of the upper laser level (N_G) and the photon density in the cavity (n_p) are as follow. Due to jitter interferences there is a delay in time for N_G and n_p system variables: $N_G(t) \to N_G(t-\tau_1)$, $n_p(t) \to n_p(t-\tau_2)$. There is no effects of time delay parameters (τ_1, τ_2) on N_G and n_p variables derivatives in time $\left(\frac{dN_G}{dt}, \frac{dn_p}{dt}\right)$. Remark: $N_G = N_G(t); n_p = n_p(t)$

$$\frac{dN_G}{dt} = p_{eff} - \frac{1}{\tau_g} \cdot N_G(t-\tau_1) - c \cdot \sigma_G \cdot n_p(t-\tau_2) \cdot N_G(t-\tau_1)$$

$$\frac{dn_p}{dt} = c \cdot \sigma_G \cdot n_p(t-\tau_2) \cdot N_G(t-\tau_1) - c \cdot \sigma_A \cdot n_p(t-\tau_2) \cdot N_A - \frac{1}{\tau_p} \cdot n_p(t-\tau_2) + r_{sp}$$

n_p—Density of photons in the laser mode, N_G—Population density of the upper laser level, τ_g—Lifetime of the upper laser level, σ_G—Emission cross section of the gain medium, c—Speed of light in vacuum, N_A—Density of the saturate-absorber atoms (in the absence of laser light it is N_{A0}), σ_A—Effective absorption cross section (it takes into account the ratio between the volumes of the active mode in the absorber and in the gain medium), τ_p—Photon cavity lifetime, r_{sp}—Rate of spontaneous emission into the laser mode, $N^0_{G,th}$—Threshold density of the upper laser level (with the unsaturated saturate absorber, by setting $\frac{dn_p}{dt} = 0$ and assuming that $r_{sp} = 0$. Assuming that $r_{sp} = 0$ because few photons generated by spontaneous emission contribute to the laser mode that is defined by the cavity mirrors; $N^0_{G,th} = \frac{1}{c \cdot \sigma_G \cdot \tau_p} + N_{A0} \cdot \frac{\sigma_A}{\sigma_G}$. By replacing N_G in equation $N_G = p_{eff} \cdot \tau_g \cdot (1 - e^{-\frac{t}{\tau_g}})$ with $N^0_{G,th}$, as $\tau_d = \tau_g \cdot \ln(\frac{\alpha}{\alpha - 1})$, where $\alpha = \frac{p_{eff} \cdot \tau_g}{N^0_{G,th}}$. τ_d—Time when the laser reaches its threshold, α—Ratio of pump power to the threshold pump power.

Remark 1 The larger the pump power is the shorter the delay time.

Remark 2 The threshold density of the upper laser level is $N^0_{G,th}$ ($\frac{dn_p}{dt} = 0; r_{sp} = 0$) and we get the time when the laser reaches it threshold (τ_d) by the equation $N_G = p_{eff} \cdot \tau_g \cdot (1 - e^{-\frac{t}{\tau_g}})$ and replacing $t \to \tau_d$; $N_G \to N^0_{G,th}$.

$$N_G = p_{eff} \cdot \tau_g \cdot (1 - e^{-\frac{t}{\tau_g}}) \Rightarrow N^0_{G,th} = p_{eff} \cdot \tau_g \cdot (1 - e^{-\frac{\tau_d}{\tau_g}}); \frac{p_{eff} \cdot \tau_g}{N^0_{G,th}}$$

$$\cdot (1 - e^{-\frac{\tau_d}{\tau_g}}) = 1; \alpha = \frac{p_{eff} \cdot \tau_g}{N^0_{G,th}} \alpha \cdot (1 - e^{-\frac{\tau_d}{\tau_g}}) = 1 \Rightarrow 1 - e^{-\frac{\tau_d}{\tau_g}}$$

$$= \frac{1}{\alpha}; e^{-\frac{\tau_d}{\tau_g}} = 1 - \frac{1}{\alpha} = \frac{\alpha - 1}{\alpha}; \ln[e^{-\frac{\tau_d}{\tau_g}}] = \ln\left[\frac{\alpha - 1}{\alpha}\right]; -\frac{\tau_d}{\tau_g} = \ln\left[\frac{\alpha - 1}{\alpha}\right]$$

$$-\frac{\tau_d}{\tau_g} = \ln[\frac{\alpha - 1}{\alpha}] \Rightarrow \tau_d = -\tau_g \cdot \ln\left[\frac{\alpha - 1}{\alpha}\right]; \tau_d = \tau_g \cdot \ln\left[\frac{\alpha}{\alpha - 1}\right]$$

Remark 3 Timing jitter differential; $d\tau_d$.

$$\tau_d = \tau_g \cdot \ln\left[\frac{\alpha}{\alpha - 1}\right] \Rightarrow d\tau_d = \tau_g \cdot \left(\frac{\alpha - 1}{\alpha}\right) \cdot \left(\frac{d\alpha \cdot (\alpha - 1) - \alpha \cdot d\alpha}{(\alpha - 1)^2}\right); d\tau_d$$

$$= \tau_g \cdot \frac{1}{\alpha \cdot (\alpha - 1)} \cdot (\alpha \cdot d\alpha - d\alpha - \alpha \cdot d\alpha)$$

$$d\tau_d = -\tau_g \cdot \frac{1}{\alpha \cdot (\alpha - 1)} \cdot d\alpha \Rightarrow d\tau_d = \tau_g \cdot \frac{1}{\alpha \cdot (1 - \alpha)}$$

$$\cdot d\alpha; d \rightleftharpoons \Delta \Rightarrow \Delta\tau_d = \tau_g \cdot \frac{1}{\alpha \cdot (1 - \alpha)} \cdot \Delta\alpha \frac{\Delta\alpha}{\alpha} \rightleftharpoons \frac{\Delta P}{P} \Rightarrow \Delta\tau_d$$

$$= \tau_g \cdot \frac{1}{(1 - \alpha)} \cdot \frac{\Delta\alpha}{\alpha}; \Delta\tau_d = \tau_g \cdot \frac{1}{(1 - \alpha)} \cdot \frac{\Delta P}{P}$$

Remark 4 A small change in pump power causes very large charge in the delay unless α is very large. It is difficult to maintain a pump power that is so large for a prolonged period of time.

Remark 5 The solution of the differential equation for the population density of the upper laser level $\frac{dN_G}{dt} = \ldots$ in the absence of lasing ($n_p = 0$) can be calculated as follow: we neglect the time delay parameters τ_1 and τ_2 then we get the differential equation:

$$\frac{dN_G}{dt} = p_{eff} - \frac{1}{\tau_g} \cdot N_G - c \cdot \sigma_G \cdot n_p \cdot N_G; n_p = 0 \Rightarrow \frac{dN_G}{dt} = p_{eff} - \frac{1}{\tau_g} \cdot N_G$$

Assumption: The population density of the upper laser level (N_G) at t = 0 is equal to zero ($N_G(t = 0) = 0$).

$$\begin{aligned}
\frac{dN_G}{dt} &= p_{eff} - \frac{1}{\tau_g} \cdot N_G \Rightarrow \tau_g \cdot \frac{dN_G}{dt} = p_{eff} \cdot \tau_g - N_G; \frac{dN_G}{p_{eff} \cdot \tau_g - N_G} \\
&= \frac{1}{\tau_g} \cdot dt \int \left(\frac{1}{p_{eff} \cdot \tau_g - N_G} \right) \cdot dN_G \\
&= \int \frac{1}{\tau_g} \cdot dt \Rightarrow -\ln(p_{eff} \cdot \tau_g - N_G) \\
&= \frac{1}{\tau_g} \cdot t + const1
\end{aligned}$$

$$\begin{aligned}
\ln(p_{eff} \cdot \tau_g - N_G) &= -\frac{1}{\tau_g} \cdot t - const1; const2 = -const1; \ln(p_{eff} \cdot \tau_g - N_G) \\
&= -\frac{1}{\tau_g} \cdot t + const2 e^{-\frac{1}{\tau_g} \cdot t + const2} = p_{eff} \cdot \tau_g - N_G; e^{-\frac{1}{\tau_g} \cdot t + const2} \\
&= e^{-\frac{1}{\tau_g} \cdot t} \cdot e^{const2}; e^{-\frac{1}{\tau_g} \cdot t} \cdot e^{const2} = p_{eff} \cdot \tau_g - N_G \\
N_G &= p_{eff} \cdot \tau_g - e^{-\frac{1}{\tau_g} \cdot t} \cdot e^{const2}
\end{aligned}$$

According to the assumption: $N_G(t = 0) = 0$.

$$\begin{aligned}
N_G(t = 0) &= p_{eff} \cdot \tau_g - e^{const2} = 0 \Rightarrow p_{eff} \cdot \tau_g = e^{const2}; \ln(p_{eff} \cdot \tau_g) \\
&= \ln(e^{const2}) \Rightarrow const2 = \ln(p_{eff} \cdot \tau_g) \\
e^{const2} &= e^{\ln(p_{eff} \cdot \tau_g)} = p_{eff} \cdot \tau_g
\end{aligned}$$

Then we get $N_G = p_{eff} \cdot \tau_g - e^{-\frac{1}{\tau_g} \cdot t} \cdot p_{eff} \cdot \tau_g$; $N_G(t) = p_{eff} \cdot \tau_g \cdot (1 - e^{-\frac{1}{\tau_g} \cdot t})$.

Q-switched laser scheme system: The Q-switched diode-pumped solid-state laser with low timing jitter scheme is uses composite pump pulses. Each composite pump pulse consists of two rectangular pulses: a long, low-power pulse followed by a short, high power pulse. The first pump pulse brings the system close to lasing, and the second one raises the upper-level population in a much faster manner and thus trigger the Q-switch pulse. Timing jitter is in reverse proportion to the slope, $\frac{dN_G}{dt}$, within the window (p_{eff}). Incorporating two pulses into the pump pulse one make the slop much higher within the lasing window, leading to a substantial reduction in timing jitter. The experimental system is constructed from a composite pulse from two separate lasers rather than one laser because the two laser arrangement makes the electronics much simpler. Two laser diodes (LDs) emitting at 808 nm with orthogonal polarization were used to generate long and short pulses, respectively. Two pulses were combined by a polarization beam splitter (PBS) unit. A variable delay generator is used to set the correct timing of the second pulse relative to the first pulse. A comparison of Q-switched pulses obtained with simple pumping when the short pulse is timed to be delayed relative to the Q-switched pulse. The composite pulse, the Q-switched pulse is timed to be triggered by the short pulse. The effect of the power of pumping on the timing jitter of a passively Q-switched laser can be also demonstrated by use of only one pumping laser diode but with constant DC injection currents of different magnitude (Fig. 4.5). The performance of the composite pumping scheme can be further improved by use of only one pump laser to provide both long and short pulses and by improvement of the electronics to yield a sharper short pulse. It is a simple way to reduce the timing jitter of a passively Q-switched laser with a composite pump pulse that is composed of a long, low-power pulse and a short, high power pulse (Fig. 4.6). The target is to obtain a reduction in timing jitter for Nd:YAG/Cr^{4+}:YAG laser.

Polarizing Beam Splitter (PBS): Polarizing beam splitter is used to split unpolarized light into polarized parts. Polarizing beam splitter is beam splitter designed to split light by polarization state rather than by wavelength or intensity. Polarizing beam splitter is used in semiconductor or photonics instrumentation to transmit p-polarized light while reflecting s-polarized light. Polarizing beam splitter is typically designed for 0^0 to 45^0 angle of incident with 90^0 separations of the beams, depending

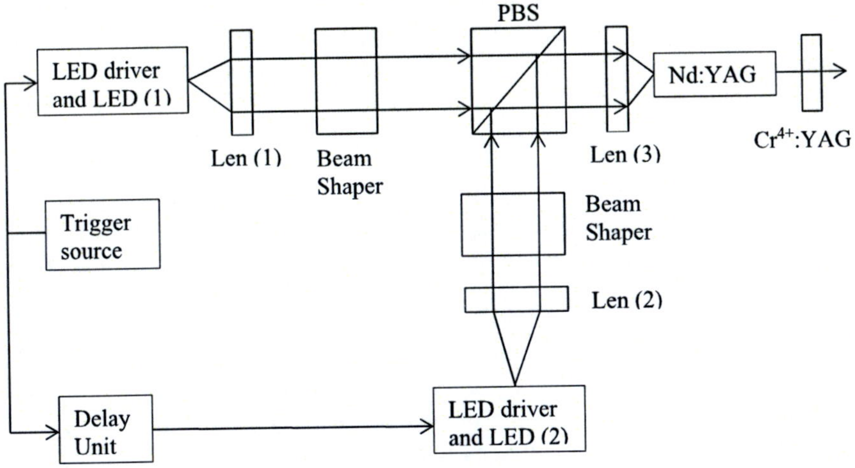

Fig. 4.5 Experimental Q-switched laser scheme system

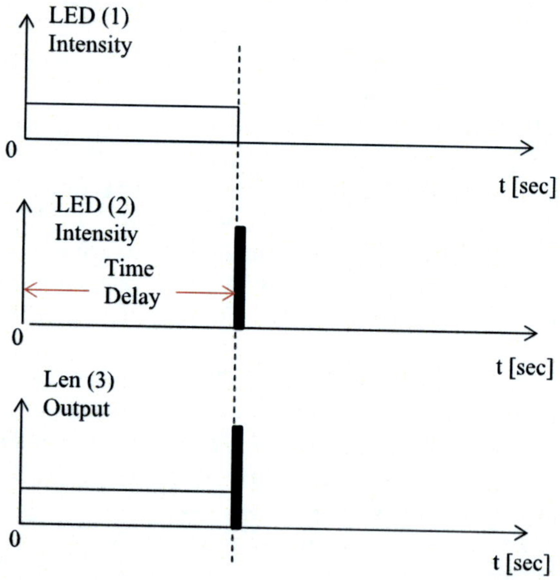

Fig. 4.6 Experimental Q-switched laser system signals in time

on the configuration. In other way we define beam splitter as an optical component used to split incident light at a designated ration into two separate beams. Additionally, beam splitter can be used in reverse to combine two different beams into a single one (our case). Beam splitter can be cube or plate.

Cube beam splitter: It is constructed using two typically right angle prisms. The hypotenuse surface of the prism is coated, and the two prisms are centered together so that they cubic shape. The light is transmitted into the coated prism, which often features a reference mark on the ground surface.

Plate beam splitter: It is consist of a thin, flat glass plate that has been coated on the first surface of the substrate. Most plate beam splitters feature an anti-reflection coating on the second surface to remove unwanted Fresnel reflections.

Beam splitter construction: It is made from two triangular glass prisms which are glued together at their base using polyester, epoxy, or urethane-based adhesives. The thickness of the resin layer is adjusted such that half of the light incident through one "port" is reflected and the other half is transmitted due to frustrated total internal reflection. There are other design configurations for beam splitter. A beam splitter that consists of a glass plate with a reflective dielectric coating on one side gives a phase shift of 0 to π, depending on the side from which it is incident. Transmitted waves have no phase shift. Reflected waves entering from the reflective side are phase-shifted by π, whereas reflected waves entering from the glass side have no phase shift. The explanation is Fresnel equations, according to which reflection causes a phase shift only when light passing through a material of low refractive index is reflected at a material of high refractive index.

Optical beam shaper: A beam shaper (beam converter) is an optical device which reshapes a light beam by modifies its spatial properties. It can be refractive and micro-optic beam shaper which creates a flat-top beam from a Gaussian input beam. Another option is to use in conjunction with a high-power laser diode. It makes both its beam radius and beam quality more symmetric with respect to two orthogonal directions. It can be a device, which based on two highly reflective mirrors and largely preserves the brightness or a device which based on micro-optical structures, containing arrays of small prisms, to perform a similar function with a smaller device.

System fixed points (equilibrium points): $\lim_{t\to\infty} N_G(t-\tau_1) = N_G(t); t \gg \tau_1 \Rightarrow t - \tau_1 \approx t$

$$\lim_{t\to\infty} n_p(t-\tau_2) = n_p(t); t \gg \tau_2 \Rightarrow t - \tau_2 \approx t; \frac{dN_G}{dt} = 0; \frac{dn_p}{dt} = 0$$

$$(*)\frac{dN_G}{dt} = 0 \Rightarrow p_{eff} - \frac{1}{\tau_g}\cdot N_G^* - c\cdot\sigma_G\cdot n_p^*\cdot N_G^* = 0$$

$$(**)\frac{dn_p}{dt} = 0 \Rightarrow c\cdot\sigma_G\cdot n_p^*\cdot N_G^* - c\cdot\sigma_A\cdot n_p^*\cdot N_A - \frac{1}{\tau_p}\cdot n_p^* + r_{sp} = 0$$

$$(*) p_{eff} - \frac{1}{\tau_g}\cdot N_G^* - c\cdot\sigma_G\cdot n_p^*\cdot N_G^* = 0 \Rightarrow n_p^*$$

$$= \frac{1}{c\cdot\sigma_G\cdot N_G^*}\cdot p_{eff} - \frac{1}{c\cdot\sigma_G\cdot\tau_g}$$

$$(*) \to (**) : c\cdot\sigma_G\cdot n_p^*\cdot N_G^* - c\cdot\sigma_A\cdot n_p^*\cdot N_A - \frac{1}{\tau_p}\cdot n_p^* + r_{sp} = 0$$

$$n_p^*\cdot\left[c\cdot\sigma_G\cdot N_G^* - c\cdot\sigma_A\cdot N_A - \frac{1}{\tau_p}\right] + r_{sp} = 0$$

$$\left[\frac{1}{c\cdot\sigma_G\cdot N_G^*}\cdot p_{eff} - \frac{1}{c\cdot\sigma_G\cdot\tau_g}\right]\cdot\left[c\cdot\sigma_G\cdot N_G^* - c\cdot\sigma_A\cdot N_A - \frac{1}{\tau_p}\right] + r_{sp} = 0$$

$$\frac{1}{c\cdot\sigma_G\cdot N_G^*}\cdot p_{eff}\cdot c\cdot\sigma_G\cdot N_G^* - \frac{1}{c\cdot\sigma_G\cdot N_G^*}\cdot p_{eff}\cdot c\cdot\sigma_A\cdot N_A$$

$$- \frac{1}{c\cdot\sigma_G\cdot N_G^*}\cdot p_{eff}\cdot\frac{1}{\tau_p} - \frac{1}{c\cdot\sigma_G\cdot\tau_g}\cdot c\cdot\sigma_G\cdot N_G^*$$

$$+ \frac{1}{c\cdot\sigma_G\cdot\tau_g}\cdot c\cdot\sigma_A\cdot N_A + \frac{1}{c\cdot\sigma_G\cdot\tau_g}\cdot\frac{1}{\tau_p} + r_{sp} = 0$$

$$p_{eff} - \frac{1}{\sigma_G\cdot N_G^*}\cdot p_{eff}\cdot\sigma_A\cdot N_A - \frac{p_{eff}}{c\cdot\sigma_G\cdot\tau_p\cdot N_G^*} - \frac{1}{\tau_g}\cdot N_G^*$$

$$+ \frac{\sigma_A\cdot N_A}{\sigma_G\cdot\tau_g} + \frac{1}{c\cdot\sigma_G\cdot\tau_g\cdot\tau_p} + r_{sp} = 0$$

$$\frac{\sigma_A\cdot N_A}{\sigma_G\cdot\tau_g} + p_{eff} + \frac{1}{c\cdot\sigma_G\cdot\tau_g\cdot\tau_p} + r_{sp} = \frac{1}{\sigma_G\cdot N_G^*}\cdot p_{eff}\cdot\sigma_A\cdot N_A$$

$$+ \frac{p_{eff}}{c\cdot\sigma_G\cdot\tau_p\cdot N_G^*} + \frac{1}{\tau_g}\cdot N_G^*$$

We define for simplicity a global parameter:

$$\Omega_1 = \frac{\sigma_A \cdot N_A}{\sigma_G \cdot \tau_g} + p_{eff} + \frac{1}{c \cdot \sigma_G \cdot \tau_g \cdot \tau_p}$$
$$+ r_{sp}\frac{1}{\sigma_G \cdot N_G^*} \cdot p_{eff} \cdot \sigma_A \cdot N_A + \frac{p_{eff}}{c \cdot \sigma_G \cdot \tau_p \cdot N_G^*} + \frac{1}{\tau_g} \cdot N_G^*$$
$$= \Omega_1 \frac{c \cdot \tau_p \cdot \tau_g \cdot p_{eff} \cdot \sigma_A \cdot N_A + \tau_g \cdot p_{eff} + c \cdot \sigma_G \cdot \tau_p \cdot (N_G^*)^2}{c \cdot \sigma_G \cdot \tau_p \cdot N_G^* \cdot \tau_g}$$
$$= \Omega_1 c \cdot \tau_p \cdot \tau_g \cdot p_{eff} \cdot \sigma_A \cdot N_A + \tau_g \cdot p_{eff} + c \cdot \sigma_G \cdot \tau_p \cdot (N_G^*)^2$$
$$= \Omega_1 \cdot c \cdot \sigma_G \cdot \tau_p \cdot N_G^* \cdot \tau_g c \cdot \sigma_G \cdot \tau_p \cdot (N_G^*)^2 - \Omega_1 \cdot c \cdot \sigma_G \cdot \tau_p \cdot \tau_g$$
$$\cdot N_G^* + c \cdot \tau_p \cdot \tau_g \cdot p_{eff} \cdot \sigma_A \cdot N_A + \tau_g \cdot p_{eff} = 0$$

We define for simplicity global parameters:

$$\Omega_2 = c \cdot \sigma_G \cdot \tau_p; \ \Omega_3 = -\Omega_1 \cdot c \cdot \sigma_G \cdot \tau_p \cdot \tau_g$$
$$\Omega_3 = -\Omega_1 \cdot c \cdot \sigma_G \cdot \tau_p \cdot \tau_g = -\left(\frac{\sigma_A \cdot N_A}{\sigma_G \cdot \tau_g} + p_{eff} + \frac{1}{c \cdot \sigma_G \cdot \tau_g \cdot \tau_p} + r_{sp}\right)$$
$$\cdot c \cdot \sigma_G \cdot \tau_p \cdot \tau_g \ \Omega_3 = -\Omega_1 \cdot c \cdot \sigma_G \cdot \tau_p \cdot \tau_g$$
$$= -(\sigma_A \cdot N_A \cdot c \cdot \tau_p + p_{eff} \cdot c \cdot \sigma_G \cdot \tau_p \cdot \tau_g + 1 + r_{sp} \cdot c \cdot \sigma_G \cdot \tau_p \cdot \tau_g)$$
$$\Omega_4 = c \cdot \tau_p \cdot \tau_g \cdot p_{eff} \cdot \sigma_A \cdot N_A + \tau_g \cdot p_{eff}; \ \Omega_2 \cdot (N_G^*)^2 + \Omega_3 \cdot N_G^*$$
$$+ \Omega_4 = 0 \ \Omega_2 \cdot (N_G^*)^2 + \Omega_3 \cdot N_G^* + \Omega_4$$
$$= 0 \Rightarrow N_G^* = \frac{-\Omega_3 \pm \sqrt{\Omega_3^2 - 4 \cdot \Omega_2 \cdot \Omega_4}}{2 \cdot \Omega_2}$$

$$\Omega_3^2 - 4 \cdot \Omega_2 \cdot \Omega_4$$
$$= (\sigma_A \cdot N_A \cdot c \cdot \tau_p + p_{eff} \cdot c \cdot \sigma_G \cdot \tau_p \cdot \tau_g + 1 + r_{sp} \cdot c \cdot \sigma_G \cdot \tau_p \cdot \tau_g)^2$$
$$- 4 \cdot c \cdot \sigma_G \cdot \tau_p \cdot c \cdot \tau_p \cdot \tau_g \cdot p_{eff} \cdot \sigma_A \cdot N_A + \tau_g \cdot p_{eff}$$

$$\Omega_3^2 - 4 \cdot \Omega_2 \cdot \Omega_4$$
$$= (\sigma_A \cdot N_A \cdot c \cdot \tau_p + p_{eff} \cdot c \cdot \sigma_G \cdot \tau_p \cdot \tau_g + 1 + r_{sp} \cdot c \cdot \sigma_G \cdot \tau_p \cdot \tau_g)^2$$
$$- 4 \cdot c^2 \cdot \sigma_G \cdot \tau_p^2 \cdot \tau_g \cdot p_{eff} \cdot \sigma_A \cdot N_A + \tau_g \cdot p_{eff}$$

N_G^* is the population density of the upper laser level—fixed points and must be real and positive value then $N_G^* \geq 0$(except specific cases when is equal to zero).

$$N_G^* \geq 0 \Rightarrow \Omega_3^2 - 4 \cdot \Omega_2 \cdot \Omega_4 > 0 \text{ and } -\Omega_3 \pm \sqrt{\Omega_3^2 - 4 \cdot \Omega_2 \cdot \Omega_4} > 0$$

$$n_p^* = \frac{1}{c \cdot \sigma_G \cdot N_G^*} \cdot p_{eff} - \frac{1}{c \cdot \sigma_G \cdot \tau_g} = \frac{1}{c \cdot \sigma_G \cdot \left(\frac{-\Omega_3 \pm \sqrt{\Omega_3^2 - 4 \cdot \Omega_2 \cdot \Omega_4}}{2 \cdot \Omega_2}\right)} \cdot p_{eff} - \frac{1}{c \cdot \sigma_G \cdot \tau_g}$$

$$n_p^* = \frac{1}{c \cdot \sigma_G \cdot N_G^*} \cdot p_{eff} - \frac{1}{c \cdot \sigma_G \cdot \tau_g} = \frac{2 \cdot \Omega_2 \cdot p_{eff}}{c \cdot \sigma_G \cdot \left(-\Omega_3 \pm \sqrt{\Omega_3^2 - 4 \cdot \Omega_2 \cdot \Omega_4}\right)} - \frac{1}{c \cdot \sigma_G \cdot \tau_g}$$

We can define our system two fixed points:

$$E^{(0)}\left(N_G^{(0)}, n_p^{(0)}\right) = \left(\frac{-\Omega_3 + \sqrt{\Omega_3^2 - 4 \cdot \Omega_2 \cdot \Omega_4}}{2 \cdot \Omega_2}, \frac{2 \cdot \Omega_2 \cdot p_{eff}}{c \cdot \sigma_G \cdot \left(-\Omega_3 + \sqrt{\Omega_3^2 - 4 \cdot \Omega_2 \cdot \Omega_4}\right)} - \frac{1}{c \cdot \sigma_G \cdot \tau_g}\right);$$

$$E^{(1)}\left(N_G^{(1)}, n_p^{(1)}\right) = \left(\frac{-\Omega_3 - \sqrt{\Omega_3^2 - 4 \cdot \Omega_2 \cdot \Omega_4}}{2 \cdot \Omega_2}, \frac{2 \cdot \Omega_2 \cdot p_{eff}}{c \cdot \sigma_G \cdot \left(-\Omega_3 - \sqrt{\Omega_3^2 - 4 \cdot \Omega_2 \cdot \Omega_4}\right)} - \frac{1}{c \cdot \sigma_G \cdot \tau_g}\right)$$

Remark System fixed points $\left(E^{(0)}\left(N_G^{(0)}, n_p^{(0)}\right), E^{(1)}\left(N_G^{(1)}, n_p^{(1)}\right)\right)$ must be non-negative and real numbers, if for specific system parameters values fixed point is complex or negative you need to ignore it (mathematical fixed point and not system physical fixed point).

Stability analysis: The standard local stability analysis about any one of the equilibrium points of the Q-switched laser system corresponding to the rate differential equation system $\left(\frac{dN_G}{dt} = \cdots ; \frac{dn_p}{dt} = \cdots\right)$ consists in adding to coordinate $[N_G, n_p]$ arbitrarily small increments of exponential form $[n_G, n_p] \cdot e^{\lambda \cdot t}$ and retaining the first order terms in N_G, n_p. The system of two homogeneous equations leads to a polynomial characteristic equation in the eigenvalues. The polynomial characteristic equations accept by set the below variables (N_G, n_p) and variables derivatives with respect to time into system rate equations [5–7]. System fixed values with arbitrarily small increments of exponential form $[n_G, n_p] \cdot e^{\lambda \cdot t}$ are: j = 0 (first fixed point), j = 1 (second fixed point), j = 2 (third fixed point), etc.

$$N_G(t) = N_G^{(j)} + n_G \cdot e^{\lambda \cdot t}; n_p(t) = n_p^{(j)} + n_p \cdot e^{\lambda \cdot t}; N_G(t - \tau_1)$$

$$= N_G^{(j)} + n_G \cdot e^{\lambda\cdot(t-\tau_1)} n_p(t-\tau_2) = n_p^{(j)} + n_p \cdot e^{\lambda\cdot(t-\tau_2)}; \frac{dN_G(t)}{dt}$$
$$= n_G \cdot \lambda \cdot e^{\lambda\cdot t}; \frac{dn_p(t)}{dt} = n_p \cdot \lambda \cdot e^{\lambda\cdot t}$$

We choose these expressions for ourselves $N_G(t), n_p(t)$ as a small displacement $[n_G, n_p]$ from the system fixed points in time t = 0.

$$N_G(t=0) = N_G^{(j)} + n_G; n_p(t=0) = n_p^{(j)} + n_p$$
$$\frac{dN_G}{dt} = p_{eff} - \frac{1}{\tau_g} \cdot N_G(t-\tau_1) - c \cdot \sigma_G \cdot n_p(t-\tau_2) \cdot N_G(t-\tau_1)$$
$$n_G \cdot \lambda \cdot e^{\lambda\cdot t} = p_{eff} - \frac{1}{\tau_g} \cdot \left(N_G^{(j)} + n_G \cdot e^{\lambda\cdot(t-\tau_1)}\right)$$
$$- c \cdot \sigma_G \cdot \left(n_p^{(j)} + n_p \cdot e^{\lambda\cdot(t-\tau_2)}\right) \cdot \left(N_G^{(j)} + n_G \cdot e^{\lambda\cdot(t-\tau_1)}\right)$$
$$n_G \cdot \lambda \cdot e^{\lambda\cdot t} = p_{eff} - \frac{1}{\tau_g} \cdot N_G^{(j)} - \frac{1}{\tau_g} \cdot n_G \cdot e^{\lambda\cdot(t-\tau_1)}$$
$$- c \cdot \sigma_G \cdot \left(n_p^{(j)} \cdot N_G^{(j)} + n_p^{(j)} \cdot n_G \cdot e^{\lambda\cdot(t-\tau_1)}\right.$$
$$\left. + N_G^{(j)} \cdot n_p \cdot e^{\lambda\cdot(t-\tau_2)} + n_p \cdot n_G \cdot e^{\lambda\cdot(t-\tau_2)} \cdot e^{\lambda\cdot(t-\tau_1)}\right)$$

Assumption: $n_p \cdot n_G \approx 0$

$$n_G \cdot \lambda \cdot e^{\lambda\cdot t} = p_{eff} - \frac{1}{\tau_g} \cdot N_G^{(j)} - \frac{1}{\tau_g} \cdot n_G \cdot e^{\lambda\cdot(t-\tau_1)}$$
$$- c \cdot \sigma_G \cdot \left(n_p^{(j)} \cdot N_G^{(j)} + n_p^{(j)} \cdot n_G \cdot e^{\lambda\cdot(t-\tau_1)} + N_G^{(j)} \cdot n_p \cdot e^{\lambda\cdot(t-\tau_2)}\right)$$
$$n_G \cdot \lambda \cdot e^{\lambda\cdot t} = p_{eff} - \frac{1}{\tau_g} \cdot N_G^{(j)} - \frac{1}{\tau_g} \cdot n_G \cdot e^{\lambda\cdot(t-\tau_1)}$$
$$- c \cdot \sigma_G \cdot n_p^{(j)} \cdot N_G^{(j)} - c \cdot \sigma_G \cdot n_p^{(j)} \cdot n_G \cdot e^{\lambda\cdot(t-\tau_1)}$$
$$- c \cdot \sigma_G \cdot N_G^{(j)} \cdot n_p \cdot e^{\lambda\cdot(t-\tau_2)}$$
$$n_G \cdot \lambda \cdot e^{\lambda\cdot t} = p_{eff} - \frac{1}{\tau_g} \cdot N_G^{(j)} - c \cdot \sigma_G \cdot n_p^{(j)} \cdot N_G^{(j)}$$
$$- \frac{1}{\tau_g} \cdot n_G \cdot e^{\lambda\cdot(t-\tau_1)} - c \cdot \sigma_G \cdot n_p^{(j)} \cdot n_G \cdot e^{\lambda\cdot(t-\tau_1)}$$
$$- c \cdot \sigma_G \cdot N_G^{(j)} \cdot n_p \cdot e^{\lambda\cdot(t-\tau_2)}$$

At fixed points: $p_{eff} - \frac{1}{\tau_g} \cdot N_G^{(j)} - c \cdot \sigma_G \cdot n_p^{(j)} \cdot N_G^{(j)} = 0.$

$$n_G \cdot \lambda \cdot e^{\lambda\cdot t} = -\frac{1}{\tau_g} \cdot n_G \cdot e^{\lambda\cdot(t-\tau_1)} - c \cdot \sigma_G \cdot n_p^{(j)} \cdot n_G \cdot e^{\lambda\cdot(t-\tau_1)}$$

$$- c \cdot \sigma_G \cdot N_G^{(j)} \cdot n_p \cdot e^{\lambda \cdot (t - \tau_2)} n_G \cdot \lambda \cdot e^{\lambda \cdot t} = -\frac{1}{\tau_g} \cdot n_G \cdot e^{\lambda \cdot t} \cdot e^{-\lambda \cdot \tau_1}$$
$$- c \cdot \sigma_G \cdot n_p^{(j)} \cdot n_G \cdot e^{\lambda \cdot t} \cdot e^{-\lambda \cdot \tau_1} - c \cdot \sigma_G \cdot N_G^{(j)} \cdot n_p \cdot e^{\lambda \cdot t} \cdot e^{-\lambda \cdot \tau_2}$$

Divide the two sides by $e^{\lambda \cdot t}$ term.

$$n_G \cdot \lambda = -\frac{1}{\tau_g} \cdot n_G \cdot e^{-\lambda \cdot \tau_1} - c \cdot \sigma_G \cdot n_p^{(j)} \cdot n_G \cdot e^{-\lambda \cdot \tau_1} - c \cdot \sigma_G \cdot N_G^{(j)} \cdot n_p \cdot e^{-\lambda \cdot \tau_2}$$

We have there possible cases: (1) $\tau_1 = \tau; \tau_2 = 0$, (2) $\tau_1 = 0; \tau_2 = \tau$, (3) $\tau_1 = \tau; \tau_2 = \tau$.

We analyze our system for the first case $\tau_1 = \tau; \tau_2 = 0$.

$$n_G \cdot \lambda = -\frac{1}{\tau_g} \cdot n_G \cdot e^{-\lambda \cdot \tau} - c \cdot \sigma_G \cdot n_p^{(j)} \cdot n_G \cdot e^{-\lambda \cdot \tau} - c \cdot \sigma_G \cdot N_G^{(j)} \cdot n_p$$
$$- n_G \cdot \lambda - \left(\frac{1}{\tau_g} + c \cdot \sigma_G \cdot n_p^{(j)} \right) \cdot n_G \cdot e^{-\lambda \cdot \tau} - c \cdot \sigma_G \cdot N_G^{(j)} \cdot n_p = 0$$

$$\frac{dn_p}{dt} = c \cdot \sigma_G \cdot n_p(t - \tau_2) \cdot N_G(t - \tau_1) - c \cdot \sigma_A \cdot n_p(t - \tau_2)$$
$$\cdot N_A - \frac{1}{\tau_p} \cdot n_p(t - \tau_2) + r_{sp}$$

$$n_p \cdot \lambda \cdot e^{\lambda \cdot t} = c \cdot \sigma_G \cdot \left(n_p^{(j)} + n_p \cdot e^{\lambda \cdot (t - \tau_2)} \right) \cdot \left(N_G^{(j)} + n_G \cdot e^{\lambda \cdot (t - \tau_1)} \right)$$
$$- c \cdot \sigma_A \cdot \left(n_p^{(j)} + n_p \cdot e^{\lambda \cdot (t - \tau_2)} \right) \cdot N_A$$
$$- \frac{1}{\tau_p} \cdot \left(n_p^{(j)} + n_p \cdot e^{\lambda \cdot (t - \tau_2)} \right) + r_{sp} \; n_p \cdot \lambda \cdot e^{\lambda \cdot t} = c \cdot \sigma_G \cdot \left(n_p^{(j)} \cdot N_G^{(j)} \right.$$
$$\left. + n_p^{(j)} \cdot n_G \cdot e^{\lambda \cdot (t - \tau_1)} + N_G^{(j)} \cdot n_p \cdot e^{\lambda \cdot (t - \tau_2)} + n_p \cdot n_G \cdot e^{\lambda \cdot (t - \tau_2)} \cdot e^{\lambda \cdot (t - \tau_1)} \right)$$
$$- c \cdot \sigma_A \cdot n_p^{(j)} \cdot N_A - c \cdot \sigma_A \cdot n_p \cdot N_A \cdot e^{\lambda \cdot (t - \tau_2)}$$
$$- \frac{1}{\tau_p} \cdot n_p^{(j)} - \frac{1}{\tau_p} \cdot n_p \cdot e^{\lambda \cdot (t - \tau_2)} + r_{sp}$$

$$n_p \cdot \lambda \cdot e^{\lambda \cdot t} = c \cdot \sigma_G \cdot n_p^{(j)} \cdot N_G^{(j)} + c \cdot \sigma_G \cdot n_p^{(j)} \cdot n_G \cdot e^{\lambda \cdot (t - \tau_1)}$$
$$+ c \cdot \sigma_G \cdot N_G^{(j)} \cdot n_p \cdot e^{\lambda \cdot (t - \tau_2)}$$
$$+ c \cdot \sigma_G \cdot n_p \cdot n_G \cdot e^{\lambda \cdot (t - \tau_2)} \cdot e^{\lambda \cdot (t - \tau_1)} - c \cdot \sigma_A \cdot n_p^{(j)} \cdot N_A$$
$$- c \cdot \sigma_A \cdot n_p \cdot N_A \cdot e^{\lambda \cdot (t - \tau_2)} - \frac{1}{\tau_p} \cdot n_p^{(j)} - \frac{1}{\tau_p} \cdot n_p \cdot e^{\lambda \cdot (t - \tau_2)} + r_{sp}$$

Assumption: $n_p \cdot n_G \approx 0$

$$n_p \cdot \lambda \cdot e^{\lambda \cdot t} = c \cdot \sigma_G \cdot n_p^{(j)} \cdot N_G^{(j)} + c \cdot \sigma_G \cdot n_p^{(j)} \cdot n_G \cdot e^{\lambda \cdot t} \cdot e^{-\lambda \cdot \tau_1}$$
$$+ c \cdot \sigma_G \cdot N_G^{(j)} \cdot n_p \cdot e^{\lambda \cdot t} \cdot e^{-\lambda \cdot \tau_2}$$

$$- c \cdot \sigma_A \cdot n_p^{(j)} \cdot N_A - c \cdot \sigma_A \cdot n_p \cdot N_A \cdot e^{\lambda \cdot t} \cdot e^{-\lambda \cdot \tau_2}$$
$$- \frac{1}{\tau_p} \cdot n_p^{(j)} - \frac{1}{\tau_p} \cdot n_p \cdot e^{\lambda \cdot t} \cdot e^{-\lambda \cdot \tau_2} + r_{sp}$$

We analyze our system for the first case $\tau_1 = \tau; \tau_2 = 0$.

$$\begin{aligned} n_p \cdot \lambda \cdot e^{\lambda \cdot t} &= c \cdot \sigma_G \cdot n_p^{(j)} \cdot N_G^{(j)} + c \cdot \sigma_G \cdot n_p^{(j)} \cdot n_G \cdot e^{\lambda \cdot t} \cdot e^{-\lambda \cdot \tau_1} \\ &+ c \cdot \sigma_G \cdot N_G^{(j)} \cdot n_p \cdot e^{\lambda \cdot t} \cdot e^{-\lambda \cdot \tau_2} \\ &- c \cdot \sigma_A \cdot n_p^{(j)} \cdot N_A - c \cdot \sigma_A \cdot n_p \cdot N_A \cdot e^{\lambda \cdot t} \cdot e^{-\lambda \cdot \tau_2} \\ &- \frac{1}{\tau_p} \cdot n_p^{(j)} - \frac{1}{\tau_p} \cdot n_p \cdot e^{\lambda \cdot t} \cdot e^{-\lambda \cdot \tau_2} + r_{sp} \end{aligned}$$

$$\begin{aligned} n_p \cdot \lambda \cdot e^{\lambda \cdot t} &= c \cdot \sigma_G \cdot n_p^{(j)} \cdot N_G^{(j)} - c \cdot \sigma_A \cdot n_p^{(j)} \cdot N_A - \frac{1}{\tau_p} \cdot n_p^{(j)} + r_{sp} \\ &+ c \cdot \sigma_G \cdot n_p^{(j)} \cdot n_G \cdot e^{\lambda \cdot t} \cdot e^{-\lambda \cdot \tau} + c \cdot \sigma_G \cdot N_G^{(j)} \cdot n_p \cdot e^{\lambda \cdot t} \\ &- c \cdot \sigma_A \cdot n_p \cdot N_A \cdot e^{\lambda \cdot t} - \frac{1}{\tau_p} \cdot n_p \cdot e^{\lambda \cdot t} \end{aligned}$$

At fixed points: $c \cdot \sigma_G \cdot n_p^{(j)} \cdot N_G^{(j)} - c \cdot \sigma_A \cdot n_p^{(j)} \cdot N_A - \frac{1}{\tau_p} \cdot n_p^{(j)} + r_{sp} = 0$

$$\begin{aligned} n_p \cdot \lambda \cdot e^{\lambda \cdot t} &= c \cdot \sigma_G \cdot n_p^{(j)} \cdot n_G \cdot e^{\lambda \cdot t} \cdot e^{-\lambda \cdot \tau} + c \cdot \sigma_G \cdot N_G^{(j)} \cdot n_p \cdot e^{\lambda \cdot t} \\ &- c \cdot \sigma_A \cdot n_p \cdot N_A \cdot e^{\lambda \cdot t} - \frac{1}{\tau_p} \cdot n_p \cdot e^{\lambda \cdot t} \end{aligned}$$

Divide the two sides by $e^{\lambda \cdot t}$ term.

$$n_p \cdot \lambda = c \cdot \sigma_G \cdot n_p^{(j)} \cdot n_G \cdot e^{-\lambda \cdot \tau} + c \cdot \sigma_G \cdot N_G^{(j)} \cdot n_p - c \cdot \sigma_A \cdot n_p \cdot N_A - \frac{1}{\tau_p} \cdot n_p$$

$$\begin{aligned} &- n_p \cdot \lambda + c \cdot \sigma_G \cdot n_p^{(j)} \cdot n_G \cdot e^{-\lambda \cdot \tau} + c \cdot \sigma_G \cdot N_G^{(j)} \cdot n_p \\ &- c \cdot \sigma_A \cdot n_p \cdot N_A - \frac{1}{\tau_p} \cdot n_p = 0 c \cdot \sigma_G \cdot n_p^{(j)} \cdot n_G \cdot e^{-\lambda \cdot \tau} \\ &- n_p \cdot \lambda + \left(c \cdot \sigma_G \cdot N_G^{(j)} - c \cdot \sigma_A \cdot N_A - \frac{1}{\tau_p} \right) \cdot n_p = 0 \end{aligned}$$

We can summarize our last results:

$$\begin{aligned} &- n_G \cdot \lambda - \left(\frac{1}{\tau_g} + c \cdot \sigma_G \cdot n_p^{(j)} \right) \cdot n_G \cdot e^{-\lambda \cdot \tau} - c \cdot \sigma_G \cdot N_G^{(j)} \\ &\cdot n_p = 0 \; c \cdot \sigma_G \cdot n_p^{(j)} \cdot n_G \cdot e^{-\lambda \cdot \tau} - n_p \cdot \lambda \end{aligned}$$

$$+\left(c \cdot \sigma_G \cdot N_G^{(j)} - c \cdot \sigma_A \cdot N_A - \frac{1}{\tau_p}\right) \cdot n_p = 0$$

The small increments Jacobian of our system is as follow:

$$\begin{pmatrix} -\left(\frac{1}{\tau_g} + c \cdot \sigma_G \cdot n_p^{(j)}\right) \cdot e^{-\lambda \cdot \tau} - \lambda & -c \cdot \sigma_G \cdot N_G^{(j)} \\ c \cdot \sigma_G \cdot n_p^{(j)} \cdot e^{-\lambda \cdot \tau} & \left(c \cdot \sigma_G \cdot N_G^{(j)} - c \cdot \sigma_A \cdot N_A - \frac{1}{\tau_p}\right) - \lambda \end{pmatrix} \cdot \begin{pmatrix} n_G \\ n_p \end{pmatrix} = 0$$

$$(A - \lambda \cdot I) = \begin{pmatrix} -\left(\frac{1}{\tau_g} + c \cdot \sigma_G \cdot n_p^{(j)}\right) \cdot e^{-\lambda \cdot \tau} - \lambda & -c \cdot \sigma_G \cdot N_G^{(j)} \\ c \cdot \sigma_G \cdot n_p^{(j)} \cdot e^{-\lambda \cdot \tau} & \left(c \cdot \sigma_G \cdot N_G^{(j)} - c \cdot \sigma_A \cdot N_A - \frac{1}{\tau_p}\right) - \lambda \end{pmatrix}$$

$$\det(A - \lambda \cdot I) = 0$$

$$\det \begin{pmatrix} -\left(\frac{1}{\tau_g} + c \cdot \sigma_G \cdot n_p^{(j)}\right) \cdot e^{-\lambda \cdot \tau} - \lambda & -c \cdot \sigma_G \cdot N_G^{(j)} \\ c \cdot \sigma_G \cdot n_p^{(j)} \cdot e^{-\lambda \cdot \tau} & \left(c \cdot \sigma_G \cdot N_G^{(j)} - c \cdot \sigma_A \cdot N_A - \frac{1}{\tau_p}\right) - \lambda \end{pmatrix} = 0$$

$$\begin{aligned}\det(A - \lambda \cdot I) = &-\left(\frac{1}{\tau_g} + c \cdot \sigma_G \cdot n_p^{(j)}\right) \cdot \left(c \cdot \sigma_G \cdot N_G^{(j)} - c \cdot \sigma_A \cdot N_A - \frac{1}{\tau_p}\right) \cdot e^{-\lambda \cdot \tau} \\ &+ \left(\frac{1}{\tau_g} + c \cdot \sigma_G \cdot n_p^{(j)}\right) \cdot e^{-\lambda \cdot \tau} \cdot \lambda \\ &- \left(c \cdot \sigma_G \cdot N_G^{(j)} - c \cdot \sigma_A \cdot N_A - \frac{1}{\tau_p}\right) \cdot \lambda + \lambda^2 \\ &+ c^2 \cdot \sigma_G^2 \cdot n_p^{(j)} \cdot N_G^{(j)} \cdot e^{-\lambda \cdot \tau}\end{aligned}$$

$$\begin{aligned}\det(A - \lambda \cdot I) = &-\left(\frac{1}{\tau_g} + c \cdot \sigma_G \cdot n_p^{(j)}\right) \cdot \left(c \cdot \sigma_G \cdot N_G^{(j)} - c \cdot \sigma_A \cdot N_A - \frac{1}{\tau_p}\right) \cdot e^{-\lambda \cdot \tau} \\ &+ (\frac{1}{\tau_g} + c \cdot \sigma_G \cdot n_p^{(j)}) \cdot e^{-\lambda \cdot \tau} \cdot \lambda \\ &- \left(c \cdot \sigma_G \cdot N_G^{(j)} - c \cdot \sigma_A \cdot N_A - \frac{1}{\tau_p}\right) \cdot \lambda + \lambda^2 + c^2 \cdot \sigma_G^2 \cdot n_p^{(j)} \cdot N_G^{(j)} \cdot e^{-\lambda \cdot \tau}\end{aligned}$$

$$\begin{aligned}\det(A - \lambda \cdot I) = &-\left(\frac{1}{\tau_g} + c \cdot \sigma_G \cdot n_p^{(j)}\right) \cdot \left(c \cdot \sigma_G \cdot N_G^{(j)} - c \cdot \sigma_A \cdot N_A - \frac{1}{\tau_p}\right) \cdot e^{-\lambda \cdot \tau} \\ &+ \left(\frac{1}{\tau_g} + c \cdot \sigma_G \cdot n_p^{(j)}\right) \cdot e^{-\lambda \cdot \tau} \cdot \lambda \\ &- \left(c \cdot \sigma_G \cdot N_G^{(j)} - c \cdot \sigma_A \cdot N_A - \frac{1}{\tau_p}\right) \cdot \lambda + \lambda^2 \\ &+ c^2 \cdot \sigma_G^2 \cdot n_p^{(j)} \cdot N_G^{(j)} \cdot e^{-\lambda \cdot \tau}\end{aligned}$$

$$\begin{aligned}D(\lambda, \tau) = &-\left(c \cdot \sigma_G \cdot N_G^{(j)} - c \cdot \sigma_A \cdot N_A - \frac{1}{\tau_p}\right) \cdot \lambda + \lambda^2 + \left[c^2 \cdot \sigma_G^2 \cdot n_p^{(j)} \cdot N_G^{(j)}\right. \\ &- \left(\frac{1}{\tau_g} + c \cdot \sigma_G \cdot n_p^{(j)}\right) \cdot \left(c \cdot \sigma_G \cdot N_G^{(j)} - c \cdot \sigma_A \cdot N_A - \frac{1}{\tau_p}\right)\end{aligned}$$

$$+\left(\frac{1}{\tau_g}+c\cdot\sigma_G\cdot n_p^{(j)}\right)\cdot\lambda\Bigg]\cdot e^{-\lambda\cdot\tau}$$

$$D(\lambda,\tau)=P_n(\lambda,\tau)+Q_m(\lambda,\tau)\cdot e^{-\lambda\cdot\tau};\,n,m\in\mathbb{N}_0;\,n>m$$

$$P_n(\lambda,\tau)=-\left(c\cdot\sigma_G\cdot N_G^{(j)}-c\cdot\sigma_A\cdot N_A-\frac{1}{\tau_p}\right)\cdot\lambda+\lambda^2$$

$$Q_m(\lambda,\tau)=c^2\cdot\sigma_G^2\cdot n_p^{(j)}\cdot N_G^{(j)}-\left(\frac{1}{\tau_g}+c\cdot\sigma_G\cdot n_p^{(j)}\right)$$
$$\cdot\left(c\cdot\sigma_G\cdot N_G^{(j)}-c\cdot\sigma_A\cdot N_A-\frac{1}{\tau_p}\right)$$
$$+\left(\frac{1}{\tau_g}+c\cdot\sigma_G\cdot n_p^{(j)}\right)\cdot\lambda$$

$$P_n(\lambda,\tau)=\sum_{k=0}^{n}p_k(\tau)\cdot\lambda^k=p_0(\tau)+p_1(\tau)\cdot\lambda$$
$$+p_2(\tau)\cdot\lambda^2;\,Q_m(\lambda,\tau)=\sum_{k=0}^{m}q_k\cdot\lambda^k=q_0(\tau)+q_1(\tau)\cdot\lambda$$

$$D(\lambda,\tau)=P_n(\lambda,\tau)+Q_m(\lambda,\tau)\cdot e^{-\lambda\cdot\tau};\,n,m\in\mathbb{N}_0;\,n=2;\,m=1;\,n>m$$

$$p_0(\tau)=0;\,p_1(\tau)=-\left(c\cdot\sigma_G\cdot N_G^{(j)}-c\cdot\sigma_A\cdot N_A-\frac{1}{\tau_p}\right);\,p_2(\tau)=1$$

$$q_0(\tau)=c^2\cdot\sigma_G^2\cdot n_p^{(j)}\cdot N_G^{(j)}-\left(\frac{1}{\tau_g}+c\cdot\sigma_G\cdot n_p^{(j)}\right)$$
$$\cdot\left(c\cdot\sigma_G\cdot N_G^{(j)}-c\cdot\sigma_A\cdot N_A-\frac{1}{\tau_p}\right);\,q_1(\tau)=\left(\frac{1}{\tau_g}+c\cdot\sigma_G\cdot n_p^{(j)}\right)$$

The homogeneous system for N_G, n_p leads to a characteristic equation for the eigenvalue λ having the form $P(\lambda,\tau)+Q(\lambda,\tau)\cdot e^{-\lambda\cdot\tau}=0;\ P(\lambda)=\sum_{k=0}^{2}a_j\cdot\lambda^k;\ Q(\lambda)=\sum_{k=0}^{1}c_j\cdot\lambda^k$. The coefficients $\{a_j(q_i,q_k,\tau),c_j(q_i,q_k,\tau)\}\in\mathbb{R}$ depend on q_i, q_k and τ. q_i, q_k are any Q-switched laser system parameters, other parameters kept as a constant.

$$a_0=0;\,a_1=-\left(c\cdot\sigma_G\cdot N_G^{(j)}-c\cdot\sigma_A\cdot N_A-\frac{1}{\tau_p}\right);\,a_2=1$$

$$c_0=c^2\cdot\sigma_G^2\cdot n_p^{(j)}\cdot N_G^{(j)}-\left(\frac{1}{\tau_g}+c\cdot\sigma_G\cdot n_p^{(j)}\right)$$
$$\cdot\left(c\cdot\sigma_G\cdot N_G^{(j)}-c\cdot\sigma_A\cdot N_A-\frac{1}{\tau_p}\right);\,c_1=\left(\frac{1}{\tau_g}+c\cdot\sigma_G\cdot n_p^{(j)}\right)$$

Unless strictly necessary, the designation of the varied arguments (q_i, q_k) will subsequently be omitted from P, Q, a_j, c_j. The coefficients a_j, c_j are continuous and differentiable functions of their arguments, and direct substitution shows that $a_0 + c_0 \neq 0$ for $q_i, q_k \in \mathbb{R}_+$; that is, $\lambda = 0$ is not of $P(\lambda, \tau) + Q(\lambda, \tau) \cdot e^{-\lambda \cdot \tau} = 0$. Furthermore $P(\lambda)$, $Q(\lambda)$ are analytic functions of λ, for which the following requirements of the analysis [5] can also be verified in the present case:

(a) If $\lambda = i \cdot \omega$; $\omega \in \mathbb{R}$, then $P(i \cdot \omega) + Q(i \cdot \omega) \neq 0$.
(b) If $|\frac{Q(\lambda)}{P(\lambda)}|$ is bounded for $|\lambda| \to \infty$; $\mathrm{Re}\lambda \geq 0$. No roots bifurcation from ∞.
(c) $F(\omega) = |P(i \cdot \omega)|^2 - |Q(i \cdot \omega)|^2$ has a finite number of zeros. Indeed, this is a polynomial in ω.
(d) Each positive root $\omega(q_i, q_k)$ of $F(\omega) = 0$ is continuous and differentiable with respect to q_i, q_k.

We assume that $P_n(\lambda, \tau)$ and $Q_m(\lambda, \tau)$ cannot have common imaginary roots. That is for any real number ω: $P_n(\lambda = i \cdot \omega, \tau) + Q_m(\lambda = i \cdot \omega, \tau) \neq 0$.

$$P_n(\lambda = i \cdot \omega, \tau) = -(c \cdot \sigma_G \cdot N_G^{(j)} - c \cdot \sigma_A \cdot N_A - \frac{1}{\tau_p}) \cdot i \cdot \omega - \omega^2$$

$$\begin{aligned} Q_m(\lambda, \tau) &= c^2 \cdot \sigma_G^2 \cdot n_p^{(j)} \cdot N_G^{(j)} \\ &\quad - \left(\frac{1}{\tau_g} + c \cdot \sigma_G \cdot n_p^{(j)}\right) \cdot \left(c \cdot \sigma_G \cdot N_G^{(j)} - c \cdot \sigma_A \cdot N_A - \frac{1}{\tau_p}\right) \\ &\quad + \left(\frac{1}{\tau_g} + c \cdot \sigma_G \cdot n_p^{(j)}\right) \cdot i \cdot \omega \end{aligned}$$

$$\begin{aligned} &P_n(\lambda = i \cdot \omega, \tau) + Q_m(\lambda = i \cdot \omega, \tau) \\ &= c^2 \cdot \sigma_G^2 \cdot n_p^{(j)} \cdot N_G^{(j)} - \left(\frac{1}{\tau_g} + c \cdot \sigma_G \cdot n_p^{(j)}\right) \cdot \left(c \cdot \sigma_G \cdot N_G^{(j)}\right. \\ &\left. - c \cdot \sigma_A \cdot N_A - \frac{1}{\tau_p}\right) - \omega^2 \\ &\quad + \left[\left(\frac{1}{\tau_g} + c \cdot \sigma_G \cdot n_p^{(j)}\right) - \left(c \cdot \sigma_G \cdot N_G^{(j)} - c \cdot \sigma_A \cdot N_A - \frac{1}{\tau_p}\right)\right] \cdot i \cdot \omega \neq 0 \end{aligned}$$

$$|P(i \cdot \omega, \tau)|^2 = (c \cdot \sigma_G \cdot N_G^{(j)} - c \cdot \sigma_A \cdot N_A - \frac{1}{\tau_p})^2 \cdot \omega^2 + \omega^4$$

$$\begin{aligned} |Q(i \cdot \omega, \tau)|^2 &= \left[c^2 \cdot \sigma_G^2 \cdot n_p^{(j)} \cdot N_G^{(j)}\right. \\ &\quad \left. - \left(\frac{1}{\tau_g} + c \cdot \sigma_G \cdot n_p^{(j)}\right) \cdot \left(c \cdot \sigma_G \cdot N_G^{(j)} - c \cdot \sigma_A \cdot N_A - \frac{1}{\tau_p}\right)\right]^2 \\ &\quad + \left(\frac{1}{\tau_g} + c \cdot \sigma_G \cdot n_p^{(j)}\right)^2 \cdot \omega^2 \end{aligned}$$

$$F(\omega, \tau) = |P(i \cdot \omega, \tau)|^2 - |Q(i \cdot \omega, \tau)|^2$$

$$= \left(c \cdot \sigma_G \cdot N_G^{(j)} - c \cdot \sigma_A \cdot N_A - \frac{1}{\tau_p}\right)^2 \cdot \omega^2 + \omega^4$$
$$- \left[c^2 \cdot \sigma_G^2 \cdot n_p^{(j)} \cdot N_G^{(j)} - \left(\frac{1}{\tau_g} + c \cdot \sigma_G \cdot n_p^{(j)}\right)\right.$$
$$\left.\cdot \left(c \cdot \sigma_G \cdot N_G^{(j)} - c \cdot \sigma_A \cdot N_A - \frac{1}{\tau_p}\right)\right]^2$$
$$- \left(\frac{1}{\tau_g} + c \cdot \sigma_G \cdot n_p^{(j)}\right)^2 \cdot \omega^2$$

$$F(\omega, \tau) = |P(i \cdot \omega, \tau)|^2 - |Q(i \cdot \omega, \tau)|^2$$
$$= -\left[c^2 \cdot \sigma_G^2 \cdot n_p^{(j)} \cdot N_G^{(j)} - \left(\frac{1}{\tau_g} + c \cdot \sigma_G \cdot n_p^{(j)}\right) \cdot \left(c \cdot \sigma_G \cdot N_G^{(j)}\right.\right.$$
$$\left.\left. - c \cdot \sigma_A \cdot N_A - \frac{1}{\tau_p}\right)\right]^2 + \left[\left(c \cdot \sigma_G \cdot N_G^{(j)} - c \cdot \sigma_A \cdot N_A - \frac{1}{\tau_p}\right)^2\right.$$
$$\left. - \left(\frac{1}{\tau_g} + c \cdot \sigma_G \cdot n_p^{(j)}\right)^2\right] \cdot \omega^2 + \omega^4$$

We define the following parameters for simplicity: Π_0, Π_2, Π_4.

$$\Pi_0 = -\left[c^2 \cdot \sigma_G^2 \cdot n_p^{(j)} \cdot N_G^{(j)} - \left(\frac{1}{\tau_g} + c \cdot \sigma_G \cdot n_p^{(j)}\right)\right.$$
$$\left.\cdot \left(c \cdot \sigma_G \cdot N_G^{(j)} - c \cdot \sigma_A \cdot N_A - \frac{1}{\tau_p}\right)\right]^2$$
$$\Pi_2 = \left(c \cdot \sigma_G \cdot N_G^{(j)} - c \cdot \sigma_A \cdot N_A - \frac{1}{\tau_p}\right)^2$$
$$- \left(\frac{1}{\tau_g} + c \cdot \sigma_G \cdot n_p^{(j)}\right)^2; \Pi_4 = 1$$

Hence $F(\omega, \tau) = 0$ implies $\sum_{k=0}^{2} \Pi_{2 \cdot k} \cdot \omega^{2 \cdot k} = 0$ and its roots are given by solving the above polynomial.

$$P_R(i \cdot \omega, \tau) = -\omega^2; P_I(i \cdot \omega, \tau)$$
$$= -\left(c \cdot \sigma_G \cdot N_G^{(j)} - c \cdot \sigma_A \cdot N_A - \frac{1}{\tau_p}\right) \cdot \omega; Q_I(i \cdot \omega, \tau)$$
$$= \left(\frac{1}{\tau_g} + c \cdot \sigma_G \cdot n_p^{(j)}\right) \cdot \omega$$

$$Q_R(i \cdot \omega, \tau) = c^2 \cdot \sigma_G^2 \cdot n_p^{(j)} \cdot N_G^{(j)}$$

$$-\left(\frac{1}{\tau_g}+c\cdot\sigma_G\cdot n_p^{(j)}\right)\cdot\left(c\cdot\sigma_G\cdot N_G^{(j)}-c\cdot\sigma_A\cdot N_A-\frac{1}{\tau_p}\right)$$

$$\sin\theta(\tau)=\frac{-P_R(i\cdot\omega,\tau)\cdot Q_I(i\cdot\omega,\tau)+P_I(i\cdot\omega,\tau)\cdot Q_R(i\cdot\omega,\tau)}{|Q(i\cdot\omega,\tau)|^2}$$

$$\cos\theta(\tau)=-\frac{P_R(i\cdot\omega,\tau)\cdot Q_R(i\cdot\omega,\tau)+P_I(i\cdot\omega,\tau)\cdot Q_I(i\cdot\omega,\tau)}{|Q(i\cdot\omega,\tau)|^2}$$

We use different parameters terminology from our last characteristics parameters definition:

$$k\to j;\ p_k(\tau)\to a_j;\ q_k(\tau)\to c_j;\ n=2;$$
$$m=1;\ n>m;\ P_n(\lambda,\tau)\to P(\lambda);\ Q_m(\lambda,\tau)\to Q(\lambda)$$

$$P(\lambda)=\sum_{j=0}^{2}a_j\cdot\lambda^j;\ Q(\lambda)=\sum_{j=0}^{1}c_j\cdot\lambda^j;\ P(\lambda)$$
$$=a_0+a_1\cdot\lambda+a_2\cdot\lambda^2;\ Q(\lambda)=c_0+c_1\cdot\lambda$$

$n, m \in \mathbb{N}_0; n > m$ and $a_j, c_j:\mathbb{R}_{+0} \to \mathbb{R}$ are continuous and differentiable function of τ such that $a_0 + c_0 \neq 0$. In the following "$-$" denotes complex and conjugate. $P(\lambda), Q(\lambda)$ are analytic functions in λ and differentiable in τ. The coefficients $a_j(p_{eff}, \tau_g, \sigma_G, \sigma_A, N_A, \tau, \ldots) \in \mathbb{R}$ and $c_j(p_{eff}, \tau_g, \sigma_G, \sigma_A, N_A, \tau, \ldots) \in \mathbb{R}$ depend on Q-switched laser system's $p_{eff}, \tau_g, \sigma_G, \sigma_A, N_A, \tau, \ldots$ values. Unless strictly necessary, the designation of the varied arguments: $(p_{eff}, \tau_g, \sigma_G, \sigma_A, N_A, \tau, \ldots)$ will subsequently be omitted from P, Q, a_j, c_j. The coefficients a_j, c_j are continuous and differentiable functions of their arguments, and direct substitution shows that $a_0 + c_0 \neq 0$.

$$a_0=0;\ c_0=c^2\cdot\sigma_G^2\cdot n_p^{(j)}\cdot N_G^{(j)}$$
$$-\left(\frac{1}{\tau_g}+c\cdot\sigma_G\cdot n_p^{(j)}\right)\cdot\left(c\cdot\sigma_G\cdot N_G^{(j)}-c\cdot\sigma_A\cdot N_A-\frac{1}{\tau_p}\right)$$
$$c^2\cdot\sigma_G^2\cdot n_p^{(j)}\cdot N_G^{(j)}-\left(\frac{1}{\tau_g}+c\cdot\sigma_G\cdot n_p^{(j)}\right)$$
$$\cdot\left(c\cdot\sigma_G\cdot N_G^{(j)}-c\cdot\sigma_A\cdot N_A-\frac{1}{\tau_p}\right)\neq 0$$

$\forall\ p_{eff}, \tau_g, \sigma_G, \sigma_A, N_A, \tau, \ldots \in \mathbb{R}_+$, i.e. $\lambda = 0$ is not a root of the characteristic equation. Furthermore $P(\lambda)$; $Q(\lambda)$ are analytic functions of λ for which the following requirements of the analysis (see Kuang, 1993, Sect. 3.4) can be varied in the present case.

(a) If $\lambda = i \cdot \omega; \omega \in \mathbb{R}$ then $P(i \cdot \omega) + Q(i \cdot \omega) \neq 0$, i.e. P and Q have no common imaginary roots. This condition was verified numerically in the entire $(p_{eff}, \tau_g, \sigma_G, \sigma_A, N_A, \tau, \ldots)$ domain of interest.

(b) $|\frac{P(\lambda)}{Q(\lambda)}|$ is bounded for $|\lambda| \to \infty; \mathrm{Re}\lambda \geq 0$. No roots bifurcation from ∞. Indeed, in the limit: $|\frac{P(\lambda)}{Q(\lambda)}| =$

$$|\frac{-(c\cdot\sigma_G\cdot N_G^{(j)}-c\cdot\sigma_A\cdot N_A-\frac{1}{\tau_p})\cdot\lambda+\lambda^2}{c^2\cdot\sigma_G^2\cdot n_p^{(j)}\cdot N_G^{(j)}-(\frac{1}{\tau_g}+c\cdot\sigma_G\cdot n_p^{(j)})\cdot(c\cdot\sigma_G\cdot N_G^{(j)}-c\cdot\sigma_A\cdot N_A-\frac{1}{\tau_p}) +(\frac{1}{\tau_g}+c\cdot\sigma_G\cdot n_p^{(j)})\cdot\lambda}|$$

(c) The following expressions exist: $F(\omega) = |P(i \cdot \omega)|^2 - |Q(i \cdot \omega)|^2$; $F(\omega) = \sum_{k=0}^{2} \Pi_{2\cdot k} \cdot \omega^{2\cdot k}$
Has at most a finite number of zeros. Indeed, this is a polynomial in ω (degree in ω^4).

(d) Each positive root $\omega(p_{eff}, \tau_g, \sigma_G, \sigma_A, N_A, \tau, \ldots)$ of $F(\omega) = 0$ is continuous and differentiable with respect to $p_{eff}, \tau_g, \sigma_G, \sigma_A, N_A, \tau, \ldots$. The condition can only be assessed numerically.

In addition, since the coefficients in P and Q are real, we have $\overline{P(-i \cdot \omega)} = P(i \cdot \omega)$ and $\overline{Q(-i \cdot \omega)} = Q(i \cdot \omega)$ thus, $\omega > 0$ maybe on eigenvalue of characteristic equations. The analysis consists in identifying the roots of the characteristic equation situated on the imaginary axis of the complex λ—plane, whereby increasing the parameters:

$p_{eff}, \tau_g, \sigma_G, \sigma_A, N_A, \tau, \ldots$ Reλ may, at the crossing, change its sign from (−) to (+), i.e. from a stable focus $E^{(*)}\left(N_G^{(*)}, n_p^{(*)}\right)$ to an unstable one, or vice versa. This feature may be further assessed by examining the sign of the partial derivatives with respect to $p_{eff}, \tau_g, \sigma_G, \sigma_A, N_A, \tau, \ldots$ and system parameters [5, 6].

$$\Lambda^{-1}(\tau) = \left(\frac{\partial \mathrm{Re}\lambda}{\partial \tau}\right)_{\lambda=i\cdot\omega}; p_{eff}, \tau_g, \sigma_G, \sigma_A, N_A, \ldots = const$$

$$\Lambda^{-1}(p_{eff}) = \left(\frac{\partial \mathrm{Re}\lambda}{\partial p_{eff}}\right)_{\lambda=i\cdot\omega}; \tau, \tau_g, \sigma_G, \sigma_A, N_A, \ldots = const$$

$$\Lambda^{-1}(\tau_g) = \left(\frac{\partial \mathrm{Re}\lambda}{\partial \tau_g}\right)_{\lambda=i\cdot\omega}; p_{eff}, \tau, \sigma_G, \sigma_A, N_A, \ldots = const$$

$$\Lambda^{-1}(\sigma_G) = \left(\frac{\partial \mathrm{Re}\lambda}{\partial \sigma_G}\right)_{\lambda=i\cdot\omega}; p_{eff}, \tau_g, \tau, \sigma_A, N_A, \ldots = const$$

$$\Lambda^{-1}(\sigma_A) = \left(\frac{\partial \mathrm{Re}\lambda}{\partial \sigma_A}\right)_{\lambda=i\cdot\omega}; p_{eff}, \tau_g, \sigma_G, \tau, N_A, \ldots = const$$

$$\Lambda^{-1}(N_A) = \left(\frac{\partial \mathrm{Re}\lambda}{\partial N_A}\right)_{\lambda=i\cdot\omega}; p_{eff}, \tau_g, \sigma_G, \sigma_A, \tau, \ldots = const$$

$$P(\lambda) = P_R(\lambda) + i \cdot P_I(\lambda); Q(\lambda) = Q_R(\lambda) + i \cdot Q_I(\lambda)$$

When writing and inserting $\lambda = i \cdot \omega$ into system's characteristic equation ω must satisfy the following equations.

$$\sin(\omega \cdot \tau) = g(\omega) = \frac{-P_R(i \cdot \omega, \tau) \cdot Q_I(i \cdot \omega, \tau) + P_I(i \cdot \omega, \tau) \cdot Q_R(i \cdot \omega, \tau)}{|Q(i \cdot \omega, \tau)|^2}$$

$$\cos(\omega \cdot \tau) = h(\omega) = -\frac{P_R(i \cdot \omega, \tau) \cdot Q_R(i \cdot \omega, \tau) + P_I(i \cdot \omega, \tau) \cdot Q_I(i \cdot \omega, \tau)}{|Q(i \cdot \omega, \tau)|^2}$$

where $|Q(i \cdot \omega, \tau)|^2 \neq 0$ in view of requirement (a) above, and $(g, h) \in \mathbb{R}$. Furthermore, it follows above $\sin(\omega \cdot \tau)$ and $\cos(\omega \cdot \tau)$ equations, that by squaring and adding the sides, ω must be a positive root of $F(\omega) = |P(i \cdot \omega)|^2 - |Q(i \cdot \omega)|^2 = 0$. Note: $F(\omega)$ is independent on τ. It is important to notice that if $\tau \notin I$(assume that $I \subseteq \mathbb{R}_{+0}$ is the set where $\omega(\tau)$ is a positive root of $F(\omega)$ and for:$\tau \notin I$;$\omega(\tau)$ is not defined. Then for all τ in I, $\omega(\tau)$ is satisfied that $F(\omega) = 0$. Then there are no positive $\omega(\tau)$ solutions for $F(\omega, \tau) = 0$, and we cannot have stability switches. For $\tau \in I$ where $\omega(\tau)$ is a positive solution of $F(\omega, \tau) = 0$, we can define the angle $\theta(\tau) \in [0, 2 \cdot \pi]$ as the solution of $\sin \theta(\tau) = \cdots$; $\cos \theta(\tau) = \cdots$

$$\sin \theta(\tau) = \frac{-P_R(i \cdot \omega) \cdot Q_I(i \cdot \omega) + P_I(i \cdot \omega) \cdot Q_R(i \cdot \omega)}{|Q(i \cdot \omega)|^2}$$

$$\cos \theta(\tau) = -\frac{P_R(i \cdot \omega) \cdot Q_R(i \cdot \omega) + P_I(i \cdot \omega) \cdot Q_I(i \cdot \omega)}{|Q(i \cdot \omega)|^2}$$

And the relation between the argument $\theta(\tau)$ and $\tau \cdot \omega(\tau)$ for $\tau \in I$ must be as describe below.

$$\omega(\tau) \cdot \tau = \theta(\tau) + 2 \cdot n \cdot \pi \; \forall n \in \mathbb{N}_0$$

Hence we can define the maps $\tau_n : I \to \mathbb{R}_{+0}$ given by $\tau_n(\tau) = \frac{\theta(\tau)+2 \cdot n \cdot \pi}{\omega(\tau)}$; $n \in \mathbb{N}_0$; $\tau \in I$. Let us introduce the functions: $I \to \mathbb{R}$; $S_n(\tau) = \tau - \tau_n(\tau)$; $\tau \in I$; $n \in \mathbb{N}_0$.

That is continuous and differentiable in τ. In the following, the subscripts $\lambda, \omega, p_{eff}, \tau_g, \sigma_G, \sigma_A, N_A, \ldots$ indicate the corresponding partial derivatives. Let us first concentrate on $\Lambda(x)$, remember in $\lambda(p_{eff}, \tau_g, \sigma_G, \sigma_A, N_A, \ldots)$ and $\omega(p_{eff}, \tau_g, \sigma_G, \sigma_A, N_A, \ldots)$, and keeping all parameters except (x) and τ. The derivation closely follows that in [BK]. Differentiating system characteristic equation $P(\lambda) + Q(\lambda) \cdot e^{-\lambda \cdot \tau} = 0$ with respect to specific parameter (x), and inverting the derivative, for convenience, one calculates:

$$x = p_{eff}, \tau_g, \sigma_G, \sigma_A, N_A, \ldots$$

$$\left(\frac{\partial \lambda}{\partial x}\right)^{-1} = \frac{-P_\lambda(\lambda, x) \cdot Q(\lambda, x) + Q_\lambda(\lambda, x) \cdot P(\lambda, x) - \tau \cdot P(\lambda, x) \cdot Q(\lambda, x)}{P_x(\lambda, x) \cdot Q(\lambda, x) - Q_x(\lambda, x) \cdot P(\lambda, x)}$$

where $P_\lambda = \frac{\partial P}{\partial \lambda}$; $Q_\lambda = \frac{\partial Q}{\partial \lambda}$; $P_x = \frac{\partial P}{\partial x}$; $Q_x = \frac{\partial Q}{\partial x}$, substituting $\lambda = i \cdot \omega$ and bearing $\overline{P(-i \cdot \omega)} = P(i \cdot \omega)$ and $\overline{Q(-i \cdot \omega)} = Q(i \cdot \omega)$ then $i \cdot P_\lambda(i \cdot \omega) = P_\omega(i \cdot \omega)$; $i \cdot Q_\lambda(i \cdot \omega) = Q_\omega(i \cdot \omega)$ and that on the surface $|P(i \cdot \omega)|^2 = |Q(i \cdot \omega)|^2$, one obtains:

$$\left(\frac{\partial \lambda}{\partial x}\right)^{-1}|_{\lambda=i\cdot\omega} = \left(\frac{i \cdot P_\omega(i \cdot \omega, x) \cdot \overline{P(i \cdot \omega, x)} + i \cdot Q_\lambda(i \cdot \omega, x) \cdot \overline{Q(i \cdot \omega, x)} - \tau \cdot |P(i \cdot \omega, x)|^2}{P_x(i \cdot \omega, x) \cdot \overline{P(i \cdot \omega, x)} - Q_x(i \cdot \omega, x) \cdot \overline{Q(i \cdot \omega, x)}}\right)$$

Upon separating into real and imaginary parts, with $P = P_R + i \cdot P_I$; $Q = Q_R + i \cdot Q_I$; $P_\omega = P_{R\omega} + i \cdot P_{I\omega}$ $Q_\omega = Q_{R\omega} + i \cdot Q_{I\omega}$; $P_x = P_{Rx} + i \cdot P_{Ix}$; $Q_x = Q_{Rx} + i \cdot Q_{Ix}$; $P^2 = P_R^2 + P_I^2$, when (x) can be any Q-switched laser system parameter $p_{eff}, \tau_g, \sigma_G, \sigma_A, N_A, \ldots$ and time delay τ etc. Where for convenience, we have dropped the arguments $(i \cdot \omega, x)$ and where

$$F_\omega = 2 \cdot [(P_{R\omega} \cdot P_R + P_{I\omega} \cdot P_I) - (Q_{R\omega} \cdot Q_R + Q_{I\omega} \cdot Q_I)];$$
$$F_x = 2 \cdot [(P_{Rx} \cdot P_R + P_{Ix} \cdot P_I) - (Q_{Rx} \cdot Q_R + Q_{Ix} \cdot Q_I)]$$

$\omega_x = -\frac{F_x}{F_\omega}$. We define U and V:

$$U = (P_R \cdot P_{I\omega} - P_I \cdot P_{R\omega}) - (Q_R \cdot Q_{I\omega} - Q_I \cdot Q_{R\omega});$$
$$V = (P_R \cdot P_{Ix} - P_I \cdot P_{Rx}) - (Q_R \cdot Q_{Ix} - Q_I \cdot Q_{Rx})$$

We choose our specific parameter as time delay $x = \tau$.

$$P_R = -\omega^2;\ P_I = -\left(c \cdot \sigma_G \cdot N_G^{(j)} - c \cdot \sigma_A \cdot N_A - \frac{1}{\tau_p}\right) \cdot \omega;$$
$$Q_I = \left(\frac{1}{\tau_g} + c \cdot \sigma_G \cdot n_p^{(j)}\right) \cdot \omega\ Q_R = c^2 \cdot \sigma_G^2 \cdot n_p^{(j)} \cdot N_G^{(j)} - \left(\frac{1}{\tau_g} + c \cdot \sigma_G \cdot n_p^{(j)}\right) \cdot \left(c \cdot \sigma_G \cdot N_G^{(j)} - c \cdot \sigma_A \cdot N_A - \frac{1}{\tau_p}\right)$$
$$P_{R\omega} = -2 \cdot \omega;\ Q_{I\omega} = \left(\frac{1}{\tau_g} + c \cdot \sigma_G \cdot n_p^{(j)}\right);$$
$$P_{I\omega} = -\left(c \cdot \sigma_G \cdot N_G^{(j)} - c \cdot \sigma_A \cdot N_A - \frac{1}{\tau_p}\right);\ Q_{R\omega} = 0$$
$$P_{R\tau} = 0;\ P_{I\tau} = 0;\ Q_{I\tau} = 0;\ Q_{R\tau} = 0;\ P_{R\omega} \cdot P_R = 2 \cdot \omega^3;\ Q_{R\omega} \cdot Q_R = 0$$
$$F_\tau = 2 \cdot [(P_{R\tau} \cdot P_R + P_{I\tau} \cdot P_I) - (Q_{R\tau} \cdot Q_R + Q_{I\tau} \cdot Q_I)] = 0;$$

$$P_I \cdot P_{R\omega} = \left(c \cdot \sigma_G \cdot N_G^{(j)} - c \cdot \sigma_A \cdot N_A - \frac{1}{\tau_p}\right) \cdot 2 \cdot \omega^2$$

$$Q_R \cdot Q_{I\omega} = \left[c^2 \cdot \sigma_G^2 \cdot n_p^{(j)} \cdot N_G^{(j)} - \left(\frac{1}{\tau_g} + c \cdot \sigma_G \cdot n_p^{(j)}\right) \cdot \left(c \cdot \sigma_G \cdot N_G^{(j)} - c \cdot \sigma_A \cdot N_A - \frac{1}{\tau_p}\right)\right] \cdot \left(\frac{1}{\tau_g} + c \cdot \sigma_G \cdot n_p^{(j)}\right)$$

$$Q_I \cdot Q_{R\omega} = 0; V = (P_R \cdot P_{I\tau} - P_I \cdot P_{R\tau}) - (Q_R \cdot Q_{I\tau} - Q_I \cdot Q_{R\tau}) = 0; F(\omega, \tau) = 0$$

Differentiating with respect to τ and we get

$$F_\omega \cdot \frac{\partial \omega}{\partial \tau} + F_\tau = 0; \ \tau \in I \Rightarrow \frac{\partial \omega}{\partial \tau} = -\frac{F_\tau}{F_\omega}; \ \Lambda^{-1}(\tau) = \mathrm{Re}\left\{\frac{-2 \cdot [U + \tau \cdot |P|^2] + i \cdot F_\omega}{F_\tau + 2 \cdot i \cdot [V + \omega \cdot |P|^2]}\right\}$$

$$sign\{\Lambda^{-1}(\tau)\} = sign\left\{\left(\frac{\partial \mathrm{Re}\lambda}{\partial \tau}\right)_{\lambda = i \cdot \omega}\right\}; \ sign\{\Lambda^{-1}(\tau)\} = sign\{F_\omega\} \cdot sign\left\{\frac{V + \frac{\partial \omega}{\partial \tau} \cdot U}{|P|^2} + \omega + \frac{\partial \omega}{\partial \tau} \cdot \tau\right\}$$

We shall presently examine the possibility of stability transitions (bifurcation) Q-switched laser system, about the equilibrium point $E^{(*)}(N_G^{(*)}, n_p^{(*)})$ as a result of a variation of delay parameterτ. The analysis consists in identifying the roots of our system characteristic equation situated on the imaginary axis of the complex λ—plane, where by increasing the delay parameterτ, Reλ may at the crossing, changes its sign from − to + , i.e. from a stable focus E^* to an unstable one, or vice versa. This feature may be further assessed by examining the sign of the partial derivatives with respect to τ.

$$\Lambda^{-1}(\tau) = \left(\frac{\partial \mathrm{Re}\lambda}{\partial \tau}\right)_{\lambda = i \cdot \omega};$$

$$p_{eff}, \tau_g, \sigma_G, \sigma_A, \ldots = const; \omega \in \mathbb{R}_+$$

$$P_R \cdot P_{I\omega} = \omega^2 \cdot \left(c \cdot \sigma_G \cdot N_G^{(j)} - c \cdot \sigma_A \cdot N_A - \frac{1}{\tau_p}\right); Q_I \cdot Q_{R\omega} = 0$$

$$Q_R \cdot Q_{I\omega} = \left[c^2 \cdot \sigma_G^2 \cdot n_p^{(j)} \cdot N_G^{(j)} - \left(\frac{1}{\tau_g} + c \cdot \sigma_G \cdot n_p^{(j)}\right) \cdot \left(c \cdot \sigma_G \cdot N_G^{(j)} - c \cdot \sigma_A \cdot N_A - \frac{1}{\tau_p}\right)\right] \cdot \left(\frac{1}{\tau_g} + c \cdot \sigma_G \cdot n_p^{(j)}\right)$$

$$U = (P_R \cdot P_{I\omega} - P_I \cdot P_{R\omega}) - (Q_R \cdot Q_{I\omega} - Q_I \cdot Q_{R\omega})$$
$$= \left\{\omega^2 \cdot \left(c \cdot \sigma_G \cdot N_G^{(j)} - c \cdot \sigma_A \cdot N_A - \frac{1}{\tau_p}\right)\right.$$
$$\left. - \left(c \cdot \sigma_G \cdot N_G^{(j)} - c \cdot \sigma_A \cdot N_A - \frac{1}{\tau_p}\right) \cdot 2 \cdot \omega^2\right\}$$
$$- \left\{\left[c^2 \cdot \sigma_G^2 \cdot n_p^{(j)} \cdot N_G^{(j)} - \left(\frac{1}{\tau_g} + c \cdot \sigma_G \cdot n_p^{(j)}\right)\right.\right.$$
$$\left.\left. \cdot \left(c \cdot \sigma_G \cdot N_G^{(j)} - c \cdot \sigma_A \cdot N_A - \frac{1}{\tau_p}\right)\right] \cdot \left(\frac{1}{\tau_g} + c \cdot \sigma_G \cdot n_p^{(j)}\right)\right\}$$

Then we get the expression for $F(\omega, \tau)$ circuit parameter values. We find those ω, τ values which fulfill $F(\omega, \tau) = 0$. We ignore negative, complex, and imaginary values of ω for specific $\tau \in [0.001 \ldots 10]$ second. We plot the stability switch diagram based on different delay values of our Q-switched laser system [5, 6].

$$sign\{\Lambda^{-1}(\tau)\} = sign\{F_\omega(\omega(\tau), \tau)\}$$
$$\cdot sign\{\tau \cdot \omega_\tau(\omega(\tau)) + \omega(\tau) + \frac{U(\omega(\tau)) \cdot \omega_\tau(\omega(\tau)) + V(\omega(\tau))}{|P(\omega(\tau))|^2}$$

Remark We know $F(\omega, \tau) = 0$ implies its roots $\omega_i(\tau)$ and finding those delays values τ which ω_i is feasible. There are τ values which give complex ω_i or imaginary number, then unable to analyze stability.F—Function is independent on τ the parameter $F(\omega, \tau) = 0$. The results: We find those ω, τ values which fulfill $F(\omega, \tau) = 0$. We ignore negative, complex, and imaginary values of ω. Next is to find those ω, τ values which fulfill $\sin\theta(\tau) = \cdots$; $\cos\theta(\tau) = \cdots$.

$$\sin(\omega \cdot \tau) = \frac{-P_R \cdot Q_I + P_I \cdot Q_R}{|Q|^2}; \cos(\omega \cdot \tau)$$
$$= -\frac{P_R \cdot Q_R + P_I \cdot Q_I}{|Q|^2}; |Q|^2 = Q_R^2 + Q_I^2$$

Finally, we plot the stability switch diagram.

$$g(\tau) = \Lambda^{-1}(\tau) = \left(\frac{\partial \mathrm{Re}\lambda}{\partial \tau}\right)_{\lambda = i \cdot \omega}$$
$$= \frac{2 \cdot \{F_\omega \cdot (V + \omega \cdot P^2) - F_\tau \cdot (U + \tau \cdot P^2)\}}{F_\tau^2 + 4 \cdot (V + \omega \cdot P^2)^2}$$
$$sign[g(\tau)] = sign[\Lambda^{-1}(\tau)] = sign\left[\left(\frac{\partial \mathrm{Re}\lambda}{\partial \tau}\right)_{\lambda = i \cdot \omega}\right]$$
$$= sign\left[\frac{2 \cdot \{F_\omega \cdot (V + \omega \cdot P^2) - F_\tau \cdot (U + \tau \cdot P^2)\}}{F_\tau^2 + 4 \cdot (V + \omega \cdot P^2)^2}\right]$$

$$F_\tau^2 + 4 \cdot (V + \omega \cdot P^2)^2 > 0; sign[\Lambda^{-1}(\tau)]$$
$$= sign\{F_\omega \cdot (V + \omega \cdot P^2) - F_\tau \cdot (U + \tau \cdot P^2)\}$$
$$sign[\Lambda^{-1}(\tau)] = sign\left\{\{[F_\omega] \cdot \left[(V + \omega \cdot P^2)\right.\right.$$
$$\left.\left. - \frac{F_\tau}{F_\omega} \cdot (U + \tau \cdot P^2)\right]\right\}; \omega_\tau = -\frac{F_\tau}{F_\omega}$$
$$\omega_\tau = \left(\frac{\partial \omega}{\partial \tau}\right)^{-1} = -\frac{\partial F/\partial \omega}{\partial F/\partial \tau}$$
$$sign[\Lambda^{-1}(\tau)] = sign\{[F_\omega]$$
$$\cdot [V + \omega_\tau \cdot U + \omega \cdot P^2 + \omega_\tau \cdot \tau \cdot P^2]\} sign[\Lambda^{-1}(\tau)]$$
$$= sign\left\{[F_\omega] \cdot [P^2] \cdot \left[\frac{V + \omega_\tau \cdot U}{P^2} + \omega + \omega_\tau \cdot \tau\right]\right\}$$
$$sign[P^2] > 0 \Rightarrow sign[\Lambda^{-1}(\tau)]$$
$$= sign\left\{[F_\omega] \cdot \left[\frac{V + \omega_\tau \cdot U}{P^2} + \omega + \omega_\tau \cdot \tau\right]\right\}$$
$$sign[\Lambda^{-1}(\tau)] = sign[F_\omega] \cdot sign\left[\frac{V + \omega_\tau \cdot U}{P^2} + \omega + \omega_\tau \cdot \tau\right]$$
$$P_{I\omega} \cdot P_I = \left(c \cdot \sigma_G \cdot N_G^{(j)} - c \cdot \sigma_A \cdot N_A - \frac{1}{\tau_p}\right)^2 \cdot \omega;$$
$$Q_{I\omega} \cdot Q_I = \left(\frac{1}{\tau_g} + c \cdot \sigma_G \cdot n_p^{(j)}\right)^2 \cdot \omega; P_{R\omega} \cdot P_R = 2 \cdot \omega^3$$
$$F_\omega = 2 \cdot [(P_{R\omega} \cdot P_R + P_{I\omega} \cdot P_I) - (Q_{R\omega} \cdot Q_R + Q_{I\omega} \cdot Q_I)]$$
$$= 2 \cdot \left[\left(2 \cdot \omega^3 + \left(c \cdot \sigma_G \cdot N_G^{(j)} - c \cdot \sigma_A \cdot N_A - \frac{1}{\tau_p}\right)^2 \cdot \omega\right)\right.$$
$$\left. - \left(\frac{1}{\tau_g} + c \cdot \sigma_G \cdot n_p^{(j)}\right)^2 \cdot \omega\right]$$

We check the sign $\Lambda^{-1}(\tau)$ according the following rule (Table 4.6):

If $sign[\Lambda^{-1}(\tau)] > 0$ then the crossing proceeds from (−) to (+), respectively (stable to unstable). If $sign[\Lambda^{-1}(\tau)] < 0$ then the crossing proceeds from (+) to (−) respectively (unstable to stable).

Table. 4.6 Q-switched laser system sign of $sign[\Lambda^{-1}(\tau)]$

$sign[F_\omega]$	$sign[\frac{V+\omega_\tau \cdot U}{P^2} + \omega + \omega_\tau \cdot \tau]$	$sign[\Lambda^{-1}(\tau)]$
±	±	+
±	±	−

4.5 Questions

1. We have a model of solid-state laser by semiconductor devices and the corresponding to the rate equations. We have two model assumptions: (1) the transverse relaxation times of the equivalent two-level models for the laser gain medium and for the saturate absorber are much faster than any other dynamics in the system (2) the changes in the laser intensity, gain, and saturate absorption on a time scale of the order of the round-trip time (T_R) in the cavity are small. The laser model of a single-mode laser with homogeneously broadband laser medium and saturate absorber is not ideal and there are interferences. We represent these interferences by time delay parameterτ. We can write our system Delay Differential Equations (DDEs):

$$\frac{dP}{dt} = \frac{2}{T_R} \cdot (g(t) - l - q(t - \tau)) \cdot P(t); \ \frac{dg}{dt} = -\frac{(g(t) - g_0)}{T_L \cdot T_R} - \frac{g(t) \cdot P(t)}{E_L}$$

$$\frac{dq}{dt} = -\frac{(q(t - \tau) - q_0)}{T_A \cdot T_R} - \frac{q(t - \tau) \cdot P(t)}{E_A}$$

T_R—Round-trip time in cavity, P—Laser power, g—Gain per round trip, l—Linear losses per round trip, q—Saturate losses per round trip, g_0—Small-signal gain per round trip, q_0—Un saturate but saturate losses per round trip, T_L—Upper state lifetime of the gain medium (normalized to the round trip time of the cavity) ; $T_L = \frac{\tau_L}{T_R}$, T_A—Absorber recovery time (normalized to the round trip time of the cavity) ; $T_A = \frac{\tau_A}{T_R}$, τ_L—Upper state lifetime of the gain medium, τ_A—Absorber recovery time, E_L—Saturation energy of the gain ; $E_L = \frac{h \cdot \nu \cdot A_{eff,L}}{2 \cdot \sigma_L}$, E_A—Saturation energy of the absorber ; $E_A = \frac{h \cdot \nu \cdot A_{eff,A}}{2 \cdot \sigma_A}$.

1.1 Find system fixed points, plot the related graphs of fixed point's coordinates as a function of system parameters ($l, T_A, T_L, T_R, E_A, E_L$).
1.2 Discuss stability and stability switching, find small increments Jacobian.
1.3 Find homogeneous system characteristic equation for the λ.
1.4 How our results change if there is no linear losses per round trip in our system ($l = 0$)? Find fixed points and discuss stability and stability switching.
1.5 How our results change if the small-signal gain per round trip is very small ($g_0 \approx 0$) and the un-saturate but saturate losses per round trip is very small ($q_0 \approx 0$)? Find fixed points and discuss stability and stability switching.

2. We have a model of solid-state laser by semiconductor devices and the corresponding to the rate equations. We control the laser dynamics by various Q-switching and continuous-wave (CW) mode-locking techniques which increase the laser peak power and obtaining ultra-short optical pulses. Solid-state lasers such as Nd:YAG or Nd:YLF, with their long upper-state lifetimes, can hardly be passively CW mode locked using saturable absorbers without Q-switching or at least Q-switched mode locking. To understand the regime of Q-switched mode locking we consider the basic system rate equations. The approximate laser

power is $P(T,t) = E_p(T) \cdot \sum_n f(t - n \cdot T_R)$with $\int_{-T_R/2}^{T_R/2} f(t - n \cdot T_R) \cdot dt = 1$where $E_p(T = n \cdot T_R)$; $E_p(T)$is the laser phase energy; $T_R$is the round trip time in the cavity; n—number of pulse. The pulse energy of the n-th pulse is changes appreciably over many cavity round trips. Function $f(t)$is the shape of the mode-locked pulses. We assume that the mode-locked pulses are much shorter than the recovery time of the absorber. We can consider that the relaxation term for the absorber in $q = \frac{q_0}{1+\frac{P}{P_A}}$; $P_A = \frac{E_A}{\tau_A}$can be neglected for the duration of the mode-locked pulses, $q = \frac{q_0}{1+\frac{P \cdot \tau_A}{E_A}}$; q_0—Un saturate but saturate losses per round trip, q—Saturate losses per round trip, $P_A$is the saturation power of the absorber, τ_A—Absorber recovery time. We take the assumption that since the absorber recovery time is assumed to be much shorter than the cavity round-trip time, the absorber is unsaturated before the arrival of the pulse. The saturation of the absorber during one pulse is $q(T,t)$; $q(T,t) = q_0 \cdot \exp\left[-\frac{E_P(T)}{E_A} \cdot \int_{-T_R/2}^{t} f(t') \cdot dt'\right]$ therefore the loss in the pulse energy per round trip is $q_p(T)$; $q_p(T) = \int_{-T_R/2}^{T_R/2} f(t) \cdot q(T,t) \cdot dt = q_0 \cdot \left(\frac{1-e^{\left[-\frac{E_P(T)}{E_A}\right]}}{E_p(T)/E_A}\right)$.

The saturable absorber essentially saturates with the pulse energy. We average the basic laser power (P) and the gain per round trip (g) rate equations of the laser over one round-trip and obtain the following two DDEs for the dynamics of the pulse energy and the gain on a coarse-grained time scale T.

$$T_R \cdot \frac{dE_p}{dT} = 2 \cdot [g(T - \tau_1) - l - q_p(E_p(T - \tau_2))] \cdot E_p(T - \tau_2)$$

$$T_R \cdot \frac{dg}{dT} = -\frac{g(T - \tau_1) - g_0}{T_L} - \frac{g(T - \tau_1) \cdot E_p(T - \tau_2)}{E_L}$$

τ_1, τ_2 are delay parameters for the gain per round trip (g) and the laser pulse energy (E_p) respectively. $g(T) \to g(T - \tau_1)$ and $E_p(T) \to E_p(T - \tau_2)$. The delay parameters are due to non-ideal averaging process for the basic rate equations.

Non ideal averaging: The saturation of the gain medium within one pulse is not completely negligible due to the small interaction cross section of the solid state laser material.

2.1 Find system fixed points, plot the related graphs of fixed points coordinates as a function of system parameters ($l, T_L, E_L, \ldots$).
2.2 Discuss stability and stability switching and find small increment Jacobian for the cases: (1) $\tau_1 = \tau$; $\tau_2 = 0$ (2) $\tau_1 = 0$; $\tau_2 = \tau$ (3) $\tau_1 = \tau$; $\tau_2 = \tau$.
2.3 Find the homogeneous system characteristic equations for the λ, cases: (1) $\tau_1 = \tau$; $\tau_2 = 0$(2) $\tau_1 = 0$; $\tau_2 = \tau$(3) $\tau_1 = \tau$; $\tau_2 = \tau$.nn
2.4 Return 2.1–2.3 if there is no linear losses per round trip in our system ($l = 0$). Explain how the results change and the behavior of the system.
2.5 Return 2.1–2.3 for the case that the small signal gain per round trip is neglected ($g_0 \approx 0$). Explain how the results change and the behavior of the system.

3. We have a model of solid-state laser by semiconductor devices and the corresponding to the rate equations. We control the laser dynamics by various Q-switching and continuous-wave (CW) mode-locking techniques which increase the laser peak power and obtaining ultrashort optical pulses. For ultrashort pulse generation; a fast saturable absorber or the interplay between a slow saturable absorber and gain saturation is usually used. Both mechanisms open a window of net gain (only the pulse itself experiences gain per round trip). This allows the system to discriminate against noise that may grow in the wings of the pulse, and therefore the pulse is kept stable against perturbations or noise. Soliton pulse shaping—the combined action of negative group velocity dispersion (GVD) and self-phase-modulation (SPM) may lead to additional pulse shortening and stabilization in actively and passively mode-locked lasers. Soliton like pulse formation in actively mode-locked lasers allows considerable pulse shortening beyond that due to the usual active mode locking. In the presence of soliton like pulse shaping even a slow saturable absorber with a recovery time much longer than the pulse width can stabilize the pulse. For the soliton the nonlinear effects due to SPM and the linear effects due to the negative GVD are in balance. In contrast, the noise or instabilities that might grow within the longer time window are not intense enough to experience the Kerr nonlinearity and are therefore spread in time. When they spread in time they experience the higher absorption due to the slowly recovering absorber after passage of the much shorter soliton like pulse. The instabilities see less gain per round trip than the soliton and will decay with time. For femtosecond pulse generation there is an extending of our laser model to include SPM and GVD. The laser by Hau's master equation of mode locking is as follow:

$$
\begin{aligned}
T_R \cdot \frac{\partial A(T,t)}{\partial T} &= \left(-i \cdot D \cdot \frac{\partial^2}{\partial t^2} + i \cdot \delta \cdot |A(T,t)|^2\right) A(T,t) \\
&\quad + \left(g - l + D_g \cdot \frac{\partial^2}{\partial t^2} - q(T,t)\right) A(T,t)
\end{aligned}
$$

$A(T,t)$ is the slowly varying field envelope of the pulse, D is the intra cavity GVD, $D_g = \frac{g}{\Omega_g^2}$ is the gain dispersion, δ is the nonlinear coefficient due to SPM, and l is the amplitude loss and g is the gain. Ω_g is the HWHM gain bandwidth.

The Hau's master equation of mode locking describes the laser dynamics on two time scales: T, which is on the order of the round trip time in the laser cavity and t, which is on the order of the pulse width and which resolves the pulse shape. In term of pulse shaping, the action of the saturable absorption $q(t)$on short time scale is as follow: $\frac{\partial q(T,t)}{\partial t} = -\frac{q-q_0}{\tau_A} - \frac{|A(T,t)|^2}{E_A} \cdot q$. Where we have assumed that the recovery time $\tau_A$of the saturable absorber is much shorter than the round-trip time of the pulse, and the absorber recovers completely between two consecutive pulses, $q(T,t) = q_0 \cdot \exp\left[-\frac{E_P(T)}{E_A} \cdot \int_{-T_R/2}^{t} f(t') \cdot dt'\right]$. The loss in pulse energy per round trip can be written as follow.

$$q_p(T) = \int_{-T_R/2}^{T_R/2} f(t) \cdot q(T,t) \cdot dt = q_0 \cdot \left(\frac{1 - \exp[-\frac{E_p(T)}{E_A}]}{\frac{E_P(T)}{E_A}} \right)$$

3.1 In term of laser by Hau's master equation of mode locking, find system fixed points ($A = A(T,t),\ A^*$). Plot the related fixed point's graphs as a function of system parameters (D, δ, g, l, D_g). Consider $\frac{\partial}{\partial T} \leftrightarrow \frac{d}{dT}$; $\frac{\partial}{\partial t} \leftrightarrow \frac{d}{dt}$. See Hint 1.

3.2 Due to non-linearity of our system there is a time delay τ in the slow varying field envelope of the pulse $A(T,t)$ in time $A(T,t) \to A(T,t-\tau)$. The time delay does not affect the derivatives of $A(T,t)$. We can rewrite our laser by Hau's master equation of mode locking and get the partial differential equation:

$$T_R \cdot \frac{\partial A(T,t)}{\partial T} = -i \cdot D \cdot \frac{\partial^2 A(T,t)}{\partial t^2} + i \cdot \delta \cdot |A(T,t-\tau)|^2 \cdot A(T,t-\tau)$$
$$+(g - l - q(T,t)) \cdot A(T,t-\tau) + D_g \cdot \frac{\partial^2 A(T,t)}{\partial t^2}$$

Discuss stability and stability switching for different values of time delay parameter τ.

3.3 Return (3.1) and (3.2) if the gain dispersion D_g is negligible ($D_g \approx 0$).

3.4 Return (3.1) and (3.2) if the system amplitude loss is negligible ($l \approx 0$).

3.5 Return (3.1) and (3.2) if the nonlinear coefficient due to SPM is negligible ($\delta \approx 0$).

Hint 1:

$$T_R \cdot \frac{\partial A(T,t)}{\partial T} = \left(-i \cdot D \cdot \frac{\partial^2}{\partial t^2} + i \cdot \delta \cdot |A(T,t)|^2\right) A(T,t)$$
$$+ \left(g - l + D_g \cdot \frac{\partial^2}{\partial t^2} - q(T,t)\right) A(T,t)$$
$$i \cdot \left[-D \cdot \frac{\partial^2 A(T,t)}{\partial t^2} + \delta \cdot |A(T,t)|^2 \cdot A(T,t)\right]$$
$$+ (g - l) \cdot A(T,t) - q(T,t) \cdot A(T,t)$$
$$+ D_g \cdot \frac{\partial^2 A(T,t)}{\partial t^2} - T_R \cdot \frac{\partial A(T,t)}{\partial T} = 0$$

We define two functions: $f_1(\frac{\partial^2 A(T,t)}{\partial t^2}, A(T,t)) = -D \cdot \frac{\partial^2 A(T,t)}{\partial t^2} + \delta \cdot |A(T,t)|^2 \cdot A(T,t)$.

Then $f_1(\frac{\partial^2 A(T,t)}{\partial t^2}, A(T,t)) = 0$ & $f_2(\frac{\partial A(T,t)}{\partial T}, \frac{\partial^2 A(T,t)}{\partial t^2}, A(T,t)) = 0$

$$f_1\left(\frac{\partial^2 A(T,t)}{\partial t^2}, A(T,t)\right) = 0 \Rightarrow -D \cdot \frac{\partial^2 A(T,t)}{\partial t^2} + \delta \cdot |A(T,t)|^2 \cdot A(T,t) = 0$$

$$D \cdot \frac{\partial^2 A(T,t)}{\partial t^2} = \delta \cdot |A(T,t)|^2 \cdot A(T,t) \Rightarrow \frac{\partial^2 A(T,t)}{\partial t^2} = \frac{1}{D} \cdot \delta \cdot |A(T,t)|^2 \cdot A(T,t)$$

Inserting into f_2 equation: $f_2(\frac{\partial A(T,t)}{\partial T}, \frac{\partial^2 A(T,t)}{\partial t^2}, A(T,t)) = 0$.

$$(g - l) \cdot A(T,t) - q(T,t) \cdot A(T,t) + D_g \cdot \frac{\partial^2 A(T,t)}{\partial t^2} - T_R \cdot \frac{\partial A(T,t)}{\partial T} = 0$$

$$(g - l) \cdot A(T,t) - q(T,t) \cdot A(T,t) + D_g \cdot \frac{1}{D} \cdot \delta \cdot |A(T,t)|^2 \cdot A(T,t) - T_R \cdot \frac{\partial A(T,t)}{\partial T} = 0$$

$$T_R \cdot \frac{\partial A(T,t)}{\partial T} = (g - l) \cdot A(T,t) - q(T,t) \cdot A(T,t) + D_g \cdot \frac{1}{D} \cdot \delta \cdot |A(T,t)|^2 \cdot A(T,t) \frac{\partial A(T,t)}{\partial T} = \frac{1}{T_R} \cdot (g - l) \cdot A(T,t) - \frac{1}{T_R} \cdot q(T,t) \cdot A(T,t) + D_g \cdot \frac{1}{D \cdot T_R} \cdot \delta \cdot |A(T,t)|^2 \cdot A(T,t)$$

4. We have a system of nanometer-vibration measurement with microchip solid-state laser. The experiment configuration is a self-mixing Doppler vibrometry and optical micro phony. The model Delay Differential Equations (DDEs) is the simulation, that is, laser with frequency-modulated optical feedback. We define three variables, $N(t)$—normalized excess population inversion, $E(t)$—normalized field amplitude and $\phi(t)$—phase of the lasing field. We define delay time parameter τ as a delay in time of the normalized excess population inversion due to non-ideal system model. We define the system model DDEs (laser with frequency-modulated optical feedback):

$$\frac{dN(t)}{dt} = K \cdot \{\omega - 1 - N(t - \tau) - [1 + 2 \cdot N(t - \tau)] \cdot E^2(t)\}$$

$$N = N(t)\ ;\ E = E(t)\ ;\ \phi = \phi(t)$$

$$\frac{dE(t)}{dt} = N(t - \tau) \cdot E(t) + m \cdot E(t - t_D) \times \cos\psi(t) + \left(\sqrt{2 \cdot \varepsilon \cdot [N(t - \tau) + 1]}\right) \cdot \xi(t)$$

$$\frac{d\phi(t)}{dt} = \left[\frac{E(t - t_D)}{E(t)}\right] \cdot \sin\psi(t)\ ;\ \psi(t) = \Delta\Omega \cdot t + \beta \cdot \sin(\Omega_m \cdot t) - \phi(t) + \phi(t - t_D) - \left(\Omega_0 + \frac{\Delta\Omega}{2}\right) \cdot t_D$$

ω—Relative pump power normalized by the threshold ($\omega = \frac{P}{P_{th}}$), K—Population to photon lifetime ration ($K = \frac{\tau}{\tau_p}$), $\phi(t)$—Phase of the lasing field, $\psi(t)$—Phase difference between the lasing and the feedback fields, $\Delta\Omega$—Normalized frequency shift, $\Delta\Omega = (\omega_i - \omega_0) \cdot \tau_p$ (τ_p—photon lifetime), m—Amplitude feedback coefficient, Ω_m—Normalized modulation frequency ($\Omega_m = \omega_m \cdot \tau_p$), t—Time normalized by the photon lifetime, t_D—Delay time normalized by the photon lifetime, ε—Spontaneous emission rate, $\xi(t)$—Gaussian white noise with zero mean. The value $\langle \xi(t) \cdot \xi(t') \rangle = \delta(t - t')$ is δcorrelated in time. $(\sqrt{2 \cdot \varepsilon \cdot [N(t) + 1]}) \cdot \xi(t)$ is the quantum spontaneous emission noise.

Assumption 1: The quantum spontaneous emission noise is very small and negligible $(\sqrt{2 \cdot \varepsilon \cdot [N(t) + 1]}) \cdot \xi(t)1$; $(\sqrt{2 \cdot \varepsilon \cdot [N(t) + 1]}) \cdot \xi(t) \to 0$.

Assumption 2: The delay time normalized by the photon lifetime $t_D$is very small ($t_D \approx 0$).

4.1 Find system fixed points, plot the related graphs of fixed point's coordinates as a function of system parameters ($K, \omega, \Delta\Omega, \beta, \ldots$).
4.2 Discuss stability and stability switching for different values of delay parameter τ. Find the small increments Jacobian.
4.3 Find the homogeneous system characteristic equation for the λ.
4.4 How our results (4.1–4.3) change if the normalized frequency shift is very small ($\Delta\Omega \approx 0$). Discuss stability and stability switching for different values of delay parameter τ.
4.5 Return (4.1–4.3) if the frequency modulation index is 100% ($\beta = 1$).

5. We have a system of nanometer-vibration measurement with microchip solid-state laser. The experiment configuration is a self-mixing Doppler vibrometry and optical micro phony. The model Delay Differential Equations (DDEs) is the simulation, that is, laser with frequency-modulated is the simulation, that is, laser with frequency-modulated optical feedback. We define three variables, $N(t)$—normalized excess population inversion, $E(t)$—normalized field amplitude and $\phi(t)$—phase of the lasing field. Due to non-ideal system model there is an additional shift in the field amplitude Ω_E; $\Omega_E \in \mathbb{R}$. The additional shift in the field amplitude Ω_E can be a positive or negative number. We define the system model DDEs:

$$\frac{dN(t)}{dt} = K \cdot \{\omega - 1 - N(t) - [1 + 2 \cdot N(t)] \cdot [E(t) + \Omega_E]^2\}$$

$$N = N(t)\,;\; E = E(t)\,;\; \phi = \phi(t)$$

$$\frac{dE(t)}{dt} = N(t) \cdot [E(t) + \Omega_E] + m \cdot [E(t - t_D) + \Omega_E] \times \cos\psi(t) + \left(\sqrt{2 \cdot \varepsilon \cdot [N(t) + 1]}\right) \cdot \xi(t)$$

$$\frac{d\phi(t)}{dt} = \left[\frac{E(t - t_D) + \Omega_E}{E(t) + \Omega_E}\right] \cdot \sin\psi(t)$$

$$\psi(t) = \Delta\Omega \cdot t + \beta \cdot \sin(\Omega_m \cdot t) - \phi(t) + \phi(t - t_D) - \left(\Omega_0 + \frac{\Delta\Omega}{2}\right) \cdot t_D$$

ω—Relative pump power normalized by the threshold ($\omega = \frac{P}{P_{th}}$), K—Population to photon lifetime ration ($K = \frac{\tau}{\tau_p}$), $\phi(t)$—Phase of the lasing field, $\psi(t)$—Phase difference between the lasing and the feedback fields, $\Delta\Omega$—Normalized frequency shift, $\Delta\Omega = (\omega_i - \omega_0) \cdot \tau_p$ (τ_p—photon lifetime), m—Amplitude feedback coefficient, Ω_m—Normalized modulation frequency ($\Omega_m = \omega_m \cdot \tau_p$), t—Time normalized by the photon lifetime, t_D—Delay time normalized by the photon lifetime, ε—Spontaneous emission rate, $\xi(t)$—Gaussian white noise with zero mean. The value $\langle \xi(t) \cdot \xi(t') \rangle = \delta(t - t')$ is δcorrelated in time. $(\sqrt{2 \cdot \varepsilon \cdot [N(t) + 1]}) \cdot \xi(t)$ is the quantum spontaneous emission noise.

Assumption 1: The quantum spontaneous emission noise is very small and negligible $(\sqrt{2 \cdot \varepsilon \cdot [N(t) + 1]}) \cdot \xi(t) \to 1$; $(\sqrt{2 \cdot \varepsilon \cdot [N(t) + 1]}) \cdot \xi(t) \to 0$.

5.1 Find system fixed points, plot the related graphs of fixed point's coordinates as a function of additional shift in the field amplitude parameter Ω_E; $\Omega_E \in \mathbb{R}$ and $(N^*, E^*, \phi^*) = f(\Omega_E)$.

5.2 Discuss stability and stability switching for different values of t_D parameter (delay time normalized by the photon lifetime). Find the small increments Jacobian.

5.3 Discuss stability and stability switching for different values of Ω_E; $\Omega_E \in \mathbb{R}$ parameter (additional shift in the field amplitude). Find the small increments Jacobian.

5.4 How our results (5.1–5.3) change if we consider that the delay time normalized by the photon lifetime is very small ($t_D \approx 0$)? Discuss stability and stability switching for different values of Ω_E; $\Omega_E \in \mathbb{R}$parameter.

5.5 How our results (5.1–5.3) change if the frequency modulation index is 100% ($\beta = 1$)? Discuss stability and stability switching for different values of Ω_E; $\Omega_E \in \mathbb{R}$parameter.

6. We have two systems of Doppler shift with a microchip solid state laser. In the first system, the number of relaxation oscillations which each mode exhibits is 4 (N = 4) and in the second system is 5 (N = 5). The differential equations which characterize our systems are as follow.

$$\frac{dN_0}{dt} = P - \frac{N_0}{\tau} - \sum_{k=1}^{N} B_k \cdot S_k \cdot \left(N_0 - \frac{N_k}{2}\right); \ \frac{dN_i}{dt} = -\frac{N_i}{\tau} + B_i \cdot N_0 \cdot S_i - N_i \cdot \sum_{k=1}^{N} B_k \cdot S_k$$

$$\frac{dE_i}{dt} = \left[B_i \cdot \left(N_0 - \frac{N_i}{2}\right) - \frac{1}{\tau_p}\right] \cdot \left(\frac{E_i}{2}\right) + \left(\frac{\kappa}{\tau_l}\right) \cdot E_{i,s} \cdot \cos[\psi_i(t) + \phi_{i,s}(t)]$$

$$\frac{d\psi_i}{dt} = \omega_i - \omega_{i,s} - \left(\frac{\kappa}{\tau_l}\right) \cdot \left(\frac{E_{i,s}}{E_i}\right) \cdot \sin[\psi_i(t) + \phi_{i,s}(t)]; \ i = 1, 2, \ldots, N$$

N_0—Space average of population inversion density, N_i—First spatial Fourier component of population inversion for the i – thmode, i—Modal index, P—Pump power, τ—Population lifetime, τ_p—Photon lifetime, κ—Effective amplitude transmission coefficient of the output mirror, τ_l—Round trip time in the cavity, ψ_i—Phase different between the two fields, $\phi_{i,s}$—Express phase fluctuations of scattered light, $\widetilde{E}_i$, E_i—Lasing field, $\widetilde{E_{i,s}}$—Scattered field, B_i—Gain, S_i—Photon density, $S_i = E_i^2$; $E_{i,s}^2(t) \propto E_i(t - t_d)$; t_d—Delay time, $\frac{E_{i,s}}{E_i} < 10^{-4}$(extremely small). In the lock-in region: $\frac{\Delta\omega}{2\cdot\pi} = \left(\frac{\kappa}{\tau_l}\right)\cdot\left(\frac{E_{i,s}}{E_i}\right) < 100\,\text{kHz}$. Where "$i$" denotes the model index (laser mode number). N—Number of relaxation oscillations which each mode may exhibit. $w = \frac{P}{P_{th,1}}$; $P_{th,1}$ is the first lasing mode pump threshold, n_0, n_i—are the space average and the first Fourier components of population inversion normalized by the mode threshold value $(N_0 - \frac{N_1}{2})_{th}$. s_i—Photon density normalized by the steady state S_1value at $w = 2$. γ_i—Gain ratio to the first lasing mode $(\gamma_i = \frac{B_i}{B})$, $m = \frac{2\cdot E_{i,s}}{\kappa\cdot E_i}$; $(\omega_i - \omega_{i,s})\cdot\tau \equiv \Omega_D$ (time is scaled by τ), $K = \frac{\tau}{\tau_p}$. ω_D— Doppler-shift frequency. We define the relaxation oscillation frequency f_{RO} as $f_{RO} = \frac{\omega_{RO}}{2\cdot\pi} = \left(\frac{1}{2\cdot\pi}\right)\cdot\sqrt{\frac{(w-1)}{\tau\cdot\tau_p}}|_{\tau=120\,\mu\text{s},\tau_p=310\,\text{ps}} = 540\,\text{kHz}$. $k\cdot v$—Angular frequency of Doppler shift and increases with increasing angular moving velocity $v$and with decreasing the scattering angle θ_s, where $\theta_s$is the angle between the laser axis and velocity vector. λ—Oscillation wavelength. Assumptions: (1) The delay time t_d is extremely small compared with modulation time scales, (2) Two systems parameters are the same except the number of relaxation oscillations which each mode exhibits (N).

6.1 Find the fixed points of the two systems and draw the related graphs of the different function between each similar fixed point coordinate. System (1) fixed point: $(N_0^*, N_i^*, E_i^*, \psi_i^*)$, System (2) fixed point: $(N_0^{**}, N_i^{**}, E_i^{**}, \psi_i^{**})$, Systems different functions: $(|N_0^* - N_0^{**}|, |N_i^* - N_i^{**}|, |E_i^* - E_i^{**}|, |\psi_i^* - \psi_i^{**}|)$.
6.2 Discuss stability and stability switching of the two systems for different parameters values. How the stability behavior changes if we move from the first system to the second one? Why?
6.3 How our results (6.1) and (6.2) are changed if the round trip time in the cavity (τ_l) is equal to the population life time (τ), $(\tau_l = \tau)$? Discuss stability and stability switching for different values of $\tau$for the two systems.
6.4 How our results (6.1) and (6.2) are changed if the round trip time in the cavity (τ_l) is equal to the photon lifetime (τ_p), $(\tau_l = \tau_p)$? Discuss stability and stability switching for different values of $\tau_l$for the two systems.
6.5 In the first system the express phase fluctuations of the scattered light $(\phi_{i,s})$ is very small and we can ignore it, $\phi_{i,s} \to \varepsilon$; $\psi_i(t) \gg \phi_{i,s}(t) \Rightarrow \psi_i(t) + \phi_{i,s}(t) \approx \psi_i(t)$. In the second system there is no change. Repeat (6.1) and (6.2) for that case.

7. We have two systems of Doppler shift with a microchip solid state laser. The express phase fluctuations of scattered light in first system $(\phi_{i,s-1}(t))$ is $m$times the express phase fluctuations of scattered light in second system $(\phi_{i,s-2}(t))$, $\phi_{i,s-1}(t) = m\cdot\phi_{i,s-1}(t)$; $m \geq 2$; $m \in \mathbb{N}$. All other parameters are the same for

the first and second systems. The differential equations which characterize our systems are as follow.

$$\frac{dN_0}{dt} = P - \frac{N_0}{\tau} - \sum_{k=1}^{N} B_k \cdot S_k \cdot \left(N_0 - \frac{N_k}{2}\right)$$

$$\frac{dN_i}{dt} = -\frac{N_i}{\tau} + B_i \cdot N_0 \cdot S_i - N_i \cdot \sum_{k=1}^{N} B_k \cdot S_k$$

$$\frac{dE_i}{dt} = \left[B_i \cdot \left(N_0 - \frac{N_i}{2}\right) - \frac{1}{\tau_p}\right] \cdot \left(\frac{E_i}{2}\right) + \left(\frac{\kappa}{\tau_l}\right) \cdot E_{i,s} \cdot \cos[\psi_i(t) + \phi_{i,s-k}(t)]$$

$$\frac{d\psi_i}{dt} = \omega_i - \omega_{i,s} - \left(\frac{\kappa}{\tau_l}\right) \cdot \left(\frac{E_{i,s}}{E_i}\right) \cdot \sin[\psi_i(t) + \phi_{i,s-k}(t)]$$

$$i = 1, 2, \ldots, N; \; k = 1, 2$$

N_0—Space average of population inversion density, N_i—First spatial Fourier component of population inversion for the i – thmode, i—Modal index, P—Pump power, τ—Population lifetime, τ_p—Photon lifetime, κ—Effective amplitude transmission coefficient of the output mirror, τ_l—Round trip time in the cavity, ψ_i—Phase different between the two fields, $\phi_{i,s-k}$—Express phase fluctuations of scattered light ($\phi_{i,s-1}(t)$ in the first system (k = 1) and $\phi_{i,s-2}(t)$in the second system (k = 2)), $\widetilde{E}_i$, E_i—Lasing field, $\widetilde{E_{i,s}}$—Scattered field, B_i—Gain, S_i—Photon density, $S_i = E_i^2$; $E_{i,s}^2(t) \propto E_i(t - t_d)$; t_d—Delay time, $\frac{E_{i,s}}{E_i} < 10^{-4}$(extremely small). In the lock-in region: $\frac{\Delta\omega}{2\cdot\pi} = \left(\frac{\kappa}{\tau_l}\right) \cdot \left(\frac{E_{i,s}}{E_i}\right) < 100\,\text{kHz}$. Where "$i$" denotes the model index (laser mode number). N—Number of relaxation oscillations which each mode may exhibit. $w = \frac{P}{P_{th,1}}$; $P_{th,1}$is the first lasing mode pump threshold, n_0, n_i—are the space average and the first Fourier components of population inversion normalized by the mode threshold value $(N_0 - \frac{N_1}{2})_{th}$. s_i—Photon density normalized by the steady state S_1value at $w = 2$. γ_i—Gain ratio to the first lasing mode $(\gamma_i = \frac{B_i}{B})$, $m = \frac{2\cdot E_{i,s}}{\kappa\cdot E_i}$; $(\omega_i - \omega_{i,s}) \cdot \tau \equiv \Omega_D$ (time is scaled by τ), $K = \frac{\tau}{\tau_p}$. ω_D— Doppler-shift frequency. We define the relaxation oscillation frequency f_{RO} as $f_{RO} = \frac{\omega_{RO}}{2\cdot\pi} = \left(\frac{1}{2\cdot\pi}\right) \cdot \sqrt{\frac{(w-1)}{\tau\cdot\tau_p}}\Big|_{\tau=120\,\mu\text{s}, \tau_p=310\,ps} = 540\,\text{kHz}$. $k \cdot v$—Angular frequency of Doppler shift and increases with increasing angular moving velocity$v$and with decreasing the scattering angle θ_s, where $\theta_s$is the angle between the laser axis and velocity vector. λ—Oscillation wavelength. Assumptions: The delay time t_d is extremely small compared with modulation time scales.

7.1 Plot the systems fixed points coordinates as a function of m parameter ($m \geq 2; m \in \mathbb{N}$). Keep all other parameters constant in your analysis.

7.2 Discuss stability and stability switching of the two systems for different m parameter values ($m \geq 2; m \in \mathbb{N}$). How the stability behavior changes if we move from the first system to the second one? Why?

7.3 How our results (7.1) and (7.2) are changed if the round trip time in the cavity (τ_l) is equal to the population life time (τ), ($\tau_l = \tau$)? Discuss stability and stability switching for different values of τ for the two systems.

7.4 How our results (7.1) and (7.2) are changed if the round trip time in the cavity (τ_l) is equal to the photon lifetime (τ_p), ($\tau_l = \tau_p$)? Discuss stability and stability switching for different values of τ_l for the two systems.

7.5 How our stability analysis results are changed if $\phi_{i,s-1}(t) = \frac{1}{m} \cdot \phi_{i,s-1}(t)$ and $m \geq 2; m \in \mathbb{N}$? Plot the related stability switching diagrams for different values of m parameter.

8. Laser feedback interferometry has become a fast-developing precision measurement modality with many kinds of lasers. By employing the frequency shifted optical feedback; microchip laser feedback interferometry has been widely researched due to its advantages of high sensitivity, simple structure, and easy alignment. Laser confocal feedback tomography combines the high sensitivity of laser frequency-shift feedback effect and the axial positioning ability of confocal microscopy. Laser feedback is a laser self-mixing interference which it is a physical phenomenon where part of the output laser is reflected or scattered by the external object returning back into the laser resonator to modulate the laser output power, phase, polarization states, and so on. Laser feedback interference occurs inside the laser resonator. The external optical feedback of $LiNdP_4O_{12}$ lasers produced using a rotating glass plate. It belongs to the class-B laser, of which the population decays slowly compared with the field. In this case, the feedback signal can be amplified by a gain factor of the ratio $\frac{\gamma_c}{\gamma_1}$, where γ_c the damping rate of the laser cavity and γ_1 is the damping rate of the population inversion. The ratio $\frac{\gamma_c}{\gamma_1}$ of a microchip laser can be high as 10^6. The ratio is only 10^3 for a laser diode (LD) laser, and much lower in a gas laser. The microchip laser has a high sensitivity compared with other kinds of lasers. In the weak feedback level, the laser power fluctuation under the effect of frequency-shifted feedback can be simulated using the modified Lang-Kobayashi equation. Due to system interferences there is a shift in time for the population inversion N(t), $N(t) \to N(t - \tau_\Delta)$; but there is not effect on the derivative of the population in time ($\frac{dN(t)}{dt}$). Our system Delay Differential Equations (DDEs) are as follow:

$$\frac{dN(t)}{dt} = \gamma \cdot (N_0 - N(t - \tau_\Delta)) - B \cdot N(t - \tau_\Delta) \cdot |E(t)|^2$$

$$\frac{dE(t)}{dt} = \frac{1}{2} \cdot (B \cdot N(t - \tau_\Delta) - \gamma_c) \cdot E(t) + \gamma_c \cdot \kappa \cdot \cos(2 \cdot \pi \cdot \Omega \cdot t - \omega \cdot \tau) \cdot E(t)$$

$N(t)$ is the population inversion, $E(t)$ is the amplitude of the laser electric field, N_0 is the inversion particle number under a small signal, γ is the decay rate of the population inversion, B is the Einstein coefficient, γ_c is the laser cavity decay rate, κ is the effective laser feedback level, ω is the optical running laser frequency, and τ is the photon round-trip time between the laser and the target.

8.1 Find system fixed points, plot the fixed points coordinate 3D graphs as a function of laser cavity decay rate (γ_c) and the decay rate of the population inversion (γ).
8.2 Discuss stability and stability switching for different values of τ_Δ parameter. If the effective laser feedback level (κ) is very low ($\kappa \to \varepsilon$), How the stability and stability switching of the system is influenced? Discuss it deeply.
8.3 Discuss (8.1) and (8.2) for the case that $\frac{dN(t)}{dt} = \ldots$ differential equation has cubed of $E(t)$ element, $\frac{dN(t)}{dt} = \ldots - B \cdot N(t - \tau_\Delta) \cdot |E(t)|^3$.
8.4 Discuss (8.1) and (8.2) for the case that $\frac{dN(t)}{dt} = \ldots$ differential equation has square root of $E(t)$ element, $\frac{dN(t)}{dt} = \ldots - B \cdot N(t - \tau_\Delta) \cdot \sqrt{|E(t)|}$.

9. We have a system of frequency shifted optical feedback measurement using a solid-state microchip laser. In the weak feedback level, the laser power fluctuation under the effect of frequency-shifted feedback is simulated using the modified Lang-Kobayashi equation (see question 8). The stationary solution of the modified Lang-Kobayashi equation can be obtained by setting the electric field E and population inversion N to be constant. Without feedback ($\kappa = 0$), the steady laser solutions are: $N_s = \frac{\gamma_c}{B}$; $I_s = |E_s|^2 = \frac{\gamma}{B} \cdot (\eta - 1)$. Where I_s, is the stationary intensity of the laser field, and $\eta = \frac{B \cdot N}{\gamma_c}$ is the normalized pumping rate. The power spectrum of the continuous pumped N_d: YVO_4 laser can be considered as slight fluctuations of the stable solutions. The electric field E(t) and the population inversion N(t) are: $N(t) = N_s + n(t)$; $E(t) = E_s + e(t)$. Combining the steady laser solutions equation and the electric field E(t) and the population inversion N(t) by Lang-Kobayashi equation gives the next differential equations (by neglecting the second-order terms):

$$\frac{dn^*(t)}{dt} = -\gamma \cdot \eta \cdot n^*(t) - 2 \cdot \gamma \cdot (\eta - 1) \cdot e(t)$$

$$\frac{de^*(t)}{dt} = \frac{1}{2} \cdot \gamma_c \cdot n^*(t) + \gamma_c \cdot \kappa \cdot \cos(2 \cdot \pi \cdot \Omega \cdot t - \omega \cdot \tau) + \gamma_c \cdot \kappa \cdot \cos(2 \cdot \pi \cdot \Omega \cdot t - \omega \cdot \tau) \cdot e^*(t)$$

where $n^*(t) = \frac{n(t)}{N_s}$; $e^*(t) = \frac{e(t)}{E_s}$ are the normalized variations of the population inversion and the electric field. The variation of laser intensity versus time can be derived accordingly, $\frac{\Delta I}{I} = \frac{2E(t) \cdot e(t)}{E^2(t)} \approx \frac{2 \cdot e(t)}{E_s} = 2 \cdot e^*(t)$. γ is the decay rate of the population inversion, γ_c is the laser cavity decay rate, κ is the effective laser feedback level, ω is the optical running laser frequency, and τ is the photon round-trip time between the laser and the target.

9.1 Find system fixed points and plot the relative coordinate 3D graphs as a function of γ_c, κ parameters.
9.2 Discuss stability and stability switching for different values of γ_c, κ parameters.
9.3 The normalized pumping rate η is double his value, how the stability of the system is changed?

9.4 The normalized pumping rate η is very small ($\eta \to \varepsilon$), how the stability of the system is changed? Find fixed points.

9.5 The normalized pumping rate η is very close to one ($\eta \to 1$), how the stability of the system is changed? Find fixed points.

10. We have a system of Q-switched diode-pumped solid-state laser. It implements and combines gain switching with passive Q switching of a miniature diode pumped solid-state laser. The simple model of a passively Q-switched laser is presented. The effective pumping rate of the upper laser level is $p_{eff} = \frac{\eta \cdot P}{h \cdot v_L \cdot V}$, where P is the electrical power input; $\eta$is the pumping efficiency, which is the product of laser-diode efficiency and optical coupling efficiency; $v_L$is the laser frequency; $V$is the laser mode volume, and $h$is the Planck constant. Then the delay differential equations (DDEs) of our system is for population density of the upper laser level (N_G) and the photon density in the cavity (n_p) are as follow. Due to jitter interferences there is a delay in time for N_G and $n_p$system variables: $N_G(t) \to N_G(t-\tau_1), n_p(t) \to n_p(t-\tau_2)$. There is no effects of time delay parameters (τ_1, τ_2) on N_G and $n_p$variables derivatives in time ($\frac{dN_G}{dt}, \frac{dn_p}{dt}$).

Remark: $N_G = N_G(t); n_p = n_p(t)$

$$\frac{dN_G}{dt} = p_{eff} - \frac{1}{\tau_g} \cdot N_G(t-\tau_1) - c \cdot \sigma_G \cdot n_p(t-\tau_2) \cdot N_G(t-\tau_1)$$
$$- \frac{1}{\tau_g} \cdot \xi \cdot \sqrt[k]{N_G(t-\tau_1)}$$

$$\frac{dn_p}{dt} = c \cdot \sigma_G \cdot n_p(t-\tau_2) \cdot N_G(t-\tau_1)$$
$$- c \cdot \sigma_A \cdot n_p(t-\tau_2) \cdot N_A - \frac{1}{\tau_p} \cdot n_p(t-\tau_2) + r_{sp}$$

nn_p—Density of photons in the laser mode, N_G—Population density of the upper laser level, τ_g—Lifetime of the upper laser level, σ_G—Emission cross section of the gain medium, c—Speed of light in vacuum, N_A—Density of the saturate-absorber atoms (in the absence of laser light it is N_{A0}), σ_A—Effective absorption cross section (it takes into account the ratio between the volumes of the active mode in the absorber and in the gain medium), ξ, k—are setting coefficients for $\frac{dN_G}{dt} = \cdots$ delay differential equation ($\xi \in \mathbb{R}; k > 2; k \in \mathbb{N}$), τ_p—Photon cavity lifetime, r_{sp}—Rate of spontaneous emission into the laser mode, $N^0_{G,th}$—Threshold density of the upper laser level (with the unsaturated saturate absorber, by setting $\frac{dn_p}{dt} = 0$ and assuming that $r_{sp} = 0$. Assuming that $r_{sp} = 0$ because few photons generated by spontaneous emission contribute to the laser mode that is defined by the cavity mirrors; $N^0_{G,th} = \frac{1}{c \cdot \sigma_G \cdot \tau_p} + N_{A0} \cdot \frac{\sigma_A}{\sigma_G}$. By replacing N_G in equation $N_G = p_{eff} \cdot \tau_g \cdot (1 - e^{-\frac{t}{\tau_g}})$ with $N^0_{G,th}$, as $\tau_d = \tau_g \cdot \ln(\frac{\alpha}{\alpha - 1})$, where $\alpha = \frac{p_{eff} \cdot \tau_g}{N^0_{G,th}}$. τ_d—Time when the laser reaches its threshold, α—Ratio of pump power to the threshold pump power.

10.1 Find system fixed points for $k = 3, 4, 5$, draw the related graph of fixed points coordinates as a function of k parameter. All other system parameters are constant.

10.2 Discuss stability and stability switching for different values of τ parameter, the case is $\tau_1 = 0; \tau_2 = \tau$.

10.3 How the results (10.1) and (10.2) are changed if $\xi = 0$? Discuss stability and stability switching for different values of τ parameter, the case is $\tau_1 = 0; \tau_2 = \tau$.

10.4 Parameter $k \to \infty$, find system fixed points and discuss stability and stability switching ($\tau_1 = 0; \tau_2 = \tau$).

10.5 Plot the fixed points coordinate graphs ($\xi = k$, setting coefficients are equal) as a function of k parameter. Discuss stability and stability switching ($\tau_1 = 0; \tau_2 = \tau$).

References

1. F.X. Kartner, L.R. Brovelli, D. Kopf, M. Kamp, I.G. Calasso, U. Keller, Control of solid state laser dynamics by semiconductor devices. Opt. Eng. **34**(7) (1995)
2. K. Otsuka, K. Abe, J.-Y. Ko, Real-time nanometer vibration measurement with a self-mixing microchip solid-state laser. Opt. Lett. **27**(15) (2002)
3. K. Otsuka, Highly sensitive measurement of Doppler-shift with a microchip solid-state laser. Jpn. J. Appl. Phys. **31**, L1546–L1548, part 2, No. 11A (1992)
4. J.B. Khurgin, F. Jin, G. Solyar, C.-C. Wang, S. Trivedi, Cost-effective low timing jitter passively Q-switched diode-pumped solid-state laser with composite pumping pulses. Appl. Opt. **41**(6) (2002)
5. Y. Kuang, *Delay Differential Equations with Applications in Population Dynamics* (Academic Press, Boston, 1993).
6. E. Beretta, Y. Kuang, Geometric stability switch criteria in delay differential systems with delay dependent parameters. SIAM J. Math. Anal. **33**, 1144–1165 (2002)
7. J. Kuang, Y. Cong, *Stability of Numerical Methods for Delay Differential Equations* Elsevier science; 1 edition (December 12, 2007).
8. E. Beretta, Y. Kuang, Geometric stability switch criteria In delay differential systems with delay dependent parameters. SIAM J.MATH. ANAL. **33**(5), 1144–1165 (Published electronically February 14, 2002).
9. Y. Kuang, *Delay Differential Equations with Applications in Population Dynamics*, Mathematics in Science and Engineering, Vol. 191 (1993 by Academic Press, Inc).
10. J. Gu, S.-C. Tam, ..., Novel use of GaAs as a passive Q- switch as well as an output coupler for diode-pumped infrared solid-state lasers, in *Proceedings of SPIE 3929, Solid State Lasers IX*, 2000.
11. U. Levy, Y. Silberberg, Second and third harmonic waves excited by focused Gaussian beams. Opt. Express 27795, **23**(21) (2015)
12. Y.A. Kuznetsov, *Elelments of Applied Bifurcation Theory* (Applied Mathematical Sciences)
13. J.K. Hale, *Dynamics and Bifurcations*. Texts in Applied Mathematics, Vol. 3
14. S.H. Strogatz, *Nonlinear Dynamics and Chaos* (Westview Press)

Chapter 5
Nd:YAG, Mid-Infrared and Q-Switched Microchip Lasers Stability Analysis

A diode-pumped Nd:YAG laser is Q-switched by a GaAs saturable absorber and intra-cavity frequency-doubled by a KTP crystal. At pump power the device produces high quality pulses at specific pulse repetition rate. We can characterize the dynamics of the pulse formation by rate equations and an energy-level model which accounts for the various energy transfer process in GaAs. High power laser diode technology brings the ability to build efficient, reliable, and compact laser sources with good beam quality. Short pulses can be generated from diode-pumped, solid-state lasers by Q-switching or model locking. Passively methods are based on nonlinear optical properties of solid-state materials. We can use efficient nonlinear wavelength conversion techniques to extend the wavelength coverage of these devices to the visible region. Frequency doubling of a Q-switched, diode-pumped laser is a method to generating compact, pulsed laser sources in the blue-green region. Second-harmonic generation is realized either inside or outside of the cavity. Advantages of external-cavity doubling are that is simplifies the cavity design and results in shorter pulses. Intra cavity frequency doubling utilizes the high peak power that is present inside the cavity and results in higher conversion efficiency. In Q-switched operation, the generated pulses are longer than those produced by a comparable Q-switched laser with no intra cavity doubling. Fully utilization of the high power inside the cavity is reachable by finding an optimal compromise between SHG and the saturable absorption that leads to Q-switching. We adjusting the spot size, the polarization, or the phase mismatch at the doubling crystal. Higher or lower second-harmonic coupling results in frequency-doubled pulses with lower peak power. The compact green-pulse source is based entirely on solid-state components. The device is a diode-pumped Nd:YAG laser that is passively Q-switched by a GaAs saturable absorber and intra-cavity frequency-doubled by a KTP crystal. One type is passive Q-switching of a diode-pump Nd:YAG laser with GaAs. Nonlinear losses in GaAs limit the obtainable pulse energies, especially at high peak powers. The frequency-doubled system operates at lower peak powers and has a performance that is similar to a comparable actively Q-switched device. KTP (KTiOPO4) is a nonlinear optical crystal, which possess excellent nonlinear, electro optical and acoustic-optical properties.

O. Aluf, *Advance Elements of Laser Circuits and Systems*,
https://doi.org/10.1007/978-3-030-64103-0_5

It is potassium titanyl phosphate KTP crystals. Potassium titanyl phosphate (KTP) is an inorganic compound with the formula KTiOPO4 and it is a white solid. KTP is a nonlinear optical material that is commonly used for frequency doubling diode pumped solid-state lasers such as Nd:YAG and other neodymium-doped lasers. Crystals of KTP are highly transparent for wavelengths between 350 and 2700 nm with a reduced transmission out to 4500 nm where the crystal is effectively opaque. It is second harmonic generation (SHG) coefficient which is about three times higher than KDP. KTP can be used as an optical parametric oscillator for near IR generation up to 4 μm. It is suited to high power operation as an optical parametric oscillator due to its high damage threshold and large crystal aperture. Estimation for the performance of Q-switched laser is obtained from a four-level rate equation model. Pulse formation with GaAs saturable absorber is analyzing by includes both two-photon absorption (TPA) and free-carrier absorption (FCA) into the model. Dynamic and stability is inspected. TPA generates free electrons and holes whereas FCA promotes electrons higher into the conduction band and holes deeper into the valence band. Combination of all these effects with the accompanying recombination process derives the set of intra-cavity rate equations. Semiconductor saturable absorber mirror (SESAM) devices are a key component of ultrafast passive mode-locked-laser sources. The key SESAM parameters such as saturation fluency, modulation depth, and non saturable losses are measured and inspected. They are important and controllable to obtain stable pulse generation for a given laser. Nonlinear reflectivity of saturable absorbers is measured for getting a high precision, wide dynamic range setup. An accurate calibration is achievable by measures a low modulation depth. The model function for the nonlinear reflectivity is based on a simple two-level traveling wave system. Spatial beam profiles, non saturable losses and high-order absorption (two-photon absorption) are included in the model. We measure nonlinear reflectivity to fit a model function to obtain the key parameters of the saturable absorber. The model function is described the nonlinear behavior of real SESAMs. A band-structure is approximated by two level systems.

The approximation model neglects intra-band relaxations as well as trapping or recombination. Effects of standing wave patterns and carrier diffusion and temperature effects in the device are not included. The traveling wave model is based on rate equations for a two-level system without relaxations. It is a good approximation for a slow saturable absorber, where the recovery time is longer then the pulse bleaching the absorber. The two-level system is characterized by set of differential equations and dynamics is inspected. Ion-doped crystalline lasers operating in the mid-IR spectral range (2–5 μm) is inspected. It is rare earth and transition metal based ionic crystals as well as color-center lasers. They are compact all-solid-state room-temperature tunable sources, belonging to the broad class of vibrionic lasers.

The efficient high-power room-temperature operation and super-broad tenability, and the possibility of generating ultrashort pulses from the novel class of chromium doped chalcogenide laser led to an analysis in vibrionic laser systems involving $3d^n$ transition-metal ions. The mid-IR lasers and spectroscopic and laser characteristics are allowing assessment of the suitability of laser materials to serve as active media in diode pumped laser systems. The process influencing formation of the inversion population in mid-IR crystalline solid-state laser is inspected and the effect of these processes on laser operation (laser threshold and efficiency). It is continuous-wave operation by assuming the four-level scheme the balance equation for the population of the upper laser level n and the rate equation for photon flux at the laser wavelength. Q-switched lasers are ideal for applications requiring the highest available peak power and/or single pulse energy. These lasers are commonly used in medical, military defense, ranging and LiDAR, scientific research, material processing and industrial applications. Passively Q-switched Nd:YAG microchip lasers can produce up to 250 μJ/pulse at 1.064 μm, with a pulse duration of 380 ps. The infrared output has been converted to 532, 355, and 266 nm with high efficiency [1, 2].

5.1 Nd:YAG Laser Passively Q-Switched with GaAs Stability Optimization

We discuss all-solid-state green-pulse laser. A diode-dumped Nd:YAG laser is Q-switched by a GaAs saturable absorber and intra-cavity frequency-doubled by a KTP crystal. The dynamics of a pulse formation can be described by rate equations and an energy-level model which accounts for the various energy-transfer processes in GaAs. High-power diode-laser technology helps as to build efficient, reliable and compact laser sources with beam quality. We generate short pulses from diode-pumped, solid-state lasers by Q-switching or model locking. The fully passive methods are based on nonlinear optical properties of solid-state materials. The efficient nonlinear wavelength conversion techniques can be utilized to extend the wavelength coverage of these devices to the visible region. Frequency doubling of a Q-switched, diode pumped laser is a convenient method of generating compact, pulsed laser sources in the blue-green region. The devices are applicable in communication, displays, chemical processing, and as pump sources for laser materials. We can realize the second-harmonic generation either inside or outside of the cavity. The external-cavity doubling has the advantages which simplifies the cavity design and results in shorter pulses. This method is applicable to laser sources that operate with high output powers, such that acceptable conversion efficiencies which can be realize. These devices typically require tens of watts of pump power. Intra-cavity frequency doubling utilizes the high peak power that is present inside the cavity and results in higher conversion efficiency. For Q-switched operation, the generated pulses are longer than those produced by a comparable Q-switched laser with no intra cavity doubling. Intensity dependent loss prevents fast lasing build-up. We fully utilize the high power inside the cavity by finding an optimal compromise between SHG and the saturable absorption that leads to Q-switching. We can get it by adjusting the spot size, the polarization, or the phase mismatch at the doubling crystal. There are high or low second-harmonic coupling results in frequency-doubled pulses with lower peak power. Actually it is a compact green-pulse source that is based on solid-state components [1]. KTP (potassium titanyl phosphate) crystal is commonly used to

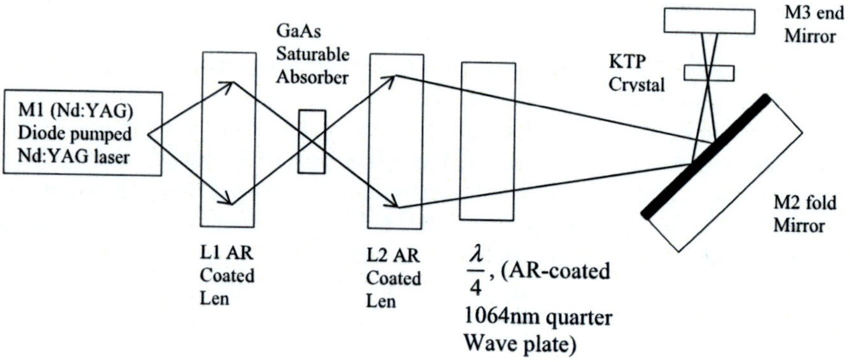

Fig. 5.1 Cavity configuration of the intra-cavity frequency doubled passively Q-switched Nd:YAG laser

frequency-double diode-pumped solid state laser. If the optical power becomes too high, KTP shows photo-induced "gray-tracks". The diode-pumped Nd:YAG laser is passively Q-switched by a GaAs saturable absorber and intra-cavity frequency doubled by a KTP crystal. Another possible is a passive Q-switching of a diode-pumped Nd:YAG laser with GaAs. Practically nonlinear losses in GaAs limit the obtaining pulse energies, mainly at high peak powers. The frequency-doubled system operates at lower peak powers and has a performance that is similar to a comparable actively Q-switched device. The cavity configuration of the intra-cavity frequency-doubled, passively Q-switched Nd:YAG laser is presented in the next figure (Fig. 5.1). First the system has partially collimated outputs from two 2 W laser diodes. They are combined by a polarizing beam splitter and focused onto the end of a 1 cm, 1.1% Nd-doped YAG crystal. The laser diodes offer high cw optical power and high brightness with unsurpassed reliability. The small emitting aperture combined with low beam divergence makes it among the highest brightness cw laser diodes available for our system.

The diodes comprised of partially coherent broad area emitters with relatively uniform emission over the emitting aperture. Operation is multi-longitudinal mode with a spectral envelope width of approximately 2 nm FWHM. The far field beam divergence in the plane perpendicular to the diode junction is nearly Gaussian while the lateral beam profile exhibits the complex pattern typical of broad area emitters. Emitting apertures can be in the range of 50–500 μm giving cw power output capability of up to 4 W with superlative reliability. There is power for 100 and 200 μmaperture devices or fiber-coupled lasers. The high efficiency of the quantum well structure combined with low thermal resistance epi-down chip mounting provides minimum junction temperature at high optical power. Low junction temperature and low thermal resistance packages extend lifetime and increase reliability. The pumping end of the Nd-YAG crystal has an anti-reflectance (AR) coating at 808 nm and a high-reflectance (HR) coating at 1064 nm (M1). The L-cavity design, with proper dichroic coatings on the end mirror (M3) and on the fold mirror (M2), allows one to extract ~90% of the green light that is generated during the double pass through the crystal. The end mirror (M3) has a radius of curvature of 10 cm and reflects 99.5% of both 1064 and 532 nm light. The fold mirror (M2) reflects approximately 99% of the fundamental light and transmits more than 90% of the second harmonic light at a 45° angle of incidence. Two anti-reflectance (AR) coated (1064 nm) 5 cm lenses are each used to form intra-cavity foci at the GaAs saturable absorber and at the 5 mm, dual AR-coated KTP crystal. The polished single-crystal GaAs wafer has a thickness of 450 μm and is oriented at Brewster's angle. The wafer 450 μm thickness produces the best compromise between pulse energy and stability. An AR-coated (1064 nm) quarter-wave plate is placed in the cavity to compensate the birefringence caused by the KTP crystal. Birefringence is the optical property of a material having a refractive index that depends on the polarization and propagation direction of light. These optically anisotropic materials are said to be birefringent. This cavity design is stable and allows considerable flexibility in positioning the elements for optimum performance. Green pulses are detected using a 1.5 GHz photodiode and a digital oscilloscope. At 3.7 W of incident pumping power 250 mW of cw green light is generated without the GaAs saturable absorber in the cavity. At the maximum pump power the pulse width is shortened, and the repetition rate decreased. This behavior is similar to what we observed with the GaAs saturable absorber without frequency doubling. We can plot the corresponding pulse energies and peak powers, as determined from the average power, repetition rate, and pulse width. The theoretical estimate for the performance of this Q-switched laser can be obtained from a four-level rate equation model. We discuss the pulse formation with the GaAs saturable absorber, and present both two-photon absorption (TPA) and free-carrier absorption (FCA) into the model. The rate equations for photon density ϕ and population inversion density N_x are as follow. Due to uncomplete of photon generation and interferences, there is a delay in time (τ_Δ) for photon density $\phi(t) \to \phi(t-\tau_\Delta)$; it is not density (ϕ) in time $\left(\frac{d\phi}{dt}\right)$.

$$\frac{d\phi}{dt} = [(2 \cdot N_x(t) \cdot \sigma \cdot L_0 - \gamma - 2 \cdot \alpha_q \cdot L_q) \cdot \phi(t-\tau_\Delta) - (K \cdot L_q^2 + B \cdot L_c) \cdot \phi^2(t-\tau_\Delta)] \cdot \frac{1}{T}$$

$$\frac{dN_x}{dt} = -2 \cdot N_x(t) \cdot \sigma \cdot c \cdot \phi(t-\tau_\Delta) - \frac{N_x(t)}{\tau} + P \cdot (N_{tot} - N_x(t)))$$

where, σ is the stimulated emission cross section, T is cavity round-trip time, P is the pump rate, N_{tot} is the density of active atoms, and L_0 is the length of the laser medium, L_c is the length of the doubling crystal, L_q is the length of the GaAs wafer. The linear losses of the cavity are included in γ, and the saturable absorption of the GaAs wafer is described by α_q. ϕ—Photon density in time. The terms quadratic in $\phi(t-\tau_\Delta)$represent SHG and TPA in GaAs, and the corresponding coupling coefficients are as describe below.

$$K = 32 \cdot \pi^2 \cdot h \cdot (\eta \cdot v)^3 \cdot c \cdot (\varepsilon_0 \cdot d \cdot L_c)^2 \cdot \left(\frac{w_0}{w_c}\right)^2 \cdot \mathrm{sinc}^2\left(\frac{\Delta k \cdot L_c}{2}\right)$$

$$B = 6 \cdot \beta \cdot h \cdot v \cdot c \cdot L_q \cdot \left(\frac{w_0}{w_q}\right)^2 \cdot \frac{1}{n_q}$$

Respectively, where v is the frequency of the fundamental field, d is the effective second order nonlinear coefficient, β is the TPA coefficient, and Δk is the wave-vector mismatch. The parameters w_0, w_c and w_q are the spot sizes of the beam at the laser rod, doubling crystal and GaAs wafer, respectively. The refractive index of GaAs appears in $B = \cdots$ because the wafer is at Brewster's angle. c is the speed of light in vacuum.

Remark Brewster's angle is an angle of incidence at which light with a particular polarization is perfectly transmitted through a transparent dielectric surface, with no reflection. When un-polarized light is incident at this angle, the light that is reflected from the surface is therefore perfectly polarized. If we have a system with a boundary between two media with different refractive indices, some of it is usually reflected. The fraction that is reflected is described by the Fresnel equations, and is dependent upon the incoming light's polarization and angle of incidence. The Fresnel equations predict that light with the p polarization (electric field polarized in the same plane as the incident ray and the surface normal at the point of incidence) will not be reflected if the angle of incidence is $\theta_B = \arctan\left(\frac{n_2}{n_1}\right)$. n_1 is the refractive index of the initial medium through which the light propagates, and n_2 is the index of the other medium. We define θ_1 as the angle of reflection (or incidence) and θ_2 is the angle of refraction then using Snell's law $n_1 \cdot \sin\theta_1 = n_2 \cdot \sin\theta_2$. We consider the incident angle, $\theta_1 = \theta_B$ at which no light is reflected, then $n_1 \cdot \sin\theta_B = n_2 \cdot \sin(90^\circ - \theta_B) = n_2 \cdot \cos\theta_B$; $\theta_B = \arctan\left(\frac{n_2}{n_1}\right)$. Gas lasers typically use a window tilted at Brewster's angle to allow the beam to leave the laser tube. Since the window reflects some s-polarized light but no p-polarized light, the round trip loss for the s polarization is higher than that of the p polarization. This causes the laser's output to be p polarized due to competition between the two modes. When the reflecting surface is absorbing, reflectivity at parallel polarization (p) goes through a non-zero minimum at the so-called pseudo-Brewster's angle [2].

The beam inside the GaAs wafer is elliptical and its area is larger, by a factor of n_q, than what the spot size suggests. The initial inversion N_{x-0} at the operating pulse repetition rate f is given by

$$N_{x-0}(f) = \frac{P \cdot N_{tot}}{\frac{1}{\tau} + P} \cdot \left(1 - \exp\left[-\left(\frac{1}{\tau} + P\right) \cdot \frac{1}{f}\right]\right)$$

At fixed points: $\frac{d\phi}{dt} = 0$; $\frac{dN_x}{dt} = 0$; $\lim_{t \to \infty} \phi(t - \tau_\Delta) = \phi(t)$; $T > 0$; $T \in \mathbb{R}$

$$\frac{d\phi}{dt} = 0 \Rightarrow \left[(2 \cdot N_x^* \cdot \sigma \cdot L_0 - \gamma - 2 \cdot \alpha_q \cdot L_q) \cdot \phi^* \right.$$
$$\left. -(K \cdot L_q^2 + B \cdot L_c) \cdot (\phi^*)^2\right] \cdot \frac{1}{T} = 0$$

$$(^*)\frac{1}{T} \neq 0 \Rightarrow (2 \cdot N_x^* \cdot \sigma \cdot L_0 - \gamma - 2 \cdot \alpha_q \cdot L_q) \cdot \phi^*$$
$$- (K \cdot L_q^2 + B \cdot L_c) \cdot (\phi^*)^2 = 0$$

$$\frac{dN_x}{dt} = 0 \Rightarrow -2 \cdot N_x^* \cdot \sigma \cdot c \cdot \phi^* - \frac{N_x^*}{\tau} + P \cdot (N_{tot} - N_x^*) = 0$$
$$- 2 \cdot N_x^* \cdot \sigma \cdot c \cdot \phi^* - \frac{N_x^*}{\tau} + P \cdot N_{tot} - P \cdot N_x^* = 0; \; P \cdot N_x^*$$
$$+ \frac{N_x^*}{\tau} + 2 \cdot N_x^* \cdot \sigma \cdot c \cdot \phi^* = P \cdot N_{tot}$$

$$(^{**})N_x^* \cdot \left(P + \frac{1}{\tau} + 2 \cdot \sigma \cdot c \cdot \phi^*\right) = P \cdot N_{tot} \Rightarrow N_x^* = \frac{P \cdot N_{tot}}{P + \frac{1}{\tau} + 2 \cdot \sigma \cdot c \cdot \phi^*}$$

$$(^{**}) \to (^*) : \left(2 \cdot \left[\frac{P \cdot N_{tot}}{P + \frac{1}{\tau} + 2 \cdot \sigma \cdot c \cdot \phi^*}\right] \cdot \sigma \cdot L_0 - \gamma - 2 \cdot \alpha_q \cdot L_q\right) \cdot \phi^*$$
$$- (K \cdot L_q^2 + B \cdot L_c) \cdot (\phi^*)^2 = 0$$

$$2 \cdot \left[\frac{P \cdot N_{tot}}{P + \frac{1}{\tau} + 2 \cdot \sigma \cdot c \cdot \phi^*}\right] \cdot \sigma \cdot L_0 \cdot \phi^* - \gamma \cdot \phi^* - 2 \cdot \alpha_q \cdot L_q \cdot \phi^*$$
$$- (K \cdot L_q^2 + B \cdot L_c) \cdot (\phi^*)^2 = 0$$

$$2 \cdot P \cdot N_{tot} \cdot \sigma \cdot L_0 \cdot \phi^* - \gamma \cdot \phi^* \cdot \left(P + \frac{1}{\tau} + 2 \cdot \sigma \cdot c \cdot \phi^*\right)$$
$$- 2 \cdot \alpha_q \cdot L_q \cdot \phi^* \cdot \left(P + \frac{1}{\tau} + 2 \cdot \sigma \cdot c \cdot \phi^*\right)$$
$$- (K \cdot L_q^2 + B \cdot L_c) \cdot (\phi^*)^2 \cdot \left(P + \frac{1}{\tau} + 2 \cdot \sigma \cdot c \cdot \phi^*\right) = 0$$

$$2 \cdot P \cdot N_{tot} \cdot \sigma \cdot L_0 \cdot \phi^* - \gamma \cdot \phi^* \cdot \left[P + \frac{1}{\tau}\right] - \gamma \cdot 2 \cdot \sigma \cdot c \cdot [\phi^*]^2$$
$$- 2 \cdot \alpha_q \cdot L_q \cdot \left[P + \frac{1}{\tau}\right] \cdot \phi^* - 4 \cdot \alpha_q \cdot L_q \cdot \sigma \cdot c \cdot [\phi^*]^2$$
$$- (K \cdot L_q^2 + B \cdot L_c) \cdot (\phi^*)^2 \cdot \left[P + \frac{1}{\tau}\right]$$
$$- (K \cdot L_q^2 + B \cdot L_c) \cdot 2 \cdot \sigma \cdot c \cdot [\phi^*]^3 = 0$$

$$\left\{2\cdot P\cdot N_{tot}\cdot\sigma\cdot L_0-\gamma\cdot\left[P+\frac{1}{\tau}\right]-2\cdot\alpha_q\cdot L_q\cdot\left[P+\frac{1}{\tau}\right]\right\}\cdot\phi^*$$
$$\{\gamma\cdot 2\cdot\sigma\cdot c+4\cdot\alpha_q\cdot L_q\cdot\sigma\cdot c$$
$$+(K\cdot L_q^2+B\cdot L_c)\cdot\left[P+\frac{1}{\tau}\right]\right\}-\cdot(\phi^*)^2$$
$$-(K\cdot L_q^2+B\cdot L_c)\cdot 2\cdot\sigma\cdot c\cdot[\phi^*]^3=0$$

$$\Bigg(\left\{2\cdot P\cdot N_{tot}\cdot\sigma\cdot L_0-\gamma\cdot\left[P+\frac{1}{\tau}\right]-2\cdot\alpha_q\cdot L_q\cdot\left[P+\frac{1}{\tau}\right]\right\}$$
$$-\left\{\gamma\cdot 2\cdot\sigma\cdot c+4\cdot\alpha_q\cdot L_q\cdot\sigma\cdot c+(K\cdot L_q^2+B\cdot L_c)\cdot\left[P+\frac{1}{\tau}\right]\right\}\cdot\phi^*$$
$$-(K\cdot L_q^2+B\cdot L_c)\cdot 2\cdot\sigma\cdot c\cdot[\phi^*]^2\Bigg)\cdot\phi^*=0$$

For simplicity we define $\xi(\phi^*)$ function:

$$\xi(\phi^*)=\left\{2\cdot P\cdot N_{tot}\cdot\sigma\cdot L_0-\gamma\cdot\left[P+\frac{1}{\tau}\right]-2\cdot\alpha_q\cdot L_q\cdot\left[P+\frac{1}{\tau}\right]\right\}$$
$$-\left\{\gamma\cdot 2\cdot\sigma\cdot c+4\cdot\alpha_q\cdot L_q\cdot\sigma\cdot c+(K\cdot L_q^2+B\cdot L_c)\cdot\left[P+\frac{1}{\tau}\right]\right\}\cdot\phi^*$$
$$-(K\cdot L_q^2+B\cdot L_c)\cdot 2\cdot\sigma\cdot c\cdot[\phi^*]^2$$

Then we get the equation $\xi(\phi^*)\cdot\phi^*=0$, first fixed point is $\phi^*=0$ then $N_x^*=\frac{P\cdot N_{tot}}{P+\frac{1}{\tau}}$. We can define the first fixed point $E^{(0)}(\phi^{(0)},N_x^{(0)})=\left(0,\frac{P\cdot N_{tot}}{P+\frac{1}{\tau}}\right)$. Additional fixed points we find numerically by $\xi(\phi^*)=0$. First we need to calculate parameters K, B.

$$B=6\cdot\beta\cdot h\cdot\nu\cdot c\cdot L_q\cdot\left(\frac{w_0}{w_q}\right)^2\cdot\frac{1}{n_q}$$

$$K=32\cdot\pi^2\cdot h\cdot(\eta\cdot\nu)^3\cdot c\cdot(\varepsilon_0\cdot d\cdot L_c)^2\cdot\left(\frac{w_0}{w_c}\right)^2\cdot\mathrm{sinc}^2\left(\frac{\Delta k\cdot L_c}{2}\right)$$

$$\alpha_q=\sigma_e\cdot(N-N^+)+\sigma_h\cdot N^++\sigma_{fc}\cdot n$$

N—is the total EL2 density, N_0^+—is the initial ionized EL2 (EL2⁺) density, N^+—is the ionized EL2 (EL2⁺) density, σ_e—is the EL2 absorption cross section, σ_h is the EL2⁺ absorption cross section, σ_{fc}—is the free-carrier absorption cross section, n—is the produce free electron, p—is the produce free holes, c—Speed of light in Vacuum $\left(c=c=3\times10^8\,\frac{\mathrm{m}}{\mathrm{s}}\right)$, h—is Planck's constant $\left(h=6.626\times10^{-34}\,\mathrm{J\,s}\right)$, $\hbar=\frac{h}{2\cdot\pi}$ is the reduced Planck constant $\left(\hbar=1.054\times10^{-34}\,\frac{\mathrm{J\,s}}{\mathrm{rad}}\right)$, ε_0 is the Vacuum permittivity

$\left(\varepsilon_0 = 8.85 \times 10^{-12}\,\frac{\mathrm{F}}{\mathrm{m}}\right)$, ν—is the frequency of the fundamental field, Δk—is the wave vector mismatch [3].

Saturable Absorber (SA): The saturable absorption of the GaAs wafer is described by parameter α_q. The SA consists of a saturable absorber layer on semi-insulating (transparent) GaAs wafer. Both sides of the SA are anti-reflect (AR) coated for the laser wavelength. The saturable absorber is a nonlinear absorber and the transmittance increases with increasing pulse energy. The typical saturation fluency of a SA is higher than that of a SAM. After a relaxation time <100 ps the SA goes back in the ground state. By using a saturable absorber (SA), a self-starting, passively mode-locked laser with a very stable pulse repetition rate can be realized. SAs mainly for fiber laser mode-locking in a wide spectral region from 1030 nm up to 2800 nm wavelength are available. The SA can be used with free-space optics or fiber coupled. The optical absorption coefficient of un-doped GaAs has mainly been affected by EL2 concentration. EL2 defects occur in bulk crystals grown LEC and the Bridgman technique, and are more easily formed under as rich conditions. Mainly contributed by two-photon-absorption and free-carrier-absorption, GaAs has the property of saturable absorption at the wavelength of 1.06 μm, which implies that it can be used as a saturable absorber for passive Q-switching. We already define the absorption coefficient of α_q of GaAs which express as $\alpha_q = \sigma_e \cdot (N - N^+) + \sigma_h \cdot N^+ + \sigma_{fc} \cdot n$, where σ_e is the cross section of EL2 absorption, N is the total density of EL2 level, N^+ is the portion of the positively charged EL2 centers, i.e., $EL2^+$, σ_h is the cross section of $EL2^+$absorption, σ_{fc} is the cross section of free carrier absorption (FCA), and n is the density of the free electrons produced by the transition from EL2 to the conduction band after the absorption of the optical energy of the incident laser beam. From the absorption coefficient $\alpha_q = \cdots$ equation we can see that, when the intensity of the incident light is low, both N^+ and n will be low and hence it can be ignored. Thus the absorption coefficient α_q equals to $\sigma_e \cdot N$, $\left(\alpha_q \approx \sigma_e \cdot N\right)$, which represents the linear absorption of GaAs at low irradiance region, and equals approximately to 1.1–1.2 cm^{-1} $\left(\alpha_q \approx 1.15\,\mathrm{cm}^{-1}\right)$ at the wavelength of 1.06 μm, which is determined only by σ_e and N. This is much lower than that of the absorption for light with the photon energy higher than the band gap energy, which is in order of 10^3 to 10^4 cm^{-1}. With the increase in the intensity of the incident laser beam, more EL2 centers will be positively charged, i.e., N^+ will increase, and thus makes the absorption coefficient lower than its linear absorption coefficient. Once the intensity of the incident laser beam becomes high enough, the decreasing of the absorption coefficient caused by the increasing of N^+ will be surpassed by the increasing caused by the free carrier absorption (FCA) when n becomes very high under the irradiation of high intensity laser beam. Then, the absorption coefficient will increase and higher than its linear absorption coefficient. This kind of the variation of absorption coefficient represents the characteristics of the saturable absorption of GaAs, i.e., its nonlinear optical properties. We can simulate the process by incorporating the rate equations [4].

Fig. 5.2 Phase mismatch for second-harmonic generation

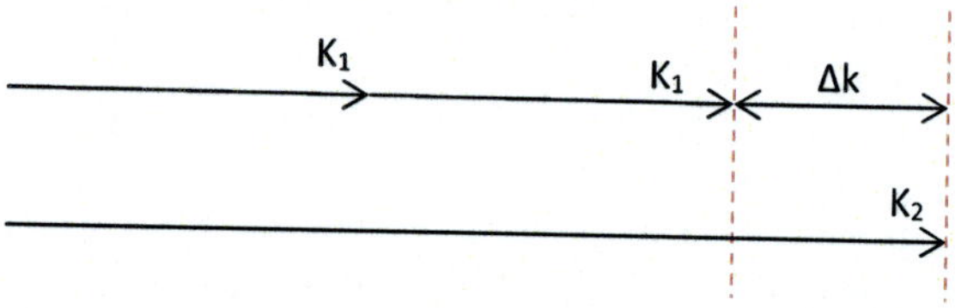

Wave-vector Mismatch (Δk): Wave-vector mismatch value is related to phase matching which is a group of techniques for achieving efficient nonlinear interactions in a medium. Many phase-sensitive nonlinear processes, for example parametric processes such as frequency doubling, sum and different frequency generation, parametric amplification and oscillation and also four-wave mixing, require phase matching to be efficient. The means ensuring that a proper phase relationship between the interacting waves is maintained along the propagation direction. Only if that condition is fulfilled, amplitude contributions from different locations to the product wave are all in phase at the end of the nonlinear crystal. Some phase-mismatch should be close to zero in order to obtain an effective nonlinear interaction. For type I phase matching of frequency doubling with collinear beams the phase mismatch is given by $\Delta k = k_2 - k_1$, where k_1 and k_2 are the wavenumbers of the fundamental and second-harmonic beam, respectively. Without chromatic dispersion, $k_2 = 2 \cdot k_1$ would hold and the phase mismatch is vanish ($k_2 = 2 \cdot k_1 \Rightarrow \Delta k = 0$). However, dispersion generally causes a non-zero phase mismatch if no special measures are taken to avoid this. The next figure demonstrated the phase mismatch for second-harmonic generation (Fig. 5.2). Due to chromatic dispersion, the wave number of the second harmonic is more than twice as large as that for the fundamental wave. This can be avoided by choosing a different polarization in a birefringent crystal (birefringent phase matching).

Δk is the wave-vector mismatch between the fundamental and harmonic wave. The wave-vector mismatch, Δk, occurs because of the natural dispersion in the refractive index that is present in all materials. A plane wave at the fundamental frequency co propagates with the phase velocity $\frac{c}{n(\omega)}$ through a nonlinear medium of length L, where $n(\omega)$ is the frequency-dependent index. A polarization wave proportional to $\chi_2 \cdot E(\omega)$ is thus generated with twice the temporal and spatial frequency of the fundamental. This polarization wave serves as an oscillatory current density that generates the electromagnetic wave at frequency $2 \cdot \omega$, propagating with phase velocity$\frac{c}{n(\omega)}$. Quantum mechanically, second-harmonic generation (SHG) involves the destruction of two photons at the fundamental frequency co and generation of one photon at the second harmonic $2 \cdot \omega$. The wave vector mismatch Δk is thus given by $\Delta k = k_{2\omega} - 2 \cdot k_{\omega}$ then $\Delta k = 4 \cdot \pi \cdot n \cdot 2 \cdot \omega - \frac{n \cdot \omega}{\lambda}$. In second-harmonic generation (SHG), the electric field of an incident light wave at frequency co interacts with a nonlinear optical medium, giving rise to an output at twice the frequency of the incident radiation (Fig. 5.3).

For the wave-vector mismatch $\Delta k \neq 0$, the generated harmonic wave gradually gets out of phase with the driving polarization as it propagates through the nonlinear

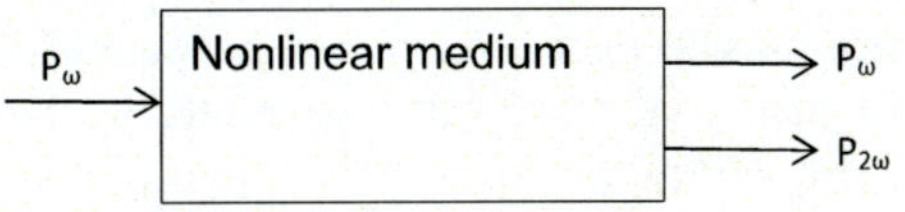

Fig. 5.3 Nonlinear medium's output (output at twice the frequency of the incident radiation)

medium. The relative phase of the interacting waves varies along the medium, and the direction of power flow oscillates, initially generating the second harmonic but then converting the second harmonic back into fundamental radiation after the relative phase reaches n. In this case, maximum conversion to the harmonic is obtained for a nonlinear medium whose length L is an odd multiple of the coherence length $L_c = \frac{\pi}{\Delta k}$. If $\Delta k = 0, n_\omega = n_{2\omega}$ the phase velocities of the fundamental and second-harmonic waves are equal and the relative phase of the driving polarization and generated field remains constant throughout the nonlinear medium. The interaction is then said to be phase-match and the harmonic power grows as the square of the length.

Affective second-order nonlinear coefficient (d): A wide range of nonlinear coefficients and various systems of notation have been used to describe optical second-harmonic generation (SHG). In laser radiation and nonlinear optical materials, accurate determination of second-order nonlinear optical coefficient is both possible and necessary with the higher optical quality that has become available. High power lasers are irregular in temporal and spatial distributions. These lasers irregular nature made absolute measurements of nonlinear optical coefficient difficult. The uncertainty in the nonlinear optical coefficients has made it difficult to perform accurate engineering calculations for nonlinear optical frequency conversion process. We can measure the nonlinear optical coefficients by the technique of phase-matched harmonic generation. Phase matched harmonic generation has an advantage of being performed under conditions that are closer to the conditions of practical applications than other techniques. Our system includes dual AR-coated KTP crystal and we discuss the value of the nonlinear coefficient of KDP. Potassium di-hydrogen phosphate (KDP) is a nonlinear crystal. The nonlinear coefficient of KDP crystal has a range of values. The values fall into two groups. The first group with lowers values has absolute scaled derived from phase-matched second-harmonic generation (SHG) with either direct measurements of KDP. The second group with higher values of nonlinear coefficients is based on an absolute scale obtained by parametric florescence measurements on $LiIO_3$. Phase-matched second-harmonic generation provide absolute or relative measurement of the effective nonlinear coefficient, and precisely controlled phase matching is required. There is a wide range of notation in nonlinear optics, and different definitions of the nonlinear optical coefficient. There is a relationship between the monochromatic plan wave treatment of harmonic generation and the focused Gaussian beam analysis. The electric field and electric polarization (real values) are expressed as products of complex amplitudes and exponentials summed with the complex conjugates of those products. The time independent electric field of a monochromatic plan wave of regular frequency ω is $E(r, t) = \frac{1}{2} \cdot E(r, \omega) \cdot e^{\{i \cdot (k \cdot r - \omega \cdot t)\}} + const$, E, r and k are vector quantities. The relationship of the vector components of the electric polarization at the harmonic frequency $\wp_i^{NL}(r, 2\cdot\omega)$ is generated by the components of the fundamental electric field, $\wp_i^{NL}(r, 2\cdot\omega) = \sum_{j,k=1}^{3} \varepsilon_0 \cdot d_{ijk}(-2\cdot\omega, \omega, \omega) \cdot E_j(r, \omega) \cdot E_k(r, \omega)$, where ε_0 is the permittivity of free space, define the second order nonlinear optical coefficients$d_{jkl}(-2\cdot\omega, \omega, \omega)$. We use the reduced notation $d_{ijk} \rightarrow d_{im}$ which allows the representation of the nonlinear optical coefficients in the customary 3×6 matrix. We reduce it to a single effective nonlinear coefficient d_{eff}, (d), dependent on nonlinear optical properties, phase matching, and crystal orientation is standard for modeling phase-matched second-harmonic generation (SHG). We need to take care on the birefringent walk off angle ρ in the calculation of effective nonlinear coefficient; with this notation the coupled equations describing harmonic generation for a monochromatic plan wave propagating in the z direction are as below.

$$\frac{d}{dt}E(z, 2\cdot\omega) = i\cdot\kappa\cdot e^{-i\cdot\Delta k\cdot z}\cdot E^2(z, \omega);\ \frac{d}{dt}E(z, \omega) = i\cdot\kappa\cdot e^{i\cdot\Delta k\cdot z}\cdot E(z, 2\cdot\omega)\cdot E^*(z, \omega)$$

where the wave vector mismatch is given by $\Delta k = k_{2\cdot\omega} - 2\cdot k_\omega$ and $\kappa = \frac{\omega\cdot d_{eff}}{n\cdot c} \overset{d=d_{eff}}{=} \frac{\omega\cdot d}{n\cdot c}$ with n the index of refraction, c the speed of light, and d_{eff}, (d), the effective nonlinear optical coefficient (in our case, $d = 3.18\,\frac{\text{pm}}{\text{V}}$). Phase matching is of critical importance in second-harmonic measurements of nonlinear coefficients. The crystal must be adjusted for optimum phase matching and phase matching must be uniform through the crystal. Crystal orientation could be confirmed with polarized alignment beams, and crystal rotation confirmed that d_{eff}, (d), was maximum.

The above equation ($\xi(\phi^*) = 0$) is a cubic equation, and we solve it numerically. The next table (Table 5.1) gives the typical values for experimental system's parameters (Four-level rate-equation model).

Remark 1 MKS is the system of units based on measuring lengths in meters, mass in kilograms, and time in seconds. MKS is generally used in engineering and beginning physics, where the so-called CGS system (based on the centimeter, gram, and second) is commonly used in theoretic physics ($\text{cm}^{-3} = 10^6\times\text{m}^{-3}$; $\text{cm}^{-2} = 10^4\times\text{m}^{-2}$; $\text{cm}^{-1} = 10^2\times\text{m}^{-1}$, $\text{cm}^3 = 10^{-6}\times\text{m}^3$; $\text{cm}^2 = 10^{-4}\times\text{m}^2$; $\text{cm} = 10^{-2}\times\text{m}$).

Remark 1.1 Effective nonlinear coefficient (d) is a coefficient for quantifying the strength of a nonlinear interaction. There are various kinds of optical nonlinearities, the strength of which can depend both on material properties and various operation conditions. Often, we use a kind of effective nonlinear coefficient to quantify such strength.

Kerr Nonlinearity in an Optical Fiber: One of the effects of the Kerr nonlinearity in an optical fiber is self-phase modulation, i.e., phase changes which are proportional to the optical intensity. In case of a single-mode fiber, the transverse intensity profile can usually be assumed to be determined by the refractive index profile of the fiber—with negligible influences of nonlinearities. The guided fiber mode has a certain

Table 5.1 Intra-cavity frequency-doubling of a Nd:YAG laser passively Q-switched with GaAs system's parameters (four-level rate-equation model)

Four-level rate-equation model parameter	Meaning	Typical value	Typical value (MKS unit)
σ	Stimulated emission cross-section	2.8×10^{-19} cm^2	2.8×10^{-23} m^2
N_{tot}	Total density of active atoms	1.5×10^{20} cm^{-3}	1.5×10^{26} m^{-3}
τ	Spontaneous fluorescence lifetime	230 μs	230×10^{-6} s
γ	Cavity round trip loss coefficient	0.002	0.002
L_0	Length of the laser rod	1 cm	10^{-2} m
L_c	Length of the doubling crystal	0.5 cm	0.5×10^{-2} m
L_q	GaAs wafer thickness	450 μm	450×10^{-6} m
T	Cavity round trip time	2 ns	2×10^{-9} s
P	Pump rate (Typical 58 s^{-1})	41–75 s^{-1}	41–75 s^{-1}
η	Plane wave impedance $\frac{377}{n_c}$	217 Ω (ohm)	217 Ω (ohm)
n_c	Refractive index of the doubling crystal	1.74	1.74
n_q	Refractive index of GaAs	3.48	3.48
d	Effective second-order nonlinear coefficient (pm = Pico metre, pm = 10^{-12} m)	$3.18 \frac{\text{pm}}{\text{V}}$	3.18×10^{-12} m/V
w_0	Spot size at the laser rod	200 μm	200×10^{-6} m
w_c	Spot size at the doubling crystal (Typical 150 μm)	100–200 μm	150×10^{-6} m
w_q	Spot size at GaAs wafer	180 μm	180×10^{-6} m
N	Total EL2 density	1.2×10^{16} cm^{-3}	1.2×10^{22} m^{-3}

(continued)

Table 5.1 (continued)

Four-level rate-equation model parameter	Meaning	Typical value	Typical value (MKS unit)
N_0^+	Initial ionized EL2 ($EL2^+$) density	1.4×10^{15} cm^{-3}	1.4×77710^{21} m^{-3}
γ_{et}	Electron-EL2 recombination coefficient	1.9×10^{-8} $\frac{cm^{-3}}{s}$	1.9×10^{-2} m^{-3}/s
γ_{hd}	Hole-EL2 recombination coefficient	3.4×10^{-11} $\frac{cm^{-3}}{s}$	3.4×10^{-5} m^{-3}/s
γ_{eh}	Direct electron-EL2 recombination coefficient	2.0×10^{-10} $\frac{cm^{-3}}{s}$	2.0×10^{-4} m^{-3}/s
σ_e	EL2 absorption cross section	1.0×10^{-16} cm^2	1.0×10^{-20} m^2
σ_h	$EL2^+$ absorption cross section	2.3×10^{-17} cm^2	2.3×10^{-21} m^2
σ_{fc}	Free-carrier absorption cross section	6.0×10^{-18} cm^2	6.0×10^{-22} m^2
β	Two-photon absorption coefficient	0.026 $\frac{cm}{MW}$	0.026×10^{-8} $\frac{m}{W}$
α_q	Saturable absorption coefficient of the GaAs wafer ($\alpha_q \approx 1.15$ cm^{-1}), see assumptions	1.1–1.2 cm^{-1}	115 m^{-1}
Δk	Wave-vector mismatch, we choose the case of no chromatic dispersion$k_2 = 2 \cdot k_1 \Rightarrow \Delta k = 0$. Phase matching of frequency doubling with collinear beams the phase mismatch $\Delta k = 0$, $n_\omega = n_{2\omega}$	0	0

effective mode area A_{eff} which is defined such that the nonlinear phase change for light within some length L of a fiber is $\varphi_{nl} = \frac{2\cdot\pi}{\lambda} \cdot n_2 \cdot \frac{P}{A_{eff}} \cdot L$ containing the nonlinear indexn_2. We can introduce an effective nonlinear coefficient, which is equal to $\frac{2\cdot\pi}{\lambda} \cdot \frac{n_2}{A_{eff}}$. Such a nonlinear coefficient (SPM coefficient) occurs e.g. in the nonlinear Schrödinger equation for the evolution of ultrashort pulses in a fiber. Its units are rad / (W m) (radians per Watt and meter). If an ultrashort pulse propagates through the fiber, the total nonlinear phase shift can be simply calculated, as long as the pulse duration is not substantially changed e.g. by the chromatic dispersion of the fiber. Otherwise, something like the nonlinear Schrödinger equation needs to be solved for that purpose.

Crystals with $\chi^{(3)}$nonlinearity: Various nonlinear crystal materials are used for nonlinear frequency conversion processes such as frequency doubling, parametric amplification or optical rectification. For calculating quantities like the output power or the power conversion efficiency, one frequently uses equations containing an effective nonlinear coefficient d_{eff} with units of pm/V. The magnitude of that coefficient depends both on material properties and on the polarization properties of the interacting light beams. In cases with noncritical phase matching, all involved light beams essentially propagate along one axis of the crystal, and the involved beams are typically linearly polarized either along another crystal axis or with an angle of 45° against such an axis. The effective nonlinearity can then relatively simply be calculated from the nonlinear tensor of the material. For example, for non-critically phase-matched frequency doubling in lithium niobate ($LiNbO_3$), one usually has light propagation in X direction, with the pump light being polarized in Y direction and the harmonic light polarized in Z direction. The nonlinear polarization is as follow:

$$\begin{pmatrix} P_X \\ P_Y \\ P_Z \end{pmatrix} = \begin{pmatrix} 0 & 0 & 0 & 0 & d_{31} & -d_{22} \\ -d_{22} & d_{22} & 0 & d_{31} & 0 & 0 \\ d_{31} & d_{31} & d_{33} & 0 & 0 & 0 \end{pmatrix} \cdot \begin{pmatrix} E_X^2 \\ E_Y^2 \\ E_Z^2 \\ 2 \cdot E_Y \cdot E_Z \\ 2 \cdot E_Z \cdot E_X \\ 2 \cdot E_X \cdot E_Y \end{pmatrix}$$

With various tensor coefficients d_{jk}, but most of these tensor coefficients are irrelevant since the pump field has only got a component E_Y while the harmonic field has only a component E_Z, which can interact only with $P_Z(P_Z = d_{31} \cdot E_Y^2)$ and find that the effective nonlinear coefficient is $d_{eff} = d_{31}$. The same result would be obtained for light propagation in Y direction, where the pump light needs to be polarized in X direction. For critical phase matching, where the beam direction is not simply along one of the crystal axes, a given phase-matching configuration, characterized by a phase-matching angle θ or φ, one often requires a formula involving one or more tensor coefficients and the angle and the relative signs of tensor components may then be relevant. The required formula depends on the crystal symmetry, which determines the structure of the nonlinear tensor. In cases with quasi-phase matching, there is another factor $\frac{\pi}{2}$ in the equation for the effective nonlinear coefficient. Its origin is that one does not have perfect phase matching in the material, i.e., periodic phase deviations to one side and the others during propagation, which effectively make the nonlinear infection somewhat weaker. The achieved effective nonlinear coefficient is often substantially higher with quasi-phase matching than with birefringent phase matching, because one can utilize a higher tensor component (e.g. d_{33} instead of d_{31} in $LiNbO_3$).

Assumption: The energy, E, of a photon is determined by its frequency ν, (frequency of the fundamental field), and Planck's constant h, $E = h \cdot \nu$, the velocity of light in vacuum, c is $3 \times 10^8\ \frac{\mathrm{m}}{\mathrm{s}}$. The wavelength, λ, of light is related to ν from the following equation $\lambda = \frac{c}{\nu}$. We choose synchronized actively Q-switched Nd:YAG lasers generating 3 W average power at 589 nm ($\lambda = 589\mathrm{nm}$), then

$$\nu = \frac{c}{\lambda} = \frac{3 \times 10^8\ \frac{\mathrm{m}}{\mathrm{s}}}{589 \times 10^{-9}\,\mathrm{m}} = 0.00509 \times 10^{17}\ \frac{1}{\mathrm{s}}$$
$$= 0.00509 \times 10^{17}\,\mathrm{Hz} = 509\,\mathrm{THz}$$

$$\nu = \frac{c}{\lambda} = \frac{3 \times 10^8\ \frac{\mathrm{m}}{\mathrm{s}}}{589 \times 10^{-9}\,\mathrm{m}} = 0.00509 \times 10^{17}\ \frac{1}{\mathrm{s}}$$
$$= 0.00509 \times 10^{17}\,\mathrm{Hz} = 509\,\mathrm{THz}.$$

Coupling coefficient B: $B = 6 \cdot \beta \cdot h \cdot \nu \cdot c \cdot L_q \cdot \left(\frac{w_0}{w_q}\right)^2 \cdot \frac{1}{n_q}$, by using MATLAB we find the value. ≫ 6∗(0.026∗1e−8)∗(6.626∗1e−34)∗(0.00509∗1e+17)∗(3∗1e+8) ∗(450∗1e−6)∗((200∗1e−6)/(180∗1e−6))^2∗(1/3.48) ans = 2.5198e−23, $B = 2.51 \times 10^{-23}$.

Coupling coefficient K: $K = 32 \cdot \pi^2 \cdot h \cdot (\eta \cdot \nu)^3 \cdot c \cdot (\varepsilon_0 \cdot d \cdot L_c)^2 \cdot \left(\frac{w_0}{w_c}\right)^2 \cdot \mathrm{sinc}^2\left(\frac{\Delta k \cdot L_c}{2}\right)$.

We consider no chromatic dispersion (or very small value, $\Delta k \to 0$), $\Delta k = 0$, then the phase velocities of the fundamental and second harmonic waves are equal and the relative phase of the driving polarization and generated field remains constant throughout the nonlinear medium.

$$\mathrm{sinc}\left(\frac{\Delta k \cdot L_c}{2}\right)|_{\Delta k \to 0} = \lim_{\Delta k \to 0} \frac{\sin\left(\Delta k \cdot \frac{L_c}{2}\right)}{\Delta k \cdot \frac{L_c}{2}}$$
$$= 1 \forall \frac{L_c}{2} \neq 0; \frac{L_c}{2} \in \mathbb{R}; \mathrm{sinc}^2\left(\frac{\Delta k \cdot L_c}{2}\right)|_{\Delta k \to 0} = 1$$

$$\mathrm{sinc}\left(\Delta k \cdot \frac{L_c}{2}\right) = \frac{\sin\left(\Delta k \cdot \frac{L_c}{2}\right)}{\Delta k \cdot \frac{L_c}{2}} = 1 - \frac{\left(\Delta k \cdot \frac{L_c}{2}\right)^2}{3!}$$

$$+\frac{(\Delta k\cdot\frac{L_c}{2})^4}{5!}-\frac{(\Delta k\cdot\frac{L_c}{2})^6}{7!}$$

$$+\cdots=\sum_{n=0}^{\infty}\frac{(-1)^n\cdot(\Delta k\cdot\frac{L_c}{2})^{2\cdot n}}{(2\cdot n+1)!}$$

$$\operatorname{sinc}\left(\Delta k\cdot\frac{L_c}{2}\right)|_{\Delta k\to 0}=\sum_{n=0}^{\infty}\frac{(-1)^n\cdot(\Delta k\cdot\frac{L_c}{2})^{2\cdot n}}{(2\cdot n+1)!}|_{\Delta k\to 0}=1;\ \ K|_{\Delta k\to 0}$$

$$=32\cdot\pi^2\cdot h\cdot(\eta\cdot v)^3\cdot c\cdot(\varepsilon_0\cdot d\cdot L_c)^2\cdot\left(\frac{w_0}{w_c}\right)^2$$

by using MATLAB we find the value.

$$(\xi(\phi^*)=0)$$

, we define some global system parameters: $\xi(\phi^*)=0\Rightarrow\Upsilon_0+\Upsilon_1\cdot\phi^*+\Upsilon_2\cdot(\phi^*)^2=0$

$$\Upsilon_0=2\cdot P\cdot N_{tot}\cdot\sigma\cdot L_0-\gamma\cdot\left[P+\frac{1}{\tau}\right]-2\cdot\alpha_q\cdot L_q\cdot\left[P+\frac{1}{\tau}\right]$$

$$\Upsilon_2=-(K\cdot L_q^2+B\cdot L_c)\cdot 2\cdot\sigma\cdot c$$

$$\Upsilon_1=-\left\{\gamma\cdot 2\cdot\sigma\cdot c+4\cdot\alpha_q\cdot L_q\cdot\sigma\cdot c+(K\cdot L_q^2+B\cdot L_c)\cdot\left[P+\frac{1}{\tau}\right]\right\}$$

$$\Upsilon_2\cdot(\phi^*)^2+\Upsilon_1\cdot\phi^*+\Upsilon_0=0$$

$$\phi^*=\frac{-\Upsilon_1\pm\sqrt{\Upsilon_1^2-4\cdot\Upsilon_0\cdot\Upsilon_2}}{2\cdot\Upsilon_2}\Rightarrow\phi^{(1)}=\frac{-\Upsilon_1+\sqrt{\Upsilon_1^2-4\cdot\Upsilon_0\cdot\Upsilon_2}}{2\cdot\Upsilon_2}$$

$$\phi^{(2)}=\frac{-\Upsilon_1-\sqrt{\Upsilon_1^2-4\cdot\Upsilon_0\cdot\Upsilon_2}}{2\cdot\Upsilon_2}$$

$$N_x^{(1)}=\frac{P\cdot N_{tot}}{P+\frac{1}{\tau}+2\cdot\sigma\cdot c\cdot\left[\frac{-\Upsilon_1+\sqrt{\Upsilon_1^2-4\cdot\Upsilon_0\cdot\Upsilon_2}}{2\cdot\Upsilon_2}\right]}$$

$$N_x^{(2)}=\frac{P\cdot N_{tot}}{P+\frac{1}{\tau}+2\cdot\sigma\cdot c\cdot\left[\frac{-\Upsilon_1-\sqrt{\Upsilon_1^2-4\cdot\Upsilon_0\cdot\Upsilon_2}}{2\cdot\Upsilon_2}\right]}$$

We can summary our system fixed points (equilibrium points) in the next table (Table 5.2).

Typical cw output power without the doubling crystal, GaAs wafer pump power is in the range of 1.65–3.7 W, the gain and the corresponding pump rate P varied

Table 5.2 Nd:YAG laser passively Q-switched with GaAs system fixed points

First fixed point	$E^{(0)}(\phi^{(0)}, N_x^{(0)}) = \left(0, \frac{P \cdot N_{tot}}{P+\frac{1}{\tau}}\right)$
Second fixed point	$E^{(1)}(\phi^{(1)}, N_x^{(1)}) = \left(\frac{-\Upsilon_1 \pm \sqrt{\Upsilon_1^2 - 4 \cdot \Upsilon_0 \cdot \Upsilon_2}}{2 \cdot \Upsilon_2}, \frac{P \cdot N_{tot}}{P + \frac{1}{\tau} + 2 \cdot \sigma \cdot c \cdot \left[\frac{-\Upsilon_1 + \sqrt{\Upsilon_1^2 - 4 \cdot \Upsilon_0 \cdot \Upsilon_2}}{2 \cdot \Upsilon_2}\right]}\right)$
Third fixed point	$E^{(2)}(\phi^{(2)}, N_x^{(2)}) = \left(\frac{-\Upsilon_1 - \sqrt{\Upsilon_1^2 - 4 \cdot \Upsilon_0 \cdot \Upsilon_2}}{2 \cdot \Upsilon_2}, \frac{P \cdot N_{tot}}{P + \frac{1}{\tau} + 2 \cdot \sigma \cdot c \cdot \left[\frac{-\Upsilon_1 - \sqrt{\Upsilon_1^2 - 4 \cdot \Upsilon_0 \cdot \Upsilon_2}}{2 \cdot \Upsilon_2}\right]}\right)$

between 0.4 cm^{-1} and 0.73 cm^{-1} and 41 s^{-1} and 75 s^{-1} (typically value 58 s^{-1}), respectively. We consider typical value of pump rate P, 58 s^{-1} in our system [5].

$$\Upsilon_0 = 2 \cdot P \cdot N_{tot} \cdot \sigma \cdot L_0 - \gamma \cdot \left[P + \frac{1}{\tau}\right] - 2 \cdot \alpha_q \cdot L_q \cdot \left[P + \frac{1}{\tau}\right]$$
$$\Upsilon_0 = 4.8255 \times 10^4$$

$$\Upsilon_1 = -\left\{\gamma \cdot 2 \cdot \sigma \cdot c + 4 \cdot \alpha_q \cdot L_q \cdot \sigma \cdot c + (K \cdot L_q^2 + B \cdot L_c) \cdot \left[P + \frac{1}{\tau}\right]\right\}$$
$$\Upsilon_1 = -1.7724 \times 10^{-15}$$

$$\Upsilon_2 = -(K \cdot L_q^2 + B \cdot L_c) \cdot 2 \cdot \sigma \cdot c; \; \Upsilon_2 = -2.1185 \times 10^{-39}$$

$$\Upsilon_2 \cdot (\phi^*)^2 + \Upsilon_1 \cdot \phi^* + \Upsilon_0 = 0 \Rightarrow \phi^{(1)} = -8.3666 \times 10^{23}; \; \phi^{(2)} = 0.0003 \times 10^{23} = 3 \times 10^{19}$$

MATLAB: a = −2.1185e−39; b = −1.7724e−15; c = 4.8255e4; p = [a b c]; r = roots(p)

r = 1.0e + 23*

− 8.3666

0.0003

Since the photon density fixed point ($\phi^{(*)}$) is a positive real number, the coordinate $\phi^{(1)} = -8.3666 \times 10^{23}$ is only mathematical solution.

We already presented the rate equations for photon density ϕ and population inversion density N_x with delay parameter τ_Δ ($\phi = \phi(t)$; $N_x = N_x(t)$).

$$\frac{d\phi}{dt} = \left[(2 \cdot N_x(t) \cdot \sigma \cdot L_0 - \gamma - 2 \cdot \alpha_q \cdot L_q) \cdot \phi(t - \tau_\Delta) - (K \cdot L_q^2 + B \cdot L_c) \cdot \phi^2(t - \tau_\Delta)\right] \cdot \frac{1}{T}$$

$$\frac{dN_x}{dt} = -2 \cdot N_x(t) \cdot \sigma \cdot c \cdot \phi(t - \tau_\Delta) - \frac{N_x(t)}{\tau} + P \cdot (N_{tot} - N_x(t)))$$

Stability analysis: The standard local stability analysis about any one of the equilibrium points of Nd:YAG laser passively Q-switched with GaAs system consists in adding to its coordinates [ϕ N_x] arbitrarily small increments of exponential terms [$\phi\, n_x$] $\cdot e^{\lambda \cdot t}$, and retaining the first order terms in $\phi\, n_x$. The system of two homogenous equations leads to a polynomial characteristic equation in the eigenvalue λ[6, 7]. The polynomial characteristic equation accepts by set the Nd:YAG laser passively Q-switched with GaAs system equations. The system fixed values with arbitrarily small increments of exponential form [$\phi\, n_x$] $\cdot e^{\lambda \cdot t}$ are; i = 0 (first fixed point), i = 1 (second fixed point), i = 2 (third fixed point), etc.

$$N_x(t) = N_x^{(i)} + n_x \cdot e^{\lambda \cdot t};\ \phi(t) = \phi^{(i)} + \phi \cdot e^{\lambda \cdot t};\ \phi(t - \tau_\Delta) = \phi^{(i)} + \phi \cdot e^{\lambda \cdot (t - \tau_\Delta)}$$

$$\frac{dN_x(t)}{dt} = n_x \cdot \lambda \cdot e^{\lambda \cdot t};\ \frac{d\phi(t)}{dt} = \phi \cdot \lambda \cdot e^{\lambda \cdot t}$$

$$\frac{d\phi}{dt} = [(2 \cdot N_x(t) \cdot \sigma \cdot L_0 - \gamma - 2 \cdot \alpha_q \cdot L_q) \cdot \phi(t - \tau_\Delta) - (K \cdot L_q^2 + B \cdot L_c) \cdot \phi^2(t - \tau_\Delta)] \cdot \frac{1}{T}$$

$$\phi \cdot \lambda \cdot e^{\lambda \cdot t} = \left[(2 \cdot (N_x^{(i)} + n_x \cdot e^{\lambda \cdot t}) \cdot \sigma \cdot L_0 - \gamma - 2 \cdot \alpha_q \cdot L_q) \cdot (\phi^{(i)} + \phi \cdot e^{\lambda \cdot (t - \tau_\Delta)}) - (K \cdot L_q^2 + B \cdot L_c) \cdot (\phi^{(i)} + \phi \cdot e^{\lambda \cdot (t - \tau_\Delta)})^2\right] \cdot \frac{1}{T}$$

$$\phi \cdot \lambda \cdot e^{\lambda \cdot t} = \left[(2 \cdot N_x^{(i)} \cdot \sigma \cdot L_0 + 2 \cdot n_x \cdot \sigma \cdot L_0 \cdot e^{\lambda \cdot t} - \gamma - 2 \cdot \alpha_q \cdot L_q) \cdot (\phi^{(i)} + \phi \cdot e^{\lambda \cdot (t - \tau_\Delta)}) - (K \cdot L_q^2 + B \cdot L_c) \cdot ([\phi^{(i)}]^2 + 2 \cdot \phi^{(i)} \cdot \phi \cdot e^{\lambda \cdot (t - \tau_\Delta)} + \phi^2 \cdot e^{2 \cdot \lambda \cdot (t - \tau_\Delta)})\right] \cdot \frac{1}{T}$$

Assumption $\phi^2 \approx 0$

$$\phi \cdot \lambda \cdot e^{\lambda \cdot t} = \{([2 \cdot N_x^{(i)} \cdot \sigma \cdot L_0 - \gamma - 2 \cdot \alpha_q \cdot L_q] + 2 \cdot n_x \cdot \sigma \cdot L_0 \cdot e^{\lambda \cdot t}) \cdot (\phi^{(i)} + \phi \cdot e^{\lambda \cdot (t - \tau_\Delta)}) - (K \cdot L_q^2 + B \cdot L_c) \cdot ([\phi^{(i)}]^2 + 2 \cdot \phi^{(i)} \cdot \phi \cdot e^{\lambda \cdot (t - \tau_\Delta)})\} \cdot \frac{1}{T}$$

$$\phi \cdot \lambda \cdot e^{\lambda \cdot t} = \{([2 \cdot N_x^{(i)} \cdot \sigma \cdot L_0 - \gamma - 2 \cdot \alpha_q \cdot L_q] + 2 \cdot n_x \cdot \sigma \cdot L_0 \cdot e^{\lambda \cdot t}) \cdot (\phi^{(i)} + \phi \cdot e^{\lambda \cdot (t - \tau_\Delta)}) - (K \cdot L_q^2 + B \cdot L_c) \cdot [\phi^{(i)}]^2 - (K \cdot L_q^2 + B \cdot L_c) \cdot 2 \cdot \phi^{(i)} \cdot \phi \cdot e^{\lambda \cdot (t - \tau_\Delta)}\} \cdot \frac{1}{T}$$

$$\phi \cdot \lambda \cdot e^{\lambda \cdot t} = \{[2 \cdot N_x^{(i)} \cdot \sigma \cdot L_0 - \gamma - 2 \cdot \alpha_q \cdot L_q] \cdot \phi^{(i)} + [2 \cdot N_x^{(i)} \cdot \sigma \cdot L_0 - \gamma - 2 \cdot \alpha_q \cdot L_q] \cdot \phi \cdot e^{\lambda \cdot (t - \tau_\Delta)} + 2 \cdot n_x \cdot \sigma \cdot L_0 \cdot e^{\lambda \cdot t} \cdot \phi^{(i)} + 2 \cdot n_x \cdot \phi \cdot \sigma \cdot L_0 \cdot e^{\lambda \cdot t} \cdot e^{\lambda \cdot (t - \tau_\Delta)} - (K \cdot L_q^2 + B \cdot L_c) \cdot [\phi^{(i)}]^2 - (K \cdot L_q^2 + B \cdot L_c) \cdot 2 \cdot \phi^{(i)} \cdot \phi \cdot e^{\lambda \cdot (t - \tau_\Delta)}\} \cdot \frac{1}{T}$$

$$\phi \cdot \lambda \cdot e^{\lambda \cdot t} = [2 \cdot N_x^{(i)} \cdot \sigma \cdot L_0 - \gamma - 2 \cdot \alpha_q \cdot L_q] \cdot \phi^{(i)} \cdot \frac{1}{T} + [2 \cdot N_x^{(i)} \cdot \sigma \cdot L_0 - \gamma - 2 \cdot \alpha_q \cdot L_q] \cdot \phi \cdot e^{\lambda \cdot (t - \tau_\Delta)} \cdot \frac{1}{T} + 2 \cdot n_x \cdot \sigma \cdot L_0 \cdot e^{\lambda \cdot t} \cdot \phi^{(i)} \cdot \frac{1}{T} + 2 \cdot n_x \cdot \phi \cdot \sigma \cdot L_0 \cdot e^{\lambda \cdot t} \cdot e^{\lambda \cdot (t - \tau_\Delta)} \cdot \frac{1}{T} - (K \cdot L_q^2 + B \cdot L_c) \cdot [\phi^{(i)}]^2 \cdot \frac{1}{T} - (K \cdot L_q^2 + B \cdot L_c) \cdot 2 \cdot \phi^{(i)} \cdot \phi \cdot e^{\lambda \cdot (t - \tau_\Delta)} \cdot \frac{1}{T}$$

Assumption $n_x \cdot \phi \approx 0$

$$\phi \cdot \lambda \cdot e^{\lambda \cdot t} = \{[2 \cdot N_x^{(i)} \cdot \sigma \cdot L_0 - \gamma - 2 \cdot \alpha_q \cdot L_q] \cdot \phi^{(i)} - (K \cdot L_q^2 + B \cdot L_c) \cdot [\phi^{(i)}]^2\} \cdot \frac{1}{T} + [2 \cdot N_x^{(i)} \cdot \sigma \cdot L_0 - \gamma - 2 \cdot \alpha_q \cdot L_q] \cdot \phi \cdot e^{\lambda \cdot (t - \tau_\Delta)} \cdot \frac{1}{T} + 2 \cdot n_x \cdot \sigma \cdot L_0 \cdot e^{\lambda \cdot t} \cdot \phi^{(i)} \cdot \frac{1}{T} - (K \cdot L_q^2 + B \cdot L_c) \cdot 2 \cdot \phi^{(i)} \cdot \phi \cdot e^{\lambda \cdot (t - \tau_\Delta)} \cdot \frac{1}{T}$$

At fixed points:
$$\{[2 \cdot N_x^{(i)} \cdot \sigma \cdot L_0 - \gamma - 2 \cdot \alpha_q \cdot L_q] \cdot \phi^{(i)} - (K \cdot L_q^2 + B \cdot L_c) \cdot [\phi^{(i)}]^2\} \cdot \frac{1}{T} = 0$$

$$\phi \cdot \lambda \cdot e^{\lambda \cdot t} = [2 \cdot N_x^{(i)} \cdot \sigma \cdot L_0 - \gamma - 2 \cdot \alpha_q \cdot L_q] \cdot \phi \cdot e^{\lambda \cdot t} \cdot e^{-\lambda \cdot \tau_\Delta} \cdot \frac{1}{T} + 2 \cdot n_x \cdot \sigma \cdot L_0 \cdot e^{\lambda \cdot t} \cdot \phi^{(i)} \cdot \frac{1}{T} - (K \cdot L_q^2 + B \cdot L_c) \cdot 2 \cdot \phi^{(i)} \cdot \phi \cdot e^{\lambda \cdot t} \cdot e^{-\lambda \cdot \tau_\Delta} \cdot \frac{1}{T}$$

Dividing the above equation by $e^{\lambda \cdot t}$ term, gives:

$$\{[2 \cdot N_x^{(i)} \cdot \sigma \cdot L_0 - \gamma - 2 \cdot \alpha_q \cdot L_q] - (K \cdot L_q^2 + B \cdot L_c) \cdot 2 \cdot \phi^{(i)}\} \cdot \frac{1}{T} \cdot \phi \cdot e^{-\lambda \cdot \tau_\Delta} - \phi \cdot \lambda + 2 \cdot \sigma \cdot L_0 \cdot \phi^{(i)} \cdot \frac{1}{T} \cdot n_x = 0$$

$$(\{[2 \cdot N_x^{(i)} \cdot \sigma \cdot L_0 - \gamma - 2 \cdot \alpha_q \cdot L_q] - (K \cdot L_q^2 + B \cdot L_c) \cdot 2 \cdot \phi^{(i)}\} \cdot \frac{1}{T} \cdot e^{-\lambda \cdot \tau_\Delta} - \lambda) \cdot \phi + 2 \cdot \frac{1}{T} \cdot \sigma \cdot L_0 \cdot \phi^{(i)} \cdot n_x = 0$$

$$\frac{dN_x}{dt} = -2 \cdot N_x(t) \cdot \sigma \cdot c \cdot \phi(t - \tau_\Delta) - \frac{N_x(t)}{\tau} + P \cdot (N_{tot} - N_x(t))$$

$$n_x \cdot \lambda \cdot e^{\lambda \cdot t} = -2 \cdot (N_x^{(i)} + n_x \cdot e^{\lambda \cdot t}) \cdot \sigma \cdot c \cdot (\phi^{(i)} + \phi \cdot e^{\lambda \cdot (t - \tau_\Delta)}) - \frac{(N_x^{(i)} + n_x \cdot e^{\lambda \cdot t})}{\tau} + P \cdot (N_{tot} - [N_x^{(i)} + n_x \cdot e^{\lambda \cdot t}])$$

$$n_x \cdot \lambda \cdot e^{\lambda \cdot t} = -2 \cdot \sigma \cdot c \cdot (N_x^{(i)} \cdot \phi^{(i)} + N_x^{(i)} \cdot \phi \cdot e^{\lambda \cdot (t - \tau_\Delta)} + \phi^{(i)} \cdot n_x \cdot e^{\lambda \cdot t} + n_x \cdot \phi \cdot e^{\lambda \cdot t} \cdot e^{\lambda \cdot (t - \tau_\Delta)}) - \frac{N_x^{(i)}}{\tau} - \frac{n_x \cdot e^{\lambda \cdot t}}{\tau} + P \cdot N_{tot} - P \cdot N_x^{(i)} - P \cdot n_x \cdot e^{\lambda \cdot t}$$

Assumption $n_x \cdot \phi \approx 0$

$$n_x \cdot \lambda \cdot e^{\lambda \cdot t} = -2 \cdot \sigma \cdot c \cdot (N_x^{(i)} \cdot \phi^{(i)} + N_x^{(i)} \cdot \phi \cdot e^{\lambda \cdot (t - \tau_\Delta)} + \phi^{(i)} \cdot n_x \cdot e^{\lambda \cdot t}) - \frac{N_x^{(i)}}{\tau} - \frac{n_x \cdot e^{\lambda \cdot t}}{\tau}$$

$$+ P \cdot N_{tot} - P \cdot N_x^{(i)} - P \cdot n_x \cdot e^{\lambda \cdot t}$$

$$\begin{aligned} n_x \cdot \lambda \cdot e^{\lambda \cdot t} &= -2 \cdot \sigma \cdot c \cdot N_x^{(i)} \cdot \phi^{(i)} - \frac{N_x^{(i)}}{\tau} + P \cdot (N_{tot} - N_x^{(i)}) \\ &\quad - 2 \cdot \sigma \cdot c \cdot N_x^{(i)} \cdot \phi \cdot e^{\lambda \cdot t} \cdot e^{-\lambda \cdot \tau_\Delta} \\ &\quad - 2 \cdot \sigma \cdot c \cdot \phi^{(i)} \cdot n_x \cdot e^{\lambda \cdot t} - \frac{n_x \cdot e^{\lambda \cdot t}}{\tau} - P \cdot n_x \cdot e^{\lambda \cdot t} \end{aligned}$$

At fixed points: $\frac{dN_x}{dt} = 0 \Rightarrow -2 \cdot \sigma \cdot c \cdot N_x^{(i)} \cdot \phi^{(i)} - \frac{N_x^{(i)}}{\tau} + P \cdot (N_{tot} - N_x^{(i)}) = 0.$

$$t \gg \tau_\Delta \Rightarrow \lim_{t \to \infty} \phi(t - \tau_\Delta) = \phi(t)$$

$$n_x \cdot \lambda = -2 \cdot \sigma \cdot c \cdot N_x^{(i)} \cdot \phi \cdot e^{-\lambda \cdot \tau_\Delta} - 2 \cdot \sigma \cdot c \cdot \phi^{(i)} \cdot n_x - \frac{n_x}{\tau} - P \cdot n_x$$

$$-2 \cdot \sigma \cdot c \cdot N_x^{(i)} \cdot \phi \cdot e^{-\lambda \cdot \tau_\Delta} - n_x \cdot \left(2 \cdot \sigma \cdot c \cdot \phi^{(i)} + \frac{1}{\tau} + P\right) - n_x \cdot \lambda = 0$$

We can summary our Nd: YAG laser passively Q-switched with GaAs system arbitrarily small increments equations:

$$\begin{aligned} &(\{[2 \cdot N_x^{(i)} \cdot \sigma \cdot L_0 - \gamma - 2 \cdot \alpha_q \cdot L_q] \\ &- (K \cdot L_q^2 + B \cdot L_c) \cdot 2 \cdot \phi^{(i)}\} \cdot \frac{1}{T} \cdot e^{-\lambda \cdot \tau_\Delta} - \lambda) \cdot \phi \\ &+ 2 \cdot \frac{1}{T} \cdot \sigma \cdot L_0 \cdot \phi^{(i)} \cdot n_x = 0 \end{aligned}$$

$$-2 \cdot \sigma \cdot c \cdot N_x^{(i)} \cdot \phi \cdot e^{-\lambda \cdot \tau_\Delta} - n_x \cdot \left(2 \cdot \sigma \cdot c \cdot \phi^{(i)} + \frac{1}{\tau} + P\right) - n_x \cdot \lambda = 0$$

We define for simplicity some functions of system parameters:

$$\begin{aligned} &g_1(N_x^{(i)}, \sigma, L_0, \gamma, \alpha_q, L_q, K, B, L_c, T) \\ &\quad = \left\{[2 \cdot N_x^{(i)} \cdot \sigma \cdot L_0 - \gamma - 2 \cdot \alpha_q \cdot L_q] \right. \\ &\quad \left. -(K \cdot L_q^2 + B \cdot L_c) \cdot 2 \cdot \phi^{(i)}\right\} \cdot \frac{1}{T} \end{aligned}$$

$$g_2(T, \sigma, L_0) = 2 \cdot \frac{1}{T} \cdot \sigma \cdot L_0 \cdot \phi^{(i)}; \; g_3(\sigma, c) = 2 \cdot \sigma \cdot c \cdot N_x^{(i)}$$

$$g_4(\sigma, c, \tau, P) = 2 \cdot \sigma \cdot c \cdot \phi^{(i)} + \frac{1}{\tau} + P$$

$$g_1 = g_1(N_x^{(i)}, \sigma, L_0, \gamma, \alpha_q, L_q, K, B, L_c, T);\ g_2 = g_2(T, \sigma, L_0)$$
$$g_3 = g_3(\sigma, c);\ g_4 = g_4(\sigma, c, \tau, P)$$

System arbitrarily small increments equations are:

$$(g_1 \cdot e^{-\lambda \cdot \tau_\Delta} - \lambda) \cdot \phi + g_2 \cdot n_x = 0;\ -g_3 \cdot \phi \cdot e^{-\lambda \cdot \tau_\Delta} - n_x \cdot g_4 - n_x \cdot \lambda = 0$$

Our eigenvalue matrix is $\begin{pmatrix} g_1 \cdot e^{-\lambda \cdot \tau_\Delta} - \lambda & g_2 \\ -g_3 \cdot e^{-\lambda \cdot \tau_\Delta} & -g_4 - \lambda \end{pmatrix} \begin{pmatrix} \phi \\ n_x \end{pmatrix} = \begin{pmatrix} 0 \\ 0 \end{pmatrix}$; $(A - \lambda \cdot I) =$
$\begin{pmatrix} g_1 \cdot e^{-\lambda \cdot \tau_\Delta} - \lambda & g_2 \\ -g_3 \cdot e^{-\lambda \cdot \tau_\Delta} & -g_4 - \lambda \end{pmatrix}$

$$\det(A - \lambda \cdot I) = 0 \Rightarrow \lambda^2 + g_4 \cdot \lambda + ([g_2 \cdot g_3 - g_1 \cdot g_4] - g_1 \cdot \lambda) \cdot e^{-\lambda \cdot \tau_\Delta} = 0$$

$$D(\lambda, \tau_\Delta) = \lambda^2 + g_4 \cdot \lambda + (-g_1 \cdot \lambda + [g_2 \cdot g_3 - g_1 \cdot g_4]) \cdot e^{-\lambda \cdot \tau_\Delta}$$

$$D(\lambda, \tau_\Delta) = P_n(\lambda, \tau_\Delta) + Q_m(\lambda, \tau_\Delta) \cdot e^{-\lambda \cdot \tau_\Delta};\ n = 2;\ m = 1;\ n > m$$

The expression for $P_n(\lambda, \tau_\Delta)$:$P_{n=2}(\lambda, \tau_\Delta) = \lambda^2 + g_4 \cdot \lambda$; $p_0(\tau_\Delta) = 0$; $p_1(\tau_\Delta) = g_4$; $p_2(\tau_\Delta) = 1$

$$P_{n=2}(\lambda, \tau_\Delta) = \sum_{k=0}^{n=2} p_k(\tau_\Delta) \cdot \lambda^k = p_0(\tau_\Delta) + p_1(\tau_\Delta) \cdot \lambda + p_2(\tau_\Delta) \cdot \lambda^2$$

The expression for $Q_m(\lambda, \tau_\Delta)$: $Q_{m=1}(\lambda, \tau_\Delta) = -g_1 \cdot \lambda + [g_2 \cdot g_3 - g_1 \cdot g_4]$

$$Q_{m=1}(\lambda, \tau_\Delta) = -g_1 \cdot \lambda + [g_2 \cdot g_3 - g_1 \cdot g_4];\ q_0(\tau_\Delta) = g_2 \cdot g_3 - g_1 \cdot g_4;\ q_1(\tau_\Delta) = -g_1$$

$$Q_{m=1}(\lambda, \tau_\Delta) = \sum_{k=0}^{m=1} q_k(\tau_\Delta) \cdot \lambda^k = q_0(\tau_\Delta) + q_1(\tau_\Delta) \cdot \lambda$$

The homogenous system for ϕ, N_x lead to characteristic equation for the eigenvalue λ having the form $P(\lambda, \tau_\Delta) + Q(\lambda, \tau_\Delta) \cdot e^{-\lambda \cdot \tau_\Delta} = 0$; $P(\lambda, \tau_\Delta) = \sum_{j=0}^{2} a_j(\tau_\Delta) \cdot \lambda^j$; $Q(\lambda, \tau_\Delta) = \sum_{j=0}^{1} c_j(\tau_\Delta) \cdot \lambda^j$ and the coefficients $\{a_j(q_i, q_k, \tau_\Delta), c_j(q_i, q_k, \tau_\Delta)\} \in \mathbb{R}$ depend on q_i, q_k and delay τ_Δ, q_i, q_k are any Nd:YAG laser passively Q-switched with GaAs system parameters, other parameters keep constant.

$$a_0 = 0; a_1 = g_4 = 2 \cdot \sigma \cdot c \cdot \phi^{(i)} + \frac{1}{\tau} + P; a_2 = 1$$

$$c_1 = -g_1 = -\{[2 \cdot N_x^{(i)} \cdot \sigma \cdot L_0 - \gamma - 2 \cdot \alpha_q \cdot L_q] - (K \cdot L_q^2 + B \cdot L_c) \cdot 2 \cdot \phi^{(i)}\} \cdot \frac{1}{T}$$

$$c_0 = g_2 \cdot g_3 - g_1 \cdot g_4 = 4 \cdot \frac{1}{T} \cdot \sigma^2 \cdot L_0 \cdot \phi^{(i)} \cdot c \cdot N_x^{(i)} - \{[2 \cdot N_x^{(i)} \cdot \sigma \cdot L_0 - \gamma - 2 \cdot \alpha_q \cdot L_q] - (K \cdot L_q^2 + B \cdot L_c) \cdot 2 \cdot \phi^{(i)}\} \cdot \frac{1}{T} \cdot \left(2 \cdot \sigma \cdot c \cdot \phi^{(i)} + \frac{1}{\tau} + P\right)$$

Unless strictly necessary, the designation of the variation arguments (q_i, q_k) will subsequently by omitted from P, Q, a_j and c_j [8, 9]. The coefficients a_j, c_j are continuous and differentiable functions of their arguments, and direct substitution shows that $a_0 + c_0 \neq 0$.

$$4 \cdot \frac{1}{T} \cdot \sigma^2 \cdot L_0 \cdot \phi^{(i)} \cdot c \cdot N_x^{(i)} - \{[2 \cdot N_x^{(i)} \cdot \sigma \cdot L_0 - \gamma - 2 \cdot \alpha_q \cdot L_q] - (K \cdot L_q^2 + B \cdot L_c) \cdot 2 \cdot \phi^{(i)}\} \cdot \frac{1}{T} \cdot \left(2 \cdot \sigma \cdot c \cdot \phi^{(i)} + \frac{1}{\tau} + P\right) \neq 0 \, \forall \, q_i, q_k \in \mathbb{R}_+$$

$\lambda = 0$, is not a $P(\lambda, \tau_\Delta) + Q(\lambda, \tau_\Delta) \cdot e^{-\lambda \cdot \tau_\Delta} = 0$. We assume that $P_n(\lambda, \tau_\Delta), Q_m(\lambda, \tau_\Delta)$ can't have common imaginary roots. That is for any real number ω: $P_n(\lambda = i \cdot \omega, \tau_\Delta), Q_m(\lambda = i \cdot \omega, \tau_\Delta) \neq 0$.

$$P_n(\lambda = i \cdot \omega, \tau_\Delta) = -\omega^2 + i \cdot g_4 \cdot \omega$$
$$Q_m(\lambda = i \cdot \omega, \tau_\Delta) = -i \cdot g_1 \cdot \omega + [g_2 \cdot g_3 - g_1 \cdot g_4]$$

$$P_n(\lambda = i \cdot \omega, \tau_\Delta) + Q_m(\lambda = i \cdot \omega, \tau_\Delta) = -\omega^2 + (g_2 \cdot g_3 - g_1 \cdot g_4) + (g_4 - g_1) \cdot i \cdot \omega$$

$$|P_n(\lambda = i \cdot \omega, \tau_\Delta)|^2 = \omega^4 + g_4^2 \cdot \omega^2; \; |Q_m(\lambda = i \cdot \omega, \tau_\Delta)|^2 = g_1^2 \cdot \omega^2 + (g_2 \cdot g_3 - g_1 \cdot g_4)^2$$

We find the expression: $F(\omega, \tau_{\Delta}) = |P(i \cdot \omega)|^2 - |Q(i \cdot \omega)|^2$

$$F(\omega, \tau_{\Delta}) = |P(i \cdot \omega)|^2 - |Q(i \cdot \omega)|^2 = \omega^4 + g_4^2 \cdot \omega^2 - g_1^2 \cdot \omega^2 - (g_2 \cdot g_3 - g_1 \cdot g_4)^2$$

$$F(\omega, \tau_{\Delta}) = |P(i \cdot \omega)|^2 - |Q(i \cdot \omega)|^2 = \omega^4 + (g_4^2 - g_1^2) \cdot \omega^2 - (g_2 \cdot g_3 - g_1 \cdot g_4)^2$$

We define the following parameters for simplicity: $\Xi_0 = -(g_2 \cdot g_3 - g_1 \cdot g_4)^2$

$$\Xi_2 = g_4^2 - g_1^2;\ \Xi_4 = 1;\ F(\omega, \tau_{\Delta}) = |P(i \cdot \omega)|^2 - |Q(i \cdot \omega)|^2 = \sum_{k=0}^{2} \Xi_{2 \cdot k} \cdot \omega^{2 \cdot k}$$

$F(\omega, \tau_{\Delta}) = \sum_{k=0}^{2} \Xi_{2 \cdot k} \cdot \omega^{2 \cdot k} = \Xi_0 + \Xi_2 \cdot \omega^2 + \Xi_4 \cdot \omega^4$. Hence $F(\omega, \tau_{\Delta}) = 0$ implies $\sum_{k=0}^{2} \Xi_{2 \cdot k} \cdot \omega^{2 \cdot k} = 0$ and its roots are given by solving the above polynomial. Furthermore, $P_R(\omega, \tau_{\Delta}) = -\omega^2$; $P_I(\omega, \tau_{\Delta}) = g_4 \cdot \omega$ and $Q_R(\omega, \tau_{\Delta}) = g_2 \cdot g_3 - g_1 \cdot g_4$; $Q_I(\omega, \tau_{\Delta}) = -g_1 \cdot \omega$. Hence

$$\sin \theta(\tau_{\Delta}) = \frac{-P_R(i\omega, \tau_{\Delta}) \cdot Q_I(i\omega, \tau_{\Delta}) + P_I(i\omega, \tau_{\Delta}) \cdot Q_R(i\omega, \tau_{\Delta})}{|Q(i\omega, \tau_{\Delta})|^2}$$

$$\cos \theta(\tau_{\Delta}) = -\frac{P_R(i\omega, \tau_{\Delta}) \cdot Q_R(i\omega, \tau_{\Delta}) + P_I(i\omega, \tau_{\Delta}) \cdot Q_I(i\omega, \tau_{\Delta})}{|Q(i\omega, \tau_{\Delta})|^2}$$

$$\sin \theta(\tau_{\Delta}) = \frac{-\omega^3 \cdot g_1 + g_4 \cdot \omega \cdot (g_2 \cdot g_3 - g_1 \cdot g_4)}{g_1^2 \cdot \omega^2 + (g_2 \cdot g_3 - g_1 \cdot g_4)^2}$$

$$\cos \theta(\tau_{\Delta}) = -\frac{-\omega^2 \cdot (g_2 \cdot g_3 - g_1 \cdot g_4) - g_4 \cdot g_1 \cdot \omega^2}{g_1^2 \cdot \omega^2 + (g_2 \cdot g_3 - g_1 \cdot g_4)^2}$$

Above expressions are continuous and differentiable in τ_{Δ} based on Lemma 1.1. Hence we use Theorem 1.2 and this proves Theorem 1.3. We use different parameters terminology from our last characteristics parameter definition $k \to j$; $p_k(\tau_{\Delta}) \to a_j$ and $q_k(\tau_{\Delta}) \to c_j$; $n = 2$; $m = 1$; $n > m$, additionally $P_n(\lambda, \tau_{\Delta}) \to P(\lambda, \tau_{\Delta})$; $Q_m(\lambda, \tau_{\Delta}) \to Q(\lambda, \tau_{\Delta})$ then $P(\lambda, \tau_{\Delta}) = \sum_{j=0}^{2} a_j \cdot \lambda^j$; $Q(\lambda, \tau_{\Delta}) = \sum_{j=0}^{1} c_j \cdot \lambda^j$; $P(\lambda) = \lambda^2 + g_4 \cdot \lambda$; $Q(\lambda) = -g_1 \cdot \lambda + [g_2 \cdot g_3 - g_1 \cdot g_4]$.

$n, m \in \mathbb{N}_0$; $n > m$ and $a_j, c_j : \mathbb{R}_{+0} \to \mathbb{R}$ are continuous and differentiable functions of τ_{Δ} such that $a_0 + c_0 \neq 0$. In the following "−" denotes complex and conjugate. $P(\lambda)$ and $Q(\lambda)$ are analytic functions in λ and differentiable in τ_{Δ}. The coefficients:

$$\{a_j(\sigma, L_0, \gamma, \alpha_q, L_q, K, B, L_c, T, \ldots) \\ \& \; c_j(\sigma, L_0, \gamma, \alpha_q, L_q, K, B, L_c, T, \ldots)\} \in \mathbb{R}$$

Depend on Nd: YAG laser passively Q-switched with GaAs system's parameters $\sigma, L_0, \gamma, \alpha_q, L_q, \tau_\Delta, \ldots$ values. Unless strictly necessary, the designation of the variation arguments, $\sigma, L_0, \gamma, \alpha_q, L_q, \tau_\Delta, \ldots$ system parameters will subsequently be omitted from P, Q, a_j, c_j. The coefficients a_j, c_j are continuous and differentiable functions of their arguments, and direct substitution shows that $a_0 + c_0 \neq 0$.

$$\frac{1}{T} \cdot \{4 \cdot \sigma^2 \cdot L_0 \cdot \phi^{(i)} \cdot c \cdot N_x^{(i)} - \{[2 \cdot N_x^{(i)} \cdot \sigma \cdot L_0 - \gamma - 2 \cdot \alpha_q \cdot L_q] \\ - (K \cdot L_q^2 + B \cdot L_c) \cdot 2 \cdot \phi^{(i)}\} \cdot (2 \cdot \sigma \cdot c \cdot \phi^{(i)} + \frac{1}{\tau} + P)\} \neq 0$$

$$\frac{1}{T} \neq 0; 4 \cdot \sigma^2 \cdot L_0 \cdot \phi^{(i)} \cdot c \cdot N_x^{(i)} - \{[2 \cdot N_x^{(i)} \cdot \sigma \cdot L_0 - \gamma - 2 \cdot \alpha_q \cdot L_q] \\ - (K \cdot L_q^2 + B \cdot L_c) \cdot 2 \cdot \phi^{(i)}\} \cdot \left(2 \cdot \sigma \cdot c \cdot \phi^{(i)} + \frac{1}{\tau} + P\right) \neq 0$$

$$\forall \sigma, L_0, \gamma, \alpha_q, L_q, \ldots \in \mathbb{R}_+$$

i.e. $\lambda = 0$ is not a root of characteristic equation. Furthermore, $P(\lambda)$, $Q(\lambda)$ are analytic function of λ for which the following requirements of the analysis (see Kuang 1993, Sect. 3.4) can also be verified in the present case [9, 10].

(a) If $\lambda = i \cdot \omega; \omega \in \mathbb{R}$ then $P(i \cdot \omega) + Q(i \cdot \omega) \neq 0$, i.e. P and Q have no common imaginary roots. This condition was verified numerically in the entire ($\sigma, L_0, \gamma, \alpha_q, L_q, \ldots$ system parameters) domain of interest.
(b) $\left|\frac{Q(\lambda)}{P(\lambda)}\right|$ is bounded for $|\lambda| \to \infty$; $\text{Re}\lambda \geq 0$. No roots bifurcation from ∞. Indeed, in the limit: $\left|\frac{Q(\lambda)}{P(\lambda)}\right| = \left|\frac{-g_1 \cdot \lambda + [g_2 \cdot g_3 - g_1 \cdot g_4]}{\lambda^2 + g_4 \cdot \lambda}\right|$.
(c) $F(\omega) = |P(i \cdot \omega)|^2 - |Q(i \cdot \omega)|^2 = \omega^4 + (g_4^2 - g_1^2) \cdot \omega^2 - (g_2 \cdot g_3 - g_1 \cdot g_4)^2$ has at most a finite number of zeros. Indeed, this is a polynomial in ω (degree in ω^4).
(d) Each positive root $\omega(\sigma, L_0, \gamma, \alpha_q, L_q, \tau_\Delta, \ldots)$ of $F(\omega) = 0$ is continuous and differentiable with respect to $\sigma, L_0, \gamma, \alpha_q, L_q, \tau_\Delta, \ldots$ parameters. This condition can only be assessed numerically.

In addition, since the coefficients in P and Q are real, we have $\overline{P(-i \cdot \omega)} = P(i \cdot \omega)$ and $\overline{Q(-i \cdot \omega)} = Q(i \cdot \omega)$ thus $\lambda = i \cdot \omega \, \forall \omega > 0$ may be on eigenvalue of characteristic equation. The analysis consists in identifying the roots of characteristic equation situated on the imaginary axis of the complex λ—plane, where by increasing the parameters $\sigma, L_0, \gamma, \alpha_q, L_q, \tau_\Delta, \ldots$ (system parameters), $\text{Re}\lambda$ may, at the crossing, change its sign from (−) to (+), i.e. from a stable focus $E^{(k)}(\phi^{(k)}, N_x^{(k)}) \, \forall k = 0, 1, 2$ to an unstable one, or vice versa. This feature may be further assessed by examining the sign of the partial derivatives with respect to $\sigma, L_0, \gamma, \alpha_q, L_q, \tau_\Delta, \ldots$ parameters.

$$\Lambda^{-1}(\sigma) = \left(\frac{\partial \mathrm{Re}\lambda}{\partial \sigma}\right)_{\lambda=i\cdot\omega}; L_0, \gamma, \alpha_q, L_q, \tau_\Delta, \ldots = const$$

$$\Lambda^{-1}(L_0) = \left(\frac{\partial \mathrm{Re}\lambda}{\partial L_0}\right)_{\lambda=i\cdot\omega}; \sigma, \gamma, \alpha_q, L_q, \tau_\Delta, \ldots = const$$

$$\Lambda^{-1}(\alpha_q) = \left(\frac{\partial \mathrm{Re}\lambda}{\partial \alpha_q}\right)_{\lambda=i\cdot\omega}; \sigma, \gamma, L_0, L_q, \tau_\Delta, \ldots = const$$

$$\Lambda^{-1}(L_q) = \left(\frac{\partial \mathrm{Re}\lambda}{\partial L_q}\right)_{\lambda=i\cdot\omega}; \sigma, \gamma, L_0, \alpha_q, \tau_\Delta, \ldots = const$$

$$\Lambda^{-1}(\tau_\Delta) = \left(\frac{\partial \mathrm{Re}\lambda}{\partial \tau_\Delta}\right)_{\lambda=i\cdot\omega}; \sigma, \gamma, L_0, \alpha_q, L_q, \ldots = const$$

where $\omega \in \mathbb{R}_+$.

Where writing $P(\lambda) = P_R(\lambda) + i \cdot P_I(\lambda)$ and $Q(\lambda) = Q_R(\lambda) + i \cdot Q_I(\lambda)$, and inserting $\lambda = i \cdot \omega$ into Nd: YAG laser passively Q-switched with GaAs system's characteristic equation, ω must satisfy the following:

$$\sin(\omega \cdot \tau_\Delta) = g(\omega) = \frac{-P_R(i\cdot\omega)\cdot Q_I(i\cdot\omega) + P_I(i\cdot\omega)\cdot Q_R(i\cdot\omega)}{|Q(i\cdot\omega)|^2}$$

$$\cos(\omega \cdot \tau_\Delta) = h(\omega) = -\frac{P_R(i\cdot\omega)\cdot Q_R(i\cdot\omega) + P_I(i\cdot\omega)\cdot Q_I(i\cdot\omega)}{|Q(i\cdot\omega)|^2}$$

where $|Q(i\cdot\omega)|^2 \neq 0$ in the view of requirement (a) above, $(g, h) \in \mathbb{R}$. Furthermore, it follows the above equations $\sin(\omega\cdot\tau_\Delta)$ and $\cos(\omega\cdot\tau_\Delta)$ that, by squaring and adding the sides, ω must be a positive root of $F(\omega) = |P(i\cdot\omega)|^2 - |Q(i\cdot\omega)|^2 = 0$. Note $F(\omega)$ is independent of τ_Δ. Now it is important to notice that if $\tau_\Delta \notin I$ (assume that $I \subseteq \mathbb{R}_{+0}$ is the set where $\omega(\tau_\Delta)$ is a positive root of $F(\omega)$ and for $\tau_\Delta \notin I$; $\omega(\tau_\Delta)$ is not define. Then for all $\tau_\Delta \in I$ is satisfies that $F(\omega, \tau_\Delta) = 0$. Then there are no positive $\omega(\tau_\Delta)$ solutions for $F(\omega, \tau_\Delta) = 0$, and we cannot have stability switch. For any $\tau_\Delta \in I$, where $\omega(\tau_\Delta)$ is a positive solution of $F(\omega, \tau_\Delta) = 0$, we can define the angle $\theta(\tau_\Delta) \in [0, 2\cdot\pi]$ as the solution of $\sin\theta(\tau_\Delta) = \frac{-P_R(i\cdot\omega)\cdot Q_I(i\cdot\omega) + P_I(i\cdot\omega)\cdot Q_R(i\cdot\omega)}{|Q(i\cdot\omega)|^2}$; $\cos\theta(\tau_\Delta) = -\frac{P_R(i\cdot\omega)\cdot Q_R(i\cdot\omega) + P_I(i\cdot\omega)\cdot Q_I(i\cdot\omega)}{|Q(i\cdot\omega)|^2}$ and the relation between the argument $\theta(\tau_\Delta)$ and $\omega(\tau_\Delta)\cdot\tau_\Delta$ for $\tau_\Delta \in I$ must be $\omega(\tau_\Delta)\cdot\tau_\Delta = \theta(\tau_\Delta) + 2\cdot n\cdot\pi \; \forall n \in \mathbb{N}_0$. Hence we can define the maps $\tau_{\Delta n} : I \to \mathbb{R}_{+0}$ given by $\tau_{\Delta n}(\tau_\Delta) = \frac{\theta(\tau_\Delta) + n\cdot 2\cdot\pi}{\omega(\tau_\Delta)}$; $n \in \mathbb{N}_0$, $\tau_\Delta \in I$. Let introduce the functions $I \to \mathbb{R}$; $S_n(\tau) = \tau_\Delta - \tau_{\Delta n}(\tau_\Delta)$, $n \in \mathbb{N}_0$, $\tau_\Delta \in I$ that are continuous and differentiable in τ. In the following, the subscripts $\lambda, \omega, L_0, \gamma, \alpha_q, L_q, \tau_\Delta, \ldots$ system parameters indicate the corresponding partial derivatives. Let us first concentrate on $\Lambda(x)$, remember in $\lambda(L_0, \gamma, \alpha_q, L_q, \tau_\Delta, \ldots \in \mathbb{R}_+)$; $\omega(L_0, \gamma, \alpha_q, L_q, \tau_\Delta, \ldots \in \mathbb{R}_+)$ and keeping all parameters except one (x) and τ_Δ. The derivation closely follows that in reference [BK]. Differentiating Nd: YAG laser passively Q-switched with GaAs system characteristic equation $P(\lambda) + Q(\lambda) \cdot e^{-\lambda\cdot\tau_\Delta} = 0$ with respect to specific parameter (x), and inverting the derivative, for convenience, one calculates, $x = \sigma, \gamma, L_0, \alpha_q, L_q, \ldots \in \mathbb{R}_+$,

$$\left(\frac{\partial\lambda}{\partial x}\right)^{-1} = \frac{-P_\lambda(\lambda, x)\cdot Q(\lambda, x) + Q_\lambda(\lambda, x)\cdot P(\lambda, x) - \tau_\Delta \cdot P(\lambda, x)\cdot Q(\lambda, x)}{P_x(\lambda, x)\cdot Q(\lambda, x) - Q_x(\lambda, x)\cdot P(\lambda, x)}$$

where $P_\lambda = \frac{\partial P}{\partial \lambda}, \ldots$ etc., Substituting $\lambda = i\cdot\omega$, and bearing $\overline{P(-i\cdot\omega)} = P(i\cdot\omega)$; $\overline{Q(-i\cdot\omega)} = Q(i\cdot\omega)$ then $i\cdot P_\lambda(i\cdot\omega) = P_\omega(i\cdot\omega)$ and $i\cdot Q_\lambda(i\cdot\omega) = Q_\omega(i\cdot\omega)$ that on the surface $|P(i\cdot\omega)|^2 = |Q(i\cdot\omega)|^2$, one obtains,

$$\left(\frac{\partial\lambda}{\partial x}\right)^{-1}|_{\lambda=i\cdot\omega} = \frac{i\cdot P_\omega(i\cdot\omega, x)\cdot\overline{P(i\cdot\omega, x)} + i\cdot Q_\lambda(i\cdot\omega, x)\cdot\overline{Q(\lambda, x)} - \tau_\Delta\cdot|P(i\cdot\omega, x)|^2}{P_x(i\cdot\omega, x)\cdot\overline{P(i\cdot\omega, x)} - Q_x(i\cdot\omega, x)\cdot\overline{Q(i\cdot\omega, x)}}$$

Upon separating into real and imaginary parts, with $P = P_R + i\cdot P_I$; $Q = Q_R + i\cdot Q_I$ and $P_\omega = P_{R\omega} + i\cdot P_{I\omega}$; $Q_\omega = Q_{R\omega} + i\cdot Q_{I\omega}$; $P_x = P_{Rx} + i\cdot P_{Ix}$; $Q_x = Q_{Rx} + i\cdot Q_{Ix}$; $P^2 = P_R^2 + P_I^2$. When (x) can be an Nd: YAG laser passively Q-switched with GaAs system parameters $L_0, \gamma, \alpha_q, L_q, \ldots$ and time delay τ_Δ etc. Where for convenience, we have dropped the arguments $(i\cdot\omega, x)$, and where $F_\omega = 2\cdot[(P_{R\omega}\cdot P_R + P_{I\omega}\cdot P_I) - (Q_{R\omega}\cdot Q_R + Q_{I\omega}\cdot Q_I)]$ and $F_x = 2\cdot[(P_{Rx}\cdot P_R + P_{Ix}\cdot P_I) - (Q_{Rx}\cdot Q_R + Q_{Ix}\cdot Q_I)]$; $\omega_x = -\frac{F_\lambda}{F_\omega}$. We define U and V:

$$U = (P_R\cdot P_{I\omega} - P_I\cdot P_{R\omega}) - (Q_R\cdot Q_{I\omega} - Q_I\cdot Q_{R\omega})$$
$$V = (P_R\cdot P_{Ix} - P_I\cdot P_{Rx}) - (Q_R\cdot Q_{Ix} - Q_I\cdot Q_{Rx})$$

We choose our specific parameter as time delay $x = \tau_\Delta$.

$$P_n(\lambda = i\cdot\omega, \tau_\Delta) = -\omega^2 + i\cdot g_4\cdot\omega;\ Q_m(\lambda = i\cdot\omega, \tau_\Delta) = -i\cdot g_1\cdot\omega + [g_2\cdot g_3 - g_1\cdot g_4]$$

$$P_R(i\cdot\omega, \tau_\Delta) = -\omega^2;\ P_I(i\cdot\omega, \tau_\Delta) = g_4\cdot\omega$$
$$Q_R(i\cdot\omega, \tau_\Delta) = g_2\cdot g_3 - g_1\cdot g_4;\ Q_I(i\cdot\omega, \tau_\Delta) = -g_1\cdot\omega$$

$$P_{R\omega}(i\cdot\omega, \tau_\Delta) = -2\cdot\omega;\ P_{I\omega}(i\cdot\omega, \tau_\Delta) = g_4$$
$$Q_{R\omega}(i\cdot\omega, \tau_\Delta) = 0;\ Q_{I\omega}(i\cdot\omega, \tau_\Delta) = -g_1$$

$$P_{R\omega} = P_{R\omega}(i\cdot\omega, \tau_\Delta);\ P_{I\omega} = P_{I\omega}(i\cdot\omega, \tau_\Delta);\ Q_{R\omega} = Q_{R\omega}(i\cdot\omega, \tau_\Delta)$$
$$Q_{I\omega} = Q_{I\omega}(i\cdot\omega, \tau_\Delta)$$

$$P_{R\tau_\Delta} = 0;\ P_{I\tau_\Delta} = 0;\ Q_{R\tau_\Delta} = 0;\ Q_{I\tau_\Delta} = 0;\ P_{R\omega} \cdot P_R = 2 \cdot \omega^3$$
$$P_{I\omega} \cdot P_I = g_4^2 \cdot \omega;\ Q_{R\omega} \cdot Q_R = 0;\ Q_{I\omega} \cdot Q_I = g_1^2 \cdot \omega$$

$$F_{\tau_\Delta} = 2 \cdot [(P_{R\tau_\Delta} \cdot P_R + P_{I\tau_\Delta} \cdot P_I) - (Q_{R\tau_\Delta} \cdot Q_R + Q_{I\tau_\Delta} \cdot Q_I)] = 0$$
$$\omega_{\tau_\Delta} = -\frac{F_{\tau_\Delta}}{F_\omega} = 0$$

$$P_R \cdot P_{I\omega} = -\omega^2 \cdot g_4;\ P_I \cdot P_{R\omega} = -g_4 \cdot 2 \cdot \omega^2;\ Q_R \cdot Q_{I\omega} = -(g_2 \cdot g_3 - g_1 \cdot g_4) \cdot g_1;\ Q_I \cdot Q_{R\omega} = 0$$

$$P_R \cdot P_{Ix}|_{x=\tau_\Delta} = 0;\ P_I \cdot P_{Rx}|_{x=\tau_\Delta} = 0;\ Q_R \cdot Q_{Ix}|_{x=\tau_\Delta} = 0;\ Q_I \cdot Q_{Rx}|_{x=\tau_\Delta} = 0$$

$$V = (P_R \cdot P_{I\tau_\Delta} - P_I \cdot P_{R\tau_\Delta}) - (Q_R \cdot Q_{I\tau_\Delta} - Q_I \cdot Q_{R\tau_\Delta}) = 0$$
$$F_\omega \cdot \frac{\partial\omega}{\partial\tau_\Delta} + F_{\tau_\Delta} = 0;\ \tau_\Delta \in I;\ \frac{\partial\omega}{\partial\tau_\Delta} = -\frac{F_{\tau_\Delta}}{F_\omega}$$

$$\Lambda^{-1}(\tau_\Delta) = \left(\frac{\partial \mathrm{Re}\lambda}{\partial\tau_\Delta}\right)_{\lambda=i\cdot\omega};\ \Lambda^{-1}(\tau_\Delta) = \mathrm{Re}\left\{\frac{-2 \cdot [U + \tau_\Delta \cdot |P|^2] + i \cdot F_\omega}{F_{\tau_\Delta} + i \cdot 2 \cdot [V + \omega \cdot |P|^2]}\right\};\ \frac{\partial\omega}{\partial\tau_\Delta} = \omega_{\tau_\Delta} = -\frac{F_{\tau_\Delta}}{F_\omega}$$

$$sign\{\Lambda^{-1}(\tau_\Delta)\} = sign\left\{\left(\frac{\partial \mathrm{Re}\lambda}{\partial\tau_\Delta}\right)_{\lambda=i\cdot\omega}\right\};\ sign\{\Lambda^{-1}(\tau_\Delta)\} = sign\{F_\omega\} \cdot sign\left\{\tau_\Delta \cdot \frac{\partial\omega}{\partial\tau_\Delta} + \omega + \frac{U \cdot \frac{\partial\omega}{\partial\tau_\Delta} + V}{|P|^2}\right\}$$

We shall presently examine the possibility of stability transitions (bifurcations) in an Nd: YAG laser passively Q-switched with GaAs system, about equilibrium points $E^{(k)}(\phi^{(k)}, N_x^{(k)})\ \forall k = 0, 1, 2$ as a result of a variation of delay parameter τ_Δ. The analysis consists in identifying the roots of our system characteristic equation situated on the imaginary axis of the complex λ-plane where by increasing the delay parameter τ_Δ, $\mathrm{Re}\lambda$ may at the crossing, change its sign from "−" to "+", i.e. from a stable focus $E^{(*)}$ to an unstable one, or vice versa. This feature may be further assessed by examining the sign of the partial derivatives with respect to τ_Δ, $\Lambda^{-1}(\tau_\Delta) = \left(\frac{\partial \mathrm{Re}\lambda}{\partial\tau_\Delta}\right)_{\lambda=i\cdot\omega}$ and $L_0, \gamma, \alpha_q, L_q, \ldots$ system parameters are constant where $\omega \in \mathbb{R}_+$. We need to plot the stability switch diagram based on different delay values of our Nd: YAG laser passively Q-switched with GaAs system. Since it is very complex function we recommend to solve it numerically rather than analytic [9, 10].

$$\Lambda^{-1}(\tau_\Delta) = \left(\frac{\partial \mathrm{Re}\lambda}{\partial \tau_\Delta}\right)_{\lambda=i\cdot\omega}; \ \Lambda^{-1}(\tau_\Delta) = \mathrm{Re}\left\{\frac{-2\cdot[U+\tau_\Delta\cdot|P|^2]+i\cdot F_\omega}{F_{\tau_\Delta}+i\cdot 2\cdot[V+\omega\cdot|P|^2]}\right\}$$

$$\Lambda^{-1}(\tau_\Delta) = \left(\frac{\partial \mathrm{Re}\lambda}{\partial \tau_\Delta}\right)_{\lambda=i\cdot\omega}; \ \Lambda^{-1}(\tau_\Delta) = \frac{2\cdot\{F_\omega\cdot(V+\omega\cdot P^2)-F_{\tau_\Delta}\cdot(U+\tau_\Delta\cdot P^2)\}}{F_{\tau_\Delta}^2+4\cdot(V+\omega\cdot P^2)^2}$$

The stability switch occurs only on those delay values (τ_Δ) which fit the equation: $\tau_\Delta = \frac{\theta_+(\tau_\Delta)}{\omega_+(\tau_\Delta)}$ and $\theta_+(\tau_\Delta)$ is the solution of $\sin\theta(\tau_\Delta) = \cdots$; $\cos\theta(\tau_\Delta) = \cdots$, when $\omega = \omega_+(\tau_\Delta)$ if only ω_+ is feasible. Additionally when all Nd: YAG laser passively Q-switched with GaAs system parameters are known and the stability switch due to various time delay values τ_Δ is describe in the following expression:

$$sign\{\wedge^{-1}(\tau_\Delta)\} = sign\{F_\omega(\omega(\tau_\Delta),\tau_\Delta)\}\cdot sign\{\tau_\Delta\cdot\omega_{\tau_\Delta}(\omega(\tau_\Delta)) + \omega(\tau_\Delta) + \frac{U(\omega(\tau_\Delta))\cdot\omega_{\tau_\Delta}(\omega(\tau_\Delta))+V(\omega(\tau_\Delta))}{|P(\omega(\tau_\Delta))|^2}\}$$

Remark we know $F(\omega,\tau_\Delta) = 0$ implies it roots $\omega_i(\tau_\Delta)$ and finding those delays values τ_Δ which ω_i is feasible. There are τ_Δ values which ω_i are complex or imaginary numbers, then unable to analyse stability.

Lemma 1.1 Assume that $\omega(\tau)$ is a positive and real root of $F(\omega,\tau) = 0$ defined for $\tau \in I$, which is continuous and differentiable. Assume further that if $\lambda = i\cdot\omega, \omega \in R$, then $P_n(i\cdot\omega,\tau)+Q_n(i\cdot\omega,\tau) \neq 0$, $\tau \in$ R hold true. The functions $S_n(\tau), n \in N_0$, are continuous and differentiable on I.

Theorem 1.2 Assume that $\omega(\tau)$ is a positive real root of $F(\omega,\tau) = 0$ defined for $\tau \in I, I \subseteq R_{+0}$, and at some $\tau^* \in I$, $S_n(\tau^*) = 0$ for some $n \in N_0$ then a pair of simple conjugate pure imaginary roots $\lambda_+(\tau^*) = i\cdot\omega(\tau^*), \lambda_-(\tau^*) = -i\cdot\omega(\tau^*)$ of $D(\lambda,\tau) = 0$ exist at $\tau = \tau^*$ which crosses the imaginary axis from left to right if $\delta(\tau^*) > 0$ and cross the imaginary axis from right to left if $\delta(\tau^*) < 0$ where

$$\delta(\tau^*) = sign\left\{\frac{d\mathrm{Re}\lambda}{d\tau}|_{\lambda=i\omega(\tau^*)}\right\} = sign\{F_\omega(\omega(\tau^*),\tau^*)\}\cdot sign\left\{\frac{dS_n(\tau)}{d\tau}|_{\tau=\tau^*}\right\}$$

Theorem 1.3 The characteristic equation has a pair of simple and conjugate pure imaginary roots $\lambda = \pm\omega(\tau^*), \omega(\tau^*)$ real at $\tau^* \in I$ if $S_n(\tau^*) = \tau^* - \tau_n(\tau^*) = 0$ for some $n \in N_0$. If $\omega(\tau^*) = \omega_+(\tau^*)$, this pair of simple conjugate pure imaginary roots crosses the imaginary axis from left to right if $\delta_+(\tau^*) > 0$ and crosses the imaginary axis from right to left if $\delta_+(\tau^*) < 0$ where $\delta_+(\tau^*) = sign\{\frac{d\mathrm{Re}\lambda}{d\tau}|_{\lambda=i\omega_+(\tau^*)}\}$; $\delta_+(\tau^*) = sign\{\frac{d\mathrm{Re}\lambda}{d\tau}|_{\lambda=i\omega_+(\tau^*)}\} = sign\left\{\frac{dS_n(\tau)}{d\tau}|_{\tau=\tau^*}\right\}$.

If $\omega(\tau^*) = \omega_-(\tau^*)$, this pair of simple conjugate pure imaginary roots cross the imaginary axis from left to right, if $\delta_-(\tau^*) > 0$ and crosses the imaginary axis from right to left. If $\delta_-(\tau^*) < 0$ where $\delta_-(\tau^*) = sign\{\frac{d\mathrm{Re}\lambda}{d\tau}|_{\lambda=i\omega_-(\tau^*)}\} = -sign\left\{\frac{dS_n(\tau)}{d\tau}|_{\tau=\tau^*}\right\}$.

If $\omega_+(\tau^*) = \omega_-(\tau^*) = \omega(\tau^*)$ then $\Delta(\tau^*) = 0$ and $sign\{\frac{d\mathrm{Re}\lambda}{d\tau}|_{\lambda=i\omega(\tau^*)}\} = 0$, the same is true when $S_n'(\tau^*) = 0$ the following result can be useful in identifying values of τ where stability switches happened. Remark: Lemma 1.1 and Theorems 1.2, 1.3: In the first and second cases we discuss delay parameter τ and in the third case we discuss delay parameter τ_Δ [8].

5.2 Q-Switched Microchip Lasers Semiconductors Saturable Absorbers Rate Equations Stability Analysis

Semiconductor saturable absorber mirror (SESAM) is a key component of ultrafast passive mode-locked laser sources. The key SESAM parameters are saturation fluence, modulation depth, and non saturable losses are measurable with a high accuracy. We can control these parameters to obtain stable pulse generation for a given laser. The model function for the nonlinear reflectivity is based on a simple two-level travelling wave system. SESAM is a mirror structure with an incorporated saturable absorber, all modes in semiconductor technology. We use SESAM for the generation of ultrafast pulses by passive mode locking of various types of lasers. A SESAM contains a semiconductor Bragg mirror and near the surface a single quantum well absorber layer (Fig. 5.4). The materials of the Bragg mirror have larger bandgap energy and no absorption occurs in that region (such SESAMs also called saturable Bragg reflectors (SBRs). We can obtain a large modulation depth for passive Q switching by using a thicker absorber layer. The suitable passivation layer on the top surface increases the device lifetime [11].

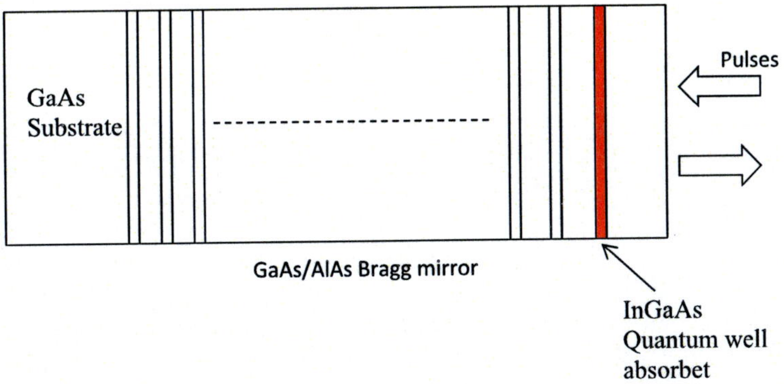

Fig. 5.4 Structure of a typical SESAM (semiconductor saturable absorber mirror)

The structure of a typical SESAM is designed to operate around 1064 nm. On a GaAs substrate a GaAs/AlGaAs bragg mirror is grown, and within the top layers, there is an InGaAs quantum well absorber layer, which may be 10 nm thick. We calculate the optical field into a SESAM with the same matrix method as applied to other types of dielectric mirrors. The optical intensity in the region where the saturable material is placed influences the modulation depth and the saturation fluence. The design of the structure also influences the bandwidth and the chromatic dispersion. The absorber layer is placed in an anti-node of the electric field and then we get a maximum saturable absorption and the smallest possible saturation fluence. If multiple absorber layers are required for a high modulation depth, they may be placed in separate anti-nodes, or possible several of them near one anti-node. Other types of semiconductor saturable absorbers can be based on quantum dots embedded in glass or on carbon nanotubes or graphene. We can tailor the macroscopic parameters of a SESAM for operation in very different regime. The macroscopic parameters of a SESAM can be the operation wavelength, of the modulation depth, the saturation fluence, and the recovery time. The most important characteristics of a SESAM for passive mode locking or Q-switching are the following: modulation depth, saturation fluence, recovery time and non saturable losses. The saturation fluence is the fluence of an incident short pulse which is required for causing significant absorption saturation. It depends on the absorber material, the wavelength, and the field penetration into the absorber structure. A "roll-over" of saturation curve (reduction in reflectivity for high fluence values) can be caused by two-photon absorption for sub-picosecond pulses or by other effects. The recovery time is the exponential time constant of absorption recovery after a saturating pulse. Typical values are between a few picoseconds and hundreds of picoseconds. The recovery time is strongly influenced by the defect density in the absorber, and possibly in nearby structures. SESAMs non saturable losses lead to device heating while not contributing to the pulse shaping. Non saturable losses tend to be higher for SESAMs with a larger modulation depth and faster recovery. There are SESAMs for high-power operation which the average output powers of well above 100w (passively mode-locked high-power laser) and the intra cavity average power of above 1 kW. There can be a substantial thermal lensing, which affects the mode properties of the laser resonator. It is recommended to optimize SESAMs for use in high-power lasers. A SESAM which generated heat must be conducted through some hundreds of micrometer of semiconductor materials which are determined by the thickness of the used wafer. A substantial local temperature increases can occur even in low-power lasers (around 1w of average output power) if the pulse repetition rate is very high ($\approx$GHz). We need to use a strong focusing of the radiation in order to achieve sufficiently strong saturation of the absorption despite the small pulse energy. High power lasers with much lower pulse repetition rate (tens of MHz) able to use of much larger beam areas and then the handle substantial absorber powers is easier. The dispersion of any sign can be engineered into a SESAM via the multilayer structure. Dispersive SESAMs serve the purpose of dispersion compensation in a laser resonator, which is in addition to the function of a passive mode locker. SESAMs are widely used for passive mode locking of lasers and for solid-state bulk and fiber lasers. They work with a wide range of laser parameters and allow for reliable self-starting mode locking for the right device and operation parameters. Another application of SESAMs is passive Q-switching (microchip lasers and fiber lasers). The key decisions using SESAMs are the selection of a suitable SESAM design and the adjustment of a numbers of laser parameters, especially the resonator mode size on the absorber. We can face problems in the form of various instabilities or SESAM damage by using inappropriate device and operation parameters. The engineering areas of nonlinear filtering and signal processing can be used SESAMs (optical fiber communications). The performance of SESAMs in solid-state lasers worked out design guidelines for application to practical laser systems, and the performance improvements in shortest pulse widths, highest average and peak power from a passively mode-locked laser, and extending the pulse repetition rate to 160 GHz. We can adjust and get stable Q-switching of compact microchip lasers with pulse durations as short as 37 ps, or pulse energies as high as 1.1 μJ at a wavelength of around 1um. We integrate the semiconductor saturable absorber into a mirror structure, which results in a device whose reflectivity increases as the incident optical intensity increases. SESAMs both the linear and nonlinear optical properties can be engineered over a wide range, allowing for more freedom in the specific laser cavity design. Semiconductor saturable absorbers are excellent solution for passive mode locking solid-state lasers because the large absorber's cross section, and therefore small saturation fluence, and is ideally suited for suppressing Q-switching instabilities. The key parameters for a saturable absorber are its wavelength range (where it absorbs), its dynamic response (how fast it recovers), and its saturation intensity and fluence (at what intensity and pulse energy density it saturated). The SESAM needs to be characterized to obtain the required parameters such as saturation fluence, modulation depth, non saturable absorption, and recovery time. There are different SESAM designs. Concern the design, a high modulation depth is suited for passive Q-switching and a smaller modulation depth for passive continuous wave (cw) mode locking. The SESAM nonlinear reflectivity R versus the incident pulse energy fluence F_p is described by three parameters: (1) the linear reflectivity R_{lin} for pulses with "zero" pulse energy fluence, (2) the reflectivity R_{ns} for "infinitely" high pulse energy fluences when all saturable absorption is bleached, and (3) the saturation fluence F_{sat}. The modulation depth ΔR and the non saturable losses ΔR_{ns} in reflectivity are defined as: $\Delta R = R_{ns} - R_{lin}$; $\Delta R_{ns} = 1 - R_{ns}$. It is implying that R_{lin} and R_{ns} are not experimentally accessible but extrapolated values from the measured data using a proper model function. The significant and stability of these parameters depend on the quality of the model function. The pulse energy fluence $F_p = \frac{E_p}{A}$ is the incident pulse energy per unit surface area. The saturation fluence F_{sat} is the fluence required to begin absorption saturation. In the case of infinitely thin absorber, the reflectivity for a pulse with fluence $F_P = F_{sat}$ is increase by $\frac{1}{e}(\approx 37\%)$ of ΔR with respect to R_{lin}. In SESAM mode-locked solid-state lasers (picosecond regime) there is a simple design guideline which prevents Q-switching instabilities. The inequality $E_p^2 > E_{sat,L} \cdot E_{sat,A} \cdot \Delta R$ describes the condition to prevent Q-switching instabilities, where E_p is the intra cavity pulse energy, $E_{sat,L}$ is the saturation energy of the laser medium, and $E_{sat,A}$ the saturation energy of the saturable absorber. Assumption is taken for a fully saturated, slow saturable absorber that fully recovers between consecutive pulses. The modified criteria to prevent Q-switching instabilities is $E_p^2 > \frac{E_{sat,A}}{\left(a_2 + \frac{1}{E_{sat,L}}\right)} \cdot \Delta R$, where a_2(scaling constant) is related to nonlinear coefficient that increases linearity with the pulse energy and with a scaling constant. The nonlinear loss coefficient is given by $q_2(E_p) = a_2 \cdot E_p$ which includes a certain roll-off at higher fluence. The roll-off (femtosecond regime) is dominated by two-photon absorption (TPA) for which $a_2 = \frac{\beta_{TPA} \cdot z_{eff}}{\tau_p \cdot A_A}$, where β_{TPA} is the two-photon absorption coefficient, z_{eff} is the effective TPA-absorber thickness, τ_p and A_A is the laser mode area on the saturable absorber. The roll-off even in the picosecond regime is observed, where TPA is negligible. Our target is to measure the most reliable nonlinear reflectivity. The experimental system is built from an incident laser beam which is focused onto a device under test (DUT), can be SESAM. The attenuators control the incident fluence. A beam-splitter feeds a small part of the incident and reflector beam onto two photodiodes. After proper calibration the applied fluence is measured by the photocurrent of the incident photodiode. The reflectivity is calculated from the ratio of the two photodiodes photocurrents. We divide system setup into five functional groups: (1) an isolated laser source, (2) power adjustment, (3) focusing system and DUT holder, (4) beam-splitter and photodiode optics, and (5) imaging system. The system setup allows us to measure the reflectivity with accuracy close to 10^{-4} over a dynamic input fluence range of more than 4 orders of magnitude. The experimental system to measure R(F) is as follow:

The experimental system (Fig. 5.5) elements are as follow: First the isolated laser source which should match the parameters (wavelength, pulse duration, repetition rate, etc.), of the laser in which the SESAM will be used. The laser needs to be stable over time and good beam quality. We need to compensate for group delay dispersion (GDD) for measurements with pulse durations below 100 fs. GDD comp element: group delay dispersion pre-compensation. An external cavity resonator between the device under test (DUT) and the laser's output coupler. It will disturb the laser source especially at the lowest attenuation of highest fluences. Therefor we use an isolator. The optical isolator prevents back reflections. A $\frac{\lambda}{2}$ computer

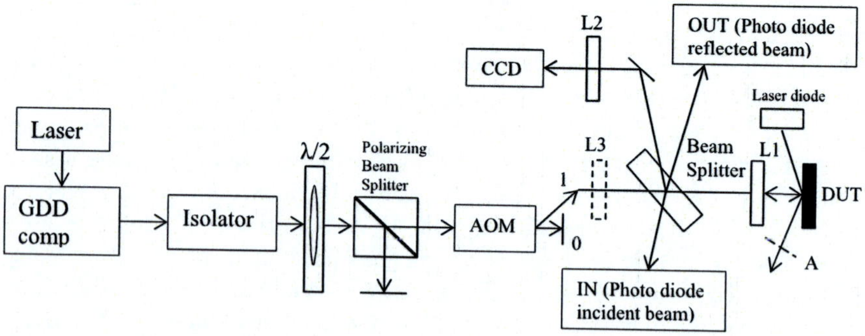

Fig. 5.5 SESAM's experimental system to measure R(F)

controlled, motorized half wave plate; PBS: Polarizing beam splitter; Attenuation $\left(\frac{\lambda}{2}+\text{PBS}\right)$: $\left(\frac{\lambda}{2}+\text{PBS}\right)$. We need a computer-controlled attenuator which operates over 4 orders of magnitude. The possible attenuator systems are: (1) motorized graded neutral density filters, (2) rotating half-wave plates in combination with a polarizing beam-splitter, (3) an Acoustic-optical modulator (AOM), or other optics. AOM: acoustic-optic modulator: chopper and attenuation: OD 0.1–3.0; 0, 1: zero and first deflection order. One option is to use a graded neutral density (ND) filters, must avoid thermal lenses inside the ND filters. An acoustic-optical modulator (AOM) is not change the shape and direction of the diffracted beam. Possible thermal effects inside the AOM caused by different RF powers applied to change the diffraction efficiency. We need to avoid strange attenuations. The half wave plate operating and at stronger attenuations reduces repeatability, and increase the sensitivity to mechanical vibrations, changes the spectrum of short pulses with broad spectra, and reduces the quality of polarization, which directly affects the calibration. By using AOM, a stronger attenuation increases amplitude noise in the diffracted beam due to electrical noise at the input side. A combination of attenuators is needed to increase the dynamic range of system setup. The half-wavelength $\left(\frac{\lambda}{2}\right)$ is used to slower coarse adjustment and the AOM for fast modulation. Then beam splitter (BS): low percentage beam splitter; L1: focusing lens; Device under test (DUT); LD: Laser diode; A: aperture; IN: photodiode incident beam; OUT: photodiode reflected beam; L2: lens which is part of imaging system; CCD: imaging laser spot on sample; L3: flip-in "illumination" lens, used for calibration of imaging system. We need to lock-in detection to detect the signals and reject the photodiodes dark currents and environmental background light. The best place to put the chopper is at a focal point of the beam. If we place the chopper in other places than the focal point of the beam the rising and falling edges of the laser power will result in measuring an average reflectivity over a fluence range instead of measuring the reflectivity at a specific fluence. We use AOM as a chopper with negligible edge effects. If we need to get larger spot sizes then a smaller duty cycles reduce the average thermal load on the sample. The higher chopping frequencies do not reduce laser noise, because the modulation induced sidebands of the lasers noise spectrum is always shifted to the detection frequency. The device under test (DUT) is mounted on a xyz-stage and a mirror mount. Reproducible surface orientation from device to device is maintained by a tilt control consisting of an alignment laser diode and an aperture. The focusing lens is implemented as a zoom lens system which allow for variable spot size. The beam splitter (BS) is an uncoated small-angle wedged glass plate. The wedge is placed close to normal incident in s-polarized beam to get the minimum sensitivity to polarization effects. Transfer optics deliver these few percent of reflected beams to large-area photo-detectors for the incident beam "IN" and reflected beam "OUT". For larger beam areas the wings of the beam profile are cut off at the edge of the diode making calibration sensitive to beam steering and lensing effects. Small focused spots on the diodes might be the origin of unnoticed local saturation or nonlinear absorption (TPA) at higher F_p and short pulse durations or deviation from linearity due to in homogeneities of the detector area. To adjust the power level at the diodes, neutral density filter wheels are placed in the transfer optics. The signal of each photo-diode is fed into a lock-in amplifier. This discriminates background light and electric DC offsets. Obviously it does not discriminate against scattered light from the setup. The scattered light is coded with the same chopping frequency and will add to the signal in the lock-in amplifiers. The diodes need to be shielded against scattered light. The imaging system (two lenses L1 and L2 with focal lengths f_1 and f_2) is used to position the sample in the focal plane and to image bad spots on the surface. The magnification is the ratio of f_2/f_1 and a typical SESAM has a homogenous flat surface, like any other mirror in the setup. The image on the CCD is not an image of the beam's spot on the SESAM surface. If we move the device under test (DUT) in the direction of the beam propagation changes the beam size on the CCD camera. It is recommended that the position of lens L1 is fixed and the device under test (DUT) is moved along the beam axis to obtain a minimum spot size on the device under test (DUT). The CCD can be placed such that the beam diameter on the CCD is minimal when the sample is placed in the focus. We put additional lens (L3) which is flipped into the beam. It increases the illuminated area on the sample and makes surface irregularities such as scratches, pitches, or dust particles visible. The focal length f_3 of L3 should be 30–50 times f_1 and placed at a distance of about f_3 in the front of L1. Len L3 has to be placed in front of the beam-splitter (BS). By calibration the image system can be used to measure the actual spot size, if the beam is collimated between the beam-splitter (BS) and the focusing lens (L1). We need to fit the model function to obtain the key parameters of the saturable absorber. It is done by measuring the nonlinear reflectivity. We need to characterize the model function $R(F_p)$ and the model function is well suited to describe to nonlinear behavior of real SESAMs. The two levels system helps us to approximate a band-structure. Assumption: neglect intra band relaxations, neglect trapping or recombination, not include effects of standing wave pattern in the device, and not include carrier diffusion and temperature effects [11–15]. The traveling wave model is based on rate equations for two levels system without relaxations, and it is a good approximation for a slow saturable absorber, where the recovery time is longer than the pulse bleaching the absorber. We can characterize the two levels system by the following equations:

$$n_1 + n_2 = n; \ \frac{dn_1}{dt} = -\frac{dn_2}{dt} = -(n_1 - n_2) \cdot \frac{\sigma}{\hbar \cdot \omega} \cdot I = -\Delta n \cdot \frac{\sigma}{\hbar \cdot \omega} \cdot I$$

$$\frac{d\Delta n}{dt} = 2 \cdot \frac{dn_1}{dt} = -\Delta n \cdot \frac{2 \cdot \sigma}{\hbar \cdot \omega} \cdot I; \alpha = \sigma \cdot \Delta n; \ \frac{d\alpha}{dt} = -\alpha \cdot \frac{2 \cdot \sigma}{\hbar \cdot \omega} \cdot I$$

n_1 and n_2 are the occupation numbers of level 1 and 2, respectively, σ is the absorption and emission cross-section, ω is the angular frequency of the optical field, α is the absorption coefficient. The microscopic definition of F_{sat} is given by

$$\frac{d\alpha}{dt} = -\frac{\alpha \cdot I}{F_{sat}}; \ F_{sat} = \frac{\hbar \cdot \omega}{2 \cdot \sigma}; \ d\alpha = -\frac{\alpha \cdot I}{F_{sat}} \cdot dt \Rightarrow \int \frac{1}{\alpha} \cdot d\alpha$$
$$= -\int \frac{I}{F_{sat}} \cdot dt \Rightarrow \ln \alpha = -\frac{I}{F_{sat}} \cdot t$$

The saturation of absorption α for times after the pulse has passed: $\alpha(F_p) = \alpha_{lin} \cdot e^{-\frac{F_p}{F_{sat}}}$. The target is to get the transmitted pulse energy through a saturable absorber; we get it by solving the following system of differential equations:

$$\frac{d\alpha(z,t)}{dt} = -\frac{\alpha(z,t) \cdot I(z,t)}{F_{sat}}; \frac{dI(z,t)}{dt} = -\alpha(z,t) \cdot I(z,t)$$

Assumption: No standing wave patters or Fabry-Perot effects in the SESAM. The reflectivity is calculated as twice the transmission through an absorber of length $\frac{L}{2}$. The deviations scale linearly with the field enhancement factor for the modulation depth and inverse linearly for the saturation fluence. We get from the model function SESAM parameters and microscopic values for material comparison or extracted SESAM design. Finally we get the internal transmission for a pulse twice through an absorber of length $\frac{L}{2}$. Parameter S is the saturation parameter $\left(S = \frac{F_p}{F_{sat}}\right)$, T_{lin} is define as the linear transmission. We replace T_{lin} by R_{lin} [11].

$$R(F_p) = \frac{F_{out}}{F_{IN}} = \frac{\int_{-\infty}^{\infty} I(L,t) \cdot dt}{\int_{-\infty}^{\infty} I(0,t) \cdot dt} = \frac{\ln\left(1 + T_{lin} \cdot \left(e^{\frac{F_p}{F_{sat}}} - 1\right)\right)}{\frac{F_p}{F_{sat}}}$$

$$= \frac{\ln(1 + T_{lin} \cdot (e^S - 1))}{S}$$

There are saturable losses (reflectivity less than 100%) causes by residual transmission losses through the Bragg mirror, scattering losses from rough interfaces, non saturable defect absorption, free carrier absorption, Auger recombination, etc. These losses are homogeneously distributed over the absorber layer or transmission losses. We evaluate these losses by a scaling factor R_{ns}.

$$R(F_p) = R_{ns} \cdot \frac{\ln\left(1 + \frac{R_{lin}}{R_{ns}} \cdot (e^S - 1)\right)}{S}$$

At higher fluences the reflectivity deceases with increasing fluence according to a second order process like two-photon absorption (TPA), leading to a roll-off in the reflectivity curve. This additional absorption coefficient increases more or less linearity with fluence and depends on the pulse duration as well. In the femtosecond regime the most significant part is due to two-photon absorption in the SESAM structure, not limited to the absorber layer only. There are other terms like thermal effects, free-carrier absorption, and other sources of induced absorption. The roll-off is taken into account by multiplying the model function $R(F_p) = \cdots$ with $e^{-\frac{F_p}{F_2}}$ term. F_2 is the fluence where SESAM reflectivity is 37% (1/e) due to induced absorption. A smaller F_2 value is corresponding to a stronger roll-off. We drive the relation between F_2 and the material parameter $a_2$nn(TPA), $q_2(E_p) = a_2 \cdot E_p$; $q_2(E_p) = \beta \cdot I \cdot z \approx \beta_{TPA} \cdot \frac{F_p}{\tau_p} \cdot z = \beta_{TPA} \cdot \frac{E_p}{\tau_p \cdot A_A} \cdot z$, β_{TPA} is TPA coefficient, τ_p is the pulse duration, and A_A is the mode area. Absorber with thickness z the reduced transmission T_2 is $T_2(E_p) = e^{-q_2(E_p)} = e^{-a_2 \cdot E_p}$. The characteristics of standing wave pattern, penetration depth of the optical field into the Bragg mirror, and different two-photon absorption cross-sections for different materials are consider for SESAM structure. We approximate these effects by an effective absorber thickness z_{eff}. Parameter z_{eff} need to be calculated numerically for specific SESAM design.

$$a_2 = \frac{\beta_{TPA} \cdot z_{eff}}{\tau_p \cdot A_A}; \; F_2 = \frac{1}{a_2 \cdot A_A} \approx \frac{\tau_p}{\beta_{TPA} \cdot z_{eff}}$$

F_p is the constant fluence for flattop spot, ω is the radius. We calculate F_p from the measured pulse energy E_p $\left(F_p = \frac{E_p}{\pi \cdot \omega^2}\right)$. In most application we use a Gaussian profile: $F_p^{Gauss}(r) = F_0 \cdot e^{-\frac{2r^2}{\omega^2}}$ where F_0 is the peak fluence and ω is $\frac{1}{e^2}$ beam radius. The pulse energy E_p is $E_p = \int F_p^{Gauss}(r) \cdot 2 \cdot \pi \cdot r \cdot dr = \frac{1}{2} \cdot F_0(\pi \cdot \omega^2)$; $F_p = \frac{1}{2} \cdot F_0 = \frac{E_p}{\pi \cdot \omega^2}$. The peak fluence in the Gaussian beam is $2 \cdot F_p$, and the saturation occurs at lower fluences. Integration over the spatial energy distribution gives: $R^{FlatTop}(x) = R_{ns} \cdot \frac{\ln\left(1+\frac{R_{lin}}{R_{ns}} \cdot (e^{\frac{x}{F_{sat}}} - 1)\right)}{\frac{x}{F_{sat}}} \times e^{-\frac{x}{F_2}}$.

The system model function describes the nonlinear behavior of real SESAMs under some approximations which describe before. Due to interferences and experimental system restrictions there are delays in time ($\tau_1, \tau_2, \tau_3, \tau_4$) for systems variables $n_1, n_2, \Delta n, \alpha$ respectively. The delays in time do not affect the derivatives in time of these variables $\left(\frac{dn_1}{dt}, \frac{dn_2}{dt}, \frac{d\Delta n}{dt}, \frac{d\alpha}{dt}\right)$, $n_1 = n_1(t)$; $n_2 = n_2(t)$; $\Delta n = \Delta n(t)$; $\alpha = \alpha(t)$. The two levels system is characterized by set of delay differential equations (DDEs).

$$\frac{dn_1}{dt} = -[n_1(t-\tau_1) - n_2(t-\tau_2)] \cdot \frac{\sigma}{\hbar \cdot \omega} \cdot I; \; \frac{dn_2}{dt} = [n_1(t-\tau_1) - n_2(t-\tau_2)] \cdot \frac{\sigma}{\hbar \cdot \omega} \cdot I$$

$$\frac{d\Delta n}{dt} = -\Delta n(t-\tau_3) \cdot \frac{2 \cdot \sigma}{\hbar \cdot \omega} \cdot I = -\Delta n(t-\tau_3) \cdot \frac{1}{F_{sat}} \cdot I$$

$$\frac{d\alpha}{dt} = -\alpha(t-\tau_4) \cdot \frac{2 \cdot \sigma}{\hbar \cdot \omega} \cdot I = \frac{d\alpha}{dt} = -\alpha(t-\tau_4) \cdot \frac{1}{F_{sat}} \cdot I; \; F_{sat} = \frac{\hbar \cdot \omega}{2 \cdot \sigma}$$

n_1—Occupation number of level 1 in time, n_2—Occupation number of level 1 in time, Δn—Occupation difference between level 1 and level 2 in time, α—Absorption coefficient in time, σ—Absorption and emission cross-section, ω—Angular frequency of the optical field, $\hbar$—Reduced Planck constant $\left(\hbar = \frac{h}{2 \cdot \pi} = 1.054 \times 10^{-34} \frac{\text{J s}}{\text{rad}} = 6.58 \times 10^{-16} \frac{\text{eV s}}{\text{rad}}\right)$. There are 16 sub cases: (1) $\tau_i = 0; i = 1, 2, 3, 4$, (2) $\tau_1 = \tau; \tau_i = 0; i = 2, 3, 4$, (3) $\tau_2 = \tau; \tau_i = 0; i = 1, 3, 4$, (4) $\tau_3 = \tau; \tau_i = 0; i = 1, 2, 4$, (4) $\tau_4 = \tau; \tau_i = 0; i = 1, 2, 3$, (5) $\tau_1 = \tau_2 = \tau; \tau_i =$

0; $i = 3, 4$, (6) …, (10) $\tau_1 = \tau_4 = \tau; \tau_i = 0; i = 2, 3$. We choose to analyze the (2) sub case, the two levels system delay differential equations (DDEs) are

$$\frac{dn_1}{dt} = -[n_1(t-\tau) - n_2(t)] \cdot \frac{\sigma}{\hbar \cdot \omega} \cdot I; \frac{dn_2}{dt} = [n_1(t-\tau) - n_2(t)] \cdot \frac{\sigma}{\hbar \cdot \omega} \cdot I$$

$$\frac{d\Delta n}{dt} = -\Delta n(t) \cdot \frac{2 \cdot \sigma}{\hbar \cdot \omega} \cdot I = -\Delta n(t) \cdot \frac{1}{F_{sat}} \cdot I$$
$$\frac{d\alpha}{dt} = -\alpha(t) \cdot \frac{2 \cdot \sigma}{\hbar \cdot \omega} \cdot I = \frac{d\alpha}{dt} = -\alpha(t) \cdot \frac{1}{F_{sat}} \cdot I; F_{sat} = \frac{\hbar \cdot \omega}{2 \cdot \sigma}$$

At fixed points: $\frac{dn_1}{dt} = 0; \frac{dn_2}{dt} = 0; \frac{d\Delta n}{dt} = 0; \frac{d\alpha}{dt} = 0; \lim_{t \to \infty} n_1(t-\tau) = n_1(t)$

$$t \to \infty \Rightarrow t \gg \tau; t - \tau \approx t$$

$$\frac{dn_1}{dt} = 0 \Rightarrow -[n_1^{(*)} - n_2^{(*)}] \cdot \frac{\sigma}{\hbar \cdot \omega} \cdot I = 0; \frac{\sigma}{\hbar \cdot \omega} \cdot I \neq 0 \Rightarrow n_1^{(*)} = n_2^{(*)}$$

$$\frac{dn_2}{dt} = 0 \Rightarrow [n_1^{(*)} - n_2^{(*)}] \cdot \frac{\sigma}{\hbar \cdot \omega} \cdot I = 0; \frac{\sigma}{\hbar \cdot \omega} \cdot I \neq 0 \Rightarrow n_1^{(*)} = n_2^{(*)}$$

$$\frac{d\Delta n}{dt} = 0 \Rightarrow -\Delta n^{(*)} \cdot \frac{2 \cdot \sigma}{\hbar \cdot \omega} \cdot I = -\Delta n^{(*)} \cdot \frac{1}{F_{sat}} \cdot I = 0$$
$$\frac{2 \cdot \sigma}{\hbar \cdot \omega} \cdot I \neq 0; \frac{1}{F_{sat}} \cdot I \neq 0; \Delta n^{(*)} = 0$$

$$\frac{d\alpha}{dt} = 0 \Rightarrow -\alpha^{(*)} \cdot \frac{2 \cdot \sigma}{\hbar \cdot \omega} \cdot I = 0; \frac{2 \cdot \sigma}{\hbar \cdot \omega} \cdot I \neq 0 \Rightarrow \alpha^{(*)} = 0$$
$$\frac{d\alpha}{dt} = 0 \Rightarrow -\alpha^{(*)} \cdot \frac{1}{F_{sat}} \cdot I = 0; \frac{1}{F_{sat}} \cdot I \neq 0; \alpha^{(*)} = 0$$

Our system fixed point: $= (n_1^{(*)}, n_1^{(*)}, 0, 0) \, \forall k = 0, 1, 2 = (n_1^{(*)}, n_1^{(*)}, 0, 0) \, \forall k = 0, 1, 2$.

Stability analysis: The standard local stability analysis about any one of the equilibrium points of Q-switched microchip lasers semiconductors saturable absorbers system consists in adding to its coordinates $[n_1 \; n_2 \; \Delta n \; \alpha]$ arbitrarily small increments of exponential terms $[n_1 \; n_2 \; \Delta n \; \alpha] \cdot e^{\lambda \cdot t}$, and retaining the first order terms in $n_1 \; n_2 \; \Delta n \; \alpha$. The system of four homogenous equations leads to a polynomial characteristic equation in the eigenvalue λ [6, 7]. The polynomial characteristic equation accepts by set the Q-switched microchip lasers semiconductors saturable absorbers system equations. The system fixed values with arbitrarily small increments of exponential form $[n_1 \; n_2 \; \Delta n \; \alpha] \cdot e^{\lambda \cdot t}$ are; i = 0 (first fixed point), i = 1 (second fixed point), i = 2 (third fixed point), etc.

$$n_1(t) = n_1^{(i)} + n_1 \cdot e^{\lambda \cdot t}; n_2(t) = n_2^{(i)} + n_2 \cdot e^{\lambda \cdot t}$$

$$\Delta n(t) = \Delta n^{(i)} + \Delta n \cdot e^{\lambda \cdot t}; \alpha(t) = \alpha^{(i)} + \alpha \cdot e^{\lambda \cdot t}$$

$$n_1(t-\tau) = n_1^{(i)} + n_1 \cdot e^{\lambda \cdot (t-\tau)}; \frac{dn_1(t)}{dt} = n_1 \cdot \lambda \cdot e^{\lambda \cdot t}; \frac{dn_2(t)}{dt} = n_2 \cdot \lambda \cdot e^{\lambda \cdot t}$$

$$\frac{d\Delta n(t)}{dt} = \Delta n \cdot \lambda \cdot e^{\lambda \cdot t}; \frac{d\alpha(t)}{dt} = \alpha \cdot \lambda \cdot e^{\lambda \cdot t}$$

$$\frac{dn_1}{dt} = -[n_1(t-\tau) - n_2(t)] \cdot \frac{\sigma}{\hbar \cdot \omega} \cdot I; n_1 \cdot \lambda \cdot e^{\lambda \cdot t}$$
$$= -[n_1^{(i)} + n_1 \cdot e^{\lambda \cdot (t-\tau)} - n_2^{(i)} - n_2 \cdot e^{\lambda \cdot t}] \cdot \frac{\sigma}{\hbar \cdot \omega} \cdot I$$

$$n_1 \cdot \lambda \cdot e^{\lambda \cdot t} = -[n_1^{(i)} - n_2^{(i)}] \cdot \frac{\sigma}{\hbar \cdot \omega} \cdot I - [n_1 \cdot e^{\lambda \cdot (t-\tau)} - n_2 \cdot e^{\lambda \cdot t}] \cdot \frac{\sigma}{\hbar \cdot \omega} \cdot I$$

At fixed points: $-[n_1^{(i)} - n_2^{(i)}] \cdot \frac{\sigma}{\hbar \cdot \omega} \cdot I = 0; n_1 \cdot \lambda \cdot e^{\lambda \cdot t} = -[n_1 \cdot e^{\lambda \cdot (t-\tau)} - n_2 \cdot e^{\lambda \cdot t}] \cdot \frac{\sigma}{\hbar \cdot \omega} \cdot I$

$$n_1 \cdot \lambda \cdot e^{\lambda \cdot t} = -[n_1 \cdot e^{-\lambda \cdot \tau} - n_2 \cdot e^{\lambda \cdot t}] \cdot e^{\lambda \cdot t} \cdot \frac{\sigma}{\hbar \cdot \omega} \cdot I$$
$$\Rightarrow -n_1 \cdot \frac{\sigma}{\hbar \cdot \omega} \cdot I \cdot e^{-\lambda \cdot \tau} - n_1 \cdot \lambda + n_2 \cdot \frac{\sigma}{\hbar \cdot \omega} \cdot I = 0$$

$$\frac{dn_2}{dt} = [n_1(t-\tau) - n_2(t)] \cdot \frac{\sigma}{\hbar \cdot \omega} \cdot I; n_2 \cdot \lambda \cdot e^{\lambda \cdot t}$$
$$= [n_1^{(i)} + n_1 \cdot e^{\lambda \cdot (t-\tau)} - n_2^{(i)} - n_2 \cdot e^{\lambda \cdot t}] \cdot \frac{\sigma}{\hbar \cdot \omega} \cdot I$$

$$n_2 \cdot \lambda \cdot e^{\lambda \cdot t} = (n_1^{(i)} - n_2^{(i)}) \cdot \frac{\sigma}{\hbar \cdot \omega} \cdot I + (n_1 \cdot e^{\lambda \cdot (t-\tau)} - n_2 \cdot e^{\lambda \cdot t}) \cdot \frac{\sigma}{\hbar \cdot \omega} \cdot I$$

At fixed points: $(n_1^{(i)} - n_2^{(i)}) \cdot \frac{\sigma}{\hbar \cdot \omega} \cdot I = 0; n_2 \cdot \lambda \cdot e^{\lambda \cdot t} = (n_1 \cdot e^{\lambda \cdot (t-\tau)} - n_2 \cdot e^{\lambda \cdot t}) \cdot \frac{\sigma}{\hbar \cdot \omega} \cdot I$

$$n_2 \cdot \lambda \cdot e^{\lambda \cdot t} = (n_1 \cdot e^{\lambda \cdot (t-\tau)} - n_2 \cdot e^{\lambda \cdot t}) \cdot \frac{\sigma}{\hbar \cdot \omega} \cdot I$$
$$\Rightarrow n_1 \cdot \frac{\sigma}{\hbar \cdot \omega} \cdot I \cdot e^{-\lambda \cdot \tau} - n_2 \cdot \frac{\sigma}{\hbar \cdot \omega} \cdot I - n_2 \cdot \lambda = 0$$

$$\frac{d\Delta n}{dt} = -\Delta n(t) \cdot \frac{2 \cdot \sigma}{\hbar \cdot \omega} \cdot I; \Delta n \cdot \lambda \cdot e^{\lambda \cdot t} = -(\Delta n^{(i)} + \Delta n \cdot e^{\lambda \cdot t}) \cdot \frac{2 \cdot \sigma}{\hbar \cdot \omega} \cdot I$$

$\Delta n \cdot \lambda \cdot e^{\lambda \cdot t} = -\Delta n^{(i)} \cdot \frac{2 \cdot \sigma}{\hbar \cdot \omega} \cdot I - \Delta n \cdot \frac{2 \cdot \sigma}{\hbar \cdot \omega} \cdot I \cdot e^{\lambda \cdot t};$
At fixed points: $-\Delta n^{(i)} \cdot \frac{2 \cdot \sigma}{\hbar \cdot \omega} \cdot I = 0$

$$\Delta n \cdot \lambda \cdot e^{\lambda \cdot t} = -\Delta n \cdot \frac{2 \cdot \sigma}{\hbar \cdot \omega} \cdot I \cdot e^{\lambda \cdot t} \Rightarrow -\Delta n \cdot \frac{2 \cdot \sigma}{\hbar \cdot \omega} \cdot I - \Delta n \cdot \lambda = 0$$

$$\frac{d\alpha}{dt} = -\alpha(t) \cdot \frac{2 \cdot \sigma}{\hbar \cdot \omega} \cdot I; \alpha \cdot \lambda \cdot e^{\lambda \cdot t} = -(\alpha^{(i)} + \alpha \cdot e^{\lambda \cdot t}) \cdot \frac{2 \cdot \sigma}{\hbar \cdot \omega} \cdot I$$

$\alpha \cdot \lambda \cdot e^{\lambda \cdot t} = -\alpha^{(i)} \cdot \frac{2 \cdot \sigma}{\hbar \cdot \omega} \cdot I - \alpha \cdot e^{\lambda \cdot t} \cdot \frac{2 \cdot \sigma}{\hbar \cdot \omega} \cdot I$
At fixed points: $-\alpha^{(i)} \cdot \frac{2 \cdot \sigma}{\hbar \cdot \omega} \cdot I = 0$

$$\alpha \cdot \lambda \cdot e^{\lambda \cdot t} = -\alpha \cdot e^{\lambda \cdot t} \cdot \frac{2 \cdot \sigma}{\hbar \cdot \omega} \cdot I \Rightarrow -\alpha \cdot \frac{2 \cdot \sigma}{\hbar \cdot \omega} \cdot I - \alpha \cdot \lambda = 0$$

We can summary our Q-switched microchip lasers semiconductors saturable absorbers system arbitrarily small increments equations:

$$-n_1 \cdot \frac{\sigma}{\hbar \cdot \omega} \cdot I \cdot e^{-\lambda \cdot \tau} - n_1 \cdot \lambda + n_2 \cdot \frac{\sigma}{\hbar \cdot \omega} \cdot I = 0$$
$$n_1 \cdot \frac{\sigma}{\hbar \cdot \omega} \cdot I \cdot e^{-\lambda \cdot \tau} - n_2 \cdot \frac{\sigma}{\hbar \cdot \omega} \cdot I - n_2 \cdot \lambda = 0$$

$$-\Delta n \cdot \frac{2 \cdot \sigma}{\hbar \cdot \omega} \cdot I - \Delta n \cdot \lambda = 0; -\alpha \cdot \frac{2 \cdot \sigma}{\hbar \cdot \omega} \cdot I - \alpha \cdot \lambda = 0$$

Our eigenvalue matrix is

$$\begin{pmatrix} -\frac{\sigma}{\hbar \cdot \omega} \cdot I \cdot e^{-\lambda \cdot \tau} - \lambda & \frac{\sigma}{\hbar \cdot \omega} \cdot I & 0 & 0 \\ \frac{\sigma}{\hbar \cdot \omega} \cdot I \cdot e^{-\lambda \cdot \tau} & -\frac{\sigma}{\hbar \cdot \omega} \cdot I - \lambda & 0 & 0 \\ 0 & 0 & -\frac{2 \cdot \sigma}{\hbar \cdot \omega} \cdot I - \lambda & 0 \\ 0 & 0 & 0 & -\frac{2 \cdot \sigma}{\hbar \cdot \omega} \cdot I - \lambda \end{pmatrix} \cdot \begin{pmatrix} n_1 \\ n_2 \\ \Delta n \\ \alpha \end{pmatrix} = 0$$

$$A - \lambda \cdot I = \begin{pmatrix} -\frac{\sigma}{\hbar \cdot \omega} \cdot I \cdot e^{-\lambda \cdot \tau} - \lambda & \frac{\sigma}{\hbar \cdot \omega} \cdot I & 0 & 0 \\ \frac{\sigma}{\hbar \cdot \omega} \cdot I \cdot e^{-\lambda \cdot \tau} & -\frac{\sigma}{\hbar \cdot \omega} \cdot I - \lambda & 0 & 0 \\ 0 & 0 & -\frac{2 \cdot \sigma}{\hbar \cdot \omega} \cdot I - \lambda & 0 \\ 0 & 0 & 0 & -\frac{2 \cdot \sigma}{\hbar \cdot \omega} \cdot I - \lambda \end{pmatrix}$$

To get the system characteristic equation: $\det(A - \lambda \cdot I) = 0$

$$\det(A - \lambda \cdot I)$$
$$= 0 \Rightarrow \left(-\frac{\sigma}{\hbar \cdot \omega} \cdot I \cdot e^{-\lambda \cdot \tau} - \lambda\right) \cdot \det \begin{pmatrix} -\frac{\sigma}{\hbar \cdot \omega} \cdot I - \lambda & 0 & 0 \\ 0 & -\frac{2 \cdot \sigma}{\hbar \cdot \omega} \cdot I - \lambda & 0 \\ 0 & 0 & -\frac{2 \cdot \sigma}{\hbar \cdot \omega} \cdot I - \lambda \end{pmatrix}$$
$$- \frac{\sigma}{\hbar \cdot \omega} \cdot I \cdot \det \begin{pmatrix} \frac{\sigma}{\hbar \cdot \omega} \cdot I \cdot e^{-\lambda \cdot \tau} & 0 & 0 \\ 0 & -\frac{2 \cdot \sigma}{\hbar \cdot \omega} \cdot I - \lambda & 0 \\ 0 & 0 & -\frac{2 \cdot \sigma}{\hbar \cdot \omega} \cdot I - \lambda \end{pmatrix} = 0$$

$$-\left(\frac{\sigma}{\hbar \cdot \omega} \cdot I \cdot e^{-\lambda \cdot \tau} + \lambda\right) \cdot \left(-\frac{\sigma}{\hbar \cdot \omega} \cdot I - \lambda\right) \cdot \left(-\frac{2 \cdot \sigma}{\hbar \cdot \omega} \cdot I - \lambda\right)$$

$$\cdot\left(-\frac{2\cdot\sigma}{\hbar\cdot\omega}\cdot I-\lambda\right)-\frac{\sigma}{\hbar\cdot\omega}\cdot I\cdot\frac{\sigma}{\hbar\cdot\omega}\cdot I\cdot e^{-\lambda\cdot\tau}$$
$$\cdot\left(-\frac{2\cdot\sigma}{\hbar\cdot\omega}\cdot I-\lambda\right)\cdot\left(-\frac{2\cdot\sigma}{\hbar\cdot\omega}\cdot I-\lambda\right)=0$$

$$\left(\frac{\sigma}{\hbar\cdot\omega}\cdot I\cdot e^{-\lambda\cdot\tau}+\lambda\right)\cdot\left(\frac{\sigma}{\hbar\cdot\omega}\cdot I+\lambda\right)\cdot\left(\frac{2\cdot\sigma}{\hbar\cdot\omega}\cdot I+\lambda\right)\cdot\left(\frac{2\cdot\sigma}{\hbar\cdot\omega}\cdot I+\lambda\right)$$
$$-\left(\frac{\sigma\cdot I}{\hbar\cdot\omega}\right)^2\cdot e^{-\lambda\cdot\tau}\cdot\left(\frac{2\cdot\sigma}{\hbar\cdot\omega}\cdot I+\lambda\right)\cdot\left(\frac{2\cdot\sigma}{\hbar\cdot\omega}\cdot I+\lambda\right)=0$$

$$\left(\frac{\sigma}{\hbar\cdot\omega}\cdot I\cdot e^{-\lambda\cdot\tau}+\lambda\right)\cdot\left(\frac{\sigma}{\hbar\cdot\omega}\cdot I+\lambda\right)\cdot\left(\frac{2\cdot\sigma}{\hbar\cdot\omega}\cdot I+\lambda\right)^2$$
$$-\left(\frac{\sigma\cdot I}{\hbar\cdot\omega}\right)^2\cdot e^{-\lambda\cdot\tau}\cdot\left(\frac{2\cdot\sigma}{\hbar\cdot\omega}\cdot I+\lambda\right)^2=0$$

$$\left(\left[\frac{\sigma}{\hbar\cdot\omega}\cdot I\right]^2\cdot e^{-\lambda\cdot\tau}+\frac{\sigma}{\hbar\cdot\omega}\cdot I\cdot e^{-\lambda\cdot\tau}\cdot\lambda+\frac{\sigma}{\hbar\cdot\omega}\cdot I\cdot\lambda+\lambda^2\right)$$
$$\cdot\left(\left[\frac{2\cdot\sigma}{\hbar\cdot\omega}\cdot I\right]^2+\frac{4\cdot\sigma}{\hbar\cdot\omega}\cdot I\cdot\lambda+\lambda^2\right)$$
$$-\left(\frac{\sigma\cdot I}{\hbar\cdot\omega}\right)^2\cdot e^{-\lambda\cdot\tau}\cdot\left(\left[\frac{2\cdot\sigma}{\hbar\cdot\omega}\cdot I\right]^2+\frac{4\cdot\sigma}{\hbar\cdot\omega}\cdot I\cdot\lambda+\lambda^2\right)=0$$

$$(\left[\frac{\sigma}{\hbar\cdot\omega}\cdot I\right]^2\cdot e^{-\lambda\cdot\tau}\cdot\left[\frac{2\cdot\sigma}{\hbar\cdot\omega}\cdot I\right]^2+\left[\frac{\sigma}{\hbar\cdot\omega}\cdot I\right]^2\cdot e^{-\lambda\cdot\tau}\cdot\frac{4\cdot\sigma}{\hbar\cdot\omega}\cdot I\cdot\lambda$$
$$+\left[\frac{\sigma}{\hbar\cdot\omega}\cdot I\right]^2\cdot e^{-\lambda\cdot\tau}\cdot\lambda^2\frac{\sigma}{\hbar\cdot\omega}\cdot I\cdot e^{-\lambda\cdot\tau}\cdot\lambda\cdot\left[\frac{2\cdot\sigma}{\hbar\cdot\omega}\cdot I\right]^2$$
$$+\frac{\sigma}{\hbar\cdot\omega}\cdot I\cdot e^{-\lambda\cdot\tau}\cdot\lambda\cdot\frac{4\cdot\sigma}{\hbar\cdot\omega}\cdot I\cdot\lambda+\frac{\sigma}{\hbar\cdot\omega}\cdot I\cdot e^{-\lambda\cdot\tau}\cdot\lambda\cdot\lambda^2$$
$$+\lambda^2\cdot\left[\frac{2\cdot\sigma}{\hbar\cdot\omega}\cdot I\right]^2+\lambda^2\cdot\frac{4\cdot\sigma}{\hbar\cdot\omega}\cdot I\cdot\lambda+\lambda^4)$$
$$-\left(\frac{\sigma\cdot I}{\hbar\cdot\omega}\right)^2\cdot e^{-\lambda\cdot\tau}\cdot\left(\left[\frac{2\cdot\sigma}{\hbar\cdot\omega}\cdot I\right]^2+\frac{4\cdot\sigma}{\hbar\cdot\omega}\cdot I\cdot\lambda+\lambda^2\right)=0$$

$$(\left[\frac{\sigma}{\hbar\cdot\omega}\cdot I\right]^2\cdot e^{-\lambda\cdot\tau}\cdot\left[\frac{2\cdot\sigma}{\hbar\cdot\omega}\cdot I\right]^2+\left[\frac{\sigma}{\hbar\cdot\omega}\cdot I\right]^2\cdot e^{-\lambda\cdot\tau}\cdot\frac{4\cdot\sigma}{\hbar\cdot\omega}\cdot I\cdot\lambda$$
$$+\left[\frac{\sigma}{\hbar\cdot\omega}\cdot I\right]^2\cdot e^{-\lambda\cdot\tau}\cdot\lambda^2\frac{\sigma}{\hbar\cdot\omega}\cdot I\cdot e^{-\lambda\cdot\tau}\cdot\lambda\cdot\left[\frac{2\cdot\sigma}{\hbar\cdot\omega}\cdot I\right]^2$$
$$+\frac{\sigma}{\hbar\cdot\omega}\cdot I\cdot e^{-\lambda\cdot\tau}\cdot\lambda\cdot\frac{4\cdot\sigma}{\hbar\cdot\omega}\cdot I\cdot\lambda+\frac{\sigma}{\hbar\cdot\omega}\cdot I\cdot e^{-\lambda\cdot\tau}\cdot\lambda\cdot\lambda^2$$

$$+\lambda^2\cdot\left[\frac{2\cdot\sigma}{\hbar\cdot\omega}\cdot I\right]^2+\lambda^2\cdot\frac{4\cdot\sigma}{\hbar\cdot\omega}\cdot I\cdot\lambda+\lambda^4)-\left(\frac{\sigma\cdot I}{\hbar\cdot\omega}\right)^2\cdot e^{-\lambda\cdot\tau}\cdot\left[\frac{2\cdot\sigma}{\hbar\cdot\omega}\cdot I\right]^2$$
$$-\left(\frac{\sigma\cdot I}{\hbar\cdot\omega}\right)^2\cdot e^{-\lambda\cdot\tau}\cdot\frac{4\cdot\sigma}{\hbar\cdot\omega}\cdot I\cdot\lambda-\left(\frac{\sigma\cdot I}{\hbar\cdot\omega}\right)^2\cdot e^{-\lambda\cdot\tau}\cdot\lambda^2=0$$

$$4\cdot\left(\frac{\sigma\cdot I}{\hbar\cdot\omega}\right)^4\cdot e^{-\lambda\cdot\tau}+4\cdot\left(\frac{\sigma\cdot I}{\hbar\cdot\omega}\right)^3\cdot e^{-\lambda\cdot\tau}\cdot\lambda+\left(\frac{\sigma\cdot I}{\hbar\cdot\omega}\right)^2\cdot e^{-\lambda\cdot\tau}\cdot\lambda^2$$
$$+4\cdot\left(\frac{\sigma\cdot I}{\hbar\cdot\omega}\right)^3\cdot e^{-\lambda\cdot\tau}\cdot\lambda+4\cdot\left(\frac{\sigma\cdot I}{\hbar\cdot\omega}\right)^2\cdot e^{-\lambda\cdot\tau}\cdot\lambda^2+\left(\frac{\sigma\cdot I}{\hbar\cdot\omega}\right)\cdot e^{-\lambda\cdot\tau}\cdot\lambda^3$$
$$+\lambda^2\cdot\left(\frac{2\cdot\sigma\cdot I}{\hbar\cdot\omega}\right)^2+\lambda^3\cdot\left(\frac{4\cdot\sigma\cdot I}{\hbar\cdot\omega}\right)+\lambda^4-4\cdot\left(\frac{\sigma\cdot I}{\hbar\cdot\omega}\right)^4\cdot e^{-\lambda\cdot\tau}$$
$$-4\cdot\left(\frac{\sigma\cdot I}{\hbar\cdot\omega}\right)^3\cdot e^{-\lambda\cdot\tau}\cdot\lambda-\left(\frac{\sigma\cdot I}{\hbar\cdot\omega}\right)^2\cdot e^{-\lambda\cdot\tau}\cdot\lambda^2=0$$

$$\lambda^2\cdot\left(\frac{2\cdot\sigma\cdot I}{\hbar\cdot\omega}\right)^2+\lambda^3\cdot\left(\frac{4\cdot\sigma\cdot I}{\hbar\cdot\omega}\right)+\lambda^4$$
$$+\left\{4\cdot\left(\frac{\sigma\cdot I}{\hbar\cdot\omega}\right)^3\cdot\lambda+4\cdot\left(\frac{\sigma\cdot I}{\hbar\cdot\omega}\right)^2\cdot\lambda^2+\left(\frac{\sigma\cdot I}{\hbar\cdot\omega}\right)\cdot\lambda^3\right\}\cdot e^{-\lambda\cdot\tau}=0$$

$$D(\lambda,\tau)=\lambda^2\cdot\left(\frac{2\cdot\sigma\cdot I}{\hbar\cdot\omega}\right)^2+\lambda^3\cdot\left(\frac{4\cdot\sigma\cdot I}{\hbar\cdot\omega}\right)$$
$$+\lambda^4+\left\{4\cdot\left(\frac{\sigma\cdot I}{\hbar\cdot\omega}\right)^3\cdot\lambda+4\cdot\left(\frac{\sigma\cdot I}{\hbar\cdot\omega}\right)^2\cdot\lambda^2+\left(\frac{\sigma\cdot I}{\hbar\cdot\omega}\right)\cdot\lambda^3\right\}\cdot e^{-\lambda\cdot\tau}$$

$$D(\lambda,\tau)=P_n(\lambda,\tau)+Q_m(\lambda,\tau)\cdot e^{-\lambda\cdot\tau};n=4;m=3;n>m$$

$$P_{n=4}(\lambda,\tau)=\lambda^2\cdot\left(\frac{2\cdot\sigma\cdot I}{\hbar\cdot\omega}\right)^2+\lambda^3\cdot\left(\frac{4\cdot\sigma\cdot I}{\hbar\cdot\omega}\right)+\lambda^4;\ Q_{m=3}(\lambda,\tau)$$
$$=4\cdot\left(\frac{\sigma\cdot I}{\hbar\cdot\omega}\right)^3\cdot\lambda+4\cdot\left(\frac{\sigma\cdot I}{\hbar\cdot\omega}\right)^2\cdot\lambda^2+\left(\frac{\sigma\cdot I}{\hbar\cdot\omega}\right)\cdot\lambda^3$$

The expression for $P_n(\lambda,\tau)$:

$$P_{n=4}(\lambda,\tau)=\sum_{k=0}^{n=4}p_k(\tau)\cdot\lambda^k=p_0(\tau)+p_1(\tau)\cdot\lambda+p_2(\tau)\cdot\lambda^2$$
$$+p_3(\tau)\cdot\lambda^3+p_4(\tau)\cdot\lambda^4$$

$$p_0(\tau)=0;\ p_1(\tau)=0;\ p_2(\tau)=\left(\frac{2\cdot\sigma\cdot I}{\hbar\cdot\omega}\right)^2;\ p_3(\tau)=\frac{4\cdot\sigma\cdot I}{\hbar\cdot\omega};\ p_4(\tau)=1$$

$$Q_{m=3}(\lambda,\tau) = \sum_{k=0}^{n=3} q_k(\tau)\cdot\lambda^k = q_0(\tau) + q_1(\tau)\cdot\lambda + q_2(\tau)\cdot\lambda^2 + q_3(\tau)\cdot\lambda^3$$

$$q_0(\tau) = 0;\ q_1(\tau) = 4\cdot\left(\frac{\sigma\cdot I}{\hbar\cdot\omega}\right)^3;\ q_2(\tau) = 4\cdot\left(\frac{\sigma\cdot I}{\hbar\cdot\omega}\right)^2;\ q_3(\tau) = \frac{\sigma\cdot I}{\hbar\cdot\omega}$$

The homogeneous system for $n_1\ n_2\ \Delta n\ \alpha$ leads to a characteristic equation for the eigenvalue λ having the form $P(\lambda)+Q(\lambda)\cdot e^{-\lambda\cdot\tau}$; $P(\lambda) = \sum_{j=0}^{4} a_j\cdot\lambda^j$; $Q(\lambda) = \sum_{j=0}^{3} c_j\cdot\lambda^j$ and the coefficients $\{a_j(q_i,q_k), c_j(q_i,q_k)\}\in\mathbb{R}$ depend on q_i, q_k but not on τ, q_i, q_k are any Q-switched microchip lasers semiconductors saturable absorbers system parameters, other parameters keep as a constant.

$$a_0 = 0;\ a_1 = 0;\ a_2 = \left(\frac{2\cdot\sigma\cdot I}{\hbar\cdot\omega}\right)^2;\ a_3 = \frac{4\cdot\sigma\cdot I}{\hbar\cdot\omega};\ a_4 = 1$$

$$c_0 = 0;\ c_1 = 4\cdot\left(\frac{\sigma\cdot I}{\hbar\cdot\omega}\right)^3;\ c_2 = 4\cdot\left(\frac{\sigma\cdot I}{\hbar\cdot\omega}\right)^2;\ c_3 = \frac{\sigma\cdot I}{\hbar\cdot\omega}$$

Unless strictly necessary, the designation of the variation argument (q_i, q_k) will subsequently be omitted from P, Q, a_j and c_j. The coefficients a_j and c_j are continuous and differential functions of their arguments, and direct substitution shows that $a_0 + c_0 = 0\,\forall q_i, q_k\in\mathbb{R}_+$. $\lambda = 0$ is not a of $P(\lambda)+Q(\lambda)\cdot e^{-\lambda\cdot\tau} = 0$. Furthermore, $P(\lambda)$, $Q(\lambda)$ are analytic functions of λ, for which the following requirements of the analysis [BK] can also be verified in the present case:

(a) If $\lambda = i\cdot\omega;\ \omega\in\mathbb{R}$ then $P(i\cdot\omega)+Q(i\cdot\omega)\neq 0$.
(b) $\left|\frac{Q(\lambda)}{P(\lambda)}\right|$ is bounded for $|\lambda|\to\infty$; $\mathrm{Re}\lambda\geq 0$. No roots bifurcation from ∞.
(c) $F(\omega) = |P(i\cdot\omega)|^2 - |Q(i\cdot\omega)|^2$ has a finite number of zeros. Indeed, this is a polynomial in ω(degree in ω^4).
(d) Each positive root $\omega(q_i,q_k)$ of $F(\omega)=0$ is continuous and differentiable with respect to q_i, q_k parameters.

Hint: We define ω parameter as the system angular frequency of the optical field and as a ω parameter in our stability analysis.

We assume that $P_n(\lambda,\tau) = P_n(\lambda)$; $Q_m(\lambda,\tau) = Q_m(\lambda)$ cannot have common imaginary roots. That is for any real number ω: $P_n(\lambda = i\cdot\omega,\tau)+Q_m(\lambda = i\cdot\omega,\tau)\neq 0$.

$$P_{n=4}(\lambda = i\cdot\omega,\tau) = \omega^4 - \omega^2\cdot\left(\frac{2\cdot\sigma\cdot I}{\hbar\cdot\omega}\right)^2 - i\cdot\omega^3\cdot\left(\frac{4\cdot\sigma\cdot I}{\hbar\cdot\omega}\right)$$

$$Q_{m=3}(\lambda = i \cdot \omega, \tau) = -4 \cdot \left(\frac{\sigma \cdot I}{\hbar \cdot \omega}\right)^2 \cdot \omega^2 + i \cdot \left[4 \cdot \left(\frac{\sigma \cdot I}{\hbar \cdot \omega}\right)^3 \cdot \omega - \left(\frac{\sigma \cdot I}{\hbar \cdot \omega}\right) \cdot \omega^3\right]$$

$$P_n(\lambda = i \cdot \omega, \tau) + Q_m(\lambda = i \cdot \omega, \tau) = \omega^4 - \omega^2 \cdot \left(\frac{2 \cdot \sigma \cdot I}{\hbar \cdot \omega}\right)^2 - i \cdot \omega^3 \cdot \left(\frac{4 \cdot \sigma \cdot I}{\hbar \cdot \omega}\right) - 4 \cdot \left(\frac{\sigma \cdot I}{\hbar \cdot \omega}\right)^2 \cdot \omega^2 + 4 \cdot \left(\frac{\sigma \cdot I}{\hbar \cdot \omega}\right)^3 \cdot i \cdot \omega - i \cdot \left(\frac{\sigma \cdot I}{\hbar \cdot \omega}\right) \cdot \omega^3 \neq 0$$

$$P_n(\lambda = i \cdot \omega, \tau) + Q_m(\lambda = i \cdot \omega, \tau) = \omega^4 - \left[\left(\frac{2 \cdot \sigma \cdot I}{\hbar \cdot \omega}\right)^2 + 4 \cdot \left(\frac{\sigma \cdot I}{\hbar \cdot \omega}\right)^2\right] \cdot \omega^2 + 4 \cdot \left(\frac{\sigma \cdot I}{\hbar \cdot \omega}\right)^3 \cdot i \cdot \omega - i \cdot \omega^3 \cdot \left[\left(\frac{4 \cdot \sigma \cdot I}{\hbar \cdot \omega}\right) + \left(\frac{\sigma \cdot I}{\hbar \cdot \omega}\right)\right] \neq 0$$

$$P_n(\lambda = i \cdot \omega, \tau) + Q_m(\lambda = i \cdot \omega, \tau) = \omega^4 - 8 \cdot \left(\frac{\sigma \cdot I}{\hbar \cdot \omega}\right)^2 \cdot \omega^2 + 4 \cdot \left(\frac{\sigma \cdot I}{\hbar \cdot \omega}\right)^3 \cdot i \cdot \omega - i \cdot \omega^3 \cdot \frac{5 \cdot \sigma \cdot I}{\hbar \cdot \omega} \neq 0$$

$$|P_n(\lambda = i \cdot \omega, \tau)|^2 = \left[\omega^4 - \omega^2 \cdot \left(\frac{2 \cdot \sigma \cdot I}{\hbar \cdot \omega}\right)^2\right]^2 + \omega^6 \cdot \left(\frac{4 \cdot \sigma \cdot I}{\hbar \cdot \omega}\right)^2$$

$$|Q_m(\lambda = i \cdot \omega, \tau)|^2 = 16 \cdot \left(\frac{\sigma \cdot I}{\hbar \cdot \omega}\right)^4 \cdot \omega^4 + \left[4 \cdot \left(\frac{\sigma \cdot I}{\hbar \cdot \omega}\right)^3 \cdot \omega - \left(\frac{\sigma \cdot I}{\hbar \cdot \omega}\right) \cdot \omega^3\right]^2$$

$$F(\omega) = |P(i \cdot \omega)|^2 - |Q(i \cdot \omega)|^2 = \left[\omega^4 - \omega^2 \cdot \left(\frac{2 \cdot \sigma \cdot I}{\hbar \cdot \omega}\right)^2\right]^2 + \omega^6 \cdot \left(\frac{4 \cdot \sigma \cdot I}{\hbar \cdot \omega}\right)^2 - 16 \cdot \left(\frac{\sigma \cdot I}{\hbar \cdot \omega}\right)^4 \cdot \omega^4 - \left[4 \cdot \left(\frac{\sigma \cdot I}{\hbar \cdot \omega}\right)^3 \cdot \omega - \left(\frac{\sigma \cdot I}{\hbar \cdot \omega}\right) \cdot \omega^3\right]^2$$

$$F(\omega) = |P(i\cdot\omega)|^2 - |Q(i\cdot\omega)|^2 = \omega^8 - 2\cdot\omega^6\cdot\left(\frac{2\cdot\sigma\cdot I}{\hbar\cdot\omega}\right)^2 + \omega^4\cdot\left(\frac{2\cdot\sigma\cdot I}{\hbar\cdot\omega}\right)^4 + \omega^6\cdot\left(\frac{4\cdot\sigma\cdot I}{\hbar\cdot\omega}\right)^2 - 16\cdot\left(\frac{\sigma\cdot I}{\hbar\cdot\omega}\right)^4\cdot\omega^4 - \left[16\cdot\left(\frac{\sigma\cdot I}{\hbar\cdot\omega}\right)^6\cdot\omega^2 - 8\cdot\left(\frac{\sigma\cdot I}{\hbar\cdot\omega}\right)^4\cdot\omega^4 + \left(\frac{\sigma\cdot I}{\hbar\cdot\omega}\right)^2\cdot\omega^6\right]$$

$$F(\omega) = |P(i\cdot\omega)|^2 - |Q(i\cdot\omega)|^2 = \omega^8 - 2\cdot\omega^6\cdot\left(\frac{2\cdot\sigma\cdot I}{\hbar\cdot\omega}\right)^2 + \omega^4\cdot\left(\frac{2\cdot\sigma\cdot I}{\hbar\cdot\omega}\right)^4 + \omega^6\cdot\left(\frac{4\cdot\sigma\cdot I}{\hbar\cdot\omega}\right)^2 - 16\cdot\left(\frac{\sigma\cdot I}{\hbar\cdot\omega}\right)^4\cdot\omega^4 - 16\cdot\left(\frac{\sigma\cdot I}{\hbar\cdot\omega}\right)^6\cdot\omega^2 + 8\cdot\left(\frac{\sigma\cdot I}{\hbar\cdot\omega}\right)^4\cdot\omega^4 - \left(\frac{\sigma\cdot I}{\hbar\cdot\omega}\right)^2\cdot\omega^6$$

$$F(\omega) = |P(i\cdot\omega)|^2 - |Q(i\cdot\omega)|^2 = \omega^8 + \omega^6\cdot\left[\left(\frac{4\cdot\sigma\cdot I}{\hbar\cdot\omega}\right)^2 - 2\cdot\left(\frac{2\cdot\sigma\cdot I}{\hbar\cdot\omega}\right)^2 - \left(\frac{\sigma\cdot I}{\hbar\cdot\omega}\right)^2\right] + \omega^4\cdot\left[\left(\frac{2\cdot\sigma\cdot I}{\hbar\cdot\omega}\right)^4 - 16\cdot\left(\frac{\sigma\cdot I}{\hbar\cdot\omega}\right)^4 + 8\cdot\left(\frac{\sigma\cdot I}{\hbar\cdot\omega}\right)^4\right] - 16\cdot\left(\frac{\sigma\cdot I}{\hbar\cdot\omega}\right)^6\cdot\omega^2$$

$$F(\omega) = |P(i\cdot\omega)|^2 - |Q(i\cdot\omega)|^2 = \omega^8 + \omega^6\cdot 7\cdot\left(\frac{\sigma\cdot I}{\hbar\cdot\omega}\right)^2 + \omega^4\cdot 8\cdot\left(\frac{\sigma\cdot I}{\hbar\cdot\omega}\right)^4 - \omega^2\cdot 16\cdot\left(\frac{\sigma\cdot I}{\hbar\cdot\omega}\right)^6$$

We define the following parameters for simplicity: $\Xi_0 = 0$; $\Xi_2 = -16\cdot\left(\frac{\sigma\cdot I}{\hbar\cdot\omega}\right)^6$

$$\Xi_4 = 8\cdot\left(\frac{\sigma\cdot I}{\hbar\cdot\omega}\right)^4;\ \Xi_6 = 7\cdot\left(\frac{\sigma\cdot I}{\hbar\cdot\omega}\right)^2;\ \Xi_8 = 1$$

$$F(\omega) = |P(i\cdot\omega)|^2 - |Q(i\cdot\omega)|^2 = \Xi_8\cdot\omega^8 + \Xi_6\cdot\omega^6 + \Xi_4\cdot\omega^4 + \Xi_2\cdot\omega^2 = \sum_{k=0}^{4}\Xi_{2\cdot k}\cdot\omega^{2\cdot k}$$

Hence $F(\omega) = 0$ implies $\sum_{k=0}^{4}\Xi_{2\cdot k}\cdot\omega^{2\cdot k} = 0$ and its roots are given by solving the above polynomial.

$$P_{n=4}(i\cdot\omega)=\omega^4-\omega^2\cdot\left(\frac{2\cdot\sigma\cdot I}{\hbar\cdot\omega}\right)^2-i\cdot\omega^3\cdot\left(\frac{4\cdot\sigma\cdot I}{\hbar\cdot\omega}\right)$$

$$Q_{m=3}(i\cdot\omega)=-4\cdot\left(\frac{\sigma\cdot I}{\hbar\cdot\omega}\right)^2\cdot\omega^2+i\cdot\left[4\cdot\left(\frac{\sigma\cdot I}{\hbar\cdot\omega}\right)^3\cdot\omega-\left(\frac{\sigma\cdot I}{\hbar\cdot\omega}\right)\cdot\omega^3\right]$$

Furthermore $P_R(i\cdot\omega)=\omega^4-\omega^2\cdot\left(\frac{2\cdot\sigma\cdot I}{\hbar\cdot\omega}\right)^2$; $P_I(i\cdot\omega)=-\omega^3\cdot\left(\frac{4\cdot\sigma\cdot I}{\hbar\cdot\omega}\right)$

$$Q_R(i\cdot\omega)=-4\cdot\left(\frac{\sigma\cdot I}{\hbar\cdot\omega}\right)^2\cdot\omega^2;\ Q_I(i\cdot\omega)=4\cdot\left(\frac{\sigma\cdot I}{\hbar\cdot\omega}\right)^3\cdot\omega-\left(\frac{\sigma\cdot I}{\hbar\cdot\omega}\right)\cdot\omega^3$$

Hence

$$\sin\theta(\tau)=\frac{-P_R(i\cdot\omega)\cdot Q_I(i\cdot\omega)+P_I(i\cdot\omega)\cdot Q_R(i\cdot\omega)}{|Q(i\cdot\omega)|^2}$$

$$\cos\theta(\tau)=-\frac{P_R(i\cdot\omega)\cdot Q_R(i\cdot\omega)+P_I(i\cdot\omega)\cdot Q_I(i\cdot\omega)}{|Q(i\cdot\omega)|^2}$$

And

$$\sin\theta(\tau)=\frac{-\left[\omega^4-\omega^2\cdot\left(\frac{2\cdot\sigma\cdot I}{\hbar\cdot\omega}\right)^2\right]\cdot\left[4\cdot\left(\frac{\sigma\cdot I}{\hbar\cdot\omega}\right)^3\cdot\omega-\left(\frac{\sigma\cdot I}{\hbar\cdot\omega}\right)\cdot\omega^3\right]+16\cdot\omega^5\cdot\left(\frac{\sigma\cdot I}{\hbar\cdot\omega}\right)^3}{16\cdot\left(\frac{\sigma\cdot I}{\hbar\cdot\omega}\right)^4\cdot\omega^4+\left[4\cdot\left(\frac{\sigma\cdot I}{\hbar\cdot\omega}\right)^3\cdot\omega-\left(\frac{\sigma\cdot I}{\hbar\cdot\omega}\right)\cdot\omega^3\right]^2}$$

$$\cos\theta(\tau)=-\frac{-\left[\omega^4-\omega^2\cdot\left(\frac{2\cdot\sigma\cdot I}{\hbar\cdot\omega}\right)^2\right]\cdot 4\cdot\left(\frac{\sigma\cdot I}{\hbar\cdot\omega}\right)^2\cdot\omega^2-\omega^3\cdot\left(\frac{4\cdot\sigma\cdot I}{\hbar\cdot\omega}\right)\cdot\left[4\cdot\left(\frac{\sigma\cdot I}{\hbar\cdot\omega}\right)^3\cdot\omega-\left(\frac{\sigma\cdot I}{\hbar\cdot\omega}\right)\cdot\omega^3\right]}{16\cdot\left(\frac{\sigma\cdot I}{\hbar\cdot\omega}\right)^4\cdot\omega^4+\left[4\cdot\left(\frac{\sigma\cdot I}{\hbar\cdot\omega}\right)^3\cdot\omega-\left(\frac{\sigma\cdot I}{\hbar\cdot\omega}\right)\cdot\omega^3\right]^2}$$

Which jointly with $F(\omega)=0\Rightarrow\sum_{k=0}^{4}\Xi_{2\cdot k}\cdot\omega^{2\cdot k}=0$, that are continuous and differentiable in τ.

It based on Lemma 1.1, hence we use Theorem 1.2 and this proves the Theorem 1.3. We use different parameters terminology from our last characteristics parameters definition $k\to j$; $p_k(\tau)\to a_j$ and $q_k(\tau)\to c_j$; $n=4$; $m=3$; $n>m$, additionally $P_n(\lambda,\tau)\to P(\lambda,\tau)$; $Q_m(\lambda,\tau)\to Q(\lambda,\tau)$ then $P(\lambda,\tau_\Delta)=\sum_{j=0}^{2}a_j\cdot\lambda^j$; $Q(\lambda,\tau_\Delta)=\sum_{j=0}^{1}c_j\cdot\lambda^j$

$$P(\lambda,\tau) = \lambda^2 \cdot \left(\frac{2\cdot\sigma\cdot I}{\hbar\cdot\omega}\right)^2 + \lambda^3 \cdot \left(\frac{4\cdot\sigma\cdot I}{\hbar\cdot\omega}\right) + \lambda^4;\ Q(\lambda,\tau)$$
$$= 4\cdot\left(\frac{\sigma\cdot I}{\hbar\cdot\omega}\right)^3 \cdot\lambda + 4\cdot\left(\frac{\sigma\cdot I}{\hbar\cdot\omega}\right)^2 \cdot\lambda^2 + \left(\frac{\sigma\cdot I}{\hbar\cdot\omega}\right)\cdot\lambda^3$$

$n, m \in \mathbb{N}_0; n > m$ and $a_j, c_j : \mathbb{R}_{+0} \to \mathbb{R}$ are continuous and differentiable functions of τ such that $a_0 + c_0 = 0$. In the following "–" denotes complex and conjugate. $P(\lambda)$ and $Q(\lambda)$ are analytic functions in λ and differentiable in τ. The coefficients $\{a_j(\sigma,\omega,I)\&\ c_j(\sigma,\omega,I)\} \in \mathbb{R}$ depend on Q-switched microchip lasers semiconductors saturable absorbers system parameters $\sigma, \omega, I, \tau, \ldots$ values. Unless strictly necessary, the designation of the variation arguments, $\sigma, \omega, I, \tau, \ldots$ system parameters will subsequently be omitted from P, Q, a_j, c_j. The coefficients a_j, c_j are continuous and differentiable functions of their arguments, and direct substitution shows that $a_0 + c_0 = 0$.

Furthermore, $P(\lambda), Q(\lambda)$ are analytic function of λ for which the following requirements of the analysis (see Kuang 1993, Sect. 3.4) can also be verified in the present case [9, 10].

(e) If $\lambda = i\cdot\omega;\ \omega \in \mathbb{R}$ then $P(i\cdot\omega) + Q(i\cdot\omega) \neq 0$, i.e. P and Q have no common imaginary roots. This condition was verified numerically in the entire (σ, ω, I, τ system parameters) domain of interest.

(f) $\left|\frac{Q(\lambda)}{P(\lambda)}\right|$ is bounded for $|\lambda| \to \infty$; $\mathrm{Re}\lambda \geq 0$. No roots bifurcation from ∞. Indeed, in the limit: $\left|\frac{Q(\lambda)}{P(\lambda)}\right| = \left|\frac{4\cdot\left(\frac{\sigma\cdot I}{\hbar\cdot\omega}\right)^3\cdot\lambda + 4\cdot\left(\frac{\sigma\cdot I}{\hbar\cdot\omega}\right)^2\cdot\lambda^2 + \left(\frac{\sigma\cdot I}{\hbar\cdot\omega}\right)\cdot\lambda^3}{\lambda^2\cdot\left(\frac{2\cdot\sigma\cdot I}{\hbar\cdot\omega}\right)^2 + \lambda^3\cdot\left(\frac{4\cdot\sigma\cdot I}{\hbar\cdot\omega}\right) + \lambda^4}\right|$.

(g) $F(\omega) = |P(i\cdot\omega)|^2 - |Q(i\cdot\omega)|^2 = \omega^8 + \omega^6\cdot 7\cdot\left(\frac{\sigma\cdot I}{\hbar\cdot\omega}\right)^2 + \omega^4\cdot 8\cdot\left(\frac{\sigma\cdot I}{\hbar\cdot\omega}\right)^4 - \omega^2\cdot 16\cdot\left(\frac{\sigma\cdot I}{\hbar\cdot\omega}\right)^6$ has at most a finite number of zeros. Indeed, this is a polynomial in ω (degree in ω^8).

(h) Each positive root $\omega(\sigma,\omega,I,\tau)$ of $F(\omega) = 0$ is continuous and differentiable with respect to σ, ω, I, τ parameters. This condition can only be assessed numerically.

In addition, since the coefficients in P and Q are real, we have $\overline{P(-i\cdot\omega)} = P(i\cdot\omega)$ and $\overline{Q(-i\cdot\omega)} = Q(i\cdot\omega)$ thus $\lambda = i\cdot\omega\,\forall\omega > 0$ may be on eigenvalue of characteristic equation. The analysis consists in identifying the roots of characteristic equation situated on the imaginary axis of the complex λ—plane, where by increasing the parameters σ, ω, I, τ(system parameters), $\mathrm{Re}\lambda$ may, at the crossing, change its sign from (−) to (+), i.e. from a stable focus $E^{(k)}(n_1^{(k)}, n_2^{(k)}, \Delta n^{(k)}, \alpha^{(k)}) = (n_1^{(*)}, n_1^{(*)}, 0, 0)\ \forall k = 0, 1, 2$ to an unstable one, or vice versa. This feature may be further assessed by examining the sign of the partial derivatives with respect to σ, ω, I, τ parameters ($\omega \in \mathbb{R}_+$).

$$\Lambda^{-1}(\sigma) = \left(\frac{\partial \mathrm{Re}\lambda}{\partial\sigma}\right)_{\lambda = i\cdot\omega};\ \omega, I, \tau = const$$

$$\Lambda^{-1}(\omega) = \left(\frac{\partial \mathrm{Re}\lambda}{\partial \omega}\right)_{\lambda=i\cdot\omega} ; \sigma, I, \tau = const$$

$$\Lambda^{-1}(I) = \left(\frac{\partial \mathrm{Re}\lambda}{\partial I}\right)_{\lambda=i\cdot\omega} ; \sigma, \omega, \tau = const$$

$$\Lambda^{-1}(\tau) = \left(\frac{\partial \mathrm{Re}\lambda}{\partial \tau}\right)_{\lambda=i\cdot\omega} ; \sigma, \omega, I = const$$

where writing $P(\lambda) = P_R(\lambda)+i\cdot P_I(\lambda)$ and $Q(\lambda) = Q_R(\lambda)+i\cdot Q_I(\lambda)$, and inserting $\lambda = i \cdot \omega$ into Q-switched microchip lasers semiconductors saturable absorbers system's characteristic equation ω must satisfy the following:

$$\sin(\omega \cdot \tau) = g(\omega) = \frac{-P_R(i\cdot\omega)\cdot Q_I(i\cdot\omega) + P_I(i\cdot\omega)\cdot Q_R(i\cdot\omega)}{|Q(i\cdot\omega)|^2}$$

$$\cos(\omega \cdot \tau) = h(\omega) = -\frac{P_R(i\cdot\omega)\cdot Q_R(i\cdot\omega) + P_I(i\cdot\omega)\cdot Q_I(i\cdot\omega)}{|Q(i\cdot\omega)|^2}$$

where $|Q(i\cdot\omega)|^2 \neq 0$ in the view of requirement (a) above, $(g, h) \in \mathbb{R}$. Furthermore, it follows the above equations $\sin(\omega\cdot\tau)$ and $\cos(\omega\cdot\tau)$ that, by squaring and adding the sides, ω must be a positive root of $F(\omega) = |P(i\cdot\omega)|^2 - |Q(i\cdot\omega)|^2 = 0$. Note $F(\omega)$ is independent of τ. Now it is important to notice that if $\tau \notin I$ (assume that $I \subseteq \mathbb{R}_{+0}$ is the set where $\omega(\tau)$ is a positive root of $F(\omega)$ and for $\tau \notin I$; $\omega(\tau)$ is not define. Then for all $\tau \in I$ is satisfies that $F(\omega, \tau) = 0$. Then there are no positive $\omega(\tau)$ solutions for $F(\omega, \tau) = 0$, and we cannot have stability switch. For any $\tau \in I$, where $\omega(\tau)$ is a positive solution of $F(\omega, \tau) = 0$, we can define the angle $\theta(\tau) \in [0, 2\cdot\pi]$ as the solution of $\sin\theta(\tau) = \frac{-P_R(i\cdot\omega)\cdot Q_I(i\cdot\omega)+P_I(i\cdot\omega)\cdot Q_R(i\cdot\omega)}{|Q(i\cdot\omega)|^2}$; $\cos\theta(\tau) = -\frac{P_R(i\cdot\omega)\cdot Q_R(i\cdot\omega)+P_I(i\cdot\omega)\cdot Q_I(i\cdot\omega)}{|Q(i\cdot\omega)|^2}$ and the relation between the argument $\theta(\tau)$ and $\omega(\tau)\cdot\tau$ for $\tau \in I$ must be $\omega(\tau)\cdot\tau = \theta(\tau)+2\cdot n\cdot\pi \; \forall n \in \mathbb{N}_0$. Hence we can define the maps $\tau_{\Delta n} : I \to \mathbb{R}_{+0}$ given by $\tau_{\Delta n}(\tau) = \frac{\theta(\tau)+n\cdot 2\cdot\pi}{\omega(\tau)}$; $n \in \mathbb{N}_0$, $\tau \in I$. Let introduce the functions $I \to \mathbb{R}$; $S_n(\tau) = \tau - \tau_{\Delta n}(\tau)$, $n \in \mathbb{N}_0$, $\tau \in I$, that are continuous and differentiable in τ. In the following, the subscripts $\lambda, \omega, \sigma, \omega, I, \tau, \ldots$ system parameters (first, ω is related to parameter in our stability analysis and second ω is related to system angular frequency of the optical field) indicate the corresponding partial derivatives. Let us first concentrate on $\Lambda(x)$, remember in $\lambda(\sigma, \omega, I, \tau, \ldots \in \mathbb{R}_+)$; $\omega(\sigma, \omega, I, \tau, \ldots \in \mathbb{R}_+)$ and keeping all parameters except one (x) and τ. The derivation closely follows that in reference [BK]. Differentiating Q-switched microchip lasers semiconductors saturable absorbers system characteristic equation $P(\lambda) + Q(\lambda)\cdot e^{-\lambda\cdot\tau} = 0$ with respect to specific parameter (x), and inverting the derivative, for convenience, one calculates, $x = \sigma, \omega, I, \tau, \ldots \in \mathbb{R}_+$,

$$\left(\frac{\partial\lambda}{\partial x}\right)^{-1} = \frac{-P_\lambda(\lambda, x)\cdot Q(\lambda, x) + Q_\lambda(\lambda, x)\cdot P(\lambda, x) - \tau\cdot P(\lambda, x)\cdot Q(\lambda, x)}{P_x(\lambda, x)\cdot Q(\lambda, x) - Q_x(\lambda, x)\cdot P(\lambda, x)}$$

where $P_\lambda = \frac{\partial P}{\partial \lambda}, \ldots$ etc., Substituting $\lambda = i \cdot \omega$, and bearing $\overline{P(-i \cdot \omega)} = P(i \cdot \omega)$; $\overline{Q(-i \cdot \omega)} = Q(i \cdot \omega)$ then $i \cdot P_\lambda(i \cdot \omega) = P_\omega(i \cdot \omega)$ and $i \cdot Q_\lambda(i \cdot \omega) = Q_\omega(i \cdot \omega)$ that on the surface $|P(i \cdot \omega)|^2 = |Q(i \cdot \omega)|^2$, one obtains,

$$\left(\frac{\partial \lambda}{\partial x}\right)^{-1}|_{\lambda = i \cdot \omega} = \frac{i \cdot P_\omega(i \cdot \omega, x) \cdot \overline{P(i \cdot \omega, x)} + i \cdot Q_\lambda(i \cdot \omega, x) \cdot \overline{Q(\lambda, x)} - \tau \cdot |P(i \cdot \omega, x)|^2}{P_x(i \cdot \omega, x) \cdot \overline{P(i \cdot \omega, x)} - Q_x(i \cdot \omega, x) \cdot \overline{Q(i \cdot \omega, x)}}$$

Upon separating into real and imaginary parts, with $P = P_R + i \cdot P_I$; $Q = Q_R + i \cdot Q_I$ and $P_\omega = P_{R\omega} + i \cdot P_{I\omega}$; $Q_\omega = Q_{R\omega} + i \cdot Q_{I\omega}$; $P_x = P_{Rx} + i \cdot P_{Ix}$; $Q_x = Q_{Rx} + i \cdot Q_{Ix}$; $P^2 = P_R^2 + P_I^2$. When (x) can be an Q-switched microchip lasers semiconductors saturable absorbers system parameters $\sigma, \omega, I, \ldots$ and time delay τ etc. Where for convenience, we have dropped the arguments $(i \cdot \omega, x)$, and where $F_\omega = 2 \cdot [(P_{R\omega} \cdot P_R + P_{I\omega} \cdot P_I) - (Q_{R\omega} \cdot Q_R + Q_{I\omega} \cdot Q_I)]$ and $F_x = 2 \cdot [(P_{Rx} \cdot P_R + P_{Ix} \cdot P_I) - (Q_{Rx} \cdot Q_R + Q_{Ix} \cdot Q_I)]$; $\omega_x = -\frac{F_x}{F_\omega}$. We define U and V:

$$U = (P_R \cdot P_{I\omega} - P_I \cdot P_{R\omega}) - (Q_R \cdot Q_{I\omega} - Q_I \cdot Q_{R\omega})$$
$$V = (P_R \cdot P_{Ix} - P_I \cdot P_{Rx}) - (Q_R \cdot Q_{Ix} - Q_I \cdot Q_{Rx})$$

We choose our specific parameter as time delay $x = \tau$.
Hint: We define the system angular frequency of the optical field as ω_{of}.

$$P_{n=4}(i \cdot \omega) = \omega^4 - \omega^2 \cdot \left(\frac{2 \cdot \sigma \cdot I}{\hbar \cdot \omega_{of}}\right)^2 - i \cdot \omega^3 \cdot \left(\frac{4 \cdot \sigma \cdot I}{\hbar \cdot \omega_{of}}\right)$$

$$Q_{m=3}(i \cdot \omega) = -4 \cdot \left(\frac{\sigma \cdot I}{\hbar \cdot \omega_{of}}\right)^2 \cdot \omega^2 + i \cdot \left[4 \cdot \left(\frac{\sigma \cdot I}{\hbar \cdot \omega_{of}}\right)^3 \cdot \omega - \left(\frac{\sigma \cdot I}{\hbar \cdot \omega_{of}}\right) \cdot \omega^3\right]$$

$$P_R(i \cdot \omega) = \omega^4 - \omega^2 \cdot \left(\frac{2 \cdot \sigma \cdot I}{\hbar \cdot \omega_{of}}\right)^2; \; P_I(i \cdot \omega) = -\omega^3 \cdot \left(\frac{4 \cdot \sigma \cdot I}{\hbar \cdot \omega_{of}}\right)$$

$$Q_R(i \cdot \omega) = -4 \cdot \left(\frac{\sigma \cdot I}{\hbar \cdot \omega_{of}}\right)^2 \cdot \omega^2; \; Q_I(i \cdot \omega) = 4 \cdot \left(\frac{\sigma \cdot I}{\hbar \cdot \omega_{of}}\right)^3 \cdot \omega - \left(\frac{\sigma \cdot I}{\hbar \cdot \omega_{of}}\right) \cdot \omega^3$$

$$P_{R\omega}(i \cdot \omega, \tau) = 4 \cdot \omega^3 - 2 \cdot \omega \cdot \left(\frac{2 \cdot \sigma \cdot I}{\hbar \cdot \omega_{of}}\right)^2; \; P_{I\omega}(i \cdot \omega, \tau)$$

$$= -3 \cdot \omega^2 \cdot \left(\frac{4 \cdot \sigma \cdot I}{\hbar \cdot \omega_{of}}\right)$$

$$Q_{R\omega}(i \cdot \omega, \tau) = -8 \cdot \left(\frac{\sigma \cdot I}{\hbar \cdot \omega_{of}}\right)^2 \cdot \omega; \; Q_{I\omega}(i \cdot \omega, \tau)$$
$$= 4 \cdot \left(\frac{\sigma \cdot I}{\hbar \cdot \omega_{of}}\right)^3 - 3 \cdot \left(\frac{\sigma \cdot I}{\hbar \cdot \omega_{of}}\right) \cdot \omega^2$$

$$P_{R\omega} = P_{R\omega}(i \cdot \omega, \tau); \; P_{I\omega} = P_{I\omega}(i \cdot \omega, \tau); \; Q_{R\omega}$$
$$= Q_{R\omega}(i \cdot \omega, \tau); \; Q_{I\omega} = Q_{I\omega}(i \cdot \omega, \tau)$$

$$P_{R\tau} = 0; \; P_{I\tau} = 0; \; Q_{R\tau} = 0; \; Q_{I\tau} = 0$$

$$P_{R\omega} \cdot P_R = \left[4 \cdot \omega^3 - 2 \cdot \omega \cdot \left(\frac{2 \cdot \sigma \cdot I}{\hbar \cdot \omega_{of}}\right)^2\right] \cdot \omega^4 - \omega^2 \cdot \left(\frac{2 \cdot \sigma \cdot I}{\hbar \cdot \omega_{of}}\right)^2$$

$$P_{I\omega} \cdot P_I = 3 \cdot \omega^5 \cdot \left(\frac{4 \cdot \sigma \cdot I}{\hbar \cdot \omega_{of}}\right)^2; \; Q_{R\omega} \cdot Q_R = 32 \cdot \left(\frac{\sigma \cdot I}{\hbar \cdot \omega_{of}}\right)^4 \cdot \omega^3$$

$$Q_{I\omega} \cdot Q_I$$
$$= \left[4 \cdot \left(\frac{\sigma \cdot I}{\hbar \cdot \omega_{of}}\right)^3 - 3 \cdot \left(\frac{\sigma \cdot I}{\hbar \cdot \omega_{of}}\right) \cdot \omega^2\right] \cdot \left[4 \cdot \left(\frac{\sigma \cdot I}{\hbar \cdot \omega_{of}}\right)^3 \cdot \omega - \left(\frac{\sigma \cdot I}{\hbar \cdot \omega_{of}}\right) \cdot \omega^3\right]$$

$$F_\tau = 2 \cdot [(P_{R\tau} \cdot P_R + P_{I\tau} \cdot P_I) - (Q_{R\tau} \cdot Q_R + Q_{I\tau} \cdot Q_I)]$$
$$= 0; \; \omega_\tau = -\frac{F_\tau}{F_\omega} = 0$$

$$P_R \cdot P_{I\omega} = -3 \cdot \omega^2 \cdot \left[\omega^4 - \omega^2 \cdot \left(\frac{2 \cdot \sigma \cdot I}{\hbar \cdot \omega_{of}}\right)^2\right] \cdot \left(\frac{4 \cdot \sigma \cdot I}{\hbar \cdot \omega_{of}}\right)$$

$$s P_I \cdot P_{R\omega} = -\omega^3 \cdot \left(\frac{4 \cdot \sigma \cdot I}{\hbar \cdot \omega_{of}}\right) \cdot \left[4 \cdot \omega^3 - 2 \cdot \omega \cdot \left(\frac{2 \cdot \sigma \cdot I}{\hbar \cdot \omega_{of}}\right)^2\right]$$

$$Q_R \cdot Q_{I\omega}$$
$$= -4 \cdot \left(\frac{\sigma \cdot I}{\hbar \cdot \omega_{of}}\right)^2 \cdot \omega^2 \cdot \left[4 \cdot \left(\frac{\sigma \cdot I}{\hbar \cdot \omega_{of}}\right)^3 - 3 \cdot \left(\frac{\sigma \cdot I}{\hbar \cdot \omega_{of}}\right) \cdot \omega^2\right]$$

$$Q_I \cdot Q_{R\omega}$$
$$4 = -8 \cdot \left(\frac{\sigma \cdot I}{\hbar \cdot \omega_{of}}\right)^2 \cdot \omega \left[4 \cdot \left(\frac{\sigma \cdot I}{\hbar \cdot \omega_{of}}\right)^3 \cdot \omega - \left(\frac{\sigma \cdot I}{\hbar \cdot \omega_{of}}\right) \cdot \omega^3\right]$$

$$P_R \cdot P_{Ix}|_{x=\tau} = 0; P_I \cdot P_{Rx}|_{x=\tau} = 0; Q_R \cdot Q_{Ix}|_{x=\tau}$$
$$= 0; Q_I \cdot Q_{Rx}|_{x=\tau} = 0$$

$$V = (P_R \cdot P_{I\tau} - P_I \cdot P_{R\tau}) - (Q_R \cdot Q_{I\tau} - Q_I \cdot Q_{R\tau}) = 0$$
$$F_\omega \cdot \frac{\partial \omega}{\partial \tau} + F_\tau = 0; \tau \in I; \frac{\partial \omega}{\partial \tau} = -\frac{F_\tau}{F_\omega}$$

$$\Lambda^{-1}(\tau) = \left(\frac{\partial \mathrm{Re}\lambda}{\partial \tau}\right)_{\lambda = i\cdot\omega}; \Lambda^{-1}(\tau)$$
$$= \mathrm{Re}\left\{\frac{-2 \cdot [U + \tau \cdot |P|^2] + i \cdot F_\omega}{F_\tau + i \cdot 2 \cdot [V + \omega \cdot |P|^2]}\right\}$$
$$\frac{\partial \omega}{\partial \tau} = \omega_\tau = -\frac{F_\tau}{F_\omega}$$

$$sign\{\Lambda^{-1}(\tau)\} = sign\left\{\left(\frac{\partial \mathrm{Re}\lambda}{\partial \tau}\right)_{\lambda = i\cdot\omega}\right\}$$
$$sign\{\Lambda^{-1}(\tau)\} = sign\{F_\omega\} \cdot sign\left\{\tau \cdot \frac{\partial \omega}{\partial \tau} + \omega + \frac{U \cdot \frac{\partial \omega}{\partial \tau} + V}{|P|^2}\right\}$$

We shall presently examine the possibility of stability transitions (bifurcations) in a Q-switched microchip lasers semiconductors saturable absorbers system, about equilibrium points $E^{(k)}(n_1^{(k)}, n_2^{(k)}, \Delta n^{(k)}, \alpha^{(k)}) = (n_1^{(*)}, n_1^{(*)}, 0, 0)\, \forall k = 0, 1, 2$ as a result of a variation of delay parameter τ. The analysis consists in identifying the roots of our system characteristic equation situated on the imaginary axis of the complex λ-plane where by increasing the delay parameter τ, $\mathrm{Re}\lambda$ may at the crossing, change its sign from "−" to "+", i.e. from a stable focus $E^{(*)}$ to an unstable one, or vice versa. This feature may be further assessed by examining the sign of the partial derivatives with respect to τ, $\Lambda^{-1}(\tau) = \left(\frac{\partial \mathrm{Re}\lambda}{\partial \tau}\right)_{\lambda = i\cdot\omega}$ and $\sigma, \omega, I, \ldots$ system parameters are constant where $\omega \in \mathbb{R}_+$. We need to plot the stability switch diagram based on different delay values of our Q-switched microchip lasers semiconductors saturable absorbers system. Since it is very complex function we recommend to solve it numerically rather than analytic [9, 10].

$$\Lambda^{-1}(\tau) = \left(\frac{\partial \mathrm{Re}\lambda}{\partial \tau}\right)_{\lambda = i\cdot\omega}; \Lambda^{-1}(\tau)$$
$$= \mathrm{Re}\left\{\frac{-2 \cdot [U + \tau \cdot |P|^2] + i \cdot F_\omega}{F_\tau + i \cdot 2 \cdot [V + \omega \cdot |P|^2]}\right\}$$

$$\Lambda^{-1}(\tau) = \left(\frac{\partial \mathrm{Re}\lambda}{\partial \tau}\right)_{\lambda = i\cdot\omega}; \Lambda^{-1}(\tau)$$

$$= \frac{2 \cdot \{F_\omega \cdot (V + \omega \cdot P^2) - F_\tau \cdot (U + \tau \cdot P^2)\}}{F_\tau^2 + 4 \cdot (V + \omega \cdot P^2)^2}$$

The stability switch occurs only on those delay values (τ) which fit the equation: $\tau = \frac{\theta_+(\tau)}{\omega_+(\tau)}$ and $\theta_+(\tau)$ is the solution of $\sin\theta(\tau) = \ldots; \cos\theta(\tau) = \ldots$, when $\omega = \omega_+(\tau)$ if only ω_+ is feasible. Additionally when all Q-switched microchip lasers semiconductors saturable absorbers system parameters are known and the stability switch due to various time delay values τ is describe in the following expression:

$$sign\{\wedge^{-1}(\tau)\} = sign\{F_\omega(\omega(\tau), \tau)\} \cdot sign \left\{\tau \cdot \omega_\tau(\omega(\tau)) + \omega(\tau) + \frac{U(\omega(\tau)) \cdot \omega_{\tau_\Delta}(\omega(\tau)) + V(\omega(\tau))}{|P(\omega(\tau))|^2}\right\}$$

Remark Lemma 1.1, Theorem 1.2 and Theorem 1.3 are described in Sect. 5.1.

5.3 Crystalline Mid-Infrared Laser Balance and Rate Equations Instability Under Delayed Variables in Time

Ion-doped crystalline laser mainly operate in the mid-IR spectral range between 2 and 5 μm. Types are rare-earth and transition-metal based ionic crystals, and color-center lasers. They are compact all-solid-state room-temperature tunable sources, belonging to class of vibrionic lasers. These lasers have the efficient high-power room-temperature operation and super-broad tenability, the possibility of generating ultrashort pulses from the novel class of chromium doped chalcogenide lasers led to an analysis in vibrionic laser systems, involving $3d^n$ transition-metal ions. They are mid-IR lasers which have spectroscopic and laser characteristics. Mainly they are continuous-wave (CW), diode-pumped and tunable lasers which are based on both rare-earth ion doped crystals (operate at fixed wavelengths and allow high-power operation) and transition-metal TM ion doped-and color-center crystals, allowing broad tuning and ultrashort pulse generation. When working in operation threshold the important parameter is bandwidth of the gain medium and tuning is done. The bandwidth definition is as the full width at half maximum relative to the gain maximum at $\lambda_0\left(\frac{\Delta\lambda}{\lambda_0} \approx \frac{\Delta\nu}{\nu_0}\right)$. The definition is the same in the wavelength and frequency domains. Then we can compare gain media with different central wavelengths. Crystalline active media have broadest gain in a certain wavelength range at room temperature, from near-infrared to mid-IR range. The number of optical cycles per pulse for ultrashort pulse generation is equal to $\left(\frac{\Delta\lambda}{\lambda_0}\right)^{-1}$. The mid-IR (MIR) wavelength region ("molecular fingerprint" region) and the range between 2 and 5 μm is characterized by a strong fundamental vibrational absorption lines of atmospheric constituents, vapors and other gases. Mid-IR radiation helps us to identify many materials by their characteristic spectra. Fingerprint region in IR spectroscopy, are the regions where all the bending vibrations are seen in spectroscopy. We can identify two different organic compounds by comparing the troughs in the right hand side of spectroscopy graph. Molecules can be in different excited states: stretch along their bands, vibrate around their center of mass and rotate around one of their axes. These excitations store energy and, energy can be expressed in wavelengths and the energy of many of these rotational and vibrational (ro-vibrational) excitations correspond to mid-IR wavelengths. We use these spectral fingerprints to identify materials using spectroscopy (by the characteristic of the energy stored in each of these excitations). Typical lasers in mid-IR region are semiconductor lasers, heterojunction lasers, and lead-salt, antimonide and quantum cascade laser sources. They require cryogenic cooling and have limited output power levels. Crystalline solid-state lasers can provide very high power levels retaining good beam quality and narrow spectral line width. Near infrared diode lasers is function as a pump sources to these lasers. We get stability, efficiency and compactness, broad spectral coverage and tuning ranges. Remote sensing and trace gas detection are possible by using room temperature diode-pumped tunable solid-state lasers. The ultra-broad gain bandwidth of laser crystals allows generation of ultrashort pulses of only a few optical cycles. These lasers fit application as mid-IR free space communications, optical frequency standards, and optical coherence tomography (OCT). Additional medical applications of mid-IR solid-state lasers are tissue cutting and welding, ophthalmology, neurosurgery, etc. we emphasis on the basics of lasers operation of tunable crystalline lasers. Laser techniques used in the visible and near-infrared regions and extended to mid-IR spectral regions. We handle request for modification the techniques and analyze the Kerr-lens mode-locking. Kerr lens mode locking is a technique of passive mode locking a laser. It use an artificial saturable absorber based on Kerr lensing in the gain medium. The effect causes a reduction in the beam size for high optical intensities by two mechanisms which act as a fast saturable absorber. The first mechanism is related to the case of hard aperture KLM, the Kerr lens reduces the optical losses at an aperture which the beam must pass in each resonator round trip. The second mechanism is in the case of soft aperture KLM, the Kerr lens leads to a better overlap of laser and pump beam, and thus to a higher gain for the peak of the pulse. The increase of gain is equivalent to decrease of losses (similar effects). Both effects increase the net round trip gain. Kerr lens mode locking has enabled the generation of the short pulses with durations down to 5fsec in Ti:sapphire lasers. It is a very fast response and that no special saturable absorber medium is required. The laser operates close to stability limit of its resonator which is disadvantage, because otherwise the Kerr lensing effect is too weak. We get difficulties for long-term stable operation and resonator design. The operation procedure for such laser starts in noisy operation mode, and after being turned on and switch to mode-locked operation only after an external trigger. KLM is called also self-mode locking since there is no request for a visible saturable absorber device. Crystalline lasers are key elements in mid-IR lasers. Laser devices key element is a chromium ion in a sapphire. A ruby laser is a three-level system and operated in the pulsed regime. Another type of mid-IR system is dysprosium doped calcium fluoride which operated both pulsed and continuous-wave (CW). It is pumped by a Xenon flash-lamp. We discuss the process in fluencing formation of the inversion population in mid-IR crystalline solid-state lasers. These processes effect on the laser operation (laser threshold and efficiency). We assume four level scheme which characterized by balance equation for the population of the upper laser level n [16].

$$\frac{dn}{dt} = F^{pump} \cdot \sigma_{GSA} \cdot (N_t - n) - F^{las} \cdot \sigma_{em} \cdot n - \frac{n}{\tau} - \alpha \cdot n^2$$

The term for pumping is $F^{pump} \cdot \sigma_{GSA} \cdot (N_t - n)$, the term for stimulated emission is $F^{las} \cdot \sigma_{em} \cdot n$, and the term for spontaneous decay is $\frac{n}{\tau}$. The term for up conversion with the up conversion macro parameter (α) is $\alpha \cdot n^2$. N_t—is the total concentration of the active ions, n—is the excited ion concentration, σ_{GSA}—is the GSA cross section at the pumping wavelength (pump absorption cross-section), σ_{em}—is the emission cross section at the laser wavelength, F^{pump}—is the pumping photon flux, F^{las}—is the intra cavity emission photon flux, and τ—is the temperature dependent lifetime of the upper laser level.

Remark ESA terms are not under the assumption that the quantum yield from the upper states to the upper laser level is equal to unity. ESA does not change the excited ion concentration.

The rate equation for the photon flux at the laser wavelength is

$$\frac{dF^{las}}{dt} = \frac{c}{l} \cdot F^{las} \cdot \sigma_{em} \cdot n \cdot l_a - \frac{F^{las}}{\tau_c} - \frac{c}{l} \cdot F^{las} \cdot \sigma_{ESA}^{las} \cdot n \cdot l_a$$

The stimulate emission term is $\frac{c}{l} \cdot F^{las} \cdot \sigma_{em} \cdot n \cdot l_a$, the loss term is $\frac{F^{las}}{\tau_c}$, and the ESA term is $\frac{c}{l} \cdot F^{las} \cdot \sigma_{ESA}^{las} \cdot n \cdot l_a$. The photon lifetime in the resonator is $\tau_c = \frac{l}{(T+L) \cdot c}$, where l is the effective cavity round-trip length, and $T + L$ are the logarithmic round-trip losses, T—output coupling, L—other losses. The threshold population n_{th} is $n_{th} = \frac{T+L}{l_a \cdot (\sigma_{em} - \sigma_{ESA}^{las})}$, where, l_a is the active medium length. We get the expression for the threshold absorbed power in longitudinally pumped configuration:

$$P_{th}^{abs} = (T + L) \cdot \frac{h \cdot \nu^{las}}{\sigma_{em} \cdot \tau} \cdot \frac{\lambda^{las}}{\lambda^{pump}} \cdot A^{pump}$$
$$\times (1 + \alpha \cdot n_{th} \cdot \tau) \cdot \left(1 - \frac{\sigma_{ESA}^{las}}{\sigma_{em}}\right)^{-1} \cdot \left(1 + \frac{\sigma_{ESA}^{pump} \cdot n_{th}}{\sigma_{GSA} \cdot (N_t - n_{th})}\right)$$

The above expression fulfills the following: $P_{th}^{abs} \sim (T + L)$; $P_{th}^{abs} \sim \frac{h \cdot \nu^{las}}{\sigma_{em} \cdot \tau}$; $P_{th}^{abs} \sim \frac{\lambda^{las}}{\lambda^{pump}}$, and $P_{th}^{abs} \sim A^{pump}$. The last three terms represent the up conversion, ESA at the laser wavelength, and ESA at the pump wavelength, respectively. The slope efficiency with respect to the absorbed pump power is

$$\eta_{slope} = \frac{\lambda^{pump}}{\lambda^{las}} \cdot \left(\frac{T}{T + L}\right) \cdot \left(1 - \frac{\sigma_{ESA}^{las}}{\sigma_{em}}\right) \cdot \left(1 + \frac{\sigma_{ESA}^{pump} \cdot n_{th}}{\sigma_{ESA}^{pump} \cdot (N_t - n_{th})}\right)^{-1}$$

where the first two terms are the stokes shift and output coupling efficiency, and the last two stand for the efficiency loss due to the ESA at the laser and pump wavelengths, respectively. The intrinsic slope efficiency is $\eta_0 = \frac{\lambda^{pump}}{\lambda^{las}} \cdot \left(1 - \frac{\sigma_{ESA}^{las}}{\sigma_{em}}\right)$.

Remark The up conversion and temperature reduction of the active ion lifetime increase the laser threshold but do not affect the slope efficiency. Our target is to keep the up conversion losses at minimum ($\alpha \cdot n_{th} \cdot \tau \ll 1$), keep overall losses $T + L$ low to reduce the threshold population. The losses and ESA affect the threshold and slope efficiency. In four-level lasers, where $n_{th} < N_t$ we can ignore the ESA at the pump wavelength [16].

Laser transitions in the mid-IR tend to be broadband. The widths of the individual states (energy units) are determined by the particular crystal-lattice interaction and inhomogeneous broadening. They are characteristic for the given ion host. The transition bandwidth $h \cdot \Delta v$ (energy units) is the sum of the widths of the upper and lower states. The transition energy $h \cdot \Delta v$ is inversely proportional to the wavelength. The relative bandwidth of the transition $\frac{\Delta v}{v} \approx \frac{\Delta\lambda}{\lambda}$ increases towards the infrared. The transition bandwidth is determining the ability of the laser material to lase and to be tuned. Additional characteristics are transition lifetime and cross-section. Both are connected with the bandwidth of the laser transition. The spontaneous emission cross section $\sigma_{em}(\lambda)$ is obtained from the fluorescence intensity signal $I(\lambda)$ by using the relation: $\sigma_{em}(\lambda) = \frac{A}{8 \cdot \pi \cdot c \cdot n^2} \cdot \frac{I(\lambda) \cdot \lambda^5}{\int I(\lambda) \cdot \lambda \cdot d\lambda}$; A—is the full spontaneous emission probability (relative decay rate) from the upper laser level, c—is the speed of light and n is the index of refraction. The formula for $\sigma_{em}(\lambda)$ does not contain assumptions; we neglect the polarization dependence of fluorescence. We get radiative decay rate by taking inverse life time at low temperature. Then we get the effective emission cross-section $\sigma_{em}(\lambda)$ by radiative decay rate. Ion-host interaction outcome is non radiative decay. The luminescence life time τ consists of purely radiative and non-radiative components, $\frac{1}{\tau} = \frac{1}{\tau_{rad}} + \frac{1}{\tau_{nr}}$.

$$\frac{1}{\tau_{rad}} = A = 8 \cdot c \cdot n^2 \cdot \int \frac{\sigma_{em}(\lambda) \cdot d\lambda}{\lambda^4} \approx 8 \cdot c \cdot n^2 \cdot \sigma_{em}(\lambda_0) \cdot \frac{1}{\lambda_0^3} \cdot \frac{\Delta\lambda}{\lambda_0}$$

We get the fact that lifetime inversely scales with emission bandwidth. The minimum absorbed pump power density at threshold is proportional to the loss coefficient and the gain saturation intensity. The incident pumping intensity (I_{th}) is inversely proportional to the pump absorption coefficient: $I_{th} \propto \frac{\alpha_{loss}}{\alpha_{abs}} \cdot \frac{h \cdot v}{\sigma(v) \cdot \tau}$, where α_{loss} include the absorption from the ground state in the case of three and quasi three level schemes. The pump intensity at threshold is proportional to the following: $I_{th} \propto \frac{1}{FOM} \cdot \frac{1}{\eta_{QE}} \cdot \frac{\Delta\lambda}{\lambda_0} \cdot \frac{n^2}{\lambda_0^4}$, where η_{QE} $\left(\eta_{QE} = \frac{\tau}{\tau_{rad}}\right)$ is the quantum yield of the transition [16]. We can summary our system differential equations:$n = n(t)$; $F^{las} = F^{las}(t)$

$$\frac{dn}{dt} = F^{pump} \cdot \sigma_{GSA} \cdot (N_t - n) - F^{las} \cdot \sigma_{em} \cdot n - \frac{n}{\tau} - \alpha \cdot n^2$$

$$\frac{dF^{las}}{dt} = \frac{c}{l} \cdot F^{las} \cdot \sigma_{em} \cdot n \cdot l_a - \frac{F^{las}}{\tau_c} - \frac{c}{l} \cdot F^{las} \cdot \sigma_{ESA}^{las} \cdot n \cdot l_a$$

At fixed points: $\frac{dn}{dt} = 0; \frac{dF^{las}}{dt} = 0$

$$(^*) F^{pump} \cdot \sigma_{GSA} \cdot (N_t - n^*) - F^{las*} \cdot \sigma_{em} \cdot n^* - \frac{n^*}{\tau} - \alpha \cdot (n^*)^2 = 0$$

$$(^{**}) \frac{c}{l} \cdot F^{las*} \cdot \sigma_{em} \cdot n^* \cdot l_a - \frac{F^{las*}}{\tau_c} - \frac{c}{l} \cdot F^{las*} \cdot \sigma_{ESA}^{las} \cdot n^* \cdot l_a = 0$$

$$(^{**}) : F^{las*} \cdot \left(\frac{c}{l} \cdot \sigma_{em} \cdot n^* \cdot l_a - \frac{1}{\tau_c} - \frac{c}{l} \cdot \sigma_{ESA}^{las} \cdot n^* \cdot l_a \right) = 0$$

Case I: $F^{las*} = 0 \Rightarrow F^{pump} \cdot \sigma_{GSA} \cdot (N_t - n^*) - \frac{n^*}{\tau} - \alpha \cdot (n^*)^2 = 0.$

$$(^{**}) \to (^*) : \alpha \cdot (n^*)^2 + \left(F^{pump} \cdot \sigma_{GSA} + \frac{1}{\tau} \right) \cdot n^* - F^{pump} \cdot \sigma_{GSA} \cdot N_t = 0.$$

Case Ia: $n^* = \frac{-(F^{pump} \cdot \sigma_{GSA} + \frac{1}{\tau}) + \sqrt{(F^{pump} \cdot \sigma_{GSA} + \frac{1}{\tau})^2 + 4 \cdot \alpha \cdot F^{pump} \cdot \sigma_{GSA} \cdot N_t}}{2 \cdot \alpha}.$

Case Ib: $n^* = \frac{-(F^{pump} \cdot \sigma_{GSA} + \frac{1}{\tau}) - \sqrt{(F^{pump} \cdot \sigma_{GSA} + \frac{1}{\tau})^2 + 4 \cdot \alpha \cdot F^{pump} \cdot \sigma_{GSA} \cdot N_t}}{2 \cdot \alpha}.$

We can summary our first and second fixed points:

$$E^0(n^{(0)}, F^{las(0)}) = \left(\frac{-(F^{pump} \cdot \sigma_{GSA} + \frac{1}{\tau}) + \sqrt{(F^{pump} \cdot \sigma_{GSA} + \frac{1}{\tau})^2 + 4 \cdot \alpha \cdot F^{pump} \cdot \sigma_{GSA} \cdot N_t}}{2 \cdot \alpha}, 0 \right)$$

$$E^1(n^{(1)}, F^{las(1)}) = \left(\frac{-(F^{pump} \cdot \sigma_{GSA} + \frac{1}{\tau}) - \sqrt{(F^{pump} \cdot \sigma_{GSA} + \frac{1}{\tau})^2 + 4 \cdot \alpha \cdot F^{pump} \cdot \sigma_{GSA} \cdot N_t}}{2 \cdot \alpha}, 0 \right)$$

Case II: $\frac{c}{l} \cdot \sigma_{em} \cdot n^* \cdot l_a - \frac{1}{\tau_c} - \frac{c}{l} \cdot \sigma_{ESA}^{las} \cdot n^* \cdot l_a = 0$

$$n^* \cdot l_a \cdot \frac{c}{l} \cdot (\sigma_{em} - \sigma_{ESA}^{las}) = \frac{1}{\tau_c} \Rightarrow n^* = \frac{l}{\tau_c \cdot l_a \cdot c \cdot (\sigma_{em} - \sigma_{ESA}^{las})}$$

$$(^{**}) \to (^*) : F^{pump} \cdot \sigma_{GSA} \cdot (N_t - n^*) - F^{las*} \cdot \sigma_{em} \cdot n^* - \frac{n^*}{\tau} - \alpha \cdot (n^*)^2 = 0$$

$$F^{las*} = F^{pump} \cdot \sigma_{GSA} \cdot (N_t - n^*) \cdot \frac{1}{\sigma_{em} \cdot n^*} - \frac{1}{\tau \cdot \sigma_{em}} - \frac{\alpha \cdot n^*}{\sigma_{em}}$$

$$F^{las*} = F^{pump} \cdot \sigma_{GSA} \cdot \left(N_t - \frac{l}{\tau_c \cdot l_a \cdot c \cdot (\sigma_{em} - \sigma_{ESA}^{las})}\right) \cdot \frac{\tau_c \cdot l_a \cdot c \cdot (\sigma_{em} - \sigma_{ESA}^{las})}{\sigma_{em} \cdot l} - \frac{1}{\tau \cdot \sigma_{em}} - \frac{\alpha}{\sigma_{em}} \cdot \frac{l}{\tau_c \cdot l_a \cdot c \cdot (\sigma_{em} - \sigma_{ESA}^{las})}$$

$$F^{las*} = \left(F^{pump} \cdot \sigma_{GSA} \cdot N_t - \frac{F^{pump} \cdot \sigma_{GSA} \cdot l}{\tau_c \cdot l_a \cdot c \cdot (\sigma_{em} - \sigma_{ESA}^{las})}\right) \cdot \frac{\tau_c \cdot l_a \cdot c \cdot (\sigma_{em} - \sigma_{ESA}^{las})}{\sigma_{em} \cdot l} - \frac{1}{\tau \cdot \sigma_{em}} - \frac{\alpha}{\sigma_{em}} \cdot \frac{l}{\tau_c \cdot l_a \cdot c \cdot (\sigma_{em} - \sigma_{ESA}^{las})}$$

$$F^{las*} = F^{pump} \cdot \sigma_{GSA} \cdot N_t \cdot \frac{\tau_c \cdot l_a \cdot c \cdot (\sigma_{em} - \sigma_{ESA}^{las})}{\sigma_{em} \cdot l} - \frac{F^{pump} \cdot \sigma_{GSA}}{\sigma_{em}} - \frac{1}{\tau \cdot \sigma_{em}} - \frac{\alpha}{\sigma_{em}} \cdot \frac{l}{\tau_c \cdot l_a \cdot c \cdot (\sigma_{em} - \sigma_{ESA}^{las})}$$

The third fixed point:

$$E^2(n^{(2)}, F^{las(2)}) = \left(\frac{l}{\tau_c \cdot l_a \cdot c \cdot (\sigma_{em} - \sigma_{ESA}^{las})}, F^{pump} \cdot \sigma_{GSA} \cdot N_t \cdot \frac{\tau_c \cdot l_a \cdot c \cdot (\sigma_{em} - \sigma_{ESA}^{las})}{\sigma_{em} \cdot l} - \frac{F^{pump} \cdot \sigma_{GSA}}{\sigma_{em}} - \frac{1}{\tau \cdot \sigma_{em}} - \frac{\alpha}{\sigma_{em}} \cdot \frac{l}{\tau_c \cdot l_a \cdot c \cdot (\sigma_{em} - \sigma_{ESA}^{las})}\right)$$

We can summary our system three fixed points:

$$E^0(n^{(0)}, F^{las(0)}) = \left(\frac{-(F^{pump} \cdot \sigma_{GSA} + \frac{1}{\tau}) + \sqrt{(F^{pump} \cdot \sigma_{GSA} + \frac{1}{\tau})^2 + 4 \cdot \alpha \cdot F^{pump} \cdot \sigma_{GSA} \cdot N_t}}{2 \cdot \alpha}, 0\right)$$

$$E^1(n^{(1)}, F^{las(1)}) = \left(\frac{-(F^{pump} \cdot \sigma_{GSA} + \frac{1}{\tau}) - \sqrt{(F^{pump} \cdot \sigma_{GSA} + \frac{1}{\tau})^2 + 4 \cdot \alpha \cdot F^{pump} \cdot \sigma_{GSA} \cdot N_t}}{2 \cdot \alpha}, 0\right)$$

$$E^2(n^{(2)}, F^{las(2)}) = \left(\frac{l}{\tau_c \cdot l_a \cdot c \cdot (\sigma_{em} - \sigma_{ESA}^{las})},\right.$$

$$F^{pump} \cdot \sigma_{GSA} \cdot N_t \cdot \frac{\tau_c \cdot l_a \cdot c \cdot (\sigma_{em} - \sigma_{ESA}^{las})}{\sigma_{em} \cdot l} - \frac{F^{pump} \cdot \sigma_{GSA}}{\sigma_{em}}$$

$$- \frac{1}{\tau \cdot \sigma_{em}} - \frac{\alpha}{\sigma_{em}} \cdot \frac{l}{\tau_c \cdot l_a \cdot c \cdot (\sigma_{em} - \sigma_{ESA}^{las})} \Bigg)$$

We need to do linearization to our system. The general form of our Crystalline Mid-infrared laser system (no delay variables in time) is

$$\frac{dn}{dt} = f_1(n, F^{las}); \frac{dF^{las}}{dt} = f_2(n, F^{las})$$

$$f_1(n, F^{las}) = F^{pump} \cdot \sigma_{GSA} \cdot (N_t - n) - F^{las} \cdot \sigma_{em} \cdot n - \frac{n}{\tau} - \alpha \cdot n^2$$

$$f_2(n, F^{las}) = \frac{c}{l} \cdot F^{las} \cdot \sigma_{em} \cdot n \cdot l_a - \frac{F^{las}}{\tau_c} - \frac{c}{l} \cdot F^{las} \cdot \sigma_{ESA}^{las} \cdot n \cdot l_a$$

System any fixed point (we have three fixed points) is n^*, F^{las*} and exist $f_1(n^*, F^{las*}) = 0$; $f_2(n^*, F^{las*}) = 0$. Let $u_1 = n - n^*; u_2 = F^{las} - F^{las*}$ denote the components of a small disturbance from the fixed point. To see whether the disturbance grows or decays, we need to derive differential equations for u_1 and u_2. The u_1 equation is $\frac{du_1}{dt} = \frac{dn}{dt}$ since n^* is a constant and by substitution $\frac{du_1}{dt} = \frac{dn}{dt} = f_1(n^* + u_1, F^{las*} + u_2)$. By using Taylor series expansion, $\frac{du_1}{dt} = \frac{dn}{dt} = f_1(n^* + u_1, F^{las*} + u_2) = f_1(n^*, F^{las*}) + u_1 \cdot \frac{\partial f_1}{\partial n} + u_2 \cdot \frac{\partial f_1}{\partial F^{las}} + O(u_1^2, u_2^2, u_1 \cdot u_2)$ and since $f_1(n^*, F^{las*}) = 0$ then $\frac{du_1}{dt} = \frac{dn}{dt} = u_1 \cdot \frac{\partial f_1}{\partial n} + u_2 \cdot \frac{\partial f_1}{\partial F^{las}} + O(u_1^2, u_2^2, u_1 \cdot u_2)$. To simplify the notation, we have written $\frac{\partial f_1}{\partial n}$ and $\frac{\partial f_1}{\partial F^{las}}$; these partial derivatives are to be evaluated at the fixed point (n^*, F^{las*}); thus they are umbers, not functions. The shorthand notation $O(u_1^2, u_2^2, u_1 \cdot u_2)$ denotes quadratic term in u_1 and u_2. Since u_1 and u_2 are small, these quadratic terms are extremely small. Similarly we find $\frac{du_2}{dt} = u_1 \cdot \frac{\partial f_2}{\partial n} + u_2 \cdot \frac{\partial f_2}{\partial F^{las}} + O(u_1^2, u_2^2, u_1 \cdot u_2)$ $\left(\frac{du_2}{dt} = \frac{dF^{las}}{dt} = u_1 \cdot \frac{\partial f_2}{\partial n} + u_2 \cdot \frac{\partial f_2}{\partial F^{las}} + O(u_1^2, u_2^2, u_1 \cdot u_2)\right)$. Hence the disturbance (u_1, u_2) evolved according to $\begin{pmatrix} \frac{du_1}{dt} \\ \frac{du_2}{dt} \end{pmatrix} = \begin{pmatrix} \frac{df_1}{dn} & \frac{df_1}{dF^{las}} \\ \frac{df_2}{dn} & \frac{df_2}{dF^{las}} \end{pmatrix} \cdot \begin{pmatrix} u_1 \\ u_2 \end{pmatrix}$ + quadratic terms.

The matrix $A = \begin{pmatrix} \frac{df_1}{dn} & \frac{df_1}{dF^{las}} \\ \frac{df_2}{dn} & \frac{df_2}{dF^{las}} \end{pmatrix}_{(n^*, F^{las*})}$ is called the Jacobian matrix at the fixed point (n^*, F^{las*}). We obtain the linearized system $\begin{pmatrix} \frac{du_1}{dt} \\ \frac{du_2}{dt} \end{pmatrix} = \begin{pmatrix} \frac{df_1}{dn} & \frac{df_1}{dF^{las}} \\ \frac{df_2}{dn} & \frac{df_2}{dF^{las}} \end{pmatrix} \cdot \begin{pmatrix} u_1 \\ u_2 \end{pmatrix}$. The linearized system gives a qualitatively correct picture of the phase portrait near (n^*, F^{las*}), as long as the fixed point for the linearized system is not one of the borderline cases (centers, degenerate nodes, stars, or non-isolated fixed points) [6, 7].

$$\frac{df_1}{dn} = -F^{pump} \cdot \sigma_{GSA} - F^{las} \cdot \sigma_{em} - \frac{1}{\tau} - 2 \cdot \alpha \cdot n; \ \frac{df_1}{dF^{las}} = -\sigma_{em} \cdot n$$

$$\frac{df_2}{dn} = \frac{c}{l} \cdot F^{las} \cdot (\sigma_{em} - \sigma_{ESA}^{las}) \cdot l_a; \ \frac{df_2}{dF^{las}} = \frac{c}{l} \cdot (\sigma_{em} - \sigma_{ESA}^{las}) \cdot n \cdot l_a - \frac{1}{\tau_c}$$

The eigenvalues of a matrix A are given by the characteristic equation $\det(A - \lambda \cdot I) = 0$, where I is the identity matrix.

$$A = \begin{pmatrix} -F^{pump} \cdot \sigma_{GSA} - F^{las*} \cdot \sigma_{em} - \frac{1}{\tau} - 2 \cdot \alpha \cdot n^* & -\sigma_{em} \cdot n^* \\ F^{las*} \cdot (\frac{c}{l} \cdot \sigma_{em} \cdot l_a - \frac{c}{l} \cdot \sigma_{ESA}^{las} \cdot l_a) & \frac{c}{l} \cdot \sigma_{em} \cdot n^* \cdot l_a - \frac{1}{\tau_c} - \frac{c}{l} \cdot \sigma_{ESA}^{las} \cdot n^* \cdot l_a \end{pmatrix}$$

The characteristic equation becomes

$$\det \begin{pmatrix} -F^{pump} \cdot \sigma_{GSA} - F^{las*} \cdot \sigma_{em} - \frac{1}{\tau} - 2 \cdot \alpha \cdot n^* - \lambda & -\sigma_{em} \cdot n^* \\ F^{las*} \cdot (\frac{c}{l} \cdot \sigma_{em} \cdot l_a - \frac{c}{l} \cdot \sigma_{ESA}^{las} \cdot l_a) & \frac{c}{l} \cdot \sigma_{em} \cdot n^* \cdot l_a - \frac{1}{\tau_c} - \frac{c}{l} \cdot \sigma_{ESA}^{las} \cdot n^* \cdot l_a - \lambda \end{pmatrix} = 0$$

We define for simplicity: $\Upsilon_1 = -F^{pump} \cdot \sigma_{GSA} - F^{las*} \cdot \sigma_{em} - \frac{1}{\tau} - 2 \cdot \alpha \cdot n^*$

$$\Upsilon_2 = F^{las*} \cdot \frac{c}{l} \cdot (\sigma_{em} - \sigma_{ESA}^{las}) \cdot l_a; \ \Upsilon_3 = \frac{c}{l} \cdot (\sigma_{em} - \sigma_{ESA}^{las}) \cdot n^* \cdot l_a - \frac{1}{\tau_c}$$

$$\det \begin{pmatrix} \Upsilon_1 - \lambda & -\sigma_{em} \cdot n^* \\ \Upsilon_2 & \Upsilon_3 - \lambda \end{pmatrix} = 0; \ (\Upsilon_1 - \lambda) \cdot (\Upsilon_3 - \lambda) + \Upsilon_2 \cdot \sigma_{em} \cdot n^* = 0$$

$$\lambda^2 - \lambda \cdot (\Upsilon_1 + \Upsilon_3) + \Upsilon_1 \cdot \Upsilon_3 + \Upsilon_2 \cdot \sigma_{em} \cdot n^* = 0$$
$$trace(A) = \Upsilon_1 + \Upsilon_3; \ \det(A) = \Upsilon_1 \cdot \Upsilon_3 + \sigma_{em} \cdot n^* \cdot \Upsilon_2$$

Then $\lambda_1 = \frac{\Upsilon_1 + \Upsilon_3 + \sqrt{(\Upsilon_1 + \Upsilon_3)^2 - 4 \cdot (\Upsilon_1 \cdot \Upsilon_3 + \sigma_{em} \cdot n^* \cdot \Upsilon_2)}}{2}$

$$\lambda_2 = \frac{\Upsilon_1 + \Upsilon_3 - \sqrt{(\Upsilon_1 + \Upsilon_3)^2 - 4 \cdot (\Upsilon_1 \cdot \Upsilon_3 + \sigma_{em} \cdot n^* \cdot \Upsilon_2)}}{2}$$

The solutions of the quadratic equation $\lambda^2 - trace(A) \cdot \lambda + \det(A) = 0$ are λ_1, λ_2. The eigenvalues depend only on the trace and determinant of the matrix A.

Classification of fixed points: If $\det(A) < 0$, the eigenvalues are real and have opposite sign; hence the fixed point is a saddle point. If $\det(A) > 0$, the eigenvalues are either real with the same sign (nodes), or complex conjugate (spirals and centers). Nodes satisfy $(trace(A))^2 - 4 \cdot \det(A) > 0$ and spirals satisfy $(trace(A))^2 - 4 \cdot \det(A) < 0$. The parabola $(trace(A))^2 - 4 \cdot \det(A) = 0$ is a borderline between nodes and spirals; star nodes and degenerate nodes live on this parabola. The stability of the nodes and spirals is determined by $trace(A)$. When $trace(A) < 0$, both eigenvalues have negative real parts, so the fixed point is stable. Unstable spirals and nodes have $trace(A) > 0$. Neutrally stable centers line on the borderline $trace(A) = 0$, where the eigenvalues are purely imaginary. If $\det(A) = 0$, at least one of the eigenvalues is zero. Then the origin is not an isolated fixed point. There is either a whole line of fixed points, or a plane of fixed points, if $A = 0$. Saddle points, nodes, and spirals are the major types of fixed points and they occur in large open regions of the ($\det(A)$, $trace(A)$) plane. Centers, stars, degenerate nodes, and non-isolated fixed points are borderline cases that occur along curves in the ($\det(A)$, $trace(A)$) plane.

Stability analysis with delay parameters in time (τ_1, τ_2): The four level schemes gives the balance equation for the population of the upper laser level (n) and since the processes influences formation is not ideal there is a delay in time τ_1 for variable $n, n(t) \rightarrow n(t - \tau_1)$ and a delay in time τ_2 for the photon flux at the laser wavelength F^{las}, $F^{las}(t) \rightarrow F^{las}(t - \tau_2)$. The derivatives of the population of the upper laser level (n) and photon flux (F^{las}) do not effected by time delay parameters τ_1, τ_2.

$$\begin{aligned} \frac{dn}{dt} &= F^{pump} \cdot \sigma_{GSA} \cdot [N_t - n(t - \tau_1)] \\ &\quad - F^{las}(t - \tau_2) \cdot \sigma_{em} \cdot n(t - \tau_1) \\ &\quad - \frac{n(t - \tau_1)}{\tau} - \alpha \cdot n^2(t - \tau_1) \end{aligned}$$

$$\begin{aligned} \frac{dF^{las}}{dt} &= \frac{c}{l} \cdot F^{las}(t - \tau_2) \cdot \sigma_{em} \cdot n(t - \tau_1) \cdot l_a \\ &\quad - \frac{F^{las}(t - \tau_2)}{\tau_c} - \frac{c}{l} \cdot F^{las}(t - \tau_2) \cdot \sigma_{ESA}^{las} \cdot n(t - \tau_1) \cdot l_a \end{aligned}$$

System fixed points (equilibrium points) are the same for the case of no delays ($\tau_1 = 0, \tau_2 = 0$) since at fixed points $\lim_{t \to \infty} n(t - \tau_1) = n(t); t \gg \tau_1 \Rightarrow t - \tau_1 \approx t$ and $\lim_{t \to \infty} F^{las}(t - \tau_2) = F^{las}(t); t \gg \tau_2 \Rightarrow t - \tau_2 \approx t$.

Stability analysis: The standard local stability analysis about any one of the equilibrium points of Crystalline Mid-infrared laser system consists in adding to its coordinated $[n \ F^{las}]$ arbitrarily small increments of exponential term $[n \ f^{las}] \cdot e^{\lambda \cdot t}$, and retaining the first order terms in $n \ f^{las}$. The system of two homogeneous equations leads to a polynomial characteristic equation in the eigenvalue λ [8, 9]. The polynomial characteristic equation accepts by set of Crystalline Mid-infrared laser system equations. The system fixed values with arbitrarily small increments of exponential form $[n \ f^{las}] \cdot e^{\lambda \cdot t}$ are; j = 0 (first fixed point), j = 1 (second fixed point), j = 2 (third fixed point), etc.

We have three cases: (1) $\tau_1 = \tau_\Delta; \tau_2 = 0$, (2) $\tau_1 = 0; \tau_2 = \tau_\Delta$, (3) $\tau_1 = \tau_\Delta; \tau_2 = \tau_\Delta$. We analyze the stability for the first case $\tau_1 = \tau_\Delta; \tau_2 = 0$. We get the following delay differential equations:

$$\frac{dn}{dt} = F^{pump} \cdot \sigma_{GSA} \cdot [N_t - n(t - \tau_\Delta)] - F^{las}(t) \cdot \sigma_{em} \cdot n(t - \tau_\Delta) - \frac{n(t - \tau_\Delta)}{\tau} - \alpha \cdot n^2(t - \tau_\Delta)$$

$$\frac{dF^{las}}{dt} = \frac{c}{l} \cdot F^{las}(t) \cdot \sigma_{em} \cdot n(t - \tau_\Delta) \cdot l_a - \frac{F^{las}(t)}{\tau_c} - \frac{c}{l} \cdot F^{las}(t) \cdot \sigma_{ESA}^{las} \cdot n(t - \tau_\Delta) \cdot l_a$$

$$n(t) = n^{(j)} + n \cdot e^{\lambda \cdot t}; F^{las}(t) = F^{las(j)} + f^{las} \cdot e^{\lambda \cdot t}; n(t - \tau_\Delta) = n^{(j)} + n \cdot e^{\lambda \cdot (t - \tau_\Delta)}$$

$$\frac{dn(t)}{dt} = n \cdot \lambda \cdot e^{\lambda \cdot t}; \frac{dF^{las}(t)}{dt} = f^{las} \cdot \lambda \cdot e^{\lambda \cdot t}$$

Implement it in the first DDE (delay differential equation):

$$n \cdot \lambda \cdot e^{\lambda \cdot t} = F^{pump} \cdot \sigma_{GSA} \cdot [N_t - (n^{(j)} + n \cdot e^{\lambda \cdot (t - \tau_\Delta)})] - (F^{las(j)} + f^{las} \cdot e^{\lambda \cdot t}) \cdot \sigma_{em} \cdot (n^{(j)} + n \cdot e^{\lambda \cdot (t - \tau_\Delta)}) - \left(\frac{n^{(j)} + n \cdot e^{\lambda \cdot (t - \tau_\Delta)}}{\tau} \right) - \alpha \cdot (n^{(j)} + n \cdot e^{\lambda \cdot (t - \tau_\Delta)})^2$$

$$n \cdot \lambda \cdot e^{\lambda \cdot t} = F^{pump} \cdot \sigma_{GSA} \cdot (N_t - n^{(j)}) - F^{pump} \cdot \sigma_{GSA} \cdot n \cdot e^{\lambda \cdot (t - \tau_\Delta)} - (F^{las(j)} + f^{las} \cdot e^{\lambda \cdot t}) \cdot \sigma_{em} \cdot n^{(j)} - (F^{las(j)} + f^{las} \cdot e^{\lambda \cdot t}) \cdot \sigma_{em} \cdot n \cdot e^{\lambda \cdot (t - \tau_\Delta)} - \frac{n^{(j)}}{\tau} - \frac{n \cdot e^{\lambda \cdot (t - \tau_\Delta)}}{\tau} - \alpha \cdot ([n^{(j)}]^2 + 2 \cdot n^{(j)} \cdot n \cdot e^{\lambda \cdot (t - \tau_\Delta)} + n^2 \cdot e^{\lambda \cdot (t - \tau_\Delta) \cdot 2})$$

$$n \cdot \lambda \cdot e^{\lambda \cdot t} = F^{pump} \cdot \sigma_{GSA} \cdot (N_t - n^{(j)}) - F^{pump} \cdot \sigma_{GSA} \cdot n \cdot e^{\lambda \cdot (t - \tau_\Delta)} - F^{las(j)} \cdot \sigma_{em} \cdot n^{(j)} - f^{las} \cdot \sigma_{em} \cdot n^{(j)} \cdot e^{\lambda \cdot t} - F^{las(j)} \cdot \sigma_{em} \cdot n \cdot e^{\lambda \cdot (t - \tau_\Delta)} - f^{las} \cdot n \cdot e^{\lambda \cdot t} \cdot \sigma_{em} \cdot e^{\lambda \cdot (t - \tau_\Delta)} - \frac{n^{(j)}}{\tau} - \frac{n \cdot e^{\lambda \cdot (t - \tau_\Delta)}}{\tau} - \alpha \cdot [n^{(j)}]^2 - 2 \cdot \alpha \cdot n^{(j)} \cdot n \cdot e^{\lambda \cdot (t - \tau_\Delta)} - \alpha \cdot n^2 \cdot e^{\lambda \cdot (t - \tau_\Delta) \cdot 2}$$

We assume $f^{las} \cdot n \approx 0; n^2 \approx 0$ then

$$n \cdot \lambda \cdot e^{\lambda \cdot t} = F^{pump} \cdot \sigma_{GSA} \cdot (N_t - n^{(j)}) - F^{pump} \cdot \sigma_{GSA} \cdot n \cdot e^{\lambda \cdot (t - \tau_\Delta)}$$
$$- F^{las(j)} \cdot \sigma_{em} \cdot n^{(j)} - f^{las} \cdot \sigma_{em} \cdot n^{(j)} \cdot e^{\lambda \cdot t} - F^{las(j)} \cdot \sigma_{em} \cdot n \cdot e^{\lambda \cdot (t - \tau_\Delta)}$$
$$- \frac{n^{(j)}}{\tau} - \frac{n \cdot e^{\lambda \cdot (t - \tau_\Delta)}}{\tau} - \alpha \cdot [n^{(j)}]^2$$
$$- 2 \cdot \alpha \cdot n^{(j)} \cdot n \cdot e^{\lambda \cdot (t - \tau_\Delta)}$$

$$n \cdot \lambda \cdot e^{\lambda \cdot t} = F^{pump} \cdot \sigma_{GSA} \cdot (N_t - n^{(j)}) - F^{las(j)} \cdot \sigma_{em} \cdot n^{(j)} - \frac{n^{(j)}}{\tau} - \alpha \cdot [n^{(j)}]^2$$
$$- F^{pump} \cdot \sigma_{GSA} \cdot n \cdot e^{\lambda \cdot t} \cdot e^{-\lambda \cdot \tau_\Delta} - f^{las} \cdot \sigma_{em} \cdot n^{(j)} \cdot e^{\lambda \cdot t}$$
$$- F^{las(j)} \cdot \sigma_{em} \cdot n \cdot e^{\lambda \cdot t} \cdot e^{-\lambda \cdot \tau_\Delta}$$
$$- \frac{n \cdot e^{\lambda \cdot t} \cdot e^{-\lambda \cdot \tau_\Delta}}{\tau} - 2 \cdot \alpha \cdot n^{(j)} \cdot n \cdot e^{\lambda \cdot t} \cdot e^{-\lambda \cdot \tau_\Delta}$$

At fixed point: $F^{pump} \cdot \sigma_{GSA} \cdot (N_t - n^{(j)}) - F^{las(j)} \cdot \sigma_{em} \cdot n^{(j)} - \frac{n^{(j)}}{\tau} - \alpha \cdot [n^{(j)}]^2 = 0$.

$$n \cdot \lambda = -F^{pump} \cdot \sigma_{GSA} \cdot n \cdot e^{-\lambda \cdot \tau_\Delta} - f^{las} \cdot \sigma_{em} \cdot n^{(j)}$$
$$- F^{las(j)} \cdot \sigma_{em} \cdot n \cdot e^{-\lambda \cdot \tau_\Delta} - \frac{n \cdot e^{-\lambda \cdot \tau_\Delta}}{\tau} - 2 \cdot \alpha \cdot n^{(j)} \cdot n \cdot e^{-\lambda \cdot \tau_\Delta}$$

$$\left[\left(-F^{pump} \cdot \sigma_{GSA} - F^{las(j)} \cdot \sigma_{em} - \frac{1}{\tau} - 2 \cdot \alpha \cdot n^{(j)}\right) \cdot e^{-\lambda \cdot \tau_\Delta} - \lambda\right]$$
$$\cdot n - f^{las} \cdot \sigma_{em} \cdot n^{(j)} = 0$$

Implement it in the second DDE (delay differential equation):

$$f^{las} \cdot \lambda \cdot e^{\lambda \cdot t} = \frac{c}{l} \cdot (F^{las(j)} + f^{las} \cdot e^{\lambda \cdot t}) \cdot \sigma_{em} \cdot (n^{(j)} + n \cdot e^{\lambda \cdot (t - \tau_\Delta)}) \cdot l_a$$
$$- \frac{F^{las(j)} + f^{las} \cdot e^{\lambda \cdot t}}{\tau_c} - \frac{c}{l} \cdot (F^{las(j)} + f^{las} \cdot e^{\lambda \cdot t})$$
$$\cdot \sigma_{ESA}^{las} \cdot (n^{(j)} + n \cdot e^{\lambda \cdot (t - \tau_\Delta)}) \cdot l_a$$

$$f^{las} \cdot \lambda \cdot e^{\lambda \cdot t} = \frac{c}{l} \cdot \sigma_{em} \cdot l_a \cdot (F^{las(j)} + f^{las} \cdot e^{\lambda \cdot t}) \cdot (n^{(j)} + n \cdot e^{\lambda \cdot (t - \tau_\Delta)})$$
$$- \frac{F^{las(j)}}{\tau_c} - \frac{f^{las} \cdot e^{\lambda \cdot t}}{\tau_c}$$
$$- \frac{c}{l} \cdot \sigma_{ESA}^{las} \cdot l_a \cdot (F^{las(j)} + f^{las} \cdot e^{\lambda \cdot t}) \cdot (n^{(j)} + n \cdot e^{\lambda \cdot (t - \tau_\Delta)})$$

Hint: we develop the expression $(F^{las(j)} + f^{las} \cdot e^{\lambda \cdot t}) \cdot (n^{(j)} + n \cdot e^{\lambda \cdot (t - \tau_\Delta)})$

$$(F^{las(j)} + f^{las} \cdot e^{\lambda \cdot t}) \cdot (n^{(j)} + n \cdot e^{\lambda \cdot (t - \tau_\Delta)})$$

$$= F^{las(j)} \cdot n^{(j)} + F^{las(j)} \cdot n \cdot e^{\lambda \cdot (t-\tau_\Delta)} + n^{(j)} \cdot f^{las} \cdot e^{\lambda \cdot t} + f^{las} \cdot n \cdot e^{\lambda \cdot t} \cdot e^{\lambda \cdot (t-\tau_\Delta)}$$

Assumption $f^{las} \cdot n \approx 0$

$$(F^{las(j)} + f^{las} \cdot e^{\lambda \cdot t}) \cdot (n^{(j)} + n \cdot e^{\lambda \cdot (t-\tau_\Delta)}) = F^{las(j)} \cdot n^{(j)} + F^{las(j)} \cdot n \cdot e^{\lambda \cdot t} \cdot e^{-\lambda \cdot \tau_\Delta} + n^{(j)} \cdot f^{las} \cdot e^{\lambda \cdot t}$$

###

$$f^{las} \cdot \lambda \cdot e^{\lambda \cdot t} = \frac{c}{l} \cdot \sigma_{em} \cdot l_a \cdot (F^{las(j)} \cdot n^{(j)} + F^{las(j)} \cdot n \cdot e^{\lambda \cdot t} \cdot e^{-\lambda \cdot \tau_\Delta} + n^{(j)} \cdot f^{las} \cdot e^{\lambda \cdot t}) - \frac{F^{las(j)}}{\tau_c} - \frac{f^{las} \cdot e^{\lambda \cdot t}}{\tau_c} - \frac{c}{l} \cdot \sigma_{ESA}^{las} \cdot l_a \cdot (F^{las(j)} \cdot n^{(j)} + F^{las(j)} \cdot n \cdot e^{\lambda \cdot t} \cdot e^{-\lambda \cdot \tau_\Delta} + n^{(j)} \cdot f^{las} \cdot e^{\lambda \cdot t})$$

$$f^{las} \cdot \lambda \cdot e^{\lambda \cdot t} = \frac{c}{l} \cdot \sigma_{em} \cdot l_a \cdot F^{las(j)} \cdot n^{(j)} + \frac{c}{l} \cdot \sigma_{em} \cdot l_a \cdot F^{las(j)} \cdot n \cdot e^{\lambda \cdot t} \cdot e^{-\lambda \cdot \tau_\Delta} + \frac{c}{l} \cdot \sigma_{em} \cdot l_a \cdot n^{(j)} \cdot f^{las} \cdot e^{\lambda \cdot t} - \frac{F^{las(j)}}{\tau_c} - \frac{f^{las} \cdot e^{\lambda \cdot t}}{\tau_c} - \frac{c}{l} \cdot \sigma_{ESA}^{las} \cdot l_a \cdot F^{las(j)} \cdot n^{(j)} - \frac{c}{l} \cdot \sigma_{ESA}^{las} \cdot l_a \cdot F^{las(j)} \cdot n \cdot e^{\lambda \cdot t} \cdot e^{-\lambda \cdot \tau_\Delta} - \frac{c}{l} \cdot \sigma_{ESA}^{las} \cdot l_a \cdot n^{(j)} \cdot f^{las} \cdot e^{\lambda \cdot t}$$

$$f^{las} \cdot \lambda \cdot e^{\lambda \cdot t} = \frac{c}{l} \cdot \sigma_{em} \cdot l_a \cdot F^{las(j)} \cdot n^{(j)} - \frac{F^{las(j)}}{\tau_c} - \frac{c}{l} \cdot \sigma_{ESA}^{las} \cdot l_a \cdot F^{las(j)} \cdot n^{(j)} + \frac{c}{l} \cdot \sigma_{em} \cdot l_a \cdot F^{las(j)} \cdot n \cdot e^{\lambda \cdot t} \cdot e^{-\lambda \cdot \tau_\Delta} + \frac{c}{l} \cdot \sigma_{em} \cdot l_a \cdot n^{(j)} \cdot f^{las} \cdot e^{\lambda \cdot t} - \frac{f^{las} \cdot e^{\lambda \cdot t}}{\tau_c} - \frac{c}{l} \cdot \sigma_{ESA}^{las} \cdot l_a \cdot F^{las(j)} \cdot n \cdot e^{\lambda \cdot t} \cdot e^{-\lambda \cdot \tau_\Delta} - \frac{c}{l} \cdot \sigma_{ESA}^{las} \cdot l_a \cdot n^{(j)} \cdot f^{las} \cdot e^{\lambda \cdot t}$$

At fixed point: $\frac{c}{l} \cdot \sigma_{em} \cdot l_a \cdot F^{las(j)} \cdot n^{(j)} - \frac{F^{las(j)}}{\tau_c} - \frac{c}{l} \cdot \sigma_{ESA}^{las} \cdot l_a \cdot F^{las(j)} \cdot n^{(j)} = 0$.

$$f^{las} \cdot \lambda \cdot e^{\lambda \cdot t} = \frac{c}{l} \cdot \sigma_{em} \cdot l_a \cdot F^{las(j)} \cdot n \cdot e^{\lambda \cdot t} \cdot e^{-\lambda \cdot \tau_\Delta} + \frac{c}{l} \cdot \sigma_{em} \cdot l_a \cdot n^{(j)} \cdot f^{las} \cdot e^{\lambda \cdot t} - \frac{f^{las} \cdot e^{\lambda \cdot t}}{\tau_c} - \frac{c}{l} \cdot \sigma_{ESA}^{las} \cdot l_a \cdot F^{las(j)} \cdot n \cdot e^{\lambda \cdot t} \cdot e^{-\lambda \cdot \tau_\Delta}$$

$$-\frac{c}{l}\cdot\sigma_{ESA}^{las}\cdot l_a\cdot n^{(j)}\cdot f^{las}\cdot e^{\lambda\cdot t}$$

$$(\sigma_{em}-\sigma_{ESA}^{las})\cdot\frac{c}{l}\cdot l_a\cdot F^{las(j)}\cdot e^{-\lambda\cdot\tau_\Delta}\cdot n+\left[(\sigma_{em}-\sigma_{ESA}^{las})\cdot\frac{c}{l}\cdot l_a\cdot n^{(j)}-\frac{1}{\tau_c}\right]\cdot f^{las}-f^{las}\cdot\lambda=0$$

We can summary our Crystalline Mid-infrared laser system arbitrarily small increments equations.

$$\left[\left(-F^{pump}\cdot\sigma_{GSA}-F^{las(j)}\cdot\sigma_{em}-\frac{1}{\tau}-2\cdot\alpha\cdot n^{(j)}\right)\cdot e^{-\lambda\cdot\tau_\Delta}-\lambda\right]\cdot n-f^{las}\cdot\sigma_{em}\cdot n^{(j)}=0$$

$$(\sigma_{em}-\sigma_{ESA}^{las})\cdot\frac{c}{l}\cdot l_a\cdot F^{las(j)}\cdot e^{-\lambda\cdot\tau_\Delta}\cdot n+\left[(\sigma_{em}-\sigma_{ESA}^{las})\cdot\frac{c}{l}\cdot l_a\cdot n^{(j)}-\frac{1}{\tau_c}\right]\cdot f^{las}-f^{las}\cdot\lambda=0$$

The small increments Jacobian of our Crystalline Mid-infrared laser system is as below:

$$\begin{pmatrix} (-F^{pump}\cdot\sigma_{GSA}-F^{las(j)}\cdot\sigma_{em}-\frac{1}{\tau}-2\cdot\alpha\cdot n^{(j)})\cdot e^{-\lambda\cdot\tau_\Delta}-\lambda & -\sigma_{em}\cdot n^{(j)} \\ (\sigma_{em}-\sigma_{ESA}^{las})\cdot\frac{c}{l}\cdot l_a\cdot F^{las(j)}\cdot e^{-\lambda\cdot\tau_\Delta} & [(\sigma_{em}-\sigma_{ESA}^{las})\cdot\frac{c}{l}\cdot l_a\cdot n^{(j)}-\frac{1}{\tau_c}]-\lambda \end{pmatrix}\cdot\begin{pmatrix} n \\ f^{las}\end{pmatrix}=0$$

$$(A-\lambda\cdot I)=\begin{pmatrix} (-F^{pump}\cdot\sigma_{GSA}-F^{las(j)}\cdot\sigma_{em}-\frac{1}{\tau}-2\cdot\alpha\cdot n^{(j)})\cdot e^{-\lambda\cdot\tau_\Delta}-\lambda & -\sigma_{em}\cdot n^{(j)} \\ (\sigma_{em}-\sigma_{ESA}^{las})\cdot\frac{c}{l}\cdot l_a\cdot F^{las(j)}\cdot e^{-\lambda\cdot\tau_\Delta} & [(\sigma_{em}-\sigma_{ESA}^{las})\cdot\frac{c}{l}\cdot l_a\cdot n^{(j)}-\frac{1}{\tau_c}]-\lambda \end{pmatrix}$$

We define for simplicity: $\Xi_1=-F^{pump}\cdot\sigma_{GSA}-F^{las(j)}\cdot\sigma_{em}-\frac{1}{\tau}-2\cdot\alpha\cdot n^{(j)}$

$$\Xi_2=(\sigma_{em}-\sigma_{ESA}^{las})\cdot\frac{c}{l}\cdot l_a\cdot F^{las(j)};\ \Xi_3=(\sigma_{em}-\sigma_{ESA}^{las})\cdot\frac{c}{l}\cdot l_a\cdot n^{(j)}-\frac{1}{\tau_c}$$

$$(A-\lambda\cdot I)=\begin{pmatrix}\Xi_1\cdot e^{-\lambda\cdot\tau_\Delta}-\lambda & -\sigma_{em}\cdot n^{(j)}\\ \Xi_2\cdot e^{-\lambda\cdot\tau_\Delta} & \Xi_3-\lambda\end{pmatrix};\ \det(A-\lambda\cdot I)=0$$

$$\det(A-\lambda\cdot I)=(\Xi_1\cdot e^{-\lambda\cdot\tau_\Delta}-\lambda)\cdot(\Xi_3-\lambda)+\Xi_2\cdot\sigma_{em}\cdot n^{(j)}\cdot e^{-\lambda\cdot\tau_\Delta}$$

$$\det(A - \lambda \cdot I) = \lambda^2 - \Xi_3 \cdot \lambda + (\Xi_1 \cdot \Xi_3 + \Xi_2 \cdot \sigma_{em} \cdot n^{(j)} - \Xi_1 \cdot \lambda) \cdot e^{-\lambda \cdot \tau_\Delta}$$

$$D(\lambda, \tau_\Delta) = \lambda^2 - \Xi_3 \cdot \lambda + (\Xi_1 \cdot \Xi_3 + \Xi_2 \cdot \sigma_{em} \cdot n^{(j)} - \Xi_1 \cdot \lambda) \cdot e^{-\lambda \cdot \tau_\Delta}$$
$$n = 2; m = 1; n > m$$

$$D(\lambda, \tau_\Delta) = P_n(\lambda, \tau_\Delta) + Q_m(\lambda, \tau_\Delta) \cdot e^{-\lambda \cdot \tau_\Delta}$$
$$n = 2; m = 1; n > m$$

The expression for $P_n(\lambda, \tau_\Delta)$: $P_{n=2}(\lambda, \tau_\Delta) = \sum_{k=0}^{2} p_k(\tau_\Delta) \cdot \lambda^k = \lambda^2 - \Xi_3 \cdot \lambda$

$$P_{n=2}(\lambda, \tau_\Delta) = \sum_{k=0}^{2} p_k(\tau_\Delta) \cdot \lambda^k = p_0(\tau_\Delta) + p_1(\tau_\Delta) \cdot \lambda + p_2(\tau_\Delta) \cdot \lambda^2$$
$$p_0(\tau_\Delta) = 0; \; p_1(\tau_\Delta) = -\Xi_3; \; p_2(\tau_\Delta) = 1$$

The expression for $Q_m(\lambda, \tau_\Delta)$: $Q_m(\lambda, \tau_\Delta) = -\Xi_1 \cdot \lambda + (\Xi_1 \cdot \Xi_3 + \Xi_2 \cdot \sigma_{em} \cdot n^{(j)})$

$$Q_{m=1}(\lambda, \tau_\Delta) = \sum_{k=0}^{1} q_k(\tau_\Delta) \cdot \lambda^k = q_0(\tau_\Delta) + q_1(\tau_\Delta) \cdot \lambda$$
$$q_0(\tau_\Delta) = \Xi_1 \cdot \Xi_3 + \Xi_2 \cdot \sigma_{em} \cdot n^{(j)}; \; q_1(\tau_\Delta) = -\Xi_1$$

The homogenous system for n F^{las} lead to a characteristic equation for the eigenvalue λ having the form $P_{n=2}(\lambda, \tau_\Delta) + Q_{m=1}(\lambda, \tau_\Delta) \cdot e^{-\lambda \cdot \tau_\Delta} = 0$; $P(\lambda, \tau_\Delta) = \sum_{j=0}^{2} a_j(\tau_\Delta) \cdot \lambda^j$; $Q(\lambda, \tau_\Delta) = \sum_{j=0}^{1} c_j(\tau_\Delta) \cdot \lambda^j$ and the coefficients $\{a_j(q_i, q_k, \tau_\Delta), c_j(q_i, q_k, \tau_\Delta)\} \in \mathbb{R}$ depend on q_i, q_k and delay τ_Δ, q_i, q_k are any Crystalline Mid-infrared laser system's parameters, other parameters keep as a constant.

$$a_0 = 0; \; a_1 = -\Xi_3; \; a_2 = 1; \; c_0 = \Xi_1 \cdot \Xi_3 + \Xi_2 \cdot \sigma_{em} \cdot n^{(j)}; \; c_1 = -\Xi_1$$

Unless strictly necessary, the designation of the variation arguments (q_i, q_k) will subsequently be omitted from P, Q, a_j and c_j. The coefficients a_j, c_j are continuous and differentiable functions of their arguments, and direct substitution shows that $a_0 + c_0 \neq 0$, $\Xi_1 \cdot \Xi_3 + \Xi_2 \cdot \sigma_{em} \cdot n^{(j)} \neq 0$; $\forall \; q_i, q_k \in \mathbb{R}_+$, $\lambda = 0$ is not a $P(\lambda, \tau_\Delta) + Q(\lambda, \tau_\Delta) \cdot e^{-\lambda \cdot \tau_\Delta} = 0$. We assume that $P_{n=2}(\lambda, \tau_\Delta), Q_{m=1}(\lambda, \tau_\Delta)$ can't have common imaginary roots. That is for any real number ω; $\omega \in \mathbb{R}$: $P_{n=2}(\lambda = i \cdot \omega, \tau_\Delta) \neq 0$; $Q_{m=1}(\lambda, \tau_\Delta) \neq 0$.

$$P_{n=2}(\lambda = i \cdot \omega, \tau_\Delta) = -\omega^2 - i \cdot \omega \cdot \Xi_3; \; Q_{m=1}(\lambda, \tau_\Delta)$$
$$= -i \cdot \omega \cdot \Xi_1 + (\Xi_1 \cdot \Xi_3 + \Xi_2 \cdot \sigma_{em} \cdot n^{(j)})$$

$$P_{n=2}(\lambda = i\cdot\omega, \tau_\Delta) + Q_{m=1}(\lambda = i\cdot\omega, \tau_\Delta) = -\omega^2 + (\Xi_1\cdot\Xi_3 + \Xi_2\cdot\sigma_{em}\cdot n^{(j)}) - i\cdot\omega\cdot(\Xi_3 + \Xi_1)$$

$$|P(i\cdot\omega,\tau_\Delta)|^2 = \omega^4 + \omega^2\cdot\Xi_3^2;\ |Q(i\cdot\omega,\tau_\Delta)|^2 = \omega^2\cdot\Xi_1^2 + (\Xi_1\cdot\Xi_3 + \Xi_2\cdot\sigma_{em}\cdot n^{(j)})^2$$

We need to find the expression: $F(\omega,\tau_\Delta) = |P(i\cdot\omega,\tau_\Delta)|^2 - |Q(i\cdot\omega,\tau_\Delta)|^2$

$$F(\omega,\tau_\Delta) = |P(i\cdot\omega,\tau_\Delta)|^2 - |Q(i\cdot\omega,\tau_\Delta)|^2 = \omega^4 + \omega^2\cdot\Xi_3^2 - \omega^2\cdot\Xi_1^2 - (\Xi_1\cdot\Xi_3 + \Xi_2\cdot\sigma_{em}\cdot n^{(j)})^2$$

$$F(\omega,\tau_\Delta) = |P(i\cdot\omega,\tau_\Delta)|^2 - |Q(i\cdot\omega,\tau_\Delta)|^2 = \omega^4 + \omega^2\cdot(\Xi_3^2 - \Xi_1^2) - (\Xi_1\cdot\Xi_3 + \Xi_2\cdot\sigma_{em}\cdot n^{(j)})^2$$

We define the following parameters: $\Upsilon_0 = -(\Xi_1\cdot\Xi_3 + \Xi_2\cdot\sigma_{em}\cdot n^{(j)})^2$; $\Upsilon_2 = \Xi_3^2 - \Xi_1^2$; $\Upsilon_4 = 1$

$$F(\omega,\tau_\Delta) = |P(i\cdot\omega,\tau_\Delta)|^2 - |Q(i\cdot\omega,\tau_\Delta)|^2 = \sum_{k=0}^{2}\Upsilon_{2\cdot k}\cdot\omega^{2\cdot k} = \Upsilon_0 + \Upsilon_2\cdot\omega^2 + \Upsilon_4\cdot\omega^4$$

Hence $F(\omega,\tau_\Delta) = 0$ implies $\sum_{k=0}^{2}\Upsilon_{2\cdot k}\cdot\omega^{2\cdot k} = 0$ and its roots are given by solving the above polynomial. Furthermore, $P_R = -\omega^2$; $P_I = -\omega\cdot\Xi_3$; $Q_R = \Xi_1\cdot\Xi_3 + \Xi_2\cdot\sigma_{em}\cdot n^{(j)}$; $Q_I = -\omega\cdot\Xi_1$.

Hence

$$\sin\theta(\tau_\Delta) = \frac{-P_R(i\cdot\omega,\tau_\Delta)\cdot Q_I(i\cdot\omega,\tau_\Delta) + P_I(i\cdot\omega,\tau_\Delta)\cdot Q_R(i\cdot\omega,\tau_\Delta)}{|Q(i\cdot\omega,\tau_\Delta)|^2}$$

$$\cos\theta(\tau_\Delta) = -\frac{P_R(i\cdot\omega,\tau_\Delta)\cdot Q_R(i\cdot\omega,\tau_\Delta) + P_I(i\cdot\omega,\tau_\Delta)\cdot Q_I(i\cdot\omega,\tau_\Delta)}{|Q(i\cdot\omega,\tau_\Delta)|^2}$$

$$\sin\theta(\tau_\Delta) = \frac{-\omega^3\cdot\Xi_1 - \omega\cdot\Xi_3\cdot(\Xi_1\cdot\Xi_3 + \Xi_2\cdot\sigma_{em}\cdot n^{(j)})}{\omega^2\cdot\Xi_1^2 + (\Xi_1\cdot\Xi_3 + \Xi_2\cdot\sigma_{em}\cdot n^{(j)})^2}$$

$$\cos\theta(\tau_\Delta) = -\frac{-\omega^2\cdot(\Xi_1\cdot\Xi_3 + \Xi_2\cdot\sigma_{em}\cdot n^{(j)}) + \omega^2\cdot\Xi_3\cdot\Xi_1}{\omega^2\cdot\Xi_1^2 + (\Xi_1\cdot\Xi_3 + \Xi_2\cdot\sigma_{em}\cdot n^{(j)})^2}$$

Above expressions are continuous and differentiable in τ_Δ based on Lemma 1.1. Hence we use Theorem 1.2 and this proves Theorem 1.3.

We use different parameters terminology from our last characteristics parameters definition $k \to j$; $p_k(\tau_\Delta) \to a_j$; $q_k(\tau_\Delta) \to c_j$; $n = 2$; $m = 1$; $n > m$ additionally $P_n(\lambda,\tau_\Delta) \to P(\lambda,\tau_\Delta)$ and $Q_m(\lambda,\tau_\Delta) \to Q(\lambda,\tau_\Delta)$ then $P(\lambda,\tau_\Delta) = \sum_{j=0}^{2} a_j\ \cdot$

$\lambda^j; Q(\lambda, \tau_\Delta) = \sum_{j=0}^{1} c_j \cdot \lambda^j$.

$$P(\lambda) = \lambda^2 - \Xi_3 \cdot \lambda; Q(\lambda) = \Xi_1 \cdot \Xi_3 + \Xi_2 \cdot \sigma_{em} \cdot n^{(j)} - \Xi_1 \cdot \lambda$$

$n, m \in \mathbb{N}_0; n > m$ and $a_j, c_j : \mathbb{R}_{+0} \to \mathbb{R}$ are continuous and differentiable functions of τ_Δ such that $a_0 + c_0 \neq 0$. In the following "–" denotes complex and conjugate. $P(\lambda)$ and $Q(\lambda)$ are analytic functions in λ and differentiable in τ_Δ.

The coefficients $\{a_j(F^{pump}, \sigma_{GSA}, \sigma_{em}, \tau, \alpha, \sigma_{ESA}^{las}, \ldots)$ & $c_j(F^{pump}, \sigma_{GSA}, \sigma_{em}, \tau, \alpha, \sigma_{ESA}^{las}, \ldots)\} \in \mathbb{R}$ depend on Crystalline Mid-infrared laser system parameters $F^{pump}, \sigma_{GSA}, \sigma_{em}, \tau, \alpha, \sigma_{ESA}^{las}, \tau_\Delta, \ldots$ values. Unless strictly necessary, the designation of the variation arguments, $F^{pump}, \sigma_{GSA}, \sigma_{em}, \tau, \alpha, \sigma_{ESA}^{las}, \tau_\Delta, \ldots$ system parameters will subsequently be omitted from P, Q, a_j, c_j. The coefficients a_j, c_j are continuous and differentiable functions of their arguments, and direct substitution shows that $a_0 + c_0 \neq 0$.

$$\begin{aligned} a_0 + c_0 = & \left[-F^{pump} \cdot \sigma_{GSA} - F^{las(j)} \cdot \sigma_{em} - \frac{1}{\tau} - 2 \cdot \alpha \cdot n^{(j)} \right] \\ & \cdot \left[(\sigma_{em} - \sigma_{ESA}^{las}) \cdot \frac{c}{l} \cdot l_a \cdot n^{(j)} - \frac{1}{\tau_c} \right] \\ & + (\sigma_{em} - \sigma_{ESA}^{las}) \cdot \frac{c}{l} \cdot l_a \cdot F^{las(j)} \cdot \sigma_{em} \cdot n^{(j)} \neq 0 \end{aligned}$$

Furthermore, $P(\lambda)$, $Q(\lambda)$ are analytic function of λ for which the following requirements of the analysis (see Kuang 1993, Sect. 3.4) can also be verified in the present case [9, 10].

(a) If $\lambda = i \cdot \omega; \omega \in \mathbb{R}$ then $P(i \cdot \omega) + Q(i \cdot \omega) \neq 0$, i.e. P and Q have no common imaginary roots. This condition was verified numerically in the entire ($F^{pump}, \sigma_{GSA}, \sigma_{em}, \tau, \alpha, \sigma_{ESA}^{las}, \tau_\Delta, \ldots$ system parameters) domain of interest.

(b) $\left|\frac{Q(\lambda)}{P(\lambda)}\right|$ is bounded for $|\lambda| \to \infty$; $\operatorname{Re}\lambda \geq 0$. No roots bifurcation from ∞. Indeed, in the limit: $\left|\frac{Q(\lambda)}{P(\lambda)}\right| = \left|\frac{\Xi_1 \cdot \Xi_3 + \Xi_2 \cdot \sigma_{em} \cdot n^{(j)} - \Xi_1 \cdot \lambda}{\lambda^2 - \Xi_3 \cdot \lambda}\right|$.

(c) $F(\omega) = |P(i \cdot \omega)|^2 - |Q(i \cdot \omega)|^2 = \omega^4 + \omega^2 \cdot (\Xi_3^2 - \Xi_1^2) - (\Xi_1 \cdot \Xi_3 + \Xi_2 \cdot \sigma_{em} \cdot n^{(j)})^2$ has at most a finite number of zeros. Indeed, this is a polynomial in ω(degree in ω^4).

(d) Each positive root $\omega(F^{pump}, \sigma_{GSA}, \sigma_{em}, \tau, \alpha, \sigma_{ESA}^{las}, \tau_\Delta, \ldots)$ of $F(\omega) = 0$ is continuous and differentiable with respect to $F^{pump}, \sigma_{GSA}, \sigma_{em}, \tau, \alpha, \sigma_{ESA}^{las}, \tau_\Delta, \ldots$ parameters. This condition can only be assessed numerically.

In addition, since the coefficients in P and Q are real, we have $\overline{P(-i \cdot \omega)} = P(i \cdot \omega)$ and $\overline{Q(-i \cdot \omega)} = Q(i \cdot \omega)$ thus $\lambda = i \cdot \omega \ \forall \ \omega > 0$ may be on eigenvalue of characteristic equation. The analysis consists in identifying the roots of characteristic equation situated on the imaginary axis of the complex λ—plane, where by increasing the parameters $F^{pump}, \sigma_{GSA}, \sigma_{em}, \tau, \alpha, \sigma_{ESA}^{las}, \tau_{\Delta}, \ldots$(system parameters), $\mathrm{Re}\lambda$ may, at the crossing, change its sign from (−) to (+), i.e. from a stable focus $E^k(n^{(k)}, F^{las(k)})\ \forall\ k = 0, 1, 2$ to an unstable one, or vice versa. This feature may be further assessed by examining the sign of the partial derivatives with respect to $F^{pump}, \sigma_{GSA}, \sigma_{em}, \tau, \alpha, \sigma_{ESA}^{las}, \tau_{\Delta}, \ldots$ parameters ($\omega \in \mathbb{R}_{+}$).

$$\Lambda^{-1}(F^{pump}) = \left(\frac{\partial \mathrm{Re}\lambda}{\partial F^{pump}}\right)_{\lambda = i\cdot\omega}; \sigma_{GSA}, \sigma_{em}, \tau, \alpha, \sigma_{ESA}^{las}, \tau_{\Delta}, \cdots = const$$

$$\Lambda^{-1}(\sigma_{GSA}) = \left(\frac{\partial \mathrm{Re}\lambda}{\partial \sigma_{GSA}}\right)_{\lambda = i\cdot\omega}; F^{pump}, \sigma_{em}, \tau, \alpha, \sigma_{ESA}^{las}, \tau_{\Delta}, \cdots = const$$

$$\Lambda^{-1}(\sigma_{ESA}^{las}) = \left(\frac{\partial \mathrm{Re}\lambda}{\partial \sigma_{ESA}^{las}}\right)_{\lambda = i\cdot\omega}; F^{pump}, \sigma_{GSA}, \sigma_{em}, \tau, \alpha, \tau_{\Delta}, \cdots = const$$

$$\Lambda^{-1}(\tau_{\Delta}) = \left(\frac{\partial \mathrm{Re}\lambda}{\partial \tau_{\Delta}}\right)_{\lambda = i\cdot\omega}; F^{pump}, \sigma_{GSA}, \sigma_{em}, \tau, \alpha, \sigma_{ESA}^{las}, \cdots = const$$

where writing $P(\lambda) = P_R(\lambda) + i\cdot P_I(\lambda)$ and $Q(\lambda) = Q_R(\lambda) + i\cdot Q_I(\lambda)$, and inserting $\lambda = i\cdot\omega$ into Crystalline Mid-infrared laser system's characteristic equation ω must satisfy the following:

$$\sin(\omega\cdot\tau_{\Delta}) = g(\omega) = \frac{-P_R(i\cdot\omega)\cdot Q_I(i\cdot\omega) + P_I(i\cdot\omega)\cdot Q_R(i\cdot\omega)}{|Q(i\cdot\omega)|^2}$$

$$\cos(\omega\cdot\tau_{\Delta}) = h(\omega) = -\frac{P_R(i\cdot\omega)\cdot Q_R(i\cdot\omega) + P_I(i\cdot\omega)\cdot Q_I(i\cdot\omega)}{|Q(i\cdot\omega)|^2}$$

where $|Q(i\cdot\omega)|^2 \neq 0$ in the view of requirement (a) above, $(g, h) \in \mathbb{R}$. Furthermore, it follows the above equations $\sin(\omega\cdot\tau_{\Delta})$ and $\cos(\omega\cdot\tau_{\Delta})$ that, by squaring and adding the sides, ω must be a positive root of $F(\omega) = |P(i\cdot\omega)|^2 - |Q(i\cdot\omega)|^2 = 0$. Note $F(\omega)$ is independent of τ_{Δ}. Now it is important to notice that if $\tau_{\Delta} \notin I$ (assume that $I \subseteq \mathbb{R}_{+0}$ is the set where $\omega(\tau_{\Delta})$ is a positive root of $F(\omega)$ and for $\tau_{\Delta} \notin I$; $\omega(\tau_{\Delta})$ is not define. Then for all $\tau_{\Delta} \in I$ is satisfies that $F(\omega, \tau_{\Delta}) = 0$. Then there are no positive $\omega(\tau_{\Delta})$ solutions for $F(\omega, \tau_{\Delta}) = 0$, and we cannot have stability switch. For any $\tau_{\Delta} \in I$, where $\omega(\tau_{\Delta})$ is a positive solution of $F(\omega, \tau_{\Delta}) = 0$, we can define the angle $\theta(\tau_{\Delta}) \in [0, 2\cdot\pi]$ as the solution of $\sin\theta(\tau_{\Delta}) = \frac{-P_R(i\cdot\omega)\cdot Q_I(i\cdot\omega)+P_I(i\cdot\omega)\cdot Q_R(i\cdot\omega)}{|Q(i\cdot\omega)|^2}$; $\cos\theta(\tau_{\Delta}) = -\frac{P_R(i\cdot\omega)\cdot Q_R(i\cdot\omega)+P_I(i\cdot\omega)\cdot Q_I(i\cdot\omega)}{|Q(i\cdot\omega)|^2}$ and the relation between the argument $\theta(\tau_{\Delta})$ and $\omega(\tau_{\Delta})\cdot\tau_{\Delta}$ for $\tau_{\Delta} \in I$ must be $\omega(\tau_{\Delta})\cdot\tau_{\Delta} = \theta(\tau_{\Delta}) + 2\cdot n\cdot\pi\ \forall\ n \in \mathbb{N}_0$. Hence we can define the maps $\tau_{\Delta,\Delta n}: I \to \mathbb{R}_{+0}$ given by $\tau_{\Delta,\Delta n}(\tau) = \frac{\theta(\tau_{\Delta})+n\cdot 2\cdot\pi}{\omega(\tau_{\Delta})}$; $n \in \mathbb{N}_0, \tau_{\Delta} \in I$. Let introduce the functions $I \to \mathbb{R}$; $S_n(\tau_{\Delta}) = \tau_{\Delta} - \tau_{\Delta,\Delta n}(\tau_{\Delta})$, $n \in \mathbb{N}_0, \tau_{\Delta} \in I$, that are continuous and differentiable in τ_{Δ}. In the following, the subscripts $\lambda, \omega, F^{pump}, \sigma_{GSA}, \sigma_{em}, \tau, \alpha, \sigma_{ESA}^{las}, \tau_{\Delta}, \ldots$ system parameters (ω is related to parameter in our stability analysis) indicate the corresponding partial derivatives. Let us first concentrate on $\Lambda(x)$, remember in $\lambda(F^{pump}, \sigma_{GSA}, \sigma_{em}, \tau, \alpha, \sigma_{ESA}^{las}, \tau_{\Delta}, \ldots \in \mathbb{R}_{+})$, $\omega(F^{pump}, \sigma_{GSA}, \sigma_{em}, \tau, \alpha, \sigma_{ESA}^{las}, \tau_{\Delta}, \ldots \in \mathbb{R}_{+})$ and keeping all parameters except one (x) and τ_{Δ}. The derivation closely follows that in reference [BK]. Differentiating Crystalline Mid-infrared laser system characteristic equation $P(\lambda)+Q(\lambda)\cdot e^{-\lambda\cdot\tau_{\Delta}} = 0$ with respect to specific parameter (x), and inverting the derivative, for convenience, one calculates, $x = F^{pump}, \sigma_{GSA}, \sigma_{em}, \tau, \alpha, \sigma_{ESA}^{las}, \tau_{\Delta}, \ldots \in \mathbb{R}_{+}$,

$$\left(\frac{\partial\lambda}{\partial x}\right)^{-1} = \frac{-P_{\lambda}(\lambda,x)\cdot Q(\lambda,x) + Q_{\lambda}(\lambda,x)\cdot P(\lambda,x) - \tau_{\Delta}\cdot P(\lambda,x)\cdot Q(\lambda,x)}{P_{x}(\lambda,x)\cdot Q(\lambda,x) - Q_{x}(\lambda,x)\cdot P(\lambda,x)}$$

where $P_{\lambda} = \frac{\partial P}{\partial \lambda}, \ldots$ etc., Substituting $\lambda = i\cdot\omega$, and bearing $\overline{P(-i\cdot\omega)} = P(i\cdot\omega)$; $\overline{Q(-i\cdot\omega)} = Q(i\cdot\omega)$ then $i\cdot P_{\lambda}(i\cdot\omega) = P_{\omega}(i\cdot\omega)$ and $i\cdot Q_{\lambda}(i\cdot\omega) = Q_{\omega}(i\cdot\omega)$ that on the surface $|P(i\cdot\omega)|^2 = |Q(i\cdot\omega)|^2$, one obtains,

$$\left(\frac{\partial\lambda}{\partial x}\right)^{-1}|_{\lambda=i\cdot\omega}$$
$$= \frac{i\cdot P_{\omega}(i\cdot\omega,x)\cdot\overline{P(i\cdot\omega,x)} + i\cdot Q_{\lambda}(i\cdot\omega,x)\cdot\overline{Q(\lambda,x)} - \tau_{\Delta}\cdot|P(i\cdot\omega,x)|^2}{P_{x}(i\cdot\omega,x)\cdot\overline{P(i\cdot\omega,x)} - Q_{x}(i\cdot\omega,x)\cdot\overline{Q(i\cdot\omega,x)}}$$

Upon separating into real and imaginary parts, with $P = P_R + i\cdot P_I$; $Q = Q_R + i\cdot Q_I$ and $P_{\omega} = P_{R\omega} + i\cdot P_{I\omega}$; $Q_{\omega} = Q_{R\omega} + i\cdot Q_{I\omega}$; $P_x = P_{Rx} + i\cdot P_{Ix}$; $Q_x = Q_{Rx} + i\cdot Q_{Ix}$; $P^2 = P_R^2 + P_I^2$. When (x) can be a Crystalline Mid-infrared laser system parameters $F^{pump}, \sigma_{GSA}, \sigma_{em}, \tau, \alpha, \sigma_{ESA}^{las}, \ldots$ and time delay τ_{Δ} etc. Where for convenience, we have dropped the arguments $(i\cdot\omega, x)$, and where $F_{\omega} = 2\cdot[(P_{R\omega}\cdot P_R + P_{I\omega}\cdot P_I) - (Q_{R\omega}\cdot Q_R + Q_{I\omega}\cdot Q_I)]$ and $F_x = 2\cdot[(P_{Rx}\cdot P_R + P_{Ix}\cdot P_I) - (Q_{Rx}\cdot Q_R + Q_{Ix}\cdot Q_I)]$; $\omega_x = -\frac{F_x}{F_{\omega}}$. We define U and V:

$$U = (P_R\cdot P_{I\omega} - P_I\cdot P_{R\omega}) - (Q_R\cdot Q_{I\omega} - Q_I\cdot Q_{R\omega})$$
$$V = (P_R\cdot P_{Ix} - P_I\cdot P_{Rx}) - (Q_R\cdot Q_{Ix} - Q_I\cdot Q_{Rx})$$

We choose our specific parameter as time delay $x = \tau_{\Delta}$.

$$P_{n=2}(i\cdot\omega) = -\omega^2 - i\cdot\omega\cdot\Xi_3;\; Q_{m=1}(i\cdot\omega)$$
$$= -i\cdot\omega\cdot\Xi_1 + (\Xi_1\cdot\Xi_3 + \Xi_2\cdot\sigma_{em}\cdot n^{(j)})$$

$$P_R(i\cdot\omega) = -\omega^2;\; P_I(i\cdot\omega) = -\omega\cdot\Xi_3;\; Q_R(i\cdot\omega)$$
$$= \Xi_1\cdot\Xi_3 + \Xi_2\cdot\sigma_{em}\cdot n^{(j)};\; Q_I(i\cdot\omega) = -\omega\cdot\Xi_1$$

$$P_{R\omega}(i\cdot\omega,\tau_{\Delta}) = -2\cdot\omega;\; P_{I\omega}(i\cdot\omega,\tau_{\Delta})$$
$$= -\Xi_3;\; Q_{R\omega}(i\cdot\omega,\tau_{\Delta}) = 0;\; Q_{I\omega}(i\cdot\omega,\tau_{\Delta}) = -\Xi_1$$

$$P_{R\omega} = P_{R\omega}(i\cdot\omega,\tau_\Delta);\ P_{I\omega} = P_{I\omega}(i\cdot\omega,\tau_\Delta)$$
$$Q_{R\omega} = Q_{R\omega}(i\cdot\omega,\tau_\Delta);\ Q_{I\omega} = Q_{I\omega}(i\cdot\omega,\tau_\Delta)$$

$$P_{R\tau_\Delta} = 0;\ P_{I\tau_\Delta} = 0;\ Q_{R\tau_\Delta} = 0;\ Q_{I\tau_\Delta} = 0;\ P_{R\omega}\cdot P_R = 2\cdot\omega^3$$
$$P_{I\omega}\cdot P_I = \omega\cdot\Xi_3^2;\ Q_{R\omega}\cdot Q_R = 0;\ Q_{I\omega}\cdot Q_I = \omega\cdot\Xi_1^2$$

$$F_{\tau_\Delta} = 2\cdot[(P_{R\tau_\Delta}\cdot P_R + P_{I\tau_\Delta}\cdot P_I) - (Q_{R\tau_\Delta}\cdot Q_R + Q_{I\tau_\Delta}\cdot Q_I)]$$
$$= 0;\ \omega_{\tau_\Delta} = -\frac{F_{\tau_\Delta}}{F_\omega} = 0$$

$$P_R\cdot P_{I\omega} = \omega^2\cdot\Xi_3;\ P_I\cdot P_{R\omega} = \Xi_3\cdot 2\cdot\omega^2$$
$$Q_R\cdot Q_{I\omega} = -(\Xi_1\cdot\Xi_3 + \Xi_2\cdot\sigma_{em}\cdot n^{(j)})\cdot\Xi_1$$

$$Q_I\cdot Q_{R\omega} = 0;\ P_R\cdot P_{Ix}|_{x=\tau_\Delta} = 0;\ P_I\cdot P_{Rx}|_{x=\tau_\Delta} = 0$$
$$Q_R\cdot Q_{Ix}|_{x=\tau_\Delta} = 0;\ Q_I\cdot Q_{Rx}|_{x=\tau_\Delta} = 0$$

$$V = (P_R\cdot P_{I\tau_\Delta} - P_I\cdot P_{R\tau_\Delta}) - (Q_R\cdot Q_{I\tau_\Delta} - Q_I\cdot Q_{R\tau_\Delta})$$
$$= 0;\ F_\omega\cdot\frac{\partial\omega}{\partial\tau_\Delta} + F_{\tau_\Delta} = 0;\ \tau_\Delta\in I;\ \frac{\partial\omega}{\partial\tau} = -\frac{F_{\tau_\Delta}}{F_\omega}$$

$$\Lambda^{-1}(\tau_\Delta) = \left(\frac{\partial \mathrm{Re}\lambda}{\partial\tau_\Delta}\right)_{\lambda=i\cdot\omega};\ \Lambda^{-1}(\tau_\Delta)$$
$$= \mathrm{Re}\left\{\frac{-2\cdot[U+\tau_\Delta\cdot|P|^2] + i\cdot F_\omega}{F_{\tau_\Delta} + i\cdot 2\cdot[V+\omega\cdot|P|^2]}\right\};\ \frac{\partial\omega}{\partial\tau_\Delta} = \omega_{\tau_\Delta} = -\frac{F_{\tau_\Delta}}{F_\omega}$$

$$sign\{\Lambda^{-1}(\tau_\Delta)\} = sign\left\{\left(\frac{\partial \mathrm{Re}\lambda}{\partial\tau_\Delta}\right)_{\lambda=i\cdot\omega}\right\};\ sign\{\Lambda^{-1}(\tau_\Delta)\}$$
$$= sign\{F_\omega\}\cdot sign\left\{\tau_\Delta\cdot\frac{\partial\omega}{\partial\tau_\Delta} + \omega + \frac{U\cdot\frac{\partial\omega}{\partial\tau_\Delta} + V}{|P|^2}\right\}$$

We shall presently examine the possibility of stability transitions (bifurcations) in a Crystalline Mid-infrared laser system, about equilibrium points $E^k(n^{(k)}, F^{las(k)})\ \forall\ k = 0,1,2$ as a result of a variation of delay parameter τ_Δ. The analysis consists in identifying the roots of our system characteristic equation situated on the imaginary axis of the complexλ-plane where by increasing the delay parameter τ_Δ, $\mathrm{Re}\,\lambda$ may at the crossing, change its sign from "–" to "+", i.e. from a stable focus $E^{(*)}$ to an unstable one, or vice versa. This feature may be further assessed by examining the sign of the partial derivatives with respect to τ_Δ, $\Lambda^{-1}(\tau_\Delta) = \left(\frac{\partial \mathrm{Re}\lambda}{\partial\tau_\Delta}\right)_{\lambda=i\cdot\omega}$ and $F^{pump}, \sigma_{GSA}, \sigma_{em}, \tau, \alpha, \sigma_{ESA}^{las}, \ldots$ system parameters are constant where $\omega \in \mathbb{R}_{+}$. We need to plot the stability switch diagram based on different delay values of our Crystalline Mid-infrared laser system. Since it is very complex function we recommend to solve it numerically rather than analytic [9, 10].

$$\Lambda^{-1}(\tau_\Delta) = \left(\frac{\partial \mathrm{Re}\lambda}{\partial \tau_\Delta}\right)_{\lambda = i\cdot\omega} ; \Lambda^{-1}(\tau_\Delta) = \mathrm{Re}\left\{\frac{-2\cdot[U + \tau_\Delta \cdot |P|^2] + i\cdot F_\omega}{F_{\tau_\Delta} + i\cdot 2\cdot [V + \omega\cdot |P|^2]}\right\}$$

$$\Lambda^{-1}(\tau_\Delta) = \left(\frac{\partial \mathrm{Re}\lambda}{\partial \tau_\Delta}\right)_{\lambda = i\cdot\omega} ; \Lambda^{-1}(\tau_\Delta) = \frac{2\cdot\{F_\omega\cdot(V + \omega\cdot P^2) - F_{\tau_\Delta}\cdot(U + \tau_\Delta\cdot P^2)\}}{F_{\tau_\Delta}^2 + 4\cdot(V + \omega\cdot P^2)^2}$$

The stability switch occurs only on those delay values (τ_Δ) which fit the equation: $\tau_\Delta = \frac{\theta_+(\tau_\Delta)}{\omega_+(\tau_\Delta)}$ and $\theta_+(\tau_\Delta)$ is the solution of $\sin\theta(\tau_\Delta) = \ldots; \cos\theta(\tau_\Delta) = \ldots$, when $\omega = \omega_+(\tau_\Delta)$ if only ω_+ is feasible. Additionally when all Crystalline Mid-infrared laser system parameters are known and the stability switch due to various time delay values τ_Δ is describe in the following expression:

$$sign\{\Lambda^{-1}(\tau_\Delta)\} = sign\{F_\omega(\omega(\tau_\Delta), \tau_\Delta)\} \cdot sign\{\tau_\Delta \cdot \omega_{\tau_\Delta}(\omega(\tau_\Delta)) + \omega(\tau_\Delta) + \frac{U(\omega(\tau_\Delta)) \cdot \omega_{\tau_\Delta}(\omega(\tau_\Delta)) + V(\omega(\tau_\Delta))}{|P(\omega(\tau_\Delta))|^2}\}$$

Remark Lemma 1.1, Theorem 1.2 and Theorem 1.3 are described in Sect. 5.1.

5.4 Questions

1. We have a system of Intra-cavity frequency-doubling of a Nd:YAG laser passively Q-switched with GaAs. The saturable absorption in the GaAs wafer needs to apply the energy-level model of Valley and Smirl for energy transfer process in GaAs. In this model, schematically, the level responsible for absorption around 1um is believed to be EL2 defect that forms a deep level 0.82 eV below the band edge. This level has a total density N, part of which (N_0^+) is positively charged. Transitions from EL2 to the conduction band absorb optical energy and produce free electrons (n), while valence to EL2$^+$ transistors produce free holes (p) and neutral EL2 donors. Two-photon absorption (TPA) generates free electrons and holes, whereas free-carrier absorption (FCA) promotes electrons higher into the conduction band and holes deeper into the valence band. Combining all these effects with the accompanying recombination process, we get the set of intra-cavity rate equations, $n = n(t)$; $p = p(t)$; $N^+ = N^+(t)$.

$$\frac{dn}{dt} = 2 \cdot \phi_q \cdot (N - N^+) \cdot \sigma_e + B \cdot \frac{\phi^2(t - \tau_\Delta)}{2} - \gamma_{et} \cdot n \cdot N^+ - \gamma_{eh} \cdot n \cdot p$$

$$\frac{dp}{dt} = 2 \cdot \phi_q \cdot N^+ \cdot \sigma_h + B \cdot \frac{\phi^2(t - \tau_\Delta)}{2} - \gamma_{hd} \cdot p \cdot (N - N^+) - \gamma_{eh} \cdot n \cdot p$$

$$\frac{dN^+}{dt} = 2 \cdot \phi_q \cdot [(N - N^+) \cdot \sigma_e + N^+ \cdot \sigma_h] - \gamma_{et} \cdot n \cdot N^+ + \gamma_{hd} \cdot p \cdot (N - N^+)$$

$$\alpha_q = \sigma_e \cdot (N - N^+) + \sigma_h \cdot N^+ + \sigma_{fc} \cdot n$$

Which need to be solved simultaneously with the rate equations for photon density ϕ and population inversion density N_x. Due to uncomplete of photon generation and interferences, there is a delay in time (τ_Δ) for photon density $\phi(t) \to \phi(t - \tau_\Delta)$. σ—is stimulated emission cross-section, N_{tot}—is the total density of active atoms, τ—is spontaneous fluorescence lifetime, γ—is the cavity round trip loss coefficient, L_0—is the length of the laser rod, L_c—is the length of the doubling crystal, L_q—is the GaAs wafer thickness, T—is the cavity round trip, P—is the pump rate, η—is the plane wave impedance $\frac{377}{n_c}$, n_c—is the refractive index of the doubling crystal, n_q—is the refractive index of GaAs, d—the effective nonlinear coefficient, w_0—is the spot size at the laser rod, w_c—is the spot size at the doubling crystal, w_q—is the spot size at GaAs wafer, N—is the total EL2 density, N^+—is the ionized EL2 ($EL2^+$) density, N_0^+—is the initial ionized EL2 ($EL2^+$) density, γ_{et}—is the electron-EL2 recombination coefficient, γ_{hd}—is the hole-EL2 recombination coefficient, γ_{eh}—is the direct electron-EL2 recombination coefficient, σ_e—is the EL2 absorption cross section, σ_h is the $EL2^+$ absorption cross section, σ_{fc}—is the free-carrier absorption cross section, β—is the two-photon absorption coefficient. Here $\phi_q = \phi(t - \tau_\Delta) \cdot \left(\frac{w_0}{w_q}\right)^2 \cdot \frac{1}{n_q}$ is the photon density at the GaAs wafer and the coefficients γ_{et}, γ_{hd} and γ_{eh} describe electron—$EL2^+$, EL2-hole, and direct electron–hole recombination process, respectively. In $\alpha_q = \cdots$ equation $n(\approx p)$ represents the number of generated electron–hole pairs. The model assumes that all the elements are homogeneously distributed throughout the cavity, which is valid under the condition that the intensity does not change substantially during a single round trip. Transverse beam shape effects including the refraction caused by free carriers in GaAs. Hint: Some system parameters are not explain in the question, see Sect. 5.1.

1.1 Find system fixed points. Draw fixed points coordinates as a function of σ_e (EL2 absorption cross section) and σ_h ($EL2^+$ absorption cross section)—3D graph.
1.2 Discuss stability and stability switching for different values of τ_Δ parameter.
1.3 Parameter L_q (length of the GaAs wafer) is very small ($L_q \to \varepsilon$), return (1.1) and (1.2). Explain how our results change.
1.4 Parameter L_c(length of the doubling crystal) is very small ($L_c \to \varepsilon$), return (1.1) and (1.2). Explain how our results change.
1.5 Discuss stability and stability switching for different values of γ_{hd} parameter values (hole—EL2 recombination coefficient).

2. The model presented the passively Q-switched lasers. By using the second-threshold criterion and numerical fitting procedure, we get the analytical function of output pulse energy. An analytical expression for the optimal output reflectivity maximized the output pulse energy of a passively Q-switched laser with a given initial transmission of the saturable absorber. The analysis of the passively Q-switched laser used the variable z, $z = \frac{1}{L} \cdot 2 \cdot \sigma \cdot n_i \cdot l$, where n_i is the initial population density in the gain medium, σ is the stimulated emission cross section of the gain medium, l is the length of the gain medium, L is the non saturable intra cavity round-trip dissipative optical loss. In the cavity Q-switched laser, n_i is proportional to the pump rate. We characterize the output pulse energy of the cavity Q-switch laser with the variable z. T_o is the initial transmission of the saturable absorber, and R is the reflectivity of the output mirror. The second threshold of passively Q-switches laser is accepted when the photon density turns upward and produces a giant pulse. The gain can saturate first and the photon density never turns upward to develop a giant pulse. The model takes the assumptions: uniform pumping of the gain medium, the intra cavity optical intensity as axially uniform, and the complete recovery of the saturable absorber. The coupled rate equations describe the model of a passively Q-switched laser. The coupled equations for three or four level gain media are as follow:

$$\frac{d\phi}{dt} = \frac{\phi(t-\tau)}{t_r} \cdot \Big[2 \cdot \sigma \cdot n(t) \cdot l - 2 \cdot \sigma_{gs} \cdot n_{gs}(t) \cdot l_s - 2 \cdot \sigma_{es} \cdot n_{es} \cdot l_s - \left(\ln\left(\frac{1}{R}\right) + L\right)\Big]$$

$$\frac{dn}{dt} = -\gamma \cdot c \cdot \sigma \cdot \phi(t-\tau) \cdot n(t); \quad \frac{dn_{gs}}{dt} = -\frac{A}{A_s} \cdot c \cdot \sigma_{gs} \cdot \phi(t-\tau) \cdot n_{gs}(t); \quad n_{gs} + n_{es} = n_{so}$$

$\phi(t) = \phi$; $n(t) = n$; $n_{gs} = n_{gs}(t)$. The parameters table is described below.

System parameter	Description
ϕ	Intra cavity photon density with respect to the effective cross-section area of the laser beam in the gain medium
n	Population density of the gain medium
l_s	Length of the saturable absorber
$\frac{A}{A_s}$	Ratio of the effective area in the gain medium and in the saturable absorber
n_{gs}	Absorber ground
n_{es}	Excited state
n_{so}	Total population densities

(continued)

(continued)

System parameter	Description
σ_{gs}, σ_{es}	GSA and ESA cross sections in the saturable absorber, respectively
R	Reflectivity of the output mirror
γ	Inversion reduction factor ($\gamma = 1$ and $\gamma = 2$ correspond to, respectively, four-level and three-level systems
$t_r = \frac{2 \cdot l'}{c}$	Round trip transit time of light in the cavity optical length l', where c is the speed of light

The passively Q-switched laser model is not ideal, and there is a shift in time in the interactivity photon density variable, $\phi(t) \to \phi(t-\tau)$. It is not affect the derivative of $\phi(t)$, $\frac{d\phi}{dt}$. τ is the delay parameter which can affect the dynamic of the system.

2.1 Find system fixed points, How the results are changed if $\tau = 0$?
2.2 Discuss stability and stability switching for different values of τ parameter.
2.3 The reflectivity of the output mirror system parameter (R) is double in his value ($R \to 2 \cdot R$), How it effect the behavior of the system?
2.4 There is a situation where, $R \to 1$, How the system dynamic change? Find fixed points and discuss stability, stability switching. What happened when $R = 1$?
2.5 Discuss the system stability switching for different values of the ratio of the effective area, $\frac{A}{A_s}$.

3. We get for the passively Q-switched lasers model (question 2), for four-level saturable absorber. $n_{gs} = n_{so} \cdot \left(\frac{n}{n_i}\right)^{\alpha}$; $\alpha = \frac{1}{\gamma} \cdot \frac{\sigma_{gs}}{\sigma} \cdot \frac{A}{A_s}$ and n_i is determined from the condition that the round-trip gain is exactly equal to the round-trip losses before the Q-switch opens. Thus $2 \cdot \sigma \cdot n_i \cdot l - 2 \cdot \sigma_{gs} \cdot n_{so} \cdot l_s - \left[\ln\left(\frac{1}{R}\right) + L\right] = 0$. The parallel between the two analyses is done by $T_0 = e^{-\sigma_{gs} \cdot n_{so} \cdot l_s}$, where T_0 is the initial transmission of the saturable absorber. We define $n_i = f(T_0)$, $n_i = \frac{\ln\left(\frac{1}{T_0^2}\right) + \ln\left(\frac{1}{R}\right) + L}{2 \cdot \sigma \cdot l}$ and the differential equation:

$$\frac{d\phi}{dn} = -\frac{l}{\gamma \cdot l'} \cdot \left[1 - \frac{(1-\beta)}{2 \cdot \sigma \cdot l \cdot n_i} \cdot \ln\left(\frac{1}{T_0^2}\right) \cdot \left(\frac{n}{n_i}\right)^{\alpha - 1} - \frac{\beta \cdot \ln\left(\frac{1}{T_0^2}\right) + \ln\left(\frac{1}{R}\right) + L}{2 \cdot \sigma \cdot l \cdot n}\right]; \beta = \frac{\sigma_{es}}{\sigma_{gs}}$$

The first derivative of ϕ with respective to n at $n = n_i$ is equal to zero, the criterion for Q-switching behavior is whether the second derivative of ϕ with respective to $n @ n = n_i$ has a positive or a negative sign. If positive, the growth curve for the photon intensity will turn increasingly upward. We take account of the influence of excited-state absorption and intra cavity focusing.

3.1 Plot the function between $\frac{d\phi}{dn}$ and T_0 parameter, $\frac{d\phi}{dn} = \xi(T_0)$, find extreme points and explore the behavior of the function.
3.2 Find the expression for $\frac{d^2\phi}{dn^2}$ and find by substitution $n = n_i$, the criterion for a giant pulse to occur and the criterion for the second threshold.
3.3 Discuss the stability of our system for different values of n_i parameter.
3.4 In the case $\beta = 0$; $\frac{A}{A_s} = \Gamma$; $\Gamma \in \mathbb{R}_+$, plot the function between $\frac{d^2\phi}{dn^2}$ and Γ parameter $\left(\frac{d^2\phi}{dn^2} = \zeta(\Gamma)\right)$. Discuss stability and stability switching for different values of Γ parameter.
3.5 The parameters α and β are determined from the cavity configuration and the physical properties of the gain medium and the saturable absorber. The initial transmission of the saturable absorber has an upper bound $T_{0-upper}$ for producing a giant pulse for a given $\ln\left(\frac{I}{R}\right) + L$, a given α, and a given value of β, $T_{0-upper} = \exp\left\{-\frac{\ln\left(\frac{1}{R}\right)+L}{2\cdot[\alpha\cdot(1-\beta)-1]}\right\}$. On the other hand, the reflectivity of the output coupler has a lower bound R_{upper} to build up a giant pulse for a given L, a given T_0, a given α, and a given β, $R_{lower} = \exp\left\{-[\alpha\cdot(1-\beta)-1]\cdot\ln\left(\frac{1}{T_0^2}\right) + L\right\}$. The parameters $T_{0-upper}$ and R_{lower} are used to express the output pulse energy and optimum output reflectivity as an analytical function, respectively. Express T_0 by R_{lower} and R by $T_{0-upper}$ in the $\frac{d^2\phi}{dn^2} = \zeta(\Gamma)$, second order differential equation. Discuss stability and stability switching for different values of $T_{0-upper}$ and R_{lower} parameters.

4. We have a device of semiconductor saturable absorber mirror (SESAM). The measured nonlinear reflectivity is used to fit a model function to obtain the key parameters of the saturable absorber. The model function is $R(F_p)$ and this model function describes the nonlinear behavior of real SESAMs. The two level system differential equation is discussed, when the reflectivity is given by $R(F_p) = c_R \cdot \frac{V_{out}}{V_{IN}}$ where, c_R is established by using a dielectric high-reflector ($R = 100\%$) as reference. Parameter c_R is a constant, and using a constant c_R and plotting $R(F_p)$ of the high-reflector does not show 100% flat curve. A real absorber device shows non-saturable losses (the device not reach 100% reflectivity) even for arbitrarily high fluences. The losses are homogeneously distributed over the absorber layer or transmission losses. These losses are accounted by a scaling factor R_{ns}.

$R(F_p) = R_{ns} \cdot \frac{\ln\left\{1+\frac{R_{lin}}{R_{ns}}\cdot(e^S-1)\right\}}{S}$, where S is the saturation parameter $\left(S = \frac{F_p}{F_{sat}}\right)$, T_{lin} is the linear transmission, F_{sat} is the saturation fluence. We can replace T_{lin} by R_{lin} in $R(F_p)$ equation. At higher fluences the reflectivity decreases with increasing fluence according to a second order process which leads to a roll-off in the reflectivity curve. The roll-off is taken into account by multiplying the model function with $e^{\left[-\frac{F_p}{F_2}\right]}$ which gives $R(F_p) = R_{ns} \cdot \frac{\ln\left\{1+\frac{R_{lin}}{R_{ns}}\cdot(e^S-1)\right\}}{S} \cdot e^{\left[-\frac{F_p}{F_2}\right]}$. A smaller F_2 value is corresponded to a stronger roll-off. F_2 is the fluence where the SESAM reflectivity is dropped to 37% $\left(\frac{1}{e}\right)$ due to induced absorption. We define a_2 as the material parameter. For a Gaussian laser beam $F_p^{Gauss}(r) = F_0 \cdot e^{\left[-\frac{2 \cdot r^2}{\omega^2}\right]}$, where F_0 is the peak fluence and ω is the $\left(\frac{1}{e^2}\right)$ beam radius. A Gaussian laser beam is focused to a small spot spreads out rapidly as it propagates away from the spot. The laser beam is kept collimated by larger diameter. The Gaussian laser beam is propagated in z direction (z – axis), r is the radial distance from the center $\left(r = \sqrt{x^2 + y^2}\right)$. The peak fluence in a Gaussian beam is $2 \cdot F_p$ which gives the outcome that saturation already occurs at lower fluences. We define $R^{FlatTop}(\chi)$, where it is a transformation for any function $R^{Gauss}(F_p)$ and $R^{FlatTop}(F_p)$ as long as lateral diffusion is neglected.

$$R^{FlatTop}(\chi) = R_{ns} \cdot \frac{\ln\left\{1 + \frac{R_{lin}}{R_{ns}} \cdot \left(e^{\frac{\chi}{F_{sat}}} - 1\right)\right\}}{\frac{\chi}{F_{sat}}} \cdot e^{\left[-\frac{\chi}{F_2},\right]}$$

χ is F_p variable $\left(S = \frac{\chi}{F_{sat}}\right)$.

4.1 Get the expression for $\frac{dR^{FlatTop}(\chi)}{d\chi}$, and plot the graph $\frac{dR^{FlatTop}(\chi)}{d\chi}$ versus χ.

4.2 $\chi = \Gamma_0 + \Gamma_1 \cdot R^{FlatTop}(\chi) + \Gamma_2 \cdot [R^{FlatTop}(\chi)]^2$; $\Gamma_0, \Gamma_1, \Gamma_2 \in \mathbb{R}_+$. Find fixed points of the system $\frac{dR^{FlatTop}(\chi)}{d\chi} = \xi(R^{FlatTop}(\chi))$ and plot fixed points coordinates as a function of F_{sat} parameter.

4.3 Discuss the stability and stability switching for different values of F_{sat} parameter.

4.4 Discuss stability and stability switching for different values of $\Gamma_0, \Gamma_1, \Gamma_2$ parameters.

4.5 There is a shifting in χ for $R^{FlatTop}$ variable $\left(R^{FlatTop}(\chi) \to R^{FlatTop}(\chi - \chi_\Delta)\right)$, it does not affect $\frac{dR^{FlatTop}(\chi)}{d\chi}$. Discuss stability and stability switching for different values of χ_Δ parameter.

5. We have a passively Q-switched laser system. The system is characterized by rate equations, the intra cavity photon density and the initial population inversion density in the diode-pumped passively Q-switched laser. We assume that the rate equations are characterized by Gaussian spatial distributions. We assume that the intra cavity photon density is a Gaussian spatial distribution during the entire formatting process of the Laser-Diode (LD) pumped passively Q-switched laser pulse. The population-inversion density at $t = 0$ is a Gaussian spatial distribution, and the ground-state population density of the saturable absorber at $t = 0$ is uniform. The intra cavity photon density $\phi(r, t)$ for TEM_{00} mode is $\phi(r, t) = \phi(0, t) \cdot e^{\left[-\frac{2 \cdot r^2}{\omega_L^2}\right]}$, where r is the radial coordinate, ω_L is the laser-mode radius, which determined by the geometry of the resonator. $r = 0$ is the laser axis, and $\phi(0, t)$ is the photon density in the laser axis. The differential equation describing $\frac{d\phi(r,t)}{dt}$ is integrated over the beam cross section to guarantee the beam Gaussian spatial distribution during the entire formatting process of the Q-switched pulse.

$$\frac{dn(r,t)}{dt} = -\gamma \cdot \sigma \cdot c \cdot \phi(r,t) \cdot n(r,t); \frac{dn_{sg}(r,t)}{dt} = -\frac{S_g}{S_s} \cdot \sigma_{gsa} \cdot c \cdot \phi(r,t) \cdot n_{sg}(r,t)$$

where, $n(r,t)$ is the population-inversion density, $n_{sg}(r,t)$ and n_{s0} are, respectively, the ground-state and total population densities of the saturable absorber. σ and l are, respectively, the simulated-emission cross section and length of the gain medium. σ_{gsa} and σ_{asa} are, respectively, the ground-state and excited-state absorption cross sections of the saturable absorber. l_s is the length of the saturable absorber, $t_r = \frac{2 \cdot l'}{c}$ is the round-trip transit time of light in the resonator of optical length l', c is the light speed in vacuum. R is the reflectivity of the output mirror, L is the remaining round-trip dissipative optical loss; S_g and S_s are, respectively, the beam cross-section area in the gain medium and in the saturable absorber; γ is the inversion-reduction factor which corresponds to the net reduction in the population inversion resulting from the stimulated emission of a single photon. We neglect the pumping and spontaneous emission during the pulse formation. The solid-state saturable absorbers are of slow recovery then the ground-state recovery of the saturable absorber during the formation of the Q-switched pulse is neglected. The pump-light intensity in the gain medium is attenuated along the longitudinal axis z, and $n(r,t)$ is a function of z. We choose the radial coordinate, r as a function of $n_{sg}(r,t), r = \sqrt{n_{sg}^K(r,t)}; K > 0; K \in \mathbb{N}$ then $\phi(r,t) = \phi(0,t) \cdot e^{\left[-\frac{2 \cdot n_{sg}^K(r,t)}{\omega_L^2}\right]}; t = t(n_{sg}, K); \phi(0,t) \in \mathbb{R}_+$ is a constant in our analysis and for specific r value ($r = r_0$), we can write our rate equations:

$$\frac{dn(r_0,t)}{dt} = -\gamma \cdot \sigma \cdot c \cdot \phi(r_0,t) \cdot n(r_0,t); \frac{dn_{sg}(r_0,t)}{dt} = -\frac{S_g}{S_s} \cdot \sigma_{gsa} \cdot c \cdot \phi(r_0,t) \cdot n_{sg}(r_0,t)$$

$$\phi(r_0, n_{sg}) = \phi(0,t) \cdot e^{\left[-\frac{2 \cdot n_{sg}^K(r_0,t)}{\omega_L^2}\right]}; t = t(n_{sg}, K); \phi(0,t) \in \mathbb{R}_+$$

5.1 Find system fixed points and discuss stability.

5.2 There is a delay in time only in $n_{sg}(r_0,t)$ variable, $n_{sg}(r_0,t) \to n_{sg}(r_0, t-\tau)$, but it is not affect the derivative of $n_{sg}(r_0,t)$ in time $\left(\frac{dn_{sg}(r_0,t)}{dt}\right)$. Discuss stability and stability switching for different values of τ parameter.

5.3 How the system dynamic is changed if we change K parameter value from $K = 1$ to $K > 1$? refer to (5.2). Discuss stability for two cases, (1) $K = 1$, (2) $K > 1$.

5.4 There is a delay in time only in $n(r_0, t)$ variable, $n(r_0, t) \rightarrow n(r_0, t - \Delta)$, but it is not affect the derivative of $n(r_0, t)$ in time $\left(\frac{dn(r_0,t)}{dt}\right)$. Discuss stability and stability switching for different values of Δ parameter.

5.5 How the system dynamic is changed if we change K parameter value from $K > 4$ to $K = 2$? refer to (5.4). Discuss stability for two cases, (1) $K > 4$, (2) $K = 2$.

6. We have a system of passively Q-switched microchip laser at specific wavelength (λ). It produces a laser beam in the eye-safe wavelength regime that produces peak power at specific pulse durations and repetition rate. An Er:Yb:glass microchip laser is passively Q-switched with a semiconductor saturable absorber mirror (SESAM). The reflectivity of a SESAM as a function of the incident pulse fluence is very important. The modulation depth ΔR is defined as the maximum reflectivity change between low and high intensity. Non saturable losses are the remaining losses at high intensity. The fluence which switch $\left(\frac{1}{e}\right)$ of the saturable losses is the saturation fluence F_{sat}. When the recovery time τ_A of the absorber is larger than the pulse duration τ_p and smaller than the pulse-to-pulse duration $\frac{1}{f_{rep}}$, $\left(\frac{1}{f_{rep}} > \tau_A > \tau_p\right)$, then the dynamic response of the absorber is not dominant in the Q switching. Shorter recovery time than the pulse duration reduces the efficiency of the laser. One system SESAM design for Q switching with pulse energies is the high-finesse anti resonant Fabry–Perot saturable absorber. The semiconductor absorber is embedded in an anti-resonant cavity between a semiconductor bottom Bragg mirror of high reflectivity and a sputtered dielectric top reflector with a transmittance T_t. The modulation depth of the AR-coated sample ΔR_{AR} and the non saturable losses $l_{ns,AR}$, the reflectivity in the saturated state $R_{ns,AR} = 1 - l_{ns,AR}$. The transmittance of the top reflector T_t, tunes adjust the intensity that enters to the absorber. Decreasing the transmittance T_t reduces the intensity on the absorber. It increases the damage fluence and reduces the non saturable losses. Other effect is decreasing the modulation depth and increasing the saturation fluence. The fluence on the absorber is F_{abs}, $(F_{abs} = \xi \cdot F_{AR})$, where F_{AR} is the fluence in the AR-coated reference sample. The low fluences in anti-resonance, ξ fraction is define as $\xi_{low} = \frac{T_t}{\left\{1+\sqrt{\frac{1-T_t}{R_{ns,AR}-\Delta R_{AR}}}\right\}^2}$. In saturated condition we get expression for ξ_{high}; which is $\xi_{high} = \frac{T_t}{\{1+\sqrt{(1-T_t)\cdot R_{ns,AR}}\}^2}$. For small non saturable losses $l_{ns,AR} \ll 1$ $(R_{ns,AR} \approx 1)$ the modulation depth is scaling with the same factor as the fluence for low intensities, $\Delta R \approx \xi_{low} \cdot \Delta R_{AR}$ where ΔR is the modulation depth of the sample with the top reflector.

6.1 Plot the graphs of two functions, $\xi_{low} = g_1(T_t)$; $\xi_{high} = g_2(T_t)$. Investigate the behavior. Are there any intersection points between the two graphs? If yes find them (consider that $R_{ns,AR}$, ΔR_{AR} are constants and $R_{ns,AR} \neq \Delta R_{AR}$).

6.2 How the graphs (6.1) change for small non saturable losses $l_{ns,AR} \ll 1$, $(R_{ns,AR} \approx 1)$? Find if exist intersection points between two graphs.

6.3 Experimental results show that $\xi_{high} = \xi_{low}^{\Xi} - 1$; $\Xi > 1$; $\Xi \in \mathbb{N}$, find the possible values T_t for specific $R_{ns,AR}$, ΔR_{AR} parameters values ($R_{ns,AR} \neq \Delta R_{AR}$). How your results change for different values of Ξ parameter? Plot the graph of T_t versus Ξ.

6.4 Find the expression for $\frac{dT_t}{dt}$ as a function of T_t, $R_{ns,AR}$, ΔR_{AR}, and Ξ. Find fixed points and explore the dynamic of the system.

6.5 For (6.4) discuss stability and stability switching for different values of Ξ parameter.

7. We have two SESAMs that have different transmittance T_t of the top reflector and the same modulation depth of the final structure. The SESAM with higher top-mirror reflectivity has a thicker absorber. The resulting laser performance is the same for SESAMs are fully saturated. The lower transmittance T_t reduces the pulse fluence on the absorber and increases the damage threshold. F_d is the damage fluence of the absorber. $F_{abs} = \frac{\xi_{high}}{T_{OC}} \cdot F_p$, where $F_p = \frac{E_p}{A_L}$ then $F_{abs} = \frac{\xi_{high}}{T_{OC}} \cdot \frac{E_p}{A_L}$, $F_{abs} = \frac{T_t}{T_{OC} \cdot \{1 + \sqrt{(1-T_t) \cdot R_{ns,AR}}\}^2} \cdot \frac{E_p}{A_L}$. F_p is the pulse fluence outside the cavity. The transmittance of the output coupler, T_{OC} is related when the SESAM is exposed to the intra cavity intensity. The pulse fluence, the pulse energy E_p, $E_p = 2 \cdot A_L \cdot F_L \cdot \Delta R \cdot \left(\frac{T_{OC}}{T_{OC} + l_{par}}\right)$ and $F_L = \frac{h \cdot \nu_L}{2 \cdot \sigma_L + 2 \cdot \sigma_L^{abs}}$ is the saturation fluence of the gain material, ν_L is the laser frequency, σ_L and σ_L^{abs} are the cross section for stimulated emission and re absorption at the laser wavelength, respectively. The fraction $\frac{T_{OC}}{(T_{OC} + l_{par}}$ is for the output coupling efficiency. The pulse duration is given by $\tau_p = \frac{7 \cdot n \cdot L}{c \cdot \Delta R}$, where n is the index of refraction of the gain material, L is the cavity length, and c is the speed of light in vacuum. The cavity round trip time is $\frac{2 \cdot n \cdot L}{c}$. The fluence on the absorber layer ($R_{ns,AR} - \Delta R_{AR} \neq 0$) is

$$F_{abs} = 2 \cdot F_L \cdot \Delta R_{AR} \cdot \frac{1}{(T_{OC} + l_{par})}$$

$$\cdot \frac{T_t^2}{\{1 + \sqrt{(1 - T_t) \cdot R_{ns,AR}}\}^2 \cdot \{1 + \sqrt{(1 - T_t) \cdot (R_{ns,AR} - \Delta R_{AR})}\}^2}.$$

The condition $F_{abs} < F_d$ avoids damage, keeping the fluence on the absorber below the damage fluence. l_{par} is the parasitic losses, it is small compared with the transmittance of the output coupler and then can be neglected. The condition for the transmittance of the top reflector to prevent the SESAM damage is $T_t < \sqrt{2 \cdot \frac{F_d \cdot T_{OC}}{F_L \cdot \Delta R_{AR}}}$. The way to achieve high modulation depth and avoiding damage for a Q switched microchip laser is by growing the semiconductor part of the SESAM with higher modulation depth which is limited by the critical thickness for strained layers and for lattice match growth. The non saturable losses reduce the modulation depth with increasing of the absorber thickness.

7.1 Plot the graph of F_{abs} as a function of T_t(transmittance), $F_{abs} = \psi_1(T_t)$ where all other parameters are constants. Explore the behavior of the function.

7.2 The transmittance of the output coupler T_{OC} is dependent on T_t (transmittance), $T_{OC} = \sqrt{T_t^2 + \Gamma}$; $\Gamma \in \mathbb{R}_+$. Plot the graph $F_{abs}|_{T_{OC}=\sqrt{T_t^2+\Gamma};\Gamma\in\mathbb{R}_+} = \psi_2(T_t)$ and discuss the behavior of the graph for different values of Γ parameter (all other parameters are constants).

7.3 Find the expression for $\frac{dF_{abs}}{dt}$, where $T_t(t) = F_{abs}^K(t)$; $K \in \mathbb{N}_+$. Discuss fixed points, How fixed points behave for different values of K parameter?

7.4 Discuss stability and stability switching (7.3) for different values of K parameter.

7.4 There is a delay in time, Δ_1 for variable $F_{abs}(t)$, $F_{abs}(t) \to F_{abs}(t - \Delta_1)$ which do not affect the derivative of $\frac{dF_{abs}}{dt}$. Discuss stability and stability switching for different values of Δ_1 parameter.

8. We have ion-doped crystalline laser which operates in the mid-IR spectral range. The bandwidth of the gain medium is the key parameter for establishing the operation threshold. The full width at half maximum is relative to the gain maximum at λ_0; $\left(\frac{\Delta\lambda}{\lambda_0} \approx \frac{\Delta\nu}{\nu_0}\right)$. We define parameter n_{ocp} as the number of optical cycles per pulse, and it is especially for ultra-short pulse generation, $n_{ocp} = \left(\frac{\Delta\lambda}{\lambda_0}\right)^{-1}$. We consider a system continuous-wave operation and assume four-level scheme, balance equation for the population of the upper laser level n: $\frac{dn}{dt} = F^{pump} \cdot \sigma_{GSA} \cdot [N_t - n(t)] - F^{las}(t) \cdot \sigma_{em} \cdot n(t) - \frac{n(t)}{\tau} - \alpha \cdot n^2(t)$.

The rate equation for the photon flux at the laser wavelength: $\frac{dF^{las}}{dt} = \frac{c}{l} \cdot F^{las}(t) \cdot \sigma_{em} \cdot n(t) \cdot l_a - \frac{F^{las}(t)}{\tau_c} - \frac{c}{l} \cdot F^{las}(t) \cdot \sigma_{ESA}^{las} \cdot n(t) \cdot l_a$. We define the intrinsic slop efficiency as η_0, where $\eta_0 = \frac{\lambda^{pump}}{\lambda^{las}} \cdot \left(1 - \frac{\sigma_{ESA}^{las}}{\sigma_{em}}\right)$. $\frac{\lambda^{las}}{\lambda_{pump}}$ is the stokes shift and we consider it as a constant, $\Gamma = \frac{\lambda^{las}}{\lambda_{pump}}$ with very small variation, $\Gamma \to \Gamma \pm \Delta\Gamma$, $\Delta\Gamma \to \varepsilon$, then we can define $\eta_0 = \frac{1}{\Gamma} \cdot \left(1 - \frac{\sigma_{ESA}^{las}}{\sigma_{em}}\right)$. Hint: All parameters are as discuss in this chapter.

8.1 Express σ_{ESA}^{las} as a function of η_0, Γ (3D graph). Find fixed points and discuss the behavior of fixed points for different values of Γ parameter.

8.2 What happened for $\Gamma \to \Gamma \pm \Delta\Gamma$; $\Delta\Gamma \in \mathbb{R}_+$; $0 < \Delta\Gamma < 1$? Investigate the behavior of the system for different values of $\Delta\Gamma$ parameter.

8.3 There is a shift in time for F^{las} (photon flux at the laser wavelength), $F^{las}(t) \to F^{las}(t - \Delta)$ but it is not affect the derivative in time of $F^{las}(t)$; $\frac{dF^{las}}{dt}$. Discuss the stability and stability switching for different values of Δ parameter.

8.4 What happened (8.3), if Δ parameter double his value, $\Delta \to 2 \cdot \Delta$? How the behavior of the system is changed?

8.5 What happened (8.3), if $\Delta \to \sqrt[3]{\Delta}$? How the stability map is changed?

9. We have crystalline and fiber Raman laser system which based on the Simulated Raman Scattering (SRS) effect in crystals and silica-based fibers. The laser system operates in the CW, nano second, and picosecond regimes with low and high repetition rates. The spontaneous Raman scattering (RS) process is a two-photon resonance, when the difference of the absorption (ν_L) and emission (ν_S) optical frequencies is equal to the frequency of an atomic or molecular vibration in Raman media $\nu_R = \nu_L - \nu_S$. Due to the two-photon

nature, RS cross-sections are very small, $\sigma \approx 10^{-30}$ cm^2. The intensity of stimulated Raman scattering is proportional to the occupation densities of the states of three interacting fields: laser pump photons (N_L), photons of the generated stokes wave (N_S), and crystal phonons (molecular vibrations) in the ground (N_g) or excited (N_e) state ($N = N_g - N_e$). The rate equation is $\frac{dN_S}{dz} \sim N_L \cdot N_S \cdot (N_g - N_e)$. In the steady-state regime, when the pump duration t_p is much longer than the Raman mode dephasing time T_R, ($t_p \gg T_R$), the intensity I_s of the stimulated Raman scattering (SRS) stokes beam passing through the Raman medium with length l is given by $I_s(l) = I_s(0) \cdot e^{[g \cdot I_L \cdot l]}$, where I_L is the intensity of the incident pump laser beam and g is the gain coefficient $g = \frac{\lambda_L \cdot \lambda_s^2 \cdot N}{\pi \cdot c \cdot \hbar \cdot n_s^2 \cdot \Delta \nu_R} \cdot \left(\frac{d\sigma}{d\Omega}\right)$. λ_L and λ_s are the wavelengths of the pump laser beam and the stokes Raman beam, respectively; c is the speed of light in vacuum, $\hbar$ is the reduced Planck constant $\left(\hbar = \frac{h}{2 \cdot \pi}\right)$, n_s is the refractive index at the stokes wavelength; $\Delta \nu_R$ is the full width at half maximum of the Raman spectral line in $\left[\frac{1}{\text{sec}}\right]$ units; and Ω is the solid scattering angle. The gain is very high for materials with a high density of scatters N, large Raman scattering integral cross-section $\frac{d\sigma}{d\Omega}$, and small Raman linewidth $\Delta \nu_R$.

9.1 The crystalline and fiber Raman laser system is not ideal, then there is a shift in z coordinate for photons of the generated stokes wave (N_S), $N_S(z) \to N_S(z - \Delta z)$ but not affect the derivative of N_S; $\left(\frac{dN_S}{dz}\right)$. Discover the dynamic of the system (fixed points, stability, and stability switching for different values of Δz parameter). Hint: consider that N_L, N_g and N_e are constants with some approximation.

9.2 The excited (N_e) state is a function of (N_S), photons of the generated stokes wave, $N_e = \xi_1(N_S) = \sqrt{\Gamma_1 \cdot N_S + \sqrt{\Gamma_2 \cdot N_S}}$; $\Gamma_1, \Gamma_2 \in \mathbb{R}_+$ and there is a shift in z parameter ($z \to z + \Delta z$) as discuss in (9.1). The dynamic differential equation for $\frac{dN_S}{dz}$ is $\frac{dN_S(z)}{dz} \sim N_L \cdot N_S(z - \Delta z) \cdot (N_g - \sqrt{\Gamma_1 \cdot N_S(z - \Delta z) + \sqrt{\Gamma_2 \cdot N_S(z - \Delta z)}})$. Find system fixed points and plot 3D graph of fixed points coordinates as a function of Γ_1, Γ_2 parameters. Hint: consider N_g and N_L as a constants.

9.3 Discuss stability and stability switching (9.2) for different values of Δz parameter.

9.4 The laser pump photons (N_L) is a function of N_S, $N_L = \xi_2(N_S) = \sqrt[3]{N_S + 1}$, then $\frac{dN_S}{dz} \sim \xi_2(N_S(z - \Delta z)) \cdot N_S(z - \Delta z) \cdot (N_g - N_e)$ and there is a shift in z parameter ($z \to z + \Delta z$) as discuss in (9.1). Find system fixed points, and discuss stability and stability switching for different values of Δz parameter. Hint: consider N_g and N_e as a constants.

9.5 g is the system gain coefficient $g = \frac{\lambda_L \cdot \lambda_s^2 \cdot N}{\pi \cdot c \cdot \hbar \cdot n_s^2 \cdot \Delta \nu_R} \cdot \left(\frac{d\sigma}{d\Omega}\right)$, $g = \sigma^{\Xi}(\Omega - \Omega_\Delta)$; $\Xi \in \mathbb{N}$; $\Xi > 1$ then $\frac{d\sigma}{d\Omega} = \left(\frac{\lambda_L \cdot \lambda_s^2 \cdot N}{\pi \cdot c \cdot \hbar \cdot n_s^2 \cdot \Delta \nu_R}\right)^{-1} \cdot \sigma^{\Xi}(\Omega - \Omega_\Delta)$; $\Omega_\Delta > 0$. Find fixed points and plot fixed points coordinates as a function of Ξ parameter. Discuss stability and stability switching for different values of Ω_Δ parameter. How the stability map is changed if we increase the value of Ξ parameter?

10. A crystalline Raman laser is based on self-Raman laser materials. It is a passive Q-switched intra-cavity Raman laser, which emits a yellow orange light beam. The laser frequency conversion is based on stimulated Raman scattering (SRS) in crystalline materials. Crystalline Raman laser generates both high average power (5 W) and high pulse energies (<1 J), with conversion efficiencies which approach the quantum limit. Intra-cavity Raman laser is based on diode-pumped solid-state laser materials, and utilizing a range of self-Raman laser materials. It can generate a variety of near infrared laser wavelengths with diode stokes optical efficiencies close to $\approx$ 20%. Spontaneous Raman scattering involves inelastic scattering of light in an optical medium, which deposits energy into an excited state of the component matter, the scattered photons have energy which reduced by the quantum of energy deposited in the process. The scattered light has frequency $\omega_{S1} = \omega_P - \omega_R$, where, ω_P is the frequency of the incident pump light, $\hbar \cdot \omega_R$ matches the energy of an electronic vibrational, or rotational energy state of the Raman medium. For crystalline Raman media $\hbar \cdot \omega_R$ matches a vibrational mode of the crystal. When the energy of the scattered photon is less than that of the incident photon, the scattered light has longer wavelength and is referred to the first stokes line. Simulated Raman scattering (SRS) is the outcome from the third-order nonlinear polarizability $P_3 = \varepsilon_0 \cdot \chi_3 \cdot E^3$ of the optical medium, where ε_0 is the dielectric constant, χ_3 is the third-order nonlinear susceptibility, and E is the amplitude of the incident optical field. The common nonlinear phenomena, two-photon absorption, stimulated Brillouin scattering are results of χ_3 (third order nonlinear susceptibility). The SRS which related to the imaginary part of χ_3 is proportional to the square of the normal-mode derivative of the molecular polarizability tensor $\left(\frac{\partial \alpha}{\partial q}\right)$ materials with high polarizability. The vibrational modes modulate the polarizability strongly as for symmetric breathing modes of covalently bound molecules, are favored for SRS. The expression for growth along the propagation direction z of the intensity $I_s(z)$ of stokes field near the threshold in steady-state is $I_s(z) = I_s(0) \cdot e^{[g_R \cdot I_p \cdot z]}$, where, I_p is the incident/pump field intensity. g_R is the steady-state Raman gain coefficient and z is the propagation distance along z-axis. The steady-state Raman gain coefficient is $g_R = \frac{8 \cdot \pi \cdot c^2 \cdot N}{\hbar \cdot \mu_S^2 \cdot \omega_S^3 \cdot \Gamma} \cdot \left(\frac{\partial \sigma}{\partial \Omega}\right)$ and the integrated Raman scattering cross section $\left(\frac{\partial \sigma}{\partial \Omega}\right)$ is $\frac{\partial \sigma}{\partial \Omega} = \frac{\omega_S^4 \cdot \mu_S}{c^4 \cdot \mu_P} \cdot \frac{\hbar}{2 \cdot m \cdot \omega_R} \cdot \left(\frac{\partial \alpha}{\partial q}\right)^2$. Where, N is the number density of Raman active molecules, μ_S and μ_P are the refractive indices at stokes and pump wavelength (λ_S and λ_P), respectively. m is the reduced mass of the molecular oscillator, and Γ is the line width of the Raman transition and it is equal to the inverse of the dephasing time T_R for the final state of the transition $\left(\Gamma = \frac{1}{T_R}\right)$. ω_S is the stokes radial frequency which is related to stokes wavelength λ_S. c is the speed of light in vacuum and $\hbar$ is the reduced Planck constant $\left(\hbar = \frac{h}{2 \cdot \pi}\right)$.$\omega_R$ is Raman radial frequency. $\frac{\partial \alpha}{\partial q}$ is the normal-mode derivative of the molecular polarizability tensor.

10.1 Plot 3D graph of the normal-mode derivative of the molecular polarizability tensor $\left(\frac{\partial \alpha}{\partial q}\right)$ as a function of ω_R and ω_S. Discuss the behavior of $\frac{\partial \alpha}{\partial q}$ for different values of ω_R and ω_S. Hint: consider all other parameters as a constants.

10.2 We know from experimental results, that Γ(linewidth of the Raman transition) is a function of α parameter $\Gamma = \sqrt{\alpha + |\alpha| \cdot \sqrt{\alpha}}$. Explore the dynamic of the system $\frac{\partial \alpha}{\partial q} \leftrightarrow \frac{d\alpha}{dq}; \frac{d\alpha}{dq} = \ldots\ldots \Gamma$, find fixed points and discuss stability.

10.3 We get the results that $\omega_S = a \cdot \omega_P; a \in \mathbb{R}_+$. Plot the graph of $\frac{d\alpha}{dq}$ for different values of a parameter.

10.4 For (10.2), $\Gamma = \sqrt{\alpha(q - q_\Delta) + |\alpha(q - q_\Delta)| \cdot \sqrt{\alpha(q - q_\Delta)}}$. Discuss stability and stability switching for different values of q_Δ parameter ($q_\Delta \in \mathbb{R}; q_\Delta > 0$).

10.5 What happened (10.4), if q_Δ parameter doubles his values $q_\Delta \rightarrow 2 \cdot q_\Delta$? How the stability map is changed?

References

1. T.T. Kajava, A.L. Gaeta, Intra-cavity frequency-doubling of a Nd:YAG laser passively Q-switched with GaAs. Opt. Commun. **137**, 93–97 (1997)
2. R. Zhou, W. Wen, Z. Cai, X. Ding, P. Wang, J. Yao, Efficient stable simultaneous CW dual-wavelength diode-end-pumped Nd:YAG laser operating at 1.319 and 1.338 μm. Chin. Opt. Lett. **3**(10), (2005)
3. R.C. Eckardt, R.L. Byer, Measurement of nonlinear optical coefficients by phase-matched harmonic generation, in *SPIE Vol. 1561 Inorganic Crystals for Optics, Electro-Optics, and Frequency Conversion* (1991).
4. J. Gu, S.-C. Tam et al., Novel use of GaAs as a passive Q-switch as well as an output coupler for diode-pumped infrared solid-state lasers, Proc. SPIE 3929, Solid State Lasers IX, 2000.
5. W. Su-Mei, Z. Qiu-Lin et al., Diode-pumped passive Q-switched 946nm Nd:YAG laser with a GaAs saturable absorber. Chin. Phys Lett. **23**(3), 619 (2006)
6. J.K. Hale, *Dynamics and Bifurcations*. Texts in Applied Mathematics, vol. 3
7. S.H. Strogatz, *Nonlinear Dynamics and Chaos* (Westview Press)
8. Y. Kuang, *Delay Differential Equations with Applications in Population Dynamics* (Academic Press, Boston, 1993).
9. E. Beretta, Y. Kuang, Geometric stability switch criteria in delay differential systems with delay dependent parameters. SIAMJ. Math. Anal. **33**, 1144–1165 (2002)
10. J. Kuang, Y. Cong, *Stability of Numerical Methods for Delay Differential Equations* (Elsevier Science, 2007)
11. M. Haiml, R. Grange, U. Keller, Optical characterization of semiconductor saturable absorbers. Appl. Phys. B **79**, 331–339 (2004)
12. G.J. Spuhler, R. Paschotta, R. Fluck, B. Braun, M. Moser, G. Zhang, E. Gini, U. Keller, Experimentally confirmed design guidelines for passive Q-switched microchip lasers using semiconductor saturable absorbers. J. Opt. Soc. Am. B/ **16**(3), (1999)
13. R. Haring, R. Paschotta, R. Fluck, E. Gini, H. Melchior, U. Keller, Passively Q-switched microchip laser at 1.5 um. J. Opt. Soc. Am. B **18**(12) (2001)
14. X. Zhang, S. Zhao, Q. Wang, Modeling of passively Q-switched lasers. J. Opt. Soc. Am. B/ **17**(7) (2000).
15. M.C. Stumpf, S.C. Zeller, A. Schlatter, T. Okuno, T. Sudmeyer, U. Keller, Compact Er:Yb:glass-laser-based supercontinuum source for high-resolution optical coherence tomography. Opt. Express **16**(14), 10572–11057 (2008)
16. I.T. Sorokina, Crystalline Mid-infrared laser. Topics Appl. Phys. **89**, 255–351 (2003)

Chapter 6
Gas Laser Systems Stability Analysis Under Parameters Variation

Lasing from molecular nitrogen is used in many scientific and industrial applications. The discharge pumped nitrogen laser, operating in a broad range of gas pressures, from several mill bars to the atmospheric pressure, and repetition rates from several hertz to several kilo hertz. It is robust source of high-power near-UV radiation. Achieving nitrogen lasing via remote excitation would pave the way to many potential applications. It is remotely initiated lasing from molecular gases by femtosecond filaments. A mid infrared femtosecond laser filament can introduce backward-directed lasing of molecular nitrogen via a resonant excitation transfer mechanism. The characterization of the filament ignited nitrogen laser, including the spatial and temporal properties of the generated UV emission and generation thresholds. A filament-assisted nitrogen laser is efficient as its conventional discharge-pumped counterpart. Mid-IR ultrashort laser pulses radically enhance fila mentation-assisted lasing of N_2 relative to ultrashort pulses in the near-IR. We need a full analysis of multiple fila mentation. Another area is a remotely pumped N_2 laser in the atmosphere and optimization of the pumping via electron-N_2 collisions and evolving long-pulse light or microwave sources in combination with the femtosecond filament. Fila mentation is modeled by numerical solving the cylindrically symmetric nonlinear Schrodinger equation, which accounts for the impact of plasma dispersion and refraction, beam diffraction, Kerr, Raman, and plasma nonlinearities. We describe a change of concentration of different neutral and ionic atomic and molecular species in the filament plasma by the following set of rate equations: $\frac{d[N_s]}{dt} = [G_s]-[L_s]$; N_s is the density of species of type-s; G_s and L_s are the relevant generation and loss rates. The rate equations are solved jointly with the equations for the electron temperature and vibrational temperature of ground-state nitrogen molecules and stability is inspected. To include the lase effects, the set of plasma kinetic equations is inspected by the rate equations for the population of the lase levels and number of generated photons, stability analysis is done. The Quasi-two level analytic model described the end pumped Alkali metal vapor laser. We consider the steady state rate equations for the number densities of the, ${}^2s_{1/2}$, ${}^2p_{3/2}$ and ${}^2p_{1/2}$, energy states for the three level laser system. We consider that the relaxation between the two upper levels, ${}^2p_{3/2}$ and ${}^2p_{1/2}$, caused by

O. Aluf, *Advance Elements of Laser Circuits and Systems*,
https://doi.org/10.1007/978-3-030-64103-0_6

collisions with additive ethane is much faster (infinity fast), by comparison with any other process in the system including stimulated emission. We get the ratio of the number densities for the upper two levels, $^2p_{3/2}$ and $^2p_{1/2}$, and its statistical equilibrium value and the mathematical description. A quasi-two level system is from which an analytic solution is extracted. The analytic model description gives expressions for the threshold pump power and the slope efficiency including intra-cavity losses. The model compares with the steady state three level models. The system of coupled differential rate equations describes the kinetics and radiation process in the system at steady state and stability analysis is done under parameters variation. A high repetition rate copper vapor lasers can be describe by a self-consistent model. We simulate the equations for the discharge pulse, laser pulse, and inter pulse afterglow period. It simultaneously integrated over many discharge cycles until a consistent solution is obtained. By modeling both an ideal and real laser we get to conclusion that the peak electron temperature and the initial density of metastable copper are factors which dominate the performance of the laser. We can describe the laser by the processes which affect these two quantities. High repetition rate (HRR) electric discharge metal vapor lasers are a source of high peak and high average power in the visible. The most sophisticated HRR copper lasers are resonantly charged electric discharge devices. These lasers are self-heated; the energy required to heat the laser tube and vaporize to copper comes from the discharge itself. Tube temperature and the copper vapor pressure are established by the insulation surrounding the discharge tube. All for a given repetition rate, buffer gas pressure and discharge pulse energy $\left(\frac{1}{2} \cdot C \cdot V^2\right)$. The copper laser is a cyclic or self-terminating laser. The upper laser level is resonant to the ground state while the lower laser level is metastable. The discharge circuit for the real laser simulates the actual circuit used for the laser. The discharge circuitry is inspected for stability under parameters variation. The gas temperature relaxes to the wall temperature with a time constant proportional to the radius of the discharge tube. We derive the time rate of change of P_g and P_{cu} inspect the stability under parameters variation. Parameters T_g, P_{cu} and P_g are thermodynamic quantities during the operation of the lasers, the temperature of the discharge tube, which determines the copper vapor pressure, can be measured by an optical pyrometer. If the gas density stays in equilibrium with the local temperature and pressure, we derive the related equations for the change in gas density due to changes in temperature and pressure and inspect the dynamic.

6.1 Nitrogen Gas Laser Filament Plasma Kinetic Equations Stability Analysis Under Parameters Variation

Nitrogen is a chemical element with symbol N and atomic number 7. Nitrogen is the lightest member of group 15 of the periodic table and is called pnictogens. Many industrially compound, such as ammonia, nitric acid, organic nitrates, and cyanides, contain nitrogen. Nitrogen is a constituent of organic compounds as diverse as Kevlar used in high-strength fabric and cyanoacrylate used in superglue. The molecular nitrogen (N_2) has some properties: inter molecular potential and the equation of state. We use Quantum mechanics (QM) calculations to determine N_2-N_2 inter molecular interaction potential. We get the anisotropic potential surface for all orientations of the pair of the nitrogen molecules in the rotation invariant form. The molecular dynamics (MD) of the N_2 molecules has been used to determine nitrogen state. The classical motion of N_2 molecules is integrated and it accounted only translational and rotational degrees of freedom. We use the molecular dynamics (MD) modeling of the nitrogen pressure to get the fifth order predictor-corrector, or Runge-kutta. The nitrogen molecule is two nitrogen atoms stuck together by a "covalent bond". They are bound together tightly and it is very difficult to separate them, except in certain chemical reactions. In our laser system a laser generation in molecular nitrogen in an argon-nitrogen gas mixture is remotely excited at a specific distance above in a femtosecond laser filament. Mid-IR laser pulses enable radical enhancement of fila mentation assisted lasing by N_2 molecules. High energies are achieved from laser pulses which generated through the second-positive-band transitions of N_2. It is corresponding to a 0.5% total conversion efficiency from mid infrared laser energy to the energy of UV lasing. Lasing from molecular nitrogen is used in many applications. The discharge-pumped nitrogen laser, operating in a broad range of gas pressures, from several millibars to the atmospheric pressure, and repetition rates from several hertz to several kilohertz. It is robust source of high-power near-UV radiation. Achieving nitrogen via remote excitation would pave the way to many potential applications. A narrow band source of stimulated emission would provide coherent, highly directional radiation for highly selective and sensitive remote spectroscopy of the atmosphere. The possibility for remote ignition of a free localized nitrogen laser is demonstrated previously in a microwave discharge. By using focused laser beams we can excite free-space nitrogen lasing. The process of femtosecond fila mentation in gases is related to remotely excited free-space nitrogen laser. Fila mentation of high-power femtosecond laser radiation in gases is connected the formation of a self-guided, high-intensity field structure, accompanied by a significant spectral broadening, super continuum generation, and creation of a plasma channel in the wake of the pulse. The femtosecond filaments can be generated at standoff distances of up to tens of kilometers and from ionized channels of up to hundreds of meters. Formation of plasma in the filament, similar to a gas discharge, initiates a chain of plasma chemical reactions in the atmosphere which lead to the appearance of a large variety of neutral and ionic species in rotationally, vibrational, and electronically excited states. We get a situation that favorable conditions for population inversion and lasing between different electronic levels in nitrogen and in oxygen is created. We can use the standoff lasing with a femtosecond filament as an advantage in comparison to scheme due to substantially less scattering and absorption losses. There is a possibility of stimulated emission from nitrogen in femtosecond filaments. The proof of lasing is based on the exponential fit of the dependence of fluorescence on the filament length. The exponential dependence is valid under the condition of a stationary population inversion only. In filament driven laser the pump is the femtosecond laser pulse and amplified spontaneous emission (ASE) is about nanosecond [1]. The laser generation is build up under the condition of a decaying population inversion which has a decay rate on the same nanoscale time scale. The filament length is calculated from the laser power assuming the direct proportionality between these values. Multiple filament formation is expected at high levels of laser power, and the assumption is that the filament length is proportional to the laser power. The generation of micro joule pulses from a cavity-free nitrogen laser which uses the plasma channel traced by a femtosecond laser filament as its active medium. In the lasing process, the electronic transitions of nitrogen are the same as in the conventional discharge-pumped nitrogen laser. In the discharge-driven scheme, the upper lasing triplet $C^3\Pi_u$ in the three-level nitrogen laser is populated by electron impact excitation from the singlet ground state $X^1\Sigma_g$. This gives rise to an efficient fluorescence in the near-UV spectral range due to radiative transitions between the vibrational manifolds of $C^3\Pi_u$ and $B^3\Pi_g$ states. The control over the density and temperature of electrons is of key importance for obtaining population inversion and lasing between these states. The impulsive RF or capacitor discharge caused to rapid joule heating of electrons due to collisions with neutrals leads to the formation of non-equilibrium energy distribution of electrons. The non-equilibrium energy distribution and the plasma concentration are maintained throughout the discharge, resulting in a sufficient amount of hot electrons—which is needed for efficient pumping. In a laser filament, the electronic energy distribution is formed on the time scale of the femtosecond laser pulse by the optical field ionization process. After the femtosecond pulse, the resultant energy distribution function evolves freely as the plasma can concentration decays. Therefore, the required $C^3\Pi_u$ excitation by electron impact can be achieved within the average lifetime of the plasma electrons. By using a single femtosecond pulse, we cannot control the electron temperature during the buildup of population inversion, nor to control the duration of the time window within which the nitrogen laser is pumped. The electric discharge pumping, which maintains hot plasma, in the femtosecond filament the optimal electronic energy distribution is governed by optical field ionization and therefore is determined by the intensity and the optical cycle duration. The effect of intensity clamping precluded the possibility of direct scaling of the temperature of electrons in a filament via an increasing of the energy pulse. The plasma density is limited by the self-consistent balance between self-focusing and plasma refraction when focusing conditions are fixed. The coherence length of ASE is determined by the radiative lifetime. We need to inspect and analyze the efficient lasing of molecular nitrogen in a femtosecond filament. In this process the excitation energy of argon atoms transfers to molecular nitrogen. Then, excited argon atoms provide a collisional pump for laser transitions of N_2 in a laser-induced filament, as same as hot electrons in a discharge-pumped nitrogen laser [1]. The analytical model is used to analyze the lasing in a laser-induced filament which includes the fila mentation dynamics and of plasma kinetics in the wake of the filament. We get the electron density and the intensity of the laser pulses. The target is to solve the system of plasma kinetics equations jointly with the equations for the electron temperature and the vibrational temperature of ground-state nitrogen molecules. The set of plasma kinetic equations is associated with the relevant rate equations for the population of the lasing levels and number of emitted photons. The change of concentration of different neutral and ionic atomic and molecular species in the filament plasma is described by set of rate equations $\frac{d[N_s]}{dt} = [G_s] - [L_s]$. Where the subscript s stands for e, Ar^+, Ar_2^+, N_2^+, N_4^+, N_3^+ and N^+, N, $N_2(A^3\Sigma_u)$, $N_2(B^3\Pi_g)$, $N_2(C^3\Pi_u)$, $Ar^*(4^3P_2)$; N_s is the density of species of type-s; and G_s, L_s are the relevant generation and loss rates. The rate equations are solved jointly with the equations for the electron temperature and the vibrational temperature of ground-state nitrogen molecules.

System No.1:

The first rate equation:

$$\begin{aligned}\frac{3}{2}\cdot k\cdot\frac{d}{dt}(N_e\cdot T_e) = &-\frac{3}{2}\cdot N_e\cdot k\cdot(T_e - T_{vibr})\cdot\upsilon_*\\ &-\frac{3}{2}\cdot N_e\cdot k\cdot(T_e - T)\cdot[\delta_{N_2}\\ &\cdot(\upsilon_{N_2}+\upsilon_{N_2^+}+\frac{1}{2}\cdot\upsilon_{N_4^+})\\ &+\delta_{A_r}\cdot(\upsilon_{A_r}+\upsilon_{A_r^+}+\frac{1}{2}\cdot\upsilon_{A_{r_2}^+})]\\ &+k_e\cdot N_e\cdot N_{A_{r*}}\cdot I_{exc}\end{aligned}$$

$-\frac{3}{2}\cdot N_e\cdot k\cdot(T_e - T_{vibr})\cdot\upsilon_*$—is the electron energy transfer to vibrational excitation of nitrogen molecules.

$-\frac{3}{2}\cdot N_e\cdot k\cdot(T_e - T)\cdot[\delta_{N_2}\cdot(\upsilon_{N_2}+\upsilon_{N_2^+}+\frac{1}{2}\cdot\upsilon_{N_4^+})+\delta_{A_r}\cdot(\upsilon_{A_r}+\upsilon_{A_r^+}+\frac{1}{2}\cdot\upsilon_{A_{r_2}^+})]$—is the electron energy transfer in elastic and Coulombic collisions to translational energy of nitrogen and argon molecules, atoms, and ions.

k is the Boltzmann constant, T, T_e, T_{vibr} are gas, electronic, and vibrational temperatures, respectively. υ_s are the frequencies of elastic collisions of electrons and corresponding species, υ_* is the frequency of inelastic collisions of electrons with N_2 in the ground state which are responsible for vibrational heating; $\delta_{N_2} = \frac{2\cdot m}{M_{N_2}}$; $\delta_{A_r} = \frac{2\cdot m}{M_{A_r}}$; k_e is the de excitation rate of excited argon atoms by electrons.

Assumption G_e and L_e are the relevant generation and loss rates, and they are very close in their values $G_e - L_e \to \varepsilon$ then $\frac{dN_e}{dt} \to \varepsilon$.

$$\begin{aligned}\frac{3}{2}\cdot k\cdot\frac{d}{dt}(N_e\cdot T_e) = \frac{3}{2}\cdot k\cdot\frac{dN_e}{dt}\cdot T_e + \frac{3}{2}\cdot k\\ \cdot N_e\cdot\frac{dT_e}{dt} \overset{\frac{dN_e}{dt}\to\varepsilon}{=} \frac{3}{2}\cdot k\cdot N_e\cdot\frac{dT_e}{dt}\end{aligned}$$

The reduced first rate equation:

$$\frac{3}{2}\cdot k\cdot N_e\cdot\frac{dT_e}{dt} = -\frac{3}{2}\cdot N_e\cdot k\cdot(T_e - T_{vibr})\cdot\upsilon_* - \frac{3}{2}\cdot N_e\cdot k$$

$$\cdot (T_e - T) \cdot [\delta_{N_2} \cdot (\upsilon_{N_2} + \upsilon_{N_2^+} + \frac{1}{2} \cdot \upsilon_{N_4^+})$$
$$+ \delta_{A_r} \cdot (\upsilon_{A_r} + \upsilon_{A_r^+} + \frac{1}{2} \cdot \upsilon_{A_{r_2}^+})] + k_e \cdot N_e \cdot N_{A_{r^*}} \cdot I_{exc}$$

The second rate equation:

$$\frac{dT_{vibr}}{dt} = \left(\frac{N_e}{N_{N_2}}\right) \cdot (T_e - T_{vibr}) \cdot \upsilon_* - (T_{vibr} - T) \times \left[\frac{1}{\tau_{VT,N_2}} + \left(\frac{N_{A_r}}{N_2}\right) / \tau_{VT,A_r}\right]$$

τ_{VT,N_2} and τ_{VT,A_r} are the vibrational relaxation times due to N_2-N_2 and N_2-A_r collisions, respectively. They are calculated by using the Landau-Teller approximation; $I_{exc} \approx 11.8eV$ stands for the argon excitation energy into $A_{r^*}(4^3P_2)$ [1]. Rigorous bounds on the rate of energy exchange between vibrational and translational degrees of freedom are described by simple classical models of diatomic molecules. The outcomes are known as the elementary approximation introduced by Landau and Teller. The initial energy of electrons in the wake of the mid-IR laser pulse is taken equal to 1 eV ($T_e = 0.66$ eV). The initial translational temperature of atoms and molecules in the gas mixture and the vibrational temperature of nitrogen molecules are set equal to $T = T_V = 290$ K. We have two variables in our systems: T_e, electronic temperature and T_{vibr}, vibrational temperature which change in time. We multiply the reduced first rate equation's two sides by $\frac{2}{3 \cdot k \cdot N_e}$.

The reduced first rate equation:

$$\frac{dT_e}{dt} = -(T_e - T_{vibr}) \cdot \upsilon_* - (T_e - T) \cdot [\delta_{N_2} \cdot (\upsilon_{N_2} + \upsilon_{N_2^+} + \frac{1}{2} \cdot \upsilon_{N_4^+}) + \delta_{A_r} \cdot (\upsilon_{A_r} + \upsilon_{A_r^+} + \frac{1}{2} \cdot \upsilon_{A_{r_2}^+})] + \frac{2}{3 \cdot k} \cdot k_e \cdot N_{A_{r^*}} \cdot I_{exc}$$

The second rate equation:

$$\frac{dT_{vibr}}{dt} = \left(\frac{N_e}{N_{N_2}}\right) \cdot (T_e - T_{vibr}) \cdot \upsilon_* - (T_{vibr} - T) \times \left[\frac{1}{\tau_{VT,N_2}} + \left(\frac{N_{A_r}}{N_2}\right) / \tau_{VT,A_r}\right]$$

At fixed points (equilibrium points): $\frac{dT_e}{dt} = 0$; $\frac{dT_{vibr}}{dt} = 0$

$$(*) \quad - (T_e^* - T_{vibr}^*) \cdot \upsilon_* - (T_e^* - T) \cdot [\delta_{N_2} \cdot (\upsilon_{N_2} + \upsilon_{N_2^+} + \frac{1}{2} \cdot \upsilon_{N_4^+}) + \delta_{A_r} \cdot (\upsilon_{A_r} + \upsilon_{A_r^+} + \frac{1}{2} \cdot \upsilon_{A_{r_2}^+})] + \frac{2}{3 \cdot k} \cdot k_e \cdot N_{A_{r^*}} \cdot I_{exc} = 0$$

$$(**)\ (\tfrac{N_e}{N_{N_2}}) \cdot (T_e^* - T_{vibr}^*) \cdot v_* - (T_{vibr}^* - T) \times [\tfrac{1}{\tau_{VT.N_2}} + (\tfrac{N_{A_r}}{N_2})/\tau_{VT,A_r}] = 0$$

$$(*) \rightarrow \begin{aligned} &(T_e^* \cdot v_* - T_{vibr}^* \cdot v_*) + (T_e^* - T) \cdot [\delta_{N_2} \cdot (v_{N_2} + v_{N_2^+} + \frac{1}{2} \cdot v_{N_4^+}) \\ &+ \delta_{A_r} \cdot (v_{A_r} + v_{A_r^+} + \frac{1}{2} \cdot v_{A_{r_2}^+})] = \frac{2}{3 \cdot k} \cdot k_e \cdot N_{A_{r*}} \cdot I_{exc} \end{aligned}$$

$$\begin{aligned} T_{vibr}^* \cdot v_* = T_e^* \cdot v_* + T_e^* \cdot [&\delta_{N_2} \cdot (v_{N_2} + v_{N_2^+} + \frac{1}{2} \cdot v_{N_4^+}) \\ &+ \delta_{A_r} \cdot (v_{A_r} + v_{A_r^+} + \frac{1}{2} \cdot v_{A_{r_2}^+})] \\ &- T \cdot [\delta_{N_2} \cdot (v_{N_2} + v_{N_2^+} + \frac{1}{2} \cdot v_{N_4^+}) \\ &+ \delta_{A_r} \cdot (v_{A_r} + v_{A_r^+} + \frac{1}{2} \cdot v_{A_{r_2}^+})] \\ &- \frac{2}{3 \cdot k} \cdot k_e \cdot N_{A_{r*}} \cdot I_{exc} \end{aligned}$$

$$\begin{aligned} T_{vibr}^* \cdot v_* = T_e^* \cdot \{&v_* + \delta_{N_2} \cdot (v_{N_2} + v_{N_2^+} + \frac{1}{2} \cdot v_{N_4^+}) \\ &+ \delta_{A_r} \cdot (v_{A_r} + v_{A_r^+} + \frac{1}{2} \cdot v_{A_{r_2}^+})\} \\ &- T \cdot [\delta_{N_2} \cdot (v_{N_2} + v_{N_2^+} + \frac{1}{2} \cdot v_{N_4^+}) \\ &+ \delta_{A_r} \cdot (v_{A_r} + v_{A_r^+} + \frac{1}{2} \cdot v_{A_{r_2}^+})] \\ &- \frac{2}{3 \cdot k} \cdot k_e \cdot N_{A_{r*}} \cdot I_{exc} \end{aligned}$$

$$\begin{aligned} T_{vibr}^* = T_e^* \cdot \frac{1}{v_*} \cdot \{&v_* + \delta_{N_2} \cdot (v_{N_2} + v_{N_2^+} + \frac{1}{2} \cdot v_{N_4^+}) \\ &+ \delta_{A_r} \cdot (v_{A_r} + v_{A_r^+} + \frac{1}{2} \cdot v_{A_{r_2}^+})\} \\ &- \frac{T}{v_*} \cdot [\delta_{N_2} \cdot (v_{N_2} + v_{N_2^+} + \frac{1}{2} \cdot v_{N_4^+}) \\ &+ \delta_{A_r} \cdot (v_{A_r} + v_{A_r^+} + \frac{1}{2} \cdot v_{A_{r_2}^+})] \\ &- \frac{2}{3 \cdot k \cdot v_*} \cdot k_e \cdot N_{A_{r*}} \cdot I_{exc} \end{aligned}$$

We define for simplicity two global parameters: Ξ_1, Ξ_2; $T_{vibr}^* = T_e^* \cdot \Xi_1 + \Xi_2$

$$\Xi_1 = \Xi_1(v_*, \delta_{N_2}, v_{N_2}, v_{N_2^+}, v_{N_4^+}, \delta_{A_r}, v_{A_r}, v_{A_r^+}, v_{A_{r_2}^+})$$

$$\Xi_2 = \Xi_2(T, v_*, \delta_{N_2}, v_{N_2}, v_{N_2^+}, v_{N_4^+}, \delta_{A_r}, v_{A_r}, v_{A_r^+}, v_{A_{r_2}^+}, k, k_e, N_{A_{r*}}, I_{exc})$$

$$\Xi_1 = \frac{1}{v_*} \cdot \{v_* + \delta_{N_2} \cdot (v_{N_2} + v_{N_2^+} + \frac{1}{2} \cdot v_{N_4^+}) + \delta_{A_r} \cdot (v_{A_r} + v_{A_r^+} + \frac{1}{2} \cdot v_{A_{r_2}^+})\}$$

$$\Xi_2 = -\frac{T}{v_*} \cdot [\delta_{N_2} \cdot (v_{N_2} + v_{N_2^+} + \frac{1}{2} \cdot v_{N_4^+}) + \delta_{A_r} \cdot (v_{A_r} + v_{A_r^+} + \frac{1}{2} \cdot v_{A_{r_2}^+})] - \frac{2}{3 \cdot k \cdot v_*} \cdot k_e \cdot N_{A_{r*}} \cdot I_{exc}$$

$$(*) \to (**)\ (\tfrac{N_e}{N_{N_2}}) \cdot (T_e^* - T_{vibr}^*) \cdot v_* - (T_{vibr}^* - T) \times [\tfrac{1}{\tau_{VT,N_2}} + (\tfrac{N_{A_r}}{N_2})/\tau_{VT,A_r}] = 0$$

$$T_e^* \cdot \frac{N_e}{N_{N_2}} \cdot v_* - T_{vibr}^* \cdot \frac{N_e}{N_{N_2}} \cdot v_* - (T_{vibr}^* - T) \times [\frac{1}{\tau_{VT,N_2}} + (\frac{N_{A_r}}{N_2})/\tau_{VT,A_r}] = 0$$

$$T_e^* \cdot \frac{N_e}{N_{N_2}} \cdot v_* = T_{vibr}^* \cdot \frac{N_e}{N_{N_2}} \cdot v_* + (T_{vibr}^* - T) \times \left[\frac{1}{\tau_{VT,N_2}} + \left(\frac{N_{A_r}}{N_2}\right)/\tau_{VT,A_r}\right]$$

$$T_e^* = T_{vibr}^* + \frac{N_{N_2}}{N_e \cdot v_*} \cdot (T_{vibr}^* - T) \times \left[\frac{1}{\tau_{VT,N_2}} + \left(\frac{N_{A_r}}{N_2}\right)/\tau_{VT,A_r}\right]$$

$$T_e^* = T_{vibr}^* + \frac{N_{N_2}}{N_e \cdot v_*} \cdot T_{vibr}^* \times \left[\frac{1}{\tau_{VT,N_2}} + \left(\frac{N_{A_r}}{N_2}\right)/\tau_{VT,A_r}\right] - \frac{N_{N_2}}{N_e \cdot v_*} \cdot T \times \left[\frac{1}{\tau_{VT,N_2}} + \left(\frac{N_{A_r}}{N_2}\right)/\tau_{VT,A_r}\right]$$

$$T_e^* = T_{vibr}^* \cdot \left\{1 + \frac{N_{N_2}}{N_e \cdot v_*} \times \left[\frac{1}{\tau_{VT,N_2}} + \left(\frac{N_{A_r}}{N_2}\right)/\tau_{VT,A_r}\right]\right\} - \frac{N_{N_2}}{N_e \cdot v_*} \cdot T \times \left[\frac{1}{\tau_{VT,N_2}} + \left(\frac{N_{A_r}}{N_2}\right)/\tau_{VT,A_r}\right]$$

We define for simplicity two global parameters: Ξ_3, Ξ_4; $T_e^* = T_{vibr}^* \cdot \Xi_3 + \Xi_4$

$$\Xi_3 = \Xi_3(N_{N_2}, N_e, v_*, \tau_{VT,N_2}, N_{A_r}, N_2, \tau_{VT,A_r})$$
$$\Xi_4 = \Xi_4(N_{N_2}, N_e, v_*, T, \tau_{VT,N_2}, N_{A_r}, N_2, \tau_{VT,A_r})$$

$$\Xi_3 = 1 + \frac{N_{N_2}}{N_e \cdot v_*} \times \left[\frac{1}{\tau_{VT,N_2}} + \left(\frac{N_{A_r}}{N_2}\right)/\tau_{VT,A_r}\right]$$

$$\Xi_4 = -\frac{N_{N_2}}{N_e \cdot v_*} \cdot T \times \left[\frac{1}{\tau_{VT,N_2}} + \left(\frac{N_{A_r}}{N_2}\right)/\tau_{VT,A_r}\right]$$

Result: $T^*_{vibr} = T^*_e \cdot \Xi_1 + \Xi_2$; $T^*_e = T^*_{vibr} \cdot \Xi_3 + \Xi_4$

$$\begin{aligned} T^*_{vibr} &= (T^*_{vibr} \cdot \Xi_3 + \Xi_4) \cdot \Xi_1 + \Xi_2 \Rightarrow T^*_{vibr} \\ &= T^*_{vibr} \cdot \Xi_3 \cdot \Xi_1 + \Xi_4 \cdot \Xi_1 + \Xi_2 \Rightarrow T^*_{vibr} \\ &= \frac{\Xi_4 \cdot \Xi_1 + \Xi_2}{1 - \Xi_3 \cdot \Xi_1} \end{aligned}$$

$$T^*_e = T^*_{vibr} \cdot \Xi_3 + \Xi_4 = \left(\frac{\Xi_4 \cdot \Xi_1 + \Xi_2}{1 - \Xi_3 \cdot \Xi_1}\right) \cdot \Xi_3 + \Xi_4$$

Our system fixed point is $E^{(*)}(T^*_e, T^*_{vibr}) = \left(\left[\frac{\Xi_4 \cdot \Xi_1 + \Xi_2}{1 - \Xi_3 \cdot \Xi_1}\right] \cdot \Xi_3 + \Xi_4, \left[\frac{\Xi_4 \cdot \Xi_1 + \Xi_2}{1 - \Xi_3 \cdot \Xi_1}\right]\right)$
The reduced two rate equations are:

$$\begin{aligned} \frac{dT_e}{dt} &= -(T_e - T_{vibr}) \cdot v_* - (T_e - T) \cdot [\delta_{N_2} \cdot (v_{N_2} + v_{N_2^+} + \frac{1}{2} \cdot v_{N_4^+}) \\ &+ \delta_{A_r} \cdot (v_{A_r} + v_{A_r^+} + \frac{1}{2} \cdot v_{A_{r_2^+}})] \\ &+ \frac{2}{3 \cdot k} \cdot k_e \cdot N_{A_{r*}} \cdot I_{exc} \end{aligned}$$

$$\begin{aligned} \frac{dT_{vibr}}{dt} &= (\frac{N_e}{N_{N_2}}) \cdot (T_e - T_{vibr}) \cdot v_* - (T_{vibr} - T) \\ &\times [\frac{1}{\tau_{VT,N_2}} + (\frac{N_{A_r}}{N_2})/\tau_{VT,A_r}] \end{aligned}$$

We define two functions: $f_1(T_e, T_{vibr})$; $f_2(T_e, T_{vibr})$

$$\begin{aligned} f_1(T_e, T_{vibr}) &= -(T_e - T_{vibr}) \cdot v_* - (T_e - T) \\ &\cdot [\delta_{N_2} \cdot (v_{N_2} + v_{N_2^+} + \frac{1}{2} \cdot v_{N_4^+}) \\ &+ \delta_{A_r} \cdot (v_{A_r} + v_{A_r^+} + \frac{1}{2} \cdot v_{A_{r_2^+}})] \\ &+ \frac{2}{3 \cdot k} \cdot k_e \cdot N_{A_{r*}} \cdot I_{exc} \end{aligned}$$

$$\begin{aligned} f_2(T_e, T_{vibr}) &= \left(\frac{N_e}{N_{N_2}}\right) \cdot (T_e - T_{vibr}) \cdot v_* - (T_{vibr} - T) \\ &\times \left[\frac{1}{\tau_{VT,N_2}} + \left(\frac{N_{A_r}}{N_2}\right)/\tau_{VT,A_r}\right] \end{aligned}$$

Our system set of differential equations: $\frac{dT_e}{dt} = f_1(T_e, T_{vibr})$; $\frac{dT_{vibr}}{dt} = f_2(T_e, T_{vibr})$

Linearization: suppose that (T_e^*, T_{vibr}^*) is a fixed point, $f_1(T_e^*, T_{vibr}^*) = 0$; $f_2(T_e^*, T_{vibr}^*) = 0$ and let $u_1 = T_e - T_e^*$; $u_2 = T_{vibr} - T_{vibr}^*$ denote the components of a small disturbance from the fixed point. To see whether the disturbance grows or decays, we need to derive differential equations for u_1 and u_2. Let's do the u_1-equation first: $\frac{du_1}{dt} = \frac{dT_e}{dt}$ since T_e^* is constant and $\frac{du_1}{dt} = \frac{dT_e}{dt} = f_1(T_e^* + u_1, T_{vibr}^* + u_2)$ by substitution, then by using Taylor series expansion $\frac{du_1}{dt} = \frac{dT_e}{dt} = f_1(T_e^*, T_{vibr}^*) + u_1 \cdot \frac{\partial f_1}{\partial T_e} + u_2 \cdot \frac{\partial f_1}{\partial T_{vibr}} + O(u_1^2, u_2^2, u_1 \cdot u_2)$ and since $f_1(T_e^*, T_{vibr}^*) = 0$ then $\frac{du_1}{dt} = \frac{dT_e}{dt} = u_1 \cdot \frac{\partial f_1}{\partial T_e} + u_2 \cdot \frac{\partial f_1}{\partial T_{vibr}} + O(u_1^2, u_2^2, u_1 \cdot u_2)$. To simplify the notation, we have written $\frac{\partial f_1}{\partial T_e}$ and $\frac{\partial f_1}{\partial T_{vibr}}$, these partial derivatives are to be evaluated at the fixed point (T_e^*, T_{vibr}^*); thus they are numbers, not functions. The shorthand notation, $O(u_1^2, u_2^2, u_1 \cdot u_2)$ denotes quadratic terms in u_1 and u_2 are small, these quadratic terms are extremely small [2, 3].

$\frac{du_2}{dt} = \frac{dT_{vibr}}{dt} = u_1 \cdot \frac{\partial f_2}{\partial T_e} + u_2 \cdot \frac{\partial f_2}{\partial T_{vibr}} + O(u_1^2, u_2^2, u_1 \cdot u_2)$. Hence the disturbance (u_1, u_2) evolves according to $\begin{pmatrix} \frac{du_1}{dt} \\ \frac{du_2}{dt} \end{pmatrix} = \begin{pmatrix} \frac{\partial f_1}{\partial T_e} & \frac{\partial f_1}{\partial T_{vibr}} \\ \frac{\partial f_2}{\partial T_e} & \frac{\partial f_2}{\partial T_{vibr}} \end{pmatrix} \cdot \begin{pmatrix} u_1 \\ u_2 \end{pmatrix}$ + quadratic terms.

The matrix $A = \begin{pmatrix} \frac{\partial f_1}{\partial T_e} & \frac{\partial f_1}{\partial T_{vibr}} \\ \frac{\partial f_2}{\partial T_e} & \frac{\partial f_2}{\partial T_{vibr}} \end{pmatrix}_{(T_e^*, T_{vibr}^*)}$ is called the Jacobian matrix at the fixed point (T_e^*, T_{vibr}^*). It is multivariable analog to $f_1'(T_e^*)$. Since the quadratic terms are tiny, we neglect them altogether. We obtain the linearized system:

$$\begin{pmatrix} \frac{du_1}{dt} \\ \frac{du_2}{dt} \end{pmatrix} = \begin{pmatrix} \frac{\partial f_1}{\partial T_e} & \frac{\partial f_1}{\partial T_{vibr}} \\ \frac{\partial f_2}{\partial T_e} & \frac{\partial f_2}{\partial T_{vibr}} \end{pmatrix} \cdot \begin{pmatrix} u_1 \\ u_2 \end{pmatrix}.$$

$$\begin{aligned} f_1 &= f_1(T_e, T_{vibr}) \\ &= -(T_e - T_{vibr}) \cdot v_* - (T_e - T) \\ &\quad \cdot [\delta_{N_2} \cdot (v_{N_2} + v_{N_2^+} + \frac{1}{2} \cdot v_{N_4^+}) \\ &\quad + \delta_{A_r} \cdot (v_{A_r} + v_{A_r^+} + \frac{1}{2} \cdot v_{A_{r_2}^+})] + \frac{2}{3 \cdot k} \cdot k_e \cdot N_{A_{r^*}} \cdot I_{exc} \end{aligned}$$

$$\begin{aligned} \frac{\partial f_1}{\partial T_e} &= -v_* - [\delta_{N_2} \cdot (v_{N_2} + v_{N_2^+} + \frac{1}{2} \cdot v_{N_4^+}) \\ &\quad + \delta_{A_r} \cdot (v_{A_r} + v_{A_r^+} + \frac{1}{2} \cdot v_{A_{r_2}^+})] \\ \frac{\partial f_1}{\partial T_{vibr}} &= v_* \end{aligned}$$

$$f_2(T_e, T_{vibr}) = (\frac{N_e}{N_{N_2}}) \cdot (T_e - T_{vibr}) \cdot v_* - (T_{vibr} - T) \times \left[\frac{1}{\tau_{VT,N_2}} + \left(\frac{N_{A_r}}{N_2}\right)/\tau_{VT,A_r}\right]$$

$$\frac{\partial f_2}{\partial T_e} = \frac{N_e}{N_{N_2}} \cdot v_*; \frac{\partial f_2}{\partial T_{vibr}} = -\frac{N_e}{N_{N_2}} \cdot v_* - \left[\frac{1}{\tau_{VT,N_2}} + \left(\frac{N_{A_r}}{N_2}\right)/\tau_{VT,A_r}\right]$$

$$A = \begin{pmatrix} -v_* - [\delta_{N_2} \cdot (v_{N_2} + v_{N_2^+} + \frac{1}{2} \cdot v_{N_4^+}) + \delta_{A_r} \cdot (v_{A_r} + v_{A_r^+} + \frac{1}{2} \cdot v_{A_{r_2}^+})] & v_* \\ \frac{N_e}{N_{N_2}} \cdot v_* & -\frac{N_e}{N_{N_2}} \cdot v_* - [\frac{1}{\tau_{VT,N_2}} + (\frac{N_{A_r}}{N_2})/\tau_{VT,A_r}] \end{pmatrix}_{(T_e^*, T_{vibr}^*)}$$

The eigenvalues of the matrix A are given by the characteristic equationdet$(A - \lambda \cdot I) = 0$, where I is the identity matrix. For our 2×2 matrix the characteristic equation becomes:

$$\det\begin{pmatrix} -v_* - [\delta_{N_2} \cdot (v_{N_2} + v_{N_2^+} + \frac{1}{2} \cdot v_{N_4^+}) + \delta_{A_r} \cdot (v_{A_r} + v_{A_r^+} + \frac{1}{2} \cdot v_{A_{r_2}^+})] - \lambda & v_* \\ \frac{N_e}{N_{N_2}} \cdot v_* & -\frac{N_e}{N_{N_2}} \cdot v_* - [\frac{1}{\tau_{VT,N_2}} + (\frac{N_{A_r}}{N_2})/\tau_{VT,A_r}] - \lambda \end{pmatrix} = 0$$

For simplicity we define two global parameters: Υ_1, Υ_2

$$\Upsilon_1 = -v_* - [\delta_{N_2} \cdot (v_{N_2} + v_{N_2^+} + \frac{1}{2} \cdot v_{N_4^+}) + \delta_{A_r} \cdot (v_{A_r} + v_{A_r^+} + \frac{1}{2} \cdot v_{A_{r_2}^+})]$$

$\Upsilon_2 = -\frac{N_e}{N_{N_2}} \cdot v_* - [\frac{1}{\tau_{VT,N_2}} + (\frac{N_{A_r}}{N_2})/\tau_{VT,A_r}]$, then we get the determinant:

$$\det\begin{pmatrix} \Upsilon_1 - \lambda & v_* \\ \frac{N_e}{N_{N_2}} \cdot v_* & \Upsilon_2 - \lambda \end{pmatrix} = 0 \Rightarrow (\Upsilon_1 - \lambda) \cdot (\Upsilon_2 - \lambda) - \frac{N_e}{N_{N_2}} \cdot v_* \cdot v_* = 0$$

Expanding the determinant yields: $\lambda^2 - \lambda \cdot (\Upsilon_1 + \Upsilon_2) + (\Upsilon_1 \cdot \Upsilon_2 - \frac{N_e}{N_{N_2}} \cdot v_*^2) = 0$

Another way to get the system characteristic equation and explore stability is by adding to coordinates $[T_e\ T_{vibr}]$ arbitrarily small increments of exponential form $[t_e\ t_{vibr}] \cdot e^{\lambda \cdot t}$, and retaining the first order terms in $T_e\ T_{vibr}$. The system of two homogeneous equations leads to a polynomial characteristics equation in the eigenvalues λ. Nitrogen gas laser filament plasma kinetic system T_e, electronic temperature and T_{vibr}, vibrational temperature fixed values with arbitrarily small increments of exponential form $[t_e\ t_{vibr}] \cdot e^{\lambda \cdot t}$ are $T_e = T_e^* + t_e \cdot e^{\lambda \cdot t}$; $T_{vibr} = T_{vibr}^* + t_{vibr} \cdot e^{\lambda \cdot t}$. We choose these expressions for our $T_e(t)$, $T_{vibr}(t)$ as small $[t_e\ t_{vibr}]$ from the system fixed points at time $t = 0$; $T_e(t = 0) = T_e^* + t_e$; $T_{vibr}(t = 0) = T_{vibr}^*$.

$$\frac{dT_e}{dt} = t_e \cdot \lambda \cdot e^{\lambda \cdot t}; \frac{dT_{vibr}}{dt} = t_{vibr} \cdot \lambda \cdot e^{\lambda \cdot t}$$

$$\begin{aligned}\frac{dT_e}{dt} &= -(T_e - T_{vibr}) \cdot \upsilon_* - (T_e - T) \cdot [\delta_{N_2} \cdot (\upsilon_{N_2} + \upsilon_{N_2^+} + \frac{1}{2} \cdot \upsilon_{N_4^+}) \\ &\quad + \delta_{A_r} \cdot (\upsilon_{A_r} + \upsilon_{A_r^+} + \frac{1}{2} \cdot \upsilon_{A_{r_2^+}})] \\ &\quad + \frac{2}{3 \cdot k} \cdot k_e \cdot N_{A_{r*}} \cdot I_{exc}\end{aligned}$$

$$\begin{aligned}t_e \cdot \lambda \cdot e^{\lambda \cdot t} &= -(T_e^* + t_e \cdot e^{\lambda \cdot t} - T_{vibr}^* - t_{vibr} \cdot e^{\lambda \cdot t}) \cdot \upsilon_* \\ &\quad - (T_e^* + t_e \cdot e^{\lambda \cdot t} - T) \cdot [\delta_{N_2} \cdot (\upsilon_{N_2} + \upsilon_{N_2^+} + \frac{1}{2} \cdot \upsilon_{N_4^+}) \\ &\quad + \delta_{A_r} \cdot (\upsilon_{A_r} + \upsilon_{A_r^+} + \frac{1}{2} \cdot \upsilon_{A_{r_2^+}})] + \frac{2}{3 \cdot k} \cdot k_e \cdot N_{A_{r*}} \cdot I_{exc}\end{aligned}$$

$$\begin{aligned}t_e \cdot \lambda \cdot e^{\lambda \cdot t} &= -(T_e^* - T_{vibr}^*) \cdot \upsilon_* - (t_e - t_{vibr}) \cdot \upsilon_* \cdot e^{\lambda \cdot t} \\ &\quad - (T_e^* - T) \cdot [\delta_{N_2} \cdot (\upsilon_{N_2} + \upsilon_{N_2^+} + \frac{1}{2} \cdot \upsilon_{N_4^+}) \\ &\quad + \delta_{A_r} \cdot (\upsilon_{A_r} + \upsilon_{A_r^+} + \frac{1}{2} \cdot \upsilon_{A_{r_2^+}})] - t_e \cdot [\delta_{N_2} \cdot (\upsilon_{N_2} + \upsilon_{N_2^+} + \frac{1}{2} \cdot \upsilon_{N_4^+}) \\ &\quad + \delta_{A_r} \cdot (\upsilon_{A_r} + \upsilon_{A_r^+} + \frac{1}{2} \cdot \upsilon_{A_{r_2^+}})] \cdot e^{\lambda \cdot t} \\ &\quad + \frac{2}{3 \cdot k} \cdot k_e \cdot N_{A_{r*}} \cdot I_{exc}\end{aligned}$$

$$\begin{aligned}t_e \cdot \lambda \cdot e^{\lambda \cdot t} &= -(T_e^* - T_{vibr}^*) \cdot \upsilon_* - (T_e^* - T) \cdot [\delta_{N_2} \cdot (\upsilon_{N_2} + \upsilon_{N_2^+} + \frac{1}{2} \cdot \upsilon_{N_4^+}) \\ &\quad + \delta_{A_r} \cdot (\upsilon_{A_r} + \upsilon_{A_r^+} + \frac{1}{2} \cdot \upsilon_{A_{r_2^+}})] \\ &\quad + \frac{2}{3 \cdot k} \cdot k_e \cdot N_{A_{r*}} \cdot I_{exc} - (t_e - t_{vibr}) \cdot \upsilon_* \cdot e^{\lambda \cdot t} - t_e \\ &\quad \cdot [\delta_{N_2} \cdot (\upsilon_{N_2} + \upsilon_{N_2^+} + \frac{1}{2} \cdot \upsilon_{N_4^+}) + \delta_{A_r} \cdot (\upsilon_{A_r} + \upsilon_{A_r^+} + \frac{1}{2} \cdot \upsilon_{A_{r_2^+}})] \cdot e^{\lambda \cdot t}\end{aligned}$$

At fixed point:

$$\begin{aligned}&- (T_e^* - T_{vibr}^*) \cdot \upsilon_* - (T_e^* - T) \cdot [\delta_{N_2} \cdot (\upsilon_{N_2} + \upsilon_{N_2^+} + \frac{1}{2} \cdot \upsilon_{N_4^+}) \\ &\quad + \delta_{A_r} \cdot (\upsilon_{A_r} + \upsilon_{A_r^+} + \frac{1}{2} \cdot \upsilon_{A_{r_2^+}})] + \frac{2}{3 \cdot k} \cdot k_e \cdot N_{A_{r*}} \cdot I_{exc} = 0\end{aligned}$$

$$\begin{aligned}t_e \cdot \lambda &= -(t_e - t_{vibr}) \cdot \upsilon_* - t_e \cdot [\delta_{N_2} \cdot (\upsilon_{N_2} + \upsilon_{N_2^+} + \frac{1}{2} \cdot \upsilon_{N_4^+}) \\ &\quad + \delta_{A_r} \cdot (\upsilon_{A_r} + \upsilon_{A_r^+} + \frac{1}{2} \cdot \upsilon_{A_{r_2^+}})]\end{aligned}$$

$$t_e \cdot \{-v_* - [\delta_{N_2} \cdot (v_{N_2} + v_{N_2^+} + \frac{1}{2} \cdot v_{N_4^+}) + \delta_{A_r} \cdot (v_{A_r} + v_{A_r^+} + \frac{1}{2} \cdot v_{A_{r_2}^+})] - \lambda\} + t_{vibr} \cdot v_* = 0$$

$$\frac{dT_{vibr}}{dt} = (\frac{N_e}{N_{N_2}}) \cdot (T_e - T_{vibr}) \cdot v_* - (T_{vibr} - T) \times [\frac{1}{\tau_{VT,N_2}} + (\frac{N_{A_r}}{N_2})/\tau_{VT,A_r}]$$

$$t_{vibr} \cdot \lambda \cdot e^{\lambda \cdot t} = (\frac{N_e}{N_{N_2}}) \cdot (T_e^* + t_e \cdot e^{\lambda \cdot t} - T_{vibr}^* - t_{vibr} \cdot e^{\lambda \cdot t}) \cdot v_* - (T_{vibr}^* + t_{vibr} \cdot e^{\lambda \cdot t} - T) \times [\frac{1}{\tau_{VT,N_2}} + (\frac{N_{A_r}}{N_2})/\tau_{VT,A_r}]$$

$$t_{vibr} \cdot \lambda \cdot e^{\lambda \cdot t} = (\frac{N_e}{N_{N_2}}) \cdot (T_e^* - T_{vibr}^*) \cdot v_* + (\frac{N_e}{N_{N_2}}) \cdot (t_e - t_{vibr}) \cdot v_* \cdot e^{\lambda \cdot t} - (T_{vibr}^* - T) \times [\frac{1}{\tau_{VT,N_2}} + (\frac{N_{A_r}}{N_2})/\tau_{VT,A_r}] - t_{vibr} \times [\frac{1}{\tau_{VT,N_2}} + (\frac{N_{A_r}}{N_2})/\tau_{VT,A_r}] \cdot e^{\lambda \cdot t}$$

$$t_{vibr} \cdot \lambda \cdot e^{\lambda \cdot t} = (\frac{N_e}{N_{N_2}}) \cdot (T_e^* - T_{vibr}^*) \cdot v_* - (T_{vibr}^* - T) \times [\frac{1}{\tau_{VT,N_2}} + (\frac{N_{A_r}}{N_2})/\tau_{VT,A_r}] + (\frac{N_e}{N_{N_2}}) \cdot (t_e - t_{vibr}) \cdot v_* \cdot e^{\lambda \cdot t} - t_{vibr} \times [\frac{1}{\tau_{VT,N_2}} + (\frac{N_{A_r}}{N_2})/\tau_{VT,A_r}] \cdot e^{\lambda \cdot t}$$

At fixed point: $(\frac{N_e}{N_{N_2}}) \cdot (T_e^* - T_{vibr}^*) \cdot v_* - (T_{vibr}^* - T) \times [\frac{1}{\tau_{VT,N_2}} + (\frac{N_{A_r}}{N_2})/\tau_{VT,A_r}] = 0$

$$t_{vibr} \cdot \lambda = \left(\frac{N_e}{N_{N_2}}\right) \cdot (t_e - t_{vibr}) \cdot v_* - t_{vibr} \times \left[\frac{1}{\tau_{VT,N_2}} + \left(\frac{N_{A_r}}{N_2}\right)/\tau_{VT,A_r}\right]$$

$$\left(\frac{N_e}{N_{N_2}}\right) \cdot t_e \cdot v_* - \left\{\left(\frac{N_e}{N_{N_2}}\right) \cdot t_{vibr} \cdot v_* + t_{vibr} \times \left[\frac{1}{\tau_{VT,N_2}} + \left(\frac{N_{A_r}}{N_2}\right)/\tau_{VT,A_r}\right]\right\} - t_{vibr} \cdot \lambda = 0$$

$$\left(\frac{N_e}{N_{N_2}}\right) \cdot v_* \cdot t_e - \left\{\left(\frac{N_e}{N_{N_2}}\right) \cdot v_* + \left[\frac{1}{\tau_{VT,N_2}} + \left(\frac{N_{A_r}}{N_2}\right)/\tau_{VT,A_r}\right]\right\}$$
$$\cdot t_{vibr} - t_{vibr} \cdot \lambda = 0$$

The small increment Jacobian of our Nitrogen gas laser filament plasma kinetic system is as follow:

$$t_e \cdot \{-v_* - [\delta_{N_2} \cdot (v_{N_2} + v_{N_2^+} + \frac{1}{2} \cdot v_{N_4^+})$$
$$+ \delta_{A_r} \cdot (v_{A_r} + v_{A_r^+} + \frac{1}{2} \cdot v_{A_{r_2}^+})] - \lambda\} + t_{vibr} \cdot v_* = 0$$

$$(\frac{N_e}{N_{N_2}}) \cdot v_* \cdot t_e - \{(\frac{N_e}{N_{N_2}}) \cdot v_* + [\frac{1}{\tau_{VT,N_2}} + (\frac{N_{A_r}}{N_2})/\tau_{VT,A_r}]\}$$
$$\cdot t_{vibr} - t_{vibr} \cdot \lambda = 0$$

$$\begin{pmatrix} -v_* - [\delta_{N_2} \cdot (v_{N_2} + v_{N_2^+} + \frac{1}{2} \cdot v_{N_4^+}) \\ +\delta_{A_r} \cdot (v_{A_r} + v_{A_r^+} + \frac{1}{2} \cdot v_{A_{r_2}^+})] - \lambda & v_* \\ (\frac{N_e}{N_{N_2}}) \cdot v_* & -\{(\frac{N_e}{N_{N_2}}) \cdot v_* + [\frac{1}{\tau_{VT,N_2}} + (\frac{N_{A_r}}{N_2})/\tau_{VT,A_r}]\} - \lambda \end{pmatrix}$$
$$\cdot \begin{pmatrix} t_e \\ t_{vibr} \end{pmatrix} = \begin{pmatrix} 0 \\ 0 \end{pmatrix}$$

$$A - \lambda \cdot I$$
$$= \begin{pmatrix} -v_* - [\delta_{N_2} \cdot (v_{N_2} + v_{N_2^+} + \frac{1}{2} \cdot v_{N_4^+}) \\ +\delta_{A_r} \cdot (v_{A_r} + v_{A_r^+} + \frac{1}{2} \cdot v_{A_{r_2}^+})] - \lambda & v_* \\ \frac{N_e}{N_{N_2}} \cdot v_* & -\frac{N_e}{N_{N_2}} \cdot v_* - [\frac{1}{\tau_{VT,N_2}} + (\frac{N_{A_r}}{N_2})/\tau_{VT,A_r}] - \lambda \end{pmatrix}$$

$$A = \begin{pmatrix} -v_* - [\delta_{N_2} \cdot (v_{N_2} + v_{N_2^+} + \frac{1}{2} \cdot v_{N_4^+}) \\ +\delta_{A_r} \cdot (v_{A_r} + v_{A_r^+} + \frac{1}{2} \cdot v_{A_{r_2}^+})] & v_* \\ \frac{N_e}{N_{N_2}} \cdot v_* & -\frac{N_e}{N_{N_2}} \cdot v_* - [\frac{1}{\tau_{VT,N_2}} + (\frac{N_{A_r}}{N_2})/\tau_{VT,A_r}] \end{pmatrix}$$

For simplicity we define two global parameters: Υ_1, Υ_2

$$\Upsilon_1 = -v_* - [\delta_{N_2} \cdot (v_{N_2} + v_{N_2^+} + \frac{1}{2} \cdot v_{N_4^+})$$
$$+ \delta_{A_r} \cdot (v_{A_r} + v_{A_r^+} + \frac{1}{2} \cdot v_{A_{r_2}^+})]$$

$\Upsilon_2 = -\frac{N_e}{N_{N_2}} \cdot v_* - [\frac{1}{\tau_{VT,N_2}} + (\frac{N_{A_r}}{N_2})/\tau_{VT,A_r}]$, then we get the determinant:

$$\det\begin{pmatrix} \Upsilon_1-\lambda & \upsilon_* \\ \frac{N_e}{N_{N_2}}\cdot\upsilon_* & \Upsilon_2-\lambda \end{pmatrix} = 0 \Rightarrow (\Upsilon_1-\lambda)\cdot(\Upsilon_2-\lambda) - \frac{N_e}{N_{N_2}}\cdot\upsilon_*\cdot\upsilon_* = 0$$

Expanding the determinant yields: $\lambda^2 - \lambda\cdot(\Upsilon_1+\Upsilon_2)+(\Upsilon_1\cdot\Upsilon_2 - \frac{N_e}{N_{N_2}}\cdot\upsilon_*^2) = 0$
We define $\wp_1 = \mathrm{trace}(A)$; $\wp_2 = \det(A)$

$$\wp_1 = \mathrm{trace}(A) = -\upsilon_* - \left[\delta_{N_2}\cdot\left(\upsilon_{N_2}+\upsilon_{N_2^+}+\frac{1}{2}\cdot\upsilon_{N_4^+}\right) + \delta_{A_r}\cdot\left(\upsilon_{A_r}+\upsilon_{A_r^+}+\frac{1}{2}\cdot\upsilon_{A_{r_2}^+}\right)\right] - \frac{N_e}{N_{N_2}}\cdot\upsilon_* - \left[\frac{1}{\tau_{VT,N_2}} + \left(\frac{N_{A_r}}{N_2}\right)/\tau_{VT,A_r}\right]$$

$$\wp_1 = \mathrm{trace}(A) = -\upsilon_*\cdot[1+\frac{N_e}{N_{N_2}}] - [\delta_{N_2}\cdot(\upsilon_{N_2}+\upsilon_{N_2^+}+\frac{1}{2}\cdot\upsilon_{N_4^+}) + \delta_{A_r}\cdot(\upsilon_{A_r}+\upsilon_{A_r^+}+\frac{1}{2}\cdot\upsilon_{A_{r_2}^+})] - [\frac{1}{\tau_{VT,N_2}} + (\frac{N_{A_r}}{N_2})/\tau_{VT,A_r}]$$

$$\wp_2 = \det(A) = \{-\upsilon_* - [\delta_{N_2}\cdot(\upsilon_{N_2}+\upsilon_{N_2^+}+\frac{1}{2}\cdot\upsilon_{N_4^+}) + \delta_{A_r}\cdot(\upsilon_{A_r}+\upsilon_{A_r^+}+\frac{1}{2}\cdot\upsilon_{A_{r_2}^+})]\}\cdot\{-\frac{N_e}{N_{N_2}}\cdot\upsilon_* - [\frac{1}{\tau_{VT,N_2}} + (\frac{N_{A_r}}{N_2})/\tau_{VT,A_r}]\} - \frac{N_e}{N_{N_2}}\cdot\upsilon_*\cdot\upsilon_*$$

$$\wp_2 = \det(A) = \{\upsilon_* + [\delta_{N_2}\cdot(\upsilon_{N_2}+\upsilon_{N_2^+}+\frac{1}{2}\cdot\upsilon_{N_4^+}) + \delta_{A_r}\cdot(\upsilon_{A_r}+\upsilon_{A_r^+}+\frac{1}{2}\cdot\upsilon_{A_{r_2}^+})]\}\cdot\{\frac{N_e}{N_{N_2}}\cdot\upsilon_* + [\frac{1}{\tau_{VT,N_2}} + (\frac{N_{A_r}}{N_2})/\tau_{VT,A_r}]\} - \frac{N_e}{N_{N_2}}\cdot\upsilon_*^2$$

Then $\lambda_1 = \frac{\wp_1+\sqrt{\wp_1^2-4\cdot\wp_2}}{2}$; $\lambda_2 = \frac{\wp_1-\sqrt{\wp_1^2-4\cdot\wp_2}}{2}$ are the solutions of the quadratic equation$\lambda^2 - \lambda\cdot(\Upsilon_1+\Upsilon_2)+(\Upsilon_1\cdot\Upsilon_2 - \frac{N_e}{N_{N_2}}\cdot\upsilon_*^2) = 0$; $\wp_1 = \Upsilon_1+\Upsilon_2$; $\wp_2 = \Upsilon_1\cdot\Upsilon_2 - \frac{N_e}{N_{N_2}}\cdot\upsilon_*^2$. The eigenvalues depend only on the trace and determinant of the matrix A. The typical situation is for the eigenvalues to be distinct: $\lambda_1 \neq \lambda_2$. In this case, the theorem of linear algebra states that the corresponding eigenvectors are linearly independent and hence span the entire plane [2, 3].

Classification of fixed points: We need to know the type and stability of all the system different fixed points on a single diagram. The axes are the trace $\wp_1$ and the determinant $\wp_2$ of the matrix A. All of the information in the diagram is implied by the following formulas: $\lambda_{1,2} = \frac{\wp_1 \pm \sqrt{\wp_1^2 - 4 \cdot \wp_2}}{2}$; $\wp_2 = \prod_{k=1}^{2} \lambda_k$; $\wp_1 = \sum_{k=1}^{2} \lambda_k$ and $(\lambda - \lambda_1) \cdot (\lambda - \lambda_2) = 0 (\lambda - \lambda_1) \cdot (\lambda - \lambda_2) = \lambda^2 - \lambda \cdot \wp_1 + \wp_2 = 0$. If $\wp_2 < 0$, the eigenvalues are real and have opposite sign; hence the fixed point is a saddle point. If $\wp_2 > 0$, the eigenvalues are either real with the same sign (nodes), or complex conjugate (spirals and centers). Nodes satisfy $\wp_1^2 - 4 \cdot \wp_2 > 0$ and spirals satisfy $\wp_1^2 - 4 \cdot \wp_2 < 0$. The parabola $\wp_1^2 - 4 \cdot \wp_2 = 0$ is the borderline between nodes and spirals; star nodes and degenerate nodes live on this parabola. The stability of the nodes and spirals is determined by $\wp_1$. When $\wp_1 < 0$, both eigenvalues have negative real parts, the fixed point is stable. Unstable spirals and nodes have $\wp_1 > 0$. Neutrally stable center live on the borderline $\wp_1 = 0$, where the eigenvalues are purely imaginary. If $\wp_2 = 0$ then at least one of the eigenvalues is zero. Then the origin is not an isolated fixed point. There is either a whole line of fixed points, or a plane of fixed points, if $A = 0$. Saddle points, nodes, and spirals are the major types of fixed points; they occur in large open regions of the $(\wp_2, \wp_1)$ plane. Centers, stars, degenerate nodes, and non-isolated fixed points are borderline cases that occur along the curves in the $(\wp_2, \wp_1)$ plane.

System No.2:

We include lase effects, and the set of plasma kinetic equations include the populations of the lasing levels and number of generated photons:

$$\frac{dn_c}{dt} = S_c - n_c \cdot (A_{CB1} + A_{CB2} + \sigma_{CB1} \cdot c \cdot n_{ph1} + \sigma_{CB2} \cdot c \cdot n_{ph2}) + n_{B1} \cdot \sigma_{B1C} \cdot c \cdot n_{ph1} + n_{B2} \cdot \sigma_{B2C} \cdot c \cdot n_{ph2}$$

$$\frac{dn_{B1}}{dt} = S_{B1} + n_c \cdot (A_{CB1} + \sigma_{CB1} \cdot c \cdot n_{ph1}) - n_{B1} \cdot \sigma_{B1C} \cdot c \cdot n_{ph1}$$

$$\frac{dn_{B2}}{dt} = S_{B2} + n_c \cdot (A_{CB2} + \sigma_{CB2} \cdot c \cdot n_{ph2}) - n_{B2} \cdot \sigma_{B2C} \cdot c \cdot n_{ph2}$$

$$\frac{dn_{ph1}}{dt} = \frac{\sigma_{CB1} \cdot (n_c - n_{B1}) \cdot n_{ph1} \cdot c}{1 + \frac{I_L}{I_{s1}}} - \frac{n_{ph1}}{\tau_{ph}}$$

$$\frac{dn_{ph2}}{dt} = \frac{\sigma_{CB2} \cdot (n_c - n_{B2}) \cdot n_{ph2} \cdot c}{1 + \frac{I_L}{I_{s2}}} - \frac{n_{ph2}}{\tau_{ph}}$$

n_c is the population of the $C^3\Pi_u$ state, n_{B1}, n_{B2} are the populations of $B^3\Pi_g$ in the ground and first excited vibrational states, S_c, S_{B1}, S_{B2} are the rates of generation and loss through all the non-radiative channels; $\sigma_{CB1}, \sigma_{CB2}, \sigma_{B1C}, \sigma_{B2C}$ are the stimulated emission cross sections; c is the speed of light; $I_{sk} = \frac{h \cdot \upsilon}{\sigma(\upsilon) \cdot \tau_{sk}}$; $k = 1, 2$ are the saturation intensities where the radiative lifetimes $\tau_s \propto A^{-1}$ and are determined from the corresponding Einstein coefficients; $I_L = n_{ph} \cdot h \cdot \upsilon \cdot c$ are the lasing intensities; and n_{ph1}, n_{ph2} are the photon number densities at specific wavelength (λ = 337nm; λ = 357 nm); τ_{ph} is a photon lifetime related with the filament length L as $\tau_{ph} = \frac{L}{c}$. The broadening of each of the laser modes is assumed equal to Δλ = 0.1nm. The pressure-induced line broadening is taken as $g(\upsilon) \approx \frac{g_0(\upsilon) \cdot 760}{p_{Ar} + p_{N_2}}$, where $g_0(\upsilon)$ is the spectral line shape at the gas pressure $p = 760$ Torr and $T = 300$ K, and p_{Ar}, p_{N_2} are the partial pressures of argon and molecular nitrogen [1]. The time dependent model is the standard model for description of chemical and plasma-chemical kinetics. The variables in the model are functions of time only, and we neglect the spatial effects (diffusion and spatial gain inhomogeneity). The plasma kinetic model is not ideal and there is a time delay for $n_c(t) \to n_c(t - \tau_D)$; $\tau_D > 0$, then the set of plasma kinetic equations: parameters $(S_c, S_{B1}, S_{B2}, \sigma_{CB1}, \sigma_{CB2}, \sigma_{B1C}, \sigma_{B2C}, \tau_D, \ldots)$

$$\begin{aligned}\frac{dn_c}{dt} &= S_c - n_c(t - \tau_D) \cdot (A_{CB1} + A_{CB2} + \sigma_{CB1} \cdot c \cdot n_{ph1} \\ &\quad + \sigma_{CB2} \cdot c \cdot n_{ph2}) + n_{B1} \cdot \sigma_{B1C} \cdot c \cdot n_{ph1} + n_{B2} \\ &\quad \cdot \sigma_{B2C} \cdot c \cdot n_{ph2}\end{aligned}$$

$$\begin{aligned}\frac{dn_{B1}}{dt} &= S_{B1} + n_c(t - \tau_D) \cdot (A_{CB1} + \sigma_{CB1} \cdot c \cdot n_{ph1}) \\ &\quad - n_{B1} \cdot \sigma_{B1C} \cdot c \cdot n_{ph1}\end{aligned}$$

$$\begin{aligned}\frac{dn_{B2}}{dt} &= S_{B2} + n_c(t - \tau_D) \cdot (A_{CB2} + \sigma_{CB2} \cdot c \cdot n_{ph2}) \\ &\quad - n_{B2} \cdot \sigma_{B2C} \cdot c \cdot n_{ph2}\end{aligned}$$

$$\frac{dn_{ph1}}{dt} = \frac{\sigma_{CB1} \cdot (n_c(t - \tau_D) - n_{B1}) \cdot n_{ph1} \cdot c}{1 + \frac{I_L}{I_{s1}}} - \frac{n_{ph1}}{\tau_{ph}}$$

$$\frac{dn_{ph2}}{dt} = \frac{\sigma_{CB2} \cdot (n_c(t - \tau_D) - n_{B2}) \cdot n_{ph2} \cdot c}{1 + \frac{I_L}{I_{s2}}} - \frac{n_{ph2}}{\tau_{ph}}$$

$$n_c = n_c(t);\, n_{B1} = n_{B1}(t)$$

$$n_{B2} = n_{B2}(t);\, n_{ph1} = n_{ph1}(t);\, n_{ph2} = n_{ph2}(t)$$

Hint: time delay τ_D does not affect the derivative of $n_c = n_c(t)$ in time, $\frac{dn_c}{dt}$.
At fixed points: $\frac{dn_c}{dt} = 0$; $\frac{dn_{B1}}{dt} = 0$; $\frac{dn_{B2}}{dt} = 0$; $\frac{dn_{ph1}}{dt} = 0$; $\frac{dn_{ph2}}{dt} = 0$

$$\lim_{t \to 0} n_c(t - \tau_D) = n_c(t);\, t \gg \tau_D;\, t - \tau_D \approx t\ \forall t \to \infty$$

$$(*)\ \begin{array}{l} S_c - n_c^* \cdot (A_{CB1} + A_{CB2} + \sigma_{CB1} \cdot c \cdot n_{ph1}^* + \sigma_{CB2} \cdot c \cdot n_{ph2}^*) \\ + n_{B1}^* \cdot \sigma_{B1C} \cdot c \cdot n_{ph1}^* + n_{B2}^* \cdot \sigma_{B2C} \cdot c \cdot n_{ph2}^* = 0 \end{array}$$

$(**)\ S_{B1} + n_c^* \cdot (A_{CB1} + \sigma_{CB1} \cdot c \cdot n_{ph1}^*) - n_{B1}^* \cdot \sigma_{B1C} \cdot c \cdot n_{ph1}^* = 0$

$(***)\ S_{B2} + n_c^* \cdot (A_{CB2} + \sigma_{CB2} \cdot c \cdot n_{ph2}^*) - n_{B2}^* \cdot \sigma_{B2C} \cdot c \cdot n_{ph2}^* = 0$

$(****)\ \frac{\sigma_{CB1}\cdot(n_c^*-n_{B1}^*)\cdot n_{ph1}^*\cdot c}{1+\frac{I_L}{I_{s1}}} - \frac{n_{ph1}^*}{\tau_{ph}} = 0;\ (*****)\ \frac{\sigma_{CB2}\cdot(n_c^*-n_{B2}^*)\cdot n_{ph2}^*\cdot c}{1+\frac{I_L}{I_{s2}}} - \frac{n_{ph2}^*}{\tau_{ph}} = 0$

Some mathematical manipulations:

$(****)\ n_{ph1}^* \cdot [\frac{\sigma_{CB1}\cdot(n_c^*-n_{B1}^*)\cdot c}{1+\frac{I_L}{I_{s1}}} - \frac{1}{\tau_{ph}}] = 0;\ (1.1)\ n_{ph1}^* = 0;\ (1.2)\ \frac{\sigma_{CB1}\cdot(n_c^*-n_{B1}^*)\cdot c}{1+\frac{I_L}{I_{s1}}} - \frac{1}{\tau_{ph}} = 0$

$n_c^* = n_{B1}^* + (1 + \frac{I_L}{I_{s1}}) \cdot \frac{1}{\tau_{ph}\cdot\sigma_{CB1}\cdot c};\ (*****);\ n_{ph2}^* \cdot [\frac{\sigma_{CB2}\cdot(n_c^*-n_{B2}^*)\cdot c}{1+\frac{I_L}{I_{s2}}} - \frac{1}{\tau_{ph}}] = 0;\ (2.1)\ n_{ph2}^* = 0$

$(2.2)\ \frac{\sigma_{CB2}\cdot(n_c^*-n_{B2}^*)\cdot c}{1+\frac{I_L}{I_{s2}}} - \frac{1}{\tau_{ph}} = 0;\ \frac{\sigma_{CB2}\cdot(n_c^*-n_{B2}^*)\cdot c}{1+\frac{I_L}{I_{s2}}} = \frac{1}{\tau_{ph}};\ n_c^* = n_{B2}^* + (1 + \frac{I_L}{I_{s2}}) \cdot \frac{1}{\sigma_{CB2}\cdot\tau_{ph}\cdot c}$

We can summary the above partial fixed points discussion in the next table (Table 6.1).

Case A, (1.1) & (2.1) $n_{ph1}^* = 0;\ n_{ph2}^* = 0$

$$(*)\ \begin{array}{l} S_c - n_c^* \cdot (A_{CB1} + A_{CB2} + \sigma_{CB1} \cdot c \cdot n_{ph1}^* + \sigma_{CB2} \cdot c \cdot n_{ph2}^*) \\ + n_{B1}^* \cdot \sigma_{B1C} \cdot c \cdot n_{ph1}^* + n_{B2}^* \cdot \sigma_{B2C} \cdot c \cdot n_{ph2}^* = 0 \end{array}$$

$$n_{ph1}^* = 0;\ n_{ph2}^* = 0 \Rightarrow S_c - n_c^* \cdot (A_{CB1} + A_{CB2}) = 0 \Rightarrow n_c^* = \frac{S_c}{A_{CB1} + A_{CB2}}$$

Table 6.1 Plasma kinetic equations (populations of the lasing levels and number of generated photons), partial fixed points discussion

Case	Description
A: (1.1) & (2.1)	$n_{ph1}^* = 0;\ n_{ph2}^* = 0$
B: (1.1) & (2.2)	$n_{ph1}^* = 0;\ n_c^* = n_{B2}^* + (1 + \frac{I_L}{I_{s2}}) \cdot \frac{1}{\sigma_{CB2}\cdot\tau_{ph}\cdot c} = n_{B2}^* + \Gamma_2$ We define for simplicity Γ_2 parameter: $\Gamma_2 = (1 + \frac{I_L}{I_{s2}}) \cdot \frac{1}{\sigma_{CB2}\cdot\tau_{ph}\cdot c};\ \Gamma_2 = \Gamma_2(I_L, I_{s2}, \sigma_{CB2}, \tau_{ph})$
C: (1.2) & (2.1)	$n_c^* = n_{B1}^* + (1 + \frac{I_L}{I_{s1}}) \cdot \frac{1}{\tau_{ph}\cdot\sigma_{CB1}\cdot c} = n_{B1}^* + \Gamma_1;\ n_{ph2}^* = 0$ We define for simplicity Γ_1 parameter: $\Gamma_1 = (1 + \frac{I_L}{I_{s1}}) \cdot \frac{1}{\tau_{ph}\cdot\sigma_{CB1}\cdot c};\ \Gamma_1 = \Gamma_1(I_L, I_{s1}, \tau_{ph}, \sigma_{CB1})$
D (1.2) & (2.2)	$n_c^* = n_{B1}^* + (1 + \frac{I_L}{I_{s1}}) \cdot \frac{1}{\tau_{ph}\cdot\sigma_{CB1}\cdot c};\ n_c^* = n_{B2}^* + (1 + \frac{I_L}{I_{s2}}) \cdot \frac{1}{\sigma_{CB2}\cdot\tau_{ph}\cdot c}$ $n_c^* = n_{B1}^* + \Gamma_1;\ n_c^* = n_{B2}^* + \Gamma_2$ Then $n_{B1}^* + (1 + \frac{I_L}{I_{s1}}) \cdot \frac{1}{\tau_{ph}\cdot\sigma_{CB1}\cdot c} = n_{B2}^* + (1 + \frac{I_L}{I_{s2}}) \cdot \frac{1}{\sigma_{CB2}\cdot\tau_{ph}\cdot c}$ $n_{B1}^* = n_{B2}^* + (1 + \frac{I_L}{I_{s2}}) \cdot \frac{1}{\sigma_{CB2}\cdot\tau_{ph}\cdot c} - (1 + \frac{I_L}{I_{s1}}) \cdot \frac{1}{\tau_{ph}\cdot\sigma_{CB1}\cdot c}$ $n_{B1}^* = n_{B2}^* + \Gamma_2 - \Gamma_1$

(**) $S_{B1} + n_c^* \cdot (A_{CB1} + \sigma_{CB1} \cdot c \cdot n_{ph1}^*) - n_{B1}^* \cdot \sigma_{B1C} \cdot c \cdot n_{ph1}^* = 0$
$n_{ph1}^* = 0; n_{ph2}^* = 0 \Rightarrow n_c^* = -\frac{S_{B1}}{A_{CB1}}$ (Conclusion: $\frac{S_c}{A_{CB1}+A_{CB2}} = -\frac{S_{B1}}{A_{CB1}}$).
(***) $S_{B2} + n_c^* \cdot (A_{CB2} + \sigma_{CB2} \cdot c \cdot n_{ph2}^*) - n_{B2}^* \cdot \sigma_{B2C} \cdot c \cdot n_{ph2}^* = 0$

$$n_{ph1}^* = 0; n_{ph2}^* = 0 \Rightarrow S_{B2} + n_c^* \cdot A_{CB2} = 0 \Rightarrow n_c^* = -\frac{S_{B2}}{A_{CB2}}$$

Conclusion: $\frac{S_c}{A_{CB1}+A_{CB2}} = -\frac{S_{B1}}{A_{CB1}} = -\frac{S_{B2}}{A_{CB2}}$

$$n_{B1}^* = n_c^* - \Gamma_1 = n_c^* - \Gamma_1(I_L, I_{s1}, \tau_{ph}, \sigma_{CB1}) = \frac{S_c}{A_{CB1} + A_{CB2}} - \Gamma_1(I_L, I_{s1}, \tau_{ph}, \sigma_{CB1})$$

$$n_{B2}^* = n_c^* - \Gamma_2 = n_c^* - \Gamma_2(I_L, I_{s2}, \sigma_{CB2}, \tau_{ph}) = \frac{S_c}{A_{CB1} + A_{CB2}} - \Gamma_2(I_L, I_{s2}, \sigma_{CB2}, \tau_{ph})$$

First fixed point:

$$E^{(0)} = \left(n_c^{(0)}, n_{B1}^{(0)}, n_{B2}^{(0)}, n_{ph1}^{(0)}, n_{ph2}^{(0)}\right) = \left(\frac{S_c}{A_{CB1} + A_{CB2}}, \frac{S_c}{A_{CB1} + A_{CB2}} - \Gamma_1(I_L, I_{s1}, \tau_{ph}, \sigma_{CB1}), \frac{S_c}{A_{CB1} + A_{CB2}} - \Gamma_2(I_L, I_{s2}, \sigma_{CB2}, \tau_{ph}), 0, 0\right)$$

Case B, (1.1) & (2.2) $n_{ph1}^* = 0; n_c^* = n_{B2}^* + \Gamma_2$

$$(*) \quad S_c - n_c^* \cdot (A_{CB1} + A_{CB2} + \sigma_{CB1} \cdot c \cdot n_{ph1}^* + \sigma_{CB2} \cdot c \cdot n_{ph2}^*) + n_{B1}^* \cdot \sigma_{B1C} \cdot c \cdot n_{ph1}^* + n_{B2}^* \cdot \sigma_{B2C} \cdot c \cdot n_{ph2}^* = 0$$

$$S_c - (n_{B2}^* + \Gamma_2) \cdot (A_{CB1} + A_{CB2} + \sigma_{CB2} \cdot c \cdot n_{ph2}^*) + n_{B2}^* \cdot \sigma_{B2C} \cdot c \cdot n_{ph2}^* = 0$$

$$S_c - (n_{B2}^* + \Gamma_2) \cdot (A_{CB1} + A_{CB2}) - (n_{B2}^* + \Gamma_2) \cdot (\sigma_{CB2} \cdot c \cdot n_{ph2}^*) + n_{B2}^* \cdot \sigma_{B2C} \cdot c \cdot n_{ph2}^* = 0$$

$$S_c - n_{B2}^* \cdot (A_{CB1} + A_{CB2}) - \Gamma_2 \cdot (A_{CB1} + A_{CB2})$$

$$- n^*_{B2} \cdot n^*_{ph2} \cdot \sigma_{CB2} \cdot c - \Gamma_2 \cdot \sigma_{CB2} \cdot c \cdot n^*_{ph2}$$
$$+ n^*_{B2} \cdot n^*_{ph2} \cdot \sigma_{B2C} \cdot c = 0$$

(B.1)
$$S_c - \Gamma_2 \cdot (A_{CB1} + A_{CB2}) + n^*_{B2} \cdot n^*_{ph2} \cdot (\sigma_{B2C} - \sigma_{CB2}) \cdot c$$
$$- \Gamma_2 \cdot \sigma_{CB2} \cdot c \cdot n^*_{ph2} - n^*_{B2} \cdot (A_{CB1} + A_{CB2}) = 0$$

(**) $S_{B1} + n^*_c \cdot (A_{CB1} + \sigma_{CB1} \cdot c \cdot n^*_{ph1}) - n^*_{B1} \cdot \sigma_{B1C} \cdot c \cdot n^*_{ph1} = 0$

$$S_{B1} + (n^*_{B2} + \Gamma_2) \cdot A_{CB1} = 0 \Rightarrow S_{B1} + n^*_{B2} \cdot A_{CB1} + \Gamma_2 \cdot A_{CB1}$$
$$= 0 \Rightarrow n^*_{B2} = -\frac{(S_{B1} + \Gamma_2 \cdot A_{CB1})}{A_{CB1}}$$

$n^*_c = n^*_{B2} + \Gamma_2 = -\frac{(S_{B1}+\Gamma_2 \cdot A_{CB1})}{A_{CB1}} + \Gamma_2; n^*_{B2} = -\frac{(S_{B1}+\Gamma_2 \cdot A_{CB1})}{A_{CB1}} \to$ (B.1):

$$S_c - \Gamma_2 \cdot (A_{CB1} + A_{CB2}) - \frac{(S_{B1} + \Gamma_2 \cdot A_{CB1})}{A_{CB1}}$$
$$\cdot n^*_{ph2} \cdot (\sigma_{B2C} - \sigma_{CB2}) \cdot c - \Gamma_2 \cdot \sigma_{CB2} \cdot c \cdot n^*_{ph2}$$
$$+ \frac{(S_{B1} + \Gamma_2 \cdot A_{CB1})}{A_{CB1}} \cdot (A_{CB1} + A_{CB2}) = 0$$

$$S_c - \Gamma_2 \cdot (A_{CB1} + A_{CB2}) + \frac{(S_{B1} + \Gamma_2 \cdot A_{CB1})}{A_{CB1}} \cdot (A_{CB1} + A_{CB2})$$
$$= \frac{(S_{B1} + \Gamma_2 \cdot A_{CB1})}{A_{CB1}} \cdot (\sigma_{B2C} - \sigma_{CB2}) \cdot c \cdot n^*_{ph2} + \Gamma_2 \cdot \sigma_{CB2} \cdot c \cdot n^*_{ph2}$$

$$[\frac{(S_{B1} + \Gamma_2 \cdot A_{CB1})}{A_{CB1}} \cdot (\sigma_{B2C} - \sigma_{CB2}) + \Gamma_2 \cdot \sigma_{CB2}] \cdot c \cdot n^*_{ph2}$$
$$= S_c - \Gamma_2 \cdot (A_{CB1} + A_{CB2}) + \frac{(S_{B1} + \Gamma_2 \cdot A_{CB1})}{A_{CB1}} \cdot (A_{CB1} + A_{CB2})$$

$$n^*_{ph2} = \frac{S_c - \Gamma_2 \cdot (A_{CB1} + A_{CB2}) + \frac{(S_{B1}+\Gamma_2 \cdot A_{CB1})}{A_{CB1}} \cdot (A_{CB1} + A_{CB2})}{[\frac{(S_{B1}+\Gamma_2 \cdot A_{CB1})}{A_{CB1}} \cdot (\sigma_{B2C} - \sigma_{CB2}) + \Gamma_2 \cdot \sigma_{CB2}] \cdot c}$$

(***) $S_{B2} + n^*_c \cdot (A_{CB2} + \sigma_{CB2} \cdot c \cdot n^*_{ph2}) - n^*_{B2} \cdot \sigma_{B2C} \cdot c \cdot n^*_{ph2} = 0$

$$S_{B2} + (-\frac{(S_{B1} + \Gamma_2 \cdot A_{CB1})}{A_{CB1}} + \Gamma_2) \cdot (A_{CB2} + \sigma_{CB2} \cdot c$$
$$\cdot \{\frac{S_c - \Gamma_2 \cdot (A_{CB1} + A_{CB2}) + \frac{(S_{B1}+\Gamma_2 \cdot A_{CB1})}{A_{CB1}} \cdot (A_{CB1} + A_{CB2})}{[\frac{(S_{B1}+\Gamma_2 \cdot A_{CB1})}{A_{CB1}} \cdot (\sigma_{B2C} - \sigma_{CB2}) + \Gamma_2 \cdot \sigma_{CB2}] \cdot c}\})$$
$$+ \frac{(S_{B1} + \Gamma_2 \cdot A_{CB1})}{A_{CB1}} \cdot \sigma_{B2C} \cdot c$$

$$\cdot \{ \frac{S_c - \Gamma_2 \cdot (A_{CB1} + A_{CB2}) + \frac{(S_{B1}+\Gamma_2 \cdot A_{CB1})}{A_{CB1}} \cdot (A_{CB1} + A_{CB2})}{[\frac{(S_{B1}+\Gamma_2 \cdot A_{CB1})}{A_{CB1}} \cdot (\sigma_{B2C} - \sigma_{CB2}) + \Gamma_2 \cdot \sigma_{CB2}] \cdot c} \} = 0$$

Remark The above mathcmatical relation must fulfill for the second fixed point.
(**) $S_{B1} + n_c^* \cdot (A_{CB1} + \sigma_{CB1} \cdot c \cdot n_{ph1}^*) - n_{B1}^* \cdot \sigma_{B1C} \cdot c \cdot n_{ph1}^* = 0$

$$n_{B1}^* = \frac{S_{B1} + n_c^* \cdot (A_{CB1} + \sigma_{CB1} \cdot c \cdot n_{ph1}^*)}{\sigma_{B1C} \cdot c \cdot n_{ph1}^*} |_{n_{ph1}^*=0} \to \infty$$

Second fixed point:

$$E^{(1)} = \left(n_c^{(1)}, n_{B1}^{(1)}, n_{B2}^{(1)}, n_{ph1}^{(1)}, n_{ph2}^{(1)}\right)$$
$$= \left(-\frac{(S_{B1} + \Gamma_2 \cdot A_{CB1})}{A_{CB1}} + \Gamma_2, n_{B1}^{(1)} \to \infty, -\frac{(S_{B1} + \Gamma_2 \cdot A_{CB1})}{A_{CB1}}, 0, \right.$$
$$\left. \frac{S_c - \Gamma_2 \cdot (A_{CB1} + A_{CB2}) + \frac{(S_{B1}+\Gamma_2 \cdot A_{CB1})}{A_{CB1}} \cdot (A_{CB1} + A_{CB2})}{\left[\frac{(S_{B1}+\Gamma_2 \cdot A_{CB1})}{A_{CB1}} \cdot (\sigma_{B2C} - \sigma_{CB2}) + \Gamma_2 \cdot \sigma_{CB2}\right] \cdot c} \right)$$

Since $n_{B1}^{(1)} \to \infty$ (population of $B^3\Pi_g$ in the ground state), the second fixed point is only "mathematical" and Not "physical".

Case C, (1.2) & (2.1) $n_c^* = n_{B1}^* + \Gamma_1; n_{ph2}^* = 0$

$$(*) \quad \begin{aligned} &S_c - n_c^* \cdot (A_{CB1} + A_{CB2} + \sigma_{CB1} \cdot c \cdot n_{ph1}^* + \sigma_{CB2} \cdot c \cdot n_{ph2}^*) \\ &+ n_{B1}^* \cdot \sigma_{B1C} \cdot c \cdot n_{ph1}^* + n_{B2}^* \cdot \sigma_{B2C} \cdot c \cdot n_{ph2}^* = 0 \end{aligned}$$

$$\begin{aligned} &S_c - (n_{B1}^* + \Gamma_1) \cdot (A_{CB1} + A_{CB2} + \sigma_{CB1} \cdot c \cdot n_{ph1}^*) \\ &+ n_{B1}^* \cdot \sigma_{B1C} \cdot c \cdot n_{ph1}^* = 0 \end{aligned}$$

$$\begin{aligned} &S_c - n_{B1}^* \cdot (A_{CB1} + A_{CB2}) - \Gamma_1 \cdot (A_{CB1} + A_{CB2}) \\ &- (n_{B1}^* + \Gamma_1) \cdot \sigma_{CB1} \cdot c \cdot n_{ph1}^* \\ &+ n_{B1}^* \cdot \sigma_{B1C} \cdot c \cdot n_{ph1}^* = 0 \end{aligned}$$

$$\begin{aligned} &S_c - n_{B1}^* \cdot (A_{CB1} + A_{CB2}) - \Gamma_1 \cdot (A_{CB1} + A_{CB2}) \\ &- n_{B1}^* \cdot \sigma_{CB1} \cdot c \cdot n_{ph1}^* - \Gamma_1 \cdot \sigma_{CB1} \cdot c \cdot n_{ph1}^* \\ &+ n_{B1}^* \cdot \sigma_{B1C} \cdot c \cdot n_{ph1}^* = 0 \end{aligned}$$

$$(C.1) \quad \begin{aligned} &S_c - \Gamma_1 \cdot (A_{CB1} + A_{CB2}) - n_{B1}^* \cdot (A_{CB1} + A_{CB2}) - \Gamma_1 \cdot \sigma_{CB1} \cdot c \cdot n_{ph1}^* \\ &+ n_{B1}^* \cdot n_{ph1}^* \cdot (\sigma_{B1C} - \sigma_{CB1}) \cdot c = 0 \end{aligned}$$

(**) $S_{B1} + n_c^* \cdot (A_{CB1} + \sigma_{CB1} \cdot c \cdot n_{ph1}^*) - n_{B1}^* \cdot \sigma_{B1C} \cdot c \cdot n_{ph1}^* = 0$

$$S_{B1} + (n^*_{B1} + \Gamma_1) \cdot (A_{CB1} + \sigma_{CB1} \cdot c \cdot n^*_{ph1}) - n^*_{B1} \cdot \sigma_{B1C} \cdot c \cdot n^*_{ph1} = 0$$

$$S_{B1} + (n^*_{B1} + \Gamma_1) \cdot A_{CB1} + (n^*_{B1} + \Gamma_1) \cdot \sigma_{CB1} \cdot c \cdot n^*_{ph1} - \sigma_{B1C} \cdot c \cdot n^*_{B1} \cdot n^*_{ph1} = 0$$

$$S_{B1} + \Gamma_1 \cdot A_{CB1} + n^*_{B1} \cdot A_{CB1} + \Gamma_1 \cdot \sigma_{CB1} \cdot c \cdot n^*_{ph1} + c \cdot (\sigma_{CB1} - \sigma_{B1C}) \cdot n^*_{B1} \cdot n^*_{ph1} = 0$$

We get two equations, two fixed coordinates: n^*_{B1}, n^*_{ph1}

(C.2) $S_c - \Gamma_1 \cdot (A_{CB1} + A_{CB2}) - n^*_{B1} \cdot (A_{CB1} + A_{CB2}) - \Gamma_1 \cdot \sigma_{CB1} \cdot c \cdot n^*_{ph1} + n^*_{B1} \cdot n^*_{ph1} \cdot (\sigma_{B1C} - \sigma_{CB1}) \cdot c = 0$

(C.3) $S_{B1} + \Gamma_1 \cdot A_{CB1} + n^*_{B1} \cdot A_{CB1} + \Gamma_1 \cdot \sigma_{CB1} \cdot c \cdot n^*_{ph1} + c \cdot (\sigma_{CB1} - \sigma_{B1C}) \cdot n^*_{B1} \cdot n^*_{ph1} = 0$

###

(***) $S_{B2} + n^*_c \cdot (A_{CB2} + \sigma_{CB2} \cdot c \cdot n^*_{ph2}) - n^*_{B2} \cdot \sigma_{B2C} \cdot c \cdot n^*_{ph2} = 0$

$$S_{B2} + n^*_c \cdot A_{CB2} = 0 \Rightarrow n^*_c = -\frac{S_{B2}}{A_{CB2}}$$

$$n^*_c = n^*_{B1} + \Gamma_1 \Rightarrow n^*_{B1} = n^*_c - \Gamma_1 = -\left(\frac{S_{B2}}{A_{CB2}} + \Gamma_1\right)$$

$n^*_{B1} = -(\frac{S_{B2}}{A_{CB2}} + \Gamma_1) \Rightarrow (C.2)$:

$$S_c - \Gamma_1 \cdot (A_{CB1} + A_{CB2}) + (\frac{S_{B2}}{A_{CB2}} + \Gamma_1) \cdot (A_{CB1} + A_{CB2}) - \Gamma_1 \cdot \sigma_{CB1} \cdot c \cdot n^*_{ph1} - (\frac{S_{B2}}{A_{CB2}} + \Gamma_1) \cdot n^*_{ph1} \cdot (\sigma_{B1C} - \sigma_{CB1}) \cdot c = 0$$

$$n^*_{ph1} = \frac{S_c - \Gamma_1 \cdot (A_{CB1} + A_{CB2}) + (\frac{S_{B2}}{A_{CB2}} + \Gamma_1) \cdot (A_{CB1} + A_{CB2})}{[\Gamma_1 \cdot \sigma_{CB1} + (\frac{S_{B2}}{A_{CB2}} + \Gamma_1) \cdot (\sigma_{B1C} - \sigma_{CB1})] \cdot c}$$

$n^*_{B1} = -(\frac{S_{B2}}{A_{CB2}} + \Gamma_1)$; $n^*_{ph1} = \ldots \Rightarrow$ (C.3):

$$S_{B1} + \Gamma_1 \cdot A_{CB1} + (-(\frac{S_{B2}}{A_{CB2}} + \Gamma_1)) \cdot A_{CB1} + \Gamma_1 \cdot \sigma_{CB1} \cdot c \cdot \{\frac{S_c - \Gamma_1 \cdot (A_{CB1} + A_{CB2}) + (\frac{S_{B2}}{A_{CB2}} + \Gamma_1) \cdot (A_{CB1} + A_{CB2})}{[\Gamma_1 \cdot \sigma_{CB1} + (\frac{S_{B2}}{A_{CB2}} + \Gamma_1) \cdot (\sigma_{B1C} - \sigma_{CB1})] \cdot c}\}$$

$$- c \cdot (\sigma_{CB1} - \sigma_{B1C}) \cdot (\frac{S_{B2}}{A_{CB2}} + \Gamma_1)$$
$$\cdot \{\frac{S_c - \Gamma_1 \cdot (A_{CB1} + A_{CB2}) + (\frac{S_{B2}}{A_{CB2}} + \Gamma_1) \cdot (A_{CB1} + A_{CB2})}{[\Gamma_1 \cdot \sigma_{CB1} + (\frac{S_{B2}}{A_{CB2}} + \Gamma_1) \cdot (\sigma_{B1C} - \sigma_{CB1})] \cdot c}\} = 0$$

Remark The above mathematical relation must fulfill for the third fixed point.
(***) $S_{B2} + n_c^* \cdot (A_{CB2} + \sigma_{CB2} \cdot c \cdot n_{ph2}^*) - n_{B2}^* \cdot \sigma_{B2C} \cdot c \cdot n_{ph2}^* = 0$

$$n_{B2}^* = \frac{S_{B2} + n_c^* \cdot (A_{CB2} + \sigma_{CB2} \cdot c \cdot n_{ph2}^*)}{\sigma_{B2C} \cdot c \cdot n_{ph2}^*}|_{n_{ph2}^*=0} \to \infty$$

Second fixed point:

$$E^{(2)} = \left(n_c^{(2)}, n_{B1}^{(2)}, n_{B2}^{(2)}, n_{ph1}^{(2)}, n_{ph2}^{(2)}\right)$$
$$= \left(-\frac{S_{B2}}{A_{CB2}}, -(\frac{S_{B2}}{A_{CB2}} + \Gamma_1\right), n_{B2}^{(1)} \to \infty,$$
$$\left.\frac{S_c - \Gamma_1 \cdot (A_{CB1} + A_{CB2}) + \left(\frac{S_{B2}}{A_{CB2}} + \Gamma_1\right) \cdot (A_{CB1} + A_{CB2})}{\left[\Gamma_1 \cdot \sigma_{CB1} + \left(\frac{S_{B2}}{A_{CB2}} + \Gamma_1\right) \cdot (\sigma_{B1C} - \sigma_{CB1})\right] \cdot c}, 0\right)$$

Since $n_{B2}^{(1)} \to \infty$ (population of $B^3\Pi_g$ in the first excited vibrational state), the third fixed point is only "mathematical" and Not "physical".

Second and third fixed points discussion: For both Second and Third fixed points, $n_c^{(1)} < 0$; $n_c^{(2)} < 0$, respectively and $n_{B1}^{(1)} < 0$ or $n_{B1}^{(2)} < 0$, respectively. Population of specific state cannot be negative; it is evidence that these fixed points are theoretical.

Case D, (1.2) & (2.2) $n_c^* = n_{B1}^* + \Gamma_1$; $n_c^* = n_{B2}^* + \Gamma_2$; $n_{B1}^* = n_{B2}^* + \Gamma_2 - \Gamma_1$

$$(*) \quad \begin{aligned} &S_c - n_c^* \cdot (A_{CB1} + A_{CB2} + \sigma_{CB1} \cdot c \cdot n_{ph1}^* + \sigma_{CB2} \cdot c \cdot n_{ph2}^*) \\ &+ n_{B1}^* \cdot \sigma_{B1C} \cdot c \cdot n_{ph1}^* + n_{B2}^* \cdot \sigma_{B2C} \cdot c \cdot n_{ph2}^* = 0 \end{aligned}$$

$$\begin{aligned} &S_c - (n_{B2}^* + \Gamma_2) \cdot (A_{CB1} + A_{CB2} + \sigma_{CB1} \cdot c \cdot n_{ph1}^* + \sigma_{CB2} \cdot c \cdot n_{ph2}^*) \\ &+ (n_{B2}^* + \Gamma_2 - \Gamma_1) \cdot \sigma_{B1C} \cdot c \cdot n_{ph1}^* \\ &+ n_{B2}^* \cdot \sigma_{B2C} \cdot c \cdot n_{ph2}^* = 0 \end{aligned}$$

$$\begin{aligned} &S_c - (n_{B2}^* + \Gamma_2) \cdot [A_{CB1} + A_{CB2}] - (n_{B2}^* + \Gamma_2) \\ &\cdot \sigma_{CB1} \cdot c \cdot n_{ph1}^* - (n_{B2}^* + \Gamma_2) \cdot \sigma_{CB2} \cdot c \cdot n_{ph2}^* \\ &+ n_{B2}^* \cdot \sigma_{B1C} \cdot c \cdot n_{ph1}^* + [\Gamma_2 - \Gamma_1] \\ &\cdot \sigma_{B1C} \cdot c \cdot n_{ph1}^* + n_{B2}^* \cdot \sigma_{B2C} \cdot c \cdot n_{ph2}^* = 0 \end{aligned}$$

$$S_c - (n_{B2}^* \cdot [A_{CB1} + A_{CB2}] + \Gamma_2 \cdot [A_{CB1} + A_{CB2}])$$

$$
\begin{aligned}
&- (n^*_{B2} \cdot \sigma_{CB1} \cdot c \cdot n^*_{ph1} + \Gamma_2 \cdot \sigma_{CB1} \cdot c \cdot n^*_{ph1}) \\
&- (n^*_{B2} \cdot \sigma_{CB2} \cdot c \cdot n^*_{ph2} + \Gamma_2 \cdot \sigma_{CB2} \cdot c \cdot n^*_{ph2}) \\
&+ n^*_{B2} \cdot \sigma_{B1C} \cdot c \cdot n^*_{ph1} + [\Gamma_2 - \Gamma_1] \cdot \sigma_{B1C} \cdot c \cdot n^*_{ph1} \\
&+ n^*_{B2} \cdot \sigma_{B2C} \cdot c \cdot n^*_{ph2} = 0
\end{aligned}
$$

$$
\begin{aligned}
S_c &- n^*_{B2} \cdot [A_{CB1} + A_{CB2}] - \Gamma_2 \cdot [A_{CB1} + A_{CB2}] \\
&- n^*_{B2} \cdot \sigma_{CB1} \cdot c \cdot n^*_{ph1} - \Gamma_2 \cdot \sigma_{CB1} \cdot c \cdot n^*_{ph1} \\
&- n^*_{B2} \cdot \sigma_{CB2} \cdot c \cdot n^*_{ph2} - \Gamma_2 \cdot \sigma_{CB2} \cdot c \cdot n^*_{ph2} \\
&+ n^*_{B2} \cdot \sigma_{B1C} \cdot c \cdot n^*_{ph1} + [\Gamma_2 - \Gamma_1] \cdot \sigma_{B1C} \cdot c \cdot n^*_{ph1} \\
&+ n^*_{B2} \cdot \sigma_{B2C} \cdot c \cdot n^*_{ph2} = 0
\end{aligned}
$$

$$
\begin{aligned}
S_c &- \Gamma_2 \cdot [A_{CB1} + A_{CB2}] - n^*_{B2} \cdot [A_{CB1} + A_{CB2}] \\
&+ n^*_{B2} \cdot n^*_{ph1} \cdot (\sigma_{B1C} - \sigma_{CB1}) \cdot c + ([\Gamma_2 - \Gamma_1] \cdot \sigma_{B1C} - \Gamma_2 \cdot \sigma_{CB1}) \\
&\cdot c \cdot n^*_{ph1} + n^*_{B2} \cdot n^*_{ph2} \cdot [\sigma_{B2C} - \sigma_{CB2}] \cdot c - \Gamma_2 \cdot \sigma_{CB2} \cdot c \cdot n^*_{ph2} = 0
\end{aligned}
$$

$$(**)\ S_{B1} + n^*_c \cdot (A_{CB1} + \sigma_{CB1} \cdot c \cdot n^*_{ph1}) - n^*_{B1} \cdot \sigma_{B1C} \cdot c \cdot n^*_{ph1} = 0$$

$$
\begin{aligned}
S_{B1} &+ (n^*_{B2} + \Gamma_2) \cdot (A_{CB1} + \sigma_{CB1} \cdot c \cdot n^*_{ph1}) \\
&- (n^*_{B2} + \Gamma_2 - \Gamma_1) \cdot \sigma_{B1C} \cdot c \cdot n^*_{ph1} = 0
\end{aligned}
$$

$$
\begin{aligned}
S_{B1} &+ \Gamma_2 \cdot A_{CB1} + n^*_{B2} \cdot A_{CB1} + n^*_{B2} \cdot n^*_{ph1} \cdot [\sigma_{CB1} - \sigma_{B1C}] \cdot c \\
&+ (\Gamma_2 \cdot \sigma_{CB1} - [\Gamma_2 - \Gamma_1] \cdot \sigma_{B1C}) \cdot c \cdot n^*_{ph1} = 0
\end{aligned}
$$

$$(***)\ S_{B2} + n^*_c \cdot (A_{CB2} + \sigma_{CB2} \cdot c \cdot n^*_{ph2}) - n^*_{B2} \cdot \sigma_{B2C} \cdot c \cdot n^*_{ph2} = 0$$

$$S_{B2} + (n^*_{B2} + \Gamma_2) \cdot (A_{CB2} + \sigma_{CB2} \cdot c \cdot n^*_{ph2}) - n^*_{B2} \cdot \sigma_{B2C} \cdot c \cdot n^*_{ph2} = 0$$

$$
\begin{aligned}
S_{B2} &+ \Gamma_2 \cdot A_{CB2} + n^*_{B2} \cdot A_{CB2} + n^*_{B2} \cdot n^*_{ph2} \cdot (\sigma_{CB2} - \sigma_{B2C}) \cdot c \\
&+ \Gamma_2 \cdot \sigma_{CB2} \cdot c \cdot n^*_{ph2} = 0
\end{aligned}
$$

We can summary the three fixed points equations: $(n^*_{B2}, n^*_{ph1}, n^*_{ph2})$

$$
\begin{aligned}
&S_c - \Gamma_2 \cdot [A_{CB1} + A_{CB2}] - n^*_{B2} \cdot [A_{CB1} + A_{CB2}] \\
\text{(D.1)}\quad &+ n^*_{B2} \cdot n^*_{ph1} \cdot (\sigma_{B1C} - \sigma_{CB1}) \cdot c + ([\Gamma_2 - \Gamma_1] \cdot \sigma_{B1C} - \Gamma_2 \cdot \sigma_{CB1}) \\
&\cdot c \cdot n^*_{ph1} + n^*_{B2} \cdot n^*_{ph2} \cdot [\sigma_{B2C} - \sigma_{CB2}] \cdot c - \Gamma_2 \cdot \sigma_{CB2} \cdot c \cdot n^*_{ph2} = 0
\end{aligned}
$$

$$
\begin{aligned}
\text{(D.2)}\quad &S_{B1} + \Gamma_2 \cdot A_{CB1} + n^*_{B2} \cdot A_{CB1} + n^*_{B2} \cdot n^*_{ph1} \cdot [\sigma_{CB1} - \sigma_{B1C}] \cdot c \\
&+ (\Gamma_2 \cdot \sigma_{CB1} - [\Gamma_2 - \Gamma_1] \cdot \sigma_{B1C}) \cdot c \cdot n^*_{ph1} = 0
\end{aligned}
$$

(D.3) $S_{B2}+\Gamma_2\cdot A_{CB2}+n^*_{B2}\cdot A_{CB2}+n^*_{B2}\cdot n^*_{ph2}\cdot(\sigma_{CB2}-\sigma_{B2C})\cdot c+\Gamma_2\cdot\sigma_{CB2}\cdot c\cdot n^*_{ph2}=0$

(D.3) $\rightarrow n^*_{ph2}=\frac{-S_{B2}-\Gamma_2\cdot A_{CB2}-n^*_{B2}\cdot A_{CB2}}{n^*_{B2}\cdot(\sigma_{CB2}-\sigma_{B2C})\cdot c+\Gamma_2\cdot\sigma_{CB2}\cdot c}$

(D.3) → (D.1):
$$\begin{aligned}&S_c-\Gamma_2\cdot[A_{CB1}+A_{CB2}]-n^*_{B2}\cdot[A_{CB1}+A_{CB2}]\\&+n^*_{B2}\cdot n^*_{ph1}\cdot(\sigma_{B1C}-\sigma_{CB1})\cdot c\\&+([\Gamma_2-\Gamma_1]\cdot\sigma_{B1C}-\Gamma_2\cdot\sigma_{CB1})\cdot c\cdot n^*_{ph1}\\&+n^*_{ph2}\cdot\{n^*_{B2}\cdot[\sigma_{B2C}-\sigma_{CB2}]\cdot c-\Gamma_2\cdot\sigma_{CB2}\cdot c\}=0\end{aligned}$$

$$\begin{aligned}&S_c-\Gamma_2\cdot[A_{CB1}+A_{CB2}]-n^*_{B2}\cdot[A_{CB1}+A_{CB2}]\\&+n^*_{B2}\cdot n^*_{ph1}\cdot(\sigma_{B1C}-\sigma_{CB1})\\&\cdot c+([\Gamma_2-\Gamma_1]\cdot\sigma_{B1C}-\Gamma_2\cdot\sigma_{CB1})\cdot c\cdot n^*_{ph1}\\&+\{\frac{-S_{B2}-\Gamma_2\cdot A_{CB2}-n^*_{B2}\cdot A_{CB2}}{n^*_{B2}\cdot(\sigma_{CB2}-\sigma_{B2C})\cdot c+\Gamma_2\cdot\sigma_{CB2}\cdot c}\}\\&\cdot\{n^*_{B2}\cdot[\sigma_{B2C}-\sigma_{CB2}]\cdot c-\Gamma_2\cdot\sigma_{CB2}\cdot c\}=0\end{aligned}$$

We get two equations with variables (n^*_{ph1}, n^*_{B2}):

(D.3) → (D.1):
$$\begin{aligned}&S_c-\Gamma_2\cdot[A_{CB1}+A_{CB2}]-n^*_{B2}\cdot[A_{CB1}+A_{CB2}]\\&+n^*_{B2}\cdot n^*_{ph1}\cdot(\sigma_{B1C}-\sigma_{CB1})\cdot c\\&+([\Gamma_2-\Gamma_1]\cdot\sigma_{B1C}-\Gamma_2\cdot\sigma_{CB1})\cdot c\cdot n^*_{ph1}\\&+\left\{\frac{-S_{B2}-\Gamma_2\cdot A_{CB2}-n^*_{B2}\cdot A_{CB2}}{n^*_{B2}\cdot(\sigma_{CB2}-\sigma_{B2C})\cdot c+\Gamma_2\cdot\sigma_{CB2}\cdot c}\right\}\\&\cdot\{n^*_{B2}\cdot[\sigma_{B2C}-\sigma_{CB2}]\cdot c-\Gamma_2\cdot\sigma_{CB2}\cdot c\}=0\end{aligned}$$

(D.2)
$$\begin{aligned}&S_{B1}+\Gamma_2\cdot A_{CB1}+n^*_{B2}\cdot A_{CB1}+n^*_{B2}\cdot n^*_{ph1}\cdot[\sigma_{CB1}-\sigma_{B1C}]\cdot c\\&+(\Gamma_2\cdot\sigma_{CB1}-[\Gamma_2-\Gamma_1]\cdot\sigma_{B1C})\cdot c\cdot n^*_{ph1}=0\end{aligned}$$

$$\begin{aligned}&S_{B1}+\Gamma_2\cdot A_{CB1}+n^*_{B2}\cdot A_{CB1}\\&+\{n^*_{B2}\cdot[\sigma_{CB1}-\sigma_{B1C}]+(\Gamma_2\cdot\sigma_{CB1}-[\Gamma_2-\Gamma_1]\\&\cdot\sigma_{B1C})\}\cdot c\cdot n^*_{ph1}=0\end{aligned}$$

$$n^*_{ph1}=\frac{-n^*_{B2}\cdot A_{CB1}-\Gamma_2\cdot A_{CB1}-S_{B1}}{\{n^*_{B2}\cdot[\sigma_{CB1}-\sigma_{B1C}]+(\Gamma_2\cdot\sigma_{CB1}-[\Gamma_2-\Gamma_1]\cdot\sigma_{B1C})\}\cdot c}$$

Inserting it to (D.3) → (D.1) equation gives

$$\begin{aligned}&S_c-\Gamma_2\cdot[A_{CB1}+A_{CB2}]-n^*_{B2}\cdot[A_{CB1}+A_{CB2}]\\&+\{n^*_{B2}\cdot(\sigma_{B1C}-\sigma_{CB1})\cdot c+([\Gamma_2-\Gamma_1]\cdot\sigma_{B1C}-\Gamma_2\cdot\sigma_{CB1})\cdot c\}\cdot n^*_{ph1}\\&+\{\frac{-S_{B2}-\Gamma_2\cdot A_{CB2}-n^*_{B2}\cdot A_{CB2}}{n^*_{B2}\cdot(\sigma_{CB2}-\sigma_{B2C})\cdot c+\Gamma_2\cdot\sigma_{CB2}\cdot c}\}\end{aligned}$$

$$\cdot \{n_{B2}^{*} \cdot [\sigma_{B2C} - \sigma_{CB2}] \cdot c - \Gamma_2 \cdot \sigma_{CB2} \cdot c\} = 0$$

$$\begin{aligned}
&S_c - \Gamma_2 \cdot [A_{CB1} + A_{CB2}] - n_{B2}^{*} \cdot [A_{CB1} + A_{CB2}] \\
&\quad + \{n_{B2}^{*} \cdot (\sigma_{B1C} - \sigma_{CB1}) \cdot c + ([\Gamma_2 - \Gamma_1] \cdot \sigma_{B1C} - \Gamma_2 \cdot \sigma_{CB1}) \cdot c\} \\
&\quad \cdot \{\frac{-n_{B2}^{*} \cdot A_{CB1} - \Gamma_2 \cdot A_{CB1} - S_{B1}}{\{n_{B2}^{*} \cdot [\sigma_{CB1} - \sigma_{B1C}] \cdot c + (\Gamma_2 \cdot \sigma_{CB1} - [\Gamma_2 - \Gamma_1] \cdot \sigma_{B1C}) \cdot c\}}\} \\
&\quad + \{\frac{-S_{B2} - \Gamma_2 \cdot A_{CB2} - n_{B2}^{*} \cdot A_{CB2}}{n_{B2}^{*} \cdot (\sigma_{CB2} - \sigma_{B2C}) \cdot c + \Gamma_2 \cdot \sigma_{CB2} \cdot c}\} \cdot \{n_{B2}^{*} \cdot [\sigma_{B2C} - \sigma_{CB2}] \\
&\quad \cdot c - \Gamma_2 \cdot \sigma_{CB2} \cdot c\} = 0
\end{aligned}$$

We define some parameters for simplicity: $A_1 = S_c - \Gamma_2 \cdot [A_{CB1} + A_{CB2}]$; $A_2 = A_{CB1} + A_{CB2}$

$$\begin{aligned}
A_3 &= (\sigma_{B1C} - \sigma_{CB1}) \cdot c \\
A_4 &= ([\Gamma_2 - \Gamma_1] \cdot \sigma_{B1C} - \Gamma_2 \cdot \sigma_{CB1}) \cdot c \\
A_5 &= -\Gamma_2 \cdot A_{CB1} - S_{B1}
\end{aligned}$$

$$\begin{aligned}
&A_6 = -S_{B2} - \Gamma_2 \cdot A_{CB2};\ [\sigma_{CB1} - \sigma_{B1C}] \cdot c = -A_3 \\
&(\Gamma_2 \cdot \sigma_{CB1} - [\Gamma_2 - \Gamma_1] \cdot \sigma_{B1C}) \cdot c = -A_4
\end{aligned}$$

$$\begin{aligned}
A_7 &= (\sigma_{CB2} - \sigma_{B2C}) \cdot c;\ A_8 = \Gamma_2 \cdot \sigma_{CB2} \cdot c \\
A_9 &= [\sigma_{B2C} - \sigma_{CB2}] \cdot c;\ A_{10} = \Gamma_2 \cdot \sigma_{CB2} \cdot c
\end{aligned}$$

$$\begin{aligned}
&A_1 - n_{B2}^{*} \cdot A_2 + (n_{B2}^{*} \cdot A_3 + A_4) \cdot (\frac{-n_{B2}^{*} \cdot A_{CB1} + A_5}{-n_{B2}^{*} \cdot A_3 - A_4}) \\
&\quad + (\frac{A_6 - n_{B2}^{*} \cdot A_{CB2}}{n_{B2}^{*} \cdot A_7 + A_8}) \cdot (n_{B2}^{*} \cdot A_9 - A_{10}) = 0
\end{aligned}$$

$$\begin{aligned}
&A_1 - n_{B2}^{*} \cdot A_2 + \frac{(n_{B2}^{*} \cdot A_3 + A_4) \cdot (-n_{B2}^{*} \cdot A_{CB1} + A_5)}{-n_{B2}^{*} \cdot A_3 - A_4} \\
&\quad + \frac{(A_6 - n_{B2}^{*} \cdot A_{CB2}) \cdot (n_{B2}^{*} \cdot A_9 - A_{10})}{n_{B2}^{*} \cdot A_7 + A_8} = 0
\end{aligned}$$

$$\begin{aligned}
&(n_{B2}^{*} \cdot A_3 + A_4) \cdot (-n_{B2}^{*} \cdot A_{CB1} + A_5) \\
&= -(n_{B2}^{*})^2 \cdot A_3 \cdot A_{CB1} + n_{B2}^{*} \cdot (A_3 \cdot A_5 - A_{CB1} \cdot A_4) + A_4 \cdot A_5
\end{aligned}$$

$$\begin{aligned}
&(A_6 - n_{B2}^{*} \cdot A_{CB2}) \cdot (n_{B2}^{*} \cdot A_9 - A_{10}) \\
&= n_{B2}^{*} \cdot (A_9 \cdot A_6 + A_{CB2} \cdot A_{10}) - A_6 \cdot A_{10} - (n_{B2}^{*})^2 \cdot A_{CB2} \cdot A_9
\end{aligned}$$

Then we get

$$
\begin{aligned}
&A_1 - n^*_{B2} \cdot A_2 \\
&\quad + \frac{-(n^*_{B2})^2 \cdot A_3 \cdot A_{CB1} + n^*_{B2} \cdot (A_3 \cdot A_5 - A_{CB1} \cdot A_4) + A_4 \cdot A_5}{-n^*_{B2} \cdot A_3 - A_4} \\
&\quad + \frac{n^*_{B2} \cdot (A_9 \cdot A_6 + A_{CB2} \cdot A_{10}) - A_6 \cdot A_{10} - (n^*_{B2})^2 \cdot A_{CB2} \cdot A_9}{n^*_{B2} \cdot A_7 + A_8} = 0
\end{aligned}
$$

$$
\begin{aligned}
&\frac{\begin{array}{l}(A_1 - n^*_{B2} \cdot A_2) \cdot (-n^*_{B2} \cdot A_3 - A_4) \cdot (n^*_{B2} \cdot A_7 + A_8) \\ \quad + \{-(n^*_{B2})^2 \cdot A_3 \cdot A_{CB1} + n^*_{B2} \cdot (A_3 \cdot A_5 - A_{CB1} \cdot A_4) + A_4 \cdot A_5\} \cdot (n^*_{B2} \cdot A_7 + A_8) \\ \quad + \{n^*_{B2} \cdot (A_9 \cdot A_6 + A_{CB2} \cdot A_{10}) - A_6 \cdot A_{10} - (n^*_{B2})^2 \cdot A_{CB2} \cdot A_9\} \cdot (-n^*_{B2} \cdot A_3 - A_4)\end{array}}{(-n^*_{B2} \cdot A_3 - A_4) \cdot (n^*_{B2} \cdot A_7 + A_8)} \\
&= 0
\end{aligned}
$$

$$
-n^*_{B2} \cdot A_3 - A_4 \neq 0 \Rightarrow n^*_{B2} \neq -\frac{A_4}{A_3};\, n^*_{B2} \cdot A_7 + A_8 \neq 0 \Rightarrow n^*_{B2} = -\frac{A_8}{A_7}
$$

$$
\begin{aligned}
&(A_1 - n^*_{B2} \cdot A_2) \cdot (-n^*_{B2} \cdot A_3 - A_4) \cdot (n^*_{B2} \cdot A_7 + A_8) \\
&\quad + \{-(n^*_{B2})^2 \cdot A_3 \cdot A_{CB1} + n^*_{B2} \cdot (A_3 \cdot A_5 - A_{CB1} \cdot A_4) \\
&\quad + A_4 \cdot A_5\} \cdot (n^*_{B2} \cdot A_7 + A_8) + \{n^*_{B2} \cdot (A_9 \cdot A_6 + A_{CB2} \cdot A_{10}) \\
&\quad - A_6 \cdot A_{10} - (n^*_{B2})^2 \cdot A_{CB2} \cdot A_9\} \cdot (-n^*_{B2} \cdot A_3 - A_4) = 0
\end{aligned}
$$

Result 1:

$$
\begin{aligned}
&(A_1 - n^*_{B2} \cdot A_2) \cdot (-n^*_{B2} \cdot A_3 - A_4) \cdot (n^*_{B2} \cdot A_7 + A_8) \\
&= (A_1 - n^*_{B2} \cdot A_2) \cdot (-[n^*_{B2}]^2 \cdot A_3 \cdot A_7 - n^*_{B2} \\
&\quad \cdot [A_3 \cdot A_8 + A_4 \cdot A_7] - A_4 \cdot A_8) \\
&= -[n^*_{B2}]^2 \cdot A_3 \cdot A_7 \cdot A_1 - n^*_{B2} \cdot [A_3 \cdot A_8 + A_4 \cdot A_7] \cdot A_1 \\
&\quad - A_4 \cdot A_8 \cdot A_1 + [n^*_{B2}]^3 \cdot A_3 \cdot A_7 \cdot A_2 + [n^*_{B2}]^2 \\
&\quad \cdot [A_3 \cdot A_8 + A_4 \cdot A_7] \cdot A_2 + n^*_{B2} \cdot A_4 \cdot A_8 \cdot A_2
\end{aligned}
$$

$$
\begin{aligned}
&(A_1 - n^*_{B2} \cdot A_2) \cdot (-n^*_{B2} \cdot A_3 - A_4) \cdot (n^*_{B2} \cdot A_7 + A_8) \\
&= (A_1 - n^*_{B2} \cdot A_2) \cdot (-[n^*_{B2}]^2 \cdot A_3 \cdot A_7 - n^*_{B2} \\
&\quad \cdot [A_3 \cdot A_8 + A_4 \cdot A_7] - A_4 \cdot A_8) \\
&= [n^*_{B2}]^3 \cdot A_3 \cdot A_7 \cdot A_2 + [n^*_{B2}]^2 \cdot \{[A_3 \cdot A_8 + A_4 \cdot A_7] \\
&\quad \cdot A_2 - A_3 \cdot A_7 \cdot A_1\} + n^*_{B2} \cdot \{A_4 \cdot A_8 \cdot A_2 \\
&\quad - [A_3 \cdot A_8 + A_4 \cdot A_7] \cdot A_1\} - A_4 \cdot A_8 \cdot A_1
\end{aligned}
$$

We define some parameters for simplicity: $\Omega_1 = A_3 \cdot A_7 \cdot A_2$; $\Omega_2 = [A_3 \cdot A_8 + A_4 \cdot A_7] \cdot A_2 - A_3 \cdot A_7 \cdot A_1$

$$\Omega_3 = A_4 \cdot A_8 \cdot A_2 - [A_3 \cdot A_8 + A_4 \cdot A_7] \cdot A_1;\ \Omega_4 = -A_4 \cdot A_8 \cdot A_1$$

$$\begin{aligned}&(A_1 - n^*_{B2} \cdot A_2) \cdot (-n^*_{B2} \cdot A_3 - A_4) \cdot (n^*_{B2} \cdot A_7 + A_8)\\ &= [n^*_{B2}]^3 \cdot \Omega_1 + [n^*_{B2}]^2 \cdot \Omega_2 + n^*_{B2} \cdot \Omega_3 + \Omega_4\end{aligned}$$

Result 2:

$$\begin{aligned}&\{-(n^*_{B2})^2 \cdot A_3 \cdot A_{CB1} + n^*_{B2} \cdot (A_3 \cdot A_5 - A_{CB1} \cdot A_4) + A_4 \cdot A_5\} \cdot (n^*_{B2} \cdot A_7 + A_8)\\ &= -(n^*_{B2})^3 \cdot A_3 \cdot A_{CB1} \cdot A_7 + (n^*_{B2})^2 \cdot [(A_3 \cdot A_5 - A_{CB1} \cdot A_4)\\ &\quad \cdot A_7 - A_3 \cdot A_{CB1} \cdot A_8] + n^*_{B2} \cdot (A_3 \cdot A_5 - A_{CB1} \cdot A_4) \cdot A_8\\ &\quad + A_4 \cdot A_5 \cdot n^*_{B2} \cdot A_7 + A_4 \cdot A_5 \cdot A_8\\ &= -(n^*_{B2})^3 \cdot A_3 \cdot A_{CB1} \cdot A_7 + (n^*_{B2})^2 \cdot [(A_3 \cdot A_5 - A_{CB1} \cdot A_4) \cdot A_7\\ &\quad - A_3 \cdot A_{CB1} \cdot A_8] + n^*_{B2} \cdot \{(A_3 \cdot A_5 - A_{CB1} \cdot A_4) \cdot A_8 + A_4 \cdot A_5 \cdot A_7\}\\ &\quad + A_4 \cdot A_5 \cdot A_8\end{aligned}$$

We define some parameters for simplicity:

$$\begin{aligned}&\Omega_5 = -A_3 \cdot A_{CB1} \cdot A_7;\ \Omega_6 = (A_3 \cdot A_5 - A_{CB1} \cdot A_4) \cdot A_7 - A_3 \cdot A_{CB1} \cdot A_8\\ &\Omega_7 = (A_3 \cdot A_5 - A_{CB1} \cdot A_4) \cdot A_8 + A_4 \cdot A_5 \cdot A_7;\ \Omega_8 = A_4 \cdot A_5 \cdot A_8\end{aligned}$$

$$\begin{aligned}&\{-(n^*_{B2})^2 \cdot A_3 \cdot A_{CB1} + n^*_{B2} \cdot (A_3 \cdot A_5 - A_{CB1} \cdot A_4)\\ &\quad + A_4 \cdot A_5\} \cdot (n^*_{B2} \cdot A_7 + A_8)\\ &= -(n^*_{B2})^3 \cdot \Omega_5 + (n^*_{B2})^2 \cdot \Omega_6 + n^*_{B2} \cdot \Omega_7 + \Omega_8\end{aligned}$$

Result 3:

$$\begin{aligned}&\{n^*_{B2} \cdot (A_9 \cdot A_6 + A_{CB2} \cdot A_{10}) - A_6 \cdot A_{10} - (n^*_{B2})^2 \cdot A_{CB2} \cdot A_9\}\\ &\quad \cdot (-n^*_{B2} \cdot A_3 - A_4) = (n^*_{B2})^3 \cdot A_{CB2} \cdot A_9 \cdot A_3 + (n^*_{B2})^2\\ &\quad \cdot A_{CB2} \cdot A_9 \cdot A_4 - (n^*_{B2})^2 \cdot (A_9 \cdot A_6 + A_{CB2} \cdot A_{10}) \cdot A_3 + n^*_{B2} \cdot A_6\\ &\quad \cdot A_{10} \cdot A_3 - n^*_{B2} \cdot (A_9 \cdot A_6 + A_{CB2} \cdot A_{10}) \cdot A_4 + A_6 \cdot A_{10} \cdot A_4\end{aligned}$$

$$\begin{aligned}&\{n^*_{B2} \cdot (A_9 \cdot A_6 + A_{CB2} \cdot A_{10}) - A_6 \cdot A_{10} - (n^*_{B2})^2 \cdot A_{CB2} \cdot A_9\}\\ &\quad \cdot (-n^*_{B2} \cdot A_3 - A_4) = (n^*_{B2})^3 \cdot A_{CB2} \cdot A_9 \cdot A_3 + (n^*_{B2})^2\\ &\quad \cdot \{A_{CB2} \cdot A_9 \cdot A_4 - (A_9 \cdot A_6 + A_{CB2} \cdot A_{10}) \cdot A_3\} + n^*_{B2}\\ &\quad \cdot \{A_6 \cdot A_{10} \cdot A_3 - (A_9 \cdot A_6 + A_{CB2} \cdot A_{10}) \cdot A_4\} + A_6 \cdot A_{10} \cdot A_4\end{aligned}$$

Table 6.2 System fixed points, results 1–3 and expressions

Result No.	Expression
1	$(A_1 - n^*_{B2} \cdot A_2) \cdot (-n^*_{B2} \cdot A_3 - A_4) \cdot (n^*_{B2} \cdot A_7 + A_8) = [n^*_{B2}]^3 \cdot \Omega_1 + [n^*_{B2}]^2 \cdot \Omega_2 + n^*_{B2} \cdot \Omega_3 + \Omega_4$
2	$\{-(n^*_{B2})^2 \cdot A_3 \cdot A_{CB1} + n^*_{B2} \cdot (A_3 \cdot A_5 - A_{CB1} \cdot A_4) + A_4 \cdot A_5\} \cdot (n^*_{B2} \cdot A_7 + A_8) = -(n^*_{B2})^3 \cdot \Omega_5 + (n^*_{B2})^2 \cdot \Omega_6 + n^*_{B2} \cdot \Omega_7 + \Omega_8$
3	$\{n^*_{B2} \cdot (A_9 \cdot A_6 + A_{CB2} \cdot A_{10}) - A_6 \cdot A_{10} - (n^*_{B2})^2 \cdot A_{CB2} \cdot A_9\} \cdot (-n^*_{B2} \cdot A_3 - A_4) = (n^*_{B2})^3 \cdot \Omega_9 + (n^*_{B2})^2 \cdot \Omega_{10} + n^*_{B2} \cdot \Omega_{11} + \Omega_{12}$

We define some parameters for simplicity:

$$\Omega_9 = A_{CB2} \cdot A_9 \cdot A_3; \Omega_{10} = A_{CB2} \cdot A_9 \cdot A_4 - (A_9 \cdot A_6 + A_{CB2} \cdot A_{10}) \cdot A_3$$
$$\Omega_{11} = A_6 \cdot A_{10} \cdot A_3 - (A_9 \cdot A_6 + A_{CB2} \cdot A_{10}) \cdot A_4; \Omega_{12} = A_6 \cdot A_{10} \cdot A_4$$

$$\begin{aligned}&\{n^*_{B2} \cdot (A_9 \cdot A_6 + A_{CB2} \cdot A_{10}) - A_6 \cdot A_{10} - (n^*_{B2})^2 \\ &\quad \cdot A_{CB2} \cdot A_9\} \cdot (-n^*_{B2} \cdot A_3 - A_4) \\ &= (n^*_{B2})^3 \cdot \Omega_9 + (n^*_{B2})^2 \cdot \Omega_{10} + n^*_{B2} \cdot \Omega_{11} + \Omega_{12}\end{aligned}$$

We can summary our three results in the next table (Table 6.2).

$$\begin{aligned}&(A_1 - n^*_{B2} \cdot A_2) \cdot (-n^*_{B2} \cdot A_3 - A_4) \cdot (n^*_{B2} \cdot A_7 + A_8) \\ &\quad + \{-(n^*_{B2})^2 \cdot A_3 \cdot A_{CB1} + n^*_{B2} \cdot (A_3 \cdot A_5 - A_{CB1} \cdot A_4) + A_4 \cdot A_5\} \\ &\quad \cdot (n^*_{B2} \cdot A_7 + A_8) + \{n^*_{B2} \cdot (A_9 \cdot A_6 + A_{CB2} \cdot A_{10}) \\ &\quad - A_6 \cdot A_{10} - (n^*_{B2})^2 \cdot A_{CB2} \cdot A_9\} \cdot (-n^*_{B2} \cdot A_3 - A_4) = 0\end{aligned}$$

$$\begin{aligned}&[n^*_{B2}]^3 \cdot \Omega_1 + [n^*_{B2}]^2 \cdot \Omega_2 + n^*_{B2} \cdot \Omega_3 + \Omega_4 - (n^*_{B2})^3 \cdot \Omega_5 \\ &\quad + (n^*_{B2})^2 \cdot \Omega_6 + n^*_{B2} \cdot \Omega_7 + \Omega_8 + (n^*_{B2})^3 \\ &\quad \cdot \Omega_9 + (n^*_{B2})^2 \cdot \Omega_{10} + n^*_{B2} \cdot \Omega_{11} + \Omega_{12} = 0\end{aligned}$$

$$\begin{aligned}&[n^*_{B2}]^3 \cdot [\Omega_1 - \Omega_5 + \Omega_9] + [n^*_{B2}]^2 \cdot [\Omega_2 + \Omega_6 + \Omega_{10}] \\ &\quad + n^*_{B2} \cdot [\Omega_3 + \Omega_4 + \Omega_7 + \Omega_8 + \Omega_{11}] + \Omega_{12} = 0\end{aligned}$$

For specific parameters values, we can solve the above equation numerically and find

The possible values for fixed point coordinate n^*_{B2} (the population of $B^3\Pi_g$ in the first excited vibrational states for $t \to \infty$). Then we get the fixed point coordinates: n^*_{ph1}, n^*_{ph2}.

Stability analysis: The standard local stability analysis about any one of the equilibrium point of the plasma kinetic system for the populations of the lasing levels and number of generated photons consists in adding to coordinate $[n_c, n_{B1}, n_{B2}, n_{ph1}, n_{ph2}]$ arbitrarily small increments of exponential form $[n_c, n_{B1}, n_{B2}, n_{ph1}, n_{ph2}] \cdot e^{\lambda \cdot t}$ and retaining the first order terms in $n_c, n_{B1}, n_{B2}, n_{ph1}, n_{ph2}$. The system of five homogeneous equations leads to a polynomial characteristic equation in the eigenvalues. The polynomial characteristic equations accept by set the below variables and variables derivative with respect to time into plasma kinetic system equations. The plasma kinetic system fixed values with arbitrarily small increments of exponential form $[n_c, n_{B1}, n_{B2}, n_{ph1}, n_{ph2}] \cdot e^{\lambda \cdot t}$ are: $j = 0$(first fixed point), $j = 1$(second fixed point), $j = 2$(third fixed point), etc. [4, 5].

$$n_c(t) = n_c^{(j)} + n_c \cdot e^{\lambda \cdot t};\ n_{B1}(t) = n_{B1}^{(j)} + n_{B1} \cdot e^{\lambda \cdot t}$$
$$n_{B2}(t) = n_{B2}^{(j)} + n_{B2} \cdot e^{\lambda \cdot t};\ n_{ph1}(t) = n_{ph1}^{(j)} + n_{ph1} \cdot e^{\lambda \cdot t}$$

$$n_{ph2}(t) = n_{ph2}^{(j)} + n_{ph2} \cdot e^{\lambda \cdot t};\ \frac{dn_c(t)}{dt} = n_c \cdot \lambda \cdot e^{\lambda \cdot t}$$
$$\frac{dn_{B1}(t)}{dt} = n_{B1} \cdot \lambda \cdot e^{\lambda \cdot t};\ \frac{dn_{B2}(t)}{dt} = n_{B2} \cdot \lambda \cdot e^{\lambda \cdot t}$$

$$\frac{dn_{ph1}(t)}{dt} = n_{ph1} \cdot \lambda \cdot e^{\lambda \cdot t}$$
$$\frac{dn_{ph2}(t)}{dt} = n_{ph2} \cdot \lambda \cdot e^{\lambda \cdot t}$$
$$n_c(t - \tau_D) = n_c^{(j)} + n_c \cdot e^{\lambda \cdot (t - \tau_D)}$$

We choose these expressions for ourselves $n_c(t), n_{B1}(t), n_{B2}(t), n_{ph1}(t), n_{ph2}(t)$ as a small displacement $[n_c, n_{B1}, n_{B2}, n_{ph1}, n_{ph2}]$ from the plasma kinetic system fixed points in time $t = 0$.

$$n_c(t = 0) = n_c^{(j)} + n_c;\ n_{B1}(t = 0) = n_{B1}^{(j)} + n_{B1}$$
$$n_{B2}(t = 0) = n_{B2}^{(j)} + n_{B2};\ n_{ph1}(t = 0) = n_{ph1}^{(j)} + n_{ph1}$$
$$n_{ph2}(t = 0) = n_{ph2}^{(j)} + n_{ph2}$$

$$\frac{dn_c}{dt} = S_c - n_c(t - \tau_D) \cdot (A_{CB1} + A_{CB2} + \sigma_{CB1} \cdot c \cdot n_{ph1} + \sigma_{CB2} \cdot c \cdot n_{ph2}) + n_{B1} \cdot \sigma_{B1C} \cdot c \cdot n_{ph1} + n_{B2} \cdot \sigma_{B2C} \cdot c \cdot n_{ph2}$$

$$n_c \cdot \lambda \cdot e^{\lambda \cdot t} = S_c - [n_c^{(j)} + n_c \cdot e^{\lambda \cdot (t - \tau_D)}] \\ \cdot (A_{CB1} + A_{CB2} + \sigma_{CB1} \cdot c \cdot [n_{ph1}^{(j)} + n_{ph1} \cdot e^{\lambda \cdot t}] \\ + \sigma_{CB2} \cdot c \cdot [n_{ph2}^{(j)} + n_{ph2} \cdot e^{\lambda \cdot t}]) + [n_{B1}^{(j)} + n_{B1} \cdot e^{\lambda \cdot t}] \\ \cdot \sigma_{B1C} \cdot c \cdot [n_{ph1}^{(j)} + n_{ph1} \cdot e^{\lambda \cdot t}] + [n_{B2}^{(j)} + n_{B2} \cdot e^{\lambda \cdot t}] \\ \cdot \sigma_{B2C} \cdot c \cdot [n_{ph2}^{(j)} + n_{ph2} \cdot e^{\lambda \cdot t}]$$

$$n_c \cdot \lambda \cdot e^{\lambda \cdot t} = S_c - [n_c^{(j)} + n_c \cdot e^{\lambda \cdot (t - \tau_D)}] \\ \cdot (A_{CB1} + A_{CB2} + \sigma_{CB1} \cdot c \cdot n_{ph1}^{(j)} + \sigma_{CB2} \cdot c \cdot n_{ph2}^{(j)} \\ + \sigma_{CB1} \cdot c \cdot n_{ph1} \cdot e^{\lambda \cdot t} + \sigma_{CB2} \cdot c \cdot n_{ph2} \cdot e^{\lambda \cdot t}) \\ + [n_{B1}^{(j)} + n_{B1} \cdot e^{\lambda \cdot t}] \cdot [n_{ph1}^{(j)} + n_{ph1} \cdot e^{\lambda \cdot t}] \cdot \sigma_{B1C} \cdot c \\ + [n_{B2}^{(j)} + n_{B2} \cdot e^{\lambda \cdot t}] \cdot [n_{ph2}^{(j)} + n_{ph2} \cdot e^{\lambda \cdot t}] \cdot \sigma_{B2C} \cdot c$$

$$n_c \cdot \lambda \cdot e^{\lambda \cdot t} = S_c - (n_c^{(j)} \cdot [A_{CB1} + A_{CB2}] + n_c^{(j)} \cdot n_{ph1}^{(j)} \\ \cdot \sigma_{CB1} \cdot c + n_c^{(j)} \cdot n_{ph2}^{(j)} \cdot \sigma_{CB2} \cdot c + n_c^{(j)} \\ \cdot \sigma_{CB1} \cdot c \cdot e^{\lambda \cdot t} \cdot n_{ph1} + n_c^{(j)} \cdot \sigma_{CB2} \cdot c \cdot e^{\lambda \cdot t} \cdot n_{ph2}) \\ - ([A_{CB1} + A_{CB2}] \cdot e^{\lambda \cdot (t - \tau_D)} \cdot n_c + \sigma_{CB1} \cdot c \cdot n_{ph1}^{(j)} \\ \cdot e^{\lambda \cdot (t - \tau_D)} \cdot n_c + \sigma_{CB2} \cdot c \cdot n_{ph2}^{(j)} \cdot e^{\lambda \cdot (t - \tau_D)} \cdot n_c \\ + \sigma_{CB1} \cdot c \cdot n_{ph1} \cdot n_c \cdot e^{\lambda \cdot t} \cdot e^{\lambda \cdot (t - \tau_D)} \\ + \sigma_{CB2} \cdot c \cdot n_{ph2} \cdot n_c \cdot e^{\lambda \cdot t} \cdot e^{\lambda \cdot (t - \tau_D)}) \\ + [n_{B1}^{(j)} \cdot n_{ph1}^{(j)} + n_{B1}^{(j)} \cdot n_{ph1} \cdot e^{\lambda \cdot t} + n_{ph1}^{(j)} \cdot n_{B1} \cdot e^{\lambda \cdot t} \\ + n_{B1} \cdot n_{ph1} \cdot e^{\lambda \cdot t} \cdot e^{\lambda \cdot t}] \cdot \sigma_{B1C} \cdot c + [n_{B2}^{(j)} \cdot n_{ph2}^{(j)} \\ + n_{B2}^{(j)} \cdot n_{ph2} \cdot e^{\lambda \cdot t} + n_{ph2}^{(j)} \cdot n_{B2} \cdot e^{\lambda \cdot t} \\ + n_{B2} \cdot n_{ph2} \cdot e^{\lambda \cdot t} \cdot e^{\lambda \cdot t}] \cdot \sigma_{B2C} \cdot c$$

We consider $n_{ph1} \cdot n_c \approx 0$; $n_{ph2} \cdot n_c \approx 0$; $n_{B1} \cdot n_{ph1} \approx 0$; $n_{B2} \cdot n_{ph2} \approx 0$ then

$$n_c \cdot \lambda \cdot e^{\lambda \cdot t} = S_c - (n_c^{(j)} \cdot [A_{CB1} + A_{CB2}] + n_c^{(j)} \cdot n_{ph1}^{(j)} \cdot \sigma_{CB1} \cdot c \\ + n_c^{(j)} \cdot n_{ph2}^{(j)} \cdot \sigma_{CB2} \cdot c + n_c^{(j)} \cdot \sigma_{CB1} \cdot c \cdot e^{\lambda \cdot t} \cdot n_{ph1} \\ + n_c^{(j)} \cdot \sigma_{CB2} \cdot c \cdot e^{\lambda \cdot t} \cdot n_{ph2}) - ([A_{CB1} + A_{CB2}] \cdot e^{\lambda \cdot (t - \tau_D)} \cdot n_c \\ + \sigma_{CB1} \cdot c \cdot n_{ph1}^{(j)} \cdot e^{\lambda \cdot (t - \tau_D)} \cdot n_c + \sigma_{CB2} \cdot c \cdot n_{ph2}^{(j)} \cdot e^{\lambda \cdot (t - \tau_D)} \cdot n_c) \\ + [n_{B1}^{(j)} \cdot n_{ph1}^{(j)} + n_{B1}^{(j)} \cdot n_{ph1} \cdot e^{\lambda \cdot t} + n_{ph1}^{(j)} \cdot n_{B1} \cdot e^{\lambda \cdot t}] \cdot \sigma_{B1C} \cdot c \\ + [n_{B2}^{(j)} \cdot n_{ph2}^{(j)} + n_{B2}^{(j)} \cdot n_{ph2} \cdot e^{\lambda \cdot t} + n_{ph2}^{(j)} \cdot n_{B2} \cdot e^{\lambda \cdot t}] \cdot \sigma_{B2C} \cdot c$$

$$\begin{aligned} n_c \cdot \lambda \cdot e^{\lambda \cdot t} &= S_c - n_c^{(j)} \cdot [A_{CB1} + A_{CB2}] - n_c^{(j)} \cdot n_{ph1}^{(j)} \cdot \sigma_{CB1} \cdot c \\ &\quad - n_c^{(j)} \cdot n_{ph2}^{(j)} \cdot \sigma_{CB2} \cdot c - n_c^{(j)} \cdot \sigma_{CB1} \cdot c \cdot e^{\lambda \cdot t} \cdot n_{ph1} \\ &\quad - n_c^{(j)} \cdot \sigma_{CB2} \cdot c \cdot e^{\lambda \cdot t} \cdot n_{ph2} - [A_{CB1} + A_{CB2}] \cdot e^{\lambda \cdot (t-\tau_D)} \cdot n_c \\ &\quad - \sigma_{CB1} \cdot c \cdot n_{ph1}^{(j)} \cdot e^{\lambda \cdot (t-\tau_D)} \cdot n_c - \sigma_{CB2} \cdot c \cdot n_{ph2}^{(j)} \cdot e^{\lambda \cdot (t-\tau_D)} \cdot n_c \\ &\quad + [n_{B1}^{(j)} \cdot n_{ph1}^{(j)}] \cdot \sigma_{B1C} \cdot c + [n_{B1}^{(j)} \cdot n_{ph1} \cdot e^{\lambda \cdot t} + n_{ph1}^{(j)} \cdot n_{B1} \cdot e^{\lambda \cdot t}] \\ &\quad \cdot \sigma_{B1C} \cdot c + [n_{B2}^{(j)} \cdot n_{ph2}^{(j)}] \cdot \sigma_{B2C} \cdot c \\ &\quad + [n_{B2}^{(j)} \cdot n_{ph2} \cdot e^{\lambda \cdot t} + n_{ph2}^{(j)} \cdot n_{B2} \cdot e^{\lambda \cdot t}] \cdot \sigma_{B2C} \cdot c \end{aligned}$$

$$\begin{aligned} n_c \cdot \lambda \cdot e^{\lambda \cdot t} &= S_c - n_c^{(j)} \cdot [A_{CB1} + A_{CB2}] - n_c^{(j)} \cdot n_{ph1}^{(j)} \cdot \sigma_{CB1} \cdot c \\ &\quad - n_c^{(j)} \cdot n_{ph2}^{(j)} \cdot \sigma_{CB2} \cdot c + [n_{B1}^{(j)} \cdot n_{ph1}^{(j)}] \cdot \sigma_{B1C} \cdot c + [n_{B2}^{(j)} \cdot n_{ph2}^{(j)}] \\ &\quad \cdot \sigma_{B2C} \cdot c - n_c^{(j)} \cdot \sigma_{CB1} \cdot c \cdot e^{\lambda \cdot t} \cdot n_{ph1} - n_c^{(j)} \cdot \sigma_{CB2} \cdot c \cdot e^{\lambda \cdot t} \cdot n_{ph2} \\ &\quad - [A_{CB1} + A_{CB2}] \cdot e^{\lambda \cdot t} \cdot e^{-\lambda \cdot \tau_D} \cdot n_c - \sigma_{CB1} \cdot c \cdot n_{ph1}^{(j)} \cdot e^{\lambda \cdot t} \cdot e^{-\lambda \cdot \tau_D} \\ &\quad \cdot n_c - \sigma_{CB2} \cdot c \cdot n_{ph2}^{(j)} \cdot e^{\lambda \cdot t} \cdot e^{-\lambda \cdot \tau_D} \cdot n_c + [n_{B1}^{(j)} \cdot n_{ph1} \cdot e^{\lambda \cdot t} \\ &\quad + n_{ph1}^{(j)} \cdot n_{B1} \cdot e^{\lambda \cdot t}] \cdot \sigma_{B1C} \cdot c + [n_{B2}^{(j)} \cdot n_{ph2} \cdot e^{\lambda \cdot t} \\ &\quad + n_{ph2}^{(j)} \cdot n_{B2} \cdot e^{\lambda \cdot t}] \cdot \sigma_{B2C} \cdot c \end{aligned}$$

$$\begin{aligned} n_c \cdot \lambda \cdot e^{\lambda \cdot t} &= S_c - n_c^{(j)} \cdot [A_{CB1} + A_{CB2} + n_{ph1}^{(j)} \cdot \sigma_{CB1} \cdot c \\ &\quad + n_{ph2}^{(j)} \cdot \sigma_{CB2} \cdot c] + n_{B1}^{(j)} \cdot n_{ph1}^{(j)} \cdot \sigma_{B1C} \cdot c + n_{B2}^{(j)} \cdot n_{ph2}^{(j)} \\ &\quad \cdot \sigma_{B2C} \cdot c - n_c^{(j)} \cdot \sigma_{CB1} \cdot c \cdot e^{\lambda \cdot t} \cdot n_{ph1} - n_c^{(j)} \cdot \sigma_{CB2} \cdot c \cdot e^{\lambda \cdot t} \\ &\quad \cdot n_{ph2} - [A_{CB1} + A_{CB2}] \cdot e^{\lambda \cdot t} \cdot e^{-\lambda \cdot \tau_D} \cdot n_c - \sigma_{CB1} \cdot c \cdot n_{ph1}^{(j)} \\ &\quad \cdot e^{\lambda \cdot t} \cdot e^{-\lambda \cdot \tau_D} \cdot n_c - \sigma_{CB2} \cdot c \cdot n_{ph2}^{(j)} \cdot e^{\lambda \cdot t} \cdot e^{-\lambda \cdot \tau_D} \cdot n_c \\ &\quad + [n_{B1}^{(j)} \cdot n_{ph1} \cdot e^{\lambda \cdot t} + n_{ph1}^{(j)} \cdot n_{B1} \cdot e^{\lambda \cdot t}] \cdot \sigma_{B1C} \cdot c \\ &\quad + [n_{B2}^{(j)} \cdot n_{ph2} \cdot e^{\lambda \cdot t} + n_{ph2}^{(j)} \cdot n_{B2} \cdot e^{\lambda \cdot t}] \cdot \sigma_{B2C} \cdot c \end{aligned}$$

At fixed point: $\frac{dn_c}{dt} = 0; \lim_{t \to 0} n_c(t - \tau_D) = n_c(t) \, \forall \, t \gg \tau_D$

$$\begin{aligned} &S_c - n_c^{(j)} \cdot [A_{CB1} + A_{CB2} + n_{ph1}^{(j)} \cdot \sigma_{CB1} \cdot c + n_{ph2}^{(j)} \cdot \sigma_{CB2} \cdot c] \\ &\quad + n_{B1}^{(j)} \cdot n_{ph1}^{(j)} \cdot \sigma_{B1C} \cdot c + n_{B2}^{(j)} \cdot n_{ph2}^{(j)} \cdot \sigma_{B2C} \cdot c = 0 \end{aligned}$$

$$\begin{aligned} n_c \cdot \lambda \cdot e^{\lambda \cdot t} &= -n_c^{(j)} \cdot \sigma_{CB1} \cdot c \cdot e^{\lambda \cdot t} \cdot n_{ph1} - n_c^{(j)} \cdot \sigma_{CB2} \cdot c \cdot e^{\lambda \cdot t} \cdot n_{ph2} \\ &\quad - [A_{CB1} + A_{CB2}] \cdot e^{\lambda \cdot t} \cdot e^{-\lambda \cdot \tau_D} \cdot n_c - \sigma_{CB1} \cdot c \cdot n_{ph1}^{(j)} \end{aligned}$$

$$\cdot e^{\lambda \cdot t} \cdot e^{-\lambda \cdot \tau_D} \cdot n_c - \sigma_{CB2} \cdot c \cdot n_{ph2}^{(j)} \cdot e^{\lambda \cdot t} \cdot e^{-\lambda \cdot \tau_D} \cdot n_c$$
$$+ [n_{B1}^{(j)} \cdot n_{ph1} \cdot e^{\lambda \cdot t} + n_{ph1}^{(j)} \cdot n_{B1} \cdot e^{\lambda \cdot t}] \cdot \sigma_{B1C} \cdot c$$
$$+ [n_{B2}^{(j)} \cdot n_{ph2} \cdot e^{\lambda \cdot t} + n_{ph2}^{(j)} \cdot n_{B2} \cdot e^{\lambda \cdot t}] \cdot \sigma_{B2C} \cdot c$$

Divide the two sides of the above equation by $e^{\lambda \cdot t}$ gives

$$n_c \cdot \lambda = -n_c^{(j)} \cdot \sigma_{CB1} \cdot c \cdot n_{ph1} - n_c^{(j)} \cdot \sigma_{CB2} \cdot c \cdot n_{ph2}$$
$$- [A_{CB1} + A_{CB2}] \cdot e^{-\lambda \cdot \tau_D} \cdot n_c - \sigma_{CB1} \cdot c \cdot n_{ph1}^{(j)} \cdot e^{-\lambda \cdot \tau_D} \cdot n_c$$
$$- \sigma_{CB2} \cdot c \cdot n_{ph2}^{(j)} \cdot e^{-\lambda \cdot \tau_D} \cdot n_c + [n_{B1}^{(j)} \cdot n_{ph1} + n_{ph1}^{(j)} \cdot n_{B1}]$$
$$\cdot \sigma_{B1C} \cdot c + [n_{B2}^{(j)} \cdot n_{ph2} + n_{ph2}^{(j)} \cdot n_{B2}] \cdot \sigma_{B2C} \cdot c$$

$$- [A_{CB1} + A_{CB2}] \cdot e^{-\lambda \cdot \tau_D} \cdot n_c - \sigma_{CB1} \cdot c \cdot n_{ph1}^{(j)} \cdot e^{-\lambda \cdot \tau_D} \cdot n_c$$
$$- \sigma_{CB2} \cdot c \cdot n_{ph2}^{(j)} \cdot e^{-\lambda \cdot \tau_D} \cdot n_c - n_c \cdot \lambda + n_{ph1}^{(j)} \cdot \sigma_{B1C} \cdot c \cdot n_{B1}$$
$$+ n_{ph2}^{(j)} \cdot \sigma_{B2C} \cdot c \cdot n_{B2} - n_c^{(j)} \cdot \sigma_{CB1} \cdot c \cdot n_{ph1} + n_{B1}^{(j)} \cdot \sigma_{B1C} \cdot c \cdot n_{ph1}$$
$$- n_c^{(j)} \cdot \sigma_{CB2} \cdot c \cdot n_{ph2} + n_{B2}^{(j)} \cdot \sigma_{B2C} \cdot c \cdot n_{ph2} = 0$$

$$\{(-[A_{CB1} + A_{CB2}] - \sigma_{CB1} \cdot c \cdot n_{ph1}^{(j)} - \sigma_{CB2} \cdot c \cdot n_{ph2}^{(j)})$$
$$\cdot e^{-\lambda \cdot \tau_D} - \lambda\} \cdot n_c + n_{ph1}^{(j)} \cdot \sigma_{B1C} \cdot c \cdot n_{B1} + n_{ph2}^{(j)} \cdot \sigma_{B2C}$$
$$\cdot c \cdot n_{B2} + [n_{B1}^{(j)} \cdot \sigma_{B1C} - n_c^{(j)} \cdot \sigma_{CB1}] \cdot c \cdot n_{ph1}$$
$$+ [n_{B2}^{(j)} \cdot \sigma_{B2C} - n_c^{(j)} \cdot \sigma_{CB2}] \cdot c \cdot n_{ph2} = 0$$

We define for simplicity some parameters: $\Xi_1 = -[A_{CB1} + A_{CB2}] - \sigma_{CB1} \cdot c \cdot n_{ph1}^{(j)} - \sigma_{CB2} \cdot c \cdot n_{ph2}^{(j)}$

$$\Xi_2 = n_{ph1}^{(j)} \cdot \sigma_{B1C} \cdot c; \ \Xi_3 = n_{ph2}^{(j)} \cdot \sigma_{B2C} \cdot c$$
$$\Xi_4 = [n_{B1}^{(j)} \cdot \sigma_{B1C} - n_c^{(j)} \cdot \sigma_{CB1}] \cdot c; \ \Xi_5 = [n_{B2}^{(j)} \cdot \sigma_{B2C} - n_c^{(j)} \cdot \sigma_{CB2}] \cdot c$$

The first arbitrarily small increments equation is $(\Xi_1 \cdot e^{-\lambda \cdot \tau_D} - \lambda) \cdot n_c + \Xi_2 \cdot n_{B1} + \Xi_3 \cdot n_{B2} + \Xi_4 \cdot n_{ph1} + \Xi_5 \cdot n_{ph2} = 0$

$$\frac{dn_{B1}}{dt} = S_{B1} + n_c(t - \tau_D) \cdot (A_{CB1} + \sigma_{CB1} \cdot c \cdot n_{ph1})$$
$$- n_{B1} \cdot \sigma_{B1C} \cdot c \cdot n_{ph1}$$

$$n_{B1} \cdot \lambda \cdot e^{\lambda \cdot t} = S_{B1} + [n_c^{(j)} + n_c \cdot e^{\lambda \cdot (t - \tau_D)}]$$
$$\cdot (A_{CB1} + \sigma_{CB1} \cdot c \cdot [n_{ph1}^{(j)} + n_{ph1} \cdot e^{\lambda \cdot t}])$$

$$- (n_{B1}^{(j)} + n_{B1} \cdot e^{\lambda \cdot t}) \cdot \sigma_{B1C} \cdot c \cdot [n_{ph1}^{(j)} + n_{ph1} \cdot e^{\lambda \cdot t}]$$

$$\begin{aligned} n_{B1} \cdot \lambda \cdot e^{\lambda \cdot t} = & S_{B1} + [n_c^{(j)} + n_c \cdot e^{\lambda \cdot (t - \tau_D)}] \cdot ([A_{CB1} + \sigma_{CB1} \cdot c \cdot n_{ph1}^{(j)}] \\ & + \sigma_{CB1} \cdot c \cdot n_{ph1} \cdot e^{\lambda \cdot t}) - [(n_{B1}^{(j)} + n_{B1} \cdot e^{\lambda \cdot t}) \cdot \sigma_{B1C} \cdot c \cdot n_{ph1}^{(j)} \\ & + (n_{B1}^{(j)} + n_{B1} \cdot e^{\lambda \cdot t}) \cdot \sigma_{B1C} \cdot c \cdot n_{ph1} \cdot e^{\lambda \cdot t}] \end{aligned}$$

$$\begin{aligned} n_{B1} \cdot \lambda \cdot e^{\lambda \cdot t} = & S_{B1} + [n_c^{(j)} + n_c \cdot e^{\lambda \cdot (t - \tau_D)}] \\ & \cdot [A_{CB1} + \sigma_{CB1} \cdot c \cdot n_{ph1}^{(j)}] + [n_c^{(j)} + n_c \cdot e^{\lambda \cdot (t - \tau_D)}] \\ & \cdot \sigma_{CB1} \cdot c \cdot n_{ph1} \cdot e^{\lambda \cdot t} - (n_{B1}^{(j)} + n_{B1} \cdot e^{\lambda \cdot t}) \cdot \sigma_{B1C} \\ & \cdot c \cdot n_{ph1}^{(j)} - (n_{B1}^{(j)} + n_{B1} \cdot e^{\lambda \cdot t}) \cdot \sigma_{B1C} \cdot c \cdot n_{ph1} \cdot e^{\lambda \cdot t} \end{aligned}$$

$$\begin{aligned} n_{B1} \cdot \lambda \cdot e^{\lambda \cdot t} = & S_{B1} + n_c^{(j)} \cdot A_{CB1} + n_c^{(j)} \cdot \sigma_{CB1} \cdot c \cdot n_{ph1}^{(j)} \\ & + A_{CB1} \cdot n_c \cdot e^{\lambda \cdot (t - \tau_D)} + \sigma_{CB1} \cdot c \cdot n_{ph1}^{(j)} \cdot n_c \cdot e^{\lambda \cdot (t - \tau_D)} \\ & + n_c^{(j)} \cdot \sigma_{CB1} \cdot c \cdot n_{ph1} \cdot e^{\lambda \cdot t} + n_c \cdot n_{ph1} \cdot e^{\lambda \cdot (t - \tau_D)} \cdot \sigma_{CB1} \cdot c \\ & \cdot e^{\lambda \cdot t} - n_{B1}^{(j)} \cdot \sigma_{B1C} \cdot c \cdot n_{ph1}^{(j)} - n_{B1} \cdot \sigma_{B1C} \cdot c \cdot n_{ph1}^{(j)} \cdot e^{\lambda \cdot t} \\ & - n_{B1}^{(j)} \cdot \sigma_{B1C} \cdot c \cdot n_{ph1} \cdot e^{\lambda \cdot t} - n_{B1} \cdot n_{ph1} \cdot e^{\lambda \cdot t} \cdot \sigma_{B1C} \cdot c \cdot e^{\lambda \cdot t} \end{aligned}$$

We consider $n_c \cdot n_{ph1} \approx 0$; $n_{B1} \cdot n_{ph1} \approx 0$ then

$$\begin{aligned} n_{B1} \cdot \lambda \cdot e^{\lambda \cdot t} = & S_{B1} + n_c^{(j)} \cdot A_{CB1} + n_c^{(j)} \cdot \sigma_{CB1} \cdot c \cdot n_{ph1}^{(j)} \\ & + A_{CB1} \cdot n_c \cdot e^{\lambda \cdot (t - \tau_D)} + \sigma_{CB1} \cdot c \cdot n_{ph1}^{(j)} \cdot n_c \cdot e^{\lambda \cdot (t - \tau_D)} \\ & + n_c^{(j)} \cdot \sigma_{CB1} \cdot c \cdot n_{ph1} \cdot e^{\lambda \cdot t} - n_{B1}^{(j)} \cdot \sigma_{B1C} \cdot c \cdot n_{ph1}^{(j)} \\ & - n_{B1} \cdot \sigma_{B1C} \cdot c \cdot n_{ph1}^{(j)} \cdot e^{\lambda \cdot t} - n_{B1}^{(j)} \cdot \sigma_{B1C} \cdot c \cdot n_{ph1} \cdot e^{\lambda \cdot t} \end{aligned}$$

$$\begin{aligned} n_{B1} \cdot \lambda \cdot e^{\lambda \cdot t} = & S_{B1} + n_c^{(j)} \cdot [A_{CB1} + \sigma_{CB1} \cdot c \cdot n_{ph1}^{(j)}] - n_{B1}^{(j)} \\ & \cdot \sigma_{B1C} \cdot c \cdot n_{ph1}^{(j)} + A_{CB1} \cdot n_c \cdot e^{\lambda \cdot (t - \tau_D)} + \sigma_{CB1} \cdot c \cdot n_{ph1}^{(j)} \\ & \cdot n_c \cdot e^{\lambda \cdot (t - \tau_D)} + n_c^{(j)} \cdot \sigma_{CB1} \cdot c \cdot n_{ph1} \cdot e^{\lambda \cdot t} \\ & - n_{B1} \cdot \sigma_{B1C} \cdot c \cdot n_{ph1}^{(j)} \cdot e^{\lambda \cdot t} - n_{B1}^{(j)} \cdot \sigma_{B1C} \cdot c \cdot n_{ph1} \cdot e^{\lambda \cdot t} \end{aligned}$$

At fixed point: $\frac{dn_{B1}}{dt} = 0; \lim_{t \to 0} n_c(t - \tau_D) = n_c(t) \, \forall t \gg \tau_D$

$$S_{B1} + n_c^{(j)} \cdot [A_{CB1} + \sigma_{CB1} \cdot c \cdot n_{ph1}^{(j)}] - n_{B1}^{(j)} \cdot \sigma_{B1C} \cdot c \cdot n_{ph1}^{(j)} = 0$$

$$n_{B1} \cdot \lambda \cdot e^{\lambda \cdot t} = A_{CB1} \cdot n_c \cdot e^{\lambda \cdot (t - \tau_D)} + \sigma_{CB1} \cdot c \cdot n_{ph1}^{(j)} \cdot n_c \cdot e^{\lambda \cdot (t - \tau_D)}$$

$$+ n_c^{(j)} \cdot \sigma_{CB1} \cdot c \cdot n_{ph1} \cdot e^{\lambda \cdot t} - n_{B1} \cdot \sigma_{B1C} \cdot c \cdot n_{ph1}^{(j)} \cdot e^{\lambda \cdot t}$$
$$- n_{B1}^{(j)} \cdot \sigma_{B1C} \cdot c \cdot n_{ph1} \cdot e^{\lambda \cdot t}$$

$$A_{CB1} \cdot n_c \cdot e^{\lambda \cdot t} \cdot e^{-\lambda \cdot \tau_D} + \sigma_{CB1} \cdot c \cdot n_{ph1}^{(j)} \cdot n_c \cdot e^{\lambda \cdot t} \cdot e^{-\lambda \cdot \tau_D}$$
$$+ n_c^{(j)} \cdot \sigma_{CB1} \cdot c \cdot n_{ph1} \cdot e^{\lambda \cdot t} - n_{B1} \cdot \sigma_{B1C} \cdot c \cdot n_{ph1}^{(j)} \cdot e^{\lambda \cdot t}$$
$$- n_{B1}^{(j)} \cdot \sigma_{B1C} \cdot c \cdot n_{ph1} \cdot e^{\lambda \cdot t} - n_{B1} \cdot \lambda \cdot e^{\lambda \cdot t} = 0$$

Divide the two sides of the above equation by $e^{\lambda \cdot t}$ gives

$$A_{CB1} \cdot n_c \cdot e^{-\lambda \cdot \tau_D} + \sigma_{CB1} \cdot c \cdot n_{ph1}^{(j)} \cdot n_c \cdot e^{-\lambda \cdot \tau_D}$$
$$+ n_c^{(j)} \cdot \sigma_{CB1} \cdot c \cdot n_{ph1} - n_{B1} \cdot \sigma_{B1C} \cdot c \cdot n_{ph1}^{(j)}$$
$$- n_{B1}^{(j)} \cdot \sigma_{B1C} \cdot c \cdot n_{ph1} - n_{B1} \cdot \lambda = 0$$

$$[A_{CB1} + \sigma_{CB1} \cdot c \cdot n_{ph1}^{(j)}] \cdot n_c \cdot e^{-\lambda \cdot \tau_D} - \sigma_{B1C} \cdot c \cdot n_{ph1}^{(j)} \cdot n_{B1}$$
$$- n_{B1} \cdot \lambda + [n_c^{(j)} \cdot \sigma_{CB1} - n_{B1}^{(j)} \cdot \sigma_{B1C}] \cdot c \cdot n_{ph1} = 0$$

We define for simplicity some parameters:

$$\Xi_6 = A_{CB1} + \sigma_{CB1} \cdot c \cdot n_{ph1}^{(j)}; \; \Xi_2 = n_{ph1}^{(j)} \cdot \sigma_{B1C} \cdot c$$
$$- \Xi_4 = [n_c^{(j)} \cdot \sigma_{CB1} - n_{B1}^{(j)} \cdot \sigma_{B1C}] \cdot c$$

The second arbitrarily small increments equation is $\Xi_6 \cdot n_c \cdot e^{-\lambda \cdot \tau_D} - \Xi_2 \cdot n_{B1} - n_{B1} \cdot \lambda - \Xi_4 \cdot n_{ph1} = 0$

$$\frac{dn_{B2}}{dt} = S_{B2} + n_c(t - \tau_D) \cdot (A_{CB2} + \sigma_{CB2} \cdot c \cdot n_{ph2})$$
$$- n_{B2} \cdot \sigma_{B2C} \cdot c \cdot n_{ph2}$$

$$n_{B2} \cdot \lambda \cdot e^{\lambda \cdot t} = S_{B2} + [n_c^{(j)} + n_c \cdot e^{\lambda \cdot (t - \tau_D)}] \cdot (A_{CB2} + \sigma_{CB2}$$
$$\cdot c \cdot [n_{ph2}^{(j)} + n_{ph2} \cdot e^{\lambda \cdot t}]) - [n_{B2}^{(j)} + n_{B2} \cdot e^{\lambda \cdot t}]$$
$$\cdot \sigma_{B2C} \cdot c \cdot [n_{ph2}^{(j)} + n_{ph2} \cdot e^{\lambda \cdot t}]$$

$$n_{B2} \cdot \lambda \cdot e^{\lambda \cdot t} = S_{B2} + [n_c^{(j)} + n_c \cdot e^{\lambda \cdot (t - \tau_D)}] \cdot ([A_{CB2} + \sigma_{CB2}$$
$$\cdot c \cdot n_{ph2}^{(j)}] + \sigma_{CB2} \cdot c \cdot n_{ph2} \cdot e^{\lambda \cdot t}) - [n_{B2}^{(j)} + n_{B2} \cdot e^{\lambda \cdot t}]$$
$$\cdot [n_{ph2}^{(j)} + n_{ph2} \cdot e^{\lambda \cdot t}] \cdot \sigma_{B2C} \cdot c$$

$$n_{B2} \cdot \lambda \cdot e^{\lambda \cdot t} = S_{B2} + n_c^{(j)} \cdot [A_{CB2} + \sigma_{CB2} \cdot c \cdot n_{ph2}^{(j)}] + n_c^{(j)} \cdot \sigma_{CB2} \cdot c \cdot n_{ph2} \cdot e^{\lambda \cdot t} + [A_{CB2} + \sigma_{CB2} \cdot c \cdot n_{ph2}^{(j)}] \cdot n_c \cdot e^{\lambda \cdot (t-\tau_D)} + e^{\lambda \cdot (t-\tau_D)} \cdot \sigma_{CB2} \cdot c \cdot n_c \cdot n_{ph2} \cdot e^{\lambda \cdot t} - [n_{B2}^{(j)} \cdot n_{ph2}^{(j)} + n_{B2}^{(j)} \cdot n_{ph2} \cdot e^{\lambda \cdot t} + n_{ph2}^{(j)} \cdot n_{B2} \cdot e^{\lambda \cdot t} + e^{\lambda \cdot t} \cdot n_{B2} \cdot n_{ph2} \cdot e^{\lambda \cdot t}] \cdot \sigma_{B2C} \cdot c$$

We consider $n_c \cdot n_{ph2} \approx 0$; $n_{B2} \cdot n_{ph2} \approx 0$ then

$$n_{B2} \cdot \lambda \cdot e^{\lambda \cdot t} = S_{B2} + n_c^{(j)} \cdot [A_{CB2} + \sigma_{CB2} \cdot c \cdot n_{ph2}^{(j)}] + n_c^{(j)} \cdot \sigma_{CB2} \cdot c \cdot n_{ph2} \cdot e^{\lambda \cdot t} + [A_{CB2} + \sigma_{CB2} \cdot c \cdot n_{ph2}^{(j)}] \cdot n_c \cdot e^{\lambda \cdot (t-\tau_D)} - [n_{B2}^{(j)} \cdot n_{ph2}^{(j)} + n_{B2}^{(j)} \cdot n_{ph2} \cdot e^{\lambda \cdot t} + n_{ph2}^{(j)} \cdot n_{B2} \cdot e^{\lambda \cdot t}] \cdot \sigma_{B2C} \cdot c$$

$$S_{B2} + n_c^{(j)} \cdot [A_{CB2} + \sigma_{CB2} \cdot c \cdot n_{ph2}^{(j)}] + n_c^{(j)} \cdot \sigma_{CB2} \cdot c \cdot n_{ph2} \cdot e^{\lambda \cdot t} + [A_{CB2} + \sigma_{CB2} \cdot c \cdot n_{ph2}^{(j)}] \cdot n_c \cdot e^{\lambda \cdot (t-\tau_D)} - [n_{B2}^{(j)} \cdot n_{ph2}^{(j)} + n_{B2}^{(j)} \cdot n_{ph2} \cdot e^{\lambda \cdot t} + n_{ph2}^{(j)} \cdot n_{B2} \cdot e^{\lambda \cdot t}] \cdot \sigma_{B2C} \cdot c - n_{B2} \cdot \lambda \cdot e^{\lambda \cdot t} = 0$$

$$S_{B2} + n_c^{(j)} \cdot [A_{CB2} + \sigma_{CB2} \cdot c \cdot n_{ph2}^{(j)}] + n_c^{(j)} \cdot \sigma_{CB2} \cdot c \cdot n_{ph2} \cdot e^{\lambda \cdot t} + [A_{CB2} + \sigma_{CB2} \cdot c \cdot n_{ph2}^{(j)}] \cdot n_c \cdot e^{\lambda \cdot (t-\tau_D)} - n_{B2}^{(j)} \cdot n_{ph2}^{(j)} \cdot \sigma_{B2C} \cdot c - n_{B2}^{(j)} \cdot \sigma_{B2C} \cdot c \cdot n_{ph2} \cdot e^{\lambda \cdot t} - n_{ph2}^{(j)} \cdot \sigma_{B2C} \cdot c \cdot n_{B2} \cdot e^{\lambda \cdot t} - n_{B2} \cdot \lambda \cdot e^{\lambda \cdot t} = 0$$

$$S_{B2} + n_c^{(j)} \cdot [A_{CB2} + \sigma_{CB2} \cdot c \cdot n_{ph2}^{(j)}] - n_{B2}^{(j)} \cdot n_{ph2}^{(j)} \cdot \sigma_{B2C} \cdot c + n_c^{(j)} \cdot \sigma_{CB2} \cdot c \cdot n_{ph2} \cdot e^{\lambda \cdot t} + [A_{CB2} + \sigma_{CB2} \cdot c \cdot n_{ph2}^{(j)}] \cdot n_c \cdot e^{\lambda \cdot (t-\tau_D)} - n_{B2}^{(j)} \cdot \sigma_{B2C} \cdot c \cdot n_{ph2} \cdot e^{\lambda \cdot t} - n_{ph2}^{(j)} \cdot \sigma_{B2C} \cdot c \cdot n_{B2} \cdot e^{\lambda \cdot t} - n_{B2} \cdot \lambda \cdot e^{\lambda \cdot t} = 0$$

At fixed point: $\frac{dn_{B2}}{dt} = 0$; $\lim_{t \to 0} n_c(t - \tau_D) = n_c(t) \, \forall t \gg \tau_D$

$$S_{B2} + n_c^{(j)} \cdot [A_{CB2} + \sigma_{CB2} \cdot c \cdot n_{ph2}^{(j)}] - n_{B2}^{(j)} \cdot n_{ph2}^{(j)} \cdot \sigma_{B2C} \cdot c = 0$$

$$n_c^{(j)} \cdot \sigma_{CB2} \cdot c \cdot n_{ph2} \cdot e^{\lambda \cdot t} + [A_{CB2} + \sigma_{CB2} \cdot c \cdot n_{ph2}^{(j)}] \cdot n_c \cdot e^{\lambda \cdot (t-\tau_D)} - n_{B2}^{(j)} \cdot \sigma_{B2C} \cdot c \cdot n_{ph2} \cdot e^{\lambda \cdot t} - n_{ph2}^{(j)} \cdot \sigma_{B2C} \cdot c \cdot n_{B2} \cdot e^{\lambda \cdot t} - n_{B2} \cdot \lambda \cdot e^{\lambda \cdot t} = 0$$

$$n_c^{(j)} \cdot \sigma_{CB2} \cdot c \cdot n_{ph2} \cdot e^{\lambda \cdot t} + [A_{CB2} + \sigma_{CB2} \cdot c \cdot n_{ph2}^{(j)}] \cdot n_c \cdot e^{\lambda \cdot t} \cdot e^{-\lambda \cdot \tau_D} - n_{B2}^{(j)} \cdot \sigma_{B2C} \cdot c \cdot n_{ph2} \cdot e^{\lambda \cdot t} - n_{ph2}^{(j)} \cdot \sigma_{B2C} \cdot c \cdot n_{B2} \cdot e^{\lambda \cdot t} - n_{B2} \cdot \lambda \cdot e^{\lambda \cdot t} = 0$$

Divide the two sides of the above equation by $e^{\lambda \cdot t}$ gives

$$n_c^{(j)} \cdot \sigma_{CB2} \cdot c \cdot n_{ph2} + [A_{CB2} + \sigma_{CB2} \cdot c \cdot n_{ph2}^{(j)}] \cdot n_c \cdot e^{-\lambda \cdot \tau_D} - n_{B2}^{(j)} \cdot \sigma_{B2C} \cdot c \cdot n_{ph2} - n_{ph2}^{(j)} \cdot \sigma_{B2C} \cdot c \cdot n_{B2} - n_{B2} \cdot \lambda = 0$$

$$[A_{CB2} + \sigma_{CB2} \cdot c \cdot n_{ph2}^{(j)}] \cdot n_c \cdot e^{-\lambda \cdot \tau_D} - n_{ph2}^{(j)} \cdot \sigma_{B2C} \cdot c \cdot n_{B2} - n_{B2} \cdot \lambda + [n_c^{(j)} \cdot \sigma_{CB2} - n_{B2}^{(j)} \cdot \sigma_{B2C}] \cdot c \cdot n_{ph2} = 0$$

We define for simplicity some parameters:

$$\Xi_7 = A_{CB2} + \sigma_{CB2} \cdot c \cdot n_{ph2}^{(j)};\ \Xi_3 = n_{ph2}^{(j)} \cdot \sigma_{B2C} \cdot c$$
$$-\Xi_5 = [n_c^{(j)} \cdot \sigma_{CB2} - n_{B2}^{(j)} \cdot \sigma_{B2C}] \cdot c$$

The third arbitrarily small increments equation is $\Xi_7 \cdot n_c \cdot e^{-\lambda \cdot \tau_D} - \Xi_3 \cdot n_{B2} - n_{B2} \cdot \lambda - \Xi_5 \cdot n_{ph2} = 0$

$$\frac{dn_{ph1}}{dt} = \frac{\sigma_{CB1} \cdot (n_c(t - \tau_D) - n_{B1}) \cdot n_{ph1} \cdot c}{1 + \frac{I_L}{I_{s1}}} - \frac{n_{ph1}}{\tau_{ph}}$$

$$n_{ph1} \cdot \lambda \cdot e^{\lambda \cdot t} = \frac{\sigma_{CB1} \cdot ([n_c^{(j)} + n_c \cdot e^{\lambda \cdot (t - \tau_D)}] - [n_{B1}^{(j)} + n_{B1} \cdot e^{\lambda \cdot t}]) \cdot [n_{ph1}^{(j)} + n_{ph1} \cdot e^{\lambda \cdot t}] \cdot c}{1 + \frac{I_L}{I_{s1}}} - \frac{[n_{ph1}^{(j)} + n_{ph1} \cdot e^{\lambda \cdot t}]}{\tau_{ph}}$$

$$n_{ph1} \cdot \lambda \cdot e^{\lambda \cdot t} = \frac{\sigma_{CB1} \cdot c}{(1 + \frac{I_L}{I_{s1}})} \cdot ([n_c^{(j)} - n_{B1}^{(j)}] + [n_c \cdot e^{-\lambda \cdot \tau_D} - n_{B1}] \cdot e^{\lambda \cdot t}) \cdot (n_{ph1}^{(j)} + n_{ph1} \cdot e^{\lambda \cdot t}) - \frac{n_{ph1}^{(j)}}{\tau_{ph}} - \frac{n_{ph1} \cdot e^{\lambda \cdot t}}{\tau_{ph}}$$

$$n_{ph1} \cdot \lambda \cdot e^{\lambda \cdot t} = \frac{\sigma_{CB1} \cdot c}{(1 + \frac{I_L}{I_{s1}})} \cdot ([n_c^{(j)} - n_{B1}^{(j)}] \cdot n_{ph1}^{(j)} + [n_c^{(j)} - n_{B1}^{(j)}] \cdot n_{ph1} \cdot e^{\lambda \cdot t} + n_{ph1}^{(j)} \cdot [n_c \cdot e^{-\lambda \cdot \tau_D} - n_{B1}] \cdot e^{\lambda \cdot t}$$

$$+ [n_c \cdot e^{-\lambda \cdot \tau_D} - n_{B1}] \cdot e^{\lambda \cdot t} \cdot n_{ph1} \cdot e^{\lambda \cdot t}) - \frac{n_{ph1}^{(j)}}{\tau_{ph}} - \frac{n_{ph1} \cdot e^{\lambda \cdot t}}{\tau_{ph}}$$

$$n_{ph1} \cdot \lambda \cdot e^{\lambda \cdot t} = \frac{\sigma_{CB1} \cdot c}{(1 + \frac{I_L}{I_{s1}})} \cdot ([n_c^{(j)} - n_{B1}^{(j)}] \cdot n_{ph1}^{(j)} + [n_c^{(j)} - n_{B1}^{(j)}] \cdot n_{ph1} \cdot e^{\lambda \cdot t} + n_{ph1}^{(j)} \cdot [n_c \cdot e^{-\lambda \cdot \tau_D} - n_{B1}] \cdot e^{\lambda \cdot t} + [n_c \cdot n_{ph1} \cdot e^{-\lambda \cdot \tau_D} - n_{B1} \cdot n_{ph1}] \cdot e^{\lambda \cdot t} \cdot e^{\lambda \cdot t}) - \frac{n_{ph1}^{(j)}}{\tau_{ph}} - \frac{n_{ph1} \cdot e^{\lambda \cdot t}}{\tau_{ph}}$$

We consider $n_c \cdot n_{ph1} \approx 0; n_{B1} \cdot n_{ph1} \approx 0$ then

$$n_{ph1} \cdot \lambda \cdot e^{\lambda \cdot t} = \frac{\sigma_{CB1} \cdot c}{(1 + \frac{I_L}{I_{s1}})} \cdot ([n_c^{(j)} - n_{B1}^{(j)}] \cdot n_{ph1}^{(j)} + [n_c^{(j)} - n_{B1}^{(j)}] \cdot n_{ph1} \cdot e^{\lambda \cdot t} + n_{ph1}^{(j)} \cdot [n_c \cdot e^{-\lambda \cdot \tau_D} - n_{B1}] \cdot e^{\lambda \cdot t}) - \frac{n_{ph1}^{(j)}}{\tau_{ph}} - \frac{n_{ph1} \cdot e^{\lambda \cdot t}}{\tau_{ph}}$$

$$n_{ph1} \cdot \lambda \cdot e^{\lambda \cdot t} = \frac{\sigma_{CB1} \cdot c}{(1 + \frac{I_L}{I_{s1}})} \cdot [n_c^{(j)} - n_{B1}^{(j)}] \cdot n_{ph1}^{(j)} + \frac{\sigma_{CB1} \cdot c}{(1 + \frac{I_L}{I_{s1}})} \cdot [n_c^{(j)} - n_{B1}^{(j)}] \cdot n_{ph1} \cdot e^{\lambda \cdot t} + \frac{\sigma_{CB1} \cdot c}{(1 + \frac{I_L}{I_{s1}})} \cdot n_{ph1}^{(j)} \cdot [n_c \cdot e^{-\lambda \cdot \tau_D} - n_{B1}] \cdot e^{\lambda \cdot t} - \frac{n_{ph1}^{(j)}}{\tau_{ph}} - \frac{n_{ph1} \cdot e^{\lambda \cdot t}}{\tau_{ph}}$$

$$n_{ph1} \cdot \lambda \cdot e^{\lambda \cdot t} = \frac{\sigma_{CB1} \cdot c}{(1 + \frac{I_L}{I_{s1}})} \cdot [n_c^{(j)} - n_{B1}^{(j)}] \cdot n_{ph1}^{(j)} - \frac{n_{ph1}^{(j)}}{\tau_{ph}} + \frac{\sigma_{CB1} \cdot c}{(1 + \frac{I_L}{I_{s1}})} \cdot [n_c^{(j)} - n_{B1}^{(j)}] \cdot n_{ph1} \cdot e^{\lambda \cdot t} + \frac{\sigma_{CB1} \cdot c}{(1 + \frac{I_L}{I_{s1}})} \cdot n_{ph1}^{(j)} \cdot [n_c \cdot e^{-\lambda \cdot \tau_D} - n_{B1}] \cdot e^{\lambda \cdot t} - \frac{n_{ph1} \cdot e^{\lambda \cdot t}}{\tau_{ph}}$$

At fixed point: $\frac{dn_{ph1}}{dt} = 0; \lim_{t \to 0} n_c(t - \tau_D) = n_c(t) \, \forall t \gg \tau_D$

$$\frac{\sigma_{CB1} \cdot c}{(1 + \frac{I_L}{I_{s1}})} \cdot [n_c^{(j)} - n_{B1}^{(j)}] \cdot n_{ph1}^{(j)} - \frac{n_{ph1}^{(j)}}{\tau_{ph}} = 0,$$

then

$$n_{ph1}\cdot\lambda\cdot e^{\lambda\cdot t}=\frac{\sigma_{CB1}\cdot c}{(1+\frac{I_L}{I_{s1}})}\cdot[n_c^{(j)}-n_{B1}^{(j)}]\cdot n_{ph1}\cdot e^{\lambda\cdot t}+\frac{\sigma_{CB1}\cdot c}{(1+\frac{I_L}{I_{s1}})}\cdot n_{ph1}^{(j)}\cdot[n_c\cdot e^{-\lambda\cdot\tau_D}-n_{B1}]\cdot e^{\lambda\cdot t}-\frac{n_{ph1}\cdot e^{\lambda\cdot t}}{\tau_{ph}}$$

Divide the two sides of the above equation by $e^{\lambda\cdot t}$ gives

$$n_{ph1}\cdot\lambda=\frac{\sigma_{CB1}\cdot c}{(1+\frac{I_L}{I_{s1}})}\cdot[n_c^{(j)}-n_{B1}^{(j)}]\cdot n_{ph1}+\frac{\sigma_{CB1}\cdot c}{(1+\frac{I_L}{I_{s1}})}\cdot n_{ph1}^{(j)}\cdot[n_c\cdot e^{-\lambda\cdot\tau_D}-n_{B1}]-\frac{n_{ph1}}{\tau_{ph}}$$

$$\frac{\sigma_{CB1}\cdot c}{(1+\frac{I_L}{I_{s1}})}\cdot[n_c^{(j)}-n_{B1}^{(j)}]\cdot n_{ph1}+\frac{\sigma_{CB1}\cdot c}{(1+\frac{I_L}{I_{s1}})}\cdot n_{ph1}^{(j)}\cdot[n_c\cdot e^{-\lambda\cdot\tau_D}-n_{B1}]-\frac{n_{ph1}}{\tau_{ph}}-n_{ph1}\cdot\lambda=0$$

$$\frac{\sigma_{CB1}\cdot c}{(1+\frac{I_L}{I_{s1}})}\cdot n_{ph1}^{(j)}\cdot n_c\cdot e^{-\lambda\cdot\tau_D}-\frac{\sigma_{CB1}\cdot c}{(1+\frac{I_L}{I_{s1}})}\cdot n_{ph1}^{(j)}\cdot n_{B1}+(\frac{\sigma_{CB1}\cdot c}{(1+\frac{I_L}{I_{s1}})}\cdot[n_c^{(j)}-n_{B1}^{(j)}]-\frac{1}{\tau_{ph}})\cdot n_{ph1}-n_{ph1}\cdot\lambda=0$$

We define for simplicity some parameters: $\Xi_8=\frac{\sigma_{CB1}\cdot c}{(1+\frac{I_L}{I_{s1}})}\cdot n_{ph1}^{(j)}$; $\Xi_9=\frac{\sigma_{CB1}\cdot c}{(1+\frac{I_L}{I_{s1}})}\cdot[n_c^{(j)}-n_{B1}^{(j)}]-\frac{1}{\tau_{ph}}$

The fourth arbitrarily small increments equation is $\Xi_8\cdot n_c\cdot e^{-\lambda\cdot\tau_D}-\Xi_8\cdot n_{B1}+\Xi_9\cdot n_{ph1}-n_{ph1}\cdot\lambda=0$

$$\frac{dn_{ph2}}{dt}=\frac{\sigma_{CB2}\cdot(n_c(t-\tau_D)-n_{B2})\cdot n_{ph2}\cdot c}{1+\frac{I_L}{I_{s2}}}-\frac{n_{ph2}}{\tau_{ph}}$$

$$n_{ph2}\cdot\lambda\cdot e^{\lambda\cdot t}=\frac{\sigma_{CB2}\cdot([n_c^{(j)}+n_c\cdot e^{\lambda\cdot(t-\tau_D)}]-[n_{B2}^{(j)}+n_{B2}\cdot e^{\lambda\cdot t}])\cdot[n_{ph2}^{(j)}+n_{ph2}\cdot e^{\lambda\cdot t}]\cdot c}{1+\frac{I_L}{I_{s2}}}-\frac{[n_{ph2}^{(j)}+n_{ph2}\cdot e^{\lambda\cdot t}]}{\tau_{ph}}$$

$$n_{ph2}\cdot\lambda\cdot e^{\lambda\cdot t}=\frac{\sigma_{CB2}\cdot c}{1+\frac{I_L}{I_{s2}}}\cdot([n_c^{(j)}-n_{B2}^{(j)}]+[n_c\cdot e^{-\lambda\cdot\tau_D}-n_{B2}]\cdot e^{\lambda\cdot t})$$

$$\cdot [n_{ph2}^{(j)} + n_{ph2} \cdot e^{\lambda \cdot t}] - \frac{n_{ph2}^{(j)}}{\tau_{ph}} - \frac{n_{ph2} \cdot e^{\lambda \cdot t}}{\tau_{ph}}$$

$$n_{ph2} \cdot \lambda \cdot e^{\lambda \cdot t} = \frac{\sigma_{CB2} \cdot c}{1 + \frac{I_L}{I_{s2}}} \cdot ([n_c^{(j)} - n_{B2}^{(j)}] \cdot n_{ph2}^{(j)} + [n_c^{(j)} - n_{B2}^{(j)}] \cdot n_{ph2}$$
$$\cdot e^{\lambda \cdot t} + n_{ph2}^{(j)} \cdot [n_c \cdot e^{-\lambda \cdot \tau_D} - n_{B2}] \cdot e^{\lambda \cdot t}$$
$$+ [n_c \cdot e^{-\lambda \cdot \tau_D} - n_{B2}] \cdot e^{\lambda \cdot t} \cdot n_{ph2} \cdot e^{\lambda \cdot t})$$
$$- \frac{n_{ph2}^{(j)}}{\tau_{ph}} - \frac{n_{ph2} \cdot e^{\lambda \cdot t}}{\tau_{ph}}$$

$$n_{ph2} \cdot \lambda \cdot e^{\lambda \cdot t} = \frac{\sigma_{CB2} \cdot c}{1 + \frac{I_L}{I_{s2}}} \cdot ([n_c^{(j)} - n_{B2}^{(j)}] \cdot n_{ph2}^{(j)} + [n_c^{(j)} - n_{B2}^{(j)}]$$
$$\cdot n_{ph2} \cdot e^{\lambda \cdot t} + n_{ph2}^{(j)} \cdot [n_c \cdot e^{-\lambda \cdot \tau_D} - n_{B2}] \cdot e^{\lambda \cdot t}$$
$$+ [n_c \cdot n_{ph2} \cdot e^{-\lambda \cdot \tau_D} - n_{B2} \cdot n_{ph2}] \cdot e^{\lambda \cdot t} \cdot e^{\lambda \cdot t})$$
$$- \frac{n_{ph2}^{(j)}}{\tau_{ph}} - \frac{n_{ph2} \cdot e^{\lambda \cdot t}}{\tau_{ph}}$$

We consider $n_c \cdot n_{ph2} \approx 0$; $n_{B2} \cdot n_{ph2} \approx 0$ then

$$n_{ph2} \cdot \lambda \cdot e^{\lambda \cdot t} = \frac{\sigma_{CB2} \cdot c}{1 + \frac{I_L}{I_{s2}}} \cdot ([n_c^{(j)} - n_{B2}^{(j)}] \cdot n_{ph2}^{(j)} + [n_c^{(j)} - n_{B2}^{(j)}]$$
$$\cdot n_{ph2} \cdot e^{\lambda \cdot t} + n_{ph2}^{(j)} \cdot [n_c \cdot e^{-\lambda \cdot \tau_D} - n_{B2}] \cdot e^{\lambda \cdot t})$$
$$- \frac{n_{ph2}^{(j)}}{\tau_{ph}} - \frac{n_{ph2} \cdot e^{\lambda \cdot t}}{\tau_{ph}}$$

$$n_{ph2} \cdot \lambda \cdot e^{\lambda \cdot t} = \frac{\sigma_{CB2} \cdot c}{1 + \frac{I_L}{I_{s2}}} \cdot [n_c^{(j)} - n_{B2}^{(j)}] \cdot n_{ph2}^{(j)} + \frac{\sigma_{CB2} \cdot c}{1 + \frac{I_L}{I_{s2}}}$$
$$\cdot [n_c^{(j)} - n_{B2}^{(j)}] \cdot n_{ph2} \cdot e^{\lambda \cdot t} + \frac{\sigma_{CB2} \cdot c}{1 + \frac{I_L}{I_{s2}}} \cdot n_{ph2}^{(j)}$$
$$\cdot [n_c \cdot e^{-\lambda \cdot \tau_D} - n_{B2}] \cdot e^{\lambda \cdot t} - \frac{n_{ph2}^{(j)}}{\tau_{ph}} - \frac{n_{ph2} \cdot e^{\lambda \cdot t}}{\tau_{ph}}$$

$$n_{ph2} \cdot \lambda \cdot e^{\lambda \cdot t} = \frac{\sigma_{CB2} \cdot c}{1 + \frac{I_L}{I_{s2}}} \cdot [n_c^{(j)} - n_{B2}^{(j)}] \cdot n_{ph2}^{(j)} - \frac{n_{ph2}^{(j)}}{\tau_{ph}} + \frac{\sigma_{CB2} \cdot c}{1 + \frac{I_L}{I_{s2}}}$$
$$\cdot [n_c^{(j)} - n_{B2}^{(j)}] \cdot n_{ph2} \cdot e^{\lambda \cdot t} + \frac{\sigma_{CB2} \cdot c}{1 + \frac{I_L}{I_{s2}}} \cdot n_{ph2}^{(j)}$$
$$\cdot [n_c \cdot e^{-\lambda \cdot \tau_D} - n_{B2}] \cdot e^{\lambda \cdot t} - \frac{n_{ph2} \cdot e^{\lambda \cdot t}}{\tau_{ph}}$$

At fixed point: $\frac{dn_{ph2}}{dt} = 0; \lim_{t \to 0} n_c(t - \tau_D) = n_c(t) \, \forall \, t \gg \tau_D$

$$\frac{\sigma_{CB2} \cdot c}{1 + \frac{I_L}{I_{s2}}} \cdot [n_c^{(j)} - n_{B2}^{(j)}] \cdot n_{ph2}^{(j)} - \frac{n_{ph2}^{(j)}}{\tau_{ph}} = 0$$

$$\begin{aligned} n_{ph2} \cdot \lambda \cdot e^{\lambda \cdot t} = & \frac{\sigma_{CB2} \cdot c}{1 + \frac{I_L}{I_{s2}}} \cdot [n_c^{(j)} - n_{B2}^{(j)}] \cdot n_{ph2} \cdot e^{\lambda \cdot t} \\ & + \frac{\sigma_{CB2} \cdot c}{1 + \frac{I_L}{I_{s2}}} \cdot n_{ph2}^{(j)} \cdot [n_c \cdot e^{-\lambda \cdot \tau_D} - n_{B2}] \cdot e^{\lambda \cdot t} \\ & - \frac{n_{ph2} \cdot e^{\lambda \cdot t}}{\tau_{ph}} \end{aligned}$$

Divide the two sides of the above equation by $e^{\lambda \cdot t}$ gives

$$\begin{aligned} n_{ph2} \cdot \lambda = & \frac{\sigma_{CB2} \cdot c}{1 + \frac{I_L}{I_{s2}}} \cdot [n_c^{(j)} - n_{B2}^{(j)}] \cdot n_{ph2} \\ & + \frac{\sigma_{CB2} \cdot c}{1 + \frac{I_L}{I_{s2}}} \cdot n_{ph2}^{(j)} \cdot [n_c \cdot e^{-\lambda \cdot \tau_D} - n_{B2}] - \frac{n_{ph2}}{\tau_{ph}} \end{aligned}$$

$$\begin{aligned} & \frac{\sigma_{CB2} \cdot c}{1 + \frac{I_L}{I_{s2}}} \cdot n_{ph2}^{(j)} \cdot e^{-\lambda \cdot \tau_D} \cdot n_c - \frac{\sigma_{CB2} \cdot c}{1 + \frac{I_L}{I_{s2}}} \cdot n_{ph2}^{(j)} \cdot n_{B2} \\ & + (\frac{\sigma_{CB2} \cdot c}{1 + \frac{I_L}{I_{s2}}} \cdot [n_c^{(j)} - n_{B2}^{(j)}] - \frac{1}{\tau_{ph}}) \cdot n_{ph2} - n_{ph2} \cdot \lambda = 0 \end{aligned}$$

We define for simplicity some parameters: $\Xi_{10} = \frac{\sigma_{CB2} \cdot c}{1 + \frac{I_L}{I_{s2}}} \cdot n_{ph2}^{(j)}$; $\Xi_{11} = \frac{\sigma_{CB2} \cdot c}{1 + \frac{I_L}{I_{s2}}} \cdot [n_c^{(j)} - n_{B2}^{(j)}] - \frac{1}{\tau_{ph}}$

The fives arbitrarily small increments equation is $\Xi_{10} \cdot e^{-\lambda \cdot \tau_D} \cdot n_c - \Xi_{10} \cdot n_{B2} + \Xi_{11} \cdot n_{ph2} - n_{ph2} \cdot \lambda = 0$

We can summary our plasma kinetic system five arbitrarily small increments equations:

$$(\Xi_1 \cdot e^{-\lambda \cdot \tau_D} - \lambda) \cdot n_c + \Xi_2 \cdot n_{B1} + \Xi_3 \cdot n_{B2} + \Xi_4 \cdot n_{ph1} + \Xi_5 \cdot n_{ph2} = 0$$
$$\Xi_6 \cdot e^{-\lambda \cdot \tau_D} \cdot n_c - \Xi_2 \cdot n_{B1} - \lambda \cdot n_{B1} - \Xi_4 \cdot n_{ph1} = 0$$

$$\Xi_7 \cdot e^{-\lambda \cdot \tau_D} \cdot n_c - \Xi_3 \cdot n_{B2} - \lambda \cdot n_{B2} - \Xi_5 \cdot n_{ph2} = 0$$
$$\Xi_8 \cdot e^{-\lambda \cdot \tau_D} \cdot n_c - \Xi_8 \cdot n_{B1} + \Xi_9 \cdot n_{ph1} - \lambda \cdot n_{ph1} = 0$$

$$\Xi_{10} \cdot e^{-\lambda \cdot \tau_D} \cdot n_c - \Xi_{10} \cdot n_{B2} + \Xi_{11} \cdot n_{ph2} - \lambda \cdot n_{ph2} = 0$$

The small increments Jacobian of our plasma kinetic system is as follow:

$$\begin{pmatrix} \Upsilon_{11} & \ldots & \Upsilon_{15} \\ \vdots & \ddots & \vdots \\ \Upsilon_{51} & \cdots & \Upsilon_{55} \end{pmatrix} \cdot \begin{pmatrix} n_c \\ n_{B1} \\ n_{B2} \\ n_{ph1} \\ n_{ph2} \end{pmatrix} = 0; \Upsilon_{11} = \Xi_1 \cdot e^{-\lambda \cdot \tau_D} - \lambda; \Upsilon_{12} = \Xi_2$$

$$\Upsilon_{13} = \Xi_3; \Upsilon_{14} = \Xi_4; \Upsilon_{15} = \Xi_5$$

$$\Upsilon_{21} = \Xi_6 \cdot e^{-\lambda \cdot \tau_D}; \Upsilon_{22} = -\Xi_2 - \lambda; \Upsilon_{23} = 0$$
$$\Upsilon_{24} = -\Xi_4; \Upsilon_{25} = 0; \Upsilon_{31} = \Xi_7 \cdot e^{-\lambda \cdot \tau_D}; \Upsilon_{32} = 0$$

$$\Upsilon_{33} = -\Xi_3 - \lambda; \Upsilon_{34} = 0; \Upsilon_{35} = -\Xi_5$$
$$\Upsilon_{41} = \Xi_8 \cdot e^{-\lambda \cdot \tau_D}; \Upsilon_{42} = -\Xi_8; \Upsilon_{43} = 0; \Upsilon_{44} = \Xi_9 - \lambda$$

$$\Upsilon_{45} = 0; \Upsilon_{51} = \Xi_{10} \cdot e^{-\lambda \cdot \tau_D}; \Upsilon_{52} = 0$$
$$\Upsilon_{53} = -\Xi_{10}; \Upsilon_{54} = 0; \Upsilon_{55} = \Xi_{11} - \lambda$$

$$A - \lambda \cdot I = \begin{pmatrix} \Upsilon_{11} & \ldots & \Upsilon_{15} \\ \vdots & \ddots & \vdots \\ \Upsilon_{51} & \cdots & \Upsilon_{55} \end{pmatrix}$$

$$\det(A - \lambda \cdot I) = \det \begin{pmatrix} \Upsilon_{11} & \ldots & \Upsilon_{15} \\ \vdots & \ddots & \vdots \\ \Upsilon_{51} & \cdots & \Upsilon_{55} \end{pmatrix}$$

$$\det(A - \lambda \cdot I) = 0$$

$$\det(A - \lambda \cdot I) = \Upsilon_{11} \cdot \det \begin{pmatrix} \Upsilon_{22} & \Upsilon_{23} & \Upsilon_{24} & \Upsilon_{25} \\ \Upsilon_{32} & \Upsilon_{33} & \Upsilon_{34} & \Upsilon_{35} \\ \Upsilon_{42} & \Upsilon_{43} & \Upsilon_{44} & \Upsilon_{45} \\ \Upsilon_{52} & \Upsilon_{53} & \Upsilon_{54} & \Upsilon_{55} \end{pmatrix}$$
$$- \Upsilon_{12} \cdot \det \begin{pmatrix} \Upsilon_{21} & \Upsilon_{23} & \Upsilon_{24} & \Upsilon_{25} \\ \Upsilon_{31} & \Upsilon_{33} & \Upsilon_{34} & \Upsilon_{35} \\ \Upsilon_{41} & \Upsilon_{43} & \Upsilon_{44} & \Upsilon_{45} \\ \Upsilon_{51} & \Upsilon_{53} & \Upsilon_{54} & \Upsilon_{55} \end{pmatrix}$$
$$+ \Upsilon_{13} \cdot \det \begin{pmatrix} \Upsilon_{21} & \Upsilon_{22} & \Upsilon_{24} & \Upsilon_{25} \\ \Upsilon_{31} & \Upsilon_{32} & \Upsilon_{34} & \Upsilon_{35} \\ \Upsilon_{41} & \Upsilon_{42} & \Upsilon_{44} & \Upsilon_{45} \\ \Upsilon_{51} & \Upsilon_{52} & \Upsilon_{54} & \Upsilon_{55} \end{pmatrix}$$

$$-\Upsilon_{14}\cdot\det\begin{pmatrix}\Upsilon_{21}&\Upsilon_{22}&\Upsilon_{23}&\Upsilon_{25}\\ \Upsilon_{31}&\Upsilon_{32}&\Upsilon_{33}&\Upsilon_{35}\\ \Upsilon_{41}&\Upsilon_{42}&\Upsilon_{43}&\Upsilon_{45}\\ \Upsilon_{51}&\Upsilon_{52}&\Upsilon_{53}&\Upsilon_{55}\end{pmatrix}$$

$$+\Upsilon_{15}\cdot\det\begin{pmatrix}\Upsilon_{21}&\Upsilon_{22}&\Upsilon_{23}&\Upsilon_{24}\\ \Upsilon_{31}&\Upsilon_{32}&\Upsilon_{33}&\Upsilon_{34}\\ \Upsilon_{41}&\Upsilon_{42}&\Upsilon_{43}&\Upsilon_{44}\\ \Upsilon_{51}&\Upsilon_{52}&\Upsilon_{53}&\Upsilon_{54}\end{pmatrix}$$

Step 1: Find the expression for $\Upsilon_{11}\cdot\det\begin{pmatrix}\Upsilon_{22}&\Upsilon_{23}&\Upsilon_{24}&\Upsilon_{25}\\ \Upsilon_{32}&\Upsilon_{33}&\Upsilon_{34}&\Upsilon_{35}\\ \Upsilon_{42}&\Upsilon_{43}&\Upsilon_{44}&\Upsilon_{45}\\ \Upsilon_{52}&\Upsilon_{53}&\Upsilon_{54}&\Upsilon_{55}\end{pmatrix}$

$$\Upsilon_{11}\cdot\det\begin{pmatrix}\Upsilon_{22}&\Upsilon_{23}&\Upsilon_{24}&\Upsilon_{25}\\ \Upsilon_{32}&\Upsilon_{33}&\Upsilon_{34}&\Upsilon_{35}\\ \Upsilon_{42}&\Upsilon_{43}&\Upsilon_{44}&\Upsilon_{45}\\ \Upsilon_{52}&\Upsilon_{53}&\Upsilon_{54}&\Upsilon_{55}\end{pmatrix}$$

$$=(\Xi_1\cdot e^{-\lambda\cdot\tau_D}-\lambda)\cdot\det\begin{pmatrix}(-\Xi_2-\lambda)&0&-\Xi_4&0\\ 0&(-\Xi_3-\lambda)&0&-\Xi_5\\ -\Xi_8&0&(\Xi_9-\lambda)&0\\ 0&-\Xi_{10}&0&(\Xi_{11}-\lambda)\end{pmatrix}$$

$$=(\Xi_1\cdot e^{-\lambda\cdot\tau_D}-\lambda)\cdot\{(-\Xi_2-\lambda)\cdot\det\begin{pmatrix}(-\Xi_3-\lambda)&0&-\Xi_5\\ 0&(\Xi_9-\lambda)&0\\ -\Xi_{10}&0&(\Xi_{11}-\lambda)\end{pmatrix}$$

$$-\Xi_4\cdot\det\begin{pmatrix}0&(-\Xi_3-\lambda)&-\Xi_5\\ -\Xi_8&0&0\\ 0&-\Xi_{10}&(\Xi_{11}-\lambda)\end{pmatrix}\Bigg)\}$$

Step 1.1: $(-\Xi_2-\lambda)\cdot\det\begin{pmatrix}(-\Xi_3-\lambda)&0&-\Xi_5\\ 0&(\Xi_9-\lambda)&0\\ -\Xi_{10}&0&(\Xi_{11}-\lambda)\end{pmatrix}$

$$(-\Xi_2-\lambda)\cdot\det\begin{pmatrix}(-\Xi_3-\lambda)&0&-\Xi_5\\ 0&(\Xi_9-\lambda)&0\\ -\Xi_{10}&0&(\Xi_{11}-\lambda)\end{pmatrix}$$

$$=(-\Xi_2-\lambda)\cdot\{(-\Xi_3-\lambda)\cdot(\Xi_9-\lambda)\cdot(\Xi_{11}-\lambda)$$
$$-\Xi_5\cdot\Xi_{10}\cdot(\Xi_9-\lambda)\}=(-\Xi_2-\lambda)\cdot\{\}$$

$$(-\Xi_2-\lambda)\cdot\det\begin{pmatrix}(-\Xi_3-\lambda) & 0 & -\Xi_5\\ 0 & (\Xi_9-\lambda) & 0\\ -\Xi_{10} & 0 & (\Xi_{11}-\lambda)\end{pmatrix}$$
$$=(-\Xi_2-\lambda)\cdot\{(-\Xi_3\cdot\Xi_9+(\Xi_3-\Xi_9\cdot)\lambda+\lambda^2)\cdot(\Xi_{11}-\lambda)$$
$$-\Xi_5\cdot\Xi_{10}\cdot\Xi_9+\Xi_5\cdot\Xi_{10}\cdot\lambda\}$$
$$=(-\Xi_2-\lambda)\cdot\{-\Xi_3\cdot\Xi_9\cdot\Xi_{11}+\Xi_3\cdot\Xi_9\cdot\lambda+\Xi_{11}\cdot(\Xi_3-\Xi_9\cdot)$$
$$\cdot\lambda-(\Xi_3-\Xi_9\cdot)\cdot\lambda^2+\Xi_{11}\cdot\lambda^2-\lambda^3-\Xi_5\cdot\Xi_{10}\cdot\Xi_9+\Xi_5\cdot\Xi_{10}\cdot\lambda\}$$

$$(-\Xi_2-\lambda)\cdot\det\begin{pmatrix}(-\Xi_3-\lambda) & 0 & -\Xi_5\\ 0 & (\Xi_9-\lambda) & 0\\ -\Xi_{10} & 0 & (\Xi_{11}-\lambda)\end{pmatrix}$$
$$=(-\Xi_2-\lambda)\cdot\{(-\Xi_3\cdot\Xi_9+(\Xi_3-\Xi_9\cdot)\lambda+\lambda^2)$$
$$\cdot(\Xi_{11}-\lambda)-\Xi_5\cdot\Xi_{10}\cdot\Xi_9+\Xi_5\cdot\Xi_{10}\cdot\lambda\}$$
$$=(-\Xi_2-\lambda)\cdot\{-\Xi_3\cdot\Xi_9\cdot\Xi_{11}-\Xi_5\cdot\Xi_{10}\cdot\Xi_9$$
$$+(\Xi_3\cdot\Xi_9+\Xi_{11}\cdot(\Xi_3-\Xi_9)+\Xi_5\cdot\Xi_{10})\cdot\lambda$$
$$+(\Xi_{11}-\Xi_3+\Xi_9)\cdot\lambda^2-\lambda^3\}$$

$$(-\Xi_2-\lambda)\cdot\det\begin{pmatrix}(-\Xi_3-\lambda) & 0 & -\Xi_5\\ 0 & (\Xi_9-\lambda) & 0\\ -\Xi_{10} & 0 & (\Xi_{11}-\lambda)\end{pmatrix}$$
$$=(-\Xi_2-\lambda)\cdot\{(-\Xi_3\cdot\Xi_9+(\Xi_3-\Xi_9\cdot)\lambda+\lambda^2)$$
$$\cdot(\Xi_{11}-\lambda)-\Xi_5\cdot\Xi_{10}\cdot\Xi_9+\Xi_5\cdot\Xi_{10}\cdot\lambda\}=(-\Xi_2-\lambda)$$
$$\cdot\{-(\Xi_3\cdot\Xi_9\cdot\Xi_{11}+\Xi_5\cdot\Xi_{10}\cdot\Xi_9)+(\Xi_3\cdot\Xi_9+\Xi_{11}$$
$$\cdot(\Xi_3-\Xi_9)+\Xi_5\cdot\Xi_{10})\cdot\lambda+(\Xi_{11}-\Xi_3+\Xi_9)\cdot\lambda^2-\lambda^3\}$$
$$=(\Xi_3\cdot\Xi_9\cdot\Xi_{11}+\Xi_5\cdot\Xi_{10}\cdot\Xi_9)\cdot\Xi_2-\Xi_2\cdot(\Xi_3\cdot\Xi_9+\Xi_{11}$$
$$\cdot[\Xi_3-\Xi_9]+\Xi_5\cdot\Xi_{10})\cdot\lambda-\Xi_2\cdot(\Xi_{11}-\Xi_3+\Xi_9)\cdot\lambda^2+\Xi_2\cdot\lambda^3$$
$$+(\Xi_3\cdot\Xi_9\cdot\Xi_{11}+\Xi_5\cdot\Xi_{10}\cdot\Xi_9)\cdot\lambda-(\Xi_3\cdot\Xi_9+\Xi_{11}\cdot(\Xi_3-\Xi_9)$$
$$+\Xi_5\cdot\Xi_{10})\cdot\lambda^2-(\Xi_{11}-\Xi_3+\Xi_9)\cdot\lambda^3+\lambda^4$$

$$(-\Xi_2-\lambda)\cdot\det\begin{pmatrix}(-\Xi_3-\lambda) & 0 & -\Xi_5\\ 0 & (\Xi_9-\lambda) & 0\\ -\Xi_{10} & 0 & (\Xi_{11}-\lambda)\end{pmatrix}$$
$$=(\Xi_3\cdot\Xi_9\cdot\Xi_{11}+\Xi_5\cdot\Xi_{10}\cdot\Xi_9)\cdot\Xi_2+\{\Xi_3\cdot\Xi_9\cdot\Xi_{11}$$
$$+\Xi_5\cdot\Xi_{10}\cdot\Xi_9-\Xi_2\cdot(\Xi_3\cdot\Xi_9+\Xi_{11}\cdot[\Xi_3-\Xi_9]$$
$$+\Xi_5\cdot\Xi_{10})\}\cdot\lambda-\{\Xi_2\cdot(\Xi_{11}-\Xi_3+\Xi_9)+\Xi_3\cdot\Xi_9$$
$$+\Xi_{11}\cdot(\Xi_3-\Xi_9)+\Xi_5\cdot\Xi_{10}\}\cdot\lambda^2+\{\Xi_2-\Xi_{11}+\Xi_3-\Xi_9\}$$

$$\cdot \lambda^3 + \lambda^4$$

We define some global parameters: $\iota_1 = (\Xi_3 \cdot \Xi_9 \cdot \Xi_{11} + \Xi_5 \cdot \Xi_{10} \cdot \Xi_9) \cdot \Xi_2$

$$\iota_2 = \Xi_3 \cdot \Xi_9 \cdot \Xi_{11} + \Xi_5 \cdot \Xi_{10} \cdot \Xi_9 - \Xi_2 \cdot (\Xi_3 \cdot \Xi_9 + \Xi_{11} \cdot [\Xi_3 - \Xi_9] + \Xi_5 \cdot \Xi_{10})$$

$$\iota_3 = -\{\Xi_2 \cdot (\Xi_{11} - \Xi_3 + \Xi_9) + \Xi_3 \cdot \Xi_9 + \Xi_{11} \cdot (\Xi_3 - \Xi_9) + \Xi_5 \cdot \Xi_{10}\};\ \iota_4 = \Xi_2 - \Xi_{11} + \Xi_3 - \Xi_9$$

$$(-\Xi_2 - \lambda) \cdot \det\begin{pmatrix} (-\Xi_3 - \lambda) & 0 & -\Xi_5 \\ 0 & (\Xi_9 - \lambda) & 0 \\ -\Xi_{10} & 0 & (\Xi_{11} - \lambda) \end{pmatrix} = \iota_1 + \iota_2 \cdot \lambda + \iota_3 \cdot \lambda^2 + \iota_4 \cdot \lambda^3 + \lambda^4$$

Step 1.2: $-\Xi_4 \cdot \det\begin{pmatrix} 0 & (-\Xi_3 - \lambda) & -\Xi_5 \\ -\Xi_8 & 0 & 0 \\ 0 & -\Xi_{10} & (\Xi_{11} - \lambda) \end{pmatrix}$

$$- \Xi_4 \cdot \det\begin{pmatrix} 0 & (-\Xi_3 - \lambda) & -\Xi_5 \\ -\Xi_8 & 0 & 0 \\ 0 & -\Xi_{10} & (\Xi_{11} - \lambda) \end{pmatrix}$$
$$= -\Xi_4 \cdot \{(\Xi_3 + \lambda) \cdot \det\begin{pmatrix} -\Xi_8 & 0 \\ 0 & (\Xi_{11} - \lambda) \end{pmatrix} - \Xi_5 \cdot \det\begin{pmatrix} -\Xi_8 & 0 \\ 0 & -\Xi_{10} \end{pmatrix}\}$$
$$= -\Xi_4 \cdot \{-(\Xi_3 + \lambda) \cdot \Xi_8 \cdot (\Xi_{11} - \lambda) - \Xi_5 \cdot \Xi_8 \cdot \Xi_{10}\}$$
$$= -\Xi_4 \cdot \{-\Xi_5 \cdot \Xi_8 \cdot \Xi_{10} - \Xi_8 \cdot \Xi_3 \cdot \Xi_{11} - \Xi_8 \cdot [\Xi_{11} - \Xi_3] \cdot \lambda + \Xi_8 \cdot \lambda^2\}$$
$$= \Xi_4 \cdot \Xi_8 \cdot (\Xi_5 \cdot \Xi_{10} + \Xi_3 \cdot \Xi_{11}) + \Xi_4 \cdot \Xi_8 \cdot [\Xi_{11} - \Xi_3] \cdot \lambda - \Xi_4 \cdot \Xi_8 \cdot \lambda^2$$

We define some global parameters: $\iota_5 = \Xi_4 \cdot \Xi_8 \cdot (\Xi_5 \cdot \Xi_{10} + \Xi_3 \cdot \Xi_{11})$; $\iota_6 = \Xi_4 \cdot \Xi_8 \cdot [\Xi_{11} - \Xi_3]$

$$\iota_7 = -\Xi_4 \cdot \Xi_8;\ -\Xi_4 \cdot \det\begin{pmatrix} 0 & (-\Xi_3 - \lambda) & -\Xi_5 \\ -\Xi_8 & 0 & 0 \\ 0 & -\Xi_{10} & (\Xi_{11} - \lambda) \end{pmatrix} = \iota_5 + \iota_6 \cdot \lambda + \iota_7 \cdot \lambda^2$$

We can summary:

$$\Upsilon_{11} \cdot \det\begin{pmatrix} \Upsilon_{22} & \Upsilon_{23} & \Upsilon_{24} & \Upsilon_{25} \\ \Upsilon_{32} & \Upsilon_{33} & \Upsilon_{34} & \Upsilon_{35} \\ \Upsilon_{42} & \Upsilon_{43} & \Upsilon_{44} & \Upsilon_{45} \\ \Upsilon_{52} & \Upsilon_{53} & \Upsilon_{54} & \Upsilon_{55} \end{pmatrix}$$

$$= (\Xi_1 \cdot e^{-\lambda \cdot \tau_D} - \lambda) \cdot \det\begin{pmatrix} (-\Xi_2 - \lambda) & 0 & -\Xi_4 & 0 \\ 0 & (-\Xi_3 - \lambda) & 0 & -\Xi_5 \\ -\Xi_8 & 0 & (\Xi_9 - \lambda) & 0 \\ 0 & -\Xi_{10} & 0 & (\Xi_{11} - \lambda) \end{pmatrix}$$

$$= (\Xi_1 \cdot e^{-\lambda \cdot \tau_D} - \lambda) \cdot \{(-\Xi_2 - \lambda) \cdot \det\begin{pmatrix} (-\Xi_3 - \lambda) & 0 & -\Xi_5 \\ 0 & (\Xi_9 - \lambda) & 0 \\ -\Xi_{10} & 0 & (\Xi_{11} - \lambda) \end{pmatrix}$$

$$- \Xi_4 \cdot \det\begin{pmatrix} 0 & (-\Xi_3 - \lambda) & -\Xi_5 \\ -\Xi_8 & 0 & 0 \\ 0 & -\Xi_{10} & (\Xi_{11} - \lambda) \end{pmatrix}\}$$

$$= (\Xi_1 \cdot e^{-\lambda \cdot \tau_D} - \lambda) \cdot \{\iota_1 + \iota_2 \cdot \lambda + \iota_3 \cdot \lambda^2 + \iota_4 \cdot \lambda^3 + \lambda^4 + \iota_5 + \iota_6 \cdot \lambda + \iota_7 \cdot \lambda^2\}$$

$$= (\Xi_1 \cdot e^{-\lambda \cdot \tau_D} - \lambda) \cdot \{\iota_1 + \iota_5 + (\iota_2 + \iota_6) \cdot \lambda + (\iota_3 + \iota_7) \cdot \lambda^2 + \iota_4 \cdot \lambda^3 + \lambda^4\}$$

$$\Upsilon_{11} \cdot \det\begin{pmatrix} \Upsilon_{22} & \Upsilon_{23} & \Upsilon_{24} & \Upsilon_{25} \\ \Upsilon_{32} & \Upsilon_{33} & \Upsilon_{34} & \Upsilon_{35} \\ \Upsilon_{42} & \Upsilon_{43} & \Upsilon_{44} & \Upsilon_{45} \\ \Upsilon_{52} & \Upsilon_{53} & \Upsilon_{54} & \Upsilon_{55} \end{pmatrix}$$

$$= (\Xi_1 \cdot e^{-\lambda \cdot \tau_D} - \lambda) \cdot \{(\iota_1 + \iota_5) + (\iota_2 + \iota_6) \cdot \lambda + (\iota_3 + \iota_7) \cdot \lambda^2 + \iota_4 \cdot \lambda^3 + \lambda^4\}$$

$$= (\iota_1 + \iota_5) \cdot \Xi_1 \cdot e^{-\lambda \cdot \tau_D} + (\iota_2 + \iota_6) \cdot \lambda \cdot \Xi_1 \cdot e^{-\lambda \cdot \tau_D} + (\iota_3 + \iota_7) \cdot \lambda^2 \cdot \Xi_1 \cdot e^{-\lambda \cdot \tau_D}$$
$$+ \iota_4 \cdot \lambda^3 \cdot \Xi_1 \cdot e^{-\lambda \cdot \tau_D} + \lambda^4 \cdot \Xi_1 \cdot e^{-\lambda \cdot \tau_D} - (\iota_1 + \iota_5) \cdot \lambda - (\iota_2 + \iota_6) \cdot \lambda^2$$
$$- (\iota_3 + \iota_7) \cdot \lambda^3 - \iota_4 \cdot \lambda^4 - \lambda^5$$

$$\Upsilon_{11} \cdot \det\begin{pmatrix} \Upsilon_{22} & \Upsilon_{23} & \Upsilon_{24} & \Upsilon_{25} \\ \Upsilon_{32} & \Upsilon_{33} & \Upsilon_{34} & \Upsilon_{35} \\ \Upsilon_{42} & \Upsilon_{43} & \Upsilon_{44} & \Upsilon_{45} \\ \Upsilon_{52} & \Upsilon_{53} & \Upsilon_{54} & \Upsilon_{55} \end{pmatrix}$$

$$= -(\iota_1 + \iota_5) \cdot \lambda - (\iota_2 + \iota_6) \cdot \lambda^2 - (\iota_3 + \iota_7) \cdot \lambda^3 - \iota_4 \cdot \lambda^4 - \lambda^5$$
$$+ \{(\iota_1 + \iota_5) \cdot \Xi_1 + (\iota_2 + \iota_6) \cdot \lambda \cdot \Xi_1 + (\iota_3 + \iota_7) \cdot \lambda^2 \cdot \Xi_1$$
$$+ \iota_4 \cdot \lambda^3 \cdot \Xi_1 + \lambda^4 \cdot \Xi_1\} \cdot e^{-\lambda \cdot \tau_D}$$

Step 2: Find the expression for $\Upsilon_{12} \cdot \det\begin{pmatrix} \Upsilon_{21} & \Upsilon_{23} & \Upsilon_{24} & \Upsilon_{25} \\ \Upsilon_{31} & \Upsilon_{33} & \Upsilon_{34} & \Upsilon_{35} \\ \Upsilon_{41} & \Upsilon_{43} & \Upsilon_{44} & \Upsilon_{45} \\ \Upsilon_{51} & \Upsilon_{53} & \Upsilon_{54} & \Upsilon_{55} \end{pmatrix}$

$$\Upsilon_{12} \cdot \det \begin{pmatrix} \Upsilon_{21} & \Upsilon_{23} & \Upsilon_{24} & \Upsilon_{25} \\ \Upsilon_{31} & \Upsilon_{33} & \Upsilon_{34} & \Upsilon_{35} \\ \Upsilon_{41} & \Upsilon_{43} & \Upsilon_{44} & \Upsilon_{45} \\ \Upsilon_{51} & \Upsilon_{53} & \Upsilon_{54} & \Upsilon_{55} \end{pmatrix}$$

$$= \Xi_2 \cdot \det \begin{pmatrix} \Xi_6 \cdot e^{-\lambda \cdot \tau_D} & 0 & -\Xi_4 & 0 \\ \Xi_7 \cdot e^{-\lambda \cdot \tau_D} & (-\Xi_3 - \lambda) & 0 & -\Xi_5 \\ \Xi_8 \cdot e^{-\lambda \cdot \tau_D} & 0 & (\Xi_9 - \lambda) & 0 \\ \Xi_{10} \cdot e^{-\lambda \cdot \tau_D} & -\Xi_{10} & 0 & (\Xi_{11} - \lambda) \end{pmatrix}$$

$$= \Xi_2 \cdot \{\Xi_6 \cdot e^{-\lambda \cdot \tau_D} \cdot \det \begin{pmatrix} (-\Xi_3 - \lambda) & 0 & -\Xi_5 \\ 0 & (\Xi_9 - \lambda) & 0 \\ -\Xi_{10} & 0 & (\Xi_{11} - \lambda) \end{pmatrix}$$

$$- \Xi_4 \cdot \det \begin{pmatrix} \Xi_7 \cdot e^{-\lambda \cdot \tau_D} & (-\Xi_3 - \lambda) & -\Xi_5 \\ \Xi_8 \cdot e^{-\lambda \cdot \tau_D} & 0 & 0 \\ \Xi_{10} \cdot e^{-\lambda \cdot \tau_D} & -\Xi_{10} & (\Xi_{11} - \lambda) \end{pmatrix}\}$$

Step 2.1: $\Xi_6 \cdot e^{-\lambda \cdot \tau_D} \cdot \det \begin{pmatrix} (-\Xi_3 - \lambda) & 0 & -\Xi_5 \\ 0 & (\Xi_9 - \lambda) & 0 \\ -\Xi_{10} & 0 & (\Xi_{11} - \lambda) \end{pmatrix}$

$$\Xi_6 \cdot e^{-\lambda \cdot \tau_D} \cdot \det \begin{pmatrix} (-\Xi_3 - \lambda) & 0 & -\Xi_5 \\ 0 & (\Xi_9 - \lambda) & 0 \\ -\Xi_{10} & 0 & (\Xi_{11} - \lambda) \end{pmatrix}$$

$$= \Xi_6 \cdot e^{-\lambda \cdot \tau_D} \cdot \{(-\Xi_3 - \lambda) \cdot \det \begin{pmatrix} (\Xi_9 - \lambda) & 0 \\ 0 & (\Xi_{11} - \lambda) \end{pmatrix}$$

$$- \Xi_5 \cdot \det \begin{pmatrix} 0 & (\Xi_9 - \lambda) \\ -\Xi_{10} & 0 \end{pmatrix}\}$$

$$= \Xi_6 \cdot e^{-\lambda \cdot \tau_D} \cdot \{(-\Xi_3 - \lambda) \cdot (\Xi_9 - \lambda) \cdot (\Xi_{11} - \lambda) - \Xi_5 \cdot \Xi_{10} \cdot (\Xi_9 - \lambda)\}$$

$$\Xi_6 \cdot e^{-\lambda \cdot \tau_D} \cdot \det \begin{pmatrix} (-\Xi_3 - \lambda) & 0 & -\Xi_5 \\ 0 & (\Xi_9 - \lambda) & 0 \\ -\Xi_{10} & 0 & (\Xi_{11} - \lambda) \end{pmatrix}$$

$$= \Xi_6 \cdot e^{-\lambda \cdot \tau_D} \cdot \{(-\Xi_3 - \lambda) \cdot (\Xi_9 - \lambda)$$

$$\cdot (\Xi_{11} - \lambda) - \Xi_5 \cdot \Xi_{10} \cdot (\Xi_9 - \lambda)\}$$

$$= \Xi_6 \cdot e^{-\lambda \cdot \tau_D} \cdot \{(-\Xi_3 \cdot \Xi_9 + [\Xi_3 - \Xi_9] \cdot \lambda + \lambda^2)$$

$$\cdot (\Xi_{11} - \lambda) - \Xi_5 \cdot \Xi_{10} \cdot \Xi_9 + \Xi_5 \cdot \Xi_{10} \cdot \lambda\}$$

$$\Xi_6 \cdot e^{-\lambda \cdot \tau_D} \cdot \det\begin{pmatrix} (-\Xi_3 - \lambda) & 0 & -\Xi_5 \\ 0 & (\Xi_9 - \lambda) & 0 \\ -\Xi_{10} & 0 & (\Xi_{11} - \lambda) \end{pmatrix}$$
$$= \Xi_6 \cdot e^{-\lambda \cdot \tau_D} \cdot \{(-\Xi_3 - \lambda) \cdot (\Xi_9 - \lambda) \cdot (\Xi_{11} - \lambda) - \Xi_5 \cdot \Xi_{10} \cdot (\Xi_9 - \lambda)\}$$
$$= \Xi_6 \cdot e^{-\lambda \cdot \tau_D} \cdot \{-\Xi_3 \cdot \Xi_9 \cdot \Xi_{11} + \Xi_3 \cdot \Xi_9 \cdot \lambda + \Xi_{11} \cdot [\Xi_3 - \Xi_9] \cdot \lambda - [\Xi_3 - \Xi_9] \cdot \lambda^2 + \Xi_{11} \cdot \lambda^2 - \lambda^3 - \Xi_5 \cdot \Xi_{10} \cdot \Xi_9 + \Xi_5 \cdot \Xi_{10} \cdot \lambda\}$$

$$\Xi_6 \cdot e^{-\lambda \cdot \tau_D} \cdot \det\begin{pmatrix} (-\Xi_3 - \lambda) & 0 & -\Xi_5 \\ 0 & (\Xi_9 - \lambda) & 0 \\ -\Xi_{10} & 0 & (\Xi_{11} - \lambda) \end{pmatrix}$$
$$= \Xi_6 \cdot e^{-\lambda \cdot \tau_D} \cdot \{(-\Xi_3 - \lambda) \cdot (\Xi_9 - \lambda) \cdot (\Xi_{11} - \lambda) - \Xi_5 \cdot \Xi_{10} \cdot (\Xi_9 - \lambda)\}$$
$$= \Xi_6 \cdot e^{-\lambda \cdot \tau_D} \cdot \{(-\Xi_3 \cdot \Xi_{11} - \Xi_5 \cdot \Xi_{10}) \cdot \Xi_9 + (\Xi_3 \cdot \Xi_9 + \Xi_{11} \cdot [\Xi_3 - \Xi_9] + \Xi_5 \cdot \Xi_{10}) \cdot \lambda + (\Xi_{11} - \Xi_3 + \Xi_9) \cdot \lambda^2 - \lambda^3\}$$

$$\Xi_6 \cdot e^{-\lambda \cdot \tau_D} \cdot \det\begin{pmatrix} (-\Xi_3 - \lambda) & 0 & -\Xi_5 \\ 0 & (\Xi_9 - \lambda) & 0 \\ -\Xi_{10} & 0 & (\Xi_{11} - \lambda) \end{pmatrix}$$
$$= \Xi_6 \cdot e^{-\lambda \cdot \tau_D} \cdot \{(-\Xi_3 - \lambda) \cdot (\Xi_9 - \lambda) \cdot (\Xi_{11} - \lambda) - \Xi_5 \cdot \Xi_{10} \cdot (\Xi_9 - \lambda)\}$$
$$= \{\Xi_6 \cdot (-\Xi_3 \cdot \Xi_{11} - \Xi_5 \cdot \Xi_{10}) \cdot \Xi_9 + \Xi_6 \cdot (\Xi_3 \cdot \Xi_9 + \Xi_{11} \cdot [\Xi_3 - \Xi_9] + \Xi_5 \cdot \Xi_{10}) \cdot \lambda + \Xi_6 \cdot (\Xi_{11} - \Xi_3 + \Xi_9) \cdot \lambda^2 - \Xi_6 \cdot \lambda^3\} \cdot e^{-\lambda \cdot \tau_D}$$

We define some global parameters: $\iota_8 = \Xi_6 \cdot (-\Xi_3 \cdot \Xi_{11} - \Xi_5 \cdot \Xi_{10}) \cdot \Xi_9$

$$\iota_9 = \Xi_6 \cdot (\Xi_3 \cdot \Xi_9 + \Xi_{11} \cdot [\Xi_3 - \Xi_9] + \Xi_5 \cdot \Xi_{10})$$
$$\iota_{10} = \Xi_6 \cdot (\Xi_{11} - \Xi_3 + \Xi_9); \; \iota_{11} = -\Xi_6$$

$$\Xi_6 \cdot e^{-\lambda \cdot \tau_D} \cdot \det\begin{pmatrix} (-\Xi_3 - \lambda) & 0 & -\Xi_5 \\ 0 & (\Xi_9 - \lambda) & 0 \\ -\Xi_{10} & 0 & (\Xi_{11} - \lambda) \end{pmatrix}$$
$$= \{\iota_8 + \iota_9 \cdot \lambda + \iota_{10} \cdot \lambda^2 + \iota_{11} \cdot \lambda^3\} \cdot e^{-\lambda \cdot \tau_D}$$

$$\text{Step 2.2: } \Xi_4 \cdot \det\begin{pmatrix} \Xi_7 \cdot e^{-\lambda\cdot\tau_D} & (-\Xi_3 - \lambda) & -\Xi_5 \\ \Xi_8 \cdot e^{-\lambda\cdot\tau_D} & 0 & 0 \\ \Xi_{10} \cdot e^{-\lambda\cdot\tau_D} & -\Xi_{10} & (\Xi_{11} - \lambda) \end{pmatrix}$$

$$\Xi_4 \cdot \det\begin{pmatrix} \Xi_7 \cdot e^{-\lambda\cdot\tau_D} & (-\Xi_3 - \lambda) & -\Xi_5 \\ \Xi_8 \cdot e^{-\lambda\cdot\tau_D} & 0 & 0 \\ \Xi_{10} \cdot e^{-\lambda\cdot\tau_D} & -\Xi_{10} & (\Xi_{11} - \lambda) \end{pmatrix}$$

$$= \Xi_4 \cdot \{\Xi_7 \cdot e^{-\lambda\cdot\tau_D} \cdot \det\begin{pmatrix} 0 & 0 \\ -\Xi_{10} & (\Xi_{11} - \lambda) \end{pmatrix}$$

$$- (-\Xi_3 - \lambda) \cdot \det\begin{pmatrix} \Xi_8 \cdot e^{-\lambda\cdot\tau_D} & 0 \\ \Xi_{10} \cdot e^{-\lambda\cdot\tau_D} & (\Xi_{11} - \lambda) \end{pmatrix}$$

$$- \Xi_5 \cdot \det\begin{pmatrix} \Xi_8 \cdot e^{-\lambda\cdot\tau_D} & 0 \\ \Xi_{10} \cdot e^{-\lambda\cdot\tau_D} & -\Xi_{10} \end{pmatrix}\}$$

$$\Xi_4 \cdot \det\begin{pmatrix} \Xi_7 \cdot e^{-\lambda\cdot\tau_D} & (-\Xi_3 - \lambda) & -\Xi_5 \\ \Xi_8 \cdot e^{-\lambda\cdot\tau_D} & 0 & 0 \\ \Xi_{10} \cdot e^{-\lambda\cdot\tau_D} & -\Xi_{10} & (\Xi_{11} - \lambda) \end{pmatrix}$$

$$= \Xi_4 \cdot \{(\Xi_3 + \lambda) \cdot \det\begin{pmatrix} \Xi_8 \cdot e^{-\lambda\cdot\tau_D} & 0 \\ \Xi_{10} \cdot e^{-\lambda\cdot\tau_D} & (\Xi_{11} - \lambda) \end{pmatrix}$$

$$- \Xi_5 \cdot \det\begin{pmatrix} \Xi_8 \cdot e^{-\lambda\cdot\tau_D} & 0 \\ \Xi_{10} \cdot e^{-\lambda\cdot\tau_D} & -\Xi_{10} \end{pmatrix}\}$$

$$\Xi_4 \cdot \det\begin{pmatrix} \Xi_7 \cdot e^{-\lambda\cdot\tau_D} & (-\Xi_3 - \lambda) & -\Xi_5 \\ \Xi_8 \cdot e^{-\lambda\cdot\tau_D} & 0 & 0 \\ \Xi_{10} \cdot e^{-\lambda\cdot\tau_D} & -\Xi_{10} & (\Xi_{11} - \lambda) \end{pmatrix}$$

$$= \Xi_4 \cdot \{(\Xi_3 + \lambda) \cdot \Xi_8 \cdot e^{-\lambda\cdot\tau_D} \cdot (\Xi_{11} - \lambda)$$

$$+ \Xi_5 \cdot \Xi_8 \cdot e^{-\lambda\cdot\tau_D} \cdot \Xi_{10}\}$$

$$\Xi_4 \cdot \det\begin{pmatrix} \Xi_7 \cdot e^{-\lambda\cdot\tau_D} & (-\Xi_3 - \lambda) & -\Xi_5 \\ \Xi_8 \cdot e^{-\lambda\cdot\tau_D} & 0 & 0 \\ \Xi_{10} \cdot e^{-\lambda\cdot\tau_D} & -\Xi_{10} & (\Xi_{11} - \lambda) \end{pmatrix}$$

$$= \{\Xi_4 \cdot (\Xi_3 + \lambda) \cdot \Xi_8 \cdot (\Xi_{11} - \lambda)$$

$$+ \Xi_4 \cdot \Xi_5 \cdot \Xi_8 \cdot \Xi_{10}\} \cdot e^{-\lambda\cdot\tau_D}$$

$$\Xi_4 \cdot \det\begin{pmatrix} \Xi_7 \cdot e^{-\lambda\cdot\tau_D} & (-\Xi_3 - \lambda) & -\Xi_5 \\ \Xi_8 \cdot e^{-\lambda\cdot\tau_D} & 0 & 0 \\ \Xi_{10} \cdot e^{-\lambda\cdot\tau_D} & -\Xi_{10} & (\Xi_{11} - \lambda) \end{pmatrix}$$

$$= \{\Xi_4 \cdot \Xi_8 \cdot (\Xi_3 \cdot \Xi_{11} + [\Xi_{11} - \Xi_3] \cdot \lambda - \lambda^2)$$

$$+ \Xi_4 \cdot \Xi_5 \cdot \Xi_8 \cdot \Xi_{10}\} \cdot e^{-\lambda \cdot \tau_D}$$

$$\Xi_4 \cdot \det\begin{pmatrix} \Xi_7 \cdot e^{-\lambda \cdot \tau_D} & (-\Xi_3 - \lambda) & -\Xi_5 \\ \Xi_8 \cdot e^{-\lambda \cdot \tau_D} & 0 & 0 \\ \Xi_{10} \cdot e^{-\lambda \cdot \tau_D} & -\Xi_{10} & (\Xi_{11} - \lambda) \end{pmatrix}$$
$$= \{(\Xi_3 \cdot \Xi_{11} + \Xi_5 \cdot \Xi_{10}) \cdot \Xi_8 \cdot \Xi_4 + \Xi_4 \cdot \Xi_8 \cdot [\Xi_{11} - \Xi_3] \cdot \lambda - \Xi_4 \cdot \Xi_8 \cdot \lambda^2\} \cdot e^{-\lambda \cdot \tau_D}$$

We define some global parameters: $\iota_{12} = (\Xi_3 \cdot \Xi_{11} + \Xi_5 \cdot \Xi_{10}) \cdot \Xi_8 \cdot \Xi_4$; $\iota_{13} = \Xi_4 \cdot \Xi_8 \cdot [\Xi_{11} - \Xi_3]$

$$\iota_{14} = -\Xi_4 \cdot \Xi_8;\ \Xi_4 \cdot \det\begin{pmatrix} \Xi_7 \cdot e^{-\lambda \cdot \tau_D} & (-\Xi_3 - \lambda) & -\Xi_5 \\ \Xi_8 \cdot e^{-\lambda \cdot \tau_D} & 0 & 0 \\ \Xi_{10} \cdot e^{-\lambda \cdot \tau_D} & -\Xi_{10} & (\Xi_{11} - \lambda) \end{pmatrix}$$
$$= \{\iota_{12} + \iota_{13} \cdot \lambda + \iota_{14} \cdot \lambda^2\} \cdot e^{-\lambda \cdot \tau_D}$$

We can summary:

$$\Upsilon_{12} \cdot \det\begin{pmatrix} \Upsilon_{21} & \Upsilon_{23} & \Upsilon_{24} & \Upsilon_{25} \\ \Upsilon_{31} & \Upsilon_{33} & \Upsilon_{34} & \Upsilon_{35} \\ \Upsilon_{41} & \Upsilon_{43} & \Upsilon_{44} & \Upsilon_{45} \\ \Upsilon_{51} & \Upsilon_{53} & \Upsilon_{54} & \Upsilon_{55} \end{pmatrix}$$
$$= \Xi_2 \cdot \{\Xi_6 \cdot e^{-\lambda \cdot \tau_D} \cdot \det\begin{pmatrix} (-\Xi_3 - \lambda) & 0 & -\Xi_5 \\ 0 & (\Xi_9 - \lambda) & 0 \\ -\Xi_{10} & 0 & (\Xi_{11} - \lambda) \end{pmatrix}$$
$$- \Xi_4 \cdot \det\begin{pmatrix} \Xi_7 \cdot e^{-\lambda \cdot \tau_D} & (-\Xi_3 - \lambda) & -\Xi_5 \\ \Xi_8 \cdot e^{-\lambda \cdot \tau_D} & 0 & 0 \\ \Xi_{10} \cdot e^{-\lambda \cdot \tau_D} & -\Xi_{10} & (\Xi_{11} - \lambda) \end{pmatrix}\Bigg\}$$
$$= \Xi_2 \cdot \{(\iota_8 + \iota_9 \cdot \lambda + \iota_{10} \cdot \lambda^2 + \iota_{11} \cdot \lambda^3) \cdot e^{-\lambda \cdot \tau_D} - (\iota_{12} + \iota_{13} \cdot \lambda + \iota_{14} \cdot \lambda^2) \cdot e^{-\lambda \cdot \tau_D}\}$$

$$\Upsilon_{12} \cdot \det\begin{pmatrix} \Upsilon_{21} & \Upsilon_{23} & \Upsilon_{24} & \Upsilon_{25} \\ \Upsilon_{31} & \Upsilon_{33} & \Upsilon_{34} & \Upsilon_{35} \\ \Upsilon_{41} & \Upsilon_{43} & \Upsilon_{44} & \Upsilon_{45} \\ \Upsilon_{51} & \Upsilon_{53} & \Upsilon_{54} & \Upsilon_{55} \end{pmatrix}$$
$$= \Xi_2 \cdot \{(\iota_8 - \iota_{12}) + (\iota_9 - \iota_{13}) \cdot \lambda + (\iota_{10} - \iota_{14}) \cdot \lambda^2 + \iota_{11} \cdot \lambda^3\} \cdot e^{-\lambda \cdot \tau_D}$$

$$\Upsilon_{12}\cdot\det\begin{pmatrix}\Upsilon_{21} & \Upsilon_{23} & \Upsilon_{24} & \Upsilon_{25}\\ \Upsilon_{31} & \Upsilon_{33} & \Upsilon_{34} & \Upsilon_{35}\\ \Upsilon_{41} & \Upsilon_{43} & \Upsilon_{44} & \Upsilon_{45}\\ \Upsilon_{51} & \Upsilon_{53} & \Upsilon_{54} & \Upsilon_{55}\end{pmatrix}$$
$$=\{\Xi_2\cdot(\iota_8-\iota_{12})+\Xi_2\cdot(\iota_9-\iota_{13})\cdot\lambda$$
$$+\Xi_2\cdot(\iota_{10}-\iota_{14})\cdot\lambda^2+\Xi_2\cdot\iota_{11}\cdot\lambda^3\}\cdot e^{-\lambda\cdot\tau_D}$$

Step 3: Find the expression for $\Upsilon_{13}\cdot\det\begin{pmatrix}\Upsilon_{21} & \Upsilon_{22} & \Upsilon_{24} & \Upsilon_{25}\\ \Upsilon_{31} & \Upsilon_{32} & \Upsilon_{34} & \Upsilon_{35}\\ \Upsilon_{41} & \Upsilon_{42} & \Upsilon_{44} & \Upsilon_{45}\\ \Upsilon_{51} & \Upsilon_{52} & \Upsilon_{54} & \Upsilon_{55}\end{pmatrix}$

$$\Xi_3\cdot\det\begin{pmatrix}\Xi_6\cdot e^{-\lambda\cdot\tau_D} & (-\Xi_2-\lambda) & -\Xi_4 & 0\\ \Xi_7\cdot e^{-\lambda\cdot\tau_D} & 0 & 0 & -\Xi_5\\ \Xi_8\cdot e^{-\lambda\cdot\tau_D} & -\Xi_8 & (\Xi_9-\lambda) & 0\\ \Xi_{10}\cdot e^{-\lambda\cdot\tau_D} & 0 & 0 & (\Xi_{11}-\lambda)\end{pmatrix}$$
$$=\Xi_3\cdot\{\Xi_6\cdot e^{-\lambda\cdot\tau_D}\cdot\det\begin{pmatrix}0 & 0 & -\Xi_5\\ -\Xi_8 & (\Xi_9-\lambda) & 0\\ 0 & 0 & (\Xi_{11}-\lambda)\end{pmatrix}$$
$$-(-\Xi_2-\lambda)\cdot\det\begin{pmatrix}\Xi_7\cdot e^{-\lambda\cdot\tau_D} & 0 & -\Xi_5\\ \Xi_8\cdot e^{-\lambda\cdot\tau_D} & (\Xi_9-\lambda) & 0\\ \Xi_{10}\cdot e^{-\lambda\cdot\tau_D} & 0 & (\Xi_{11}-\lambda)\end{pmatrix}$$
$$-\Xi_4\cdot\det\begin{pmatrix}\Xi_7\cdot e^{-\lambda\cdot\tau_D} & 0 & -\Xi_5\\ \Xi_8\cdot e^{-\lambda\cdot\tau_D} & -\Xi_8 & 0\\ \Xi_{10}\cdot e^{-\lambda\cdot\tau_D} & 0 & (\Xi_{11}-\lambda)\end{pmatrix}\}$$

$$\det\begin{pmatrix}0 & 0 & -\Xi_5\\ -\Xi_8 & (\Xi_9-\lambda) & 0\\ 0 & 0 & (\Xi_{11}-\lambda)\end{pmatrix}$$
$$=-\Xi_5\cdot\det\begin{pmatrix}-\Xi_8 & (\Xi_9-\lambda)\\ 0 & 0\end{pmatrix}=0$$

$$\Xi_3\cdot\det\begin{pmatrix}\Xi_6\cdot e^{-\lambda\cdot\tau_D} & (-\Xi_2-\lambda) & -\Xi_4 & 0\\ \Xi_7\cdot e^{-\lambda\cdot\tau_D} & 0 & 0 & -\Xi_5\\ \Xi_8\cdot e^{-\lambda\cdot\tau_D} & -\Xi_8 & (\Xi_9-\lambda) & 0\\ \Xi_{10}\cdot e^{-\lambda\cdot\tau_D} & 0 & 0 & (\Xi_{11}-\lambda)\end{pmatrix}$$
$$=\Xi_3\cdot\{(\Xi_2+\lambda)\cdot\det\begin{pmatrix}\Xi_7\cdot e^{-\lambda\cdot\tau_D} & 0 & -\Xi_5\\ \Xi_8\cdot e^{-\lambda\cdot\tau_D} & (\Xi_9-\lambda) & 0\\ \Xi_{10}\cdot e^{-\lambda\cdot\tau_D} & 0 & (\Xi_{11}-\lambda)\end{pmatrix}$$

$$- \Xi_4 \cdot \det\begin{pmatrix} \Xi_7 \cdot e^{-\lambda \cdot \tau_D} & 0 & -\Xi_5 \\ \Xi_8 \cdot e^{-\lambda \cdot \tau_D} & -\Xi_8 & 0 \\ \Xi_{10} \cdot e^{-\lambda \cdot \tau_D} & 0 & (\Xi_{11} - \lambda) \end{pmatrix}\}$$

Step 3.1: $(\Xi_2 + \lambda) \cdot \det\begin{pmatrix} \Xi_7 \cdot e^{-\lambda \cdot \tau_D} & 0 & -\Xi_5 \\ \Xi_8 \cdot e^{-\lambda \cdot \tau_D} & (\Xi_9 - \lambda) & 0 \\ \Xi_{10} \cdot e^{-\lambda \cdot \tau_D} & 0 & (\Xi_{11} - \lambda) \end{pmatrix}$

$$(\Xi_2 + \lambda) \cdot \det\begin{pmatrix} \Xi_7 \cdot e^{-\lambda \cdot \tau_D} & 0 & -\Xi_5 \\ \Xi_8 \cdot e^{-\lambda \cdot \tau_D} & (\Xi_9 - \lambda) & 0 \\ \Xi_{10} \cdot e^{-\lambda \cdot \tau_D} & 0 & (\Xi_{11} - \lambda) \end{pmatrix}$$

$$= (\Xi_2 + \lambda) \cdot \{\Xi_7 \cdot e^{-\lambda \cdot \tau_D} \cdot \det\begin{pmatrix} (\Xi_9 - \lambda) & 0 \\ 0 & (\Xi_{11} - \lambda) \end{pmatrix}$$

$$- \Xi_5 \cdot \det\begin{pmatrix} \Xi_8 \cdot e^{-\lambda \cdot \tau_D} & (\Xi_9 - \lambda) \\ \Xi_{10} \cdot e^{-\lambda \cdot \tau_D} & 0 \end{pmatrix}\}$$

$$= (\Xi_2 + \lambda) \cdot \{\Xi_7 \cdot e^{-\lambda \cdot \tau_D} \cdot (\Xi_9 - \lambda) \cdot (\Xi_{11} - \lambda)$$

$$+ \Xi_5 \cdot \Xi_{10} \cdot e^{-\lambda \cdot \tau_D} \cdot (\Xi_9 - \lambda)\}$$

$$(\Xi_2 + \lambda) \cdot \det\begin{pmatrix} \Xi_7 \cdot e^{-\lambda \cdot \tau_D} & 0 & -\Xi_5 \\ \Xi_8 \cdot e^{-\lambda \cdot \tau_D} & (\Xi_9 - \lambda) & 0 \\ \Xi_{10} \cdot e^{-\lambda \cdot \tau_D} & 0 & (\Xi_{11} - \lambda) \end{pmatrix}$$

$$= (\Xi_2 + \lambda) \cdot \{\Xi_7 \cdot e^{-\lambda \cdot \tau_D} \cdot (\Xi_9 \cdot \Xi_{11} - [\Xi_9 + \Xi_{11}]$$

$$\cdot \lambda + \lambda^2) + \Xi_9 \cdot \Xi_5 \cdot \Xi_{10} \cdot e^{-\lambda \cdot \tau_D} - \lambda \cdot \Xi_5 \cdot \Xi_{10} \cdot e^{-\lambda \cdot \tau_D}\}$$

$$(\Xi_2 + \lambda) \cdot \det\begin{pmatrix} \Xi_7 \cdot e^{-\lambda \cdot \tau_D} & 0 & -\Xi_5 \\ \Xi_8 \cdot e^{-\lambda \cdot \tau_D} & (\Xi_9 - \lambda) & 0 \\ \Xi_{10} \cdot e^{-\lambda \cdot \tau_D} & 0 & (\Xi_{11} - \lambda) \end{pmatrix}$$

$$= (\Xi_2 + \lambda) \cdot \{\Xi_7 \cdot \Xi_9 \cdot \Xi_{11} \cdot e^{-\lambda \cdot \tau_D} - \Xi_7 \cdot [\Xi_9 + \Xi_{11}]$$

$$\cdot \lambda \cdot e^{-\lambda \cdot \tau_D} + \Xi_7 \cdot \lambda^2 \cdot e^{-\lambda \cdot \tau_D} + \Xi_9 \cdot \Xi_5 \cdot \Xi_{10}$$

$$\cdot e^{-\lambda \cdot \tau_D} - \lambda \cdot \Xi_5 \cdot \Xi_{10} \cdot e^{-\lambda \cdot \tau_D}\}$$

$$(\Xi_2 + \lambda) \cdot \det\begin{pmatrix} \Xi_7 \cdot e^{-\lambda \cdot \tau_D} & 0 & -\Xi_5 \\ \Xi_8 \cdot e^{-\lambda \cdot \tau_D} & (\Xi_9 - \lambda) & 0 \\ \Xi_{10} \cdot e^{-\lambda \cdot \tau_D} & 0 & (\Xi_{11} - \lambda) \end{pmatrix}$$

$$= (\Xi_2 + \lambda) \cdot \{(\Xi_7 \cdot \Xi_9 \cdot \Xi_{11} + \Xi_9 \cdot \Xi_5 \cdot \Xi_{10})$$

$$- (\Xi_7 \cdot [\Xi_9 + \Xi_{11}] + \Xi_5 \cdot \Xi_{10}) \cdot \lambda + \Xi_7 \cdot \lambda^2\} \cdot e^{-\lambda \cdot \tau_D}$$

$$(\Xi_2+\lambda)\cdot\det\begin{pmatrix}\Xi_7\cdot e^{-\lambda\cdot\tau_D} & 0 & -\Xi_5\\ \Xi_8\cdot e^{-\lambda\cdot\tau_D} & (\Xi_9-\lambda) & 0\\ \Xi_{10}\cdot e^{-\lambda\cdot\tau_D} & 0 & (\Xi_{11}-\lambda)\end{pmatrix}$$
$$=\{\Xi_2\cdot(\Xi_7\cdot\Xi_9\cdot\Xi_{11}+\Xi_9\cdot\Xi_5\cdot\Xi_{10})$$
$$+[(\Xi_7\cdot\Xi_9\cdot\Xi_{11}+\Xi_9\cdot\Xi_5\cdot\Xi_{10})-\Xi_2\cdot(\Xi_7\cdot[\Xi_9+\Xi_{11}]+\Xi_5\cdot\Xi_{10})]$$
$$\cdot\lambda+[\Xi_2\cdot\Xi_7-(\Xi_7\cdot[\Xi_9+\Xi_{11}]+\Xi_5\cdot\Xi_{10})]\cdot\lambda^2+\Xi_7\cdot\lambda^3\}\cdot e^{-\lambda\cdot\tau_D}$$

We define some global parameters: $\iota_{15}=\Xi_2\cdot(\Xi_7\cdot\Xi_9\cdot\Xi_{11}+\Xi_9\cdot\Xi_5\cdot\Xi_{10})$

$$\iota_{16}=(\Xi_7\cdot\Xi_9\cdot\Xi_{11}+\Xi_9\cdot\Xi_5\cdot\Xi_{10})-\Xi_2\cdot(\Xi_7\cdot[\Xi_9+\Xi_{11}]+\Xi_5\cdot\Xi_{10})$$

$$\iota_{17}=\Xi_2\cdot\Xi_7-(\Xi_7\cdot[\Xi_9+\Xi_{11}]+\Xi_5\cdot\Xi_{10});\ \iota_{18}=\Xi_7$$

$$(\Xi_2+\lambda)\cdot\det\begin{pmatrix}\Xi_7\cdot e^{-\lambda\cdot\tau_D} & 0 & -\Xi_5\\ \Xi_8\cdot e^{-\lambda\cdot\tau_D} & (\Xi_9-\lambda) & 0\\ \Xi_{10}\cdot e^{-\lambda\cdot\tau_D} & 0 & (\Xi_{11}-\lambda)\end{pmatrix}$$
$$=\{\iota_{15}+\iota_{16}\cdot\lambda+\iota_{17}\cdot\lambda^2+\iota_{18}\cdot\lambda^3\}\cdot e^{-\lambda\cdot\tau_D}$$

Step 3.2: $\Xi_4\cdot\det\begin{pmatrix}\Xi_7\cdot e^{-\lambda\cdot\tau_D} & 0 & -\Xi_5\\ \Xi_8\cdot e^{-\lambda\cdot\tau_D} & -\Xi_8 & 0\\ \Xi_{10}\cdot e^{-\lambda\cdot\tau_D} & 0 & (\Xi_{11}-\lambda)\end{pmatrix}$

$$\Xi_4\cdot\det\begin{pmatrix}\Xi_7\cdot e^{-\lambda\cdot\tau_D} & 0 & -\Xi_5\\ \Xi_8\cdot e^{-\lambda\cdot\tau_D} & -\Xi_8 & 0\\ \Xi_{10}\cdot e^{-\lambda\cdot\tau_D} & 0 & (\Xi_{11}-\lambda)\end{pmatrix}$$
$$=\Xi_4\cdot\{\Xi_7\cdot e^{-\lambda\cdot\tau_D}\cdot\det\begin{pmatrix}-\Xi_8 & 0\\ 0 & (\Xi_{11}-\lambda)\end{pmatrix}$$
$$-\Xi_5\cdot\det\begin{pmatrix}\Xi_8\cdot e^{-\lambda\cdot\tau_D} & -\Xi_8\\ \Xi_{10}\cdot e^{-\lambda\cdot\tau_D} & 0\end{pmatrix}\}$$
$$=\Xi_4\cdot\{\Xi_7\cdot e^{-\lambda\cdot\tau_D}\cdot(-\Xi_8)\cdot(\Xi_{11}-\lambda)-\Xi_5\cdot\Xi_8\cdot\Xi_{10}\cdot e^{-\lambda\cdot\tau_D}\}$$

$$\Xi_4\cdot\det\begin{pmatrix}\Xi_7\cdot e^{-\lambda\cdot\tau_D} & 0 & -\Xi_5\\ \Xi_8\cdot e^{-\lambda\cdot\tau_D} & -\Xi_8 & 0\\ \Xi_{10}\cdot e^{-\lambda\cdot\tau_D} & 0 & (\Xi_{11}-\lambda)\end{pmatrix}$$
$$=\Xi_4\cdot\{\Xi_7\cdot e^{-\lambda\cdot\tau_D}\cdot\det\begin{pmatrix}-\Xi_8 & 0\\ 0 & (\Xi_{11}-\lambda)\end{pmatrix}$$
$$-\Xi_5\cdot\det\begin{pmatrix}\Xi_8\cdot e^{-\lambda\cdot\tau_D} & -\Xi_8\\ \Xi_{10}\cdot e^{-\lambda\cdot\tau_D} & 0\end{pmatrix}\}$$

$$= \{-\Xi_4 \cdot \Xi_8 \cdot (\Xi_7 \cdot \Xi_{11} + \Xi_5 \cdot \Xi_{10}) + \Xi_4 \cdot \Xi_7 \cdot \Xi_8 \cdot \lambda\} \cdot e^{-\lambda \cdot \tau_D}$$

We define some global parameters: $\iota_{19} = -\Xi_4 \cdot \Xi_8 \cdot (\Xi_7 \cdot \Xi_{11} + \Xi_5 \cdot \Xi_{10})$; $\iota_{20} = \Xi_4 \cdot \Xi_7 \cdot \Xi_8$

$$\Xi_4 \cdot \det\begin{pmatrix} \Xi_7 \cdot e^{-\lambda \cdot \tau_D} & 0 & -\Xi_5 \\ \Xi_8 \cdot e^{-\lambda \cdot \tau_D} & -\Xi_8 & 0 \\ \Xi_{10} \cdot e^{-\lambda \cdot \tau_D} & 0 & (\Xi_{11} - \lambda) \end{pmatrix}$$

$$= \Xi_4 \cdot \{\Xi_7 \cdot e^{-\lambda \cdot \tau_D} \cdot \det\begin{pmatrix} -\Xi_8 & 0 \\ 0 & (\Xi_{11} - \lambda) \end{pmatrix}$$

$$- \Xi_5 \cdot \det\begin{pmatrix} \Xi_8 \cdot e^{-\lambda \cdot \tau_D} & -\Xi_8 \\ \Xi_{10} \cdot e^{-\lambda \cdot \tau_D} & 0 \end{pmatrix}\} = \{\iota_{19} + \iota_{20} \cdot \lambda\} \cdot e^{-\lambda \cdot \tau_D}$$

We can summary:

$$\Upsilon_{13} \cdot \det\begin{pmatrix} \Upsilon_{21} & \Upsilon_{22} & \Upsilon_{24} & \Upsilon_{25} \\ \Upsilon_{31} & \Upsilon_{32} & \Upsilon_{34} & \Upsilon_{35} \\ \Upsilon_{41} & \Upsilon_{42} & \Upsilon_{44} & \Upsilon_{45} \\ \Upsilon_{51} & \Upsilon_{52} & \Upsilon_{54} & \Upsilon_{55} \end{pmatrix}$$

$$== \Xi_3 \cdot \{(\Xi_2 + \lambda) \cdot \det\begin{pmatrix} \Xi_7 \cdot e^{-\lambda \cdot \tau_D} & 0 & -\Xi_5 \\ \Xi_8 \cdot e^{-\lambda \cdot \tau_D} & (\Xi_9 - \lambda) & 0 \\ \Xi_{10} \cdot e^{-\lambda \cdot \tau_D} & 0 & (\Xi_{11} - \lambda) \end{pmatrix}$$

$$- \Xi_4 \cdot \det\begin{pmatrix} \Xi_7 \cdot e^{-\lambda \cdot \tau_D} & 0 & -\Xi_5 \\ \Xi_8 \cdot e^{-\lambda \cdot \tau_D} & -\Xi_8 & 0 \\ \Xi_{10} \cdot e^{-\lambda \cdot \tau_D} & 0 & (\Xi_{11} - \lambda) \end{pmatrix}\}$$

$$= \Xi_3 \cdot \{(\iota_{15} + \iota_{16} \cdot \lambda + \iota_{17} \cdot \lambda^2 + \iota_{18} \cdot \lambda^3) \cdot e^{-\lambda \cdot \tau_D} - (\iota_{19} + \iota_{20} \cdot \lambda) \cdot e^{-\lambda \cdot \tau_D}\}$$

$$\Upsilon_{13} \cdot \det\begin{pmatrix} \Upsilon_{21} & \Upsilon_{22} & \Upsilon_{24} & \Upsilon_{25} \\ \Upsilon_{31} & \Upsilon_{32} & \Upsilon_{34} & \Upsilon_{35} \\ \Upsilon_{41} & \Upsilon_{42} & \Upsilon_{44} & \Upsilon_{45} \\ \Upsilon_{51} & \Upsilon_{52} & \Upsilon_{54} & \Upsilon_{55} \end{pmatrix}$$

$$== \Xi_3 \cdot \{(\Xi_2 + \lambda) \cdot \det\begin{pmatrix} \Xi_7 \cdot e^{-\lambda \cdot \tau_D} & 0 & -\Xi_5 \\ \Xi_8 \cdot e^{-\lambda \cdot \tau_D} & (\Xi_9 - \lambda) & 0 \\ \Xi_{10} \cdot e^{-\lambda \cdot \tau_D} & 0 & (\Xi_{11} - \lambda) \end{pmatrix}$$

$$- \Xi_4 \cdot \det\begin{pmatrix} \Xi_7 \cdot e^{-\lambda \cdot \tau_D} & 0 & -\Xi_5 \\ \Xi_8 \cdot e^{-\lambda \cdot \tau_D} & -\Xi_8 & 0 \\ \Xi_{10} \cdot e^{-\lambda \cdot \tau_D} & 0 & (\Xi_{11} - \lambda) \end{pmatrix}\}$$

$$= \Xi_3 \cdot \{(\iota_{15} - \iota_{19}) + (\iota_{16} - \iota_{20}) \cdot \lambda + \iota_{17} \cdot \lambda^2 + \iota_{18} \cdot \lambda^3\} \cdot e^{-\lambda \cdot \tau_D}$$

$$\Upsilon_{13} \cdot \det\begin{pmatrix} \Upsilon_{21} & \Upsilon_{22} & \Upsilon_{24} & \Upsilon_{25} \\ \Upsilon_{31} & \Upsilon_{32} & \Upsilon_{34} & \Upsilon_{35} \\ \Upsilon_{41} & \Upsilon_{42} & \Upsilon_{44} & \Upsilon_{45} \\ \Upsilon_{51} & \Upsilon_{52} & \Upsilon_{54} & \Upsilon_{55} \end{pmatrix}$$

$$== \Xi_3 \cdot \{(\Xi_2 + \lambda) \cdot \det\begin{pmatrix} \Xi_7 \cdot e^{-\lambda \cdot \tau_D} & 0 & -\Xi_5 \\ \Xi_8 \cdot e^{-\lambda \cdot \tau_D} & (\Xi_9 - \lambda) & 0 \\ \Xi_{10} \cdot e^{-\lambda \cdot \tau_D} & 0 & (\Xi_{11} - \lambda) \end{pmatrix}$$

$$- \Xi_4 \cdot \det\begin{pmatrix} \Xi_7 \cdot e^{-\lambda \cdot \tau_D} & 0 & -\Xi_5 \\ \Xi_8 \cdot e^{-\lambda \cdot \tau_D} & -\Xi_8 & 0 \\ \Xi_{10} \cdot e^{-\lambda \cdot \tau_D} & 0 & (\Xi_{11} - \lambda) \end{pmatrix}\}$$

$$= \{\Xi_3 \cdot (\iota_{15} - \iota_{19}) + \Xi_3 \cdot (\iota_{16} - \iota_{20}) \cdot \lambda$$
$$+ \Xi_3 \cdot \iota_{17} \cdot \lambda^2 + \Xi_3 \cdot \iota_{18} \cdot \lambda^3\} \cdot e^{-\lambda \cdot \tau_D}$$

Step 4: Find the expression for $\Upsilon_{14} \cdot \det\begin{pmatrix} \Upsilon_{21} & \Upsilon_{22} & \Upsilon_{23} & \Upsilon_{25} \\ \Upsilon_{31} & \Upsilon_{32} & \Upsilon_{33} & \Upsilon_{35} \\ \Upsilon_{41} & \Upsilon_{42} & \Upsilon_{43} & \Upsilon_{45} \\ \Upsilon_{51} & \Upsilon_{52} & \Upsilon_{53} & \Upsilon_{55} \end{pmatrix}$

$$\Upsilon_{14} \cdot \det\begin{pmatrix} \Upsilon_{21} & \Upsilon_{22} & \Upsilon_{23} & \Upsilon_{25} \\ \Upsilon_{31} & \Upsilon_{32} & \Upsilon_{33} & \Upsilon_{35} \\ \Upsilon_{41} & \Upsilon_{42} & \Upsilon_{43} & \Upsilon_{45} \\ \Upsilon_{51} & \Upsilon_{52} & \Upsilon_{53} & \Upsilon_{55} \end{pmatrix}$$

$$= \Xi_4 \cdot \det\begin{pmatrix} \Xi_6 \cdot e^{-\lambda \cdot \tau_D} & (-\Xi_2 - \lambda) & 0 & 0 \\ \Xi_7 \cdot e^{-\lambda \cdot \tau_D} & 0 & (-\Xi_3 - \lambda) & -\Xi_5 \\ \Xi_8 \cdot e^{-\lambda \cdot \tau_D} & -\Xi_8 & 0 & 0 \\ \Xi_{10} \cdot e^{-\lambda \cdot \tau_D} & 0 & -\Xi_{10} & (\Xi_{11} - \lambda) \end{pmatrix}$$

$$\Xi_4 \cdot \det\begin{pmatrix} \Xi_6 \cdot e^{-\lambda \cdot \tau_D} & (-\Xi_2 - \lambda) & 0 & 0 \\ \Xi_7 \cdot e^{-\lambda \cdot \tau_D} & 0 & (-\Xi_3 - \lambda) & -\Xi_5 \\ \Xi_8 \cdot e^{-\lambda \cdot \tau_D} & -\Xi_8 & 0 & 0 \\ \Xi_{10} \cdot e^{-\lambda \cdot \tau_D} & 0 & -\Xi_{10} & (\Xi_{11} - \lambda) \end{pmatrix}$$

$$= \Xi_4 \cdot \{\Xi_6 \cdot e^{-\lambda \cdot \tau_D} \cdot \det\begin{pmatrix} 0 & (-\Xi_3 - \lambda) & -\Xi_5 \\ -\Xi_8 & 0 & 0 \\ 0 & -\Xi_{10} & (\Xi_{11} - \lambda) \end{pmatrix}$$

$$- (-\Xi_2 - \lambda) \cdot \det\begin{pmatrix} \Xi_7 \cdot e^{-\lambda \cdot \tau_D} & (-\Xi_3 - \lambda) & -\Xi_5 \\ \Xi_8 \cdot e^{-\lambda \cdot \tau_D} & 0 & 0 \\ \Xi_{10} \cdot e^{-\lambda \cdot \tau_D} & -\Xi_{10} & (\Xi_{11} - \lambda) \end{pmatrix}\}$$

Step 4.1: $\Xi_6 \cdot e^{-\lambda \cdot \tau_D} \cdot \det\begin{pmatrix} 0 & (-\Xi_3 - \lambda) & -\Xi_5 \\ -\Xi_8 & 0 & 0 \\ 0 & -\Xi_{10} & (\Xi_{11} - \lambda) \end{pmatrix}$

$$\Xi_6 \cdot e^{-\lambda \cdot \tau_D} \cdot \det\begin{pmatrix} 0 & (-\Xi_3 - \lambda) & -\Xi_5 \\ -\Xi_8 & 0 & 0 \\ 0 & -\Xi_{10} & (\Xi_{11} - \lambda) \end{pmatrix}$$
$$= \Xi_6 \cdot e^{-\lambda \cdot \tau_D} \cdot \{-(-\Xi_3 - \lambda) \cdot \det\begin{pmatrix} -\Xi_8 & 0 \\ 0 & (\Xi_{11} - \lambda) \end{pmatrix}$$
$$- \Xi_5 \cdot \det\begin{pmatrix} -\Xi_8 & 0 \\ 0 & -\Xi_{10} \end{pmatrix}\}$$
$$= \Xi_6 \cdot e^{-\lambda \cdot \tau_D} \cdot \{(\Xi_3 + \lambda) \cdot (-\Xi_8) \cdot (\Xi_{11} - \lambda) - \Xi_5 \cdot \Xi_8 \cdot \Xi_{10}\}$$

$$\Xi_6 \cdot e^{-\lambda \cdot \tau_D} \cdot \det\begin{pmatrix} 0 & (-\Xi_3 - \lambda) & -\Xi_5 \\ -\Xi_8 & 0 & 0 \\ 0 & -\Xi_{10} & (\Xi_{11} - \lambda) \end{pmatrix}$$
$$= \Xi_6 \cdot e^{-\lambda \cdot \tau_D} \cdot \{-(-\Xi_3 - \lambda) \cdot \det\begin{pmatrix} -\Xi_8 & 0 \\ 0 & (\Xi_{11} - \lambda) \end{pmatrix}$$
$$- \Xi_5 \cdot \det\begin{pmatrix} -\Xi_8 & 0 \\ 0 & -\Xi_{10} \end{pmatrix}\}$$
$$= \Xi_6 \cdot e^{-\lambda \cdot \tau_D} \cdot \{(-\Xi_8) \cdot (\Xi_3 \cdot \Xi_{11} + (\Xi_{11} - \Xi_3) \cdot \lambda - \lambda^2)$$
$$- \Xi_5 \cdot \Xi_8 \cdot \Xi_{10}\}$$

$$\Xi_6 \cdot e^{-\lambda \cdot \tau_D} \cdot \det\begin{pmatrix} 0 & (-\Xi_3 - \lambda) & -\Xi_5 \\ -\Xi_8 & 0 & 0 \\ 0 & -\Xi_{10} & (\Xi_{11} - \lambda) \end{pmatrix}$$
$$= \Xi_6 \cdot e^{-\lambda \cdot \tau_D} \cdot \{-(-\Xi_3 - \lambda) \cdot \det\begin{pmatrix} -\Xi_8 & 0 \\ 0 & (\Xi_{11} - \lambda) \end{pmatrix}$$
$$- \Xi_5 \cdot \det\begin{pmatrix} -\Xi_8 & 0 \\ 0 & -\Xi_{10} \end{pmatrix}\}$$
$$= \Xi_6 \cdot e^{-\lambda \cdot \tau_D} \cdot \{-\Xi_8 \cdot \Xi_3 \cdot \Xi_{11} + \Xi_8 \cdot (-\Xi_{11} + \Xi_3)$$
$$\cdot \lambda + \Xi_8 \cdot \lambda^2 - \Xi_5 \cdot \Xi_8 \cdot \Xi_{10}\}$$

$$\Xi_6 \cdot e^{-\lambda \cdot \tau_D} \cdot \det\begin{pmatrix} 0 & (-\Xi_3 - \lambda) & -\Xi_5 \\ -\Xi_8 & 0 & 0 \\ 0 & -\Xi_{10} & (\Xi_{11} - \lambda) \end{pmatrix}$$

$$= \Xi_6 \cdot e^{-\lambda \cdot \tau_D} \cdot \{-(-\Xi_3 - \lambda) \cdot \det\begin{pmatrix} -\Xi_8 & 0 \\ 0 & (\Xi_{11} - \lambda) \end{pmatrix}$$

$$- \Xi_5 \cdot \det\begin{pmatrix} -\Xi_8 & 0 \\ 0 & -\Xi_{10} \end{pmatrix}\} = \Xi_6 \cdot \{-\Xi_8 \cdot (\Xi_3 \cdot \Xi_{11}$$

$$+ \Xi_5 \cdot \Xi_{10}) + \Xi_8 \cdot (-\Xi_{11} + \Xi_3) \cdot \lambda + \Xi_8 \cdot \lambda^2\} \cdot e^{-\lambda \cdot \tau_D}$$

$$\Xi_6 \cdot e^{-\lambda \cdot \tau_D} \cdot \det\begin{pmatrix} 0 & (-\Xi_3 - \lambda) & -\Xi_5 \\ -\Xi_8 & 0 & 0 \\ 0 & -\Xi_{10} & (\Xi_{11} - \lambda) \end{pmatrix}$$

$$= \Xi_6 \cdot e^{-\lambda \cdot \tau_D} \cdot \{-(-\Xi_3 - \lambda) \cdot \det\begin{pmatrix} -\Xi_8 & 0 \\ 0 & (\Xi_{11} - \lambda) \end{pmatrix}$$

$$- \Xi_5 \cdot \det\begin{pmatrix} -\Xi_8 & 0 \\ 0 & -\Xi_{10} \end{pmatrix}\} = \{-\Xi_8 \cdot \Xi_6 \cdot (\Xi_3 \cdot \Xi_{11} + \Xi_5 \cdot \Xi_{10})$$

$$+ \Xi_8 \cdot \Xi_6 \cdot (-\Xi_{11} + \Xi_3) \cdot \lambda + \Xi_6 \cdot \Xi_8 \cdot \lambda^2\} \cdot e^{-\lambda \cdot \tau_D}$$

We define some global parameters: $\iota_{21} = -\Xi_8 \cdot \Xi_6 \cdot (\Xi_3 \cdot \Xi_{11} + \Xi_5 \cdot \Xi_{10})$; $\iota_{22} = \Xi_8 \cdot \Xi_6 \cdot (-\Xi_{11} + \Xi_3)$

$$\iota_{23} = \Xi_6 \cdot \Xi_8$$

$$\Xi_6 \cdot e^{-\lambda \cdot \tau_D} \cdot \det\begin{pmatrix} 0 & (-\Xi_3 - \lambda) & -\Xi_5 \\ -\Xi_8 & 0 & 0 \\ 0 & -\Xi_{10} & (\Xi_{11} - \lambda) \end{pmatrix}$$

$$= \Xi_6 \cdot e^{-\lambda \cdot \tau_D} \cdot \{-(-\Xi_3 - \lambda) \cdot \det\begin{pmatrix} -\Xi_8 & 0 \\ 0 & (\Xi_{11} - \lambda) \end{pmatrix}$$

$$- \Xi_5 \cdot \det\begin{pmatrix} -\Xi_8 & 0 \\ 0 & -\Xi_{10} \end{pmatrix}\}$$

$$= (\iota_{21} + \iota_{22} \cdot \lambda + \iota_{23} \cdot \lambda^2) \cdot e^{-\lambda \cdot \tau_D}$$

Step 4.2: $(-\Xi_2 - \lambda) \cdot \det\begin{pmatrix} \Xi_7 \cdot e^{-\lambda \cdot \tau_D} & (-\Xi_3 - \lambda) & -\Xi_5 \\ \Xi_8 \cdot e^{-\lambda \cdot \tau_D} & 0 & 0 \\ \Xi_{10} \cdot e^{-\lambda \cdot \tau_D} & -\Xi_{10} & (\Xi_{11} - \lambda) \end{pmatrix}$

$$(-\Xi_2 - \lambda) \cdot \det\begin{pmatrix} \Xi_7 \cdot e^{-\lambda \cdot \tau_D} & (-\Xi_3 - \lambda) & -\Xi_5 \\ \Xi_8 \cdot e^{-\lambda \cdot \tau_D} & 0 & 0 \\ \Xi_{10} \cdot e^{-\lambda \cdot \tau_D} & -\Xi_{10} & (\Xi_{11} - \lambda) \end{pmatrix}$$

$$= (-\Xi_2 - \lambda) \cdot \{\Xi_7 \cdot e^{-\lambda \cdot \tau_D} \cdot \det\begin{pmatrix} 0 & 0 \\ -\Xi_{10} & (\Xi_{11} - \lambda) \end{pmatrix}$$

$$- (-\Xi_3 - \lambda) \cdot \det\begin{pmatrix} \Xi_8 \cdot e^{-\lambda \cdot \tau_D} & 0 \\ \Xi_{10} \cdot e^{-\lambda \cdot \tau_D} & (\Xi_{11} - \lambda) \end{pmatrix}$$

$$- \Xi_5 \cdot \det\begin{pmatrix} \Xi_8 \cdot e^{-\lambda \cdot \tau_D} & 0 \\ \Xi_{10} \cdot e^{-\lambda \cdot \tau_D} & -\Xi_{10} \end{pmatrix}\}$$

$$\Xi_7 \cdot e^{-\lambda \cdot \tau_D} \cdot \det\begin{pmatrix} 0 & 0 \\ -\Xi_{10} & (\Xi_{11} - \lambda) \end{pmatrix} = 0$$

$$(-\Xi_2 - \lambda) \cdot \det\begin{pmatrix} \Xi_7 \cdot e^{-\lambda \cdot \tau_D} & (-\Xi_3 - \lambda) & -\Xi_5 \\ \Xi_8 \cdot e^{-\lambda \cdot \tau_D} & 0 & 0 \\ \Xi_{10} \cdot e^{-\lambda \cdot \tau_D} & -\Xi_{10} & (\Xi_{11} - \lambda) \end{pmatrix}$$
$$= (-\Xi_2 - \lambda) \cdot \{(\Xi_3 + \lambda) \cdot \det\begin{pmatrix} \Xi_8 \cdot e^{-\lambda \cdot \tau_D} & 0 \\ \Xi_{10} \cdot e^{-\lambda \cdot \tau_D} & (\Xi_{11} - \lambda) \end{pmatrix}$$
$$- \Xi_5 \cdot \det\begin{pmatrix} \Xi_8 \cdot e^{-\lambda \cdot \tau_D} & 0 \\ \Xi_{10} \cdot e^{-\lambda \cdot \tau_D} & -\Xi_{10} \end{pmatrix}\}$$

$$(-\Xi_2 - \lambda) \cdot \det\begin{pmatrix} \Xi_7 \cdot e^{-\lambda \cdot \tau_D} & (-\Xi_3 - \lambda) & -\Xi_5 \\ \Xi_8 \cdot e^{-\lambda \cdot \tau_D} & 0 & 0 \\ \Xi_{10} \cdot e^{-\lambda \cdot \tau_D} & -\Xi_{10} & (\Xi_{11} - \lambda) \end{pmatrix}$$
$$= (-\Xi_2 - \lambda) \cdot \{(\Xi_3 + \lambda) \cdot \Xi_8 \cdot e^{-\lambda \cdot \tau_D} \cdot (\Xi_{11} - \lambda)$$
$$- \Xi_5 \cdot \Xi_8 \cdot e^{-\lambda \cdot \tau_D} \cdot (-\Xi_{10})\}$$

$$(-\Xi_2 - \lambda) \cdot \det\begin{pmatrix} \Xi_7 \cdot e^{-\lambda \cdot \tau_D} & (-\Xi_3 - \lambda) & -\Xi_5 \\ \Xi_8 \cdot e^{-\lambda \cdot \tau_D} & 0 & 0 \\ \Xi_{10} \cdot e^{-\lambda \cdot \tau_D} & -\Xi_{10} & (\Xi_{11} - \lambda) \end{pmatrix}$$
$$= (-\Xi_2 - \lambda) \cdot \{(\Xi_3 + \lambda) \cdot \Xi_8 \cdot (\Xi_{11} - \lambda)$$
$$+ \Xi_5 \cdot \Xi_8 \cdot \Xi_{10}\} \cdot e^{-\lambda \cdot \tau_D}$$

$$(-\Xi_2 - \lambda) \cdot \det\begin{pmatrix} \Xi_7 \cdot e^{-\lambda \cdot \tau_D} & (-\Xi_3 - \lambda) & -\Xi_5 \\ \Xi_8 \cdot e^{-\lambda \cdot \tau_D} & 0 & 0 \\ \Xi_{10} \cdot e^{-\lambda \cdot \tau_D} & -\Xi_{10} & (\Xi_{11} - \lambda) \end{pmatrix}$$
$$= (-\Xi_2 - \lambda) \cdot \{\Xi_8 \cdot (\Xi_3 \cdot \Xi_{11} + \Xi_5 \cdot \Xi_{10}) + \Xi_8$$
$$\cdot [\Xi_{11} - \Xi_3] \cdot \lambda - \Xi_8 \cdot \lambda^2\} \cdot e^{-\lambda \cdot \tau_D}$$

$$(-\Xi_2 - \lambda) \cdot \det\begin{pmatrix} \Xi_7 \cdot e^{-\lambda \cdot \tau_D} & (-\Xi_3 - \lambda) & -\Xi_5 \\ \Xi_8 \cdot e^{-\lambda \cdot \tau_D} & 0 & 0 \\ \Xi_{10} \cdot e^{-\lambda \cdot \tau_D} & -\Xi_{10} & (\Xi_{11} - \lambda) \end{pmatrix}$$
$$= -\{\Xi_2 \cdot \Xi_8 \cdot (\Xi_3 \cdot \Xi_{11} + \Xi_5 \cdot \Xi_{10}) + [\Xi_2 \cdot \Xi_8 \cdot [\Xi_{11} - \Xi_3]$$
$$+ \Xi_8 \cdot (\Xi_3 \cdot \Xi_{11} + \Xi_5 \cdot \Xi_{10})] \cdot \lambda + (\Xi_8 \cdot [\Xi_{11} - \Xi_3] - \Xi_2 \cdot \Xi_8)$$

$$\cdot \lambda^2 - \Xi_8 \cdot \lambda^3\} \cdot e^{-\lambda \cdot \tau_D}$$

$$(-\Xi_2 - \lambda) \cdot \det\begin{pmatrix} \Xi_7 \cdot e^{-\lambda \cdot \tau_D} & (-\Xi_3 - \lambda) & -\Xi_5 \\ \Xi_8 \cdot e^{-\lambda \cdot \tau_D} & 0 & 0 \\ \Xi_{10} \cdot e^{-\lambda \cdot \tau_D} & -\Xi_{10} & (\Xi_{11} - \lambda) \end{pmatrix}$$
$$= \{-\Xi_2 \cdot \Xi_8 \cdot (\Xi_3 \cdot \Xi_{11} + \Xi_5 \cdot \Xi_{10}) - (\Xi_2 \cdot \Xi_8 \cdot [\Xi_{11} - \Xi_3] + \Xi_8 \cdot [\Xi_3 \cdot \Xi_{11} + \Xi_5 \cdot \Xi_{10}]) \cdot \lambda - (\Xi_8 \cdot [\Xi_{11} - \Xi_3] - \Xi_2 \cdot \Xi_8) \cdot \lambda^2 + \Xi_8 \cdot \lambda^3\} \cdot e^{-\lambda \cdot \tau_D}$$

We define some global parameters: $\iota_{24} = -\Xi_2 \cdot \Xi_8 \cdot (\Xi_3 \cdot \Xi_{11} + \Xi_5 \cdot \Xi_{10})$

$$\iota_{25} = -(\Xi_2 \cdot \Xi_8 \cdot [\Xi_{11} - \Xi_3] + \Xi_8 \cdot [\Xi_3 \cdot \Xi_{11} + \Xi_5 \cdot \Xi_{10}])$$
$$\iota_{26} = -(\Xi_8 \cdot [\Xi_{11} - \Xi_3] - \Xi_2 \cdot \Xi_8); \iota_{27} = \Xi_8$$

$$(-\Xi_2 - \lambda) \cdot \det\begin{pmatrix} \Xi_7 \cdot e^{-\lambda \cdot \tau_D} & (-\Xi_3 - \lambda) & -\Xi_5 \\ \Xi_8 \cdot e^{-\lambda \cdot \tau_D} & 0 & 0 \\ \Xi_{10} \cdot e^{-\lambda \cdot \tau_D} & -\Xi_{10} & (\Xi_{11} - \lambda) \end{pmatrix}$$
$$= (\iota_{24} + \iota_{25} \cdot \lambda + \iota_{26} \cdot \lambda^2 + \iota_{27} \cdot \lambda^3) \cdot e^{-\lambda \cdot \tau_D}$$

We can summary:

$$\Upsilon_{14} \cdot \det\begin{pmatrix} \Upsilon_{21} & \Upsilon_{22} & \Upsilon_{23} & \Upsilon_{25} \\ \Upsilon_{31} & \Upsilon_{32} & \Upsilon_{33} & \Upsilon_{35} \\ \Upsilon_{41} & \Upsilon_{42} & \Upsilon_{43} & \Upsilon_{45} \\ \Upsilon_{51} & \Upsilon_{52} & \Upsilon_{53} & \Upsilon_{55} \end{pmatrix}$$
$$= \Xi_4 \cdot \{\Xi_6 \cdot e^{-\lambda \cdot \tau_D} \cdot \det\begin{pmatrix} 0 & (-\Xi_3 - \lambda) & -\Xi_5 \\ -\Xi_8 & 0 & 0 \\ 0 & -\Xi_{10} & (\Xi_{11} - \lambda) \end{pmatrix}$$
$$- (-\Xi_2 - \lambda) \cdot \det\begin{pmatrix} \Xi_7 \cdot e^{-\lambda \cdot \tau_D} & (-\Xi_3 - \lambda) & -\Xi_5 \\ \Xi_8 \cdot e^{-\lambda \cdot \tau_D} & 0 & 0 \\ \Xi_{10} \cdot e^{-\lambda \cdot \tau_D} & -\Xi_{10} & (\Xi_{11} - \lambda) \end{pmatrix}\}$$
$$= \Xi_4 \cdot \{(\iota_{21} + \iota_{22} \cdot \lambda + \iota_{23} \cdot \lambda^2) \cdot e^{-\lambda \cdot \tau_D} - (\iota_{24} + \iota_{25} \cdot \lambda + \iota_{26} \cdot \lambda^2 + \iota_{27} \cdot \lambda^3) \cdot e^{-\lambda \cdot \tau_D}\}$$

$$\Upsilon_{14} \cdot \det\begin{pmatrix} \Upsilon_{21} & \Upsilon_{22} & \Upsilon_{23} & \Upsilon_{25} \\ \Upsilon_{31} & \Upsilon_{32} & \Upsilon_{33} & \Upsilon_{35} \\ \Upsilon_{41} & \Upsilon_{42} & \Upsilon_{43} & \Upsilon_{45} \\ \Upsilon_{51} & \Upsilon_{52} & \Upsilon_{53} & \Upsilon_{55} \end{pmatrix}$$

$$= \Xi_4 \cdot \{\Xi_6 \cdot e^{-\lambda \cdot \tau_D} \cdot \det \begin{pmatrix} 0 & (-\Xi_3 - \lambda) & -\Xi_5 \\ -\Xi_8 & 0 & 0 \\ 0 & -\Xi_{10} & (\Xi_{11} - \lambda) \end{pmatrix}$$

$$- (-\Xi_2 - \lambda) \cdot \det \begin{pmatrix} \Xi_7 \cdot e^{-\lambda \cdot \tau_D} & (-\Xi_3 - \lambda) & -\Xi_5 \\ \Xi_8 \cdot e^{-\lambda \cdot \tau_D} & 0 & 0 \\ \Xi_{10} \cdot e^{-\lambda \cdot \tau_D} & -\Xi_{10} & (\Xi_{11} - \lambda) \end{pmatrix} \}$$

$$= \{\Xi_4 \cdot (\iota_{21} - \iota_{24}) + \Xi_4 \cdot (\iota_{22} - \iota_{25}) \cdot \lambda + \Xi_4 \cdot (\iota_{23} - \iota_{26}) \cdot \lambda^2$$
$$- \Xi_4 \cdot \iota_{27} \cdot \lambda^3\} \cdot e^{-\lambda \cdot \tau_D}$$

<u>Step 5</u>: Find the expression for $\Upsilon_{15} \cdot \det \begin{pmatrix} \Upsilon_{21} & \Upsilon_{22} & \Upsilon_{23} & \Upsilon_{24} \\ \Upsilon_{31} & \Upsilon_{32} & \Upsilon_{33} & \Upsilon_{34} \\ \Upsilon_{41} & \Upsilon_{42} & \Upsilon_{43} & \Upsilon_{44} \\ \Upsilon_{51} & \Upsilon_{52} & \Upsilon_{53} & \Upsilon_{54} \end{pmatrix}$

$$\Upsilon_{15} \cdot \det \begin{pmatrix} \Upsilon_{21} & \Upsilon_{22} & \Upsilon_{23} & \Upsilon_{24} \\ \Upsilon_{31} & \Upsilon_{32} & \Upsilon_{33} & \Upsilon_{34} \\ \Upsilon_{41} & \Upsilon_{42} & \Upsilon_{43} & \Upsilon_{44} \\ \Upsilon_{51} & \Upsilon_{52} & \Upsilon_{53} & \Upsilon_{54} \end{pmatrix}$$

$$= \Xi_5 \cdot \det \begin{pmatrix} \Xi_6 \cdot e^{-\lambda \cdot \tau_D} & (-\Xi_2 - \lambda) & 0 & -\Xi_4 \\ \Xi_7 \cdot e^{-\lambda \cdot \tau_D} & 0 & (-\Xi_3 - \lambda) & 0 \\ \Xi_8 \cdot e^{-\lambda \cdot \tau_D} & -\Xi_8 & 0 & (\Xi_9 - \lambda) \\ \Xi_{10} \cdot e^{-\lambda \cdot \tau_D} & 0 & -\Xi_{10} & 0 \end{pmatrix}$$

$$\Upsilon_{15} \cdot \det \begin{pmatrix} \Upsilon_{21} & \Upsilon_{22} & \Upsilon_{23} & \Upsilon_{24} \\ \Upsilon_{31} & \Upsilon_{32} & \Upsilon_{33} & \Upsilon_{34} \\ \Upsilon_{41} & \Upsilon_{42} & \Upsilon_{43} & \Upsilon_{44} \\ \Upsilon_{51} & \Upsilon_{52} & \Upsilon_{53} & \Upsilon_{54} \end{pmatrix}$$

$$= \Xi_5 \cdot \det \begin{pmatrix} \Xi_6 \cdot e^{-\lambda \cdot \tau_D} & (-\Xi_2 - \lambda) & 0 & -\Xi_4 \\ \Xi_7 \cdot e^{-\lambda \cdot \tau_D} & 0 & (-\Xi_3 - \lambda) & 0 \\ \Xi_8 \cdot e^{-\lambda \cdot \tau_D} & -\Xi_8 & 0 & (\Xi_9 - \lambda) \\ \Xi_{10} \cdot e^{-\lambda \cdot \tau_D} & 0 & -\Xi_{10} & 0 \end{pmatrix}$$

$$= \Xi_5 \cdot \{\Xi_6 \cdot e^{-\lambda \cdot \tau_D} \cdot \det \begin{pmatrix} 0 & (-\Xi_3 - \lambda) & 0 \\ -\Xi_8 & 0 & (\Xi_9 - \lambda) \\ 0 & -\Xi_{10} & 0 \end{pmatrix}$$

$$- (-\Xi_2 - \lambda) \cdot \det \begin{pmatrix} \Xi_7 \cdot e^{-\lambda \cdot \tau_D} & (-\Xi_3 - \lambda) & 0 \\ \Xi_8 \cdot e^{-\lambda \cdot \tau_D} & 0 & (\Xi_9 - \lambda) \\ \Xi_{10} \cdot e^{-\lambda \cdot \tau_D} & -\Xi_{10} & 0 \end{pmatrix}$$

$$-(-\Xi_4)\cdot\det\begin{pmatrix}\Xi_7\cdot e^{-\lambda\cdot\tau_D} & 0 & (-\Xi_3-\lambda)\\ \Xi_8\cdot e^{-\lambda\cdot\tau_D} & -\Xi_8 & 0\\ \Xi_{10}\cdot e^{-\lambda\cdot\tau_D} & 0 & -\Xi_{10}\end{pmatrix}\}$$

$$\Xi_6\cdot e^{-\lambda\cdot\tau_D}\cdot\det\begin{pmatrix}0 & (-\Xi_3-\lambda) & 0\\ -\Xi_8 & 0 & (\Xi_9-\lambda)\\ 0 & -\Xi_{10} & 0\end{pmatrix}$$

$$=\Xi_6\cdot e^{-\lambda\cdot\tau_D}\cdot\{-(-\Xi_3-\lambda)\cdot\det\begin{pmatrix}-\Xi_8 & (\Xi_9-\lambda)\\ 0 & 0\end{pmatrix}\}=0$$

$$\Upsilon_{15}\cdot\det\begin{pmatrix}\Upsilon_{21} & \Upsilon_{22} & \Upsilon_{23} & \Upsilon_{24}\\ \Upsilon_{31} & \Upsilon_{32} & \Upsilon_{33} & \Upsilon_{34}\\ \Upsilon_{41} & \Upsilon_{42} & \Upsilon_{43} & \Upsilon_{44}\\ \Upsilon_{51} & \Upsilon_{52} & \Upsilon_{53} & \Upsilon_{54}\end{pmatrix}$$

$$=\Xi_5\cdot\det\begin{pmatrix}\Xi_6\cdot e^{-\lambda\cdot\tau_D} & (-\Xi_2-\lambda) & 0 & -\Xi_4\\ \Xi_7\cdot e^{-\lambda\cdot\tau_D} & 0 & (-\Xi_3-\lambda) & 0\\ \Xi_8\cdot e^{-\lambda\cdot\tau_D} & -\Xi_8 & 0 & (\Xi_9-\lambda)\\ \Xi_{10}\cdot e^{-\lambda\cdot\tau_D} & 0 & -\Xi_{10} & 0\end{pmatrix}$$

$$=\Xi_5\cdot\{-(-\Xi_2-\lambda)\cdot\det\begin{pmatrix}\Xi_7\cdot e^{-\lambda\cdot\tau_D} & (-\Xi_3-\lambda) & 0\\ \Xi_8\cdot e^{-\lambda\cdot\tau_D} & 0 & (\Xi_9-\lambda)\\ \Xi_{10}\cdot e^{-\lambda\cdot\tau_D} & -\Xi_{10} & 0\end{pmatrix}$$

$$-(-\Xi_4)\cdot\det\begin{pmatrix}\Xi_7\cdot e^{-\lambda\cdot\tau_D} & 0 & (-\Xi_3-\lambda)\\ \Xi_8\cdot e^{-\lambda\cdot\tau_D} & -\Xi_8 & 0\\ \Xi_{10}\cdot e^{-\lambda\cdot\tau_D} & 0 & -\Xi_{10}\end{pmatrix}\}$$

$$\Upsilon_{15}\cdot\det\begin{pmatrix}\Upsilon_{21} & \Upsilon_{22} & \Upsilon_{23} & \Upsilon_{24}\\ \Upsilon_{31} & \Upsilon_{32} & \Upsilon_{33} & \Upsilon_{34}\\ \Upsilon_{41} & \Upsilon_{42} & \Upsilon_{43} & \Upsilon_{44}\\ \Upsilon_{51} & \Upsilon_{52} & \Upsilon_{53} & \Upsilon_{54}\end{pmatrix}$$

$$=\Xi_5\cdot\det\begin{pmatrix}\Xi_6\cdot e^{-\lambda\cdot\tau_D} & (-\Xi_2-\lambda) & 0 & -\Xi_4\\ \Xi_7\cdot e^{-\lambda\cdot\tau_D} & 0 & (-\Xi_3-\lambda) & 0\\ \Xi_8\cdot e^{-\lambda\cdot\tau_D} & -\Xi_8 & 0 & (\Xi_9-\lambda)\\ \Xi_{10}\cdot e^{-\lambda\cdot\tau_D} & 0 & -\Xi_{10} & 0\end{pmatrix}$$

$$=\Xi_5\cdot\{(\Xi_2+\lambda)\cdot\det\begin{pmatrix}\Xi_7\cdot e^{-\lambda\cdot\tau_D} & (-\Xi_3-\lambda) & 0\\ \Xi_8\cdot e^{-\lambda\cdot\tau_D} & 0 & (\Xi_9-\lambda)\\ \Xi_{10}\cdot e^{-\lambda\cdot\tau_D} & -\Xi_{10} & 0\end{pmatrix}$$

$$+\Xi_4\cdot\det\begin{pmatrix}\Xi_7\cdot e^{-\lambda\cdot\tau_D} & 0 & (-\Xi_3-\lambda)\\ \Xi_8\cdot e^{-\lambda\cdot\tau_D} & -\Xi_8 & 0\\ \Xi_{10}\cdot e^{-\lambda\cdot\tau_D} & 0 & -\Xi_{10}\end{pmatrix}\}$$

Step 5.1: $(\Xi_2 + \lambda) \cdot \det\begin{pmatrix} \Xi_7 \cdot e^{-\lambda \cdot \tau_D} & (-\Xi_3 - \lambda) & 0 \\ \Xi_8 \cdot e^{-\lambda \cdot \tau_D} & 0 & (\Xi_9 - \lambda) \\ \Xi_{10} \cdot e^{-\lambda \cdot \tau_D} & -\Xi_{10} & 0 \end{pmatrix}$

$$(\Xi_2 + \lambda) \cdot \det\begin{pmatrix} \Xi_7 \cdot e^{-\lambda \cdot \tau_D} & (-\Xi_3 - \lambda) & 0 \\ \Xi_8 \cdot e^{-\lambda \cdot \tau_D} & 0 & (\Xi_9 - \lambda) \\ \Xi_{10} \cdot e^{-\lambda \cdot \tau_D} & -\Xi_{10} & 0 \end{pmatrix}$$
$$= (\Xi_2 + \lambda) \cdot \{\Xi_7 \cdot e^{-\lambda \cdot \tau_D} \cdot \det\begin{pmatrix} 0 & (\Xi_9 - \lambda) \\ -\Xi_{10} & 0 \end{pmatrix}$$
$$- (-\Xi_3 - \lambda) \cdot \det\begin{pmatrix} \Xi_8 \cdot e^{-\lambda \cdot \tau_D} & (\Xi_9 - \lambda) \\ \Xi_{10} \cdot e^{-\lambda \cdot \tau_D} & 0 \end{pmatrix}\}$$

$$(\Xi_2 + \lambda) \cdot \det\begin{pmatrix} \Xi_7 \cdot e^{-\lambda \cdot \tau_D} & (-\Xi_3 - \lambda) & 0 \\ \Xi_8 \cdot e^{-\lambda \cdot \tau_D} & 0 & (\Xi_9 - \lambda) \\ \Xi_{10} \cdot e^{-\lambda \cdot \tau_D} & -\Xi_{10} & 0 \end{pmatrix}$$
$$= (\Xi_2 + \lambda) \cdot \{\Xi_7 \cdot e^{-\lambda \cdot \tau_D} \cdot \Xi_{10} \cdot (\Xi_9 - \lambda) - (-\Xi_3 - \lambda)$$
$$\cdot (-\Xi_{10} \cdot e^{-\lambda \cdot \tau_D}) \cdot (\Xi_9 - \lambda)\}$$

$$(\Xi_2 + \lambda) \cdot \det\begin{pmatrix} \Xi_7 \cdot e^{-\lambda \cdot \tau_D} & (-\Xi_3 - \lambda) & 0 \\ \Xi_8 \cdot e^{-\lambda \cdot \tau_D} & 0 & (\Xi_9 - \lambda) \\ \Xi_{10} \cdot e^{-\lambda \cdot \tau_D} & -\Xi_{10} & 0 \end{pmatrix}$$
$$= (\Xi_2 + \lambda) \cdot \{\Xi_7 \cdot \Xi_{10} \cdot (\Xi_9 - \lambda) \cdot e^{-\lambda \cdot \tau_D}$$
$$- (\Xi_3 + \lambda) \cdot \Xi_{10} \cdot (\Xi_9 - \lambda) \cdot e^{-\lambda \cdot \tau_D}\}$$

$$(\Xi_2 + \lambda) \cdot \det\begin{pmatrix} \Xi_7 \cdot e^{-\lambda \cdot \tau_D} & (-\Xi_3 - \lambda) & 0 \\ \Xi_8 \cdot e^{-\lambda \cdot \tau_D} & 0 & (\Xi_9 - \lambda) \\ \Xi_{10} \cdot e^{-\lambda \cdot \tau_D} & -\Xi_{10} & 0 \end{pmatrix}$$
$$= (\Xi_2 + \lambda) \cdot \{\Xi_7 \cdot \Xi_{10} \cdot (\Xi_9 - \lambda) - (\Xi_3 + \lambda)$$
$$\cdot \Xi_{10} \cdot (\Xi_9 - \lambda)\} \cdot e^{-\lambda \cdot \tau_D}$$

$$(\Xi_2 + \lambda) \cdot \det\begin{pmatrix} \Xi_7 \cdot e^{-\lambda \cdot \tau_D} & (-\Xi_3 - \lambda) & 0 \\ \Xi_8 \cdot e^{-\lambda \cdot \tau_D} & 0 & (\Xi_9 - \lambda) \\ \Xi_{10} \cdot e^{-\lambda \cdot \tau_D} & -\Xi_{10} & 0 \end{pmatrix}$$
$$= (\Xi_2 + \lambda) \cdot \{\Xi_7 \cdot \Xi_{10} \cdot \Xi_9 - \Xi_7 \cdot \Xi_{10} \cdot \lambda - \Xi_{10} \cdot (\Xi_3 \cdot \Xi_9$$
$$+ [\Xi_9 - \Xi_3] \cdot \lambda - \lambda^2)\} \cdot e^{-\lambda \cdot \tau_D}$$

$$(\Xi_2+\lambda)\cdot\det\begin{pmatrix}\Xi_7\cdot e^{-\lambda\cdot\tau_D} & (-\Xi_3-\lambda) & 0\\ \Xi_8\cdot e^{-\lambda\cdot\tau_D} & 0 & (\Xi_9-\lambda)\\ \Xi_{10}\cdot e^{-\lambda\cdot\tau_D} & -\Xi_{10} & 0\end{pmatrix}$$
$$=(\Xi_2+\lambda)\cdot\{\Xi_7\cdot\Xi_{10}\cdot\Xi_9-\Xi_7\cdot\Xi_{10}\cdot\lambda-\Xi_{10}\cdot\Xi_3\cdot\Xi_9$$
$$-\Xi_{10}\cdot[\Xi_9-\Xi_3]\cdot\lambda+\Xi_{10}\cdot\lambda^2\}\cdot e^{-\lambda\cdot\tau_D}$$

$$(\Xi_2+\lambda)\cdot\det\begin{pmatrix}\Xi_7\cdot e^{-\lambda\cdot\tau_D} & (-\Xi_3-\lambda) & 0\\ \Xi_8\cdot e^{-\lambda\cdot\tau_D} & 0 & (\Xi_9-\lambda)\\ \Xi_{10}\cdot e^{-\lambda\cdot\tau_D} & -\Xi_{10} & 0\end{pmatrix}$$
$$=(\Xi_2+\lambda)\cdot\{(\Xi_7-\Xi_3)\cdot\Xi_{10}\cdot\Xi_9-(\Xi_7\cdot\Xi_{10}+\Xi_{10}$$
$$\cdot[\Xi_9-\Xi_3])\cdot\lambda+\Xi_{10}\cdot\lambda^2\}\cdot e^{-\lambda\cdot\tau_D}$$

$$(\Xi_2+\lambda)\cdot\det\begin{pmatrix}\Xi_7\cdot e^{-\lambda\cdot\tau_D} & (-\Xi_3-\lambda) & 0\\ \Xi_8\cdot e^{-\lambda\cdot\tau_D} & 0 & (\Xi_9-\lambda)\\ \Xi_{10}\cdot e^{-\lambda\cdot\tau_D} & -\Xi_{10} & 0\end{pmatrix}$$
$$=\{\Xi_2\cdot(\Xi_7-\Xi_3)\cdot\Xi_{10}\cdot\Xi_9-\Xi_2\cdot(\Xi_7\cdot\Xi_{10}$$
$$+\Xi_{10}\cdot[\Xi_9-\Xi_3])\cdot\lambda+\Xi_2\cdot\Xi_{10}\cdot\lambda^2+(\Xi_7-\Xi_3)$$
$$\cdot\Xi_{10}\cdot\Xi_9\cdot\lambda-(\Xi_7\cdot\Xi_{10}+\Xi_{10}\cdot[\Xi_9-\Xi_3])$$
$$\cdot\lambda^2+\Xi_{10}\cdot\lambda^3\}\cdot e^{-\lambda\cdot\tau_D}$$

$$(\Xi_2+\lambda)\cdot\det\begin{pmatrix}\Xi_7\cdot e^{-\lambda\cdot\tau_D} & (-\Xi_3-\lambda) & 0\\ \Xi_8\cdot e^{-\lambda\cdot\tau_D} & 0 & (\Xi_9-\lambda)\\ \Xi_{10}\cdot e^{-\lambda\cdot\tau_D} & -\Xi_{10} & 0\end{pmatrix}$$
$$=\{\Xi_2\cdot(\Xi_7-\Xi_3)\cdot\Xi_{10}\cdot\Xi_9+[(\Xi_7-\Xi_3)\cdot\Xi_{10}\cdot\Xi_9$$
$$-\Xi_2\cdot(\Xi_7\cdot\Xi_{10}+\Xi_{10}\cdot[\Xi_9-\Xi_3])]\cdot\lambda+[\Xi_2\cdot\Xi_{10}$$
$$-(\Xi_7\cdot\Xi_{10}+\Xi_{10}\cdot[\Xi_9-\Xi_3])]\cdot\lambda^2+\Xi_{10}\cdot\lambda^3\}\cdot e^{-\lambda\cdot\tau_D}$$

We define some global parameters: $\iota_{28}=\Xi_2\cdot(\Xi_7-\Xi_3)\cdot\Xi_{10}\cdot\Xi_9$; $\iota_{31}=\Xi_{10}$

$$\iota_{29}=(\Xi_7-\Xi_3)\cdot\Xi_{10}\cdot\Xi_9-\Xi_2\cdot(\Xi_7\cdot\Xi_{10}+\Xi_{10}\cdot[\Xi_9-\Xi_3])$$
$$\iota_{30}=\Xi_2\cdot\Xi_{10}-(\Xi_7\cdot\Xi_{10}+\Xi_{10}\cdot[\Xi_9-\Xi_3])$$

$$(\Xi_2+\lambda)\cdot\det\begin{pmatrix}\Xi_7\cdot e^{-\lambda\cdot\tau_D} & (-\Xi_3-\lambda) & 0\\ \Xi_8\cdot e^{-\lambda\cdot\tau_D} & 0 & (\Xi_9-\lambda)\\ \Xi_{10}\cdot e^{-\lambda\cdot\tau_D} & -\Xi_{10} & 0\end{pmatrix}$$
$$=\{\iota_{28}+\iota_{29}\cdot\lambda+\iota_{30}\cdot\lambda^2+\iota_{31}\cdot\lambda^3\}\cdot e^{-\lambda\cdot\tau_D}$$

Step 5.2: $\Xi_4 \cdot \det\begin{pmatrix} \Xi_7 \cdot e^{-\lambda \cdot \tau_D} & 0 & (-\Xi_3 - \lambda) \\ \Xi_8 \cdot e^{-\lambda \cdot \tau_D} & -\Xi_8 & 0 \\ \Xi_{10} \cdot e^{-\lambda \cdot \tau_D} & 0 & -\Xi_{10} \end{pmatrix}$

$$\Xi_4 \cdot \det\begin{pmatrix} \Xi_7 \cdot e^{-\lambda \cdot \tau_D} & 0 & (-\Xi_3 - \lambda) \\ \Xi_8 \cdot e^{-\lambda \cdot \tau_D} & -\Xi_8 & 0 \\ \Xi_{10} \cdot e^{-\lambda \cdot \tau_D} & 0 & -\Xi_{10} \end{pmatrix}$$

$$= \Xi_4 \cdot \{\Xi_7 \cdot e^{-\lambda \cdot \tau_D} \cdot \det\begin{pmatrix} -\Xi_8 & 0 \\ 0 & -\Xi_{10} \end{pmatrix}$$

$$+ (-\Xi_3 - \lambda) \cdot \det\begin{pmatrix} \Xi_8 \cdot e^{-\lambda \cdot \tau_D} & -\Xi_8 \\ \Xi_{10} \cdot e^{-\lambda \cdot \tau_D} & 0 \end{pmatrix}\}$$

$$\Xi_4 \cdot \det\begin{pmatrix} \Xi_7 \cdot e^{-\lambda \cdot \tau_D} & 0 & (-\Xi_3 - \lambda) \\ \Xi_8 \cdot e^{-\lambda \cdot \tau_D} & -\Xi_8 & 0 \\ \Xi_{10} \cdot e^{-\lambda \cdot \tau_D} & 0 & -\Xi_{10} \end{pmatrix}$$
$$= \Xi_4 \cdot \{\Xi_7 \cdot \Xi_8 \cdot \Xi_{10} + (-\Xi_3 - \lambda) \cdot \Xi_8 \cdot \Xi_{10}\} \cdot e^{-\lambda \cdot \tau_D}$$

$$\Xi_4 \cdot \det\begin{pmatrix} \Xi_7 \cdot e^{-\lambda \cdot \tau_D} & 0 & (-\Xi_3 - \lambda) \\ \Xi_8 \cdot e^{-\lambda \cdot \tau_D} & -\Xi_8 & 0 \\ \Xi_{10} \cdot e^{-\lambda \cdot \tau_D} & 0 & -\Xi_{10} \end{pmatrix}$$
$$= \{\Xi_4 \cdot \Xi_8 \cdot \Xi_{10} \cdot [\Xi_7 - \Xi_3] - \Xi_4 \cdot \Xi_8 \cdot \Xi_{10} \cdot \lambda\} \cdot e^{-\lambda \cdot \tau_D}$$

We define some global parameters: $\iota_{32} = \Xi_4 \cdot \Xi_8 \cdot \Xi_{10} \cdot [\Xi_7 - \Xi_3]$; $\iota_{33} = -\Xi_4 \cdot \Xi_8 \cdot \Xi_{10}$

$$\Xi_4 \cdot \det\begin{pmatrix} \Xi_7 \cdot e^{-\lambda \cdot \tau_D} & 0 & (-\Xi_3 - \lambda) \\ \Xi_8 \cdot e^{-\lambda \cdot \tau_D} & -\Xi_8 & 0 \\ \Xi_{10} \cdot e^{-\lambda \cdot \tau_D} & 0 & -\Xi_{10} \end{pmatrix}$$
$$= \{\iota_{32} + \iota_{33} \cdot \lambda\} \cdot e^{-\lambda \cdot \tau_D}$$

We can summary:

$$\Upsilon_{15} \cdot \det\begin{pmatrix} \Upsilon_{21} & \Upsilon_{22} & \Upsilon_{23} & \Upsilon_{24} \\ \Upsilon_{31} & \Upsilon_{32} & \Upsilon_{33} & \Upsilon_{34} \\ \Upsilon_{41} & \Upsilon_{42} & \Upsilon_{43} & \Upsilon_{44} \\ \Upsilon_{51} & \Upsilon_{52} & \Upsilon_{53} & \Upsilon_{54} \end{pmatrix}$$
$$= \Xi_5 \cdot \det\begin{pmatrix} \Xi_6 \cdot e^{-\lambda \cdot \tau_D} & (-\Xi_2 - \lambda) & 0 & -\Xi_4 \\ \Xi_7 \cdot e^{-\lambda \cdot \tau_D} & 0 & (-\Xi_3 - \lambda) & 0 \\ \Xi_8 \cdot e^{-\lambda \cdot \tau_D} & -\Xi_8 & 0 & (\Xi_9 - \lambda) \\ \Xi_{10} \cdot e^{-\lambda \cdot \tau_D} & 0 & -\Xi_{10} & 0 \end{pmatrix}$$

$$= \Xi_5 \cdot \{\{\iota_{28} + \iota_{29} \cdot \lambda + \iota_{30} \cdot \lambda^2 + \iota_{31} \cdot \lambda^3\} \cdot e^{-\lambda \cdot \tau_D} + (\iota_{32} + \iota_{33} \cdot \lambda) \cdot e^{-\lambda \cdot \tau_D}\}$$

$$\Upsilon_{15} \cdot \det\begin{pmatrix} \Upsilon_{21} & \Upsilon_{22} & \Upsilon_{23} & \Upsilon_{24} \\ \Upsilon_{31} & \Upsilon_{32} & \Upsilon_{33} & \Upsilon_{34} \\ \Upsilon_{41} & \Upsilon_{42} & \Upsilon_{43} & \Upsilon_{44} \\ \Upsilon_{51} & \Upsilon_{52} & \Upsilon_{53} & \Upsilon_{54} \end{pmatrix}$$

$$= \Xi_5 \cdot \det\begin{pmatrix} \Xi_6 \cdot e^{-\lambda \cdot \tau_D} & (-\Xi_2 - \lambda) & 0 & -\Xi_4 \\ \Xi_7 \cdot e^{-\lambda \cdot \tau_D} & 0 & (-\Xi_3 - \lambda) & 0 \\ \Xi_8 \cdot e^{-\lambda \cdot \tau_D} & -\Xi_8 & 0 & (\Xi_9 - \lambda) \\ \Xi_{10} \cdot e^{-\lambda \cdot \tau_D} & 0 & -\Xi_{10} & 0 \end{pmatrix}$$

$$= \Xi_5 \cdot \{(\Xi_2 + \lambda) \cdot \det\begin{pmatrix} \Xi_7 \cdot e^{-\lambda \cdot \tau_D} & (-\Xi_3 - \lambda) & 0 \\ \Xi_8 \cdot e^{-\lambda \cdot \tau_D} & 0 & (\Xi_9 - \lambda) \\ \Xi_{10} \cdot e^{-\lambda \cdot \tau_D} & -\Xi_{10} & 0 \end{pmatrix}$$

$$+ \Xi_4 \cdot \det\begin{pmatrix} \Xi_7 \cdot e^{-\lambda \cdot \tau_D} & 0 & (-\Xi_3 - \lambda) \\ \Xi_8 \cdot e^{-\lambda \cdot \tau_D} & -\Xi_8 & 0 \\ \Xi_{10} \cdot e^{-\lambda \cdot \tau_D} & 0 & -\Xi_{10} \end{pmatrix}\}$$

$$= \{\Xi_5 \cdot [\iota_{28} + \iota_{32}] + \Xi_5 \cdot [\iota_{29} + \iota_{33}] \cdot \lambda + \Xi_5 \cdot \iota_{30} \cdot \lambda^2$$
$$+ \Xi_5 \cdot \iota_{31} \cdot \lambda^3\} \cdot e^{-\lambda \cdot \tau_D}$$

We can summary steps 1–5 results (Table 6.3):

$$\det(A - \lambda \cdot I) = (\Xi_1 \cdot e^{-\lambda \cdot \tau_D} - \lambda) \cdot \det\begin{pmatrix} \Upsilon_{22} & \Upsilon_{23} & \Upsilon_{24} & \Upsilon_{25} \\ \Upsilon_{32} & \Upsilon_{33} & \Upsilon_{34} & \Upsilon_{35} \\ \Upsilon_{42} & \Upsilon_{43} & \Upsilon_{44} & \Upsilon_{45} \\ \Upsilon_{52} & \Upsilon_{53} & \Upsilon_{54} & \Upsilon_{55} \end{pmatrix}$$

$$- \Upsilon_{12} \cdot \det\begin{pmatrix} \Upsilon_{21} & \Upsilon_{23} & \Upsilon_{24} & \Upsilon_{25} \\ \Upsilon_{31} & \Upsilon_{33} & \Upsilon_{34} & \Upsilon_{35} \\ \Upsilon_{41} & \Upsilon_{43} & \Upsilon_{44} & \Upsilon_{45} \\ \Upsilon_{51} & \Upsilon_{53} & \Upsilon_{54} & \Upsilon_{55} \end{pmatrix}$$

$$+ \Upsilon_{13} \cdot \det\begin{pmatrix} \Upsilon_{21} & \Upsilon_{22} & \Upsilon_{24} & \Upsilon_{25} \\ \Upsilon_{31} & \Upsilon_{32} & \Upsilon_{34} & \Upsilon_{35} \\ \Upsilon_{41} & \Upsilon_{42} & \Upsilon_{44} & \Upsilon_{45} \\ \Upsilon_{51} & \Upsilon_{52} & \Upsilon_{54} & \Upsilon_{55} \end{pmatrix}$$

$$- \Upsilon_{14} \cdot \det\begin{pmatrix} \Upsilon_{21} & \Upsilon_{22} & \Upsilon_{23} & \Upsilon_{25} \\ \Upsilon_{31} & \Upsilon_{32} & \Upsilon_{33} & \Upsilon_{35} \\ \Upsilon_{41} & \Upsilon_{42} & \Upsilon_{43} & \Upsilon_{45} \\ \Upsilon_{51} & \Upsilon_{52} & \Upsilon_{53} & \Upsilon_{55} \end{pmatrix}$$

Table 6.3 Small increments Jacobian of plasma kinetic system determinants, results 1–5 and expressions

Step No.	Determinant	Expression
1	$\Upsilon_{11}\cdot\det\begin{pmatrix}\Upsilon_{22} & \Upsilon_{23} & \Upsilon_{24} & \Upsilon_{25}\\ \Upsilon_{32} & \Upsilon_{33} & \Upsilon_{34} & \Upsilon_{35}\\ \Upsilon_{42} & \Upsilon_{43} & \Upsilon_{44} & \Upsilon_{45}\\ \Upsilon_{52} & \Upsilon_{53} & \Upsilon_{54} & \Upsilon_{55}\end{pmatrix}$	$-(\iota_1+\iota_5)\cdot\lambda-(\iota_2+\iota_6)\cdot\lambda^2-(\iota_3+\iota_7)\cdot\lambda^3-\iota_4\cdot\lambda^4-\lambda^5$ $+\{(\iota_1+\iota_5)\cdot\Xi_1+(\iota_2+\iota_6)\cdot\lambda\cdot\Xi_1+(\iota_3+\iota_7)\cdot\lambda^2\cdot\Xi_1$ $+\iota_4\cdot\lambda^3\cdot\Xi_1+\lambda^4\cdot\Xi_1\}\cdot e^{-\lambda\cdot\tau_D}$
2	$\Upsilon_{12}\cdot\det\begin{pmatrix}\Upsilon_{21} & \Upsilon_{23} & \Upsilon_{24} & \Upsilon_{25}\\ \Upsilon_{31} & \Upsilon_{33} & \Upsilon_{34} & \Upsilon_{35}\\ \Upsilon_{41} & \Upsilon_{43} & \Upsilon_{44} & \Upsilon_{45}\\ \Upsilon_{51} & \Upsilon_{53} & \Upsilon_{54} & \Upsilon_{55}\end{pmatrix}$	$\{\Xi_2\cdot(\iota_8-\iota_{12})+\Xi_2\cdot(\iota_9-\iota_{13})\cdot\lambda+\Xi_2\cdot(\iota_{10}-\iota_{13})\cdot\lambda^2$ $+\Xi_2\cdot\iota_{11}\cdot\lambda^3\}\cdot e^{-\lambda\cdot\tau_D}$
3	$\Upsilon_{13}\cdot\det\begin{pmatrix}\Upsilon_{21} & \Upsilon_{22} & \Upsilon_{24} & \Upsilon_{25}\\ \Upsilon_{31} & \Upsilon_{32} & \Upsilon_{34} & \Upsilon_{35}\\ \Upsilon_{41} & \Upsilon_{42} & \Upsilon_{44} & \Upsilon_{45}\\ \Upsilon_{51} & \Upsilon_{52} & \Upsilon_{54} & \Upsilon_{55}\end{pmatrix}$	$\{\Xi_3\cdot(\iota_{15}-\iota_{19})+\Xi_3\cdot(\iota_{16}-\iota_{20})\cdot\lambda+\Xi_3\cdot\iota_{17}\cdot\lambda^2$ $+\Xi_3\cdot\iota_{18}\cdot\lambda^3\}\cdot e^{-\lambda\cdot\tau_D}$
4	$\Upsilon_{14}\cdot\det\begin{pmatrix}\Upsilon_{21} & \Upsilon_{22} & \Upsilon_{23} & \Upsilon_{25}\\ \Upsilon_{31} & \Upsilon_{32} & \Upsilon_{33} & \Upsilon_{35}\\ \Upsilon_{41} & \Upsilon_{42} & \Upsilon_{43} & \Upsilon_{45}\\ \Upsilon_{51} & \Upsilon_{52} & \Upsilon_{53} & \Upsilon_{55}\end{pmatrix}$	$\{\Xi_4\cdot(\iota_{21}-\iota_{24})+\Xi_4\cdot(\iota_{22}-\iota_{25})\cdot\lambda$ $+\Xi_4\cdot(\iota_{23}-\iota_{26})\cdot\lambda^2-\Xi_4\cdot\iota_{27}\cdot\lambda^3\}\cdot e^{-\lambda\cdot\tau_D}$

(continued)

Table 6.3 (continued)

Step No.	Determinant	Expression
5	$\Upsilon_{15} \cdot \det \begin{pmatrix} \Upsilon_{21} & \Upsilon_{22} & \Upsilon_{23} & \Upsilon_{24} \\ \Upsilon_{31} & \Upsilon_{32} & \Upsilon_{33} & \Upsilon_{34} \\ \Upsilon_{41} & \Upsilon_{42} & \Upsilon_{43} & \Upsilon_{44} \\ \Upsilon_{51} & \Upsilon_{52} & \Upsilon_{53} & \Upsilon_{54} \end{pmatrix}$	$\{\Xi_5 \cdot [\iota_{28} + \iota_{32}] + \Xi_5 \cdot [\iota_{29} + \iota_{33}] \cdot \lambda + \Xi_5 \cdot \iota_{30} \cdot \lambda^2 + \Xi_5 \cdot \iota_{31} \cdot \lambda^3\} \cdot e^{-\lambda \cdot \tau_D}$

$$+\Upsilon_{15}\cdot\det\begin{pmatrix}\Upsilon_{21} & \Upsilon_{22} & \Upsilon_{23} & \Upsilon_{24}\\ \Upsilon_{31} & \Upsilon_{32} & \Upsilon_{33} & \Upsilon_{34}\\ \Upsilon_{41} & \Upsilon_{42} & \Upsilon_{43} & \Upsilon_{44}\\ \Upsilon_{51} & \Upsilon_{52} & \Upsilon_{53} & \Upsilon_{54}\end{pmatrix}$$

$$\begin{aligned}\det(A-\lambda\cdot I) = {} & [-(\iota_1+\iota_5)\cdot\lambda-(\iota_2+\iota_6)\cdot\lambda^2\\ & -(\iota_3+\iota_7)\cdot\lambda^3-\iota_4\cdot\lambda^4-\lambda^5+\{(\iota_1+\iota_5)\cdot\Xi_1\\ & +(\iota_2+\iota_6)\cdot\lambda\cdot\Xi_1+(\iota_3+\iota_7)\cdot\lambda^2\cdot\Xi_1+\iota_4\cdot\lambda^3\\ & \cdot\Xi_1+\lambda^4\cdot\Xi_1\}\cdot e^{-\lambda\cdot\tau_D}]-[\{\Xi_2\cdot(\iota_8-\iota_{12})+\Xi_2\\ & \cdot(\iota_9-\iota_{13})\cdot\lambda+\Xi_2\cdot(\iota_{10}-\iota_{13})\cdot\lambda^2+\Xi_2\cdot\iota_{11}\cdot\lambda^3\}\cdot e^{-\lambda\cdot\tau_D}]\\ & +[\{\Xi_3\cdot(\iota_{15}-\iota_{19})+\Xi_3\cdot(\iota_{16}-\iota_{20})\cdot\lambda+\Xi_3\cdot\iota_{17}\cdot\lambda^2\\ & +\Xi_3\cdot\iota_{18}\cdot\lambda^3\}\cdot e^{-\lambda\cdot\tau_D}]-[\{\Xi_4\cdot(\iota_{21}-\iota_{24})+\Xi_4\cdot(\iota_{22}-\iota_{25})\\ & \cdot\lambda+\Xi_4\cdot(\iota_{23}-\iota_{26})\cdot\lambda^2-\Xi_4\cdot\iota_{27}\cdot\lambda^3\}\cdot e^{-\lambda\cdot\tau_D}]\\ & +[\{\Xi_5\cdot[\iota_{28}+\iota_{32}]+\Xi_5\cdot[\iota_{29}+\iota_{33}]\cdot\lambda+\Xi_5\cdot\iota_{30}\cdot\lambda^2\\ & +\Xi_5\cdot\iota_{31}\cdot\lambda^3\}\cdot e^{-\lambda\cdot\tau_D}]\end{aligned}$$

$$\begin{aligned}\det(A-\lambda\cdot I) = {} & -(\iota_1+\iota_5)\cdot\lambda-(\iota_2+\iota_6)\cdot\lambda^2-(\iota_3+\iota_7)\\ & \cdot\lambda^3-\iota_4\cdot\lambda^4-\lambda^5+\{(\iota_1+\iota_5)\cdot\Xi_1+(\iota_2+\iota_6)\cdot\lambda\cdot\Xi_1\\ & +(\iota_3+\iota_7)\cdot\lambda^2\cdot\Xi_1+\iota_4\cdot\lambda^3\cdot\Xi_1+\lambda^4\cdot\Xi_1-\Xi_2\\ & \cdot(\iota_8-\iota_{12})-\Xi_2\cdot(\iota_9-\iota_{13})\cdot\lambda-\Xi_2\cdot(\iota_{10}-\iota_{13})\cdot\lambda^2\\ & -\Xi_2\cdot\iota_{11}\cdot\lambda^3+\Xi_3\cdot(\iota_{15}-\iota_{19})+\Xi_3\cdot(\iota_{16}-\iota_{20})\cdot\lambda\\ & +\Xi_3\cdot\iota_{17}\cdot\lambda^2+\Xi_3\cdot\iota_{18}\cdot\lambda^3-\Xi_4\cdot(\iota_{21}-\iota_{24})-\Xi_4\\ & \cdot(\iota_{22}-\iota_{25})\cdot\lambda-\Xi_4\cdot(\iota_{23}-\iota_{26})\cdot\lambda^2+\Xi_4\cdot\iota_{27}\cdot\lambda^3\\ & +\Xi_5\cdot[\iota_{28}+\iota_{32}]+\Xi_5\cdot[\iota_{29}+\iota_{33}]\cdot\lambda+\Xi_5\cdot\iota_{30}\\ & \cdot\lambda^2+\Xi_5\cdot\iota_{31}\cdot\lambda^3\}\cdot e^{-\lambda\cdot\tau_D}\end{aligned}$$

$$\begin{aligned}\det(A-\lambda\cdot I) = {} & -(\iota_1+\iota_5)\cdot\lambda-(\iota_2+\iota_6)\cdot\lambda^2-(\iota_3+\iota_7)\\ & \cdot\lambda^3-\iota_4\cdot\lambda^4-\lambda^5+\{(\iota_1+\iota_5)\cdot\Xi_1-\Xi_2\cdot(\iota_8-\iota_{12})\\ & +\Xi_3\cdot(\iota_{15}-\iota_{19})-\Xi_4\cdot(\iota_{21}-\iota_{24})+\Xi_5\cdot(\iota_{28}+\iota_{32})\\ & +[(\iota_2+\iota_6)\cdot\Xi_1-\Xi_2\cdot(\iota_9-\iota_{13})+\Xi_3\cdot(\iota_{16}-\iota_{20})-\Xi_4\\ & \cdot(\iota_{22}-\iota_{25})+\Xi_5\cdot(\iota_{29}+\iota_{33})]\cdot\lambda+[(\iota_3+\iota_7)\cdot\Xi_1-\Xi_2\\ & \cdot(\iota_{10}-\iota_{13})+\Xi_3\cdot\iota_{17}-\Xi_4\cdot(\iota_{23}-\iota_{26})+\Xi_5\cdot\iota_{30}]\cdot\lambda^2\\ & +[\iota_4\cdot\Xi_1-\Xi_2\cdot\iota_{11}+\Xi_3\cdot\iota_{18}+\Xi_4\cdot\iota_{27}+\Xi_5\cdot\iota_{31}]\\ & \cdot\lambda^3+\lambda^4\cdot\Xi_1\}\cdot e^{-\lambda\cdot\tau_D}\end{aligned}$$

$$D(\lambda,\tau_D) = -(\iota_1+\iota_5)\cdot\lambda - (\iota_2+\iota_6)\cdot\lambda^2 - (\iota_3+\iota_7)\cdot\lambda^3 - \iota_4\cdot\lambda^4 - \lambda^5 + \{(\iota_1+\iota_5)\cdot\Xi_1 - \Xi_2\cdot(\iota_8-\iota_{12}) + \Xi_3\cdot(\iota_{15}-\iota_{19}) - \Xi_4\cdot(\iota_{21}-\iota_{24}) + \Xi_5\cdot(\iota_{28}+\iota_{32}) + [(\iota_2+\iota_6)\cdot\Xi_1 - \Xi_2\cdot(\iota_9-\iota_{13}) + \Xi_3\cdot(\iota_{16}-\iota_{20}) - \Xi_4\cdot(\iota_{22}-\iota_{25}) + \Xi_5\cdot(\iota_{29}+\iota_{33})]\cdot\lambda + [(\iota_3+\iota_7)\cdot\Xi_1 - \Xi_2\cdot(\iota_{10}-\iota_{13}) + \Xi_3\cdot\iota_{17} - \Xi_4\cdot(\iota_{23}-\iota_{26}) + \Xi_5\cdot\iota_{30}]\cdot\lambda^2 + [\iota_4\cdot\Xi_1 - \Xi_2\cdot\iota_{11} + \Xi_3\cdot\iota_{18} + \Xi_4\cdot\iota_{27} + \Xi_5\cdot\iota_{31}]\cdot\lambda^3 + \lambda^4\cdot\Xi_1\}\cdot e^{-\lambda\cdot\tau_D}$$

$$D(\lambda,\tau_D) = P_n(\lambda,\tau_D) + Q_m(\lambda,\tau_D)\cdot e^{-\lambda\cdot\tau_D}$$
$$n, m \in \mathbb{N}_0;\ n=5;\ m=4;\ n>m$$

$$P_n(\lambda,\tau_D) = -(\iota_1+\iota_5)\cdot\lambda - (\iota_2+\iota_6)\cdot\lambda^2 - (\iota_3+\iota_7)\cdot\lambda^3 - \iota_4\cdot\lambda^4 - \lambda^5;\ n=5$$

$$Q_m(\lambda,\tau_D) = (\iota_1+\iota_5)\cdot\Xi_1 - \Xi_2\cdot(\iota_8-\iota_{12}) + \Xi_3\cdot(\iota_{15}-\iota_{19}) - \Xi_4\cdot(\iota_{21}-\iota_{24}) + \Xi_5\cdot(\iota_{28}+\iota_{32}) + [(\iota_2+\iota_6)\cdot\Xi_1 - \Xi_2\cdot(\iota_9-\iota_{13}) + \Xi_3\cdot(\iota_{16}-\iota_{20}) - \Xi_4\cdot(\iota_{22}-\iota_{25}) + \Xi_5\cdot(\iota_{29}+\iota_{33})]\cdot\lambda + [(\iota_3+\iota_7)\cdot\Xi_1 - \Xi_2\cdot(\iota_{10}-\iota_{13}) + \Xi_3\cdot\iota_{17} - \Xi_4\cdot(\iota_{23}-\iota_{26}) + \Xi_5\cdot\iota_{30}]\cdot\lambda^2 + [\iota_4\cdot\Xi_1 - \Xi_2\cdot\iota_{11} + \Xi_3\cdot\iota_{18} + \Xi_4\cdot\iota_{27} + \Xi_5\cdot\iota_{31}]\cdot\lambda^3 + \Xi_1\cdot\lambda^4;\ m=4$$

$$P_n(\lambda,\tau_D) = \sum_{k=0}^{n=5} p_k(\tau_D)\cdot\lambda^k = p_0(\tau_D) + p_1(\tau_D)\cdot\lambda + p_2(\tau_D)\cdot\lambda^2 + p_3(\tau_D)\cdot\lambda^3 + p_4(\tau_D)\cdot\lambda^4 + p_5(\tau_D)\cdot\lambda^5$$

$$p_0 = p_0(\tau_D);\ p_1 = p_1(\tau_D);\ p_2 = p_2(\tau_D)$$
$$p_3 = p_3(\tau_D);\ p_4 = p_4(\tau_D);\ p_5 = p_5(\tau_D)$$

$$p_0 = 0;\ p_1 = -(\iota_1+\iota_5);\ p_2 = -(\iota_2+\iota_6)$$
$$p_3 = -(\iota_3+\iota_7);\ p_4 = -\iota_4;\ p_5 = -1$$

$$Q_m(\lambda, \tau_D) = \sum_{k=0}^{m=4} q_k(\tau_D) \cdot \lambda^k$$
$$= q_0(\tau_D) + q_1(\tau_D) \cdot \lambda + q_2(\tau_D) \cdot \lambda^2 + q_3(\tau_D) \cdot \lambda^3 + q_4(\tau_D) \cdot \lambda^4$$

$$q_0 = q_0(\tau_D); q_1{=}q_1(\tau_D); q_2 = q_2(\tau_D)$$
$$q_3 = q_3(\tau_D); q_4 = q_4(\tau_D)$$

$$q_0 = (\iota_1 + \iota_5) \cdot \Xi_1 - \Xi_2 \cdot (\iota_8 - \iota_{12}) + \Xi_3 \cdot (\iota_{15} - \iota_{19}) - \Xi_4 \cdot (\iota_{21} - \iota_{24}) + \Xi_5 \cdot (\iota_{28} + \iota_{32})$$

$$q_1 = (\iota_2 + \iota_6) \cdot \Xi_1 - \Xi_2 \cdot (\iota_9 - \iota_{13}) + \Xi_3 \cdot (\iota_{16} - \iota_{20}) - \Xi_4 \cdot (\iota_{22} - \iota_{25}) + \Xi_5 \cdot (\iota_{29} + \iota_{33})$$

$$q_2 = (\iota_3 + \iota_7) \cdot \Xi_1 - \Xi_2 \cdot (\iota_{10} - \iota_{13}) + \Xi_3 \cdot \iota_{17} - \Xi_4 \cdot (\iota_{23} - \iota_{26}) + \Xi_5 \cdot \iota_{30}$$

$$q_3 = \iota_4 \cdot \Xi_1 - \Xi_2 \cdot \iota_{11} + \Xi_3 \cdot \iota_{18} + \Xi_4 \cdot \iota_{27} + \Xi_5 \cdot \iota_{31}; q_4 = \Xi_1$$

The homogeneous system for $n_c, n_{B1}, n_{B2}, n_{ph1}, n_{ph2}$ leads to a characteristic equation for the eigenvalue λ having the form $D(\lambda, \tau_D) = P(\lambda, \tau_D) + Q(\lambda, \tau_D) \cdot e^{-\lambda \cdot \tau_D} = 0$; and $P(\lambda) = \sum_{j=0}^{5} a_j \cdot \lambda^j$; $Q(\lambda) = \sum_{j=0}^{4} c_j \cdot \lambda^j$. The coefficients $\{a_j(q_i, q_k, \tau_D), c_j(q_i, q_k, \tau_D)\} \in \mathbb{R}$ depend on q_i, q_k and delay τ_D. q_i, q_k are any plasma kinetic system's parameters, other parameters kept as a constant [4, 5].

$$a_0 = 0; a_1{=} - (\iota_1 + \iota_5); a_2{=} - (\iota_2 + \iota_6)$$
$$a_3{=} - (\iota_3 + \iota_7); a_4{=} - \iota_4; a_5{=} - 1$$

$$c_0 = (\iota_1 + \iota_5) \cdot \Xi_1 - \Xi_2 \cdot (\iota_8 - \iota_{12}) + \Xi_3 \cdot (\iota_{15} - \iota_{19}) - \Xi_4 \cdot (\iota_{21} - \iota_{24}) + \Xi_5 \cdot (\iota_{28} + \iota_{32})$$

$$c_1 = (\iota_2 + \iota_6) \cdot \Xi_1 - \Xi_2 \cdot (\iota_9 - \iota_{13}) + \Xi_3 \cdot (\iota_{16} - \iota_{20}) - \Xi_4 \cdot (\iota_{22} - \iota_{25}) + \Xi_5 \cdot (\iota_{29} + \iota_{33})$$

$$c_2 = (\iota_3 + \iota_7) \cdot \Xi_1 - \Xi_2 \cdot (\iota_{10} - \iota_{13}) + \Xi_3 \cdot \iota_{17} - \Xi_4 \cdot (\iota_{23} - \iota_{26}) + \Xi_5 \cdot \iota_{30}$$

$$c_3 = \iota_4 \cdot \Xi_1 - \Xi_2 \cdot \iota_{11} + \Xi_3 \cdot \iota_{18} + \Xi_4 \cdot \iota_{27} + \Xi_5 \cdot \iota_{31}; \; c_4 = \Xi_1$$

Unless strictly necessary, the designation of the varied arguments (q_i, q_k) will subsequently be omitted from P, Q, a_j and c_j. The coefficients a_j, c_j are continuous and differentiable functions of their arguments, and direct substitution shows that $a_0 + c_0 \neq 0$ for $q_i, q_k \in \mathbb{R}_+$; that is, $\lambda = 0$ is not of $P(\lambda) + Q(\lambda) \cdot e^{-\lambda \cdot \tau_D} = 0$. Furthermore, $P(\lambda)$, $Q(\lambda)$ are analytic functions of λ, for which the following requirements of the analysis (Kuang J and Cong Y 2005; Kuang Y 1993) can also be verified in the present case:

(a) If $\lambda = i \cdot \omega; \omega \in \mathbb{R}$, then $P(i \cdot \omega) + Q(i \cdot \omega) \neq 0$.
(b) If $|\frac{Q(\lambda)}{P(\lambda)}|$ is bounded for $|\lambda| \to \infty$; $\mathrm{Re}\lambda \geq 0$. No roots bifurcation from ∞.
(c) $F(\omega) = |P(i \cdot \omega)|^2 - |Q(i \cdot \omega)|^2$ has a finite number of zeros. Indeed, this is a polynomial in ω.
(d) Each positive root $\omega(q_i, q_k)$ of $F(\omega) = 0$ is continuous and differentiable with respect to q_i, q_k.

We assume that $P_n(\lambda, \tau_D)$ and $Q_m(\lambda, \tau_D)$ cannot have common imaginary roots. That is for any real numberω: $P_n(\lambda = i \cdot \omega, \tau_D) + Q_m(\lambda = i \cdot \omega, \tau_D) \neq 0$ and

$$P_n(\lambda = i \cdot \omega, \tau_D) = p_0(\tau_D) + p_1(\tau_D) \cdot i \cdot \omega - p_2(\tau_D) \cdot \omega^2 - p_3(\tau_D) \cdot i \cdot \omega^3 + p_4(\tau_D) \cdot \omega^4 + p_5(\tau_D) \cdot i \cdot \omega^5$$

$$P_n(\lambda = i \cdot \omega, \tau_D) = p_0(\tau_D) - p_2(\tau_D) \cdot \omega^2 + p_4(\tau_D) \cdot \omega^4 + [p_1(\tau_D) \cdot \omega - p_3(\tau_D) \cdot \omega^3 + p_5(\tau_D) \cdot \omega^5] \cdot i$$

$$P_n(\lambda = i \cdot \omega, \tau_D) = (\iota_2 + \iota_6) \cdot \omega^2 - \iota_4 \cdot \omega^4 + [-(\iota_1 + \iota_5) \cdot \omega + (\iota_3 + \iota_7) \cdot \omega^3 - \omega^5] \cdot i$$

$$Q_m(\lambda = i \cdot \omega, \tau_D) = q_0(\tau_D) + q_1(\tau_D) \cdot i \cdot \omega - q_2(\tau_D) \cdot \omega^2 - q_3(\tau_D) \cdot i \cdot \omega^3 + q_4(\tau_D) \cdot \omega^4$$

$$Q_m(\lambda = i \cdot \omega, \tau_D) = q_0(\tau_D) - q_2(\tau_D) \cdot \omega^2 + q_4(\tau_D) \cdot \omega^4 + [q_1(\tau_D) \cdot \omega - q_3(\tau_D) \cdot \omega^3] \cdot i$$

$$Q_m(\lambda = i \cdot \omega, \tau_D) = [(\iota_1 + \iota_5) \cdot \Xi_1 - \Xi_2 \cdot (\iota_8 - \iota_{12}) + \Xi_3 \cdot (\iota_{15} - \iota_{19}) - \Xi_4 \cdot (\iota_{21} - \iota_{24}) + \Xi_5 \cdot (\iota_{28} + \iota_{32})] - [(\iota_3 + \iota_7)$$

$$\cdot \Xi_1 - \Xi_2 \cdot (\iota_{10} - \iota_{13}) + \Xi_3 \cdot \iota_{17} - \Xi_4 \cdot (\iota_{23} - \iota_{26}) + \Xi_5 \cdot \iota_{30}] \cdot \omega^2 + \Xi_1 \cdot \omega^4 + \{[(\iota_2 + \iota_6) \cdot \Xi_1 - \Xi_2 \cdot (\iota_9 - \iota_{13}) + \Xi_3 \cdot (\iota_{16} - \iota_{20}) - \Xi_4 \cdot (\iota_{22} - \iota_{25}) + \Xi_5 \cdot (\iota_{29} + \iota_{33})] \cdot \omega - [\iota_4 \cdot \Xi_1 - \Xi_2 \cdot \iota_{11} + \Xi_3 \cdot \iota_{18} + \Xi_4 \cdot \iota_{27} + \Xi_5 \cdot \iota_{31}] \cdot \omega^3\} \cdot i$$

$$P_n(\lambda = i \cdot \omega, \tau_D) + Q_m(\lambda = i \cdot \omega, \tau_D) = (\iota_2 + \iota_6) \cdot \omega^2 - \iota_4 \cdot \omega^4 + \{-(\iota_1 + \iota_5) \cdot \omega + (\iota_3 + \iota_7) \cdot \omega^3 - \omega^5\} \cdot i + [(\iota_1 + \iota_5) \cdot \Xi_1 - \Xi_2 \cdot (\iota_8 - \iota_{12}) + \Xi_3 \cdot (\iota_{15} - \iota_{19}) - \Xi_4 \cdot (\iota_{21} - \iota_{24}) + \Xi_5 \cdot (\iota_{28} + \iota_{32})] - [(\iota_3 + \iota_7) \cdot \Xi_1 - \Xi_2 \cdot (\iota_{10} - \iota_{13}) + \Xi_3 \cdot \iota_{17} - \Xi_4 \cdot (\iota_{23} - \iota_{26}) + \Xi_5 \cdot \iota_{30}] \cdot \omega^2 + \Xi_1 \cdot \omega^4 + \{[(\iota_2 + \iota_6) \cdot \Xi_1 - \Xi_2 \cdot (\iota_9 - \iota_{13}) + \Xi_3 \cdot (\iota_{16} - \iota_{20}) - \Xi_4 \cdot (\iota_{22} - \iota_{25}) + \Xi_5 \cdot (\iota_{29} + \iota_{33})] \cdot \omega - [\iota_4 \cdot \Xi_1 - \Xi_2 \cdot \iota_{11} + \Xi_3 \cdot \iota_{18} + \Xi_4 \cdot \iota_{27} + \Xi_5 \cdot \iota_{31}] \cdot \omega^3\} \cdot i \neq 0$$

$$P_n(\lambda = i \cdot \omega, \tau_D) + Q_m(\lambda = i \cdot \omega, \tau_D) = (\iota_2 + \iota_6) \cdot \omega^2 - \iota_4 \cdot \omega^4 + [(\iota_1 + \iota_5) \cdot \Xi_1 - \Xi_2 \cdot (\iota_8 - \iota_{12}) + \Xi_3 \cdot (\iota_{15} - \iota_{19}) - \Xi_4 \cdot (\iota_{21} - \iota_{24}) + \Xi_5 \cdot (\iota_{28} + \iota_{32})] - [(\iota_3 + \iota_7) \cdot \Xi_1 - \Xi_2 \cdot (\iota_{10} - \iota_{13}) + \Xi_3 \cdot \iota_{17} - \Xi_4 \cdot (\iota_{23} - \iota_{26}) + \Xi_5 \cdot \iota_{30}] \cdot \omega^2 + \Xi_1 \cdot \omega^4 + \{-(\iota_1 + \iota_5) \cdot \omega + (\iota_3 + \iota_7) \cdot \omega^3 - \omega^5 + [(\iota_2 + \iota_6) \cdot \Xi_1 - \Xi_2 \cdot (\iota_9 - \iota_{13}) + \Xi_3 \cdot (\iota_{16} - \iota_{20}) - \Xi_4 \cdot (\iota_{22} - \iota_{25}) + \Xi_5 \cdot (\iota_{29} + \iota_{33})] \cdot \omega - [\iota_4 \cdot \Xi_1 - \Xi_2 \cdot \iota_{11} + \Xi_3 \cdot \iota_{18} + \Xi_4 \cdot \iota_{27} + \Xi_5 \cdot \iota_{31}] \cdot \omega^3\} \cdot i \neq 0$$

$$|P(i \cdot \omega, \tau_D)|^2 = [(\iota_2 + \iota_6) \cdot \omega^2 - \iota_4 \cdot \omega^4]^2 + [-(\iota_1 + \iota_5) \cdot \omega + (\iota_3 + \iota_7) \cdot \omega^3 - \omega^5]^2$$

$$|Q(i \cdot \omega, \tau_D)|^2 = \{[(\iota_1 + \iota_5) \cdot \Xi_1 - \Xi_2 \cdot (\iota_8 - \iota_{12}) + \Xi_3 \cdot (\iota_{15} - \iota_{19}) - \Xi_4 \cdot (\iota_{21} - \iota_{24}) + \Xi_5 \cdot (\iota_{28} + \iota_{32})] - [(\iota_3 + \iota_7) \cdot \Xi_1 - \Xi_2 \cdot (\iota_{10} - \iota_{13}) + \Xi_3 \cdot \iota_{17} - \Xi_4 \cdot (\iota_{23} - \iota_{26}) + \Xi_5 \cdot \iota_{30}] \cdot \omega^2 + \Xi_1 \cdot \omega^4\}^2 + \{[(\iota_2 + \iota_6) \cdot \Xi_1 - \Xi_2 \cdot (\iota_9 - \iota_{13}) + \Xi_3 \cdot (\iota_{16} - \iota_{20}) - \Xi_4 \cdot (\iota_{22} - \iota_{25}) + \Xi_5 \cdot (\iota_{29} + \iota_{33})] \cdot \omega - [\iota_4 \cdot \Xi_1 - \Xi_2 \cdot \iota_{11} + \Xi_3 \cdot \iota_{18} + \Xi_4 \cdot \iota_{27} + \Xi_5 \cdot \iota_{31}] \cdot \omega^3\}^2$$

Table 6.4 Plasma kinetic system's stability analysis $F(\omega, \tau_D)$ elements

Expression with ω	Equivalent expression
$[(\iota_2+\iota_6)\cdot\omega^2-\iota_4\cdot\omega^4]^2$	$(\iota_2+\iota_6)^2\cdot\omega^4-2\cdot(\iota_2+\iota_6)\cdot\iota_4\cdot\omega^6+\iota_4^2\cdot\omega^8$
$[-(\iota_1+\iota_5)\cdot\omega+(\iota_3+\iota_7)\cdot\omega^3-\omega^5]^2$	$(\iota_1+\iota_5)^2\cdot\omega^2-2\cdot(\iota_1+\iota_5)\cdot(\iota_3+\iota_7)\cdot\omega^4$ $+[2\cdot(\iota_1+\iota_5)+(\iota_3+\iota_7)^2]\cdot\omega^6-2\cdot(\iota_3+\iota_7)\cdot\omega^8+\omega^{10}$
$[q_0(\tau_D)-q_2(\tau_D)\cdot\omega^2+q_4(\tau_D)\cdot\omega^4]^2$	$q_0^2(\tau_D)-2\cdot q_0(\tau_D)\cdot q_2(\tau_D)\cdot\omega^2$ $+[q_0(\tau_D)\cdot q_4(\tau_D)+q_2^2(\tau_D)+q_0(\tau_D)\cdot q_4(\tau_D)]\cdot\omega^4$ $-2\cdot q_2(\tau_D)\cdot q_4(\tau_D)\cdot\omega^6+q_4^2(\tau_D)\cdot\omega^8$
$[q_1(\tau_D)\cdot\omega-q_3(\tau_D)\cdot\omega^3]^2$	$q_1^2(\tau_D)\cdot\omega^2-2\cdot q_1(\tau_D)\cdot q_3(\tau_D)\cdot\omega^4+q_3^2(\tau_D)\cdot\omega^6$

$$\begin{aligned}F(\omega,\tau_D) &= |P(i\cdot\omega,\tau_D)|^2-|Q(i\cdot\omega,\tau_D)|^2\\ &= [(\iota_2+\iota_6)\cdot\omega^2-\iota_4\cdot\omega^4]^2+[-(\iota_1+\iota_5)\cdot\omega\\ &\quad+(\iota_3+\iota_7)\cdot\omega^3-\omega^5]^2-\{[q_0(\tau_D)\\ &\quad-q_2(\tau_D)\cdot\omega^2+q_4(\tau_D)\cdot\omega^4]^2\\ &\quad+[q_1(\tau_D)\cdot\omega-q_3(\tau_D)\cdot\omega^3]^2\}\end{aligned}$$

The plasma kinetic system's stability analysis $F(\omega, \tau_D)$ elements are (Table 6.4):

$$\begin{aligned}F(\omega,\tau_D) &= |P(i\cdot\omega,\tau_D)|^2-|Q(i\cdot\omega,\tau_D)|^2\\ &= (\iota_2+\iota_6)^2\cdot\omega^4-2\cdot(\iota_2+\iota_6)\cdot\iota_4\cdot\omega^6\\ &\quad+\iota_4^2\cdot\omega^8+(\iota_1+\iota_5)^2\cdot\omega^2-2\cdot(\iota_1+\iota_5)\\ &\quad\cdot(\iota_3+\iota_7)\cdot\omega^4+[2\cdot(\iota_1+\iota_5)+(\iota_3+\iota_7)^2]\\ &\quad\cdot\omega^6-2\cdot(\iota_3+\iota_7)\cdot\omega^8+\omega^{10}-q_0^2(\tau_D)\\ &\quad+2\cdot q_0(\tau_D)\cdot q_2(\tau_D)\cdot\omega^2-[q_0(\tau_D)\cdot q_4(\tau_D)\\ &\quad+q_2^2(\tau_D)+q_0(\tau_D)\cdot q_4(\tau_D)]\cdot\omega^4\\ &\quad+2\cdot q_2(\tau_D)\cdot q_4(\tau_D)\cdot\omega^6-q_4^2(\tau_D)\cdot\omega^8\\ &\quad-q_1^2(\tau_D)\cdot\omega^2+2\cdot q_1(\tau_D)\cdot q_3(\tau_D)\cdot\omega^4\\ &\quad-q_3^2(\tau_D)\cdot\omega^6\end{aligned}$$

$$\begin{aligned}F(\omega,\tau_D) &= |P(i\cdot\omega,\tau_D)|^2-|Q(i\cdot\omega,\tau_D)|^2\\ &= -q_0^2(\tau_D)+[(\iota_1+\iota_5)^2+2\cdot q_0(\tau_D)\cdot q_2(\tau_D)\\ &\quad-q_1^2(\tau_D)]\cdot\omega^2+\{(\iota_2+\iota_6)^2-2\cdot(\iota_1+\iota_5)\\ &\quad\cdot(\iota_3+\iota_7)-[q_0(\tau_D)\cdot q_4(\tau_D)+q_2^2(\tau_D)\\ &\quad+q_0(\tau_D)\cdot q_4(\tau_D)]+2\cdot q_1(\tau_D)\cdot q_3(\tau_D)\}\end{aligned}$$

$$\cdot \omega^4 + \{[2 \cdot (\iota_1 + \iota_5) + (\iota_3 + \iota_7)^2] - q_3^2(\tau_D) + 2 \cdot q_2(\tau_D) \cdot q_4(\tau_D)\} \cdot \omega^6 - \{2 \cdot (\iota_2 + \iota_6) \cdot \iota_4 - \iota_4^2 + 2 \cdot (\iota_3 + \iota_7) + q_4^2(\tau_D)\} \cdot \omega^8 + \omega^{10}$$

We define the following parameters for simplicity: Π_0, Π_2, Π_4, $\Pi_6 . \Pi_8$, Π_{10}

$$\Pi_0 = -q_0^2(\tau_D);\ \Pi_2 = (\iota_1 + \iota_5)^2 + 2 \cdot q_0(\tau_D) \cdot q_2(\tau_D) - q_1^2(\tau_D)$$

$$\Pi_4 = (\iota_2 + \iota_6)^2 - 2 \cdot (\iota_1 + \iota_5) \cdot (\iota_3 + \iota_7) - [q_0(\tau_D) \cdot q_4(\tau_D) + q_2^2(\tau_D) + q_0(\tau_D) \cdot q_4(\tau_D)] + 2 \cdot q_1(\tau_D) \cdot q_3(\tau_D)$$

$$\Pi_6 = [2 \cdot (\iota_1 + \iota_5) + (\iota_3 + \iota_7)^2] - q_3^2(\tau_D) + 2 \cdot q_2(\tau_D) \cdot q_4(\tau_D)$$

$$\Pi_8 = -\{2 \cdot (\iota_2 + \iota_6) \cdot \iota_4 - \iota_4^2 + 2 \cdot (\iota_3 + \iota_7) + q_4^2(\tau_D)\} \Pi_{10} = 1$$

Hence $F(\omega, \tau_D) = 0$ implies $\sum_{k=0}^{5} \Pi_{2 \cdot k} \cdot \omega^{2 \cdot k} = 0$ and its roots are given by solving the above polynomial.

$$P_R(i\omega, \tau_D) = (\iota_2 + \iota_6) \cdot \omega^2 - \iota_4 \cdot \omega^4;\ P_I(i\omega, \tau_D) = -(\iota_1 + \iota_5) \cdot \omega + (\iota_3 + \iota_7) \cdot \omega^3 - \omega^5$$

$$Q_R(i\omega, \tau_D) = q_0(\tau_D) - q_2(\tau_D) \cdot \omega^2 + q_4(\tau_D) \cdot \omega^4$$
$$Q_I(i\omega, \tau_D) = q_1(\tau_D) \cdot \omega - q_3(\tau_D) \cdot \omega^3$$

$$Q_R(i\omega, \tau_D) = [(\iota_1 + \iota_5) \cdot \Xi_1 - \Xi_2 \cdot (\iota_8 - \iota_{12}) + \Xi_3 \cdot (\iota_{15} - \iota_{19}) - \Xi_4 \cdot (\iota_{21} - \iota_{24}) + \Xi_5 \cdot (\iota_{28} + \iota_{32})] - [(\iota_3 + \iota_7) \cdot \Xi_1 - \Xi_2 \cdot (\iota_{10} - \iota_{13}) + \Xi_3 \cdot \iota_{17} - \Xi_4 \cdot (\iota_{23} - \iota_{26}) + \Xi_5 \cdot \iota_{30}] \cdot \omega^2 + \Xi_1 \cdot \omega^4$$

$$Q_I(i\omega, \tau_D) = [(\iota_2 + \iota_6) \cdot \Xi_1 - \Xi_2 \cdot (\iota_9 - \iota_{13}) + \Xi_3 \cdot (\iota_{16} - \iota_{20}) - \Xi_4 \cdot (\iota_{22} - \iota_{25}) + \Xi_5 \cdot (\iota_{29} + \iota_{33})] \cdot \omega - [\iota_4 \cdot \Xi_1 - \Xi_2 \cdot \iota_{11} + \Xi_3 \cdot \iota_{18} + \Xi_4 \cdot \iota_{27} + \Xi_5 \cdot \iota_{31}] \cdot \omega^3$$

$$\sin\theta(\tau_D) = \frac{-P_R(i\omega,\tau_D)\cdot Q_I(i\omega,\tau_D) + P_I(i\omega,\tau_D)\cdot Q_R(i\omega,\tau_D)}{|Q(i\omega,\tau_D)|^2}$$

$$\cos\theta(\tau_D) = -\frac{P_R(i\omega,\tau_D)\cdot Q_R(i\omega,\tau_D) + P_I(i\omega,\tau_D)\cdot Q_I(i\omega,\tau_D)}{|Q(i\omega,\tau_D)|^2}$$

We use different terminology from our last characteristics parameters definition:

$$k \to j;\ p_k(\tau_D) \to a_j;\ q_k(\tau_D) \to c_j$$
$$n, m \in \mathbb{N}_0;\ n = 5;\ m = 4;\ n > m$$

$$P_n(\lambda,\tau_D) \to P(\lambda);\ Q_m(\lambda,\tau_D) \to Q(\lambda)$$
$$P(\lambda) = \sum_{j=0}^{5} a_j\cdot\lambda^j;\ Q(\lambda) = \sum_{j=0}^{4} c_j\cdot\lambda^j$$

$$P(\lambda) = a_0 + a_1\cdot\lambda + a_2\cdot\lambda^2 + a_3\cdot\lambda^3 + a_4\cdot\lambda^4 + a_5\cdot\lambda^5$$
$$Q(\lambda) = c_0 + c_1\cdot\lambda + c_2\cdot\lambda^2 + c_3\cdot\lambda^3 + c_4\cdot\lambda^4$$

$n, m \in \mathbb{N}_0; n > m$ and $a_j, c_j : \mathbb{R}_{0+} \to \mathbb{R}$ are continuous and differentiable function of τ_D such that $a_0 + c_0 \neq 0$. In the following "–" denoted complex and conjugate. $P(\lambda)$, $Q(\lambda)$ are analytic functions in λ and differentiable in τ_D. The coefficients $a_j(S_c, S_{B1}, S_{B2}, \sigma_{CB1}, \sigma_{CB2}, \tau_D, \ldots) \in \mathbb{R}$ and $c_j(S_c, S_{B1}, S_{B2}, \sigma_{CB1}, \sigma_{CB2}, \tau_D, \ldots) \in \mathbb{R}$ depend on plasma kinetic system's parameters, $S_c, S_{B1}, S_{B2}, \sigma_{CB1}, \sigma_{CB2}, \tau_D, \ldots$ values. Unless strictly necessary, the designation of the varied arguments: $S_c, S_{B1}, S_{B2}, \sigma_{CB1}, \sigma_{CB2}, \tau_D, \ldots$ will subsequently be omitted from P, Q, a_j, c_j. The coefficients a_j, c_j are continuous and differentiable functions of their arguments, and direct substitution shows that $a_0 + c_0 \neq 0$[4, 5].

$$a_0 = 0$$
$$c_0 = (\iota_1 + \iota_5)\cdot\Xi_1 - \Xi_2\cdot(\iota_8 - \iota_{12}) + \Xi_3 \cdot(\iota_{15} - \iota_{19}) - \Xi_4\cdot(\iota_{21} - \iota_{24}) + \Xi_5\cdot(\iota_{28} + \iota_{32})$$

$$a_0 + c_0 = (\iota_1 + \iota_5)\cdot\Xi_1 - \Xi_2\cdot(\iota_8 - \iota_{12}) + \Xi_3\cdot(\iota_{15} - \iota_{19}) - \Xi_4\cdot(\iota_{21} - \iota_{24}) + \Xi_5\cdot(\iota_{28} + \iota_{32}) \neq 0$$

$\forall\, S_c, S_{B1}, S_{B2}, \sigma_{CB1}, \sigma_{CB2}, \tau_D, \ldots \in \mathbb{R}_+$ I.e. $\lambda = 0$ is not a root of the characteristic equation. Furthermore $P(\lambda)$, $Q(\lambda)$ are analytic functions of λ for which the following requirements of the analysis (see Kuang, 1993, Sect. 3.4) can also be verified in the present case.

(a) If $\lambda = i \cdot \omega; \omega \in \mathbb{R}$ then $P(i \cdot \omega) + Q(i \cdot \omega) \neq 0$, i.e. P and Q have no common imaginary roots. This condition was verified numerically in the entire $S_c, S_{B1}, S_{B2}, \sigma_{CB1}, \sigma_{CB2}, \tau_D, \ldots$ domain of interest.
(b) $|\frac{P(\lambda)}{Q(\lambda)}|$ is bounded for $|\lambda| \to \infty$; $\mathrm{Re}\lambda \geq 0$. No roots bifurcation from ∞. Indeed, in the limit: $|\frac{Q(\lambda)}{P(\lambda)}| = |\frac{c_0+c_1\cdot\lambda+c_2\cdot\lambda^2+c_3\cdot\lambda^3+c_4\cdot\lambda^4}{a_0+a_1\cdot\lambda+a_2\cdot\lambda^2+a_3\cdot\lambda^3+a_4\cdot\lambda^4+a_5\cdot\lambda^5}|$.
(c) The following expressions exist: $F(\omega) = |P(i \cdot \omega)|^2 - |Q(i \cdot \omega)|^2$ $F(\omega, \tau_D) = |P(i \cdot \omega, \tau_D)|^2 - |Q(i \cdot \omega, \tau_D)|^2 = \sum_{k=0}^{5} \Pi_{2\cdot k} \cdot \omega^{2\cdot k}$ has at most a finite number of zeros. Indeed, this is a polynomial in ω (degree in ω^{10}).
(d) Each positive root $\omega(S_c, S_{B1}, S_{B2}, \sigma_{CB1}, \sigma_{CB2}, \tau_D, \ldots)$ of $F(\omega) = 0$ is continuous and differentiable with respect to $S_c, S_{B1}, S_{B2}, \sigma_{CB1}, \sigma_{CB2}, \tau_D, \ldots$ and the condition can only be assessed numerically.

In addition, since the coefficients in P and Q are real, we have $\overline{P(-i \cdot \omega)} = P(i \cdot \omega)$ and $\overline{Q(-i \cdot \omega)} = Q(i \cdot \omega)$ thus, $\omega > 0$ may be on eigenvalue of characteristic equations. The analysis consists in identifying the roots of the characteristic equation situated on the imaginary axis of the complex λ plane, whereby increasing the parameters: $S_c, S_{B1}, S_{B2}, \sigma_{CB1}, \sigma_{CB2}, \tau_D, \ldots$, $\mathrm{Re}\lambda$ may, at the crossing, change its sign from (-) to (+). i.e. from a stable focus $E^{(j)} = (n_c^{(j)}, n_{B1}^{(j)}, n_{B2}^{(j)}, n_{ph1}^{(j)}, n_{ph2}^{(j)})$; $j = 0, 1, 2$ to an unstable one, or vice versa. This feature may be further assessed by examining the sign of the partial derivatives with respect to $S_c, S_{B1}, S_{B2}, \sigma_{CB1}, \sigma_{CB2}, \tau_D, \ldots$ and any system parameters.

$$\Lambda^{-1}(S_c) = \left(\frac{\partial \mathrm{Re}\lambda}{\partial S_c}\right)_{\lambda=i\cdot\omega} ; S_{B1}, S_{B2}, \sigma_{CB1}, \sigma_{CB2}, \tau_D, \ldots = const$$

$$\Lambda^{-1}(S_{B1}) = \left(\frac{\partial \mathrm{Re}\lambda}{\partial S_{B1}}\right)_{\lambda=i\cdot\omega} ; S_c, S_{B2}, \sigma_{CB1}, \sigma_{CB2}, \tau_D, \ldots = const$$

$$\Lambda^{-1}(S_{B2}) = \left(\frac{\partial \mathrm{Re}\lambda}{\partial S_{B2}}\right)_{\lambda=i\cdot\omega} ; S_c, S_{B1}, \sigma_{CB1}, \sigma_{CB2}, \tau_D, \ldots = const$$

$$\Lambda^{-1}(\sigma_{CB1}) = \left(\frac{\partial \mathrm{Re}\lambda}{\partial \sigma_{CB1}}\right)_{\lambda=i\cdot\omega} ; S_{B1}, S_{B2}, S_c, \sigma_{CB2}, \tau_D, \ldots = const$$

$$\Lambda^{-1}(\sigma_{CB2}) = \left(\frac{\partial \mathrm{Re}\lambda}{\partial \sigma_{CB2}}\right)_{\lambda=i\cdot\omega} ; S_{B1}, S_{B2}, S_c, \sigma_{CB1}, \tau_D, \ldots = const$$

$$\Lambda^{-1}(\tau_D) = \left(\frac{\partial \mathrm{Re}\lambda}{\partial \tau_D}\right)_{\lambda=i\cdot\omega} ; S_{B1}, S_{B2}, S_c, \sigma_{CB1}, \sigma_{CB2}, \ldots = const$$

$P(\lambda) = P_R(\lambda) + i \cdot P_I(\lambda)$; $Q(\lambda) = Q_R(\lambda) + i \cdot Q_I(\lambda)$, When writing and inserting $\lambda = i \cdot \omega$ into plasma kinetic system's characteristic equation ω must satisfy the following equations:

$$\sin(\omega \cdot \tau_D) = g(\omega) = \frac{-P_R(i\omega, \tau_D) \cdot Q_I(i\omega, \tau_D) + P_I(i\omega, \tau_D) \cdot Q_R(i\omega, \tau_D)}{|Q(i\omega, \tau_D)|^2}$$

$$\cos(\omega \cdot \tau_D) = h(\omega) = -\frac{P_R(i\omega, \tau_D) \cdot Q_R(i\omega, \tau_D) + P_I(i\omega, \tau_D) \cdot Q_I(i\omega, \tau_D)}{|Q(i\omega, \tau_D)|^2}$$

where $|Q(i\omega, \tau_D)|^2 \neq 0$ in view of requirement (a) above, and $(g, h) \in \mathbb{R}$. Furthermore, it follows above $\sin(\omega \cdot \tau_D)$ and $\cos(\omega \cdot \tau_D)$ equation that, by squaring and adding the sides, ω must be a positive root of $F(\omega) = |P(i \cdot \omega)|^2 - |Q(i \cdot \omega)|^2 = 0$. Note: $F(\omega)$ is independent on τ_D. Now it is important to notice that if $\tau_D \notin I$(assume that $I \subseteq \mathbb{R}_{+0}$ is the set where $\omega(\tau_D)$ is a positive root of $F(\omega)$ and for, $\tau_D \notin I, \omega(\tau_D)$ is not defined. Then for all τ_D in I, $\omega(\tau_D)$ is satisfied that $F(\omega, \tau_D) = 0$. Then there are no positive $\omega(\tau_D)$ solutions for $F(\omega, \tau_D) = 0$, and we cannot have stability switches. For $\tau_D \in I$ where $\omega(\tau_D)$ is a positive solution of $F(\omega, \tau_D) = 0$, we can define the angle $\theta(\tau_D) \in [0, 2 \cdot \Pi]$ as the solution of $\sin\theta(\tau_D) = \ldots$ and $\cos\theta(\tau_D) = \ldots$; the relation between the arguments $\theta(\tau_D)$ and $\tau_D \cdot \omega(\tau_D)$ for $\tau_D \in I$ must be describing below. $\tau_D \cdot \omega(\tau_D) = \theta(\tau_D) + 2 \cdot n \cdot \Pi \, \forall n \in \mathbb{N}_0$ Hence we can define the maps: $\tau_{D(n)} : I \to \mathbb{R}_{+0}$, is given by $\tau_{D(n)}(\tau_D) = \frac{\theta(\tau_D)+2 \cdot n \cdot \Pi}{\omega(\tau_D)}; n \in \mathbb{N}_0; \tau_D \in I$. Let us introduce the function $I \to \mathbb{R}$; $S_n(\tau_D) = \tau_D - \tau_{D(n)}(\tau_D); \tau_D \in I; n \in \mathbb{N}_0$ that is continuous and differentiable in τ_D. In the following, the subscripts $\lambda, \omega, S_c, S_{B1}, S_{B2}, \sigma_{CB1}, \sigma_{CB2}, \ldots$ indicate the corresponding partial derivatives. Let us first concentrate on $\Lambda(x)$, remember in $\lambda(S_c, S_{B1}, S_{B2}, \sigma_{CB1}, \sigma_{CB2}, \ldots)$ and $\omega(S_c, S_{B1}, S_{B2}, \sigma_{CB1}, \sigma_{CB2}, \ldots)$, and keeping all parameters except one (x) and τ_D. The derivation closely follows that in reference [BK]. Differentiating plasma kinetic system's characteristic equation $P(\lambda) + Q(\lambda) \cdot e^{-\lambda \cdot \tau_D} = 0$ with respect to specific parameter (x), and inverting the derivative, for convenience, one calculates: $x = S_c, S_{B1}, S_{B2}, \sigma_{CB1}, \sigma_{CB2}, \tau_D, \ldots$

$$\left(\frac{\partial \lambda}{\partial x}\right)^{-1} = \frac{-P_\lambda(\lambda, x) \cdot Q(\lambda, x) + Q_\lambda(\lambda, x) \cdot P(\lambda, x) - \tau_D \cdot P(\lambda, x) \cdot Q(\lambda, x)}{P_x(\lambda, x) \cdot Q(\lambda, x) - Q_x(\lambda, x) \cdot P(\lambda, x)}$$

where $P_\lambda = \frac{\partial P}{\partial \lambda}, \ldots$ etc., substituting $\lambda = i \cdot \omega$ and bearing $\overline{P(-i \cdot \omega)} = P(i \cdot \omega)$; $\overline{Q(-i \cdot \omega)} = Q(i \cdot \omega)$ then $i \cdot P_\lambda(i \cdot \omega) = P_\omega(i \cdot \omega); i \cdot Q_\lambda(i \cdot \omega) = Q_\omega(i \cdot \omega)$ and that on the surface $|P(i\omega)|^2 = |Q(i\omega)|^2$, one obtain: $(\frac{\partial \lambda}{\partial x})^{-1}|_{\lambda = i \cdot \omega} = (\frac{i \cdot P_\omega(i \cdot \omega, x) \cdot \overline{P(i \cdot \omega, x)} + i \cdot Q_\lambda(i \cdot \omega, x) \cdot \overline{Q(\lambda, x)} - \tau_D \cdot |P(i \cdot \omega, x)|^2}{P_x(i \cdot \omega, x) \cdot \overline{P(i \cdot \omega, x)} - Q_x(i \cdot \omega, x) \cdot \overline{Q(i \cdot \omega, x)}})$

Upon separating into real and imaginary parts, with $P = P_R + i \cdot P_I$; $Q = Q_R + i \cdot Q_I$ and

$P_\omega = P_{R\omega} + i \cdot P_{I\omega}$; $Q_\omega = Q_{R\omega} + i \cdot Q_{I\omega}$; $P_x = P_{Rx} + i \cdot P_{Ix}$; $Q_x = Q_{Rx} + i \cdot Q_{Ix}$; $P^2 = P_R^2 + P_I^2$, When (x) can be any plasma kinetic system's parameters $S_c, S_{B1}, S_{B2}, \sigma_{CB1}, \sigma_{CB2}, \ldots$ and time delay τ_D etc. Where for convenience, we have dropped the arguments $(i \cdot \omega, x)$, and where $F_\omega = 2 \cdot [(P_{R\omega} \cdot P_R + P_{I\omega} \cdot P_I) - (Q_{R\omega} \cdot Q_R + Q_{I\omega} \cdot Q_I)]$;$F_x = 2 \cdot [(P_{Rx} \cdot P_R + P_{Ix} \cdot P_I) - (Q_{Rx} \cdot Q_R + Q_{Ix} \cdot Q_I)]$

and $\omega_x = -\frac{F_x}{F_\omega}$. We define U and V: $U = (P_R \cdot P_{I\omega} - P_I \cdot P_{R\omega}) - (Q_R \cdot Q_{I\omega} - Q_I \cdot Q_{R\omega})$, $V = (P_R \cdot P_{Ix} - P_I \cdot P_{Rx}) - (Q_R \cdot Q_{Ix} - Q_I \cdot Q_{Rx})$, we choose our specific parameter as time delay$x = \tau_D$: $P_R = (\iota_2 + \iota_6) \cdot \omega^2 - \iota_4 \cdot \omega^4$; $P_I = -(\iota_1 + \iota_5) \cdot \omega + (\iota_3 + \iota_7) \cdot \omega^3 - \omega^5$

$$\begin{aligned} Q_R = {} & [(\iota_1 + \iota_5) \cdot \Xi_1 - \Xi_2 \cdot (\iota_8 - \iota_{12}) + \Xi_3 \cdot (\iota_{15} - \iota_{19}) \\ & - \Xi_4 \cdot (\iota_{21} - \iota_{24}) + \Xi_5 \cdot (\iota_{28} + \iota_{32})] - [(\iota_3 + \iota_7) \cdot \Xi_1 \\ & - \Xi_2 \cdot (\iota_{10} - \iota_{13}) + \Xi_3 \cdot \iota_{17} - \Xi_4 \cdot (\iota_{23} - \iota_{26}) \\ & + \Xi_5 \cdot \iota_{30}] \cdot \omega^2 + \Xi_1 \cdot \omega^4 \end{aligned}$$

We define for simplicity the following global parameters:

$$\begin{aligned} \vartheta_2 = {} & (\iota_1 + \iota_5) \cdot \Xi_1 - \Xi_2 \cdot (\iota_8 - \iota_{12}) + \Xi_3 \cdot (\iota_{15} - \iota_{19}) \\ & - \Xi_4 \cdot (\iota_{21} - \iota_{24}) + \Xi_5 \cdot (\iota_{28} + \iota_{32}) \end{aligned}$$

$$\begin{aligned} \vartheta_1 = {} & (\iota_3 + \iota_7) \cdot \Xi_1 - \Xi_2 \cdot (\iota_{10} - \iota_{13}) + \Xi_3 \cdot \iota_{17} \\ & - \Xi_4 \cdot (\iota_{23} - \iota_{26}) + \Xi_5 \cdot \iota_{30} \end{aligned}$$

$$\begin{aligned} Q_R = {} & \vartheta_2 - \vartheta_1 \cdot \omega^2 + \Xi_1 \cdot \omega^4 \Rightarrow Q_{R\omega} \\ & = -2 \cdot \vartheta_1 \cdot \omega + 4 \cdot \Xi_1 \cdot \omega^3 \end{aligned}$$

$$\begin{aligned} Q_I = {} & [(\iota_2 + \iota_6) \cdot \Xi_1 - \Xi_2 \cdot (\iota_9 - \iota_{13}) + \Xi_3 \cdot (\iota_{16} - \iota_{20}) \\ & - \Xi_4 \cdot (\iota_{22} - \iota_{25}) + \Xi_5 \cdot (\iota_{29} + \iota_{33})] \cdot \omega - [\iota_4 \cdot \Xi_1 - \Xi_2 \cdot \iota_{11} \\ & + \Xi_3 \cdot \iota_{18} + \Xi_4 \cdot \iota_{27} + \Xi_5 \cdot \iota_{31}] \cdot \omega^3 \end{aligned}$$

We define for simplicity the following global parameters:

$$\begin{aligned} \vartheta_3 = {} & (\iota_2 + \iota_6) \cdot \Xi_1 - \Xi_2 \cdot (\iota_9 - \iota_{13}) + \Xi_3 \cdot (\iota_{16} - \iota_{20}) \\ & - \Xi_4 \cdot (\iota_{22} - \iota_{25}) + \Xi_5 \cdot (\iota_{29} + \iota_{33}) \end{aligned}$$

$$\vartheta_4 = \iota_4 \cdot \Xi_1 - \Xi_2 \cdot \iota_{11} + \Xi_3 \cdot \iota_{18} + \Xi_4 \cdot \iota_{27} + \Xi_5 \cdot \iota_{31}$$

$$Q_I = \vartheta_3 \cdot \omega - \vartheta_4 \cdot \omega^3 \Rightarrow Q_{I\omega} = \vartheta_3 - 3 \cdot \vartheta_4 \cdot \omega^2$$

$$P_{R\omega} = 2 \cdot (\iota_2 + \iota_6) \cdot \omega - 4 \cdot \iota_4 \cdot \omega^3$$
$$P_{I\omega} = -(\iota_1 + \iota_5) + 3 \cdot (\iota_3 + \iota_7) \cdot \omega^2 - 5 \cdot \omega^4$$

$$Q_{R\omega} = -2 \cdot [(\iota_3 + \iota_7) \cdot \Xi_1 - \Xi_2 \cdot (\iota_{10} - \iota_{13}) + \Xi_3 \cdot \iota_{17} - \Xi_4 \cdot (\iota_{23} - \iota_{26}) + \Xi_5 \cdot \iota_{30}] \cdot \omega + 4 \cdot \Xi_1 \cdot \omega^3$$

$$Q_{I\omega} = [(\iota_2 + \iota_6) \cdot \Xi_1 - \Xi_2 \cdot (\iota_9 - \iota_{13}) + \Xi_3 \cdot (\iota_{16} - \iota_{20}) - \Xi_4 \cdot (\iota_{22} - \iota_{25}) + \Xi_5 \cdot (\iota_{29} + \iota_{33})] - 3 \cdot [\iota_4 \cdot \Xi_1 - \Xi_2 \cdot \iota_{11} + \Xi_3 \cdot \iota_{18} + \Xi_4 \cdot \iota_{27} + \Xi_5 \cdot \iota_{31}] \cdot \omega^2$$

$$P_{R\tau_D} = 0; P_{I\tau_D} = 0; Q_{R\tau_D} = 0; Q_{I\tau_D} = 0$$
$$F_{\tau_D} = 0; V = (P_R \cdot P_{I\tau_D} - P_I \cdot P_{R\tau_D}) - (Q_R \cdot Q_{I\tau_D} - Q_I \cdot Q_{R\tau_D}) = 0$$

$F_\omega = \ldots.$ Elements:

$$P_{R\omega} \cdot P_R = [2 \cdot (\iota_2 + \iota_6) \cdot \omega - 4 \cdot \iota_4 \cdot \omega^3] \cdot [(\iota_2 + \iota_6) \cdot \omega^2 - \iota_4 \cdot \omega^4]$$
$$= 2 \cdot (\iota_2 + \iota_6)^2 \cdot \omega^3 - 6 \cdot \iota_4 \cdot (\iota_2 + \iota_6) \cdot \omega^5 + 4 \cdot \iota_4 \cdot \iota_4 \cdot \omega^7$$

$$P_{I\omega} \cdot P_I = \{-(\iota_1 + \iota_5) + 3 \cdot (\iota_3 + \iota_7) \cdot \omega^2 - 5 \cdot \omega^4\} \cdot \{-(\iota_1 + \iota_5) \cdot \omega + (\iota_3 + \iota_7) \cdot \omega^3 - \omega^5\}$$
$$= (\iota_1 + \iota_5)^2 \cdot \omega - 4 \cdot (\iota_1 + \iota_5) \cdot (\iota_3 + \iota_7) \cdot \omega^3 + \{(\iota_3 + \iota_7)^2 + 2 \cdot (\iota_1 + \iota_5)\} \cdot 3 \cdot \omega^5 - 8 \cdot (\iota_3 + \iota_7) \cdot \omega^7 + 5 \cdot \omega^9$$

$$Q_{R\omega} \cdot Q_R = \{-2 \cdot [(\iota_3 + \iota_7) \cdot \Xi_1 - \Xi_2 \cdot (\iota_{10} - \iota_{13}) + \Xi_3 \cdot \iota_{17} - \Xi_4 \cdot (\iota_{23} - \iota_{26}) + \Xi_5 \cdot \iota_{30}] \cdot \omega + 4 \cdot \Xi_1 \cdot \omega^3\}$$
$$\cdot \{[(\iota_1 + \iota_5) \cdot \Xi_1 - \Xi_2 \cdot (\iota_8 - \iota_{12}) + \Xi_3 \cdot (\iota_{15} - \iota_{19}) - \Xi_4 \cdot (\iota_{21} - \iota_{24}) + \Xi_5 \cdot (\iota_{28} + \iota_{32})] - [(\iota_3 + \iota_7) \cdot \Xi_1 - \Xi_2 \cdot (\iota_{10} - \iota_{13}) + \Xi_3 \cdot \iota_{17} - \Xi_4 \cdot (\iota_{23} - \iota_{26}) + \Xi_5 \cdot \iota_{30}] \cdot \omega^2 + \Xi_1 \cdot \omega^4\}$$

$$\vartheta_1 = (\iota_3 + \iota_7) \cdot \Xi_1 - \Xi_2 \cdot (\iota_{10} - \iota_{13}) + \Xi_3 \cdot \iota_{17} - \Xi_4 \cdot (\iota_{23} - \iota_{26}) + \Xi_5 \cdot \iota_{30}$$

$$\vartheta_2 = (\iota_1 + \iota_5) \cdot \Xi_1 - \Xi_2 \cdot (\iota_8 - \iota_{12}) + \Xi_3 \cdot (\iota_{15} - \iota_{19})$$

$$- \Xi_4 \cdot (\iota_{21} - \iota_{24}) + \Xi_5 \cdot (\iota_{28} + \iota_{32})$$

$$\begin{aligned} Q_{R\omega} \cdot Q_R &= \{-2 \cdot \vartheta_1 \cdot \omega + 4 \cdot \Xi_1 \cdot \omega^3\} \cdot \{\vartheta_2 - \vartheta_1 \cdot \omega^2 + \Xi_1 \cdot \omega^4\} \\ &= -2 \cdot \vartheta_1 \cdot \vartheta_2 \cdot \omega + (2 \cdot \vartheta_1^2 + 4 \cdot \Xi_1 \cdot \vartheta_2) \cdot \omega^3 \\ &\quad - 6 \cdot \vartheta_1 \cdot \Xi_1 \cdot \omega^5 + 4 \cdot \Xi_1 \cdot \Xi_1 \cdot \omega^7 \end{aligned}$$

$$\begin{aligned} Q_{I\omega} \cdot Q_I = &\{[(\iota_2 + \iota_6) \cdot \Xi_1 - \Xi_2 \cdot (\iota_9 - \iota_{13}) + \Xi_3 \cdot (\iota_{16} - \iota_{20}) \\ &- \Xi_4 \cdot (\iota_{22} - \iota_{25}) + \Xi_5 \cdot (\iota_{29} + \iota_{33})] - 3 \\ &\cdot [\iota_4 \cdot \Xi_1 - \Xi_2 \cdot \iota_{11} + \Xi_3 \cdot \iota_{18} + \Xi_4 \cdot \iota_{27} + \Xi_5 \cdot \iota_{31}] \cdot \omega^2\} \\ &\cdot \{[(\iota_2 + \iota_6) \cdot \Xi_1 - \Xi_2 \cdot (\iota_9 - \iota_{13}) + \Xi_3 \cdot (\iota_{16} - \iota_{20}) - \Xi_4 \\ &\cdot (\iota_{22} - \iota_{25}) + \Xi_5 \cdot (\iota_{29} + \iota_{33})] \cdot \omega - [\iota_4 \cdot \Xi_1 - \Xi_2 \cdot \iota_{11} \\ &+ \Xi_3 \cdot \iota_{18} + \Xi_4 \cdot \iota_{27} + \Xi_5 \cdot \iota_{31}] \cdot \omega^3\} \end{aligned}$$

$$\begin{aligned} \vartheta_3 = &(\iota_2 + \iota_6) \cdot \Xi_1 - \Xi_2 \cdot (\iota_9 - \iota_{13}) + \Xi_3 \cdot (\iota_{16} - \iota_{20}) \\ &- \Xi_4 \cdot (\iota_{22} - \iota_{25}) + \Xi_5 \cdot (\iota_{29} + \iota_{33}) \end{aligned}$$

$$\vartheta_4 = \iota_4 \cdot \Xi_1 - \Xi_2 \cdot \iota_{11} + \Xi_3 \cdot \iota_{18} + \Xi_4 \cdot \iota_{27} + \Xi_5 \cdot \iota_{31}$$

$$\begin{aligned} Q_{I\omega} \cdot Q_I &= (\vartheta_3 - 3 \cdot \vartheta_4 \cdot \omega^2) \cdot (\vartheta_3 \cdot \omega - \vartheta_4 \cdot \omega^3) \\ &= \vartheta_3^2 \cdot \omega - 4 \cdot \vartheta_4 \cdot \vartheta_3 \cdot \omega^3 + 3 \cdot \vartheta_4 \cdot \vartheta_4 \cdot \omega^5 \end{aligned}$$

$U = \ldots.$Elements:

$$\begin{aligned} P_R \cdot P_{I\omega} &= \{(\iota_2 + \iota_6) \cdot \omega^2 - \iota_4 \cdot \omega^4\} \cdot \{-(\iota_1 + \iota_5) + 3 \cdot (\iota_3 + \iota_7) \cdot \omega^2 - 5 \cdot \omega^4\} \\ &= -(\iota_1 + \iota_5) \cdot (\iota_2 + \iota_6) \cdot \omega^2 + \{3 \cdot (\iota_2 + \iota_6) \cdot (\iota_3 + \iota_7) + \iota_4 \cdot (\iota_1 + \iota_5)\} \cdot \omega^4 \\ &\quad - \{(\iota_2 + \iota_6) \cdot 5 + 3 \cdot \iota_4 \cdot (\iota_3 + \iota_7)\} \cdot \omega^6 + \iota_4 \cdot 5 \cdot \omega^8 \end{aligned}$$

$$\begin{aligned} P_I \cdot P_{R\omega} &= \{-(\iota_1 + \iota_5) \cdot \omega + (\iota_3 + \iota_7) \cdot \omega^3 - \omega^5\} \\ &\quad \cdot \{2 \cdot (\iota_2 + \iota_6) \cdot \omega - 4 \cdot \iota_4 \cdot \omega^3\} \\ &= -2 \cdot (\iota_1 + \iota_5) \cdot (\iota_2 + \iota_6) \cdot \omega^2 + 2 \cdot \{2 \cdot (\iota_1 + \iota_5) \\ &\quad \cdot \iota_4 + (\iota_3 + \iota_7) \cdot (\iota_2 + \iota_6)\} \cdot \omega^4 - 2 \\ &\quad \cdot \{2 \cdot (\iota_3 + \iota_7) \cdot \iota_4 + (\iota_2 + \iota_6)\} \cdot \omega^6 + 4 \cdot \iota_4 \cdot \omega^8 \end{aligned}$$

$$\begin{aligned} Q_R \cdot Q_{I\omega} &= (\vartheta_2 - \vartheta_1 \cdot \omega^2 + \Xi_1 \cdot \omega^4) \cdot (\vartheta_3 - 3 \cdot \vartheta_4 \cdot \omega^2) \\ &= \vartheta_2 \cdot \vartheta_3 - (3 \cdot \vartheta_2 \cdot \vartheta_4 + \vartheta_1 \cdot \vartheta_3) \cdot \omega^2 \\ &\quad + (3 \cdot \vartheta_1 \cdot \vartheta_4 + \Xi_1 \cdot \vartheta_3) \cdot \omega^4 - 3 \cdot \Xi_1 \cdot \vartheta_4 \cdot \omega^6 \end{aligned}$$

Table 6.5 Plasma kinetic system's stability analysis F_{ω}, U elements

F_{ω}, U elements	Expression
$P_{R\omega} \cdot P_R$	$2 \cdot (\iota_2 + \iota_6)^2 \cdot \omega^3 - 6 \cdot \iota_4 \cdot (\iota_2 + \iota_6) \cdot \omega^5 + 4 \cdot \iota_4 \cdot \iota_4 \cdot \omega^7$
$P_{I\omega} \cdot P_I$	$(\iota_1 + \iota_5)^2 \cdot \omega - 4 \cdot (\iota_1 + \iota_5) \cdot (\iota_3 + \iota_7) \cdot \omega^3 + \{(\iota_3 + \iota_7)^2 + 2 \cdot (\iota_1 + \iota_5)\} \cdot 3 \cdot \omega^5$ $- 8 \cdot (\iota_3 + \iota_7) \cdot \omega^7 + 5 \cdot \omega^9$
$Q_{R\omega} \cdot Q_R$	$-2 \cdot \vartheta_1 \cdot \vartheta_2 \cdot \omega + (2 \cdot \vartheta_1^2 + 4 \cdot \Xi_1 \cdot \vartheta_2) \cdot \omega^3 - 6 \cdot \vartheta_1 \cdot \Xi_1 \cdot \omega^5$ $+ 4 \cdot \Xi_1 \cdot \Xi_1 \cdot \omega^7$
$Q_{I\omega} \cdot Q_I$	$\vartheta_3^2 \cdot \omega - 4 \cdot \vartheta_4 \cdot \vartheta_3 \cdot \omega^3 + 3 \cdot \vartheta_4 \cdot \vartheta_4 \cdot \omega^5$
$P_R \cdot P_{I\omega}$	$-(\iota_1 + \iota_5) \cdot (\iota_2 + \iota_6) \cdot \omega^2 + \{3 \cdot (\iota_2 + \iota_6) \cdot (\iota_3 + \iota_7) + \iota_4 \cdot (\iota_1 + \iota_5)\} \cdot \omega^4$ $- \{(\iota_2 + \iota_6) \cdot 5 + 3 \cdot \iota_4 \cdot (\iota_3 + \iota_7)\} \cdot \omega^6 + \iota_4 \cdot 5 \cdot \omega^8$
$P_I \cdot P_{R\omega}$	$-2 \cdot (\iota_1 + \iota_5) \cdot (\iota_2 + \iota_6) \cdot \omega^2 + 2 \cdot \{2 \cdot (\iota_1 + \iota_5) \cdot \iota_4 + (\iota_3 + \iota_7) \cdot (\iota_2 + \iota_6)\} \cdot \omega^4$ $- 2 \cdot \{2 \cdot (\iota_3 + \iota_7) \cdot \iota_4 + (\iota_2 + \iota_6)\} \cdot \omega^6 + 4 \cdot \iota_4 \cdot \omega^8$
$Q_R \cdot Q_{I\omega}$	$\vartheta_2 \cdot \vartheta_3 - (3 \cdot \vartheta_2 \cdot \vartheta_4 + \vartheta_1 \cdot \vartheta_3) \cdot \omega^2 + (3 \cdot \vartheta_1 \cdot \vartheta_4 + \Xi_1 \cdot \vartheta_3) \cdot \omega^4 - 3 \cdot \Xi_1 \cdot \vartheta_4 \cdot \omega^6$
$Q_I \cdot Q_{R\omega}$	$-2 \cdot \vartheta_1 \cdot \vartheta_3 \cdot \omega^2 + 2 \cdot (2 \cdot \vartheta_3 \cdot \Xi_1 + \vartheta_4 \cdot \vartheta_1) \cdot \omega^4 - 4 \cdot \vartheta_4 \cdot \Xi_1 \cdot \omega^6$

$$
\begin{aligned}
Q_I \cdot Q_{R\omega} &= (\vartheta_3 \cdot \omega - \vartheta_4 \cdot \omega^3) \cdot (-2 \cdot \vartheta_1 \cdot \omega + 4 \cdot \Xi_1 \cdot \omega^3) \\
&= -2 \cdot \vartheta_1 \cdot \vartheta_3 \cdot \omega^2 + 2 \cdot (2 \cdot \vartheta_3 \cdot \Xi_1 + \vartheta_4 \cdot \vartheta_1) \cdot \omega^4 - 4 \cdot \vartheta_4 \cdot \Xi_1 \cdot \omega^6
\end{aligned}
$$

We can summary our last results in the next table (Table 6.5).

$F(\omega, \tau_D) = 0$, differentiating with respect to τ_D and we get $F_{\omega} \cdot \frac{\partial \omega}{\partial \tau_D} + F_{\tau_D} = 0$; $\tau_D \in I \Rightarrow \frac{\partial \omega}{\partial \tau_D} = -\frac{F_{\tau_D}}{F_{\omega}}$

$$
\Lambda^{-1}(\tau_D) = \left(\frac{\partial \mathrm{Re}\lambda}{\partial \tau_D}\right)_{\lambda = i\omega}
$$

$$
\frac{\partial \omega}{\partial \tau_D} = \omega_{\tau_D} = -\frac{F_{\tau_D}}{F_{\omega}};\ \Lambda^{-1}(\tau_D) = \mathrm{Re}\left\{\frac{-2 \cdot [U + \tau_D \cdot |P|^2 + i \cdot F_{\omega}}{F_{\tau_D} + i \cdot 2 \cdot [V + \omega \cdot |P|^2]}\right\}
$$

$$
\mathrm{sign}\{\Lambda^{-1}(\tau_D)\} = \mathrm{sign}\left\{\left(\frac{\partial \mathrm{Re}\lambda}{\partial \tau_D}\right)_{\lambda = i\omega}\right\}
$$

$$
\mathrm{sign}\{\Lambda^{-1}(\tau_D)\} = \mathrm{sign}\{F_{\omega}\} \cdot \mathrm{sign}\left\{\frac{V + \frac{\partial \omega}{\partial \tau_D} \cdot U}{|P|^2} + \omega + \frac{\partial \omega}{\partial \tau_D} \cdot \tau_D\right\}
$$

We shall presently examine the possibility of stability transitions (bifurcations) plasma kinetic system, about the equilibrium points $E^{(j)} = (n_c^{(j)}, n_{B1}^{(j)}, n_{B2}^{(j)}, n_{ph1}^{(j)}, n_{ph2}^{(j)})$; $j = 0, 1, 2, \ldots$. The analysis consists in identifying the roots of our system characteristic equation situated on the imaginary axis of the complex λ-plane, where by increasing the delay parameterτ_D. Reλ, may at the crossing, changes its sign from – to +, i.e. from a stable focus $E^{(*)}$ to an unstable one, or vice versa. This feature may be further assessed by examining the sign of the partial derivatives with respect to τ_D [4, 5].

$$\Lambda^{-1}(\tau_D) = \left(\frac{\partial \mathrm{Re}\lambda}{\partial \tau_D}\right)_{\lambda=i\omega};\ S_{B1}, S_{B2}, S_c, \sigma_{CB1}, \sigma_{CB2}, \ldots = const$$

$$\begin{aligned} U &= (P_R \cdot P_{I\omega} - P_I \cdot P_{R\omega}) - (Q_R \cdot Q_{I\omega} - Q_I \cdot Q_{R\omega}) \\ &= -(\iota_1 + \iota_5) \cdot (\iota_2 + \iota_6) \cdot \omega^2 + \{3 \cdot (\iota_2 + \iota_6) \cdot (\iota_3 + \iota_7) \\ &\quad + \iota_4 \cdot (\iota_1 + \iota_5)\} \cdot \omega^4 - \{(\iota_2 + \iota_6) \cdot 5 + 3 \cdot \iota_4 \cdot (\iota_3 + \iota_7)\} \\ &\quad \cdot \omega^6 + \iota_4 \cdot 5 \cdot \omega^8 - [-2 \cdot (\iota_1 + \iota_5) \cdot (\iota_2 + \iota_6) \cdot \omega^2 \\ &\quad + 2 \cdot \{2 \cdot (\iota_1 + \iota_5) \cdot \iota_4 + (\iota_3 + \iota_7) \cdot (\iota_2 + \iota_6)\} \cdot \omega^4 \\ &\quad - 2 \cdot \{2 \cdot (\iota_3 + \iota_7) \cdot \iota_4 + (\iota_2 + \iota_6)\} \cdot \omega^6 + 4 \cdot \iota_4 \cdot \omega^8] \\ &\quad - [\vartheta_2 \cdot \vartheta_3 - (3 \cdot \vartheta_2 \cdot \vartheta_4 + \vartheta_1 \cdot \vartheta_3) \cdot \omega^2 \\ &\quad + (3 \cdot \vartheta_1 \cdot \vartheta_4 + \Xi_1 \cdot \vartheta_3) \cdot \omega^4 - 3 \cdot \Xi_1 \cdot \vartheta_4 \cdot \omega^6 \\ &\quad - \{-2 \cdot \vartheta_1 \cdot \vartheta_3 \cdot \omega^2 + 2 \cdot (2 \cdot \vartheta_3 \cdot \Xi_1 + \vartheta_4 \cdot \vartheta_1) \cdot \omega^4 - 4 \cdot \vartheta_4 \cdot \Xi_1 \cdot \omega^6\}] \end{aligned}$$

$$\begin{aligned} U &= (P_R \cdot P_{I\omega} - P_I \cdot P_{R\omega}) - (Q_R \cdot Q_{I\omega} - Q_I \cdot Q_{R\omega}) \\ &= -(\iota_1 + \iota_5) \cdot (\iota_2 + \iota_6) \cdot \omega^2 + [3 \cdot (\iota_2 + \iota_6) \cdot (\iota_3 + \iota_7) \\ &\quad + \iota_4 \cdot (\iota_1 + \iota_5)] \cdot \omega^4 - [(\iota_2 + \iota_6) \cdot 5 + 3 \cdot \iota_4 \cdot (\iota_3 + \iota_7)] \\ &\quad \cdot \omega^6 + \iota_4 \cdot 5 \cdot \omega^8 + 2 \cdot (\iota_1 + \iota_5) \cdot (\iota_2 + \iota_6) \cdot \omega^2 - 2 \\ &\quad \cdot [2 \cdot (\iota_1 + \iota_5) \cdot \iota_4 + (\iota_3 + \iota_7) \cdot (\iota_2 + \iota_6)] \cdot \omega^4 \\ &\quad + 2 \cdot [2 \cdot (\iota_3 + \iota_7) \cdot \iota_4 - (\iota_2 + \iota_6)] \cdot \omega^6 - 4 \cdot \iota_4 \cdot \omega^8 \\ &\quad - \vartheta_2 \cdot \vartheta_3 + (3 \cdot \vartheta_2 \cdot \vartheta_4 + \vartheta_1 \cdot \vartheta_3) \cdot \omega^2 - (3 \cdot \vartheta_1 \cdot \vartheta_4 + \Xi_1 \cdot \vartheta_3) \\ &\quad \cdot \omega^4 + 3 \cdot \Xi_1 \cdot \vartheta_4 \cdot \omega^6 - 2 \cdot \vartheta_1 \cdot \vartheta_3 \cdot \omega^2 \\ &\quad + 2 \cdot (2 \cdot \vartheta_3 \cdot \Xi_1 + \vartheta_4 \cdot \vartheta_1) \cdot \omega^4 - 4 \cdot \vartheta_4 \cdot \Xi_1 \cdot \omega^6 \end{aligned}$$

$$\begin{aligned} U &= (P_R \cdot P_{I\omega} - P_I \cdot P_{R\omega}) - (Q_R \cdot Q_{I\omega} - Q_I \cdot Q_{R\omega}) \\ &= -\vartheta_2 \cdot \vartheta_3 + \{2 \cdot (\iota_1 + \iota_5) \cdot (\iota_2 + \iota_6) + (3 \cdot \vartheta_2 \cdot \vartheta_4 + \vartheta_1 \cdot \vartheta_3) \\ &\quad - 2 \cdot \vartheta_1 \cdot \vartheta_3 - (\iota_1 + \iota_5) \cdot (\iota_2 + \iota_6)\} \cdot \omega^2 + \{[3 \cdot (\iota_2 + \iota_6) \\ &\quad \cdot (\iota_3 + \iota_7) + \iota_4 \cdot (\iota_1 + \iota_5)] - 2 \cdot [2 \cdot (\iota_1 + \iota_5) \cdot \iota_4 + (\iota_3 + \iota_7) \\ &\quad \cdot (\iota_2 + \iota_6)] - (3 \cdot \vartheta_1 \cdot \vartheta_4 + \Xi_1 \cdot \vartheta_3) + 2 \cdot (2 \cdot \vartheta_3 \cdot \Xi_1 + \vartheta_4 \cdot \vartheta_1)\} \\ &\quad \cdot \omega^4 + \{2 \cdot [2 \cdot (\iota_3 + \iota_7) \cdot \iota_4 - (\iota_2 + \iota_6)] - [(\iota_2 + \iota_6) \cdot 5 + 3 \cdot \iota_4 \\ &\quad \cdot (\iota_3 + \iota_7)] + 3 \cdot \Xi_1 \cdot \vartheta_4 - 4 \cdot \vartheta_4 \cdot \Xi_1\} \cdot \omega^6 + \{\iota_4 \cdot 5 - 4 \cdot \iota_4\} \cdot \omega^8 \end{aligned}$$

We define for simplicity the following global parameters:

$$A_0 = -\vartheta_2 \cdot \vartheta_3$$
$$A_2 = 2 \cdot (\iota_1 + \iota_5) \cdot (\iota_2 + \iota_6) + (3 \cdot \vartheta_2 \cdot \vartheta_4 + \vartheta_1 \cdot \vartheta_3) - 2 \cdot \vartheta_1 \cdot \vartheta_3 - (\iota_1 + \iota_5) \cdot (\iota_2 + \iota_6)$$

$$A_4 = [3 \cdot (\iota_2 + \iota_6) \cdot (\iota_3 + \iota_7) + \iota_4 \cdot (\iota_1 + \iota_5)] - 2 \cdot [2 \cdot (\iota_1 + \iota_5) \cdot \iota_4 + (\iota_3 + \iota_7) \cdot (\iota_2 + \iota_6)] - (3 \cdot \vartheta_1 \cdot \vartheta_4 + \Xi_1 \cdot \vartheta_3) + 2 \cdot (2 \cdot \vartheta_3 \cdot \Xi_1 + \vartheta_4 \cdot \vartheta_1)$$

$$A_6 = 2 \cdot [2 \cdot (\iota_3 + \iota_7) \cdot \iota_4 - (\iota_2 + \iota_6)] - [(\iota_2 + \iota_6) \cdot 5 + 3 \cdot \iota_4 \cdot (\iota_3 + \iota_7)] + 3 \cdot \Xi_1 \cdot \vartheta_4 - 4 \cdot \vartheta_4 \cdot \Xi_1$$

$$A_8 = \iota_4 \cdot 5 - 4 \cdot \iota_4$$
$$U = (P_R \cdot P_{I\omega} - P_I \cdot P_{R\omega}) - (Q_R \cdot Q_{I\omega} - Q_I \cdot Q_{R\omega}) = \sum_{k=0}^{4} A_{2k} \cdot \omega^{2 \cdot k}$$

$$F_\omega = 2 \cdot [(P_{R\omega} \cdot P_R + P_{I\omega} \cdot P_I) - (Q_{R\omega} \cdot Q_R + Q_{I\omega} \cdot Q_I)] = 2 \cdot [(2 \cdot (\iota_2 + \iota_6)^2 \cdot \omega^3 - 6 \cdot \iota_4 \cdot (\iota_2 + \iota_6) \cdot \omega^5 + 4 \cdot \iota_4 \cdot \iota_4 \cdot \omega^7 + (\iota_1 + \iota_5)^2 \cdot \omega - 4 \cdot (\iota_1 + \iota_5) \cdot (\iota_3 + \iota_7) \cdot \omega^3 + \{(\iota_3 + \iota_7)^2 + 2 \cdot (\iota_1 + \iota_5)\} \cdot 3 \cdot \omega^5 - 8 \cdot (\iota_3 + \iota_7) \cdot \omega^7 + 5 \cdot \omega^9) - (-2 \cdot \vartheta_1 \cdot \vartheta_2 \cdot \omega + (2 \cdot \vartheta_1^2 + 4 \cdot \Xi_1 \cdot \vartheta_2) \cdot \omega^3 - 6 \cdot \vartheta_1 \cdot \Xi_1 \cdot \omega^5 + 4 \cdot \Xi_1 \cdot \Xi_1 \cdot \omega^7 + \vartheta_3^2 \cdot \omega - 4 \cdot \vartheta_4 \cdot \vartheta_3 \cdot \omega^3 + 3 \cdot \vartheta_4 \cdot \vartheta_4 \cdot \omega^5)]$$

$$F_\omega = 2 \cdot [(P_{R\omega} \cdot P_R + P_{I\omega} \cdot P_I) - (Q_{R\omega} \cdot Q_R + Q_{I\omega} \cdot Q_I)] = 2 \cdot [2 \cdot (\iota_2 + \iota_6)^2 \cdot \omega^3 - 6 \cdot \iota_4 \cdot (\iota_2 + \iota_6) \cdot \omega^5 + 4 \cdot \iota_4 \cdot \iota_4 \cdot \omega^7 + (\iota_1 + \iota_5)^2 \cdot \omega - 4 \cdot (\iota_1 + \iota_5) \cdot (\iota_3 + \iota_7) \cdot \omega^3 + \{(\iota_3 + \iota_7)^2 + 2 \cdot (\iota_1 + \iota_5)\} \cdot 3 \cdot \omega^5 - 8 \cdot (\iota_3 + \iota_7) \cdot \omega^7 + 5 \cdot \omega^9 + 2 \cdot \vartheta_1 \cdot \vartheta_2 \cdot \omega - (2 \cdot \vartheta_1^2 + 4 \cdot \Xi_1 \cdot \vartheta_2) \cdot \omega^3 + 6 \cdot \vartheta_1 \cdot \Xi_1 \cdot \omega^5 - 4 \cdot \Xi_1 \cdot \Xi_1 \cdot \omega^7 - \vartheta_3^2 \cdot \omega + 4 \cdot \vartheta_4 \cdot \vartheta_3 \cdot \omega^3 - 3 \cdot \vartheta_4 \cdot \vartheta_4 \cdot \omega^5]$$

$$F_\omega = 2 \cdot [(P_{R\omega} \cdot P_R + P_{I\omega} \cdot P_I) - (Q_{R\omega} \cdot Q_R + Q_{I\omega} \cdot Q_I)]$$

$$= 2 \cdot [(\iota_1 + \iota_5)^2 + 2 \cdot \vartheta_1 \cdot \vartheta_2 - \vartheta_3^2] \cdot \omega + 2 \cdot [2 \cdot (\iota_2 + \iota_6)^2 - 4 \cdot (\iota_1 + \iota_5) \cdot (\iota_3 + \iota_7) - (2 \cdot \vartheta_1^2 + 4 \cdot \Xi_1 \cdot \vartheta_2) + 4 \cdot \vartheta_4 \cdot \vartheta_3] \cdot \omega^3 + 2 \cdot [6 \cdot \vartheta_1 \cdot \Xi_1 - 6 \cdot \iota_4 \cdot (\iota_2 + \iota_6) + \{(\iota_3 + \iota_7)^2 + 2 \cdot (\iota_1 + \iota_5)\} \cdot 3 - 3 \cdot \vartheta_4 \cdot \vartheta_4] \cdot \omega^5 + 2 \cdot [4 \cdot \iota_4 \cdot \iota_4 - 8 \cdot (\iota_3 + \iota_7) - 4 \cdot \Xi_1 \cdot \Xi_1] \cdot \omega^7 + 10 \cdot \omega^9$$

We define for simplicity the following global parameters:

$$B_1 = 2 \cdot [(\iota_1 + \iota_5)^2 + 2 \cdot \vartheta_1 \cdot \vartheta_2 - \vartheta_3^2]$$
$$B_7 = 2 \cdot [4 \cdot \iota_4 \cdot \iota_4 - 8 \cdot (\iota_3 + \iota_7) - 4 \cdot \Xi_1 \cdot \Xi_1];\ B_9 = 10$$

$$B_3 = 2 \cdot [2 \cdot (\iota_2 + \iota_6)^2 - 4 \cdot (\iota_1 + \iota_5) \cdot (\iota_3 + \iota_7) - (2 \cdot \vartheta_1^2 + 4 \cdot \Xi_1 \cdot \vartheta_2) + 4 \cdot \vartheta_4 \cdot \vartheta_3]$$

$$B_5 = 2 \cdot [6 \cdot \vartheta_1 \cdot \Xi_1 - 6 \cdot \iota_4 \cdot (\iota_2 + \iota_6) + \{(\iota_3 + \iota_7)^2 + 2 \cdot (\iota_1 + \iota_5)\} \cdot 3 - 3 \cdot \vartheta_4 \cdot \vartheta_4]$$

$$F_\omega = 2 \cdot [(P_{R\omega} \cdot P_R + P_{I\omega} \cdot P_I) - (Q_{R\omega} \cdot Q_R + Q_{I\omega} \cdot Q_I)] = \sum_{k=0}^{4} B_{2 \cdot k+1} \cdot \omega^{2 \cdot k+1}$$

Then we get the expression for $F_\omega = \sum_{k=0}^{4} B_{2 \cdot k+1} \cdot \omega^{2 \cdot k+1}$ plasma kinetic system parameter values. We find those ω, τ_D values which fulfill $F_\omega(\omega, \tau_D) = 0$. We ignore negative, complex, and imaginary values of ω for specific τ_D values. $\tau_D \in [0.001 \ldots 10]$, we can express by 3D function $F_\omega(\omega, \tau_D) = 0$. We plot the stability switch diagram based on different delay values of our plasma kinetic system.

$$\Lambda^{-1}(\tau_D) = \left(\frac{\partial \mathrm{Re}\lambda}{\partial \tau_D}\right)_{\lambda = i\omega} = \mathrm{Re}\left\{\frac{-2 \cdot [U + \tau_D \cdot |P|^2] + i \cdot F_\omega}{F_{\tau_D} + 2i \cdot [V + \omega \cdot |P|^2]}\right\}$$

$$\Lambda^{-1}(\tau_D) = \left(\frac{\partial \mathrm{Re}\lambda}{\partial \tau_D}\right)_{\lambda = i\omega} = \frac{2 \cdot \{F_\omega \cdot (V + \omega \cdot P^2) - F_{\tau_D} \cdot (U + \tau_D \cdot P^2)\}}{F_{\tau_D}^2 + 4 \cdot (V + \omega \cdot P^2)^2}$$

The stability switch occurs only on those delay values (τ_D) which fit the equation: $\tau_D = \frac{\theta_+(\tau_D)}{\omega_+(\tau_D)}$ and $\theta_+(\tau_D)$ is the solution of $\sin \theta(\tau_D) = \ldots.$ and $\cos \theta(\tau_D) = \ldots.$

when $\omega = \omega_{+}(\tau_D)$ If only ω_{+} is feasible. Additionally, when all plasma kinetic system's parameters are known and the stability switch due to various time delay values τ_D is described in the following expression:

$$\begin{aligned} \text{sign}\{\Lambda^{-1}(\tau_D)\} &= \text{sign}\{F_\omega(\omega(\tau_D), \tau_D)\} \\ &\quad \cdot \text{sign}\{\tau_D \cdot \omega_{\tau_D}(\omega(\tau_D)) + \omega(\tau_D) \\ &\quad + \frac{U(\omega(\tau_D)) \cdot \omega_{\tau_D}(\omega(\tau_D)) + V(\omega(\tau_D))}{|P(\omega(\tau_D))|^2}\} \end{aligned}$$

Remark we know $F(\omega, \tau_D) = 0$ implies its roots $\omega_i(\tau_D)$ and finding those delays values τ_D which ω_i is feasible. There are τ_D values which give complex ω_i or imaginary number, then unable to analyze stability. $F(\omega, \tau_D)$, function is independent on τ_D the parameter $F(\omega, \tau_D) = 0$. The results: We find those ω, τ_D values which fulfill $F(\omega, \tau_D) = 0$. We ignore negative, complex, and imaginary values of ω. Next is to find those ω, τ_D values which fulfill $\sin\theta(\tau_D) = \ldots.$ and $\cos\theta(\tau_D) = \ldots.$; $\sin(\omega \cdot \tau_D) = \frac{-P_R \cdot Q_I + P_I \cdot Q_R}{|Q|^2}$; $\cos(\omega \cdot \tau_D) = -\frac{(P_R \cdot Q_R + P_I \cdot Q_I)}{|Q|^2} |Q|^2 = Q_R^2 + Q_I^2$. Finally we plot the stability switch diagram $g(\tau_D) = \Lambda^{-1}(\tau_D) = (\frac{\partial \text{Re}\lambda}{\partial \tau_D})_{\lambda = i\omega}$

$$\begin{aligned} g(\tau_D) &= \Lambda^{-1}(\tau_D) \\ &= (\frac{\partial \text{Re}\lambda}{\partial \tau_D})_{\lambda = i\omega} \\ &= \frac{2 \cdot \{F_\omega \cdot (V + \omega \cdot P^2) - F_{\tau_D} \cdot (U + \tau_D \cdot P^2)\}}{F_{\tau_D}^2 + 4 \cdot (V + \omega \cdot P^2)^2} \end{aligned}$$

$$\begin{aligned} \text{sign}[g(\tau_D)] &= \text{sign}[\Lambda^{-1}(\tau_D)] \\ &= \text{sign}\left[\left(\frac{\partial \text{Re}\lambda}{\partial \tau_D}\right)_{\lambda = i\omega}\right] \\ &= \text{sign}\left[\frac{2 \cdot \{F_\omega \cdot (V + \omega \cdot P^2) - F_{\tau_D} \cdot (U + \tau_D \cdot P^2)\}}{F_{\tau_D}^2 + 4 \cdot (V + \omega \cdot P^2)^2}\right] \end{aligned}$$

$$\begin{aligned} &F_{\tau_D}^2 + 4 \cdot (V + \omega \cdot P^2)^2 > 0 \Rightarrow \text{sign}[\Lambda^{-1}(\tau_D)] \\ &= \text{sign}[F_\omega \cdot (V + \omega \cdot P^2) - F_{\tau_D} \cdot (U + \tau_D \cdot P^2)] \end{aligned}$$

$$\begin{aligned} \text{sign}[\Lambda^{-1}(\tau_D)] &= \text{sign}\{[F_\omega] \cdot [(V + \omega \cdot P^2) \\ &\quad - \frac{F_{\tau_D}}{F_\omega} \cdot (U + \tau_D \cdot P^2)]\} \\ \omega_{\tau_D} &= -\frac{F_{\tau_D}}{F_\omega}; \; \omega_{\tau_D} = (\frac{\partial \omega}{\partial \tau_D})^{-1} = -\frac{\partial F / \partial \omega}{\partial F / \partial \tau_D} \end{aligned}$$

$$\begin{aligned} \text{sign}[\Lambda^{-1}(\tau_D)] &= \text{sign}\{[F_\omega] \cdot [P^2] \cdot [\frac{V + \omega_{\tau_D} \cdot U}{P^2} \\ &\quad + \omega + \omega_{\tau_D} \cdot \tau_D]\} \; ; \; \text{sign}[P^2] > 0 \end{aligned}$$

Table 6.6 Plasma kinetic system sign of sign[$\Lambda^{-1}(\tau_D)$]

sign[F_ω]	sign[$\frac{V+\omega_{\tau_D}\cdot U}{P^2}+\omega+\omega_{\tau_D}\cdot\tau_D$]	sign[$\Lambda^{-1}(\tau_D)$]
$\pm$	$\pm$	+
$\pm$	$\mp$	−

$$\text{sign}[\Lambda^{-1}(\tau_D)] = \text{sign}\{[F_\omega]\cdot[\frac{V+\omega_{\tau_D}\cdot U}{P^2} + \omega + \omega_{\tau_D}\cdot\tau_D]\}$$

$$\text{sign}[\Lambda^{-1}(\tau_D)] = \text{sign}[F_\omega]\cdot\text{sign}[\frac{V+\omega_{\tau_D}\cdot U}{P^2} + \omega + \omega_{\tau_D}\cdot\tau_D]$$

$$F_\omega = 2\cdot[(P_{R\omega}\cdot P_R + P_{I\omega}\cdot P_I) - (Q_{R\omega}\cdot Q_R + Q_{I\omega}\cdot Q_I)] = \sum_{k=0}^{4} B_{2\cdot k+1}\cdot\omega^{2\cdot k+1}$$

We check the sign of $\Lambda^{-1}(\tau_D)$ according to the following rule (Table 6.6):

If, sign[$\Lambda^{-1}(\tau_D)$] > 0, then the crossing proceeds from (−) to (+) respectively (stable to unstable). If, sign[$\Lambda^{-1}(\tau_D)$] < 0, then the crossing proceeds from (+) to (−) respectively (unstable to stable).

6.2 A Quasi-Two Level Analytic Model for Metal Vapor Laser System Stability Analysis

A quasi-two level analytic model for end pumped Alkali metal vapor laser describes precisely the mechanism between two upper levels, $^2P_{3/2}$ and $^2P_{1/2}$, caused by collision with additive ethane which is much faster (infinity) by comparison with any other process in the system including stimulated emission. The Alkali metals (lithium, sodium, potassium, rubidium and cesium) are "visible hydrogen", readily ionized, and strongly interacting with laser light. It is systems which describe the microscopic inter conversion mechanisms between photon (laser), chemical, electrical and thermal energy. There are many practical applications such as sodium lamps, thermionic converters, magneto hydrodynamic devices and lasers ("lithium water fall"). Alkali-vapor lasers (rubidium) have a very high efficiency and the flowing-gas versions have a very high energy. The optical pumping of Alkali-metal vapors

Fig. 6.1 Diode pumping along the laser axis (flowing gas through an Alkali-metal laser)

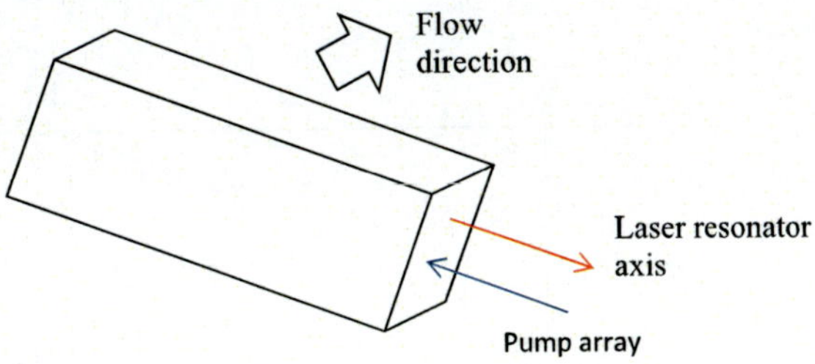

has only a single valence electron in their outer shells. Alkali metals have pair of low-lying energy levels of their single valence electron. Flowing gas through an Alkali-metal laser is enable the generation of much higher powers. Diode pumping can be either along the laser axis or transverse to the laser axis. Operating at higher power levels requires that Alkali vapor be flowed transversely through the pump volume, and waste heat is removed outside of the laser cavity. There are two diode pumping structures, one is along the laser axis and the second is transverse to the laser axis, when flowing gas through an Alkali-metal laser. The diode pumping along the laser axis structure is as follow (Fig. 6.1).

The quasi-two level analytic model takes the assumption of steady state rate equations for the number of densities of the, ${}^2S_{1/2}$, ${}^2P_{3/2}$ and ${}^2P_{1/2}$, energy states for the three level laser system. The ratio of the number densities for the upper two levels, ${}^2P_{3/2}$ and ${}^2P_{1/2}$, is given by its statistical equilibrium value. We extract analytical solution from the quasi-two system. We get the threshold pump power and the slope efficiency including intra-cavity losses [6].

The diode pumping transverse to the laser axis structure is as follow (Fig. 6.2).

The diode pumped Alkali metal vapor laser (DPAL) is a single narrow bond diodes partially narrow bonded low power diode arrays, and surrogate pump sources such as Ti-Sapphire as pump sources. The three-level laser devices are pumped from the Alkali metal ${}^2S_{1/2}$ ground state to the lowest excited ${}^2P_{3/2}$ state, the D_2 transition. Rapid collision induced spin-orbit relaxation with additive ethane, (200 Torr) populated the lower ${}^2P_{1/2}$ state. Lasing then occurs between ${}^2P_{1/2}$ and the ${}^2S_{1/2}$ ground state, the D_1 transition (Fig. 6.3). The difference in energy between the ${}^2P_{3/2}$ and ${}^2P_{1/2}$ fine structure spin-orbit components is 554 cm^{-1} (C_s), 237 cm^{-1} (R_b), 57.7 cm^{-1} (K). The small energy defects lead to intrinsic quantum efficiencies (>95%) and to

Fig. 6.2 Diode pumping transverse to the laser axis (flowing gas through on Alkali-metal laser)

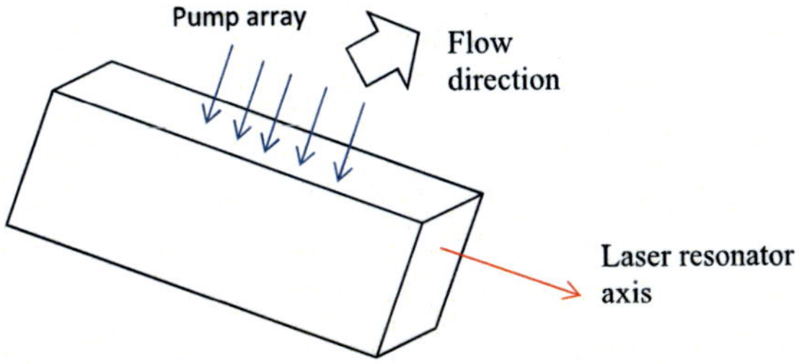

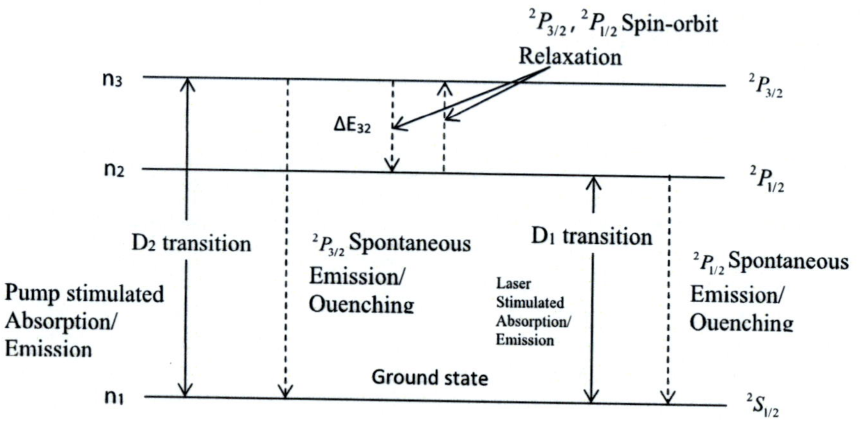

Fig. 6.3 Schematic energy level diagram of Alkali metal vapor laser

the possibility of very efficient laser systems. A quasi-two level analytic model can be applied to longitudinal laser pumped Alkali metal vapor lasers. The model includes with modeling three level solid state laser systems that are pumped longitudinally with diodes or other lasers. An analytic DPAL model extracts an analytical solution for the DPAL from the rate equations. The energy balance methodology is used longitudinally averaged number densities, or (LANDs), its formulation. By using LAND's formulation, we allow double passing of the pump beam and the flexibility to include intra-cavity losses. The pump and laser beams are single frequency and centered at the peaks of the absorption and emission cross-sections. The pressure broad ended absorption and emission lines are assumed to be especially homogeneous. We specify in the model the intensity and number densities [6]. The diode pumped Alkali metal vapor laser (DPAL) model and pump and laser beam geometry includes two mirrors (M1 and M2) and laser Gain cell. We consider that the collision induced mixing or relaxation between $^2P_{3/2}$ and $^2P_{1/2}$ states is much faster by comparison with any other process in the system including stimulated emission. The mathematical description is of a quasi-two level system from which an analytical solution is extracted. The optical schematics for DPAL and the geometry for single or double pass longitudinal pumping are described in the next figure (Fig. 6.4). It includes the right and left running laser waves.

Where I_p is the intensity of the input pump beam, I_+ and I_- are right and left running intra-cavity laser intensities. n_1 is the lower level and n_2, n_3 are the two upper levels.

$$f = \frac{1}{1 + 2 \cdot e^{-\left(\frac{\Delta E_{32}}{k_B \cdot T}\right)}};\ n_1 = n_L;\ n_2 = f \cdot n_U;\ n_3 = (1 - f) \cdot n_U$$

$$n_U = n_2 + n_3;\ n_{tot} = n_U + n_L$$

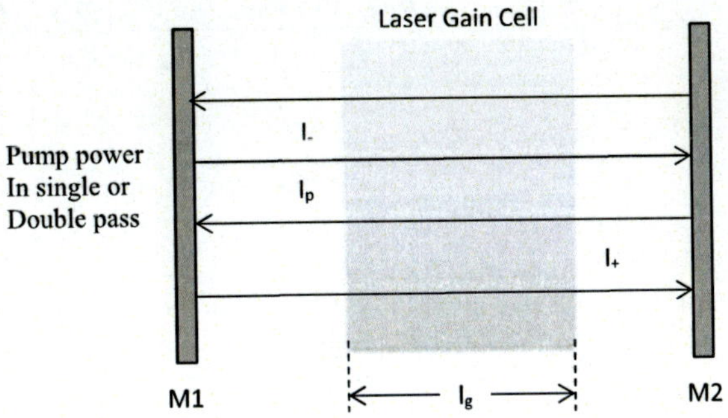

Fig. 6.4 Optical schematic for diode pumped Alkali metal vapor laser (DPAL)

where T is the temperature in degrees Kelvin, k_B is the Boltzmann constant and $\Delta E_{3,2}$ is the energy different between level 3 and level 2.n_{tot} is the total Alkali atom number density.

$$\frac{n_3}{n_2} = 2 \cdot e^{-[\frac{\Delta E_{32}}{k_B \cdot T}]}$$

The two upper levels n_2 and n_3 are decayed with a common life time since by assumption they are coupled by infinity collision induced processes.n_1, n_L—Density of lower level population ($^2S_{1/2}$ ground state), n_2—Density of the mid-level population ($^2P_{1/2}$ state),n_3—Density of the upper level population ($^2P_{3/2}$ state), n_U—Total density of the mid and upper states, f- Scaling parameter for the density of population in the mid and upper state ($0 \leq f \leq 1$), n_{tot}—Total density of all mode states ($n_{tot} = \sum_{k=1}^{3} n_k$). Parameters σ_{32} and σ_{21} are the line center simulated emission/absorption cross sections, h is Planck's constant, υ_p and υ_L are the line center frequencies of the pump and laser transitions, and τ is the effective lifetime of the upper level manifold. The system of coupled differential equations is as follow. They describe the kinetics and radiation processes in the system at steady state [6].

$$\begin{aligned} \frac{dn_L}{dt} &= \frac{I_p}{h \cdot \upsilon_p} \cdot \sigma_{31} \cdot [(1-f) \cdot n_U - 2 \cdot n_L] \\ &+ \frac{(I_+ + I_-)}{h \cdot \upsilon_L} \cdot \sigma_{21} \cdot (f \cdot n_U - n_L) + \frac{n_U}{\tau} \end{aligned}$$

$$\frac{dn_U}{dt} = -\frac{dn_L}{dt}; n_{tot} = n_U + n_L$$

At any plane z, of the laser cell, the number densities of the upper and lower levels are in steady state with the local pump and laser radiation fields present. Solving the above differential equations for n_U and n_L, and the expressions we get are

$$n_U = \xi_1(I_p, I_+, I_-)$$

$$= -\left(\frac{\tau \cdot n_{tot} \cdot (2 \cdot \sigma_{31} \cdot I_p \cdot v_L + \sigma_{21} \cdot (I_+ + I_-) \cdot v_p)}{(f-3) \cdot \sigma_{31} \cdot \tau \cdot I_p \cdot v_L - [(f+1) \cdot \sigma_{21} \cdot \tau \cdot (I_+ + I_-) + h \cdot v_L] \cdot v_p}\right)$$

$$n_L = \xi_2(I_p, I_+, I_-)$$
$$= \frac{n_{tot} \cdot [(f-1) \cdot \sigma_{31} \cdot \tau \cdot I_p \cdot v_L - (f \cdot \sigma_{21} \cdot \tau \cdot (I_+ + I_-) + h \cdot v_L) \cdot v_p]}{(f-3) \cdot \sigma_{31} \cdot \tau \cdot I_p \cdot v_L - [(f+1) \cdot \sigma_{21} \cdot \tau \cdot (I_+ + I_-) + h \cdot v_L] \cdot v_p}$$

The three rate equations for I_p (intensity of the input pump beam), I_+ and I_- (right and left running intra-cavity laser intensities) are as follow:

$$\frac{dI_p}{dz} = \sigma_{31} \cdot [(1-f) \cdot n_U - 2 \cdot n_L] \cdot I_p$$
$$\frac{dI_+}{dz} = \sigma_{21} \cdot (f \cdot n_U - n_L) \cdot I_+; \quad \frac{dI_-}{dz} = -\sigma_{21} \cdot (f \cdot n_U - n_L) \cdot I_-$$

And we can write the above rate equations by using $\xi_1(I_p, I_+, I_-)$ and $\xi_2(I_p, I_+, I_-)$ functions.

$$\frac{dI_p}{dz} = \sigma_{31} \cdot [(1-f) \cdot \xi_1(I_p, I_+, I_-) - 2 \cdot \xi_2(I_p, I_+, I_-)]$$
$$\xi_1 = \xi_1(I_p, I_+, I_-); \; \xi_2 = \xi_2(I_p, I_+, I_-)$$

$$\frac{dI_+}{dz} = \sigma_{21} \cdot (f \cdot \xi_1(I_p, I_+, I_-) - \xi_2(I_p, I_+, I_-)) \cdot I_+$$
$$\frac{dI_-}{dz} = -\sigma_{21} \cdot (f \cdot \xi_1(I_p, I_+, I_-) - \xi_2(I_p, I_+, I_-)) \cdot I_-$$

At fixed points (equilibrium points): $\frac{dI_p}{dz} = 0; \frac{dI_+}{dz} = 0; \frac{dI_-}{dz} = 0$

$$\frac{dI_p}{dz} = 0 \Rightarrow \sigma_{31} \cdot [(1-f) \cdot \xi_1 - 2 \cdot \xi_2] = 0 \Rightarrow \sigma_{31} \neq 0$$
$$(1-f) \cdot \xi_1^* = 2 \cdot \xi_2^* \Rightarrow \frac{\xi_1^*}{\xi_2^*} = \frac{2}{(1-f)}$$

$$\frac{\xi_1^*}{\xi_2^*} = -\left(\frac{\tau \cdot n_{tot} \cdot (2 \cdot \sigma_{31} \cdot I_p^* \cdot v_L + \sigma_{21} \cdot (I_+^* + I_-^*) \cdot v_p)}{n_{tot} \cdot [(f-1) \cdot \sigma_{31} \cdot \tau \cdot I_p^* \cdot v_L - (f \cdot \sigma_{21} \cdot \tau \cdot (I_+^* + I_-^*) + h \cdot v_L) \cdot v_p]}\right)$$
$$= \frac{2}{(1-f)}$$

$$\frac{\tau \cdot n_{tot} \cdot (2 \cdot \sigma_{31} \cdot I_p^* \cdot v_L + \sigma_{21} \cdot (I_+^* + I_-^*) \cdot v_p)}{n_{tot} \cdot [(f-1) \cdot \sigma_{31} \cdot \tau \cdot I_p^* \cdot v_L - (f \cdot \sigma_{21} \cdot \tau \cdot (I_+^* + I_-^*) + h \cdot v_L) \cdot v_p]}$$
$$= \frac{2}{(f-1)}$$

$$\frac{\tau \cdot n_{tot} \cdot (2 \cdot \sigma_{31} \cdot I_p^* \cdot \upsilon_L + \sigma_{21} \cdot (I_+^* + I_-^*) \cdot \upsilon_p) \cdot (f-1)}{n_{tot} \cdot [(f-1) \cdot \sigma_{31} \cdot \tau \cdot I_p^* \cdot \upsilon_L - (f \cdot \sigma_{21} \cdot \tau \cdot (I_+^* + I_-^*) + h \cdot \upsilon_L) \cdot \upsilon_p]}$$
$$= 2$$

$$\tau \cdot n_{tot} \cdot (2 \cdot \sigma_{31} \cdot I_p^* \cdot \upsilon_L + \sigma_{21} \cdot (I_+^* + I_-^*) \cdot \upsilon_p) \cdot (f-1)$$
$$= 2 \cdot n_{tot} \cdot [(f-1) \cdot \sigma_{31} \cdot \tau \cdot I_p^* \cdot \upsilon_L - (f \cdot \sigma_{21} \cdot \tau \cdot (I_+^* + I_-^*)$$
$$+ h \cdot \upsilon_L) \cdot \upsilon_p]$$

$$\tau \cdot n_{tot} \cdot 2 \cdot \sigma_{31} \cdot I_p^* \cdot \upsilon_L \cdot (f-1) + \tau \cdot n_{tot} \cdot \sigma_{21}$$
$$\cdot \upsilon_p \cdot (I_+^* + I_-^*) \cdot (f-1)$$
$$= 2 \cdot n_{tot} \cdot (f-1) \cdot \sigma_{31} \cdot \tau \cdot I_p^* \cdot \upsilon_L - 2 \cdot n_{tot} \cdot f \cdot \sigma_{21}$$
$$\cdot \tau \cdot (I_+^* + I_-^*) \cdot \upsilon_p - 2 \cdot n_{tot} \cdot h \cdot \upsilon_L \cdot \upsilon_p$$

$$[2 \cdot \tau \cdot n_{tot} \cdot \sigma_{31} - 2 \cdot \tau \cdot n_{tot} \cdot \sigma_{31}] \cdot (f-1) \cdot \upsilon_L \cdot I_p^*$$
$$+ (3 \cdot f - 1) \cdot (I_+^* + I_-^*) \cdot \upsilon_p \cdot \tau \cdot n_{tot} \cdot \sigma_{21}$$
$$= -2 \cdot n_{tot} \cdot h \cdot \upsilon_L \cdot \upsilon_p$$

$$[2 \cdot \tau \cdot n_{tot} \cdot \sigma_{31} - 2 \cdot \tau \cdot n_{tot} \cdot \sigma_{31}] \cdot (f-1) \cdot \upsilon_L \cdot I_p^*$$
$$= 0 \Rightarrow (3 \cdot f - 1) \cdot (I_+^* + I_-^*) \cdot \upsilon_p \cdot \tau \cdot n_{tot} \cdot \sigma_{21}$$
$$= -2 \cdot n_{tot} \cdot h \cdot \upsilon_L \cdot \upsilon_p$$

$$I_+^* + I_-^* = -\frac{2 \cdot n_{tot} \cdot h \cdot \upsilon_L \cdot \upsilon_p}{(3 \cdot f - 1) \cdot \upsilon_p \cdot \tau \cdot n_{tot} \cdot \sigma_{21}}$$

$$\frac{dI_+}{dz} = 0 \Rightarrow \sigma_{21} \cdot (f \cdot \xi_1^* - \xi_2^*) \cdot I_+^* = 0; \sigma_{21} \neq 0$$

$(f \cdot \xi_1^* - \xi_2^*) \cdot I_+^* = 0$; **Case 1** $f \cdot \xi_1^* - \xi_2^* = 0 \Rightarrow \frac{\xi_1^*}{\xi_2^*} = \frac{1}{f}$; **Case 2** $I_+^* = 0$

Case 1 $\frac{\xi_1^*}{\xi_2^*} = \frac{2}{(1-f)}$; $\frac{\xi_1^*}{\xi_2^*} = \frac{1}{f}$; $\frac{2}{(1-f)} = \frac{1}{f} \Rightarrow f = \frac{1}{3}$

$$f = \frac{1}{1 + 2 \cdot e^{-(\frac{\Delta E_{32}}{k_B \cdot T})}} = \frac{1}{3} \Rightarrow 1 + 2 \cdot e^{-(\frac{\Delta E_{32}}{k_B \cdot T})}$$
$$= 3 \Rightarrow e^{-(\frac{\Delta E_{32}}{k_B \cdot T})} = 1 \Rightarrow \frac{\Delta E_{32}}{k_B \cdot T} = 0; k_B \cdot T \neq 0; \Delta E_{32} = 0$$

Meaning: $\Delta E_{32} = 0 \Rightarrow \frac{n_3}{n_2} = 2 \Rightarrow n_3 = 2 \cdot n_2$; First fixed point: $E^{(0)} = (I_p^{(0)}, I_+^{(0)}, I_-^{(0)})$

Case 2 $I_+^* = 0 \Rightarrow I_-^* = -\frac{2\cdot n_{tot}\cdot h\cdot v_L\cdot v_p}{(3\cdot f-1)\cdot v_p\cdot\tau\cdot n_{tot}\cdot\sigma_{21}}$

Second Fixed point: $E^{(1)} = (I_p^{(1)}, I_+^{(1)}, I_-^{(1)}) = (I_p^{(1)}, 0, -\frac{2\cdot n_{tot}\cdot h\cdot v_L\cdot v_p}{(3\cdot f-1)\cdot v_p\cdot\tau\cdot n_{tot}\cdot\sigma_{21}})$

$\frac{dI_-}{dz} = 0 \Rightarrow -\sigma_{21}\cdot(f\cdot\xi_1^* - \xi_2^*)\cdot I_-$; $\sigma_{21} \neq 0$; **Case 3** $f\cdot\xi_1^* - \xi_2^* = 0 \Rightarrow f\cdot\xi_1^* = \xi_2^* \Rightarrow \frac{\xi_1^*}{\xi_2^*} = \frac{1}{f}$

$$\frac{\xi_1^*}{\xi_2^*} = -\left(\frac{\tau\cdot n_{tot}\cdot(2\cdot\sigma_{31}\cdot I_p^*\cdot v_L + \sigma_{21}\cdot(I_+^* + I_-^*)\cdot v_p)}{n_{tot}\cdot[(f-1)\cdot\sigma_{31}\cdot\tau\cdot I_p^*\cdot v_L - (f\cdot\sigma_{21}\cdot\tau\cdot(I_+^* + I_-^*) + h\cdot v_L)\cdot v_p]}\right) = \frac{1}{f}$$

$$\frac{f\cdot\tau\cdot n_{tot}\cdot(2\cdot\sigma_{31}\cdot I_p^*\cdot v_L + \sigma_{21}\cdot(I_+^* + I_-^*)\cdot v_p)}{n_{tot}\cdot[(f-1)\cdot\sigma_{31}\cdot\tau\cdot I_p^*\cdot v_L - (f\cdot\sigma_{21}\cdot\tau\cdot(I_+^* + I_-^*) + h\cdot v_L)\cdot v_p]} = -1$$

$$f\cdot\tau\cdot n_{tot}\cdot(2\cdot\sigma_{31}\cdot I_p^*\cdot v_L + \sigma_{21}\cdot(I_+^* + I_-^*)\cdot v_p) = -n_{tot}\cdot[(f-1)\cdot\sigma_{31}\cdot\tau\cdot I_p^*\cdot v_L - (f\cdot\sigma_{21}\cdot\tau\cdot(I_+^* + I_-^*) + h\cdot v_L)\cdot v_p]$$

$$f\cdot\tau\cdot n_{tot}\cdot 2\cdot\sigma_{31}\cdot I_p^*\cdot v_L + f\cdot\tau\cdot n_{tot}\cdot\sigma_{21}\cdot(I_+^* + I_-^*)\cdot v_p = -n_{tot}\cdot[(f-1)\cdot\sigma_{31}\cdot\tau\cdot I_p^*\cdot v_L - f\cdot\sigma_{21}\cdot\tau\cdot(I_+^* + I_-^*)\cdot v_p - h\cdot v_L\cdot v_p]$$

$$f\cdot\tau\cdot n_{tot}\cdot 2\cdot\sigma_{31}\cdot I_p^*\cdot v_L + f\cdot\tau\cdot n_{tot}\cdot\sigma_{21}\cdot(I_+^* + I_-^*)\cdot v_p = -n_{tot}\cdot(f-1)\cdot\sigma_{31}\cdot\tau\cdot I_p^*\cdot v_L + n_{tot}\cdot f\cdot\sigma_{21}\cdot\tau\cdot(I_+^* + I_-^*)\cdot v_p + n_{tot}\cdot h\cdot v_L\cdot v_p$$

$$f\cdot\tau\cdot n_{tot}\cdot 2\cdot\sigma_{31}\cdot I_p^*\cdot v_L + n_{tot}\cdot(f-1)\cdot\sigma_{31}\cdot\tau\cdot I_p^*\cdot v_L = n_{tot}\cdot f\cdot\sigma_{21}\cdot\tau\cdot(I_+^* + I_-^*)\cdot v_p - f\cdot\tau\cdot n_{tot}\cdot\sigma_{21}\cdot(I_+^* + I_-^*)\cdot v_p + n_{tot}\cdot h\cdot v_L\cdot v_p$$

$n_{tot}\cdot f\cdot\tau\cdot\sigma_{21}\cdot(I_+^* + I_-^*)\cdot v_p - n_{tot}\cdot f\cdot\tau\cdot\sigma_{21}\cdot(I_+^* + I_-^*)\cdot v_p = 0$ then

$$[3\cdot f - 1]\cdot I_p^*\cdot\tau\cdot v_L\cdot\sigma_{31}\cdot n_{tot} = n_{tot}\cdot h\cdot v_L\cdot v_p \Rightarrow I_p^* = \frac{h\cdot v_p}{[3\cdot f - 1]\cdot\tau\cdot\sigma_{31}}$$

$$\frac{dI_p}{dz} = 0 \Rightarrow \sigma_{31}\cdot[(1-f)\cdot\xi_1 - 2\cdot\xi_2] = 0 \Rightarrow \sigma_{31} \neq 0$$

$$(1-f)\cdot\xi_1^* = 2\cdot\xi_2^* \Rightarrow \frac{\xi_1^*}{\xi_2^*} = \frac{2}{(1-f)}$$

$$\frac{dI_-}{dz} = 0 \Rightarrow \frac{\xi_1^*}{\xi_2^*} = \frac{1}{f};\ \frac{dI_p}{dz} = 0 \Rightarrow \frac{2}{(1-f)}$$

$$\frac{1}{f} = \frac{2}{(1-f)} \Rightarrow 1 = 3\cdot f \Rightarrow f = \frac{1}{3}$$

$$f = \frac{1}{1+2\cdot e^{-(\frac{\Delta E_{32}}{k_B\cdot T})}} = \frac{1}{3} \Rightarrow 1+2\cdot e^{-(\frac{\Delta E_{32}}{k_B\cdot T})}$$

$$= 3 \Rightarrow e^{-(\frac{\Delta E_{32}}{k_B\cdot T})} = 1 \Rightarrow \frac{\Delta E_{32}}{k_B\cdot T} = 0;\ k_B\cdot T \neq 0;\ \Delta E_{32} = 0$$

Meaning: $\Delta E_{32} = 0 \Rightarrow \frac{n_3}{n_2} = 2 \Rightarrow n_3 = 2\cdot n_2;\ I_p^* = \frac{h\cdot v_p}{[3\cdot f-1]\cdot\tau\cdot\sigma_{31}};\ \frac{dI_p}{dz} = 0 \Rightarrow \frac{\xi_1^*}{\xi_2^*} = \frac{2}{(1-f)}$

$$\frac{\xi_1^*}{\xi_2^*} = -\left(\frac{\tau\cdot n_{tot}\cdot(2\cdot\sigma_{31}\cdot I_p^*\cdot v_L + \sigma_{21}\cdot(I_+^* + I_-^*)\cdot v_p)}{n_{tot}\cdot[(f-1)\cdot\sigma_{31}\cdot\tau\cdot I_p^*\cdot v_L - (f\cdot\sigma_{21}\cdot\tau\cdot(I_+^* + I_-^*) + h\cdot v_L)\cdot v_p]}\right) = \frac{2}{(1-f)}$$

$$\frac{\xi_1^*(I_p^*)}{\xi_2^*(I_p^*)} = -\left(\frac{\tau\cdot n_{tot}\cdot(2\cdot\sigma_{31}\cdot[\frac{h\cdot v_p}{[3\cdot f-1]\cdot\tau\cdot\sigma_{31}}]\cdot v_L + \sigma_{21}\cdot(I_+^* + I_-^*)\cdot v_p)}{n_{tot}\cdot[(f-1)\cdot\sigma_{31}\cdot\tau\cdot[\frac{h\cdot v_p}{[3\cdot f-1]\cdot\tau\cdot\sigma_{31}}]\cdot v_L - (f\cdot\sigma_{21}\cdot\tau\cdot(I_+^* + I_-^*) + h\cdot v_L)\cdot v_p]}\right) = \frac{2}{(1-f)}$$

$$\frac{\tau\cdot n_{tot}\cdot(2\cdot[\frac{h\cdot v_p}{[3\cdot f-1]\cdot\tau}]\cdot v_L + \sigma_{21}\cdot(I_+^* + I_-^*)\cdot v_p)}{n_{tot}\cdot[(f-1)\cdot\tau\cdot[\frac{h\cdot v_p}{[3\cdot f-1]\cdot\tau}]\cdot v_L - (f\cdot\sigma_{21}\cdot\tau\cdot(I_+^* + I_-^*) + h\cdot v_L)\cdot v_p]} = \frac{2}{(f-1)}$$

$$\frac{n_{tot}\cdot 2\cdot[\frac{h\cdot v_p}{[3\cdot f-1]}]\cdot v_L + \tau\cdot n_{tot}\cdot\sigma_{21}\cdot(I_+^* + I_-^*)\cdot v_p}{[n_{tot}\cdot(f-1)\cdot[\frac{h\cdot v_p}{[3\cdot f-1]}]\cdot v_L - n_{tot}\cdot(f\cdot\sigma_{21}\cdot\tau\cdot(I_+^* + I_-^*) + h\cdot v_L)\cdot v_p]} = \frac{2}{(f-1)}$$

$$n_{tot}\cdot 2\cdot[\frac{h\cdot v_p}{[3\cdot f-1]}]\cdot v_L + \tau\cdot n_{tot}\cdot\sigma_{21}\cdot(I_+^* + I_-^*)\cdot v_p$$

$$= [n_{tot} \cdot (f-1) \cdot [\frac{h \cdot v_p}{[3 \cdot f - 1]}] \cdot v_L - n_{tot} \cdot (f \cdot \sigma_{21} \cdot \tau \cdot (I_+^* + I_-^*) + h \cdot v_L) \cdot v_p] \cdot \frac{2}{(f-1)}$$

$$n_{tot} \cdot 2 \cdot [\frac{h \cdot v_p}{[3 \cdot f - 1]}] \cdot v_L + \tau \cdot n_{tot} \cdot \sigma_{21} \cdot (I_+^* + I_-^*) \cdot v_p = n_{tot} \cdot [\frac{h \cdot v_p}{[3 \cdot f - 1]}] \cdot v_L \cdot 2 - n_{tot} \cdot f \cdot \sigma_{21} \cdot \tau \cdot (I_+^* + I_-^*) \cdot v_p \cdot \frac{2}{(f-1)} - n_{tot} \cdot h \cdot v_L \cdot v_p \cdot \frac{2}{(f-1)}$$

$$\tau \cdot n_{tot} \cdot \sigma_{21} \cdot (I_+^* + I_-^*) \cdot v_p + n_{tot} \cdot f \cdot \sigma_{21} \cdot \tau \cdot (I_+^* + I_-^*) \cdot v_p \cdot \frac{2}{(f-1)} = n_{tot} \cdot [\frac{h \cdot v_p}{[3 \cdot f - 1]}] \cdot v_L \cdot 2 - n_{tot} \cdot h \cdot v_L \cdot v_p \cdot \frac{2}{(f-1)} - n_{tot} \cdot 2 \cdot [\frac{h \cdot v_p}{[3 \cdot f - 1]}] \cdot v_L$$

$$[1 + f \cdot \frac{2}{(f-1)}] \cdot (I_+^* + I_-^*) \cdot v_p \cdot n_{tot} \cdot \sigma_{21} \cdot \tau = n_{tot} \cdot [\frac{h \cdot v_p}{[3 \cdot f - 1]}] \cdot v_L \cdot 2 - n_{tot} \cdot 2 \cdot [\frac{h \cdot v_p}{[3 \cdot f - 1]}] \cdot v_L - n_{tot} \cdot h \cdot v_L \cdot v_p \cdot \frac{2}{(f-1)}$$

$$[1 + f \cdot \frac{2}{(f-1)}] \cdot (I_+^* + I_-^*) \cdot v_p \cdot n_{tot} \cdot \sigma_{21} \cdot \tau = -n_{tot} \cdot h \cdot v_L \cdot v_p \cdot \frac{2}{(f-1)}$$

$$(I_+^* + I_-^*) = -[\frac{n_{tot} \cdot h \cdot v_L \cdot v_p \cdot \frac{2}{(f-1)}}{[1 + f \cdot \frac{2}{(f-1)}] \cdot v_p \cdot n_{tot} \cdot \sigma_{21} \cdot \tau}]$$

$$(I_+^* + I_-^*) = -[\frac{h \cdot v_L \cdot v_p \cdot \frac{2}{(f-1)}}{[1 + f \cdot \frac{2}{(f-1)}] \cdot v_p \cdot \sigma_{21} \cdot \tau}]$$

$$(I_+^* + I_-^*) = \frac{h \cdot v_L \cdot 2}{[1 - 3 \cdot f] \cdot \sigma_{21} \cdot \tau}$$

Third Fixed point: $E^{(2)} = (I_p^{(2)}, I_+^{(2)}, I_-^{(2)}) = (\frac{h \cdot v_p}{[3 \cdot f - 1] \cdot \tau \cdot \sigma_{31}}, I_+^*, \frac{h \cdot v_L \cdot 2}{[1 - 3 \cdot f] \cdot \sigma_{21} \cdot \tau} - I_+^*)$
We can summary our system three fixed points: $E^{(0)} = (I_p^{(0)}, I_+^{(0)}, I_-^{(0)})$

$$E^{(1)} = (I_p^{(1)}, I_+^{(1)}, I_-^{(1)}) = (I_p^{(1)}, 0, -\frac{2 \cdot n_{tot} \cdot h \cdot \upsilon_L \cdot \upsilon_p}{(3 \cdot f - 1) \cdot \upsilon_p \cdot \tau \cdot n_{tot} \cdot \sigma_{21}})$$

$$E^{(2)} = (I_p^{(2)}, I_+^{(2)}, I_-^{(2)}) = (\frac{h \cdot \upsilon_p}{[3 \cdot f - 1] \cdot \tau \cdot \sigma_{31}}, I_+^*, \frac{h \cdot \upsilon_L \cdot 2}{[1 - 3 \cdot f] \cdot \sigma_{21} \cdot \tau} - I_+^*)$$

Stability analysis, we define three functions:

$$g_1(I_p, I_+, I_-) = \sigma_{31} \cdot [(1 - f) \cdot \xi_1(I_p, I_+, I_-) - 2 \cdot \xi_2(I_p, I_+, I_-)] = \sigma_{31} \cdot [(1 - f) \cdot \xi_1 - 2 \cdot \xi_2]$$

$$g_2(I_p, I_+, I_-) = \sigma_{21} \cdot (f \cdot \xi_1(I_p, I_+, I_-) - \xi_2(I_p, I_+, I_-)) \cdot I_+ = \sigma_{21} \cdot (f \cdot \xi_1 - \xi_2) \cdot I_+$$

$$g_3(I_p, I_+, I_-) = -\sigma_{21} \cdot (f \cdot \xi_1(I_p, I_+, I_-) - \xi_2(I_p, I_+, I_-)) \cdot I_- = -\sigma_{21} \cdot (f \cdot \xi_1 - \xi_2) \cdot I_-$$

Our system set of differential equations: $$\frac{dI_p}{dz} = g_1(I_p, I_+, I_-) \quad \frac{dI_+}{dz} = g_2(I_p, I_+, I_-) \quad \frac{dI_-}{dz} = g_3(I_p, I_+, I_-)$$

Linearization: suppose that (I_p^*, I_+^*, I_-^*) is a fixed point, $g_1(I_p^*, I_+^*, I_-^*) = 0$; $g_2(I_p^*, I_+^*, I_-^*) = 0$ $g_3(I_p^*, I_+^*, I_-^*) = 0$ and let $u_1 = I_p - I_p^*$; $u_2 = I_+ - I_+^*$; $u_3 = I_- - I_-^*$ denote the components of a small disturbance from the fixed point. To see whether the disturbance grows or decays, we need to derive differential equations for u_1, u_2 and u_3. Let's do the u_1-equation first: $\frac{du_1}{dt} = \frac{dI_p}{dt}$ since I_p^* is constant and $\frac{du_1}{dt} = \frac{dI_p}{dt} = g_1(I_p^* + u_1, I_+^* + u_2, I_-^* + u_3)$ by substitution, then by using Taylor series expansion $\frac{du_1}{dt} = \frac{dI_p}{dt} = g_1(I_p^*, I_+^*, I_-^*) + \ldots$ $\frac{du_1}{dt} = \frac{dI_p}{dt} = g_1(I_p^*, I_+^*, I_-^*) + u_1 \cdot \frac{\partial g_1}{\partial I_p} + u_2 \cdot \frac{\partial g_1}{\partial I_+} + u_3 \cdot \frac{\partial g_1}{\partial I_-} + O(u_1^2, u_2^2, u_3^2, u_1 \cdot u_2 \cdot u_3)$ and since $g_1(I_p^*, I_+^*, I_-^*) = 0$ then $\frac{du_1}{dt} = \frac{dI_p}{dt} = u_1 \cdot \frac{\partial g_1}{\partial I_p} + u_2 \cdot \frac{\partial g_1}{\partial I_+} + u_3 \cdot \frac{\partial g_1}{\partial I_-} + O(u_1^2, u_2^2, u_3^2, u_1 \cdot u_2 \cdot u_3)$. To simplify the notation, we have written $\frac{\partial g_1}{\partial I_p}, \frac{\partial g_1}{\partial I_+}$ and $\frac{\partial g_1}{\partial I_-}$, these partial derivatives are to be evaluated at the fixed point (I_p^*, I_+^*, I_-^*); thus they are numbers, not functions. The shorthand notation, $O(u_1^2, u_2^2, u_3^2, u_1 \cdot u_2 \cdot u_3)$ denotes quadratic terms in u_1, u_2

and u_3 are small, these quadratic terms are extremely small [2, 3]. Similarly we find

$$\frac{du_2}{dt} = \frac{dI_+}{dt} = u_1 \cdot \frac{\partial g_2}{\partial I_p} + u_2 \cdot \frac{\partial g_2}{\partial I_+} + u_3 \cdot \frac{\partial g_2}{\partial I_-} + O(u_1^2, u_2^2, u_3^2, u_1 \cdot u_2 \cdot u_3)$$

$$\frac{du_3}{dt} = \frac{dI_-}{dt} = u_1 \cdot \frac{\partial g_3}{\partial I_p} + u_2 \cdot \frac{\partial g_3}{\partial I_+} + u_3 \cdot \frac{\partial g_3}{\partial I_-} + O(u_1^2, u_2^2, u_3^2, u_1 \cdot u_2 \cdot u_3)$$

Hence the disturbance (u_1, u_2, u_3) evolves according to

$$\begin{pmatrix} \frac{du_1}{dt} \\ \frac{du_2}{dt} \\ \frac{du_3}{dt} \end{pmatrix} = \begin{pmatrix} \frac{\partial g_1}{\partial I_p} & \frac{\partial g_1}{\partial I_+} & \frac{\partial g_1}{\partial I_-} \\ \frac{\partial g_2}{\partial I_p} & \frac{\partial g_2}{\partial I_+} & \frac{\partial g_2}{\partial I_-} \\ \frac{\partial g_3}{\partial I_p} & \frac{\partial g_3}{\partial I_+} & \frac{\partial g_3}{\partial I_-} \end{pmatrix} \cdot \begin{pmatrix} u_1 \\ u_2 \\ u_3 \end{pmatrix} + \text{quadratic term and the matrix}$$

$$A = \begin{pmatrix} \frac{\partial g_1}{\partial I_p} & \frac{\partial g_1}{\partial I_+} & \frac{\partial g_1}{\partial I_-} \\ \frac{\partial g_2}{\partial I_p} & \frac{\partial g_2}{\partial I_+} & \frac{\partial g_2}{\partial I_-} \\ \frac{\partial g_3}{\partial I_p} & \frac{\partial g_3}{\partial I_+} & \frac{\partial g_3}{\partial I_-} \end{pmatrix}_{(I_p^*, I_+^*, I_-^*)}$$ is called the Jacobian matrix at the fixed point (I_p^*, I_+^*, I_-^*).

Since the quadratic terms are tiny, it's tempting to neglect them altogether. If we do that, we obtain the linearized system.

$$\begin{pmatrix} \frac{du_1}{dt} \\ \frac{du_2}{dt} \\ \frac{du_3}{dt} \end{pmatrix} = \begin{pmatrix} \frac{\partial g_1}{\partial I_p} & \frac{\partial g_1}{\partial I_+} & \frac{\partial g_1}{\partial I_-} \\ \frac{\partial g_2}{\partial I_p} & \frac{\partial g_2}{\partial I_+} & \frac{\partial g_2}{\partial I_-} \\ \frac{\partial g_3}{\partial I_p} & \frac{\partial g_3}{\partial I_+} & \frac{\partial g_3}{\partial I_-} \end{pmatrix} \cdot \begin{pmatrix} u_1 \\ u_2 \\ u_3 \end{pmatrix}$$

Remark The linearized system gives a qualitatively correct picture of the phase portrait near (I_p^*, I_+^*, I_-^*) as long as the fixed point for the linearized system is not one of the borderline cases. The linearized system predicts a saddle, node, or a spiral, and then it is the real fixed point classification for the original nonlinear system.

$$g_1(I_p, I_+, I_-) = \sigma_{31} \cdot [(1-f) \cdot \xi_1(I_p, I_+, I_-) - 2 \cdot \xi_2(I_p, I_+, I_-)] = \sigma_{31} \cdot [(1-f) \cdot \xi_1 - 2 \cdot \xi_2]$$

$$\frac{\partial g_1}{\partial I_p} = \sigma_{31} \cdot [(1-f) \cdot \frac{\partial \xi_1}{\partial I_p} - 2 \cdot \frac{\xi_2}{I_p}]$$

$$\frac{\partial \xi_1}{\partial I_p} = \frac{\partial \xi_1(I_p, I_+, I_-)}{\partial I_p} = \frac{\partial n_U}{\partial I_p}; \frac{\partial \xi_2}{\partial I_p} = \frac{\partial \xi_2(I_p, I_+, I_-)}{\partial I_p} = \frac{\partial n_L}{\partial I_p}$$

$$n_U = \xi_1(I_p, I_+, I_-) = -\left(\frac{\tau \cdot n_{tot} \cdot (2 \cdot \sigma_{31} \cdot I_p \cdot v_L + \sigma_{21} \cdot (I_+ + I_-) \cdot v_p)}{(f-3) \cdot \sigma_{31} \cdot \tau \cdot I_p \cdot v_L - [(f+1) \cdot \sigma_{21} \cdot \tau \cdot (I_+ + I_-) + h \cdot v_L] \cdot v_p}\right)$$

$$n_L = \xi_2(I_p, I_+, I_-) = \frac{n_{tot} \cdot [(f-1) \cdot \sigma_{31} \cdot \tau \cdot I_p \cdot v_L - (f \cdot \sigma_{21} \cdot \tau \cdot (I_+ + I_-) + h \cdot v_L) \cdot v_p]}{(f-3) \cdot \sigma_{31} \cdot \tau \cdot I_p \cdot v_L - [(f+1) \cdot \sigma_{21} \cdot \tau \cdot (I_+ + I_-) + h \cdot v_L] \cdot v_p}$$

We define new variables: $\psi_1 = \psi_1(I_p, I_+, I_-)$; $\psi_2 = \psi_2(I_p, I_+, I_-)$; $\psi_3 = \psi_3(I_p, I_+, I_-)$

$$\psi_1 = \psi_1(I_p, I_+, I_-) = (f-3) \cdot \sigma_{31} \cdot \tau \cdot I_p \cdot v_L - [(f+1) \cdot \sigma_{21} \cdot \tau \cdot (I_+ + I_-) + h \cdot v_L] \cdot v_p$$

$$\psi_2 = \psi_2(I_p, I_+, I_-) = \tau \cdot n_{tot} \cdot (2 \cdot \sigma_{31} \cdot I_p \cdot v_L + \sigma_{21} \cdot (I_+ + I_-) \cdot v_p)$$

$$n_U = \xi_1(I_p, I_+, I_-) = -\frac{\psi_2}{\psi_1}$$

$$\psi_3 = \psi_3(I_p, I_+, I_-) = n_{tot} \cdot [(f-1) \cdot \sigma_{31} \cdot \tau \cdot I_p \cdot v_L - (f \cdot \sigma_{21} \cdot \tau \cdot (I_+ + I_-) + h \cdot v_L) \cdot v_p]$$

$$\frac{\partial \xi_1}{\partial I_p} = \frac{\partial n_U}{\partial I_p} = -\frac{\frac{\partial \psi_2}{\partial I_p} \cdot \psi_1 - \frac{\partial \psi_1}{\partial I_p} \cdot \psi_2}{\psi_1^2}$$

$$\frac{\partial \psi_2}{\partial I_p} = \tau \cdot n_{tot} \cdot 2 \cdot \sigma_{31} \cdot v_L; \frac{\partial \psi_1}{\partial I_p} = (f-3) \cdot \sigma_{31} \cdot \tau \cdot v_L$$

$$n_L = \xi_2(I_p, I_+, I_-) = \frac{\psi_3}{\psi_1}; \frac{\partial \xi_2}{\partial I_p} = \frac{\partial n_L}{\partial I_p} = \frac{\partial \xi_2(I_p, I_+, I_-)}{\partial I_p} = \frac{\frac{\partial \psi_3}{\partial I_p} \cdot \psi_1 - \frac{\partial \psi_1}{\partial I_p} \cdot \psi_3}{\psi_1^2}$$

$$\frac{\partial \psi_3}{\partial I_p} = n_{tot} \cdot (f-1) \cdot \sigma_{31} \cdot \tau \cdot v_L$$

$$\frac{\partial g_1}{\partial I_p} = \sigma_{31} \cdot [(1-f) \cdot \frac{\partial \xi_1}{\partial I_p} - 2 \cdot \frac{\xi_2}{I_p}]$$

$$= \sigma_{31} \cdot [-(1-f) \cdot (\frac{\frac{\partial \psi_2}{\partial I_p} \cdot \psi_1 - \frac{\partial \psi_1}{\partial I_p} \cdot \psi_2}{\psi_1^2})$$

$$- 2 \cdot (\frac{\frac{\partial \psi_3}{\partial I_p} \cdot \psi_1 - \frac{\partial \psi_1}{\partial I_p} \cdot \psi_3}{\psi_1^2})]$$

$$\frac{\partial g_1}{\partial I_+} = \sigma_{31} \cdot [(1-f) \cdot \frac{\partial \xi_1}{\partial I_+} - 2 \cdot \frac{\partial \xi_2}{\partial I_+}]$$

$$\frac{\partial \xi_1}{\partial I_+} = \frac{\partial \xi_1(I_p, I_+, I_-)}{\partial I_+} = \frac{\partial n_U}{\partial I_+}$$

$$\frac{\partial \xi_2}{\partial I_+} = \frac{\partial \xi_2(I_p, I_+, I_-)}{\partial I_+} = \frac{\partial n_L}{\partial I_+}$$

$$\frac{\partial \xi_1}{\partial I_+} = \frac{\partial n_U}{\partial I_+} = -\frac{\frac{\partial \psi_2}{\partial I_+} \cdot \psi_1 - \frac{\partial \psi_1}{\partial I_+} \cdot \psi_2}{\psi_1^2}$$

$$\frac{\partial \psi_2}{\partial I_+} = \tau \cdot n_{tot} \cdot \sigma_{21} \cdot v_p; \frac{\partial \psi_1}{\partial I_+} = -(f+1) \cdot \sigma_{21} \cdot \tau \cdot v_p$$

$$\frac{\partial \psi_3}{\partial I_+} = -n_{tot} \cdot f \cdot \sigma_{21} \cdot \tau \cdot v_p$$

$$\frac{\partial \xi_2}{\partial I_+} = \frac{\partial n_L}{\partial I_+} = \frac{\partial \xi_2(I_p, I_+, I_-)}{\partial I_+}$$

$$= \frac{\frac{\partial \psi_3}{\partial I_+} \cdot \psi_1 - \frac{\partial \psi_1}{\partial I_+} \cdot \psi_3}{\psi_1^2}$$

$$\frac{\partial g_1}{\partial I_-} = \sigma_{31} \cdot [(1-f) \cdot \frac{\partial \xi_1}{\partial I_-} - 2 \cdot \frac{\partial \xi_2}{\partial I_-}]$$

$$\frac{\partial \xi_1}{\partial I_-} = \frac{\partial n_U}{\partial I_-} = -\frac{\frac{\partial \psi_2}{\partial I_-} \cdot \psi_1 - \frac{\partial \psi_1}{\partial I_-} \cdot \psi_2}{\psi_1^2}$$

$$\frac{\partial \psi_1}{\partial I_-} = -(f+1) \cdot \sigma_{21} \cdot \tau \cdot v_p$$

$$\frac{\partial \psi_2}{\partial I_-} = \tau \cdot n_{tot} \cdot \sigma_{21} \cdot v_p; \frac{\partial \psi_3}{\partial I_-} = -n_{tot} \cdot f \cdot \sigma_{21} \cdot \tau \cdot v_p$$

$$\frac{\partial \xi_2}{\partial I_-} = \frac{\partial n_L}{\partial I_-} = \frac{\partial \xi_2(I_p, I_+, I_-)}{\partial I_-}$$

$$= \frac{\frac{\partial \psi_3}{\partial I_-} \cdot \psi_1 - \frac{\partial \psi_1}{\partial I_-} \cdot \psi_3}{\psi_1^2}$$

Table 6.7 Quasi-two level analytic model for metal vapor laser system's Jacobian

$\frac{\partial\psi_1}{\partial I_p}$	$(f-3)\cdot\sigma_{31}\cdot\tau\cdot\upsilon_L$	$\frac{\partial\psi_1}{\partial I_+}$	$-(f+1)\cdot\sigma_{21}\cdot\tau\cdot\upsilon_p$	$\frac{\partial\psi_1}{\partial I_-}$	$-(f+1)\cdot\sigma_{21}\cdot\tau\cdot\upsilon_p$
$\frac{\partial\psi_2}{\partial I_p}$	$\tau\cdot n_{tot}\cdot 2\cdot\sigma_{31}\cdot\upsilon_L$	$\frac{\partial\psi_2}{\partial I_+}$	$\tau\cdot n_{tot}\cdot\sigma_{21}\cdot\upsilon_p$	$\frac{\partial\psi_2}{\partial I_-}$	$\tau\cdot n_{tot}\cdot\sigma_{21}\cdot\upsilon_p$
$\frac{\partial\psi_3}{\partial I_p}$	$n_{tot}\cdot(f-1)\cdot\sigma_{31}\cdot\tau\cdot\upsilon_L$	$\frac{\partial\psi_3}{\partial I_+}$	$-n_{tot}\cdot f\cdot\sigma_{21}\cdot\tau\cdot\upsilon_p$	$\frac{\partial\psi_3}{\partial I_-}$	$-n_{tot}\cdot f\cdot\sigma_{21}\cdot\tau\cdot\upsilon_p$

We can summary our last results in the following table (Table 6.7):
Elements: $\left(\frac{\partial\psi_1}{\partial I_p},\frac{\partial\psi_1}{\partial I_+},\frac{\partial\psi_1}{\partial I_-},\frac{\partial\psi_2}{\partial I_p},\frac{\partial\psi_2}{\partial I_+},\frac{\partial\psi_2}{\partial I_-},\frac{\partial\psi_3}{\partial I_p},\frac{\partial\psi_3}{\partial I_+},\frac{\partial\psi_3}{\partial I_-}\right)$.

$$\begin{aligned} g_2 &= g_2(I_p, I_+, I_-) \\ &= \sigma_{21}\cdot(f\cdot\xi_1(I_p,I_+,I_-)-\xi_2(I_p,I_+,I_-))\cdot I_+ \\ &= \sigma_{21}\cdot(f\cdot\xi_1-\xi_2)\cdot I_+ \end{aligned}$$

$$\frac{\partial g_2}{\partial I_p}=\sigma_{21}\cdot(f\cdot\frac{\partial\xi_1}{\partial I_p}-\frac{\partial\xi_2}{\partial I_p})\cdot I_+$$
$$\frac{\partial g_2}{\partial I_+}=\sigma_{21}\cdot(f\cdot\frac{\partial\xi_1}{\partial I_+}-\frac{\partial\xi_2}{\partial I_+})\cdot I_+ +\sigma_{21}\cdot(f\cdot\xi_1-\xi_2)$$
$$\frac{\partial g_2}{\partial I_-}=\sigma_{21}\cdot(f\cdot\frac{\partial\xi_1}{\partial I_-}-\frac{\partial\xi_2}{\partial I_-})\cdot I_+$$
$$\frac{\partial g_3}{\partial I_p}=-\sigma_{21}\cdot(f\cdot\frac{\xi_1}{\partial I_p}-\frac{\xi_2}{\partial I_p})\cdot I_-$$
$$\frac{\partial g_3}{\partial I_+}=-\sigma_{21}\cdot(f\cdot\frac{\partial\xi_1}{\partial I_+}-\frac{\partial\xi_2}{\partial I_+})\cdot I_-$$
$$\frac{\partial g_3}{\partial I_-}=-\sigma_{21}\cdot(f\cdot\frac{\partial\xi_1}{\partial I_-}-\frac{\partial\xi_2}{\partial I_-})\cdot I_- -\sigma_{21}\cdot(f\cdot\xi_1-\xi_2)$$

We can summary our last results in the following table (Table 6.8):
The system Jacobian matrix at the fixed point (I_p^*, I_+^*, I_-^*) is

Table 6.8 Quasi-two level analytic model system's Jacobian elements: $\partial\xi_i/\partial I_{p,+,-}; i=1,2$

$\frac{\partial\xi_1}{\partial I_p}$	$-\frac{\frac{\partial\psi_2}{\partial I_p}\cdot\psi_1-\frac{\partial\psi_1}{\partial I_p}\cdot\psi_2}{\psi_1^2}$	$\frac{\partial\xi_1}{\partial I_-}$	$-\frac{\frac{\partial\psi_2}{\partial I_-}\cdot\psi_1-\frac{\partial\psi_1}{\partial I_-}\cdot\psi_2}{\psi_1^2}$	$\frac{\partial\xi_2}{\partial I_+}$	$\frac{\frac{\partial\psi_3}{\partial I_+}\cdot\psi_1-\frac{\partial\psi_1}{\partial I_+}\cdot\psi_3}{\psi_1^2}$
$\frac{\partial\xi_1}{\partial I_+}$	$-\frac{\frac{\partial\psi_2}{\partial I_+}\cdot\psi_1-\frac{\partial\psi_1}{\partial I_+}\cdot\psi_2}{\psi_1^2}$	$\frac{\partial\xi_2}{\partial I_p}$	$\frac{\frac{\partial\psi_3}{\partial I_p}\cdot\psi_1-\frac{\partial\psi_1}{\partial I_p}\cdot\psi_3}{\psi_1^2}$	$\frac{\partial\xi_2}{\partial I_-}$	$\frac{\frac{\partial\psi_3}{\partial I_-}\cdot\psi_1-\frac{\partial\psi_1}{\partial I_-}\cdot\psi_3}{\psi_1^2}$

$$A = \begin{pmatrix} \sigma_{31}\cdot[(1-f)\cdot\frac{\partial\xi_1}{\partial I_p} - 2\cdot\frac{\xi_2}{I_p}] & \sigma_{31}\cdot[(1-f)\cdot\frac{\partial\xi_1}{\partial I_+} - 2\cdot\frac{\partial\xi_2}{\partial I_+}] & \sigma_{31}\cdot[(1-f)\cdot\frac{\partial\xi_1}{\partial I_-} - 2\cdot\frac{\partial\xi_2}{\partial I_-}] \\ \sigma_{21}\cdot(f\cdot\frac{\partial\xi_1}{\partial I_p} - \frac{\partial\xi_2}{\partial I_p})\cdot I_+ & \begin{matrix}\sigma_{21}\cdot(f\cdot\frac{\partial\xi_1}{\partial I_+} - \frac{\partial\xi_2}{\partial I_+})\cdot I_+ \\ +\sigma_{21}\cdot(f\cdot\xi_1 - \xi_2)\end{matrix} & \sigma_{21}\cdot(f\cdot\frac{\partial\xi_1}{\partial I_-} - \frac{\partial\xi_2}{\partial I_-})\cdot I_+ \\ -\sigma_{21}\cdot(f\cdot\frac{\xi_1}{\partial I_p} - \frac{\xi_2}{\partial I_p})\cdot I_- & -\sigma_{21}\cdot(f\cdot\frac{\partial\xi_1}{\partial I_+} - \frac{\partial\xi_2}{\partial I_+})\cdot I_- & \begin{matrix}-\sigma_{21}\cdot(f\cdot\frac{\partial\xi_1}{\partial I_-} - \frac{\partial\xi_2}{\partial I_-})\cdot I_- \\ -\sigma_{21}\cdot(f\cdot\xi_1 - \xi_2)\end{matrix} \end{pmatrix}_{(I_p^*, I_+^*, I_-^*)}$$

$$A - \lambda\cdot I$$
$$= \begin{pmatrix} \sigma_{31}\cdot[(1-f)\cdot\frac{\partial\xi_1}{\partial I_p} - 2\cdot\frac{\xi_2}{I_p}] - \lambda & \sigma_{31}\cdot[(1-f)\cdot\frac{\partial\xi_1}{\partial I_+} - 2\cdot\frac{\partial\xi_2}{\partial I_+}] & \sigma_{31}\cdot[(1-f)\cdot\frac{\partial\xi_1}{\partial I_-} - 2\cdot\frac{\partial\xi_2}{\partial I_-}] \\ \sigma_{21}\cdot(f\cdot\frac{\partial\xi_1}{\partial I_p} - \frac{\partial\xi_2}{\partial I_p})\cdot I_+ & \begin{matrix}\sigma_{21}\cdot(f\cdot\frac{\partial\xi_1}{\partial I_+} - \frac{\partial\xi_2}{\partial I_+})\cdot I_+ \\ +\sigma_{21}\cdot(f\cdot\xi_1 - \xi_2) - \lambda\end{matrix} & \sigma_{21}\cdot(f\cdot\frac{\partial\xi_1}{\partial I_-} - \frac{\partial\xi_2}{\partial I_-})\cdot I_+ \\ -\sigma_{21}\cdot(f\cdot\frac{\xi_1}{\partial I_p} - \frac{\xi_2}{\partial I_p})\cdot I_- & -\sigma_{21}\cdot(f\cdot\frac{\partial\xi_1}{\partial I_+} - \frac{\partial\xi_2}{\partial I_+})\cdot I_- & \begin{matrix}-\sigma_{21}\cdot(f\cdot\frac{\partial\xi_1}{\partial I_-} - \frac{\partial\xi_2}{\partial I_-})\cdot I_- \\ -\sigma_{21}\cdot(f\cdot\xi_1 - \xi_2) - \lambda\end{matrix} \end{pmatrix}_{(I_p^*, I_+^*, I_-^*)}$$

For simplicity we define the following global Jacobian elements:

$$\Upsilon_{11} = \sigma_{31}\cdot[(1-f)\cdot\frac{\partial\xi_1}{\partial I_p} - 2\cdot\frac{\xi_2}{I_p}]|_{(I_p^*, I_+^*, I_-^*)}$$

$$\Upsilon_{12} = \sigma_{31}\cdot[(1-f)\cdot\frac{\partial\xi_1}{\partial I_+} - 2\cdot\frac{\partial\xi_2}{\partial I_+}]|_{(I_p^*, I_+^*, I_-^*)}$$

$$\Upsilon_{13} = \sigma_{31}\cdot[(1-f)\cdot\frac{\partial\xi_1}{\partial I_-} - 2\cdot\frac{\partial\xi_2}{\partial I_-}]|_{(I_p^*, I_+^*, I_-^*)}$$

$$\Upsilon_{21} = \sigma_{21}\cdot(f\cdot\frac{\partial\xi_1}{\partial I_p} - \frac{\partial\xi_2}{\partial I_p})\cdot I_+|_{(I_p^*, I_+^*, I_-^*)}$$

$$\Upsilon_{22} = \sigma_{21}\cdot(f\cdot\frac{\partial\xi_1}{\partial I_+} - \frac{\partial\xi_2}{\partial I_+})\cdot I_+ + \sigma_{21}\cdot(f\cdot\xi_1 - \xi_2)|_{(I_p^*, I_+^*, I_-^*)}$$

$$\Upsilon_{23} = \sigma_{21}\cdot(f\cdot\frac{\partial\xi_1}{\partial I_-} - \frac{\partial\xi_2}{\partial I_-})\cdot I_{+(I_p^*, I_+^*, I_-^*)}$$

$$\Upsilon_{31} = -\sigma_{21}\cdot(f\cdot\frac{\xi_1}{\partial I_p} - \frac{\xi_2}{\partial I_p})\cdot I_-|_{(I_p^*, I_+^*, I_-^*)}$$

$$\Upsilon_{32} = -\sigma_{21}\cdot(f\cdot\frac{\partial\xi_1}{\partial I_+} - \frac{\partial\xi_2}{\partial I_+})\cdot I_-|_{(I_p^*, I_+^*, I_-^*)}$$

$$\Upsilon_{33} = -\sigma_{21}\cdot(f\cdot\frac{\partial\xi_1}{\partial I_-} - \frac{\partial\xi_2}{\partial I_-})\cdot I_- - \sigma_{21}\cdot(f\cdot\xi_1 - \xi_2)|_{(I_p^*, I_+^*, I_-^*)}$$

$$A - \lambda \cdot I = \begin{pmatrix} \Upsilon_{11} - \lambda & \Upsilon_{12} & \Upsilon_{13} \\ \Upsilon_{21} & \Upsilon_{22} - \lambda & \Upsilon_{23} \\ \Upsilon_{31} & \Upsilon_{32} & \Upsilon_{33} - \lambda \end{pmatrix}_{(I_p^*, I_+^*, I_-^*)}$$

Expanding thc dctcrminant yields:

$$\begin{aligned} \det(A - \lambda \cdot I) = (\Upsilon_{11} - \lambda) \cdot \det\begin{pmatrix} \Upsilon_{22} - \lambda & \Upsilon_{23} \\ \Upsilon_{32} & \Upsilon_{33} - \lambda \end{pmatrix} \\ - \Upsilon_{12} \cdot \det\begin{pmatrix} \Upsilon_{21} & \Upsilon_{23} \\ \Upsilon_{31} & \Upsilon_{33} - \lambda \end{pmatrix} \\ + \Upsilon_{13} \cdot \det\begin{pmatrix} \Upsilon_{21} & \Upsilon_{22} - \lambda \\ \Upsilon_{31} & \Upsilon_{32} \end{pmatrix} \end{aligned}$$

$$\begin{aligned} \det(A - \lambda \cdot I) = (\Upsilon_{11} - \lambda) \cdot [(\Upsilon_{22} - \lambda) \cdot (\Upsilon_{33} - \lambda) - \Upsilon_{32} \cdot \Upsilon_{23}] \\ - \Upsilon_{12} \cdot [\Upsilon_{21} \cdot (\Upsilon_{33} - \lambda) - \Upsilon_{31} \cdot \Upsilon_{23}] \\ + \Upsilon_{13} \cdot [\Upsilon_{21} \cdot \Upsilon_{32} - \Upsilon_{31} \cdot (\Upsilon_{22} - \lambda)] \end{aligned}$$

$$\begin{aligned} \det(A - \lambda \cdot I) = (\Upsilon_{11} - \lambda) \cdot [(\Upsilon_{22} \cdot \Upsilon_{33} - \Upsilon_{32} \cdot \Upsilon_{23}) \\ - \lambda \cdot (\Upsilon_{22} + \Upsilon_{33}) + \lambda^2] - \Upsilon_{12} \cdot [(\Upsilon_{21} \cdot \Upsilon_{33} \\ - \Upsilon_{31} \cdot \Upsilon_{23}) - \Upsilon_{21} \cdot \lambda] + \Upsilon_{13} \cdot [(\Upsilon_{21} \cdot \Upsilon_{32} \\ - \Upsilon_{31} \cdot \Upsilon_{22}) + \Upsilon_{31} \cdot \lambda] \end{aligned}$$

$$\begin{aligned} \det(A - \lambda \cdot I) = \Upsilon_{11} \cdot (\Upsilon_{22} \cdot \Upsilon_{33} - \Upsilon_{32} \cdot \Upsilon_{23}) - \lambda \cdot \Upsilon_{11} \\ \cdot (\Upsilon_{22} + \Upsilon_{33}) + \Upsilon_{11} \cdot \lambda^2 - \lambda \cdot (\Upsilon_{22} \cdot \Upsilon_{33} - \Upsilon_{32} \cdot \Upsilon_{23}) \\ + \lambda^2 \cdot (\Upsilon_{22} + \Upsilon_{33}) - \lambda^3 - \Upsilon_{12} \cdot (\Upsilon_{21} \cdot \Upsilon_{33} - \Upsilon_{31} \cdot \Upsilon_{23}) \\ + \Upsilon_{12} \cdot \Upsilon_{21} \cdot \lambda + \Upsilon_{13} \cdot (\Upsilon_{21} \cdot \Upsilon_{32} - \Upsilon_{31} \cdot \Upsilon_{22}) + \Upsilon_{13} \cdot \Upsilon_{31} \cdot \lambda \end{aligned}$$

$$\begin{aligned} \det(A - \lambda \cdot I) = \{\Upsilon_{11} \cdot (\Upsilon_{22} \cdot \Upsilon_{33} - \Upsilon_{32} \cdot \Upsilon_{23}) - \Upsilon_{12} \\ \cdot (\Upsilon_{21} \cdot \Upsilon_{33} - \Upsilon_{31} \cdot \Upsilon_{23}) + \Upsilon_{13} \cdot (\Upsilon_{21} \cdot \Upsilon_{32} \\ - \Upsilon_{31} \cdot \Upsilon_{22})\} + \lambda \cdot \{\Upsilon_{12} \cdot \Upsilon_{21} + \Upsilon_{13} \cdot \Upsilon_{31} \\ - \Upsilon_{11} \cdot (\Upsilon_{22} + \Upsilon_{33}) - (\Upsilon_{22} \cdot \Upsilon_{33} - \Upsilon_{32} \cdot \Upsilon_{23})\} \\ + \lambda^2 \cdot (\Upsilon_{11} + \Upsilon_{22} + \Upsilon_{33}) - \lambda^3 \end{aligned}$$

We define for simplicity the following global parameters:

$$\begin{aligned} \Xi_0 = \Upsilon_{11} \cdot (\Upsilon_{22} \cdot \Upsilon_{33} - \Upsilon_{32} \cdot \Upsilon_{23}) - \Upsilon_{12} \cdot (\Upsilon_{21} \cdot \Upsilon_{33} \\ - \Upsilon_{31} \cdot \Upsilon_{23}) + \Upsilon_{13} \cdot (\Upsilon_{21} \cdot \Upsilon_{32} - \Upsilon_{31} \cdot \Upsilon_{22}) \end{aligned}$$

$$\Xi_1 = \Upsilon_{12} \cdot \Upsilon_{21} + \Upsilon_{13} \cdot \Upsilon_{31} - \Upsilon_{11} \cdot (\Upsilon_{22} + \Upsilon_{33}) - (\Upsilon_{22} \cdot \Upsilon_{33} - \Upsilon_{32} \cdot \Upsilon_{23})$$

$$\Xi_2 = \Upsilon_{11} + \Upsilon_{22} + \Upsilon_{33}; \; \Xi_3 = -1$$

$$\det(A - \lambda \cdot I) = \Xi_0 + \lambda \cdot \Xi_1 + \lambda^2 \cdot \Xi_2 + \lambda^3 \cdot \Xi_3$$

$$\det(A - \lambda \cdot I) = \sum_{k=0}^{3} \Xi_k \cdot \lambda^k$$

$$\det(A - \lambda \cdot I) = 0 \Rightarrow \sum_{k=0}^{3} \Xi_k \cdot \lambda^k = 0$$

$$\Xi_0 + \lambda \cdot \Xi_1 + \lambda^2 \cdot \Xi_2 + \lambda^3 \cdot \Xi_3 = 0$$

Eigenvalue stability discussion: Our Quasi-two level analytic model system involving N variables (N > 2, N = 3, arbitrarily small increments), the characteristic equation is of degree N = 3 and must often be solved numerically. Expect in some particular cases, such an equation has (N = 3) distinct roots that can be real or complex. These values are eigenvalues of the (3 × 3) Jacobian matrix (A). The general rule is that the Quasi-two level analytic model system is stable if there is no eigenvalue with positive real part. It is sufficient that one eigenvalue is positive for the steady state to be unstable. Our three variables (I_p, I_+, I_-) system has three eigenvalues (three system's arbitrarily small increments). The type of behavior can be characterized as a function of the position of these eigenvalues in the Re/Im plane. Five non-generated cases can be distinguished: (1) the three eigenvalues are real and negative (stable steady state), (2) the three eigenvalues are real, at least one of them is positive (unstable steady state), (3) and (4) two eigenvalues are complex conjugates with a negative real part and other eigenvalue real is negative (stable steady state), two cases can be distinguished depending on the relative value of the real part of the complex eigenvalues and of the real one, (5) two eigenvalues are complex conjugates with a negative real part and other eigenvalue real is positive (unstable steady state) [7, 8].

$$\det(A - \lambda \cdot I) = 0 \Rightarrow \sum_{k=0}^{3} \Xi_k \cdot \lambda^k = 0$$

$$\Xi_0 + \lambda \cdot \Xi_1 + \lambda^2 \cdot \Xi_2 + \lambda^3 \cdot \Xi_3 = 0$$

6.3 A Self-consistent Model Copper Vapor Laser (CVL) Circuity Stability Analysis Under Parameters Variation

A self-consistent model copper vapor laser (CVL) is very helpful for characterization of the copper laser and its implementation. It uses vapors of copper as the lasing medium in a 3-level laser. It can produce green laser light (510.6 nm) and yellow laser light (578.2 nm). The peak power of the laser pulse can be between 50kw and 5×10^3 kw. The pulse width is typically from 5 nsec to 60 nsec and the pulse repetition frequencies are typically between 2 kHz to 100 kHz. We can characterize the average power of copper vapor laser in the range of 25 w to 2 kw. To a create vapor it is necessary to get extremely high temperature, about 1500 °C. Coper compound vapors increase the complexity of the pump signal applied to the device. Typically, two engineering pulses are required, the first to dissociate vapor molecules, and the second to cause the dissociated ions to lase. We can reduce the operating temperature by the use of copper nitrate or copper acetylacetonate. By using those coppers we get vapors with peak laser output power at 180 °C and 40 °C, respectively. If we want to use a copper vapor lasers that are cw (continuous wave) we need to use singly ionized species of Cu. These lasers can provide average UV powers of several mW mainly for spectroscopy applications. The copper vapor laser makes use of copper vapors as the lasing medium. Generally we use copper bromide, copper chloride, and copper iodide to implement copper vapor laser (CVL). The laser head of the copper vapor laser consists of a refractory ceramic tube that contains a low-pressure buffer gas, neon, and copper pellets. A pulsed electrical discharge is between electrodes at the tube ends, and the temperature of the laser head increases to $\approx$ 1450 C degree, and producing copper vapor at low temperature. The excited electrons collide with vaporized copper atoms and excite these atoms to the upper ${}^2P_{3/2}$ and ${}^2P_{1/2}$ laser levels. The electrons in the ${}^2P_{3/2}$ level decay to the lower ${}^2D_{5/2}$ laser level to produce green light, and the ${}^2P_{1/2}$ decay to ${}^2D_{3/2}$ level to produce yellow laser light. Copper vapor laser has good beam quality, excellent power stability and along operational life time. The main applications are laser cutting and precision micromachining [9]. When producing copper vapor laser it is recommended to use several mbars of neon which added as a buffer gas to prevent window contamination and loss of copper. The overall well plug efficiency of copper vapor laser is ~1% which is highest for visible gas lasers. The copper vapor laser tube is sealed with flat glass windows; the rear mirror is a total reflector, with 90% transmission chosen for the output windows. Beams from copper vapor lasers can be from 10 mm diameter to ~50 mm diameter, beam divergence for a stable type resonator is 3 mradians to 5 mradians. In the visible part of the spectrum, beam focusing is recommended to use glass optics. The copper vapor laser structure is present below (Fig. 6.5).

We describe the equations for the discharge pulse, laser pulse, and inter pulse afterglow period. The simulation analysis is over many discharge cycles until a consistent solution is obtained. We present a consistent initial conditions and scaling characteristics of the copper laser. Ideal and real laser are simulate and a conclusion is accepted that the peak electron temperature and the initial density of metastable

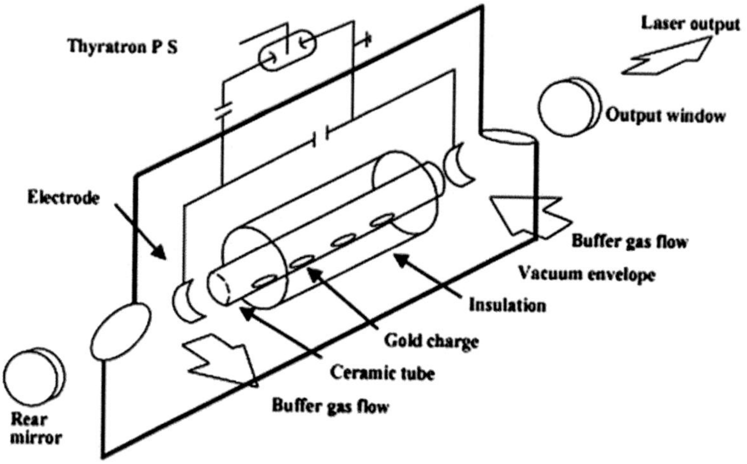

Fig. 6.5 Copper vapor laser (CVL) structure

copper are the factors which dominate the performance of the laser. These processes affect these two quantities. We produce a high peak and high average power in the visible by using high repetition rate (HRR) electric discharge metal vapor laser. The most sophisticated HRR copper lasers are resonantly charged electric discharge devices. These lasers are self-heated and the heating energy of the laser tube and vaporize the copper comes from the discharge itself. The copper laser is a cyclic or self-terminating laser. The upper laser level is resonant to the ground state while the lower laser level is metastable. Atoms which making the laser transition, accumulate in the lower level and caused to gain to become negative. The choice of repetition rate immediately, sets threshold pumping requirements. The choice of repetition rate determined the initial conditions for the discharge pulse. The initial electron density, plasma tube impedance, and electron temperature are a function of inter pulse processes. In an ideal HRR copper vapor laser, quantities such as charging voltage, capacitance, metal density, and repetition rate are independent be uniquely specified. In an ideal laser, we specify thermodynamic quantities, access any tube temperature, and uniformly deposit energy throughout the active volume. The discharge circuit of an ideal laser is free of inductance. In real HRR laser, the tube temperature and metal density are functions of the discharge input power, and then the charging voltage, capacitance, repetition rate, and metal density are not independent. The buffer gas pressure can be specified exterior to the tube but not accurately within the tube due to transient discharge heating. The laser's materials define an upper limit on the tube temperature, and inductance in the discharge circuit limits the rate at which current and voltage appear across the tube. These limitations mask the physical process which occur in a real HRR laser and make it difficult to de convolve the effect of changing a single variable. The analysis is divided into two sections. The first section is the modeling of an ideal laser and the second section examines and models the real

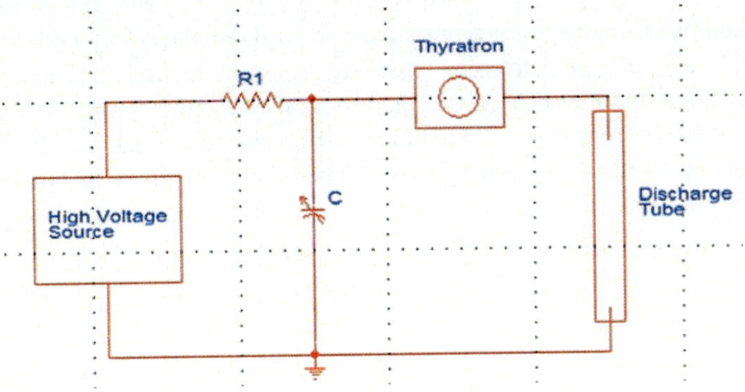

Fig. 6.6 Ideal copper vapor laser (CVL) discharge circuit

laser. The experiment is related to VENUS laser (15w device, 6hHz). The VENUS laser discharge tube is made of alumina and heat region. The laser power on metal density (tube temperature) is a function of the peak electron temperature. The peak electron temperature is a function of the peak voltage across the discharge tube, and it determines the optimum tube temperature and maximum laser power. The dependence of laser power on repetition rate is a function of the initial metastable density, thus the value is a function of the dominant mode of relaxation which is electron collision or diffusion. The model simulates a discharge pulse, laser pulse, and inter-pulse afterglow in a high repetition rate copper laser. The initial conditions are chosen and a number of cycles are computed, the outcome is consistent pulses. The voltage across the laser tube, and current through laser tube are a function of the plasma impedance and the particulars of the discharge circuit. Detailed discharge circuitry is an integral part of the model. There are two groups of discharge circuits which describe the ideal and real lasers [9].

Ideal laser discharge circuit: The ideal laser discharge circuit consists of a capacitor initially charged to a voltage V_0 and a Thyratron with a switching time τ_s and discharge tube (Fig. 6.6). A thyratron is a type of gas-filled tube which used as a high-power electrical switch and controlled rectifier. The Thyratrons handle currents are much greater than hard-vacuum tube's currents. The working details of Thyratron are as follow: An electron multiplication occurs when the gas becomes ionized, and Townsend discharge phenomenon is happened. The Townsend discharge or Townsend avalanche is a gas ionization process where free electrons are accelerated by an electric field, collide with gas molecules, and finally free additional electron. Then these electrons are in turn accelerated and free additional electrons. We get avalanche multiplication and electrical conduction through the gas is happened. The phenomenon occurs when there is a source of free electrons and a significant electric field. Thyratron, gas-filled discharge chamber grid controls the starting of a current and thus provides a trigger effect. The grid potential is negative with respect to the cathode and prevents electrons from flowing to the plate and exciting a discharge. The grid potential is raised enough to start electrons flowing from the cathode then discharge happened. As free electrons stream toward the plate, they collide with gas molecules, freeing other electrons and ionizing the gas within the discharge chamber. When a sufficient number of ions and electrons are present, a "short" occurs, and a large current flows from the cathode to the plate, causing discharge. The discharge stops when the anode voltage has been sufficiently lowered. A discharge tube is an electron tube which contains gas or vapor at low pressure and through which conduction takes place when a high voltage is applied. The ionization is induced by an electric field and the gas molecules emit light as they return to the ground state.

The time rate of change of voltage across the tube is $\frac{dV}{dt} = V_0 \cdot f(\tau_s) - \frac{V}{R_d \cdot C}$, where $f(\tau_s)$ is the Thyratron voltage turn on function, and R_d, $\left(R_d = \frac{l \cdot m_e \cdot v}{n_e \cdot A \cdot e^2}\right)$, is the instantaneous discharge impedance. Where, m_e is the mass of the electron, v is the electron collision frequency, and A is the effective cross-section area of the discharge tube.

Real laser discharge circuit: The discharge circuit for the real laser simulates the actual circuit which use in VENUS laser (Fig. 6.7). The model discharge circuit elements are capacitor C_1 which is the storage capacitor resonantly charged to voltage V_0 by a DC power supply and inductanceL_c. Capacitor C_2 is the peaking capacitor. The Thyratron switch (S_1) is modeled by a reverse bias voltage and removed with a switching time τ_s. R_c is the parallel charging resistance. The Thyratron behaves like a diode with current flowing through it in only one direction. The inductances L_1 and L_2 represent the lumped circuit values [9].

Where current I_1 flows through L_1 and current I_2 flows through L_2, voltage V_1 is the voltage across capacitor C_1 and V_2 is the voltage across capacitor C_2. The voltage across the discharge tube is V_d, $\left(V_d = I_2(t - \tau_2) \cdot \frac{R_c \cdot R_d}{R_c + R_d}\right)$, and the current that flows through the discharge tube is I_d, $I_d = I_2(t - \tau_2) \cdot \frac{R_c}{R_c + R_d}$. The real model differential equations circuitry which represents the actual behavior of our system is as follow:

$$\frac{dI_1}{dt} = \frac{(V_1 - V_2) + V_0 \cdot f(\tau_s)}{L_1}$$

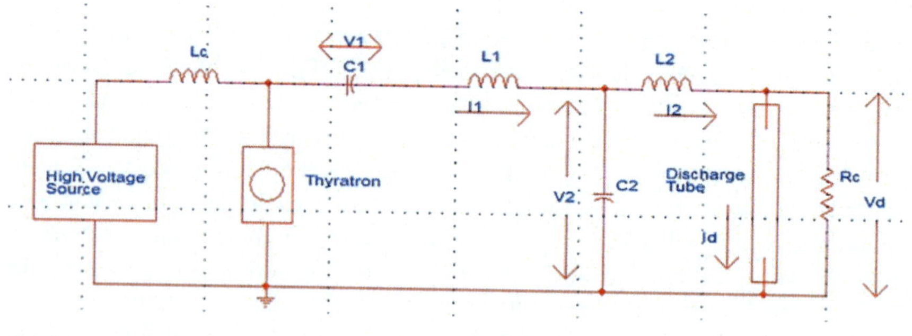

Fig. 6.7 Real copper vapor laser (CVL) discharge circuit

$$\frac{dI_2}{dt} = \frac{V_2 - V_d}{L_2}; \frac{dV_1}{dt} = -\frac{I_1(t-\tau_1)}{C_1}$$

$$\frac{dV_2}{dt} = \frac{I_1(t-\tau_1) - I_2(t-\tau_2)}{C_2}$$

Remark The real laser discharge circuit lumped inductances L_1 and L_2 are not ideal and there is time delays τ_1, τ_2 of currents I_1 and I_2 that flow through them, respectively. τ_1 and τ_2 are circuit lumped inductances (L_1, L_2) delay in time which causes by electromagnetic interferences.

Equilibrium points (fixed points): $\lim_{t\to\infty} I_1(t-\tau_1)|_{t\gg\tau_1} = I_1(t)$; $\lim_{t\to\infty} I_2(t-\tau_2) = I_2(t)$

$$\frac{dI_1}{dt} = 0; \frac{dI_2}{dt} = 0; \frac{dV_1}{dt} = 0; \frac{dV_2}{dt} = 0$$

$$\frac{(V_1^* - V_2^*) + V_0 \cdot f(\tau_s)}{L_1} = 0; \frac{V_2^* - V_d}{L_2} = 0 \Rightarrow V_2^* = V_d$$

$$\frac{(V_1^* - V_2^*) + V_0 \cdot f(\tau_s)}{L_1} = 0 \Rightarrow \frac{(V_1^* - V_d) + V_0 \cdot f(\tau_s)}{L_1} = 0$$

$$(V_1^* - V_d) + V_0 \cdot f(\tau_s) = 0$$

$$V_1^* = V_d - V_0 \cdot f(\tau_s); -\frac{I_1^*}{C_1} = 0 \Rightarrow I_1^* = 0$$

$$\frac{I_1^* - I_2^*}{C_2} = 0 \Rightarrow I_1^* = I_2^* = 0$$

System fixed points: $E^*(I_1^*, I_2^*, V_1^*, V_2^*) = (0, 0, V_d - V_0 \cdot f(\tau_s), V_d)$

Stability analysis: The standard local stability analysis about any one of the equilibrium point of the copper vapor laser (CVL) circuity consists in adding to coordinate $[I_1, I_2, V_1, V_2]$ arbitrarily small increments of exponential form $[i_1, i_2, v_1, v_2] \cdot e^{\lambda \cdot t}$ and retaining the first order terms in I_1, I_2, V_1, V_2. The system of four homogeneous equations leads to a polynomial characteristic equation in the eigenvalues. The polynomial characteristic equations accept by set the below variables and variables derivative with respect to time into copper vapor laser (CVL) circuity equations. The copper vapor laser (CVL) circuity fixed values with arbitrarily small increments of exponential form $[i_1, i_2, v_1, v_2] \cdot e^{\lambda \cdot t}$ are: $j = 0$(first fixed point), $j = 1$(second fixed point), $j = 2$(third fixed point), etc. [4, 5].

$$I_1(t) = I_1^{(j)} + i_1 \cdot e^{\lambda \cdot t}; I_2(t) = I_2^{(j)} + i_2 \cdot e^{\lambda \cdot t}$$

$$V_1(t) = V_1^{(j)} + v_1 \cdot e^{\lambda \cdot t}; V_2(t) = V_2^{(j)} + v_2 \cdot e^{\lambda \cdot t}$$

$$I_1(t-\tau_1) = I_1^{(j)} + i_1 \cdot e^{\lambda\cdot(t-\tau_1)};\ I_2(t-\tau_2) = I_2^{(j)} + i_2 \cdot e^{\lambda\cdot(t-\tau_2)}$$

$$\frac{dI_1(t)}{dt} = i_1 \cdot \lambda \cdot e^{\lambda\cdot t};\ \frac{dI_2(t)}{dt} = i_2 \cdot \lambda \cdot e^{\lambda\cdot t}$$
$$\frac{dV_1(t)}{dt} = v_1 \cdot \lambda \cdot e^{\lambda\cdot t};\ \frac{dV_2(t)}{dt} = v_2 \cdot \lambda \cdot e^{\lambda\cdot t}$$

We choose these expressions for ourselves $I_1(t), I_2(t), V_1(t), V_2(t)$ as a small displacement $[i_1, i_2, v_1, v_2]$ from the copper vapor laser (CVL) circuity fixed points in time $t = 0$.

$$I_1(t=0) = I_1^{(j)} + i_1;\ I_2(t=0) = I_2^{(j)} + i_2$$
$$V_1(t=0) = V_1^{(j)} + v_1;\ V_2(t=0) = V_2^{(j)} + v_2$$

$$\frac{dI_1}{dt} = \frac{(V_1 - V_2) + V_0 \cdot f(\tau_s)}{L_1}$$
$$i_1 \cdot \lambda \cdot e^{\lambda\cdot t} = \frac{(V_1^{(j)} + v_1 \cdot e^{\lambda\cdot t} - V_2^{(j)} - v_2 \cdot e^{\lambda\cdot t}) + V_0 \cdot f(\tau_s)}{L_1}$$

$$i_1 \cdot \lambda \cdot e^{\lambda\cdot t} = \frac{(V_1^{(j)} - V_2^{(j)}) + V_0 \cdot f(\tau_s)}{L_1} + \frac{(v_1 \cdot e^{\lambda\cdot t} - v_2 \cdot e^{\lambda\cdot t})}{L_1}$$

At fixed point: $\frac{dI_1}{dt} = 0 \Rightarrow \frac{(V_1^{(j)} - V_2^{(j)}) + V_0\cdot f(\tau_s)}{L_1} = 0;\ i_1 \cdot \lambda \cdot e^{\lambda\cdot t} = \frac{1}{L_1} \cdot v_1 \cdot e^{\lambda\cdot t} - \frac{1}{L_1} \cdot v_2 \cdot e^{\lambda\cdot t}$

$$-i_1 \cdot \lambda + \frac{1}{L_1} \cdot v_1 - \frac{1}{L_1} \cdot v_2 = 0;\ \frac{dI_2}{dt} = \frac{V_2 - V_d}{L_2}$$
$$i_2 \cdot \lambda \cdot e^{\lambda\cdot t} = \frac{V_2^{(j)} + v_2 \cdot e^{\lambda\cdot t} - V_d}{L_2}$$
$$i_2 \cdot \lambda \cdot e^{\lambda\cdot t} = \frac{V_2^{(j)} - V_d}{L_2} + \frac{v_2 \cdot e^{\lambda\cdot t}}{L_2}$$

At fixed point: $\frac{dI_2}{dt} = 0 \Rightarrow \frac{V_2^{(j)} - V_d}{L_2} = 0;\ i_2\cdot\lambda\cdot e^{\lambda\cdot t} = \frac{1}{L_2}\cdot v_2\cdot e^{\lambda\cdot t};\ -i_2\cdot\lambda + \frac{1}{L_2}\cdot v_2 = 0$
We have four sub cases for (1) $\tau_1 = 0;\ \tau_2 = 0$(2) $\tau_1 = \tau;\ \tau_2 = 0$(3) $\tau_1 = 0;\ \tau_2 = \tau$ (4) $\tau_1 = \tau;\ \tau_2 = \tau$. We choose to analyze the second case $\tau_1 = \tau;\ \tau_2 = 0$.

$$\frac{dV_1}{dt} = -\frac{I_1(t-\tau)}{C_1};\ v_1 \cdot \lambda \cdot e^{\lambda\cdot t} = -\frac{I_1^{(j)} + i_1 \cdot e^{\lambda\cdot(t-\tau)}}{C_1}$$
$$v_1 \cdot \lambda \cdot e^{\lambda\cdot t} = -\frac{I_1^{(j)}}{C_1} - \frac{i_1 \cdot e^{\lambda\cdot t} \cdot e^{-\lambda\cdot\tau}}{C_1}$$

At fixed point: $\frac{dV_1}{dt} = 0 \Rightarrow -\frac{I_1^{(j)}}{C_1} = 0;\ v_1 \cdot \lambda = -\frac{i_1 \cdot e^{-\lambda \cdot \tau}}{C_1};\ -\frac{i_1 \cdot e^{-\lambda \cdot \tau}}{C_1} - v_1 \cdot \lambda = 0$

$$\frac{dV_2}{dt} = \frac{I_1(t-\tau) - I_2(t)}{C_2}$$

$$v_2 \cdot \lambda \cdot e^{\lambda \cdot t} = \frac{I_1^{(j)} + i_1 \cdot e^{\lambda \cdot (t-\tau)} - I_2^{(j)} - i_2 \cdot e^{\lambda \cdot t}}{C_2}$$

$v_2 \cdot \lambda \cdot e^{\lambda \cdot t} = \frac{I_1^{(j)} - I_2^{(j)}}{C_2} + \frac{i_1 \cdot e^{\lambda \cdot (t-\tau)} - i_2 \cdot e^{\lambda \cdot t}}{C_2}$, At fixed point: $\frac{dV_2}{dt} = 0 \Rightarrow \frac{I_1^{(j)} - I_2^{(j)}}{C_2} = 0$

$$v_2 \cdot \lambda \cdot e^{\lambda \cdot t} = \frac{i_1 \cdot e^{\lambda \cdot t} \cdot e^{-\lambda \cdot \tau} - i_2 \cdot e^{\lambda \cdot t}}{C_2}$$

$$\Rightarrow \frac{e^{-\lambda \cdot \tau}}{C_2} \cdot i_1 - \frac{1}{C_2} \cdot i_2 - v_2 \cdot \lambda = 0$$

We can summary our copper vapor laser (CVL) circuity four arbitrarily small increments equations:

$$-i_1 \cdot \lambda + \frac{1}{L_1} \cdot v_1 - \frac{1}{L_1} \cdot v_2 = 0;\ -i_2 \cdot \lambda + \frac{1}{L_2} \cdot v_2 = 0$$

$$-\frac{i_1 \cdot e^{-\lambda \cdot \tau}}{C_1} - v_1 \cdot \lambda = 0;\ \frac{e^{-\lambda \cdot \tau}}{C_2} \cdot i_1 - \frac{1}{C_2} \cdot i_2 - v_2 \cdot \lambda = 0$$

The small increments Jacobian of our copper vapor laser (CVL) circuity is as follow:

$$\begin{pmatrix} \Upsilon_{11} & \dots & \Upsilon_{14} \\ \vdots & \ddots & \vdots \\ \Upsilon_{41} & \cdots & \Upsilon_{44} \end{pmatrix} \cdot \begin{pmatrix} i_1 \\ i_2 \\ v_1 \\ v_2 \end{pmatrix} = 0$$

$$\Upsilon_{11} = -\lambda;\ \Upsilon_{12} = 0;\ \Upsilon_{13} = \frac{1}{L_1};\ \Upsilon_{14} = -\frac{1}{L_1};\ \Upsilon_{21} = 0;\ \Upsilon_{22} = -\lambda$$

$$\Upsilon_{23} = 0;\ \Upsilon_{24} = \frac{1}{L_2};\ \Upsilon_{31} = -\frac{1}{C_1} \cdot e^{-\lambda \cdot \tau};\ \Upsilon_{32} = 0$$

$$\Upsilon_{33} = -\lambda;\ \Upsilon_{34} = 0;\ \Upsilon_{41} = \frac{1}{C_2} \cdot e^{-\lambda \cdot \tau};\ \Upsilon_{42} = -\frac{1}{C_2}$$

$$\Upsilon_{43} = 0;\ \Upsilon_{44} = -\lambda$$

$$A - \lambda \cdot I = \begin{pmatrix} \Upsilon_{11} & \ldots & \Upsilon_{14} \\ \vdots & \ddots & \vdots \\ \Upsilon_{41} & \cdots & \Upsilon_{44} \end{pmatrix}$$

$$\det(A - \lambda \cdot I) = \det\begin{pmatrix} \Upsilon_{11} & \ldots & \Upsilon_{14} \\ \vdots & \ddots & \vdots \\ \Upsilon_{41} & \cdots & \Upsilon_{44} \end{pmatrix}$$

$$\det(A - \lambda \cdot I) = 0$$

$$\begin{aligned} \det(A - \lambda \cdot I) = \Upsilon_{11} \cdot \det\begin{pmatrix} \Upsilon_{22} & \Upsilon_{23} & \Upsilon_{24} \\ \Upsilon_{32} & \Upsilon_{33} & \Upsilon_{34} \\ \Upsilon_{42} & \Upsilon_{43} & \Upsilon_{44} \end{pmatrix} \\ - \Upsilon_{12} \cdot \det\begin{pmatrix} \Upsilon_{21} & \Upsilon_{23} & \Upsilon_{24} \\ \Upsilon_{31} & \Upsilon_{33} & \Upsilon_{34} \\ \Upsilon_{41} & \Upsilon_{43} & \Upsilon_{44} \end{pmatrix} \\ + \Upsilon_{13} \cdot \det\begin{pmatrix} \Upsilon_{21} & \Upsilon_{22} & \Upsilon_{24} \\ \Upsilon_{31} & \Upsilon_{32} & \Upsilon_{34} \\ \Upsilon_{41} & \Upsilon_{42} & \Upsilon_{44} \end{pmatrix} \\ - \Upsilon_{14} \cdot \det\begin{pmatrix} \Upsilon_{21} & \Upsilon_{22} & \Upsilon_{23} \\ \Upsilon_{31} & \Upsilon_{32} & \Upsilon_{33} \\ \Upsilon_{41} & \Upsilon_{42} & \Upsilon_{43} \end{pmatrix} \end{aligned}$$

$$\begin{aligned} \Upsilon_{12} = 0 \Rightarrow \det(A - \lambda \cdot I) = \Upsilon_{11} \cdot \det\begin{pmatrix} \Upsilon_{22} & \Upsilon_{23} & \Upsilon_{24} \\ \Upsilon_{32} & \Upsilon_{33} & \Upsilon_{34} \\ \Upsilon_{42} & \Upsilon_{43} & \Upsilon_{44} \end{pmatrix} \\ + \Upsilon_{13} \cdot \det\begin{pmatrix} \Upsilon_{21} & \Upsilon_{22} & \Upsilon_{24} \\ \Upsilon_{31} & \Upsilon_{32} & \Upsilon_{34} \\ \Upsilon_{41} & \Upsilon_{42} & \Upsilon_{44} \end{pmatrix} \\ - \Upsilon_{14} \cdot \det\begin{pmatrix} \Upsilon_{21} & \Upsilon_{22} & \Upsilon_{23} \\ \Upsilon_{31} & \Upsilon_{32} & \Upsilon_{33} \\ \Upsilon_{41} & \Upsilon_{42} & \Upsilon_{43} \end{pmatrix} \end{aligned}$$

$$\begin{aligned} \det\begin{pmatrix} \Upsilon_{22} & \Upsilon_{23} & \Upsilon_{24} \\ \Upsilon_{32} & \Upsilon_{33} & \Upsilon_{34} \\ \Upsilon_{42} & \Upsilon_{43} & \Upsilon_{44} \end{pmatrix} = \Upsilon_{22} \cdot \det\begin{pmatrix} \Upsilon_{33} & \Upsilon_{34} \\ \Upsilon_{43} & \Upsilon_{44} \end{pmatrix} \\ - \Upsilon_{23} \cdot \det\begin{pmatrix} \Upsilon_{32} & \Upsilon_{34} \\ \Upsilon_{42} & \Upsilon_{44} \end{pmatrix} \\ + \Upsilon_{24} \cdot \det\begin{pmatrix} \Upsilon_{32} & \Upsilon_{33} \\ \Upsilon_{42} & \Upsilon_{43} \end{pmatrix} \end{aligned}$$

$$\Upsilon_{23} = 0 \Rightarrow \det\begin{pmatrix} \Upsilon_{22} & \Upsilon_{23} & \Upsilon_{24} \\ \Upsilon_{32} & \Upsilon_{33} & \Upsilon_{34} \\ \Upsilon_{42} & \Upsilon_{43} & \Upsilon_{44} \end{pmatrix}$$

$$= \Upsilon_{22} \cdot \det\begin{pmatrix} \Upsilon_{33} & \Upsilon_{34} \\ \Upsilon_{43} & \Upsilon_{44} \end{pmatrix} + \Upsilon_{24} \cdot \det\begin{pmatrix} \Upsilon_{32} & \Upsilon_{33} \\ \Upsilon_{42} & \Upsilon_{43} \end{pmatrix}$$

$$\det\begin{pmatrix} \Upsilon_{22} & \Upsilon_{23} & \Upsilon_{24} \\ \Upsilon_{32} & \Upsilon_{33} & \Upsilon_{34} \\ \Upsilon_{42} & \Upsilon_{43} & \Upsilon_{44} \end{pmatrix} = -\lambda \cdot \det\begin{pmatrix} -\lambda & 0 \\ 0 & -\lambda \end{pmatrix}$$

$$+ \frac{1}{L_2} \cdot \det\begin{pmatrix} 0 & -\lambda \\ -\frac{1}{C_2} & 0 \end{pmatrix}$$

$$= -\lambda^3 - \frac{1}{C_2 \cdot L_2} \cdot \lambda$$

$$\det\begin{pmatrix} \Upsilon_{21} & \Upsilon_{22} & \Upsilon_{24} \\ \Upsilon_{31} & \Upsilon_{32} & \Upsilon_{34} \\ \Upsilon_{41} & \Upsilon_{42} & \Upsilon_{44} \end{pmatrix} = \Upsilon_{21} \cdot \det\begin{pmatrix} \Upsilon_{32} & \Upsilon_{34} \\ \Upsilon_{42} & \Upsilon_{44} \end{pmatrix}$$

$$- \Upsilon_{22} \cdot \det\begin{pmatrix} \Upsilon_{31} & \Upsilon_{34} \\ \Upsilon_{41} & \Upsilon_{44} \end{pmatrix}$$

$$+ \Upsilon_{24} \cdot \det\begin{pmatrix} \Upsilon_{31} & \Upsilon_{32} \\ \Upsilon_{41} & \Upsilon_{42} \end{pmatrix}$$

$$\Upsilon_{21} = 0 \Rightarrow \det\begin{pmatrix} \Upsilon_{21} & \Upsilon_{22} & \Upsilon_{24} \\ \Upsilon_{31} & \Upsilon_{32} & \Upsilon_{34} \\ \Upsilon_{41} & \Upsilon_{42} & \Upsilon_{44} \end{pmatrix}$$

$$= -\Upsilon_{22} \cdot \det\begin{pmatrix} \Upsilon_{31} & \Upsilon_{34} \\ \Upsilon_{41} & \Upsilon_{44} \end{pmatrix} + \Upsilon_{24} \cdot \det\begin{pmatrix} \Upsilon_{31} & \Upsilon_{32} \\ \Upsilon_{41} & \Upsilon_{42} \end{pmatrix}$$

$$\det\begin{pmatrix} \Upsilon_{21} & \Upsilon_{22} & \Upsilon_{24} \\ \Upsilon_{31} & \Upsilon_{32} & \Upsilon_{34} \\ \Upsilon_{41} & \Upsilon_{42} & \Upsilon_{44} \end{pmatrix} = \lambda \cdot \det\begin{pmatrix} -\frac{1}{C_1} \cdot e^{-\lambda \cdot \tau} & 0 \\ \frac{1}{C_2} \cdot e^{-\lambda \cdot \tau} & -\lambda \end{pmatrix}$$

$$+ \frac{1}{L_2} \cdot \det\begin{pmatrix} -\frac{1}{C_1} \cdot e^{-\lambda \cdot \tau} & 0 \\ \frac{1}{C_2} \cdot e^{-\lambda \cdot \tau} & -\frac{1}{C_2} \end{pmatrix}$$

$$\det\begin{pmatrix} \Upsilon_{21} & \Upsilon_{22} & \Upsilon_{24} \\ \Upsilon_{31} & \Upsilon_{32} & \Upsilon_{34} \\ \Upsilon_{41} & \Upsilon_{42} & \Upsilon_{44} \end{pmatrix} = \lambda^2 \cdot \frac{1}{C_1} \cdot e^{-\lambda \cdot \tau} + \frac{1}{C_1 \cdot C_2 \cdot L_2}$$

$$\cdot e^{-\lambda \cdot \tau} = (\lambda^2 \cdot \frac{1}{C_1} + \frac{1}{C_1 \cdot C_2 \cdot L_2}) \cdot e^{-\lambda \cdot \tau}$$

$$\det\begin{pmatrix} \Upsilon_{21} & \Upsilon_{22} & \Upsilon_{23} \\ \Upsilon_{31} & \Upsilon_{32} & \Upsilon_{33} \\ \Upsilon_{41} & \Upsilon_{42} & \Upsilon_{43} \end{pmatrix} = \Upsilon_{21} \cdot \det\begin{pmatrix} \Upsilon_{32} & \Upsilon_{33} \\ \Upsilon_{42} & \Upsilon_{43} \end{pmatrix} - \Upsilon_{22} \cdot \det\begin{pmatrix} \Upsilon_{31} & \Upsilon_{33} \\ \Upsilon_{41} & \Upsilon_{43} \end{pmatrix} + \Upsilon_{23} \cdot \det\begin{pmatrix} \Upsilon_{31} & \Upsilon_{32} \\ \Upsilon_{41} & \Upsilon_{42} \end{pmatrix}$$

$$\Upsilon_{21} = 0; \Upsilon_{23} = 0 \Rightarrow \det\begin{pmatrix} \Upsilon_{21} & \Upsilon_{22} & \Upsilon_{23} \\ \Upsilon_{31} & \Upsilon_{32} & \Upsilon_{33} \\ \Upsilon_{41} & \Upsilon_{42} & \Upsilon_{43} \end{pmatrix} = -\Upsilon_{22} \cdot \det\begin{pmatrix} \Upsilon_{31} & \Upsilon_{33} \\ \Upsilon_{41} & \Upsilon_{43} \end{pmatrix} = \lambda \cdot \det\begin{pmatrix} -\frac{1}{C_1} \cdot e^{-\lambda \cdot \tau} & -\lambda \\ \frac{1}{C_2} \cdot e^{-\lambda \cdot \tau} & 0 \end{pmatrix}$$

$$\det\begin{pmatrix} \Upsilon_{21} & \Upsilon_{22} & \Upsilon_{23} \\ \Upsilon_{31} & \Upsilon_{32} & \Upsilon_{33} \\ \Upsilon_{41} & \Upsilon_{42} & \Upsilon_{43} \end{pmatrix} = \frac{1}{C_2} \cdot \lambda^2 \cdot e^{-\lambda \cdot \tau}$$

$$\det(A - \lambda \cdot I) = \Upsilon_{11} \cdot \det\begin{pmatrix} \Upsilon_{22} & \Upsilon_{23} & \Upsilon_{24} \\ \Upsilon_{32} & \Upsilon_{33} & \Upsilon_{34} \\ \Upsilon_{42} & \Upsilon_{43} & \Upsilon_{44} \end{pmatrix} + \Upsilon_{13} \cdot \det\begin{pmatrix} \Upsilon_{21} & \Upsilon_{22} & \Upsilon_{24} \\ \Upsilon_{31} & \Upsilon_{32} & \Upsilon_{34} \\ \Upsilon_{41} & \Upsilon_{42} & \Upsilon_{44} \end{pmatrix} - \Upsilon_{14} \cdot \det\begin{pmatrix} \Upsilon_{21} & \Upsilon_{22} & \Upsilon_{23} \\ \Upsilon_{31} & \Upsilon_{32} & \Upsilon_{33} \\ \Upsilon_{41} & \Upsilon_{42} & \Upsilon_{43} \end{pmatrix}$$

$$\det(A - \lambda \cdot I) = -\lambda \cdot \left(-\lambda^3 - \frac{1}{C_2 \cdot L_2} \cdot \lambda\right) + \frac{1}{L_1} \cdot \left(\lambda^2 \cdot \frac{1}{C_1} + \frac{1}{C_1 \cdot C_2 \cdot L_2}\right) \cdot e^{-\lambda \cdot \tau} + \frac{1}{L_1} \cdot \frac{1}{C_2} \cdot \lambda^2 \cdot e^{-\lambda \cdot \tau}$$

$$\det(A - \lambda \cdot I) = \lambda^4 + \frac{1}{C_2 \cdot L_2} \cdot \lambda^2 + \left(\lambda^2 \cdot \frac{1}{C_1 \cdot L_1} + \frac{1}{C_1 \cdot C_2 \cdot L_2 \cdot L_1}\right)$$

$$\cdot e^{-\lambda \cdot \tau} + \frac{1}{L_1} \cdot \frac{1}{C_2} \cdot \lambda^2 \cdot e^{-\lambda \cdot \tau}$$

$$\begin{aligned} \det(A - \lambda \cdot I) &= \lambda^4 + \frac{1}{C_2 \cdot L_2} \cdot \lambda^2 \\ &+ \left[\frac{1}{L_1} \cdot \left(\frac{1}{C_1} + \frac{1}{C_2} \right) \cdot \lambda^2 \right. \\ &\left. + \frac{1}{C_1 \cdot C_2 \cdot L_2 \cdot L_1} \right] \cdot e^{-\lambda \cdot \tau} \end{aligned}$$

$$D(\lambda, \tau) = P_n(\lambda, \tau) + Q_m(\lambda, \tau) \cdot e^{-\lambda \cdot \tau}$$
$$n, m \in \mathbb{N}_0;\ n = 4;\ m = 2;\ n > m$$

$$P_n(\lambda, \tau) = \lambda^4 + \frac{1}{C_2 \cdot L_2} \cdot \lambda^2; n = 4$$
$$Q_m(\lambda, \tau) = \frac{1}{L_1} \cdot \left(\frac{1}{C_1} + \frac{1}{C_2} \right) \cdot \lambda^2 + \frac{1}{C_1 \cdot C_2 \cdot L_2 \cdot L_1}; m = 2$$

We can summary our copper vapor laser (CVL) circuity determinants small increments Jacobian results in the next table (Table 6.9)

$$\begin{aligned} P_n(\lambda, \tau) &= \sum_{k=0}^{n=4} p_k(\tau) \cdot \lambda^k \\ &= p_0(\tau) + p_1(\tau) \cdot \lambda + p_2(\tau) \cdot \lambda^2 \\ &\quad + p_3(\tau) \cdot \lambda^3 + p_4(\tau) \cdot \lambda^4 \end{aligned}$$

Table 6.9 Copper vapor laser (CVL) circuity determinants (small increments Jacobian results)

Copper vapor laser (CVL) circuity determinants	Expression
$\det\begin{pmatrix} \Upsilon_{22} & \Upsilon_{23} & \Upsilon_{24} \\ \Upsilon_{32} & \Upsilon_{33} & \Upsilon_{34} \\ \Upsilon_{42} & \Upsilon_{43} & \Upsilon_{44} \end{pmatrix}$	$-\lambda^3 - \frac{1}{C_2 \cdot L_2} \cdot \lambda$
$\det\begin{pmatrix} \Upsilon_{21} & \Upsilon_{22} & \Upsilon_{24} \\ \Upsilon_{31} & \Upsilon_{32} & \Upsilon_{34} \\ \Upsilon_{41} & \Upsilon_{42} & \Upsilon_{44} \end{pmatrix}$	$(\lambda^2 \cdot \frac{1}{C_1} + \frac{1}{C_1 \cdot C_2 \cdot L_2}) \cdot e^{-\lambda \cdot \tau}$
$\det\begin{pmatrix} \Upsilon_{21} & \Upsilon_{22} & \Upsilon_{23} \\ \Upsilon_{31} & \Upsilon_{32} & \Upsilon_{33} \\ \Upsilon_{41} & \Upsilon_{42} & \Upsilon_{43} \end{pmatrix}$	$\frac{1}{C_2} \cdot \lambda^2 \cdot e^{-\lambda \cdot \tau}$

$$p_0 = p_0(\tau);\ p_1 = p_1(\tau);\ p_2 = p_2(\tau);\ p_3 = p_3(\tau)$$

$$p_4 = p_4(\tau);\ p_0 = 0;\ p_1 = 0;\ p_2 = \frac{1}{C_2 \cdot L_2};\ p_3 = 0;\ p_4 = 1$$

$$Q_m(\lambda, \tau) = \sum_{k=0}^{m=2} q_k(\tau) \cdot \lambda^k$$
$$= q_0(\tau) + q_1(\tau) \cdot \lambda + q_2(\tau) \cdot \lambda^2$$
$$q_0 = q_0(\tau);\ q_1 = q_1(\tau);\ q_2 = q_2(\tau)$$

$$q_0 = \frac{1}{C_1 \cdot C_2 \cdot L_2 \cdot L_1};\ q_1 = 0$$
$$q_2 = \frac{1}{L_1} \cdot \left(\frac{1}{C_1} + \frac{1}{C_2}\right)$$

The homogeneous system for I_1, I_2, V_1, V_2 leads to a characteristic equation for the eigenvalue λ having the form $D(\lambda, \tau) = P(\lambda, \tau) + Q(\lambda, \tau) \cdot e^{-\lambda \cdot \tau} = 0$; and $P(\lambda) = \sum_{j=0}^{4} a_j \cdot \lambda^j$; $Q(\lambda) = \sum_{j=0}^{2} c_j \cdot \lambda^j$. The coefficients $\{a_j(q_i, q_k, \tau), c_j(q_i, q_k, \tau)\} \in \mathbb{R}$ depend on q_i, q_k and delayτ. q_i, q_k are any Copper vapor laser (CVL) circuity's parameters, other parameters kept as a constant [4, 5].

$$P(\lambda) = \sum_{j=0}^{4} a_j \cdot \lambda^j$$
$$= a_0 + a_1 \cdot \lambda + a_2 \cdot \lambda^2 + a_3 \cdot \lambda^3 + a_4 \cdot \lambda^4$$
$$Q(\lambda) = \sum_{j=0}^{2} c_j \cdot \lambda^j = c_0 + c_1 \cdot \lambda + c_2 \cdot \lambda^2$$

$$a_0 = 0;\ a_1 = 0;\ a_2 = \frac{1}{C_2 \cdot L_2};\ a_3 = 0$$
$$a_4 = 1;\ c_0 = \frac{1}{C_1 \cdot C_2 \cdot L_2 \cdot L_1};\ c_1 = 0;\ c_2 = \frac{1}{L_1} \cdot \left(\frac{1}{C_1} + \frac{1}{C_2}\right)$$

Unless strictly necessary, the designation of the varied arguments (q_i, q_k) will subsequently be omitted from P, Q, a_j and c_j. The coefficients a_j, c_j are continuous and differentiable functions of their arguments, and direct substitution shows that $a_0 + c_0 \neq 0$ for $q_i, q_k \in \mathbb{R}_+$; that is, $\lambda = 0$ is not of $P(\lambda) + Q(\lambda) \cdot e^{-\lambda \cdot \tau} = 0$. Furthermore, $P(\lambda)$, $Q(\lambda)$ are analytic functions of λ, for which the following requirements of the analysis (Kuang J and Cong Y 2005; Kuang Y 1993) can also be verified in the present case:

(a) If $\lambda = i \cdot \omega;\ \omega \in \mathbb{R}$, then $P(i \cdot \omega) + Q(i \cdot \omega) \neq 0$.
(b) If $|\frac{Q(\lambda)}{P(\lambda)}|$ is bounded for $|\lambda| \to \infty$; $\mathrm{Re}\lambda \geq 0$. No roots bifurcation from ∞.

(c) $F(\omega) = |P(i \cdot \omega)|^2 - |Q(i \cdot \omega)|^2$ has a finite number of zeros. Indeed, this is a polynomial in ω.

(d) Each positive root $\omega(q_i, q_k)$ of $F(\omega) = 0$ is continuous and differentiable with respect to q_i, q_k.

We assume that $P_n(\lambda, \tau)$ and $Q_m(\lambda, \tau)$ cannot have common imaginary roots. That is for any real numberω: $P_n(\lambda = i \cdot \omega, \tau) + Q_m(\lambda = i \cdot \omega, \tau) \neq 0$ and

$$P_n(\lambda = i \cdot \omega, \tau) = p_0(\tau) - p_2(\tau) \cdot \omega^2 + p_4(\tau) \cdot \omega^4 + [p_1(\tau) \cdot \omega - p_3(\tau) \cdot \omega^3] \cdot i$$

$$P_n(\lambda = i \cdot \omega, \tau) = -\frac{1}{C_2 \cdot L_2} \cdot \omega^2 + \omega^4$$

$$Q_m(\lambda = i \cdot \omega, \tau) = q_0(\tau) + q_1(\tau) \cdot i \cdot \omega - q_2(\tau) \cdot \omega^2$$

$$Q_m(\lambda = i \cdot \omega, \tau) = \frac{1}{C_1 \cdot C_2 \cdot L_2 \cdot L_1} - \frac{1}{L_1} \cdot \left(\frac{1}{C_1} + \frac{1}{C_2}\right) \cdot \omega^2$$

$$P_n(\lambda = i \cdot \omega, \tau) + Q_m(\lambda = i \cdot \omega, \tau) = -\frac{1}{C_2 \cdot L_2} \cdot \omega^2 + \omega^4 + \frac{1}{C_1 \cdot C_2 \cdot L_2 \cdot L_1} - \frac{1}{L_1} \cdot (\frac{1}{C_1} + \frac{1}{C_2}) \cdot \omega^2 \neq 0$$

$$P_n(\lambda = i \cdot \omega, \tau) + Q_m(\lambda = i \cdot \omega, \tau) = \frac{1}{C_1 \cdot C_2 \cdot L_2 \cdot L_1} - \left[\frac{1}{C_2 \cdot L_2} + \frac{1}{L_1} \cdot \left(\frac{1}{C_1} + \frac{1}{C_2}\right)\right] \cdot \omega^2 + \omega^4 \neq 0$$

$$|P(i \cdot \omega, \tau)|^2 = [-\frac{1}{C_2 \cdot L_2} \cdot \omega^2 + \omega^4]^2 = \frac{1}{(C_2 \cdot L_2)^2} \cdot \omega^4 - 2 \cdot \frac{1}{C_2 \cdot L_2} \cdot \omega^6 + \omega^8$$

$$|Q(i \cdot \omega, \tau)|^2 = [\frac{1}{C_1 \cdot C_2 \cdot L_2 \cdot L_1} - \frac{1}{L_1} \cdot (\frac{1}{C_1} + \frac{1}{C_2}) \cdot \omega^2]^2 = \frac{1}{(C_1 \cdot C_2 \cdot L_2 \cdot L_1)^2} - 2 \cdot \frac{1}{(C_1 \cdot C_2 \cdot L_2 \cdot L_1)} \cdot \frac{1}{L_1} \cdot (\frac{1}{C_1} + \frac{1}{C_2}) \cdot \omega^2 + \frac{1}{L_1^2} \cdot (\frac{1}{C_1} + \frac{1}{C_2})^2 \cdot \omega^4$$

$$F(\omega, \tau) = |P(i \cdot \omega, \tau)|^2 - |Q(i \cdot \omega, \tau)|^2$$

$$= \frac{1}{(C_2 \cdot L_2)^2} \cdot \omega^4 - 2 \cdot \frac{1}{C_2 \cdot L_2} \cdot \omega^6 + \omega^8$$
$$- \frac{1}{(C_1 \cdot C_2 \cdot L_2 \cdot L_1)^2} + 2 \cdot \frac{1}{(C_1 \cdot C_2 \cdot L_2 \cdot L_1)} \cdot \frac{1}{L_1}$$
$$\cdot \left(\frac{1}{C_1} + \frac{1}{C_2} \right) \cdot \omega^2 - \frac{1}{L_1^2} \cdot \left(\frac{1}{C_1} + \frac{1}{C_2} \right)^2 \cdot \omega^4$$

$$F(\omega, \tau) = |P(i \cdot \omega, \tau)|^2 - |Q(i \cdot \omega, \tau)|^2$$
$$= - \frac{1}{(C_1 \cdot C_2 \cdot L_2 \cdot L_1)^2} + 2 \cdot \frac{1}{(C_1 \cdot C_2 \cdot L_2 \cdot L_1)}$$
$$\cdot \frac{1}{L_1} \cdot (\frac{1}{C_1} + \frac{1}{C_2}) \cdot \omega^2 + \left[\frac{1}{(C_2 \cdot L_2)^2} - \frac{1}{L_1^2} \cdot \left(\frac{1}{C_1} + \frac{1}{C_2} \right)^2 \right]$$
$$\cdot \omega^4 - 2 \cdot \frac{1}{C_2 \cdot L_2} \cdot \omega^6 + \omega^8$$

We define the following parameters for simplicity: $\Pi_0, \Pi_2, \Pi_4, \Pi_6.\Pi_8$

$$\Pi_0 = - \frac{1}{(C_1 \cdot C_2 \cdot L_2 \cdot L_1)^2}$$
$$\Pi_2 = 2 \cdot \frac{1}{(C_1 \cdot C_2 \cdot L_2 \cdot L_1)} \cdot \frac{1}{L_1} \cdot \left(\frac{1}{C_1} + \frac{1}{C_2} \right)$$
$$\Pi_4 = \frac{1}{(C_2 \cdot L_2)^2} - \frac{1}{L_1^2} \cdot \left(\frac{1}{C_1} + \frac{1}{C_2} \right)^2$$
$$\Pi_6 = -2 \cdot \frac{1}{C_2 \cdot L_2}; \; \Pi_8 = 1$$

Hence $F(\omega, \tau) = 0$ implies $\sum_{k=0}^{4} \Pi_{2 \cdot k} \cdot \omega^{2 \cdot k} = 0$ and its roots are given by solving the above polynomial.

$$P_R(i\omega, \tau) = - \frac{1}{C_2 \cdot L_2} \cdot \omega^2 + \omega^4$$
$$P_I(i\omega, \tau) = 0; \; Q_R(i\omega, \tau) = \frac{1}{C_1 \cdot C_2 \cdot L_2 \cdot L_1} - \frac{1}{L_1} \cdot (\frac{1}{C_1} + \frac{1}{C_2}) \cdot \omega^2$$

$$Q_I(i\omega, \tau) = 0$$
$$\sin \theta(\tau) = \frac{-P_R(i\omega, \tau) \cdot Q_I(i\omega, \tau) + P_I(i\omega, \tau) \cdot Q_R(i\omega, \tau)}{|Q(i\omega, \tau)|^2}$$

$$\cos \theta(\tau) = - \frac{P_R(i\omega, \tau) \cdot Q_R(i\omega, \tau) + P_I(i\omega, \tau) \cdot Q_I(i\omega, \tau)}{|Q(i\omega, \tau)|^2}$$

We use different terminology from our last characteristics parameters definition:

$$k \to j;\ p_k(\tau) \to a_j;\ q_k(\tau) \to c_j$$
$$n, m \in \mathbb{N}_0;\ n = 4;\ m = 2;\ n > m$$

$$P_n(\lambda, \tau) \to P(\lambda);\ Q_m(\lambda, \tau) \to Q(\lambda)$$
$$P(\lambda) = \sum_{j=0}^{4} a_j \cdot \lambda^j;\ Q(\lambda) = \sum_{j=0}^{2} c_j \cdot \lambda^j$$

$$P(\lambda) = a_0 + a_1 \cdot \lambda + a_2 \cdot \lambda^2 + a_3 \cdot \lambda^3 + a_4 \cdot \lambda^4$$
$$Q(\lambda) = c_0 + c_1 \cdot \lambda + c_2 \cdot \lambda^2$$

$n, m \in \mathbb{N}_0; n > m$ and $a_j, c_j : \mathbb{R}_{0+} \to \mathbb{R}$ are continuous and differentiable function of τ such that $a_0 + c_0 \neq 0$. In the following "–" denoted complex and conjugate. $P(\lambda)$, $Q(\lambda)$ are analytic functions in λ and differentiable in τ. The coefficients$a_j(L_1, L_2, C_1, C_2, V_0, \tau, \ldots) \in \mathbb{R}$ and $c_j(L_1, L_2, C_1, C_2, V_0, \tau, \ldots) \in \mathbb{R}$ depend on Copper vapor laser (CVL) circuity's parameters, $L_1, L_2, C_1, C_2, V_0, \tau, \ldots$ values. Unless strictly necessary, the designation of the varied arguments: $L_1, L_2, C_1, C_2, V_0, \tau, \ldots$ will subsequently be omitted from P, Q, a_j, c_j. The coefficients a_j, c_j are continuous and differentiable functions of their arguments, and direct substitution shows that $a_0 + c_0 \neq 0$[4, 5].

$$a_0 = 0;\ c_0 = \frac{1}{C_1 \cdot C_2 \cdot L_2 \cdot L_1}$$
$$a_0 + c_0 = \frac{1}{C_1 \cdot C_2 \cdot L_2 \cdot L_1} \neq 0$$

$\forall L_1, L_2, C_1, C_2, V_0, \tau, \ldots \in \mathbb{R}_+$ I.e. $\lambda = 0$ is not a root of the characteristic equation. Furthermore $P(\lambda)$, $Q(\lambda)$ are analytic functions of λ for which the following requirements of the analysis (see Kuang, 1993, Sect. 3.4) can also be verified in the present case.

(a) If $\lambda = i \cdot \omega;\ \omega \in \mathbb{R}$ then $P(i \cdot \omega) + Q(i \cdot \omega) \neq 0$, i.e. P and Q have no common imaginary roots. This condition was verified numerically in the entire $L_1, L_2, C_1, C_2, V_0, \tau, \ldots$ domain of interest.
(b) $|\frac{P(\lambda)}{Q(\lambda)}|$ is bounded for $|\lambda| \to \infty$; $\mathrm{Re}\lambda \geq 0$. No roots bifurcation from ∞. Indeed, in the limit: $|\frac{Q(\lambda)}{P(\lambda)}| = |\frac{c_0 + c_1 \cdot \lambda + c_2 \cdot \lambda^2}{a_0 + a_1 \cdot \lambda + a_2 \cdot \lambda^2 + a_3 \cdot \lambda^3 + a_4 \cdot \lambda^4}|$.
(c) The following expressions exist: $F(\omega) = |P(i \cdot \omega)|^2 - |Q(i \cdot \omega)|^2$ $F(\omega, \tau) = |P(i \cdot \omega, \tau)|^2 - |Q(i \cdot \omega, \tau)|^2 = \sum_{k=0}^{4} \Pi_{2 \cdot k} \cdot \omega^{2 \cdot k}$ has at most a finite number of zeros. Indeed, this is a polynomial in ω (degree in ω^8).
(d) Each positive root $\omega(L_1, L_2, C_1, C_2, V_0, \tau, \ldots)$ of $F(\omega) = 0$ is continuous and differentiable with respect to $L_1, L_2, C_1, C_2, V_0, \tau, \ldots$ and the condition can only be assessed numerically.

In addition, since the coefficients in P and Q are real, we have $\overline{P(-i \cdot \omega)} = P(i \cdot \omega)$ and $\overline{Q(-i \cdot \omega)} = Q(i \cdot \omega)$ thus, $\omega > 0$ may be on eigenvalue of characteristic equations. The analysis consists in identifying the roots of the characteristic equation situated on the imaginary axis of the complex λ plane, whereby increasing the parameters: $L_1, L_2, C_1, C_2, V_0, \tau, \ldots$, Re$\lambda$ may, at the crossing, change its sign from (−) to (+). i.e. from a stable focus $E^*(I_1^*, I_2^*, V_1^*, V_2^*) = (0, 0, V_d - V_0 \cdot f(\tau_s), V_d)$ to an unstable one, or vice versa. This feature may be further assessed by examining the sign of the partial derivatives with respect to $L_1, L_2, C_1, C_2, V_0, \tau, \ldots$ and any system parameters.

$$\Lambda^{-1}(L_1) = (\frac{\partial \text{Re}\lambda}{\partial L_1})_{\lambda = i \cdot \omega};\ L_2, C_1, C_2, V_0, V_d, \tau, \ldots = const$$

$$\Lambda^{-1}(L_2) = (\frac{\partial \text{Re}\lambda}{\partial L_2})_{\lambda = i \cdot \omega};\ L_1, C_1, C_2, V_0, V_d, \tau, \ldots = const$$

$$\Lambda^{-1}(C_1) = (\frac{\partial \text{Re}\lambda}{\partial C_1})_{\lambda = i \cdot \omega};\ L_1, L_2, C_2, V_0, V_d, \tau, \ldots = const$$

$$\Lambda^{-1}(C_2) = (\frac{\partial \text{Re}\lambda}{\partial C_2})_{\lambda = i \cdot \omega};\ L_1, L_2, C_1, V_0, V_d, \tau, \ldots = const$$

$$\Lambda^{-1}(V_0) = (\frac{\partial \text{Re}\lambda}{\partial V_0})_{\lambda = i \cdot \omega};\ L_1, L_2, C_1, C_2, V_d, \tau, \ldots = const$$

$$\Lambda^{-1}(\tau) = (\frac{\partial \text{Re}\lambda}{\partial \tau})_{\lambda = i \cdot \omega};\ L_1, L_2, C_1, C_2, V_0, V_d, \ldots = const$$

$P(\lambda) = P_R(\lambda) + i \cdot P_I(\lambda)$; $Q(\lambda) = Q_R(\lambda) + i \cdot Q_I(\lambda)$, When writing and inserting $\lambda = i \cdot \omega$ into Copper vapor laser (CVL) circuity's characteristic equation ω must satisfy the following equations:

$$\sin(\omega \cdot \tau) = g(\omega) = \frac{-P_R(i\omega, \tau) \cdot Q_I(i\omega, \tau) + P_I(i\omega, \tau) \cdot Q_R(i\omega, \tau)}{|Q(i\omega, \tau)|^2}$$

$$\cos(\omega \cdot \tau) = h(\omega) = -\frac{P_R(i\omega, \tau) \cdot Q_R(i\omega, \tau) + P_I(i\omega, \tau) \cdot Q_I(i\omega, \tau)}{|Q(i\omega, \tau)|^2}$$

where $|Q(i\omega, \tau)|^2 \neq 0$ in view of requirement (a) above, and $(g, h) \in \mathbb{R}$. Furthermore, it follows above $\sin(\omega \cdot \tau)$ and $\cos(\omega \cdot \tau)$ equation that, by squaring and adding the sides, ω must be a positive root of $F(\omega) = |P(i \cdot \omega)|^2 - |Q(i \cdot \omega)|^2 = 0$. Note: $F(\omega)$ is independent on τ. Now it is important to notice that if $\tau \notin I$(assume that $I \subseteq \mathbb{R}_{+0}$ is the set where $\omega(\tau)$ is a positive root of $F(\omega)$ and for, $\tau \notin I$, $\omega(\tau)$ is not defined. Then for all τ in I, $\omega(\tau)$ is satisfied that $F(\omega, \tau) = 0$. Then there are no positive $\omega(\tau)$ solutions for $F(\omega,\tau)=0$, and we cannot have stability switches. For $\tau \in I$ where $\omega(\tau)$ is a positive solution of $F(\omega,\tau)=0$, we can define the angle $\theta(\tau) \in [0, 2\cdot\Pi]$ as the solution of $\sin\theta(\tau) = \ldots$ and $\cos\theta(\tau) = \ldots$; the relation between the arguments $\theta(\tau)$ and $\tau\cdot\omega(\tau)$ for $\tau \in I$ must be describing below. $\tau\cdot\omega(\tau) = \theta(\tau) + 2\cdot n\cdot\Pi \; \forall n \in \mathbb{N}_0$ Hence we can define the maps:

$\tau_n : I \to \mathbb{R}_{+0}$, is given by $\tau_n(\tau) = \frac{\theta(\tau)+2\cdot n\cdot\Pi}{\omega(\tau)}$; $n \in \mathbb{N}_0$; $\tau \in I$. Let us introduce the function $I \to \mathbb{R}$; $S_n(\tau) = \tau - \tau_n(\tau)$; $\tau \in I$; $n \in \mathbb{N}_0$ that is continuous and differentiable in τ. In the following, the subscripts $\lambda, \omega, L_1, L_2, C_1, C_2, V_0, V_d, \ldots$ indicate the corresponding partial derivatives. Let us first concentrate on $\Lambda(x)$, remember in $\lambda(L_1, L_2, C_1, C_2, V_0, V_d, \ldots)$ and $\omega(L_1, L_2, C_1, C_2, V_0, V_d, \ldots)$, and keeping all parameters except one (x) and τ. The derivation closely follows that in reference [BK]. Differentiating Copper vapor laser (CVL) circuity's characteristic equation $P(\lambda) + Q(\lambda)\cdot e^{-\lambda\cdot\tau} = 0$ with respect to specific parameter (x), and inverting the derivative, for convenience, one calculates: $x = L_1, L_2, C_1, C_2, V_0, V_d, \tau, \ldots$

$$(\frac{\partial\lambda}{\partial x})^{-1} = \frac{-P_\lambda(\lambda,x)\cdot Q(\lambda,x) + Q_\lambda(\lambda,x)\cdot P(\lambda,x) - \tau\cdot P(\lambda,x)\cdot Q(\lambda,x)}{P_x(\lambda,x)\cdot Q(\lambda,x) - Q_x(\lambda,x)\cdot P(\lambda,x)}$$

where $P_\lambda = \frac{\partial P}{\partial\lambda}, \ldots$ etc., substituting $\lambda = i\cdot\omega$ and bearing $\overline{P(-i\cdot\omega)} = P(i\cdot\omega)$; $\overline{Q(-i\cdot\omega)} = Q(i\cdot\omega)$ then $i\cdot P_\lambda(i\cdot\omega) = P_\omega(i\cdot\omega)$; $i\cdot Q_\lambda(i\cdot\omega) = Q_\omega(i\cdot\omega)$ and that on the surface $|P(i\omega)|^2 = |Q(i\omega)|^2$, one obtain: $(\frac{\partial\lambda}{\partial x})^{-1}|_{\lambda=i\cdot\omega} = (\frac{i\cdot P_\omega(i\cdot\omega,x)\cdot\overline{P(i\cdot\omega,x)} + i\cdot Q_\lambda(i\cdot\omega,x)\cdot\overline{Q(\lambda,x)} - \tau\cdot|P(i\cdot\omega,x)|^2}{P_x(i\cdot\omega,x)\cdot\overline{P(i\cdot\omega,x)} - Q_x(i\cdot\omega,x)\cdot\overline{Q(i\cdot\omega,x)}})$ Upon separating into real and imaginary parts, with $P = P_R + i\cdot P_I$; $Q = Q_R + i\cdot Q_I$ and $P_\omega = P_{R\omega} + i\cdot P_{I\omega}$; $Q_\omega = Q_{R\omega} + i\cdot Q_{I\omega}$; $P_x = P_{Rx} + i\cdot P_{Ix}$; $Q_x = Q_{Rx} + i\cdot Q_{Ix}$; $P^2 = P_R^2 + P_I^2$, When (x) can be any Copper vapor laser (CVL) circuity's parameters $L_1, L_2, C_1, C_2, V_0, V_d, \ldots$ and time delay τ etc. Where for convenience, we have dropped the arguments $(i\cdot\omega, x)$, and where $F_\omega = 2\cdot[(P_{R\omega}\cdot P_R + P_{I\omega}\cdot P_I) - (Q_{R\omega}\cdot Q_R + Q_{I\omega}\cdot Q_I)]F_x = 2\cdot[(P_{Rx}\cdot P_R + P_{Ix}\cdot P_I) - (Q_{Rx}\cdot Q_R + Q_{Ix}\cdot Q_I)]$

And $\omega_x = -\frac{F_x}{F_\omega}$. We define U and V: $U = (P_R\cdot P_{I\omega} - P_I\cdot P_{R\omega}) - (Q_R\cdot Q_{I\omega} - Q_I\cdot Q_{R\omega})$, $V = (P_R\cdot P_{Ix} - P_I\cdot P_{Rx}) - (Q_R\cdot Q_{Ix} - Q_I\cdot Q_{Rx})$, we choose our specific parameter as time delay$x = \tau$; $P_R = -\frac{1}{C_2\cdot L_2}\cdot\omega^2 + \omega^4$; $P_I = 0$; $Q_R = \frac{1}{C_1\cdot C_2\cdot L_2\cdot L_1} - \frac{1}{L_1}\cdot(\frac{1}{C_1} + \frac{1}{C_2})\cdot\omega^2$; $Q_I = 0$:

$$Q_{R\omega} = -2\cdot\frac{1}{L_1}\cdot(\frac{1}{C_1} + \frac{1}{C_2})\cdot\omega;\; Q_{I\omega} = 0$$

$$P_{R\omega} = -2\cdot\frac{1}{C_2\cdot L_2}\cdot\omega + 4\cdot\omega^3;\; P_{I\omega} = 0$$

$$P_{R\tau} = 0;\; P_{I\tau} = 0;\; Q_{R\tau} = 0;\; Q_{I\tau} = 0;\; F_\tau = 0$$
$$V = (P_R\cdot P_{I\tau} - P_I\cdot P_{R\tau}) - (Q_R\cdot Q_{I\tau} - Q_I\cdot Q_{R\tau}) = 0$$

$F_\omega = \ldots$. Elements: $P_{R\omega} \cdot P_R = [-2 \cdot \frac{1}{C_2 \cdot L_2} \cdot \omega + 4 \cdot \omega^3] \cdot [-\frac{1}{C_2 \cdot L_2} \cdot \omega^2 + \omega^4]$

$$P_{R\omega} \cdot P_R = [-2 \cdot \frac{1}{C_2 \cdot L_2} \cdot \omega + 4 \cdot \omega^3] \cdot [-\frac{1}{C_2 \cdot L_2} \cdot \omega^2 + \omega^4]$$
$$= 2 \cdot \frac{1}{(C_2 \cdot L_2)^2} \cdot \omega^3 - 6 \cdot \frac{1}{C_2 \cdot L_2} \cdot \omega^5 + 4 \cdot \omega^7$$

We define the following parameters for simplicity: $A_3 = 2 \cdot \frac{1}{(C_2 \cdot L_2)^2}$; $A_5 = -6 \cdot \frac{1}{C_2 \cdot L_2}$; $A_7 = 4$

$$P_{R\omega} \cdot P_R = \sum_{k=1}^{3} A_{1+2\cdot k} \cdot \omega^{1+2\cdot k};\ P_{I\omega} \cdot P_I = 0;\ Q_{I\omega} \cdot Q_I = 0$$

$$Q_{R\omega} \cdot Q_R = \left[-2 \cdot \frac{1}{L_1} \cdot \left(\frac{1}{C_1} + \frac{1}{C_2}\right) \cdot \omega\right] \cdot \left[\frac{1}{C_1 \cdot C_2 \cdot L_2 \cdot L_1} - \frac{1}{L_1} \cdot \left(\frac{1}{C_1} + \frac{1}{C_2}\right) \cdot \omega^2\right]$$

$$Q_{R\omega} \cdot Q_R = -2 \cdot \frac{1}{L_1} \cdot \left(\frac{1}{C_1} + \frac{1}{C_2}\right) \cdot \frac{1}{C_1 \cdot C_2 \cdot L_2 \cdot L_1} \cdot \omega + 2 \cdot \frac{1}{L_1^2} \cdot (\frac{1}{C_1} + \frac{1}{C_2})^2 \cdot \omega^3$$

We define the following parameters for simplicity: $B_1 = -2 \cdot \frac{1}{L_1} \cdot (\frac{1}{C_1} + \frac{1}{C_2}) \cdot \frac{1}{C_1 \cdot C_2 \cdot L_2 \cdot L_1}$

$$B_3 = 2 \cdot \frac{1}{L_1^2} \cdot (\frac{1}{C_1} + \frac{1}{C_2})^2;\ Q_{R\omega} \cdot Q_R = \sum_{k=1}^{2} B_{2\cdot k-1} \cdot \omega^{2\cdot k-1}$$

$U = \ldots$. Elements: $P_R \cdot P_{I\omega} = 0$; $P_I \cdot P_{R\omega} = 0$; $Q_R \cdot Q_{I\omega} = 0$; $Q_I \cdot Q_{R\omega} = 0$
$F(\omega, \tau) = 0$, differentiating with respect to τ and we get $F_\omega \cdot \frac{\partial \omega}{\partial \tau} + F_\tau = 0$; $\tau \in I \Rightarrow \frac{\partial \omega}{\partial \tau} = -\frac{F_\tau}{F_\omega}$

$$\Lambda^{-1}(\tau) = (\frac{\partial \mathrm{Re}\lambda}{\partial \tau})_{\lambda = i\omega};\ \frac{\partial \omega}{\partial \tau} = \omega_\tau = -\frac{F_\tau}{F_\omega}$$
$$\Lambda^{-1}(\tau) = \mathrm{Re}\{\frac{-2 \cdot [U + \tau \cdot |P|^2 + i \cdot F_\omega}{F_\tau + i \cdot 2 \cdot [V + \omega \cdot |P|^2]}\}$$

$$\mathrm{sign}\{\Lambda^{-1}(\tau)\} = \mathrm{sign}\{(\frac{\partial \mathrm{Re}\lambda}{\partial \tau})_{\lambda = i\omega}\}$$

$$\mathrm{sign}\{\Lambda^{-1}(\tau)\} = \mathrm{sign}\{F_\omega\} \cdot \mathrm{sign}\{\frac{V + \frac{\partial \omega}{\partial \tau} \cdot U}{|P|^2} + \omega + \frac{\partial \omega}{\partial \tau} \cdot \tau\}$$

We shall presently examine the possibility of stability transitions (bifurcations) Copper vapor laser (CVL) circuity, about the equilibrium point $E^*(I_1^*, I_2^*, V_1^*, V_2^*) = (0, 0, V_d - V_0 \cdot f(\tau_s), V_d)$. The analysis consists in identifying the roots of our system characteristic equation situated on the imaginary axis of the complex λ-plane, where by increasing the delay parameter τ. Reλ, may at the crossing, changes its sign from – to +, i.e. from a stable focus $E^{(*)}$ to an unstable one, or vice versa. This feature may be further assessed by examining the sign of the partial derivatives with respect to τ[4, 5].

$$\Lambda^{-1}(\tau) = (\frac{\partial \mathrm{Re}\lambda}{\partial \tau})_{\lambda = i\cdot\omega};\ L_1, L_2, C_1, C_2, V_0, V_d, \ldots = const$$

$$U = (P_R \cdot P_{I\omega} - P_I \cdot P_{R\omega}) - (Q_R \cdot Q_{I\omega} - Q_I \cdot Q_{R\omega}) = 0$$
$$V = (P_R \cdot P_{I\tau} - P_I \cdot P_{R\tau}) - (Q_R \cdot Q_{I\tau} - Q_I \cdot Q_{R\tau}) = 0$$

$$F_\omega = 2 \cdot [(\sum_{k=1}^{3} A_{1+2\cdot k} \cdot \omega^{1+2\cdot k}) - (\sum_{k=1}^{2} B_{2\cdot k-1} \cdot \omega^{2\cdot k-1})]$$
$$\frac{\partial \omega}{\partial \tau} = \omega_\tau = -\frac{F_\tau}{F_\omega}|_{F_\tau = 0, F_\omega \neq 0} = -\frac{F_\tau}{F_\omega} = 0$$

$$F_\tau = 2 \cdot [(P_{R\tau} \cdot P_R + P_{I\tau} \cdot P_I) - (Q_{R\tau} \cdot Q_R + Q_{I\tau} \cdot Q_I)] = 0$$

$$\mathrm{sign}\{\Lambda^{-1}(\tau)\} = \mathrm{sign}\{F_\omega\} \cdot \mathrm{sign}\{\frac{V + \frac{\partial\omega}{\partial\tau} \cdot U}{|P|^2} + \omega + \frac{\partial \omega}{\partial \tau} \cdot \tau\}$$

$$\mathrm{sign}\{\Lambda^{-1}(\tau)\} = \mathrm{sign}\{2 \cdot [(\sum_{k=1}^{3} A_{1+2\cdot k} \cdot \omega^{1+2\cdot k}) - (\sum_{k=1}^{2} B_{2\cdot k-1} \cdot \omega^{2\cdot k-1})]\} \cdot \mathrm{sign}\{\omega\}$$

$$\mathrm{sign}\{\Lambda^{-1}(\tau)\} = \mathrm{sign}\{[(\sum_{k=1}^{3} A_{1+2\cdot k} \cdot \omega^{1+2\cdot k}) - (\sum_{k=1}^{2} B_{2\cdot k-1} \cdot \omega^{2\cdot k-1})]\} \cdot \mathrm{sign}\{\omega\}$$

We get the expression for $F_\omega = 2\cdot[(\sum_{k=1}^{3} A_{1+2\cdot k}\cdot\omega^{1+2\cdot k}) - (\sum_{k=1}^{2} B_{2\cdot k-1}\cdot\omega^{2\cdot k-1})]$ Copper vapor laser (CVL) circuity parameter values. We find those ω, τ values which fulfill $F_\omega(\omega, \tau) = 0$. We ignore negative, complex, and imaginary values of ω for specific τ values. $\tau \in [0.001 \ldots 10]$, we can express by 3D function $F_\omega(\omega, \tau) = 0$. We plot the stability switch diagram based on different delay values of our Copper vapor laser (CVL) circuity.

$$\Lambda^{-1}(\tau) = (\frac{\partial \mathrm{Re}\lambda}{\partial \tau})_{\lambda = i\omega}$$
$$= \mathrm{Re}\{\frac{-2 \cdot [U + \tau \cdot |P|^2] + i \cdot F_\omega}{F_\tau + 2 \cdot i \cdot [V + \omega \cdot |P|^2]}\}$$

$$\Lambda^{-1}(\tau) = (\frac{\partial \mathrm{Re}\lambda}{\partial \tau})_{\lambda = i\omega}$$
$$= \frac{2 \cdot \{F_\omega \cdot (V + \omega \cdot P^2) - F_\tau \cdot (U + \tau \cdot P^2)\}}{F_\tau^2 + 4 \cdot (V + \omega \cdot P^2)^2}$$

The stability switch occurs only on those delay values (τ) which fit the equation: $\tau = \frac{\theta_+(\tau)}{\omega_+(\tau)}$ and $\theta_+(\tau)$ is the solution of $\sin\theta(\tau) = \ldots.$ and $\cos\theta(\tau) = \ldots.$ when $\omega = \omega_+(\tau)$ If only ω_+ is feasible. Additionally, when all Copper vapor laser (CVL) circuity's parameters are known and the stability switch due to various time delay values τ is described in the following expression:

$$\mathrm{sign}\{\Lambda^{-1}(\tau)\} = \mathrm{sign}\{F_\omega(\omega(\tau), \tau)\}$$
$$\cdot \mathrm{sign}\{\tau \cdot \omega_\tau(\omega(\tau)) + \omega(\tau)$$
$$+ \frac{U(\omega(\tau)) \cdot \omega_\tau(\omega(\tau)) + V(\omega(\tau))}{|P(\omega(\tau))|^2}\}$$

Remark we know $F(\omega, \tau) = 0$ implies its roots $\omega_i(\tau)$ and finding those delays values τ which ω_i is feasible. There are τ values which give complex ω_i or imaginary number, then unable to analyze stability. $F(\omega, \tau)$, function is independent on τ the paramester $F(\omega, \tau) = 0$. The results: We find those ω, τ values which fulfill $F(\omega, \tau) = 0$. We ignore negative, complex, and imaginary values of ω. Next is to find those ω, τ values which fulfill $\sin\theta(\tau) = \ldots.$ and $\cos\theta(\tau) = \ldots.$; $\sin(\omega \cdot \tau) = \frac{-P_R \cdot Q_I + P_I \cdot Q_R}{|Q|^2}$; $\cos(\omega \cdot \tau) = -\frac{(P_R \cdot Q_R + P_I \cdot Q_I)}{|Q|^2} |Q|^2 = Q_R^2 + Q_I^2$. Finally we plot the stability switch diagram $g(\tau) = \Lambda^{-1}(\tau) = (\frac{\partial \mathrm{Re}\lambda}{\partial \tau})_{\lambda = i\omega}$

$$g(\tau) = \Lambda^{-1}(\tau)$$
$$= \left(\frac{\partial \mathrm{Re}\lambda}{\partial \tau}\right)_{\lambda = i\omega}$$
$$= \frac{2 \cdot \{F_\omega \cdot (V + \omega \cdot P^2) - F_\tau \cdot (U + \tau \cdot P^2)\}}{F_\tau^2 + 4 \cdot (V + \omega \cdot P^2)^2}$$

$$\mathrm{sign}[g(\tau)] = \mathrm{sign}[\Lambda^{-1}(\tau)]$$

$$= \text{sign}[(\frac{\partial \text{Re}\lambda}{\partial \tau})_{\lambda=i\omega}]$$
$$= \text{sign}[\frac{2 \cdot \{F_\omega \cdot (V + \omega \cdot P^2) - F_\tau \cdot (U + \tau \cdot P^2)\}}{F_\tau^2 + 4 \cdot (V + \omega \cdot P^2)^2}]$$

$$F_\tau^2 + 4 \cdot (V + \omega \cdot P^2)^2 > 0 \Rightarrow \text{sign}[\Lambda^{-1}(\tau)]$$
$$= \text{sign}[F_\omega \cdot (V + \omega \cdot P^2) - F_\tau \cdot (U + \tau \cdot P^2)]$$

$$\text{sign}[\Lambda^{-1}(\tau)] = \text{sign}\{[F_\omega] \cdot [(V + \omega \cdot P^2)$$
$$- \frac{F_\tau}{F_\omega} \cdot (U + \tau \cdot P^2)]\}$$
$$\omega_\tau = -\frac{F_\tau}{F_\omega}; \omega_\tau = (\frac{\partial \omega}{\partial \tau})^{-1} = -\frac{\partial F / \partial \omega}{\partial F / \partial \tau}$$

$$\text{sign}[\Lambda^{-1}(\tau)] = \text{sign}\{[F_\omega] \cdot [P^2] \cdot [\frac{V + \omega_\tau \cdot U}{P^2}$$
$$+ \omega + \omega_\tau \cdot \tau]\} \text{ ; sign}[P^2] > 0$$

$$\text{sign}[\Lambda^{-1}(\tau)] = \text{sign}\{[F_\omega] \cdot [\frac{V + \omega_\tau \cdot U}{P^2} + \omega + \omega_\tau \cdot \tau]\}$$

$$\text{sign}[\Lambda^{-1}(\tau)] = \text{sign}[F_\omega] \cdot \text{sign}[\frac{V + \omega_\tau \cdot U}{P^2} + \omega + \omega_\tau \cdot \tau]$$

$$F_\omega = 2 \cdot [(P_{R\omega} \cdot P_R + P_{I\omega} \cdot P_I) - (Q_{R\omega} \cdot Q_R + Q_{I\omega} \cdot Q_I)]$$
$$= 2 \cdot [(\sum_{k=1}^{3} A_{1+2\cdot k} \cdot \omega^{1+2\cdot k}) - (\sum_{k=1}^{2} B_{2\cdot k-1} \cdot \omega^{2\cdot k-1})]$$

We check the sign of $\Lambda^{-1}(\tau)$ according to the following rule: (Table 6.10)

If, $\text{sign}[\Lambda^{-1}(\tau)] > 0$, then the crossing proceeds from (-) to (+) respectively (stable to unstable). If, $\text{sign}[\Lambda^{-1}(\tau)] < 0$, then the crossing proceeds from (+) to (-) respectively (unstable to stable).

Table 6.10 Copper vapor laser (CVL) circuity sign of $\text{sign}[\Lambda^{-1}(\tau)]$

$\text{sign}[F_\omega]$	$\text{sign}[\frac{V+\omega_\tau \cdot U}{P^2} + \omega + \omega_\tau \cdot \tau]$	$\text{sign}[\Lambda^{-1}(\tau)]$
$\pm$	$\pm$	+
$\pm$	$\mp$	−

6.4 A Self-consistent Model Copper Vapor Laser (CVL) Electron Density Upper and Lower Laser Levels Stability Analysis Under Parameters Variation

A copper vapor laser makes use of copper vapor as a lasing medium. It emits yellow and green laser light at 578.2 nm and 510.6 nm of the electromagnetic spectrum, respectively. Laser is made using pure metal vapor produced from elemental copper is hard to make because of the high temperature required to form vapor. This is the reason why compounds like chloride, copper iodide, and copper bromide are used and they form vapors at low temperatures. The laser head of copper vapor laser consists of a refractory ceramic tube that contains a low-pressure buffer gas, neon, and copper pellets. A pulsed electrical discharge between electrodes at the tube ends is happened. The temperature of the laser head increased to about 1450 C degree, thereby producing copper vapor at low pressure. The excited electrons collide with vaporized copper atoms and excite these atoms to the upper ${}^2P_{3/2}$ and ${}^2P_{1/2}$ laser level. The electrons in the ${}^2P_{3/2}$ level decay to the lower ${}^2D_{5/2}$ laser level to produce green light, and the ${}^2P_{1/2}$ decay to ${}^2D_{3/2}$ level to produce yellow laser light. The features of the laser are good power stability, good beam quality, and a long operational lifetime. The copper laser has a wide range of applications due to his very good beam characteristics. Some of the applications are high speed photography, dermatological applications, and pump for dye lasers. The copper vapor laser (CVL) is very applicable to engineering lasers. Diverse applications exist because of an inherent beam characteristic and a natural choice for high precision micromachining of features down to 1 μm. Laser is an acronym for Light Amplification by the Stimulated Emission of Radiation. The emission of radiation is stimulated when atoms are excited to an energy state above their ground state. The copper vapor laser (CVL) uses vaporized copper atoms as the lasing medium. The self- consistent model for high repetition rate copper vapor laser (CVL) simulates a discharge pulse, laser pulse, an inter-pulse afterglow in a high repetition rate. We need to choose initial conditions and number of cycles for computing until consistent pulses are obtained [9]. The model includes various species as follow (Table 6.1).

The helium metastable state is the 2^3S level. Lamped radiative state (He^*, Cu^{**}) is induced for both helium and copper. These states act as a buffer for recombining ions. Result atom in an excited state is happened when an electron and ion recombine. This state is He^* or Cu^{**}. Subsequent processes determine what distribution of excited states results from electron-ion recombination. The subsequent processes which involving these states are spontaneous emission, electron excitation or relaxation, etc. If electron-ion recombination preferentially populates the metastable ${}^2D$ lower laser level, then pumping requirements to reach threshold are increased for the next laser pulse. System model thermodynamic quantities are T_g, P_{Cu}, and P_g. During the operation of the laser, the temperature of the discharge tube, which determines the copper vapor pressure, can be measured by an optical pyrometer. The optical pyrometer is a non-contact type temperature measuring device. It works on

Table 6.11 Copper vapor laser (CVL) model species

Copper vapor laser (CVL) model species	Specie meaning
n_e	Electrons
Cu	^2S ground state copper
Cu_m	^2D metastable copper (lower laser level)
Cu^*	^2P copper (upper laser level)
Cu^{**}	Lumped radiative state in copper
Cu^+	Atomic copper ion
H_e	Ground state helium
H_{em}	Metastable state helium
He^*	Lumped radiative state in helium
He^+	Atomic helium ion
He_2^+	Molecular helium ion
T_e	Electron temperature
T_g	Gas temperature
P_g	Buffer gas pressure
P_{Cu}	Copper vapor pressure
I_i, V_i	Discharge current and voltage for circuit elements i
R	Copper vapor laser (CVL) mirror reflectivity
Σ	Laser intensity

the principle of matching the brightness of an object to the brightness of the filament which is placed inside the pyrometer. Mainly the optical pyrometer is used for measuring the temperature of the furnaces, molten metals, and other overhead material or liquids. Furnace is a device used for high-temperature heating. The heat energy to fuel a furnace is supplied directly by fuel combustion, by electricity such as the electric arc furnace, or through induction heating in induction furnaces. The pyrometer is cylindrical inside which the lens is placed on one end and the eyepiece on the other end. An eyepiece, or ocular lens, is a type of lens that is attached to a variety of optical devices such as telescopes and microscopes. The objective lens or mirror collects light and brings it to focus creating an image. The eyepiece is placed near the focal point of the objective to magnify this image. The amount of magnification depends on the focal length of the eyepiece. The lamp is kept between the eyepiece and the lens. The filter is placed in front of the eyepiece. By using filter we get monochromatic light. The lamp has the filament which is connected to the battery, ammeter and rheostat. A rheostat is a variable resistor which is used to control current. It is able to vary the resistance in a circuit without interruption. The rheostats have to carry a significant current therefore they mostly constructed as a wire wound resistors. Resistive wire is wound around an insulating ceramic core and the wiper slides over the windings. There are some types of rheostats, rotary type, multi-gang types, slide rheostats, and linear rheostats [9]. The optical pyrometer

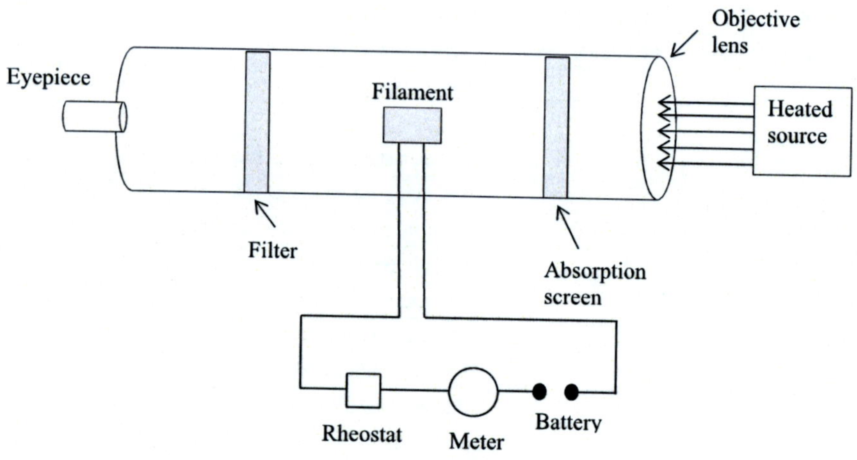

Fig. 6.8 Optical pyrometer structure

(Fig. 6.8) consists the lens which are focused the radiated energy from the heated object and target it on the electric filament lamp. The intensity of the filament depends on the current passes through it. The adjustable current is passed through the lamp. The magnitude of the current is adjusted until the brightness of the filament is similar to the brightness of the object. When the brightness of the filament and the brightness of the object are same, then the outline of the filament is completely disappeared. The optical pyrometer has advantages of high accuracy, temperature measurement without contacting the heated body. Typically pyrometer is used for measuring the temperature above 700 °C. The pyrometer accuracy depends on the adjustment of the filament current (Table 6.11).

In our system the buffer gas pressure is measured external to the laser tube. The temperature of the discharge tube remains constant, as does the external buffer gas pressure. The gas temperature within the tube is changed by many hundreds of degrees during the discharge pulse. An increase is gas temperature increases the local gas pressures P_g and P_{Cu} at the same rate, but the relaxation of P_g and P_{Cu} proceeds at different rates. The equilibrium vapor pressure for P_{Cu} is the pressure determined by the wall temperature. The changes in the gas temperature, the time constant for relaxation of P_{Cu} to its equilibrium value is $\frac{\Lambda}{v_d}$, where the radial diffusion length of the discharge tube is Λ and v_d is the speed at which copper atoms diffuse to the wall. The equilibrium pressure for the buffer gas is the pressure measured external to the tube. Exclusive of changes in the gas temperature, the time constant for relaxation of P_g to its equilibrium value is $\frac{l}{2 \cdot v_s}$; where the length of the discharge tube is L and v_s is the sound speed. The gas temperature relaxes to the wall temperature with a time constant proportional to the radius of the discharge tube. We can express the time rate of change of P_g and P_{Cu}:

$$\frac{dP_g}{dt} = \frac{dT_g}{dt} \cdot \frac{P_g(t-\tau)}{T_g} - [P_g(t-\tau) - P_{ex}] \cdot \frac{2 \cdot v_2}{l}$$
$$\frac{dP_{cu}}{dt} = \frac{dT_g}{dt} \cdot \frac{P_{cu}}{T_g} - [P_{cu} - P_{vp}] \cdot \frac{v_d}{R}$$

$$P_{cu} = P_{cu}(t);\ P_g = P_g(t);\ T_g = T_g(t)$$

Temperature increasing elements: $\frac{dT_g}{dt} \cdot \frac{P_g(t-\tau)}{T_g}, \frac{dT_g}{dt} \cdot \frac{P_{cu}}{T_g}$.

Relaxation elements: $[P_g(t-\tau) - P_{ex}] \cdot \frac{2 \cdot v_2}{l}, [P_{cu} - P_{vp}] \cdot \frac{v_d}{R}$.

P_g- Buffer gas pressure, T_g- Gas temperature, P_{cu}- Copper vapor pressure, P_{ex}- External buffer gas pressure, P_{vp}- Copper vapor pressure based on the wall temperature, R- Mirror reflectivity [9]. Time delay τ is related to the operational interferences of system discharge tube which affect the thermodynamic quantity P_g. The time delay τ is related to buffer gas pressure. ***Remark*** The time delay parameter τ does not affect the derivative of the buffer gas pressure in time $\left(\frac{dP_g}{dt}\right)$. We assume that the gas density stays in equilibrium with the local temperature and pressure, and we express the changes in gas density due to changes in temperature and pressure by the following differential equations:

$$\frac{\partial n_g}{\partial t} = -\frac{n_g}{T_g} \cdot \frac{dT_g}{dt} + \frac{n_g}{P_g(t-\tau)} \cdot \frac{dP_g}{dt}$$
$$= -[P_g(t-\tau) - P_{ex}] \cdot \frac{n_g \cdot 2 \cdot v_s}{l \cdot P_g(t-\tau)}$$

$$\frac{\partial n_{cu}}{\partial t} = -\frac{n_{cu}}{T_g} \cdot \frac{dT_g}{dt} + \frac{n_{cu}}{P_{cu}} \cdot \frac{dP_{cu}}{dt}$$
$$= -[P_{cu} - P_{vp}] \cdot \frac{n_{cu} \cdot v_d}{\Lambda \cdot P_{cu}}$$

where n_g and n_{cu} are the densities of the buffer gas and copper atoms, respectively. We can consider $\frac{\partial n_g}{\partial t} \leftrightarrow \frac{dn_g}{dt}; \frac{\partial n_{cu}}{\partial t} \leftrightarrow \frac{dn_{cu}}{dt}$, and then we can claim that the above equations:

$$\frac{dn_g}{dt} = -[P_g(t-\tau) - P_{ex}] \cdot \frac{n_g \cdot 2 \cdot v_s}{l \cdot P_g(t-\tau)}$$
$$\frac{dn_{cu}}{dt} = -[P_{cu} - P_{vp}] \cdot \frac{n_{cu} \cdot v_d}{\Lambda \cdot P_{cu}}$$

We assume in our system that P_{ex} (external buffer gas pressure), and P_{cu}(copper vapor pressure) are constant in time, then $\frac{dP_{ex}}{dt} \to 0; \frac{dP_{cu}}{dt} \to 0$.

$$\frac{dP_{cu}}{dt} \to 0 \Rightarrow \frac{dT_g}{dt} \cdot \frac{P_{cu}}{T_g} - [P_{cu} - P_{vp}] \cdot \frac{v_d}{R} = 0; \frac{dT_g}{dt} = [P_{cu} - P_{vp}] \cdot \frac{v_d \cdot T_g}{R \cdot P_{cu}}$$

Submitting $\frac{dT_g}{dt} = [P_{cu} - P_{vp}] \cdot \frac{v_d \cdot T_g}{R \cdot P_{cu}}$ in $\frac{dP_g}{dt} = \ldots$ equation gives:

$$\frac{dP_g}{dt} = [P_{cu} - P_{vp}] \cdot \frac{v_d}{R \cdot P_{cu}} \cdot P_g(t-\tau) - [P_g(t-\tau) - P_{ex}] \cdot \frac{2 \cdot v_2}{l}$$

We can summary our system delay differential equations:

$$\frac{dP_g}{dt} = [P_{cu} - P_{vp}] \cdot \frac{v_d}{R \cdot P_{cu}} \cdot P_g(t-\tau) - [P_g(t-\tau) - P_{ex}] \cdot \frac{2 \cdot v_2}{l}$$

$$\frac{dn_g}{dt} = -[P_g(t-\tau) - P_{ex}] \cdot \frac{n_g \cdot 2 \cdot v_s}{l \cdot P_g(t-\tau)}$$

$$\frac{dn_{cu}}{dt} = -[P_{cu} - P_{vp}] \cdot \frac{n_{cu} \cdot v_d}{\Lambda \cdot P_{cu}}$$

Our system variables are P_g, n_g, n_{cu} and all other parameters are constant. At fixed point $\frac{dP_g}{dt} = 0$; $\frac{dn_g}{dt} = 0$; $\frac{dn_{cu}}{dt} = 0$; $\lim_{t\to\infty} P_g(t-\tau) = P_g(t); t \gg \tau$

$$\frac{dP_g}{dt} = 0 \Rightarrow [P_{cu} - P_{vp}] \cdot \frac{v_d}{R \cdot P_{cu}} \cdot P_g^* - [P_g^* - P_{ex}] \cdot \frac{2 \cdot v_2}{l} = 0$$

$$\frac{dn_g}{dt} = 0 \Rightarrow -[P_g^* - P_{ex}] \cdot \frac{n_g^* \cdot 2 \cdot v_s}{l \cdot P_g^*} = 0$$

$$\frac{dn_{cu}}{dt} = 0 \Rightarrow -[P_{cu} - P_{vp}] \cdot \frac{n_{cu}^* \cdot v_d}{\Lambda \cdot P_{cu}} = 0$$

$$P_{cu} \neq P_{vp} \Rightarrow n_{cu}^* = 0$$

$$[P_{cu} - P_{vp}] \cdot \frac{v_d}{R \cdot P_{cu}} \cdot P_g^* - [P_g^* - P_{ex}] \cdot \frac{2 \cdot v_2}{l} = 0$$

$$\Rightarrow [P_{cu} - P_{vp}] \cdot \frac{v_d}{R \cdot P_{cu}} \cdot P_g^* = P_g^* \cdot \frac{2 \cdot v_2}{l} - P_{ex} \cdot \frac{2 \cdot v_2}{l}$$

$$P_{ex} \cdot \frac{2 \cdot v_2}{l} = \left(\frac{2 \cdot v_2}{l} - [P_{cu} - P_{vp}] \cdot \frac{v_d}{R \cdot P_{cu}}\right) \cdot P_g^*$$

$$\Rightarrow P_g^* = \frac{P_{ex} \cdot \frac{2 \cdot v_2}{l}}{\frac{2 \cdot v_2}{l} - [P_{cu} - P_{vp}] \cdot \frac{v_d}{R \cdot P_{cu}}}$$

$-[P_g^* - P_{ex}] \cdot \frac{n_g^* \cdot 2 \cdot v_s}{l \cdot P_g^*} = 0$ then we have two cases:

(a)
$$P_g^* = P_{ex} \Rightarrow \frac{P_{ex} \cdot \frac{2 \cdot v_2}{l}}{\frac{2 \cdot v_2}{l} - [P_{cu} - P_{vp}] \cdot \frac{v_d}{R \cdot P_{cu}}} = P_{ex}$$
$$\frac{2 \cdot v_2}{l} = \frac{2 \cdot v_2}{l} - [P_{cu} - P_{vp}] \cdot \frac{v_d}{R \cdot P_{cu}} \Rightarrow P_{cu} = P_{vp}$$

(b) $P_g^* \neq P_{ex}; n_g^* = 0$

We can summary our system fixed points:

$$E^{(0)} = (P_g^{(0)}, n_g^{(0)}, n_{cu}^{(0)}) = \left(\frac{P_{ex} \cdot \frac{2 \cdot v_2}{l}}{\frac{2 \cdot v_2}{l} - [P_{cu} - P_{vp}] \cdot \frac{v_d}{R \cdot P_{cu}}}, 0, 0 \right)$$

$$E^{(1)} = (P_g^{(1)}, n_g^{(1)}, n_{cu}^{(1)}) = (P_{ex}, n_g^*, 0) \, \forall \, P_{cu} = P_{vp}$$

Stability analysis: The standard local stability analysis about any one of the equilibrium point of the copper vapor laser (CVL) rate equations consists in adding to coordinate $[P_g, n_g, n_{cu}]$ arbitrarily small increments of exponential form $[p_g, n_g, n_{cu}] \cdot e^{\lambda \cdot t}$ and retaining the first order terms in P_g, n_g, n_{cu}. The system of three homogeneous equations leads to a polynomial characteristic equation in the eigenvalues. The polynomial characteristic equations accept by set the below variables and variables derivative with respect to time into copper vapor laser (CVL) rate equations. The copper vapor laser (CVL) rate equations fixed values with arbitrarily small increments of exponential form $[p_g, n_g, n_{cu}] \cdot e^{\lambda \cdot t}$ are: $j = 0$ (first fixed point), $j = 1$ (second fixed point), $j = 2$(third fixed point), etc. [4, 5].

$$P_g(t) = P_g^{(j)} + p_g \cdot e^{\lambda \cdot t}; n_g(t) = n_g^{(j)} + n_g \cdot e^{\lambda \cdot t}$$
$$n_{cu}(t) = n_{cu}^{(j)} + n_{cu} \cdot e^{\lambda \cdot t}; P_g(t - \tau) = P_g^{(j)} + p_g \cdot e^{\lambda \cdot (t - \tau)}$$

$$\frac{dP_g(t)}{dt} = p_g \cdot \lambda \cdot e^{\lambda \cdot t}$$
$$\frac{dn_g(t)}{dt} = n_g \cdot \lambda \cdot e^{\lambda \cdot t}; \frac{dn_{cu}(t)}{dt} = n_{cu} \cdot \lambda \cdot e^{\lambda \cdot t}$$

We choose these expressions for ourselves $P_g(t), n_g(t), V_1(t), n_{cu}(t)$ as a small displacement $[p_g, n_g, n_{cu}]$ from the copper vapor laser (CVL) rate equations fixed points in time $t = 0$.

$$P_g(t = 0) = P_g^{(j)} + p_g; n_g(t = 0) = n_g^{(j)} + n_g$$
$$n_{cu}(t = 0) = n_{cu}^{(j)} + n_{cu}$$

$$\frac{dP_g}{dt} = [P_{cu} - P_{vp}] \cdot \frac{v_d}{R \cdot P_{cu}} \cdot P_g(t - \tau)$$

$$- [P_g(t-\tau) - P_{ex}] \cdot \frac{2 \cdot v_2}{l}$$

$$p_g \cdot \lambda \cdot e^{\lambda \cdot t} = [P_{cu} - P_{vp}] \cdot \frac{v_d}{R \cdot P_{cu}} \cdot [P_g^{(j)} + p_g \cdot e^{\lambda \cdot (t-\tau)}]$$
$$- [P_g^{(j)} + p_g \cdot e^{\lambda \cdot (t-\tau)} - P_{ex}] \cdot \frac{2 \cdot v_2}{l}$$

$$p_g \cdot \lambda \cdot e^{\lambda \cdot t} = [P_{cu} - P_{vp}] \cdot \frac{v_d}{R \cdot P_{cu}} \cdot P_g^{(j)}$$
$$+ [P_{cu} - P_{vp}] \cdot \frac{v_d}{R \cdot P_{cu}} \cdot p_g \cdot e^{\lambda \cdot (t-\tau)}$$
$$- [P_g^{(j)} - P_{ex}] \cdot \frac{2 \cdot v_2}{l} - p_g \cdot e^{\lambda \cdot (t-\tau)} \cdot \frac{2 \cdot v_2}{l}$$

$$p_g \cdot \lambda \cdot e^{\lambda \cdot t} = [P_{cu} - P_{vp}] \cdot \frac{v_d}{R \cdot P_{cu}} \cdot P_g^{(j)} - [P_g^{(j)} - P_{ex}]$$
$$\cdot \frac{2 \cdot v_2}{l} + [P_{cu} - P_{vp}] \cdot \frac{v_d}{R \cdot P_{cu}} \cdot p_g \cdot e^{\lambda \cdot (t-\tau)}$$
$$- p_g \cdot e^{\lambda \cdot (t-\tau)} \cdot \frac{2 \cdot v_2}{l}$$

At fixed points: $[P_{cu} - P_{vp}] \cdot \frac{v_d}{R \cdot P_{cu}} \cdot P_g^{(j)} - [P_g^{(j)} - P_{ex}] \cdot \frac{2 \cdot v_2}{l} = 0$

$$\left([P_{cu} - P_{vp}] \cdot \frac{v_d}{R \cdot P_{cu}} - \frac{2 \cdot v_2}{l}\right) \cdot p_g \cdot e^{-\lambda \cdot \tau} - p_g \cdot \lambda = 0$$

$$\frac{dn_g}{dt} = -[P_g(t-\tau) - P_{ex}] \cdot \frac{n_g \cdot 2 \cdot v_s}{l \cdot P_g(t-\tau)}$$

$$n_g \cdot \lambda \cdot e^{\lambda \cdot t} = -[P_g^{(j)} + p_g \cdot e^{\lambda \cdot (t-\tau)} - P_{ex}] \cdot \frac{[n_g^{(j)} + n_g \cdot e^{\lambda \cdot t}] \cdot 2 \cdot v_s}{l \cdot [P_g^{(j)} + p_g \cdot e^{\lambda \cdot (t-\tau)}]}$$

$$n_g \cdot \lambda \cdot e^{\lambda \cdot t} = -[P_g^{(j)} + p_g \cdot e^{\lambda \cdot (t-\tau)} - P_{ex}]$$
$$\cdot \frac{[n_g^{(j)} + n_g \cdot e^{\lambda \cdot t}] \cdot 2 \cdot v_s}{l \cdot [P_g^{(j)} + p_g \cdot e^{\lambda \cdot (t-\tau)}]}$$

First we analyze the expression: $\frac{[n_g^{(j)} + n_g \cdot e^{\lambda \cdot t}] \cdot 2 \cdot v_s}{l \cdot [P_g^{(j)} + p_g \cdot e^{\lambda \cdot (t-\tau)}]}$

$$\frac{[n_g^{(j)} + n_g \cdot e^{\lambda \cdot t}] \cdot 2 \cdot v_s}{l \cdot [P_g^{(j)} + p_g \cdot e^{\lambda \cdot (t-\tau)}]} \cdot \frac{[P_g^{(j)} - p_g \cdot e^{\lambda \cdot (t-\tau)}]}{[P_g^{(j)} - p_g \cdot e^{\lambda \cdot (t-\tau)}]}$$

$$= \frac{2 \cdot v_s \cdot [n_g^{(j)} \cdot P_g^{(j)} - n_g^{(j)} \cdot p_g \cdot e^{\lambda \cdot (t-\tau)} + n_g \cdot e^{\lambda \cdot t} \cdot P_g^{(j)} - n_g \cdot p_g \cdot e^{\lambda \cdot (t-\tau)} \cdot e^{\lambda \cdot t}]}{l \cdot [(P_g^{(j)})^2 - p_g^2 \cdot e^{2 \cdot \lambda \cdot (t-\tau)}]}$$

Assumptions $n_g \cdot p_g \approx 0; p_g^2 \approx 0$

$$\frac{[n_g^{(j)} + n_g \cdot e^{\lambda \cdot t}] \cdot 2 \cdot v_s}{l \cdot [P_g^{(j)} + p_g \cdot e^{\lambda \cdot (t-\tau)}]} \cdot \frac{[P_g^{(j)} - p_g \cdot e^{\lambda \cdot (t-\tau)}]}{[P_g^{(j)} - p_g \cdot e^{\lambda \cdot (t-\tau)}]} = \frac{2 \cdot v_s \cdot [n_g^{(j)} \cdot P_g^{(j)} - n_g^{(j)} \cdot p_g \cdot e^{\lambda \cdot (t-\tau)} + n_g \cdot e^{\lambda \cdot t} \cdot P_g^{(j)}]}{l \cdot (P_g^{(j)})^2}$$

$$\frac{[n_g^{(j)} + n_g \cdot e^{\lambda \cdot t}] \cdot 2 \cdot v_s}{l \cdot [P_g^{(j)} + p_g \cdot e^{\lambda \cdot (t-\tau)}]} \cdot \frac{[P_g^{(j)} - p_g \cdot e^{\lambda \cdot (t-\tau)}]}{[P_g^{(j)} - p_g \cdot e^{\lambda \cdot (t-\tau)}]} = \frac{2 \cdot v_s}{l} \cdot [\frac{n_g^{(j)} \cdot P_g^{(j)}}{(P_g^{(j)})^2} - \frac{n_g^{(j)} \cdot p_g \cdot e^{\lambda \cdot (t-\tau)}}{(P_g^{(j)})^2} + \frac{n_g \cdot e^{\lambda \cdot t} \cdot P_g^{(j)}}{(P_g^{(j)})^2}]$$

$$\frac{[n_g^{(j)} + n_g \cdot e^{\lambda \cdot t}] \cdot 2 \cdot v_s}{l \cdot [P_g^{(j)} + p_g \cdot e^{\lambda \cdot (t-\tau)}]} \cdot \frac{[P_g^{(j)} - p_g \cdot e^{\lambda \cdot (t-\tau)}]}{[P_g^{(j)} - p_g \cdot e^{\lambda \cdot (t-\tau)}]} = \frac{2 \cdot v_s}{l} \cdot [\frac{n_g^{(j)}}{P_g^{(j)}} - \frac{n_g^{(j)} \cdot p_g \cdot e^{\lambda \cdot (t-\tau)}}{(P_g^{(j)})^2} + \frac{n_g \cdot e^{\lambda \cdot t}}{P_g^{(j)}}]$$

$$\frac{[n_g^{(j)} + n_g \cdot e^{\lambda \cdot t}] \cdot 2 \cdot v_s}{l \cdot [P_g^{(j)} + p_g \cdot e^{\lambda \cdot (t-\tau)}]} = \frac{2 \cdot v_s}{l} \cdot [\frac{n_g^{(j)}}{P_g^{(j)}} - \frac{n_g^{(j)} \cdot p_g \cdot e^{\lambda \cdot (t-\tau)}}{(P_g^{(j)})^2} + \frac{n_g \cdot e^{\lambda \cdot t}}{P_g^{(j)}}]$$

$$n_g \cdot \lambda \cdot e^{\lambda \cdot t} = -[P_g^{(j)} + p_g \cdot e^{\lambda \cdot (t-\tau)} - P_{ex}] \cdot \frac{2 \cdot v_s}{l} \cdot [\frac{n_g^{(j)}}{P_g^{(j)}} - \frac{n_g^{(j)} \cdot p_g \cdot e^{\lambda \cdot (t-\tau)}}{(P_g^{(j)})^2} + \frac{n_g \cdot e^{\lambda \cdot t}}{P_g^{(j)}}]$$

$$n_g \cdot \lambda \cdot e^{\lambda \cdot t} = -[(P_g^{(j)} - P_{ex}) + p_g \cdot e^{\lambda \cdot (t-\tau)}] \cdot \frac{2 \cdot v_s}{l} \cdot [\frac{n_g^{(j)}}{P_g^{(j)}} - \frac{n_g^{(j)} \cdot p_g \cdot e^{\lambda \cdot (t-\tau)}}{(P_g^{(j)})^2} + \frac{n_g \cdot e^{\lambda \cdot t}}{P_g^{(j)}}]$$

$$n_g \cdot \lambda \cdot e^{\lambda \cdot t} = -(P_g^{(j)} - P_{ex}) \cdot \frac{2 \cdot v_s}{l} \cdot \frac{n_g^{(j)}}{P_g^{(j)}} + (P_g^{(j)} - P_{ex}) \cdot \frac{2 \cdot v_s}{l} \cdot \frac{n_g^{(j)} \cdot p_g \cdot e^{\lambda \cdot (t-\tau)}}{(P_g^{(j)})^2} - (P_g^{(j)} - P_{ex})$$

$$\cdot \frac{2 \cdot v_s}{l} \cdot \frac{n_g \cdot e^{\lambda \cdot t}}{P_g^{(j)}} - p_g \cdot e^{\lambda \cdot (t-\tau)} \cdot \frac{2 \cdot v_s}{l} \cdot \frac{n_g^{(j)}}{P_g^{(j)}}$$
$$+ p_g \cdot e^{\lambda \cdot (t-\tau)} \cdot \frac{2 \cdot v_s}{l} \cdot \frac{n_g^{(j)} \cdot p_g \cdot e^{\lambda \cdot (t-\tau)}}{(P_g^{(j)})^2}$$
$$- p_g \cdot e^{\lambda \cdot (t-\tau)} \cdot \frac{2 \cdot v_s}{l} \cdot \frac{n_g \cdot e^{\lambda \cdot t}}{P_g^{(j)}}$$

$$n_g \cdot \lambda \cdot e^{\lambda \cdot t} = -(P_g^{(j)} - P_{ex}) \cdot \frac{2 \cdot v_s}{l} \cdot \frac{n_g^{(j)}}{P_g^{(j)}} + (P_g^{(j)} - P_{ex})$$
$$\cdot \frac{2 \cdot v_s}{l} \cdot \frac{n_g^{(j)} \cdot p_g \cdot e^{\lambda \cdot (t-\tau)}}{(P_g^{(j)})^2} - (P_g^{(j)} - P_{ex})$$
$$\cdot \frac{2 \cdot v_s}{l} \cdot \frac{n_g \cdot e^{\lambda \cdot t}}{P_g^{(j)}} - p_g \cdot e^{\lambda \cdot (t-\tau)} \cdot \frac{2 \cdot v_s}{l} \cdot \frac{n_g^{(j)}}{P_g^{(j)}}$$
$$+ p_g \cdot p_g \cdot e^{\lambda \cdot (t-\tau)} \cdot \frac{2 \cdot v_s}{l} \cdot \frac{n_g^{(j)} \cdot e^{\lambda \cdot (t-\tau)}}{(P_g^{(j)})^2}$$
$$- p_g \cdot n_g \cdot e^{\lambda \cdot (t-\tau)} \cdot \frac{2 \cdot v_s}{l} \cdot \frac{e^{\lambda \cdot t}}{P_g^{(j)}}$$

Assumption $p_g \cdot p_g \approx 0; \; p_g \cdot n_g \approx 0$

$$n_g \cdot \lambda \cdot e^{\lambda \cdot t} = -(P_g^{(j)} - P_{ex}) \cdot \frac{2 \cdot v_s}{l} \cdot \frac{n_g^{(j)}}{P_g^{(j)}} + (P_g^{(j)} - P_{ex})$$
$$\cdot \frac{2 \cdot v_s}{l} \cdot \frac{n_g^{(j)}}{(P_g^{(j)})^2} \cdot p_g \cdot e^{\lambda \cdot (t-\tau)} - \frac{2 \cdot v_s}{l}$$
$$\cdot \frac{n_g^{(j)}}{P_g^{(j)}} \cdot p_g \cdot e^{\lambda \cdot (t-\tau)} - (P_g^{(j)} - P_{ex}) \cdot \frac{2 \cdot v_s}{l}$$
$$\cdot \frac{1}{P_g^{(j)}} \cdot n_g \cdot e^{\lambda \cdot t}$$

$$n_g \cdot \lambda \cdot e^{\lambda \cdot t} = -(P_g^{(j)} - P_{ex}) \cdot \frac{2 \cdot v_s}{l} \cdot \frac{n_g^{(j)}}{P_g^{(j)}} + [(P_g^{(j)} - P_{ex})$$
$$\cdot \frac{1}{P_g^{(j)}} - 1] \cdot \frac{n_g^{(j)}}{P_g^{(j)}} \cdot \frac{2 \cdot v_s}{l} \cdot p_g \cdot e^{\lambda \cdot (t-\tau)}$$
$$- (P_g^{(j)} - P_{ex}) \cdot \frac{2 \cdot v_s}{l} \cdot \frac{1}{P_g^{(j)}} \cdot n_g \cdot e^{\lambda \cdot t}$$

At fixed point: $-(P_g^{(j)} - P_{ex}) \cdot \frac{2 \cdot v_s}{l} \cdot \frac{n_g^{(j)}}{P_g^{(j)}} = 0$

$$n_g \cdot \lambda \cdot e^{\lambda \cdot t} = [(P_g^{(j)} - P_{ex}) \cdot \frac{1}{P_g^{(j)}} - 1] \cdot \frac{n_g^{(j)}}{P_g^{(j)}} \cdot \frac{2 \cdot v_s}{l} \cdot p_g \cdot e^{\lambda \cdot (t-\tau)} - (P_g^{(j)} - P_{ex}) \cdot \frac{2 \cdot v_s}{l} \cdot \frac{1}{P_g^{(j)}} \cdot n_g \cdot e^{\lambda \cdot t}$$

$$n_g \cdot \lambda = [(P_g^{(j)} - P_{ex}) \cdot \frac{1}{P_g^{(j)}} - 1] \cdot \frac{n_g^{(j)}}{P_g^{(j)}} \cdot \frac{2 \cdot v_s}{l} \cdot p_g \cdot e^{-\lambda \cdot \tau} - (P_g^{(j)} - P_{ex}) \cdot \frac{2 \cdot v_s}{l} \cdot \frac{1}{P_g^{(j)}} \cdot n_g$$

$$[(P_g^{(j)} - P_{ex}) \cdot \frac{1}{P_g^{(j)}} - 1] \cdot \frac{n_g^{(j)}}{P_g^{(j)}} \cdot \frac{2 \cdot v_s}{l} \cdot p_g \cdot e^{-\lambda \cdot \tau} - (P_g^{(j)} - P_{ex}) \cdot \frac{2 \cdot v_s}{l} \cdot \frac{1}{P_g^{(j)}} \cdot n_g - n_g \cdot \lambda = 0$$

$$\frac{dn_{cu}}{dt} = -[P_{cu} - P_{vp}] \cdot \frac{n_{cu} \cdot v_d}{\Lambda \cdot P_{cu}}; n_{cu} \cdot \lambda \cdot e^{\lambda \cdot t}$$
$$= -[P_{cu} - P_{vp}] \cdot \frac{(n_{cu}^{(j)} + n_{cu} \cdot e^{\lambda \cdot t}) \cdot v_d}{\Lambda \cdot P_{cu}}$$

$$n_{cu} \cdot \lambda \cdot e^{\lambda \cdot t} = -[P_{cu} - P_{vp}] \cdot n_{cu}^{(j)} \cdot \frac{v_d}{\Lambda \cdot P_{cu}} - [P_{cu} - P_{vp}] \cdot \frac{v_d}{\Lambda \cdot P_{cu}} \cdot n_{cu} \cdot e^{\lambda \cdot t}$$

At fixed point: $-[P_{cu} - P_{vp}] \cdot n_{cu}^{(j)} \cdot \frac{v_d}{\Lambda \cdot P_{cu}} = 0; -[P_{cu} - P_{vp}] \cdot \frac{v_d}{\Lambda \cdot P_{cu}} \cdot n_{cu} - n_{cu} \cdot \lambda = 0$

We can summary our copper vapor laser (CVL) rate equations three arbitrarily small increments equations:

$$([P_{cu} - P_{vp}] \cdot \frac{v_d}{R \cdot P_{cu}} - \frac{2 \cdot v_2}{l}) \cdot p_g \cdot e^{-\lambda \cdot \tau} - p_g \cdot \lambda = 0$$

$$[(P_g^{(j)} - P_{ex}) \cdot \frac{1}{P_g^{(j)}} - 1] \cdot \frac{n_g^{(j)}}{P_g^{(j)}} \cdot \frac{2 \cdot v_s}{l} \cdot p_g \cdot e^{-\lambda \cdot \tau} - (P_g^{(j)} - P_{ex}) \cdot \frac{2 \cdot v_s}{l} \cdot \frac{1}{P_g^{(j)}} \cdot n_g - n_g \cdot \lambda = 0$$

$$-[P_{cu} - P_{vp}] \cdot \frac{v_d}{\Lambda \cdot P_{cu}} \cdot n_{cu} - n_{cu} \cdot \lambda = 0$$

The small increments Jacobian of our copper vapor laser (CVL) rate equations is as follow:

$$\begin{pmatrix} ([P_{cu}-P_{vp}]\cdot\frac{v_d}{R\cdot P_{cu}}-\frac{2\cdot v_2}{l})\cdot e^{-\lambda\cdot\tau}-\lambda & 0 & 0 \\ [(P_g^{(j)}-P_{ex})\cdot\frac{1}{P_g^{(j)}}-1]\cdot\frac{n_g^{(j)}}{P_g^{(j)}}\cdot\frac{2\cdot v_s}{l}\cdot e^{-\lambda\cdot\tau} & -(P_g^{(j)}-P_{ex})\cdot\frac{2\cdot v_s}{l}\cdot\frac{1}{P_g^{(j)}}-\lambda & 0 \\ 0 & 0 & -[P_{cu}-P_{vp}]\cdot\frac{v_d}{\Lambda\cdot P_{cu}}-\lambda \end{pmatrix}$$

$$\cdot\begin{pmatrix} p_g \\ n_g \\ n_{cu} \end{pmatrix}=0$$

$$A-\lambda\cdot I=$$

$$\begin{pmatrix} ([P_{cu}-P_{vp}]\cdot\frac{v_d}{R\cdot P_{cu}} -\frac{2\cdot v_2}{l})\cdot e^{-\lambda\cdot\tau}-\lambda & 0 & 0 \\ \cdot\frac{n_g^{(j)}}{P_g^{(j)}}\cdot\frac{2\cdot v_s}{l}\cdot e^{-\lambda\cdot\tau} & -(P_g^{(j)}-P_{ex})\cdot\frac{2\cdot v_s}{l}\cdot\frac{1}{P_g^{(j)}}-\lambda & 0 \\ 0 & 0 & -[P_{cu}-P_{vp}]\cdot\frac{v_d}{\Lambda\cdot P_{cu}} -\lambda \end{pmatrix}$$

$$\det(A-\lambda\cdot I)=$$

$$\det\begin{pmatrix} ([P_{cu}-P_{vp}]\cdot\frac{v_d}{R\cdot P_{cu}} -\frac{2\cdot v_2}{l})\cdot e^{-\lambda\cdot\tau}-\lambda & 0 & 0 \\ \cdot\frac{n_g^{(j)}}{P_g^{(j)}}\cdot\frac{2\cdot v_s}{l}\cdot e^{-\lambda\cdot\tau} & -(P_g^{(j)}-P_{ex})\cdot\frac{2\cdot v_s}{l}\cdot\frac{1}{P_g^{(j)}}-\lambda & 0 \\ 0 & 0 & -[P_{cu}-P_{vp}]\cdot\frac{v_d}{\Lambda\cdot P_{cu}} -\lambda \end{pmatrix}$$

$$\det(A-\lambda\cdot I)=0$$

$$\det(A-\lambda\cdot I) = \{([P_{cu}-P_{vp}]\cdot\frac{v_d}{R\cdot P_{cu}}-\frac{2\cdot v_2}{l})\cdot e^{-\lambda\cdot\tau}-\lambda\} \cdot\det\begin{pmatrix} -(P_g^{(j)}-P_{ex})\cdot\frac{2\cdot v_s}{l}\cdot\frac{1}{P_g^{(j)}}-\lambda & 0 \\ 0 & -[P_{cu}-P_{vp}]\cdot\frac{v_d}{\Lambda\cdot P_{cu}}-\lambda \end{pmatrix}$$

$$\det(A-\lambda\cdot I)=\{([P_{cu}-P_{vp}]\cdot\frac{v_d}{R\cdot P_{cu}}-\frac{2\cdot v_2}{l})\cdot e^{-\lambda\cdot\tau}-\lambda\} \cdot\{[-(P_g^{(j)}-P_{ex})\cdot\frac{2\cdot v_s}{l}\cdot\frac{1}{P_g^{(j)}}-\lambda]$$

$$\cdot [-[P_{cu} - P_{vp}] \cdot \frac{v_d}{\Lambda \cdot P_{cu}} - \lambda]\}$$

We define for simplicity global parameters: $\Upsilon_1 = [P_{cu} - P_{vp}] \cdot \frac{v_d}{R \cdot P_{cu}} - \frac{2 \cdot v_2}{l}$

$$\Upsilon_2 = -(P_g^{(j)} - P_{ex}) \cdot \frac{2 \cdot v_s}{l} \cdot \frac{1}{P_g^{(j)}}$$

$$\Upsilon_3 = -[P_{cu} - P_{vp}] \cdot \frac{v_d}{\Lambda \cdot P_{cu}}$$

$$\det(A - \lambda \cdot I) = (\Upsilon_1 \cdot e^{-\lambda \cdot \tau} - \lambda) \cdot ([\Upsilon_2 - \lambda] \cdot [\Upsilon_3 - \lambda]) \\ = (\Upsilon_1 \cdot e^{-\lambda \cdot \tau} - \lambda) \cdot (\Upsilon_2 \cdot \Upsilon_3 - [\Upsilon_2 + \Upsilon_3] \cdot \lambda + \lambda^2)$$

$$\det(A - \lambda \cdot I) = (\Upsilon_1 \cdot e^{-\lambda \cdot \tau} - \lambda) \cdot (\Upsilon_2 \cdot \Upsilon_3 - [\Upsilon_2 + \Upsilon_3] \cdot \lambda + \lambda^2) \\ = \Upsilon_2 \cdot \Upsilon_3 \cdot \Upsilon_1 \cdot e^{-\lambda \cdot \tau} - [\Upsilon_2 + \Upsilon_3] \cdot \Upsilon_1 \cdot e^{-\lambda \cdot \tau} \cdot \lambda \\ + \lambda^2 \cdot \Upsilon_1 \cdot e^{-\lambda \cdot \tau} - \Upsilon_2 \cdot \Upsilon_3 \cdot \lambda + [\Upsilon_2 + \Upsilon_3] \cdot \lambda^2 - \lambda^3$$

$$\det(A - \lambda \cdot I) = -\Upsilon_2 \cdot \Upsilon_3 \cdot \lambda + [\Upsilon_2 + \Upsilon_3] \cdot \lambda^2 - \lambda^3 \\ + (\Upsilon_2 \cdot \Upsilon_3 \cdot \Upsilon_1 - [\Upsilon_2 + \Upsilon_3] \cdot \Upsilon_1 \cdot \lambda + \lambda^2 \cdot \Upsilon_1) \cdot e^{-\lambda \cdot \tau}$$

$$D(\lambda, \tau) = P_n(\lambda, \tau) + Q_m(\lambda, \tau) \cdot e^{-\lambda \cdot \tau} \\ n, m \in \mathbb{N}_0;\ n = 3;\ m = 2;\ n > m$$

$$P_n(\lambda, \tau) = -\Upsilon_2 \cdot \Upsilon_3 \cdot \lambda + [\Upsilon_2 + \Upsilon_3] \cdot \lambda^2 - \lambda^3 \\ n = 3;\ Q_m(\lambda, \tau) = \Upsilon_2 \cdot \Upsilon_3 \cdot \Upsilon_1 - [\Upsilon_2 + \Upsilon_3] \cdot \Upsilon_1 \cdot \lambda + \lambda^2 \cdot \Upsilon_1;\ m = 2$$

$$P_n(\lambda, \tau) = \sum_{k=0}^{n=3} p_k(\tau) \cdot \lambda^k \\ = p_0(\tau) + p_1(\tau) \cdot \lambda + p_2(\tau) \cdot \lambda^2 + p_3(\tau) \cdot \lambda^3$$

$$p_0 = p_0(\tau);\ p_1 = p_1(\tau);\ p_2 = p_2(\tau) \\ p_3 = p_3(\tau);\ p_0 = 0;\ p_1 = -\Upsilon_2 \cdot \Upsilon_3 \\ p_2 = \Upsilon_2 + \Upsilon_3;\ p_3 = -1$$

$$Q_m(\lambda, \tau) = \sum_{k=0}^{m=2} q_k(\tau) \cdot \lambda^k \\ = q_0(\tau) + q_1(\tau) \cdot \lambda + q_2(\tau) \cdot \lambda^2$$

$$q_0 = q_0(\tau);\ q_1 = q_1(\tau);\ q_2 = q_2(\tau)$$

$$Q_m(\lambda, \tau) = \sum_{k=0}^{m=2} q_k(\tau) \cdot \lambda^k = q_0(\tau) + q_1(\tau) \cdot \lambda + q_2(\tau) \cdot \lambda^2$$

$$q_0 = q_0(\tau);\ q_1 = q_1(\tau);\ q_2 = q_2(\tau)$$

$$q_0 = \Upsilon_2 \cdot \Upsilon_3 \cdot \Upsilon_1;\ q_1 = -[\Upsilon_2 + \Upsilon_3] \cdot \Upsilon_1;\ q_2 = \Upsilon_1$$

The homogeneous system for P_g, n_g, V_1, n_{cu} leads to a characteristic equation for the eigenvalue λ having the form $D(\lambda, \tau) = P(\lambda, \tau) + Q(\lambda, \tau) \cdot e^{-\lambda \cdot \tau} = 0$; and $P(\lambda) = \sum_{j=0}^{3} a_j \cdot \lambda^j$; $Q(\lambda) = \sum_{j=0}^{2} c_j \cdot \lambda^j$. The coefficients $\{a_j(q_i, q_k, \tau), c_j(q_i, q_k, \tau)\} \in \mathbb{R}$ depend on q_i, q_k and delayτ. q_i, q_k are any copper vapor laser (CVL) rate equations parameters, other parameters kept as a constant [4, 5].

$$P(\lambda) = \sum_{j=0}^{3} a_j \cdot \lambda^j = a_0 + a_1 \cdot \lambda + a_2 \cdot \lambda^2 + a_3 \cdot \lambda^3$$

$$Q(\lambda) = \sum_{j=0}^{2} c_j \cdot \lambda^j = c_0 + c_1 \cdot \lambda + c_2 \cdot \lambda^2$$

$$a_0 = 0;\ a_1 = -\Upsilon_2 \cdot \Upsilon_3;\ a_2 = \Upsilon_2 + \Upsilon_3$$

$$a_3 = -1;\ c_0 = \Upsilon_2 \cdot \Upsilon_3 \cdot \Upsilon_1;\ c_1 = -[\Upsilon_2 + \Upsilon_3] \cdot \Upsilon_1;\ c_2 = \Upsilon_1$$

Unless strictly necessary, the designation of the varied arguments (q_i, q_k) will subsequently be omitted from P, Q, a_j and c_j. The coefficients a_j, c_j are continuous and

differentiable functions of their arguments, and direct substitution shows that $a_0 + c_0 \neq 0$ for $q_i, q_k \in \mathbb{R}_+$; that is, $\lambda = 0$ is not of $P(\lambda) + Q(\lambda) \cdot e^{-\lambda \cdot \tau} = 0$. Furthermore, $P(\lambda)$, $Q(\lambda)$ are analytic functions of λ, for which the following requirements of the analysis (Kuang J and Cong Y 2005; Kuang Y 1993) can also be verified in the present case:

(a) If $\lambda = i \cdot \omega$; $\omega \in \mathbb{R}$, then $P(i \cdot \omega) + Q(i \cdot \omega) \neq 0$.
(b) If $|\frac{Q(\lambda)}{P(\lambda)}|$ is bounded for $|\lambda| \to \infty$; $\mathrm{Re}\lambda \geq 0$. No roots bifurcation from ∞.
(c) $F(\omega) = |P(i \cdot \omega)|^2 - |Q(i \cdot \omega)|^2$ has a finite number of zeros. Indeed, this is a polynomial in ω.
(d) Each positive root $\omega(q_i, q_k)$ of $F(\omega) = 0$ is continuous and differentiable with respect to q_i, q_k.

We assume that $P_n(\lambda, \tau)$ and $Q_m(\lambda, \tau)$ cannot have common imaginary roots. That is for any real numberω: $P_n(\lambda = i \cdot \omega, \tau) + Q_m(\lambda = i \cdot \omega, \tau) \neq 0$ and

$$P_n(\lambda = i \cdot \omega, \tau) = p_0(\tau) - p_2(\tau) \cdot \omega^2 + [p_1(\tau) \cdot \omega - p_3(\tau) \cdot \omega^3] \cdot i$$

$$P_n(\lambda = i \cdot \omega, \tau) = -[\Upsilon_2 + \Upsilon_3] \cdot \omega^2 + [\omega^3 - \Upsilon_2 \cdot \Upsilon_3 \cdot \omega] \cdot i$$
$$Q_m(\lambda = i \cdot \omega, \tau) = q_0(\tau) + q_1(\tau) \cdot i \cdot \omega - q_2(\tau) \cdot \omega^2$$

$$Q_m(\lambda = i \cdot \omega, \tau) = \Upsilon_2 \cdot \Upsilon_3 \cdot \Upsilon_1 - \Upsilon_1 \cdot \omega^2 - [\Upsilon_2 + \Upsilon_3] \cdot \Upsilon_1 \cdot \omega \cdot i$$

$$\begin{aligned} &P_n(\lambda = i \cdot \omega, \tau) + Q_m(\lambda = i \cdot \omega, \tau) \\ &= -[\Upsilon_2 + \Upsilon_3] \cdot \omega^2 + [\omega^3 - \Upsilon_2 \cdot \Upsilon_3 \cdot \omega] \cdot i \\ &\quad + \Upsilon_2 \cdot \Upsilon_3 \cdot \Upsilon_1 - \Upsilon_1 \cdot \omega^2 - [\Upsilon_2 + \Upsilon_3] \\ &\quad \cdot \Upsilon_1 \cdot i \cdot \omega \neq 0 \end{aligned}$$

$$\begin{aligned} &P_n(\lambda = i \cdot \omega, \tau) + Q_m(\lambda = i \cdot \omega, \tau) \\ &= -[\Upsilon_2 + \Upsilon_3] \cdot \omega^2 + \Upsilon_2 \cdot \Upsilon_3 \cdot \Upsilon_1 - \Upsilon_1 \cdot \omega^2 \\ &\quad - [\Upsilon_2 + \Upsilon_3] \cdot \Upsilon_1 \cdot i \cdot \omega + [\omega^3 - \Upsilon_2 \cdot \Upsilon_3 \cdot \omega] \cdot i \neq 0 \end{aligned}$$

$$\begin{aligned} |P(i \cdot \omega, \tau)|^2 &= [\Upsilon_2 + \Upsilon_3]^2 \cdot \omega^4 + [\omega^3 - \Upsilon_2 \cdot \Upsilon_3 \cdot \omega]^2 \\ &= [\Upsilon_2 + \Upsilon_3]^2 \cdot \omega^4 + \omega^6 - 2 \cdot \Upsilon_2 \cdot \Upsilon_3 \cdot \omega^4 \\ &\quad + (\Upsilon_2 \cdot \Upsilon_3)^2 \cdot \omega^2 \end{aligned}$$

$$\begin{aligned} |P(i \cdot \omega, \tau)|^2 &= [\Upsilon_2 + \Upsilon_3]^2 \cdot \omega^4 + [\omega^3 - \Upsilon_2 \cdot \Upsilon_3 \cdot \omega]^2 \\ &= \omega^6 + ([\Upsilon_2 + \Upsilon_3]^2 - 2 \cdot \Upsilon_2 \cdot \Upsilon_3) \cdot \omega^4 \\ &\quad + (\Upsilon_2 \cdot \Upsilon_3)^2 \cdot \omega^2 \end{aligned}$$

$$|Q(i \cdot \omega, \tau)|^2 = [\Upsilon_2 \cdot \Upsilon_3 \cdot \Upsilon_1 - \Upsilon_1 \cdot \omega^2]^2 + [\Upsilon_2 + \Upsilon_3]^2 \cdot \Upsilon_1^2 \cdot \omega^2$$

$$\begin{aligned} |Q(i \cdot \omega, \tau)|^2 &= [\Upsilon_2 \cdot \Upsilon_3 \cdot \Upsilon_1]^2 + ([\Upsilon_2 + \Upsilon_3]^2 \cdot \Upsilon_1^2 \\ &\quad - 2 \cdot \Upsilon_2 \cdot \Upsilon_3 \cdot \Upsilon_1^2) \cdot \omega^2 + \Upsilon_1^2 \cdot \omega^4 \end{aligned}$$

$$\begin{aligned} F(\omega, \tau) &= |P(i \cdot \omega, \tau)|^2 - |Q(i \cdot \omega, \tau)|^2 \\ &= \omega^6 + ([\Upsilon_2 + \Upsilon_3]^2 - 2 \cdot \Upsilon_2 \cdot \Upsilon_3) \cdot \omega^4 \\ &\quad + (\Upsilon_2 \cdot \Upsilon_3)^2 \cdot \omega^2 - [\Upsilon_2 \cdot \Upsilon_3 \cdot \Upsilon_1]^2 \\ &\quad - ([\Upsilon_2 + \Upsilon_3]^2 \cdot \Upsilon_1^2 - 2 \cdot \Upsilon_2 \cdot \Upsilon_3 \cdot \Upsilon_1^2) \cdot \omega^2 - \Upsilon_1^2 \cdot \omega^4 \end{aligned}$$

$$F(\omega, \tau) = |P(i \cdot \omega, \tau)|^2 - |Q(i \cdot \omega, \tau)|^2$$

$$= -[\Upsilon_2 \cdot \Upsilon_3 \cdot \Upsilon_1]^2 + \{(\Upsilon_2 \cdot \Upsilon_3)^2 - ([\Upsilon_2 + \Upsilon_3]^2 \cdot \Upsilon_1^2 - 2 \cdot \Upsilon_2 \cdot \Upsilon_3 \cdot \Upsilon_1^2)\} \cdot \omega^2 + ([\Upsilon_2 + \Upsilon_3]^2 - 2 \cdot \Upsilon_2 \cdot \Upsilon_3 - \Upsilon_1^2) \cdot \omega^4 + \omega^6$$

We define the following parameters for simplicity: $\Pi_0, \Pi_2, \Pi_4, \Pi_6$

$$\Pi_0 = -[\Upsilon_2 \cdot \Upsilon_3 \cdot \Upsilon_1]^2$$
$$\Pi_2 = (\Upsilon_2 \cdot \Upsilon_3)^2 - ([\Upsilon_2 + \Upsilon_3]^2 \cdot \Upsilon_1^2 - 2 \cdot \Upsilon_2 \cdot \Upsilon_3 \cdot \Upsilon_1^2)$$

$$\Pi_4 = [\Upsilon_2 + \Upsilon_3]^2 - 2 \cdot \Upsilon_2 \cdot \Upsilon_3 - \Upsilon_1^2; \; \Pi_6 = 1$$

Hence $F(\omega, \tau) = 0$ implies $\sum_{k=0}^{3} \Pi_{2 \cdot k} \cdot \omega^{2 \cdot k} = 0$ and its roots are given by solving the above polynomial.

$$P_R(i\omega, \tau) = -[\Upsilon_2 + \Upsilon_3] \cdot \omega^2$$
$$P_I(i\omega, \tau) = \omega^3 - \Upsilon_2 \cdot \Upsilon_3 \cdot \omega$$
$$Q_R(i\omega, \tau) = \Upsilon_2 \cdot \Upsilon_3 \cdot \Upsilon_1 - \Upsilon_1 \cdot \omega^2$$
$$Q_I(i\omega, \tau) = -[\Upsilon_2 + \Upsilon_3] \cdot \Upsilon_1 \cdot \omega$$

$$\sin\theta(\tau) = \frac{-P_R(i\omega, \tau) \cdot Q_I(i\omega, \tau) + P_I(i\omega, \tau) \cdot Q_R(i\omega, \tau)}{|Q(i\omega, \tau)|^2}$$

$$\cos\theta(\tau) = -\frac{P_R(i\omega, \tau) \cdot Q_R(i\omega, \tau) + P_I(i\omega, \tau) \cdot Q_I(i\omega, \tau)}{|Q(i\omega, \tau)|^2}$$

We use different terminology from our last characteristics parameters definition:

$$k \to j; \; p_k(\tau) \to a_j; \; q_k(\tau) \to c_j$$
$$n, m \in \mathbb{N}_0; \; n = 3; \; m = 2; \; n > m$$

$$P_n(\lambda, \tau) \to P(\lambda); \; Q_m(\lambda, \tau) \to Q(\lambda)$$
$$P(\lambda) = \sum_{j=0}^{3} a_j \cdot \lambda^j; \; Q(\lambda) = \sum_{j=0}^{2} c_j \cdot \lambda^j$$

$$P(\lambda) = a_0 + a_1 \cdot \lambda + a_2 \cdot \lambda^2 + a_3 \cdot \lambda^3$$
$$Q(\lambda) = c_0 + c_1 \cdot \lambda + c_2 \cdot \lambda^2$$

$n, m \in \mathbb{N}_0; n > m$ and $a_j, c_j : \mathbb{R}_{0+} \to \mathbb{R}$ are continuous and differentiable function of τ such that $a_0 + c_0 \neq 0$. In the following "–" denoted complex and conjugate. $P(\lambda)$, $Q(\lambda)$ are analytic functions in λ and differentiable in τ. The coefficients$a_j(P_{cu}, P_{vp}, v_d, R, v_2, l, P_{ex}, v_s, \Lambda, \tau) \in \mathbb{R}$ and $c_j(P_{cu}, P_{vp}, v_d, R, v_2, l, P_{ex}, v_s, \Lambda, \tau) \in \mathbb{R}$ depend on copper vapor laser (CVL) rate equations parameters, $P_{cu}, P_{vp}, v_d, R, v_2, l, P_{ex}, v_s, \Lambda, \tau$ values. Unless strictly necessary, the designation of the varied arguments: $P_{cu}, P_{vp}, v_d, R, v_2, l, P_{ex}, v_s, \Lambda, \tau$ will subsequently be omitted from P, Q, a_j, c_j. The coefficients a_j, c_j are continuous and differentiable functions of their arguments, and direct substitution shows that $a_0 + c_0 \neq 0$ [4, 5].

$$a_0 = 0; c_0 = \Upsilon_2 \cdot \Upsilon_3 \cdot \Upsilon_1$$

$$a_0 + c_0 = \Upsilon_2 \cdot \Upsilon_3 \cdot \Upsilon_1 = \prod_{k=1}^{3} \Upsilon_k \neq 0$$

$\forall P_{cu}, P_{vp}, v_d, R, v_2, l, P_{ex}, v_s, \Lambda, \tau \in \mathbb{R}_+$ I.e. $\lambda = 0$ is not a root of the characteristic equation. Furthermore $P(\lambda)$, $Q(\lambda)$ are analytic functions of λ for which the following requirements of the analysis (see Kuang, 1993, Sect. 3.4) can also be verified in the present case.

(a) If $\lambda = i \cdot \omega; \omega \in \mathbb{R}$ then $P(i \cdot \omega) + Q(i \cdot \omega) \neq 0$, i.e. P and Q have no common imaginary roots. This condition was verified numerically in the entire $P_{cu}, P_{vp}, v_d, R, v_2, l, P_{ex}, v_s, \Lambda, \tau$ domain of interest.
(b) $|\frac{P(\lambda)}{Q(\lambda)}|$ is bounded for $|\lambda| \to \infty$; Re$\lambda \geq 0$. No roots bifurcation from ∞. Indeed, in the limit: $|\frac{Q(\lambda)}{P(\lambda)}| = |\frac{c_0+c_1\cdot\lambda+c_2\cdot\lambda^2}{a_0+a_1\cdot\lambda+a_2\cdot\lambda^2+a_3\cdot\lambda^3}|$.
(c) The following expressions exist: $F(\omega) = |P(i \cdot \omega)|^2 - |Q(i \cdot \omega)|^2$ $F(\omega, \tau) = |P(i \cdot \omega, \tau)|^2 - |Q(i \cdot \omega, \tau)|^2 = \sum_{k=0}^{3} \Pi_{2\cdot k} \cdot \omega^{2\cdot k}$ has at most a finite number of zeros. Indeed, this is a polynomial in ω (degree in ω^6).
(d) Each positive root $\omega(P_{cu}, P_{vp}, v_d, R, v_2, l, P_{ex}, v_s, \Lambda, \tau)$ of $F(\omega) = 0$ is continuous and differentiable with respect to $P_{cu}, P_{vp}, v_d, R, v_2, l, P_{ex}, v_s, \Lambda, \tau$ and the condition can only be assessed numerically.

In addition, since the coefficients in P and Q are real, we have $\overline{P(-i \cdot \omega)} = P(i \cdot \omega)$ and $\overline{Q(-i \cdot \omega)} = Q(i \cdot \omega)$ thus, $\omega > 0$ may be on eigenvalue of characteristic equations. The analysis consists in identifying the roots of the characteristic equation situated on the imaginary axis of the complex λ plane, whereby increasing the parameters: $P_{cu}, P_{vp}, v_d, R, v_2, l, P_{ex}, v_s, \Lambda, \tau$, Re$\lambda$ may, at the crossing, change its sign from (-) to (+). i.e. from a stable focus $E^{(*)} = (P_g^{(*)}, n_g^{(*)}, n_{cu}^{(*)})$ to an unstable one, or vice versa. This feature may be further assessed by examining the sign of the partial derivatives with respect to $P_{cu}, P_{vp}, v_d, R, v_2, l, P_{ex}, v_s, \Lambda, \tau$ and any system parameters.

$$\Lambda^{-1}(P_{cu}) = \left(\frac{\partial \text{Re}\lambda}{\partial P_{cu}}\right)_{\lambda=i\cdot\omega} ; P_{vp}, v_d, R, v_2, l, P_{ex}, v_s, \Lambda, \tau = const$$

$$\Lambda^{-1}(P_{vp}) = \left(\frac{\partial \text{Re}\lambda}{\partial P_{vp}}\right)_{\lambda=i\cdot\omega} ;\ P_{cu}, v_d, R, v_2, l, P_{ex}, v_s, \Lambda, \tau = const$$

$$\Lambda^{-1}(v_d) = \left(\frac{\partial \text{Re}\lambda}{\partial v_d}\right)_{\lambda=i\cdot\omega} ;\ P_{cu}, P_{vp}, R, v_2, l, P_{ex}, v_s, \Lambda, \tau = const$$

$$\Lambda^{-1}(R) = \left(\frac{\partial \text{Re}\lambda}{\partial R}\right)_{\lambda=i\cdot\omega} ;\ P_{cu}, P_{vp}, v_d, v_2, l, P_{ex}, v_s, \Lambda, \tau = const$$

$$\Lambda^{-1}(v_2) = \left(\frac{\partial \text{Re}\lambda}{\partial v_2}\right)_{\lambda=i\cdot\omega} ;\ P_{cu}, P_{vp}, v_d, R, l, P_{ex}, v_s, \Lambda, \tau = const$$

$$\Lambda^{-1}(\tau) = \left(\frac{\partial \text{Re}\lambda}{\partial \tau}\right)_{\lambda=i\cdot\omega} ;\ P_{cu}, P_{vp}, v_d, R, v_2, l, P_{ex}, v_s, \Lambda = const$$

$P(\lambda) = P_R(\lambda) + i \cdot P_I(\lambda)$; $Q(\lambda) = Q_R(\lambda) + i \cdot Q_I(\lambda)$, When writing and inserting $\lambda = i \cdot \omega$ into copper vapor laser (CVL) rate equations characteristic equation ω must satisfy the following equations:

$$\sin(\omega\cdot\tau) = g(\omega) = \frac{-P_R(i\omega,\tau)\cdot Q_I(i\omega,\tau) + P_I(i\omega,\tau)\cdot Q_R(i\omega,\tau)}{|Q(i\omega,\tau)|^2}$$

$$\cos(\omega\cdot\tau) = h(\omega) = -\frac{P_R(i\omega,\tau)\cdot Q_R(i\omega,\tau) + P_I(i\omega,\tau)\cdot Q_I(i\omega,\tau)}{|Q(i\omega,\tau)|^2}$$

where $|Q(i\omega,\tau)|^2 \neq 0$ in view of requirement (a) above, and $(g,h) \in \mathbb{R}$. Furthermore, it follows above $\sin(\omega\cdot\tau)$ and $\cos(\omega\cdot\tau)$ equation that, by squaring and adding the sides, ω must be a positive root of $F(\omega) = |P(i\cdot\omega)|^2 - |Q(i\cdot\omega)|^2 = 0$. Note: $F(\omega)$ is independent on τ. Now it is important to notice that if $\tau \notin I$(assume that $I \subseteq \mathbb{R}_{+0}$ is the set where $\omega(\tau)$ is a positive root of $F(\omega)$ and for, $\tau \notin I$, $\omega(\tau)$ is not defined. Then for all τ in I, $\omega(\tau)$ is satisfied that $F(\omega,\tau) = 0$. Then there are no positive $\omega(\tau)$ solutions for $F(\omega,\tau) = 0$, and we cannot have stability switches. For $\tau \in I$ where $\omega(\tau)$ is a positive solution of $F(\omega,\tau) = 0$, we can define the angle $\theta(\tau) \in [0, 2\cdot\Pi]$ as the solution of $\sin\theta(\tau) = \ldots$ and $\cos\theta(\tau) = \ldots$; the relation between the arguments $\theta(\tau)$ and $\tau\cdot\omega(\tau)$ for $\tau \in I$ must be describing below. $\tau\cdot\omega(\tau) = \theta(\tau) + 2\cdot n\cdot\Pi\ \forall n \in \mathbb{N}_0$ Hence we can define the maps:

$\tau_n : I \to \mathbb{R}_{+0}$, is given by $\tau_n(\tau) = \frac{\theta(\tau)+2\cdot n\cdot\Pi}{\omega(\tau)}$; $n \in \mathbb{N}_0$; $\tau \in I$. Let us introduce the function $I \to \mathbb{R}$; $S_n(\tau) = \tau - \tau_n(\tau)$; $\tau \in I$; $n \in \mathbb{N}_0$ that is continuous and differentiable in τ. In the following, the subscripts $\lambda, \omega, P_{cu}, P_{vp}, v_d, R, v_2, l, P_{ex}, v_s, \Lambda$ indicate the corresponding partial derivatives. Let us first concentrate on $\Lambda(x)$, remember in $\lambda(P_{cu}, P_{vp}, v_d, R, v_2, l, P_{ex}, v_s, \Lambda, \ldots)$ and $\omega(P_{cu}, P_{vp}, v_d, R, v_2, l, P_{ex}, v_s, \Lambda, \ldots)$, and keeping all parameters except one (x) and τ. The derivation closely follows that in reference [BK]. Differentiating copper vapor laser (CVL) rate equations characteristic equation $P(\lambda) + Q(\lambda) \cdot e^{-\lambda \cdot \tau} = 0$ with respect to specific parameter (x), and inverting the derivative, for convenience, one calculates: $x = P_{cu}, P_{vp}, v_d, R, v_2, l, \Lambda, \tau, \ldots$

$$\left(\frac{\partial \lambda}{\partial x}\right)^{-1} = \frac{-P_\lambda(\lambda, x) \cdot Q(\lambda, x) + Q_\lambda(\lambda, x) \cdot P(\lambda, x) - \tau \cdot P(\lambda, x) \cdot Q(\lambda, x)}{P_x(\lambda, x) \cdot Q(\lambda, x) - Q_x(\lambda, x) \cdot P(\lambda, x)}$$

where $P_\lambda = \frac{\partial P}{\partial \lambda}, \ldots$ etc., substituting $\lambda = i \cdot \omega$ and bearing $\overline{P(-i \cdot \omega)} = P(i \cdot \omega)$; $\overline{Q(-i \cdot \omega)} = Q(i \cdot \omega)$ then $i \cdot P_\lambda(i \cdot \omega) = P_\omega(i \cdot \omega)$; $i \cdot Q_\lambda(i \cdot \omega) = Q_\omega(i \cdot \omega)$ and that on the surface $|P(i\omega)|^2 = |Q(i\omega)|^2$, one obtain: $(\frac{\partial \lambda}{\partial x})^{-1}|_{\lambda = i \cdot \omega} = (\frac{i \cdot P_\omega(i \cdot \omega, x) \cdot \overline{P(i \cdot \omega, x)} + i \cdot Q_\lambda(i \cdot \omega, x) \cdot \overline{Q(\lambda, x)} - \tau \cdot |P(i \cdot \omega, x)|^2}{P_x(i \cdot \omega, x) \cdot \overline{P(i \cdot \omega, x)} - Q_x(i \cdot \omega, x) \cdot \overline{Q(i \cdot \omega, x)}})$

Upon separating into real and imaginary parts, with $P = P_R + i \cdot P_I$; $Q = Q_R + i \cdot Q_I$ and $P_\omega = P_{R\omega} + i \cdot P_{I\omega}$; $Q_\omega = Q_{R\omega} + i \cdot Q_{I\omega}$; $P_x = P_{Rx} + i \cdot P_{Ix}$; $Q_x = Q_{Rx} + i \cdot Q_{Ix}$; $P^2 = P_R^2 + P_I^2$, When (x) can be any copper vapor laser (CVL) rate equations parameters $P_{cu}, P_{vp}, v_d, R, v_2, l, \Lambda, \ldots$ and time delay τ etc. Where for convenience, we have dropped the arguments $(i \cdot \omega, x)$, and where $F_\omega = 2 \cdot [(P_{R\omega} \cdot P_R + P_{I\omega} \cdot P_I) - (Q_{R\omega} \cdot Q_R + Q_{I\omega} \cdot Q_I)]$; $F_x = 2 \cdot [(P_{Rx} \cdot P_R + P_{Ix} \cdot P_I) - (Q_{Rx} \cdot Q_R + Q_{Ix} \cdot Q_I)]$ and $\omega_x = -\frac{F_x}{F_\omega}$. We define U and V: $U = (P_R \cdot P_{I\omega} - P_I \cdot P_{R\omega}) - (Q_R \cdot Q_{I\omega} - Q_I \cdot Q_{R\omega})$,

$V = (P_R \cdot P_{Ix} - P_I \cdot P_{Rx}) - (Q_R \cdot Q_{Ix} - Q_I \cdot Q_{Rx})$, we choose our specific parameter as time delay$x = \tau$; $P_R = -[\Upsilon_2 + \Upsilon_3] \cdot \omega^2$; $P_I = \omega^3 - \Upsilon_2 \cdot \Upsilon_3 \cdot \omega$; $Q_R = \Upsilon_2 \cdot \Upsilon_3 \cdot \Upsilon_1 - \Upsilon_1 \cdot \omega^2$; $Q_I = -[\Upsilon_2 + \Upsilon_3] \cdot \Upsilon_1 \cdot \omega$:

$$P_{R\omega} = -2 \cdot [\Upsilon_2 + \Upsilon_3] \cdot \omega;\ P_{I\omega} = 3 \cdot \omega^2 - \Upsilon_2 \cdot \Upsilon_3$$
$$Q_{R\omega} = -2 \cdot \Upsilon_1 \cdot \omega;\ Q_{I\omega} = -[\Upsilon_2 + \Upsilon_3] \cdot \Upsilon_1$$

$$P_{R\tau} = 0;\ P_{I\tau} = 0;\ Q_{R\tau} = 0;\ Q_{I\tau} = 0;\ F_\tau = 0$$
$$V = (P_R \cdot P_{I\tau} - P_I \cdot P_{R\tau}) - (Q_R \cdot Q_{I\tau} - Q_I \cdot Q_{R\tau}) = 0$$

$F_\omega = \ldots$. Elements: $P_{R\omega} \cdot P_R = 2 \cdot [\Upsilon_2 + \Upsilon_3]^2 \cdot \omega^3$; $P_{I\omega} \cdot P_I = (3 \cdot \omega^2 - \Upsilon_2 \cdot \Upsilon_3) \cdot (\omega^3 - \Upsilon_2 \cdot \Upsilon_3 \cdot \omega)$

$$Q_{I\omega} \cdot Q_I = [\Upsilon_2 + \Upsilon_3]^2 \cdot \Upsilon_1^2 \cdot \omega$$
$$Q_{R\omega} \cdot Q_R = -2 \cdot \Upsilon_1 \cdot \omega \cdot (\Upsilon_2 \cdot \Upsilon_3 \cdot \Upsilon_1 - \Upsilon_1 \cdot \omega^2)$$

$U = \ldots$. Elements: $P_R \cdot P_{I\omega} = -(\Upsilon_2 + \Upsilon_3) \cdot \omega^2 \cdot (3 \cdot \omega^2 - \Upsilon_2 \cdot \Upsilon_3)$; $P_I \cdot P_{R\omega} = -2 \cdot (\omega^3 - \Upsilon_2 \cdot \Upsilon_3 \cdot \omega) \cdot (\Upsilon_2 + \Upsilon_3) \cdot \omega$ $Q_R \cdot Q_{I\omega} = -(\Upsilon_2 \cdot \Upsilon_3 \cdot \Upsilon_1 - \Upsilon_1 \cdot \omega^2) \cdot (\Upsilon_2 + \Upsilon_3) \cdot \Upsilon_1$; $Q_I \cdot Q_{R\omega} = (\Upsilon_2 + \Upsilon_3) \cdot 2 \cdot \Upsilon_1^2 \cdot \omega^2$

$F(\omega, \tau) = 0$, differentiating with respect to τ and we get $F_\omega \cdot \frac{\partial \omega}{\partial \tau} + F_\tau = 0$; $\tau \in I \Rightarrow \frac{\partial \omega}{\partial \tau} = -\frac{F_\tau}{F_\omega}$

$$\Lambda^{-1}(\tau) = (\frac{\partial \text{Re}\lambda}{\partial \tau})_{\lambda=i\omega}; \ \frac{\partial \omega}{\partial \tau} = \omega_\tau = -\frac{F_\tau}{F_\omega}$$
$$\Lambda^{-1}(\tau) = \text{Re}\{\frac{-2 \cdot [U + \tau \cdot |P|^2 + i \cdot F_\omega}{F_\tau + i \cdot 2 \cdot [V + \omega \cdot |P|^2]}\}$$

$$\text{sign}\{\Lambda^{-1}(\tau)\} = \text{sign}\{(\frac{\partial \text{Re}\lambda}{\partial \tau})_{\lambda=i\omega}\}$$

$$\text{sign}\{\Lambda^{-1}(\tau)\} = \text{sign}\{F_\omega\} \cdot \text{sign}\{\frac{V + \frac{\partial \omega}{\partial \tau} \cdot U}{|P|^2} + \omega + \frac{\partial \omega}{\partial \tau} \cdot \tau\}$$

We shall presently examine the possibility of stability transitions (bifurcations) copper vapor laser (CVL) rate equations, about the equilibrium point $E^{(*)} = (P_g^{(*)}, n_g^{(*)}, n_{cu}^{(*)})$. The analysis consists in identifying the roots of our system characteristic equation situated on the imaginary axis of the complex λ-plane, where by increasing the delay parameter τ. Reλ, may at the crossing, changes its sign from – to +, i.e. from a stable focus $E^{(*)}$ to an unstable one, or vice versa. This feature may be further assessed by examining the sign of the partial derivatives with respect to τ [4, 5].

$$\Lambda^{-1}(\tau) = (\frac{\partial \text{Re}\lambda}{\partial \tau})_{\lambda=i \cdot \omega}; \ P_{cu}, P_{vp}, v_d, R, v_2, l, P_{ex}, v_s, \Lambda = const$$

$$V = (P_R \cdot P_{I\tau} - P_I \cdot P_{R\tau}) - (Q_R \cdot Q_{I\tau} - Q_I \cdot Q_{R\tau}) = 0$$

$$\begin{aligned} U &= (P_R \cdot P_{I\omega} - P_I \cdot P_{R\omega}) - (Q_R \cdot Q_{I\omega} - Q_I \cdot Q_{R\omega}) \\ &= -(\Upsilon_2 + \Upsilon_3) \cdot \omega^2 \cdot (3 \cdot \omega^2 - \Upsilon_2 \cdot \Upsilon_3) \\ &\quad + 2 \cdot (\omega^3 - \Upsilon_2 \cdot \Upsilon_3 \cdot \omega) \cdot (\Upsilon_2 + \Upsilon_3) \cdot \omega \\ &\quad + (\Upsilon_2 \cdot \Upsilon_3 \cdot \Upsilon_1 - \Upsilon_1 \cdot \omega^2) \cdot (\Upsilon_2 + \Upsilon_3) \\ &\quad \cdot \Upsilon_1 + (\Upsilon_2 + \Upsilon_3) \cdot 2 \cdot \Upsilon_1^2 \cdot \omega^2 \end{aligned}$$

$$\begin{aligned} F_\omega &= 2 \cdot [2 \cdot [\Upsilon_2 + \Upsilon_3]^2 \cdot \omega^3 + (3 \cdot \omega^2 - \Upsilon_2 \cdot \Upsilon_3) \\ &\quad \cdot (\omega^3 - \Upsilon_2 \cdot \Upsilon_3 \cdot \omega) + 2 \cdot \Upsilon_1 \cdot \omega \cdot (\Upsilon_2 \cdot \Upsilon_3 \cdot \Upsilon_1 - \Upsilon_1 \cdot \omega^2) \\ &\quad - [\Upsilon_2 + \Upsilon_3]^2 \cdot \Upsilon_1^2 \cdot \omega]; \ \frac{\partial \omega}{\partial \tau} = \omega_\tau = -\frac{F_\tau}{F_\omega}|_{F_\tau=0, F_\omega \neq 0} = -\frac{F_\tau}{F_\omega} = 0 \end{aligned}$$

$$F_\tau = 2 \cdot [(P_{R\tau} \cdot P_R + P_{I\tau} \cdot P_I) - (Q_{R\tau} \cdot Q_R + Q_{I\tau} \cdot Q_I)] = 0$$

$$\begin{aligned} \text{sign}\{\Lambda^{-1}(\tau)\} = \text{sign}\{F_\omega\} \cdot \text{sign}\{&\frac{V + \frac{\partial \omega}{\partial \tau} \cdot U}{|P|^2} \\ &+ \omega + \frac{\partial \omega}{\partial \tau} \cdot \tau\} \end{aligned}$$

We find those ω, τ values which fulfill $F_{\omega}(\omega, \tau) = 0$. We ignore negative, complex, and imaginary values of ω for specific τ values. $\tau \in [0.001 \ldots 10]$, we can express by 3D function $F_{\omega}(\omega, \tau) = 0$. We plot the stability switch diagram based on different delay values of our Copper vapor laser (CVL) circuity.

$$\Lambda^{-1}(\tau) = \left(\frac{\partial \mathrm{Re}\lambda}{\partial \tau}\right)_{\lambda=i\omega} = \mathrm{Re}\left\{\frac{-2 \cdot [U + \tau \cdot |P|^2] + i \cdot F_{\omega}}{F_{\tau} + 2 \cdot i \cdot [V + \omega \cdot |P|^2]}\right\}$$

$$\Lambda^{-1}(\tau) = (\frac{\partial \mathrm{Re}\lambda}{\partial \tau})_{\lambda=i\omega} = \frac{2 \cdot \{F_{\omega} \cdot (V + \omega \cdot P^2) - F_{\tau} \cdot (U + \tau \cdot P^2)\}}{F_{\tau}^2 + 4 \cdot (V + \omega \cdot P^2)^2}$$

The stability switch occurs only on those delay values (τ) which fit the equation: $\tau = \frac{\theta_{+}(\tau)}{\omega_{+}(\tau)}$ and $\theta_{+}(\tau)$ is the solution of $\sin\theta(\tau) = \ldots.$ and $\cos\theta(\tau) = \ldots.$ when $\omega = \omega_{+}(\tau)$ If only ω_{+} is feasible. Additionally, when all copper vapor laser (CVL) rate equations parameters are known and the stability switch due to various time delay values τ is described in the following expression:

$$\mathrm{sign}\{\Lambda^{-1}(\tau)\} = \mathrm{sign}\{F_{\omega}(\omega(\tau), \tau)\} \cdot \mathrm{sign}\left\{\tau \cdot \omega_{\tau}(\omega(\tau)) + \omega(\tau) + \frac{U(\omega(\tau)) \cdot \omega_{\tau}(\omega(\tau)) + V(\omega(\tau))}{|P(\omega(\tau))|^2}\right\}$$

Remark we know $F(\omega, \tau) = 0$ implies its roots $\omega_i(\tau)$ and finding those delays values τ which ω_i is feasible. There are τ values which give complex ω_i or imaginary number, then unable to analyze stability. $F(\omega, \tau)$, function is independent on τ the parameter $F(\omega, \tau) = 0$. The results: We find those ω, τ values which fulfill $F(\omega, \tau) = 0$. We ignore negative, complex, and imaginary values of ω. Next is to find those ω, τ values which fulfill $\sin\theta(\tau) = \ldots.$ and $\cos\theta(\tau) = \ldots.$; $\sin(\omega \cdot \tau) = \frac{-P_R \cdot Q_I + P_I \cdot Q_R}{|Q|^2}$; $\cos(\omega \cdot \tau) = -\frac{(P_R \cdot Q_R + P_I \cdot Q_I)}{|Q|^2}$ $|Q|^2 = Q_R^2 + Q_I^2$. Finally we plot the stability switch diagram $g(\tau) = \Lambda^{-1}(\tau) = (\frac{\partial \mathrm{Re}\lambda}{\partial \tau})_{\lambda=i\omega}$

$$g(\tau) = \Lambda^{-1}(\tau) = \left(\frac{\partial \mathrm{Re}\lambda}{\partial \tau}\right)_{\lambda=i\omega} = \frac{2 \cdot \{F_{\omega} \cdot (V + \omega \cdot P^2) - F_{\tau} \cdot (U + \tau \cdot P^2)\}}{F_{\tau}^2 + 4 \cdot (V + \omega \cdot P^2)^2}$$

$$\mathrm{sign}[g(\tau)] = \mathrm{sign}[\Lambda^{-1}(\tau)] = \mathrm{sign}\left[\left(\frac{\partial \mathrm{Re}\lambda}{\partial \tau}\right)_{\lambda=i\omega}\right] = \mathrm{sign}\left[\frac{2 \cdot \{F_{\omega} \cdot (V + \omega \cdot P^2) - F_{\tau} \cdot (U + \tau \cdot P^2)\}}{F_{\tau}^2 + 4 \cdot (V + \omega \cdot P^2)^2}\right]$$

$$F_{\tau}^2 + 4 \cdot (V + \omega \cdot P^2)^2 > 0 \Rightarrow \mathrm{sign}[\Lambda^{-1}(\tau)]$$

Table 6.12 Copper vapor laser (CVL) rate equations sign of sign$[\Lambda^{-1}(\tau)]$

sign$[F_\omega]$	sign$[\frac{V+\omega_\tau \cdot U}{P^2} + \omega + \omega_\tau \cdot \tau]$	sign$[\Lambda^{-1}(\tau)]$
$\pm$	$\pm$	+
$\pm$	$\mp$	−

$$= \text{sign}[F_\omega \cdot (V + \omega \cdot P^2) - F_\tau \cdot (U + \tau \cdot P^2)]$$

$$\text{sign}[\Lambda^{-1}(\tau)] = \text{sign}\left\{[F_\omega] \cdot \left[(V + \omega \cdot P^2) - \frac{F_\tau}{F_\omega} \cdot (U + \tau \cdot P^2)\right]\right\}$$

$$\omega_\tau = -\frac{F_\tau}{F_\omega}; \omega_\tau = \left(\frac{\partial \omega}{\partial \tau}\right)^{-1} = -\frac{\partial F / \partial \omega}{\partial F / \partial \tau}$$

$$\text{sign}[\Lambda^{-1}(\tau)] = \text{sign}\left\{[F_\omega] \cdot [P^2] \cdot \left[\frac{V + \omega_\tau \cdot U}{P^2} + \omega + \omega_\tau \cdot \tau\right]\right\}; \text{sign}[P^2] > 0$$

$$\text{sign}[\Lambda^{-1}(\tau)] = \text{sign}\left\{[F_\omega] \cdot \left[\frac{V + \omega_\tau \cdot U}{P^2} + \omega + \omega_\tau \cdot \tau\right]\right\}$$

$$\text{sign}[\Lambda^{-1}(\tau)] = \text{sign}[F_\omega] \cdot \text{sign}\left[\frac{V + \omega_\tau \cdot U}{P^2} + \omega + \omega_\tau \cdot \tau\right]$$

$$F_\omega = 2 \cdot [2 \cdot [\Upsilon_2 + \Upsilon_3]^2 \cdot \omega^3 + (3 \cdot \omega^2 - \Upsilon_2 \cdot \Upsilon_3) \cdot (\omega^3 - \Upsilon_2 \cdot \Upsilon_3 \cdot \omega) + 2 \cdot \Upsilon_1 \cdot \omega \cdot (\Upsilon_2 \cdot \Upsilon_3 \cdot \Upsilon_1 - \Upsilon_1 \cdot \omega^2) - [\Upsilon_2 + \Upsilon_3]^2 \cdot \Upsilon_1^2 \cdot \omega]$$

We check the sign of $\Lambda^{-1}(\tau)$ according to the following rule: (Table 6.12)

If, sign$[\Lambda^{-1}(\tau)] > 0$, then the crossing proceeds from (-) to (+) respectively (stable to unstable). If, sign$[\Lambda^{-1}(\tau)] < 0$, then the crossing proceeds from (+) to (−) respectively (unstable to stable).

6.5 Questions

1. Optical system with diode pumped Alkali metal vapor laser (DPAL), Where I_p is the intensity of the input pump beam, I_+ and I_- are right and left running intra-cavity laser intensities. n_1 is the lower level and n_2, n_3 are the two upper levels.

$$f = \frac{1}{1 + 2 \cdot e^{-\left(\frac{\Delta E_{32}}{k_B \cdot T}\right)}}; n_1 = n_L; n_2 = f \cdot n_U$$

$$n_3 = (1-f)\cdot n_U; n_U = n_2 + n_3; n_{tot} = n_U + n_L$$

Where T is the temperature in degrees Kelvin, k_B is the Boltzmann constant and $\Delta E_{3,2}$ is the energy different between level 3 and level 2. n_{tot} is the total Alkali atom number density, $\frac{n_3}{n_2} = 2\cdot e^{-[\frac{\Delta E_{32}}{k_B\cdot T}]}$. The diode pumped Alkali metal vapor laser (DPAL) model and pump and laser beam geometry includes two mirrors (M1 and M2) and laser Gain cell. We consider that the collision induced mixing or relaxation between $^2P_{3/2}$ and $^2P_{1/2}$ states is much faster by comparison with any other process in the system including stimulated emission. The mathematical description is of a quasi-two level system from which an analytical solution is extracted. The optical schematics for DPAL and the geometry for single or double pass longitudinal pumping are described our system. It includes the right and left running laser waves. Due to non-ideal mirrors (M1 and M2) in Optical schematic for diode pumped Alkali metal vapor laser (DPAL), there are shifting in z, $(\Delta_{z_1}, \Delta_{z_2})$ for the right and left running intra-cavity laser intensities (I_+, I_-), respectively. The shifts in z do not affect the derivatives of I_+, I_-in z $(\frac{dI_+}{dz}, \frac{dI_-}{dz})$. The non-ideal mirrors M1 and M2 is caused by mirror's surfaces and non-uniformity reflection (all other parameters as describe in subchapter 6.2).

$$\frac{dI_p(z)}{dz} = \sigma_{31}\cdot[(1-f)\cdot n_U - 2\cdot n_L]\cdot I_p(z)$$

$$\frac{dI_+(z)}{dz} = \sigma_{21}\cdot(f\cdot n_U - n_L)\cdot I_+(z-\Delta_{z_1})$$

$$\frac{dI_-(z)}{dz} = -\sigma_{21}\cdot(f\cdot n_U - n_L)\cdot I_-(z-\Delta_{z_2}); n_U = \xi_1(I_p, I_+(z-\Delta_{z_1}), I_-(z-\Delta_{z_2}))$$

$$n_L = \xi_2(I_p, I_+(z-\Delta_{z_1}), I_-(z-\Delta_{z_2})); n_U = \xi_1(I_p, I_+(z-\Delta_{z_1}), I_-(z-\Delta_{z_2}))$$

$$n_U = \xi_1(I_p, I_+, I_-) = -\left(\frac{\tau\cdot n_{tot}\cdot(2\cdot\sigma_{31}\cdot I_p\cdot v_L + \sigma_{21}\cdot(I_+ + I_-)\cdot v_p)}{(f-3)\cdot\sigma_{31}\cdot\tau\cdot I_p\cdot v_L - [(f+1)\cdot\sigma_{21}\cdot\tau\cdot(I_+ + I_-) + h\cdot v_L]\cdot v_p}\right)$$

$$n_L = \xi_2(I_p, I_+, I_-) = \frac{n_{tot}\cdot[(f-1)\cdot\sigma_{31}\cdot\tau\cdot I_p\cdot v_L - (f\cdot\sigma_{21}\cdot\tau\cdot(I_+ + I_-) + h\cdot v_L)\cdot v_p]}{(f-3)\cdot\sigma_{31}\cdot\tau\cdot I_p\cdot v_L - [(f+1)\cdot\sigma_{21}\cdot\tau\cdot(I_+ + I_-) + h\cdot v_L]\cdot v_p}$$

1.1 Find system fixed points. Draw fixed points coordinates as a function of f (Scaling parameter for the density of population in the mid and upper state, $0 \le f \le 1$), and σ_{31} - 3D graph.
1.2 Discuss stability and stability switching for different values of Δ_{z_1}, $\Delta_{z_2} \to 0$.
1.3 Discuss stability and stability switching for different values of Δ_{z_2}, $\Delta_{z_1} \to 0$.
1.4 Parameter σ_{21} (line center simulated emission/absorption cross sections) is Very small, $\sigma_{21} \to \varepsilon$, return (1.1) and (1.2). Explain how our results change.

1.5 Parameter σ_{21} (line center simulated emission/absorption cross sections) doubles his value, $\sigma_{21-\text{new}} = 2 \cdot \sigma_{21}$, Explain how our results change?

2. A three-level analytic DPAL model is applied to pulsed potassium and quasi cw rubidium laser demonstrations. This model is adequate to describe the entire output laser dynamics observed for the broad range of gain cell conditions. Application of the model to gain cells with helium buffer gases requires a modification to the alkali concentration and is further interpreted as a thermal effect. The model uses longitudinally averaged number densities to predict cw or temporally evolving output intensities, including conditions where the spin-orbit relaxation rate limits performance. Determining the number of absorbed pump photons via the longitudinally averaged pump intensity, Ω(t), requires the solution of a transcendental equation, we take it in our analysis as a constant $\Omega(t) = \Omega$. Where the populations in the ground $^2S_{1/2}$ state, $n_1(t)$, pumped $^2P_{3/2}$ state, $n_3(t)$, and upper laser $^2P_{1/2}, n_2(t)$, are implicitly a function of the pump intensity and specified by the standard laser rate equations: $n_1 = n_1(t); n_2 = n_2(t); n_3 = n_3(t)$

$$\frac{dn_1}{dt} = \sigma_{31} \cdot (n_3 - 2 \cdot n_1) \cdot \frac{\Omega}{h \cdot v_p} + \sigma_{21} \cdot (n_2 - n_1) \cdot \frac{\psi}{h \cdot v_L} + n_2 \cdot \Gamma_{21} + n_3 \cdot \Gamma_{31}$$

$$\frac{dn_2}{dt} = -\sigma_{21} \cdot (n_2 - n_1) \cdot \frac{\psi}{h \cdot v_L} + \gamma_{mix} \cdot (n_3 - 2 \cdot e^{-\theta} \cdot n_2) - n_2 \cdot \Gamma_{21}$$

$$\frac{dn_3}{dt} = -\sigma_{31} \cdot (n_3 - 2 \cdot n_1) \cdot \frac{\Omega}{h \cdot v_p} + \gamma_{mix} \cdot (n_3 - 2 \cdot e^{-\theta} \cdot n_2) - n_3 \cdot \Gamma_{31}$$

The material and cavity parameters are defined as: σ_{31} is the stimulated emission cross-section for the pump transition, σ_{21} is the stimulated emission cross-section for the lasing transition, $\Gamma_{21} = 3.61 \times 10^7 \text{s}^{-1}$ (assumes no quenching), $\Gamma_{31} = 3.81 \times 10^7 \text{s}^{-1}$ (assumes no quenching), $\theta = \frac{\Delta E_{21}}{k \cdot T} = 0.91$; $T = 376$ K, Rubidium concentration, n, $n = n_1 + n_2 + n_3 = \sum_{i=1}^{3} n_i = 0.15 - 2.01 \times 10^{13} \frac{\text{atoms}}{\text{cm}^3}$, γ_{mix} is spin-orbit mixing rate, typically is equal to $4.56 - 9.14 \times 10^9 \text{s}^{-1}$ (depending on gas mixture). The average intra-cavity DPAL laser intensity, ψ, is amplified via stimulated emission from the spontaneous noise and limited by output coupling and cavity losses:
$\frac{d\psi}{dt} = 0 = \{r \cdot T^4 \cdot e^{(\sigma_{21} \cdot (n_2 - n_1) \cdot 2 \cdot l_g} - 1\} \cdot \frac{\psi}{\tau_{RT}} + \frac{n_2 \cdot c^2 \cdot \sigma_{21} \cdot h \cdot v_L}{l_g}$, where r is the output coupler reflectivity, l_g is the gain length (12.7cm), $\tau_{RT} = \frac{2 \cdot l_g}{c} = 2.5$nsec, c is the speed of light in vacuum.

2.1 Find system fixed points (equilibrium points), Plot system fixed points (1D) as a function of σ_{31} (n_1, n_2 and n_3 are system variables).
2.2 Discuss system stability and stability switching for different values of σ_{31} and σ_{21} parameters.
2.3 Discuss stability and stability switching, where $n_3(t) \to n_3(t - \Delta)$ is delay in time (not effect $\frac{dn_3}{dt}$) for different values of Δ, delay parameter.

2.4 Parameter $\theta = \frac{\Delta E_{21}}{k \cdot T}$ is very small, $\Delta E_{21} \to 0 \Rightarrow \theta \to 0$, find system fixed points and discuss stability and stability switching for different values of σ_{31} parameter.

2.5 The system gain length (l_g) triple his value, $l_g \to 3 \cdot l_g$, How the system dynamic is changed? Discuss stability.

3. We have a Nitrogen gas laser filament plasma kinetic system. The set of plasma kinetic equations is associated with the relevant rate equations for the population of the lasing levels and number of emitted photons. The change of concentration of different neutral and ionic atomic and molecular species in the filament plasma is described by set of rate equations $\frac{d[N_s]}{dt} = [G_s] - [L_s]$. Where the subscript s stands for $e, Ar^+, Ar_2^+, N_2^+, N_4^+, N_3^+$ and $N^+, N, N_2(A^3\Sigma_u), N_2(B^3\Pi_g), N_2(C^3\Pi_u), Ar^*(4^3P_2)$; N_s is the density of species of type- s; and G_s,L_s are the relevant generation and loss rates. The rate equations are solved jointly with the equations for the electron temperature and the vibrational temperature of ground-state nitrogen molecules.

$$\frac{3}{2} \cdot k \cdot N_e \cdot \frac{dT_e}{dt} = -\frac{3}{2} \cdot N_e \cdot k \cdot (T_e - T_{vibr}) \cdot v_* \\ - \frac{3}{2} \cdot N_e \cdot k \cdot (T_e - T) \cdot [\delta_{N_2} \cdot (v_{N_2} + v_{N_2^+} + \frac{1}{2} \cdot v_{N_4^+}) \\ + \delta_{A_r} \cdot (v_{A_r} + v_{A_r^+} + \frac{1}{2} \cdot v_{A_{r_2}^+})] + k_e \cdot N_e \cdot N_{A_r*} \cdot I_{exc}$$

$$\frac{dT_{vibr}}{dt} = (\frac{N_e}{N_{N_2}}) \cdot (T_e - T_{vibr}) \cdot v_* - (T_{vibr} - T) \times [\frac{1}{\tau_{VT,N_2}} + (\frac{N_{A_r}}{N_2})/\tau_{VT,A_r}]$$

All system parameters and variables as describe in subchapter (6.1), T, T_e, T_{vibr} are gas, electronic, and vibrational temperatures, respectively. The concentration of system gas is not balanced and $T_e \approx T_{vibr}$, and additionally there is a new behavior for T parameter, $T \to T_\Xi + \Omega \cdot \sqrt{T}$; $T_\Xi \in \mathbb{N}$; $\Omega \in \mathbb{R}$.

3.1 Find system fixed points and discuss how the fixed points coordinate are changed for different values of T_Ξ and Ω parameters.

3.2 Do linearization to system $f_1(T_e, T_{vibr})$; $f_2(T_e, T_{vibr})$ equations and get the system eigenvalues of the matrix A.

3.3 Discuss stability and stability switching for different values of T_Ξ and Ω parameters.

3.4 $T \to T_\Xi + \Omega \cdot \sqrt{T}$; $T_\Xi \in \mathbb{N}$; $\Omega \in \mathbb{R}$; $T_\Xi = 0$, How the dynamic of the system is changed? Find fixed points and discuss stability.

3.5 How our system dynamic is changed for $T_\Xi \gg \Omega$, discuss stability.

4. A laser initiation RF technique create and sustain large-volume, high-pressure air and nitrogen plasmas. The technique utilizes a laser-initiated, 15 mTorr partial tetrakis ethylene seed plasma with a 75 Torr background gas pressure to achieve high-pressure air/nitrogen plasma breakdown and reduce the RF power requirement needed to sustain the plasma. The target is to create a power-efficient, large-volume

atmospheric plasma, breakdown of atmospheric-pressure air is assisted by a laser-produced seed plasma. The phase and attenuation constants of the plasma is

$$\beta_p = \frac{\omega}{c} \cdot \{\frac{1}{2} \cdot (1 - \frac{\omega_p^2}{\omega^2 + v^2}) + \frac{1}{2} \cdot [(1 - \frac{\omega_p^2}{\omega^2 + v^2})^2 + (\frac{\omega_p^2}{\omega^2 + v^2} \cdot \frac{v}{\omega})^2]^{\frac{1}{2}}\}^{\frac{1}{2}}$$

$$\alpha_p = \frac{\omega}{c} \cdot \{-\frac{1}{2} \cdot (1 - \frac{\omega_p^2}{\omega^2 + v^2}) + \frac{1}{2} \cdot [(1 - \frac{\omega_p^2}{\omega^2 + v^2})^2 + (\frac{\omega_p^2}{\omega^2 + v^2} \cdot \frac{v}{\omega})^2]^{\frac{1}{2}}\}^{\frac{1}{2}}$$

where, ω is the frequency of the electromagnetic wave, c is the speed of light in the free space, and the electron plasma frequency is $\omega_p = \sqrt{\frac{n_e \cdot e^2}{\varepsilon_0 \cdot m}\,\frac{textrad}{sec}}$. The plasma interferometry objective is to accurately diagnose the line-averaged v and $n_e$during the entire plasma pulse. The plasma characteristics are v, ω_p, n_e and T_e. The phase constant is β_p and the attenuation constant is α_p of the wave. We know that due to system non linearity there is a dependent of β_p parameter on ω (frequency of the electromagnetic wave), $\beta_p(\omega)$ and α_p on ω, $\alpha_p(\omega)$. We define our system by two differential equations: $\frac{d\beta_p}{d\omega} = f_1(\beta_p, \alpha_p)$; $\frac{d\alpha_p}{d\omega} = f_2(\beta_p, \alpha_p)$.

4.1 Find system fixed points ($\frac{d\beta_p}{d\omega} = 0$; $\frac{d\alpha_p}{d\omega} = 0$).
4.2 Discuss stability and stability switching for different values of ω_p, plasma parameter.
4.3 How the system dynamic is changed if the plasma frequency (ω_p) is equal to the frequency of the electromagnetic wave ($\omega_p \approx \omega$)?
4.4 Find system fixed points and discuss stability if $\omega_p = \Omega + \omega$; $\Omega \in \mathbb{R}$, Hint: Ω parameter can be negative or positive and fulfil, $\omega_p > 0$; $\omega > 0$.
4.5 How the system dynamics is changed if $\omega_p \gg \omega$ and if $\omega_p \ll \omega$?

5. We have a self-consistent model copper vapor laser (CVL) system. Our system is represented by a related CVL discharge circuit (see subchapter 6.3, Fig. 6.7). The model discharge circuit elements are capacitor C_1 which is the storage capacitor resonantly charged to voltage V_0 by a DC power supply and inductance L_c. Capacitor C_2 is the peaking capacitor. The Thyratron switch (S_1) is modeled by a reverse bias voltage and removed with a switching time τ_s. R_c is the parallel charging resistance. The Thyratron behaves like a diode with current flowing through it in only one direction. The inductances L_1 and L_2 represent the lumped circuit values. Where current I_1 flows through L_1 and current I_2 flows through L_2, voltage V_1 is the voltage across capacitor C_1 and V_2 is the voltage across capacitor C_2. The voltage across the discharge tube is V_d, and the current that flows through the discharge tube is I_d, $I_d = I_2(t - \tau_2) \cdot \frac{R_c}{R_c + R_d}$. The real model differential equations circuitry which represents the actual behavior of our system is as follow:

$$\frac{dI_1}{dt} = \frac{(V_1 - V_2) + V_0 \cdot f(\tau_s)}{L_1}; \frac{dI_2}{dt} = \frac{V_2 - V_d}{L_2}$$

$$\frac{dV_1}{dt} = -\frac{I_1(t - \tau_1)}{C_1}; \frac{dV_2}{dt} = \frac{I_1(t - \tau_1) - I_2(t - \tau_2)}{C_2}$$

$f(\tau_s)$ is the Thyratron voltage turn on function, function $f(\tau_s) = \sum_{k=1}^{2} I_k(t-\tau_k) + \frac{dI_2}{dt}$.

5.1 Find system (CVL discharge circuit) fixed points. How the value of $\frac{dI_2}{dt}$ influences the fixed points coordinates?
5.2 Discuss stability and stability switching for $\tau_1 = \tau;\ \tau_2 = 0,\ x = \tau$.
5.3 Discuss stability and stability switching for $\tau_1 = 0;\ \tau_2 = \tau,\ x = \tau$.
5.4 Return 5.1 – 5.3 for $I_2(t-\tau_2) \to 0$, at fixed point $\lim_{t\to\infty} I_2(t-\tau_2)|_{t\gg\tau_2} = I_2(t)$.
5.5 Return 5.1 – 5.3 for $I_1(t-\tau_1) \to 0$, at fixed point $\lim_{t\to\infty} I_1(t-\tau_2)|_{t\gg\tau_1} = I_1(t)$.

6. We have a self-consistent model copper vapor laser (CVL) system. Our system is represented by a related CVL discharge circuit (see subchapter 6.3, Fig. 6.7). The model discharge circuit elements are capacitor C_1 which is the storage capacitor resonantly charged to voltage V_0 by a DC power supply and inductance L_c. Capacitor C_2 is the peaking capacitor. The Thyratron switch (S_1) is modeled by a reverse bias voltage and removed with a switching time τ_s. R_c is the parallel charging resistance. The Thyratron behaves like a diode with current flowing through it in only one direction. The inductances L_1 and L_2 represent the lumped circuit values. Where current I_1 flows through L_1 and current I_2 flows through L_2, voltage V_1 is the voltage across capacitor C_1 and V_2 is the voltage across capacitor C_2. The voltage across the discharge tube is V_d, and the current that flows through the discharge tube is I_d, $I_d = I_2(t-\tau_2)\cdot\frac{R_c}{R_c+R_d}$. The real model differential equations circuitry which represents the actual behavior of our system is as follow:

$$\frac{dI_1}{dt} = \frac{(V_1 - V_2) + V_0\cdot f(\tau_s)}{L_1};\ \frac{dI_2}{dt} = \frac{V_2 - V_d}{L_2}$$
$$\frac{dV_1}{dt} = -\frac{I_1(t-\tau_1)}{C_1};\ \frac{dV_2}{dt} = \frac{I_1(t-\tau_1) - I_2(t-\tau_2)}{C_2}$$

$f(\tau_s)$ is the Thyratron voltage turn on function, function $f(\tau_s) = \sum_{k=1}^{2} I_k^2(t-\tau_k) + \sqrt[\Gamma]{I_2};\ \Gamma \in \mathbb{N}$.

6.1 Find system (CVL discharge circuit) fixed points. Plot fixed points coordinates.
6.2 Discuss stability and stability switching for different values of Γ parameter ($\tau_1 = 0,\ \tau_2 = 0$).
6.3 Discuss stability and stability switching for different values of τ parameter ($\tau_1 = \tau,\ \tau_2 = 0$).
6.4 Discuss stability and stability switching for different values of τ parameter ($\tau_1 = 0,\ \tau_2 = \tau$).
6.5 Discuss stability and stability switching for different values of τ parameter ($\tau_1 = \sqrt{\tau},\ \tau_2 = \tau$).

7. We have a heat and mass transfer in pulsed laser alloying system. Typical process monitoring is based on the detection of electromagnetic radiation from the zone of laser action. We get information on the various surface phenomena and the

material removal (vapor and plasma flows, droplets of melt, etc.), but no information from the inside of irradiated metal is available. By analysis of the treatment results after the end of energy action, we reconstruct mainly integral parameters, but not the dynamics of the process. The calculation of temperature fields on the base of linear mathematical models is crucial. The main difficulties appear during the simulation of phase transformations of melting and evaporation. The mathematical models are nonlinear. The heat transfer phenomena are described by the dynamics of phase boundaries in pulsed laser action. The two phase fronts model: movement of two phase fronts, melting and evaporation. The evaporation phenomenon into the model prevents the rise of temperature up to nonrealistic values and makes it possible to consider comparatively high values of energy density flows. We assume that the energy flow is absorbed on the irradiated surface; convection and radiation mechanisms of heat losses from both the sides of the slab are taking care and the melting is determined. The model includes both melting and evaporation phenomena. There is a wide range of energy density flow and pulse duration values. The mathematical model is describes by the following equations: $T(x, t = 0) = T_0$

$$\frac{\partial^2 T_1}{\partial x^2} = \frac{1}{a_1} \cdot \frac{\partial T_1}{\partial t}; \; S_1(t) < x < S_2(t); \; t_m < t < 0$$

$$q_0(t) - \alpha_g \cdot [T_1(x, t) - T_g] - \sigma \cdot [\varepsilon_1 \cdot T_1^4(x, t) - \varepsilon_g \cdot T_g^4] = -\lambda_1 \cdot \frac{\partial T_1}{\partial x} + \rho_1 \cdot L_v \cdot \frac{dS_1}{dt}$$

$$\frac{dS_1}{dt} = \frac{V_\#}{\sqrt{T_1(S_1(t), t)}} \cdot e^{[-\frac{T_\#}{T_1(S_1(t), t)}]}; \; x = S_1(t)$$

$$\lambda_1 \cdot \frac{\partial T_1}{\partial x} = \lambda_2 \cdot \frac{\partial T_2}{\partial x} - \rho_2 \cdot L_m \cdot \frac{dS_2}{dt}; \; T_1 = T_2 = T_m; \; x = S_2(t); \; S_2(T_m) = 0$$

$$\frac{\partial^2 T_2}{\partial x^2} = \frac{1}{a_2} \cdot \frac{\partial T_2}{\partial t}; \; S_2(t) < x < L$$

$$-\lambda_2 \cdot \frac{\partial T_2}{\partial x} = \alpha_f \cdot (T_2 - T_f) + \sigma \cdot (\varepsilon_2 \cdot T_2^4 - \varepsilon_f \cdot T_f^4); \; x = L$$

Where $T_1(x, t)$ temperature of liquid phase; $T_2(x, t)$ temperature of solid phase; x, t are distance and time; $S_1(t)$, $S_2(t)$ positions of evaporation and melting phase boundaries, respectively; $q_0(t)$ is the absorbed energy density flux; a_1, a_2 are thermal diffusivity of liquid and solid phase, respectively; λ_1, λ_2 are thermal conductivity of liquid and solid phase; ρ_1, ρ_2 are densities of liquid and solid phase, respectively; L_v is specific heat of evaporation; L_m is latent heat of melting; α_g, α_f are convection heat losses coefficient of irradiated and rear surfaces of the slab; ε_1, ε_2 are emissivity of irradiated and rear surfaces of the slab; ε_g, ε_f are emissivity of the environment near the irradiated and rear surfaces of the slab; T_m is the temperature of melting; t_m is the starting time for melting; L is the thickness of the slab; T_0 is the initial temperature. $V_\#$, $T_\#$ are constants which determined by the Herz-Knudsen law of evaporation:

$$V_{\#} = \frac{P_v}{2 \cdot \rho_1 \cdot \sqrt{\frac{2 \cdot \pi \cdot k}{m}}} \cdot e^{[\frac{L_v}{T_v \cdot (\frac{k}{m})}]}; \; T_{\#} = \frac{L_v}{(k/m)}$$

Where $k = 1.38 \times 10^{-23} [\frac{J}{K}]$, Boltzmann constant; m atomic mass of slab material; T_v is the boiling temperature corresponding to the pressure P_v. Hint: we can replace the partial derivative by regular derivative, $\frac{\partial}{\partial x} \leftrightarrow \frac{d}{dx}$; $\frac{\partial}{\partial t} \leftrightarrow \frac{d}{dt}$. The variables of our system are T_1, T_2, S_1, S_2 and all other parameters.

7.1 Try to find system fixed points in each distance interval which is characterize by nonlinear partial differential equations.
7.2 Discuss stability and stability switching for values of α_g parameter.
7.3 Discuss stability and stability switching for values of α_f parameter.
7.4 Assume that $\frac{\partial T_1}{\partial x} = \sqrt{S_2}$; $\frac{\partial T_2}{\partial x} = \sqrt[\Gamma]{S_2}$; $\Gamma \in \mathbb{N}$, how the dynamic of the system is changed? Discuss stability and stability switching for different values of Γ parameter.
7.5 Variable $T_1(x, t)$ is delayed in time, $T_1(x, t - \tau)$, discuss stability and stability switching for different values of τparameter. ***Remark*** $T_1(x, t)$ derivative in distance, $\frac{\partial T_1}{\partial x}$ do not effected by τ parameter.

8. Laser cladding is a technique to coat structural materials with a wear and/or corrosion resistance material without altering the mechanical properties of the bulk. This process is suited to coat medium to small areas with a moderate thickness coating. The advantages include easy control of the coating thickness, minimal heat load to the substrate, low roughness, excellent adhesion between the coating and substrate, minimal dilution of the coating's material by the substrate's material, and fine grained micro structures with superior hardness, wear and corrosion properties which result from the high cooling rates involved. Laser cladding is performed by either by melting a power preplaced on the substrate or by blowing the powder into the molten pool produced by the laser beam. There is an influence of the process parameters on the cladding quality and concluded that dilution depends on the specific energy and on the injected powder mass flow (q). The angel α depends on the dilution, the spot diameter, the scanning speed (v), the powder feed rate, the catchment efficiency and the powder's material density (ρ). A simple theoretical correlation between the cladding parameters and the dimensions and shape of the clad track that is valid for low dilution. The model: we consider the cladding process is stationary, the area (A) of the cross section of the clad track is given by the mass balance: $A = \frac{q}{\rho \cdot v} \cdot \eta$, where q is the powder feed rate, ρ is the density of the powder's material, v is the scanning speed and η is the catchment efficiency of the powder. The variation of the track width (w) with the scanning speed is expressed by $w = a - b \cdot v$, where, a, b are constants that depend on the laser power, the beam diameter, the powder mass flow rate and the powder particle velocity. The profile of the transverse cross section of the liquid track (S) is $S : y = f(x)$ in two dimensional frames. The position of the contact line between three different single-phase regions, which corresponds to the points $(\frac{w}{2})$ and $(-\frac{w}{2})$, and we can use the width values with no powder injection. The shape of the interface at $x = \pm\frac{w}{2}$ determines the contact

angle between the solid and the liquid $\tan(\alpha) = \frac{df}{dx}|_{x=\frac{w}{2}})$. This shape is determined by the normal component of the force balance over the interface which is given by $\Delta p(x) + \Delta(\mu\overline{\overline{S}} \cdot \overline{n} \cdot \overline{n}) = \gamma \cdot [1 + (\frac{df}{dx})^2]^{-\frac{3}{2}} \cdot \frac{d^2 f}{dx^2}$

Where, ΔQ is the jump at the interface of a property Q, p is the pressure, m is the viscosity, $\mu\overline{\overline{S}}$ is the viscous stress tensor (constant) and $\overline{n}$ is the unit normal pointing out of liquid. g is the surface tension times the mean curvature of S. The gravitational forces are $\Delta p(x) = \Delta p$ and the variation of γ with temperature (T) and the concentration (C) which lead to the Marangoni effect. If $\overline{\overline{S}} = 0$ s (static case), the balance nonlinear differential equation states that the interface S has a constant curvature radius. We define two variables for our system $y_1 = \frac{df}{dx}$; $y_2 = \frac{dy_1}{dx}$ and all other balance nonlinear differential equations are constant.

8.1 Find system fixed points, and plot as a function of viscous stress tensor $\mu\overline{\overline{S}}$.
8.2 Discuss stability and stability switching for different values of gravitational forces parameter Δp.
8.3 We have y_1 variable shifting in x coordinate, $y_1(x) \to y_1(x - \Delta x)$, Investigate stability and stability switching for different values of Δx parameter.
8.4 We know that the gravitational forces parameter Δp is a function of f function, $\Delta p = \sqrt{f + \frac{df}{dx}}$, find system fixed points and discuss stability.
8.5 We know that the gravitational forces parameter Δp is a function of f function, $\Delta p = \sqrt{f^2 + (\frac{df}{dx})^{\Xi}}$; $\Xi \in \mathbb{N}$, find system fixed points and discuss stability and stability switching for different values of Ξ parameter.

9. We have a copper vapor laser (CVL) which characterize by set of delay differential equations (DDEs), rate equations. The self- consistent model for high repetition rate copper vapor laser (CVL) simulates a discharge pulse, laser pulse, an inter-pulse afterglow in a high repetition rate. System model thermodynamic quantities are T_g, P_{Cu}, and P_g. The radial diffusion length of the discharge tube is Λ and v_d is the speed at which copper atoms diffuse to the wall. Exclusive of changes in the gas temperature, the time constant for relaxation of P_g to its equilibrium value is $\frac{l}{2 \cdot v_s}$; where the length of the discharge tube is $l$and v_s is the sound speed. P_g—Buffer gas pressure, T_g—Gas temperature, P_{cu}—Copper vapor pressure, P_{ex}—External buffer gas pressure, P_{vp}—Copper vapor pressure based on the wall temperature, R—Mirror reflectivity. The time delays τ_1, τ_2 and τ_3 are related to buffer gas pressure, and densities of the buffer gas and copper atoms, respectively. ***Remark*** The time delays parameters τ_1, τ_2 and τ_3 do not affect the derivatives $\frac{dP_g}{dt}$, $\frac{dn_g}{dt}$, $\frac{dn_{cu}}{dt}$.

$$\frac{dP_g}{dt} = [P_{cu} - P_{vp}] \cdot \frac{v_d}{R \cdot P_{cu}} \cdot P_g(t - \tau_1) - [P_g(t - \tau_1) - P_{ex}] \cdot \frac{2 \cdot v_2}{l}$$
$$\frac{dn_g}{dt} = -\sqrt{[P_g(t - \tau_1) - P_{ex}]} \cdot \frac{n_g(t - \tau_2) \cdot 2 \cdot v_s}{l \cdot P_g(t - \tau_1)}; \quad \frac{dn_{cu}}{dt} = -[P_{cu} - P_{vp}] \cdot \frac{n_{cu}(t - \tau_3) \cdot v_d}{\Lambda \cdot P_{cu}}$$

9.1 Find system fixed points, and draw coordinates as a function of P_{ex} parameter.
9.2 Discuss stability and stability switching (for different values of τ parameter) for $\tau_1 = 0$, $\tau_2 = \tau$ and $\tau_3 = 0$.

9.3 9.3 Discuss stability and stability switching (for different values of τ parameter) for $\tau_1 = \sqrt{\tau},\ \tau_2 = 0$ and $\tau_3 = \tau$.

9.4 Discuss stability and stability switching (for different values of τ parameter) for $\tau_1 = \sqrt{\tau},\ \tau_2 = \sqrt[3]{\tau}$ and $\tau_3 = \sqrt[4]{\tau}$.

9.5 The external buffer gas pressure P_{ex} is very small, $P_{ex} \to 0$, How it influence the dynamic of the system? Find fixed points and discuss stability for $\tau_1 = 0,\ \tau_2 = \tau$ and $\tau_3 = 0$.

10. We have a copper vapor laser (CVL) which characterize by set of delay differential equations (DDEs), rate equations. The self- consistent model for high repetition rate copper vapor laser (CVL) simulates a discharge pulse, laser pulse, an inter-pulse afterglow in a high repetition rate. System model thermodynamic quantities are T_g, P_{Cu}, and P_g. The radial diffusion length of the discharge tube is Λ and v_d is the speed at which copper atoms diffuse to the wall. Exclusive of changes in the gas temperature, the time constant for relaxation of P_g to its equilibrium value is $\frac{l}{2 \cdot v_s}$; where the length of the discharge tube is l and v_s is the sound speed. P_g—Buffer gas pressure, T_g—Gas temperature, P_{cu}—Copper vapor pressure, P_{ex}—External buffer gas pressure, P_{vp}—Copper vapor pressure based on the wall temperature, R—Mirror reflectivity. The time delays τ_1, τ_2 and τ_3 are related to buffer gas pressure, and densities of the buffer gas and copper atoms, respectively, Γ parameter $1 > \Gamma > 0; \Gamma \in \mathbb{R}$. Remark: The time delays parameters τ_1, τ_2 and τ_3 do not affect the derivatives $\frac{dP_g}{dt}, \frac{dn_g}{dt}, \frac{dn_{cu}}{dt}$.

$$\frac{dP_g}{dt} = [P_{cu} - P_{vp}] \cdot \frac{v_d}{R \cdot P_{cu}} \cdot P_g(t - \tau_1) - [P_g(t - \tau_1) - P_{ex}] \cdot \frac{2 \cdot v_2}{l}$$

$$\frac{dn_g}{dt} = -[P_g(t - \tau_1) - \sqrt{P_{ex}}] \cdot \frac{n_g(t - \tau_2) \cdot 2 \cdot v_s}{l \cdot P_g(t - \tau_1)}; \ \frac{dn_{cu}}{dt} = -[P_{cu} - P_{vp}] \cdot \sqrt[(1+\Gamma)]{\frac{n_{cu}(t - \tau_3) \cdot v_d}{\Lambda \cdot P_{cu}}}$$

10.1 Find system fixed points, and draw coordinates as a function of Γ parameter.

10.2 Discuss stability and stability switching (for different values of τ parameter) for $\tau_1 = 0,\ \tau_2 = 0$ and $\tau_3 = \tau$.

10.3 Discuss stability and stability switching (for different values of τ parameter) for $\tau_1 = 0,\ \tau_2 = \sqrt{\tau}$ and $\tau_3 = 0$.

10.4 Discuss stability and stability switching (for different values of Γ Parameter) for $\tau_1 = \sqrt{\tau},\ \tau_2 = \sqrt[\Gamma]{\tau}; 1 > \Gamma > 0; \Gamma \in \mathbb{R}$ and $\tau_3 = \sqrt[4]{\tau}$.

10.5 The external buffer gas pressure P_{ex} is very small, $P_{ex} \to 0$, How it influence the dynamic of the system? Find fixed points and discuss stability for $\tau_1 = \tau,\ \tau_2 = 0$ and $\tau_3 = 0$.

References

1. K. Daniel, A. Skimantas, et al., Free-space nitrogen gas laser driven by a femtosecond filament. Phys. Rev. **A 86**, 033831 (2012)

2. J.K. Hale, *Dynamics and Bifurcations*. Texts in Applied Mathematics, vol. 3
3. S.H. Strogatz, *Nonlinear Dynamics and Chaos*. Westview Press
4. Y. Kuang, *Delay Differential Equations with Applications in Population Dynamics* (Academic Press, Boston, 1993)
5. E. Betta, Y. Kuang, Geometric stability switch criteria in delay differential systems with delay dependent parameters. SIAM J. Math. Anal. **33**, 1144–1165 (2002)
6. G. Hager, J. McIver, D. Hostutler, G. Pitz, G. Perram, A Quasi-two level analytic model for end pumped Alkali metal vapor laser, in *Proceedings of SPIE*, vol. 7005 700528-1
7. J. Guckenheimer, Nonlinear oscillations, dynamical systems, and bifurcations of vector fields, in *Applied Mathematical Sciences*, vol. 42
8. S. Wiggins, Introduction to applied nonlinear dynamical systems and chaos, in *Text in Applied Mathematics* (Hardcover)
9. M.J. Kushner, A self-consistent model for high repetition rate copper vapor lasers. IEEE J. Quantum Electronics **QE-17**(8) (1981)

Chapter 7
Dual-Wavelength Laser Systems Stability Analysis Under Parameters Variation (I)

Dual wavelength lasers are not common in scientific facilities but some applications need to use them. Dual wave lasers demonstrate simultaneous dual wavelength lasing and outcome of operating in multiple wavelength is possible. The related sub topics to dual wavelength lasers are dual wavelength laser Raman, simultaneous dual wavelength laser, dual wavelength laser tunable, dual wavelength laser speckle contrast imaging, dual wavelength laser operation, dual wavelength laser diode, dual wavelength laser interference, and dual wavelength laser output power. Dual wavelength Ti:Sapphire laser is used in many scientific and industrial applications. Ti:Sapphire laser system simultaneously generates sequence of femtosecond pulses at two independent wavelength regions. Rate-equation characterize the four energy levels of dual wavelength Ti:Sapphire laser. There is a separation of the wavelengths that increased with the pump power which is thermal lensing in Ti:Sapphire laser. Ti:Sapphire laser can generates a laser beam at more than one wavelength simultaneously. It is non-linear frequency changing via dual-wavelength laser operation of Ti:Sapphire. The dual wavelength tunable laser is characterized by rate equations for dynamical behavior over time and stability analysis. We split the Ti:Sapphire laser cavity to two sub resonator and one wavelength is established by both resonator. Ti:Sapphire sub resonators (A and B) are inspected for their operation and stability. We can get dual wavelength emission laser from Vertical External Cavity Surface Emitting Laser (VECSEL). It is simultaneously dual-wavelength emission, an inhomogeneous optical pumping of the non-identical quantum wells of the active region. The VECSEL rate equations include a strong time-delay feedback. The dynamic of dual-wavelength lasing VECSEL is analyzed and steady-state stability conditions is inspected. There are many applications that use laser sources which emit coaxial beams at two different wavelengths. The dual-wavelength VECSEL model is characterized and a stability analysis for the factors that impede simultaneous steady-state dual-wavelength emission. The VECSEL laser outcome can be stable or self-modulated optical output. Erbium-doped fiber laser is Er-doped fiber laser (EDFL) which operates in a particular regime where coherent oscillation of ASE occurs by feedback. The Erbium doped fiber laser (EDFLs) are used as sources for

O. Aluf, *Advance Elements of Laser Circuits and Systems*,
https://doi.org/10.1007/978-3-030-64103-0_7

coherent light signal generation, while EDFAs are used as wave-wave amplifiers for coherent light signal generation. Erbium-Doped fiber amplifier (EDFA) is an optical amplifier used in the C-band and L-band, where loss of telecom optical fibers becomes lowest in the entire optical communication wavelength bands. An EDFA is used to compensate the loss of an optical fiber in long-distance optical communication. An EDFA can amplify multiple optical signals simultaneously and can be easily integrated with WDM solution. Erbium-doped fiber laser exhibits quasiperiodic route to chaos and operates simultaneously at two wavelengths. The EDFL start from a CW state for high pumping rates and the system becomes T, 2T, 3T, …, nT periodics and chaotic for decreasing pumping ratios. There are two fundamental frequencies and the nT- periodic (n = 1, 2, 3, …) regimes are related to a frequency locking of the low frequency on a subharmonic of the high frequency. The model is based on the rate equations and the dynamic is analyzed and simulate. An intra cavity sum-frequency mixing in a diode and pump Q-switched Nd:YVO_4 dual-wavelength laser emits a compact high power yellow pulses laser beam. The gain match for simultaneous dual-wavelength emission Q-switched is optimized by a three mirror configuration forming two separate laser cavities. The Nd:YVO_4 crystal is good for diode-pumped lasers with dual-wavelength emission. It has high absorption over a wide pump-wavelength bandwidth. The model exhibits the dynamical behavior of a dual-wavelength Q-switched laser. The dynamic and stability of the laser model is inspected and analyze.

7.1 Dual-Wavelength Operation of a Ti:Sapphire Laser Stability Analysis Under Delay Variables in Time

A four-energy-level system of dual-wavelength Ti:Sapphire laser is characterized by rate-equation. Two sub-resonators structure gives dual wavelength laser. The Ti:Sapphire laser system simultaneously generates sequence of femtosecond pulses at two independent wavelength regions. It shows a high degree of synchronization and reduced relative timing jitter. Laser beam which oscillates at dual wavelength shows separation of the wavelengths that increased with the pump power (thermal lensing in Ti:Sapphire). The dual wavelengths laser source is very applicable in differential absorption lidar (DIAL), laser probing of the atmosphere, non-linear frequency conversion, and laser multistage ionization of atoms and molecules. Differential absorption lidar (DIAL) is a laser remote-sensing technique used for range-resolved measurements of atmospheric gas concentrations. It's provides species specific concentrations of gases with range resolution. The LIDAR technique also uses as an optical tool for distance and speed measurement, or for the determination of gas parameters. Differential LIDAR uses two different optical wavelengths and provides a self-calibrating method for the depth profile of gas concentration. A ground-based differential absorption LIDAR (light detection and ranging) system

(DIAL) uses an optical parametric oscillator (OPO). The optical parametric oscillator (OPO) is injection seeded with the output of a confocal filter cavity at frequencies generated by an electro-optic phase modulator (EOM) from a fixed-frequency external cavity diode laser (ECDL). Laser probing of the gas used in optical interferometry to study the propagation of femtosecond laser pulses in gases. The propagation in a gas jet gives details on the temporal structure of the laser pulse and the phase and resulting electron-density map. We get the gas ionization dynamic profile and characteristics. If we want to know the behavior of gas density, we use an array of laser beams, which are used to probe discrete points within the volume. The fluctuations in gas density are inspected, and therefore refractive index, and resulting beam deflection, which is detected by a neutral density wedge. Nonlinear frequency conversion is the conversion of input light to light of other frequencies by using optical nonlinearities. We get an efficient frequency conversion at sufficiently high optical intensities. By using non-linear crystal we can modify and change the laser's wavelength to that required for our system [1]. The principles of nonlinear frequency conversion is based on the fact that light travels through a material interacts with it and the electric field from the incident light drives material dipoles and causing them to oscillate at it travels through the material. The light forces dipoles to oscillate with a nonlinear response; reemitted light contains additional frequencies, harmonics. The type of nonlinear response is dependent on the structure of the material. The multistage ionization limits the increase in the efficiency and emission power of self-mode-locking lasers. The lasers which have outcome of two different wavelengths simultaneously are using dyes, alexandrite, and crystals containing F color centers. Ti:Sapphire laser generates a laser beam at more than one wavelength simultaneously. Non-linear frequency changing via dual-wavelength laser operation of Ti:Sapphire is used in many scientific experimental facilities. The unique of a pulsed dual-wavelength Ti:Sapphire laser is the pump source which is a Q-switched nanosecond laser pulse, the changeable operation wavelengths, and the different gains for different wavelengths. We characterize the dual wavelength tunable laser by the rate equations, and inspect the Ti:Sapphire dual wavelength results and nonlinear dynamic. The lower energy level of a Ti:Sapphire crystal is split into an energy band with closely spaced energy levels. The Ti:Sapphire laser can operate at two or more different-wavelengths. A single resonator with specific wavelength of the laser, the laser oscillates at one wavelength. By splitting the laser cavity to two sub-resonators, the one wavelength is established by both resonators. Outcome of two laser wavelengths is result of balance between the gain and loss on specific wavelength. The energy levels scheme of a Ti:Sapphire crystal contains an upper state (E_3) and two states (E_2^A and E_2^B) which are the lower energy band (Fig. 7.1).

There are two sub-resonators, A and B. The sub-resonator A causes a transition of a part of the population inversion from E_3 state to E_2^A state and the laser oscillates at wavelength λ_A. Simultaneously, the sub-resonator B causes a transition of the other part of the population inversion from the E_3 state to the E_2^B state and the laser oscillates at wavelength λ_B. The lasers at two wavelengths deplete the population inversion at the same time (homogeneous broadening). The rate equations of tunable solid-state lasers, a quasi-four energy level system simulate the dual-wavelength operation of

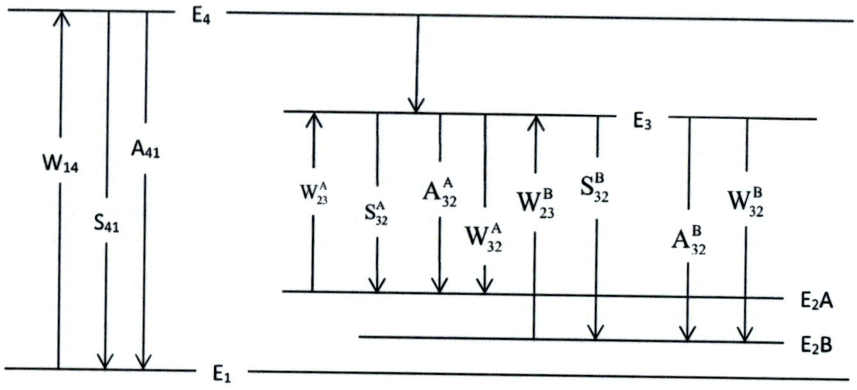

Fig. 7.1 Energy levels in the dual-wavelength operation of Ti:Sapphire laser

a Ti:Sapphire laser. The subscripts A and B refer to the sub-resonators A and B, respectively [1]. We define the following system parameters list (Table 7.1).

The system rate equations for the two separate lower energy levels are

Table 7.1 Dual-wavelength Ti:Sapphire laser system parameters list

Dual-wavelength Ti:Sapphire laser system parameter	Description
Φ	Photon density
ν	Center frequency of laser pulse in sub-resonator
σ	Stimulated-emission cross-section
g	Small-signal gain coefficient
A_{mn}	Rate of spontaneous transition between energy levels E_m and E_n
S_{mn}	Non-radiative emission rate between energy levels E_m and E_n
W_{mn}	Stimulate transition probability between energy levels E_m and E_n
l	Length of Ti:Sapphire crystal
L	Length of sub-resonators
V	Velocity of light in Ti:Sapphire crystal
δ	Single-pass loss in sub-resonator
$\tau_R = \frac{L}{\delta_c}$	Photon lifetime in sub-resonator
n_i	Population density at every energy level E_i i = 1, 2, 3, 4
n_{tot}	Total population density

$$\frac{d\Phi_A}{dt} = \left(n_3 - \frac{g_3}{g_2} \cdot n_2\right) \cdot \frac{A_{32}^A}{n_{v_A}} \cdot g(v, v_A) \cdot \Phi_A(t - \tau_A) \cdot \frac{l}{L_A} - \frac{\Phi_A(t - \tau_A)}{\tau_{RA}}$$

$$\frac{d\Phi_B}{dt} = \left(n_3 - \frac{g_3}{g_2} \cdot n_2\right) \cdot \frac{A_{32}^B}{n_{v_B}} \cdot g(v, v_B) \cdot \Phi_B(t - \tau_B) \cdot \frac{l}{L_B} - \frac{\Phi_B(t - \tau_B)}{\tau_{RB}}$$

$$\frac{dn_4}{dt} = n_1 \cdot W_{14} - n_4 \cdot (S_{43} + S_{41} + A_{41})$$

$$\frac{dn_3}{dt} = -\left(n_3 - \frac{g_3}{g_2} \cdot n_2\right) \cdot \frac{A_{32}^A}{n_{v_A}} \cdot g(v, v_A) \cdot \Phi_A(t - \tau_A) - \left(n_3 - \frac{g_3}{g_2} \cdot n_2\right) \cdot \frac{A_{32}^B}{n_{v_B}} \cdot g(v, v_B) \cdot \Phi_B(t - \tau_B) - n_3 \cdot (A_{32}^A + S_{32}^A) - n_3 \cdot (A_{32}^B + S_{32}^B) + n_4 \cdot S_{43}$$

$$\frac{dn_1}{dt} = n_3 \cdot (S_{21}^A + S_{21}^B) - n_1 \cdot W_{14}; \sum_{k=1}^{4} n_k = n_{tot}$$

Our system variables are $\Phi_A, \Phi_B, n_3, n_4, n_1$; $\Phi_A = \Phi_A(t)$, $\Phi_B = \Phi_B(t)$, $n_3 = n_3(t)$, $n_4 = n_4(t)$, $n_1 = n_1(t)$.

We consider that all other parameters in our system are constant.

$$\sum_{k=1}^{4} n_k = n_{tot} \Rightarrow n_2 = n_{tot} - n_1 - n_3 - n_4 = n_{tot} - \sum_{\substack{k=1\\k\neq 2}}^{4} n_k$$

We can reorganize our system DDEs as

$$\frac{d\Phi_A}{dt} = \left(n_3 - \frac{g_3}{g_2} \cdot \left[n_{tot} - \sum_{\substack{k=1\\k\neq 2}}^{4} n_k\right]\right) \cdot \frac{A_{32}^A}{n_{v_A}} \cdot g(v, v_A) \cdot \Phi_A(t - \tau_A) \cdot \frac{l}{L_A} - \frac{\Phi_A(t - \tau_A)}{\tau_{RA}}$$

$$\frac{d\Phi_B}{dt} = \left(n_3 - \frac{g_3}{g_2} \cdot \left[n_{tot} - \sum_{\substack{k=1\\k\neq 2}}^{4} n_k\right]\right) \cdot \frac{A_{32}^B}{n_{v_B}} \cdot g(v, v_B) \cdot \Phi_B(t - \tau_B) \cdot \frac{l}{L_B} - \frac{\Phi_B(t - \tau_B)}{\tau_{RB}}$$

$$\frac{dn_4}{dt} = n_1 \cdot W_{14} - n_4 \cdot (S_{43} + S_{41} + A_{41}); \frac{dn_1}{dt} = n_3 \cdot (S_{21}^A + S_{21}^B) - n_1 \cdot W_{14}$$

$$\frac{dn_3}{dt} = -\left(n_3 - \frac{g_3}{g_2}\cdot\left[n_{tot} - \sum_{\substack{k=1\\k\neq 2}}^{4} n_k\right]\right)\cdot\frac{A_{32}^{A}}{n_{\nu_A}}\cdot g(\nu,\nu_A)\cdot\Phi_A(t-\tau_A)$$
$$-\left(n_3 - \frac{g_3}{g_2}\cdot\left[n_{tot} - \sum_{\substack{k=1\\k\neq 2}}^{4} n_k\right]\right)\cdot\frac{A_{32}^{B}}{n_{\nu_B}}\cdot g(\nu,\nu_B)\cdot\Phi_B(t-\tau_B)$$
$$-n_3\cdot[A_{32}^{A}+S_{32}^{A}+A_{32}^{B}+S_{32}^{B}]+n_4\cdot S_{43}$$

Parameters τ_A, τ_B are photon densities (Φ_A, Φ_B) delays in time, respectively. The photon densities (Φ_A, Φ_B) delays in time is caused by Dual-wavelength Ti:Sapphire optical interferences. The derivatives $\frac{d\Phi_A}{dt}, \frac{d\Phi_B}{dt}$ are not affected by τ_A, τ_B time delays.

At fixed points, $\frac{d\Phi_A}{dt}=0$; $\frac{d\Phi_B}{dt}=0$; $\frac{dn_4}{dt}=0$; $\frac{dn_1}{dt}=0$; $\frac{dn_3}{dt}=0$; $\lim_{t\to\infty}\Phi_A(t-\tau_A)=\Phi_A(t)\ \forall\ t\gg\tau_A$ $\lim_{t\to\infty}\Phi_B(t-\tau_B)=\Phi_B(t)\ \forall\ t\gg\tau_B$. We define for simplicity $\Gamma^* = n_3^* - \frac{g_3}{g_2}\cdot[n_{tot}-\sum_{\substack{k=1\\k\neq 2}}^{4} n_k^*]$, where Γ^* is fixed global variable. $\Gamma^* = n_3^* - \frac{g_3}{g_2}\cdot[n_{tot}-\sum_{\substack{k=1\\k\neq 2}}^{4} n_k^*] = n_3^* - \frac{g_3}{g_2}\cdot[n_{tot}-(n_1^*+n_3^*+n_4^*)]$.

$$\Gamma^* = \Gamma^*(n_1^*, n_3^*, n_4^*) = \left(1+\frac{g_3}{g_2}\right)\cdot n_3^* - \frac{g_3}{g_2}\cdot n_{tot} + \frac{g_3}{g_2}\cdot n_1^* + \frac{g_3}{g_2}\cdot n_4^*$$

$$\frac{d\Phi_A}{dt} = 0 \Rightarrow \Gamma^*\cdot\frac{A_{32}^{A}}{n_{\nu_A}}\cdot g(\nu,\nu_A)\cdot\Phi_A^*\cdot\frac{l}{L_A} - \frac{\Phi_A^*}{\tau_{RA}} = 0$$
$$\left[\Gamma^*\cdot\frac{A_{32}^{A}}{n_{\nu_A}}\cdot g(\nu,\nu_A)\cdot\frac{l}{L_A} - \frac{1}{\tau_{RA}}\right]\cdot\Phi_A^* = 0$$

Case A: $\Phi_A^* = 0$

$$\frac{d\Phi_B}{dt} = 0 \Rightarrow \left[\Gamma^*\cdot\frac{A_{32}^{B}}{n_{\nu_B}}\cdot g(\nu,\nu_B)\cdot\frac{l}{L_B} - \frac{1}{\tau_{RB}}\right]\cdot\Phi_B^* = 0, \text{Case A1}: \Phi_B^* = 0$$
$$\frac{dn_4}{dt} = 0 \Rightarrow n_1^*\cdot W_{14} - n_4^*\cdot(S_{43}+S_{41}+A_{41}) = 0$$
$$\frac{dn_1}{dt} = 0 \Rightarrow n_3^*\cdot(S_{21}^{A}+S_{21}^{B}) - n_1^*\cdot W_{14} = 0$$

$$n_1^* = n_4^*\cdot\frac{1}{W_{14}}\cdot(S_{43}+S_{41}+A_{41});\ n_1^* = n_3^*\cdot\frac{1}{W_{14}}\cdot(S_{21}^{A}+S_{21}^{B})$$
$$n_3^* = n_4^*\cdot\frac{(S_{43}+S_{41}+A_{41})}{(S_{21}^{A}+S_{21}^{B})}$$

$$\frac{dn_3}{dt} = 0 \Rightarrow -\Gamma^*\cdot\frac{A_{32}^{A}}{n_{\nu_A}}\cdot g(\nu,\nu_A)\cdot\Phi_A^* - \Gamma^*\cdot\frac{A_{32}^{B}}{n_{\nu_B}}\cdot g(\nu,\nu_B)\cdot\Phi_B^*$$

$$- n_3^* \cdot [A_{32}^A + S_{32}^A + A_{32}^B + S_{32}^B] + n_4^* \cdot S_{43} = 0$$

Case A, A1: $\Phi_A^* = 0;\ \Phi_B^* = 0;\ n_3^* = n_4^* \cdot \frac{S_{43}}{A_{32}^A + S_{32}^A + A_{32}^B + S_{32}^B}$

$$n_3^* = n_4^* \cdot \frac{S_{43}}{A_{32}^A + S_{32}^A + A_{32}^B + S_{32}^B}$$

$$n_3^* = n_4^* \cdot \frac{(S_{43} + S_{41} + A_{41})}{(S_{21}^A + S_{21}^B)} \Rightarrow \frac{S_{43}}{A_{32}^A + S_{32}^A + A_{32}^B + S_{32}^B}$$

$$= \frac{(S_{43} + S_{41} + A_{41})}{(S_{21}^A + S_{21}^B)}$$

First fixed point:

$$E^{(0)} = (\Phi_A^{(0)}, \Phi_B^{(0)}, n_1^{(0)}, n_3^{(0)}, n_4^{(0)})$$

$$= \left(0, 0, n_4^* \cdot \frac{1}{W_{14}} \cdot (S_{43} + S_{41} + A_{41}), n_4^* \cdot \frac{S_{43}}{A_{32}^A + S_{32}^A + A_{32}^B + S_{32}^B}, n_4^*\right)$$

$$n_1^{(0)} = f_1(n_4^*);\ f_1(n_4^*) = n_4^* \cdot \frac{1}{W_{14}} \cdot (S_{43} + S_{41} + A_{41})$$

$$n_3^{(0)} = f_2(n_4^*);\ f_2(n_4^*) = n_4^* \cdot \frac{S_{43}}{A_{32}^A + S_{32}^A + A_{32}^B + S_{32}^B}$$

$$E^{(0)} = (\Phi_A^{(0)}, \Phi_B^{(0)}, n_1^{(0)}, n_3^{(0)}, n_4^{(0)}) = (0, 0, f_1(n_4^*), f_2(n_4^*), n_4^*)$$

Case A, A2: $\Gamma^* \cdot \frac{A_{32}^B}{n_{\nu_B}} \cdot g(\nu, \nu_B) \cdot \frac{l}{L_B} - \frac{1}{\tau_{RB}} = 0 \Rightarrow \Gamma^* \cdot \frac{A_{32}^B}{n_{\nu_B}} \cdot g(\nu, \nu_B) \cdot \frac{l}{L_B} = \frac{1}{\tau_{RB}};\ \Gamma^* = \frac{L_B \cdot n_{\nu_B}}{\tau_{RB} \cdot g(\nu, \nu_B) \cdot A_{32}^B \cdot l}$

$$\Phi_A^* = 0;\ \Phi_B^* \neq 0$$

$$\Gamma^* = \left(1 + \frac{g_3}{g_2}\right) \cdot n_3^* - \frac{g_3}{g_2} \cdot n_{tot} + \frac{g_3}{g_2} \cdot n_1^* + \frac{g_3}{g_2} \cdot n_4^*$$

$$= \frac{L_B \cdot n_{\nu_B}}{\tau_{RB} \cdot g(\nu, \nu_B) \cdot A_{32}^B \cdot l}$$

$$\frac{dn_3}{dt} = 0 \Rightarrow -\Gamma^* \cdot \frac{A_{32}^B}{n_{\nu_B}} \cdot g(\nu, \nu_B) \cdot \Phi_B^*$$

$$- n_3^* \cdot [A_{32}^A + S_{32}^A + A_{32}^B + S_{32}^B] + n_4^* \cdot S_{43} = 0$$

$$- \frac{L_B \cdot n_{\nu_B}}{\tau_{RB} \cdot g(\nu, \nu_B) \cdot A_{32}^B \cdot l} \cdot \frac{A_{32}^B}{n_{\nu_B}} \cdot g(\nu, \nu_B) \cdot \Phi_B^*$$

$$- n_3^* \cdot [A_{32}^A + S_{32}^A + A_{32}^B + S_{32}^B] + n_4^* \cdot S_{43} = 0$$

$$\Phi_B^* = \frac{n_4^* \cdot S_{43} - n_3^* \cdot [A_{32}^A + S_{32}^A + A_{32}^B + S_{32}^B]}{\frac{L_B \cdot n_{\nu_B}}{\tau_{RB} \cdot g(\nu,\nu_B) \cdot A_{32}^B \cdot l} \cdot \frac{A_{32}^B}{n_{\nu_B}} \cdot g(\nu,\nu_B)}$$

We get three fixed points equations:

$$\left(1 + \frac{g_3}{g_2}\right) \cdot n_3^* - \frac{g_3}{g_2} \cdot n_{tot} + \frac{g_3}{g_2} \cdot n_1^* + \frac{g_3}{g_2} \cdot n_4^* = \frac{L_B \cdot n_{\nu_B}}{\tau_{RB} \cdot g(\nu,\nu_B) \cdot A_{32}^B \cdot l}$$

$$\frac{dn_4}{dt} = 0 \Rightarrow n_1^* = n_4^* \cdot \frac{1}{W_{14}} \cdot (S_{43} + S_{41} + A_{41})$$

$$\frac{dn_1}{dt} = 0 \Rightarrow n_1^* = n_3^* \cdot \frac{1}{W_{14}} \cdot (S_{21}^A + S_{21}^B);\ n_3^* = n_4^* \cdot \frac{(S_{43} + S_{41} + A_{41})}{(S_{21}^A + S_{21}^B)}$$

$$\left(1 + \frac{g_3}{g_2}\right) \cdot n_4^* \cdot \frac{(S_{43} + S_{41} + A_{41})}{(S_{21}^A + S_{21}^B)} - \frac{g_3}{g_2} \cdot n_{tot} + \frac{g_3}{g_2} \cdot n_4^* \cdot \frac{1}{W_{14}} \cdot (S_{43} + S_{41} + A_{41})$$
$$+ \frac{g_3}{g_2} \cdot n_4^* = \frac{L_B \cdot n_{\nu_B}}{\tau_{RB} \cdot g(\nu,\nu_B) \cdot A_{32}^B \cdot l}$$

$$\left[\left(1 + \frac{g_3}{g_2}\right) \cdot \frac{(S_{43} + S_{41} + A_{41})}{(S_{21}^A + S_{21}^B)} + \frac{g_3}{g_2} \cdot \frac{1}{W_{14}} \cdot (S_{43} + S_{41} + A_{41}) + \frac{g_3}{g_2}\right] \cdot n_4^*$$
$$= \frac{L_B \cdot n_{\nu_B}}{\tau_{RB} \cdot g(\nu,\nu_B) \cdot A_{32}^B \cdot l} + \frac{g_3}{g_2} \cdot n_{tot}$$

$$n_4^{(1)} = \frac{\frac{L_B \cdot n_{\nu_B}}{\tau_{RB} \cdot g(\nu,\nu_B) \cdot A_{32}^B \cdot l} + \frac{g_3}{g_2} \cdot n_{tot}}{\left(1 + \frac{g_3}{g_2}\right) \cdot \frac{(S_{43}+S_{41}+A_{41})}{(S_{21}^A+S_{21}^B)} + \frac{g_3}{g_2} \cdot \frac{1}{W_{14}} \cdot (S_{43} + S_{41} + A_{41}) + \frac{g_3}{g_2}}$$

$$n_3^{(1)} = \left[\frac{\frac{L_B \cdot n_{\nu_B}}{\tau_{RB} \cdot g(\nu,\nu_B) \cdot A_{32}^B \cdot l} + \frac{g_3}{g_2} \cdot n_{tot}}{\left(1 + \frac{g_3}{g_2}\right) \cdot \frac{(S_{43}+S_{41}+A_{41})}{(S_{21}^A+S_{21}^B)} + \frac{g_3}{g_2} \cdot \frac{1}{W_{14}} \cdot (S_{43} + S_{41} + A_{41}) + \frac{g_3}{g_2}}\right]$$
$$\cdot \left[\frac{(S_{43} + S_{41} + A_{41})}{(S_{21}^A + S_{21}^B)}\right]$$

$$n_1^{(1)} = \left[\frac{\frac{L_B \cdot n_{\nu_B}}{\tau_{RB} \cdot g(\nu,\nu_B) \cdot A_{32}^B \cdot l} + \frac{g_3}{g_2} \cdot n_{tot}}{\left(1 + \frac{g_3}{g_2}\right) \cdot \frac{(S_{43}+S_{41}+A_{41})}{(S_{21}^A+S_{21}^B)} + \frac{g_3}{g_2} \cdot \frac{1}{W_{14}} \cdot (S_{43} + S_{41} + A_{41}) + \frac{g_3}{g_2}}\right]$$
$$\cdot \left[\frac{(S_{43} + S_{41} + A_{41})}{W_{14}}\right]$$

$$\Phi_B^{(1)} = \frac{\left\{\left[\frac{\frac{L_B \cdot n_{\nu_B}}{\tau_{RB}\cdot g(\nu,\nu_B)\cdot A_{32}^B\cdot l} + \frac{g_3}{g_2}\cdot n_{tot}}{(1+\frac{g_3}{g_2})\cdot\frac{(S_{43}+S_{41}+A_{41})}{(S_{21}^A+S_{21}^B)} + \frac{g_3}{g_2}\cdot\frac{1}{W_{14}}\cdot(S_{43}+S_{41}+A_{41}) + \frac{g_3}{g_2}}\right]\cdot S_{43} - \left[\frac{\frac{L_B \cdot n_{\nu_B}}{\tau_{RB}\cdot g(\nu,\nu_B)\cdot A_{32}^B\cdot l} + \frac{g_3}{g_2}\cdot n_{tot}}{(1+\frac{g_3}{g_2})\cdot\frac{(S_{43}+S_{41}+A_{41})}{(S_{21}^A+S_{21}^B)} + \frac{g_3}{g_2}\cdot\frac{1}{W_{14}}\cdot(S_{43}+S_{41}+A_{41}) + \frac{g_3}{g_2}}\right]\cdot\left[\frac{(S_{43}+S_{41}+A_{41})}{(S_{21}^A+S_{21}^B)}\right]\cdot\left[A_{32}^A+S_{32}^A+A_{32}^B+S_{32}^B\right]\right\}}{\frac{L_B\cdot n_{\nu_B}}{\tau_{RB}\cdot g(\nu,\nu_B)\cdot A_{32}^B\cdot l}\cdot\frac{A_{32}^B}{n_{\nu_B}}\cdot g(\nu,\nu_B)}$$

Second fixed point: $E^{(1)} = (\Phi_A^{(1)}, \Phi_B^{(1)}, n_1^{(1)}, n_3^{(1)}, n_4^{(1)}) = (0, \Phi_B^{(1)}, n_1^{(1)}, n_3^{(1)}, n_4^{(1)})$.
Case B: $\Phi_B^* = 0$; $\frac{d\Phi_B}{dt} = 0 \Rightarrow [\Gamma^*\cdot\frac{A_{32}^B}{n_{\nu_B}}\cdot g(\nu,\nu_B)\cdot\frac{l}{L_B} - \frac{1}{\tau_{RB}}]\cdot\Phi_B^* = 0$

$$\frac{d\Phi_A}{dt} = 0 \Rightarrow \Gamma^*\cdot\frac{A_{32}^A}{n_{\nu_A}}\cdot g(\nu,\nu_A)\cdot\Phi_A^*\cdot\frac{l}{L_A} - \frac{\Phi_A^*}{\tau_{RA}} = 0$$

$$\left[\Gamma^*\cdot\frac{A_{32}^A}{n_{\nu_A}}\cdot g(\nu,\nu_A)\cdot\frac{l}{L_A} - \frac{1}{\tau_{RA}}\right]\cdot\Phi_A^* = 0$$

Case B, B1: $\Phi_A^* = 0$ then $\Phi_A^* = 0$; $\Phi_B^* = 0$; $n_3^* = n_4^*\cdot\frac{S_{43}}{A_{32}^A+S_{32}^A+A_{32}^B+S_{32}^B}$ and the fixed point is identical as in Case A, A1, and we don't count it as a new fixed point.

$$E^{(0)} = (\Phi_A^{(0)}, \Phi_B^{(0)}, n_1^{(0)}, n_3^{(0)}, n_4^{(0)}) = \left(0, 0, n_4^*\cdot\frac{1}{W_{14}}\cdot(S_{43}+S_{41}+A_{41}), n_4^*\cdot\frac{S_{43}}{A_{32}^A+S_{32}^A+A_{32}^B+S_{32}^B}, n_4^*\right)$$

$$n_1^{(0)} = f_1(n_4^*);\ f_1(n_4^*) = n_4^*\cdot\frac{1}{W_{14}}\cdot(S_{43}+S_{41}+A_{41})$$

$$n_3^{(0)} = f_2(n_4^*);\ f_2(n_4^*) = n_4^*\cdot\frac{S_{43}}{A_{32}^A+S_{32}^A+A_{32}^B+S_{32}^B}$$

$$E^{(0)} = (\Phi_A^{(0)}, \Phi_B^{(0)}, n_1^{(0)}, n_3^{(0)}, n_4^{(0)}) = (0, 0, f_1(n_4^*), f_2(n_4^*), n_4^*)$$

Case B, B2: $\Gamma^*\cdot\frac{A_{32}^A}{n_{\nu_A}}\cdot g(\nu,\nu_A)\cdot\frac{l}{L_A} - \frac{1}{\tau_{RA}} = 0 \Rightarrow \Gamma^* = \frac{\frac{1}{\tau_{RA}}}{\frac{A_{32}^A}{n_{\nu_A}}\cdot g(\nu,\nu_A)\cdot\frac{l}{L_A}} = \frac{n_{\nu_A}\cdot L_A}{\tau_{RA}\cdot A_{32}^B\cdot g(\nu,\nu_B)\cdot l}$

$$\Phi_A^* \neq 0;\ \Phi_B^* = 0$$

$$\Gamma^* = \left(1+\frac{g_3}{g_2}\right)\cdot n_3^* - \frac{g_3}{g_2}\cdot n_{tot} + \frac{g_3}{g_2}\cdot n_1^* + \frac{g_3}{g_2}\cdot n_4^*$$

$$= \frac{n_{\nu_A} \cdot L_A}{\tau_{RA} \cdot A_{32}^B \cdot g(\nu, \nu_B) \cdot l}$$

$$\frac{dn_3}{dt} = 0 \Rightarrow -\Gamma^* \cdot \frac{A_{32}^A}{n_{\nu_A}} \cdot g(\nu, \nu_A) \cdot \Phi_A^* - n_3^* \cdot [A_{32}^A + S_{32}^A + A_{32}^B + S_{32}^B] + n_4^* \cdot S_{43} = 0$$

$$-\left(\frac{n_{\nu_A} \cdot L_A}{\tau_{RA} \cdot A_{32}^B \cdot g(\nu, \nu_B) \cdot l}\right) \cdot \frac{A_{32}^A}{n_{\nu_A}} \cdot g(\nu, \nu_A) \cdot \Phi_A^* - n_3^* \cdot [A_{32}^A + S_{32}^A + A_{32}^B + S_{32}^B] + n_4^* \cdot S_{43} = 0$$

$$\left(\frac{n_{\nu_A} \cdot L_A}{\tau_{RA} \cdot A_{32}^B \cdot g(\nu, \nu_B) \cdot l}\right) \cdot \frac{A_{32}^A}{n_{\nu_A}} \cdot g(\nu, \nu_A) \cdot \Phi_A^* = -n_3^* \cdot [A_{32}^A + S_{32}^A + A_{32}^B + S_{32}^B] + n_4^* \cdot S_{43}$$

$$\Phi_A^* = \frac{-n_3^* \cdot [A_{32}^A + S_{32}^A + A_{32}^B + S_{32}^B] + n_4^* \cdot S_{43}}{\left(\frac{n_{\nu_A} \cdot L_A}{\tau_{RA} \cdot A_{32}^B \cdot g(\nu, \nu_B) \cdot l}\right) \cdot \frac{A_{32}^A}{n_{\nu_A}} \cdot g(\nu, \nu_A)}$$

We get three fixed points equations:

$$\left(1 + \frac{g_3}{g_2}\right) \cdot n_3^* - \frac{g_3}{g_2} \cdot n_{tot} + \frac{g_3}{g_2} \cdot n_1^* + \frac{g_3}{g_2} \cdot n_4^* = \frac{n_{\nu_A} \cdot L_A}{\tau_{RA} \cdot A_{32}^B \cdot g(\nu, \nu_B) \cdot l}$$

$$\frac{dn_4}{dt} = 0 \Rightarrow n_1^* = n_4^* \cdot \frac{1}{W_{14}} \cdot (S_{43} + S_{41} + A_{41})$$

$$\frac{dn_1}{dt} = 0 \Rightarrow n_1^* = n_3^* \cdot \frac{1}{W_{14}} \cdot (S_{21}^A + S_{21}^B);\ n_3^* = n_4^* \cdot \frac{(S_{43} + S_{41} + A_{41})}{(S_{21}^A + S_{21}^B)}$$

$$\left(1 + \frac{g_3}{g_2}\right) \cdot n_4^* \cdot \frac{(S_{43} + S_{41} + A_{41})}{(S_{21}^A + S_{21}^B)} - \frac{g_3}{g_2} \cdot n_{tot} + \frac{g_3}{g_2} \cdot n_4^* \cdot \frac{1}{W_{14}} \cdot (S_{43} + S_{41} + A_{41}) + \frac{g_3}{g_2} \cdot n_4^* = \frac{n_{\nu_A} \cdot L_A}{\tau_{RA} \cdot A_{32}^B \cdot g(\nu, \nu_B) \cdot l}$$

$$\left[\left(1 + \frac{g_3}{g_2}\right) \cdot \frac{(S_{43} + S_{41} + A_{41})}{(S_{21}^A + S_{21}^B)} + \frac{g_3}{g_2} \cdot \frac{1}{W_{14}} \cdot (S_{43} + S_{41} + A_{41}) + \frac{g_3}{g_2}\right] \cdot n_4^* = \frac{n_{\nu_A} \cdot L_A}{\tau_{RA} \cdot A_{32}^B \cdot g(\nu, \nu_B) \cdot l} + \frac{g_3}{g_2} \cdot n_{tot}$$

$$n_4^{(2)} = \frac{\frac{n_{\nu_A} \cdot L_A}{\tau_{RA} \cdot A_{32}^B \cdot g(\nu, \nu_B) \cdot l} + \frac{g_3}{g_2} \cdot n_{tot}}{(1 + \frac{g_3}{g_2}) \cdot \frac{(S_{43} + S_{41} + A_{41})}{(S_{21}^A + S_{21}^B)} + \frac{g_3}{g_2} \cdot \frac{1}{W_{14}} \cdot (S_{43} + S_{41} + A_{41}) + \frac{g_3}{g_2}}$$

$$n_3^{(2)} = \frac{\frac{n_{\nu_A} \cdot L_A}{\tau_{RA} \cdot A_{32}^B \cdot g(\nu, \nu_B) \cdot l} + \frac{g_3}{g_2} \cdot n_{tot}}{\left(1 + \frac{g_3}{g_2}\right) \cdot \frac{(S_{43}+S_{41}+A_{41})}{(S_{21}^A+S_{21}^B)} + \frac{g_3}{g_2} \cdot \frac{1}{W_{14}} \cdot (S_{43} + S_{41} + A_{41}) + \frac{g_3}{g_2} \cdot \frac{(S_{43} + S_{41} + A_{41})}{(S_{21}^A + S_{21}^B)}}$$

$$n_1^{(2)} = \frac{\frac{n_{\nu_A} \cdot L_A}{\tau_{RA} \cdot A_{32}^B \cdot g(\nu, \nu_B) \cdot l} + \frac{g_3}{g_2} \cdot n_{tot}}{\left(1 + \frac{g_3}{g_2}\right) \cdot \frac{(S_{43}+S_{41}+A_{41})}{(S_{21}^A+S_{21}^B)} + \frac{g_3}{g_2} \cdot \frac{1}{W_{14}} \cdot (S_{43} + S_{41} + A_{41}) + \frac{g_3}{g_2} \cdot \frac{(S_{43} + S_{41} + A_{41})}{W_{14}}}$$

$$\Phi_A^{(2)} = \frac{\left\{ -\left[\frac{\frac{n_{\nu_A} \cdot L_A}{\tau_{RA} \cdot A_{32}^B \cdot g(\nu, \nu_B) \cdot l} + \frac{g_3}{g_2} \cdot n_{tot}}{\left(1 + \frac{g_3}{g_2}\right) \cdot \frac{(S_{43}+S_{41}+A_{41})}{(S_{21}^A+S_{21}^B)} + \frac{g_3}{g_2} \cdot \frac{1}{W_{14}} \cdot (S_{43} + S_{41} + A_{41}) + \frac{g_3}{g_2} \cdot \frac{(S_{43} + S_{41} + A_{41})}{(S_{21}^A + S_{21}^B)}} \right] \cdot [A_{32}^A + S_{32}^A + A_{32}^B + S_{32}^B] + \left[\frac{\frac{n_{\nu_A} \cdot L_A}{\tau_{RA} \cdot A_{32}^B \cdot g(\nu, \nu_B) \cdot l} + \frac{g_3}{g_2} \cdot n_{tot}}{\left(1 + \frac{g_3}{g_2}\right) \cdot \frac{(S_{43}+S_{41}+A_{41})}{(S_{21}^A+S_{21}^B)} + \frac{g_3}{g_2} \cdot \frac{1}{W_{14}} \cdot (S_{43} + S_{41} + A_{41}) + \frac{g_3}{g_2}} \right] \cdot S_{43} \right\}}{\left(\frac{n_{\nu_A} \cdot L_A}{\tau_{RA} \cdot A_{32}^B \cdot g(\nu, \nu_B) \cdot l}\right) \cdot \frac{A_{32}^A}{n_{\nu_A}} \cdot g(\nu, \nu_A)}$$

Third fixed point: $E^{(2)} = (\Phi_A^{(2)}, \Phi_B^{(2)}, n_1^{(2)}, n_3^{(2)}, n_4^{(2)}) = (\Phi_A^{(2)}, 0, n_1^{(2)}, n_3^{(2)}, n_4^{(2)})$.
We can summary our system fixed points (Fig. 7.2).

$$E^{(0)} = (\Phi_A^{(0)}, \Phi_B^{(0)}, n_1^{(0)}, n_3^{(0)}, n_4^{(0)})$$
$$= \left(0, 0, n_4^* \cdot \frac{1}{W_{14}} \cdot (S_{43} + S_{41} + A_{41}), n_4^* \cdot \frac{S_{43}}{A_{32}^A + S_{32}^A + A_{32}^B + S_{32}^B}, n_4^*\right)$$

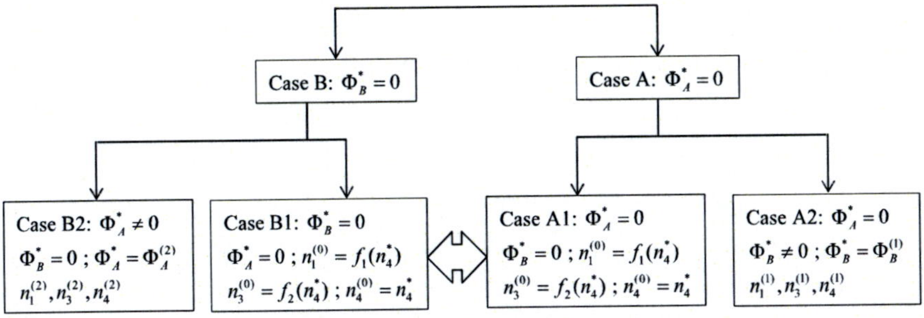

Fig. 7.2 Dual-wavelength operation of Ti:Sapphire laser system fixed points

$$E^{(1)} = (\Phi_A^{(1)}, \Phi_B^{(1)}, n_1^{(1)}, n_3^{(1)}, n_4^{(1)}) = (0, \Phi_B^{(1)}, n_1^{(1)}, n_3^{(1)}, n_4^{(1)})$$

$$E^{(2)} = (\Phi_A^{(2)}, \Phi_B^{(2)}, n_1^{(2)}, n_3^{(2)}, n_4^{(2)}) = (\Phi_A^{(2)}, 0, n_1^{(2)}, n_3^{(2)}, n_4^{(2)})$$

Stability analysis: The standard local stability analysis about any one of the equilibrium point of the Dual-wavelength operation of Ti:Sapphire laser system consists in adding to coordinate $[\Phi_A, \Phi_B, n_1, n_3, n_4]$ arbitrarily small increments of exponential form $[\phi_A, \phi_B, n_1, n_3, n_4] \cdot e^{\lambda \cdot t}$ and retaining the first order terms in $\Phi_A, \Phi_B, n_1, n_3, n_4$. The system of five homogeneous equations leads to a polynomial characteristic equation in the eigenvalues. The polynomial characteristic equations accept by set the below variables and variables derivative with respect to time into Dual-wavelength system equations. The Dual-wavelength system fixed values with arbitrarily small increments of exponential form $[\phi_A, \phi_B, n_1, n_3, n_4] \cdot e^{\lambda \cdot t}$ are: $j = 0$ (first fixed point), $j = 1$ (second fixed point), $j = 2$ (third fixed point), etc. [2, 3].

$$\Phi_A(t) = \Phi_A^{(j)} + \phi_A \cdot e^{\lambda \cdot t}; \ \Phi_B(t) = \Phi_B^{(j)} + \phi_B \cdot e^{\lambda \cdot t}$$
$$n_1(t) = n_1^{(j)} + n_1 \cdot e^{\lambda \cdot t}; \ n_3(t) = n_3^{(j)} + n_3 \cdot e^{\lambda \cdot t}$$

$$n_4(t) = n_4^{(j)} + n_4 \cdot e^{\lambda \cdot t}; \ \frac{d\Phi_A(t)}{dt} = \phi_A \cdot \lambda \cdot e^{\lambda \cdot t}$$
$$\frac{d\Phi_B(t)}{dt} = \phi_B \cdot \lambda \cdot e^{\lambda \cdot t}; \ \frac{dn_1(t)}{dt} = n_1 \cdot \lambda \cdot e^{\lambda \cdot t}$$

$$\frac{dn_3(t)}{dt} = n_3 \cdot \lambda \cdot e^{\lambda \cdot t}; \ \frac{dn_4(t)}{dt} = n_4 \cdot \lambda \cdot e^{\lambda \cdot t}$$
$$\Phi_A(t - \tau_A) = \Phi_A^{(j)} + \phi_A \cdot e^{\lambda \cdot (t - \tau_A)}; \ \Phi_B(t - \tau_B) = \Phi_B^{(j)} + \phi_B \cdot e^{\lambda \cdot (t - \tau_B)}$$

We choose these expressions for ourselves $\Phi_A(t), \Phi_B(t), n_1(t), n_3(t), n_4(t)$ as a small displacement $[\phi_A, \phi_B, n_1, n_3, n_4]$ from the Dual-wavelength operation of Ti:Sapphire laser system fixed points in time $t = 0$, $\Phi_A(t = 0) = \Phi_A^{(j)} + \phi_A$; $\Phi_B(t = 0) = \Phi_B^{(j)} + \phi_B$; $n_1(t = 0) = n_1^{(j)} + n_1$.

1. $$\frac{d\Phi_A}{dt} = \left(n_3 - \frac{g_3}{g_2} \cdot \left[n_{tot} - \sum_{\substack{k=1 \\ k \neq 2}}^{4} n_k \right] \right) \cdot \frac{A_{32}^A}{n_{v_A}} \cdot g(v, v_A) \cdot \Phi_A(t - \tau_A) \cdot \frac{l}{L_A} - \frac{\Phi_A(t - \tau_A)}{\tau_{RA}}$$

$$\phi_A \cdot \lambda \cdot e^{\lambda \cdot t} = \left(n_3^{(j)} + n_3 \cdot e^{\lambda \cdot t} - \frac{g_3}{g_2} \cdot \left[n_{tot} - \sum_{\substack{k=1 \\ k \neq 2}}^{4} (n_k^{(j)} + n_k \cdot e^{\lambda \cdot t}) \right] \right)$$

$$\cdot \frac{A_{32}^A}{n_{\nu_A}} \cdot g(\nu, \nu_A) \cdot (\Phi_A^{(j)} + \phi_A \cdot e^{\lambda \cdot (t - \tau_A)}) \cdot \frac{l}{L_A}$$

$$- \frac{1}{\tau_{RA}} \cdot (\Phi_A^{(j)} + \phi_A \cdot e^{\lambda \cdot (t - \tau_A)})$$

$$\phi_A \cdot \lambda \cdot e^{\lambda \cdot t} = \left(\left[n_3^{(j)} - \frac{g_3}{g_2} \cdot \left(n_{tot} - \sum_{\substack{k=1 \\ k \neq 2}}^{4} n_k^{(j)} \right) \right] + \left[\frac{g_3}{g_2} \cdot \sum_{\substack{k=1 \\ k \neq 2}}^{4} n_k + n_3 \right] \cdot e^{\lambda \cdot t} \right)$$

$$\cdot \frac{A_{32}^A}{n_{\nu_A}} \cdot g(\nu, \nu_A) \cdot (\Phi_A^{(j)} + \phi_A \cdot e^{\lambda \cdot t} \cdot e^{-\lambda \cdot \tau_A}) \cdot \frac{l}{L_A}$$

$$- \frac{1}{\tau_{RA}} \cdot \Phi_A^{(j)} - \frac{1}{\tau_{RA}} \cdot \phi_A \cdot e^{\lambda \cdot t} \cdot e^{-\lambda \cdot \tau_A}$$

$$\phi_A \cdot \lambda \cdot e^{\lambda \cdot t} = \left[n_3^{(j)} - \frac{g_3}{g_2} \cdot \left(n_{tot} - \sum_{\substack{k=1 \\ k \neq 2}}^{4} n_k^{(j)} \right) \right] \cdot \frac{A_{32}^A}{n_{\nu_A}} \cdot g(\nu, \nu_A) \cdot (\Phi_A^{(j)}$$

$$+ \phi_A \cdot e^{\lambda \cdot t} \cdot e^{-\lambda \cdot \tau_A}) \cdot \frac{l}{L_A}$$

$$+ \left[\frac{g_3}{g_2} \cdot \sum_{\substack{k=1 \\ k \neq 2}}^{4} n_k + n_3 \right] \cdot e^{\lambda \cdot t} \cdot \frac{A_{32}^A}{n_{\nu_A}} \cdot g(\nu, \nu_A) \cdot (\Phi_A^{(j)}$$

$$+ \phi_A \cdot e^{\lambda \cdot t} \cdot e^{-\lambda \cdot \tau_A}) \cdot \frac{l}{L_A} - \frac{1}{\tau_{RA}} \cdot \Phi_A^{(j)} - \frac{1}{\tau_{RA}} \cdot \phi_A \cdot e^{\lambda \cdot t} \cdot e^{-\lambda \cdot \tau_A}$$

$$\phi_A \cdot \lambda \cdot e^{\lambda \cdot t} = \left[n_3^{(j)} - \frac{g_3}{g_2} \cdot \left(n_{tot} - \sum_{\substack{k=1 \\ k \neq 2}}^{4} n_k^{(j)} \right) \right] \cdot \frac{A_{32}^A}{n_{\nu_A}} \cdot g(\nu, \nu_A) \cdot \Phi_A^{(j)} \cdot \frac{l}{L_A}$$

$$+ \left[n_3^{(j)} - \frac{g_3}{g_2} \cdot \left(n_{tot} - \sum_{\substack{k=1 \\ k \neq 2}}^{4} n_k^{(j)} \right) \right] \cdot \frac{A_{32}^A}{n_{\nu_A}} \cdot g(\nu, \nu_A) \cdot \phi_A \cdot e^{\lambda \cdot t} \cdot e^{-\lambda \cdot \tau_A}$$

$$\cdot \frac{l}{L_A} + \left[\frac{g_3}{g_2} \cdot \sum_{\substack{k=1 \\ k \neq 2}}^{4} n_k + n_3 \right] \cdot e^{\lambda \cdot t} \cdot \frac{A_{32}^A}{n_{\nu_A}} \cdot g(\nu, \nu_A) \cdot \Phi_A^{(j)} \cdot \frac{l}{L_A}$$

$$+\left[\frac{g_3}{g_2}\cdot\sum_{\substack{k=1\\k\neq 2}}^{4}n_k+n_3\right]\cdot e^{\lambda\cdot t}\cdot\frac{A_{32}^{A}}{n_{\nu_A}}\cdot g(\nu,\nu_A)\cdot\phi_A\cdot e^{\lambda\cdot t}\cdot e^{-\lambda\cdot\tau_A}\cdot\frac{l}{L_A}$$
$$-\frac{1}{\tau_{RA}}\cdot\Phi_A^{(j)}-\frac{1}{\tau_{RA}}\cdot\phi_A\cdot e^{\lambda\cdot t}\cdot e^{-\lambda\cdot\tau_A}$$

$$\phi_A\cdot\lambda\cdot e^{\lambda\cdot t}=\left[n_3^{(j)}-\frac{g_3}{g_2}\cdot\left(n_{tot}-\sum_{\substack{k=1\\k\neq 2}}^{4}n_k^{(j)}\right)\right]\cdot\frac{A_{32}^{A}}{n_{\nu_A}}\cdot g(\nu,\nu_A)\cdot\Phi_A^{(j)}\cdot\frac{l}{L_A}$$
$$-\frac{1}{\tau_{RA}}\cdot\Phi_A^{(j)}+\left[n_3^{(j)}-\frac{g_3}{g_2}\cdot\left(n_{tot}-\sum_{\substack{k=1\\k\neq 2}}^{4}n_k^{(j)}\right)\right]\cdot\frac{A_{32}^{A}}{n_{\nu_A}}\cdot g(\nu,\nu_A)$$
$$\cdot\phi_A\cdot e^{\lambda\cdot t}\cdot e^{-\lambda\cdot\tau_A}\cdot\frac{l}{L_A}+\left[\frac{g_3}{g_2}\cdot\sum_{\substack{k=1\\k\neq 2}}^{4}n_k+n_3\right]\cdot e^{\lambda\cdot t}$$
$$\cdot\frac{A_{32}^{A}}{n_{\nu_A}}\cdot g(\nu,\nu_A)\cdot\Phi_A^{(j)}\cdot\frac{l}{L_A}$$
$$+\left[\frac{g_3}{g_2}\cdot\sum_{\substack{k=1\\k\neq 2}}^{4}n_k+n_3\right]\cdot e^{\lambda\cdot t}\cdot\frac{A_{32}^{A}}{n_{\nu_A}}\cdot g(\nu,\nu_A)$$
$$\cdot\phi_A\cdot e^{\lambda\cdot t}\cdot e^{-\lambda\cdot\tau_A}\cdot\frac{l}{L_A}-\frac{1}{\tau_{RA}}\cdot\phi_A\cdot e^{\lambda\cdot t}\cdot e^{-\lambda\cdot\tau_A}$$

At fixed point $[n_3^{(j)}-\frac{g_3}{g_2}\cdot(n_{tot}-\sum_{\substack{k=1\\k\neq 2}}^{4}n_k^{(j)})]\cdot\frac{A_{32}^{A}}{n_{\nu_A}}\cdot g(\nu,\nu_A)\cdot\Phi_A^{(j)}\cdot\frac{l}{L_A}-\frac{1}{\tau_{RA}}\cdot\Phi_A^{(j)}=0$

$$\phi_A\cdot\lambda=\left[n_3^{(j)}-\frac{g_3}{g_2}\cdot\left(n_{tot}-\sum_{\substack{k=1\\k\neq 2}}^{4}n_k^{(j)}\right)\right]\cdot\frac{A_{32}^{A}}{n_{\nu_A}}\cdot g(\nu,\nu_A)\cdot\phi_A\cdot e^{-\lambda\cdot\tau_A}\cdot\frac{l}{L_A}$$
$$+\left[\frac{g_3}{g_2}\cdot\sum_{\substack{k=1\\k\neq 2}}^{4}n_k+n_3\right]\cdot\frac{A_{32}^{A}}{n_{\nu_A}}\cdot g(\nu,\nu_A)\cdot\Phi_A^{(j)}\cdot\frac{l}{L_A}$$
$$+\left[\frac{g_3}{g_2}\cdot\sum_{\substack{k=1\\k\neq 2}}^{4}n_k+n_3\right]\cdot\frac{A_{32}^{A}}{n_{\nu_A}}\cdot g(\nu,\nu_A)$$
$$\cdot\phi_A\cdot e^{\lambda\cdot t}\cdot e^{-\lambda\cdot\tau_A}\cdot\frac{l}{L_A}-\frac{1}{\tau_{RA}}\cdot\phi_A\cdot e^{-\lambda\cdot\tau_A}$$

$$\phi_A \cdot \lambda = \left[n_3^{(j)} - \frac{g_3}{g_2} \cdot \left(n_{tot} - \sum_{\substack{k=1 \\ k \neq 2}}^{4} n_k^{(j)} \right) \right] \cdot \frac{A_{32}^A}{n_{\nu_A}} \cdot g(\nu, \nu_A) \cdot \phi_A \cdot e^{-\lambda \cdot \tau_A} \cdot \frac{l}{L_A}$$
$$+ \left[\frac{g_3}{g_2} \cdot \sum_{\substack{k=1 \\ k \neq 2}}^{4} n_k + n_3 \right] \cdot \frac{A_{32}^A}{n_{\nu_A}} \cdot g(\nu, \nu_A) \cdot \Phi_A^{(j)} \cdot \frac{l}{L_A}$$
$$+ \left[\frac{g_3}{g_2} \cdot \sum_{\substack{k=1 \\ k \neq 2}}^{4} n_k \cdot \phi_A + n_3 \cdot \phi_A \right] \cdot \frac{A_{32}^A}{n_{\nu_A}} \cdot g(\nu, \nu_A)$$
$$\cdot e^{\lambda \cdot t} \cdot e^{-\lambda \cdot \tau_A} \cdot \frac{l}{L_A} - \frac{1}{\tau_{RA}} \cdot \phi_A \cdot e^{-\lambda \cdot \tau_A}$$

Assumption: $n_k \cdot \phi_A \approx 0$; $n_3 \cdot \phi_A \approx 0$

$$\phi_A \cdot \lambda = \left[n_3^{(j)} - \frac{g_3}{g_2} \cdot \left(n_{tot} - \sum_{\substack{k=1 \\ k \neq 2}}^{4} n_k^{(j)} \right) \right] \cdot \frac{A_{32}^A}{n_{\nu_A}} \cdot g(\nu, \nu_A) \cdot \phi_A \cdot e^{-\lambda \cdot \tau_A} \cdot \frac{l}{L_A}$$
$$+ \left[\frac{g_3}{g_2} \cdot \sum_{\substack{k=1 \\ k \neq 2}}^{4} n_k + n_3 \right] \cdot \frac{A_{32}^A}{n_{\nu_A}} \cdot g(\nu, \nu_A) \cdot \Phi_A^{(j)} \cdot \frac{l}{L_A} - \frac{1}{\tau_{RA}} \cdot \phi_A \cdot e^{-\lambda \cdot \tau_A}$$

We consider that n_{tot} is slightly changes and take it as a constant ($\sum_{\substack{k=1 \\ k \neq 2}}^{4} n_k = n_1 + n_3 + n_4$).

$$\phi_A \cdot \lambda = \left[n_3^{(j)} - \frac{g_3}{g_2} \cdot \left(n_{tot} - \sum_{\substack{k=1 \\ k \neq 2}}^{4} n_k^{(j)} \right) \right] \cdot \frac{A_{32}^A}{n_{\nu_A}} \cdot g(\nu, \nu_A) \cdot \phi_A \cdot e^{-\lambda \cdot \tau_A} \cdot \frac{l}{L_A}$$
$$+ \left[\frac{g_3}{g_2} \cdot (n_1 + n_3 + n_4) + n_3 \right] \cdot \frac{A_{32}^A}{n_{\nu_A}} \cdot g(\nu, \nu_A) \cdot \Phi_A^{(j)} \cdot \frac{l}{L_A}$$
$$- \frac{1}{\tau_{RA}} \cdot \phi_A \cdot e^{-\lambda \cdot \tau_A}$$

$$\left\{ \left[n_3^{(j)} - \frac{g_3}{g_2} \cdot \left(n_{tot} - \sum_{\substack{k=1 \\ k \neq 2}}^{4} n_k^{(j)} \right) \right] \cdot \frac{A_{32}^A}{n_{\nu_A}} \cdot g(\nu, \nu_A) \cdot \frac{l}{L_A} - \frac{1}{\tau_{RA}} \right\} \cdot \phi_A \cdot e^{-\lambda \cdot \tau_A}$$
$$- \phi_A \cdot \lambda + \frac{g_3}{g_2} \cdot \frac{A_{32}^A}{n_{\nu_A}} \cdot g(\nu, \nu_A) \cdot \Phi_A^{(j)} \cdot \frac{l}{L_A} \cdot n_1$$

$$+\left(1+\frac{g_3}{g_2}\right)\cdot\frac{A_{32}^A}{n_{\nu_A}}\cdot g(\nu,\nu_A)\cdot\Phi_A^{(j)}\cdot\frac{l}{L_A}\cdot n_3$$
$$+\frac{g_3}{g_2}\cdot\frac{A_{32}^A}{n_{\nu_A}}\cdot g(\nu,\nu_A)\cdot\Phi_A^{(j)}\cdot\frac{l}{L_A}\cdot n_4=0$$

We define for simplicity some global parameters:

$$\Omega_1=\left[n_3^{(j)}-\frac{g_3}{g_2}\cdot\left(n_{tot}-\sum_{\substack{k=1\\k\neq 2}}^{4}n_k^{(j)}\right)\right]\cdot\frac{A_{32}^A}{n_{\nu_A}}\cdot g(\nu,\nu_A)\cdot\frac{l}{L_A}-\frac{1}{\tau_{RA}}$$
$$\Omega_2=\frac{g_3}{g_2}\cdot\frac{A_{32}^A}{n_{\nu_A}}\cdot g(\nu,\nu_A)\cdot\Phi_A^{(j)}\cdot\frac{l}{L_A}$$
$$\Omega_3=\left(1+\frac{g_3}{g_2}\right)\cdot\frac{A_{32}^A}{n_{\nu_A}}\cdot g(\nu,\nu_A)\cdot\Phi_A^{(j)}\cdot\frac{l}{L_A}$$
$$\Omega_1\cdot\phi_A\cdot e^{-\lambda\cdot\tau_A}-\phi_A\cdot\lambda+\Omega_2\cdot n_1+\Omega_3\cdot n_3+\Omega_2\cdot n_4=0$$

First arbitrarily small increments equation: $\Omega_1\cdot\phi_A\cdot e^{-\lambda\cdot\tau_A}-\phi_A\cdot\lambda+\Omega_2\cdot n_1+\Omega_3\cdot n_3+\Omega_2\cdot n_4=0$.

$$2.\ \frac{d\Phi_B}{dt}=\left(n_3-\frac{g_3}{g_2}\cdot\left[n_{tot}-\sum_{\substack{k=1\\k\neq 2}}^{4}n_k\right]\right)\cdot\frac{A_{32}^B}{n_{\nu_B}}\cdot g(\nu,\nu_B)\cdot\Phi_B(t-\tau_B)\cdot\frac{l}{L_B}-\frac{\Phi_B(t-\tau_B)}{\tau_{RB}}$$

$$\phi_B\cdot\lambda\cdot e^{\lambda\cdot t}=\left(n_3^{(j)}+n_3\cdot e^{\lambda\cdot t}-\frac{g_3}{g_2}\cdot\left[n_{tot}-\sum_{\substack{k=1\\k\neq 2}}^{4}(n_k^{(j)}+n_k\cdot e^{\lambda\cdot t})\right]\right)\cdot\frac{A_{32}^B}{n_{\nu_B}}$$
$$\cdot g(\nu,\nu_B)\cdot(\Phi_B^{(j)}+\phi_B\cdot e^{\lambda\cdot(t-\tau_B)})\cdot\frac{l}{L_B}$$
$$-\frac{1}{\tau_{RB}}\cdot(\Phi_B^{(j)}+\phi_B\cdot e^{\lambda\cdot(t-\tau_B)})$$

$$\phi_B\cdot\lambda\cdot e^{\lambda\cdot t}=\left(\left[n_3^{(j)}-\frac{g_3}{g_2}\cdot\left(n_{tot}-\sum_{\substack{k=1\\k\neq 2}}^{4}n_k^{(j)}\right)\right]+\left[\frac{g_3}{g_2}\cdot\sum_{\substack{k=1\\k\neq 2}}^{4}n_k+n_3\right]\cdot e^{\lambda\cdot t}\right)$$
$$\cdot\frac{A_{32}^B}{n_{\nu_B}}\cdot g(\nu,\nu_B)\cdot(\Phi_B^{(j)}+\phi_B\cdot e^{\lambda\cdot t}\cdot e^{-\lambda\cdot\tau_B})\cdot\frac{l}{L_B}$$

$$- \frac{1}{\tau_{RB}} \cdot \Phi_B^{(j)} - \frac{1}{\tau_{RB}} \cdot \phi_B \cdot e^{\lambda \cdot t} \cdot e^{-\lambda \cdot \tau_B}$$

$$\phi_B \cdot \lambda \cdot e^{\lambda \cdot t} = \left[n_3^{(j)} - \frac{g_3}{g_2} \cdot \left(n_{tot} - \sum_{\substack{k=1 \\ k \neq 2}}^{4} n_k^{(j)} \right) \right] \cdot \frac{A_{32}^B}{n_{\nu_B}} \cdot g(\nu, \nu_B) \cdot (\Phi_B^{(j)}$$
$$+ \phi_B \cdot e^{\lambda \cdot t} \cdot e^{-\lambda \cdot \tau_B}) \cdot \frac{l}{L_B}$$
$$+ \left[\frac{g_3}{g_2} \cdot \sum_{\substack{k=1 \\ k \neq 2}}^{4} n_k + n_3 \right] \cdot e^{\lambda \cdot t} \cdot \frac{A_{32}^A}{n_{\nu_B}} \cdot g(\nu, \nu_B) \cdot (\Phi_B^{(j)}$$
$$+ \phi_B \cdot e^{\lambda \cdot t} \cdot e^{-\lambda \cdot \tau_B}) \cdot \frac{l}{L_B} - \frac{1}{\tau_{RB}} \cdot \Phi_B^{(j)} - \frac{1}{\tau_{RB}} \cdot \phi_B \cdot e^{\lambda \cdot t} \cdot e^{-\lambda \cdot \tau_B}$$

$$\phi_B \cdot \lambda \cdot e^{\lambda \cdot t} = \left[n_3^{(j)} - \frac{g_3}{g_2} \cdot \left(n_{tot} - \sum_{\substack{k=1 \\ k \neq 2}}^{4} n_k^{(j)} \right) \right] \cdot \frac{A_{32}^B}{n_{\nu_B}} \cdot g(\nu, \nu_B) \cdot \Phi_B^{(j)} \cdot \frac{l}{L_B}$$
$$+ \left[n_3^{(j)} - \frac{g_3}{g_2} \cdot \left(n_{tot} - \sum_{\substack{k=1 \\ k \neq 2}}^{4} n_k^{(j)} \right) \right] \cdot \frac{A_{32}^B}{n_{\nu_B}} \cdot g(\nu, \nu_B) \cdot \phi_B$$
$$\cdot e^{\lambda \cdot t} \cdot e^{-\lambda \cdot \tau_B} \cdot \frac{l}{L_B}$$
$$+ \left[\frac{g_3}{g_2} \cdot \sum_{\substack{k=1 \\ k \neq 2}}^{4} n_k + n_3 \right] \cdot e^{\lambda \cdot t} \cdot \frac{A_{32}^B}{n_{\nu_B}} \cdot g(\nu, \nu_B) \cdot \Phi_B^{(j)} \cdot \frac{l}{L_B}$$
$$+ \left[\frac{g_3}{g_2} \cdot \sum_{\substack{k=1 \\ k \neq 2}}^{4} n_k + n_3 \right] \cdot e^{\lambda \cdot t} \cdot \frac{A_{32}^B}{n_{\nu_B}} \cdot g(\nu, \nu_B) \cdot \phi_B \cdot e^{\lambda \cdot t} \cdot e^{-\lambda \cdot \tau_B} \cdot \frac{l}{L_B}$$
$$- \frac{1}{\tau_{RB}} \cdot \Phi_B^{(j)} - \frac{1}{\tau_{RB}} \cdot \phi_B \cdot e^{\lambda \cdot t} \cdot e^{-\lambda \cdot \tau_B}$$

$$\phi_B \cdot \lambda \cdot e^{\lambda \cdot t} = \left[n_3^{(j)} - \frac{g_3}{g_2} \cdot \left(n_{tot} - \sum_{\substack{k=1 \\ k \neq 2}}^{4} n_k^{(j)} \right) \right] \cdot \frac{A_{32}^B}{n_{\nu_B}} \cdot g(\nu, \nu_B) \cdot \Phi_B^{(j)} \cdot \frac{l}{L_B}$$
$$- \frac{1}{\tau_{RB}} \cdot \Phi_B^{(j)} + \left[n_3^{(j)} - \frac{g_3}{g_2} \cdot \left(n_{tot} - \sum_{\substack{k=1 \\ k \neq 2}}^{4} n_k^{(j)} \right) \right] \cdot \frac{A_{32}^B}{n_{\nu_B}} \cdot g(\nu, \nu_B) \cdot \phi_B$$

$$\cdot e^{\lambda \cdot t} \cdot e^{-\lambda \cdot \tau_B} \cdot \frac{l}{L_B} + \left[\frac{g_3}{g_2} \cdot \sum_{\substack{k=1 \\ k \neq 2}}^{4} n_k + n_3 \right] \cdot e^{\lambda \cdot t} \cdot \frac{A_{32}^B}{n_{v_B}} \cdot g(v, v_B) \cdot \Phi_B^{(j)}$$

$$\cdot \frac{l}{L_B} + \left[\frac{g_3}{g_2} \cdot \sum_{\substack{k=1 \\ k \neq 2}}^{4} n_k + n_3 \right] \cdot e^{\lambda \cdot t} \cdot \frac{A_{32}^B}{n_{v_B}} \cdot g(v, v_B) \cdot \phi_B \cdot e^{\lambda \cdot t}$$

$$\cdot e^{-\lambda \cdot \tau_B} \cdot \frac{l}{L_B} - \frac{1}{\tau_{RB}} \cdot \phi_B \cdot e^{\lambda \cdot t} \cdot e^{-\lambda \cdot \tau_B}$$

At fixed point $[n_3^{(j)} - \frac{g_3}{g_2} \cdot (n_{tot} - \sum_{\substack{k=1 \\ k \neq 2}}^{4} n_k^{(j)})] \cdot \frac{A_{32}^B}{n_{v_B}} \cdot g(v, v_B) \cdot \Phi_B^{(j)} \cdot \frac{l}{L_B} - \frac{1}{\tau_{RB}} \cdot \Phi_B^{(j)} = 0$

$$\phi_B \cdot \lambda = \left[n_3^{(j)} - \frac{g_3}{g_2} \cdot \left(n_{tot} - \sum_{\substack{k=1 \\ k \neq 2}}^{4} n_k^{(j)} \right) \right] \cdot \frac{A_{32}^B}{n_{v_B}} \cdot g(v, v_B) \cdot \phi_B \cdot e^{-\lambda \cdot \tau_B} \cdot \frac{l}{L_B}$$

$$+ \left[\frac{g_3}{g_2} \cdot \sum_{\substack{k=1 \\ k \neq 2}}^{4} n_k + n_3 \right] \cdot \frac{A_{32}^B}{n_{v_B}} \cdot g(v, v_B) \cdot \Phi_B^{(j)} \cdot \frac{l}{L_B}$$

$$+ \left[\frac{g_3}{g_2} \cdot \sum_{\substack{k=1 \\ k \neq 2}}^{4} n_k + n_3 \right] \cdot \frac{A_{32}^B}{n_{v_B}} \cdot g(v, v_B) \cdot \phi_B$$

$$\cdot e^{\lambda \cdot t} \cdot e^{-\lambda \cdot \tau_B} \cdot \frac{l}{L_B} - \frac{1}{\tau_{RB}} \cdot \phi_B \cdot e^{-\lambda \cdot \tau_B}$$

$$\phi_B \cdot \lambda = \left[n_3^{(j)} - \frac{g_3}{g_2} \cdot \left(n_{tot} - \sum_{\substack{k=1 \\ k \neq 2}}^{4} n_k^{(j)} \right) \right] \cdot \frac{A_{32}^B}{n_{v_B}} \cdot g(v, v_B) \cdot \phi_B \cdot e^{-\lambda \cdot \tau_B} \cdot \frac{l}{L_B}$$

$$+ \left[\frac{g_3}{g_2} \cdot \sum_{\substack{k=1 \\ k \neq 2}}^{4} n_k + n_3 \right] \cdot \frac{A_{32}^B}{n_{v_B}} \cdot g(v, v_B) \cdot \Phi_B^{(j)} \cdot \frac{l}{L_B}$$

$$+ \left[\frac{g_3}{g_2} \cdot \sum_{\substack{k=1 \\ k \neq 2}}^{4} n_k \cdot \phi_B + n_3 \cdot \phi_B \right] \cdot \frac{A_{32}^B}{n_{v_B}} \cdot g(v, v_B)$$

$$\cdot e^{\lambda \cdot t} \cdot e^{-\lambda \cdot \tau_B} \cdot \frac{l}{L_B} - \frac{1}{\tau_{RB}} \cdot \phi_B \cdot e^{-\lambda \cdot \tau_B}$$

Assumption: $n_k \cdot \phi_B \approx 0$; $n_3 \cdot \phi_B \approx 0$

$$\phi_B \cdot \lambda = \left[n_3^{(j)} - \frac{g_3}{g_2} \cdot \left(n_{tot} - \sum_{\substack{k=1\\k\neq 2}}^{4} n_k^{(j)} \right) \right] \cdot \frac{A_{32}^B}{n_{\nu_B}} \cdot g(\nu, \nu_B) \cdot \phi_B \cdot e^{-\lambda \cdot \tau_B} \cdot \frac{l}{L_B}$$
$$+ \left[\frac{g_3}{g_2} \cdot \sum_{\substack{k=1\\k\neq 2}}^{4} n_k + n_3 \right] \cdot \frac{A_{32}^B}{n_{\nu_B}} \cdot g(\nu, \nu_B) \cdot \Phi_B^{(j)} \cdot \frac{l}{L_B} - \frac{1}{\tau_{RB}} \cdot \phi_B \cdot e^{-\lambda \cdot \tau_B}$$

We consider that n_{tot} is slightly changes and take it as a constant ($\sum_{\substack{k=1\\k\neq 2}}^{4} n_k = n_1 + n_3 + n_4$).

$$\phi_B \cdot \lambda = \left[n_3^{(j)} - \frac{g_3}{g_2} \cdot \left(n_{tot} - \sum_{\substack{k=1\\k\neq 2}}^{4} n_k^{(j)} \right) \right] \cdot \frac{A_{32}^B}{n_{\nu_B}} \cdot g(\nu, \nu_B) \cdot \phi_B \cdot e^{-\lambda \cdot \tau_B} \cdot \frac{l}{L_B}$$
$$+ \left[\frac{g_3}{g_2} \cdot (n_1 + n_3 + n_4) + n_3 \right] \cdot \frac{A_{32}^B}{n_{\nu_B}} \cdot g(\nu, \nu_B) \cdot \Phi_B^{(j)} \cdot \frac{l}{L_B}$$
$$- \frac{1}{\tau_{RB}} \cdot \phi_B \cdot e^{-\lambda \cdot \tau_B}$$

$$\left\{ \left[n_3^{(j)} - \frac{g_3}{g_2} \cdot \left(n_{tot} - \sum_{\substack{k=1\\k\neq 2}}^{4} n_k^{(j)} \right) \right] \cdot \frac{A_{32}^B}{n_{\nu_B}} \cdot g(\nu, \nu_B) \cdot \frac{l}{L_B} - \frac{1}{\tau_{RB}} \right\} \cdot \phi_B \cdot e^{-\lambda \cdot \tau_B}$$
$$- \phi_B \cdot \lambda + \frac{g_3}{g_2} \cdot \frac{A_{32}^B}{n_{\nu_B}} \cdot g(\nu, \nu_B) \cdot \Phi_B^{(j)} \cdot \frac{l}{L_B} \cdot n_1 + \left(1 + \frac{g_3}{g_2} \right)$$
$$\cdot \frac{A_{32}^B}{n_{\nu_B}} \cdot g(\nu, \nu_B) \cdot \Phi_B^{(j)}$$
$$\cdot \frac{l}{L_B} \cdot n_3 + \frac{g_3}{g_2} \cdot \frac{A_{32}^B}{n_{\nu_B}} \cdot g(\nu, \nu_B) \cdot \Phi_B^{(j)} \cdot \frac{l}{L_B} \cdot n_4 = 0$$

We define for simplicity some global parameters:

$$\Omega_4 = \left[n_3^{(j)} - \frac{g_3}{g_2} \cdot \left(n_{tot} - \sum_{\substack{k=1\\k\neq 2}}^{4} n_k^{(j)} \right) \right] \cdot \frac{A_{32}^B}{n_{\nu_B}} \cdot g(\nu, \nu_B) \cdot \frac{l}{L_B} - \frac{1}{\tau_{RB}}$$
$$\Omega_5 = \frac{g_3}{g_2} \cdot \frac{A_{32}^B}{n_{\nu_B}} \cdot g(\nu, \nu_B) \cdot \Phi_B^{(j)} \cdot \frac{l}{L_B}$$
$$\Omega_6 = \left(1 + \frac{g_3}{g_2} \right) \cdot \frac{A_{32}^B}{n_{\nu_B}} \cdot g(\nu, \nu_B) \cdot \Phi_B^{(j)} \cdot \frac{l}{L_B}$$

$$\Omega_4 \cdot \phi_B \cdot e^{-\lambda \cdot \tau_B} - \phi_B \cdot \lambda + \Omega_5 \cdot n_1 + \Omega_6 \cdot n_3 + \Omega_5 \cdot n_4 = 0$$

Second arbitrarily small increments equation: $\Omega_4 \cdot \phi_B \cdot e^{-\lambda \cdot \tau_B} - \phi_B \cdot \lambda + \Omega_5 \cdot n_1 + \Omega_6 \cdot n_3 + \Omega_5 \cdot n_4 = 0$.

3. $$\frac{dn_1}{dt} = n_3 \cdot (S_{21}^A + S_{21}^B) - n_1 \cdot W_{14} \Rightarrow n_1 \cdot \lambda \cdot e^{\lambda \cdot t} = (n_3^{(j)} + n_3 \cdot e^{\lambda \cdot t}) \cdot (S_{21}^A + S_{21}^B) - (n_1^{(j)} + n_1 \cdot e^{\lambda \cdot t}) \cdot W_{14}$$

$$n_3^{(j)} \cdot (S_{21}^A + S_{21}^B) - n_1^{(j)} \cdot W_{14} + n_3 \cdot (S_{21}^A + S_{21}^B) \cdot e^{\lambda \cdot t} - n_1 \cdot W_{14} \cdot e^{\lambda \cdot t} - n_1 \cdot \lambda \cdot e^{\lambda \cdot t} = 0$$

At fixed point: $n_3^{(j)} \cdot (S_{21}^A + S_{21}^B) - n_1^{(j)} \cdot W_{14} = 0; -n_1 \cdot W_{14} - n_1 \cdot \lambda + n_3 \cdot (S_{21}^A + S_{21}^B) = 0$.

Third arbitrarily small increments equation: $-n_1 \cdot W_{14} - n_1 \cdot \lambda + n_3 \cdot (S_{21}^A + S_{21}^B) = 0$.

4. $$\frac{dn_3}{dt} = -\left(n_3 - \frac{g_3}{g_2} \cdot \left[n_{tot} - \sum_{\substack{k=1 \\ k \neq 2}}^{4} n_k\right]\right) \cdot \frac{A_{32}^A}{n_{\nu_A}} \cdot g(\nu, \nu_A) \cdot \Phi_A(t - \tau_A) - \left(n_3 - \frac{g_3}{g_2} \cdot \left[n_{tot} - \sum_{\substack{k=1 \\ k \neq 2}}^{4} n_k\right]\right) \cdot \frac{A_{32}^B}{n_{\nu_B}} \cdot g(\nu, \nu_B) \cdot \Phi_B(t - \tau_B) - n_3 \cdot [A_{32}^A + S_{32}^A + A_{32}^B + S_{32}^B] + n_4 \cdot S_{43}$$

$$n_3 \cdot \lambda \cdot e^{\lambda \cdot t} = -\left(n_3 - \frac{g_3}{g_2} \cdot \left[n_{tot} - \sum_{\substack{k=1 \\ k \neq 2}}^{4} n_k\right]\right) \cdot \frac{A_{32}^A}{n_{\nu_A}} \cdot g(\nu, \nu_A) \cdot \Phi_A(t - \tau_A) - \left(n_3 - \frac{g_3}{g_2} \cdot \left[n_{tot} - \sum_{\substack{k=1 \\ k \neq 2}}^{4} n_k\right]\right) \cdot \frac{A_{32}^B}{n_{\nu_B}} \cdot g(\nu, \nu_B) \cdot \Phi_B(t - \tau_B) - (n_3^{(j)} + n_3 \cdot e^{\lambda \cdot t}) \cdot [A_{32}^A + S_{32}^A + A_{32}^B + S_{32}^B] + (n_4^{(j)} + n_4 \cdot e^{\lambda \cdot t}) \cdot S_{43}$$

First step, arbitrarily expression: $\left(n_3 - \frac{g_3}{g_2} \cdot \left[n_{tot} - \sum_{\substack{k=1 \\ k \neq 2}}^{4} n_k\right]\right) \cdot \frac{A_{32}^A}{n_{\nu_A}} \cdot g(\nu, \nu_A) \cdot \Phi_A(t - \tau_A)$

$$\left(n_3^{(j)} + n_3 \cdot e^{\lambda \cdot t} - \frac{g_3}{g_2} \cdot \left[n_{tot} - \sum_{\substack{k=1 \\ k \neq 2}}^{4} (n_k^{(j)} + n_k \cdot e^{\lambda \cdot t})\right]\right) \cdot \frac{A_{32}^A}{n_{\nu_A}}$$
$$\cdot g(\nu, \nu_A) \cdot (\Phi_A^{(j)} + \phi_A \cdot e^{\lambda \cdot (t - \tau_A)})$$

$$\left[n_3^{(j)} + n_3 \cdot e^{\lambda \cdot t} - \frac{g_3}{g_2} \cdot \left(n_{tot} - \sum_{\substack{k=1 \\ k \neq 2}}^{4} n_k^{(j)}\right) + \frac{g_3}{g_2} \cdot \sum_{\substack{k=1 \\ k \neq 2}}^{4} n_k \cdot e^{\lambda \cdot t}\right] \cdot \frac{A_{32}^A}{n_{\nu_A}}$$
$$\cdot g(\nu, \nu_A) \cdot (\Phi_A^{(j)} + \phi_A \cdot e^{\lambda \cdot t} \cdot e^{-\lambda \cdot \tau_A})$$

$$\left\{\left[n_3^{(j)} - \frac{g_3}{g_2} \cdot \left(n_{tot} - \sum_{\substack{k=1 \\ k \neq 2}}^{4} n_k^{(j)}\right)\right] + \frac{g_3}{g_2} \cdot \sum_{\substack{k=1 \\ k \neq 2}}^{4} n_k \cdot e^{\lambda \cdot t} + n_3 \cdot e^{\lambda \cdot t}\right\} \cdot \frac{A_{32}^A}{n_{\nu_A}}$$
$$\cdot g(\nu, \nu_A) \cdot (\Phi_A^{(j)} + \phi_A \cdot e^{\lambda \cdot t} \cdot e^{-\lambda \cdot \tau_A})$$

$$\left\{\left[n_3^{(j)} - \frac{g_3}{g_2} \cdot \left(n_{tot} - \sum_{\substack{k=1 \\ k \neq 2}}^{4} n_k^{(j)}\right)\right]\right.$$
$$\left. + \frac{g_3}{g_2} \cdot \sum_{\substack{k=1 \\ k \neq 2}}^{4} n_k \cdot e^{\lambda \cdot t} + n_3 \cdot e^{\lambda \cdot t}\right\} \cdot \frac{A_{32}^A}{n_{\nu_A}} \cdot g(\nu, \nu_A) \cdot \Phi_A^{(j)}$$
$$+ \left\{\left[n_3^{(j)} - \frac{g_3}{g_2} \cdot \left(n_{tot} - \sum_{\substack{k=1 \\ k \neq 2}}^{4} n_k^{(j)}\right)\right] \cdot \phi_A\right.$$
$$\left. + \frac{g_3}{g_2} \cdot \sum_{\substack{k=1 \\ k \neq 2}}^{4} n_k \cdot \phi_A \cdot e^{\lambda \cdot t} + n_3 \cdot \phi_A \cdot e^{\lambda \cdot t}\right\} \cdot \frac{A_{32}^A}{n_{\nu_A}} \cdot g(\nu, \nu_A) \cdot e^{\lambda \cdot t} \cdot e^{-\lambda \cdot \tau_A}$$

Assumption: $n_k \cdot \phi_A \approx 0$; $n_3 \cdot \phi_A \approx 0$

$$\left[n_3^{(j)} - \frac{g_3}{g_2} \cdot \left(n_{tot} - \sum_{\substack{k=1 \\ k \neq 2}}^{4} n_k^{(j)}\right)\right] \cdot \frac{A_{32}^A}{n_{\nu_A}} \cdot g(\nu, \nu_A) \cdot \Phi_A^{(j)}$$

$$+\left[\frac{g_3}{g_2}\cdot\sum_{\substack{k=1\\k\neq 2}}^{4}n_k\cdot e^{\lambda\cdot t}+n_3\cdot e^{\lambda\cdot t}\right]\cdot\frac{A_{32}^{A}}{n_{\nu_A}}\cdot g(\nu,\nu_A)\cdot\Phi_A^{(j)}$$

$$+\left[n_3^{(j)}-\frac{g_3}{g_2}\cdot\left(n_{tot}-\sum_{\substack{k=1\\k\neq 2}}^{4}n_k^{(j)}\right)\right]\cdot\frac{A_{32}^{A}}{n_{\nu_A}}\cdot g(\nu,\nu_A)\cdot\phi_A\cdot e^{\lambda\cdot t}\cdot e^{-\lambda\cdot\tau_A}$$

We consider that n_{tot} is slightly changes and take it as a constant $(\sum_{\substack{k=1\\k\neq 2}}^{4}n_k=n_1+n_3+n_4)$.

$$\left[n_3^{(j)}-\frac{g_3}{g_2}\cdot\left(n_{tot}-\sum_{\substack{k=1\\k\neq 2}}^{4}n_k^{(j)}\right)\right]\cdot\frac{A_{32}^{A}}{n_{\nu_A}}\cdot g(\nu,\nu_A)\cdot\Phi_A^{(j)}$$

$$+\left[n_3^{(j)}-\frac{g_3}{g_2}\cdot\left(n_{tot}-\sum_{\substack{k=1\\k\neq 2}}^{4}n_k^{(j)}\right)\right]\cdot\frac{A_{32}^{A}}{n_{\nu_A}}\cdot g(\nu,\nu_A)\cdot\phi_A\cdot e^{\lambda\cdot t}\cdot e^{-\lambda\cdot\tau_A}$$

$$+\frac{g_3}{g_2}\cdot\frac{A_{32}^{A}}{n_{\nu_A}}\cdot g(\nu,\nu_A)\cdot\Phi_A^{(j)}\cdot e^{\lambda\cdot t}\cdot n_1$$

$$+\left(\frac{g_3}{g_2}+1\right)\cdot e^{\lambda\cdot t}\cdot\frac{A_{32}^{A}}{n_{\nu_A}}\cdot g(\nu,\nu_A)\cdot\Phi_A^{(j)}\cdot n_3+\frac{g_3}{g_2}\cdot\frac{A_{32}^{A}}{n_{\nu_A}}\cdot g(\nu,\nu_A)\cdot\Phi_A^{(j)}\cdot e^{\lambda\cdot t}\cdot n_4$$

At fixed points: $[n_3^{(j)}-\frac{g_3}{g_2}\cdot(n_{tot}-\sum_{\substack{k=1\\k\neq 2}}^{4}n_k^{(j)})]\cdot\frac{A_{32}^{A}}{n_{\nu_A}}\cdot g(\nu,\nu_A)\cdot\Phi_A^{(j)}=0$.

$$\left[n_3^{(j)}-\frac{g_3}{g_2}\cdot\left(n_{tot}-\sum_{\substack{k=1\\k\neq 2}}^{4}n_k^{(j)}\right)\right]\cdot\frac{A_{32}^{A}}{n_{\nu_A}}\cdot g(\nu,\nu_A)\cdot\phi_A\cdot e^{\lambda\cdot t}\cdot e^{-\lambda\cdot\tau_A}$$

$$+\frac{g_3}{g_2}\cdot\frac{A_{32}^{A}}{n_{\nu_A}}\cdot g(\nu,\nu_A)\cdot\Phi_A^{(j)}\cdot e^{\lambda\cdot t}\cdot n_1$$

$$+\left(\frac{g_3}{g_2}+1\right)\cdot e^{\lambda\cdot t}\cdot\frac{A_{32}^{A}}{n_{\nu_A}}\cdot g(\nu,\nu_A)\cdot\Phi_A^{(j)}\cdot n_3$$

$$+\frac{g_3}{g_2}\cdot\frac{A_{32}^{A}}{n_{\nu_A}}\cdot g(\nu,\nu_A)\cdot e^{\lambda\cdot t}\cdot\Phi_A^{(j)}\cdot n_4$$

We divide it by $e^{\lambda\cdot t}$ and get

$$\left[n_3^{(j)} - \frac{g_3}{g_2} \cdot \left(n_{tot} - \sum_{\substack{k=1\\k\neq 2}}^{4} n_k^{(j)}\right)\right] \cdot \frac{A_{32}^A}{n_{\nu_A}} \cdot g(\nu, \nu_A) \cdot \phi_A \cdot e^{-\lambda \cdot \tau_A}$$

$$+ \frac{g_3}{g_2} \cdot \frac{A_{32}^A}{n_{\nu_A}} \cdot g(\nu, \nu_A) \cdot \Phi_A^{(j)} \cdot n_1$$

$$+ \left(\frac{g_3}{g_2} + 1\right) \cdot \frac{A_{32}^A}{n_{\nu_A}} \cdot g(\nu, \nu_A) \cdot \Phi_A^{(j)} \cdot n_3 + \frac{g_3}{g_2} \cdot \frac{A_{32}^A}{n_{\nu_A}} \cdot g(\nu, \nu_A) \cdot \Phi_A^{(j)} \cdot n_4$$

Second step, arbitrarily expression: $\left(n_3 - \frac{g_3}{g_2} \cdot \left[n_{tot} - \sum_{\substack{k=1\\k\neq 2}}^{4} n_k\right]\right) \cdot \frac{A_{32}^B}{n_{\nu_B}} \cdot g(\nu, \nu_B) \cdot \Phi_B(t - \tau_B)$.

$$\left(n_3^{(j)} + n_3 \cdot e^{\lambda \cdot t} - \frac{g_3}{g_2} \cdot \left[n_{tot} - \sum_{\substack{k=1\\k\neq 2}}^{4} (n_k^{(j)} + n_k \cdot e^{\lambda \cdot t})\right]\right) \cdot \frac{A_{32}^B}{n_{\nu_B}}$$
$$\cdot g(\nu, \nu_B) \cdot (\Phi_B^{(j)} + \phi_B \cdot e^{\lambda \cdot (t - \tau_B)})$$

$$\left[n_3^{(j)} + n_3 \cdot e^{\lambda \cdot t} - \frac{g_3}{g_2} \cdot \left(n_{tot} - \sum_{\substack{k=1\\k\neq 2}}^{4} n_k^{(j)}\right) + \frac{g_3}{g_2} \cdot \sum_{\substack{k=1\\k\neq 2}}^{4} n_k \cdot e^{\lambda \cdot t}\right] \cdot \frac{A_{32}^B}{n_{\nu_B}}$$
$$\cdot g(\nu, \nu_B) \cdot (\Phi_B^{(j)} + \phi_B \cdot e^{\lambda \cdot t} \cdot e^{-\lambda \cdot \tau_B})$$

$$\left\{\left[n_3^{(j)} - \frac{g_3}{g_2} \cdot \left(n_{tot} - \sum_{\substack{k=1\\k\neq 2}}^{4} n_k^{(j)}\right)\right] + \frac{g_3}{g_2} \cdot \sum_{\substack{k=1\\k\neq 2}}^{4} n_k \cdot e^{\lambda \cdot t} + n_3 \cdot e^{\lambda \cdot t}\right\}$$
$$\cdot \frac{A_{32}^B}{n_{\nu_B}} \cdot g(\nu, \nu_B) \cdot (\Phi_B^{(j)} + \phi_B \cdot e^{\lambda \cdot t} \cdot e^{-\lambda \cdot \tau_B})$$

$$\left\{\left[n_3^{(j)} - \frac{g_3}{g_2} \cdot \left(n_{tot} - \sum_{\substack{k=1\\k\neq 2}}^{4} n_k^{(j)}\right)\right] + \frac{g_3}{g_2} \cdot \sum_{\substack{k=1\\k\neq 2}}^{4} n_k \cdot e^{\lambda \cdot t} + n_3 \cdot e^{\lambda \cdot t}\right\}$$
$$\cdot \frac{A_{32}^B}{n_{\nu_B}} \cdot g(\nu, \nu_B) \cdot \Phi_B^{(j)} + \left\{\left[n_3^{(j)} - \frac{g_3}{g_2} \cdot \left(n_{tot} - \sum_{\substack{k=1\\k\neq 2}}^{4} n_k^{(j)}\right)\right] \cdot \phi_B\right.$$
$$\left. + \frac{g_3}{g_2} \cdot \sum_{\substack{k=1\\k\neq 2}}^{4} n_k \cdot \phi_B \cdot e^{\lambda \cdot t} + n_3 \cdot \phi_B \cdot e^{\lambda \cdot t}\right\} \cdot \frac{A_{32}^B}{n_{\nu_B}}$$

$$\cdot g(\nu, \nu_B) \cdot e^{\lambda \cdot t} \cdot e^{-\lambda \cdot \tau_B}$$

Assumption: $n_k \cdot \phi_B \approx 0; n_3 \cdot \phi_B \approx 0$

$$\left[n_3^{(j)} - \frac{g_3}{g_2} \cdot \left(n_{tot} - \sum_{\substack{k=1 \\ k \neq 2}}^{4} n_k^{(j)}\right)\right] \cdot \frac{A_{32}^B}{n_{\nu_B}} \cdot g(\nu, \nu_B) \cdot \Phi_B^{(j)}$$
$$+ \left[\frac{g_3}{g_2} \cdot \sum_{\substack{k=1 \\ k \neq 2}}^{4} n_k \cdot e^{\lambda \cdot t} + n_3 \cdot e^{\lambda \cdot t}\right] \cdot \frac{A_{32}^B}{n_{\nu_B}} \cdot g(\nu, \nu_B) \cdot \Phi_B^{(j)}$$
$$+ \left[n_3^{(j)} - \frac{g_3}{g_2} \cdot \left(n_{tot} - \sum_{\substack{k=1 \\ k \neq 2}}^{4} n_k^{(j)}\right)\right] \cdot \frac{A_{32}^B}{n_{\nu_B}} \cdot g(\nu, \nu_B) \cdot \phi_B \cdot e^{\lambda \cdot t} \cdot e^{-\lambda \cdot \tau_B}$$

We consider that n_{tot} is slightly changes and take it as a constant $(\sum_{\substack{k=1 \\ k \neq 2}}^{4} n_k = n_1 + n_3 + n_4)$.

$$\left[n_3^{(j)} - \frac{g_3}{g_2} \cdot \left(n_{tot} - \sum_{\substack{k=1 \\ k \neq 2}}^{4} n_k^{(j)}\right)\right] \cdot \frac{A_{32}^B}{n_{\nu_B}} \cdot g(\nu, \nu_B) \cdot \Phi_B^{(j)}$$
$$+ \left[n_3^{(j)} - \frac{g_3}{g_2} \cdot \left(n_{tot} - \sum_{\substack{k=1 \\ k \neq 2}}^{4} n_k^{(j)}\right)\right] \cdot \frac{A_{32}^B}{n_{\nu_B}} \cdot g(\nu, \nu_B) \cdot \phi_B \cdot e^{\lambda \cdot t} \cdot e^{-\lambda \cdot \tau_B}$$
$$+ \frac{g_3}{g_2} \cdot \frac{A_{32}^B}{n_{\nu_B}} \cdot g(\nu, \nu_B) \cdot \Phi_B^{(j)} \cdot e^{\lambda \cdot t} \cdot n_1 + \left(\frac{g_3}{g_2} + 1\right) \cdot e^{\lambda \cdot t} \cdot \frac{A_{32}^B}{n_{\nu_B}} \cdot g(\nu, \nu_B)$$
$$\cdot \Phi_B^{(j)} \cdot n_3 + \frac{g_3}{g_2} \cdot \frac{A_{32}^B}{n_{\nu_B}} \cdot g(\nu, \nu_B) \cdot \Phi_B^{(j)} \cdot e^{\lambda \cdot t} \cdot n_4$$

At fixed points: $[n_3^{(j)} - \frac{g_3}{g_2} \cdot (n_{tot} - \sum_{\substack{k=1 \\ k \neq 2}}^{4} n_k^{(j)})] \cdot \frac{A_{32}^B}{n_{\nu_B}} \cdot g(\nu, \nu_B) \cdot \Phi_B^{(j)} = 0.$

$$\left[n_3^{(j)} - \frac{g_3}{g_2} \cdot \left(n_{tot} - \sum_{\substack{k=1 \\ k \neq 2}}^{4} n_k^{(j)}\right)\right] \cdot \frac{A_{32}^B}{n_{\nu_B}} \cdot g(\nu, \nu_B) \cdot \phi_B \cdot e^{\lambda \cdot t} \cdot e^{-\lambda \cdot \tau_B}$$
$$+ \frac{g_3}{g_2} \cdot \frac{A_{32}^B}{n_{\nu_B}} \cdot g(\nu, \nu_B) \cdot \Phi_B^{(j)} \cdot e^{\lambda \cdot t} \cdot n_1$$
$$+ \left(\frac{g_3}{g_2} + 1\right) \cdot e^{\lambda \cdot t} \cdot \frac{A_{32}^B}{n_{\nu_B}} \cdot g(\nu, \nu_B) \cdot \Phi_B^{(j)} \cdot n_3$$

$$+\frac{g_3}{g_2}\cdot\frac{A_{32}^B}{n_{\nu_B}}\cdot g(\nu,\nu_B)\cdot\Phi_B^{(j)}\cdot e^{\lambda\cdot t}\cdot n_4$$

We divide it by $e^{\lambda\cdot t}$ and get

$$\left[n_3^{(j)}-\frac{g_3}{g_2}\cdot\left(n_{tot}-\sum_{\substack{k=1\\k\neq 2}}^{4}n_k^{(j)}\right)\right]\cdot\frac{A_{32}^B}{n_{\nu_B}}\cdot g(\nu,\nu_B)\cdot\phi_B\cdot e^{-\lambda\cdot\tau_B}$$
$$+\frac{g_3}{g_2}\cdot\frac{A_{32}^B}{n_{\nu_B}}\cdot g(\nu,\nu_B)\cdot\Phi_B^{(j)}\cdot n_1$$
$$+\left(\frac{g_3}{g_2}+1\right)\cdot\frac{A_{32}^B}{n_{\nu_B}}\cdot g(\nu,\nu_B)\cdot\Phi_B^{(j)}\cdot n_3+\frac{g_3}{g_2}\cdot\frac{A_{32}^B}{n_{\nu_B}}\cdot g(\nu,\nu_B)\cdot\Phi_B^{(j)}\cdot n_4$$

Third step, arbitrarily expression: $-(n_3^{(j)}+n_3\cdot e^{\lambda\cdot t})\cdot[A_{32}^A+S_{32}^A+A_{32}^B+S_{32}^B]$ $+(n_4^{(j)}+n_4\cdot e^{\lambda\cdot t})\cdot S_{43}$.

$$-n_3^{(j)}\cdot[A_{32}^A+S_{32}^A+A_{32}^B+S_{32}^B]-[A_{32}^A+S_{32}^A+A_{32}^B+S_{32}^B]\cdot n_3\cdot e^{\lambda\cdot t}$$
$$+n_4^{(j)}\cdot S_{43}+n_4\cdot S_{43}\cdot e^{\lambda\cdot t}$$

$$-n_3^{(j)}\cdot[A_{32}^A+S_{32}^A+A_{32}^B+S_{32}^B]+n_4^{(j)}\cdot S_{43}-[A_{32}^A+S_{32}^A+A_{32}^B$$
$$+S_{32}^B]\cdot n_3\cdot e^{\lambda\cdot t}+n_4\cdot S_{43}\cdot e^{\lambda\cdot t}$$

At fixed point: $-n_3^{(j)}\cdot[A_{32}^A+S_{32}^A+A_{32}^B+S_{32}^B]+n_4^{(j)}\cdot S_{43}=0$.

$$-[A_{32}^A+S_{32}^A+A_{32}^B+S_{32}^B]\cdot n_3\cdot e^{\lambda\cdot t}+n_4\cdot S_{43}\cdot e^{\lambda\cdot t}$$

We divide it by $e^{\lambda\cdot t}$ and get $-[A_{32}^A+S_{32}^A+A_{32}^B+S_{32}^B]\cdot n_3+n_4\cdot S_{43}$.
We can summary our third arbitrarily equation's three steps:

$$\frac{dn_3}{dt}=-\left(n_3-\frac{g_3}{g_2}\cdot\left[n_{tot}-\sum_{\substack{k=1\\k\neq 2}}^{4}n_k\right]\right)\cdot\frac{A_{32}^A}{n_{\nu_A}}\cdot g(\nu,\nu_A)\cdot\Phi_A(t-\tau_A)$$
$$-\left(n_3-\frac{g_3}{g_2}\cdot\left[n_{tot}-\sum_{\substack{k=1\\k\neq 2}}^{4}n_k\right]\right)\cdot\frac{A_{32}^B}{n_{\nu_B}}\cdot g(\nu,\nu_B)\cdot\Phi_B(t-\tau_B)$$
$$-n_3\cdot[A_{32}^A+S_{32}^A+A_{32}^B+S_{32}^B]+n_4\cdot S_{43}$$

$$n_3 \cdot \lambda = -\left\{\left[n_3^{(j)} - \frac{g_3}{g_2} \cdot \left(n_{tot} - \sum_{\substack{k=1 \\ k\neq 2}}^{4} n_k^{(j)}\right)\right] \cdot \frac{A_{32}^A}{n_{\nu_A}} \cdot g(\nu, \nu_A) \cdot \phi_A \right.$$

$$\cdot e^{-\lambda \cdot \tau_A} + \frac{g_3}{g_2} \cdot \frac{A_{32}^A}{n_{\nu_A}} \cdot g(\nu, \nu_A) \cdot \Phi_A^{(j)} \cdot n_1$$

$$+ \left(\frac{g_3}{g_2} + 1\right) \cdot \frac{A_{32}^A}{n_{\nu_A}} \cdot g(\nu, \nu_A) \cdot \Phi_A^{(j)} \cdot n_3$$

$$\left. + \frac{g_3}{g_2} \cdot \frac{A_{32}^A}{n_{\nu_A}} \cdot g(\nu, \nu_A) \cdot \Phi_A^{(j)} \cdot n_4 \right\}$$

$$-\left\{\left[n_3^{(j)} - \frac{g_3}{g_2} \cdot \left(n_{tot} - \sum_{\substack{k=1 \\ k\neq 2}}^{4} n_k^{(j)}\right)\right] \cdot \frac{A_{32}^B}{n_{\nu_B}} \cdot g(\nu, \nu_B) \cdot \phi_B \cdot e^{-\lambda \cdot \tau_B} \right.$$

$$+ \frac{g_3}{g_2} \cdot \frac{A_{32}^B}{n_{\nu_B}} \cdot g(\nu, \nu_B) \cdot \Phi_B^{(j)} \cdot n_1$$

$$+ \left(\frac{g_3}{g_2} + 1\right) \cdot \frac{A_{32}^B}{n_{\nu_B}} \cdot g(\nu, \nu_B) \cdot \Phi_B^{(j)} \cdot n_3$$

$$\left. + \frac{g_3}{g_2} \cdot \frac{A_{32}^B}{n_{\nu_B}} \cdot g(\nu, \nu_B) \cdot \Phi_B^{(j)} \cdot n_4 \right\}$$

$$- [A_{32}^A + S_{32}^A + A_{32}^B + S_{32}^B] \cdot n_3 + n_4 \cdot S_{43}$$

For simplicity we define some global parameters:

$$\Gamma_1 = \left[n_3^{(j)} - \frac{g_3}{g_2} \cdot \left(n_{tot} - \sum_{\substack{k=1 \\ k\neq 2}}^{4} n_k^{(j)}\right)\right] \cdot \frac{A_{32}^A}{n_{\nu_A}} \cdot g(\nu, \nu_A)$$

$$\Gamma_2 = \frac{g_3}{g_2} \cdot \frac{A_{32}^A}{n_{\nu_A}} \cdot g(\nu, \nu_A) \cdot \Phi_A^{(j)}$$

$$\Gamma_3 = \left(\frac{g_3}{g_2} + 1\right) \cdot \frac{A_{32}^A}{n_{\nu_A}} \cdot g(\nu, \nu_A) \cdot \Phi_A^{(j)}$$

$$\Gamma_4 = \left[n_3^{(j)} - \frac{g_3}{g_2} \cdot \left(n_{tot} - \sum_{\substack{k=1 \\ k\neq 2}}^{4} n_k^{(j)}\right)\right] \cdot \frac{A_{32}^B}{n_{\nu_B}} \cdot g(\nu, \nu_B)$$

$$\Gamma_5 = \frac{g_3}{g_2} \cdot \frac{A_{32}^B}{n_{\nu_B}} \cdot g(\nu, \nu_B) \cdot \Phi_B^{(j)}$$

$$\Gamma_6 = \left(\frac{g_3}{g_2} + 1\right) \cdot \frac{A_{32}^B}{n_{\nu_B}} \cdot g(\nu, \nu_B) \cdot \Phi_B^{(j)}$$

$$\Gamma_7 = A_{32}^A + S_{32}^A + A_{32}^B + S_{32}^B$$

$$\begin{aligned} n_3 \cdot \lambda = & -(\Gamma_1 \cdot \phi_A \cdot e^{-\lambda \cdot \tau_A} + \Gamma_2 \cdot n_1 + \Gamma_3 \cdot n_3 + \Gamma_2 \cdot n_4) \\ & - (\Gamma_4 \cdot \phi_B \cdot e^{-\lambda \cdot \tau_B} + \Gamma_5 \cdot n_1 + \Gamma_6 \cdot n_3 + \Gamma_5 \cdot n_4) \\ & - \Gamma_7 \cdot n_3 + n_4 \cdot S_{43} \end{aligned}$$

Forth arbitrarily small increments equation:

$$\begin{aligned} & - \Gamma_1 \cdot \phi_A \cdot e^{-\lambda \cdot \tau_A} - \Gamma_4 \cdot \phi_B \cdot e^{-\lambda \cdot \tau_B} - (\Gamma_2 + \Gamma_5) \cdot n_1 - (\Gamma_3 + \Gamma_6 + \Gamma_7) \cdot n_3 \\ & - n_3 \cdot \lambda + (S_{43} - \Gamma_2 - \Gamma_5) \cdot n_4 = 0 \end{aligned}$$

5. $\frac{dn_4}{dt} = n_1 \cdot W_{14} - n_4 \cdot (S_{43} + S_{41} + A_{41})$

$$n_4 \cdot \lambda \cdot e^{\lambda \cdot t} = (n_1^{(j)} + n_1 \cdot e^{\lambda \cdot t}) \cdot W_{14} - (n_4^{(j)} + n_4 \cdot e^{\lambda \cdot t}) \cdot (S_{43} + S_{41} + A_{41})$$

$$\begin{aligned} n_4 \cdot \lambda \cdot e^{\lambda \cdot t} = & (n_1^{(j)} \cdot W_{14} + W_{14} \cdot n_1 \cdot e^{\lambda \cdot t}) - [n_4^{(j)} \cdot (S_{43} + S_{41} + A_{41}) \\ & + (S_{43} + S_{41} + A_{41}) \cdot n_4 \cdot e^{\lambda \cdot t}] \end{aligned}$$

$$\begin{aligned} n_4 \cdot \lambda \cdot e^{\lambda \cdot t} = & n_1^{(j)} \cdot W_{14} - n_4^{(j)} \cdot (S_{43} + S_{41} + A_{41}) + W_{14} \cdot n_1 \cdot e^{\lambda \cdot t} \\ & - (S_{43} + S_{41} + A_{41}) \cdot n_4 \cdot e^{\lambda \cdot t} \end{aligned}$$

At fixed points: $n_1^{(j)} \cdot W_{14} - n_4^{(j)} \cdot (S_{43} + S_{41} + A_{41}) = 0$.

$$W_{14} \cdot n_1 - (S_{43} + S_{41} + A_{41}) \cdot n_4 - n_4 \cdot \lambda = 0; \Gamma_8 = S_{43} + S_{41} + A_{41}$$

Fifth arbitrarily small increments equation: $W_{14} \cdot n_1 - \Gamma_8 \cdot n_4 - n_4 \cdot \lambda = 0$.
We can summary our system global parameters list: $\Gamma_6 = \left(\frac{g_3}{g_2} + 1\right) \cdot \Gamma_5 \cdot \frac{g_2}{g_3}$

$$\Omega_1 = \left[n_3^{(j)} - \frac{g_3}{g_2} \cdot \left(n_{tot} - \sum_{\substack{k=1 \\ k \neq 2}}^{4} n_k^{(j)} \right) \right] \cdot \frac{A_{32}^A}{n_{v_A}} \cdot g(v, v_A) \cdot \frac{l}{L_A} - \frac{1}{\tau_{RA}}$$

$$\Omega_2 = \frac{g_3}{g_2} \cdot \frac{A_{32}^A}{n_{v_A}} \cdot g(v, v_A) \cdot \Phi_A^{(j)} \cdot \frac{l}{L_A}$$

$$\Omega_3 = \left(1 + \frac{g_3}{g_2}\right) \cdot \frac{A_{32}^A}{n_{v_A}} \cdot g(v, v_A) \cdot \Phi_A^{(j)} \cdot \frac{l}{L_A}$$

$$\Omega_4 = \left[n_3^{(j)} - \frac{g_3}{g_2} \cdot \left(n_{tot} - \sum_{\substack{k=1 \\ k \neq 2}}^{4} n_k^{(j)} \right) \right] \cdot \frac{A_{32}^B}{n_{\nu_B}} \cdot g(\nu, \nu_B) \cdot \frac{l}{L_B} - \frac{1}{\tau_{RB}}$$

$$\Omega_5 = \frac{g_3}{g_2} \cdot \frac{A_{32}^B}{n_{\nu_B}} \cdot g(\nu, \nu_B) \cdot \Phi_B^{(j)} \cdot \frac{l}{L_B}$$

$$\Omega_6 = \left(1 + \frac{g_3}{g_2}\right) \cdot \frac{A_{32}^B}{n_{\nu_B}} \cdot g(\nu, \nu_B) \cdot \Phi_B^{(j)} \cdot \frac{l}{L_B}$$

$$\Gamma_1 = \left[n_3^{(j)} - \frac{g_3}{g_2} \cdot \left(n_{tot} - \sum_{\substack{k=1 \\ k \neq 2}}^{4} n_k^{(j)} \right) \right] \cdot \frac{A_{32}^A}{n_{\nu_A}} \cdot g(\nu, \nu_A)$$

$$\Gamma_2 = \frac{g_3}{g_2} \cdot \frac{A_{32}^A}{n_{\nu_A}} \cdot g(\nu, \nu_A) \cdot \Phi_A^{(j)}$$

$$\Gamma_3 = \left(\frac{g_3}{g_2} + 1\right) \cdot \frac{A_{32}^A}{n_{\nu_A}} \cdot g(\nu, \nu_A) \cdot \Phi_A^{(j)}$$

$$\Gamma_4 = \left[n_3^{(j)} - \frac{g_3}{g_2} \cdot \left(n_{tot} - \sum_{\substack{k=1 \\ k \neq 2}}^{4} n_k^{(j)} \right) \right] \cdot \frac{A_{32}^B}{n_{\nu_B}} \cdot g(\nu, \nu_B)$$

$$\Gamma_5 = \frac{g_3}{g_2} \cdot \frac{A_{32}^B}{n_{\nu_B}} \cdot g(\nu, \nu_B) \cdot \Phi_B^{(j)}$$

$$\Gamma_6 = \left(\frac{g_3}{g_2} + 1\right) \cdot \frac{A_{32}^B}{n_{\nu_B}} \cdot g(\nu, \nu_B) \cdot \Phi_B^{(j)}$$

$$\Gamma_7 = A_{32}^A + S_{32}^A + A_{32}^B + S_{32}^B$$

$$\Omega_3 = \left(\frac{g_3}{g_2} + 1\right) \cdot \Omega_2 \cdot \frac{g_2}{g_3}$$

$$\Omega_6 = \left(\frac{g_3}{g_2} + 1\right) \cdot \Omega_5 \cdot \frac{g_2}{g_3}$$

$$\Gamma_3 = \left(\frac{g_3}{g_2} + 1\right) \cdot \Gamma_2 \cdot \frac{g_2}{g_3}$$

The next table presents the System arbitrarily small increments equations (Table 7.2).

The small increments Jacobian of our Dual-wavelength Ti:Sapphire laser system is as follow:

Table 7.2 Dual-wavelength Ti:Sapphire laser system arbitrarily small increments equations

First arbitrarily small increments equation	$\Omega_1 \cdot \phi_A \cdot e^{-\lambda \cdot \tau_A} - \phi_A \cdot \lambda + \Omega_2 \cdot n_1 + \Omega_3 \cdot n_3 + \Omega_2 \cdot n_4 = 0$
Second arbitrarily small increments equation	$\Omega_4 \cdot \phi_B \cdot e^{-\lambda \cdot \tau_B} - \phi_B \cdot \lambda + \Omega_5 \cdot n_1 + \Omega_6 \cdot n_3 + \Omega_5 \cdot n_4 = 0$
Third arbitrarily small increments equation	$-n_1 \cdot W_{14} - n_1 \cdot \lambda + n_3 \cdot (S_{21}^A + S_{21}^B) = 0$
Forth arbitrarily small increments equation	$-\Gamma_1 \cdot \phi_A \cdot e^{-\lambda \cdot \tau_A} - \Gamma_4 \cdot \phi_B \cdot e^{-\lambda \cdot \tau_B} - (\Gamma_2 + \Gamma_5) \cdot n_1$ $- (\Gamma_3 + \Gamma_6 + \Gamma_7) \cdot n_3 - n_3 \cdot \lambda + (S_{43} - \Gamma_2 - \Gamma_5) \cdot n_4 = 0$
Fifth arbitrarily small increments equation	$W_{14} \cdot n_1 - \Gamma_8 \cdot n_4 - n_4 \cdot \lambda = 0$

$$\begin{pmatrix} \Upsilon_{11} & \dots & \Upsilon_{15} \\ \vdots & \ddots & \vdots \\ \Upsilon_{51} & \cdots & \Upsilon_{55} \end{pmatrix} \cdot \begin{pmatrix} \phi_A \\ \phi_B \\ n_1 \\ n_3 \\ n_4 \end{pmatrix} = 0; \ \Upsilon_{11} = \Omega_1 \cdot e^{-\lambda \cdot \tau_A} - \lambda$$

$$\Upsilon_{12} = 0; \ \Upsilon_{13} = \Omega_2; \ \Upsilon_{14} = \Omega_3; \ \Upsilon_{15} = \Omega_2$$

$$\Upsilon_{21} = 0; \ \Upsilon_{22} = \Omega_4 \cdot e^{-\lambda \cdot \tau_B} - \lambda; \ \Upsilon_{23} = \Omega_5; \ \Upsilon_{24} = \Omega_6$$
$$\Upsilon_{25} = \Omega_5; \ \Upsilon_{31} = 0; \ \Upsilon_{32} = 0$$

$$\Upsilon_{33} = -W_{14} - \lambda; \ \Upsilon_{34} = S_{21}^A + S_{21}^B; \ \Upsilon_{35} = 0$$
$$\Upsilon_{41} = -\Gamma_1 \cdot e^{-\lambda \cdot \tau_A}; \ \Upsilon_{42} = -\Gamma_4 \cdot e^{-\lambda \cdot \tau_B}$$

$$\Upsilon_{43} = -(\Gamma_2 + \Gamma_5); \ \Upsilon_{44} = -(\Gamma_3 + \Gamma_6 + \Gamma_7) - \lambda$$
$$\Upsilon_{45} = S_{43} - \Gamma_2 - \Gamma_5; \ \Upsilon_{51} = 0; \ \Upsilon_{52} = 0$$

$$\Upsilon_{53} = W_{14}; \ \Upsilon_{54} = 0; \ \Upsilon_{55} = -\Gamma_8 - \lambda$$

$$A - \lambda \cdot I = \begin{pmatrix} \Upsilon_{11} & \dots & \Upsilon_{15} \\ \vdots & \ddots & \vdots \\ \Upsilon_{51} & \cdots & \Upsilon_{55} \end{pmatrix}; \ \det(A - \lambda \cdot I) = \det \begin{pmatrix} \Upsilon_{11} & \dots & \Upsilon_{15} \\ \vdots & \ddots & \vdots \\ \Upsilon_{51} & \cdots & \Upsilon_{55} \end{pmatrix}$$

$$\det(A - \lambda \cdot I) = 0$$

$$\det(A - \lambda \cdot I) = \Upsilon_{11} \cdot \det\begin{pmatrix} \Upsilon_{22} & \Upsilon_{23} & \Upsilon_{24} & \Upsilon_{25} \\ \Upsilon_{32} & \Upsilon_{33} & \Upsilon_{34} & \Upsilon_{35} \\ \Upsilon_{42} & \Upsilon_{43} & \Upsilon_{44} & \Upsilon_{45} \\ \Upsilon_{52} & \Upsilon_{53} & \Upsilon_{54} & \Upsilon_{55} \end{pmatrix} - \Upsilon_{12} \cdot \det\begin{pmatrix} \Upsilon_{21} & \Upsilon_{23} & \Upsilon_{24} & \Upsilon_{25} \\ \Upsilon_{31} & \Upsilon_{33} & \Upsilon_{34} & \Upsilon_{35} \\ \Upsilon_{41} & \Upsilon_{43} & \Upsilon_{44} & \Upsilon_{45} \\ \Upsilon_{51} & \Upsilon_{53} & \Upsilon_{54} & \Upsilon_{55} \end{pmatrix}$$

$$+ \Upsilon_{13} \cdot \det\begin{pmatrix} \Upsilon_{21} & \Upsilon_{22} & \Upsilon_{24} & \Upsilon_{25} \\ \Upsilon_{31} & \Upsilon_{32} & \Upsilon_{34} & \Upsilon_{35} \\ \Upsilon_{41} & \Upsilon_{42} & \Upsilon_{44} & \Upsilon_{45} \\ \Upsilon_{51} & \Upsilon_{52} & \Upsilon_{54} & \Upsilon_{55} \end{pmatrix} - \Upsilon_{14} \cdot \det\begin{pmatrix} \Upsilon_{21} & \Upsilon_{22} & \Upsilon_{23} & \Upsilon_{25} \\ \Upsilon_{31} & \Upsilon_{32} & \Upsilon_{33} & \Upsilon_{35} \\ \Upsilon_{41} & \Upsilon_{42} & \Upsilon_{43} & \Upsilon_{45} \\ \Upsilon_{51} & \Upsilon_{52} & \Upsilon_{53} & \Upsilon_{55} \end{pmatrix}$$

$$+ \Upsilon_{15} \cdot \det\begin{pmatrix} \Upsilon_{21} & \Upsilon_{22} & \Upsilon_{23} & \Upsilon_{24} \\ \Upsilon_{31} & \Upsilon_{32} & \Upsilon_{33} & \Upsilon_{34} \\ \Upsilon_{41} & \Upsilon_{42} & \Upsilon_{43} & \Upsilon_{44} \\ \Upsilon_{51} & \Upsilon_{52} & \Upsilon_{53} & \Upsilon_{54} \end{pmatrix}$$

Step 1: Find the expression for $\Upsilon_{11} \cdot \det\begin{pmatrix} \Upsilon_{22} & \Upsilon_{23} & \Upsilon_{24} & \Upsilon_{25} \\ \Upsilon_{32} & \Upsilon_{33} & \Upsilon_{34} & \Upsilon_{35} \\ \Upsilon_{42} & \Upsilon_{43} & \Upsilon_{44} & \Upsilon_{45} \\ \Upsilon_{52} & \Upsilon_{53} & \Upsilon_{54} & \Upsilon_{55} \end{pmatrix}$.

$$(\Omega_1 \cdot e^{-\lambda \cdot \tau_A} - \lambda) \cdot \det\begin{pmatrix} \Upsilon_{22} & \Upsilon_{23} & \Upsilon_{24} & \Upsilon_{25} \\ \Upsilon_{32} & \Upsilon_{33} & \Upsilon_{34} & \Upsilon_{35} \\ \Upsilon_{42} & \Upsilon_{43} & \Upsilon_{44} & \Upsilon_{45} \\ \Upsilon_{52} & \Upsilon_{53} & \Upsilon_{54} & \Upsilon_{55} \end{pmatrix}$$
$$= (\Omega_1 \cdot e^{-\lambda \cdot \tau_A} - \lambda)$$
$$\cdot \det\begin{pmatrix} \Omega_4 \cdot e^{-\lambda \cdot \tau_B} - \lambda & \Omega_5 & \Omega_6 & \Omega_5 \\ 0 & -W_{14} - \lambda & S_{21}^A + S_{21}^B & 0 \\ -\Gamma_4 \cdot e^{-\lambda \cdot \tau_B} & -(\Gamma_2 + \Gamma_5) & -(\Gamma_3 + \Gamma_6 + \Gamma_7) - \lambda & S_{43} - \Gamma_2 - \Gamma_5 \\ 0 & W_{14} & 0 & -\Gamma_8 - \lambda \end{pmatrix}$$

$$(\Omega_1 \cdot e^{-\lambda \cdot \tau_A} - \lambda) \cdot \det\begin{pmatrix} \Upsilon_{22} & \Upsilon_{23} & \Upsilon_{24} & \Upsilon_{25} \\ \Upsilon_{32} & \Upsilon_{33} & \Upsilon_{34} & \Upsilon_{35} \\ \Upsilon_{42} & \Upsilon_{43} & \Upsilon_{44} & \Upsilon_{45} \\ \Upsilon_{52} & \Upsilon_{53} & \Upsilon_{54} & \Upsilon_{55} \end{pmatrix}$$
$$= (\Omega_1 \cdot e^{-\lambda \cdot \tau_A} - \lambda)$$
$$\cdot \det\begin{pmatrix} \Omega_4 \cdot e^{-\lambda \cdot \tau_B} - \lambda & \Omega_5 & \Omega_6 & \Omega_5 \\ 0 & -W_{14} - \lambda & S_{21}^A + S_{21}^B & 0 \\ -\Gamma_4 \cdot e^{-\lambda \cdot \tau_B} & -(\Gamma_2 + \Gamma_5) & -(\Gamma_3 + \Gamma_6 + \Gamma_7) - \lambda & S_{43} - \Gamma_2 - \Gamma_5 \\ 0 & W_{14} & 0 & -\Gamma_8 - \lambda \end{pmatrix}$$
$$= (\Omega_1 \cdot e^{-\lambda \cdot \tau_A} - \lambda) \cdot \{(\Omega_4 \cdot e^{-\lambda \cdot \tau_B} - \lambda)$$

$$\cdot \det\begin{pmatrix} -(W_{14}+\lambda) & (S_{21}^{A}+S_{21}^{B}) & 0 \\ -(\Gamma_2+\Gamma_5) & -(\Gamma_3+\Gamma_6+\Gamma_7)-\lambda & S_{43}-\Gamma_2-\Gamma_5 \\ W_{14} & 0 & -(\Gamma_8+\lambda) \end{pmatrix}$$

$$\left. -\Gamma_4 \cdot e^{-\lambda \cdot \tau_B} \cdot \det\begin{pmatrix} \Omega_5 & \Omega_6 & \Omega_5 \\ -(W_{14}+\lambda) & (S_{21}^{A}+S_{21}^{B}) & 0 \\ W_{14} & 0 & -(\Gamma_8+\lambda) \end{pmatrix}\right\}$$

Step 1.1: $(\Omega_4 \cdot e^{-\lambda \cdot \tau_B} - \lambda)$

$$\cdot \det\begin{pmatrix} -(W_{14}+\lambda) & (S_{21}^{A}+S_{21}^{B}) & 0 \\ -(\Gamma_2+\Gamma_5) & -(\Gamma_3+\Gamma_6+\Gamma_7)-\lambda & S_{43}-\Gamma_2-\Gamma_5 \\ W_{14} & 0 & -(\Gamma_8+\lambda) \end{pmatrix}.$$

$$\begin{aligned}
&(\Omega_4 \cdot e^{-\lambda \cdot \tau_B} - \lambda) \\
&\quad \cdot \det\begin{pmatrix} -(W_{14}+\lambda) & (S_{21}^{A}+S_{21}^{B}) & 0 \\ -(\Gamma_2+\Gamma_5) & -(\Gamma_3+\Gamma_6+\Gamma_7)-\lambda & S_{43}-\Gamma_2-\Gamma_5 \\ W_{14} & 0 & -(\Gamma_8+\lambda) \end{pmatrix} \\
&= (\Omega_4 \cdot e^{-\lambda \cdot \tau_B} - \lambda) \cdot \{-(W_{14}+\lambda) \\
&\quad \cdot \det\begin{pmatrix} -(\Gamma_3+\Gamma_6+\Gamma_7)-\lambda & (S_{43}-\Gamma_2-\Gamma_5) \\ 0 & -(\Gamma_8+\lambda) \end{pmatrix} \\
&\quad \left. -(S_{21}^{A}+S_{21}^{B}) \cdot \det\begin{pmatrix} -(\Gamma_2+\Gamma_5) & (S_{43}-\Gamma_2-\Gamma_5) \\ W_{14} & -(\Gamma_8+\lambda) \end{pmatrix}\right\} \\
&= (\Omega_4 \cdot e^{-\lambda \cdot \tau_B} - \lambda) \cdot \{-(W_{14}+\lambda) \\
&\quad \cdot ([\Gamma_3+\Gamma_6+\Gamma_7]+\lambda) \cdot (\Gamma_8+\lambda) \\
&\quad - (S_{21}^{A}+S_{21}^{B}) \cdot [(\Gamma_2+\Gamma_5) \cdot (\Gamma_8+\lambda) \\
&\quad - W_{14} \cdot (S_{43}-\Gamma_2-\Gamma_5)]\}
\end{aligned}$$

Step 1.1.1: $-(W_{14}+\lambda) \cdot ([\Gamma_3+\Gamma_6+\Gamma_7]+\lambda) \cdot (\Gamma_8+\lambda)$.

$$\begin{aligned}
&-(W_{14}+\lambda) \cdot ([\Gamma_3+\Gamma_6+\Gamma_7]+\lambda) \cdot (\Gamma_8+\lambda) \\
&= -\{W_{14} \cdot [\Gamma_3+\Gamma_6+\Gamma_7] \cdot \Gamma_8 \\
&\quad + \lambda \cdot (W_{14} \cdot [\Gamma_3+\Gamma_6+\Gamma_7] + \Gamma_8 \cdot [W_{14} \\
&\quad + \Gamma_3+\Gamma_6+\Gamma_7]) + \lambda^2 \cdot (W_{14}+\Gamma_3 \\
&\quad + \Gamma_6+\Gamma_7+\Gamma_8) + \lambda^3\}
\end{aligned}$$

We define some parameters for simplicity: $\Xi_1 = W_{14} \cdot [\Gamma_3+\Gamma_6+\Gamma_7] \cdot \Gamma_8$

$$\Xi_2 = W_{14} \cdot [\Gamma_3 + \Gamma_6 + \Gamma_7] + \Gamma_8 \cdot [W_{14} + \Gamma_3 + \Gamma_6 + \Gamma_7]$$
$$\Xi_3 = W_{14} + \Gamma_3 + \Gamma_6 + \Gamma_7 + \Gamma_8$$

$$-(W_{14} + \lambda) \cdot ([\Gamma_3 + \Gamma_6 + \Gamma_7] + \lambda) \cdot (\Gamma_8 + \lambda) = -\{\Xi_1 + \lambda \cdot \Xi_2 + \lambda^2 \cdot \Xi_3 + \lambda^3\}$$

Step 1.1.2: $(S_{21}^A + S_{21}^B) \cdot [(\Gamma_2 + \Gamma_5) \cdot (\Gamma_8 + \lambda) - W_{14} \cdot (S_{43} - \Gamma_2 - \Gamma_5)]$.

$$\begin{aligned}
&(S_{21}^A + S_{21}^B) \cdot [(\Gamma_2 + \Gamma_5) \cdot (\Gamma_8 + \lambda) - W_{14} \cdot (S_{43} - \Gamma_2 - \Gamma_5)] \\
&\quad = (S_{21}^A + S_{21}^B) \cdot [\{(\Gamma_2 + \Gamma_5) \cdot \Gamma_8 \\
&\qquad - W_{14} \cdot (S_{43} - \Gamma_2 - \Gamma_5)\} + (\Gamma_2 + \Gamma_5) \cdot \lambda] \\
&\quad = (S_{21}^A + S_{21}^B) \cdot \{(\Gamma_2 + \Gamma_5) \cdot \Gamma_8 - W_{14} \cdot (S_{43} - \Gamma_2 - \Gamma_5)\} \\
&\qquad + (S_{21}^A + S_{21}^B) \cdot (\Gamma_2 + \Gamma_5) \cdot \lambda
\end{aligned}$$

We define some parameters for simplicity: $\Xi_4 = (S_{21}^A + S_{21}^B) \cdot \{(\Gamma_2 + \Gamma_5) \cdot \Gamma_8 - W_{14} \cdot (S_{43} - \Gamma_2 - \Gamma_5)\}$

$$\Xi_5 = (S_{21}^A + S_{21}^B) \cdot (\Gamma_2 + \Gamma_5)$$

$$(S_{21}^A + S_{21}^B) \cdot [(\Gamma_2 + \Gamma_5) \cdot (\Gamma_8 + \lambda) - W_{14} \cdot (S_{43} - \Gamma_2 - \Gamma_5)] = \Xi_4 + \Xi_5 \cdot \lambda$$

$$\begin{aligned}
&(\Omega_4 \cdot e^{-\lambda \cdot \tau_B} - \lambda) \cdot \det \begin{pmatrix} -(W_{14} + \lambda) & (S_{21}^A + S_{21}^B) & 0 \\ -(\Gamma_2 + \Gamma_5) & -(\Gamma_3 + \Gamma_6 + \Gamma_7) - \lambda & S_{43} - \Gamma_2 - \Gamma_5 \\ W_{14} & 0 & -(\Gamma_8 + \lambda) \end{pmatrix} \\
&\quad = (\Omega_4 \cdot e^{-\lambda \cdot \tau_B} - \lambda) \cdot \{-(W_{14} + \lambda) \cdot ([\Gamma_3 + \Gamma_6 + \Gamma_7] + \lambda) \cdot (\Gamma_8 + \lambda) \\
&\qquad - (S_{21}^A + S_{21}^B) \cdot [(\Gamma_2 + \Gamma_5) \cdot (\Gamma_8 + \lambda) - W_{14} \cdot (S_{43} - \Gamma_2 - \Gamma_5)]\} \\
&\quad = (\Omega_4 \cdot e^{-\lambda \cdot \tau_B} - \lambda) \cdot \{-(\Xi_1 + \lambda \cdot \Xi_2 + \lambda^2 \cdot \Xi_3 + \lambda^3) - (\Xi_4 + \Xi_5 \cdot \lambda)\} \\
&\quad = (\Omega_4 \cdot e^{-\lambda \cdot \tau_B} - \lambda) \cdot \{-(\Xi_1 + \Xi_4) - (\Xi_2 + \Xi_5) \cdot \lambda - \lambda^2 \cdot \Xi_3 - \lambda^3\}
\end{aligned}$$

$$\begin{aligned}
&(\Omega_4 \cdot e^{-\lambda \cdot \tau_B} - \lambda) \cdot \det \begin{pmatrix} -(W_{14} + \lambda) & (S_{21}^A + S_{21}^B) & 0 \\ -(\Gamma_2 + \Gamma_5) & -(\Gamma_3 + \Gamma_6 + \Gamma_7) - \lambda & S_{43} - \Gamma_2 - \Gamma_5 \\ W_{14} & 0 & -(\Gamma_8 + \lambda) \end{pmatrix} \\
&\quad = \lambda \cdot (\Xi_1 + \Xi_4) + \lambda^2 \cdot (\Xi_2 + \Xi_5) + \lambda^3 \cdot \Xi_3 + \lambda^4 \\
&\qquad - [(\Xi_1 + \Xi_4) \cdot \Omega_4 + (\Xi_2 + \Xi_5) \cdot \Omega_4 \cdot \lambda \\
&\qquad + \Xi_3 \cdot \Omega_4 \cdot \lambda^2 + \Omega_4 \cdot \lambda^3] \cdot e^{-\lambda \cdot \tau_B}
\end{aligned}$$

Step 1.2: $\Gamma_4 \cdot e^{-\lambda \cdot \tau_B} \cdot \det\begin{pmatrix} \Omega_5 & \Omega_6 & \Omega_5 \\ -(W_{14}+\lambda) & (S_{21}^A + S_{21}^B) & 0 \\ W_{14} & 0 & -(\Gamma_8+\lambda) \end{pmatrix}$.

$$\Gamma_4 \cdot e^{-\lambda \cdot \tau_B} \cdot \det\begin{pmatrix} \Omega_5 & \Omega_6 & \Omega_5 \\ -(W_{14}+\lambda) & (S_{21}^A + S_{21}^B) & 0 \\ W_{14} & 0 & -(\Gamma_8+\lambda) \end{pmatrix}$$
$$= \Gamma_4 \cdot e^{-\lambda \cdot \tau_B} \cdot \left\{ \Omega_5 \cdot \det\begin{pmatrix} (S_{21}^A + S_{21}^B) & 0 \\ 0 & -(\Gamma_8+\lambda) \end{pmatrix} \right.$$
$$- \Omega_6 \cdot \det\begin{pmatrix} -(W_{14}+\lambda) & 0 \\ W_{14} & -(\Gamma_8+\lambda) \end{pmatrix}$$
$$\left. + \Omega_5 \cdot \det\begin{pmatrix} -(W_{14}+\lambda) & (S_{21}^A + S_{21}^B) \\ W_{14} & 0 \end{pmatrix} \right\}$$

$$\Gamma_4 \cdot e^{-\lambda \cdot \tau_B} \cdot \det\begin{pmatrix} \Omega_5 & \Omega_6 & \Omega_5 \\ -(W_{14}+\lambda) & (S_{21}^A + S_{21}^B) & 0 \\ W_{14} & 0 & -(\Gamma_8+\lambda) \end{pmatrix}$$
$$= \Gamma_4 \cdot e^{-\lambda \cdot \tau_B} \cdot \left\{ \Omega_5 \cdot \det\begin{pmatrix} (S_{21}^A + S_{21}^B) & 0 \\ 0 & -(\Gamma_8+\lambda) \end{pmatrix} \right.$$
$$\left. -\Omega_6 \cdot \det\begin{pmatrix} -(W_{14}+\lambda) & 0 \\ W_{14} & -(\Gamma_8+\lambda) \end{pmatrix} + \Omega_5 \cdot \det\begin{pmatrix} -(W_{14}+\lambda) & (S_{21}^A + S_{21}^B) \\ W_{14} & 0 \end{pmatrix} \right\}$$

Step 1.2.1: $\Omega_5 \cdot \det\begin{pmatrix} (S_{21}^A + S_{21}^B) & 0 \\ 0 & -(\Gamma_8+\lambda) \end{pmatrix}$.

$$\Omega_5 \cdot \det\begin{pmatrix} (S_{21}^A + S_{21}^B) & 0 \\ 0 & -(\Gamma_8+\lambda) \end{pmatrix} = -\Omega_5 \cdot (S_{21}^A + S_{21}^B) \cdot (\Gamma_8+\lambda)$$
$$= -\Omega_5 \cdot (S_{21}^A + S_{21}^B) \cdot \Gamma_8$$
$$- \Omega_5 \cdot (S_{21}^A + S_{21}^B) \cdot \lambda$$

Step 1.2.2: $\Omega_6 \cdot \det\begin{pmatrix} -(W_{14}+\lambda) & 0 \\ W_{14} & -(\Gamma_8+\lambda) \end{pmatrix}$.

$$\Omega_6 \cdot \det\begin{pmatrix} -(W_{14}+\lambda) & 0 \\ W_{14} & -(\Gamma_8+\lambda) \end{pmatrix} = \Omega_6 \cdot (W_{14}+\lambda) \cdot (\Gamma_8+\lambda)$$
$$= \Omega_6 \cdot W_{14} \cdot \Gamma_8 + \Omega_6 \cdot (W_{14}+\Gamma_8) \cdot \lambda$$
$$+ \Omega_6 \cdot \lambda^2$$

Step 1.2.3: $\Omega_5 \cdot \det\begin{pmatrix} -(W_{14}+\lambda) & (S_{21}^A + S_{21}^B) \\ W_{14} & 0 \end{pmatrix} = -\Omega_5 \cdot W_{14} \cdot (S_{21}^A + S_{21}^B)$.

$$\Gamma_4 \cdot e^{-\lambda\cdot\tau_B} \cdot \det\begin{pmatrix} \Omega_5 & \Omega_6 & \Omega_5 \\ -(W_{14}+\lambda) & (S_{21}^A + S_{21}^B) & 0 \\ W_{14} & 0 & -(\Gamma_8+\lambda) \end{pmatrix}$$
$$= \Gamma_4 \cdot e^{-\lambda\cdot\tau_B} \cdot \{-\Omega_5 \cdot (S_{21}^A + S_{21}^B) \cdot \Gamma_8 - \Omega_5 \cdot (S_{21}^A + S_{21}^B) \cdot \lambda$$
$$- [\Omega_6 \cdot W_{14} \cdot \Gamma_8 + \Omega_6 \cdot (W_{14} + \Gamma_8) \cdot \lambda + \Omega_6 \cdot \lambda^2] - \Omega_5 \cdot W_{14} \cdot (S_{21}^A + S_{21}^B)\}$$
$$= \Gamma_4 \cdot e^{-\lambda\cdot\tau_B} \cdot \{-\Omega_5 \cdot (S_{21}^A + S_{21}^B) \cdot \Gamma_8$$
$$- \Omega_5 \cdot (S_{21}^A + S_{21}^B) \cdot \lambda - \Omega_6 \cdot W_{14} \cdot \Gamma_8$$
$$- \Omega_6 \cdot (W_{14} + \Gamma_8) \cdot \lambda - \Omega_6 \cdot \lambda^2 - \Omega_5 \cdot W_{14} \cdot (S_{21}^A + S_{21}^B)\}$$

$$\Gamma_4 \cdot e^{-\lambda\cdot\tau_B} \cdot \det\begin{pmatrix} \Omega_5 & \Omega_6 & \Omega_5 \\ -(W_{14}+\lambda) & (S_{21}^A + S_{21}^B) & 0 \\ W_{14} & 0 & -(\Gamma_8+\lambda) \end{pmatrix}$$
$$= \Gamma_4 \cdot e^{-\lambda\cdot\tau_B} \cdot \{-\Omega_5 \cdot (S_{21}^A + S_{21}^B) \cdot \Gamma_8 - \Omega_6 \cdot W_{14} \cdot \Gamma_8$$
$$- \Omega_5 \cdot W_{14} \cdot (S_{21}^A + S_{21}^B) - [\Omega_5 \cdot (S_{21}^A + S_{21}^B)$$
$$+ \Omega_6 \cdot (W_{14} + \Gamma_8)] \cdot \lambda - \Omega_6 \cdot \lambda^2\}$$

We define some parameters for simplicity: $\Xi_7 = [\Omega_5 \cdot (S_{21}^A + S_{21}^B) + \Omega_6 \cdot (W_{14} + \Gamma_8)]$

$$\Xi_6 = -\Omega_5 \cdot (S_{21}^A + S_{21}^B) \cdot \Gamma_8 - \Omega_6 \cdot W_{14} \cdot \Gamma_8 - \Omega_5 \cdot W_{14} \cdot (S_{21}^A + S_{21}^B)$$

$$\Gamma_4 \cdot e^{-\lambda\cdot\tau_B} \cdot \det\begin{pmatrix} \Omega_5 & \Omega_6 & \Omega_5 \\ -(W_{14}+\lambda) & (S_{21}^A + S_{21}^B) & 0 \\ W_{14} & 0 & -(\Gamma_8+\lambda) \end{pmatrix}$$
$$= \Gamma_4 \cdot e^{-\lambda\cdot\tau_B} \cdot (\Xi_6 - \Xi_7 \cdot \lambda - \Omega_6 \cdot \lambda^2)$$

We can implement 1.1 and 1.2 into step 1 expression:

$$(\Omega_1 \cdot e^{-\lambda\cdot\tau_A} - \lambda) \cdot \det\begin{pmatrix} \Upsilon_{22} & \Upsilon_{23} & \Upsilon_{24} & \Upsilon_{25} \\ \Upsilon_{32} & \Upsilon_{33} & \Upsilon_{34} & \Upsilon_{35} \\ \Upsilon_{42} & \Upsilon_{43} & \Upsilon_{44} & \Upsilon_{45} \\ \Upsilon_{52} & \Upsilon_{53} & \Upsilon_{54} & \Upsilon_{55} \end{pmatrix}$$
$$= (\Omega_1 \cdot e^{-\lambda\cdot\tau_A} - \lambda)$$

$$\cdot \det\begin{pmatrix} \Omega_4 \cdot e^{-\lambda\cdot\tau_B} - \lambda & \Omega_5 & \Omega_6 & \Omega_5 \\ 0 & -W_{14} - \lambda & S_{21}^A + S_{21}^B & 0 \\ -\Gamma_4 \cdot e^{-\lambda\cdot\tau_B} & -(\Gamma_2 + \Gamma_5) & -(\Gamma_3 + \Gamma_6 + \Gamma_7) - \lambda & S_{43} - \Gamma_2 - \Gamma_5 \\ 0 & W_{14} & 0 & -\Gamma_8 - \lambda \end{pmatrix}$$

$$= (\Omega_1 \cdot e^{-\lambda\cdot\tau_A} - \lambda) \cdot \Big\{(\Omega_4 \cdot e^{-\lambda\cdot\tau_B} - \lambda)$$

$$\cdot \det\begin{pmatrix} -(W_{14} + \lambda) & (S_{21}^A + S_{21}^B) & 0 \\ -(\Gamma_2 + \Gamma_5) & -(\Gamma_3 + \Gamma_6 + \Gamma_7) - \lambda & S_{43} - \Gamma_2 - \Gamma_5 \\ W_{14} & 0 & -(\Gamma_8 + \lambda) \end{pmatrix}$$

$$-\Gamma_4 \cdot e^{-\lambda\cdot\tau_B} \cdot \det\begin{pmatrix} \Omega_5 & \Omega_6 & \Omega_5 \\ -(W_{14} + \lambda) & (S_{21}^A + S_{21}^B) & 0 \\ W_{14} & 0 & -(\Gamma_8 + \lambda) \end{pmatrix}\Bigg\}$$

$$= (\Omega_1 \cdot e^{-\lambda\cdot\tau_A} - \lambda) \cdot \{\lambda \cdot (\Xi_1 + \Xi_4) + \lambda^2 \cdot (\Xi_2 + \Xi_5)$$
$$+ \lambda^3 \cdot \Xi_3 + \lambda^4 - [(\Xi_1 + \Xi_4) \cdot \Omega_4 + (\Xi_2 + \Xi_5) \cdot \Omega_4 \cdot \lambda$$
$$+ \Xi_3 \cdot \Omega_4 \cdot \lambda^2 + \Omega_4 \cdot \lambda^3] \cdot e^{-\lambda\cdot\tau_B}$$
$$- \Gamma_4 \cdot e^{-\lambda\cdot\tau_B} \cdot (\Xi_6 - \Xi_7 \cdot \lambda - \Omega_6 \cdot \lambda^2)\}$$

$$(\Omega_1 \cdot e^{-\lambda\cdot\tau_A} - \lambda) \cdot \det\begin{pmatrix} \Upsilon_{22} & \Upsilon_{23} & \Upsilon_{24} & \Upsilon_{25} \\ \Upsilon_{32} & \Upsilon_{33} & \Upsilon_{34} & \Upsilon_{35} \\ \Upsilon_{42} & \Upsilon_{43} & \Upsilon_{44} & \Upsilon_{45} \\ \Upsilon_{52} & \Upsilon_{53} & \Upsilon_{54} & \Upsilon_{55} \end{pmatrix}$$
$$= (\Omega_1 \cdot e^{-\lambda\cdot\tau_A} - \lambda) \cdot \{\lambda \cdot (\Xi_1 + \Xi_4) + \lambda^2 \cdot (\Xi_2 + \Xi_5) + \lambda^3 \cdot \Xi_3 + \lambda^4$$
$$- \{[(\Xi_1 + \Xi_4) \cdot \Omega_4 + \Xi_6 \cdot \Gamma_4] + [(\Xi_2 + \Xi_5) \cdot \Omega_4 - \Xi_7 \cdot \Gamma_4] \cdot \lambda$$
$$+ [\Xi_3 \cdot \Omega_4 - \Omega_6 \cdot \Gamma_4] \cdot \lambda^2 + \Omega_4 \cdot \lambda^3\} \cdot e^{-\lambda\cdot\tau_B}\}$$

We choose the case: $\tau_A = \tau_\Delta$; $\tau_B = 0 \Rightarrow e^{-\lambda\cdot\tau_B} = 1$ then

$$(\Omega_1 \cdot e^{-\lambda\cdot\tau_\Delta} - \lambda) \cdot \det\begin{pmatrix} \Upsilon_{22} & \Upsilon_{23} & \Upsilon_{24} & \Upsilon_{25} \\ \Upsilon_{32} & \Upsilon_{33} & \Upsilon_{34} & \Upsilon_{35} \\ \Upsilon_{42} & \Upsilon_{43} & \Upsilon_{44} & \Upsilon_{45} \\ \Upsilon_{52} & \Upsilon_{53} & \Upsilon_{54} & \Upsilon_{55} \end{pmatrix}$$
$$= (\Omega_1 \cdot e^{-\lambda\cdot\tau_\Delta} - \lambda) \cdot \{\lambda \cdot (\Xi_1 + \Xi_4) + \lambda^2 \cdot (\Xi_2 + \Xi_5) + \lambda^3 \cdot \Xi_3 + \lambda^4$$
$$- [(\Xi_1 + \Xi_4) \cdot \Omega_4 + \Xi_6 \cdot \Gamma_4] - [(\Xi_2 + \Xi_5) \cdot \Omega_4 - \Xi_7 \cdot \Gamma_4] \cdot \lambda$$
$$- [\Xi_3 \cdot \Omega_4 - \Omega_6 \cdot \Gamma_4] \cdot \lambda^2 - \Omega_4 \cdot \lambda^3\}$$

$$(\Omega_1 \cdot e^{-\lambda\cdot\tau_\Delta} - \lambda) \cdot \det\begin{pmatrix} \Upsilon_{22} & \Upsilon_{23} & \Upsilon_{24} & \Upsilon_{25} \\ \Upsilon_{32} & \Upsilon_{33} & \Upsilon_{34} & \Upsilon_{35} \\ \Upsilon_{42} & \Upsilon_{43} & \Upsilon_{44} & \Upsilon_{45} \\ \Upsilon_{52} & \Upsilon_{53} & \Upsilon_{54} & \Upsilon_{55} \end{pmatrix}$$

$$= (\Omega_1 \cdot e^{-\lambda \cdot \tau_\Delta} - \lambda) \cdot \{-[(\Xi_1 + \Xi_4) \cdot \Omega_4 + \Xi_6 \cdot \Gamma_4] + [(\Xi_1 + \Xi_4) - (\Xi_2 + \Xi_5) \cdot \Omega_4 + \Xi_7 \cdot \Gamma_4] \cdot \lambda + [\Xi_2 + \Xi_5 - \Xi_3 \cdot \Omega_4 + \Omega_6 \cdot \Gamma_4] \cdot \lambda^2 + [\Xi_3 - \Omega_4] \cdot \lambda^3 + \lambda^4\}$$

We define some parameters for simplicity: $\Xi_8 = -[(\Xi_1 + \Xi_4) \cdot \Omega_4 + \Xi_6 \cdot \Gamma_4]$

$$\Xi_9 = (\Xi_1 + \Xi_4) - (\Xi_2 + \Xi_5) \cdot \Omega_4 + \Xi_7 \cdot \Gamma_4$$
$$\Xi_{10} = \Xi_2 + \Xi_5 - \Xi_3 \cdot \Omega_4 + \Omega_6 \cdot \Gamma_4$$
$$\Xi_{11} = \Xi_3 - \Omega_4$$

$$(\Omega_1 \cdot e^{-\lambda \cdot \tau_\Delta} - \lambda) \cdot \det \begin{pmatrix} \Upsilon_{22} & \Upsilon_{23} & \Upsilon_{24} & \Upsilon_{25} \\ \Upsilon_{32} & \Upsilon_{33} & \Upsilon_{34} & \Upsilon_{35} \\ \Upsilon_{42} & \Upsilon_{43} & \Upsilon_{44} & \Upsilon_{45} \\ \Upsilon_{52} & \Upsilon_{53} & \Upsilon_{54} & \Upsilon_{55} \end{pmatrix}$$
$$= (\Omega_1 \cdot e^{-\lambda \cdot \tau_\Delta} - \lambda) \cdot \{\Xi_8 + \Xi_9 \cdot \lambda + \Xi_{10} \cdot \lambda^2 + \Xi_{11} \cdot \lambda^3 + \lambda^4\}$$
$$= -(\Xi_8 \cdot \lambda + \Xi_9 \cdot \lambda^2 + \Xi_{10} \cdot \lambda^3 + \Xi_{11} \cdot \lambda^4 + \lambda^5) + (\Xi_8 \cdot \Omega_1 + \Xi_9 \cdot \Omega_1 \cdot \lambda + \Xi_{10} \cdot \Omega_1 \cdot \lambda^2 + \Xi_{11} \cdot \Omega_1 \cdot \lambda^3 + \Omega_1 \cdot \lambda^4) \cdot e^{-\lambda \cdot \tau_\Delta}$$

Step 2: Find the expression for $\Upsilon_{12} \cdot \det \begin{pmatrix} \Upsilon_{21} & \Upsilon_{23} & \Upsilon_{24} & \Upsilon_{25} \\ \Upsilon_{31} & \Upsilon_{33} & \Upsilon_{34} & \Upsilon_{35} \\ \Upsilon_{41} & \Upsilon_{43} & \Upsilon_{44} & \Upsilon_{45} \\ \Upsilon_{51} & \Upsilon_{53} & \Upsilon_{54} & \Upsilon_{55} \end{pmatrix}$.

$$\Upsilon_{12} = 0 \Rightarrow \Upsilon_{12} \cdot \det \begin{pmatrix} \Upsilon_{21} & \Upsilon_{23} & \Upsilon_{24} & \Upsilon_{25} \\ \Upsilon_{31} & \Upsilon_{33} & \Upsilon_{34} & \Upsilon_{35} \\ \Upsilon_{41} & \Upsilon_{43} & \Upsilon_{44} & \Upsilon_{45} \\ \Upsilon_{51} & \Upsilon_{53} & \Upsilon_{54} & \Upsilon_{55} \end{pmatrix} = 0$$

Step 3: Find the expression for $\Upsilon_{13} \cdot \det \begin{pmatrix} \Upsilon_{21} & \Upsilon_{22} & \Upsilon_{24} & \Upsilon_{25} \\ \Upsilon_{31} & \Upsilon_{32} & \Upsilon_{34} & \Upsilon_{35} \\ \Upsilon_{41} & \Upsilon_{42} & \Upsilon_{44} & \Upsilon_{45} \\ \Upsilon_{51} & \Upsilon_{52} & \Upsilon_{54} & \Upsilon_{55} \end{pmatrix}$; $\Upsilon_{13} = \Omega_2$.

$$\Omega_2 \cdot \det \begin{pmatrix} \Upsilon_{21} & \Upsilon_{22} & \Upsilon_{24} & \Upsilon_{25} \\ \Upsilon_{31} & \Upsilon_{32} & \Upsilon_{34} & \Upsilon_{35} \\ \Upsilon_{41} & \Upsilon_{42} & \Upsilon_{44} & \Upsilon_{45} \\ \Upsilon_{51} & \Upsilon_{52} & \Upsilon_{54} & \Upsilon_{55} \end{pmatrix}$$

$$= \Omega_2 \cdot \det\begin{pmatrix} 0 & (\Omega_4 \cdot e^{-\lambda \cdot \tau_B} - \lambda) & \Omega_6 & \Omega_5 \\ 0 & 0 & (S_{21}^A + S_{21}^B) & 0 \\ -\Gamma_1 \cdot e^{-\lambda \cdot \tau_A} & -\Gamma_4 \cdot e^{-\lambda \cdot \tau_B} & \begin{matrix} -(\Gamma_3 + \Gamma_6 \\ +\Gamma_7) - \lambda \end{matrix} & S_{43} - \Gamma_2 - \Gamma_5 \\ 0 & 0 & 0 & -\Gamma_8 - \lambda \end{pmatrix}$$

$$\Omega_2 \cdot \det\begin{pmatrix} \Upsilon_{21} & \Upsilon_{22} & \Upsilon_{24} & \Upsilon_{25} \\ \Upsilon_{31} & \Upsilon_{32} & \Upsilon_{34} & \Upsilon_{35} \\ \Upsilon_{41} & \Upsilon_{42} & \Upsilon_{44} & \Upsilon_{45} \\ \Upsilon_{51} & \Upsilon_{52} & \Upsilon_{54} & \Upsilon_{55} \end{pmatrix}$$

$$= \Omega_2 \cdot (-\Gamma_1 \cdot e^{-\lambda \cdot \tau_A}) \cdot \det\begin{pmatrix} (\Omega_4 \cdot e^{-\lambda \cdot \tau_B} - \lambda) & \Omega_6 & \Omega_5 \\ 0 & (S_{21}^A + S_{21}^B) & 0 \\ 0 & 0 & -\Gamma_8 - \lambda \end{pmatrix}$$

$$\Omega_2 \cdot \det\begin{pmatrix} \Upsilon_{21} & \Upsilon_{22} & \Upsilon_{24} & \Upsilon_{25} \\ \Upsilon_{31} & \Upsilon_{32} & \Upsilon_{34} & \Upsilon_{35} \\ \Upsilon_{41} & \Upsilon_{42} & \Upsilon_{44} & \Upsilon_{45} \\ \Upsilon_{51} & \Upsilon_{52} & \Upsilon_{54} & \Upsilon_{55} \end{pmatrix}$$

$$= \Omega_2 \cdot (-\Gamma_1 \cdot e^{-\lambda \cdot \tau_A}) \cdot (\Omega_4 \cdot e^{-\lambda \cdot \tau_B} - \lambda) \cdot \det\begin{pmatrix} (S_{21}^A + S_{21}^B) & 0 \\ 0 & -(\Gamma_8 + \lambda) \end{pmatrix}$$

We choose the case: $\tau_A = \tau_\Delta;\ \tau_B = 0 \Rightarrow e^{-\lambda \cdot \tau_B} = 1$ then

$$\Omega_2 \cdot \det\begin{pmatrix} \Upsilon_{21} & \Upsilon_{22} & \Upsilon_{24} & \Upsilon_{25} \\ \Upsilon_{31} & \Upsilon_{32} & \Upsilon_{34} & \Upsilon_{35} \\ \Upsilon_{41} & \Upsilon_{42} & \Upsilon_{44} & \Upsilon_{45} \\ \Upsilon_{51} & \Upsilon_{52} & \Upsilon_{54} & \Upsilon_{55} \end{pmatrix}$$

$$= \Omega_2 \cdot (-\Gamma_1 \cdot e^{-\lambda \cdot \tau_\Delta}) \cdot (\Omega_4 - \lambda) \cdot \det\begin{pmatrix} (S_{21}^A + S_{21}^B) & 0 \\ 0 & -(\Gamma_8 + \lambda) \end{pmatrix}$$

$$= \Omega_2 \cdot \Gamma_1 \cdot e^{-\lambda \cdot \tau_\Delta} \cdot (\Omega_4 - \lambda) \cdot (S_{21}^A + S_{21}^B) \cdot (\Gamma_8 + \lambda)$$

$$= (S_{21}^A + S_{21}^B) \cdot \Omega_2 \cdot \Gamma_1 \cdot (\Omega_4 - \lambda) \cdot (\Gamma_8 + \lambda) \cdot e^{-\lambda \cdot \tau_\Delta}$$

$$= [(S_{21}^A + S_{21}^B) \cdot \Omega_2 \cdot \Gamma_1 \cdot \Omega_4 \cdot \Gamma_8 + (S_{21}^A + S_{21}^B) \cdot \Omega_2 \cdot \Gamma_1 \cdot (\Omega_4 - \Gamma_8) \cdot \lambda - (S_{21}^A + S_{21}^B) \cdot \Omega_2 \cdot \Gamma_1 \cdot \lambda^2] \cdot e^{-\lambda \cdot \tau_\Delta}$$

We define some parameters for simplicity: $\Xi_{12} = (S_{21}^A + S_{21}^B) \cdot \Omega_2 \cdot \Gamma_1 \cdot \Omega_4 \cdot \Gamma_8$

$$\Xi_{13} = (S_{21}^A + S_{21}^B) \cdot \Omega_2 \cdot \Gamma_1 \cdot (\Omega_4 - \Gamma_8);\ \Xi_{14} = (S_{21}^A + S_{21}^B) \cdot \Omega_2 \cdot \Gamma_1$$

$$\Omega_2 \cdot \det\begin{pmatrix} \Upsilon_{21} & \Upsilon_{22} & \Upsilon_{24} & \Upsilon_{25} \\ \Upsilon_{31} & \Upsilon_{32} & \Upsilon_{34} & \Upsilon_{35} \\ \Upsilon_{41} & \Upsilon_{42} & \Upsilon_{44} & \Upsilon_{45} \\ \Upsilon_{51} & \Upsilon_{52} & \Upsilon_{54} & \Upsilon_{55} \end{pmatrix} = [\Xi_{12} + \Xi_{13} \cdot \lambda - \Xi_{14} \cdot \lambda^2] \cdot e^{-\lambda \cdot \tau_\Delta}$$

<u>Step 4</u>: Find the expression for $\Upsilon_{14} \cdot \det\begin{pmatrix} \Upsilon_{21} & \Upsilon_{22} & \Upsilon_{23} & \Upsilon_{25} \\ \Upsilon_{31} & \Upsilon_{32} & \Upsilon_{33} & \Upsilon_{35} \\ \Upsilon_{41} & \Upsilon_{42} & \Upsilon_{43} & \Upsilon_{45} \\ \Upsilon_{51} & \Upsilon_{52} & \Upsilon_{53} & \Upsilon_{55} \end{pmatrix}$; $\Upsilon_{14} = \Omega_3$.

$$\Upsilon_{14} \cdot \det\begin{pmatrix} \Upsilon_{21} & \Upsilon_{22} & \Upsilon_{23} & \Upsilon_{25} \\ \Upsilon_{31} & \Upsilon_{32} & \Upsilon_{33} & \Upsilon_{35} \\ \Upsilon_{41} & \Upsilon_{42} & \Upsilon_{43} & \Upsilon_{45} \\ \Upsilon_{51} & \Upsilon_{52} & \Upsilon_{53} & \Upsilon_{55} \end{pmatrix}$$

$$= \Omega_3 \cdot \det\begin{pmatrix} 0 & (\Omega_4 \cdot e^{-\lambda \cdot \tau_B} - \lambda) & \Omega_5 & \Omega_5 \\ 0 & 0 & -(W_{14} + \lambda) & 0 \\ -\Gamma_1 \cdot e^{-\lambda \cdot \tau_A} & -\Gamma_4 \cdot e^{-\lambda \cdot \tau_B} & -(\Gamma_2 + \Gamma_5) & S_{43} - \Gamma_2 - \Gamma_5 \\ 0 & 0 & W_{14} & -(\Gamma_8 + \lambda) \end{pmatrix}$$

$$\Upsilon_{14} \cdot \det\begin{pmatrix} \Upsilon_{21} & \Upsilon_{22} & \Upsilon_{23} & \Upsilon_{25} \\ \Upsilon_{31} & \Upsilon_{32} & \Upsilon_{33} & \Upsilon_{35} \\ \Upsilon_{41} & \Upsilon_{42} & \Upsilon_{43} & \Upsilon_{45} \\ \Upsilon_{51} & \Upsilon_{52} & \Upsilon_{53} & \Upsilon_{55} \end{pmatrix}$$

$$= \Omega_3 \cdot (-\Gamma_1 \cdot e^{-\lambda \cdot \tau_A}) \cdot \det\begin{pmatrix} (\Omega_4 \cdot e^{-\lambda \cdot \tau_B} - \lambda) & \Omega_5 & \Omega_5 \\ 0 & -(W_{14} + \lambda) & 0 \\ 0 & W_{14} & -(\Gamma_8 + \lambda) \end{pmatrix}$$

$$\Upsilon_{14} \cdot \det\begin{pmatrix} \Upsilon_{21} & \Upsilon_{22} & \Upsilon_{23} & \Upsilon_{25} \\ \Upsilon_{31} & \Upsilon_{32} & \Upsilon_{33} & \Upsilon_{35} \\ \Upsilon_{41} & \Upsilon_{42} & \Upsilon_{43} & \Upsilon_{45} \\ \Upsilon_{51} & \Upsilon_{52} & \Upsilon_{53} & \Upsilon_{55} \end{pmatrix}$$

$$= \Omega_3 \cdot (-\Gamma_1 \cdot e^{-\lambda \cdot \tau_A}) \cdot (\Omega_4 \cdot e^{-\lambda \cdot \tau_B} - \lambda) \cdot \det\begin{pmatrix} -(W_{14} + \lambda) & 0 \\ W_{14} & -(\Gamma_8 + \lambda) \end{pmatrix}$$

$$\Upsilon_{14} \cdot \det\begin{pmatrix} \Upsilon_{21} & \Upsilon_{22} & \Upsilon_{23} & \Upsilon_{25} \\ \Upsilon_{31} & \Upsilon_{32} & \Upsilon_{33} & \Upsilon_{35} \\ \Upsilon_{41} & \Upsilon_{42} & \Upsilon_{43} & \Upsilon_{45} \\ \Upsilon_{51} & \Upsilon_{52} & \Upsilon_{53} & \Upsilon_{55} \end{pmatrix}$$

$$= \Omega_3 \cdot (-\Gamma_1 \cdot e^{-\lambda \cdot \tau_A}) \cdot (\Omega_4 \cdot e^{-\lambda \cdot \tau_B} - \lambda) \cdot (W_{14} + \lambda) \cdot (\Gamma_8 + \lambda)$$

We choose the case: $\tau_A = \tau_\Delta$; $\tau_B = 0 \Rightarrow e^{-\lambda \cdot \tau_B} = 1$ then

$$\Upsilon_{14} \cdot \det \begin{pmatrix} \Upsilon_{21} & \Upsilon_{22} & \Upsilon_{23} & \Upsilon_{25} \\ \Upsilon_{31} & \Upsilon_{32} & \Upsilon_{33} & \Upsilon_{35} \\ \Upsilon_{41} & \Upsilon_{42} & \Upsilon_{43} & \Upsilon_{45} \\ \Upsilon_{51} & \Upsilon_{52} & \Upsilon_{53} & \Upsilon_{55} \end{pmatrix}$$
$$= -\Omega_3 \cdot \Gamma_1 \cdot e^{-\lambda \cdot \tau_\Delta} \cdot (\Omega_4 - \lambda) \cdot (W_{14} + \lambda) \cdot (\Gamma_8 + \lambda)$$

$$\Upsilon_{14} \cdot \det \begin{pmatrix} \Upsilon_{21} & \Upsilon_{22} & \Upsilon_{23} & \Upsilon_{25} \\ \Upsilon_{31} & \Upsilon_{32} & \Upsilon_{33} & \Upsilon_{35} \\ \Upsilon_{41} & \Upsilon_{42} & \Upsilon_{43} & \Upsilon_{45} \\ \Upsilon_{51} & \Upsilon_{52} & \Upsilon_{53} & \Upsilon_{55} \end{pmatrix}$$
$$= -\Omega_3 \cdot \Gamma_1 \cdot e^{-\lambda \cdot \tau_\Delta} \cdot \{\Gamma_8 \Omega_4 W_{14} + (\Gamma_8 \cdot [\Omega_4 - W_{14}] + \Omega_4 W_{14}) \cdot \lambda$$
$$+ (\Omega_4 - W_{14} - \Gamma_8) \cdot \lambda^2 - \lambda^3\}$$

$$\Upsilon_{14} \cdot \det \begin{pmatrix} \Upsilon_{21} & \Upsilon_{22} & \Upsilon_{23} & \Upsilon_{25} \\ \Upsilon_{31} & \Upsilon_{32} & \Upsilon_{33} & \Upsilon_{35} \\ \Upsilon_{41} & \Upsilon_{42} & \Upsilon_{43} & \Upsilon_{45} \\ \Upsilon_{51} & \Upsilon_{52} & \Upsilon_{53} & \Upsilon_{55} \end{pmatrix}$$
$$= \{-\Omega_3 \cdot \Gamma_1 \cdot \Gamma_8 \cdot \Omega_4 \cdot W_{14} - \Omega_3 \cdot \Gamma_1 \cdot (\Gamma_8 \cdot [\Omega_4 - W_{14}] + \Omega_4 \cdot W_{14}) \cdot \lambda$$
$$- \Omega_3 \cdot \Gamma_1 \cdot (\Omega_4 - W_{14} - \Gamma_8) \cdot \lambda^2 + \Omega_3 \cdot \Gamma_1 \cdot \lambda^3\} \cdot e^{-\lambda \cdot \tau_\Delta}$$

We define some parameters for simplicity: $\Xi_{15} = -\Omega_3 \cdot \Gamma_1 \cdot \Gamma_8 \cdot \Omega_4 \cdot W_{14}$

$$\Xi_{16} = \Omega_3 \cdot \Gamma_1 \cdot (\Gamma_8 \cdot [\Omega_4 - W_{14}] + \Omega_4 \cdot W_{14})$$
$$\Xi_{17} = \Omega_3 \cdot \Gamma_1 \cdot (\Omega_4 - W_{14} - \Gamma_8); \; \Xi_{18} = \Omega_3 \cdot \Gamma_1$$

$$\Upsilon_{14} \cdot \det \begin{pmatrix} \Upsilon_{21} & \Upsilon_{22} & \Upsilon_{23} & \Upsilon_{25} \\ \Upsilon_{31} & \Upsilon_{32} & \Upsilon_{33} & \Upsilon_{35} \\ \Upsilon_{41} & \Upsilon_{42} & \Upsilon_{43} & \Upsilon_{45} \\ \Upsilon_{51} & \Upsilon_{52} & \Upsilon_{53} & \Upsilon_{55} \end{pmatrix} = (\Xi_{15} - \Xi_{16} \cdot \lambda - \Xi_{17} \cdot \lambda^2 + \Xi_{18} \cdot \lambda^3) \cdot e^{-\lambda \cdot \tau_\Delta}$$

Step 5: Find the expression for $\Upsilon_{15} \cdot \det \begin{pmatrix} \Upsilon_{21} & \Upsilon_{22} & \Upsilon_{23} & \Upsilon_{24} \\ \Upsilon_{31} & \Upsilon_{32} & \Upsilon_{33} & \Upsilon_{34} \\ \Upsilon_{41} & \Upsilon_{42} & \Upsilon_{43} & \Upsilon_{44} \\ \Upsilon_{51} & \Upsilon_{52} & \Upsilon_{53} & \Upsilon_{54} \end{pmatrix}$.

$$\Upsilon_{15} \cdot \det \begin{pmatrix} \Upsilon_{21} & \Upsilon_{22} & \Upsilon_{23} & \Upsilon_{24} \\ \Upsilon_{31} & \Upsilon_{32} & \Upsilon_{33} & \Upsilon_{34} \\ \Upsilon_{41} & \Upsilon_{42} & \Upsilon_{43} & \Upsilon_{44} \\ \Upsilon_{51} & \Upsilon_{52} & \Upsilon_{53} & \Upsilon_{54} \end{pmatrix}$$

$$= \Omega_2 \cdot \det \begin{pmatrix} 0 & (\Omega_4 \cdot e^{-\lambda \cdot \tau_B} - \lambda) & \Omega_5 & \Omega_6 \\ 0 & 0 & -(W_{14} + \lambda) & (S_{21}^A + S_{21}^B) \\ -\Gamma_1 \cdot e^{-\lambda \cdot \tau_A} & -\Gamma_4 \cdot e^{-\lambda \cdot \tau_B} & -(\Gamma_2 + \Gamma_5) & -[(\Gamma_3 + \Gamma_6 + \Gamma_7) + \lambda] \\ 0 & 0 & W_{14} & 0 \end{pmatrix}$$

$$\Upsilon_{15} \cdot \det \begin{pmatrix} \Upsilon_{21} & \Upsilon_{22} & \Upsilon_{23} & \Upsilon_{24} \\ \Upsilon_{31} & \Upsilon_{32} & \Upsilon_{33} & \Upsilon_{34} \\ \Upsilon_{41} & \Upsilon_{42} & \Upsilon_{43} & \Upsilon_{44} \\ \Upsilon_{51} & \Upsilon_{52} & \Upsilon_{53} & \Upsilon_{54} \end{pmatrix}$$

$$= \Omega_2 \cdot (-\Gamma_1 \cdot e^{-\lambda \cdot \tau_A}) \cdot \det \begin{pmatrix} (\Omega_4 \cdot e^{-\lambda \cdot \tau_B} - \lambda) & \Omega_5 & \Omega_6 \\ 0 & -(W_{14} + \lambda) & (S_{21}^A + S_{21}^B) \\ 0 & W_{14} & 0 \end{pmatrix}$$

$$\Upsilon_{15} \cdot \det \begin{pmatrix} \Upsilon_{21} & \Upsilon_{22} & \Upsilon_{23} & \Upsilon_{24} \\ \Upsilon_{31} & \Upsilon_{32} & \Upsilon_{33} & \Upsilon_{34} \\ \Upsilon_{41} & \Upsilon_{42} & \Upsilon_{43} & \Upsilon_{44} \\ \Upsilon_{51} & \Upsilon_{52} & \Upsilon_{53} & \Upsilon_{54} \end{pmatrix}$$

$$= \Omega_2 \cdot (-\Gamma_1 \cdot e^{-\lambda \cdot \tau_A}) \cdot (\Omega_4 \cdot e^{-\lambda \cdot \tau_B} - \lambda) \cdot \det \begin{pmatrix} -(W_{14} + \lambda) & (S_{21}^A + S_{21}^B) \\ W_{14} & 0 \end{pmatrix}$$

We choose the case: $\tau_A = \tau_\Delta$; $\tau_B = 0 \Rightarrow e^{-\lambda \cdot \tau_B} = 1$ then

$$\Upsilon_{15} \cdot \det \begin{pmatrix} \Upsilon_{21} & \Upsilon_{22} & \Upsilon_{23} & \Upsilon_{24} \\ \Upsilon_{31} & \Upsilon_{32} & \Upsilon_{33} & \Upsilon_{34} \\ \Upsilon_{41} & \Upsilon_{42} & \Upsilon_{43} & \Upsilon_{44} \\ \Upsilon_{51} & \Upsilon_{52} & \Upsilon_{53} & \Upsilon_{54} \end{pmatrix}$$

$$= \Omega_2 \cdot (-\Gamma_1 \cdot e^{-\lambda \cdot \tau_\Delta}) \cdot (\Omega_4 - \lambda) \cdot \det \begin{pmatrix} -(W_{14} + \lambda) & (S_{21}^A + S_{21}^B) \\ W_{14} & 0 \end{pmatrix}$$

$$\Upsilon_{15} \cdot \det \begin{pmatrix} \Upsilon_{21} & \Upsilon_{22} & \Upsilon_{23} & \Upsilon_{24} \\ \Upsilon_{31} & \Upsilon_{32} & \Upsilon_{33} & \Upsilon_{34} \\ \Upsilon_{41} & \Upsilon_{42} & \Upsilon_{43} & \Upsilon_{44} \\ \Upsilon_{51} & \Upsilon_{52} & \Upsilon_{53} & \Upsilon_{54} \end{pmatrix}$$

$$= \Omega_2 \cdot \Gamma_1 \cdot W_{14} \cdot (S_{21}^A + S_{21}^B) \cdot (\Omega_4 - \lambda) \cdot e^{-\lambda \cdot \tau_\Delta}$$

$$\Upsilon_{15} \cdot \det \begin{pmatrix} \Upsilon_{21} & \Upsilon_{22} & \Upsilon_{23} & \Upsilon_{24} \\ \Upsilon_{31} & \Upsilon_{32} & \Upsilon_{33} & \Upsilon_{34} \\ \Upsilon_{41} & \Upsilon_{42} & \Upsilon_{43} & \Upsilon_{44} \\ \Upsilon_{51} & \Upsilon_{52} & \Upsilon_{53} & \Upsilon_{54} \end{pmatrix}$$

$$= [\Omega_2 \cdot \Gamma_1 \cdot W_{14} \cdot (S_{21}^A + S_{21}^B) \cdot \Omega_4 - \Omega_2 \cdot \Gamma_1 \cdot W_{14} \cdot (S_{21}^A + S_{21}^B) \cdot \lambda] \cdot e^{-\lambda \cdot \tau_\Delta}$$

We define some parameters for simplicity: $\Xi_{19} = \Omega_2 \cdot \Gamma_1 \cdot W_{14} \cdot (S_{21}^A + S_{21}^B) \cdot \Omega_4$

$$\Xi_{20} = \Omega_2 \cdot \Gamma_1 \cdot W_{14} \cdot (S_{21}^A + S_{21}^B);\ \Xi_{19} = \Omega_4 \cdot \Xi_{20}$$

$$\Upsilon_{15} \cdot \det\begin{pmatrix} \Upsilon_{21} & \Upsilon_{22} & \Upsilon_{23} & \Upsilon_{24} \\ \Upsilon_{31} & \Upsilon_{32} & \Upsilon_{33} & \Upsilon_{34} \\ \Upsilon_{41} & \Upsilon_{42} & \Upsilon_{43} & \Upsilon_{44} \\ \Upsilon_{51} & \Upsilon_{52} & \Upsilon_{53} & \Upsilon_{54} \end{pmatrix} = [\Xi_{19} - \Xi_{20} \cdot \lambda] \cdot e^{-\lambda \cdot \tau_\Delta}$$

We can summary our system additional parameters in the next table (Table 7.3). We can summary steps 1 to 5 in the next table (Table 7.4).

$$\det(A - \lambda \cdot I) = \Upsilon_{11} \cdot \det\begin{pmatrix} \Upsilon_{22} & \Upsilon_{23} & \Upsilon_{24} & \Upsilon_{25} \\ \Upsilon_{32} & \Upsilon_{33} & \Upsilon_{34} & \Upsilon_{35} \\ \Upsilon_{42} & \Upsilon_{43} & \Upsilon_{44} & \Upsilon_{45} \\ \Upsilon_{52} & \Upsilon_{53} & \Upsilon_{54} & \Upsilon_{55} \end{pmatrix} - \Upsilon_{12} \cdot \det\begin{pmatrix} \Upsilon_{21} & \Upsilon_{23} & \Upsilon_{24} & \Upsilon_{25} \\ \Upsilon_{31} & \Upsilon_{33} & \Upsilon_{34} & \Upsilon_{35} \\ \Upsilon_{41} & \Upsilon_{43} & \Upsilon_{44} & \Upsilon_{45} \\ \Upsilon_{51} & \Upsilon_{53} & \Upsilon_{54} & \Upsilon_{55} \end{pmatrix}$$
$$+ \Upsilon_{13} \cdot \det\begin{pmatrix} \Upsilon_{21} & \Upsilon_{22} & \Upsilon_{24} & \Upsilon_{25} \\ \Upsilon_{31} & \Upsilon_{32} & \Upsilon_{34} & \Upsilon_{35} \\ \Upsilon_{41} & \Upsilon_{42} & \Upsilon_{44} & \Upsilon_{45} \\ \Upsilon_{51} & \Upsilon_{52} & \Upsilon_{54} & \Upsilon_{55} \end{pmatrix} - \Upsilon_{14} \cdot \det\begin{pmatrix} \Upsilon_{21} & \Upsilon_{22} & \Upsilon_{23} & \Upsilon_{25} \\ \Upsilon_{31} & \Upsilon_{32} & \Upsilon_{33} & \Upsilon_{35} \\ \Upsilon_{41} & \Upsilon_{42} & \Upsilon_{43} & \Upsilon_{45} \\ \Upsilon_{51} & \Upsilon_{52} & \Upsilon_{53} & \Upsilon_{55} \end{pmatrix}$$

Table 7.3 Dual-wavelength Ti:Sapphire laser system additional parameters

Parameter	Expression	Parameter	
Ξ_1	$W_{14} \cdot [\Gamma_3 + \Gamma_6 + \Gamma_7] \cdot \Gamma_8$	Ξ_{11}	$\Xi_3 - \Omega_4$
Ξ_2	$W_{14} \cdot [\Gamma_3 + \Gamma_6 + \Gamma_7] + \Gamma_8 \cdot [W_{14} + \Gamma_3 + \Gamma_6 + \Gamma_7]$	Ξ_{12}	$(S_{21}^A + S_{21}^B) \cdot \Omega_2 \cdot \Gamma_1 \cdot \Omega_4 \cdot \Gamma_8$
Ξ_3	$W_{14} + \Gamma_3 + \Gamma_6 + \Gamma_7 + \Gamma_8$	Ξ_{13}	$(S_{21}^A + S_{21}^B) \cdot \Omega_2 \cdot \Gamma_1 \cdot (\Omega_4 - \Gamma_8)$
Ξ_4	$(S_{21}^A + S_{21}^B) \cdot \{(\Gamma_2 + \Gamma_5) \cdot \Gamma_8 - W_{14} \cdot (S_{43} - \Gamma_2 - \Gamma_5)\}$	Ξ_{14}	$(S_{21}^A + S_{21}^B) \cdot \Omega_2 \cdot \Gamma_1$
Ξ_5	$(S_{21}^A + S_{21}^B) \cdot (\Gamma_2 + \Gamma_5)$	Ξ_{15}	$-\Omega_3 \cdot \Gamma_1 \cdot \Gamma_8 \cdot \Omega_4 \cdot W_{14}$
Ξ_6	$-\Omega_5 \cdot (S_{21}^A + S_{21}^B) \cdot \Gamma_8 - \Omega_6 \cdot W_{14} \cdot \Gamma_8 - \Omega_5 \cdot W_{14} \cdot (S_{21}^A + S_{21}^B)$	Ξ_{16}	$\Omega_3 \cdot \Gamma_1 \cdot (\Gamma_8 \cdot [\Omega_4 - W_{14}] + \Omega_4 \cdot W_{14})$
Ξ_7	$[\Omega_5 \cdot (S_{21}^A + S_{21}^B) + \Omega_6 \cdot (W_{14} + \Gamma_8)]$	Ξ_{17}	$\Omega_3 \cdot \Gamma_1 \cdot (\Omega_4 - W_{14} - \Gamma_8)$
Ξ_8	$-[(\Xi_1 + \Xi_4) \cdot \Omega_4 + \Xi_6 \cdot \Gamma_4]$	Ξ_{18}	$\Omega_3 \cdot \Gamma_1$
Ξ_9	$(\Xi_1 + \Xi_4) - (\Xi_2 + \Xi_5) \cdot \Omega_4 + \Xi_7 \cdot \Gamma_4$	Ξ_{19}	$\Omega_2 \cdot \Gamma_1 \cdot W_{14} \cdot (S_{21}^A + S_{21}^B) \cdot \Omega_4$
Ξ_{10}	$\Xi_2 + \Xi_5 - \Xi_3 \cdot \Omega_4 + \Omega_6 \cdot \Gamma_4$	Ξ_{20}	$\Omega_2 \cdot \Gamma_1 \cdot W_{14} \cdot (S_{21}^A + S_{21}^B)$

Table 7.4 Dual-wavelength Ti:Sapphire laser system small increments Jacobian Matrix's determinant elements

Matrix's determinant element	Expression
$\Upsilon_{11} \cdot \det\begin{pmatrix} \Upsilon_{22} & \Upsilon_{23} & \Upsilon_{24} & \Upsilon_{25} \\ \Upsilon_{32} & \Upsilon_{33} & \Upsilon_{34} & \Upsilon_{35} \\ \Upsilon_{42} & \Upsilon_{43} & \Upsilon_{44} & \Upsilon_{45} \\ \Upsilon_{52} & \Upsilon_{53} & \Upsilon_{54} & \Upsilon_{55} \end{pmatrix}$	$-(\Xi_8 \cdot \lambda + \Xi_9 \cdot \lambda^2 + \Xi_{10} \cdot \lambda^3 + \Xi_{11} \cdot \lambda^4 + \lambda^5)$ $+(\Xi_8 \cdot \Omega_1 + \Xi_9 \cdot \Omega_1 \cdot \lambda + \Xi_{10} \cdot \Omega_1 \cdot \lambda^2 + \Xi_{11} \cdot \Omega_1 \cdot \lambda^3 + \Omega_1 \cdot \lambda^4) \cdot e^{-\lambda \cdot \tau_\Delta}$
$\Upsilon_{12} \cdot \det\begin{pmatrix} \Upsilon_{21} & \Upsilon_{23} & \Upsilon_{24} & \Upsilon_{25} \\ \Upsilon_{31} & \Upsilon_{33} & \Upsilon_{34} & \Upsilon_{35} \\ \Upsilon_{41} & \Upsilon_{43} & \Upsilon_{44} & \Upsilon_{45} \\ \Upsilon_{51} & \Upsilon_{53} & \Upsilon_{54} & \Upsilon_{55} \end{pmatrix}$	$\Upsilon_{12} = 0 \Rightarrow 0$
$\Upsilon_{13} \cdot \det\begin{pmatrix} \Upsilon_{21} & \Upsilon_{22} & \Upsilon_{24} & \Upsilon_{25} \\ \Upsilon_{31} & \Upsilon_{32} & \Upsilon_{34} & \Upsilon_{35} \\ \Upsilon_{41} & \Upsilon_{42} & \Upsilon_{44} & \Upsilon_{45} \\ \Upsilon_{51} & \Upsilon_{52} & \Upsilon_{54} & \Upsilon_{55} \end{pmatrix}$	$[\Xi_{12} + \Xi_{13} \cdot \lambda - \Xi_{14} \cdot \lambda^2] \cdot e^{-\lambda \cdot \tau_\Delta}$
$\Upsilon_{14} \cdot \det\begin{pmatrix} \Upsilon_{21} & \Upsilon_{22} & \Upsilon_{23} & \Upsilon_{25} \\ \Upsilon_{31} & \Upsilon_{32} & \Upsilon_{33} & \Upsilon_{35} \\ \Upsilon_{41} & \Upsilon_{42} & \Upsilon_{43} & \Upsilon_{45} \\ \Upsilon_{51} & \Upsilon_{52} & \Upsilon_{53} & \Upsilon_{55} \end{pmatrix}$	$(\Xi_{15} - \Xi_{16} \cdot \lambda - \Xi_{17} \cdot \lambda^2 + \Xi_{18} \cdot \lambda^3) \cdot e^{-\lambda \cdot \tau_\Delta}$

(continued)

Table 7.4 (continued)

Matrix's determinant element	Expression
$\Upsilon_{15} \cdot \det\begin{pmatrix} \Upsilon_{21} & \Upsilon_{22} & \Upsilon_{23} & \Upsilon_{24} \\ \Upsilon_{31} & \Upsilon_{32} & \Upsilon_{33} & \Upsilon_{34} \\ \Upsilon_{41} & \Upsilon_{42} & \Upsilon_{43} & \Upsilon_{44} \\ \Upsilon_{51} & \Upsilon_{52} & \Upsilon_{53} & \Upsilon_{54} \end{pmatrix}$	$[\Xi_{19} - \Xi_{20} \cdot \lambda] \cdot e^{-\lambda \cdot \tau_{\Delta}}$

$$+\Upsilon_{15}\cdot\det\begin{pmatrix}\Upsilon_{21} & \Upsilon_{22} & \Upsilon_{23} & \Upsilon_{24}\\ \Upsilon_{31} & \Upsilon_{32} & \Upsilon_{33} & \Upsilon_{34}\\ \Upsilon_{41} & \Upsilon_{42} & \Upsilon_{43} & \Upsilon_{44}\\ \Upsilon_{51} & \Upsilon_{52} & \Upsilon_{53} & \Upsilon_{54}\end{pmatrix}$$

$$\Upsilon_{12}=0\Rightarrow\det(A-\lambda\cdot I)$$

$$=\Upsilon_{11}\cdot\det\begin{pmatrix}\Upsilon_{22} & \Upsilon_{23} & \Upsilon_{24} & \Upsilon_{25}\\ \Upsilon_{32} & \Upsilon_{33} & \Upsilon_{34} & \Upsilon_{35}\\ \Upsilon_{42} & \Upsilon_{43} & \Upsilon_{44} & \Upsilon_{45}\\ \Upsilon_{52} & \Upsilon_{53} & \Upsilon_{54} & \Upsilon_{55}\end{pmatrix}+\Upsilon_{13}\cdot\det\begin{pmatrix}\Upsilon_{21} & \Upsilon_{22} & \Upsilon_{24} & \Upsilon_{25}\\ \Upsilon_{31} & \Upsilon_{32} & \Upsilon_{34} & \Upsilon_{35}\\ \Upsilon_{41} & \Upsilon_{42} & \Upsilon_{44} & \Upsilon_{45}\\ \Upsilon_{51} & \Upsilon_{52} & \Upsilon_{54} & \Upsilon_{55}\end{pmatrix}$$

$$-\Upsilon_{14}\cdot\det\begin{pmatrix}\Upsilon_{21} & \Upsilon_{22} & \Upsilon_{23} & \Upsilon_{25}\\ \Upsilon_{31} & \Upsilon_{32} & \Upsilon_{33} & \Upsilon_{35}\\ \Upsilon_{41} & \Upsilon_{42} & \Upsilon_{43} & \Upsilon_{45}\\ \Upsilon_{51} & \Upsilon_{52} & \Upsilon_{53} & \Upsilon_{55}\end{pmatrix}+\Upsilon_{15}\cdot\det\begin{pmatrix}\Upsilon_{21} & \Upsilon_{22} & \Upsilon_{23} & \Upsilon_{24}\\ \Upsilon_{31} & \Upsilon_{32} & \Upsilon_{33} & \Upsilon_{34}\\ \Upsilon_{41} & \Upsilon_{42} & \Upsilon_{43} & \Upsilon_{44}\\ \Upsilon_{51} & \Upsilon_{52} & \Upsilon_{53} & \Upsilon_{54}\end{pmatrix}$$

We choose the case: $\tau_A=\tau_\Delta;\ \tau_B=0\Rightarrow e^{-\lambda\cdot\tau_B}=1$ then

$$\begin{aligned}\det(A-\lambda\cdot I)&=-(\Xi_8\cdot\lambda+\Xi_9\cdot\lambda^2+\Xi_{10}\cdot\lambda^3+\Xi_{11}\cdot\lambda^4+\lambda^5)\\&+(\Xi_8\cdot\Omega_1+\Xi_9\cdot\Omega_1\cdot\lambda+\Xi_{10}\cdot\Omega_1\cdot\lambda^2\\&+\Xi_{11}\cdot\Omega_1\cdot\lambda^3+\Omega_1\cdot\lambda^4)\cdot e^{-\lambda\cdot\tau_\Delta}\\&+[\Xi_{12}+\Xi_{13}\cdot\lambda-\Xi_{14}\cdot\lambda^2]\cdot e^{-\lambda\cdot\tau_\Delta}\\&-(\Xi_{15}-\Xi_{16}\cdot\lambda-\Xi_{17}\cdot\lambda^2+\Xi_{18}\cdot\lambda^3)\cdot e^{-\lambda\cdot\tau_\Delta}\\&+[\Xi_{19}-\Xi_{20}\cdot\lambda]\cdot e^{-\lambda\cdot\tau_\Delta}\end{aligned}$$

$$\begin{aligned}\det(A-\lambda\cdot I)&=-\Xi_8\cdot\lambda-\Xi_9\cdot\lambda^2-\Xi_{10}\cdot\lambda^3-\Xi_{11}\cdot\lambda^4-\lambda^5\\&+\{\Xi_8\cdot\Omega_1+\Xi_{12}-\Xi_{15}+\Xi_{19}\\&+(\Xi_9\cdot\Omega_1+\Xi_{13}+\Xi_{16}-\Xi_{20})\cdot\lambda\\&+(\Xi_{10}\cdot\Omega_1-\Xi_{14}+\Xi_{17})\cdot\lambda^2\\&+(\Xi_{11}\cdot\Omega_1-\Xi_{18})\cdot\lambda^3+\Omega_1\cdot\lambda^4\}\cdot e^{-\lambda\cdot\tau_\Delta}\end{aligned}$$

We define some parameters for simplicity: $\Xi_{21}=\Xi_8\cdot\Omega_1+\Xi_{12}-\Xi_{15}+\Xi_{19}$

$$\Xi_{22}=\Xi_9\cdot\Omega_1+\Xi_{13}+\Xi_{16}-\Xi_{20}$$
$$\Xi_{23}=\Xi_{10}\cdot\Omega_1-\Xi_{14}+\Xi_{17}$$
$$\Xi_{24}=\Xi_{11}\cdot\Omega_1-\Xi_{18}$$

$$\begin{aligned}\det(A-\lambda\cdot I)&=-\Xi_8\cdot\lambda-\Xi_9\cdot\lambda^2-\Xi_{10}\cdot\lambda^3-\Xi_{11}\cdot\lambda^4-\lambda^5\\&+(\Xi_{21}+\Xi_{22}\cdot\lambda+\Xi_{23}\cdot\lambda^2+\Xi_{24}\cdot\lambda^3+\Omega_1\cdot\lambda^4)\cdot e^{-\lambda\cdot\tau_\Delta}\end{aligned}$$

$$D(\lambda, \tau_\Delta) = -\Xi_8 \cdot \lambda - \Xi_9 \cdot \lambda^2 - \Xi_{10} \cdot \lambda^3 - \Xi_{11} \cdot \lambda^4 - \lambda^5 + (\Xi_{21} + \Xi_{22} \cdot \lambda + \Xi_{23} \cdot \lambda^2 + \Xi_{24} \cdot \lambda^3 + \Omega_1 \cdot \lambda^4) \cdot e^{-\lambda \cdot \tau_\Delta}$$

$$D(\lambda, \tau_\Delta) = P_n(\lambda, \tau_\Delta) + Q_m(\lambda, \tau_\Delta) \cdot e^{-\lambda \cdot \tau_\Delta}; n, m \in \mathbb{N}_0; n = 5; m = 4; n > m$$

$$P_n(\lambda, \tau_\Delta) = -\Xi_8 \cdot \lambda - \Xi_9 \cdot \lambda^2 - \Xi_{10} \cdot \lambda^3 - \Xi_{11} \cdot \lambda^4 - \lambda^5; n = 5$$

$$Q_m(\lambda, \tau_\Delta) = \Xi_{21} + \Xi_{22} \cdot \lambda + \Xi_{23} \cdot \lambda^2 + \Xi_{24} \cdot \lambda^3 + \Omega_1 \cdot \lambda^4; m = 4$$

$$P_n(\lambda, \tau_\Delta) = \sum_{k=0}^{n=5} p_k(\tau_\Delta) \cdot \lambda^k = p_0(\tau_\Delta) + p_1(\tau_\Delta) \cdot \lambda + p_2(\tau_\Delta) \cdot \lambda^2 + p_3(\tau_\Delta) \cdot \lambda^3 + p_4(\tau_\Delta) \cdot \lambda^4 + p_5(\tau_\Delta) \cdot \lambda^5$$

$$p_0(\tau_\Delta) = 0; p_1(\tau_\Delta) = -\Xi_8; p_2(\tau_\Delta) = -\Xi_9$$
$$p_3(\tau_\Delta) = -\Xi_{10}; p_4(\tau_\Delta) = -\Xi_{11}; p_5(\tau_\Delta) = -1$$

$$Q_m(\lambda, \tau_\Delta) = \sum_{k=0}^{m=4} q_k(\tau_\Delta) \cdot \lambda^k = q_0(\tau_\Delta) + q_1(\tau_\Delta) \cdot \lambda + q_2(\tau_\Delta) \cdot \lambda^2 + q_3(\tau_\Delta) \cdot \lambda^3 + q_4(\tau_\Delta) \cdot \lambda^4$$

$$q_0(\tau_\Delta) = \Xi_{21}; q_1(\tau_\Delta) = \Xi_{22}; q_2(\tau_\Delta) = \Xi_{23}; q_3(\tau_\Delta) = \Xi_{24}; q_4(\tau_\Delta) = \Omega_1$$

The homogeneous system for $\Phi_A, \Phi_B, n_1.n_3, n_4$ leads to a characteristic equation for the eigenvalue λ having the form $D(\lambda, \tau_\Delta) = P(\lambda, \tau_\Delta) + Q(\lambda, \tau_\Delta) \cdot e^{-\lambda \cdot \tau_\Delta} = 0$; and $P(\lambda) = \sum_{j=0}^{5} a_j \cdot \lambda^j$; $Q(\lambda) = \sum_{j=0}^{4} c_j \cdot \lambda^j$. The coefficients $\{a_j(q_i, q_k, \tau_\Delta), c_j(q_i, q_k, \tau_\Delta)\} \in \mathbb{R}$ depend on q_i, q_k and delay τ_Δ. q_i, q_k are any Dual-wavelength Ti:Sapphire laser system's parameters, other parameters kept as a constant [2, 3].

$$a_0 = 0; a_1 = -\Xi_8; a_2 = -\Xi_9; a_3 = -\Xi_{10}; a_4 = -\Xi_{11}; a_5 = -1$$

$$c_0 = \Xi_{21}; c_1 = \Xi_{22}; c_2 = \Xi_{23}; c_3 = \Xi_{24}; c_4 = \Omega_1$$

Unless strictly necessary, the designation of the varied arguments (q_i, q_k) will subsequently be omitted from P, Q, a_j and c_j. The coefficients a_j, c_j are continuous and differentiable functions of their arguments, and direct substitution shows that $a_0 + c_0 \neq 0$ for $q_i, q_k \in \mathbb{R}_+$; that is, $\lambda = 0$ is not of $P(\lambda) + Q(\lambda) \cdot e^{-\lambda \cdot \tau_\Delta} = 0$. Furthermore, $P(\lambda)$, $Q(\lambda)$ are analytic functions of λ, for which the following requirements of the analysis [2] can also be verified in the present case:

1. If $\lambda = i \cdot \omega$; $\omega \in \mathbb{R}$, then $P(i \cdot \omega) + Q(i \cdot \omega) \neq 0$.
2. If $\left|\frac{Q(\lambda)}{P(\lambda)}\right|$ is bounded for $|\lambda| \to \infty$; $\operatorname{Re} \lambda \geq 0$. No roots bifurcation from ∞.
3. $F(\omega) = |P(i \cdot \omega)|^2 - |Q(i \cdot \omega)|^2$ has a finite number of zeros. Indeed, this is a polynomial in ω.
4. Each positive root $\omega(q_i, q_k)$ of $F(\omega) = 0$ is continuous and differentiable with respect to q_i, q_k.

We assume that $P_n(\lambda, \tau_\Delta)$ and $Q_m(\lambda, \tau_\Delta)$ cannot have common imaginary roots. That is for any real number ω: $P_n(\lambda = i \cdot \omega, \tau_\Delta) + Q_m(\lambda = i \cdot \omega, \tau_\Delta) \neq 0$ and

$$P_n(\lambda = i \cdot \omega, \tau_\Delta) = p_0(\tau_\Delta) + p_1(\tau_\Delta) \cdot i \cdot \omega - p_2(\tau_\Delta) \cdot \omega^2 - p_3(\tau_\Delta) \cdot i \cdot \omega^3 + p_4(\tau_\Delta) \cdot \omega^4 + p_5(\tau_\Delta) \cdot i \cdot \omega^5$$

$$P_n(\lambda = i \cdot \omega, \tau_\Delta) = p_0(\tau_\Delta) - p_2(\tau_\Delta) \cdot \omega^2 + p_4(\tau_\Delta) \cdot \omega^4 + [p_1(\tau_\Delta) \cdot \omega - p_3(\tau_\Delta) \cdot \omega^3 + p_5(\tau_\Delta) \cdot \omega^5] \cdot i$$

$$P_n(\lambda = i \cdot \omega, \tau_\Delta) = \Xi_9 \cdot \omega^2 - \Xi_{11} \cdot \omega^4 + [-\Xi_8 \cdot \omega + \Xi_{10} \cdot \omega^3 - \omega^5] \cdot i$$

$$Q_m(\lambda = i \cdot \omega, \tau_\Delta) = q_0(\tau_\Delta) + q_1(\tau_\Delta) \cdot i \cdot \omega - q_2(\tau_\Delta) \cdot \omega^2 - q_3(\tau_\Delta) \cdot i \cdot \omega^3 + q_4(\tau_\Delta) \cdot \omega^4$$

$$Q_m(\lambda = i \cdot \omega, \tau_\Delta) = q_0(\tau_\Delta) - q_2(\tau_\Delta) \cdot \omega^2 + q_4(\tau_\Delta) \cdot \omega^4 + [q_1(\tau_\Delta) \cdot \omega - q_3(\tau_\Delta) \cdot \omega^3] \cdot i$$

$$Q_m(\lambda = i \cdot \omega, \tau_\Delta) = \Xi_{21} - \Xi_{23} \cdot \omega^2 + \Omega_1 \cdot \omega^4 + [\Xi_{22} \cdot \omega - \Xi_{24} \cdot \omega^3] \cdot i$$

$$\begin{aligned} P_n(\lambda = i \cdot \omega, \tau_\Delta) + Q_m(\lambda = i \cdot \omega, \tau_\Delta) = {} & \Xi_9 \cdot \omega^2 - \Xi_{11} \cdot \omega^4 \\ & + [-\Xi_8 \cdot \omega + \Xi_{10} \cdot \omega^3 - \omega^5] \cdot i \\ & + \Xi_{21} - \Xi_{23} \cdot \omega^2 + \Omega_1 \cdot \omega^4 \\ & + [\Xi_{22} \cdot \omega - \Xi_{24} \cdot \omega^3] \cdot i \neq 0 \end{aligned}$$

$$\begin{aligned} P_n(\lambda = i \cdot \omega, \tau_\Delta) + Q_m(\lambda = i \cdot \omega, \tau_\Delta) = {} & \Xi_{21} + (\Xi_9 - \Xi_{23}) \cdot \omega^2 + (\Omega_1 - \Xi_{11}) \\ & \cdot \omega^4 + [(\Xi_{22} - \Xi_8) \cdot \omega + (\Xi_{10} \\ & - \Xi_{24}) \cdot \omega^3 - \omega^5] \cdot i \neq 0 \end{aligned}$$

Table 7.5 Dual-wavelength Ti:Sapphire laser system's stability analysis $F(\omega, \tau_\Delta)$ elements

Expression with ω	Equivalent expression
$(\Xi_9 \cdot \omega^2 - \Xi_{11} \cdot \omega^4)^2$	$\Xi_9^2 \cdot \omega^4 - 2 \cdot \Xi_9 \cdot \Xi_{11} \cdot \omega^6 + \Xi_{11}^2 \cdot \omega^8$
$(-\Xi_8 \cdot \omega + \Xi_{10} \cdot \omega^3 - \omega^5)^2$	$\Xi_8^2 \cdot \omega^2 - 2 \cdot \Xi_{10} \cdot \Xi_8 \cdot \omega^4$ $+ (2 \cdot \Xi_8 + \Xi_{10}^2) \cdot \omega^6 - 2 \cdot \Xi_{10} \cdot \omega^8 + \omega^{10}$
$(\Xi_{21} - \Xi_{23} \cdot \omega^2 + \Omega_1 \cdot \omega^4)^2$	$\Xi_{21}^2 - 2 \cdot \Xi_{21} \cdot \Xi_{23} \cdot \omega^2 + (2 \cdot \Xi_{21} \cdot \Omega_1 + \Xi_{23}^2) \cdot \omega^4$ $- 2 \cdot \Xi_{23} \cdot \Omega_1 \cdot \omega^6 + \Omega_1^2 \cdot \omega^8$
$(\Xi_{22} \cdot \omega - \Xi_{24} \cdot \omega^3)^2$	$\Xi_{22}^2 \cdot \omega^2 - 2 \cdot \Xi_{22} \cdot \Xi_{24} \cdot \omega^4 + \Xi_{24}^2 \cdot \omega^6$

$$|P(i \cdot \omega, \tau_\Delta)|^2 = (\Xi_9 \cdot \omega^2 - \Xi_{11} \cdot \omega^4)^2 + (-\Xi_8 \cdot \omega + \Xi_{10} \cdot \omega^3 - \omega^5)^2$$

$$|Q(i \cdot \omega, \tau_\Delta)|^2 = (\Xi_{21} - \Xi_{23} \cdot \omega^2 + \Omega_1 \cdot \omega^4)^2 + (\Xi_{22} \cdot \omega - \Xi_{24} \cdot \omega^3)^2$$

$$\begin{aligned} F(\omega, \tau_\Delta) &= |P(i \cdot \omega, \tau_\Delta)|^2 - |Q(i \cdot \omega, \tau_\Delta)|^2 \\ &= (\Xi_9 \cdot \omega^2 - \Xi_{11} \cdot \omega^4)^2 + (-\Xi_8 \cdot \omega + \Xi_{10} \cdot \omega^3 - \omega^5)^2 \\ &\quad - (\Xi_{21} - \Xi_{23} \cdot \omega^2 + \Omega_1 \cdot \omega^4)^2 - (\Xi_{22} \cdot \omega - \Xi_{24} \cdot \omega^3)^2 \end{aligned}$$

The Dual-wavelength Ti:Sapphire laser system's stability analysis $F(\omega, \tau_\Delta)$ elements are (Table 7.5).

$$\begin{aligned} F(\omega, \tau_\Delta) &= |P(i \cdot \omega, \tau_\Delta)|^2 - |Q(i \cdot \omega, \tau_\Delta)|^2 \\ &= \Xi_9^2 \cdot \omega^4 - 2 \cdot \Xi_9 \cdot \Xi_{11} \cdot \omega^6 + \Xi_{11}^2 \cdot \omega^8 \\ &\quad + \Xi_8^2 \cdot \omega^2 - 2 \cdot \Xi_{10} \cdot \Xi_8 \cdot \omega^4 + (2 \cdot \Xi_8 \\ &\quad + \Xi_{10}^2) \cdot \omega^6 - 2 \cdot \Xi_{10} \cdot \omega^8 + \omega^{10} \\ &\quad - [\Xi_{21}^2 - 2 \cdot \Xi_{21} \cdot \Xi_{23} \cdot \omega^2 + (2 \cdot \Xi_{21} \cdot \Omega_1 + \Xi_{23}^2) \cdot \omega^4 \\ &\quad - 2 \cdot \Xi_{23} \cdot \Omega_1 \cdot \omega^6 + \Omega_1^2 \cdot \omega^8] \\ &\quad - [\Xi_{22}^2 \cdot \omega^2 - 2 \cdot \Xi_{22} \cdot \Xi_{24} \cdot \omega^4 + \Xi_{24}^2 \cdot \omega^6] \end{aligned}$$

$$\begin{aligned} F(\omega, \tau_\Delta) &= |P(i \cdot \omega, \tau_\Delta)|^2 - |Q(i \cdot \omega, \tau_\Delta)|^2 \\ &= \Xi_9^2 \cdot \omega^4 - 2 \cdot \Xi_9 \cdot \Xi_{11} \cdot \omega^6 + \Xi_{11}^2 \cdot \omega^8 \\ &\quad + \Xi_8^2 \cdot \omega^2 - 2 \cdot \Xi_{10} \cdot \Xi_8 \cdot \omega^4 \\ &\quad + (2 \cdot \Xi_8 + \Xi_{10}^2) \cdot \omega^6 - 2 \cdot \Xi_{10} \cdot \omega^8 + \omega^{10} \\ &\quad - \Xi_{21}^2 + 2 \cdot \Xi_{21} \cdot \Xi_{23} \cdot \omega^2 - (2 \cdot \Xi_{21} \cdot \Omega_1 + \Xi_{23}^2) \cdot \omega^4 \\ &\quad + 2 \cdot \Xi_{23} \cdot \Omega_1 \cdot \omega^6 - \Omega_1^2 \cdot \omega^8 \\ &\quad - \Xi_{22}^2 \cdot \omega^2 + 2 \cdot \Xi_{22} \cdot \Xi_{24} \cdot \omega^4 - \Xi_{24}^2 \cdot \omega^6 \end{aligned}$$

$$\begin{aligned}F(\omega,\tau_\Delta) &= |P(i\cdot\omega,\tau_\Delta)|^2 - |Q(i\cdot\omega,\tau_\Delta)|^2\\&= -\Xi_{21}^2 + [\Xi_8^2 + 2\cdot\Xi_{21}\cdot\Xi_{23} - \Xi_{22}^2]\cdot\omega^2\\&\quad + [\Xi_9^2 - 2\cdot\Xi_{10}\cdot\Xi_8 - (2\cdot\Xi_{21}\cdot\Omega_1 + \Xi_{23}^2) + 2\cdot\Xi_{22}\cdot\Xi_{24}]\cdot\omega^4\\&\quad + [(2\cdot\Xi_8 + \Xi_{10}^2) - 2\cdot\Xi_9\cdot\Xi_{11} + 2\cdot\Xi_{23}\cdot\Omega_1 - \Xi_{24}^2]\cdot\omega^6\\&\quad + [\Xi_{11}^2 - 2\cdot\Xi_{10} - \Omega_1^2]\cdot\omega^8 + \omega^{10}\end{aligned}$$

We define the following parameters for simplicity: $\Pi_0, \Pi_2, \Pi_4, \Pi_6.\Pi_8, \Pi_{10}$

$$\Pi_0 = -\Xi_{21}^2;\ \Pi_2 = \Xi_8^2 + 2\cdot\Xi_{21}\cdot\Xi_{23} - \Xi_{22}^2$$
$$\Pi_4 = \Xi_9^2 - 2\cdot\Xi_{10}\cdot\Xi_8 - (2\cdot\Xi_{21}\cdot\Omega_1 + \Xi_{23}^2) + 2\cdot\Xi_{22}\cdot\Xi_{24}$$

$$\Pi_6 = (2\cdot\Xi_8 + \Xi_{10}^2) - 2\cdot\Xi_9\cdot\Xi_{11} + 2\cdot\Xi_{23}\cdot\Omega_1 - \Xi_{24}^2$$
$$\Pi_8 = \Xi_{11}^2 - 2\cdot\Xi_{10} - \Omega_1^2;\ \Pi_{10} = 1$$

Hence $F(\omega,\tau_\Delta) = 0$ implies $\sum_{k=0}^{5}\Pi_{2\cdot k}\cdot\omega^{2\cdot k} = 0$ and its roots are given by solving the above polynomial.

$$P_R(i\omega,\tau_\Delta) = \Xi_9\cdot\omega^2 - \Xi_{11}\cdot\omega^4;\ P_I(i\omega,\tau_\Delta) = -\Xi_8\cdot\omega + \Xi_{10}\cdot\omega^3 - \omega^5$$

$$Q_R(i\omega,\tau_\Delta) = \Xi_{21} - \Xi_{23}\cdot\omega^2 + \Omega_1\cdot\omega^4;\ Q_I(i\omega,\tau_\Delta) = \Xi_{22}\cdot\omega - \Xi_{24}\cdot\omega^3$$

$$\sin\theta(\tau_\Delta) = \frac{-P_R(i\omega,\tau_\Delta)\cdot Q_I(i\omega,\tau_\Delta) + P_I(i\omega,\tau_\Delta)\cdot Q_R(i\omega,\tau_\Delta)}{|Q(i\omega,\tau_\Delta)|^2}$$

$$\cos\theta(\tau_\Delta) = -\frac{P_R(i\omega,\tau_\Delta)\cdot Q_R(i\omega,\tau_\Delta) + P_I(i\omega,\tau_\Delta)\cdot Q_I(i\omega,\tau_\Delta)}{|Q(i\omega,\tau_\Delta)|^2}$$

We use different terminology from our last characteristics parameters definition:

$$k \to j;\ p_k(\tau_\Delta) \to a_j;\ q_k(\tau_\Delta) \to c_j;\ n, m \in \mathbb{N}_0;\ n = 5;\ m = 4;\ n > m$$

$$P_n(\lambda,\tau_\Delta) \to P(\lambda);\ Q_m(\lambda,\tau_\Delta) \to Q(\lambda)$$
$$P(\lambda) = \sum_{j=0}^{5} a_j\cdot\lambda^j;\ Q(\lambda) = \sum_{j=0}^{4} c_j\cdot\lambda^j$$

$$P(\lambda) = a_0 + a_1\cdot\lambda + a_2\cdot\lambda^2 + a_3\cdot\lambda^3 + a_4\cdot\lambda^4 + a_5\cdot\lambda^5$$
$$Q(\lambda) = c_0 + c_1\cdot\lambda + c_2\cdot\lambda^2 + c_3\cdot\lambda^3 + c_4\cdot\lambda^4$$

$n, m \in \mathbb{N}_0; n > m$ and $a_j, c_j: \mathbb{R}_{0+} \to \mathbb{R}$ are continuous and differentiable function of τ_Δ such that $a_0 + c_0 \neq 0$. In the following "—" denoted complex and conjugate. $P(\lambda)$, $Q(\lambda)$ are analytic functions in λ and differentiable in τ_Δ. The coefficients $a_j(A_{32}^A, L_A, l, W_{14}, n_{v_A}, n_{v_B}, \tau_\Delta, \ldots) \in \mathbb{R}$ and $c_j(A_{32}^A, A_{32}^B, L_A, L_B, l, W_{14}, n_{v_A}, n_{v_B}, \tau_\Delta, \ldots) \in \mathbb{R}$ depend on Dual-wavelength Ti:Sapphire laser system's parameters, $A_{32}^A, A_{32}^B, L_A, L_B, l, W_{14}, n_{v_A}, n_{v_B}, \tau_\Delta, \ldots$ values. Unless strictly necessary, the designation of the varied arguments: $A_{32}^A, A_{32}^B, L_A, L_B, l, W_{14}, n_{v_A}, n_{v_B}, \tau_\Delta, \ldots$ will subsequently be omitted from P, Q, a_j, c_j. The coefficients a_j, c_j are continuous and differentiable functions of their arguments, and direct substitution shows that $a_0 + c_0 \neq 0$ [2, 3].

$$a_0 = 0; c_0 = \Xi_{21} = \Xi_8 \cdot \Omega_1 + \Xi_{12} - \Xi_{15} + \Xi_{19}$$
$$\Xi_8 = -[(\Xi_1 + \Xi_4) \cdot \Omega_4 + \Xi_6 \cdot \Gamma_4]$$

$$\Xi_{12} = (S_{21}^A + S_{21}^B) \cdot \Omega_2 \cdot \Gamma_1 \cdot \Omega_4 \cdot \Gamma_8; \Xi_{15} = -\Omega_3 \cdot \Gamma_1 \cdot \Gamma_8 \cdot \Omega_4 \cdot W_{14}$$
$$\Xi_{19} = \Omega_2 \cdot \Gamma_1 \cdot W_{14} \cdot (S_{21}^A + S_{21}^B) \cdot \Omega_4$$

$$c_0 = (-[(\Xi_1 + \Xi_4) \cdot \Omega_4 + \Xi_6 \cdot \Gamma_4]) \cdot \Omega_1 + (S_{21}^A + S_{21}^B) \cdot \Omega_2 \cdot \Gamma_1 \cdot \Omega_4 \cdot \Gamma_8$$
$$+ \Omega_3 \cdot \Gamma_1 \cdot \Gamma_8 \cdot \Omega_4 \cdot W_{14} + \Omega_2 \cdot \Gamma_1 \cdot W_{14} \cdot (S_{21}^A + S_{21}^B) \cdot \Omega_4$$

$$\Xi_1 = W_{14} \cdot [\Gamma_3 + \Gamma_6 + \Gamma_7] \cdot \Gamma_8; \Xi_4$$
$$= (S_{21}^A + S_{21}^B) \cdot \{(\Gamma_2 + \Gamma_5) \cdot \Gamma_8 - W_{14} \cdot (S_{43} - \Gamma_2 - \Gamma_5)\}$$

$$\Xi_6 = -\Omega_5 \cdot (S_{21}^A + S_{21}^B) \cdot \Gamma_8 - \Omega_6 \cdot W_{14} \cdot \Gamma_8 - \Omega_5 \cdot W_{14} \cdot (S_{21}^A + S_{21}^B)$$

$$c_0 = (-[(W_{14} \cdot [\Gamma_3 + \Gamma_6 + \Gamma_7] \cdot \Gamma_8 + (S_{21}^A + S_{21}^B) \cdot \{(\Gamma_2 + \Gamma_5) \cdot \Gamma_8$$
$$- W_{14} \cdot (S_{43} - \Gamma_2 - \Gamma_5)\}) \cdot \Omega_4 + (-\Omega_5 \cdot (S_{21}^A + S_{21}^B) \cdot \Gamma_8 - \Omega_6 \cdot W_{14} \cdot \Gamma_8$$
$$- \Omega_5 \cdot W_{14} \cdot (S_{21}^A + S_{21}^B)) \cdot \Gamma_4]) \cdot \Omega_1 + (S_{21}^A + S_{21}^B) \cdot \Omega_2 \cdot \Gamma_1 \cdot \Omega_4 \cdot \Gamma_8$$
$$+ \Omega_3 \cdot \Gamma_1 \cdot \Gamma_8 \cdot \Omega_4 \cdot W_{14} + \Omega_2 \cdot \Gamma_1 \cdot W_{14} \cdot (S_{21}^A + S_{21}^B) \cdot \Omega_4$$

$$a_0 + c_0 = (-[(W_{14} \cdot [\Gamma_3 + \Gamma_6 + \Gamma_7] \cdot \Gamma_8 + (S_{21}^A + S_{21}^B) \cdot \{(\Gamma_2 + \Gamma_5) \cdot \Gamma_8$$
$$- W_{14} \cdot (S_{43} - \Gamma_2 - \Gamma_5)\}) \cdot \Omega_4 + (-\Omega_5 \cdot (S_{21}^A + S_{21}^B) \cdot \Gamma_8 - \Omega_6 \cdot W_{14} \cdot \Gamma_8$$
$$- \Omega_5 \cdot W_{14} \cdot (S_{21}^A + S_{21}^B)) \cdot \Gamma_4]) \cdot \Omega_1 + (S_{21}^A + S_{21}^B) \cdot \Omega_2 \cdot \Gamma_1 \cdot \Omega_4 \cdot \Gamma_8$$
$$+ \Omega_3 \cdot \Gamma_1 \cdot \Gamma_8 \cdot \Omega_4 \cdot W_{14} + \Omega_2 \cdot \Gamma_1 \cdot W_{14} \cdot (S_{21}^A + S_{21}^B) \cdot \Omega_4 \neq 0$$

$\forall A_{32}^A, A_{32}^B, L_A, L_B, l, W_{14}, n_{v_A}, n_{v_B}, \tau_\Delta, \ldots \in \mathbb{R}_+$ I.e. $\lambda = 0$ is not a root of the characteristic equation. Furthermore $P(\lambda)$, $Q(\lambda)$ are analytic functions of λ for which the following requirements of the analysis (see [2], Sect. 3.4) can also be verified in the present case.

1. If $\lambda = i \cdot \omega; \omega \in \mathbb{R}$ then $P(i \cdot \omega) + Q(i \cdot \omega) \neq 0$, i.e. P and Q have no common imaginary roots. This condition was verified numerically in the entire $A_{32}^{A}, L_A, l, W_{14}, n_{v_A}, n_{v_B}, \tau_\Delta, \ldots$ domain of interest.
2. $\left|\frac{P(\lambda)}{Q(\lambda)}\right|$ is bounded for $|\lambda| \to \infty$; $\mathrm{Re}\,\lambda \geq 0$. No roots bifurcation from ∞. Indeed, in the limit: $\left|\frac{Q(\lambda)}{P(\lambda)}\right| = \left|\frac{c_0+c_1\cdot\lambda+c_2\cdot\lambda^2+c_3\cdot\lambda^3+c_4\cdot\lambda^4}{a_0+a_1\cdot\lambda+a_2\cdot\lambda^2+a_3\cdot\lambda^3+a_4\cdot\lambda^4+a_5\cdot\lambda^5}\right|$.
3. The following expressions exist: $F(\omega) = |P(i \cdot \omega)|^2 - |Q(i \cdot \omega)|^2$. $F(\omega, \tau_\Delta) = |P(i \cdot \omega, \tau_\Delta)|^2 - |Q(i \cdot \omega, \tau_\Delta)|^2 = \sum_{k=0}^{5} \Pi_{2\cdot k} \cdot \omega^{2\cdot k}$ has at most a finite number of zeros. Indeed, this is a polynomial in ω (degree in ω^{10}).
4. Each positive root $\omega(A_{32}^{A}, L_A, l, W_{14}, n_{v_A}, n_{v_B}, \tau_\Delta, \ldots)$ of $F(\omega) = 0$ is continuous and differentiable with respect to $A_{32}^{A}, L_A, l, W_{14}, n_{v_A}, n_{v_B}, \tau_\Delta, \ldots$ and the condition can only be assessed numerically.

In addition, since the coefficients in P and Q are real, we have $\overline{P(-i \cdot \omega)} = P(i \cdot \omega)$ and $\overline{Q(-i \cdot \omega)} = Q(i \cdot \omega)$ thus, $\omega > 0$ may be on eigenvalue of characteristic equations. The analysis consists in identifying the roots of the characteristic equation situated on the imaginary axis of the complex λ plane, whereby increasing the parameters: $A_{32}^{A}, L_A, l, W_{14}, n_{v_A}, n_{v_B}, \tau_\Delta, \ldots$, $\mathrm{Re}\lambda$ may, at the crossing, change its sign from (−) to (+). i.e. from a stable focus $E^{(j)} = (\Phi_A^{(j)}, \Phi_B^{(j)}, n_1^{(j)}, n_3^{(j)}, n_4^{(j)}); j = 0, 1, 2$ to an unstable one, or vice versa. This feature may be further assessed by examining the sign of the partial derivatives with respect to $A_{32}^{A}, L_A, l, W_{14}, n_{v_A}, n_{v_B}, \tau_\Delta, \ldots$ and any system parameters.

$$\Lambda^{-1}(A_{32}^{A}) = \left(\frac{\partial \mathrm{Re}\lambda}{\partial A_{32}^{A}}\right)_{\lambda=i\cdot\omega}; A_{32}^{B}, L_A, L_B, l, W_{14}, n_{v_A}, n_{v_B}, \tau_\Delta, \ldots = const$$

$$\Lambda^{-1}(L_A) = \left(\frac{\partial \mathrm{Re}\lambda}{\partial L_A}\right)_{\lambda=i\cdot\omega}; A_{32}^{A}, A_{32}^{B}, L_B, l, W_{14}, n_{v_A}, n_{v_B}, \tau_\Delta, \ldots = const$$

$$\Lambda^{-1}(W_{14}) = \left(\frac{\partial \mathrm{Re}\lambda}{\partial W_{14}}\right)_{\lambda=i\cdot\omega}; A_{32}^{A}, A_{32}^{B}, L_A, L_B, l, n_{v_A}, n_{v_B}, \tau_\Delta, \ldots = const$$

$$\Lambda^{-1}\left(n_{v_A}\right) = \left(\frac{\partial \mathrm{Re}\lambda}{\partial n_{v_A}}\right)_{\lambda=i\cdot\omega}; A_{32}^{A}, A_{32}^{B}, L_A, L_B, l, W_{14}, n_{v_B}, \tau_\Delta, \ldots = const$$

$$\Lambda^{-1}(n_{v_B}) = \left(\frac{\partial \mathrm{Re}\lambda}{\partial n_{v_B}}\right)_{\lambda=i\cdot\omega}; A_{32}^{A}, A_{32}^{B}, L_A, L_B, l, W_{14}, n_{v_A}, \tau_\Delta, \ldots = const$$

$$\Lambda^{-1}(\tau_\Delta) = \left(\frac{\partial \mathrm{Re}\lambda}{\partial \tau_\Delta}\right)_{\lambda=i\cdot\omega}; A_{32}^{A}, A_{32}^{B}, L_A, L_B, l, W_{14}, n_{v_A}, n_{v_B}, \ldots = const$$

$P(\lambda) = P_R(\lambda) + i \cdot P_I(\lambda)$; $Q(\lambda) = Q_R(\lambda) + i \cdot Q_I(\lambda)$, When writing and inserting $\lambda = i \cdot \omega$ into Dual-wavelength Ti:Sapphire laser system's characteristic equation ω must satisfy the following equations:

$$\sin(\omega \cdot \tau_\Delta) = g(\omega) = \frac{-P_R(i\omega, \tau_\Delta) \cdot Q_I(i\omega, \tau_\Delta) + P_I(i\omega, \tau_\Delta) \cdot Q_R(i\omega, \tau_\Delta)}{|Q(i\omega, \tau_\Delta)|^2}$$

$$\cos(\omega \cdot \tau_\Delta) = h(\omega) = -\frac{P_R(i\omega, \tau_\Delta) \cdot Q_R(i\omega, \tau_\Delta) + P_I(i\omega, \tau_\Delta) \cdot Q_I(i\omega, \tau_\Delta)}{|Q(i\omega, \tau_\Delta)|^2}$$

where $|Q(i\omega, \tau_\Delta)|^2 \neq 0$ in view of requirement (1) above, and $(g, h) \in \mathbb{R}$. Furthermore, it follows above $\sin(\omega \cdot \tau_\Delta)$ and $\cos(\omega \cdot \tau_\Delta)$ equation that, by squaring and adding the sides, ω must be a positive root of $F(\omega) = |P(i \cdot \omega)|^2 - |Q(i \cdot \omega)|^2 = 0$. Note: $F(\omega)$ is independent on τ_Δ. Now it is important to notice that if $\tau_\Delta \notin I$ (assume that $I \subseteq \mathbb{R}_{+0}$ is the set where $\omega(\tau_\Delta)$ is a positive root of $F(\omega)$ and for, $\tau_\Delta \notin I, \omega(\tau_\Delta)$ is not defined. Then for all τ_Δ in $I, \omega(\tau_\Delta)$ is satisfied that $F(\omega, \tau_\Delta) = 0$. Then there are no positive $\omega(\tau_\Delta)$ solutions for $F(\omega, \tau_\Delta) = 0$, and we cannot have stability switches. For $\tau_\Delta \in I$ where $\omega(\tau_\Delta)$ is a positive solution of $F(\omega, \tau_\Delta) = 0$, we can define the angle $\theta(\tau_\Delta) \in [0, 2 \cdot \pi]$ as the solution of $\sin\theta(\tau_\Delta) = \ldots$ and $\cos\theta(\tau_\Delta) = \ldots$; the relation between the arguments $\theta(\tau_\Delta)$ and $\tau_\Delta \cdot \omega(\tau_\Delta)$ for $\tau_\Delta \in I$ must be describing below. $\tau_\Delta \cdot \omega(\tau_\Delta) = \theta(\tau_\Delta) + 2 \cdot n \cdot \pi \; \forall n \in \mathbb{N}_0$ Hence we can define the maps: $\tau_{\Delta(n)} : I \to \mathbb{R}_{+0}$, is given by $\tau_{\Delta(n)}(\tau_\Delta) = \frac{\theta(\tau_\Delta)+2 \cdot n \cdot \pi}{\omega(\tau_\Delta)}; n \in \mathbb{N}_0; \tau_\Delta \in I$. Let us introduce the function $I \to \mathbb{R}; S_n(\tau_\Delta) = \tau_\Delta - \tau_{\Delta(n)}(\tau_\Delta) \; ; \tau_\Delta \in I; n \in \mathbb{N}_0$ that is continuous and differentiable in τ_Δ. In the following, the subscripts $\lambda, \omega, A_{32}^A, L_A, l, W_{14}, n_{v_A}, n_{v_B}, \ldots$ indicate the corresponding partial derivatives. Let us first concentrate on $\Lambda(x)$, remember in $\lambda(A_{32}^A, L_A, l, W_{14}, n_{v_A}, n_{v_B}, \ldots)$ and $\omega(A_{32}^A, L_A, l, W_{14}, n_{v_A}, n_{v_B}, \ldots)$, and keeping all parameters except one (x) and τ_Δ. The derivation closely follows that in reference [BK]. Differentiating Dual-wavelength Ti:Sapphire laser system's characteristic equation $P(\lambda) + Q(\lambda) \cdot e^{-\lambda \cdot \tau_D} = 0$ with respect to specific parameter (x), and inverting the derivative, for convenience, one calculates: $x = A_{32}^A, L_A, l, W_{14}, n_{v_A}, n_{v_B}, \tau_\Delta, \ldots$

$$\left(\frac{\partial \lambda}{\partial x}\right)^{-1} = \frac{-P_\lambda(\lambda, x) \cdot Q(\lambda, x) + Q_\lambda(\lambda, x) \cdot P(\lambda, x) - \tau_\Delta \cdot P(\lambda, x) \cdot Q(\lambda, x)}{P_x(\lambda, x) \cdot Q(\lambda, x) - Q_x(\lambda, x) \cdot P(\lambda, x)}$$

where $P_\lambda = \frac{\partial P}{\partial \lambda}, \ldots$ etc., substituting $\lambda = i \cdot \omega$ and bearing $\overline{P(-i \cdot \omega)} = P(i \cdot \omega)$; $\overline{Q(-i \cdot \omega)} = Q(i \cdot \omega)$ then $i \cdot P_\lambda(i \cdot \omega) = P_\omega(i \cdot \omega) \; ; i \cdot Q_\lambda(i \cdot \omega) = Q_\omega(i \cdot \omega)$ and that on the surface $|P(i\omega)|^2 = |Q(i\omega)|^2$, one obtain: $(\frac{\partial \lambda}{\partial x})^{-1}|_{\lambda = i \cdot \omega} = (\frac{i \cdot P_\omega(i \cdot \omega, x) \cdot \overline{P(i \cdot \omega, x)} + i \cdot Q_\lambda(i \cdot \omega, x) \cdot \overline{Q(\lambda, x)} - \tau_\Delta \cdot |P(i \cdot \omega, x)|^2}{P_x(i \cdot \omega, x) \cdot \overline{P(i \cdot \omega, x)} - Q_x(i \cdot \omega, x) \cdot \overline{Q(i \cdot \omega, x)}})$.

Upon separating into real and imaginary parts, with $P = P_R + i \cdot P_I$; $Q = Q_R + i \cdot Q_I$ and $P_\omega = P_{R\omega} + i \cdot P_{I\omega}$; $Q_\omega = Q_{R\omega} + i \cdot Q_{I\omega}$; $P_x = P_{Rx} + i \cdot P_{Ix}$; $Q_x = Q_{Rx} + i \cdot Q_{Ix}$; $P^2 = P_R^2 + P_I^2$, When (x) can be any Dual-wavelength Ti:Sapphire laser system's parameters $A_{32}^A, L_A, l, W_{14}, n_{v_A}, n_{v_B}, \ldots$ and time delay τ_Δ etc. Where for convenience, we have dropped the arguments $(i \cdot \omega, x)$, and where $F_\omega = 2 \cdot [(P_{R\omega} \cdot P_R + P_{I\omega} \cdot P_I) - (Q_{R\omega} \cdot Q_R + Q_{I\omega} \cdot Q_I)]$; $F_x = 2 \cdot [(P_{Rx} \cdot P_R + P_{Ix} \cdot P_I) - (Q_{Rx} \cdot Q_R + Q_{Ix} \cdot Q_I)]$ and $\omega_x = -\frac{F_x}{F_\omega}$. We define U and V: $U = (P_R \cdot P_{I\omega} - P_I \cdot P_{R\omega}) - (Q_R \cdot Q_{I\omega} - Q_I \cdot Q_{R\omega})$, $V = (P_R \cdot P_{Ix} - P_I \cdot P_{Rx}) - (Q_R \cdot Q_{Ix} - Q_I \cdot Q_{Rx})$, we choose our specific parameter as time delay $x = \tau_\Delta$: $P_R = \Xi_9 \cdot \omega^2 - \Xi_{11} \cdot \omega^4$; $P_I =$

$-\Xi_8\cdot\omega+\Xi_{10}\cdot\omega^3-\omega^5$; $Q_R=\Xi_{21}-\Xi_{23}\cdot\omega^2+\Omega_1\cdot\omega^4$; $Q_I=\Xi_{22}\cdot\omega-\Xi_{24}\cdot\omega^3$; $P_{R\omega}=2\cdot\Xi_9\cdot\omega-4\cdot\Xi_{11}\cdot\omega^3$; $P_{I\omega}=-\Xi_8+3\cdot\Xi_{10}\cdot\omega^2-5\cdot\omega^4$; $Q_{I\omega}=\Xi_{22}-3\cdot\Xi_{24}\cdot\omega^2$; $Q_{R\omega}=-2\cdot\Xi_{23}\cdot\omega+4\cdot\Omega_1\cdot\omega^3$; $P_{R\tau_\Delta}=0$; $P_{I\tau_\Delta}=0$; $Q_{R\tau_\Delta}=0$; $Q_{I\tau_\Delta}=0$; $F_{\tau_\Delta}=0$; $P_R\cdot P_{I\tau_\Delta}-P_I\cdot P_{R\tau_\Delta}=0$; $Q_R\cdot Q_{I\tau_\Delta}-Q_I\cdot Q_{R\tau_\Delta}=0$; $V=(P_R\cdot P_{I\tau_\Delta}-P_I\cdot P_{R\tau_\Delta})-(Q_R\cdot Q_{I\tau_\Delta}-Q_I\cdot Q_{R\tau_\Delta})=0$.

$F_\omega=\ldots$ Elements:

$$\begin{aligned}P_{R\omega}\cdot P_R&=(2\cdot\Xi_9\cdot\omega-4\cdot\Xi_{11}\cdot\omega^3)\cdot(\Xi_9\cdot\omega^2-\Xi_{11}\cdot\omega^4)\\&=2\cdot\Xi_9^2\cdot\omega^3-6\cdot\Xi_{11}\cdot\Xi_9\cdot\omega^5+4\cdot\Xi_{11}^2\cdot\omega^7\end{aligned}$$

$$\begin{aligned}P_{I\omega}\cdot P_I&=(-\Xi_8+3\cdot\Xi_{10}\cdot\omega^2-5\cdot\omega^4)\cdot(-\Xi_8\cdot\omega+\Xi_{10}\cdot\omega^3-\omega^5)\\&=\Xi_8^2\cdot\omega-4\cdot\Xi_{10}\cdot\Xi_8\cdot\omega^3\\&\quad+3\cdot(\Xi_{10}^2+2\cdot\Xi_8)\cdot\omega^5-8\cdot\Xi_{10}\cdot\omega^7+5\cdot\omega^9\end{aligned}$$

$$\begin{aligned}Q_{R\omega}\cdot Q_R&=(-2\cdot\Xi_{23}\cdot\omega+4\cdot\Omega_1\cdot\omega^3)\cdot(\Xi_{21}-\Xi_{23}\cdot\omega^2+\Omega_1\cdot\omega^4)\\&=-2\cdot\Xi_{23}\Xi_{21}\cdot\omega+2\cdot(\Xi_{23}^2+2\cdot\Omega_1\cdot\Xi_{21})\cdot\omega^3\\&\quad-2\cdot(\Xi_{23}\cdot\Omega_1+2\cdot\Omega_1\cdot\Xi_{23})\cdot\omega^5+4\cdot\Omega_1^2\cdot\omega^7\end{aligned}$$

$$\begin{aligned}Q_{I\omega}\cdot Q_I&=(\Xi_{22}-3\cdot\Xi_{24}\cdot\omega^2)\cdot(\Xi_{22}\cdot\omega-\Xi_{24}\cdot\omega^3)\\&=\Xi_{22}^2\cdot\omega-4\cdot\Xi_{24}\cdot\Xi_{22}\cdot\omega^3+3\cdot\Xi_{24}^2\cdot\omega^5\end{aligned}$$

$U=\ldots$ Elements:

$$\begin{aligned}P_R\cdot P_{I\omega}&=(\Xi_9\cdot\omega^2-\Xi_{11}\cdot\omega^4)\cdot(-\Xi_8+3\cdot\Xi_{10}\cdot\omega^2-5\cdot\omega^4)\\&=-\Xi_9\cdot\Xi_8\cdot\omega^2+(3\cdot\Xi_9\cdot\Xi_{10}+\Xi_{11}\cdot\Xi_8)\cdot\omega^4\\&\quad-(3\cdot\Xi_{11}\cdot\Xi_{10}+5\cdot\Xi_9)\cdot\omega^6+5\cdot\Xi_{11}\cdot\omega^8\end{aligned}$$

$$\begin{aligned}P_I\cdot P_{R\omega}&=(-\Xi_8\cdot\omega+\Xi_{10}\cdot\omega^3-\omega^5)\cdot(2\cdot\Xi_9\cdot\omega-4\cdot\Xi_{11}\cdot\omega^3)\\&=-2\cdot\Xi_8\cdot\Xi_9\cdot\omega^2+2\cdot(2\cdot\Xi_8\cdot\Xi_{11}+\Xi_{10}\cdot\Xi_9)\cdot\omega^4\\&\quad-2\cdot(2\cdot\Xi_{10}\cdot\Xi_{11}+\Xi_9)\cdot\omega^6+4\cdot\Xi_{11}\cdot\omega^8\end{aligned}$$

$$\begin{aligned}Q_R\cdot Q_{I\omega}&=(\Xi_{21}-\Xi_{23}\cdot\omega^2+\Omega_1\cdot\omega^4)\cdot(\Xi_{22}-3\cdot\Xi_{24}\cdot\omega^2)\\&=\Xi_{21}\cdot\Xi_{22}-(3\cdot\Xi_{21}\cdot\Xi_{24}+\Xi_{23}\cdot\Xi_{22})\cdot\omega^2+(3\cdot\Xi_{23}\cdot\Xi_{24}\\&\quad+\Omega_1\cdot\Xi_{22})\cdot\omega^4-3\cdot\Omega_1\cdot\Xi_{24}\cdot\omega^6\end{aligned}$$

Table 7.6 Dual-wavelength Ti:Sapphire laser system's stability analysis F_ω, U elements

F_ω, U elements	Expression
$P_{R\omega}\cdot P_R$	$2\cdot\Xi_9^2\cdot\omega^3-6\cdot\Xi_{11}\cdot\Xi_9\cdot\omega^5+4\cdot\Xi_{11}^2\cdot\omega^7$
$P_{I\omega}\cdot P_I$	$\Xi_8^2\cdot\omega-4\cdot\Xi_{10}\cdot\Xi_8\cdot\omega^3+3\cdot(\Xi_{10}^2+2\cdot\Xi_8)\cdot\omega^5-8\cdot\Xi_{10}\cdot\omega^7+5\cdot\omega^9$
$Q_{R\omega}\cdot Q_R$	$-2\cdot\Xi_{23}\Xi_{21}\cdot\omega+2\cdot(\Xi_{23}^2+2\cdot\Omega_1\cdot\Xi_{21})\cdot\omega^3$ $-2\cdot(\Xi_{23}\cdot\Omega_1+2\cdot\Omega_1\cdot\Xi_{23})\cdot\omega^5+4\cdot\Omega_1^2\cdot\omega^7$
$Q_{I\omega}\cdot Q_I$	$\Xi_{22}^2\cdot\omega-4\cdot\Xi_{24}\cdot\Xi_{22}\cdot\omega^3+3\cdot\Xi_{24}^2\cdot\omega^5$
$P_R\cdot P_{I\omega}$	$-\Xi_9\cdot\Xi_8\cdot\omega^2+(3\cdot\Xi_9\cdot\Xi_{10}+\Xi_{11}\cdot\Xi_8)\cdot\omega^4$ $-(3\cdot\Xi_{11}\cdot\Xi_{10}+5\cdot\Xi_9)\cdot\omega^6+5\cdot\Xi_{11}\cdot\omega^8$
$P_I\cdot P_{R\omega}$	$-2\cdot\Xi_8\cdot\Xi_9\cdot\omega^2+2\cdot(2\cdot\Xi_8\cdot\Xi_{11}+\Xi_{10}\cdot\Xi_9)\cdot\omega^4$ $-2\cdot(2\cdot\Xi_{10}\cdot\Xi_{11}+\Xi_9)\cdot\omega^6+4\cdot\Xi_{11}\cdot\omega^8$
$Q_R\cdot Q_{I\omega}$	$\Xi_{21}\cdot\Xi_{22}-(3\cdot\Xi_{21}\cdot\Xi_{24}+\Xi_{23}\cdot\Xi_{22})\cdot\omega^2+(3\cdot\Xi_{23}\cdot\Xi_{24}+\Omega_1\cdot\Xi_{22})\cdot\omega^4$ $-3\cdot\Omega_1\cdot\Xi_{24}\cdot\omega^6$
$Q_I\cdot Q_{R\omega}$	$-2\cdot\Xi_{22}\cdot\Xi_{23}\cdot\omega^2+2\cdot(2\cdot\Xi_{22}\cdot\Omega_1+\Xi_{24}\cdot\Xi_{23})\cdot\omega^4-4\cdot\Xi_{24}\cdot\Omega_1\cdot\omega^6$

$$Q_I\cdot Q_{R\omega}=(\Xi_{22}\cdot\omega-\Xi_{24}\cdot\omega^3)\cdot(-2\cdot\Xi_{23}\cdot\omega+4\cdot\Omega_1\cdot\omega^3)$$
$$=-2\cdot\Xi_{22}\cdot\Xi_{23}\cdot\omega^2+2\cdot(2\cdot\Xi_{22}\cdot\Omega_1+\Xi_{24}\cdot\Xi_{23})\cdot\omega^4$$
$$-4\cdot\Xi_{24}\cdot\Omega_1\cdot\omega^6$$

We can summary our last results in the next table (Table 7.6).

$F(\omega,\tau_\Delta)=0$, differentiating with respect to τ_Δ and we get $F_\omega\cdot\frac{\partial\omega}{\partial\tau_\Delta}+F_{\tau_\Delta}=0$; $\tau_\Delta\in I\Rightarrow\frac{\partial\omega}{\partial\tau_\Delta}=-\frac{F_{\tau_\Delta}}{F_\omega}$

$$\Lambda^{-1}(\tau_\Delta)=\left(\frac{\partial\mathrm{Re}\lambda}{\partial\tau_\Delta}\right)_{\lambda=i\omega};\ \frac{\partial\omega}{\partial\tau_\Delta}=\omega_{\tau_\Delta}=-\frac{F_{\tau_\Delta}}{F_\omega}$$
$$\Lambda^{-1}(\tau_\Delta)=\mathrm{Re}\left\{\frac{-2\cdot[U+\tau_\Delta\cdot|P|^2+i\cdot F_\omega]}{F_{\tau_\Delta}+i\cdot 2\cdot[V+\omega\cdot|P|^2]}\right\}$$
$$\mathrm{sign}\{\Lambda^{-1}(\tau_\Delta)\}=\mathrm{sign}\left\{\left(\frac{\partial\mathrm{Re}\lambda}{\partial\tau_\Delta}\right)_{\lambda=i\omega}\right\}$$
$$\mathrm{sign}\{\Lambda^{-1}(\tau_\Delta)\}=\mathrm{sign}\{F_\omega\}\cdot\mathrm{sign}\{\frac{V+\frac{\partial\omega}{\partial\tau_\Delta}\cdot U}{|P|^2}+\omega+\frac{\partial\omega}{\partial\tau_\Delta}\cdot\tau_\Delta\}$$

We shall presently examine the possibility of stability transitions (bifurcations) Dual-wavelength Ti:Sapphire laser system, about the equilibrium points, $E^{(j)}$; $j=0,1,2,\ldots,$

$E^{(j)} = (\Phi_A^{(j)}, \Phi_B^{(j)}, n_1^{(j)}, n_3^{(j)}, n_4^{(j)})$; $j = 0, 1, 2, \ldots$. The analysis consists in identifying the roots of our system characteristic equation situated on the imaginary axis of the complex λ-plane, where by increasing the delay parameter τ_Δ. Reλ, may at the crossing, changes its sign from – to +, i.e. from a stable focus $E^{(*)}$ to an unstable one, or vice versa. This feature may be further assessed by examining the sign of the partial derivatives with respect to τ_Δ [2, 3].

$$\Lambda^{-1}(\tau_\Delta) = \left(\frac{\partial \mathrm{Re}\lambda}{\partial \tau_\Delta}\right)_{\lambda = i\cdot\omega}; A_{32}^A, A_{32}^B, L_A, L_B, l, W_{14}, n_{v_A}, n_{v_B}, \ldots = const$$

$$\begin{aligned} U &= (P_R \cdot P_{I\omega} - P_I \cdot P_{R\omega}) - (Q_R \cdot Q_{I\omega} - Q_I \cdot Q_{R\omega}) \\ &= -\Xi_{21} \cdot \Xi_{22} + [2 \cdot \Xi_8 \cdot \Xi_9 + (3 \cdot \Xi_{21} \cdot \Xi_{24} + \Xi_{23} \cdot \Xi_{22}) \\ &\quad - 2 \cdot \Xi_{22} \cdot \Xi_{23} - \Xi_9 \cdot \Xi_8] \cdot \omega^2 + [(3 \cdot \Xi_9 \cdot \Xi_{10} + \Xi_{11} \cdot \Xi_8) \\ &\quad - 2 \cdot (2 \cdot \Xi_8 \cdot \Xi_{11} + \Xi_{10} \cdot \Xi_9) - (3 \cdot \Xi_{23} \cdot \Xi_{24} + \Omega_1 \cdot \Xi_{22}) \\ &\quad + 2 \cdot (2 \cdot \Xi_{22} \cdot \Omega_1 + \Xi_{24} \cdot \Xi_{23})] \cdot \omega^4 + [2 \cdot (2 \cdot \Xi_{10} \cdot \Xi_{11} + \Xi_9) \\ &\quad + 3 \cdot \Omega_1 \cdot \Xi_{24} - 4 \cdot \Xi_{24} \cdot \Omega_1 - (3 \cdot \Xi_{11} \cdot \Xi_{10} + 5 \cdot \Xi_9)] \cdot \omega^6 \\ &\quad + [5 \cdot \Xi_{11} - 4 \cdot \Xi_{11}] \cdot \omega^8 \end{aligned}$$

We define for simplicity the following global parameters: $A_0 = -\Xi_{21} \cdot \Xi_{22}$

$$A_2 = 2 \cdot \Xi_8 \cdot \Xi_9 + (3 \cdot \Xi_{21} \cdot \Xi_{24} + \Xi_{23} \cdot \Xi_{22}) - 2 \cdot \Xi_{22} \cdot \Xi_{23} - \Xi_9 \cdot \Xi_8$$

$$\begin{aligned} A_4 &= (3 \cdot \Xi_9 \cdot \Xi_{10} + \Xi_{11} \cdot \Xi_8) - 2 \cdot (2 \cdot \Xi_8 \cdot \Xi_{11} + \Xi_{10} \cdot \Xi_9) \\ &\quad - (3 \cdot \Xi_{23} \cdot \Xi_{24} + \Omega_1 \cdot \Xi_{22}) + 2 \cdot (2 \cdot \Xi_{22} \cdot \Omega_1 + \Xi_{24} \cdot \Xi_{23}) \end{aligned}$$

$$\begin{aligned} A_6 &= 2 \cdot (2 \cdot \Xi_{10} \cdot \Xi_{11} + \Xi_9) + 3 \cdot \Omega_1 \cdot \Xi_{24} - 4 \cdot \Xi_{24} \cdot \Omega_1 \\ &\quad - (3 \cdot \Xi_{11} \cdot \Xi_{10} + 5 \cdot \Xi_9);\ A_8 = 5 \cdot \Xi_{11} - 4 \cdot \Xi_{11} \end{aligned}$$

$$U = (P_R \cdot P_{I\omega} - P_I \cdot P_{R\omega}) - (Q_R \cdot Q_{I\omega} - Q_I \cdot Q_{R\omega}) = \sum_{k=0}^{4} A_{2k} \cdot \omega^{2\cdot k}$$

$$\begin{aligned} F_\omega &= 2 \cdot [(P_{R\omega} \cdot P_R + P_{I\omega} \cdot P_I) - (Q_{R\omega} \cdot Q_R + Q_{I\omega} \cdot Q_I)] \\ &= 2 \cdot [\Xi_8^2 \cdot \omega + (2 \cdot \Xi_9^2 - 4 \cdot \Xi_{10} \cdot \Xi_8) \cdot \omega^3 \\ &\quad + (3 \cdot (\Xi_{10}^2 + 2 \cdot \Xi_8) - 6 \cdot \Xi_{11} \cdot \Xi_9) \cdot \omega^5 + (4 \cdot \Xi_{11}^2 - 8 \cdot \Xi_{10}) \cdot \omega^7 \\ &\quad + 5 \cdot \omega^9 + (2 \cdot \Xi_{23} \Xi_{21} - \Xi_{22}^2) \cdot \omega \\ &\quad + (4 \cdot \Xi_{24} \cdot \Xi_{22} - 2 \cdot (\Xi_{23}^2 + 2 \cdot \Omega_1 \cdot \Xi_{21})) \cdot \omega^3 \\ &\quad + (2 \cdot (\Xi_{23} \cdot \Omega_1 + 2 \cdot \Omega_1 \cdot \Xi_{23}) - 3 \cdot \Xi_{24}^2) \cdot \omega^5 - 4 \cdot \Omega_1^2 \cdot \omega^7] \end{aligned}$$

$$
\begin{aligned}
F_{\omega} &= 2 \cdot [(P_{R\omega} \cdot P_R + P_{I\omega} \cdot P_I) - (Q_{R\omega} \cdot Q_R + Q_{I\omega} \cdot Q_I)] \\
&= 2 \cdot [(\Xi_8^2 + 2 \cdot \Xi_{23}\Xi_{21} - \Xi_{22}^2) \cdot \omega \\
&\quad + \{2 \cdot \Xi_9^2 - 4 \cdot \Xi_{10} \cdot \Xi_8 + 4 \cdot \Xi_{24} \cdot \Xi_{22} - 2 \cdot (\Xi_{23}^2 + 2 \cdot \Omega_1 \cdot \Xi_{21})\} \cdot \omega^3 \\
&\quad + \{3 \cdot (\Xi_{10}^2 + 2 \cdot \Xi_8) - 6 \cdot \Xi_{11} \cdot \Xi_9 \\
&\quad + 2 \cdot (\Xi_{23} \cdot \Omega_1 + 2 \cdot \Omega_1 \cdot \Xi_{23}) - 3 \cdot \Xi_{24}^2\} \cdot \omega^5 \\
&\quad + \{4 \cdot \Xi_{11}^2 - 8 \cdot \Xi_{10} - 4 \cdot \Omega_1^2\} \cdot \omega^7 + 5 \cdot \omega^9]
\end{aligned}
$$

We define for simplicity the following global parameters: $B_1 = 2 \cdot (\Xi_8^2 + 2 \cdot \Xi_{23}\Xi_{21} - \Xi_{22}^2)$

$$B_3 = 2 \cdot \{2 \cdot \Xi_9^2 - 4 \cdot \Xi_{10} \cdot \Xi_8 + 4 \cdot \Xi_{24} \cdot \Xi_{22} - 2 \cdot (\Xi_{23}^2 + 2 \cdot \Omega_1 \cdot \Xi_{21})\}$$

$$B_5 = \{3 \cdot (\Xi_{10}^2 + 2 \cdot \Xi_8) - 6 \cdot \Xi_{11} \cdot \Xi_9 + 2 \cdot (\Xi_{23} \cdot \Omega_1 + 2 \cdot \Omega_1 \cdot \Xi_{23}) - 3 \cdot \Xi_{24}^2\}$$

$$B_7 = 2\{4 \cdot \Xi_{11}^2 - 8 \cdot \Xi_{10} - 4 \cdot \Omega_1^2\};\ B_9 = 10$$

$$F_{\omega} = 2 \cdot [(P_{R\omega} \cdot P_R + P_{I\omega} \cdot P_I) - (Q_{R\omega} \cdot Q_R + Q_{I\omega} \cdot Q_I)] = \sum_{k=0}^{4} B_{2 \cdot k+1} \cdot \omega^{2 \cdot k+1}$$

Then we get the expression for $F_{\omega} = \sum_{k=0}^{4} B_{2 \cdot k+1} \cdot \omega^{2 \cdot k+1}$, Dual-wavelength Ti:Sapphire laser system parameter values. We find those ω, τ_{Δ} values which fulfill $F_{\omega}(\omega, \tau_{\Delta}) = 0$. We ignore negative, complex, and imaginary values of ω for specific τ_{Δ} values. $\tau_{\Delta} \in [0.001 \ldots 10]$, we can express by 3D function $F_{\omega}(\omega, \tau_{\Delta}) = 0$. We plot the stability switch diagram based on different delay values of our Dual-wavelength Ti:Sapphire laser system.

$$\Lambda^{-1}(\tau_{\Delta}) = \left(\frac{\partial \mathrm{Re}\lambda}{\partial \tau_{\Delta}}\right)_{\lambda = i\omega} = \mathrm{Re}\left\{\frac{-2 \cdot [U + \tau_{\Delta} \cdot |P|^2] + i \cdot F_{\omega}}{F_{\tau_{\Delta}} + 2i \cdot [V + \omega \cdot |P|^2]}\right\}$$

$$\Lambda^{-1}(\tau_{\Delta}) = \left(\frac{\partial \mathrm{Re}\lambda}{\partial \tau_{\Delta}}\right)_{\lambda = i\omega} = \frac{2 \cdot \{F_{\omega} \cdot (V + \omega \cdot P^2) - F_{\tau_{\Delta}} \cdot (U + \tau_{\Delta} \cdot P^2)\}}{F_{\tau_{\Delta}}^2 + 4 \cdot (V + \omega \cdot P^2)^2}$$

The stability switch occurs only on those delay values (τ_{Δ}) which fit the equation: $\tau_{\Delta} = \frac{\theta_+(\tau_{\Delta})}{\omega_+(\tau_{\Delta})}$ and $\theta_+(\tau_{\Delta})$ is the solution of $\sin\theta(\tau_{\Delta}) = \ldots$. and $\cos\theta(\tau_{\Delta}) = \ldots$. when $\omega = \omega_+(\tau_{\Delta})$ If only ω_+ is feasible. Additionally, when all Dual-wavelength Ti:Sapphire laser system's parameters are known and the stability switch due to various time delay values τ_{Δ} is described in the following expression:

$$\operatorname{sign}\{\Lambda^{-1}(\tau_\Delta)\} = \operatorname{sign}\{F_\omega(\omega(\tau_\Delta), \tau_\Delta)\} \cdot \operatorname{sign}\Big\{\tau_\Delta \cdot \omega_{\tau_\Delta}(\omega(\tau_\Delta)) + \omega(\tau_\Delta) + \frac{U(\omega(\tau_\Delta)) \cdot \omega_{\tau_\Delta}(\omega(\tau_\Delta)) + V(\omega(\tau_\Delta))}{|P(\omega(\tau_\Delta))|^2}\Big\}$$

Remark: we know $F(\omega, \tau_\Delta) = 0$ implies its roots $\omega_i(\tau_\Delta)$ and finding those delays values τ_Δ which ω_i is feasible. There are τ_Δ values which give complex ω_i or imaginary number, then unable to analyze stability. $F(\omega, \tau_\Delta)$, function is independent on τ_Δ the parameter $F(\omega, \tau_\Delta) = 0$. The results: We find those ω, τ_Δ values which fulfill $F(\omega, \tau_\Delta) = 0$. We ignore negative, complex, and imaginary values of ω. Next is to find those ω, τ_Δ values which fulfill $\sin\theta(\tau_\Delta) = \ldots.$ and $\cos\theta(\tau_\Delta) = \ldots.$; $\sin(\omega \cdot \tau_\Delta) = \frac{-P_R \cdot Q_I + P_I \cdot Q_R}{|Q|^2}$; $\cos(\omega \cdot \tau_\Delta) = -\frac{(P_R \cdot Q_R + P_I \cdot Q_I)}{|Q|^2}$ $|Q|^2 = Q_R^2 + Q_I^2$. Finally we plot the stability switch diagram $g(\tau_\Delta) = \Lambda^{-1}(\tau_\Delta) = (\frac{\partial \mathrm{Re}\lambda}{\partial \tau_\Delta})_{\lambda = i\omega}$.

$$g(\tau_\Delta) = \Lambda^{-1}(\tau_\Delta) = \left(\frac{\partial \mathrm{Re}\lambda}{\partial \tau_\Delta}\right)_{\lambda = i\omega} = \frac{2 \cdot \{F_\omega \cdot (V + \omega \cdot P^2) - F_{\tau_\Delta} \cdot (U + \tau_\Delta \cdot P^2)\}}{F_{\tau_\Delta}^2 + 4 \cdot (V + \omega \cdot P^2)^2}$$

$$\operatorname{sign}[g(\tau_\Delta)] = \operatorname{sign}[\Lambda^{-1}(\tau_\Delta)] = \operatorname{sign}[\left(\frac{\partial \mathrm{Re}\lambda}{\partial \tau_\Delta}\right)_{\lambda = i\omega}] = \operatorname{sign}\left[\frac{2 \cdot \{F_\omega \cdot (V + \omega \cdot P^2) - F_{\tau_\Delta} \cdot (U + \tau_\Delta \cdot P^2)\}}{F_{\tau_\Delta}^2 + 4 \cdot (V + \omega \cdot P^2)^2}\right]$$

$$F_{\tau_\Delta}^2 + 4 \cdot (V + \omega \cdot P^2)^2 > 0 \Rightarrow \operatorname{sign}[\Lambda^{-1}(\tau_\Delta)] = \operatorname{sign}[F_\omega \cdot (V + \omega \cdot P^2) - F_{\tau_\Delta} \cdot (U + \tau_\Delta \cdot P^2)]$$

$$\operatorname{sign}[\Lambda^{-1}(\tau_\Delta)] = \operatorname{sign}\left\{[F_\omega] \cdot \left[(V + \omega \cdot P^2) - \frac{F_{\tau_\Delta}}{F_\omega} \cdot (U + \tau_\Delta \cdot P^2)\right]\right\}$$

$$\omega_{\tau_\Delta} = -\frac{F_{\tau_\Delta}}{F_\omega};\ \omega_{\tau_\Delta} = \left(\frac{\partial \omega}{\partial \tau_\Delta}\right)^{-1} = -\frac{\partial F / \partial \omega}{\partial F / \partial \tau_\Delta}$$

$$\operatorname{sign}[\Lambda^{-1}(\tau_\Delta)] = \operatorname{sign}\left\{[F_\omega] \cdot [P^2] \cdot \left[\frac{V + \omega_{\tau_\Delta} \cdot U}{P^2} + \omega + \omega_{\tau_\Delta} \cdot \tau_\Delta\right]\right\}$$

$$\operatorname{sign}[P^2] > 0$$

$$\operatorname{sign}[\Lambda^{-1}(\tau_\Delta)] = \operatorname{sign}\left\{[F_\omega] \cdot \left[\frac{V + \omega_{\tau_\Delta} \cdot U}{P^2} + \omega + \omega_{\tau_\Delta} \cdot \tau_\Delta\right]\right\}$$

$$\operatorname{sign}[\Lambda^{-1}(\tau_\Delta)] = \operatorname{sign}[F_\omega] \cdot \operatorname{sign}\left[\frac{V + \omega_{\tau_\Delta} \cdot U}{P^2} + \omega + \omega_{\tau_\Delta} \cdot \tau_\Delta\right]$$

Table 7.7 Dual-wavelength Ti:Sapphire laser system sign of $\text{sign}[\Lambda^{-1}(\tau_\Delta)]$

$\text{sign}[F_\omega]$	$\text{sign}\left[\frac{V+\omega_{\tau_\Delta}\cdot U}{P^2}+\omega+\omega_{\tau_\Delta}\cdot\tau_\Delta\right]$	$\text{sign}[\Lambda^{-1}(\tau_\Delta)]$
$\pm$	$\pm$	+
$\pm$	$\mp$	−

$$F_\omega = 2\cdot[(P_{R\omega}\cdot P_R + P_{I\omega}\cdot P_I) - (Q_{R\omega}\cdot Q_R + Q_{I\omega}\cdot Q_I)] = \sum_{k=0}^{4} B_{2\cdot k+1}\cdot\omega^{2\cdot k+1}$$

We check the sign of $\Lambda^{-1}(\tau_\Delta)$ according to the following rule (Table 7.7).

If, $\text{sign}[\Lambda^{-1}(\tau_\Delta)] > 0$, then the crossing proceeds from (−) to (+) respectively (stable to unstable). If, $\text{sign}[\Lambda^{-1}(\tau_\Delta)] < 0$, then the crossing proceeds from (+) to (−) respectively (unstable to stable).

7.2 Dual-Wavelength Emission from Vertical External-Cavity Surface-Emitting Laser Stability Analysis Under Delay Parameters in Time

A vertical external-cavity surface-emitting laser (VECSEL) is a semiconductor laser based on a surface emitting semiconductor gain chip and a laser resonator. VECSELs generate high optical power in diffraction limited beams and have wavelength versatility. VECSEL unit is constructed from a semiconductor gain chip and an external laser resonator and additionally there are arrangements for pumping and cooling. The semiconductor device, gain chip contains a single semiconductor Bragg mirror and the active/gain region with several quantum wells (QWs). The semiconductor structure total thickness is around few micrometers and it is amounted on heat sink substrate. The laser resonator and external mirror is one functional VECSEL unit. The semiconductor chip laser mode size is defined by the external resonator. The external resonator is folded with flat or curved mirrors and additional optical elements. The additional optical elements can be optical filters (single frequency operation, wavelength tuning), a nonlinear crystal (intra-cavity frequency doubling), saturable absorber (passive mode locking). The VECSEL resonator can be monolithic with a micro lens. The monolithic resonator is being contacted with the gain chip on one side and has an output coupler mirror coating on the other surface [4]. The schematic structure of a VECSEL with a semiconductor gain chip and an external laser resonator is described (Fig. 7.3).

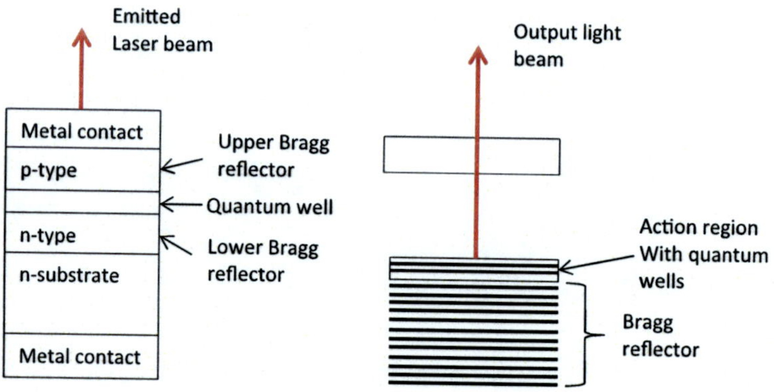

Fig. 7.3 Schematic structure of a VECSEL with a semiconductor gain chip and an external laser resonator

A VECSEL thin film Bragg reflector (Bragg mirror) is constructed from a multilayer-stack of alternate high-index and low-index films, which characterized by one quarter wavelength thick. The geometrical structure thickness of the high index and low index films are $t_H = \frac{\lambda}{4 \cdot n_H}; t_L = \frac{\lambda}{4 \cdot n_L}$, respectively. n_H and n_L are the indices of refraction of high and low index films, respectively and λ is the center wavelength of the Bragg mirror. On every interface in the stack a part of the incident beam is reflected. There is a phase shift of 180° which related to reflected parts if the incident light goes from low index medium in a high index medium. The relative phase difference of all reflected beams is zero or a multiple of 360° then they interfere constructively. The intensity of the incident light beam is decreases during his travel through the quarter-wave stack and at the same time the reflected light increases, if the absorptance A of the stack is negligible (Fig. 7.4).

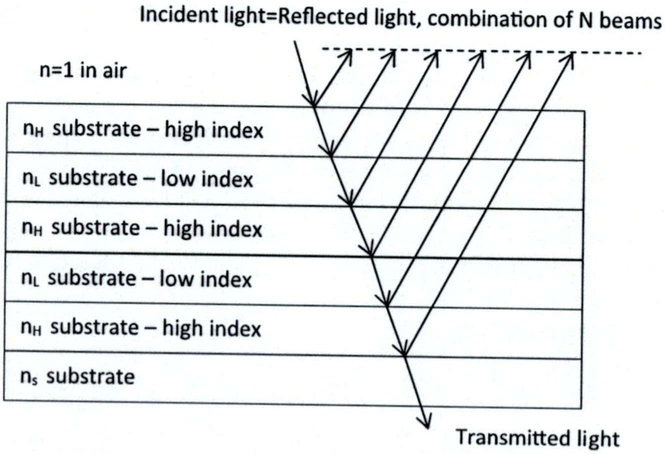

Fig. 7.4 VECSEL thin film Bragg reflector with multilayer-stack

The lasing occurs in VECSEL quantum well. Quantum well is hetero structure in which a thin layer of one semiconductor is sandwiched between two layers of a different semiconductor material by forming a heterojunction. The demand is that the two semiconductors have different energy gaps and different refractive indices (VECSEL). We choose the material those electrons available for conduction in the middle layer has lower energy than those in the outer layers, creating energy well (dip) that confines the electrons in the middle layer. Differences in the energy gap permit spatial confinement of electrons and holes injected into the middle layer. A quantum well laser is a laser diode unit in which the active region of the device is narrow that quantum confinement occurs. The wavelength of the light emitted by a quantum well laser is determined by the thickness of the active region rather than the bandgap of the material. Shorter wavelengths are obtained from quantum well lasers compare to conventional laser diode by using a particular semiconductor material. The efficiency of a quantum well laser is greater than a conventional laser diode due to the stepwise form of its density of states function. A VECSEL can be designed for simultaneous dual wavelength emission. The operation mechanism is inhomogeneous optical pumping of the non-identical quantum wells of the active region. The simultaneous dual-wavelength emission VECSEL is characterized by rate equations and exhibits a strong time-delayed feedback. VECSEL tendency to self-pulsation due to Q-switching at high pump power is avoidable by accurate modeling and epitaxial growth of the active region of the laser. VECSELs have features as circular non-diffraction limited beams, high output power, and intra cavity nonlinear frequency conversion. Many industrial applications need laser source which emits coaxial beam at two different wavelengths. The two-color laser is applied to dual-wavelength interferometry, and generation of middle infrared and far-infrared radiation. Dual-wavelength lasing source systems utilize an array of lasers with physically separated gain media for each wavelength. We can get dual-wavelength emission with a wide-band semiconductor amplifier and with a super luminescent diode located inside an external cavity containing optical grating and a slit mirror. The term "optical grating" is synonym with diffraction grating or multiple slit. Optical gratings are periodical structures for diffraction of light. Other relevant devices are characterized by vertical coupled-cavity geometry. Simultaneous dual-wavelength emissions from optically pumped VECSEL can gives a single transverse mode continues wave (CW) at specific two different wavelengths. The simultaneous dual-wavelength VECSEL demonstrated an inhomogeneous pumping of the active region containing non-identical quantum wells (QWs). A single-cavity quantum well (QW) hetero structure is designed for dual-wavelength emission. The active region is divided into three sections by blocking layers. We enhance the achievable gain coefficient by placing the QWs at the antinodes of their own standing waves. Quantum well L and S, $QW_{L,S}$ are located corresponds to the standing wave at $\lambda_{L,S}$. $QW_{L,S}$ are near the nodes of the cavity mode of λ_S to minimize optical absorption of short-wavelength emission in the deeper QWs. The blocking layers are constructed from wide bandgap AlAs, are transparent to the pump light and laser emission, but impenetrable for charge carriers. The distributed Bragg reflector (DBR), made of alternating GaAs and AlAs layers, and the window layer is the same of a conventional VECSEL at these wavelengths. The field distribution and geometry of the device are accepted by solving the boundary eigenvalue problem for the Helmholts equation within the one-dimensional transfer matrix. The VECSEL pump beam is incident to the absorber region through the output window layer. A one-dimensional diffusion equation is solved with the assumption of no electric bias applied to VECSEL, and then there is no carrier drift contribution. The carrier distribution in the absorber region is characterized by the second order differential equation: $n = n(x)$, $D_a \cdot \frac{d^2 n}{dx^2} - \frac{n}{\tau} + G = 0$; where n is the carrier density, D_a is the ambipolar diffusion coefficient, and τ is the ambipolar carrier lifetime. It is related to the diffusion length by $L_a = \sqrt{D_a \cdot \tau}$, due to spontaneous recombination. The carrier generation rate is $G = \frac{P_{in}}{h \cdot c} \cdot \frac{\alpha \cdot \lambda_p}{\pi \cdot r^2} \cdot e^{-\alpha \cdot x} = G_0 \cdot e^{-\alpha \cdot x}$; $G_0 = \frac{P_{in}}{h \cdot c} \cdot \frac{\alpha \cdot \lambda_p}{\pi \cdot r^2}$; where P_{in} is the pump power at the onset (x = 0) of the absorber region, r is the spot radius, λ_p is the pump wavelength, α is the absorption coefficient, h is Planck's constant, and c is the speed of light in vacuum. The general solution we get is $n(X) = C_1 \cdot \sinh(X) + C_2 \cdot \cosh(X) + n_G \cdot e^{-a \cdot X}$, where $X = \frac{x}{L_a}$; $a = \alpha \cdot L_a$. And The general solution is consistent within a barrier region between two quantum wells, QWs or between a QW and a blocking layer [4]. We can characterize dual-wavelength VECSEL by rate equations. The laser consists of two cavities, a sub-cavity with a low reflectivity mirror (close to zero) and an external cavity with a high feedback mirror (close to one). The sub-cavity is formed between the semiconductor DBR (Distributed Bragg Reflector) and the top of the diamond heat spreader in spite of its Anti-Reflect (AR) coated surface. By inspecting the rate equations for dual-wavelength emission, we reduce the number of rate equations. It is done by the assumption that the gain coefficient of the quantum wells (QWs) of each section is lumped to a single "equivalent" quantum well (QW) per section. Parameter n_{QW} in the QWs within each section is constant with high probably. The outcome is that the optical field confinement factor of the equivalent QW equals the number of the QWs in that section times the confinement factor of one QW. The whole active region contains three equivalents QWs. Minority (very small) influence is related to the contributions of spontaneous emission to the lasing modes and effects or random generation-recombination process. The rate equations for dual-wavelength emission are:

$$\frac{dS_1}{dt} = v_g \cdot \left[\sum_{j=1}^{3} \Gamma_{1j} \cdot g_{1j} - \alpha_{in} + \frac{1}{L_{in}} \cdot \ln(r_{DBR}(\lambda_S) \cdot |r_{ef}(\lambda_S)|) \right] \cdot S_1$$

$$\frac{dS_2}{dt} = v_g \cdot \left[\sum_{j=1}^{3} \Gamma_{2j} \cdot g_{2j} - \alpha_{in} + \frac{1}{L_{in}} \cdot \ln(r_{DBR}(\lambda_L) \cdot |r_{ef}(\lambda_L)|) \right] \cdot S_2$$

S_1 and S_2 are the photon densities at wavelength λ_S and λ_L, respectively, Γ_{ij} is the confinement factor of the ith optical field (i = 1 for λ_S and i = 2 for λ_L) in the jth equivalent QW, g_{ij} is the gain coefficient for the jth QW at λ_S and λ_L, α_{in} is the loss factor inside the sub cavity, v_g is the group velocity of optical fields, L_{in} is the sub cavity length, r_{DBR} ($r_{DBR}(\lambda_S), r_{DBR}(\lambda_L)$) is the reflectivity of the DBR mirror, and r_{ef} ($r_{ef}(\lambda_S), r_{ef}(\lambda_L)$) is the effective reflectivity of the external mirror and defined as

$$r_{ef} = r_f - r_{ext} \cdot (1 - r_f^2) \cdot \sum_{m=1}^{\infty} (r_f \cdot r_{ext})^{m-1} \cdot \sqrt{\frac{S(t - m \cdot \tau_{ext})}{S(t)}} \cdot e^{-j \cdot \theta_m}$$

r_f is the sub cavity facet reflectivity and r_{ext} is the external mirror reflectivity, τ_{ext} is the round-trip time in the external cavity, and θ_m, $\theta_m = m \cdot \omega \cdot \tau_{ext} + \varphi(t) - \varphi(t - m \cdot \tau_{ext})$, where ω is the angular frequency and φ is the time-varying phase deviation from the steady state of the solitary laser [4]. The rate equations for dual-wavelength emission with high feedback reflectivity $\left(\chi = r_{ext} \cdot \frac{(1-r_f^2)}{r_f} \gg 1\right)$, are

$$\frac{dS_1}{dt} = v_g \cdot \left[\sum_{j=1}^{3} \Gamma_{1j} \cdot g_{1j} - \alpha_{s1} + \frac{1}{2 \cdot L_{in}} \cdot \ln\left(\frac{S_1(t - \tau_{ext})}{S_1(t)} \right) \right] \cdot S_1(t);\ S_1 = S_1(t)$$

$$\frac{dS_2}{dt} = v_g \cdot \left[\sum_{j=1}^{3} \Gamma_{2j} \cdot g_{2j} - \alpha_{s2} + \frac{1}{2 \cdot L_{in}} \cdot \ln\left(\frac{S_2(t - \tau_{ext})}{S_2(t)} \right) \right] \cdot S_2(t);\ S_2 = S_2(t)$$

α_{s1} and α_{s2} are the total loss coefficients, $\alpha_{s1,s2} = \alpha_{in} - (\frac{1}{L_{in}}) \cdot \ln[r_{ext} \cdot (1 - r_f^2) \cdot r_{DBR}(\lambda_{S,L})]$.

At fixed points $\frac{dS_1}{dt} = 0$; $\frac{dS_2}{dt} = 0$; $lim_{t\to\infty} S_1(t - \tau_{ext}) = S_1(t)$; $\lim_{t\to\infty} S_2(t - \tau_{ext}) = S_2(t)\ \forall\, t \gg \tau_{ext}$.

$$v_g \cdot \left[\sum_{j=1}^{3} \Gamma_{1j} \cdot g_{1j} - \alpha_{s1} + \frac{1}{2 \cdot L_{in}} \cdot \ln\left(\frac{S_1^*}{S_1^*} \right) \right] \cdot S_1^* = 0$$

$$v_g \neq 0;\ \ln\left(\frac{S_1^*}{S_1^*} \right) = \ln(1) = 0$$

$$v_g \cdot \left[\sum_{j=1}^{3} \Gamma_{2j} \cdot g_{2j} - \alpha_{s2} + \frac{1}{2 \cdot L_{in}} \cdot \ln\left(\frac{S_2^*}{S_2^*} \right) \right] \cdot S_2^* = 0$$

$$v_g \neq 0;\ \ln\left(\frac{S_2^*}{S_2^*} \right) = \ln(1) = 0$$

We get the system fixed points equations:

$$\left[\sum_{j=1}^{3} \Gamma_{1j} \cdot g_{1j} - \alpha_{s1} \right] \cdot S_1^* = 0 \Rightarrow \sum_{j=1}^{3} \Gamma_{1j} \cdot g_{1j} \neq \alpha_{s1};\ S_1^* = 0$$

$$\left[\sum_{j=1}^{3}\Gamma_{2j}\cdot g_{2j}-\alpha_{s2}\right]\cdot S_2^*=0\Rightarrow\sum_{j=1}^{3}\Gamma_{2j}\cdot g_{2j}\neq\alpha_{s2};\ S_2^*=0$$

Remark: Special case is when $\sum_{j=1}^{3}\Gamma_{1j}\cdot g_{1j}=\alpha_{s1}$ then S_1^* can be any value, $S_1^*>0$; $S_1^*\in\mathbb{R}$ or/and $\sum_{j=1}^{3}\Gamma_{2j}\cdot g_{2j}=\alpha_{s2}$ then S_2^* can be any value, $S_2^*>0$; $S_2^*\in\mathbb{R}$.

Dual-wavelength emission VECSEL system fixed point: $E^{(*)}=(S_1^*,S_2^*)=(0,0)$.

Assumption: τ_{ext} is the round-trip time in the external cavity is not the same for the photon densities S_1 and S_2. We consider τ_{ext-1} and τ_{ext-2} for photon densities S_1 and S_2, respectively. The rate equations for dual-wavelength emission with high feedback reflectivity $\left(\chi=r_{ext-1,2}\cdot\frac{(1-r_f^2)}{r_f}\gg 1\right)$, are

$$\frac{dS_1}{dt}=v_g\cdot\left[\sum_{j=1}^{3}\Gamma_{1j}\cdot g_{1j}-\alpha_{s1}+\frac{1}{2\cdot L_{in}}\cdot\ln\left(\frac{S_1(t-\tau_{ext-1})}{S_1(t)}\right)\right]\cdot S_1(t)$$
$$S_1=S_1(t)$$

$$\frac{dS_2}{dt}=v_g\cdot\left[\sum_{j=1}^{3}\Gamma_{2j}\cdot g_{2j}-\alpha_{s2}+\frac{1}{2\cdot L_{in}}\cdot\ln\left(\frac{S_2(t-\tau_{ext-2})}{S_2(t)}\right)\right]\cdot S_2(t)$$
$$S_2=S_2(t)$$

Stability analysis: The standard local stability analysis about any one of the equilibrium point of the dual-wavelength emission VECSEL system consists in adding to coordinate $[S_1,S_2]$ arbitrarily small increments of exponential form $[s_1,s_2]\cdot e^{\lambda\cdot t}$ and retaining the first order terms in S_1,S_2. The system of two homogeneous equations leads to a polynomial characteristic equation in the eigenvalues. The polynomial characteristic equations accept by set the below variables and variables derivative with respect to time into dual-wavelength emission VECSEL system rate equations (DDEs). The dual-wavelength emission VECSEL system equations fixed values with arbitrarily small increments of exponential form $[s_1,s_2]\cdot e^{\lambda\cdot t}$ are: $j=0$ (first fixed point), $j=1$ (second fixed point), $j=2$ (third fixed point), etc. [2, 3].

$$S_1(t)=S_1^{(j)}+s_1\cdot e^{\lambda\cdot t};\ S_2(t)=S_2^{(j)}+s_2\cdot e^{\lambda\cdot t}$$
$$S_1(t-\tau_{ext-1})=S_1^{(j)}+s_1\cdot e^{\lambda\cdot(t-\tau_{ext-1})}$$

$$S_2(t-\tau_{ext-2}) = S_2^{(j)} + s_2 \cdot e^{\lambda\cdot(t-\tau_{ext-2})};\ \frac{dS_1(t)}{dt} = s_1 \cdot \lambda \cdot e^{\lambda\cdot t}$$

$$\frac{dS_2(t)}{dt} = s_2 \cdot \lambda \cdot e^{\lambda\cdot t}$$

We choose these expressions for ourselves $S_1(t), S_2(t)$ as a small displacement $[s_1, s_2]$ from the dual-wavelength emission VECSEL system rate equations fixed points in time $t = 0$, $S_1(t=0) = S_1^{(j)} + s_1$; $S_2(t=0) = S_2^{(j)} + s_2$. There are four possible cases: (1) $\tau_{ext-1} = 0, \tau_{ext-2} = 0$; (2) $\tau_{ext-1} = \tau, \tau_{ext-2} = 0$; (3) $\tau_{ext-1} = 0, \tau_{ext-2} = \tau$, (4) $\tau_{ext-1} = \tau, \tau_{ext-2} = \tau$. We analyze the second case, where $\tau_{ext-1} = \tau, \tau_{ext-2} = 0$.

The rate equations for dual-wavelength emission with high feedback reflectivity $\left(\chi = r_{ext-1,2} \cdot \frac{(1-r_f^2)}{r_f} \gg 1\right)$, are

$$\frac{dS_1}{dt} = v_g \cdot \left[\sum_{j=1}^{3} \Gamma_{1j} \cdot g_{1j} - \alpha_{s1} + \frac{1}{2 \cdot L_{in}} \cdot \ln\left(\frac{S_1(t-\tau)}{S_1(t)}\right)\right] \cdot S_1(t);\ S_1 = S_1(t)$$

$$\frac{dS_2}{dt} = v_g \cdot \left[\sum_{j=1}^{3} \Gamma_{2j} \cdot g_{2j} - \alpha_{s2}\right] \cdot S_2(t);\ S_2 = S_2(t)$$

Stability analysis:

$$\frac{dS_1}{dt} = v_g \cdot \left[\sum_{j=1}^{3} \Gamma_{1j} \cdot g_{1j} - \alpha_{s1} + \frac{1}{2 \cdot L_{in}} \cdot \ln\left(\frac{S_1(t-\tau)}{S_1(t)}\right)\right] \cdot S_1(t)$$

$$s_1 \cdot \lambda \cdot e^{\lambda\cdot t} = v_g \cdot \left[\sum_{j=1}^{3} \Gamma_{1j} \cdot g_{1j} - \alpha_{s1} + \frac{1}{2 \cdot L_{in}} \cdot \ln\left(\frac{S_1^{(j)} + s_1 \cdot e^{\lambda\cdot(t-\tau)}}{S_1^{(j)} + s_1 \cdot e^{\lambda\cdot t}}\right)\right] \cdot (S_1^{(j)} + s_1 \cdot e^{\lambda\cdot t})$$

$$\frac{S_1^{(j)} + s_1 \cdot e^{\lambda\cdot(t-\tau)}}{S_1^{(j)} + s_1 \cdot e^{\lambda\cdot t}} = \left(\frac{S_1^{(j)} + s_1 \cdot e^{\lambda\cdot(t-\tau)}}{S_1^{(j)} + s_1 \cdot e^{\lambda\cdot t}}\right) \cdot \left(\frac{S_1^{(j)} - s_1 \cdot e^{\lambda\cdot t}}{S_1^{(j)} - s_1 \cdot e^{\lambda\cdot t}}\right)$$

$$= \frac{\{[S_1^{(j)}]^2 - S_1^{(j)} \cdot s_1 \cdot e^{\lambda\cdot t} + s_1 \cdot e^{\lambda\cdot(t-\tau)} \cdot S_1^{(j)} - s_1^2 \cdot e^{\lambda\cdot(t-\tau)} \cdot e^{\lambda\cdot t}\}}{[S_1^{(j)}]^2 - s_1^2 \cdot e^{2\cdot\lambda\cdot t}}$$

Assumption: $s_1^2 \approx 0$ then $\frac{S_1^{(j)} + s_1 \cdot e^{\lambda\cdot(t-\tau)}}{S_1^{(j)} + s_1 \cdot e^{\lambda\cdot t}} = \frac{[S_1^{(j)}]^2 - S_1^{(j)} \cdot s_1 \cdot e^{\lambda\cdot t} + s_1 \cdot e^{\lambda\cdot(t-\tau)} \cdot S_1^{(j)}}{[S_1^{(j)}]^2}$

$$\frac{S_1^{(j)} + s_1 \cdot e^{\lambda \cdot (t-\tau)}}{S_1^{(j)} + s_1 \cdot e^{\lambda \cdot t}} = \frac{[S_1^{(j)}]^2 - S_1^{(j)} \cdot s_1 \cdot e^{\lambda \cdot t} + s_1 \cdot e^{\lambda \cdot (t-\tau)} \cdot S_1^{(j)}}{[S_1^{(j)}]^2}$$
$$= 1 - \frac{1}{S_1^{(j)}} \cdot s_1 \cdot e^{\lambda \cdot t} + \frac{1}{S_1^{(j)}} \cdot s_1 \cdot e^{\lambda \cdot (t-\tau)}$$

$$s_1 \cdot \lambda \cdot e^{\lambda \cdot t} = v_g \cdot \left[\sum_{j=1}^{3} \Gamma_{1j} \cdot g_{1j} - \alpha_{s1} + \frac{1}{2 \cdot L_{in}} \cdot \ln\left(1 - \frac{1}{S_1^{(j)}} \cdot s_1 \cdot e^{\lambda \cdot t} + \frac{1}{S_1^{(j)}} \cdot s_1 \cdot e^{\lambda \cdot (t-\tau)}\right)\right] \cdot (S_1^{(j)} + s_1 \cdot e^{\lambda \cdot t})$$

$$s_1 \cdot \lambda \cdot e^{\lambda \cdot t} = v_g \cdot \left[\sum_{j=1}^{3} \Gamma_{1j} \cdot g_{1j} - \alpha_{s1} \right] \cdot S_1^{(j)} + \left[v_g \cdot \frac{1}{2 \cdot L_{in}} \cdot \ln\left(1 - \frac{1}{S_1^{(j)}} \cdot s_1 \cdot e^{\lambda \cdot t} + \frac{1}{S_1^{(j)}} \cdot s_1 \cdot e^{\lambda \cdot (t-\tau)}\right)\right] \cdot S_1^{(j)} + v_g \cdot \left[\sum_{j=1}^{3} \Gamma_{1j} \cdot g_{1j} - \alpha_{s1} \right] \cdot s_1 \cdot e^{\lambda \cdot t} + \left[v_g \cdot \frac{1}{2 \cdot L_{in}} \cdot \ln\left(1 - \frac{1}{S_1^{(j)}} \cdot s_1 \cdot e^{\lambda \cdot t} + \frac{1}{S_1^{(j)}} \cdot s_1 \cdot e^{\lambda \cdot (t-\tau)}\right)\right] \cdot s_1 \cdot e^{\lambda \cdot t}$$

At fixed points: $\frac{dS_1}{dt} = 0 \Rightarrow v_g \cdot [\sum_{j=1}^{3} \Gamma_{1j} \cdot g_{1j} - \alpha_{s1} + \frac{1}{2 \cdot L_{in}} \cdot \ln\left(\frac{S_1^*}{S_1^*}\right)] \cdot S_1^* = 0; \ln\left(\frac{S_1^*}{S_1^*}\right) = 0.$

$$\frac{dS_1}{dt} = 0 \Rightarrow v_g \cdot \left[\sum_{j=1}^{3} \Gamma_{1j} \cdot g_{1j} - \alpha_{s1} \right] \cdot S_1^* = 0 \Rightarrow v_g \cdot \left[\sum_{j=1}^{3} \Gamma_{1j} \cdot g_{1j} - \alpha_{s1} \right] \cdot S_1^{(j)}$$

$$s_1 \cdot \lambda \cdot e^{\lambda \cdot t} = \left[v_g \cdot \frac{1}{2 \cdot L_{in}} \cdot \ln\left(1 - \frac{1}{S_1^{(j)}} \cdot s_1 \cdot e^{\lambda \cdot t} + \frac{1}{S_1^{(j)}} \cdot s_1 \cdot e^{\lambda \cdot (t-\tau)}\right)\right] \cdot S_1^{(j)} + v_g \cdot \left[\sum_{j=1}^{3} \Gamma_{1j} \cdot g_{1j} - \alpha_{s1} \right] \cdot s_1 \cdot e^{\lambda \cdot t} + \left[v_g \cdot \frac{1}{2 \cdot L_{in}} \cdot \ln\left(1 - \frac{1}{S_1^{(j)}} \cdot s_1 \cdot e^{\lambda \cdot t} + \frac{1}{S_1^{(j)}} \cdot s_1 \cdot e^{\lambda \cdot (t-\tau)}\right)\right] \cdot s_1 \cdot e^{\lambda \cdot t}$$

$$A = \frac{1}{S_1^{(j)}} \cdot s_1 \cdot e^{\lambda \cdot t}; \; B = \frac{1}{S_1^{(j)}} \cdot s_1 \cdot e^{\lambda \cdot (t-\tau)}$$

$$\ln\left(1 - \frac{1}{S_1^{(j)}} \cdot s_1 \cdot e^{\lambda \cdot t} + \frac{1}{S_1^{(j)}} \cdot s_1 \cdot e^{\lambda \cdot (t-\tau)}\right) = \ln([1 - A] + B)$$

$$\ln([1 - A] + B) = \ln[1 - A] + \ln\left[1 + \frac{B}{1 - A}\right]$$

$$\ln[1 - A] = \ln\left[1 - \frac{1}{S_1^{(j)}} \cdot s_1 \cdot e^{\lambda \cdot t}\right]$$

$$\ln\left[1 + \frac{B}{1 - A}\right] = \ln\left[1 + \frac{\frac{1}{S_1^{(j)}} \cdot s_1 \cdot e^{\lambda \cdot (t-\tau)}}{1 - \frac{1}{S_1^{(j)}} \cdot s_1 \cdot e^{\lambda \cdot t}}\right]$$

$$\ln[1 - A] = -\sum_{n=1}^{\infty} \frac{A^n}{n} = -A - \frac{A^2}{2} - \frac{A^3}{3} - \cdots$$

$$\ln[1 - A] = -\sum_{n=1}^{\infty} \frac{A^n}{n}$$

$$= -\frac{1}{S_1^{(j)}} \cdot s_1 \cdot e^{\lambda \cdot t} - \frac{[\frac{1}{S_1^{(j)}} \cdot s_1 \cdot e^{\lambda \cdot t}]^2}{2} - \frac{[\frac{1}{S_1^{(j)}} \cdot s_1 \cdot e^{\lambda \cdot t}]^3}{3} - \cdots$$

Assumption: $\forall\, k \geq 2; k \in \mathbb{N} \,\exists\, s_1^k \to 0$

$$\ln[1 - A] = -\sum_{n=1}^{\infty} \frac{A^n}{n} \approx -\frac{1}{S_1^{(j)}} \cdot s_1 \cdot e^{\lambda \cdot t}$$

$$\ln\left[1 + \frac{B}{1 - A}\right] = \sum_{n=1}^{\infty} (-1)^{n+1} \cdot \frac{1}{n} \cdot \left[\frac{B}{1 - A}\right]^n$$

$$= \frac{B}{1 - A} - \frac{1}{2} \cdot \left[\frac{B}{1 - A}\right]^2 + \frac{1}{3} \cdot \left[\frac{B}{1 - A}\right]^3 - \cdots$$

$$\frac{B}{1 - A} = \frac{\frac{1}{S_1^{(j)}} \cdot s_1 \cdot e^{\lambda \cdot (t-\tau)}}{1 - \frac{1}{S_1^{(j)}} \cdot s_1 \cdot e^{\lambda \cdot t}} = \left[\frac{\frac{1}{S_1^{(j)}} \cdot s_1 \cdot e^{\lambda \cdot (t-\tau)}}{1 - \frac{1}{S_1^{(j)}} \cdot s_1 \cdot e^{\lambda \cdot t}}\right] \cdot \left[\frac{1 + \frac{1}{S_1^{(j)}} \cdot s_1 \cdot e^{\lambda \cdot t}}{1 + \frac{1}{S_1^{(j)}} \cdot s_1 \cdot e^{\lambda \cdot t}}\right]$$

$$= \frac{\frac{1}{S_1^{(j)}} \cdot s_1 \cdot e^{\lambda \cdot (t-\tau)} + \left(\frac{1}{S_1^{(j)}}\right)^2 \cdot s_1^2 \cdot e^{\lambda \cdot (t-\tau)} \cdot e^{\lambda \cdot t}}{1 - \left(\frac{1}{S_1^{(j)}}\right)^2 \cdot s_1^2 \cdot e^{2 \cdot \lambda \cdot t}}$$

Assumption: $s_1^2 \to 0$; $\frac{B}{1-A} \approx \frac{1}{S_1^{(j)}} \cdot s_1 \cdot e^{\lambda \cdot (t-\tau)}$; $\forall\, k \geq 2; k \in \mathbb{N} \,\exists\, s_1^k \to 0$ then $\left[\frac{B}{1-A}\right]^k \to 0$

$$-\frac{1}{2}\cdot\left[\frac{B}{1-A}\right]^2+\frac{1}{3}\cdot\left[\frac{B}{1-A}\right]^3-\cdots\to 0$$

$$\ln\left[1+\frac{B}{1-A}\right]\approx\frac{1}{S_1^{(j)}}\cdot s_1\cdot e^{\lambda\cdot(t-\tau)}$$

$$\ln[1-A]\approx-\frac{1}{S_1^{(j)}}\cdot s_1\cdot e^{\lambda\cdot t}$$

$$\begin{aligned}\ln\left(1-\frac{1}{S_1^{(j)}}\cdot s_1\cdot e^{\lambda\cdot t}+\frac{1}{S_1^{(j)}}\cdot s_1\cdot e^{\lambda\cdot(t-\tau)}\right)&=\ln([1-A]+B)\\&=\ln[1-A]+\ln\left[1+\frac{B}{1-A}\right]\\&\approx-\frac{1}{S_1^{(j)}}\cdot s_1\cdot e^{\lambda\cdot t}+\frac{1}{S_1^{(j)}}\cdot s_1\cdot e^{\lambda\cdot(t-\tau)}\\&\approx\frac{1}{S_1^{(j)}}\cdot s_1\cdot e^{\lambda\cdot t}\cdot(e^{-\lambda\cdot\tau}-1)\end{aligned}$$

Back to the original expression:

$$\begin{aligned}s_1\cdot\lambda\cdot e^{\lambda\cdot t}=&\left[v_g\cdot\frac{1}{2\cdot L_{in}}\cdot\ln\left(1-\frac{1}{S_1^{(j)}}\cdot s_1\cdot e^{\lambda\cdot t}+\frac{1}{S_1^{(j)}}\cdot s_1\cdot e^{\lambda\cdot(t-\tau)}\right)\right]\cdot S_1^{(j)}\\&+v_g\cdot\left[\sum_{j=1}^{3}\Gamma_{1j}\cdot g_{1j}-\alpha_{s1}\right]\cdot s_1\cdot e^{\lambda\cdot t}\\&+\left[v_g\cdot\frac{1}{2\cdot L_{in}}\cdot\ln\left(1-\frac{1}{S_1^{(j)}}\cdot s_1\cdot e^{\lambda\cdot t}+\frac{1}{S_1^{(j)}}\cdot s_1\cdot e^{\lambda\cdot(t-\tau)}\right)\right]\cdot s_1\cdot e^{\lambda\cdot t}\end{aligned}$$

$$\begin{aligned}s_1\cdot\lambda\cdot e^{\lambda\cdot t}=&\,v_g\cdot\frac{1}{2\cdot L_{in}}\cdot\frac{1}{S_1^{(j)}}\cdot s_1\cdot e^{\lambda\cdot t}\cdot(e^{-\lambda\cdot\tau}-1)\cdot S_1^{(j)}\\&+v_g\cdot\left[\sum_{j=1}^{3}\Gamma_{1j}\cdot g_{1j}-\alpha_{s1}\right]\cdot s_1\cdot e^{\lambda\cdot t}\\&+v_g\cdot\frac{1}{2\cdot L_{in}}\cdot\frac{1}{S_1^{(j)}}\cdot e^{\lambda\cdot t}\cdot(e^{-\lambda\cdot\tau}-1)\cdot s_1^2\cdot e^{\lambda\cdot t}\end{aligned}$$

$$\begin{aligned}s_1^2\to 0\Rightarrow s_1\cdot\lambda=&\,v_g\cdot\frac{1}{2\cdot L_{in}}\cdot\frac{1}{S_1^{(j)}}\cdot s_1\cdot(e^{-\lambda\cdot\tau}-1)\cdot S_1^{(j)}\\&+v_g\cdot\left[\sum_{j=1}^{3}\Gamma_{1j}\cdot g_{1j}-\alpha_{s1}\right]\cdot s_1\end{aligned}$$

$$\left\{v_g \cdot \frac{1}{2 \cdot L_{in}} \cdot (e^{-\lambda \cdot \tau} - 1) + v_g \cdot \left[\left[\sum_{j=1}^{3} \Gamma_{1j} \cdot g_{1j} - \alpha_{s1}\right]\right]\right\} \cdot s_1 - s_1 \cdot \lambda = 0$$

$$\frac{dS_2}{dt} = v_g \cdot \left[\sum_{j=1}^{3} \Gamma_{2j} \cdot g_{2j} - \alpha_{s2}\right] \cdot S_2(t); \; S_2 = S_2(t)$$

$$s_2 \cdot \lambda \cdot e^{\lambda \cdot t} = v_g \cdot \left[\sum_{j=1}^{3} \Gamma_{2j} \cdot g_{2j} - \alpha_{s2}\right] \cdot (S_2^{(j)} + s_2 \cdot e^{\lambda \cdot t})$$

$$s_2 \cdot \lambda \cdot e^{\lambda \cdot t} = v_g \cdot \left[\sum_{j=1}^{3} \Gamma_{2j} \cdot g_{2j} - \alpha_{s2}\right] \cdot S_2^{(j)} + v_g \cdot \left[\sum_{j=1}^{3} \Gamma_{2j} \cdot g_{2j} - \alpha_{s2}\right] \cdot s_2 \cdot e^{\lambda \cdot t}$$

At fixed points: $v_g \cdot [\sum_{j=1}^{3} \Gamma_{2j} \cdot g_{2j} - \alpha_{s2}] \cdot S_2^{(j)} = 0$.

$$s_2 \cdot \lambda = v_g \cdot \left[\sum_{j=1}^{3} \Gamma_{2j} \cdot g_{2j} - \alpha_{s2}\right] \cdot s_2 \Rightarrow v_g \cdot \left[\sum_{j=1}^{3} \Gamma_{2j} \cdot g_{2j} - \alpha_{s2}\right] \cdot s_2 - s_2 \cdot \lambda = 0$$

The System arbitrarily small increments equations:

$$\left\{v_g \cdot \frac{1}{2 \cdot L_{in}} \cdot (e^{-\lambda \cdot \tau} - 1) + v_g \cdot \left[\left[\sum_{j=1}^{3} \Gamma_{1j} \cdot g_{1j} - \alpha_{s1}\right]\right]\right\} \cdot s_1 - s_1 \cdot \lambda = 0$$

$$v_g \cdot \left[\sum_{j=1}^{3} \Gamma_{2j} \cdot g_{2j} - \alpha_{s2}\right] \cdot s_2 - s_2 \cdot \lambda = 0$$

The small increments Jacobian of our dual-wavelength emission VECSEL system is as follow:

$$\begin{pmatrix} v_g \cdot \frac{1}{2 \cdot L_{in}} \cdot (e^{-\lambda \cdot \tau} - 1) + v_g \cdot \left[\sum_{j=1}^{3} \Gamma_{1j} \cdot g_{1j} - \alpha_{s1} \right] - \lambda & 0 \\ 0 & v_g \cdot \left[\sum_{j=1}^{3} \Gamma_{2j} \cdot g_{2j} - \alpha_{s2} \right] - \lambda \end{pmatrix} \cdot \begin{pmatrix} s_1 \\ s_2 \end{pmatrix} = 0$$

$$A - \lambda \cdot I = \begin{pmatrix} v_g \cdot \frac{1}{2 \cdot L_{in}} \cdot (e^{-\lambda \cdot \tau} - 1) + v_g \cdot \left[\sum_{j=1}^{3} \Gamma_{1j} \cdot g_{1j} - \alpha_{s1} \right] - \lambda & 0 \\ 0 & v_g \cdot \left[\sum_{j=1}^{3} \Gamma_{2j} \cdot g_{2j} - \alpha_{s2} \right] - \lambda \end{pmatrix}$$

$$\det(A - \lambda \cdot I) = \det \begin{pmatrix} v_g \cdot \frac{1}{2 \cdot L_{in}} \cdot (e^{-\lambda \cdot \tau} - 1) + v_g \cdot \left[\sum_{j=1}^{3} \Gamma_{1j} \cdot g_{1j} - \alpha_{s1} \right] - \lambda & 0 \\ 0 & v_g \cdot \left[\sum_{j=1}^{3} \Gamma_{2j} \cdot g_{2j} - \alpha_{s2} \right] - \lambda \end{pmatrix}$$

$$\det(A - \lambda \cdot I) = 0 \Rightarrow \left\{ v_g \cdot \frac{1}{2 \cdot L_{in}} \cdot (e^{-\lambda \cdot \tau} - 1) + v_g \cdot \left[\sum_{j=1}^{3} \Gamma_{1j} \cdot g_{1j} - \alpha_{s1} \right] - \lambda \right\} \cdot \left\{ v_g \cdot \left[\sum_{j=1}^{3} \Gamma_{2j} \cdot g_{2j} - \alpha_{s2} \right] - \lambda \right\} = 0$$

$$\left\{ v_g \cdot \frac{1}{2 \cdot L_{in}} \cdot (e^{-\lambda \cdot \tau} - 1) + v_g \cdot \left[\sum_{j=1}^{3} \Gamma_{1j} \cdot g_{1j} - \alpha_{s1} \right] - \lambda \right\} \cdot \left\{ v_g \cdot \left[\sum_{j=1}^{3} \Gamma_{2j} \cdot g_{2j} - \alpha_{s2} \right] - \lambda \right\}$$
$$= v_g \cdot \frac{1}{2 \cdot L_{in}} \cdot (e^{-\lambda \cdot \tau} - 1) \cdot v_g \cdot \left[\sum_{j=1}^{3} \Gamma_{2j} \cdot g_{2j} - \alpha_{s2} \right] - v_g \cdot \frac{1}{2 \cdot L_{in}} \cdot (e^{-\lambda \cdot \tau} - 1) \cdot \lambda$$

$$+ v_g \cdot \left[\sum_{j=1}^{3} \Gamma_{1j} \cdot g_{1j} - \alpha_{s1} \right] \cdot v_g \cdot \left[\sum_{j=1}^{3} \Gamma_{2j} \cdot g_{2j} - \alpha_{s2} \right]$$

$$- v_g \cdot \left[\sum_{j=1}^{3} \Gamma_{1j} \cdot g_{1j} - \alpha_{s1} \right] \cdot \lambda$$

$$- v_g \cdot \left[\sum_{j=1}^{3} \Gamma_{2j} \cdot g_{2j} - \alpha_{s2} \right] \cdot \lambda + \lambda^2 = 0$$

$$\left\{ v_g^2 \cdot \frac{1}{2 \cdot L_{in}} \cdot \left[\sum_{j=1}^{3} \Gamma_{2j} \cdot g_{2j} - \alpha_{s2} \right] - v_g \cdot \frac{1}{2 \cdot L_{in}} \cdot \lambda \right\} \cdot e^{-\lambda \cdot \tau}$$

$$+ v_g^2 \cdot \left[\sum_{j=1}^{3} \Gamma_{1j} \cdot g_{1j} - \alpha_{s1} - \frac{1}{2 \cdot L_{in}} \right] \cdot \left[\sum_{j=1}^{3} \Gamma_{2j} \cdot g_{2j} - \alpha_{s2} \right]$$

$$+ v_g \cdot \left\{ \frac{1}{2 \cdot L_{in}} - \left[\sum_{j=1}^{3} \Gamma_{1j} \cdot g_{1j} - \alpha_{s1} \right] \right.$$

$$\left. - \left[\sum_{j=1}^{3} \Gamma_{2j} \cdot g_{2j} - \alpha_{s2} \right] \right\} \cdot \lambda + \lambda^2 = 0$$

We define global parameters for simplicity: $\Xi_0 = v_g^2 \cdot \frac{1}{2 \cdot L_{in}} \cdot [\sum_{j=1}^{3} \Gamma_{2j} \cdot g_{2j} - \alpha_{s2}]$

$$\Xi_1 = -v_g \cdot \frac{1}{2 \cdot L_{in}}$$

$$\Xi_2 = v_g^2 \cdot \left[\sum_{j=1}^{3} \Gamma_{1j} \cdot g_{1j} - \alpha_{s1} - \frac{1}{2 \cdot L_{in}} \right] \cdot \left[\sum_{j=1}^{3} \Gamma_{2j} \cdot g_{2j} - \alpha_{s2} \right]$$

$$\Xi_3 = v_g \cdot \left\{ \frac{1}{2 \cdot L_{in}} - \left[\sum_{j=1}^{3} \Gamma_{1j} \cdot g_{1j} - \alpha_{s1} \right] - \left[\sum_{j=1}^{3} \Gamma_{2j} \cdot g_{2j} - \alpha_{s2} \right] \right\}$$

$$\Xi_2 + \Xi_3 \cdot \lambda + \lambda^2 + (\Xi_0 + \Xi_1 \cdot \lambda) \cdot e^{-\lambda \cdot \tau} = 0$$

$$\det(A - \lambda \cdot I) = \Xi_2 + \Xi_3 \cdot \lambda + \lambda^2 + (\Xi_0 + \Xi_1 \cdot \lambda) \cdot e^{-\lambda \cdot \tau}$$
$$D(\lambda, \tau) = \Xi_2 + \Xi_3 \cdot \lambda + \lambda^2 + (\Xi_0 + \Xi_1 \cdot \lambda) \cdot e^{-\lambda \cdot \tau}$$

$$D(\lambda, \tau) = P_n(\lambda, \tau) + Q_m(\lambda, \tau) \cdot e^{-\lambda \cdot \tau}$$
$$n, m \in \mathbb{N}_0; n = 2; m = 1; n > m$$

$$P_n(\lambda,\tau) = \Xi_2 + \Xi_3 \cdot \lambda + \lambda^2;\, n = 2;\, Q_m(\lambda,\tau) = \Xi_0 + \Xi_1 \cdot \lambda;\, m = 1$$

$$P_n(\lambda,\tau) = \sum_{k=0}^{n=2} p_k(\tau) \cdot \lambda^k = p_0(\tau) + p_1(\tau) \cdot \lambda + p_2(\tau) \cdot \lambda^2$$

$$p_0(\tau) = \Xi_2 = v_g^2 \cdot \left[\sum_{j=1}^{3} \Gamma_{1j} \cdot g_{1j} - \alpha_{s1} - \frac{1}{2 \cdot L_{in}}\right] \cdot \left[\sum_{j=1}^{3} \Gamma_{2j} \cdot g_{2j} - \alpha_{s2}\right]$$

$$p_1(\tau) = \Xi_3 = v_g \cdot \left\{\frac{1}{2 \cdot L_{in}} - \left[\sum_{j=1}^{3} \Gamma_{1j} \cdot g_{1j} - \alpha_{s1}\right] - \left[\sum_{j=1}^{3} \Gamma_{2j} \cdot g_{2j} - \alpha_{s2}\right]\right\}$$

$$p_2(\tau) = 1$$

$$Q_m(\lambda,\tau) = \sum_{k=0}^{m=1} q_k(\tau) \cdot \lambda^k = q_0(\tau) + q_1(\tau) \cdot \lambda$$

$$q_0(\tau) = \Xi_0 = v_g^2 \cdot \frac{1}{2 \cdot L_{in}} \cdot \left[\sum_{j=1}^{3} \Gamma_{2j} \cdot g_{2j} - \alpha_{s2}\right];\, q_1(\tau) = \Xi_1 = -v_g \cdot \frac{1}{2 \cdot L_{in}}$$

The homogeneous system for S_1, S_2 leads to a characteristic equation for the eigenvalue λ having the form $D(\lambda,\tau) = P(\lambda,\tau) + Q(\lambda,\tau) \cdot e^{-\lambda \cdot \tau} = 0$; and $P(\lambda) = \sum_{j=0}^{2} a_j \cdot \lambda^j$; $Q(\lambda) = \sum_{j=0}^{1} c_j \cdot \lambda^j$. The coefficients $\{a_j(q_i, q_k, \tau), c_j(q_i, q_k, \tau)\} \in \mathbb{R}$ depend on q_i, q_k and delay τ. q_i, q_k are any Dual-wavelength emission VECSEL system's parameters, other parameters kept as a constant [2, 3].

$$a_0 = \Xi_2 = v_g^2 \cdot \left[\sum_{j=1}^{3} \Gamma_{1j} \cdot g_{1j} - \alpha_{s1} - \frac{1}{2 \cdot L_{in}}\right] \cdot \left[\sum_{j=1}^{3} \Gamma_{2j} \cdot g_{2j} - \alpha_{s2}\right]$$

$$a_1 = \Xi_3 = v_g \cdot \left\{\frac{1}{2 \cdot L_{in}} - \left[\sum_{j=1}^{3} \Gamma_{1j} \cdot g_{1j} - \alpha_{s1}\right] - \left[\sum_{j=1}^{3} \Gamma_{2j} \cdot g_{2j} - \alpha_{s2}\right]\right\}$$

$$a_2 = 1$$

$$c_0 = \Xi_0 = v_g^2 \cdot \frac{1}{2 \cdot L_{in}} \cdot \left[\sum_{j=1}^{3} \Gamma_{2j} \cdot g_{2j} - \alpha_{s2}\right];\, c_1 = \Xi_1 = -v_g \cdot \frac{1}{2 \cdot L_{in}}$$

Unless strictly necessary, the designation of the varied arguments (q_i, q_k) will subsequently be omitted from P, Q, a_j and c_j. The coefficients a_j, c_j are continuous and differentiable functions of their arguments, and direct substitution shows that $a_0 + c_0 \neq 0$ for $q_i, q_k \in \mathbb{R}_+$; that is, $\lambda = 0$ is not of $P(\lambda) + Q(\lambda) \cdot e^{-\lambda \cdot \tau} = 0$. Furthermore, $P(\lambda)$, $Q(\lambda)$ are analytic functions of λ, for which the following requirements of the analysis (Kuang and Cong 2005) [2] can also be verified in the present case:

1. If $\lambda = i \cdot \omega$; $\omega \in \mathbb{R}$, then $P(i \cdot \omega) + Q(i \cdot \omega) \neq 0$.
2. If $\left|\frac{Q(\lambda)}{P(\lambda)}\right|$ is bounded for $|\lambda| \to \infty$; $\operatorname{Re}\lambda \geq 0$. No roots bifurcation from ∞.
3. $F(\omega) = |P(i \cdot \omega)|^2 - |Q(i \cdot \omega)|^2$ has a finite number of zeros. Indeed, this is a polynomial in ω.
4. Each positive root $\omega(q_i, q_k)$ of $F(\omega) = 0$ is continuous and differentiable with respect to q_i, q_k.

We assume that $P_n(\lambda, \tau)$ and $Q_m(\lambda, \tau)$ cannot have common imaginary roots. That is for any real number ω: $P_n(\lambda = i \cdot \omega, \tau) + Q_m(\lambda = i \cdot \omega, \tau) \neq 0$ and

$$\begin{aligned} P_n(\lambda = i \cdot \omega, \tau) &= p_0(\tau) + p_1(\tau) \cdot i \cdot \omega - p_2(\tau) \cdot \omega^2 \\ &= p_0(\tau) - p_2(\tau) \cdot \omega^2 + p_1(\tau) \cdot i \cdot \omega \end{aligned}$$

$$\begin{aligned} P_n(\lambda = i \cdot \omega, \tau) &= \Xi_2 - \omega^2 + \Xi_3 \cdot \omega \cdot i \\ Q_m(\lambda = i \cdot \omega, \tau) &= q_0(\tau) + q_1(\tau) \cdot i \cdot \omega \\ Q_m(\lambda = i \cdot \omega, \tau) &= \Xi_0 + \Xi_1 \cdot \omega \cdot i \end{aligned}$$

$$\begin{aligned} P_n(\lambda = i \cdot \omega, \tau) + Q_m(\lambda = i \cdot \omega, \tau) &= q_0(\tau) + p_0(\tau) - p_2(\tau) \cdot \omega^2 \\ &\quad + [p_1(\tau) + q_1(\tau)] \cdot i \cdot \omega \neq 0 \end{aligned}$$

$$P_n(\lambda = i \cdot \omega, \tau) + Q_m(\lambda = i \cdot \omega, \tau) = \Xi_0 + \Xi_2 - \omega^2 + [\Xi_3 + \Xi_1] \cdot i \cdot \omega \neq 0$$

$$\begin{aligned} |P(i \cdot \omega, \tau)|^2 &= (\Xi_2 - \omega^2)^2 + (\Xi_3 \cdot \omega)^2 \\ &= \Xi_2^2 + (\Xi_3^2 - 2 \cdot \Xi_2) \cdot \omega^2 + \omega^4; \ |Q(i \cdot \omega, \tau)|^2 \\ &= \Xi_0^2 + \Xi_1^2 \cdot \omega^2 \end{aligned}$$

$$\begin{aligned} F(\omega, \tau) &= |P(i \cdot \omega, \tau)|^2 - |Q(i \cdot \omega, \tau)|^2 \\ &= \Xi_2^2 - \Xi_0^2 + (\Xi_3^2 - 2 \cdot \Xi_2 - \Xi_1^2) \cdot \omega^2 + \omega^4 \end{aligned}$$

We define the following parameters for simplicity: Π_0, Π_2, Π_4

$$\Pi_0 = \Xi_2^2 - \Xi_0^2; \ \Pi_2 = \Xi_3^2 - 2 \cdot \Xi_2 - \Xi_1^2; \ \Pi_4 = 1$$

Hence $F(\omega,\tau)=0$ implies $\sum_{k=0}^{2}\Pi_{2\cdot k}\cdot\omega^{2\cdot k}=0$ and its roots are given by solving the above polynomial. $P_R(i\omega,\tau)=\Xi_2-\omega^2$; $P_I(i\omega,\tau)=\Xi_3\cdot\omega$; $Q_R(i\omega,\tau)=\Xi_0$; $Q_I(i\omega,\tau)=\Xi_1\cdot\omega$

$$\sin\theta(\tau)=\frac{-P_R(i\omega,\tau)\cdot Q_I(i\omega,\tau)+P_I(i\omega,\tau)\cdot Q_R(i\omega,\tau)}{|Q(i\omega,\tau)|^2}$$

$$\cos\theta(\tau)=-\frac{P_R(i\omega,\tau)\cdot Q_R(i\omega,\tau)+P_I(i\omega,\tau)\cdot Q_I(i\omega,\tau)}{|Q(i\omega,\tau)|^2}$$

We use different terminology from our last characteristics parameters definition:

$$k\to j;\ p_k(\tau)\to a_j;\ q_k(\tau)\to c_j;\ n,m\in\mathbb{N}_0;\ n=2;\ m=1;\ n>m$$

$$P_n(\lambda,\tau)\to P(\lambda);\ Q_m(\lambda,\tau)\to Q(\lambda);\ P(\lambda)=\sum_{j=0}^{2}a_j\cdot\lambda^j;\ Q(\lambda)=\sum_{j=0}^{1}c_j\cdot\lambda^j$$

$$P(\lambda)=a_0+a_1\cdot\lambda+a_2\cdot\lambda^2;\ Q(\lambda)=c_0+c_1\cdot\lambda$$

$n,m\in\mathbb{N}_0;n>m$ and $a_j,c_j:\mathbb{R}_{0+}\to\mathbb{R}$ are continuous and differentiable function of τ such that $a_0+c_0\neq 0$. In the following "—" denoted complex and conjugate. $P(\lambda)$, $Q(\lambda)$ are analytic functions in λ and differentiable in τ. The coefficients $a_j(v_g,\Gamma_{1j},\alpha_{s1},L_{in},\Gamma_{2j},\alpha_{s2},\tau,\ldots)\in\mathbb{R}$ and $c_j(v_g,\Gamma_{1j},\alpha_{s1},L_{in},\Gamma_{2j},\alpha_{s2},\tau,\ldots)\in\mathbb{R}$ depend on Dual-wavelength emission VECSEL system's parameters, $v_g,\Gamma_{1j},\alpha_{s1},L_{in},\Gamma_{2j},\alpha_{s2},\tau,\ldots$ values. Unless strictly necessary, the designation of the varied arguments: $v_g,\Gamma_{1j},\alpha_{s1},L_{in},\Gamma_{2j},\alpha_{s2},\tau,\ldots$ will subsequently be omitted from P, Q, a_j, c_j. The coefficients a_j, c_j are continuous and differentiable functions of their arguments, and direct substitution shows that $a_0+c_0\neq 0$ [2, 3].

$$\begin{aligned}a_0+c_0&=\Xi_2+\Xi_0\\&=v_g^2\cdot\left[\sum_{j=1}^{3}\Gamma_{1j}\cdot g_{1j}-\alpha_{s1}-\frac{1}{2\cdot L_{in}}\right]\cdot\left[\sum_{j=1}^{3}\Gamma_{2j}\cdot g_{2j}\right.\\&\quad\left.-\alpha_{s2}\right]+v_g^2\cdot\frac{1}{2\cdot L_{in}}\cdot\left[\sum_{j=1}^{3}\Gamma_{2j}\cdot g_{2j}-\alpha_{s2}\right]\\&=v_g^2\cdot\left(\left[\sum_{j=1}^{3}\Gamma_{1j}\cdot g_{1j}-\alpha_{s1}-\frac{1}{2\cdot L_{in}}\right]\right.\\&\quad\left.+\frac{1}{2\cdot L_{in}}\right)\cdot\left[\sum_{j=1}^{3}\Gamma_{2j}\cdot g_{2j}-\alpha_{s2}\right]\end{aligned}$$

$$= v_g^2 \cdot \left[\sum_{j=1}^{3} \Gamma_{1j} \cdot g_{1j} - \alpha_{s1} \right] \cdot \left[\sum_{j=1}^{3} \Gamma_{2j} \cdot g_{2j} - \alpha_{s2} \right] \neq 0$$

$\forall\, v_g, \Gamma_{1j}, \alpha_{s1}, L_{in}, \Gamma_{2j}, \alpha_{s2}, \tau, \ldots \in \mathbb{R}_+$ I.e. $\lambda = 0$ is not a root of the characteristic equation. Furthermore $P(\lambda)$, $Q(\lambda)$ are analytic functions of λ for which the following requirements of the analysis (see Kuang 2, Sect. 3.4) can also be verified in the present case.

1. If $\lambda = i \cdot \omega;\, \omega \in \mathbb{R}$ then $P(i \cdot \omega) + Q(i \cdot \omega) \neq 0$, i.e. P and Q have no common imaginary roots. This condition was verified numerically in the entire $\forall\, v_g, \Gamma_{1j}, \alpha_{s1}, L_{in}, \Gamma_{2j}, \alpha_{s2}, \tau, \ldots$ domain of interest.
2. $\left|\frac{P(\lambda)}{Q(\lambda)}\right|$ is bounded for $|\lambda| \to \infty$; $\mathrm{Re}\lambda \geq 0$. No roots bifurcation from ∞. Indeed, in the limit: $\left|\frac{Q(\lambda)}{P(\lambda)}\right| = \left|\frac{c_0 + c_1 \cdot \lambda}{a_0 + a_1 \cdot \lambda + a_2 \cdot \lambda^2}\right|$.
3. The following expressions exist: $F(\omega) = |P(i \cdot \omega)|^2 - |Q(i \cdot \omega)|^2$. $F(\omega, \tau) = |P(i \cdot \omega, \tau)|^2 - |Q(i \cdot \omega, \tau)|^2 = \sum_{k=0}^{2} \Pi_{2\cdot k} \cdot \omega^{2\cdot k}$ has at most a finite number of zeros. Indeed, this is a polynomial in ω (degree in ω^4).
4. Each positive root $\omega(v_g, \Gamma_{1j}, \alpha_{s1}, L_{in}, \Gamma_{2j}, \alpha_{s2}, \tau, \ldots)$ of $F(\omega) = 0$ is continuous and differentiable with respect to $v_g, \Gamma_{1j}, \alpha_{s1}, L_{in}, \Gamma_{2j}, \alpha_{s2}, \tau, \ldots$ and the condition can only be assessed numerically.

In addition, since the coefficients in P and Q are real, we have $\overline{P(-i \cdot \omega)} = P(i \cdot \omega)$ and $\overline{Q(-i \cdot \omega)} = Q(i \cdot \omega)$ thus, $\omega > 0$ may be on eigenvalue of characteristic equations. The analysis consists in identifying the roots of the characteristic equation situated on the imaginary axis of the complex λ plane, whereby increasing the parameters: $v_g, \Gamma_{1j}, \alpha_{s1}, L_{in}, \Gamma_{2j}, \alpha_{s2}, \tau, \ldots$, $\mathrm{Re}\lambda$ may, at the crossing, change its sign from $(-)$ to $(+)$. i.e. from a stable focus $E^{(j)} = (S_1^{(j)}, S_2^{(j)});\ j = 0, 1, 2$ to an unstable one, or vice versa. This feature may be further assessed by examining the sign of the partial derivatives with respect to $v_g, \Gamma_{1j}, \alpha_{s1}, L_{in}, \Gamma_{2j}, \alpha_{s2}, \tau, \ldots$ and any system parameters.

$$\Lambda^{-1}(v_g) = \left(\frac{\partial \mathrm{Re}\lambda}{\partial v_g}\right)_{\lambda = i \cdot \omega} ;\ \Gamma_{1j}, \alpha_{s1}, L_{in}, \Gamma_{2j}, \alpha_{s2}, \tau, \ldots = const$$

$$\Lambda^{-1}(\Gamma_{1j}) = \left(\frac{\partial \mathrm{Re}\lambda}{\partial \Gamma_{1j}}\right)_{\lambda = i \cdot \omega} ;\ v_g, \alpha_{s1}, L_{in}, \Gamma_{2j}, \alpha_{s2}, \tau, \ldots = const$$

$$\Lambda^{-1}(L_{in}) = \left(\frac{\partial \mathrm{Re}\lambda}{\partial L_{in}}\right)_{\lambda = i \cdot \omega} ;\ v_g, \Gamma_{1j}, \alpha_{s1}, \Gamma_{2j}, \alpha_{s2}, \tau, \ldots = const$$

$$\Lambda^{-1}(\Gamma_{2j}) = \left(\frac{\partial \mathrm{Re}\lambda}{\partial \Gamma_{2j}}\right)_{\lambda = i \cdot \omega} ;\ v_g, \Gamma_{1j}, \alpha_{s1}, L_{in}, \alpha_{s2}, \tau, \ldots = const$$

$$\Lambda^{-1}(\alpha_{s2}) = \left(\frac{\partial \mathrm{Re}\lambda}{\partial \alpha_{s2}}\right)_{\lambda = i \cdot \omega} ;\ v_g, \Gamma_{1j}, \alpha_{s1}, L_{in}, \Gamma_{2j}, \alpha_{s2}, \tau, \ldots = const$$

$$\Lambda^{-1}(\tau) = \left(\frac{\partial \mathrm{Re}\lambda}{\partial \tau}\right)_{\lambda=i\cdot\omega} ; v_g, \Gamma_{1j}, \alpha_{s1}, L_{in}, \Gamma_{2j}, \alpha_{s2}, \ldots = const$$

$P(\lambda) = P_R(\lambda) + i \cdot P_I(\lambda)$; $Q(\lambda) = Q_R(\lambda) + i \cdot Q_I(\lambda)$, When writing and inserting $\lambda = i \cdot \omega$ into Dual-wavelength emission VECSEL system's characteristic equation ω must satisfy the following equations:

$$\sin(\omega \cdot \tau) = g(\omega) = \frac{-P_R(i\omega, \tau) \cdot Q_I(i\omega, \tau) + P_I(i\omega, \tau) \cdot Q_R(i\omega, \tau)}{|Q(i\omega, \tau)|^2}$$

$$\cos(\omega \cdot \tau) = h(\omega) = -\frac{P_R(i\omega, \tau) \cdot Q_R(i\omega, \tau) + P_I(i\omega, \tau) \cdot Q_I(i\omega, \tau)}{|Q(i\omega, \tau)|^2}$$

where $|Q(i\omega, \tau)|^2 \neq 0$ in view of requirement (1) above, and $(g, h) \in \mathbb{R}$. Furthermore, it follows above $\sin(\omega \cdot \tau)$ and $\cos(\omega \cdot \tau)$ equation that, by squaring and adding the sides, ω must be a positive root of $F(\omega) = |P(i \cdot \omega)|^2 - |Q(i \cdot \omega)|^2 = 0$. Note: $F(\omega)$ is independent on τ. Now it is important to notice that if $\tau \notin I$ (assume that $I \subseteq \mathbb{R}_{+0}$ is the set where $\omega(\tau)$ is a positive root of $F(\omega)$ and for, $\tau \notin I$, $\omega(\tau)$ is not defined. Then for all τ in I, $\omega(\tau)$ is satisfied that $F(\omega, \tau) = 0$. Then there are no positive $\omega(\tau)$ solutions for $F(\omega, \tau) = 0$, and we cannot have stability switches. For $\tau \in I$ where $\omega(\tau)$ is a positive solution of $F(\omega, \tau) = 0$, we can define the angle $\theta(\tau) \in [0, 2 \cdot \pi]$ as the solution of $\sin\theta(\tau) = \ldots$ and $\cos\theta(\tau) = \ldots$; the relation between the arguments $\theta(\tau)$ and $\tau \cdot \omega(\tau)$ for $\tau \in I$ must be describing below. $\tau \cdot \omega(\tau) = \theta(\tau) + 2 \cdot n \cdot \pi \; \forall n \in \mathbb{N}_0$ Hence we can define the maps:

$\tau_n : I \to \mathbb{R}_{+0}$, is given by $\tau_n(\tau) = \frac{\theta(\tau)+2\cdot n\cdot\pi}{\omega(\tau)}$; $n \in \mathbb{N}_0$; $\tau \in I$. Let us introduce the function $I \to \mathbb{R}$; $S_n(\tau) = \tau - \tau_n(\tau)$; $\tau \in I$; $n \in \mathbb{N}_0$ that is continuous and differentiable in τ. In the following, the subscripts $\lambda, \omega, v_g, \Gamma_{1j}, \alpha_{s1}, L_{in}, \Gamma_{2j}, \alpha_{s2}, \ldots$ indicate the corresponding partial derivatives. Let us first concentrate on $\Lambda(x)$, remember in $\lambda(v_g, \Gamma_{1j}, \alpha_{s1}, L_{in}, \Gamma_{2j}, \alpha_{s2}, \ldots)$ and $\omega(v_g, \Gamma_{1j}, \alpha_{s1}, L_{in}, \Gamma_{2j}, \alpha_{s2}, \ldots)$, and keeping all parameters except one (x) and τ. The derivation closely follows that in reference [BK]. Differentiating Dual-wavelength emission VECSEL system's characteristic equation $P(\lambda) + Q(\lambda) \cdot e^{-\lambda\cdot\tau} = 0$ with respect to specific parameter (x), and inverting the derivative, for convenience, one calculates: $x = v_g, \Gamma_{1j}, \alpha_{s1}, L_{in}, \Gamma_{2j}, \alpha_{s2}, \tau, \ldots$

$$\left(\frac{\partial\lambda}{\partial x}\right)^{-1} = \frac{-P_\lambda(\lambda, x) \cdot Q(\lambda, x) + Q_\lambda(\lambda, x) \cdot P(\lambda, x) - \tau \cdot P(\lambda, x) \cdot Q(\lambda, x)}{P_x(\lambda, x) \cdot Q(\lambda, x) - Q_x(\lambda, x) \cdot P(\lambda, x)}$$

where $P_\lambda = \frac{\partial P}{\partial \lambda}, \ldots$ etc., substituting $\lambda = i \cdot \omega$ and bearing $\overline{P(-i \cdot \omega)} = P(i \cdot \omega)$; $\overline{Q(-i \cdot \omega)} = Q(i \cdot \omega)$ then $i \cdot P_\lambda(i \cdot \omega) = P_\omega(i \cdot \omega)$; $i \cdot Q_\lambda(i \cdot \omega) = Q_\omega(i \cdot \omega)$ and that on the surface $|P(i\omega)|^2 = |Q(i\omega)|^2$, one obtain: $(\frac{\partial\lambda}{\partial x})^{-1}|_{\lambda=i\cdot\omega} = (\frac{i\cdot P_\omega(i\cdot\omega,x)\cdot\overline{P(i\cdot\omega,x)}+i\cdot Q_\lambda(i\cdot\omega,x)\cdot\overline{Q(\lambda,x)}-\tau\cdot|P(i\cdot\omega,x)|^2}{P_x(i\cdot\omega,x)\cdot\overline{P(i\cdot\omega,x)}-Q_x(i\cdot\omega,x)\cdot\overline{Q(i\cdot\omega,x)}})$.

Upon separating into real and imaginary parts, with $P = P_R + i \cdot P_I$; $Q = Q_R + i \cdot Q_I$ and $P_\omega = P_{R\omega} + i \cdot P_{I\omega}$; $Q_\omega = Q_{R\omega} + i \cdot Q_{I\omega}$; $P_x = P_{Rx} + i \cdot P_{Ix}$; $Q_x =$

$Q_{Rx} + i \cdot Q_{Ix}$; $P^2 = P_R^2 + P_I^2$, When (x) can be any Dual-wavelength emission VECSEL system's parameters $v_g, \Gamma_{1j}, \alpha_{s1}, L_{in}, \Gamma_{2j}, \alpha_{s2}, \ldots$ and time delay τ etc. Where for convenience, we have dropped the arguments $(i \cdot \omega, x)$, and where $F_\omega = 2 \cdot [(P_{R\omega} \cdot P_R + P_{I\omega} \cdot P_I) - (Q_{R\omega} \cdot Q_R + Q_{I\omega} \cdot Q_I)]$; $F_x = 2 \cdot [(P_{Rx} \cdot P_R + P_{Ix} \cdot P_I) - (Q_{Rx} \cdot Q_R + Q_{Ix} \cdot Q_I)]$ and $\omega_x = -\frac{F_x}{F_\omega}$. We define U and V: $U = (P_R \cdot P_{I\omega} - P_I \cdot P_{R\omega}) - (Q_R \cdot Q_{I\omega} - Q_I \cdot Q_{R\omega})$, $V = (P_R \cdot P_{Ix} - P_I \cdot P_{Rx}) - (Q_R \cdot Q_{Ix} - Q_I \cdot Q_{Rx})$, we choose our specific parameter as time delay $x = \tau$: $P_R = \Xi_2 - \omega^2$; $P_I = \Xi_3 \cdot \omega$; $Q_R = \Xi_0$; $Q_I = \Xi_1 \cdot \omega$; $P_{R\omega} = -2 \cdot \omega$; $P_{I\omega} = \Xi_3$; $Q_{I\omega} = \Xi_1$.

$$Q_{R\omega} = 0;\ P_{R\tau} = 0;\ P_{I\tau} = 0;\ Q_{R\tau} = 0;\ Q_{I\tau} = 0$$
$$F_\tau = 0;\ P_R \cdot P_{I\tau} - P_I \cdot P_{R\tau} = 0$$

$$Q_R \cdot Q_{I\tau} - Q_I \cdot Q_{R\tau} = 0$$
$$V = (P_R \cdot P_{I\tau} - P_I \cdot P_{R\tau}) - (Q_R \cdot Q_{I\tau} - Q_I \cdot Q_{R\tau}) = 0$$

$F_\omega = \ldots$ Elememts:

$$P_{R\omega} \cdot P_R = -2 \cdot \omega \cdot (\Xi_2 - \omega^2) = 2 \cdot \omega^3 - 2 \cdot \omega \cdot \Xi_2$$
$$P_{I\omega} \cdot P_I = \Xi_3^2 \cdot \omega;\ Q_{R\omega} \cdot Q_R = 0$$
$$Q_{I\omega} \cdot Q_I = \Xi_1^2 \cdot \omega$$

$U = \ldots$ Elememts:

$$P_R \cdot P_{I\omega} = (\Xi_2 - \omega^2) \cdot \Xi_3;\ P_I \cdot P_{R\omega} = -2 \cdot \Xi_3 \cdot \omega^2$$
$$Q_R \cdot Q_{I\omega} = \Xi_0 \cdot \Xi_1;\ Q_I \cdot Q_{R\omega} = 0$$

We can summary our last results in the next table (Table 7.8).

$F(\omega, \tau) = 0$, differentiating with respect to τ and we get $F_\omega \cdot \frac{\partial \omega}{\partial \tau} + F_\tau = 0$; $\tau \in I \Rightarrow \frac{\partial \omega}{\partial \tau} = -\frac{F_\tau}{F_\omega}$

Table 7.8 Dual-wavelength emission VECSEL system's stability analyses F_ω, U elements

F_ω, U elements	Expression	F_ω, U elements	Expression
$P_{R\omega} \cdot P_R$	$2 \cdot \omega^3 - 2 \cdot \omega \cdot \Xi_2$	$P_R \cdot P_{I\omega}$	$(\Xi_2 - \omega^2) \cdot \Xi_3$
$P_{I\omega} \cdot P_I$	$\Xi_3^2 \cdot \omega$	$P_I \cdot P_{R\omega}$	$-2 \cdot \Xi_3 \cdot \omega^2$
$Q_{R\omega} \cdot Q_R$	0	$Q_R \cdot Q_{I\omega}$	$\Xi_0 \cdot \Xi_1$
$Q_{I\omega} \cdot Q_I$	$\Xi_1^2 \cdot \omega$	$Q_I \cdot Q_{R\omega}$	0

$$\Lambda^{-1}(\tau) = \left(\frac{\partial \mathrm{Re}\lambda}{\partial \tau}\right)_{\lambda = i\omega}; \quad \frac{\partial \omega}{\partial \tau} = \omega_{\tau} = -\frac{F_{\tau}}{F_{\omega}}$$

$$\Lambda^{-1}(\tau) = \mathrm{Re}\left\{\frac{-2 \cdot [U + \tau \cdot |P|^2 + i \cdot F_{\omega}]}{F_{\tau} + i \cdot 2 \cdot [V + \omega \cdot |P|^2]}\right\}$$

$$\mathrm{sign}\{\Lambda^{-1}(\tau)\} = \mathrm{sign}\left\{\left(\frac{\partial \mathrm{Re}\lambda}{\partial \tau}\right)_{\lambda = i\omega}\right\}$$

$$\mathrm{sign}\{\Lambda^{-1}(\tau)\} = \mathrm{sign}\{F_{\omega}\} \cdot \mathrm{sign}\left\{\frac{V + \frac{\partial \omega}{\partial \tau} \cdot U}{|P|^2} + \omega + \frac{\partial \omega}{\partial \tau} \cdot \tau\right\}$$

We shall presently examine the possibility of stability transitions (bifurcations) Dual-wavelength emission VECSEL system, about the equilibrium points, $E^{(j)}$; $j = 0, 1, 2, \ldots$,

$E^{(j)} = (S_1^{(j)}, S_2^{(j)})$; $j = 0, 1, 2$. The analysis consists in identifying the roots of our system characteristic equation situated on the imaginary axis of the complex λ-plane, where by increasing the delay parameter τ. Reλ, may at the crossing, changes its sign from − to +, i.e. from a stable focus $E^{(*)}$ to an unstable one, or vice versa. This feature may be further assessed by examining the sign of the partial derivatives with respect to τ [2, 3].

$$\Lambda^{-1}(\tau) = \left(\frac{\partial \mathrm{Re}\lambda}{\partial \tau}\right)_{\lambda = i \cdot \omega}; \; v_g, \Gamma_{1j}, \alpha_{s1}, L_{in}, \Gamma_{2j}, \alpha_{s2}, \ldots = const$$

$$\begin{aligned} U &= (P_R \cdot P_{I\omega} - P_I \cdot P_{R\omega}) - (Q_R \cdot Q_{I\omega} - Q_I \cdot Q_{R\omega}) \\ &= (\Xi_2 - \omega^2) \cdot \Xi_3 + 2 \cdot \Xi_3 \cdot \omega^2 - \Xi_0 \cdot \Xi_1 \end{aligned}$$

$$\begin{aligned} U &= (P_R \cdot P_{I\omega} - P_I \cdot P_{R\omega}) - (Q_R \cdot Q_{I\omega} - Q_I \cdot Q_{R\omega}) \\ &= \Xi_2 \cdot \Xi_3 - \Xi_0 \cdot \Xi_1 + \Xi_3 \cdot \omega^2 \end{aligned}$$

$$U = (P_R \cdot P_{I\omega} - P_I \cdot P_{R\omega}) - (Q_R \cdot Q_{I\omega} - Q_I \cdot Q_{R\omega}) = \sum_{k=0}^{1} A_{2k} \cdot \omega^{2 \cdot k}$$

$$A_0 = \Xi_2 \cdot \Xi_3 - \Xi_0 \cdot \Xi_1; \; A_2 = \Xi_3$$

$$\begin{aligned} F_{\omega} &= 2 \cdot [(P_{R\omega} \cdot P_R + P_{I\omega} \cdot P_I) - (Q_{R\omega} \cdot Q_R + Q_{I\omega} \cdot Q_I)] \\ &= 2 \cdot (\Xi_3^2 - 2 \cdot \Xi_2 - \Xi_1^2) \cdot \omega + 4 \cdot \omega^3 \end{aligned}$$

$$\begin{aligned} F_{\omega} &= 2 \cdot [(P_{R\omega} \cdot P_R + P_{I\omega} \cdot P_I) - (Q_{R\omega} \cdot Q_R + Q_{I\omega} \cdot Q_I)] \\ &= \sum_{k=0}^{1} B_{2 \cdot k+1} \cdot \omega^{2 \cdot k+1}; \; B_1 = 2 \cdot (\Xi_3^2 - 2 \cdot \Xi_2 - \Xi_1^2); \; B_3 = 4 \end{aligned}$$

Then we get the expression for $F_\omega = \sum_{k=0}^{1} B_{2\cdot k+1} \cdot \omega^{2\cdot k+1}$, Dual-wavelength emission VECSEL system parameter values. We find those ω, τ values which fulfill $F_\omega(\omega, \tau) = 0$. We ignore negative, complex, and imaginary values of ω for specific τ values. $\tau \in [0.001 \ldots 10]$, we can express by 3D function $F_\omega(\omega, \tau) = 0$. We plot the stability switch diagram based on different delay values of our Dual-wavelength emission VECSEL system.

$$\Lambda^{-1}(\tau) = \left(\frac{\partial \mathrm{Re}\lambda}{\partial \tau}\right)_{\lambda=i\omega} = \mathrm{Re}\left\{\frac{-2\cdot[U+\tau\cdot|P|^2]+i\cdot F_\omega}{F_\tau + 2i\cdot[V+\omega\cdot|P|^2]}\right\}$$

$$\Lambda^{-1}(\tau) = \left(\frac{\partial \mathrm{Re}\lambda}{\partial \tau}\right)_{\lambda=i\omega} = \frac{2\cdot\{F_\omega\cdot(V+\omega\cdot P^2) - F_\tau\cdot(U+\tau\cdot P^2)\}}{F_\tau^2 + 4\cdot(V+\omega\cdot P^2)^2}$$

The stability switch occurs only on those delay values (τ) which fit the equation: $\tau = \frac{\theta_+(\tau)}{\omega_+(\tau)}$ and $\theta_+(\tau)$ is the solution of $\sin\theta(\tau) = \ldots.$ and $\cos\theta(\tau) = \ldots.$ when $\omega = \omega_+(\tau)$ If only ω_+ is feasible. Additionally, when all Dual-wavelength emission VECSEL system's parameters are known and the stability switch due to various time delay values τ is described in the following expression:

$$\mathrm{sign}\{\Lambda^{-1}(\tau)\} = \mathrm{sign}\{F_\omega(\omega(\tau),\tau)\}\cdot\mathrm{sign}\left\{\tau\cdot\omega_\tau(\omega(\tau)) + \omega(\tau) + \frac{U(\omega(\tau))\cdot\omega_\tau(\omega(\tau)) + V(\omega(\tau))}{|P(\omega(\tau))|^2}\right\}$$

Remark: we know $F(\omega, \tau) = 0$ implies its roots $\omega_i(\tau)$ and finding those delays values τ which ω_i is feasible. There are τ values which give complex ω_i or imaginary number, then unable to analyze stability. $F(\omega, \tau)$, function is independent on τ the parameter $F(\omega, \tau) = 0$. The results: We find those ω, τ values which fulfill $F(\omega, \tau) = 0$. We ignore negative, complex, and imaginary values of ω . Next is to find those ω, τ values which fulfill $\sin\theta(\tau) = \ldots.$ and $\cos\theta(\tau) = \ldots.$; $\sin(\omega\cdot\tau) = \frac{-P_R\cdot Q_I + P_I\cdot Q_R}{|Q|^2}$; $\cos(\omega\cdot\tau) = -\frac{(P_R\cdot Q_R + P_I\cdot Q_I)}{|Q|^2}$ $|Q|^2 = Q_R^2 + Q_I^2$. Finally we plot the stability switch diagram $g(\tau) = \Lambda^{-1}(\tau) = (\frac{\partial \mathrm{Re}\lambda}{\partial \tau})_{\lambda=i\omega}$

$$g(\tau) = \Lambda^{-1}(\tau) = \left(\frac{\partial \mathrm{Re}\lambda}{\partial \tau}\right)_{\lambda=i\omega} = \frac{2\cdot\{F_\omega\cdot(V+\omega\cdot P^2) - F_\tau\cdot(U+\tau\cdot P^2)\}}{F_\tau^2 + 4\cdot(V+\omega\cdot P^2)^2}$$

$$\mathrm{sign}[g(\tau)] = \mathrm{sign}[\Lambda^{-1}(\tau)] = \mathrm{sign}\left[\left(\frac{\partial \mathrm{Re}\lambda}{\partial \tau}\right)_{\lambda=i\omega}\right] = \mathrm{sign}\left[\frac{2\cdot\{F_\omega\cdot(V+\omega\cdot P^2) - F_\tau\cdot(U+\tau\cdot P^2)\}}{F_\tau^2 + 4\cdot(V+\omega\cdot P^2)^2}\right]$$

$$F_\tau^2 + 4\cdot(V+\omega\cdot P^2)^2 > 0 \Rightarrow \mathrm{sign}[\Lambda^{-1}(\tau)] = \mathrm{sign}[F_\omega\cdot(V+\omega\cdot P^2) - F_\tau\cdot(U+\tau\cdot P^2)]$$

Table 7.9 Dual-wavelength emission VECSEL system sign of sign[$\Lambda^{-1}(\tau)$]

sign[F_ω]	sign[$\frac{V+\omega_\tau \cdot U}{P^2} + \omega + \omega_\tau \cdot \tau$]	sign[$\Lambda^{-1}(\tau)$]
$\pm$	$\pm$	+
$\pm$	$\mp$	−

$$\text{sign}[\Lambda^{-1}(\tau)] = \text{sign}\left\{[F_\omega] \cdot \left[(V + \omega \cdot P^2) - \frac{F_\tau}{F_\omega} \cdot (U + \tau \cdot P^2)\right]\right\}$$

$$\omega_\tau = -\frac{F_\tau}{F_\omega}; \omega_\tau = \left(\frac{\partial \omega}{\partial \tau}\right)^{-1} = -\frac{\partial F/\partial \omega}{\partial F/\partial \tau}$$

$$\text{sign}[\Lambda^{-1}(\tau)] = \text{sign}\left\{[F_\omega] \cdot [P^2] \cdot \left[\frac{V + \omega_\tau \cdot U}{P^2} + \omega + \omega_\tau \cdot \tau\right]\right\}; \text{sign}[P^2] > 0$$

$$\text{sign}[\Lambda^{-1}(\tau)] = \text{sign}\left\{[F_\omega] \cdot \left[\frac{V + \omega_\tau \cdot U}{P^2} + \omega + \omega_\tau \cdot \tau\right]\right\}$$

$$\text{sign}[\Lambda^{-1}(\tau)] = \text{sign}[F_\omega] \cdot \text{sign}\left[\frac{V + \omega_\tau \cdot U}{P^2} + \omega + \omega_\tau \cdot \tau\right]$$

$$F_\omega = 2 \cdot [(P_{R\omega} \cdot P_R + P_{I\omega} \cdot P_I) - (Q_{R\omega} \cdot Q_R + Q_{I\omega} \cdot Q_I)] = \sum_{k=0}^{1} B_{2\cdot k+1} \cdot \omega^{2\cdot k+1}$$

We check the sign of $\Lambda^{-1}(\tau)$ according to the following rule (Table 7.9).

If, sign[$\Lambda^{-1}(\tau)$] > 0, then the crossing proceeds from (−) to (+) respectively (stable to unstable). If, sign[$\Lambda^{-1}(\tau)$] < 0, then the crossing proceeds from (+) to (−) respectively (unstable to stable).

7.3 A Quasi-periodic in Erbium-Doped Fiber Laser Nonlinearity and Stability Analysis

Er-doped fiber lasers (EDFLs) is a EDFA (Erbium-doped fiber amplifier) which operate in the particular regime where coherent oscillation of ASE occurs due to feedbacks mean. EDFLs are used as sources for coherent light signal generation, while EDFAs are used as wave-wave amplifiers for coherent light signal generation. EDFLs can be pumped with compact, efficient laser diodes. They are compatible with differ fibers and fiber optic components used in communications; they have a negligible coupling losses. Fiber wave guiding and splicing alleviate any mechanical alignment of parts and provide environment stability. The tunable EDFL configuration uses an all-fiber ring laser cavity. Wavelength selectivity is achieved by using a tunable transmission filter. Basically Erbium (chemical symbol: Er) is a chemical element which belong to the group of metals. It is used also in the form of the trivalent ion Er^{3+} as the laser-active dopant of gain media based on various host materials (crystal and glasses). Erbium-doped glasses are mostly of silicate and phosphate types, are used both for bulk lasers and fiber lasers and amplifiers. Erbium-Doped Fiber Amplifier (EDFA) is an optical amplifier used in the C-band and L-band, where loss of telecom optical fibers becomes lowest in the entire optical telecommunication wavelength bands. EDFA is commonly used to compensate the loss of an optical fiber in long-distance optical communication. EDFA can amplify multiple optical signals simultaneously, and thus can be easily combined with WDM (wavelength division multiplexing) systems. EDFAs are used as a booster, inline, and pre-amplifier in an optical transmission line. The booster amplifier is placed after the transmitter to increase the optical power launched to the transmission line. The inline amplifiers are placed in the transmission line, compensating the attenuation induced by the optical fiber. The pre-amplifier is placed before the receiver, such that sufficient optical power is launched to the receiver. A typical distance between each of the EDFAs is several tens of kilometers. When an EDFA is pumped at 1480 nm, Er ion doped in the fiber absorbs the pump light and is excited to an excited state. There is a quasiperiodic route to chaos in erbium-doped fiber laser operating simultaneously at 1.55 and 1.536 μm. It is starting from a CW state for high pumping rates the system become T-periodic, 2T-periodic, 3T-periodic and chaotic for decreasing pumping ratios. The low-frequency spectra show two fundamental frequencies. The nT-periodic regimes correspond to a frequency locking of the low frequency on a subharmonic of the high frequency. The group of rare-earth doped fibers is focused on their potential applications as amplifiers or optical sources for telecommunication networks [5]. Commercially optical amplifier are at 1.55 μm and they demonstrated a large variety of dynamical behaviors including static and dynamics polarization effects, antiphase and chaotic dynamics under autonomous or pump modulation conditions. They are characterized by highly multimode operation due to the large inhomogeneous linewidth, and the fiber lasers exhibit cooperative effects between longitudinal modes leading to a simple modeling of the dynamics by low dimensional equations. Erbium-doped fiber laser (EDFL) is preferred because of its spontaneous different dynamical behaviors. EDFL can operate in a CW, sinusoidal or self-pulsing regime, depending on some parameters: the ion pairs concentration (x), the pumping ration ($r = \frac{P}{P_{th}}$), and the photon lifetime in the cavity. The ion pair's concentration is responsible for the dynamics. For low x-values, EDFL is always CW, whereas for large x, the laser is self-pulsing for any pumping ratio achievable experimentally and for intermediate x-values, the EDFL behavior changes continuously from self-pulsing to CW operation via a Hopf bifurcation. The ion clusters (ion pairs) are able to explain qualitatively the whole dynamics observed when operating around 1.55 μm. There is a antiphase and chaotic dynamics in EDFL having an ion pair concentration of about $\approx$ 18%, when operating simultaneously at 1.55 and 1.536 μm. Experimentally it is approve that quasiperiodic route to chaos in erbium-doped fiber laser operate simultaneously at 1.55 and 1.536 μm (Fig. 7.5).

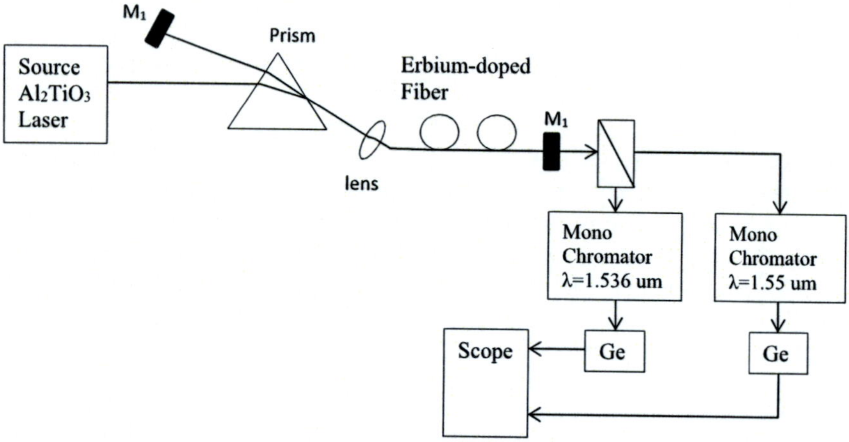

Fig. 7.5 Experimental set up system for two-wavelength operation

The experimental system includes the following elements: A Titane Sapphire laser which is used to optically pump the erbium-doped fiber. The output power is around 810 nm and about 800 mW which allows pumping ratios up to 20 times the threshold. The geometrical characteristics of the fiber are core diameter Φ_c, cutoff wavelength λ_c, and length (l). Typical doping concentrations are: 100 ppm Er^{3+}, 1000 ppm Al^{3+} and 145,000 ppm Ge^{4+}. The ion pair concentration is 7.5% ($x \simeq 7.5\%$). The ion pair concentration is responsible of the pulsed behavior of erbium-doped fiber lasers. We do not operate the laser pumping at 514.5 nm to avoid the photochromism resulting from the high Ge^{4+} concentration. The prism is inside the cavity in order to allow a spectral tune-ability together with a spatial separation between the pump beam and the laser beam. The spectral tune-ability is achieved with a tilt of mirror M1. We get an oscillation between the mirror M1 ($R = 100\%@1.5\,\mu\text{m}$) and the mirror M2 ($R = 80\%@1.5\,\mu\text{m}$). The output fiber end is directly butted on the output mirror. A collimating lens is placed after the output mirror. The laser beam is then split by a non-polarizing cube. The signals are analyzed through mono chromators adjusted respectively at 1.536 and 1.55 μm and detected with two germanium photodetectors. The output signal from the scope is analyzed by appropriate PC [5]. The erbium laser model simulation is for the case of laser operation around 1.55 μm. The model is based on existence of two atomic systems: isolated ions and ion pairs coupled via the laser field. The diagram of the energy levels adopted for two systems (an isolated ion and an ion pair) is described (Fig. 7.6). An isolated ion is modeled as a two-level system while ion pairs are assumed three level systems [6].

The system rate equations described by four differential nonlinear coupled equations.

$$\frac{dd}{dt} = \Lambda - a_2 \cdot (1 + d) - 2 \cdot I \cdot d$$

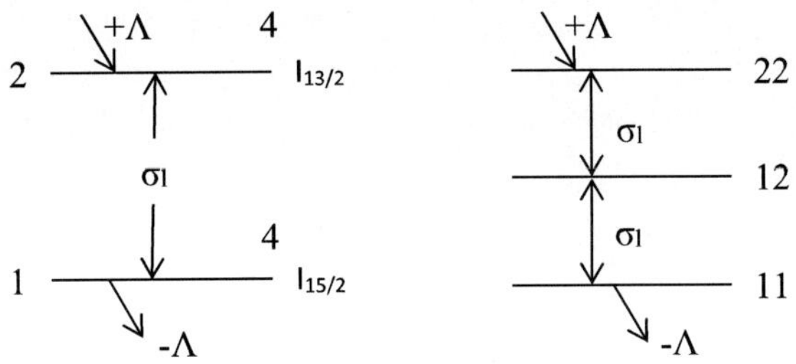

Fig. 7.6 Schematic energy levels for an isolated ion and an ion pair

$$\frac{dd_+}{dt} = a_{12}\cdot(1-d_+) - \frac{a_{22}}{2}\cdot(d_+ + d_-) + y\cdot I\cdot(2-3\cdot d_+)$$

$$\frac{dd_-}{dt} = \Lambda - a_{12}\cdot(1-d_+) - \frac{a_{22}}{2}\cdot(d_+ + d_-) + y\cdot I\cdot d_-$$

$$\frac{dI}{dt} = -I + (1-2\cdot x)\cdot A\cdot I\cdot d + x\cdot y\cdot A\cdot I\cdot d_-$$

$$d = d(t);\ d_+ = d_+(t);\ d_- = d_-(t);\ I = I(t)$$

where $d = n_2 - n_1$; $d_\pm = n_{22} \pm n_{11}$; $a_2 = \frac{\tau_1}{\tau_2}$; $a_{ij} = \frac{\tau_1}{\tau_{ij}}$; $A = \sigma_l \cdot N_0 \cdot \tau_1$; I, is the normalized laser intensity. Λ is the pumping rate, x is the ion pair concentration, y is the ratio between the absorption cross-sections of an ion pair and an isolated ion, n_k is the population of level $|k\rangle$ for the isolated ions and n_{kl} is the population of level $|kl\rangle$ for the ion pairs, N_0 is the erbium concentration, τ_1 is the photon lifetime and τ_k is the lifetime of level $|k\rangle$. Equation $\frac{dd}{dt} = \ldots$, describe the evolution of the population inversion for isolated ions. Equations $\frac{dd_+}{dt} = \ldots.$ and $\frac{dd_-}{dt} = \ldots$ are associated respectively with the sum (difference) of the populations of the upper and lower levels of ion pairs. Equation $\frac{dI}{dt} = \ldots$, describe the field dynamics. At fixed points (equilibrium points), $\frac{dd}{dt} = 0$; $\frac{dd_+}{dt} = 0$.

$$\frac{dd_-}{dt} = 0;\ \frac{dI}{dt} = 0,\ \Lambda - a_2\cdot(1+d^*) - 2\cdot I^*\cdot d^* = 0$$

$$a_{12}\cdot(1-d_+^*) - \frac{a_{22}}{2}\cdot(d_+^* + d_-^*) + y\cdot I^*\cdot(2-3\cdot d_+^*) = 0$$

$$\Lambda - a_{12}\cdot(1-d_+^*) - \frac{a_{22}}{2}\cdot(d_+^* + d_-^*) + y\cdot I^*\cdot d_-^* = 0$$

$$-I^* + (1-2\cdot x)\cdot A\cdot I^*\cdot d^* + x\cdot y\cdot A\cdot I^*\cdot d_-^* = 0$$

$$-I^* + (1-2\cdot x)\cdot A\cdot I^*\cdot d^* + x\cdot y\cdot A\cdot I^*\cdot d_-^* = 0$$
$$\Rightarrow [-1 + (1-2\cdot x)\cdot A\cdot d^* + x\cdot y\cdot A\cdot d_-^*]\cdot I^* = 0$$

Case A: $I^* = 0$ then

$$\Lambda - a_2 \cdot (1 + d^*) = 0; a_{12} \cdot (1 - d_+^*) - \frac{a_{22}}{2} \cdot (d_+^* + d_-^*) = 0$$
$$\Lambda - a_{12} \cdot (1 - d_+^*) - \frac{a_{22}}{2} \cdot (d_+^* + d_-^*) = 0$$

$$\Lambda - a_2 \cdot (1 + d^*) = 0 \Rightarrow d^* = \frac{\Lambda}{a_2} - 1; \frac{a_{22}}{2} \cdot (d_+^* + d_-^*) = a_{12} \cdot (1 - d_+^*)$$
$$\frac{a_{22}}{2} \cdot (d_+^* + d_-^*) = \Lambda - a_{12} \cdot (1 - d_+^*)$$

$$a_{12} \cdot (1 - d_+^*) = \Lambda - a_{12} \cdot (1 - d_+^*) \Rightarrow d_+^* = 1 - \frac{\Lambda}{2 \cdot a_{12}}$$

$$\begin{aligned} d_-^* &= 2 \cdot \frac{a_{12}}{a_{22}} \cdot (1 - d_+^*) - d_+^* \\ &= 2 \cdot \frac{a_{12}}{a_{22}} - 2 \cdot \frac{a_{12}}{a_{22}} \cdot d_+^* - d_+^* \\ &= 2 \cdot \frac{a_{12}}{a_{22}} - \left(2 \cdot \frac{a_{12}}{a_{22}} + 1\right) \cdot d_+^* \\ &= 2 \cdot \frac{a_{12}}{a_{22}} - \left(2 \cdot \frac{a_{12}}{a_{22}} + 1\right) \cdot \left(1 - \frac{\Lambda}{2 \cdot a_{12}}\right) \end{aligned}$$

$$d_-^* = 2 \cdot \frac{a_{12}}{a_{22}} - \left(2 \cdot \frac{a_{12}}{a_{22}} + 1\right) \cdot \left(1 - \frac{\Lambda}{2 \cdot a_{12}}\right) = \left(\frac{1}{a_{22}} + \frac{1}{2 \cdot a_{12}}\right) \cdot \Lambda - 1$$

First fixed point: $E^{(0)} = (d^{(0)}, d_+^{(0)}, d_-^{(0)}, I^{(0)}) = (\frac{\Lambda}{a_2} - 1, 1 - \frac{\Lambda}{2 \cdot a_{12}}, (\frac{1}{a_{22}} + \frac{1}{2 \cdot a_{12}}) \cdot \Lambda - 1, 0)$.

Case B: $-1 + (1 - 2 \cdot x) \cdot A \cdot d^* + x \cdot y \cdot A \cdot d_-^* = 0 \Rightarrow d^* = \frac{1 - x \cdot y \cdot A \cdot d_-^*}{(1 - 2 \cdot x) \cdot A}$

$$\begin{aligned} \Lambda - a_2 \cdot (1 + d^*) - 2 \cdot I^* \cdot d^* = 0 \Rightarrow I^* &= \frac{\Lambda - a_2 \cdot (1 + d^*)}{2 \cdot d^*} \\ &= \frac{\Lambda - a_2 - a_2 \cdot d^*}{2 \cdot d^*} \\ &= \frac{1}{2 \cdot d^*} \cdot (\Lambda - a_2) - \frac{a_2}{2} \end{aligned}$$

1. $$\begin{aligned} &a_{12} \cdot (1 - d_+^*) - \frac{a_{22}}{2} \cdot (d_+^* + d_-^*) \\ &+ y \cdot \left(\frac{\Lambda - a_2 \cdot (1 + d^*)}{2 \cdot d^*}\right) \cdot (2 - 3 \cdot d_+^*) = 0 \end{aligned}$$

$$a_{12} - a_{12} \cdot d_+^* - \frac{a_{22}}{2} \cdot d_+^* - \frac{a_{22}}{2} \cdot d_-^*$$

$$+y\cdot\left(\frac{\Lambda}{2\cdot d^*}-\frac{a_2}{2\cdot d^*}-\frac{a_2}{2}\right)\cdot(2-3\cdot d_+^*)=0$$

$$a_{12}-a_{12}\cdot d_+^*-\frac{a_{22}}{2}\cdot d_+^*-\frac{a_{22}}{2}\cdot d_-^*+y\cdot\left(\frac{\Lambda}{d^*}-\frac{3\cdot\Lambda}{2\cdot d^*}\cdot d_+^*\right.$$
$$\left.-\frac{a_2}{d^*}+\frac{3\cdot a_2}{2\cdot d^*}\cdot d_+^*-a_2+\frac{3\cdot a_2}{2}\cdot d_+^*\right)=0$$

$$a_{12}-a_{12}\cdot d_+^*-\frac{a_{22}}{2}\cdot d_+^*-\frac{a_{22}}{2}\cdot d_-^*+\frac{y\cdot\Lambda}{d^*}-\frac{y\cdot 3\cdot\Lambda}{2\cdot d^*}\cdot d_+^*$$
$$-\frac{y\cdot a_2}{d^*}+\frac{y\cdot 3\cdot a_2}{2\cdot d^*}\cdot d_+^*-y\cdot a_2+\frac{y\cdot 3\cdot a_2}{2}\cdot d_+^*=0$$

$$a_{12}-a_{12}\cdot d_+^*-\frac{a_{22}}{2}\cdot d_+^*-\frac{a_{22}}{2}\cdot d_-^*+\frac{1}{d^*}\cdot\left[y\cdot\Lambda-\frac{y\cdot 3\cdot\Lambda}{2}\cdot d_+^*\right.$$
$$\left.-y\cdot a_2+\frac{y\cdot 3\cdot a_2}{2}\cdot d_+^*\right]-y\cdot a_2+\frac{y\cdot 3\cdot a_2}{2}\cdot d_+^*=0$$

$$\frac{1}{d^*}\cdot\left[y\cdot(\Lambda-a_2)-\frac{y\cdot 3\cdot\Lambda}{2}\cdot d_+^*+\frac{y\cdot 3\cdot a_2}{2}\cdot d_+^*\right]$$
$$=y\cdot a_2-\frac{y\cdot 3\cdot a_2}{2}\cdot d_+^*+\frac{a_{22}}{2}\cdot d_-^*+\frac{a_{22}}{2}\cdot d_+^*+a_{12}\cdot d_+^*-a_{12}$$

$$\frac{1}{d^*}=\frac{y\cdot a_2-\frac{y\cdot 3\cdot a_2}{2}\cdot d_+^*+\frac{a_{22}}{2}\cdot d_-^*+\frac{a_{22}}{2}\cdot d_+^*+a_{12}\cdot d_+^*-a_{12}}{y\cdot(\Lambda-a_2)-\frac{y\cdot 3\cdot\Lambda}{2}\cdot d_+^*+\frac{y\cdot 3\cdot a_2}{2}\cdot d_+^*}$$

$$d^*=\frac{y\cdot(\Lambda-a_2)-\frac{y\cdot 3\cdot\Lambda}{2}\cdot d_+^*+\frac{y\cdot 3\cdot a_2}{2}\cdot d_+^*}{y\cdot a_2-\frac{y\cdot 3\cdot a_2}{2}\cdot d_+^*+\frac{a_{22}}{2}\cdot d_-^*+\frac{a_{22}}{2}\cdot d_+^*+a_{12}\cdot d_+^*-a_{12}}$$

$$\frac{1-x\cdot y\cdot A\cdot d_-^*}{(1-2\cdot x)\cdot A}=\frac{y\cdot(\Lambda-a_2)-\frac{y\cdot 3\cdot\Lambda}{2}\cdot d_+^*+\frac{y\cdot 3\cdot a_2}{2}\cdot d_+^*}{y\cdot a_2-\frac{y\cdot 3\cdot a_2}{2}\cdot d_+^*+\frac{a_{22}}{2}\cdot d_-^*+\frac{a_{22}}{2}\cdot d_+^*+a_{12}\cdot d_+^*-a_{12}}$$

2. $\Lambda-a_{12}\cdot(1-d_+^*)-\frac{a_{22}}{2}\cdot(d_+^*+d_-^*)+y\cdot\left(\frac{\Lambda-a_2\cdot(1+d^*)}{2\cdot d^*}\right)\cdot d_-^*=0$

$$\Lambda-a_{12}+a_{12}\cdot d_+^*-\frac{a_{22}}{2}\cdot d_+^*-\frac{a_{22}}{2}\cdot d_-^*+y\cdot\left(\frac{\Lambda}{2\cdot d^*}-\frac{a_2}{2\cdot d^*}-\frac{a_2}{2}\right)\cdot d_-^*=0$$

$$\Lambda-a_{12}+a_{12}\cdot d_+^*-\frac{a_{22}}{2}\cdot d_+^*-\frac{a_{22}}{2}\cdot d_-^*+y\cdot\frac{\Lambda}{2\cdot d^*}\cdot d_-^*$$
$$-y\cdot\frac{a_2}{2\cdot d^*}\cdot d_-^*-y\cdot\frac{a_2}{2}\cdot d_-^*=0$$

$$\frac{y}{2 \cdot d^*} \cdot (\Lambda - a_2) \cdot d_-^* = \frac{1}{2} \cdot (y \cdot a_2 + a_{22}) \cdot d_-^* + \left(\frac{a_{22}}{2} - a_{12}\right) \cdot d_+^* - \Lambda + a_{12}$$

$$y \cdot (\Lambda - a_2) \cdot d_-^* = 2 \cdot d^* \cdot \left[\frac{1}{2} \cdot (y \cdot a_2 + a_{22}) \cdot d_-^* + \left(\frac{a_{22}}{2} - a_{12}\right) \cdot d_+^* - \Lambda + a_{12}\right]$$

$$d^* = \frac{y \cdot (\Lambda - a_2) \cdot d_-^*}{2 \cdot d^* \cdot [\frac{1}{2} \cdot (y \cdot a_2 + a_{22}) \cdot d_-^* + (\frac{a_{22}}{2} - a_{12}) \cdot d_+^* - \Lambda + a_{12}]}$$

$$\frac{1 - x \cdot y \cdot A \cdot d_-^*}{(1 - 2 \cdot x) \cdot A} = \frac{y \cdot (\Lambda - a_2) \cdot d_-^*}{2 \cdot d^* \cdot [\frac{1}{2} \cdot (y \cdot a_2 + a_{22}) \cdot d_-^* + (\frac{a_{22}}{2} - a_{12}) \cdot d_+^* - \Lambda + a_{12}]}$$

We can summary our last results for second, third... Fixed points d_+^* and d_-^* coordinate

$$\frac{1 - x \cdot y \cdot A \cdot d_-^*}{(1 - 2 \cdot x) \cdot A} = \frac{y \cdot (\Lambda - a_2) - \frac{y \cdot 3 \cdot \Lambda}{2} \cdot d_+^* + \frac{y \cdot 3 \cdot a_2}{2} \cdot d_+^*}{y \cdot a_2 - \frac{y \cdot 3 \cdot a_2}{2} \cdot d_+^* + \frac{a_{22}}{2} \cdot d_-^* + \frac{a_{22}}{2} \cdot d_+^* + a_{12} \cdot d_+^* - a_{12}}$$

$$\frac{1 - x \cdot y \cdot A \cdot d_-^*}{(1 - 2 \cdot x) \cdot A} = \frac{y \cdot (\Lambda - a_2) \cdot d_-^*}{2 \cdot d^* \cdot [\frac{1}{2} \cdot (y \cdot a_2 + a_{22}) \cdot d_-^* + (\frac{a_{22}}{2} - a_{12}) \cdot d_+^* - \Lambda + a_{12}]}$$

It is recommended to solve the above two equations for d_+^* and d_-^* coordinate numerically rather than analytic. Then we can get system second, third... fixed points.

$$\begin{aligned} E^{(k)} &= (d^{(k)}, d_+^{(k)}, d_-^{(k)}, I^{(k)}) \\ &= \left(\frac{1 - x \cdot y \cdot A \cdot d_-^*}{(1 - 2 \cdot x) \cdot A}, d_+^*, d_-^*, \frac{1}{2 \cdot d^*} \cdot (\Lambda \right. \\ &\quad \left. - a_2) - \frac{a_2}{2}\right); \quad k > 0; k \in \mathbb{N}; k = 1, 2, 3, \ldots \end{aligned}$$

$$\begin{aligned} E^{(k)} &= (d^{(k)}, d_+^{(k)}, d_-^{(k)}, I^{(k)}) \\ &= \left(\frac{1 - x \cdot y \cdot A \cdot d_-^*}{(1 - 2 \cdot x) \cdot A}, d_+^*, d_-^*, \frac{(1 - 2 \cdot x) \cdot A \cdot (\Lambda - a_2)}{2 \cdot (1 - x \cdot y \cdot A \cdot d_-^*)} - \frac{a_2}{2}\right) \\ &\forall k > 0; k \in \mathbb{N}; k = 1, 2, 3, \ldots \end{aligned}$$

Stability analysis: The standard local stability analysis about any one of the equilibrium point of the Er-doped fiber laser (EDFL) system consists in adding to coordinate $[d, d_+, d_-, I]$ arbitrarily small increments of exponential form $[d, d_+, d_-, I] \cdot e^{\lambda \cdot t}$ and retaining the first order terms in d, d_+, d_-, I. The system of four homogeneous equations leads to a polynomial characteristic equation in the eigenvalues. The polynomial characteristic equations accept by set the below variables and variables derivative with respect to time into Er-doped fiber laser (EDFL) system equations. The Er-doped fiber laser (EDFL) system fixed values with arbitrarily small increments of exponential form $[d, d_+, d_-, I] \cdot e^{\lambda \cdot t}$ are: $j = 0$ (first fixed point), $j = 1$ (second fixed point), $j = 2$ (third fixed point), etc. [2, 3].

$$d(t) = d^{(j)} + d \cdot e^{\lambda \cdot t};\ d_+(t) = d_+^{(j)} + d_+ \cdot e^{\lambda \cdot t};\ d_-(t) = d_-^{(j)} + d_- \cdot e^{\lambda \cdot t}$$
$$I(t) = I^{(j)} + I \cdot e^{\lambda \cdot t}$$

$$\frac{dd(t)}{dt} = \lambda \cdot d \cdot e^{\lambda \cdot t};\ \frac{dd_+(t)}{dt} = \lambda \cdot d_+ \cdot e^{\lambda \cdot t};\ \frac{dd_-(t)}{dt} = \lambda \cdot d_- \cdot e^{\lambda \cdot t}$$
$$\frac{dI(t)}{dt} = \lambda \cdot I \cdot e^{\lambda \cdot t}$$

$$\frac{dd}{dt} = \Lambda - a_2 \cdot (1 + d) - 2 \cdot I \cdot d$$

$$\lambda \cdot d \cdot e^{\lambda \cdot t} = \Lambda - a_2 \cdot (1 + d^{(j)} + d \cdot e^{\lambda \cdot t}) - 2 \cdot (I^{(j)} + I \cdot e^{\lambda \cdot t}) \cdot (d^{(j)} + d \cdot e^{\lambda \cdot t})$$

$$\lambda \cdot d \cdot e^{\lambda \cdot t} = \Lambda - (a_2 + a_2 \cdot d^{(j)} + a_2 \cdot d \cdot e^{\lambda \cdot t}) - 2 \cdot (I^{(j)} \cdot d^{(j)} + I^{(j)} \cdot d \cdot e^{\lambda \cdot t} + d^{(j)} \cdot I \cdot e^{\lambda \cdot t} + I \cdot d \cdot e^{\lambda \cdot t} \cdot e^{\lambda \cdot t})$$

Assumption: $I \cdot d \approx 0$

$$\lambda \cdot d \cdot e^{\lambda \cdot t} = \Lambda - (a_2 + a_2 \cdot d^{(j)} + a_2 \cdot d \cdot e^{\lambda \cdot t}) - 2 \cdot (I^{(j)} \cdot d^{(j)} + I^{(j)} \cdot d \cdot e^{\lambda \cdot t} + d^{(j)} \cdot I \cdot e^{\lambda \cdot t})$$

$$\lambda \cdot d \cdot e^{\lambda \cdot t} = \Lambda - a_2 \cdot (1 + d^{(j)}) - 2 \cdot I^{(j)} \cdot d^{(j)} - a_2 \cdot d \cdot e^{\lambda \cdot t} - 2 \cdot I^{(j)} \cdot d \cdot e^{\lambda \cdot t} - 2 \cdot d^{(j)} \cdot I \cdot e^{\lambda \cdot t}$$

At fixed points: $\Lambda - a_2 \cdot (1 + d^{(j)}) - 2 \cdot I^{(j)} \cdot d^{(j)} = 0; -(a_2 + 2 \cdot I^{(j)}) \cdot d - \lambda \cdot d - 2 \cdot d^{(j)} \cdot I = 0$.

$$\frac{dd_+}{dt} = a_{12} \cdot (1 - d_+) - \frac{a_{22}}{2} \cdot (d_+ + d_-) + y \cdot I \cdot (2 - 3 \cdot d_+)$$

$$\lambda \cdot d_+ \cdot e^{\lambda \cdot t} = a_{12} \cdot (1 - [d_+^{(j)} + d_+ \cdot e^{\lambda \cdot t}]) - \frac{a_{22}}{2} \cdot (d_+^{(j)} + d_+ \cdot e^{\lambda \cdot t} + d_-^{(j)} + d_- \cdot e^{\lambda \cdot t}) + y \cdot (I^{(j)} + I \cdot e^{\lambda \cdot t}) \cdot (2 - 3 \cdot [d_+^{(j)} + d_+ \cdot e^{\lambda \cdot t}])$$

$$\lambda \cdot d_+ \cdot e^{\lambda \cdot t} = a_{12} \cdot (1 - d_+^{(j)}) - a_{12} \cdot d_+ \cdot e^{\lambda \cdot t} - \frac{a_{22}}{2} \cdot (d_+^{(j)} + d_-^{(j)}) - \frac{a_{22}}{2} \cdot (d_+ \cdot e^{\lambda \cdot t} + d_- \cdot e^{\lambda \cdot t}) + y \cdot (2 \cdot I^{(j)} - 3 \cdot I^{(j)} \cdot [d_+^{(j)} + d_+ \cdot e^{\lambda \cdot t}] + 2 \cdot I \cdot e^{\lambda \cdot t}$$

$$- I \cdot e^{\lambda \cdot t} \cdot 3 \cdot [d_+^{(j)} + d_+ \cdot e^{\lambda \cdot t}])$$

$$\lambda \cdot d_+ \cdot e^{\lambda \cdot t} = a_{12} \cdot (1 - d_+^{(j)}) - a_{12} \cdot d_+ \cdot e^{\lambda \cdot t} - \frac{a_{22}}{2} \cdot (d_+^{(j)} + d_-^{(j)})$$
$$- \frac{a_{22}}{2} \cdot (d_+ \cdot e^{\lambda \cdot t} + d_- \cdot e^{\lambda \cdot t})$$
$$+ y \cdot (2 \cdot I^{(j)} - 3 \cdot I^{(j)} \cdot d_+^{(j)} - 3 \cdot I^{(j)} \cdot d_+ \cdot e^{\lambda \cdot t} + 2 \cdot I \cdot e^{\lambda \cdot t}$$
$$- I \cdot e^{\lambda \cdot t} \cdot 3 \cdot d_+^{(j)} - I \cdot d_+ \cdot e^{\lambda \cdot t} \cdot 3 \cdot e^{\lambda \cdot t})$$

$$\lambda \cdot d_+ \cdot e^{\lambda \cdot t} = a_{12} \cdot (1 - d_+^{(j)}) - \frac{a_{22}}{2} \cdot (d_+^{(j)} + d_-^{(j)}) + y \cdot I^{(j)} \cdot (2 - 3 \cdot d_+^{(j)})$$
$$- a_{12} \cdot d_+ \cdot e^{\lambda \cdot t} - \frac{a_{22}}{2} \cdot (d_+ \cdot e^{\lambda \cdot t} + d_- \cdot e^{\lambda \cdot t}) + y \cdot (2 \cdot I$$
$$- I \cdot 3 \cdot d_+^{(j)} - I \cdot d_+ \cdot e^{\lambda \cdot t} \cdot 3 - 3 \cdot I^{(j)} \cdot d_+) \cdot e^{\lambda \cdot t}$$

Assumption: $I \cdot d_+ \approx 0$

$$\lambda \cdot d_+ \cdot e^{\lambda \cdot t} = a_{12} \cdot (1 - d_+^{(j)}) - \frac{a_{22}}{2} \cdot (d_+^{(j)} + d_-^{(j)}) + y \cdot I^{(j)} \cdot (2 - 3 \cdot d_+^{(j)})$$
$$- a_{12} \cdot d_+ \cdot e^{\lambda \cdot t} - \frac{a_{22}}{2} \cdot (d_+ + d_-) \cdot e^{\lambda \cdot t} + y \cdot (2 \cdot I - I \cdot 3 \cdot d_+^{(j)}$$
$$- 3 \cdot I^{(j)} \cdot d_+) \cdot e^{\lambda \cdot t}$$

At fixed points: $a_{12} \cdot (1 - d_+^{(j)}) - \frac{a_{22}}{2} \cdot (d_+^{(j)} + d_-^{(j)}) + y \cdot I^{(j)} \cdot (2 - 3 \cdot d_+^{(j)}) = 0$.

$$- a_{12} \cdot d_+ - \frac{a_{22}}{2} \cdot (d_+ + d_-) + y \cdot 2 \cdot I - I \cdot y \cdot 3 \cdot d_+^{(j)}$$
$$- 3 \cdot y \cdot I^{(j)} \cdot d_+ - \lambda \cdot d_+ = 0$$

$$-(a_{12} + \frac{a_{22}}{2} + 3 \cdot y \cdot I^{(j)}) \cdot d_+ - \lambda \cdot d_+ - \frac{a_{22}}{2} \cdot d_- + y \cdot (2 - 3 \cdot d_+^{(j)}) \cdot I = 0$$

$$\frac{dd_-}{dt} = \Lambda - a_{12} \cdot (1 - d_+) - \frac{a_{22}}{2} \cdot (d_+ + d_-) + y \cdot I \cdot d_-$$

$$\lambda \cdot d_- \cdot e^{\lambda \cdot t} = \Lambda - a_{12} \cdot (1 - [d_+^{(j)} + d_+ \cdot e^{\lambda \cdot t}])$$
$$- \frac{a_{22}}{2} \cdot (d_+^{(j)} + d_+ \cdot e^{\lambda \cdot t} + d_-^{(j)} + d_- \cdot e^{\lambda \cdot t})$$
$$+ y \cdot (I^{(j)} + I \cdot e^{\lambda \cdot t}) \cdot (d_-^{(j)} + d_- \cdot e^{\lambda \cdot t})$$

$$\lambda \cdot d_- \cdot e^{\lambda \cdot t} = \Lambda - a_{12} \cdot (1 - d_+^{(j)}) + a_{12} \cdot d_+ \cdot e^{\lambda \cdot t} - \frac{a_{22}}{2} \cdot (d_+^{(j)} + d_-^{(j)})$$
$$- \frac{a_{22}}{2} \cdot (d_+ + d_-) \cdot e^{\lambda \cdot t}$$

$$+ y \cdot (I^{(j)} \cdot d_-^{(j)} + I^{(j)} \cdot d_- \cdot e^{\lambda \cdot t} + d_-^{(j)} \cdot I \cdot e^{\lambda \cdot t} + I \cdot d_- \cdot e^{\lambda \cdot t} \cdot e^{\lambda \cdot t})$$

Assumption: $I \cdot d_- \approx 0$

$$\lambda \cdot d_- \cdot e^{\lambda \cdot t} = \Lambda - a_{12} \cdot (1 - d_+^{(j)}) + a_{12} \cdot d_+ \cdot e^{\lambda \cdot t} - \frac{a_{22}}{2} \cdot (d_+^{(j)} + d_-^{(j)})$$
$$- \frac{a_{22}}{2} \cdot (d_+ + d_-) \cdot e^{\lambda \cdot t}$$
$$+ y \cdot I^{(j)} \cdot d_-^{(j)} + y \cdot (I^{(j)} \cdot d_- + d_-^{(j)} \cdot I) \cdot e^{\lambda \cdot t}$$

$$\lambda \cdot d_- \cdot e^{\lambda \cdot t} = \Lambda - a_{12} \cdot (1 - d_+^{(j)}) - \frac{a_{22}}{2} \cdot (d_+^{(j)} + d_-^{(j)})$$
$$+ y \cdot I^{(j)} \cdot d_-^{(j)} + a_{12} \cdot d_+ \cdot e^{\lambda \cdot t}$$
$$- \frac{a_{22}}{2} \cdot (d_+ + d_-) \cdot e^{\lambda \cdot t} + y \cdot (I^{(j)} \cdot d_- + d_-^{(j)} \cdot I) \cdot e^{\lambda \cdot t}$$

At fixed points: $\Lambda - a_{12} \cdot (1 - d_+^{(j)}) - \frac{a_{22}}{2} \cdot (d_+^{(j)} + d_-^{(j)}) + y \cdot I^{(j)} \cdot d_-^{(j)} = 0$.

$$\lambda \cdot d_- = a_{12} \cdot d_+ - \frac{a_{22}}{2} \cdot (d_+ + d_-) + y \cdot (I^{(j)} \cdot d_- + d_-^{(j)} \cdot I)$$

$$\left(a_{12} - \frac{a_{22}}{2}\right) \cdot d_+ + \left(y \cdot I^{(j)} - \frac{a_{22}}{2}\right) \cdot d_- - \lambda \cdot d_- + y \cdot d_-^{(j)} \cdot I = 0$$

$$\frac{dI}{dt} = -I + (1 - 2 \cdot x) \cdot A \cdot I \cdot d + x \cdot y \cdot A \cdot I \cdot d_-$$

$$\lambda \cdot I \cdot e^{\lambda \cdot t} = -(I^{(j)} + I \cdot e^{\lambda \cdot t}) + (1 - 2 \cdot x) \cdot A \cdot (I^{(j)}$$
$$+ I \cdot e^{\lambda \cdot t}) \cdot (d^{(j)} + d \cdot e^{\lambda \cdot t})$$
$$+ x \cdot y \cdot A \cdot (I^{(j)} + I \cdot e^{\lambda \cdot t}) \cdot (d_-^{(j)} + d_- \cdot e^{\lambda \cdot t})$$

$$\lambda \cdot I \cdot e^{\lambda \cdot t} = -(I^{(j)} + I \cdot e^{\lambda \cdot t}) + (1 - 2 \cdot x) \cdot A \cdot (I^{(j)} \cdot d^{(j)}$$
$$+ I^{(j)} \cdot d \cdot e^{\lambda \cdot t} + d^{(j)} \cdot I \cdot e^{\lambda \cdot t} + I \cdot d \cdot e^{\lambda \cdot t} \cdot e^{\lambda \cdot t})$$
$$+ x \cdot y \cdot A \cdot (I^{(j)} \cdot d_-^{(j)} + I^{(j)} \cdot d_- \cdot e^{\lambda \cdot t} + d_-^{(j)} \cdot I \cdot e^{\lambda \cdot t}$$
$$+ I \cdot d_- \cdot e^{\lambda \cdot t} \cdot e^{\lambda \cdot t})$$

At fixed points: $I \cdot d \approx 0$; $I \cdot d_- \approx 0$.

$$\lambda \cdot I \cdot e^{\lambda \cdot t} = -(I^{(j)} + I \cdot e^{\lambda \cdot t}) + (1 - 2 \cdot x) \cdot A \cdot (I^{(j)} \cdot d^{(j)})$$
$$+ (1 - 2 \cdot x) \cdot A \cdot (I^{(j)} \cdot d + d^{(j)} \cdot I) \cdot e^{\lambda \cdot t}$$
$$+ x \cdot y \cdot A \cdot (I^{(j)} \cdot d_-^{(j)}) + x \cdot y \cdot A \cdot (I^{(j)} \cdot d_- + d_-^{(j)} \cdot I) \cdot e^{\lambda \cdot t}$$

$$\lambda \cdot I \cdot e^{\lambda \cdot t} = -I^{(j)} + (1 - 2 \cdot x) \cdot A \cdot I^{(j)} \cdot d^{(j)} + x \cdot y \cdot A \cdot I^{(j)} \cdot d_{-}^{(j)} \\ - I \cdot e^{\lambda \cdot t} + (1 - 2 \cdot x) \cdot A \cdot (I^{(j)} \cdot d + d^{(j)} \cdot I) \cdot e^{\lambda \cdot t} \\ + x \cdot y \cdot A \cdot (I^{(j)} \cdot d_{-} + d_{-}^{(j)} \cdot I) \cdot e^{\lambda \cdot t}$$

At fixed points: $-I^{(j)} + (1 - 2 \cdot x) \cdot A \cdot I^{(j)} \cdot d^{(j)} + x \cdot y \cdot A \cdot I^{(j)} \cdot d_{-}^{(j)} = 0$.

$$\lambda \cdot I = -I + (1 - 2 \cdot x) \cdot A \cdot (I^{(j)} \cdot d + d^{(j)} \cdot I) \\ + x \cdot y \cdot A \cdot (I^{(j)} \cdot d_{-} + d_{-}^{(j)} \cdot I)$$

$$(1 - 2 \cdot x) \cdot A \cdot I^{(j)} \cdot d + x \cdot y \cdot A \cdot I^{(j)} \cdot d_{-} + [(1 - 2 \cdot x) \cdot A \cdot d^{(j)} \\ + x \cdot y \cdot A \cdot d_{-}^{(j)} - 1] \cdot I - \lambda \cdot I = 0$$

We can summary our system four arbitrarily equations:

$$-(a_2 + 2 \cdot I^{(j)}) \cdot d - \lambda \cdot d - 2 \cdot d^{(j)} \cdot I = 0$$

$$-\left(a_{12} + \frac{a_{22}}{2} + 3 \cdot y \cdot I^{(j)}\right) \cdot d_{+} - \lambda \cdot d_{+} - \frac{a_{22}}{2} \cdot d_{-} \\ + y \cdot (2 - 3 \cdot d_{+}^{(j)}) \cdot I = 0$$

$$\left(a_{12} - \frac{a_{22}}{2}\right) \cdot d_{+} + \left(y \cdot I^{(j)} - \frac{a_{22}}{2}\right) \cdot d_{-} - \lambda \cdot d_{-} + y \cdot d_{-}^{(j)} \cdot I = 0$$

$$(1 - 2 \cdot x) \cdot A \cdot I^{(j)} \cdot d + x \cdot y \cdot A \cdot I^{(j)} \cdot d_{-} + [(1 - 2 \cdot x) \cdot A \cdot d^{(j)} \\ + x \cdot y \cdot A \cdot d_{-}^{(j)} - 1] \cdot I - \lambda \cdot I = 0$$

We define for simplicity the following global parameters: $\Upsilon_1 = -(a_2 + 2 \cdot I^{(j)})$

$$\Upsilon_2 = -\left(a_{12} + \frac{a_{22}}{2} + 3 \cdot y \cdot I^{(j)}\right); \Upsilon_3 = y \cdot (2 - 3 \cdot d_{+}^{(j)}) \\ \Upsilon_4 = a_{12} - \frac{a_{22}}{2}; \Upsilon_5 = y \cdot I^{(j)} - \frac{a_{22}}{2}$$

$$\Upsilon_6 = (1 - 2 \cdot x) \cdot A \cdot I^{(j)}; \Upsilon_7 = x \cdot y \cdot A \cdot I^{(j)} \\ \Upsilon_8 = (1 - 2 \cdot x) \cdot A \cdot d^{(j)} + x \cdot y \cdot A \cdot d_{-}^{(j)} - 1$$

We can summary our system four arbitrarily equations ($\Upsilon_k; k = 1, 2, \ldots, 8$):

$$\Upsilon_1 \cdot d - \lambda \cdot d - 2 \cdot d^{(j)} \cdot I = 0; \Upsilon_2 \cdot d_{+} - \lambda \cdot d_{+} - \frac{a_{22}}{2} \cdot d_{-} + \Upsilon_3 \cdot I = 0$$

$$\Upsilon_4 \cdot d_+ + \Upsilon_5 \cdot d_- - \lambda \cdot d_- + y \cdot d_-^{(j)} \cdot I = 0$$
$$\Upsilon_6 \cdot d + \Upsilon_7 \cdot d_- + \Upsilon_8 \cdot I - \lambda \cdot I = 0$$

Er-doped fiber laser (EDFL) system arbitrarily increment matrix is

$$\begin{bmatrix} (\Upsilon_1 - \lambda) & 0 & 0 & -2 \cdot d^{(j)} \\ 0 & (\Upsilon_2 - \lambda) & -\frac{a_{22}}{2} & \Upsilon_3 \\ 0 & \Upsilon_4 & (\Upsilon_5 - \lambda) & y \cdot d_-^{(j)} \\ \Upsilon_6 & 0 & \Upsilon_7 & (\Upsilon_8 - \lambda) \end{bmatrix} \cdot \begin{bmatrix} d \\ d_+ \\ d_- \\ I \end{bmatrix} = 0$$

$$A - \lambda \cdot I = \begin{bmatrix} (\Upsilon_1 - \lambda) & 0 & 0 & -2 \cdot d^{(j)} \\ 0 & (\Upsilon_2 - \lambda) & -\frac{a_{22}}{2} & \Upsilon_3 \\ 0 & \Upsilon_4 & (\Upsilon_5 - \lambda) & y \cdot d_-^{(j)} \\ \Upsilon_6 & 0 & \Upsilon_7 & (\Upsilon_8 - \lambda) \end{bmatrix}$$

$$\det(A - \lambda \cdot I) = 0$$

$$\det(A - \lambda \cdot I) = (\Upsilon_1 - \lambda) \cdot \det\begin{pmatrix} (\Upsilon_2 - \lambda) & -\frac{a_{22}}{2} & \Upsilon_3 \\ \Upsilon_4 & (\Upsilon_5 - \lambda) & y \cdot d_-^{(j)} \\ 0 & \Upsilon_7 & (\Upsilon_8 - \lambda) \end{pmatrix}$$
$$+ 2 \cdot d^{(j)} \cdot \det\begin{pmatrix} 0 & (\Upsilon_2 - \lambda) & -\frac{a_{22}}{2} \\ 0 & \Upsilon_4 & (\Upsilon_5 - \lambda) \\ \Upsilon_6 & 0 & \Upsilon_7 \end{pmatrix}$$

Step 1: $\det\begin{pmatrix} (\Upsilon_2 - \lambda) & -\frac{a_{22}}{2} & \Upsilon_3 \\ \Upsilon_4 & (\Upsilon_5 - \lambda) & y \cdot d_-^{(j)} \\ 0 & \Upsilon_7 & (\Upsilon_8 - \lambda) \end{pmatrix}$.

$$\det\begin{pmatrix} (\Upsilon_2 - \lambda) & -\frac{a_{22}}{2} & \Upsilon_3 \\ \Upsilon_4 & (\Upsilon_5 - \lambda) & y \cdot d_-^{(j)} \\ 0 & \Upsilon_7 & (\Upsilon_8 - \lambda) \end{pmatrix}$$
$$= (\Upsilon_2 - \lambda) \cdot \det\begin{pmatrix} (\Upsilon_5 - \lambda) & y \cdot d_-^{(j)} \\ \Upsilon_7 & (\Upsilon_8 - \lambda) \end{pmatrix}$$
$$- \Upsilon_4 \cdot \det\begin{pmatrix} -\frac{a_{22}}{2} & \Upsilon_3 \\ \Upsilon_7 & (\Upsilon_8 - \lambda) \end{pmatrix}$$

$$\det\begin{pmatrix} (\Upsilon_2 - \lambda) & -\frac{a_{22}}{2} & \Upsilon_3 \\ \Upsilon_4 & (\Upsilon_5 - \lambda) & y \cdot d_-^{(j)} \\ 0 & \Upsilon_7 & (\Upsilon_8 - \lambda) \end{pmatrix}$$
$$= (\Upsilon_2 - \lambda) \cdot [(\Upsilon_5 - \lambda) \cdot (\Upsilon_8 - \lambda)$$

$$- \Upsilon_7 \cdot y \cdot d_-^{(j)}] + \Upsilon_4 \cdot \left[\frac{a_{22}}{2} \cdot (\Upsilon_8 - \lambda) + \Upsilon_7 \cdot \Upsilon_3\right]$$
$$= -\lambda^3 + \lambda^2 \cdot (\Upsilon_8 + \Upsilon_2 + \Upsilon_5) + \lambda \cdot \Big(\Upsilon_7 \cdot y \cdot d_-^{(j)}$$
$$- \Upsilon_8 \cdot \Upsilon_5 - \Upsilon_2 \cdot \Upsilon_5 - \Upsilon_2 \cdot \Upsilon_8 - \Upsilon_4 \cdot \frac{a_{22}}{2}\Big)$$
$$+ \Upsilon_2 \cdot \Upsilon_8 \cdot \Upsilon_5 - \Upsilon_2 \cdot \Upsilon_7 \cdot y \cdot d_-^{(j)}$$
$$+ \Upsilon_4 \cdot \frac{a_{22}}{2} \cdot \Upsilon_8 + \Upsilon_4 \cdot \Upsilon_7 \cdot \Upsilon_3$$

We define for simplicity global parameters: $\Xi_1 = \Upsilon_8 + \Upsilon_2 + \Upsilon_5$

$$\Xi_2 = \Upsilon_7 \cdot y \cdot d_-^{(j)} - \Upsilon_8 \cdot \Upsilon_5 - \Upsilon_2 \cdot \Upsilon_5 - \Upsilon_2 \cdot \Upsilon_8 - \Upsilon_4 \cdot \frac{a_{22}}{2}$$

$$\Xi_3 = \Upsilon_2 \cdot \Upsilon_8 \cdot \Upsilon_5 - \Upsilon_2 \cdot \Upsilon_7 \cdot y \cdot d_-^{(j)} + \Upsilon_4 \cdot \frac{a_{22}}{2} \cdot \Upsilon_8 + \Upsilon_4 \cdot \Upsilon_7 \cdot \Upsilon_3$$

$$\det\begin{pmatrix} (\Upsilon_2 - \lambda) & -\frac{a_{22}}{2} & \Upsilon_3 \\ \Upsilon_4 & (\Upsilon_5 - \lambda) & y \cdot d_-^{(j)} \\ 0 & \Upsilon_7 & (\Upsilon_8 - \lambda) \end{pmatrix} = -\lambda^3 + \lambda^2 \cdot \Xi_1 + \lambda \cdot \Xi_2 + \Xi_3$$

Step 2: $\det\begin{pmatrix} 0 & (\Upsilon_2 - \lambda) & -\frac{a_{22}}{2} \\ 0 & \Upsilon_4 & (\Upsilon_5 - \lambda) \\ \Upsilon_6 & 0 & \Upsilon_7 \end{pmatrix}$.

$$\det\begin{pmatrix} 0 & (\Upsilon_2 - \lambda) & -\frac{a_{22}}{2} \\ 0 & \Upsilon_4 & (\Upsilon_5 - \lambda) \\ \Upsilon_6 & 0 & \Upsilon_7 \end{pmatrix} = \Upsilon_6 \cdot \det\begin{pmatrix} (\Upsilon_2 - \lambda) & -\frac{a_{22}}{2} \\ \Upsilon_4 & (\Upsilon_5 - \lambda) \end{pmatrix}$$
$$= \Upsilon_6 \cdot \left[(\Upsilon_2 - \lambda) \cdot (\Upsilon_5 - \lambda) + \Upsilon_4 \cdot \frac{a_{22}}{2}\right]$$

$$\det\begin{pmatrix} 0 & (\Upsilon_2 - \lambda) & -\frac{a_{22}}{2} \\ 0 & \Upsilon_4 & (\Upsilon_5 - \lambda) \\ \Upsilon_6 & 0 & \Upsilon_7 \end{pmatrix} = \Upsilon_6 \cdot \lambda^2 - \Upsilon_6 \cdot (\Upsilon_2 + \Upsilon_5) \cdot \lambda$$
$$+ \Upsilon_6 \cdot \Upsilon_4 \cdot \frac{a_{22}}{2} + \Upsilon_6 \cdot \Upsilon_2 \cdot \Upsilon_5$$

We define for simplicity global parameters: $\Xi_4 = \Upsilon_6 \cdot (\Upsilon_2 + \Upsilon_5)$; $\Xi_5 = \Upsilon_6 \cdot \Upsilon_4 \cdot \frac{a_{22}}{2} + \Upsilon_6 \cdot \Upsilon_2 \cdot \Upsilon_5$

$$\det\begin{pmatrix} 0 & (\Upsilon_2 - \lambda) & -\frac{a_{22}}{2} \\ 0 & \Upsilon_4 & (\Upsilon_5 - \lambda) \\ \Upsilon_6 & 0 & \Upsilon_7 \end{pmatrix} = \Upsilon_6 \cdot \lambda^2 - \Xi_4 \cdot \lambda + \Xi_5$$

We can summary steps 1 and 2:

$$\det\begin{pmatrix} (\Upsilon_2-\lambda) & -\frac{a_{22}}{2} & \Upsilon_3 \\ \Upsilon_4 & (\Upsilon_5-\lambda) & y\cdot d_{-}^{(j)} \\ 0 & \Upsilon_7 & (\Upsilon_8-\lambda) \end{pmatrix} = -\lambda^3+\lambda^2\cdot\Xi_1+\lambda\cdot\Xi_2+\Xi_3$$

$$\det\begin{pmatrix} 0 & (\Upsilon_2-\lambda) & -\frac{a_{22}}{2} \\ 0 & \Upsilon_4 & (\Upsilon_5-\lambda) \\ \Upsilon_6 & 0 & \Upsilon_7 \end{pmatrix} = \Upsilon_6\cdot\lambda^2-\Xi_4\cdot\lambda+\Xi_5$$

$$\det(A-\lambda\cdot I) = (\Upsilon_1-\lambda)\cdot(-\lambda^3+\lambda^2\cdot\Xi_1+\lambda\cdot\Xi_2+\Xi_3) + 2\cdot d^{(j)}\cdot(\Upsilon_6\cdot\lambda^2-\Xi_4\cdot\lambda+\Xi_5)$$

$$\det(A-\lambda\cdot I) = -\lambda^3\cdot\Upsilon_1+\lambda^2\cdot\Xi_1\cdot\Upsilon_1+\lambda\cdot\Xi_2\cdot\Upsilon_1+\Xi_3\cdot\Upsilon_1 + \lambda^4-\lambda^3\cdot\Xi_1-\lambda^2\cdot\Xi_2-\lambda\cdot\Xi_3+2\cdot d^{(j)}\cdot\Upsilon_6\cdot\lambda^2 - 2\cdot d^{(j)}\cdot\Xi_4\cdot\lambda+2\cdot d^{(j)}\cdot\Xi_5$$

$$\det(A-\lambda\cdot I) = \lambda^4-\lambda^3\cdot(\Upsilon_1+\Xi_1)+\lambda^2\cdot(\Xi_1\cdot\Upsilon_1-\Xi_2+2\cdot d^{(j)}\cdot\Upsilon_6) + \lambda\cdot(\Xi_2\cdot\Upsilon_1-\Xi_3-2\cdot d^{(j)}\cdot\Xi_4) + \Xi_3\cdot\Upsilon_1+2\cdot d^{(j)}\cdot\Xi_5$$

We define for simplicity global parameters: $\Pi_4 = 1$; $\Pi_3 = -(\Upsilon_1+\Xi_1)$

$$\Pi_2 = \Xi_1\cdot\Upsilon_1-\Xi_2+2\cdot d^{(j)}\cdot\Upsilon_6;\ \Pi_1 = \Xi_2\cdot\Upsilon_1-\Xi_3-2\cdot d^{(j)}\cdot\Xi_4$$
$$\Pi_0 = \Xi_3\cdot\Upsilon_1+2\cdot d^{(j)}\cdot\Xi_5$$

$$\det(A-\lambda\cdot I) = \lambda^4\cdot\Pi_4+\lambda^3\cdot\Pi_3+\lambda^2\cdot\Pi_2+\lambda\cdot\Pi_1+\Pi_0 = \sum_{k=0}^{4}\lambda^k\cdot\Pi_k$$

Eigenvalue stability discussion: Our Er-doped fiber laser (EDFL) system involving N variables (N > 2, N = 4, arbitrarily small increments), the characteristic equation is of degree N = 4 and must often be solved numerically. Expect in some particular cases, such an equation has (N = 4) distinct roots that can be real or complex. These values are eigenvalues of the (4 × 4) Jacobian matrix (A). The general rule is that the Er-doped fiber laser (EDFL) system is stable if there is no eigenvalue with positive real part. It is sufficient that one eigenvalue is positive for the steady state to be unstable. Our four variables (d, d_+, d_-, I) system has four eigenvalues (four system's arbitrarily small increments). The type of behavior can be characterized as a function of the position of these eigenvalues in the Re/Im plane. Five non-generated cases can be distinguished: (1) the four eigenvalues are real and negative (stable steady state), (2) the four eigenvalues are real, at least one of them is positive (unstable steady state), (3) and (4) three eigenvalues are complex conjugates with a negative real part and other eigenvalue real is negative (stable steady state), two cases can be distinguished depending on the relative value of the real part of the complex eigenvalues and of the real one, (5) three eigenvalues are complex conjugates with a negative real part and other eigenvalue real is positive (unstable steady state) [7, 8].

7.4 A Diode-Pumped Q-Switched Nd:YVO_4 Yellow Laser Analysis Under Delay Variables in Time

A compact high power yellow pulsed laser use an intra cavity sum-frequency mixing in a diode-end-pumped Q-switched Nd:YVO_4 dual-wavelength laser. A three-mirror configuration forming two separate laser cavities is used to optimize the gain match for simultaneous dual-wavelength emission in Q-switched operation. A diode-pumped Nd-doped lasers is a compact all-solid-state sources in the blue, green, and red spectral regions by use of intra cavity frequency doubling. These sources cover the region from 550 to 650 nm. The laser sources in the yellow-orange spectrum are used mainly for medical applications. The system consists of pulse dye lasers, copper-vapor lasers, and kryptonion lasers, on all-solid-state laser is another possible source. Krypton laser belongs to the gas lasers family, which use rare gases as the lasing medium. It is referred to as a krypton ion laser. The stimulated emission process in ion lasers occurs between the two energy states of the ion. Argon is one of the most common gases to be used in gas lasers, as it generates white light. The unique feature of krypton lasers is that with the use of proper mirrors it will lease on four sharp spectral lines of red, blue, yellow and green color. The krypton laser is capable of emitting lights of a number of wavelengths (as many as 10), the most significant one being the ones in the visible spectrum of the electromagnetic spectrum. The system is based on solid-state laser techniques for generating yellow lasers include sum-frequency mixing (SFM) of a flash-pumped Q-switched Nd:YAG dual-wavelength laser and frequency doubling of an intra cavity Raman-shifted Nd:YAG laser. A sum and difference frequency generation is a nonlinear process generating beams with the sum or difference of the frequencies of the input beams. Crystal materials lacking inversion symmetry exhibits $\chi^{(2)}$ nonlinearity. By using a nonlinear crystal we get the sum frequency generation (SFG) or difference frequency generation (DFG), where two pump beams generate another beam with the sum or difference of the optical frequencies of the pump beams. A possible model of the self-sum frequency mixing (self-SFM) laser is generated by a single crystal and characterized by the spatial distribution of the pump and circulating fundamental lasers with arbitrary beam waists. The beam waist (or beam focus) of a laser beam is the location along the propagation direction where the beam radius has a minimum. The waist radius is the beam radius at this location. There are several ways to improve the self-SFM laser output. A radiation with a high frequency density is generated by sum frequency mixing the output of two frequency modulated lasers. The radiation is useful for efficient excitation of a Doppler broadened atomic vapor. The Nd:YVO$_4$ crystal is the preferred material for diode-pumped lasers with dual-wavelength emission. It is characterized by high absorption over a wide pump-wavelength bandwidth and a large stimulated-emission cross section at both $^4F_{3/2}$–$^4I_{11/2}$ and the $^4F_{3/2}$–$^4I_{13/2}$ transitions [9]. Neodymium Doped Yttrium Aluminum Garnet—Nd:YAG laser crystal is the most popular lasing medial for solid-state lasers. It possesses a combination of properties uniquely favorable for laser operations. There are Nd:YAG crystals with diffusion bonded undoped YAG endcaps. Nd:YAG crystals have a good fluorescence lifetime and thermal conductivity, as well as a robust nature. It is suitable for high-power continuous wave (CW), high-intensity Q-switched and single-mode operations. In Nd:YAG crystal, the ratio of the stimulated-emission cross section between the $^4F_{3/2}$–$^4I_{11/2}$ and the $^4F_{3/2}$–$^4I_{13/2}$ transitions is approximately 5/5.1, whereas this ratio is approximately 2/2.1 for Nd:YVO$_4$ crystal. There is a compact, efficient scheme for generating a 593 nm laser based on intra-cavity SFM (sum-frequency mixing) of a diode-pumped Q-switched Nd:YVO$_4$ dual-wavelength laser. There is a spatial and temporal overlap of the two different wavelengths in SFM. It is an overlapping collinear cavities for simultaneous 1064 and 1342 nm emission from a Q-switched Nd:YVO$_4$ laser and empty an intra cavity β-barium borate crystal to obtain SFM 593 nm output. There is an option for Y cavity, Q-switched dual-wavelength Nd:YAG lasing and offset of the open times of the Q switches for the different wavelength emissions and the lasers are characterized by relatively good temporal and spatial overlap between the two wavelengths. The schematic of intra cavity SFM in the diode-end-pumped Q-switched Nd:YVO$_4$ dual-wavelength laser at 1064 and 1342 nm is presented (Fig. 7.7). AO, acousto-optic; HR, highly reflective; HT, highly transmitting; BBO, β-barium borate. The Nd^{3+} concentration of the laser crystal is 0.5 and its length is 6 mm. The pump source is 15 W fiber-coupled laser diode with a core diameter of 0.8 mm and a numerical aperture of 0.18. The fiber output is focused into the crystal, and the pump spot size is ~0.35 mm. The input mirror, M1, is a 1-m radius-of-curvature concave mirror with anti-reflection coating

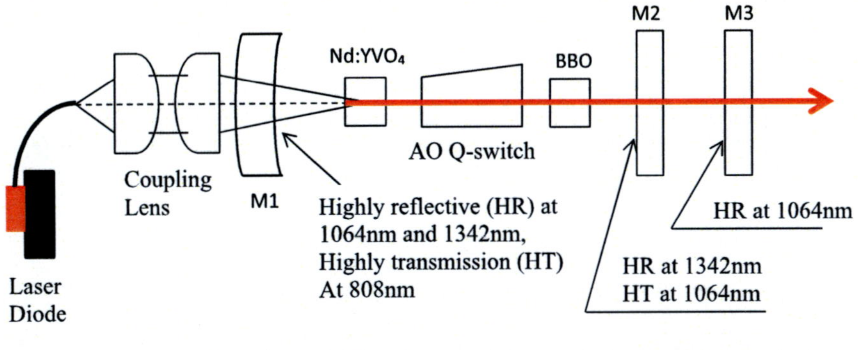

Fig. 7.7 Schematic of intra cavity SFM in the diode-end-pumped Q-switched Nd:YVO$_4$ dual-wavelength laser at 1064 and 1342 nm

at the pump wavelength on the entrance face (R < 0.2%), high-reflection coating at both lasing wavelengths (R > 99.8%), and high-transmission coating at the pump wavelength on the other surface (T > 90%). The mirror M1 is not optimum because of limited mirror availability. The optimum mirror M1 has a high-reflection coating at 593 nm. One side of flat mirror M2 is coated to be highly reflecting at 1342 nm (R > 99.8%) and highly transmitting at 1064 nm (T > 95%). The other side of mirror M2 is anti-reflective at 1064 nm (R < 0.2%). Flat mirror M3 has one side coated to be highly reflecting at 1064 nm and highly transmitting at 593 nm (T > 90%). The other side of mirror M3 is anti-reflective at 593 nm. The 20 mm long Q switcher has anti-reflection coatings at 1064 and 1342 nm on both faces and is driven at a 41 MHz center frequency with 3 W of RF power. The cavity length between M1 and M2 is 7 cm for 1342 nm oscillation. The cavity length between M1 and M3 is 15 cm for 1064 nm emission.

The dynamical model of our system described the behavior of a dual-wavelength Q-switched laser. The model is presented by three delay differential equations (DDEs).

$$\frac{dN}{dt} = R_p - c \cdot N \cdot [\sigma_1 \cdot \phi_1(t - \tau_1) + \sigma_2 \cdot \phi_2(t - \tau_2)] - \frac{N}{\tau_f}$$

$$N = N(t); \phi_1 = \phi_1(t); \phi_2 = \phi_2(t)$$

$$\frac{d\phi_1}{dt} = \frac{l_{cr}}{l_1} \cdot c \cdot \sigma_1 \cdot N - \frac{\phi_1(t - \tau_1)}{\tau_{c1}} - \eta_{SFM} \cdot \phi_1(t - \tau_1) \cdot \phi_2(t - \tau_2)$$

$$\frac{d\phi_2}{dt} = \frac{l_{cr}}{l_2} \cdot c \cdot \sigma_2 \cdot \phi_2(t - \tau_2) \cdot N - \frac{1}{\tau_{c2}} \cdot \phi_2(t - \tau_2) - \eta_{SFM} \cdot \phi_1(t - \tau_1) \cdot \phi_2(t - \tau_2)$$

$$\sum_{k=1}^{2} \sigma_k \cdot \phi_k(t - \tau_k) = \sigma_1 \cdot \phi_1(t - \tau_1) + \sigma_2 \cdot \phi_2(t - \tau_2)$$

$$\prod_{k=1}^{2} \phi_k(t - \tau_k) = \phi_1(t - \tau_1) \cdot \phi_2(t - \tau_2)$$

We can write our Diode-pumped Q-switched Nd:YVO_4 yellow laser system DDEs:

$$\frac{dN}{dt} = R_p - c \cdot N \cdot \left[\sum_{k=1}^{2} \sigma_k \cdot \phi_k(t - \tau_k)\right] - \frac{N}{\tau_f}$$

$$\frac{d\phi_1}{dt} = \frac{l_{cr}}{l_1} \cdot c \cdot \sigma_1 \cdot N - \frac{\phi_1(t - \tau_1)}{\tau_{c1}} - \eta_{SFM} \cdot \prod_{k=1}^{2} \phi_k(t - \tau_k)$$

Table 7.10 Diode-pumped Q-switched Nd:YVO4 yellow laser system parameters for $\lambda_1 = 1342$ nm and $\lambda_2 = 1064$ nm

Parameter @ $\lambda_1 = 1342$ nm	Parameter @ $\lambda_2 = 1064$ nm	Description
ϕ_1	ϕ_2	Photon intensity
σ_1	σ_2	Stimulated-emission cross section
l_1	l_2	Cavity length
N	N	Population inversion density
R_p	R_p	Average pump intensity
τ_f	τ_f	Emission lifetime
l_{cr}	l_{cr}	Crystal length
τ_{c1}	τ_{c2}	Effective photon decay time as a result of all the linear losses
η_{SFM}	η_{SFM}	Effective conversion rate as a result of the intra cavity SFM (sum-frequency mixing)

$$\frac{d\phi_2}{dt} = \frac{l_{cr}}{l_2} \cdot c \cdot \sigma_2 \cdot \phi_2(t-\tau_2) \cdot N - \frac{1}{\tau_{c2}} \cdot \phi_2(t-\tau_2) - \eta_{SFM} \cdot \prod_{k=1}^{2} \phi_k(t-\tau_k)$$

The parameters of our system are described in the next table (Table 7.10), where the subscripts 1 and 2 for $\lambda_1 = 1342$ nm and $\lambda_2 = 1064$ nm, respectively.

c is the speed of light in vacuum. The photon densities at time $\phi_1(t)$ and $\phi_2(t)$ for $\lambda_1 = 1342$ nm and $\lambda_2 = 1064$ nm, respectively is shifted in time $\phi_1(t) \to \phi_1(t-\tau_1)$, $\phi_2(t) \to \phi_2(t-\tau_2)$ but it is not affect the derivatives in time $\frac{d\phi_1}{dt}, \frac{d\phi_2}{dt}$. The shifting in time ($\tau_1$ and τ_2 for $\lambda_1 = 1342$ nm and $\lambda_2 = 1064$ nm, respectively) of these variables is due to parasitic effects in our system. The optimum temporal overlap between the two wavelengths is happened when $\frac{d\phi_1}{dt} = \frac{d\phi_2}{dt}$. When $\frac{\sigma_1}{l_1} = \frac{\sigma_2}{l_2}$ and $\tau_{c1} = \tau_{c2}$ then $\frac{d\phi_1}{dt} = \frac{d\phi_2}{dt}$, parameter τ_c is varied by the fine adjustment of the cavity alignment. The cavity lengths satisfy the relationship $\frac{\sigma_1}{\sigma_2} = \frac{l_1}{l_2}$ for optimum temporal overlap. By increasing the cavity length we lower the average photon density. The result is that the competing ability of the transition with larger emission cross section (σ) is reduced.

At fixed points $\frac{dN}{dt} = 0$; $\frac{d\phi_1}{dt} = 0$; $\frac{d\phi_2}{dt} = 0$; $\lim_{t\to\infty} \phi_1(t-\tau_1) = \phi_1(t)\ \forall\, t \gg \tau_1$.

$$\lim_{t\to\infty} \phi_1(t-\tau_2) = \phi_2(t)\ \forall\, t \gg \tau_2;\ N = N(t);\ \phi_1 = \phi_1(t);\ \phi_2 = \phi_2(t)$$

$$\frac{dN}{dt} = 0 \Rightarrow R_p - c \cdot N^* \cdot [\sigma_1 \cdot \phi_1^* + \sigma_2 \cdot \phi_2^*] - \frac{N^*}{\tau_f} = 0$$

$$\Rightarrow R_p - N^* \cdot \left(c \cdot [\sigma_1 \cdot \phi_1^* + \sigma_2 \cdot \phi_2^*] + \frac{1}{\tau_f}\right) = 0$$

$$N^* \cdot \left(c \cdot [\sigma_1 \cdot \phi_1^* + \sigma_2 \cdot \phi_2^*] + \frac{1}{\tau_f}\right) = R_p \Rightarrow N^* = \frac{R_p}{c \cdot [\sigma_1 \cdot \phi_1^* + \sigma_2 \cdot \phi_2^*] + \frac{1}{\tau_f}}$$

$$\frac{d\phi_1}{dt} = 0 \Rightarrow \frac{l_{cr}}{l_1} \cdot c \cdot \sigma_1 \cdot N^* - \frac{\phi_1^*}{\tau_{c1}} - \eta_{SFM} \cdot \phi_1^* \cdot \phi_2^* = 0$$

$$\frac{l_{cr}}{l_1} \cdot c \cdot \sigma_1 \cdot \left[\frac{R_p}{c \cdot [\sigma_1 \cdot \phi_1^* + \sigma_2 \cdot \phi_2^*] + \frac{1}{\tau_f}}\right] - \frac{\phi_1^*}{\tau_{c1}} - \eta_{SFM} \cdot \phi_1^* \cdot \phi_2^* = 0$$

$$\frac{d\phi_2}{dt} = 0 \Rightarrow \frac{l_{cr}}{l_2} \cdot c \cdot \sigma_2 \cdot \phi_2^* \cdot N^* - \frac{1}{\tau_{c2}} \cdot \phi_2^* - \eta_{SFM} \cdot \phi_1^* \cdot \phi_2^* = 0$$

$$\frac{l_{cr}}{l_2} \cdot c \cdot \sigma_2 \cdot \phi_2^* \cdot \left[\frac{R_p}{c \cdot [\sigma_1 \cdot \phi_1^* + \sigma_2 \cdot \phi_2^*] + \frac{1}{\tau_f}}\right] - \frac{1}{\tau_{c2}} \cdot \phi_2^* - \eta_{SFM} \cdot \phi_1^* \cdot \phi_2^* = 0$$

We get two fixed points coordinate equations: (ϕ_1^*, ϕ_2^*).

1. $$\frac{l_{cr}}{l_1} \cdot c \cdot \sigma_1 \cdot \left[\frac{R_p}{c \cdot [\sigma_1 \cdot \phi_1^* + \sigma_2 \cdot \phi_2^*] + \frac{1}{\tau_f}}\right] - \frac{\phi_1^*}{\tau_{c1}} - \eta_{SFM} \cdot \phi_1^* \cdot \phi_2^* = 0$$

2. $$\frac{l_{cr}}{l_2} \cdot c \cdot \sigma_2 \cdot \phi_2^* \cdot \left[\frac{R_p}{c \cdot [\sigma_1 \cdot \phi_1^* + \sigma_2 \cdot \phi_2^*] + \frac{1}{\tau_f}}\right] - \frac{1}{\tau_{c2}} \cdot \phi_2^* - \eta_{SFM} \cdot \phi_1^* \cdot \phi_2^* = 0$$

$$\phi_2^* \cdot \left(\frac{l_{cr}}{l_2} \cdot c \cdot \sigma_2 \cdot \left[\frac{R_p}{c \cdot [\sigma_1 \cdot \phi_1^* + \sigma_2 \cdot \phi_2^*] + \frac{1}{\tau_f}}\right] - \frac{1}{\tau_{c2}} - \eta_{SFM} \cdot \phi_1^*\right) = 0$$

Case A: $\phi_2^* = 0$

$$\frac{l_{cr}}{l_1} \cdot c \cdot \sigma_1 \cdot \left[\frac{R_p}{c \cdot [\sigma_1 \cdot \phi_1^* + \sigma_2 \cdot \phi_2^*] + \frac{1}{\tau_f}}\right] - \frac{\phi_1^*}{\tau_{c1}} - \eta_{SFM} \cdot \phi_1^* \cdot \phi_2^*|_{\phi_2^*=0}$$
$$= \frac{l_{cr}}{l_1} \cdot c \cdot \sigma_1 \cdot \left[\frac{R_p}{c \cdot \sigma_1 \cdot \phi_1^* + \frac{1}{\tau_f}}\right] - \frac{\phi_1^*}{\tau_{c1}} = 0$$

$$\frac{l_{cr}}{l_1} \cdot c \cdot \sigma_1 \cdot \left[\frac{R_p}{c \cdot \sigma_1 \cdot \phi_1^* + \frac{1}{\tau_f}}\right] - \frac{\phi_1^*}{\tau_{c1}} = 0$$

$$\Rightarrow \frac{\frac{l_{cr}}{l_1} \cdot c \cdot \sigma_1 \cdot R_p \cdot \tau_{c1} - \left(c \cdot \sigma_1 \cdot \phi_1^* + \frac{1}{\tau_f}\right) \cdot \phi_1^*}{\left(c \cdot \sigma_1 \cdot \phi_1^* + \frac{1}{\tau_f}\right) \cdot \tau_{c1}} = 0$$

$$\left(c \cdot \sigma_1 \cdot \phi_1^* + \frac{1}{\tau_f}\right) \cdot \tau_{c1} \neq 0 \Rightarrow \tau_{c1} \neq 0 \text{ and}$$

$$c \cdot \sigma_1 \cdot \phi_1^* + \frac{1}{\tau_f} \neq 0 \Rightarrow \phi_1^* \neq -\frac{1}{c \cdot \sigma_1 \cdot \tau_f}$$

$$\frac{l_{cr}}{l_1} \cdot c \cdot \sigma_1 \cdot R_p \cdot \tau_{c1} - \left(c \cdot \sigma_1 \cdot \phi_1^* + \frac{1}{\tau_f}\right) \cdot \phi_1^* = 0$$

$$\frac{l_{cr}}{l_1} \cdot c \cdot \sigma_1 \cdot R_p \cdot \tau_{c1} - \left(c \cdot \sigma_1 \cdot [\phi_1^*]^2 + \frac{1}{\tau_f} \cdot \phi_1^*\right) = 0$$

$$c \cdot \sigma_1 \cdot [\phi_1^*]^2 + \frac{1}{\tau_f} \cdot \phi_1^* - \frac{l_{cr}}{l_1} \cdot c \cdot \sigma_1 \cdot R_p \cdot \tau_{c1} = 0$$

$$\phi_1^* = \frac{-\frac{1}{\tau_f} \pm \sqrt{\frac{1}{\tau_f^2} + 4 \cdot \frac{l_{cr}}{l_1} \cdot c^2 \cdot \sigma_1^2 \cdot R_p \cdot \tau_{c1}}}{2 \cdot c \cdot \sigma_1}$$

$$\text{Case A}_1\text{: } \phi_1^* = \frac{-\frac{1}{\tau_f} + \sqrt{\frac{1}{\tau_f^2} + 4 \cdot \frac{l_{cr}}{l_1} \cdot c^2 \cdot \sigma_1^2 \cdot R_p \cdot \tau_{c1}}}{2 \cdot c \cdot \sigma_1}; \quad \sqrt{\frac{1}{\tau_f^2} + 4 \cdot \frac{l_{cr}}{l_1} \cdot c^2 \cdot \sigma_1^2 \cdot R_p \cdot \tau_{c1}} > \frac{1}{\tau_f}$$

$$N^* = \frac{R_p}{c \cdot [\sigma_1 \cdot \phi_1^* + \sigma_2 \cdot \phi_2^*] + \frac{1}{\tau_f}} \Big|_{\phi_2^*=0}$$

$$= \frac{R_p}{c \cdot \sigma_1 \cdot \phi_1^* + \frac{1}{\tau_f}}$$

$$= \frac{R_p}{c \cdot \sigma_1 \cdot \left[\frac{-\frac{1}{\tau_f} + \sqrt{\frac{1}{\tau_f^2} + 4 \cdot \frac{l_{cr}}{l_1} \cdot c^2 \cdot \sigma_1^2 \cdot R_p \cdot \tau_{c1}}}{2 \cdot c \cdot \sigma_1}\right] + \frac{1}{\tau_f}}$$

$$N^*|_{\phi_2^*=0} = \frac{R_p}{-\frac{1}{2 \cdot \tau_f} + \frac{1}{2} \cdot \sqrt{\frac{1}{\tau_f^2} + 4 \cdot \frac{l_{cr}}{l_1} \cdot c^2 \cdot \sigma_1^2 \cdot R_p \cdot \tau_{c1}} + \frac{1}{\tau_f}}$$

$$= \frac{2 \cdot R_p}{\frac{1}{\tau_f} + \sqrt{\frac{1}{\tau_f^2} + 4 \cdot \frac{l_{cr}}{l_1} \cdot c^2 \cdot \sigma_1^2 \cdot R_p \cdot \tau_{c1}}}$$

First fixed point:

$E^{(0)}(N^{(0)}, \phi_1^{(0)}, \phi_2^{(0)})$

$$= \left(\frac{2 \cdot R_p}{\frac{1}{\tau_f} + \sqrt{\frac{1}{\tau_f^2} + 4 \cdot \frac{l_{cr}}{l_1} \cdot c^2 \cdot \sigma_1^2 \cdot R_p \cdot \tau_{c1}}}, \frac{-\frac{1}{\tau_f} + \sqrt{\frac{1}{\tau_f^2} + 4 \cdot \frac{l_{cr}}{l_1} \cdot c^2 \cdot \sigma_1^2 \cdot R_p \cdot \tau_{c1}}}{2 \cdot c \cdot \sigma_1}, 0 \right)$$

Case A$_2$: $\phi_1^* = \frac{-\frac{1}{\tau_f} - \sqrt{\frac{1}{\tau_f^2} + 4 \cdot \frac{l_{cr}}{l_1} \cdot c^2 \cdot \sigma_1^2 \cdot R_p \cdot \tau_{c1}}}{2 \cdot c \cdot \sigma_1}$; $\phi_1^* < 0$.

<u>Remark</u>: It is a mathematical fixed point, since Photon intensity cannot be negative.

$$N^*|_{\phi_2^*=0} = \frac{R_p}{-\frac{1}{2 \cdot \tau_f} - \frac{1}{2} \cdot \sqrt{\frac{1}{\tau_f^2} + 4 \cdot \frac{l_{cr}}{l_1} \cdot c^2 \cdot \sigma_1^2 \cdot R_p \cdot \tau_{c1}} + \frac{1}{\tau_f}}$$
$$= \frac{2 \cdot R_p}{\frac{1}{\tau_f} - \sqrt{\frac{1}{\tau_f^2} + 4 \cdot \frac{l_{cr}}{l_1} \cdot c^2 \cdot \sigma_1^2 \cdot R_p \cdot \tau_{c1}}}$$

Second fixed point:

$$E^{(1)}(N^{(1)}, \phi_1^{(1)}, \phi_2^{(1)})$$
$$= \left(\frac{2 \cdot R_p}{\frac{1}{\tau_f} - \sqrt{\frac{1}{\tau_f^2} + 4 \cdot \frac{l_{cr}}{l_1} \cdot c^2 \cdot \sigma_1^2 \cdot R_p \cdot \tau_{c1}}}, \frac{-\frac{1}{\tau_f} - \sqrt{\frac{1}{\tau_f^2} + 4 \cdot \frac{l_{cr}}{l_1} \cdot c^2 \cdot \sigma_1^2 \cdot R_p \cdot \tau_{c1}}}{2 \cdot c \cdot \sigma_1}, 0 \right)$$

Case B: $\frac{l_{cr}}{l_2} \cdot c \cdot \sigma_2 \cdot \left[\frac{R_p}{c \cdot [\sigma_1 \cdot \phi_1^* + \sigma_2 \cdot \phi_2^*] + \frac{1}{\tau_f}} \right] - \frac{1}{\tau_{c2}} - \eta_{SFM} \cdot \phi_1^* = 0$

$$\frac{l_{cr}}{l_2} \cdot c \cdot \sigma_2 \cdot \frac{R_p}{\left(c \cdot [\sigma_1 \cdot \phi_1^* + \sigma_2 \cdot \phi_2^*] + \frac{1}{\tau_f} \right)} = \eta_{SFM} \cdot \phi_1^* + \frac{1}{\tau_{c2}}$$
$$\Rightarrow \frac{l_{cr}}{l_2} \cdot c \cdot \sigma_2 \cdot R_p = \left(\eta_{SFM} \cdot \phi_1^* + \frac{1}{\tau_{c2}} \right) \cdot \left(c \cdot [\sigma_1 \cdot \phi_1^* + \sigma_2 \cdot \phi_2^*] + \frac{1}{\tau_f} \right)$$
$$\frac{l_{cr}}{l_2} \cdot c \cdot \sigma_2 \cdot R_p = \left(\eta_{SFM} \cdot \phi_1^* + \frac{1}{\tau_{c2}} \right) \cdot \left(\left[c \cdot \sigma_1 \cdot \phi_1^* + \frac{1}{\tau_f} \right] + c \cdot \sigma_2 \cdot \phi_2^* \right)$$
$$\frac{l_{cr}}{l_2} \cdot c \cdot \sigma_2 \cdot R_p = \left(\eta_{SFM} \cdot \phi_1^* + \frac{1}{\tau_{c2}} \right) \cdot \left(c \cdot \sigma_1 \cdot \phi_1^* + \frac{1}{\tau_f} \right) + \left(\eta_{SFM} \cdot \phi_1^* + \frac{1}{\tau_{c2}} \right) \cdot c \cdot \sigma_2 \cdot \phi_2^*$$
$$\left(\eta_{SFM} \cdot \phi_1^* + \frac{1}{\tau_{c2}} \right) \cdot c \cdot \sigma_2 \cdot \phi_2^* = \frac{l_{cr}}{l_2} \cdot c \cdot \sigma_2 \cdot R_p$$

$$-\left(\eta_{SFM}\cdot\phi_1^*+\frac{1}{\tau_{c2}}\right)\cdot\left(c\cdot\sigma_1\cdot\phi_1^*+\frac{1}{\tau_f}\right)$$

$$\phi_2^*|_{\text{CaseB}}=\frac{\frac{l_{cr}}{l_2}\cdot c\cdot\sigma_2\cdot R_p-\left(\eta_{SFM}\cdot\phi_1^*+\frac{1}{\tau_{c2}}\right)\cdot\left(c\cdot\sigma_1\cdot\phi_1^*+\frac{1}{\tau_f}\right)}{\left(\eta_{SFM}\cdot\phi_1^*+\frac{1}{\tau_{c2}}\right)\cdot c\cdot\sigma_2}$$

$$\frac{dN}{dt}=0\Rightarrow N^*=\frac{R_p}{c\cdot[\sigma_1\cdot\phi_1^*+\sigma_2\cdot\phi_2^*]+\frac{1}{\tau_f}}$$

$$N^*=\frac{R_p}{c\cdot[\sigma_1\cdot\phi_1^*+\sigma_2\cdot\phi_2^*|_{\text{CaseB}}]+\frac{1}{\tau_f}}$$

$$N^*=\frac{R_p}{c\cdot\left[\sigma_1\cdot\phi_1^*+\sigma_2\cdot\left\{\frac{\frac{l_{cr}}{l_2}\cdot c\cdot\sigma_2\cdot R_p-(\eta_{SFM}\cdot\phi_1^*+\frac{1}{\tau_{c2}})\cdot(c\cdot\sigma_1\cdot\phi_1^*+\frac{1}{\tau_f})}{(\eta_{SFM}\cdot\phi_1^*+\frac{1}{\tau_{c2}})\cdot c\cdot\sigma_2}\right\}\right]+\frac{1}{\tau_f}}$$

$$\frac{d\phi_1}{dt}=0\Rightarrow\frac{l_{cr}}{l_1}\cdot c\cdot\sigma_1\cdot N^*-\frac{\phi_1^*}{\tau_{c1}}-\eta_{SFM}\cdot\phi_1^*\cdot\phi_2^*|_{\text{Case B}}=0$$

$$\frac{l_{cr}}{l_1}\cdot c\cdot\sigma_1\cdot N^*-\frac{\phi_1^*}{\tau_{c1}}-\eta_{SFM}\cdot\phi_1^*$$
$$\cdot\left\{\frac{\frac{l_{cr}}{l_2}\cdot c\cdot\sigma_2\cdot R_p-\left(\eta_{SFM}\cdot\phi_1^*+\frac{1}{\tau_{c2}}\right)\cdot\left(c\cdot\sigma_1\cdot\phi_1^*+\frac{1}{\tau_f}\right)}{\left(\eta_{SFM}\cdot\phi_1^*+\frac{1}{\tau_{c2}}\right)\cdot c\cdot\sigma_2}\right\}=0$$

We get two fixed points coordinate equations: (ϕ_1^*, N^*), it is recommended to solve it numerically to find the third and fourth fixed coordinates ϕ_1^*, N^*. Then find $\phi_2^*|_{\text{Case B}}$.

Stability analysis: The standard local stability analysis about any one of the equilibrium point of the Diode-pumped Q-switched Nd:YVO$_4$ yellow laser system consists in adding to coordinate $[N, \phi_1, \phi_2]$ arbitrarily small increments of exponential form $[n, \phi_1, \phi_2]\cdot e^{\lambda\cdot t}$ and retaining the first order terms in N, ϕ_1, ϕ_2. The system of three homogeneous equations leads to a polynomial characteristic equation in the eigenvalues. The polynomial characteristic equations accept by set the below variables and variables derivative with respect to time into Diode-pumped Q-switched Nd:YVO$_4$ yellow laser system equations. The Diode-pumped Q-switched Nd:YVO$_4$ yellow laser system fixed values with arbitrarily small increments of exponential form $[n, \phi_1, \phi_2]\cdot e^{\lambda\cdot t}$ are: $j=0$ (first fixed point), $j=1$ (second fixed point), $j=2$ (third fixed point), etc. [2, 3].

$$N(t)=N^{(j)}+n\cdot e^{\lambda\cdot t};\ \phi_1(t)=\phi_1^{(j)}+\phi_1\cdot e^{\lambda\cdot t}$$
$$\phi_2(t)=\phi_2^{(j)}+\phi_2\cdot e^{\lambda\cdot t};\ \phi_1(t-\tau_1)=\phi_1^{(j)}+\phi_1\cdot e^{\lambda\cdot(t-\tau_1)}$$

$$\phi_2(t-\tau_2) = \phi_2^{(j)} + \phi_2 \cdot e^{\lambda\cdot(t-\tau_2)}; \frac{dN(t)}{dt} = n\cdot\lambda\cdot e^{\lambda\cdot t}$$
$$\frac{d\phi_1(t)}{dt} = \phi_1\cdot\lambda\cdot e^{\lambda\cdot t}; \frac{d\phi_2(t)}{dt} = \phi_2\cdot\lambda\cdot e^{\lambda\cdot t}$$

We choose these expressions for ourselves $N(t), \phi_1(t), \phi_2(t)$ as a small displacement $[n, \phi_1, \phi_2]$ from the Diode-pumped Q-switched Nd:YVO_4 yellow laser system fixed points in time $t = 0$. $N(t=0) = N^{(j)} + n$; $\phi_1(t=0) = \phi_1^{(j)} + \phi_1$; $\phi_2(t=0) = \phi_2^{(j)} + \phi_2$. We have three possible cases: (1) $\tau_1 = \tau; \tau_2 = 0$ (2) $\tau_1 = 0; \tau_2 = \tau$ (3) $\tau_1 = \tau; \tau_2 = \tau$, and we choose to analyze the first case, $\tau_1 = \tau; \tau_2 = 0$.

$$\frac{dN}{dt} = R_p - c\cdot N\cdot[\sigma_1\cdot\phi_1(t-\tau_1) + \sigma_2\cdot\phi_2(t-\tau_2)] - \frac{N}{\tau_f}$$

$$n\cdot\lambda\cdot e^{\lambda\cdot t} = R_p - c\cdot(N^{(j)} + n\cdot e^{\lambda\cdot t})\cdot[\sigma_1\cdot(\phi_1^{(j)} + \phi_1\cdot e^{\lambda\cdot(t-\tau)}) + \sigma_2\cdot(\phi_2^{(j)} + \phi_2\cdot e^{\lambda\cdot t})] - \frac{N^{(j)} + n\cdot e^{\lambda\cdot t}}{\tau_f}$$

$$n\cdot\lambda\cdot e^{\lambda\cdot t} = R_p - c\cdot[\sigma_1\cdot(N^{(j)} + n\cdot e^{\lambda\cdot t})\cdot(\phi_1^{(j)} + \phi_1\cdot e^{\lambda\cdot(t-\tau)}) + \sigma_2\cdot(N^{(j)} + n\cdot e^{\lambda\cdot t})\cdot(\phi_2^{(j)} + \phi_2\cdot e^{\lambda\cdot t})] - \frac{N^{(j)}}{\tau_f} - \frac{n\cdot e^{\lambda\cdot t}}{\tau_f}$$

$$\begin{aligned} n\cdot\lambda\cdot e^{\lambda\cdot t} = R_p - c\cdot[\sigma_1\cdot(N^{(j)}\cdot\phi_1^{(j)} + N^{(j)}\cdot\phi_1\cdot e^{\lambda\cdot(t-\tau)} + \phi_1^{(j)}\cdot n\cdot e^{\lambda\cdot t} \\ + n\cdot\phi_1\cdot e^{\lambda\cdot t}\cdot e^{\lambda\cdot(t-\tau)}) \\ + \sigma_2\cdot(N^{(j)}\cdot\phi_2^{(j)} + N^{(j)}\cdot\phi_2\cdot e^{\lambda\cdot t} + \phi_2^{(j)}\cdot n\cdot e^{\lambda\cdot t} \\ + n\cdot\phi_2\cdot e^{\lambda\cdot t}\cdot e^{\lambda\cdot t})] - \frac{N^{(j)}}{\tau_f} - n\cdot\frac{1}{\tau_f}\cdot e^{\lambda\cdot t} \end{aligned}$$

Assumption: $n\cdot\phi_1 \approx 0; n\cdot\phi_2 \approx 0$

$$\begin{aligned} n\cdot\lambda\cdot e^{\lambda\cdot t} = R_p - c\cdot[\sigma_1\cdot N^{(j)}\cdot\phi_1^{(j)} + \sigma_1\cdot(N^{(j)}\cdot\phi_1\cdot e^{\lambda\cdot(t-\tau)} + \phi_1^{(j)}\cdot n\cdot e^{\lambda\cdot t}) \\ + \sigma_2\cdot N^{(j)}\cdot\phi_2^{(j)} + \sigma_2\cdot(N^{(j)}\cdot\phi_2\cdot e^{\lambda\cdot t} \\ + \phi_2^{(j)}\cdot n\cdot e^{\lambda\cdot t})] - \frac{N^{(j)}}{\tau_f} - n\cdot\frac{1}{\tau_f}\cdot e^{\lambda\cdot t} \end{aligned}$$

$$\begin{aligned} n\cdot\lambda\cdot e^{\lambda\cdot t} = R_p - c\cdot N^{(j)}\cdot[\sigma_1\cdot\phi_1^{(j)} + \sigma_2\cdot\phi_2^{(j)}] - \frac{N^{(j)}}{\tau_f} \\ - [c\cdot\sigma_1\cdot(N^{(j)}\cdot\phi_1\cdot e^{\lambda\cdot(t-\tau)} + \phi_1^{(j)}\cdot n\cdot e^{\lambda\cdot t}) \\ + c\cdot\sigma_2\cdot(N^{(j)}\cdot\phi_2\cdot e^{\lambda\cdot t} + \phi_2^{(j)}\cdot n\cdot e^{\lambda\cdot t})] - n\cdot\frac{1}{\tau_f}\cdot e^{\lambda\cdot t} \end{aligned}$$

At fixed point: $\frac{dN}{dt} = 0 \Rightarrow R_p - c \cdot N^{(j)} \cdot [\sigma_1 \cdot \phi_1^{(j)} + \sigma_2 \cdot \phi_2^{(j)}] - \frac{N^{(j)}}{\tau_f} = 0$

$$n \cdot \lambda \cdot e^{\lambda \cdot t} = -[c \cdot \sigma_1 \cdot N^{(j)} \cdot \phi_1 \cdot e^{\lambda \cdot (t-\tau)} + c \cdot \sigma_1 \cdot \phi_1^{(j)} \cdot n \cdot e^{\lambda \cdot t} + c \cdot \sigma_2 \cdot N^{(j)} \cdot \phi_2 \cdot e^{\lambda \cdot t} + c \cdot \sigma_2 \cdot \phi_2^{(j)} \cdot n \cdot e^{\lambda \cdot t}] - n \cdot \frac{1}{\tau_f} \cdot e^{\lambda \cdot t}$$

$$-c \cdot \sigma_1 \cdot N^{(j)} \cdot e^{\lambda \cdot (t-\tau)} \cdot \phi_1 - c \cdot \sigma_1 \cdot \phi_1^{(j)} \cdot e^{\lambda \cdot t} \cdot n - c \cdot \sigma_2 \cdot N^{(j)} \cdot e^{\lambda \cdot t} \cdot \phi_2 - c \cdot \sigma_2 \cdot \phi_2^{(j)} \cdot e^{\lambda \cdot t} \cdot n - \frac{1}{\tau_f} \cdot e^{\lambda \cdot t} \cdot n - n \cdot \lambda \cdot e^{\lambda \cdot t} = 0$$

$$-\left(c \cdot [\sigma_1 \cdot \phi_1^{(j)} + \sigma_2 \cdot \phi_2^{(j)}] + \frac{1}{\tau_f}\right) \cdot n - n \cdot \lambda - c \cdot \sigma_1 \cdot N^{(j)} \cdot e^{-\lambda \cdot \tau} \cdot \phi_1 - c \cdot \sigma_2 \cdot N^{(j)} \cdot \phi_2 = 0$$

$$\frac{d\phi_1}{dt} = \frac{l_{cr}}{l_1} \cdot c \cdot \sigma_1 \cdot N - \frac{\phi_1(t - \tau_1)}{\tau_{c1}} - \eta_{SFM} \cdot \phi_1(t - \tau_1) \cdot \phi_2(t - \tau_2)$$

$$\phi_1 \cdot \lambda \cdot e^{\lambda \cdot t} = \frac{l_{cr}}{l_1} \cdot c \cdot \sigma_1 \cdot (N^{(j)} + n \cdot e^{\lambda \cdot t}) - \frac{(\phi_1^{(j)} + \phi_1 \cdot e^{\lambda \cdot (t-\tau)})}{\tau_{c1}} - \eta_{SFM} \cdot (\phi_1^{(j)} + \phi_1 \cdot e^{\lambda \cdot (t-\tau)}) \cdot (\phi_2^{(j)} + \phi_2 \cdot e^{\lambda \cdot t})$$

$$\phi_1 \cdot \lambda \cdot e^{\lambda \cdot t} = \frac{l_{cr}}{l_1} \cdot c \cdot \sigma_1 \cdot N^{(j)} + n \cdot \frac{l_{cr}}{l_1} \cdot c \cdot \sigma_1 \cdot e^{\lambda \cdot t} - \frac{\phi_1^{(j)}}{\tau_{c1}} - \frac{\phi_1 \cdot e^{\lambda \cdot (t-\tau)}}{\tau_{c1}} - \eta_{SFM} \cdot (\phi_1^{(j)} \cdot \phi_2^{(j)} + \phi_1^{(j)} \cdot \phi_2 \cdot e^{\lambda \cdot t} + \phi_2^{(j)} \cdot \phi_1 \cdot e^{\lambda \cdot (t-\tau)} + \phi_1 \cdot \phi_2 \cdot e^{\lambda \cdot t} \cdot e^{\lambda \cdot (t-\tau)})$$

Assumption: $\phi_1 \cdot \phi_2 \approx 0$

$$\phi_1 \cdot \lambda \cdot e^{\lambda \cdot t} = \frac{l_{cr}}{l_1} \cdot c \cdot \sigma_1 \cdot N^{(j)} + n \cdot \frac{l_{cr}}{l_1} \cdot c \cdot \sigma_1 \cdot e^{\lambda \cdot t} - \frac{\phi_1^{(j)}}{\tau_{c1}} - \frac{\phi_1 \cdot e^{\lambda \cdot (t-\tau)}}{\tau_{c1}} - \eta_{SFM} \cdot (\phi_1^{(j)} \cdot \phi_2^{(j)} + \phi_1^{(j)} \cdot \phi_2 \cdot e^{\lambda \cdot t} + \phi_2^{(j)} \cdot \phi_1 \cdot e^{\lambda \cdot (t-\tau)})$$

$$\phi_1 \cdot \lambda \cdot e^{\lambda \cdot t} = \frac{l_{cr}}{l_1} \cdot c \cdot \sigma_1 \cdot N^{(j)} - \frac{\phi_1^{(j)}}{\tau_{c1}} - \eta_{SFM} \cdot \phi_1^{(j)} \cdot \phi_2^{(j)} + n \cdot \frac{l_{cr}}{l_1} \cdot c \cdot \sigma_1 \cdot e^{\lambda \cdot t} - \frac{\phi_1 \cdot e^{\lambda \cdot (t-\tau)}}{\tau_{c1}} - \eta_{SFM} \cdot (\phi_1^{(j)} \cdot \phi_2 \cdot e^{\lambda \cdot t} + \phi_2^{(j)} \cdot \phi_1 \cdot e^{\lambda \cdot (t-\tau)})$$

At fixed points: $\frac{l_{cr}}{l_1} \cdot c \cdot \sigma_1 \cdot N^{(j)} - \frac{\phi_1^{(j)}}{\tau_{c1}} - \eta_{SFM} \cdot \phi_1^{(j)} \cdot \phi_2^{(j)} = 0.$

$$\phi_1 \cdot \lambda \cdot e^{\lambda \cdot t} = n \cdot \frac{l_{cr}}{l_1} \cdot c \cdot \sigma_1 \cdot e^{\lambda \cdot t} - \frac{\phi_1 \cdot e^{\lambda \cdot t} \cdot e^{-\lambda \cdot \tau}}{\tau_{c1}} - \eta_{SFM} \cdot (\phi_1^{(j)} \cdot \phi_2 + \phi_2^{(j)} \cdot \phi_1 \cdot e^{-\lambda \cdot \tau}) \cdot e^{\lambda \cdot t}$$

$$n \cdot \frac{l_{cr}}{l_1} \cdot c \cdot \sigma_1 - \frac{\phi_1 \cdot e^{-\lambda \cdot \tau}}{\tau_{c1}} - \eta_{SFM} \cdot \phi_1^{(j)} \cdot \phi_2 - \eta_{SFM} \cdot \phi_2^{(j)} \cdot \phi_1 \cdot e^{-\lambda \cdot \tau} - \phi_1 \cdot \lambda = 0$$

$$\frac{l_{cr}}{l_1} \cdot c \cdot \sigma_1 \cdot n - \left(\frac{1}{\tau_{c1}} + \eta_{SFM} \cdot \phi_2^{(j)}\right) \cdot e^{-\lambda \cdot \tau} \cdot \phi_1 - \lambda \cdot \phi_1 - \eta_{SFM} \cdot \phi_1^{(j)} \cdot \phi_2 = 0$$

$$\frac{d\phi_2}{dt} = \frac{l_{cr}}{l_2} \cdot c \cdot \sigma_2 \cdot \phi_2(t - \tau_2) \cdot N - \frac{1}{\tau_{c2}} \cdot \phi_2(t - \tau_2) - \eta_{SFM} \cdot \phi_1(t - \tau_1) \cdot \phi_2(t - \tau_2)$$

$$\phi_2 \cdot \lambda \cdot e^{\lambda \cdot t} = \frac{l_{cr}}{l_2} \cdot c \cdot \sigma_2 \cdot (\phi_2^{(j)} + \phi_2 \cdot e^{\lambda \cdot t}) \cdot (N^{(j)} + n \cdot e^{\lambda \cdot t}) - \frac{1}{\tau_{c2}} \cdot (\phi_2^{(j)} + \phi_2 \cdot e^{\lambda \cdot t}) - \eta_{SFM} \cdot (\phi_1^{(j)} + \phi_1 \cdot e^{\lambda \cdot (t - \tau)}) \cdot (\phi_2^{(j)} + \phi_2 \cdot e^{\lambda \cdot t})$$

$$\phi_2 \cdot \lambda \cdot e^{\lambda \cdot t} = \frac{l_{cr}}{l_2} \cdot c \cdot \sigma_2 \cdot (\phi_2^{(j)} \cdot N^{(j)} + \phi_2^{(j)} \cdot n \cdot e^{\lambda \cdot t} + N^{(j)} \cdot \phi_2 \cdot e^{\lambda \cdot t} + \phi_2 \cdot n \cdot e^{\lambda \cdot t} \cdot e^{\lambda \cdot t}) - \frac{1}{\tau_{c2}} \cdot \phi_2^{(j)} - \frac{1}{\tau_{c2}} \cdot \phi_2 \cdot e^{\lambda \cdot t} - \eta_{SFM} \cdot (\phi_1^{(j)} \cdot \phi_2^{(j)} + \phi_1^{(j)} \cdot \phi_2 \cdot e^{\lambda \cdot t} + \phi_2^{(j)} \cdot \phi_1 \cdot e^{\lambda \cdot (t - \tau)} + \phi_1 \cdot \phi_2 \cdot e^{\lambda \cdot t} \cdot e^{\lambda \cdot (t - \tau)})$$

At fixed points: $\phi_2 \cdot n \approx 0$; $\phi_1 \cdot \phi_2 \approx 0$.

$$\phi_2 \cdot \lambda \cdot e^{\lambda \cdot t} = \frac{l_{cr}}{l_2} \cdot c \cdot \sigma_2 \cdot (\phi_2^{(j)} \cdot N^{(j)} + \phi_2^{(j)} \cdot n \cdot e^{\lambda \cdot t} + N^{(j)} \cdot \phi_2 \cdot e^{\lambda \cdot t}) - \frac{1}{\tau_{c2}} \cdot \phi_2^{(j)} - \frac{1}{\tau_{c2}} \cdot \phi_2 \cdot e^{\lambda \cdot t} - \eta_{SFM} \cdot (\phi_1^{(j)} \cdot \phi_2^{(j)} + \phi_1^{(j)} \cdot \phi_2 \cdot e^{\lambda \cdot t} + \phi_2^{(j)} \cdot \phi_1 \cdot e^{\lambda \cdot (t - \tau)})$$

$$\phi_2 \cdot \lambda \cdot e^{\lambda \cdot t} = \frac{l_{cr}}{l_2} \cdot c \cdot \sigma_2 \cdot \phi_2^{(j)} \cdot N^{(j)} + \frac{l_{cr}}{l_2} \cdot c \cdot \sigma_2 \cdot \phi_2^{(j)} \cdot n \cdot e^{\lambda \cdot t} + \frac{l_{cr}}{l_2} \cdot c \cdot \sigma_2 \cdot N^{(j)} \cdot \phi_2 \cdot e^{\lambda \cdot t} - \frac{1}{\tau_{c2}} \cdot \phi_2^{(j)} - \frac{1}{\tau_{c2}} \cdot \phi_2 \cdot e^{\lambda \cdot t} - \eta_{SFM} \cdot \phi_1^{(j)} \cdot \phi_2^{(j)} - \eta_{SFM} \cdot (\phi_1^{(j)} \cdot \phi_2 \cdot e^{\lambda \cdot t} + \phi_2^{(j)} \cdot \phi_1 \cdot e^{\lambda \cdot (t-\tau)})$$

$$\phi_2 \cdot \lambda \cdot e^{\lambda \cdot t} = \frac{l_{cr}}{l_2} \cdot c \cdot \sigma_2 \cdot \phi_2^{(j)} \cdot N^{(j)} - \frac{1}{\tau_{c2}} \cdot \phi_2^{(j)} - \eta_{SFM} \cdot \phi_1^{(j)} \cdot \phi_2^{(j)} + \frac{l_{cr}}{l_2} \cdot c \cdot \sigma_2 \cdot \phi_2^{(j)} \cdot n \cdot e^{\lambda \cdot t} + \frac{l_{cr}}{l_2} \cdot c \cdot \sigma_2 \cdot N^{(j)} \cdot \phi_2 \cdot e^{\lambda \cdot t} - \frac{1}{\tau_{c2}} \cdot \phi_2 \cdot e^{\lambda \cdot t} - \eta_{SFM} \cdot \phi_1^{(j)} \cdot \phi_2 \cdot e^{\lambda \cdot t} - \eta_{SFM} \cdot \phi_2^{(j)} \cdot \phi_1 \cdot e^{\lambda \cdot (t-\tau)}$$

At fixed points: $\frac{l_{cr}}{l_2} \cdot c \cdot \sigma_2 \cdot \phi_2^{(j)} \cdot N^{(j)} - \frac{1}{\tau_{c2}} \cdot \phi_2^{(j)} - \eta_{SFM} \cdot \phi_1^{(j)} \cdot \phi_2^{(j)} = 0$.

$$\frac{l_{cr}}{l_2} \cdot c \cdot \sigma_2 \cdot \phi_2^{(j)} \cdot n \cdot e^{\lambda \cdot t} + \frac{l_{cr}}{l_2} \cdot c \cdot \sigma_2 \cdot N^{(j)} \cdot \phi_2 \cdot e^{\lambda \cdot t} - \frac{1}{\tau_{c2}} \cdot \phi_2 \cdot e^{\lambda \cdot t} - \eta_{SFM} \cdot \phi_1^{(j)} \cdot \phi_2 \cdot e^{\lambda \cdot t} - \eta_{SFM} \cdot \phi_2^{(j)} \cdot \phi_1 \cdot e^{\lambda \cdot t} \cdot e^{-\lambda \cdot \tau} - \phi_2 \cdot \lambda \cdot e^{\lambda \cdot t} = 0$$

$$\frac{l_{cr}}{l_2} \cdot c \cdot \sigma_2 \cdot \phi_2^{(j)} \cdot n - \eta_{SFM} \cdot \phi_2^{(j)} \cdot e^{-\lambda \cdot \tau} \cdot \phi_1 + \left(\frac{l_{cr}}{l_2} \cdot c \cdot \sigma_2 \cdot N^{(j)} - \frac{1}{\tau_{c2}} - \eta_{SFM} \cdot \phi_1^{(j)}\right) \cdot \phi_2 - \phi_2 \cdot \lambda = 0$$

We can summary our Diode-pumped Q-switched Nd:YVO_4 yellow laser system arbitrarily small increments equations:

$$-\left(c \cdot [\sigma_1 \cdot \phi_1^{(j)} + \sigma_2 \cdot \phi_2^{(j)}] + \frac{1}{\tau_f}\right) \cdot n - n \cdot \lambda - c \cdot \sigma_1 \cdot N^{(j)} \cdot e^{-\lambda \cdot \tau} \cdot \phi_1 - c \cdot \sigma_2 \cdot N^{(j)} \cdot \phi_2 = 0$$

$$\frac{l_{cr}}{l_1} \cdot c \cdot \sigma_1 \cdot n - \left(\frac{1}{\tau_{c1}} + \eta_{SFM} \cdot \phi_2^{(j)}\right) \cdot e^{-\lambda \cdot \tau} \cdot \phi_1 - \lambda \cdot \phi_1 - \eta_{SFM} \cdot \phi_1^{(j)} \cdot \phi_2 = 0$$

$$\frac{l_{cr}}{l_2} \cdot c \cdot \sigma_2 \cdot \phi_2^{(j)} \cdot n - \eta_{SFM} \cdot \phi_2^{(j)} \cdot e^{-\lambda \cdot \tau} \cdot \phi_1 + \left(\frac{l_{cr}}{l_2} \cdot c \cdot \sigma_2 \cdot N^{(j)} - \frac{1}{\tau_{c2}} - \eta_{SFM} \cdot \phi_1^{(j)}\right) \cdot \phi_2 - \phi_2 \cdot \lambda = 0$$

We define for simplicity some global parameters: $\Xi_1 = -\left(c \cdot [\sigma_1 \cdot \phi_1^{(j)} + \sigma_2 \cdot \phi_2^{(j)}] + \frac{1}{\tau_f}\right)$

$$\Xi_2 = c \cdot \sigma_1 \cdot N^{(j)};\ \Xi_3 = c \cdot \sigma_2 \cdot N^{(j)};\ \Xi_4 = \frac{l_{cr}}{l_1} \cdot c \cdot \sigma_1$$

$$\Xi_5 = \frac{1}{\tau_{c1}} + \eta_{SFM} \cdot \phi_2^{(j)};\ \Xi_6 = \eta_{SFM} \cdot \phi_1^{(j)}$$

$$\Xi_7 = \frac{l_{cr}}{l_2} \cdot c \cdot \sigma_2 \cdot \phi_2^{(j)};\ \Xi_8 = \eta_{SFM} \cdot \phi_2^{(j)}$$

$$\Xi_9 = \frac{l_{cr}}{l_2} \cdot c \cdot \sigma_2 \cdot N^{(j)} - \frac{1}{\tau_{c2}} - \eta_{SFM} \cdot \phi_1^{(j)}$$

$$\Xi_1 \cdot n - n \cdot \lambda - \Xi_2 \cdot e^{-\lambda \cdot \tau} \cdot \phi_1 - \Xi_3 \cdot \phi_2 = 0$$

$$\Xi_4 \cdot n - \Xi_5 \cdot e^{-\lambda \cdot \tau} \cdot \phi_1 - \lambda \cdot \phi_1 - \Xi_6 \cdot \phi_2 = 0$$

$$\Xi_7 \cdot n - \Xi_8 \cdot e^{-\lambda \cdot \tau} \cdot \phi_1 + \Xi_9 \cdot \phi_2 - \phi_2 \cdot \lambda = 0$$

$$\begin{pmatrix} (\Xi_1 - \lambda) & -\Xi_2 \cdot e^{-\lambda \cdot \tau} & -\Xi_3 \\ \Xi_4 & -(\Xi_5 \cdot e^{-\lambda \cdot \tau} + \lambda) & -\Xi_6 \\ \Xi_7 & -\Xi_8 \cdot e^{-\lambda \cdot \tau} & (\Xi_9 - \lambda) \end{pmatrix} \cdot \begin{pmatrix} n \\ \phi_1 \\ \phi_2 \end{pmatrix} = 0$$

$$A - \lambda \cdot I = \begin{pmatrix} (\Xi_1 - \lambda) & -\Xi_2 \cdot e^{-\lambda \cdot \tau} & -\Xi_3 \\ \Xi_4 & -(\Xi_5 \cdot e^{-\lambda \cdot \tau} + \lambda) & -\Xi_6 \\ \Xi_7 & -\Xi_8 \cdot e^{-\lambda \cdot \tau} & (\Xi_9 - \lambda) \end{pmatrix};\ \det(A - \lambda \cdot I) = 0$$

$$\det(A - \lambda \cdot I) = \det\begin{pmatrix} (\Xi_1 - \lambda) & -\Xi_2 \cdot e^{-\lambda \cdot \tau} & -\Xi_3 \\ \Xi_4 & -(\Xi_5 \cdot e^{-\lambda \cdot \tau} + \lambda) & -\Xi_6 \\ \Xi_7 & -\Xi_8 \cdot e^{-\lambda \cdot \tau} & (\Xi_9 - \lambda) \end{pmatrix}$$

$$= (\Xi_1 - \lambda) \cdot \det\begin{pmatrix} -(\Xi_5 \cdot e^{-\lambda \cdot \tau} + \lambda) & -\Xi_6 \\ -\Xi_8 \cdot e^{-\lambda \cdot \tau} & (\Xi_9 - \lambda) \end{pmatrix} + \Xi_2 \cdot e^{-\lambda \cdot \tau} \cdot \det\begin{pmatrix} \Xi_4 & -\Xi_6 \\ \Xi_7 & (\Xi_9 - \lambda) \end{pmatrix} - \Xi_3 \cdot \det\begin{pmatrix} \Xi_4 & -(\Xi_5 \cdot e^{-\lambda \cdot \tau} + \lambda) \\ \Xi_7 & -\Xi_8 \cdot e^{-\lambda \cdot \tau} \end{pmatrix}$$

$$\det(A-\lambda\cdot I)=\det\begin{pmatrix}(\Xi_1-\lambda) & -\Xi_2\cdot e^{-\lambda\cdot\tau} & -\Xi_3\\ \Xi_4 & -(\Xi_5\cdot e^{-\lambda\cdot\tau}+\lambda) & -\Xi_6\\ \Xi_7 & -\Xi_8\cdot e^{-\lambda\cdot\tau} & (\Xi_9-\lambda)\end{pmatrix}$$
$$=(\Xi_1-\lambda)\cdot[-(\Xi_5\cdot e^{-\lambda\cdot\tau}+\lambda)\cdot(\Xi_9-\lambda)-\Xi_8\cdot e^{-\lambda\cdot\tau}\cdot\Xi_6]$$
$$+\Xi_2\cdot e^{-\lambda\cdot\tau}\cdot[\Xi_4\cdot(\Xi_9-\lambda)+\Xi_7\cdot\Xi_6]$$
$$-\Xi_3\cdot[-\Xi_8\cdot\Xi_4\cdot e^{-\lambda\cdot\tau}+\Xi_7\cdot(\Xi_5\cdot e^{-\lambda\cdot\tau}+\lambda)]$$

$$\det(A-\lambda\cdot I)=\det\begin{pmatrix}(\Xi_1-\lambda) & -\Xi_2\cdot e^{-\lambda\cdot\tau} & -\Xi_3\\ \Xi_4 & -(\Xi_5\cdot e^{-\lambda\cdot\tau}+\lambda) & -\Xi_6\\ \Xi_7 & -\Xi_8\cdot e^{-\lambda\cdot\tau} & (\Xi_9-\lambda)\end{pmatrix}$$
$$=(\Xi_1-\lambda)\cdot[(-\Xi_5\cdot\Xi_9-\Xi_8\cdot\Xi_6+\Xi_5\cdot\lambda)\cdot e^{-\lambda\cdot\tau}-\lambda\cdot\Xi_9+\lambda^2]$$
$$+\Xi_2\cdot e^{-\lambda\cdot\tau}\cdot[(\Xi_4\cdot\Xi_9+\Xi_7\cdot\Xi_6)-\Xi_4\cdot\lambda]$$
$$-\Xi_3\cdot[(\Xi_7\cdot\Xi_5-\Xi_8\cdot\Xi_4)\cdot e^{-\lambda\cdot\tau}+\Xi_7\cdot\lambda]$$

$$\det(A-\lambda\cdot I)=\det\begin{pmatrix}(\Xi_1-\lambda) & -\Xi_2\cdot e^{-\lambda\cdot\tau} & -\Xi_3\\ \Xi_4 & -(\Xi_5\cdot e^{-\lambda\cdot\tau}+\lambda) & -\Xi_6\\ \Xi_7 & -\Xi_8\cdot e^{-\lambda\cdot\tau} & (\Xi_9-\lambda)\end{pmatrix}$$
$$=\Xi_1\cdot(-\Xi_5\cdot\Xi_9-\Xi_8\cdot\Xi_6+\Xi_5\cdot\lambda)\cdot e^{-\lambda\cdot\tau}$$
$$-\lambda\cdot\Xi_1\cdot\Xi_9+\Xi_1\cdot\lambda^2-(-\Xi_5\cdot\Xi_9-\Xi_8\cdot\Xi_6$$
$$+\Xi_5\cdot\lambda)\cdot\lambda\cdot e^{-\lambda\cdot\tau}+\lambda^2\cdot\Xi_9-\lambda^3$$
$$+(\Xi_4\cdot\Xi_9+\Xi_7\cdot\Xi_6)\cdot\Xi_2\cdot e^{-\lambda\cdot\tau}-\Xi_4\cdot\Xi_2\cdot\lambda\cdot e^{-\lambda\cdot\tau}$$
$$-\Xi_3\cdot(\Xi_7\cdot\Xi_5-\Xi_8\cdot\Xi_4)\cdot e^{-\lambda\cdot\tau}-\Xi_3\cdot\Xi_7\cdot\lambda$$

$$\det(A-\lambda\cdot I)=\det\begin{pmatrix}(\Xi_1-\lambda) & -\Xi_2\cdot e^{-\lambda\cdot\tau} & -\Xi_3\\ \Xi_4 & -(\Xi_5\cdot e^{-\lambda\cdot\tau}+\lambda) & -\Xi_6\\ \Xi_7 & -\Xi_8\cdot e^{-\lambda\cdot\tau} & (\Xi_9-\lambda)\end{pmatrix}$$
$$=(-\Xi_1\cdot\Xi_5\cdot\Xi_9-\Xi_1\cdot\Xi_8\cdot\Xi_6+\Xi_1\cdot\Xi_5\cdot\lambda)\cdot e^{-\lambda\cdot\tau}$$
$$-\lambda\cdot\Xi_1\cdot\Xi_9+\Xi_1\cdot\lambda^2+\Xi_5\cdot\Xi_9\cdot\lambda\cdot e^{-\lambda\cdot\tau}$$
$$+\Xi_8\cdot\Xi_6\cdot\lambda\cdot e^{-\lambda\cdot\tau}-\Xi_5\cdot\lambda^2\cdot e^{-\lambda\cdot\tau}+\lambda^2\cdot\Xi_9-\lambda^3$$
$$+(\Xi_4\cdot\Xi_9+\Xi_7\cdot\Xi_6)\cdot\Xi_2\cdot e^{-\lambda\cdot\tau}-\Xi_4\cdot\Xi_2\cdot\lambda\cdot e^{-\lambda\cdot\tau}$$
$$-\Xi_3\cdot(\Xi_7\cdot\Xi_5-\Xi_8\cdot\Xi_4)\cdot e^{-\lambda\cdot\tau}-\Xi_3\cdot\Xi_7\cdot\lambda$$

$$\det(A-\lambda\cdot I)=\det\begin{pmatrix}(\Xi_1-\lambda) & -\Xi_2\cdot e^{-\lambda\cdot\tau} & -\Xi_3\\ \Xi_4 & -(\Xi_5\cdot e^{-\lambda\cdot\tau}+\lambda) & -\Xi_6\\ \Xi_7 & -\Xi_8\cdot e^{-\lambda\cdot\tau} & (\Xi_9-\lambda)\end{pmatrix}$$
$$=[-\Xi_1\cdot\Xi_5\cdot\Xi_9-\Xi_1\cdot\Xi_8\cdot\Xi_6+(\Xi_4\cdot\Xi_9+\Xi_7\cdot\Xi_6)\cdot\Xi_2$$

$$
\begin{aligned}
&- \Xi_3 \cdot (\Xi_7 \cdot \Xi_5 - \Xi_8 \cdot \Xi_4) + (\Xi_1 \cdot \Xi_5 + \Xi_5 \cdot \Xi_9 \\
&+ \Xi_8 \cdot \Xi_6 - \Xi_4 \cdot \Xi_2) \cdot \lambda - \Xi_5 \cdot \lambda^2] \cdot e^{-\lambda \cdot \tau} \\
&- (\Xi_1 \cdot \Xi_9 + \Xi_3 \cdot \Xi_7) \cdot \lambda + (\Xi_1 + \Xi_9) \cdot \lambda^2 - \lambda^3
\end{aligned}
$$

We define for simplicity some global parameters:

$$
\begin{aligned}
\Upsilon_1 = &-\Xi_1 \cdot \Xi_5 \cdot \Xi_9 - \Xi_1 \cdot \Xi_8 \cdot \Xi_6 + (\Xi_4 \cdot \Xi_9 + \Xi_7 \cdot \Xi_6) \cdot \Xi_2 \\
&- \Xi_3 \cdot (\Xi_7 \cdot \Xi_5 - \Xi_8 \cdot \Xi_4)
\end{aligned}
$$

$$
\begin{aligned}
&\Upsilon_2 = \Xi_1 \cdot \Xi_5 + \Xi_5 \cdot \Xi_9 + \Xi_8 \cdot \Xi_6 - \Xi_4 \cdot \Xi_2; \ \Upsilon_3 = -\Xi_5 \\
&\Upsilon_4 = -(\Xi_1 \cdot \Xi_9 + \Xi_3 \cdot \Xi_7); \ \Upsilon_5 = \Xi_1 + \Xi_9
\end{aligned}
$$

$$
\begin{aligned}
\det(A - \lambda \cdot I) &= \det \begin{pmatrix} (\Xi_1 - \lambda) & -\Xi_2 \cdot e^{-\lambda \cdot \tau} & -\Xi_3 \\ \Xi_4 & -(\Xi_5 \cdot e^{-\lambda \cdot \tau} + \lambda) & -\Xi_6 \\ \Xi_7 & -\Xi_8 \cdot e^{-\lambda \cdot \tau} & (\Xi_9 - \lambda) \end{pmatrix} \\
&= \Upsilon_4 \cdot \lambda + \Upsilon_5 \cdot \lambda^2 - \lambda^3 + (\Upsilon_1 + \Upsilon_2 \cdot \lambda + \Upsilon_3 \cdot \lambda^2) \cdot e^{-\lambda \cdot \tau}
\end{aligned}
$$

$$
\det(A - \lambda \cdot I) = \Upsilon_4 \cdot \lambda + \Upsilon_5 \cdot \lambda^2 - \lambda^3 + (\Upsilon_1 + \Upsilon_2 \cdot \lambda + \Upsilon_3 \cdot \lambda^2) \cdot e^{-\lambda \cdot \tau}
$$

$$
D(\lambda, \tau) = \Upsilon_4 \cdot \lambda + \Upsilon_5 \cdot \lambda^2 - \lambda^3 + (\Upsilon_1 + \Upsilon_2 \cdot \lambda + \Upsilon_3 \cdot \lambda^2) \cdot e^{-\lambda \cdot \tau}
$$

$$
D(\lambda, \tau) = P_n(\lambda, \tau) + Q_m(\lambda, \tau) \cdot e^{-\lambda \cdot \tau}; n, m \in \mathbb{N}_0; n = 3; m = 2; n > m
$$

$$
\begin{aligned}
&P_n(\lambda, \tau) = \Upsilon_4 \cdot \lambda + \Upsilon_5 \cdot \lambda^2 - \lambda^3; n = 3 \\
&Q_m(\lambda, \tau) = \Upsilon_1 + \Upsilon_2 \cdot \lambda + \Upsilon_3 \cdot \lambda^2; m = 2
\end{aligned}
$$

$$
P_n(\lambda, \tau) = \sum_{k=0}^{n=3} p_k(\tau) \cdot \lambda^k = p_0(\tau) + p_1(\tau) \cdot \lambda + p_2(\tau) \cdot \lambda^2 + p_3(\tau) \cdot \lambda^3
$$

$$
p_0(\tau) = 0; \ p_1(\tau) = \Upsilon_4; \ p_2(\tau) = \Upsilon_5; \ p_3(\tau) = -1
$$

$$
\begin{aligned}
&Q_m(\lambda, \tau) = \sum_{k=0}^{m=2} q_k(\tau) \cdot \lambda^k = q_0(\tau) + q_1(\tau) \cdot \lambda + q_2(\tau) \cdot \lambda^2; \ q_0(\tau) = \Upsilon_1 \\
&q_1(\tau) = \Upsilon_2; \ q_2(\tau) = \Upsilon_3
\end{aligned}
$$

The homogeneous system for N, ϕ_1, ϕ_2 leads to a characteristic equation for the eigenvalue λ having the form $D(\lambda, \tau) = P(\lambda, \tau) + Q(\lambda, \tau) \cdot e^{-\lambda \cdot \tau} = 0$; and $P(\lambda) =$

$\sum_{j=0}^{3} a_j \cdot \lambda^j$; $Q(\lambda) = \sum_{j=0}^{2} c_j \cdot \lambda^j$. The coefficients $\{a_j(q_i, q_k, \tau), c_j(q_i, q_k, \tau)\} \in \mathbb{R}$ depend on q_i, q_k and delay τ. q_i, q_k are any Diode-pumped Q-switched Nd:YVO$_4$ yellow laser system's parameters, other parameters kept as a constant [2, 3].

$$a_0 = 0; a_1 = \Upsilon_4; a_2 = \Upsilon_5; a_3 = -1; c_0 = \Upsilon_1; c_1 = \Upsilon_2; c_2 = \Upsilon_3$$

Unless strictly necessary, the designation of the varied arguments (q_i, q_k) will subsequently be omitted from P, Q, a_j and c_j. The coefficients a_j, c_j are continuous and differentiable functions of their arguments, and direct substitution shows that $a_0 + c_0 \neq 0$ for $q_i, q_k \in \mathbb{R}_+$; that is, $\lambda = 0$ is not of $P(\lambda) + Q(\lambda) \cdot e^{-\lambda \cdot \tau} = 0$. Furthermore, $P(\lambda)$, $Q(\lambda)$ are analytic functions of λ, for which the following requirements of the analysis (Kuang and Cong 2005) [2] can also be verified in the present case:

1. If $\lambda = i \cdot \omega; \omega \in \mathbb{R}$, then $P(i \cdot \omega) + Q(i \cdot \omega) \neq 0$.
2. If $\left|\frac{Q(\lambda)}{P(\lambda)}\right|$ is bounded for $|\lambda| \to \infty$; $\operatorname{Re} \lambda \geq 0$. No roots bifurcation from ∞.
3. $F(\omega) = |P(i \cdot \omega)|^2 - |Q(i \cdot \omega)|^2$ has a finite number of zeros. Indeed, this is a polynomial in ω.
4. Each positive root $\omega(q_i, q_k)$ of $F(\omega) = 0$ is continuous and differentiable with respect to q_i, q_k.

We assume that $P_n(\lambda, \tau)$ and $Q_m(\lambda, \tau)$ cannot have common imaginary roots. That is for any real number ω: $P_n(\lambda = i \cdot \omega, \tau) + Q_m(\lambda = i \cdot \omega, \tau) \neq 0$ and

$$\begin{aligned} P_n(\lambda = i \cdot \omega, \tau) &= p_0(\tau) + p_1(\tau) \cdot i \cdot \omega - p_2(\tau) \cdot \omega^2 - p_3(\tau) \cdot i \cdot \omega^3 \\ &= p_0(\tau) - p_2(\tau) \cdot \omega^2 + (p_1(\tau) \cdot \omega - p_3(\tau) \cdot \omega^3) \cdot i \end{aligned}$$

$$P_n(\lambda = i \cdot \omega, \tau) = p_0(\tau) - p_2(\tau) \cdot \omega^2 + (p_1(\tau) \cdot \omega - p_3(\tau) \cdot \omega^3) \cdot i$$
$$P_n(\lambda = i \cdot \omega, \tau) = -\Upsilon_5 \cdot \omega^2 + (\Upsilon_4 \cdot \omega + \omega^3) \cdot i$$

$$Q_m(\lambda = i \cdot \omega, \tau) = q_0(\tau) + q_1(\tau) \cdot i \cdot \omega - q_2(\tau) \cdot \omega^2$$
$$Q_m(\lambda = i \cdot \omega, \tau) = q_0(\tau) - q_2(\tau) \cdot \omega^2 + q_1(\tau) \cdot i \cdot \omega$$

$$Q_m(\lambda = i \cdot \omega, \tau) = \Upsilon_1 - \Upsilon_3 \cdot \omega^2 + \Upsilon_2 \cdot i \cdot \omega$$

$$\begin{aligned} P_n(\lambda = i \cdot \omega, \tau) + Q_m(\lambda = i \cdot \omega, \tau) = {} & p_0(\tau) - p_2(\tau) \cdot \omega^2 \\ & + (p_1(\tau) \cdot \omega - p_3(\tau) \cdot \omega^3) \cdot i \\ & + q_0(\tau) - q_2(\tau) \cdot \omega^2 + q_1(\tau) \cdot i \cdot \omega \neq 0 \end{aligned}$$

$$\begin{aligned} P_n(\lambda = i \cdot \omega, \tau) + Q_m(\lambda = i \cdot \omega, \tau) = {} & \Upsilon_1 - (\Upsilon_5 + \Upsilon_3) \cdot \omega^2 \\ & + [(\Upsilon_4 + \Upsilon_2) \cdot \omega + \omega^3] \cdot i \neq 0 \end{aligned}$$

$$|P(i \cdot \omega, \tau)|^2 = \Upsilon_5^2 \cdot \omega^4 + (\Upsilon_4 \cdot \omega + \omega^3)^2 = \Upsilon_4^2 \cdot \omega^2 + (\Upsilon_5^2 + 2 \cdot \Upsilon_4) \cdot \omega^4 + \omega^6$$

$$|Q(i \cdot \omega, \tau)|^2 = (\Upsilon_1 - \Upsilon_3 \cdot \omega^2)^2 + \Upsilon_2^2 \cdot \omega^2$$
$$= \Upsilon_1^2 + (\Upsilon_2^2 - 2 \cdot \Upsilon_1 \cdot \Upsilon_3) \cdot \omega^2 + \Upsilon_3^2 \cdot \omega^4$$

$$F(\omega, \tau) = |P(i \cdot \omega, \tau)|^2 - |Q(i \cdot \omega, \tau)|^2$$
$$= -\Upsilon_1^2 + (\Upsilon_4^2 - \Upsilon_2^2 + 2 \cdot \Upsilon_1 \cdot \Upsilon_3) \cdot \omega^2 + (\Upsilon_5^2 + 2 \cdot \Upsilon_4 - \Upsilon_3^2) \cdot \omega^4 + \omega^6$$

We define the following parameters for simplicity: $\Pi_0, \Pi_2, \Pi_4, \Pi_6$

$$\Pi_0 = -\Upsilon_1^2;\ \Pi_2 = \Upsilon_4^2 - \Upsilon_2^2 + 2 \cdot \Upsilon_1 \cdot \Upsilon_3;\ \Pi_4 = \Upsilon_5^2 + 2 \cdot \Upsilon_4 - \Upsilon_3^2;\ \Pi_6 = 1$$

Hence $F(\omega, \tau) = 0$ implies $\sum_{k=0}^{3} \Pi_{2 \cdot k} \cdot \omega^{2 \cdot k} = 0$ and its roots are given by solving the above polynomial. $P_R(i\omega, \tau) = -\Upsilon_5 \cdot \omega^2$; $P_I(i\omega, \tau) = \Upsilon_4 \cdot \omega + \omega^3$; $Q_R(i\omega, \tau) = \Upsilon_1 - \Upsilon_3 \cdot \omega^2$; $Q_I(i\omega, \tau) = \Upsilon_2 \cdot \omega$.

$$\sin \theta(\tau) = \frac{-P_R(i\omega, \tau) \cdot Q_I(i\omega, \tau) + P_I(i\omega, \tau) \cdot Q_R(i\omega, \tau)}{|Q(i\omega, \tau)|^2}$$

$$\cos \theta(\tau) = -\frac{P_R(i\omega, \tau) \cdot Q_R(i\omega, \tau) + P_I(i\omega, \tau) \cdot Q_I(i\omega, \tau)}{|Q(i\omega, \tau)|^2}$$

We use different terminology from our last characteristics parameters definition:

$$k \to j;\ p_k(\tau) \to a_j;\ q_k(\tau) \to c_j;\ n, m \in \mathbb{N}_0;\ n = 3;\ m = 2;\ n > m$$

$$P_n(\lambda, \tau) \to P(\lambda);\ Q_m(\lambda, \tau) \to Q(\lambda);\ P(\lambda) = \sum_{j=0}^{3} a_j \cdot \lambda^j$$

$$Q(\lambda) = \sum_{j=0}^{2} c_j \cdot \lambda^j$$

$$P(\lambda) = a_0 + a_1 \cdot \lambda + a_2 \cdot \lambda^2 + a_3 \cdot \lambda^3;\ Q(\lambda) = c_0 + c_1 \cdot \lambda + c_2 \cdot \lambda^2$$

$$P(\lambda) = \Upsilon_4 \cdot \lambda + \Upsilon_5 \cdot \lambda^2 - \lambda^3;\ Q(\lambda) = \Upsilon_1 + \Upsilon_2 \cdot \lambda + \Upsilon_3 \cdot \lambda^2$$

$n, m \in \mathbb{N}_0; n > m$ and $a_j, c_j : \mathbb{R}_{0+} \to \mathbb{R}$ are continuous and differentiable function of τ such that $a_0 + c_0 \neq 0$. In the following "—" denoted complex and conjugate. $P(\lambda)$, $Q(\lambda)$ are analytic functions in λ and differentiable in τ. The coefficients $a_j(R_p, \sigma_1, \tau_f, l_{cr}, l_1, \eta_{SFM}, \tau_{c1}, \tau, \ldots) \in \mathbb{R}$ and $c_j(R_p, \sigma_1, \tau_f, l_{cr}, l_1, \eta_{SFM}, \tau_{c1}, \tau, \ldots) \in \mathbb{R}$ depend on Diode-pumped Q-switched

Nd:YVO4 yellow laser system's parameters, $R_p, \sigma_1, \tau_f, l_{cr}, l_1, \eta_{SFM}, \tau_{c1}, \tau, \ldots$ values. Unless strictly necessary, the designation of the varied arguments: $R_p, \sigma_1, \tau_f, l_{cr}, l_1, \eta_{SFM}, \tau_{c1}, \tau, \ldots$ will subsequently be omitted from P, Q, a_j, c_j. The coefficients a_j, c_j are continuous and differentiable functions of their arguments, and direct substitution shows that $a_0 + c_0 \neq 0$ [2, 3].

$$\begin{aligned} a_0 + c_0 &= \Upsilon_1 \\ &= -\Xi_1 \cdot \Xi_5 \cdot \Xi_9 - \Xi_1 \cdot \Xi_8 \cdot \Xi_6 + (\Xi_4 \cdot \Xi_9 + \Xi_7 \cdot \Xi_6) \cdot \Xi_2 \\ &\quad - \Xi_3 \cdot (\Xi_7 \cdot \Xi_5 - \Xi_8 \cdot \Xi_4) \neq 0 \end{aligned}$$

$\forall\, R_p, \sigma_1, \tau_f, l_{cr}, l_1, \eta_{SFM}, \tau_{c1}, \tau, \ldots \in \mathbb{R}_+$ I.e. $\lambda = 0$ is not a root of the characteristic equation. Furthermore $P(\lambda)$, $Q(\lambda)$ are analytic functions of λ for which the following requirements of the analysis (see Kuang [2], Sect. 3.4) can also be verified in the present case.

1. If $\lambda = i \cdot \omega; \omega \in \mathbb{R}$ then $P(i \cdot \omega) + Q(i \cdot \omega) \neq 0$, i.e. P and Q have no common imaginary roots. This condition was verified numerically in the entire $\forall\, R_p, \sigma_1, \tau_f, l_{cr}, l_1, \eta_{SFM}, \tau_{c1}, \tau, \ldots$ domain of interest.
2. $\left|\frac{P(\lambda)}{Q(\lambda)}\right|$ is bounded for $|\lambda| \to \infty$; $\operatorname{Re}\lambda \geq 0$. No roots bifurcation from ∞. Indeed, in the limit: $\left|\frac{Q(\lambda)}{P(\lambda)}\right| = \left|\frac{c_0 + c_1 \cdot \lambda + c_2 \cdot \lambda^2}{a_0 + a_1 \cdot \lambda + a_2 \cdot \lambda^2 + a_3 \cdot \lambda^3}\right|$.
3. The following expressions exist: $F(\omega) = |P(i \cdot \omega)|^2 - |Q(i \cdot \omega)|^2$. $F(\omega, \tau) = |P(i \cdot \omega, \tau)|^2 - |Q(i \cdot \omega, \tau)|^2 = \sum_{k=0}^{3} \Pi_{2 \cdot k} \cdot \omega^{2 \cdot k}$ has at most a finite number of zeros. Indeed, this is a polynomial in ω (degree in ω^6).
4. Each positive root $\omega(R_p, \sigma_1, \tau_f, l_{cr}, l_1, \eta_{SFM}, \tau_{c1}, \tau, \ldots)$ of $F(\omega) = 0$ is continuous and differentiable with respect to $R_p, \sigma_1, \tau_f, l_{cr}, l_1, \eta_{SFM}, \tau_{c1}, \tau, \ldots$ and the condition can only be assessed numerically.

In addition, since the coefficients in P and Q are real, we have $\overline{P(-i \cdot \omega)} = P(i \cdot \omega)$ and $\overline{Q(-i \cdot \omega)} = Q(i \cdot \omega)$ thus, $\omega > 0$ may be on eigenvalue of characteristic equations. The analysis consists in identifying the roots of the characteristic equation situated on the imaginary axis of the complex λ plane, whereby increasing the parameters: $R_p, \sigma_1, \tau_f, l_{cr}, l_1, \eta_{SFM}, \tau_{c1}, \tau, \ldots$, $\operatorname{Re}\lambda$ may, at the crossing, change its sign from (−) to (+). i.e. from a stable focus $E^{(j)} = (N^{(j)}, \phi_1^{(j)}, \phi_2^{(j)})$; $j = 0, 1, 2$ to an unstable one, or vice versa. This feature may be further assessed by examining the sign of the partial derivatives with respect to $R_p, \sigma_1, \tau_f, l_{cr}, l_1, \eta_{SFM}, \tau_{c1}, \tau, \ldots$ and any system parameters.

$$\Lambda^{-1}(R_p) = \left(\frac{\partial \operatorname{Re}\lambda}{\partial R_p}\right)_{\lambda = i \cdot \omega}; \sigma_1, \tau_f, l_{cr}, l_1, \eta_{SFM}, \tau_{c1}, \tau, \ldots = const$$

$$\Lambda^{-1}(\sigma_1) = \left(\frac{\partial \operatorname{Re}\lambda}{\partial \sigma_1}\right)_{\lambda = i \cdot \omega}; R_p, \tau_f, l_{cr}, l_1, \eta_{SFM}, \tau_{c1}, \tau, \ldots = const$$

$$\Lambda^{-1}(\tau_f) = \left(\frac{\partial \mathrm{Re}\lambda}{\partial \tau_f}\right)_{\lambda=i\cdot\omega};\sigma_1, R_p, l_{cr}, l_1, \eta_{SFM}, \tau_{c1}, \tau, \ldots = const$$

$$\Lambda^{-1}(l_{cr}) = \left(\frac{\partial \mathrm{Re}\lambda}{\partial l_{cr}}\right)_{\lambda=i\cdot\omega};\sigma_1, \tau_f, R_p, l_1, \eta_{SFM}, \tau_{c1}, \tau, \ldots = const$$

$$\Lambda^{-1}(\eta_{SFM}) = \left(\frac{\partial \mathrm{Re}\lambda}{\partial \eta_{SFM}}\right)_{\lambda=i\cdot\omega};\sigma_1, \tau_f, l_{cr}, l_1, R_p, \tau_{c1}, \tau, \ldots = const$$

$$\Lambda^{-1}(\tau) = \left(\frac{\partial \mathrm{Re}\lambda}{\partial \tau}\right)_{\lambda=i\cdot\omega};R_p, \sigma_1, \tau_f, l_{cr}, l_1, \eta_{SFM}, \tau_{c1}, \ldots = const$$

$P(\lambda) = P_R(\lambda) + i \cdot P_I(\lambda)$; $Q(\lambda) = Q_R(\lambda) + i \cdot Q_I(\lambda)$, When writing and inserting $\lambda = i \cdot \omega$ into Diode-pumped Q-switched Nd:YVO_4 yellow laser system's characteristic equation ω must satisfy the following equations:

$$\sin(\omega \cdot \tau) = g(\omega) = \frac{-P_R(i\omega,\tau) \cdot Q_I(i\omega,\tau) + P_I(i\omega,\tau) \cdot Q_R(i\omega,\tau)}{|Q(i\omega,\tau)|^2}$$

$$\cos(\omega \cdot \tau) = h(\omega) = -\frac{P_R(i\omega,\tau) \cdot Q_R(i\omega,\tau) + P_I(i\omega,\tau) \cdot Q_I(i\omega,\tau)}{|Q(i\omega,\tau)|^2}$$

where $|Q(i\omega,\tau)|^2 \neq 0$ in view of requirement (1) above, and $(g, h) \in \mathbb{R}$. Furthermore, it follows above $\sin(\omega \cdot \tau)$ and $\cos(\omega \cdot \tau)$ equation that, by squaring and adding the sides, ω must be a positive root of $F(\omega) = |P(i \cdot \omega)|^2 - |Q(i \cdot \omega)|^2 = 0$. Note: $F(\omega)$ is independent on τ. Now it is important to notice that if $\tau \notin I$ (assume that $I \subseteq \mathbb{R}_{+0}$ is the set where $\omega(\tau)$ is a positive root of $F(\omega)$ and for, $\tau \notin I$, $\omega(\tau)$ is not defined. Then for all τ in I, $\omega(\tau)$ is satisfied that $F(\omega,\tau) = 0$. Then there are no positive $\omega(\tau)$ solutions for $F(\omega,\tau) = 0$, and we cannot have stability switches. For $\tau \in I$ where $\omega(\tau)$ is a positive solution of $F(\omega,\tau) = 0$, we can define the angle $\theta(\tau) \in [0, 2 \cdot \pi]$ as the solution of $\sin\theta(\tau) = \ldots$ and $\cos\theta(\tau) = \ldots$; the relation between the arguments $\theta(\tau)$ and $\tau \cdot \omega(\tau)$ for $\tau \in I$ must be describing below. $\tau \cdot \omega(\tau) = \theta(\tau) + 2 \cdot n \cdot \pi \ \forall n \in \mathbb{N}_0$ Hence we can define the maps: $\tau_n : I \to \mathbb{R}_{+0}$, is given by $\tau_n(\tau) = \frac{\theta(\tau)+2\cdot n\cdot\pi}{\omega(\tau)}$; $n \in \mathbb{N}_0$; $\tau \in I$. Let us introduce the function $I \to \mathbb{R}$; $S_n(\tau) = \tau - \tau_n(\tau)$; $\tau \in I$; $n \in \mathbb{N}_0$ that is continuous and differentiable in τ. In the following, the subscripts $\lambda, \omega, R_p, \sigma_1, \tau_f, l_{cr}, l_1, \eta_{SFM}, \tau_{c1}, \ldots$ indicate the corresponding partial derivatives. Let us first concentrate on $\Lambda(x)$, remember in $\lambda(R_p, \sigma_1, \tau_f, l_{cr}, l_1, \eta_{SFM}, \tau_{c1}, \ldots)$ and $\omega(R_p, \sigma_1, \tau_f, l_{cr}, l_1, \eta_{SFM}, \tau_{c1}, \ldots)$, and keeping all parameters except one (x) and τ. The derivation closely follows that in reference [BK]. Differentiating Diode-pumped Q-switched Nd:YVO_4 yellow laser system's characteristic equation $P(\lambda) + Q(\lambda) \cdot e^{-\lambda\cdot\tau} = 0$ with respect to specific parameter (x), and inverting the derivative, for convenience, one calculates: $x = R_p, \sigma_1, \tau_f, l_{cr}, l_1, \eta_{SFM}, \tau_{c1}, \tau, \ldots$

$$\left(\frac{\partial\lambda}{\partial x}\right)^{-1} = \frac{-P_\lambda(\lambda,x)\cdot Q(\lambda,x) + Q_\lambda(\lambda,x)\cdot P(\lambda,x) - \tau\cdot P(\lambda,x)\cdot Q(\lambda,x)}{P_x(\lambda,x)\cdot Q(\lambda,x) - Q_x(\lambda,x)\cdot P(\lambda,x)}$$

where $P_\lambda = \frac{\partial P}{\partial\lambda}, \ldots$ etc., substituting $\lambda = i\cdot\omega$ and bearing $\overline{P(-i\cdot\omega)} = P(i\cdot\omega)$; $\overline{Q(-i\cdot\omega)} = Q(i\cdot\omega)$ then $i\cdot P_\lambda(i\cdot\omega) = P_\omega(i\cdot\omega)$; $i\cdot Q_\lambda(i\cdot\omega) = Q_\omega(i\cdot\omega)$ and that on the surface $|P(i\omega)|^2 = |Q(i\omega)|^2$, one obtain: $(\frac{\partial\lambda}{\partial x})^{-1}|_{\lambda=i\cdot\omega} = \left(\frac{i\cdot P_\omega(i\cdot\omega,x)\cdot\overline{P(i\cdot\omega,x)} + i\cdot Q_\lambda(i\cdot\omega,x)\cdot\overline{Q(\lambda,x)} - \tau\cdot|P(i\cdot\omega,x)|^2}{P_x(i\cdot\omega,x)\cdot\overline{P(i\cdot\omega,x)} - Q_x(i\cdot\omega,x)\cdot\overline{Q(i\cdot\omega,x)}}\right)$.

Upon separating into real and imaginary parts, with $P = P_R + i\cdot P_I$; $Q = Q_R + i\cdot Q_I$ and $P_\omega = P_{R\omega} + i\cdot P_{I\omega}$; $Q_\omega = Q_{R\omega} + i\cdot Q_{I\omega}$; $P_x = P_{Rx} + i\cdot P_{Ix}$; $Q_x = Q_{Rx} + i\cdot Q_{Ix}$; $P^2 = P_R^2 + P_I^2$, When (x) can be any Diode-pumped Q-switched Nd:YVO$_4$ yellow laser system's parameters $R_p, \sigma_1, \tau_f, l_{cr}, l_1, \eta_{SFM}, \tau_{c1}, \ldots$ and time delay τ etc. Where for convenience, we have dropped the arguments $(i\cdot\omega, x)$, and where $F_\omega = 2\cdot[(P_{R\omega}\cdot P_R + P_{I\omega}\cdot P_I) - (Q_{R\omega}\cdot Q_R + Q_{I\omega}\cdot Q_I)]$; $F_x = 2\cdot[(P_{Rx}\cdot P_R + P_{Ix}\cdot P_I) - (Q_{Rx}\cdot Q_R + Q_{Ix}\cdot Q_I)]$ and $\omega_x = -\frac{F_x}{F_\omega}$. We define U and V: $U = (P_R\cdot P_{I\omega} - P_I\cdot P_{R\omega}) - (Q_R\cdot Q_{I\omega} - Q_I\cdot Q_{R\omega})$, $V = (P_R\cdot P_{Ix} - P_I\cdot P_{Rx}) - (Q_R\cdot Q_{Ix} - Q_I\cdot Q_{Rx})$, we choose our specific parameter as time delay $x = \tau$: $P_R = -\Upsilon_5\cdot\omega^2$; $P_I = \Upsilon_4\cdot\omega + \omega^3$; $Q_R = \Upsilon_1 - \Upsilon_3\cdot\omega^2$ $Q_I = \Upsilon_2\cdot\omega$; $P_{R\omega} = -2\cdot\Upsilon_5\cdot\omega$; $P_{I\omega} = \Upsilon_4 + 3\cdot\omega^2$; $Q_{I\omega} = \Upsilon_2$; $Q_{R\omega} = -2\cdot\Upsilon_3\cdot\omega$; $P_{R\tau} = 0$; $P_{I\tau} = 0$; $Q_{R\tau} = 0$.

$$Q_{I\tau} = 0;\ F_\tau = 0;\ P_R\cdot P_{I\tau} - P_I\cdot P_{R\tau} = 0$$
$$Q_R\cdot Q_{I\tau} - Q_I\cdot Q_{R\tau} = 0$$
$$V = (P_R\cdot P_{I\tau} - P_I\cdot P_{R\tau}) - (Q_R\cdot Q_{I\tau} - Q_I\cdot Q_{R\tau}) = 0$$

$F_\omega = \ldots$ Elements:

$$P_{R\omega}\cdot P_R = 2\cdot\Upsilon_5^2\cdot\omega^3$$
$$P_{I\omega}\cdot P_I = (\Upsilon_4 + 3\cdot\omega^2)\cdot(\Upsilon_4\cdot\omega + \omega^3) = \Upsilon_4^2\cdot\omega + 4\cdot\Upsilon_4\cdot\omega^3 + 3\cdot\omega^5$$

$$Q_{R\omega}\cdot Q_R = -2\cdot\Upsilon_3\cdot\omega\cdot(\Upsilon_1 - \Upsilon_3\cdot\omega^2) = -2\cdot\Upsilon_1\cdot\Upsilon_3\cdot\omega + 2\cdot\Upsilon_3^2\cdot\omega^3$$
$$Q_{I\omega}\cdot Q_I = \Upsilon_2^2\cdot\omega$$

$U = \ldots$ Elements:

$$P_R\cdot P_{I\omega} = -(\Upsilon_5\cdot\Upsilon_4\cdot\omega^2 + \Upsilon_5\cdot 3\cdot\omega^4)$$
$$P_I\cdot P_{R\omega} = (\Upsilon_4\cdot\omega + \omega^3)\cdot(-2\cdot\Upsilon_5\cdot\omega) = -2\cdot\Upsilon_5\cdot\Upsilon_4\cdot\omega^2 - 2\cdot\Upsilon_5\cdot\omega^4$$

$$Q_R\cdot Q_{I\omega} = \Upsilon_2\cdot\Upsilon_1 - \Upsilon_2\cdot\Upsilon_3\cdot\omega^2;\ Q_I\cdot Q_{R\omega} = -2\cdot\Upsilon_2\cdot\Upsilon_3\cdot\omega^2$$

We can summary our last results in the next table (Table 7.11).

Table 7.11 Diode-pumped Q-switched Nd:YVO_4 yellow laser system's stability analyses F_ω, U elements

F_ω, U elements	Expression	F_ω, U elements	Expression
$P_{R\omega} \cdot P_R$	$2 \cdot \Upsilon_5^2 \cdot \omega^3$	$P_R \cdot P_{I\omega}$	$-(\Upsilon_5 \cdot \Upsilon_4 \cdot \omega^2 + \Upsilon_5 \cdot 3 \cdot \omega^4)$
$P_{I\omega} \cdot P_I$	$\Upsilon_4^2 \cdot \omega + 4 \cdot \Upsilon_4 \cdot \omega^3 + 3 \cdot \omega^5$	$P_I \cdot P_{R\omega}$	$-2 \cdot \Upsilon_5 \cdot \Upsilon_4 \cdot \omega^2 - 2 \cdot \Upsilon_5 \cdot \omega^4$
$Q_{R\omega} \cdot Q_R$	$-2 \cdot \Upsilon_1 \cdot \Upsilon_3 \cdot \omega + 2 \cdot \Upsilon_3^2 \cdot \omega^3$	$Q_R \cdot Q_{I\omega}$	$\Upsilon_2 \cdot \Upsilon_1 - \Upsilon_2 \cdot \Upsilon_3 \cdot \omega^2$
$Q_{I\omega} \cdot Q_I$	$\Upsilon_2^2 \cdot \omega$	$Q_I \cdot Q_{R\omega}$	$-2 \cdot \Upsilon_2 \cdot \Upsilon_3 \cdot \omega^2$

$F(\omega, \tau) = 0$, differentiating with respect to τ and we get $F_\omega \cdot \frac{\partial \omega}{\partial \tau} + F_\tau = 0; \tau \in I \Rightarrow \frac{\partial \omega}{\partial \tau} = -\frac{F_\tau}{F_\omega}$

$$\Lambda^{-1}(\tau) = \left(\frac{\partial \mathrm{Re}\lambda}{\partial \tau}\right)_{\lambda = i\omega}; \frac{\partial \omega}{\partial \tau} = \omega_\tau = -\frac{F_\tau}{F_\omega}$$

$$\Lambda^{-1}(\tau) = \mathrm{Re}\left\{\frac{-2 \cdot [U + \tau \cdot |P|^2 + i \cdot F_\omega}{F_\tau + i \cdot 2 \cdot [V + \omega \cdot |P|^2]}\right\}$$

$$\mathrm{sign}\{\Lambda^{-1}(\tau)\} = \mathrm{sign}\left\{\left(\frac{\partial \mathrm{Re}\lambda}{\partial \tau}\right)_{\lambda = i\omega}\right\}$$

$$\mathrm{sign}\{\Lambda^{-1}(\tau)\} = \mathrm{sign}\{F_\omega\} \cdot \mathrm{sign}\left\{\frac{V + \frac{\partial \omega}{\partial \tau} \cdot U}{|P|^2} + \omega + \frac{\partial \omega}{\partial \tau} \cdot \tau\right\}$$

We shall presently examine the possibility of stability transitions (bifurcations) Diode-pumped Q-switched Nd:YVO_4 yellow laser system, about the equilibrium points, $E^{(j)}; j = 0, 1, 2, \ldots$, $E^{(j)} = (N^{(j)}, \phi_1^{(j)}, \phi_2^{(j)}); j = 0, 1, 2$. The analysis consists in identifying the roots of our system characteristic equation situated on the imaginary axis of the complex λ-plane, where by increasing the delay parameter τ. Reλ, may at the crossing, changes its sign from – to +, i.e. from a stable focus $E^{(*)}$ to an unstable one, or vice versa. This feature may be further assessed by examining the sign of the partial derivatives with respect to τ [2, 3].

$$\Lambda^{-1}(\tau) = \left(\frac{\partial \mathrm{Re}\lambda}{\partial \tau}\right)_{\lambda = i \cdot \omega}; R_p, \sigma_1, \tau_f, l_{cr}, l_1, \eta_{SFM}, \tau_{c1}, \ldots = const$$

$$\begin{aligned} U &= (P_R \cdot P_{I\omega} - P_I \cdot P_{R\omega}) - (Q_R \cdot Q_{I\omega} - Q_I \cdot Q_{R\omega}) \\ &= [-\Upsilon_5 \cdot \Upsilon_4 \cdot \omega^2 - \Upsilon_5 \cdot 3 \cdot \omega^4 + 2 \cdot \Upsilon_5 \cdot \Upsilon_4 \cdot \omega^2 + 2 \cdot \Upsilon_5 \cdot \omega^4] \\ &\quad - [\Upsilon_2 \cdot \Upsilon_1 - \Upsilon_2 \cdot \Upsilon_3 \cdot \omega^2 + 2 \cdot \Upsilon_2 \cdot \Upsilon_3 \cdot \omega^2] \end{aligned}$$

$$U = -\Upsilon_2 \cdot \Upsilon_1 + (\Upsilon_5 \cdot \Upsilon_4 - \Upsilon_2 \cdot \Upsilon_3) \cdot \omega^2 - \Upsilon_5 \cdot \omega^4 = \sum_{k=0}^{2} A_{2k} \cdot \omega^{2\cdot k}$$

$$A_4 = -\Upsilon_5$$

$$U = (P_R \cdot P_{I\omega} - P_I \cdot P_{R\omega}) - (Q_R \cdot Q_{I\omega} - Q_I \cdot Q_{R\omega}) = \sum_{k=0}^{2} A_{2k} \cdot \omega^{2\cdot k}$$

$$A_0 = -\Upsilon_2 \cdot \Upsilon_1; A_2 = \Upsilon_5 \cdot \Upsilon_4 - \Upsilon_2 \cdot \Upsilon_3$$

$$\begin{aligned} F_\omega &= 2 \cdot [(P_{R\omega} \cdot P_R + P_{I\omega} \cdot P_I) - (Q_{R\omega} \cdot Q_R + Q_{I\omega} \cdot Q_I)] \\ &= 2 \cdot [(2 \cdot \Upsilon_5^2 \cdot \omega^3 + \Upsilon_4^2 \cdot \omega + 4 \cdot \Upsilon_4 \cdot \omega^3 + 3 \cdot \omega^5) \\ &\quad - (-2 \cdot \Upsilon_1 \cdot \Upsilon_3 \cdot \omega + 2 \cdot \Upsilon_3^2 \cdot \omega^3 + \Upsilon_2^2 \cdot \omega)] \end{aligned}$$

$$\begin{aligned} F_\omega &= 2 \cdot [(P_{R\omega} \cdot P_R + P_{I\omega} \cdot P_I) - (Q_{R\omega} \cdot Q_R + Q_{I\omega} \cdot Q_I)] \\ &= 2 \cdot (\Upsilon_4^2 + 2 \cdot \Upsilon_1 \cdot \Upsilon_3 - \Upsilon_2^2) \cdot \omega \\ &\quad + 4 \cdot (\Upsilon_5^2 + 2 \cdot \Upsilon_4 - \Upsilon_3^2) \cdot \omega^3 + 6 \cdot \omega^5 \end{aligned}$$

$$\begin{aligned} F_\omega &= 2 \cdot [(P_{R\omega} \cdot P_R + P_{I\omega} \cdot P_I) - (Q_{R\omega} \cdot Q_R + Q_{I\omega} \cdot Q_I)] \\ &= \sum_{k=0}^{2} B_{2\cdot k+1} \cdot \omega^{2\cdot k+1} \end{aligned}$$

$$B_1 = 2 \cdot (\Upsilon_4^2 + 2 \cdot \Upsilon_1 \cdot \Upsilon_3 - \Upsilon_2^2)$$

$$B_3 = 4 \cdot (\Upsilon_5^2 + 2 \cdot \Upsilon_4 - \Upsilon_3^2); \; B_5 = 6$$

Then we get the expression for $F_\omega = \sum_{k=0}^{2} B_{2\cdot k+1} \cdot \omega^{2\cdot k+1}$, Diode-pumped Q-switched Nd:YVO$_4$ yellow laser system parameter values. We find those ω, τ values which fulfill $F_\omega(\omega, \tau) = 0$. We ignore negative, complex, and imaginary values of ω for specific τ values. $\tau \in [0.001 \ldots 10]$, we can express by 3D function $F_\omega(\omega, \tau) = 0$. We plot the stability switch diagram based on different delay values of our Diode-pumped Q-switched Nd:YVO$_4$ yellow laser system.

$$\Lambda^{-1}(\tau) = \left(\frac{\partial \mathrm{Re}\lambda}{\partial \tau}\right)_{\lambda = i\omega} = \mathrm{Re}\left\{\frac{-2 \cdot [U + \tau \cdot |P|^2] + i \cdot F_\omega}{F_\tau + 2i \cdot [V + \omega \cdot |P|^2]}\right\}$$

$$\Lambda^{-1}(\tau) = \left(\frac{\partial \mathrm{Re}\lambda}{\partial \tau}\right)_{\lambda = i\omega} = \frac{2 \cdot \{F_\omega \cdot (V + \omega \cdot P^2) - F_\tau \cdot (U + \tau \cdot P^2)\}}{F_\tau^2 + 4 \cdot (V + \omega \cdot P^2)^2}$$

The stability switch occurs only on those delay values (τ) which fit the equation: $\tau = \frac{\theta_+(\tau)}{\omega_+(\tau)}$ and $\theta_+(\tau)$ is the solution of $\sin\theta(\tau) = \ldots$. and $\cos\theta(\tau) = \ldots$. when

$\omega = \omega_+(\tau)$ If only ω_+ is feasible. Additionally, when all Diode-pumped Q-switched Nd:YVO_4 yellow laser system's parameters are known and the stability switch due to various time delay values τ is described in the following expression:

$$\text{sign}\{\Lambda^{-1}(\tau)\} = \text{sign}\{F_\omega(\omega(\tau),\tau)\} \cdot \text{sign}\left\{\tau \cdot \omega_\tau(\omega(\tau)) + \omega(\tau) + \frac{U(\omega(\tau)) \cdot \omega_\tau(\omega(\tau)) + V(\omega(\tau))}{|P(\omega(\tau))|^2}\right\}$$

Remark: we know $F(\omega,\tau) = 0$ implies its roots $\omega_i(\tau)$ and finding those delays values τ which ω_i is feasible. There are τ values which give complex ω_i or imaginary number, then unable to analyze stability. $F(\omega,\tau)$, function is independent on τ the parameter $F(\omega,\tau) = 0$. The results: We find those ω,τ values which fulfill $F(\omega,\tau) = 0$. We ignore negative, complex, and imaginary values of ω. Next is to find those ω,τ values which fulfill $\sin\theta(\tau) = \ldots.$ and $\cos\theta(\tau) = \ldots.$; $\sin(\omega\cdot\tau) = \frac{-P_R\cdot Q_I + P_I\cdot Q_R}{|Q|^2}$; $\cos(\omega\cdot\tau) = -\frac{(P_R\cdot Q_R + P_I\cdot Q_I)}{|Q|^2}$ $|Q|^2 = Q_R^2 + Q_I^2$. Finally we plot the stability switch diagram $g(\tau) = \Lambda^{-1}(\tau) = (\frac{\partial \text{Re}\lambda}{\partial\tau})_{\lambda=i\omega}$

$$g(\tau) = \Lambda^{-1}(\tau) = \left(\frac{\partial \text{Re}\lambda}{\partial\tau}\right)_{\lambda=i\omega} = \frac{2\cdot\{F_\omega\cdot(V+\omega\cdot P^2) - F_\tau\cdot(U+\tau\cdot P^2)\}}{F_\tau^2 + 4\cdot(V+\omega\cdot P^2)^2}$$

$$\text{sign}[g(\tau)] = \text{sign}[\Lambda^{-1}(\tau)] = \text{sign}[\left(\frac{\partial \text{Re}\lambda}{\partial\tau}\right)_{\lambda=i\omega}] = \text{sign}\left[\frac{2\cdot\{F_\omega\cdot(V+\omega\cdot P^2) - F_\tau\cdot(U+\tau\cdot P^2)\}}{F_\tau^2 + 4\cdot(V+\omega\cdot P^2)^2}\right]$$

$$F_\tau^2 + 4\cdot(V+\omega\cdot P^2)^2 > 0 \Rightarrow \text{sign}[\Lambda^{-1}(\tau)] = \text{sign}[F_\omega\cdot(V+\omega\cdot P^2) - F_\tau\cdot(U+\tau\cdot P^2)]$$

$$\text{sign}[\Lambda^{-1}(\tau)] = \text{sign}\left\{[F_\omega]\cdot\left[(V+\omega\cdot P^2) - \frac{F_\tau}{F_\omega}\cdot(U+\tau\cdot P^2)\right]\right\};\ \omega_\tau = -\frac{F_\tau}{F_\omega}$$

$$\omega_\tau = \left(\frac{\partial\omega}{\partial\tau}\right)^{-1} = -\frac{\partial F/\partial\omega}{\partial F/\partial\tau}$$

$$\text{sign}[\Lambda^{-1}(\tau)] = \text{sign}\left\{[F_\omega]\cdot[P^2]\cdot\left[\frac{V+\omega_\tau\cdot U}{P^2} + \omega + \omega_\tau\cdot\tau\right]\right\};\ \text{sign}[P^2] > 0$$

$$\text{sign}[\Lambda^{-1}(\tau)] = \text{sign}\left\{[F_\omega]\cdot\left[\frac{V+\omega_\tau\cdot U}{P^2} + \omega + \omega_\tau\cdot\tau\right]\right\}$$

$$\text{sign}[\Lambda^{-1}(\tau)] = \text{sign}[F_\omega]\cdot\text{sign}\left[\frac{V+\omega_\tau\cdot U}{P^2} + \omega + \omega_\tau\cdot\tau\right]$$

$$F_\omega = 2\cdot[(P_{R\omega}\cdot P_R + P_{I\omega}\cdot P_I) - (Q_{R\omega}\cdot Q_R + Q_{I\omega}\cdot Q_I)]$$

Table 7.12 Diode-pumped Q-switched Nd:YVO4 yellow laser system sign of $\mathrm{sign}[\Lambda^{-1}(\tau)]$

$\mathrm{sign}[F_{\omega}]$	$\mathrm{sign}\left[\frac{V+\omega_{\tau}\cdot U}{P^2}+\omega+\omega_{\tau}\cdot\tau\right]$	$\mathrm{sign}[\Lambda^{-1}(\tau)]$
$\pm$	$\pm$	+
$\pm$	$\mp$	−

$$=\sum_{k=0}^{2} B_{2\cdot k+1}\cdot\omega^{2\cdot k+1}$$

We check the sign of $\Lambda^{-1}(\tau)$ according to the following rule (Table 7.12).

If, $\mathrm{sign}[\Lambda^{-1}(\tau)] > 0$, then the crossing proceeds from (−) to (+) respectively (stable to unstable). If, $\mathrm{sign}[\Lambda^{-1}(\tau)] < 0$, then the crossing proceeds from (+) to (−) respectively (unstable to stable).

7.5 Questions

1. In the four-level system of a Ti:sapphire laser, $n_2 \approx n_4 \approx 0$, the population inversion density $\Delta n \approx n_3 - \frac{g_3}{g_2}\cdot n_2 \approx n_3 = N$. We consider that $A_{32} \gg S_{32}$ (the transition from level E_3 to E_2 level is dominating and non-radiative transitions are neglected), $S_{43} \gg S_{41}$ then $S_{43} \gg S_{41}$ (the population in level E_4 decays immediately to level E_3, non-radiative), and $\frac{dn_4}{dt} \approx 0$ (the population of the pump level de-excites rapidly to the upper laser level E_3. Additionally, $n_4\cdot S_{43} = n_1\cdot W_{14} = W_p$, W_p is the pumping rate. Furthermore, $A_{32} = A_{32}^A + A_{32}^B$; $S_{32} = S_{32}^A + S_{32}^B$ and $\tau_0 = \frac{1}{A_{32}}$ (self-emission lifetime). $\sigma_A = \frac{A_{32}^A\cdot g(\nu,\nu_A)}{n_{\nu_A}\cdot\nu}$; $\sigma_B = \frac{A_{32}^B\cdot g(\nu,\nu_B)}{n_{\nu_B}\cdot\nu}$, we can summarize our system differential equations:

$$\frac{dN}{dt} = W_p - (\sigma_A\cdot\Phi_A + \sigma_B\cdot\Phi_B)\cdot\nu\cdot N(t-\tau_N) - \frac{N(t-\tau_N)}{\tau_0}$$

$$\frac{d\Phi_A}{dt} = \sigma_A\cdot\nu\cdot N(t-\tau_N)\cdot\Phi_A\cdot\frac{l}{L_A} - \frac{\Phi_A}{\tau_{R_A}};\ \frac{d\Phi_B}{dt} = \sigma_B\cdot\nu\cdot N(t-\tau_N)\cdot\Phi_B\cdot\frac{l}{L_B} - \frac{\Phi_B}{\tau_{R_B}}$$

N is the population inversion density (PID) and due to unideal population inversion process, there is shifting in time, $N(t)\to N(t-\tau_N)$ but it is not affect the derivative $\frac{dN}{dt}$; $N = N(t)$; $\Phi_A = \Phi_A(t)$; $\Phi_B = \Phi_B(t)$.

1.1 Find system fixed points for $\tau_N = 0$, and discuss stability and stability switching for different system parameter values.

1.2 Discuss stability and stability switching for different values of τ_N parameter ($\tau_N \neq 0$; $\tau_N > 0$).
1.3 $N(t) \to N(t-\tau_N^k)$; $k > 1$; $k \in \mathbb{N}$, Discuss stability and stability switching for different values of k parameter.
1.4 The four-level system of Ti:sapphire laser self-emission life time $\tau_0 = \Gamma \cdot \tau_N$; $\Gamma \in \mathbb{N}$, Discuss stability and stability switching for different values of τ_N and Γ parameters.
1.5 Repeat (1.4) for $\tau_0 = \sqrt{\Gamma \cdot \sqrt{\tau_N}}$; $\Gamma \in \mathbb{N}$ and discus system dynamical behavior.

2. We have a four-level system of a Ti:sapphire laser, same as in (1), but we do the following scaling in the variables: $N \to n = \frac{N}{N_0}$; $\Phi_A \to \varphi_A = \frac{\Phi_A}{N_0}$; $\Phi_B \to \varphi_B = \frac{\Phi_B}{N_0}$, and $W_p \to \omega_p = \frac{W_p}{N_0}$, where N_0 is the concentration of Ti^{3+}. The normalized rate equations of a pulsed dual-wavelength laser are $n = n(t)$; $\varphi_A = \varphi_A(t)$; $\varphi_B = \varphi_B(t)$.

$$\frac{dn}{dt} = \omega_p - (\sigma_A \cdot \varphi_A + \sigma_B \cdot \varphi_B) \cdot v \cdot N_0 - \frac{n(t-\tau_n)}{\tau_0}$$

$$\frac{d\varphi_A}{dt} = \sigma_A \cdot v \cdot n(t-\tau_n) \cdot N_0 \cdot \varphi_A \cdot \frac{l}{L_A} - \frac{\varphi_A}{\tau_{R_A}}; \frac{d\varphi_B}{dt}$$
$$= \sigma_B \cdot v \cdot n(t-\tau_n) \cdot N_0 \cdot \varphi_B \cdot \frac{l}{L_B} - \frac{\varphi_B}{\tau_{R_B}}$$

where, n is the normalized population inversion. Due to uncomplete population inversion there is a shift in time (delay) in n, $n(t) \to n(t-\tau_n)$, but it is not affect the derivative of n with time, $\frac{dn}{dt}$.

2.1 Find system fixed points and how they changed for different values of ω_p parameter.
2.2 Discuss stability and stability switching for different values of τ_n parameter.
2.3 We assume that the pumping rate is a Gaussian $\omega_p = \omega_p(t)$, $\omega_p(t) = \frac{2 \cdot E_p}{\sqrt{2 \cdot \pi} \cdot T_0 \cdot h \cdot v \cdot N_0} \cdot (1 - e^{-\alpha_p \cdot l}) \cdot e^{-\frac{2 \cdot t^2}{T_0^2}}$, where E_p is the pumping energy, T_0 is the pulse width of the pump laser, and α_p is the absorption efficiency of the Ti:sapphire crystal at the wavelength of the pump laser. Analyze the system dynamic for $t \to \infty$, find system fixed points and discuss stability and stability switching for different values of τ_n and σ_A, σ_B parameters.
2.4 The absorption efficiency α_p is dependent on the normalized population inversion, n, $\alpha_p = \ln[\frac{\Gamma}{n(t-\tau_n)}]$; $\Gamma \in \mathbb{R}$; $\Gamma > 0$, Hint: $e^{-\alpha_p \cdot l} = e^{-l \cdot \log[\frac{\Gamma}{n(t-\tau_n)}]}$, $e^{-\alpha_p \cdot l} = e^{-l \cdot \ln[\frac{\Gamma}{n(t-\tau_n)}]} = e^{l \cdot \ln[\frac{\Gamma}{n(t-\tau_n)}]^{-1}} = e^{\ln[\frac{n(t-\tau_n)}{\Gamma}]^l} = [\frac{n(t-\tau_n)}{\Gamma}]^l$. And we consider that The pumping rate is not attenuated for time t, $\forall\ t \geq$

t_0, $e^{-\frac{2 \cdot t^2}{T_0^2}} = e^{-\frac{2 \cdot t_0^2}{T_0^2}}$, Find system fixed points and how they changed for different values of t_0 parameter.

2.5 Discuss stability and stability switching (2.4) for different values of τ_n parameter.

3. We have a system of dual-wavelength VECSELs. Inhomogeneous pumping of the active region contains non identical Quantum Wells (QWs). The carrier density across the QW layer to be continuous, $(C_1^{(j)} - C_1^{(j+1)}) \cdot \tanh(X_{QW}^{(j)}) = C_2^{(j+1)} - C_2^{(j)}$, where $C_{1,2}^{(j)}$ are constants which are related to the j-th barrier region, $X_{QW}^{(j)}$ is the position of the j-th QW layer. We have two rate equations which determined the carrier balance in the continuous state above the QW and in the QW below lasing threshold reads, respectively, $n^{(j)} = n^{(j)}(t); n_{QW}^{(j)} = n_{QW}^{(j)}(t)$.

$$\frac{dn^{(j)}}{dt} = \frac{1}{t_{QW}^{(j)}} \cdot (J_-^{(j)} - J_+^{(j)}) - \frac{n^{(j)}}{\tau_c} + \frac{n_{QW}^{(j)}}{\tau_e}; \frac{dn_{QW}^{(j)}}{dt}$$
$$= \frac{n^{(j)}}{\tau_c} - n_{QW}^{(j)} \cdot (\frac{1}{\tau_e} + \frac{1}{\tau_r})$$

$$J_{-,+}^{(j)} = -\frac{D_a}{L_a} \cdot \frac{dn^{(j)}}{dX}|_{X_{QW}^{(j)} \mp 0}; J_-^{(j)} - J_+^{(j)} = \xi_0 \cdot n^{(j)}; \xi_0 = \frac{t_{QW}^{(j)}}{\tau_c} \cdot (\frac{\tau_e}{\tau_e + \tau_r})$$

Then $J_-^{(j)} - J_+^{(j)} = \xi_0 \cdot n^{(j)} = \frac{t_{QW}^{(j)}}{\tau_c} \cdot (\frac{\tau_e}{\tau_e+\tau_r}) \cdot n^{(j)}$.
Remark: j index is for the Quantum Well (QW) layer number.
$J_{-,+}^{(j)}$ is the density of carrier flux into the continuous state above the quantum confined states of the j-th QW from the left and right hand sides (− and +), $t_{QW}^{(j)}$ is the QW thickness, and $n_{QW}^{(j)}$ is the carrier density in the QW; $n^{(j)} = n(X_{QW}^{(j)})$ is the carrier density above the QW, τ_c is the capture time for a charge carrier from the continuous state into the QW, and τ_e is the escape time describing the reverse process, $\frac{1}{\tau_r} = A + B \cdot n_{QW}^{(j)} + C \cdot (n_{QW}^{(j)})^2$ where A, B, and C are the monomolecular, bimolecular, and Auger recombination coefficients.

3.1 Draw the graph for $X_{QW}^{(j)}$ versus the ration between the differences related to the first barrier region and the second barrier region $X_{QW}^{(j)} = \tanh^{-1}[\frac{C_2^{(j+1)} - C_2^{(j)}}{C_1^{(j)} - C_1^{(j+1)}}]$. Is there any intersection that $X_{QW}^{(j)} = 0$?

3.2 Find system fixed points and discuss stability for different values of τ_c, τ_e, and τ_r.

3.3 When the carrier balance is not perfect in our system in the continuous state above the QW and in the QW below lasing threshold reads, respectively then $n^{(j)}(t) \to n^{(j)}(t - \tau_1); n_{QW}^{(j)}(t) \to n_{QW}^{(j)}(t - \tau_2)$, but it doesn't affect the derivatives $\frac{dn^{(j)}}{dt}, \frac{dn_{QW}^{(j)}}{dt}$. We have four cases: (1) $\tau_1 = \tau; \tau_2 = 0$, (2) $\tau_1 = 0; \tau_2 = \tau$, (3) $\tau_1 = \tau; \tau_2 = \tau^2$. Discuss stability and stability switching for different values of τ parameter in the three cases.

3.4 Return (3.3) for $\tau_c \gg \tau_e$, explain how the system dynamic is changed.

3.5 Return (3.3) for $\tau_e \gg \tau_r$, explain how the system dynamic is changed.

4. The rate equations in (3), $\frac{dn^{(j)}}{dt} = \frac{1}{t_{QW}^{(j)}} \cdot (J_-^{(j)} - J_+^{(j)}) - \frac{n^{(j)}}{\tau_c} + \frac{n_{QW}^{(j)}}{\tau_e}$; $\frac{dn_{QW}^{(j)}}{dt} = \frac{n^{(j)}}{\tau_c} - n_{QW}^{(j)} \cdot (\frac{1}{\tau_e} + \frac{1}{\tau_r})$ and the recursive equation, $(C_1^{(j)} - C_1^{(j+1)}) \cdot \tanh(X_{QW}^{(j)}) = C_2^{(j+1)} - C_2^{(j)}$, and $J_-^{(j)} - J_+^{(j)} = \xi_0 \cdot n^{(j)} = \frac{t_{QW}^{(j)}}{\tau_c} \cdot (\frac{\tau_e}{\tau_e+\tau_r}) \cdot n^{(j)}$ establish the relationship between $C_{1,2}^{(j)}$ related to the barrier layers adjacent to the j-th quantum well (QW).

$$\begin{pmatrix} C_1^{(j+1)} \\ C_2^{(j+1)} \end{pmatrix} = M \cdot \begin{pmatrix} C_1^{(j)} \\ C_2^{(j)} \end{pmatrix} + \xi \cdot n_G \cdot e^{-a \cdot X_{QW}^{(j)}} \cdot \begin{pmatrix} \cosh(X_{QW}^{(j)}) \\ \sinh(X_{QW}^{(j)}) \end{pmatrix}$$

$$M = \begin{pmatrix} 1 + \frac{1}{2} \cdot \xi \cdot \sinh(2 \cdot X_{QW}^{(j)}) & \xi \cdot \cosh^2(X_{QW}^{(j)}) \\ -\xi \cdot \sinh^2(X_{QW}^{(j)}) & 1 - \frac{1}{2} \cdot \xi \cdot \sinh(2 \cdot X_{QW}^{(j)}) \end{pmatrix}$$

$$\xi = \frac{\xi_0 \cdot L_a}{D_a} = \frac{\xi_0 \cdot L_a}{D_a}; \xi_0 = \frac{t_{QW}^{(j)}}{\tau_c} \cdot \left(\frac{\tau_e}{\tau_e + \tau_r} \right)$$

4.1 Draw a graph values of $C_1^{(j)}$ and $C_2^{(j)}$, $1 \le j < K; K > 0; K \in \mathbb{N}; j \in \mathbb{N}; j = 1, 2, 3, \ldots$ start from some initial values, $X_{QW}^{(j=1)}, C_1^{(j=1)}, C_2^{(j=1)}$ and parameters, ξ, n_G, a. Are the values of $C_1^{(j)}$ and $C_2^{(j)}$ converge to specific values for $j \to K \; \forall \; K \gg 1$?

4.2 Discuss the convergence in (4.1) for different values of a parameter, $(e^{-a \cdot X_{QW}^{(j)}})$.
How the system dynamic is changed for $a = 0$? Draw the relative graph.

4.3 Parameter $\xi = 0$, How the results (4.1 and 4.2) are changed? Draw the relative graph.

4.4 Discuss (4.1 and 4.2) if $X_{QW}^{(j)}$ is a geometric series.

4.5 Discuss (4.1 and 4.2) if $X_{QW}^{(j)}$ is an algebraic series.

5. We have an EDFL in dual wavelength system which is described as a two-mode laser system. The existence of a frequency is lower than that of the relaxation oscillations is characteristic of a two-mode laser. The laser is modeled with four equations: two for the population's inversion and two for the corresponding laser fields. The two sub-systems are coupled with a cross-saturation parameter. The dynamical behavior of an ion is described with two equations. The nonlinear system for the dual wavelength EDFL with ion pairs differential equations:

$$\frac{dd_1}{dt} = \Lambda - a_2 \cdot (1 + d_1) - 2 \cdot d_1 \cdot (I_1 + \beta \cdot I_2)$$

$$\frac{dd_2}{dt} = \gamma \cdot \Lambda - a_2 \cdot (1 + d_2) - 2 \cdot d_2 \cdot (I_2 + \beta \cdot I_1)$$

$$\frac{dd_+}{dt} = a_{12} \cdot (1 - d_+) - \frac{a_{22}}{2} \cdot (d_+ + d_-) + y \cdot (I_1 + I_2) \cdot (2 - 3 \cdot d_+)$$

$$\frac{dd_-}{dt} = \Lambda - a_{12} \cdot (1 - d_+) - \frac{a_{22}}{2} \cdot (d_+ + d_-) + y \cdot (I_1 + I_2) \cdot d_-$$

$$\frac{dI_1}{dt} = I_1 \cdot (-1 + A \cdot (1 - 2 \cdot x) \cdot (d_1 + \beta \cdot d_2) + A \cdot x \cdot y \cdot d_-)$$

$$\frac{dI_2}{dt} = I_2 \cdot (-1 + A \cdot (1 - 2 \cdot x) \cdot (d_2 + \beta \cdot d_1) + A \cdot x \cdot y \cdot d_-)$$

The system parameters are described in the next Table.

Variable/Parameter	Description
I_1, I_2	Two normalized laser intensities
d_1, d_2	Normalized population inversions
β	Cross saturation parameter
γ ($\gamma = 1$ for perfect symmetry between the two modes – antiphase effect vanish)	A dichroism in the pumping process in order to take into account the small anisotropies of the laser
Λ	Pumping rate
x	Ion pair concentration
y	Ratio between the absorption cross-sections of an ion pair and isolated ion
$A = \sigma_l \cdot N_0 \cdot \tau_l$	N_0—Erbium concentration l—fiber length, τ_l [s]—photon lifetime σ_l [$cm^3\ s^{-1}$]—Absorption cross section for the isolated ion
$d_+ = n_{22} + n_{11}$	Summation of populations in levels $\|11\rangle$ and $\|22\rangle$. n_{22} is the population of level $\|22\rangle$ and n_{11} is the population in level $\|11\rangle$
$d_+ = n_{22} - n_{11}$	Difference between the population of level $\|22\rangle$ and the population of level $\|11\rangle$
$a_{ij} = \frac{\tau_l}{\tau_{ij}}$	The ratio between the photon lifetime (τ_l) and the lifetime of level $\|ij\rangle$
$a_2 = \frac{\tau_l}{\tau_2}$	The ratio between the photon lifetime (τ_l) and the life time of level $\|2\rangle$, (τ_2)

5.1 Find system fixed points and draw there coordinates as a function of x and y parameters.

5.2 Discuss stability and stability switching for different values of Λ parameter.

5.3 Due to populations of levels $|kl\rangle$ shifting in time (parasitic system effects), we get variables d_+, d_- shifting in time $d_+(t) \to d_+(t-\tau_+); d_-(t) \to d_-(t-\tau_-)$ but it is not affect the derivatives $\frac{dd_+}{dt}, \frac{dd_-}{dt}$. Discuss stability and stability switching for different values of τ_+ and τ_- shifting parameters.

5.4 How the system dynamic is changed if the photon lifetime is much smaller compare to the lifetime of levels $|ij\rangle$, $\tau_+, \tau_- > 0;\ \forall\ \tau_l \ll \tau_{ij}\ \exists\, a_{ij} = \frac{\tau_l}{\tau_{ij}} \to \varepsilon, (0<\varepsilon \ll 1)$.

5.5 Find system fixed points and stability switching for different values of τ_+ and τ_- shifting parameters.

Remark: The net gains for the two frequencies are different because of small anisotropies induced by twists or bends on the fiber, and to the position of the mirror M1.

6. We have a system of Diode-pumped Q-switched Nd:YVO_4 yellow laser with intra cavity sum-frequency mixing. The system is an intra cavity SFM in the diode-end-pumped Nd:YVO_2 dual-wavelength laser at 1064 and 1342 nm. The model of the dynamic behavior of a dual-wavelength Q-switched laser is described by set of DDEs (Delay Differential Equations), $N = N(t); \phi_1 = \phi_1(t); \phi_2 = \phi_2(t)$.

$$\frac{dN}{dt} = R_p \cdot c \cdot N(t-\tau_N) \cdot \left(\sum_{k=1}^{2} \sigma_k \cdot \phi_k\right) - \frac{N(t-\tau_N)}{\tau_f}$$

$$\frac{d\phi_1}{dt} = \frac{l_{cr}}{l_1} \cdot c \cdot \sigma_1 \cdot N(t-\tau_N) - \frac{\phi_1}{\tau_{c_1}} - \eta_{SFM} \cdot \phi_1 \cdot \phi_2$$

$$\frac{d\phi_2}{dt} = \frac{l_{cr}}{l_2} \cdot c \cdot \sigma_2 \cdot \phi_2 \cdot N(t-\tau_N) - \frac{\phi_2}{\tau_{c_2}} - \eta_{SFM} \cdot \phi_1 \cdot \phi_2$$

The subscripts 1 and 2 denote ϕ, σ, l and τ_c at $\lambda_1 = 1342$ nm and $\lambda_2 = 1064$ nm, respectively. N is the population inversion density and system parasitic effects cause to shift in time for $N(t) \to N(t-\tau_N)$ but it not affect $\frac{dN}{dt}$. R_p is the average pump intensity, τ_f is the emission lifetime, $\phi_i; i = 1, 2$ are the photon densities at 1064 nm and 1342 nm, respectively, $\sigma_k; k = 1, 2$ are the stimulated emission cross sections, l_{cr} is the crystal length, $\tau_{c_i}; i = 1, 2$ are the effective photon time as a result of all the linear losses, and η_{SFM} is the effective conversion rate as a result of the intra cavity SFM.

6.1 Find system fixed points, and discuss stability and stability switching for different values of τ_N parameter.

6.2 What happened when the emission lifetime,τ_f is much more bigger than the population inversion density ($\tau_f \gg \tau_N$)? Discuss the system dynamic.

6.3 How the system dynamic is changed for $\tau_{c_1} \gg \phi_1$? Discuss stability and stability switching for different values of τ_N parameter. ϕ_1 is the photon density at $\lambda_1 = 1342$ nm.
6.4 How the system dynamic is changed for $\tau_{c_2} \gg \phi_2$? Discuss stability and stability switching for different values of τ_N parameter. ϕ_2 is the photon density at $\lambda_2 = 1064$ nm.
6.5 Discuss system dynamic and fixed points for $\tau_{c_1} \gg \phi_1$ and $\tau_{c_2} \gg \phi_2$, Case A: $\tau_N = 0$, Case B: $\tau_N > 0$.

7. Erbium-doped fiber lasers are characterized by cw, sinusoidal and self-pulsing operation. The obtained regimes depend on three control parameters: ion-pair concentration, photon lifetime, and pumping rate. The system model described the active medium as a mixture of isolated ions and ion pairs. We describe our system by adapted laser rate equations, four first order coupled equations. The linear stability analysis demonstrates the existence of self-pulsing for a finite range of pumping rates. There are two classes of ions: (1) isolated ions and (2) ion pairs. The quenching process is associated with ion pairs and establishes the dynamical behavior of a laser. The laser operates in cw regime without pairs and their presence lead to a self-pulsing instability. The isolated ion is described as a two-level system and the ion pair as a three-level system. The system rate equations describe the fast relaxation time associated with the electronic polarizations, large homogeneous broadening due to collisions with photons. The dynamical behavior of an isolated ion is described by the rate equation:

$$\frac{dn_1}{dt} = -\Lambda + a_2 \cdot n_2 + i_l \cdot (n_2 - n_1(t - \tau))$$
$$\frac{dn_2}{dt} = \Lambda - a_2 \cdot n_2 - i_l \cdot (n_2 - n_1(t - \tau))$$

$$a_2 = \frac{\gamma_2}{\gamma_l}; n_1 = \frac{N_1}{(1 - 2 \cdot x) \cdot N_0}; n_2 = \frac{N_2}{(1 - 2 \cdot x) \cdot N_0}$$

$$i_l = \frac{\sigma_l \cdot I_l}{\gamma_l}; \sum_{k=1}^{2} N_k = (1 - 2x) \cdot N_0$$

N_0 is the erbium concentration, γ_2 and γ_l are the relaxation rates, respectively, of the upper level of the laser transition and of the laser field, σ_l is the absorption cross section for the isolated ion, I_l is the photon flux of the laser field, and Λ is the pumping rate. The time is normalized with respect to the photon lifetime $\tau_l = \frac{1}{\gamma_l}$. Due to parasitic effect there is shifting in time for the population in level 1, $n_1(t - \tau)$, but it not affect the derivative $\frac{dn_1}{dt}$.

7.1 Find system fixed points and draw the relative graphs as a function of x parameter.
7.2 Discuss stability and stability switching for different values of τ parameter.

7.3 $\gamma_l \gg \gamma_2 \Rightarrow a_2 \to \varepsilon; 0 < \varepsilon \ll 1$, how the system dynamic is changed? Find fixed points and discuss stability switching for different values of τ parameter.

7.4 Find the expression for x parameter as a function of fixed points coordinates.

7.5 $\tau = 0$, discuss stability and stability switching for different values of x parameter.

8. Erbium-doped fiber lasers are characterized by cw, sinusoidal and self-pulsing operation. The obtained regimes depend on three control parameters: ion-pair concentration, photon lifetime, and pumping rate. The system model described the active medium as a mixture of isolated ions and ion pairs. We describe our system by adapted laser rate equations, four first order coupled equations. The linear stability analysis demonstrates the existence of self-pulsing for a finite range of pumping rates. There are two classes of ions: (1) isolated ions and (2) ion pairs. The quenching process is associated with ion pairs and establishes the dynamical behavior of a laser. The laser operates in cw regime without pairs and their presence lead to a self-pulsing instability. The isolated ion is described as a two-level system and the ion pair as a three-level system. The system rate equations describe the fast relaxation time associated with the electronic polarizations, large homogeneous broadening due to collisions with photons. The three-level scheme is adopted for the ion pairs and is described by three rate equations:

$$\frac{dn_{11}}{dt} = -\Lambda + a_{12} \cdot n_{12} + y \cdot i_l \cdot (n_{12} - n_{11})$$

$$\frac{dn_{22}}{dt} = \Lambda - a_{22} \cdot n_{22} - y \cdot i_l \cdot (n_{22} - n_{12})$$

$$\frac{dn_{12}}{dt} = -a_{12} \cdot n_{12} + a_{22} \cdot n_{22} - y \cdot i_l \cdot (n_{12} - n_{11}) + y \cdot i_l \cdot (n_{22} - n_{12})$$

$$y = \frac{\sigma_l'}{\sigma_l}; a_{ij} = \frac{\gamma_{ij}}{\gamma_l}; n_{ij} = \frac{N_{ij}}{x \cdot N_0}; n_{11} + n_{12} + n_{22} = 1$$

σ_l' is the absorption cross section at the lasing wavelength for a pair, γ_{ij} is the relaxation rate of level $|ij\rangle$, σ_l is the absorption cross section for the isolated ion, γ_l is the relaxation rate of the laser field.

8.1 Find system fixed points and discuss stability switching for different values of N_0 parameter (N_0 is the erbium concentration).

8.2 $a_{12} = \frac{\gamma_{12}}{\gamma_l}; \gamma_l \gg \gamma_{12}$, How our system dynamic is changed? Find fixed points and discuss stability.

8.3 $a_{22} = \frac{\gamma_{22}}{\gamma_l}; \gamma_l \gg \gamma_{22}$, How our system dynamic is changed? Find fixed points and discuss stability.

8.4 Find the expression of parameter x as a function of system parameters and fixed points coordinates. Draw the relative graphs.
8.5 $N_0 \gg N_{11}$; $x = 1$, Find system fixed points and discuss stability and stability switching for different values of Λ parameter.

9. Erbium-doped fiber lasers are characterized by cw, sinusoidal and self-pulsing operation. The obtained regimes depend on three control parameters: ion-pair concentration, photon lifetime, and pumping rate. The laser field interacts with two systems: (1) the fraction $1 - 2 \cdot x$ of isolated ions and (2) the fraction x of ion pairs. The dynamics of the laser intensity is described by the differential equation, $\frac{di_l}{dt} = -i_l + (1 - 2 \cdot x) \cdot A \cdot i_l \cdot n_i + x \cdot B \cdot i_l \cdot n_-$; $A = \frac{\sigma_l \cdot N_0}{\gamma_l}$; $B = \frac{\sigma_l' \cdot N_0}{\gamma_l}$; $y = \frac{B}{A} = \frac{\sigma_l'}{\sigma_l}$. The dynamical behavior of the system is described by four coupled first order differential equations:

$$\frac{dn_i}{dt} = 2 \cdot \Lambda - a_2 \cdot (1 + n_i) - 2 \cdot i_l \cdot n_i$$

$$\frac{dn_+}{dt} = a_{12} \cdot (1 - n_+) - \frac{1}{2} \cdot a_{22} \cdot (n_+ + n_-) + y \cdot i_l \cdot (2 - 3 \cdot n_+)$$

$$\frac{dn_-}{dt} = 2 \cdot \Lambda - a_{12} \cdot (1 - n_+) - \frac{1}{2} \cdot a_{22} \cdot (n_+ + n_-) - y \cdot i_l \cdot n_-$$

$$\frac{di_l}{dt} = -i_l + (1 - 2 \cdot x) \cdot A \cdot i_l \cdot n_i + x \cdot B \cdot i_l \cdot n_-$$

σ_l' is the absorption cross section at the lasing wavelength for a pair, γ_{ij} is the relaxation rate of level $|ij\rangle$, σ_l is the absorption cross section for the isolated ion, γ_l is the relaxation rate of the laser field, N_0 is the erbium concentration, Λ is the pumping rate, $a_2 = \frac{\gamma_2}{\gamma_l}$, γ_2 and γ_l are the relaxation rates, respectively, of the upper level of the laser transition and of the laser field, $n_+ = n_{22} + n_{11}$; $n_- = n_{22} - n_{11}$.

9.1 Find system fixed points and draw the function of x versus fixed point's coordinates and system parameters.
9.2 Discuss stability and stability switching for different values of Λ parameter.
9.3 $y = \frac{B}{A} = \frac{\sigma_l'}{\sigma_l}$; $\sigma_l \gg \sigma_l'$; $y \to \varepsilon$; $0 \le \varepsilon \ll 1$, How the system dynamic is changed? Find system fixed points and discuss stability, stability switching for different values of x parameter.
9.4 $i_l(t) \to i_l(t - \tau_l)$, Discuss stability and stability switching for different values of τ_l parameter.
9.5 $n_+(t) \to n_+(t - \tau_+)$, Discuss stability and stability switching for different values of τ_+ parameter.

10. We have a uniform multi-wavelength fiber laser which is based on hybrid Raman and Erbium-doped fiber. There is a gain competition effects in the

fiber Raman amplification (FRA) and EDF amplification. The FRA gain mechanism is suppresses the gain competition efficiency and makes the present multi-wavelength laser stable at room temperature. The hybrid gain medium increased the lasing bandwidth (compared with a pure EDF laser) and the power conversion efficiency (compared with a pure fiber Raman laser). The instability of multi wavelength lasing is mainly due to the gain competition effect. There is a gain competition (between two signal channels) for the FRA and EDFA parts of the fiber laser. The parameters of the signals and pumps are defined. Signal #A (fixed wavelength, λ_A) is injected to port (a) for both FRA and EDFA. Signal #B is at wavelength λ_B (tuned from 1525 to 1580 nm) is used to study the gain competition. The gains for signal #A when signal #B is off ($G_{A,B-off}$) and on ($G_{A,B-on}$) are measured. The gain difference $\Delta G_A(\Delta\lambda) = G_{A,B-off}(\lambda_A; \Delta\lambda) - G_{A,B-on}(\lambda_A; \Delta\lambda)$ for signal #A in EDFA and in FRA as wavelength spacing $\Delta\lambda = \lambda_B - \lambda_A$ varies.

10.1 Define the functions $G_{A,B-off}(\lambda_A; \Delta\lambda)$ and $G_{A,B-on}(\lambda_A; \Delta\lambda)$.
10.2 Write the differential equation $\frac{\Delta G_A(\Delta\lambda)}{\Delta\Delta\lambda} = \ldots.; \Delta \leftrightarrow d \Rightarrow \frac{dG_A(\Delta\lambda)}{d\Delta\lambda} =$ .....
10.3 Find fixed points and discuss stability.
10.4 $G_A(\Delta\lambda) \rightarrow G_A(\Delta\lambda - \tau_\lambda)$, discuss stability and stability switching for different values of τ_λ parameter.
10.5 We switch between signal #A and signal #B, what happened? How the system dynamic is changed? Discuss stability.

References

1. F. Song, J.Q. Yao, D.W. Zhou, J.Y. Qiao, G.Y. Zhang, J.G. Tian, Rate-equation theory and experimental research on dual-wavelength operation of a Ti: sapphire laser. Appl. Phys. B **72**, 605–610 (2001)
2. Y. Kuang, *Delay Differential Equations with Applications in Population Dynamics* (Academic Press, Boston, 1993)
3. E. Beretta, Y. Kuang, Geometric stability switch criteria in delay differential systems with delay dependent parameters. SIAM J. Math. Anal. **33**, 1144–1165 (2002)
4. Y.A. Morozov, L. Tomi, H. Antti, P. Markus, Simultaneous dual-wavelength emission from vertical external-cavity surface-emitting laser: a numerical modeling. IEEE J. Quantum Electron. **42**(10) (2006)
5. S. Francois, L. Marc, M. Guy, L. Patrice, P.-L. Francois, Quasi-periodic route to chaos in erbium-doped fiber laser. IEEE J. Quantum Electron. **31**(3) (1995)
6. F. Sanchez, P. Le Boudec, P.-L. Francois, G. Stephan, Effects of ion pairs the dynamics of erbium-doped fiber lasers. Phys. Rev. A **48**(3) (1993)
7. J. Guckenheimer, Nonlinear oscillations, dynamical systems, and bifurcations of vector fields. Appl. Math. Sci. **42**
8. S. Wiggins, *Introduction to Applied Nonlinear Dynamical Systems and Chaos*. Text in Applied Mathematics (Hardcover)
9. Y.-F. Chen, S.W. Tsai, Diode-pumped Q-switched Nd:YVO_4 yellow laser with intra-cavity sum-frequency mixing. Opt. Lett. **27**(6) (2002)

Chapter 8
Dual-Wavelength Laser Systems Stability Analysis Under Parameters Variation (II)

The dynamic responses of the photon and carrier densities in an asymmetric dual quantum well laser exhibits wavelength switching. The lasers consist of two quantum wells of different energy gap, which are separated by a high and/or thick barrier layer. The barrier layer blocks carrier transport between the wells and the rate of the transport becomes comparable to the rate of the radiative recombination. The laser two different quantum wells which are isolated by a high and/or thick barrier layer results in an inhomogeneous carrier injection into the two wells. The pulse response at a voltage of the dual-wavelength lasing shows that the emission of the shorter wavelength light precedes that of the longer wavelength light by a nanosecond. The asymmetric dual quantum well (ADQW) laser diode consists of two quantum wells, well 1 and well 2, of different emission wavelengths, λ_1 and λ_2, respectively ($\lambda_2 < \lambda_1$). The two wells are located in the core of a single optical waveguide. The well 2 of a wider band gap is located on the p-type side. We control the height and thickness of the barrier layer in a way that holes are injected into the two wells in homogeneously, and electrons homogeneously. When defining the component of the hole current which is directly injected into well 1, equal to zero, the set of rate equations for hole densities n_1, n_2 in well 1 and well 2, respectively, and the photon densities $S^{\lambda 1}$, $S^{\lambda 2}$ of the lasing modes at, λ_1 and λ_2, then we can inspect the dynamic and stability of our system. A self-pulsing effects in T_m^{3+}-doped silica fiber laser is inspected and operate near 2 μm. There are various self-pulsing regimes which are observed for a range of pumping rates when the fiber is end-pumped with a high power Nd:YAG laser operating at 1.319 μm in a linear bidirectional cavity. We can define a variety of nonlinear phenomena, ranging from self-pulsing to self-Q-switching and to a modulated quasi-CW wave. The phenomenon of pair-induced quenching (PIQ) is inspected and it is compatible with excited-state absorption (ESA). The model of T_m^{3+}-doped silica fiber laser is based on the rate equations for a three-level system. The laser system dynamic and stability is analyzed by the cavity photon density $C_p(t)$ and through rate equations. Terahertz (THz) quantum cascade laser (QCL) is based on GaN/AlGaN quantum wells and emits at two widely separated wavelengths 33 and 52 μm simultaneously in a single active region. The large LO-phonon energy

O. Aluf, *Advance Elements of Laser Circuits and Systems*,
https://doi.org/10.1007/978-3-030-64103-0_8

(~90 meV), the ultrafast resonant phonon depopulation of the lower radiative levels, suppression of the electrons that escape to the continuum states and selective carrier injection and extraction lead to an enhancement in the operating temperature of the structure, for temperature around 256 °K. Rate equations models describe the coupled electronic and electromagnetic system and certain structural parameters are defined for dynamic behavior and stability inspection.

8.1 Asymmetric Dual Quantum Well Lasers and Dual-Wavelength Lasing Stability Analysis Under Parameters Variation and Delay Variables in Time

We perform analysis of dynamic response of asymmetric dual quantum well lasers. The dynamic responses of the photon and carrier densities in an asymmetric dual quantum well laser exhibits wavelength switching. The laser consists of two different quantum wells isolated by a high and/or thick barrier layer that results in an inhomogeneous carrier injection into the two wells. The pulse response at a voltage of the dual-wavelength lasing shows that the emission of the shorter wavelength light precedes that of the longer wavelength light by a nanosecond. The monolithic laser diodes (LDs) changes the wavelength λ with a wide variable range $\Delta\lambda$ and/or emit multiple-wavelength light are required for future optoelectronic applications. An asymmetric dual quantum well (ADQW) laser diode (LD) is demonstrated the discrete switching of $\Delta\lambda = 13$ nm and $\Delta\lambda = 50$ nm, quasi-continuous wavelength tuning $\Delta\lambda = 22$ nm, when $\lambda \approx 800$ nm. The ADQW LD consists of two quantum wells, well 1 and well 2, of different emission wavelength, λ_1 and λ_2, ($\lambda_2 < \lambda_1$), respectively. The two wells are located in the core of a single optical waveguide, where well 2 of a wider band gap are located on the p-type side. In the case that well 2 are located on the n-type side, roles of the electrons and holes are interchanged (Fig. 8.1).

We control the height and thickness of the barrier layer and the holes are injected into the two wells in homogeneously, and electrons homogeneously. By using various types of ADQW structure, it is easier to inject electrons homogeneously than holes. The electron mobility is 20 times larger than that of holes. The holes are injected from the p-type cladding layer and first they trapped in well 2, and then thermally activated to be transferred over the barrier into well 1. By blocking this transfer with the high and/or thick barrier, the rate of the transfer becomes comparable with the recombination rate. It is constructed with normal multiple-quantum well (MQW) LDs, for which barrier layers are too low and thin to inject carriers uniformly across many wells. For a low and thin barrier in an ADQW LD, most injected carrier would occupy well 1 only, and the carrier density n_2 of well 2 would not reach the threshold density at a reasonable value of the injection current. For electrons, the sufficient electron densities are homogeneously distributed in both wells. In the time that sufficient electron densities are obtained in both wells, the analysis is focused on

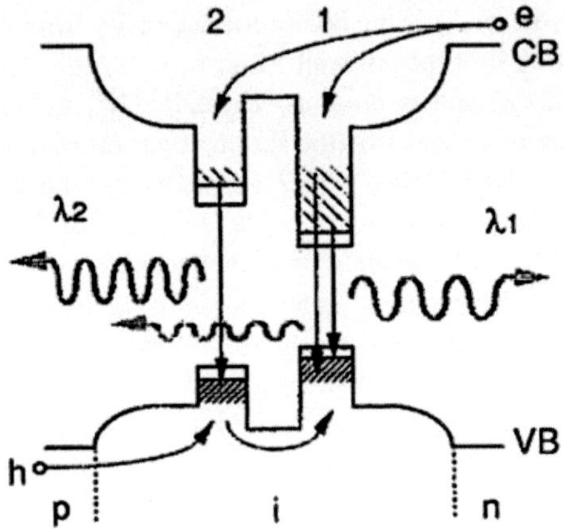

Fig. 8.1 Schematic band diagram of an asymmetric dual quantum well (ADQW) LD

holes and photons, for which the rate equations are defined. The component of the hole current is directly injected into well 1 is zero, than the set of rate equations for hole densities n_1, n_2 in well 1, well 2, and the photon densities $s^{\lambda_1}, s^{\lambda_2}$ of the lasing modes at λ_1, λ_2 is described [1]. Due to variation structure of our asymmetric dual quantum well laser there is shift in time in the hole densities $n_1(t) \to n_1(t-\tau_1)$, $n_2(t) \to n_2(t-\tau_2)$ and for photon densities $s^{\lambda_1}(t) \to s^{\lambda_1}(t-\tau_{ph1})$; and $s^{\lambda_2}(t) \to s^{\lambda_2}(t-\tau_{ph2})$. The shifting in time does not affect the hole and photon densities derivative in time ($\frac{dn_1}{dt}, \frac{dn_2}{dt}, \frac{ds^{\lambda_1}}{dt}, \frac{ds^{\lambda_2}}{dt}$). The component of the hole current which is directly injected into well 1 is zero, the set of rate equations for hole densities n_1, n_2 in well 1, well 2, respectively, and the photon densities $s^{\lambda_1}, s^{\lambda_2}$ of the lasing modes at λ_1 and λ_2 are

$$\frac{dn_1}{dt} = \frac{J_{21}(n_1(t-\tau_1), n_2(t-\tau_2))}{e \cdot d_1} - G_1^{\lambda_1}(n_1(t-\tau_1)) \cdot s^{\lambda_1}(t-\tau_{ph1}) - G_1^{\lambda_2}(n_1(t-\tau_1)) \cdot s^{\lambda_2}(t-\tau_{ph2}) - \frac{n_1(t-\tau_1)}{\tau_{n_1}}$$

$$\frac{dn_2}{dt} = \frac{J - J_{21}(n_1(t-\tau_1), n_2(t-\tau_2))}{e \cdot d_2} - G_2^{\lambda_2}(n_2(t-\tau_2)) \cdot s^{\lambda_2}(t-\tau_{ph2}) - \frac{n_2(t-\tau_2)}{\tau_{n_2}}$$

$$\frac{ds^{\lambda_1}}{dt} = G_1^{\lambda_1}(n_1(t-\tau_1)) \cdot s^{\lambda_1}(t-\tau_{ph1}) - \frac{s^{\lambda_1}(t-\tau_{ph1})}{\tau^{\lambda_1}} + \beta_1^{\lambda_1} \cdot \frac{n_1(t-\tau_1)}{\tau_{n_1}}$$

$$\frac{ds^{\lambda_2}}{dt} = \left[G_1^{\lambda_2}(n_1(t-\tau_1)) + G_2^{\lambda_2}(n_2(t-\tau_2))\right] \cdot s^{\lambda_2}(t-\tau_{ph2}) - \frac{s^{\lambda_2}(t-\tau_{ph2})}{\tau^{\lambda_2}} + \beta_1^{\lambda_2} \cdot \frac{n_1(t-\tau_1)}{\tau_{n_1}} + \beta_2^{\lambda_2} \cdot \frac{n_2(t-\tau_2)}{\tau_{n_2}}$$

where, J_{21} is the hole-current density from well 2 to 1, τ_{n_1} and τ_{n_2} are the recombination lifetimes in well 1 and well 2, respectively, and τ^{λ_1} and τ^{λ_2} are the lifetimes of the λ_1 and λ_2 photons. The $G_1^{\lambda_1}$, $G_1^{\lambda_2}$, and $G_2^{\lambda_2}$ ($\beta_1^{\lambda_1}$, $\beta_1^{\lambda_2}$, and $\beta_2^{\lambda_2}$) denote the optical gain coefficients (the spontaneous emission factors) of λ_1 from well 1, λ_2 from well 1, and λ_2 from well 2, respectively. Assumption: $J_{21} \simeq e \cdot v \cdot n_2$, where v is the effective velocity of holes moving over the barrier. The barrier must be high and/or thick enough and v becomes much slower than that in normal MQW LDs. $\beta_1^{\lambda_2}$ and $\beta_2^{\lambda_2}$ are constant, independent of the light intensity. $\beta_1^{\lambda_1}$ of ADQW LDs is expected to be an increasing function of s^{λ_2}, λ_2 photons induce the simultaneous emission of the λ_1 photon and another quantum of energy $\hbar \cdot \omega_{21} = \hbar \cdot \omega_2 - \hbar \cdot \omega_1$. The quantum can be either a phonon (Raman process) or a photon (electronic Raman process). The electronic Raman process is enhanced as $\hbar \cdot \omega_{21}$ approach the inter sub-band level separation in well 1. Both types of the Raman processes are modify $\beta_1^{\lambda_1}$ into the form $\beta_1^{\lambda_1} = \beta_{10}^{\lambda_1} + \zeta^{\lambda_1} \cdot s^{\lambda_2}$, where the first term in the right hand side is the usual term, and the second represents the Raman processes. It is difficult to evaluate the coefficient ζ^{λ_1} from the first principles, since it required detailed information on level broadening, phase matching conditions. The way is to present ζ^{λ_1} as a fitting parameter. In ADQW LDs, the λ_2 light is emitted from the LD and λ_2 is much different from the λ_1 light. In usual LDs the spontaneous emission factor reduces the relaxation oscillation of the λ_1 light. The optical gains are defined as $G_1^{\lambda_1}(n_1) = g_a \cdot (n_1 - f_a)$ and $G_2^{\lambda_2}(n_2) = g_b \cdot (n_2 - f_b)$, where g_a and g_b are constants, and f_a, f_b are the inversion required to overcome the bulk losses. The assumption is $G_1^{\lambda_2}(n_1) = \alpha \cdot G_1^{\lambda_1}(n_1)$, where, α is a constant. The threshold hole densities n_{1th} and n_{2th} are $G_1^{\lambda_1}(n_{1th}) = \frac{1}{\tau^{\lambda_1}}$ and $G_2^{\lambda_2}(n_{2th}) = \frac{1}{\tau^{\lambda_2}} - G_1^{\lambda_2}(n_{1th})$, $G_1^{\lambda_1}(n_1) = \frac{n_1 - n_{1th}}{n_{1th}} \cdot \left(\frac{1}{\tau^{\lambda_1}} + g_a \cdot f_a\right) + \frac{1}{\tau^{\lambda_1}}$; $G_2^{\lambda_2}(n_2) = \frac{n_2 - n_{2th}}{n_{2th}} \cdot \left(\frac{1}{\tau^{\lambda_2}} - G_1^{\lambda_2}(n_{1th}) + g_b \cdot f_b\right) + \frac{1}{\tau^{\lambda_2}}$. It is recommended to the Runge–Kutta method to solve numerically the set of rate equations for hole densities n_1, n_2 in well 1, well 2, respectively, and the photon densities s^{λ_1}, s^{λ_2} of the lasing modes at λ_1 and λ_2. The Runge–Kutta methods are a family of implicit and explicit iterative methods, which include the well-known routine called the Euler Method, used in temporal discretization for the approximate solutions of ordinary differential equations and judiciously uses the information on the 'slope' at more than one point to extrapolate the solution to the future time step. The Runge–Kutta calculates four different slopes and uses them as weighted averages. The injection current J is assumed to be a rectangular pulse of height J_p and width $3 \cdot \tau_{n_1}$, starting at t = 0 [1]. The typical values of parameters used are described in next table (Table 8.1).

$$G_1^{\lambda_1}(n_1(t - \tau_1)) = \frac{n_1(t - \tau_1) - n_{1th}}{n_{1th}} \cdot \left(\frac{1}{\tau^{\lambda_1}} + g_a \cdot f_a\right) + \frac{1}{\tau^{\lambda_1}}$$

$$J_{21} = J_{21}(n_2(t - \tau_2)) \simeq e \cdot v \cdot n_2(t - \tau_2)$$

$$G_2^{\lambda_2}(n_2(t - \tau_2)) = \frac{n_2(t - \tau_2) - n_{2th}}{n_{2th}} \cdot \left(\frac{1}{\tau^{\lambda_2}} - G_1^{\lambda_2}(n_{1th}) + g_b \cdot f_b\right) + \frac{1}{\tau^{\lambda_2}}$$

Table 8.1 Asymmetric dual quantum well lasers and dual-wavelength lasing system typical parameters values

System parameter	Typical value
ξ	0
$\frac{d_2}{d_1}$	1.35
$\frac{J_{th}^{\lambda_2}}{J_{th}^{\lambda_1}}$	2
$\frac{n_{2th}}{n_{1th}}$	0.5
τ_{n_1}, τ_{n_2}	10^{-9} s
$\tau^{\lambda_1}, \tau^{\lambda_2}$	10^{-12} s
$\tau_{n_1} \cdot g_a \cdot f_a, \tau_{n_2} \cdot g_b \cdot f_b$	10
α	0.4
$\beta_{10}^{\lambda_1}, \beta_1^{\lambda_2}, \beta_2^{\lambda_2}$	5×10^{-3}
$\zeta^{\lambda_1} \cdot n_{2th}$	1

$$G_1^{\lambda_2}(n_1(t-\tau_1)) = \alpha \cdot G_1^{\lambda_1}(n_1(t-\tau_1))$$
$$= \alpha \cdot \left[\frac{n_1(t-\tau_1) - n_{1th}}{n_{1th}} \cdot \left(\frac{1}{\tau^{\lambda_1}} + g_a \cdot f_a\right) + \frac{1}{\tau^{\lambda_1}}\right]$$

We can write our system full differential equations:

1. $$\frac{dn_1}{dt} = \frac{e \cdot v \cdot n_2(t-\tau_2)}{e \cdot d_1} - \left[\frac{n_1(t-\tau_1) - n_{1th}}{n_{1th}} \cdot \left(\frac{1}{\tau^{\lambda_1}} + g_a \cdot f_a\right) + \frac{1}{\tau^{\lambda_1}}\right]$$
$$\cdot s^{\lambda_1}\left(t - \tau_{ph1}\right) - \alpha \cdot \left[\frac{n_1(t-\tau_1) - n_{1th}}{n_{1th}} \cdot \left(\frac{1}{\tau^{\lambda_1}} + g_a \cdot f_a\right) + \frac{1}{\tau^{\lambda_1}}\right]$$
$$\cdot s^{\lambda_2}\left(t - \tau_{ph2}\right) - \frac{n_1(t-\tau_1)}{\tau_{n_1}}$$

2. $$\frac{dn_2}{dt} = \frac{J - e \cdot v \cdot n_2(t-\tau_2)}{e \cdot d_2}$$
$$- \left[\frac{n_2(t-\tau_2) - n_{2th}}{n_{2th}} \cdot \left(\frac{1}{\tau^{\lambda_2}} - G_1^{\lambda_2}(n_{1th}) + g_b \cdot f_b\right) + \frac{1}{\tau^{\lambda_2}}\right]$$
$$\cdot s^{\lambda_2}(t - \tau_{ph2}) - \frac{n_2(t-\tau_2)}{\tau_{n_2}}$$

3. $$\frac{ds^{\lambda_1}}{dt} = \left[\frac{n_1(t-\tau_1) - n_{1th}}{n_{1th}} \cdot \left(\frac{1}{\tau^{\lambda_1}} + g_a \cdot f_a\right) + \frac{1}{\tau^{\lambda_1}}\right] \cdot s^{\lambda_1}(t - \tau_{ph1})$$
$$- \frac{s^{\lambda_1}(t-\tau_{ph1})}{\tau^{\lambda_1}} + \beta_1^{\lambda_1} \cdot \frac{n_1(t-\tau_1)}{\tau_{n_1}}$$

4. $$\frac{ds^{\lambda_2}}{dt} = \left\{\alpha \cdot \left[\frac{n_1(t-\tau_1) - n_{1th}}{n_{1th}} \cdot \left(\frac{1}{\tau^{\lambda_1}} + g_a \cdot f_a\right) + \frac{1}{\tau^{\lambda_1}}\right]\right.$$

$$+\left[\frac{n_2(t-\tau_2)-n_{2th}}{n_{2th}}\cdot\left(\frac{1}{\tau^{\lambda_2}}-G_1^{\lambda_2}(n_{1th})+g_b\cdot f_b\right)+\frac{1}{\tau^{\lambda_2}}\right]\Bigg\}$$

$$\cdot s^{\lambda_2}(t-\tau_{ph2})-\frac{s^{\lambda_2}(t-\tau_{ph2})}{\tau^{\lambda_2}}+\beta_1^{\lambda_2}\cdot\frac{n_1(t-\tau_1)}{\tau_{n_1}}+\beta_2^{\lambda_2}\cdot\frac{n_2(t-\tau_2)}{\tau_{n_2}}$$

At fixed points:

$$\frac{dn_1}{dt}=0;\ \frac{dn_2}{dt}=0;\ \frac{ds^{\lambda_1}}{dt}=0$$
$$\frac{ds^{\lambda_2}}{dt}=0;\ \lim_{t\to\infty}n_1(t-\tau_1)=n_1(t)\,\forall t\gg\tau_1$$

$$\lim_{t\to\infty}n_1(t-\tau_2)=n_2(t)\,\forall t\gg\tau_2;\ \lim_{t\to\infty}s^{\lambda_1}\left(t-\tau_{ph1}\right)$$
$$=s^{\lambda_1}(t);\ \lim_{t\to\infty}s^{\lambda_2}\left(t-\tau_{ph2}\right)$$
$$=s^{\lambda_2}(t)\,\forall t\gg\tau_{ph1}\ \&\ t\gg\tau_{ph2}$$

$$\frac{dn_1}{dt}=0\Rightarrow\frac{v\cdot n_2^*}{d_1}-\left[\frac{n_1^*-n_{1th}}{n_{1th}}\cdot\left(\frac{1}{\tau^{\lambda_1}}+g_a\cdot f_a\right)+\frac{1}{\tau^{\lambda_1}}\right]\cdot[s^{\lambda_1}]^*$$
$$-\alpha\cdot\left[\frac{n_1^*-n_{1th}}{n_{1th}}\cdot\left(\frac{1}{\tau^{\lambda_1}}+g_a\cdot f_a\right)+\frac{1}{\tau^{\lambda_1}}\right]\cdot[s^{\lambda_2}]^*-\frac{n_1^*}{\tau_{n_1}}=0$$

$$\frac{v\cdot n_2^*}{d_1}-n_1^*\cdot\left[s^{\lambda_1}\right]^*\cdot\frac{\left(\frac{1}{\tau^{\lambda_1}}+g_a\cdot f_a\right)}{n_{1th}}$$
$$+g_a\cdot f_a\cdot\left[s^{\lambda_1}\right]^*-\alpha\cdot n_1^*\cdot\left[s^{\lambda_2}\right]^*\cdot\frac{\left(\frac{1}{\tau^{\lambda_1}}+g_a\cdot f_a\right)}{n_{1th}}$$
$$+\alpha\cdot g_a\cdot f_a\cdot\left[s^{\lambda_2}\right]^*-\frac{n_1^*}{\tau_{n_1}}=0$$

$$\frac{v\cdot n_2^*}{d_1}-(1+\alpha)\cdot\frac{\left(\frac{1}{\tau^{\lambda_1}}+g_a\cdot f_a\right)}{n_{1th}}\cdot n_1^*\cdot\left[s^{\lambda_2}\right]^*+(\alpha+1)\cdot g_a\cdot f_a\cdot\left[s^{\lambda_1}\right]^*-\frac{n_1^*}{\tau_{n_1}}=0$$

$$\frac{dn_2}{dt}=0\Rightarrow\frac{J-e\cdot v\cdot n_2^*}{e\cdot d_2}$$
$$-\left[\frac{n_2^*-n_{2th}}{n_{2th}}\cdot\left(\frac{1}{\tau^{\lambda_2}}-G_1^{\lambda_2}(n_{1th})+g_b\cdot f_b\right)+\frac{1}{\tau^{\lambda_2}}\right]\cdot[s^{\lambda_2}]^*$$
$$-\frac{n_2^*}{\tau_{n_2}}=0$$

$$\frac{J-e\cdot v\cdot n_2^*}{e\cdot d_2}-\left[\left(n_2^*\cdot\frac{1}{n_{2th}}-1\right)\cdot\left(\frac{1}{\tau^{\lambda_2}}-G_1^{\lambda_2}(n_{1th})+g_b\cdot f_b\right)+\frac{1}{\tau^{\lambda_2}}\right]$$

$$\cdot [s^{\lambda_2}]^* - \frac{n_2^*}{\tau_{n_2}} = 0$$

We define $\Gamma_1 = \left(\frac{1}{\tau^{\lambda_2}} - G_1^{\lambda_2}(n_{1th}) + g_b \cdot f_b\right)$, Γ_1 is a global parameter then

$$\frac{J - e \cdot v \cdot n_2^*}{e \cdot d_2} - \left[\left(n_2^* \cdot \frac{1}{n_{2th}} - 1\right) \cdot \Gamma_1 + \frac{1}{\tau^{\lambda_2}}\right] \cdot \left[s^{\lambda_2}\right]^* - \frac{n_2^*}{\tau_{n_2}} = 0$$

$$\frac{J}{e \cdot d_2} - \frac{v}{d_2} \cdot n_2^* - n_2^* \cdot \left[s^{\lambda_2}\right]^* \cdot \frac{\Gamma_1}{n_{2th}} - \left(\frac{1}{\tau^{\lambda_2}} - \Gamma_1\right) \cdot \left[s^{\lambda_2}\right]^* - \frac{n_2^*}{\tau_{n_2}} = 0$$

$$(*)\frac{J}{e \cdot d_2} - n_2^* \cdot \left[s^{\lambda_2}\right]^* \cdot \frac{\Gamma_1}{n_{2th}} - \left(\frac{1}{\tau^{\lambda_2}} - \Gamma_1\right) \cdot \left[s^{\lambda_2}\right]^* - \left(\frac{1}{\tau_{n_2}} + \frac{v}{d_2}\right) \cdot n_2^* = 0$$

We define $\Gamma_2 = (1 + \alpha) \cdot \frac{\left(\frac{1}{\tau^{\lambda_1}} + g_a \cdot f_a\right)}{n_{1th}}$, $\Gamma_3 = (\alpha + 1) \cdot g_a \cdot f_a$ global parameters then

$$\frac{v \cdot n_2^*}{d_1} - (1 + \alpha) \cdot \frac{\left(\frac{1}{\tau^{\lambda_1}} + g_a \cdot f_a\right)}{n_{1th}} \cdot n_1^* \cdot \left[s^{\lambda_2}\right]^* + (\alpha + 1) \cdot g_a \cdot f_a \cdot \left[s^{\lambda_1}\right]^* - \frac{n_1^*}{\tau_{n_1}} = 0$$

$$(**)\frac{v \cdot n_2^*}{d_1} - \Gamma_2 \cdot n_1^* \cdot \left[s^{\lambda_2}\right]^* + \Gamma_3 \cdot \left[s^{\lambda_1}\right]^* - \frac{n_1^*}{\tau_{n_1}} = 0$$

$$\frac{ds^{\lambda_1}}{dt} = 0 \Rightarrow \left[\frac{n_1^* - n_{1th}}{n_{1th}} \cdot \left(\frac{1}{\tau^{\lambda_1}} + g_a \cdot f_a\right) + \frac{1}{\tau^{\lambda_1}}\right] \cdot \left[s^{\lambda_1}\right]^* - \frac{\left[s^{\lambda_1}\right]^*}{\tau^{\lambda_1}} + \beta_1^{\lambda_1} \cdot \frac{n_1^*}{\tau_{n_1}} = 0$$

$$(***)\frac{n_1^*}{n_{1th}} \cdot \left(\frac{1}{\tau^{\lambda_1}} + g_a \cdot f_a\right) \cdot \left[s^{\lambda_1}\right]^* - \left(g_a \cdot f_a + \frac{1}{\tau^{\lambda_1}}\right) \cdot \left[s^{\lambda_1}\right]^* + \beta_1^{\lambda_1} \cdot \frac{n_1^*}{\tau_{n_1}} = 0$$

$$\frac{ds^{\lambda_2}}{dt} = 0 \Rightarrow \left\{\alpha \cdot \left[\frac{n_1^* - n_{1th}}{n_{1th}} \cdot \left(\frac{1}{\tau^{\lambda_1}} + g_a \cdot f_a\right) + \frac{1}{\tau^{\lambda_1}}\right] + \left[\frac{n_2^* - n_{2th}}{n_{2th}} \cdot \left(\frac{1}{\tau^{\lambda_2}} - G_1^{\lambda_2}(n_{1th}) + g_b \cdot f_b\right) + \frac{1}{\tau^{\lambda_2}}\right]\right\} \cdot [s^{\lambda_2}]^* - \frac{[s^{\lambda_2}]^*}{\tau^{\lambda_2}} + \beta_1^{\lambda_2} \cdot \frac{n_1^*}{\tau_{n_1}} + \beta_2^{\lambda_2} \cdot \frac{n_2^*}{\tau_{n_2}} = 0$$

Since $\Gamma_1 = \left(\frac{1}{\tau^{\lambda_2}} - G_1^{\lambda_2}(n_{1th}) + g_b \cdot f_b\right)$ then

$$\left\{\alpha \cdot \left[\frac{n_1^* - n_{1th}}{n_{1th}} \cdot \left(\frac{1}{\tau^{\lambda_1}} + g_a \cdot f_a\right) + \frac{1}{\tau^{\lambda_1}}\right] + \left[\frac{n_2^* - n_{2th}}{n_{2th}} \cdot \Gamma_1 + \frac{1}{\tau^{\lambda_2}}\right]\right\} \cdot \left[s^{\lambda_2}\right]^* - \frac{\left[s^{\lambda_2}\right]^*}{\tau^{\lambda_2}} + \beta_1^{\lambda_2} \cdot \frac{n_1^*}{\tau_{n_1}} + \beta_2^{\lambda_2} \cdot \frac{n_2^*}{\tau_{n_2}} = 0$$

$$\left\{\frac{n_1^*}{n_{1th}}\cdot\alpha\cdot\left(\frac{1}{\tau^{\lambda_1}}+g_a\cdot f_a\right)+\frac{n_2^*}{n_{2th}}\cdot\Gamma_1+\left(\frac{1}{\tau^{\lambda_2}}-\Gamma_1-\alpha\cdot g_a\cdot f_a\right)\right\}\cdot\left[s^{\lambda_2}\right]^*$$
$$-\frac{\left[s^{\lambda_2}\right]^*}{\tau^{\lambda_2}}+\beta_1^{\lambda_2}\cdot\frac{n_1^*}{\tau_{n_1}}+\beta_2^{\lambda_2}\cdot\frac{n_2^*}{\tau_{n_2}}=0$$

We define for simplicity new parameters: $\Gamma_4=\alpha\cdot\left(\frac{1}{\tau^{\lambda_1}}+g_a\cdot f_a\right)$; $\Gamma_5=\frac{1}{\tau^{\lambda_2}}-\Gamma_1-\alpha\cdot g_a\cdot f_a$.

$$(****)\left(\frac{n_1^*}{n_{1th}}\cdot\Gamma_4+\frac{n_2^*}{n_{2th}}\cdot\Gamma_1+\Gamma_5\right)\cdot\left[s^{\lambda_2}\right]^*$$
$$-\frac{\left[s^{\lambda_2}\right]^*}{\tau^{\lambda_2}}+\beta_1^{\lambda_2}\cdot\frac{n_1^*}{\tau_{n_1}}+\beta_2^{\lambda_2}\cdot\frac{n_2^*}{\tau_{n_2}}=0$$

We can summary our Asymmetric dual quantum well lasers and dual-wavelength Lasing system fixed points equations:

$$(*)\frac{J}{e\cdot d_2}-n_2^*\cdot\left[s^{\lambda_2}\right]^*\cdot\frac{\Gamma_1}{n_{2th}}-\left(\frac{1}{\tau^{\lambda_2}}-\Gamma_1\right)\cdot\left[s^{\lambda_2}\right]^*-\left(\frac{1}{\tau_{n_2}}+\frac{v}{d_2}\right)\cdot n_2^*=0$$

$$(**)\frac{v\cdot n_2^*}{d_1}-\Gamma_2\cdot n_1^*\cdot\left[s^{\lambda_2}\right]^*+\Gamma_3\cdot\left[s^{\lambda_1}\right]^*-\frac{n_1^*}{\tau_{n_1}}=0$$

$$(***)\frac{n_1^*}{n_{1th}}\cdot\left(\frac{1}{\tau^{\lambda_1}}+g_a\cdot f_a\right)\cdot\left[s^{\lambda_1}\right]^*-\left(g_a\cdot f_a+\frac{1}{\tau^{\lambda_1}}\right)\cdot\left[s^{\lambda_1}\right]^*+\beta_1^{\lambda_1}\cdot\frac{n_1^*}{\tau_{n_1}}=0$$

$$(****)\left(\frac{n_1^*}{n_{1th}}\cdot\Gamma_4+\frac{n_2^*}{n_{2th}}\cdot\Gamma_1+\Gamma_5\right)\cdot\left[s^{\lambda_2}\right]^*$$
$$-\frac{\left[s^{\lambda_2}\right]^*}{\tau^{\lambda_2}}+\beta_1^{\lambda_2}\cdot\frac{n_1^*}{\tau_{n_1}}+\beta_2^{\lambda_2}\cdot\frac{n_2^*}{\tau_{n_2}}=0$$

It is recommended to solve it numerically rather than analytic to find system fixed points coordinate, $E^{(j)}\left(n_1^{(j)},n_2^{(j)},\left[s^{\lambda_1}\right]^{(j)},\left[s^{\lambda_2}\right]^{(j)}\right)\forall j=0,1,2,\ldots$.

Stability analysis: We have four possible cases to inspect our system stability, (1) $\tau_1=\tau,\tau_2=0,\tau_{ph1}=0,\tau_{ph2}=0$, (2) $\tau_1=0,\tau_2=\tau,\tau_{ph1}=0,\tau_{ph2}=0$, (3) $\tau_1=0,\tau_2=0,\tau_{ph1}=\tau,\tau_{ph2}=0$, (4) $\tau_1=0,\tau_2=0,\tau_{ph1}=0,\tau_{ph2}=\tau$. We choose to inspect the first case $\tau_1=\tau,\tau_2=0,\tau_{ph1}=0,\tau_{ph2}=0$. $n_1=n_1(t)$; $n_2=n_2(t)$; $s^{\lambda_1}=s^{\lambda_1}(t)$; $s^{\lambda_2}=s^{\lambda_2}(t)$.

$$1.\ \frac{dn_1}{dt}=\frac{e\cdot v\cdot n_2}{e\cdot d_1}-\left[\frac{n_1(t-\tau)-n_{1th}}{n_{1th}}\cdot\left(\frac{1}{\tau^{\lambda_1}}+g_a\cdot f_a\right)+\frac{1}{\tau^{\lambda_1}}\right]\cdot s^{\lambda_1}$$
$$-\alpha\cdot\left[\frac{n_1(t-\tau)-n_{1th}}{n_{1th}}\cdot\left(\frac{1}{\tau^{\lambda_1}}+g_a\cdot f_a\right)+\frac{1}{\tau^{\lambda_1}}\right]\cdot s^{\lambda_2}-\frac{n_1(t-\tau)}{\tau_{n_1}}$$

2. $$\frac{dn_2}{dt} = \frac{J - e \cdot v \cdot n_2}{e \cdot d_2} - \left[\frac{n_2 - n_{2th}}{n_{2th}} \cdot \left(\frac{1}{\tau^{\lambda_2}} - G_1^{\lambda_2}(n_{1th}) + g_b \cdot f_b\right) + \frac{1}{\tau^{\lambda_2}}\right] \cdot s^{\lambda_2} - \frac{n_2}{\tau_{n_2}}$$

3. $$\frac{ds^{\lambda_1}}{dt} = \left[\frac{n_1(t-\tau) - n_{1th}}{n_{1th}} \cdot \left(\frac{1}{\tau^{\lambda_1}} + g_a \cdot f_a\right) + \frac{1}{\tau^{\lambda_1}}\right] \cdot s^{\lambda_1} - \frac{s^{\lambda_1}}{\tau^{\lambda_1}} + \beta_1^{\lambda_1} \cdot \frac{n_1(t-\tau)}{\tau_{n_1}}$$

4. $$\frac{ds^{\lambda_2}}{dt} = \left\{\alpha \cdot \left[\frac{n_1(t-\tau) - n_{1th}}{n_{1th}} \cdot \left(\frac{1}{\tau^{\lambda_1}} + g_a \cdot f_a\right) + \frac{1}{\tau^{\lambda_1}}\right] + \left[\frac{n_2 - n_{2th}}{n_{2th}} \cdot \left(\frac{1}{\tau^{\lambda_2}} - G_1^{\lambda_2}(n_{1th}) + g_b \cdot f_b\right) + \frac{1}{\tau^{\lambda_2}}\right]\right\} \cdot s^{\lambda_2} - \frac{s^{\lambda_2}}{\tau^{\lambda_2}} + \beta_1^{\lambda_2} \cdot \frac{n_1(t-\tau)}{\tau_{n_1}} + \beta_2^{\lambda_2} \cdot \frac{n_2}{\tau_{n_2}}$$

Stability analysis: The standard local stability analysis about any one of the equilibrium point of the Asymmetric dual quantum well lasers and dual-wavelength Lasing system consists in adding to coordinate $[n_1, n_2, s^{\lambda_1}, s^{\lambda_2}]$ arbitrarily small increments of exponential form $\left[n_1, n_2, s^{\lambda_1}, s^{\lambda_2}\right] \cdot e^{\lambda \cdot t}$ and retaining the first order terms in $n_1, n_2, s^{\lambda_1}, s^{\lambda_2}$. The system of four homogeneous equations leads to a polynomial characteristic equation in the eigenvalues. The polynomial characteristic equations accept by set the below variables and variables derivative with respect to time into Asymmetric dual quantum well lasers and dual-wavelength Lasing system equations. The Asymmetric dual quantum well lasers and dual-wavelength Lasing system fixed values with arbitrarily small increments of exponential form $\left[n_1, n_2, s^{\lambda_1}, s^{\lambda_2}\right] \cdot e^{\lambda \cdot t}$ are: $j = 0$ (first fixed point), $j = 1$ (second fixed point), $j = 2$ (third fixed point), etc. [2, 3].

$$n_1(t) = n_1^{(j)} + n_1 \cdot e^{\lambda \cdot t}; n_2(t) = n_2^{(j)} + n_2 \cdot e^{\lambda \cdot t}; s^{\lambda_1}(t) = \left[s^{\lambda_1}\right]^{(j)} + s^{\lambda_1} \cdot e^{\lambda \cdot t}; s^{\lambda_2}(t) = \left[s^{\lambda_2}\right]^{(j)} + s^{\lambda_2} \cdot e^{\lambda \cdot t}$$

$$\frac{dn_1(t)}{dt} = n_1 \cdot \lambda \cdot e^{\lambda \cdot t}; \frac{dn_2(t)}{dt} = n_2 \cdot \lambda \cdot e^{\lambda \cdot t}$$
$$\frac{ds^{\lambda_1}(t)}{dt} = s^{\lambda_1} \cdot \lambda \cdot e^{\lambda \cdot t}; \frac{ds^{\lambda_2}(t)}{dt} = s^{\lambda_2} \cdot \lambda \cdot e^{\lambda \cdot t}$$

$$n_1(t-\tau) = n_1^{(j)} + n_1 \cdot e^{\lambda \cdot (t-\tau)}; \Gamma_6 = \frac{1}{\tau^{\lambda_1}} + g_a \cdot f_a$$

$$\Gamma_4 = \alpha \cdot \left(\frac{1}{\tau^{\lambda_1}} + g_a \cdot f_a\right); \Gamma_4 = \alpha \cdot \Gamma_6$$

We choose these expressions for ourselves $n_1(t), n_2(t), s^{\lambda_1}(t), s^{\lambda_2}(t)$ as a small displacement $\left[n_1, n_2, s^{\lambda_1}, s^{\lambda_2}\right]$ from the Asymmetric dual quantum well lasers and dual-wavelength Lasing system fixed points in time $t = 0$.

$$n_1(t=0) = n_1^{(j)} + n_1; n_2(t=0) = n_2^{(j)} + n_2$$
$$s^{\lambda_1}(t=0) = \left[s^{\lambda_1}\right]^{(j)} + s^{\lambda_1}; s^{\lambda_2}(t=0)$$
$$= \left[s^{\lambda_2}\right]^{(j)} + s^{\lambda_2}$$

$$\frac{dn_1}{dt} = \frac{e \cdot v \cdot n_2}{e \cdot d_1} - \left[\frac{n_1(t-\tau) - n_{1th}}{n_{1th}} \cdot \Gamma_6 + \frac{1}{\tau^{\lambda_1}}\right] \cdot s^{\lambda_1}$$
$$- \alpha \cdot \left[\frac{n_1(t-\tau) - n_{1th}}{n_{1th}} \cdot \Gamma_6 + \frac{1}{\tau^{\lambda_1}}\right] \cdot s^{\lambda_2} - \frac{n_1(t-\tau)}{\tau_{n_1}}$$

$$n_1 \cdot \lambda \cdot e^{\lambda \cdot t} = \frac{e \cdot v \cdot (n_2^{(j)} + n_2 \cdot e^{\lambda \cdot t})}{e \cdot d_1}$$
$$- \left[\frac{(n_1^{(j)} + n_1 \cdot e^{\lambda \cdot (t-\tau)}) - n_{1th}}{n_{1th}} \cdot \Gamma_6 + \frac{1}{\tau^{\lambda_1}}\right] \cdot ([s^{\lambda_1}]^{(j)} + s^{\lambda_1} \cdot e^{\lambda \cdot t})$$
$$- \alpha \cdot \left[\frac{(n_1^{(j)} + n_1 \cdot e^{\lambda \cdot (t-\tau)}) - n_{1th}}{n_{1th}} \cdot \Gamma_6 + \frac{1}{\tau^{\lambda_1}}\right] \cdot ([s^{\lambda_2}]^{(j)} + s^{\lambda_2} \cdot e^{\lambda \cdot t})$$
$$- \frac{(n_1^{(j)} + n_1 \cdot e^{\lambda \cdot (t-\tau)})}{\tau_{n_1}}$$

$$n_1 \cdot \lambda \cdot e^{\lambda \cdot t} = \frac{e \cdot v \cdot n_2^{(j)}}{e \cdot d_1} + \frac{e \cdot v \cdot n_2 \cdot e^{\lambda \cdot t}}{e \cdot d_1}$$
$$- \left[\frac{n_1^{(j)} - n_{1th}}{n_{1th}} \cdot \Gamma_6 + \frac{n_1 \cdot e^{\lambda \cdot (t-\tau)}}{n_{1th}} \cdot \Gamma_6 + \frac{1}{\tau^{\lambda_1}}\right]$$
$$\cdot ([s^{\lambda_1}]^{(j)} + s^{\lambda_1} \cdot e^{\lambda \cdot t})$$
$$- \alpha \cdot \left[\frac{n_1^{(j)} - n_{1th}}{n_{1th}} \cdot \Gamma_6 + \frac{n_1 \cdot e^{\lambda \cdot (t-\tau)}}{n_{1th}} \cdot \Gamma_6 + \frac{1}{\tau^{\lambda_1}}\right]$$
$$\cdot ([s^{\lambda_2}]^{(j)} + s^{\lambda_2} \cdot e^{\lambda \cdot t}) - \frac{n_1^{(j)}}{\tau_{n_1}} - \frac{n_1 \cdot e^{\lambda \cdot (t-\tau)}}{\tau_{n_1}}$$

$$n_1 \cdot \lambda \cdot e^{\lambda \cdot t} = \frac{e \cdot v \cdot n_2^{(j)}}{e \cdot d_1} + \frac{e \cdot v \cdot n_2 \cdot e^{\lambda \cdot t}}{e \cdot d_1}$$

$$
\begin{aligned}
&-\left(\frac{n_1^{(j)}-n_{1th}}{n_{1th}}\cdot\Gamma_6+\frac{1}{\tau^{\lambda_1}}\right)\cdot[s^{\lambda_1}]^{(j)}\\
&-\left(\frac{n_1^{(j)}-n_{1th}}{n_{1th}}\cdot\Gamma_6+\frac{1}{\tau^{\lambda_1}}\right)\cdot s^{\lambda_1}\cdot e^{\lambda\cdot t}-\frac{n_1\cdot e^{\lambda\cdot(t-\tau)}}{n_{1th}}\cdot\Gamma_6\cdot[s^{\lambda_1}]^{(j)}\\
&-\frac{e^{\lambda\cdot(t-\tau)}}{n_{1th}}\cdot\Gamma_6\cdot n_1\cdot s^{\lambda_1}\cdot e^{\lambda\cdot t}-\alpha\cdot\left\{\left(\frac{n_1^{(j)}-n_{1th}}{n_{1th}}\cdot\Gamma_6+\frac{1}{\tau^{\lambda_1}}\right)\cdot[s^{\lambda_2}]^{(j)}\right.\\
&+\left(\frac{n_1^{(j)}-n_{1th}}{n_{1th}}\cdot\Gamma_6+\frac{1}{\tau^{\lambda_1}}\right)\cdot s^{\lambda_2}\cdot e^{\lambda\cdot t}+\frac{n_1\cdot e^{\lambda\cdot(t-\tau)}}{n_{1th}}\cdot\Gamma_6\cdot[s^{\lambda_2}]^{(j)}\\
&\left.+\frac{e^{\lambda\cdot(t-\tau)}}{n_{1th}}\cdot\Gamma_6\cdot n_1\cdot s^{\lambda_2}\cdot e^{\lambda\cdot t}\right\}-\frac{n_1^{(j)}}{\tau_{n_1}}-\frac{n_1\cdot e^{\lambda\cdot(t-\tau)}}{\tau_{n_1}}
\end{aligned}
$$

Assumption: $n_1\cdot s^{\lambda_1}\approx 0;\, n_1\cdot s^{\lambda_2}\approx 0$

$$
\begin{aligned}
n_1\cdot\lambda\cdot e^{\lambda\cdot t}=&\frac{e\cdot v\cdot n_2^{(j)}}{e\cdot d_1}+\frac{e\cdot v\cdot n_2\cdot e^{\lambda\cdot t}}{e\cdot d_1}-\left(\frac{n_1^{(j)}-n_{1th}}{n_{1th}}\cdot\Gamma_6+\frac{1}{\tau^{\lambda_1}}\right)\cdot[s^{\lambda_1}]^{(j)}\\
&-\left(\frac{n_1^{(j)}-n_{1th}}{n_{1th}}\cdot\Gamma_6+\frac{1}{\tau^{\lambda_1}}\right)\cdot s^{\lambda_1}\cdot e^{\lambda\cdot t}-\frac{n_1\cdot e^{\lambda\cdot(t-\tau)}}{n_{1th}}\cdot\Gamma_6\cdot[s^{\lambda_1}]^{(j)}\\
&-\alpha\cdot\left\{\left(\frac{n_1^{(j)}-n_{1th}}{n_{1th}}\cdot\Gamma_6+\frac{1}{\tau^{\lambda_1}}\right)\cdot[s^{\lambda_2}]^{(j)}\right.\\
&+\left(\frac{n_1^{(j)}-n_{1th}}{n_{1th}}\cdot\Gamma_6+\frac{1}{\tau^{\lambda_1}}\right)\cdot s^{\lambda_2}\cdot e^{\lambda\cdot t}\\
&\left.+\frac{n_1\cdot e^{\lambda\cdot(t-\tau)}}{n_{1th}}\cdot\Gamma_6\cdot[s^{\lambda_2}]^{(j)}\right\}-\frac{n_1^{(j)}}{\tau_{n_1}}-\frac{n_1\cdot e^{\lambda\cdot(t-\tau)}}{\tau_{n_1}}
\end{aligned}
$$

$$
\begin{aligned}
n_1\cdot\lambda\cdot e^{\lambda\cdot t}=&\frac{e\cdot v\cdot n_2^{(j)}}{e\cdot d_1}-\left(\frac{n_1^{(j)}-n_{1th}}{n_{1th}}\cdot\Gamma_6+\frac{1}{\tau^{\lambda_1}}\right)\cdot[s^{\lambda_1}]^{(j)}\\
&-\alpha\cdot\left(\frac{n_1^{(j)}-n_{1th}}{n_{1th}}\cdot\Gamma_6+\frac{1}{\tau^{\lambda_1}}\right)\cdot[s^{\lambda_2}]^{(j)}-\frac{n_1^{(j)}}{\tau_{n_1}}\\
&+\frac{e\cdot v\cdot n_2\cdot e^{\lambda\cdot t}}{e\cdot d_1}-\left(\frac{n_1^{(j)}-n_{1th}}{n_{1th}}\cdot\Gamma_6+\frac{1}{\tau^{\lambda_1}}\right)\cdot s^{\lambda_1}\cdot e^{\lambda\cdot t}\\
&-\frac{n_1\cdot e^{\lambda\cdot(t-\tau)}}{n_{1th}}\cdot\Gamma_6\cdot[s^{\lambda_1}]^{(j)}\\
&-\alpha\cdot\left(\frac{n_1^{(j)}-n_{1th}}{n_{1th}}\cdot\Gamma_6+\frac{1}{\tau^{\lambda_1}}\right)\cdot s^{\lambda_2}\cdot e^{\lambda\cdot t}
\end{aligned}
$$

$$-\alpha \cdot \frac{n_1 \cdot e^{\lambda \cdot (t-\tau)}}{n_{1th}} \cdot \Gamma_6 \cdot [s^{\lambda_2}]^{(j)} - \frac{n_1 \cdot e^{\lambda \cdot (t-\tau)}}{\tau_{n_1}}$$

At fixed points:

$$\frac{e \cdot v \cdot n_2^{(j)}}{e \cdot d_1} - \left(\frac{n_1^{(j)} - n_{1th}}{n_{1th}} \cdot \Gamma_6 + \frac{1}{\tau^{\lambda_1}} \right) \cdot [s^{\lambda_1}]^{(j)}$$
$$- \alpha \cdot \left(\frac{n_1^{(j)} - n_{1th}}{n_{1th}} \cdot \Gamma_6 + \frac{1}{\tau^{\lambda_1}} \right) \cdot [s^{\lambda_2}]^{(j)} - \frac{n_1^{(j)}}{\tau_{n_1}} = 0$$

$$-n_1 \cdot \lambda - n_1 \cdot \left\{ (1+\alpha) \cdot \frac{1}{n_{1th}} \cdot \Gamma_6 \cdot [s^{\lambda_1}]^{(j)} + \frac{1}{\tau_{n_1}} \right\} \cdot e^{-\lambda \cdot \tau}$$
$$+ \frac{e \cdot v \cdot n_2}{e \cdot d_1} - \left(\frac{n_1^{(j)} - n_{1th}}{n_{1th}} \cdot \Gamma_6 + \frac{1}{\tau^{\lambda_1}} \right) \cdot s^{\lambda_1}$$
$$- \alpha \cdot \left(\frac{n_1^{(j)} - n_{1th}}{n_{1th}} \cdot \Gamma_6 + \frac{1}{\tau^{\lambda_1}} \right) \cdot s^{\lambda_2} = 0$$

$$\frac{dn_2}{dt} = \frac{J - e \cdot v \cdot n_2}{e \cdot d_2} - \left[\frac{n_2 - n_{2th}}{n_{2th}} \cdot \left(\frac{1}{\tau^{\lambda_2}} - G_1^{\lambda_2}(n_{1th}) \right.\right.$$
$$\left.\left. + g_b \cdot f_b \right) + \frac{1}{\tau^{\lambda_2}} \right] \cdot s^{\lambda_2} - \frac{n_2}{\tau_{n_2}}$$

$$n_2 \cdot \lambda \cdot e^{\lambda \cdot t} = \frac{J - e \cdot v \cdot (n_2^{(j)} + n_2 \cdot e^{\lambda \cdot t})}{e \cdot d_2}$$
$$- \left[\frac{(n_2^{(j)} + n_2 \cdot e^{\lambda \cdot t}) - n_{2th}}{n_{2th}} \cdot \left(\frac{1}{\tau^{\lambda_2}} - G_1^{\lambda_2}(n_{1th}) \right.\right.$$
$$\left.\left. + g_b \cdot f_b \right) + \frac{1}{\tau^{\lambda_2}} \right] \cdot ([s^{\lambda_2}]^{(j)} + s^{\lambda_2} \cdot e^{\lambda \cdot t}) - \frac{(n_2^{(j)} + n_2 \cdot e^{\lambda \cdot t})}{\tau_{n_2}}$$

$$n_2 \cdot \lambda \cdot e^{\lambda \cdot t} = \frac{J - e \cdot v \cdot n_2^{(j)}}{e \cdot d_2} - \frac{e \cdot v \cdot n_2 \cdot e^{\lambda \cdot t}}{e \cdot d_2}$$
$$- \left[\left(\frac{n_2^{(j)} - n_{2th}}{n_{2th}} + \frac{n_2 \cdot e^{\lambda \cdot t}}{n_{2th}} \right) \cdot \left(\frac{1}{\tau^{\lambda_2}} - G_1^{\lambda_2}(n_{1th}) \right.\right.$$
$$\left.\left. + g_b \cdot f_b \right) + \frac{1}{\tau^{\lambda_2}} \right] \cdot ([s^{\lambda_2}]^{(j)} + s^{\lambda_2} \cdot e^{\lambda \cdot t})$$
$$- \frac{n_2^{(j)}}{\tau_{n_2}} - \frac{n_2 \cdot e^{\lambda \cdot t}}{\tau_{n_2}}$$

We already know that $\Gamma_1 = \frac{1}{\tau^{\lambda_2}} - G_1^{\lambda_2}(n_{1th}) + g_b \cdot f_b$.

$$n_2 \cdot \lambda \cdot e^{\lambda \cdot t} = \frac{J - e \cdot v \cdot n_2^{(j)}}{e \cdot d_2} - \frac{e \cdot v \cdot n_2 \cdot e^{\lambda \cdot t}}{e \cdot d_2} - \left[\left(\frac{n_2^{(j)} - n_{2th}}{n_{2th}} + \frac{n_2 \cdot e^{\lambda \cdot t}}{n_{2th}}\right) \cdot \Gamma_1 + \frac{1}{\tau^{\lambda_2}}\right] \cdot ([s^{\lambda_2}]^{(j)} + s^{\lambda_2} \cdot e^{\lambda \cdot t}) - \frac{n_2^{(j)}}{\tau_{n_2}} - \frac{n_2 \cdot e^{\lambda \cdot t}}{\tau_{n_2}}$$

$$n_2 \cdot \lambda \cdot e^{\lambda \cdot t} = \frac{J - e \cdot v \cdot n_2^{(j)}}{e \cdot d_2} - \frac{e \cdot v \cdot n_2 \cdot e^{\lambda \cdot t}}{e \cdot d_2} - \left[\left(\frac{n_2^{(j)} - n_{2th}}{n_{2th}} \cdot \Gamma_1 + \frac{1}{\tau^{\lambda_2}}\right) + \frac{n_2 \cdot e^{\lambda \cdot t}}{n_{2th}} \cdot \Gamma_1\right] \cdot ([s^{\lambda_2}]^{(j)} + s^{\lambda_2} \cdot e^{\lambda \cdot t}) - \frac{n_2^{(j)}}{\tau_{n_2}} - \frac{n_2 \cdot e^{\lambda \cdot t}}{\tau_{n_2}}$$

$$\begin{aligned} n_2 \cdot \lambda \cdot e^{\lambda \cdot t} = {} & \frac{J - e \cdot v \cdot n_2^{(j)}}{e \cdot d_2} - \frac{e \cdot v \cdot n_2 \cdot e^{\lambda \cdot t}}{e \cdot d_2} \\ & - \left\{\left(\frac{n_2^{(j)} - n_{2th}}{n_{2th}} \cdot \Gamma_1 + \frac{1}{\tau^{\lambda_2}}\right) \cdot [s^{\lambda_2}]^{(j)} \right. \\ & + \left(\frac{n_2^{(j)} - n_{2th}}{n_{2th}} \cdot \Gamma_1 + \frac{1}{\tau^{\lambda_2}}\right) \cdot s^{\lambda_2} \cdot e^{\lambda \cdot t} \\ & \left. + \frac{n_2 \cdot e^{\lambda \cdot t}}{n_{2th}} \cdot \Gamma_1 \cdot [s^{\lambda_2}]^{(j)} + \frac{e^{\lambda \cdot t}}{n_{2th}} \cdot \Gamma_1 \cdot n_2 \cdot s^{\lambda_2} \cdot e^{\lambda \cdot t}\right\} \\ & - \frac{n_2^{(j)}}{\tau_{n_2}} - \frac{n_2 \cdot e^{\lambda \cdot t}}{\tau_{n_2}} \end{aligned}$$

Assumption: $n_2 \cdot s^{\lambda_2} \approx 0$

$$\begin{aligned} n_2 \cdot \lambda \cdot e^{\lambda \cdot t} = {} & \frac{J - e \cdot v \cdot n_2^{(j)}}{e \cdot d_2} - \frac{e \cdot v \cdot n_2 \cdot e^{\lambda \cdot t}}{e \cdot d_2} \\ & - \left(\frac{n_2^{(j)} - n_{2th}}{n_{2th}} \cdot \Gamma_1 + \frac{1}{\tau^{\lambda_2}}\right) \cdot \left[s^{\lambda_2}\right]^{(j)} - \left(\frac{n_2^{(j)} - n_{2th}}{n_{2th}} \cdot \Gamma_1 \right. \\ & \left. + \frac{1}{\tau^{\lambda_2}}\right) \cdot s^{\lambda_2} \cdot e^{\lambda \cdot t} - \frac{n_2 \cdot e^{\lambda \cdot t}}{n_{2th}} \cdot \Gamma_1 \cdot \left[s^{\lambda_2}\right]^{(j)} \\ & - \frac{n_2^{(j)}}{\tau_{n_2}} - \frac{n_2 \cdot e^{\lambda \cdot t}}{\tau_{n_2}} \end{aligned}$$

$$n_2 \cdot \lambda \cdot e^{\lambda \cdot t} = \frac{J - e \cdot v \cdot n_2^{(j)}}{e \cdot d_2} - \left(\frac{n_2^{(j)} - n_{2th}}{n_{2th}} \cdot \Gamma_1 + \frac{1}{\tau^{\lambda_2}} \right) \cdot \left[s^{\lambda_2} \right]^{(j)} - \frac{n_2^{(j)}}{\tau_{n_2}} - \frac{e \cdot v \cdot n_2 \cdot e^{\lambda \cdot t}}{e \cdot d_2} - \left(\frac{n_2^{(j)} - n_{2th}}{n_{2th}} \cdot \Gamma_1 + \frac{1}{\tau^{\lambda_2}} \right) \cdot s^{\lambda_2} \cdot e^{\lambda \cdot t} - \frac{n_2 \cdot e^{\lambda \cdot t}}{n_{2th}} \cdot \Gamma_1 \cdot \left[s^{\lambda_2} \right]^{(j)} - \frac{n_2 \cdot e^{\lambda \cdot t}}{\tau_{n_2}}$$

At fixed points:

$$\frac{J - e \cdot v \cdot n_2^{(j)}}{e \cdot d_2} - \left(\frac{n_2^{(j)} - n_{2th}}{n_{2th}} \cdot \Gamma_1 + \frac{1}{\tau^{\lambda_2}} \right) \cdot \left[s^{\lambda_2} \right]^{(j)} - \frac{n_2^{(j)}}{\tau_{n_2}} = 0$$

$$- n_2 \cdot \lambda - n_2 \cdot \left\{ \frac{e \cdot v}{e \cdot d_2} + \frac{1}{n_{2th}} \cdot \Gamma_1 \cdot \left[s^{\lambda_2} \right]^{(j)} + \frac{1}{\tau_{n_2}} \right\} - \left(\frac{n_2^{(j)} - n_{2th}}{n_{2th}} \cdot \Gamma_1 + \frac{1}{\tau^{\lambda_2}} \right) \cdot s^{\lambda_2} = 0$$

$$\frac{ds^{\lambda_1}}{dt} = \left[\frac{n_1(t - \tau) - n_{1th}}{n_{1th}} \cdot \left(\frac{1}{\tau^{\lambda_1}} + g_a \cdot f_a \right) + \frac{1}{\tau^{\lambda_1}} \right] \cdot s^{\lambda_1} - \frac{s^{\lambda_1}}{\tau^{\lambda_1}} + \beta_1^{\lambda_1} \cdot \frac{n_1(t - \tau)}{\tau_{n_1}}$$

$$s^{\lambda_1} \cdot \lambda \cdot e^{\lambda \cdot t} = \left[\frac{\left(n_1^{(j)} + n_1 \cdot e^{\lambda \cdot (t - \tau)} \right) - n_{1th}}{n_{1th}} \cdot \left(\frac{1}{\tau^{\lambda_1}} + g_a \cdot f_a \right) + \frac{1}{\tau^{\lambda_1}} \right] \cdot \left(\left[s^{\lambda_1} \right]^{(j)} + s^{\lambda_1} \cdot e^{\lambda \cdot t} \right) - \frac{\left(\left[s^{\lambda_1} \right]^{(j)} + s^{\lambda_1} \cdot e^{\lambda \cdot t} \right)}{\tau^{\lambda_1}} + \beta_1^{\lambda_1} \cdot \frac{\left(n_1^{(j)} + n_1 \cdot e^{\lambda \cdot (t - \tau)} \right)}{\tau_{n_1}}$$

We already know that $\Gamma_6 = \frac{1}{\tau^{\lambda_1}} + g_a \cdot f_a$.

$$s^{\lambda_1} \cdot \lambda \cdot e^{\lambda \cdot t} = \left[\left(\frac{n_1^{(j)} - n_{1th}}{n_{1th}} \cdot \Gamma_6 + \frac{1}{\tau^{\lambda_1}} \right) \right.$$

$$+\frac{n_1\cdot e^{\lambda\cdot(t-\tau)}}{n_{1th}}\cdot\Gamma_6\Bigg]\cdot\left([s^{\lambda_1}]^{(j)}+s^{\lambda_1}\cdot e^{\lambda\cdot t}\right)$$
$$-\frac{[s^{\lambda_1}]^{(j)}}{\tau^{\lambda_1}}-\frac{s^{\lambda_1}\cdot e^{\lambda\cdot t}}{\tau^{\lambda_1}}+\beta_1^{\lambda_1}\cdot\frac{n_1^{(j)}}{\tau_{n_1}}+\beta_1^{\lambda_1}\cdot\frac{n_1\cdot e^{\lambda\cdot(t-\tau)}}{\tau_{n_1}}$$

$$\begin{aligned}s^{\lambda_1}\cdot\lambda\cdot e^{\lambda\cdot t}&=\left(\frac{n_1^{(j)}-n_{1th}}{n_{1th}}\cdot\Gamma_6+\frac{1}{\tau^{\lambda_1}}\right)\cdot\left[s^{\lambda_1}\right]^{(j)}\\&+\left(\frac{n_1^{(j)}-n_{1th}}{n_{1th}}\cdot\Gamma_6+\frac{1}{\tau^{\lambda_1}}\right)\cdot s^{\lambda_1}\cdot e^{\lambda\cdot t}\\&+\frac{n_1\cdot e^{\lambda\cdot(t-\tau)}}{n_{1th}}\cdot\Gamma_6\cdot\left[s^{\lambda_1}\right]^{(j)}\\&+\frac{e^{\lambda\cdot(t-\tau)}}{n_{1th}}\cdot\Gamma_6\cdot n_1\cdot s^{\lambda_1}\cdot e^{\lambda\cdot t}-\frac{\left[s^{\lambda_1}\right]^{(j)}}{\tau^{\lambda_1}}\\&-\frac{s^{\lambda_1}\cdot e^{\lambda\cdot t}}{\tau^{\lambda_1}}+\beta_1^{\lambda_1}\cdot\frac{n_1^{(j)}}{\tau_{n_1}}+\beta_1^{\lambda_1}\cdot\frac{n_1\cdot e^{\lambda\cdot(t-\tau)}}{\tau_{n_1}}\end{aligned}$$

Assumption: $n_1\cdot s^{\lambda_1}\approx 0$

$$\begin{aligned}s^{\lambda_1}\cdot\lambda\cdot e^{\lambda\cdot t}&=\left(\frac{n_1^{(j)}-n_{1th}}{n_{1th}}\cdot\Gamma_6+\frac{1}{\tau^{\lambda_1}}\right)\cdot\left[s^{\lambda_1}\right]^{(j)}\\&+\left(\frac{n_1^{(j)}-n_{1th}}{n_{1th}}\cdot\Gamma_6+\frac{1}{\tau^{\lambda_1}}\right)\cdot s^{\lambda_1}\cdot e^{\lambda\cdot t}\\&+\frac{n_1\cdot e^{\lambda\cdot(t-\tau)}}{n_{1th}}\cdot\Gamma_6\cdot\left[s^{\lambda_1}\right]^{(j)}\\&-\frac{\left[s^{\lambda_1}\right]^{(j)}}{\tau^{\lambda_1}}-\frac{s^{\lambda_1}\cdot e^{\lambda\cdot t}}{\tau^{\lambda_1}}\\&+\beta_1^{\lambda_1}\cdot\frac{n_1^{(j)}}{\tau_{n_1}}+\beta_1^{\lambda_1}\cdot\frac{n_1\cdot e^{\lambda\cdot(t-\tau)}}{\tau_{n_1}}\end{aligned}$$

$$\begin{aligned}s^{\lambda_1}\cdot\lambda\cdot e^{\lambda\cdot t}&=\left(\frac{n_1^{(j)}-n_{1th}}{n_{1th}}\cdot\Gamma_6+\frac{1}{\tau^{\lambda_1}}\right)\cdot\left[s^{\lambda_1}\right]^{(j)}\\&-\frac{\left[s^{\lambda_1}\right]^{(j)}}{\tau^{\lambda_1}}+\beta_1^{\lambda_1}\cdot\frac{n_1^{(j)}}{\tau_{n_1}}\\&+\left(\frac{n_1^{(j)}-n_{1th}}{n_{1th}}\cdot\Gamma_6+\frac{1}{\tau^{\lambda_1}}\right)\cdot s^{\lambda_1}\cdot e^{\lambda\cdot t}\\&+\frac{n_1\cdot e^{\lambda\cdot(t-\tau)}}{n_{1th}}\cdot\Gamma_6\cdot\left[s^{\lambda_1}\right]^{(j)}-\frac{s^{\lambda_1}\cdot e^{\lambda\cdot t}}{\tau^{\lambda_1}}\end{aligned}$$

$$+ \beta_1^{\lambda_1} \cdot \frac{n_1 \cdot e^{\lambda \cdot (t-\tau)}}{\tau_{n_1}}$$

At fixed points:

$$\left(\frac{n_1^{(j)} - n_{1th}}{n_{1th}} \cdot \Gamma_6 + \frac{1}{\tau^{\lambda_1}} \right) \cdot \left[s^{\lambda_1} \right]^{(j)} - \frac{\left[s^{\lambda_1} \right]^{(j)}}{\tau^{\lambda_1}} + \beta_1^{\lambda_1} \cdot \frac{n_1^{(j)}}{\tau_{n_1}} = 0$$

$$s^{\lambda_1} \cdot \lambda \cdot e^{\lambda \cdot t} = \left(\frac{n_1^{(j)} - n_{1th}}{n_{1th}} \cdot \Gamma_6 + \frac{1}{\tau^{\lambda_1}} \right) \cdot s^{\lambda_1} \cdot e^{\lambda \cdot t}$$
$$+ \frac{n_1 \cdot e^{\lambda \cdot (t-\tau)}}{n_{1th}} \cdot \Gamma_6 \cdot \left[s^{\lambda_1} \right]^{(j)} - \frac{s^{\lambda_1} \cdot e^{\lambda \cdot t}}{\tau^{\lambda_1}}$$
$$+ \beta_1^{\lambda_1} \cdot \frac{n_1 \cdot e^{\lambda \cdot (t-\tau)}}{\tau_{n_1}}$$

$$\left[\frac{1}{n_{1th}} \cdot \Gamma_6 \cdot \left[s^{\lambda_1} \right]^{(j)} + \beta_1^{\lambda_1} \cdot \frac{1}{\tau_{n_1}} \right] \cdot n_1 \cdot e^{-\lambda \cdot \tau}$$
$$+ \frac{n_1^{(j)} - n_{1th}}{n_{1th}} \cdot \Gamma_6 \cdot s^{\lambda_1} - s^{\lambda_1} \cdot \lambda = 0$$

$$\frac{ds^{\lambda_2}}{dt} = \left\{ \alpha \cdot \left[\frac{n_1(t-\tau) - n_{1th}}{n_{1th}} \cdot \left(\frac{1}{\tau^{\lambda_1}} + g_a \cdot f_a \right) + \frac{1}{\tau^{\lambda_1}} \right] \right.$$
$$\left. + \left[\frac{n_2 - n_{2th}}{n_{2th}} \cdot \left(\frac{1}{\tau^{\lambda_2}} - G_1^{\lambda_2}(n_{1th}) + g_b \cdot f_b \right) + \frac{1}{\tau^{\lambda_2}} \right] \right\} \cdot s^{\lambda_2}$$
$$- \frac{s^{\lambda_2}}{\tau^{\lambda_2}} + \beta_1^{\lambda_2} \cdot \frac{n_1(t-\tau)}{\tau_{n_1}} + \beta_2^{\lambda_2} \cdot \frac{n_2}{\tau_{n_2}}$$

We already know that $\Gamma_6 = \frac{1}{\tau^{\lambda_1}} + g_a \cdot f_a$; $\Gamma_1 = \frac{1}{\tau^{\lambda_2}} - G_1^{\lambda_2}(n_{1th}) + g_b \cdot f_b$.

$$\frac{ds^{\lambda_2}}{dt} = \left\{ \alpha \cdot \left[\frac{n_1(t-\tau) - n_{1th}}{n_{1th}} \cdot \Gamma_6 + \frac{1}{\tau^{\lambda_1}} \right] + \left[\frac{n_2 - n_{2th}}{n_{2th}} \cdot \Gamma_1 + \frac{1}{\tau^{\lambda_2}} \right] \right\} \cdot s^{\lambda_2}$$
$$- \frac{s^{\lambda_2}}{\tau^{\lambda_2}} + \beta_1^{\lambda_2} \cdot \frac{n_1(t-\tau)}{\tau_{n_1}} + \beta_2^{\lambda_2} \cdot \frac{n_2}{\tau_{n_2}}$$

$$s^{\lambda_2} \cdot \lambda \cdot e^{\lambda \cdot t} = \left\{ \alpha \cdot \left[\frac{\left(n_1^{(j)} + n_1 \cdot e^{\lambda \cdot (t-\tau)} \right) - n_{1th}}{n_{1th}} \cdot \Gamma_6 + \frac{1}{\tau^{\lambda_1}} \right] \right.$$
$$\left. + \left[\frac{\left(n_2^{(j)} + n_2 \cdot e^{\lambda \cdot t} \right) - n_{2th}}{n_{2th}} \cdot \Gamma_1 + \frac{1}{\tau^{\lambda_2}} \right] \right\} \cdot \left(\left[s^{\lambda_2} \right]^{(j)} + s^{\lambda_2} \cdot e^{\lambda \cdot t} \right)$$

$$- \frac{\left(\left[s^{\lambda_2}\right]^{(j)} + s^{\lambda_2} \cdot e^{\lambda \cdot t}\right)}{\tau^{\lambda_2}} + \beta_1^{\lambda_2} \cdot \frac{\left(n_1^{(j)} + n_1 \cdot e^{\lambda \cdot (t-\tau)}\right)}{\tau_{n_1}} + \beta_2^{\lambda_2} \cdot \frac{\left(n_2^{(j)} + n_2 \cdot e^{\lambda \cdot t}\right)}{\tau_{n_2}}$$

$$s^{\lambda_2} \cdot \lambda \cdot e^{\lambda \cdot t} = \left\{\alpha \cdot \left[\left(\frac{n_1^{(j)} - n_{1th}}{n_{1th}} \cdot \Gamma_6 + \frac{1}{\tau^{\lambda_1}}\right) + \frac{n_1 \cdot e^{\lambda \cdot (t-\tau)}}{n_{1th}} \cdot \Gamma_6\right] + \left[\left(\frac{n_2^{(j)} - n_{2th}}{n_{2th}} \cdot \Gamma_1 + \frac{1}{\tau^{\lambda_2}}\right) + \frac{n_2 \cdot e^{\lambda \cdot t}}{n_{2th}} \cdot \Gamma_1\right]\right\} \cdot \left(\left[s^{\lambda_2}\right]^{(j)} + s^{\lambda_2} \cdot e^{\lambda \cdot t}\right) - \frac{\left[s^{\lambda_2}\right]^{(j)}}{\tau^{\lambda_2}} - \frac{s^{\lambda_2} \cdot e^{\lambda \cdot t}}{\tau^{\lambda_2}} + \beta_1^{\lambda_2} \cdot \frac{n_1^{(j)}}{\tau_{n_1}} + \beta_1^{\lambda_2} \cdot \frac{n_1 \cdot e^{\lambda \cdot (t-\tau)}}{\tau_{n_1}} + \beta_2^{\lambda_2} \cdot \frac{n_2^{(j)}}{\tau_{n_2}} + \beta_2^{\lambda_2} \cdot \frac{n_2 \cdot e^{\lambda \cdot t}}{\tau_{n_2}}$$

$$s^{\lambda_2} \cdot \lambda \cdot e^{\lambda \cdot t} = \alpha \cdot \left[\left(\frac{n_1^{(j)} - n_{1th}}{n_{1th}} \cdot \Gamma_6 + \frac{1}{\tau^{\lambda_1}}\right) + \frac{n_1 \cdot e^{\lambda \cdot (t-\tau)}}{n_{1th}} \cdot \Gamma_6\right] \cdot \left(\left[s^{\lambda_2}\right]^{(j)} + s^{\lambda_2} \cdot e^{\lambda \cdot t}\right) + \left[\left(\frac{n_2^{(j)} - n_{2th}}{n_{2th}} \cdot \Gamma_1 + \frac{1}{\tau^{\lambda_2}}\right) + \frac{n_2 \cdot e^{\lambda \cdot t}}{n_{2th}} \cdot \Gamma_1\right] \cdot \left(\left[s^{\lambda_2}\right]^{(j)} + s^{\lambda_2} \cdot e^{\lambda \cdot t}\right) - \frac{\left[s^{\lambda_2}\right]^{(j)}}{\tau^{\lambda_2}} - \frac{s^{\lambda_2} \cdot e^{\lambda \cdot t}}{\tau^{\lambda_2}} + \beta_1^{\lambda_2} \cdot \frac{n_1^{(j)}}{\tau_{n_1}} + \beta_1^{\lambda_2} \cdot \frac{n_1 \cdot e^{\lambda \cdot (t-\tau)}}{\tau_{n_1}} + \beta_2^{\lambda_2} \cdot \frac{n_2^{(j)}}{\tau_{n_2}} + \beta_2^{\lambda_2} \cdot \frac{n_2 \cdot e^{\lambda \cdot t}}{\tau_{n_2}}$$

$$s^{\lambda_2} \cdot \lambda \cdot e^{\lambda \cdot t} = \alpha \cdot \left\{\left(\frac{n_1^{(j)} - n_{1th}}{n_{1th}} \cdot \Gamma_6 + \frac{1}{\tau^{\lambda_1}}\right) \cdot [s^{\lambda_2}]^{(j)} + \left(\frac{n_1^{(j)} - n_{1th}}{n_{1th}} \cdot \Gamma_6 + \frac{1}{\tau^{\lambda_1}}\right) \cdot s^{\lambda_2} \cdot e^{\lambda \cdot t} + \frac{n_1 \cdot e^{\lambda \cdot (t-\tau)}}{n_{1th}} \cdot \Gamma_6 \cdot [s^{\lambda_2}]^{(j)} + \frac{e^{\lambda \cdot (t-\tau)}}{n_{1th}} \cdot \Gamma_6 \cdot n_1 \cdot s^{\lambda_2} \cdot e^{\lambda \cdot t}\right\} + \left(\frac{n_2^{(j)} - n_{2th}}{n_{2th}} \cdot \Gamma_1 + \frac{1}{\tau^{\lambda_2}}\right) \cdot [s^{\lambda_2}]^{(j)} + \left(\frac{n_2^{(j)} - n_{2th}}{n_{2th}} \cdot \Gamma_1 + \frac{1}{\tau^{\lambda_2}}\right) \cdot s^{\lambda_2} \cdot e^{\lambda \cdot t} + \frac{n_2 \cdot e^{\lambda \cdot t}}{n_{2th}} \cdot \Gamma_1 \cdot [s^{\lambda_2}]^{(j)} + \frac{e^{\lambda \cdot t}}{n_{2th}} \cdot \Gamma_1 \cdot n_2 \cdot s^{\lambda_2} \cdot e^{\lambda \cdot t} - \frac{[s^{\lambda_2}]^{(j)}}{\tau^{\lambda_2}} - \frac{s^{\lambda_2} \cdot e^{\lambda \cdot t}}{\tau^{\lambda_2}} + \beta_1^{\lambda_2} \cdot \frac{n_1^{(j)}}{\tau_{n_1}} + \beta_1^{\lambda_2} \cdot \frac{n_1 \cdot e^{\lambda \cdot (t-\tau)}}{\tau_{n_1}} + \beta_2^{\lambda_2} \cdot \frac{n_2^{(j)}}{\tau_{n_2}} + \beta_2^{\lambda_2} \cdot \frac{n_2 \cdot e^{\lambda \cdot t}}{\tau_{n_2}}$$

Assumption: $n_1 \cdot s^{\lambda_2} \approx 0; n_2 \cdot s^{\lambda_2} \approx 0$

$$s^{\lambda_2} \cdot \lambda \cdot e^{\lambda \cdot t} = \alpha \cdot \left\{ \left(\frac{n_1^{(j)} - n_{1th}}{n_{1th}} \cdot \Gamma_6 + \frac{1}{\tau^{\lambda_1}} \right) \cdot [s^{\lambda_2}]^{(j)} + \left(\frac{n_1^{(j)} - n_{1th}}{n_{1th}} \cdot \Gamma_6 + \frac{1}{\tau^{\lambda_1}} \right) \cdot s^{\lambda_2} \cdot e^{\lambda \cdot t} + \frac{n_1 \cdot e^{\lambda \cdot (t-\tau)}}{n_{1th}} \cdot \Gamma_6 \cdot [s^{\lambda_2}]^{(j)} \right\} + \left(\frac{n_2^{(j)} - n_{2th}}{n_{2th}} \cdot \Gamma_1 + \frac{1}{\tau^{\lambda_2}} \right) \cdot [s^{\lambda_2}]^{(j)} + \left(\frac{n_2^{(j)} - n_{2th}}{n_{2th}} \cdot \Gamma_1 + \frac{1}{\tau^{\lambda_2}} \right) \cdot s^{\lambda_2} \cdot e^{\lambda \cdot t} + \frac{n_2 \cdot e^{\lambda \cdot t}}{n_{2th}} \cdot \Gamma_1 \cdot [s^{\lambda_2}]^{(j)} - \frac{[s^{\lambda_2}]^{(j)}}{\tau^{\lambda_2}} - \frac{s^{\lambda_2} \cdot e^{\lambda \cdot t}}{\tau^{\lambda_2}} + \beta_1^{\lambda_2} \cdot \frac{n_1^{(j)}}{\tau_{n_1}} + \beta_1^{\lambda_2} \cdot \frac{n_1 \cdot e^{\lambda \cdot (t-\tau)}}{\tau_{n_1}} + \beta_2^{\lambda_2} \cdot \frac{n_2^{(j)}}{\tau_{n_2}} + \beta_2^{\lambda_2} \cdot \frac{n_2 \cdot e^{\lambda \cdot t}}{\tau_{n_2}}$$

$$s^{\lambda_2} \cdot \lambda \cdot e^{\lambda \cdot t} = \alpha \cdot \left(\frac{n_1^{(j)} - n_{1th}}{n_{1th}} \cdot \Gamma_6 + \frac{1}{\tau^{\lambda_1}} \right) \cdot [s^{\lambda_2}]^{(j)} + \left(\frac{n_2^{(j)} - n_{2th}}{n_{2th}} \cdot \Gamma_1 + \frac{1}{\tau^{\lambda_2}} \right) \cdot [s^{\lambda_2}]^{(j)} - \frac{[s^{\lambda_2}]^{(j)}}{\tau^{\lambda_2}} + \beta_1^{\lambda_2} \cdot \frac{n_1^{(j)}}{\tau_{n_1}} + \beta_2^{\lambda_2} \cdot \frac{n_2^{(j)}}{\tau_{n_2}} + \alpha \cdot \left(\frac{n_1^{(j)} - n_{1th}}{n_{1th}} \cdot \Gamma_6 + \frac{1}{\tau^{\lambda_1}} \right) \cdot s^{\lambda_2} \cdot e^{\lambda \cdot t} + \alpha \cdot \frac{n_1 \cdot e^{\lambda \cdot (t-\tau)}}{n_{1th}} \cdot \Gamma_6 \cdot [s^{\lambda_2}]^{(j)} + \left(\frac{n_2^{(j)} - n_{2th}}{n_{2th}} \cdot \Gamma_1 + \frac{1}{\tau^{\lambda_2}} \right) \cdot s^{\lambda_2} \cdot e^{\lambda \cdot t} + \frac{n_2 \cdot e^{\lambda \cdot t}}{n_{2th}} \cdot \Gamma_1 \cdot [s^{\lambda_2}]^{(j)} - \frac{s^{\lambda_2} \cdot e^{\lambda \cdot t}}{\tau^{\lambda_2}} + \beta_1^{\lambda_2} \cdot \frac{n_1 \cdot e^{\lambda \cdot (t-\tau)}}{\tau_{n_1}} + \beta_2^{\lambda_2} \cdot \frac{n_2 \cdot e^{\lambda \cdot t}}{\tau_{n_2}}$$

At fixed points:

$$\alpha \cdot \left(\frac{n_1^{(j)} - n_{1th}}{n_{1th}} \cdot \Gamma_6 + \frac{1}{\tau^{\lambda_1}} \right) \cdot [s^{\lambda_2}]^{(j)} + \left(\frac{n_2^{(j)} - n_{2th}}{n_{2th}} \cdot \Gamma_1 + \frac{1}{\tau^{\lambda_2}} \right) \cdot [s^{\lambda_2}]^{(j)} - \frac{[s^{\lambda_2}]^{(j)}}{\tau^{\lambda_2}} + \beta_1^{\lambda_2} \cdot \frac{n_1^{(j)}}{\tau_{n_1}} + \beta_2^{\lambda_2} \cdot \frac{n_2^{(j)}}{\tau_{n_2}} = 0$$

$$s^{\lambda_2} \cdot \lambda \cdot e^{\lambda \cdot t} = \alpha \cdot \left(\frac{n_1^{(j)} - n_{1th}}{n_{1th}} \cdot \Gamma_6 + \frac{1}{\tau^{\lambda_1}} \right) \cdot s^{\lambda_2} \cdot e^{\lambda \cdot t} + \alpha \cdot \frac{n_1 \cdot e^{\lambda \cdot (t-\tau)}}{n_{1th}} \cdot \Gamma_6 \cdot [s^{\lambda_2}]^{(j)} + \left(\frac{n_2^{(j)} - n_{2th}}{n_{2th}} \cdot \Gamma_1 + \frac{1}{\tau^{\lambda_2}} \right) \cdot s^{\lambda_2} \cdot e^{\lambda \cdot t} + \frac{n_2 \cdot e^{\lambda \cdot t}}{n_{2th}} \cdot \Gamma_1 \cdot [s^{\lambda_2}]^{(j)} - \frac{s^{\lambda_2} \cdot e^{\lambda \cdot t}}{\tau^{\lambda_2}} + \beta_1^{\lambda_2} \cdot \frac{n_1 \cdot e^{\lambda \cdot (t-\tau)}}{\tau_{n_1}} + \beta_2^{\lambda_2} \cdot \frac{n_2 \cdot e^{\lambda \cdot t}}{\tau_{n_2}}$$

$$\left\{\beta_1^{\lambda_2}\cdot\frac{1}{\tau_{n_1}}+\alpha\cdot\frac{1}{n_{1th}}\cdot\Gamma_6\cdot[s^{\lambda_2}]^{(j)}\right\}\cdot e^{-\lambda\cdot\tau}\cdot n_1+\left\{\beta_2^{\lambda_2}\cdot\frac{1}{\tau_{n_2}}+\frac{1}{n_{2th}}\cdot\Gamma_1\cdot[s^{\lambda_2}]^{(j)}\right\}\cdot n_2$$
$$+\left\{\alpha\cdot\left(\frac{n_1^{(j)}-n_{1th}}{n_{1th}}\cdot\Gamma_6+\frac{1}{\tau^{\lambda_1}}\right)+\left(\frac{n_2^{(j)}-n_{2th}}{n_{2th}}\cdot\Gamma_1+\frac{1}{\tau^{\lambda_2}}\right)-\frac{1}{\tau^{\lambda_2}}\right\}\cdot s^{\lambda_2}$$
$$-s^{\lambda_2}\cdot\lambda=0$$

We can summary our system arbitrarily small increment equations:

1. $$-n_1\cdot\lambda-n_1\cdot\left\{(1+\alpha)\cdot\frac{1}{n_{1th}}\cdot\Gamma_6\cdot[s^{\lambda_1}]^{(j)}+\frac{1}{\tau_{n_1}}\right\}\cdot e^{-\lambda\cdot\tau}$$
$$+\frac{e\cdot v}{e\cdot d_1}\cdot n_2-\left(\frac{n_1^{(j)}-n_{1th}}{n_{1th}}\cdot\Gamma_6+\frac{1}{\tau^{\lambda_1}}\right)\cdot s^{\lambda_1}$$
$$-\alpha\cdot\left(\frac{n_1^{(j)}-n_{1th}}{n_{1th}}\cdot\Gamma_6+\frac{1}{\tau^{\lambda_1}}\right)\cdot s^{\lambda_2}=0$$

2. $$-n_2\cdot\lambda-n_2\cdot\left\{\frac{e\cdot v}{e\cdot d_2}+\frac{1}{n_{2th}}\cdot\Gamma_1\cdot[s^{\lambda_2}]^{(j)}+\frac{1}{\tau_{n_2}}\right\}$$
$$-\left(\frac{n_2^{(j)}-n_{2th}}{n_{2th}}\cdot\Gamma_1+\frac{1}{\tau^{\lambda_2}}\right)\cdot s^{\lambda_2}=0$$

3. $$\left[\frac{1}{n_{1th}}\cdot\Gamma_6\cdot[s^{\lambda_1}]^{(j)}+\beta_1^{\lambda_1}\cdot\frac{1}{\tau_{n_1}}\right]\cdot n_1\cdot e^{-\lambda\cdot\tau}$$
$$+\frac{n_1^{(j)}-n_{1th}}{n_{1th}}\cdot\Gamma_6\cdot s^{\lambda_1}-s^{\lambda_1}\cdot\lambda=0$$

4. $$\left\{\beta_1^{\lambda_2}\cdot\frac{1}{\tau_{n_1}}+\alpha\cdot\frac{1}{n_{1th}}\cdot\Gamma_6\cdot[s^{\lambda_2}]^{(j)}\right\}\cdot e^{-\lambda\cdot\tau}\cdot n_1+\left\{\beta_2^{\lambda_2}\cdot\frac{1}{\tau_{n_2}}+\frac{1}{n_{2th}}\cdot\Gamma_1\cdot[s^{\lambda_2}]^{(j)}\right\}\cdot n_2$$
$$+\left\{\alpha\cdot\left(\frac{n_1^{(j)}-n_{1th}}{n_{1th}}\cdot\Gamma_6+\frac{1}{\tau^{\lambda_1}}\right)+\left(\frac{n_2^{(j)}-n_{2th}}{n_{2th}}\cdot\Gamma_1+\frac{1}{\tau^{\lambda_2}}\right)-\frac{1}{\tau^{\lambda_2}}\right\}\cdot s^{\lambda_2}$$
$$-s^{\lambda_2}\cdot\lambda=0$$

We define in the next table (Table 8.2) some global parameters for simplicity. We can summary our system arbitrarily small increment equations:

$$-n_1\cdot\lambda-n_1\cdot\Xi_1\cdot e^{-\lambda\cdot\tau}+\Xi_2\cdot n_2-\Xi_3\cdot s^{\lambda_1}-\Xi_4\cdot s^{\lambda_2}=0$$
$$-n_2\cdot\lambda-n_2\cdot\Xi_5-\Xi_6\cdot s^{\lambda_2}=0$$

$$\Xi_7\cdot n_1\cdot e^{-\lambda\cdot\tau}+\Xi_8\cdot s^{\lambda_1}-s^{\lambda_1}\cdot\lambda=0$$
$$\Xi_9\cdot e^{-\lambda\cdot\tau}\cdot n_1+\Xi_{10}\cdot n_2+\Xi_{11}\cdot s^{\lambda_2}-s^{\lambda_2}\cdot\lambda=0$$

Table 8.2 Asymmetric dual quantum well lasers and dual-wavelength Lasing system global parameters

Parameter	Expression
Ξ_1	$(1+\alpha)\cdot\frac{1}{n_{1th}}\cdot\Gamma_6\cdot\left[s^{\lambda_1}\right]^{(j)}+\frac{1}{\tau_{n_1}}$
Ξ_2	$\frac{e\cdot v}{e\cdot d_1}$
Ξ_3	$\frac{n_1^{(j)}-n_{1th}}{n_{1th}}\cdot\Gamma_6+\frac{1}{\tau^{\lambda_1}}$
Ξ_4	$\alpha\cdot\left(\frac{n_1^{(j)}-n_{1th}}{n_{1th}}\cdot\Gamma_6+\frac{1}{\tau^{\lambda_1}}\right)$
Ξ_5	$\frac{e\cdot v}{e\cdot d_2}+\frac{1}{n_{2th}}\cdot\Gamma_1\cdot\left[s^{\lambda_2}\right]^{(j)}+\frac{1}{\tau_{n_2}}$
Ξ_6	$\frac{n_2^{(j)}-n_{2th}}{n_{2th}}\cdot\Gamma_1+\frac{1}{\tau^{\lambda_2}}$
Ξ_7	$\frac{1}{n_{1th}}\cdot\Gamma_6\cdot\left[s^{\lambda_1}\right]^{(j)}+\beta_1^{\lambda_1}\cdot\frac{1}{\tau_{n_1}}$
Ξ_8	$\frac{n_1^{(j)}-n_{1th}}{n_{1th}}\cdot\Gamma_6$
Ξ_9	$\beta_1^{\lambda_2}\cdot\frac{1}{\tau_{n_1}}+\alpha\cdot\frac{1}{n_{1th}}\cdot\Gamma_6\cdot\left[s^{\lambda_2}\right]^{(j)}$
Ξ_{10}	$\beta_2^{\lambda_2}\cdot\frac{1}{\tau_{n_2}}+\frac{1}{n_{2th}}\cdot\Gamma_1\cdot\left[s^{\lambda_2}\right]^{(j)}$
Ξ_{11}	$\alpha\cdot\left(\frac{n_1^{(j)}-n_{1th}}{n_{1th}}\cdot\Gamma_6+\frac{1}{\tau^{\lambda_1}}\right)+\left(\frac{n_2^{(j)}-n_{2th}}{n_{2th}}\cdot\Gamma_1+\frac{1}{\tau^{\lambda_2}}\right)-\frac{1}{\tau^{\lambda_2}}$

$$\begin{pmatrix} -\lambda-\Xi_1\cdot e^{-\lambda\cdot\tau} & \Xi_2 & -\Xi_3 & -\Xi_4 \\ 0 & -\lambda-\Xi_5 & 0 & -\Xi_6 \\ \Xi_7\cdot e^{-\lambda\cdot\tau} & 0 & \Xi_8-\lambda & 0 \\ \Xi_9\cdot e^{-\lambda\cdot\tau} & \Xi_{10} & 0 & \Xi_{11}-\lambda \end{pmatrix}\cdot\begin{pmatrix} n_1 \\ n_2 \\ s^{\lambda_1} \\ s^{\lambda_2} \end{pmatrix}=0$$

$$A-\lambda\cdot I=\begin{pmatrix} -\lambda-\Xi_1\cdot e^{-\lambda\cdot\tau} & \Xi_2 & -\Xi_3 & -\Xi_4 \\ 0 & -\lambda-\Xi_5 & 0 & -\Xi_6 \\ \Xi_7\cdot e^{-\lambda\cdot\tau} & 0 & \Xi_8-\lambda & 0 \\ \Xi_9\cdot e^{-\lambda\cdot\tau} & \Xi_{10} & 0 & \Xi_{11}-\lambda \end{pmatrix}$$

$$\det(A-\lambda\cdot I)=0$$

$$\det(A-\lambda\cdot I)=\det\begin{pmatrix} (-\lambda-\Xi_1\cdot e^{-\lambda\cdot\tau}) & \Xi_2 & -\Xi_3 & -\Xi_4 \\ 0 & (-\lambda-\Xi_5) & 0 & -\Xi_6 \\ \Xi_7\cdot e^{-\lambda\cdot\tau} & 0 & (\Xi_8-\lambda) & 0 \\ \Xi_9\cdot e^{-\lambda\cdot\tau} & \Xi_{10} & 0 & (\Xi_{11}-\lambda) \end{pmatrix}$$

$$= \left(-\lambda - \Xi_1 \cdot e^{-\lambda \cdot \tau}\right) \cdot \det\begin{pmatrix} (-\lambda - \Xi_5) & 0 & -\Xi_6 \\ 0 & (\Xi_8 - \lambda) & 0 \\ \Xi_{10} & 0 & (\Xi_{11} - \lambda) \end{pmatrix}$$

$$- \Xi_2 \cdot \det\begin{pmatrix} 0 & 0 & -\Xi_6 \\ \Xi_7 \cdot e^{-\lambda \cdot \tau} & (\Xi_8 - \lambda) & 0 \\ \Xi_9 \cdot e^{-\lambda \cdot \tau} & 0 & (\Xi_{11} - \lambda) \end{pmatrix}$$

$$- \Xi_3 \cdot \det\begin{pmatrix} 0 & (-\lambda - \Xi_5) & -\Xi_6 \\ \Xi_7 \cdot e^{-\lambda \cdot \tau} & 0 & 0 \\ \Xi_9 \cdot e^{-\lambda \cdot \tau} & \Xi_{10} & (\Xi_{11} - \lambda) \end{pmatrix}$$

$$+ \Xi_4 \cdot \det\begin{pmatrix} 0 & (-\lambda - \Xi_5) & 0 \\ \Xi_7 \cdot e^{-\lambda \cdot \tau} & 0 & (\Xi_8 - \lambda) \\ \Xi_9 \cdot e^{-\lambda \cdot \tau} & \Xi_{10} & 0 \end{pmatrix}$$

Step 1:

$$\det\begin{pmatrix} (-\lambda - \Xi_5) & 0 & -\Xi_6 \\ 0 & (\Xi_8 - \lambda) & 0 \\ \Xi_{10} & 0 & (\Xi_{11} - \lambda) \end{pmatrix}$$

$$\det\begin{pmatrix} (-\lambda - \Xi_5) & 0 & -\Xi_6 \\ 0 & (\Xi_8 - \lambda) & 0 \\ \Xi_{10} & 0 & (\Xi_{11} - \lambda) \end{pmatrix}$$
$$= (-\lambda - \Xi_5) \cdot \det\begin{pmatrix} (\Xi_8 - \lambda) & 0 \\ 0 & (\Xi_{11} - \lambda) \end{pmatrix}$$
$$- \Xi_6 \cdot \det\begin{pmatrix} 0 & (\Xi_8 - \lambda) \\ \Xi_{10} & 0 \end{pmatrix}$$

$$\det\begin{pmatrix} (-\lambda - \Xi_5) & 0 & -\Xi_6 \\ 0 & (\Xi_8 - \lambda) & 0 \\ \Xi_{10} & 0 & (\Xi_{11} - \lambda) \end{pmatrix}$$
$$= (-\lambda - \Xi_5) \cdot (\Xi_8 - \lambda) \cdot (\Xi_{11} - \lambda)$$
$$+ \Xi_6 \cdot \Xi_{10} \cdot (\Xi_8 - \lambda)$$

$$\det\begin{pmatrix} (-\lambda - \Xi_5) & 0 & -\Xi_6 \\ 0 & (\Xi_8 - \lambda) & 0 \\ \Xi_{10} & 0 & (\Xi_{11} - \lambda) \end{pmatrix}$$
$$= [-\Xi_5 \cdot \Xi_8 \cdot \Xi_{11} + \Xi_6 \cdot \Xi_{10} \cdot \Xi_8]$$
$$+ [-\Xi_8 \cdot \Xi_{11} + \Xi_5 \cdot \Xi_{11} + \Xi_5 \cdot \Xi_8 - \Xi_6 \cdot \Xi_{10}] \cdot \lambda$$

$$+ [\Xi_{11} + \Xi_8 - \Xi_5] \cdot \lambda^2 - \lambda^3$$

$$(-\lambda - \Xi_1 \cdot e^{-\lambda \cdot \tau}) \cdot \det \begin{pmatrix} (-\lambda - \Xi_5) & 0 & -\Xi_6 \\ 0 & (\Xi_8 - \lambda) & 0 \\ \Xi_{10} & 0 & (\Xi_{11} - \lambda) \end{pmatrix}$$
$$= (-\lambda - \Xi_1 \cdot e^{-\lambda \cdot \tau}) \cdot ([-\Xi_5 \cdot \Xi_8 \cdot \Xi_{11} + \Xi_6 \cdot \Xi_{10} \cdot \Xi_8]$$
$$+ [-\Xi_8 \cdot \Xi_{11} + \Xi_5 \cdot \Xi_{11} + \Xi_5 \cdot \Xi_8 - \Xi_6 \cdot \Xi_{10}] \cdot \lambda$$
$$+ [\Xi_{11} + \Xi_8 - \Xi_5] \cdot \lambda^2 - \lambda^3)$$

$$(-\lambda - \Xi_1 \cdot e^{-\lambda \cdot \tau}) \cdot \det \begin{pmatrix} (-\lambda - \Xi_5) & 0 & -\Xi_6 \\ 0 & (\Xi_8 - \lambda) & 0 \\ \Xi_{10} & 0 & (\Xi_{11} - \lambda) \end{pmatrix}$$
$$= (-\lambda - \Xi_1 \cdot e^{-\lambda \cdot \tau}) \cdot ([-\Xi_5 \cdot \Xi_8 \cdot \Xi_{11} + \Xi_6 \cdot \Xi_{10} \cdot \Xi_8]$$
$$+ [-\Xi_8 \cdot \Xi_{11} + \Xi_5 \cdot \Xi_{11} + \Xi_5 \cdot \Xi_8 - \Xi_6 \cdot \Xi_{10}] \cdot \lambda$$
$$+ [\Xi_{11} + \Xi_8 - \Xi_5] \cdot \lambda^2 - \lambda^3)$$
$$= -\lambda \cdot [-\Xi_5 \cdot \Xi_8 \cdot \Xi_{11} + \Xi_6 \cdot \Xi_{10} \cdot \Xi_8]$$
$$- [-\Xi_8 \cdot \Xi_{11} + \Xi_5 \cdot \Xi_{11} + \Xi_5 \cdot \Xi_8 - \Xi_6 \cdot \Xi_{10}] \cdot \lambda^2$$
$$- [\Xi_{11} + \Xi_8 - \Xi_5] \cdot \lambda^3 + \lambda^4 + \{-[-\Xi_5 \cdot \Xi_8 \cdot \Xi_{11} + \Xi_6 \cdot \Xi_{10} \cdot \Xi_8] \cdot \Xi_1$$
$$- [-\Xi_8 \cdot \Xi_{11} + \Xi_5 \cdot \Xi_{11} + \Xi_5 \cdot \Xi_8 - \Xi_6 \cdot \Xi_{10}] \cdot \Xi_1 \cdot \lambda$$
$$- \Xi_1 \cdot [\Xi_{11} + \Xi_8 - \Xi_5] \cdot \lambda^2 + \Xi_1 \cdot \lambda^3\} \cdot e^{-\lambda \cdot \tau}$$

We define for simplicity some global parameters: $\Upsilon_1 = -\Xi_5 \cdot \Xi_8 \cdot \Xi_{11} + \Xi_6 \cdot \Xi_{10} \cdot \Xi_8$

$$\Upsilon_2 = -\Xi_8 \cdot \Xi_{11} + \Xi_5 \cdot \Xi_{11} + \Xi_5 \cdot \Xi_8 - \Xi_6 \cdot \Xi_{10}; \Upsilon_3$$
$$= \Xi_{11} + \Xi_8 - \Xi_5; \Upsilon_6 = \Xi_1 \cdot [\Xi_{11} + \Xi_8 - \Xi_5]$$

$$\Upsilon_4 = -[-\Xi_5 \cdot \Xi_8 \cdot \Xi_{11} + \Xi_6 \cdot \Xi_{10} \cdot \Xi_8] \cdot \Xi_1; \Upsilon_5$$
$$= [-\Xi_8 \cdot \Xi_{11} + \Xi_5 \cdot \Xi_{11} + \Xi_5 \cdot \Xi_8 - \Xi_6 \cdot \Xi_{10}] \cdot \Xi_1$$

$$\left(-\lambda - \Xi_1 \cdot e^{-\lambda \cdot \tau}\right) \cdot \det \begin{pmatrix} (-\lambda - \Xi_5) & 0 & -\Xi_6 \\ 0 & (\Xi_8 - \lambda) & 0 \\ \Xi_{10} & 0 & (\Xi_{11} - \lambda) \end{pmatrix}$$
$$= -\lambda \cdot \Upsilon_1 - \Upsilon_2 \cdot \lambda^2 - \Upsilon_3 \cdot \lambda^3 + \lambda^4$$
$$+ \left(\Upsilon_4 - \Upsilon_5 \cdot \lambda - \Upsilon_6 \cdot \lambda^2 + \Xi_1 \cdot \lambda^3\right) \cdot e^{-\lambda \cdot \tau}$$

Step 2:

$$\Xi_2 \cdot \det\begin{pmatrix} 0 & 0 & -\Xi_6 \\ \Xi_7 \cdot e^{-\lambda\cdot\tau} & (\Xi_8 - \lambda) & 0 \\ \Xi_9 \cdot e^{-\lambda\cdot\tau} & 0 & (\Xi_{11} - \lambda) \end{pmatrix}$$

$$\begin{aligned} &\Xi_2 \cdot \det\begin{pmatrix} 0 & 0 & -\Xi_6 \\ \Xi_7 \cdot e^{-\lambda\cdot\tau} & (\Xi_8 - \lambda) & 0 \\ \Xi_9 \cdot e^{-\lambda\cdot\tau} & 0 & (\Xi_{11} - \lambda) \end{pmatrix} \\ &= \Xi_2 \cdot (-\Xi_6) \cdot \det\begin{pmatrix} \Xi_7 \cdot e^{-\lambda\cdot\tau} & (\Xi_8 - \lambda) \\ \Xi_9 \cdot e^{-\lambda\cdot\tau} & 0 \end{pmatrix} \\ &= \Xi_2 \cdot \Xi_6 \cdot \Xi_9 \cdot e^{-\lambda\cdot\tau} \cdot (\Xi_8 - \lambda) \\ &= (\Xi_2 \cdot \Xi_6 \cdot \Xi_9 \cdot \Xi_8 - \Xi_2 \cdot \Xi_6 \cdot \Xi_9 \cdot \lambda) \cdot e^{-\lambda\cdot\tau} \end{aligned}$$

We define for simplicity some global parameters: $\Upsilon_7 = \Xi_2 \cdot \Xi_6 \cdot \Xi_9 \cdot \Xi_8$; $\Upsilon_8 = \Xi_2 \cdot \Xi_6 \cdot \Xi_9$

$$\Xi_2 \cdot \det\begin{pmatrix} 0 & 0 & -\Xi_6 \\ \Xi_7 \cdot e^{-\lambda\cdot\tau} & (\Xi_8 - \lambda) & 0 \\ \Xi_9 \cdot e^{-\lambda\cdot\tau} & 0 & (\Xi_{11} - \lambda) \end{pmatrix} = (\Upsilon_7 - \Upsilon_8 \cdot \lambda) \cdot e^{-\lambda\cdot\tau}$$

Step 3:

$$\Xi_3 \cdot \det\begin{pmatrix} 0 & (-\lambda - \Xi_5) & -\Xi_6 \\ \Xi_7 \cdot e^{-\lambda\cdot\tau} & 0 & 0 \\ \Xi_9 \cdot e^{-\lambda\cdot\tau} & \Xi_{10} & (\Xi_{11} - \lambda) \end{pmatrix}$$

$$\begin{aligned} &\Xi_3 \cdot \det\begin{pmatrix} 0 & (-\lambda - \Xi_5) & -\Xi_6 \\ \Xi_7 \cdot e^{-\lambda\cdot\tau} & 0 & 0 \\ \Xi_9 \cdot e^{-\lambda\cdot\tau} & \Xi_{10} & (\Xi_{11} - \lambda) \end{pmatrix} \\ &= \Xi_3 \cdot \left\{ -(-\lambda - \Xi_5) \cdot \det\begin{pmatrix} \Xi_7 \cdot e^{-\lambda\cdot\tau} & 0 \\ \Xi_9 \cdot e^{-\lambda\cdot\tau} & (\Xi_{11} - \lambda) \end{pmatrix} \right. \\ &\left. -\Xi_6 \cdot \det\begin{pmatrix} \Xi_7 \cdot e^{-\lambda\cdot\tau} & 0 \\ \Xi_9 \cdot e^{-\lambda\cdot\tau} & \Xi_{10} \end{pmatrix} \right\} \end{aligned}$$

$$\begin{aligned} &\Xi_3 \cdot \det\begin{pmatrix} 0 & (-\lambda - \Xi_5) & -\Xi_6 \\ \Xi_7 \cdot e^{-\lambda\cdot\tau} & 0 & 0 \\ \Xi_9 \cdot e^{-\lambda\cdot\tau} & \Xi_{10} & (\Xi_{11} - \lambda) \end{pmatrix} \\ &= \Xi_3 \cdot \{(\lambda + \Xi_5) \cdot (\Xi_{11} - \lambda) \cdot \Xi_7 - \Xi_6 \cdot \Xi_7 \cdot \Xi_{10}\} \cdot e^{-\lambda\cdot\tau} \end{aligned}$$

$$\Xi_3 \cdot \det\begin{pmatrix} 0 & (-\lambda - \Xi_5) & -\Xi_6 \\ \Xi_7 \cdot e^{-\lambda\cdot\tau} & 0 & 0 \\ \Xi_9 \cdot e^{-\lambda\cdot\tau} & \Xi_{10} & (\Xi_{11} - \lambda) \end{pmatrix}$$
$$= \{\Xi_3 \cdot (\Xi_5 \cdot \Xi_{11} - \Xi_6 \cdot \Xi_{10}) \cdot \Xi_7$$
$$+ \lambda \cdot (\Xi_{11} - \Xi_5) \cdot \Xi_3 \cdot \Xi_7 - \lambda^2 \cdot \Xi_3 \cdot \Xi_7\} \cdot e^{-\lambda\cdot\tau}$$

We define for simplicity some global parameters:

$$\Upsilon_9 = \Xi_3 \cdot (\Xi_5 \cdot \Xi_{11} - \Xi_6 \cdot \Xi_{10}) \cdot \Xi_7$$

$$\Upsilon_{10} = (\Xi_{11} - \Xi_5) \cdot \Xi_3 \cdot \Xi_7;\ \Upsilon_{11} = \Xi_3 \cdot \Xi_7$$

$$\Xi_3 \cdot \det\begin{pmatrix} 0 & (-\lambda - \Xi_5) & -\Xi_6 \\ \Xi_7 \cdot e^{-\lambda\cdot\tau} & 0 & 0 \\ \Xi_9 \cdot e^{-\lambda\cdot\tau} & \Xi_{10} & (\Xi_{11} - \lambda) \end{pmatrix} = \left(\Upsilon_9 + \lambda \cdot \Upsilon_{10} - \lambda^2 \cdot \Upsilon_{11}\right) \cdot e^{-\lambda\cdot\tau}$$

Step 4:

$$\Xi_4 \cdot \det\begin{pmatrix} 0 & (-\lambda - \Xi_5) & 0 \\ \Xi_7 \cdot e^{-\lambda\cdot\tau} & 0 & (\Xi_8 - \lambda) \\ \Xi_9 \cdot e^{-\lambda\cdot\tau} & \Xi_{10} & 0 \end{pmatrix}$$

$$\Xi_4 \cdot \det\begin{pmatrix} 0 & (-\lambda - \Xi_5) & 0 \\ \Xi_7 \cdot e^{-\lambda\cdot\tau} & 0 & (\Xi_8 - \lambda) \\ \Xi_9 \cdot e^{-\lambda\cdot\tau} & \Xi_{10} & 0 \end{pmatrix}$$
$$= -\Xi_4 \cdot (-\lambda - \Xi_5) \cdot \det\begin{pmatrix} \Xi_7 \cdot e^{-\lambda\cdot\tau} & (\Xi_8 - \lambda) \\ \Xi_9 \cdot e^{-\lambda\cdot\tau} & 0 \end{pmatrix}$$

$$\Xi_4 \cdot \det\begin{pmatrix} 0 & (-\lambda - \Xi_5) & 0 \\ \Xi_7 \cdot e^{-\lambda\cdot\tau} & 0 & (\Xi_8 - \lambda) \\ \Xi_9 \cdot e^{-\lambda\cdot\tau} & \Xi_{10} & 0 \end{pmatrix}$$
$$= -\Xi_4 \cdot \Xi_9 \cdot (\lambda + \Xi_5) \cdot (\Xi_8 - \lambda) \cdot e^{-\lambda\cdot\tau}$$

$$\Xi_4 \cdot \det\begin{pmatrix} 0 & (-\lambda - \Xi_5) & 0 \\ \Xi_7 \cdot e^{-\lambda\cdot\tau} & 0 & (\Xi_8 - \lambda) \\ \Xi_9 \cdot e^{-\lambda\cdot\tau} & \Xi_{10} & 0 \end{pmatrix}$$
$$= \{-\Xi_4 \cdot \Xi_9 \cdot \Xi_8 \cdot \Xi_5 - \lambda \cdot \Xi_4 \cdot \Xi_9 \cdot [\Xi_8 - \Xi_5]$$
$$+ \Xi_4 \cdot \Xi_9 \cdot \lambda^2\} \cdot e^{-\lambda\cdot\tau}$$

We define for simplicity some global parameters:

$$\Upsilon_{12} = -\Xi_4 \cdot \Xi_9 \cdot \Xi_8 \cdot \Xi_5; \Upsilon_{13} = \Xi_4 \cdot \Xi_9 \cdot [\Xi_8 - \Xi_5]$$

$$\Upsilon_{14} = \Xi_4 \cdot \Xi_9; \Xi_4 \cdot \det\begin{pmatrix} 0 & (-\lambda - \Xi_5) & 0 \\ \Xi_7 \cdot e^{-\lambda\cdot\tau} & 0 & (\Xi_8 - \lambda) \\ \Xi_9 \cdot e^{-\lambda\cdot\tau} & \Xi_{10} & 0 \end{pmatrix}$$
$$= \left(\Upsilon_{12} - \lambda \cdot \Upsilon_{13} + \Upsilon_{14} \cdot \lambda^2\right) \cdot e^{-\lambda\cdot\tau}$$

We can summary system determinant elements (Steps 1–4) in Table 8.3.

$$\det(A - \lambda \cdot I) = \det\begin{pmatrix} \left(-\lambda - \Xi_1 \cdot e^{-\lambda\cdot\tau}\right) & \Xi_2 & -\Xi_3 & -\Xi_4 \\ 0 & (-\lambda - \Xi_5) & 0 & -\Xi_6 \\ \Xi_7 \cdot e^{-\lambda\cdot\tau} & 0 & (\Xi_8 - \lambda) & 0 \\ \Xi_9 \cdot e^{-\lambda\cdot\tau} & \Xi_{10} & 0 & (\Xi_{11} - \lambda) \end{pmatrix}$$
$$= \left\{-\lambda \cdot \Upsilon_1 - \Upsilon_2 \cdot \lambda^2 - \Upsilon_3 \cdot \lambda^3 + \lambda^4 + \left(\Upsilon_4 - \Upsilon_5 \cdot \lambda - \Upsilon_6 \cdot \lambda^2 + \Xi_1 \cdot \lambda^3\right) \cdot e^{-\lambda\cdot\tau}\right\}$$
$$- (\Upsilon_7 - \Upsilon_8 \cdot \lambda) \cdot e^{-\lambda\cdot\tau} - \left(\Upsilon_9 + \lambda \cdot \Upsilon_{10} - \lambda^2 \cdot \Upsilon_{11}\right) \cdot e^{-\lambda\cdot\tau}$$
$$+ \left(\Upsilon_{12} - \lambda \cdot \Upsilon_{13} + \Upsilon_{14} \cdot \lambda^2\right) \cdot e^{-\lambda\cdot\tau}$$

$$\det(A - \lambda \cdot I) = \det\begin{pmatrix} \left(-\lambda - \Xi_1 \cdot e^{-\lambda\cdot\tau}\right) & \Xi_2 & -\Xi_3 & -\Xi_4 \\ 0 & (-\lambda - \Xi_5) & 0 & -\Xi_6 \\ \Xi_7 \cdot e^{-\lambda\cdot\tau} & 0 & (\Xi_8 - \lambda) & 0 \\ \Xi_9 \cdot e^{-\lambda\cdot\tau} & \Xi_{10} & 0 & (\Xi_{11} - \lambda) \end{pmatrix}$$
$$= -\lambda \cdot \Upsilon_1 - \Upsilon_2 \cdot \lambda^2 - \Upsilon_3 \cdot \lambda^3 + \lambda^4$$
$$+ [(\Upsilon_4 - \Upsilon_7 - \Upsilon_9 + \Upsilon_{12}) + (\Upsilon_8 - \Upsilon_5 - \Upsilon_{10} - \Upsilon_{13}) \cdot \lambda$$
$$+ (\Upsilon_{11} - \Upsilon_6 + \Upsilon_{14}) \cdot \lambda^2 + \Xi_1 \cdot \lambda^3] \cdot e^{-\lambda\cdot\tau}$$

We define for simplicity some global parameters:

$$\Omega_0 = \Upsilon_4 - \Upsilon_7 - \Upsilon_9 + \Upsilon_{12}$$

$$\Omega_1 = \Upsilon_8 - \Upsilon_5 - \Upsilon_{10} - \Upsilon_{13}; \Omega_2 = \Upsilon_{11} - \Upsilon_6 + \Upsilon_{14}$$

$$\det(A - \lambda \cdot I) = \det\begin{pmatrix} \left(-\lambda - \Xi_1 \cdot e^{-\lambda\cdot\tau}\right) & \Xi_2 & -\Xi_3 & -\Xi_4 \\ 0 & (-\lambda - \Xi_5) & 0 & -\Xi_6 \\ \Xi_7 \cdot e^{-\lambda\cdot\tau} & 0 & (\Xi_8 - \lambda) & 0 \\ \Xi_9 \cdot e^{-\lambda\cdot\tau} & \Xi_{10} & 0 & (\Xi_{11} - \lambda) \end{pmatrix}$$
$$= -\lambda \cdot \Upsilon_1 - \Upsilon_2 \cdot \lambda^2 - \Upsilon_3 \cdot \lambda^3 + \lambda^4$$

Table 8.3 Asymmetric dual quantum well lasers and dual-wavelength Lasing system determinant elements

Determinant element	Expression
$(-\lambda - \Xi_1 \cdot e^{-\lambda\cdot\tau}) \cdot \det\begin{pmatrix} (-\lambda - \Xi_5) & 0 & -\Xi_6 \\ 0 & (\Xi_8 - \lambda) & 0 \\ \Xi_{10} & 0 & (\Xi_{11} - \lambda) \end{pmatrix}$	$-\lambda \cdot \Upsilon_1 - \Upsilon_2 \cdot \lambda^2 - \Upsilon_3 \cdot \lambda^3 + \lambda^4 + \left(\Upsilon_4 - \Upsilon_5 \cdot \lambda - \Upsilon_6 \cdot \lambda^2 + \Xi_1 \cdot \lambda^3\right) \cdot e^{-\lambda\cdot\tau}$
$\Xi_2 \cdot \det\begin{pmatrix} 0 & 0 & -\Xi_6 \\ \Xi_7 \cdot e^{-\lambda\cdot\tau} & (\Xi_8 - \lambda) & 0 \\ \Xi_9 \cdot e^{-\lambda\cdot\tau} & 0 & (\Xi_{11} - \lambda) \end{pmatrix}$	$(\Upsilon_7 - \Upsilon_8 \cdot \lambda) \cdot e^{-\lambda\cdot\tau}$
$\Xi_3 \cdot \det\begin{pmatrix} 0 & (-\lambda - \Xi_5) & -\Xi_6 \\ \Xi_7 \cdot e^{-\lambda\cdot\tau} & 0 & 0 \\ \Xi_9 \cdot e^{-\lambda\cdot\tau} & \Xi_{10} & (\Xi_{11} - \lambda) \end{pmatrix}$	$\left(\Upsilon_9 + \lambda \cdot \Upsilon_{10} - \lambda^2 \cdot \Upsilon_{11}\right) \cdot e^{-\lambda\cdot\tau}$
$\Xi_4 \cdot \det\begin{pmatrix} 0 & (-\lambda - \Xi_5) & 0 \\ \Xi_7 \cdot e^{-\lambda\cdot\tau} & 0 & (\Xi_8 - \lambda) \\ \Xi_9 \cdot e^{-\lambda\cdot\tau} & \Xi_{10} & 0 \end{pmatrix}$	$\left(\Upsilon_{12} - \lambda \cdot \Upsilon_{13} + \Upsilon_{14} \cdot \lambda^2\right) \cdot e^{-\lambda\cdot\tau}$

$$+\left[\Omega_0 + \Omega_1 \cdot \lambda + \Omega_2 \cdot \lambda^2 + \Xi_1 \cdot \lambda^3\right] \cdot e^{-\lambda \cdot \tau}$$

$$\det(A - \lambda \cdot I) = -\lambda \cdot \Upsilon_1 - \Upsilon_2 \cdot \lambda^2 - \Upsilon_3 \cdot \lambda^3 + \lambda^4 + \left[\Omega_0 + \Omega_1 \cdot \lambda + \Omega_2 \cdot \lambda^2 + \Xi_1 \cdot \lambda^3\right] \cdot e^{-\lambda \cdot \tau}$$

$$D(\lambda, \tau) = -\lambda \cdot \Upsilon_1 - \Upsilon_2 \cdot \lambda^2 - \Upsilon_3 \cdot \lambda^3 + \lambda^4 + \left[\Omega_0 + \Omega_1 \cdot \lambda + \Omega_2 \cdot \lambda^2 + \Xi_1 \cdot \lambda^3\right] \cdot e^{-\lambda \cdot \tau}$$

$$D(\lambda, \tau) = P_n(\lambda, \tau) + Q_m(\lambda, \tau) \cdot e^{-\lambda \cdot \tau}; n, m \in \mathbb{N}_0;\ n = 4;\ m = 3;\ n > m$$

$$P_n(\lambda, \tau) = -\lambda \cdot \Upsilon_1 - \Upsilon_2 \cdot \lambda^2 - \Upsilon_3 \cdot \lambda^3 + \lambda^4; n = 4$$
$$Q_m(\lambda, \tau) = \Omega_0 + \Omega_1 \cdot \lambda + \Omega_2 \cdot \lambda^2 + \Xi_1 \cdot \lambda^3; m = 3$$

$$P_n(\lambda, \tau) = \sum_{k=0}^{n=4} p_k(\tau) \cdot \lambda^k = p_0(\tau) + p_1(\tau) \cdot \lambda + p_2(\tau) \cdot \lambda^2 + p_3(\tau) \cdot \lambda^3 + p_4(\tau) \cdot \lambda^4$$

$$p_0(\tau) = 0;\ p_1(\tau) = -\Upsilon_1;\ p_2(\tau) = -\Upsilon_2;\ p_3(\tau) = -\Upsilon_3;\ p_4(\tau) = 1$$

$$Q_m(\lambda, \tau) = \sum_{k=0}^{m=3} q_k(\tau) \cdot \lambda^k = q_0(\tau) + q_1(\tau) \cdot \lambda + q_2(\tau) \cdot \lambda^2 + q_3(\tau) \cdot \lambda^3$$
$$q_0(\tau) = \Omega_0 q_1(\tau) = \Omega_1;\ q_2(\tau) = \Omega_2;\ q_3(\tau) = \Xi_1$$

The homogeneous system for $n_1, n_2, s^{\lambda_1}, s^{\lambda_2}$ leads to a characteristic equation for the eigenvalue λ having the form $D(\lambda, \tau) = P(\lambda, \tau) + Q(\lambda, \tau) \cdot e^{-\lambda \cdot \tau} = 0$; and $P(\lambda) = \sum_{j=0}^{4} a_j \cdot \lambda^j$; $Q(\lambda) = \sum_{j=0}^{3} c_j \cdot \lambda^j$. The coefficients $\{a_j(q_i, q_k, \tau), c_j(q_i, q_k, \tau)\} \in \mathbb{R}$ depend on q_i, q_k and delay τ. q_i, q_k are any Asymmetric dual quantum well lasers and dual-wavelength Lasing system's parameters, other parameters kept as a constant [2, 3].

$$a_0 = 0;\ a_1 = -\Upsilon_1;\ a_2 = -\Upsilon_2;\ a_3 = -\Upsilon_3;\ a_4 = 1$$
$$c_0 = \Omega_0;\ c_1 = \Omega_1;\ c_2 = \Omega_2;\ c_3 = \Xi_1$$

Unless strictly necessary, the designation of the varied arguments (q_i, q_k) will subsequently be omitted from P, Q, a_j and c_j. The coefficients a_j, c_j are continuous and differentiable functions of their arguments, and direct substitution shows that $a_0 + c_0 \neq 0$ for $q_i, q_k \in \mathbb{R}_+$; that is, $\lambda = 0$ is not of $P(\lambda) + Q(\lambda) \cdot e^{-\lambda \cdot \tau} =$

0. Furthermore, $P(\lambda)$, $Q(\lambda)$ are analytic functions of λ, for which the following requirements of the analysis (Kuang and Cong 2005) [2] can also be verified in the present case:

1. If $\lambda = i \cdot \omega$; $\omega \in \mathbb{R}$, then $P(i \cdot \omega) + Q(i \cdot \omega) \neq 0$.
2. If $|\frac{Q(\lambda)}{P(\lambda)}|$ is bounded for $|\lambda| \to \infty$; $\text{Re}\lambda \geq 0$. No roots bifurcation from ∞.
3. $F(\omega) = |P(i \cdot \omega)|^2 - |Q(i \cdot \omega)|^2$ has a finite number of zeros. Indeed, this is a polynomial in ω.
4. Each positive root $\omega(q_i, q_k)$ of $F(\omega) = 0$ is continuous and differentiable with respect to q_i, q_k.

We assume that $P_n(\lambda, \tau)$ and $Q_m(\lambda, \tau)$ cannot have common imaginary roots. That is for any real number ω: $P_n(\lambda = i \cdot \omega, \tau) + Q_m(\lambda = i \cdot \omega, \tau) \neq 0$ and

$$P_n(\lambda = i \cdot \omega, \tau) = p_0(\tau) + p_1(\tau) \cdot i \cdot \omega - p_2(\tau) \cdot \omega^2 - p_3(\tau) \cdot i \cdot \omega^3 + p_4(\tau) \cdot \omega^4$$

$$P_n(\lambda = i \cdot \omega, \tau) = p_0(\tau) - p_2(\tau) \cdot \omega^2 + p_4(\tau) \cdot \omega^4 + \left(p_1(\tau) \cdot \omega - p_3(\tau) \cdot \omega^3\right) \cdot i$$

$$P_n(\lambda = i \cdot \omega, \tau) = \Upsilon_2 \cdot \omega^2 + \omega^4 + \left(-\Upsilon_1 \cdot \omega + \Upsilon_3 \cdot \omega^3\right) \cdot i$$

$$Q_m(\lambda = i \cdot \omega, \tau) = q_0(\tau) + q_1(\tau) \cdot i \cdot \omega - q_2(\tau) \cdot \omega^2 - q_3(\tau) \cdot i \cdot \omega^3$$

$$Q_m(\lambda = i \cdot \omega, \tau) = q_0(\tau) - q_2(\tau) \cdot \omega^2 + \left(q_1(\tau) \cdot \omega - q_3(\tau) \cdot \omega^3\right) \cdot i$$

$$Q_m(\lambda = i \cdot \omega, \tau) = \Omega_0 - \Omega_2 \cdot \omega^2 + \left(\Omega_1 \cdot \omega - \Xi_1 \cdot \omega^3\right) \cdot i$$

$$\begin{aligned} P_n(\lambda = i \cdot \omega, \tau) + Q_m(\lambda = i \cdot \omega, \tau) &= p_0(\tau) - p_2(\tau) \cdot \omega^2 + p_4(\tau) \cdot \omega^4 \\ &+ \left(p_1(\tau) \cdot \omega - p_3(\tau) \cdot \omega^3\right) \cdot i + q_0(\tau) - q_2(\tau) \cdot \omega^2 \\ &+ \left(q_1(\tau) \cdot \omega - q_3(\tau) \cdot \omega^3\right) \cdot i \neq 0 \end{aligned}$$

$$\begin{aligned} P_n(\lambda = i \cdot \omega, \tau) + Q_m(\lambda = i \cdot \omega, \tau) &= \Upsilon_2 \cdot \omega^2 + \omega^4 + \left(-\Upsilon_1 \cdot \omega + \Upsilon_3 \cdot \omega^3\right) \cdot i \\ &+ \Omega_0 - \Omega_2 \cdot \omega^2 + \left(\Omega_1 \cdot \omega - \Xi_1 \cdot \omega^3\right) \cdot i \neq 0 \end{aligned}$$

$$\begin{aligned} P_n(\lambda = i \cdot \omega, \tau) + Q_m(\lambda = i \cdot \omega, \tau) &= \Omega_0 + (\Upsilon_2 - \Omega_2) \cdot \omega^2 + \omega^4 \\ &+ \left\{(\Omega_1 - \Upsilon_1) \cdot \omega + (\Upsilon_3 - \Xi_1) \cdot \omega^3\right\} \cdot i \neq 0 \end{aligned}$$

$$\begin{aligned} |P(i \cdot \omega, \tau)|^2 &= \left(\Upsilon_2 \cdot \omega^2 + \omega^4\right)^2 + \left(-\Upsilon_1 \cdot \omega + \Upsilon_3 \cdot \omega^3\right)^2 \\ &= \Upsilon_2^2 \cdot \omega^4 + 2 \cdot \Upsilon_2 \cdot \omega^6 \\ &\quad + \omega^8 + \Upsilon_1^2 \cdot \omega^2 - 2 \cdot \Upsilon_1 \cdot \Upsilon_3 \cdot \omega^4 \end{aligned}$$

$$+\Upsilon_3^2 \cdot \omega^6$$

$$\begin{aligned}|P(i\cdot\omega,\tau)|^2 &= \left(\Upsilon_2\cdot\omega^2+\omega^4\right)^2+\left(-\Upsilon_1\cdot\omega+\Upsilon_3\cdot\omega^3\right)^2\\ &= \Upsilon_1^2\cdot\omega^2+\left(\Upsilon_2^2-2\cdot\Upsilon_1\cdot\Upsilon_3\right)\cdot\omega^4\\ &\quad+\left(2\cdot\Upsilon_2+\Upsilon_3^2\right)\cdot\omega^6+\omega^8\end{aligned}$$

$$\begin{aligned}|Q(i\cdot\omega,\tau)|^2 &= \left(\Omega_0-\Omega_2\cdot\omega^2\right)^2+\left(\Omega_1\cdot\omega-\Xi_1\cdot\omega^3\right)^2\\ &= \Omega_0^2-2\cdot\Omega_0\cdot\Omega_2\cdot\omega^2+\Omega_2^2\cdot\omega^4+\Omega_1^2\cdot\omega^2\\ &\quad-2\cdot\Omega_1\cdot\Xi_1\cdot\omega^4+\Xi_1^2\cdot\omega^6\end{aligned}$$

$$\begin{aligned}|Q(i\cdot\omega,\tau)|^2 &= \left(\Omega_0-\Omega_2\cdot\omega^2\right)^2+\left(\Omega_1\cdot\omega-\Xi_1\cdot\omega^3\right)^2\\ &= \Omega_0^2+\left(\Omega_1^2-2\cdot\Omega_0\cdot\Omega_2\right)\cdot\omega^2\\ &\quad+\left(\Omega_2^2-2\cdot\Omega_1\cdot\Xi_1\right)\cdot\omega^4+\Xi_1^2\cdot\omega^6\end{aligned}$$

$$\begin{aligned}F(\omega,\tau) &= |P(i\cdot\omega,\tau)|^2-|Q(i\cdot\omega,\tau)|^2\\ &= \Upsilon_1^2\cdot\omega^2+\left(\Upsilon_2^2-2\cdot\Upsilon_1\cdot\Upsilon_3\right)\cdot\omega^4\\ &\quad+\left(2\cdot\Upsilon_2+\Upsilon_3^2\right)\cdot\omega^6+\omega^8-\{\Omega_0^2+\left(\Omega_1^2-2\cdot\Omega_0\cdot\Omega_2\right)\cdot\omega^2\\ &\quad+\left(\Omega_2^2-2\cdot\Omega_1\cdot\Xi_1\right)\cdot\omega^4+\Xi_1^2\cdot\omega^6\}\end{aligned}$$

$$\begin{aligned}F(\omega,\tau) &= |P(i\cdot\omega,\tau)|^2-|Q(i\cdot\omega,\tau)|^2\\ &= -\Omega_0^2+\left[\Upsilon_1^2-\Omega_1^2+2\cdot\Omega_0\cdot\Omega_2\right]\cdot\omega^2\\ &\quad+\left[\Upsilon_2^2-2\cdot\Upsilon_1\cdot\Upsilon_3-\Omega_2^2+2\cdot\Omega_1\cdot\Xi_1\right]\cdot\omega^4\\ &\quad+\left[2\cdot\Upsilon_2+\Upsilon_3^2-\Xi_1^2\right]\cdot\omega^6+\omega^8\end{aligned}$$

We define the following parameters for simplicity: $\Pi_0, \Pi_2, \Pi_4, \Pi_6, \Pi_8$.

$$\Pi_0=-\Omega_0^2;\ \Pi_2=\Upsilon_1^2-\Omega_1^2+2\cdot\Omega_0\cdot\Omega_2$$
$$\Pi_4=\Upsilon_2^2-2\cdot\Upsilon_1\cdot\Upsilon_3-\Omega_2^2+2\cdot\Omega_1\cdot\Xi_1$$

$\Pi_6=2\cdot\Upsilon_2+\Upsilon_3^2-\Xi_1^2$; $\Pi_8=1$. Hence $F(\omega,\tau)=0$ implies $\sum_{k=0}^{4}\Pi_{2\cdot k}\cdot\omega^{2\cdot k}=0$ and its roots are given by solving the above polynomial.

$$\begin{aligned}P_R(i\omega,\tau) &= \Upsilon_2\cdot\omega^2+\omega^4;\ P_I(i\omega,\tau)\\ &= -\Upsilon_1\cdot\omega+\Upsilon_3\cdot\omega^3;\ Q_R(i\omega,\tau)\\ &= \Omega_0-\Omega_2\cdot\omega^2;\ Q_I(i\omega,\tau)=\Omega_1\cdot\omega-\Xi_1\cdot\omega^3\end{aligned}$$

$$\sin\theta(\tau) = \frac{-P_R(i\omega,\tau)\cdot Q_I(i\omega,\tau) + P_I(i\omega,\tau)\cdot Q_R(i\omega,\tau)}{|Q(i\omega,\tau)|^2}$$

$$\cos\theta(\tau) = -\frac{P_R(i\omega,\tau)\cdot Q_R(i\omega,\tau) + P_I(i\omega,\tau)\cdot Q_I(i\omega,\tau)}{|Q(i\omega,\tau)|^2}$$

$$k\to j;\ p_k(\tau)\to a_j;\ q_k(\tau)\to c_j;\ n,m\in \mathbb{N}_0;\ n=4;\ m=3;\ n>m$$

$$P_n(\lambda,\tau)\to P(\lambda);\ Q_m(\lambda,\tau)\to Q(\lambda);\ P(\lambda)=\sum_{j=0}^{4} a_j\cdot\lambda^j;\ Q(\lambda)=\sum_{j=0}^{3} c_j\cdot\lambda^j$$

$$P(\lambda)=a_0+a_1\cdot\lambda+a_2\cdot\lambda^2+a_3\cdot\lambda^3+a_4\cdot\lambda^4;\ Q(\lambda)=c_0+c_1\cdot\lambda+c_2\cdot\lambda^2+c_3\cdot\lambda^3$$

$$P(\lambda)=-\lambda\cdot\Upsilon_1-\Upsilon_2\cdot\lambda^2-\Upsilon_3\cdot\lambda^3+\lambda^4;\ Q(\lambda)$$
$$=\Omega_0+\Omega_1\cdot\lambda+\Omega_2\cdot\lambda^2+\Xi_1\cdot\lambda^3$$

$n,m\in\mathbb{N}_0; n>m$ and $a_j, c_j:\mathbb{R}_{0+}\to\mathbb{R}$ are continuous and differentiable function of τ such that $a_0+c_0\neq 0$. In the following "–" denoted complex and conjugate. $P(\lambda)$, $Q(\lambda)$ are analytic functions in λ and differentiable in τ. The coefficients $a_j\left(\tau_{n_1},\tau_{n_2},\beta_1^{\lambda_1},\beta_1^{\lambda_2},\beta_2^{\lambda_2},\tau,\ldots\right)\in\mathbb{R}$ and $c_j\left(\tau_{n_1},\tau_{n_2},\beta_1^{\lambda_1},\beta_1^{\lambda_2},\beta_2^{\lambda_2},\tau,\ldots\right)\in\mathbb{R}$ depend on Asymmetric dual quantum well lasers and dual-wavelength Lasing system's parameters, $\tau_{n_1},\tau_{n_2},\beta_1^{\lambda_1},\beta_1^{\lambda_2},\beta_2^{\lambda_2},\tau,\ldots$ values. Unless strictly necessary, the designation of the varied arguments: $\tau_{n_1},\tau_{n_2},\beta_1^{\lambda_1},\beta_1^{\lambda_2},\beta_2^{\lambda_2},\tau,\ldots$ will subsequently be omitted from P, Q, a_j, c_j. The coefficients a_j, c_j are continuous and differentiable functions of their arguments, and direct substitution shows that $a_0+c_0\neq 0$ [2, 3].

$$a_0+c_0=\Omega_0=\Upsilon_4-\Upsilon_7-\Upsilon_9+\Upsilon_{12}\neq 0$$

$\forall\,\tau_{n_1},\tau_{n_2},\beta_1^{\lambda_1},\beta_1^{\lambda_2},\beta_2^{\lambda_2},\tau,\ldots\in\mathbb{R}_+$ I.e. $\lambda=0$ is not a root of the characteristic equation. Furthermore $P(\lambda)$, $Q(\lambda)$ are analytic functions of λ for which the following requirements of the analysis (see [2], Sect. 3.4) can also be verified in the present case.

1. If $\lambda=i\cdot\omega;\ \omega\in\mathbb{R}$ then $P(i\cdot\omega)+Q(i\cdot\omega)\neq 0$, i.e. P and Q have no common imaginary roots. This condition was verified numerically in the entire $\forall\,\tau_{n_1},\tau_{n_2},\beta_1^{\lambda_1},\beta_1^{\lambda_2},\beta_2^{\lambda_2},\tau,\ldots$ domain of interest.
2. $|\frac{P(\lambda)}{Q(\lambda)}|$ is bounded for $|\lambda|\to\infty$; $\mathrm{Re}\,\lambda\geq 0$. No roots bifurcation from ∞. Indeed, in the limit: $|\frac{Q(\lambda)}{P(\lambda)}|=|\frac{c_0+c_1\cdot\lambda+c_2\cdot\lambda^2+c_3\cdot\lambda^3}{a_0+a_1\cdot\lambda+a_2\cdot\lambda^2+a_3\cdot\lambda^3+a_4\cdot\lambda^4}|$.
3. The following expressions exist: $F(\omega)=|P(i\cdot\omega)|^2-|Q(i\cdot\omega)|^2$.
$F(\omega,\tau)=|P(i\cdot\omega,\tau)|^2-|Q(i\cdot\omega,\tau)|^2=\sum_{k=0}^{4}\Pi_{2\cdot k}\cdot\omega^{2\cdot k}$ has at most a finite number of zeros. Indeed, this is a polynomial in ω (degree in ω^8).

4. Each positive root $\omega\left(\tau_{n_1}, \tau_{n_2}, \beta_1^{\lambda_1}, \beta_1^{\lambda_2}, \beta_2^{\lambda_2}, \tau, \ldots\right)$ of $F(\omega) = 0$ is continuous and differentiable with respect to $\tau_{n_1}, \tau_{n_2}, \beta_1^{\lambda_1}, \beta_1^{\lambda_2}, \beta_2^{\lambda_2}, \tau, \ldots$ and the condition can only be assessed numerically.

In addition, since the coefficients in P and Q are real, we have $\overline{P(-i \cdot \omega)} = P(i \cdot \omega)$ and $\overline{Q(-i \cdot \omega)} = Q(i \cdot \omega)$ thus, $\omega > 0$ may be on eigenvalue of characteristic equations. The analysis consists in identifying the roots of the characteristic equation situated on the imaginary axis of the complex λ plane, whereby increasing the parameters: $\tau_{n_1}, \tau_{n_2}, \beta_1^{\lambda_1}, \beta_1^{\lambda_2}, \beta_2^{\lambda_2}, \tau, \ldots$, $\mathrm{Re}\lambda$ may, at the crossing, change its sign from (−) to (+). i.e. from a stable focus $E^{(j)}\left(n_1^{(j)}, n_2^{(j)}, \left[s^{\lambda_1}\right]^{(j)}, \left[s^{\lambda_2}\right]^{(j)}\right) \forall j = 0, 1, 2, \ldots$ to an unstable one, or vice versa. This feature may be further assessed by examining the sign of the partial derivatives with respect to $\tau_{n_1}, \tau_{n_2}, \beta_1^{\lambda_1}, \beta_1^{\lambda_2}, \beta_2^{\lambda_2}, \tau, \ldots$ and any system parameters.

$$\Lambda^{-1}\left(\tau_{n_1}\right) = \left(\frac{\partial \mathrm{Re}\lambda}{\partial \tau_{n_1}}\right)_{\lambda = i\cdot\omega}; \tau_{n_2}, \beta_1^{\lambda_1}, \beta_1^{\lambda_2}, \beta_2^{\lambda_2}, \tau, \ldots = const$$

$$\Lambda^{-1}\left(\tau_{n_2}\right) = \left(\frac{\partial \mathrm{Re}\lambda}{\partial \tau_{n_2}}\right)_{\lambda = i\cdot\omega}; \tau_{n_1}, \beta_1^{\lambda_1}, \beta_1^{\lambda_2}, \beta_2^{\lambda_2}, \tau, \ldots = const$$

$$\Lambda^{-1}\left(\beta_1^{\lambda_1}\right) = \left(\frac{\partial \mathrm{Re}\lambda}{\partial \beta_1^{\lambda_1}}\right)_{\lambda = i\cdot\omega}; \tau_{n_1}, \tau_{n_2}, \beta_1^{\lambda_1}, \beta_1^{\lambda_2}, \beta_2^{\lambda_2}, \tau, \ldots = const$$

$$\Lambda^{-1}\left(\beta_1^{\lambda_2}\right) = \left(\frac{\partial \mathrm{Re}\lambda}{\partial \beta_1^{\lambda_2}}\right)_{\lambda = i\cdot\omega}; \tau_{n_1}, \tau_{n_2}, \beta_1^{\lambda_1}, \beta_2^{\lambda_2}, \tau, \ldots = const$$

$$\Lambda^{-1}\left(\beta_2^{\lambda_2}\right) = \left(\frac{\partial \mathrm{Re}\lambda}{\partial \beta_2^{\lambda_2}}\right)_{\lambda = i\cdot\omega}; \tau_{n_1}, \tau_{n_2}, \beta_1^{\lambda_1}, \beta_1^{\lambda_2}, \tau, \ldots = const$$

$$\Lambda^{-1}(\tau) = \left(\frac{\partial \mathrm{Re}\lambda}{\partial \tau}\right)_{\lambda = i\cdot\omega}; \tau_{n_1}, \tau_{n_2}, \beta_1^{\lambda_1}, \beta_1^{\lambda_2}, \beta_2^{\lambda_2}, \ldots = const$$

$P(\lambda) = P_R(\lambda) + i \cdot P_I(\lambda)$; $Q(\lambda) = Q_R(\lambda) + i \cdot Q_I(\lambda)$, When writing and inserting $\lambda = i \cdot \omega$ into Asymmetric dual quantum well lasers and dual-wavelength Lasing system's characteristic equation ω must satisfy the following equations:

$$\sin(\omega \cdot \tau) = g(\omega) = \frac{-P_R(i\omega, \tau) \cdot Q_I(i\omega, \tau) + P_I(i\omega, \tau) \cdot Q_R(i\omega, \tau)}{|Q(i\omega, \tau)|^2}$$

$$\cos(\omega \cdot \tau) = h(\omega) = -\frac{P_R(i\omega, \tau) \cdot Q_R(i\omega, \tau) + P_I(i\omega, \tau) \cdot Q_I(i\omega, \tau)}{|Q(i\omega, \tau)|^2}$$

where $|Q(i\omega, \tau)|^2 \neq 0$ in view of requirement (1) above, and $(g, h) \in \mathbb{R}$. Furthermore, it follows above $\sin(\omega \cdot \tau)$ and $\cos(\omega \cdot \tau)$ equation that, by squaring and adding the sides, ω must be a positive root of $F(\omega) = |P(i \cdot \omega)|^2 - |Q(i \cdot \omega)|^2 = 0$. Note: $F(\omega)$ is independent on τ. Now it is important to notice that if $\tau \notin I$ (assume that $I \subseteq \mathbb{R}_{+0}$ is the set) where $\omega(\tau)$ is a positive root of $F(\omega)$ and for, $\tau \notin I$, $\omega(\tau)$ is not defined. Then for all τ in I, $\omega(\tau)$ is satisfied that $F(\omega, \tau) = 0$. Then there are no positive $\omega(\tau)$ solutions for $F(\omega, \tau) = 0$, and we cannot have stability switches. For $\tau \in I$ where $\omega(\tau)$ is a positive solution of $F(\omega, \tau) = 0$, we can define the angle $\theta(\tau) \in [0, 2 \cdot \pi]$ as the solution of $\sin\theta(\tau) = \ldots$ and $\cos\theta(\tau) = \ldots$; the relation between the arguments $\theta(\tau)$ and $\tau \cdot \omega(\tau)$ for $\tau \in I$ must be describing below. $\tau \cdot \omega(\tau) = \theta(\tau) + 2 \cdot n \cdot \pi \, \forall n \in \mathbb{N}_0$. Hence we can define the maps: $\tau_n : I \to \mathbb{R}_{+0}$, is given by $\tau_n(\tau) = \frac{\theta(\tau)+2\cdot n\cdot\pi}{\omega(\tau)}$; $n \in \mathbb{N}_0$; $\tau \in I$. Let us introduce the function $I \to \mathbb{R}$; $S_n(\tau) = \tau - \tau_n(\tau)$; $\tau \in I$; $n \in \mathbb{N}_0$ that is continuous and differentiable in τ. In the following, the subscripts $\lambda, \omega, \tau_{n_1}, \tau_{n_2}, \beta_1^{\lambda_1}, \beta_1^{\lambda_2}, \beta_2^{\lambda_2}, \ldots$ indicate the corresponding partial derivatives. Let us first concentrate on $\Lambda(x)$, remember in $\lambda\left(\tau_{n_1}, \tau_{n_2}, \beta_1^{\lambda_1}, \beta_1^{\lambda_2}, \beta_2^{\lambda_2}, \ldots\right)$ and $\omega\left(\tau_{n_1}, \tau_{n_2}, \beta_1^{\lambda_1}, \beta_1^{\lambda_2}, \beta_2^{\lambda_2}, \ldots\right)$, and keeping all parameters except one (x) and τ. The derivation closely follows that in reference [BK]. Differentiating Asymmetric dual quantum well lasers and dual-wavelength Lasing system's characteristic equation $P(\lambda) + Q(\lambda) \cdot e^{-\lambda \cdot \tau} = 0$ with respect to specific parameter (x), and inverting the derivative, for convenience, one calculates: $x = \tau_{n_1}, \tau_{n_2}, \beta_1^{\lambda_1}, \beta_1^{\lambda_2}, \beta_2^{\lambda_2}, \tau, \ldots$

$$\left(\frac{\partial\lambda}{\partial x}\right)^{-1} = \frac{-P_\lambda(\lambda, x) \cdot Q(\lambda, x) + Q_\lambda(\lambda, x) \cdot P(\lambda, x) - \tau \cdot P(\lambda, x) \cdot Q(\lambda, x)}{P_x(\lambda, x) \cdot Q(\lambda, x) - Q_x(\lambda, x) \cdot P(\lambda, x)}$$

where $P_\lambda = \frac{\partial P}{\partial \lambda}, \ldots$ etc., substituting $\lambda = i \cdot \omega$ and bearing $\overline{P(-i \cdot \omega)} = P(i \cdot \omega)$; $\overline{Q(-i \cdot \omega)} = Q(i \cdot \omega)$ then $i \cdot P_\lambda(i \cdot \omega) = P_\omega(i \cdot \omega)$; $i \cdot Q_\lambda(i \cdot \omega) = Q_\omega(i \cdot \omega)$ and that on the surface $|P(i\omega)|^2 = |Q(i\omega)|^2$, one obtain:

$$\left(\frac{\partial\lambda}{\partial x}\right)^{-1}|_{\lambda=i\cdot\omega} = \left(\frac{i \cdot P_\omega(i \cdot \omega, x) \cdot \overline{P(i \cdot \omega, x)} + i \cdot Q_\lambda(i \cdot \omega, x) \cdot \overline{Q(\lambda, x)} - \tau \cdot |P(i \cdot \omega, x)|^2}{P_x(i \cdot \omega, x) \cdot \overline{P(i \cdot \omega, x)} - Q_x(i \cdot \omega, x) \cdot \overline{Q(i \cdot \omega, x)}}\right)$$

Upon separating into real and imaginary parts, with $P = P_R + i \cdot P_I$; $Q = Q_R + i \cdot Q_I$ and $P_\omega = P_{R\omega} + i \cdot P_{I\omega}$; $Q_\omega = Q_{R\omega} + i \cdot Q_{I\omega}$; $P_x = P_{Rx} + i \cdot P_{Ix}$; $Q_x = Q_{Rx} + i \cdot Q_{Ix}$; $P^2 = P_R^2 + P_I^2$, When (x) can be any Asymmetric dual quantum well lasers and dual-wavelength Lasing system's parameters $\tau_{n_1}, \tau_{n_2}, \beta_1^{\lambda_1}, \beta_1^{\lambda_2}, \beta_2^{\lambda_2}, \ldots$ and time delay τ etc. Where for convenience, we have dropped the arguments $(i \cdot \omega, x)$, and where $F_\omega = 2 \cdot [(P_{R\omega} \cdot P_R + P_{I\omega} \cdot P_I) - (Q_{R\omega} \cdot Q_R + Q_{I\omega} \cdot Q_I)]$; $F_x = 2 \cdot [(P_{Rx} \cdot P_R + P_{Ix} \cdot P_I) - (Q_{Rx} \cdot Q_R + Q_{Ix} \cdot Q_I)]$ and $\omega_x = -\frac{F_x}{F_\omega}$. We define U and V: $U = (P_R \cdot P_{I\omega} - P_I \cdot P_{R\omega}) - (Q_R \cdot Q_{I\omega} - Q_I \cdot Q_{R\omega})$, $V = (P_R \cdot P_{Ix} - P_I \cdot P_{Rx}) - (Q_R \cdot Q_{Ix} - Q_I \cdot Q_{Rx})$, we choose our specific parameter as time delay $x = \tau$: $P_R = \Upsilon_2 \cdot \omega^2 + \omega^4$; $P_I = -\Upsilon_1 \cdot \omega + \Upsilon_3 \cdot \omega^3$ $Q_R = \Omega_0 - \Omega_2 \cdot \omega^2$; $Q_I = \Omega_1 \cdot \omega - \Xi_1 \cdot \omega^3$; $P_{R\omega} = 2 \cdot \Upsilon_2 \cdot \omega + 4 \cdot \omega^3$; $P_{I\omega} = -\Upsilon_1 + 3 \cdot \Upsilon_3 \cdot \omega^2$; $Q_{R\omega} = -2 \cdot \Omega_2 \cdot \omega$ $Q_{I\omega} = \Omega_1 - 3 \cdot \Xi_1 \cdot \omega^2$; $P_{R\tau} = 0$; $P_{I\tau} = 0$; $Q_{R\tau} = 0$; $Q_{I\tau} = 0$; $F_\tau = 0$; $P_R \cdot P_{I\tau} - P_I \cdot P_{R\tau} = 0$; $Q_R \cdot Q_{I\tau} - Q_I \cdot Q_{R\tau} = 0$

$$V = (P_R \cdot P_{I\tau} - P_I \cdot P_{R\tau}) - (Q_R \cdot Q_{I\tau} - Q_I \cdot Q_{R\tau}) = 0.$$

$$\begin{aligned} F_\omega &= \ldots \text{ Elements } P_{R\omega} \cdot P_R \\ &= \left(2 \cdot \Upsilon_2 \cdot \omega + 4 \cdot \omega^3\right) \cdot \left(\Upsilon_2 \cdot \omega^2 + \omega^4\right) \\ &= 2 \cdot \Upsilon_2 \cdot \Upsilon_2 \cdot \omega^3 + 6 \cdot \Upsilon_2 \cdot \omega^5 + 4 \cdot \omega^7 \end{aligned}$$

$$\begin{aligned} P_{I\omega} \cdot P_I &= \left(-\Upsilon_1 + 3 \cdot \Upsilon_3 \cdot \omega^2\right) \cdot \left(-\Upsilon_1 \cdot \omega + \Upsilon_3 \cdot \omega^3\right) \\ &= \Upsilon_1^2 \cdot \omega - 4 \cdot \Upsilon_3 \cdot \Upsilon_1 \cdot \omega^3 \\ &\quad + 3 \cdot \Upsilon_3 \cdot \Upsilon_3 \cdot \omega^5 \end{aligned}$$

$$Q_{R\omega} \cdot Q_R = -2 \cdot \Omega_2 \cdot \omega \cdot \left(\Omega_0 - \Omega_2 \cdot \omega^2\right) = -2 \cdot \Omega_2 \cdot \Omega_0 \cdot \omega + 2 \cdot \Omega_2^2 \cdot \omega^3$$

$$\begin{aligned} Q_{I\omega} \cdot Q_I &= \left(\Omega_1 - 3 \cdot \Xi_1 \cdot \omega^2\right) \cdot \left(\Omega_1 \cdot \omega - \Xi_1 \cdot \omega^3\right) \\ &= \Omega_1^2 \cdot \omega - 4 \cdot \Xi_1 \cdot \Omega_1 \cdot \omega^3 + 3 \cdot \Xi_1^2 \cdot \omega^5 \end{aligned}$$

$U = \ldots$ Elements:

$$\begin{aligned} P_R \cdot P_{I\omega} &= \left(\Upsilon_2 \cdot \omega^2 + \omega^4\right) \cdot \left(-\Upsilon_1 + 3 \cdot \Upsilon_3 \cdot \omega^2\right) \\ &= -\Upsilon_1 \cdot \Upsilon_2 \cdot \omega^2 + (3 \cdot \Upsilon_2 \cdot \Upsilon_3 - \Upsilon_1) \cdot \omega^4 + 3 \cdot \Upsilon_3 \cdot \omega^6 \end{aligned}$$

$$\begin{aligned} P_I \cdot P_{R\omega} &= \left(-\Upsilon_1 \cdot \omega + \Upsilon_3 \cdot \omega^3\right) \cdot \left(2 \cdot \Upsilon_2 \cdot \omega + 4 \cdot \omega^3\right) \\ &= -2 \cdot \Upsilon_1 \cdot \Upsilon_2 \cdot \omega^2 + 2 \cdot (\Upsilon_3 \cdot \Upsilon_2 - 2 \cdot \Upsilon_1) \cdot \omega^4 + 4 \cdot \Upsilon_3 \cdot \omega^6 \end{aligned}$$

$$\begin{aligned} Q_R \cdot Q_{I\omega} &= \left(\Omega_0 - \Omega_2 \cdot \omega^2\right) \cdot \left(\Omega_1 - 3 \cdot \Xi_1 \cdot \omega^2\right) \\ &= \Omega_0 \cdot \Omega_1 - (3 \cdot \Omega_0 \cdot \Xi_1 + \Omega_2 \cdot \Omega_1) \cdot \omega^2 + 3 \cdot \Omega_2 \cdot \Xi_1 \cdot \omega^4 \end{aligned}$$

$$\begin{aligned} Q_I \cdot Q_{R\omega} &= \left(\Omega_1 \cdot \omega - \Xi_1 \cdot \omega^3\right) \cdot (-2 \cdot \Omega_2 \cdot \omega) \\ &= -2 \cdot \Omega_2 \cdot \Omega_1 \cdot \omega^2 + 2 \cdot \Xi_1 \cdot \Omega_2 \cdot \omega^4 \end{aligned}$$

We can summary our last results in the next table (Table 8.4).

$F(\omega, \tau) = 0$, differentiating with respect to τ and we get $F_\omega \cdot \frac{\partial \omega}{\partial \tau} + F_\tau = 0$; $\tau \in I \Rightarrow \frac{\partial \omega}{\partial \tau} = -\frac{F_\tau}{F_\omega}$.

$$\Lambda^{-1}(\tau) = \left(\frac{\partial \mathrm{Re}\lambda}{\partial \tau}\right)_{\lambda = i\omega}; \quad \frac{\partial \omega}{\partial \tau} = \omega_\tau = -\frac{F_\tau}{F_\omega};$$

Table 8.4 Asymmetric dual quantum well lasers and dual-wavelength Lasing system's stability analyses F_ω, U elements

F_ω, U elements	Expression	F_ω, U elements	Expression
$P_{R\omega}\cdot P_R$	$2\cdot\Upsilon_2\cdot\Upsilon_2\cdot\omega^3+6\cdot\Upsilon_2\cdot\omega^5+4\cdot\omega^7$	$P_R\cdot P_{I\omega}$	$-\Upsilon_1\cdot\Upsilon_2\cdot\omega^2+(3\cdot\Upsilon_2\cdot\Upsilon_3-\Upsilon_1)\cdot\omega^4+3\cdot\Upsilon_3\cdot\omega^6$
$P_{I\omega}\cdot P_I$	$\Upsilon_1^2\cdot\omega-4\cdot\Upsilon_3\cdot\Upsilon_1\cdot\omega^3+3\cdot\Upsilon_3\cdot\Upsilon_3\cdot\omega^5$	$P_I\cdot P_{R\omega}$	$-2\cdot\Upsilon_1\cdot\Upsilon_2\cdot\omega^2+2\cdot(\Upsilon_3\cdot\Upsilon_2-2\cdot\Upsilon_1)\cdot\omega^4+4\cdot\Upsilon_3\cdot\omega^6$
$Q_{R\omega}\cdot Q_R$	$-2\cdot\Omega_2\cdot\Omega_0\cdot\omega+2\cdot\Omega_2^2\cdot\omega^3$	$Q_R\cdot Q_{I\omega}$	$\Omega_0\cdot\Omega_1-(3\cdot\Omega_0\cdot\Xi_1+\Omega_2\cdot\Omega_1)\cdot\omega^2+3\cdot\Omega_2\cdot\Xi_1\cdot\omega^4$
$Q_{I\omega}\cdot Q_I$	$\Omega_1^2\cdot\omega-4\cdot\Xi_1\cdot\Omega_1\cdot\omega^3+3\cdot\Xi_1^2\cdot\omega^5$	$Q_I\cdot Q_{R\omega}$	$-2\cdot\Omega_2\cdot\Omega_1\cdot\omega^2+2\cdot\Xi_1\cdot\Omega_2\cdot\omega^4$

$$\Lambda^{-1}(\tau)=\mathrm{Re}\left\{\frac{-2\cdot[U+\tau\cdot|P|^2+i\cdot F_\omega]}{F_\tau+i\cdot 2\cdot[V+\omega\cdot|P|^2]}\right\}$$

$$\mathrm{sign}\{\Lambda^{-1}(\tau)\}=\mathrm{sign}\left\{\left(\frac{\partial\mathrm{Re}\lambda}{\partial\tau}\right)_{\lambda=i\omega}\right\}$$

$$\mathrm{sign}\{\Lambda^{-1}(\tau)\}=\mathrm{sign}\{F_\omega\}\cdot\mathrm{sign}\left\{\frac{V+\frac{\partial\omega}{\partial\tau}\cdot U}{|P|^2}+\omega+\frac{\partial\omega}{\partial\tau}\cdot\tau\right\}$$

We shall presently examine the possibility of stability transitions (bifurcations) Asymmetric dual quantum well lasers and dual-wavelength Lasing system, about the equilibrium points, $E^{(j)}$; $j=0,1,2,\ldots$, $E^{(j)}\left(n_1^{(j)},n_2^{(j)},[s^{\lambda_1}]^{(j)},[s^{\lambda_2}]^{(j)}\right)\forall j=0,1,2,\ldots$. The analysis consists in identifying the roots of our system characteristic equation situated on the imaginary axis of the complex λ-plane, where by increasing the delay parameter τ. Reλ , may at the crossing, changes its sign from − to + , i.e. from a stable focus $E^{(*)}$ to an unstable one, or vice versa. This feature may be further assessed by examining the sign of the partial derivatives with respect to τ [2, 3].

$$\Lambda^{-1}(\tau)=\left(\frac{\partial\mathrm{Re}\lambda}{\partial\tau}\right)_{\lambda=i\cdot\omega};\ \tau_{n_1},\tau_{n_2},\beta_1^{\lambda_1},\beta_1^{\lambda_2},\beta_2^{\lambda_2},\ldots=const$$

$$\begin{aligned}U&=(P_R\cdot P_{I\omega}-P_I\cdot P_{R\omega})-(Q_R\cdot Q_{I\omega}-Q_I\cdot Q_{R\omega})\\&=-\Upsilon_1\cdot\Upsilon_2\cdot\omega^2+(3\cdot\Upsilon_2\cdot\Upsilon_3-\Upsilon_1)\cdot\omega^4+3\cdot\Upsilon_3\cdot\omega^6\\&\quad-\{-2\cdot\Upsilon_1\cdot\Upsilon_2\cdot\omega^2+2\cdot(\Upsilon_3\cdot\Upsilon_2-2\cdot\Upsilon_1)\cdot\omega^4+4\cdot\Upsilon_3\cdot\omega^6\}\\&\quad-(\Omega_0\cdot\Omega_1-(3\cdot\Omega_0\cdot\Xi_1+\Omega_2\cdot\Omega_1)\cdot\omega^2\\&\quad+3\cdot\Omega_2\cdot\Xi_1\cdot\omega^4-\{-2\cdot\Omega_2\cdot\Omega_1\cdot\omega^2+2\cdot\Xi_1\cdot\Omega_2\cdot\omega^4\})\end{aligned}$$

$$\begin{aligned}U&=(P_R\cdot P_{I\omega}-P_I\cdot P_{R\omega})-(Q_R\cdot Q_{I\omega}-Q_I\cdot Q_{R\omega})\\&=-\Omega_0\cdot\Omega_1+(3\cdot\Omega_0\cdot\Xi_1+\Upsilon_1\cdot\Upsilon_2-\Omega_2\cdot\Omega_1)\cdot\omega^2\end{aligned}$$

$$+ (\Upsilon_2 \cdot \Upsilon_3 + 3 \cdot \Upsilon_1 - \Omega_2 \cdot \Xi_1) \cdot \omega^4 - \Upsilon_3 \cdot \omega^6$$

$$U = \sum_{k=0}^{3} A_{2k} \cdot \omega^{2 \cdot k}; A_0 = -\Omega_0 \cdot \Omega_1; A_2 = 3 \cdot \Omega_0 \cdot \Xi_1 + \Upsilon_1 \cdot \Upsilon_2 - \Omega_2 \cdot \Omega_1$$

$$A_4 = \Upsilon_2 \cdot \Upsilon_3 + 3 \cdot \Upsilon_1 - \Omega_2 \cdot \Xi_1$$

$$A_6 = -\Upsilon_3$$

$$U = (P_R \cdot P_{I\omega} - P_I \cdot P_{R\omega}) - (Q_R \cdot Q_{I\omega} - Q_I \cdot Q_{R\omega}) = \sum_{k=0}^{3} A_{2k} \cdot \omega^{2 \cdot k}$$

$$\begin{aligned} F_\omega &= 2 \cdot [(P_{R\omega} \cdot P_R + P_{I\omega} \cdot P_I) - (Q_{R\omega} \cdot Q_R + Q_{I\omega} \cdot Q_I)] \\ &= 2 \cdot [(2 \cdot \Upsilon_2 \cdot \Upsilon_2 \cdot \omega^3 + 6 \cdot \Upsilon_2 \cdot \omega^5 + 4 \cdot \omega^7 \\ &\quad + \Upsilon_1^2 \cdot \omega - 4 \cdot \Upsilon_3 \cdot \Upsilon_1 \cdot \omega^3 + 3 \cdot \Upsilon_3 \cdot \Upsilon_3 \cdot \omega^5) \\ &\quad - (-2 \cdot \Omega_2 \cdot \Omega_0 \cdot \omega + 2 \cdot \Omega_2^2 \cdot \omega^3 + \Omega_1^2 \cdot \omega \\ &\quad - 4 \cdot \Xi_1 \cdot \Omega_1 \cdot \omega^3 + 3 \cdot \Xi_1^2 \cdot \omega^5)] \end{aligned}$$

$$\begin{aligned} F_\omega &= 2 \cdot [(P_{R\omega} \cdot P_R + P_{I\omega} \cdot P_I) - (Q_{R\omega} \cdot Q_R + Q_{I\omega} \cdot Q_I)] \\ &= 2 \cdot \left(\Upsilon_1^2 + 2 \cdot \Omega_2 \cdot \Omega_0 - \Omega_1^2\right) \cdot \omega \\ &\quad + 2 \cdot \left(2 \cdot \Upsilon_2 \cdot \Upsilon_2 - 4 \cdot \Upsilon_3 \cdot \Upsilon_1 + 4 \cdot \Xi_1 \cdot \Omega_1 - 2 \cdot \Omega_2^2\right) \cdot \omega^3 \\ &\quad + 2 \cdot \left(6 \cdot \Upsilon_2 + 3 \cdot \Upsilon_3 \cdot \Upsilon_3 - 3 \cdot \Xi_1^2\right) \cdot \omega^5 + 8 \cdot \omega^7 \end{aligned}$$

$$\begin{aligned} B_1 &= 2 \cdot \left(\Upsilon_1^2 + 2 \cdot \Omega_2 \cdot \Omega_0 - \Omega_1^2\right) \\ B_3 &= 2 \cdot \left(2 \cdot \Upsilon_2 \cdot \Upsilon_2 - 4 \cdot \Upsilon_3 \cdot \Upsilon_1 + 4 \cdot \Xi_1 \cdot \Omega_1 - 2 \cdot \Omega_2^2\right) \end{aligned}$$

$$B_5 = 2 \cdot \left(6 \cdot \Upsilon_2 + 3 \cdot \Upsilon_3 \cdot \Upsilon_3 - 3 \cdot \Xi_1^2\right); B_7 = 8; F_\omega = \sum_{k=0}^{3} B_{2 \cdot k+1} \cdot \omega^{2 \cdot k+1}$$

$$F_\omega = 2 \cdot [(P_{R\omega} \cdot P_R + P_{I\omega} \cdot P_I) - (Q_{R\omega} \cdot Q_R + Q_{I\omega} \cdot Q_I)] = \sum_{k=0}^{3} B_{2 \cdot k+1} \cdot \omega^{2 \cdot k+1}$$

Then we get the expression for $F_\omega = \sum_{k=0}^{3} B_{2 \cdot k+1} \cdot \omega^{2 \cdot k+1}$, Asymmetric dual quantum well lasers and dual-wavelength Lasing system parameter values. We find those ω, τ values which fulfill $F_\omega(\omega, \tau) = 0$. We ignore negative, complex, and imaginary values of ω for specific τ values. $\tau \in [0.001 \ldots 10]$, we can express by 3D function $F_\omega(\omega, \tau) = 0$. We plot the stability switch diagram based on different delay values of our Asymmetric dual quantum well lasers and dual-wavelength Lasing system.

$$\Lambda^{-1}(\tau) = \left(\frac{\partial \mathrm{Re}\lambda}{\partial \tau}\right)_{\lambda=i\omega} = \mathrm{Re}\left\{\frac{-2\cdot\left[U+\tau\cdot|P|^2\right]+i\cdot F_\omega}{F_\tau+2i\cdot\left[V+\omega\cdot|P|^2\right]}\right\}$$

$$\Lambda^{-1}(\tau) = \left(\frac{\partial \mathrm{Re}\lambda}{\partial \tau}\right)_{\lambda=i\omega} = \frac{2\cdot\left\{F_\omega\cdot\left(V+\omega\cdot P^2\right)-F_\tau\cdot\left(U+\tau\cdot P^2\right)\right\}}{F_\tau^2+4\cdot\left(V+\omega\cdot P^2\right)^2}$$

The stability switch occurs only on those delay values (τ) which fit the equation: $\tau = \frac{\theta_+(\tau)}{\omega_+(\tau)}$ and $\theta_+(\tau)$ is the solution of $\sin\theta(\tau) = \ldots.$ and $\cos\theta(\tau) = \ldots.$ when $\omega = \omega_+(\tau)$ If only ω_+ is feasible. Additionally, when all Asymmetric dual quantum well lasers and dual-wavelength Lasing system's parameters are known and the stability switch due to various time delay values τ is described in the following expression:

$$\mathrm{sign}\left\{\Lambda^{-1}(\tau)\right\} = \mathrm{sign}\{F_\omega(\omega(\tau),\tau)\}\cdot\mathrm{sign}\left\{\tau\cdot\omega_\tau(\omega(\tau))+\omega(\tau) + \frac{U(\omega(\tau))\cdot\omega_\tau(\omega(\tau))+V(\omega(\tau))}{|P(\omega(\tau))|^2}\right\}$$

Remark: we know $F(\omega,\tau)=0$ implies its roots $\omega_i(\tau)$ and finding those delays values τ which ω_i is feasible. There are τ values which give complex ω_i or imaginary number, then unable to analyze stability. $F(\omega,\tau)$, function is independent on τ the parameter $F(\omega,\tau)=0$. The results: We find those ω,τ values which fulfill $F(\omega,\tau)=0$. We ignore negative, complex, and imaginary values of ω. Next is to find those ω,τ values which fulfill $\sin\theta(\tau)=\ldots.$ and $\cos\theta(\tau)=\ldots.$; $\sin(\omega\cdot\tau) = \frac{-P_R\cdot Q_I+P_I\cdot Q_R}{|Q|^2}$; $\cos(\omega\cdot\tau) = -\frac{(P_R\cdot Q_R+P_I\cdot Q_I)}{|Q|^2}|Q|^2 = Q_R^2+Q_I^2$. Finally we plot the stability switch diagram$g(\tau)=\Lambda^{-1}(\tau)=\left(\frac{\partial\mathrm{Re}\lambda}{\partial\tau}\right)_{\lambda=i\omega}$.

$$g(\tau)=\Lambda^{-1}(\tau) = \left(\frac{\partial \mathrm{Re}\lambda}{\partial \tau}\right)_{\lambda=i\omega} = \frac{2\cdot\left\{F_\omega\cdot\left(V+\omega\cdot P^2\right)-F_\tau\cdot\left(U+\tau\cdot P^2\right)\right\}}{F_\tau^2+4\cdot\left(V+\omega\cdot P^2\right)^2}$$

$$\mathrm{sign}[g(\tau)] = \mathrm{sign}\left[\Lambda^{-1}(\tau)\right] = \mathrm{sign}\left[\left(\frac{\partial \mathrm{Re}\lambda}{\partial \tau}\right)_{\lambda=i\omega}\right] = \mathrm{sign}\left[\frac{2\cdot\left\{F_\omega\cdot\left(V+\omega\cdot P^2\right)-F_\tau\cdot\left(U+\tau\cdot P^2\right)\right\}}{F_\tau^2+4\cdot\left(V+\omega\cdot P^2\right)^2}\right]$$

$$F_\tau^2+4\cdot\left(V+\omega\cdot P^2\right)^2>0 \Rightarrow \mathrm{sign}\left[\Lambda^{-1}(\tau)\right] = \mathrm{sign}\left[F_\omega\cdot\left(V+\omega\cdot P^2\right)-F_\tau\cdot\left(U+\tau\cdot P^2\right)\right]$$

$$\mathrm{sign}\left[\Lambda^{-1}(\tau)\right] = \mathrm{sign}\left\{[F_\omega]\cdot\left[\left(V+\omega\cdot P^2\right)-\frac{F_\tau}{F_\omega}\cdot\left(U+\tau\cdot P^2\right)\right]\right\}$$

$$\omega_\tau = -\frac{F_\tau}{F_\omega};\ \omega_\tau = \left(\frac{\partial\omega}{\partial\tau}\right)^{-1} = -\frac{\partial F/\partial\omega}{\partial F/\partial\tau}$$

Table 8.5 Asymmetric dual quantum well lasers and dual-wavelength Lasing system sign of $\mathrm{sign}\left[\Lambda^{-1}(\tau)\right]$

$\mathrm{sign}[F_\omega]$	$\mathrm{sign}\left[\frac{V+\omega_\tau \cdot U}{P^2}+\omega+\omega_\tau\cdot\tau\right]$	$\mathrm{sign}\left[\Lambda^{-1}(\tau)\right]$
+/−	+/−	+
+/−	−/+	−

$$\mathrm{sign}\left[\Lambda^{-1}(\tau)\right]=\mathrm{sign}\left\{[F_\omega]\cdot\left[P^2\right]\cdot\left[\frac{V+\omega_\tau\cdot U}{P^2}+\omega+\omega_\tau\cdot\tau\right]\right\};\ \mathrm{sign}\left[P^2\right]>0$$

$$\mathrm{sign}\left[\Lambda^{-1}(\tau)\right]=\mathrm{sign}\left\{[F_\omega]\cdot\left[\frac{V+\omega_\tau\cdot U}{P^2}+\omega+\omega_\tau\cdot\tau\right]\right\}$$

$$\mathrm{sign}\left[\Lambda^{-1}(\tau)\right]=\mathrm{sign}[F_\omega]\cdot\mathrm{sign}\left[\frac{V+\omega_\tau\cdot U}{P^2}+\omega+\omega_\tau\cdot\tau\right]$$

$$F_\omega=2\cdot[(P_{R\omega}\cdot P_R+P_{I\omega}\cdot P_I)-(Q_{R\omega}\cdot Q_R+Q_{I\omega}\cdot Q_I)]=\sum_{k=0}^{3}B_{2\cdot k+1}\cdot\omega^{2\cdot k+1}$$

We check the sign of $\Lambda^{-1}(\tau)$ according to the following rule (Table 8.5).

If, $\mathrm{sign}\left[\Lambda^{-1}(\tau)\right]>0$, then the crossing proceeds from (−) to (+) respectively (stable to unstable). If, $\mathrm{sign}\left[\Lambda^{-1}(\tau)\right]<0$, then the crossing proceeds from (+) to (−) respectively (unstable to stable).

8.2 Tm^{3+}-Doped Silica Fibre Lasers Nonlinearity and Stability Analysis Under Parameters Variation

The self-pulsing of the Tm^{3+}-doped silica fiber laser is operated near 2 μm. There are various self-pulsing regimes for a range of pumping rates where the fiber is end-pumped with a high power Nd:YAG laser which operated at 1.319 μm in a linear bidirectional cavity. There are a rich variety of nonlinear phenomena, ranging from self-pulsing to self-Q-switching and to a modulated quasi-cw wave. The Tm^{3+}-doped silica glasses and fiber are prepared from mesoporous silica glass by glass phase-separation technology. The Judd–Ofelt parameters and radiation properties of Tm^{3+}-doped silica glasses are evaluated according to the recorded absorption spectrum. Tm^{3+}-doped large-core double-cladding fiber with a core NA of 0.08 and fabricated by the rod-in-tube method. The fiber has a Tm^{3+} concentration of 10,300 ppm by weight and excellent homogeneity in the fiber core. There is a phenomenon of pair-induced quenching (PIQ) which is compatible with excited-state absorption (ESA). There is influence of thulium concentration on ion-pair production. In highly doped rare-earth glasses the ions can end up close to each other that, rather than emitting and absorption energy from the core glasses, they start to pass energy between themselves; this effectively removes them from externally observed energy transfer processes and is almost as if they were not in the glass-reducing the effective doping concentration. This effect can occur when two ions are co-located, and is known as pair-induced quenching, or PIQ. In fact in low-concentration erbium-doped silicate fibers (below 1000 p.p.m) a residual absorption at 980 nm cannot be saturated. The usual models for up conversion of Er/sup 3+/cannot match this behavior nor explain a fluorescence lifetime independent of pump power and erbium concentration. The Excited-state Absorption (ESA) is the absorption of light by ions or atoms in an excited electronic state, rather than in the electronic ground state. The ESA is a common problem for broadband gain media such as transition-metal-doped crystal, but less so for rare-earth-doped crystals with their relatively narrow-bandwidth transitions. ESA is more usual for laser ions with multiple electronic levels, such as erbium or thulium, whereas it is not possible for ytterbium. Rare-earth-doped optical fiber lasers are characterized by their potential applications as optical sources or amplifiers and in fields of telecommunication networks, sensing, spectroscopy, medicine and LIDAR. The rate-earth-doped fiber lasers have a large variety of dynamical behaviors, static and dynamic polarization effects, antiphase and chaotic dynamics under autonomous or pump modulation conditions. They have a highly longitudinal multimode operation due to the large inhomogeneous linewidth, fiber lasers exhibit co-operative effects between longitudinal modes; and we get a simple modeling of the dynamics by low dimensional equations [4]. The conjunction of the inherent nonlinear character of both the optical fiber and the light amplification process makes it suited for analysis of nonlinear dynamics in optical systems. Due to the optical-guiding characteristics of their amplifying medium, fiber lasers have a cavity length of the order of tens of meters, orders of magnitude higher than in most other lasers. It ensures that a large number of longitudinal modes experience gain and coexist inside the cavity, coupled through gain sharing. The fiber laser usually operates in a strongly longitudinal multimode regime. The dynamics of multimode lasers include antiphase behavior and self-pulsing organized collective oscillation. There is both antiphase dynamics in self-pulsing, and a quasi-periodic route to chaos in $Er^{3\pm}$ doped fiber laser operating simultaneously at 1.536 and 1.55 μm wave-lengths. We use a CW state for high pumping rates and the system becomes T-periodic, 2T-periodic, 3T-periodic and chaotic for decreasing pumping ratio, T is represented the fundamental periodicity of the pulses, independently of their magnitude. There is a stable self-pulsing in a linear cavity Er^{3+}-doped fiber laser at 1550 nm with a unidirectional pump from a 980 nm laser diode for continuous-wave pumping. When the pump laser is externally chopped at low frequencies, which is typically in 20–200 Hz. The dynamics of a high power Yb^{3+}-doped double-clad fiber laser is done in a various optical configurations, operating in 1.08 μm wavelength. The fiber is side-pumped with a high power laser diode using the V-groove technique. Different self-pulsing regimes exist and attributed to third-order nonlinear effects. There is also an influence of the cavity losses on the dynamical behavior of our system. The system behavior typical guidelines:

1. Self-pulsing behavior in the fiber laser is due to the presence of ion-pairs or clusters.
2. Bidirectional propagation in "high-loss cavity" and Brillouin-scattering effects in the fiber laser.
3. Ion-pairs (ion clusters) are responsible for the performances of both heavily doped erbium-fiber amplifiers or super fluorescent fiber sources and self-pulsing instability in erbium-doped fiber lasers (EDFLs) at $\lambda = 1.55\,\mu m$.
4. For low ion-pair concentrations ($x \leq 5\%$) the EDFL operates continuously, where x is the proportion of ion-pairs in the doped fiber, whereas for high ion-pair concentration ($x > 5\%$) the laser is self-pulsing.

The physical mechanism of saturable absorption is a quenching effect which it is accrued between two neighboring ions in the $^4I_{13/2}$ state of Er^{3+}: one ion transfers its energy to the other, producing one up-converted $^4I_{9/2}$ ion and one ground-state ion. The characteristic time is associated with the process is some μsec, and the up-converted ion quickly decays to the $^4I_{13/2}$ state. We get the loss of one excited erbium ion. The quenching effect in negligible for low ion-pair concentrations which allow CW operation of the EDFL. For high ion-pair concentrations an efficient saturable absorption effect which occurs leading to self-pulsing behavior. We get output relaxation oscillations of a high power free-running diode-pumped Tm^{3+}-doped silica fiber laser. The laser operates on the $^3H_4 \rightarrow {}^3H_6$ quasi-three-level transition and operates efficiency from 1.9 to 2.0 μm. By evaluation of the $^3F_4 \rightarrow {}^3H_4$ fluorescence we get that after 1.98 μm pumping of a separate laser, the $({}^3H_4, {}^3H_4) \rightarrow ({}^3H_6, {}^3F_4)$ up conversion process is significant in heavily Tm^{3+}-doped silica fibers. The process causes the saturable absorption which is indicated by the presence of the relaxation oscillations. There is a model that describes the ion-pair dynamics, relevant to the Tm^{3+}-doped silica system. For large emission-to-absorption cross-section ratios, which are relevant to Tm^{3+}-doped silica and for pump rates for which stable output is predicted, the oscillations are weakly damped before the steady state is reached. The emission behavior of CW diode pumped neodymium doped double clad fiber lasers is with and without a semiconductor saturable absorber mirror (SESAM). Without the SESAM the CW-pumped multi-longitudinal mode lasers start oscillating in a pulsed fashion. By maximizing the emission rate, the lasing system favors pulsed emission (self-pulsing) over CW emission. The simultaneous oscillation of a large number of modes in homogeneously broadened laser is not alone due to spatial hole burning but due to spontaneous mode-locking spanning a certain frequency range regardless of the number of modes contained in this range. The SESAM modulation's self-pulsing goes over into a self-locked mode of emission. The conditions to achieve this type of emission behavior are inferred from the self-pulsing state. In the onset dynamics in erbium-doped fiber lasers, the lasing light have a build-up time when the pump is on, and it existed in the EDFL whether the laser is CW or self-pulsing. Self-pulsing is a transient phenomenon in CW lasers and it takes place at the beginning of laser action. As the pump is switched on, the gain in the active medium rises and exceeds the steady-state value. There is a mechanism that the number of photons in the cavity increases, depleting the gain below the steady-state value. After several strong peaks, the amplitude of pulsation reduces, and the system behaves as a linear oscillator with damping. Then the pulsation decays; this is the beginning of the CW operation. The build-up time is from 1 ms to more than 10 ms. The onset behavior is characterized by delayed time, the frequency of the relaxation oscillations, and the pump power profile in the resonator. It determined the loss and gain coefficients of the laser. We target on a passively Q-switched fiber laser system with erbium-doped fiber (EDF) and samarium-doped. A linear stability analysis gives an unstable steady-state region for an erbium-doped fiber of high gain coefficient and a samarium-doped fiber of moderate absorption. The ubiquitous erbium-doped fiber amplifier (EDFA) is typically doped with rare earth elements, which are erbium and ytterbium. The obtaining of effective signal amplification with those rare-earth ions requires advanced materials. The heavy-metal and alkali/alkaline earth elements are lead, bismuth, gallium, lithium, potassium, and barium in an oxide glass doped with trivalent samarium (Sm) rate-earth ions. There are an optical properties of trivalent samarium doped silica glass fiber which this material has a narrow fluorescence of 2.2 nm f.w.h.m at a wavelength of 650 nm. The performance of the laser in continuous, Q-switched and self-mode-locked operation is very important to the operation of the laser. In the unstable steady-state region of erbium-doped fiber and a samarium-doped fiber, a self-pulsing appears and giant pulses of good Q-switched light is obtained with increasing ratio of the cross section of samarium-doped fiber to that of erbium-doped fiber. There is a bi-stable behavior between two distinct periodic orbits in a self-pulsing dual-wavelength erbium-doped fiber laser which operated simultaneously at 1.536 and 1.548 μm. The interval between the two wavelengths is around 12 nm and be varied by adjustment of a Bragg grating. When stable output intensity is required, there is a need to stabilize the laser. The laser stabilization is achieved with a Bragg grating or in a unidirectional ring cavity. The unidirectional optical isolator allows the suppression of Brillouin back scattering. A self-Q-switched $T_m{}^{3+}$-doped fiber lasers is capable of generating powerful nanosecond optical pulses at wavelengths around 2 μm. Our system includes a 2 μm $T_m{}^{3+}$-doped silica fiber laser, pumped by a CW-Nd:YAG laser at a wavelength of 1.319 μm. It is characterized by parameters such as the input and the output powers from which the laser threshold, slop and launch efficiency are defined for fibers of various lengths. The $T_m{}^{3+}$-doped silica fiber has a $T_m{}^{3+}$ concentration of 1.1 wt% (7000 ppm), a core diameter of 17 μm, a numerical aperture (NA) of 0.25 and an optimized length of 3.1 m for maximum power which is obtained in cw mode of operation. The $T_m^{3+}-{}^3H_4$ upper laser level has lifetime of $\approx$ 500 μs. The fiber ends are cleaved with a vibrating diamond blade to ensure flatness and incidence at right angles for minimization of losses. The source for the fiber laser is a high power continuous wave 1.319 μm Nd:YAG laser, operating in the fundamental transverse TEM_{00} mode at the maximum output. This laser is pumped by a (CW) high-pressure krypton arc lamp. A cooling system avoids instabilities and thermal problems (circulating water in the crystal chamber). A warm-up time establish the stable laser conditions. The pump irradiance is varied by the use of a plane polarizer, while keeping the lamp current of the Nd:YAG laser constant, then the current variations not change the pump beam characteristics but change the launch efficiency. The pump light is focused into the fiber with a NA IR microscope objective with calculated 70% transmittance at 2 μm wavelength, and about 85% transmittance at the 1.319 μm wavelength. The cleaved ends of the fiber are located on x-y-z differential micrometer translation stages. It allowed fine and coarse alignment. The pump light is propagated only inside the core of the fiber (assumption), because only light launched into the cladding is absorbed in the outer jacket layer of the fiber. The pump beam misalignment affects the fiber laser performance and the absorption characteristics of the fiber. There are x-y-z micrometer translation stages which continuously adjusted, in order to obtain maximum fluorescence at 2 μm, and thus ensure the maximum amount of pump power launched into the core. The maximum fluorescence is achieved by observation of the blue emission from the fiber due to pump excited-state absorption (ESA). The blue up-conversion can become weaker because the stimulated emission process reduces pump exited-state absorption (PESA). For high efficiency use of pump power we choose a mirror with high reflectivity (>90%) at the fiber laser wavelength and it butted onto the output end of the fiber to provide feedback. Another mirror which characterize by a highly reflective at the fiber laser wavelength and highly transmissivity at the pump wavelength (≈80%) is placed before the input of the fiber. It formed part of the fiber cavity arrangement to guide the fiber laser radiation into a 300 lines/mm grating mono chromatic. The mono chromatic output is detected by an InAs photodiode which is cooled by liquid nitrogen and has a typical response time <10 ns. A Ge filter with (93%) transmittance at the fiber laser wavelength is used to block the pump radiation, then the beam is send through the mono-chromatic (≈ 2 μm wavelength beam). For laser oscillation observation and self-pulsing at specific wavelength the InAs photodiode is linked to digital oscilloscope. The coincidence between the pump energy and the energy between $^3H_4$ and $^3F_4$ levels in Tm gives rise to undesirable pump ESA. ESA becomes weaker under lasing condition. Very important function is the fiber laser output power versus input pump power. Tm^{3+}-doped silica fiber laser model based on pair-induced quenching: The model simulated the pump transmittance and the laser dynamics. The model is based on the time evolution of the Tm^{3+} ion excited-state populations. For typical fiber laser conditions, most of the pump input light is absorbed and it efficiency activated ion-pairs. A quenching affect is occurred between two excited neighboring ions (ion-pair). This effect results in one up converted ion, which relaxes to the $^3H_4$ level and one ground-state ion. Then the energy transfer leads to the loss of one potential simulated photon [4]. The quenching effect is a saturable absorber or an additional dynamical loss mechanism. Energy transfer lifetime is of the order of a few micro-seconds. If the ion-pairs contain two ions in the excited Meta stable $^3H_4$ level then we get efficient absorption mechanism and the excited ions are sufficiently dense for the occurrence of the quenching effect. The energy levels of the Tm^{3+} ion is involved in laser action when pumped with 1.319 μm light (Fig. 8.2).

It happened for specific laser source with pump power [watts] above a lower limit value. The quenching effect between two excited neighboring ions (ion-pair) in Tm^{3+} shows that the pump light takes ions from ground state $^3H_6$ to state $^3H_5$ (lifetime ≈ 0.007 μs). There is a fast decay into $^3H_4$ state, the upper lasing level. State $^3H_4$ is metastable (lifetime $\tau_{31} \approx 334.7\,\mu s$). We assume that in Er-ion-pair effect the

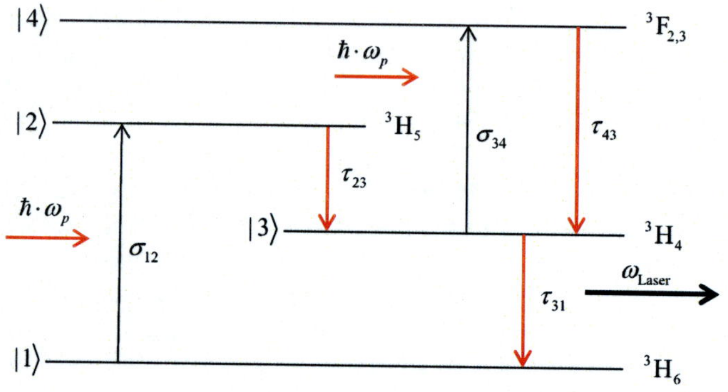

Fig. 8.2 Energy levels of the Tm^{3+} ion for pumping with 1.319 μm light

ratio of emission to absorption cross section (γ) is equal to one ($\gamma = 1$). For $T_{m\text{-ions}}$ is in wavelength range 1.99 μm, $\gamma \approx 7$. The $T_{m\text{-ion-pair}}$ effect on laser dynamics is depending on the emission to absorption ratio γ. The Er-ion-pair interaction model defines the possibility that the $T_{m\text{-ions}}$ are separated into two groups (isolated ions and ion-pairs). The isolated-ions model is described by a two-level system consisting of the ground and upper laser level population densities N_1, and N_2, respectively (ESA in neglected under lasing condition). T_m-ion-pairs model is described by a three-level system, and characterize by ground state (no ion excited), one-ion excited state and ion-pair excited state (both ions excited). The population densities of the three ion-pair states are N_{11}, N_{12}, and N_{22}, respectively. The ion-pair fraction in the total ion concentration N_0 is x. We can define that $\sum_{k=1}^{2} N_k = (1 - 2 \cdot x) \cdot N_0$; $N_{11} + N_{12} + N_{22} = x \cdot N_0$, then $N_0 = \frac{1}{(1-2 \cdot x)} \cdot \sum_{k=1}^{2} N_k$; $N_{11} + N_{12} + N_{22} = x \cdot \left[\frac{1}{(1-2 \cdot x)} \cdot \sum_{k=1}^{2} N_k\right]$. The cavity photon density is $C_p(t)$ which described the laser dynamic behavior. We can characterize our Tm^{3+}-doped silica fiber laser system by four non-linear differential equations:

$$\frac{dn_2}{dt} = R_p \cdot (1 - n_2) - C_p \cdot r_c \cdot \left[(\gamma + 1) \cdot n_2 - 1\right] - \frac{n_2}{\tau_2}$$

$$\frac{dn_{11}}{dt} = -R_p \cdot n_{11} + C_p \cdot \alpha \cdot r_c \cdot \left[\gamma - (\gamma + 1) \cdot n_{11} - \gamma \cdot n_{11}\right] + \frac{(1 - n_{11} - n_{22})}{\tau_2}$$

$$\frac{dn_{22}}{dt} = R_p \cdot (1 - n_{11} - n_{22}) + C_p \cdot \alpha \cdot r_c \times \left[1 - n_{11} - (\gamma + 1) \cdot n_{22}\right] - \frac{n_{22}}{\tau_{22}}$$

$$\frac{dC_p}{dt} = C_p \cdot r_c \cdot N_0 \cdot \{(1 - 2 \cdot x) \cdot \left[(\gamma + 1) \cdot n_2 - 1\right] + \alpha \cdot x \cdot (\gamma - 1 - \gamma \cdot n_{11} + n_{22})\} - \frac{C_p}{\tau_c}$$

Table 8.6 Tm^{3+}-doped silica fiber laser system's parameters

Parameter	
P_p	Power pump
A_{eff}	Effective mode field area at pump wavelength
$h \cdot \nu_p$	Pump photon energy
σ_p	Absorption cross section at pump
σ_a	Absorption cross section at lasing wavelength
$\gamma = \frac{\sigma_e}{\sigma_a}$	The ratio of emission to absorption cross section at lasing wavelength
c	Light velocity in vacuum
n	Refractive index of the laser medium
τ_2	Lifetime of isolated ion upper level
τ_{22}	Lifetime of ion-pair excited state
τ_c	Lifetime of photons in the cavity

$$\gamma = \frac{\sigma_e}{\sigma_a};\ R_p = \frac{P_p}{h \cdot \nu_p \cdot A_{eff}} \cdot \sigma_p;\ r_c = \frac{c}{n} \cdot \sigma_a$$

The normalized ion densities are $n_i = \frac{N_i}{(1-2 \cdot x) \cdot N_0}\ \forall i = 1, 2;\ n_{ij} = \frac{N_{ij}}{x \cdot N_0}\ \forall ij = 11, 12, 22$.

The parameters of the system are described in the next table (Table 8.6).

We take the following system assumptions:

1. The lifetime of non-ion excited state is equal to τ_2.
2. Due to the interaction in ion-pairs, the emission and absorption cross sections of paired ion are usually different from isolated ions.
3. The paired ions have the same $\gamma = \frac{\sigma_e}{\sigma_a}$ value as that of isolated ions.
4. The absorption cross section of paired ions is $\alpha \cdot \sigma_a$.
5. Ion-pairs is not sufficiently pumped and act as additional absorber [the lifetime of the ion-pair excited level τ_{22} is very short (few μsec)].
6. Additional absorber can make the laser system unstable depending on whether the steady-state solution is stable or not.

At steady state (equilibrium), $\frac{dn_2}{dt} = 0;\ \frac{dn_{11}}{dt} = 0;\ \frac{dn_{22}}{dt} = 0;\ \frac{dC_p}{dt} = 0$.

$$\frac{dn_2}{dt} = 0 \Rightarrow R_p \cdot (1 - n_2^*) - C_p^* \cdot r_c \cdot [(\gamma + 1) \cdot n_2^* - 1] - \frac{n_2^*}{\tau_2} = 0$$

$$C_p^* = \frac{R_p \cdot (1 - n_2^*) - \frac{n_2^*}{\tau_2}}{r_c \cdot [(\gamma + 1) \cdot n_2^* - 1]}$$

$$\frac{dn_{11}}{dt} = 0 \Rightarrow -R_p \cdot n_{11}^* + C_p^* \cdot \alpha \cdot r_c \cdot [\gamma - (\gamma + 1) \cdot n_{11}^* - \gamma \cdot n_{11}^*]$$

$$+\frac{\left(1-n_{11}^{*}-n_{22}^{*}\right)}{\tau_{2}}=0$$

$$-R_{p}\cdot n_{11}^{*}+C_{p}^{*}\cdot\alpha\cdot r_{c}\cdot\left[\gamma-(\gamma+1)\cdot n_{11}^{*}-\gamma\cdot n_{11}^{*}\right]+\frac{\left(1-n_{11}^{*}-n_{22}^{*}\right)}{\tau_{2}}=0$$

$$\frac{dn_{22}}{dt}=0\Rightarrow R_{p}\cdot\left(1-n_{11}^{*}-n_{22}^{*}\right)+C_{p}^{*}\cdot\alpha\cdot r_{c}\times[1-n_{11}^{*}$$
$$-(\gamma+1)\cdot n_{22}^{*}]-\frac{n_{22}^{*}}{\tau_{22}}=0$$

$$\frac{dC_{p}}{dt}=0\Rightarrow C_{p}^{*}\cdot r_{c}\cdot N_{0}\cdot\{(1-2\cdot x)\cdot\left[(\gamma+1)\cdot n_{2}^{*}-1\right]$$
$$+\alpha\cdot x\cdot\left(\gamma-1-\gamma\cdot n_{11}^{*}+n_{22}^{*}\right)\}-\frac{C_{p}^{*}}{\tau_{c}}=0$$

$$C_{p}^{*}\cdot r_{c}\cdot N_{0}\cdot\{(1-2\cdot x)\cdot\left[(\gamma+1)\cdot n_{2}^{*}-1\right]$$
$$+\alpha\cdot x\cdot\left(\gamma-1-\gamma\cdot n_{11}^{*}+n_{22}^{*}\right)\}-\frac{C_{p}^{*}}{\tau_{c}}=0$$

$$C_{p}^{*}\cdot[r_{c}\cdot N_{0}\cdot\{(1-2\cdot x)\cdot\left[(\gamma+1)\cdot n_{2}^{*}-1\right]$$
$$+\alpha\cdot x\cdot\left(\gamma-1-\gamma\cdot n_{11}^{*}+n_{22}^{*}\right)\}-\frac{1}{\tau_{c}}]=0$$

Case A: $C_{p}^{*}=0$

$$R_{p}\cdot\left(1-n_{11}^{*}-n_{22}^{*}\right)-\frac{n_{22}^{*}}{\tau_{22}}=0;-R_{p}\cdot n_{11}^{*}+\frac{\left(1-n_{11}^{*}-n_{22}^{*}\right)}{\tau_{2}}=0$$

$$R_{p}\cdot\left(1-n_{2}^{*}\right)-\frac{n_{2}^{*}}{\tau_{2}}=0\Rightarrow R_{p}-R_{p}\cdot n_{2}^{*}-\frac{n_{2}^{*}}{\tau_{2}}=0$$
$$n_{2}^{*}\cdot\left(R_{p}+\frac{1}{\tau_{2}}\right)=R_{p}\Rightarrow n_{2}^{*}=\frac{R_{p}}{R_{p}+\frac{1}{\tau_{2}}}$$

$$R_{p}\cdot\left(1-n_{11}^{*}-n_{22}^{*}\right)-\frac{n_{22}^{*}}{\tau_{22}}=0\Rightarrow n_{11}^{*}=1-n_{22}^{*}\cdot\frac{1}{R_{p}}\cdot\left(R_{p}+\frac{1}{\tau_{22}}\right)$$
$$n_{11}^{*}=1-n_{22}^{*}\cdot\left(1+\frac{1}{\tau_{22}\cdot R_{p}}\right)$$

$$-R_{p}\cdot n_{11}^{*}+\frac{\left(1-n_{11}^{*}-n_{22}^{*}\right)}{\tau_{2}}=0\Rightarrow-R_{p}\cdot n_{11}^{*}+\frac{\left(1-n_{11}^{*}-n_{22}^{*}\right)}{\tau_{2}}=0$$
$$-R_{p}\cdot n_{11}^{*}+\frac{1}{\tau_{2}}-\frac{n_{11}^{*}}{\tau_{2}}-\frac{n_{22}^{*}}{\tau_{2}}=0$$

$n_{11}^* \cdot \left(R_p + \frac{1}{\tau_2}\right) = \frac{1}{\tau_2} \cdot (1 - n_{22}^*) \Rightarrow n_{11}^* = \frac{\frac{1}{\tau_2}\cdot(1-n_{22}^*)}{R_p+\frac{1}{\tau_2}}$, then we can compare $n_{11}^* = \ldots$

$$1 - n_{22}^* \cdot \left(1 + \frac{1}{\tau_{22} \cdot R_p}\right) = \frac{\frac{1}{\tau_2} \cdot (1 - n_{22}^*)}{R_p + \frac{1}{\tau_2}} \Rightarrow \left(R_p + \frac{1}{\tau_2}\right) - n_{22}^* \cdot \left(1 + \frac{1}{\tau_{22} \cdot R_p}\right) \cdot \left(R_p + \frac{1}{\tau_2}\right) = \frac{1}{\tau_2} \cdot (1 - n_{22}^*)$$

$$n_{22}^* \cdot \left[\frac{1}{\tau_2} - \left(1 + \frac{1}{\tau_{22} \cdot R_p}\right) \cdot \left(R_p + \frac{1}{\tau_2}\right)\right] = \frac{1}{\tau_2} - \left(R_p + \frac{1}{\tau_2}\right)$$

$$n_{22}^* = \frac{\frac{1}{\tau_2} - \left(R_p + \frac{1}{\tau_2}\right)}{\frac{1}{\tau_2} - \left(1 + \frac{1}{\tau_{22} \cdot R_p}\right) \cdot \left(R_p + \frac{1}{\tau_2}\right)}$$

$$n_{11}^* = 1 - n_{22}^* \cdot \left(1 + \frac{1}{\tau_{22} \cdot R_p}\right) = 1 - \left(\frac{\frac{1}{\tau_2} - \left(R_p + \frac{1}{\tau_2}\right)}{\frac{1}{\tau_2} - \left(1 + \frac{1}{\tau_{22} \cdot R_p}\right) \cdot \left(R_p + \frac{1}{\tau_2}\right)}\right) \cdot \left(1 + \frac{1}{\tau_{22} \cdot R_p}\right)$$

$n_{11}^* = 1 - \left(\frac{\left[\frac{1}{\tau_2} - \left(R_p + \frac{1}{\tau_2}\right)\right] \cdot \left(1 + \frac{1}{\tau_{22} \cdot R_p}\right)}{\frac{1}{\tau_2} - \left(1 + \frac{1}{\tau_{22} \cdot R_p}\right) \cdot \left(R_p + \frac{1}{\tau_2}\right)}\right)$. The first system fixed point is

$$E^{(0)} = \left(n_2^{(0)}, n_{11}^{(0)}, n_{22}^{(0)}, C_p^{(0)}\right) = \left(\frac{R_p}{R_p + \frac{1}{\tau_2}}, 1 - \left(\frac{\left[\frac{1}{\tau_2} - \left(R_p + \frac{1}{\tau_2}\right)\right] \cdot \left(1 + \frac{1}{\tau_{22} \cdot R_p}\right)}{\frac{1}{\tau_2} - \left(1 + \frac{1}{\tau_{22} \cdot R_p}\right) \cdot \left(R_p + \frac{1}{\tau_2}\right)}\right), \frac{\frac{1}{\tau_2} - \left(R_p + \frac{1}{\tau_2}\right)}{\frac{1}{\tau_2} - \left(1 + \frac{1}{\tau_{22} \cdot R_p}\right) \cdot \left(R_p + \frac{1}{\tau_2}\right)}, 0\right)$$

Case B: $C_p^* \neq 0$; $r_c \cdot N_0 \cdot \{(1 - 2 \cdot x) \cdot [(\gamma + 1) \cdot n_2^* - 1] + \alpha \cdot x \cdot (\gamma - 1 - \gamma \cdot n_{11}^* + n_{22}^*)\} - \frac{1}{\tau_c} = 0$.

$$n_2^* = \frac{\frac{\frac{1}{r_c \cdot N_0 \cdot \tau_c} - \alpha \cdot x \cdot (\gamma - 1 - \gamma \cdot n_{11}^* + n_{22}^*)}{(1 - 2 \cdot x)} + 1}{\gamma + 1}$$

$$R_p \cdot (1 - n_2^*) - C_p^* \cdot r_c \cdot [(\gamma + 1) \cdot n_2^* - 1] - \frac{n_2^*}{\tau_2} = 0$$

$$n_2^* \cdot \left[\frac{1}{\tau_2} + R_p + C_p^* \cdot r_c \cdot (\gamma + 1)\right] = R_p + C_p^* \cdot r_c; n_2^* = \frac{R_p + C_p^* \cdot r_c}{\frac{1}{\tau_2} + R_p + C_p^* \cdot r_c \cdot (\gamma + 1)}$$

We get three fixed points $n_{11}^{(*)}, n_{22}^{(*)}, C_p^{(*)}$ equations and it is easy to solve it numerically rather than analytically for fixed points coordinates, $E^{(k)} = \left(n_2^{(k)}, n_{11}^{(k)}, n_{22}^{(k)}, C_p^{(k)}\right); k \geq 1; k \in \mathbb{N}$.

1. $-R_p \cdot n_{11}^* + C_p^* \cdot \alpha \cdot r_c \cdot [\gamma - (\gamma + 1) \cdot n_{11}^* - \gamma \cdot n_{11}^*] + \frac{(1 - n_{11}^* - n_{22}^*)}{\tau_2} = 0$
2. $R_p \cdot (1 - n_{11}^* - n_{22}^*) + C_p^* \cdot \alpha \cdot r_c \times [1 - n_{11}^* - (\gamma + 1) \cdot n_{22}^*] - \frac{n_{22}^*}{\tau_{22}} = 0$
3. $\frac{\left(\frac{\frac{1}{r_c \cdot N_0 \cdot \tau_c} - \alpha \cdot x \cdot (\gamma - 1 - \gamma \cdot n_{11}^* + n_{22}^*)}{(1 - 2 \cdot x)} + 1\right)}{\gamma + 1} = \frac{R_p + C_p^* \cdot r_c}{R_p + C_p^* \cdot r_c \cdot (\gamma + 1) + \frac{1}{\tau_2}}$

Stability analysis: The standard local stability analysis about any one of the equilibrium point of the Tm^{3+}-doped silica fiber laser system consists in adding to coordinate $[n_2, n_{11}, n_{22}, C_p]$ arbitrarily small increments of exponential form $[n_2, n_{11}, n_{22}, c_p] \cdot e^{\lambda \cdot t}$ and retaining the first order terms in n_2, n_{11}, n_{22}, C_p. The system of four homogeneous equations leads to a polynomial characteristic equation in the eigenvalues. The polynomial characteristic equations accept by set the below variables and variables derivative with respect to time into Tm^{3+}-doped silica fiber laser system equations. Tm^{3+}-doped silica fiber laser system fixed values with arbitrarily small increments of exponential form $[n_2, n_{11}, n_{22}, c_p] \cdot e^{\lambda \cdot t}$ are: $j = 0$ (first fixed point), $j = 1$ (second fixed point), $j = 2$ (third fixed point), etc. [2, 3].

$$n_2(t) = n_2^{(j)} + n_2 \cdot e^{\lambda \cdot t}; \; n_{11}(t) = n_{11}^{(j)} + n_{11} \cdot e^{\lambda \cdot t}$$
$$n_{22}(t) = n_{22}^{(j)} + n_{22} \cdot e^{\lambda \cdot t}; \; C_p(t) = C_p^{(j)} + c_p \cdot e^{\lambda \cdot t}$$

$$\frac{dn_2(t)}{dt} = n_2 \cdot \lambda \cdot e^{\lambda \cdot t}; \; \frac{dn_{11}(t)}{dt} = n_{11} \cdot \lambda \cdot e^{\lambda \cdot t}$$
$$\frac{dn_{22}(t)}{dt} = n_{22} \cdot \lambda \cdot e^{\lambda \cdot t}; \; \frac{dC_p(t)}{dt} = c_p \cdot \lambda \cdot e^{\lambda \cdot t}$$

We choose these expressions for ourselves $n_2(t), n_{11}(t), n_{22}(t), C_p(t)$ as a small displacement $[n_2, n_{11}, n_{22}, c_p]$ from the Tm^{3+}-doped silica fiber laser system fixed points in time $t = 0$, $n_2(t=0) = n_2^{(j)} + n_2$; $n_{11}(t=0) = n_{11}^{(j)} + n_{11}$; $n_{22}(t=0) = n_{22}^{(j)} + n_{22}$; $C_p(t=0) = C_p^{(j)} + c_p$.

$$\frac{dn_2}{dt} = R_p \cdot (1 - n_2) - C_p \cdot r_c \cdot \left[(\gamma + 1) \cdot n_2 - 1\right] - \frac{n_2}{\tau_2}$$

$$\begin{aligned} n_2 \cdot \lambda \cdot e^{\lambda \cdot t} = {} & R_p \cdot \left(1 - n_2^{(j)} - n_2 \cdot e^{\lambda \cdot t}\right) \\ & - \left(C_p^{(j)} + c_p \cdot e^{\lambda \cdot t}\right) \cdot r_c \cdot \left[(\gamma + 1) \cdot \left(n_2^{(j)} + n_2 \cdot e^{\lambda \cdot t}\right) - 1\right] \\ & - \frac{\left(n_2^{(j)} + n_2 \cdot e^{\lambda \cdot t}\right)}{\tau_2} \end{aligned}$$

$$\begin{aligned} n_2 \cdot \lambda \cdot e^{\lambda \cdot t} = {} & R_p \cdot \left(1 - n_2^{(j)}\right) - R_p \cdot n_2 \cdot e^{\lambda \cdot t} \\ & - \left(C_p^{(j)} + c_p \cdot e^{\lambda \cdot t}\right) \cdot r_c \cdot [\left\{(\gamma + 1) \cdot n_2^{(j)} - 1\right\} \\ & + (\gamma + 1) \cdot n_2 \cdot e^{\lambda \cdot t}] - \frac{n_2^{(j)}}{\tau_2} - \frac{n_2 \cdot e^{\lambda \cdot t}}{\tau_2} \end{aligned}$$

$$\begin{aligned} n_2 \cdot \lambda \cdot e^{\lambda \cdot t} = {} & R_p \cdot \left(1 - n_2^{(j)}\right) - R_p \cdot n_2 \cdot e^{\lambda \cdot t} - r_c \cdot [C_p^{(j)} \cdot \left\{(\gamma + 1) \cdot n_2^{(j)} - 1\right\} \\ & + C_p^{(j)} \cdot (\gamma + 1) \cdot n_2 \cdot e^{\lambda \cdot t} + c_p \cdot \left\{(\gamma + 1) \cdot n_2^{(j)} - 1\right\} \cdot e^{\lambda \cdot t} \\ & + e^{\lambda \cdot t} \cdot (\gamma + 1) \cdot c_p \cdot n_2 \cdot e^{\lambda \cdot t}] - \frac{n_2^{(j)}}{\tau_2} - \frac{n_2 \cdot e^{\lambda \cdot t}}{\tau_2} \end{aligned}$$

Assumption: $c_p \cdot n_2 \approx 0$

$$\begin{aligned} n_2 \cdot \lambda \cdot e^{\lambda \cdot t} = {} & R_p \cdot (1 - n_2^{(j)}) - R_p \cdot n_2 \cdot e^{\lambda \cdot t} - r_c \cdot [C_p^{(j)} \cdot \{(\gamma + 1) \cdot n_2^{(j)} - 1\} \\ & + C_p^{(j)} \cdot (\gamma + 1) \cdot n_2 \cdot e^{\lambda \cdot t} + c_p \cdot \{(\gamma + 1) \cdot n_2^{(j)} \\ & - 1\} \cdot e^{\lambda \cdot t}] - \frac{n_2^{(j)}}{\tau_2} - \frac{n_2 \cdot e^{\lambda \cdot t}}{\tau_2} \end{aligned}$$

$$\begin{aligned} n_2 \cdot \lambda \cdot e^{\lambda \cdot t} = {} & R_p \cdot (1 - n_2^{(j)}) - R_p \cdot n_2 \cdot e^{\lambda \cdot t} - r_c \cdot C_p^{(j)} \cdot \{(\gamma + 1) \cdot n_2^{(j)} - 1\} \\ & - r_c \cdot C_p^{(j)} \cdot (\gamma + 1) \cdot n_2 \cdot e^{\lambda \cdot t} - r_c \cdot c_p \cdot \{(\gamma + 1) \cdot n_2^{(j)} \\ & - 1\} \cdot e^{\lambda \cdot t} - \frac{n_2^{(j)}}{\tau_2} - \frac{n_2 \cdot e^{\lambda \cdot t}}{\tau_2} \end{aligned}$$

$$\begin{aligned} n_2 \cdot \lambda \cdot e^{\lambda \cdot t} = {} & R_p \cdot (1 - n_2^{(j)}) - r_c \cdot C_p^{(j)} \cdot \{(\gamma + 1) \cdot n_2^{(j)} - 1\} \\ & - \frac{n_2^{(j)}}{\tau_2} - R_p \cdot n_2 \cdot e^{\lambda \cdot t} - r_c \cdot C_p^{(j)} \cdot (\gamma + 1) \cdot n_2 \cdot e^{\lambda \cdot t} \end{aligned}$$

$$- r_c \cdot c_p \cdot \{(\gamma + 1) \cdot n_2^{(j)} - 1\} \cdot e^{\lambda \cdot t} - \frac{n_2 \cdot e^{\lambda \cdot t}}{\tau_2}$$

At fixed points:

$$R_p \cdot \left(1 - n_2^{(j)}\right) - r_c \cdot C_p^{(j)} \cdot \left\{(\gamma + 1) \cdot n_2^{(j)} - 1\right\} - \frac{n_2^{(j)}}{\tau_2} = 0$$

$$-n_2 \cdot \lambda - \left[R_p + r_c \cdot C_p^{(j)} \cdot (\gamma + 1) + \frac{1}{\tau_2}\right] \cdot n_2 - r_c \cdot \left[(\gamma + 1) \cdot n_2^{(j)} - 1\right] \cdot c_p = 0$$

We define for simplicity the global parameters:

$$\Xi_1 = R_p + r_c \cdot C_p^{(j)} \cdot (\gamma + 1) + \frac{1}{\tau_2}$$

$$\Xi_2 = r_c \cdot \left[(\gamma + 1) \cdot n_2^{(j)} - 1\right]; -n_2 \cdot \lambda - \Xi_1 \cdot n_2 - \Xi_2 \cdot c_p = 0$$

$$\frac{dn_{11}}{dt} = -R_p \cdot n_{11} + C_p \cdot \alpha \cdot r_c \cdot [\gamma - (\gamma + 1) \cdot n_{11} - \gamma \cdot n_{11}] + \frac{(1 - n_{11} - n_{22})}{\tau_2}$$

$$\begin{aligned} n_{11} \cdot \lambda \cdot e^{\lambda \cdot t} = & -R_p \cdot (n_{11}^{(j)} + n_{11} \cdot e^{\lambda \cdot t}) + [C_p^{(j)} + c_p \cdot e^{\lambda \cdot t}] \cdot \alpha \cdot r_c \cdot [\gamma \\ & - (\gamma + 1) \cdot (n_{11}^{(j)} + n_{11} \cdot e^{\lambda \cdot t}) - \gamma \cdot (n_{11}^{(j)} + n_{11} \cdot e^{\lambda \cdot t})] \\ & + \frac{(1 - [n_{11}^{(j)} + n_{11} \cdot e^{\lambda \cdot t}] - [n_{22}^{(j)} + n_{22} \cdot e^{\lambda \cdot t}])}{\tau_2} \end{aligned}$$

$$\begin{aligned} n_{11} \cdot \lambda \cdot e^{\lambda \cdot t} = & -R_p \cdot n_{11}^{(j)} - R_p \cdot n_{11} \cdot e^{\lambda \cdot t} + [C_p^{(j)} + c_p \cdot e^{\lambda \cdot t}] \cdot \alpha \cdot r_c \cdot [\{\gamma \\ & - (\gamma + 1) \cdot n_{11}^{(j)} - \gamma \cdot n_{11}^{(j)}\} - (2 \cdot \gamma + 1) \cdot n_{11} \cdot e^{\lambda \cdot t}] \\ & + \frac{(1 - n_{11}^{(j)} - n_{22}^{(j)})}{\tau_2} - \frac{(n_{11} + n_{22})}{\tau_2} \cdot e^{\lambda \cdot t} \end{aligned}$$

$$\begin{aligned} n_{11} \cdot \lambda \cdot e^{\lambda \cdot t} = & -R_p \cdot n_{11}^{(j)} - R_p \cdot n_{11} \cdot e^{\lambda \cdot t} + \alpha \cdot r_c \cdot C_p^{(j)} \cdot [\gamma \\ & - (\gamma + 1) \cdot n_{11}^{(j)} - \gamma \cdot n_{11}^{(j)}] - \alpha \cdot r_c \cdot C_p^{(j)} \cdot (2 \cdot \gamma + 1) \cdot n_{11} \cdot e^{\lambda \cdot t} \\ & + \alpha \cdot r_c \cdot c_p \cdot e^{\lambda \cdot t} \cdot [\gamma - (\gamma + 1) \cdot n_{11}^{(j)} - \gamma \cdot n_{11}^{(j)}] \\ & - \alpha \cdot r_c \cdot e^{\lambda \cdot t} \cdot (2 \cdot \gamma + 1) \cdot c_p \cdot n_{11} \cdot e^{\lambda \cdot t} \\ & + \frac{(1 - n_{11}^{(j)} - n_{22}^{(j)})}{\tau_2} - \frac{(n_{11} + n_{22})}{\tau_2} \cdot e^{\lambda \cdot t} \end{aligned}$$

Assumption: $c_p \cdot n_{11} \approx 0$

$$n_{11} \cdot \lambda \cdot e^{\lambda \cdot t} = -R_p \cdot n_{11}^{(j)} - R_p \cdot n_{11} \cdot e^{\lambda \cdot t} + \alpha \cdot r_c \cdot C_p^{(j)} \cdot [\gamma$$

$$- (\gamma + 1) \cdot n_{11}^{(j)} - \gamma \cdot n_{11}^{(j)}] - \alpha \cdot r_c \cdot C_p^{(j)} \cdot (2 \cdot \gamma + 1) \cdot n_{11} \cdot e^{\lambda \cdot t}$$
$$+ \alpha \cdot r_c \cdot c_p \cdot e^{\lambda \cdot t} \cdot [\gamma - (\gamma + 1) \cdot n_{11}^{(j)} - \gamma \cdot n_{11}^{(j)}]$$
$$+ \frac{(1 - n_{11}^{(j)} - n_{22}^{(j)})}{\tau_2} - \frac{(n_{11} + n_{22})}{\tau_2} \cdot e^{\lambda \cdot t}$$

$$n_{11} \cdot \lambda \cdot e^{\lambda \cdot t} = -R_p \cdot n_{11}^{(j)} + \alpha \cdot r_c \cdot C_p^{(j)} \cdot [\gamma - (\gamma + 1) \cdot n_{11}^{(j)}$$
$$- \gamma \cdot n_{11}^{(j)}] + \frac{(1 - n_{11}^{(j)} - n_{22}^{(j)})}{\tau_2} - R_p \cdot n_{11} \cdot e^{\lambda \cdot t}$$
$$- \alpha \cdot r_c \cdot C_p^{(j)} \cdot (2 \cdot \gamma + 1) \cdot n_{11} \cdot e^{\lambda \cdot t} + \alpha \cdot r_c \cdot c_p \cdot e^{\lambda \cdot t} \cdot [\gamma$$
$$- (\gamma + 1) \cdot n_{11}^{(j)} - \gamma \cdot n_{11}^{(j)}] - \frac{(n_{11} + n_{22})}{\tau_2} \cdot e^{\lambda \cdot t}$$

At fixed points:

$$-R_p \cdot n_{11}^{(j)} + \alpha \cdot r_c \cdot C_p^{(j)} \cdot \left[\gamma - (\gamma + 1) \cdot n_{11}^{(j)} - \gamma \cdot n_{11}^{(j)}\right] + \frac{\left(1 - n_{11}^{(j)} - n_{22}^{(j)}\right)}{\tau_2} = 0$$

$$- n_{11} \cdot \lambda - \left[R_p + \alpha \cdot r_c \cdot C_p^{(j)} \cdot (2 \cdot \gamma + 1) + \frac{1}{\tau_2}\right] \cdot n_{11} - \frac{n_{22}}{\tau_2}$$
$$+ \alpha \cdot r_c \cdot \left[\gamma - (\gamma + 1) \cdot n_{11}^{(j)} - \gamma \cdot n_{11}^{(j)}\right] \cdot c_p = 0$$

We define global parameters for simplicity:

$$\Xi_3 = R_p + \alpha \cdot r_c \cdot C_p^{(j)} \cdot (2 \cdot \gamma + 1) + \frac{1}{\tau_2}$$

$$\Xi_4 = \alpha \cdot r_c \cdot \left[\gamma - (\gamma + 1) \cdot n_{11}^{(j)} - \gamma \cdot n_{11}^{(j)}\right]$$
$$- n_{11} \cdot \lambda - \Xi_3 \cdot n_{11} - \frac{n_{22}}{\tau_2} + \Xi_4 \cdot c_p = 0$$

$$\frac{dn_{22}}{dt} = R_p \cdot (1 - n_{11} - n_{22}) + C_p \cdot \alpha \cdot r_c \times \left[1 - n_{11} - (\gamma + 1) \cdot n_{22}\right] - \frac{n_{22}}{\tau_{22}}$$

$$n_{22} \cdot \lambda \cdot e^{\lambda \cdot t} = R_p \cdot \left(1 - n_{11}^{(j)} - n_{11} \cdot e^{\lambda \cdot t} - n_{22}^{(j)} - n_{22} \cdot e^{\lambda \cdot t}\right)$$
$$+ \left(C_p^{(j)} + c_p \cdot e^{\lambda \cdot t}\right) \cdot \alpha \cdot r_c \times [1 - n_{11}^{(j)} - n_{11} \cdot e^{\lambda \cdot t}$$
$$- (\gamma + 1) \cdot \left(n_{22}^{(j)} + n_{22} \cdot e^{\lambda \cdot t}\right)] - \frac{\left(n_{22}^{(j)} + n_{22} \cdot e^{\lambda \cdot t}\right)}{\tau_{22}}$$

$$n_{22} \cdot \lambda \cdot e^{\lambda \cdot t} = R_p \cdot \left[1 - n_{11}^{(j)} - n_{22}^{(j)}\right] - R_p \cdot [n_{11} + n_{22}] \cdot e^{\lambda \cdot t}$$

$$+ \left(C_p^{(j)} + c_p \cdot e^{\lambda \cdot t}\right) \cdot \alpha \cdot r_c \times [1 - n_{11}^{(j)} - (\gamma + 1) \cdot n_{22}^{(j)}$$
$$- \{n_{11} + (\gamma + 1) \cdot n_{22}\} \cdot e^{\lambda \cdot t}] - \frac{n_{22}^{(j)}}{\tau_{22}} - \frac{n_{22} \cdot e^{\lambda \cdot t}}{\tau_{22}}$$

$$n_{22} \cdot \lambda \cdot e^{\lambda \cdot t} = R_p \cdot [1 - n_{11}^{(j)} - n_{22}^{(j)}] - R_p \cdot [n_{11} + n_{22}] \cdot e^{\lambda \cdot t}$$
$$+ \alpha \cdot r_c \times \{C_p^{(j)} \cdot \{1 - n_{11}^{(j)} - (\gamma + 1) \cdot n_{22}^{(j)}\}$$
$$- C_p^{(j)} \cdot \{n_{11} + (\gamma + 1) \cdot n_{22}\} \cdot e^{\lambda \cdot t}$$
$$+ c_p \cdot e^{\lambda \cdot t} \cdot \{1 - n_{11}^{(j)} - (\gamma + 1) \cdot n_{22}^{(j)}\}$$
$$- e^{\lambda \cdot t} \cdot \{c_p \cdot n_{11} + (\gamma + 1) \cdot c_p \cdot n_{22}\} \cdot e^{\lambda \cdot t}\} - \frac{n_{22}^{(j)}}{\tau_{22}} - \frac{n_{22} \cdot e^{\lambda \cdot t}}{\tau_{22}}$$

Assumption: $c_p \cdot n_{11} \approx 0; c_p \cdot n_{22} \approx 0$

$$n_{22} \cdot \lambda \cdot e^{\lambda \cdot t} = R_p \cdot [1 - n_{11}^{(j)} - n_{22}^{(j)}] - R_p \cdot [n_{11} + n_{22}] \cdot e^{\lambda \cdot t}$$
$$+ \alpha \cdot r_c \times \{C_p^{(j)} \cdot \{1 - n_{11}^{(j)} - (\gamma + 1) \cdot n_{22}^{(j)}\}$$
$$- C_p^{(j)} \cdot \{n_{11} + (\gamma + 1) \cdot n_{22}\} \cdot e^{\lambda \cdot t}$$
$$+ c_p \cdot e^{\lambda \cdot t} \cdot \{1 - n_{11}^{(j)} - (\gamma + 1) \cdot n_{22}^{(j)}\}\} - \frac{n_{22}^{(j)}}{\tau_{22}} - \frac{n_{22} \cdot e^{\lambda \cdot t}}{\tau_{22}}$$

$$n_{22} \cdot \lambda \cdot e^{\lambda \cdot t} = R_p \cdot [1 - n_{11}^{(j)} - n_{22}^{(j)}] + \alpha \cdot r_c \times C_p^{(j)} \cdot [1 - n_{11}^{(j)}$$
$$- (\gamma + 1) \cdot n_{22}^{(j)}] - \frac{n_{22}^{(j)}}{\tau_{22}} - \alpha \cdot r_c \times C_p^{(j)} \cdot [n_{11} + (\gamma + 1) \cdot n_{22}] \cdot e^{\lambda \cdot t}$$
$$+ \alpha \cdot r_c \times c_p \cdot e^{\lambda \cdot t} \cdot [1 - n_{11}^{(j)} - (\gamma + 1) \cdot n_{22}^{(j)}] - \frac{n_{22} \cdot e^{\lambda \cdot t}}{\tau_{22}}$$
$$- R_p \cdot [n_{11} + n_{22}] \cdot e^{\lambda \cdot t}$$

At fixed points:

$$R_p \cdot \left[1 - n_{11}^{(j)} - n_{22}^{(j)}\right] + \alpha \cdot r_c \times C_p^{(j)} \cdot \left[1 - n_{11}^{(j)} - (\gamma + 1) \cdot n_{22}^{(j)}\right] - \frac{n_{22}^{(j)}}{\tau_{22}} = 0$$

$$- n_{22} \cdot \lambda \cdot e^{\lambda \cdot t} - \alpha \cdot r_c \times C_p^{(j)} \cdot \left[n_{11} + (\gamma + 1) \cdot n_{22}\right] \cdot e^{\lambda \cdot t}$$
$$+ \alpha \cdot r_c \times c_p \cdot e^{\lambda \cdot t} \cdot \left[1 - n_{11}^{(j)} - (\gamma + 1) \cdot n_{22}^{(j)}\right]$$
$$- \frac{n_{22} \cdot e^{\lambda \cdot t}}{\tau_{22}} - R_p \cdot [n_{11} + n_{22}] \cdot e^{\lambda \cdot t} = 0$$

$$- \left[\alpha \cdot r_c \cdot C_p^{(j)} + R_p\right] \cdot n_{11} - \left[\alpha \cdot r_c \cdot C_p^{(j)} \times (\gamma + 1)\right.$$

$$+\frac{1}{\tau_{22}}+R_p\Bigg]\cdot n_{22}-n_{22}\cdot\lambda+\alpha\cdot r_c\times[1-n_{11}^{(j)}$$
$$-(\gamma+1)\cdot n_{22}^{(j)}]\cdot c_p=0$$

We define global parameters for simplicity:

$$\Xi_5=-\left[\alpha\cdot r_c\cdot C_p^{(j)}+R_p\right]$$

$$\Xi_6=\alpha\cdot r_c\cdot C_p^{(j)}\times(\gamma+1)+\frac{1}{\tau_{22}}+R_p$$
$$\Xi_7=\alpha\cdot r_c\times\left[1-n_{11}^{(j)}-(\gamma+1)\cdot n_{22}^{(j)}\right]$$

$$\Xi_5\cdot n_{11}-\Xi_6\cdot n_{22}-n_{22}\cdot\lambda+\Xi_7\cdot c_p=0$$

$$\frac{dC_p}{dt}=C_p\cdot r_c\cdot N_0\cdot\{(1-2\cdot x)\cdot\left[(\gamma+1)\cdot n_2-1\right]$$
$$+\alpha\cdot x\cdot(\gamma-1-\gamma\cdot n_{11}+n_{22})\}-\frac{C_p}{\tau_c}$$

$$c_p\cdot\lambda\cdot e^{\lambda\cdot t}=(C_p^{(j)}+c_p\cdot e^{\lambda\cdot t})\cdot r_c\cdot N_0\cdot\{(1-2\cdot x)\cdot[(\gamma+1)\cdot(n_2^{(j)}$$
$$+n_2\cdot e^{\lambda\cdot t})-1]+\alpha\cdot x\cdot(\gamma-1-\gamma\cdot[n_{11}^{(j)}+n_{11}\cdot e^{\lambda\cdot t}]$$
$$+n_{22}^{(j)}+n_{22}\cdot e^{\lambda\cdot t})\}-\frac{(C_p^{(j)}+c_p\cdot e^{\lambda\cdot t})}{\tau_c}$$

$$c_p\cdot\lambda\cdot e^{\lambda\cdot t}=(C_p^{(j)}+c_p\cdot e^{\lambda\cdot t})\cdot r_c\cdot N_0\cdot\{(1-2\cdot x)\cdot[(\gamma+1)\cdot n_2^{(j)}$$
$$-1+(\gamma+1)\cdot n_2\cdot e^{\lambda\cdot t}+\alpha\cdot x\cdot[\gamma-1-\gamma\cdot n_{11}^{(j)}+n_{22}^{(j)}]$$
$$+\alpha\cdot x\cdot[n_{22}-\gamma\cdot n_{11}]\cdot e^{\lambda\cdot t}\}-\frac{C_p^{(j)}}{\tau_c}-\frac{c_p\cdot e^{\lambda\cdot t}}{\tau_c}$$

$$c_p\cdot\lambda\cdot e^{\lambda\cdot t}=r_c\cdot N_0\cdot\{C_p^{(j)}\cdot(1-2\cdot x)\cdot[(\gamma+1)\cdot n_2^{(j)}$$
$$-1+(\gamma+1)\cdot n_2\cdot e^{\lambda\cdot t}]+C_p^{(j)}\cdot\alpha\cdot x\cdot[\gamma-1-\gamma\cdot n_{11}^{(j)}+n_{22}^{(j)}]$$
$$+C_p^{(j)}\cdot\alpha\cdot x\cdot[n_{22}-\gamma\cdot n_{11}]\cdot e^{\lambda\cdot t}+e^{\lambda\cdot t}\cdot(1-2\cdot x)\cdot[c_p\cdot\{(\gamma$$
$$+1)\cdot n_2^{(j)}-1\}+(\gamma+1)\cdot c_p\cdot n_2\cdot e^{\lambda\cdot t}]$$
$$+c_p\cdot e^{\lambda\cdot t}\cdot\alpha\cdot x\cdot[\gamma-1-\gamma\cdot n_{11}^{(j)}+n_{22}^{(j)}]$$
$$+e^{\lambda\cdot t}\cdot\alpha\cdot x\cdot[c_p\cdot n_{22}-\gamma\cdot c_p\cdot n_{11}]\cdot e^{\lambda\cdot t}\}-\frac{C_p^{(j)}}{\tau_c}-\frac{c_p\cdot e^{\lambda\cdot t}}{\tau_c}$$

Assumption: $c_p\cdot n_2\approx 0$; $c_p\cdot n_{22}\approx 0$; $c_p\cdot n_{11}\approx 0$

$$
\begin{aligned}
c_p \cdot \lambda \cdot e^{\lambda \cdot t} = r_c \cdot N_0 \cdot \{C_p^{(j)} \cdot (1 - 2 \cdot x) \cdot [(\gamma + 1) \cdot n_2^{(j)} \\
- 1 + (\gamma + 1) \cdot n_2 \cdot e^{\lambda \cdot t}] + C_p^{(j)} \cdot \alpha \cdot x \cdot [\gamma - 1 - \gamma \cdot n_{11}^{(j)} + n_{22}^{(j)}] \\
+ C_p^{(j)} \cdot \alpha \cdot x \cdot [n_{22} - \gamma \cdot n_{11}] \cdot e^{\lambda \cdot t} \\
+ e^{\lambda \cdot t} \cdot (1 - 2 \cdot x) \cdot [c_p \cdot \{(\gamma + 1) \cdot n_2^{(j)} - 1\}] \\
+ c_p \cdot e^{\lambda \cdot t} \cdot \alpha \cdot x \cdot [\gamma - 1 - \gamma \cdot n_{11}^{(j)} + n_{22}^{(j)}]\} \\
- \frac{C_p^{(j)}}{\tau_c} - \frac{c_p \cdot e^{\lambda \cdot t}}{\tau_c}
\end{aligned}
$$

$$
\begin{aligned}
c_p \cdot \lambda \cdot e^{\lambda \cdot t} = r_c \cdot N_0 \cdot C_p^{(j)} \cdot (1 - 2 \cdot x) \cdot [(\gamma + 1) \cdot n_2^{(j)} \\
- 1 + (\gamma + 1) \cdot n_2 \cdot e^{\lambda \cdot t}] + r_c \cdot N_0 \cdot C_p^{(j)} \cdot \alpha \cdot x \cdot [\gamma \\
- 1 - \gamma \cdot n_{11}^{(j)} + n_{22}^{(j)}] + r_c \cdot N_0 \cdot C_p^{(j)} \cdot \alpha \cdot x \cdot [n_{22} \\
- \gamma \cdot n_{11}] \cdot e^{\lambda \cdot t} + r_c \cdot N_0 \cdot e^{\lambda \cdot t} \cdot (1 - 2 \cdot x) \cdot [c_p \cdot \{(\gamma + 1) \cdot n_2^{(j)} - 1\}] \\
+ r_c \cdot N_0 \cdot c_p \cdot e^{\lambda \cdot t} \cdot \alpha \cdot x \cdot [\gamma - 1 - \gamma \cdot n_{11}^{(j)} + n_{22}^{(j)}] \\
- \frac{C_p^{(j)}}{\tau_c} - \frac{c_p \cdot e^{\lambda \cdot t}}{\tau_c}
\end{aligned}
$$

$$
\begin{aligned}
c_p \cdot \lambda \cdot e^{\lambda \cdot t} = r_c \cdot N_0 \cdot C_p^{(j)} \cdot (1 - 2 \cdot x) \cdot \{(\gamma + 1) \cdot n_2^{(j)} - 1\} \\
+ r_c \cdot N_0 \cdot C_p^{(j)} \cdot (1 - 2 \cdot x) \cdot (\gamma + 1) \cdot n_2 \cdot e^{\lambda \cdot t} \\
+ r_c \cdot N_0 \cdot C_p^{(j)} \cdot \alpha \cdot x \cdot [\gamma - 1 - \gamma \cdot n_{11}^{(j)} + n_{22}^{(j)}] \\
+ r_c \cdot N_0 \cdot C_p^{(j)} \cdot \alpha \cdot x \cdot [n_{22} - \gamma \cdot n_{11}] \cdot e^{\lambda \cdot t} \\
+ r_c \cdot N_0 \cdot e^{\lambda \cdot t} \cdot (1 - 2 \cdot x) \cdot [c_p \cdot \{(\gamma + 1) \cdot n_2^{(j)} - 1\}] \\
+ r_c \cdot N_0 \cdot c_p \cdot e^{\lambda \cdot t} \cdot \alpha \cdot x \cdot [\gamma - 1 - \gamma \cdot n_{11}^{(j)} + n_{22}^{(j)}] \\
- \frac{C_p^{(j)}}{\tau_c} - \frac{c_p \cdot e^{\lambda \cdot t}}{\tau_c}
\end{aligned}
$$

$$
\begin{aligned}
c_p \cdot \lambda \cdot e^{\lambda \cdot t} = r_c \cdot N_0 \cdot C_p^{(j)} \cdot \{(1 - 2 \cdot x) \cdot \{(\gamma + 1) \cdot n_2^{(j)} - 1\} \\
+ \alpha \cdot x \cdot [\gamma - 1 - \gamma \cdot n_{11}^{(j)} + n_{22}^{(j)}]\} - \frac{C_p^{(j)}}{\tau_c} \\
+ r_c \cdot N_0 \cdot C_p^{(j)} \cdot (1 - 2 \cdot x) \cdot (\gamma + 1) \cdot n_2 \cdot e^{\lambda \cdot t} \\
+ r_c \cdot N_0 \cdot C_p^{(j)} \cdot \alpha \cdot x \cdot [n_{22} - \gamma \cdot n_{11}] \cdot e^{\lambda \cdot t} \\
+ r_c \cdot N_0 \cdot e^{\lambda \cdot t} \cdot (1 - 2 \cdot x) \cdot [c_p \cdot \{(\gamma + 1) \cdot n_2^{(j)} - 1\}] \\
+ r_c \cdot N_0 \cdot c_p \cdot e^{\lambda \cdot t} \cdot \alpha \cdot x \cdot [\gamma - 1 - \gamma \cdot n_{11}^{(j)} + n_{22}^{(j)}] \\
- \frac{c_p \cdot e^{\lambda \cdot t}}{\tau_c}
\end{aligned}
$$

At fixed points:

$$r_c \cdot N_0 \cdot C_p^{(j)} \cdot \{(1-2\cdot x)\cdot\{(\gamma+1)\cdot n_2^{(j)} - 1\} + \alpha\cdot x\cdot[\gamma - 1 - \gamma\cdot n_{11}^{(j)} + n_{22}^{(j)}]\} - \frac{C_p^{(j)}}{\tau_c} = 0$$

$$\begin{aligned} c_p\cdot\lambda\cdot e^{\lambda\cdot t} &= r_c\cdot N_0\cdot C_p^{(j)}\cdot(1-2\cdot x)\cdot(\gamma+1)\cdot n_2\cdot e^{\lambda\cdot t} \\ &+ r_c\cdot N_0\cdot C_p^{(j)}\cdot\alpha\cdot x\cdot[n_{22}-\gamma\cdot n_{11}]\cdot e^{\lambda\cdot t} \\ &+ r_c\cdot N_0\cdot e^{\lambda\cdot t}\cdot(1-2\cdot x)\cdot[c_p\cdot\{(\gamma+1)\cdot n_2^{(j)} - 1\}] \\ &+ r_c\cdot N_0\cdot c_p\cdot e^{\lambda\cdot t}\cdot\alpha\cdot x\cdot[\gamma-1-\gamma\cdot n_{11}^{(j)} + n_{22}^{(j)}] - \frac{c_p\cdot e^{\lambda\cdot t}}{\tau_c} \end{aligned}$$

$$\begin{aligned} & r_c\cdot N_0\cdot C_p^{(j)}\cdot(1-2\cdot x)\cdot(\gamma+1)\cdot n_2 - r_c\cdot N_0\cdot C_p^{(j)}\cdot\alpha\cdot x\cdot\gamma\cdot n_{11} \\ &+ r_c\cdot N_0\cdot C_p^{(j)}\cdot\alpha\cdot x\cdot n_{22} + \Big\{r_c\cdot N_0\cdot(1-2\cdot x)\cdot\{(\gamma+1)\cdot n_2^{(j)} - 1\} \\ &+ r_c\cdot N_0\cdot\alpha\cdot x\cdot[\gamma-1-\gamma\cdot n_{11}^{(j)} + n_{22}^{(j)}] - \frac{1}{\tau_c}\Big\}\cdot c_p - c_p\cdot\lambda = 0 \end{aligned}$$

We define global parameters for simplicity:

$$\Xi_8 = r_c\cdot N_0\cdot C_p^{(j)}\cdot(1-2\cdot x)\cdot(\gamma+1)$$

$$\Xi_9 = r_c\cdot N_0\cdot C_p^{(j)}\cdot\alpha\cdot x\cdot\gamma;\ \Xi_{10} = r_c\cdot N_0\cdot C_p^{(j)}\cdot\alpha\cdot x$$

$$\begin{aligned} \Xi_{11} &= r_c\cdot N_0\cdot(1-2\cdot x)\cdot\Big\{(\gamma+1)\cdot n_2^{(j)} - 1\Big\} \\ &+ r_c\cdot N_0\cdot\alpha\cdot x\cdot\Big[\gamma-1-\gamma\cdot n_{11}^{(j)} + n_{22}^{(j)}\Big] - \frac{1}{\tau_c} \end{aligned}$$

$$\Xi_8\cdot n_2 - \Xi_9\cdot n_{11} + \Xi_{10}\cdot n_{22} + \Xi_{11}\cdot c_p - c_p\cdot\lambda = 0$$

We can summary our system arbitrary small increment equations:

$$-n_2\cdot\lambda - \Xi_1\cdot n_2 - \Xi_2\cdot c_p = 0;\ -n_{11}\cdot\lambda - \Xi_3\cdot n_{11} - \frac{n_{22}}{\tau_2} + \Xi_4\cdot c_p = 0$$

$$\begin{aligned} &\Xi_5\cdot n_{11} - \Xi_6\cdot n_{22} - n_{22}\cdot\lambda + \Xi_7\cdot c_p = 0 \\ &\Xi_8\cdot n_2 - \Xi_9\cdot n_{11} + \Xi_{10}\cdot n_{22} + \Xi_{11}\cdot c_p - c_p\cdot\lambda = 0 \end{aligned}$$

The small increments Jacobian of our Tm^{3+}-doped silica fiber laser system is as follow:

$$\begin{pmatrix} (-\lambda - \Xi_1) & 0 & 0 & -\Xi_2 \\ 0 & (-\lambda - \Xi_3) & -\frac{1}{\tau_2} & \Xi_4 \\ 0 & \Xi_5 & (-\Xi_6 - \lambda) & \Xi_7 \\ \Xi_8 & -\Xi_9 & \Xi_{10} & (\Xi_{11} - \lambda) \end{pmatrix} \cdot \begin{pmatrix} n_2 \\ n_{11} \\ n_{22} \\ c_p \end{pmatrix} = 0$$

$$A - \lambda \cdot I = \begin{pmatrix} (-\lambda - \Xi_1) & 0 & 0 & -\Xi_2 \\ 0 & (-\lambda - \Xi_3) & -\frac{1}{\tau_2} & \Xi_4 \\ 0 & \Xi_5 & (-\Xi_6 - \lambda) & \Xi_7 \\ \Xi_8 & -\Xi_9 & \Xi_{10} & (\Xi_{11} - \lambda) \end{pmatrix}$$

$$\det(A - \lambda \cdot I) = \det \begin{pmatrix} (-\lambda - \Xi_1) & 0 & 0 & -\Xi_2 \\ 0 & (-\lambda - \Xi_3) & -\frac{1}{\tau_2} & \Xi_4 \\ 0 & \Xi_5 & (-\Xi_6 - \lambda) & \Xi_7 \\ \Xi_8 & -\Xi_9 & \Xi_{10} & (\Xi_{11} - \lambda) \end{pmatrix}$$

$$\det(A - \lambda \cdot I) = 0$$

$$\det \begin{pmatrix} (-\lambda - \Xi_1) & 0 & 0 & -\Xi_2 \\ 0 & (-\lambda - \Xi_3) & -\frac{1}{\tau_2} & \Xi_4 \\ 0 & \Xi_5 & (-\Xi_6 - \lambda) & \Xi_7 \\ \Xi_8 & -\Xi_9 & \Xi_{10} & (\Xi_{11} - \lambda) \end{pmatrix}$$

$$= (-\lambda - \Xi_1) \cdot \det \begin{pmatrix} (-\lambda - \Xi_3) & -\frac{1}{\tau_2} & \Xi_4 \\ \Xi_5 & (-\Xi_6 - \lambda) & \Xi_7 \\ -\Xi_9 & \Xi_{10} & (\Xi_{11} - \lambda) \end{pmatrix}$$

$$+ \Xi_2 \cdot \det \begin{pmatrix} 0 & (-\lambda - \Xi_3) & -\frac{1}{\tau_2} \\ 0 & \Xi_5 & (-\Xi_6 - \lambda) \\ \Xi_8 & -\Xi_9 & \Xi_{10} \end{pmatrix}$$

Step 1:

$$\det \begin{pmatrix} (-\lambda - \Xi_3) & -\frac{1}{\tau_2} & \Xi_4 \\ \Xi_5 & (-\Xi_6 - \lambda) & \Xi_7 \\ -\Xi_9 & \Xi_{10} & (\Xi_{11} - \lambda) \end{pmatrix}$$

$$\det \begin{pmatrix} (-\lambda - \Xi_3) & -\frac{1}{\tau_2} & \Xi_4 \\ \Xi_5 & (-\Xi_6 - \lambda) & \Xi_7 \\ -\Xi_9 & \Xi_{10} & (\Xi_{11} - \lambda) \end{pmatrix}$$

$$= (-\lambda - \Xi_3) \cdot \det \begin{pmatrix} (-\Xi_6 - \lambda) & \Xi_7 \\ \Xi_{10} & (\Xi_{11} - \lambda) \end{pmatrix}$$

$$+ \frac{1}{\tau_2} \cdot \det \begin{pmatrix} \Xi_5 & \Xi_7 \\ -\Xi_9 & (\Xi_{11} - \lambda) \end{pmatrix} + \Xi_4 \cdot \det \begin{pmatrix} \Xi_5 & (-\Xi_6 - \lambda) \\ -\Xi_9 & \Xi_{10} \end{pmatrix}$$

Step 1.1:

$$(-\lambda - \Xi_3) \cdot \det\begin{pmatrix} (-\Xi_6 - \lambda) & \Xi_7 \\ \Xi_{10} & (\Xi_{11} - \lambda) \end{pmatrix}$$

$$(-\lambda - \Xi_3) \cdot \det\begin{pmatrix} (-\Xi_6 - \lambda) & \Xi_7 \\ \Xi_{10} & (\Xi_{11} - \lambda) \end{pmatrix} = (-\lambda - \Xi_3) \cdot \{(-\Xi_6 - \lambda) \cdot (\Xi_{11} - \lambda) - \Xi_{10} \cdot \Xi_7\}$$

$$(-\lambda - \Xi_3) \cdot \det\begin{pmatrix} (-\Xi_6 - \lambda) & \Xi_7 \\ \Xi_{10} & (\Xi_{11} - \lambda) \end{pmatrix} = (-\lambda - \Xi_3) \cdot \left\{-\Xi_6 \cdot \Xi_{11} - \Xi_{10} \cdot \Xi_7 + (\Xi_6 - \Xi_{11}) \cdot \lambda + \lambda^2\right\}$$

$$\begin{aligned}(-\lambda - \Xi_3) \cdot \det\begin{pmatrix} (-\Xi_6 - \lambda) & \Xi_7 \\ \Xi_{10} & (\Xi_{11} - \lambda) \end{pmatrix} &= \Xi_6 \cdot \Xi_{11} \cdot \lambda + \Xi_{10} \cdot \Xi_7 \cdot \lambda - (\Xi_6 - \Xi_{11}) \cdot \lambda^2 - \lambda^3 \\ &+ \Xi_6 \cdot \Xi_{11} \cdot \Xi_3 + \Xi_{10} \cdot \Xi_7 \cdot \Xi_3 - (\Xi_6 - \Xi_{11}) \cdot \Xi_3 \cdot \lambda - \Xi_3 \cdot \lambda^2\end{aligned}$$

$$\begin{aligned}(-\lambda - \Xi_3) \cdot \det\begin{pmatrix} (-\Xi_6 - \lambda) & \Xi_7 \\ \Xi_{10} & (\Xi_{11} - \lambda) \end{pmatrix} &= (\Xi_6 \cdot \Xi_{11} + \Xi_{10} \cdot \Xi_7) \cdot \Xi_3 + [\Xi_6 \cdot \Xi_{11} + \Xi_{10} \cdot \Xi_7 - (\Xi_6 - \Xi_{11}) \cdot \Xi_3] \cdot \lambda \\ &- [\Xi_6 - \Xi_{11} + \Xi_3] \cdot \lambda^2 - \lambda^3\end{aligned}$$

Step 1.2:

$$\frac{1}{\tau_2} \cdot \det\begin{pmatrix} \Xi_5 & \Xi_7 \\ -\Xi_9 & (\Xi_{11} - \lambda) \end{pmatrix} = \frac{1}{\tau_2} \cdot [\Xi_5 \cdot (\Xi_{11} - \lambda) + \Xi_9 \cdot \Xi_7]$$

$$\frac{1}{\tau_2} \cdot \det\begin{pmatrix} \Xi_5 & \Xi_7 \\ -\Xi_9 & (\Xi_{11} - \lambda) \end{pmatrix} = \frac{1}{\tau_2} \cdot (\Xi_5 \cdot \Xi_{11} + \Xi_9 \cdot \Xi_7) - \frac{1}{\tau_2} \cdot \Xi_5 \cdot \lambda$$

Step 1.3:

$$\Xi_4 \cdot \det\begin{pmatrix} \Xi_5 & (-\Xi_6 - \lambda) \\ -\Xi_9 & \Xi_{10} \end{pmatrix} = \Xi_4 \cdot [(\Xi_5 \cdot \Xi_{10} - \Xi_9 \cdot \Xi_6) - \Xi_9 \cdot \lambda]$$

$$\Xi_4 \cdot \det\begin{pmatrix} \Xi_5 & (-\Xi_6 - \lambda) \\ -\Xi_9 & \Xi_{10} \end{pmatrix} = \Xi_4 \cdot (\Xi_5 \cdot \Xi_{10} - \Xi_9 \cdot \Xi_6) - \Xi_4 \cdot \Xi_9 \cdot \lambda$$

$$\det\begin{pmatrix} (-\lambda - \Xi_3) & -\frac{1}{\tau_2} & \Xi_4 \\ \Xi_5 & (-\Xi_6 - \lambda) & \Xi_7 \\ -\Xi_9 & \Xi_{10} & (\Xi_{11} - \lambda) \end{pmatrix}$$
$$= (\Xi_6 \cdot \Xi_{11} + \Xi_{10} \cdot \Xi_7) \cdot \Xi_3 + [\Xi_6 \cdot \Xi_{11} + \Xi_{10} \cdot \Xi_7 - (\Xi_6 - \Xi_{11}) \cdot \Xi_3] \cdot \lambda$$
$$- [\Xi_6 - \Xi_{11} + \Xi_3] \cdot \lambda^2 - \lambda^3 + \frac{1}{\tau_2} \cdot (\Xi_5 \cdot \Xi_{11} + \Xi_9 \cdot \Xi_7)$$
$$- \frac{1}{\tau_2} \cdot \Xi_5 \cdot \lambda + \Xi_4 \cdot (\Xi_5 \cdot \Xi_{10} - \Xi_9 \cdot \Xi_6) - \Xi_4 \cdot \Xi_9 \cdot \lambda$$

$$\det\begin{pmatrix} (-\lambda - \Xi_3) & -\frac{1}{\tau_2} & \Xi_4 \\ \Xi_5 & (-\Xi_6 - \lambda) & \Xi_7 \\ -\Xi_9 & \Xi_{10} & (\Xi_{11} - \lambda) \end{pmatrix}$$
$$= \{(\Xi_6 \cdot \Xi_{11} + \Xi_{10} \cdot \Xi_7) \cdot \Xi_3 + \frac{1}{\tau_2} \cdot (\Xi_5 \cdot \Xi_{11} + \Xi_9 \cdot \Xi_7)$$
$$+ \Xi_4 \cdot (\Xi_5 \cdot \Xi_{10} - \Xi_9 \cdot \Xi_6)\} + \{\Xi_6 \cdot \Xi_{11} + \Xi_{10} \cdot \Xi_7 - (\Xi_6 - \Xi_{11}) \cdot \Xi_3$$
$$- \frac{1}{\tau_2} \cdot \Xi_5 - \Xi_4 \cdot \Xi_9\} \cdot \lambda - [\Xi_6 - \Xi_{11} + \Xi_3] \cdot \lambda^2 - \lambda^3$$

We define for simplicity global parameters:

$$\Upsilon_1 = (\Xi_6 \cdot \Xi_{11} + \Xi_{10} \cdot \Xi_7) \cdot \Xi_3 + \frac{1}{\tau_2} \cdot (\Xi_5 \cdot \Xi_{11} + \Xi_9 \cdot \Xi_7)$$
$$+ \Xi_4 \cdot (\Xi_5 \cdot \Xi_{10} - \Xi_9 \cdot \Xi_6)$$

$$\Upsilon_2 = \Xi_6 \cdot \Xi_{11} + \Xi_{10} \cdot \Xi_7 - (\Xi_6 - \Xi_{11}) \cdot \Xi_3$$
$$- \frac{1}{\tau_2} \cdot \Xi_5 - \Xi_4 \cdot \Xi_9; \; \Upsilon_3 = \Xi_6 - \Xi_{11} + \Xi_3$$

$$\det\begin{pmatrix} (-\lambda - \Xi_3) & -\frac{1}{\tau_2} & \Xi_4 \\ \Xi_5 & (-\Xi_6 - \lambda) & \Xi_7 \\ -\Xi_9 & \Xi_{10} & (\Xi_{11} - \lambda) \end{pmatrix} = \Upsilon_1 + \Upsilon_2 \cdot \lambda - \Upsilon_3 \cdot \lambda^2 - \lambda^3$$

$$(-\lambda - \Xi_1) \cdot \det\begin{pmatrix} (-\lambda - \Xi_3) & -\frac{1}{\tau_2} & \Xi_4 \\ \Xi_5 & (-\Xi_6 - \lambda) & \Xi_7 \\ -\Xi_9 & \Xi_{10} & (\Xi_{11} - \lambda) \end{pmatrix}$$
$$= (-\lambda - \Xi_1) \cdot \left(\Upsilon_1 + \Upsilon_2 \cdot \lambda - \Upsilon_3 \cdot \lambda^2 - \lambda^3\right)$$

$$(-\lambda - \Xi_1) \cdot \det\begin{pmatrix} (-\lambda - \Xi_3) & -\frac{1}{\tau_2} & \Xi_4 \\ \Xi_5 & (-\Xi_6 - \lambda) & \Xi_7 \\ -\Xi_9 & \Xi_{10} & (\Xi_{11} - \lambda) \end{pmatrix}$$

$$= -\Xi_1 \cdot \Upsilon_1 - (\Upsilon_1 + \Xi_1 \cdot \Upsilon_2) \cdot \lambda + (\Xi_1 \cdot \Upsilon_3 - \Upsilon_2) \cdot \lambda^2 + (\Upsilon_3 + \Xi_1) \cdot \lambda^3 + \lambda^4$$

Step 2:

$$\Xi_2 \cdot \det\begin{pmatrix} 0 & (-\lambda - \Xi_3) & -\frac{1}{\tau_2} \\ 0 & \Xi_5 & (-\Xi_6 - \lambda) \\ \Xi_8 & -\Xi_9 & \Xi_{10} \end{pmatrix}$$

$$\Xi_2 \cdot \det\begin{pmatrix} 0 & (-\lambda - \Xi_3) & -\frac{1}{\tau_2} \\ 0 & \Xi_5 & (-\Xi_6 - \lambda) \\ \Xi_8 & -\Xi_9 & \Xi_{10} \end{pmatrix}$$
$$= \Xi_2 \cdot \Xi_8 \cdot \det\begin{pmatrix} (-\lambda - \Xi_3) & -\frac{1}{\tau_2} \\ \Xi_5 & (-\Xi_6 - \lambda) \end{pmatrix}$$
$$= \Xi_2 \cdot \Xi_8 \cdot \left\{(-\lambda - \Xi_3) \cdot (-\Xi_6 - \lambda) + \Xi_5 \cdot \frac{1}{\tau_2}\right\}$$

$$\Xi_2 \cdot \det\begin{pmatrix} 0 & (-\lambda - \Xi_3) & -\frac{1}{\tau_2} \\ 0 & \Xi_5 & (-\Xi_6 - \lambda) \\ \Xi_8 & -\Xi_9 & \Xi_{10} \end{pmatrix}$$
$$= \Xi_2 \cdot \Xi_8 \cdot \det\begin{pmatrix} (-\lambda - \Xi_3) & -\frac{1}{\tau_2} \\ \Xi_5 & (-\Xi_6 - \lambda) \end{pmatrix}$$
$$= \Xi_2 \cdot \Xi_8 \cdot \left\{(\lambda + \Xi_3) \cdot (\Xi_6 + \lambda) + \Xi_5 \cdot \frac{1}{\tau_2}\right\}$$
$$= \Xi_2 \cdot \Xi_8 \cdot \left\{\Xi_5 \cdot \frac{1}{\tau_2} + \Xi_3 \cdot \Xi_6 + (\Xi_6 + \Xi_3) \cdot \lambda + \lambda^2\right\}$$

$$\Xi_2 \cdot \det\begin{pmatrix} 0 & (-\lambda - \Xi_3) & -\frac{1}{\tau_2} \\ 0 & \Xi_5 & (-\Xi_6 - \lambda) \\ \Xi_8 & -\Xi_9 & \Xi_{10} \end{pmatrix}$$
$$= \Xi_2 \cdot \Xi_8 \cdot \det\begin{pmatrix} (-\lambda - \Xi_3) & -\frac{1}{\tau_2} \\ \Xi_5 & (-\Xi_6 - \lambda) \end{pmatrix}$$
$$= \Xi_2 \cdot \Xi_8 \cdot \left\{(\lambda + \Xi_3) \cdot (\Xi_6 + \lambda) + \Xi_5 \cdot \frac{1}{\tau_2}\right\}$$
$$= \Xi_2 \cdot \Xi_8 \cdot \left(\Xi_5 \cdot \frac{1}{\tau_2} + \Xi_3 \cdot \Xi_6\right) + \Xi_2 \cdot \Xi_8 \cdot (\Xi_6 + \Xi_3) \cdot \lambda + \Xi_2 \cdot \Xi_8 \cdot \lambda^2$$

We define for simplicity global parameters:

$$\Upsilon_4 = \Xi_2 \cdot \Xi_8 \cdot \left(\Xi_5 \cdot \frac{1}{\tau_2} + \Xi_3 \cdot \Xi_6 \right)$$

$$\Upsilon_5 = \Xi_2 \cdot \Xi_8 \cdot (\Xi_6 + \Xi_3);\ \Upsilon_6 = \Xi_2 \cdot \Xi_8$$

$$\Xi_2 \cdot \det \begin{pmatrix} 0 & (-\lambda - \Xi_3) & -\frac{1}{\tau_2} \\ 0 & \Xi_5 & (-\Xi_6 - \lambda) \\ \Xi_8 & -\Xi_9 & \Xi_{10} \end{pmatrix} = \Upsilon_4 + \Upsilon_5 \cdot \lambda + \Upsilon_6 \cdot \lambda^2$$

We can summary our last results (steps 1 and 2) in Table 8.7:

$$\det \begin{pmatrix} (-\lambda - \Xi_1) & 0 & 0 & -\Xi_2 \\ 0 & (-\lambda - \Xi_3) & -\frac{1}{\tau_2} & \Xi_4 \\ 0 & \Xi_5 & (-\Xi_6 - \lambda) & \Xi_7 \\ \Xi_8 & -\Xi_9 & \Xi_{10} & (\Xi_{11} - \lambda) \end{pmatrix}$$
$$= -\Xi_1 \cdot \Upsilon_1 - (\Upsilon_1 + \Xi_1 \cdot \Upsilon_2) \cdot \lambda + (\Xi_1 \cdot \Upsilon_3 - \Upsilon_2) \cdot \lambda^2$$
$$+ (\Upsilon_3 + \Xi_1) \cdot \lambda^3 + \lambda^4 + \Upsilon_4 + \Upsilon_5 \cdot \lambda + \Upsilon_6 \cdot \lambda^2$$

$$\det \begin{pmatrix} (-\lambda - \Xi_1) & 0 & 0 & -\Xi_2 \\ 0 & (-\lambda - \Xi_3) & -\frac{1}{\tau_2} & \Xi_4 \\ 0 & \Xi_5 & (-\Xi_6 - \lambda) & \Xi_7 \\ \Xi_8 & -\Xi_9 & \Xi_{10} & (\Xi_{11} - \lambda) \end{pmatrix}$$
$$= -\Xi_1 \cdot \Upsilon_1 - (\Upsilon_1 + \Xi_1 \cdot \Upsilon_2) \cdot \lambda + (\Xi_1 \cdot \Upsilon_3 - \Upsilon_2) \cdot \lambda^2$$
$$+ (\Upsilon_3 + \Xi_1) \cdot \lambda^3 + \lambda^4 + \Upsilon_4 + \Upsilon_5 \cdot \lambda + \Upsilon_6 \cdot \lambda^2$$

Table 8.7 Tm^{3+}-doped silica fiber laser system's determinant elements

Determinant	Expression
$(-\lambda - \Xi_1) \cdot \det \begin{pmatrix} (-\lambda - \Xi_3) & -\frac{1}{\tau_2} & \Xi_4 \\ \Xi_5 & (-\Xi_6 - \lambda) & \Xi_7 \\ -\Xi_9 & \Xi_{10} & (\Xi_{11} - \lambda) \end{pmatrix}$	$-\Xi_1 \cdot \Upsilon_1 - (\Upsilon_1 + \Xi_1 \cdot \Upsilon_2) \cdot \lambda + (\Xi_1 \cdot \Upsilon_3 - \Upsilon_2) \cdot \lambda^2 + (\Upsilon_3 + \Xi_1) \cdot \lambda^3 + \lambda^4$
$\Xi_2 \cdot \det \begin{pmatrix} 0 & (-\lambda - \Xi_3) & -\frac{1}{\tau_2} \\ 0 & \Xi_5 & (-\Xi_6 - \lambda) \\ \Xi_8 & -\Xi_9 & \Xi_{10} \end{pmatrix}$	$\Upsilon_4 + \Upsilon_5 \cdot \lambda + \Upsilon_6 \cdot \lambda^2$

$$\det\begin{pmatrix} (-\lambda - \Xi_1) & 0 & 0 & -\Xi_2 \\ 0 & (-\lambda - \Xi_3) & -\frac{1}{\tau_2} & \Xi_4 \\ 0 & \Xi_5 & (-\Xi_6 - \lambda) & \Xi_7 \\ \Xi_8 & -\Xi_9 & \Xi_{10} & (\Xi_{11} - \lambda) \end{pmatrix}$$
$$= \Upsilon_4 - \Xi_1 \cdot \Upsilon_1 + (\Upsilon_5 - \Upsilon_1 - \Xi_1 \cdot \Upsilon_2) \cdot \lambda + (\Xi_1 \cdot \Upsilon_3 - \Upsilon_2 + \Upsilon_6) \cdot \lambda^2 + (\Upsilon_3 + \Xi_1) \cdot \lambda^3 + \lambda^4$$

$$\det(A - \lambda \cdot I) = \Upsilon_4 - \Xi_1 \cdot \Upsilon_1 + (\Upsilon_5 - \Upsilon_1 - \Xi_1 \cdot \Upsilon_2) \cdot \lambda + (\Xi_1 \cdot \Upsilon_3 - \Upsilon_2 + \Upsilon_6) \cdot \lambda^2 + (\Upsilon_3 + \Xi_1) \cdot \lambda^3 + \lambda^4$$

We define for simplicity global parameters:

$$E_0 = \Upsilon_4 - \Xi_1 \cdot \Upsilon_1; E_1 = \Upsilon_5 - \Upsilon_1 - \Xi_1 \cdot \Upsilon_2$$

$$E_2 = \Xi_1 \cdot \Upsilon_3 - \Upsilon_2 + \Upsilon_6; E_3 = \Upsilon_3 + \Xi_1; E_4 = 1; \det(A - \lambda \cdot I) = \sum_{k=0}^{4} E_k \cdot \lambda^k$$

$$\det(A - \lambda \cdot I) = 0 \Rightarrow \sum_{k=0}^{4} E_k \cdot \lambda^k = 0$$

Eigenvalue stability discussion: Our Tm^{3+}-doped silica fiber laser system involving N variables (N > 2, N = 4, arbitrarily small increments), the characteristic equation is of degree N = 4 and must often be solved numerically. Expect in some particular cases, such an equation has (N = 4) distinct roots that can be real or complex. These values are eigenvalues of the (4 × 4) Jacobian matrix (A). The general rule is that the Tm^{3+}-doped silica fiber laser system is stable if there is no eigenvalue with positive real part. It is sufficient that one eigenvalue is positive for the steady state to be unstable. Our four variables (n_2, n_{11}, n_{22}, C_p) system has four eigenvalues (four system's arbitrarily small increments). The type of behavior can be characterized as a function of the position of these eigenvalues in the Re/Im plane. Five non-generated cases can be distinguished: (1) the four eigenvalues are real and negative (stable steady state), (2) the four eigenvalues are real, at least one of them is positive (unstable steady state), (3) and (4) three eigenvalues are complex conjugates with a negative real part and other eigenvalue real is negative (stable steady state), two cases can be distinguished depending on the relative value of the real part of the complex eigenvalues and of the real one, (5) three eigenvalues are complex conjugates with a negative real part and other eigenvalue real is positive (unstable steady state) [5, 6].

8.3 Terahertz Dual-Wavelength Quantum Cascade Laser Stability Analysis Under Parameters Variations and Delay Variables in Time

Quantum cascade (QC) lasers are semiconductor devices in which radiation is generated by electronic transitions within an artificial crystal, hetero structure crystal. This hetero structure consists of alternating layers of two semiconductor materials, with the layer thickness determining the electronic states inside the crystal, and thereby the frequency of the emitted light as well as the electrical transport. The operation of QC lasers in the range of frequencies 1–10 THz is demonstrated by employing super lattice active material and developing a waveguide concept based on interface modes called surface Plasmon. One implementation of THz QCLs into advanced photonic structures is allowing the emission of laser light from the surface of the semiconductor, and into external cavity set-ups from improved tune ability. The terahertz (THz) frequency quantum cascade laser (QCL) is a source of THz radiation offering high power, high spectral purity and moderate tune ability. It is particularly suited for THz frequency imaging across a range of disciplines. There are a number of imaging approaches targeted to application areas including multiple-frequency transmission and diffuse reflection imaging for the spectral mapping of targets; as well as coherent approaches based on the self-mixing phenomenon in THz QCLs for long-range imaging, three-dimensional imaging, materials analysis, and high-resolution inverse synthetic aperture radar imaging. Quantum-cascade lasers can be based on a resonant-phonon depopulation scheme, and use a semi-insulation surface-Plasmon waveguide, a typical QCL resonant-phonon parameters are $f \simeq 4.4$ THz, $\lambda \simeq 68\,\mu$m, $P_{peak} = 248$ mW (pulsed mode), $P_{cw} = 138$ mW at heatsink temperature of 10 K. There is a hetero generous active region in THz QCL which is based on frequency combs. The important parameters are dynamic range, spectral bandwidth and output power. A narrow and stable inter mode beat note indicate frequency comb operation up to a spectral bandwidth of 1.1 THz, while a dispersion-dominated regime a bandwidth up to 1.94 THz at a center frequency of 3 THz. A self-detected dual-comb setup is used to verify the frequency comb nature of the lasers. There is an alternative way to produce terahertz radiation by frequency down-conversion in a nonlinear optical crystal using infrared or visible pump lasers [7–9]. This approach offers broad spectral tune ability at room temperature. It requires powerful laser pumps and a complicated optical set-up, resulting in bulky and un-widely sources. It is done by a monolithically integrated device which is designed to combine the advantages of electrically pumped semiconductor lasers and nonlinear optical sources [10–12]. In this subchapter we define and analyze Terahertz (THz) quantum cascade laser (QCL), based on GaN/AIGaN quantum wells which emits at two widely separated wavelengths 33 and 52 μm simultaneously in a single active region. The large LO-phonon energy (~90 meV), the ultrafast resonant phonon depopulation of the lower radiative levels, suppression of the electrons that escape to the continuum states and selective carrier injection and extraction all together lead to a considerable enhancement in the operating temperature of the structure. The terahertz frequency range, is the frequencies from 1 to 10 THz, or wavelengths from 30 to 300 μm (it is between ranges which covered by RF and microwave and photonics, Infrared and visible). THz potential application is spectroscopy where all molecules, biological, organic, and inorganic have vibrational and rotational spectra in the terahertz range. They are short enough for sub-millimeter imaging, and long enough to penetrate a large variety of materials. The Terahertz quantum cascade laser (QCL) is based on the GaAs/AlGaAs quantum wells and span the 1.2–5 THz range with maximum temperatures near 170 and 120 K for pulsed and continuous wave emissions, respectively. There are a longitudinal optical (LO) phonons with energies ($\hbar \cdot \omega_{L_0} \approx 36\,\text{meV}$) which is relatively close to the room temperature thermal energy $K_B \cdot T$ (≈26 meV), and cause to limitation. The thermally activated LO-phonon scatterings and back fillings (≈25 °) cause a considerable reduction in device population inversion and optical gain. The material system with large LO-phonon energy is needed for high-temperature operation of THz QCLs. Another system is GaN/AlGaN hetero structures where LO-phonon energy is ≈ 91 meV. Conventional QCLs which is based on GaAs/AlGaAs material system does not emit in the range around the LO-phonon energies and there is a spectral gap between 30 and 40 μm. In many applications, differential spectroscopy free-space wavelength division multiplexing, frequency mixing, trace gas sensing, vapor sensing of organic molecules or liquids, and complex sensing applications, a multi wavelength light source is needed. The absence of re-absorption for widely spaced photon energies effects on an inter sub band devices ideal for multi-wavelength operation. Structures with multi-peak gain spectra and multiple wavelength emissions is done using mid-IR spectral region. Some structures work at extremely low operational temperatures or high magnetic fields, and terahertz frequencies. Second-harmonic generation (SHG) in quantum-cascade lasers are the way to access multi-wavelength operation. An optimized structure design is utilized the super symmetric quantum mechanics techniques, linear and SHG output powers of about 0.5 W and 300 μm is demonstrated, respectively. The high difference between linear and nonlinear powers results in poor efficiency of the second harmonic power for typical applications. The quantum cascade laser (QCL) active region is based on GaN/AlGaN material system with a resonant-phonon-assisted depopulation scheme, which emits at two widely separated wavelengths (33 μm, 52 μm), and is located in THz frequency range. Our system is described by the conduction band structure of the designed QCL active region and features. Also it is characterized by the rate equations. The quantum cascade structure is based on a $GaN/Al_{0.46}Ga_{0.54}N$ material system, which contains 50 repetitions of GaN multiple quantum well periods. Each period consists of four wells for achievement of gain and four wells for the injection process. Two and half periods of the active region and wave functions are simulated. Differences of lattice constants at hetero junctions of nitride quantum-well (QW) devices are sources of stresses in a crystal lattice, and piezo electric polarization. The piezoelectric polarization with the spontaneous polarization in III nitride quantum wells causes electric fields, which influence the quantum structure. An electric field is applied to have the first radiative transition ($4 \rightarrow 3$) and second one ($2 \rightarrow 1$). These radiations are separated by following the electrons through an injector mini band followed by their injection into level 4 of the next period. (1) First radiative transition $(4 \to 3)$, $\lambda_1 = 33\,\mu\text{m}$; $\Delta E_{43} = 38\,\text{meV}$; $\nu_1 = 9.1\,\text{THz}$, (2) Second radiative transition $(2 \to 1)$, $\lambda_2 = 52\,\mu\text{m}$; $\Delta E_{21} = 24\,\text{meV}$; $\nu_2 = 5.8\,\text{THz}$. There is a sensitive issue when designing terahertz QCLs which is the difficulty of selective population of the upper lasing level and selective depopulation of the lower lasing level. For the system structure the wave functions have a small overlaps where all transitions are diagonal. It suppresses the electron thermal backfilling and enhanced the temperature characteristics. System dipole matrix elements are subsequently the optical gains are reduced; the appropriate electron injections and extractions are key issue. The system lower/upper radiative level is almost depopulated (populated) without depopulation (population) of the upper (lower) radiative level. The extraction of carriers from the lower level is much more rapid than repopulation by carriers scattered from the upper level then the population inversion condition between the two involved radiative sub bonds for gain attainment is satisfied. An energy spacing equivalent to the LO-phonon energy ($\approx$91.9 meV) is chosen next to the lower radiative levels to extract electrons via an ultra-fast LO-phonon scattering mechanism. The energy spacing creates a large barrier in front of the thermally activated phonon scatterings and back fillings. It plays a prominent role in the enhancement of the temperature characteristics of GaN quantum wells. The probability of thermal activation of the electrons to the continuum states increases as a confined level approaches the top of the barriers. There is a large energy spacing above the lasing states ($\sim$350 mV), which make it difficult for electrons to be thermally activated and escape to the continuum states, which is happened also at high temperatures. The system structure works also at higher temperature. The rate equation model is for a coupled electronic and electromagnetic system, and includes structural system parameters. The schematic energy diagram of the QCL structure and the electron transport and the photon generation process is presented (Fig. 8.3), there are a various transition lifetimes which describe the rate equation analysis of a five level QCL. The five levels QCL are 4, 3, 2, and 1 which denote

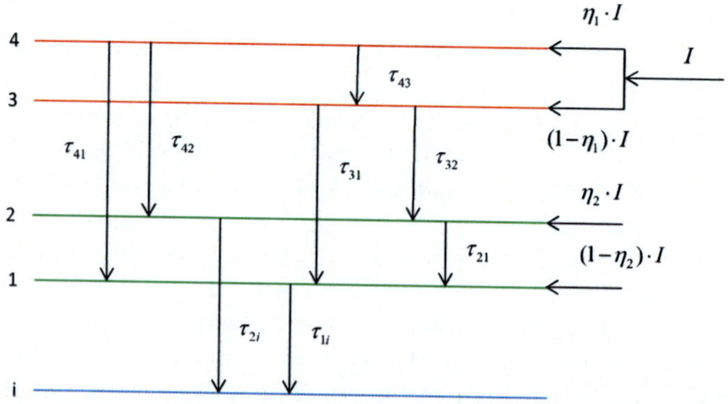

Fig. 8.3 Schematic of a level diagram and various transition lifetimes which describe the rate equation analysis of five level QCL

the radiative levels. Level i is the injector state. I is the active current, η_1 and η_2 are injection efficiencies and τ_{43}, τ_{42}, τ_{41}, τ_{32}, τ_{31}, τ_{21}, τ_{2i}, and τ_{1i} are the lifetimes due to non-radiative transitions, $4 \to 3, 4 \to 2, 4 \to 1, 3 \to 2, 3 \to 1, 2 \to 1, 2 \to i$, and $1 \to i$, respectively [13].

The dynamic electron and photon populations is determined as a function of various radiative and non-radiative scattering lifetimes and the source term I, which is the current flowing through the quantum cascade structure. The polarization effects include a large two-dimensional electron gas (2DEG) at the interface between AlGaN/GaN hetero-junctions. The induced polarization charge is enough to cause the electron flowing. States 4 and 3 are the upper and the lower radiative levels of the first optical transition at $\lambda_1 = 33\,\mu\text{m}$ with electron injection rates of η_1 and $1 - \eta_1$, respectively. Levels 2 and 1 are related to the second optical transition (at $\lambda_2 = 52\,\mu\text{m}$). i is the reservoir (injector) state. It is recommended that the total outflow rate from levels 4 and 3, $\frac{1}{\tau_{42}} + \frac{1}{\tau_{41}} + \frac{1}{\tau_{32}} + \frac{1}{\tau_{31}}$ is equal to flowing current injected into levels 2 and 1 with respective rates of η_2 and $1 - \eta_2$, respectively [13]. The injection efficiency η is the fraction of the current injected into upper radiative levels. The THz QCL rate equations are

$$\frac{dn_4}{dt} = \frac{\eta_1 \cdot I \cdot N_p}{|e|} - \left(\frac{1}{\tau_{41}} + \frac{1}{\tau_{43}} + \frac{1}{\tau_{42}} + \frac{1}{\tau_{sp1}} \cdot \frac{\Gamma_1}{N_p}\right) \cdot n_4 - (n_4 - n_3) \cdot \frac{\Gamma_1}{N_p} \cdot \int \frac{n_{ph}(\hbar \cdot \omega_1)}{\tau_{sp}(\hbar \cdot \omega_1)} \cdot \hbar \cdot d\omega$$

$$\frac{dn_3}{dt} = \frac{(1 - \eta_1) \cdot I \cdot N_p}{|e|} - \left(\frac{1}{\tau_{32}} + \frac{1}{\tau_{31}}\right) \cdot n_3 + \left(\frac{1}{\tau_{43}} + \frac{1}{\tau_{sp1}} \cdot \frac{\Gamma_1}{N_p}\right) \cdot n_4 + (n_4 - n_3) \cdot \frac{\Gamma_1}{N_p} \cdot \int \frac{n_{ph}(\hbar \cdot \omega_1)}{\tau_{sp}(\hbar \cdot \omega_1)} \cdot \hbar \cdot d\omega$$

$$\frac{dn_2}{dt} = \frac{\eta_2 \cdot I \cdot N_p}{|e|} - \left(\frac{1}{\tau_{21}} + \frac{1}{\tau_{2i}} + \frac{1}{\tau_{sp2}} \cdot \frac{\Gamma_2}{N_p}\right) \cdot n_2 - (n_2 - n_1) \cdot \frac{\Gamma_2}{N_p} \cdot \int \frac{n_{ph}(\hbar \cdot \omega_2)}{\tau_{sp}(\hbar \cdot \omega_2)} \cdot \hbar \cdot d\omega$$

$$\frac{dn_1}{dt} = \frac{(1 - \eta_2) \cdot I \cdot N_p}{|e|} - \frac{n_1}{\tau_{1i}} + \frac{n_4}{\tau_{41}} + \frac{n_2}{\tau_{21}} + \frac{n_2}{\tau_{sp2}} \cdot \frac{\Gamma_2}{N_p} + (n_2 - n_1) \cdot \frac{\Gamma_2}{N_p} \cdot \int \frac{n_{ph}(\hbar \cdot \omega_2)}{\tau_{sp}(\hbar \cdot \omega_2)} \cdot \hbar \cdot d\omega$$

$$\frac{dn_{ph}(\hbar \cdot \omega_1)}{dt} = (n_4 - n_3) \cdot \frac{\Gamma_1}{N_p} \cdot \frac{n_{ph}(\hbar \cdot \omega_1)}{\tau_{sp}(\hbar \cdot \omega_1)} - \frac{n_{ph}(\hbar \cdot \omega_1)}{\tau_{ph1}}$$

$$\frac{dn_{ph}(\hbar \cdot \omega_2)}{dt} = (n_2 - n_1) \cdot \frac{\Gamma_2}{N_p} \cdot \frac{n_{ph}(\hbar \cdot \omega_2)}{\tau_{sp}(\hbar \cdot \omega_2)} - \frac{n_{ph}(\hbar \cdot \omega_2)}{\tau_{ph2}}$$

The next table is described the THz dual-wavelength QCL system parameters (Table 8.8).

τ_{ph1}, τ_{ph2} are photon lifetimes in the cavity, and $\tau_{ph1} = \frac{n_{eff1}}{c\cdot(\alpha_{w_1}+\alpha_m)}$, $\tau_{ph2} = \frac{n_{eff2}}{c\cdot(\alpha_{w_2}+\alpha_m)}$, $\frac{1}{\tau_{ph1}} = \frac{c\cdot(\alpha_{w_1}+\alpha_m)}{n_{eff1}}$; $\frac{1}{\tau_{ph2}} = \frac{c\cdot(\alpha_{w_2}+\alpha_m)}{n_{eff2}}$. The energy spaces, ΔE_{32} and ΔE_{1i} are equal to LO-phonon energy which make electrons scatter via an ultrafast electron-LO-phonon mechanism that cancels the formation of population inversion between those levels. There is a need to calculate the scattering times between an initial sub band i and the final sub band f. Although radiative sub bands are very close together, the electron LO-phonon scattering is the dominant scattering mechanism because of low sub band populations ($<5 \times 10^{10}\,\text{cm}^{-2}$), and its rate is about an order of magnitude greater than the electron–electron scattering rate. We neglect

Table 8.8 THz dual-wavelength QCL system parameters

Parameter	Description
I	Electron current in the direction of transport, flowing through the device
N_p	Number of repeated modules in the QCL
Γ_1, Γ_2	The fraction of the optical mode that overlaps with the entire active region (confinement overlap factor), Γ_1 is the parameter at $\lambda_1 = 33\,\mu\text{m}$, Γ_2 is the parameter at $\lambda_2 = 52\,\mu\text{m}$
n_4, n_3, n_2 and n_1	Electron populations in sub bands 4, 3, 2, and 1, respectively
$n_{ph}(\hbar\cdot\omega_1), n_{ph}(\hbar\cdot\omega_2)$	Number of photons per unit energy (J^{-1}) generated in all available cavity modes, due to both spontaneous and stimulated optical transitions
$\tau_{sp1}\cdot\frac{N_p}{\Gamma_1}, \tau_{sp2}\cdot\frac{N_p}{\Gamma_2}$	Spontaneous emission lifetime in all available cavity modes
$\tau_{sp}(\hbar\cdot\omega_1)\cdot\frac{N_p}{\Gamma_1}, \tau_{sp}(\hbar\cdot\omega_2)\cdot\frac{N_p}{\Gamma_2}$	Spontaneous emission lifetime in the lasing mode
τ_{ph1}, τ_{ph2}	Photon lifetime in the cavity
$\alpha_{w_1}, \alpha_{w_2}$	Waveguide losses (m^{-1})
α_m	Mirror losses (m^{-1})
c	Speed of light in vacuum
n_{eff1}, n_{eff2}	Modal effective refractive index
η_1, η_2	Injection efficiencies
$\tau_{43}, \tau_{42}, \tau_{41}, \tau_{32}, \tau_{31}, \tau_{21}, \tau_{2i}$, and τ_{1i}	lifetimes due to non-radiative transitions, $4\rightarrow 3, 4\rightarrow 2, 4\rightarrow 1, 3\rightarrow 2, 3\rightarrow 1, 2\rightarrow 1, 2\rightarrow i$, and $1\rightarrow i$, respectively
e	Electron charge, (symbol e), fundamental physical constant expressing the naturally occurring unit of electric charge, equal to $1.602176634\times 10^{-19}$ C

Table 8.9 Threshold condition, the steady state sub band population expressions

Parameter	Description	Expression
n_{1th}	Steady state sub band population 1 (threshold condition)	$n_{1th} = I_{th2} \cdot \frac{N_p}{\lvert e\rvert} \cdot \left[\frac{\tau_{1i} \cdot (\tau_{2i} + (1-\eta_{th2}) \cdot \tau_{21}}{(\tau_{2i} + \tau_{21})}\right]$
n_{2th}	Steady state sub band population 2 (threshold condition)	$n_{2th} = I_{th2} \cdot \frac{\eta_{th2} \cdot N_p}{\lvert e\rvert} \cdot \left[\frac{\tau_{2i} \cdot \tau_{21}}{(\tau_{2i} + \tau_{21})}\right]$
n_{3th}	Steady state sub band population 3 (threshold condition)	$n_{3th} = I_{th1} \cdot \frac{N_p}{\lvert e\rvert} \cdot \left[\frac{\tau_3 \cdot (\tau_{42} + (1-\eta_{th1}) \cdot \tau_{43})}{(\tau_{42} + \tau_{43})}\right]$
n_{4th}	Steady state sub band population 4 (threshold condition)	$n_{4th} = I_{th1} \cdot \frac{\eta_{th1} \cdot N_p}{\lvert e\rvert} \cdot \left[\frac{\tau_{42} \cdot \tau_{43}}{(\tau_{42} + \tau_{43})}\right]$

the electron–electron scattering and only considered the electron-LO-phonon interactions. For the calculated lifetimes, i' is the injector state of the previous stage, which injects electrons into level 4, and $\tau_{i'4}, \tau_{i'3}$ are corresponding lifetimes due to $i' \to 4$ and $i' \to 3$ transitions, respectively. The calculated transition lifetimes for our analysis are $\tau_{i'4} = 0.3\,\text{ps}$; $\tau_{i'3} = 1.96\,\text{ps}$; $\tau_{43} = 0.24\,\text{ps}$; $\tau_{42} = 41\,\text{ps}$ and $\tau_{32} = 0.14\,\text{ps}$; $\tau_{31} = 12\,\text{ps}$; $\tau_{21} = 0.23\,\text{ps}$; $\tau_{2i} = 7\,\text{ps}$; $\tau_{1i} = 0.096\,\text{ps}$. The large quantities of $\tau_{i'3}$, τ_{31}, τ_{42}, and τ_{2i} are due to the low overlapping between the relevant electron wave functions leads to two prominent features: high injection efficiency, and high selectivity of electron extraction. Below or near lasing thresholds, $n_{ph}(\hbar \cdot \omega_1)$ and $n_{ph}(\hbar \cdot \omega_2)$ does not have a considerable value and τ_{sp1}, τ_{sp2}, $\tau_{sp}(\hbar \cdot \omega_1)$, and $\tau_{sp}(\hbar \cdot \omega_2)$ are several orders of magnitude greater than non-radiative lifetimes. The threshold condition, the steady state sub band population expressions are described (Table 8.9).

The expressions for $n_{ph}(\hbar \cdot \omega_1)$ and $n_{ph}(\hbar \cdot \omega_2)$ are

$$n_{ph}(\hbar \cdot \omega_1) = \frac{N_p}{|e|} \cdot \tau_{ph1} \cdot \left[\frac{(\eta_{th1} \cdot \tau_{42} \cdot \tau_{43} - \tau_3 \cdot \tau_{42} - (1 - \eta_{th1}) \cdot \tau_3 \cdot \tau_{43})}{(\tau_3 + \tau_{42}) \cdot \tau_{43}} \times (I - I_{th1}) + I \cdot (\eta - \eta_{th1})\right]$$

$$n_{ph}(\hbar \cdot \omega_2) = \frac{N_p}{|e|} \cdot \tau_{ph2} \cdot \left[\frac{(\eta_{th2} \cdot \tau_{2i} \cdot \tau_{21} - \tau_{1i} \cdot \tau_{2i} - (1 - \eta_{th2}) \cdot \tau_{1i} \cdot \tau_{21})}{(\tau_{1i} + \tau_{2i}) \cdot \tau_{21}} \times (I - I_{th2}) + I \cdot (\eta - \eta_{th2})\right]$$

$$\int \frac{n_{ph}(\hbar \cdot \omega_1)}{\tau_{sp}(\hbar \cdot \omega_1)} \cdot \hbar \cdot d\omega = \int \frac{\hbar}{\tau_{sp}(\hbar \cdot \omega_1)} \cdot \frac{N_p}{|e|} \cdot \tau_{ph1} \cdot \left[\frac{(\eta_{th1} \cdot \tau_{42} \cdot \tau_{43} - \tau_3 \cdot \tau_{42} - (1 - \eta_{th1}) \cdot \tau_3 \cdot \tau_{43})}{(\tau_3 + \tau_{42}) \cdot \tau_{43}} \times (I - I_{th1}) + I \cdot (\eta - \eta_{th1})\right] \cdot d\omega$$

$$\int \frac{n_{ph}(\hbar \cdot \omega_2)}{\tau_{sp}(\hbar \cdot \omega_2)} \cdot \hbar \cdot d\omega = \int \frac{\hbar}{\tau_{sp}(\hbar \cdot \omega_2)} \cdot \frac{N_p}{|e|} \cdot \tau_{ph2}$$

$$\cdot \left[\frac{(\eta_{th2} \cdot \tau_{2i} \cdot \tau_{21} - \tau_{1i} \cdot \tau_{2i} - (1 - \eta_{th2}) \cdot \tau_{1i} \cdot \tau_{21})}{(\tau_{1i} + \tau_{2i}) \cdot \tau_{21}} \right. \\ \left. \times (I - I_{th2}) + I \cdot (\eta - \eta_{th2}) \right] \cdot d\omega$$

We define two functions of system parameters: $f_1(N_p, \tau_{ph1}, \ldots)$; $f_2(N_p, \tau_{ph1}, \ldots)$.

$$f_1(N_p, \tau_{ph1}, \ldots) = \frac{\hbar}{\tau_{sp}(\hbar \cdot \omega_1)} \cdot \frac{N_p}{|e|} \cdot \tau_{ph1} \\ \cdot \left[\frac{(\eta_{th1} \cdot \tau_{42} \cdot \tau_{43} - \tau_3 \cdot \tau_{42} - (1 - \eta_{th1}) \cdot \tau_3 \cdot \tau_{43})}{(\tau_3 + \tau_{42}) \cdot \tau_{43}} \right. \\ \left. \times (I - I_{th1}) + I \cdot (\eta - \eta_{th1}) \right]$$

$$f_2(N_p, \tau_{ph1}, \ldots) = \frac{\hbar}{\tau_{sp}(\hbar \cdot \omega_2)} \cdot \frac{N_p}{|e|} \cdot \tau_{ph2} \\ \cdot \left[\frac{(\eta_{th2} \cdot \tau_{2i} \cdot \tau_{21} - \tau_{1i} \cdot \tau_{2i} - (1 - \eta_{th2}) \cdot \tau_{1i} \cdot \tau_{21})}{(\tau_{1i} + \tau_{2i}) \cdot \tau_{21}} \right. \\ \left. \times (I - I_{th2}) + I \cdot (\eta - \eta_{th2}) \right]$$

$$\int \frac{n_{ph}(\hbar \cdot \omega_1)}{\tau_{sp}(\hbar \cdot \omega_1)} \cdot \hbar \cdot d\omega = \int f_1(N_p, \tau_{ph1}, \ldots) \cdot d\omega = f_1(N_p, \tau_{ph1}, \ldots) \cdot \omega$$
$$f_1 = f_1(N_p, \tau_{ph1}, \ldots)$$

$$\int \frac{n_{ph}(\hbar \cdot \omega_2)}{\tau_{sp}(\hbar \cdot \omega_2)} \cdot \hbar \cdot d\omega = \int f_2(N_p, \tau_{ph1}, \ldots) \cdot d\omega = f_2(N_p, \tau_{ph1}, \ldots) \cdot \omega$$
$$f_2 = f_2(N_p, \tau_{ph1}, \ldots)$$

We can rewrite our system differential equations: $\frac{dn_4}{dt} = \ldots$; $\frac{dn_3}{dt} = \ldots$; $\frac{dn_2}{dt} = \ldots$

$$\frac{dn_4}{dt} = \frac{\eta_1 \cdot I \cdot N_p}{|e|} - \left(\frac{1}{\tau_{41}} + \frac{1}{\tau_{43}} + \frac{1}{\tau_{42}} + \frac{1}{\tau_{sp1}} \cdot \frac{\Gamma_1}{N_p} \right) \cdot n_4 \\ - (n_4 - n_3) \cdot \frac{\Gamma_1}{N_p} \cdot f_1(N_p, \tau_{ph1}, \ldots) \cdot \omega$$

$$\frac{dn_3}{dt} = \frac{(1 - \eta_1) \cdot I \cdot N_p}{|e|} - \left(\frac{1}{\tau_{32}} + \frac{1}{\tau_{31}} \right) \cdot n_3 + \left(\frac{1}{\tau_{43}} + \frac{1}{\tau_{sp1}} \cdot \frac{\Gamma_1}{N_p} \right) \cdot n_4 \\ + (n_4 - n_3) \cdot \frac{\Gamma_1}{N_p} \cdot f_1(N_p, \tau_{ph1}, \ldots) \cdot \omega$$

$$\frac{dn_2}{dt} = \frac{\eta_2 \cdot I \cdot N_p}{|e|} - \left(\frac{1}{\tau_{21}} + \frac{1}{\tau_{2i}} + \frac{1}{\tau_{sp2}} \cdot \frac{\Gamma_2}{N_p} \right) \cdot n_2$$

$$- (n_2 - n_1) \cdot \frac{\Gamma_2}{N_p} \cdot f_2(N_p, \tau_{ph1}, \ldots) \cdot \omega$$

$$\frac{dn_1}{dt} = \frac{(1-\eta_2) \cdot I \cdot N_p}{|e|} - \frac{n_1}{\tau_{1i}} + \frac{n_4}{\tau_{41}} + \frac{n_2}{\tau_{21}} + \frac{n_2}{\tau_{sp2}} \cdot \frac{\Gamma_2}{N_p}$$
$$+ (n_2 - n_1) \cdot \frac{\Gamma_2}{N_p} \cdot f_2(N_p, \tau_{ph1}, \ldots) \cdot \omega$$

The five levels QCL (quantum cascade laser) system has very important feature, the injection efficiencies. In the ideal case $\eta = 1$ and the value is related to either thermal activation from the injector to the continuum, higher excited states, which have been nearly canceled in this structure, or direct injection from the injector to the lower states. The approximations for injection efficiencies: $\eta_1 \approx \frac{R_{i'4}}{R_{i'4}+R_{i'3}}$; $\eta_2 \approx \frac{R_{32}+R_{42}}{R_{32}+R_{42}+R_{31}}$. These approximations results in $\eta_1 \approx 0.91$; $\eta_2 \approx 0.98$, where sample R represents the transition rate $(\frac{1}{\tau})$ between the involved sub bands. In η_1, η_2 expressions, the numerators are the net injection rates into the upper radiative levels and the denominators are the total outflow rates. If the population inversion is established in the system, each of the optical transitions leads to a net stimulated emission into a cavity mode and correspondingly, the optical power would be added to the field as it propagated through the system.

At fixed points:

$$\frac{dn_4}{dt} = 0; \frac{dn_3}{dt} = 0; \frac{dn_2}{dt} = 0; \frac{dn_1}{dt} = 0; \frac{dn_{ph}(\hbar \cdot \omega_1)}{dt} = 0; \frac{dn_{ph}(\hbar \cdot \omega_2)}{dt} = 0$$

$$\frac{dn_4}{dt} = 0 \Rightarrow \frac{\eta_1 \cdot I \cdot N_p}{|e|} - \left(\frac{1}{\tau_{41}} + \frac{1}{\tau_{43}} + \frac{1}{\tau_{42}} + \frac{1}{\tau_{sp1}} \cdot \frac{\Gamma_1}{N_p}\right) \cdot n_4^*$$
$$- \left(n_4^* - n_3^*\right) \cdot \frac{\Gamma_1}{N_p} \cdot f_1(N_p, \tau_{ph1}, \ldots) \cdot \omega = 0$$

$$\frac{dn_3}{dt} = 0 \Rightarrow \frac{(1-\eta_1) \cdot I \cdot N_p}{|e|} - \left(\frac{1}{\tau_{32}} + \frac{1}{\tau_{31}}\right) \cdot n_3^* + \left(\frac{1}{\tau_{43}} + \frac{1}{\tau_{sp1}} \cdot \frac{\Gamma_1}{N_p}\right) \cdot n_4^*$$
$$+ \left(n_4^* - n_3^*\right) \cdot \frac{\Gamma_1}{N_p} \cdot f_1(N_p, \tau_{ph1}, \ldots) \cdot \omega = 0$$

$$\frac{dn_2}{dt} = 0 \Rightarrow \frac{\eta_2 \cdot I \cdot N_p}{|e|} - \left(\frac{1}{\tau_{21}} + \frac{1}{\tau_{2i}} + \frac{1}{\tau_{sp2}} \cdot \frac{\Gamma_2}{N_p}\right) \cdot n_2^*$$
$$- \left(n_2^* - n_1^*\right) \cdot \frac{\Gamma_2}{N_p} \cdot f_2(N_p, \tau_{ph1}, \ldots) \cdot \omega = 0$$

$$\frac{dn_1}{dt} = 0 \Rightarrow \frac{(1-\eta_2) \cdot I \cdot N_p}{|e|} - \frac{n_1^*}{\tau_{1i}} + \frac{n_4^*}{\tau_{41}} + \frac{n_2^*}{\tau_{21}} + \frac{n_2^*}{\tau_{sp2}} \cdot \frac{\Gamma_2}{N_p}$$
$$+ \left(n_2^* - n_1^*\right) \cdot \frac{\Gamma_2}{N_p} \cdot f_2(N_p, \tau_{ph1}, \ldots) \cdot \omega = 0$$

$$\frac{dn_{ph}(\hbar\cdot\omega_1)}{dt}=0\Rightarrow\left(n_4^*-n_3^*\right)\cdot\frac{\Gamma_1}{N_p}\cdot\frac{n_{ph}^*(\hbar\cdot\omega_1)}{\tau_{sp}(\hbar\cdot\omega_1)}-\frac{n_{ph}^*(\hbar\cdot\omega_1)}{\tau_{ph1}}=0$$

$$\frac{dn_{ph}(\hbar\cdot\omega_2)}{dt}=0\Rightarrow\left(n_2^*-n_1^*\right)\cdot\frac{\Gamma_2}{N_p}\cdot\frac{n_{ph}^*(\hbar\cdot\omega_2)}{\tau_{sp}(\hbar\cdot\omega_2)}-\frac{n_{ph}^*(\hbar\cdot\omega_2)}{\tau_{ph2}}=0$$

First fixed point's equation:

$$\frac{\eta_1\cdot I\cdot N_p}{|e|}-\left(\frac{1}{\tau_{41}}+\frac{1}{\tau_{43}}+\frac{1}{\tau_{42}}+\frac{1}{\tau_{sp1}}\cdot\frac{\Gamma_1}{N_p}\right)\cdot n_4^* \\ -\left(n_4^*-n_3^*\right)\cdot\frac{\Gamma_1}{N_p}\cdot f_1\left(N_p,\tau_{ph1},\ldots\right)\cdot\omega=0$$

$$\frac{\eta_1\cdot I\cdot N_p}{|e|}-\left(\frac{1}{\tau_{41}}+\frac{1}{\tau_{43}}+\frac{1}{\tau_{42}}+\frac{1}{\tau_{sp1}}\cdot\frac{\Gamma_1}{N_p}\right)\cdot n_4^* \\ -n_4^*\cdot\frac{\Gamma_1}{N_p}\cdot f_1\left(N_p,\tau_{ph1},\ldots\right)\cdot\omega+n_3^*\cdot\frac{\Gamma_1}{N_p}\cdot f_1\left(N_p,\tau_{ph1},\ldots\right)\cdot\omega=0$$

$$\frac{\eta_1\cdot I\cdot N_p}{|e|}+n_3^*\cdot\frac{\Gamma_1}{N_p}\cdot f_1\left(N_p,\tau_{ph1},\ldots\right)\cdot\omega \\ =n_4^*\cdot\frac{\Gamma_1}{N_p}\cdot f_1\left(N_p,\tau_{ph1},\ldots\right)\cdot\omega+\left(\frac{1}{\tau_{41}}+\frac{1}{\tau_{43}}+\frac{1}{\tau_{42}}+\frac{1}{\tau_{sp1}}\cdot\frac{\Gamma_1}{N_p}\right)\cdot n_4^*$$

$$\left[\frac{\Gamma_1}{N_p}\cdot f_1\left(N_p,\tau_{ph1},\ldots\right)\cdot\omega+\left(\frac{1}{\tau_{41}}+\frac{1}{\tau_{43}}+\frac{1}{\tau_{42}}+\frac{1}{\tau_{sp1}}\cdot\frac{\Gamma_1}{N_p}\right)\right]\cdot n_4^* \\ =\frac{\eta_1\cdot I\cdot N_p}{|e|}+n_3^*\cdot\frac{\Gamma_1}{N_p}\cdot f_1\left(N_p,\tau_{ph1},\ldots\right)\cdot\omega$$

$$n_4^*=\frac{\frac{\eta_1\cdot I\cdot N_p}{|e|}+n_3^*\cdot\frac{\Gamma_1}{N_p}\cdot f_1\left(N_p,\tau_{ph1},\ldots\right)\cdot\omega}{\frac{\Gamma_1}{N_p}\cdot f_1\left(N_p,\tau_{ph1},\ldots\right)\cdot\omega+\left(\frac{1}{\tau_{41}}+\frac{1}{\tau_{43}}+\frac{1}{\tau_{42}}+\frac{1}{\tau_{sp1}}\cdot\frac{\Gamma_1}{N_p}\right)}$$

Second fixed point's equation:

$$\frac{(1-\eta_1)\cdot I\cdot N_p}{|e|}-\left(\frac{1}{\tau_{32}}+\frac{1}{\tau_{31}}\right)\cdot n_3^*+\left(\frac{1}{\tau_{43}}+\frac{1}{\tau_{sp1}}\cdot\frac{\Gamma_1}{N_p}\right)\cdot n_4^* \\ +\left(n_4^*-n_3^*\right)\cdot\frac{\Gamma_1}{N_p}\cdot f_1\left(N_p,\tau_{ph1},\ldots\right)\cdot\omega=0$$

$$\frac{(1-\eta_1)\cdot I\cdot N_p}{|e|}-\left(\frac{1}{\tau_{32}}+\frac{1}{\tau_{31}}\right)\cdot n_3^*+\left(\frac{1}{\tau_{43}}+\frac{1}{\tau_{sp1}}\cdot\frac{\Gamma_1}{N_p}\right)\cdot n_4^* \\ +n_4^*\cdot\frac{\Gamma_1}{N_p}\cdot f_1\left(N_p,\tau_{ph1},\ldots\right)\cdot\omega-n_3^*\cdot\frac{\Gamma_1}{N_p}\cdot f_1\left(N_p,\tau_{ph1},\ldots\right)\cdot\omega=0$$

$$\frac{(1-\eta_1)\cdot I\cdot N_p}{|e|}-\left(\frac{1}{\tau_{32}}+\frac{1}{\tau_{31}}\right)\cdot n_3^*+\left[\left(\frac{1}{\tau_{43}}+\frac{1}{\tau_{sp1}}\cdot\frac{\Gamma_1}{N_p}\right)\right.$$
$$\left.+\frac{\Gamma_1}{N_p}\cdot f_1\left(N_p,\tau_{ph1},\ldots\right)\cdot\omega\right]\cdot n_4^*$$
$$-n_3^*\cdot\frac{\Gamma_1}{N_p}\cdot f_1\left(N_p,\tau_{ph1},\ldots\right)\cdot\omega=0$$

$$\frac{(1-\eta_1)\cdot I\cdot N_p}{|e|}-\left[\left(\frac{1}{\tau_{32}}+\frac{1}{\tau_{31}}\right)+\frac{\Gamma_1}{N_p}\cdot f_1\left(N_p,\tau_{ph1},\ldots\right)\cdot\omega\right]\cdot n_3^*$$
$$+\left[\left(\frac{1}{\tau_{43}}+\frac{1}{\tau_{sp1}}\cdot\frac{\Gamma_1}{N_p}\right)+\frac{\Gamma_1}{N_p}\cdot f_1\left(N_p,\tau_{ph1},\ldots\right)\cdot\omega\right]\cdot n_4^*=0$$

$$\frac{(1-\eta_1)\cdot I\cdot N_p}{|e|}-\left[\left(\frac{1}{\tau_{32}}+\frac{1}{\tau_{31}}\right)+\frac{\Gamma_1}{N_p}\cdot f_1\left(N_p,\tau_{ph1},\ldots\right)\cdot\omega\right]\cdot n_3^*$$
$$+\left[\left(\frac{1}{\tau_{43}}+\frac{1}{\tau_{sp1}}\cdot\frac{\Gamma_1}{N_p}\right)+\frac{\Gamma_1}{N_p}\cdot f_1\left(N_p,\tau_{ph1},\ldots\right)\cdot\omega\right]$$
$$\cdot\left[\frac{\frac{\eta_1\cdot I\cdot N_p}{|e|}+n_3^*\cdot\frac{\Gamma_1}{N_p}\cdot f_1\left(N_p,\tau_{ph1},\ldots\right)\cdot\omega}{\frac{\Gamma_1}{N_p}\cdot f_1\left(N_p,\tau_{ph1},\ldots\right)\cdot\omega+\left(\frac{1}{\tau_{41}}+\frac{1}{\tau_{43}}+\frac{1}{\tau_{42}}+\frac{1}{\tau_{sp1}}\cdot\frac{\Gamma_1}{N_p}\right)}\right]=0$$

$$\frac{(1-\eta_1)\cdot I\cdot N_p}{|e|}-\left[\left(\frac{1}{\tau_{32}}+\frac{1}{\tau_{31}}\right)+\frac{\Gamma_1}{N_p}\cdot f_1\left(N_p,\tau_{ph1},\ldots\right)\cdot\omega\right]\cdot n_3^*$$
$$+\left\{\frac{\left[\left(\frac{1}{\tau_{43}}+\frac{1}{\tau_{sp1}}\cdot\frac{\Gamma_1}{N_p}\right)+\frac{\Gamma_1}{N_p}\cdot f_1\left(N_p,\tau_{ph1},\ldots\right)\cdot\omega\right]}{\frac{\Gamma_1}{N_p}\cdot f_1\left(N_p,\tau_{ph1},\ldots\right)\cdot\omega+\left(\frac{1}{\tau_{41}}+\frac{1}{\tau_{43}}+\frac{1}{\tau_{42}}+\frac{1}{\tau_{sp1}}\cdot\frac{\Gamma_1}{N_p}\right)}\right\}$$
$$\cdot\left[\frac{\eta_1\cdot I\cdot N_p}{|e|}+n_3^*\cdot\frac{\Gamma_1}{N_p}\cdot f_1\left(N_p,\tau_{ph1},\ldots\right)\cdot\omega\right]=0$$

We define for simplicity some global parameters:

$$\Xi_1=\frac{(1-\eta_1)\cdot I\cdot N_p}{|e|}$$

$$\Xi_2=\left(\frac{1}{\tau_{32}}+\frac{1}{\tau_{31}}\right)+\frac{\Gamma_1}{N_p}\cdot f_1\left(N_p,\tau_{ph1},\ldots\right)\cdot\omega$$
$$\Xi_3=\frac{\left[\left(\frac{1}{\tau_{43}}+\frac{1}{\tau_{sp1}}\cdot\frac{\Gamma_1}{N_p}\right)+\frac{\Gamma_1}{N_p}\cdot f_1\left(N_p,\tau_{ph1},\ldots\right)\cdot\omega\right]}{\frac{\Gamma_1}{N_p}\cdot f_1\left(N_p,\tau_{ph1},\ldots\right)\cdot\omega+\left(\frac{1}{\tau_{41}}+\frac{1}{\tau_{43}}+\frac{1}{\tau_{42}}+\frac{1}{\tau_{sp1}}\cdot\frac{\Gamma_1}{N_p}\right)}$$

$$\Xi_4=\frac{\eta_1\cdot I\cdot N_p}{|e|};\ \Xi_5=\frac{\Gamma_1}{N_p}\cdot f_1\left(N_p,\tau_{ph1},\ldots\right)\cdot\omega;\ \Xi_2=\left(\frac{1}{\tau_{32}}+\frac{1}{\tau_{31}}\right)+\Xi_5$$

$$\Xi_3 = \frac{\left[\left(\frac{1}{\tau_{43}} + \frac{1}{\tau_{sp1}} \cdot \frac{\Gamma_1}{N_p}\right) + \Xi_5\right]}{\Xi_5 + \left(\frac{1}{\tau_{41}} + \frac{1}{\tau_{43}} + \frac{1}{\tau_{42}} + \frac{1}{\tau_{sp1}} \cdot \frac{\Gamma_1}{N_p}\right)}$$

$$\Xi_1 - \Xi_2 \cdot n_3^* + \Xi_3 \cdot \left[\Xi_4 + n_3^* \cdot \Xi_5\right] = 0 \Rightarrow n_3^* = \frac{\Xi_1 + \Xi_3 \cdot \Xi_4}{\Xi_2 - \Xi_3 \cdot \Xi_5}$$

$$n_4^* = \frac{\frac{\eta_1 \cdot I \cdot N_p}{|e|} + n_3^* \cdot \frac{\Gamma_1}{N_p} \cdot f_1(N_p, \tau_{ph1}, \ldots) \cdot \omega}{\frac{\Gamma_1}{N_p} \cdot f_1(N_p, \tau_{ph1}, \ldots) \cdot \omega + \left(\frac{1}{\tau_{41}} + \frac{1}{\tau_{43}} + \frac{1}{\tau_{42}} + \frac{1}{\tau_{sp1}} \cdot \frac{\Gamma_1}{N_p}\right)} = \frac{\frac{\eta_1 \cdot I \cdot N_p}{|e|} + n_3^* \cdot \Xi_5}{\Xi_5 + \left(\frac{1}{\tau_{41}} + \frac{1}{\tau_{43}} + \frac{1}{\tau_{42}} + \frac{1}{\tau_{sp1}} \cdot \frac{\Gamma_1}{N_p}\right)}$$

$$n_4^* = \frac{\frac{\eta_1 \cdot I \cdot N_p}{|e|} + n_3^* \cdot \frac{\Gamma_1}{N_p} \cdot f_1(N_p, \tau_{ph1}, \ldots) \cdot \omega}{\frac{\Gamma_1}{N_p} \cdot f_1(N_p, \tau_{ph1}, \ldots) \cdot \omega + \left(\frac{1}{\tau_{41}} + \frac{1}{\tau_{43}} + \frac{1}{\tau_{42}} + \frac{1}{\tau_{sp1}} \cdot \frac{\Gamma_1}{N_p}\right)} = \frac{\frac{\eta_1 \cdot I \cdot N_p}{|e|} + \left[\frac{\Xi_1 + \Xi_3 \cdot \Xi_4}{\Xi_2 - \Xi_3 \cdot \Xi_5}\right] \cdot \Xi_5}{\Xi_5 + \left(\frac{1}{\tau_{41}} + \frac{1}{\tau_{43}} + \frac{1}{\tau_{42}} + \frac{1}{\tau_{sp1}} \cdot \frac{\Gamma_1}{N_p}\right)}$$

We define for simplicity the following global parameters:

$$\Xi_6 = \frac{\eta_1 \cdot I \cdot N_p}{|e|};\ \Xi_7 = \frac{1}{\tau_{41}} + \frac{1}{\tau_{43}} + \frac{1}{\tau_{42}} + \frac{1}{\tau_{sp1}} \cdot \frac{\Gamma_1}{N_p}$$

$$n_4^* = \frac{\Xi_6 + \left[\frac{\Xi_1 + \Xi_3 \cdot \Xi_4}{\Xi_2 - \Xi_3 \cdot \Xi_5}\right] \cdot \Xi_5}{\Xi_5 + \Xi_7}$$

Third fixed point's equation:

$$\frac{\eta_2 \cdot I \cdot N_p}{|e|} - \left(\frac{1}{\tau_{21}} + \frac{1}{\tau_{2i}} + \frac{1}{\tau_{sp2}} \cdot \frac{\Gamma_2}{N_p}\right) \cdot n_2^* - (n_2^* - n_1^*) \cdot \frac{\Gamma_2}{N_p} \cdot f_2(N_p, \tau_{ph1}, \ldots) \cdot \omega = 0$$

$$\frac{\eta_2 \cdot I \cdot N_p}{|e|} - \left(\frac{1}{\tau_{21}} + \frac{1}{\tau_{2i}} + \frac{1}{\tau_{sp2}} \cdot \frac{\Gamma_2}{N_p}\right) \cdot n_2^* - n_2^* \cdot \frac{\Gamma_2}{N_p} \cdot f_2(N_p, \tau_{ph1}, \ldots) \cdot \omega + n_1^* \cdot \frac{\Gamma_2}{N_p} \cdot f_2(N_p, \tau_{ph1}, \ldots) \cdot \omega = 0$$

$$\left[\left(\frac{1}{\tau_{21}} + \frac{1}{\tau_{2i}} + \frac{1}{\tau_{sp2}} \cdot \frac{\Gamma_2}{N_p}\right) + \frac{\Gamma_2}{N_p} \cdot f_2(N_p, \tau_{ph1}, \ldots) \cdot \omega\right] \cdot n_2^*$$

$$= \frac{\eta_2 \cdot I \cdot N_p}{|e|} + n_1^* \cdot \frac{\Gamma_2}{N_p} \cdot f_2(N_p, \tau_{ph1}, \ldots) \cdot \omega$$

$$n_2^* = \frac{\frac{\eta_2 \cdot I \cdot N_p}{|e|} + n_1^* \cdot \frac{\Gamma_2}{N_p} \cdot f_2(N_p, \tau_{ph1}, \ldots) \cdot \omega}{\left(\frac{1}{\tau_{21}} + \frac{1}{\tau_{2i}} + \frac{1}{\tau_{sp2}} \cdot \frac{\Gamma_2}{N_p}\right) + \frac{\Gamma_2}{N_p} \cdot f_2(N_p, \tau_{ph1}, \ldots) \cdot \omega} = \frac{\Xi_8 + n_1^* \cdot \Xi_9}{\Xi_{10} + \Xi_9}$$

We define for simplicity some global parameters:

$$\Xi_8 = \frac{\eta_2 \cdot I \cdot N_p}{|e|}; \ \Xi_9 = \frac{\Gamma_2}{N_p} \cdot f_2(N_p, \tau_{ph1}, \ldots) \cdot \omega$$

$$\Xi_{10} = \frac{1}{\tau_{21}} + \frac{1}{\tau_{2i}} + \frac{1}{\tau_{sp2}} \cdot \frac{\Gamma_2}{N_p}; \ n_2^* = \frac{\Xi_8 + n_1^* \cdot \Xi_9}{\Xi_{10} + \Xi_9}$$

Forth fixed point's equation:

$$\frac{(1-\eta_2) \cdot I \cdot N_p}{|e|} - \frac{n_1^*}{\tau_{1i}} + \frac{n_4^*}{\tau_{41}} + \frac{n_2^*}{\tau_{21}} + \frac{n_2^*}{\tau_{sp2}} \cdot \frac{\Gamma_2}{N_p} + (n_2^* - n_1^*) \cdot \frac{\Gamma_2}{N_p} \cdot f_2(N_p, \tau_{ph1}, \ldots) \cdot \omega = 0$$

$$\frac{(1-\eta_2) \cdot I \cdot N_p}{|e|} - \frac{n_1^*}{\tau_{1i}} + \frac{n_4^*}{\tau_{41}} + \frac{n_2^*}{\tau_{21}} + \frac{n_2^*}{\tau_{sp2}} \cdot \frac{\Gamma_2}{N_p} + n_2^* \cdot \frac{\Gamma_2}{N_p} \cdot f_2(N_p, \tau_{ph1}, \ldots) \cdot \omega - n_1^* \cdot \frac{\Gamma_2}{N_p} \cdot f_2(N_p, \tau_{ph1}, \ldots) \cdot \omega = 0$$

$$\frac{(1-\eta_2) \cdot I \cdot N_p}{|e|} - n_1^* \cdot \left[\frac{1}{\tau_{1i}} + \frac{\Gamma_2}{N_p} \cdot f_2(N_p, \tau_{ph1}, \ldots) \cdot \omega\right] + \frac{n_4^*}{\tau_{41}} + n_2^* \cdot \left[\frac{1}{\tau_{21}} + \frac{1}{\tau_{sp2}} \cdot \frac{\Gamma_2}{N_p} + \frac{\Gamma_2}{N_p} \cdot f_2(N_p, \tau_{ph1}, \ldots) \cdot \omega\right] = 0$$

$$n_1^* = \frac{\frac{(1-\eta_2) \cdot I \cdot N_p}{|e|} + \frac{n_4^*}{\tau_{41}} + n_2^* \cdot \left[\frac{1}{\tau_{21}} + \frac{1}{\tau_{sp2}} \cdot \frac{\Gamma_2}{N_p} + \frac{\Gamma_2}{N_p} \cdot f_2(N_p, \tau_{ph1}, \ldots) \cdot \omega\right]}{\frac{1}{\tau_{1i}} + \frac{\Gamma_2}{N_p} \cdot f_2(N_p, \tau_{ph1}, \ldots) \cdot \omega}$$

We define for simplicity some global parameters:

$$\Xi_{11} = \frac{(1-\eta_2) \cdot I \cdot N_p}{|e|}; \ \Xi_{12} = \frac{1}{\tau_{21}} + \frac{1}{\tau_{sp2}} \cdot \frac{\Gamma_2}{N_p}$$

$$n_1^* = \frac{\Xi_{11} + \frac{n_4^*}{\tau_{41}} + n_2^* \cdot [\Xi_{12} + \Xi_9]}{\frac{1}{\tau_{1i}} + \Xi_9}$$

$$n_1^* = \frac{\Xi_{11} + \frac{1}{\tau_{41}} \cdot \left[\frac{\Xi_6 + \left[\frac{\Xi_1 + \Xi_3 \cdot \Xi_4}{\Xi_2 - \Xi_3 \cdot \Xi_5}\right] \cdot \Xi_5}{\Xi_5 + \Xi_7}\right] + \left[\frac{\Xi_8 + n_1^* \cdot \Xi_9}{\Xi_{10} + \Xi_9}\right] \cdot [\Xi_{12} + \Xi_9]}{\frac{1}{\tau_{1i}} + \Xi_9}$$

Fifth fixed point's equation:

$$\left(n_4^* - n_3^*\right) \cdot \frac{\Gamma_1}{N_p} \cdot \frac{n_{ph}^*(\hbar \cdot \omega_1)}{\tau_{sp}(\hbar \cdot \omega_1)} - \frac{n_{ph}^*(\hbar \cdot \omega_1)}{\tau_{ph1}} = 0$$

$$n_{ph}^*(\hbar \cdot \omega_1) \cdot \left[\left(n_4^* - n_3^*\right) \cdot \frac{\Gamma_1}{N_p} \cdot \frac{1}{\tau_{sp}(\hbar \cdot \omega_1)} - \frac{1}{\tau_{ph1}}\right] = 0$$

Case A: $n_{ph}^*(\hbar \cdot \omega_1) = 0$; Case B: $\left(n_4^* - n_3^*\right) \cdot \frac{\Gamma_1}{N_p} \cdot \frac{1}{\tau_{sp}(\hbar \cdot \omega_1)} - \frac{1}{\tau_{ph1}} = 0$

$$n_4^* - n_3^* = \frac{\frac{1}{\tau_{ph1}}}{\frac{\Gamma_1}{N_p} \cdot \frac{1}{\tau_{sp}(\hbar \cdot \omega_1)}} = \frac{N_p \cdot \tau_{sp}(\hbar \cdot \omega_1)}{\tau_{ph1} \cdot \Gamma_1}; n_4^* - n_3^* = \frac{N_p \cdot \tau_{sp}(\hbar \cdot \omega_1)}{\tau_{ph1} \cdot \Gamma_1}$$

$$\left[\frac{\frac{\eta_1 \cdot I \cdot N_p}{|e|} + \left[\frac{\Xi_1 + \Xi_3 \cdot \Xi_4}{\Xi_2 - \Xi_3 \cdot \Xi_5}\right] \cdot \Xi_5}{\Xi_5 + \left(\frac{1}{\tau_{41}} + \frac{1}{\tau_{43}} + \frac{1}{\tau_{42}} + \frac{1}{\tau_{sp1}} \cdot \frac{\Gamma_1}{N_p}\right)}\right] - \left[\frac{\Xi_1 + \Xi_3 \cdot \Xi_4}{\Xi_2 - \Xi_3 \cdot \Xi_5}\right] = \frac{N_p \cdot \tau_{sp}(\hbar \cdot \omega_1)}{\tau_{ph1} \cdot \Gamma_1}$$

Sixth fixed point's equation: $\left(n_2^* - n_1^*\right) \cdot \frac{\Gamma_2}{N_p} \cdot \frac{n_{ph}^*(\hbar \cdot \omega_2)}{\tau_{sp}(\hbar \cdot \omega_2)} - \frac{n_{ph}^*(\hbar \cdot \omega_2)}{\tau_{ph2}} = 0$

$$\left[\left(n_2^* - n_1^*\right) \cdot \frac{\Gamma_2}{N_p} \cdot \frac{1}{\tau_{sp}(\hbar \cdot \omega_2)} - \frac{1}{\tau_{ph2}}\right] \cdot n_{ph}^*(\hbar \cdot \omega_2) = 0$$

Case C: $n_{ph}^*(\hbar \cdot \omega_2) = 0$; Case D: $\left(n_2^* - n_1^*\right) \cdot \frac{\Gamma_2}{N_p} \cdot \frac{1}{\tau_{sp}(\hbar \cdot \omega_2)} - \frac{1}{\tau_{ph2}} = 0$

$$n_2^* - n_1^* = \frac{\frac{1}{\tau_{ph2}}}{\frac{\Gamma_2}{N_p} \cdot \frac{1}{\tau_{sp}(\hbar \cdot \omega_2)}} = \frac{N_p \cdot \tau_{sp}(\hbar \cdot \omega_2)}{\tau_{ph2} \cdot \Gamma_2}$$

$$\left[\frac{\Xi_8 + n_1^* \cdot \Xi_9}{\Xi_{10} + \Xi_9}\right] - \left[\frac{\Xi_{11} + \frac{1}{\tau_{41}} \cdot \left[\frac{\Xi_6 + \left[\frac{\Xi_1 + \Xi_3 \cdot \Xi_4}{\Xi_2 - \Xi_3 \cdot \Xi_5}\right] \cdot \Xi_5}{\Xi_5 + \Xi_7}\right] + \left[\frac{\Xi_8 + n_1^* \cdot \Xi_9}{\Xi_{10} + \Xi_9}\right] \cdot [\Xi_{12} + \Xi_9]}{\frac{1}{\tau_{1i}} + \Xi_9}\right]$$

$$= \frac{N_p \cdot \tau_{sp}(\hbar \cdot \omega_2)}{\tau_{ph2} \cdot \Gamma_2}$$

We can summary our system fixed points coordinates:

$$n_2^* = \frac{\Xi_8 + n_1^* \cdot \Xi_9}{\Xi_{10} + \Xi_9}; n_3^* = \frac{\Xi_1 + \Xi_3 \cdot \Xi_4}{\Xi_2 - \Xi_3 \cdot \Xi_5}; n_4^* = \frac{\Xi_6 + \left[\frac{\Xi_1 + \Xi_3 \cdot \Xi_4}{\Xi_2 - \Xi_3 \cdot \Xi_5}\right] \cdot \Xi_5}{\Xi_5 + \Xi_7}$$

$$n_1^* = \frac{\Xi_{11} + \frac{1}{\tau_{41}} \cdot \left[\frac{\Xi_6 + \left[\frac{\Xi_1 + \Xi_3 \cdot \Xi_4}{\Xi_2 - \Xi_3 \cdot \Xi_5}\right] \cdot \Xi_5}{\Xi_5 + \Xi_7}\right] + \left[\frac{\Xi_8 + n_1^* \cdot \Xi_9}{\Xi_{10} + \Xi_9}\right] \cdot [\Xi_{12} + \Xi_9]}{\frac{1}{\tau_{1i}} + \Xi_9}$$

$$n_{ph}^*(\hbar \cdot \omega_1) = 0 \,\text{or}\, n_{ph}^*(\hbar \cdot \omega_1) \neq 0; n_4^* - n_3^* = \frac{N_p \cdot \tau_{sp}(\hbar \cdot \omega_1)}{\tau_{ph1} \cdot \Gamma_1}.$$

$$n_{ph}^*(\hbar \cdot \omega_2) = 0 \,\text{or}\, n_{ph}^*(\hbar \cdot \omega_2) \neq 0; n_2^* - n_1^* = \frac{N_p \cdot \tau_{sp}(\hbar \cdot \omega_2)}{\tau_{ph2} \cdot \Gamma_2}.$$

We can summary our system fixed points:

$$E^{(0)} = \left(n_1^{(0)}, n_2^{(0)}, n_3^{(0)}, n_4^{(0)}, n_{ph}^{(0)}(\hbar \cdot \omega_1), n_{ph}^{(0)}(\hbar \cdot \omega_2)\right) = \left(n_1^*, n_2^*, n_3^*, n_4^*, 0, 0\right)$$

$$\begin{aligned} E^{(1)} &= \left(n_1^{(1)}, n_2^{(1)}, n_3^{(1)}, n_4^{(1)}, n_{ph}^{(1)}(\hbar \cdot \omega_1), n_{ph}^{(1)}(\hbar \cdot \omega_2)\right) \\ &= \left(n_1^*, n_2^*, n_3^*, n_4^*, n_{ph}^*(\hbar \cdot \omega_1) \neq 0; n_4^* - n_3^* = \frac{N_p \cdot \tau_{sp}(\hbar \cdot \omega_1)}{\tau_{ph1} \cdot \Gamma_1}, 0\right) \end{aligned}$$

$$\begin{aligned} E^{(2)} &= \left(n_1^{(2)}, n_2^{(2)}, n_3^{(2)}, n_4^{(2)}, n_{ph}^{(2)}(\hbar \cdot \omega_1), n_{ph}^{(2)}(\hbar \cdot \omega_2)\right) \\ &= \left(n_1^*, n_2^*, n_3^*, n_4^*, 0, n_{ph}^*(\hbar \cdot \omega_2) \neq 0; n_2^* - n_1^* = \frac{N_p \cdot \tau_{sp}(\hbar \cdot \omega_2)}{\tau_{ph2} \cdot \Gamma_2}\right) \end{aligned}$$

$$\begin{aligned} E^{(3)} &= \left(n_1^{(3)}, n_2^{(3)}, n_3^{(3)}, n_4^{(3)}, n_{ph}^{(3)}(\hbar \cdot \omega_1), n_{ph}^{(3)}(\hbar \cdot \omega_2)\right) \\ &= \left(n_1^*, n_2^*, n_3^*, n_4^*, n_{ph}^*(\hbar \cdot \omega_1) \neq 0; n_4^* - n_3^*\right. \\ &= \frac{N_p \cdot \tau_{sp}(\hbar \cdot \omega_1)}{\tau_{ph1} \cdot \Gamma_1}, n_{ph}^*(\hbar \cdot \omega_2) \neq 0; n_2^* - n_1^* \\ &= \left.\frac{N_p \cdot \tau_{sp}(\hbar \cdot \omega_2)}{\tau_{ph2} \cdot \Gamma_2}\right) \end{aligned}$$

Stability analysis: The standard local stability analysis about any one of the equilibrium point of the THz dual-wavelength QCL system consists in adding to coordinate $[n_1, n_2, n_3, n_4, n_{ph}(\hbar \cdot \omega_1), n_{ph}(\hbar \cdot \omega_2)]$ arbitrarily small increments of exponential form $\left[n_1, n_2, n_3, n_4, n_{ph}(\hbar \cdot \omega_1), n_{ph}(\hbar \cdot \omega_2)\right] \cdot e^{\lambda \cdot t}$ and retaining the first order terms in $n_1, n_2, n_3, n_4, n_{ph}(\hbar \cdot \omega_1), n_{ph}(\hbar \cdot \omega_2)$. The system of six homogeneous equations leads to a polynomial characteristic equation in the eigenvalues. The polynomial characteristic equations accept by set the below variables and variables derivative with respect to time into THz dual-wavelength QCL system equations. THz dual-wavelength QCL system fixed values with arbitrarily small increments of exponential form $[n_1, n_2, n_3, n_4, n_{ph}(\hbar \cdot \omega_1), n_{ph}(\hbar \cdot \omega_2)] \cdot e^{\lambda \cdot t}$ are: $j = 0$ (first fixed point), $j = 1$ (second fixed point), $j = 2$ (third fixed point), etc. [2, 3].

$$n_1(t) = n_1^{(j)} + n_1 \cdot e^{\lambda \cdot t}; n_2(t) = n_2^{(j)} + n_2 \cdot e^{\lambda \cdot t}; n_3(t) = n_3^{(j)} + n_3 \cdot e^{\lambda \cdot t}$$
$$n_4(t) = n_4^{(j)} + n_4 \cdot e^{\lambda \cdot t}$$

$$n_{ph}(\hbar \cdot \omega_1)(t) = n_{ph}^{(j)}(\hbar \cdot \omega_1) + n_{ph}(\hbar \cdot \omega_1) \cdot e^{\lambda \cdot t}; n_{ph}(\hbar \cdot \omega_2)(t) = n_{ph}^{(j)}(\hbar \cdot \omega_2) + n_{ph}(\hbar \cdot \omega_2) \cdot e^{\lambda \cdot t}$$

$$\frac{dn_1(t)}{dt} = n_1 \cdot \lambda \cdot e^{\lambda \cdot t}; \frac{dn_2(t)}{dt} = n_2 \cdot \lambda \cdot e^{\lambda \cdot t}; \frac{dn_3(t)}{dt} = n_3 \cdot \lambda \cdot e^{\lambda \cdot t}$$
$$\frac{dn_4(t)}{dt} = n_4 \cdot \lambda \cdot e^{\lambda \cdot t}$$

$$\frac{dn_{ph}(\hbar \cdot \omega_1)(t)}{dt} = n_{ph}(\hbar \cdot \omega_1) \cdot \lambda \cdot e^{\lambda \cdot t}; \frac{dn_{ph}(\hbar \cdot \omega_2)(t)}{dt} = n_{ph}(\hbar \cdot \omega_2) \cdot \lambda \cdot e^{\lambda \cdot t}$$

We choose these expressions for ourselves $n_1(t), n_2(t), n_3(t), n_4(t), n_{ph}(\hbar \cdot \omega_1)(t)$, $n_1(t), n_2(t), n_3(t), n_4(t), n_{ph}(\hbar \cdot \omega_1)(t), n_{ph}(\hbar \cdot \omega_2)(t)$ as a small displacement $[n_1, n_2, n_3, n_4, n_{ph}(\hbar \cdot \omega_1), n_{ph}(\hbar \cdot \omega_2)]$ from the THz dual-wavelength QCL system fixed points in time $t = 0, n_1(t = 0) = n_1^{(j)} + n_1; n_2(t = 0) = n_2^{(j)} + n_2; n_3(t = 0) = n_3^{(j)} + n_3, n_4(t = 0) = n_4^{(j)} + n_4; n_{ph}(\hbar \cdot \omega_1)(t = 0) = n_{ph}^{(j)}(\hbar \cdot \omega_1) + n_{ph}(\hbar \cdot \omega_1)$; $n_{ph}(\hbar \cdot \omega_2)(t = 0) = n_{ph}^{(j)}(\hbar \cdot \omega_2) + n_{ph}(\hbar \cdot \omega_2)$.

$$\frac{dn_1}{dt} = \frac{(1 - \eta_2) \cdot I \cdot N_p}{|e|} - \frac{n_1}{\tau_{1i}} + \frac{n_4}{\tau_{41}} + \frac{n_2}{\tau_{21}} + \frac{n_2}{\tau_{sp2}} \cdot \frac{\Gamma_2}{N_p} + (n_2 - n_1) \cdot \frac{\Gamma_2}{N_p} \cdot f_2(N_p, \tau_{ph1}, \ldots) \cdot \omega$$

$$n_1 \cdot \lambda \cdot e^{\lambda \cdot t} = \frac{(1 - \eta_2) \cdot I \cdot N_p}{|e|} - \left(\frac{n_1^{(j)} + n_1 \cdot e^{\lambda \cdot t}}{\tau_{1i}}\right) + \frac{\left(n_4^{(j)} + n_4 \cdot e^{\lambda \cdot t}\right)}{\tau_{41}} + \left(\frac{n_2^{(j)} + n_2 \cdot e^{\lambda \cdot t}}{\tau_{21}}\right) + \left(\frac{n_2^{(j)} + n_2 \cdot e^{\lambda \cdot t}}{\tau_{sp2}}\right) \cdot \frac{\Gamma_2}{N_p} + (n_2^{(j)} + n_2 \cdot e^{\lambda \cdot t} - n_1^{(j)} - n_1 \cdot e^{\lambda \cdot t}) \cdot \frac{\Gamma_2}{N_p} \cdot f_2(N_p, \tau_{ph1}, \ldots) \cdot \omega$$

$$n_1 \cdot \lambda \cdot e^{\lambda \cdot t} = \frac{(1-\eta_2) \cdot I \cdot N_p}{|e|} - \frac{n_1^{(j)}}{\tau_{1i}} + \frac{n_4^{(j)}}{\tau_{41}} + \frac{n_2^{(j)}}{\tau_{21}} + \frac{n_2^{(j)}}{\tau_{sp2}} \cdot \frac{\Gamma_2}{N_p}$$
$$+ \left(n_2^{(j)} - n_1^{(j)}\right) \cdot \frac{\Gamma_2}{N_p} \cdot f_2\left(N_p, \tau_{ph1}, \ldots\right) \cdot \omega$$
$$- \frac{n_1 \cdot e^{\lambda \cdot t}}{\tau_{1i}} + \frac{n_4 \cdot e^{\lambda \cdot t}}{\tau_{41}} + \frac{n_2 \cdot e^{\lambda \cdot t}}{\tau_{21}} + \frac{n_2 \cdot e^{\lambda \cdot t}}{\tau_{sp2}} \cdot \frac{\Gamma_2}{N_p}$$
$$+ (n_2 - n_1) \cdot \frac{\Gamma_2}{N_p} \cdot f_2\left(N_p, \tau_{ph1}, \ldots\right) \cdot \omega \cdot e^{\lambda \cdot t}$$

At fixed points:

$$\frac{(1-\eta_2) \cdot I \cdot N_p}{|e|} - \frac{n_1^{(j)}}{\tau_{1i}} + \frac{n_4^{(j)}}{\tau_{41}} + \frac{n_2^{(j)}}{\tau_{21}} + \frac{n_2^{(j)}}{\tau_{sp2}} \cdot \frac{\Gamma_2}{N_p}$$
$$+ \left(n_2^{(j)} - n_1^{(j)}\right) \cdot \frac{\Gamma_2}{N_p} \cdot f_2\left(N_p, \tau_{ph1}, \ldots\right) \cdot \omega = 0$$

$$-\left[\frac{1}{\tau_{1i}} + \frac{\Gamma_2}{N_p} \cdot f_2\left(N_p, \tau_{ph1}, \ldots\right) \cdot \omega\right] \cdot n_1 - n_1 \cdot \lambda$$
$$+\left[\frac{1}{\tau_{21}} + \frac{1}{\tau_{sp2}} \cdot \frac{\Gamma_2}{N_p} + \frac{\Gamma_2}{N_p} \cdot f_2\left(N_p, \tau_{ph1}, \ldots\right) \cdot \omega\right] \cdot n_2 + \frac{1}{\tau_{41}} \cdot n_4 = 0$$

$$\frac{dn_2}{dt} = \frac{\eta_2 \cdot I \cdot N_p}{|e|} - \left(\frac{1}{\tau_{21}} + \frac{1}{\tau_{2i}} + \frac{1}{\tau_{sp2}} \cdot \frac{\Gamma_2}{N_p}\right) \cdot n_2$$
$$- (n_2 - n_1) \cdot \frac{\Gamma_2}{N_p} \cdot f_2\left(N_p, \tau_{ph1}, \ldots\right) \cdot \omega$$

$$n_2 \cdot \lambda \cdot e^{\lambda \cdot t} = \frac{\eta_2 \cdot I \cdot N_p}{|e|} - \left(\frac{1}{\tau_{21}} + \frac{1}{\tau_{2i}} + \frac{1}{\tau_{sp2}} \cdot \frac{\Gamma_2}{N_p}\right) \cdot \left(n_2^{(j)} + n_2 \cdot e^{\lambda \cdot t}\right)$$
$$- \left(n_2^{(j)} + n_2 \cdot e^{\lambda \cdot t} - n_1^{(j)} - n_1 \cdot e^{\lambda \cdot t}\right) \cdot \frac{\Gamma_2}{N_p} \cdot f_2\left(N_p, \tau_{ph1}, \ldots\right) \cdot \omega$$

$$n_2 \cdot \lambda \cdot e^{\lambda \cdot t} = \frac{\eta_2 \cdot I \cdot N_p}{|e|} - \left(\frac{1}{\tau_{21}} + \frac{1}{\tau_{2i}} + \frac{1}{\tau_{sp2}} \cdot \frac{\Gamma_2}{N_p}\right) \cdot n_2^{(j)}$$
$$- \left(n_2^{(j)} - n_1^{(j)}\right) \cdot \frac{\Gamma_2}{N_p} \cdot f_2\left(N_p, \tau_{ph1}, \ldots\right) \cdot \omega$$
$$- \left(\frac{1}{\tau_{21}} + \frac{1}{\tau_{2i}} + \frac{1}{\tau_{sp2}} \cdot \frac{\Gamma_2}{N_p}\right) \cdot n_2 \cdot e^{\lambda \cdot t} - (n_2 \cdot e^{\lambda \cdot t}$$
$$- n_1 \cdot e^{\lambda \cdot t}) \cdot \frac{\Gamma_2}{N_p} \cdot f_2\left(N_p, \tau_{ph1}, \ldots\right) \cdot \omega$$

At fixed points:

$$\frac{\eta_2 \cdot I \cdot N_p}{|e|} - \left(\frac{1}{\tau_{21}} + \frac{1}{\tau_{2i}} + \frac{1}{\tau_{sp2}} \cdot \frac{\Gamma_2}{N_p}\right) \cdot n_2^{(j)}$$
$$- \left(n_2^{(j)} - n_1^{(j)}\right) \cdot \frac{\Gamma_2}{N_p} \cdot f_2(N_p, \tau_{ph1}, \ldots) \cdot \omega = 0$$

$$\frac{\Gamma_2}{N_p} \cdot f_2(N_p, \tau_{ph1}, \ldots) \cdot \omega \cdot n_1 - \left[\left(\frac{1}{\tau_{21}} + \frac{1}{\tau_{2i}} + \frac{1}{\tau_{sp2}} \cdot \frac{\Gamma_2}{N_p}\right)\right.$$
$$\left. + \frac{\Gamma_2}{N_p} \cdot f_2(N_p, \tau_{ph1}, \ldots) \cdot \omega\right] \cdot n_2 - n_2 \cdot \lambda = 0$$

$$\frac{dn_3}{dt} = \frac{(1-\eta_1) \cdot I \cdot N_p}{|e|} - \left(\frac{1}{\tau_{32}} + \frac{1}{\tau_{31}}\right) \cdot n_3 + \left(\frac{1}{\tau_{43}} + \frac{1}{\tau_{sp1}} \cdot \frac{\Gamma_1}{N_p}\right) \cdot n_4$$
$$+ (n_4 - n_3) \cdot \frac{\Gamma_1}{N_p} \cdot f_1(N_p, \tau_{ph1}, \ldots) \cdot \omega$$

$$n_3 \cdot \lambda \cdot e^{\lambda \cdot t} = \frac{(1-\eta_1) \cdot I \cdot N_p}{|e|} - \left(\frac{1}{\tau_{32}} + \frac{1}{\tau_{31}}\right) \cdot \left(n_3^{(j)} + n_3 \cdot e^{\lambda \cdot t}\right)$$
$$+ \left(\frac{1}{\tau_{43}} + \frac{1}{\tau_{sp1}} \cdot \frac{\Gamma_1}{N_p}\right) \cdot \left(n_4^{(j)} + n_4 \cdot e^{\lambda \cdot t}\right)$$
$$+ \left(n_4^{(j)} + n_4 \cdot e^{\lambda \cdot t} - n_3^{(j)} - n_3 \cdot e^{\lambda \cdot t}\right) \cdot \frac{\Gamma_1}{N_p} \cdot f_1(N_p, \tau_{ph1}, \ldots) \cdot \omega$$

$$n_3 \cdot \lambda \cdot e^{\lambda \cdot t} = \frac{(1-\eta_1) \cdot I \cdot N_p}{|e|} - \left(\frac{1}{\tau_{32}} + \frac{1}{\tau_{31}}\right) \cdot n_3^{(j)} + \left(\frac{1}{\tau_{43}} + \frac{1}{\tau_{sp1}} \cdot \frac{\Gamma_1}{N_p}\right) \cdot n_4^{(j)}$$
$$+ \left(n_4^{(j)} - n_3^{(j)}\right) \cdot \frac{\Gamma_1}{N_p} \cdot f_1(N_p, \tau_{ph1}, \ldots) \cdot \omega - \left(\frac{1}{\tau_{32}} + \frac{1}{\tau_{31}}\right) \cdot n_3 \cdot e^{\lambda \cdot t}$$
$$- n_3 \cdot e^{\lambda \cdot t} \cdot \frac{\Gamma_1}{N_p} \cdot f_1(N_p, \tau_{ph1}, \ldots) \cdot \omega + \left(\frac{1}{\tau_{43}} + \frac{1}{\tau_{sp1}} \cdot \frac{\Gamma_1}{N_p}\right) \cdot n_4 \cdot e^{\lambda \cdot t}$$
$$+ n_4 \cdot e^{\lambda \cdot t} \cdot \frac{\Gamma_1}{N_p} \cdot f_1(N_p, \tau_{ph1}, \ldots) \cdot \omega$$

At fixed points:

$$\frac{(1-\eta_1) \cdot I \cdot N_p}{|e|} - \left(\frac{1}{\tau_{32}} + \frac{1}{\tau_{31}}\right) \cdot n_3^{(j)} + \left(\frac{1}{\tau_{43}} + \frac{1}{\tau_{sp1}} \cdot \frac{\Gamma_1}{N_p}\right) \cdot n_4^{(j)}$$
$$+ \left(n_4^{(j)} - n_3^{(j)}\right) \cdot \frac{\Gamma_1}{N_p} \cdot f_1(N_p, \tau_{ph1}, \ldots) \cdot \omega = 0$$

$$- \left[\left(\frac{1}{\tau_{32}} + \frac{1}{\tau_{31}}\right) + \frac{\Gamma_1}{N_p} \cdot f_1(N_p, \tau_{ph1}, \ldots) \cdot \omega\right] \cdot n_3 - n_3 \cdot \lambda$$
$$+ \left[\left(\frac{1}{\tau_{43}} + \frac{1}{\tau_{sp1}} \cdot \frac{\Gamma_1}{N_p}\right) + \frac{\Gamma_1}{N_p} \cdot f_1(N_p, \tau_{ph1}, \ldots) \cdot \omega\right] \cdot n_4 = 0$$

$$\frac{dn_4}{dt} = \frac{\eta_1 \cdot I \cdot N_p}{|e|} - \left(\frac{1}{\tau_{41}} + \frac{1}{\tau_{43}} + \frac{1}{\tau_{42}} + \frac{1}{\tau_{sp1}} \cdot \frac{\Gamma_1}{N_p}\right) \cdot n_4 - (n_4 - n_3) \cdot \frac{\Gamma_1}{N_p} \cdot f_1(N_p, \tau_{ph1}, \ldots) \cdot \omega$$

$$n_4 \cdot \lambda \cdot e^{\lambda \cdot t} = \frac{\eta_1 \cdot I \cdot N_p}{|e|} - \left(\frac{1}{\tau_{41}} + \frac{1}{\tau_{43}} + \frac{1}{\tau_{42}} + \frac{1}{\tau_{sp1}} \cdot \frac{\Gamma_1}{N_p}\right) \cdot \left(n_4^{(j)} + n_4 \cdot e^{\lambda \cdot t}\right) - \left(n_4^{(j)} + n_4 \cdot e^{\lambda \cdot t} - n_3^{(j)} - n_3 \cdot e^{\lambda \cdot t}\right) \cdot \frac{\Gamma_1}{N_p} \cdot f_1(N_p, \tau_{ph1}, \ldots) \cdot \omega$$

$$n_4 \cdot \lambda \cdot e^{\lambda \cdot t} = \frac{\eta_1 \cdot I \cdot N_p}{|e|} - \left(\frac{1}{\tau_{41}} + \frac{1}{\tau_{43}} + \frac{1}{\tau_{42}} + \frac{1}{\tau_{sp1}} \cdot \frac{\Gamma_1}{N_p}\right) \cdot n_4^{(j)} - \left(n_4^{(j)} - n_3^{(j)}\right) \cdot \frac{\Gamma_1}{N_p} \cdot f_1(N_p, \tau_{ph1}, \ldots) \cdot \omega - \left(\frac{1}{\tau_{41}} + \frac{1}{\tau_{43}} + \frac{1}{\tau_{42}} + \frac{1}{\tau_{sp1}} \cdot \frac{\Gamma_1}{N_p}\right) \cdot n_4 \cdot e^{\lambda \cdot t} - \left(n_4 \cdot e^{\lambda \cdot t} - n_3 \cdot e^{\lambda \cdot t}\right) \cdot \frac{\Gamma_1}{N_p} \cdot f_1(N_p, \tau_{ph1}, \ldots) \cdot \omega$$

At fixed points:

$$\frac{\eta_1 \cdot I \cdot N_p}{|e|} - \left(\frac{1}{\tau_{41}} + \frac{1}{\tau_{43}} + \frac{1}{\tau_{42}} + \frac{1}{\tau_{sp1}} \cdot \frac{\Gamma_1}{N_p}\right) \cdot n_4^{(j)} - \left(n_4^{(j)} - n_3^{(j)}\right) \cdot \frac{\Gamma_1}{N_p} \cdot f_1(N_p, \tau_{ph1}, \ldots) \cdot \omega = 0$$

$$\frac{\Gamma_1}{N_p} \cdot f_1(N_p, \tau_{ph1}, \ldots) \cdot \omega \cdot n_3 - \left[\left(\frac{1}{\tau_{41}} + \frac{1}{\tau_{43}} + \frac{1}{\tau_{42}} + \frac{1}{\tau_{sp1}} \cdot \frac{\Gamma_1}{N_p}\right) + \frac{\Gamma_1}{N_p} \cdot f_1(N_p, \tau_{ph1}, \ldots) \cdot \omega\right] \cdot n_4 - n_4 \cdot \lambda = 0$$

$$\frac{dn_{ph}(\hbar \cdot \omega_1)}{dt} = (n_4 - n_3) \cdot \frac{\Gamma_1}{N_p} \cdot \frac{n_{ph}(\hbar \cdot \omega_1)}{\tau_{sp}(\hbar \cdot \omega_1)} - \frac{n_{ph}(\hbar \cdot \omega_1)}{\tau_{ph1}}$$

$$n_{ph}(\hbar \cdot \omega_1) \cdot \lambda \cdot e^{\lambda \cdot t} = \left(n_4^{(j)} + n_4 \cdot e^{\lambda \cdot t} - n_3^{(j)} - n_3 \cdot e^{\lambda \cdot t}\right) \cdot \frac{\Gamma_1}{N_p} \cdot \frac{\left(n_{ph}^{(j)}(\hbar \cdot \omega_1) + n_{ph}(\hbar \cdot \omega_1) \cdot e^{\lambda \cdot t}\right)}{\tau_{sp}(\hbar \cdot \omega_1)} - \frac{\left(n_{ph}^{(j)}(\hbar \cdot \omega_1) + n_{ph}(\hbar \cdot \omega_1) \cdot e^{\lambda \cdot t}\right)}{\tau_{ph1}}$$

$$n_{ph}(\hbar\cdot\omega_1)\cdot\lambda\cdot e^{\lambda\cdot t} = \left(n_4^{(j)} - n_3^{(j)}\right)\cdot\frac{\Gamma_1}{N_p}\cdot\frac{\left(n_{ph}^{(j)}(\hbar\cdot\omega_1) + n_{ph}(\hbar\cdot\omega_1)\cdot e^{\lambda\cdot t}\right)}{\tau_{sp}(\hbar\cdot\omega_1)} + \left(n_4\cdot e^{\lambda\cdot t} - n_3\cdot e^{\lambda\cdot t}\right)\cdot\frac{\Gamma_1}{N_p}\cdot\frac{\left(n_{ph}^{(j)}(\hbar\cdot\omega_1) + n_{ph}(\hbar\cdot\omega_1)\cdot e^{\lambda\cdot t}\right)}{\tau_{sp}(\hbar\cdot\omega_1)} - \frac{n_{ph}^{(j)}(\hbar\cdot\omega_1)}{\tau_{ph1}} - \frac{n_{ph}(\hbar\cdot\omega_1)\cdot e^{\lambda\cdot t}}{\tau_{ph1}}$$

$$n_{ph}(\hbar\cdot\omega_1)\cdot\lambda\cdot e^{\lambda\cdot t} = \left(n_4^{(j)} - n_3^{(j)}\right)\cdot\frac{\Gamma_1}{N_p}\cdot\frac{n_{ph}^{(j)}(\hbar\cdot\omega_1)}{\tau_{sp}(\hbar\cdot\omega_1)} - \frac{n_{ph}^{(j)}(\hbar\cdot\omega_1)}{\tau_{ph1}} + \left(n_4^{(j)} - n_3^{(j)}\right)\cdot\frac{\Gamma_1}{N_p}\cdot\frac{n_{ph}(\hbar\cdot\omega_1)\cdot e^{\lambda\cdot t}}{\tau_{sp}(\hbar\cdot\omega_1)} + \left(n_4\cdot e^{\lambda\cdot t} - n_3\cdot e^{\lambda\cdot t}\right)\cdot\frac{\Gamma_1}{N_p}\cdot\frac{\left(n_{ph}^{(j)}(\hbar\cdot\omega_1) + n_{ph}(\hbar\cdot\omega_1)\cdot e^{\lambda\cdot t}\right)}{\tau_{sp}(\hbar\cdot\omega_1)} - \frac{n_{ph}(\hbar\cdot\omega_1)\cdot e^{\lambda\cdot t}}{\tau_{ph1}}$$

At fixed points:

$$\left(n_4^{(j)} - n_3^{(j)}\right)\cdot\frac{\Gamma_1}{N_p}\cdot\frac{n_{ph}^{(j)}(\hbar\cdot\omega_1)}{\tau_{sp}(\hbar\cdot\omega_1)} - \frac{n_{ph}^{(j)}(\hbar\cdot\omega_1)}{\tau_{ph1}} = 0$$

$$n_{ph}(\hbar\cdot\omega_1)\cdot\lambda\cdot e^{\lambda\cdot t} = \left(n_4^{(j)} - n_3^{(j)}\right)\cdot\frac{\Gamma_1}{N_p}\cdot\frac{n_{ph}(\hbar\cdot\omega_1)\cdot e^{\lambda\cdot t}}{\tau_{sp}(\hbar\cdot\omega_1)} + (n_4 - n_3)\cdot e^{\lambda\cdot t}\cdot\frac{\Gamma_1}{N_p}\cdot\frac{\left(n_{ph}^{(j)}(\hbar\cdot\omega_1) + n_{ph}(\hbar\cdot\omega_1)\cdot e^{\lambda\cdot t}\right)}{\tau_{sp}(\hbar\cdot\omega_1)} - \frac{n_{ph}(\hbar\cdot\omega_1)\cdot e^{\lambda\cdot t}}{\tau_{ph1}}$$

$$n_{ph}(\hbar\cdot\omega_1)\cdot\lambda\cdot e^{\lambda\cdot t} = (n_4^{(j)} - n_3^{(j)})\cdot\frac{\Gamma_1}{N_p}\cdot\frac{n_{ph}(\hbar\cdot\omega_1)\cdot e^{\lambda\cdot t}}{\tau_{sp}(\hbar\cdot\omega_1)} + [(n_4 - n_3)\cdot n_{ph}^{(j)}(\hbar\cdot\omega_1) + (n_4 - n_3)\cdot n_{ph}(\hbar\cdot\omega_1)\cdot e^{\lambda\cdot t}]\cdot e^{\lambda\cdot t}\cdot\frac{\Gamma_1}{N_p\cdot\tau_{sp}(\hbar\cdot\omega_1)} - \frac{n_{ph}(\hbar\cdot\omega_1)\cdot e^{\lambda\cdot t}}{\tau_{ph1}}$$

Assumption: $(n_4 - n_3) \cdot n_{ph}(\hbar \cdot \omega_1) \approx 0$

$$n_{ph}(\hbar \cdot \omega_1) \cdot \lambda \cdot e^{\lambda \cdot t} = \left(n_4^{(j)} - n_3^{(j)}\right) \cdot \frac{\Gamma_1}{N_p} \cdot \frac{n_{ph}(\hbar \cdot \omega_1) \cdot e^{\lambda \cdot t}}{\tau_{sp}(\hbar \cdot \omega_1)} + (n_4 - n_3) \cdot n_{ph}^{(j)}(\hbar \cdot \omega_1) \cdot e^{\lambda \cdot t} \cdot \frac{\Gamma_1}{N_p \cdot \tau_{sp}(\hbar \cdot \omega_1)} - \frac{n_{ph}(\hbar \cdot \omega_1) \cdot e^{\lambda \cdot t}}{\tau_{ph1}}$$

$$- n_3 \cdot n_{ph}^{(j)}(\hbar \cdot \omega_1) \cdot \frac{\Gamma_1}{N_p \cdot \tau_{sp}(\hbar \cdot \omega_1)} + n_4 \cdot n_{ph}^{(j)}(\hbar \cdot \omega_1) \cdot \frac{\Gamma_1}{N_p \cdot \tau_{sp}(\hbar \cdot \omega_1)} + \left[\left(n_4^{(j)} - n_3^{(j)}\right) \cdot \frac{\Gamma_1}{N_p \cdot \tau_{sp}(\hbar \cdot \omega_1)} - \frac{1}{\tau_{ph1}}\right] \cdot n_{ph}(\hbar \cdot \omega_1) - n_{ph}(\hbar \cdot \omega_1) \cdot \lambda = 0$$

$$\frac{dn_{ph}(\hbar \cdot \omega_2)}{dt} = (n_2 - n_1) \cdot \frac{\Gamma_2}{N_p} \cdot \frac{n_{ph}(\hbar \cdot \omega_2)}{\tau_{sp}(\hbar \cdot \omega_2)} - \frac{n_{ph}(\hbar \cdot \omega_2)}{\tau_{ph2}}$$

$$n_{ph}(\hbar \cdot \omega_2) \cdot \lambda \cdot e^{\lambda \cdot t} = \left(n_2^{(j)} + n_2 \cdot e^{\lambda \cdot t} - n_1^{(j)} - n_1 \cdot e^{\lambda \cdot t}\right) \cdot \frac{\Gamma_2}{N_p} \cdot \left[\frac{n_{ph}^{(j)}(\hbar \cdot \omega_2) + n_{ph}(\hbar \cdot \omega_2) \cdot e^{\lambda \cdot t}}{\tau_{sp}(\hbar \cdot \omega_2)}\right] - \left[\frac{n_{ph}^{(j)}(\hbar \cdot \omega_2) + n_{ph}(\hbar \cdot \omega_2) \cdot e^{\lambda \cdot t}}{\tau_{ph2}}\right]$$

$$n_{ph}(\hbar \cdot \omega_2) \cdot \lambda \cdot e^{\lambda \cdot t} = \left(n_2^{(j)} - n_1^{(j)}\right) \cdot \frac{\Gamma_2}{N_p} \cdot \left[\frac{n_{ph}^{(j)}(\hbar \cdot \omega_2) + n_{ph}(\hbar \cdot \omega_2) \cdot e^{\lambda \cdot t}}{\tau_{sp}(\hbar \cdot \omega_2)}\right] + (n_2 - n_1) \cdot \frac{\Gamma_2}{N_p} \cdot \left[\frac{n_{ph}^{(j)}(\hbar \cdot \omega_2) + n_{ph}(\hbar \cdot \omega_2) \cdot e^{\lambda \cdot t}}{\tau_{sp}(\hbar \cdot \omega_2)}\right] \cdot e^{\lambda \cdot t} - \frac{n_{ph}^{(j)}(\hbar \cdot \omega_2)}{\tau_{ph2}} - \frac{n_{ph}(\hbar \cdot \omega_2) \cdot e^{\lambda \cdot t}}{\tau_{ph2}}$$

$$n_{ph}(\hbar \cdot \omega_2) \cdot \lambda \cdot e^{\lambda \cdot t} = \left(n_2^{(j)} - n_1^{(j)}\right) \cdot \frac{\Gamma_2}{N_p} \cdot \frac{n_{ph}^{(j)}(\hbar \cdot \omega_2)}{\tau_{sp}(\hbar \cdot \omega_2)} + \left(n_2^{(j)} - n_1^{(j)}\right) \cdot \frac{\Gamma_2}{N_p} \cdot \frac{n_{ph}(\hbar \cdot \omega_2) \cdot e^{\lambda \cdot t}}{\tau_{sp}(\hbar \cdot \omega_2)} + (n_2 - n_1) \cdot \frac{\Gamma_2}{N_p} \cdot \frac{n_{ph}^{(j)}(\hbar \cdot \omega_2)}{\tau_{sp}(\hbar \cdot \omega_2)} \cdot e^{\lambda \cdot t}$$

$$+ (n_2 - n_1) \cdot n_{ph}(\hbar \cdot \omega_2) \cdot \frac{\Gamma_2}{N_p} \cdot \frac{e^{\lambda \cdot t}}{\tau_{sp}(\hbar \cdot \omega_2)} \cdot e^{\lambda \cdot t}$$
$$- \frac{n_{ph}^{(j)}(\hbar \cdot \omega_2)}{\tau_{ph2}} - \frac{n_{ph}(\hbar \cdot \omega_2) \cdot e^{\lambda \cdot t}}{\tau_{ph2}}$$

Assumption: $(n_2 - n_1) \cdot n_{ph}(\hbar \cdot \omega_2) \approx 0$

$$n_{ph}(\hbar \cdot \omega_2) \cdot \lambda \cdot e^{\lambda \cdot t} = \left(n_2^{(j)} - n_1^{(j)}\right) \cdot \frac{\Gamma_2}{N_p} \cdot \frac{n_{ph}^{(j)}(\hbar \cdot \omega_2)}{\tau_{sp}(\hbar \cdot \omega_2)} - \frac{n_{ph}^{(j)}(\hbar \cdot \omega_2)}{\tau_{ph2}}$$
$$+ (n_2 - n_1) \cdot \frac{\Gamma_2}{N_p} \cdot \frac{n_{ph}^{(j)}(\hbar \cdot \omega_2)}{\tau_{sp}(\hbar \cdot \omega_2)} \cdot e^{\lambda \cdot t} - \frac{n_{ph}(\hbar \cdot \omega_2) \cdot e^{\lambda \cdot t}}{\tau_{ph2}}$$
$$+ \left(n_2^{(j)} - n_1^{(j)}\right) \cdot \frac{\Gamma_2}{N_p} \cdot \frac{n_{ph}(\hbar \cdot \omega_2) \cdot e^{\lambda \cdot t}}{\tau_{sp}(\hbar \cdot \omega_2)}$$

At fixed points:

$$\left(n_2^{(j)} - n_1^{(j)}\right) \cdot \frac{\Gamma_2}{N_p} \cdot \frac{n_{ph}^{(j)}(\hbar \cdot \omega_2)}{\tau_{sp}(\hbar \cdot \omega_2)} - \frac{n_{ph}^{(j)}(\hbar \cdot \omega_2)}{\tau_{ph2}} = 0$$

$$- n_1 \cdot \frac{\Gamma_2}{N_p} \cdot \frac{n_{ph}^{(j)}(\hbar \cdot \omega_2)}{\tau_{sp}(\hbar \cdot \omega_2)} + n_2 \cdot \frac{\Gamma_2}{N_p} \cdot \frac{n_{ph}^{(j)}(\hbar \cdot \omega_2)}{\tau_{sp}(\hbar \cdot \omega_2)}$$
$$+ \left[\left(n_2^{(j)} - n_1^{(j)}\right) \cdot \frac{\Gamma_2}{N_p} \cdot \frac{1}{\tau_{sp}(\hbar \cdot \omega_2)} - \frac{1}{\tau_{ph2}}\right] \cdot n_{ph}(\hbar \cdot \omega_2)$$
$$- n_{ph}(\hbar \cdot \omega_2) \cdot \lambda = 0$$

We can summary our system arbitrarily small increments of exponential form equations:

$$-\left[\frac{1}{\tau_{1i}} + \frac{\Gamma_2}{N_p} \cdot f_2(N_p, \tau_{ph1}, \ldots) \cdot \omega\right] \cdot n_1 - n_1 \cdot \lambda$$
$$+ \left[\frac{1}{\tau_{21}} + \frac{1}{\tau_{sp2}} \cdot \frac{\Gamma_2}{N_p} + \frac{\Gamma_2}{N_p} \cdot f_2(N_p, \tau_{ph1}, \ldots) \cdot \omega\right] \cdot n_2 + \frac{1}{\tau_{41}} \cdot n_4 = 0$$

$$\frac{\Gamma_2}{N_p} \cdot f_2(N_p, \tau_{ph1}, \ldots) \cdot \omega \cdot n_1 - \left[\frac{1}{\tau_{21}} + \frac{1}{\tau_{2i}} + \frac{1}{\tau_{sp2}} \cdot \frac{\Gamma_2}{N_p}\right.$$
$$\left. + \frac{\Gamma_2}{N_p} \cdot f_2(N_p, \tau_{ph1}, \ldots) \cdot \omega\right] \cdot n_2 - n_2 \cdot \lambda = 0$$

$$-\left[\left(\frac{1}{\tau_{32}} + \frac{1}{\tau_{31}}\right) + \frac{\Gamma_1}{N_p} \cdot f_1(N_p, \tau_{ph1}, \ldots) \cdot \omega\right] \cdot n_3 - n_3 \cdot \lambda$$

$$+\left[\frac{1}{\tau_{43}}+\frac{1}{\tau_{sp1}}\cdot\frac{\Gamma_1}{N_p}+\frac{\Gamma_1}{N_p}\cdot f_1\left(N_p,\tau_{ph1},\ldots\right)\cdot\omega\right]\cdot n_4=0$$

$$\frac{\Gamma_1}{N_p}\cdot f_1\left(N_p,\tau_{ph1},\ldots\right)\cdot\omega\cdot n_3-\left[\frac{1}{\tau_{41}}+\frac{1}{\tau_{43}}+\frac{1}{\tau_{42}}+\frac{1}{\tau_{sp1}}\cdot\frac{\Gamma_1}{N_p}\right.$$
$$\left.+\frac{\Gamma_1}{N_p}\cdot f_1\left(N_p,\tau_{ph1},\ldots\right)\cdot\omega\right]\cdot n_4-n_4\cdot\lambda=0$$

$$-n_3\cdot n_{ph}^{(j)}(\hbar\cdot\omega_1)\cdot\frac{\Gamma_1}{N_p\cdot\tau_{sp}(\hbar\cdot\omega_1)}+n_4\cdot n_{ph}^{(j)}(\hbar\cdot\omega_1)\cdot\frac{\Gamma_1}{N_p\cdot\tau_{sp}(\hbar\cdot\omega_1)}$$
$$+\left[\left(n_4^{(j)}-n_3^{(j)}\right)\cdot\frac{\Gamma_1}{N_p\cdot\tau_{sp}(\hbar\cdot\omega_1)}-\frac{1}{\tau_{ph1}}\right]\cdot n_{ph}(\hbar\cdot\omega_1)-n_{ph}(\hbar\cdot\omega_1)\cdot\lambda=0$$

$$-n_1\cdot\frac{\Gamma_2}{N_p}\cdot\frac{n_{ph}^{(j)}(\hbar\cdot\omega_2)}{\tau_{sp}(\hbar\cdot\omega_2)}+n_2\cdot\frac{\Gamma_2}{N_p}\cdot\frac{n_{ph}^{(j)}(\hbar\cdot\omega_2)}{\tau_{sp}(\hbar\cdot\omega_2)}$$
$$+\left[(n_2^{(j)}-n_1^{(j)})\cdot\frac{\Gamma_2}{N_p}\cdot\frac{1}{\tau_{sp}(\hbar\cdot\omega_2)}-\frac{1}{\tau_{ph2}}\right]\cdot n_{ph}(\hbar\cdot\omega_2)$$
$$-n_{ph}(\hbar\cdot\omega_2)\cdot\lambda=0$$

We define some global parameters for simplicity (Table 8.10).

Table 8.10 THz dual-wavelength QCL system global parameters

Parameter	Expression
Ξ_1	$-\left[\frac{1}{\tau_{1i}}+\frac{\Gamma_2}{N_p}\cdot f_2\left(N_p,\tau_{ph1},\ldots\right)\cdot\omega\right]$
Ξ_2	$\frac{1}{\tau_{21}}+\frac{1}{\tau_{sp2}}\cdot\frac{\Gamma_2}{N_p}+\frac{\Gamma_2}{N_p}\cdot f_2\left(N_p,\tau_{ph1},\ldots\right)\cdot\omega$
Ξ_3	$\frac{\Gamma_2}{N_p}\cdot f_2\left(N_p,\tau_{ph1},\ldots\right)\cdot\omega$
Ξ_4	$\frac{1}{\tau_{21}}+\frac{1}{\tau_{2i}}+\frac{1}{\tau_{sp2}}\cdot\frac{\Gamma_2}{N_p}+\frac{\Gamma_2}{N_p}\cdot f_2\left(N_p,\tau_{ph1},\ldots\right)\cdot\omega$
Ξ_5	$-\left[\left(\frac{1}{\tau_{32}}+\frac{1}{\tau_{31}}\right)+\frac{\Gamma_1}{N_p}\cdot f_1\left(N_p,\tau_{ph1},\ldots\right)\cdot\omega\right]$
Ξ_6	$\frac{1}{\tau_{43}}+\frac{1}{\tau_{sp1}}\cdot\frac{\Gamma_1}{N_p}+\frac{\Gamma_1}{N_p}\cdot f_1\left(N_p,\tau_{ph1},\ldots\right)\cdot\omega$
Ξ_7	$\frac{\Gamma_1}{N_p}\cdot f_1\left(N_p,\tau_{ph1},\ldots\right)\cdot\omega$
Ξ_8	$\frac{1}{\tau_{41}}+\frac{1}{\tau_{43}}+\frac{1}{\tau_{42}}+\frac{1}{\tau_{sp1}}\cdot\frac{\Gamma_1}{N_p}+\frac{\Gamma_1}{N_p}\cdot f_1\left(N_p,\tau_{ph1},\ldots\right)\cdot\omega$
Ξ_9	$n_{ph}^{(j)}(\hbar\cdot\omega_1)\cdot\frac{\Gamma_1}{N_p\cdot\tau_{sp}(\hbar\cdot\omega_1)}$
Ξ_{10}	$\left(n_4^{(j)}-n_3^{(j)}\right)\cdot\frac{\Gamma_1}{N_p\cdot\tau_{sp}(\hbar\cdot\omega_1)}-\frac{1}{\tau_{ph1}}$
Ξ_{11}	$\frac{\Gamma_2}{N_p}\cdot\frac{n_{ph}^{(j)}(\hbar\cdot\omega_2)}{\tau_{sp}(\hbar\cdot\omega_2)}$
Ξ_{12}	$\left(n_2^{(j)}-n_1^{(j)}\right)\cdot\frac{\Gamma_2}{N_p}\cdot\frac{1}{\tau_{sp}(\hbar\cdot\omega_2)}-\frac{1}{\tau_{ph2}}$

$$\Xi_1 \cdot n_1 - n_1 \cdot \lambda + \Xi_2 \cdot n_2 + \frac{1}{\tau_{41}} \cdot n_4 = 0;\ \Xi_3 \cdot n_1 - \Xi_4 \cdot n_2 - n_2 \cdot \lambda = 0$$
$$\Xi_5 \cdot n_3 - n_3 \cdot \lambda + \Xi_6 \cdot n_4 = 0$$

$$\Xi_7 \cdot n_3 - \Xi_8 \cdot n_4 - n_4 \cdot \lambda = 0$$
$$-n_3 \cdot \Xi_9 + n_4 \cdot \Xi_9 + \Xi_{10} \cdot n_{ph}(\hbar \cdot \omega_1) - n_{ph}(\hbar \cdot \omega_1) \cdot \lambda = 0$$

$$-n_1 \cdot \Xi_{11} + n_2 \cdot \Xi_{11} + \Xi_{12} \cdot n_{ph}(\hbar \cdot \omega_2) - n_{ph}(\hbar \cdot \omega_2) \cdot \lambda = 0$$

The small increments Jacobian of our THz dual-wavelength QCL system is as follow:

$$\begin{pmatrix} \Upsilon_{11} & \ldots & \Upsilon_{16} \\ \vdots & \ddots & \vdots \\ \Upsilon_{61} & \cdots & \Upsilon_{66} \end{pmatrix} \cdot \begin{pmatrix} n_1 \\ n_2 \\ n_3 \\ n_4 \\ n_{ph}(\hbar \cdot \omega_1) \\ n_{ph}(\hbar \cdot \omega_2) \end{pmatrix} = 0;\ \Upsilon_{11} = \Xi_1 - \lambda;\ \Upsilon_{12} = \Xi_2$$
$$\Upsilon_{13} = 0;\ \Upsilon_{14} = \frac{1}{\tau_{41}};\ \Upsilon_{15} = \Upsilon_{16} = 0$$

$$\Upsilon_{21} = \Xi_3;\ \Upsilon_{22} = -\Xi_4 - \lambda;\ \Upsilon_{23} = \Upsilon_{24} = \Upsilon_{25} = \Upsilon_{26} = 0$$
$$\Upsilon_{31} = \Upsilon_{32} = 0;\ \Upsilon_{33} = \Xi_5 - \lambda;\ \Upsilon_{34} = \Xi_6$$

$$\Upsilon_{35} = \Upsilon_{36} = 0;\ \Upsilon_{41} = \Upsilon_{42} = 0;\ \Upsilon_{43} = \Xi_7;\ \Upsilon_{44} = -\Xi_8 - \lambda$$
$$\Upsilon_{45} = \Upsilon_{46} = 0;\ \Upsilon_{51} = \Upsilon_{52} = 0$$

$$\Upsilon_{53} = -\Xi_9;\ \Upsilon_{54} = \Xi_9;\ \Upsilon_{55} = \Xi_{10} - \lambda;\ \Upsilon_{56} = 0;\ \Upsilon_{61} = -\Xi_{11}$$
$$\Upsilon_{62} = \Xi_{11};\ \Upsilon_{63} = \Upsilon_{64} = \Upsilon_{65} = 0;\ \Upsilon_{66} = \Xi_{12} - \lambda$$

$$A - \lambda \cdot I = \begin{pmatrix} \Upsilon_{11} & \ldots & \Upsilon_{16} \\ \vdots & \ddots & \vdots \\ \Upsilon_{61} & \cdots & \Upsilon_{66} \end{pmatrix};\ \det(A - \lambda \cdot I) = \det \begin{pmatrix} \Upsilon_{11} & \ldots & \Upsilon_{16} \\ \vdots & \ddots & \vdots \\ \Upsilon_{61} & \cdots & \Upsilon_{66} \end{pmatrix}$$
$$\det(A - \lambda \cdot I) = 0$$

$$\det(A - \lambda \cdot I) = \det \begin{pmatrix} \Upsilon_{11} & \ldots & \Upsilon_{16} \\ \vdots & \ddots & \vdots \\ \Upsilon_{61} & \cdots & \Upsilon_{66} \end{pmatrix} = \Upsilon_{11} \cdot \det \begin{pmatrix} \Upsilon_{22} & \ldots & \Upsilon_{26} \\ \vdots & \ddots & \vdots \\ \Upsilon_{62} & \cdots & \Upsilon_{66} \end{pmatrix}$$

$$-\Upsilon_{12}\cdot\det\begin{pmatrix}\Upsilon_{2k} & \ldots & \Upsilon_{2k}\\ \vdots & \ddots & \vdots\\ \Upsilon_{6k} & \cdots & \Upsilon_{6k}\end{pmatrix}|_{k=1,3,4,5,6}$$

$$-\Upsilon_{14}\cdot\det\begin{pmatrix}\Upsilon_{2k} & \ldots & \Upsilon_{2k}\\ \vdots & \ddots & \vdots\\ \Upsilon_{6k} & \cdots & \Upsilon_{6k}\end{pmatrix}|_{k=1,2,3,5,6}$$

All other elements of the matrix first line are zero since $\Upsilon_{13}=0;\ \Upsilon_{15}=\Upsilon_{16}=0.$

$$\Upsilon_{13}\cdot\det\begin{pmatrix}\Upsilon_{2k} & \ldots & \Upsilon_{2k}\\ \vdots & \ddots & \vdots\\ \Upsilon_{6k} & \cdots & \Upsilon_{6k}\end{pmatrix}|_{k=1,2,4,5,6}=0$$

$$\Upsilon_{15}\cdot\det\begin{pmatrix}\Upsilon_{2k} & \ldots & \Upsilon_{2k}\\ \vdots & \ddots & \vdots\\ \Upsilon_{6k} & \cdots & \Upsilon_{6k}\end{pmatrix}|_{k=1,2,3,4,6}=0$$

$$\Upsilon_{16}\cdot\det\begin{pmatrix}\Upsilon_{21} & \ldots & \Upsilon_{25}\\ \vdots & \ddots & \vdots\\ \Upsilon_{61} & \cdots & \Upsilon_{65}\end{pmatrix}=0$$

$$(*)\Upsilon_{11}\cdot\det\begin{pmatrix}\Upsilon_{22} & \ldots & \Upsilon_{26}\\ \vdots & \ddots & \vdots\\ \Upsilon_{62} & \cdots & \Upsilon_{66}\end{pmatrix}$$

$$=(\Xi_1-\lambda)\cdot(-\Xi_4-\lambda)\cdot\det\begin{pmatrix}(\Xi_5-\lambda) & \Xi_6 & 0 & 0\\ \Xi_7 & (-\Xi_8-\lambda) & 0 & 0\\ -\Xi_9 & \Xi_9 & (\Xi_{10}-\lambda) & 0\\ 0 & 0 & 0 & (\Xi_{12}-\lambda)\end{pmatrix}$$

$$\Upsilon_{11}\cdot\det\begin{pmatrix}\Upsilon_{22} & \ldots & \Upsilon_{26}\\ \vdots & \ddots & \vdots\\ \Upsilon_{62} & \cdots & \Upsilon_{66}\end{pmatrix}=\left[-\Xi_1\cdot\Xi_4+(\Xi_4-\Xi_1)\cdot\lambda+\lambda^2\right]\cdot\{(\Xi_5-\lambda)$$

$$\cdot\det\begin{pmatrix}(-\Xi_8-\lambda) & 0 & 0\\ \Xi_9 & (\Xi_{10}-\lambda) & 0\\ 0 & 0 & (\Xi_{12}-\lambda)\end{pmatrix}$$

$$-\Xi_6\cdot\det\begin{pmatrix}\Xi_7 & 0 & 0\\ -\Xi_9 & (\Xi_{10}-\lambda) & 0\\ 0 & 0 & (\Xi_{12}-\lambda)\end{pmatrix}\Bigg\}$$

Step A:

$$(\Xi_5-\lambda)\cdot\det\begin{pmatrix}(-\Xi_8-\lambda) & 0 & 0\\ \Xi_9 & (\Xi_{10}-\lambda) & 0\\ 0 & 0 & (\Xi_{12}-\lambda)\end{pmatrix}$$

$$(\Xi_5-\lambda)\cdot\det\begin{pmatrix}(-\Xi_8-\lambda) & 0 & 0\\ \Xi_9 & (\Xi_{10}-\lambda) & 0\\ 0 & 0 & (\Xi_{12}-\lambda)\end{pmatrix}$$
$$=(\Xi_5-\lambda)\cdot(-\Xi_8-\lambda)\cdot(\Xi_{10}-\lambda)\cdot(\Xi_{12}-\lambda)$$
$$=\left[-\Xi_5\cdot\Xi_8+(\Xi_8-\Xi_5)\cdot\lambda+\lambda^2\right]\left[\Xi_{10}\cdot\Xi_{12}-(\Xi_{10}+\Xi_{12})\cdot\lambda+\lambda^2\right]$$

$$(\Xi_5-\lambda)\cdot\det\begin{pmatrix}(-\Xi_8-\lambda) & 0 & 0\\ \Xi_9 & (\Xi_{10}-\lambda) & 0\\ 0 & 0 & (\Xi_{12}-\lambda)\end{pmatrix}$$
$$=-\Xi_5\cdot\Xi_8\cdot\Xi_{10}\cdot\Xi_{12}+[\Xi_5\cdot\Xi_8\cdot(\Xi_{10}+\Xi_{12})+(\Xi_8-\Xi_5)\cdot\Xi_{10}\cdot\Xi_{12}]\cdot\lambda$$
$$+[\Xi_{10}\cdot\Xi_{12}-\Xi_5\cdot\Xi_8-(\Xi_8-\Xi_5)\cdot(\Xi_{10}+\Xi_{12})]\cdot\lambda^2$$
$$+[(\Xi_8-\Xi_5)-(\Xi_{10}+\Xi_{12})]\cdot\lambda^3+\lambda^4$$

We define some global parameters for simplicity:

$$A_0=-\Xi_5\cdot\Xi_8\cdot\Xi_{10}\cdot\Xi_{12}$$

$$A_1=\Xi_5\cdot\Xi_8\cdot(\Xi_{10}+\Xi_{12})+(\Xi_8-\Xi_5)\cdot\Xi_{10}\cdot\Xi_{12}$$
$$A_2=\Xi_{10}\cdot\Xi_{12}-\Xi_5\cdot\Xi_8-(\Xi_8-\Xi_5)\cdot(\Xi_{10}+\Xi_{12})$$

$$A_3=(\Xi_8-\Xi_5)-(\Xi_{10}+\Xi_{12});\ A_4=1$$

$$(\Xi_5-\lambda)\cdot\det\begin{pmatrix}(-\Xi_8-\lambda) & 0 & 0\\ \Xi_9 & (\Xi_{10}-\lambda) & 0\\ 0 & 0 & (\Xi_{12}-\lambda)\end{pmatrix}$$
$$=\sum_{k=0}^{4}A_k\cdot\lambda^k=A_0+A_1\cdot\lambda+A_2\cdot\lambda^2+A_3\cdot\lambda^3+A_4\cdot\lambda^4$$

Step B:

$$\Xi_6\cdot\det\begin{pmatrix}\Xi_7 & 0 & 0\\ -\Xi_9 & (\Xi_{10}-\lambda) & 0\\ 0 & 0 & (\Xi_{12}-\lambda)\end{pmatrix}$$

$$\Xi_6 \cdot \det\begin{pmatrix} \Xi_7 & 0 & 0 \\ -\Xi_9 & (\Xi_{10}-\lambda) & 0 \\ 0 & 0 & (\Xi_{12}-\lambda) \end{pmatrix} = \Xi_6 \cdot \Xi_7 \cdot (\Xi_{10}-\lambda) \cdot (\Xi_{12}-\lambda)$$

$$= \Xi_6 \cdot \Xi_7 \cdot [\Xi_{10} \cdot \Xi_{12} - (\Xi_{10} + \Xi_{12}) \cdot \lambda + \lambda^2]$$

$$\Xi_6 \cdot \det\begin{pmatrix} \Xi_7 & 0 & 0 \\ -\Xi_9 & (\Xi_{10}-\lambda) & 0 \\ 0 & 0 & (\Xi_{12}-\lambda) \end{pmatrix} = \Xi_6 \cdot \Xi_7 \cdot \Xi_{10} \cdot \Xi_{12} - \Xi_6 \cdot \Xi_7 \cdot (\Xi_{10} + \Xi_{12}) \cdot \lambda + \Xi_6 \cdot \Xi_7 \cdot \lambda^2$$

We define some global parameters for simplicity:

$$B_0 = \Xi_6 \cdot \Xi_7 \cdot \Xi_{10} \cdot \Xi_{12}$$

$$B_1 = -\Xi_6 \cdot \Xi_7 \cdot (\Xi_{10} + \Xi_{12});\ B_2 = \Xi_6 \cdot \Xi_7$$

$$\Xi_6 \cdot \det\begin{pmatrix} \Xi_7 & 0 & 0 \\ -\Xi_9 & (\Xi_{10}-\lambda) & 0 \\ 0 & 0 & (\Xi_{12}-\lambda) \end{pmatrix} = \sum_{k=0}^{2} B_k \cdot \lambda^k = B_0 + B_1 \cdot \lambda + B_2 \cdot \lambda^2$$

##

$$\Upsilon_{11} \cdot \det\begin{pmatrix} \Upsilon_{22} & \dots & \Upsilon_{26} \\ \vdots & \ddots & \vdots \\ \Upsilon_{62} & \cdots & \Upsilon_{66} \end{pmatrix}$$

$$= \left[-\Xi_1 \cdot \Xi_4 + (\Xi_4 - \Xi_1) \cdot \lambda + \lambda^2\right] \cdot \left\{(\Xi_5 - \lambda) \cdot \det\begin{pmatrix} (-\Xi_8-\lambda) & 0 & 0 \\ \Xi_9 & (\Xi_{10}-\lambda) & 0 \\ 0 & 0 & (\Xi_{12}-\lambda) \end{pmatrix} - \Xi_6 \cdot \det\begin{pmatrix} \Xi_7 & 0 & 0 \\ -\Xi_9 & (\Xi_{10}-\lambda) & 0 \\ 0 & 0 & (\Xi_{12}-\lambda) \end{pmatrix}\right\}$$

$$= \left[-\Xi_1 \cdot \Xi_4 + (\Xi_4 - \Xi_1) \cdot \lambda + \lambda^2\right] \cdot \left[\sum_{k=0}^{4} A_k \cdot \lambda^k - \sum_{k=0}^{2} B_k \cdot \lambda^k\right]$$

$$\Upsilon_{11}\cdot\det\begin{pmatrix}\Upsilon_{22} & \ldots & \Upsilon_{26}\\ \vdots & \ddots & \vdots\\ \Upsilon_{62} & \cdots & \Upsilon_{66}\end{pmatrix} = \left[-\Xi_1\cdot\Xi_4 + (\Xi_4-\Xi_1)\cdot\lambda+\lambda^2\right]$$

$$\cdot\left[\sum_{k=0}^{4}A_k\cdot\lambda^k - \sum_{k=0}^{2}B_k\cdot\lambda^k\right]$$

$$= [-\Xi_1\cdot\Xi_4 + (\Xi_4-\Xi_1)\cdot\lambda+\lambda^2]$$

$$\cdot[A_0 - B_0 + (A_1-B_1)\cdot\lambda$$

$$+ (A_2-B_2)\cdot\lambda^2 + A_3\cdot\lambda^3 + A_4\cdot\lambda^4)]$$

$$= [-\Xi_1\cdot\Xi_4 + (\Xi_4-\Xi_1)\cdot\lambda+\lambda^2]$$

$$\cdot\left[\sum_{k=0}^{2}(A_k-B_k)\cdot\lambda^k + \sum_{k=3}^{4}A_k\cdot\lambda^k\right]$$

$$\Upsilon_{11}\cdot\det\begin{pmatrix}\Upsilon_{22} & \ldots & \Upsilon_{26}\\ \vdots & \ddots & \vdots\\ \Upsilon_{62} & \cdots & \Upsilon_{66}\end{pmatrix}$$

$$= [-\Xi_1\cdot\Xi_4 + (\Xi_4-\Xi_1)\cdot\lambda+\lambda^2]\cdot\left[\sum_{k=0}^{2}(A_k-B_k)\cdot\lambda^k + \sum_{k=3}^{4}A_k\cdot\lambda^k\right]$$

$$= [-\Xi_1\cdot\Xi_4 + (\Xi_4-\Xi_1)\cdot\lambda+\lambda^2]\cdot[(A_0-B_0)$$

$$+ (A_1-B_1)\cdot\lambda + (A_2-B_2)\cdot\lambda^2 + A_3\cdot\lambda^3 + A_4\cdot\lambda^4)]$$

$$= -\Xi_1\cdot\Xi_4\cdot(A_0-B_0) + [(\Xi_4-\Xi_1)\cdot(A_0-B_0) - \Xi_1\cdot\Xi_4\cdot(A_1-B_1)]\cdot\lambda$$

$$+ [(\Xi_4-\Xi_1)\cdot(A_1-B_1) + (A_0-B_0) - \Xi_1\cdot\Xi_4\cdot(A_2-B_2)]\cdot\lambda^2$$

$$+ [(\Xi_4-\Xi_1)\cdot(A_2-B_2) - \Xi_1\cdot\Xi_4\cdot A_3 + (A_1-B_1)]\cdot\lambda^3$$

$$+ [(\Xi_4-\Xi_1)\cdot A_3 - \Xi_1\cdot\Xi_4\cdot A_4 + (A_2-B_2)]\cdot\lambda^4$$

$$+ [(\Xi_4-\Xi_1)\cdot A_4 + A_3]\cdot\lambda^5 + A_4\cdot\lambda^6$$

We define for simplicity some global parameters:

$$C_0 = -\Xi_1\cdot\Xi_4\cdot(A_0-B_0)$$

$$C_1 = (\Xi_4-\Xi_1)\cdot(A_0-B_0) - \Xi_1\cdot\Xi_4\cdot(A_1-B_1)$$

$$C_2 = (\Xi_4-\Xi_1)\cdot(A_1-B_1) + (A_0-B_0) - \Xi_1\cdot\Xi_4\cdot(A_2-B_2)$$

$$C_3 = (\Xi_4-\Xi_1)\cdot(A_2-B_2) - \Xi_1\cdot\Xi_4\cdot A_3 + (A_1-B_1)$$

$$C_4 = (\Xi_4-\Xi_1)\cdot A_3 - \Xi_1\cdot\Xi_4\cdot A_4 + (A_2-B_2)$$

$$C_5 = (\Xi_4 - \Xi_1) \cdot A_4 + A_3;\ C_6 = A_4$$

$$\Upsilon_{11} \cdot \det\begin{pmatrix} \Upsilon_{22} & \ldots & \Upsilon_{26} \\ \vdots & \ddots & \vdots \\ \Upsilon_{62} & \cdots & \Upsilon_{66} \end{pmatrix} = \sum_{k=0}^{6} C_k \cdot \lambda^k$$
$$= C_0 + C_1 \cdot \lambda + C_2 \cdot \lambda^2 + C_3 \cdot \lambda^3 + C_4 \cdot \lambda^4 + C_5 \cdot \lambda^5 + C_6 \cdot \lambda^6$$

$$(**)\Upsilon_{12} \cdot \det\begin{pmatrix} \Upsilon_{2k} & \ldots & \Upsilon_{2k} \\ \vdots & \ddots & \vdots \\ \Upsilon_{6k} & \cdots & \Upsilon_{6k} \end{pmatrix}|_{k=1,3,4,5,6}$$
$$= \Xi_2 \cdot \Xi_3 \cdot \det\begin{pmatrix} \Xi_5 - \lambda & \Xi_6 & 0 & 0 \\ \Xi_7 & -\Xi_8 - \lambda & 0 & 0 \\ -\Xi_9 & \Xi_9 & \Xi_{10} - \lambda & 0 \\ 0 & 0 & 0 & \Xi_{12} - \lambda \end{pmatrix}$$
$$= \Xi_2 \cdot \Xi_3 \cdot \left\{ (\Xi_5 - \lambda) \cdot \det\begin{pmatrix} (-\Xi_8 - \lambda) & 0 & 0 \\ \Xi_9 & (\Xi_{10} - \lambda) & 0 \\ 0 & 0 & \Xi_{12} - \lambda \end{pmatrix} \right.$$
$$\left. - \Xi_6 \cdot \det\begin{pmatrix} \Xi_7 & 0 & 0 \\ -\Xi_9 & (\Xi_{10} - \lambda) & 0 \\ 0 & 0 & (\Xi_{12} - \lambda) \end{pmatrix} \right\}$$

$$\Upsilon_{12} \cdot \det\begin{pmatrix} \Upsilon_{2k} & \ldots & \Upsilon_{2k} \\ \vdots & \ddots & \vdots \\ \Upsilon_{6k} & \cdots & \Upsilon_{6k} \end{pmatrix}|_{k=1,3,4,5,6}$$
$$= \Xi_2 \cdot \Xi_3 \cdot \left\{ (\Xi_5 - \lambda) \cdot \det\begin{pmatrix} (-\Xi_8 - \lambda) & 0 & 0 \\ \Xi_9 & (\Xi_{10} - \lambda) & 0 \\ 0 & 0 & (\Xi_{12} - \lambda) \end{pmatrix} \right.$$
$$\left. - \Xi_6 \cdot \det\begin{pmatrix} \Xi_7 & 0 & 0 \\ -\Xi_9 & (\Xi_{10} - \lambda) & 0 \\ 0 & 0 & (\Xi_{12} - \lambda) \end{pmatrix} \right\}$$
$$= \Xi_2 \cdot \Xi_3 \cdot \{(\Xi_5 - \lambda) \cdot (-\Xi_8 - \lambda) \cdot (\Xi_{10} - \lambda) \cdot (\Xi_{12} - \lambda)$$
$$- \Xi_6 \cdot \Xi_7 \cdot (\Xi_{10} - \lambda) \cdot (\Xi_{12} - \lambda)\}$$

$$\Upsilon_{12} \cdot \det\begin{pmatrix} \Upsilon_{2k} & \ldots & \Upsilon_{2k} \\ \vdots & \ddots & \vdots \\ \Upsilon_{6k} & \cdots & \Upsilon_{6k} \end{pmatrix}|_{k=1,3,4,5,6}$$
$$= \Xi_2 \cdot \Xi_3 \cdot \{(\Xi_5 - \lambda) \cdot (-\Xi_8 - \lambda) \cdot (\Xi_{10} - \lambda) \cdot (\Xi_{12} - \lambda)$$
$$- \Xi_6 \cdot \Xi_7 \cdot (\Xi_{10} - \lambda) \cdot (\Xi_{12} - \lambda)\} = \Xi_2 \cdot \Xi_3 \cdot \{(-\Xi_5 \cdot \Xi_8$$
$$+ [\Xi_8 - \Xi_5] \cdot \lambda + \lambda^2) \cdot (\Xi_{10} \cdot \Xi_{12} - [\Xi_{10} + \Xi_{12}] \cdot \lambda + \lambda^2)$$
$$- \Xi_6 \cdot \Xi_7 \cdot (\Xi_{10} \cdot \Xi_{12} - [\Xi_{10} + \Xi_{12}] \cdot \lambda + \lambda^2)\}$$

$$\Upsilon_{12} \cdot \det\begin{pmatrix} \Upsilon_{2k} & \ldots & \Upsilon_{2k} \\ \vdots & \ddots & \vdots \\ \Upsilon_{6k} & \cdots & \Upsilon_{6k} \end{pmatrix}|_{k=1,3,4,5,6}$$
$$= \Xi_2 \cdot \Xi_3 \cdot \{(-\Xi_5 \cdot \Xi_8 + [\Xi_8 - \Xi_5] \cdot \lambda$$
$$+ \lambda^2) \cdot (\Xi_{10} \cdot \Xi_{12} - [\Xi_{10} + \Xi_{12}] \cdot \lambda + \lambda^2)$$
$$- \Xi_6 \cdot \Xi_7 \cdot (\Xi_{10} \cdot \Xi_{12} - [\Xi_{10} + \Xi_{12}] \cdot \lambda + \lambda^2)\}$$
$$= \Xi_2 \cdot \Xi_3 \cdot \{-\Xi_5 \cdot \Xi_8 \cdot \Xi_{10} \cdot \Xi_{12} + \Xi_5 \cdot \Xi_8 \cdot [\Xi_{10} + \Xi_{12}] \cdot \lambda - \Xi_5 \cdot \Xi_8 \cdot \lambda^2$$
$$+ [\Xi_8 - \Xi_5] \cdot \Xi_{10} \cdot \Xi_{12} \cdot \lambda - [\Xi_8 - \Xi_5] \cdot [\Xi_{10} + \Xi_{12}] \cdot \lambda^2$$
$$+ [\Xi_8 - \Xi_5] \cdot \lambda^3 + \Xi_{10} \cdot \Xi_{12} \cdot \lambda^2 - [\Xi_{10} + \Xi_{12}] \cdot \lambda^3 + \lambda^4$$
$$- \Xi_6 \cdot \Xi_7 \cdot \Xi_{10} \cdot \Xi_{12} + \Xi_6 \cdot \Xi_7 \cdot [\Xi_{10} + \Xi_{12}] \cdot \lambda - \Xi_6 \cdot \Xi_7 \cdot \lambda^2\}$$

$$\Upsilon_{12} \cdot \det\begin{pmatrix} \Upsilon_{2k} & \ldots & \Upsilon_{2k} \\ \vdots & \ddots & \vdots \\ \Upsilon_{6k} & \cdots & \Upsilon_{6k} \end{pmatrix}|_{k=1,3,4,5,6}$$
$$= \Xi_2 \cdot \Xi_3 \cdot \{-\Xi_5 \cdot \Xi_8 \cdot \Xi_{10} \cdot \Xi_{12}$$
$$- \Xi_6 \cdot \Xi_7 \cdot \Xi_{10} \cdot \Xi_{12}$$
$$+ (\Xi_5 \cdot \Xi_8 \cdot [\Xi_{10} + \Xi_{12}] + [\Xi_8 - \Xi_5] \cdot \Xi_{10} \cdot \Xi_{12}$$
$$+ \Xi_6 \cdot \Xi_7 \cdot [\Xi_{10} + \Xi_{12}]) \cdot \lambda$$
$$+ (\Xi_{10} \cdot \Xi_{12} - \Xi_5 \cdot \Xi_8 - [\Xi_8 - \Xi_5] \cdot [\Xi_{10} + \Xi_{12}] - \Xi_6 \cdot \Xi_7) \cdot \lambda^2$$
$$+ ([\Xi_8 - \Xi_5] - [\Xi_{10} + \Xi_{12}]) \cdot \lambda^3 + \lambda^4\}$$

$$\Upsilon_{12} \cdot \det\begin{pmatrix} \Upsilon_{2k} & \ldots & \Upsilon_{2k} \\ \vdots & \ddots & \vdots \\ \Upsilon_{6k} & \cdots & \Upsilon_{6k} \end{pmatrix}|_{k=1,3,4,5,6}$$
$$= -\Xi_2 \cdot \Xi_3 \cdot (\Xi_5 \cdot \Xi_8 \cdot \Xi_{10} \cdot \Xi_{12} + \Xi_6 \cdot \Xi_7 \cdot \Xi_{10} \cdot \Xi_{12})$$
$$+ \Xi_2 \cdot \Xi_3 \cdot (\Xi_5 \cdot \Xi_8 \cdot [\Xi_{10} + \Xi_{12}] + [\Xi_8 - \Xi_5] \cdot \Xi_{10} \cdot \Xi_{12}$$

$$+ \Xi_6 \cdot \Xi_7 \cdot [\Xi_{10} + \Xi_{12}]) \cdot \lambda$$
$$+ \Xi_2 \cdot \Xi_3 \cdot (\Xi_{10} \cdot \Xi_{12}$$
$$- \Xi_5 \cdot \Xi_8 - [\Xi_8 - \Xi_5] \cdot [\Xi_{10} + \Xi_{12}] - \Xi_6 \cdot \Xi_7) \cdot \lambda^2$$
$$+ \Xi_2 \cdot \Xi_3 \cdot ([\Xi_8 - \Xi_5] - [\Xi_{10} + \Xi_{12}]) \cdot \lambda^3 + \Xi_2 \cdot \Xi_3 \cdot \lambda^4$$

We define for simplicity some global parameters:

$$D_0 = -\Xi_2 \cdot \Xi_3 \cdot (\Xi_5 \cdot \Xi_8 \cdot \Xi_{10} \cdot \Xi_{12} + \Xi_6 \cdot \Xi_7 \cdot \Xi_{10} \cdot \Xi_{12})$$

$$D_1 = \Xi_2 \cdot \Xi_3 \cdot (\Xi_5 \cdot \Xi_8 \cdot [\Xi_{10} + \Xi_{12}] + [\Xi_8 - \Xi_5] \cdot \Xi_{10} \cdot \Xi_{12} + \Xi_6 \cdot \Xi_7 \cdot [\Xi_{10} + \Xi_{12}])$$

$$D_2 = \Xi_2 \cdot \Xi_3 \cdot (\Xi_{10} \cdot \Xi_{12} - \Xi_5 \cdot \Xi_8 - [\Xi_8 - \Xi_5] \cdot [\Xi_{10} + \Xi_{12}] - \Xi_6 \cdot \Xi_7)$$

$$D_3 = \Xi_2 \cdot \Xi_3 \cdot ([\Xi_8 - \Xi_5] - [\Xi_{10} + \Xi_{12}]);\ D_4 = \Xi_2 \cdot \Xi_3$$

$$\Upsilon_{12} \cdot \det\begin{pmatrix} \Upsilon_{2k} & \dots & \Upsilon_{2k} \\ \vdots & \ddots & \vdots \\ \Upsilon_{6k} & \cdots & \Upsilon_{6k} \end{pmatrix}|_{k=1,3,4,5,6} = \sum_{k=0}^{4} D_k \cdot \lambda^k = D_0 + D_1 \cdot \lambda + D_2 \cdot \lambda^2 + D_3 \cdot \lambda^3 + D_4 \cdot \lambda^4$$

$$(***)\Upsilon_{14} \cdot \det\begin{pmatrix} \Upsilon_{2k} & \dots & \Upsilon_{2k} \\ \vdots & \ddots & \vdots \\ \Upsilon_{6k} & \cdots & \Upsilon_{6k} \end{pmatrix}|_{k=1,2,3,5,6}$$

$$\Upsilon_{14} \cdot \det\begin{pmatrix} \Upsilon_{2k} & \dots & \Upsilon_{2k} \\ \vdots & \ddots & \vdots \\ \Upsilon_{6k} & \cdots & \Upsilon_{6k} \end{pmatrix}|_{k=1,2,3,5,6}$$
$$= \frac{1}{\tau_{41}} \cdot \left\{ \Xi_3 \cdot \det\begin{pmatrix} 0 & (\Xi_5 - \lambda) & 0 & 0 \\ 0 & \Xi_7 & 0 & 0 \\ 0 & -\Xi_9 & (\Xi_{10} - \lambda) & 0 \\ \Xi_{11} & 0 & 0 & (\Xi_{12} - \lambda) \end{pmatrix} \right.$$
$$\left. -(-\Xi_4 - \lambda) \cdot \det\begin{pmatrix} 0 & (\Xi_5 - \lambda) & 0 & 0 \\ 0 & \Xi_7 & 0 & 0 \\ 0 & -\Xi_9 & (\Xi_{10} - \lambda) & 0 \\ -\Xi_{11} & 0 & 0 & (\Xi_{12} - \lambda) \end{pmatrix} \right\}$$

Step A:

$$\det\begin{pmatrix} 0 & (\Xi_5-\lambda) & 0 & 0 \\ 0 & \Xi_7 & 0 & 0 \\ 0 & -\Xi_9 & (\Xi_{10}-\lambda) & 0 \\ \Xi_{11} & 0 & 0 & (\Xi_{12}-\lambda) \end{pmatrix}$$

$$\det\begin{pmatrix} 0 & (\Xi_5-\lambda) & 0 & 0 \\ 0 & \Xi_7 & 0 & 0 \\ 0 & -\Xi_9 & (\Xi_{10}-\lambda) & 0 \\ \Xi_{11} & 0 & 0 & (\Xi_{12}-\lambda) \end{pmatrix}$$
$$= -(\Xi_5-\lambda)\cdot\det\begin{pmatrix} 0 & 0 & 0 \\ 0 & (\Xi_{10}-\lambda) & 0 \\ \Xi_{11} & 0 & (\Xi_{12}-\lambda) \end{pmatrix} = 0$$

Step B:

$$\det\begin{pmatrix} 0 & (\Xi_5-\lambda) & 0 & 0 \\ 0 & \Xi_7 & 0 & 0 \\ 0 & -\Xi_9 & (\Xi_{10}-\lambda) & 0 \\ -\Xi_{11} & 0 & 0 & (\Xi_{12}-\lambda) \end{pmatrix}$$

$$\det\begin{pmatrix} 0 & (\Xi_5-\lambda) & 0 & 0 \\ 0 & \Xi_7 & 0 & 0 \\ 0 & -\Xi_9 & (\Xi_{10}-\lambda) & 0 \\ -\Xi_{11} & 0 & 0 & (\Xi_{12}-\lambda) \end{pmatrix}$$
$$= -(\Xi_5-\lambda)\cdot\det\begin{pmatrix} 0 & 0 & 0 \\ 0 & (\Xi_{10}-\lambda) & 0 \\ -\Xi_{11} & 0 & (\Xi_{12}-\lambda) \end{pmatrix} = 0$$

$$\Upsilon_{14}\cdot\det\begin{pmatrix} \Upsilon_{2k} & \dots & \Upsilon_{2k} \\ \vdots & \ddots & \vdots \\ \Upsilon_{6k} & \cdots & \Upsilon_{6k} \end{pmatrix}|_{k=1,2,3,5,6} = 0$$

We can summary our last results in the next table (Table 8.11).

$$\det(A-\lambda\cdot I) = \det\begin{pmatrix} \Upsilon_{11} & \dots & \Upsilon_{16} \\ \vdots & \ddots & \vdots \\ \Upsilon_{61} & \cdots & \Upsilon_{66} \end{pmatrix}$$

Table 8.11 THz dual-wavelength QCL system determinant elements

Determinant element	Expression
$\Upsilon_{11}\cdot\det\begin{pmatrix}\Upsilon_{22} & \ldots & \Upsilon_{26}\\ \vdots & \ddots & \vdots\\ \Upsilon_{62} & \cdots & \Upsilon_{66}\end{pmatrix}$	$\sum_{k=0}^{6} C_k\cdot\lambda^k = C_0 + C_1\cdot\lambda + C_2\cdot\lambda^2 + C_3\cdot\lambda^3 + C_4\cdot\lambda^4 + C_5\cdot\lambda^5 + C_6\cdot\lambda^6$
$\Upsilon_{12}\cdot\det\begin{pmatrix}\Upsilon_{2k} & \ldots & \Upsilon_{2k}\\ \vdots & \ddots & \vdots\\ \Upsilon_{6k} & \cdots & \Upsilon_{6k}\end{pmatrix}\Big\vert_{k=1,3,4,5,6}$	$\sum_{k=0}^{4} D_k\cdot\lambda^k = D_0 + D_1\cdot\lambda + D_2\cdot\lambda^2 + D_3\cdot\lambda^3 + D_4\cdot\lambda^4$
$\Upsilon_{14}\cdot\det\begin{pmatrix}\Upsilon_{2k} & \ldots & \Upsilon_{2k}\\ \vdots & \ddots & \vdots\\ \Upsilon_{6k} & \cdots & \Upsilon_{6k}\end{pmatrix}\Big\vert_{k=1,2,3,5,6}$	0

$$
\begin{aligned}
&= \Upsilon_{11}\cdot\det\begin{pmatrix}\Upsilon_{22} & \ldots & \Upsilon_{26}\\ \vdots & \ddots & \vdots\\ \Upsilon_{62} & \cdots & \Upsilon_{66}\end{pmatrix}\\
&\quad - \Upsilon_{12}\cdot\det\begin{pmatrix}\Upsilon_{2k} & \ldots & \Upsilon_{2k}\\ \vdots & \ddots & \vdots\\ \Upsilon_{6k} & \cdots & \Upsilon_{6k}\end{pmatrix}\Big\vert_{k=1,3,4,5,6}\\
&\quad - \Upsilon_{14}\cdot\det\begin{pmatrix}\Upsilon_{2k} & \ldots & \Upsilon_{2k}\\ \vdots & \ddots & \vdots\\ \Upsilon_{6k} & \cdots & \Upsilon_{6k}\end{pmatrix}\Big\vert_{k=1,2,3,5,6}\\
&= \sum_{k=0}^{6} C_k\cdot\lambda^k - \sum_{k=0}^{4} D_k\cdot\lambda^k
\end{aligned}
$$

$$
\begin{aligned}
\det(A-\lambda\cdot I) &= \det\begin{pmatrix}\Upsilon_{11} & \ldots & \Upsilon_{16}\\ \vdots & \ddots & \vdots\\ \Upsilon_{61} & \cdots & \Upsilon_{66}\end{pmatrix}\\
&= C_0 + C_1\cdot\lambda + C_2\cdot\lambda^2 + C_3\cdot\lambda^3\\
&\quad + C_4\cdot\lambda^4 + C_5\cdot\lambda^5 + C_6\cdot\lambda^6\\
&\quad - \left(D_0 + D_1\cdot\lambda + D_2\cdot\lambda^2 + D_3\cdot\lambda^3 + D_4\cdot\lambda^4\right)
\end{aligned}
$$

$$\det(A - \lambda \cdot I) = \det\begin{pmatrix} \Upsilon_{11} & \ldots & \Upsilon_{16} \\ \vdots & \ddots & \vdots \\ \Upsilon_{61} & \cdots & \Upsilon_{66} \end{pmatrix}$$
$$= (C_0 - D_0) + (C_1 - D_1) \cdot \lambda$$
$$+ (C_2 - D_2) \cdot \lambda^2 + (C_3 - D_3) \cdot \lambda^3$$
$$+ (C_4 - D_4) \cdot \lambda^4 + C_5 \cdot \lambda^5 + C_6 \cdot \lambda^6$$

We define for simplicity new global parameters:

$$E_0 = C_0 - D_0; \; E_1 = C_1 - D_1$$

$$E_2 = C_2 - D_2; \; E_3 = C_3 - D_3; \; E_4 = C_4 - D_4; \; E_5 = C_5; \; E_6 = C_6$$

$$\det(A - \lambda \cdot I) = \det\begin{pmatrix} \Upsilon_{11} & \ldots & \Upsilon_{16} \\ \vdots & \ddots & \vdots \\ \Upsilon_{61} & \cdots & \Upsilon_{66} \end{pmatrix}$$
$$= \sum_{k=0}^{6} E_k \cdot \lambda^k; \; \det(A - \lambda \cdot I) = 0 \Rightarrow \sum_{k=0}^{6} E_k \cdot \lambda^k = 0$$

Eigenvalue stability discussion: Our THz dual-wavelength QCL system involving N variables (N > 2, N = 6, arbitrarily small increments), the characteristic equation is of degree N = 6 and must often be solved numerically. Expect in some particular cases, such an equation has (N = 6) distinct roots that can be real or complex. These values are eigenvalues of the (6 × 6) Jacobian matrix (A). The general rule is that the THz dual-wavelength QCL system is stable if there is no eigenvalue with positive real part. It is sufficient that one eigenvalue is positive for the steady state to be unstable. Our six variables $(n_1, n_2, n_3, n_4, n_{ph}(\hbar \cdot \omega_1), n_{ph}(\hbar \cdot \omega_2))$ system has six eigenvalues (six system's arbitrarily small increments). The type of behavior can be characterized as a function of the position of these eigenvalues in the Re/Im plane. Five non-generated cases can be distinguished: (1) the six eigenvalues are real and negative (stable steady state), (2) the six eigenvalues are real, at least one of them is positive (unstable steady state), (3) and (4) five eigenvalues are complex conjugates with a negative real part and other eigenvalue real is negative (stable steady state), two cases can be distinguished depending on the relative value of the real part of the complex eigenvalues and of the real one, (5) five eigenvalues are complex conjugates with a negative real part and other eigenvalue real is positive (unstable steady state) [5, 6].

8.4 Questions

1. We have a system a system of asymmetric dual quantum well laser and we analyze the dynamic responses of the photon and carrier densities. The laser consists of two different quantum wells isolated by a high and/or thick barrier layer that result in an inhomogeneous carrier injection into the two wells. We write our system differential equations for the component of the hole current which is directly injected into well 1 (zero), the rate equations for hole densities n_1 and n_2 in well 1, well 2, respectively. The rate equations for the photon densities $S^{\lambda 1}$, $S^{\lambda 2}$ of the lasing modes at λ_1 and λ_2 are

$$\frac{dn_1}{dt} = \frac{J_{21}(n_1(t), n_2(t))}{e \cdot d_1} - G_1^{\lambda_1}(n_1(t)) \cdot s^{\lambda_1}(t) - G_1^{\lambda_2}(n_1(t)) \cdot s^{\lambda_2}(t) - \frac{n_1(t)}{\tau_{n_1}}$$

$$\frac{dn_2}{dt} = \frac{J - J_{21}(n_1(t), n_2(t))}{e \cdot d_2} - G_2^{\lambda_2}(n_2(t)) \cdot s^{\lambda_2}(t) - \frac{n_2(t)}{\tau_{n_2}}$$

$$\frac{ds^{\lambda_1}}{dt} = G_1^{\lambda_1}(n_1(t)) \cdot s^{\lambda_1}(t) - \frac{s^{\lambda_1}(t)}{\tau^{\lambda_1}} + \beta_1^{\lambda_1} \cdot \frac{n_1(t)}{\tau_{n_1}}$$

$$\frac{ds^{\lambda_2}}{dt} = \left[G_1^{\lambda_2}(n_1(t)) + G_2^{\lambda_2}(n_2(t))\right] \cdot s^{\lambda_2}(t) - \frac{s^{\lambda_2}(t)}{\tau^{\lambda_2}} + \beta_1^{\lambda_2} \cdot \frac{n_1(t)}{\tau_{n_1}} + \beta_2^{\lambda_2} \cdot \frac{n_2(t)}{\tau_{n_2}}$$

If we define the threshold hole densities n_{1th} and n_{2th} by $G_1^{\lambda_1}(n_{1th}) = \frac{1}{\tau^{\lambda 1}}$ and $G_2^{\lambda_2}(n_{2th}) = \frac{1}{\tau^{\lambda 2}} - G_1^{\lambda_2}(n_{1th})$ and the optical gains expressions:

$$G_1^{\lambda_1}(n_1) = \frac{n_1 - n_{1th}}{n_{1th}} \cdot \left(\frac{1}{\tau^{\lambda_1}} + g_a \cdot f_a\right) + \frac{1}{\tau^{\lambda_1}}$$

$$G_2^{\lambda_2}(n_2) = \frac{n_2 - n_{2th}}{n_{2th}} \cdot \left(\frac{1}{\tau^{\lambda_2}} - G_1^{\lambda_2}(n_{1th}) + g_b \cdot f_b\right) + \frac{1}{\tau^{\lambda_2}}$$

All system parameters as discuss in (8.3), the injection current J is assumed to be a rectangular pulse of height J_p and width $3 \cdot \tau_{n_1}$ starting at $t = 0$.

1.1 Find system fixes points (equilibrium points) if $\frac{1}{\tau^{\lambda 1}}, \frac{1}{\tau^{\lambda 2}} \gg g_a \cdot f_a, g_b \cdot f_b$, respectively.
Hint: $\tau^{\lambda_1}, \tau^{\lambda_2} \approx 10^{-12}$ s; $\frac{1}{\tau^{\lambda 1}}, \frac{1}{\tau^{\lambda 2}} \approx 10^{12}$; $\tau_{n_1} \cdot g_a \cdot f_a \approx 10^{-7}$; $\tau_{n_1} \approx 10^{-9}$ s, $\tau_{n_2} \cdot g_b \cdot f_b \approx 10^{-7}$; $\tau_{n_2} \approx 10^{-9}$ s; $g_a \cdot f_a \approx 100$; $g_b \cdot f_b \approx 100$.

1.2 Discuss stability and stability switching for different values of n_{1th} and n_{2th}.
1.3 $n_1 > n_{1th}; n_2 > n_{2th}; \frac{n_1 - n_{1th}}{n_{1th}} \in [0.1, 0.9]; \frac{n_2 - n_{2th}}{n_{2th}} \in [0.2, 0.8]$, Find system fixed points within the bands (0.1 … .0.9; 0.2 … .0.8), and discuss stability.
1.4 $n_1 \gg n_{1th}; n_2 \gg n_{2th}$, Find system fixed points and discuss stability.
1.5 $\frac{d_2}{d_1} = \Gamma; \Gamma > 1; \Gamma \in \mathbb{R}$, Find system fixed points coordinates as a function of Γ parameter, and plot the relative graph. Discuss stability and stability switching for different values of Γ parameter.

2. We have a system a system of asymmetric dual quantum well laser and we analyze the dynamic responses of the photon and carrier densities. The laser consists of two different quantum wells isolated by a high and/or thick barrier layer that result in an inhomogeneous carrier injection into the two wells. We write our system differential equations for the component of the hole current which is directly injected into well 1 (zero), the rate equations for hole densities n_1 and n_2 in well 1, well 2, respectively. As usual $\beta_1^{\lambda_1}$ and $\beta_1^{\lambda_2}$ are constant, independent of light intensity. On the other hand $\beta_1^{\lambda_1}$ of ADQW LDs is expected to be an increasing function of s^{λ_2}. The λ_2 photons induce the simultaneous emission of the λ_1 photon and another quantum of energy $\hbar \cdot \omega_{21} = \hbar \cdot \omega_2 - \hbar \cdot \omega_1$. The quantum can be either a phonon (Raman process) or a photon (electronic Raman process). The electronic Raman process becomes resonantly enhanced as $\hbar \cdot \omega_{21}$ approaches the inter sub band level separation in well 1. Both types of the Raman process modify $\beta_1^{\lambda_1}$ into $\beta_1^{\lambda_1} = \beta_{10}^{\lambda_1} + \zeta^{\lambda_1} \cdot s^{\lambda_2}$, where the first term in the right-hand side is the usual term, and the second represents the Raman process. The estimation of ζ^{λ_1}(fitting parameter) is done by knowing the level broadening, phase matching conditions, etc. The rate equations for the photon densities $S^{\lambda 1}$, $S^{\lambda 2}$ of the lasing modes at λ_1 and λ_2 are

$$\frac{dn_1}{dt} = \frac{J_{21}(n_1(t), n_2(t))}{e \cdot d_1} - G_1^{\lambda_1}(n_1(t)) \cdot s^{\lambda_1}(t) - G_1^{\lambda_2}(n_1(t)) \cdot s^{\lambda_2}(t) - \frac{n_1(t)}{\tau_{n_1}}$$

$$\frac{dn_2}{dt} = \frac{J - J_{21}(n_1(t), n_2(t))}{e \cdot d_2} - G_2^{\lambda_2}(n_2(t)) \cdot s^{\lambda_2}(t) - \frac{n_2(t)}{\tau_{n_2}}$$

$$\frac{ds^{\lambda_1}}{dt} = G_1^{\lambda_1}(n_1(t)) \cdot s^{\lambda_1}(t) - \frac{s^{\lambda_1}(t)}{\tau^{\lambda_1}} + \left(\beta_{10}^{\lambda_1} + \zeta^{\lambda_1} \cdot s^{\lambda_2}\right) \cdot \frac{n_1(t)}{\tau_{n_1}}$$

$$\frac{ds^{\lambda_2}}{dt} = \left[G_1^{\lambda_2}(n_1(t)) + G_2^{\lambda_2}(n_2(t))\right] \cdot s^{\lambda_2}(t) - \frac{s^{\lambda_2}(t)}{\tau^{\lambda_2}} + \beta_1^{\lambda_2} \cdot \frac{n_1(t)}{\tau_{n_1}} + \beta_2^{\lambda_2} \cdot \frac{n_2(t)}{\tau_{n_2}}$$

If we define the threshold hole densities n_{1th} and n_{2th} by $G_1^{\lambda_1}(n_{1th}) = \frac{1}{\tau^{\lambda_1}}$ and $G_2^{\lambda_2}(n_{2th}) = \frac{1}{\tau^{\lambda_2}} - G_1^{\lambda_2}(n_{1th})$ and the optical gains expressions:

$$G_1^{\lambda_1}(n_1) = \frac{n_1 - n_{1th}}{n_{1th}} \cdot \left(\frac{1}{\tau^{\lambda_1}} + g_a \cdot f_a \right) + \frac{1}{\tau^{\lambda_1}}$$

$$G_2^{\lambda_2}(n_2) = \frac{n_2 - n_{2th}}{n_{2th}} \cdot \left(\frac{1}{\tau^{\lambda_2}} - G_1^{\lambda_2}(n_{1th}) + g_b \cdot f_b \right) + \frac{1}{\tau^{\lambda_2}}$$

All system parameters as discuss in (8.3), the injection current J is assumed to be a rectangular pulse of height J_p and width $3 \cdot \tau_{n_1}$ starting at $t = 0$.

2.1 Find system fixed points (equilibrium points) and plot as a function of $\beta_{10}^{\lambda_1}$ and ζ^{λ_1} parameters.
2.2 Discuss stability and stability switching for different values of ζ^{λ_1} parameter.
2.3 $n_1 \gg n_{1th}$, Find system fixed points and discuss stability and stability switching for different values of $\beta_{10}^{\lambda_1}$ parameter.
2.4 $\frac{d_2}{d_1} = \sqrt[k]{\Gamma}; \Gamma > 1; \Gamma \in \mathbb{R}; k \in \mathbb{N}$, Find system fixed points coordinates as a function of k parameter, and plot the relative graph. Discuss stability and stability switching for different values of Γ and k parameters.
2.5 $\beta_1^{\lambda_2} = f\left(\beta_1^{\lambda_1}\right)$, Find system fixed points and discuss stability for the cases: (1) $f\left(\beta_1^{\lambda_1}\right)$ is a quadratic function of $\beta_1^{\lambda_1}$ parameter, (2) $f\left(\beta_1^{\lambda_1}\right)$ is a Surjective function of $\beta_1^{\lambda_1}$ parameter, (3) $f\left(\beta_1^{\lambda_1}\right)$ is a composition function of quadratic function, g ,$f \circ g = f\left(g\left(\beta_1^{\lambda_1}\right)\right)$,$f$ is Surjective function.

Hint: A function f from a set X to a set Y is surjective (also known as onto, or a surjection), if for every element y in the codomain Y of f , there is at least one element x in the domain X of f such that f(x) = y. It is not required that x be unique; the function f may map one or more elements of X to the same element of Y.

3. Typical applications for ~2 μm, ..., 3 μm Laser beam are laser microscopy, environmental monitoring, LIDAR, and as optical-parametric-oscillation (OPO) pump sources. Tm^{3+}-doped fiber is suitable for producing ~2 μm laser emission due to its several unique advantages. First Tm^{3+}-doped fiber has a strong absorption spectrum that has good overlap with the emission spectrum of commercially available AlGaAs laser diodes. Second, the specific energy-level structure of Tm^{3+} ions provides the Tm^{3+}-doped fiber laser (TDFL) quantum efficiency close to two through the cross relaxation (CR) process. Thirdly, the Tm^{3+}-doped fiber has a very broad emission band (~400 nm), offering great wavelength tenability for the fiber laser. To get high-power output, Tm^{3+} ions are often doped into silica glass fiber to construct ~2 μm fiber lasers. The silicate glass fiber has maximum phonon energy as large as 1100 cm^{-1} (compared to 550 cm^{-1} for fluorides), which limits its infrared transparency wavelength less than ~2.2μm. There is a reason accounting for the self-pulsing formation in TDFLs through theoretical analysis and numerical simulation. For Tm^{3+} ions, the energy-level structure diagram includes several important transition and energy-transfer processes. The rate equations for the local population densities of these levels are as follows.

$$\frac{dN_4}{dt} = R(z,t)\cdot(N_1 - N_4) - k_{4212}\cdot N_1\cdot N_4 + k_{2124}\cdot N_2^2 - \frac{N_4}{\tau_4} + \sigma_{sa}\cdot c\cdot\phi(z,t)\cdot(N_3 - N_4)$$

$$\frac{dN_3}{dt} = k_{2123}\cdot N_2^2 + \beta_{43}\cdot\frac{N_4}{\tau_4} - \frac{N_3}{\tau_3} - \sigma_{sa}\cdot c\cdot\phi(z,t)\cdot(N_3 - N_4)$$

$$\frac{dN_2}{dt} = 2\cdot k_{4212}\cdot N_1\cdot N_4 - 2\cdot(k_{2124} + k_{2123})\cdot N_2^2 + \beta_{42}\cdot\frac{N_4}{\tau_4} + \beta_{32}\cdot\frac{N_3}{\tau_3} - \frac{N_2}{\tau_2} - c\cdot\phi(z,t)\cdot\sigma_e\cdot\left(N_2 - \frac{g_2}{g_1}\cdot N_1\right) + c\cdot\phi(z,t)\cdot\sigma_{ga}\cdot(N_1 - N_2)$$

$$\frac{d\phi}{dt} = c\cdot\phi(z,t)\cdot\sigma_e\cdot\left(N_2 - \frac{g_2}{g_1}\cdot N_1\right) + m\cdot\frac{N_2}{\tau_2} - \sigma_{ga}\cdot c\cdot\phi(z,t)\cdot(N_1 - N_2) - r_c\cdot\phi(z,t)$$

$$N_1 = N_{tot} - \sum_{k=2}^{4} N_k;\ R(z,t) = R(z=0,t)\cdot e^{-\alpha_p\cdot z}$$

N_i values are the populations of four energy manifolds $^3H_6$, $^3F_4$, $^3F_5$, $^3H_4$, and N_{tot} is the total density of Tm^{3+} ions. R is the pump rate, and ϕ is the average photon density of the laser field, σ_e is the stimulated emission cross section of signal light, σ_{ga} and σ_{sa} are the absorption cross sections of ground state and excited state, respectively. Functions g_1 and g_2 are the degeneracies of the upper and lower laser levels, τ_i is the level lifetimes of four manifolds, and r_c is the signal photon decay rate. β_{ij} are branch ratios from the i to j level, m is the ratio of laser modes to total spontaneous emission modes. The coefficients k_{ijkl} describe the energy transfer process; k_{4212} and k_{3212} are the cross relaxation constants, and k_{2124} and k_{2123} are the up-conversion constants. The coefficient α_p is the pump absorption of the fiber, which is calculated by $\alpha_p = \sigma_{ap}\cdot N_{tot}$, where σ_{ap} is the pump absorption cross section. In our analysis, the phonon-assisted ESA process of $^3F_4$, $^3H_5 \rightarrow {}^3H_6$, $^3H_4$ and ground state absorption (GSA) through $^3F_4$, $^3H_5 \rightarrow {}^3H_6$, $^3F_3$ energy transfer process are considered.

3.1 Find system fixed points (equilibrium points).
3.2 Discuss stability and stability switching for different values of σ_{ga} and σ_{sa} parameters.
3.3 $z \rightarrow \infty$ (very big value), Find system fixed points and discuss stability.
3.4 $r_c \rightarrow \varepsilon$ (signal photon decay rate) is very small, how the system dynamic is influenced? Find fixed points and discuss stability.
3.5 Functions g_1 and g_2 are the degeneracies of the upper and lower laser levels $g_1 = \sqrt[\psi]{g_2}$; $\psi \in \mathbb{N}$; $\psi > 1$, Find system fixed points coordinate and plots as a

function of ψ parameter. Discuss stability and stability switching for different values of ψ parameter.

4. We have a system of Tm^{3+} ions in Alkali Germanate Glass. Germanate glasses are of particular interest for their excellent optical properties as well as their abnormal structural changes that appear with the addition of modifiers, giving rise to the so-called *germanate anomaly*. This anomaly refers to the non-monotonic compositional scaling of properties exhibited by alkali germanate glasses and has been studied with various spectroscopy techniques. We inspect T_m-doped alkali germanate glass and absorption, emission and luminescence decay. When pumped with 0.792 μm diodes (low phonon energies) there is T_m:germanate promise as fiber laser. There are low non radiative quenching and pulsed laser performance with quantum efficiency of 1.69. T_m lasers uses materials co-doped with Yb and Er and ≈0.95 μm pump diode. Yb absorbs the pump photons, transfers the energy to Er, which subsequently transfers the energy to the T_m upper laser manifold. In ideal case, every energy transfer process has unity efficiency; the ratio of photon energies limits the efficiency to ≈ 0.5. Another approach pumps T_m directly in the $^3H_4$ manifold at 0.79 μm. If each pump photon produces only one T_m atom in the upper laser manifold, the $^3F_4$, then the ratio of pump to laser photon energies limits the efficiency to ≈0.4. T_m pumped by a 0.79 μm diode laser undergoes self-quenching. It produces two T_m atoms in the upper laser manifold. The quantum efficiency is ideally 2.0 and the efficiency limit increases to ≈0.8. Concern the absorption of the $^3F_4$ manifold in T_m:germanate glass, it exhibits a double peak structure which is not seen in T_m:silica or T_m:ZBLAN glasses. It happened in $^1G_4$ manifold, T_m:ZBLAN, but not for T_m:silica. There is a localized symmetry about the T_m ions in germanate glasses. The absorption cross sections are similar to those measured for T_m-doped lead germanate glass and have similar values for the cross section. The absorption parameters of T_m:germanate glass are for manifolds $^1D_2$, $^1G_4$, $^3F_2$, $^3F_3$, $^3H_4$, $^3H_5$, and $^3F_4$. τ_2 and τ_4 are the related system lifetimes. The equations for the rate of change of the population of the $^3F_4$ and $^3H_4$ manifolds are ($n_1 = n_1(t)$, $n_4 = n_4(t)$, $n_2 = n_2(t)$):

$$\frac{dn_2(t)}{dt} = -\frac{n_2(t-\tau_\Delta)}{\tau_2} + 2 \cdot p_{41} \cdot n_1(t) \cdot n_4(t) - 2 \cdot p_{22} \cdot n_2^2(t-\tau_\Delta)$$

$$\frac{dn_4(t)}{dt} = -\frac{n_4(t)}{\tau_4} - p_{41} \cdot n_1(t) \cdot n_4(t) + p_{22} \cdot n_2^2(t-\tau_\Delta)$$

$$C_{T_m} \cdot N_s = n_1 + n_2 + n_4$$

The population in T_m-T_m energy transfer processes level 2, is shifted in time by τ_Δ parameter, due to un-perfect energy transfer process. $^3H_6$, $^3F_4$ and $^3H_4$ manifolds are designed by the numbers 1, 2 and 4, respectively. The energy transfer parameters in the system differential equations are presented. The p_{41} parameter is the self-quenching process and p_{22} is its reverse process. All energy transfer process can have a forward and reverse process. In our analysis, pumping the $^3F_4$ manifold, p_{22}

is the forward up conversion process and p_{41} is the reverse self-quenching process. Since there must be conservation of T_m excitations, then $C_{T_m} \cdot N_s = n_1 + n_2 + n_4$, where C_{T_m} is the percent thulium concentration and N_s is the total density of sites available for occupation by the T_m ions. It can be used to eliminate n_1 from the system equations.

4.1 Find system fixed points and plot the related graphs as a function of p_{41} and p_{22} system parameters.
4.2 Discuss stability and stability switching for different values of τ_Δ parameter.
4.3 Discuss the system dynamic if p_{22} parameter \left(reverse process) is very small $p_{22} \to \varepsilon$. Find stability switching for different values of τ_Δ parameter.
4.4 Discuss the system dynamic if p_{41} parameter (self-quenching process) is very small $p_{41} \to \varepsilon$. Find stability switching for different values of τ_Δ parameter.
4.5 Discuss stability and stability switching for different values of C_{T_m} and N_s parameters.

5. We have a system of Tm^{3+} ions in Alkali Germanate Glass. We inspect T_m-doped alkali germanate glass and absorption, emission and luminescence decay. When pumped with 0.792 μm diodes (low phonon energies) there is T_m:germanate promise as fiber laser. There are low non radiative quenching and pulsed laser performance with quantum efficiency of 1.69. T_m lasers uses materials co-doped with Yb and Er and ≈0.95 μm pump diode. Yb absorbs the pump photons, transfers the energy to Er, which subsequently transfers the energy to the T_m upper laser manifold. In ideal case, every energy transfer process has unity efficiency; the ratio of photon energies limits the efficiency to ≈0.5. Another approach pumps T_m directly in the $^3H_4$ manifold at 0.79 μm. If each pump photon produces only one T_m atom in the upper laser manifold, the $^3F_4$, then the ratio of pump to laser photon energies limits the efficiency to ≈0.4, see (4). Since the majority of excitations reside in the $^3F_4$ pumping level, then $n_2 \gg n_4$, and all terms involving the terms $n_4^2(t)$ and $n_2(t) \cdot n_4(t)$ can be neglected compared to $n_2^2(t)$ and $n_4(t)$. τ_2 and τ_4 are the related system lifetimes. The reduced equations for the rate of change of the population of the $^3F_4$ and $^3H_4$ manifolds are ($n_1 = n_1(t), n_4 = n_4(t), n_2 = n_2(t)$):

$$\frac{dn_2(t)}{dt} = -\frac{n_2\left(t - \tau_{\Delta_1}\right)}{\tau_2} - 2 \cdot p_{22} \cdot n_2^2\left(t - \tau_{\Delta_1}\right)$$

$$\frac{dn_4(t)}{dt} = -\left(\frac{1}{\tau_4} + p_{41} \cdot C_{T_m} \cdot N_s\right) \cdot n_4\left(t - \tau_{\Delta_2}\right) + p_{22} \cdot n_2^2\left(t - \tau_{\Delta_1}\right)$$

The population in T_m-T_m energy transfer processes levels 2 and 4 are shifted in time by τ_{Δ_1} andτ_{Δ_2}, respectively. It is due to un-perfect energy transfer process.

$^3H_6$, $^3F_4$ and $^3H_4$ manifolds are designed by the numbers 1, 2 and 4, respectively.

The energy transfer parameters in the system differential equations are presented. The p_{41}parameter is the self-quenching process and p_{22} is its reverse process. All energy transfer process can have a forward and reverse process. In our analysis, pumping the $^3F_4$ manifold, p_{22} is the forward up conversion process and p_{41} is the reverse self-quenching process. Since there must be conservation of T_m excitations, then $C_{T_m} \cdot N_s = n_1 + n_2 + n_4$, where C_{T_m} is the percent thulium concentration and N_s is the total density of sites available for occupation by the T_m ions. It can be used to eliminate n_1 from the system equations.

5.1 Find system fixed points and plot the related graphs as a function of C_{T_m} and N_s parameters.
5.2 Discuss stability and stability switching for the following cases: (1) $\tau_{\Delta_1} = \tau_\Delta, \tau_{\Delta_2} = 0$, (2) $\tau_{\Delta_1} = 0, \tau_{\Delta_2} = \tau_\Delta$.
Hint: the parameter value for stability switching is τ_Δ.
5.3 Discuss stability and stability switching for the following cases: (1) $\tau_{\Delta_1} = \tau_\Delta$, $\tau_{\Delta_2} = \sqrt{\tau_\Delta}$, (2) $\tau_{\Delta_1} = \sqrt[3]{\tau_\Delta}$, $\tau_{\Delta_2} = \tau_\Delta$.
Hint: the parameter value for stability switching is τ_Δ.
5.4 Discuss the system dynamic if p_{22} parameter (reverse process) is very small $p_{22} \to \varepsilon$. Find stability switching for different values of τ_Δ parameter.
5.5 Discuss the system dynamic if p_{41} parameter (self-quenching process) is very small $p_{41} \to \varepsilon$. Find stability switching for different values of τ_Δ parameter.

6. We have a system of Tm^{3+} ions in Alkali Germanate Glass. We inspect T_m-doped alkali germanate glass and absorption, emission and luminescence decay. When pumped with 0.792 μm diodes (low phonon energies) there is T_m:germanate promise as fiber laser. There are low non radiative quenching and pulsed laser performance with quantum efficiency of 1.69. T_m lasers uses materials co-doped with Yb and Er and ≈ 0.95 μm pump diode. Yb absorbs the pump photons, transfers the energy to Er, which subsequently transfers the energy to the T_m upper laser manifold. In ideal case, every energy transfer process has unity efficiency; the ratio of photon energies limits the efficiency to ≈0.5. Another approach pumps T_m directly in the $^3H_4$ manifold at 0.79 μm. If each pump photon produces only one T_m atom in the upper laser manifold, the $^3F_4$, then the ratio of pump to laser photon energies limits the efficiency to ≈ 0.4, see (4). Since the majority of excitations reside in the $^3F_4$pumping level, then $n_2 \gg n_4$, and all terms involving the terms $n_4^2(t)$ and $n_2(t) \cdot n_4(t)$ can be neglected compared to $n_2^2(t)$ and $n_4(t)$. τ_2 and τ_4 are the related system lifetimes. The reduced equations for the rate of change of the population of the $^3F_4$ and $^3H_4$ manifolds are ($n_1 = n_1(t), n_4 = n_4(t), n_2 = n_2(t)$):

$$\frac{dn_2(t)}{dt} = -\frac{n_2(t)}{\tau_2} - 2 \cdot p_{22} \cdot n_2^2(t)$$

$$\frac{dn_4(t)}{dt} = -\left(\frac{1}{\tau_4} + p_{41} \cdot C_{T_m} \cdot N_s\right) \cdot n_4(t) + p_{22} \cdot n_2^2(t)$$

$^3H_6$, $^3F_4$ and $^3H_4$ manifolds are designed by the numbers 1, 2 and 4, respectively. The energy transfer parameters in the system differential equations are presented. The p_{41}parameter is the self-quenching process and p_{22} is its reverse process. All energy transfer process can have a forward and reverse process. In our analysis, pumping the $^3F_4$ manifold, p_{22} is the forward up conversion process and p_{41} is the reverse self-quenching process. Since there must be conservation of T_m excitations, then $C_{T_m} \cdot N_s = n_1 + n_2 + n_4$, where C_{T_m} is the percent thulium concentration and N_s is the total density of sites available for occupation by the T_m ions. It can be used to eliminate n_1 from the system equations. The solution for differential equation $\frac{dn_2(t)}{dt} = -\frac{n_2(t)}{\tau_2} - 2 \cdot p_{22} \cdot n_2^2(t)$ is $n_2(t) = \frac{n_{20} \cdot e^{-\frac{t}{\tau_2}}}{1 + 2 \cdot n_{20} \cdot \tau_2 \cdot p_{22} \cdot \left(1 - e^{-\frac{t}{\tau_2}}\right)}$, where n_{20} represents the initial population density for $t = 0$ at the initiation of the pulse.

6.1 Plot the function $n_2(t)$ and check for each time values ($t = t_1, t_2, t_3, \ldots, t_k; k \in \mathbb{N}$) if there are extremum points in the graph.
6.2 Find the expression for $\frac{\partial n_2(t)}{\partial \tau_2}$ and plot the graph $\frac{\partial n_2(t)}{\partial \tau_2} = f_1(\tau_2)$ for specific time values ($t = t_1, t_2, t_3, \ldots, t_k; k \in \mathbb{N}$) and $t_1 < t_2 < t_3 < \cdots < t_k; k \in \mathbb{N}$.
6.3 Find the expression to $\frac{\partial n_2(t)}{\partial n_{20}}$ and plot the graph $\frac{\partial n_2(t)}{\partial n_{20}} = f_2(n_{20})$ for specific time values ($t = t_1, t_2, t_3, \ldots, t_k; k \in \mathbb{N}$) and $t_1 < t_2 < t_3 < \cdots < t_k; k \in \mathbb{N}$.
6.4 We want to check if for specific time values $t = t_1, t_2, t_3, \ldots, t_k; k \in \mathbb{N}$ exist $\frac{\partial n_2(t)}{\partial \tau_2} = \frac{\partial n_2(t)}{\partial n_{20}}$. If it exists, find the expression for $t = t_1, t_2, t_3, \ldots, t_k; f_1(\tau_2) = f_2(n_{20})$.
6.5 Find the expression for $\frac{\partial n_2(t)}{\partial \tau_2} + \frac{\partial n_2(t)}{\partial n_{20}} = f(\tau_2, n_{20})$, plot 3D graph for $f(\tau_2, n_{20})$ versus τ_2 and n_{20}, discrete time values $t_1 < t_2 < t_3 < \cdots < t_k; k \in \mathbb{N}$.

7. We have a system of THz dual-wavelength quantum cascade laser (QCL) which is based on GaN active region. The rate equation model is for coupled electronic and electromagnetic system when certain structure parameters are obtained. The thermally activated phonon scattering mechanism can be suppressed by a large LO-phonon-energy. The conduction band diagram can be described by a simple schematic of a level diagram and various transition lifetimes to describe the rate equation analysis of a five level QCL 4, 3, 2, and 1 denote the radiative levels (see Sect. 8.3). Dynamic electron and photon populations is determined as a function of various radiative and non-radiative scattering lifetimes and the source term I, which is the current flowing through the quantum cascade structure. The rate equation are written for the derivative of n_4, n_3, n_2 and n_1 which are the electron populations in sub-bands 4, 3, 2, and 1, respectively. Additionally the rate equation include the derivative of the number of photon per unit energy (J^{-1}), $n_{ph}(\hbar \cdot \omega_{1,2})$ which is generated in all available cavity modes, due to both spontaneous and stimulated optical transitions. When the population inversion is established in the system then each of the optical transitions leads to a net stimulated emission into a cavity mode and correspondingly, the optical power would be added to the field as it propagates through the system. The propagation gain (material gain) is defined as variation of the photon flux divided by the number of photons per unit length of the structure, and it described with$g(\omega)$ due to radiative transition between the upper and the lower sub-bands u and l, respectively, $g(\omega) = \frac{\pi \cdot e^2 \cdot \omega}{\varepsilon_0 \cdot n_{eff} \cdot c} \cdot |Z_{ul}|^2 \cdot \Delta N_{3D} \cdot \frac{\Delta E/(2 \cdot \pi)}{(E_u - E_l + \hbar \cdot \omega)^2 + \left(\frac{\Delta E}{2 \cdot \pi}\right)^2}$. The peak gain is obtained at the resonance condition $\omega = \omega_{ul} = \frac{E_u - E_l}{\hbar}$ and is given by $g(\omega_{ul}) = \frac{2 \cdot e^2}{\hbar \cdot \varepsilon_0 \cdot n_{eff} \cdot c} \cdot |Z_{ul}|^2 \cdot \Delta N_{3D} \cdot \frac{\nu_{ul}}{\Delta \nu}$.

7.1 Plot $g(\omega)$ function versus ω, and as a function of ΔE. Find extremum points in your graphs.
7.2 Find $\frac{\partial g(\omega)}{\partial \omega} = f_1(\omega)$ and $\frac{\partial g(\omega)}{\partial \Delta E} = f_2(\Delta E)$, and plot the relative graphs.
7.3 Prove that $g(\omega)$ function has a term that represents the broadening in the gain spectrum, which is approximated by a normalized Lorentzian function, where ΔE is the full-width at half-maximum (FWHM).
7.4 Discuss the case for threshold condition in which the optical gain becomes equal to the losses$g(\omega_{ul}) = \alpha_m + \alpha_w$.
7.5 Find the expression for $\frac{\partial g(\omega)}{\partial \omega} + \frac{\partial g(\omega)}{\partial \Delta E} = f(\omega, \Delta E)$, plot the 3D graph for$f(\omega, \Delta E)$ versus ω and ΔE. Find extremum values in the graph.
Hint: ΔN_{3D} is the 3D population inversion density computed over the volume of all QCL modules. $v_{ul} = \frac{\omega_{ul}}{2 \cdot \pi}$, Z_{ul} is the dipole matrix element which is defined by $Z_{ul} = \int_{-\infty}^{\infty} \psi_u(z) \cdot z \cdot \psi_1(z) \cdot dz$.

8. We have a system of THz dual-wavelength quantum cascade laser (QCL) which is based on GaN active region. The rate equation model is for coupled electronic and electromagnetic system when certain structure parameters are obtained. The thermally activated phonon scattering mechanism can be suppressed by a large LO-phonon-energy. The conduction band diagram can be described by a simple schematic of a level diagram and various transition lifetimes to describe the rate equation analysis of a five level QCL 4, 3, 2, and 1 denote the radiative levels (see Sect. 8.3). The system includes some utilized parameters, η_{eff} is constant for both wavelengths because of the flat refractive index spectrum of GaN at long wavelengths. Where a typical FWHM of 3 meV is for $\lambda = 44.3\ \mu m$. The mirror and waveguide losses are chosen, where $\alpha_m = 1{-}2\ cm^{-1}$ and $\alpha_w = 10{-}20\ cm^{-1}$ for terahertz frequencies, α_{w2} is assumed to be larger than α_{w1} because of the larger carrier absorption losses in longer wavelengths. The level population densities and the relevant population inversions below and above the lasing threshold are established. The population of individual sub bands increases linearly with the current above the threshold. $N_4 - N_3$ and $N_2 - N_1$ quantities do not vary from their threshold values ($N_i; i = 1, 2, 3, 4$ are the population densities of the radiative levels and the relevant population inversion densities). This is a general property of all lasers under stationary conditions (gain clamping). The power coupled out from the laser is $P(\hbar \cdot \omega_1)$ and $P(\hbar \cdot \omega_2)$, the expressions are

$$
\begin{aligned}
P(\hbar \cdot \omega_1) &= \hbar \cdot \omega_0 \cdot n_{ph}(\hbar \cdot \omega_1) \cdot \alpha_m \cdot \frac{c}{n_{eff1}} \\
&= \frac{N_p \cdot \hbar \cdot \omega_0}{|e|} \cdot \left[\frac{\alpha_m}{\alpha_{w1} + \alpha_m} \right] \\
&\quad \times \left[\frac{(\eta_{th1} \cdot \tau_{42} \cdot \tau_{43} - \tau_3 \cdot \tau_{42} - (1 - \eta_{th1}) \cdot \tau_3 \cdot \tau_{43})}{(\tau_3 + \tau_{42}) \cdot \tau_{43}} \right. \\
&\quad \left. \cdot (I - I_{th1}) + I \cdot (\eta - \eta_{th1}) \right]
\end{aligned}
$$

$$P(\hbar\cdot\omega_2) = \hbar\cdot\omega_0\cdot n_{ph}(\hbar\cdot\omega_2)\cdot\alpha_m\cdot\frac{c}{n_{eff2}}$$
$$= \frac{N_p\cdot\hbar\cdot\omega_0}{|e|}\cdot\left[\frac{\alpha_m}{\alpha_{w2}+\alpha_m}\right]$$
$$\times\left[\frac{(\eta_{th2}\cdot\tau_{2i}\cdot\tau_{21}-\tau_{1i}\cdot\tau_{2i}-(1-\eta_{th2})\cdot\tau_{1i}\cdot\tau_{21})}{(\tau_{1i}+\tau_{2i})\cdot\tau_{21}}\right.$$
$$\left.\cdot(I-I_{th2})+I\cdot(\eta-\eta_{th2})\right]$$

η_{th} is the injection efficiency at the threshold condition. In the power versus current curves together with the gain spectra, the maximum output powers of $P_{1\max} = 20\,\text{mW}$ and $P_{2\max} = 25\,\text{mW}$ are obtained. The peak gains and output powers are completely comparable together. The system constant and typical utilized parameters value are $T = 265\,\text{K}$, $n_{eff1,2} = 2.22$, $N_p = 50$, $m^* = 0.22\cdot m_0$, $\gamma_{43} = 2.8\,\text{meV}$, $\gamma_{21} = 3.2\,\text{meV}$, $\alpha_m = 2\,\text{cm}^{-1}$, $\alpha_{w1} = 15\,\text{cm}^{-1}$, $\alpha_{w2} = 18\,\text{cm}^{-1}$.

8.1 Plot $P(\hbar\cdot\omega_1)$ as a function of τ_{43} parameter, find extremum points.
8.2 Find the typical function ($\eta_{th1} = f(\eta_{th2})$) which $\frac{\partial P(\hbar\cdot\omega_1)}{\partial\tau_{43}} = \frac{\partial P(\hbar\cdot\omega_2)}{\partial\tau_{2i}}$ and plot the function f for system typical parameters. Find extremum points.
8.3 Find constrain on η if $\frac{\partial P(\hbar\cdot\omega_1)}{\partial I} < \frac{\partial^2 P(\hbar\cdot\omega_2)}{\partial I^2}$, Discuss how system parameters are influence constrain on η.
8.4 $\frac{\partial P(\hbar\cdot\omega_1)}{\partial I} < \left[\frac{\partial^2 P(\hbar\cdot\omega_2)}{\partial I^2}\right]^k$; $k\in\mathbb{N}$, Discuss constrain on η as a function of k parameter, plot the related graphs.
8.5 $\frac{\partial^2 P(\hbar\cdot\omega_2)}{\partial I^2}, \left[\frac{\partial^2 P(\hbar\cdot\omega_2)}{\partial I^2}\right]^2, \ldots, \left[\frac{\partial^2 P(\hbar\cdot\omega_2)}{\partial I^2}\right]^k$; $k\in\mathbb{N}$, Try to characterize the series and find the expression $f(\eta)$, which transfer one element to the other. Plot the function $f(\eta)$.

9. Quantum cascade laser (QCLs) fabricated the first one-dimensional periodic potential multilayer by periodically varying the composition during epitaxial growth (supper lattice). QCLs use supper lattice lead to quantum confinement and tunneling phenomena, the basic process in QCLs operation. A QCL is made of a periodic repetition of active sections, which consist of tunnel-coupled quantum wells and injector, where a mini-band is formed. The main particularity in QCL is that instead of using bulk semiconductor materials in their optically active region, a periodic series of thin layers are used, that is, super lattice, consisting of a number of quantum-well-barrier system equally spaced, introducing a multiple quantum well (MQW) leading to one-dimensional confinement allowing an electric potential variation (band splitting) that results in a number of discrete electronic sub-bands. QCLs are usually three-level lasers, and the wave function formation is faster than scattering between states. Each sub-band i, is considered to have n_i electrons, and a scatter between the initial and the final sub-band, f, will have a scattering rate, W_{if}, and lifetime τ_{if}. If no other sub-bands are populated, the rate equations are described as extraction and injection of electron, respectively. The values are both equal in steady-state conditions, where time derivatives are zero. Hence, the general rate equation

for electrons in a sub-band i within an N level system is where $I = I_{ee} = I_{ie}$. At low temperature absorption is near zero: hence, $\tau_{32} > \tau_{21} \Rightarrow W_{21} > W_{32}$, and $n_3 > n_2$ leading to the existence of population inversion ($n_1 = n_1(t)$, $n_2 = n_2(t), n_3 = n_3(t)$).

$$\frac{dn_1(t)}{dt} = \frac{n_2(t - \tau_{\Delta_2})}{\tau_{21}} + \frac{n_3(t)}{\tau_{31}} - \frac{n_1(t - \tau_{\Delta_1})}{\tau_{13}} - \frac{n_1(t - \tau_{\Delta_1})}{\tau_{12}} - I_{ee}$$

$$\frac{dn_2(t)}{dt} = \frac{n_3(t)}{\tau_{32}} + \frac{n_1(t - \tau_{\Delta_1})}{\tau_{12}} - \frac{n_2(t - \tau_{\Delta_2})}{\tau_{21}} - \frac{n_1(t - \tau_{\Delta_1})}{\tau_{23}}$$

$$\frac{dn_3(t)}{dt} = I_{ie} + \frac{n_1(t - \tau_{\Delta_1})}{\tau_{13}} + \frac{n_2(t - \tau_{\Delta_2})}{\tau_{23}} - \frac{n_3(t)}{\tau_{31}} - \frac{n_3(t)}{\tau_{23}}$$

$\tau_{\Delta_1}, \tau_{\Delta_2}$ are time delays in the number of electrons in sub-band i at time t due to uncomplete structure of three-level inter sub band transition and scattering in QCLs. I_{ee} and I_{ie}a re the extraction and injection of electrons, respectively, whose values are both equal in steady-state conditions, where time derivatives are zero.

9.1 Find system fixed points coordinates and plot each coordinate as a function of I_{ee} and I_{ie} (3D graph).

9.2 Discuss stability and stability switching for different values of τ parameter, cases: (1) $\tau_{\Delta_1} = \tau$; $\tau_{\Delta_2} = 0$, (2) $\tau_{\Delta_1} = 0$; $\tau_{\Delta_2} = \tau$.

9.3 The general rate equation for electrons in a sub-band i within an N level system is $\frac{dn_i}{dt} = S_{f=1}^{N} \cdot n_j(t - \tau_j) \cdot \frac{1}{\tau_{fi}} - n_i(t - \tau_i) \cdot S_{f=1}^{N} \cdot \frac{1}{\tau_{if}} + I \cdot (d_{i0} - d_{iN})$, where $I = I_{ee} = I_{ie}$. At low temperature, absorption is near to zero: $\frac{n_3}{\tau_{32}} = \frac{n_2}{\tau_{21}}$ hence $\tau_{32} > \tau_{21} \Rightarrow W_{21} > W_{32}$ and $n_3 > n_2$ leading to the existence of population inversion.
Discuss stability and stability switching (different values of τ parameter) for the following cases: (1) $\tau_j = \tau, \tau_i = 0$, (2) $\tau_j = 0$, $\tau_i = \tau$.

9.4 What is the difference between τ_{if} and τ_{if} parameters? Plot 3D graph of system fixed points as a function of τ_{if} and τ_{if} parameters.

9.5 How the dynamic of the system is changed if $d_{i0} = d_{iN}$, Find fixed points and discuss stability.

10. We have a system of dual-wavelength distributed feedback (DFB) quantum cascade laser (QCL) which emits light at several wavelengths in the mid-infrared (mid-IR) spectrum. There are possible two structures, one is "Neighbour" DFB, which two single-mode DFB QCLs are fabricated next to each other, with minimal lateral distance, to allow efficient beam-coupling into multi-pass gas cells. In addition, the minimal distance allows either laser to be used as an integrated heater for the other, allowing extending the tuning range of its neighbor without any electrical cross-talk. The second structure used Vernier effects to realize a switchable DFB laser, with two target wavelengths which are distant by about 300 cm^{-1}. These devices are lasers sources for

Tunable diode laser absorption spectroscopy applications targeting simultaneous detection of multiple gasses, with distant spectral features, in compact and mobile setups. Quantum cascade laser is unipolar semiconductor device, based on inter sub band transitions, emitting in the mid-IR and THz parts of the electromagnetic spectrum. It offer excellent tailoring of its spectral gain by band structure engineering of the active region, as well as by stacking cascades of different gain profiles. In addition, it offers the possibility of precise mode control via distributed feedback (DFB) and distributed Bragg reflector (DBR) gratings. Their performance in the mid-IR range, in which one can access the fundamental ro-vibrational transitions of many pollutant and greenhouse gas molecules, makes them the preferred choice for laser based trace gas sensing applications. It is very important the design of digitalized dual-frequency grating. The first order Bragg reflection condition for light occurs when the wavelength is twice as large as the inverse of the spatial frequency component of the grating (k_{grating}). It is similar to sampled grating Vernier QCLs, two DFB mirrors for each of the front and back sections of the laser. Each mirror section consists of two reflectivity modes. One of the modes is kept aligned in frequency in both DFB mirrors and the other mode of one mirror is offset in frequency, with respect to the same mode in the other mirror. The mode which will be favored for lasing is the aligned one. By tuning the effective refractive index, n_{eff} using a heater close to the front section, we get red shift the reflectivity of the front mirror to align the offset modes and, therefore, switch the lasing mode. To implement two reflectivity modes on the grating, there is a need to combine two periodic functions $f(x) = \sin\left(\frac{x}{\Lambda_1}\right) + \sin\left(\frac{x}{\Lambda_2}\right)$, where $\Lambda_i = \frac{1}{2 \cdot n_{eff} \cdot f_i}$; $i = 1, 2$ is the grating periodicity to have Bragg reflection condition at frequency $f_i = \frac{10000}{\lambda_i}$ (frequency f_i is in cm^{-1}, wavelength λ in μ). The same function with slight frequency shift of one of the target frequencies is applied to the DFB grating of the other mirror section $f'(x) = \sin\left(\frac{x}{\Lambda'_1}\right) + \sin\left(\frac{x}{\Lambda'_2}\right)$, where Λ'_1 is kept the same as Λ_1, and Λ'_2 is slightly offset by δf as $\Lambda'_2 = \frac{1}{2 \cdot n_{eff} \cdot (f_2 + \delta f)}$.

10.1 Discuss the analogue profile resulting from the functions $f(x)$ and $f'(x)$.

10.2 Discuss the digitalizing of functions $f(x)$ and $f'(x)$ to 0 (for $f(x)$ or $f'(x) < 0$) and 1 (for $f(x)$ or $f'(x) > 0$) sequences, which correspond to etched and un-etched strips of the laser ridge InGaAs cladding, with respective effective modal refractive indices n_e for InP and n_u for un-etched InGaAs layer.

10.3 Plot function $f'(x)$ versus offset parameter δf, find extremum points.

10.4 Find δf parameter expression for $f(x) = f'(x)$, fixed x value. Find extremum points for different values of x variable.

10.5 Discuss the behavior of $g(x) = f(x) \cdot f'(x)$ function, plot $g(\delta f)$ for different values of x variable.

References

1. I. Sotomitsu, S. Arika, Theoretical analysis of dynamic response of asymmetric dual quantum well lasers. Appl. Phys. Lett. **61**(9) (1992)
2. Y. Kuang, *Delay Differential Equations with Applications in Population Dynamics* (Academic Press, Boston, 1993)
3. E. Beretta, Y. Kuang, Geometric stability switch criteria in delay differential systems with delay dependent parameters. SIAM J. Math. Anal. **33**, 1144–1165 (2002)
4. A.F. El-sherif, T.A. King, Dynamics and self-pulsing effects in Tm^{3+}-doped silica fibre lasers. Opt. Commun. **208**, 381–389 (2002)
5. J. Guckenheimer, Nonlinear oscillations, dynamical systems, and bifurcations of vector fields, in *Applied Mathematical Sciences*, vol 42
6. S. Wiggins, Introduction to applied nonlinear dynamical systems and chaos, in *Text in Applied Mathematics* (Hardcover)
7. A. Tredicucci, Terahertz quantum cascade lasers. NEST scientific report 2007–2009
8. K. Vijayraghaven, R.W. Adams, Terahertz sources based on cerenkov difference-frequency generation in quantum cascade lasers. Appl. Phys. Lett. **100**, 251104 (2012)
9. P. Dean, A. Valavanis, J. Keeley, Terahertz imaging using quantum cascade lasers—a review of systems and applications. J. Phys. D Appl. Phys. **47**, 374008 (22pp) (2014)
10. B.S. Williams, S. Kumar, Q. Hu, J.L. Reno, High-power terahertz quantum-cascade lasers. Electron. Lett. **42**(2)
11. M. Rosch, M. Beck, Heterogeneous terahertz quantum cascade lasers exceeding 1.9 THz spectral bandwidth and featuring dual comb operation. Nanophotonics **7**(1), 237–242 (2018)
12. M.A. Belkin, F. Capasso, Terahertz quantum-cascade-laser source based on intracavity difference-frequency generation. Nat. Photonics **1** (2007)
13. B. Mirzaei, A. Rostami, H. Baghban, Terahertz dual-wavelength quantum cascade laser based on GaN active region. Opt. Laser Technol. **44**, 378–383 (2012)

Chapter 9
Laser Circuits and Systems Bifurcation Behaviors Investigation, Comparison and Conclusions

Light **A**mplification by **S**timulated **E**mission of **R**adiation is demonstrated to be a viable technique for producing intense beams of single-color light. Travelling through space like a pencil beam, laser light is highly ordered and provides an invaluable tool in the world of science and technology because its properties are easy to control and manipulate to cut and drill, for instance, or to provide information and insights about virtually any object it interacts with. Lasers are increasingly important in surface science, both for the diagnostic evaluation and the processing of surfaces, for example, higher harmonic generation for diagnosis and the widespread use of laser surface micro structuring and annealing for processing. The physics behind such applications might be described in some cases by simple heating and melting processes. It can include much more complex phenomena such as plasma generation or elementary collective surface excitations. Laser application in surface science and technology provides an overview of the different techniques discusses the principles behind them and gives a concise description of laser-induced and laser-detected processes on surfaces. The developments in the field such as nonlinear surface spectroscopies and the interactions of ultrashort pulses with materials help to implement laser sources in many applications. The laser source is special in that it emits a coherent source of light—the wave source will always have the same frequency, waveform, and phase difference. This allows a myriad of applications for laser beams, including among others, manipulating very small (atom-sized) objects. Like any electromagnetic wave, a laser beam carries energy and momentum, which it can transfer to a particle in its path. This interaction is substantially impacted by objects' refractive index, which is a measure of how fast light propagates through the material. By carefully considering the differences in refractive index between an object and its surrounding medium, a laser can be tuned to physically hold and move objects, acting like a very precise tweezer. Optical tweezers have been used to trap a multitude of objects, including viruses, bacteria, metal particles, and even strands of DNA. There is great potential for applications in the life sciences, with some studies suggesting that optical tweezers might be pivotal in the detection and treatment of diseases such as cancer. Scientists, lab technicians, engineers, and industrial

O. Aluf, *Advance Elements of Laser Circuits and Systems*,
https://doi.org/10.1007/978-3-030-64103-0_9

technicians regularly utilize lasers to perform a wide range of important tasks. They measure distances, both short and long, with lasers, giving astronomers, geographers, and surveyors much more accurate figures than were available before the invention of these devices. They also use lasers to drill, weld, cut, and mark all sorts of materials; to study microscopic objects, including molecules. One of the most important scientific uses for lasers is that of an advanced measuring tool. The science point out that the strange mirror has revealed the distance from Earth to the moon. A figure that was accurate to within the length of a person's finger. This moon mirror was neither mysterious nor magical, though it would have seemed. Measuring distance by means of lasers and mirrors works just as well on Earth as it does in space. Lasers can measure enormous distances with great accuracy. A laser beam travels at a constant speed (the speed of light). The time it takes a laser beam to travel from its source, reflect off an object, and return to the source, will indicate the exact distance between the source and the object. The photons of laser light are collimated. They all travel from the laser in the same direction. Laser light is coherent, meaning that all the waves in a laser beam have the same wave pattern. These properties make laser light more intense than ordinary light and allow it to travel long distances. Ordinary light waves scatter in all directions from their source. Also, an ordinary beam of light is incoherent, meaning that it contains waves of many different patterns, which tend to interfere with one another.

Laser light is highly directional, or collimated. This means that all the photons travel in the same direction. They tend to stay together rather than spread apart as the photons in ordinary light do. The farther a beam of ordinary light travels, the more it spreads out and gets dimmer. On the other hand, collimated laser light can travel a great distance without losing very much of its energy and brightness. Ordinary light could never have made it to the moon mirror and back, whereas laser light did quite easily. Laser light is coherent, or very organized. This means that the light waves are lined up with each other and moving along in step, almost like a regiment of soldiers marching in a parade. By contrast, ordinary light is incoherent. Its waves become mixed up as they move along. So, the laser light is special. It is concentrated, directional, and organized. The three qualities combine to make the laser an extremely powerful and useful tool in materials processing, the industrial manipulation of metals, plastics, wood, ceramics, cloth, and other materials for making a wide range of products. Lasers are used to cut, drill, weld, heat-treat, and otherwise alter both metals and nonmetals. Lasers can drill tiny holes in turbine blades more quickly and less expensively than mechanical drills. Lasers have several advantages over conventional techniques of cutting materials. Lasers make cuts with better edge quality than most mechanical cutters. The edges of metal parts cut by a laser rarely need to be filed or polished because the laser makes such a clean cut. One of the more important uses of the laser drill in industry is in the production of copper wire. The wire is formed by forcing copper metal into a small round hole that a laser has drilled into a diamond. The hard diamond acts like a mold, and the much softer copper squeezes out the other end in the form of wire. Furthermore, the drilling process took several hours, so a worker could drill only two or three holes in a workday. In contrast, a laser beam drills holes in diamonds at the speed of light.

Lasers are effective in boring holes in very soft materials. Some of these materials are easily stretched or torn by ordinary methods. Similarly, lasers are used to drill tiny holes in the soft plastic valves of spray cans (such as those of hair spray or glass cleaner). One such laser can punch over a thousand valve holes in one minute. Another industrial application of lasers is welding. The advantage of the laser over normal welding methods is similar to its advantage in other industrial areas. The laser is hotter, faster, more accurate, and also safer because the welder does not have to go near the hot metal. Laser welding is worked on both large and small scales. The Laser Science and Technology (LST) bring latest advance in advanced laser sources, laser materials processing, laser micro/Nano manufacturing, and laser spectroscopic technologies. The main topics in laser physics are Solid state lasers, Fiber lasers, Nonlinear laser optics, Advanced beam shaping and guiding, Advanced laser micro-processing, Ultrafast laser nanofabrication, Laser-based welding, Laser metal deposition, Laser induced breakdown spectroscopy, Nonlinear spectroscopy.

The laser generates light energy by converting the energy states of a material. The energy level of an atom is a function of its temperature. Its lowest energy level is called its "ground state. Laser cutting, laser welding, laser marking or laser drilling—lasers are used in a wide variety of processes in many different industrial sectors to optimize manufacturing processes and to ensure the high quality of components. Light can be produced by atomic processes, and it is these processes which are responsible for the generation of laser light. The process is at atomic energy levels and how changes in these energy levels can lead to the production of laser light. A number of simplifications made regarding the concept of the atom. An atom consists of a small dense nucleus and one or more electrons in motion about the nucleus. The relationship between the electrons and the nucleus is described in terms of energy levels. Quantum mechanics predicts that these energy levels are discrete. The electrons normally occupy the lowest available energy levels. When this is the case, the atom is said to be in its ground state. However, electrons can occupy higher energy levels, leaving some of the lower energy states vacant or sparsely populated. One way that electrons and atoms can change from one energy state to another is by the absorption or emission of light energy, via a process called a radiative transition. An electron can absorb energy from a variety of external sources. From the point of view of laser action, two methods of supplying energy to the electrons are of prime importance. The first of these is the transfer of all the energy of a photon directly to an orbital electron. The increase in the energy of the electron causes it to "jump" to a higher energy level; the atom is then said to be in an "excited" state. An electron can accept only the precise amount of energy that is needed to move it from one allowable energy level to another. Only photons of the exact energy acceptable to the electron can be absorbed. Photons of slightly more (or slightly less) energy will not be absorbed. Another means often used to excite electrons is an electrical discharge. In this technique, the energy is supplied by collisions with electrons which have been accelerated by an electric field. The result of either type of excitation is that through the absorption of energy, an electron has been placed in a higher energy level than it originally resided. As a result, the atom of which it is a part is said to be excited. The nature of all matter is such that atomic and molecular structures tend to exist in the lowest energy state possible. Thus, an excited electron in a higher energy level will soon attempt to DE-EXCITE itself by any of several means. Some of the energy may be converted to heat. Another means of de-excitation is the spontaneous emission of a photon. The photon released by an atom as it is de-excited will have a total energy exactly equal to the difference in energy between the excited and lower energy levels. This release of a photon is called spontaneous emission. The exciting force is not of a unique energy, so that the electrons may be excited to any one of several allowable levels. A photon released from an excited atom could, upon interacting with a second, similarly excited atom, trigger the second atom into de-exciting itself with the release of another photon. The photon released by the second atom would be identical in frequency, energy, direction, and phase with the triggering photon, and the triggering photon would continue on its way, unchanged. Where there was one photon now there are two. These two photons could then proceed to trigger more through the process of stimulated emission. If an appropriate medium contains a great many excited atoms and de-excitation occurs only by spontaneous emission, the light output will be random and approximately equal in all directions. The process of stimulated emission, however, can cause an amplification of the number of photons traveling in a particular direction—a photon cascade. A preferential direction is established by placing mirrors at the ends of an optical cavity. Thus the number of photons traveling along the axis of the two mirrors increases greatly and Light Amplification by the Stimulated Emission of Radiation may occur. If enough amplification occurs, LASER beam is created. The process of stimulated emission will not produce a very efficient or even noticeable amplification of light unless a condition called "population inversion" occurs. If only a few atoms of several million are in an excited state, the chances of stimulated emission occurring are small. The greater the percentage of atoms in an excited state cause the greater the probability of stimulated emission. In the normal state of matter the population of electrons will be such that most of the electrons reside in the ground or lowest levels, leaving the upper levels somewhat depopulated. When electrons are excited and fill these upper levels to the extent that there are more atoms excited than not excited, the population is said to be inverted. A generalized laser consists of a lasing medium, a "pumping" system and an optical cavity. The laser material must have a metastable state in which the atoms or molecules can be trapped after receiving energy from the pumping system. The laser components are pumping systems, optical cavity, laser media, and time modes of operation. Pumping system: The pumping system imparts energy to the atoms or molecules of the lasing medium enabling them to be raised to an excited "metastable state" creating a population inversion. Optical pumping uses photons provided by a source such as a Xenon gas flash lamp or another laser to transfer energy to the lasing material. The optical source must provide photons which correspond to the allowed transition levels of the lasing material. Collision pumping relies on the transfer of energy to the lasing material by collision with the atoms (or molecules) of the lasing material. Energies which correspond to the allowed transitions must be provided. This is often done by electrical discharge in a pure gas or gas mixture in a tube. Chemical pumping systems use the binding energy released in chemical reactions to state. Optical cavity: An optical cavity is required to provide the amplification desired in the laser and to select the photons which are traveling in the desired direction. As the first atom or molecule in the metastable state of the inverted population decays, it triggers via stimulated emission, the decay of another atom or molecule in the metastable state. If the photons are traveling in a direction which leads to the walls of the lasing material, which is usually in the form of a rod or tube, they are lost and the amplification process terminates. They may actually be reflected at the wall of the rod or tube, but sooner or later they will be lost in the system and will not contribute to the beam. If, one of the decaying atoms or molecules releases a photon parallel to the axis of the lasing material, it can trigger the emission of another photon and both will be reflected by the mirror on the end of the lasing rod or tube. The reflected photons then pass back through the material triggering further emissions along exactly the same path which are reflected by the mirrors on the ends of the lasing material. As this amplification process continues, a portion of the radiation will always escape through the partially reflecting mirror. When the amount of amplification or gain through this process exceeds the losses in the cavity, laser oscillation is occurring. In this way, a narrow concentrated beam of coherent light is formed. The mirrors on the laser optical cavity must be precisely aligned for light beams parallel to the axis. The optical cavity itself, i.e., the lasing medium material must not be a strong absorber of the light energy. Laser Media: Lasers are commonly designated by the type of lasing material employed. There are four types which are: solid state, gas, dye, and semiconductor. Solid State Lasers employ a lasing material distributed in a solid matrix. One example is the Neodymium:YAG laser (Nd:YAG). The term: YAG is an abbreviation for the crystal: Yttrium Aluminum Garnet which serves as the host for the Neodymium ions. This laser emits an infrared beam at the wavelength of 1.064 μm ($\mu m = 10^{-6}$ m). Accessory devices that may be internal or external to the cavity may be used to convert the output to visible or ultraviolet wavelength. Gas Lasers use a gas or a mixture of gases within a tube. The most common gas laser uses a mixture of helium and neon (HeNe), with a primary output of 632.8 nm ($nm = 10^{-9}$ m) which is a visible red color. All gas lasers are quite similar in construction and behavior. For example, the carbon dioxide (CO_2) gas laser radiates at 10.6 μm in the far-infrared spectrum. Argon and krypton gas lasers operate with multiple frequency emissions principally in the visible spectra. The main emission wavelengths of an argon laser are 488 and 514 nm. Dye Lasers use a laser medium that is usually a complex organic dye in liquid solution or suspension. The most striking feature of these lasers is their tune ability. Dye lasers commonly employ optical pumping although some types have used chemical reaction pumping. The most commonly used dye is Rhodamine 6G which provides tune ability over 200 nm bandwidth in the red portion (620 nm) of the spectrum. Semiconductor Lasers (sometimes referred to as diode lasers) are not to be confused with solid state lasers. Semiconductor devices consist of two layers of semiconductor material sandwiched together. These lasers are generally very small physically, and individually of only modest power. However, they may be built into larger arrays. The most common diode laser is the Gallium Arsenide diode laser with a central emission of 840 nm. Time Modes of Operation: The different time modes of operation of a laser are distinguished by the rate at which energy is delivered. Continuous Wave (CW) lasers operate with a stable average beam power. In higher power systems, one is able to adjust the power. In low power gas lasers, such as HeNe, the power level is fixed by design and performance usually degrades with long term use. Single Pulsed (normal mode) lasers generally have pulse durations of a few hundred microseconds to a few milliseconds. This mode of operation is sometimes referred to as long pulse or normal mode. Single Pulsed Q-Switched lasers are the result of an intra-cavity delay (Q-switch cell) which allows the laser media to store a maximum of potential energy. Then, under optimum gain conditions, emission occurs in single pulses; typically of 10^{-8} s time domain. These pulses will have high peak powers often in the range from 10^6 to 10^9 W peak. Repetitively Pulsed or scanning lasers generally involve the operation of pulsed laser performance operating at a fixed (or variable) pulse rates which may range from a few pulses per second to as high as 20,000 pulses per second. The direction of a CW laser can be scanned rapidly using optical scanning systems to produce the equivalent of a repetitively pulsed output at a given location. Mode Locked lasers operate as a result of the resonant modes of the optical cavity which can affect the characteristics of the output beam. When the phases of different frequency modes are synchronized, i.e., "locked together," the different modes will interfere with one another to generate a beat effect. The result is a laser output which is observed as regularly spaced pulsations. Lasers operating in this mode-locked fashion, usually produce a train of regularly spaced pulses, each having a duration of 10^{-15} (femto) to 10^{-12} (pico)s. A mode-locked laser can deliver extremely high peak powers than the same laser operating in the Q-switched mode. These pulses will have enormous peak powers often in the range from 10^{12} W peak. The typical laser types are Helium Neon Laser, Argon, Krypton, and Xenon Ion Lasers, Carbon Dioxide Laser, ND:YAG Laser Systems, Excimer Lasers, Semiconductor Diode Lasers, Dye Lasers, and Argon or Nitrogen laser. Helium Neon Laser: The CW system is the helium neon (HeNe) gas mixture. Although its first successful operation was at an infrared wavelength of 1.15 μm, the HeNe laser is most well-known operating at the red 633 nm transition. Some HeNe lasers also can emit operate at other wavelengths (594 nm, 612 nm, 543 nm). Some earlier HeNe lasers were excited by radio frequency (RF) discharge but virtually all HeNe lasers today are driven by a small DC discharge between electrodes in the laser tube. The HeNe laser operates by an excitation of the helium atoms from the ground state. This energy excess is coupled to an unexcited neon atom by a collisional process with the net result of an inversion in the neon atom population, thus allowing laser action to begin. Power levels available from the HeNe laser ranges from a fraction of a milliwatt to about 75 mW in the largest available systems. The HeNe laser is noted for its high-frequency stability and TEM(oo) (single mode) operation. The HeNe laser is one of the most widely used laser in existence today. Its pencil-thin beam is used in surveying work, to align pipelines, as a sawing guide in sawmills, and is also used to "align" patients in medical X-ray units, just to name a few of its many applications. It is also used in many retail scanners, lecture hall pointers and display devices. In addition, holograms are often made using the coherent light of HeNe lasers. Argon, Krypton, and Xenon Ion Lasers: The family of ion lasers utilize argon, krypton, xenon, and neon gases to provides a source for over 35 different laser frequencies, ranging from the near ultraviolet (neon at 0.322 μm) to the near-infrared (krypton at 0.799 μm). It is possible to mix the gases, for example, argon and krypton, to produce either single frequency or simultaneous emission at ten different wavelengths, ranging from the violet through the red end of the spectrum. The basic design of an ion gas laser is similar to the HeNe. The major difference is that the electrical current flowing in the laser tube will be 10–20 A; sufficient to ionize the gas. Population inversion is obtained only in the ionized state of the gas. An important feature of these lasers is the very stable (0.2%) high output power of up to 20 W/CW. Commercial models will normally have a wavelength selector (a prism) within the cavity to allow for operation at any one of the wavelengths available. In addition, approximately single frequency operation can be achieved by placing an etalon inside the optical resonator cavity. Argon ion lasers produce the highest visible power levels and have up to 10 lasing wavelengths in the blue-green portion of the spectrum. These lasers are normally rated by the power level (typically 1–10 W) produced by all of the six major visible wavelengths from 458 to 514 nm. The most prominent argon wavelengths are the 514 and 488 nm lines. Wavelengths in the ultraviolet spectrum at 351 and 364 nm are available by changing resonator mirrors. To dissipate the large amount of generated heat, the larger argon ion laser tubes are water cooled. Although some lasers have separate heat exchangers, most use tap water. Simple pulsed versions of argon ion lasers also are available. Since the duty cycle ("on" time divided by the time between pulses) is low, the heat energy generated is small, and usually only convective cooling is needed. The average power output may be as high as several Watts, thought the peak powers can be as high as several kilowatts. Pulse widths are approximately five to fifty microseconds, with repetition rates as high as 60 Hz. Carbon Dioxide Laser: The carbon dioxide laser is the most efficient and powerful of all CW laser devices. Continuous powers have been reported above 30 kW at the far infrared 10.6 μm wavelength. An electrical discharge is initiated in a plasma tube containing carbon dioxide gas. CO_2 molecules are excited by electron collisions to higher vibrational levels, from which they decay to the metastable vibrational level occurs; which has a lifetime of approximately 2×10^{-3} s at low pressure of a few Torr. Establishing a population inversion between certain vibrational levels leads to lasing transitions at 10.6 μm, while a population inversion between other vibrational levels can result in lasing transitions at 9.6 μm. Although lasing can be obtained in a plasma tube containing CO_2 gas alone, various gases usually added, including N_2, He, Xe, CO_2 and H_2O. Such additives are used to increase the operating efficiency of CO_2 lasers. The most common gas composition in CO_2 lasers is a mixture of He, N_2 and CO_2. Carbon dioxide lasers are capable of producing tremendous amounts of output power, primarily because of the high efficiency of about 30%, as compared to less than 0.1% for most HeNe lasers. The principal difference between the CO_2 and other gas lasers is that the optics must be coated, or made of special materials, to be reflective or transmissivity at the far infrared wavelength of 10.6 μm. The output mirror can be made of germanium, which, if cooled, has very low loss at 10.6 μm. There are three common laser cavity configurations of the CO_2 laser. The first is the gas discharge tube encountered with the discussion of the HeNe laser. Secondly is the axial gas flow, where the gas mixture is pumped into one end of the tube and taken out the other. The gas flow allows for the replacement of the CO_2 molecules depleted (disassociated CO_2 molecules) by the electrical discharge. Nitrogen is added to the CO_2 to increase the efficiency of the pumping process and transfers energy by collisions. Associated effects enhance the de-excitation process. Helium is added to the mixture to further increase the efficiency of the process of pumping and stimulated emissions. The third method is the transverse gas flow. This technique can produce CO_2 laser emissions at power levels approaching 25 kW. The CO_2 laser has a strong emission wavelength at 10.6 μm. There is another strong line at 9.6 μm and a multitude of lines between 9 and 11 μm. CO_2 lasers are highly efficient (10–30%), give high output powers (used for welding and cutting), and applications out-of-doors can take advantage of low transmission loss atmospheric window at about 10 μm. ND:YAG Laser Systems: One of the most widely used laser sources for moderate to high power uses a neodymium doped crystal Yttrium Aluminum Garnet (YAG), commonly designated Nd:YAG. In addition, other hosts can be used with Nd, such as calcium tungstate and glass. The Nd:YAG laser is optically pumped either by tungsten or krypton pump lamps and is capable of CW outputs approaching 2000 W at the 1.06 μm wavelength. The ends of the crystal, which is usually in the form of a rod, are lapped, polished, and may be coated to provide the cavity mirrors. Nd:YAG lasers belong to the class of solid state lasers. Solid state lasers occupy a unique place in laser development. The first operational laser medium was a crystal of pink ruby (a sapphire crystal doped with chromium); since that time, the term "solid state laser" usually has been used to describe a laser whose active medium is a crystal doped with an impurity ion. Solid state lasers are rugged, simple to maintain, and capable of generating high powers. Although solid state lasers offer some unique advantages over gas lasers, crystals are not ideal cavities or perfect laser media. Real crystals contain refractive index variations that distort the wave front and mode structure of the laser. High power operation causes thermal expansion of the crystal that alters the effective cavity dimensions and thus changes the modes. The laser crystals are cooled by forced air or liquids, particularly for high repetition rates. The most striking aspect of solid state lasers is that the output is usually not continuous, but consists of a large number of often separated power bursts. Normal mode and Q-Switched solid-state lasers are often designed for a high repetition-rate operation. Usually the specific parameters of operation are dictated by the application. Pulsed YAG lasers operating 1 Hz at 150 J per pulse are used in metal removal applications. As the repetition rate increases, the allowable exit energy per pulse necessarily decreases. Systems are in operation, for example, which produce up to ten Joules per pulse at a repetition rate of 10 Hz. A similar laser, operated in the Q-Switched mode, could produce a one megawatt per pulse at a rate up to ten pulses a minute. Excimer Lasers: High power ultraviolet (UV) lasers have been the desire of many in the laser applications community for over twenty-five years. Theoretically, such a laser could produce a focused beam of sub-micrometer size and, therefore, be useful in laser microsurgery and industrial microlithography. Also, photochemical processes which are dependent upon the shorter UV wavelength would be possible at significantly greater speeds because of the enormous UV photon flux presented by a laser beam. Excimer lasers operate using reactive gases such as chlorine and fluorine mixed with inert gases such as argon, krypton or xenon. The various gas combinations, when electrically excited, produce a pseudo molecule (called a "dimer") with an energy level configuration that causes the generation of a specific laser wavelength emission which falls in the UV spectrum. The reliability of excimer lasers has made significant strides over the past several years. Now, systems operating at average powers from 50 to 100 W are commercially available. A typical excimer operates in a repetitively pulsed mode of 30–40 ns pulses at pulse rates up to 50 Hz with pulse energies of 1–2 J/pulse. Some systems use x-rays to preionize the excimer laser's gas mixture so-as-to enhance lasing efficiency and increase the overall output power. Excimer lasers are used either as a specific UV source or, in many cases, to serve as a "pumping" or exciting source to generate visible laser emissions. The excimer's UV output is directed into a tunable dye laser or Raman shifter module and converted into a modestly high power visible frequency emission. Excimer lasers are making the transition from the lab to the production area for a few unique uses in industry or in the operating room for exploratory surgical applications. Semiconductor Diode Lasers: The semiconductor or diode injection laser is another type of solid state laser. The energy level scheme is constructed by charge carriers in the semiconductor. They may be pumped optically or by electron beam bombardment, but most commonly, they are pumped by an externally applied current. Although all of these devices operate in the near infrared spectral region, visible laser diodes are being made today. A useful feature is that many are tunable by varying the applied current, changing temperature, or by applying an external magnetic field. Laser diodes are used extensively for communications, in compact disc players, retail scanners, printer, and are beginning to be used in ophthalmology. Semiconductor lasers are used in distance detectors and remote sensing systems, rangefinders, and for voice and data communications. Many of the diode lasers may be operated on a continuous wave basis. The most common diode uses a gallium-arsenide junction which emits a fan shaped infrared beam at 840 nm. Dye Lasers: Dye Lasers are true tunable laser. Using different organic dyes, a dye laser is capable of producing emission from the ultraviolet to near infrared. Most are operated in the visible with tunable emissions of red, yellow, green, or blue laser emission at almost any wavelength. The more common organic dye lasers are optically pumped. The most common dye used is Rhodamine-6G in solution. Such lasers may either be flash lamp pumped, or more commonly pumped with another laser such as an Argon or Nitrogen laser. To obtain CW reliable operation the dye is made to flow through a thin cell. Using the appropriate dye solutions, an argon-ion laser as a pump, and a prism, the dye laser is tunable across most of the visible spectrum. Tunable dye lasers are now widely used in high resolution atomic and molecular spectroscopy. The main Laser Beam Parameters are common to the beams emitted from all laser types and are the factors which, when combined together, distinguish laser outputs from other sources of electromagnetic radiation. The laser beam properties are (1) A nearly single frequency operation of low bandwidth, (2) A beam with a Gaussian beam intensity profile, (3) A beam of small divergence, (4) A beam of enormous intensity, (5) A beam which maintains a high degree of temporal and spatial coherence, (6) A beam that is, in many laser devices, highly plane polarized,

(7) A beam with enormous electromagnetic field strengths. Single Frequency Operation (Mono-chromacity): The frequency of any electro-magnetic wave is related to the number of cycles the electric or magnetic field undergoes each second. A completely coherent, monochromatic wave oscillates exactly at a constant frequency. Most laser systems display a narrow multi frequency characteristic. This frequency spread is, however, very narrow when compared to the average laser frequency. In most lasers, the frequency degeneracy is solely dependent upon the quantum transition characteristics of the active media, and the geometry of the laser resonator cavity (Fabry-Perot). In this sense, the laser media may be considered as a high number of isolated light generators placed between two mirrors. The electromagnetic field developed between the mirrors may be regarded as a superposition of plane waves at each of the slightly different frequencies which the laser media generates and allows oscillating. These different frequencies are termed the "modes" of the laser resonator. The off-axis modes result from plane waves propagating at an angle with respect to the axis of the resonator. These different modes are produced by diffraction effects in the Fabry-Perot cavity. The lowest order axial mode is designated as the TEM(oo) mode. This mode has the lowest diffraction losses and often will be the predominant mode of oscillation. For each transverse mode, there will be many longitudinal modes which can oscillate; hence the output of a multimode laser will actually contain a superposition of plane waves oscillating at many discrete frequencies. However, as previously mentioned, this frequency spread will be very small. In each laser, there will be specific "allowed" frequencies of the resonator cavity (Fabry-Perot modes). For most cases, the average wavelength at which the laser oscillates is sufficient to describe it's operation. If more precision is needed, then the frequency spread or bandwidth is given. Depending on the type of laser, bandwidths range typically from 10^{-4} to 10^{-9} times the average frequency of the laser; although bandwidths as low as 0.1 Hz. have been reported for stabilized gas lasers. The wave nature of light most often allows adequate description of the output of the laser, and for most cases it will be sufficient to use geometrical optics to describe the output as a beam with well-defined edges and some beam divergence. The beam is emitted from the laser with a beam diameter (alpha) and a beam divergence (PHI), as though it came from a small point source far behind the laser output aperture. Point Source Emission: The emission from most lasers can be considered as emanating from a "virtual point source" located within or behind the laser device. A "virtual point source" is one which really doesn't exist, but the properties of the emitted beam are such that there appears to be a sourcat this position. Thus, the virtual point source is located two meters behind the exit mirror. Gaussian distribution of the Beam: The intensity profile across a TEM(oo) laser beam will be in the form of a bell-shaped (Gaussian) distribution. The decrease in intensity at the edge of the beam is the result of diffraction effects produced at the edges. The spatial intensity distribution of this mode may be expressed by equation: (see printed copy) where R is the radius and W is a constant which defines the mean radius and is commonly referred to as the "spot size." At this point the intensity has fallen to e^{-2} of the peak intensity at the center of distribution. The edges f the laser beam is not well defined. If one were to measure the energy or power per unit area point by point across the center of

the output aperture, a Gaussian beam distribution is defined. The peak intensity is in the center of the beam and approaches zero as one move from the center. This shape is maintained as the beam propagates through space subject to broadening and distortion by atmospheric effects. Important points on the distribution curve are the e^{-1} and e^{-2} intensity points since they are used as standard quantities to define the laser beam divergence parameter. (The e is the natural number associated with the natural logarithm and is equal to: $e = 2.7183$). The e^{-1} point is where the intensity is reduced by the factor or approximately 63% of the energy (or power) is contained within the aperture of diameter (a) centered in the beam. In most all laser safety and compliance standards, the output aperture, a, and the beam divergence, (PHI), are defined relative to the e^{-1} points. The total laser energy (Q(t)) or beam power (PHI) is defined as that which is collected from the entire beam (total values). Many manufactures specifications use the e^{-2} i points to define beam divergence. In this case, e^{-2} i = 0.1353, or the total power (energy) is: $100\% - 0.1353 \times 100\% = 86.47\%$ or approximately 86% of the total energy/power is within the e^{-2} aperture. In some cases, they specify relative to the 90% point. Note that the beam divergence is larger at the e^{-2} or 90% point. Hazard calculations are sensitive to the beam divergence and conversions from e^{-2} points to e^{-2} power points are often performed on beam sizes. The beam diameters at the two points are related: Departure from the Gaussian distribution arises when independent oscillation occurs within the resonator at higher order modes. Some gas lasers may be designed to have sufficient gain to support simultaneous oscillation in many different transverse modes. Mode selection may often be accomplished by slight adjustment of the mirror alignments. With this technique, one can observe the different complex intensity distributions of each mode. The lowest order TEM(oo) mode with the nearly Gaussian intensity distribution has the lowest cavity losses and hence will generally be the dominant mode of oscillation. Optically pumped solid-state lasers such as the normal mode Nd:YAG laser usually display a randomly varying mode output. Thermal gradients in the optical media (i.e., the crystal) caused by non-uniform absorption of the pump light give rise to lens effects in the crystal which change during the pumping cycle. The result is a sporadic switching of transverse modes during the laser pulse. The time average is generally a bell-shaped distribution which is dependent upon the optical purity of the laser crystal, the pumping scheme, and the level at which the system is operated above lasing threshold. Some pumping schemes produce pronounced "hot spots" in the intensity distributions. For long range transmission, atmospheric effects can also produce intensity variations by a factor of ten over localized regions of the beam. Such non-uniformities distribution make it difficult to specify the cross sectional area of the beam. As a result, an average value of beam radius must be chosen. Typically, this is often (1) the half-power point; (2) the e^{-1} power point; or (3) the e^{-2} power point. In the case of optically pumped solid-state lasers, the size of the beam cross section is generally a function of the pumping level of the laser. In general, the higher the pumping level, the wider the beam size. Only when pulsed lasers are operated near threshold, or in special cavity conditions, the zero order modes (lowest beam spread) predominate. Beam Divergence: Beam divergence is a very important laser parameter and is often expressed in units of m-radians. The symmetry of the laser beam allows the geometry to be reduced to the two dimensions of a plane. The angle (phi), in radians can be related to degrees by noting that for a full circle, phi is 360°. For smaller angle, the arc length(s) intercepted along the circumference of the circle can be used to define the angle as the minimum beam divergence, called the diffraction limited beam divergence, is related by an equation. This concept is expanded to three dimensions by introducing the concept of solid angle. The solid angle (OMEGA) is expressed in units of steradians (sr) and is determined by using the area cut out of a surface of a sphere divided by the square of the distance to that surface; that is: (see printed copy). For a sphere, the solid angle may be opened up to include the entire sphere surface area ($A = 4$ pi R^2), therefore, the output of a typical laser will be confined to less than 10^{-6} sr. Intensity of Laser Emission: In many applications, the most important laser beam characteristic is the enormous intensity of the beam. Intensity is related to the beam power the cross sectional area and the manner in which the beam spreads from one point in space to the next. Power, by definition, is the time-rate at which work is done; specifically, it is the rate at which energy is used or produced. Energy relates the ability to do work. As with other forms of energy (e.g., chemical, mechanical, electrical), electromagnetic energy (light energy) is a conserved quantity. The intensity of the laser is usually expressed by the IRRADIANCE (power/area) of the beam. This is determined by dividing the average value of beam power by the average value of the beam cross section. Irradiance units are expressed in Watts per square centimeter. In pulsed laser operation, instantaneous (peak) Irradiances in excess of 100,000 W/cm^2 are quite easily generated in an unfocused high energy pulsed solid state laser pulse. If this output were contained within a typical beam divergence of 20 mrad and focused by only moderate power optics, the Irradiance at the focal plane would be increased at least one-hundred fold. A CW laser is rated in Watts and a pulsed laser is normally rated according to the total energy (Joules) per pulse. Pulsed outputs are also expressed as a RADIANT EXPOSURE in units of Joules per square centimeter. In order to determine the peak power of pulsed laser, it is necessary to know the pulse shape and duration. The peak power may be closely approximated by assuming a triangular pulse shape and dividing the energy per pulse by the pulse duration at half power. Average power is an important factor for high PRF lasers when determining the laser classification and maximum permissible exposure levels. The radiometric units of RADIANCE and INTEGRATED RADIANCE are used to describe the diffuse reflection of a continuous wave or pulsed laser beam. Radiance is expressed, by definition, as the Irradiance per unit solid angle (Watts per square centimeter per steradian). Integrated Radiance is expressed as the Radiant Exposure per unit solid angle (Joules per square centimeter per steradian). The unit of solid angle is defined such that all space about a point source (i.e., the source of light) will encompass 4pi sr. Focused Laser Beams: The beam from an ideal laser, i.e., a laser which emits a coherent wave, can be considered as a diffraction-limited beam. In this case, divergence of the beam is limited to the effects of diffraction at the beam edges. The emission from such a laser will display a far-field diffraction pattern at a distance (see printed copy) where a is the diameter of the emergent laser radiation. The TEM(oo) beam from a typical helium neon laser will display a 0.5–1.0

mrad beam spread at a distance of 1.0–2.0 m from the laser. Due to the high degree of coherence of a laser beam, it is theoretically possible to focus the beam to the diffraction limit of the wavelength of light. Typically, however, the laser will have a finite beam spread and can be expressed by the simple equations of geometrical optics. The spot diameter (d) is given by the simple equation: d = f phi where: d = spot diameter at focus f = focal length of lens phi = laser beam divergence (radians). As the spot diameter approaches the wavelength of light, the spot becomes diffraction-limited. For example, the beam from a highly coherent single transverse mode (TEM(oo)) gas laser will produce a Gaussian intensity pattern when focused. This distribution may be described mathematically by an equation where it is considered that the beam energy will be contained in a diameter defined at the e^{-2} power point. Therefore, the smallest possible spot size of a focused laser beam will approach the dimensions the wavelength of light which is being focused. Combining the equations above can yield an expression for the spot area. Thus, the Irradiance (power per unit area) of a focused laser beam will vary inversely with the square of the focal length of the lens and with the square of the beam divergence angle. Hence, these two factors have dramatic effects on the power distribution at the focal plane of the lens. Consequently, either a reduction in the focal length of the lens used to focus the beam or a reduction in the beam spread by a factor of ten will produce a one-hundred fold increase in the irradiance at the focal plane of the lens. Simultaneous reduction of both by a factor of ten would increase the Irradiance at the focal plane by a factor of 10^4. In practice, however, it is usually the beam divergence value that limits the focal spot diameter. This is especially true with pulsed laser systems. To achieve high power outputs, the laser crystal is usually pumped well over threshold; consequently, the beam will contain a conglomerate of high order "off-axis" modes which subsequently increase the beam size. Typical beam divergence values for gas lasers (helium-neon, argon, etc.) will be about one milliradian (1 mrad = 3.44 min of arc). Solid-state ruby and neodymium lasers generally have a higher beam spread (1–30 mrad), due primarily to the high beam divergence associated with the random multimode operation of such devices. Scanning Lasers: Some laser applications employ electro-mechanical or electro-optical scanner units to allow a raster-scan capability to the beam. In this way, the beam can be scanned over a large area (such as in a laser print maker) or over a small area (such as a laser UPC label reader) in a repeated geometry. Coherence: The coherency of a laser beam relates to the constancy of the spatial and temporal variations in the radiation wave-fronts. A high degree of coherence implies a constant phase different between two points on a series of equal-amplitude wave-fronts (spatial coherence), and in a correlation in time between the same points on different wave-fronts (temporal coherence). The two coherence terms are a part of the overall four-dimensional coherence function which completely describes the degree of coherency of the beam. If the laser beam is considered as a plane wave traveling in one direction, it will be spatially coherent due to the perpendicularity of the wave-fronts in the direction of propagation. Also, due to the monochromatic nature of the laser light, the beam will be temporally coherent; that is, it will display a fixed-phase relation between a part of the beam emitted at one time and portion emitted at another. Should the wavelength (or frequency) change, then the temporal coherency would degrade. Due to the wave nature of light and the corresponding diffraction effects produced by finite apertures, the image a point source of light (provided by any real optical system) is not actually a point. Such a distribution has a bright central area surrounded by light and dark rings. When an optical system's resolution is only limited by the diffraction effects, it is said to be diffraction limited. Even in this condition, a point is "spread out" as it is imaged. The more defects and aberrations introduced by the optical system, the more the spreading. Each lens images a point with some spreading and the manner which it behaves is defined as the point spread function. Since an optical system spreads a point source image, there is a corresponding limit to its resolution or ability to separate two points close together. This is especially important when looking at stars through a telescope. There are at least two criteria used to define the resolution of two points close together. A simple one, and the one most commonly used, is the Rayleigh criterion. This states that when the peak of one Airy disk is over the first dark ring of the other, the points are resolved. This is normally defined in terms of the apparent angle between the two points. Lasers are often referred to as coherent sources, but in fact, they really only partially coherent. Only absolutely monochromatic or single frequency waves are truly coherent; however, lasers are so close, relative to anything else, that a loose definition may seem justified. The degree to which two waves are coherent determines how well they interfere when brought together at some point in space. One frequently encounters the terms spatial and temporal coherence. Temporal coherence effects are those which arise from the finiteness of the spectral band. An increase in fringe visibility with a decrease in source size is a measure of the spatial coherence. An important measure of coherence is the coherence length, dL, which can be conceptually related to the duration of an uninterrupted wave-train. Even in the beam from an "ideal" laser, there will be random fluctuations in the phase difference of the electromagnetic fields at two separate points on a wave-front. The distance between points on the wave-front for which the average of this phase different is equal to radians is generally defined as the lateral coherence distance. Recombination of the light samples from points separated by a distance equal to, or less than, this amount can produce interference fringes. The distance is a classical measure of the spatial coherence of a light beam. The temporal coherence is a measure of the length of time that the beam is truly monochromatic. This may be considered as the time during which the amplitude of the electromagnetic field will remain constant at a given point in space while the phase varies linearly with time. During this time, the beam will travel a length $dL = c\ dT$ defined as the coherence length (where $c = 3 \times 10^8$ m/s, the velocity of light). Thus the coherence time is the time required for light to travel the coherence length in the direction of travel of the beam. The frequency bandwidth is a measure of temporal coherence. Thus a frequency stabilized HeNe gas laser ($d(v) = 3$–5 Hz) will have a coherence time of several hundred milliseconds and a corresponding coherence length of 10^5 km. The high spectral purity of gas lasers, the coherence lengths of pulsed ruby lasers are in the order of 15 m with corresponding coherence times in the order of only 100 ns. Polarization of the Laser Output: The polarization of most lasers is directly related to the nature of the resonator. Many high power gas lasers are built with Brewster's angle windows on both ends of the gas discharge tube. Such windows present virtually no losses to a beam which has a linear polarization component lying in the plane of incident. Hence the output will be linearly polarized in this plane. In some solid-state crystal lasers the ruby laser, the output will be linearly polarized. This is a result of the birefringent nature of the crystal in which the slower "ordinary" polarized photons will have a longer time to interact with the excited chromium ions, thereby favoring a polarized output in this plane. This is generally only true for ruby crystals operating near lasing threshold unless Brewster's angles are fabricated on the ends of the crystal. This latter practice is often necessary for very high power Q-switched laser systems. In diode lasers, linear polarized light is observed. This may be attributed to the linear symmetry of the junction region. Electrical Field Strength: The electromagnetic theory of light depicts a light wave as having instantaneous electric and magnetic fields which oscillate at the same frequency. The electrical (E) and magnetic (H) fields are fixed at right angles and are mutually perpendicular to the direction of propagation of the wave. Of particular importance in the description of laser beam interactions is the magnitude of the electric field associated with the beam. The electric field (E) in volts per centimeter associated with a light beam in a vacuum (or air) of average power (PHI) in Watts, spread over a cross-sectional area (A) in cm^2. When a laser beam with a field of comparable magnitude enters a transparent structure, an instantaneous massive redistribution of the electric system of the material can occur due to the interaction of the fields. At the present, the interaction of these enormous electromagnetic fields is not fully understood, to be sure. The production of free electrons, ionized atoms, and X-rays has been detected in the reaction association with the interaction of high power laser beams. Light from conventional thermal sources is emitted over a wide spectral band. The polarizations of the photons are distributed over all possible states of polarization and leave the source in all possible directions (Lambertian source). In contrast, a laser source has a very narrow spectral linewidth even in comparison to special, narrow band thermal sources; the photons may have, essentially, the same polarization and they are highly directional as they leave the laser cavity. Conventional optical sources can most certainly constitute a hazard to the human eye and/or skin, particularly close up and when focused. One's first introduction to optical physics might well have been using a magnifying glass to focus the sun's rays on dry leaves to start a fire. However, even a relatively small laser is capable of producing power/energy distributions much greater those conventional sources. In addition, the hazard can exist at very long ranges due to the highly directional nature of the laser output. A conventional thermal source will emit light into a sphere or hemisphere. The power/energy per unit area (intensity) may be large at the source; however, the intensity at the observer falls off rapidly as the observer moves away from the source. The intensity at the observer can be dramatically increased by using optics to reduce the divergence, making a searchlight; however, the effect is limited by the size of the source. The output of the laser has a very small divergence, typically less than 1 mrad (1 mrad = 0.0573°), and the intensity decreases very slowly as the distance to the observer increases. It would take a very powerful thermal source to put as much power into as tight a beam as offered by even the smaller lasers. If one were to insert a very narrow bandpass filter into the searchlight (in order to approximate the spectral purity of monochromatic nature of the laser output), the laser would be brighter than any thermal source by an enormous factor, where brightness is defined as the power output per steradian of solid angle. To illustrate the relative brightness of the laser over its narrow band, one notes that the sun emits, at its surface, approximately 10^4 W/cm^2/sr/μm and lasers can produce greater than 10^{10} W/cm^2/sr/μm in single pulse. Therefore, it is not difficult for a laser to be a million times brighter that the sun. Indeed, a laser can not only burn dry leaves, but some are used to weld metal. The most significant factor is not total power, but rather the power per unit area, where the laser may be focused to an extremely small spot (approximately a wavelength in diameter). A one milliwatt laser focused to a one micrometer spot will produce a focused irradiance greater than 1×10^5 W/cm^2.

Basic Laser Operation: The term LASER is an acronym. It stands for Light Amplification by Stimulated Emission of Radiation. Thus the laser is a device which produces and amplifies light. The light which the laser produces is unique, for it is characterized by properties which are very desirable, but almost impossible to obtain by any means other than the laser.

Energy Levels: Light can be produced by atomic processes, and it is these processes which are responsible for the generation of laser light. Let's look first at atomic energy levels and then see how changes in these energy levels can lead to the production of laser light. A number of simplifications will be made regarding the concept of the atom. It can be assumed, for the purposes of this discussion, that an atom consists of a small dense nucleus and one or more electrons in motion about the nucleus. The relationship between the electrons and the nucleus is described in terms of energy levels. Quantum mechanics predicts that these energy levels are discrete.

Radiative Transitions: The electrons normally occupy the lowest available energy levels. When this is the case, the atom is said to be in its ground state. However, electrons can occupy higher energy levels, leaving some of the lower energy states vacant or sparsely populated. One way that electrons and atoms can change from one energy state to another is by the absorption or emission of light energy, via a process called a radiative transition.

Absorption: An electron can absorb energy from a variety of external sources. From the point of view of laser action, two methods of supplying energy to the electrons are of prime importance. The first of these is the transfer of all the energy of a photon directly to an orbital electron. The increase in the energy of the electron causes it to "jump" to a higher energy level; the atom is then said to be in an "excited" state. It is important to note that an electron can accept only the precise amount of energy that is needed to move it from one allowable energy level to another. Only photons of the exact energy acceptable to the electron can be absorbed. Photons of slightly more (or slightly less) energy will not be absorbed. Another means often used to excite electrons is an electrical discharge. In this technique, the energy is supplied by collisions with electrons which have been accelerated by an electric field. The result of either type of excitation is that through the absorption of energy, an electron has been placed in a higher energy level than it originally resided. As a result, the atom of which it is a part is said to be excited.

Spontaneous Emission: The nature of all matter is such that atomic and molecular structures tend to exist in the lowest energy state possible. Thus, an excited electron in a higher energy level will soon attempt to DE-EXCITE itself by any of several means. Some of the energy may be converted to heat. Another means of de-excitation is the spontaneous emission of a photon. The photon released by an atom as it is de-excited will have a total energy exactly equal to the difference in energy between the excited and lower energy levels. This release of a photon is called spontaneous emission. One example of spontaneous emission is the common neon sign. Atoms of neon are excited by an electrical discharge through the tube. They de-excite themselves by spontaneously emitting photons of visible light.

Stimulated Emission: Excited atom trigger the second atom into de-exciting itself with the release of another photon. The photon released by the second atom would be identical in frequency, energy, direction, and phase with the triggering photon, and the triggering photon would continue on its way, unchanged. Where there was one photon now there are two. These two photons could then proceed to trigger more through the process of stimulated emission. If an appropriate medium contains a great many excited atoms and de-excitation occurs only by spontaneous emission, the light output will be random and approximately equal in all directions. The process of stimulated emission, however, can cause an amplification of the number of photons traveling in a particular direction—a photon cascade. A preferential direction is established by placing mirrors at the ends of an optical cavity. Thus the number of photons traveling along the axis of the two mirrors increases greatly and Light Amplification by the Stimulated Emission of Radiation may occur. If enough amplification occurs, LASER beam is created.

Population Inversion: The process of stimulated emission will not produce a very efficient or even noticeable amplification of light unless a condition called "population inversion" occurs. If only a few atoms of several million are in an excited state, the chances of stimulated emission occurring are small. The greater the percentage of atoms in an excited state then the greater the probability of stimulated emission. In the normal state of matter the population of electrons will be such that most of the electrons reside in the ground or lowest levels, leaving the upper levels somewhat depopulated. When electrons are excited and fill these upper levels to the extent that there are more atoms excited than not excited, the population is said to be inverted.

Laser Components: A generalized laser consists of a lasing medium, a "pumping" system and an optical cavity. The laser material must have a metastable state in which the atoms or molecules can be trapped after receiving energy from the pumping system.

Laser components are as follow: (1) Pumping Systems: The pumping system imparts energy to the atoms or molecules of the lasing medium enabling them to be raised to an excited "metastable state" creating a population inversion. Optical pumping uses photons provided by a source such as a Xenon gas flash lamp or another laser to transfer energy to the lasing material. The optical source must provide photons which correspond to the allowed transition levels of the lasing material. Collision pumping relies on the transfer of energy to the lasing material by collision with the atoms (or molecules) of the lasing material. Energies which correspond to the allowed transitions must be provided. This is often done by electrical discharge in a pure gas or gas mixture in a tube. Chemical pumping systems use the binding energy released in chemical reactions to state, (2) Optical Cavity: An optical cavity is required to provide the amplification desired in the laser and to select the photons which are traveling in the desired direction. As the first atom or molecule in the metastable state of the inverted population decays, it triggers via stimulated emission, the decay of another atom or molecule in the metastable state. If the photons are traveling in a direction which leads to the walls of the lasing material, which is usually in the form of a rod or tube, they are lost and the amplification process terminates. They may actually be reflected at the wall of the rod or tube, but sooner or later they will be lost in the system and will not contribute to the beam. If, on the other hand, one of the decaying atoms or molecules releases a photon parallel to the axis of the lasing material, it can trigger the emission of another photon and both will be reflected by the mirror on the end of the lasing rod or tube. The reflected photons then pass back through the material triggering further emissions along exactly the same path which are reflected by the mirrors on the ends of the lasing material. As this amplification process continues, a portion of the radiation will always escape through the partially reflecting mirror. When the amount of amplification or gain through this process exceeds the losses in the cavity, laser oscillation is said to occur. In this way, a narrow concentrated beam of coherent light is formed. The mirrors on the laser optical cavity must be precisely aligned for light beams parallel to the axis. The optical cavity itself, i.e., the lasing medium material must not be a strong absorber of the light energy, (3) Laser Media: Lasers are commonly designated by the type of lasing material employed. There are four types which are: solid state, gas, dye, and semiconductor. Solid State Lasers employ a lasing material distributed in a solid matrix. One example is the Neodymium:YAG laser (Nd:YAG). The term: YAG is an abbreviation for the crystal: Yttrium Aluminum Garnet which serves as the host for the Neodymium ions. This laser emits an infrared beam at the wavelength of 1.064 μm ($\mu m = 10^{-6}$ m). Accessory devices that may be internal or external to the cavity may be used to convert the output to visible or ultraviolet wavelength. Gas Lasers use a gas or a mixture of gases within a tube. The most common gas laser uses a mixture of helium and neon (HeNe), with a primary output of 632.8 nm (nm $= 10^{-9}$ m) which is a visible red color. It was first developed in 1961 and has proved to be the forerunner of a whole family of gas lasers. All gas lasers are quite similar in construction and behavior. For example, the carbon dioxide (CO_2) gas laser radiates at 10.6 μm in the far-infrared spectrum. Argon and krypton gas lasers operate with multiple frequency emissions principally in the visible spectra. The main emission wavelengths of an argon laser are 488 and 514 nm. Dye Lasers use a laser medium that is usually a complex organic dye in liquid solution or suspension. The most striking feature of these lasers is their "tunability." Proper choice of the dye and its concentration allows the production of laser light over a broad range of wavelengths in or near the visible spectrum. Dye lasers commonly employ optical pumping although some types have used chemical reaction pumping. The most commonly used dye is Rhodamine 6G which provides tunability over 200 nm bandwidth in the red portion (620 nm) of the spectrum. Semiconductor Lasers (sometimes referred to as diode lasers) are not to be confused with solid state lasers. Semiconductor devices consist of two layers of semiconductor material sandwiched together. These lasers are generally very small physically, and individually of only modest power. However, they may be built into larger arrays. The most common diode laser is the Gallium Arsenide diode laser with a central emission of 840 nm, (4) Time Modes of Operation: The different time modes of operation of a laser are distinguished by the rate at which energy is delivered. Continuous Wave (CW) lasers operate with a stable average beam power. In higher power systems, one is able to adjust the power. In low power gas lasers, such as HeNe, the power level is fixed by design and performance usually degrades with long term use. Single Pulsed (normal mode) lasers generally have pulse durations of a few hundred microseconds to a few milliseconds. This mode of operation is sometimes referred to as long pulse or normal mode. Single Pulsed Q-Switched lasers are the result of an intra cavity delay (Q-switch cell) which allows the laser media to store a maximum of potential energy. Then, under optimum gain conditions, emission occurs in single pulses; typically of 10^{-8} s time domain. These pulses will have high peak powers often in the range from 10^6 to 10^9 W peak.

Repetitively Pulsed or scanning lasers generally involve the operation of pulsed laser performance operating at a fixed (or variable) pulse rates which may range from a few pulses per second to as high as 20,000 pulses per second. The direction of a CW laser can be scanned rapidly using optical scanning systems to produce the equivalent of a repetitively pulsed output at a given location. Mode Locked lasers operate as a result of the resonant modes of the optical cavity which can affect the characteristics of the output beam. When the phases of different frequency modes are synchronized, i.e., "locked together," the different modes will interfere with one another to generate a beat effect. The result is a laser output which is observed as regularly spaced pulsations. Lasers operating in this mode-locked fashion, usually produce a train of regularly spaced pulses, each having a duration of 10^{-15} (femto) to 10^{-12} (pico)s. A mode-locked laser can deliver extremely high peak powers than the same laser operating in the Q-switched mode. These pulses will have enormous peak powers often in the range from 10^{12} W peak. Typical lasers which exist are Helium Neon Laser, Argon, Krypton, and Xenon Ion Lasers, Carbon Dioxide Laser, ND:YAG Laser Systems, Excimer Lasers, and Semiconductor Diode Lasers.

Plasma physics is a very important scientific area to Laser physics. Plasma is a collection of charge particles such as ions and electrons as well as neutral particles. The energy of these particles is mainly dependent on the temperature of the plasma. Plasma, in physics, an electrically conducting medium in which there are roughly equal numbers of positively and negatively charged particles, produced when the atoms in a gas become ionized. It is sometimes referred to as the fourth state of matter, distinct from the solid, liquid, and gaseous states. The negative charge is usually carried by electrons, each of which has one unit of negative charge. The positive charge is typically carried by atoms or molecules that are missing those same electrons. In some rare but interesting cases, electrons missing from one type of atom or molecule become attached to another component, resulting in plasma containing both positive and negative ions. The most extreme case of this type occurs when small but macroscopic dust particles become charged in a state referred to as dusty plasma.

The uniqueness of the plasma state is due to the importance of electric and magnetic forces that act on a plasma in addition to such forces as gravity that affect all forms of matter. Since these electromagnetic forces can act at large distances, plasma will act collectively much like a fluid even when the particles seldom collide with one another. Nearly all the visible matter in the universe exists in the plasma state, occurring predominantly in this form in the Sun and stars and in interplanetary and interstellar space. Auroras, lightning, and welding arcs are also plasmas; plasmas exist in neon and fluorescent tubes, in the crystal structure of metallic solids, and in many other phenomena and objects. The Earth itself is immersed in a tenuous plasma called the solar wind and is surrounded by a dense plasma called the ionosphere. A plasma may be produced in the laboratory by heating a gas to an extremely high temperature, which causes such vigorous collisions between its atoms and molecules that electrons are ripped free, yielding the requisite electrons and ions. A similar process occurs inside stars. In space the dominant plasma formation process is photoionization, wherein photons from sunlight or starlight are absorbed by an existing gas, causing electrons to be emitted. Since the Sun and stars shine continuously, virtually all the matter becomes ionized in such cases, and the plasma is said to be fully ionized. This need not be the case, however, for plasma may be only partially ionized. A completely ionized hydrogen plasma, consisting solely of electrons and protons (hydrogen nuclei), is the most elementary plasma. Plasma can occur by subjecting matter to very high temperatures, to radiation or high voltages, as in a lightning bolt. At low temperatures atoms lock together to form solids, such as a crystal. Higher temperatures loosen the bonds between atoms, bringing them into a liquid state. At even higher temperatures, the bonds between atoms loosen further, turning substances into gases. Extremely high temperatures, such as that of the sun, strip some or all of the electrons away from atoms, creating a "soup" of atomic nuclei, ions and electrons; this is the plasma state. Like gases and unlike solids, plasmas drift and flow freely; if enclosed, plasmas expand to fill the container. Like gases, plasmas have density and pressure. In deep space, plasmas can be extremely thin and tenuous, averaging about one atom per cubic centimeter; by contrast, the plasma in the Sun's core is 10 times denser than lead. Because they consist of free-flowing electrically charged particles, plasmas have several unique characteristics. In most plasma, protons and electrons occur in equal numbers, making it electrically neutral; however, because they flow freely, plasmas are affected by electric and magnetic fields in ways not seen in other forms of matter. These fields can influence plasmas over great distances, pinching, warping and shaping them, as in the great, twisting flares seen on the surface of the Sun. A thermal plasma is one in which the electrons and ions are at the same temperature as their surroundings, such as in the stars; by comparison, non-thermal plasmas are pockets of energetic, charged particles in an otherwise "cool" environment. An example of this is the artificial plasmas used by the food service industry to sterilize fresh produce. In this process, a jet of plasma kills bacteria; because only small amounts of plasma are needed, its atoms mix with room-temperature air and cool quickly.

Plasma diagnostics are a pool of methods, instruments, and experimental techniques used to measure properties of a plasma, such as plasma components' density, distribution function over energy (temperature), their spatial profiles and dynamics, which enable to derive plasma parameters.

Plasma diagnostic tools are an essential element towards the proper understanding and development of technological plasmas. Knowledge of the particle densities and energies in the bulk and at boundaries, the electrical potentials and the spatial and temporal evolution of these parameters allow technologists to operate plasmas in the most efficient way and allow the intrinsic plasma processes to be tailored to suit a particular application.

There are many different diagnostic tools that can be used, depending on the type of plasma under investigation and the specific information that is required. Here, we have chosen to highlight four techniques frequently used in both academia and industrial settings. The first of these is the interpretation of the driving current and voltage waveforms. These measurements do not affect the plasma and can yield useful information on the major processes in the discharge. The second is electrical probing which, by their nature, are intrusive, since their presence affects the plasma under investigation. Their use is usually confined to low-pressure and low-temperature plasmas in which the heat flux will not destroy the integrity of the probe. The third area is mass spectrometry, which is most often performed at the substrate or plasma boundaries and may in many cases not affect the plasma unduly. The fourth diagnostic method discussed, optical emission spectroscopy, is non-perturbing; however, interpretation of spectral response is often difficult in low-pressure plasmas where the species are not in local thermodynamic equilibrium. The field of plasma diagnostics is a large and varied one, and here we wish only to present the basics, applicable to the type of discharges used in the treatment of textiles, allowing readers to find more detailed and specific information relevant to their applications. Plasma diagnostics are based on a multitude of different physical processes with wavelengths in the range from sub-mm to tens of cm. Many different techniques are being employed for measuring the spatial profile and evolution of various plasma parameters. Although most of them are already well established, plasma diagnostics is still a very challenging and vivid discipline. On the one hand this is caused by the always-continuing effort to attain a better spatial and temporal resolution, to reach higher accuracies and to measure with more spatial channels. On the other hand diagnostic techniques based on more subtle physical processes than used in the routine diagnostics are continuously being developed. Langmuir Probes in Collisional-Dominated Regime: There is certain controversy with respect to the application of Langmuir probes to the highly collisional atmospheric plasma when the electron mean free path is much less than Debye length. The argument about applicability of the Langmuir probe technique for the plasma diagnostics in collision-dominated conditions is based on the fact that the presence of collisions in the Debye sheath does not cancel the Boltzmann distribution for plasma electrons in the sheath (for probe potentials in the vicinity and lower than the floating potential). As such, the probe can be used to measure the electron temperature. To evaluate the electron density from the probe current–voltage characteristic, one needs to invoke a model of the ion and electron transport across the sheath. In contrast with a collision less sheath in which particle transport across the sheath is determined by the flux from the plasma, the transport across the collisional sheath is due to diffusion. In addition, the model should be coupled with the plasma model. Typically, the local thermodynamic equilibrium (LTE) is reached at a pressure of about 1 atm for most gases and any arc currents except for inert gases. In helium (~1 atm), the arc electron temperature exceeds the gas temperature by a factor of two for arc currents up to 100 A and LTE is reached only for arc currents larger than 200 A. LTE allows simplifying the analysis of Langmuir electrostatic probe operation significantly due to the availability of an additional relation between plasma density and temperature. Langmuir Probe Diagnostics: The applicability of Langmuir probe technique for highly collision plasma of atmospheric anodic arc producing SWCNT remains the subject of active ongoing investigation. The complication of use of Langmuir probes in nanostructures producing arc is the very fast condensed contamination of the probe with the synthesized carbon nanostructures leading to the uncontrollable increasing of the collecting probe area and short circuit of the probe with surrounding bodies. In this respect, probe having a fast-moving shutter, providing exposure times to the plasma environment in the millisecond range, was shown to be the effective technique for plasma diagnostics. Three modifications of single electrostatic probes were used for arc plasma studies. The first two types of regular metallic probes for measurements were relatively far from the discharge (the distance to arc center $r > 15$ mm). In this case, probe contamination with the synthesized nanostructures was relatively low; therefore, shutter was not necessary. The circular probe oriented perpendicular to radial plasma flow expanding for the gap is demonstrated, while another surrounding probe used large-area prolonged cylinder covering the arc axis, oriented coaxially to the axis and equipped with opening for arc video recording. The single electrostatic probe with shutter was put in vicinity of the inter electrode gap ($r = 8$ mm). The probe was equipped with electrically controlled shutter providing exposure time to the plasma environment in milliseconds range. Fast shutter was utilized to limit the material contamination to the probe and to prevent its fast deposition by various carbon species synthesized in arc. Such deposition may lead to uncontrollable growth of collecting area and produce inadequate data. The probe consists of a copper foil collector with 3 mm width installed inside the ceramic tube which has outside diameter of 6.3 mm. The probe opening is about 5×3 mm^2. The cylindrical shutter made of molybdenum sheet was closely fit to the outside surface of the ceramic tube and was able to slide along it, so that the collector was exposed to the plasma solely during the period of time when shutter was open. Shutter was controlled using electrically driven solenoid. The collector surface containing nanostructures deposited on it during the exposition to plasma flux was analyzed using SEM. A 1 kHz voltage sweeping voltage is applied to the probes for measurements of V-I characteristics. A wide range of gas pressures were considered, while the main focus is concerned with relatively high-pressure range of several hundred Torr corresponding to the synthesis conditions of SWCNT and graphene. Interferometry: One of the commonly found plasma diagnostic is the interferometer. Light from a laser beam acquires a phase shift as it passes through a plasma medium, which is proportional to the density along the path ($\int$ n dl), compared to a reference leg, which does not go through the discharge. The wavelength of the laser must be carefully chosen to optimize the expected phase shift range and the optical characteristics of the beam path, including the plasma (e.g., refraction). A second wavelength is normally needed to isolate the potential effects due to vibration or changes in path length. Currently the most common wavelength used on these applications is near 10 μm (e.g., CO_2 laser). A polarimeter is sometimes integrated into this system, and can support the interferometer measurement by mitigating potential fringe skips. It is most commonly used toroidally where the magnetic field component of the Faraday rotation is simply the main toroidal field. An interferometer system (and possibly the polarimeter) is well-suited for a power plant, as its footprint is relatively small, and can be run in steady-state conditions. However, stability issues must be addressed, including maintaining alignment for the duration of the pulse. An automatic alignment feedback system must be integrated into the diagnostic over its long beam path (e.g., $> \sim 100$ m for ITER) and its large number of mirrors. There is a use of BEM (boundary element method) and NN (Neural networks) System to Estimate Density Distribution of Plasma. Non-destructive measurement techniques are essential to plasma diagnostics. There is an inverse analysis to estimate a density distribution of plasma by measuring scattered electromagnetic fields. It is assumed that the plasma distribution is represented by some parameters e.g., radius, center, and maximum density. The electromagnetic field is scattered by the plasma, and it is observed at some measurement points on the boundary. The purpose is to estimate the parameters from these observed data. Since the vessel acts as a resonator in this model, electromagnetic field is given as a sum of various Eigen modes of resonator, which is formed the complicated distribution. As a result, it is predicted that a ratio of a change in measurements to a change in source distribution seems large (i.e., good sensitivity). On the other hand, it seems to be difficult to estimate the parameters due to the complicated field. Neural networks (NN) are widely used in many fields for an application of inverse problem. As another choice, a genetic algorithm is also a famous method. Considering the cost for calculation, the science adopts the multi-layered NN for the inverse analysis. The multi-layered NN needs a solver for direct problem to generate the teacher's data. As this application, adopt boundary element method (BEM) which includes the domain integral for inhomogeneous plasma analysis. Furthermore, in order to improve the convergence of NN, there are two techniques, a normalization technique for input data and a dividing section technique. It will be clarified that combining of these techniques provide us a rapid convergence and lead the availability of the estimation system. Applications and Configurations of RF/Microwave Plasmas for Materials Processing: Electrically induced plasmas are formed by imposing a sufficiently strong electric field to ionize a gas, which means separating the electrons and the rest of the molecule, which forms an ion. Once such separation takes place, the ions, being chemically active species, can be utilized to create a physical or chemical impact on the material that is being processed. In general, electrically induced plasmas can be formed with both direct current (DC) and alternating current (AC) electric fields. DC plasmas have been used in the industry, but they suffer several disadvantages compared with RF/microwave plasmas, chief among them being electrode erosion and secondary electron emissions. The presence of separated active ions is what makes plasma useful; however, plasma is, indeed, a mixture of ions, free electrons, and intact gas atoms and molecules. This would give various plasmas their specific properties. This is, at least in part, the reason that different applicators with various configurations and frequency ranges have been developed. Plasmas formed via RF/microwave means can be classified according to several criteria: Operating pressure: Depending on application and required properties, industrial plasmas are formed at pressures from fractions of milliTorr to atmospheric pressure (760 Torr = 101.3 kPA). Thermodynamic equilibrium: Thermal plasmas, where there is equilibrium between gas and ions, and non-equilibrium (cold) plasmas, where electron temperature is higher than the ion temperature.

Gas temperature: Low temperature plasma ($T_{gas} < 2000$ K) and high temperature plasma ($T_{gas} > 2000$ K). Electron temperature: Is a measure of plasma's electron kinetic energy in kelvins. The electron temperature is often expressed in terms of electron-volt (eV). Each kelvin (1 K) corresponds to 8.617×10^{-5} eV; this factor is the ratio of the Boltzmann constant to the elementary charge. Plasma density or number density, η_e: Number of free electron per cubic centimeter, expressed typically as a large number with exponents of 10 and then cm^{-3} (e.g., 3.4×10^{14} cm^{-3}). Both the number density and electron temperature are measured by plasma diagnostic devices such as the Langmuir probe. Degree of ionization, α: $\eta_i/(\eta_i + \eta_n)$, where η_i is the number density of ions, and η_n is the number density of neutral atoms.

Electrical conductivity and skin depth: A plasma body's electrical conductivity (S/m) and degree of RF/microwave field penetration into the plasma. Damage: When the plasma, for various reasons, damages the material being processed. When too high a plasma temperature or electron bombardment damages a silicon wafer being etched [1]. Apart from the constraints on the resolution, dynamic range and accuracy, there are complications that should be taken into account when developing plasma diagnostics. Because of the high temperatures and densities of present-day fusion plasmas, only diagnostic techniques can be employed that have no physical contact with the plasma, except for probes that are usually applied at the very plasma edge. Hence, the plasma must be diagnosed either by analyzing the radiation and particles emitted by the plasma itself (passive diagnostics) or by probing the plasma with electromagnetic waves or particle beams (active diagnostics). None of these techniques should perturb the plasma. Especially in the larger fusion devices it is important that the diagnostics are insensitive to the hostile environment (high heat loads, neutron and gamma-radiation) which can lead to thermal and mechanical stresses as well as to a large number of radiation induced effects. They must be well screened for the high electromagnetic stray fields around these devices. Magnetic diagnostics operate in the frequency range from about 100 Hz up to several MHz. This is the frequency range in which many typical plasma processes are active, like MHD instabilities. Magnetic diagnostics are indispensable for the operation of magnetic confinement devices. They are used for measuring basic plasma parameters as the plasma current, position, shape and pressure, as well as for detecting plasma instabilities. Magnetic diagnostics make use of the electromagnetic waves emitted by the plasma and are therefore passive. The simplest magnetic diagnostic is the pick-up coil whose integrated voltage output is a measure for the magnetic field strength. Combinations of pick-up coils are generally used to determine the plasma position and shape.

Another magnetic diagnostic is the Rogowski coil, which is a solenoid wound in such a way around a poloidal cross section of the plasma that its integrated output voltage is proportional to the plasma current enclosed by the coil. Voltage loops are used to measure the loop voltage and, hence, if the plasma current is known from a Rogowski coil, also the ohm input power. Diamagnetic loops are used to yield a value for the total energy content of the plasma (plasma pressure). Microwave diagnostics are indicated by mm and sub-mm diagnostics. They are in the frequency range from 1 GHz to 3 THz. Many powerful and widely applied diagnostics like reflectometry, electron cyclotron emission (ECE) and absorption (ECA) and interferometry/polarimetry belong to this group. Apart from ECE all diagnostics in this group are active. Probes are active diagnostics in direct contact with the plasma. Therefore, they can only be applied at the very plasma edge. The most well-known is the Langmuir probe. The simplest probe consists of an isolated metallic pin inserted into the plasma. Only the tip is not isolated. By applying a voltage to the tip, a current is drawn from the plasma. From the exponential part of the I-V characteristic, the electron temperature at the edge can be deduced. Other plasma parameters like the electron density and the plasma potential can also be determined from probe measurements. Sophisticated analyzers with electron repelled grids often measure parameters of the ion population. The logarithmic slope of the ion collection current is proportional to the ion temperature at the plasma edge. Bolometers are used to measure the radiation losses from the plasma in a wide wavelength range. Bolometers are also sensitive to particle losses. Wide-angle bolometers yield a value for the total radiation losses from the plasma. We can categorizes plasma in two types as cold and hot plasma, which compelled plasma physicist to develop different kinds of diagnostics for measuring accurately the parameters of plasma for achieving the goals as well as to better understand the physics of plasma. This involves the knowledge of a wide variety of branches both theoretical as well as experimental like the other sub-discipline of physics. The techniques developed for diagnosing the properties of plasma is known as plasma diagnostics. The main objective of the plasma diagnostics is to deduce information about the state of the plasma from practical observations of physical processes and their effects. Accurate and reliable measurements of the plasma condition in the various plasma experiments are an important objective for making a significant progress in the field. The plasma diagnostics can be categorized in various ways. First of all one can consider cold and hot plasma. Material probes such as the most common Langmuir probe can be inserted in the cold plasma. In the hot plasma ($T_e \geq$ few KeV) the use of material probes is limited to the extreme edge of the plasma ($T_e \leq 50$ eV) and the non-invasive techniques must be used to diagnose other plasma regions. These include passive methods, which detects radiation or particles emitted spontaneously by the plasma, where as in active methods radiations or particles produced by external sources are used to probe the plasma.

There are some typical electric and magnetic probes for plasma diagnostics.

Langmuir probe: It is simply a small metallic electrode in the form of wire inserted into the plasma. A variable biasing voltage is applied on this electrode, which may be positive or negative with respect to the plasma. The current collected by the probe with respect to the applied voltage provides information about the plasma condition.

The probe is a very simple device but has very complicated theory. This is because probe acts as a boundary and the equations governing the plasma motion change its characteristics. Due to the quasi-neutral behavior of plasma a sheath is formed around the probe, which can sustain a large electric field. The characteristics of the Langmuir probe can be easily understood by plotting the I-V curve. I is the current flowing in the probe when a biasing voltage V is applied to it with respect to the plasma or the body of the vessel containing plasma. The probe current is given by $I = I_0 \cdot \exp(\frac{e \cdot V}{k \cdot T})$, taking the ln, we get $\ln(I) = \ln(I_0) + \frac{e \cdot V}{k \cdot T}$. The $\ln(I - V)$ curve can provide information about the plasma temperature from the slope of the straight line. Since probe current mainly depend on electron current it can provide better information about electron temperature in comparison to ion temperature. A simple relation can approximate probe current in the absence of disturbance in the plasma as $J = n \cdot e \cdot v$, where J is the current density on the probe, n is the plasma density and v is the velocity of the electron corresponding to plasma temperature.

Double probe: Double Langmuir probes can be used in plasma, where reference point is not available. The example of this type of plasma is RF discharge, which is an electrode less plasma. Double probe (electrode) is used simultaneously in the plasma. The biasing voltages on the probe are varied from negative to positive potential for getting I-V characteristics of the probe. This provides ion saturation current in the negative (I_{1+}) as well as in positive direction (I_{2+}). This curve passes through the origin ($V = 0, I = 0$), where current is linearly varying with applied voltage. The rate of change of current with respect to voltage in the linear range can be expressed as $\frac{dI}{dV} = \frac{e}{k \cdot T_e} \cdot \left(\frac{I_{1+} \cdot I_{2+}}{I_{1+} + I_{2+}}\right)$. This shows that once the plasma temperature $k \cdot T_e$ is known plasma density can be estimated from the amplitude of ion saturation current.

Magnetic probe: There are cases which strong currents are present in the plasma, particularly in the tokamak plasma. Presence of plasma current generates a magnetic field around the plasma current. Plasma current in the magnetized plasma generates different types of magneto hydrodynamic (MHD) instability, which is observed as a fluctuation superimposed on the magnetic field generated due to plasma current. Presence of a search coil which is a small area solenoid having only few turns, near the plasma current can measure any fluctuation in the magnetic field. An analysis of voltage induced in the search coil provides information about the MHD instability present in the plasma. If the search coil has n number of turns on a small area a, then the induced voltage in the search coil can be given by $V = n \cdot \frac{d\phi}{dt}$ $V = n \cdot \frac{d\phi}{dt} = n \cdot a \cdot \frac{d(\delta B)}{dt}$, where δB is the small fluctuation in the magnetic field around the plasma. This provides a different signal, which should be integrated for getting direct information about the fluctuation in magnetic field. If $R \cdot C$ is the time constant of an integrator, then amplitude of the fluctuating magnetic field can be obtained from the induced voltage on the search coil and can be written as $\delta B = R \cdot C \cdot \frac{V}{a \cdot n}$. Before using the search coil for measurement, it should be properly shielded. Another important point is that search coil should not be in contact with the hot plasma. It is useful particularly for measuring either local magnetic field or its fluctuations in the tokamak or in laser produced plasma.

Rogowski coil: We can measure the amplitude of current flowing in the plasma by integrating the induced magnetic field around the plasma column using a long solenoid known as Rogowski coil. Amplitude of the total current can be given by $I = \int B \cdot dl$. The plasma current will induce a voltage in Rogowski coil, which will be proportional to the rate of change of plasma current with time. The expression for the induced voltage in Rogowski coil in term of plasma can be given by $V = \frac{\mu_0 \cdot N \cdot A}{2 \cdot \pi \cdot a} \cdot \frac{dI_p}{dt}$, where I_p is the plasma current, a is the radius of plasma column, N is the number of turns in the Rogowski coil, and A is its area of cross section. After integrating the Rogowski coil signal by a suitable active electronic or passive integrator one we can get information about the time evolution of the plasma current. The final expression for the plasma current can be given by $I_p = \frac{2 \cdot \pi \cdot a}{\mu_0 \cdot N \cdot A} \cdot R \cdot C \cdot V$, $R \cdot C$ is the time constant of the integrator. For better integrating of Rogowski coil signal, integrating time must be nearly 10 times more than the total duration of the plasma current. In this case measured plasma current remains independent of position of the plasma column inside the Rogowski coil. To avoid any electrostatic pick up on the signal Rogowski coil must be shielded properly. Coil winding must return from the end of initial winding (bifilar winding), otherwise the presence of any other magnetic field passing in the same or opposite direction of the plasma current can generate an electrical signal considering Rogowski solenoid as one turn coil. These factors are important for designing a Rogowski coil to provide nearly true shape of plasma current in time. This is an useful diagnostics for a tokamak machine.

We inspect laser systems and circuits which involve many elements, as a dynamical system where a fixed rule describes the time dependent of specific laser system or circuit variable in a geometrical space. Examples include the mathematical models that describe laser system or circuit. At any time a dynamical system has a state given by a set of real numbers (a vector) which can be represented by a specific variable in an appropriate state space (a geometrical manifold). Small changes in the state of the laser system or circuit create small changes in the numbers.

The evolution rule of the laser dynamical system is a fixed rule that describes what future states follow from the current state. The rule is deterministic, for a given time interval only one future state follows from the current state. A laser system and circuit dynamical system is a phase (or state) space endowed with a family of smooth evolution functions that for any element of time (t), map a point of the phase space back into the phase space. Laser systems can be described by numbers. The state vector is a numerical description of the current configuration of a system. For example, solid state laser source can be described using some numbers: its dynamical variables $(X_1, X_2, X_3, \ldots)$ and $(Y_1, Y_2, Y_3, \ldots)$. Once we know these numbers $X_1, X_2, X_3, \ldots$ and $Y_1, Y_2, Y_3, \ldots$ the system variables trajectories are completely determined. The group of numbers $(X_1, X_2, X_3, \ldots, Y_1, Y_2, Y_3, \ldots)$ is a vector which completely describes the state of our laser system and hence is called the state vector system. There are two main behaviors which are related to laser systems: (1) the system gravitates toward a fixed point, or (2) the system blowup. There are additional cases of oscillators related to laser system. We assume $f_i; i = 1, 2, 3, \ldots$ are differentiable with continuous derivatives. The vectors X_i are the state of the laser dynamical system, and the functions $f_i; i = 1, 2, 3, \ldots$ tell us how the system moves. In special circumstances, however, the system does not move. The system can be stuck (we will say fixed) in a special state; we call these states fixed points of the dynamical laser system. Not all fixed points are the same. We call some stable and others unstable. Is the specific fixed point Y_i^*, X_i^* stable or unstable? The answer is, neither. To see that it is not stable, consider any points X_{i0}, Y_{i0} near (but not equal to) X_i^*, Y_i^* (i = 1, 2, 3...). At $t \to \infty$ the laser system never approaches (X_i^*, Y_i^*). Further, X_i^*, Y_i^* are not unstable. To be unstable, points near X_i^*, Y_i^* must be sent "far" away from X_i^*, Y_i^*. Clearly, if we start at certain distance from X_i^*, Y_i^* the system does not get any farther away. Stable fixed points give excellent information about the fate of a dynamical system. In our analysis we investigate laser circuit's bifurcation and dynamical behavior. Bifurcation behavior in our laser system is the study of changes in the qualitative or topological structure of laser system, vector fields, and the solutions of a family of differential equations. A bifurcation occurs when a small smooth change made to the parameter values (the bifurcation parameters) of a laser system causes a sudden topological change in behavior. Bifurcations occur in both our continuous laser systems (ODEs, DDEs, and PDEs) and discrete systems (described by maps). We can inspect in our laser system two principal bifurcation classes: local bifurcation, which our laser system can be analyzed through changes in the local stability properties of equilibria, periodic orbits or other invariant sets as parameters cross through critical thresholds. Global bifurcations, which often occur in laser system, happened when larger invariant sets of the system collide with each other, or with equilibria of the system. They cannot be detected only by stability analysis (fixed points). A local bifurcation occurs when a system parameter change causes the stability of an equilibrium (or fixed point) to change. In continuous system, this corresponds to the real part of an eigenvalue of an equilibrium passing through zero. Global bifurcations occur when larger invariant sets, such as periodic orbits, collide with equilibria. This causes changes in the topology of the trajectories in the phase space which cannot be confined to a small neighborhood (local bifurcation). The changes in system topology extend out to an arbitrarily large distance. In our analysis we pay attention to the co-dimension of a bifurcation which is the number of parameters which must be varied for the bifurcation to occur. The co-dimension of the parameter set for which the bifurcation occurs within the full space of laser system parameters. The simple case for stability analysis is when there are no time delay elements in our laser system ($\tau_1 = 0, \tau_2 = 0, \ \tau_3 = 0, \ldots; \ \Delta_1 = 0, \Delta_2 = 0, \Delta_3 = 0, \ldots; \ \Delta_{\mu_i} = 0; \ i = 1, 2.3, \ldots$). If our laser system involving N_a variables ($N_a > 2$), the characteristic equation is of degree N_a $\left(\sum_{k=0}^{N_a} \lambda^k \cdot \Pi_k = 0\right)$ and must often be solved numerically. Expect in some particular cases, such an equation has N_a distinct roots that can be real or complex. These values are the eigenvalues of the $N_a \times N_a$ Jacobian matrix (A). The general rule is that the Steady State (SS) is stable if there is no eigenvalue with positive real part. It is sufficient that one eigenvalue is positive for the steady state to be unstable. Our N_a-variables laser system has N_a eigenvalues. The type of behavior can be characterized as a function of the position of these eigenvalues in the Re/Im plane. Five non-degenerated cases can be distinguished: (1) the N_a eigenvalues are real and negative (stable steady state), (2) the N_a eigenvalues are real, $N_a - 1$ of them are negative (unstable steady state), (3) and (4) two eigenvalues are complex conjugates with a negative real part and the other eigenvalues are real and negative (stable steady state), two cases can be distinguished depending on the relative value of the real part of the complex eigenvalues and of the real one, (5) two eigenvalues are complex conjugates with a negative real part and at least one eigenvalue is positive (unstable steady state). The next case is when there are delay elements in our laser system and we can't neglect them. In that case, our laser system stability analysis is related to two main cases: first case, laser system is characterized by a set of variables $(X_1, X_2, X_3, \ldots.)$ or/and variables $((Y_1, Y_2, Y_3, \ldots))$. Due to instabilities or parasitic effects there are differences in time delays with respect to system variables

$$X_1(t) \to X_1(t-\tau_1);\ X_2(t) \to X_2(t-\tau_2);\ X_i(t) \to X_i(t-\tau_i)$$
$$Y_1(t) \to Y_1(t-\tau_1)$$
$$Y_2(t) \to Y_2(t-\tau_2);\ Y_i(t) \to Y_i(t-\tau_i);\ i = 1, 2, 3, \ldots$$
$$\tau_k \neq \tau_l;\ k \neq l;\ \tau_i \in \mathbb{R}_+.$$

Sometimes the delay in time is related to laser system or circuit variables derivatives: $\frac{dX_1(t)}{dt} \to \frac{dX_1(t-\Delta_1)}{dt}$; $\frac{dX_2(t)}{dt} \to \frac{dX_2(t-\Delta_2)}{dt}$; $\frac{dX_i(t)}{dt} \to \frac{dX_i(t-\Delta_i)}{dt}$; $i = 1, 2, 3, \ldots$, We assume $\Delta_k \neq \Delta_l; k \neq l; \Delta_i \in \mathbb{R}_+$. The stability of a given steady state is determined by the graphs of some function of $\tau_1, \tau_2, \tau_3, \ldots$ or/and $\Delta_1, \Delta_2, \Delta_3, \ldots$ which can be expressed explicitly and thus can be easily depicted by software. We need to look at one such function and locate the zeros. The stability switching is due to different values of delay parameters $\tau_1, \tau_2, \tau_3, \ldots$ or/and $\Delta_1, \Delta_2, \Delta_3, \ldots$. Second case, laser circuits and systems include many elements. Laser systems or circuits have parasitic effects, a delay in time $\Delta_{\mu_i}; i = 1, 2.3, \ldots$. The delays are on the variables $Y(t) \to Y(t - \Delta_{\mu_i}); i = 1, 2.3, \ldots$. We consider $X_{\Delta_{\mu_i}} \to \varepsilon$ (neglect other variables). We inspect the stability behavior and stability switching under variation of delay parameters $\Delta_{\mu_i}; i = 1, 2.3, \ldots$. In our laser systems which many elements, we take care in our analysis and calculation the mutual interaction between elements within the system. If we minimize our laser system to specific case where there is only one time delay parameter τ then the general geometric criterion: The occurrence of any possible stability switching resulting from the increase of value of the time delay τ for our laser system characteristic equation.

$$D(\lambda, \tau_1, \tau_2, \tau_3, \ldots) = \det(A - \lambda \cdot I);\ D(\lambda, \tau_1, \tau_2, \tau_3, \ldots) = 0$$
$$D(\lambda, \tau) = P_n(\lambda, \tau) + Q_m(\lambda, \tau) \cdot e^{-\lambda \cdot \tau}$$
$$P_n(\lambda, \tau) = \sum_{k=0}^{n} p_k(\tau) \cdot \lambda^k;\ Q_m(\lambda, \tau) = \sum_{k=0}^{m} q_k(\tau) \cdot \lambda^k;\ n, m \in \mathbb{N}_0;\ n > m$$

$p_k(\cdot), q_k(\cdot) : \mathbb{R}_{+0} \to \mathbb{R}$ are continuous and differentiable functions of τ.

$$P_n(\lambda = 0, \tau) + Q_m(\lambda = 0, \tau) = p_0(\tau) + q_0(\tau) \neq 0 \forall \tau \in \mathbb{R}_{+0}$$

$P_n(\lambda, \tau)$, $Q_m(\lambda, \tau)$ are analytic functions in λ and differentiable in τ for which we assume ($P_n(\lambda, \tau) \to P$, $Q_m(\lambda, \tau) \to Q$):

(I) If $\lambda = i \cdot \omega$; $\omega \in \mathbb{R}$ then $P_n(i \cdot \omega, \tau) + Q_m(i \cdot \omega, \tau) \neq 0$; $\tau \in \mathbb{R}$.
(II) $\lim \sup\{|Q_m(\lambda, \tau)/P_n(\lambda, \tau)|:|\lambda| \to \infty, \operatorname{Re} \lambda \geq 0\} < 1$ for any τ.
(III) $F(\omega, \tau) = |P_n(i \cdot \omega, \tau)|^2 - |Q_m(i \cdot \omega, \tau)|^2$ for each τ has at most a finite number of real zeros.
(IV) Each positive root $\omega(\tau)$ of $F(\omega, \tau) = 0$ is continuous and differentiable in τ whenever it exists.

In addition, since the coefficients in P and Q are real, we have $\overline{P(-i \cdot \omega)} = P(i \cdot \omega)$, and $\overline{Q(-i \cdot \omega)} = Q(i \cdot \omega)$ thus $\lambda = i \cdot \omega, \omega > 0$ may be on eigenvalue of characteristic equation. The analysis consists in identifying the roots of characteristic equation situated on the imaginary axis of the complex λ-plane, where by increasing the parameters of laser system, Re λ may, at the crossing Change its sign from (−) to (+), i.e. from a stable focus $E^{(0)}$ to an unstable one, or vice versa. This feature may be further assessed by examining the sign of the partial derivatives with respect to τ and laser system parameters [2, 3].

$$\wedge^{-1}(\tau) = \left(\frac{\partial \operatorname{Re} \lambda}{\partial \tau}\right)_{\lambda = i \cdot \omega}, \text{ Other laser system parameters} = const \text{ where } \omega \in \mathbb{R}_+.$$

Hence $F(\omega, \tau) = 0$ implies $\sum_{k=0}^{N} \Xi_{2 \cdot k} \cdot \omega^{2 \cdot k} = 0$; $N \in \mathbb{Z}_+$ when writing $P(\lambda) = P_R(\lambda) + i \cdot P_I(\lambda)$ and $Q(\lambda) = Q_R(\lambda) + i \cdot Q_I(\lambda)$, and inserting $\lambda = i \cdot \omega$ into laser system's characteristic equation, ω must satisfy the following:

$$\sin \omega \cdot \tau = g(\omega) = \frac{-P_R(i \cdot \omega) \cdot Q_I(i \cdot \omega) + P_I(i \cdot \omega) \cdot Q_R(i \cdot \omega)}{|Q(i \cdot \omega)|^2}$$

$$\cos \omega \cdot \tau = h(\omega) = -\frac{P_R(i \cdot \omega) \cdot Q_R(i \cdot \omega) + P_I(i \cdot \omega) \cdot Q_I(i \cdot \omega)}{|Q(i \cdot \omega)|^2}$$

where $|Q(i \cdot \omega)|^2 \neq 0$ in view of requirement (a) above, $(g, h) \in R$. Furthermore, it follows above $\sin \omega \cdot \tau$ and $\cos \omega \cdot \tau$ equations that, by squaring and adding the sides, ω must be a positive root of $F(\omega) = |P(i \cdot \omega)|^2 - |Q(i \cdot \omega)|^2 = 0$. Note: $F(\omega)$ is dependent of τ. Now it is important to notice that if $\tau \notin I$ (assume that $I \subseteq R_{+0}$ is the set where $\omega(\tau)$ is a positive root of $F(\omega)$ and for $\tau \notin I$, $\omega(\tau)$ is not define. Then for all τ in I $\omega(\tau)$ is satisfies that $F(\omega, \tau) = 0$). Then there are no positive $\omega(\tau)$ solutions for $F(\omega, \tau) = 0$, and we cannot have stability switches. For any $\tau \in I$, where $\omega(\tau)$ is a positive solution of $F(\omega, \tau) = 0$, we can define the angle $\theta(\tau) \in [0, 2 \cdot \pi]$ as the solution of the below equations:

$$\sin \theta(\tau) = \frac{-P_R(i \cdot \omega) \cdot Q_I(i \cdot \omega) + P_I(i \cdot \omega) \cdot Q_R(i \cdot \omega)}{|Q(i \cdot \omega)|^2}$$

$$\cos\theta(\tau) = -\frac{P_R(i\cdot\omega)\cdot Q_R(i\cdot\omega) + P_I(i\cdot\omega)\cdot Q_I(i\cdot\omega)}{|Q(i\cdot\omega)|^2}$$

And the relation between the argument $\theta(\tau)$ and $\omega(\tau)\cdot\tau$ for $\tau \in I$ must be $\omega(\tau)\cdot\tau = \theta(\tau) + n\cdot 2\cdot\pi \quad \forall n \in \mathbb{N}_0$. Hence we can define the maps $\tau_n : I \to R_{+0}$ given by $\tau_n(\tau) = \frac{\theta(\tau)+n\cdot 2\cdot\pi}{\omega(\tau)}$; $n \in \mathbb{N}_0, \tau \in I$. Let us introduce the functions $I \to R$; $S_n(\tau) = \tau - \tau_n(\tau), \tau \in I, n \in \mathbb{N}_0$ that are continuous and differentiable in τ. In the following, the subscripts $\lambda, \omega, a_1, a_2, \ldots$ and laser system or circuit parameters indicate the corresponding partial derivatives. Let us first concentrate on, $\wedge(x)$ remember in λ(parameters); ω(parameters), and keeping all parameters except one (x) and τ. The derivation closely follows that in reference [BK]. Differentiating laser system characteristic equation $P(\lambda) + Q(\lambda)\cdot e^{-\lambda\cdot\tau} = 0$ with respect to specific parameter (x), and inverting the derivative, for convenience, one calculates:

Remark $x =$ Laser system specific parameter

$$\left(\frac{\partial\lambda}{\partial x}\right)^{-1} = \frac{-P_\lambda(\lambda,x)\cdot Q(\lambda,x) + Q_\lambda(\lambda,x)\cdot P(\lambda,x) - \tau\cdot P(\lambda,x)\cdot Q(\lambda,x)}{P_x(\lambda,x)\cdot Q(\lambda,x) - Q_x(\lambda,x)\cdot P(\lambda,x)}$$

where $P_\lambda = \frac{\partial P}{\partial\lambda}, \ldots$. etc., Substituting $\lambda = i\cdot\omega$, and bearing $\overline{P(-i\cdot\omega)} = P(i\cdot\omega)$, $\overline{Q(-i\cdot\omega)} = Q(i\cdot\omega)$ then $i\cdot P_\lambda(i\cdot\omega) = P_\omega(i\cdot\omega)$ and $i\cdot Q_\lambda(i\cdot\omega) = Q_\omega(i\cdot\omega)$ that on the surface $|P(i\cdot\omega)|^2 = |Q(i\cdot\omega)|^2$, one obtains

$$\left.\left(\frac{\partial\lambda}{\partial x}\right)^{-1}\right|_{\lambda=i\cdot\omega} = \left(\frac{i\cdot P_\omega(i\cdot\omega,x)\cdot\overline{P(i\cdot\omega,x)} + i\cdot Q_\lambda(i\cdot\omega,x)\cdot\overline{Q(\lambda,x)} - \tau\cdot|P(i\cdot\omega,x)|^2}{P_x(i\cdot\omega,x)\cdot\overline{P(i\cdot\omega,x)} - Q_x(i\cdot\omega,x)\cdot\overline{Q(i\cdot\omega,x)}}\right)$$

Upon separating into real and imaginary parts, with $P = P_R + i\cdot P_I$; $Q = Q_R + i\cdot Q_I$ $P_\omega = P_{R\omega} + i\cdot P_{I\omega}$; $Q_\omega = Q_{R\omega} + i\cdot Q_{I\omega}$; $P_x = P_{Rx} + i\cdot P_{Ix}$; $Q_x = Q_{Rx} + i\cdot Q_{Ix}$.

$P^2 = P_R^2 + P_I^2$. When (x) can be any laser system parameters and time delay τ etc.,.Where for convenience, we have dropped the arguments $(i\cdot\omega, x)$, and where $F_\omega = 2\cdot[(P_{R\omega}\cdot P_R + P_{I\omega}\cdot P_I) - (Q_{R\omega}\cdot Q_R + Q_{I\omega}\cdot Q_I)]$; $F_x = 2\cdot[(P_{Rx}\cdot P_R + P_{Ix}\cdot P_I) - (Q_{Rx}\cdot Q_R + Q_{Ix}\cdot Q_I)]$.

$\omega_x = -F_x/F_\omega$. We define U and V: $U = (P_R\cdot P_{I\omega} - P_I\cdot P_{R\omega}) - (Q_R\cdot Q_{I\omega} - Q_I\cdot Q_{R\omega})$.

$V = (P_R\cdot P_{Ix} - P_I\cdot P_{Rx}) - (Q_R\cdot Q_{Ix} - Q_I\cdot Q_{Rx})$. We choose our specific parameter as time delay x = τ.

Differentiating with respect to τ and we get $F_\omega\cdot\frac{\partial\omega}{\partial\tau} + F_\tau = 0$; $\tau \in I \Rightarrow \frac{\partial\omega}{\partial\tau} = -\frac{F_\tau}{F_\omega}$

$$\wedge^{-1}(\tau) = \left(\frac{\partial\,\mathrm{Re}\,\lambda}{\partial\tau}\right)_{\lambda=i\cdot\omega};\ \wedge^{-1}(\tau) = \mathrm{Re}\left\{\frac{-2\cdot[U + \tau\cdot|P|^2] + i\cdot F_\omega}{F_\tau + i\cdot 2\cdot[V + \omega\cdot|P|^2]}\right\}$$

$$\frac{\partial\omega}{\partial\tau} = \omega_\tau = -\frac{F_\tau}{F_\omega}$$

$$sign\{\wedge^{-1}(\tau)\} = sign\left\{\left(\frac{\partial \operatorname{Re}\lambda}{\partial\tau}\right)_{\lambda=i\cdot\omega}\right\}$$

$$sign\{\wedge^{-1}(\tau)\} = sign\{F_\omega\}\cdot sign\left\{\tau\cdot\frac{\partial\omega}{\partial\tau} + \omega + \frac{U\cdot\frac{\partial\omega}{\partial\tau} + V}{|P|^2}\right\}$$

We shall presently examine the possibility of stability transitions (bifurcations) in laser system, about the equilibrium point as a result of a variation of delay parameterτ. The analysis consists in identifying the roots of our system characteristic equation situated on the imaginary axis of the complex λ-plane where by increasing the delay parameter τ, Re λ may at the crossing, change its sign from − to +, i.e. from a stable focus $E^{(*)}$ to an unstable one, or vice versa. This feature may be further assessed by examining the sign of the partial derivatives with respect to τ, $\wedge^{-1}(\tau) = \left(\frac{\partial \operatorname{Re}\lambda}{\partial\tau}\right)_{\lambda=i\cdot\omega}$.

References

1. A.J.H. Donne, Introduction to plasma diagnostics. Trans. Fusion Sci. Technol. **49**, 349–356 (2006)
2. Measured close lightning leader step electric field derivative waveforms. J. Geophys. Res. **116** (2011). https://doi.org/10.1029/2010JD015249
3. A Basic Primer in Lightning Effects and Protection, weighing-systems.com. Recovered 8 Sept 2015

Appendix A
Laser X-ray Guiding System

Ideal plasma contains equal number of electron and ions. m_e—Electron mass, m_i—Ion mass, electron charge (−e), Ion charge (+e). Kinetic temperature of particle (s), kinetic temperature of particle (s) $T_s = \frac{1}{3} \cdot m_s \cdot \langle V_s^2 \rangle$, particle (s) velocity ensemble average $\langle V_s^2 \rangle$. We measure kinetic temperature in eV, 1 Joules $= 6.24 \times 10^{18}$ eV. Quasi-neutrally: $n_i \simeq n_e \simeq n$. Number density n_s, $n_s = \{\frac{\text{number of particles}}{\text{cubic meter}}\}$. Thermal speed $V_{ts} = \sqrt{\frac{2 \cdot T}{m_s}}$. Ion thermal speed $V_{ti} \simeq V_{te} \cdot \sqrt{\frac{m_e}{m_i}}$; {Ion thermal speed} $\ll$ {electron thermal speed}. n, T as a function of location in the plasma. Plasma: group of charged particles, there is interaction between each two particles through electromagnetic potential. The basic parameter that characterize plasma is Debye length λ_D; $\frac{1}{\lambda_D^2} = \sum_j \frac{4 \cdot n_j \cdot q_j^2}{k_B \cdot T_j}$.

nj—density of particle j, qj—electric charge of particle j, Tj—temperature of particle j. $\lambda_D = \sqrt{\frac{k_B \cdot T_e}{4 \cdot \pi \cdot n_e \cdot e^2}} \approx 7.43 \times 10^2 \cdot \left(\frac{T_e(\text{eV})}{n_e(\text{cm}^{-3})} \right)$ cm. The Debye length (λ_D) gives a measurement of the distance over which the electromagnetic potentials due to a single charge are screened by the surrounding charges. ne ~ cm^{-3}, Te ~ [eV]. Typical parameters for the plasmas of interest as X-ray sources: Electron density $n_e = 10^{21}\,\text{cm}^{-3}$, Electron temperature $T_e = 500\,\text{eV}$, and Debye length $\lambda_D \sim 50\text{A}^0$; $1\text{A}^0 = 10^{-10}$ m. We can take the assumption and ignore the interaction inside the plasma between particles with scale length less than Debye length (λ_D) {Plasma scale length} $\gg$ {Debye length}. The condition $\nu_{e,90} \ll \omega_{pe}$ must exist, when the time it takes electron to scatter by 900 from the original direction (cause by contradictions) is $\frac{1}{\nu_{e,90}}$ and the plasma frequency $\omega_{pe} = \sqrt{\frac{4 \cdot \pi \cdot n_e \cdot e^2}{m_e}}$ which is the typical frequency of plasma electron's oscillations. The condition which state that the number of particles in Debye sphere is greater than one (ND > 1); $\frac{\omega_{pe}}{\nu_{e,90}} \approx (\frac{4\pi}{3}) \cdot n_e \cdot \lambda_D^3 = N_D$. We fulfill the conditions on Te and ne and get $N_D \approx 600$ by the equation:

$$N_D = \frac{4}{3} \cdot \pi \cdot \lambda_D^3 \cdot n_e \approx 1.72 \times 10^9 \cdot (T_e(\text{eV}))^{\frac{3}{2}} \cdot n_e^{-\frac{1}{2}}.$$

O. Aluf, *Advance Elements of Laser Circuits and Systems*,
https://doi.org/10.1007/978-3-030-64103-0

Laser-Plasma Self-focusing Mechanisms:

There are some basic mechanisms which are related to Laser-plasma self-focusing (PF = Ponderomotive Force, TF = Thermal pressure):

1. Ponderomotive force self-focusing.
2. Thermal self-focusing.
3. Small-scale self-focusing.
4. Relativistic self-focusing.

Fila-mentation = perturbations or non-uniformities in the laser beam cause local changes in the refractive index of the medium.

Self-focusing and fila mentation: self-focusing of optical laser beams in nonlinear media (plasma) involve the electric field (E) independent term (η_0) and the field dependent term (η_2) so that $\eta = \eta_0 + \eta_2(E^2)$. It happened because the index of refraction η of the optical plasma involve in both η_0 and η_2. The whole beam is focused and fila-mentation when the incident laser beam break-up into number of small channels which them-selves focus. The fila-mentation instability causes perturbations in the intensity profile of an incident laser beam to grow in amplitude. The beam is break up into intense filaments. In laser-plasmas there are a variety of mechanisms that give rise to an intensity dependent refractive index and produce ponderomotive force (PF), thermal pressure (TF) and relativistic effects. The ponderomotive force and thermal pressure mechanisms change the refractive index by expelling the plasma from the high intensity region and are active in laser-fusion plasmas. The relativistic affects the refractive index by increasing the electron mass from the quiver velocity in the electromagnetic wave. It is a very important mechanism with short pulses at ultra-intense interactions (beside the fact that is unimportant at the intensity ranges used in laser-fusion applications).

Optical power: At certain power level, critical power (P_{cr}) the diffraction spreading of the optical beam in the media is balanced by nonlinear refraction (the laser beam propagates without divergence maintaining its original diameter). If $P > P_{cr}$ the laser beam is focused. λ_0—laser wavelength, $\lambda_\perp$—Modulation wavelength.

$$k_\perp = \frac{2\pi}{\lambda_\perp};\ k_0 = \frac{2\pi}{\lambda_0};\ E^2 = \left(\frac{\lambda_0}{\lambda_\perp}\right)^2 \cdot \frac{\eta_0}{\eta_2};\ I = \frac{E^2}{8\pi}\cdot\frac{c}{\eta};\ A = \frac{\pi}{4}\cdot\lambda_\perp^2$$

$P = I \cdot A = \left(\frac{E^2 \cdot c}{8 \cdot \pi \cdot \eta}\right) \cdot \left(\frac{\pi}{4} \cdot \lambda_\perp^2\right) = \frac{c \cdot \lambda_0^2}{32 \cdot \eta_2 \cdot \eta}$. Weakly nonlinear media, $\eta = \eta_0 \Rightarrow$ $P = \frac{c \cdot \lambda_0^2}{32 \cdot \eta_2 \cdot \eta_0}$.

$$\eta = \eta_0 + \eta_2(E^2)$$

η_0 Index of refraction, electric field E independent term.
η_2 Index of refraction, field dependent term.
η Index of refraction.

1. Ponderomotive force self-focusing

Ponderomotive force self-focusing, initially incident light beam that is slightly more intense in one region by an amount δI. The ponderomotive force F_p pushes the electrons away from the region of highest laser intensity leaving behind a region of lowered electron density.

$$F_p = -\frac{n_e}{n_c} \cdot \frac{\nabla(\langle E^2 \rangle)}{8\pi}; \; n = \sqrt{\left(1 - \frac{n_e}{n_c}\right)}$$

Since the index of refraction η of plasma depends on a local electron density n_e, the depletion of electrons at the place where, δI is a large raise the index of refraction there, slow down the phase velocity of the light wave, and causes the wave-fronts to curve. As a result, the light is focused toward the original region of enhanced intensity, since the light wave energy propagates normal to the curved wave-front. Thus, an initial section of the beam with a slightly higher intensity than its surroundings will become more intense as the beam propagates. The instability will saturate when the focusing of the intensity "hot spot" is balanced by diffraction, at which point the filament intensity may be orders of magnitude higher than the initial perturbation.

2. Thermal self-focusing

The thermal self-focusing is similar, except that the density perturbations arise because of temperature gradients associated with heating by a non-uniform laser field. Refraction of laser light acts to enhance the density perturbation. The characteristic minimum scale-size for thermal pressure is the electron conduction mean free path λ_{mfp}. The ponderomotive farce and thermal pressure mechanism occur simultaneously, although their relative strengths vary depending on the density and temperature of the plasma. The time scale for ponderomotive force self-focusing is determined by the ion acoustic transit time across the channel $T_i = \frac{r_0}{c_s}$; r_0—radius of the intensity perturbation. For thermal pressure (TP) self-focusing, the growth time is determined by the larger of the ion acoustic transit time across the channel or the thermal conduction transit time, $T_{cr} = \frac{n_e \cdot r_0^2}{\kappa}$ where κ is the electron thermal conductivity. The ion acoustic transit time will be larger when

$$\frac{1.35 \times 10^{-5} \cdot \frac{n_e}{n_c} \cdot r_0(\mu\text{m}) \cdot \ln(\Lambda) \cdot Z \cdot (1 + 0.24 \cdot Z) \cdot \sqrt{Z+1}}{[T_e(\text{KeV})]^2 \cdot [\lambda_0(\mu\text{m})]^2 \cdot [(Z + 0.24) \cdot \sqrt{A}]} < 1$$

$\ln(\Lambda)$ Coulomb logarithmCoulomb logarithm.

In general the ion acoustic transit time will be larger except for cool plasmas, short wavelength laser and large perturbation.

Laser self-focusing nonlinearity falls into two categories:

1. The channel forms and is able to remain more or less intact while it propagates to the critical density.

2. The channel is unable to retain its shape over substantial distances after it forms, due to the disruptive effects of diffraction or refraction in the plasma, or stimulated scattering.

In either case, self-focusing has major implications for processes such as electron-heating and laser absorption. The final intensity and hot-electron temperatures inside the channel can be large. Channels propagating up to n_c can increase density profile deformation and resonance absorption at n_c. Additional result of such channel is DC magnetic fields around the outside of the channel, due to $\Delta n_e \times \Delta T_e$ effect. The channel like structures explain the localized hot-electron emission of x-ray images recorded. The high intensity channels can induce other harmful laser-plasma instabilities such as SRS (or TPD) or SBS. It may also affect the ablation pressure uniformity [1–3].

3. Small-scale self-focusing

Laser-beam self-focusing is a dominant mechanism in laser-plasma interaction. The instability can be seen as "enhanced" breakup of the interaction beam, via small-scale self-focusing of the intensity "hot spots" of incident beam. The instability spatial growth rate is given by the classical filamentation theory. The estimation of the spatial growth rate ($\alpha_{k\perp}$) of an intensity modulation of wave number ($k_\perp$) can be given by the expression

$$\alpha_{k\perp}(\mathrm{cm}^{-1}) = \left[\alpha_{pm} \cdot \left(\frac{k_\perp^2}{k_0}\right) + \alpha_{TF}^2 - \left(\frac{k_\perp^2}{2k_0}\right)^2\right]^{\frac{1}{2}}$$

$$k_\perp = \frac{2\pi}{\lambda_\perp};\ k_0 = \frac{2\pi}{\lambda_0};\ \lambda_0[\mathrm{cm}];\ \lambda_\perp[\mathrm{cm}]$$

$\lambda\perp$ Modulation wavelength.
$\lambda 0$ Laser wavelength.
α_{PF} Ponderomotive grow rate.
α_{TF} Thermal pressure grow rate

$$\alpha_{PF}(\mathrm{cm}^{-1}) = \frac{2.9 \times 10^{-9} \cdot n \cdot I_0(\mathrm{W\ cm}^{-2}) \cdot \lambda_0(\mu\mathrm{m})}{\left(1 + \frac{1}{Z}\right) \cdot T_e(\mathrm{eV})};\ n = \frac{n_e}{n_c}$$

$$\alpha_{TF}(\mathrm{cm}^{-1}) = \left(\frac{13.2 \cdot \ln\Lambda}{\lambda_0(\mu\mathrm{m})}\right) \cdot \left[\frac{Z^2 \cdot n^3 \cdot I_0(\mathrm{W\ cm}^{-2})}{\left(1 + \frac{1}{Z}\right) \cdot \Phi \cdot T_e^5(\mathrm{eV})}\right]^{\frac{1}{2}}$$

Z Average ionization potential.
$\ln\Lambda$ Coulomb logarithm.

$$\Phi = (Z + 0.24) \cdot (1 + 0.24 \cdot Z)$$

I_0 Zeroth order (unperturbed) intensity.
n_e Electron plasma density.
T_e Electron temperature.

The ratio of ponderomotive (PF) to thermal term (TF):

$$\frac{PF}{TF} = 1.05 \times 10^{-6} \cdot \frac{[\Phi^{\frac{1}{2}} \cdot \lambda_0^2(\text{cm}) \cdot T_e^2(\text{eV})]}{[Z \cdot \ln\Lambda \cdot \lambda_\perp(\mu\text{m}) \cdot n]}$$

4. Relativistic self-focusing

Relativistic self-focusing—the motion of the plasma electrons in the field of ultra-intense laser-pulse becomes relativistic. The relativistic expression for refractive index of plasma is $\eta = \sqrt{1 - \frac{\omega_p^2}{\gamma \cdot \omega^2}}$. The laser intensity will be higher at the center of laser spot; the value of the refractive index will be maximal on the laser axis. It causes to pulse self-focusing. In the mildly relativistic regime (very under dense plasma) the refractive index η is $\eta = 1 - \frac{1}{2} \cdot \frac{\omega_p^2}{\omega^2} + \frac{1}{8} \cdot \frac{\omega_p^2}{\omega^2} \cdot a_L^2\ a_L^2 = \left(\frac{2 \cdot r_e}{\pi \cdot m \cdot c^3}\right) \cdot \lambda^2 \cdot I$. The relation $a_L^2 = \left(\frac{2 \cdot r_e}{\pi \cdot m \cdot c^3}\right) \cdot \lambda^2 \cdot I$ is between normalized vector potential and intensity. r_e-Classical electron radius $\left(r_e = \frac{e^2}{m \cdot c^2}\right)$. We can write the refraction index in the form $\eta = \eta_0 + \eta_2 \cdot I$; $\eta_2 = \frac{1}{4} \cdot \frac{\omega_p^2}{\omega^2} \cdot \left(\frac{r_e}{\pi \cdot m \cdot c^3}\right) \cdot \lambda^2$.

Consequently, one obtains: $P_c = \frac{\lambda^2}{2\pi \cdot \eta_0 \cdot \eta_2} = \frac{2 \cdot m_e^2 \cdot c^5}{e^2} \cdot \frac{n_c}{n_e}$; $P_c = 17.4 \cdot \frac{n_c}{n_e}$ GW.

$$I \cdot \lambda^2 > 10^{18}\ \text{W cm}^{-2}\mu\text{m}^2 \text{ for } P > P_c \simeq 17.4 \cdot \frac{n_c}{n_e}\ \text{GW}.$$

$$V_{os} = \frac{e \cdot E_L}{m_e \cdot \omega};\ \gamma = \left(1 - \frac{V_{os}^2}{c^2}\right)^{-\frac{1}{2}}$$

$$a_0 = \frac{\gamma \cdot V_{os}}{c} = \frac{e \cdot E_L}{m_e \cdot \omega \cdot c} = 8.5 \times 10^{-10} \cdot \lambda(\mu\text{m}) \cdot I^{\frac{1}{2}}\left(\frac{\text{W}}{\text{cm}^2}\right)$$

$$\gamma = \sqrt{1 + a_0^2}$$

Instability in plasma due to shooting intense laser pulses:

The propagation of intense laser through plasmas is an important concern in the laser driven inertial confinement fusion, the laser-plasma accelerators, the x-ray laser and other effects. The amplitude of laser can be describe by the normalize vector potential $\{a, \phi\}$, where $a = \frac{e \cdot A}{m_0 \cdot c^2}$; $\phi = \frac{e \cdot \Phi}{m_0 \cdot c^2}$. While $a \ll 1$, the propagation equation of laser is linear. When a increase such as density wave effects, relativistic mass corrections. While a > 1 we have highly nonlinear problems because of the relativistic effects. When the plasma is fully ionized there are main two nonlinear effects: relativistic mass increase of the electron and laser's diffraction by electron density waves.

1. {Amplitude of vector potential increases} → {Vibration velocity of electron increase} → {the mass of electron is changed to $\gamma \cdot m$; $\gamma = \frac{1}{\sqrt{1-\frac{v^2}{c^2}}}$}.
2. {Intense laser propagate} → {electron in the plasma vibrate (parallel to electric field and perpendicular to propagation of the laser)} → {vibration elections pushed by Lorentz force (produce by Lorentz magnet filed of laser)} → {electron density oscillation form density waves} → {laser is diffracted by density waves}

Density wave (electron) is in the relativistic region, density fluctuation ~ $a \cdot a$, and $f_{\text{electron-density wave}} \simeq 2 \cdot f_{\text{laser}}$.

Both non-linear effects cause corrections of the refract index of the plasma and self-focusing of the intense laser pulse.

U	is the normalized fluid momentum $u = \frac{p}{m_o \cdot c}$.
m_0	Electron rest mass.
p	Fluid momentum.
n	Electron density.
n_0	Ambient density.

$$\gamma = \sqrt{1+u^2};\ \rho = \frac{n}{\gamma \cdot n_0};\ \rho^{(0)} = \frac{n^{(0)}}{n_0};\ n_0 = 1.2 \times 10^{16}\,\text{cm}^{-3}$$

Modulation instability is associated with propagation of intense laser pulses in a partially stripped, preformed plasma channel, modulation instabilities are caused by the interplay between (anomalous) group velocity dispersion and self-phase modulation. The analysis is based on a systematic approach that includes finite-perturbation-length effects, nonlinearities, group velocity dispersion, and transverse effects. To properly include the radial variation of the laser field and plasma channel, the source-dependent expansion method for analyzing the wave equation must be employed. Matched equilibria for a laser beam propagating in a plasma channel. Modulation of a uniform (matched) laser beam equilibrium in a plasma channel leads to a coupled pair of differential equations for the perturbed spot size and laser field amplitude. A general dispersion relation is derived and solved. Surface plots of the spatial growth rate as a function of laser beam power.

Brillouin instability, Stimulate Brillouin Scattering (SBS) is a very important phenomenon because it transfers almost all energy to the scattered light and reaches a maximum on backward scattering. In the case of backward SBS, it causes laser's reflection and energy loss. During the implosion of ICF, the Brillouin reflectivity can vary from 10% (short wave incident laser) to more than 40% (long wave laser). In addition the Brillouin instability related closely with the fila-mentation instability. The SBS takes place in all the under-dense region of plasmas. For Brillouin instability of an intense laser with $a^2 \sim 0.1-0.5$, the intensity region is Weak-Relativistic Region (nonlinear effects can be processed by the series expansion with respect to a^2.

Raman Instability, The stimulated Raman scattering (SRS) is one of the parametric instabilities, which develop in the under dense region of the laser produced plasmas (below quarter critical density). SRS can be characterized as a decay of incident pumping electromagnetic wave in another electromagnetic wave and in a fast forward going electrostatic Langmuir. The generation of hot electrons is an important issue, because it causes the target preheating and prevents the efficient compression of the fuel capsule. In the fast ignition scheme an ultra-short ($\approx$10 ps) and ultra-intense heating laser pulse follow another longer ($\approx$100 ps) hole boring pulse or it is applied through a guiding cone. The idea is to separate the processes of implosion and heating of the fuel. The both pulses may interact with the corona plasma surrounding the core earlier than they reach it. One of the possible interactions is SRS, which causes an energy loss of the primary electromagnetic wave and decreases the energy conversion from the laser pulse to the fuel capsule. Then the Raman reflectivity becomes a very important parameter. The Raman scattered electromagnetic wave can propagate in the backward and the forward direction with respect to the direction of motion of the laser pulse. In both the Raman backscattering (SRS-B) and the forward Raman scattering (SRS-F) the electron plasma waves travel in the forward direction, but their velocities may differ considerably. Slower SRS-B electrostatic wave in the laser produced plasmas usually dominates over SRS-F, whole phase velocity is often comparable with the speed of light. With a high phase velocity of SRS-F electrostatic wave a small Landau damping rate is connected and vice versa, in the well under dense region of the plasma corona or in the high temperature plasma, where the SRS-B electrostatic wave is strongly Landau damped, SRS-F can become comparable with SRS-B in spite of its small growth rate. The phase velocity of SRS-B electrostatic wave is closer to the thermal velocity then in the case of SRS-F. Hence, the large amplitude SRS-B plasma wave can interact with a significantly large amount of electrons, which can get trapped in the wave. SRS can couple to other plasma instabilities, which leads to a cascading, whereas the deformation of electron distribution function and the occurrence of population of the trapped accelerated electrons gives rise to new nonlinear modes of which the impinging laser may also be scattered.

The nonlinear interaction of intense laser pulses in plasmas:

The interaction of ultra-high power laser beams with plasmas is rich of in a variety of wave-particle phenomena.

1. The interaction of ultra-high power laser with plasmas cause variety of wave-particle phenomena.
2. Laser power high enough cause electron oscillation (quiver) velocity to become highly relativistic.

Laser plasma process:

1. Relativistic optical guiding of the laser beam.

2. Excitation of coherent radiation at harmonics of the fundamental laser frequency.
3. Generation of large amplitude plasma waves (wake fields)—Hot plasma.
4. Frequency shifts induced in the laser pulse by plasma waves.
5. Frequency amplification using an ionization front. Ionization front may be used to amplify the frequency of the laser pulse.
6. Single particle acceleration in a laser pulse. Single particle may be accelerated by the passage of a laser pulse.

Intense laser pulse heats plasmas then self-consistence interaction. We analyze nonlinear 1D model. The "Quasi-static" cold fluid plasma response is discussed.

Set of coupled nonlinear equations: (1) vector potential of the radiation field (2) electrostatic potential of the plasmas.

$$\left(\frac{\partial^2}{\partial z^2} - \frac{1}{c^2} \cdot \frac{\partial^2}{\partial t^2}\right) a = k_p^2 \cdot \frac{n}{n_0} \cdot \beta_\perp = k_p^2 \cdot \frac{n}{n_0} \cdot \frac{a}{\gamma}$$

$$\beta_\perp = \frac{a}{\gamma}; \quad \frac{\partial^2 \phi}{\partial z^2} = k_p^2 \cdot \left(\frac{n}{n_0} - 1\right)$$

$$a(z,t) = \frac{|e| \cdot A}{m_0 \cdot c^2}; \quad \phi(z,t) = \frac{|e| \cdot \Phi}{m_0 \cdot c^2}; \quad k_p = \frac{\omega_p}{c}$$

$$\omega_p = \sqrt{\frac{4 \cdot \pi \cdot |e|^2 \cdot n_0}{m_0}}; \quad k_p = \frac{1}{c} \cdot \sqrt{\frac{4 \cdot \pi \cdot |e|^2 \cdot n_0}{m_0}}$$

There are two cases: $a \ll 1$ the propagating equation of laser is linear, $a > 1$ the propagation of the laser becomes highly non-linear because of the relativistic effects (many non-linear effects will appear when a increases, such as density wave effect, relativistic mass correction, etc). In full ionized plasmas, there are two dominating nonlinear effects on the laser propagation: one is the relativistic mass increase of the electron $m \to m \cdot \gamma = m \cdot \frac{1}{\sqrt{1-\frac{v^2}{c^2}}}$ (cause by high vibration velocity of electron), where the Lorentz factor $\gamma = \frac{1}{\sqrt{1-\frac{v^2}{c^2}}}$ (coefficient determined by the velocity), other is the laser's diffraction by electron density waves. When the intense laser propagates, electron in the plasma first vibrates parallel to the electric field and perpendicular to the propagation direction of the laser. Then the vibrating electrons are pushed by the Lorentz force produced by the magnet field of the laser. Electron density oscillations parallel to the propagating direction of the laser form density waves. Finally the laser is diffracted by density waves (density fluctuation is proportional to $a \cdot a$ and its frequency is twice as the laser frequency and its order is the same as relativistic effect's order). In the relativistic region, both non-linear effects cause corrections of the refract index of the plasma by modifying the dispersion relation of the laser. The correction is very important for the shaping and self-focusing of the intense laser pulse (Fig. A.1). additionally, we have some mechanisms:

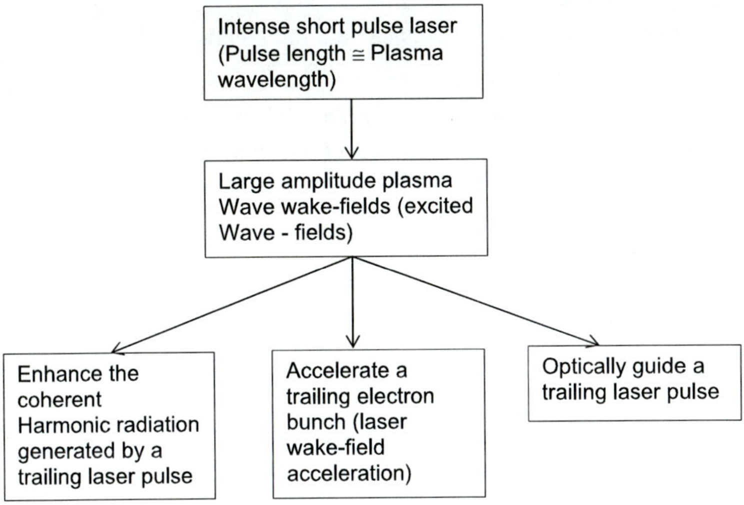

Fig. A.1 Intense short laser pulse, large amplitude plasma and outcomes

1. The nonlinear equation Examine various laser-plasma interaction phenomena.
2. Laser plasma instability limits the laser propagation distance.
3. Relativistic quiver motion of electron by laser field cause relativistic optical guiding.
4. {short laser pulses—pulse length} < {plasma wave length}.

Plasma wavelength $\lambda_p = 1$ mm is for typical plasma density $n_p = 10^{15}\,\text{cm}^{-3}$. To excite a large amplitude wake-fields the proton Burch length need to be $\sigma_z \approx \lambda_p = 1$ mm. Interaction of Ultra-high power laser with plasmas **causes** a variety of wave-particle phenomena. Laser power high enough **causes** electron oscillation (quiver) velocity to become highly relativistic.

$$\text{The plasma wavelength } \lambda_p = \frac{2\cdot\pi}{k_p} = \sqrt{\frac{\pi}{r_e\cdot n_e}}$$

$$\text{Acceleration length } L_{acc} \approx \frac{\lambda_p}{1-\beta_p} = 2\cdot\gamma_p^2\cdot\lambda_p;\ 2\cdot\gamma_p^2 = \frac{1}{1-\beta_p}$$

$$L_{acc}[\text{cm}] \approx 2\cdot\frac{n_c}{n_e}\cdot\sqrt{\frac{\pi}{r\cdot_e n_e}} = \frac{2}{\lambda_0^2}\cdot\left(\frac{\pi}{r_e\cdot n_e}\right)^{\frac{3}{2}} \approx \frac{7.4\times 10^{27}}{\lambda_0^2[\mu\text{m}]\cdot n_e^{\frac{3}{2}}[\text{cm}^{-3}]}$$

$n(z,t)$ is the plasma density, n_0 is the ambient plasma density.

$\beta_\perp = \frac{V_\perp}{c} = \frac{a}{\gamma};\ \gamma = \frac{1}{\sqrt{1-\beta_z^2-\beta_\perp^2}} = \frac{\sqrt{1+a^2}}{\sqrt{1-\beta_z^2}}$; $\beta_\perp$—normalized transverse plasma fluid velocity. γ—Relativistic mass factor. The transverse canonical momentum is invariant and prior to the laser pulse interaction the plasma is assumed stationary. n, β_z, γ—fulfill the relativistic fluid equations, continuity equation and equation of

motion from the first two moments of the Vlasov equation: $\frac{\partial n}{\partial t} + c \cdot \frac{\partial (n \cdot \beta_z)}{\partial z} = 0$
$\frac{d\beta_z}{dt} = -\frac{1}{2\cdot\gamma^2} \cdot \left(c \cdot \frac{\partial}{\partial z} + \beta_z \cdot \frac{\partial}{\partial t}\right)a^2 + \frac{c}{\gamma} \cdot \left(1 - \beta_z^2\right) \cdot \frac{\partial \phi}{\partial z}$.

The electron relativistic continuity equation: $\frac{\partial n}{\partial t} + c \cdot \frac{\partial (n \cdot \beta_z)}{\partial z} = 0$.

n—Electron density [m^{-3}], e—charge of electron [Coulomb], v_z—electron velocity vector in z direction [m/s], J—electron current density vector [A/m^2], ρ—electron charge density [Coulomb/m^3].

$$\rho = n \cdot e;\ \beta_z = \frac{v_z}{c};\ J = \rho \cdot v_z = n \cdot e \cdot v_z$$

$$\frac{\partial n}{\partial t} + c \cdot \frac{\partial (n \cdot \beta_z)}{\partial z} = 0 \Rightarrow \frac{\partial}{\partial t}\left[\frac{1}{e} \cdot \rho\right] + cn\frac{\partial}{\partial z}\left[\frac{1}{c} \cdot v_z\right]$$

$$\frac{\partial}{\partial t}\left[\frac{1}{e} \cdot \rho\right] + c \cdot n \cdot \frac{\partial}{\partial z}\left[\frac{1}{c} \cdot v_z\right] \Rightarrow \frac{1}{e} \cdot \frac{\partial \rho}{\partial t} + c \cdot \frac{1}{c} \cdot n \cdot \frac{\partial}{\partial z}\left[\frac{J}{n \cdot e}\right] = 0$$

$$\Rightarrow \frac{1}{e} \cdot \frac{\partial \rho}{\partial t} + \frac{1}{e} \cdot \frac{\partial J}{\partial z} = 0$$

$$\frac{1}{e} \cdot \frac{\partial \rho}{\partial t} + \frac{1}{e} \cdot \frac{\partial J}{\partial z} = 0 \Rightarrow \frac{\partial \rho}{\partial t} + \frac{\partial J}{\partial z} = 0;\ \text{Is continuity equation}$$

$$\frac{d\gamma}{dt} = c \cdot \beta_z \cdot \frac{\partial \phi}{\partial z} + \frac{1}{2 \cdot \gamma} \cdot \frac{\partial a^2}{\partial t};\ \frac{d}{dt} = \frac{\partial}{\partial t} + c \cdot \beta_z \cdot \frac{\partial}{\partial z}$$

The term a^2 contains the ponderomotive force; thermal effects may be neglected provided the electron quiver velocity which is much greater than the electron thermal velocity and the ions are assumed to be stationary. Algebraic transformation: laboratory frame independent space and the variables (z, t) **To** Independent variables (ξ, τ) where $\xi = z - c \cdot \beta_t \cdot t;\ \tau = t;\ \beta_t = \frac{v_t}{c}, \beta_t = \frac{v_t}{c} = 1$ is the normalized transformation velocity. $(z, t) \to (\xi, \tau);\ \frac{\partial}{\partial z} = \frac{\partial}{\partial \xi};\ \frac{\partial}{\partial t} = \frac{\partial}{\partial \tau} - c \cdot \beta_t \cdot \frac{\partial}{\partial \xi}$.

We get the following equations:

$$\left(\frac{1}{\gamma_t^2} \cdot \frac{\partial^2}{\partial \xi^2} + \frac{2 \cdot \beta_t}{c} \cdot \frac{\partial^2}{\partial \xi \partial t} - \frac{1}{c^2} \cdot \frac{\partial^2}{\partial t^2}\right)a = \frac{k_p^2}{\gamma} \cdot \frac{n}{n_0} \cdot a$$

$$\frac{\partial^2 \phi}{\partial \xi^2} = k_p^2 \cdot \left(\frac{n}{n_0} - 1\right);\ \frac{\partial}{\partial \xi}[n \cdot (\beta_t - \beta_z)] = \frac{1}{c} \cdot \frac{\partial n}{\partial \tau}$$

$$\frac{\partial}{\partial \xi}[\gamma \cdot (1 - \beta_t \cdot \beta_z) - \phi] = -\frac{1}{c} \cdot \frac{\partial}{\partial \tau}(\gamma \cdot \beta_z)$$

$\gamma_t^2 = \frac{1}{1-\beta_t^2};\ \gamma = \frac{\sqrt{1+a^2}}{\sqrt{1-\beta_z^2}}$. We get a complete set of fully nonlinear, relativistic, cold fluid equations which describe the 1D laser-plasma interaction. The 1D model is valid as long as the radiation spot size is large compared to the plasma wavelength $\left(r_s \gg \lambda_p = \frac{2\pi}{k_p}\right)$. We choose $v_t = c$, since it lies between the group velocity and phase velocity of the laser pulse. The laser pulse is propagated in the positive z direction and cross the s = 0 plane at t = 0. The $\xi = 0$ plane defines the front of the pulse which extends into the $\xi \leq 0$ region. The important section of ξ, τ frame is the region where $\xi \leq 0$ since for $\xi \geq 0$ we have$a = 0; n = n_0; \beta_z = 0;$ $\gamma = 1$ The quasi-static approximation: In a frame moving at the speed of light, the plasma fluid sees a nearly steady-state (s.s) radiation field after the transients have decayed away. The speed of light frame ($\beta_t = 1$) under certain conditions a quasi-static state will exist in the macroscopic plasma quantities n, β_z, γ. If the laser pulse is sufficiently short, the fields a, ϕ which drive the plasma are expected to change little during a transit time of the plasma through the laser pulse. The envelope of a changes on a characteristic time $\tau_e \sim 2 \cdot \gamma \cdot |\frac{n_0}{n}| \cdot \frac{\omega}{\omega_p} \cdot \frac{1}{\omega_p}$ where ω is the laser frequency. We assume that $\omega \gg \omega_p$, the radiation envelope changes on a time scale which is long compared to a plasma period. If the laser pulse duration $\tau_L \ll \tau_e$ then the quasi-static approximation is valid. The validity of the 1D model requires that the laser beam diffraction time (transverse spreading time) $\tau_d = \frac{\pi \cdot r_s^2}{\lambda \cdot c}$ be long compare to τ_e. This condition is satisfied as long as $r_s \gg \lambda_p$. The quasi-static approximation: $|\frac{1}{c} \cdot \frac{\partial}{\partial \tau} \int_{\xi}^{0} n \cdot d\xi'| \ll n_0$; $|\frac{1}{c} \cdot \frac{\partial}{\partial \tau} \int_{\xi}^{0} (\gamma \cdot \beta_z) \cdot d\xi'| \ll 1$ and we get $n \cdot (1 - \beta_z) = n_0;\ \gamma \cdot (1 - \beta_z) - \phi = 1$. The coupled laser field equations:

$$\left(\frac{2}{c} \cdot \frac{\partial}{\partial \xi} - \frac{1}{c^2} \cdot \frac{\partial}{\partial \tau}\right) \frac{\partial a}{\partial \tau} = k_p^2 \cdot \frac{a}{1 + \phi}$$

$$\frac{\partial^2 \phi}{\partial \xi^2} + k_p^2 \cdot \frac{1}{(1 + \phi)^2} \cdot \left(1 + \frac{\phi}{2}\right) \cdot \phi = k_p^2 \cdot \frac{1}{(1 + \phi)^2} \cdot \frac{a^2}{2}$$

The plasma quantities in terms of the fields:

$$\frac{n}{n_0} = 1 + \frac{1}{2} \cdot \left[\frac{(1 + a^2)}{(1 + \phi)^2} - 1\right];\ \gamma = \frac{[1 + a^2 + (1 + \phi)^2]}{[2 \cdot (1 + \phi)]}$$

$$\beta_z = \frac{[1 + a^2 - (1 + \phi)^2]}{[1 + a^2 + (1 + \phi)^2]}$$

The fields and plasma quantities are fully nonlinear and hold for arbitrary polarization of the laser pulse. Parallel to the main laser beam which we shoot to the plasma there is X-ray laser which heat the plasma. We consider X-ray laser normalized vector potential $\{a_{x-ray}, \phi_{x-ray}\}$; $a_{x-ray} = \frac{e \cdot A_{x-ray}}{m_0 \cdot c^2}$ $\phi_{x-ray} = \frac{e \cdot \Phi_{x-ray}}{m_0 \cdot c^2}$. We consider x-ray laser beam intensity is not strong enough $a_{x-ray} \ll 1$ (cannot cause to optical guiding). The x-ray propagation equation is as follow:

$$\left(\frac{\partial^2}{\partial z^2} - \frac{1}{c^2} \cdot \frac{\partial^2}{\partial t^2}\right) a_{x-ray} = k_p^2 \cdot \frac{n}{n_0} \cdot \beta_\perp = k_p^2 \cdot \frac{n}{n_0} \cdot \frac{a_{x-ray}}{\gamma}$$

$$\beta_\perp = \frac{a_{x-ray}}{\gamma};\ \frac{\partial^2 \phi_{x-ray}}{\partial z^2} = k_p^2 \cdot \left(\frac{n}{n_0} - 1\right)$$

We consider two-dimensional homogenous plasma and ignore the electron thermal motion in the direction of laser polarization. Let

$$\nabla_{\perp} = \left\{\frac{\partial}{\partial x}, \frac{\partial}{\partial y}\right\};\ \nabla_{\perp}^2 = \frac{\partial^2}{\partial x^2} + \frac{\partial^2}{\partial y^2}$$

$$\beta = \left\{\frac{v_x}{c}, \frac{v_y}{c}, \frac{v_z}{c}\right\};\ \nabla^2 = \frac{\partial^2}{\partial z^2} + \nabla_{\perp}^2$$

The incident lasers propagate in z direction and is polarized parallel to x axis; the electron's temperature in the plasma is T, then $k \cdot T \ll m_0 \cdot v_x^2$. The conservation of the transverse canonical momentum leads to $\frac{\gamma \cdot v_x}{c} \approx a_x$.

The Lorentz force has the following form $v \times (\nabla \times a) \approx \frac{1}{2 \cdot \gamma} \cdot \nabla(a \cdot a)$. The equation of electrons becomes:

$$\frac{\partial}{\partial ct}(\gamma\beta) + (\beta \cdot \nabla)(\gamma\beta) = \nabla(\phi + \phi_{x-ray}) - \frac{1}{2 \cdot \gamma} \cdot \nabla([a + a_{x-ray}] \cdot [a + a_{x-ray}]) - \nabla \frac{P_e}{n_0 \cdot m_0 \cdot c^2}$$

Instead P_e, the normalized electron thermal-pressure,

$$P_e \to P = \frac{P_e}{m_0 \cdot c^2};\ P = \frac{P_e}{m_0 \cdot c^2} = \frac{(n_0 + n) \cdot \theta}{m_0 \cdot c^2}$$

$$\theta = k \cdot T;\ P_e = (n_0 + n) \cdot \theta$$

Plasma frequency:

Plasma frequency is the electron plasma frequency—electrostatic oscillation frequency. There are different values to different particles. The most important is the plasma frequency of relatively fast electron. ω_p—Electrostatic oscillation frequency of particle which caused by small charge separation. The density of charge particles is created in leading face of slab. Quasi-neutral position is δx. The x directed electric field inside the slab $E_x = -\frac{\sigma}{\varepsilon_0} = -\frac{e \cdot n \cdot \delta x}{\varepsilon_0};\ \sigma = e \cdot n \cdot \delta x;\ \omega_p^2 = \frac{n \cdot e^2}{\varepsilon_0 \cdot m}$. The charge particle density is $\sigma = e \cdot n \cdot \delta x$. The Newton's law of individual particle inside slab $m \cdot \frac{d^2 \delta x}{dt^2} = e \cdot E_x = -m \cdot \omega_p^2 \cdot \delta x;\ \delta x = (\delta x)_0 \cdot \cos(\omega_p \cdot t)$.

We can observe the plasma oscillation if we inspect the plasma system for period of time τ which is bigger than plasma period $\tau > \tau_p = \frac{1}{\omega_p}$. There is no behavior of plasma for length-scales $L < v_t \cdot \tau_p$. Particles will be in the system before they finish the plasma oscillation. Debye length $\lambda_D = \sqrt{\frac{T}{m}} \cdot \omega_p^{-1}$ and it is not effect by mass and is true for different particles $\lambda_D = \sqrt{\frac{\varepsilon_0 \cdot T}{n \cdot e^2}}$. The system can be plasmatic only if $\frac{\lambda_D}{L} \ll 1;\ \frac{\tau_p}{\tau} \ll 1$when τ is typical time scale, L is the length scale. The distance λ_D is equivalent to τ_p. The equilibrium potential is perturbed by amount $\delta\rho$, by small localized charge density $\delta\rho_{ext}$.

Table A.1 Plasma parameter Λ for strong and weakly coupled plasma

Description	$\Lambda \ll 1 (\Gamma \gg 1)$	$\Lambda \gg 1 (\Gamma \ll 1)$
Coupling	Strongly coupled plasma	Weakly coupled plasma
Debye sphere	Sparsely populated, corresponds to a strongly coupled plasma	Densely populated, corresponds to a weakly coupled plasma
Electrostatic influence	Almost continuously	Occasional
Typical characteristic	Cold and dense	**Hot and diffuse**

$$\delta\rho = \delta\rho_{ext} + e\cdot(\delta n_i - \delta n_e) = \delta\rho_{ext} - 2\cdot e^2\cdot n_0\cdot\frac{\delta\phi}{T}$$

$$\text{Poisson's equation: } \nabla^2\delta\phi = -\frac{\delta\rho}{\varepsilon_0} = -\left(\frac{\delta\rho_{ext} - 2\cdot e^2\cdot n_0\cdot\frac{\delta\phi}{T}}{\varepsilon_0}\right)$$

$\left(\nabla^2 - \frac{2}{\lambda_D^2}\right)\delta\phi = -\frac{\delta\rho_{ext}}{\varepsilon_0}$, perturbing charge density consists of a point charge q located at the origin ($\delta\rho_{ext} = q\cdot\delta(r)$). $\delta\phi(r) = \frac{q}{4\cdot\pi\cdot\varepsilon_0\cdot r}\cdot e^{-\frac{\sqrt{2}\cdot r}{\lambda_D}}$, the coulomb potential of the perturbing point of charge q is shielded on distance larger than Debye length. By shielding cloud approximate radius λ_D its consisting charge of the opposite sign.

Debye shielding: plasmas generally do not contain strong electric fields in their rest frames. The shielding of an external electric field from the interior of plasma can be viewed as a result of high plasma conductivity. Shielding is a Dielectric phenomenon. Polarization of the plasma medium is associated redistribution of space charge which prevents penetration by an external electric field. The length scale = Debye length. For Quasi-neutral plasma close to thermal equilibrium, the particles density by Maxwell-Boltzmann law:

$n_s = n_0\cdot e^{-\frac{e_S\cdot\phi(r)}{T}}$, n_0, T—constant. Quasi neutrally ($e_i = -e_e = e$) requires the equilibrium potential to be constant. Plasma parameter: Average distance between particles $r_d = n^{-\frac{1}{3}} = \frac{1}{\sqrt[3]{n}}$; distance of closest approach $r_c = \frac{e^2}{4\cdot\pi\cdot\varepsilon_0\cdot T}$. Distance r_c which the coulomb energy of one charge particle in the field disappear $U(r_c, v_t) = 0$ which $U(r_c, v) = \frac{1}{2}\cdot m\cdot v^2 - \frac{e^2}{4\cdot\pi\cdot\varepsilon_0\cdot T}$. The plasma parameter is a dimensionless number, denoted by capital Lambda, Λ (Table A.1). The plasma parameter is usually interpreted to be the argument of the Coulomb logarithm, which is the ratio of the maximum impact parameter to the classical distance of closest approach in Coulomb scattering. In this case, the plasma parameter is given by $\Lambda = 4\cdot\pi\cdot n\cdot\lambda_D^3$. n is the number density of electrons, λ_D is the Debye length. This definition of the plasma parameter is called the Debye number (N_D, $N_D = \frac{(\varepsilon_0\cdot k\cdot T_e)^{\frac{3}{2}}}{q_e^2\cdot n_e^{\frac{1}{2}}}$). The plasma parameter is defined as $N_D = \frac{4\cdot\pi}{3}\cdot n\cdot\lambda_D^3$. Debye length: $\lambda_D = \sqrt{\frac{\varepsilon_0\cdot k\cdot T_e}{n_e\cdot q_e^2}}$. Γ—Plasma coupling parameter, $\Gamma = \frac{\text{coulomb energy}}{\text{thermal energy}}$. E_c—Coulomb energy (per particle), $E_c = \frac{q_e^2}{4\cdot\pi\cdot\varepsilon_0\cdot\langle r\rangle}$; $\Gamma = \frac{q_e^2}{4\cdot\pi\cdot\varepsilon_0\cdot k\cdot T_e}\cdot\sqrt[3]{\frac{4\cdot\pi\cdot n_e}{3}}$; $\Gamma \sim \Lambda^{-\frac{2}{3}}$; $\frac{r_d}{r_c} = \frac{1}{\sqrt[3]{n}}\cdot\frac{4\cdot\pi\cdot\varepsilon_0\cdot T}{e^2}$. We can characterize the weakly coupled plasma by Fokker-planck equation.

$$\Lambda = 4 \cdot \pi \cdot n \cdot \lambda_D^3;\ r_c = \frac{e^2}{4 \cdot \pi \cdot \varepsilon_0 \cdot T};\ \lambda_D = \sqrt{\frac{\varepsilon_0 \cdot T}{n \cdot e^2}}$$

$$\Lambda = \frac{\lambda_D}{r_c} = \frac{4 \cdot \pi \cdot \varepsilon_0 \cdot T}{e^2} \cdot \sqrt{\frac{\varepsilon_0 \cdot T}{n \cdot e^2}} = \frac{4 \cdot \pi \cdot \varepsilon_0^{\frac{3}{2}} \cdot T^{\frac{3}{2}}}{e^3 \cdot \sqrt{n}}$$

$$\Lambda = \frac{\lambda_D}{r_c} = \frac{1}{\sqrt{4 \cdot \pi}} \cdot \left(\frac{r_d}{r_c}\right)^{\frac{3}{2}}$$

High power short pulses laser and X-ray in Hot plasma, differential equations of electrons:

The differential equation of electrons:

$$\frac{\partial}{\partial ct}(\gamma \cdot \beta) + (\beta \cdot \nabla)(\gamma \cdot \beta) = \nabla(\phi + \phi_{x-ray}) - \frac{1}{2 \cdot \gamma} \cdot \nabla([a + a_{x-ray}] \cdot [a + a_{x-ray}]) - \nabla \frac{P_e}{n_0 \cdot m_0 \cdot c^2}$$

We define $a' = a + a_{x-ray}$; where a' is a summation of vector potential of the radiation field main laser (a) PLUS the vector potential of the radiation field of the X-laser. We define $\phi' = \phi + \phi_{x-ray}$; where ϕ' is a summation of the scalar electrostatic potential of the plasma (cause by main laser beam) and ϕ_{x-ray} is the scalar electrostatic potential of the plasma cause by X-ray laser. β is the ratio of electron velocity $\{v_x, v_y, v_z\}$to the speed of light (c), $\beta = \left\{\frac{v_x}{c}, \frac{v_y}{c}, \frac{v_z}{c}\right\}$ $\beta = \beta_x \cdot \hat{x} + \beta_y \cdot \hat{y} + \beta_z \cdot \hat{z}$; $\beta^2 = \beta_x^2 + \beta_y^2 + \beta_z^2$; $\beta = \sqrt{\beta_x^2 + \beta_y^2 + \beta_z^2}$. γ is Lorentz factor (relativistic mass factor). $\gamma = \frac{1}{\sqrt{1-\beta^2}}$; $\gamma \cdot \beta = \frac{\beta}{\sqrt{1-\beta^2}}$.

The equation of electron becomes:

$$\frac{\partial}{\partial ct}(\gamma \cdot \beta) + (\beta \cdot \nabla)(\gamma \cdot \beta) = \nabla(\phi') - \frac{1}{2 \cdot \gamma} \cdot \nabla(a' \cdot a') - \nabla \frac{P_e}{n_0 \cdot m_0 \cdot c^2}$$

Equation left side, element (1):

$$\gamma \cdot \beta = \frac{\beta}{\sqrt{1-\beta^2}} = \frac{\beta_x}{\sqrt{1-\beta^2}} \cdot \hat{x} + \frac{\beta_y}{\sqrt{1-\beta^2}} \cdot \hat{y} + \frac{\beta_z}{\sqrt{1-\beta^2}} \cdot \hat{z}$$

$$\frac{\partial}{\partial ct}(\gamma \cdot \beta) = \frac{1}{c} \cdot \frac{\partial}{\partial t}(\gamma \cdot \beta)$$
$$= \frac{1}{c} \cdot \frac{\partial}{\partial t}\left(\frac{\beta_x}{\sqrt{1-\beta^2}} \cdot \hat{x} + \frac{\beta_y}{\sqrt{1-\beta^2}} \cdot \hat{y} + \frac{\beta_z}{\sqrt{1-\beta^2}} \cdot \hat{z}\right)$$

$$\frac{\partial}{\partial ct}(\gamma\cdot\beta)=\frac{1}{c}\cdot\frac{\partial}{\partial t}(\gamma\cdot\beta)$$
$$=\frac{1}{c}\cdot\left(\hat{x}\cdot\frac{\partial}{\partial t}\left[\frac{\beta_x}{\sqrt{1-\beta^2}}\right]+\hat{y}\cdot\frac{\partial}{\partial t}\left[\frac{\beta_y}{\sqrt{1-\beta^2}}\right]+\hat{z}\cdot\frac{\partial}{\partial t}\left[\frac{\beta_z}{\sqrt{1-\beta^2}}\right]\right)$$

$$\frac{\partial}{\partial t}\left[\frac{\beta_x}{\sqrt{1-\beta^2}}\right]=\frac{\frac{\partial\beta_x}{\partial t}\cdot\sqrt{1-\beta^2}-\beta_x\cdot\frac{\partial}{\partial t}(\sqrt{1-\beta^2})}{1-\beta^2}$$
$$=\frac{1}{\sqrt{1-\beta^2}}\cdot\frac{\partial\beta_x}{\partial t}-\frac{\beta_x}{1-\beta^2}\cdot\frac{\partial}{\partial t}(\sqrt{1-\beta^2})$$

$$\frac{\partial}{\partial t}(\sqrt{1-\beta^2})=\frac{\partial}{\partial t}(1-\beta^2)^{\frac{1}{2}}=\frac{1}{2}\cdot(1-\beta^2)^{-\frac{1}{2}}\cdot\frac{\partial}{\partial t}(1-\beta^2)$$

$$\frac{\partial}{\partial t}(\sqrt{1-\beta^2})=\frac{1}{2}\cdot(1-\beta^2)^{-\frac{1}{2}}\cdot\frac{\partial}{\partial t}(1-\beta_x^2-\beta_y^2-\beta_z^2)$$
$$=-(1-\beta^2)^{-\frac{1}{2}}\cdot\left(\beta_x\cdot\frac{\partial\beta_x}{\partial t}+\beta_y\cdot\frac{\partial\beta_y}{\partial t}+\beta_z\cdot\frac{\partial\beta_z}{\partial t}\right)$$

$$\frac{\partial}{\partial t}\left[\frac{\beta_x}{\sqrt{1-\beta^2}}\right]=\frac{1}{\sqrt{1-\beta^2}}\cdot\frac{\partial\beta_x}{\partial t}+\frac{\beta_x}{1-\beta^2}\cdot(1-\beta^2)^{-\frac{1}{2}}\cdot\left(\beta_x\cdot\frac{\partial\beta_x}{\partial t}+\beta_y\cdot\frac{\partial\beta_y}{\partial t}+\beta_z\cdot\frac{\partial\beta_z}{\partial t}\right)$$

And in the same way we get the following:

$$\frac{\partial}{\partial t}\left[\frac{\beta_y}{\sqrt{1-\beta^2}}\right]=\frac{1}{\sqrt{1-\beta^2}}\cdot\frac{\partial\beta_y}{\partial t}+\frac{\beta_y}{1-\beta^2}\cdot(1-\beta^2)^{-\frac{1}{2}}\cdot\left(\beta_x\cdot\frac{\partial\beta_x}{\partial t}+\beta_y\cdot\frac{\partial\beta_y}{\partial t}+\beta_z\cdot\frac{\partial\beta_z}{\partial t}\right)$$

$$\frac{\partial}{\partial t}\left[\frac{\beta_z}{\sqrt{1-\beta^2}}\right]=\frac{1}{\sqrt{1-\beta^2}}\cdot\frac{\partial\beta_z}{\partial t}+\frac{\beta_z}{1-\beta^2}\cdot(1-\beta^2)^{-\frac{1}{2}}\cdot\left(\beta_x\cdot\frac{\partial\beta_x}{\partial t}+\beta_y\cdot\frac{\partial\beta_y}{\partial t}+\beta_z\cdot\frac{\partial\beta_z}{\partial t}\right)$$

Finally we get

$$\frac{\partial}{\partial ct}(\gamma\cdot\beta)=\frac{1}{c}\cdot\frac{\partial}{\partial t}(\gamma\cdot\beta)$$

$$= \frac{1}{c} \cdot \left(\hat{x} \cdot \frac{\partial}{\partial t} \left[\frac{\beta_x}{\sqrt{1-\beta^2}} \right] + \hat{y} \cdot \frac{\partial}{\partial t} \left[\frac{\beta_y}{\sqrt{1-\beta^2}} \right] + \hat{z} \cdot \frac{\partial}{\partial t} \left[\frac{\beta_z}{\sqrt{1-\beta^2}} \right] \right)$$

$$\frac{\partial}{\partial ct}(\gamma \cdot \beta) = \frac{1}{c} \cdot \frac{\partial}{\partial t}(\gamma \cdot \beta)$$

$$= \frac{1}{c} \cdot \left(\hat{x} \cdot \left[\frac{1}{\sqrt{1-\beta^2}} \cdot \frac{\partial \beta_x}{\partial t} + \frac{\beta_x}{1-\beta^2} \cdot (1-\beta^2)^{-\frac{1}{2}} \cdot \left(\beta_x \cdot \frac{\partial \beta_x}{\partial t} + \beta_y \cdot \frac{\partial \beta_y}{\partial t} + \beta_z \cdot \frac{\partial \beta_z}{\partial t} \right) \right] \right.$$
$$+ \hat{y} \cdot \left[\frac{1}{\sqrt{1-\beta^2}} \cdot \frac{\partial \beta_y}{\partial t} + \frac{\beta_y}{1-\beta^2} \cdot (1-\beta^2)^{-\frac{1}{2}} \cdot \left(\beta_x \cdot \frac{\partial \beta_x}{\partial t} + \beta_y \cdot \frac{\partial \beta_y}{\partial t} + \beta_z \cdot \frac{\partial \beta_z}{\partial t} \right) \right]$$
$$\left. + \hat{z} \cdot \left[\frac{1}{\sqrt{1-\beta^2}} \cdot \frac{\partial \beta_z}{\partial t} + \frac{\beta_z}{1-\beta^2} \cdot (1-\beta^2)^{-\frac{1}{2}} \cdot \left(\beta_x \cdot \frac{\partial \beta_x}{\partial t} + \beta_y \cdot \frac{\partial \beta_y}{\partial t} + \beta_z \cdot \frac{\partial \beta_z}{\partial t} \right) \right] \right)$$

Equation left side, element (2):

$$(\beta \cdot \nabla) = \beta_x \frac{\partial}{\partial x} + \beta_y \frac{\partial}{\partial y} + \beta_z \frac{\partial}{\partial z}$$
$$(\beta \cdot \nabla)(\gamma \cdot \beta) = \beta_x \cdot \frac{\partial (\gamma \cdot \beta)}{\partial x} + \beta_y \cdot \frac{\partial (\gamma \cdot \beta)}{\partial y} + \beta_z \cdot \frac{\partial (\gamma \cdot \beta)}{\partial z}$$

$$\frac{\partial (\gamma \cdot \beta)}{\partial x} = \frac{\partial}{\partial x} \frac{\beta}{\sqrt{1-\beta^2}}$$
$$= \frac{\partial}{\partial x} \left[\frac{\beta_x \cdot \hat{x} + \beta_y \cdot \hat{y} + \beta_z \cdot \hat{z}}{\sqrt{1-\beta^2}} \right]$$
$$= \hat{x} \cdot \frac{\partial}{\partial x} \left[\frac{\beta_x}{\sqrt{1-\beta^2}} \right] + \hat{y} \cdot \frac{\partial}{\partial x} \left[\frac{\beta_y}{\sqrt{1-\beta^2}} \right] + \hat{z} \cdot \frac{\partial}{\partial x} \frac{\beta_z}{\sqrt{1-\beta^2}}$$

$$\frac{\partial (\gamma \cdot \beta)}{\partial y} = \frac{\partial}{\partial y} \frac{\beta}{\sqrt{1-\beta^2}}$$
$$= \frac{\partial}{\partial y} \left[\frac{\beta_x \cdot \hat{x} + \beta_y \cdot \hat{y} + \beta_z \cdot \hat{z}}{\sqrt{1-\beta^2}} \right]$$

$$= \hat{x} \cdot \frac{\partial}{\partial y}\left[\frac{\beta_x}{\sqrt{1-\beta^2}}\right] + \hat{y} \cdot \frac{\partial}{\partial y}\left[\frac{\beta_y}{\sqrt{1-\beta^2}}\right] + \hat{z} \cdot \frac{\partial}{\partial y}\frac{\beta_z}{\sqrt{1-\beta^2}}$$

$$\frac{\partial(\gamma \cdot \beta)}{\partial z} = \frac{\partial}{\partial z}\frac{\beta}{\sqrt{1-\beta^2}}$$

$$= \frac{\partial}{\partial z}\left[\frac{\beta_x \cdot \hat{x} + \beta_y \cdot \hat{y} + \beta_z \cdot \hat{z}}{\sqrt{1-\beta^2}}\right]$$

$$= \hat{x} \cdot \frac{\partial}{\partial z}\left[\frac{\beta_x}{\sqrt{1-\beta^2}}\right] + \hat{y} \cdot \frac{\partial}{\partial z}\left[\frac{\beta_y}{\sqrt{1-\beta^2}}\right] + \hat{z} \cdot \frac{\partial}{\partial z}\frac{\beta_z}{\sqrt{1-\beta^2}}$$

$$(\beta \cdot \nabla)(\gamma \cdot \beta)$$

$$= \beta_x \cdot \left(\hat{x} \cdot \frac{\partial}{\partial x}\left[\frac{\beta_x}{\sqrt{1-\beta^2}}\right] + \hat{y} \cdot \frac{\partial}{\partial x}\left[\frac{\beta_y}{\sqrt{1-\beta^2}}\right] + \hat{z} \cdot \frac{\partial}{\partial x}\left[\frac{\beta_z}{\sqrt{1-\beta^2}}\right]\right)$$

$$+ \beta_y \cdot \left(\hat{x} \cdot \frac{\partial}{\partial y}\left[\frac{\beta_x}{\sqrt{1-\beta^2}}\right] + \hat{y} \cdot \frac{\partial}{\partial y}\left[\frac{\beta_y}{\sqrt{1-\beta^2}}\right] + \hat{z} \cdot \frac{\partial}{\partial y}\frac{\beta_z}{\sqrt{1-\beta^2}}\right)$$

$$+ \beta_z \cdot \left(\hat{x} \cdot \frac{\partial}{\partial z}\left[\frac{\beta_x}{\sqrt{1-\beta^2}}\right] + \hat{y} \cdot \frac{\partial}{\partial z}\left[\frac{\beta_y}{\sqrt{1-\beta^2}}\right] + \hat{z} \cdot \frac{\partial}{\partial z}\frac{\beta_z}{\sqrt{1-\beta^2}}\right)$$

$$(\beta \cdot \nabla)(\gamma \cdot \beta)$$

$$= \hat{x} \cdot \left(\beta_x \cdot \frac{\partial}{\partial x}\left[\frac{\beta_x}{\sqrt{1-\beta^2}}\right] + \beta_y \cdot \frac{\partial}{\partial y}\left[\frac{\beta_x}{\sqrt{1-\beta^2}}\right] + \beta_z \cdot \frac{\partial}{\partial z}\left[\frac{\beta_x}{\sqrt{1-\beta^2}}\right]\right)$$

$$+ \hat{y} \cdot \left(\beta_x \cdot \frac{\partial}{\partial x}\left[\frac{\beta_y}{\sqrt{1-\beta^2}}\right] + \beta_y \cdot \frac{\partial}{\partial y}\left[\frac{\beta_y}{\sqrt{1-\beta^2}}\right] + \beta_z \cdot \frac{\partial}{\partial z}\left[\frac{\beta_y}{\sqrt{1-\beta^2}}\right]\right)$$

$$+ \hat{z} \cdot \left(\beta_x \cdot \frac{\partial}{\partial x}\left[\frac{\beta_z}{\sqrt{1-\beta^2}}\right] + \beta_y \cdot \frac{\partial}{\partial y}\frac{\beta_z}{\sqrt{1-\beta^2}} + \beta_z \cdot \frac{\partial}{\partial z}\frac{\beta_z}{\sqrt{1-\beta^2}}\right)$$

$$\frac{\partial}{\partial x}\left[\frac{\beta_x}{\sqrt{1-\beta^2}}\right] = \frac{\frac{\partial \beta_x}{\partial x} \cdot \sqrt{1-\beta^2} - \beta_x \cdot \frac{\partial}{\partial x}(\sqrt{1-\beta^2})}{1-\beta^2}$$

$$= \frac{\partial \beta_x}{\partial x} \cdot \frac{1}{\sqrt{1-\beta^2}} - \frac{\beta_x}{1-\beta^2} \cdot \frac{\partial}{\partial x}(\sqrt{1-\beta^2})$$

$$\frac{\partial}{\partial x}(\sqrt{1-\beta^2}) = \frac{\partial}{\partial x}(1-\beta^2)^{\frac{1}{2}}$$

$$= \frac{1}{2} \cdot (1-\beta^2)^{-\frac{1}{2}} \cdot \frac{\partial}{\partial x}(1-\beta^2)$$

$$= \frac{1}{2} \cdot (1-\beta^2)^{-\frac{1}{2}} \cdot \frac{\partial}{\partial x}(1-\beta_x^2-\beta_y^2-\beta_z^2)$$

$$\frac{\partial}{\partial x}\left(\sqrt{1-\beta^2}\right) = -(1-\beta^2)^{-\frac{1}{2}} \cdot \left[\beta_x \cdot \frac{\partial \beta_x}{\partial x} + \beta_y \cdot \frac{\partial \beta_y}{\partial x} + \beta_z \cdot \frac{\partial \beta_z}{\partial x}\right]$$

$$\frac{\partial}{\partial x}\left[\frac{\beta_x}{\sqrt{1-\beta^2}}\right] = \frac{\partial \beta_x}{\partial x} \cdot \frac{1}{\sqrt{1-\beta^2}} + \frac{\beta_x}{1-\beta^2} \cdot (1-\beta^2)^{-\frac{1}{2}} \cdot \left[\beta_x \cdot \frac{\partial \beta_x}{\partial x} + \beta_y \cdot \frac{\partial \beta_y}{\partial x} + \beta_z \cdot \frac{\partial \beta_z}{\partial x}\right]$$

Equation right side, element (1): $\nabla(\phi') = \hat{x} \cdot \frac{\partial \phi'}{\partial x} + \hat{y} \cdot \frac{\partial \phi'}{\partial y} + \hat{z} \cdot \frac{\partial \phi'}{\partial z}$

Equation right side, element (2): $\frac{1}{2 \cdot \gamma} \cdot \nabla(a' \cdot a')$.

The scalar or dot products of vector a' and a', where ϕ is the angle ($\phi \leq 180°$) between vectors. In our case $\phi = 0$ ($\cos(\phi) = 1$). Additionally $a' = \{a'_x, a'_y, a'_z\}$.

$$a'_x = a'_x(x, y, z);\; a'_y = a'_y(x, y, z);\; a'_z = a'_z(x, y, z)$$

$$a' \cdot a' = a'_x \cdot a'_x + a'_y \cdot a'_y + a'_z \cdot a'_z$$

$$\nabla = \hat{x}\frac{\partial}{\partial x} + \hat{y}\frac{\partial}{\partial y} + \hat{z}\frac{\partial}{\partial z}$$

$$\nabla(a' \cdot a') = \left(\hat{x}\frac{\partial}{\partial x} + \hat{y}\frac{\partial}{\partial y} + \hat{z}\frac{\partial}{\partial z}\right)(a'_x \cdot a'_x + a'_y \cdot a'_y + a'_z \cdot a'_z)$$

$$\begin{aligned}\nabla(a' \cdot a') = {} & \hat{x}\frac{\partial}{\partial x}[a'_x \cdot a'_x] + \hat{x}\frac{\partial}{\partial x}[a'_y \cdot a'_y] \\ & + \hat{x}\frac{\partial}{\partial x}[a'_z \cdot a'_z] + \hat{y}\frac{\partial}{\partial y}[a'_x \cdot a'_x] \\ & + \hat{y}\frac{\partial}{\partial y}[a'_y \cdot a'_y] + \hat{y}\frac{\partial}{\partial y}[a'_z \cdot a'_z] + \hat{z}\frac{\partial}{\partial z}[a'_x \cdot a'_x] \\ & + \hat{z}\frac{\partial}{\partial z}[a'_y \cdot a'_y] + \hat{z}\frac{\partial}{\partial z}[a'_z \cdot a'_z]\end{aligned}$$

$$\begin{aligned}\nabla(a' \cdot a') = {} & 2 \cdot \hat{x} \cdot \frac{\partial a'_x}{\partial x} \cdot a'_x + 2 \cdot \hat{x} \cdot \frac{\partial a'_y}{\partial x} \cdot a'_y \\ & + 2 \cdot \hat{x} \cdot \frac{\partial a'_z}{\partial x} \cdot a'_z + 2 \cdot \hat{y} \cdot \frac{\partial a'_x}{\partial y} \cdot a'_x \\ & + 2 \cdot \hat{y} \cdot \frac{\partial a'_y}{\partial y} \cdot a'_y + 2 \cdot \hat{y} \cdot \frac{\partial a'_z}{\partial y} \cdot a'_z + 2 \cdot \hat{z} \cdot \frac{\partial a'_x}{\partial z} \cdot a'_x \\ & + 2 \cdot \hat{z} \cdot \frac{\partial a'_y}{\partial z} \cdot a'_y + 2 \cdot \hat{z} \cdot \frac{\partial a'_z}{\partial z} \cdot a'_z\end{aligned}$$

$$\nabla(a' \cdot a') = 2 \cdot \hat{x} \cdot \left[\frac{\partial a'_x}{\partial x} \cdot a'_x + \frac{\partial a'_y}{\partial x} \cdot a'_y + \frac{\partial a'_z}{\partial x} \cdot a'_z\right]$$

$$+2\cdot\hat{y}\cdot\left[\frac{\partial a'_x}{\partial y}\cdot a'_x+\frac{\partial a'_y}{\partial y}\cdot a'_y+\frac{\partial a'_z}{\partial y}\cdot a'_z\right]$$
$$+2\cdot\hat{z}\cdot\left[\frac{\partial a'_x}{\partial z}\cdot a'_x+\frac{\partial a'_y}{\partial z}\cdot a'_y+\frac{\partial a'_z}{\partial z}\cdot a'_z\right]$$

Equation right side, element (3): $\nabla\frac{P_e}{n_0\cdot m_0\cdot c^2}$

$$\nabla\frac{P_e}{n_0\cdot m_0\cdot c^2}=\left(\hat{x}\frac{\partial}{\partial x}+\hat{y}\frac{\partial}{\partial y}+\hat{z}\frac{\partial}{\partial z}\right)\left(\frac{P_e}{n_0\cdot m_0\cdot c^2}\right)$$
$$=\left(\hat{x}\frac{\partial}{\partial x}+\hat{y}\frac{\partial}{\partial y}+\hat{z}\frac{\partial}{\partial z}\right)\left(\frac{(n_0+n)\cdot\theta}{n_0\cdot m_0\cdot c^2}\right)$$

Instead P_e, the normalized electron thermal-pressure,

$$P_e\to P=\frac{P_e}{m_0\cdot c^2};\ P=\frac{P_e}{m_0\cdot c^2}=\frac{(n_0+n)\cdot\theta}{m_0\cdot c^2}$$
$$\theta=k\cdot T;\ P_e=(n_0+n)\cdot\theta$$

$$\nabla\frac{P_e}{n_0\cdot m_0\cdot c^2}=\left(\hat{x}\frac{\partial}{\partial x}+\hat{y}\frac{\partial}{\partial y}+\hat{z}\frac{\partial}{\partial z}\right)\left(\frac{(n_0+n)\cdot k\cdot T}{n_0\cdot m_0\cdot c^2}\right)$$
$$n=n(x,y,z)$$

$$\nabla\frac{P_e}{n_0\cdot m_0\cdot c^2}=\left(\hat{x}\frac{\partial}{\partial x}+\hat{y}\frac{\partial}{\partial y}+\hat{z}\frac{\partial}{\partial z}\right)\left(\frac{k\cdot T}{m_0\cdot c^2}+\frac{k\cdot T}{n_0\cdot m_0\cdot c^2}\cdot n\right)$$

$$\nabla\frac{P_e}{n_0\cdot m_0\cdot c^2}=\left(\hat{x}\frac{\partial}{\partial x}+\hat{y}\frac{\partial}{\partial y}+\hat{z}\frac{\partial}{\partial z}\right)\left(\frac{k\cdot T}{m_0\cdot c^2}\right)$$
$$+\left(\hat{x}\frac{\partial}{\partial x}+\hat{y}\frac{\partial}{\partial y}+\hat{z}\frac{\partial}{\partial z}\right)\left(\frac{k\cdot T}{n_0\cdot m_0\cdot c^2}\cdot n\right)$$

$$\left(\hat{x}\frac{\partial}{\partial x}+\hat{y}\frac{\partial}{\partial y}+\hat{z}\frac{\partial}{\partial z}\right)\left(\frac{k\cdot T}{m_0\cdot c^2}\right)\to\varepsilon$$
$$\nabla\frac{P_e}{n_0\cdot m_0\cdot c^2}=\left(\hat{x}\frac{\partial}{\partial x}+\hat{y}\frac{\partial}{\partial y}+\hat{z}\frac{\partial}{\partial z}\right)\left(\frac{k\cdot T}{n_0\cdot m_0\cdot c^2}\cdot n\right)$$
$$\nabla\frac{P_e}{n_0\cdot m_0\cdot c^2}=\hat{x}\frac{k\cdot T}{n_0\cdot m_0\cdot c^2}\cdot\frac{\partial n}{\partial x}+\hat{y}\frac{k\cdot T}{n_0\cdot m_0\cdot c^2}\cdot\frac{\partial n}{\partial y}$$
$$+\hat{z}\frac{k\cdot T}{n_0\cdot m_0\cdot c^2}\cdot\frac{\partial n}{\partial z}$$

We separate our main electrons differential equation to three differential equations ($\hat{x}\cdots=\hat{x}\cdots;\ \hat{y}\cdots=\hat{y}\cdots;\ \hat{z}\cdots=\hat{z}\cdots$):

1. $\hat{x} \cdots = \hat{x} \cdots$

$$\frac{\partial}{\partial ct}(\gamma \cdot \beta) + (\beta \cdot \nabla)(\gamma \cdot \beta) = \nabla(\phi') - \frac{1}{2 \cdot \gamma} \cdot \nabla(a' \cdot a') - \nabla \frac{P_e}{n_0 \cdot m_0 \cdot c^2}$$

$$\begin{aligned} \frac{\partial}{\partial ct}(\gamma \cdot \beta) &= \frac{1}{c} \cdot \frac{\partial}{\partial t}(\gamma \cdot \beta) \\ &= \frac{1}{c} \cdot \left(\hat{x} \cdot \left[\frac{1}{\sqrt{1-\beta^2}} \cdot \frac{\partial \beta_x}{\partial t} + \frac{\beta_x}{1-\beta^2} \cdot (1-\beta^2)^{-\frac{1}{2}} \right.\right. \\ &\left.\left. \cdot \left(\beta_x \frac{\partial \beta_x}{\partial t} + \beta_y \frac{\partial \beta_y}{\partial t} + \beta_z \frac{\partial \beta_z}{\partial t} \right) \right] \right) \end{aligned}$$

$$\begin{aligned} (\beta \cdot \nabla)(\gamma \cdot \beta) = \hat{x} \cdot & \left(\beta_x \cdot \frac{\partial}{\partial x}\left[\frac{\beta_x}{\sqrt{1-\beta^2}} \right] + \beta_y \cdot \frac{\partial}{\partial y}\left[\frac{\beta_x}{\sqrt{1-\beta^2}} \right] \right. \\ & \left. + \beta_z \cdot \frac{\partial}{\partial z}\left[\frac{\beta_x}{\sqrt{1-\beta^2}} \right] \right) \end{aligned}$$

$$\nabla(\phi') = \hat{x} \cdot \frac{\partial \phi'}{\partial x}$$

$$\nabla(a' \cdot a') = 2 \cdot \hat{x} \cdot \left[\frac{\partial a'_x}{\partial x} \cdot a'_x + \frac{\partial a'_y}{\partial x} \cdot a'_y + \frac{\partial a'_z}{\partial x} \cdot a'_z \right]$$

$$\nabla \frac{P_e}{n_0 \cdot m_0 \cdot c^2} = \hat{x} \frac{k \cdot T}{n_0 \cdot m_0 \cdot c^2} \cdot \frac{\partial n}{\partial x}$$

We get the electron <u>first</u> differential equation:

$$\begin{aligned} & \frac{1}{c} \cdot \left[\frac{1}{\sqrt{1-\beta^2}} \cdot \frac{\partial \beta_x}{\partial t} + \frac{\beta_x}{1-\beta^2} \cdot (1-\beta^2)^{-\frac{1}{2}} \cdot \left(\beta_x \frac{\partial \beta_x}{\partial t} + \beta_y \frac{\partial \beta_y}{\partial t} + \beta_z \frac{\partial \beta_z}{\partial t} \right) \right] \\ & + \left(\beta_x \cdot \frac{\partial}{\partial x}\left[\frac{\beta_x}{\sqrt{1-\beta^2}} \right] + \beta_y \cdot \frac{\partial}{\partial y}\left[\frac{\beta_x}{\sqrt{1-\beta^2}} \right] + \beta_z \cdot \frac{\partial}{\partial z}\left[\frac{\beta_x}{\sqrt{1-\beta^2}} \right] \right) \\ & = \frac{\partial \phi'}{\partial x} - \sqrt{1-\beta^2} \cdot \left[\frac{\partial a'_x}{\partial x} \cdot a'_x + \frac{\partial a'_y}{\partial x} \cdot a'_y + \frac{\partial a'_z}{\partial x} \cdot a'_z \right] - \frac{k \cdot T}{n_0 \cdot m_0 \cdot c^2} \cdot \frac{\partial n}{\partial x} \end{aligned}$$

2. $\hat{y} \cdot \ldots = \hat{y} \cdot \ldots$

$$\frac{\partial}{\partial ct}(\gamma \cdot \beta) + (\beta \cdot \nabla)(\gamma \cdot \beta) = \nabla(\phi') - \frac{1}{2 \cdot \gamma} \cdot \nabla(a' \cdot a') - \nabla \frac{P_e}{n_0 \cdot m_0 \cdot c^2}$$

$$\frac{\partial}{\partial ct}(\gamma \cdot \beta) = \frac{1}{c} \cdot \hat{y} \cdot \left[\frac{1}{\sqrt{1-\beta^2}} \cdot \frac{\partial \beta_y}{\partial t} + \frac{\beta_y}{1-\beta^2} \cdot (1-\beta^2)^{-\frac{1}{2}} \cdot \left(\beta_x \frac{\partial \beta_x}{\partial t} + \beta_y \frac{\partial \beta_y}{\partial t} + \beta_z \frac{\partial \beta_z}{\partial t} \right) \right]$$

$$(\beta \cdot \nabla)(\gamma \cdot \beta) = \hat{y} \cdot \left(\beta_x \cdot \frac{\partial}{\partial x}\left[\frac{\beta_y}{\sqrt{1-\beta^2}} \right] + \beta_y \cdot \frac{\partial}{\partial y}\left[\frac{\beta_y}{\sqrt{1-\beta^2}} \right] + \beta_z \cdot \frac{\partial}{\partial z}\left[\frac{\beta_y}{\sqrt{1-\beta^2}} \right] \right)$$

$$\nabla(\phi') = \hat{y} \cdot \frac{\partial \phi'}{\partial y}$$

$$\nabla(a' \cdot a') = 2 \cdot \hat{y} \cdot \left[\frac{\partial a'_x}{\partial y} \cdot a'_x + \frac{\partial a'_y}{\partial y} \cdot a'_y + \frac{\partial a'_z}{\partial y} \cdot a'_z \right]$$

$$\nabla \frac{P_e}{n_0 \cdot m_0 \cdot c^2} = \hat{y} \frac{k \cdot T}{n_0 \cdot m_0 \cdot c^2} \cdot \frac{\partial n}{\partial y}$$

We get the electron second differential equation:

$$\frac{1}{c} \cdot \left[\frac{1}{\sqrt{1-\beta^2}} \cdot \frac{\partial \beta_y}{\partial t} + \frac{\beta_y}{1-\beta^2} \cdot (1-\beta^2)^{-\frac{1}{2}} \cdot \left(\beta_x \frac{\partial \beta_x}{\partial t} + \beta_y \frac{\partial \beta_y}{\partial t} + \beta_z \frac{\partial \beta_z}{\partial t} \right) \right]$$
$$+ \beta_x \cdot \frac{\partial}{\partial x}\left[\frac{\beta_y}{\sqrt{1-\beta^2}} \right] + \beta_y \cdot \frac{\partial}{\partial y}\left[\frac{\beta_y}{\sqrt{1-\beta^2}} \right] + \beta_z \cdot \frac{\partial}{\partial z}\left[\frac{\beta_y}{\sqrt{1-\beta^2}} \right]$$
$$= \frac{\partial \phi'}{\partial y} - \sqrt{1-\beta^2} \cdot \left[\frac{\partial a'_x}{\partial y} \cdot a'_x + \frac{\partial a'_y}{\partial y} \cdot a'_y + \frac{\partial a'_z}{\partial y} \cdot a'_z \right] - \frac{k \cdot T}{n_0 \cdot m_0 \cdot c^2} \cdot \frac{\partial n}{\partial y}$$

3. $\hat{z} \cdot \cdots = \hat{z} \cdot \cdots$

$$\frac{\partial}{\partial ct}(\gamma \cdot \beta) + (\beta \cdot \nabla)(\gamma \cdot \beta) = \nabla(\phi') - \frac{1}{2 \cdot \gamma} \cdot \nabla(a' \cdot a') - \nabla \frac{P_e}{n_0 \cdot m_0 \cdot c^2}$$

$$\frac{\partial}{\partial ct}(\gamma \cdot \beta) = \frac{1}{c} \cdot \frac{\partial}{\partial t}(\gamma \beta)$$
$$= \frac{1}{c} \cdot \hat{z} \cdot \left[\frac{1}{\sqrt{1-\beta^2}} \cdot \frac{\partial \beta_z}{\partial t} + \frac{\beta_z}{1-\beta^2} \cdot (1-\beta^2)^{-\frac{1}{2}} \right.$$

$$\cdot\left(\beta_x\frac{\partial\beta_x}{\partial t}+\beta_y\frac{\partial\beta_y}{\partial t}+\beta_z\frac{\partial\beta_z}{\partial t}\right)\Bigg]$$

$$(\beta\cdot\nabla)(\gamma\cdot\beta)=\hat{z}\cdot\left(\beta_x\cdot\frac{\partial}{\partial x}\left[\frac{\beta_z}{\sqrt{1-\beta^2}}\right]+\beta_y\cdot\frac{\partial}{\partial y}\frac{\beta_z}{\sqrt{1-\beta^2}}+\beta_z\cdot\frac{\partial}{\partial z}\frac{\beta_z}{\sqrt{1-\beta^2}}\right)$$

$$\nabla(\phi')=\hat{z}\cdot\frac{\partial\phi'}{\partial z}$$

$$\nabla(a'\cdot a')=2\cdot\hat{z}\cdot\left[\frac{\partial a'_x}{\partial z}\cdot a'_x+\frac{\partial a'_y}{\partial z}\cdot a'_y+\frac{\partial a'_z}{\partial z}\cdot a'_z\right]$$

$$\nabla\frac{P_e}{n_0\cdot m_0\cdot c^2}=\hat{z}\frac{k\cdot T}{n_0\cdot m_0\cdot c^2}\cdot\frac{\partial n}{\partial z}$$

We get the electron third differential equation:

$$\frac{1}{c}\cdot\left[\frac{1}{\sqrt{1-\beta^2}}\cdot\frac{\partial\beta_z}{\partial t}+\frac{\beta_z}{1-\beta^2}\cdot(1-\beta^2)^{-\frac{1}{2}}\cdot\left(\beta_x\frac{\partial\beta_x}{\partial t}+\beta_y\frac{\partial\beta_y}{\partial t}+\beta_z\frac{\partial\beta_z}{\partial t}\right)\right]$$
$$+\left(\beta_x\cdot\frac{\partial}{\partial x}\left[\frac{\beta_z}{\sqrt{1-\beta^2}}\right]+\beta_y\cdot\frac{\partial}{\partial y}\frac{\beta_z}{\sqrt{1-\beta^2}}+\beta_z\cdot\frac{\partial}{\partial z}\frac{\beta_z}{\sqrt{1-\beta^2}}\right)$$
$$=\frac{\partial\phi'}{\partial z}-\sqrt{1-\beta^2}\cdot\left[\frac{\partial a'_x}{\partial z}\cdot a'_x+\frac{\partial a'_y}{\partial z}\cdot a'_y+\frac{\partial a'_z}{\partial z}\cdot a'_z\right]-\frac{k\cdot T}{n_0\cdot m_0\cdot c^2}\cdot\frac{\partial n}{\partial z}$$

We can summery our electron three differential equations (Table A.2):

1. $$\frac{1}{c}\cdot\left[\frac{1}{\sqrt{1-\beta^2}}\cdot\frac{\partial\beta_x}{\partial t}+\frac{\beta_x}{1-\beta^2}\cdot(1-\beta^2)^{-\frac{1}{2}}\cdot\left(\beta_x\frac{\partial\beta_x}{\partial t}+\beta_y\frac{\partial\beta_y}{\partial t}+\beta_z\frac{\partial\beta_z}{\partial t}\right)\right]$$
$$+\left(\beta_x\cdot\frac{\partial}{\partial x}\left[\frac{\beta_x}{\sqrt{1-\beta^2}}\right]+\beta_y\cdot\frac{\partial}{\partial y}\left[\frac{\beta_x}{\sqrt{1-\beta^2}}\right]+\beta_z\cdot\frac{\partial}{\partial z}\left[\frac{\beta_x}{\sqrt{1-\beta^2}}\right]\right)$$
$$=\frac{\partial\phi'}{\partial x}-\sqrt{1-\beta^2}\cdot\left[\frac{\partial a'_x}{\partial x}\cdot a'_x+\frac{\partial a'_y}{\partial x}\cdot a'_y+\frac{\partial a'_z}{\partial x}\cdot a'_z\right]-\frac{k\cdot T}{n_0\cdot m_0\cdot c^2}\cdot\frac{\partial n}{\partial x}$$

2. $$\frac{1}{c}\cdot\left[\frac{1}{\sqrt{1-\beta^2}}\cdot\frac{\partial\beta_y}{\partial t}+\frac{\beta_y}{1-\beta^2}\cdot(1-\beta^2)^{-\frac{1}{2}}\cdot\left(\beta_x\frac{\partial\beta_x}{\partial t}+\beta_y\frac{\partial\beta_y}{\partial t}+\beta_z\frac{\partial\beta_z}{\partial t}\right)\right]$$
$$+\beta_x\cdot\frac{\partial}{\partial x}\left[\frac{\beta_y}{\sqrt{1-\beta^2}}\right]+\beta_y\cdot\frac{\partial}{\partial y}\left[\frac{\beta_y}{\sqrt{1-\beta^2}}\right]+\beta_z\cdot\frac{\partial}{\partial z}\left[\frac{\beta_y}{\sqrt{1-\beta^2}}\right]$$

Table A.2 Derivatives $\left(\frac{\partial}{\partial x}, \frac{\partial}{\partial y}, \frac{\partial}{\partial z}\right)$ of $\frac{\beta_x}{\sqrt{1-\beta^2}}, \frac{\beta_y}{\sqrt{1-\beta^2}}, \frac{\beta_z}{\sqrt{1-\beta^2}}$

Item	Expression
$\frac{\partial}{\partial x}\left[\frac{\beta_x}{\sqrt{1-\beta^2}}\right]$	$\frac{\partial}{\partial x}\left[\frac{\beta_x}{\sqrt{1-\beta^2}}\right] = \frac{\partial\beta_x}{\partial x}\cdot\frac{1}{\sqrt{1-\beta^2}} + \frac{\beta_x}{1-\beta^2}\cdot(1-\beta^2)^{-\frac{1}{2}}\cdot[\beta_x\cdot\frac{\partial\beta_x}{\partial x}+\beta_y\cdot\frac{\partial\beta_y}{\partial x}+\beta_z\cdot\frac{\partial\beta_z}{\partial x}]$
$\frac{\partial}{\partial y}\left[\frac{\beta_x}{\sqrt{1-\beta^2}}\right]$	$\frac{\partial}{\partial y}\left[\frac{\beta_x}{\sqrt{1-\beta^2}}\right] = \frac{\partial\beta_x}{\partial y}\cdot\frac{1}{\sqrt{1-\beta^2}} + \frac{\beta_x}{1-\beta^2}\cdot(1-\beta^2)^{-\frac{1}{2}}\cdot[\beta_x\cdot\frac{\partial\beta_x}{\partial y}+\beta_y\cdot\frac{\partial\beta_y}{\partial y}+\beta_z\cdot\frac{\partial\beta_z}{\partial y}]$
$\frac{\partial}{\partial z}\left[\frac{\beta_x}{\sqrt{1-\beta^2}}\right]$	$\frac{\partial}{\partial z}\left[\frac{\beta_x}{\sqrt{1-\beta^2}}\right] = \frac{\partial\beta_x}{\partial z}\cdot\frac{1}{\sqrt{1-\beta^2}} + \frac{\beta_x}{1-\beta^2}\cdot(1-\beta^2)^{-\frac{1}{2}}\cdot[\beta_x\cdot\frac{\partial\beta_x}{\partial z}+\beta_y\cdot\frac{\partial\beta_y}{\partial z}+\beta_z\cdot\frac{\partial\beta_z}{\partial z}]$
$\frac{\partial}{\partial x}\left[\frac{\beta_y}{\sqrt{1-\beta^2}}\right]$	$\frac{\partial}{\partial x}\left[\frac{\beta_y}{\sqrt{1-\beta^2}}\right] = \frac{\partial\beta_y}{\partial x}\cdot\frac{1}{\sqrt{1-\beta^2}} + \frac{\beta_y}{1-\beta^2}\cdot(1-\beta^2)^{-\frac{1}{2}}\cdot[\beta_x\cdot\frac{\partial\beta_x}{\partial x}+\beta_y\cdot\frac{\partial\beta_y}{\partial x}+\beta_z\cdot\frac{\partial\beta_z}{\partial x}]$
$\frac{\partial}{\partial y}\left[\frac{\beta_y}{\sqrt{1-\beta^2}}\right]$	$\frac{\partial}{\partial y}\left[\frac{\beta_y}{\sqrt{1-\beta^2}}\right] = \frac{\partial\beta_y}{\partial y}\cdot\frac{1}{\sqrt{1-\beta^2}} + \frac{\beta_y}{1-\beta^2}\cdot(1-\beta^2)^{-\frac{1}{2}}\cdot[\beta_x\cdot\frac{\partial\beta_x}{\partial y}+\beta_y\cdot\frac{\partial\beta_y}{\partial y}+\beta_z\cdot\frac{\partial\beta_z}{\partial y}]$
$\frac{\partial}{\partial z}\left[\frac{\beta_y}{\sqrt{1-\beta^2}}\right]$	$\frac{\partial}{\partial z}\left[\frac{\beta_y}{\sqrt{1-\beta^2}}\right] = \frac{\partial\beta_y}{\partial z}\cdot\frac{1}{\sqrt{1-\beta^2}} + \frac{\beta_y}{1-\beta^2}\cdot(1-\beta^2)^{-\frac{1}{2}}\cdot[\beta_x\cdot\frac{\partial\beta_x}{\partial z}+\beta_y\cdot\frac{\partial\beta_y}{\partial z}+\beta_z\cdot\frac{\partial\beta_z}{\partial z}]$
$\frac{\partial}{\partial x}\left[\frac{\beta_z}{\sqrt{1-\beta^2}}\right]$	$\frac{\partial}{\partial x}\left[\frac{\beta_z}{\sqrt{1-\beta^2}}\right] = \frac{\partial\beta_z}{\partial x}\cdot\frac{1}{\sqrt{1-\beta^2}} + \frac{\beta_z}{1-\beta^2}\cdot(1-\beta^2)^{-\frac{1}{2}}\cdot[\beta_x\cdot\frac{\partial\beta_x}{\partial x}+\beta_y\cdot\frac{\partial\beta_y}{\partial x}+\beta_z\cdot\frac{\partial\beta_z}{\partial x}]$
$\frac{\partial}{\partial y}\frac{\beta_z}{\sqrt{1-\beta^2}}$	$\frac{\partial}{\partial y}\left[\frac{\beta_z}{\sqrt{1-\beta^2}}\right] = \frac{\partial\beta_z}{\partial y}\cdot\frac{1}{\sqrt{1-\beta^2}} + \frac{\beta_z}{1-\beta^2}\cdot(1-\beta^2)^{-\frac{1}{2}}\cdot[\beta_x\cdot\frac{\partial\beta_x}{\partial y}+\beta_y\cdot\frac{\partial\beta_y}{\partial y}+\beta_z\cdot\frac{\partial\beta_z}{\partial y}]$
$\frac{\partial}{\partial z}\frac{\beta_z}{\sqrt{1-\beta^2}}$	$\frac{\partial}{\partial z}\left[\frac{\beta_z}{\sqrt{1-\beta^2}}\right] = \frac{\partial\beta_z}{\partial z}\cdot\frac{1}{\sqrt{1-\beta^2}} + \frac{\beta_z}{1-\beta^2}\cdot(1-\beta^2)^{-\frac{1}{2}}\cdot[\beta_x\cdot\frac{\partial\beta_x}{\partial z}+\beta_y\cdot\frac{\partial\beta_y}{\partial z}+\beta_z\cdot\frac{\partial\beta_z}{\partial z}]$

$$= \frac{\partial\phi'}{\partial y} - \sqrt{1-\beta^2}\cdot\left[\frac{\partial a'_x}{\partial y}\cdot a'_x + \frac{\partial a'_y}{\partial y}\cdot a'_y + \frac{\partial a'_z}{\partial y}\cdot a'_z\right] - \frac{k\cdot T}{n_0\cdot m_0\cdot c^2}\cdot\frac{\partial n}{\partial y}$$

3. $$\frac{1}{c}\cdot\left[\frac{1}{\sqrt{1-\beta^2}}\cdot\frac{\partial\beta_z}{\partial t} + \frac{\beta_z}{1-\beta^2}\cdot(1-\beta^2)^{-\frac{1}{2}}\cdot\left(\beta_x\frac{\partial\beta_x}{\partial t}+\beta_y\frac{\partial\beta_y}{\partial t}+\beta_z\frac{\partial\beta_z}{\partial t}\right)\right]$$
$$+\left(\beta_x\cdot\frac{\partial}{\partial x}\left[\frac{\beta_z}{\sqrt{1-\beta^2}}\right]+\beta_y\cdot\frac{\partial}{\partial y}\frac{\beta_z}{\sqrt{1-\beta^2}}+\beta_z\cdot\frac{\partial}{\partial z}\frac{\beta_z}{\sqrt{1-\beta^2}}\right)$$
$$= \frac{\partial\phi'}{\partial z} - \sqrt{1-\beta^2}\cdot\left[\frac{\partial a'_x}{\partial z}\cdot a'_x + \frac{\partial a'_y}{\partial z}\cdot a'_y + \frac{\partial a'_z}{\partial z}\cdot a'_z\right] - \frac{k\cdot T}{n_0\cdot m_0\cdot c^2}\cdot\frac{\partial n}{\partial z}$$

We can summery the Laser-Plasma and X-ray interaction—main parameters (Table A.3).

Table A.3 Laser-plasma and X-ray interaction—main parameters

Symbol	Description
m_e, m	Electron mass
m_0	Electron rest mass
m_i	Ion massIon mass
T_s	Kinetic temperature of particle (s)
V_s	Particle (s) velocity
$\langle V_s^2 \rangle$	Particle (s) velocity Ensemble average
n_s	Number of particles (s) per cubic meter [density of particle (s)]
V_{ts}	Thermal speed of particle (s)
V_{ti}	Ion thermal speed
V_{te}	Electron thermal speed
λ_D	Debye length
n_j	Density of particle (j) in plasma (can be electron or ion)
q_j	Electric charge of particle (j) in plasma (can be electron or ion)
T_j	Temperature of particle (j) in plasma (can be electron or ion)
e, q_e	Electron charge
T_e	Electron kinetic temperature
n_e	Electron density (electron plasma density)
k_B, k	Boltzman constant
ω_{pe}, ω_p	Plasma frequency of the ionized gaz
$v_{e,90}$	The velocity which electron is scattered by 90° from the original direction (caused by contradictions inside the plasma)
$\frac{1}{v_{e,90}}$	The time it takes electron to scatter by 90° from the original direction (caused by contradictions)
N_D	Number of particles on Debye sphere (Debye number)
η	Index of refraction of the optical plasma
η_0	Index of refraction, electric field (E) independent term
η_2	Index of refraction, electric field (E) dependent term
Γ	Plasma coupling parameter
Γ_s	Plasma coupling parameter for multi component plasma, species (s)
P_{cr}	Critical power which the diffraction spreading of the optical beam in media is balanced by nonlinear refraction
λ_0	Laser wavelength
$\lambda_\perp$	Modulation wavelength
P	Laser power
F_P	Ponderomotive force
$\alpha_{k\perp}$	Spatial growth rate
$k_\perp$	Intensity modulation wave number

(continued)

Table A.3 (continued)

Symbol	Description
Z	Average ionization potential Average ionization potential
$\ln(\Lambda)$	Coulomb logarithm
Λ, ε_p	Plasma parameter $\left(\varepsilon_p = \frac{1}{\Lambda}\right)$
I_0	Zero order (unperturbed) intensity (Laser)
r_e	Classical electron radius
r_0	Radius of the intensity perturbation
κ	Electron thermal conductivity
T_i	Ion acoustic transit time
T_{cr}	Thermal conduction transit timeThermal conduction transit time
λ_{mfp}	Electron conduction mean free path
n_c	Channel particle density
α_{PF}	Ponderomotive grow rate term
α_{TF}	Thermal pressure grow rate term Thermal pressure grow rate term
λ_0	Laser wavelength
k_0	Laser wavenumber
α_{pm}	Plasma growth rate
ω	Laser frequency
γ	Lorentz factor (relativistic mass factor)Lorentz factor (relativistic mass factor)
I	Laser intensityLaser intensity
β	Ratio of v, electron velocity (v_x, v_y, v_z) to speed of light (c) $\beta = \left\{\frac{v_x}{c}, \frac{v_y}{c}, \frac{v_z}{c}\right\}$
$n, n(z, t)$	Plasma density (density of electron)
n_0	Ambient plasma density (local average of n)Ambient plasma density (local average of n)
$n^{(0)}, n^{(0)}(r)$	Initial plasma density (density of electron) prior to the laser interaction (it is function of radial position)
$\rho^{(0)}, \rho^{(0)}(r)$	Initial plasma density profile prior to the laser interaction (it is function of radial position)
A, A(z, t)	Transverse vector (laser)
$\Phi(z, t), \Phi$	Scalar potentials
k_p	Plasma wavenumber
$\{a, \phi\}$	Normalized vector potential
a	Vector potential of the radiation field of the laser
$\phi, \phi(r)$	Scalar electrostatic potential of the plasmas
$n(z, t)$	Plasma density
v	Electron velocity
λ_p	Plasma wavelength

(continued)

Table A.3 (continued)

Symbol	Description
L_{acc}	Acceleration length
$\beta_{\perp}$	Normalized transverse plasma fluid velocity
n, β_z, γ	Plasma fluid quantities (macroscopic plasma quantities)
β_t	Normalized transformation theory
r_s	Radiation spot size
a, ϕ	Fields which drive the plasma
τ_L	Laser pulse duration
τ_e	Characteristic time
τ_d	Laser beam diffraction time (transverse spreading time)
a_L	Laser complex field
$\|a_L\|$	Characteristic spatial variation in the laser envelope
ϕ_s	Slow part of the scalar potential ϕ
ϕ_f	Rapidly varying part of ϕ
L	Envelope scale length (length scale)
V_E	Erosion velocity back through the body of the pulse
Z_R	Vacuum Rayleigh length
$\frac{\Delta\omega}{\omega}$	Frequency variation within the pulse
$\|\Delta\omega\|$	Frequency shift approaches the laser frequency
ω_0	Frequency of the pump (ionizing) laser
$\omega(z)$	Frequency of the secondary pulse laser
$\omega(0)$	Initial frequency of the secondary laser pulse
d	Spatial gradient length of the ionization front
z	Interaction length
ω_s	Frequency of the reflected radiation
a_f	Final value of the laser's vector potential
ε_f	Electron final energy (initially at rest)
ω_i	Frequency of the incident radiation
L_{op}	Optimal pulse length
ψ	Real and denotes the slowly varying phase shift
σ_z	Proton Burch length
T, T_e	Electron temperature in the plasma
P	Normalized electron thermal-pressure
P_e	Electron thermal-pressure
a_{x-ray}	Vector potential of the radiation field of the x-laser
ϕ_{x-ray}	Scalar electrostatic potential of the plasmas cause by x-ray laser
A_{x-ray}	Transverse vector (x-laser)

(continued)

Table A.3 (continued)

Symbol	Description
Φ_{x-ray}	Scalar potentials (x-laser)
$\beta_{x-ray\perp}$	Normalized transverse plasma fluid velocity cause by x-laser
τ	Time scale
Λ	Plasma parameter
r_d	Average distance between particles
r_c	Particles distance of closest approach
r_L	Gaussian radial profile
$\delta\rho$	Total perturbed charge density
$\delta\phi$	The amount of equilibrium perturbed potential
$\delta\rho_{ext}$	Small localized charge density
δn_i	Perturbed ion density
δn_e	Perturbed electron density
τ_p	Plasma period
ε_0	Permittivity of free space
$\langle r\rangle$	Typical inter particle distance
E_c	Coulomb energy (per particle)
ρ	Electron charge density [Coulomb/m^3]
J	Current density vector
v_z	Electron velocity vector in z direction
u	Normalized fluid momentum
p	Fluid momentum
α	Degree of plasma ionization
n_i	Number density of ions
n_n	Number density of neutral atoms
<Z>	Average charge state of the ions

References

[1] K.J. Gaffney, H.N. Chapman, Imaging atomic structure and dynamics with ultrafast X-ray scattering. Science **316** (2007)
[2] C. Bostedt, J.D. Bozek et al., Ultra-fast and ultra-intense x-ray sciences: first results from the Linac coherent light source free-electron laser. J. Phys. B At. Mol. Opt. Phys. **46** (2013)
[3] R.C. Elton, *X-ray Lasers*, 1st edn. (Academic Press, Cambridge, 1990)

Appendix B
Plasma Diagnostics Fundamental

Plasma is a form of matter in which many of the electrons wonder around freely among the nuclei of the atoms. Plasma has been called the fourth state of matter, the other three being solid, liquid and gas. Normally, the electrons in a solid, liquid or gaseous sample of matter stay with the same atomic nucleus. Some electrons can move from atom to atom if an electrical current flows in a solid or liquid, but the motion occurs as short jumps by individual electrons between adjacent nuclei. In plasma, a significant number of electrons have such high energy level that no nucleus can hold them. An atom that has lost some of its electrons, thereby attaining an electric charge, is an ion. When a gas is subjected to heat or an electric field, some of its atoms become ions, and the gas is said to be ionized. An ionized gas, unlike a gas in its normal condition, can conduct electrical current to a limited extent. If the heat or electric field becomes extreme, many of the atoms become ions. The resulting super-ionized gas is plasma, which can conduct a large and sustained electric current. The behavior and properties of plasmas are very useful in many applications include electric lamps, lasers, medical devices, energy converters, water purifiers, and flat-panel video displays. Laser plasma interaction is a very investigate area among many research centers. Basically, laser is a coherent and focused beam of photons, coherent, in this context, mean that it is all one wavelength, unlike ordinary light which showers on us in many wavelengths. The acronym laser stands for "light amplification by stimulated emission of radiation. Lasers work as a result of resonant effects. The output of a laser is a coherent electromagnetic field. In a coherent beam of electromagnetic energy, all the waves have the same frequency and phase. In a basic laser, a chamber called a cavity is designed to internally reflect infrared (IR), visible-light, or ultraviolet (UV) waves and they reinforce each other. The cavity can contain gases, liquids, or solids. The choice of cavity material determines the wavelength of the output. At each end of the cavity, there is a mirror. One mirror is totally reflective, allowing none of the energy to pass through; the other mirror is partially reflective, allowing approximately 5 percent of the energy to pass through. Pumping is the energy which introduced into the cavity from an external source. As a result of pumping, an electromagnetic field appears inside the laser cavity at the natural (resonant) frequency of the atoms of the material that fills the cavity. The waves reflect

O. Aluf, *Advance Elements of Laser Circuits and Systems*,
https://doi.org/10.1007/978-3-030-64103-0

back and forth between the mirrors. The length of the cavity is such that the reflected and re-reflected wave fronts reinforce each other in phase at the natural frequency of the cavity substance. Electromagnetic waves at this resonant frequency emerge from the end of the cavity having the partially-reflective mirror. The output appears as a continuous beam, or as a series of brief, intense pulses. Lasers have found a tremendous variety of uses in electronics, computer hardware, medicine, and experimental science. High intensity laser produced plasmas are capable of accelerating particles to high energies over very short distances. Laser beam is able to drive the accelerating structure, called a wake field, over a distance greater than the length over which it would normally propagate because it is constrained from defocusing by the wake field. As beam propagates it changes from being circular in cross section to being slightly elliptical, which indicated that the electron beam is initially generated behind the laser beam but as it accelerates moves forward, it gains energy from the electric field of the laser. Broadband radiation, meanwhile, is emitted transverse to the motion of the electrons when they are initially trapped and accelerated. Additionally popular experiment is the physics of neighboring laser produced plasmas, such as those generated by the multiple beams heating system in ICF (Inertial Confinement Fusion) experiments. Each laser-target interaction produces a plume of plasma with an azimuthal magnetic field. The magnetic reconnection between the magnetic fields of adjacent plasma plumes results in extraneous heating of a collision region far separated from the laser sources that could affect the illumination and symmetry of ICF targets. There is other area of the production and transport of high current, relativistic electron beams for use in fast ignition ICF. The long pulse drive lasers compress a DT fuel capsule, and then a high intensity laser pulse is used to produce an e-beam that provides sufficient energy to ignite the fuel. In idealized plasma consisting of an equal number of electrons, with mass m_e and charge (−e), when (e) denotes the magnitude of the electron charge, and ions, with mass m_i and charge (+e). The system has attained thermal equilibrium, T_s denote a kinetic temperature measured in energy units (Joules), $T_s = \frac{1}{3} \cdot m_s \cdot \langle V_s^2 \rangle$. V is a particle speed, and the angular brackets denote an ensemble average. The kinetic temperature of species (s) is essentially the average kinetic energy of particles of this species. In plasma physics, kinetic temperature is invariably measured in electron-volts when one joule is equivalent to 6.24×10^{18} eV. Quasi-neutrality demands that $n_i \simeq n_e \equiv n$, where n_s is the number density (number of particles per cubic meter) of species (s). Assuming that both ions and electrons are characterized by the same T, and we estimate typical particle speeds via the so-called thermal speed v_{th}, $v_{th} = \sqrt{\frac{2 \cdot T}{m_s}}$. The ion thermal speed is usually far smaller than the electron thermal speed v_{ti}, $v_{ti} \simeq \sqrt{\frac{m_e}{m_i}} \cdot v_{te}$. Parameters n and T are generally functions of position in plasma. We can define the average distance between particles r_d, $r_d = \frac{1}{\sqrt[3]{n}}$ and the distance of closest approach $r_c \equiv \frac{e^2}{4 \cdot \pi \cdot \varepsilon_0 \cdot T}$. Basically, plasma is a state of matter in which an ionized gaseous substance becomes highly electrically conductive to the point that long-range electric and magnetic fields dominate the behavior of the matter. Plasma mostly does not naturally exist on earth's surface under normal conditions, and must be artificially generated from neutral gases. Plasma is an electrically neutral medium of unbound positive and negative particles (over charge of a plasma is roughly zero). Although these particles are unbound, they are not "free" in the sense of not experiencing focus. Moving charges particles generate an electric current within a magnetic field, and any movement of a charges plasma particle affects and is affected by the fields created by the other charges. There are three factors which define a plasma: (1) The plasma approximation—the plasma approximation applies when the plasma parameter Λ, representing the number of charge carriers within a sphere (Debye sphere whose radius is the Debye screening length) surrounding a given charged particles, is sufficiently high as to shield the electrostatic influence of the particle outside of the sphere. (2) Bulk interaction—The Debye screening length is short compared to the physical size of the plasma. This criterion means that interactions in the bulk of the plasma are more important than those at its edges, where boundary effects may take place. When this criterion is satisfied the plasma is quasi neutral. (3) Plasma frequency—The electron plasma frequency which measuring plasma oscillations of the electrons is large compared to the electron-neutral collision frequency (measuring frequency of collisions between electrons and neutral particles). When this condition is valid, electrostatic interactions dominate over the process of ordinary gas kinetics. Plasma exists if there is ionization. The term "plasma density" refers to the "electron density", that is, the number of free electrons per unit volume. The degree of ionization of a plasma is the proportion of atoms that have lost or gained electrons, and is controlled mostly by the temperature. A partially ionized gas in which as little as 1% of the particles are ionized and can have the characteristics of plasma (response to magnetic fields and high electrical conductivity). We define the degree of ionization as (α) parameter, $\alpha = \frac{n_i}{n_i + n_n}$, where n_i is the number density of ions and n_n is the number density of neutral atoms. The electron density is related to this by the average charge state $\langle z \rangle$ of the ions through $n_e = \langle z \rangle \cdot n_i$, where n_e is the number density of electrons. Plasma temperature is commonly measured in Kelvins or electron-volts and is a measure of the thermal kinetic energy per particle. High temperatures are usually needed to sustain ionization, which is a defining feature of plasma. The degree of plasma ionization is determined by the electron temperature relative to the ionization energy. At low temperatures, ions and electrons tend to recombine into bound states—atoms—and the plasma will eventually become a gas. Based on the relative temperatures of the electrons, ions and neutrals, plasmas are classified as "thermal" or "non-thermal". Thermal plasmas have electrons and the heavy particles at the same temperature. They are in thermal equilibrium with each other. Non thermal plasmas on the other hand have the ions and neutrals at a much lower temperature, whereas electrons are much "hotter" ($T_e \gg T_n$). A plasma is sometimes referred to as being "Hot" if it is nearly fully ionized, or "Cold" if only a small fraction ($\approx 1\%$) of the gas molecules are ionized. Plasmas utilized in "plasma technology" are usually cold plasmas in the sense that only a small fraction of the gas molecules are ionized. The plasmas are very good electrical conductors and the electric potential play an important role. "Plasma potential" or the "space potential" is the potential as it exists on average in t he space between charged particles. If an electrode is inserted into plasma, its potential lies considerably below the plasma potential due to Debye sheath. The good electrical conductivity of plasmas makes their electric fields very small. The term "quasi-neutrality" states that the density of negative charges is approximately equal to the density of positive charges over large volumes of the plasma ($n_e = \langle z \rangle \cdot n_i$), the scale of the Debye length there can be charge imbalance. If double layers are formed, the charge separation can extend some tens of Debye lengths. The magnitude of the potentials and electric fields must be determined by means other than simply finding the net charge density. We assume that the electrons satisfy the Boltzmann relation $n_e \propto e^{\frac{e \cdot \Phi}{k_B \cdot T_e}}$. When differentiating this relation gives a means to calculate the electric field from the density: $\overrightarrow{E} = \left(\frac{k_B \cdot T_e}{e}\right) \cdot \left(\frac{\nabla n_e}{n_e}\right)$. Debye screening prevents electric fields from directly affecting the plasma over large distances. The existence of charged particles causes the plasma to generate, and be affected by, magnetic fields. Magnetized plasma is related to plasma with a magnetic field strong enough to influence the motion of the charged particles. Is that a particle on average completes at least one gyration around the magnetic field before making a collision $\frac{\omega_{ce}}{\nu_{coll}} > 1$, where ω_{ce} is the "electron gyro-frequency" and ν_{coll} is the "electron collision rate". Can be a situation where the electrons are magnetized while the ions are not. Magnetized plasmas are anisotropic and their properties in the direction parallel to the magnetic field are different from those appendicular to it. The electric field associated with a plasma moving in a magnetic field is given by $E = -v \times B$ (where E is the electric field, v is the velocity, and B is the magnetic field). The electric fields in plasmas are usually small due to the high conductivity. Plasma diagnostics are methods, instruments, and experimental techniques used to measure properties of plasma, such as plasma component's density, distribution function over energy (temperature) [1]. We get plasma parameters by their spatial profiles and dynamics. One way to do the inspection is by using probes. There are some probes in use to do the plasma inspection like Langmuir probe, Ball-pen probe and magnetic (B-dot) probe [2–4]. We use also a conventional Faraday cup for measurements of ion (or electron) flows from plasma boundaries and for mass spectrometry. Measurements with electric probes, called Langmuir probes are the most often procedures for low-temperature plasmas. Langmuir probe enables us to measure the current-voltage characteristics directly inside the plasma by collecting a certain part of the charge carriers according to the bias applied it. By analyzing the characteristics, we can study some of the plasma properties. We can measure the parameters spatially by installing the probe to the demanded position. On the other hand, by putting the probe into the plasma, we inevitably cause a perturbation in the plasma, which is even greater with the biased probes. Therefore highly persistent materials like wolfram or platinum are used for the probe construction to present interactions with the plasma. The method is based on the measurements and estimation of the current-voltage characteristics (probe characteristics) of a circuit consisting of two electrodes both placed in the plasma. The two basic cases are single probe method and double probe method. In single probe method both the electrodes surface areas differ by several orders of magnitude and one of the electrodes is often the probe itself. In double probe method the surface areas of both the electrodes in the plasma are approximately equal and considerably smaller than the dimensions of the vessel containing plasma. The single probe method is represented by the electrostatic Langmuir probe. The three of the most used geometric shapes of Langmuir probes are spherical probe, cylindrical probe and planar (flat) probe. The theoretical description is more simple for the spherical and planar probe and the cylindrical probe is the least complicated to construct. All parts of the probe have to be small in comparison with the dimensions of the plasma vessel in order to decrease the influence on the plasma surrounding the probe. The probe has to be designed carefully to protect it from damage and to ensure that it gives the correct I-V values. The probe tip is usually made of tungsten rod or wire 0, 1–1 mm in diameter. The wire is placed in a ceramic tube that insulates it from plasma except for a short exposed tip (≈2–10 mm long). The probe tip should be centered in the ceramic tube without touching it as there might occur, electrical contact with the conductive coating occurred because of the sputtering of the tip material. This system is encased in a vacuum vessel. The vacuum seal is preferred to be at the outside end of the probe rather than at the end immersed in the plasma, because a leak might occur. The probe diagnostics has important advantageous attributes like relatively simple probe construction and necessary experimental set-up for the probe measurements. From the measured data it is possible to determine a large number of plasma parameters (electron and ion temperature, electron velocity distribution function, plasma (space) potential, etc.). We can measure these parameters spatially. The disadvantages are a complicate measurements and data processing. The probe perturbs the surrounding plasma by the drain of the charge carriers to the probe. The presence of the probe might initiate some inhomogeneity in the plasma. There are effects of photo-emission or secondary emissions of the electrons from the probe surface and there are obstacles to include in the probe characteristics evaluation. Heavy particles may influence the probe surface during the measurement which leads to change of the surface properties like creating of thin film on the surface. Also there are difficulties of data interpretation. The probe construction must be in the direct contact with plasma. There is restriction of low-temperature plasma regarding the applications. Practically plasma physics use Langmuir probes in low temperature plasmas, few electron volts, to measure the plasma density, electron temperature, and the plasma potential. It consists of a bare wire or metal disk, which is inserted into plasma and electrically biased with respect to a reference electrode to collect electron and/or positive ion currents, the potential of the plasma at the location of the probe, could be determined by measuring the potential on the probe relative to one of the electrodes. The potential of the plasma at the location of the probe is known as the plasma potential or space potential and designated as V_P. The floating potential V_f of the probe is generally not the same as the plasma potential. A probe that is electrically floating collects no net current from the plasma. Its potential rises and falls to weather potential is necessary to maintain zero net current. In a typical plasma, the electrons, because of their smaller mass, have significantly higher thermal speeds than the positive ions (even if the electrons and ions are at the same temperature). Electrons have a higher temperature than the positive ions. A floating probe will tend initially to draw a higher electron current because the electrons reach the probe faster than the more massive ions. We can consider that plasma is electrically neutral, and the electron and ion densities are very nearly equal. Due to the fact that the net current to the floating probe must be zero, the probe floats to a negative potential relative to the plasma. The further collection of electrons is retarded and ion collection is enhanced. Thus, the floating potential is less than the plasma potential ($V_f < V_P$). An ion is an atom or molecule that has a non-zero net electrical charge, its total number of electrons is not equal to its total number of protons. A cation is a positively-charged ion, while an anion is negatively charged. Due to the fact that cations and anions have opposite electric charges, they attract each other and readily form ionic compounds, such as salts. Ions consisting of only a single atom are atomic or monatomic ions. If they consist of two or more atoms, then they are called molecular ions or polyatomic ions. In physical ionization of a gas, "ion pairs" are created by ion impact, and each pair consists of a free electron and a positive ion.

Definition: Plasma potential is the potential of the plasma with respect to the walls of the device (at a given location in the plasma). The swifter electrons tend to escape to the device walls first then leaving the plasma with a slight excess of positive space charge. The result is that V_P is a few volts positive with respect to the walls. The main part of the plasma (Bulk) is "Quasi-neutral" (electron density ≈ ion density) then the potential different between the main part of the plasma and the wall is concentrated in a thin layer or sheath near the wall. The electrons are due to the gradient of the plasma potential which cause to electric field. The discharge is done through ionization. The values of V_P and V_f are not the same if you try to measure it. By using Langmuir probe, we obtain the current-voltage (I-V) characteristics of the probe as the applied bias voltage V_B is swept from a negative to a positive potential. Practically, electrons and ions are not mono energetic and often have very different temperatures. This is the reason why probe sometimes collects only ion current, sometimes only electron current, and sometimes both. If the ion and electron current contributions are separated then it is easy to analyze the full I-V characteristics. The Langmuir probe gets two kinds of currents (Ion current and electron current). A simple way to produce plasma is to use a vessel (tube) in which a plasma is formed in a low pressure gas (<1 Torr) by applying several hundred volts across the cathode and anode. We insert a cylindrical wire probe into the discharge to measure the properties of the plasma. Another way is to use a chamber with rows of permanent magnet of opposite polarity which enhance the ionization efficiency. We produce the plasma by electron impact ionization of argon atoms by electrons that are thermionically emitted and accelerated from the hot tungsten (W) filament. The chamber includes hot wire filament which emits electrons. The Langmuir probe is inserted inside the plasma chamber to inspect the plasma parameters (Fig B.1).

Plasma chamber (vessel):

Langmuir probe I-V characteristics, total current $I(V_B) = I_e(V_B) + I_i(V_B)$. Because the electron current is much larger than the ion current, it is necessary to bias the probe to very negative voltages to even see the ion current (Table B.1). The probe bias for which $I(V_B = V_f) = I_e + I_i = 0$ is the probe's floating potential which occurs at negative voltage ($V_f < 0$). The floating potential can be calculated as the bias voltage at which

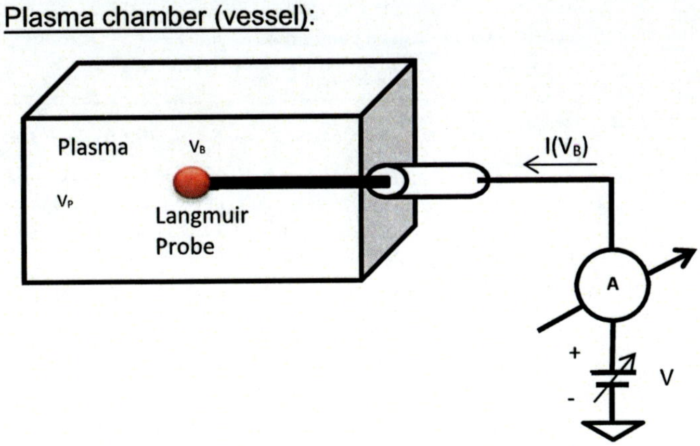

Fig. B.1 Plasma chamber and Langmuir probe system

Table B.1 V_B, V_P conditions and Langmuir probe ion current

Condition	Status	$I_i(V_B)$ equal to
$V_B < V_P$	Probe collects the ions saturation current I_{is} (positive ions)	$-I_{is}$
$V_B = V_P$ or $V_B > V_P$	The bias voltage V_B reaches V_P and ions begin to be repelled by the probe	$-I_{is} \cdot e^{\left[\frac{e\cdot(V_P - V_B)}{K\cdot T_i}\right]}$
$V_B \gg V_P$	All positive ions are repelled, and the ion current to the probe vanished $I_{is} = 0$	$V_B \gg V_P \Rightarrow V_P - V_B \approx -V_B$ $e^{\left[\frac{e\cdot(V_P - V_B)}{K\cdot T_i}\right]} \to \varepsilon$ $I_i(V_B) \to 0$

$$I_e(V_f) + I_i(V_f) = 0 \text{ then}$$

$$I_{es} \cdot e^{\left[\frac{e\cdot(V_f - V_P)}{K\cdot T_e}\right]} = I_{is} \Rightarrow V_f = V_P + \left(\frac{K \cdot T_e}{e}\right) \cdot \ln\left(0.6 \times \sqrt{\frac{2 \cdot \pi \cdot m_e}{m_i}}\right).$$

Langmuir probe ion current:

Parameters definition: e—Electron's charge, K, k_B—Boltzman constant, T_i—ion temperature, T_e—electron temperature, I_{is}—ion saturation current, n_i—ion density, $v_{i,th}$—ion thermal speed, m_i—ion mass, A_{probe}—probe collecting area, n_e—electron density.

$$v_{i,th} = \sqrt{\frac{8 \cdot K \cdot T_i}{\pi \cdot m_i}}; \; I_{is} = \frac{1}{4} \cdot e \cdot n_i \cdot v_{i,th} \cdot A_{probe} = \frac{1}{4} \cdot e \cdot n_i \cdot \sqrt{\frac{8 \cdot K \cdot T_i}{\pi \cdot m_i}} \cdot A_{probe}$$

$$A_{probe} = 2 \cdot \pi \cdot \left(\frac{d_{probe}}{2}\right)^2$$

Table B.2 Ion saturation current equations and conditions

$T_e > T_i;\ T_e \le T_i$	$T_e \gg T_i;\ \frac{T_e}{T_i} \gg 1$
$I_{is} = \frac{1}{4} \cdot e \cdot n_i \cdot v_{i,th} \cdot A_{probe}$ $v_{i,th} = \sqrt{\frac{8 \cdot K \cdot T_i}{\pi \cdot m_i}}$	$I_{is} = I_{Bohm} = 0.6 \times e \cdot n_i \cdot u_B \cdot A_{probe};\ u_B = \sqrt{\frac{K \cdot T_e}{m_i}}$ Note: The factor 0.6 is due to the reduction in the density of the ions in the pre-sheath. This is the region over which the ions are accelerated up to the Bohm speed u_B
Ion current is determined by the ion temperature T_i	Ion current is determined by the electron temperature T_e
$I_{is} \sim \sqrt{\frac{8 \cdot K \cdot T_i}{\pi \cdot m_i}}$	$I_{is} \sim \sqrt{\frac{K \cdot T_e}{m_i}}$

Table B.3 Positive and negative biased electrodes versus results

Status	Result
Positive biased electrode	An electron shielding cloud surrounding it
Negative biased electrode	1. A positive space charge cloud 2. The characteristic shielding distance of the potential disturbance is the electron Debye length $\lambda_{De} = \sqrt{\frac{\varepsilon_0 \cdot K \cdot T_e}{e^2 \cdot n_e}}$ 3. Both the electron and ion densities decrease as the particles approach the probe (not at the same rate)

$$I_{is} = e \cdot n_i \cdot \pi \cdot \sqrt{\frac{2 \cdot K \cdot T_i}{\pi \cdot m_i}} \cdot \left(\frac{d_{probe}}{2}\right)^2$$
$$= e \cdot n_i \cdot \sqrt{\frac{2 \cdot \pi \cdot K \cdot T_i}{m_i}} \cdot \left(\frac{d_{probe}}{2}\right)^2$$

If $T_e \gg T_i;\ \frac{T_e}{T_i} \gg 1$ then the ion saturation current is not determined by the ion thermal speed, but rather by the Bohm ion current I_{Bohm}. We can summery ion saturation current equations in the following table (Table B.2).

The formation of a sheath around a negatively biased probe is the physical reason for the dependence $I_{is} \sim \sqrt{\frac{K \cdot T_e}{m_i}}$ for $T_e \gg T_i;\ \frac{T_e}{T_i} \gg 1$.

{Electrode potential $\neq$ Local plasma potential} then electrons and ions distribute themselves spatially around the electrode in order to limit, or shield, the effect of this potential on the main part of the plasma (Bulk plasma) (Table B.3).

In the vicinity of a negative biased probe the electron density decreases because electrons are repelled by the probe. The ions are accelerated toward the probe, and ion density decrease due to the continuity of the current density. If the ion density exceeds the electron density ($n_i > n_e$) at the sheath edge then we get a positive space charge sheath. If the ions approach the sheath with a speed exceeding the

Table B.4 Electron current equations and conditions

Condition	Status	$I_e(V_B)$ equal to
$V_B \leq V_P$	Electrons are partially repelled by the probe and electron current decreases exponentially with decreasing V (Maxwellian electron velocity distribution)	$I_{es} \cdot e^{\left\{-e \cdot \frac{(V_P - V_B)}{K \cdot T_e}\right\}}$
$V_B \gg V_P$ $V_B > V_P$	Probe collects electron saturation current I_{es}	I_{es}
$V_B \ll V_P$	All electrons are repelled, $I_e(V_B) \to \varepsilon$	$V_B \ll V_P \Rightarrow V_P - V_B \approx V_P$ $e^{\left\{-e \cdot \frac{(V_P - V_B)}{K \cdot T_e}\right\}} \to \varepsilon;\ I_e(V_B) \to 0$

Bohm velocity $u_B = \sqrt{\frac{K \cdot T_e}{m_i}}$ then the ions density decreases more slowly than the electron density.

Langmuir probe electron current:

The electron saturation current I_{es} is given by $I_{es} = \frac{1}{4} \cdot e \cdot n_e \cdot v_{e,th} \cdot A_{probe}$; $v_{e,th} = \sqrt{\frac{8 \cdot K \cdot T_e}{\pi \cdot m_e}}$.

n_e—Electron density, $v_{e,th}$—Electron thermal speed, m_e—electron mass. If $n_e = n_i$ and $m_e \ll m_i$ then $I_{es} \gg I_{is}$(Table B.4).

Electron and ion temperatures measure by electro volt (eV) unit: The basic definition of temperature is $T = \frac{dU}{dS}$, the change in internal energy (dU) with respect to entropy (dS) holding volume and particle number constant. If the velocities of a plasma (group of electrons) follows a Maxwell-Boltzmann distribution, then the electron temperature is defined as the temperature of that distribution. The Maxwell-Boltzmann distribution with three degrees of freedom is defined as $\langle E \rangle = (\frac{3}{2}) \cdot \langle k_B \cdot T \rangle$. The SI unit of temperature is the kelvin, but the electron temperature is expressed in terms of the energy electron volt (eV). Each kelvin degree corresponds to 8.617×10^{-5} eV approximately. Each eV is equivalent to 11,605 °K. It can be calculated by the relation $E = k_B \cdot T$. The electron temperature of plasma (T_e) can be several orders higher than the temperature of ions ($T_e \gg T_i$). The reasons are first, many plasma sources heat the electrons more strongly than the ions. Secondary ions are much heavier than electrons, and energy transfer in a two-body collision is much more efficient if the masses are similar. In low pressure plasma discharges can be two distinct Maxwellian distributions of electrons, a cold and hot distribution with temperatures T_{ec} and T_{eh} respectively. We can define the electron current as $I_e(V_B) = I_{ec}(V_B) + I_{eh}(V_B)$. The n_{ec}, n_{eh} are the respective densities of cold and hot components in total electron density n_e with $n_e = n_{ec} + n_{eh}$. We define a new parameters $f_{eh} = \frac{n_{eh}}{n_e} = \frac{n_{eh}}{n_{ec} + n_{eh}}$ and $f_{ec} = \frac{n_{ec}}{n_e} = \frac{n_{ec}}{n_{ec} + n_{eh}}$. The appropriate electron temperature T_e is the harmonic average of T_{ec} and T_{eh}; $\frac{1}{T_e} = (\frac{n_{ec}}{n_e}) \cdot \frac{1}{T_{ec}} + (\frac{n_{eh}}{n_e}) \cdot \frac{1}{T_{eh}}$. We can produce plasmas in hot-filament discharges by using the effect of the ionizing (primary) electrons. The probe analysis can be done by the presence of these energetic primary electrons and we can model it as an isotropic mono-magnetic distribution.

We define primary electron current contribution as $I_{ep}(V_B)$ and $I_e(V_B)$ is the electron current contribution from the bulk electrons, $I_{et}(V_B) = I_e(V_B) + I_{ep}(V_B)$. $I_{et}(V_B)$—Total electron current, n_{ep}—density of primary electrons, v_{ep} is the speed of primary electrons with energy E_p; $v_{ep} = \sqrt{\frac{2 \cdot E_p}{m_e}}$. Electron sheaths are commonly found near Langmuir probes collecting the electron saturation current. The common assumption is that the probe collects the random flux of electrons incident on the sheath. Some assumptions imply that there is no electron pre sheath and that the flux collected is due to a velocity space truncation of the electron velocity distribution function (EVDF). At floating boundaries the sheath is ion rich, an ion sheath, providing a thin positive space charge layer that limits the electron losses to the boundary. The fact is that not all sheaths need to be ion rich. Sheaths near small electrodes, such those around Langmuir probes, can be electron rich (electron sheaths) when the electrode is biased positive with respect to the plasma potential. There are requirements of global current balance and electron sheaths are possible only near electrodes that are small enough that the ratio of their area (A_E) to the plasma chamber wall area (A_w) satisfies $\frac{A_E}{A_w} < \sqrt{\frac{2.3 \times m_e}{m_i}}$ where A_E and A_w are the effective surface areas for collecting charged particles at the electrode and the wall, respectively. Another way to inspect plasmas is by using ball-pen probe. A bell-pen probe is a modified Langmuir probe used to measure the plasma potential V_P in magnetized plasmas. The ball-pen probe balances the electron and ion saturation current (I_{es} and I_{is}) and the result is that its floating potential V_f is equal to the plasma potential V_P. Electrons have much smaller gyro radius than ions, a moving ceramic shield can be used to screen off an adjustable part of the electron current from the probe collector. Gyro radius is also known as radius of gyration, Larmor radius or cyclotron radius is the radius of the circular motion of a charged particle in the presence of a uniform magnetic field. If a Langmuir probe is inserted into plasma, its potential is not equal to the plasma potential V_P because a Debye sheath forms, but instead it equals to the floating potential V_f. We define the difference between the plasma potential and the floating potential as $V_P - V_f = \Xi \cdot T_e$ where the coefficient Ξ is given by the ratio of the electron and ion saturation current density j_{es}, j_{is} and collecting areas for electron and ions A_e, A_i. We define the coefficient Ξ as $\Xi = \ln\left(\frac{A_e \cdot j_{es}}{A_i \cdot j_{is}}\right) = \ln R$; $R = \frac{A_e \cdot j_{es}}{A_i \cdot j_{is}}$. The ball-pen probe modifies the collecting areas for electrons and ions in such a way that the ration R is equal to one ($A_e \cdot j_{es} = A_i \cdot j_{is}$). Consequently, $R = 1 \Rightarrow \Xi = 0$ and the floating potential of the ball-pen probe becomes equal to the plasma potential regardless of the electron temperature ($V_f = V_P$). One important use of Ball-pen probe (BPP) is to make profile measurements of plasma potential, electron temperature and radial electric field on the Mega Amp Spherical Tokamak (MAST). There is no big similarity between the potential profile measured by the BPP (Ball-pen probe) and the floating potential both in polarity and profile shape. We can measure the electron temperature T_e by combining the BPP potential and the floating potential V_f. Electron temperature measurement can be comparable to Thomson Scattering (TS) diagnostic. The unique feature of ball pen probe (BPP) is the diagnostic technique capable of measuring the plasma potential V_P, withstanding particle and heat fluxes from the plasma at radii up to and within the separatix. There is a recommended structure of Mega Amp Spherical Tokamak which includes some ball pen probes (BPPs) and some Langmuir probes (LPs). Some of them are floating pins whilst other are externally biased. The floating potential of a probe in a plasma, which arises by the constraint of anti-polarity is given by $V_f = V_P - \Xi \cdot T_e$ where V_f is the floating voltage of the probe, V_P is the plasma potential, T_e is the electron temperature in eV and $\Xi = \ln\left(\frac{I_e}{I_i}\right)$; $I_e = A_e \cdot j_{es}$; $I_i = A_i \cdot j_{is}$, Ξ is the logarithmic ratio of electron to ion currents flowing to the probe. In magnetized plasma the dominant flux of electrons and ions to an exposed probe, such as a Langmuir probe, happens along magnetic field lines. If we neglect the effect of secondary electron emission, Ξ_{LP} is given in the case of a Langmuir probe $\Xi_{LP} = \ln\left(\frac{I_e}{I_i}\right) = -\frac{1}{2} \cdot \ln\{2 \cdot \pi \cdot \frac{m_e}{m_i} \cdot (1 + \frac{T_i}{T_e})\}$. Practically Ball-pen probe reduces the value of Ξ to zero by equalizing the ratio of electron to ion currents at the collector. It is done by shielding the collector from fluxes directed along the magnetic field line such that cross-field transport is required for the net collection of ions and electrons. There is no current model exists to predict Ξ_{BPP} and must be treated as a free parameter. The Ξ_{BPP} can be measure directly by applying an external voltage sweep to the Ball-pen probe (BPP). The BPP signal cannot be a suppressed version of the Langmuir probe (LP). Practically BPP shows regions of opposite polarity to the floating potential. The floating potential should always be more negative than the BPP signal. The estimate electron temperature T_e when using Ball-pin probe and Langmuir probe is given by $T_e \approx \frac{V_{BPP} - V_f}{\Xi_{LP} - \Xi_{BPP}}$. We get it by subtracting the floating potential from the BPP, V_{BPP} from the floating potential of the Langmuir probe (LP). If we include the effect of secondary electron emission Ξ_{LP} is given by $\Xi_{LP} = -\frac{1}{2} \cdot \ln\{2 \cdot \pi \cdot \frac{m_e}{m_i} \cdot (1 + \frac{T_i}{T_e}) \cdot \frac{1}{(1-\delta_e)^2}\}$ where δ_e is the ratio of secondary electrons emitted per primary incident electron. The value of δ_e can be calculated for a graphite probe tip from the semi-empirical formula $\delta_e = 2.72^2 \cdot \frac{E}{300} \cdot e^{-2 \cdot \sqrt{\frac{E}{300}}}$ where E is the incident electron energy. The third probe type which can inspect the plasma is magnetic (B-dot) probe. The differential magnetic probe (B-dot probe) is used as diagnostic of fast transient effects in exploding plasmas. The B-dot probe designs means are reduce stray pickup, eliminate electrostatic pickup, reduce physical size, and increase magnetic signals while maintaining a high bandwidth. It is three-axis, high-frequency magnetic (B-dot) probe which supporting electronics to measure transient events in plasmas. Such phenomena include exploding plasmas, magnetic shocks, rapidly expanding diamagnetic cavities in vacuum or surrounding plasma environments, nonlinear Alfven waves, and many other nonlinear phenomena. To measure these events we need a probe that is sensitive to three components of the field, respond quicker than the time scales of interest, size smaller than the Larmor radius, and is sufficiently sensitive to measure week magnetic fields. The main characteristics of the B-dot probe are sensitivity and response. They are determined by the probe's internal resistance, internal capacitance, self-inductance, and noise levels. The B-dot probe follows the magnetic pickup coil design. It utilizes Faraday's law to detect magnetic field fluctuations. The probe has three key elements, the probe tip, the probe shaft and enclosure, the differential amplifier. The probe tip has three sets

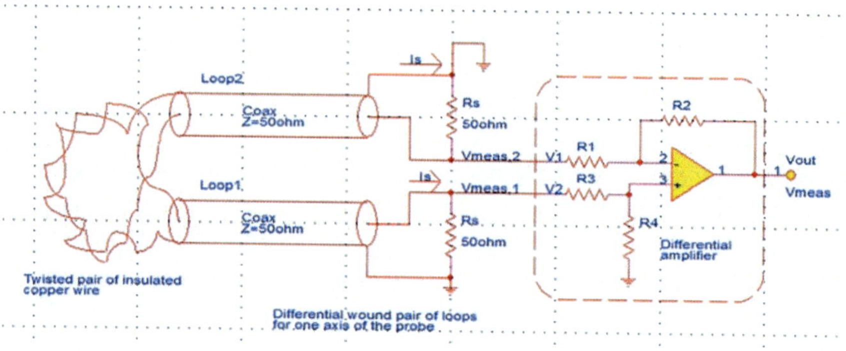

Fig B.2 Differential wound pair of loops for one axis of the probe system

of wire loops sensitive to the three components of the magnetic field at a specific point. The probe shaft one enclosure is designed to transmit the signals to the differential amplifier, protect the sensory elements from plasma saturation current, and reduce stray pickup. A high-frequency differential amplifier mitigates much of the electrostatic pick up from the plasma. In most cases probe tip consists of a 2 × 2 mm cylindrical Vespel. Vespel is used because of its high thermal capability, machinability, and relative magnetic permeability near unity at high frequencies. The B-dot probe's core is wined around by twisted pair of insulated copper wire for five times or more. This creates a differential pair of five turn loops, loop 1 and loop 2, or ten total turns along each axis. As a result, magnetic fields induce equal but opposite voltage in loop 1 and loop 2, while induced electrostatic voltage have the same polarity. By subtracting the loop 1 and loop 2 signals, the electrostatic components are subtracted out, while the magnetic components are doubled. Finally we get two loops in our B-dot probe (loop 1 and loop 2). The leads for loop 1 are cut to length and soldered to one coax cable. The loop 2 leads are soldered to another coax cable in reverse direction of loop 1. If we have some B-probe axis, the process is repeated for each axis and connects the loops in an arrangement that forms the differential pair (Fig. B.2).

$$V_{out} = \frac{(R_1 + R_2)}{(R_3 + R_4)} \cdot \frac{R_4}{R_1} \cdot V_2 - \frac{R_2}{R_1} \cdot V_1$$

$$R_1 = R_3 \text{ and } R_2 = R_4 \Rightarrow V_{out} = \frac{R_2}{R_1} \cdot (V_2 - V_1)$$

We can choose $\frac{R_1}{R_2} = \frac{R_3}{R_4}$ for minimum offset error due to input bias current. We define the g as the gain of the differential amplifier $\left(g = \frac{R_2}{R_1}\right)$ for $R_1 = R_3$ and $R_2 = R_4$.

$$V_1 = V_{meas,2};\quad V_2 = V_{meas,1};\quad V_{out} = \frac{R_2}{R_1} \cdot (V_{meas,1} - V_{meas,2})$$

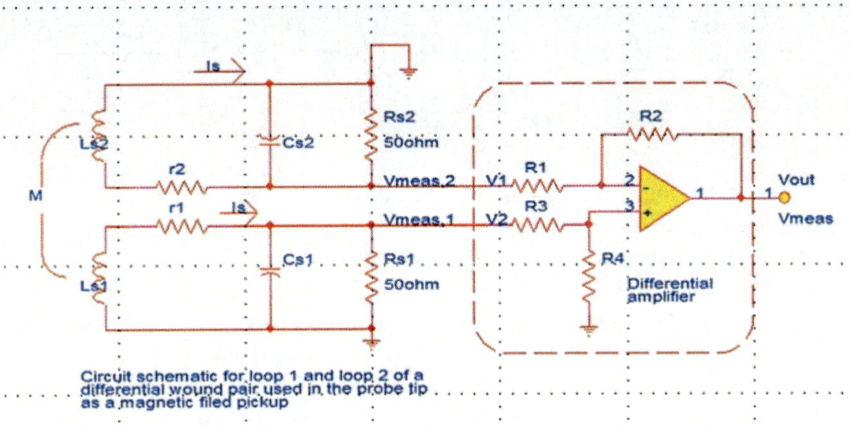

Fig B.3 Circuit schematic for loop 1 and loop 2 of a differential wound pair used in the probe tip as a magnetic field pickup

$$g = \frac{R_2}{R_1}; \;\; V_{meas} = V_{out}; \;\; V_{meas} = g \cdot (V_{meas,1} - V_{meas,2})$$

If the connections to the coax cables are not properly shielded and isolated from the plasma, then the connections can pick up plasma current and/or magnetic field fluctuations that contaminate the desired signals. When a probe is placed in plasma, there is a potential difference between the probe and the plasma, which shows up as a capacitive effect on the probe. This capacitive effect is substantially identical on both loop1 and loop 2, which shows up as a common mode potential $V_{electrostatic}$. Since the electrostatic potential is identical on both loops, the differential amplifier subtracts out the common induced voltage to create the final $V_{meas}(t)$ signal.

$$V_{meas}(t) = g \cdot (V_{meas,1}(t) - V_{meas,2}(t)); \;\; V_{meas,1}(t) = -V_{meas,2}(t)$$
$$V_{meas}(t) = 2 \cdot g \cdot V_{meas,1}(t)$$

The pickup loops of the probe consist of a pair of twisted wires turned five times or more around a plastic core. The only difference between the two set of loops, loop 1 and loop 2, is the location of ground and all other parameters are approximately identical. We can draw the circuit schematic for loops (loop 1 and loop 2) of a differentially wound pair used in the probe tip as a magnetic field pickup (Fig. B.3).

$$L_s = L_{s1} = L_{s2}; \;\; r = r_1 = r_2; \;\; C_s = C_{s1} = C_{s2}; \;\; R_s = R_{s1} = R_{s2}$$

We assume that the current through loop 1 is equal to the current through loop 2, and the circuit equation is as follow:

$$a \cdot n \cdot \frac{dB(t)}{dt} = \left(1 + \frac{r}{R_s}\right) \cdot V_{meas,1}(t) + \left(\frac{L_s + M}{R_s} + r \cdot C_s\right) \cdot \frac{dV_{meas,1}(t)}{dt} + (L_s + M) \cdot C_s \cdot \frac{d^2 V_{meas,1}(t)}{dt^2}$$

a—area of a single turn, n—number of turns in a loop, r—internal resistance of the loop, R_s—load resistance of the loop, L_s—self-inductance of the loop, M—mutual inductance between two loops, C_s—internal capacitance of the loop, N—total number of turns between the two loops, g—gain of the differential amplifier. The magnetic field B(t) created by either the Helmholtz coil or wire acts as a voltage source for the circuit. In the case of Helmholtz coil, this term is the mutual-inductance between the Helmholtz coil and the loop. Helmholtz coils is for getting a fairly uniform magnetic field, pairs of circular coils on a common axis with equal currents flowing in the same sense. There is a separation needed to give the most uniform central field. This separation is equal to the radius of the coils. V_{meas} is proportional to the first time derivative of the magnetic field. We define other two parameters (ρ_s, τ_s) where $\rho_s = \frac{r}{R_s}$; $\tau_s = \frac{L_s+M}{R_s}$. We can get the magnetic field B(t) from a probe signal $V_{meas}(t)$ by discretely integrate the signal according to the model given by the circuit equation.

$$a \cdot N \cdot g \cdot \frac{dB(t)}{dt} = \left(1 + \tau_s \frac{d}{dt}\right) V_{meas}(t)$$

$$a \cdot N \cdot g \cdot \frac{dB(t)}{dt} = V_{meas}(t) + \tau_s \frac{dV_{meas}(t)}{dt}$$

$$\frac{dB(t)}{dt} = \frac{1}{a \cdot N \cdot g} \cdot V_{meas}(t) + \frac{1}{a \cdot N \cdot g} \cdot \tau_s \frac{dV_{meas}(t)}{dt}$$

$$B(t) = \frac{1}{a \cdot N \cdot g} \cdot \int V_{meas}(t) \cdot dt + \frac{1}{a \cdot N \cdot g} \cdot \tau_s \cdot V_{meas}(t)$$

$$V_{meas}(t) = 2 \cdot g \cdot V_{meas,1}(t); \; a \cdot N \cdot \frac{dB(t)}{dt} = 2 \cdot V_{meas,1}(t) + 2 \cdot \tau_s \frac{dV_{meas,1}(t)}{dt}$$

$$\frac{dB(t)}{dt} = \frac{2}{a \cdot N} \cdot V_{meas,1}(t) + \frac{2 \cdot \tau_s}{a \cdot N} \cdot \frac{dV_{meas,1}(t)}{dt}$$

We define a new variable $X(t) = \frac{dV_{meas,1}(t)}{dt}$. $\frac{dB(t)}{dt} = \frac{2}{a \cdot N} \cdot V_{meas,1}(t) + \frac{2 \cdot \tau_s}{a \cdot N} \cdot \frac{dV_{meas,1}(t)}{dt}$

$$\frac{dB(t)}{dt} = \frac{2}{a \cdot N} \cdot V_{meas,1}(t) + \frac{2 \cdot \tau_s}{a \cdot N} \cdot X(t); \; \frac{dX(t)}{dt} = \frac{d^2 V_{meas,1}(t)}{dt^2}$$

$$a \cdot n \cdot \frac{dB(t)}{dt} = \left(1 + \frac{r}{R_s}\right) \cdot V_{meas,1}(t) + \left(\frac{L_s + M}{R_s} + r \cdot C_s\right) \cdot X(t) + (L_s + M) \cdot C_s \cdot \frac{dX(t)}{dt}$$

Combine the two above differential equations gives the following differential equation:

$$a \cdot n \cdot \left(\frac{2}{a \cdot N} \cdot V_{meas,1}(t) + \frac{2 \cdot \tau_s}{a \cdot N} \cdot X(t) \right) = \left(1 + \frac{r}{R_s} \right) \cdot V_{meas,1}(t) + \left(\frac{L_s + M}{R_s} + r \cdot C_s \right) \cdot X(t) + (L_s + M) \cdot C_s \cdot \frac{dX(t)}{dt}$$

$$(L_s + M) \cdot C_s \cdot \frac{dX(t)}{dt} = \frac{n \cdot 2}{N} \cdot V_{meas,1}(t) + \frac{n \cdot 2 \cdot \tau_s}{N} \cdot X(t) - \left(1 + \frac{r}{R_s} \right) \cdot V_{meas,1}(t) - \left(\frac{L_s + M}{R_s} + r \cdot C_s \right) \cdot X(t)$$

$$(L_s + M) \cdot C_s \cdot \frac{dX(t)}{dt} = \left[\frac{n \cdot 2}{N} - \left(1 + \frac{r}{R_s} \right) \right] \cdot V_{meas,1}(t) + \left[\frac{n \cdot 2 \cdot \tau_s}{N} - \left(\frac{L_s + M}{R_s} + r \cdot C_s \right) \right] \cdot X(t)$$

$$\frac{dX(t)}{dt} = \frac{\left[\frac{n \cdot 2}{N} - \left(1 + \frac{r}{R_s} \right) \right]}{(L_s + M) \cdot C_s} \cdot V_{meas,1}(t) + \frac{\left[\frac{n \cdot 2 \cdot \tau_s}{N} - \left(\frac{L_s + M}{R_s} + r \cdot C_s \right) \right]}{(L_s + M) \cdot C_s} \cdot X(t)$$

We can summery our B-probe system differential equations:

$$\frac{dV_{meas,1}(t)}{dt} = X(t)$$

$$\frac{dX(t)}{dt} = \frac{\left[\frac{n \cdot 2}{N} - \left(1 + \frac{r}{R_s} \right) \right]}{(L_s + M) \cdot C_s} \cdot V_{meas,1}(t) + \frac{\left[\frac{n \cdot 2 \cdot \tau_s}{N} - \left(\frac{L_s + M}{R_s} + r \cdot C_s \right) \right]}{(L_s + M) \cdot C_s} \cdot X(t)$$

Due to vessel's plasma interferences effects there is a delay in the measured prove variables: $V_{meas,1}(t) \to V_{meas,1}(t - \tau_1)$; $X(t) \to X(t - \tau_2)$ and no effect on variable derivatives. We get the following delay differential equations:

$$\frac{dV_{meas,1}(t)}{dt} = X(t - \tau_2)$$

$$\frac{dX(t)}{dt} = \frac{\left[\frac{n\cdot 2}{N} - \left(1 + \frac{r}{R_s}\right)\right]}{(L_s + M)\cdot C_s} \cdot V_{meas,1}(t-\tau_1) + \frac{\left[\frac{n\cdot 2\cdot \tau_s}{N} - \left(\frac{L_s+M}{R_s} + r\cdot C_s\right)\right]}{(L_s + M)\cdot C_s} \cdot X(t-\tau_2)$$

We can check our B-dot probe plasma system stability for the following cases: (A) $\tau_1 = \tau;\ \ \tau_2 = 0$, (B) $\tau_1 = 0;\ \ \tau_2 = \tau$, (C) $\tau_1 = \tau;\ \ \tau_2 = \tau$.

We choose to inspect the first case where (A) $\tau_1 = \tau;\ \ \tau_2 = 0$. It is reader exercise to check the other two cases.

$$\frac{dV_{meas,1}(t)}{dt} = X(t)$$

$$\frac{dX(t)}{dt} = \frac{\left[\frac{n\cdot 2}{N} - \left(1 + \frac{r}{R_s}\right)\right]}{(L_s + M)\cdot C_s} \cdot V_{meas,1}(t-\tau) + \frac{\left[\frac{n\cdot 2\cdot \tau_s}{N} - \left(\frac{L_s+M}{R_s} + r\cdot C_s\right)\right]}{(L_s + M)\cdot C_s} \cdot X(t)$$

Stability analysis: The standard local stability analysis about any one of the equilibrium points of B-dot probe system consists in adding to coordinate $[V_{meas,1}, X]$ arbitrarily small increments of exponential form $[v_{meas,1}, x]\cdot e^{\lambda\cdot t}$ and retaining the first order terms in $V_{meas,1}, X$. The system of two homogeneous equations leads to a polynomial characteristic equation in the eigenvalues. The polynomial characteristic equations accept by set the below variables derivative with respect to time into B-dot probe system equations. B-dot probe system fixed values with arbitrarily small increments of exponential form $[v_{meas,1}, x]\cdot e^{\lambda\cdot t}$ are: j = 0 (first fixed point), j = 1 (second fixed point), j = 2 (third fixed point), etc., at fixed points $\lim_{t\to\infty} V_{meas,1}(t-\tau_1) = V_{meas,1}(t)$

$$t \gg \tau_1;\ \ \lim_{t\to\infty} X(t-\tau_2) = X(t);\ \ t \gg \tau_2;\ \ \frac{dV_{meas,1}(t)}{dt} = 0$$

$$\frac{dX(t)}{dt} = 0 \text{ then } V^*_{meas,1} = 0;\ \ X^* = 0.$$

$$X(t) = X^{(j)} + x\cdot e^{\lambda\cdot t};\ \ V_{meas,1}(t) = V^{(j)}_{meas,1} + v_{meas,1}\cdot e^{\lambda\cdot t}$$

$$V_{meas,1}(t-\tau) = V^{(j)}_{meas,1} + v_{meas,1}\cdot e^{\lambda\cdot t}\cdot e^{-\lambda\cdot\tau}$$

$\frac{dX(t)}{dt} = \lambda\cdot x\cdot e^{\lambda\cdot t};\ \frac{dV_{meas,1}(t)}{dt} = \lambda\cdot v_{meas,1}\cdot e^{\lambda\cdot t}$. We choose these expressions for ourselves $X(t), V_{meas,1}(t)$ as a small displacement $[v_{meas,1}, x]$ from the B-dot probe system fixed points in time t = 0; $X(t=0) = X^{(j)} + x;\ \ V_{meas,1}(t=0) = V^{(j)}_{meas,1} + v_{meas,1}$. For $\lambda < 0,\ t > 0$ the selected fixed point is stable otherwise $\lambda > 0,\ t > 0$ is unstable. Our B-dot probe system tends to the selected fixed point exponentially for $\lambda < 0$, $t > 0$ otherwise go away from the selected fixed point exponentially. Eigenvalue λ parameter is established if the fixed point is stable or unstable; additionality, his absolute value $|\lambda|$ establishes the speed of flow toward or away from the selected fixed point. The speed of flow toward or away from the selected fixed point for B-dot probe system variables and variables derivatives with respect to time are as follow. We define $V_{meas,1}(t-\tau) = V_{meas,1}^{(j)} + v_{meas,1} \cdot e^{\lambda \cdot t} \cdot e^{-\lambda \cdot \tau}$ then we get two delayed differential equations with respect to coordinate $[V_{meas,1}, X]$ arbitrarily small increments of exponential $[v_{meas,1}, x] \cdot e^{\lambda \cdot t}$. We consider no delay effects on $\frac{dV_{meas,1}(t)}{dt}$, $\frac{dX(t)}{dt}$. We get the following equations:

$\frac{dV_{meas,1}(t)}{dt} = X(t) \Rightarrow \lambda \cdot v_{meas,1} \cdot e^{\lambda \cdot t} = X^{(j)} + x \cdot e^{\lambda \cdot t}$ At fixed point $X^{(j)} = 0 \Rightarrow -\lambda \cdot v_{meas,1} + x = 0$.

$$\frac{dX(t)}{dt} = \frac{\left[\frac{n \cdot 2}{N} - \left(1 + \frac{r}{R_s}\right)\right]}{(L_s + M) \cdot C_s} \cdot V_{meas,1}(t - \tau) + \frac{\left[\frac{n \cdot 2 \cdot \tau_s}{N} - \left(\frac{L_s + M}{R_s} + r \cdot C_s\right)\right]}{(L_s + M) \cdot C_s} \cdot X(t)$$

$$\lambda \cdot x \cdot e^{\lambda \cdot t} = \frac{\left[\frac{n \cdot 2}{N} - \left(1 + \frac{r}{R_s}\right)\right]}{(L_s + M) \cdot C_s} \cdot \left[V_{meas,1}^{(j)} + v_{meas,1} \cdot e^{\lambda \cdot t} \cdot e^{-\lambda \cdot \tau}\right] + \frac{\left[\frac{n \cdot 2 \cdot \tau_s}{N} - \left(\frac{L_s + M}{R_s} + r \cdot C_s\right)\right]}{(L_s + M) \cdot C_s} \cdot \left[X^{(j)} + x \cdot e^{\lambda \cdot t}\right]$$

$$\lambda \cdot x \cdot e^{\lambda \cdot t} = \frac{\left[\frac{n \cdot 2}{N} - \left(1 + \frac{r}{R_s}\right)\right]}{(L_s + M) \cdot C_s} \cdot V_{meas,1}^{(j)} + \frac{\left[\frac{n \cdot 2 \cdot \tau_s}{N} - \left(\frac{L_s + M}{R_s} + r \cdot C_s\right)\right]}{(L_s + M) \cdot C_s} \cdot X^{(j)} + \frac{\left[\frac{n \cdot 2}{N} - \left(1 + \frac{r}{R_s}\right)\right]}{(L_s + M) \cdot C_s} \cdot v_{meas,1} \cdot e^{\lambda \cdot t} \cdot e^{-\lambda \cdot \tau} + \frac{\left[\frac{n \cdot 2 \cdot \tau_s}{N} - \left(\frac{L_s + M}{R_s} + r \cdot C_s\right)\right]}{(L_s + M) \cdot C_s} \cdot x \cdot e^{\lambda \cdot t}$$

At fixed point $\frac{\left[\frac{n \cdot 2}{N} - \left(1 + \frac{r}{R_s}\right)\right]}{(L_s + M) \cdot C_s} \cdot V_{meas,1}^{(j)} + \frac{\left[\frac{n \cdot 2 \cdot \tau_s}{N} - \left(\frac{L_s + M}{R_s} + r \cdot C_s\right)\right]}{(L_s + M) \cdot C_s} \cdot X^{(j)} = 0$

$$\frac{\left[\frac{n \cdot 2}{N} - \left(1 + \frac{r}{R_s}\right)\right]}{(L_s + M) \cdot C_s} \cdot v_{meas,1} \cdot e^{-\lambda \cdot \tau} + \frac{\left[\frac{n \cdot 2 \cdot \tau_s}{N} - \left(\frac{L_s + M}{R_s} + r \cdot C_s\right)\right]}{(L_s + M) \cdot C_s} \cdot x - \lambda \cdot x = 0$$

We can summery our last results: $-\lambda \cdot v_{meas,1} + x = 0$

$$\frac{\left[\frac{n\cdot 2}{N}-\left(1+\frac{r}{R_s}\right)\right]}{(L_s+M)\cdot C_s}\cdot v_{meas,1}\cdot e^{-\lambda\cdot\tau}+\frac{\left[\frac{n\cdot 2\cdot\tau_s}{N}-\left(\frac{L_s+M}{R_s}+r\cdot C_s\right)\right]}{(L_s+M)\cdot C_s}\cdot x-\lambda\cdot x=0$$

The small increments Jacobian of our B-dot probe system is as follows:

$$\begin{pmatrix} -\lambda & 1 \\ \frac{\left[\frac{n\cdot 2}{N}-\left(1+\frac{r}{R_s}\right)\right]}{(L_s+M)\cdot C_s}\cdot e^{-\lambda\cdot\tau} & \frac{\left[\frac{n\cdot 2\cdot\tau_s}{N}-\left(\frac{L_s+M}{R_s}+r\cdot C_s\right)\right]}{(L_s+M)\cdot C_s}-\lambda \end{pmatrix}\cdot\begin{pmatrix} v_{meas,1} \\ x \end{pmatrix}=0$$

$$|A-\lambda\cdot I|=\begin{pmatrix} -\lambda & 1 \\ \frac{\left[\frac{n\cdot 2}{N}-\left(1+\frac{r}{R_s}\right)\right]}{(L_s+M)\cdot C_s}\cdot e^{-\lambda\cdot\tau} & \frac{\left[\frac{n\cdot 2\cdot\tau_s}{N}-\left(\frac{L_s+M}{R_s}+r\cdot C_s\right)\right]}{(L_s+M)\cdot C_s}-\lambda \end{pmatrix}$$

$$\det|A-\lambda\cdot I|=0$$

$$\lambda^2-\lambda\cdot\frac{\left[\frac{n\cdot 2\cdot\tau_s}{N}-\left(\frac{L_s+M}{R_s}+r\cdot C_s\right)\right]}{(L_s+M)\cdot C_s}-\frac{\left[\frac{n\cdot 2}{N}-\left(1+\frac{r}{R_s}\right)\right]}{(L_s+M)\cdot C_s}\cdot e^{-\lambda\cdot\tau}=0$$

We study the occurrence of any possible stability switching, resulting from the increase of the value of the time delay τ for general characteristic equation $D(\tau)$. If we choose τ as a parameter, then the expression: $D(\lambda,\tau)=P_n(\lambda,\tau)+Q_m(\lambda,\tau)\cdot e^{-\lambda\cdot\tau}$; $n,m\in\mathbb{N}_0$; $n>m$. $n=2$; $m=0$; $P_n(\lambda,\tau)=\sum_{k=0}^{n=2}p_k(\tau)\cdot\lambda^k=p_0+p_1\cdot\lambda+p_2\cdot\lambda^2$; $Q_m(\lambda,\tau)=\sum_{k=0}^{m=0}q_k(\tau)\cdot\lambda^k=q_0$.

$$p_0=0;\ \ p_1=-\frac{\left[\frac{n\cdot 2\cdot\tau_s}{N}-\left(\frac{L_s+M}{R_s}+r\cdot C_s\right)\right]}{(L_s+M)\cdot C_s}$$

$$p_2=1;\ \ q_0=-\frac{\left[\frac{n\cdot 2}{N}-\left(1+\frac{r}{R_s}\right)\right]}{(L_s+M)\cdot C_s}$$

The homogeneous system for $V_{meas,1}, X$ leads to a characteristic equation for the eigenvalue λ having the form $P(\lambda,\tau)+Q(\lambda,\tau)\cdot e^{-\lambda\cdot\tau}=0$; $P(\lambda)=\sum_{j=0}^{2}a_j\cdot\lambda^j$; $Q(\lambda)=\sum_{j=0}^{0}c_j\cdot\lambda^j$. The coefficients $\{a_j(q_i,q_k,\tau),c_j(q_i,q_k,\tau)\}\in\mathbb{R}$ depend on q_i,q_k and delay τ. q_i,q_k are any B-dot probe parameters, other parameters kept as a constant.

$$a_0=0;\ \ a_1=-\frac{\left[\frac{n\cdot 2\cdot\tau_s}{N}-\left(\frac{L_s+M}{R_s}+r\cdot C_s\right)\right]}{(L_s+M)\cdot C_s}$$

$$a_2=1;\ \ c_0=-\frac{\left[\frac{n\cdot 2}{N}-\left(1+\frac{r}{R_s}\right)\right]}{(L_s+M)\cdot C_s}$$

Unless strictly necessary; the designation of the varied arguments (q_i, q_k) will subsequently be omitted from P, Q, aj, cj. The coefficients aj, cj are continuous and differentiable functions of their arguments, and direct substitution shows that $a_0 + c_0 \neq 0 - \frac{\left[\frac{n \cdot 2}{N} - \left(1 + \frac{r}{R_s}\right)\right]}{(L_s + M) \cdot C_s} \neq 0$ for $q_i, q_k \in \mathbb{R}_+$; that is, $\lambda = 0$ is not of $P(\lambda) + Q(\lambda) \cdot e^{-\lambda \cdot \tau} = 0$. Furthermore, $P(\lambda)$, $Q(\lambda)$ are analytic functions of λ, for which the following requirements of the analysis can be verified in the present case.

1. If $\lambda = i \cdot \omega$, $\omega \in \mathbb{R}$, then $P(i \cdot \omega) + Q(i \cdot \omega) \neq 0$.
2. If $|\frac{Q(\lambda)}{P(\lambda)}|$ is bounded for $|\lambda| \to \infty$; $\text{Re}\lambda \geq 0$. No roots bifurcation from ∞.
3. $F(\omega) = |P(i \cdot \omega)|^2 - |Q(i \cdot \omega)|^2$ has a finite number of zeros. Indeed, this is a polynomial in ω.
4. Each positive root $\omega(q_i, q_k)$ of $F(\omega) = 0$ is continuous and differentiable with respect to q_i, q_k.

We assume that $P_n(\lambda, \tau)$ and $Q_m(\lambda, \tau)$ cannot have common imaginary roots. That is for any real number ω; $P_n(\lambda = i \cdot \omega, \tau) + Q_m(\lambda = i \cdot \omega, \tau) \neq 0$.

$$P_n(\lambda = i \cdot \omega, \tau) = -\omega^2 - i \cdot \omega \cdot \frac{\left[\frac{n \cdot 2 \cdot \tau_s}{N} - \left(\frac{L_s + M}{R_s} + r \cdot C_s\right)\right]}{(L_s + M) \cdot C_s}$$

$$Q_m(\lambda = i \cdot \omega, \tau) = -\frac{\left[\frac{n \cdot 2}{N} - \left(1 + \frac{r}{R_s}\right)\right]}{(L_s + M) \cdot C_s}$$

$$|P(i \cdot \omega, \tau)|^2 = \omega^4 + \omega^2 \cdot \left(\frac{\left[\frac{n \cdot 2 \cdot \tau_s}{N} - \left(\frac{L_s + M}{R_s} + r \cdot C_s\right)\right]}{(L_s + M) \cdot C_s}\right)^2$$

$$|Q(i \cdot \omega, \tau)|^2 = \left(\frac{\left[\frac{n \cdot 2}{N} - \left(1 + \frac{r}{R_s}\right)\right]}{(L_s + M) \cdot C_s}\right)^2$$

$$\begin{aligned} F(\omega, \tau) &= |P(i \cdot \omega, \tau)|^2 - |Q(i \cdot \omega, \tau)|^2 \\ &= \omega^4 + \omega^2 \cdot \left(\frac{\left[\frac{n \cdot 2 \cdot \tau_s}{N} - \left(\frac{L_s + M}{R_s} + r \cdot C_s\right)\right]}{(L_s + M) \cdot C_s}\right)^2 \\ &\quad - \left(\frac{\left[\frac{n \cdot 2}{N} - \left(1 + \frac{r}{R_s}\right)\right]}{(L_s + M) \cdot C_s}\right)^2 \end{aligned}$$

We define the following parameters for simplicity: $\Pi_0 = -\left(\frac{\left[\frac{n \cdot 2}{N} - \left(1 + \frac{r}{R_s}\right)\right]}{(L_s + M) \cdot C_s}\right)^2$

$\Pi_2 = \left(\frac{\left[\frac{n \cdot 2 \cdot \tau_s}{N} - \left(\frac{L_s + M}{R_s} + r \cdot C_s \right) \right]}{(L_s + M) \cdot C_s} \right)^2$; $\Pi_4 = 1$. Hence $\sum_{k=0}^{2} \Pi_{2 \cdot k} \cdot \omega^{2 \cdot k} = 0$ and its roots are given by solving the above polynomial.

$$P_R(i\omega, \tau) = -\omega^2$$

$$P_I(i\omega, \tau) = -\omega \cdot \frac{\left[\frac{n \cdot 2 \cdot \tau_s}{N} - \left(\frac{L_s + M}{R_s} + r \cdot C_s \right) \right]}{(L_s + M) \cdot C_s}$$

$$Q_R(i\omega, \tau) = -\frac{\left[\frac{n \cdot 2}{N} - \left(1 + \frac{r}{R_s} \right) \right]}{(L_s + M) \cdot C_s}$$

$$Q_I(i\omega, \tau) = 0$$

$$\sin\theta(\tau) = \frac{-P_R(i \cdot \omega, \tau) \cdot Q_I(i \cdot \omega, \tau) + P_I(i \cdot \omega, \tau) \cdot Q_R(i \cdot \omega, \tau)}{|Q(i \cdot \omega, \tau)|^2}$$

$$\cos\theta(\tau) = -\frac{P_R(i \cdot \omega, \tau) \cdot Q_R(i \cdot \omega, \tau) + P_I(i \cdot \omega, \tau) \cdot Q_I(i \cdot \omega, \tau)}{|Q(i \cdot \omega, \tau)|^2}$$

We use different parameters terminology from our last characteristics parameters definition: $k \to j$; $p_k(\tau) \to a_j$; $q_k(\tau) \to c_j$; $n = 2$; $m = 0$; $n > m$; $P_n(\lambda, \tau) \to P(\lambda)$, $Q_m(\lambda, \tau) \to Q(\lambda)$; $P(\lambda) = \sum_{j=0}^{2} a_j \cdot \lambda^j$; $Q(\lambda) = \sum_{j=0}^{0} c_j \cdot \lambda^j$; $P(\lambda) = a_0 + a_1 \cdot \lambda + a_2 \cdot \lambda^2$; $Q(\lambda) = c_0$ $n, m \in \mathbb{N}_0$; $n > m$ and $a_j, c_j : \mathbb{R}_{0+} \to \mathbb{R}$ are continuous and differential function of τ such that $a_0 + c_0 \neq 0$. In the following "–" denotes complex and conjugate. $P(\lambda)$, $Q(\lambda)$ are analytic functions in λ and differentiable in τ. The coefficients $a_j(V_{meas,1}, X) \in \mathbb{R}$ and $c_j(V_{meas,1}, X) \in \mathbb{R}$ depend on B-dot probe system $V_{meas,1}$, X values. Unless strictly necessary, the designation of the varied arguments: $(n, N, r, R_s, L_s, M, C_s, \tau, \ldots)$ will subsequently be omitted from P, Q, a_j, c_j. The coefficients a_j, c_j are continuous and differentiable functions of their arguments. Direct substitutions show that $a_0 + c_0 \neq 0$.

$$a_0 = 0; \; c_0 = -\frac{\left[\frac{n \cdot 2}{N} - \left(1 + \frac{r}{R_s} \right) \right]}{(L_s + M) \cdot C_s}; \; a_0 + c_0 = -\frac{\left[\frac{n \cdot 2}{N} - \left(1 + \frac{r}{R_s} \right) \right]}{(L_s + M) \cdot C_s} \neq 0$$

$$\frac{n \cdot 2}{N} \neq \left(1 + \frac{r}{R_s} \right)$$

$\forall n, N, r, R_s, L_s, M, C_s, \tau, \ldots \in \mathbb{R}_+$ i.e. $\lambda = 0$ is not a root of the characteristic equation. Furthermore $P(\lambda)$, $Q(\lambda)$ are analytic functions of λ for which the following requirements of the analysis can also be verified in the present case.

1. If $\lambda = i \cdot \omega$; $\omega \in \mathbb{R}$ then $P(i \cdot \omega) + Q(i \cdot \omega) \neq 0$, i.e. P and Q have no common imaginary roots. This condition was verified numerically in the entire $(n, N, r, R_s, L_s, M, C_s, \tau, \ldots)$ domain of interest.

2. $|\frac{P(\lambda)}{Q(\lambda)}|$ is bounded for $|\lambda| \to \infty$; $\text{Re}\lambda \geq 0$. No roots bifurcation from ∞. Indeed, in the limit:

$$\left|\frac{P(\lambda)}{Q(\lambda)}\right| = \left|\frac{\lambda^2 - \lambda \cdot \frac{\left[\frac{n \cdot 2 \cdot \tau_S}{N} - \left(\frac{L_s + M}{R_s} + r \cdot C_s\right)\right]}{(L_s + M) \cdot C_s}}{-\frac{\left[\frac{n \cdot 2}{N} - \left(1 + \frac{r}{R_S}\right)\right]}{(L_s + M) \cdot C_s}}\right|.$$

3. The following expressions exist: $F(\omega, \tau) = |P(i \cdot \omega, \tau)|^2 - |Q(i \cdot \omega, \tau)|^2 = \sum_{k=0}^{2} \Pi_{2 \cdot k} \cdot \omega^{2 \cdot k}$ has at most a finite number of zeros. Indeed, this is a polynomial in ω (degree in ω^4).
4. Each positive root $\omega(n, N, r, R_s, L_s, M, C_s, \tau, \ldots)$ of $F(\omega) = 0$ is continuous and differentiable with respect to $n, N, r, R_s, L_s, M, C_s, \tau, \ldots$ The condition can only be assessed numerically.

In addition, since the coefficients in P and Q are real, we have $\overline{P(-i \cdot \omega)} = P(i \cdot \omega) \overline{Q(-i \cdot \omega)} = Q(i \cdot \omega)$ thus, $\omega > 0$ maybe on eigenvalue of characteristic equations. The analysis consists in identifying the roots of the characteristic equation situated on the imaginary axis of the complex λ-plane, whereby increasing the parameters: $n, N, r, R_s, L_s, M, C_s, \tau, \ldots$ Re λ may, at crossing, change its sign from (−) to (+), i.e. from a stable focus $V^*_{meas,1} = 0$; $X^* = 0$ to an unstable one, or vice versa. This feature may be further assessed by examining the sign of the partial derivatives with respect to $n, N, r, R_s, L_s, M, C_s, \tau, \ldots$ and system parameters.

$$\Lambda^{-1}(\tau) = \left(\frac{\partial \text{Re}\lambda}{\partial \tau}\right)_{\lambda = i \cdot \omega}; \quad n, N, r, R_s, L_s, M, C_s, \ldots = \text{const}$$

$$\Lambda^{-1}(\tau) = \left(\frac{\partial \text{Re}\lambda}{\partial C_s}\right)_{\lambda = i \cdot \omega}; \quad n, N, r, R_s, L_s, M, \tau, \ldots = \text{const}$$

$$\Lambda^{-1}(\tau) = \left(\frac{\partial \text{Re}\lambda}{\partial M}\right)_{\lambda = i \cdot \omega}; \quad n, N, r, R_s, L_s, \tau, C_s, \ldots = \text{const}$$

$$\Lambda^{-1}(\tau) = \left(\frac{\partial \text{Re}\lambda}{\partial L_s}\right)_{\lambda = i \cdot \omega}; \quad n, N, r, R_s, \tau, M, C_s, \ldots = \text{const}$$

$$\Lambda^{-1}(\tau) = \left(\frac{\partial \text{Re}\lambda}{\partial R_s}\right)_{\lambda = i \cdot \omega}; \quad n, N, r, \tau, L_s, M, C_s, \ldots = \text{const}$$

$$\Lambda^{-1}(\tau) = \left(\frac{\partial \text{Re}\lambda}{\partial r}\right)_{\lambda = i \cdot \omega}; \quad n, N, \tau, R_s, L_s, M, C_s, \ldots = \text{const}$$

$$\Lambda^{-1}(\tau) = \left(\frac{\partial \text{Re}\lambda}{\partial N}\right)_{\lambda = i \cdot \omega}; \quad n, \tau, r, R_s, L_s, M, C_s, \ldots = \text{const}$$

$$\Lambda^{-1}(\tau) = \left(\frac{\partial \text{Re}\lambda}{\partial n}\right)_{\lambda = i \cdot \omega}; \quad \tau, N, r, R_s, L_s, M, C_s, \ldots = \text{const}$$

$P(\lambda) = P_R(\lambda) + i \cdot P_I(\lambda)$; $Q(\lambda) = Q_R(\lambda) + i \cdot Q_I(\lambda)$ when writing and inserting $\lambda = i \cdot \omega$ into B-dot probe system's characteristic equation ω must satisfy the following equations.

$$\sin(\omega \cdot \tau) = g(\omega) = \frac{-P_R(i \cdot \omega) \cdot Q_I(i \cdot \omega) + P_I(i \cdot \omega) \cdot Q_R(i \cdot \omega)}{|Q(i \cdot \omega)|^2}$$

$$\cos(\omega \cdot \tau) = h(\omega) = -\frac{P_R(i \cdot \omega) \cdot Q_R(i \cdot \omega) + P_I(i \cdot \omega) \cdot Q_I(i \cdot \omega)}{|Q(i \cdot \omega)|^2}$$

where $|Q(i \cdot \omega)|^2 \neq 0$ in view of requirement (a) above, and $g, h \in \mathbb{R}$. Furthermore, it follows above $\sin(\omega \cdot \tau)$ and $\cos(\omega \cdot \tau)$ equations that, by squaring and adding the sides, ω must be a positive root of $F(\omega, \tau) = |P(i \cdot \omega, \tau)|^2 - |Q(i \cdot \omega, \tau)|^2 = 0$. Note: $F(\omega)$ is dependent on τ. Now it is important to notice that if $\tau \notin I$ (assume that $I \subseteq \mathbb{R}_{+0}$ is the set where $\omega(\tau)$ is a positive root of $F(\omega)$ and for, $\tau \notin I, \omega(\tau)$ is not defined). Then for all τ in I, $\omega(\tau)$ is satisfied that $F(\omega, \tau) = 0$. Then there are no positive $\omega(\tau)$ solutions for $F(\omega, \tau) = 0$, and we cannot have stability switches. For $\tau \in I$ where $\omega(\tau)$ is a positive solution of $F(\omega, \tau) = 0$, we can define the angle $\theta(\tau) \in [0, 2 \cdot \pi]$ as the solution of $\sin\theta(\tau) = \cdots$ $\cos\theta(\tau) = \cdots$

$$\sin\theta(\tau) = \frac{-P_R(i \cdot \omega) \cdot Q_I(i \cdot \omega) + P_I(i \cdot \omega) \cdot Q_R(i \cdot \omega)}{|Q(i \cdot \omega)|^2}$$

$$\cos\theta(\tau) = -\frac{P_R(i \cdot \omega) \cdot Q_R(i \cdot \omega) + P_I(i \cdot \omega) \cdot Q_I(i \cdot \omega)}{|Q(i \cdot \omega)|^2}$$

And the relation between the argument $\theta(\tau)$ and $\tau \cdot \omega(\tau)$ for $\tau \in I$ must be as describe below. $\omega(\tau) \cdot \tau = \theta(\tau) + 2 \cdot n \cdot \pi \, \forall n \in \mathbb{N}_0$, Hence we can define the maps $\tau_n : I \to \mathbb{R}_{+0}$ given by $\tau_n(\tau) = \frac{\theta(\tau)+2 \cdot n \cdot \pi}{\omega(\tau)}$; $n \in \mathbb{N}_0$; $\tau \in I$. Let introduce the functions: $I \to \mathbb{R}$, $S_n(\tau) = \tau - \tau_n(\tau)$; $\tau \in I$; $n \in \mathbb{N}_0$ that is continuous and differentiable in τ. In the following, the subscripts $\lambda, \omega, n, N, r, R_s, L_s, M, C_s, \ldots$ indicate the corresponding partial derivatives. Let us concentrate on $\Lambda(x)$, remember in $\lambda(n, N, r, R_s, L_s, M, C_s, \ldots)$ and $\omega(n, N, r, R_s, L_s, M, C_s, \ldots)$, and keeping all parameters except one (x) and τ. The derivation closely follows that in references. Differentiating B-dot probe system characteristic equation $P(\lambda) + Q(\lambda) \cdot e^{-\lambda \cdot \tau} = 0$ with respect to specific parameter (x), and inverting the derivative, for convenience, one calculates: $x = n, N, r, R_s, L_s, M, C_s, \ldots$

$$\left(\frac{\partial \lambda}{\partial x}\right)^{-1} = \frac{-P_\lambda(\lambda, x) \cdot Q(\lambda, x) + Q_\lambda(\lambda, x) \cdot P(\lambda, x) - \tau \cdot P(\lambda, x) \cdot Q(\lambda, x)}{P_x(\lambda, x) \cdot Q(\lambda, x) - Q_x(\lambda, x) \cdot P(\lambda, x)}$$

where $P_\lambda = \frac{\partial P}{\partial \lambda}$; $Q_\lambda = \frac{\partial Q}{\partial \lambda}$, ….etc. Substituting $\lambda = i \cdot \omega$ and bearing $\overline{P(-i \cdot \omega)} = P(i \cdot \omega)$ and $\overline{Q(-i \cdot \omega)} = Q(i \cdot \omega), i \cdot P_\lambda(i \cdot \omega) = P_\omega(i \cdot \omega)$; $i \cdot Q_\lambda(i \cdot \omega) = Q_\omega(i \cdot \omega)$ and that on the surface $|P(i \cdot \omega)|^2 = |Q(i \cdot \omega)|^2$, one obtains:

$$\left(\frac{\partial\lambda}{\partial x}\right)^{-1}\Big|_{\lambda=i\cdot\omega}$$
$$= \left(\frac{i\cdot P_\omega(i\cdot\omega,x)\cdot\overline{P(i\cdot\omega,x)}+i\cdot Q_\lambda(i\cdot\omega,x)\cdot\overline{Q(\lambda,x)}-\tau\cdot|P(i\cdot\omega,x)|^2}{P_x(i\cdot\omega,x)\cdot\overline{P(i\cdot\omega,x)}-Q_x(i\cdot\omega,x)\cdot\overline{Q(i\cdot\omega,x)}}\right.$$

Upon separating into real and imaginary parts, with $P = P_R + i\cdot P_I$; $Q = Q_R + i\cdot Q_I$.

$$P_\omega = P_{R\omega}+i\cdot P_{I\omega};\ Q_\omega = Q_{R\omega}+i\cdot Q_{I\omega};\ P_x = P_{Rx}+i\cdot P_{Ix}$$
$$Q_x = Q_{Rx}+i\cdot Q_{Ix};\ P^2 = P_R^2+P_I^2$$

When (x) can be any B-dot probe parameter's $n, N, r, R_s, L_s, M, C_s, \ldots$ and time delay τ, etc. Where for convenience, we have dropped the arguments $(i\cdot\omega, x)$, and where $F_\omega = 2\cdot[(P_{R\omega}\cdot P_R + P_{I\omega}\cdot P_I)-(Q_{R\omega}\cdot Q_R + Q_{I\omega}\cdot Q_I)]$; $F_x = 2\cdot[(P_{Rx}\cdot P_R + P_{Ix}\cdot P_I)-(Q_{Rx}\cdot Q_R + Q_{Ix}\cdot Q_I)]$ $\omega_x = -\frac{F_x}{F_\omega}$ and we define U and V:

$$U = (P_R\cdot P_{I\omega}-P_I\cdot P_{R\omega})-(Q_R\cdot Q_{I\omega}-Q_I\cdot Q_{R\omega})$$
$$V = (P_R\cdot P_{Ix}-P_I\cdot P_{Rx})-(Q_R\cdot Q_{Ix}-Q_I\cdot Q_{Rx})$$

We choose our specific parameter as time delay $x=\tau$.

$$P_R = -\omega^2;\ P_I = -\omega\cdot\frac{\left[\frac{n\cdot 2\cdot\tau_s}{N}-\left(\frac{L_s+M}{R_s}+r\cdot C_s\right)\right]}{(L_s+M)\cdot C_s}$$
$$Q_R = -\frac{\left[\frac{n\cdot 2}{N}-\left(1+\frac{r}{R_s}\right)\right]}{(L_s+M)\cdot C_s};\ Q_I = 0$$
$$P_{R\omega} = -2\cdot\omega;\ P_{I\omega} = -\frac{\left[\frac{n\cdot 2\cdot\tau_s}{N}-\left(\frac{L_s+M}{R_s}+r\cdot C_s\right)\right]}{(L_s+M)\cdot C_s}$$
$$Q_{R\omega}=0;\ Q_{I\omega}=0$$

$$P_{R\tau}=0;\ P_{I\tau}=0;\ Q_{R\tau}=0;\ Q_{I\tau}=0;\ \omega_\tau = -\frac{F_\tau}{F_\omega}$$

$$P_{R\omega}\cdot P_R = 2\cdot\omega^3;\ Q_{R\omega}\cdot Q_R = 0;\ F_\tau = 0$$
$$P_R\cdot P_{I\omega} = \omega^2\cdot\frac{\left[\frac{n\cdot 2\cdot\tau_s}{N}-\left(\frac{L_s+M}{R_s}+r\cdot C_s\right)\right]}{(L_s+M)\cdot C_s}$$

$$P_I \cdot P_{R\omega} = 2 \cdot \omega^2 \cdot \frac{\left[\frac{n \cdot 2 \cdot \tau_s}{N} - \left(\frac{L_s + M}{R_s} + r \cdot C_s\right)\right]}{(L_s + M) \cdot C_s}$$
$$Q_R \cdot Q_{I\omega} = 0; \;\; Q_I \cdot Q_{R\omega} = 0$$

$$V = (P_R \cdot P_{I\tau} - P_I \cdot P_{R\tau}) - (Q_R \cdot Q_{I\tau} - Q_I \cdot Q_{R\tau}) = 0; \;\; F(\omega, \tau) = 0$$

Differentiating with respect to τ and we get

$$F_\omega \cdot \frac{\partial \omega}{\partial \tau} + F_\tau = 0; \;\; \tau \in I \Rightarrow \frac{\partial \omega}{\partial \tau} = -\frac{F_\tau}{F_\omega}$$
$$\Lambda^{-1}(\tau) = \left(\frac{\partial \mathrm{Re}\lambda}{\partial \tau}\right)_{\lambda = i \cdot \omega}; \;\; \frac{\partial \omega}{\partial \tau} = \omega_\tau = -\frac{F_\tau}{F_\omega}$$
$$\Lambda^{-1}(\tau) = \mathrm{Re}\left\{\frac{-2 \cdot [U + \tau \cdot |P|^2] + i \cdot F_\omega}{F_\tau + i \cdot 2 \cdot [V + \omega \cdot |P|^2]}\right\}$$
$$\mathrm{sign}\{\Lambda^{-1}(\tau)\} = \mathrm{sign}\{\left(\frac{\partial \mathrm{Re}\lambda}{\partial \tau}\right)_{\lambda = i \cdot \omega}\}$$
$$\mathrm{sign}[\Lambda^{-1}(\tau)] = \mathrm{sign}\{F_\omega\} \cdot \mathrm{sign}\left\{\frac{V + \frac{\partial \omega}{\partial \tau} \cdot U}{|P|^2} + \omega + \frac{\partial \omega}{\partial \tau} \cdot \tau\right\}$$

We shall presently examine the possibility of stability transitions (bifurcations) of B-dot probe system, about the equilibrium point $V^*_{meas,1} = 0; \;\; X^* = 0$ as a result of a variation of delay parameter τ. The analysis consists in identifying the roots of our B-dot probe system characteristic equation situated on the imaginary axis of the complex λ-plane, where by increasing the delay parameter τ, Reλ may at the crossing, changes its sign from "−" to "+", i.e. from a stable focus $E^{(*)}$ to an unstable one, or vice versa. This feature may be further assessed by examining the sign of the partial derivatives with respect to τ.

$$\Lambda^{-1}(\tau) = \left(\frac{\partial \mathrm{Re}\lambda}{\partial \tau}\right)_{\lambda = i \cdot \omega}; \;\; n, N, r, R_s, L_s, M, C_s, \ldots = \mathrm{const}$$

$$V = 0; \;\; U = (P_R \cdot P_{I\omega} - P_I \cdot P_{R\omega}) - (Q_R \cdot Q_{I\omega} - Q_I \cdot Q_{R\omega})$$
$$= -\omega^2 \cdot \frac{\left[\frac{n \cdot 2 \cdot \tau_s}{N} - \left(\frac{L_s + M}{R_s} + r \cdot C_s\right)\right]}{(L_s + M) \cdot C_s}$$

We get the expression for $F(\omega, \tau)$ B-dot probe parameter values. We find those ω, τ values which fulfill $F(\omega, \tau) = 0$. We ignore negative, complex, and imaginary values of ω for specific τ values. $\tau \in [0.001 \ldots 10]$ Second, we can be express by 3D function $F(\omega, \tau) = 0$. We plot the stability switch diagram based on different delay values of our B-dot probe system.

$$\Lambda^{-1}(\tau) = \left(\frac{\partial \mathrm{Re}\lambda}{\partial \tau}\right)_{\lambda=i\cdot\omega} = \frac{2\cdot\{F_{\omega}\cdot(V+\omega\cdot P^2) - F_{\tau}\cdot(U+\tau\cdot P^2)\}}{F_{\tau}^2 + 4\cdot(V+\omega\cdot P^2)^2}$$

The stability switch occurs only on those delay values (τ) which fit the equation: $\tau = \frac{\theta_+(\tau)}{\omega_+(\tau)}$ and $\theta_+(\tau)$ is the solution of $\sin\theta(\tau) = \ldots$; $\cos\theta(\tau) = \ldots$ when $\omega = \omega_+(\tau)$ if only ω_+ is feasible. Additionally, when all B-dot probe parameters are known and the stability switch due to various time delays τ is described in the following expression:

$$\mathrm{sign}\{\Lambda^{-1}(\tau)\} = \mathrm{sign}\{F_{\omega}(\omega(\tau),\tau)\}\cdot\mathrm{sign}\left\{\tau\cdot\omega_{\tau}(\omega(\tau)) + \omega(\tau) + \frac{U(\omega(\tau))\cdot\omega_{\tau}(\omega(\tau)) + V(\omega(\tau))}{|P(\omega(\tau))|^2}\right\}$$

Remark: we know $F(\omega,\tau) = 0$ implies roots $\omega_i(\tau)$ and finding those delays values τ which ω_i is feasible. There are τ values which give complex ω_i or imaginary number, then unable to analyze stability. F function is independent on τ parameter $F(\omega,\tau) = 0$.

The results: We find those ω, τ values which fulfill $F(\omega,\tau) = 0$. We ignore negative, complex, and imaginary values of ω.

$$g(\tau) = \Lambda^{-1}(\tau) = \left(\frac{\partial \mathrm{Re}\lambda}{\partial \tau}\right)_{\lambda=i\cdot\omega} = \frac{2\cdot[F_{\omega}\cdot(V+\omega\cdot P^2) - F_{\tau}\cdot(U+\tau\cdot P^2)]}{F_{\tau}^2 + 4\cdot(V+\omega\cdot P^2)^2}$$

$$\mathrm{sign}[g(\tau)] = \mathrm{sign}[\Lambda^{-1}(\tau)] = \mathrm{sign}\left[\left(\frac{\partial \mathrm{Re}\lambda}{\partial \tau}\right)_{\lambda=i\cdot\omega}\right] = \mathrm{sign}\left\{\frac{2\cdot[F_{\omega}\cdot(V+\omega\cdot P^2) - F_{\tau}\cdot(U+\tau\cdot P^2)]}{F_{\tau}^2 + 4\cdot(V+\omega\cdot P^2)^2}\right\}$$

$$F_{\tau}^2 + 4\cdot(V+\omega\cdot P^2)^2 > 0 \Rightarrow \mathrm{sign}[\Lambda^{-1}(\tau)] = \mathrm{sign}[F_{\omega}\cdot(V+\omega\cdot P^2) - F_{\tau}\cdot(U+\tau\cdot P^2)]$$

$$\mathrm{sign}[\Lambda^{-1}(\tau)] = \mathrm{sign}\left\{\mathrm{F}_{\omega}\cdot\left[(V+\omega\cdot P^2) - \frac{F_{\tau}}{F_{\omega}}\cdot(U+\tau\cdot P^2)\right]\right\}$$

$$\omega_{\tau} = -\frac{F_{\tau}}{F_{\omega}}$$

$$\mathrm{sign}[\Lambda^{-1}(\tau)] = \mathrm{sign}\{\mathrm{F}_{\omega}\cdot[V+\omega\cdot P^2 + \omega_{\tau}\cdot U + \omega_{\tau}\cdot\tau\cdot P^2]\}$$

$$\mathrm{sign}[\Lambda^{-1}(\tau)] = \mathrm{sign}\{\mathrm{F}_{\omega}\}\cdot\mathrm{sign}\{V + \omega_{\tau}\cdot U + \omega_{\tau}\cdot\tau\cdot P^2 + \omega\cdot P^2\}$$

$$\operatorname{sign}[\Lambda^{-1}(\tau)] = \operatorname{sign}\{F_\omega\} \cdot \operatorname{sign}\left\{\frac{1}{P^2} \cdot \left[\frac{V + \omega_\tau \cdot U}{P^2} + \omega_\tau \cdot \tau + \omega\right]\right\}$$
$$= \operatorname{sign}\{F_\omega\} \cdot \operatorname{sign}\left\{\frac{1}{P^2}\right\} \cdot \operatorname{sign}\left\{\frac{V + \omega_\tau \cdot U}{P^2} + \omega_\tau \cdot \tau + \omega\right\}$$

$$\operatorname{sign}\left\{\frac{1}{P^2}\right\} > 0 \Rightarrow \operatorname{sign}[\Lambda^{-1}(\tau)] = \operatorname{sign}\{F_\omega\} \cdot \operatorname{sign}\left\{\frac{V + \omega_\tau \cdot U}{P^2} + \omega_\tau \cdot \tau + \omega\right\}$$

$$P_{I\omega} \cdot P_I = \omega \cdot \frac{\left[\frac{n \cdot 2 \cdot \tau_s}{N} - \left(\frac{L_s + M}{R_s} + r \cdot C_s\right)\right]^2}{(L_s + M)^2 \cdot C_s^2}$$
$$Q_{R\omega} \cdot Q_R = 0; \; Q_{I\omega} \cdot Q_I = 0; \; V = 0$$

$$\omega_\tau = -\frac{F_\tau}{F_\omega}; \; F_\tau = 2 \cdot [(P_{R\tau} \cdot P_R + P_{I\tau} \cdot P_I) - (Q_{R\tau} \cdot Q_R + Q_{I\tau} \cdot Q_I)] = 0; \; \omega_\tau = 0$$

$$U = (P_R \cdot P_{I\omega} - P_I \cdot P_{R\omega}) - (Q_R \cdot Q_{I\omega} - Q_I \cdot Q_{R\omega})$$
$$= -\omega^2 \cdot \frac{\left[\frac{n \cdot 2 \cdot \tau_s}{N} - \left(\frac{L_s + M}{R_s} + r \cdot C_s\right)\right]}{(L_s + M) \cdot C_s}$$

$$P^2 = P_I^2 + P_R^2 = \omega^2 \cdot \left\{\frac{\left[\frac{n \cdot 2 \cdot \tau_s}{N} - \left(\frac{L_s + M}{R_s} + r \cdot C_s\right)\right]^2}{(L_s + M)^2 \cdot C_s^2} + \omega^2\right\}$$

$$F_\omega = 2 \cdot [(P_{R\omega} \cdot P_R + P_{I\omega} \cdot P_I) - (Q_{R\omega} \cdot Q_R + Q_{I\omega} \cdot Q_I)]$$
$$= 2 \cdot \omega \cdot \left[2 \cdot \omega^2 + \frac{\left[\frac{n \cdot 2 \cdot \tau_s}{N} - \left(\frac{L_s + M}{R_s} + r \cdot C_s\right)\right]^2}{(L_s + M)^2 \cdot C_s^2}\right]$$

$$\operatorname{sign}[\Lambda^{-1}(\tau)] = \operatorname{sign}\left\{2 \cdot \omega \cdot \left[2 \cdot \omega^2 + \frac{\left[\frac{n \cdot 2 \cdot \tau_s}{N} - \left(\frac{L_s + M}{R_s} + r \cdot C_s\right)\right]^2}{(L_s + M)^2 \cdot C_s^2}\right]\right\} \cdot \operatorname{sign}\{\omega\}$$

We check the sign of $\Lambda^{-1}(\tau)$ according the following rule:

If $\operatorname{sign}[\Lambda^{-1}(\tau)] > 0$, then the crossing proceeds from (−) to (+) respectively (stable to unstable). If $\operatorname{sign}[\Lambda^{-1}(\tau)] < 0$, then the crossing proceeds from (+) to (−) respectively (unstable to stable) (Table B.5).

Table B.5 sign of $\Lambda^{-1}(\tau)$ function

$\text{sign}\{F_\omega\}$	$\text{sign}\left\{\frac{V+\omega_\tau \cdot U}{P^2} + \omega_\tau \cdot \tau + \omega\right\}$	$\text{sign}[\Lambda^{-1}(\tau)]$
+/−	+/−	+
+/−	−/+	−

References

[1] I.H. Hutchinson, *Principles of Plasma Diagnostics* (Cambridge University Press, Cambridge, 2005)

[2] R.L. Merlino, Understanding Langmuir probe current-voltage characteristics. Am. J. Phys. **75**(12) (2007)

[3] J. Adamek, C. Ionita, R. Schrittwieser, J. Stockel, M. Tichy, G. Van Oost, Direct measurements of the electron temperature by a ball-pen/Langmuir probe, in *32nd EPS Conference of Plasma Physics Tarragona*, 27-June–1 July 2005, ECA, vol 29c, p 5.081 (2005)

[4] S. Meshkani, M. Ghoranneviss, A. Salar Elahi, M. Lafouti, Design and fabrication of competitive Langmuir ball-pen probe (LBP) for Tokamak. J Fusion Energ. **34**, 394–397 (2015)

Appendix C
Laser Beam Shaping, Jitter and Crosstalk

Laser beam shaping is the process of redistributing the irradiance and phase of a beam of optical radiation. Irradiance is the power (radiant flux) received by a surface per unit area. The SI unit of irradiant is watt per square meter $\left(\frac{\mathrm{W}}{\mathrm{m}^2}\right)$. Spectral irradiance is the irradiance of a surface per unit frequency or wavelength, depending on whether the spectrum is taken as a function of frequency or wavelength. The two forms have different dimensions: spectral irradiance of a frequency spectrum is measured in watts per square meter per Hertz $\left(\frac{\mathrm{W}}{\mathrm{m}^2\,\mathrm{Hz}}\right)$, while spectral irradiance of a wavelength spectrum is measured in watts per square meter per meter $\left(\frac{\mathrm{W}}{\mathrm{m}^3}\right)$, or more commonly watts per square meter per nanometer $\left(\frac{\mathrm{W}}{\mathrm{m}^2\,\mathrm{nm}}\right)$. The beam shape is defined by the irradiance distribution and the phase of the shaped beam which is a major factor in determining the propagation properties of the beam profile. The applications of beam shaping are laser/material processing, laser/material interaction, laser weapons, optical data/image processing, lithography, printing, and laser art patterns. There are techniques for producing shaped beams that are uniform over a specified cross-section. The shaped beam cross section may be arbitrary, including rectangular, circular, triangular, hexagonal, and ring shaped. Laser beam shaping techniques can be divided into three broad classes [1–3]. The first technique is aperture of the beam, the beam is expanded and an aperture is used to select a suitably flat portion of the beam. Resulting irradiance can be imaged with magnification to control the size of the output beam. This technique is not lossless. There is a desire that the beam shaping operation conserve energy. If the input beam's irradiance is not suitably smooth, not possible to find an aperture size and position that give the desired result. The other way is when a Gaussian laser beam encounters a circular aperture and the fraction of the power passing through is equal to $1-e^{-\left\{\frac{2a^2}{w^2}\right\}}$; a—radius of the aperture, w—radial distance from the beam's center to the point where the beam intensity is 0.135 of the intensity at the center of the beam (Fig. C.1). we define r as the radius at some point in the beam and I_0 as the beam intensity at the center of the beam (r = 0). Then the beam intensity as a function of radius is $I(r) = I_0 \cdot e^{\left(-\frac{2 \cdot r^2}{w^2}\right)}$, The power as a function of size of the aperture is computed from $\frac{P(a)}{P_0} = \int_0^a I(r) \cdot \frac{2 \cdot \pi \cdot r \cdot dr}{P_0} = 1 - e^{\left(-\frac{2 \cdot a^2}{w^2}\right)}$.

O. Aluf, *Advance Elements of Laser Circuits and Systems*,
https://doi.org/10.1007/978-3-030-64103-0

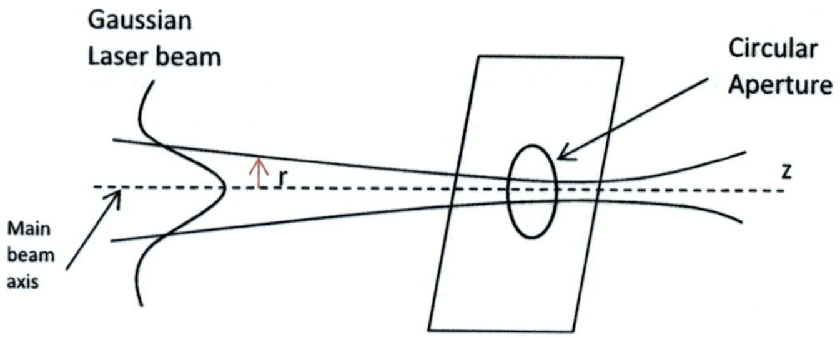

Fig. C.1 Gaussian laser beam and circular aperture system

Fig C.2 Mode locked semiconductor laser—intense and short optical pulses

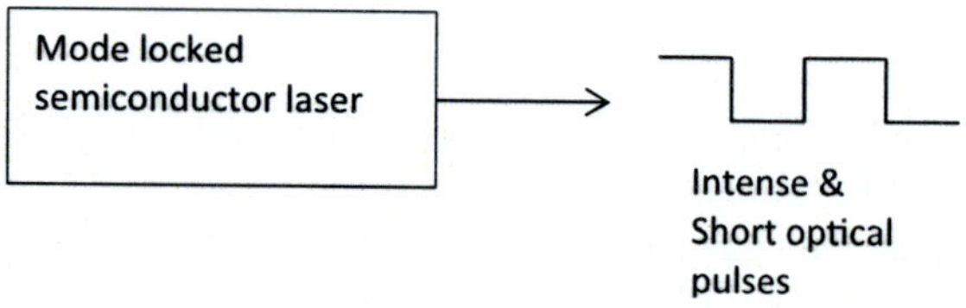

The 0.135 comes from the fact that at the $1/e^2$ point of the beam, the intensity is down to 0.135 of the intensity in the center of the beam. If the aperture a ($a = 3 \cdot w$), it transmits 99% of the beam. When an electromagnetic wave encounters an aperture, diffraction will occur. In the far field of the aperture, there are fringes, rings, and other artifacts of diffraction superimposed on the geometric optics result of a Gaussian beam with the edges clipped off.

The second major technique for beam shaping is field mapping. Fields mappers transform the input field into the desired field in a controlled manner. Basic case is the mapping of a single mode Gaussian beam into a beam with a uniform irradiance. They are uniformly distributed in the output plane. The ray bending defines a wave front that can be associated with an optical phase element and the field mappers can be made effectively lossless. The field mapping approach to beam shaping is applicable to well defined single mode laser beams. The third technique of beam shapers is beam integrators (beam homogenizers). The input beam is broken up into beam lets by a lens let array and superimposed in the output plane by a primary lens. Integrator is for the fact that the output pattern is a sum of diffraction patterns determined by the lens let apertures. Beam integrators are suited to multimode lasers with a relatively low degree of spatial coherence and they can design effectively lossless.

Mode locked semiconductor lasers: Typical semiconductor laser is a diode laser which can be used for generating ultrashort pulses with various techniques of mode locking or with gain switching (Fig. C.2). Typically, pulses with duration between 0.5 and 5 ps and pulse repetition rates between 1 GHz and hundreds of GHz are generated with mode locking. The main application of ultrafast diode lasers is in optical fiber communications systems, where such lasers function as pulse, sources of fast data transmitters or for all optical signal processing. There are some techniques for mode locking of diode lasers, active mode locking, passive mode locking, and hybrid mode locking. Active mode locking can be accomplished with an optical modulator in the laser resonator. This is an electro absorption modulator in the form of an un-pumped region with some modulated voltage, or an amplifying section where the drive current is modulated. Passive mode locking relies on a saturate absorber in the resonator, which is an un-pumped section of the device. We need to apply an electrical bias for adjusting the absorber properties. The recovery time of that king of absorber is fairly long. Short recovery times are achieved by implanting nitrogen ions from one facet, and the crystal has defects, where carriers can recombine. The absorber is often placed at a resonator end, but it can be placed somewhere within the resonator so that different pulses can meet in the absorber, colliding pulse mode locking. Hybrid mode locking is a combination of active and passive mode locking and it is possible to control the pulse timing externally, as in an actively mode-locked laser, while achieving shorter pulse duration as in a passively mode-locked device. For pulse repetition rates roughly below 10 GHz an external cavity setup is usually required. The laser resonator usually has the semiconductor chip on one end and is completed with bulk-optical elements, at least a collimation lens (GRIN lens) and an end mirror. Other techniques are to incorporate the semiconductor chip into a ring laser resonator made of optical single-mode fiber. The resonator is typically much longer, and allows the use of fiber-optic components. The semiconductor device may than is a fiber-coupled semiconductor optical amplifier (SOA). External cavity lasers have various advantages:

1. Pulse repetition rate can be chosen in a wide range and can be tuned by moving the end mirror.
2. It is possible to insert an optical filter for fixing the emission wavelength or use a diffraction grating as the end mirror.
3. The external-cavity devices have a potential for lower laser noise even for higher pulse repetition.

There are some limitations for mode-locked diode laser and one of them is the timing jitter and the noise of other pulse parameters which are higher than for other mode-locked lasers. This is partly consequence of the low power level. Timing jitter is the phenomenon that pulse trains which generated in mode-locked laser, exhibit some deviations of the temporal pulse positions from those in a perfectly periodic pulse train. Similarly, pulse from Q switched lasers exhibit timing laser. There are some kinds of timing errors: the deviations between the temporal pulse positions and those of perfectly regular clock ticks, the deviations between the temporal pulse positions and those of the ticks of a real (noisy) oscillator, the deviations of the pulse-to-pulse spacing from the average pulse period.

Typical applications: optical sampling, clocking, spectroscopy, high speed communication systems using optical time-division multiplexing (OTDM)—short pulse widths and low timing jitter.

Transmission rate 160 Gbps (640 Gbps), timing jitter <500 fs (100 fs). There is a gain dynamics which is responsible for the relaxation oscillations in semiconductor lasers and that leads to amplitude-timing jitter coupling. The orthogonal fluctuations are mutually dependent when driven by the noise sources. There are different mechanisms couple fluctuations into timing jitter: the projection of spontaneous emission noise onto timing jitter (quantum limited jitter), amplitude-timing jitter coupling due to gain fluctuations, jitter mediated by emission frequency fluctuations (Gordon-Haus jitter), changes in group refractive index, etc. we look into the mode-locking problem through Haus' master equation and pulse iterative models. Mainly in soliton communications, noise induced Gordon-Haus timing jitter in arbitrary dispersion-managed transmission systems and is reduced by the power-enhancement factor required to support a dispersion managed solitons provided the path-average soliton period which is much greater than the dispersion-map period. A soliton is a self-reinforcing solitary wave packet that maintains its shape while it propagates at a constant velocity. Solitons are caused by a cancellation of nonlinear and dispersive effects in the medium. Soliton based transmission systems is limited in part due to Gordon-Haus timing jitter which arises from the inherent amplified spontaneous emission (ASE) noise that accompanies coherent amplification. The jitter is a direct result of the nonlinear interaction of the noise field with the propagating soliton and sets a fundamental transmission limit. The target is to significantly suppress the ASE noise level in order to enhance systems performance. The jitter reduction is given by the power enhancement factor which is required to support a dispersion-managed soliton. In the Gordon-Haus analysis we can neglect the periodic intensity fluctuations due to fiber loss and amplification. The periodic fluctuations only introduce small perturbations to the average pulse dynamics. The noise field generated from the spontaneous emission of the amplifier gain through the function $S(Z,T)$. The evolution of the slowly varying envelope of the electric field is given by the nonlinear Schrodinger equation:

$$S(Z,T) = i \cdot \frac{\partial Q}{\partial Z} + \frac{1}{2} \cdot \sigma(Z) \cdot \frac{\partial^2 Q}{\partial T^2} + |Q|^2 \cdot Q; \; Z_0 = \frac{2 \cdot \pi \cdot c}{\lambda_0^2 \cdot D_{ave}} \cdot \left(\frac{T_0}{1.76}\right)^2$$

White noise process condition: $S(Z,T) \ll 1$.

Z—Variable of physical distance divided by the dispersion length Z_0 corresponding to the path average dispersion value D_{ave}, Z_0—Dispersion length corresponding to the path average dispersion value D_{ave}, T—physical time normalized by $\frac{T_0}{1.76}$, T_0—width at half maximum intensity of the pulse. Q(Z, T)—normalized function by the fundamental soliton peak field power $|E_0|^2 = \frac{\lambda_0 \cdot A_{eff}}{2 \cdot \pi \cdot n_2 \cdot Z_0}$, $|E_0|^2$—peak field power of fundamental soliton, D_{ave}—path average dispersion value, n_2—nonlinear coefficient of the fiber, A_{eff}—effective cross sectional area of the fiber, λ_0—carrier's free-space wavelength, c—speed of light, $\sigma(Z)$—dispersion map (it alternates between the anomalous and normal dispersion regimes in a piecewise constant, periodic manner). We can consider that the dispersion map function $\sigma(Z)$ is periodic then $\sigma(Z) = \sigma(Z + P)$, P—dispersion map period, $\langle\sigma\rangle = 1$—dispersion averaged over one period. The laser pulse shape and corresponding free parameters as center-frequency and center-position dynamics which is driven by the noise field. We define the normalized function by the fundamental soliton peak field power Q(Z,T) depends on free parameters $(A, \beta, \eta, C, \Omega, \phi)$ which are depend upon Z (Variable of physical distance divided by the dispersion length Z_0 corresponding to the path average dispersion value D_{ave}), $A = A(Z)$, $\beta = \beta(Z)$, $\eta = \eta(Z)$, $C = C(Z)$, $\Omega = \Omega(Z)$, $\phi = \phi(Z)$.

$$Q(Z, T) = A \cdot \sqrt{\eta} \cdot \text{sech}\{\eta \cdot (T - C)\} \cdot e^{[i \cdot (\Omega \cdot (T-C) + \beta \cdot (T-C)^2 + \frac{\phi}{2})]}$$

A—Power enhancement factor, η—Amplitude (and width), β—Quadratic chirp, C—center position, Ω—Center-frequency, ϕ—Phase dynamic. The power enhancement factor A required to support a periodic, dispersion-managed soliton solution is determined for an arbitrary dispersion map using shooting methods for ODE. We can represent the function which describes the evolution of the slowly varying envelope of the electric field (modified nonlinear Schrodinger equation) by set of coupled differential equations. The projection of noise on each of the evolution parameters is defined as S_x; $x = A, \beta, \eta, C, \Omega, \phi$ and we take it as a constant in our analysis (very small variation). If $S(Z, T) = 0$, there is an absence of noise and the evolution equations for the amplitude and chirp reduce, $\Omega = 0$; $C = 0$ so that no center-frequency and center-position shifts occur.

$$\frac{d\eta}{dZ} = -2 \cdot \sigma(Z) \cdot \beta(Z) \cdot \eta(Z) - \frac{6 \cdot [\eta(z)]^3}{\pi^2} \cdot S_\beta$$
$$\frac{d\Omega}{dZ} = -S_C; \quad \frac{dC}{dZ} = \sigma(Z) \cdot \Omega(Z) + S_\Omega$$

$$\frac{\pi^2}{2} \cdot \frac{d\beta}{dZ} = \sigma(Z) \cdot ([\eta(Z)]^4 - \pi^2 \cdot [\beta(Z)]^2) - A^2 \cdot [\eta(Z)]^3 + 3 \cdot [\eta(Z)]^3 \cdot S_\eta$$

Assumptions: $S_\beta, S_\eta, S_C, S_\Omega$ are constants (minority variation for different values of Z variable. $\frac{d\sigma(Z)}{dZ} \to \varepsilon$; $\frac{dA}{dZ} \to \varepsilon$, $\sigma = \sigma(Z)$, A are constants. Due to parasitic effects in our soliton communication system there is a physical distance shifting Z_τ in the interaction element $\beta(Z) \cdot \eta(Z) \to \beta(Z - Z_\tau) \cdot \eta(Z - Z_\tau)$ which related to amplitude (and width), η derivative in physical distance Z differential equation.

$$\frac{d\eta}{dZ} = -2 \cdot \sigma(Z) \cdot \beta(Z - Z_\tau) \cdot \eta(Z - Z_\tau) - \frac{6 \cdot [\eta(z)]^3}{\pi^2} \cdot S_\beta$$
$$\frac{d\Omega}{dZ} = -S_C; \quad \frac{dC}{dZ} = \sigma(Z) \cdot \Omega(Z) + S_\Omega$$

$$\frac{\pi^2}{2} \cdot \frac{d\beta}{dZ} = \sigma(Z) \cdot ([\eta(Z)]^4 - \pi^2 \cdot [\beta(Z)]^2) - A^2 \cdot [\eta(Z)]^3 + 3 \cdot [\eta(Z)]^3 \cdot S_\eta$$

At fixed points (equilibrium points): $\frac{d\eta}{dZ} = 0$; $\frac{d\Omega}{dZ} = 0 \Rightarrow S_C = 0$; $\frac{dC}{dZ} = 0$; $\frac{d\beta}{dZ} = 0$.

$$\lim_{Z\to\infty} \cdot\eta(Z - Z_\tau) = \eta(Z);\ \ Z \gg Z_\tau;\ \ \lim_{Z\to\infty} \cdot\beta(Z - Z_\tau) = \beta(Z);\ \ Z \gg Z_\tau$$

Remark: The physical distance Z is very big $Z \to \infty$ at fixed points.

$$-2\cdot\sigma\cdot\beta^*\cdot\eta^* - \frac{6\cdot[\eta^*]^3}{\pi^2}\cdot S_\beta = 0;\ \ \sigma\cdot\Omega^* + S_\Omega = 0 \Rightarrow \Omega^* = -\frac{S_\Omega}{\sigma}$$

$$\sigma\cdot([\eta^*]^4 - \pi^2\cdot[\beta^*]^2) + [\eta^*]^3\cdot(3\cdot S_\eta - A^2) = 0$$

$$-2\cdot\sigma\cdot\beta^*\cdot\eta^* - \frac{6\cdot[\eta^*]^3}{\pi^2}\cdot S_\beta = 0 \Rightarrow \beta^* = -\frac{3\cdot[\eta^*]^2}{\pi^2\cdot\sigma}\cdot S_\beta$$

$$-\eta^*\cdot\left\{\sigma\cdot\beta^* + \frac{3\cdot[\eta^*]^2}{\pi^2}\cdot S_\beta\right\} = 0 \Rightarrow \eta^{(0)} = 0$$

$$\eta^{(1)} = \sqrt{-\frac{\pi^2\cdot\sigma\cdot\beta^*}{3\cdot S_\beta}} = i\cdot\sqrt{\frac{\pi^2\cdot\sigma\cdot\beta^*}{3\cdot S_\beta}}$$

Remark: The amplitude (and width) η^* can't be negative or complex.

$$\eta^{(0)} = 0 \Rightarrow -\sigma\cdot\pi^2\cdot[\beta^{(0)}]^2 = 0;\ \ \sigma\cdot\pi^2 \neq 0 \Rightarrow \beta^{(0)} = 0$$

First fixed point: $E^{(0)}(\eta^{(0)}, \beta^{(0)}, C^{(0)}, \Omega^{(0)}) = (0, 0, C^{(0)}, -\frac{S_\Omega}{\sigma})$.
Second fixed point: $\beta^* = -\frac{3\cdot[\eta^*]^2}{\pi^2\cdot\sigma}\cdot S_\beta;\ \ \Omega^* = -\frac{S_\Omega}{\sigma}$.

$$\sigma\cdot[\eta^*]^4 - \sigma\cdot\pi^2\cdot\frac{9\cdot[\eta^*]^4}{\pi^4\cdot\sigma^2}\cdot S_\beta^2 + [\eta^*]^3\cdot(3\cdot S_\eta - A^2) = 0$$

$$[\eta^*]^3\cdot\left\{\sigma\cdot\eta^*\cdot\left(1 - \pi^2\cdot\frac{9}{\pi^4\cdot\sigma^2}\cdot S_\beta^2\right) + (3\cdot S_\eta - A^2)\right\} = 0$$

$$\eta^* \neq 0;\ \ \eta^* = \frac{A^2 - 3\cdot S_\eta}{\sigma\cdot\left(1 - \pi^2\cdot\frac{9}{\pi^4\cdot\sigma^2}\cdot S_\beta^2\right)}$$

$$C^{(0)} \in \mathbb{R}_+;\ \ C^{(1)} \in \mathbb{R}_+;\ \ \Omega^{(0)} = \Omega^{(1)} = -\frac{S_\Omega}{\sigma}$$

$$E^{(1)}(\eta^{(1)}, \beta^{(1)}, C^{(1)}, \Omega^{(1)}) = \left(\frac{A^2 - 3\cdot S_\eta}{\sigma\cdot\left(1 - \pi^2\cdot\frac{9}{\pi^4\cdot\sigma^2}\cdot S_\beta^2\right)}, -\frac{3\cdot[\eta^*]^2}{\pi^2\cdot\sigma}\cdot S_\beta, C^{(1)}, -\frac{S_\Omega}{\sigma}\right).$$

Stability analysis: The standard local stability analysis about any one of the equilibrium points of evolution of the slowly varying envelope of the electric field set of couples differential equations consists in adding to its coordinated $[\eta, \beta, C, \Omega]$ arbitrarily small increments of exponential terms $[\eta, \beta, C, \Omega] \cdot e^{\lambda \cdot Z}$, and retaining the first order terms in η, β, C, Ω. The system of four homogeneous equations leads to a polynomial characteristic equation in the eigenvalue λ. The polynomial characteristic equation accepts by set of system equations. The evolution of the slowly varying envelope of the electric field set of couple's differential equations fixed values with arbitrarily small increments of exponential form $[\eta, \beta, C, \Omega] \cdot e^{\lambda \cdot Z}$ are i = 0 (first fixed point), i = 1 (second fixed point), etc.

$$\eta(Z) = \eta^{(i)} + \eta \cdot e^{\lambda \cdot Z}; \ \beta(Z) = \beta^{(i)} + \beta \cdot e^{\lambda \cdot Z}$$
$$C(Z) = C^{(i)} + c \cdot e^{\lambda \cdot Z}; \ \Omega(Z) = \Omega^{(i)} + \Omega \cdot e^{\lambda \cdot Z}$$

$$\beta(Z - Z_\tau) = \beta^{(i)} + \beta \cdot e^{\lambda \cdot (Z - Z_\tau)}; \ \eta(Z - Z_\tau) = \eta^{(i)} + \eta \cdot e^{\lambda \cdot (Z - Z_\tau)}$$

$$\frac{d\eta}{dZ} = \eta \cdot \lambda \cdot e^{\lambda \cdot Z}; \ \frac{d\Omega}{dZ} = \Omega \cdot \lambda \cdot e^{\lambda \cdot Z}; \ \frac{dC}{dZ} = c \cdot \lambda \cdot e^{\lambda \cdot Z}; \ \frac{d\beta}{dZ} = \beta \cdot \lambda \cdot e^{\lambda \cdot Z}$$

$$\frac{d\eta}{dZ} = -2 \cdot \sigma(Z) \cdot \beta(Z - Z_\tau) \cdot \eta(Z - Z_\tau) - \frac{6 \cdot [\eta(z)]^3}{\pi^2} \cdot S_\beta$$

$$\eta \cdot \lambda \cdot e^{\lambda \cdot Z} = -2 \cdot \sigma \cdot (\beta^{(i)} + \beta \cdot e^{\lambda \cdot (Z - Z_\tau)}) \cdot (\eta^{(i)} + \eta \cdot e^{\lambda \cdot (Z - Z_\tau)}) - \frac{6 \cdot [\eta^{(i)} + \eta \cdot e^{\lambda \cdot Z}]^3}{\pi^2} \cdot S_\beta$$

$$\eta \cdot \lambda \cdot e^{\lambda \cdot Z} = -2 \cdot \sigma \cdot (\beta^{(i)} \cdot \eta^{(i)} + \beta^{(i)} \cdot \eta \cdot e^{\lambda \cdot (Z - Z_\tau)} + \eta^{(i)} \cdot \beta \cdot e^{\lambda \cdot (Z - Z_\tau)} + \beta \cdot \eta \cdot e^{\lambda \cdot (Z - Z_\tau)} \cdot e^{\lambda \cdot (Z - Z_\tau)}) - \frac{6 \cdot [(\eta^{(i)})^2 + 2 \cdot \eta^{(i)} \cdot \eta \cdot e^{\lambda \cdot Z} + \eta^2 \cdot e^{2 \cdot \lambda \cdot Z}] \cdot [\eta^{(i)} + \eta \cdot e^{\lambda \cdot Z}]}{\pi^2} \cdot S_\beta$$

$$\eta \cdot \lambda \cdot e^{\lambda \cdot Z} = -2 \cdot \sigma \cdot (\beta^{(i)} \cdot \eta^{(i)} + \beta^{(i)} \cdot \eta \cdot e^{\lambda \cdot (Z - Z_\tau)} + \eta^{(i)} \cdot \beta \cdot e^{\lambda \cdot (Z - Z_\tau)} + \beta \cdot \eta \cdot e^{\lambda \cdot (Z - Z_\tau)} \cdot e^{\lambda \cdot (Z - Z_\tau)}) - \frac{6 \cdot [(\eta^{(i)})^3 + (\eta^{(i)})^2 \cdot \eta \cdot e^{\lambda \cdot Z} + 2 \cdot (\eta^{(i)})^2 \cdot \eta \cdot e^{\lambda \cdot Z} + 2 \cdot \eta^{(i)} \cdot \eta^2 \cdot e^{2 \cdot \lambda \cdot Z} + \eta^2 \cdot e^{2 \cdot \lambda \cdot Z} \cdot (\eta^{(i)} + \eta \cdot e^{\lambda \cdot Z})]}{\pi^2} \cdot S_\beta$$

$$\beta \cdot \eta \to \varepsilon; \ \eta^2 \to \varepsilon$$

$$\eta \cdot \lambda \cdot e^{\lambda \cdot Z} = -2 \cdot \sigma \cdot (\beta^{(i)} \cdot \eta^{(i)} + \beta^{(i)} \cdot \eta \cdot e^{\lambda \cdot (Z - Z_\tau)} + \eta^{(i)} \cdot \beta \cdot e^{\lambda \cdot (Z - Z_\tau)})$$

$$-\frac{6\cdot[(\eta^{(i)})^3+(\eta^{(i)})^2\cdot\eta\cdot e^{\lambda\cdot Z}+2\cdot(\eta^{(i)})^2\cdot\eta\cdot e^{\lambda\cdot Z}]}{\pi^2}\cdot S_\beta$$

$$\begin{aligned}\eta\cdot\lambda\cdot e^{\lambda\cdot Z}&=-2\cdot\sigma\cdot\beta^{(i)}\cdot\eta^{(i)}-\frac{6\cdot(\eta^{(i)})^3}{\pi^2}\cdot S_\beta\\&\quad-2\cdot\sigma\cdot(\beta^{(i)}\cdot\eta\cdot e^{\lambda\cdot(Z-Z_\tau)}+\eta^{(i)}\cdot\beta\cdot e^{\lambda\cdot(Z-Z_\tau)})\\&\quad-\frac{6\cdot[(\eta^{(i)})^2\cdot\eta\cdot e^{\lambda\cdot Z}+2\cdot(\eta^{(i)})^2\cdot\eta\cdot e^{\lambda\cdot Z}]}{\pi^2}\cdot S_\beta\end{aligned}$$

At fixed point $-2\cdot\sigma\cdot\beta^{(i)}\cdot\eta^{(i)}-\frac{6\cdot(\eta^{(i)})^3}{\pi^2}\cdot S_\beta=0$

$$\begin{aligned}\eta\cdot\lambda\cdot e^{\lambda\cdot Z}&=-2\cdot\sigma\cdot(\beta^{(i)}\cdot\eta\cdot e^{-\lambda\cdot Z_\tau}+\eta^{(i)}\cdot\beta\cdot e^{-\lambda\cdot Z_\tau})\cdot e^{\lambda\cdot Z}\\&\quad-\frac{18\cdot(\eta^{(i)})^2\cdot\eta}{\pi^2}\cdot S_\beta\cdot e^{\lambda\cdot Z}\end{aligned}$$

$$\begin{aligned}&-\eta\cdot\lambda-\left[2\cdot\sigma\cdot\beta^{(i)}\cdot e^{-\lambda\cdot Z_\tau}-\frac{18\cdot(\eta^{(i)})^2}{\pi^2}\cdot S_\beta\right]\cdot\eta\\&-2\cdot\sigma\cdot\eta^{(i)}\cdot\beta\cdot e^{-\lambda\cdot Z_\tau}=0\end{aligned}$$

$\frac{d\Omega}{dZ}=-S_C;\ \Omega\cdot\lambda\cdot e^{\lambda\cdot Z}=-S_C$. At fixed point $S_C=0;\ -\Omega\cdot\lambda=0$
$\frac{dC}{dZ}=\sigma\cdot\Omega(Z)+S_\Omega;\ c\cdot\lambda\cdot e^{\lambda\cdot Z}=\sigma\cdot\Omega^{(i)}+S_\Omega+\sigma\cdot\Omega\cdot e^{\lambda\cdot Z}$. At fixed point $\sigma\cdot\Omega^{(i)}+S_\Omega=0$.

$$-c\cdot\lambda+\sigma\cdot\Omega=0$$

$$\frac{\pi^2}{2}\cdot\frac{d\beta}{dZ}=\sigma\cdot([\eta(Z)]^4-\pi^2\cdot[\beta(Z)]^2)+(3\cdot S_\eta-A^2)\cdot[\eta(Z)]^3$$

$$\begin{aligned}[\eta(Z)]^4&=[\eta^{(i)}+\eta\cdot e^{\lambda\cdot Z}]^4=[\eta^{(i)}+\eta\cdot e^{\lambda\cdot Z}]^2\cdot[\eta^{(i)}+\eta\cdot e^{\lambda\cdot Z}]^2\\&=([\eta^{(i)}]^2+2\cdot\eta^{(i)}\cdot\eta\cdot e^{\lambda\cdot Z}\\&\quad+\eta^2\cdot e^{2\cdot\lambda\cdot Z})\cdot([\eta^{(i)}]^2+2\cdot\eta^{(i)}\cdot\eta\cdot e^{\lambda\cdot Z}+\eta^2\cdot e^{2\cdot\lambda\cdot Z})\end{aligned}$$

$$\eta^2\to\varepsilon$$

$$\begin{aligned}[\eta(Z)]^4&=[\eta^{(i)}+\eta\cdot e^{\lambda\cdot Z}]^4\\&=([\eta^{(i)}]^2+2\cdot\eta^{(i)}\cdot\eta\cdot e^{\lambda\cdot Z})\cdot([\eta^{(i)}]^2+2\cdot\eta^{(i)}\cdot\eta\cdot e^{\lambda\cdot Z})=[\eta^{(i)}]^4\\&\quad+4\cdot[\eta^{(i)}]^2\cdot\eta^{(i)}\cdot\eta\cdot e^{\lambda\cdot Z}+4\cdot[\eta^{(i)}]^2\cdot\eta^2\cdot e^{2\cdot\lambda\cdot Z}\end{aligned}$$

$$\eta^2 \to \varepsilon$$

$$\begin{aligned}[\eta(Z)]^4 &= [\eta^{(i)} + \eta \cdot e^{\lambda \cdot Z}]^4 \\ &= ([\eta^{(i)}]^2 + 2 \cdot \eta^{(i)} \cdot \eta \cdot e^{\lambda \cdot Z}) \cdot ([\eta^{(i)}]^2 + 2 \cdot \eta^{(i)} \cdot \eta \cdot e^{\lambda \cdot Z}) \\ &= [\eta^{(i)}]^4 + 4 \cdot [\eta^{(i)}]^3 \cdot \eta \cdot e^{\lambda \cdot Z}\end{aligned}$$

$$\begin{aligned}[\eta(Z)]^3 &= [\eta^{(i)} + \eta \cdot e^{\lambda \cdot Z}]^3 \\ &= [[\eta^{(i)}]^2 + 2 \cdot \eta^{(i)} \cdot \eta \cdot e^{\lambda \cdot Z} + \eta^2 \cdot e^{2 \cdot \lambda \cdot Z}] \cdot [\eta^{(i)} \\ &\quad + \eta \cdot e^{\lambda \cdot Z}]\end{aligned}$$

$$\eta^2 \to \varepsilon$$

$$\begin{aligned}[\eta(Z)]^3 &= [\eta^{(i)} + \eta \cdot e^{\lambda \cdot Z}]^3 \\ &= [[\eta^{(i)}]^2 + 2 \cdot \eta^{(i)} \cdot \eta \cdot e^{\lambda \cdot Z}] \cdot [\eta^{(i)} + \eta \cdot e^{\lambda \cdot Z}] \\ &= [\eta^{(i)}]^3 + 3 \cdot [\eta^{(i)}]^2 \cdot \eta \cdot e^{\lambda \cdot Z} + 2 \cdot \eta^{(i)} \cdot \eta^2 \cdot e^{2 \cdot \lambda \cdot Z}\end{aligned}$$

$$\eta^2 \to \varepsilon$$

$$[\eta(Z)]^3 = [\eta^{(i)} + \eta \cdot e^{\lambda \cdot Z}]^3 = [\eta^{(i)}]^3 + 3 \cdot [\eta^{(i)}]^2 \cdot \eta \cdot e^{\lambda \cdot Z}$$

$$[\beta(Z)]^2 = [\beta^{(i)} + \beta \cdot e^{\lambda \cdot Z}]^2 = [\beta^{(i)}]^2 + 2 \cdot \beta^{(i)} \cdot \beta \cdot e^{\lambda \cdot Z} + \beta^2 \cdot e^{2 \cdot \lambda \cdot Z}$$

$$\beta^2 \to \varepsilon$$

$$[\beta(Z)]^2 = [\beta^{(i)} + \beta \cdot e^{\lambda \cdot Z}]^2 = [\beta^{(i)}]^2 + 2 \cdot \beta^{(i)} \cdot \beta \cdot e^{\lambda \cdot Z}$$

$$\frac{\pi^2}{2} \cdot \frac{d\beta}{dZ} = \sigma \cdot ([\eta(Z)]^4 - \pi^2 \cdot [\beta(Z)]^2) + (3 \cdot S_\eta - A^2) \cdot [\eta(Z)]^3$$

$$\begin{aligned}\frac{\pi^2}{2} \cdot \beta \cdot \lambda \cdot e^{\lambda \cdot Z} = \sigma \cdot ([\eta^{(i)}]^4 &+ 4 \cdot [\eta^{(i)}]^3 \cdot \eta \cdot e^{\lambda \cdot Z} \\ &- \pi^2 \cdot \{[\beta^{(i)}]^2 + 2 \cdot \beta^{(i)} \cdot \beta \cdot e^{\lambda \cdot Z}\}) \\ &+ (3 \cdot S_\eta - A^2) \cdot \{[\eta^{(i)}]^3 \\ &+ 3 \cdot [\eta^{(i)}]^2 \cdot \eta \cdot e^{\lambda \cdot Z}\}\end{aligned}$$

$$\frac{\pi^2}{2} \cdot \beta \cdot \lambda \cdot e^{\lambda \cdot Z} = \sigma \cdot [\eta^{(i)}]^4 + \sigma \cdot 4 \cdot [\eta^{(i)}]^3 \cdot \eta \cdot e^{\lambda \cdot Z}$$

$$- \sigma \cdot \pi^2 \cdot [\beta^{(i)}]^2 - \sigma \cdot \pi^2 \cdot 2 \cdot \beta^{(i)} \cdot \beta \cdot e^{\lambda \cdot Z} + (3 \cdot S_\eta - A^2) \cdot [\eta^{(i)}]^3 + (3 \cdot S_\eta - A^2) \cdot 3 \cdot [\eta^{(i)}]^2 \cdot \eta \cdot e^{\lambda \cdot Z}$$

$$\frac{\pi^2}{2} \cdot \beta \cdot \lambda \cdot e^{\lambda \cdot Z} = \sigma \cdot ([\eta^{(i)}]^4 - \pi^2 \cdot [\beta^{(i)}]^2) + (3 \cdot S_\eta - A^2) \cdot [\eta^{(i)}]^3 + \sigma \cdot 4 \cdot [\eta^{(i)}]^3 \cdot \eta \cdot e^{\lambda \cdot Z} - \sigma \cdot \pi^2 \cdot 2 \cdot \beta^{(i)} \cdot \beta \cdot e^{\lambda \cdot Z} + (3 \cdot S_\eta - A^2) \cdot 3 \cdot [\eta^{(i)}]^2 \cdot \eta \cdot e^{\lambda \cdot Z}$$

At fixed point $\sigma \cdot ([\eta^{(i)}]^4 - \pi^2 \cdot [\beta^{(i)}]^2) + (3 \cdot S_\eta - A^2) \cdot [\eta^{(i)}]^3 = 0$

$$\frac{\pi^2}{2} \cdot \beta \cdot \lambda \cdot e^{\lambda \cdot Z} = \sigma \cdot 4 \cdot [\eta^{(i)}]^3 \cdot \eta \cdot e^{\lambda \cdot Z} - \sigma \cdot \pi^2 \cdot 2 \cdot \beta^{(i)} \cdot \beta \cdot e^{\lambda \cdot Z} + (3 \cdot S_\eta - A^2) \cdot 3 \cdot [\eta^{(i)}]^2 \cdot \eta \cdot e^{\lambda \cdot Z}$$

$$- \frac{\pi^2}{2} \cdot \beta \cdot \lambda + \sigma \cdot 4 \cdot [\eta^{(i)}]^3 \cdot \eta - \sigma \cdot \pi^2 \cdot 2 \cdot \beta^{(i)} \cdot \beta + (3 \cdot S_\eta - A^2) \cdot 3 \cdot [\eta^{(i)}]^2 \cdot \eta = 0$$

$$- \frac{\pi^2}{2} \cdot \beta \cdot \lambda - \sigma \cdot \pi^2 \cdot 2 \cdot \beta^{(i)} \cdot \beta + \{\sigma \cdot 4 \cdot [\eta^{(i)}]^3 + (3 \cdot S_\eta - A^2) \cdot 3 \cdot [\eta^{(i)}]^2\} \cdot \eta = 0$$

We can summery our system arbitrarily small increments equations:

$$-\eta \cdot \lambda - \left[2 \cdot \sigma \cdot \beta^{(i)} \cdot e^{-\lambda \cdot Z_\tau} - \frac{18 \cdot (\eta^{(i)})^2}{\pi^2} \cdot S_\beta\right] \cdot \eta - 2 \cdot \sigma \cdot \eta^{(i)} \cdot \beta \cdot e^{-\lambda \cdot Z_\tau} = 0$$

$$- \Omega \cdot \lambda = 0; \quad -c \cdot \lambda + \sigma \cdot \Omega = 0$$

$$\left\{-\frac{\pi^2}{2} \cdot \lambda - \sigma \cdot \pi^2 \cdot 2 \cdot \beta^{(i)}\right\} \cdot \beta + \{\sigma \cdot 4 \cdot [\eta^{(i)}]^3 + (3 \cdot S_\eta - A^2) \cdot 3 \cdot [\eta^{(i)}]^2\} \cdot \eta = 0$$

$$\frac{2}{\pi^2} \cdot \left\{-\frac{\pi^2}{2} \cdot \lambda - \sigma \cdot \pi^2 \cdot 2 \cdot \beta^{(i)}\right\} \cdot \beta + \{\sigma \cdot 4 \cdot [\eta^{(i)}]^3 + (3 \cdot S_\eta - A^2) \cdot 3 \cdot [\eta^{(i)}]^2\} \cdot \eta = 0$$

$$\{-\lambda - 4 \cdot \sigma \cdot \beta^{(i)}\} \cdot \beta + \frac{2}{\pi^2} \cdot \{\sigma \cdot 4 \cdot [\eta^{(i)}]^3 + (3 \cdot S_\eta - A^2) \cdot 3 \cdot [\eta^{(i)}]^2\} \cdot \eta = 0$$

The small increment Jacobian of our system is as follow:

$$\begin{pmatrix} \Upsilon_{11} & \ldots & \Upsilon_{14} \\ \vdots & \ddots & \vdots \\ \Upsilon_{41} & \cdots & \Upsilon_{44} \end{pmatrix} \cdot \begin{pmatrix} \eta \\ \Omega \\ c \\ \beta \end{pmatrix} = 0$$

$$\Upsilon_{11} = -\lambda - \left[2 \cdot \sigma \cdot \beta^{(i)} \cdot e^{-\lambda \cdot Z_\tau} - \frac{18 \cdot (\eta^{(i)})^2}{\pi^2} \cdot S_\beta\right]$$

$$\Upsilon_{12} = 0; \;\; \Upsilon_{13} = 0$$

$$\Upsilon_{14} = -2 \cdot \sigma \cdot \eta^{(i)} \cdot e^{-\lambda \cdot Z_\tau}; \;\; \Upsilon_{21} = 0; \;\; \Upsilon_{22} = -\lambda; \;\; \Upsilon_{23} = 0$$

$$\Upsilon_{24} = 0; \;\; \Upsilon_{31} = 0; \;\; \Upsilon_{32} = \sigma; \;\; \Upsilon_{33} = -\lambda$$

$$\Upsilon_{34} = 0; \;\; \Upsilon_{41} = \frac{2}{\pi^2} \cdot \{\sigma \cdot 4 \cdot [\eta^{(i)}]^3 + (3 \cdot S_\eta - A^2) \cdot 3 \cdot [\eta^{(i)}]^2\}$$

$$\Upsilon_{42} = 0; \;\; \Upsilon_{43} = 0; \;\; \Upsilon_{44} = -\lambda - 4 \cdot \sigma \cdot \beta^{(i)}$$

$$A - \lambda \cdot I| = \begin{pmatrix} \Upsilon_{11} & \ldots & \Upsilon_{14} \\ \vdots & \ddots & \vdots \\ \Upsilon_{41} & \cdots & \Upsilon_{44} \end{pmatrix}; \;\; \det|A - \lambda \cdot I| = 0$$

$$\begin{aligned} D(\lambda, Z_\tau) = \lambda^4 + \lambda^3 \cdot \left(4 \cdot \sigma \cdot \beta^{(i)} - \frac{18 \cdot [\eta^{(i)}]^2}{\pi^2} \cdot S_\beta\right) \\ - \lambda^2 \cdot 4 \cdot \sigma \cdot \beta^{(i)} \cdot \frac{18 \cdot [\eta^{(i)}]^2}{\pi^2} \cdot S_\beta \\ + (\lambda^3 \cdot 2 \cdot \sigma \cdot \beta^{(i)} + \lambda^2 \cdot [8 \cdot \sigma^2 \cdot [\beta^{(i)}]^2 \\ + 2 \cdot \sigma \cdot \eta^{(i)} \cdot \{4 \cdot \sigma \cdot [\eta^{(i)}]^3 \\ + (3 \cdot S_\eta - A) \cdot 3 \cdot [\eta^{(i)}]^2\} \cdot \frac{2}{\pi^2}]) \cdot e^{-\lambda \cdot Z_\tau} \end{aligned}$$

$$D(\lambda, Z_\tau) = P_n(\lambda, Z_\tau) + Q_m(\lambda, Z_\tau) \cdot e^{-\lambda \cdot Z_\tau}; \;\; n, m \in \mathbb{N}_0; \;\; n > m; \;\; n = 4; \;\; m = 3$$

$$\begin{aligned} P_n(\lambda, Z_\tau) = \lambda^4 + \lambda^3 \cdot \left(4 \cdot \sigma \cdot \beta^{(i)} - \frac{18 \cdot [\eta^{(i)}]^2}{\pi^2} \cdot S_\beta\right) \\ - \lambda^2 \cdot 4 \cdot \sigma \cdot \beta^{(i)} \cdot \frac{18 \cdot [\eta^{(i)}]^2}{\pi^2} \cdot S_\beta \end{aligned}$$

$$Q_m(\lambda, Z_\tau) = \lambda^3 \cdot 2 \cdot \sigma \cdot \beta^{(i)} + \lambda^2 \cdot [8 \cdot \sigma^2 \cdot [\beta^{(i)}]^2 + 2 \cdot \sigma \cdot \eta^{(i)} \cdot \{4 \cdot \sigma \cdot [\eta^{(i)}]^3$$

$$+(3 \cdot S_\eta - A) \cdot 3 \cdot [\eta^{(i)}]^2\} \cdot \frac{2}{\pi^2}\Bigg]$$

$$P_n(\lambda, Z_\tau) = \sum_{k=0}^{n=4} P_k(Z_\tau) \cdot \lambda^k$$
$$= P_0(Z_\tau) + P_1(Z_\tau) \cdot \lambda + P_2(Z_\tau) \cdot \lambda^2 + P_3(Z_\tau) \cdot \lambda^3 + P_4(Z_\tau) \cdot \lambda^4$$

$$P_0(Z_\tau) = 0; \;\; P_1(Z_\tau) = 0; \;\; P_2(Z_\tau) = -4 \cdot \sigma \cdot \beta^{(i)} \cdot \frac{18 \cdot [\eta^{(i)}]^2}{\pi^2} \cdot S_\beta$$
$$P_3(Z_\tau) = 4 \cdot \sigma \cdot \beta^{(i)} - \frac{18 \cdot [\eta^{(i)}]^2}{\pi^2} \cdot S_\beta; \;\; P_4(Z_\tau) = 1$$

$$Q_m(\lambda, Z_\tau) = \sum_{k=0}^{m=3} q_k(Z_\tau) \cdot \lambda^k = q_0(Z_\tau) + q_1(Z_\tau) \cdot \lambda + q_2(Z_\tau) \cdot \lambda^2 + q_3(Z_\tau) \cdot \lambda^3$$

$$q_0(Z_\tau) = 0; \;\; q_1(Z_\tau) = 0$$
$$q_2(Z_\tau) = 8 \cdot \sigma^2 \cdot [\beta^{(i)}]^2 + 2 \cdot \sigma \cdot \eta^{(i)} \cdot \{4 \cdot \sigma \cdot [\eta^{(i)}]^3 + (3 \cdot S_\eta - A) \cdot 3 \cdot [\eta^{(i)}]^2\} \cdot \frac{2}{\pi^2}$$
$$q_3(Z_\tau) = 2 \cdot \sigma \cdot \beta^{(i)}$$

We can omit Z_τ variable from the above P_k, q_k elements.

$$P_0 = 0; \;\; P_1 = 0; \;\; P_2 = -4 \cdot \sigma \cdot \beta^{(i)} \cdot \frac{18 \cdot [\eta^{(i)}]^2}{\pi^2} \cdot S_\beta$$
$$P_3 = 4 \cdot \sigma \cdot \beta^{(i)} - \frac{18 \cdot [\eta^{(i)}]^2}{\pi^2} \cdot S_\beta; \;\; P_4 = 1$$

$$q_0 = 0; \;\; q_1 = 0$$
$$q_2 = 8 \cdot \sigma^2 \cdot [\beta^{(i)}]^2 + 2 \cdot \sigma \cdot \eta^{(i)} \cdot \{4 \cdot \sigma \cdot [\eta^{(i)}]^3 + (3 \cdot S_\eta - A) \cdot 3 \cdot [\eta^{(i)}]^2\} \cdot \frac{2}{\pi^2}$$
$$q_3 = 2 \cdot \sigma \cdot \beta^{(i)}$$

The homogeneous system for η, β, C, Ω leads to a characteristic equation for the eigenvalue λ having the form $D(\lambda, Z_\tau) = P(\lambda, Z_\tau) + Q(\lambda, Z_\tau) \cdot e^{-\lambda \cdot Z_\tau}$; $P(\lambda, Z_\tau) = \sum_{j=0}^{4} a_j \cdot \lambda^j$.

$Q(\lambda, Z_\tau) = Q(\lambda, Z_\tau) = \sum_{j=0}^{3} c_j \cdot \lambda^j$. The coefficients $\{a_j(q_i, q_k, Z_\tau), c_j(q_i, q_k, Z_\tau)\} \in \mathbb{R}$ depend on q_i, q_k and physical distance shifting Z_τ. q_i, q_k are any slowly varying envelope of the electric field set of couples differential equation's parameters, other parameters kept as a constant.

$$a_0 = 0; \;\; a_1 = 0; \;\; a_2 = -4 \cdot \sigma \cdot \beta^{(i)} \cdot \frac{18 \cdot [\eta^{(i)}]^2}{\pi^2} \cdot S_\beta$$

$$a_3 = 4 \cdot \sigma \cdot \beta^{(i)} - \frac{18 \cdot [\eta^{(i)}]^2}{\pi^2} \cdot S_\beta; \; a_4 = 1$$

$$c_0 = 0; \; c_1 = 0$$

$$c_2 = 8 \cdot \sigma^2 \cdot [\beta^{(i)}]^2 + 2 \cdot \sigma \cdot \eta^{(i)} \cdot \{4 \cdot \sigma \cdot [\eta^{(i)}]^3 + (3 \cdot S_\eta - A) \cdot 3 \cdot [\eta^{(i)}]^2\} \cdot \frac{2}{\pi^2}$$

$$c_3 = 2 \cdot \sigma \cdot \beta^{(i)}$$

Unless strictly necessary, the designation of the varied arguments (q_i, q_k) will subsequently be omitted from P, Q, a_j, and c_j. The coefficients a_j, c_j are continuous and differentiable functions of their arguments and direct substitution show that $a_0 + c_0 \neq 0$ (not in our case) for $q_i, q_k \in \mathbb{R}_+$; that is, $\lambda = 0$ is not of $P(\lambda) + Q(\lambda) \cdot e^{-\lambda \cdot Z_\tau} = 0$. Furthermore, $P(\lambda)$, $Q(\lambda)$ are analytic functions of λ, for which the following requirements of the analysis can also be verified in the present case:

1. If $\lambda = i \cdot \omega$; $\omega \in \mathbb{R}$ then $P(i \cdot \omega) + Q(i \cdot \omega) \neq 0$.
2. If $|\frac{Q(\lambda)}{P(\lambda)}|$ is bounded for $|\lambda| \to \infty$; $\mathrm{Re}\lambda \geq 0$. No roots bifurcation from ∞.
3. $F(\omega) = |P(i \cdot \omega)|^2 - |Q(i \cdot \omega)|^2$ has a finite number of zeros. Indeed, this is a polynomial in ω.
4. Each positive root $\omega(q_i, q_k)$ of $F(\omega) = 0$i s continuous and differentiable with respect to q_i, q_k.

We assume that $P_n(\lambda, Z_\tau)$ and $Q_m(\lambda, Z_\tau)$ cannot have common imaginary roots. That is for any real number ω: $P_n(\lambda = i \cdot \omega, Z_\tau) + Q_m(\lambda = i \cdot \omega, Z_\tau) \neq 0$.

$$\begin{aligned} P_n(\lambda = i \cdot \omega, Z_\tau) &= \omega^4 + \omega^2 \cdot 4 \cdot \sigma \cdot \beta^{(i)} \cdot \frac{18 \cdot [\eta^{(i)}]^2}{\pi^2} \cdot S_\beta \\ &\quad - i \cdot \omega^3 \cdot \left(4 \cdot \sigma \cdot \beta^{(i)} - \frac{18 \cdot [\eta^{(i)}]^2}{\pi^2} \cdot S_\beta\right) \end{aligned}$$

$$\begin{aligned} Q_m(\lambda = i \cdot \omega, Z_\tau) &= -i \cdot \omega^3 \cdot 2 \cdot \sigma \cdot \beta^{(i)} - \omega^2 \cdot \Big[8 \cdot \sigma^2 \cdot [\beta^{(i)}]^2 \\ &\quad + 2 \cdot \sigma \cdot \eta^{(i)} \cdot \{4 \cdot \sigma \cdot [\eta^{(i)}]^3 \\ &\quad + (3 \cdot S_\eta - A) \cdot 3 \cdot [\eta^{(i)}]^2\} \cdot \frac{2}{\pi^2}\Big] \end{aligned}$$

$$\begin{aligned} &P_n(\lambda = i \cdot \omega, Z_\tau) + Q_m(\lambda = i \cdot \omega, Z_\tau) \\ &= \omega^4 + \omega^2 \cdot \Big\{4 \cdot \sigma \cdot \beta^{(i)} \cdot \frac{18 \cdot [\eta^{(i)}]^2}{\pi^2} \cdot S_\beta \\ &\quad - \Big[8 \cdot \sigma^2 \cdot [\beta^{(i)}]^2 + 2 \cdot \sigma \cdot \eta^{(i)} \cdot \Big\{4 \cdot \sigma \cdot [\eta^{(i)}]^3 \\ &\quad + (3 \cdot S_\eta - A) \cdot 3 \cdot [\eta^{(i)}]^2\Big\} \cdot \frac{2}{\pi^2}\Big]\Big\} \\ &\quad - i \cdot \omega^3 \cdot \Big\{4 \cdot \sigma \cdot \beta^{(i)} \end{aligned}$$

$$-\frac{18 \cdot [\eta^{(i)}]^2}{\pi^2} \cdot S_\beta + 2 \cdot \sigma \cdot \beta^{(i)}\Bigg\}$$

$$|P(i \cdot \omega, Z_\tau)|^2 = \omega^8 + \omega^6 \cdot \Bigg\{2 \cdot 4 \cdot \sigma \cdot \beta^{(i)} \cdot \frac{18 \cdot [\eta^{(i)}]^2}{\pi^2} \cdot S_\beta + \left(4 \cdot \sigma \cdot \beta^{(i)} - \frac{18 \cdot [\eta^{(i)}]^2}{\pi^2} \cdot S_\beta\right)\Bigg\} + \omega^4 \cdot 4 \cdot \sigma \cdot \beta^{(i)} \cdot \frac{18 \cdot [\eta^{(i)}]^2}{\pi^2} \cdot S_\beta$$

$$|Q(i \cdot \omega, Z_\tau)|^2 = \omega^6 \cdot 2 \cdot \sigma \cdot \beta^{(i)} + \omega^4 \cdot \Big[8 \cdot \sigma^2 \cdot [\beta^{(i)}]^2 + 2 \cdot \sigma \cdot \eta^{(i)} \cdot \{4 \cdot \sigma \cdot [\eta^{(i)}]^3 + (3 \cdot S_\eta - A) \cdot 3 \cdot [\eta^{(i)}]^2\} \cdot \frac{2}{\pi^2}\Big]$$

$$\begin{aligned} F(\omega, Z_\tau) &= |P(i \cdot \omega, Z_\tau)|^2 - |Q(i \cdot \omega, Z_\tau)|^2 \\ &= \omega^8 + \omega^6 \cdot \Bigg\{2 \cdot 4 \cdot \sigma \cdot \beta^{(i)} \cdot \frac{18 \cdot [\eta^{(i)}]^2}{\pi^2} \cdot S_\beta + \Big(4 \cdot \sigma \cdot \beta^{(i)} \\ &\quad - \frac{18 \cdot [\eta^{(i)}]^2}{\pi^2} \cdot S_\beta\Big)\Bigg\} + \omega^4 \cdot 4 \cdot \sigma \cdot \beta^{(i)} \cdot \frac{18 \cdot [\eta^{(i)}]^2}{\pi^2} \cdot S_\beta \\ &\quad - \omega^6 \cdot 2 \cdot \sigma \cdot \beta^{(i)} - \omega^4 \cdot \Big[8 \cdot \sigma^2 \cdot [\beta^{(i)}]^2 \\ &\quad + 2 \cdot \sigma \cdot \eta^{(i)} \cdot \{4 \cdot \sigma \cdot [\eta^{(i)}]^3 + (3 \cdot S_\eta - A) \cdot 3 \cdot [\eta^{(i)}]^2\} \cdot \frac{2}{\pi^2}\Big] \end{aligned}$$

$$\begin{aligned} F(\omega, Z_\tau) &= |P(i \cdot \omega, Z_\tau)|^2 - |Q(i \cdot \omega, Z_\tau)|^2 \\ &= \omega^8 + \omega^6 \cdot \Bigg\{2 \cdot 4 \cdot \sigma \cdot \beta^{(i)} \cdot \frac{18 \cdot [\eta^{(i)}]^2}{\pi^2} \cdot S_\beta \\ &\quad + \left(4 \cdot \sigma \cdot \beta^{(i)} - \frac{18 \cdot [\eta^{(i)}]^2}{\pi^2} \cdot S_\beta\right) - 2 \cdot \sigma \cdot \beta^{(i)}\Bigg\} \\ &\quad + \omega^4 \cdot \Bigg\{4 \cdot \sigma \cdot \beta^{(i)} \cdot \frac{18 \cdot [\eta^{(i)}]^2}{\pi^2} \cdot S_\beta \\ &\quad - \Big[8 \cdot \sigma^2 \cdot [\beta^{(i)}]^2 + 2 \cdot \sigma \cdot \eta^{(i)} \cdot \Big\{4 \cdot \sigma \cdot [\eta^{(i)}]^3 \\ &\quad + (3 \cdot S_\eta - A) \cdot 3 \cdot [\eta^{(i)}]^2\Big\} \cdot \frac{2}{\pi^2}\Big]\Bigg\} \end{aligned}$$

We define the following parameters for simplicity: $\Pi_0, \Pi_2, \Pi_4, \Pi_8$. $\Pi_0 = 0$; $\Pi_2 = 0$.

$$\Pi_4 = 4 \cdot \sigma \cdot \beta^{(i)} \cdot \frac{18 \cdot [\eta^{(i)}]^2}{\pi^2} \cdot S_\beta - \Big[8 \cdot \sigma^2 \cdot [\beta^{(i)}]^2 + 2 \cdot \sigma \cdot \eta^{(i)}$$

$$\cdot\{4\cdot\sigma\cdot[\eta^{(i)}]^3+(3\cdot S_\eta-A)\cdot 3\cdot[\eta^{(i)}]^2\}\cdot\frac{2}{\pi^2}\Bigg]$$

$$\Pi_6=8\cdot\sigma\cdot\beta^{(i)}\cdot\frac{18\cdot[\eta^{(i)}]^2}{\pi^2}\cdot S_\beta+4\cdot\sigma\cdot\beta^{(i)}$$
$$-\frac{18\cdot[\eta^{(i)}]^2}{\pi^2}\cdot S_\beta-2\cdot\sigma\cdot\beta^{(i)};\ \Pi_8=1$$

Hence $F(\omega,Z_\tau)=0$ implies $\sum_{k=0}^{4}\Pi_{2\cdot k}\cdot\omega^{2\cdot k}=0$ are given by solving the polynomial.

$$P_R(i\cdot\omega,Z_\tau)=\omega^4+\omega^2\cdot 4\cdot\sigma\cdot\beta^{(i)}\cdot\frac{18\cdot[\eta^{(i)}]^2}{\pi^2}\cdot S_\beta;\ P_I(i\cdot\omega,Z_\tau)$$
$$=-\omega^3\cdot\left(4\cdot\sigma\cdot\beta^{(i)}-\frac{18\cdot[\eta^{(i)}]^2}{\pi^2}\cdot S_\beta\right)$$

$$Q_R(i\cdot\omega,Z_\tau)=-\omega^2\cdot\Big[8\cdot\sigma^2\cdot[\beta^{(i)}]^2+2\cdot\sigma\cdot\eta^{(i)}\cdot\big\{4\cdot\sigma\cdot[\eta^{(i)}]^3$$
$$+(3\cdot S_\eta-A)\cdot 3\cdot[\eta^{(i)}]^2\big\}\cdot\frac{2}{\pi^2}\Big]$$

$$Q_I(i\cdot\omega,Z_\tau)=-\omega^3\cdot 2\cdot\sigma\cdot\beta^{(i)}$$

$$\sin\theta(Z_\tau)=\frac{-P_R(i\cdot\omega,Z_\tau)\cdot Q_I(i\cdot\omega,Z_\tau)+P_I(i\cdot\omega,Z_\tau)\cdot Q_R(i\cdot\omega,Z_\tau)}{|Q(i\cdot\omega,Z_\tau)|^2}$$

$$\cos\theta(Z_\tau)=-\frac{P_R(i\cdot\omega,Z_\tau)\cdot Q_R(i\cdot\omega,Z_\tau)+P_I(i\cdot\omega,Z_\tau)\cdot Q_I(i\cdot\omega,Z_\tau)}{|Q(i\cdot\omega,Z_\tau)|^2}$$

We use different parameters terminology from our last characteristics parameters definition: $k\to j;\ p_k(Z_\tau)\to a_j;\ q_k(Z_\tau)\to c_j;\ n=4;\ m=3$.

$$P_n(\lambda,Z_\tau)\to P(\lambda);\ Q_n(\lambda,Z_\tau)\to Q(\lambda)$$
$$P(\lambda)=\sum_{j=0}^{4}a_j\cdot\lambda^j;\ Q(\lambda)=\sum_{j=0}^{3}c_j\cdot\lambda^j$$

$$P(\lambda)=a_0+a_1\cdot\lambda+a_2\cdot\lambda^2+a_3\cdot\lambda^3+a_4\cdot\lambda^4$$
$$Q(\lambda)=c_0+c_1\cdot\lambda+c_2\cdot\lambda^2+c_3\cdot\lambda^3$$

$n,m\in\mathbb{N}_0;\ n>m;\ a_j,c_j:\mathbb{R}_{+0}\to\mathbb{R}$ are continuous and differentiable function of Z_τ. In the following "–" denotes complex and conjugate. $P(\lambda)$, $Q(\lambda)$ are analytic functions in λ and differentiable in Z_τ. The coefficients $a_j(\eta,\beta,C,\Omega,Z_\tau,\ldots)\in\mathbb{R};\ c_j(\eta,\beta,C,\Omega,Z_\tau,\ldots)\in\mathbb{R}$ depend on system $\eta,\beta,C,\Omega,Z_\tau,\ldots$ values. Unless strictly necessary the designation of the varied arguments: $(\eta, \beta, C, \Omega, Z_\tau, \ldots)$ will subsequently be omitted from P, Q, a_j, c_j. The coefficients a_j, c_j are continuous and differentiable functions of their arguments $a_0 + c_0 = 0$.

$\forall \eta, \beta, C, \Omega, Z_\tau, \ldots \in \mathbb{R}_+$ i.e. $\lambda = 0$ is not a root of the characteristic equation. Furthermore $P(\lambda)$, $Q(\lambda)$ are analytic functions of λ for which the following requirements of the analysis can also be verified in the present case.

1. If $\lambda = i \cdot \omega$; $\omega \in \mathbb{R}$ then $P(i \cdot \omega) + Q(i \cdot \omega) \neq 0$, i.e. P and Q have no common imaginary roots. This condition can be verified numerically in the entire $(\eta, \beta, C, \Omega, Z_\tau, \ldots)$ domain of interest.
2. $|\frac{P(\lambda)}{Q(\lambda)}|$ is bounded for $|\lambda| \to \infty$; $\mathrm{Re}\lambda \geq 0$. No roots bifurcation from ∞. Indeed, in the limit: $|\frac{Q(\lambda)}{P(\lambda)}| = |\frac{c_0 + c_1 \cdot \lambda + c_2 \cdot \lambda^2 + c_3 \cdot \lambda^3}{a_0 + a_1 \cdot \lambda + a_2 \cdot \lambda^2 + a_3 \cdot \lambda^3 + a_4 \cdot \lambda^4}|$.
3. The following expressions exist: $F(\omega) = |P(i \cdot \omega)|^2 - |Q(i \cdot \omega)|^2$

$F(\omega, Z_\tau) = |P(i \cdot \omega, Z_\tau)|^2 - |Q(i \cdot \omega, Z_\tau)|^2 = \sum_{k=0}^{4} \Pi_{2 \cdot k} \cdot \omega^{2 \cdot k}$ Has at most a finite number of zeros. Indeed, this is a polynomial in ω (degree in ω^8).

4. Each positive root $\omega(\eta, \beta, C, \Omega, Z_\tau, \ldots)$ of $F(\omega) = 0$ is continuous and differentiable with respect to $\eta, \beta, C, \Omega, Z_\tau, \ldots$. The condition can only be assessed numerically.

In addition, since the coefficients in P and Q are real, we have $\overline{P(-i \cdot \omega)} = P(i \cdot \omega)$ $\overline{Q(-i \cdot \omega)} = Q(i \cdot \omega)$ thus, $\omega > 0$ maybe on eigenvalue of characteristic equations. The analysis consists in identifying the roots of the characteristic equation situated on the imaginary axis of the complex λ-plane, whereby increasing the parameters: $\eta, \beta, C, \Omega, Z_\tau, \ldots$ $\mathrm{Re}\lambda$ may, at the crossing change its sign from $(-)$ to $(+)$, i.e. from stable focus $E^{(0)}(\eta^{(0)}, \beta^{(0)}, C^{(0)}, \Omega^{(0)}) = (0, 0, C^{(0)}, -\frac{S_\Omega}{\sigma})$ or $E^{(1)}(\eta^{(1)}, \beta^{(1)}, C^{(1)}, \Omega^{(1)}) = \left(\frac{A^2 - 3 \cdot S_\eta}{\sigma \cdot \left(1 - \pi^2 \cdot \frac{9}{\pi^4 \cdot \sigma^2} \cdot S_\beta^2\right)}, -\frac{3 \cdot [\eta^*]^2}{\pi^2 \cdot \sigma} \cdot S_\beta, C^{(1)}, -\frac{S_\Omega}{\sigma}\right)$ to an unstable one, or vice versa. This feature may be further assessed by examining the sign of the partial derivatives with respect to $\eta, \beta, C, \Omega, Z_\tau, \ldots$ and system parameters.

$$\Lambda^{-1}(Z_\tau) = \left(\frac{\partial \mathrm{Re}\lambda}{\partial Z_\tau}\right)_{\lambda = i \cdot \omega} ; \ \eta, \beta, C, \Omega, \ldots = const$$

$$\Lambda^{-1}(\eta) = \left(\frac{\partial \mathrm{Re}\lambda}{\partial \eta}\right)_{\lambda = i \cdot \omega} ; \ Z_\tau, \beta, C, \Omega, \ldots = const$$

$$\Lambda^{-1}(\beta) = \left(\frac{\partial \mathrm{Re}\lambda}{\partial \beta}\right)_{\lambda = i \cdot \omega} ; \ \eta, Z_\tau, C, \Omega, \ldots = const$$

$$\Lambda^{-1}(C) = \left(\frac{\partial \mathrm{Re}\lambda}{\partial C}\right)_{\lambda = i \cdot \omega} ; \ \eta, \beta, C, \Omega, \ldots = const$$

$$\Lambda^{-1}(\Omega) = \left(\frac{\partial \mathrm{Re}\lambda}{\partial \Omega}\right)_{\lambda = i \cdot \omega} ; \ \eta, \beta, C, Z_\tau, \ldots = const$$

$$P(\lambda) = P_R(\lambda) + i \cdot P_I(\lambda);\ \ Q(\lambda) = Q_R(\lambda) + i \cdot Q_I(\lambda)$$

When writing and inserting $\lambda = i \cdot \omega$ into slowly varying envelope of the electric field set of couples differential equations system's characteristic equation ω must satisfy the following equations.

$$\sin(\omega \cdot Z_\tau) = g(\omega) = \frac{-P_R(i \cdot \omega) \cdot Q_I(i \cdot \omega) + P_I(i \cdot \omega) \cdot Q_R(i \cdot \omega)}{|Q(i \cdot \omega)|^2}$$

$$\cos(\omega \cdot Z_\tau) = h(\omega) = -\frac{P_R(i \cdot \omega) \cdot Q_R(i \cdot \omega) + P_I(i \cdot \omega) \cdot Q_I(i \cdot \omega)}{|Q(i \cdot \omega)|^2}$$

where $|Q(i \cdot \omega)|^2 \neq 0$ in view of requirement (a) above, and $(h, g) \in \mathbb{R}$. Furthermore, it follow the above $\sin(\omega \cdot Z_\tau)$ and $\cos(\omega \cdot Z_\tau)$ equations that, by squaring and adding the sides, ω must be a positive root of $F(\omega) = |P(i \cdot \omega)|^2 - |Q(i \cdot \omega)|^2 = 0$. Note: $F(\omega)$ is dependent on Z_τ. Now it is important to notice that if $Z_\tau \notin I$ (assume that $I \subseteq \mathbb{R}_{+0}$ is the set where $\omega(Z_\tau)$ is a positive root of $F(\omega)$ and for, $Z_\tau \notin I$, $\omega(Z_\tau)$ is not defined). Then for all Z_τ in I, $\omega(Z_\tau)$ is satisfied that $F(\omega, Z_\tau) = 0$. Then there are no positive $\omega(Z_\tau)$ solutions for $F(\omega, Z_\tau) = 0$, and we cannot have stability switches. For $Z_\tau \in I$ where $\omega(Z_\tau)$ is a positive solution of $F(\omega, Z_\tau) = 0$, we can define the angle $\theta(Z_\tau) \in [0, 2 \cdot \pi]$ as the solution of $\sin(\omega \cdot Z_\tau) = \cdots;\ \cos(\omega \cdot Z_\tau) = \cdots$

$$\sin\theta(Z_\tau) = \frac{-P_R(i \cdot \omega) \cdot Q_I(i \cdot \omega) + P_I(i \cdot \omega) \cdot Q_R(i \cdot \omega)}{|Q(i \cdot \omega)|^2}$$

$$\cos\theta(Z_\tau) = -\frac{P_R(i \cdot \omega) \cdot Q_R(i \cdot \omega) + P_I(i \cdot \omega) \cdot Q_I(i \cdot \omega)}{|Q(i \cdot \omega)|^2}$$

And the relation between the argument $\theta(Z_\tau)$ and $Z_\tau \cdot \omega(Z_\tau)$ for $Z_\tau \in I$ must be as describe as: $\omega(Z_\tau) \cdot Z_\tau = \theta(Z_\tau) + 2 \cdot n \cdot \pi\ \forall n \in \mathbb{N}_0$. Hence we can define the maps $(Z_\tau)_n : I \to \mathbb{R}_{+0}$ given by $(Z_\tau)_n(Z_\tau) = \frac{\theta(Z_\tau)+2 \cdot n \cdot \pi}{\omega(Z_\tau)}$; $n \in \mathbb{N}_0$; $Z_\tau \in I$. Let us introduce the functions: $I \to \mathbb{R}$; $S_n(Z_\tau) = Z_\tau - (Z_\tau)_n(Z_\tau)$; $Z_\tau \in I$; $n \in \mathbb{N}_0$ that is continuous and differentiable in Z_τ. In the following, the subscripts $\lambda, \omega, \eta, \beta, C, \Omega, \ldots$ indicate the corresponding partial derivatives. Let us first concentrate on $\Lambda(x)$, remember in $\lambda(\eta, \beta, C, \Omega, \ldots)$ and $\omega(\eta, \beta, C, \Omega, \ldots)$, and keeping all parameters except one (x) and Z_τ. The derivation closely follows that in reference [BK]. Differentiating system characteristic equation $P(\lambda) + Q(\lambda) \cdot e^{-\lambda \cdot Z_\tau} = 0$ with respect to specific parameter (x) and inverting the derivative, for convenience, one calculates: $x = \eta, \beta, C, \Omega, \ldots$.

$$\left(\frac{\partial\lambda}{\partial x}\right)^{-1} = \frac{-P_\lambda(\lambda, x) \cdot Q(\lambda, x) + Q_\lambda(\lambda, x) \cdot P(\lambda, x) - Z_\tau \cdot P(\lambda, x) \cdot Q(\lambda, x)}{P_x(\lambda, x) \cdot Q(\lambda, x) - Q_x(\lambda, x) \cdot P(\lambda, x)}$$

where $P_{\lambda}(\lambda, x) = \frac{\partial P}{\partial \lambda}$; $Q_{\lambda}(\lambda, x) = \frac{\partial Q}{\partial \lambda}$; $P_x(\lambda, x) = \frac{\partial P}{\partial x}$; $Q_x(\lambda, x) = \frac{\partial Q}{\partial x}$, substituting $\lambda = i \cdot \omega$ and bearing $\overline{P(-i \cdot \omega)} = P(i \cdot \omega)$; $\overline{Q(-i \cdot \omega)} = Q(i \cdot \omega)$; $i \cdot P_{\lambda}(i \cdot \omega) = P_{\omega}(i \cdot \omega)$; $i \cdot Q_{\lambda}(i \cdot \omega) = Q_{\omega}(i \cdot \omega)$ and that on the surface $|P(i \cdot \omega)|^2 = |Q(i \cdot \omega)|^2$, one obtain:

$$\left(\frac{\partial \lambda}{\partial x}\right)^{-1} |_{\lambda = i \cdot \omega}$$

$$= \frac{i \cdot P_{\omega}(i \cdot \omega, x) \cdot \overline{P(i \cdot \omega, x)} + i \cdot Q_{\lambda}(i \cdot \omega, x) \cdot \overline{Q(\lambda, x)} - Z_{\tau} \cdot |P(i \cdot \omega, x)|^2}{P_x(i \cdot \omega, x) \cdot \overline{P(i \cdot \omega, x)} - Q_x(i \cdot \omega, x) \cdot \overline{Q(i \cdot \omega, x)}}$$

Upon separating into real and imaginary parts, with $P = P_R + i \cdot P_I$; $Q = Q_R + i \cdot Q_I$ $P_{\omega} = P_{R\omega} + i \cdot P_{I\omega}$; $Q_{\omega} = Q_{R\omega} + i \cdot Q_{I\omega}$; $P_x = P_{Rx} + i \cdot P_{Ix}$; $Q_x = Q_{Rx} + i \cdot Q_{Ix}$; $P^2 = P_R^2 + P_I^2$.

When (x) can be any system's parameter $\eta, \beta, C, \Omega, \ldots$ and physical distance shifting Z_{τ} etc. where for convenience, we have dropped the arguments $(i \cdot \omega, x)$ and where

$$F_{\omega} = 2 \cdot [(P_{R\omega} \cdot P_R + P_{I\omega} \cdot P_I) - (Q_{R\omega} \cdot Q_R + Q_{I\omega} \cdot Q_I)]$$
$$F_x = 2 \cdot [(P_{Rx} \cdot P_R + P_{Ix} \cdot P_I) - (Q_{Rx} \cdot Q_R + Q_{Ix} \cdot Q_I)].$$

$\omega_x = -\frac{F_x}{F_{\omega}}$. We define U and V: $U = (P_R \cdot P_{I\omega} - P_I \cdot P_{R\omega}) - (Q_R \cdot Q_{I\omega} - Q_I \cdot Q_{R\omega})$ $V = (P_R \cdot P_{Ix} - P_I \cdot P_{Rx}) - (Q_R \cdot Q_{Ix} - Q_I \cdot Q_{Rx})$. We choose our specific parameter as physical distance shifting $x = Z_{\tau}$.

$$P_R(i \cdot \omega, Z_{\tau}) = \omega^4 + \omega^2 \cdot 4 \cdot \sigma \cdot \beta^{(i)} \cdot \frac{18 \cdot [\eta^{(i)}]^2}{\pi^2} \cdot S_{\beta}$$
$$P_I(i \cdot \omega, Z_{\tau}) = -\omega^3 \cdot \left(4 \cdot \sigma \cdot \beta^{(i)} - \frac{18 \cdot [\eta^{(i)}]^2}{\pi^2} \cdot S_{\beta}\right)$$

$$Q_R(i \cdot \omega, Z_{\tau}) = -\omega^2 \cdot \left[8 \cdot \sigma^2 \cdot [\beta^{(i)}]^2 + 2 \cdot \sigma \cdot \eta^{(i)} \cdot \{4 \cdot \sigma \cdot [\eta^{(i)}]^3 + (3 \cdot S_{\eta} - A) \cdot 3 \cdot [\eta^{(i)}]^2\} \cdot \frac{2}{\pi^2}\right]$$

$$Q_I(i \cdot \omega, Z_{\tau}) = -\omega^3 \cdot 2 \cdot \sigma \cdot \beta^{(i)}$$

$$P_{R\omega} = 4 \cdot \omega^3 + 2 \cdot \omega \cdot 4 \cdot \sigma \cdot \beta^{(i)} \cdot \frac{18 \cdot [\eta^{(i)}]^2}{\pi^2} \cdot S_{\beta}$$
$$P_{I\omega} = -3 \cdot \omega^2 \cdot \left(4 \cdot \sigma \cdot \beta^{(i)} - \frac{18 \cdot [\eta^{(i)}]^2}{\pi^2} \cdot S_{\beta}\right)$$

$$Q_{R\omega} = -2 \cdot \omega \cdot [8 \cdot \sigma^2 \cdot [\beta^{(i)}]^2 + 2 \cdot \sigma \cdot \eta^{(i)} \cdot \{4 \cdot \sigma \cdot [\eta^{(i)}]^3$$

$$+(3 \cdot S_\eta - A) \cdot 3 \cdot [\eta^{(i)}]^2\} \cdot \frac{2}{\pi^2}\Bigg]$$

$$Q_{I\omega} = -3 \cdot \omega^2 \cdot 2 \cdot \sigma \cdot \beta^{(i)}; \quad P_{RZ_\tau} = 0; \quad P_{IZ_\tau} = 0; \quad Q_{RZ_\tau} = 0; \quad Q_{IZ_\tau} = 0$$

$$\begin{aligned}
P_{R\omega} \cdot P_R &= \left\{4 \cdot \omega^3 + 2 \cdot \omega \cdot 4 \cdot \sigma \cdot \beta^{(i)} \cdot \frac{18 \cdot [\eta^{(i)}]^2}{\pi^2} \cdot S_\beta\right\} \\
&\quad \cdot \left\{\omega^4 + \omega^2 \cdot 4 \cdot \sigma \cdot \beta^{(i)} \cdot \frac{18 \cdot [\eta^{(i)}]^2}{\pi^2} \cdot S_\beta\right\} \\
&= \omega^3 \cdot \left\{4 \cdot \omega^2 + 8 \cdot \sigma \cdot \beta^{(i)} \cdot \frac{18 \cdot [\eta^{(i)}]^2}{\pi^2} \cdot S_\beta\right\} \\
&\quad \cdot \left\{\omega^2 + 4 \cdot \sigma \cdot \beta^{(i)} \cdot \frac{18 \cdot [\eta^{(i)}]^2}{\pi^2} \cdot S_\beta\right\} \\
&= \omega^3 \cdot \left\{4 \cdot \omega^4 + 16 \cdot \omega^2 \cdot \sigma \cdot \beta^{(i)} \cdot \frac{18 \cdot [\eta^{(i)}]^2}{\pi^2} \cdot S_\beta\right. \\
&\quad + \omega^2 \cdot 8 \cdot \sigma \cdot \beta^{(i)} \cdot \frac{18 \cdot [\eta^{(i)}]^2}{\pi^2} \cdot S_\beta \\
&\quad \left. +32 \cdot \sigma^2 \cdot [\beta^{(i)}]^2 \cdot \frac{18^2 \cdot [\eta^{(i)}]^4}{\pi^4} \cdot S_\beta^2\right\} \\
&= 4 \cdot \omega^3 \cdot \left\{\omega^4 + 6 \cdot \omega^2 \cdot \sigma \cdot \beta^{(i)} \cdot \frac{18 \cdot [\eta^{(i)}]^2}{\pi^2} \cdot S_\beta\right. \\
&\quad \left. +8 \cdot \sigma^2 \cdot [\beta^{(i)}]^2 \cdot \frac{18^2 \cdot [\eta^{(i)}]^4}{\pi^4} \cdot S_\beta^2\right\}
\end{aligned}$$

$$\begin{aligned}
P_{R\omega} \cdot P_R = 4 \cdot \omega^3 \cdot &\left\{\omega^4 + 6 \cdot \omega^2 \cdot \sigma \cdot \beta^{(i)} \cdot \frac{18 \cdot [\eta^{(i)}]^2}{\pi^2} \cdot S_\beta\right. \\
&\left. +8 \cdot \sigma^2 \cdot [\beta^{(i)}]^2 \cdot \frac{18^2 \cdot [\eta^{(i)}]^4}{\pi^4} \cdot S_\beta^2\right\}
\end{aligned}$$

$$\begin{aligned}
Q_{R\omega} \cdot Q_R &= \Big\{-2 \cdot \omega \cdot \Big[8 \cdot \sigma^2 \cdot [\beta^{(i)}]^2 + 2 \cdot \sigma \cdot \eta^{(i)} \\
&\quad \cdot \{4 \cdot \sigma \cdot [\eta^{(i)}]^3 + (3 \cdot S_\eta - A) \cdot 3 \cdot [\eta^{(i)}]^2\} \cdot \frac{2}{\pi^2}\Big]\Big\} \\
&\quad \Big\{\cdot - \omega^2 \cdot \Big[8 \cdot \sigma^2 \cdot [\beta^{(i)}]^2 + 2 \cdot \sigma \cdot \eta^{(i)} \\
&\quad \cdot \{4 \cdot \sigma \cdot [\eta^{(i)}]^3 + (3 \cdot S_\eta - A) \cdot 3 \cdot [\eta^{(i)}]^2\} \cdot \frac{2}{\pi^2}\Big]\Big\} \\
&= 2 \cdot \omega^3 \cdot \Big[8 \cdot \sigma^2 \cdot [\beta^{(i)}]^2 + 2 \cdot \sigma \cdot \eta^{(i)} \\
&\quad \cdot \{4 \cdot \sigma \cdot [\eta^{(i)}]^3 + (3 \cdot S_\eta - A) \cdot 3 \cdot [\eta^{(i)}]^2\} \cdot \frac{2}{\pi^2}\Big]^2
\end{aligned}$$

$$Q_{R\omega} \cdot Q_R = 2 \cdot \omega^3 \cdot \left[8 \cdot \sigma^2 \cdot [\beta^{(i)}]^2 + 2 \cdot \sigma \cdot \eta^{(i)} \cdot \{4 \cdot \sigma \cdot [\eta^{(i)}]^3 + (3 \cdot S_\eta - A) \cdot 3 \cdot [\eta^{(i)}]^2\} \cdot \frac{2}{\pi^2}\right]^2$$

$$Q_R \cdot Q_{I\omega} = \left\{-\omega^2 \cdot \left[8 \cdot \sigma^2 \cdot [\beta^{(i)}]^2 + 2 \cdot \sigma \cdot \eta^{(i)} \cdot \{4 \cdot \sigma \cdot [\eta^{(i)}]^3 + (3 \cdot S_\eta - A) \cdot 3 \cdot [\eta^{(i)}]^2\} \cdot \frac{2}{\pi^2}\right]\right\} \cdot \{-3 \cdot \omega^2 \cdot 2 \cdot \sigma \cdot \beta^{(i)}\}$$

$$Q_R \cdot Q_{I\omega} = 3 \cdot \omega^4 \cdot 2 \cdot \sigma \cdot \beta^{(i)} \cdot \left[8 \cdot \sigma^2 \cdot [\beta^{(i)}]^2 + 2 \cdot \sigma \cdot \eta^{(i)} \cdot \{4 \cdot \sigma \cdot [\eta^{(i)}]^3 + (3 \cdot S_\eta - A) \cdot 3 \cdot [\eta^{(i)}]^2\} \cdot \frac{2}{\pi^2}\right]$$

$$Q_I \cdot Q_{R\omega} = -\omega^3 \cdot 2 \cdot \sigma \cdot \beta^{(i)} \cdot \left\{-2 \cdot \omega \cdot \left[8 \cdot \sigma^2 \cdot [\beta^{(i)}]^2 + 2 \cdot \sigma \cdot \eta^{(i)} \cdot \{4 \cdot \sigma \cdot [\eta^{(i)}]^3 + (3 \cdot S_\eta - A) \cdot 3 \cdot [\eta^{(i)}]^2\} \cdot \frac{2}{\pi^2}\right]\right\}$$

$$Q_I \cdot Q_{R\omega} = \omega^4 \cdot 4 \cdot \sigma \cdot \beta^{(i)} \cdot \left[8 \cdot \sigma^2 \cdot [\beta^{(i)}]^2 + 2 \cdot \sigma \cdot \eta^{(i)} \cdot \{4 \cdot \sigma \cdot [\eta^{(i)}]^3 + (3 \cdot S_\eta - A) \cdot 3 \cdot [\eta^{(i)}]^2\} \cdot \frac{2}{\pi^2}\right]$$

We can summery the intermediate results in the following table (Table C.1):

$$V = (P_R \cdot P_{IZ_\tau} - P_I \cdot P_{RZ_\tau}) - (Q_R \cdot Q_{IZ_\tau} - Q_I \cdot Q_{RZ_\tau}) = 0; \;\; F(\omega, Z_\tau) = 0$$

Differentiating $F(\omega, Z_\tau)$ with respect to Z_τ and we get $F_\omega \cdot \frac{\partial \omega}{\partial Z_\tau} + F_{Z_\tau} = 0;\;\; Z_\tau \in I$

$$\frac{\partial \omega}{\partial Z_\tau} = -\frac{F_{Z_\tau}}{F_\omega};\;\; \Lambda^{-1}(Z_\tau) = \left(\frac{\partial \mathrm{Re}\lambda}{\partial Z_\tau}\right)_{\lambda = i\cdot\omega};\;\; \frac{\partial \omega}{\partial Z_\tau} = \omega_{Z_\tau} = -\frac{F_{Z_\tau}}{F_\omega}$$

$$\Lambda^{-1}(Z_\tau) = \mathrm{Re}\left\{\frac{-2 \cdot [U + Z_\tau \cdot |P|^2] + i \cdot F_\omega}{F_{Z_\tau} + i \cdot 2 \cdot [V + \omega \cdot |P|^2]}\right\}$$

$$sign\{\Lambda^{-1}(Z_\tau)\} = sign\left\{\left(\frac{\partial \mathrm{Re}\lambda}{\partial Z_\tau}\right)_{\lambda = i\cdot\omega}\right\}$$

$$sign[\Lambda^{-1}(Z_\tau)] = sign\{F_\omega\} \cdot sign\left\{\frac{V + \frac{\partial \omega}{\partial Z_\tau} \cdot U}{|P|^2} + \omega + \frac{\partial \omega}{\partial Z_\tau} \cdot Z_\tau\right\}$$

We shall presently examine the possibility of stability transitions (bifurcations) slowly varying envelope of the electric field set of couple's differential equation system, about the equilibrium points $E^{(0)}(\eta^{(0)}, \beta^{(0)}, C^{(0)}, \Omega^{(0)}) = (0, 0, C^{(0)}, -\frac{S_\Omega}{\sigma})$

Table C.1 $P_{R\omega} \cdot P_R$, $Q_{R\omega} \cdot Q_R$, $Q_R \cdot Q_{I\omega}$, and $Q_I \cdot Q_{R\omega}$ outcomes and expressions

Outcome	Expression
$P_{R\omega} \cdot P_R$	$4 \cdot \omega^3 \cdot \left\{\omega^4 + 6 \cdot \omega^2 \cdot \sigma \cdot \beta^{(i)} \cdot \frac{18 \cdot [\eta^{(i)}]^2}{\pi^2} \cdot S_\beta + 8 \cdot \sigma^2 \cdot [\beta^{(i)}]^2 \cdot \frac{18^2 \cdot [\eta^{(i)}]^4}{\pi^4} \cdot S_\beta^2\right\}$
$Q_{R\omega} \cdot Q_R$	$2 \cdot \omega^3 \cdot \left[8 \cdot \sigma^2 \cdot [\beta^{(i)}]^2 + 2 \cdot \sigma \cdot \eta^{(i)} \cdot \{4 \cdot \sigma \cdot [\eta^{(i)}]^3 + (3 \cdot S_\eta - A) \cdot 3 \cdot [\eta^{(i)}]^2\} \cdot \frac{2}{\pi^2}\right]^2$
$Q_R \cdot Q_{I\omega}$	$3 \cdot \omega^4 \cdot 2 \cdot \sigma \cdot \beta^{(i)} \cdot \left[8 \cdot \sigma^2 \cdot [\beta^{(i)}]^2 + 2 \cdot \sigma \cdot \eta^{(i)} \cdot \{4 \cdot \sigma \cdot [\eta^{(i)}]^3 + (3 \cdot S_\eta - A) \cdot 3 \cdot [\eta^{(i)}]^2\} \cdot \frac{2}{\pi^2}\right]$
$Q_I \cdot Q_{R\omega}$	$\omega^4 \cdot 4 \cdot \sigma \cdot \beta^{(i)} \cdot \left[8 \cdot \sigma^2 \cdot [\beta^{(i)}]^2 + 2 \cdot \sigma \cdot \eta^{(i)} \cdot \{4 \cdot \sigma \cdot [\eta^{(i)}]^3 + (3 \cdot S_\eta - A) \cdot 3 \cdot [\eta^{(i)}]^2\} \cdot \frac{2}{\pi^2}\right]$

or $E^{(1)}(\eta^{(1)}, \beta^{(1)}, C^{(1)}, \Omega^{(1)}) = \left(\frac{A^2-3\cdot S_\eta}{\sigma\cdot\left(1-\pi^2\cdot\frac{9}{\pi^4\cdot\sigma^2}\cdot S_\beta^2\right)}, -\frac{3\cdot[\eta^*]^2}{\pi^2\cdot\sigma}\cdot S_\beta, C^{(1)}, -\frac{S_\Omega}{\sigma}\right)$ as a result of a variation of physical distance shifting Z_τ. The analysis consists in identifying the roots of our system characteristic equation situated on the imaginary axis of the complex λ-plane where by increasing the physical distance shifting Z_τ, Reλ may at the crossing, changes its sign from "−" to "+", i.e. from a stable focus $E^{(*)}$ to an unstable one, or vice versa. This feature may be further assessed by examining the sign of the partial derivatives with respect to Z_τ.

$$\Lambda^{-1}(Z_\tau) = \left(\frac{\partial \mathrm{Re}\lambda}{\partial Z_\tau}\right)_{\lambda=i\cdot\omega};\ \eta, \beta, C, \Omega, \ldots = const$$

$$U = (P_R \cdot P_{I\omega} - P_I \cdot P_{R\omega}) - (Q_R \cdot Q_{I\omega} - Q_I \cdot Q_{R\omega})$$

$$P_R \cdot P_{I\omega} = -3\cdot\omega^4\cdot\left\{\omega^2 + 4\cdot\sigma\cdot\beta^{(i)}\cdot\frac{18\cdot[\eta^{(i)}]^2}{\pi^2}\cdot S_\beta\right\} \cdot\left(4\cdot\sigma\cdot\beta^{(i)} - \frac{18\cdot[\eta^{(i)}]^2}{\pi^2}\cdot S_\beta\right)$$

$$P_I \cdot P_{R\omega} = -\omega^3\cdot\left(4\cdot\sigma\cdot\beta^{(i)} - \frac{18\cdot[\eta^{(i)}]^2}{\pi^2}\cdot S_\beta\right)\cdot\left\{4\cdot\omega^3 +2\cdot\omega\cdot 4\cdot\sigma\cdot\beta^{(i)}\cdot\frac{18\cdot[\eta^{(i)}]^2}{\pi^2}\cdot S_\beta\right\}$$

$$P_I \cdot P_{R\omega} = -\omega^4\cdot\left(4\cdot\sigma\cdot\beta^{(i)} - \frac{18\cdot[\eta^{(i)}]^2}{\pi^2}\cdot S_\beta\right)\cdot 4 \cdot\left\{\omega^2 + 2\cdot\sigma\cdot\beta^{(i)}\cdot\frac{18\cdot[\eta^{(i)}]^2}{\pi^2}\cdot S_\beta\right\}$$

$$U = \left(-3\cdot\omega^4\cdot\left\{\omega^2 + 4\cdot\sigma\cdot\beta^{(i)}\cdot\frac{18\cdot[\eta^{(i)}]^2}{\pi^2}\cdot S_\beta\right\}\cdot\left(4\cdot\sigma\cdot\beta^{(i)} - \frac{18\cdot[\eta^{(i)}]^2}{\pi^2}\cdot S_\beta\right)\right.$$
$$+\omega^4\cdot\left(4\cdot\sigma\cdot\beta^{(i)} - \frac{18\cdot[\eta^{(i)}]^2}{\pi^2}\cdot S_\beta\right)$$
$$\left.\cdot 4\cdot\left\{\omega^2 + 2\cdot\sigma\cdot\beta^{(i)}\cdot\frac{18\cdot[\eta^{(i)}]^2}{\pi^2}\cdot S_\beta\right\}\right)$$
$$-\left(3\cdot\omega^4\cdot 2\cdot\sigma\cdot\beta^{(i)}\cdot\left[8\cdot\sigma^2\cdot[\beta^{(i)}]^2 + 2\cdot\sigma\cdot\eta^{(i)}\right.\right.$$
$$\left.\cdot\{4\cdot\sigma\cdot[\eta^{(i)}]^3 + (3\cdot S_\eta - A)\cdot 3\cdot[\eta^{(i)}]^2\}\cdot\frac{2}{\pi^2}\right]$$
$$-\omega^4\cdot 4\cdot\sigma\cdot\beta^{(i)}\cdot\left[8\cdot\sigma^2\cdot[\beta^{(i)}]^2 + 2\cdot\sigma\cdot\eta^{(i)}\right.$$
$$\left.\left.\cdot\{4\cdot\sigma\cdot[\eta^{(i)}]^3 + (3\cdot S_\eta - A)\cdot 3\cdot[\eta^{(i)}]^2\}\cdot\frac{2}{\pi^2}\right]\right)$$

Then we get the expression for $F(\omega, Z_\tau)$ system parameter values. We find those ω, Z_τ values which fulfill $F(\omega, Z_\tau) = 0$. We ignore negative, complex, and imaginary values of ω for specific Z_τ values. For specific Z_τ values, we can be express by 3D function $F(\omega, Z_\tau) = 0$. We plot the stability switch diagram based on different physical distance shifting values.

$$\Lambda^{-1}(Z_\tau) = \left(\frac{\partial \mathrm{Re}\lambda}{\partial Z_\tau}\right)_{\lambda=i\cdot\omega} = \mathrm{Re}\left\{\frac{-2\cdot[U + Z_\tau\cdot|P|^2] + i\cdot F_\omega}{F_{Z_\tau} + i\cdot 2\cdot[V + \omega\cdot|P|^2]}\right\}$$

$$\Lambda^{-1}(Z_\tau) = \left(\frac{\partial \mathrm{Re}\lambda}{\partial Z_\tau}\right)_{\lambda=i\cdot\omega} = \frac{2\cdot\{F_\omega\cdot(V+\omega\cdot P^2) - F_{Z_\tau}\cdot(U + Z_\tau\cdot P^2)\}}{F_{Z_\tau}^2 + 4\cdot(V+\omega\cdot P^2)^2}$$

The stability switch occurs only on those physical distance shifting values (Z_τ) which fit the equation: $Z_\tau = \frac{\theta_+(Z_\tau)}{\omega_+(Z_\tau)}$ and $\theta_+(Z_\tau)$ is the solution of $\sin\theta(Z_\tau) = \cdots$; $\cos\theta(Z_\tau) = \cdots$, when $\omega = \omega_+(Z_\tau)$ if only ω_+ is feasible. Additionally, when all system parameters are known and the stability switch due to various physical distance shifting values Z_τ is describe in the following expression:

$$sign\{\Lambda^{-1}(Z_\tau)\} = sign\{F_\omega(\omega(Z_\tau), Z_\tau)\}\cdot sign\{Z_\tau\cdot\omega_{Z_\tau}(\omega(Z_\tau)) + \omega(Z_\tau) + \frac{U(\omega(Z_\tau))\cdot\omega_{Z_\tau}(\omega(Z_\tau)) + V(\omega(Z_\tau))}{|P(\omega(Z_\tau))|^2}$$

Remark: we know $F(\omega, Z_\tau) = 0$ implies its roots $\omega_i(Z_\tau)$ and finding those physical distance shifting Z_τ values which ω_i is feasible. There are Z_τ values which give complex ω_i or imaginary number, then unable to analyze stability. F function is independent on Z_τ the parameter $F(\omega, Z_\tau) = 0$.

The results: We find those ω, Z_τ values which fulfill $F(\omega, Z_\tau) = 0$. We ignore negative, complex, and imaginary values of ω.

Next is to find those ω, Z_τ values which fulfill $\sin\theta(Z_\tau) = \cdots$; $\cos\theta(Z_\tau) = \cdots$

$$\sin(\omega\cdot Z_\tau) = \frac{-P_R\cdot Q_I + P_I\cdot Q_R}{|Q|^2}$$

$$\cos(\omega\cdot Z_\tau) = -\frac{(P_R\cdot Q_R + P_I\cdot Q_I)}{|Q|^2};\ |Q|^2 = Q_R^2 + Q_I^2$$

Finally, we plot the stability switch diagram $g(Z_\tau) = \Lambda^{-1}(Z_\tau) = (\frac{\partial \mathrm{Re}\lambda}{\partial Z_\tau})_{\lambda=i\cdot\omega}$

$$g(Z_\tau) = \Lambda^{-1}(Z_\tau) = \left(\frac{\partial \mathrm{Re}\lambda}{\partial Z_\tau}\right)_{\lambda=i\cdot\omega} = \frac{2\cdot\{F_\omega\cdot(V+\omega\cdot P^2) - F_{Z_\tau}\cdot(U + Z_\tau\cdot P^2)\}}{F_{Z_\tau}^2 + 4\cdot(V+\omega\cdot P^2)^2}$$

$$F_{Z_\tau}^2 + 4\cdot(V+\omega\cdot P^2)^2 > 0$$

Table C.2 Sign of $\Lambda^{-1}(Z_\tau)$ expression

$sign[F_\omega]$	$sign\left[\frac{V+\omega_{Z_\tau}\cdot U}{P^2}+\omega+\omega_{Z_\tau}\cdot Z_\tau\right]$	$\mathrm{sign}[\Lambda^{-1}(Z_\tau)]$
+/−	+/−	+
+/−	−/+	−

$$\mathrm{sign}[\Lambda^{-1}(Z_\tau)] = sign\{F_\omega \cdot (V + \omega \cdot P^2) - F_{Z_\tau} \cdot (U + Z_\tau \cdot P^2)\}$$

$$\mathrm{sign}[\Lambda^{-1}(Z_\tau)] = sign\left\{[F_\omega] \cdot \left[(V + \omega \cdot P^2) - \frac{F_{Z_\tau} \cdot (U + Z_\tau \cdot P^2)}{F_\omega}\right]\right\}$$

$$\omega_{Z_\tau} = -\frac{F_{Z_\tau}}{F_\omega}$$

$$\omega_{Z_\tau} = \left(\frac{\partial \omega}{\partial Z_\tau}\right)^{-1} = -\frac{\partial F/\partial \omega}{\partial F/\partial Z_\tau}$$

$$\mathrm{sign}[\Lambda^{-1}(Z_\tau)] = sign\left\{[F_\omega] \cdot \left[\frac{1}{P^2}\right] \cdot \left[\frac{V}{P^2} + \omega - \frac{F_{Z_\tau}}{F_\omega} \cdot \frac{(U + Z_\tau \cdot P^2)}{P^2}\right]\right\}$$

$$\mathrm{sign}[\Lambda^{-1}(Z_\tau)] = sign\left\{[F_\omega] \cdot \left[\frac{1}{P^2}\right] \cdot \left[\frac{V}{P^2} + \omega + \omega_{Z_\tau} \cdot \left(\frac{U}{P^2} + Z_\tau\right)\right]\right\}$$

$$\mathrm{sign}[\Lambda^{-1}(Z_\tau)] = sign\left\{[F_\omega] \cdot \left[\frac{1}{P^2}\right] \cdot \left[\frac{V}{P^2} + \omega + \frac{\omega_{Z_\tau} \cdot U}{P^2} + \omega_{Z_\tau} \cdot Z_\tau\right]\right\}$$

$$\mathrm{sign}[\Lambda^{-1}(Z_\tau)] = sign\left\{[F_\omega] \cdot \left[\frac{1}{P^2}\right] \cdot \left[\frac{V + \omega_{Z_\tau} \cdot U}{P^2} + \omega + \omega_{Z_\tau} \cdot Z_\tau\right]\right\}$$

$$sign\left[\frac{1}{P^2}\right] > 0 \Rightarrow \mathrm{sign}[\Lambda^{-1}(Z_\tau)] = sign\left\{[F_\omega] \cdot \left[\frac{V + \omega_{Z_\tau} \cdot U}{P^2} + \omega + \omega_{Z_\tau} \cdot Z_\tau\right]\right\}$$

$$\mathrm{sign}[\Lambda^{-1}(Z_\tau)] = sign[F_\omega] \cdot sign\left[\frac{V + \omega_{Z_\tau} \cdot U}{P^2} + \omega + \omega_{Z_\tau} \cdot Z_\tau\right]$$

$$F_\omega = 2 \cdot [(P_{R\omega} \cdot P_R + P_{I\omega} \cdot P_I) - (Q_{R\omega} \cdot Q_R + Q_{I\omega} \cdot Q_I)]$$

We check the sign of $\Lambda^{-1}(Z_\tau)$ according to the following rule:

If the $\mathrm{sign}[\Lambda^{-1}(Z_\tau)] > 0$ then the crossing proceeds from (−) to (+) respectively (stable to unstable). If the $\mathrm{sign}[\Lambda^{-1}(Z_\tau)] < 0$ then the crossing proceeds from (+) to (−) respectively (unstable to stable) (Table C.2).

References

[1] F.M. Dickey, *Laser Beam shaping: Theory and Techniques*, 2nd edn. (CRC Press, Boca Raton, 2014)
[2] F.M. Dickey, T.E. Lizotte, *Laser Beam Shaping Applications (Optical Science and Engineering)*, 2nd edn. (CRC Press, Boca Raton, 2017)
[3] J. Nathan kutz, P.K.A. Wai, Gordon-Haus timing jitter reduction in dispersion-managed soliton communications. IEEE Photonics Technol. Lett. **10**(5) (1998)

Appendix D
Plasma Mirror System—Modeling, Characterization, Optic Design, Operation and Working Principles

Plasma mirror (PM) is a self-induced ultrafast optical shutter. Its operation is based on the ultrafast ionization of intense laser pulses. A laser pulses is focused onto a transparent bulk target, and while the low intensity pre pulses and pedestal traverses the target, the rising edge of the main pulse with its high intensity generates a highly reflective flat plasma layer. The main pulse cleaned from the pedestal and pre pulses is secularly reflected off the target thus the contrast of the reflected beam is enhanced. Plasma mirror (PM) rapidly changing its optical properties from almost perfectly transmissivity to highly reflective. An intense laser pulses are focused onto an anti-reflective coated substrate, which ionizes and forms an over critical plasma surface at the leading edge of the main pulse and separates the reflected high intensity peak from the pulse preceding low intensity background. The fluence on the plasma mirror (PM) has to be adapted to the initial contrast of the laser system then we get high reflectivity and proper triggering of the PM substrate. If the fluence is too high, the plasma forms very early and thus reflects off unwanted signal. A long expansion of the plasma surface prior to the reflection of the peak pulse induces wave front distortions and therefore reduced focus ability of the reflected beam. If the fluence is chosen too low, the plasma mirror (PM) ionizes too late, which in turn reduces the overall reflectivity and energy throughput of the system. If our system is a conventional CPA laser system with intrinsic contrast then the plasma mirrors (PMs) should be operated in the range of 10 to 100 sJ/cm^2. Practically it is recommended to optimize the PM reflectivity up to 80%. The contrast enhancement can be achieved by setting the ratio of the plasma reflectivity and the reflectivity of the anti-reflective coating $\frac{R_{plasma}}{R_{AR}} \approx 10^2$. If we use several cascading PMs and use multiple reflections then the ratio is increased. The plasma mirror principle is that the preceding, low intensity part of the pulse is transmitted through the plasma mirror substrate, whereas the main pulse ionizes the surface and reflects off the plasma. Plasma mirror system is structured from a plasma mirror on AR (Anti Reflect) coated glass substrate (Fig. D.1). In many experimental systems we shot an intense ultrashort pulsed laser beam with a high pulse contrast (pulse contrast is define as the ratio between peak intensity and the intensity of amplified spontaneous emission (ASE), and existing pre-pulses). We

O. Aluf, *Advance Elements of Laser Circuits and Systems*,
https://doi.org/10.1007/978-3-030-64103-0

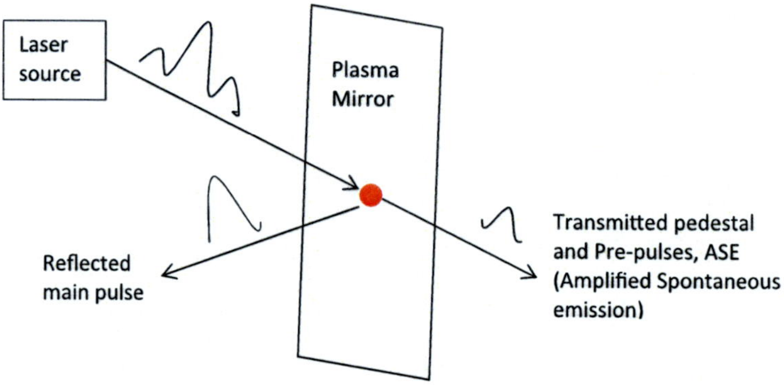

Fig. D.1 Plasma mirror (PM) and laser source system

already emphasis that the plasma mirror (PM) system is based on the technique of self-induced plasma shuttering. By using this technique we ensure that the contrast of a high intensity ultrashort laser pulse is enhanced. We can describe the operation of plasma mirror as follow: The ultrashort laser pulse is focused onto a glass substrate with an anti-reflection (AR) coating, and the ionization is triggered only by the main pulse but not the pre-pulse, which is transmitted through the sample. The outcome is that only the main pulse is reflected and the contrast will be enhanced. Basically the plasma mirror is operated with an undistorted reflected pulse and we can guarantee it by fulfill the condition $C_s \cdot \Delta t < \lambda_{laser}$, where C_s is the sound speed, Δt is the time elapsed between the time of plasma formation and the peak of the pulse. Our target is to find the optimal intensity value which yielding a reflectivity higher than specific value without changing the spatial quality of the beam. If we use in our system two PMs we get larger contrast enhancement. If the plasma mirror has a large incident angle then there is a high sensitivity to laser polarization [1, 2].

Plasma mirror (PM) suppress the pedestal and pre pulses of high-energy ultrashort pulse lasers (<1 ps). The laser light switches the ultrashort shutter at the striking edge of the main pulse and the pulse is cleaned from any leading pedestal. The operation of the PM is based on the ultrafast ionization of the larger light. The laser pulse strikes a flat transparent target, at a determined fluence. The fluence is sufficiently low for the leading pedestal and pre pulses to be transmitted through the target, but high enough for the striking edge of the main pulse to ionize it. A very fast ionization causes the plasma layer (dense and flat) to be triggered on the surface of the target. The ionization takes place within a few optical cycles, and we get an expansion of the thin plasma layer which is negligible during the laser pulse. Some energy of the main pulse is absorb by the target material for the creation of the plasma layer. The reflectivity of the triggered plasma layer is high. There is an energy loss of the main reflected pulse [3]. We can estimate the energy loss of the main reflected pulse by the overall reflectivity of the PM which is the ratio of the total energy of the main pulse before and after the plasma mirror (PM). The pre pulse traverses the target without any reflective plasma formation. The pre pulse is not suppressed perfectly.

The process we can observe in the plasma mirror system is that a laser pulse is focused onto a transparent bulk target. The fluence is adjusted then the leading pedestal and pre pulses traverse the target while the striking edge of the main pulse ionizes it and instantaneously generates a highly reflective flat plasma layer on the target's surface. The main pulse and the post pulse cleaned from the pedestal and pre pulse are secularly reflected off the plasma mirror. The PM is dedicated to the suppression of the pre pulse. The reflectivity of the PM is dependent on the fluence of the applied laser pulse. Below the threshold fluence, no breakdown occurs. The target remains transparent with its low Fresnel reflectivity. Pulses above the threshold ionize the target to the breakdown level, and create the reflective plasma layer. There is a loss by the absorbed energy that is necessary for the generation of the plasma, since no reflection occurs until the breakdown. This loss is a higher fraction of a low energy pulse than of a high-energy pulse. The reflectivity increases with increasing fluence of the incident laser beam. We get the highest reflectivity by the laser fluence which is near the limit. At that time the pre pulse still cannot trigger any plasma formation itself. The most important features of the PM are the contrast enhancement and the peak reflectivity which are strongly dependent on the polarization. There is a nonlinear dependence of the reflectivity on the laser fluence. The spatial profile of the reflected beam differs from that of the incident beam. If the focused beam has a Gaussian spatial energy distribution on the PM surface then the reflectivity of the PM is much higher in the center of the focal spot than at the boundary. We get a state that the reflected beam's spatial energy distribution is steeper than that of the original Gaussian beam. The peak reflectivity characterizes the PM better than the total or overall reflectivity. The PM spatially filters the beam and we can see a low intensity ripples in the focal plane which is related to imperfections of the beam. By the operation of the PM, it acts as a pin mirror placed in the center of the focal spot. If we use a high intensity laser then we can get rough profile that limits focus ability. By shooting short pulses to the PM, the plasma expands and the density gradient declines. The expansion of the plasma can be calculated from the pulse width and the expansion velocity, which equals approximately the plasma sound speed. If the pulse is with the appropriate fluence the striking edge of the pulse creates the plasma within a few optical cycles. During plasma formation, the absorption dominates, and the reflectivity gradually increases from its initial value. We get a reflected pulse which has steeper rising edge and shorter pulse duration. We get a significant contrast improvement by using PM. We can create nearly solid-density plasma by ultrashort high-intensity laser pulses which are focused on a solid target, and the emitting of short X-ray bursts (pulse width of the X-ray pulse falls into Pico sec range). The main pulse interacts with the lower density plasma and we get a longer pulse width of the emitted X-ray bursts. A pulse with an artificially increased ASE is cleaned by a PM and focused onto a target to generate X-ray bursts. The plasma mirror (PM) is beneficial to the generation of high density laser plasmas. It reduces the pedestal intensity without mark able spatial or temporal degradation of the laser pulse. The PM operation improves the contrast of high-power lasers. The plasma evolves over time, the plasma mirror surface contaminant layer is undergo a quasi-neutral self-similar plasma expansion, where C_s is the ion sound speed $\left(C_s = \sqrt{\frac{\gamma \cdot Z \cdot k \cdot T_e}{m_i}}\right)$. It is the speed of the longitudinal waves resulting from the mass of the ions and the pressure of the electrons.

$$C_s = \sqrt{\frac{\gamma \cdot Z \cdot k \cdot T_e}{m_i}} = 9.79 \times 10^5 \left(\sqrt{\frac{\gamma \cdot Z \cdot T_e}{\mu}}\right) \frac{\text{cm}}{\text{s}}$$

γ—is the adiabatic index, m_i—is the ion mass, Z—is the charge state, k—is the Boltzmann's constant, $\mu = \frac{m_i}{m_p}$, m_p—is the proton mass, n_e—is the electron plasma density, n_{e0}—is the unperturbed electron number density, for hydrogen plasma $\left(1 \times 10^{22}\ \text{cm}^{-3}\text{–}6 \times 10^{22}\ \text{cm}^{-3}\right)$, $n_e(x, t) = Z \cdot n_i(x, t) = n_{e0} \cdot e^{-\frac{x}{C_s \cdot t}}$. The laser radiation pressure is estimated to be almost two orders of magnitude lower than the final thermal pressure of the plasma. The plasma retains a mono exponentially decaying density profile $\frac{\partial n_e}{\partial x} = \frac{n_{e0}}{I_{ss}}$, characterized by the plasma density scale length I_{ss}. The distortion of the wave front at the PM surface is determined by the introduction of phase shifts, $\delta\varphi$, by non-uniform critical surface expansion in the quasi near field of the beam. The maximum phase shift introduced to two rays which incident at an angle θ_i, on a surface with a height perturbation, δx is giving by $\delta\varphi = \frac{2 \cdot \pi}{\lambda} \cdot 2 \cdot \delta_x \cdot \cos\theta_i$. A phase shift of zero corresponds to a perfectly smooth surface, while phases shift of π cause a complete destructive interference of the rays. This leads to a $\frac{\pi}{2}$ phase shift being the arbitrary choice in perturbation height, which separates a rough surface from a smooth one (Rayleigh criterion). Regarding the experimental system which include plasma mirror elements, plasma mirrors self-induced plasma shuttering are the final tools to clean laser pulses from pre pulses. Ultrashort laser pulses are generally preceded by pre pulses originating either from the incomplete compression of the chirped pulses or from the long ASE of the amplifier (pedestal). Pedestals can be detrimental for the investigations of high intensity laser-matter interactions, because they generate pre plasmas even if the intensity of the pedestal is several orders of magnitude less than that of the main pulse. In this case the main laser pulse interacts not with the solid target but with the preformed plasma. The principle of plasma mirror is that in case the intensity of the laser pulse which falling onto a transparent solid material is chosen and only the leading edge of the main ultrashort pulse is above the threshold for plasma production, the pedestal of lower intensity will be transmitted. Thus the part of the pulse reflected from the plasma which does not have time to expand during the short pulse duration can be separated resulting in a contrast improvement up to several orders of magnitude. There is a mechanism of cleaning the pulses of titanium Sapphire and OPA (optical parametric amplifier) laser systems by plasma mirrors (PMs). Planar plasma mirrors (PPMs) suppress laser pulse amplified spontaneous emission (ASE) and possible pre-pulses inherently present in intense pulses produced through the chirped pulse amplification (CPA) technique. The operation of PPMs is as follow: plasma is created on the surface of a solid (optical glass or plastic substrate), which is otherwise transparent to the laser light. Reflection of the laser pulse occurs near the

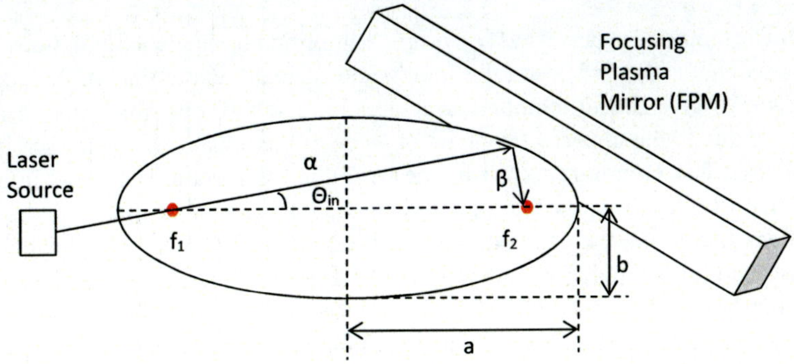

Fig. D.2 Schematic which describes the operation of ellipsoidal focusing plasma mirror (FPM)

plasma critical density. We define the laser frequency as the plasma electron oscillation frequency which is related to electron density. The laser intensity on the optic surface is selected such that the intensity of ASE/Pre-pulses preceding the main pulse is lower than the substrate ionization intensity threshold and is transmitted through the substrate. After plasma formation at threshold intensity, on the rising edge of the laser pulse, the remainder of the pulse is reflected, resulting in a pulse with a higher temporal intensity contrast which is defined as the ratio of the peak intensity to the ASE pedestal intensity. In the focusing plasma mirror (FPM) the reflecting surface of the optic substrate is curved. Employing an ellipsoidal geometry with two foci enables demagnification of a focal spot to be achieved. The geometry of FPM optic shape is ellipsoidal. The ellipse possesses two foci positions, f_1 and f_2, located on the major axis, of equal distance from the shape center, enabling point-to-point re-imaging. The degree of elliptical eccentricity is e, and a reduction or enlargement in the image size at one focus is obtained when an object is located at the other. The magnification, m, is equal to the ratio of lengths $\frac{\beta}{\alpha}$; where α and β are the distances from the mirror surface to f_1 and f_2, respectively. This ratio is changed as a function of the beam incident angle, θ_{in} with respect to the major axis and express as $m = \frac{(1+e^2)-2\cdot e\cdot \cos\theta_{in}}{1-e^2}$. Where $e = \sqrt{1 - \frac{b^2}{a^2}}$, with a and b being the semi-major and semi-minor axis lengths, respectively. This geometry enables a conventional off-axis parabola (OAP) to be aligned such that its focal spot coincides with position f_1. The light is diverged beyond f_1, and is reflected by the plasma which generates on the optic surface and is then brought to focus at position f_2 (Fig. D.2). The OPA focal spot, at f_1, is de-magnified at f_2 by a factor that depends on the selected FPM geometry and θ_{in} [3, 4].

Laser pulses of femtosecond to picosecond duration with very high peak powers, up to the Petta watt (PW) are an integrated part of every scientific facility. The intense light fields cause to any target rapid ionization, and electrons in the resulting plasma oscillate at relativistic velocities. These ultrashort and ultra-intense laser pulses are obtained with relativity small size sub picosecond laser systems by chirped pulse amplification (CPA) technique. The techniques are the temporal pedestal and pre pulses. The temporal pedestal has two origins: amplified spontaneous emission (ASE) in nanosecond scale, and spectral clipping and residual aberrations of the spectral phase, which give rise to temporal structures on a sub 100 ps time scale. The pre pulses are due to the imperfect contrast of some of the optical elements used in this type of laser. The temporal contrast at time t is defined as the ratio of the peak intensity of the main laser pulse, and the intensity of the background at time t (pedestal or pre pulse). Increasing the peak intensity requires a parallel increase of the temporal contrast then the pre pulse and ASE remain at the same intensity level. We achieve the increasing of the temporal contrast by improving the laser chain, or by cleaning the ultrashort pulse from its pedestal and pre pulse at the end of the chain. The main methods are: (1) non-linear process, such as frequency doubling at the output of the laser, cleans pulse injection, nonlinear Sagnac interferometer, and optical parametric CPA (OPCPA), (2) electro optic methods, which can remove Nano second pre pulses for systems with a regenerative amplifier, (3) self-induced plasma shuttering (plasma mirror). When using self-induced plasma shuttering (plasma mirror) the laser beam is focused on a low-reflectivity vacuum-dielectric interface, so that most of pedestal and pre pulses energy transmitted. When the laser intensity increases in time, electrons are excited in the dielectric medium by non-linear mechanisms such as multi photon absorption. If the electron density exceeds the critical density at the laser wavelength the reflectivity suddenly increases, the plasma mirror (PM) is "triggered", and the laser beam is reflected. The triggering only occurs at the very beginning of the main pulse, and the reflected beam thus has an improved contrast ration. For early triggering the pedestal and pre pulses be partly reflected, and the motion of the PM surface before and during the main pulse might distort the laser beam wave front, and a good subsequent focusing is prevented. The incident fluence is the most crucial parameter to make an efficient PM. Mainly we use high power Ti-Sapphire laser which based on the CPA technique.

Modeling of the plasma mirror (PM): The model is divided to two sub models, the first sub model describes the optical switching of the PM and the second sub model describes the expansion into vacuum. The first sub model is called "population-propagation model" and describes the ionization of the target with a population equation, and the propagation of the laser field in this in homogeneously ionized medium, using the Helmholtz wave-equation. The second sub model is a hydrodynamic model based on a Lagrangian fluid code, and is required to describe the regime where the hydrodynamics must be included because the plasma expands into vacuum. Thus the equations which describe the ionization of the target and the propagation of the field are equivalent in these two sub models.

The population-propagation sub model: The plan wave travels from vacuum impinging on the target. The target is initially a pure dielectric, with no electron in the conduction band. Electric field $E(z, t)$ is inside the target and the target gets ionized by strong field. We calculate the reflectivity of the plasma by knowing $E(z, t)$. The plane wave impinges (zero incidence angle) on the medium with a dielectric constant $\varepsilon(z)$, a part of the field energy, given by the complex reflection coefficient will be reflected back in vacuum ($z < 0$), and thus produce standing waves on the vacuum side by interference with the incident beam. The continuity of the electric field and its z-derivative at the vacuum-target boundary leads to: $E_0(1+r) = E(z=0^+)$; $i \cdot k_0 \cdot E_0(1-r) = \frac{dE(z=0^+)}{dz}$, E_0 is the amplitude of the incoming electric field, and $k_0 = \frac{\omega}{c}$. For a strongly-inhomogeneous medium like a plasma mirror (PM), the reflection coefficient cannot be simply obtained from the Fresnel equations, and is unknown. We get a solution by solving the Helmholtz equation backward, from a region far inside the target to the target surface. We take into account the initial condition at a point z = L in this region. The electric field is a plane wave propagating in the positive direction of the z-axis. Function $\psi(z, t)$ is define by $E(z, t) = A \cdot \psi(z, t)$ and $|\psi(z = L)| = 1$. The initial condition $\psi(z)|_{z=L}$ is as follow: $\psi(z = L) = e^{i \cdot k_1 \cdot L}$ and $\psi'(z = L) = i \cdot k_1 \cdot e^{i \cdot k_1 \cdot L}$ ($\psi'(z) = \frac{d\psi}{dz}$ and $k_1 = \frac{n_1 \cdot \omega}{c}$). We obtain the reflection coefficient: $r = \frac{i \cdot k_0 \cdot \psi(z=0) - \frac{d\psi(z=0)}{dz}}{i \cdot k_0 \cdot \psi(z=0) + \frac{d\psi(z=0)}{dz}}$. The amplitude of the incident electric field at the surface is $A = E_0 \cdot \frac{1+r}{\psi(z=0)}$. By knowing the electric field, we calculate the ionization rate W, and the excitation density $N(z)$ at time $t + dt$. The value of $N(z, t + dt)$ brings us the dielectric function using $\varepsilon(z, t + dt)$ and we determine the electric field at time $t + dt$. We evaluate the field propagation by using the quasi-stationary approximation by the assumption that the electric field at any point is harmonic in time, at the frequency of the external driving field. In the quasi-stationary regime, the propagation of an S-polarized electromagnetic wave, impinging on a medium is defined by its dielectric function $\varepsilon(z, t)$; $z > 0$, with an angle θ and described by the Helmholtz equation:

$$\frac{\partial^2 E(z, t)}{\partial z^2} + \frac{\omega^2}{c^2} \cdot [\varepsilon(z, t) - \sin^2(\theta)] \cdot E(z, t) = 0$$

where the angular frequency of the incident pulse is ω, and c is the speed of light in vacuum. $\varepsilon(z, t)$ is the dielectric function of the plasma and is given by the Drude model:

$\varepsilon(z, t) = n_1^2 - \frac{e^2 \cdot N(z,t)}{\varepsilon_0 \cdot m^* \cdot (\omega^2 - \frac{i \cdot \omega}{\tau})}$, n_1 is the initial refractive index of the solid, e is the electron charge, m is the effective mass, and ε_0 is the vacuum permittivity, τ is the mean electric collision time, and is assumed to be a constant in the model. $N(z, t)$ is the instantaneous conduction electron density at point z. It is varies in time because of the non-linear excitation of the dielectric by the high-intensity incident laser pulse. The dielectric constant $\varepsilon(z, t)$ is time dependent. Once some electrons are in the conduction band, they absorb some energy from the incident field (collision time $\tau \neq 0$) then $N(z, t)$ depends on z. The laser field in the dielectric is inhomogeneous. If we inspect the conduction electron density before the main laser pulse is assume to be zero $N(z, t = 0) = 0$, and the temporal evolution is given by the rate equation: $\frac{\partial N}{\partial t} = W \cdot (N_0 - N)$. Where, W is the electron excitation rate from the valence band to the conduction band, and N_0 is the bound electron density. The excitation is both due to multiphoton absorption and electronic avalanche, n is the minimum number of photons required to cross the band gap and σ_n is the corresponding cross section [5, 6]. W is given by

$$W = \sigma_n \cdot |E(z,t)|^{2 \cdot n} + \beta \cdot \frac{N}{N_0} \cdot |E(z,t)|^2$$

The hydrodynamic sub model: Plasma mirror response in high intensity regime and long interaction time is described by hydrodynamic simulation, which describes the plasma expansion, and accounts for multi-ionized atomic states and collisional processes in the high temperature plasma. There are hydrodynamic equations which simulate the plasma mirror response. The equation of state describes the temperature, pressure, mean ionization (z) and electron density for each internal energy and mass density. By knowing the temperature and density, the elastic electron-ion collision frequency is available. We get the energy source due to laser field by the Helmholtz equation model solution. The collision rate is calculated by knowing the electron-ion collisions. The expansion of the plasma during the laser pulse leads to a decrease of the plasma density and hence of the electron-ion collision rate. We inspect the measured temporal evolution of the reflected beam phase at different fluences with the model results. We use the model to calculate the phase shift induced by the PM on the reflected beam. This phase shift originates mostly from the motion of the critical density surface due to the plasma expansion in vacuum. We define a time t_e that after the plasma starts to expand and it decreases with the incident fluence. The expansion after time t_e, is with constant velocity.

Relativistic plasma mirror: Many scientific facilities include ultra-intense laser pulses source which might be capable of producing plasma mirrors through ultra-relativistic oscillating electrons. Mainly there are two kinds of plasma mirrors with relativistic factor $\gamma \sim 100{-}300$ and they are a tool for producing frequency upshifts $4 \cdot \gamma^2$ of reflected laser pulses as well as compressing and focusing of those pulses. The combination of these effects would results in dramatic light intensification toward the vacuum breakdown intensities (Schwinger limit). High power laser can envision lasers capable of producing vacuum breakdown intensities. The probability of an electron-positron pair creation from vacuum by a constant field E or electromagnetic wave of amplitude E is proportional to $e^{-\frac{\pi \cdot E_c}{E}}$, where $E_c = \frac{m^2 \cdot c^3}{e \cdot \hbar} = 1.32 \times 10^{16} \frac{\text{V}}{\text{cm}}$, c is the speed of light in vacuum, m is the electron mass, e is the positron charge, and $\hbar$ is the Planck constant. The density (power density) of a linearly polarized laser pulse having a maximal electric field E_c is $I_c = \frac{c \cdot E_c^2}{8 \cdot \pi}$, $I_c = \frac{c \cdot E_c^2}{8 \cdot \pi} = 2.3 \times 10^{29} \frac{\text{W}}{\text{cm}^2}$. The critical electric field E_c is to be obtained by means of compression and focusing laser pulses to a spot of linear sizes about the laser wavelength λ, the energy located within a spot is $\varepsilon \sim \frac{E_c^2 \cdot \lambda^3}{8 \cdot \pi} \sim 8\,\text{MJ} \frac{\lambda^3}{\mu\text{m}^3}$ (MJ = Mega Joule at half-micron). The extremely high intensities are achievable by shorter laser wavelengths. We define the mirror intensity reflection coefficient as R which includes the longitudinal compression factor, and then the total intensity amplification factor is $\frac{I_f}{I_s} \sim R \cdot (\frac{2 \cdot \gamma \cdot D_s}{\lambda_s})^2$, D_s is the diameter of the pulse which is focused to the spot of diameter about the laser wavelength in the mirror frame $\lambda_s' \approx \frac{\lambda_s}{2 \cdot \gamma}$. It gives additional intensity amplification factor $(\frac{2 \cdot \gamma \cdot D_s}{\lambda_s})^2$. γ is the relativistic factor, λ_s is the laser wavelength. We use the ultra-relativistic mirror in the plasma spike in electron concentration produced by the nearly breaking Langmuir wave driven by another laser pulse of frequency ω_d

in a highly under dense plasma of electron plasma frequency $\omega_p \ll \omega_d$. ω_p is the plasma frequency, ω_d is the frequency of another laser pulse. The intensity reflection coefficient of that mirror is $R \sim 2 \cdot \gamma \cdot (\frac{\omega_d}{\omega_s})^2$. The total intensity amplification factor is $\frac{I_f}{I_s} \sim 8 \cdot \gamma^3 \cdot (\frac{\omega_d}{\omega_s})^2 \cdot (\frac{D_s}{\lambda_s})^2$. Another kind of relativistic plasma mirror considers the production of ultrashort highly focused laser pulses by reflection of ultra-intense laser pulses from concave plasma surface. The ultra-relativistic electron motion at the over dense plasma surface causes an efficient generation of high harmonics. Coherent focusing of harmonics in the very small focal volume is the way to get extremely high intensities. We can take the assumption that the width of the electron concentration spike in the plasma wake wave is much smaller than the seed pulse wavelength, in the wake reference frame. We have some difficulties based on the fact that the seed pulse wavelength gets much smaller in the wave reference frame (Doppler effect), while the plasma wavelength gets much larger there and the thermal motion tends to make the spike in electron concentration smoother, reducing the over barrier reflection. The random bends of the mirror spoil the focus ability of reflected light. The bends can be associated with random fluctuations of the plasma concentration and the driving pulse intensity or incident laser intensity for the concave plasma surface. We discuss the seed pulse reflection from the density spike in the near-breaking plasma wave. The one-dimensional cold-plasma model equation for electrostatic wake field potential ϕ is measured in units of $\frac{m \cdot c^2}{e}$. The on-dimensional cold-plasma model equation for electrostatic wake field potential ϕ is as follow:

$$k_p^{-2} \cdot \frac{\partial^2 \phi}{\partial \xi^2} = \frac{n}{n_0} - 1 = \gamma^2 \cdot \left\{ \beta \cdot \left(\sqrt{1 - \frac{1 + a^2}{\gamma^2 \cdot (1 + \phi)^2}} \right)^{-1} - 1 \right\}$$

ϕ is the electrostatic wake field potential, a is the amplitude of vector potential of driving laser field in units of $\frac{m \cdot c^2}{e}$, $\xi = z - v_d \cdot t$, v_d is the driving laser group velocity, $\gamma = \frac{1}{\sqrt{1-\beta^2}}$, $\beta = \frac{v_d}{c}$, n and n_0 are the current and initial electron plasma concentrations, respectively, $k_p = \frac{\omega_p}{c}$; $\omega_p = \sqrt{\frac{4 \cdot \pi \cdot e^2 \cdot n_0}{m}}$. The axial electric field of the wake is given by $E_z \equiv E = -E_0 \cdot \frac{1}{k_p} \cdot \frac{\partial \phi}{\partial \xi}$, where $E_0 = \frac{m \cdot \omega_p \cdot c}{e}$ is the non-relativistic wave breaking field. We analyze the dynamic of our one-dimensional cold-plasma model system. Consider $\frac{\partial^2 \phi}{\partial \xi^2} \leftrightarrow \frac{d^2 \phi}{d \xi^2}$, then we get the second and differential equation:

$$k_p^{-2} \cdot \frac{d^2 \phi}{d \xi^2} = \gamma^2 \cdot \left\{ \beta \cdot \frac{1}{\sqrt{1 - \frac{(1+a^2)}{\gamma^2 \cdot (1+\phi)^2}}} - 1 \right\}; \quad \frac{d\varphi}{d\xi} = \frac{d^2 \phi}{d \xi^2}$$

$\varphi = \frac{d\phi}{d\xi}$, then we get

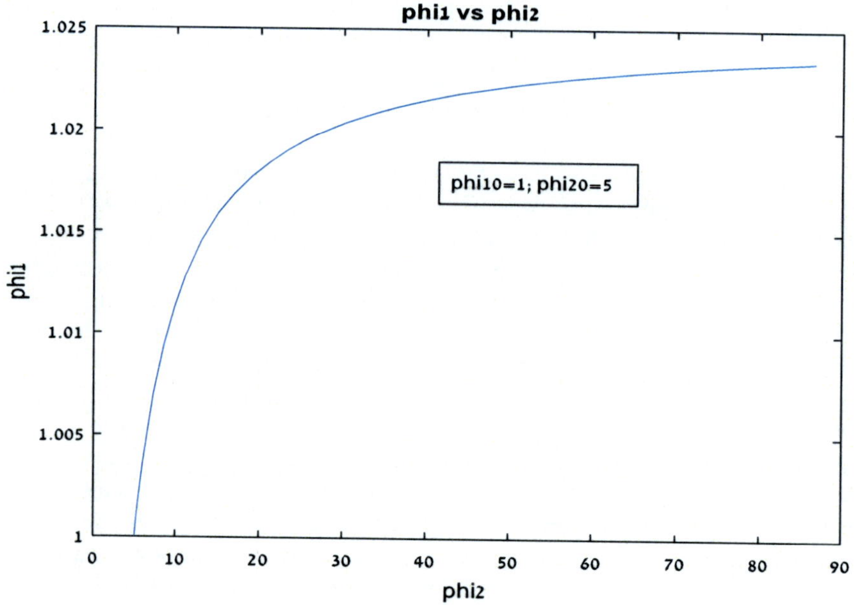

Fig. D.3 One-dimensional cold-plasma model φ versus ϕ phase space

$$k_p^{-2} \cdot \frac{d\varphi}{d\xi} = \gamma^2 \cdot \left\{ \beta \cdot \frac{1}{\sqrt{1 - \frac{(1+a^2)}{\gamma^2 \cdot (1+\phi)^2}}} - 1 \right\}$$

$$\frac{d\varphi}{d\xi} = k_p^2 \cdot \gamma^2 \cdot \left\{ \beta \cdot \frac{1}{\sqrt{1 - \frac{(1+a^2)}{\gamma^2 \cdot (1+\phi)^2}}} - 1 \right\}$$

We plot the phase space, φ versus ϕ (Fig. D.3) by using MATLAB script (Table D.1).

$$\omega_p = 1.174 \times 10^8; \;\; k_p = 0.391; \;\; \beta = 1; \;\; a = 1; \;\; \gamma = 100$$

```
function k = coldplasma1 (ksi,x)
k=zeros (2,1);
kp=0.391;
beta=1;
a=1;
gamma=100;
k(1)= kp.^2*gamma.^2*(beta*(1./(sqrt(1-(1+a.^2).
    /(gamma.^2*(1+x(2)).^2))))-1);
k(2)=x(1);
function p=coldplasma2 (phi10,phi20)
[ksi,x]=ode45 (@ coldplasma1, [0,80],[phi10,phi20],[]);
plot(x(:,2),x(:,1))% phi1 vs phi2
```

Table D.1 One-dimensional cold-plasma model parameter/variable (MATLAB)

One-dimensional cold-plasma model parameter/variable	MATLAB
φ (ξ derivative of electrostatic wake field potential), $\varphi = \frac{d\phi}{d\xi}$	x(1), phi1
ϕ (electrostatic wake field potential)	x(2), phi2
ξ ($\xi = z - v_d \cdot t$, v_d is the driving laser group velocity)	ksi
$k_p^2 \cdot \gamma^2 \cdot \left\{\beta \cdot \frac{1}{\sqrt{1-\frac{(1+a^2)}{\gamma^2\cdot(1+\phi)^2}}} - 1\right\}$	k(1)
φ	k(2)
φ_0 (initial value of φ)	phi10
ϕ_0 (initial value of ϕ)	phi20
k_p $\left(k_p = \frac{\omega_p}{c};\ \omega_p = \sqrt{\frac{4\cdot\pi\cdot e^2\cdot n_0}{m}}\right)$	kp
β $\left(\beta = \frac{v_d}{c}\right)$	beta
a (amplitude of vector potential of driving laser field in units of $\frac{m\cdot c^2}{e}$)	a
γ (relativistic factor)	gamma

```
xlabel('phi2');ylabel('phi1');
title ('phi1 vs phi2');
```

$n_0 = 10^{17}\ \text{cm}^{-3}$ (Initial density of plasma); $m = 9.1 \times 10^{-31}$ kg (Electron mass); $e = +1.602 \times 10^{-19}$ C (Positron's charge).

Hint: The coulomb (symbol: C) is the (SI) unit of electric charge. It is the charge transported by a constant current of 1 A in one second (1 C = 1 A × 1 s).

$$k_p = \frac{\omega_p}{c};\ \omega_p = \sqrt{\frac{4\cdot\pi\cdot e^2\cdot n_0}{m}};\ k_p = \frac{1}{c}\cdot\sqrt{\frac{4\cdot\pi\cdot e^2\cdot n_0}{m}}$$

At fixed point (equilibrium point): $\frac{d\phi}{d\xi} = 0;\ \frac{d\varphi}{d\xi} = 0$

$$\frac{d\phi}{d\xi} = 0 \Rightarrow \varphi^* = 0$$

$$\frac{d\varphi}{d\xi} \Longrightarrow k_p^2\cdot\gamma^2\cdot\left\{\beta\cdot\frac{1}{\sqrt{1-\frac{(1+a^2)}{\gamma^2\cdot(1+\phi^*)^2}}} - 1\right\} = 0$$

$$k_p^2\cdot\gamma^2 \neq 0$$

$$\beta\cdot\frac{1}{\sqrt{1-\frac{(1+a^2)}{\gamma^2\cdot(1+\phi^*)^2}}} - 1 = 0 \Rightarrow \phi^* = \sqrt{\frac{1+a^2}{(1-\beta^2)\cdot\gamma^2}} - 1$$

Hint: we consider that $\phi^* > 0$, and then we take only the positive root of $\frac{1+a^2}{(1-\beta^2)\cdot\gamma^2}$ expression. Our system fixed point is $E^{(*)}(\phi^{(*)}, \varphi^{(*)}) = E^{(0)}(\phi^{(0)}, \varphi^{(0)})$.

$E^{(0)}(\phi^{(0)}, \varphi^{(0)}) = \left(\sqrt{\frac{1+a^2}{(1-\beta^2)\cdot\gamma^2}} - 1, 0\right)$. We choose our one-dimensional cold-plasma model parameters as follow: $\gamma = 100$ (which corresponds to plasma electron concentration, $n \approx 10^{17}\ \text{cm}^{-3}$), $a = 1$ (amplitude of vector potential of driving laser field), $n_0 = 10^{17}\ \text{cm}^{-3}$ (initial density of the plasma). $v_d = c = 3 \times 10^8\ \frac{\text{m}}{\text{s}}$ (Driving laser group velocity) then $\beta = 1$.

Hint: The group velocity of light in a medium is defined as the inverse of the derivative of the wave number with respect to the angular frequency: $v_d = (\frac{\partial k}{\partial \omega})^{-1}$

$$v_d = \left(\frac{\partial k}{\partial \omega}\right)^{-1} = c \cdot \left(\frac{\partial}{\partial \omega}[\omega \cdot n(\omega)]\right)^{-1} = \frac{c}{n(\omega) + \omega \cdot \frac{\partial n}{\partial \omega}} = \frac{c}{n_d(\omega)}$$

$n(\omega)$ is the refractive index, $n_d(\omega)$ is the group index and the wave number k is the change in spectral phase per unit length. The group velocity is the velocity with which the envelope of a pulse propagates in a medium. We can integrate the one-dimensional cold-plasma model equation (a = 0) over ξ and get the following equation ($x = \phi + 1$):

$$k_p^{-2} \cdot \left(\frac{\partial x}{\partial \xi}\right)^2 = \left(\frac{E}{E_0}\right)^2 = 2 \cdot \gamma^2 \cdot [\beta \cdot (x^2 - \gamma^{-2})^{\frac{1}{2}} - x + C]$$

where, C is the integration constant. $x \to \gamma^{-1}$ and we get wave breaking and $x_{\text{min}} = \gamma^{-1}$ then $\gamma^2 \cdot \{\beta \cdot [1 - \frac{1+a^2}{\gamma^2\cdot(1+\phi)^2}]^{-\frac{1}{2}} - 1\} \to \infty$. At the threshold of Langmuir wave breaking, $C = \frac{1}{\gamma}$ and x oscillates between $x_{\text{min}} = \gamma^{-1}$ and $x_{\text{max}} = \frac{2\cdot\gamma - 1}{\gamma}$ (the maximal electric field reached at $x = 1$, $\phi = 0$). The maximal electric field is $E_{\text{max}} = \sqrt{2} \cdot \sqrt{\gamma - 1} \cdot E_0$. The scalar potential ϕ (in units of $\frac{\text{m}\,\text{c}^2}{\text{e}}$) is $\phi = \frac{1-v\cdot\beta}{\sqrt{1-v^2}} \cdot [1 + \frac{3}{2} \cdot \beta_T^2 \cdot \frac{(1-v)}{(1-\frac{v}{\beta})^2}] - 1 - \frac{3}{2} \cdot \beta_T^2$ and it satisfies to the Poison equation: $k_p^{-2} \cdot \frac{\partial^2 \phi}{\partial \xi^2} = -k_p^{-1} \cdot \frac{\partial(\frac{E}{E_0})}{\partial \xi} = \frac{n}{n_0} - 1 = \frac{v}{\beta - v}$, v is the hydrodynamic electron velocity normalized to the speed of light, $\beta_T = \sqrt{\frac{T}{m\cdot c}}$, and $\gamma = \frac{1}{\sqrt{1-\beta^2}}$ is the relativistic factor corresponding to the plasma wave phase velocity. The effects of the wake field electrostatic force and thermal electron pressure are included. Thermal pressure does not allow the electron density to rise to infinity. The maximum electron velocity v_0 at wave breaking is found from $\frac{\partial \phi}{\partial v} = 0$. We can find the electron velocity and density profiles at the maximum by the substitution $\phi \approx \phi(v = v_0) + \phi'' \cdot \frac{\Delta v^2}{2}$. It results in $\Delta v \approx \Delta v_0 \cdot (\frac{|\Delta\xi|}{\Delta\xi_p})$, and $\frac{n}{n_0} \approx \Delta v_0^{-1} \cdot (1 - \frac{|\Delta\xi|}{\Delta\xi_p})$ where $|\Delta\xi| < \Delta\xi_p$.

$$\Delta v_0 = \beta - v_0 = 3^{\frac{1}{4}} \cdot \frac{\beta_T^{1/2}}{\gamma^{3/2}};\quad \Delta\xi_p = 3^{\frac{3}{8}} \cdot 2 \cdot \left(\frac{\beta_T}{\gamma}\right)^{3/4} \cdot k_p^{-1}$$

Condition $\beta_T \cdot \gamma < 1$ is the case of plasma mirror (PM). If $\beta_T \cdot \gamma > 1$ we get $\Delta \nu_0 = 3 \cdot \beta_T^2$ and $\Delta \xi_p = (\frac{3}{2})^{3/4} \cdot \beta_T^{3/2} \cdot k_p^{-1}$.

Plasma mirror implementation on LFEX laser: We can reduce the laser pulse pedestal level, by implementing a plasma mirror (PM) device on LFEX laser, we can get high shot-to-shot pointing and focusing reliability and stable results. The pedestal reduction can be seen by transverse interferometry of the target surface, showing absence of preformed plasma before the main pulse arrival, while in absence of PM, under dense plasma expansion is observed before the main pulse arrival. We focus the LFEX beam on an Au block target and generate a fast electron spectrum. There is a fast electron temperature reduction within the range of interest for fast ignition in presence of PM device. The fast electron spectrum is strongly determined by the presence of preformed plasma. The PM is constituted by a transparent dielectric usually BK glass, with anti-reflective (AR) coating. Only laser fluxes above the breakdown threshold for the AR coating generate plasma on the PM surface. When the plasma reaches the critical density ($n_0 = 1.1 \times 10^{21}$ cm^{-3} for 1.053 mm LFEX laser), the laser pulse cannot propagate and is reflected in the same fashion as a classic metal mirror. The pre-pulse flux on the PM is lower than the breakdown threshold and is then transmitted, increasing the pulse contrast. The reflectivity of the PMs is a fundamental parameter and strongly depends on the laser pulse duration, polarization, wavelength and intensity on the PM. We need to characterize the PM reflectivity versus laser energy flux, using s-polarized LFEX laser with reduced pulse energy. We can get equivalent condition for high power interaction by varying the distance between the PM and target chamber center (TCC), where the LFEX beams are focused. The distance between the PM and TCC is varied, corresponding to laser energy flux on PM respectively. The beams is reflected onto a diffusive white board, and imaged by means of IR CCD camera. We measure the PM reflectivity from the ration between the energy-normalized integrated signal of the PM-reflected beam and the one from a fully reflective mirror. The reflected laser energy and the reflected beam spatial profile are dependent on the laser pulse duration as well as laser flux on the PM. At high laser fluxes, the large plasma expansion velocity acts as a moving mirror on the laser pulse. For pulse duration above specific value, a large plasma expansion velocity may induce a sensitive phase shift in the laser pulse with detrimental effects on the reflected beam spatial profile. The laser energy flux on target is chosen value to maximize the reflectivity while keeping a uniform beam spatial profile. In experimental setup a spherical PM is used, with anti-reflective (AR) coating on both surfaces. The spherical shape represents the best compromise between the easiness of alignment and the prevention of important optical aberrations. The PM is aligned together with the target on a single platform mounted on a kinematic base. We are first alignment, and focusing of the PM on the target position in the alignment station. By knowing the skeleton spot size before and after the PM, we estimate the astigmatic aberration induced by the PM. The broadening of a point like image s induced by astigmatism is $\sigma = \frac{M+1}{M} \cdot (1 - \sin \theta_i) \cdot D$, where M is the magnification of the optical system (M = 1), D is the aperture of the beam on the focusing optics, and θ_i is the incident angle of the beam on the focusing optics. The experimental target is to evaluate the performances of the PM in realistic laser-plasma conditions. Laser generated fast electron spectra are dependent on the laser intensity as well as on the pre-plasma scale length. In fast ignition, a key parameter to achieve efficient fuel heating is the hot electron temperature. We need to stop and energy release in the compressed plasma. In low contrast laser pulses the pedestal generates large scale-length plasma, leading to high electron temperatures. The pre-plasma scale length is enhanced by the presence of the cone, limiting the lateral plasma expansion and increasing the plasma scale-length along the laser axis. A high-contrast laser is strongly required in order to keep a reduced hot electron temperature and enhance the fuel heating. By focusing the LFEX beam, with and without PM on the cubic Au targets is possible to evaluate the fast electron beam temperature by means of two magnetic spectrometers, looking respectively at the target front and rear surface (without and with PM). In high contrast laser plasma interaction (with PM), the electron temperature demonstrates a significant reduction compared to the corresponding low contrast case, in absence of PM.

Water jet plasma mirror: Staging laser plasma accelerators (LPAs) is describe in order to maintain a high overall accelerating gradient while reaching beam energies of specific value and above. We minimize the distance that is needed for laser in coupling in order to achieve this high overall accelerating gradient. A plasma mirror is the final coupling optic reducing the coupling distance from tens of meters, using a conventional optic. Both a planar water jet and a nitrocellulose foil are used as reflecting surfaces and characterized. Laser plasma X-ray generation has a target of an in vacuum flowing steam of liquid water droplets. The plasma is produced by this target and the related its extreme ultra violet (EUV) emission spectrum. Plasma mirror is based on a laminar water film with low speed and is characterized for use as an ultrahigh intensity optical reflector. The use of flowing water as a target surface results in each laser seeing a new interaction surface and avoids the need for mechanical scanning of the target surface. The breakdown of water does not produce contaminating debris that can be deleterious to vacuum chamber conditions and optics, such as the case of using conventional solid targets. Laser plasma accelerators (LPAs) achieved electron beams with percent level energy, small divergence. The energy gain of single stage, W, scales inversely with the plasma density, n^{-1}. Using normalized laser intensities of $a_0 > 1$, as is typical with LPAs, the laser depletion length scales as $n^{-3/2}$. For each stage, a high energy gain requires a low density which in turn lengthens the accelerator. Multiple stages of LPAs could achieve 1 TeV energy gain over a shorter distance. A staged LPA can potentially run at higher repetition rates due to the power laser energy requirements. In a multi-stage LPA, reducing the coupling distance is important in reducing the total length of the accelerator. The size of the laser beam on the final coupling optic should be large enough to keep the fluence below the damage threshold of the optic. A complicated transport system is needed to couple the electrons across a specific distance. Plasma mirrors (PMs) are developed with high power ultrafast lasers. A laser pulse with peak intensity of the order 10^{15}–10^{17} W/cm^2 can ionize most targets creating supercritical plasma during the rise of the main pulse of the laser. At such densities, a sheet of plasma is act as a reflecting surface to the rest of the pulse and has been used to improve the pulse contrast in multi-TW class lasers. Pre pulses and background light are not intense enough to "trigger" the mirror and will not be reflected. Since PMs operate at much higher intensities than conventional optics, they might be used as a final coupling optic in staging LPAs and reduce the coupling distance. Due to the nonlinearity of PMs, they are harder to operate than conventional optics. The PM should have high reflectivity, maintain the quality of the laser profile, operate at high repetition rates, and not produce enough debris to damage other optics in the accelerator system. By using solid target we get a better vacuum and a more easily controllable surface profile. It needs to be mechanically scanned due to the local destruction of the surface on every shot and can produce debris. With 1–10 Hz laser systems, the scanning should not prove a problem; and with a kHz system, this becomes a target design and achievement. A planar liquid jet target solves the repetition rate problem as the surface is renewed automatically. Ethylene glycol's high viscosity and low vapor pressure make it favorable in maintaining a flat surface. It can be damage the optics due to the carbon containing compounds. Water does not contain carbon, and will not eject any damaging debris. The low viscosity of water poses a challenge in creating a planar liquid jet, though the use of guiding structures and low flow speeds can mitigate this problem. The water jet system is made using a specific diameter glass capillary glued in a copper tube bent to an angle of specific degree. The water is flowed onto a guiding structure consisting of a thick metal shim with a hole cut into it. A mesh is glued to the opposite side of the water flow to stabilize the film at low pressure. The film is accessible through the holes in the mesh. Using fs pulse in one of the probe lines of the TW Ti:Sapphire TREX laser system, reflectivity of the water jet PM specific backing pressure is measured by collecting the reflected beam on a CCD camera and calibrating the CCD camera with respect to input laser power. The reflectivity of intro cellulose foil is measured using the same technique at various pressures. We can use XPW system to enhance the pulse contrast of the input pulse. The beam is focused to a spot using an achromatic lens with a specific focus. The reflectivity curves are similar at similar pressures and it is not limited by the target used as the PM. The triggering occurs at specific intensity. The remaining air inside the chamber is being ionized by the main pulse creating small plasma mirrors that scatter the pulse. This is done by higher reflectivity of the solid target and the reflected spectrum of the laser pulse. The reflected beam is seen to be blue shifted at the highest intensity, which is characteristic of the ionization of air. This blue shifting is not observed below the leveling off of the reflectivity. Additional testing using the foil in milli torr backing pressure is done to illuminate this problem. Below a specific value of the reflected spot (pressure), bubbles begin to form near the outlet of the water jet and flow across the film. We can summery that the staging technology defines the total length of LPA systems. By reducing the coupling distance of an LPA, we maintain the high accelerating gradients which define our system. Plasma mirrors (PMs) are used as the final coupling optic for the laser. We inspect the planar water jet and solid target plasma mirrors (PMs) in the system. A solid target makes operation at a high-vacuum easier and gets a potentially better reflectivity. The planar water jets allow high repetition rate systems. We get the maximum reflectivity value by using both types of PMs. We can increase the reflectivity by reducing the backing pressure in the chamber. A differential pumping stage reduces the amount of air that the laser must pass through before reaching the water jet. Many experimental tasks are done with a solid target in millimeter backing pressure. The reflected beam is similar to the input beam by using the water jet and re-focused images. Possible case is when the solid targets eject enough contaminants to damage optics in the LPA system, which is preventing their use [7].

High contrast plasma mirror: Plasma optics is used to improve temporal pulse contrast and have an impact on the field of high-power laser-solid density interaction. We need to get a plasma density gradient in the high intensity focus which it related to the regimes of harmonic generation and ion acceleration. Generally plasma optics is used for fundamental laser reflection in the sub-relativistic intensity regime (10^{15}–10^{16} W cm^{-2}) with high reflectivity (~70%) and good focus ability. We need to check if plasma optics can be used for applications in the relativistic intensity regime (>10^{18} W cm^{-2}). High contrast laser pulses resulting in a regime of operation (High contrast plasma mirror (HCPM)). In this regime, pulses with contrast ration >10^{10} are used to form the PM surface at 10^{19} W cm^{-2}, displaying excellent spatial filtering, reflected near-field beam profile of the fundamental beam are reflectivity of ≈60%. Efficient second harmonic generation with exceptional beam quality routes to achieving the highest focusable harmonic intensity. By using plasma optics we can manipulate ultra-intense laser beams both spatially and temporally. Additionally, we get ultrafast frequency up shifting without detrimental effects due to group velocity dispersion (GVD). It reduced focus ability which frequently occur when nonlinear crystals are used for frequency conversion. At intensities approaching the breakdown threshold for the bulk materials/crystals, spatial filtering and second harmonic generation are a critical issue. Arbitrarily high intensities are applicable for very high-intensity lasers. We rely on the reflection of light off an over dense plasma layer, instead of an optical coating (optics which can withstand arbitrarily high intensities). Plasma optics like plasma mirrors (PMs) are used at laser intensities of 5×10^{15} W cm^{-2} to increase the contrast between the peak and the picosecond pedestal of high-intensity laser pulses. The rapid transition from low reflectivity (Anti-reflection (AR)-coated surface) to highly reflective plasma surface is resulted in an optical switch with a rise time of several hundred femtoseconds. High power laser systems laser pre pulse and amplified spontaneous emission are set to be below threshold intensity and they are transmitted through the PM. All is happened while the peak of the pulse is reflected. The contrast enhancement (CE_{PM}) is defined as the ration of plasma reflectivity (R_{PM}) to that of the AR coated optics (R_{AR}): $CE_{PM} = \frac{R_{PM}}{R_{AR}}$. The key constraint on the quality of the reflected radiation is the quality of the reflective plasma surface generated. The intensity of the pulse on target must stay below a certain value to avoid significant distortion of the smooth plasma surface. The pulses obeying the constraint $C_s \cdot \Delta t < \lambda_{laser}$, where the expansion velocity of the plasma is C_s, Δt is the time between plasma formation and the peak of the laser pulse and λ_{laser} is the laser wavelength, and there is an efficient specular reflection of the incident light. Another area in plasma optics is the efficient generation of high-order harmonics of the fundamental laser frequency from the plasma generated at the surface of a solid density target. Harmonic generation is due to two distinct processes: (1) Coherent wake emission (CWE); and (2) Relativistic ally oscillating plasma surface (ROS). Relativistic oscillating plasma harmonics have demonstrated diffraction-limited XUV beams at a wavelength of 40nm and generation of harmonics extending as high as the 3000th order of the fundamental laser frequency corresponding to specific photon energies. There is a significant spatial filtering of the incident laser beam, dispersion-free generation of high-quality second harmonic of the laser radiation with an efficiency of the order of >5% and a near-Gaussian beam profile. Plasma optics is route to overcome the limitations of conventional optics for spatial filtering, frequency doubling and contrast enhancement for ultra-high intensity laser systems. It gives insights into the process of high order harmonic generation. There are some reflection behaviors of the fundamental laser beam off the target under different conditions. It is very critical the position of the laser focus relative to the target surface, thus probing both the reflection in the near—and far-fields of the laser. The laser systems use cascaded PMs to maintain an acceptable contrast for mirror formation far beyond the regimes using double PMs. When the front surface of the target is positioned in the light focus (far-field of the laser beam), significant spatial filtering of the reflected light is observed. Higher spatial frequencies of the beam, that are located further out in the wings of the focal region, are reflected less efficiently than those spatial frequencies focused in the center of the beam. Typical high-power laser systems have a flat-top near-field laser beam profile for maximum energy extraction from the amplifier chain. In the far-field (focal plane) this flat-top profile corresponds to an Airy profile. The higher frequency contributions to the beam profile are suppressed and the resulting reflected beam is significantly smoothed. Only those frequency components focused within the second minimum of the focal pattern are reflected efficiently on the PM. The key to the operation of the plasma optics in the HCPM regime is the intensity on the target is sufficiently low prior to switch-on intensity for reflection. The plasma surface is not deformed due to pre plasma expansion, resulting in specular reflection. For high intensity (>10^{18} W cm^{-2}), the pulse must already have a significant contrast (>10^{10}) and by the time of arrival of the main pulse the target surface is essentially still at solid density, and not a tenuous plasma with a long density scale length, $L > \lambda_{laser}$, compared to the laser wavelength. In both, the low and high contrast case, the laser beam is cleaned using the PM and the near and far fields of the cleaned beam look identical. In both cases the PM is switched on only a few picoseconds before the arrival of the main pulse and not by the nanosecond pedestal. The shape of the pre pulse is determined by the timing of the fast pockets cell is preserved but the intensity of it is reduced by a factor given by the ratio of the reflectivity of the plasma divided by the reflectivity of the un ionized PM-substrate. A striking difference is observed in the reflected beam profile from the relativistic interaction on the main target for the low and the high contrast case. For low contrast the reflected beam profile breaks up completely and no beaming in the direction of the specular reflection. In the high contrast case, the behavior of the high intensity PM is very different. The beam is reflected off a well-defined surface with very short scale-length pre plasma and that despite the high-focused intensity of the laser beam the contrast is high enough to prevent an early ionization of the target. The different behavior for high and low nanosecond contrast is attributed to different plasma switch on times. In the low contrast case the nanosecond pedestal ionizes the target leading to a reflection of the beam off long scale-length pre plasma. This leads to a break-up of the reflected beam and a loss of specular reflection. In the high-contrast case, the plasma is switched on by the foot of the main pulse not giving the target enough time to expand resulting in specular reflection. The interaction on the target is more sensitive to the intensity of the pre pulse than that on the conventional PM. The interaction intensity at the target is three to four orders of magnitude higher than at the PM, while the contrast is only two orders better. The pre pulse level is low enough to prevent early ionization on the PM even for low contrast case. Possibility is when the main target leading to the destruction of the reflected beam under these conditions. Two implications: (1) A PM can operate properly at intensities much higher than regular intensities and provided the contrast of the laser beam incident on the PM, and (2) the reflected fundamental of the driving laser from a solid target can be used to monitor the quality of the surface at the time when the main laser pulse is incident on it. Good reproduction of the near-field indicates an interaction with a well-defined, steep density gradient, while beam breakup implies that long scale length pre plasma is formed. By monitoring the reflected fundamental light in, we can insight into the state of the target at the instance of the interaction with the main laser beam. A surface that is "clean" enough to reflect the fundamental laser beam is the absolute pre requisite for the observation of near diffraction limited harmonic beams. The efficient generation of the second harmonic and its high beam quality is important for the high-intensity laser-solid interactions. A second harmonic generation with nonlinear crystals (potassium dihydrogen phosphate (KDP) and beta barium borate (BBO)) is a vital issue in our system. Crystals are not ideal for the shortest pulses frequency doubling. Crystals lengthen the pulse of the frequency doubled beam. It happened because they do not have sufficient bandwidth and due to GVD (Group velocity dispersion) walk-off between the fundamental and harmonic. For high power femtosecond laser beams which have intensities of 2–4 TW cm^{-2} in the expanded beam, nonlinear effects both in time and space become intolerable even for ultra-thin crystals. The second harmonic is predicted to reach reflectivity $R_{2\omega} > \frac{1}{2} \cdot R_{\omega}$ for $a_0 > 3$ (harmonics from solid targets). The high efficiency can be demonstrated without degradation of the beam quality. Since the harmonic generation is a surface effect, there also is no pulse stretching due to dispersive effects or bandwidth limitations. The near-field beam profile of, and the conversion efficiency using HCPM help us to inspect the properties of the second harmonic emission as an alternative means of achieving $2 \cdot \omega$ operation. A lower bound for the conversion efficiency is estimated by comparing the signals measured for the fundamental and the second harmonic and the filters used in the measurements. There are some beam profiles of fundamental and second harmonic which establish the interaction of the laser with the HCPM (high contrast plasma mirror). Firstly, the divergence of the fundamental and the $2 \cdot \omega$ beams are very different and secondly, the beam profile of the second harmonic is much smoother and more Gaussian-like than that of the fundamental. The divergence of the $2 \cdot \omega$ beam corresponds to a diffraction limited second harmonic beam being emitted from a source the same size as the laser focal spot. The $2 \cdot \omega$ beam must be focusable to near-diffraction limit for its wavelength. Even at the lower bound of the conversion efficiency the peak intensity that can be achieved at $2 \cdot \omega$ is likely to exceed that of a beam frequency doubled in a crystal. The divergence of the frequency doubled beam is typically similar to that of the fundamental and consequently non-diffraction limited. The shorter pulse duration (femtosecond pulses) is taken into account using HCPM appears to be favorable in terms of the peak intensity that is achievable. Since the fundamental and the second harmonic have a very different origin then the beam profiles are different. The fundamental is the reflection of the incident laser beam off the over-critical plasma density surface formed on the target. The second harmonic is generated during the relativistic interaction of the driving laser with the target. The difference between the fundamental and the second harmonic is obvious when considering what kind of intensity distribution in the focus results in the experimentally measured beam profiles after the beam has expanded from the target. We can calculate the focal distributions from the measured beam profiles via Fourier transform and consider the different wavelengths of fundamental and second harmonic. For the fundamental, there is a substantial reflectivity of the mirror at intensities of 10^{17} W cm^{-2}. The side maxima of the focal distribution at least partially reflect the incident light. There is no emission from the second harmonic originating from the side maxima of the focal distribution. The source distribution is nearly Gaussian with some extra energy in the wings smearing out cross the first minimum in the fundamental distribution. The second harmonic is generated via the relativistic oscillating surface mechanism. The intensity in the central part of the focus is sufficiently high to generate harmonics, whereas the side lobes with intensities below the relativistic limit show negligible conversion efficiency. The central peak in the source is not narrowed despite the nonlinearity of the generation process. The transverse motion of the electrons in the oscillating surface results in an oscillating mirror of about the same width as the focal spot. There is a stable operation of plasma optics under high contrast conditions at intensities of 2×10^{19} W cm^{-2}. It is three orders of magnitude more intense than in regular PM operation. This regime of operation, the HCPM is capable of reproducing the beam profile of the incident laser beam very well when operated in the near-field, and functions like an efficient spatial filter in the far-field. An ultra-high contrast operation is scalable to lasers with highest peak intensities by cascading multiple PMs and provides a means of spatial filtering that in contrast to the conventional method with pinholes, is scalable to arbitrarily high intensities. The observation of dispersion free, diffraction limited second harmonic generation offers an advantageous method for frequency doubling ultra-short or ultra-high power laser pulses. This method is not limited to the generation of the second harmonic, even though the beam profile of higher harmonics [8].

Plasma mirrors for ultrahigh-intensity optics: We can describe specular reflection as a fundamental process of optics. Specular reflection (regular reflection) is the mirror like reflection of waves, such as light, from a surface. In this process, each incident ray is reflected at the same angle to the surface normal as the incident ray, but on the opposing side of the surface normal in the plane formed by incident and reflected rays. An image reflected by the surface is reproduced in mirror-like (specular) fashion. Each incident ray the angle of incident equals the angle of reflection, and the incident, normal, and reflected directions are coplanar. The behavior of specular reflection is contrasted with diffuse reflection, in which light is scattered away from the surface in a range of directions rather than just one. The behavior of light is related to its reflection off of plane (flat) mirrors. Mirrors are typically smooth surfaces, at microscopic levels. They offer each individual ray of light the same surface orientation. Mirrors are not the only types of objects which light reflects off of. A light beam can be described as a bundle of individual light rays which are traveling parallel to each other. Each individual light ray of the bundle follows the law of reflection. If the bundle of light rays is incident upon a smooth surface, then the light rays reflect and remain concentrated in a bundle upon leaving the surface. If the surface is microscopically rough, the light rays will reflect and diffuse in many different directions. For each type of reflection, each individual ray follows the law of reflection. The roughness of the material, each individual ray meets a surface which has a different orientation. The normal line at the point of incident is different for different rays. When the individual rays reflect off the rough surface according to the law of reflection, they scatter in different directions. The rays of light are incident upon the surface in a concentrated bundle and are diffused upon reflection. There are many applications of this distinction between specular and diffuse reflection. There are many potential useful phenomena by using modern ultrahigh-intensity lasers. When a pulse form such a laser hits an optically polished surface, it generates dense plasma which acts as a mirror [plasma mirror (PM)]. Plasma mirrors (PMs) reflect the remainder of the incident beam and act as active optical elements. The target is to improve the temporal contrast of femtosecond pulses, and demonstrate that high-order harmonics of the laser frequency which can be generated through two distinct mechanisms. It is done by using a set of three consecutive PMs in different regimes. The task is to develop a system of laser-driven atto second sources for use in fields from materials science to molecular biology. Plasmas are considered as unstable and hardly controllable media. During an ultrashort laser pulse (typically <100 fs), this plasma expands by a small fraction of the light wavelength, and thus behaves as a high-flatness mirror which leading to high-intensity specular reflection. It avoids the complications associated with the nonlinear propagation of the intense laser light in plasma. It characterized the high-intensity laser-plasma interaction. By giving a large change in reflectivity that occurs when an initially solid target is ionized and converted into plasma, PMs are used as laser-triggered ultrafast optical switches. We use a shorten nanosecond pulses from CO_2 lasers down to the 100 fs range, to produce a time gate for the temporal characterization of ultrashort pulses and to improve the temporal contrast of ultrashort laser pulses. Another use of PMs is in the relativistic interaction regime which is above a few 10^{18} W cm^{-2} at visible wavelengths, where the quivering motion of electrons in the laser field involves velocities of the order of speed of light. In this regime, PMs are driven in a highly nonlinear regime, leading to the generation of very high-order harmonics of the incident light in the spectrum of the reflected beam, up to the extreme ultraviolet and X-ray ranges. These harmonics are associated with intense trains of atto second or even septo second pulses; whereas isolated atto second bursts are expected if intense few-cycle-long laser pulses are used. It is alternative to high-order harmonic generation (HHG) in gases to obtain atto second pulses with higher pulse and photon energies. For main pulse intensity which exceeds 10^{17}–10^{18} W cm^{-2}, the nanosecond light background surrounding this pulse, although typically 10^6–10^8 times weaker, is intense enough to strongly ionize any solid target well before the main pulse. There is a subsequent expansion of the resulting plasma, the main pulse, as short as it may be, ends up interacting with a long plasma density gradient with no specular reflection. An ultrafast high-dynamics optical switch is used by double-plasma-mirror (DPM). It improves the temporal contrast by four orders of magnitude. We get high-contrast pulses which are used to reveal the basic mechanisms of HHG in the specular reflection of a PM driven at ultrahigh intensities, up to the relativistic regime. Plasma mirrors (PMs) are used in both optical elements and objects. We can improve the temporal contrast of high-power lasers by using a method, which consists of reflecting the laser beam on a dielectric target with an initially low reflectivity. If we choose specific beam focusing, and the temporal pedestal is not intense enough to affect this target, then much more intense main laser pulse includes a strong ionization, through nonlinear excitation processes. Then plasma is created within femtoseconds at the rising edge of the main pulse. The target reflectivity increases when the plasma becomes dense enough to screen the incident laser field, that is, when the electron density, n, exceeds the critical density, $n_c = \frac{m \cdot \varepsilon_0 \cdot \omega_L^2}{e^2}$, characteristic of the laser frequency ω_L, where e and m are the electron charge and mass and ε_0 is the vacuum dielectric constant. The final result is a contrast enhancement ahead of the main pulse, with a magnitude that simply depends on the ratio of the plasma and initial target reflectivity. We want to maximize the plasma reflectivity then the demand is s-polarized laser beam that is, with its electric field along the target's surface. The plasma absorbs the laser energy, inverse Bremsstrahlung due to electron-ion collisions. The final reflectivity, R_f, is approximately $\approx$70%. We can get contrast improvement, $\frac{R_f}{R_i}$ by using anti reflection coated optical flat glass. The typical reflectivity is 0.3%, $R_i \approx 0.3\%$. The pedestal is following the main pulse and is strongly decreased within a few picoseconds after the peak. It is owing to the hydrodynamics expansion of the plasma on the 10 ps time's scale, which leads to its diffuse reflection by the PMs. This expansion is negligible on the 100 fs time scale. Then the quality of the laser focal spot is not degraded by double plasma mirror (DPM). We can implement that plasma based technique at the output of any laser chain, scaled to arbitrarily high laser powers, and cascaded in N stages, to gain 2N orders of magnitude in temporal contrast at a repetition rate of up to 1 Hz. When PMs are driven in s-polarization (moderate intensities), the laser-driven dynamics is mostly along the target surface, then plasma electrons behave as free electrons, quivering in the laser field. The plasma electrons are driven across a steep electrostatic potential at the plasma-vacuum interface when the focused intensity is gradually increased up to the relativistic interaction regime, or when the laser field is p- rather than s-polarized. This leads to a much richer and complex physics of the interaction, and to the generation of high-order harmonics of the incident frequency in the spectrum of the reflected beam. A collimated and smooth extreme-ultraviolet beam is diverged about three times less than the fundamental beam. We get a good phase-front quality and a good transverse coherence of the harmonic source in the plane of the target surface. Plasmas can be used on ultrashort time scales as active optical components with excellent optical quality which exist also at ultrahigh laser intensities, and thus exploited to produce new light sources with good properties (shorter wavelengths and sub-femtosecond durations). This leads to better-defined interaction conditions and simplifying the physics and the interpretation of high-intensity systems. There are many models that characterize high-order harmonic generation (HHG) from plasmas. There are two generation mechanisms, coherent wake emission (CWE) and the relativistic oscillating mirror process. These two mechanisms lead to harmonics with very different properties. The mechanism for high-order harmonic generation is by reflection of a laser beam from an over dense plasma, even at moderate laser intensities (down to $I \cdot \lambda^2 \approx 4 \times 10^{15}$ W cm^{-2} μm^2). In this mechanism, a transient phase matching between the electromagnetic field and oscillations within a density gradient leads to the emission of harmonics up to the plasma frequency. These plasma oscillations are periodically excited in the wake of atto second electron bunches which sweep across the density gradient. This process leads to a train of unevenly spaced chirped atto second pulses, and hence to broadened and chirped harmonics. The mechanism of coherent wake emission (CWE) is described as follow. In p-polarization, the laser electric field has a large component, $E_\perp$, along the target normal. When it points inwards, some electrons near the surface are pulled out of the plasma. When $E_\perp$ decreases, then space-charge electrostatic fields drive some of these electrons back into the plasma. They no longer feel the laser field owing to plasma screening. These electrons thus propagate deeply into the plasma. We get a significant absorption of the laser pulse energy by steep density-gradient plasmas; it is an effect of "vacuum heating" or Brunel absorption. The Brunel electrons are involved in harmonic generation through the coherent wake emission process. The analytically model and particles-in-cells (PIC) simulations reveal that Brunel electrons that move further into vacuum come back to the plasma with larger velocities, and thus overtake the electrons that had smaller excursions. This trajectory crossing causes them to form an atto second bunch that penetrates the plasma with a new bunch once every optical cycle. These bunches of charge push the plasma background electrons, driving up electronic plasma oscillations throughout a range of densities in the steep gradient, n(x), at the surface of the plasma. An ensemble of linear plasma oscillations with space-dependent frequencies, $\omega_p(x) = \omega_L \cdot \sqrt{\frac{n}{n_c}}$ is coherently triggered. The wave vector k of this plasma wave has a time-varying direction. The wave, initially purely electrostatic and longitudinal (E || k) when triggered by Brunel electrons, then progressively acquires a partial transverse character, and can radiate light. This light emission is triggered one every optical cycle by returning Brunel electrons, the emitted spectrum consists of harmonics of the incident frequency. The harmonics extend as the maximum plasma frequency, $\omega_p^{\max}$, of the target, and hence up to the order $p_{CWE} = \sqrt{\frac{n_{\max}}{n_c}}$, where $n_{\max}$ is the maximum plasma density. The typical values of $n_{\max}$, is of a few hundred times n_c. This spectrum reaches into the extreme-ultraviolet regime. The maximum harmonic frequency is corresponded to the expected maximum plasma frequency, $\omega_p^{\max}$, for fully ionized solid-density targets. We need for this mode of conversion efficiency then a phase-matching between the plasma oscillations and the outgoing electromagnetic modes at the plasma vacuum interface is required. If k is parallel to the surface then we get the phase-matching. The fact is that k is gradually rotating, which occurs transiently, in a few-hundred-atto second window following excitation by each electron burst. Extreme ultraviolet light is emitted in the form of a train of atto second pulses. We can compare it with the reflection of light pulses on "chirped mirrors" which used in conventional optics. Each frequency, ω, originates from a different depth in the density gradient, where $\omega_p(x) = \omega$, and these atto second pulses have a positive chirp progressively drifting to higher frequencies. Harmonics produce by CWE is the way to inspect the dynamics of plasma mirrors (PMs) driven at high intensity, and especially of Brunel electrons. By using particles-in-cells (PIC) simulation we get the prediction that CWE has a typical conversion efficiency of 10^{-4}. The mechanism of relativistic oscillating mirror process is described as follow. When the laser intensity is increased beyond a few 10^{18} W cm^{-2} at visible wavelengths, the laser-driven motion of the plasma surface becomes relativistic, leading to strong Doppler shifts of the laser light. As this oscillating surface chases the retreating optical phase fronts, it compresses the reflected electromagnetic field, distorting it from its original sinusoid. As this phase distortion repeats itself with the periodicity of the driving laser field, harmonics of the incident frequency then appear in the reflected beam. This harmonic comb is associated with the train of atto second pulses, each pulse corresponding to an outward excursion of the plasma surface. It is "relativistic oscillating mirror" (ROM) process [9]. The interaction of relativistic intense ($I \cdot \lambda^2 > 1.3 \times 10^{18}$ W cm^{-2} μm^2) laser pulses with a near step-like plasma density profile is results in relativistic oscillations of the reflection point. This process results in efficient conversion of the incident laser to high harmonic spectrum. The beam quality is near diffraction limited and consequently that very high focused intensities can be achieved (Ultra-intense X-ray interactions). The advent of ultrahigh power femtosecond lasers open the demand to optical components based on plasma. Plasma mirror is a type of optical component formed when an intense femtosecond laser ionizes a solid surface. Plasma mirror reflects the main part of a laser pulse and is used as active optical elements to manipulate its temporal and spatial properties. The pressures exerted by the laser deform the mirror surface. The pressures affect the reflected beam and complicating or preventing, the use of plasma mirrors at ultrahigh intensities. The analytical model in laser-induced deformation of a plasma mirror characterized the dynamic of the system. If we compare it to the coherent wake emission (CWE), the maximum harmonic order that can be obtain by the relativistic oscillating mirror process is strongly depends on the laser intensity and can exceed the plasma frequency. The high contrast pulses obtained from the DPM (double-plasma-mirror) is focused at the highest intensities, using a parabolic mirror with a $\frac{f}{4}$ numerical aperture. We need to inspect the spectral properties of coherent wake emission and relativistic oscillating mirror (ROM) harmonics. There is a feature which is related to upper HHG spectrum, ROM harmonics, those with p>15, are about half the spectral width of those generated by CWE on any material. At the lower intensity, broad CWE harmonics are up to order $p_{CWE} = \sqrt{\frac{n\text{max}}{n_c}}$. At the higher intensity, narrow harmonics appear for orders above p_{CWE}, which PIC simulations allow as to clearly attribute to the ROM process. If the individual harmonics are all Fourier-transform limited, CWE harmonics would have to be generated during a shorter time period than ROM harmonics, to account for these differences in spectral width. By using PIC simulations, we get a conclusion that these differences in spectral width are due to the very different chirps of the individual harmonics generated by these two mechanisms. If we consider CWE, in the non-relativistic regime, the peak velocity of Brunel electrons varies in time, according to the intensity envelope of the laser pulse. As a result, the delay between the formation of an electron bunch by returning Brunel electrons at $n \approx n_c$, and the generation of the associated atto second pulse at $n \gg n_c$, is time dependent, and the period between successive atto second pulses in the train generated by this process shifts gradually. Then CWE harmonics are not Fourier-transform limited, negatively chirped, and this accounts for their large spectral width. The chirp of the harmonics from ROM is much smaller, better periodicity of the associated atto second pulse train; impose by the regular oscillations of the plasma surface. The CWE and ROM mechanisms are characterized by very different dependences on laser intensities. The intensities dependence the harmonic spectrum from a plastic target, are measured by varying the distance between the target surface and the best focus of the laser beam.

CWE is a quasi-linear conversion process which the overall efficiency is dependent only weakly on laser intensity. There is a characterization of ROM mechanism, a relativistic process that is driven near its intensity threshold, highly nonlinear. Many schemes exist which put plasmas to work as high-intensity optical elements, such as in the case of the plasma-based optical switches. "Plasma optics" is predicted to further amplify ultra-intense lasers, to tailor or to temporally compress ultrashort pulses down to the few femtosecond range, and to generate ultrashort coherent radiation from the terahertz to X-ray range. Relativistic flying mirrors in plasmas are as thin dense electron (or electron-ion) layers accelerated by high-intensity electromagnetic waves to velocities close to the speed of light in vacuum. The reflection of an electromagnetic wave form the relativistic mirror results in its energy and frequency changing. In a counter-propagation configuration, the frequency of the reflected wave is multiplied by the factor proportional to the Lorentz factor squared. There are many sources of ultrashort X-ray pulses in the atto second range which the expected intensity reach the level at which effects predicted by nonlinear electrodynamics [10].

References

[1] G.G. Scott, V. Bagnoud, C. Brabetz et al., Optimization of plasma mirror reflectivity and optical quality using double laser pulses. New J. Phys. **17**, 033027 (2015)

[2] I.B. Foldes, A. Barna, D. Csati, F.L. Szucs, S. Szatmari, Plasma mirror effect with a short-pulse KrF laser. J. Phys. Conf. Ser. **244**, 032004 (2010)

[3] R. Wilson, M. King et al., Developing of focusing plasma mirrors for ultra intense laser-driven particle and radiation sources. Quantum Beam Sci. **2**, 1 (2018)
[4] J.L. Ferreira, S.F. da Silva, I. da S. Rego, A multi magnetic mirror machine for plasma production with electron cyclotron resonance. Revista de Fisica de Fisica Aplicada e Instrumentacao **17**(2) (2004)
[5] M.W. Kunz, A.A. Schekochihin, J.M. Stone, Firehose and mirror instabilities in a collision less shearing plasma. Phys. Rev. Lett. PRL **112**, 205003 (2014)
[6] A.A. Solodov, V.M. Malkin, N.J. Fisch, Limits for light intensification by reflection from relativistic plasma mirrors. Phys. Plasmas **13**, 093102 (2006)
[7] D. Panasenko, A.J. Gonsalves et al., Development of water jet plasma mirror for staging of laser plasma accelerators, in *Proceedings of PAC09*, Vancouver, BC, Canada (WE6RFP077), pp. 2976–2978 (2010)
[8] R. Horlein, B. Dromey et al., High contrast plasma mirror: spatial filtering and second harmonic generation at 1019 W cm^{-2}. New J. Phys. **10**, 083002 (2008)
[9] S.V. Bulanov, T.Zh. Esirkepov, M. Kando, J. Koga, Relativistic mirrors in laser plasmas (analytical methods). Plasma Sources Sci. Technol. **25**(5), 1015
[10] C. Thaury, F. Quere, J.P. Geindre et al., Plasma mirrors for ultrahigh intensity optics. Nat. Phys. **3** (2007)

Appendix E
High Power Laser/Target Diagnostic System Optical Elements

Optical diagnostic elements in high power laser system are aimed to monitor and explore system outcomes when operate. The optical probe pulses explore laser-target parameter space. It is real-time information on laser plasma interaction and exploring the interaction mechanism. By implementing diagnostic system, we can optimize and analyze system performances and characteristics. It is needed for scaling and control our system overall parameters. The synchronization of the optical probe beams with the driving pulse and mitigate any jitter issues. There are two probing methods, first transverse probing (interferometry or polarimetry) and second longitudinal probing methods, like frequency domain holography and frequency domain shadowgraph. The main system optical elements are elliptical mirrors (EMs), off axis parabolic mirror (OAP), Spherical mirror (SM), Chirp mirror (CM), Echelon mirror (ECM), Optical delay line (ODL), Glass wedges (GW), Optical silicon window (OSW), Electro-optic crystal (EOC), Quarter wave plate (QWP), Polarization beam splitter (PBS), and Optical relay imaging system.

Elliptical mirror (EM): Elliptical flat mirrors are suitable for applications which require bending/folding light at precise angles with minimum wave front distortion. The elliptical mirror has a shape of ellipse that has two focus (foci) which lie on the major (longest) axis, spaced equally each side of the center. The mirror resides inside of an ellipse and any light ray leaving one focus will always pass through the other. Every light ray leaving F1 reflects off the inside of the elliptical mirror and bounces back, passing through the other focus F2. When a light ray bounces off the elliptical mirror, it likes reflected off a flat mirror which is tangential to the ellipse at that point. There are some typical elliptical mirrors coating materials: broadband dielectric, protected aluminum, protected silver, and protected gold. A broadband dielectric coated mirrors ofer high reflectance in the visible range (400–750 nm) or in the NIR range (750–1100 nm). It offers smaller operating wavelength ranges than metallic coatings and characterized by higher reflectance in the ranges than their metallic counterparts. The protected-aluminum-coated mirrors are good solution for many broadband applications. A SiO_2 coating is used to protect the delicate aluminum coating. It reflectance is close to the reflectance of a bare aluminum coating. It characterize by reflectance greater than 90% over specific spectral range. The silver-coated

O. Aluf, *Advance Elements of Laser Circuits and Systems*,
https://doi.org/10.1007/978-3-030-64103-0

mirrors have the highest reflectance of all metal-coated mirrors in the visible wavelength range. Silver offers high reflectance from 450 nm to 20 μm. Mirror anti oxidation is overcoat SiO_2. The overcoat protects silver from tarnishing, and high humidity environments. The silver coated mirrors require additional operational activities due their susceptibility to damage from environmental conditions and improper handling. Protected Gold-coated elliptical mirrors are ideal for use over the entire IR range, with an average reflectance in excess of >95% for 800 nm–20 μm. For purposes of damage resistance and easy to clean the gold-coated elliptical mirror has a protective SiO_2 overcoat. Elliptical mirrors are typically used at 45° where a circular cross section of a given size is required. Elliptical mirrors are used as secondary mirrors in Newton telescopes or mounting in a tube where maximum aperture is required such as periscopes. To get a greater precision it is recommended to use an elliptical mirror with flatness better than $\frac{\lambda}{10}$. Elliptical flat mirrors are designed to present a circular profile when oriented at 45°. The top surfaces are coated with metallic mirror coating and rear surface is fine ground for maximum durability. The elliptical mirror surface is accurate to $\frac{\lambda}{2}$ and they are ideally suited for systems where alignment time is an important consideration. In comparison with a parabolic mirror based system, an elliptical mirror based system is potentially promising in the aberration control of incident beam when the aperture of mirror is enlarged to adapt a large stage or specimen container at a small beam shading ratio. Parabolic and elliptical mirrors are almost the same when their apertures are similar in sizes. The size of a point-like illuminator in an elliptical mirror based system won't be changes with the aperture of the focusing mirror, and this kind of point-like illuminator can be easily obtained by a general small aperture objective. A focusing formula is established for an elliptical mirror with wide open aperture to highlight the potential application of an elliptical mirror. Apodization factor is of great importance for the discovery of the transformation between inbound and outbound waves on a mirror or lens. Apodization is an optical filtering technique ("removing the foot"). It is the technical term for changing the shape of a mathematical function, an electrical signal, an optical transmission or a mechanical structure. In optics, it is primarily used to remove Airy disks caused by diffraction around an intensity peak, improving the focus. The reflection of an elliptical mirror, it is theoretically different from that of a parabolic mirror because the incident wave is no longer a plane wave. The reflection structure on an elliptical mirror is present (Fig. E.1). S_m is the real surface of an elliptical mirror, S_{w1} and S_{w2} are two wave front surfaces which corresponding to inbound and outbound spherical waves with centers F_1 and F_2, respectively. Their focal lengths are $a + c$ and $a - c$, respectively. Angle α is the open aperture of a point-like source at focal point F_1. Angle θ is the focusing angle at focal point F_2 where there is a scanning stage or specimen container. The point diffractive spherical wave S_{w1} moves into spherical wave S_{w2} after reflection of an elliptical mirror. Then the wave front surface outbound S_{w2} focuses at F_2. This propagation is passing by M which is an incident point on real mirror surface S_m.

The optical path from first focal point (F_1) to the second focal point (F_2) gives the value of a parameter $\left(a = \frac{1}{2} \cdot \{|F_1 - M| + |M - F_2|\}\right)$. The mirror function is

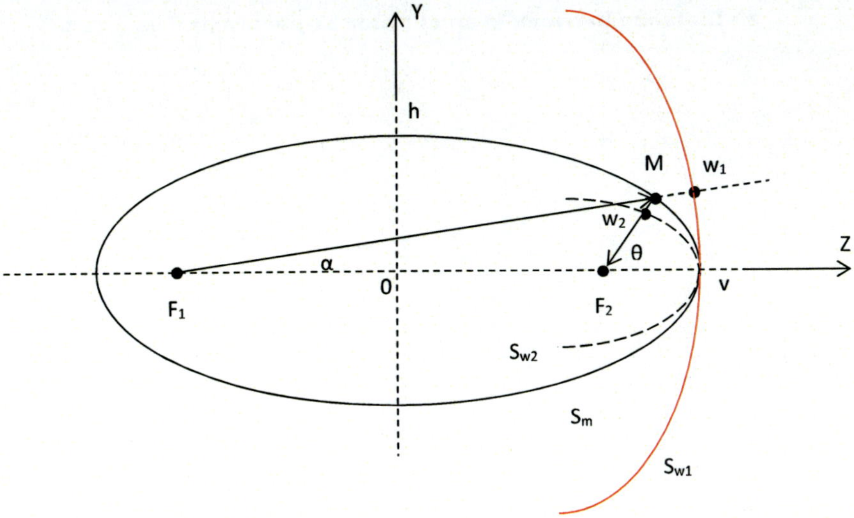

Fig. E.1 Reflections structure of elliptical mirror

$\frac{(x^2+y^2)}{b^2}+\frac{(z-c)^2}{a^2}=1$, a is the long axes of an elliptical mirror ($a=|0-\text{v}|$), b is the short axes of an elliptical mirror ($b=|0-\text{h}|$), $c=|0-\text{F}_1|=|0-\text{F}_2|=\sqrt{a^2-b^2}$. Apodization factor $l(\theta)$ defines the amplitude ratio of inbound and outbound beams on a mirror at focusing angel θ and for ideal mirror, the energy loss caused by reflection and absorption is negligible. Then $l^2(\theta)=\frac{A_1^2}{A_2^2}=\frac{\delta S_{w2}}{\delta S_{w1}}$, the energy is kept constant in an infinitesimal area. A_1 and A_2 are amplitudes of inbound and outbound beams. δS_{w1} and δS_{w2} are the infinitesimal surface elements on waves S_{w1} and S_{w2}, respectively. f_1 and f_2 are the focal lengths of S_{w1} and S_{w2} [1].

$$\delta S_{w1}=2\cdot\pi\cdot f_1^2\cdot\sin(\alpha)\cdot d\alpha;\ \ \delta S_{w2}=2\cdot\pi\cdot f_2^2\cdot\sin(\theta)\cdot d\theta$$

$$f_1=a+c=a+\sqrt{a^2-b^2}$$

$$f_2=a-c=a-\sqrt{a^2-b^2}$$

$$l(\theta)=\frac{a+c}{a-c}\cdot\sqrt{\frac{\sin(\alpha)\cdot d\alpha}{\sin(\theta)\cdot d\theta}}=\left(\frac{a+\sqrt{a^2-b^2}}{a-\sqrt{a^2-b^2}}\right)\cdot\sqrt{\frac{\sin(\alpha)\cdot d\alpha}{\sin(\theta)\cdot d\theta}}$$

$$\alpha=tg^{-1}\left[\frac{(z-2\cdot\sqrt{a^2-b^2})\cdot tg\theta}{z}\right]$$

$$z=\frac{a^2\cdot\sqrt{a^2-b^2}\cdot tg^2(\theta)+a\cdot b^2\cdot\sqrt{1+tg^2(\theta)}}{a^2\cdot tg^2(\theta)+b^2}+\sqrt{a^2-b^2}$$

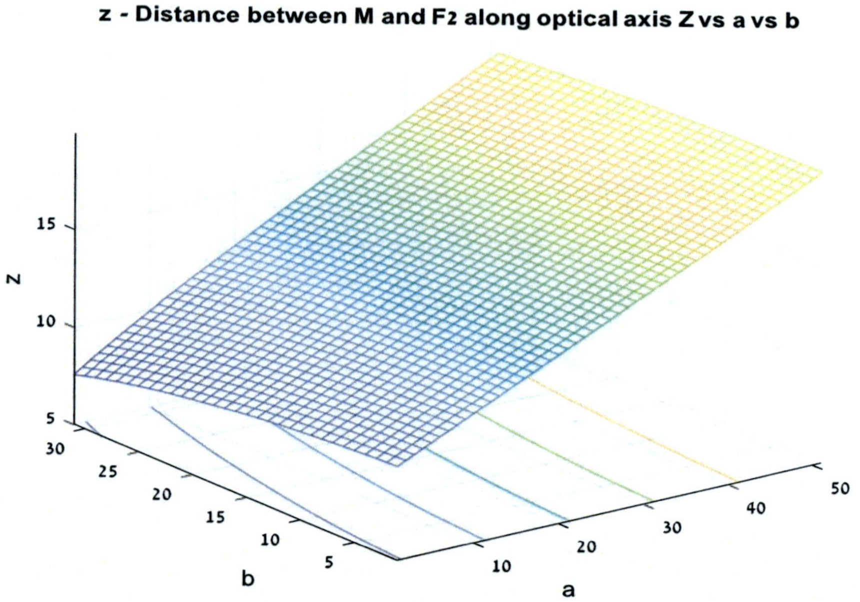

Fig. E.2 Elliptical mirror z-distance between M and F_2 along optical axis as a function of a and b

Variable z is the distance between M and F_2 along optical axis Z. In the real mirror reflection, the optical path from F_1 to F_2 is $|F_1 - M| + |M - F_2|$ which is related to the theoretical optical path $a = \frac{1}{2} \cdot \{|F_1 - w_1| + |w_2 - F_2|\}$; $2 \cdot a = f_1 + f_2 = |F_1 - w_1| + |w_2 - F_2|$. First we plot the distance z (between M and F_2 along optical axis Z) as a function of a (long axes of an elliptical mirror) and b (short axes of an elliptical mirror), $\theta = 45°$; $90° > \theta > 0°$ (Fig. E.2).

$$z = \frac{a^2 \cdot \sqrt{a^2 - b^2} \cdot tg^2(\theta) + a \cdot b^2 \cdot \sqrt{1 + tg^2(\theta)}}{a^2 \cdot tg^2(\theta) + b^2} + \sqrt{a^2 - b^2}; \ tg45° = 1$$

$$z = \frac{a^2 \cdot \sqrt{a^2 - b^2} + a \cdot b^2 \cdot \sqrt{2}}{a^2 + b^2} + \sqrt{a^2 - b^2}$$

$$a = 5 \ldots 10\,\text{m}; \ b = 1 \ldots 4\,\text{m}\ (\text{m} = \text{meter})$$

MATLAB:

```
[a,b]=meshgrid (5:.1:10,1:.1:4);
M=a.^2+b.^2;
n=sqrt(a.^2-b.^2);
Q=(((a.^2).*n+a.*(b.^2)*sqrt(2))./m)+n;
meshc (Q); % this is z vs a vs b for Teta=45 degree
ylabel('b'); xlabel('a');zlabel('z');
title ('z - Distance between M and F2 along optical axis Z vs a vs b');
```

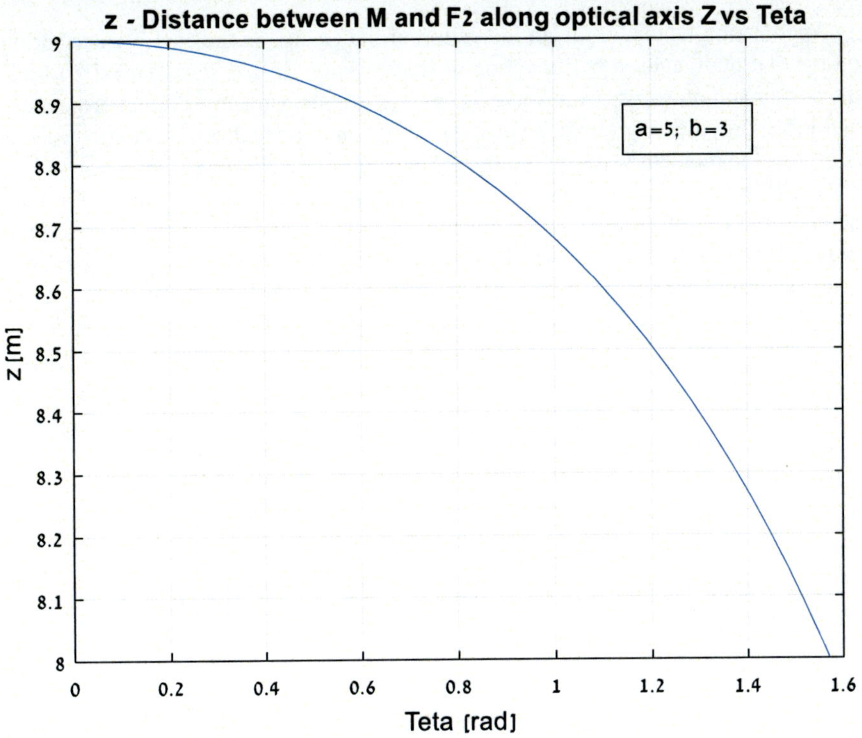

Fig. E.3 Elliptical mirror z-distance between M and F_2 along optical axis as a function of θ (a = 5; b = 3)

Second we plot the distance z (between M and F_2 along optical axis Z) as a function of θ (focusing angle, $90° > \theta > 0°$), $a = 5;\ b = 3;\ \sqrt{a^2 - b^2} = \sqrt{25 - 9} = 4$ (Fig. E.3).

$$z = \frac{100 \cdot tg^2(\theta) + 45 \cdot \sqrt{1 + tg^2(\theta)}}{25 \cdot tg^2(\theta) + 9} + 4$$

MATLAB:

```
x=0:0.01:(pi/2);
a1=(tan(x)).^2;
a2=sqrt(1+a1);
Q=(100*a1+45*a2)./(25*a1+9)+4;
plot (x,Q), grid on
ylabel('z'); xlabel('Teta [rad]');
title ('z - Distance between M and F2 along optical axis Z vs Teta');
```

Apodization factor $l(\theta)$ is an indication of the change in the energy distribution on the exit pupil caused by reflection or transmission. If $0 < \theta \ll 1 \Rightarrow l(\theta) \approx 1$, the conversion of energy from wave S_{w1} to S_{w2} is approximately a direct projection, and the energy density remains constant before and after reflection. If $l(\theta) > 1$ or $l(\theta) < 1$, the incident wave within the infinitesimal area will be compress or stretch after reflection or transmission, the energy density will increase or decrease. For both parabolic and elliptic mirrors, $l(\theta)$ increase from one toward two when θ increases up to 1.5 rad $\left(\approx \frac{\pi}{2}\right)$. If $l(\theta)$ increases then more energy is distributes as high frequency in a mirror system rather than in a system of lenses, $l(\theta)$ of an elliptical mirror is slightly sharper than that of a parabolic mirror. Then we get a compressed scanning spot. We get sharper curve of $l(\theta)$ if $c = \sqrt{a^2 - b^2}$ is close to a $(b \to 0)$. The apodization factor $l(\theta)$ of an elliptical mirror is changed not only with θ, and with a, b, and c. A smaller c value means focal point F_1 and F_2 are close one to another. If $c = \sqrt{a^2 - b^2}$ is small enough then elliptical mirror is turn into a sphere-like mirror. We can see tendency of energy conversion caused by the mirror reflection related to $l(\theta)$ for a different values of c.

$$l(\theta) = 1 \Rightarrow \left(\frac{a + \sqrt{a^2 - b^2}}{a - \sqrt{a^2 - b^2}}\right) \cdot \sqrt{\frac{\sin(\alpha) \cdot d\alpha}{\sin(\theta) \cdot d\theta}} = 1 \Rightarrow \alpha = \theta$$

The deepness of elliptical mirror is described by b or c [2–5].

Off axis parabolic mirror (OAP): An off-axis parabolic (OAP) mirror is a side section of a parent parabolic mirror which has the ability to focus collimated light without introducing spherical aberration. Collimated light that is incident to an OAP mirror is focused to a point. An OAP mirror has an advantage in that it allows more interactive space around the focal point without disrupting the beam if we compare it to centered parabolic mirror. Depending on which section of a parabolic shape an OAP mirror is replicating, the angle between the focal point and the central ray axis can be large or small. For right operation, it is important to keep the incident beam parallel to the optical axis; any angular displacement will produce chromatic aberration. Commercial off axis parabolic mirrors have additional bare gold, protected gold, protected Aluminum, UV Enhanced Aluminum, and protected silver coatings. By using OAP mirror substrate we get low scatter OAP mirrors which decrease scatter in the visible spectrum. It is designed to direct and focus incident collimated light at a specific angle with minimal scatter loss. Off-axis parabolic mirrors (OAPs) are very useful tool in many scientific optical designs. Off-axis parabolic (OAP) mirrors combine the achromatic and diffraction limited imaging properties of a parabolic mirror with the ability to deviate the light path off-axis, which is useful in many imaging systems. A parabolic mirror takes light from a point source which is located at the focus and creates a collimated beam. A source with spherical wave fronts placed at the parabolic focus is converted into a beam with plane wave fronts. We can describe the reverse operation when plane wave fronts incident on the parabolic mirror are focused at the focal point. An OAP mirror can produce a diffraction limited image without chromatic effects. A complete parabolic mirror would focus a collimated beam at its focal point, which is often not useful because it overlaps with part of the incoming beam. Obstructing part of the incoming beam help to access the focal point. If only a portion of the parabolic surface is used, the beam will focus off-axis at a more accessible location. There are important design issues if we integrate off-axis parabolic mirror in our system. First, a collimated beam incident upon the OAP from the off-axis focal angle will not produce a diffraction limited image. A spherical wave incident on-axis will not produce a collimated beam. It is an orientation issue. Second, OAPs only achieve diffraction limited imaging when focusing a collimated beam. They produce a perfectly collimated beam from a spherical wave. OAPs have terrible imaging quality when used at finite conjugates and they can be particular useful when used as a part of a relay system, or similar operation to transition between the focal plane and pupil plane. Many optical instruments include parabolic mirror which is the most common type of aspherical mirrors. They are free from spherical aberrations, and thus focus the parallel beam to a point or point source to infinity. Aberration phenomenon in lens is when different rays do not meet after the lens in one focal point. It occurs when all incoming light rays end up focusing at different points after passing through a spherical surface. Spherical aberration can affect resolution and clarity, making it hard to obtain sharp images. Spherical aberration is a type of aberration found in optical systems that use elements with spherical surfaces. Light rays that strike a spherical surface off-center are refracted of reflected more or less than those that strike close to the center. This deviation reduces the quality of images produced by optical systems. In many optical systems it is not necessary to use aperture that is completely rotationally symmetric. In some applications the central part of the mirror obscures the beam path and therefor is not good. For such systems, off-axis mirrors offer many advantages versus the traditional paraboloids. The main applications of off-axis parabolic mirrors are as follow: Target simulators, collimators, MTF measuring system and other optics test devices, spectroscopic and FTIR systems, Radiometers, Beam expanders, and Laser divergence measuring systems. The key advantages of off-axis parabolic mirrors are, minimize system sizes, minimize system weight, both "wedged" and equip-thick mirror are available, and minimize system cost. The advantages give an option to maximize of system efficiency [6, 7]. The structure specification of off-axis parabolic (OAP) mirror is as follow (Fig. E.4).

Parent focal length (PFL) is the focal length of the parent paraboloid. It defines the shape of the surface as $Z = \frac{R^2}{4 \cdot \mathrm{PFL}}$, where R is the radial distance from vertex and Z is surface Sagitta. Surface Sagitta (SAG) of a OAP circular arc (An arc (symbol: $\frown$) is a closed segment of a differentiable curve) is the distance from the center of the arc to the center of its base (Fig. E.5). It used to find the depth of a spherical mirror or lens [Sagitta (Latin) = Arrow (English)].

Slant focal length (SFL) is the distance between OAP mechanical center and parabola focus. We can calculate it from PFL and vice versa. Optical centerline is the line parallel to parent parabola optical axis and coming through the mechanical center of OAP. Zonal radius (ZR) is the distance between parent parabola optical

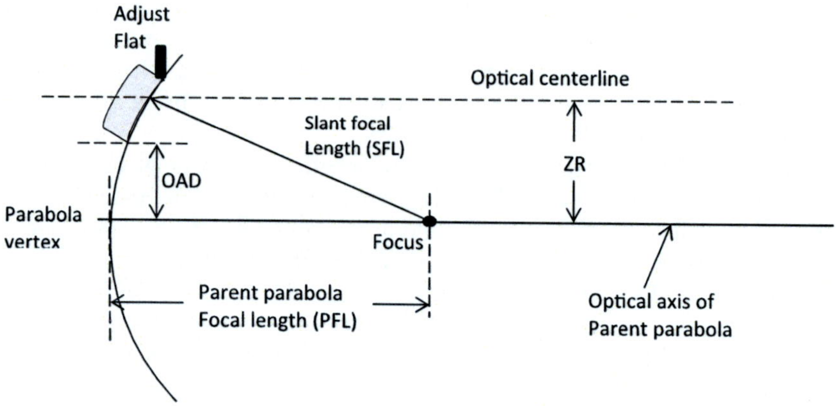

Fig E.4 Structure specification of off-axis parabolic (OAP) mirror

Fig. E.5 Surface Sagitta (SAG) of a OAP circular arc

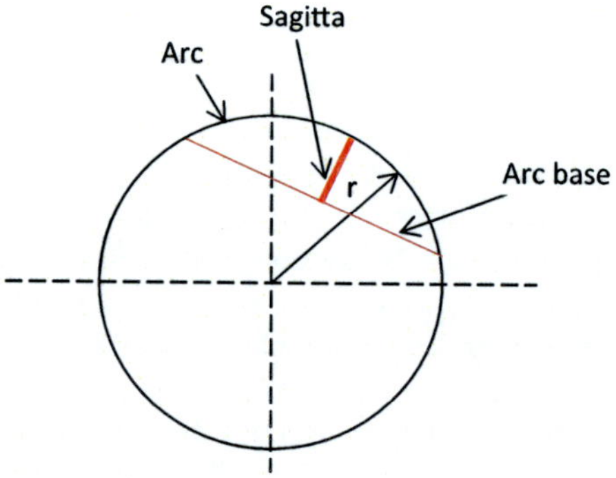

axis and optical centerline of the OAP. Off-axis distance (OAD) is the distance from parent parabola. Parent parabola is the simplest function of a family of parabolas (y = x ^ 2). It is calculated from ZR and vice versa. We can adjust the flat of the mirror through mechanical adjustment. It is perpendicular to parent parabola optical axis and to the OAP centerline and helps to adjust OAP in the optical system. If we want to specify our off-axis parabola (OAP) mirror, we need to specify main five parameters: PFL (or SFL), ZR (or OAD), CA (Clear aperture), SA (Surface accuracy), and SQ (Surface quality). There are auxiliary parameters such as mechanical size and thickness, material, and coating type. Typically, OAPs are made by polishing and slicing-up large on-axis parent paraboloids. This method has restrictions on the range of available combinations of focal lengths and off-axis distances. Another method is diamond turning which has the disadvantages of limitations for substrate materials (metals), lower surface roughness, and accuracy. The most recent and advance method to made OAPs is computer-controlled spot-correction polishing process. It combines advantages of conventional polishing (smooth surface and possibility to use common glasses) and diamond turning which is OAP production without polishing

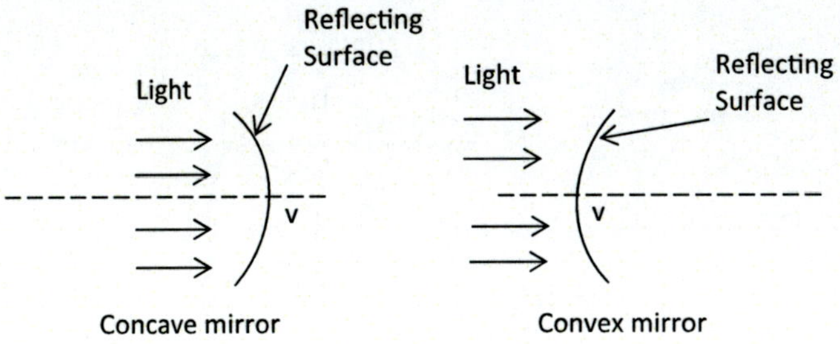

Fig E.6 Spherical mirrors: concave, and convex

of full paraboloid. When we use off-axis parabolic mirrors, we need to mount it and allow precise positioning of optical elements within setup or instrument. Precision holders are needed for that purpose. Every holder is available in both manual and motor-driven variants.

Spherical mirror (SM): A spherical mirror is a mirror which has the shape of a piece cut out of a spherical surface. There are two types of spherical mirrors: Concave, and Convex. Concave mirrors are shaving mirrors and make up mirrors. These types of mirrors magnify objects placed close to them (Fig. E.6).

The normal to the center of the mirror is called the principle axis. The mirror is assumed to be rotationally symmetric about this axis. It is possible to present a three dimensional mirror in a two-dimensional diagram, without loss of generality. The point v at which the principal axis touches the surface of the mirror is called the vertex. The point c, on the principal axis, which is equidistant from all points on the reflecting surface of the mirror, is called the center of curvature. The distance along the principle axis from point c to point v is called the radius of curvature of the mirror, and is denoted by R. The rays striking a concave mirror parallel to its principal axis, but not too far away from this axis are reflected by the mirror such that they all pass through the same point F on the principal axis. The point, which is lies between the center of curvature and the vertex, is called the focal point, or focus, of the mirror. The distance along the principal axis from the focus to the vertex is called the focal length of the mirror, and is denoted by f (Fig. E.7).

In ideal spherical mirror all light-rays which strike a mirror parallel to its principal axis are brought to a focus at the same point F. As rays from a distant object depart further from the principal axis of a concave mirror they are brought to a focus ever closer to the mirror. The lack of perfect focusing of a spherical mirror is called spherical aberration. The approximation in which we neglect spherical aberration is called the paraxial approximation. Image formation with paraxial approximation is called paraxial optics. A mirror whose reflecting surface is the surface of revolution of a parabola is a type of mirror which does not suffer from spherical aberration. Thus, a ray travelling parallel to the principal axis of a parabolic mirror is brought to a focus at the same point F, no matter how far the ray is from the axis. The path of a light-ray is completely reversible, and it follows that a light source placed at the focus F of a parabolic mirror yields a perfectly parallel beam of light, after the light

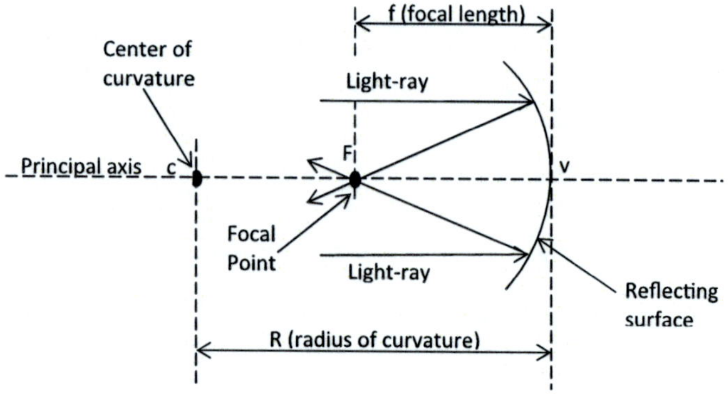

Fig. E.7 Spherical mirror (concave) geometry and important parameters

has reflected off the surface of the mirror. In contrast, parabolic mirrors are more difficult, and, therefore, more expensive, to make than spherical mirrors. Parabolic mirrors mainly used in situations where the spherical aberration of a conventional spherical mirror would be a serious obstacle. More general, curved mirrors come in two basic types: Curved mirrors which converge parallel to incident rays of light and those that diverge parallel incident rays of light. A spherical mirror is the easiest shape to analyze. It is a complete sphere, but a spherical cap a piece sliced from a larger imaginary sphere with a single cut. Most spherical mirrors are spherical caps. We start by tracing a line from the center of curvature of the sphere through the geometric center of the spherical cap. If we extend it to infinity in both directions, this imaginary line is called the principal axis or optical axis of the mirror. Any line through the center of curvature of a sphere is an axis of symmetry for the sphere, and only one of these is a line of symmetry for the spherical cap. Rays of light parallel to the principle axis of a concave mirror will appear to converge on a point in front of the mirror somewhere between the mirror's pole and its center of curvature. Positions in the space around a spherical mirror are described using the principal axis like the axis of a coordinate system. The pole serves as the origin. Locations in front of a spherical mirror (or a plane mirror) are assigned positive coordinate values, those behind, negative. The distance from the pole to the center of curvature is called the radius of curvature (r). If the focal length is (f) then the local length of a spherical mirror is approximately half its radius of curvature $\left(f \approx \frac{r}{2}\right)$. If the mirror is with an opposite curvature then one where the reflecting surface is on the "outside" then we have a convex mirror (Fig. E.8). Convex mirrors are diverging mirrors. Instead of converging onto a point in front of the mirror, rays of light parallel to the principal axis appear to diverge from a point behind the mirror. It is the focal point (focus) location of the mirror even through it disagrees with the original concept of the focus as a place where things meet up. Locations in front of a diverging mirror have opposite position value, since points in front of any mirror are always positive. The distance from the pole to the center of curvature is the radius of curvature (r), but it is negative. The distance from the pole to the focus is the focal length (f), but it

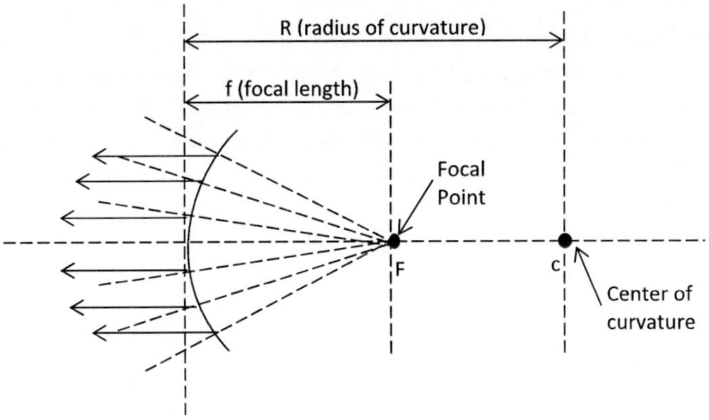

Fig. E.8 Spherical mirror (convex) geometry and important parameters

is negative. With two sign switches, the rule that focal length is half the radius of curvature exists and $f \approx \frac{r}{2}$ [8, 9].

Chirp mirror (CM): A chirped mirror is a dielectric mirror with chirped spaces which are spaces of varying depth designed to reflect varying wavelengths of lights between the dielectric layers. A dielectric mirror (Bragg mirror) is a type of mirror composed of multiple thin layers of dielectric material (substrate of glass or some other optical material). Mainly dielectric mirrors used to produce ultra-high reflectivity mirrors. Dielectric mirror reflects a broad spectrum of light, such as the entire visible range or the spectrum of the T-Sapphire laser. Typical applications are laser cavity end mirrors, hot and cold mirrors, thin-film beam splitters, and the coatings on modem mirror shades. We use chirped mirrors for applications like lasers to reflect a wider range of light wavelengths than ordinary dielectric mirrors, and compensation for the dispersion of wavelengths which create by optical elements. The dielectric mirror is made of transparent materials that are uniformly layered at a depth of quarter the wavelength of light the dielectric mirror is designed to reflect. The dielectric mirror is transparent to other wavelengths of light except those in very narrow band around the wavelength it is designed to reflect. A chirped mirror is targeted to reflect a wider range of frequencies. The chirped mirror is constructed by layers with different lengths. Mainly ten layers with a depth designed to reflect a certain wavelength of light, another layers with greater depth to reflect a longer wavelength of light, and so on for other wavelength of light the mirror is designed to reflect. The final outcome of chirped mirror is a mirror that reflects a whole range of light wavelengths rather than single narrow band of wavelengths. The mirror structure caused to light reflected from deeper layers of mirror to travel a longer distance than the light that reflects off the surface layers. A chirped mirror is designed to change the relative times of wave fronts of differ wavelength reflecting from it. We can use the chirped mirror to disperse a pulse of light of different wavelengths that arrives all at one time or to tighten a pulse of light where different wavelengths arrive dispersed in time. The chirped mirrors are used to compensate group delay variations when using

Ti:Sapphire lasers which employing Kerr-lens mode locking. These mirrors are used to correct residual variations of group delay after a grating compressor is inserted into the system when using chirped pulse amplification. The chirped pulse amplification (CPA) is a technique for amplifying an ultrashort laser pulse up to the petawatt level ($\approx 10^{15}$ W) with the laser pulse being stretched out temporally and spectrally prior to amplification. CPA is used by all of the highest-power lasers systems, with the exception of the $\approx$500 TW. In CPA an ultrashort laser pulse is stretched out in time prior to introducing it to the gain medium using a pair of gratings which arranged then the low frequency component of the laser pulse travels a shorter path than the high frequency component does. After passing the grating pair, the laser pulse is positively chirped (high-frequency component lags behind the low-frequency component) and the outcome is longer pulse duration than the original. Then the stretched pulse is introduced to the gain medium and amplified by the factor of $\approx 10^6$. There are negative and positive dispersions. A negative dispersion is when light with higher frequencies (shorter wavelengths) takes less time to travel through the device than the light with lower frequencies (longer wavelengths), and positive dispersion is the other way around. The dispersion of an optical device is a function $\tau(\omega)$, where τ is the time delay experienced by a frequency component ω. We define $\phi(\omega) = \frac{2 \cdot \pi \cdot \tau(\omega) \cdot c}{\lambda(\omega)}$, where c is the speed of light and λ is the wavelength. By using double-chirped mirrors we achieve simultaneously a high reflectivity and dispersion compensation over an extended bandwidth compared with these of standard quarter-wave Bragg mirrors. The mirrors are fabricated by ion beam sputtering. Mainly we use double-chirped mirrors (DCMs) for dispersion compensation in ultrashort pulse laser sources. The structure is multilayer interference coating by exact coupled-mode equations. We need to control the harmonic spectral range and the spectral phase when we need to generate high-order harmonics in rare gases. Multilayer extreme-ultraviolet chirped mirrors are optimized and designed to compensate for the intrinsic harmonic chirp. The chirped mirror (dielectric dispersive mirror) has a spatial variation of the layer thickness values. We use it as dispersion compensation in mode-locked lasers. We get a broader reflection bandwidth than ordinary Bragg mirrors. The Bragg wavelength is not constant and varies within the structure along the propagation direction. Light at different wavelengths penetrates to a different extent into the mirror structure and we get a different group delay. There are strong oscillations of the group delay and group delay dispersion. The double-chirped mirror can be used with semiconductor mirrors and generate a much higher amount of dispersion in smaller bandwidth. They used to compensate the dispersion in a mode-locked laser with a single compact device when a long pulse duration is required a high amount of anomalous dispersion for soliton mode locking. Pulse reflection at a dielectric mirror leads to a broadening from the pulses and the reflected signal is modified. Different wavelengths pass the coating layers with different speed and the pulse itself is delayed. The pulse energy is affected by this phenomenon, which leads to a reduced energy level. To overcome these effects, dispersion-controlled mirrors are designed. The coating design is adapted to the required spectral performance. The next schematic represents the chirped mirror with negative group delay dispersion (GDD), $\lambda_1 < \lambda_2 < \lambda_3$ (Fig. E.9).

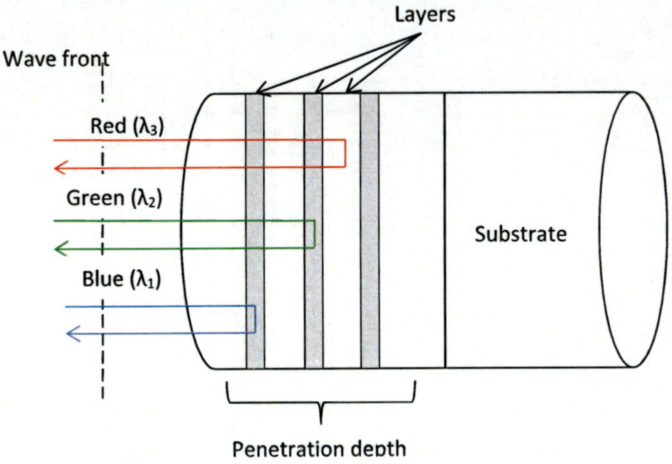

Fig. E.9 Chirped mirror structure and various wavelengths of light penetration depth

It described the chirped mirror structure and various wavelengths of light penetration depth [10, 11].

Group delay dispersion (GDD), second-order dispersion, of an optical element is the derivative of the group delay with respect to the angular frequency, or a second derivative of the change in spectral phase $\left(D_2(\omega) = \frac{\partial T_g}{\partial \omega} = \frac{\partial^2 \varphi}{\partial \omega^2}\right)$. It is usually specified in fs^2 or ps^2. If an optical element has only second order dispersion (frequency-independent D_2 value), its effect on an optical pulse or signal is characterized by the change of the spectral phase $\left(\Delta\varphi(\omega) = \frac{D_2}{2} \cdot (\omega - \omega_0)^2\right)$, where ω_0 is the angular frequency at the center of the spectrum. The group delay dispersion (GDD) is referred to optical element or to the length of a medium. The GDD per unit length (units of $\frac{s^2}{m}$) is the group velocity dispersion (GVD). Chirped mirrors provide accurate means of controlling the group delay dispersion (GDD), the second derivative of the phase shift on reflection with respect to the angular frequency to support the compression of coherent continua to pulses lasting only few optical cycles. The frequency dependence of the group delay is characterized by a broadband pulse upon reflection of a chirped mirror (CM) and is controlled by means of the penetration depth of the different spectral components in the multilayer (Fig. E.9). we suppress the GDD-fluctuations by an anti-reflection (AR) coating or a narrow band-stop filter which deposited on top of the mirror. We fulfill these requirements by AR-coating. Group delay (GD) and Group delay dispersion (GDD) of the mirror do not need to be low order polynomial functions of frequency in order to support the compensation of sub-two-cycle pulses. The GDD introduced by chirped mirrors (CMs) exhibit extremely large oscillations (order of fs^2) as a function of frequency, without impairing on the shape of the reflected pulse. We can design of single chirped mirrors (CMs) with a broadband larger than one optical octave and supporting the compression of sub-two-cycle pulses. A CafCA (Compression after Compression Approach) system is

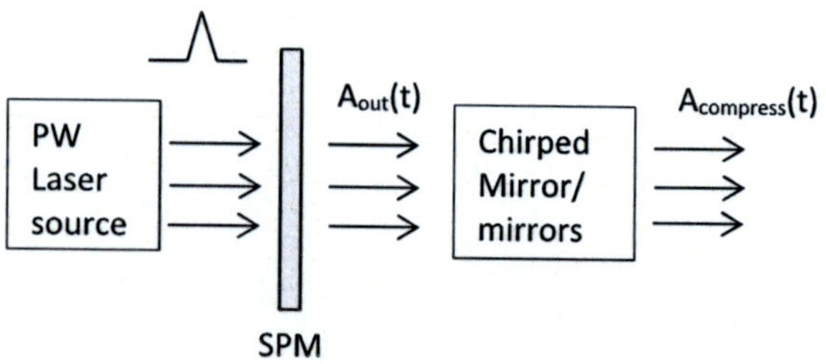

Fig. E.10 CafCA (Compression after Compression Approach) system

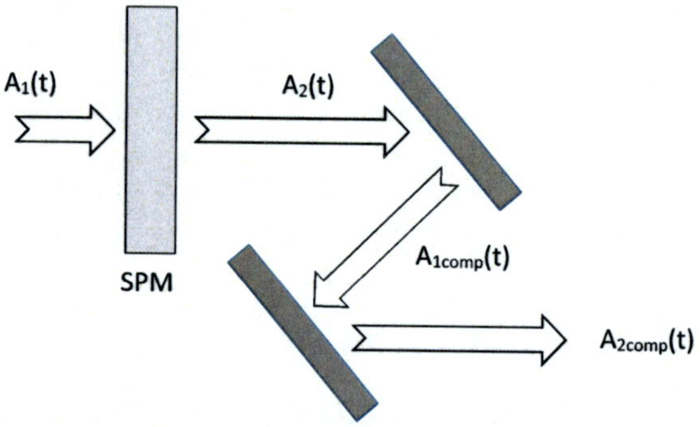

Fig. E.11 Second harmonic generation (SHG) and pulse compression system

constructed from Pettawatt (PW) laser source, self-phase modulation (SPM) medium, and a chirped mirror (CM) (Fig. E.10).

It is CafCA system which influence of residual spectral phase of small-scale self-focusing suppression in powerful laser beams.

$$A_{compress}(t) = F\left\{e^{-\frac{i\cdot\alpha\cdot\omega^2}{2}} \cdot F^{-1}\{A_{out}(t)\}\right\}$$

$F\{\}$ Fourier transform,
$F^{-1}\{\}$ Inverse Fourier transforms.

The KDP (non-linear crystal) crystal is for second harmonic generation and for self-compression after compressor. It is a system of pulse compression after SPM in a specific material. The KDP crystal controls the temporal intensity profile of the pulse and has the advantage of Aperture up to 400 nm. Additionally KDP crystal has a good relation between linear dispersion properties, second and third order nonlinear coefficients [12, 13]. The second harmonic generation (SHG) and pulse recompression technique is as follow (Fig. E.11).

$$A_{2comp}(t) = F^{-1}\left\{e^{\frac{i\cdot\alpha\cdot\omega^2}{2}} \cdot F\{A_2(t)\}\right\}$$

$e^{\frac{i\cdot\alpha\cdot\omega^2}{2}}$ Quadratic phase correction.

There is a variety of methods for compressing or shortening optical pulses. By compressing we reduce the pulse duration (typically in the picosecond or femtosecond region)—Ultrashort pulses. There are two main compression methods: Linear pulse compression and nonlinear pulse compression. In Linear compression, pulses are chirped and their duration is reduced by removing this chirp (flattening the spectral phase). De chirping is done by sending the pulses through an optical element with a suitable amount of chromatic dispersion. An optical element can be a pair of diffraction gratings, prism pair, an optical fiber, a chirped mirror, a chirped fiber Bragg grating or a volume Bragg grating. The optical bandwidth of the pulses sets the smallest pulse duration. In nonlinear pulse compression, the optical bandwidth is increased, typically with a nonlinear interaction [self-phase modulation (SPM)]. This leads to chirped pulses with duration larger than the original pulse duration. The pulse duration can be reduced by linear dispersive compression, which removes or decreases the chirp. Compression is using a chirped QPM structure (Quasi-phase matching). Quasi-phase matching is a technique for achieving similar results to those with phase matching of nonlinear interactions (nonlinear frequency conversion). We use a material with spatially modulated nonlinear properties. A phase mismatch is happened over some propagation distance. It reverses the nonlinear interaction at positions where otherwise the interaction would take place with the wrong direction of conversion. The main benefits of QPM are: a very wide range of nonlinear interactions, utilizing a larger nonlinear coefficient than is accessible with birefringent phase matching, reduced tendency for photo refractive problems of periodically poled crystals. The main limitations of QPM are: Fabrication of periodically poled crystals with high and reliably confirmed quality is possible only with certain crystal materials, periodic poling can be applied only to crystals with limited thickness which excludes large aperture devices for very high power levels, the need of many poling periods, light at additional wavelengths is generated by a parasitic higher-order processes [14, 15].

Echelon mirror (ECM): An Echelon mirror is constructed by a terraforming unit and acts as a solar collector and increasing the solar energy collected by solar collectors in all surrounding squares by +1. Echelon mirror produces the same base energy as solar collector and increases the production of adjacent solar collector by one. Multiple mirrors can affect a single solar collector which can result in that collector to produce ten or more energy. Echelon mirrors cause significant amounts of eco-damage. Echelon mirror is used for measuring ultrashort laser pulses as well as ultrafast responses of materials using the same optical setup (High power laser facilities). The Echelon mirror produces a spatially encoded time delay for the probe pulse where by both the probe and pump pulses are focused on samples with small spot size. We can demonstrate single-shot time-frequency imaging spectroscopy by using Echelon mirror. Mainly we use the optical Kerr gate apparatus and mapped the time-frequency images of ultrashort laser pulses and evaluation the chirp characteristics with the phase-retrieved procedure on a single-shot basis. Another reflective Echelon mirror use is to demonstrate single-shot measurements of terahertz (THz) electric-field waveforms. It produces multiple probe pulses with different time-delays. The polarization rotation of the probe pulses, due to the electro-optic effect induced by the THz electric field generated from grating-coupled $LiNbO_3$. Imaging onto a two-dimensional complementary metal-oxide-semiconductor-camera is done. An Echelon mirror is used for real-time time-frequency imaging which is for the auto correlation traces of ultrashort laser pulses. The Echelon mirror is employed to generate spatially encoded time delays for the probe pulses. We use the frequency-resolved optical gating (FROG) technique with the Echelon mirror, the time-frequency images of ultrashort laser pulses are presented. Frequency-resolved optical gating (FROG) is a general method for measuring the spectral phase of ultrashort laser pulses, which range from sub femtosecond to about a nanosecond in length. FROG measures an "auto-spectrogram" of the pulse, in which the pulse gates is in a nonlinear-optical medium and the resulting gated piece of the pulse is then spectrally resolved as a function of the delay between the two pulses. FROG is the technique to measure ultrashort laser pulses and it is competitive method to auto correlation which gives estimation for the pulse length. FROG is a spectrally resolved auto correlation, which gives the use of a phase-retrieval algorithm to retrieve the precise pulse intensity and phase versus time. We can evaluate the chirp characteristics of the laser pulses by using a phase-retrieval procedure on a single shot basis. When using a dual-Echelon technique, the transmitted pulse passing through the Echelons causes spatially dependent chirps that distort the temporal profile. There are main two popular techniques to characterize the ultrashort laser pulses, first the spectral phase interferometry for direct electric field reconstruction (SPIDER) and second, the frequency-resolved optical gating (FROG). The use of an Echelon mirror with a micro step structure provides temporal delays for the probe laser pulses. We can get images of the Echelon mirror by using a conventional laser microscope. The Echelon mirror is mostly fabricated on a Ni block by using a precise micro-machining technique, and it generates a spatially encoded time delay for the probe pulses. The Echelon mirror is an integral part of a spatially encoded time delay optics and reflective grating. The grating separates polychromatic (multiple wavelength) light into its component wavelengths. It is done by diffraction; diffraction is a process in which light incident to a surface with dimensions similar in size to its wavelength and is dispersed at certain angles. The diffraction angle is dependent on the wavelength of light. A polychromatic light has a separate diffraction angle for each individual wavelength. The difference in diffraction angle is what separates light into its component wavelengths. In a transmission grating the diffracted light is passed through at an angle equal to the diffraction angle. In refractive gratings, the light is first diffracted by the grating and then reflected by the coating at an angle equal to the diffraction angle. The reflective and transmission gratings fulfill the diffraction equation. Diffraction grating is an optical component with a periodic structure that splits and diffracts light into several beams travelling in different directions. The directions of these beams depend on the spacing of the grating and the wavelength of the light, and grating operated as the dispersive element. Popular use of gratings is in mono chromatics and spectrometer. Gratings have ridges or rulings on their surface and can be transmissivity or reflective. The Echelon mirror plays a role of a reflective grating with a periodicity of d; $\left(d = \sqrt{(d')^2 + h^2}\right)$, where d', h are the Echelon mirror reflective

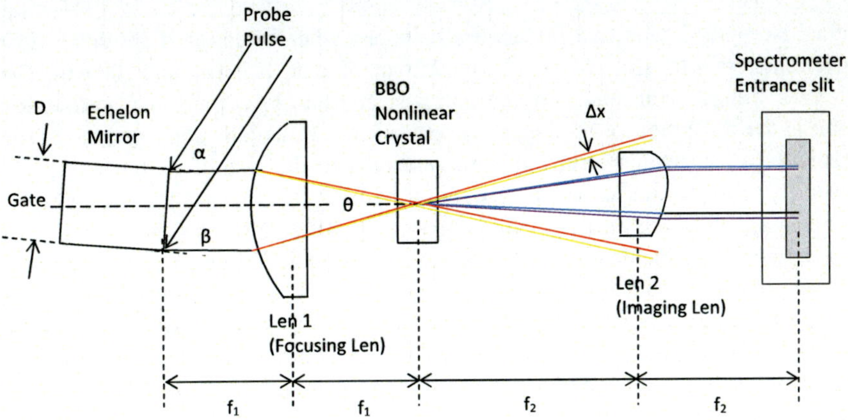

Fig. E.12 Echelon mirror system for obtaining time frequency 2D auto correlation trace of ultrashort laser pulses

grating parameters. We define the incident and diffraction angles of the probe pulse against the normal of the echelon surface as α and β, respectively. The diffraction condition is written as: $d \cdot (\sin\alpha + \sin\beta) = m \cdot \lambda; m = 0, \pm1, \pm2, \ldots$, where λ is the wavelength of the probe and m is the diffraction order and it is an integer which can be positive or negative, d is the Echelon mirror separation of reflective grating. The probe pulse has a finite spectral width $\Delta\lambda$ due to the ultrashort pulse duration then the diffracted light from each stair is spectrally dispersed with increasing distance from the Echelon mirror. Focusing and imaging lenses (f_1 and f_2) are placed between the Echelon and spectrometer and we get a clear image of the second harmonic (SH) auto correlation signal generated at a nonlinear crystal at the slit. We suppress the spectral dispersion at the entrance slit of a spectrometer and the Echelon mirror behaves as a spatial encoding optics for the time delay. The following system is for obtaining time frequency 2D auto correlation trace of ultrashort laser pulses (Fig. E.12).

The system includes the following optical elements: Lens one (1, focusing Len) and two (2, imaging Len) to suppress the spectral dispersion, Echelon mirror, Probe pulse source with finite spectral width $\Delta\lambda$, BBO nonlinear crystal (beta – $B_a B_2 O_4$) and spectrometer. We use Echelon mirror as a spatial encoding optics for the time delay and f_1, f_2 are the focal lengths of the focusing lens, respectively. The spot size Δx is bigger than the diffraction limit, and then the diffraction limit of the spot size is neglected. θ is the intersection angle between the gate and probe pulses at the nonlinear crystal (define the time resolution of the system). Intersecting angle for the probe pulse that is diffracted from the edge of the Echelon mirror is $\theta \approx \frac{D}{2 \cdot f_1}$, and time resolution is decreased by $\approx \frac{\Delta x \cdot \theta}{c}$, c is the speed of light. The intersecting angle for the probe pulse that is diffracted around the center is close to zero, and then the time resolution around the center is as short as the pulse duration. The slit is an opening that controls how much light enters the spectrometer. The width of the slit affects resolutions, the narrower of the slit, the higher resolution. Narrower slits decrease signal strength. These two factors are balanced when selecting a slit size. Most spectrometers offer slit sizes of 14 μm, 25 μm, 50 μm, 100 μm, and 200 μm. The high resolution spectrometers have small slits such as 14 μm or 25 μm. Diffraction limit is the resolution of an optical system reaches a fundamental maximum. The resolution of a given instrument is proportional to the size of its objective, and inversely proportional to the wavelength of the light being observed. The ability of an imaging system to resolve detail is limited by diffraction. The reason is the fact that a plane wave incident on a circular lens or mirror is diffracted. The light is not focused to a point but forms an Airy disk having a central spot size in the focal plane with radius to first null of $d = 1.22 \cdot \frac{\lambda}{NA}$, where λ is the wavelength of the light and *NA* is numerical aperture. If we have larger aperture of the lens then we get smaller wavelength and finer the resolution of an imaging system. The resolution of an optical imaging system can be limited by factors such as imperfections in the lenses or misalignment. An optical system with resolution performance at the instrument's theoretical has limit which is diffraction-limited. It is difficult to observe of sub-wavelength structures with microscopes because of the Abbe diffraction limit. Light with wavelength λ, traveling in a medium with refractive index n and covering to spot with half-angle θ will have a minimum resolvable distance of $d = \frac{\lambda}{2 \cdot n \cdot \sin\theta} = \frac{\lambda}{2 \cdot NA}$. Multiplication $n \cdot \sin\theta$ is called the numerical aperture (NA). To increase the resolution, shorter wavelength can be used as UV and X-ray microscopes. Lens has an absolute upper performance limit. The limitation is controlled by the working $\frac{f}{\#}$, $\left(\frac{f}{\#} = \frac{EFL}{D_{EP}}\left[\frac{\text{Lens Iris}}{\text{Aperture}}\right]\right)$, of lens and the wavelengths of light that pass through the lens. Iris diaphragm is a type of adjustable diaphragm. A diaphragm is a thin opaque structure with opening (aperture) at its center. The diaphragm stops the passage of light, except for the light passing through the aperture. It is known as the diffraction limit, and this limitation is given in line-pairs/mm and determines the theoretical maximum resolving power of lens. The diffraction limit formula is

$$\text{Diffraction limit}\left[\frac{\text{lp}}{\text{mm}}\right] \simeq \frac{1000\,\frac{\mu\text{m}}{\text{mm}}}{\frac{EFL}{D_{EP}} \cdot \lambda[\mu\text{m}]};\quad \text{lp} = \text{line pairs}$$

where EFL is the ratio of the effective focal length of the lens and D_{EP} is the effective aperture diameter. Parameters EFL and D_{EP} control lens's parameters like overall light throughout, depth of filed, and the ability to produce contrast at a given resolution [16, 17].

Every lens has diffraction limited. This limit is the point where two Airy patterns are no longer distinguishable from each other. The Airy disk (Airy disc) and Airy pattern are descriptions of the best-focused spot of light that a perfect lens with a circular aperture can make and it is limited by the diffraction of light. The diffraction pattern resulting from a uniformly illuminated, circular aperture has a bright central region (Airy disk) which with the series of concentric rings around is called Airy pattern. The circular aperture plane and observable plane/point system is presented below (Fig. E.13).

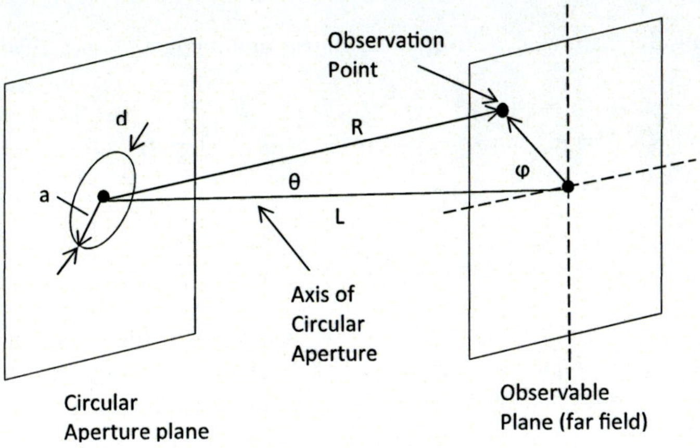

Fig. E.13 Circular aperture plane and observable plane/point system

Fig. E.14 A beam transferring from one material to another system

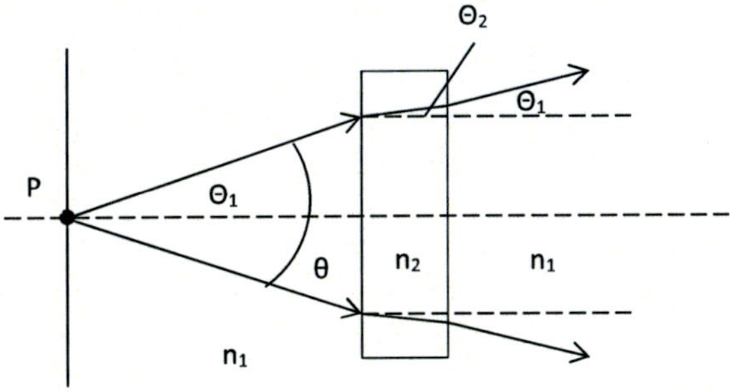

We can observe the Airy pattern when $R > \frac{a^2}{\lambda}$, where R is the observation distance, a is the radius of aperture, λ is the wavelength of radiated light, k is the wave number $\left(k = \frac{2 \cdot \pi}{\lambda}\right)$, θ is the angle of observation (angle between the axis of the circular aperture and the line between aperture center and observation point), L is the distance between aperture center and observable plane center. We can define the following mathematical connection: $k \cdot a \cdot \sin\theta = \frac{2 \cdot \pi \cdot a}{\lambda} \cdot \frac{q}{R} = \frac{\pi \cdot q}{\lambda \cdot N}$, where q is the radial distance from the optics axis in the observation or focal plane and N is the f number of the system, d is the diameter of the aperture $\left(N = \frac{R}{d};\ d = 2 \cdot a;\ k \cdot a \cdot \sin\theta = \frac{\pi \cdot q}{\lambda} \cdot \frac{d}{R}\right)$. φ is the angle between the observable plane x-axis and the line between observable plane center and observation point. The numerical aperture (NA) of an optical system is a number (dimensionless) that characterized the range of angles over which the system can accept or emit light. Numerical Aperture (NA) is constant for a beam transferring from one material to another (Fig. E.14). The numerical aperture of an optical system (objective lens) is defined $NA = n \cdot \sin\theta$, where n is the index of refraction of the medium in which the lens works, and θ is the maximal half-angle of the cone of light that can enter or exit the lens (real marginal ray in the system).

By using Snell's law, $n_1 \cdot \sin\theta_1 = n_2 \cdot \sin\theta_2$ and it equal to numerical aperture (NA),

$$NA = n_1 \cdot \sin\theta_1 = n_2 \cdot \sin\theta_2;\ \ \xi_1(n_1, \theta_1) = n_1 \cdot \sin\theta_1$$
$$\xi_2(n_2, \theta_2) = n_2 \cdot \sin\theta_2$$

$$NA = \xi_1(n_1, \theta_1);\ \ NA = \xi_2(n_2, \theta_2)$$

P is the point of light source and θ_1 is the half angle θ, $\left(\theta_1 = \frac{\theta}{2}\right)$, which is the maximum cone of light angle that can enter or exit the lens, θ_2 is the half-angle change which is related to n_2. BBO is a Barium Borate nonlinear crystal and it used when we need harmonic generation operations, optical parametric oscillators. Electro optical applications use BBO in the near infrared to the deep ultraviolet spectrum. BBO crystals are characterized by large nonlinear coefficients, high threshold for laser damage, and low thermos-optic coefficient. By using BBO crystal in Nd:YAG lasers (355 nm, 266 nm, and 213 nm, respectively), third, fourth, and fifth harmonics are generated. Additional to generate the second and third harmonics of Ti:Sapphire amplifiers (400 nm and 266.7 nm). By using dye lasers we get a variety of sum frequency mixing schemes. We use BBO to generate tunable output in the visible to near infrared range and implement it in Q-switching and high power UV laser systems [18].

Optical delay line (ODL): An optical delay line (ODL) enable computer-controlled variation of the optical path length. Mainly an ODL system includes the following elements: a DC servo stage, controller, and retroreflector optics with mounts. Additionally, can be drop-in irises with mounting positions for alignment. A DC servo stages feature long travel ranges and driven at high speeds compared to stepper motor stages. Its makes ODL systems ideal for high repetition rate scientific systems. High-speed scans minimize the effects of slow changes in the system, such as thermal drifts. In Fiber-optics, fiber-optic delay lines are used to produce defined fiber-optic transmission delays to compensate for delay differences in optical fiber systems. Optical delay line modules are used to produce defined optical transmission delays and to compensate for delay differences in optical fibers. In fiber sensor technology, optical coherence tomography, and fiber-optic interferometers, the transmission time of the optical signal needs to be adjusted or delayed among the different optical paths. Fine adjustment of the delay time is carried out via phase shifters and fiber stretcher modules. Fiber-optic phase shifter or fiber stretchers are used to finely adjust the optical phase and delay time. As by definition, optical delay line is an optical setup which is used to introduce a time delay for a light beam. In many optical devices and applications there is a need for time delay for a light beam. Laboratories interferometers need a variable delay line. Adjusting the optical phase is done by small changes of propagation distance of the order of one optical wavelength. Wave meters and optical tomography (OCT) require larger variable time delays. Variable delays are required for coherent receivers in optical fiber communications applications. An auto correlator requires a variable time delay for controlling the relative timing between two ultrashort pulses. An electro-optic sampling devices require variable time delays. In microwave photonic devices there is a need for time delays of the order of a microwave oscillation period. In pump-probe measurements, we use a variable time delay for measuring responses to a pulse after a variable amount of time. Data buffer units in optical computing systems realized with time delays. In scientific facilities there is use with both variable optical delay line and a fixed optical delay. Variable delays are periodically scanning through range of delays. Most optical delay lines work on the principle that the pulses are sent through an optical arrangement with a variable path length. Auto correlators and cross correlators devices contain an optical delay line and by moving two mirrors, which are both mounted on some movable part (movable retroreflector), and the pulse timing can be controlled. In some cases the delay is adjusted manually for reaching a high precision. In fiber optical communication, we use delay lines with fiber-optic connectors which contain optical single-mode fiber which is wound up to a coil, and fixed time delay is achieved. Group delays are calculated by the formula $\tau_g = \frac{L \cdot n_g}{c}$, where L is the fiber length, n_g is the group index, and c is the velocity of light in vacuum. Group delay (τ_g) and phase delay (τ_ϕ) are frequency dependent $\left(\tau_g(\omega) = -\frac{d\phi(\omega)}{d\omega};\ \tau_\phi(\omega) = -\frac{\phi(\omega)}{\omega}\right)$. Group delay is the time delay of amplitude envelopes of the various sinusoidal components of the optical signal through CFBGs and is a function of frequency for each component. Phase delay is the time delay of the phase as opposed to the time delay of the amplitude envelop. The group index (group refractive index), n_g is defined as the ratio of the speed of light in vacuum (c) to the group velocity in medium, $n_g = \frac{c}{v_g} = c \cdot \frac{\partial k}{\partial \omega} = \frac{\partial}{\partial \omega}[\omega \cdot n(\omega)] = n(\omega) + \omega \cdot \frac{\partial n}{\partial \omega}$. To calculate the group index we need to have the values of the refractive index at the wavelength of our system and the frequency dependence. We use the group index value n_g to calculate time delays for ultrashort pulses propagating in a medium and the resonator free space range which includes a dispersive medium. Mainly, the group index of a homogenous optical medium is larger than the refractive index. Fiber stretcher devices exhibit variable time delay; in the way that piece of fiber is stretched using a piezo transducer. Phase shifter devices provide a delay range which is more than one wavelength. We can switch between two different time delays if the polarization direction of light propagating in a birefringent medium is manipulated. A variable optical delay lines (ODLs) are characterized by path length, delay time, resolution, accuracy, speed, usable wave length range, insertion loss, etc. Many delay lines are connected to computer or other electronic control platform with digital bus system for controlling and commands. Optical coherence tomography is a high-resolution optical imaging technique which applied to transparent or semi-transparent objects of limited depth. It gives the capability of high-resolution (sub-micro meters). The working principles of optical coherence tomography (OCT) is in two transverse dimensions, which image resolution is obtained by scanning a tightly focused light beam over the sample and performing measurements on back-scattered light. Some optical delay lines (ODLs) use chromatic dispersion to obtain a substantial wavelength dependence of the delay time. A long optical fiber is used also as a dispersive delay line for ultrafast pulses. We use pulse stretchers for chirped-pulse amplification. Pulse stretchers devices are for

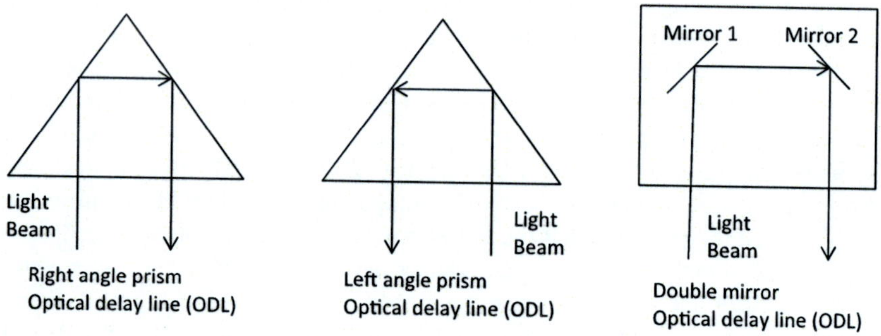

Fig. E.15 Double mirror optical delay line and right/left angle prism optical delay line

stretching ultrashort light pulses. We increase the duration of an ultrashort pulse by a large factor and a tunable pulse durations are achieved by tunable pulse stretchers. Optical delay lines are also used for microwave applications (microwave photonics). It utilizes photonics in combination with microwaves utilizing properties of optical delay lines (ODLs) for microwave delay lines devices. The process is first convert microwave signals to optical signals, then sends them through a fiber-optic delay line and finally converts them back to microwave signals with a fast photodetector. We can achieve a longer time delays with fiber low propagation losses. When using system of multiple pulsed lasers in an experimental facility (pump-probe, stimulated Raman spectroscopy, STED microscope), we need to control the timing between the two pulses [19, 20]. The typical solution for controlling the time delay between two pulses is an optical delay line. It is implemented basically by double mirror optical delay line or Right/Left angle prism optical delay line (Fig. E.15).

First, Double mirror ODL delays the beam light until it reflects from the second mirror, and second, a Right/Left prism used as a delay line mirror, since the angle of incident is more than the critical angle for total internal reflectance. The single prism can be mounted on a translation stage. It is also possible to use two individual mirrors on kinematic mounts on a translation stage or a cube mirrors or retroreflectors which send the incoming light back in a parallel path. We need to set up the translation stage parallel to the beam patch such that the beam direction does not change as the delay line is translated. A variable optical delay lines are implemented in optical coherence tomography applications. They are implemented by using two linearly chirped fibers Bragg gratings cascaded in reverse order. A fiber Bragg grating (FBG) is a type of distributed Bragg reflector constructed in a short segment of optical fiber that reflects specific wavelengths of light and transmits all other. This is a refractive index of the fiber core, which generates a wavelength specific dielectric mirror. We can use fiber Bragg grating as an inline optical filter which block specific wavelengths and act as wavelength specific reflector. An optical Bragg grating is a transparent device with a periodic variation of the refractive index, and we get a large reflectance in specific wavelength range (bandwidth) which fulfills the Bragg condition: $\frac{2\cdot\pi}{\Lambda} = 2 \cdot \frac{2\cdot\pi\cdot n}{\lambda} \cdot \cos\theta$, where λ is the vacuum wavelength of light, n

is the average refractive index of the medium, θ is the propagation angle in the medium relative to the direction normal to the grating, and Λ is the grating period. By fulfilling the condition, wave number of the grating matches the difference of the wave numbers of the incident and reflected waves. Other wavelengths which do not fulfill the condition are only weakly affected by the Bragg grating except for some side lobes in the reflection spectrum. Under other angles of incidence the reflection can disappear. Weak index modulation around the Bragg wavelength causes total reflection for long grating. Bragg grating is functioned as an optical fiber due to the fact that wavelengths are dependent on reflection and transmission. Optical coherence tomography (OCT) is used for biomedical imaging. It obtained non-invasive cross-sectional images and transparent media with a high resolution. Micro-scale, non-invasive cross-sectional imaging of material is done by using ultra-high resolution optical coherence tomography (UHR–OCT) with low noise. The technique facilitated visualization of some material in medium. Additional there are various types of THz imaging systems for computed tomography (CT) THz with continuous wave. A high speed optical delay lines (ODLs) are developed in achieving the real time imaging. We use ODL to characterize the ultrafast laser pulses and path length-resolved measurement of reflectivity and providing along optical delay at high repetition rate. The basic ODL uses a mirror mounted on a linear oscillator and fulfill a large optical delay by using a stage with long traveling length. For pulse stretching or compression we use a non-linearity in the optical delay, Fourier domain optical delay line, and rapid scanning optical delay line. By using it we get a high-speed axial scanning suitable for real-time imaging. A rapid scanning optical delay line (RSODL) system elements are a diffraction grating, focusing lens, and a scanning mirror. Scanning mirrors provide a large optical delay and high speed axial scanning which characterized by a linear phase ramp in the frequency domain. A fiber-based ODL is compactness and low power loss. We utilize an optical fiber coiled around a piezo electric transducer (PZT) and real-time imaging is achieved. A chromatic dispersion compensator in optical communication systems is used chirped fiber Bragg grating (CFBG). A fiber Bragg grating (FBG) has grating period Λ along its whole length, the grating period of a CFBG varies along its length. Linearly chirped FBGs exhibit linear group delay property that used for performing dispersion compensation. The spectrum of the CFBG is more boarded compared than that of a FBG. In a CFBG, different wavelength components of an incident light are reflected at different positions along the grating which fulfill the local grating period: $\lambda_B(z) = 2 \cdot n_{eff} \cdot \Lambda(z)$ where, $\lambda_B(z)$ is the reflection wavelength (Bragg wavelength) at a position z along the gratings, n_{eff} is the effective refractive index of the core mode, and $\Lambda(z)$ is the local grating period (Fig. E.16).

Chirped fiber Bragg grating (CFBG) structure and position is defined by the distributive reflection characteristic. The shorter wavelength component λ_1 (blue) is reflected earlier than the longer ones $\lambda_2, \lambda_3, \ldots, \lambda_k, \ldots, \lambda_{N-1}$ and λ_N (red), where, $\lambda_1 < \lambda_2 < \lambda_3 < \cdots < \lambda_k < \cdots < \lambda_N$ for $1 \leq k \leq N$. The distributive reflection properties of the CFBG provide a broad reflection spectrum and the linear group delay. The linear group delay gives variable time delay in our chirped fiber Bragg grating. The CFBG-based delay line is used for optical coherence tomography (OCT)

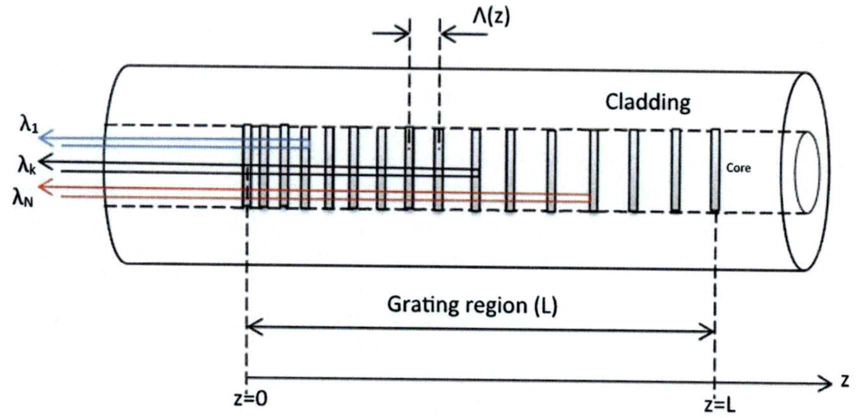

Fig. E.16 Linearly chirped fiber Bragg grating (CFBG) system

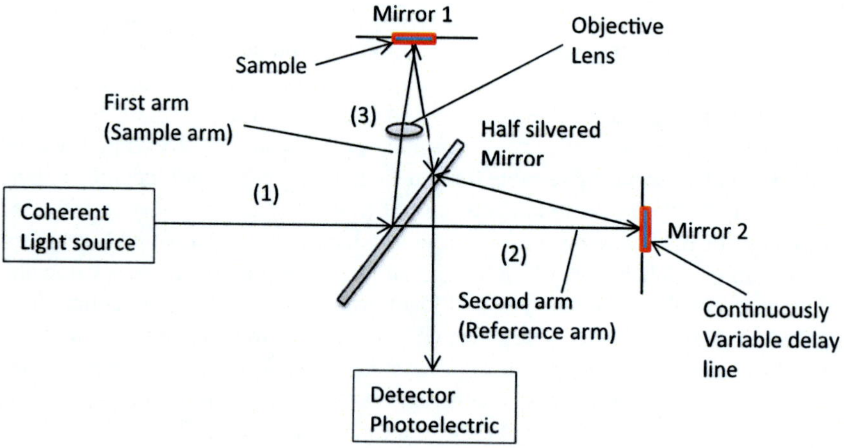

Fig. E.17 Michelson interferometer OCT system with half silvered mirror

applications. Michelson interferometer uses optical coherence tomography (OCT) system. Optical coherence tomography (OCT) is a non-invasive imaging check which uses light waves to take cross-section pictures of specific object. It is an emerging technology for performing high-resolution cross-section imaging. OCT is analogous to ultra sound imaging, except that it used light instead of sound. Michelson interferometer is an optical coherence tomography (OCT) system with a sample in one arm and a continuously variable delay line in the other arm (Fig. E.17). The Michelson interferometer is a common configuration for optical interferometry [21, 22]. It uses a beam splitter; a light source is split into two arms. Each of those light beams is reflected back toward the beam splitter which combines their amplitudes using the superposition principle. The result is some interference pattern which inspected by photoelectric detector. The Michelson interferometer produces interference fringes by splitting a beam of monochromatic light; one beam strikes a fixed mirror and the other a movable mirror. When the reflected beams are brought back together, an interference pattern results are received. Half silver mirror or beam splitter is an optical device that splits a beam of light in two (In practice, the reflective layer absorbs some light). A half silvered mirror is a sheet of glass or plastic with a transparently thin coating of metal (usually aluminum deposited from aluminum vapor).

Incident light (1), 50% transmitter (2), and 50% reflected light (3). The thickness of the deposit is controlled so half part of the light is transmitted, and the remainder is reflected. Other popular beam splitters are a cube of two triangular glass prisms, dichroic mirrored prism, etc. An objective lens is used to collimate and focus the beam on the sample (first arm) in our system. A simple collimator is used in the reference arm. Rapid scanning optical delay line (RSODL) has different bulk optic elements in both arms then undesirable artifacts appear in the OCT image unless the dispersion imbalance between the arms is canceled. The time domain interferometric reflectometer with a rapidly scanning optical delay line (RSODL) is based on a non-parallel diffracting grating (DG) pair. The light wave in the reference channel is focused on a galvo mirror in a sub-mm static spot, which allows implementation of fast micro-electro-mechanical system scan optics. The diffraction grating (DG) pair can be operated as a non-dispersive element that provides dynamic group delay of a reference light wave. The diffraction grating (DG) pair system is capable of tuning the RSODL dispersion from negative to positive values. The system dispersion imbalance is compensated by extra optical components which deployed in the reference arm or the OCT image is post-processed to remove the artifacts. We use a pair CFBGs cascaded reversely for the dispersion cancellation scheme and output light with constant group delay for every wavelength (Fig. E.17). Regarding the chirping, different wavelength components $\lambda_1, \ldots, \lambda_k, \ldots, \lambda_N$ for $1 \leq k \leq N$ of an incident beam are reflected at different locations $z_1, z_2, \ldots, z_k, \ldots, z_N$ in both CFBGs which measured from the shorter period side. The λ_1 wavelength component travels a distance z_1 in CFBG1 and a distance $L - z_1$ in CFBG2. The λ_k, $(1 \leq k \leq N)$ wavelength component travel a distance z_k in CFBG1 and a distance $L - z_k$ in CFBG2. Thus, the total traveling distance in both gratings becomes L, where both CFBGs are identical and have the same length L. The same thing happened with the wavelength component λ_k, $(1 \leq k \leq N)$. Therefore, for any wavelength component the typical optical path length (OPL) in both gratings is $2 \cdot n_{eff} \cdot L$; where round trip factor is (2). An optical path length (OPL) or optical distance (OD) is the product of the geometric length of the path followed by light through a gratings system, and the index of refraction of the medium through which it propagates. The delay occurred by traveling through the fiber between the gratings is neglected. All wavelength components have the same OPLs within the gratings. One of the CFBG is stretched and elongated then the reflection points in the gratings are shifted increasing the delay. The amount of the shift is dependent on the chirping ratio and is much larger than the stain applied on the fiber. The CFBG is stretched uniformly and the chirping ratio and the dispersive properties are not affected. We enable wavelength independent variable optical delay. The wavelength independent optical delay property is preserved

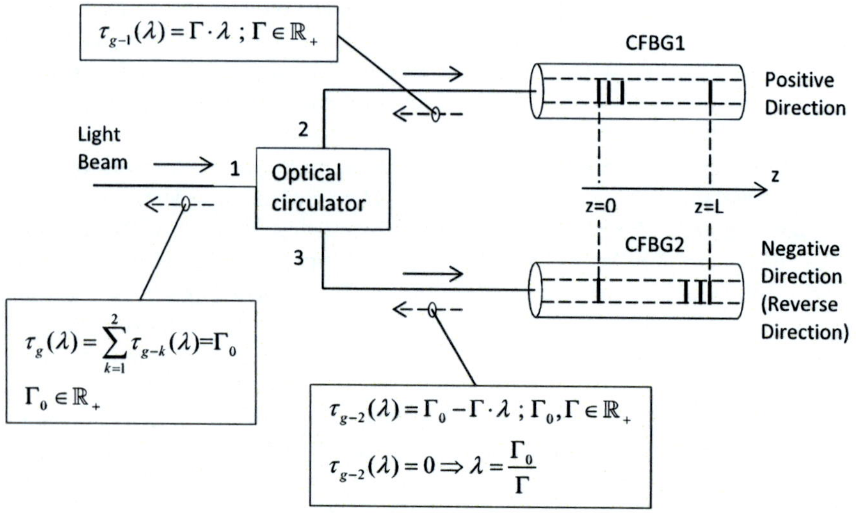

Fig. E.18 A pair CFBGs cascaded reversely for the dispersion cancellation scheme

under stain and the gain in reflection position is happened. The enhance shift in the reflection point is the most key feature for our system. We use the following elements for our dispersion cancellation system: two CFBGs which are cascaded reversely for the dispersion cancellation and optical (three ports) circulator [23] (Fig. E.18).

The optical circulator is used to direct the optical signal from one port to another port and in one direction only. It prevents the optical signal from propagating in unintended direction. In a 3-port optical circulator an optical signal is transmitted from port 1 to port 2, another optical signal, is transmitted from port 2 to port 3 and, finally, a third optical signal is transmitted from port 3 to port 1. In practice, one or two ports are used as inputs and the third port is used as the output. The light beam enters port 1 it is emitted from port 2, but if some of the emitted light is reflected back to the circulator, it does not come out of port 1 but instead exists from port 3. The fiber optic circulator is used to separate optical signals that travel in opposite directions in an optical fiber. Optical circulator is non-reciprocal optic, which means that changes in the properties of light passing through the device are not reversed when the light passes through in the opposite direction. Operation analysis of the three port optical circulator is as follow: The configuration of a three-port optical circulator is described below (Fig. E.19).

The three-port optical circulator are a beam splitting polarizer (A), a reflection prism (B), two birefringent crystal (C), a Faraday rotator (D), and a half-wave plate (E). Our dispersion cancellation system's dispersion curves show wavelength-independent of the delay line. If the CFBG is linearly chirped then it has a linear group delay. When a light is launched from the opposite direction of the identical CFBG, the slope of the dispersion curve remains the same ($\tau_{g-1}(\lambda) = \Gamma \cdot \lambda; \; \Gamma \in \mathbb{R}_+$) but the sign change and we get ($\tau_{g-2}(\lambda) \propto -\Gamma \cdot \lambda; \; \Gamma \in \mathbb{R}_+$). Thus, the result is when

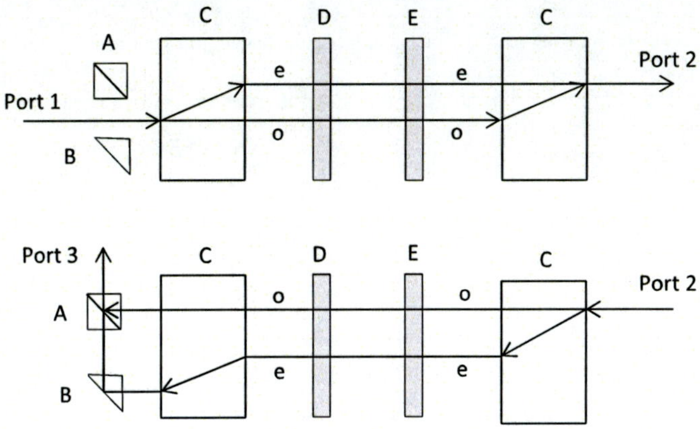

Fig E.19 Configuration of a three-port optical circulator

the light is launched into two identical CFBGs cascaded in the opposite direction, the dispersion associated with the CFBGs is canceled with each other and the outcome is flat $\left(\tau_g(\lambda) = \sum_{k=1}^{2} \tau_{g-k}(\lambda) = \Gamma_0;\ \Gamma_0 \in \mathbb{R}_+\right)$ when a CFBG is stretched, the related group delay curve is shifted to longer wavelength while the slop remain the same, not changed. Then stretching a grating does not affect the dispersion cancellation. When a gratings is stretched, the flat group delay curve moves up proportional to the amplified delay. The delay is a function of grating parameters. Identical gratings (CFBG1 and CFBG2) help to get a perfect dispersion cancellation. There are some methods to get a large optical delay from a small physical displacement in optical delay line (ODL). We define the following parameters for the pair of CFBGs dispersion cancelation system (Table E.1).

The wavelength of the light beam which reflected at a position z along a CFBG has position dependent grating $\Lambda(z) = \Lambda_0 + \beta \cdot z$ is $\lambda_B(z) = 2 \cdot n_{eff} \cdot \Lambda(z) = 2 \cdot n_{eff} \cdot (\Lambda_0 + \beta \cdot z)$. The CFBG grating pitch elongation is $\Lambda'(z') = \alpha \cdot \Lambda_0 + \beta \cdot z'$. We know $\beta = (\Lambda(z) - \Lambda_0) \cdot \frac{1}{z}$ then $\Lambda'(z') = \Lambda_0 \cdot (\alpha - \frac{z'}{z}) + \Lambda(z) \cdot \frac{z'}{z}; \alpha = \frac{L+a}{L};\ \alpha = 1 + \varepsilon$, the strain ε applied on the fiber is defined as $\frac{a}{L}$, $\left(\varepsilon = \frac{a}{L}\right)$. The strain on the fiber reduces the effective refractive index by a factor $(1 - p_e \cdot \varepsilon)$ due to elasto-optic coefficient p_e then $n'_{eff} = n_{eff} \cdot (1 - p_e \cdot \varepsilon)$. The Elasto optic effect in optical fiber is the change in the refractive index of an optical fiber caused by variation in the length of the fiber core in response to mechanical stress. Thus the pitch and effective index are modified and we get the new Bragg wavelength of a strained CFBG, $\lambda_{B'}(z') = 2 \cdot n'_{eff} \cdot \Lambda'(z') = 2 \cdot n_{eff} \cdot (1 - p_e \cdot \varepsilon) \cdot (\alpha \cdot \Lambda_0 + \beta \cdot z')$. When a wavelength component λ of light is reflected at a position z in an un-stretched grating, it will be reflected at a distance z' when the grating is stretched. The expression for the system gratings optical path length is $OPL = n'_{eff} \cdot z' = n_{eff} \cdot (1 - p_e \cdot \varepsilon) \cdot \frac{1}{\beta} \cdot \left[\left(\frac{\lambda_{B'}(z')}{2 \cdot n_{eff} \cdot (1 - p_e \cdot \varepsilon)}\right) - \alpha \cdot \Lambda_0\right]$ then $OPL = n'_{eff} \cdot z' = n_{eff} \cdot (1 - p_e \cdot \varepsilon) \cdot z'$. The Bragg wavelength $\lambda_{B'}(z') = \lambda_B(z)$ then $OPL = n_{eff} \cdot (1 - p_e \cdot \varepsilon) \cdot \frac{1}{\beta} \cdot \left[\left(\frac{\Lambda_0 + \beta \cdot z}{1 - p_e \cdot \varepsilon}\right) - (1 + \varepsilon) \cdot \Lambda_0\right]$ and we can express it

Table E.1 Parameters for the pair of CFBGs dispersion cancelation system

Dispersion cancelation system parameter	Description
Λ_0	A minimum (maximum) pitch for a positively (negatively) chirped CFBG
λ_0	A minimum Bragg wavelength of the CFBG
$\beta\left[\frac{\text{nm}}{\text{mm}}\right]$	Positive (negative) chirping ratio
L	Total traveling distance in both gratings
a	Elongation length
$\alpha;\ \alpha = 1 + \frac{a}{L}$	Elongation ratio
$\varepsilon;\ \varepsilon = \frac{a}{L}$	Fiber optic strain
p_e	Elasto-optic coefficient
n_{eff}	Fiber effective refractive index
n'_{eff}	Reduced fiber effective refractive index
$\Lambda(z)$	Position dependent grating pitch of the light beam reflected at a position z along the CFBG
z	Position z value along the CFBG
$\lambda_B(z)$	Wavelength of the light beam reflected at a position z along the CFBG
$\Lambda'(z')$	Elongated grating pitch when stretch the CFBG of a length L by an amount of an elongation length a
z'	New position z' value when stretch the CFBG of a length L an amount of an elongation length a
$\lambda_{B'}(z')$	Modified Bragg wavelength of the stained CFBG
γ	Delayed amplified factor, caused by stretching a CFBG by a length a
OPL	System gratings optical path length
Λ_m	Maximum grating period of the CFBG
λ_m	Maximum Bragg wavelength of the CFBG
$\Delta\lambda,\ \Delta\lambda = \lambda_m - \lambda_0$	Spectral bandwidth of the grating

(stretched OPL) as $OPL = n'_{eff} \cdot z' = n_{eff} \cdot z - \frac{n_{eff} \cdot \Lambda_0}{\beta} \cdot (1 - p_e) \cdot \varepsilon + O(\varepsilon^2)$. We can get the optical path length difference (OPD, $OPD = n'_{eff} \cdot z' - n_{eff} \cdot z$) by stretching the CFBG by the difference between un-stretched OPL $n_{eff} \cdot z$ and the stretched OPL, $OPL \approx n_{eff} \cdot z - \frac{n_{eff} \cdot \Lambda_0}{\beta} \cdot (1 - p_e) \cdot \varepsilon$ and neglecting the second order $O(\varepsilon^2)$, the round trip optical path length difference (OPD) is $2 \cdot OPD \approx -2 \cdot \frac{n_{eff} \cdot \Lambda_0}{\beta} \cdot (1 - p_e) \cdot \varepsilon$. The chirping ratio $\beta\left[\frac{\text{nm}}{\text{mm}}\right]$ is expressed with grating parameters: $\beta = \frac{\Lambda_m - \Lambda_0}{L} = \frac{\lambda_m - \lambda_0}{2 \cdot n_{eff} \cdot L} = \frac{\Delta\lambda}{2 \cdot n_{eff} \cdot L}$; $\Delta\lambda = \lambda_m - \lambda_0$, where Λ_m and λ_m are the maximum grating period and maximum Bragg wavelength of the CFBG, respectively. The spectral bandwidth of the grating is $\Delta\lambda = \lambda_m - \lambda_0$. We can express the round trip optical path length difference (OPD) as $2 \cdot OPD \approx -2 \cdot n_{eff} \cdot \frac{\lambda_0}{\Delta\lambda} \cdot (1 - p_e) \cdot a$. The equation states

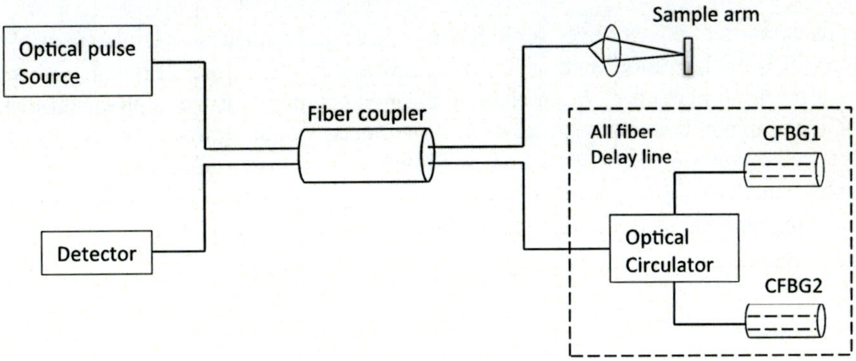

Fig. E.20 Experimental system of all-fiber optical delay line (AFODL)

which stretching the CFBG by a length a is the delayed amplified factor, caused by stretching a CFBG by a length a, $\gamma \approx n_{eff} \cdot (1 - p_e) \cdot \frac{\lambda_0}{\Delta\lambda}$. When the fiber is stretched by a length a, the induced optical delay is increased by a factor γ. For $\gamma \gg 1$ then $0 < \Delta\lambda \ll 1$ and the stretching is very small, large OPD. When we applied stain, CFBG is reversed (CFBG2 instead of CFBG1) the sign becomes positive in OPD expression. The amplification factor γ is depended on the grating parameters, we get the maximum delay. A big value of amplified delay is accepted by CFBG which his bandwidth is small. We get a constant amplification factor γ for the amplified optical delay, and the factor γ is linearly proportional to the elongation length a [23]. The experimental system of all-fiber optical delay line (AFODL) uses optical coherence tomography (OCT) images since the resolution of OCT is higher at lower wavelength (Fig E.20). The gratings demonstrate the dispersion properties of the delay line. The CFBGs units are characterized by a high reflectivity (>90%) and wide spectral bandwidth. The schematic of the OCT system includes fiber coupler unit which divides and directed the source power into the sample and reference arms and re-combine them after being reflected from the respective arms. The interference signal is obtained by recombining the signals from the sample and reference arm is detected with a photo detector, and we get the details about the sample. The optical delay is measured by stretching the grating, and the fiber delay line is used as a sample in the fiber based OCT system with conventional flat mirror mounted on the translational stage as the delay element in the reference arm. Stretching of the CFBG is done by mounting the grating on the translational stage with fiber-chucks. The translation stage on which mirror is mounted provides a variable delay. The images of the samples are obtained using the fiber grating based fiber delay line in the OCT system.

Fiber optic couplers are used to split or combine the light contained in optical fibers. It couples light from one or several input fibers to one or several output fibers. Light from an input fiber is appeared at one or more outputs, with a power distribution which dependent on the wavelength and polarization. It is possible also to couple light from free space into a fiber. The fibers are thermally tapered and fuse that their cores come into contact over some length of a few centimeters. It is also possible to do it by polarization maintaining fibers. Couplers which use side-polish fibers give access to fiber core. When using multi-mode pump fibers we implement fiber-optic pump combiners and pump-signal combiners. Planar light wave circuits with branching waveguides are function as couplers. If we use a fiber which supports only a single mode per polarization direction for specific wavelength (single-mode), coupling performances are restricted. The restrictions are:

1. Two light beam inputs need to be mutually coherent for good coupling.
2. Excess loss when combining two or more inputs of the same optical frequency into a single polarization output.

Dichroic couplers can combine two inputs of different wavelengths into one output without losses. Wavelength-sensitive couplers are used in wavelength division multiplexing (WDM) units and combine several input channels with different wavelengths, or separate channels. When using multimode fibers, the powers of two mutually incoherent beams are combined without power loss. Fiber optic couplers function in a limited range of wavelength (specific bandwidth), since the coupling strength is dependent on wavelength. The coupling occurs over a certain length. Broadband fiber couplers are used in fiber interferometers (optical coherence tomography, OCT). Multimode fiber couplers are also used for combining the radiation of several laser diodes and sending them into inner cladding of double-clad fiber.

We can use an ultrafast optical delay line which based on reversible frequency conversion by use of a time-prism pair. By using electro-optic phase modulators, we provide the time-prism phase profile. We use an optical delay lines for applications such as optical coherence tomography (OCT), optical metrology, interferometry, and telecommunications. For high scanning rates we use acousto-optic modulators or deflectors. A grating-based dispersive optics and scanning mirrors are used to obtain high scanning rates. By using a time-prism pair we get an all-fiber, programmable, ultrafast optical delay line. An aberration correction technique is to eliminate nonlinearity in the phase profile of each time prism is done and improve the performance of the system. A time prism converts the frequency of an incoming electric field. Dispersion following the prism results in a displacement of the electric field in time. Another time prism is used to convert the frequency back to its original value, with the resultant field being the original electric field delayed in time. We can describe an analogous operation to a spatial prism pair, which changes the $\vec{k}$ vector on the incoming field, resulting in a displacement in space ($\vec{k} \to \vec{k} + \vec{k}'$). The second prism converts the $\vec{k}_{new} = \vec{k} + \vec{k}'$ vector back its original value $\vec{k}_{new} = \vec{k} + \vec{k}' \to \vec{k}$. The result field is the original electric field translated in space. After a time prism introduces a delay $\Delta\tau = \Delta\lambda \cdot D = \pi \cdot \frac{\lambda^2}{2 \cdot c} \cdot \frac{V}{V_\pi} \cdot f_m \cdot D$, where, f_m is the modulation frequency, V is the peak-to-peak drive voltage, V_π is the voltage required to obtain a π phase shift, D is the fiber dispersion, $D\left[\frac{\text{ps}}{\text{nm}}\right]$, c is the speed of light in vacuum, λ is the center wavelength of the incident pulse train, and $\Delta\lambda$ is the corresponding wavelength shift that is due to phase modulation. The $\Delta\lambda$ wavelength shift is $\Delta\lambda = \pi \cdot \frac{\lambda^2}{2 \cdot c} \cdot \frac{V}{V_\pi} \cdot f_m$; $\Delta\lambda = \left(\frac{\lambda^2}{2 \cdot \pi \cdot c}\right) \cdot \Delta\omega$ then $\Delta\omega = \pi^2 \cdot \left(\frac{V}{V_\pi}\right) \cdot f_m$, where

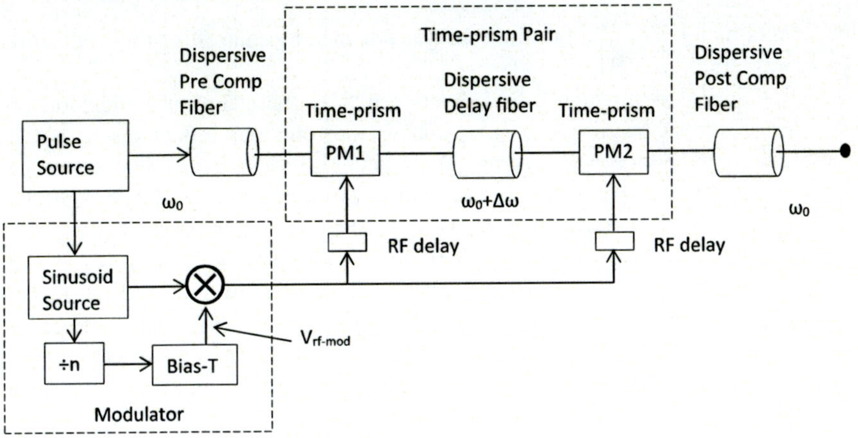

Fig. E.21 Time-prism pair with RF modulator system

$\Delta\omega$ is the corresponding radial frequency shift that is due to phase modulation. A time prism imposes a linear phase in time onto an incoming electric field similar to spatial prism does in space. Drive voltage of the time prism is analogous to the apex angle of a spatial prism. When the phase slope is changed the frequency $\left(k = \frac{2\cdot\pi}{\lambda}\right)$ is changed and hence the delay (spatial transition), $\Delta\lambda \propto \lambda^2$; $\Delta\lambda \propto \left(\frac{2\cdot\pi}{k}\right)^2$, where k is the spatial angular frequency (wavenumber) of the wave, describing how many oscillations it completes per unit of space. Parameter k is the magnitude of the wave vector. To convert the frequency of the optical field to its original value, we need to impose a phase $-\Delta\omega \cdot t$, which accomplished by phase modulator driven with the same sinusoid but π out of phase. The experimental system architecture which includes Time-prism pair, Dispersive Pre comp and Post comp fibers, RF modulator and RF delays units is described below (Fig. E.21). PM1 (frequency upshifter) is phase modulator No.1 and PM2 (frequency down shifter) is phase modulator No.2.

The pulsed source is generated by pulse-carving cw light and the time lens compensation. Pulses are upshifted in frequency by the first time prism (PM1). A time delay is presented by dispersive delay (dispersive compensation fiber). The second time prism (PM2) is a phase modulator driven with the same sinusoid but π out of phase and used to down shift the frequencies back to their original values. The system includes dispersion pre compensation and post compensation to eliminate the pulse broadening effects caused by the dispersive delay and to achieve symmetrical operations for the frequency shifters. We get pulse delay without the diverse effect of the pulse distortion. There is a drive voltage to the phase modulators which vary the frequency shift and therefore the delay. The drive voltage to the phase modulators is adjusted through a broadband RF modulator. The RF modulator controls the drive voltage to the optical phase modulators then the time delay is $\Delta\tau = \Delta\lambda \cdot D \propto V \propto V_{rf-\text{mod}}$. Where, $V_{rf-\text{mod}}$ is the modulation input into the RF modulator. The ultrafast time delay is obtained by controlling the modulation waveform, $V_{rf-\text{mod}}$ [24].

A long optical path which is folded between two spherical or aspherical mirrors provides an output beam which is separated from previous reflections with high number of passes between the mirrors. This system is used as dispersion less optical delay line (ODL) for use in filtering or storage of information modulated onto the light beam. A small perturbing mirror is inserted to give a series of offset ellipses, or on or both of mirrors with astigmatic to give a Lissajous pattern of spots on each mirror. A Lissajous curve or Lissajous pattern is the spot of an optical system of parametric equations: $x = A \cdot \sin(\omega_A \cdot t + \phi_A)$; $y = B \cdot \sin(\omega_B \cdot t)$ which describes complex harmonic motion. The appearance of the spot is highly sensitive to the ratio $\frac{\omega_A}{\omega_B}$. If $\omega_A = \omega_B$ then spot has an ellipse pattern. If $\omega_A = \omega_B$ and $\phi_A = \frac{\pi}{2}$ radians then the spot has a circles pattern. If $\omega_A = \omega_B$ and $\phi_A = 0$ then the spot is a lines. If $\omega_A = \omega_B$ and $\phi_A = \frac{\pi}{4}$ then the pattern is parabola. Other $\frac{\omega_A}{\omega_B}$ ratios produce complicated spot curves. If the $\frac{\omega_A}{\omega_B}$ is rational number these spot curves are very close in their shape. The ratio $\frac{\omega_A}{\omega_B}$ determines the number of "lobes" of the spot pattern. We can create long optical path for optical delay line (ODL) by reflecting the beam of light repeatedly between two spherical mirrors without interference between adjacent beams. The converging power of the spherical mirrors reduces the diffraction losses. We use astigmatic mirrors to cause the spot that follows a Lissajous pattern. It is a small perturbing mirror which cause the ellipse on each spherical mirror to shift and follow a set of new ellipses each different from the previous one. We get a mirror area with clear separation of the input and output spots. Astigmatic mirror is a mirror where rays that propagate in two perpendicular planes have different foci. If we use an astigmatic mirror to form an image, the vertical and horizontal lines are sharp focus at two different distances. There are two distinct of astigmatism, the first is a third-order aberration and the second is when the optical system is not symmetric about the optical axis. A system of off-axis path creates a long optical paths folded between two spherical mirrors. The light is injected through one mirror at a specific angle and position, and is reflected back and forth between the two mirrors. It traces out elliptical paths on the mirrors, and after many reflections the optical beam leaves the cavity. If the ray wave front radius and spot size correspond to the resonant modes between the mirrors, the diameter of the ray stays the same after many reflections and the diffraction losses is very small. The maximum delay is limited by the spacing of the mirrors and number of spots of finite size that be spaced around an ellipse within the diameter of the mirror without overlapping spots. The applications to off-axis cavity delay line can be the measurement of optic all loss (need long path lengths and many travels for interface measurements) and communication fields (need long time delays and multiple output taps).

Optical communication systems and phase array antennas need optical circuit which provides a time delay signal modulated on optical carriers. It can be constructed from low-cost CMOS compatible LPCVD planar waveguide which gives continuously tunable optical delay line. Photonics devices can provide a broadband tunable RF phase shifter. It is possible to use an optical ring-resonator all-pass filter and get variable delays. The ideal optical ring-resonator filter (all-pass filter) has large group

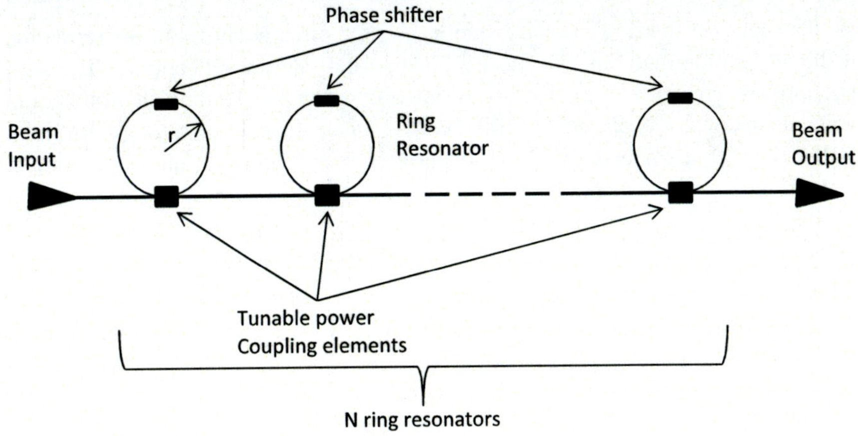

Fig. E.22 Schematic of N-stage ring resonator filter

delay which the modulated signal is delayed by the group delay at resonance and a unity magnitude response. We can tune the delay by the amount of power coupling the ring. Typical N-rings; $N \geq 1$; $N \in \mathbb{N}$ is connected to the waveguide by Mach-zehnder interferometer (MZI) based on variable couplers. Heaters are located on the ring for tuning the resonance frequency, and on one of the branches of MZI ring coupler is for adjusting the power coupling to the ring. The Mach-zehnder interferometer is a device which determined the relative phase shift variation between two collimated beams derived by splitting light from a single source. It measures the phase shifts between the two beams caused by a sample or change in length of one of the path. Mach-zehnder interferometers are used in electro-optic modulators, electronic devices for fiber-optic communication [25]. It incorporated in monolithic integrated circuits with high-bandwidth electro-optic amplitude and phase responses over a multiple-gigahertz frequency range.

An optical ring resonator is a set of waveguides which at least one is a closed loop coupled to light input and output. When light of the resonator wavelength is passed through the loop from input waveguide, the intensity is builds up over multiple round-trips due to constructive interference and is output to the bus waveguide which is detector waveguide. The optical ring resonator functions as a filter due to the fact that only a select few wavelengths are at resonance within the loop. Two or more ring waveguides are coupled to each other to form add/drop optical filter (Fig. E.22).

The light traveling through the waveguides in an optical ring resonator remain within the waveguides due to the ray optics phenomenon, total internal reflection (TIR). TIR optical phenomenon is when a ray of light strikes the boundary of a medium and fails to refract through the boundary. When total internal reflection (TIR) happened no light is able to pass through the boundary relative to the incident light ray. In optical ring resonator, the total internal reflection conditions must fulfil the light traveling through the waveguides which not be allowed to escape. The optical path length difference (OPD) of a single-ring resonator is $OPD = 2 \cdot \pi \cdot r \cdot n_{eff}$; where,

r is the radius of the ring resonator and n_{eff} is the effective index of the refraction of the waveguide material. We must fulfil the total internal reflection (TIR) when n_{eff} must be greater than the index of refraction of the surrounding environment in which the resonator is placed. The following resonance condition must be fulfilling: $OPD = m \cdot \lambda_m$, where λ_m is the resonant wavelength and m is the mode number of the ring resonator, then $2 \cdot \pi \cdot r \cdot n_{eff} = m \cdot \lambda_m \Rightarrow \lambda_m = 2 \cdot \pi \cdot r \cdot \frac{n_{eff}}{m}$. The result is that in order for light to interfere constructively inside the ring resonator, the circumference of the ring $(2 \cdot \pi \cdot r)$ must be an integer multiple of the wavelength of the light $\left(2 \cdot \pi \cdot r = \frac{m}{n_{eff}} \cdot \lambda_m\right)$. The mode number ($m$) must be a positive integer for resonant to happened. The incident light contains multiple wavelengths, but only the resonant wavelengths able to pass through the ring resonator. The quality factor of an optical ring resonator is $Q = m \cdot F = m \cdot \frac{v_f}{\delta v}$; $F = \frac{v_f}{\delta v}$; where, F is the finesse of the ring resonator, v_f is the free spectral range, and δv is the full-width half-max of the transmission spectra. We use the quality factor (Q) to determine the spectral range of the resonance condition for ring resonator. We can estimate the amount of losses in the resonator by the quality factor (Q), large losses are accepted for low Q factor. The free spectral rang (v_f) is the spacing in optical frequency or wavelength between two successive reflected or transmitted optical intensity maxima or minima of the transmission spectra. Full width of half maximum (FWHM) is an expression of the extent of function which is the difference between the two extreme values of the independent variable at which the dependent variable is equal to half of its maximum value. The group delay (τ_g) of one round trip in the ring is $\tau_g = \frac{1}{f_{FSR}} = \frac{2 \cdot \pi \cdot r \cdot n_g}{c}$, where $2 \cdot \pi \cdot r$ is the circumference, n_g is the group index of the waveguide, and c is the speed of light in vacuum. The normalized group delay τ_{g-n} which is the sum of the individual normalized group delays τ_{g-ni} for the lossless waveguide is given by $\tau_{g-n} = \sum_{i=1}^{N} \tau_{g-ni}(\Omega) = -\sum_{i=1}^{N} \frac{d\Phi_i(\Omega)}{d\Omega}$; $\tau_{g-n} = \sum_{i=1}^{N} \frac{1-\Gamma_i^2}{1+\Gamma_i^2 - 2 \cdot \Gamma_i \cdot \cos(\Omega + \phi_i)}$, where $\Omega = \omega \cdot \tau_g$ is the normalized angular frequency; $\Phi_i(\Omega)$ is the phase response of each individual filter; $\Gamma_i = \sqrt{1-\kappa_i}$ is the bar port transfer of the waveguide to ring coupler; κ_i is its power coupling coefficient, and ϕ_i is an additional phase of the ring that set by the ring heater. The normalized group delay (τ_{g-n}) is multiplied by (τ_g) to obtain the absolute delay. We get the broadband group delay response by adjusting the coupling coefficient κ_i and the extra phase shift ϕ_i of each single stage. By using ring resonator s we get compact long delays, the delay is fixed and is limited in bandwidth. The parameters which limit the bandwidth are high finesse, high order group delay dispersion. If the delay line operation is away from the resonance peak then we reduce data distortion and increase bandwidth in resonance. This technique is achieved by using stimulated Brillouin scattering (SBS) and thermally tuned silica based micro rings. Brillouin scattering is an effect caused by the $\chi^{(3)}$ nonlinearity of a medium, part of the nonlinearity which is related to acoustic phonons. An incident photon is converted into a scattered photon of lower energy, and propagating in the backward direction, and a phonon. The coupling of optical fields an acoustic waves happened by electrostriction. The frequency of the reflected beam is lower than that of the incident beam; the frequency difference

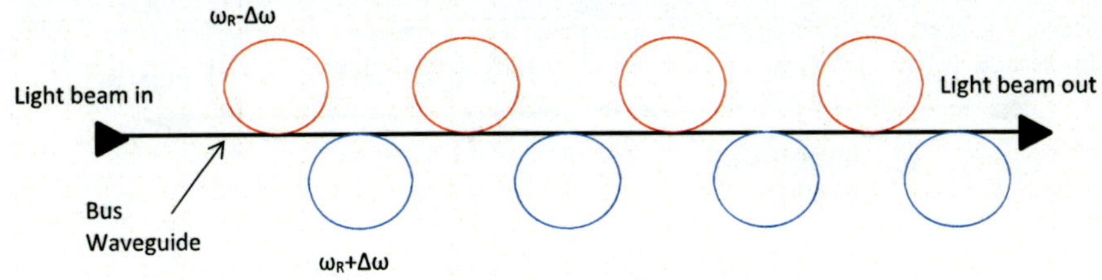

Fig. E.23 Tunable delay line structure with overall central frequency ω_r

ν_B corresponds to the frequency of limited phonons. It is Brillouin frequency shift and we set it by a phase matching requirement. In optical fiber, Brillouin scattering occurs only in backward direction and the Brillouin frequency shift is dependent on the material composition, temperature and pressure of the medium. Stimulated Brillouin scattering (BSB) is the frequency uncounted when narrow-band optical-signals are amplified in a fiber amplifier or propagated through a passive fiber. The silica is based on large radius rings (few hundred microns up to a millimeter). The delay system uses compact silicon-based micro ring resonators for enabling high bandwidth and continuous tenability based on the balanced side-coupled integrated spaced sequence of resonators (SCISSOR). The balanced SCISSOR consists of two sets of identical racetrack shaped micro ring resonators, which is opposite sides of the bus waveguides. The racetrack micro rings are characterized by a bend radius and the straight sections, for a specific overall circumference. The device is planned to work in the Transverse Electric (TE) polarization. We get the delay spectra if all the rings are at the same resonant frequency. This delay has a narrow bandwidth and we can expand the band width if we shift the resonant frequencies of the resonators by a small amount $\Delta\omega$ from the central resonance frequency ω_r; half of the rings are blue shifted and the other half are red shifted. The outcome is reducing the third and higher orders of group delay dispersion, wide spectral regions with flat and ripple-free group delay. The resonances of the rings are changed symmetrically about the central resonant frequency ω_r. The operating frequency of the overall structure remains constant when the group delay is varied. The spectral width of the structure broadens, increasing its bandwidth, as the resonance frequency detuning from ω_r is increased. The tunable delay line structure with overall central frequency ω_r is described below (Fig. E.23). the resonances of the red upper rings are red shifted by increasing the refractive index of the ring. The resonances of the blue lower rings are blue shifted by decreasing the refractive index of the rings [26]. The relationship between the delay and resonance detuning is given by $\Delta T(\omega) = \sum_{n=1}^{N} \frac{k^2 \cdot \tau}{1+\rho^2-2\cdot\cos(\omega-\omega_n)}$; $\rho = \sqrt{1-\kappa^2}$, where τ is the round trip time, and κ is the coupling coefficient.

An optical tunable delay line is based on a side-coupled integrated spaced sequence of resonator (SCISSOR) structure; the pairs of resonances are tuned in opposite directions around the signal frequency. The balanced SCISSOR fulfils the deleterious effects of group delay dispersion and provides both wide band-width and continuously tunable long delay without distortion. Tunable optical delay

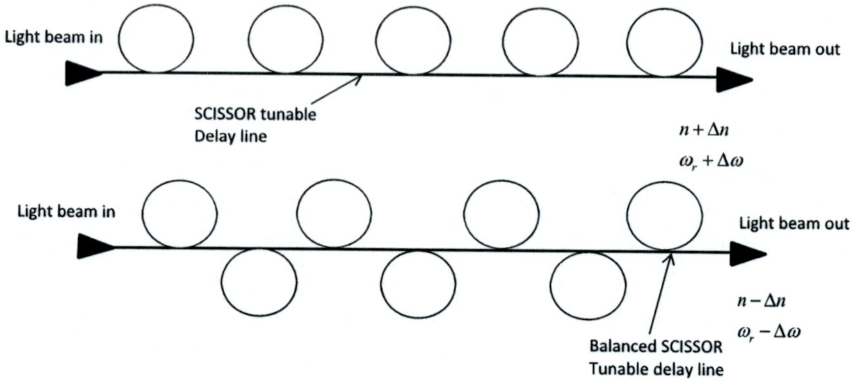

Fig. E.24 SCISSOR tunable delay line and balanced SCISSOR tunable delay line architectures

lines are applicable in the areas of optical communications, data processing, and microwave photonics. We can get a large tunable delay which based on the existence of a sharp resonance. Implementation of compensate for group delay dispersion (GDD) of coupled resonators and get attain tunable optical delay lines which characterized by a wide bandwidth, a long tunable delay, and a low distortion. Delay lines are based on coupled resonators and exist in two basic designs: Coupled resonator optical waveguides (CROWs) and side-coupled integrated spaced sequence of resonators (SCISSORs). CROW delay lines exhibits lower delays compare to SCISSOR structures. The phase shift in a SCISSOR structure with N rings is $\tan\left(\frac{\Phi}{N}\right) = \frac{\kappa^2\cdot\sin(\omega-\omega_r)\cdot\tau}{(1+\rho^2)\cdot\cos(\omega-\omega_r)\cdot\tau-2\cdot\rho}$, where, τ is a round trip time, $\omega_r = 2\cdot\pi\cdot m\cdot\frac{1}{\tau}$ is the resonance angular frequency, κ is the coupling coefficient, and $\rho = \sqrt{1-\kappa^2}$. The group or time delay in a SCISSOR structure is

$$\begin{aligned} T_d(\omega) &= T_{d0} + N\cdot\sum_{i=1}^{\infty}\left[\beta_{1+2\cdot i}\cdot(-1)^{i+2}\cdot(\omega-\omega_r)^{2\cdot i}\right] \\ &= T_{d0} - N\cdot\beta_3\cdot(\omega-\omega_r)^2 + N\cdot\beta_5\cdot(\omega-\omega_r)^4 - \ldots \end{aligned}$$

where, $T_{d0} = \frac{N\cdot\tau\cdot(1+\rho)}{(1-\rho)}$ is the on-resonance delay, $\beta_3 = \frac{\tau^3\cdot\rho\cdot(1+\rho)}{(1-\rho)^3}$ and $\beta_5 = \frac{\tau^5\cdot\rho\cdot(1+\rho)\cdot(1+10\cdot\rho+\rho^2)}{12\cdot(1-\rho)^5}$ are the terms for the higher-order GDD per-ring. The SCISSOR tunable delay line and Balanced SCISSOR tunable delay line architectures are presented in the next figure (Fig. E.24).

We can get tunable delay from SCISSOR structure by varying the resonant frequency ω_r. By shifting the resonant frequency up and down by the amount of $\Delta\omega \sim (\frac{\Delta n}{n})\cdot\omega$, where, Δn is the change in the ring's effective index; causes a change in the delay time ΔT_{d0}, $\Delta T_{d0} = N\cdot\beta_3\cdot\Delta\omega^2$. An off-resonance position of the signal, the GDD becomes a linear function of the bandwidth, $\delta T_d(B) = 2\cdot N\cdot\beta_3\cdot\Delta\omega\cdot B$, and a restriction on the attainable tunable DBP, $\Delta T_{d0}\cdot B \le \frac{\Delta\omega}{2\cdot B} \sim (\frac{\Delta n}{n})\cdot(\frac{\omega}{2\cdot B})$. If we tune the coupling coefficient κ then the delay varied. We can get tenability and at the same time to expand the bandwidth of a SCISSOR structure by taking the advantage that the third and fifth order dispersion have opposite sign ($T_d(\omega) = T_{d0} - N \cdot \beta_3 \cdot (\omega - \omega_r)^2 + N \cdot \beta_5 \cdot (\omega - \omega_r)^4 - \cdots$). The balanced SCISSOR tunable delay line structure, one half of rings have their resonant frequencies shifted up by a small amount $\Delta\omega$ relative to the central signal frequency ω_0, while the resonance frequency of the other half are shifted down by the small amount ($\omega_{1,2} = \omega_r \pm \Delta\omega$). The up and down shifted rings are located on opposite sides of the same bus and implement the shift using the thermal or the carrier induced index change $\pm\Delta n$ on the two sides of the central bus. We can express the group dispersion (GD) of balanced SCISSOR by,

$$T_d(\omega) = T_{d0} - N \cdot \Delta\omega^2 \cdot (\beta_3 - \beta_5 \cdot \Delta\omega^2) - N \cdot (\beta_3 - 6 \cdot \beta_5 \cdot \Delta\omega^2) \times (\omega - \omega_r)^2 + N \cdot (\beta_5 - 15 \cdot \beta_7 \cdot \Delta\omega^2) \cdot (\omega - \omega_r)^4 + \cdots$$

The term quadratic in $(\omega - \omega_r)$, vanishes for a detuning $\Delta\omega_0 = \sqrt{\frac{\beta_3}{6 \cdot \beta_5}} \approx \frac{1-\rho}{\tau \cdot \sqrt{6}}$. If we change $\Delta\omega$ then tunable delay is achieved.

$$\frac{dT_d}{d\Delta\omega}|_{\Delta\omega_0} = -\frac{4}{3} \cdot N \cdot \beta_3 \cdot \Delta\omega_0 = -\frac{4}{3 \cdot \sqrt{6}} \cdot T_d \cdot \frac{\rho \cdot \tau}{(1 - \rho)}$$

$$\Delta T_d = -\frac{2}{3 \cdot \sqrt{3}} \cdot \left(\frac{\rho \cdot \omega \cdot \tau}{1 - \rho}\right) \cdot \left(\frac{\Delta n}{n}\right) \cdot T_d$$

It is very interesting to get the value of the maximum group dispersion (GD) per ring as a function of the index change for specific effective index [27].

Glass wedges (GW): Precision laser quality around wedge windows is ideal for applications where direct back reflections would be a problem. These elements are similar to the precision flat windows but have a deliberate wedge of some minutes. The glass wedge allows for dispersion fine tuning and optimal pulse compression in combination with dispersion compensating mirror. The dispersion characteristics of dispersion compensating mirror is especially designed to work together with specific glass materials and thus a glass wedge can be used to fine-tune the dispersion and compress pulses for the desired application. The dispersion compensating mirror is used for the compensation of positive dispersion affecting femtosecond laser pulses. We can get compress pulses which fit the lasers spectral bandwidth. Precision flat windows are used for interferometer flats, laser windows, beam splitter substrates, and parallel plates. Precision flat mirrors are designed to be used as dielectric laser mirrors or metal coated mirrors. It is very applicable for optical path folding applications various imaging systems, and laser applications. An optical flat is a precisely polished flat surface, used as a reference against which the flatness of an unknown surface is compared. When an optical flat's polished surface is placed in contact with a surface, dark and light bands are formed when viewed with monochromatic light. The bands are interference fringes and their shape gives a visual representation of the flatness of the surface being examined. The surface flatness is indicated by the amount of curve and spacing between the interference fringes. Straight, parallel, and spaced interference fringes indicate that the surface flatness is equal to or higher than that of the reference surface. Interference fringe is a bright or dark band caused by beams of light that are in phase or out of phase with one another. Light waves and similar wave propagation, when superimposed, add their crests if they meet in the same phase and the waves are both increasing or both decreasing and the troughs cancel the crests if they are out of phase. It is constructive and destructive interference, respectively. If a beam of monochromatic light is passed through two narrow slits, the two resulting light beams are directed to a flat screen on which, instead of forming two patches of overlapping light, they will form interference fringes, a pattern of evenly spaced alternating bright and dark bands. An optical interferometers function by virtue of the interference fringes that they produce. Wedge prisms are used for deviate the laser beam set angle, and two wedge prisms are used together for beam steering applications. A single wedge prism has the ability to deviate the angle of an incident beam and two wedge prisms are used as an anamorphic pair of beam shaping and to correct the elliptical shape of diode outputs. Beam deviation is a degrees and diopters, when one diopter is 1cm of deviation at a distance of 1m from the prism. The wedge prism is a prism with a shallow angle between its input and output surface. The refraction at the surfaces causes the prism to detect light by a fixed angle. Rays of light passing through the prism (wedge prism in air) are deflected by the angle δ, which is given by $\delta \approx (n-1) \cdot \alpha$, where n is the index of refraction of the prism material, and α is the angle between the prism's surfaces. Optical beam steering is about changing the direction of the main laser beam of a radiation pattern. In optical systems beam steering is accomplished by changing the refractive index of the medium through which the beam is transmitted or by the use of mirrors, prisms, lenses, and rotating diffraction gratings [28, 29]. It is important in ultrashort laser facility to retrieve the phase of ultrashort laser pulses by a chirped mirror and glass wedges compressor. The compression system is functioned as a diagnostic tool and there is no need for complementary tools. An ultrashort pulses system is characterized by nonlinear auto correlation diagnostics. There are several methods for the reconstruction of the amplitude and phase of the pulses by combination of autocorrelation and spectral measurements. Frequency resolved optical grating (FROG) gives spectrally resolving an auto correlation or cross-correlation signal. FROG variants rely on spectral resolving some time-gated signal. Other methods are technique of spectral phase interferometry for direct electric field construction (SPIDER). There methods based on interferometry in the spectral domain. The spectrum of a given pulse is made to interfere with a frequency-shifter replica of it, and the resulting spectral interferogram is analyzed. The intensity profile of the optical spectrum of ultrashort pulse is measured by spectrograph. By using Fourier transform spectral interferometry, we retrieve the spectral phase. The principle of spectral interferometry is that two pulses are interferometically combined, and the optical spectrum of the combination is recorded. One of the pulses is delayed (time delay τ_p). If the electric fields of the two pulses are $E_1(\omega)$ and $E_2(\omega)$, the additional time delay translates into a phase factor $e^{i \cdot \omega \cdot \tau_p}$ for one of the pulses. The combine fields is

$E(\omega) = E_1(\omega) + E_2(\omega) \cdot e^{-i \cdot \omega \cdot t}$, the optical spectrum is modulated and the oscillating term has a phase $\varphi(\omega) = \omega \cdot \tau_p + \varphi_1(\omega) - \varphi_2(\omega)$ which emphasis that there is an oscillation which is determine by the time delay (τ_p) and the phase of which is influenced by the difference of the spectral phases of the two pulses. Interferogram combines light from two different sources. We use the same laser light source and split the light into two beams. One is the reference beam (comparison wave front) and the other is the best beam, which is passed through the optical system to be tested. The two laser beams are combined together to make an interferogram. It is possible that the pulse characterization which is based on phase scanning, multiphoton intra pulse interference phase scan (MIIPS) is used. It apply spectral phases to the pulse to be characterized and measuring the resulting second-harmonic generation (SHG) signal. By compression at a given wavelength we get group delay dispersion (GDD), the GDD of the pulse allows the reconstruction of the unknown phase. The MIIPS technique gives a phase scan which is performed on the pulse to be measured. The technique is implemented by using a standard chirped mirror compressor, the chirp mirrors ensure that the pulse is negatively chirped and adding glass continuously and finally the pulse becomes as short as possible. The ultrashort laser pulse has complex spectral amplitude $\widetilde{U}(\omega) = |\widetilde{U}(\omega)| \cdot e^{i \cdot \phi(\omega)}$. The pulse goes through a piece of transparent glass and then a second harmonic generation (SHG) crystal. The SHG spectral power is a function of the thickness and is expressed in the next expression:

$S(\omega, z) = |\int [\int \widetilde{U}(\Omega) \cdot e^{i \cdot z \cdot k(\Omega)} \cdot e^{i \cdot \Omega \cdot t} \cdot d\Omega]^2 \cdot e^{-i \cdot \omega \cdot t} \cdot dt|^2$, where, z is the thickness of the glass and $k(\Omega)$ its frequency-dependent phase per unit length or wavenumber. The SHG is performed the time-dependent fields is squared, and an inverse Fourier transform gives the SHG spectrum. The SHG process is squaring the electric field in time, assuming an instantaneous and wavelength-independent nonlinearity. The dispersion-scanned SHG traces of representative spectra are used. For each wavelength λ_0 in the SHG spectrum, finding the insertion that maximized the signal. The SHG for specific wavelength is determined by the spectral power and the phase at twice that wavelength in the fundamental field. We need to retrieve the phase in our system. The spectral power density for different phase curves, merit function of the RMS error between the measured and simulated scan, Frequency resolved optical grating (FROG) is given by $G = \sqrt{\frac{1}{N_i \cdot N_j} \cdot \sum_{i,j} [S_{meas}(\omega_i, z_j) - \mu \cdot S_{sim}(\omega_i, z_j)]^2}$, where $S_{meas}(\omega_i, z_j)$ and $S_{sim}(\omega_i, z_j)$ are the measured and simulated scans, respectively, and μ is the factor that minimizes the error. The μ factor can be calculated by differentiating the error with respect to μ, is given by $\mu = \frac{\sum_{i,j} S_{meas}(\omega_i, z_j) \cdot S_{sim}(\omega_i, z_j)}{\sum_{i,j} [S_{sim}(\omega_i, z_j)]^2}$. In the case of ultra-broadband few-cycle pulses, the SHG signal is described by model $S(\omega, z)$ of the spectrum which is multiplied by the spectral filter and the measured signal is given by $S_{meas}(\omega, z) = S_{ideal}(\omega, z) \cdot R(\omega)$, where $R(\omega)$ is the spectral filter and $S_{ideal}(\omega, z)$ denotes the ideal, flat response process. The integral of the trace over the thickness parameter is given by $M(\omega) = \int_{-\infty}^{+\infty} S(\omega, z) \cdot dz$ and it is not dependent on the original spectral phase of the pulse, $\phi(\omega)$. By allowing the error function to be minimized for each wavelength, with overall error weighted function of all errors, the factor that minimizes the error for each frequency component is

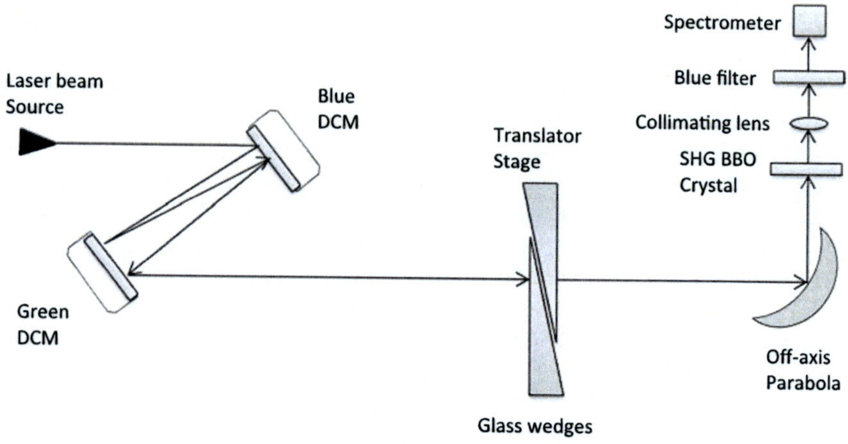

Fig. E.25 Ultrashort laser pulses chirped mirrors and glass wedges system

given by $\mu_i = \frac{\sum_j S_{meas}(\omega_i, z_j) \cdot S_{sim}(\omega_i, z_j)}{\sum_j [S_{sim}(\omega_i, z_j)]^2}$ and the overall error is

$$G = \sqrt{\frac{1}{N_i \cdot N_j} \cdot \sum_{i,j} [S_{meas}(\omega_i, z_j) - \mu_i \cdot S_{sim}(\omega_i, z_j)]^2}.$$

The minimizing factor μ_i gives the complete filter response. The ultrashort laser pulses chirped mirrors and glass wedges system consists the following elements: an ultrafast oscillator, four double chirped mirror pair, Anti-reflect (AR) coated glass wedges (α angle), off-axis aluminum coated parabola and (β-Barium borate) BBO crystal. The chirped mirrors are matched pair to minimize GDD (group delay dispersion) oscillations and come in two types (Fig. E.25).

The DCM is double chirped mirrors which made in matched pairs to minimize phase ringing and SHG (second harmonic generation) is thick BBO crystal. The laser is femtosecond laser source. The aluminum off-axis parabola 90° reflects the coming laser beam to the SHG BBO crystal. The system characterized ultrafast laser pulses based on iterative phase retrieval from dispersion scans by using chirped mirrors, wedges and a standard SHG crystal (BBO crystal) [28].

Optical silicon window (OSW): Optical silicon window is mono crystalline silicon (Si) windows. It comes in uncoated and an AR (Anti Reflect) coating versions. The AR (Anti Reflect) coating version, AR coating is on both sides. Silicon offers high thermally conductivity and low density, making it suitable for laser windows. Since silicon has a strong absorption band, it is not suitable for the use with CO_2 laser transmission application. The precision window for silicon (Si) in the wavelength range 1.2–8 μm. Infrared (IR) windows are optical windows designed for optimal performance in the infrared (IR) spectrums. IR windows are ideal for use in FTIR spectroscopy, thermal imaging, FLIR, and medical systems. Mainly the infrared window is in multiple anti reflecting coating or substrate. The silicon windows are ideal for infrared (IR) applications with good transmittance from 1.2 to 7 μm. Silicon has a low 2.329 $\frac{g}{cm^3}$ density which is half the density of the Germanium, which makes it ideal for weight sensitive applications. It is also characterize by broadband 3–12 μm anti-reflection coating. Silicon optical material is available for use in the NIR (1 μm) to 6 μm and the optical quality silicon is doped (5–40 Ω cm) to prevent absorption bands within the transmission waveband. Silicon material has a lower refractive index compare to germanium. It has a lower density that makes for less weighty optical systems. Silicon material is ideal for use as windows in the 3–5 μm waveband and as substrate for optical fibers. The waveband (3–5 μm) is crucial for the detection of sources radiation at a black body temperature (≈700 °K). Optical window (OW) in an optical system provides a clear aperture to transmit the desired radiation and to keep the two environments separate. It keeps gases present in the two environments separate and atmospheric differences (temperature, pressure, etc.,). There are critical key design topics when using optical window, mounting of plan windows and domes, structural considerations, and optical performance degradation. Additionally key design issues are mechanical and thermal distortions and sealing technique. There are several optical window materials and their important properties are transmission range, index of refraction, etc. Optical windows usually consist of two flat surfaces and have negligible effect on collimated light and effect non-collimated beams by introducing aberrations. A typical window performance includes the allowable deterioration of a plane wave front through the entire aperture of the window. We can evaluate the approximate optical path difference (OPD) created, round window deformed by a uniform load ΔP_W on its face. The unsupported aperture diameter is A_W and the optical path difference (OPD) is $OPD = 0.00889 \cdot (n-1)^2 \cdot (\Delta P_W)^2 \cdot A_W^6 \cdot \frac{1}{E_G^2 \cdot t_W^5}$, where, E_G is the Young's modulus and n is the refractive index of the specific window material. If the window is subjected to acceleration a_G, then $\Delta P_W = a_G \cdot \rho_G \cdot t_W$ and $OPD = 0.00889 \cdot (n-1)^2 \cdot (a_G \cdot \rho_G \cdot t_W)^2 \cdot A_W^6 \cdot \frac{1}{E_G^2 \cdot t_W^5}$. Where, ρ_G is the window material density and t_W is the thickness of the window. Typical optical window mounting technique includes of a round glass element held in place by adhesive which is injected through holes at the window in place. The housing for the window has a threaded auteur diameter for mounting. The clearances between the window and the housing usually are large enough and easy to assemble. It is possible to use structure that provides a clamping force to hold the window in place. A typical minimum thickness t_W equation for a circular plane window with unsupported aperture diameter A_W with a uniform differential pressure ΔP_W is $t_W = \frac{1}{2} \cdot A_W \cdot \sqrt{K_W \cdot f_s \cdot \frac{\Delta P_W}{S_F}}$. The factor of safety f_s is typically 4, and K_W is a support condition constant that is ≈ 1.25 if the window is unclamped and $K_W \approx 0.75$ if it is clamped around the entire aperture. The optical path difference (OPD) is

$$OPD = 0.00889 \cdot (n-1)^2 \cdot \left(a_G \cdot \rho_G \cdot \frac{1}{2} \cdot A_W \cdot \sqrt{K_W} \cdot f_s \cdot \left[\frac{a_G \cdot \rho_G \cdot t_W}{S_F}\right]\right)^2 \cdot A_W^6 \cdot \frac{1}{E_G^2 \cdot t_W^5}$$

The material fracture strength is S_F. The typical minimum value for fracture strength S_F of infrared window silicon material is $S_F = 120$ MPa ; $S_F = 1.74 \times 10^4 \frac{\text{lb}}{\text{in}^2}$. The pascal (SI derived units), 1 Pa $= 1 \frac{\text{N}}{\text{m}^2} = 1 \frac{\text{kg}}{\text{m s}^2} = 1 \frac{\text{J}}{\text{m}^3}$; MPa $= 10^6 \times$Pa (Mega Pascal), where N is the newton, m is the meter, kg is the kilogram, s is the second, and J is the joule. One pascal is the pressure exerted by a force of magnitude one newton perpendicularly upon an area of one square meter. $\frac{\text{lb}}{\text{in}^2}$ is pound per square inch $\left(1 \text{ Pa} \simeq 0.000145 \frac{\text{lb}}{\text{in}^2}\right)$. These values are dependent on the quality of the silicon surface, fabrication method, silicon purity, type of test, and size of the silicon sample. Other important factors when designing optical windows (OWs) are related to high loads (surface imperfections, internal stresses, and the nature of the loads which is static or dynamic). The fracture toughness of the glass is a very important parameter. Fracture toughness in a material (silicon) is its resistance to crack growth when under tensile stress. The glass fails when the stress intensity factor K_I due to crack, exceeds the fracture toughness of the material, K_{IC}. The stress intensity factor is calculated by Griffith's law: $K_I = Y \cdot \sigma \cdot \sqrt{a}$, where Y is a crack geometry factor, σ is the nominal stress, and a is the crack, or flaw size. Choosing a material for a window is very critical. The material must be able to transmit over the desired wavelengths, strong, and easy to coat. Silicon is similar to Germanium; it transmits well in the infrared, but has a high $\frac{dn}{dT}$ value. Silicon has a high transmittance in the mid-wave infrared range and is half the density of Germanium. Silicon is much harder than Germanium making is less brittle. Optical window coating is a very important part of the design. The coating is specified based on the operating environment and the desired wavelength range of the system. For systems which work in harsh environments a diamond-like coating is appropriate to protect the window from dust and moisture. In optical systems which work over a wide range of wavelengths a broadband coating is required. For high transmission over a wide spectral range optical window (OW) the complexity is much higher. The operating environments for optical windows (OWs) can range from large pressure and temperature differentials to abrasive dust and debris. Several materials are suitable for optical windows (OWs). The material substrate must withstand the operation environment and transmit the desired wavelengths. There are some design recommendations for using optical windows (OWs):

1. Include only windows in the optical design of an optical system which have impact on system performance.
2. Mechanical mounting which stands the demand that the window is sealed from the outside environment and that thermal and pressure differentials are fit.
3. The window is sandwiched between two o-rings and the top plate bolts to the window cell causing the o-rings to compress and seal the optical system.
4. The windows must be mounted in a low stress state to reduce the surface distortion caused by thermal and pressure differentials.

The silicon material for an optical window has the following parameters: 3.4223 index at 5 μm, wavelength range 1.2–15 μm, Young's modulus 131 GPa, CTE $2.6 \times 10^{-6} \frac{1}{\text{C}}$, thermal conductivity 163.3 $\frac{\text{W}}{\text{m K}}$ (thermal conductivity is measured in watts per meter-kelvin), and Poisson's ratio 0.266. Other typical optical windows

(OWs) materials are: N-Bk7 (its transmission range is limited to the visible and near infrared), Sapphire (it has a very large transmission range and good for windows and domes from the UV to the IR), ZnSe (it is an infrared transparent material and soft material which degrades quickly when subjected to particle-impacting influence), CLEARTRAN (it is a water clear from of zinc sulfide, similar to zinc selenide and it transmits from UV to the far IR. It is good to multi spectral applications that require a single window to transmit all portfolios of wavelengths), Germanium (it transmits very good the infrared but poor thermal properties, high $\frac{dn}{dT}$ value and it becomes nearly opaque at temperature above 100 °C and completely non-transmissivity at 200 °C), and CaF2 (it is calcium fluoride with transmits a wide range of wavelengths from the UV to the IR and used in cryogenically cooled thermal imaging systems, good water chemical and heat resistance, a relatively high CTE). Cryogenics is the production and behavior of materials at very low temperatures. In high power laser and scientific facilities, optical windows (OWs) are exposed to axial temperature gradients which caused by absorption of an incident heat flux. Axial temperature gradient affects the heat transfer, momentum transfer and energy conversion mechanisms. When it exposed to an incident heat flux, a plane parallel circular window becomes a concentric meniscus. In an optical system, a monocentric element is one whose surfaces are concentric. The power of the window when exposed to an axial temperature gradient is $P = \frac{1}{f} = \frac{(n-1)}{n} \cdot (\frac{\alpha}{k})^2 \cdot t \cdot q^2$, where, f is the focal length, n is the window refractive index, t is the thickness, q is the incident heat flux per unit area, α is the window CTE (common table expression) and k is the window material's thermal conductivity. The common table expression (CTE) specifies a temporary result set. Optical system windows are exposed to pressure differentials such as airborne FLIR and cryogenic systems. The circular window exposed to a uniform pressure differential, and then the optical path difference (OPD) through the window is $OPD = (8.89 \times 10^{-3}) \cdot \frac{(n-1) \cdot (\Delta P)^2 \cdot d^2}{E^2 \cdot h^5}$, where, n is the refractive index of the window material, ΔP is the pressure differential, d is the window diameter, E is the material Young's modulus and h is the window thickness. If the window is exposed to acceleration loading then $\Delta P = \frac{a \cdot \rho \cdot h}{g}$; $\left(OPD = (8.89 \times 10^{-3}) \cdot (n-1) \cdot \frac{a^2 \cdot \rho^2}{g^2 \cdot E^2 \cdot h^3} \cdot d^2\right)$, where, a is the acceleration, ρ is the window density and g is the acceleration due to gravity. The optical window (OW) shift due to a plane parallel plate is presented in the next figure (Fig. E.26). The optical window (OW) is diverging the beam and the focus of the system is affected. In the case of optical system with cryogenically cooled focal plane then the last element of the system is a window that seals the Dewar. The optical window (OW) causes the focus of the system to shift and the approximate focal shift is (Δz) in terms of the index (n) and the thickness (t) of the optical window (OW). The focal shift (Δz) is $\Delta z = (\frac{n-1}{n}) \cdot t$, where, $\frac{n-1}{n} < 1$ and $n = 1 \Rightarrow \Delta z = 0$, and $n \to \infty$ ($n \gg 1$) $\Rightarrow \Delta z \approx t$. The diverging or covering beams cause to aberrations in the optical window (OW). Optical windows (OWs) induce spherical aberration, coma and astigmatism, transverse, and longitudinal chromatic aberrations [30].

Electro-optic crystal (EOC): Electro-optical Crystals are widely used for electro-optic modulators and electro-optic deflectors. An electro-optic effect is a change in the optical properties of a material in response to an electric field that varies

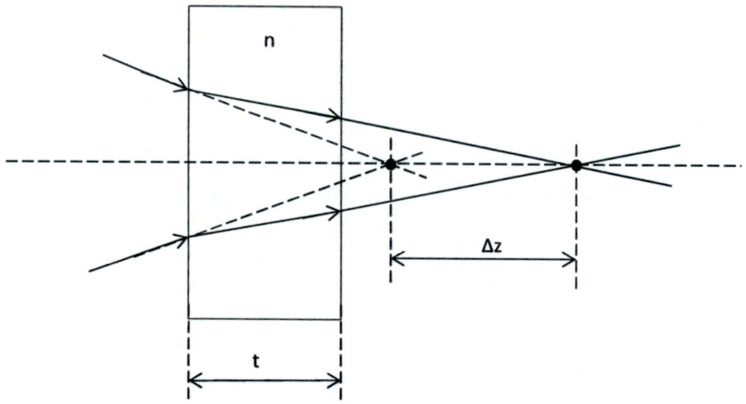

Fig. E.26 Optical window (OW) shift due to a plane parallel plate

slowly compared with the frequency of light. Electro optic modulator (EOM) is an optical device that a signal-controlled element presents the electro optic effect and we use it to modulate the beam of light. The beam is affected by frequency, phase, amplitude or polarization of the beam. It is also possible to use laser-controlled modulator for many scientific applications. The electro optic effect happened when the material (crystal) refractive index is changed by a DC voltage or low-frequency electric field. By applying electric field to crystal the material molecules change their shape, orientation and position. Crystals have the characteristic that refractive index is changed by the strength of the electric field. We expose the crystal to electric field than the light travel more slowly through it. If the time which takes to light to pass through the crystal is longer, than the phase of the light leaving the crystal is bigger. We can change the laser light that leave the crystal by electric field. We create an electric field surrounding the crystal by two parallel plate's capacitor across the crystal. Electro-optic deflector is the most efficient mechanism for changing the angle of the laser beam. Electric field in an electro-optic material (crystal) caused to refractive index gradient proportional to the applied signal voltage. By using an electro-optic crystal (EOC) we get an electro optic deflector with high speed and high efficiency. A time varying spatial slope of the refractive index is induced in the crystal by the electric standing wave. The result is ultrafast reflection of the laser beam. The change of the refractive index is caused by electromagnetic forces that perturb the position, orientation or shape of the atom or molecule structure in the crystal. Electro optic deflection (EOD) refracts a laser beam by introducing a phase delay across the cross section of the laser beam. There are two types of electro optic effects: first, is when the refractive index varied linearly with the electric field and second, it is varied quadratic ally with the field of strength ($\propto E^2$). It is referred to the Kerr effect or quadratic electro-optic effect (QEO effect). Typical crystals which their refractive index is changed linearly with the electric field are $LiNbO_3$, $LiTaO_3$ and $KTiOPO_4$. The crystal EOD which causes by index gradient which is perpendicular to the direction of the propagation of the laser beam is present (Fig. E.27).

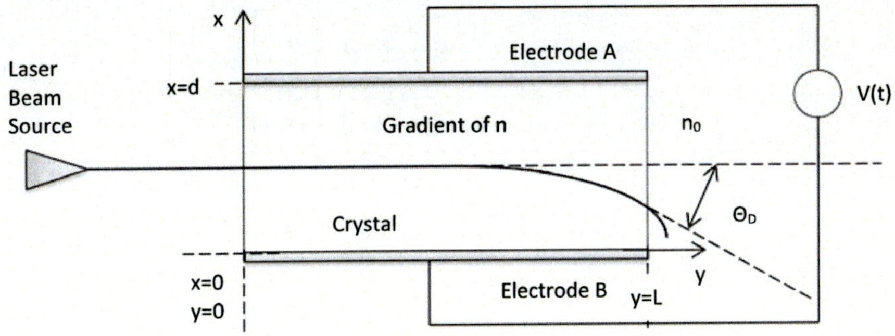

Fig. E.27 Crystal EOD which causes by index gradient which is perpendicular to the direction of the propagation of the laser beam

V(t) is the AC voltage source which is connected to electrodes A and B, n is the gradient refractive index of the crystal, n_0 is the refractive index of the crystal surrounding medium, and Θ_D is the laser beam deflection beam angle. The angle of deflection Θ_D is proportional to the length of the crystal, where, n denotes the refractive index of the crystal and Δn is the difference between refractive index of the crystal and the surrounding material. An electro optic deflection (EOD) is based on refraction by an index gradient that exist perpendicular to the direction of the propagation of the laser beam. The laser beam is cumulatively refracted as it propagated through the crystal. Deflections are attributed to a specific distribution of the electric field perpendicular to the beam propagation. The distribution which related to the inside of the crystal causes to electrons injected into the crystal from the electrodes A and B. It is superimpose of space-charge effect and Kerr effect. If the laser beam enters the crystal along the optical axis, the angle of deflection, due to a voltage V(t), V = V(t) is equal to Θ_D: $\theta_D = -0.153 \cdot n^3 \cdot \varepsilon_0^2 \cdot \varepsilon_r^2 \cdot \frac{V^2}{d^3} \cdot L$, where, ε_0 denotes the permittivity of vacuum, ε_r is the relative permittivity of the crystal, n is the refractive index of the crystal at V = 0, L denotes the propagation length (mostly crystal length) of the laser beam in the crystal and d denotes the thickness of the crystal. The negative sign in the θ_D equation indicates that the laser beam is deflected towards the cathode. By controlling the voltage V, we can control the beam deflection angle (θ_D). Another way to deflect the light beam is by an acousto-optic deflector (AOD) which is based on a periodically changing refractive index n in an optically transparent material (mainly TeO_2 or PbM_0O_4 crystals) (Fig. E.28). It induced by propagating sound waves in the material (crystal). The changing refractive index is the result of refraction and compression of the material, inducing a changing density of the material. The periodically changing refractive index, n, is function as an optical grating, moving at the speed of sound in the crystal which diffracts a laser beam traveling through the material.

It consist acousto-optic crystal, Acoustic transducer which is typically a piezo-electric element, AC voltage source, Acoustic absorber, and laser beam source. The acousto-optic crystal is glued to the transducer which emits acoustic wave to the

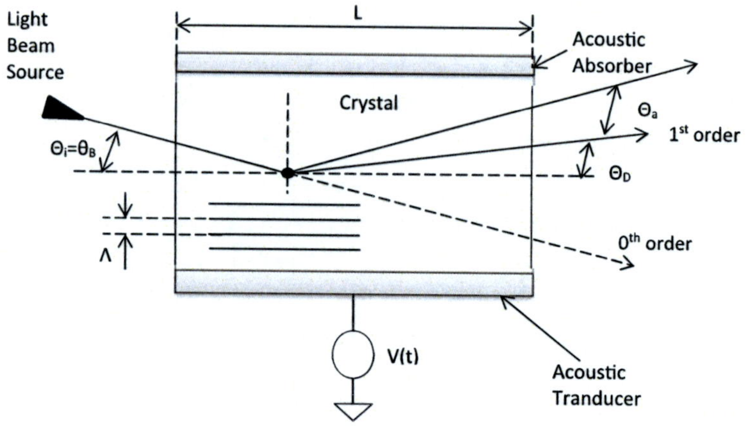

Fig. E.28 Typical configuration of an Acoustic optic deflector (AOD) system

crystal. The transducer launches acoustic waves, of frequency f_a (80 MHz–1 GHz), into the crystal production grating. At opposite end, acoustic absorber is glued to the crystal and avoids reflections of the acoustic wave back into the crystal. The interaction length is L and the laser beam is incident at the Bragg angle $\left(\theta_i = \theta_B \approx \frac{\lambda}{2\cdot\Lambda}\right)$, where Λ is the grating pitch of the light beam reflected. The acoustic wavelength, first order refraction of the light is observed and higher order diffraction orders effects by destructive interference. The angle of deflection Θ_d outside the crystal of the 1st order diffraction depends on the wavelength λ of the laser light in vacuum, the frequency f_a of the acoustic wave, n is the refractive index of the unperturbed material and the velocity V_a of the acoustic wave in crystal. The angle of deflection is θ_d; $\left(\theta_d = \frac{\lambda\cdot f_a}{2\cdot n\cdot v_a}\right)$. When the frequency of the sound wave is varied, incident angle θ_i of the laser beam is fixed at Bragg angle $\theta_i = \theta_B \approx \frac{\lambda}{2\cdot\Lambda}$ than the direction of the diffracted laser beam varies. We change the acoustic frequency from f_a to $f_a+\Delta f$ $(f_a \rightarrow f_a+\Delta f)$, where Δf_a denotes frequency bandwidth of the driver, and changes the deflection angle (θ_d). The frequency ranges from 40 MHz to 500 MHz which dependent on the crystal material. The maximum deflection angle θ_a of an Acoustic optic deflector (AOD) is $\theta_a = \frac{\lambda\cdot\Delta f_a}{2\cdot n\cdot v_a}$. The frequency bandwidth Δf_a affects the optical efficiency of the deflection. When the incident angle equals the deflection angle then the Bragg angle $(\theta_B = \theta_i = \theta_d)$, the diffraction efficiency is maximum. When the acoustic f_a is increased, the condition is not fully fulfilled and the Bragg condition is increasingly not fulfil. Typically the diffraction efficiency is greater than 50–60%. The time duration to fill the crystal by acoustic wave is finite; there is a "switch" from one acoustic frequency value to the other acoustic frequency value. It is the time to "switch" from specific angular direction of the deflected beam to the next one. It is the response time τ_a (transit time or random-access time) of the acoustic optical deflection (AOD) and it dependent on the acoustic velocity v_a and on D (beam diameter or the optical aperture of the AOD), we define $\tau_a = \frac{D}{v_a}$, the

Table E.2 Definitions of tensors (up to order 2)

Description	Type of tensor
A tensor of a single number	Tensor of order zero/ scalar (array of dimensional zero)
A point, or points, or vary continuously from point-to-point which define a vector filed	A tensor of order one/ vector
Tensor which is a two dimensional array (matrices physical properties)	Tensor of order 2

response time τ_a, is a measure of the deflection angle velocity of the AOD. A small response time $\tau_a \rightarrow \varepsilon$ implies high deflection angle velocity [31]. Oxygen octahedral ferroelectrics are materials which used in crystals. Oxygen octahedral ferroelectrics are materials family which have good electro-optic, nonlinear-optic, and acoustic and acoust-optic properties. These properties are characterized by common polarization potential tensors and common refractive index dispersion parameter. Oxygen octahedral ferroelectrics common refractive index dispersion behavior give the result that in second harmonic generation (SHG) applications there is connection between crystal birefringence and the refractive index at the fundamental for collinear phase matching. The harmonic power which the only important material quantities is the refractive index at the fundamental, the crystal symmetry. All oxygen octahedral ferroelectrics have the same tensors which related to quadratic electro-optic g coefficients and the nonlinear optic δ coefficients to static and optical-frequency polarization-potential tensors. The result is static and optical-frequency polarization potential s which are the same. The oxygen-octahedral ferroelectrics materials have nearly the same dispersion parameter ration $\frac{\varepsilon_0}{S_0}, \frac{\varepsilon_0}{S_0} = 6 \pm 0.5 \times 10^{-14}$ eV m^2, where ε_0 is an average inter band-oscillator energy in eV (It is not the regular definition "the permittivity of free space") and S_0 is an average inter band-oscillator strength defined by a single-term. A tensor is a geometric object that maps a multi-linear geometric vector, scalars and other tensor to a resulting tensor. Tensors are used in optical systems and provide formulation and solutions to optical systems. A tensor is a mathematical entity structure which interacts with other mathematical entities. The tensor must fulfil transformation rule in the case of transformation. We can present a tensor as a generalized matrix. It can be a 1-D matrix (a vector is such a tensor), a 3-D matrix (cube of numbers), and 0-D matrix (a single number), or a higher dimensional structure. Rank is the dimension of the tensor. Practically, tensors are simply arrays of numbers (or functions), that transform according to specific rules under a change of coordinates. Tensors characterize the properties of $BaTiO_3$ crystals and other optical crystals. A tensor is defined at a single point or collection of isolated points of space or space time, and be defined over a continuum points. The element of the tensor is function of position and the tensor terms is a tensor field. The tensor is defined at every point within a region of space or space-time or a collection of isolated of points. The definitions of tensors (up to order 2) can be as follow (Table E.2).

An n-th-rank tensor in m-dimensional space is a mathematical object that has n indices and m^n components which fulfil transformation rules. The related simple

Table E.3 Elastic and dielectric response variables and description

Elastic and dielectric response variables	Description
S_{kl}	Elastic strain tensor
E_i	Vector of the electric field strength
T_{kl}	Elastic stress tensor
ε_{ij}^{S}	Clamped (zero strain) dielectric constant
ε_{ij}^{T}	Un clamped (zero stress) dielectric constant
c_{ijkl}^{E} and s_{ijkl}^{E}	Elastic stiffness and compliance tensors at E = 0
e_{kij}	Piezo electric stress tensor
d_{kij}	Strain tensor

crystal perovskite materials are $SrTiO_3$, and $KNbO_3$. These crystals have large values of spontaneous polarization and high packing density of oxygen octahedral. The perovskite material is a calcium titanium oxide mineral, and is characterized by the chemical formula $CaTiO_3$. The perovskite is formed of calcium, titanium and oxygen forms of $CaTiO_3$ and it has the generic form ABX3 with the same crystallographic structure. Perovskite materials typical applications are sensors, catalyst electrodes, fuel cells, solar cells, lasers, and memory devices and spintronics applications. Barium titanate has large electro optic coefficients and used as photorefractive material. The effective electro-optic coefficients which give details on the optical indicatrix changes under the influence of the electric field grating is estimated by knowing the refractive indices, clamped (stain-free) electro-optic coefficients e_{ijk}, the elasto-optic (pockels) tensor at constant electric field p_{ijkl}^{E}, and the tensor of the elastic contents at constant electric field c_{ijkl}^{E}. $BaTiO_3$ crystals have various properties like elastic, elasto-optic, dielectric, piezoelectric and electro-optics. In many scientific industrial applications it is important to know the response of a piezoelectric crystal to external electrical and mechanical fields. The response is described by set of independent variables (mechanical and electrical variables). The piezoelectric crystal elastic and dielectric response is described by two different sets of equations. The first set is $T_{ij} = c_{ijkl}^{E} \cdot S_{kl} - e_{kij} \cdot E_k$ and $D_i = e_{ijk} \cdot S_{jk} + \varepsilon_0 \cdot \varepsilon_{ij}^{S} \cdot E_j$, the second set $S_{ij} = s_{ijkl}^{E} \cdot T_{kl} + d_{kij} \cdot E_k$ and $D_i = d_{ijk} \cdot T_{jk} + \varepsilon_0 \cdot \varepsilon_{ij}^{T} \cdot E_j$. The set of independent variables are as follow in the next table (Table E.3).

Additionally the piezoelectric crystal system fulfil $c_{ijkl}^{D} - c_{ijkl}^{E} = \frac{e_{mij} \cdot e_{nkl}}{\varepsilon_0 \cdot \varepsilon_{mn}^{S}}$; $\varepsilon_0 \cdot (\varepsilon_{ii}^{T} - \varepsilon_{ii}^{S}) = e_{ilm} \cdot d_{ilm}$ and $e_{ijk} = c_{lmjk}^{E} \cdot d_{ilm}$. Crystal $BaTiO_3$ has specific dielectric, elastic, and piezoelectric properties. In the case of acoustic waves which propagate along [101] direction, the velocities of these two waves (polarized in 010 plane) are $\rho \cdot v_{QL}^2 = A + B + \sqrt{(A - B)^2 + 4 \cdot C}$ and $\rho \cdot v_{QT}^2 = A + B - \sqrt{(A - B)^2 + 4 \cdot C}$, QL stands for quasi-longitudinal and QT stands for quasi-transverse polarization. The crystal mass density is ρ, and the coefficients A, B, and C are related to elastic

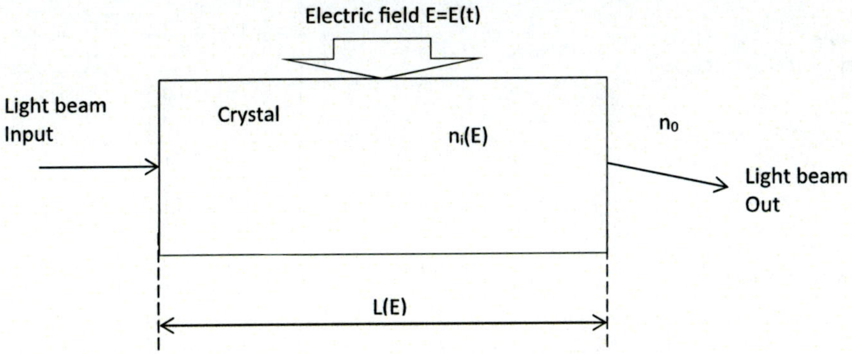

Fig. E.29 Crystal and laser light beam system under electrical filed influence

stiffness constants. The piezoelectric stress coefficients are e_{ijk}. The expressions for coefficients A, B, and C, are $A = c^E_{1111} + c^E_{1313} + \frac{(e_{311}+e_{131})^2}{\varepsilon_0\cdot(\varepsilon^S_{11}+\varepsilon^S_{33})}$ and $B = c^E_{3333} + c^E_{1313} + \frac{(e_{333}+e_{131})^2}{\varepsilon_0\cdot(\varepsilon^S_{11}+\varepsilon^S_{33})}$; $C = c^E_{1133} + c^E_{1313} + \frac{(e_{333}+e_{131})\cdot(e_{311}+e_{131})}{\varepsilon_0\cdot(\varepsilon^S_{11}+\varepsilon^S_{33})}$. $BaTiO_3$ crystal material tensors which is in use for electro-optical and acousto-optical systems are r_{ijk} (electro-optic coefficients) and p_{ijkl} (elasto-optic coefficients). When purely longitudinal waves are propagated along the crystallographic axes of $BaTiO_3$ the change of the inverse of the dielectric constant at specific optical frequencies is $\Delta(\frac{1}{n^2})_{ii} = r^S_{iik}\cdot E_k + p^E_{iill}\cdot S_{ll}$, r^S_{iik} is the electro-optic strain-free coefficient, p^E_{iill} is the elasto-optic coefficients at constant electric field. The symmetrical strain tensor is S_{kl}; which described the elastic deformation. A longitudinal sound wave propagating along the x axis induces no electric field in the crystal. The soundwave is propagated along z axis a longitudinal electric field and caused by the piezoelectric coupling through the e_{333} tensor element. The effective elasto-optic tensor components are p^*_{ii33}. The mathematical connection between p^*_{ii33} and elasto-optic tensor at zero electric field is $p^*_{ii33} = p^E_{ii33} - \frac{r^S_{ii3}\cdot e_{333}}{\varepsilon_0\cdot\varepsilon^S_{33}}$. Remark: ε_0 is the permittivity of free space in the analysis above, dielectric constant. The total optical phase change of an optical beam $\Delta\phi$ in a crystal influenced by an electric field $\Delta\phi = f(E)$ is $\Delta\phi = \Delta[\frac{2\cdot\pi}{\lambda_0}\cdot(n_i - n_0)\cdot L] = \frac{2\cdot\pi}{\lambda_0}\cdot[L\cdot\Delta n_i + (n_i - n_0)\cdot\Delta L]$, and $\phi = \frac{2\cdot\pi}{\lambda_0}\cdot(n_i - n_0)\cdot L$ is the optical phase of the light beam, where, n_i is the refractive index of the crystal, n_0 is the refractive index of the surrounding medium (air or other index matching index), and λ_0 is the wavelength of the light in vacuum, L is the length of the geometrical path of the beam inside crystal. n_i and L are parameters which related to the crystal and are changed under the influence of the electric field E(t), $\Delta n_i = \Delta n_i(E)$; $\Delta L = \Delta L(E)$.Δ, stand for change of specific crystal parameter, $n_i \to n_i + \Delta n_i$; $L \to L + \Delta L$ (Fig. E.29).

The speed of light slows down in a crystal dielectric relative to the speed in vacuum. The wavelength for an electromagnetic or light wave of a given frequency is smaller in a dielectric than it in vacuum: $\lambda_0 = \frac{c_0}{f}$; $c_0 = 3\times 10^8\ \frac{\mathrm{m}}{\mathrm{s}}$ (free space), $\lambda = \frac{c_0}{f\cdot\sqrt{\varepsilon_r}}$; $\lambda < \lambda_0$; $\sqrt{\varepsilon_r} > 1$. $c_0 = \frac{1}{\sqrt{\mu_0\cdot\varepsilon_0}} = 3\times 10^8\ \frac{\mathrm{m}}{\mathrm{s}}$; $\frac{c}{c_0} = \varepsilon_r$; $c = \frac{1}{\sqrt{\mu_0\cdot\varepsilon_0\cdot\varepsilon_r}} = \frac{1}{\sqrt{\mu_0\cdot\varepsilon}}$

Table E.4 Typical values of crystal dielectric content

Material	ε_r
Vacuum	1
Air	≈1.0006
Silicon	11.7
Calcium copper titanium	25 × 10^4

(in crystal medium). ε_0, is the permittivity of free space $\left(\varepsilon_0 = 8.854 \times 10^{-12} \left[\frac{\mathrm{F}}{\mathrm{m}}\right]\right)$, this is permittivity in vacuum. ε_r is the relative permittivity or dielectric constant. The permittivity of a medium (crystal) is expressed as the product of the dielectric constant and the free space permittivity $\varepsilon = \varepsilon_r \cdot \varepsilon_0$, ε_r is always greater than or equal to one ($\varepsilon_r \geq 1$), the outcome is that the E-field is always reduced relative to what the E-field could be in free space. The permittivity is a measure of this reduction; typical values of crystal dielectric content are presented in the next table (Table E.4).

If the medium is water then the dielectric constant start with $\varepsilon_r = 88$ at 0 °C and down to $\varepsilon_r = 34.5$ at 200 °C. Complex photo-elastic constants are tensors whose components define the effect of individual strain (elasto-optic coefficients) or stress (piezo-optic coefficients) tensor terms. The dielectric constant (ε_r) and refractive index (n_i) are functions of frequency, hence wavelength. The frequency wavelength variation of refractive index is called dispersion. Dispersion is an important property for optical design and transmission of information (pulse spreading). Other important optical properties are derived from the changes in refractive index with temperature, $n_i = f_1(T)$, which is thermo-optic coefficient, stress (force per unit area) or stain (resulting compression/stretching, deformation), $n_i = f_2(\sigma = \frac{F}{A})$, piezo-optic or elasto-optic coefficients, applied field $n_i = f_3(E)$, electro-optic or piezo-optic coefficients. Crystal dielectric constants are a second order tensor with three principal values, and the coefficients are defined as the tensor properties [32].

A liquid-crystal electro-optic modulating device is mostly identical to surface-stabilized ferroelectric liquid-crystal device. It used ferroelectric chiral smectics and takes the advantaged of the electro-clinic (EC) effect. There is a good achievable optical intensity modulation that can be observed. The response is linear with respect to the applied electric field and it good for modulator applications. The electro-clinic (EC) effect is the tilt of the optical axis of a liquid crystal in the plane perpendicular to an applied electric field. It is characterized by electro-clinic coefficients, the material property that estimates the linear coupling between tilt and electric field, in liquid-crystal. It is done by modeling the molecular properties (shape, electric dipole moment, and polarizability). The electro-clinic (EC) effect occurs when the liquid crystal is chiral (chiral is asymmetric in such a way that the structure and its mirror are not superimposable). Liquid crystals have properties between liquids and solids. The liquid crystal is chiral when the mirror image of the molecule cannot be superimposed on the molecule. A molecule is chiral if it is not superimposable on its mirror image. Most chiral molecules have lack of plane of symmetry or a center of symmetry. Basically chirality is a property of asymmetry. Liquid crystal molecules are chiral if they are distinguishable from their mirror image. A mirror image of

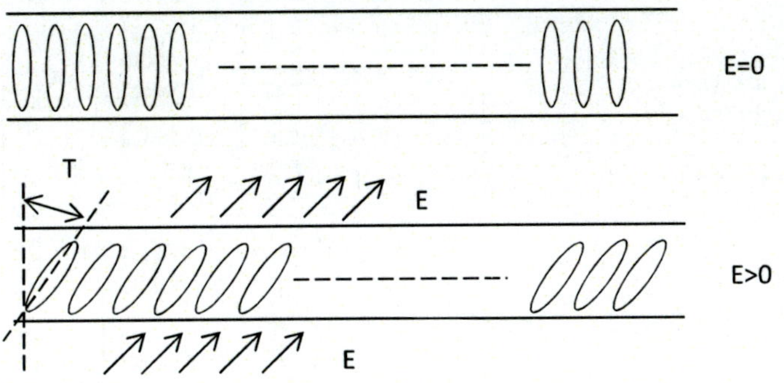

Fig. E.30 Tilt of the optical axis of a liquid crystal in the plane perpendicular to an applied electric field

achiral molecules cannot be distinguished from the molecule object (Fig E.30).

$$e_c = \frac{k}{(T - T_{AC*})}; \ \theta = e_c \cdot E = \frac{k \cdot E}{(T - T_{AC*})}$$

A chiral object (molecule) and its mirror image are called enantiomorphs (enantiomers). A non-chiral object (molecule) is called achiral (amphichiral) and can be superposed on its mirror image. Ferroelectric smectics are integrated part of liquid-crystal applications domain. It is surface stabilized ferroelectric liquid crystal (SSFLC) device. It is utilized any chiral tilted smectic phase which is ferroelectric. Spontaneous polarization happened in chiral tilted smectic phases. In non-tilted smectic phases build up from chiral molecules, an induced tilt on applying an external electric field. It is electro-clinic (EC) effect which the smectic layers are parallel to the confining glass plates. In non-tilted smectic phases (smectic A) which build up from chiral molecules by applying external electric field we get molecules induced tilt. It is electro-clinic effect on smectic layers. Practically the electro-clinic effect is small and need sensitive phase detection methods. The electro-clinical (EC) effect is general and good for all substances (good for wide range of the A phase). The electro-clinic effect behavior happened in number of substances. The liquid crystal is placed between two indium-tin-oxide-coated glass plates and electrode etching for overlapping area. The applied electric field E across the chiral smectic is above the C-A transition. The way to measure the electric field induced tilt of the optic axis is by apply simultaneously a DC voltage and AC voltage ($V_{ac} \ll V_{dc}$). The DC voltage (V_{dc}) induced a tilt and the AC voltage (V_{ac}) causes a modulation of that tilt and as a result modulation of the light intensity with AC frequency. The amplitude of modulation is proportional to the slope of $\phi(V_{dc})$ is curve. $\vec{n}_0$ and $\vec{n}(E)$ are the positions of the director ϕ without and with electric field E, φ is the angle of rotation of the director due to the electric field. The director $\vec{n}_0$ swings out to be $\vec{n}(E)$ and points new direction $\vec{n}(E)$ under electric field influence. The light intensity is I and it varied as $\sin^2(2 \cdot \phi)$, $I \propto \sin^2(2 \cdot \phi)$, where ϕ is the deviation of the optic axis from one of the polarizer directions. ϕ is a function of the applied

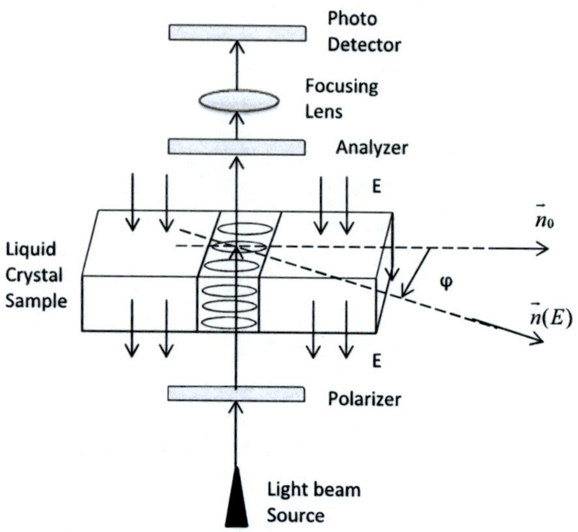

Fig. E.31 Structure of liquid crystal optic axis change on applying an electric field

electric field $\phi = \phi(E)$, which caused by electro-clinic (EC) effect (Fig. E.31). The intensity modulation which caused by V_{ac} is the differential of the light intensity I, ΔI, $\Delta I = \frac{dI}{d\phi} \cdot \frac{d\phi}{dE} \cdot \Delta E \sim \sin(4\phi) \cdot \frac{d\phi}{dE} \cdot \Delta E$ and $\frac{dI}{d\phi} \propto \sin(4 \cdot \phi)$

$$I \propto \sin^2(2\phi) \Rightarrow \frac{dI}{d\phi} \propto 2 \cdot \sin(2 \cdot \phi) \cdot \cos(2 \cdot \phi) = \sin(4 \cdot \phi).$$

The electromagnetic effect is a linear dependence of the change in the optic axis direction ϕ on the electric filed E. The direct application is light intensity modulation. To get the maximum intensity variation: $\Delta I \sim \sin(4\phi) \cdot \frac{d\phi}{dE} \cdot \Delta E \Rightarrow \frac{d\Delta I}{dt} \sim \frac{d}{dt}\left[\sin(4\phi) \cdot \frac{d\phi}{dE} \cdot \Delta E\right]$, $\frac{d\Delta I}{dt} \sim \frac{d}{dt}\left[\sin(4 \cdot \phi) \cdot \frac{d\phi}{dE} \cdot \Delta E\right] = 4 \cdot \cos(4 \cdot \phi) \cdot \frac{d\phi}{dE} \cdot \Delta E$. The extremum point is when $\frac{d\Delta I}{dt} = 0$ then $4 \cdot \cos(4 \cdot \phi) \cdot \frac{d\phi}{dE} \cdot \Delta E = 0$; $\frac{d\phi}{dE} \cdot \Delta E \neq 0$; $\cos(4 \cdot \phi) = 0 \Rightarrow 4 \cdot \phi = \pi \cdot (\frac{1}{2} + k)$; $k = 0, 1, 2, \ldots$ and $\phi_{extre} = \frac{\pi}{4} \cdot (\frac{1}{2} + k); k = 0 \Rightarrow \phi_{extre} = \frac{\pi}{8} = 22.5°$ (we get largest optical intensity at $\phi_{extre} = \frac{\pi}{8} = 22.5°$). To get the largest optical amplitude (intensity), the liquid crystal sample is oriented with its optic axis (the director) at an angle of $\phi_{extre} = \frac{\pi}{8} = 22.5°$ to the polarizer of the system. The crossed-polarizer setting is $\phi_{extre} = \frac{\pi}{8} = 22.5°$ then the response is linear and the absolute value of intensity variation (ΔI) is maximized. We can define the modulation depth as $\frac{\Delta I}{I} \sim \frac{\sin(4\phi)}{\sin^2(2 \cdot \phi)} \sim \frac{1}{\tan(2 \cdot \phi)}$ [33].

In high power laser facilities we monitor the phase retardation (PR) of an infrared femtosecond probe pulse. By using electro-optic (EO) crystals we detect propagation THz pulses. It is a very good way to inspect THz pulses shape in sub picosecond time domain. Typical EO crystals are $LiTaO_3$, $LiNbO_3$, and ZnTe. Tera Hertz (THz) pulses are dominant in the sub picosecond domain. A nearly flat frequency response, low dispersion for specific electro-optic (EO) crystal is detected in the far-infrared region.

The fractional lagging behind of waves; by lagging behind a portion of a wave length the corresponding phases, full amplitude, are kept back and retarded. There are some methods for optical phase retardation (PR) measurement with the wave plate. The methods are based on laser frequency splitting or laser feedback. Optical phase retardation elements (OPREs) widely used in scientific instruments such as heterodynes interferometer, polarizing microscope, optical communications and filters, optical attenuator and storage, etc. Retardation plates (wave plates), produce a phase shift, and the principle are based on polarization. Retardation plates are optical elements that create a phase shift in the transmitted light with birefringence crystalline quartz. A phase shift of 90° (l/4 plate, quarter wave plate, or circular polarizer) and phase shift of 180° (l/2 plate, or half wave plate) are used in many optical probe systems. Most wave plates are equipped with an anti-reflect (AR) coating which is optimizing to specific one wavelength. The optical phase retardation system includes electro-optic (EO) modulators wave plate, Fresnel prisms, etc. When light beam directed through wave plate (mostly quarter-wave plate), the result beams are ordinary light (o-light) and an extraordinary light (e-light). Their polarized directions are mutually perpendicular. There are fast and slow axis, the larger phase velocity is fast axis, and the smaller velocity is slow axis. There is a phase difference (ϕ) which is related to o-light and e-light belong to different phase velocities in two axis directions. The phase difference ϕ, $\phi = \frac{2\cdot\pi}{\lambda} \cdot |n_o - n_e| \cdot d$, where, d is the thickness of the wave plate, λ is the light beam wavelength, n_o and n_e are the refractive indices of o-light and e-light, respectively. Most popular phase retardation is $\phi = 2\cdot n \cdot \pi + \phi_0$, periodic in ϕ value. There are some typical phase retardation measurement methods, polarization rotation, Interference, Spectroscopy, Nonlinear optical, Time domain, Intra-cavity, and phase comparison methods. In the polarization rotation methods (PRM), light beam is passed through the polarizer, the compensator, the measured element and the analyzer. The rotation angle of the analyzer/polarizer, change of light intensity gives the phase retardation of the measured element. In interference method, element's phase retardation is measured by measuring the optical path difference (OPD). It is based on interference principle, or by knowing the interference beam's light intensity signal. In spectroscopy method we detect a series of extreme values in the transmission spectra of the light passing through the analyzer. The phase retardations are related to certain wave-lengths of a wave plate. In nonlinear optical method, we measure phase retardation which based on second harmonic generation (SHG) from thin film using two beams at the fundamental frequency. A time domain method is used for characterizing retardation plates and the phase retardation is determined by measuring the ring-down signal of an F-P cavity in time domain. Intra-cavity polarization method is good for small optical anisotropies. It is done by using intra-cavity polarization characteristics in a ring laser. The phase comparison method is based on comparing the phase difference between reference signal and test signal. A phase comparator is used to detect the phase retardation. Additional optical phase retardation measurement methods are laser frequency-splitting, polarization flipping, and orthogonal phase difference methods. We measure the system for wave plates by laser frequency-splitting method. The wave plate is placed in the resonant cavity of a half-intra-cavity laser. The divergent refractive indices in fast axis and slow axis of the wave plate and the optical lengths of laser cavity for o-light and e-light are different. One resonant frequency is split into two components with a certain frequency difference between each other. The laser resonance condition frequency is $\nu_q = \frac{c}{2 \cdot n \cdot l} \cdot q$, where, n is the refractive index of the medium in laser cavity, l is the length of laser cavity, q is a positive integer, and c is the light velocity in vacuum. The frequency difference $\Delta\nu$ between o-light and e-light is $\Delta L = \frac{\Delta\nu}{\Lambda} \cdot \frac{\lambda}{2}$; $L = n \cdot l$; $\Lambda = \frac{c}{2 \cdot L}$, where, ΔL is the optical phase difference (OPD) then $\Delta\nu = 2 \cdot \Delta L \cdot \frac{\Lambda}{\lambda}$; $\Delta\nu = \Delta L \cdot \frac{c}{\lambda \cdot L}$, where, Λ is the longitudinal mode spacing between two adjacent laser longitudinal modes. The phase difference ϕ is expressed as $\phi = \frac{2 \cdot \pi}{\lambda} \cdot \Delta L$; $\phi = \pi \cdot \frac{\Delta\nu}{\Lambda}$. The phase difference ϕ is the phase retardation of the wave plate. For calculation of the phase retardation, we need to know the longitudinal mode spacing Λ and the frequency difference $\Delta\nu$ in our system. By measuring small phase retardation or half-wave and full-wave plates, frequency splitting lock-in occurs. Frequency splitting lock-in phenomenon leads to zero frequency difference then we cannot measure the half-wave or full-wave plates. We compare the adjacent longitudinal mode spacing and phase offset with an additional quarter-wave plate. Phase offset increases the phase retardation and then increase the frequency difference. We compare the adjacent longitudinal mode spacing and the measured phase retardation is $\phi = \left[\frac{\Delta_1 - \Delta_2}{\Delta_1 + \Delta_2} + 1\right] \cdot \pi = \frac{2 \cdot \Delta_1}{\Delta_1 + \Delta_2} \times 180°$, where, Δ_1 and Δ_2 are the two adjacent longitudinal mode spacing the arbitrary phase retardation is measured and calculated. The polarization flipping method is based on the polarization flipping phenomenon which exists in laser feedback with a moving feedback mirror. The two sides of the flipping point are corresponding to the two eigenstates of one laser mode, respectively. The position of polarization flipping or duty cycle of the two polarization states in one period of laser intensity modulation is linear with the phase retardation of birefringence element. The phase retardation is calculated by knowing the duty cycle. In orthogonal phase difference method the initial state of laser output is single longitudinal mode, fundamental transverse mode and linearly polarized. It is a weaker optical feedback level method which the feedback mirror is uncoated. The polarization flipping phenomenon is not happened in that specific feedback level. The result phase difference is inspected by the phase retardation of the measured birefringent element. By using electro = optic (EO) effect (pockels effect), we detect the THz pulses. The electric THz field influences the refractive index ellipsoid of the electro-optic (EO) crystal. The refractive index n_i, $n_i = n_i(E_{THz})$, where i is the index of the principal axes x, y, z. The THz beam is propagated along the z direction different phase shifts, ϕ_x and ϕ_y are the results for the electric field components of electric fields E_x and E_y, respectively. The electric field strength E_x and E_y are related to optical probe beam which propagated along the z direction. The differential phase retardation (PR), $\delta\phi$, which caused by the probe beam due to the THz field over a distance dz is equal to $\delta\phi = \phi_y - \phi_x$. The experiment system includes laser probe beam source and THz pulse source, polarizer, electro-optic (EO) crystal ($LiTaO_3$, $LiNbO_3$, and ZnTe), analyzer, and photo detector (Fig. E.32).

Different orientations of the electro-optic (EO) crystals are related to the E_{THz} polarization. The pase retardation (PR) caused by the THz field is dependent on

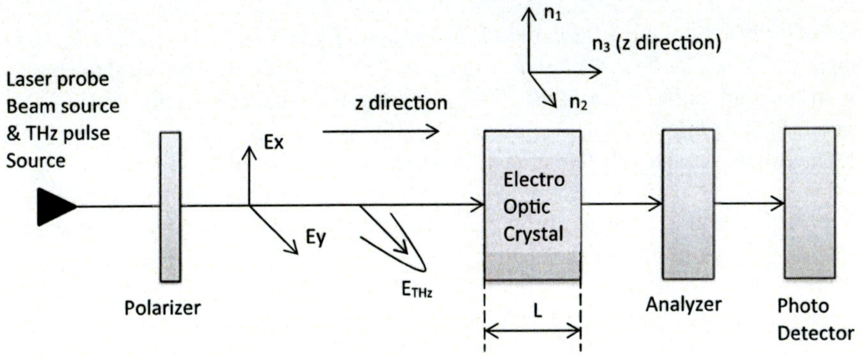

Fig. E.32 Experiment system (laser probe beam source and THz pulse source, polarizer, electro-optic (EO) crystal ($LiTaO_3$, $LiNbO_3$, and ZnTe), analyzer, and photo detector)

Table E.5 b-cut and c-cut crystals, $\delta\phi_{THz}(\Delta T)$, C_b, C_c, and $\delta\phi_{nat}$

	b-cut crystal	c-cut crystal
$\delta\phi_{THz}(\Delta T)$, differential phase retardation	$\frac{1}{2}\cdot\frac{\omega}{c}\cdot[n_0^3\cdot r_{13}-n_e^3\cdot r_{33}]\cdot E_{THz}(\Delta T)\cdot dz$	$\frac{\omega}{c}\cdot n_o^3\cdot r_{22}\cdot E_{THz}(\Delta T)\cdot dz$
C_b	$\frac{1}{2}\cdot\frac{\omega}{c}\cdot[n_0^3\cdot r_{13}-n_e^3\cdot r_{33}]$	NA
C_c	NA	$\frac{\omega}{c}\cdot n_o^3\cdot r_{22}$
$\delta\phi_{nat}$, static phase retardation	$\frac{\omega}{c}\cdot(n_e-n_o)\cdot dz$	0

crystal orientation. The differential phase retardation (PR) is $\delta\phi_{THz}(\Delta T)$, where C_b is related to b-cut crystals family (Table E.5).

$$\delta\phi_{THz}(\Delta T)=\frac{1}{2}\cdot\frac{\omega}{c}\cdot[n_0^3\cdot r_{13}-n_e^3\cdot r_{33}]\cdot E_{THz}(\Delta T)\cdot dz=C_b\cdot E_{THz}(\Delta T)\cdot dz$$

$$C_b=\frac{1}{2}\cdot\frac{\omega}{c}\cdot[n_0^3\cdot r_{13}-n_e^3\cdot r_{33}]$$

where, ΔT is the relative time delay between the THz pulse and the probe pulse, r_{13} and r_{33} are components of the electro-optic (EO) tensor, C_b is a constant appropriate for the b-cut crystal orientation, ω is the optical probe beam frequency and c is the velocity of light in vacuum. n_o and n_e are the refractive indices of o-light and e-light, respectively. The different phase retardation (PR) which is related to o-light and e-light. It is related to two phase velocities in two axis directions. The optical probe pulse and the THz pulse propagate along the z axis. The optical probe beam is linearly polarized and the THz field strength E_{THz} is parallel to y-axis. The specific private case of b-cut crystal, $n_o=n_1=n_3$ and $n_e=n_2$, and for c-cut crystal $n_o=n_1=n_2$ and $n_e=n_3$. Additionally there is a static phase retardation (PR), $\delta\phi_{nat}$ due to the b-cut crystal natural birefringence: $\delta\phi_{nat}=\frac{\omega}{c}\cdot(n_e-n_o)\cdot dz$. It is background signal. In the case of c-cut crystal the differential PR is $\delta\phi_{THz}(\Delta T) = \frac{\omega}{c} \cdot n_o^3 \cdot r_{22} \cdot E_{THz}(\Delta T) \cdot dz$; $\delta\phi_{THz}(\Delta T) = C_c \cdot E_{THz}(\Delta T) \cdot dz$ and $C_c = \frac{\omega}{c} \cdot n_o^3 \cdot r_{22}$. For c-cut crystal the probe beam does not suffer a static PR, ($\delta\phi_{nat} = 0$), but a smaller C-factor and smaller overall PR magnitude $\delta\phi_{THz}$ happened [34, 35].

Quarter wave plate (QWP): A quarter-wave plate consists of an adjusted thickness of a birefringent material such that the light beam is associated with the larger index of refraction and is retreated by 90° in phase, quarter wavelength with respect to that associated with the smaller index. The quarter-wave plate material geometric is that the optic axis is parallel to the front and back plates of the plate. If we shot a linearly polarized light beam to the plate, then it will divide into two components with different indices of refraction. The scientific application of a quarter-wave plate is to convert linearly polarized light to circularly polarize light and vice versa. It is done by adjusting the plane of the incident light, which makes 45° angle with the optic axis. The outcome is equal amplitude o-wave and e-wave. When the o-wave is slower, then it falls behind by 90° in phase, and produces circularity polarized light. When a linearly polarized light is incident on a quarter-wave plate 45° to the optic axis, then the light is divided into two equal electric field components. One is retarded by quarter wavelength by the plate. It produces circularly polarized light and the incident circularly polarized light is changed to linearly polarized light. The quarter wave plate has two perpendicular axes, x and y, each with a different index of refraction. Light polarized along one axis, E_x, goes out with a phase shifted by a quarter wave relative to the phase of beam light polarized along the axis, E_y. Any plane polarized light is decomposed into components, E_x and E_y, polarized along these axes. If the quarter wave plate optical axes are oriented at 45° with respect to the plane polarization of an incoming beam light, then the incoming beam light has equal components, E_x and E_y, along the optical axes of the quarter wave plate. The phase of one of these components gets shifted by one quarter wave $\left(\frac{\lambda}{4}\right)$ with respect to the phase of the other component. By adding two equal perpendicular components, E_x and E_y, and there is a phase difference of a quarter wave between two components, the result vector, E, sweeps out a circle (circular polarization). Other oriented degrees (quarter-wave plate) which are not equal to 45°, to the polarization of the incoming beam of light then the incoming electric filed components, E_x and E_y, are not equal, when one component is shifted with respect to the other. By adding two unequal perpendicular electric field components, E_x and E_y, and there is a difference between the two components (E_x and E_y) then the result vector, E, sweeps out an elilipse (elliptic polarization). Mainly we use wave plates which are transparent plates with a defined amount of birefringence when the target result beam is a modifying polarization of light. A wave plate has a slow axis and fast axis, both perpendicular to the beam of direction, and each other. The most useful types of wave-plates are quarter-wave plates ($\frac{\lambda}{4}$ plates) and half-wave plates ($\frac{\lambda}{2}$ plates), where the difference of phase delays between the two linear polarization directions is $\frac{\pi}{2}$ or π, respectively. They are corresponding to propagation phase shifts over a distance of $\frac{\lambda}{4}$ or $\frac{\lambda}{2}$, respectively. When the light beam is linearly polarized, and the polarization direction is along one of the axes of the wave plate, the polarization

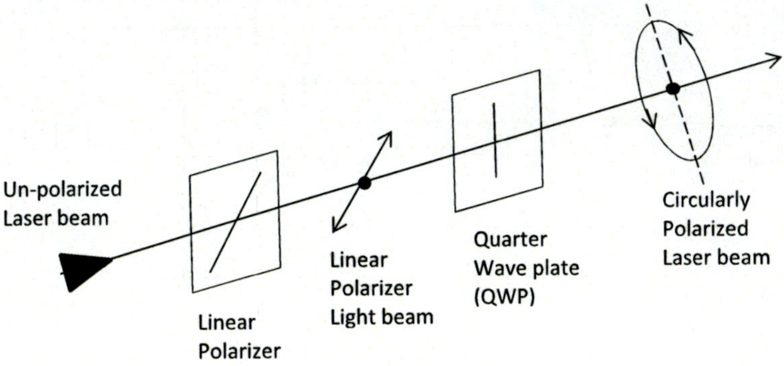

Fig. E.33 Quarter wave plate (QWP) light beam system

remains unchanged. If we want to get a single-frequency operation (twisted-mode technique) within a laser resonator then we place two quarter wave plates around the gain window. If we want to reduce depolarization loss then we insert a half-wave plate between a laser crystal and a resonator end mirror. The combination of a half-wave plate and a polarizer gives output coupler with adjustable transmission. The most popular wave plate's materials are SiO_2 which is crystalline quartz that exhibits a wide wavelength range with very high transparency, Calcite ($CaCO_3$), magnesium fluoride (MgF_2), Sapphire (Al_2O_3), a silicate material, and birefringent polymers (Fig. E.33). Polarization applications use quarter- and half-wave plates in achromatic, super achromatic, zero-order, low-order, or multi-order versions. In addition, we use Fresnel Rhombs, Liquid Crystal Retarders (LCR) come mainly in two sizes, Empty Liquid Crystal Cells, Pockels Cells, Rotating Retarders, and Soleil-Babinet Compensators.

Phase retarders are dependent on frequency. Phase retarder components are used in laser systems and laser facilities and convert a linearly polarized beam into a circularly polarized beam. The way to convert linear polarization to circular polarization, a total one-quarter wave, or $\frac{\lambda}{4}$ retardation (900) is used. We can use a single $\frac{\lambda}{4}$ retarder or two $\frac{\lambda}{8}$ retarders. The major phase retarder components are mirrors and used in high power CO_2 laser systems. The mirrors are treated with a special coating. The effect of $\frac{\lambda}{4}$ phase retardation is to cause the electric field direction to "spin" with the frequency of the radiation. There are some kinds of phase retarders, copper phase retarders; prism phase retarders (operate over a broad band of wavelength). One kind of quarter-wave plate (QWP) is terahertz achromatic quarter-wave plate. Achromatic wave plates use two birefringent crystalline materials which provide a constant phase shift independent on wavelength. Achromatic wave plates are constructed by aligning the fast axis of a birefringent quartz plate with the slow axis of another birefringent plate, can be made from magnesium fluoride or UV sapphire, resulting in a relative phase delay of $\frac{\lambda}{4}$ or $\frac{\lambda}{2}$ between the two axes. Quarter-wave plates (QWP) are mainly used to change linearly polarized light into circularly polarized light. The terahertz achromatic wave plate is constructed from six birefringent quartz plates

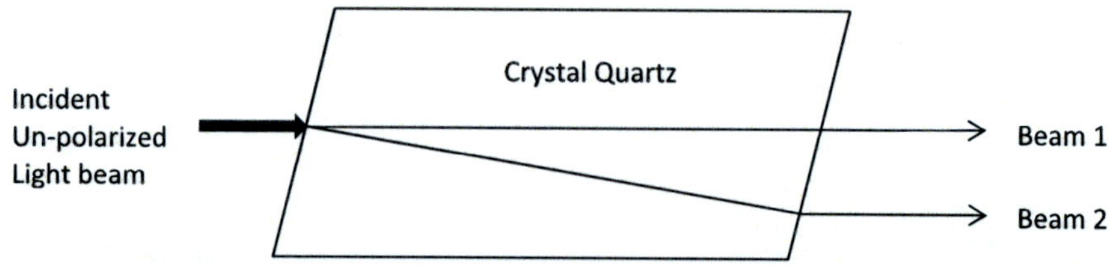

Fig. E.34 Birefringent wave plates separation of incident un-polarized light beam into parallel and orthogonal components

which stacked together. Birefringent wave plates are made from birefringent materials (crystal quartz) which have slightly different indices of refraction for light polarized in different orientations. They separate incident un-polarized light beam into its parallel and orthogonal components (Fig. E.34). There are fast axis and slow axis, light polarized along the fast axis distinguishes a lower index of refraction and travels faster through wave plates than the light polarized along the slow axis. The fast axis is indicated by a small flat spot or dot on the fast axis diameter of an unmounted wave plate. The word "Retardation" describes the phase shift between the polarization component projected along the fast axis and the component projected along the slow axis.

We specify retardation in units of degrees, waves, or nanometers. Full wave retardation is equivalent to 360°, or number of nanometers at specific wavelength. Typically, we state tolerance on retardation in degrees, natural or decimal fractions of a full wave, or nanometers ($\frac{\lambda}{4} \pm \frac{\lambda}{400}$, $\frac{\lambda}{2} \pm 0.005 \cdot \lambda$, $\frac{\lambda}{4} \pm 1°$, 780 nm ± 5 nm). The most used retardation values are $\frac{\lambda}{4}$, $\frac{\lambda}{2}$, λ. By using prism, prism internal refraction causes a phase shift between components; a compensating wave plate able to restore the desired polarization. In multiple order wave plates, the total retardation is the desired retardation plus an integer. Multiple order wave plates are designed with only a single birefringent material, which relatively thick, which gives easy system integration. In zero order wave plates, the total retardation is the desired value without excess. Quartz zero order wave plates consist of two multiple order quartz wave plates with their axes crossed. The effective retardation is the difference between the two multiple order quartz wave plates. Achromatic wave plates consist of two different materials that eliminate chromatic dispersion. Typical achromatic lenses are made from two types of glass. They are to achieve a desired focal length while minimizing or removing chromatic aberration. Achromatic wave plates operate on the same basic principle. In achromatic wave plates there is compensation of two materials, then for several spectral wavelengths there is different Retardant value (λ) (Fig. E.35).

In many scientific optical systems there is a need to change the existing polarization. Laser beam which is horizontally polarized and we want to change to vertical polarization. It is done by $\frac{\lambda}{2}$ wave plate which axes oriented 45°. Rotating the wave plate axis an angle θ from the incident polarization causes to 2θ rotation of the existing polarization. Inserting or rotating a $\frac{\lambda}{2}$ wave plate reconfigures the system optical setup with no need for realignment. The orientation of linear polarizer and $\frac{\lambda}{4}$

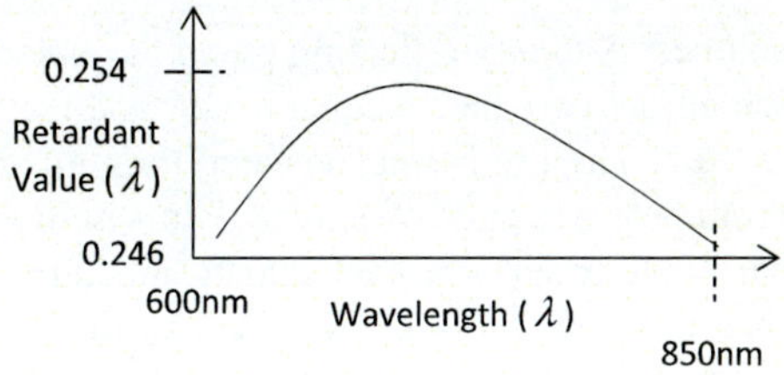

Fig. E.35 Achromatic wave plates, retardant value (λ) versus wavelengths

wave plate establish the transformation between linearly polarized light to circularly polarize light, and vice versa. If linearly polarized light enters a $\frac{\lambda}{4}$ wave plate at any angle beside 45° ($\theta \neq 45°$), then outcome of elliptically polarized beam light happened. If linearly polarized light enters a $\frac{\lambda}{4}$ wave plate at 45°, $\theta = 45°$, then we get linear polarization at 45° to the wave plate's axis. A linear polarizer and a $\frac{\lambda}{4}$ wave plate create an optical isolation system where light polarized by the linear polarizer passes through the $\frac{\lambda}{4}$ wave plate without attenuation and it is transformed to circular polarization. By using a polarizing beam splitter (linear polarizer in the optical isolation) the returning light into an alternate path is redirected without attenuation. Wave plates, retardation plates or phase shifters are made from birefringence materials. What happened when a laser beam light heats the wave plate? The velocities of the extraordinary and ordinary light rays through the birefringent materials vary inversely with their refractive indices. The difference in velocities causes to a phase difference when the two beams recombine. An incident linearly polarized beam phase difference (α) when the two beams recombine is $\alpha = 2 \cdot \pi \cdot d \cdot \frac{(n_e - n_o)}{\lambda}$, where, α is the phase difference, d is the thickness of wave plate, n_e and n_o are refractive indices of extraordinary and ordinary rays, respectively, λ is the wavelength. Low order wave plate properties are much better than the multi-order wave plate because of its thinner thickness. A zero-order wave plate is constructed of two multi-order wave plates with their axis crossed. The effect of two plates counteracting each other performs a zero-order wave plate. The true zero order wave plate has a thin thickness which makes it good in high temperature. Terahertz polarimetry measured the phase retardation. The achromatic wave plate is characterized by a big frequency bandwidth ($\nu_{\text{max}} \approx 7 \cdot \nu_{\text{min}}$) and is applicable to terahertz time domain scientific systems. The terahertz (THz) frequency band is located between the microwaves and infrared light bands. We need to characterize the THz beams by divergence, spatial or frequency modes, and characteristic profile. Polarimetry techniques use polarization converters and wave plates. The functionality of polarization converters and wave plates (WPs) is dependent on wavelength. Birefringent wave plates are made from birefringent materials for specific wavelength value [36]. Total reflection achromatic dephasing properties are used in Fresnel rhombs or total reflection prisms. A Fresnel rhomb is an optical prism that introduces a 90° phase difference between two perpendicular components of polarization, by means of two total internal reflections. Fresnel Rhomb Retarders act like broadband wave plates providing uniform λ/4 or λ/2 retardant over a wider range of wavelengths than possible with birefringent wave plates. They can replace retardation plates for broadband, multi-line or tunable laser sources. A totally reflecting prism has one of its angle equal to 90° and each of the remaining two angles equal to 45° and the phenomenon of total internal reflection is used. The ray is thus deviated through an angle of 90°. Total internal reflection (complete reflection of a ray of light within a medium) such as water or glass from the surrounding surfaces back into the medium occurs if the angle of incidence is greater than a certain limiting angle, called the critical angle. Total Internal Reflection of Light is used in many optical instruments like telescopes, microscopes, binoculars, spectroscopes, periscopes, etc. If we combine two wave plates (WPs) of different materials then a phenomenon of partial cancelation between the dispersion of the two materials happened. THz achromatic quartz (TAQ) quarter-wave plate (QWP) design combined a standard half-wave plate (HWP) and a standard quarter-wave plate (QWP). The outcome is extended bandwidth of the resulting retardation plate in the whole visible range. There is a partial cancellation of the change of retardation from each plate with respect to the frequency. Quartz material is transparent in the THz range and is characterized by strong birefringent, typical values of ordinary and extraordinary refractive indices are $n_o = 2.108$ and $n_e = 2.156$ respectively at 1THz. The quartz plates are stacked together and are cut parallel to the optical axis. We characterize each plate by Jones matrix J_i (i is plate index number).

$$J = \prod_i J_i = J_1 \cdot J_2 \cdot \ldots = \begin{pmatrix} A & B \\ -B^* & A^* \end{pmatrix}$$

$$A = A(\delta_i, \theta_i) = \cos\left(\frac{\delta_i}{2}\right) + i \cdot \cos(2 \cdot \theta_i) \cdot \sin\left(\frac{\delta_i}{2}\right)$$

$$B = B(\delta_i, \theta_i) = i \cdot \sin(2 \cdot \theta_i) \cdot \sin\left(\frac{\delta_i}{2}\right); \; B^* = -i \cdot \sin(2 \cdot \theta_i) \cdot \sin\left(\frac{\delta_i}{2}\right); \; B = -B^*$$

Polarized light is represented by a Jones vector, and linear optical elements are represented by Jones matrices (J_i). We find the emerging light which crosses an optical element (quartz plate) and is polarized, by the products of the Jones matrix of the optical element (quartz plate) and the Jones vector of incident light. The Jones vector describes the polarization of light in free space or homogeneous isotropic non-attenuating medium, where the light can be properly described as transverse waves. Functions A and B are dependent on two parameters, dephasing δ_i and orientation θ_i with respect to the optical axis. The dephasing δ_i is equal to $e_i \cdot (n_e - n_o) \cdot \frac{\nu}{c}$; $\left(\delta_i = e_i \cdot (n_e - n_o) \cdot \frac{\nu}{c}\right)$.

The retardation dephasing δ expression is $\tan^2\left(\frac{\delta}{2}\right) = \frac{|\text{Im}A|^2 + |\text{Im}B|^2}{|\text{Re}A|^2 + |\text{Re}B|^2}$ and $\text{Im}A = \cos(2 \cdot \theta_i) \cdot \sin\left(\frac{\delta_i}{2}\right)$; $\text{Re}A = \cos\left(\frac{\delta_i}{2}\right)$; $\text{Im}B = \sin(2 \cdot \theta_i) \cdot \sin\left(\frac{\delta_i}{2}\right)$; $\text{Re}B = 0$.

$$\tan\left(\frac{\delta}{2}\right) = \pm\sqrt{\frac{|\text{Im}A|^2 + |\text{Im}B|^2}{|\text{Re}A|^2 + |\text{Re}B|^2}}; \; \frac{\delta}{2} = \arctan\left\{\pm\sqrt{\frac{|\text{Im}A|^2 + |\text{Im}B|^2}{|\text{Re}A|^2 + |\text{Re}B|^2}}\right\}$$

$$\delta = 2 \cdot \arctan\left\{\pm\sqrt{\frac{|\mathrm{Im}A|^2 + |\mathrm{Im}B|^2}{|\mathrm{Re}A|^2 + |\mathrm{Re}B|^2}}\right\}.$$

Typical quasi-optic components are phase shifter, modulators, attenuators, polarizers and wave plates. A tunable phase shifter can be based on optically or electrically controlled carrier concentration in quantum-well structures. A Nematic Liquid Crystal (NLC) exhibits relatively large birefringence and a small extinction coefficient at frequency around 1THz. The tunable room-temperature THz phase shifter is capable of more than 360° of phase shift at 1THz. The device is based on magnetically controlled birefringence in a sandwiched dual NLC cell. The device is operated as a THz quarter-wave $\left(\frac{\lambda}{4}\right)$ plate. The THz wave propagation and polarization directions are z and y, respectively. The cell is constructed by sandwiching E7 (Merck) layer between two fused silica substrates. A square-wave AC voltage at 1 kHz is applied to the NLC $\frac{\lambda}{4}$ plate electrodes. If we don't supply the biasing field, the NLC molecules are in the initially stable state and perpendicular to the substrates. When the applied bias voltage is increased beyond the threshold voltage, $V > V_{th}$; $V_{th} = \pi \cdot \left(\frac{L}{d}\right) \times \sqrt{\frac{k_3}{\varepsilon_a \cdot \varepsilon_0}}$ then the E7 molecules in the bulk of the cell reoriented toward the applied electric field (positive dielectric anisotropy), where, L is the distance between two electrodes, d is the thickness of the NLC layer, k_3 is the bend elastic constant [37], $\varepsilon_a = \varepsilon_{\parallel} - \varepsilon_{\perp}$ is the dielectric anisotropy, and ε_0 is the electric permittivity of free space. In case of nematic liquid crystals with a positive dielectric anisotropy, $\varepsilon_a = \varepsilon_{\parallel} - \varepsilon_{\perp} > 0$, where, $\varepsilon_{\parallel}$ and $\varepsilon_{\perp}$ are the electric permittivity measured, respectively, parallel and perpendicular to the macroscopic molecular ordering (the direction n), the probing electric field E of a sufficient intensity causes a distortion of the director. The phase shift $\delta(V)$ experienced by the THz beam which is transmitted through the cell bias at a voltage (V), $\delta(V) = \int_0^d \frac{2 \cdot \pi \cdot f}{c} \cdot \Delta n_{eff}(V, z) \cdot dz$, where, f is the frequency of the THz wave, c is the speed of light in vacuum, and $\Delta n_{eff}(V, z)$ is the change in the effective birefringence for the NLC at position z along the propagation direction of the THz beam. The effective birefringence is $\Delta n_{eff} = \frac{1}{\sqrt{\frac{\cos^2\theta}{n_o^2} + \frac{\sin^2\theta}{n_e^2}}} - n_o$, where, n_o and n_e are ordinary and extraordinary indices of refraction of NLC and θ is the reorientation angle of NLC molecules from the original orientation. For $V > V_{th}$ the angle θ at any point z in the cell is computed by $\frac{z}{d} = \frac{V_{th}}{\pi \cdot V} \cdot \int_0^\theta \sqrt{\frac{1 + q \cdot \sin^2\theta}{\sin^2\theta_m - \sin^2\theta}} \cdot d\theta$, where $q = \frac{(k_1 - k_3)}{k_3}$ and $k_1 = 11.1 \times 10^{-12}$ N. k_1 is the splay elastic constant of NLC. The angle θ_m is the maximum reorientation angle located at $z = \frac{d}{2}$ then $\frac{V}{V_{th}} = \frac{2}{\pi} \cdot \int_0^{\theta_m} \sqrt{\frac{1 + q \cdot \sin^2\theta}{\sin^2\theta_m - \sin^2\theta}} \cdot d\theta$. Then we get the profile of molecular orientation in the cell for a given applied voltage (V) [38].

Polarization beam splitter (PBS): Many scientific laser systems include polarizing beam splitters (PBSs). Polarizing beam splitters are used to split un-polarized light into two polarized parts. The polarization beam splitters split light by polarization state rather than by wavelength or intensity. Semiconductor and photonic instruments use polarizing beam splitter element to transmit p-polarized light while reflecting s-polarized light. Polarizing beam splitters are typically designed for 0° or 45° incident with a 90° separation of the beams. There are many kind of polarizing beam splitters elements including plate, cube, or lateral displacement. Plate beam splitters are characterized by size, and type of laser light beam (UV, visible or infrared wavelengths). Cube beam splitters are good for applications requiring durability or simplified mounting or system integration. Lateral displacement beam splitters are designed to split the incident light beam into two displaced parallel beams. The polarization beam splitters are available for scientific laser wavelengths or wavelength ranges. Polarization beam splitters (PBSs) are designed to split un-polarized light at a specific Reflection/Transmission (R/T) ratio with unspecified polarization tendencies. They split light into reflected S-polarized and transmitted P-polarized beams. The beam splitter material is a light transmitting glass with mirror effect. The beam splitter glass has an optical thin-film coating on one side that reflects and transmits light waves that fall upon it at a specific ratio. The core material of beam splitter glass can be monolithic glass, laminated glass or toughened/tempered glass. While the main use of optical beam splitter is to split incident light at a designated ratio into two separate beams, it can be used in reverse to combine two different beams into a single beam. Cube beam splitters are constructed using two typically right angle prisms. The hypotenuse surface of one prism is coated, and the two prisms are centered together and form a cubic shape. The plate beam splitters are constructed a thin, flat glass plate that has been coated on the first surface of the substrate. Mostly, plate beam splitters have an anti-reflection coating on the second surface to remove unwanted Fresnel reflections. Additional beam splitter is a diffractive optical element (DOE) which used to split a single laser beam into several beams, each with the characteristics of the original beams, beside the power and angle of propagation. A beam splitter can generate either a one dimensional beam array ($1 \times N$) or two dimensional beam's matrix ($M \times N$), depending on the diffractive pattern on the element. Another kind of laser beam splitter is multisport diffractive optical element (DOE), a multi-line array, where instead of a $1 \times N$ array of spots, we use a $1 \times N$ array of lines, whose length is determined for specific design and application.

We can implement polarization beam splitter (PBS) by using a binary blazed grating coupler. It use a two-layer grating coupler and can directly couple the normally incident light from fiber into two separate waveguides according to their polarization states and in parallel splitting them. By using binary blazed grating couplers we get high coupling efficiency and a good extinction ratio. For evaluation of vertical coupler with transverse-electric polarized incident light, we calculated the coupling efficiency, which is defined as the ratio between the powers coupled to the fundamental mode of the chip waveguide. It carried by the fundamental mode of the input optical fiber. Optical systems such as polarization-based imaging systems, magneto-optic data storage in optical information processing and optical switching, routing, and isolating in optical communication use polarization beam splitters (PBSs). Conventional PBS use typical natural birefringent of crystals such as Glan-Thomson prisms, Nicol prisms, and Wollaston prisms. Glan-Thomson polarizers consist of two cemented prisms mode from the highest optical grade calcite. Un polarized light enters the polarizer, then is split at the intersection of the two crystals allowing s-polarized (e-ray) light to continue and p-polarized light (o-ray) to be reflected. Due to the large field of view, these polarizers can be used for highly converting or diverging beams. The maximum optical intensity of Glan-Thompson polarizes is limited by the cemented prism interface. Nicol prism polarizer is an optical device made from calcite crystal and used to produce analyze plane polarized light. When an ordinary ray of light is passed through a calcite crystal, it is broken up into two rays. The first ray is ordinary ray which is polarized and has its vibrations perpendicular to the principle section of the crystal. The second ray is the extra ordinary which is polarized and its vibration parallel to the principle section. In Nicol prism ordinary ray is eliminated extra ordinary ray, which is plane polarized, is transmitted through the prism. Wollaston prisms split un-polarized incident light into two orthogonally polarized light outputs. Wollaston prisms are commonly used in rotation mounts. The Wollaston prisms are fabricated from select-grade calcite prism pairs, cemented together, and assembled for convenient mounting. It separates randomly polarized and un-polarized light into linearly polarized light beams. It provides the widest deviation angle of all beam-displacing polarizers. Another way to implement PBS is by using the polarization properties of multilayer dielectric coatings. High extinction ratio is accepted from many layers and a narrow wavelength bandwidth. Polarization beam splitter which based on negative refraction character of photonic crystal is widely used in many scientific systems and facilities. Polarization beam splitters exhibits feature which the polarized light is separated in free space (medium) and needs oblique incidence for the total internal reflection condition or Brewster angle. Some PBSs are used integrated waveguide technology to separate light between waveguides and need couplers to couple light into waveguides first, based on the birefringent of materials, which require a long distance to separate two polarizations. Additional PBS is based on the strong polarization dependence feature of a binary blazed grating coupler which functional as a PBS and a coupler. It also used as a polarimetry unit to detect the polarization state of incident light from the fiber by comparing the energies in the two waveguides. A polarimetry unit is widely used in many scientific instrument and measure the angle of rotation caused by passing polarized light through an optically active substrate. PBS allows planar coupling which ease the design. The PBS is constructed from two-layer polarization-sensitive SWG (sub wavelength grating) couplers. Subwavelength grating (SWG) structures are an essential tool in silicon photonics, enabling the synthesis of metamaterials with a controllable refractive index. It is recommended to tilt the grating elements to gain control over the anisotropy of the metamaterial. The PBS upper layer is designed for TE (Transverse Electric) mode coupling, while the lower layer is for the TM (Transverse Magnetic) mode. The incident light with both polarizations is coupled separately into two waveguides. The light in the two waveguides is in the TE waveguide mode, while a two-layer grating coupler structure splits the input light into TE and TM waveguide modes. A two-layer grating coupler as a PBS is constructed from upper grating coupler for the TE mode (layer A), spacer between two couplers (layer B), lower grating coupler for the TM mode (layer C), SiO_2 spacer (layer D), and Si substrate (layer E) (Fig. E.36).

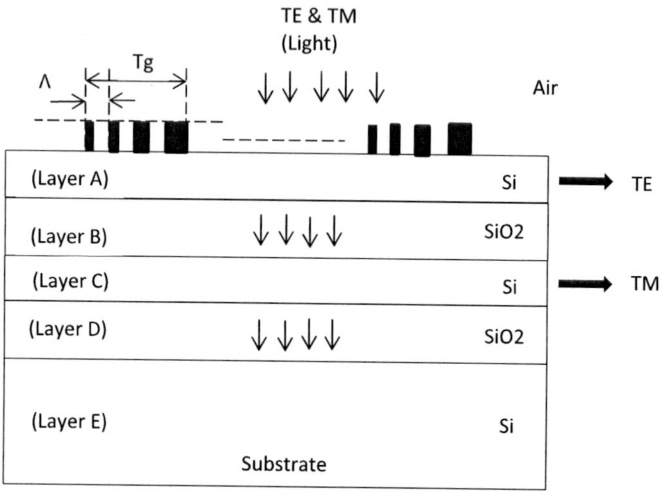

Fig. E.36 Polarization beam splitter (PBS) two-layer grating coupler

A grating coupler is a function of polarization. The TE and TM modes cannot be maximally coupled into one waveguide at the same angle of incidence by a grating coupler. There is a difference in effective index between the fundamental TE and TM waveguide modes then the polarization is dependent on the conventional symmetrical grating coupler. The coupling efficiency difference of the grating coupler is described by the phase matching equation: $T_g \cdot (N \pm n \cdot \sin\theta) = \lambda$, where, λ is the wavelength in vacuum, T_g is the grating period, N is the effective refractive index of the waveguide for the propagating mode, n is the refractive index of input space, and θ is the input beam angle. The effective refractive index (N) of the waveguide for the propagating mode (TE or TM) is $N = \frac{\lambda}{T_g} \mp n \cdot \sin\theta$. The effective refractive indices N for TE and TM waveguide modes are different and the TE and TM modes get their maximal coupling efficiencies at different input angle θ. For specific difference of angle (θ) only one polarized light can be coupled into the waveguide efficiently, while other polarized light is transmitted out of the substrate. Usually the grating coupler is for normal incidence which the maximum achievable coupling efficiency of a symmetric grating is not enough. For higher coupling efficiency and good extinction ratio, binary blazed gratings are used. The ratio of the power of a plane-polarized beam that is transmitted through a polarized placed in its path with its polarizing axis parallel to the beam's plane, as compared with the transmitted power when the polarizer's axis is perpendicular to the beam's plane. The extinction ratio is used to describe the efficiency with which the transmitted optical power is modulated over the fiber-optic transport and related also to our two-layer grating coupler as a PBS. The binary blazed grating is composed of subwavelength pillars with uniform height. The binary blazed grating exhibits higher coupling efficiency than the conventional blazed grating. A grating coupler with two layers is which each layer of the grating coupler is optimized individually for one polarization state. The coupling efficiencies for both polarizations reach their maximum. The binary blazed grating is used to modulate the grating widths to imitate the phase shift of a conventional blazed grating. The binary blazed grating shows linear phase shift along one grating period then the effective refractive indices (ERIs) are vary linearly in one grating period. The effective refractive indices (ERIs), n_{eff}^{TE} and n_{eff}^{TM}, of binary grating with a localized sub wavelength structure consist of ridges of material n_g with material n_1 in between and we get the n_{eff}^{TE} and n_{eff}^{TM} expressions for TE and TM polarization, respectively, where f is the fill factor, which is defined as the ratio of pillar width to grating sub period (Λ).

$$n_{eff}^{TE} = \sqrt{f \cdot n_g^2 + (1-f) \cdot n_1^2};\ n_{eff}^{TM} = \frac{1}{\sqrt{\frac{f}{n_g^2} + \frac{(1-f)}{n_1^2}}}$$

Typically $n_1 = 1;\ n_g = 3.7$ and we plot $n_{eff}^{TE} = n_{eff}^{TE}(f)|_{n_1=1;\ n_g=3.7}$ and $n_{eff}^{TM} = n_{eff}^{TM}(f)|_{n_1=1;\ n_g=3.7}$.

MATLAB:

```
n1=1;ng=3.7;
f=0:0.01:1;
plot(f,sqrt(f*ng.^2+(1-f)*n1^2),'r');hold;
plot(f,1./sqrt((f./(ng.^2))+(1-f)./(n1.^2)),'b');hold;
xlabel('Fill factor [f]');ylabel('[n^TE]eff, [n^TM]eff ');
title ('ERI of TE and TM modes as a function of fill factor (f) ');
```

It is possible to control the width of each pillar or the fill factor (f) to get the desired refractive index distribution. There is a different on the ERI (effective refractive indices) distributions behavior for TE and TM modes (birefringence) [39] (Fig. E.37).

Example: The input field is TE or TM mode with $\lambda = 1.55\ \mu m$ in vacuum. The binary blaze gratings are 0.3 μm deep on a 0.3 μm thick Si waveguide. The period of the upper grating coupler is $T_g(\text{up}) = 0.5\ \mu m$. The period of the lower grating coupler is $T_g(\text{low}) = 0.6\ \mu m$. The upper layer grating is optimized for TE mode coupling, and the lower grating is optimized for TM mode coupling [40].

Another kind of PBSs (polarization beam splitters) are implemented the anisotropic spectral reflectivity (ASR) characteristic of high spatial frequency multilayer binary gratings. Using ASR effect with a coupled wave analysis is result with an optical element that is transparent for TM polarization and reflective for TE polarization at an arbitrary incidence angle and operational wavelength. Reflectance anisotropy of objects arises in variety of ways, and being viewing and illuminating angles. We define the ASR characteristics of the device and these ASR devices combine many characteristics, such as compactness, low insertion loss, high efficiency, and broad angular and spectral bandwidth operations. A polarizing beam splitter (PBS) is an essential element in numerous optical information processing and imaging systems and it separate two orthogonal polarized light beams. By different polarization states, we increase the information bandwidth and reduce the cross talk

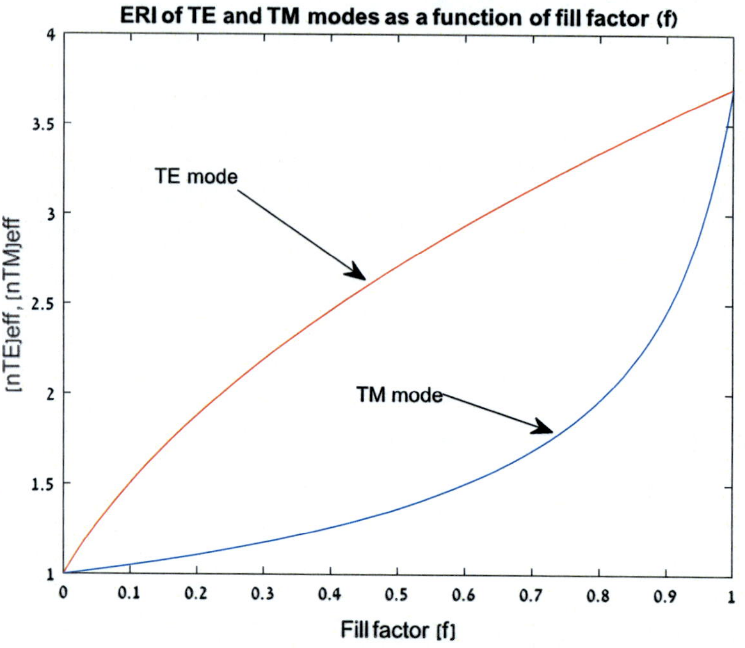

Fig. E.37 ERIs ($n_{eff}^{TE}, n_{eff}^{TM}$) of TE and TM modes as a function of fill factor (f)

Table E.6 Conventional PBSs, Wollaston prism and PBS cubes and description

Conventional PBS	Description
Wollaston prism	It requires a large thickness to generate enough walk-off distance between the two orthogonal polarizations owing to the intrinsically small birefringence of natural anisotropic materials
PBS cubes	It based on polarization selectivity of multilayer structures. It easy to fabricate, but it provide good extinction ratios only in a narrow angular bandwidth for a limited spectral range

between different channels. Integrate PBS unit in optical scientific applications gives high extinction ratios tolerate a wide angular bandwidth for high-resolution imaging, a broad wavelength range for operation with broadband sources, and a compact size for effective packaging. Conventional PBS is based on either natural crystal birefringence or polarization selectivity of multilayer structures. The next table describes the conventional PBSs, Wollaston prism and PBS cubes and description (Table E.6).

The PBS device that uses the unique properties of anisotropic spectral reflectivity (ASR) characteristics of a high-spatial-frequency multilayer binary grating is widely used in many scientific systems. The ASR mechanism is based on combining the effects of the form birefringence of a high-spatial-frequency grating with the resonant reflectivity of a multilayer structure. The ASR is designed to operate with normally incident light and it acts as a highly efficient polarization-selective mirror. There are some main important design topics for PBS ASR device:

- The principle of the ASR effect of a high-spatial-frequency multilayer binary grating.
- The effective-medium theory (EMT) and the high-spatial-frequency multilayer binary grating possess such characteristics.
- The design methodology of the PBS that employs ASR properties.
- The angular and the wavelength dependence of the ASR PBS.
- The effects of various fabrication errors on performance of the PBS.
- The fabrication techniques to make PBS.
- PBS design, the polarization extinction ratio and efficiency for operation with waves of wide angular bandwidth.

We consider a multilayer structure formed on a substrate by deposition of alternating layers of isotropic dielectric materials with high and low indices of refraction n_h and n_l, respectively. It exhibits high reflectivity in a wide spectral bandwidth. The thickness of each layer is corresponded to a quarter-wave optical thickness for a selected wavelength. A high reflectivity spectral band is centered at that specific wavelength. It is a principle of anisotropic spectral reflectivity (ASR). A typical structure (quarter-wave) is constructed of two isotropic materials, Si and SiO_2, and the number of layers is n_{layers}. We consider layers refractive indices are $n_{\mathrm{Si}} = 3.48$ and $n_{\mathrm{SiO_2}} = 1.44$ for an incident wavelength of $\lambda = 1.523\,\mu\mathrm{m}$. It shows a high-reflectivity spectral band near that wavelength. We can increase the reflectivity of quarter-wave layered structure by increasing the value of the ratio $\frac{n_h}{n_l}$ and the total number of layers (n_{layers}).

Additionally a larger value of the ratio $\frac{n_h}{n_l}$ increases the spectral bandwidth of high reflectivity (Fig. E.38).

For an optical field which the angle of incident is normal the multilayer structure (isotropic dielectric materials) shows identical reflectivity spectra for any two orthogonal linear polarizations and is due to the symmetry of the structure for a normally incident wave. In that case the structure cannot be used to separate normally incident fields by polarization. If we want to get a multilayer structure that possess reflectivity spectral bands centered at different wavelengths for the two orthogonal polarizations at normal incident then we use birefringent materials which replace one of the isotropic materials that are used to form a multilayer structure (Fig. E.39). The thickness of the $LiNbO_3$ layer (anisotropic material) is as the quarter-wave optical thickness corresponding to the refractive index for the ordinary wave. The outcome of $LiNbO_3$ layers is that the spectral reflectivity bands are separated. It is anisotropic spectral reflectivity (ASR) phenomenon and since the natural anisotropic materials possess small birefringence, this separation is small. By implementing a high-spatial-frequency multilayer binary grating, the separation between reflectivity spectral bands is increased owing to the exceptionally large anisotropic which obtained with a form-birefringent nanostructure. In high spatial frequency gratings which are constructed of isotropic dielectric materials, we see form-birefringence-effects. The two orthogonally polarized optical fields (TE field and TM field) show

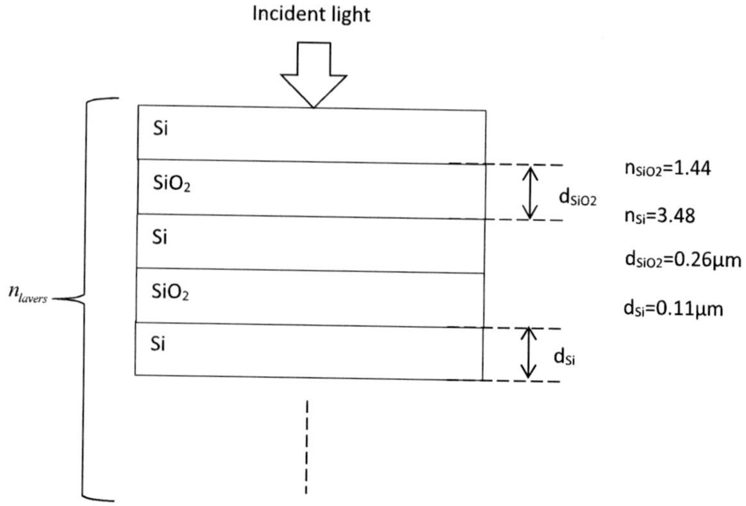

Fig. E.38 Schematic of n_{layers} quarter-wave structure constructed from two isotropic materials, Si and SiO_2

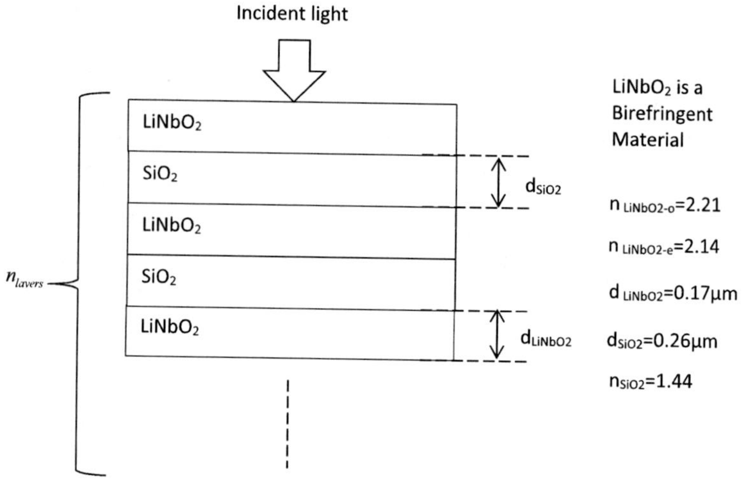

Fig. E.39 Schematic of n_{layers} quarter-wave structure constructed from isotropic material (SiO_2) and anisotropic material ($LiNbO_3$)

different boundary conditions, resulting in distinct effective indices of refraction. The TE field is parallel to the grating grooves and the TM field is perpendicular to the grating grooves. Gratings can be classified into several different types according to the profile of the grooves.

Typical three gratings are Blazed Holographic Gratings (BHG)—Sawtooth grooves, Holographic gratings (HG)—sinusoidal grooves, Holographic gratings (HG)—sinusoidal grooves, and Laminar gratings (Rectangular grooves). The Blazed Holographic gratings (BHG) exhibit high diffraction efficiency for specific wavelengths in a range from UV to visible light. The blaze wavelength which corresponding to peak diffraction efficiency is determined by the blaze angle and they are widely used in spectroscopes for UV and visible light. The Holographic gratings (HG) are characterized by relatively high diffraction efficiency at a broad range of wavelengths. The groove depth establishes the wavelength corresponding to the peak diffraction efficiency. The diffraction efficiency is around half that of blazed holographic gratings (BHG) and the value of groove depth period is large. Holographic gratings (HG) are used for wide wavelength regions and near intra red range. In Laminar gratings the diffraction efficiency for even orders of light is lower than that of BHG and HG. The wavelength corresponding to the peak diffraction efficiency is determined by the groove depth. The Laminar gratings are used in the soft X-ray region and large reflectance is accepted. The TE and TM fields light propagate through the grating structure and need a relative phase difference similar to that obtained in natural anisotropic materials. It is similar because the sub wavelength structure is designed such that only the zeroth diffraction order is propagated, while all the higher diffraction orders become evanescent. A high-spatial-frequency grating at the boundary between two isotropic materials is considered as an equivalent thin film of anisotropic material. The effective indices $n_{TE}^{(2)}$ and $n_{TM}^{(2)}$ of refraction of a surface-relief high spatial frequency binary gratings for normally incident TE and TM polarized optical waves, are evaluated by the second order Effective Medium Theory (EMT), $n_{TE}^{(2)} = n_{TE}^{(2)}(\Lambda, \lambda, F, n_I, n_{III})$ and $n_{TM}^{(2)} = n_{TM}^{(2)}(\Lambda, \lambda, F, n_I, n_{III})$:

$$n_{TE}^{(2)} = \sqrt{\left[n_{TE}^{(0)}\right]^2 + \frac{1}{3} \cdot \left(\frac{\Lambda}{\lambda}\right)^2 \cdot \pi^2 \cdot F^2 \cdot (1 - F)^2 \cdot (n_{III}^2 - n_I^2)^2}$$

$$n_{TM}^{(2)} = \sqrt{\left[n_{TM}^{(0)}\right]^2 + \frac{1}{3} \cdot \left(\frac{\Lambda}{\lambda}\right)^2 \cdot \pi^2 \cdot F^2 \cdot (1 - F)^2 \cdot \left(\frac{1}{n_{III}^2} - \frac{1}{n_I^2}\right)^2 \cdot \left[n_{TE}^{(0)}\right]^2 \cdot \left[n_{TM}^{(0)}\right]^6}$$

The above $n_{TE}^{(2)}$ and $n_{TM}^{(2)}$ equations parameters typical values and description is presented in the next table (Table E.7).In the current case the high-spatial-frequency grating is formed in a multilayer structure and made of two isotropic materials, where the composition becomes an artificial anisotropic multilayer structure which possess ASR (anisotropic spectral reflectivity) characteristics (Figs. E.40, E.41, E.42 and E.43). It can be fabricated by etching a high spatial frequency binary grating directly into a multilayer mirror structure. In the current example two isotropic material (Si, SiO_2) are used for construction the multilayer structures. The SiO_2 and Si materials are characterized by fabrication compatibility and low absorption coefficients in the near-infrared region which ensure a low insertion loss of the device. The calculated

Table E.7 $n_{TE}^{(2)}$ and $n_{TM}^{(2)}$ equations parameters typical values and description

Parameter	Description	Typical value
F	The duty cycle of the grating defined by F, $\left(F = \frac{\omega}{\Lambda}\right)$	0.5 Remark: operation as a form-birefringent zeroth diffraction order grating
ω	The width of the binary grating	$F = \frac{\omega}{\Lambda} \Rightarrow \omega = F \cdot \Lambda$ $\omega = 0.5 \times 0.6\,\mu\text{m} = 0.3\,\mu\text{m}$
Λ	The grating period	$0.6\,\mu\text{m}$ Remark: Operation as a form-birefringent zeroth Diffraction order grating
λ	It is the wavelength of the incident wave	$1.523\,\mu\text{m}$
n_I	Indices of air	1
n_{III}	Indices of grating material	Two isotropic materials used to construct the multilayer structures, SiO_2 and S_i. $n_{III-SiO_2} = 1.44$; $n_{III-Si} = 3.48$
$n_{TE}^{(0)}$	Effective indices of refraction for TE wave, provided by the zero-order EMT	$n_{TE}^{(0)} = \sqrt{F \cdot n_{III}^2 + (1-F) \cdot n_I^2}$ $n_{TE-SiO_2}^{(0)} = 1.239$ (calculated) $n_{TE-Si}^{(0)} = 2.56$ (calculated)
$n_{TM}^{(0)}$	Effective indices of refraction for TM wave provided by the zero order EMT	$n_{TM}^{(0)} = \sqrt{\frac{n_{III}^2 \cdot n_I^2}{[F \cdot n_I^2 + (1-F) \cdot n_{III}^2]}}$ $n_{TM-SiO_2}^{(0)} = 1.161$ (calculated) $n_{TM-Si}^{(0)} = 1.359$ (calculated)

effective refractive indices for the two materials:

$$n_{TE}^{(2)} = n_{TE}^{(2)}(\lambda)|_{F=0.5;\; n_I=1;\; n_{III}=\left\{\begin{smallmatrix}1.44\\3.48\end{smallmatrix}\right\};\; \Lambda=0.6\,\mu\text{m}}$$
$$n_{TM}^{(2)} = n_{TM}^{(2)}(\lambda)|_{F=0.5;\; n_I=1;\; n_{III}=\left\{\begin{smallmatrix}1.44\\3.48\end{smallmatrix}\right\};\; \Lambda=0.6\,\mu\text{m}}$$

$$n_{TE}^{(2)} = n_{TE}^{(2)}(n_{III})|_{F=0.5;\; n_I=1;\; \lambda=1.523\,\mu\text{m};\; \Lambda=0.6\,\mu\text{m}}$$
$$n_{TM}^{(2)} = n_{TM}^{(2)}(n_{III})|_{F=0.5;\; n_I=1;\; \lambda=1.523\,\mu\text{m};\; \Lambda=0.6\,\mu\text{m}}$$

The effective indices of both multilayer structure materials (Si and SiO_2) are larger for TH polarization than for TM polarization ($n_{TE-Si}^{(2)} > n_{TM-Si}^{(2)}$; $n_{TE-SiO_2}^{(2)} > n_{TM-SiO_2}^{(2)}$). This indicates in the spectral domain the reflectivity band for TE polarization are centered at a longer wavelength as compared with that for TM polarization. There is a large ASR effect which the separation between the two reflectivity spectral bands has increased over that possible with natural anisotropic materials. The ASR

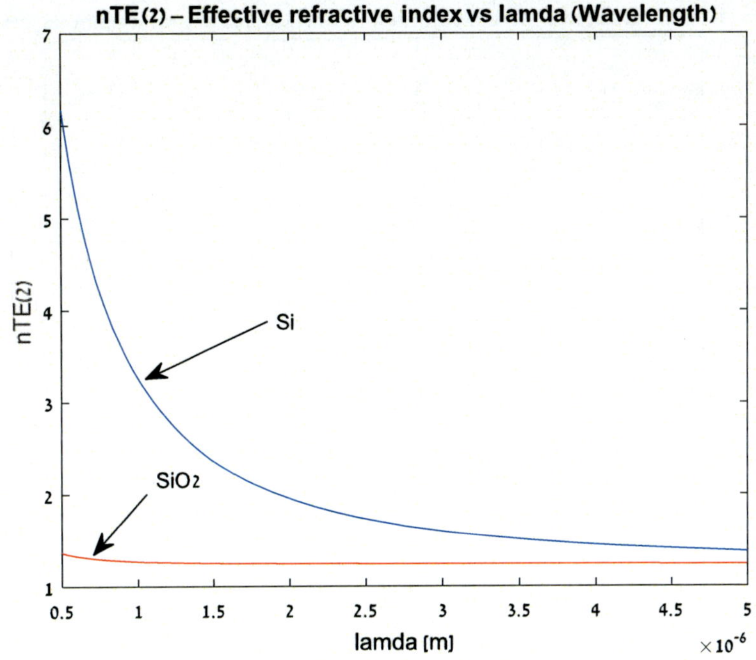

Fig. E.40 nTE(2)—Effective refractive index versus λ (Wavelength)

characteristic of a multilayer from birefringent binary grating is important for realization of the ASR PBS. The ASR property can create a compact PBS and the optical wave propagation in a stratified medium is characterized by second order EMT. By inspection of the spectral reflectivity of several layer grating we see the result that near a wavelength of 1.523 μm ($\lambda = 1.523\,\mu$m) the grating is transparent for TM polarization and reflective for TE polarization. Thus the two orthogonal polarization components of the light wave are separated to propagate in opposite directions with efficiency ≈100% [41].

Graph $n_{TE}^{(2)} = n_{TE}^{(2)}(\lambda)\big|_{F=0.5;\ n_I=1;\ n_{III}=\left\{\begin{smallmatrix}1.44\\3.48\end{smallmatrix}\right\};\ \Lambda=0.6\,\mu\text{m}}$:

MATLAB

```
lamda=0.5e-6:1e-8:5e-6; % lamda is wavelength of the incident wave
F=0.5;nIII=1.44;Lamda=0.6e-6;nTE0=1.239;nI=1; % nIII=1.44 for SiO2
a=nTE0.^2+(1/3).*(Lamda./lamda).^2.*(pi.^2).*(F.^2).*((1-F).^2).*(nIII.^2-
nI.^2).^2;
nTE2=sqrt(a); % Equation for effective refractive indices for SiO2
plot(lamda,nTE2,'r'),grid;hold;
lamda=0.5e-6:1e-8:5e-6; % lamda is wavelength of the incident wave
F=0.5;nIII=3.48;Lamda=0.6e-6;nTE0=1.239;nI=1; % nIII=3.48 for Si
```

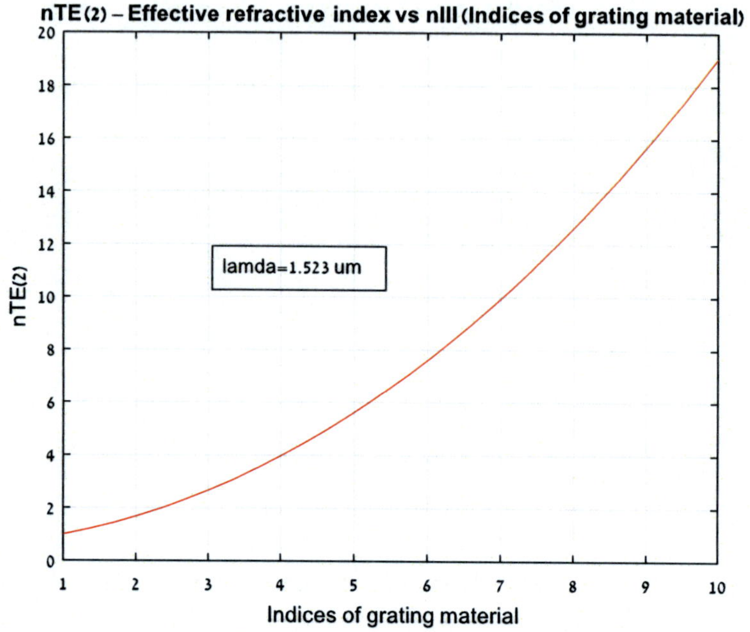

Fig. E.41 nTE(2)—Effective refractive index versus n_{III} (indices of grating material) for $\lambda = 1.523\ \mu m$

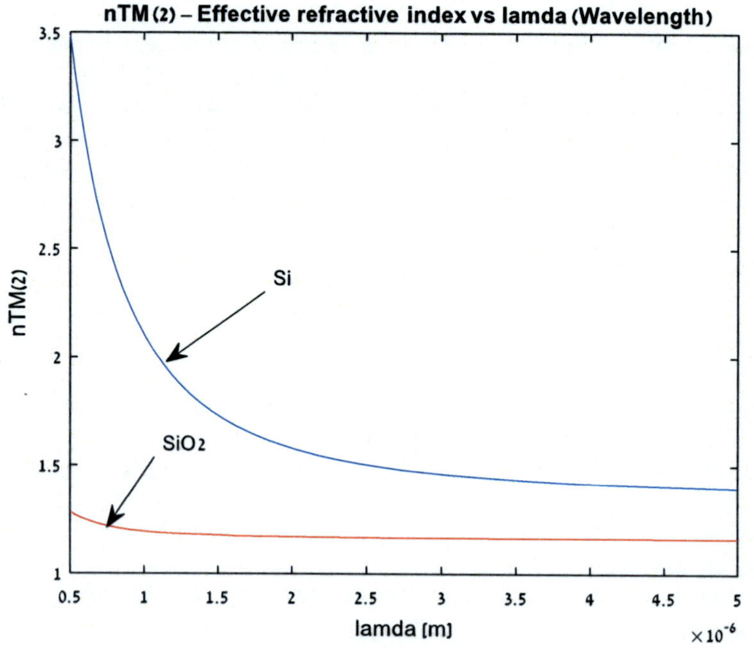

Fig. E.42 nTM(2)—Effective refractive index versus λ (wavelength)

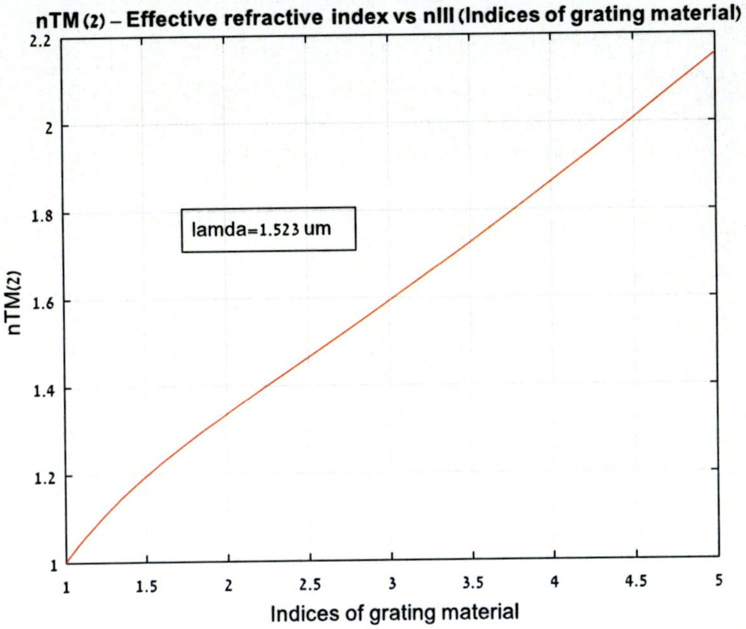

Fig. E.43 nTM(2)—Effective refractive index versus n_{III} (indices of grating material) for $\lambda =$ 1.523 μm

```
a=nTE0.^2+(1/3).*(Lamda./lamda).^2.*(pi.^2).*(F.^2).*((1-F).^2).*(nIII.^2-
nI.^2).^2;
nTE2=sqrt(a); % Equation for effective refractive indices for Si
plot(lamda,nTE2,'b'),grid;
xlabel('lamda [m]'); ylabel('nTE(2)');

title ('nTE(2)—Effective refractive index vs lamda (Wavelength)');
```

Graph $n_{TE}^{(2)} = n_{TE}^{(2)}(n_{III})|_{F=0.5;\ n_I=1;\ \lambda=1.523\,\mu m;\ \Lambda=0.6\,\mu m}$

MATLAB

```
nIII=1:0.01:10; % Indices of grating material
F=0.5;Lamda=0.6e-6;lamda=1.523e-6;nI=1; % nIII=1..10
nTE0=sqrt(F.*(nIII.^2)+(1-F).* (nI.^2)); % Effective indices of refraction for TE
wave by the zero-order EMT
a=nTE0.^2+(1/3).*(Lamda./lamda).^2.*(pi.^2).*(F.^2).*((1-F).^2).*(nIII.^2-
nI.^2).^2;
nTE2=sqrt(a); % Equation for effective refractive indices
plot(nIII,nTE2,'r'),grid;
xlabel(' Indices of grating material'); ylabel('nTE(2)');
title ('nTE(2)—Effective refractive index vs nIII (Indices of grating material)');
```

Graph $n_{TM}^{(2)} = n_{TM}^{(2)}(\lambda)|_{F=0.5;\ n_I=1;\ n_{III}=\left\{\begin{smallmatrix}1.44\\3.48\end{smallmatrix}\right\};\ \Lambda=0.6\mu m}$

MATLAB

```
lamda=0.5e-6:1e-8:5e-6; % lamda is wavelength of the incident wave
F=0.5;nIII=1.44;Lamda=0.6e-6;nTM0=1.161;nTE0=1.239;nI=1; % nIII=1.44
for SiO2
a=nTM0.^2+(1/3).*(Lamda./lamda).^2.*(pi.^2).*(F.^2).*((1-F).^2).*(1./(nIII.^2)-
1./(nI.^2)).^2.*(nTE0.^2).*(nTM0.^6);
nTM2=sqrt(a); % Equation for effective refractive indices for SiO2
plot(lamda,nTM2,'r'),grid;hold;
lamda=0.5e-6:1e-8:5e-6; % lamda is wavelength of the incident wave
F=0.5;nIII=3.48;Lamda=0.6e-6;nTM0=1.359;nTE0=2.56;nI=1; % nIII=3.48 for
Si
a=nTM0.^2+(1/3).*(Lamda./lamda).^2.*(pi.^2).*(F.^2).*((1-F).^2).*(1./(nIII.^2)-
1./(nI.^2)).^2.*(nTE0.^2).*(nTM0.^6);
nTM2=sqrt(a); % Equation for effective refractive indices for Si
plot(lamda,nTM2,'b'),grid;
xlabel('lamda [m]'); ylabel('nTM(2)');
title ('nTM(2)—Effective refractive index vs lamda (Wavelength)');
```

Graph $n_{TM}^{(2)} = n_{TM}^{(2)}(n_{III})|_{F=0.5;\ n_I=1;\ \lambda=1.523\,\mu m\ ;\ \Lambda=0.6\,\mu m}$

MATLAB

```
nIII=1:0.01:5; % Indices of grating material
F=0.5;Lamda=0.6e-6;lamda=1.523e-6;nI=1; % nIII=1..10
nTE0=sqrt(F.*(nIII.^2)+(1-F).* (nI.^2)); % Effective indices of refraction for TE
wave by the zero-order EMT
nTM0=sqrt(((nIII.^2).*(nI.^2))./(F.*(nI.^2)+(1-F).*(nIII.^2)));
a=nTM0.^2+(1/3).*(Lamda./lamda).^2.*(pi.^2).*(F.^2).*((1-F).^2).*(1./(nIII.^2)-
1./(nI.^2)).^2.*(nTE0.^2).*(nTM0.^6);
nTM2=sqrt(a); % Equation for effective refractive indices
plot(nIII,nTM2,'r'),grid;
xlabel(' Indices of grating material'); ylabel('nTM(2)');
title ('nTM(2)—Effective refractive index vs nIII (Indices of grating material)');
```

One application that used a polarization beam splitter is a laser frequency stabilization system by polarization spectroscopy of a reflecting reference cavity. The linear polarizer or Brewster plate is placed inside the reference cavity and the reflected light acquires a frequency-dependent elliptical polarization. The dispersion shaped resonances can be detected by a polarization analyzer. We get an error signal for electronic frequency stabilization. There are many perturbations that affect the cavity of the laser. We can get a better stability of a single-mode laser by locking its frequency to some passive reference cavity. There are number of method for laser frequency locking and one of them utilized a reference cavity with internal linear polarizer or Brewster plate and monitors changes in the polarization of the reflected light. A Brewster plate is a plate of an optical material with

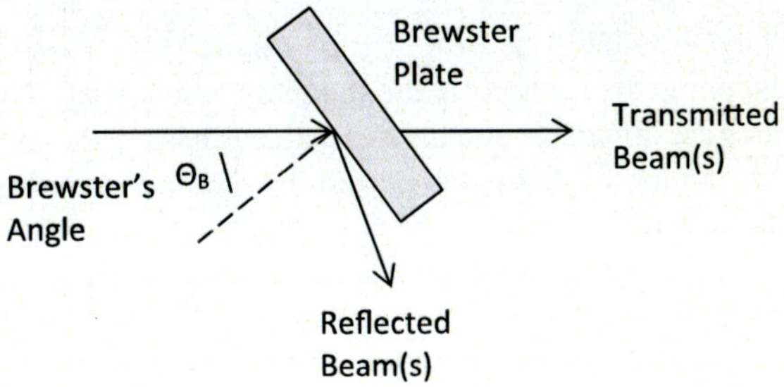

Fig. E.44 Brewster plate, inserted into a laser beam at Brewster's angle

coplanar surfaces, it is inserted into a polarized laser beam under Brewster's angle and minimum reflection losses occur for p—polarization (Fig. E.44). Brewster angle id the angle of incidence and it is calculated from the refractive index of the plate, $\theta_B \approx \arctan\left[\frac{n_{plate}}{n_{air}}\right]|_{n_{air}=1} = \arctan[n_{plate}]$.

For light beam which hit Brewster plate, the reflection loss becomes minimal for p—polarized light (Fig. E.44). actually the reflection losses of p—polarized and s—polarized light at each surface of the Brewster plate is dependent on the angle of incidence. Mainly we use Brewster plates and inserted it into laser resonators of bulk lasers. It represents the polarization-dependent losses. We get laser with a stable linear polarization. The direction is corresponded to p—polarization at the Brewster plate. The Brewster plate causes to small losses. Insertion of a Brewster place decreases the laser efficiency even if the Brewster plate losses for p—polarization are small. In high power laser facilities the Brewster plate exhibits thermal lensing. If we use mode-lock laser non linearity and chromatic dispersion phenomenon are dominant. Thermal lensing is a lensing effect induced by temperature gradients. It is very dominant in high power laser systems, the heating of the gain medium (laser crystal) can causes a significant thermal lens by some mechanisms. The first mechanism: The gain medium (laser crystal) is heated on the beam axis, compared with the outer regions and causes to transverse gradient of the refractive index $\left(\frac{dn}{dT}\right)$. The second mechanism: Additional index changes is caused by thermally induced mechanism stress (photo elastic effect), which is characterized by photo elastic coefficients (ρ_{ij}) and thermal expansion coefficient (α). The third mechanism: The mechanical stress caused by rising temperature leading to bulging of the end faces of the gain medium, and lensing happened. It is more critical in short laser crystals. The thermal lensing effect can be deduced by using a convex (defocusing) laser mirror next to the laser crystal. Dioptric power of the thermal lens is optical power (refractive power, focusing power, or convergence power) and is the degree to which a lens, mirror, or other optical system converges or diverges light. It is the reciprocal of the focal length $p = \frac{1}{f}$ for specific device. By calculating the dissipated power we can get the dioptric power of the thermal lens. It is part of the absorbed pump power. In lasers which quenching and up conversion process happened the dioptric power calculation is more difficult. The thermal lensing effect depends on the pumping and cooling geometry. For an isotropic laser gain medium in which there is a uniform pump intensity and a purely radial heat flow the thermal lens is with the dioptric (inverse of focal length), $\frac{1}{f} = \left(\frac{\frac{dn}{dT}}{2 \cdot \kappa \cdot A}\right) \cdot P_{heat}$, where P_{heat} is the dissipated power, A is the pump area, and κ is the thermal conductivity. The length of the gain medium does not influence, and the thermal lens has no aberrations. For a non-homogeneous distribution of dissipated power, the thermal lens exhibit aberrations. Mostly, laser crystals have $\frac{dn}{dT} > 0$ of the order of a few times $10^{-6}\,K^{-1}$. Nd:YLF laser crystal has a negative temperature coefficient, and defocusing lens happened if it does not over compensated by stress effects. Raman crystal material $Ba(NO_3)_2$, barium nitride, has a strongly negative $\frac{dn}{dT}$ combined with a low thermal conductivity, leading to strong defocusing. We can reduce the thermal lensing effect by using a gain medium with low power losses, high thermal conductivity, and small thermo-optical coefficients. Choosing pump and laser wavelength which gives a small quantum detect reduce the thermal lensing strength. By using cryogenic cooling of the gain medium (cryogenic lasers), we get less thermal lensing effects. Optimizing the shape of the pump beam and the geometry of the laser head reduces the thermal lensing effect. Additionally, thermal lensing occurs in Q switches, optical modulators, Raman-active crystals, and dielectric mirrors. The thermal lensing is thermally induced birefringence which lead to depolarization loss and to polarization-dependent astigmatism. In high power laser cutting head the position of a focus is dependent on the optical power level. In our laser frequency stabilization system a simple analyzer for elliptical polarization provides dispersion-shaped resonances, which give the error signal for the servo loop. Then the laser locked to the center of a cavity resonance without any need for modulation techniques. There is an automatic re-locking even after larger accidental frequency jumps. Generally, locking schemes monitor the light intensity transmitted by the reference cavity. The transmitted intensity versus frequency exhibits a periodic resonance maxima or Frings. The laser frequency is servo-locked to the side of one fringe by using a fast differencing technique. The system is out of lock state when accidental frequency jump of as little as half as fringe width. By using a reference cavity of low finesse then the laser frequency becomes sensitive to small changes in the finesse or to drifts in the photo detectors or associated amplifiers. If the frequency of the laser or reference cavity is modulated, and the amplitude modulation of the transmitted light is monitored with a phase sensitive detector then the signal corresponds to the derivative of the Airy function permits locking to the fringe center. The signal drops to zero a few fringe widths away from resonance and it recovers from large frequency jumps difficult. The stabilization method is related to frequency stabilization method and laser frequency stabilization system. Linearly polarized light beam from a tunable single mode laser is reflected by a confocal reference cavity used off-axis then a small angle between incident and reflected beam avoids feedback into the laser cavity (Fig. E.45).

An optical cavity, resonating cavity is an arrangement of mirrors (M1 and M2) that forms a standing wave cavity resonator. The optical resonator, the cavity consist of two mirrors coated with partially reflective coating when the length of the cavity can vary and the result is an output light beam of greater intensity than the original beam and of a single frequency. There are differences between confocal cavity and concentric cavity. The linear polarizer inside the cavity is rotated then its transmission

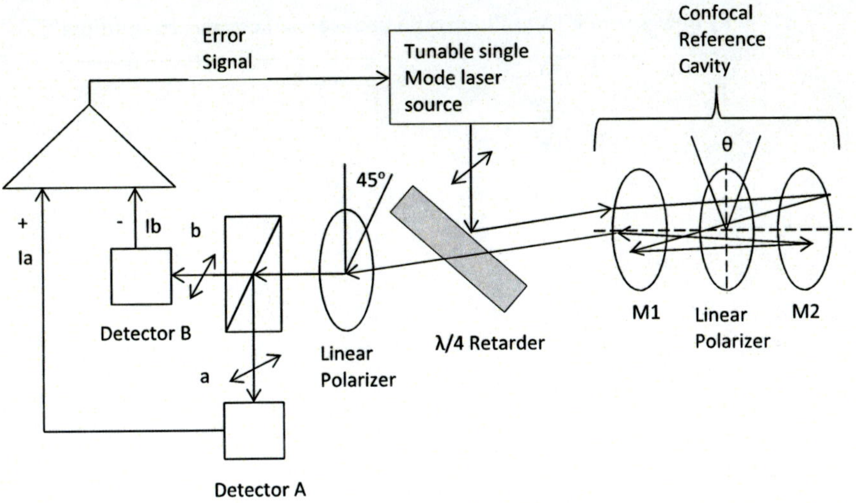

Fig. E.45 Diagram of laser frequency stabilization system

axis forms an angle θ with the polarization axis of the incident light beam. The incoming light is decomposed into two orthogonal linearly polarized components with the electric field vector parallel and perpendicular to the transmission axis of the intra cavity polarizer. Their filed amplitudes in the plane wave are $E_{\parallel}^{(i)} = E^{(i)} \cdot \cos\theta$ and $E_{\perp}^{(i)} = E^{(i)} \cdot \sin\theta$, where $E^{(i)}$ the amplitude of the incident beam is. The parallel component, $E_{\parallel}^{(i)}$, sees a cavity of low loss and has a frequency-dependent-phase shift in reflection. The perpendicular component, $E_{\perp}^{(i)}$, is reflected by mirror M1, serves as a reference. Any relative phase change between the two reflected components cause to resulting beam elliptically polarized. The complex amplitude of the reflected wave is

$$E_{\parallel}^{(r)} = E_{\parallel}^{(i)} \cdot \left[\sqrt{R_1} - \frac{T_1}{\sqrt{R_1}} \cdot \frac{R \cdot e^{i\cdot\delta}}{(1 - R \cdot e^{i\cdot\delta})}\right]$$
$$= E_{\parallel}^{(i)} \cdot \left[\sqrt{R_1} - \frac{T_1 \cdot R}{\sqrt{R_1}} \cdot \left\{\frac{\cos\delta - R + i \cdot \sin\delta}{(1 - R)^2 + 4 \cdot R \cdot \sin^2(\frac{1}{2} \cdot \delta)}\right\}\right]$$

Confocal reference cavity mirror M1: R_1 and T_1 are the reflectivity and transmissivity of the cavity entrance mirror (M1). R is the amplitude ratio between successive roundtrips($R < 1$). It determines the cavity finesse $F = \frac{\pi\cdot\sqrt{R}}{(1-R)}$ (Fig. E.46).

MATLAB:

```
R=0:0.01:1; % Amplitude ratio between successive roundtrips
F=(pi*sqrt(R))./(1-R); % Cavity finesse
plot(R,F,'r'),grid;
```

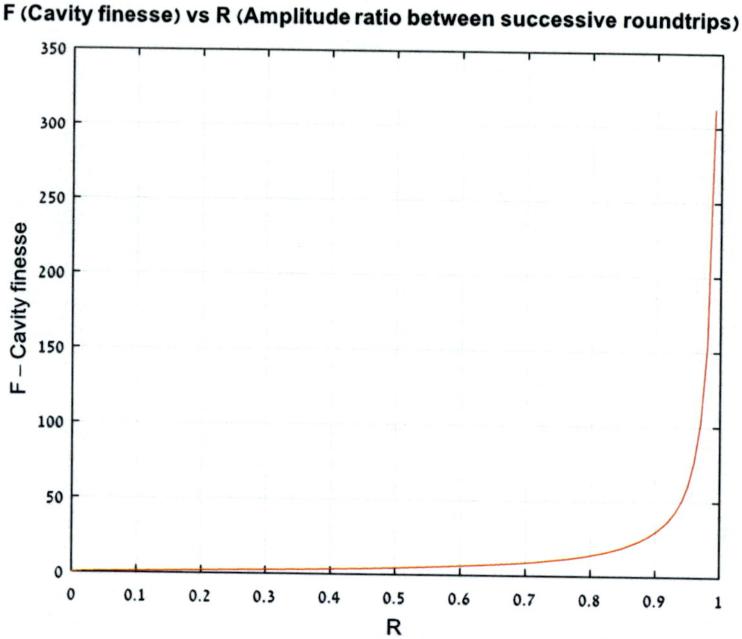

Fig. E.46 F (Cavity finesse) versus R (Amplitude ratio between successive roundtrips)

xlabel('R'); ylabel('F—Cavity finesse');
title ('F (Cavity finesse) vs R (Amplitude ratio between successive roundtrips)');

The finesse of an optical cavity (resonator) is defined as its free spectral range divided by the bandwidth of its resonances. We consider the optical cavity bandwidth as the full width at half-maximum. It is determined by the resonator (cavity) losses and is independent of the cavity length. We define the fraction ρ of the circulating power which left after one round trip (A fraction $(1-\rho)$ of the power is lost) when there is no incident field from outside the cavity (resonator), then the finesse parameter is

$$F = \frac{\pi}{2 \cdot \arcsin\left[\frac{(1-\sqrt{\rho})}{2 \cdot \sqrt[4]{\rho}}\right]} \approx \frac{\pi}{1-\sqrt{\rho}} \approx \frac{2 \cdot \pi}{(1-\rho)}$$

The approximation is true for low round-trip losses (<10%) and only for high finesse values. The ratio R (amplitude ratio between successive roundtrips) is related for any attenuation by the internal polarizer and for other losses, and two extra reflections which are required for one round trip in a confocal resonator used off-axis [42]. The amplitude of the reflected perpendicular component $(E_{\perp}^{(r)})$ can be approximate by $E_{\perp}^{(r)} = E_{\perp}^{(i)} \cdot \sqrt{R_1}$ and at exact resonance $(\delta = 2 \cdot m \cdot \pi)$ both reflection coefficient are real and the reflected wave components remain in phase.

Even through the reflected beam polarization axis is rotated from the original direction, it remains linearly polarized. The parallel component acquires a phase shift relative to the perpendicular component which is far from resonance and owing to the imaginary part of $E_{\parallel}^{(r)}$. The reflected beam acquires an elliptical polarization. The sign of the detuning from resonance established the handedness of the polarization ellipse. The reflected light goes to an analyzer assembly consisting of a $\frac{\lambda}{4}$ retarder and a linear polarization beam splitter and we detect the elliptically. The quarter-wave plate converts linearly polarized light into circularly polarized light and vice versa. The quarter-wave plate is used also to produce elliptically polarization. Ideal quarter wave plate does not absorb any light and the speed of the light through the quarter wave plate is depending on the orientation of the electric field vector relative to plate. Retarder plates (wave plates) are transparent plates with specific birefringence. We use it for manipulating the polarization state of light beams. The retarding (delaying) is related to one component of polarization with respect to its orthogonal component. Retarders, it has a meaning of transmit light and modify its polarization state without attenuation, deviation, or displacing the beam. In un polarized light, wave plates are equivalent to window and they are both flat optical components through which light passes. The fast axis of the retarders is rotated by 45° relative to the polarization axis of beam splitter output a. The PBS (Polarized Beam Splitter) has two outputs light beam with intensities I_a and I_b, respectively. The PBS two outputs are monitored by two photo detectors (A and B). The two photo detectors (A and B) are connected to a differential amplifier which give output of error signal. Analyzer scheme and function: The elliptically polarized light is superposition of two counter rotating circularly polarized components of different amplitudes. The $\frac{\lambda}{4}$ retarder transforms the circular components into orthogonal linearly polarized waves, which are separated by the beam splitter. For linearly polarized incoming light, the two circular components have equal intensities. The equal intensities help to balance the sensitivity of the two photo detectors. The signal $I_a - I_b$ is dependent on the magnitude and handedness of the elliptically, but not on the azimuth angle. Difference signal calculation ($I_a - I_b$): we assume that the fast axis of the $\frac{\lambda}{4}$ retarder is parallel to the polarization axis of the intra-cavity polarizer. The amplitudes of the reflected beam after passing through the retarder and polarization beam splitter is $E_{a,b} = \frac{1}{2} \cdot \begin{pmatrix} 1 & \pm 1 \\ \pm 1 & 1 \end{pmatrix} \cdot \begin{pmatrix} 1 & 0 \\ 0 & i \end{pmatrix} \cdot \begin{pmatrix} E_{\parallel}^{(r)} \\ E_{\perp}^{(r)} \end{pmatrix}$; $E_{a,b} = \frac{1}{2} \cdot \begin{pmatrix} 1 & \pm i \\ \pm 1 & i \end{pmatrix} \cdot \begin{pmatrix} E_{\parallel}^{(r)} \\ E_{\perp}^{(r)} \end{pmatrix}$,

then $E_{a,b} = \frac{1}{2} \cdot \begin{pmatrix} E_{\parallel}^{(r)} \pm i \cdot E_{\perp}^{(r)} \\ \pm E_{\parallel}^{(r)} + i \cdot E_{\perp}^{(r)} \end{pmatrix}$. The intensities at the two outputs I_a, I_b are $I_{a,b} = \frac{1}{2} \cdot ce \cdot |E_{a,b}|^2$ and $I_{a,b} = \frac{1}{2} \cdot ce \cdot |E_{a,b}|^2 = \frac{1}{2} \cdot ce \cdot |\frac{1}{2} \cdot (E_{\parallel}^{(r)} \pm i \cdot E_{\perp}^{(r)})|^2$. The difference signal ($I^{(i)}$ is the intensity of the incident beam) is$I_a - I_b = I^{(i)} \cdot 2 \cdot \cos\theta \cdot \sin\theta \cdot [\frac{T_1 \cdot R \cdot \sin(\delta)}{(1-R)^2 \cdot 4 \cdot R \cdot \sin^2(\frac{1}{2} \cdot \delta)}]$; $I^{(i)} = \frac{1}{2} \cdot ce \cdot |E^{(i)}|^2$. Hint: At exact resonance $\delta = 2 \cdot m \cdot \pi$ the signal function is an ideal error signal for servo locking of a laser frequency. The maximum signal intensity is accepted for $\theta = 45°$ then $2 \cdot \cos\theta \cdot \sin\theta = \sin(2 \cdot \theta)|_{\theta=45°} = 1$. The total intensity of the reflected light at resonance is very small $\theta \approx 0$, and operation at smaller angles θ offers a better signal to noise ratio (SNR). In that case the laser intensity fluctuations are the dominant source of noise.

Remark: Birefringence due to stress in the dielectric mirror coatings can happened in our system but it is not part of our analysis. Residual birefringence produce line symmetries which is not good and need to be compensated [42].

Optical relay imaging system: Relay lens or optical relay imaging system is a lens, lenses, group of lenses that invert an image and extend the optical tube. Typical applications to implement an optical relay imaging system are refracting telescopes, endoscopes, periscopes, etc. The target is to extend the length of the system, and the purpose is inverting the image. Typically relaying imaging system includes one or more conventional lenses or achromatic doublets, or a long cylindrical gradient-index of refraction lens (a GRIN lens). An achromatic lens or achromatic is a lens that is designed to limit the effects of chromatic and spherical aberration (this type of aberration found in optical systems that use elements with spherical surfaces). Achromatic lenses are corrected to bring two wavelengths into focus on the same plane (typically red and blue). A gradient index of refraction lens (a GRIN lens) is a lens behavior with optical effects produced by a gradient of the refractive index of a material. The gradual variation is used to produce lenses with flat surfaces, or lenses with better aberrations compare to the typical spherical lenses. The refraction gradient of Gradient-index lenses can be spherical, axial, or radial. The relay lenses operate by producing intermediate planes of focus. Additional application which we implement optical relay imaging system is high power laser facilities laser beam delivery systems. In high power laser [0.1 PW, 1 PW, and 10 PW, (PW = Peta-watt, 1 PW = 10^{15} W)] systems we need to deliver the laser beam to specific experimental areas. It is done by relay imaging system which includes optical elements like flat mirrors, parabolic mirrors, etc. The target of using relay imaging system in high power facilities is to focus the output laser beam to the experimental areas. We implements some technologies to get our target in high power laser beam delivery system, relay imaging, plasma mirror, polarization control, adaptive optics, back reflection isolation, and focusing optics. The technologies are related to specific beam parameters and the various properties to implement at the interaction points (experimental areas). The relay imaging technology has impact on the pointing stability of laser beam delivery system. The beams stabilization laser system need to stabilize the beam in the desired target position and beam direction. Our target is to get a very precise, fast and reliable beam alignment, active beam position and beam direction control, compensation of laser beam pointing, precise motion and vibration control, automated adjustment of laser beams, and fast delivery of laser beam to changing applications. The plasma mirror technology has impact on laser beam temporal contrast. We need to get a temporal and spatial cleaning of high-brightness laser beam pulses. There is a way which utilizes nonlinear modulation of the beam in the Fourier-plane leading both to directional and to temporal modulation. It gives more energy-scalable than that based on crossed polarizers and offers better contrast improvement compared to the plasma mirror techniques [43]. The polarization control technology has impact on the process induced on target. The scanning multi-component pulsed laser deposition (PLD) method realized uniform depositions of the desired coatings by a modified pulse laser deposition process, preferably with a femto-second laser system. Multi-component coatings are deposited onto substrates via laser induced ablation of segmented targets. The pulsed laser deposition (PLD) method enables the achievement of uniform, large-area coatings. The adaptive optics technology impacts on the quality of focus. We must ensure that quality is not compromised within the laser beam environment. It needs to fulfil the desire for beam accuracy. The laser beam quality defines aspects of the beam illumination pattern in the beam delivery system. It is related to laser beams propagation and transformation properties. The back reflection isolation technology impacts the laser existence. The laser back reflection can result in unwanted fluctuations of emitted power, or even burn-out of the cavity optics. Near the focal plane of the focusing lens, a large fraction of any reflected energy is directed back towards the laser cavity in the form of a fairly parallel beam. Many types of laser technology suffer from inherent sensitivity to back-reflected light, which can cause an unstable process, disruptive automatic shut downs, or even catastrophic failure of the laser. We define laser beam back reflection at the percentage (%) of power reflected back in relation to forward power at a particular point in a light path. An optical isolator is a passive magneto-optic device that allows light to travel in one direction, and protect a laser source from back reflections or signals that may occur after the isolator. Back reflections can damage a laser source or cause it to mode hop, amplitude modulates, or frequency shift. In high power laser systems, back reflections can cause instabilities and power spikes. The isolation from back reflecting or opposing beams is if a target is not in place during a system firing. We consider the worst case, which is 100% back reflection. If no isolation where in place, such a pulse would eventually damage or destroy costly components in the system. Finally the focusing optics which we choose for our laser beam delivery system influences on every experiment results. The layout of the beam transportation for the high power laser system (HPLS) output is very important and it includes design for all technologies, for both arms of HPLS. The high power laser transport lines include arms of the beam transport which includes a plasma mirror system, an adaptive optics systems, a system to generate circular polarization and relay imaging systems. The propagation distances after compression of the high power laser beam lines are mainly in the range of several tens of meters. We can get the effect of the beam spatial phase on the propagation which can lead to different modifications of the laser beam. The energy distribution could develop some high spatial frequencies modulations which can damage the transport optics. Some spatio-spectral coupling can cause to local changes of the pulse duration and the beam pointing stability can be decreased. The solution is relay imaging system which avoids these drawbacks, into the beam transport path. We use achromatic and stigmatic optical lay-out in our system. Relay systems are used for two mainly purposes: (1) to provide the proper orientation of the image, (2) to transfer the beam light from one region to another. In some cases the distance between the object and the final image is large and, in addition, the diameter of the lenses must not become excessive. These conditions require a series of relay lenses resulting in a periscope system. A simple relay system

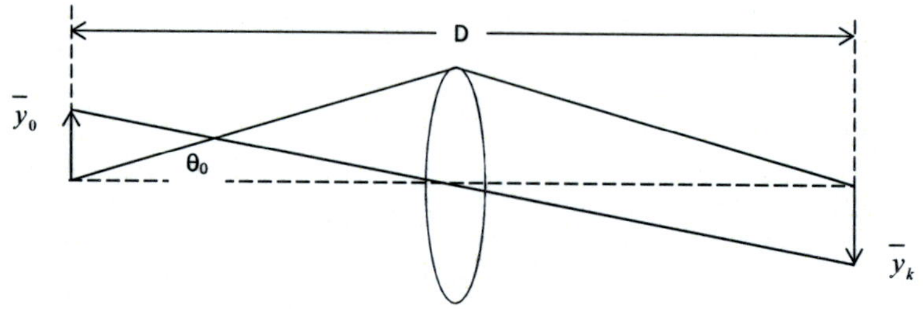

Fig. E.47 A single relay system

needs to transfer light from an object plane ($\bar{y}_0$) to the image plane ($\bar{y}_k$). The two planes are separated by the distance D. The magnification should be (−1) and the diameter of the objective is determined by the angle θ_0. The type of lens to use for the objective depends upon the image quality required. The simple approach is to assume that the relay lens consists of two telescope objectives. The objectives can be placed so that the light would be parallel between them (Fig. E.47).

We need to evaluate the visual performance with the telescope objectives and it is done by first suppose that the system is a visual system with a 10X eyepiece used to view the image. We must concern the effect of Petzval curvature [44]. The value of P for a thin lens $P = -\frac{\phi}{n}$ and the value of P for our specific relay lens is the $P = -\frac{4}{D \cdot n_e}$, where n_e is the effective index of the doublets which used for the relay lens. There are other types of lenses for relay objectives. In the case of two triplet objectives, the value of n_e in the equation $P = -\frac{\phi}{n_e}$ could be raised. The triplet would not be as well corrected on the axis as the doublets. It is not recommended to correct the Petzval curvature of the relay lenses until they start to introduce a contribution which is one half, or at most equal to, that of the eyepiece. The telescope lenses are made from ordinary glass with an amount of secondary color $\text{TA}ch_{F-D} = \frac{y_1}{2200}$ and the relay lens (Fig. E.47) have secondary color given by the equation $\text{TA}ch_{F-D} = \frac{2 \cdot y_1}{2200} = \frac{\theta_0 \cdot D}{2200}$. We can see that the secondary blur at the focal plane of the eyepiece increase as θ_0 or D is increased. If we choose $\theta_0 = 0.17$; $D = 89$ cm then the radius of the secondary blurs $\text{TA}ch_{F-D} = \frac{\theta_0 \cdot D}{2200} = 0.687 \times 10^{-4}$ m. The secondary color is usually a series problem in relay systems. If the distance D must be maintained, then the secondary color can most easily be reduced by making θ_0 smaller. We can reduce the secondary color by separating two doublets. As long as there is parallel light beam between the two lenses, the space between the lenses (d) can be considered free space. If this distance is d, the secondary color is given by the equation $\text{TA}ch_{F-D} = \frac{\theta_0 \cdot (D-d)}{2200}$. For larger d, the focal lengths of the lenses are reduced and there is more field curvature. If we try to reduce the secondary color without sacrificing clear aperture then we get more field curvature. A unit power relay system, there are advantages in using two identical doublets with parallel light between them. The doublets are air-spaced, and there are eight glass air surfaces. We can combine the positive elements of the doublets into a single lens surrounded by two negative lenses. There is an option to combine the two negative lenses and surround the combination with two positive

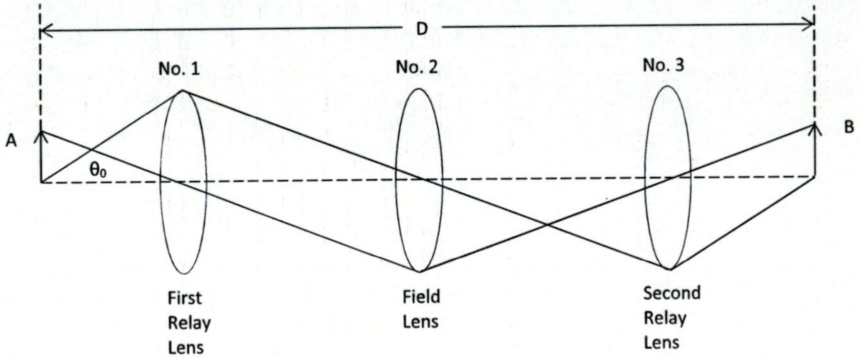

Fig. E.48 A double-relay lens system

elements. If we use positive lens, they can be made into a single equip convex lens. The structure of triplet relay lens is possible but we must take to our consideration higher order spherical aberration than the two doublets. The use of a triplet of this type is done when the zonal spherical aberration is tolerable. Some relay systems are limited in the diameter of lens which may be used when the lens must be confined to a given diameter. In the double-relay lens system (Fig. E.48), the first and second relay lenses must have just one-half the focal length of the single-relay lens, then each add twice the Petzval contribution. The two lenses do not introduce any more secondary color, and each relay transfers by the distance $\frac{D}{2}$, and each has half the secondary color and the total gives the final result. We can characterize the double relay system by $\mathrm{TA}ch_{F-D} = \frac{\theta_0 \cdot D}{2200}$. The double relay system includes additional lens in the intermediate focal plane (field lens). The field lens function is to image the chief ray passing through the center of the first relay lens at the center of the second relay lens. This field lens has a focal length equal to $\frac{D}{8}$, and negative field curvature. Doubling up the relay system in order to reduce the diameter of the lenses is characterized by Petzval field curvature.

The relay lenses introduce four times as much as field curvature, and the field lens adds as much as one of the relays. The double-relay lens introduces six times as much field curvature as the single-relay system of equal length and numerical aperture, when the secondary color is not changed. By using relay lenses, we get an image at the end of the field more blurred, and stop at the field lens which gives smaller field lens. Improving the quality at the edge of the field is done by abandon the doublet relay lenses and uses a lens with reduced field curvature. It is possible to introduce compound filed lenses (triplet), to reduce the Petzval curvature of the filed lens. If the field lens is located between the two unit power relay systems then it is symmetrical. The lens is two inverted telephoto objectives with parallel light between them. Relay systems are limited by problems of field curvature and secondary color. We use more than one relay and the relay and field lenses should be kept as large as practicable. The size of the relay or field lenses should not be reduced. The secondary color depends on the over-all distance of relay and the value of θ_0. We reduce θ_0 value or resort to special lens materials. We use Doublet relay lenses system

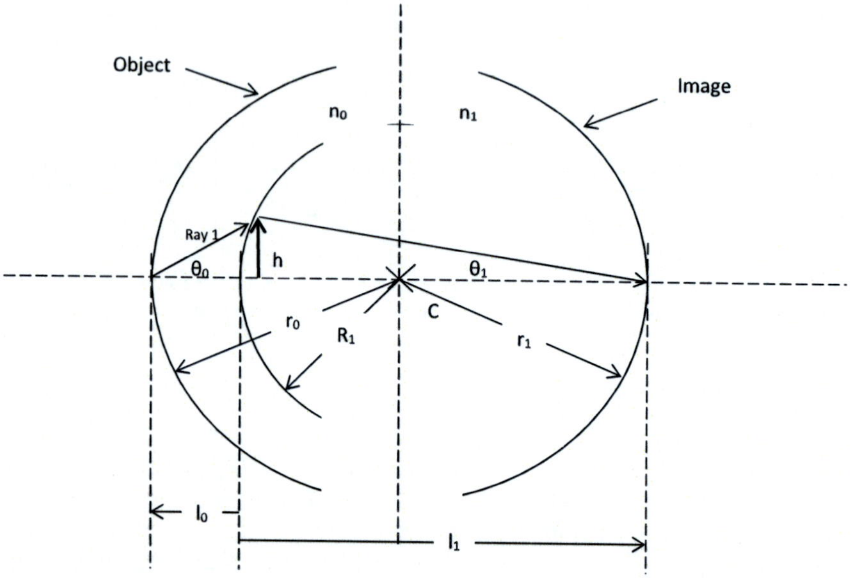

Fig. E.49 Petzval field curvature system, one spherical surface and a spherical object surface co-centered with it

for complicated types. By reducing the field curvature, the result will be a design with increased secondary color and chromatic variation of aberrations. The Petzval curvature (Petzval field curvature) described the optical aberration in which a flat object normal to the optical axis cannot be brought properly into focus on a flat image plane. We can also concern to the non-flat object past the hyper focal distance. The Petzval field curvature shows that even thin lenses have aberration which causes the image plane to curve. The sign of the curvature is opposite for surfaces of positive and negative power. The correction is done by having both positive and negative surfaces in an optical system. For a flat detector, the Petzval curvature must be corrected by balancing the positive and negative power surfaces [45]. We present an optical system of one spherical surface, with a spherical object surface co-centered with it (Fig. E.49).

We sign the center of our system by "C" and any axis drawn through the center "C" is equivalent. Any "paraxial" rays traced along the alternate rays traced along these alternate axes produce an image at the same distance from "C". The image surface is a sphere with the center at "C". For the tracing ray 1 through the surface we get $n_1 \cdot \theta_1 = n_0 \cdot \theta_0 - h \cdot (\frac{n_1 - n_0}{R_1})$ substituting $\theta_0 = -\frac{h}{l_0}$; $\theta_1 = -\frac{h}{l_1}$, it gives $\frac{n_1}{l_1} = \frac{n_0}{l_0} + (\frac{n_1 - n_0}{R_1})$. We need to get the equations for the curvature of the object surface, the curvature of the image surface, and the power of the lens surface. $l_0 = R_1 - r_0$; $l_1 = R_1 - r_1$, by using the sign convention of measuring from the vertex, we get $\frac{n_1}{R_1 - r_1} - \frac{n_0}{R_1 - r_0} = \frac{n_1 - n_0}{R_1}$ and

$$n_1 \cdot (R_1 - r_0) \cdot R_1 - n_0 \cdot (R_1 - r_1) \cdot R_1 = (n_1 - n_0) \cdot (R_1 - r_1) \cdot (R_1 - r_0)$$

$$R_1 \cdot n_1 \cdot R_1 - r_0 \cdot n_1 \cdot R_1 - R_1 \cdot n_0 \cdot R_1 + r_1 \cdot n_0 \cdot R_1 = (n_1 - n_0) \cdot (R_1^2 - R_1 \cdot r_0 - r_1 \cdot R_1 + r_1 \cdot r_0)$$

$$R_1 \cdot n_1 \cdot R_1 - r_0 \cdot n_1 \cdot R_1 - R_1 \cdot n_0 \cdot R_1 + r_1 \cdot n_0 \cdot R_1 = -n_0 \cdot R_1^2 + n_0 \cdot R_1 \cdot r_0 + n_0 \cdot r_1 \cdot R_1 - n_0 \cdot r_1 \cdot r_0 + n_1 \cdot R_1^2 - n_1 \cdot R_1 \cdot r_0 - n_1 \cdot r_1 \cdot R_1 + n_1 \cdot r_1 \cdot r_0$$

$-n_0 \cdot R_1 \cdot r_0 + n_1 \cdot r_1 \cdot R_1 + r_1 \cdot r_0 \cdot (n_0 - n_1) = 0$, We multiply by $\frac{1}{n_0 \cdot n_1 \cdot R_1 \cdot r_1 \cdot r_0}$ and we get $-\frac{1}{n_1 \cdot r_1} + \frac{1}{n_0 \cdot r_0} + \frac{(n_0 - n_1)}{n_0 \cdot n_1 \cdot R_1} = 0$, we define the power of the lens surface $K_1 = \frac{n_1 - n_0}{R_1}$.

Then $-\frac{1}{n_1 \cdot r_1} + \frac{1}{n_0 \cdot r_0} - \frac{1}{n_0 \cdot n_1} \cdot K_1 = 0 \Rightarrow \frac{1}{n_1 \cdot r_1} - \frac{1}{n_0 \cdot r_0} + \frac{1}{n_0 \cdot n_1} \cdot K_1 = 0$.

$\frac{1}{n_0 \cdot n_1} \cdot K_1 = \frac{1}{n_0 \cdot r_0} - \frac{1}{n_1 \cdot r_1}$. The above equations help to calculate the change in curvature from the image surface to the object surface as a function of the power of the lens surface. Assumption: The image surface for lens No.1 is the object surface for lens No.2. We chain this calculation through several surfaces. Applying this equation to the next surface give $\frac{1}{n_1 \cdot n_2} \cdot K_2 = \frac{1}{n_1 \cdot r_1} - \frac{1}{n_2 \cdot r_2}$ and by substitution for $\frac{1}{n_1 \cdot r_1}$ in the first equation:

$$\frac{1}{n_1 \cdot n_2} \cdot K_2 + \frac{1}{n_2 \cdot r_2} = \frac{1}{n_1 \cdot r_1} \Rightarrow \frac{1}{n_0 \cdot n_1} \cdot K_1 = \frac{1}{n_0 \cdot r_0} - \frac{1}{n_1 \cdot n_2} \cdot K_2 - \frac{1}{n_2 \cdot r_2}$$

$\frac{1}{n_1} \cdot \left(\frac{1}{n_0} \cdot K_1 + \frac{1}{n_2} \cdot K_2\right) = \frac{1}{n_0 \cdot r_0} - \frac{1}{n_2 \cdot r_2}$. The general equation for K surfaces is $\frac{1}{n_0 \cdot r_0} - \frac{1}{n_k \cdot r_k} = \sum_{j=1}^{k} \frac{K_j}{n_{j-1} \cdot n_j}$. For flat initial object plane ($r_0 \to \infty$), $r_0 \to \infty \Rightarrow \frac{1}{n_0 \cdot r_0} \to 0$ then $-\frac{1}{n_k \cdot r_k} = \sum_{j=1}^{k} \frac{K_j}{n_{j-1} \cdot n_j} \Rightarrow \frac{1}{r_k} = -n_k \cdot \sum_{j=1}^{k} \frac{K_j}{n_{j-1} \cdot n_j}$ which is Petzval curvature.

Result: Positive power surfaces ($K_j > 0;\ j = 1, 2, \ldots, k$) cause the image plane to curve negatively, the center of curvature to the left of the surface. Negative power surfaces ($K_j < 0;\ j = 1, 2, \ldots, k$) cause it to curve positively. The condition for flat image plane is that the surface powers sum to zero $\sum_{j=1}^{k} \frac{K_j}{n_{j-1} \cdot n_j} \to 0 \Rightarrow \frac{1}{r_k} \to 0 \Rightarrow r_k \to \infty;\ \forall k$. The image plane curvature exists for paraxial rays. We use the Petzval curvature for various lens design. The multiple lens systems with net positive power the negative lenses contribute relatively less to the total power than the positive lenses. By generating multiple-lens power formula which tracing a ray parallel to the axis through the system and exhibit it by the ray heights (not lens spacing), we get $K = K_1 + \frac{h_2}{h_1} \cdot K_2 + \frac{h_3}{h_1} \cdot K_3 + \cdots$ and the optical system fulfils: Arranging the negative lenses, where the parallel ray has a low height from the axis, gives minimization of the negative contribution to the total power and balance the

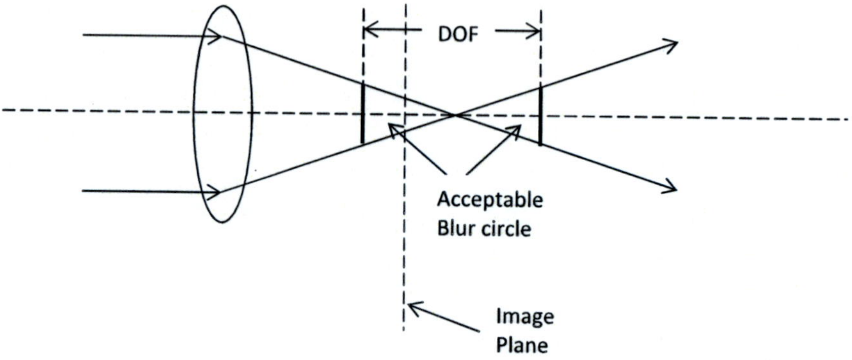

Fig. E.50 Optical system with the depth of focus (DOF) for image plane which is close to flat

Petzval sum [46, 47]. We define the Depth of focus (DOF) of the optic system for image plane which is close to flat (Fig. E.50).

The acceptable blur circle is related to the pixel-size, for a digital detector (Bayer-pattern). Example: We have optic system with lens and the parameters as follow: $n = 1.79$; $R_1 = -85$; $R_2 = -26.4$; $f = 78$. The surface power $K_1 = \frac{n-1}{R_1} = \frac{1.79-1}{-85} \approx -0.00929$ and $K_2 = \frac{1-n}{R_2} = \frac{1-1.79}{-26.4} \approx 0.0299$ then $P_{sum} = \frac{K_1}{n} + \frac{K_2}{n} = \frac{1}{n} \cdot \sum_{j=1}^{2} K_j$ and

$$P_{sum} = \frac{K_1}{n} + \frac{K_2}{n} = \frac{1}{n} \cdot \sum_{j=1}^{2} K_j = \frac{1}{1.79} \cdot (-0.00929 + 0.0299) = 11.51 \times 10^{-3}$$

The Petzval sum is positive ($P_{sum} > 0$) which indicate a negative image surface curvature (curved towards the front of the system). The lens aperture is kept small enough such that the pictures are acceptable. The assumption for thin lens: $P_{sum} = \frac{1}{n} \cdot \sum_{j=1}^{2} K_j \simeq \frac{K}{n}$ and there is no option to affect the P_{sum} value by bending the lens (Re-distributing the surface power of a thin lens while keeping to total power constant). The assumption for thick lens: $K = \sum_{j=1}^{2} K_j - K_1 \cdot K_2 \cdot d$; $P_{sum} = \frac{1}{n} \cdot \sum_{j=1}^{2} K_j = \frac{1}{n} \cdot (K + K_1 \cdot K_2 \cdot d)$ then Petzval sum is reduced if $d > 0$ and $\prod_{j=1}^{2} K_j = K_1 \cdot K_2 < 0$.

In many high power laser facilities there is a use of high-energy laser amplifiers. Applications in high-energy laser amplifiers need transport system (TS) which the laser beam propagate in relay imaging multi-pass layouts. Ray tracing analysis uses ABCD matrix formalism. A compensation for low order aberrations is needed. There are system parameter variation for typical system characteristics, such as number of round-trips, focal length and optics sizes. We need to compensate of defocus caused by a thermal lens and astigmatism caused by non-normal incidence on the imaging optics. We need to avoid intensity spikes leading to damages of optics in multi-pass laser amplifiers. The method to analyze optical systems with multiple

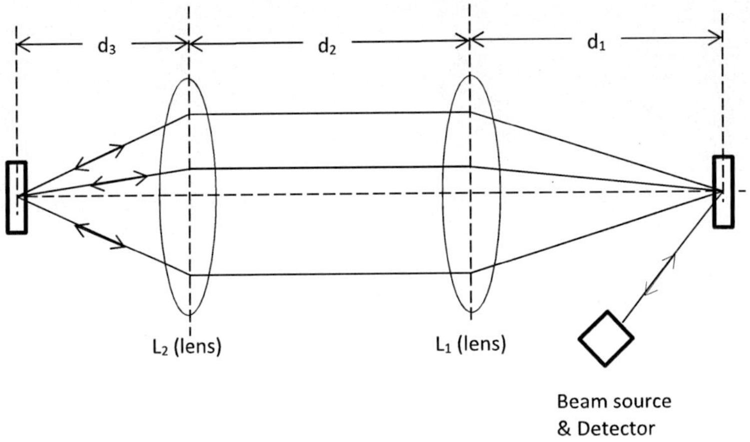

Fig. E.51 An imaging system with three round trips based on lenses

round-trips and repeated use of the imaging optics by using ABCD matrix formalism gives good analysis results. These imaging elements are modeled as ideal lenses, and ray tracing simulations, spherical optics. The system inspection and analysis is categorized into two fundamental types. First, the defocus, caused by thermal lensing and secondly, astigmatism, caused by tilted incidence on the imaging elements. In both cases, the generalized design parameters and the number of round-trips are varied. There are corrections to the system parameters that compensate for these unwanted effects. We can describe optical system which contain a translations of length d and imaging elements with focal length f by the matrices T for translation and L for the imaging element, $T = \begin{pmatrix} 1 & d \\ 0 & 1 \end{pmatrix}$; $L = \begin{pmatrix} 1 & 0 \\ -\frac{1}{f} & 1 \end{pmatrix}$. The round-trip in an imaging multi pass amplifier can be described by free space propagation and ideal imaging elements. The matrix representing one round-trip in a system of n elements is G_n; $G_n = T_{n+1} \cdot L_n \cdot T_n \cdot L_{n-1} \cdot \ldots \cdot L_1 \cdot T_1$; $G_n = T_1 \cdot \prod_{k=1}^{n} T_{k+1} \cdot L_k$. An imaging system for several round-trips restores both the spatial and the angular distribution after each round-trip. The optical system matrix equals either the unity matrix I or in the case of a symmetric profile, the negative unity matrix $(-I)$. By a constant small angle add-on per pass of the system to separate the different beam passes geometrically, the alternating direction of the angular add on prevents a summation. It is done by introducing a little tilt of the flat optics in the image plane. We can build the image system by solving the matrix equation $G_n = I$ (I is unity matrix), the results is for $n = 1, 2, 3, \ldots$. Systems with four elements and more are more sophistical for solution by our method. The algorithm is perfect for systems with less than four imaging elements. System of one imaging element ($n = 1$): It is one imaging element per one round-trip, then the focusing or de-focusing effect of the single optic cannot be mitigated and alter the collimation state of the input beam. System of two imaging elements ($n = 2$): we cannot get a solution that gives a

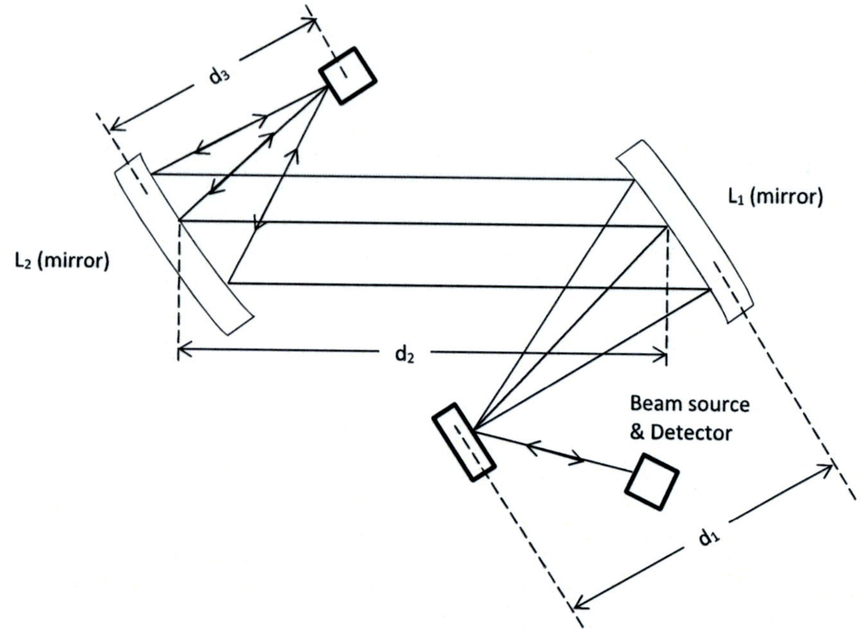

Fig. E.52 An imaging system with three round trips based on mirrors

positive unity matrix, and a solution is for negative unity matrix. The double passing back and forth in our optical system results in a unity matrix. It is a narrower sense a special case for four imaging elements, then the solution $d_1 = \frac{f_1 \cdot f_2 \cdot (f_1 + f_2) - d_3 \cdot f_1^2}{f_2^2}$; $d_2 = \sum_{k=1}^{2} f_k$; $f_1 = f_2 = f \Rightarrow d_1 = d_2 - d_3$. Where, d_1, d_2 and d_3 are the corresponding free space propagation distances, and f_1, f_2 are the focal lengths of the imaging elements. It is used for a double-passed 4f-telescope for a multiply imaging layout. The type I optical system is characterize by three round-trip and is divided to a system with three round trips based on lenses (Fig. E.51), and three round trips based on mirrors (Fig. E.52).

The system is characterized by double passed to fulfill the constraints. Since $d_1, d_2, d_3 > 0$; $d_i \in \mathbb{R} \forall i = 1, 2, 3, \ldots$ then f_1, $f_2 > 0$ (at least positive). If we want to implement our method (algorithm) for system with three imaging elements, there are many degrees of freedom with respect to the four different distances to get the solution for each distance by using matrix model. For a symmetrical system, the layout is mirrored at the second imaging element. We can use imaging optical system constrains $f_3 = f_1$; $d_4 = d_1$ and $d_3 = d_2$ to reduce the degrees of freedom. It requires two different imaging elements and the ABCD model—results constrains: $d_1 = \frac{f_1^2 + 2 \cdot f_1 \cdot f_2}{2 \cdot f_2}$; $d_2 = f_1 + 2 \cdot f_2$, both f_1, $f_2 > 0$ (positive) then we get a positive distances. The results for $\frac{1}{f_1}$ give the imaging equation: $\frac{1}{f_1} = \frac{1}{d_1} + \frac{1}{d_2}$ the system generated an image of the original beam on the second optical element. The beam is imaged back to the original image plane by the first element. The diameter of the

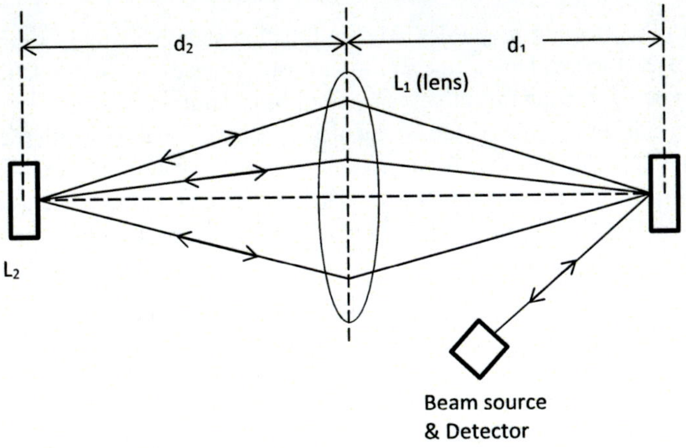

Fig. E.53 Type *II* imaging system with three round-trips based on lenses

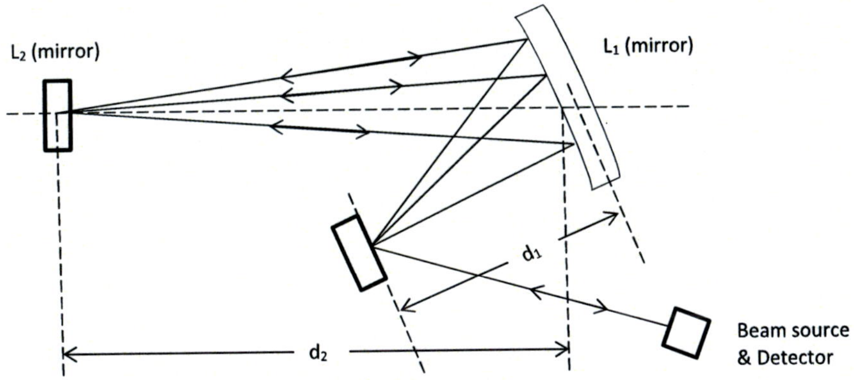

Fig. E.54 Type *II* imaging system with three round-trips based on mirrors

second element is large for supporting the beam itself, and the individual round-trips are not spatially separated on this optics. It is type *II* system with three round-trips based on lenses imaging system (Fig. E.53) and based on mirrors imaging elements (Fig. E.54).

If $d_3 = 0$ and half the focal length of the second element to consider the double pass on this element, then type *I* systems are identical to type *II* systems. In many types of bulk lasers, the thermal load originating from the laser process generates a temperature profile in the laser medium due to thermal expansion and the temperature dependence of the refractive index. It is the effect of thermal lensing which increase intensities on amplifier optics and cause to optical damages. The effect is a focusing or defocusing of the beam by the thermal lens and can cause damage. In the imaging multi-pass schemes the laser active medium is envisioned to be situated in the primary image plane. The actual beam shape in the plane is constant irrespective of a thermal lens due to the imaging properties. The collimation state is altered, adding up the focusing of the thermal lens with every round-trip. A means of compensation within the system counteracting this effect is taken to consideration.

The thermal lens is represented by the matrix L_t as an ideal lens with focal length f_t. For type I system, this lens is passed once per half round trip represented by G_2. The modified matrix G_2^* is $G_2^* = G_2 \cdot L_t$. The compensation in G_2^* is achieved when all entries are zero, except for the main axis. The corrected distances d_1, d_2 and d_3 due to thermal lensing effect is

$$d_1 = f_1 + \frac{f_t}{2} \cdot \left[1 - \sqrt{1 + \frac{4 \cdot f_1^2 \cdot (d_3 - f_2)}{f_t \cdot f_2^2}}\right];\ d_2 = \frac{f_1 \cdot f_2 + (f_1 + f_2) \cdot (f_t - d_1)}{f_t + f_1 - d_1}$$

$$d_3 = f_2 + \frac{f_2^2 \cdot (f_1 - d_1)}{f_1^2} \cdot \left(1 + \frac{f_1 - d_1}{f_t}\right)$$

It is compensating solution, and other solutions for $f_t \to \infty$ are omitted since the systems integrating the thermal lens as an additional imaging element are taking into account. For standard layout $d_1 = f_1;\ d_2 = f_1 + f_2 - \frac{f_1^2}{f_t};\ d_3 = f_2$, the spherical part of a thermal lens in a type I imaging system is compensated by adjusting the distance between the imaging optics. The focal length of the thermal lens is compensated as long as $f_t \geq \frac{f_1^2}{f_1 + f_2}$ otherwise, d_2 is negative. Practically, the setup f_t is much bigger than f_1 and avoids intensity spikes on the imaging elements. The modified matrix for a type II imaging system is $G_3^* = L_t \cdot G_3 \cdot L_t$ (the thermal lens is passed in the beginning and at the end of each round-trip). To compensate for a thermal lens, G_3^* needs to be equal to I (unity matrix) which gives $d_1 = f_1 + \frac{f_t}{2} \cdot \left[1 - \sqrt{1 - \frac{2 \cdot f_1^2}{f_2 \cdot f_t}}\right];\ d_2 = f_1 + f_2 \cdot \left[1 + \sqrt{1 - \frac{2 \cdot f_1^2}{f_2 \cdot f_t}}\right]$. As for type I system, all solutions that do not converge to the original type II solution for $f_t \to \infty$ are omitted; though the length of both arms is changed. The lens equation $\frac{1}{f_1} = \frac{1}{d_1} + \frac{1}{d_2}$ is valid for the new solution. The major effect of the thermal lens can be seen as a virtual change of the focal length of L_2 [48].

Micro lens let-array-based compact relay systems are used in various optical configurations that include an appropriate baffle computation to eliminate ghost images. There is a Trial-off between compactness of the system and a tiling effect which is corresponding to image. It is ray-tracing optimization with a sufficient signal-to-noise ratio (SNR). The most compact configuration to image gray-scale images is the 5f-based system. Micro lens are formed using photolithographic techniques based on semiconductor processing technology. We get excellent uniformity in the shape and position of each micro lens. This method produces micro lens array where the spot size of each micro lens is nearly diffraction limited. The micro lens is a small lens with good optical quality. By implementing micro lens with two flat and parallel interfaces, we get focusing by a variation of refractive index across the lens (gradient-index (GRIN) lenses). Other king of micro-lenses is focused by a variation of refractive index across the lens and by the surface shape. The micro-Fresnel lenses focus light by refraction in a set of concentric curved surfaces. Micro-lens arrays contain many lenses formed in a one dimensional or two dimensional arrays on a substrate. Single micro-lens is used to couple light to optical fiber while micro lens array mainly used to increase the light collection efficiency. Combinations of micro lenses are used for imaging applications. Micro lens let-array-based imaging systems are used for optical scanners and copies, three-dimensional integral photography, printers, and photolithography. Scientific optical imaging applications require extremely compact, lightweight relay systems. In the case of two lenses which satisfy 1:1 image relay conditions, the difference among them lies in the overall length of the optical system and its field pf view (FOV). For all systems, each pair of lenses is of the same focal length f and the same diameter D. The overall length is OAL (optical array logic) of such a system defined as the distance from the object to the final image plane is given by $\mathrm{OAL} = \frac{2 \cdot x^2}{|x| - |f|}$, where x is the distance from the object to the first lens in each pair. The minimum of the function (OAL = ...) takes place at $|x| = 2 \cdot |f|$, which gives $\mathrm{OAL} = \frac{2 \cdot x^2}{|x| - |f|}\Big|_{|x| = 2 \cdot |f|} = \frac{2 \cdot x^2}{|f|}$. The most compact possible arrangement is the $2f - 4f - 2f$, referred to as the $2f$ system, in which the object is located at a distance $2f$ in front of the first lens and an intermediary image is formed at a distance $2f$ after the first lens with a magnification of (-1), which is consequently imaged at a distance $2f$ after the second lens with a magnification of (-1). The total magnification of the system is (+1). The full field of view (FOV) is defined as 100% vignetting at its edge and is given by $\mathrm{FOV} = (M - 1) \cdot D$, where M denotes the system type ($2f$ system, $M = 2$, and for full FOV, it is equal to the diameter of the lens D, $M = D$ then for $\mathrm{FOV} = (D - 1) \cdot D$). Optical array logic (OAL) is a logic system suitable for the digital optical computing scheme. Complicated processing can be described and executed by the optical array logic (OAL) scheme. Compared with symbolic substitution, all processing in optical array logic is described by logical operations for pixels within a neighborhood area on the two images. We define $\mathrm{OAL} = g(x);\ g(x) = \frac{2 \cdot x^2}{|x| - |f|}$ where, $f = 0.5\,\mathrm{mm} = 5 \times 10^{-4}\,\mathrm{m}$ and plot the graph (Fig. E.55).

MATLAB

```
x=-5:0.1:5; % x is the distance from the
     object to the first lens in each pair
f=0.5; % f is f-number of optical system
y=(2*x.*x)./(abs(x)-abs(f)); % y is the OAL (Optical Array Logic)
plot(x,y);
xlabel('x') ;ylabel('g(x)');
title ('g(x), OAL vs x');
```

The 1:1 imaging with a single pair of micro lenses concept cannot be simply extended to array of lenses because of the formation of ghost images. To prevent ghost images, we use appropriate baffle arrays in combination with the arrays of lenses. For $2f$ system and determination the location and size of the appropriate baffles, we place two lenses L_1 and L_2 which limit equally the amount of light

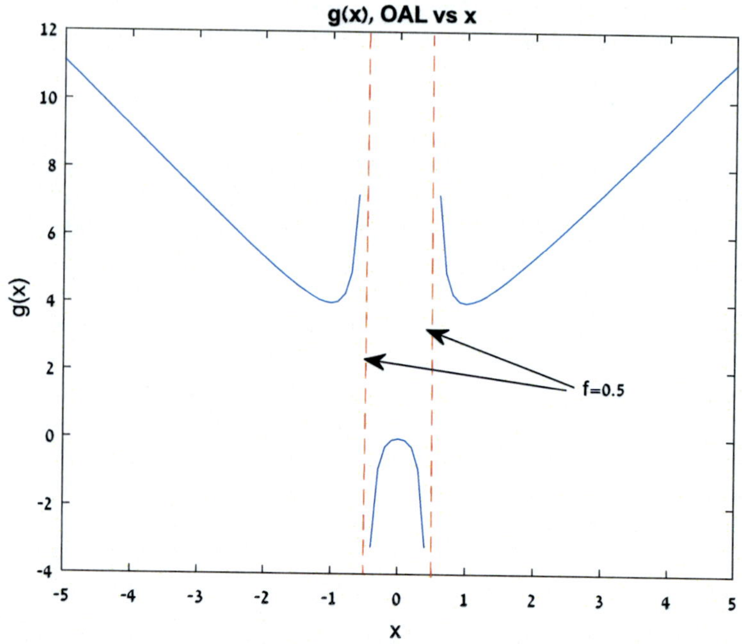

Fig. E.55 g(x), OAL (optical array logic) versus x (distance from the object to the first lens in each pair)

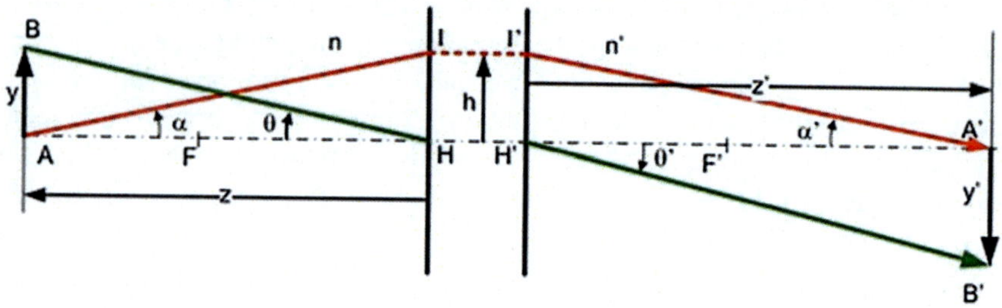

Fig. E.56 Optical system and magnification for two principal points H and H'

entering the system from a considered point object on axis. One of the lenses (L_1 or L_2) can be the Aperture Stop (AS), the other one being the window. We use Descartes imaging equations, and get location and size of the exit pupil for the $2f$ system which is $\frac{4}{3}|f|$ after the second lens and three times smaller than the diameter of L_1, respectively. The location of the entrance window is computed as the image of L_2 through L_1. A property of optical imaging with a stack of two micro lens let array is the sampling of object by each pair of micro lens lets in the stack, where each pair operates over a limited FOV, (FOV $= (M - 1) \cdot D$). The Descartes conjugate equation and magnification for two principal points H and H' is described in the next figure (Fig. E.56). The principal point H and H' are taken to have for origin $\overline{HA} = z$ and $\overline{H'A'} = z'$ then $\overline{FA} = z - f$ and $\overline{F'A'} = z' - f'$. The system equation is $\frac{n'}{z'} = \frac{n}{z} + \frac{n'}{f'} = \frac{n}{z} + C \cdot v$.

The ray (green) issued from B and going through H forms an angle with the axis. It image going from H' forms an angle θ' such that $n \cdot \theta = n' \cdot \theta'$. Since $\theta = \frac{y}{z};\ \theta' = \frac{y}{z'}$ the transversal magnification g_y: $g_y = \frac{\overline{A'B'}}{\overline{AB}} = \frac{y'}{y} = \frac{nz'}{n'z};\ g_y = \frac{f}{f-z} = 1 - \frac{z'}{f'}$. One value of z (or z') is corresponded only one value of g_y and conversely. The angular magnification g_α: $g_\alpha = \frac{\alpha'}{\alpha} = \frac{z}{z'}$ and to an axial magnification g_z: $g_z = \frac{dz'}{dz} = \frac{n}{n'} \cdot \frac{(z')^2}{z^2}$. Ray-trace system with noise: The noise is associated with each pixel in the detector. The probability of a ray reaching a pixel is small with respect to one, and the probability density of noise associated with a pixel reduces to a Poisson distribution. The standard deviation of the noise is equal to the square root of the number of rays per pixel n in the image plane. The SNR in the imaging process, which varies as the square root of n, enables to select a minimum number of rays emitted per pixel in object space. η—is the efficiency of ray tracing defined as the ratio of the total number of rays reaching the detector plane over the total number of the rays x emitted from the object and N is the total number of pixels in the object and thus in the image. The SNR as a function of x, SNR(dB)(x) isSNR(dB)$(x) = 10 \cdot \log_{10}\{\sqrt{\frac{x}{N}} \cdot \eta\}$, to quantify the SNR as a function of the number of rays per pixel $\left(n = \frac{x}{N}\right)$. The efficiency ($\eta$) varies with both the optical configuration and specifies of the optical ray-tracing approach. The efficiency (η) mainly increases from approximately 5% in the $2f$ configuration to 40% in the $7f$ configuration, respectively. The effect of lens let due to the micro lenses in the array is observed as an oscillating curve in the subtracted auto correlation functions. The $2f$ system is most compact, the lens let makes the configuration unsuitable to image gray-scale images. Lens let appears to be negligible for the $4f$ system and less pronounced for the $5f$, $6f$, and $7f$ systems. The propagation of noise from the images to the auto correlation function is negligible. By implementing a thin-lens approximation, the compactness of each configuration, defined as the OAL from the light source to the image plan, and is given by OAL $= \frac{2 \cdot M^2}{(M-1)} \cdot |f|$, where M denotes the system type, and f is the focal length. The OAL (optical array logic) is scaled linearly with focal length [49].

In many industrial and scientific optical applications there is a need to design a spherical focal surface by using close-packed relay optics. By using close-packing circular finite conjugate optics, we create a spherical focal surface. The method to generate the circle distributions is based on a distorted icosahedral geodesic. An icosahedron is a solid in three dimensions with flat polygonal faces, straight edges and sharp corners or vertices, with 20 faces. A convex polyhedron is the convex hull of infinity many points, not all on the same plane. Cubes and pyramids are examples of convex polyhedral. There are infinitely many non-similar shapes of icosahedra, and part of them is more symmetrical than the others. The regular icosahedron is the one of the platonic solids, whose faces is 20 equilateral triangles. A spherical focal surface has the advantages of high degrees of symmetry, minimized variations in a circle separations, and generation of configurations with N circles. N is the number of vertices on the geodesic. These properties are used for making a continuous focal

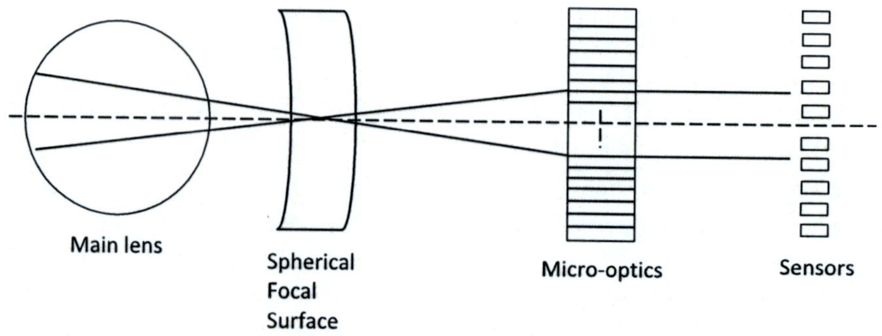

Fig. E.57 Scheme for spherical focal lenses—cross section of optical model

surface. The icosahedral packings in the 3D Euclidean space E^3 is produced by projecting and mapping of the fragments of a multidimensional icosahedral packing, which is accompanied by some insignificant distortions. The origin of coordinates and the method of projection on a collection of concentric spheres, a variety of sphere-or rod-shaped structures are produced with the relative position of the subunits close to the icosahedral arrangement. Isomorphic imaging systems typically correct geometric aberrations with the right design of optical elements then the image is captured on a flat sensor. A flat focal surface converts a spherically distributed field to a flat field and correction of field curvature. A larger field ranges compensate for curvature by manipulating the focal surface shape rather than the optical elements. In a focal imaging system, all rays passing through point x_0 in the object plane pass through point x_i in the image plane. The visibility between points in the object plane and points in the image plane is $v(x_0, y_0, x_i, y_i) = \delta(x_i - M \cdot x_0, y_i - M \cdot y_0)$, where, M is the magnification. The systems which implement point-to-point visibilities are isomorphic systems. The isomorphism of focal imaging applies only between planes in 3D space. Multiplex systems enable higher system performance relative to data efficiency, resolution, depth of field, and other parameters. There are various curved sensors to account for field curvature. The pixel elements can be connected via flexible interconnects and mounted on a flexible membrane. The substrate is then deformed by pressing with a shaped rod. A system with a flat sensor pattern is overlaid onto a flexible substrate are good for small-scale, and low resolution images. Monocentric imaging systems are characterized by wide field, high pixel count imaging with the availability of a spherical focal surface. A focal surface is approximated by incorporation of a monocentric lens as the primary optic in a multiscale design. By utilizing an array of secondary optics with overlapping fields (micro-optics) to capture the spherical image which produced by a monocentric lens. A spherical focal surface with no gaps is realized from an array of flat sensors (Fig. E.57).

The field covered by each set of micro-optics is small, and then the amount of curvature correction needed to image each sub field onto a flat local plane is minimal. The distribution of pores on pollen grains is very important for optical system design.

The target is to pack a given number of points on the surface of a sphere and get a minimum chord length between two points. We pack a given number of equal circles on the sphere, and the diameter of the circles is maximized without overlap. The minimum chord length determines the diameter of the circles, and the solution maximized the packing density (ρ), defined by $\rho = \frac{N \cdot a}{A}$, where, N is the number of circles, a is the surface of the sphere enclosed by a single circle, and A is the total surface area of the sphere. The technique takes to consideration the points as equal point charges floating on the surface of the sphere and iteratively they repel one another until a stable minimum energy configuration is reached [50].

References

[1] J. Liu, J. Tan, T. Wilson, C. Zhong, Rigorous theory on elliptical mirror focusing for point scanning microscopy. Opt. Express **20**(6) (2012)
[2] G.E. Ice, J.-S. Chung, J.Z. Tischler, Elliptical x-ray microprobe mirrors by differential deposition. Rev. Sci. Instrum. **71**(7) (2000)
[3] M. Carrara, M.R. Cacan, M.J. Leamy, M. Ruzzene, A. Erturk, Dramatic enhancement of structure-borne wave energy harvesting using an elliptical acoustic mirror. Appl. Phys. Lett. **100**, 204105 (2012)
[4] C.M. Kewish, L. Assoufid, A.T. Macrander, J. Qian, Wave-optical simulation of hard-X-ray nano focusing by precisely figured elliptical mirrors. Appl. Opt. **46**(11) (2007)
[5] J. Liu, C. Zhong, J. Tan, T. Wang, T. Wilson, Elliptical mirror based imaging with aperture angle greater than π/2. Opt. Express **20**(17) (2012)
[6] M. Avendano-Alejo, R. Diaz-Uribe, Testing a fast off-axis parabola mirror by using tilted null screens. Appl. Opt. **45**(12) (2006)
[7] M. Avendano-Alejo, V.I. Moreno-Oliva, M. Campos-Garcia, R. Diaze-Uribe, Quantitative evaluation of an off-axis parabolic mirror by using a tilted null screen. Appl. Opt. **48**(5) (2009)
[8] D. Herriott, H. Kogelnik, R. Kompfner, Off-axis path in spherical mirror interferometers. Appl. Opt. **3**(4) (1964)
[9] H.M. Ozaktas, D. Mendlovic, Fractional Fourier transform as a tool for analyzing beam propagation and spherical resonators. Opt. Lett. **19**(21) (1994)
[10] F.X. Kartner, N. Matuschek, T. Schibli, U. Keller, Design and fabrication of double-chirped mirrors. Opt. Lett. **22**(11) (1997)
[11] N. Matuschek, F.X. Kartner, U. Keller, Theory of double-chirped mirrors. IEEE J. Sel. Top. Quantum Electron. **4**(2) (1998)
[12] A.-S. Morlens, P. Balcou, P. Zeitoun, C. Valentin, Compression of attosecond harmonic pulses by extreme-ultraviolet chirped mirrors. Opt. Lett. **30**(12) (2005)
[13] J. Hebling, E.J. Mayer, J. Kuhl, Chirped-mirror dispersion-compensated femtosecond optical parametric oscillation. Opt. Lett. **20**(8) (1995)
[14] R. Szipocs, K. Ferencz, C. Spielmann, F. Krausz, Chirped multilayer coatings for broadband dispersion control in femtosecond lasers. Opt. Lett. **19**(3) (1994)
[15] P. Dombi, V.S. Yakovlev, K. O'Keeffe, T. Fuji, M. Lezius, G. Tempea, Pulse compression with time-domain optimized chirped mirrors. Opt. Express **13**(26) (2005)

[16] Y. Minami, Y. Hayashi, J. Takeda, I. Katayama, Single-shot measurement of a terahertz electric-field waveform using a reflective Echelon mirror. Appl. Phys. Lett. **103**, 051103 (2013)
[17] B.K. Ofori-Okai, P. Sivarajah, W. Ronny Huang, K.A. Nelson, THz generation using reflective stair-step Echelon. Opt. Express **24**(5), 5057–5068 (2015)
[18] I. Katayama, H. Sakaibara, J. Takeda, Real-time time-frequency imaging of ultrashort laser pulses using Echelon mirror. Jpn. J. Appl. Phys. **50**(102701), 1–5 (2011)
[19] O. Llopis, P.H. Merrer, H. Brahimi, K. Saleh, P. Lacroix, Phase noise measurement of a narrow linewidth CW laser using delay line approaches. Opt. Lett. **36**(14) (2011)
[20] D.W. Sukow, J.R. Gardner, D.J. Gauthier, Statistics of power-dropout events in semiconductor lasers with time-delayed optical feedback. Phys. Rev. A **56**(5) (1997)
[21] K.P. Jackson, S.A. Newton et al., Optical fiber delay-line signal processing. IEEE Trans. Microwave Theor. Tech. **MTT-33**(3) (1985)
[22] D.R. Herriott, H.J. Schulte, Folded optical delay lines. Appl. Opt. **4**(8) (1965)
[23] E.S. Choi, J. Na, S.Y. Ryu, G. Mudhana, B.H. Lee, All-fiber variable optical delay line for applications in optical coherence tomography: feasibility study for a novel delay line. Opt. Express **13**(4) (2005)
[24] J. van Howe, C. Xu, Ultrafast optical delay line by use of a time-prism pair. Opt. Lett. **30**(1) (2005)
[25] C.G.H. Roeloffzen, L. Zhuang, R.G. Heideman, A. Borreman, W. van Etten, Ring resonator-based tunable optical delay line in LPCVD waveguide technology, in *Proceedings Symposium IEEE/LEOS Benelux Chapter* (2005)
[26] J. Cardenas, M.A. Faster et al., Wide-bandwidth continuously tunable optical delay line using silicon micro-ring resonators. Opt. Express **18**(25) (2010)
[27] J.B. Khurgin, P.A. Morton, Tunable wideband optical delay line based on balanced coupled resonator structures. Opt. Lett. **34**(17) (2009)
[28] M. Miranda, T. Fordell, C. Arnold, A.L. Huillier, H. Crespo, Simultaneous compression and characterization of ultrashort laser pulses using chirped mirrors and glass wedges. Opt. Express **20**(1) (2011)
[29] G.S. Sarkisov, Shearing interferometer with an air wedge for the electron density diagnostics in a dense plasma. Instrum. Exp. Tech. **39**(5) (1996)
[30] L. Yin, J. Zhang, P.M. Fauchet, G.P. Agrawal, Optical switching using nonlinear polarization rotation inside silicon waveguides. Opt. Lett. **34**(4) (2009)
[31] G.R.B.E. Romer, P. Bechtold, Electro-optic and acousto-optic laser beam scanners. Phys. Procedia **56**, 29–39 (2014)
[32] M. Zgonik, P. Bernasconi, M. Duelli, R. Schlesser et al., Dielectric, elastic, piezoelectric, electro-optic, and elasto-optic tensors of BaTiO3 crystal
[33] G. Andersson, I. Dahl, P. Keller, W. Kuczynski, S.T. Lagerwall, K. Skarp, B. Stebler, Sub microsecond electro-optic switching in the liquid-crystal smectic A phase: the soft-mode ferroelectric effect. Appl. Phys. Lett. **51**(9) (1987)

[34] C. Winnewisser, P. Uhd Jepsen, M. Schall, V. Schyja, H. Helm, Electro-optic detection of THz radiation in LiTaO3, LiNbO3 and ZnTe. Appl. Phys. Lett. **70**(23) (1997)
[35] Z. Peng, T. Yidong, L. Weixin, C. Wenxue, Methods for optical retardation measurement: a review. Sci. China Tech. Sci. **56**(5) (2013)
[36] J.-B. Masson, G. Gallot, Terahertz achromatic quarter-wave plate. Opt. Lett. **31**(2) (2006)
[37] G. Czechowski, J. Jadzyn, Singular elastic properties of mixtures with the re-entrant nematic phase. Acta Phys. Pol. A **106**(4) (2004)
[38] C.-F. Hsieh, R.-P. Pan et al., Voltage-controlled liquid-crystal terahertz phase shifter and quarter-wave plate. Opt. Lett. **31**(8) (2006)
[39] J. Feng, Z. Zhou, Polarization beam splitter using a binary blazed grating coupler. Opt. Lett. **32**(12) (2007)
[40] L. Yu, L. Liu, Z. Zhou, X. Wang, High efficiency binary blazed grating coupler for perfectly-vertical and near-vertical coupling in chip level optical interconnection. Opt. Commun. **355**, 161–166 (2015)
[41] R.-C. Tyan, A.A. Salvekar, H.-P. Chou, C.-C. Cheng, A. Scherer, P.-C. Sun, F. Xu, Y. Fainman, Design fabrication and characterization of form-birefringent multilayer polarizing beam splitter. J. Opt. Soc. Am. A **14**(7) (1997)
[42] T.W. Hansch, B. Couillaud, Laser frequency stabilization by polarization spectroscopy of a reflecting reference cavity. Opt. Commun. **35**(3) (1980)
[43] S. Szatmari, R. Dajka, A. Barna, B. Gilicze, I.B. Foldes, Improvement of the temporal and spatial contrast of high-brightness laser beams. Laser Phys. Lett. **13**, 075301 (5pp) (2016)
[44] R.D. Sigler, Compound Schmidt telescope design with nonzero Petzval curvatures. Appl. Opt. **14**(9), 2302–2305 (1975)
[45] T.-Y. Lim, Y.-S. Kim, S. Park, Achromatic and thermal design of an optical system with corrected Petzval curvature on a three-dimensional glass chart. Curr. Opt. Photonics **1**(4), 278–388 (2017)
[46] J.U. Lee, S.M. Yu, Analytic design procedure of three mirror telescope corrected for spherical aberration, Coma, Astigmatism, and Petzval field curvature. J. Opt. Soc. Korea **13**(2), 184–192 (2009)
[47] J. Choi, T.H. Kim, H.J. Kong, J.U. Lee, Third-order aberration correction by using an iterative process and its applications to a Petzval lens with an external entrance pupil. J. Korean Phys. Soc. **47**(4), 631–635 (2005)
[48] J. Korner, J. Hein, M. Christoph Kaluza, Compact aberration-free relay-imaging multi-pass layouts for high energy laser amplifiers. Appl. Sci. **6**, 353 (2016)
[49] V. Shaoulov, J.P. Rolland, Design and assessment of micro lens let array relay optics. Appl. Opt. **42**(34) (2003)
[50] H.S. Son, D.L. Marks, J. Hahn, J. Kim, D.J. Brady, Design of a spherical focal surface using close-packed relay optics. Opt. Express **19**(17) (2011)

Bibliography

1. Fundamentals of electrostatic discharge. Compliance Magazine, 1 May 2015. Retrieved 25 June 2015
2. S. Younger et al., Scientific collaborations between Los Alamos and Arzamas-16 using explosive-driven flux compression generators. Los Alamos Science (24), 48–71. Retrieved 2009-10-24 (1996)
3. C.E. Baum, Reminiscences of high-power electromagnetics. IEEE Trans. Electromagn. Compat. **49**(2), 211–218 (2007)
4. C.E. Baum, From the electromagnetic pulse to high-power electromagnetics. Proc. IEEE **80**(6), 789–817 (1992)
5. Report Meta-R-320, The early-time (E1) high-altitude electromagnetic pulse (HEMP) and Its impact on the U.S. power grid
6. C.G. Brown Jr., A. Throop, D. Eder, J. Kimbrough, Electromagnetic pulses at short-pulse laser facilities. J. Phys. Conf. Ser. **112**, 032025 (2008). https://doi.org/10.1088/1742-6596/112/3/032025
7. C.G. Brown Jr., E. Bond, T. Clancy, S. Dangi, D.C. Eder, W. Ferguson, J. Kimbrough, A. Throop, Assessment and mitigation of electromagnetic pulse (EMP) impacts at short-pulse laser facilities. J. Phys. Conf. Ser. **244**, 032001 (2010). https://doi.org/10.1088/1742-6596/244/3/032001
8. D.C. Eder, R.W. Anderson et al., Assessment and mitigation of radiation EMP, Debris & Shrapnel impacts at megajoule-class laser facilities. J. Phys. Conf. Ser. **244**, 032018 (2010). https://doi.org/10.1088/1742-6596/244/3/032018
9. C.N. Danson, P.A. Brummitt et al., Vulcan Petawatt—an ultra-high intensity interaction facility. Nucl. Fusion **44**, S239–S246 (2004)
10. W. Lu, C. Huang, M. Zhou, M. Tzoufras, F.S. Tsung, W.B. Mori, T. Katsouleas, A nonlinear theory for multidimensional relativistic plasma wave wake fields. Phys. Plasma **13**, 056709 (2006)
11. P. Mora, T.M. Antonsen, Kinetic modelling of intense, short laser pulses propagating in tenuous plasmas. Phys. Plasmas **4**(1) (1997)
12. M. Lobert, E. d'Humieres, M. Grech, C. Ruyer, X. Davoine, L. Gremillet, Modeling of radiative and quantum electrodynamics effects in PIC simulations of ultra-relativistic laser-plasma interaction. Phys. Plasm-ph (2013)
13. D.F. Gordon, B. Hafizi, M.H. Helle, Solution of relativistic quantum optics problems using clusters of graphical processing units. J. Comput. Phys. **267**, 50–62 (2014)
14. P. Mora, Plasma expansion into a vacuum. Phys. Rev. Lett. **90**(18) (2003)
15. A. Poye, S. Hulin, M. Bailly-Grandvaux, J.L. Dubois, J. Ribolzi, D. Raffestin, M. Bardon, F. Lubrano-Lavaderci, E. D'Humieres, J.J. Santos, Ph Nicolai, V. Tikhonchuk, Physics of giant electromagnetic pulse generation in short pulse laser experiments. Phys. Rev. E **91**, 043106 (2015)

O. Aluf, *Advance Elements of Laser Circuits and Systems*,
https://doi.org/10.1007/978-3-030-64103-0

16. D. Kocon, D. Klir, J. Krasa, T. Lastovicka, L. Pribyl, R. Vrana, Operating semiconductor timepix detector with optical readout in an extremely hostile environment of laser plasma acceleration experiment, in *Proceedings of IBIC2013*, Oxford, UK
17. J. Krasa, P. Parys, L. Velardi, A. Velyhan, L. Ryc, D. Delle side, V. Nassisi, Time of flight spectra for mapping of charge density of ions produced by laser. Laser Particle Beams **32**, 15–20 (2014)
18. M. De Marco, J. Krasa, J. Cikhardt, M. Pfeifer, E. Krousky, D. Margarone, *Measurement of Electromagnetic Pulses Generated During Interactions of High Power Lasers with Solid Targets*. Published by IOP Publishing for Sissa Medialab.
19. J. Krasa, D. Klir, A. Velyhan, K. Rezac, J. Cikhardt, L. Ryc, E. Krousky, *Generation of Fast Neutrons Through Deuteron Acceleration at the PALS Laser Facility*. Published by IOP Publishing for Sissa Medialab
20. J. Cikhardt, J. Krasa, M. De Marco, M. Pfeifer, A. Velyhan, E. Krousky, Measurement of the target current by inductive probe during laser interaction on terawatt laser system PALS. Rev. Sci. Instrum. **85**, 103507 (2014)
21. M. De Marco, J. Cikhardt, J. Krasa, A. Velyhan, M. Pfeifer, E. Krousky, D. Klir, K. Rezac, J. Limpouch, D. Margarone, J. Ullschmied, Electromagnetic pulses produced by expanding laser produced Au plasma. Nukleonika **60**(2), 239–243 (2015)
22. H. Haug, S.W. Koch, *Quantum Theory of the Optical and Electronic Properties of Semiconductors*, 5th edn. (World Scientific, Singapore, 2008)
23. S.M. Sze, K.K. Ng, *Physics of Semiconductor Devices* (Wiley Interscience, Hoboken, Hardcover—Oct 27, 2006)
24. C.-T. Sah, *Fundamentals of Solid State Electronics* (World Scientific, Singapore)
25. W.Y. Yang, W. Cao, T.-S. Chung, J. Morris, *Applied Numerical Methods Using MATLAB* (Wiley, Hoboken, 2005)
26. A. Band, J. Unguris, *Optically Isolated Current-to-Voltage Converter for an Electron Optics System* (Electron Physics Group National Institute of Standards and Technology, Gaithersburg, Maryland), p. 20899
27. A.C. Flower, M.J. Mcguinnless, A description of the Lorenz attractor at high prandtl number. Physica 5D 149–182 (1982)
28. H.R. Dullin, S. Schmidt, P.H. Richter, S.K. Grossmann, Extended phase diagram of the Lorenz model (2005)
29. G. van der Schrier, L.R.M. Mass, The diffusionless Lorenz equations; Shil'nikov bifurcations and reduction to an explicit map. Physica D **141**, 19–36 (2000)
30. L. Marton, *Advanced Electronics and Electron Physics*, vol 35 (Academic Press Inc, Cambridge, 1974)
31. M.H. Rashid, *Power Electronics Handbook* (Academic Press, Cambridge, 2001)
32. C. Galup-Montoro, M.C. Schneider, *MOSFET Modeling for Circuit Analysis and Design* (World Scientific, Singapore)
33. K. Gopalsamy, *Stability and Oscillation in Delay Differential Equations of Population Dynamics* (Kluwer Academic Publishers, Amsterdam)
34. Y. Kuznetsov, *Elements of Applied Bifurcation Theory* [Paperback] (Springer, Berlin, 2010)
35. E.A. Butcher, H. Ma, E. Bueler, V. Averina, Z. Szabo, Stability of linear time periodic delay differential equations via Chebyshev polynomials. Int. J. Numer. Meth. Eng. **59**, 895–922 (2004)
36. T. Insperger, G. Stepan, Stability of the damped Mathieu equation with time delay. J. Dyn. Syst. Meas. Control **125** (2003)
37. N.K. Garg, B.P. Mann, N.H. Kim, M.H. Kurdi, Stability of a time delayed system with parametric excitation. J. Dyn. Syst. Meas. Control **129** (2007)
38. A. Band, J. Unguris, *Optically Isolated Current to Voltage Converter for an Electron Optics System* (National Institute of Standards and Technology Gaithersburg, Maryland), p. 20899
39. S. Wieczorek, T.B. Simpson, B. Krauskopf, D. Lenstra, Bifurcation transitions in an optically injected diode laser: theory and experiment. Opt. Commun. **215**, 125–134 (2003)

40. T. Pereira, M.S. Baptista, M.B. Reyes, I.L. Caldas, J.C. Sartorelli, J. Kurths, A scenario for torus T2 destruction via a global bifurcation. Chaos Solitons Fractals (2007)
41. A. Algaba, M. Merino, Taken Bogdanov bifurcation of periodic orbits and Arnold's tongues in a three dimensional electronic model. Int. J. Bifurcat. Chaos **11**(2), 513–531 (2001)
42. G. Nicholas Benes, A.M. Barry, T.J. Kaper, M.A. Kramer, J. Burke, An elementary model of torus canards. Chaos **21**, 023131 (2011)
43. H.W. Broer, H. Hanbmann, J. You, Umbilical torus bifurcations in Hamiltonian systems. J. Differ. Eqn. **222**, 233–262 (2006)
44. F. AMDJADI, Torus doubling cascade in problems with symmetries, in *Proceedings of Institute of Mathematics of NAS of Ukraine 2004*, vol 50, Part 3 (2004), pp. 1221–1227
45. M. Aguiar, N. Stollenwerk, B.W. Kooi, Torus bifurcation, isolas and chaotic attractors in a simple dengue fever model with ADE and temporary cross immunity. Int. J. Comput. Math. 1–18 (2008)
46. X. Yu, H. Jia, D. Cun, Complex dynamic caused by Torus bifurcation in power systems. Trans. Tianjin Univ. **12**(3), 186–192 (2006). ISSN 1006-4982
47. K. Chowdhury, A. Roy Chowdhury, An analytic approach to Torus bifurcation in a quasi periodically forced doffing oscillator. Nonlinear Math. Phys. **4**(3–4), 358–363 (1997)
48. E. Allgower, U.L.F. Garbotz, K. Georg, Matrix-free numerical torus bifurcation of periodic orbits. Rendiconti di Matematica, Serie VII **25**, 33–52 (2005)
49. J. Guckenheimer, Singular hopf bifurcation in systems with two slow variables. SIAM J. Appl. Dyn. Syst. **7**(4), 1355–1377
50. S.-I. Chu, D.A. Telnov, Beyond the Floquet theorem: generalized Floquet formalisms and quasi-energy methods for atomic and molecular multi-photon processes in intense laser fields. Phys. Rep. **390**, 1–131 (2004)
51. C.A. Klausmeier, Floquet theory: a useful tool for understanding non-equilibrium dynamics. Theor. Ecol. **1**, 153–161 (2008)
52. G. Moore, Floquet theory as a computational tool. SIAM J. Numer. Anal. **42**(6), 2522–2568
53. E. Di Cera, P.E. Phillipson, J. Wyman, Limit-cycle oscillations and chaos in reaction networks subject to conservation of mass. Proc. Natl. Acad. Sci. USA **86**, 142–146 (1989)
54. K. Briggs, A precise calculation of the Feigenbaum constants. Math. Comput. **57**(195), 435–439 (1991)
55. C.A. Jones, N.O. Weiss, F. Cattaneo, Non-linear dynamos: a complex generalization of the Lorenz equations. Physica **14D**, 161–176 (1985)
56. V. Anishchenko, S. Nikolaev, G. Strelkova, Oscillators of quasiperiodic oscillations. Two-dimensional Torus doubling bifurcation, in *2005 International Symposium on Nonlinear theory and its applications (NOLTA2005) Bruges*, Belgium, Oct 18–21, 2005
57. W.L. Keith, R.H. Rand, Dynamics of a system exhibiting the global bifurcation of a limit cycle at infinity. Int. J. Non-linear Mech. **20**(4), 325–338 (1985)
58. Z. Jing, D. Xu, Y. Chang, L. Chen, Bifurcation Chaos and system collapse in a three node power system. Electr. Power Energ. Syst. **25**, 443–461 (2003)
59. A.M. Harb, M.S. Widyan, Controlling Chaos and bifurcation of sub-synchronous resonance in power system. Nonlinear Anal. Model. Control **7**(2), 15–36 (2002)
60. Y. Komatsu, Y. Uwate, Y. Nishio, Synchronization in Chaotic circuits coupled by mutual inductors, in *2006 RISP International Workshop on Nonlinear Circuits and Signal Processing (NCSP06)*
61. I. Belykh, M. Hasler, M. Lauret, H. Nijmeijer, Synchronization and graph topology. Int. J. Bifurcat. Chaos **15**(11), 3423–3433 (2005)
62. S.K. Han, D.E. Postnov, Chaotic bursting as chaotic itinerancy in coupled neural oscillators. Chaos **13**(3) (2003)
63. T. Ueta, G. Chen, On synchronization and control of coupled Wilson-Cowan neural oscillators. Int. J. Bifurcation Chaos **13**(1) 163–175 (2003)
64. J. Doyne Farmer, I.I. Satija, Renormalization of the quasi-periodic transition to chaos for arbitrary winding numbers. Phys. Rev. A **31**(5) (1985)

65. S.K. Mazumder, A.H. Nayfeh, D. Boroyevich, An investigation into the fast and slow scale instabilities of a single phase bidirectional boost converter. IEEE Trans. Power Electron. **18**(4) (2003)
66. K.D. Edoh, J. Lorenz, Lyapunov-type numbers for poincare maps. Bol. Soc. Paran. Mat. Essaya (3s.) **24**(1–2), 89–98 (2006)
67. Z. Zhu, Z. Liu, Strange non-chaotic attractors of Chua's circuit with quasi-periodic excitation. Int. J. Bifurcation Chaos **7**(1), 227–238 (1997)
68. R. Artuso, G. Casati, D.L. Shepelyansky, Breakdown of universality in renormalization dynamics for critical invariant torus. Europhys. Lett. **15**(4), 381–386 (1991)
69. S. De, S. Banerjee, A.R. Roy, Occurrence of chaos via torus breakdown in a piecewise smooth non-invertible map, in *National Conference on Nonlinear systems & Dynamics*, Saha Institute of Nuclear Ohysics, Kolkata, India, 5–7 Mar 2009
70. C. Gan, Q. Wang, M. Perc, Torus breakdown and noise-induced dynamics in the randomly driven Morse oscillator.. J. Phys. A Math. Theor. **43**, 125102 (13pp) (2010)
71. I. Manimehan, K. Thamilmaran, G. Chidambaram, P. Philominathan, Transition from Torus to Chaos in a piecewise-linear forced parallel LCR circuit, in *National Conference on Nonlinear Systems & Dynamics*
72. A. Stefanski, T. Kapitaniak, J. Brindley, V. Astakhov, Torus on-off intermittency in coupled Lorenz systems. Phys. Rev. E **57**(1) (1998)
73. A. Agliari, G. Italo, R. Dieci, L. Gardini, Homoclinic tangles associated with closed invariant curves in families of 2D maps. Grazer Math. Ber. Bericht Nr. **350**, 1–14 (2006). ISSN 1016-7692
74. G. Looss, D.D. Joseph, *Elementary Stability and Bifurcation Theory* (Springer, New York, 1980, 1990)
75. L. Perko, *Differential Equations and Dynamical Systems* (Springer, New York, 2001, 1996, 1991)
76. C. Hobbs, H. Osinga (2008) Bifurcation of the global stable set of a planar endomorphism near a cusp singularity. Int. J. Bifurcation Chaos **18**(8), 2207–2222. ISSN 0218-1274
77. J.E. Franke, A.-A. Yakubu, Multiple attractors via CUSP bifurcation in periodically varying environments. J. Diff. Equat. **11**(4–5), 365–377 (2005)
78. J. Harlim, W.F. Langford, The Cusp-Hopf bifurcation. Int. J. Bifurcation Chaos **17**(8), 2547–2570 (2007)
79. S. Yu, J. Lu, G. Chen, Multifolded torus chaotic attractors: design and implementation. CHAOS **17**, 013118 (2007)
80. L. Yang, Floquet theory: a stroboscopic approach to study stability of a limit—cycle solution. PACS numbers: 82.40.Ck, 47.54.+r, 89.75.Kd
81. R. Kihc, A practical guide for studying Chua's circuits. World Sci. Ser. Nonlinear Sci. Ser. A **71**
82. E. Zeraoulia, J.C. Sprott, Robust chaos and its applications. World Sci. Ser. Nonlinear Sci. Ser. A **79**
83. A.M. Samoilenko, O. Stanzhytskyi, Qualitative and asymptotic analysis of differential equations with random perturbations. World Sci. Ser. Nonlinear Sci. Ser. A **78**
84. O. Aluf, *Optoisolation Circuits Nonlinearity Dynamic and Chaos—Application in Engineering* (World Scientific, Singapore, 2012)
85. I. Farmakis, M. Moskowitz, *Fixed Point Theorems and Their Applications* (World Scientific, Singapore, 2013)
86. M.W. Hirsch, S. Smale, R.L. Devaney, *Differential Equations, Dynamical Systems, and Introduction to Chaos*, 3rd edn. (Academic Press, Cambridge, 2012)
87. W. Yu, J. Cao, Hopf bifurcation and stability of periodic solutions for Van der pol equation with time delay. Nonlinear Anal. **62**, 141–165 (2005)
88. T. Chakraborty, R.H. Rand, The transition from phase locking to drift in a system of two weakly coupled Van der pol oscillators. Int. J. Non-linear Mech. **23**(5/6), 369–376 (1988)
89. W. Weckesser, Canards and horseshoes in the forced Van der pol equation. 3:50 WSPC/Trim size: 9in x 6in for proceedings, Oct 29, 2003

90. R. Mettin, U. Parlitz, W. Lauterborn, Bifurcation structure of the driven Van der pol oscillator. Int. J. Bifurcation Chaos **3**(6), 1529–1555 (1993)
91. H. Shirahama, C.-U. Choe, K. Fukushima, Conversion of stability observed in Van der pol oscillator by unstable time-delayed controller, in *the 23rd International Technical Conference on Circuits/Systems, Computers and Communications* (ITC-CSCC 2008) (1993)
92. J. Guckenheimer, Dynamic of the Van der pol equation. IEEE Trans. Circ. Syst. **CAS-27**(11) (1980)
93. A. Granas, J. Dugundji, *Fixed Point Theory*, 2003 edn. (Springer, Berlin, 2003)
94. K.C. Border, *Fixed Point Theorems with Application to Economics and Game Theory*, Rep sub edn. (Cambridge University Press, Cambridge, 1989)
95. R.P. Agarwal, M. Meehan, D. O'Regan, *Fixed Point Theory and Applications*, 1 edn. (Cambridge University Press, Cambridge, 2009)
96. V.I. Istratescu, *Fixed Point Theory: An Introduction*, 1981 edn. (Springer, Berlin, 2001)
97. D.P. Atherton, *Stability of Nonlinear System*. Electronic & Electrical Engineering Research Studies (Wiley, Hoboken, 1981)
98. D. Xue, Y.Q. Chen, D.P. Atherton, *Linear Feedback Control: Analysis and Design with MATLAB*. Advanced in Design and Control, 1st edn. (Society for Industrial and Applied Mathematics, USA, 2009)
99. D.P. Atherton, *Nonlinear Control Engineering*, stu sub edn. (Chapman & Hall, London, 1982)
100. Y. Bar-yam, *Dynamics of Complex Systems*. Studies in Nonlinearity, 1st edn. (Westview Press, USA, 1997)
101. W.-H. Steeb, *The Nonlinear Workbook: Chaos, Fractals, Cellular Automata, Generic Algorithms, Gene Expression Programming, Support Vector Machine, Wavelets, Hidden...Java and Symbolic C++ Programs*, 6th edn. (World Scientific, Singapore, 2014)
102. S. Sternberg, *Curvature in Mathematics and Physics* (Dover Publications, USA, 2012)
103. S. Sternberg, *Dynamical Systems* (Dover Publications, USA, 2010)
104. E.R. Scheinerman, *Invitation to Dynamical Systems*, Reprint edn. (Dover Publications, USA, 2012)
105. R. Abraha, *Dynamics the Geometry of Behavior Part 2: Chaotic Behavior*. Visual Mathematics Library (Aerial Pr, 1983)
106. R. Abraham, *Dynamics the Geometry of Behavior: Periodic Behavior*. Visual Mathematics Library (Aerial Pr, 1982)
107. H. Ho-Ching Iu, A.L. Fitch, *Development of Memristor Based Circuit* (World Scientific, Singapore, 2013)
108. B. Wing-kuen Ling, H. Ho-ching Iu, *Control of Chaos in Nonlinear Circuits and Systems* (World Scientific, Singapore, 2008)
109. R. Kilic, *A Practical Guide for Studying Chua's Circuits* (World Scientific, Singapore, 2010)
110. Qiying, *Limit Theorems for Nonlinear Cointegrating Regression* (World Scientific, Singapore, 2015)
111. T. Tatarinova, A. Schumitzky, *Nonlinear Mixture Models* (World Scientific, Singapore, 2015)
112. C. Diks, *Nonlinear Time Series Analysis* (World Scientific, Singapore, 1999)
113. C. Letellier, R. Gilmore, *Topology and Dynamics of Chaos* (World Scientific, Singapore, 2013)
114. Q. Wang, *Limit Theorems for Nonlinear Cointegrating Regression* (World Scientific, Singapore, 2015)
115. P. Sibani, H.J. Jensen, *Stochastic Dynamics of Complex Systems* (World Scientific, Singapore, 2013)
116. J.C. Sprott, *Elegant Chaos* (World Scientific, Singapore, 2010)
117. E. Zeraoulia, J.C. Sprott, *Robust Chaos and Its Applications* (World Scientific, Singapore, 2011)
118. E. Zeraoulia, J.C. Sprott, *Frontiers in the Study of Chaotic Dynamical Systems with Open Problems* (World Scientific, Singapore, 2011)
119. W.G. Hoover, C.G. Hoover, *Simulation and Control of Chaotic Nonequilibrium Systems* (World Scientific, Singapore, 2015)

120. V.G. Ivancevic, D.J. Reid, *Complexity and Control* (World Scientific, Singapore, 2014)
121. G. Nicolis, C. Nicolis, *Foundations of Complex Systems* (World Scientific, Singapore, 2012)
122. M. Cencini, F. Cecconi, A. Vulpiani, *Chaos* (World Scientific, Singapore, 2009)
123. M. Bier, B. Teusink, B.N. Kholodenko, H.V. Westerhoff, Control analysis of glycolytic oscillations. Biophys. Chem. **62**, 15–24 (1996)
124. E.E. Selkov, Self-oscillations in glycolysis. Eur. J. Biochem. **4**, 79–86 (1968)
125. F.A. Chandra, G. Buzi, J.C. Doyle, Glycolytic oscillations and limits on robust efficiency. Science **333** (2011)
126. F. Takens, Detecting strange attractors in turbulence, in *Dynamic Systems and Turbulence, Lecture Notes in Mathematics 898* (1981), pp. 366–381
127. D.T. Kaplan, L. Glass, *Understanding Nonlinear Dynamics* (Springer, New York, 1995)
128. E. Ott, *Chaos in Dynamical Systems* (Cambridge University Press, Cambridge, 1993)
129. D. Ruelle, *Chaotic Evolution and Strange Attractors* (Cambridge University Press, Cambridge, 1989)
130. J.L. McCauley, *Chaos, Dynamics, and Fractals: An Algorithmic Approach to Deterministic Chaos* (Cambridge University Press, Cambridge, 1993)
131. G.P. Williams, *Chaos Theory Tamed* (Joseph Henry Press, Washington D.C., 1997)
132. S.H.M.J. Houben, J.M.L. Maubach, R.M.M. Mattheij, An accelerated Poincare-map method for autonomous oscillators. Appl. Math. Comp. **140**, 2–3 (2003)
133. W. Just, H. Kantz, Some considerations on Poincare maps for chaotic flows. J. Phys. A Math. Gen. **33**, 163 (2000)
134. T. Parker, L. Chua, *Practical Numerical Algorithms for Chaotic Systems* (Springer, New York, 2000)
135. H.G. Schuster, *Deterministic Chaos* (VCH, Weinheim, 1989)
136. R. Efrem, Numerical approximation of Poincare maps. Romai. J. **4**, 101–106 (2008)
137. P. Cvitanovi, R. Artuso, R. Mainieri, G. Tanner, G. Vattay, *Chaos: Classical and Quantum* (Niels Bohr institute, Copenhagen, 2007)
138. M. Henon, On the numerical computation of Poincare maps. J. Phys. D **5**(2–3), 412–414 (1982)
139. K.T. Alligood, T. Sauer, J.A. Yorke, *Chaos: An Introduction to Dynamical Systems* (Springer, New York Inc, 1996)
140. D. Ilhem, K. Amel, One-dimensional and two dimensional dynamics of cubic maps, in *Discrete Dynamics in Nature and Society* (2006), pp. 1–13
141. E.N. Lorentz, Deterministic non periodic flow. J. Atmos Sci. **20**, 130–141 (1963)
142. E.D. Courant, High-energy inverse free-electron-laser accelerator. Phys. Rev. A **32**(5) (1985)
143. Y. Eli, Inhibited spontaneous emission in solid-state physics and electronics. Phys. Rev. Lett. **58**(20) (1987)
144. U. Keller, Ultrafast all-solid-state laser technology., Appl. Phys. B **58**, 347–363 (1994)
145. T. Masaki, T. Noboru, K. Tadash, I. Humio, Analytical and experimental studies on the characteristics of composite solid-state laser rods in diode-end-pumped geometry. IEEE J. Sel. Top. Quantum Electron. **3**(1) (1997)
146. K. Christian, M. Sonja, Z. Patrick, W. Harald, Stable solid-state source of single photons. Phy. Rev. Lett. **85**(2) (2000)
147. U. Keller, Ultrafast solid-state laser oscillators: a success story for the last 20 years with no end in sight. Appl. Phys. B **100**, 15–28 (2010)
148. P. Wang, S.-H. Zhou, K.K. Lee, Y.C. Chen, Picosecond laser pulse generation in a monolithic self-Q-switched solid-state laser. Opt. Commun. **114**, 439–441 (1995)
149. A. Baltuska, Z. Wei, M.S. Pshenichnikov, D.A. Wiersma, S. Robert, All solid-state cavity-dumped sub-5-fs laser. Appl. Phys. B **65**, 175–188 (1997)
150. G. Rashid, *Interaction of Mid-infrared Parametric Waves in Laser Plasmas* (World Scientific, Singapore, 2017)
151. M. Ghioni, A. Lacaita, S. Cova, G. Ripamonti, 20ps resolution single-photon solid-state detector, in *OSA Proceedings on Picosecond Electronics and Optoelectronics* (1989)

152. A.P. Kazantsev, S.G. Rautian, G.I. Surdutovich, Theory of a gas laser with nonlinear absorption. Sov. Phys. Jetp **27**(5) (1968)
153. C. Eck, H. Garcke, P. Knabner, *Mathematical Modeling*. Springer Undergraduate Mathematics Series, 1st edn. (Springer, Berlin, 2017)
154. K. Hironobu, A. Nobutada, K. Chikara, S. Shimpey, M. Tatsuo, Development of 200w high-performance copper vapor laser with 6cm diameter, 300cm length. J. Nucl. Sci. Technol. **31**(1), 34–47 (1994)
155. S.H. Park, D.C. Lee, J. Heo, Energy transfer between Er3+ and Pr3+ in chalcogenide glasses for dual-wavelength fiber-optics amplifiers. J. Appl. Phys. **91**(11) (2002)
156. H. Andreas, M. Richard, F. Jerome, External cavity quantum cascade laser. Semicond. Sci. Technol. **25**, 083001 (14pp) (2010)
157. Z. Xingyu, Z. Shengzhi, W. Qingpu, Z. Qidi, S. Lianke, Z. Shaojun, Optimization of Cr4+-doped saturable-absorber Q-switched lasers. IEEE J. Quantum Electron. **33**(12) (1997)
158. B.M. Walsh, *Review of Tm and Ho Materials; Spectroscopy and Lasers* (NASA Langley Research Center)
159. H. Jian-Jang, C.C. Yang, D.-W. Huang, Carrier capture competition between two different quantum wells in dual-wavelength semiconductor lasers. IEEE Photonics Technol. Lett. **8**(6) (1996)
160. B.F. Kuntsevich, A.N. Pisarchik, Synchronization effects in a dual-wavelength class-B laser with modulated losses. Phys. Rev. E **64**, 046221
161. B.M. Walsh, N.P. Barnes, P. Mulugeta, Y. Jirang, U.N. Singh, J. Appl. Phys. **95**(7) (2004)
162. U. Keller, Recent developments in compact ultrafast lasers. Nature **424** (2003)
163. H. Xu, T. Okino, K. Yamanouchi, Tracing ultrafast hydrogen migration in allene in intense laser fields by triple-ion coincidence momentum imaging. J. Chem. Phys. **131**, 151102 (2009)
164. H. Xu, C. Marceau, K. Nakai, T. Okino, S.-L. Chin, K. Yamanouchi, Communication: two stages of ultrafast hydrogen migration in methanol driven by intense laser fields. J. Chem. Phys. **133**, 071103 (2010)
165. J. Faure, Y. Glinec et al.,A laser-plasma accelerator producing mono energetic electron beams. Nature **431** (2004)
166. N. . Thipparapu, C. Guo, A.A. Umnikov, P. Barua, A. Taranta, J.K. Sahu, Bismuth-doped all-fiber mode-locked laser operating at 1340nm. Opt. Lett. **42**(24) (2017)
167. A. Dolatabady, N. Granpayeh, Plasmonic magnetic sensor based on graphene mounted on a magneto-optic grating. IEEE Trans. Magn. (2018)
168. A.K. Maini, *Lasers and Optoelectronics; Fundamental Devices and Applications*, 1st edn. (Wiley, Hoboken, 2013)
169. L. Jie, W. Chen, Y. Li, Rate-equation-based VCSEL thermal model and simulation. J. Zhejiang Univ. Sci. A **7**(12), 1968–1972 (2006)
170. Y. Yi, R. Yu, L. Zhengjia, Rate-equation-based thermal VCSEL model and its equivalent circuit application in transceiver module design. J. Opt. Commun. **23**(6), 205–208 (2002)
171. M. Bruensteinet, G.C. Papen, Extraction of VCSEL rate-equation parameters for low-bias system simulation. IEEE J. Sel. Top. Quantum Electron. **5**(3) (1999)
172. T.D. Arber, K. Bennett et al., Contemporary particle-in-cell approach to laser-plasma modeling. Plasma Phys. Control. Fusion **57**, 113001 (26pp) (2015)
173. G.S. Sarkisov, V.Yu. Bychenkov, V.N. Novikov, V.T. Tikhonchuk, Self-focusing channel formation and high-energy ion generation in interaction of an intense short laser pulse with He jet. Phys. Rev. E **59**(6) (1999)
174. B. Ersfeld, R. Bonifacio, S. Chen, M.R. Islam, P.W. Smorenburg, D.A. Jaroszynski, The ion channel free-electron laser with varying betatron amplitude. New J. Phys. **16**, 093025 (2014)
175. A.V. Arefiev, V.N. Khudik, M. Schollmeier, Enhancement of laser-driven electron acceleration in an ion channel. Phys. Plasmas (2013)
176. X. Davoine, F. Fiuza, R.A. Fonseca, W.B. Mori, L.O. Silva, Beam quality requirements for the ion-channel laser. Phys. Plasmas (2014)
177. S.K. Rajouria, H.K. Malik, V.K. Tripathi, P. Kumar, Step density model of laser sustained ion channel and coulomb explosion. Phys. Plasmas **22**, 023104 (2015)

178. Z. Ouyang, S. Zhang, Nonlinear study of an ion-channel guiding free-electron laser. Phys. Plasmas **22**, 043111 (2015)
179. H. Kupershmidt, A. Ron, Ion-channel laser theory revised, in *Laser and Electro-Optics society Annual Meeting* (1993)
180. K.-R. Chen, T.C. Katsouleas, J.M. Dawson, On the amplification mechanism of the ion-channel laser. IEEE Trans. Plasma Sci. **18**(5) (1990)
181. Y. Yao, X. Chen, Y. Dai, S. Xie, Dual-wavelength erbium-doped fiber laser with a simple linear cavity and its application in microwave generation. IEEE Photonics Technol. Lett. **18**(1) (2006)
182. Z. Luo, M. Zhou, J. Weng, G.Huang, H. Xu, C. Ye, Z. Cai, Graphene-based passively Q-switched dual-wavelength erbium-doped fiber laser. Opt. Lett. **35**(21) (2010)
183. X. Liu, X. Yang, F. Lu, J. Ng, X. Zhou, C. Lu, Stable and uniform dual-wavelength erbium-doped fiber laser based on fiber Bragg gratings and photonic crystal fiber. Opt. Express **13**(1) (2005)
184. B. Wu, P. Jiang, D. Yang, T. Chen, J. Kong, Y. Shen, Compact dual-wavelength Nd:GdVO4 laser working at 1063 and 1065 nm. Opt. Express **17**(8) (2009)
185. P. Pellandini, R.P. Stanley, R. Houdre, U. Oesterle, M. llegems, Dual-wavelength laser emission from a coupled semiconductor microcavity. Appl. Phys. Lett. **71**(7) (1997)
186. Y.F. Chen, S.W. Tsai, S.C. Wang, Y.C. Huang, T.C. Lin, B.C. Wong, Efficient generation of continuous-wave yellow light by single-pass sum-frequency mixing of a diode-pumped Nd:YVO4 dual-wavelength laser with periodically poled lithium niobate. Opt. Lett. **27**(20) (2002)
187. Y.-F. Chen, CW dual-wavelength operation of a diode-end-pumped Nd:YVO4 laser. Appl. Phys. B **70**, 475–478 (2000)
188. L. Wilmer, J.B. Boffard, *Laser for Scientists and Engineers*, 1st edn. (World Scientific Publishing Company, Singapore, 2017)
189. E. Kukk, K. Motomura, H. Fukuzawa, K. Nagaya, K. Veda, Molecular dynamics of XFEL-induced photo-dissociation revealed by Ion-Ion coincidence measurements. Appl. Sci. **7**, 531 (2017)
190. Yu.A. Kuznetsov, H.G.E. Meijer, B. Al Hdaibat, W. Govaerts, Improved homoclinic predictor for bogdanov-takens bifurcation. Int. J. Bifurcation Chaos **24**(04) (2014)
191. V. De Witte, W. Govaerts, Y.A. Kuznetsov, M. Friedman, Interactive initialization and continuation of homoclinic and heteroclinic orbits in MATLAB. ACM Trans. Math. Softw. **38**(3) Article 18 (2012)
192. P.K. Chu, X.P. Lu, *Low Temperature Plasma Technology* (CRC Press Taylor & Francis Group, UK, 2013)
193. R. Michalzik, VCSELs: *Fundamentals, Technology and Applications of Vertical-Cavity Surface-Emitting Lasers* (Springer, Berlin, 2013)
194. M. Peterka, J. Adamek, P. Kudrna, M. Tichy, Ball-pen probe—a useful tool for measuring the plasma potential in magnetized plasma, in *39th EPS Conference & 16th International Congress on Plasma Physics*, p. 147
195. J. Adamek, M. Peterka, T. Gyergyek, P. Kudrna, M. Tichy, Diagnostic of magnetized low temperature plasma by ball-pen probe. Nukleonika **57**(2), 297–300 (2012)
196. R. Schrittwieser, C. Ionita et al., Probe methods for direct measurements of the plasma potential. Rom. J. Phys. **50**(7–8), 723–739 (2005)
197. S.D. Baalrud, B. Scheiner, B. Yee, M. Hopkins, E. Barnat, Extensions and applications of the Bohm criterion. Plasma Phys. Control. Fusion 57 (2015)
198. B. Scheiner, S.D. Baalrud, B.T. Yee, M.M. Hopkins, E.V. Barnet, Theory of the electron sheath and pre-sheath. Phys. Plasmas **22**, 123520 (2015)
199. J.P. Sheehan, I.D. kaganovich, H. Wang, D. Sydorenko, Y. Raitses, N. Hershkowitz, Effects of emitted electron temperature on the plasma sheath. Phys. Plasmas **21**, 063502 (2014)
200. K.-U. Riemann, The Bohm criterion and sheath formation. J. Phys. D Appl. Phys. **24**, 493–518 (1991)
201. N. Hershkowitz, Sheaths: more complicated than you think. Phys. Plasmas **12**, 055502 (2005)

202. N.R. Walkden, J. Adamek et al., Profile measurements in the plasma edge of MAST using a ball pen probe. Rev. Sci. Instrum. **86**(2), 1–10 (2015)
203. E.T. Everson, P. Pribyl, C.G. Constantin, A. Zylstra, D. Schaeffer, Design construction and calibration of a three-axis high frequency magnetic probe (B-dot probe) as a diagnostic for exploding plasmas. Rev. Sci. Instrum. **80**, 113505 (2009)
204. H.L. Kennedy, Optimal digital design of steerable (FIR and IIR) differentiators with the flatness of polynomial filters and the isotropy of Gaussian filters. J. Electr. Imaging (JEI) (2018)
205. R. Haight, A.V. Carr, *Industrial Applications of Ultrafast Lasers* (World Scientific Publishing Co pte Ltd, Singapore, 2018)
206. S. Chen, Y. Zhou, F. Baronio, D. Mihalache, Special types of elastic resonant soliton solutions of the kadomtsev-petviashvili II equation. Rom. Rep. Phys. **70**, 102 (2018)
207. A. Isar, Dynamics of quantum interferometric power in Gaussian open systems. Rom. Rep. Phys. **70**, 103 (2018)
208. I. Ghiu, Entanglement versus quantum degree of polarization. Rom. Rep. Phys. **70**, 104 (2018)
209. S. Micluta-Campeanu, M.C. Raportaru, A.I. Nicolin, V. Baran, Fingerprints of classical chaotic dynamics in quantum behavior. Rom. Rep. Phys. **70**, 105 (2018)
210. S.V. Sazonov, Optical solitons in systems of two-level atoms. Rom. Rep. Phys. **70**, 401 (2018)
211. M. Meisterhans, A. Coillet et al., Silicon micro ring within a fiber laser cavity for high-repetition-rate pulse train generation. Rom. Rep. Phys. **70**, 405 (2018)
212. A. Motes, Understanding and simulation free-space laser beam propagation. *AM Photonics*, 3rd edn. (2014)
213. M. Wordemann, *Structure Light Field; Application in Optical Trapping, Manipulation, and Organization* (Springer Theses) (Springer, Berlin, 2014)
214. Q. Zhang, J.R. Leger, Focus shaping using cylindrical vector beams. Opt. Express **10**(7) (2002)
215. J. Mulet, J. Mork, Analysis of timing jitter external cavity mode locked semiconductor lasers. IEEE J. Quantum Electron. **42**(3) (2006)
216. M. Ams, G.D. Marshall, D.J. Spence, M.J. Withford, Slit beam shaping method for femtosecond laser direct write fabrication of symmetric waveguides in bulk glasses. Opt. Express **13**(15) (2005)
217. J.B. Schlager, B.E. Callicoatt, R.P. Mirin, N.A. Sanford, Passively mode-locked waveguide laser with low residual jitter. IEEE Photonics Technol. Lett. **14**(9) (2002)
218. H. Yoon, B.E. Bateman, B.N. Agrawal, Laser beam jitter control using recursive least squares adaptive filters. J. Dyn. Syst. Meas. Control **133** (2011)
219. P. Mangeol, U. Bockelmann, Interference and crosstalk in double optical tweezers using a single laser source. Rev. Sci. Instrum. **79**, 083103 (2008)
220. A. McGurn, *Nano Photonics*. Springer Series in Optical Sciences, 1st edn. (Springer, Berlin, 2018)
221. H.P. Freund, T.M. Antonsen Jr., *Principles of Free Electron Lasers*, 3rd edn. (Springer, Berlin, 2018)
222. J. Nathan Kutz, P.K.A. Wai, Gordon-Haus timing jitter reduction in dispersion-managed soliton communications. IEEE Photonics Technol. Lett. **10**(5) (1998)
223. Yu.S. Kivshar, M. Haelterman, Ph. Emplit, J.-P. Hamaide, Gordon-Haus effect on dark solitons. Opt. Lett. **19**(1) (1994)
224. W. Chen, Y. Song, K. Jung, M. Hu, C. Wang, J. Kim, Few-femtosecond timing jitter from a picosecond all polarization maintaining Yb-fiber laser. Opt. Express **24**(2) (2016)
225. A. Nolasco pino, G.P. Agrawal, J. Ferreira da Rocha, Effect of soliton interaction on timing jitter in communication systems. **16**(4) (1998)
226. C. Kim, S. Bae, K. Kieu, J. Kim, Sub-femtosecond timing jitter, all-fiber, CNT-mode-locked Er-laser at telecom wavelength. Opt. Express **21**(22) (2013)
227. C. Kim, K. Jung, K. Kieu, J. Kim, Low timing jitter and intensity noise from a soliton Er-fiber laser mode-locked by a fiber-taper carbon nanotube saturable absorber. Opt. Express **20**(28) (2012)

228. R.-J. Essiambre, G.P. Agrawal, Timing jitter of ultrashort solitons in high-speed communication systems. I. General formulation and application to dispersion-decreasing fibers. J. Opt. Soc. Am. B **14**(2) (1997)
229. A.M. Abobaker, K. Nakkeeran, A.B. Moubissi, Dispersion managed optical fiber systems with zero Hamiltonian, in *Proceeding of the International Conference on Computer and Communication Engineering* (2008)
230. T.I. Lakoba, G.P. Agrawal, Optimization of the average-dispersion range for Long-Haul dispersion-managed soliton system. J. Lightwave Technol. **18**(11) (2000)
231. C.R. Menyuk, Non-Gaussian corrections to the Gordon-Haus distribution resulting from soliton interactions. Opt. Lett. **20**(3) (1995)
232. C.J. Mckinstrie, J. Santhanam, G.P. Agrawal, Gordon-Haus timing jitter in dispersion-managed systems with lumped amplification: analytical approach. J. opt. Soc. Am. B **19**(4) (2002)
233. R.-J. Essiambre, G.P. Agrawal, Timing jitter of ultrashort solitons in high-speed communication systems. II. Control of jitter by periodic optical phase conjunction. J. Opt. Soc. Am. B **14**(2) (1997)
234. J. Santhanam, C.J. Mckinstrie, T.I. Lakoba, G.P. Agrawal, Effects of pre compensation and post compensation on timing jitter in dispersion-managed systems. Opt. Lett. **26**(15) (2001)
235. M.A. Ablowitz, P.A. Clarkson, *Soliton Nonlinear Evolution Equations and Inverse Scattering*. London Mathematical Society Lecture Note Series, 1st edn. (Cambridge University Press, Cambridge, 1992)
236. A.S. Ko, N. Kushnir, N. Bokla, G. Kosovan, Creation of pseudo-random sequences based on chaos for forming of wideband signal. Inf. Telecommun. Sci. **9**(2) (2017)
237. S.P. Kuo, *Plasma Physics in Active Wave Ionosphere Interaction* (World Scientific Publishing Co Pte Ltd, Singapore, 2018)
238. P. Dahan, P. Malits, Effect of symmetry on the electronic properties of arbitrarily shaped quantum rings in a magnetic field. Phys. E **56**, 165–171 (2014)
239. B. Al-Hdaibat, W. Govaerts, D.L. Van kekem, Yu.A. Kuznetsov, Remarks on homoclinic structure in Bogdanov-Takens map. J. Diff. Equat. Appl. (2017)
240. B. Al-Hdaibat, W. Govaerts, YuA Kuznetsov, H.G.E. Meijer, Initialization of homoclinic solutions near Bogdanov-Takens points: Lindstedt-Poincare compared with regular perturbation method. SIAM J. Appl. Dyn. Syst. **15**(2), 952–980 (2016)
241. C. Wang, X.-M. Chen, K.-W. Sun, J. Ren, Negative differential thermal conductance and heat amplification in a non equilibrium triangle-coupled spin-boson system at strong coupling. Phys. Rev. A **97**, 052112 (2018)
242. M. Oane, D. Ticos, C. Ristoscu, M. Badiceanu, A. Buca, I.N. Mihailescu, Analytical and semi-analytical heat transfer models for strong laser-optical windows weak interaction using integral transform technique. Rom. J. Phys. **63**, 501 (2018)
243. E. Shwartz, S. Shwartz, Difference-frequency generation of optical radiation from two-color x-ray pulses. Opt. Express **23**(6), 7472 (2015)
244. D. Borodin, A. Schori, F. Zontone, S. Shwartz, X-ray photon pairs with highly suppressed background. Phys. Rev. A **94**, 013843 (2016)
245. M.hdi Abedi-Varaki, S. Jafari, Self-focusing and de-focusing of intense left and right-hand polarized laser pulse in hot magnetized plasma: Laser out-put power and laser spot-size. Optik **142**, 360–369 (2017)
246. P. Sprangle, E. Esarey, J. Krall, G. Joyce, Propagation and guiding of intense laser pulses in plasmas. Phys. Rev. Lett. **69**(15) (1992)
247. J.T. Verdeyen, *Laser Electronics*, 3rd edn. (Pearson, London, 1995)
248. S.O. Kasap, *Optoelectronics & Photonics; Principles & Practices*, 2nd edn. (Pearson, London, 2012)
249. K. Lau, *Ultra-High Frequency Linear Fiber Optic Systems, Springer Series in Optical Sciences 159*, Softcover reprint of hardcover 2nd edn. 2011 edn. (Springer, Berlin, 2013)
250. G. Giovannetti, L. Landini, M.F. Santarelli, V. Positano, A fast and accurate simulator for the design of birdcage coils in MRI. Magn. Reson. Mater. Phys. Biol. Med. **15**, 36–44 (2002)

251. A. Zadok, *Dynamic Optical Filters: Implementation, Characterization and Applications* (Lambert Academic Publishing, 2009)
252. P. Shum, H. Ghafouri-shiraz, S.F. Yu, Analysis of a DPSK soliton transmission system. Opt. Laser Technol. **29**(7), 411–414 (1997)
253. J. Martin-Regalado, F. Prati, M. San Miguel, N. B. Abraham, Polarization properties of vertical-cavity surface-emitting laser. IEEE J. Quantum Electron. **33**(5) (1997)
254. M. Bruensteiner, G.C. Papen, Extraction of VCSEL rate equation parameters for low-bias system simulation. IEEE J. Sel. Top. Quantum Electron. **5**(3) (1999)
255. V. De Witte, W. Govaerts, Yu.A. Kuznetsov, H.G.E. Meijer, Analysis of bifurcations of limit cycles with Lyapunov exponents and numerical normal forms. Phys. D **269**, 126–141 (2014)
256. I.C.E. Turcu, F. Negoita et al., High field physics and QED experiments at ELI-NP. Rom. Rep. Phys. **68**, s145–s231 (2016)
257. M. Apostol, *Magnetic and Electric Resonance* (Cambridge Scholars Publishing, Cambridge, 2018)
258. Z. Liang, P.C. Lauterbur, *Principles of Magnetic Resonance Imaging: A Signal Processing Perspective* (Wiley-IEEE Press, 1999)
259. L. Reichwein, J. Thomas, A. Pukhov, Two-dimensional structures of electron bunches in relativistic plasma cavities. Phys. Rev. E **98**, 013201 (2018)
260. A. Argyros, S.G. Leon-Saval, J. Pla, A. Docherty, Anti resonant reflection and inhibited coupling in hollow-core square lattice optical fibers. Opt. Exp. **16**(8), 5642–5648 (2008)
261. M. Masruri, S. Balascuta, I. Dancus, Hollow core inhibited coupling fibers design for femtosecond pulse spectral broadening in multi petawatt laser-induced plasma diagnostics, in *Conference on Lasers and Electro-optics Pacific Rim (CLEO-PR)* (2017)
262. Y. Kumagai, H. Fukuzawa et al., Following the birth of a nano plasma produced by an ultrashort hard-X-ray laser in xenon clusters. Phys. Rev. X **8**, 031034 (2018)
263. J. Tang, Y.-R. Liu et al., Flexible thermo-optic variable attenuator based on long-range surface plasmon-polariton waveguide. Micromachines **9**, 369 (2018)
264. C. Maroli, V. Petrillo, I. Drebot, L. Serafini, B. Terzic, G.A. Krafft, Compensation of non-linear bandwidth broadening by laser chirping in Thomson sources. J. Appl. Phys. **124**, 063105 (2018)
265. B. Di Bartolo, *Classical Theory of Electromagnetism*, 3rd edn. (World Scientific Publishing Company, Singapore, 2018)
266. W. Baumjohann, R.A. Treumann, *Basic Space Plasma Physics*, Revised edn. (Imperial College Pr, 2012)
267. Y. Lim, *Problems and Solutions on Electromagnetism* (World Scientific Publishing Company, Singapore, 1993)
268. Y. Liu, *Advances of Atoms and Molecules in Strong Laser Fields* (World Scientific Publishing Company (WSPC), Singapore, 2015)
269. B. Friedrich, D. Herschbach, Alignment and trapping of molecules in intense laser fields. Phys. Rev. Lett. **74**(23) (1995)
270. C. Guo, M. Li, J.P. Nibarger, G.N. Gibson, Single and double ionization of diatomic molecules in strong laser fields. Phys. Rev. A **58**(6) (1998)
271. J.J. Larsen, K. Hald, N. Bjerre, H. Stapelfeldt, Three dimensional alignment of molecules using elliptically polarized laser fields. Phys. Rev. Lett. **85**(12) (2000)
272. R.-F. Lu, P.-Y. Zhang, K.-L. Han, Attosecond-resolution quantum dynamics calculations for atoms and molecules in strong laser fields. Phys. Rev. E **77**, 066701 (2008)
273. B. Friedrich, D. Herschbach, Polarization of molecules induced by intense non resonant laser fields. J. Phys. Chem. **99**, 15686–15693 (1995)
274. L. Cai, J. Marango, B. Friedrich, Time-dependent alignment and orientation of molecules in combined electrostatic and pulsed non resonant laser fields. Phys. Rev. Lett. **86**(5) (2001)
275. H. Ohmura, N. Saito, M. Tachiya, Selective ionization of oriented non polar molecules with asymmetric structure by phase-controlled two-color laser fields. Phys. Rev. Lett. **PRL 96**, 173001 (2006)

276. T. Kato, H. Kono, Time-dependent multi configuration theory for electric dynamics of molecules in intense laser fields: a description in terms of numerical orbital functions. J. Chem. Phys. **128**, 184102 (2008)
277. T. Brabec, F. Krausz, Intense few-cycle laser fields: Frontiers of nonlinear optics. Rev. Modern Phys. **72**(2) (2000)
278. C. Born, M. Sorel, S. Yu, Linear and nonlinear mode interactions in a semiconductor ring laser. IEEE J. Quantum Electron. **41**(3) (2005)
279. L. Gelens, S. Beri, G. Van der Sande et al., Optical injection in semiconductor ring lasers: backfire dynamics. Opt. Express **16**(15) (2008)
280. S.-K. Nam, K.-B. Kim, Evolution of electron beam in the tapered planar wiggler, in *Proceedings of the 2004 FEL Conference*, pp. 49–52
281. N.S. Ginzburg, N.Yu. Peskov, Nonlinearity theory of a free electron laser with a helical wiggler and an axial guide magnetic field. Phys. Rev. Spec. Top. Accelerators Beams **16**, 090701 (2013)
282. Y.M. Penkin, V.A. Katrich, M.V. Nesterenko, S.L. Berdnik, V.M. Dakhov, *Electromagnetic Fields Excited in Volumes with Spherical Boundaries*. Lecture Notes in Electrical Engineering, 1st edn. (Springer, Berlin, 2019)
283. E.I. Nefyodov, S.M. Smolskiy, *Electromagnetic Fields and Waves: Microwave and Mmwave Engineering with Generalized Macroscopic Electrodynamics*. Textbooks in Telecommunication Engineering, 1st edn. (Springer, Berlin, 2019)
284. I. Pomerantz, E. McCary et al., Ultrashort pulsed neutron source. Phys. Rev. Lett. **PRL 113**, 184801 (2014)
285. M. Borghesi, J. Fuchs, S.V. Bulanov, A.J. Mackinnon, P.K. Patel, M. Roth, Fast ion generation by high-intensity laser irradiation of solid targets and applications. Fusion Sci. Technol. **49** (2006)
286. G. Mourou, B. Brocklesby, T. Tajima, J. Limpert, The future is fibre accelerators. Nat. Photonics **7** (2013)
287. P. Jacquet, A. Pailloux, Laser absorption spectroscopy for xenon monitoring in the cover gas of sodium cooled fast reactors. J. Anal. At. Spectrum. **28**, 1298–1302 (2013)
288. N. Guler, P. Volegov, A. Favalli et al., Neutron imaging with the short-pulse laser driven neutron source at the trident laser facility. J. Appl. Phys. **120**(15) (2016)
289. M. Titberidze, F. Gruner, A.R. Maier et al., Novel femtosecond level synchronization of titanium sapphire laser and relativistic electron beams, in *Proceedings of IBIC2014*, CA, USA
290. G.J. Tserevelakis, S. Psycharakis et al., Femtosecond laser nano-surgery of sub-cellular structures in HeLa cells by employing third harmonic generation imaging modality as diagnostic tool. Biophotenics 1–8 (2011)
291. S. Mondal, M. Shirozhan et al., Surface plasma attosource beamlines at ELI-ALPS. J. Opt. Soc. Am. B **35**(5) (2018)
292. O. Auciello, *Plasma Diagnostics: Discharge Parameters and Chemistry* (Academic Press, Cambridge, 2013)
293. I.H. Hutchinson, *Principles of Plasma Diagnostics by I.H. Hutchinson (2005-07-14)* (Cambridge University Press, Cambridge, 1874)
294. R.H. Huddlestone, S.L. Leonard, *Plasma Diagnostic Techniques*, 1st edn. (Academic Press, Cambridge, 1965)
295. H.-J. HartfuB, T. Geist, *Fusion Plasma Diagnostics with mm-Waves: An Introduction*, 1st edn. (Wiley-VCH, Germany, 2013)
296. M.A. Heald, C.B. Wharton, *Plasma Diagnostics with Microwaves*. Wiley Series in Plasma Physics, 1st edn. (Wiley, Hoboken, 1965)
297. M.F. Zhukov, A.A. Ovsyannikov, *Plasma Diagnostics*, Box edn. (Cambridge International Science Publishing Ltd, UK, 2000)
298. K. Muraoka, M. Maeda, *Laser-Aided Diagnostics of Plasmas and Gases*. Series in Plasma Physics, 1st edn. (CRC Press, Boca Raton, 2000)
299. I. Podgornyi, *Topics in Plasma Diagnostics*, 1st edn. (Springer, Berlin, 1971)

300. S. Takeda, *Microwave Reflection Techniques for Dense Plasma Diagnostics* (classic reprint) (Forgotten Books, London, 2017)
301. E. Oks, *Diagnostics of Laboratory and Astrophysical Plasmas Using Spectral Line Shapes of One-, Two-, and Three-Electron System* (World Scientific Publishing Company, Singapore, 2017)
302. O. Auciello, D.L. Flamm, *Plasma Diagnostics: Discharge Parameters and Chemistry*, vol 1. (Academic Press, Cambridge, 1989)
303. M. Umehara, K. Yamada, *Differential Geometry of Curves and Surfaces* (World Scientific, Singapore, 2017)
304. W.D. Evans, A. Gogatishvill, B. Opic, *Weighted Inequalities Involving ρ-Quasi Concave Operators* (World Scientific, Singapore, 2018)
305. O. Shulika, I. Sukhoivanov, *Advanced Lasers: Laser Physics and Technology for Applied and Fundamental Science.* Springer Series in Optical Sciences (Springer, Berlin, 2015)
306. S. Ma, Application of extended geometrical criterion to population model with two time delays. J. Math. Res. **10**(3) (2018)
307. J. Zegenhagen, A. Kazimirov, *The X-ray Standing Wave Technique* (World Scientific, Singapore, 2013)
308. M.A. Baqir, A. Farmani, T. Fatima et al., Nanoscale, tunable and highly sensitive biosensor utilizing hyperbolic metamaterials in the near-infrared range. Appl. Opt. **57**(33) (2018)
309. A. Ashkin, *Optical Trapping and Manipulation of Neutral Particles Using Laser* (World Scientific Publishing Company, Singapore, 2006)
310. A.W. Chao, *Reviews of Accelerator Science and Technology* (World Scientific, Singapore, 2013)
311. L. Teofili, M. Migliorati et al., Design of new proton synchrotron booster absorber scraper (PSBAS) in the framework of the large hadron collider injection upgrade (LIU) project. IOP Conf. Ser. J. Phys. Conf. Ser. **1067**, 082027 (2018)
312. U. Keller, K.J. Weingarten, F.X. Kartner, D. Kopf et al., Semiconductor satrable absorber mirrors (SESAM's) for femtosecond to nanosecond pulse generation in solid-state lasers. IEEE J. Sel. Top. Quantum Electron. **2**(3) (1996)
313. X. Zhang, S. Zhao, Q. Wang, Modeling of passively Q-switched lasers. J. Opt. Soc. Am. B **17**(7) (2000)
314. L. Lamberti, M.-T. Lin, C. Furlong, C. Sciammarella, P.L. Reu, M.A. Sutton, Advancement of optical methods & digital image correlation in experimental mechanics, in *Proceedings of the 2018 Annual Conference an Experimental ... Society for Experimental Mechanics Series*, vol 3 (Springer, Berlin, 2019)
315. P.A. Ribeiro, M. Raposo, *Optics, Photonics and Laser Technology.* Springer Series in Optical Sciences Book 2018, 1st edn. (Springer, Berlin, 2018)
316. F.J. Duarte, *Coherence and Ultrashort Pulse Laser Emission* (Intech Open, 2010)
317. M.C. Veale, P. Seller, M. Wilson, E. Liotti, HEXITEC: a high-energy X-ray spectroscopic imaging detector for synchrotron applications. Synchrotron Radiat. News **31**(6) (2018)
318. S.-M. Jung, S. Min, Stability of the diffusion equation with a source. J. Func. Spaces **2018**, Article ID 1216901, 8 p
319. M. Bertolotti, C. Sibilia, A.M. Guzman, *Evanescent Waves in Optics: An Introduction to Plasmonics.* Springer Series in Optical Sciences, 1st edn. (Springer, Berlin, 2017)
320. A. Agrawal, T. Benson, R.M. De La Rue, G.A. Wurtz, *Recent Trends in Computational Photonics.* Springer Series in Optical Sciences Book 204, 1st edn. (Springer, Berlin, 2017)
321. J. Zhu, J. Zhu et al., Status and development of high-power laser facilities at the NLHPLP. High Power Laser Sci. Eng. **6**(eSS), 23 p (2018)
322. A. Zangwill, *Modern Electrodynamics*, 1st edn. (Cambridge University Press, Cambridge, 2012)
323. J. Jang-Yu Hsu, *Visual and Computational Plasma Physics* (Wspc, 2014)
324. F. Chen, *Introduction to Plasma Physics and Controlled Fusion*, 3rd edn. (Springer, Berlin, 2016)

325. D. Vick, M. Kado, H. Yamamoto et al., Lateral energy transport in laser-produced plasmas. Phys. Rev. E **52**(6) (1995)
326. A.G. Markelz, Terahertz dielectric sensitivity to biomolecular structure and function. IEEE J. Sel. Top. Quantum Electron. **14**(1) (2008)
327. M. Rosu-Hamzescu, C. Polonschii et al., High speed CMOS acquisition system based on FPGA embedded image processing for electro-optical measurements. Rev. Sci. Instrum. **89**, 065103 (2018)
328. D. Ursescu, G. Cheriaux et al., Laser beam delivery at ELI-NP. Rom. Rep. Phys. **68**(Supplement), S11–S36 (2016)
329. E. Petrescu, C. Cirtoaje, O. Danila, Dynamic behavior of nematic liquid crystal mixtures with quantum dots in electric fields. Beilstein J. Nanotechnol. **9**, 399–406 (2018)
330. P. Srisungsitthisunti, M. Zamfirescu et al., Real-time adaptive optimization of laser induced nano ripples by laser pulse shaping, in *Proceedings of SPIE-The International Society for Optical Engineering*, vol 8967896704-2 (2014)
331. T. Takanashi, K. Nakamura et al., Ultrafast coulomb explosion of a diiodomethane molecule induced by an X-ray free electron laser pulse. Phys. Chem. Chem. Phys. (2017)
332. Y. Li, X. Guo, X. Zou, X. Yi, Y. Leng, High temporal contrast frontend with optical parametric amplifier for high powerful femtosecond laser. Opt. Laser Technol. **57**, 165–168 (2014)
333. H. Lin, C. Wang, C.P. Liu, Ultrashort hard X-ray pulse driven by cylindrical vector beam. Opt. Commun. **292**, 101–105 (2013)
334. C. Wang, X.-M. Chen, K.-W. Sun, J. Ren, Negative differential thermal conductance and heat amplification in a non-equilibrium triangle-coupled spin-boson system at strong coupling. Phys. Rev. A **97**, 052112 (2018)
335. C.T. Zhou, S.Z. Wu, H.B. Cai et al., Hot electron transport and heating in dense plasma core by hollow guiding. Laser Particle Beams **28**, 563–570 (2010)
336. S. Ataman, A. Preda, R. Ionicioiu, Phase sensitivity of a Mach-Zehnder in interferometer with single-intensity and difference-intensity detection. Phys. Rev. A **98**, 043856 (2018)
337. S. Ataman, Vacuum birefringence detection in all-optical scenarios. Phys. Rev. A **97**, 063811 (2018)
338. I. Pickova, M. Tichy, COIL-chemical oxygen Iodine laser, in *WDS'06 Proceedings of Contributed Papers*, Part II (2006), pp. 81–85
339. A.D. Palla, J.W. Zimmerman et al., Oxygen discharge and post-discharge kinetics experiments and modeling for the electric oxygen-iodine laser system. J. Phys. Chem. A **111**, 6713–6721 (2007)
340. R. Huang, R. Zhang, R.N. Zare, Nature of the red emission in the chemical oxygen iodine laser system. Chem. Phys. Lett. **170**(5), 6 (1990)
341. D.L. Carroll, W.C. Solomon, Electricoil: an advanced chemical Iodine laser concept. Int. Soc. Opt. Eng. **4184**, 40–44 (2000)
342. G. Brederlow, R. Brodmann et al., Performance of the Asterix III high power Iodine laser. IEEE J. Quantum Electron. **QE-16**(2) (1980)
343. L.M. Garcia-Raffi, L.J. Salmeron-Contreras, Nonlinear waves in a chain of magnetically coupled pendula, in *Proceeding of Meetings on Acoustics*, vol 34 (2018), p. 045037
344. H.B. Zhuo, Z.L. Chen et al., Collimation of energetic electrons from a laser-target interaction by a magnetized target back plasma performed by a long-pulse laser. Phys. Rev. Lett. **PRL 112**, 215003 (2014)
345. J. Jong-Yu Hsu, *Visual and Computational Plasma Physics* (Wspc, 2014)
346. A.A. Harms, D.R. Kingdon, K.F. Schoepf, G.H. Miley, *Principles of Fusion energy: An Introduction to Fusion Energy for Students of Science and Engineering* (Wspc, 2000)
347. R. Fitzpatrick, *Plasma Physics: An Introduction*, 1st edn. (CRC Press, Boca Raton, 2014)
348. R.J. Goldston, *Introduction to Plasma Physics*. Plasma Physics Series, 1st edn. (CRC Press, Boca Raton, 1995)
349. J.P. Freidberg, *Plasma Physics and Fusion Energy*, 1st edn. (Cambridge University Press, Cambridge, 2008)

350. U.S. Inan, *Principles of Plasma Physics for Engineers and Scientists*, Har/Psc edn. (Cambridge University Press, Cambridge, 2011)
351. P.M. Bellan, *Fundamentals of Plasma Physics*, 1st edn. (Cambridge University Press, Cambridge, 2008)
352. A. Li, *Double-Prism Multi-mode Scanning: Principles and Technology*. Springer Series in Optical Sciences Book 216, 1st edn. (Springer, Berlin, 2018)
353. S.K. Turitsyn, J.E. Prilepsky et al., Nonlinear Fourier transform for optical data processing and transmission: advances and perspectives. Optica **4**(3) (2017)
354. M.C. Veale, C. Angelsen et al., Cadmium zinc telluride pixel detectors for high-intensity X-ray imaging at free electron lasers. J. Phys. D Appl. Phys. (2018)
355. A. Marsh, *Mathematics for Physics: An Illustrated Handbook*, 1st edn. (World Scientific Publishing Co Pte, Singapore, 2018)
356. R.G. Ungureanu, G.V. Cojocaru, R.A. Banici, D. Ursescu, Phase measurement in long chirped pulses with spectral phase jumps. Opt. Express **22**(13) (2014)
357. J.A. Bittencout, *Fundamentals of Plasma Physics*, 3rd edn. (Springer, Berlin, 2017)
358. F.F. Chen, *Introduction to Plasma Physics*, 1974 edn. (Springer, Berlin, 2012)
359. R.G. Jahn, *Physics of Electric Propulsion* (Dover Publications, USA, 2006)
360. F.F. Chen, *Introduction to Plasma Physics and Controlled Fusion: Volume 1: Plasma Physics* (Springer, Berlin, 2010)
361. H. Wang, *Mathematics for Physics*, 1st edn. (WSPC, 2017)
362. J.P. Freidberg, *Ideal MHD*, Revised edn. (Cambridge University Press, Cambridge, 2014)
363. J. Wesson, *Tokamaks*. International Series of Monographs on Physics, 4th edn. (Oxford University Press, Oxford, 2011)
364. W. Rudin, *Principles of Mathematical Analysis*. International Series in Pure & Applied Mathematics, 3rd edn. (Example Product Manufacturer, 2013)
365. M. Ulmeanu, P. Petkov, D. Ursescu et al., Pattern formation on silicon by laser-initiated liquid-assisted colloidal lithography nanotechnology **26**, 455303 (9pp) (2015)
366. J. Belfi, N. Beverini, G. Carelli et al., Analysis of 90 day operation of the GINGERINO gyroscope. Appl. Opt. **57**(20) (2018)
367. K. Otsuka, Multimode laser dynamics. Prog. Quantum Electron. **23**, 97–129 (1999)
368. K. Otsuka, P. Mandel, E.A. Viktorov, Breakup of cw multimode oscillations and low-frequency instability in a microchip solid-state laser by high-density pumping. Phys. Rev. A **56**(4) (1997)
369. K. Otsuka, *Nonlinear Dynamics in Optical Complex Systems*, 2000 edn. (Springer, Berlin, 2000)
370. L.C. Evans, *Partial Differential Equations*. Graduate Studies in Mathematics, 2nd edn. (American Mathematical Society, 2010)
371. A. Alastuey, M. Clusel, M. Magro, P. Pujol, *Physics and Mathematical Tools: Methods and Examples*, Reprint edn. (WSPC, 2015)
372. K.S. Thome, R.D. Blandford, *Modern Classical Physics: Optics, Fluids, Plasmas, Elasticity, Relativity, and Statistical Physics* (Princeton University Press, Princeton, 2017)
373. M. Petrini, G. Pradisi, A. Zaffaroni, *A Guide to Mathematical Methods for Physics: with Problems and Solutions*. Essential Textbooks in Physics (World Scientific publishing Europe Ltd, 2017)
374. Kit-Wing, A complete solution guide to principles of mathematical analysis. 978-988-78797-1-8 (2018)
375. G. Cristoforetti, A. Colaitis et al., Experimental observation of parametric instabilities at laser intensities relevant for shock ignition. EPL **117**, 35001 (2017)
376. S. Sujecki, L. Sojka, Comparative modeling of infrared fiber lasers. Photonics **5**, 48 (2018)
377. S.H. Strogatz, M. Dichter, *Nonlinear Dynamics and Chaos with Student Solutions Manual*, 2nd edn. (CRC Press, Boca Raton, 2016)
378. H. Kleinert, *Particles and Quantum Fields* (World Scientific Publishing Company, Singapore, 2016)
379. K. Zhu, H. Chen et al., Frequency-shifted optical feedback measurement technologies using a solid-state microchip laser. Appl. Sci. **9**, 109 (2019)

380. G.B. Arfken, H.J. Weber, F.E. Harris, *Mathematical Methods for Physics: A Comprehensive Guide*, 7th edn. (Academic Press, Cambridge, 2012)
381. H. Fukuzawa, S.K. Son, K. Motomura et al., Deep inner-shell multiphoton ionization by intense X-ray free-electron laser pulses. Phys. Rev. Lett. **PRL 110**, 173005 (2013)
382. M. Zeng, O. Tesileanu, High flux electron beams from laser wakefield accelerators driven by petawatt laser. Plasma Sci. Technol. (2017)
383. D. Griffiths, *Introduction to Elementary Particles*, 2nd edn. (Wiley-VCH, 2008)
384. V. Parameswaran Nair, *Concepts in Particle Physics: A Concise Introduction to the Standard Model* (World Scientific Publishing Co Pte Ltd, Singapore, 2018)
385. Z. Gang, A.P.L. Robinson, X.Q. Yan, A.V. Arefiev, Highly collimated electron acceleration by longitudinal laser fields in a hollow-core target. Plasma Phys. Controlled Fusion (2018)
386. J. Alvarez Ruiz, A. Rivera, K. Mima et al., Plasma-wall interaction in laser inertial fusion reactors: novel proposals for radiation tests of first wall materials. Plasma Phys. Control. Fusion **54**, 124051 (7pp) (2012)
387. J. Alvarez, K. Mima, K.A. Tanaka et al., Ultra intense lasers as a promising research tool for fusion material testing: production of ions, X-rays and neutrons. Plasma Fusion Res. Regul. Art. **8**, 3404055 (2013)
388. P. Popivanov, A. Slavova, *Nonlinear Waves: A Geometrical Approach*. Series on Analysis, Applications and Computation Book 9 (World Scientific Publishing Company, Singapore, 2018)
389. L. Lu, Efficient computation of the sinc function by two-points Taylor expansion. Int. J. Adv. Comput. Technol. (IJACT) **4**(20) (2012)
390. Y. Zeng, Z. Tang, Reversing the sign of the effective nonlinear coefficient of a nearly isotropic kerr photonic crystal via the local-field effect. Opt. Mater. Express **8**(8), 2256 (2018)
391. S. Wiggins, *Introduction to Applied Nonlinear Dynamical Systems and Chaos*. Texts in Applied Mathematics, 2nd edn. (Springer, Berlin, 2003)
392. D.K. Arrowsmith, *An Introduction to Dynamical Systems*, 1st edn. (Cambridge University Press, Cambridge, 1990)
393. D. Arrowsmith, C.M. Place, *Dynamical Systems: Differential Equations, Maps, and Chaotic Behavior*. Chapman Hall/CRC Mathematics Series, 1st edn. (Chapman and Hall/CRC, UK, 1992)
394. X.P. Lu, S. Reuter, M. Laroussi, D.W. Liu, *Non Equilibrium Atmospheric Pressure Plasma Jets: Fundamentals, Diagnostic, and Medical Applications*, 1st edn. (CRC Press, Boca Raton, 2019)
395. D. Haertle, M. Jazbinsek, G. Montemezzani, P. Gunter, Nonlinear optical coefficients and phase-matching conditions in Sn2P2S6. Opt. Express **13**(10), 3765–3776 (2005)
396. R. Devaney, R.L. Devaney, *An Introduction to Chaotic Dynamical Systems*, 2nd edn. (CRC Press, Boca Raton, 2003)
397. F. Verhulst, *Nonlinear Differential Equations and Dynamical System* (Universitext) (Springer, Berlin, 2nd rev, and expanded ed. 1996, 2nd printing 2006)
398. M. Desouky, A.M. Mahmoud, M.A. Swillam, Silicon based mid-IR super absorber using hyperbolic metamaterial. Sci. Rep. **8**, 2036 (2018)
399. F. Mazenc, M. Malisoff, H. Ozbay, Stability and robustness analysis for switched systems with time-varying delays. SIAM J. Control Optim. **56**(1), 158–182 (2018)
400. E. Kukk, H. Myllynen, K. Nagaya et al., Coulomb implosion of tetrabromothiophene observed under multiphoton ionization by free-electron laser soft-x-ray pulses. Phys. Rev. A **99**, 023411 (2019)
401. R. Brown, *A Modern Introduction to Dynamical Systems* (Oxford University Press, Oxford, 2018)
402. M. Brin, G. Stuck, *Introduction to Dynamical Systems*, 1st edn. (Cambridge University Press, Cambridge, 2002)
403. J. David Jackson, *Classical Electrodynamics*, 3rd edn. (Wiley, Hoboken, 1998)
404. H. Mehrivash, M.H. Shafiei, Robust model predictive control of discrete-time delayed positive systems, in *5th International Conference on Control, Instrumentation, and Automation (ICCIA)* (2017)

405. M.A. Lieberman, A.J. Lichtenberg, *Principles of Plasma Discharges and Materials Processing*, 2nd edn. (Wiley-Inter Science, 2005)
406. Md. Saiful Islam, J. Sultana et al., Low loss and low dispersion hybrid core photonic crystal fiber for terahertz propagation. Photonic Netw. Commun. (2018)
407. D. Panasenko, A.J. Shu et al., Demonstration of a plasma mirror based on a laminar flow water film. J. Appl. Phys. **108**, 044913 (2010)
408. M. Misina, Y. Setsuhara, S. Miyake, Electron cyclotron resonance plasma enhanced direct current sputtering discharge with magnetic-mirror plasma confinement. J. Vac. Sci. Technol. A **15**(4) (1997)
409. R. Vallee, M. Bernier et al., *New Perspectives for Mid-infrared Fiber Lasers* (Advanced Photonics Congress, OSA, 2018)
410. R.A. Struble, *Nonlinear Differential Equations*. Dover Books on Mathematics, Reprint edn. (Dover Publishing, USA, 2018)
411. T.S. Georgescu, I. Dancus, D. Ursescu, Free space variable optical attenuator using frustrated total internal reflection with 70dB dynamic range. Appl. Opt. **57**(34) (2018)
412. F. Verhulst, *Nonlinear Differential Equations and Dynamical Systems* (Universitext) (Springer, Berlin, 2nd rev. and expanded ed. 1996., 2nd printing, 2006)
413. H.T. Davis, *Introduction to Nonlinear Differential and Integral Equations*. Dover Books on Mathematics, 1st edn. (Dover Publications, USA, 2010)
414. R. Mazrooei-Sebdani, Z. Eskandari, H.G.E. Meijer, Numerical bifurcation analysis of double +1 multiplier in Z3—symmetric maps. J. Diff. Equat. Appl. **24**(9) (2018)
415. A.E. Siegman, *Lasers*, Revised edn. (Univ Science Books, 1986)
416. W. Koechner, M. Bass, *Solid-State Lasers: A Graduate Text*. Advanced Texts in Physics (Springer, Berlin, 2003)
417. W. Koechner, *Solid-State Laser Engineering*. Springer Series in Optical Sciences (Springer, Berlin, 2006)
418. X. Luo, *Engineering Optics 2.0: A Revolution in Optical Theories, Materials, Devices and Systems*, 1st edn. (Springer, Berlin, 2019)
419. R. Grimshaw, *Nonlinear Ordinary Differential Equations*. Applied Mathematics and Engineering Science Texts, 1st edn. (CRC Press, Boca Raton, 1991)
420. D. Jordan, P. Smith, *Nonlinear Ordinary Differential Equations: An Introduction for Scientists and Engineers*. Oxford Texts in Applied and Engineering Mathematics, 4th edn. (Oxford University Press, Oxford, 2007)
421. J.A. Sanders, F. Verhulst, J. Murdock, *Averaging Methods in Nonlinear Dynamical Systems*. Applied Mathematical Sciences, 2nd edn. (Springer, Berlin, 2007)
422. G.R. Fowles, *Introduction to Modern Optics*. Dover Books on Physics, 2nd edn. (Dover Publications, USA, 1989)
423. G.S. He, S.H. Liu, *Advanced Nonlinear Optics*, 2nd revised edn. (World Scientific Publishing Company, Singapore, 2018)
424. E. Hecht, *Optics*, 5th edn. (Pearson, UK, 2016)
425. M.G. Kuzyk, *Nonlinear Optics: A Student's Perspective: With Python Problems and Examples* (Create Space Independent Publishing Platform, 2017)
426. A. McClelland, M. Mankin, *Optical Measurements for Scientists and Engineers: A Practical Guide*, 1st edn. (Cambridge University Press, Cambridge, 2018)
427. M. Burla, D. Marpaung et al., On-chip CMOS compatible reconfigurable optical delay line with separate carrier tuning for microwave photonic signal processing. Opt. Express **19**(22) (2011)
428. J. Gleick, *Chaos: Making a New Science*, Anniversary, Reprint edn. (Penguin Books, UK, 2008)
429. H.J. Lee, *Fundamentals of Theoretical Plasma Physics: Mathematical Description of Plasma Waves* (World Scientific Publishing Co, Singapore, 2019)
430. A. Weiner, *Ultrafast Optics*, 1st edn. (Wiley, Hoboken, 2009)
431. N. Bloembergen, *Nonlinear Optics*, 4th edn. (World Scientific Pub co Inc, Singapore, 1996)

432. T. Numai, *Fundamentals of Semiconductor Lasers*. Springer Series in Optical Sciences, 2nd edn. (Springer, Berlin, 2015)
433. Y.F. Chen, Y.P. Lan, H.L. Chang, Analytical model for design criteria of passively Q-switched lasers. IEEE J. Quantum Electron. **37**(3), (2001)
434. E.C. Stancu, A. Quade, K.-D. Weltmann, Polystyrene surface modification for serum-free cell culture using an atmospheric pressure dielectric barrier discharge. Rom. Rep. Phys. **71**, 409 (2019)
435. V.A. Popescu, Interrogation methods for Bragg fiber-based plasmonic sensors. Rom. Rep. Phys. **71**, 408 (2019)
436. T. Dreischuh, L. Gurdev, O. Vankov, E. Toncheva, L. Avramov, D. Stoyanov, Peculiarities of single-sided CW laser beam imaging of small-depth in homogeneities in highly-scattering media. Rom. Rep. Phys. **71**, 407 (2019)
437. E. Grigore, C. Ruset, M. Gherendi, M. Firdaouss, C. Hernandez, Investigation of the He content within W coatings by using glow discharge optical emission spectrometry. Rom. Rep. Phys. **71**, 410 (2019)
438. F. Kapsalidis, M. Shahmohammadi, M.J. Suess et al., Dual wavelength DFB quantum cascade lasers: sources for multi-species trace gas spectroscopy. Appl. Phys. B **124**, 107 (2018)
439. M.J. Suess, P. Morten Hundt et al., Dual-section DFB-QCLs for multi-species trace Gas analysis. Photonics **3**, 24 (2016)
440. M. Shahmohammadi, F. Kapsalidis, M.J. Suess et al., Multi-wavelength distributed feedback quantum cascade lasers for broadband trace gas spectroscopy. Semicond. Sci. Technol. **34**, 083001 (14pp) (2019)
441. J. Zhou (ed.), Complex sciences, in *First International Conference, Complex 2009*, Shanghai, China, Feb 2009, Revised papers, Part 1, (Springer, Berlin, 2009)
442. F.M. Atay, *Complex Time-Delay systems, Theory and Application* (Springer, Berlin, 2010)
443. P. Hovel, *Control of Complex Nonlinear Systems with Delay* (Springer, Berlin, 2010)
444. M. Lakshmanan, D.V. Senthilkumar, *Dynamics of Nonlinear Time-Delay Systems* (Springer, Berlin, 2010)
445. A.C.J. Luo, J.-Q. Sun, *Complex Systems, Fractionally, Time-Delay and Synchronization* (Springer, Berlin, 2011)
446. A.C.J. Luo, H. Merdan, *Mathematical Modeling and Applications in Nonlinear Dynamics* (Springer, Berlin, 2016)
447. D. Volchenkov, X. Leoncini, *Regularity and Stochasticity of Nonlinear Dynamical Systems* (Springer, Berlin, 2018)
448. D. Pieroux, P. Mandel, Bifurcation diagram of a complex delay-differential equation with cubic nonlinearity. Phys. Rev. E **67**, 056213 (2003)
449. A. Matsumoto, Frenc Szidarovszky, Delay differential neoclassical growth model. J. Econ. Behav. Organ. **78**, 272–289 (2011)
450. M. Wolfrum, S. Yanchuk, P. Hovel, E. Scholl, Complex dynamics in delay-differential equations with large delay. Eur. Phys. J. Spec. Top. **191**, 91–103 (2010)

Index

O. Aluf, *Advance Elements of Laser Circuits and Systems*,
https://doi.org/10.1007/978-3-030-64103-0

D

E

F

G

H

I

J

N

O

Q

R

S

T

W

X

Y